PHANEROZOIC TIME

ERA	PERIOD	EPOCH	AGE (MILLIONS OF YEARS AGO)

HOLOCENE (0.1)
PLEISTOCENE (2)
PLIOCENE (5)
QUATERNARY

HOMO SAPIENS: 500,000 YEARS AGO
ORIGIN OF *HOMO:* 2 MILLION YEARS AGO
NORTH AND SOUTH AMERICA CONNECTED: 3.6 TO 3.1 MILLION YEARS AGO
FORMATION OF DESERTS: 4 MILLION YEARS AGO TO PRESENT
GLACIATION IN ARCTIC: 4 MILLION YEARS AGO
ORIGIN OF HOMINIDS: 5.5 MILLION YEARS AGO
GLACIATION IN ANTARCTICA: 13 MILLION YEARS AGO
WORLDWIDE FORESTS RECEDE: 15 MILLION YEARS AGO TO PRESENT

0

CENOZOIC

TERTIARY

MIOCENE — EARLIEST HOMINOIDS
OLIGOCENE — SOUTH AMERICA SEPARATES FROM ANTARCTICA
EOCENE — ANTHROPOID PRIMATES
50 — AUSTRALIA SEPARATES FROM ANTARCTICA
PALEOCENE — EARLY INSECTIVORES AND PRIMATES

FOURTH MAJOR EXTINCTION EVENT: DINOSAURS
    AND AMMONITES EXTINCT

100 — FINAL SEPARATION OF SOUTH AMERICA AND AFRICA
— ANGIOSPERMS DOMINANT WORLDWIDE

CRETACEOUS

— FIRST FOSSIL ANGIOSPERM

MESOZOIC

150 — *ARCHAEOPTERYX:* FIRST FOSSIL BIRD

JURASSIC

200 — MAMMALS APPEAR
— DINOSAURS APPEAR
— FIRST FOSSIL MAMMALS

TRIASSIC

250 — DISAPPEARANCE OF TRILOBITES
— APPEARANCE OF THERAPSID REPTILES

PERMIAN

— CONIFERS APPEAR

300 — OLDEST FOSSIL REPTILES
— OLDEST FOSSIL INSECTS

CARBONIFEROUS

— FORESTS APPEAR AND DOMINATE THE LAND

350 — AMPHIBIANS APPEAR ON LAND
— THIRD MAJOR EXTINCTION EVENT
— AGE OF FISHES

PALEOZOIC

DEVONIAN

400 — PLANTS COLONIZE THE LAND

— FIRST JAWED FISHES

SILURIAN

— FIRST FOSSIL PLANTS

450 — SECOND MAJOR EXTINCTION EVENT

ORDOVICIAN

— DIVERSIFICATION OF MOLLUSKS
— OLDEST FOSSIL CRUSTACEANS AND ONYCHOPHORA
500 — FIRST MAJOR EXTINCTION EVENT
— EXPLOSIVE EVOLUTION OF PHYLA

CAMBRIAN

— EVOLUTION OF LANCELETS AND LAMPREYS

550 —

— EVOLUTION OF EXTERNAL PHYLA

600

BIOLOGY

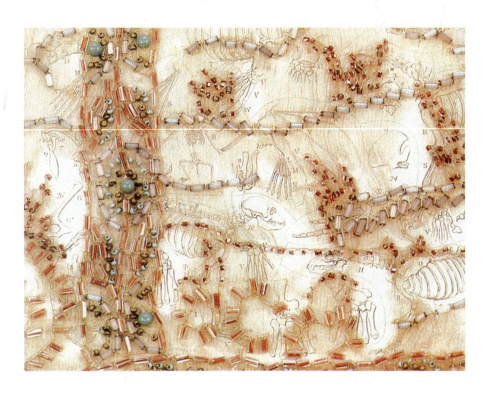

BIOLOGY

PETER H. RAVEN

Director, Missouri Botanical Garden;
Engelmann Professor of Botany,
Washington University, St. Louis, Missouri

GEORGE B. JOHNSON

Professor of Biology,
Washington University, St. Louis, Missouri

Illustration program developed by William C. Ober, Crozet, Virginia

TIMES MIRROR/MOSBY COLLEGE PUBLISHING

ST. LOUIS • TORONTO • SANTA CLARA • 1986

Executive Editor: Donald G. Mason
Editorial Assistant: June Schaffer Heath
Project Editor: Karen A. Edwards
Manuscript Editors: Linda Duncan, John P. Rowan, Suzanne Seeley

Art Director: Kay Michael Kramer
Designer: William A. Seabright
Production: Gail Morey Hudson, Susan Trail, Susan Lane, Kathy Burmann

Illustrators: William C. Ober and Ronald J. Ervin
          with Molly K. Ryan and George J. Venable
Photo Editor: Margaret B. Ober

Cover photo: "Hybrids" 1978 by Olivia Parker
Back cover photo: "Columbine" 1980 by Olivia Parker
Halftitle photo: "Residue" 1981 by Olivia Parker
Title page photo: "Contact" 1978 by Olivia Parker

Copyright © 1986 by Times Mirror/Mosby College Publishing

A division of The C.V. Mosby Company
11830 Westline Industrial Drive
St. Louis, Missouri 63146

Printed in the United States of America

Library of Congress Cataloging in Publication Data

Raven, Peter H.
  Biology.

  Includes index.
  1. Biology. I. Johnson, George B. (George Brooks),
1942-   . II. Title. [DNLM: 1. Biology.
QH 308.2 R253b]
QH308.2.R38 1986   574   85-10635
ISBN 0-8016-4091-1

C/VH/VH 9 8 7 6 5 4 3 2 1   03/A/317

PREFACE

Teaching biology in the 1980s is a greater challenge than ever before. The last decade has seen the beginnings of fundamental change in biological science. Even in its early stages, this revolution in biology is having a visible impact on the world around us. Announcements appear daily about such matters as test-tube babies, artificial insulin through genetic engineering, and new weapons against malaria and cancer. The science of biology is moving rapidly, and our teaching of it must keep up.

A survey of existing undergraduate introductory biology texts—a survey that every teacher of the subject undertakes almost every year—quickly reveals three facts. What first becomes evident is that few if any of the texts fully incorporate new advances in biology. Passing reference to new advances can sometimes be found in "gee-whiz" essays appended to the latest editions, but there is no thorough integration into the texts of the great progress biology has made in the last decade. It is as though each text were written or revised by means of a look at the other texts. The discussion of how the mammalian kidney functions is a simple and familiar example. Biologists proposed in the 1950s that the mammalian renal tubule functions as an active countercurrent multiplier, and this proposal rapidly found its way into undergraduate texts. The hypothesis has proved to be incorrect, however, and has been abandoned by renal physiologists for 15 years in favor of more realistic mechanisms—and yet the obsolete hypothesis remains in most biology texts, presented as gospel.

The second fact revealed by looking at available introductory texts is that, although most pay lip service to evolution as the organizing principle of biology and often assign it one or more chapters, none of the texts is actually organized consistently around evolutionary principles. Instead, most employ some variant of the "levels-of-organization" approach that achieved popularity in the 1950s. Basic

chemistry is taught first, then cells and their biology, then single-celled organisms, then metazoans, then community biology and evolution. This levels-of-organization approach has a major drawback: it organizes the enormous amount of information that students must be taught along arbitrary lines, rather than presenting it within an organization that devolves naturally from basic principles. The relationship between the biological phenomena characteristic of each level of organization, the fact that they have had an evolutionary background, and the organizing principles that link them—all of these tend to be lost in this approach, to the disadvantage of both the instructor and the student.

The third fact revealed is that few of these texts pay much attention to pedagogical issues. Space is at a premium in a biology text, since there is a great deal to cover, and few texts budget much space to features that facilitate learning. Nor do many texts take care to tie illustrations properly to the discussions they are intended to illustrate. These are important deficiencies because for any text to teach successfully, especially at the introductory level, it must be an effective learning tool for the student *outside the classroom*.

All of this means that teaching biology in the 1980s requires a new kind of textbook, one that fully integrates modern information, organizes its material from an evolutionary perspective, and presents the subject in a fashion that facilitates both teaching by the professor and learning by the student. This, in brief, is the strategy taken by *Biology* to meet the challenge.

DEVELOPMENT OF THIS TEXT

The truest test of any textbook should be how effectively it complements the instructor's objectives and how successfully it can be used by the student. Few successful texts are written in isolation. The more contact authors have with other scientists, on the one hand, and with instructors and students on the other hand, the more the text will reflect the current state of biology, and the more useful and usable it will be. The integration of these many sources of information by authors, orchestrated by the publisher, is referred to as "development" of a text.

Biology represents the most carefully developed introductory biology text for majors ever published. Quite simply, more steps were taken in the prepublication stage of this text than with any other competing book, to ensure both the highest quality of content and the greatest utility to instructor and students. The following demonstrates why we think you have never had a chance to use a book as carefully reviewed and as market sensitive as *Biology*.

The project was launched with **formal market research:** a detailed, eight-page questionnaire was sent to over 3400 introductory biology instructors nationwide. We synthesized the extensive information gathered in the survey into a detailed prospectus and table of contents, which was critiqued in a **qualifying review** by 17 instructors to help us refine the proposed contents and organization. The first draft was reviewed in total or in major portions by a panel of 36 reviewers. In a product **focus group session** held after completion of the first draft, eight of the first-draft reviewers met with us and the editors for 2 days to advise us in detail on the revision of the manuscript. Using the extensive first-draft reviews and focus group comments, we developed a second draft, which was class-tested by students in two schools: Northern State College in Aberdeen, South Dakota, and St. Louis Community College at Meramec. The second draft was also critiqued in detail for writing style, organization, and clarity by two professional writers, Linda Duncan and John Sturman. Class testing and professional writers' comments provided us with extensive direction in pro-

ducing a third draft, which was reviewed in depth by 27 reviewers, one third from the original panel and the remainder new to the project. Using the comments from this panel's review, we developed a refined fourth and final draft. To ensure that there were no inadvertent errors, we asked seven additional reviewers to comment on this final draft. Extensive care was exercised by us and by the publisher in the actual production process to maximize the effectiveness of the design and to minimize errors.

WHAT IS DIFFERENT ABOUT THIS TEXT AND WHY

Biology contains many distinctive features and, in fact, a number of unique ones. Many of these resulted from the extensive feedback we received from scientists and instructors throughout the country. Five themes may be readily distinguished: more modern information, consistent application of an evolutionary perspective, outstanding illustration, improved teaching and learning aids, and more usable supplements.

NEW INSIGHTS

In writing this text, we have attempted to modernize the basic core of knowledge presented to the student, and to recast certain major sections of biology along lines that could scarcely have been suspected just a few years ago. Major new insights have changed how we view the structure of the eukaryotic gene, the ways in which genetic information is processed during development, the nature of neuronal organization, and the mode of evolutionary change. In some cases this has necessitated the addition of entirely new chapters, such as Chapter 18 (Gene Technology) and Chapter 54 (The Future of the Biosphere), but more typically it has involved the recasting of traditional material in new contexts. Thus Chapter 9 (Membrane Transport) presents an account of membrane transport across biological membranes. This is a subject treated in all texts, but our chapter explains transport processes in terms of specific transmembrane protein "channels," imparting to the student a crisp and clear mechanical idea of how transport works, rather than explaining the process with a "black box" description that is difficult to grasp or remember.

Both molecular and organismal topics have undergone significant updating. Chapter 17 provides an example of the degree to which our knowledge of molecular genetics has changed in recent years. It includes a discussion of multigene families (we now know that relatively few eukaryotic genes are present only once on a chromosome), of how new eukaryotic genes evolve by the reshuffling of their segments, of the molecular basis of meiotic recombination, and of the molecular processes that lead to cancer.

The scope of new information integrated into this text extends far beyond molecular biology. Consider as routine a topic as Mendelian genetics. It is traditional to teach that Mendel did not discover linkage because the seven traits he studied were not linked (close to each other on a chromosome); in fact, two of the traits *are* closely linked—Mendel just never happened to test that combination. Other areas of traditional wisdom that have undergone significant alteration include very fundamental aspects of classification. Thus it is traditional to teach that ctenophores are related to cnidarians, when in fact they prove not to be.

A great deal of progress has been made in our understanding of evolution. It is traditional, for example, to teach that extinct apes such as *Ramapithecus* are ancestors to humans, but we now know that they are not on the line of human descent. Our closest living relatives are now thought to be pigmy chimpanzees. Progress in our understanding of evolution has also included the discovery of previously undescribed organisms such as *Prochloron,* the bacterium now thought to have given rise to the chloroplasts of green plants, and *Pelomyxa palustris,* the most primitive eukaryote known (it even lacks mitosis). In some cases, entirely new phyla, such as the Loricifera, are described.

Nor has our understanding of vertebrates gone without significant change. We have already mentioned that our understanding of how the vertebrate kidney functions has been radically revised. Knowledge in many other areas has undergone similar change. We have come to appreciate, for example, the key role of neural crest cells in vertebrate development; they are now considered the first uniquely vertebrate stage of development. The way in which nerve impulses travel down neurons is now understood to involve voltage-sensitive ion channels through the neuron cell membrane— "gates" that open in response to the arrival of the impulse. Down (not the outdated "Down's") syndrome is now known to be caused by a gene that seems identical to or closely associated with an oncogene—a gene that in other circumstances causes cancer.

Many other areas in the text might be mentioned. Far from being limited to the advances in molecular biology that happened last week, the updating of information has been pervasive, affecting all parts of our text.

EVOLUTIONARY PERSPECTIVE

Our text is presented as a scientific "narrative," whose beginning is the origin of living things, and whose storyline is the evolutionary history of life on earth. The first several chapters are devoted to the origin and early history of life. They provide an introduction to basic chemistry, because an adequate understanding of biology is impossible without a minimal chemical background. We have, however, deliberately avoided overdevelopment of chemistry, limiting our treatment to what is necessary for a clear understanding of what follows. The text then addresses the evolution of cellular organization. Here the merit of an evolutionary approach as a teaching device can be clearly seen. Consider as an example cell movement (Chapter 7), taught here as a characteristic property of cells, based on a few simple mechanisms that evolved early in the history of life. Similarly, photosynthesis in green plants (Chapter 12) is presented not as a complex process different from that of bacteria, but rather as a logical evolutionary extension of bacterial photosynthesis, a new device added to an older machine that had already been successful for billions of years. The recasting of photosynthesis in an evolutionary framework renders it far easier to understand, and teaches far more about its fundamental nature.

Heredity is the mechanism by which evolution occurs. Parts Three and Four deal with the mechanisms of heredity and evolution. The details of how heredity and evolution work, at both the microlevel and macrolevel, are now better understood than ever before. Using descriptions of the basic properties of cells covered in the previous sections, these chapters develop in the student an understanding of the process of evolution, the mechanisms of heredity and adaptation. The placement of the detailed treatment of evolution here, early in the text, is of critical importance to the teaching of biology, since it permits all the biological diversity that follows in the text to be considered within an evolutionary framework. These chapters on genetics and evolution form the core of the book, the foundation on which all that follows will stand.

Parts Five through Eight are devoted to an indepth examination of biological diversity—the parade of life on earth that is evolution's product. We start with a detailed treatment of viruses, which are best considered not truly alive so much as fragments of organisms. These are followed by the simplest true organisms: bacteria, simply organized cells that are thought to most closely resemble early life forms. We then examine plants, invertebrate animals, and vertebrates. The treatment of terrestrial plants and of vertebrates is extensive because of their particular interest. Six chapters are devoted to plants, including a novel chapter (30) stressing flowering plants as the dominant and most important group of land plants. Eleven chapters are devoted to the biology of vertebrates, stressing mammals and particularly humans. Here, also, the benefit of learning evolutionary principles before studying diversity can be clearly seen. Each major function is presented in an evolutionary context, examining the degree to which diversity of physiological function within animal or plant groups reflects different evolutionary solutions to particular environmental challenges.

The extensive treatment of diversity provides a basis for Part Nine, an elaboration of the ecological context of which all organisms form a part. Insight is provided into the functioning of biomes through a discussion of the dynamics of ecosystems and populations, and through a description of community ecology as it varies from place to place. In the last chapter we consider a subject oddly absent in most competing texts: the challenge of the world's future, endangered in many ways by human activities. Of the many lessons that we may teach students, these are among the most important.

Biological diversity, then, is taught in this text as a natural consequence of key biological principles of heredity and evolution. It is for this reason that a solid core of genetics and evolution is presented early in the text, rather than being tagged onto the end of the book as though it were an afterthought.

ILLUSTRATION

In the illustration program for *Biology,* we have departed somewhat from the usual practice of having the authors provide their own photographs and illustrations. We and the editors felt that the artwork, a critical element in any biology text, would have greater continuity and would be of higher quality if one person were responsible for overseeing the entire art program, from photo research to planning and execution of illustrations. With this kind of organization, illustrations could be produced with specific micrographs or emphasis in mind, the illustrations would be of compatible styles, and the photographs would be of consistent quality. Thus 98% of all line art was created specifically for *Biology.*

A professional photo research/illustration team of five worked closely with us and the editors during the development of the text to search for the photographs and create the line illustrations. Bill Ober, the Illustration Coordinator and an artist in his own right, was ably assisted in his efforts by a remarkable

group of artists. Ron Ervin, formerly Chief Medical Illustrator at the University of Virginia and now a free-lance medical illustrator, has done a fine job with full-color paintings of such diverse subjects as prehistoric animals and plant anatomy. George Venable, who is with the Entomology Department of the Smithsonian Institution's National Museum of Natural History, has provided astonishing illustrations for the arthropod chapter, plus illustrations of mammals and birds throughout the book. Molly Ryan, also of the Smithsonian, has been a valuable asset as a pen-and-ink illustrator. Illustrations of invertebrates are her specialty, and her knowledge of this group has been very helpful indeed. Maggie Ober, Bill's wife, coordinated the search for photographic materials. David Hunter, Jeff Smiling, and Dianne Ruscher provided Bill and Maggie with invaluable technical assistance.

TEACHING AND LEARNING AIDS
Getting the science correct and up to date is only one phase of developing a biology text. To be an effective teaching tool, it must present the science in a way that can be clearly taught and easily learned. With this in mind, we have included a variety of different teaching and learning aids.

General Approach
Readability. A text is only as easy to use as it is easy to read. This text has a "discovery" approach and a lively writing style intended to engage the interest of students. Simple analogies are employed extensively as an aid to understanding. The level of difficulty has been carefully controlled to correspond to the actual abilities of today's undergraduates.

Vocabulary. The large number of terms encountered in studying biology presents a real stumbling block for many students. This book lessens the problem by addressing and reinforcing a student's vocabulary development at three levels. First, no term is used without first being clearly defined. A student does not have to be familiar with the terminology of biology to use this book. Second, terms that are particularly important are printed in **boldface** at the point they are first used and defined. Finally, a complete and up-to-date alphabetical glossary of terms is presented at the end of the book.

Illustration. Because illustrations are so critical to a student's learning and enjoyment of biology, we have used full-color photographs generously throughout the text. Almost all line art in *Biology* is original, done under the direct supervision of the authors, and planned to illustrate and reinforce specific

points in the text. Consequently, every illustration is directly related to the text narrative and specifically cited in the text when the illustrated point is discussed. Great care has been taken in page make-up to place illustrations as close as possible to the text discussion. All illustrations and captions have been independently and carefully reviewed for accuracy and clarity.

Individual Chapter Elements

Chapter Overviews. Each chapter is introduced by a brief paragraph that tells the student what the chapter is about and why the subject is important. Not a summary, the overview allows students to place the chapter in perspective as they begin it.

For Review. As students progress through the text, they acquire an increasingly large vocabulary of terms and concepts, many of which play an important role in later chapters. To aid students in reviewing particularly important concepts, we have included a unique pedagogical device: each chapter is preceded by a short list of terms and concepts from earlier chapters that will be critical to understanding the present chapter, with an indication of where students should look to review any terms or concepts that are unfamiliar.

Concept Summaries. Another unique learning element that we have employed in each chapter is brief concept summaries. Throughout the text, important discussions and key points are succinctly summarized, with the summaries set off from the text and in boldface for emphasis. These concept summaries recap concisely the essential points that a student should learn.

Boxed Essays. To engage the student's interest and make the learning more exciting and fun, most chapters contain one, and often several, boxed essays on topics of special interest. Only a few paragraphs long, these essays provide an opportunity to briefly examine specialized topics for which there would otherwise be no space in the text.

Chapter Summaries. At the end of each chapter, a numbered summary provides a quick review of key concepts in the chapter.

Self-Quiz. A list of objective questions follows each chapter, intended to spot-check key information in the chapter. These questions can be used by students to test the degree to which they have read the chapter with sufficient attention and retention. The answers are presented in an appendix at the back of the book, so that students may check their performance.

Thought Questions. A brief list of thought-provoking questions and problems follows each chapter. These questions are not simply review, but rather are intended to challenge the student to think about the lessons of the chapter. Where chapters treat material such as genetics, in which solving problems plays an important role in learning, more problems are presented.

For Further Reading. Each chapter is followed by a short annotated list of books and articles, to which interested students can refer for additional information. Many of the articles are from *Scientific American* or other sources at a level appropriate to students with a limited science background.

SUPPLEMENTS

Few introductory courses attempt to cover the enormous amount of material that biology texts typically address. In the range of topics covered, introductory biology is almost unique. For this reason, we have provided a complete package of supplements for use with this text, to aid both student and instructor in dealing with what must sometimes seem an immense amount of material. As with the text itself, producing supplements of extraordinary quality and utility was, from the outset, a primary objective for us and the publisher. Too often, emphasis and investment in these components is based on quantity, not utility. All supplements accompanying *Biology* that are to be used with students, from the test items to the study guide, have been reviewed by many of the same instructors who critiqued the text. Additionally, much attention has been given throughout to providing elements and features in these supplements that were requested by both inexperienced and experienced instructors.

For the Student

BIOLOGY STUDY GUIDE, written by David Stetler of Virginia Polytechnic Institute and State University, Ronald M. Taylor of Lansing Community College and Margaret Gould Burke of the University of North Dakota. A direct companion to the text, the study guide provides students with significant additional study aids, including expanded chapter overviews, learning objectives, capsule summaries of key points, vocabulary reviews, additional self-study review questions, and other features intended to support motivated self-study of the textual materials.

BIOLOGY LABORATORY MANUAL, written by Darrell Vodopich and Randall Moore of Baylor University. The 35 laboratory exercises in the manual illustrate basic concepts and focus on experiments that will actually work when attempted. The emphasis in the laboratory manual is on manageability of equipment and content, with extensive visual support through numerous illustrations.

For the Instructor

INSTRUCTOR'S MANUAL, prepared by the authors and Vincent Zenger of Washington University, provides text adopters with substantial support in preparing for and teaching introductory biology with this text. The manual contains suggested course outlines, extensive sources of supplementary materials and additional resources such as film and computer software, 106 overhead transparency masters (these include some art not found in the text) to supplement the acetates available with the text; suggested learning objectives for each chapter; and chapter-by-chapter notes containing the following sections:

Chapter Outline
  "What's different in this chapter, and why." Detailed notes that compare the treatment in each chapter with that of other leading textbooks, to aid instructors who have used other books in the past in converting their courses and lecture notes.

  Chapter Objectives. Notes to the instructor highlighting important concepts covered in the chapter.

  Lecture Notes. Detailed suggestions for teaching the chapter, including suggestions for the use of overhead transparencies and other supplemental materials.

  Answers to Thought Questions in Text. A brief guide to the discussion that these questions are intended to engender, and specific answers to the review questions and problems.

OVERHEAD TRANSPARENCY ACETATES of 96 of the text's most important two-color and four-color drawings are available to instructors from the publisher for use as teaching aids. These transparencies were selected with the assistance of a number of teachers of introductory biology, to provide the maximum utility to the instructor. The instructor's manual contains suggestions for the use of these transparencies.

A printed *TEST BANK,* prepared by the authors and Karen M. Nein, provides an extensive battery of over 2100 objective text items which may be used by instructors as a powerful instructional tool. Each chapter has an average of 40 questions (10 true-false and 30 multiple choice). For each question, in addition to the answer, we have identified the subject tested, given an approximate difficulty rating, and indicated the type of question (factual or conceptual) and the text page on which the question's information appears.

QUESTBANK, a computerized version of the test bank, is available to instructors from the publisher on disks compatible with the IBM-PC or the Apple IIe. The computerized QUESTBANK contains approximately half of the questions in the printed test bank. QUESTBANK also provides instructors with the opportunity to add questions of their own, as well as to modify or delete questions already on file.

Biology is such a dynamic field today that there will be changes in many of its most exciting areas even while you are using this book. We hope that in putting it together as we have, we have provided you with a firm foundation for understanding these changes and evaluating them in context. Being able to follow the unfolding of a rapidly changing science such as biology is one of the most exciting prospects available to any student. Not only is it intellectually challenging, it also can and should be an enormous amount of fun.

A FEW WORDS OF THANKS

This book had its genesis in the request of another publisher, Bob Worth, that we review the Curtis *Biology* text for him and recommend how it should be revised. This review sparked our interest and convinced us that there was a real need for a kind of text different from those currently available.

During the course of development of a book of this size, many people within a publishing company help the authors, in a hundred different ways, and this book has been no exception. We hope that each of these people will accept our gratitude and thanks, particularly a publisher who never let us lose sight of our intended audience, an editor who showed us no mercy with deadlines but worked himself even harder, a copy editor who would not be bullied but could be convinced, and a production staff whose only concern was to do the book right.

Four people outside of Times Mirror/Mosby were of particular help to us. Bill and Maggie Ober were responsible for the art program and the photograph acquisition, respectively. They did a beautiful job, as you can see by thumbing through this book. Bill is a practicing physician, and his own hands-on knowledge of anatomy and broad knowledge of biology have added focus and originality to his considerable talents as an artist. Persistence is Maggie's special virtue—many a photograph is present in this book only because she refused to give up chasing it. Bill Seabright designed the book and laid out each page. He is a perfectionist with a great sense of style and an eye for what looks good. Much of the beauty of this book reflects his clear vision. Karel Liem teaches freshman biology at Harvard University, and his physiology lectures are famous for their clarity and excitement. He was unstinting in his efforts to help us develop a first-rate physiology section. He lent us his personal lecture notes, reviewed all drafts of chapters 40 through 49, and in many instances pointed out subjects such as kidney

function on which other texts were woefully out of date. These chapters reflect in detail his expertise; they cannot begin to reflect his charm, kindness, and generosity.

We also give special thanks to Dr. Peter C. Hoch, Peter Raven's Research Associate, for his expert help in organizing and reviewing many parts of the manuscript, and to Ms. Jo Ann Collins for her help with many details concerning the manuscript.

This book was written and produced over 3 years, which saw the birth of several children in our families, and during all of that time we had the constant support of our wives, Tamra Raven and Barbara Johnson, who had to put up with absent husbands too often. The phrase "It's almost done" now brings only shrieks of laughter in our homes. Tamra and Barbara know better.

Good reviewers are a critical element in developing a successful text. We would like to thank the more than 100 excellent and dedicated reviewers who commented on various drafts of the manuscript and suggested countless improvements. As authors, we called personally on many of our colleagues to share with us their knowledge through critiques of larger or smaller sections of this work in its various stages.

The editors likewise prevailed on the many instructors of biology listed on the next page to read and comment on the different drafts of this text. We

Vernon Ahmadjian
Clark University

Bruce Allen
Missouri Botanical Garden

Thomas E. Bowman
U.S. National Museum of Natural History
Smithsonian Institution

Winslow Briggs
Carnegie Institution of Washington

Allan Campbell
Stanford University

Hampton L. Carson
University of Hawaii

Steven Chambers
U.S. Fish and Wildlife Service

Marshall Crosby
Missouri Botanical Garden

Charles C. Doane
Arizona State University

John Fincham
Cambridge University

Victoria Finnerty
Emory University

Sturla Fridriksson
Agricultural Research Institute
Reykjavik, Iceland

Arthur C. Gibson
University of California-Los Angeles

G.R. Harbison
Woods Hole Oceanographic Institution

Daniel Hartl
Washington University

Peter C. Hoch
Missouri Botanical Garden

Paul L. Illg
University of Washington

Kenneth Y. Kaneshiro
University of Hawaii

Karel Liem
Harvard University

Malcolm McKenna
American Museum of Natural History

Robert Magill
Missouri Botanical Garden

David L. Pawson
U.S. National Museum of Natural History
Smithsonian Institution

B.O. Phinney
University of California-Los Angeles

David Pilbeam
Harvard University

Stanley B. Prusiner
University of California-San Francisco

Alan Solem
Field Museum of Natural History

Charles H. Southwick
University of Colorado

Robert W. Sussman
Washington University

Ian Tattersall
American Museum of Natural History

Alan R. Templeton
Washington University

Harrison Tordoff
Bell Museum of Natural History
University of Minnesota

Joseph Varner
Washington University

C.R. Woese
University of Illinois

Thomas Woolsey
Washington University

are pleased to acknowledge our debt to them for their unstinting effort, often in the face of almost impossible schedules. We especially thank two groups of reviewers: the members of the Focus Group—Lester Bazinet, Erik Bonde, Richard Boohar, Margaret Gould Burke, Matt Dakin, Lawrence Friedman, Frank Nordlie, and Ronald Taylor; and the instructors who class-tested the manuscript—David Campbell and his students at St. Louis Community College at Meramec, and John Naughten and Dan Tallman and their students at Northern State College in Aberdeen, South Dakota. In many ways a successful textbook represents the combined efforts of authors, illustrators, reviewers, and publisher. This text is a testimony to that effective cooperation.

Peter H. Raven

George B. Johnson

Michael Adams
Louisiana State University

Marilyn Bachmann
Iowa State University

Steve Barnhart
Santa Rosa Junior College

William Barstow
University of Georgia

William Baxter
Bowling Green State University

Lester Bazinet
Community College of Philadelphia

Cecile Boehmer
Alice Lloyd College, Kentucky

Erik Bonde
University of Colorado at Boulder

Richard Boohar
University of Nebraska at Lincoln

Stephen Bromley
Michigan State University

Margaret Gould Burke
University of North Dakota

David Campbell
St. Louis Community College at Meramec

Robert Cleland
University of Washington

Richard Collins
Louisana State University

Matt Dakin
University of Southwestern Louisiana

David W. Deamer
University of California, Davis

A.B. Dickinson
Memorial University of Newfoundland

Marvin Druger
Syracuse University

Barbara Evans
University of Melbourne

Susan Feldkamp
University of Wyoming

David L. Fox
University of Tennessee at Knoxville

Douglas G. Fratianne
Ohio State University

Lawrence D. Friedman
University of Missouri at St. Louis

Lawrence Gilbert
University of Texas at Austin

Elliott Goldstein
Arizona State University

Stuart Goldstein
University of Minnesota

Jonathan Goldthwaite
Boston College

L.C. Graham
University of Manitoba

Shirley Hanrahan
University of the Witwatersrand South Africa

Brian Hazlett
University of Michigan

Robert J. Huskey
University of Virginia

Malcolm Jollie
Northern Illinois University

Russell Jones
University of California, Berkeley

Samuel Kirkwood
University of Minnesota

David Knox
University of the Witwatersrand South Africa

Meredith Lane
University of Colorado at Boulder

Charles Leavell
Fullerton College, California

Peter Leveque
Santa Rosa Junior College

Joe Leverich
St. Louis University

Joseph Levine
Boston College

O.A.M. Lewis
University of Cape Town South Africa

Edward McCrady
University of North Carolina at Greensboro

Lynn Margulis
Boston University

William Mason
Auburn University

Robert Mellor
University of Arizona

Ralph Meyer
University of Cincinnati

Stanley Miller
University of California, San Diego

Michael Moore
Arizona State University

John Naughten
Northern State College Aberdeen, South Dakota

Frank Nordlie
University of Florida

Joseph O'Kelley
University of Alabama

C.O. Patterson
Texas A&M University

William Rayburn
Washington State University

Ian K. Ross
University of California, Santa Barbara

William Rumbach
Central Florida Community College

G.A.M. Scott
Monash University Australia

G.G.E. Scudder
University of British Columbia

Peter Shugarman
University of Southern California

David Smyth
Monash University Australia

Jerry Snider
University of Cincinnati

Irwin Spear
University of Texas at Austin

Diana Stein
Mount Holyoke College

Gerald Summers
University of Missouri at Columbia

Daryl Sweeney
University of Illinois at Urbana-Champaign

Stan Szarek
Arizona State University

Dan Tallman
Northern State College Aberdeen, South Dakota

Ronald Taylor
Lansing Community College

Roy Umber
University of Wyoming

James W. Warren
Monash University Australia

T. Weaver
Montana State University

David Woodruff
University of California, San Diego

Charles Yarish
University of Connecticut

John Zimmerman
Kansas State University

CONTENTS

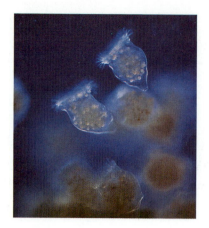

PART TWO

BIOLOGY OF THE CELL

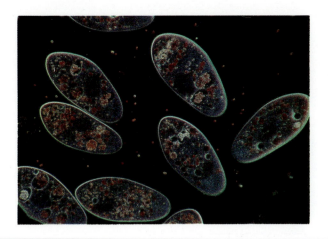

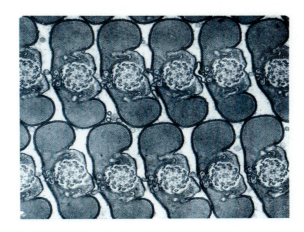

PART FIVE

BIOLOGY OF SIMPLE ORGANISMS

PART SIX

BIOLOGY OF PLANTS

Life on earth is inextricably tied to water.
It is ever present in the world about us in
varying forms—the moving oceans covering
much of the earth's surface, icebergs as large
as small cities, fog, rain, and morning dew.
Other planets are not as rich in water; on the
surface of Mars there is practically none.
Perhaps this is why life has not evolved on
other planets.

The footprints you see on the opposite page were made in Africa 3.7 million years ago. A mother and child walking on the beach might leave such tracks. But these tracks are not human. They were made by our ancestors. Preserved in volcanic ash, these tracks record the passage of two individuals of the genus *Australopithecus,* the group from which our genus, *Homo,* evolved. They were found in 1978 at Laetoli, in the central Rift Valley of East Africa, by a team of workers headed by Mary Leakey. The tracks continue 24 meters before disappearing, and by looking at them, we journey into our past.

The traces left by some of our ancestors seem a fitting place to begin our consideration of biology, which is the study of living things and how they have evolved. In this introduction to biology you will encounter many kinds of information, from the structure of molecules to the architecture of teeth. Before we begin, however, it is useful to take a moment to consider just what a biologist is.

WHAT IS A BIOLOGIST?

Perhaps the best way for us to answer the question "What is a biologist?" is to describe some of the kinds of things that biologists do. Mary Leakey, who discovered the footprints at Laetoli, is a biologist. Her specialized branch of biology is **anthropology,** the study of human evolution. Other biologists study other things. Indeed, biologists study so many different things that at first the common thread in what they do may not be evident.

Some scientists have suggested that our ancestors might have resembled the pensive orangutan in Figure 1-1. In 1983 jaw fragments of an extinct ape 17 million years old were discovered in northern Kenya. They resemble other fossils of a genus of extinct apes called *Sivapithecus* that are the ancestors of the orangutan. The fragments are so old that some investigators suspect *Sivapithecus* may be an ancestor of humans as well. Others—the clear majority now—believe that we are more closely related to the African apes (chimpanzees and gorillas). They base their deduction on biochemical and structural features. Biologists who study the characteristics of the group of mammals to which we belong, the primates, are called **primate biologists.**

The mountain in Figure 1-2 is Cerro de la Neblina, the Mountain of the Mist, rising high above the jungle of southernmost Venezuela. It is the remnant of an ancient plateau that has eroded away over millions of years, leaving this mountain as the highest point in South America outside of the Andes. For this vast period of time the creatures living on its upper slopes have been isolated from the jungle below and from other peaks weathered out

FIGURE 1-1

Orangutan

FIGURE 1-2

Mountain of the Mist

FIGURE 1-4

Cigarette smoker

of the same ancient plateau. Evolution has taken a different path on this isolated mountaintop, and the plants and animals there are unlike those found anywhere else on earth. Biologists who attempt to identify and catalogue this diversity are called **taxonomists.**

The plant in Figure 1-3 is a petunia. It appears to be a normal petunia in every respect, except that it is growing in a test tube. This plant was grown from cells that were extracted from another plant and suspended in tissue culture, where for many cell generations they simply grew and divided as undifferentiated cells. The experimental biologists who induced one of these cells to differentiate into a developing plant are called **molecular geneticists** or **developmental biologists.**

The person in Figure 1-4 who is peering out through the cloud of smoke is smoking a cigarette. Smoking is an addiction that may kill him. Over 95% of the people who will die of lung cancer in the United States this year smoke cigarettes. Cancers are initiated by chemicals such as those which this smoker is drawing into his lungs. Once inside the cells lining the lungs, these hazardous chemicals initiate cancer by damaging growth-restraining genes. Often associated with viruses living in human chromosomes, these genes act in normal cells to shut down the activities of a variety of other genes that induce cell growth. When chemicals damage the regulating genes, the growth-promoting activities are unleashed, with cancer the result. Scientists who study the molecular events leading to cancer are called **cancer biologists** or **oncologists.**

The insect in Figure 1-5 is a mosquito of the genus *Anopheles.* It carries within its saliva a small parasite, *Plasmodium.* If this mosquito were to bite you, about 1000 reproductive

FIGURE 1-3

*Petunia growing in
test tube*

FIGURE 1-6

White Cliffs of Dover

FIGURE 1-5

Anopheles *mosquito*

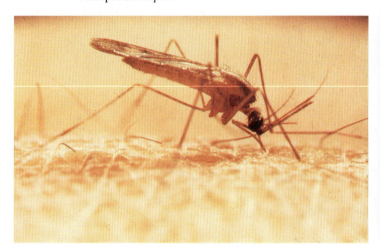

cells of *Plasmodium* would enter your bloodstream immediately, travel to your liver within 30 minutes, and multiply rapidly there. Over time each cell would give rise to thousands of parasitic cells. You would have the disease called malaria. It is one of the most common infectious diseases on earth, and one of the most deadly. At any one time about 200 million people are infected with malaria, and about 2 million of them die each year. Understanding the role that the mosquito plays in spreading malaria has made it possible to eradicate the disease in many areas. Investigators who study parasites such as *Plasmodium* are called **parasitologists.**

The White Cliffs of Dover, seen in Figure 1-6, are constructed almost entirely of fossil shells of minute marine animals called forams. Their shells are largely composed of calcium carbonate. Forams have been common in the seas of the earth for hundreds of millions of years. They are useful as indicators of past times and past conditions, because their shells are readily preserved in the sediments formed beneath the seas where they live. Foram fossils are particularly interesting because they provide useful clues in finding oil-bearing strata. Scientists who study fossils are called **paleontologists.**

FIGURE 1-7

Pocket mouse

A peculiar feature of the pocket mouse, *Perognathus* (Figure 1-7), is that although it is common in the deserts of North America, it never drinks. Even if you were to place these mice by the coolest stream, they still would not drink. Yet they have blood and excrete urine. What is their source of water? This mystery was solved by scientists who learned that pocket mice obtain their water chemically from dry plant seeds. Biologists who study how the body functions are called **physiologists.**

FIGURE 1-9
Mexico City

FIGURE 1-8
Polluted lake

The lake in Figure 1-8 is dead. Nothing lives in it. Polluted by industrial waste to the extent that no organism could survive, this lake represents the ultimate biological lesson, the bottom line in our relationship with our environment. Scientists who investigate this relationship and the many complex interactions in functional, living ecosystems are called **ecologists.**

The enormous sprawl of houses in Figure 1-9 constitutes part of one of the world's largest cities, Mexico City, built on the site of the ancient and beautiful lake city of the Aztecs, Tenochtitlán. The population of Mexico City will have reached over 18 million by 1986, and it is projected to be 26 million by the year 2000. The congestion shown here is just one effect of this enormous density of people. Human populations are increasing rapidly throughout the less developed countries of the world and are becoming more concentrated in their major cities. Such rates of increase and redistribution are frightening. Those who monitor this population nightmare are called **demographers.**

The hopeful-looking baby ducks following Konrad Lorenz in Figure 1-10 think he is their mother. He is the first "animal" they saw when they were born. As a result, the ducks have become imprinted with his image and will always recognize him as their parent. Lorenz and other students of animal behavior are called **ethologists.**

All of these different kinds of scientists are biologists. They study living organisms, or ones that lived in the past, in an attempt to understand their origins, how they cope with the stresses and strains of living, and how their diversity can be used to improve human existence.

FIGURE 1-10

*Ducks following
Konrad Lorenz*

WHY IS BIOLOGY IMPORTANT TO YOU?

For many students, biology is an important subject because it serves as an introduction to their careers as physicians, agriculturists, dentists, ecologists, veterinarians, industrial molecular biologists, or any of a wide range of other fields of biology. If you are such a person, this book will provide you with your first taste of a field of inquiry whose results you will use for the rest of your life. Biology is also an important subject for the larger numbers of students who will not go on to careers in the field, simply because biology will affect their futures in many ways. Because the activities of biologists are altering the world around us, an understanding of biology is increasingly becoming necessary for any educated person. In the following sections we will discuss some of the ways in which the activities of biologists are having important effects on society.

The Battle Against Disease

In the early 1980s a new and vicious disease, **acquired immune deficiency syndrome (AIDS),** was first reported, initially among homosexuals and later among a wider cross section of the American population. Affected persons die of a bewildering variety of infections, but they all have one thing in common: they have no resistance to infection. Their immune systems are inoperative. Most AIDS victims die. Because AIDS is infectious, it is potentially a very dangerous disease.

Biologists all over the world began work to determine the cause of AIDS. It was not long before the infectious agent was identified by laboratories in France and the United States. Surprisingly, the culprit proved to be a cancer virus, one of a group of human viruses called T cell lymphoma viruses (*h*uman *T* cell *l*ymphoma *v*irus, or HTLV) (Figure 1-11).

Biologists had already learned a great deal about cancer, and in particular about the leukemia caused by HTLV. HTLV initiates malignant growth, the uncontrolled cell proliferation that we call cancer, among human T cells, one of two types of cells that are responsible for the human body's resistance to disease. Normal T cells attack and kill invading cells within the body. They constantly monitor the surfaces of all cells they encounter, checking for the presence of the specific combination of surface proteins unique to that human individual. Any foreign cells, lacking the correct combination, are attacked and killed. Few people have the same combinations of these "self"-identifying proteins, which is why organ transplants are rejected unless T cells are suppressed with drugs. In HTLV-infected leukemia patients the T cells grow and divide uncontrollably. If the progress of the leukemia is unchecked, and it is difficult and often impossible to do so, the condition is fatal.

Cancer biologists (called oncologists, after the Greek word for cancer) learned that a form of HTLV is also present in the T cells of AIDS victims, a form that functions differently from the usual HTLV. Although cells infected with this new kind of HTLV do not exhibit malignant growth, the infection has a side effect with disastrous consequences. Proteins that the virus manufactures, using the machinery of the human body, redirect the functions of the T cell to producing HTLVs. The infected T cell, after producing many viruses, eventually ruptures and dies. Because HTLV is infectious, all of the T cells eventually become infected. Finally, with no T cells, the body is unable to marshal a defense against infection.

Diseases such as AIDS can appear at any time. In the case of AIDS, however, researchers are beginning to be able to fight back. As we will discuss in Chapter 18, biologists are stitching or inserting the specific HTLV genes responsible for specifying the virus exterior (and no other HTLV genes) into an otherwise harmless human virus (Figure 1-12). They then plan to use this engineered virus to "trick" the human body into developing a defense against HTLV without ever actually experiencing the disease. When humans are inoculated with the engineered virus, a class of infection-resisting cells called B cells begins to produce proteins called antibodies that bind to the exterior surface of any invading cell or virus, acting as a signal directing the body to destroy the invader. In this case, because

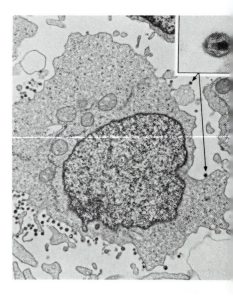

FIGURE 1-11

The virus responsible for acquired immune deficiency syndrome (AIDS), isolated in 1985. The AIDS viruses home in on the infection-resisting T cells of their victims, destroying a person's ability to resist disease. The T cell in the main photograph is releasing numerous AIDS viruses, the dark particles on the periphery of the cell. When free from the cell, a virus particle is able to infect other T cells. The inset shows an enlargement of one such AIDS virus particle. Few AIDS victims survive.

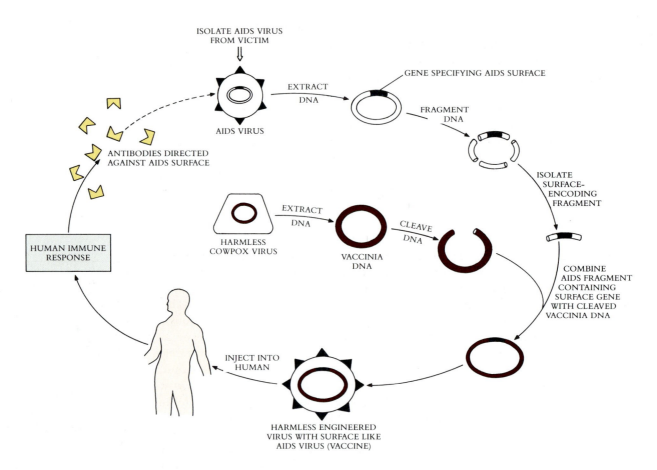

ISOLATE AIDS VIRUS
FROM VICTIM

EXTRACT
DNA

GENE SPECIFYING AIDS SURFACE

FRAGMENT
DNA

AIDS VIRUS

ANTIBODIES DIRECTED
AGAINST AIDS SURFACE

ISOLATE
SURFACE-
ENCODING
FRAGMENT

HUMAN IMMUNE
RESPONSE

EXTRACT
DNA

HARMLESS
COWPOX VIRUS

VACCINIA
DNA

CLEAVE
DNA

COMBINE
AIDS FRAGMENT
CONTAINING
SURFACE GENE
WITH CLEAVED
VACCINIA DNA

INJECT INTO
HUMAN

HARMLESS ENGINEERED
VIRUS WITH SURFACE LIKE
AIDS VIRUS (VACCINE)

FIGURE 1-12

How scientists are attempting to construct a vaccine for AIDS. Of the many genes of AIDS virus, one is selected that encodes a surface feature of the virus. All of the other AIDS genes are discarded. This one gene is not in itself harmful; it simply specifies the shape of the AIDS virus protein coat. This one gene is inserted into the deoxyribonucleic acid (DNA) of a harmless cowpox virus, with the result that the cowpox virus eventually possesses a protein coat with surface features specified by the added AIDS gene. Humans infected with the altered cowpox virus do not become ill—the cowpox virus is harmless—but they do develop antibodies directed against the infecting virus surface. Because the surface is that of an AIDS virus, the new antibodies would serve to protect the infected person against any subsequent exposure to AIDS. Any AIDS virus particle that infects a person would be destroyed by these antibodies.

the engineered virus has the exterior of HTLV, the body produces antibodies directed against HTLV, providing immunity to AIDS infection. When anti-HTLV antibodies are present, any AIDS-causing virus that enters the bloodstream will be attacked by the body's immune system. Although the engineered virus is not yet available for use (the experiments perfecting it are going on right now), the prognosis for the future is bright.

AIDS is not the only disease against which biologists have fought. Smallpox, once one of the scourges of humankind, has been reduced to history. In 1978 the last case of smallpox was recorded, and the virus that is responsible for this disease now exists only in the laboratory, where cultures are maintained as insurance in the event that a vaccine would need to be prepared if the disease reappeared. Chapter 25 discusses the elimination of this major killer. The elimination of other diseases has proven more difficult. For example, only now has the problem of malaria begun to yield to persistent effort. A gene triggering the antibody response to the *Plasmodium* parasite has been isolated. As with HTLV, stitching this gene into another harmless virus is expected to create an agent that will render inoculated persons immune to malaria.

Preserving Our Heritage

The ultimate biological tragedy is extinction. When the last member of a plant or animal species dies, the world will never see it again. Extinction is forever. Extinction has played an important role in evolution, with older forms making way for new ones. Of all the species that have ever inhabited earth, only a small percentage live now. Extinction has many different causes. For example, at the end of the Cretaceous Period, 65 million years ago, dinosaurs and many other living forms became extinct. Scientists have had many theories about why this event occurred so abruptly. Many of them now think that the extinction occurred when a large asteroid, perhaps 10 kilometers across, collided with earth, throwing up a dense cloud of dust that darkened the skies and lowered temperatures drastically worldwide for a year or more. This occurrence and similar factors that may have been important in causing extinctions in the past are reviewed in Chapter 23.

It is possible, however, that the greatest mass extinction of species in the history of life on earth is going on right now. It is not being caused by the stars, but rather by human beings. Sometimes extinction caused by humans has been deliberate. The eradication of smallpox was, in effect, an extinction event, one purposefully engineered by biologists to promote human well-being. Most extinction events are not deliberate, however. They occur as a consequence of the spread of the human population and of the destruction of natural habitats that seems inevitably to be associated with this spread. Especially in the tropics, the current rate of extinction is appalling. By the end of the century, hundreds of thousands of species, perhaps amounting to 1 out of 10 on earth, may have become extinct.

Could we not simply collect the endangered species and put them all in zoos and botanical gardens? No. To understand why, consider the cheetah. It is among the most beautiful of the large cats, long, languid, and very fast—the essence of the word *feline*. Cheetahs and their close relatives were once distributed widely over the world, including the United States. As human populations have expanded, however, their range has shrunk, and they are now found only in East and South Africa—and in many of the world's zoos. It is possible that in your lifetime they will be extinct, not just shot by hunters in the wild, but gone from zoos as well, living only in pictures and in your memory.

The cheetah problem first became evident in Oregon, in one of the most successful cheetah-breeding colonies in the world. In May 1982, a seemingly healthy female was introduced to the colony. She died 6 weeks later. On autopsy, biologists found that she had been infected with a virus of domestic house cats called feline infectious peritonitis (FIP). In house cats this is not a serious disease; perhaps 1 domestic cat in 100 dies when infected with this virus—but this cheetah died of the infection. A check of the other 40-odd cheetahs in the colony revealed that all had been exposed to the virus and possessed antibodies directed against it. Soon clinical signs of the disease could be seen in over 90% of them. To date, 18 have died, including the three in Figure 1-13. Others were shipped to other zoos and are being traced now.

The reason for the epidemic was pinpointed by Steve O'Brien, head of the National Cancer Institute's viral genetics section. O'Brien found that cheetahs, both in zoos and in the wild, are genetically almost identical to one another, as if they were all identical twins. A cheetah taken from the Oregon colony and another captured from South Africa would be much more similar to one another genetically than you would be to a brother or sister.

Genetic similarity is not in itself necessarily bad. The problem with the Oregon cheetahs stems from the lack of variation in the genes which encode the proteins that their bodies use to identify their own tissues. Skin grafts between two different mammals rarely "take," because they possess different combinations of these proteins. Skin grafts between different cheetahs always take. These "self"-recognition proteins are the ones recognized by T cells and B cells, telling the T and B cells which cells are normal parts of the body and which are foreign. Like a key to a door, however, the great weakness of using a specific combination of proteins as a "self"-recognition system is that anyone with the correct key may gain entry. Thus a virus whose surface possesses the appropriate combination of recognition proteins would not be recognized as foreign by the body's T cells and B cells. Unfortunately, viruses infecting humans occasionally acquire the human genes specifying these proteins. Human populations, however, possess hundreds of different "self"-recognition protein genes, and in different persons different combinations are used, the choice of which is random. Many thousands of combinations are possible. It is very unlikely that anyone you know has exactly the same combination you do. A virus might successfully circumvent one person's defenses but not those of another person.

It seems likely that the FIP domestic cat virus that infected the Oregon colony had acquired a genetic capacity to circumvent the "self"-recognition defenses of the first af-

FIGURE 1-13

Jezabel and her two kittens, Greensleeves and Whiskers, were infected with the FIP virus a few days after this photograph was taken and died soon after. The majority of cheetahs infected with FIP have not survived.

HOW BIOLOGISTS DO THEIR WORK

THE CONSENT

Late in November, on a single night
Not even near to freezing, the ginkgo trees
That stand along the walk drop all their leaves
In one consent, and neither to rain nor to wind
But as though to time alone: the golden and green
Leaves litter the lawn today, that yesterday
Had spread aloft their fluttering fans of light.

What signal from the stars? What senses took it in?
What in those wooden motives so decided
To strike their leaves, to down their leaves,
Rebellion or surrender? and if this
Can happen thus, what race shall be exempt?
What use to learn the lessons taught by time,
If a star at any time may tell us: Now.

HOWARD NEMEROV

What is bothering the poet Howard Nemerov is that life is influenced by forces he cannot control, or even identify. It is the job of biologists to solve puzzles such as the one he poses—to identify and try to understand those things which influence life.

Nemerov asks, "Why do ginkgo trees drop all their leaves at once?" To find an answer to questions such as this, biologists and other scientists pose *possible* answers and try to see which answers are false. Tests of alternative possibilities are called **experiments.** To learn why the ginkgo trees drop

all their leaves simultaneously, a scientist would first formulate several alternative possible answers, called hypothetical answers, or **hypotheses:**

Hypothesis 1: Ginkgo trees possess an internal clock that times the release of leaves to match the season. On the day Nemerov describes, this clock sends a "drop" signal (perhaps a chemical) to all the leaves at the same time.

Hypothesis 2: The individual leaves of ginkgo trees are each able to sense day length, and when in the fall the days get short enough, each leaf responds independently by falling.

Hypothesis 3: A strong wind arose the night before Nemerov made his observation, blowing all the leaves off the ginkgo trees.

Next the scientist attempts to eliminate one or more of the hypotheses by conducting an experiment. In this case one might cover some of the leaves so that these leaves cannot use light to sense day length. If hypothesis 2 is true, then the covered leaves should not fall when the others do, since they are not receiving the same information. Suppose, however, that despite the covering of some of the leaves, all the leaves fall together. This eliminates hypothesis 2 as a possibility. Either of the other hypotheses, and many others, remain possibilities.

This simple experiment with ginkgos serves to point out the essence of scientific progress: science

flicted individual. As the virus spread through the Oregon compound, all of the other cheetahs, being genetically identical, were also vulnerable. Because all of the world's cheetahs appear to be genetically nearly identical, they are probably all at risk for this lethal virus. If the virus spreads through zoos or, worse, through the remaining natural populations, the species may be doomed.

The problem faced by cheetahs is a consequence of their genetic similarity. For many other rare animals, the dangers associated with genetic uniformity can still be avoided. Biologists that manage the breeding programs in the world's zoos are actively seeking breeding individuals from very different genetic backgrounds, in order to preserve in the zoo population as much of the natural genetic variability as possible. In large measure this program of outbreeding appears to be a success. In only a few cases, such as with Speke's gazelle of Somalia, has it proven impossible to obtain additional individuals from the wild. Because all the world's zoos contain Speke's gazelles descended from only a few individuals, they are all similar genetically and, like the cheetahs, are at risk of viral infection. In almost all other cases it has proven possible to avoid the danger. Maintaining the genetic diversity of our captive populations, of our agricultural breeds, and of the seed banks used

does not prove that certain explanations are true, but rather that some possibilities are not. Hypotheses that are not consistent with experimental results are rejected. Hypotheses that are not proven false by an experiment or series of experiments are provisionally accepted—but may be rejected in the future when more information becomes available, if they are not consistent with the new information. Just as a computer finds the correct path through a maze by trying and eliminating false paths, so a scientist gropes toward reality by eliminating false possibilities.

A hypothesis that stands the test of time, often tested and never rejected, is called a **theory.** In biology the hypothesis that tiny organelles, called mitochondria, within your cells are the descendants of bacteria is a theory. It is not certain that this idea is correct, but the overwhelming weight of evidence supports the hypothesis and most biologists accept it as "proven." There is no absolute truth in science, however, only varying degrees of uncertainty, and the possibility always remains that future evidence will cause a theory to be revised. A scientist's acceptance of a theory is always provisional.

Some theories are so strongly supported that the likelihood of their being rejected in the future is very small. The theory of evolution, for example, is so broadly supported by different lines of inquiry that

most biologists accept it with as much certainty as they do the theory of gravity. The theory of evolution provides the conceptual framework that unifies biology as a science.

As you proceed through this text, you will encounter a great deal of information, often coupled to explanations. These explanations are hypotheses that have stood the test of experiment. Many will continue to do so; others will be revised. Biology, like all healthy sciences, is in a constant state of flux, with new ideas bubbling up and replacing old ones.

in raising crops is among the most important tasks facing biologists. Our biological heritage, a portion of which we have taken from nature for our use, is a trust that we hold for future generations. Like a garden, this trust requires constant tending.

An Overcrowded World

One corollary of our rapid population growth has been the degradation of the environment. Many of the effects of this growth are evident to any observer: the industrial pollution of lakes and rivers, polluted air and smog, and spills and "dumps" of poisonous industrial wastes. These are obvious evils, but ones that are not easily combatted because of the many conflicting forces that act together to produce them. Other, more subtle dangers are perhaps even more threatening, however. Paramount among them is the rapid destruction of the natural resources of the tropics. The degradation occurring there now will forever alter the face of the earth and affect future generations.

A major component of this problem is overpopulation. When the authors of this text took freshman biology, the world's population was less than 3 billion. As you take fresh-

FIGURE 1-14

The world's population is growing rapidly. In the next 40 years it will almost double. The growth rate is considerably slower in the highly developed nations of the temperate zone than it is in the less developed nations of the third world. Compare the projected population increase of the United States with that of third world countries. A far greater percent of third world populations consists of children, who will in the coming years bear children themselves.

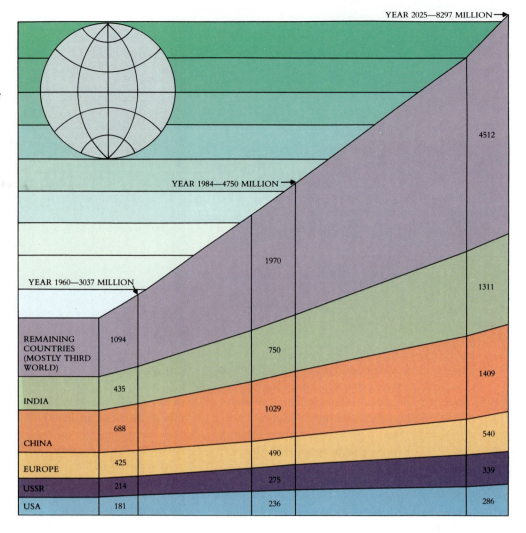

man biology, it stands at more than 5 billion. When your children take freshman biology, it will be in excess of 7 billion. In this short time span the population of the world will have more than doubled.

In the less developed, mainly tropical areas of the world, the rate of population growth is alarmingly high (Figure 1-14). The population of India, about 800 million—three times that of the United States—is estimated to double within the next 32 years. This projected increase means that India will have to produce twice as much corn, garner twice as much power, and build twice as many houses, just to keep its people in as good an economic situation as they are in now! And then the population will grow even more. With little to slow population growth, it is perhaps not surprising that the resources of the tropical areas of the world are strained—and beginning to fail under the strain.

About half of the world's people live in the tropics or subtropics, less than a quarter in developed countries, and the remainder mostly in China. By the year 2000, three out of five people in the world will live in tropical or subtropical regions. The result is ever-increasing pressure on the resources of these areas. The forests are being cut and the land devoted to short-term farming (Figure 1-15). This presents a serious problem, since tropical soils are not, for the most part, suitable for sustainable agricultural productivity. Tropical soils are characteristically poor in minerals, and what little minerals they contain are generally tied up in trees and other vegetation. When these trees are cut and burned to make way for crops, their minerals temporarily enrich the soil; the limited amounts of nutrients at any one site are soon exhausted, however, and the farmer must move on, after only a few years, having done permanent damage to the soils.

FIGURE 1-15

The slash-and-burn agriculture commonly practiced in the tropics has a devastating effect on the quality of the soil. Most of the inorganic ions and other nutrients of the forest are present in trees, and when these trees are burned like this, most nutrients are lost, carried away by rainwater. The larger the area cleared, the greater the runoff and damage. Few of these soils will recover in your lifetime.

FIGURE 1-16

Some of the most dramatic products of the Green Revolution have been varieties of rice that are more resistant to disease and that produce a higher proportion of grain per acre. Rice is the principal crop of the tropics, and the increase in productivity has had a significant impact on agricultural productivity there. Raising rice involves flooding fields where it is planted. Much of tropical Asia is hilly, and rice is grown in terraces cut into hillsides.

Even assuming that population growth can be brought under control, more than a third of the people living in the tropics and subtropics exist in a state of absolute poverty. About half of them are malnourished. If these people are ever to have any acceptable standard of living, modern science must find a way to farm these areas successfully. To do so will take a major investment in biological research. This problem is one of the greatest challenges facing biologists today. If the problem is not solved, a large and growing proportion of the world's population is going to be increasingly hungry, and the wealthy nations will certainly be increasingly affected by the consequences.

One promising avenue of biological research that may help to feed an increasingly hungry world is the creation of new crop plants. By collecting many different varieties of the principal crops of the tropics and then breeding them to select for desirable traits, biologists often have been able to improve their yields. For example, corn, wheat, and rice (Figure 1-16) now produce higher yields than were thought possible at the end of World War II—with the help of abundant fertilizer. The advent of these high-yield crops has been referred to as the **Green Revolution,** since they have significantly improved the ability of tropical countries to feed themselves.

Efforts continue in the laboratory as well as on the farm. Molecular biologists, using the techniques described in Chapter 18, have been learning to transfer genes from one plant to another. One way in which this has been possible has been to carry the genes piggyback on viruses. The hope is to assemble within a crop plant a battery of additional genes that will provide it with a desirable combination of new traits. Among these would be greater resistance to pests, increased yield, and a faster rate of growth.

One of the most difficult and challenging transfers that is being attempted is the incorporation into crop plants of the genes from clover and other legumes that are responsible for nitrogen fixation. Legume plants interact with soil bacteria to take nitrogen gas from the air and combine it with hydrogen so that nitrogen is available for incorporation into living organisms. Crop plants cannot do this and so must be fertilized with nitrogen-rich compounds. If this long-term research effort succeeds, the result would be a crop that would not need fertilization with nitrogen but instead would actually enrich the soil in which it grows. Few human endeavors could be expected to make as large a contribution to the future well-being of humankind.

YOUR STUDY OF BIOLOGY

In its broadest sense biology is the study of living things. Life, however, does not take the form of a uniform green slime covering the surface of the earth; rather, it consists of a diverse array of living forms. A biologist tries to understand the sources of this diversity and in many cases to harness particular life forms to perform useful tasks. Even the narrowest study of a seemingly unimportant life form is a study in biological diversity, one more brushstroke in the painting that biologists have labored over for centuries.

As a science, biology is devoted to understanding biological diversity and its consequences. From centuries of observation and inquiry (Figure 1-17), one organizing principle has emerged: biological diversity reflects *history,* a record of success, failure, and change extending back to a period soon after the formation of the earth. The weeding out of failures and the rewarding of success by increased reproduction is called natural selection, and the pattern of changes that result from this process is called evolution. The theory of evolution will form the backbone of your study of biological science, just as the theory of the covalent bond is the backbone of the study of chemistry or the theory of quantum mechanics that of physics. It is a thread that runs through everything that you will learn in this text. Evolution is the essence of the science of biology.

FIGURE 1-17

In the 3.7 million years between when these two sets of footprints were made, all of human evolution has occurred. Life has existed on earth for 3.5 billion years.

LAETOLI   MOON

Overview

Organisms are chemical machines; to understand them, we must start by considering chemistry. Organisms are composed of chemicals called molecules, which are collections of smaller units, called atoms, bound to one another. All of the atoms now in the universe are thought to have been formed long ago, as the universe itself was formed. Every carbon atom in your body was created in a star. The forces that hold atoms together in molecules are called chemical bonds; such bonds often are created by two or more atoms sharing the electrons that orbit around them. Sometimes molecules reshuffle their atoms or trade atoms with one another. These processes are called chemical reactions. The molecule that is most important to the evolution of life is water. The electrons that orbit water molecules are distributed asymmetrically. This asymmetry leads directly to many of the properties that are characteristic of the chemistry of living organisms.

In a small hut in the Pine Barrens of New Jersey, in the predawn darkness of a summer morning in 1962, two research scientists from Bell Laboratories first listened to the beginnings of creation. These scientists were helping to design receiving stations for the early communications satellites, their job being to find ways to reduce background interference caused by cities and other sources of radio waves. To do this, they designed a special horn-shaped antenna that allowed them to pinpoint most sources of interference. However, they were unable to identify the source of one low-strength background emission at a frequency of 3 Hertz (Hz). Wherever in the sky they pointed their antenna, the same 3 Hz signal arrived at the same low strength. The scientists dismantled and rebuilt their antenna, to no avail. The signal that they were picking up, although it was very weak, was real. What they were hearing was the residual thermal energy left over from an enormous explosion that is thought to have marked the beginning of the universe about 20 billion years ago.

With this explosion began the process of evolution, which ultimately led to the origin and diversification of life on earth. When viewed from the perspective of 20 billion years, life within our solar system is a recent development. To understand its origins, we need to consider events that took place much earlier. The same processes that led to the evolution of life were responsible for the evolution of molecules. Thus our study of life on earth begins at a seemingly odd place, with physics and chemistry. As chemical machines ourselves, we must understand chemistry to begin to understand our origins.

ATOMS

Nothing survives of any universe that might have existed before the "big bang." The explosion that we think created our universe would have been so violent that even the atoms themselves would have been torn apart, all matter reduced to subatomic particles shooting through space at velocities so fast that the collisions between them destroyed even these elementary particles. Matter as we know it is thought to have come into being as the universe cooled.

Over 100 distinct kinds of subatomic particles are known. Most associations between them are fleeting, the duration of their lives measured in millionths of a second. Three subatomic associations, however, are stable and are responsible for the three subatomic particles: **protons, neutrons,** and **electrons.** The nature and duration of the associations between these particular particles, associations called **atoms,** ultimately lead to the properties of the universe that we experience.

Atoms are very small, and it is no simple task to determine their structure. It was only early in this century that experiments were carried out suggesting the first vague outlines of atomic structure. By 1913 Niels Bohr was able to propose that a cloud of electrons orbits around every atom, like a miniature solar system. The center of each atom is a small, very dense **nucleus,** formed of the other two stable subatomic particles, protons and neutrons.

Within the nucleus, the cluster of protons and neutrons is held together by a kind of force that works only over very short subatomic distances. Each proton carries a positive (+) charge. The number of these charged protons **(atomic number)** determines the chemical character of the atom, because it dictates the number of electrons orbiting the nucleus and available for chemical activity. Neutrons are similar to protons in mass, but, as their name implies, they possess no charge. The **atomic mass** of an atom is equal to the sum of the masses of its protons and neutrons. Atoms that occur naturally on earth contain from 1 to 92 protons and up to 146 neutrons (Figure 2-1). Atoms that have the same number of protons but different numbers of neutrons are called **isotopes.** Isotopes of an atom differ in atomic mass but have similar chemical properties.

The positive charges in the nucleus of an atom are counterbalanced by negatively (−) charged electrons orbiting around the atomic nucleus at varying distances. The negative charge of one electron exactly balances the positive charge of one proton. Thus atoms with the same number of protons and electrons have no net charge. An atom in which the number of protons in the nucleus is the same as the number of orbiting electrons is known as a neutral atom.

Electrons have very little mass (only 1/1840 of the mass of a proton). Of all the mass contributing to your weight, the portion that is contributed by electrons is less than the mass of your eyelashes. Electrons are maintained in their orbits by their attraction to the positively charged nucleus. Sometimes this attraction is overcome by other forces, and one or more electrons are lost by an atom. Sometimes, too, atoms may gain additional elec-

FIGURE 2-1

A *The smallest atom is hydrogen (atomic mass, 1), whose nucleus consists of a single proton.*
B *Hydrogen also has two naturally occurring isotopic forms, which possess neutrons as well as the single proton in the nucleus: deuterium (one neutron) and tritium (two neutrons).*
C *A nucleus that contains two protons is not an isotope of hydrogen, but rather the next element in the periodic table, helium. The largest naturally occurring atom is uranium (atomic mass, 238), whose nucleus contains 92 protons and 146 neutrons.*

A NUCLEUS OF HYDROGEN

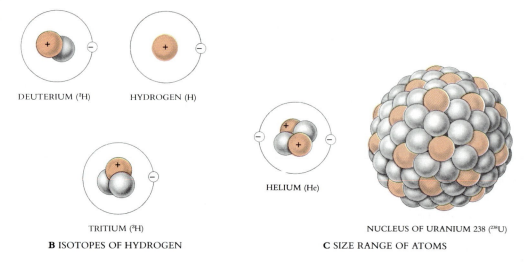

DEUTERIUM (^{2}H) HYDROGEN (H)

TRITIUM (^{3}H)
B ISOTOPES OF HYDROGEN

HELIUM (He)

NUCLEUS OF URANIUM 238 (^{238}U)
C SIZE RANGE OF ATOMS

trons. Atoms in which the number of electrons does not equal the number of protons are known as **ions,** which do carry an electrical charge. For example, an atom of sodium (Na) that has lost an electron becomes a positively charged sodium ion (Na^+).

> **An atom is a core nucleus of protons and neutrons surrounded by a cloud of electrons. The number of its electrons largely determines the chemical properties of an atom.**

Energy in the Atom: Electron Orbitals

The key to the chemical behavior of atoms lies in the arrangement of the electrons in their orbits. It is convenient to visualize individual electrons as following discrete circular orbits around a central nucleus, as in the Bohr model of the atom. Such a simple picture is not realistic, however; it is not possible to locate the position of any individual electron precisely at any given time. Theory, in fact, tells us that a particular electron can be anywhere at a given instant, from close to the nucleus to infinitely far away from it.

However, a particular electron is not equally likely to be located at all positions. Some locations are much more probable than others. In other words, it is possible to say where an electron is *most likely* to be. The volume of space around a nucleus where an electron is most likely to be found is called the **orbital** of that electron (Figure 2-2). Some electron orbitals near the nucleus are spherical orbitals (*s* orbitals), whereas others are dumbbell-shaped orbitals (*p* orbitals) (Figure 2-3). Still other orbitals, more distant from the nucleus, may have different shapes.

Because of their high orbiting velocities, electrons possess enormous energy. Not all electrons, however, possess the same energy. Although all electrons within a given orbital possess the same energy, different orbitals often possess different energies. The electrons in the outer orbitals, for example, move faster than those in the orbitals nearer the nucleus. Consequently, these outer electrons possess more energy than the inner ones.

It is convenient to think of the arrangement of electron orbitals around the nucleus in terms of their **energy levels.** These energy levels can be depicted simply as concentric, two-dimensional rings. These simple circular schematics are far easier to conceptualize than are three-dimensional orbitals. It is important to keep in mind as we use these energy-level schematics, however, that actual electron orbitals often have complex three-dimensional shapes. In an energy-level schematic of an atom (Figure 2-4) the nucleus is represented as a small circle, with the number of protons and neutrons indicated. The electron energy lev-

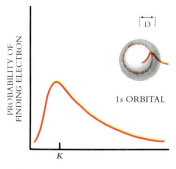

PROBABILITY OF FINDING ELECTRON

1s ORBITAL

DISTANCE (D) FROM NUCLEUS →

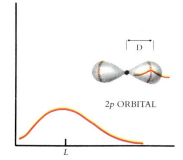

2p ORBITAL

DISTANCE (D) FROM NUCLEUS →

FIGURE 2-2

The simplest electron orbital is an s *orbital, which has a spherical shape. In an* s *orbital the probability of finding an electron orbiting the nucleus is greatest at a fixed distance from the nucleus—a sphere.*
A somewhat more complex orbital is the p *orbital, which has a dumbbell shape. In a* p *orbital the electrons are most likely to be found in two donut-shaped rings on opposite sides of the nucleus.*

1s ORBITAL

FIRST ELECTRON SHELL—ENERGY LEVEL *K*

2s ORBITAL

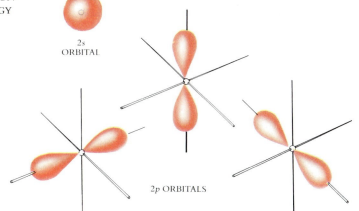

2p ORBITALS

ORBITALS FOR ENERGY LEVEL *L*: ONE SPHERICAL ORBITAL (2s) AND THREE DUMBBELL-SHAPED ORBITALS (2p)

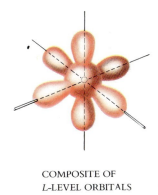

COMPOSITE OF L-LEVEL ORBITALS

FIGURE 2-3

Electron orbitals. The lowest energy level, nearest the nucleus, is level K. *It is occupied by a single s orbital, referred to as 1s. The next highest energy level,* L, *is occupied by four orbitals, one s orbital (referred to as the 2s orbital) and three p orbitals (each referred to as a 2p orbital). Viewed simultaneously, the four L level orbitals compactly fill the space around the nucleus, like two pyramids set base to base.*

FIGURE 2-4

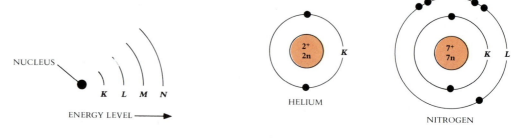

Electron energy levels for some familiar
elements. The two electrons of helium,
indicated as black dots, fill the K energy
level. Nitrogen has seven electrons, two
in the K energy level occupying 1s
orbitals and the rest in L level orbitals.
All five of these L energy level
electrons are represented as dots on the
same circle, since they have the same
energy; when two electrons occupy the
same orbital, they are grouped on the
circle together as a pair. The K energy
level has only one orbital and so may
posse:s no more than two electrons;
the outermost L and M energy levels
possess four orbitals and so may possess
a maximum of eight electrons.

ELEMENT	ATOMIC NUMBER	NUMBER OF ELECTRONS IN EACH ENERGY LEVEL		
		FIRST (K)	SECOND (L)	THIRD (M)
Hydrogen (H)	1	1	—	—
Helium (He)	2	2	—	—
Carbon (C)	6	2	4	—
Nitrogen (N)	7	2	5	—
Oxygen (O)	8	2	6	—
Neon (Ne)	10	2	8	—
Sodium (Na)	11	2	8	1
Phosphorus (P)	15	2	8	5
Sulfur (S)	16	2	8	6
Chlorine (Cl)	17	2	8	7

els are depicted as concentric rings. In helium, for example, there is a single s orbital containing two electrons. Since these two electrons are in the same orbital, they have the same energy level. They are thus located on the same ring in the energy-level schematic.

All electrons possessing the same energy are placed on the same ring in the energy-level schematic. Although no one orbital can contain more than two electrons, an atom can possess several different orbitals that are equal distance from the nucleus. All of the electrons in these orbitals will have the same energy. In terms of the energy-level schematic all of these electrons are located on the same ring. By convention, two electrons from the same orbital are placed together as an electron pair.

The Periodic Table

There are 92 naturally occurring kinds of atoms, called **elements.** Each has a different number of protons and a different arrangement of electrons. When the nineteenth-century Russian chemist Dmitri Mendeleev arranged the known elements in a table according to their atomic number (Figure 2-5), he discovered one of the great generalizations of all science. Mendeleev found that the entries in the table exhibited a repeating pattern of chemical properties, in recurring groups of eight elements.

The eight-element periodicity that Mendeleev found is based on the interactions that occur between the electrons of the outer energy levels of the different elements. These interactions, in turn, are the basis for the differing chemical properties of the elements. The maximum number of electrons that an outer energy level can possess is eight; the chemical behavior that we observe reflects how many of the eight positions are filled. Elements possessing all eight electrons at their outer energy level are not reactive (neon, argon, xenon, krypton). In sharp contrast, elements with one fewer than the maximum number of eight electrons at their outer energy level, such as chlorine, bromine, and fluorine, are highly reactive. Also very reactive are those elements with only one electron in their outer energy level, such as sodium, lithium, or potassium.

All atoms tend to fill their outer energy levels with the maximum number of electrons.

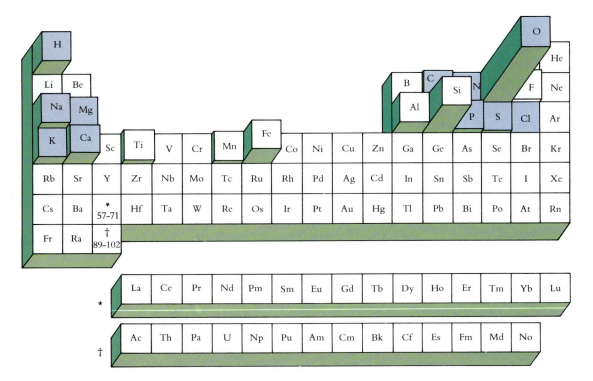

Mendeleev's **periodic table** thus leads to a very useful generalization, the **octet rule** or **rule of eight** (Greek, *octo,* eight): atoms tend to establish completely full outer energy levels. Most chemical behavior can be predicted quite accurately on the basis of this simple rule and from the tendency of atoms to balance positive and negative charges.

MOLECULES

Much of the core of the earth is thought to consist of iron, nickel, and other heavy elements. The crust of the earth is quite different in composition from that of the bulk of the planet (Table 2-1), being composed primarily of the lighter elements. By weight, 74.3% of the earth's crust (its land, oceans, and atmosphere) consists of oxygen or silicon. Most of these atoms are combined in stable associations called **molecules.**

The combining activity of atoms in forming molecules reflects the chemical forces mentioned earlier: (1) the tendency of electrons to occur in pairs, (2) the tendency of atoms to balance positive and negative charges, and (3) the tendency of outer energy levels to satisfy the octet rule. Only those elements which possess equal numbers of protons and electrons (and hence have no net charge), no unpaired electrons, and full outer electron energy levels can exist as free atoms. This is so because only elements that satisfy these requirements do not experience chemical forces that lead them to combine. Atoms with this set of characteristics are called **noble elements,** since they do not associate with other elements. The noble elements are all gases, and most are rare in the earth's crust. All other elements are usually found on earth in molecular combinations that satisfy the three fundamental chemical tendencies just enumerated.

NATURE OF THE CHEMICAL BOND

An atom that does not carry an electrical charge and that is not a noble element can satisfy the octet rule in one of three ways:

1. It can gain one or two electrons from another atom.
2. It can lose one or two electrons to another atom.
3. It can share one or more electron pairs with another atom.

TABLE 2-1   THE MOST COMMON ELEMENTS ON EARTH AND THEIR DISTRIBUTION
IN THE HUMAN BODY

ELEMENT	SYMBOL	ATOMIC NUMBER	APPROXIMATE PERCENT OF EARTH'S CRUST BY WEIGHT	PERCENT OF HUMAN BODY BY WEIGHT
Oxygen	O	8	46.6	65.0
Silicon	Si	14	27.7	Trace
Aluminum	Al	13	6.5	Trace
Iron	Fe	26	5.0	Trace
Calcium	Ca	20	3.6	1.5
Sodium	Na	11	2.8	0.2
Potassium	K	19	2.6	0.4
Magnesium	Mg	12	2.1	0.1
Hydrogen	H	1	0.14	9.5
Manganese	Mn	25	0.1	Trace
Fluorine	F	9	0.07	Trace
Phosphorus	P	15	0.07	1.0
Carbon	C	6	0.03	18.5
Sulfur	S	16	0.03	0.3
Chlorine	Cl	17	0.01	0.2
Vanadium	V	23	0.01	Trace
Chromium	Cr	24	0.01	Trace
Copper	Cu	29	0.01	Trace
Nitrogen	N	7	Trace	3.3
Boron	B	5	Trace	Trace
Cobalt	Co	27	Trace	Trace
Zinc	Zn	30	Trace	Trace
Selenium	Se	34	Trace	Trace
Molybdenum	Mo	42	Trace	Trace
Tin	Sn	50	Trace	Trace
Iodine	I	53	Trace	Trace

These changes in electron distribution act to hold atoms together. When one atom loses an electron from its outer energy level, it becomes a positive ion. The atom that gains the electron becomes a negative ion. The two atoms are then held together by attraction of opposite charges, forming an **ionic bond** between them. An ionic bond is one kind of chemical bond by which atoms are held together in molecules. A different kind of chemical bond results when two atoms share one or more electron pairs, an attraction called a **covalent bond.**

Ionic and covalent bonds are the two principal kinds of chemical bonds that unite atoms into molecules. Other weaker chemical bonds will be discussed later in this chapter; weak chemical bonds form looser associations between molecules.

A chemical bond is a force holding two atoms together. The force can result from the attraction of opposite charges, as in an ionic bond, or from the sharing of one or more pairs of electrons, as in a covalent bond. Other weaker kinds of bonds also occur.

Ionic Bonds

Common table salt, sodium chloride (NaCl), is a lattice of ions in which atoms are held together by ionic bonds. Sodium (Na) has 11 electrons. Two of these are at the inner energy level (Figure 2-6, level K), eight are at the next energy level (level L), and one is at the outer energy level (level M). Because of this distribution of electrons in the sodium atom, two

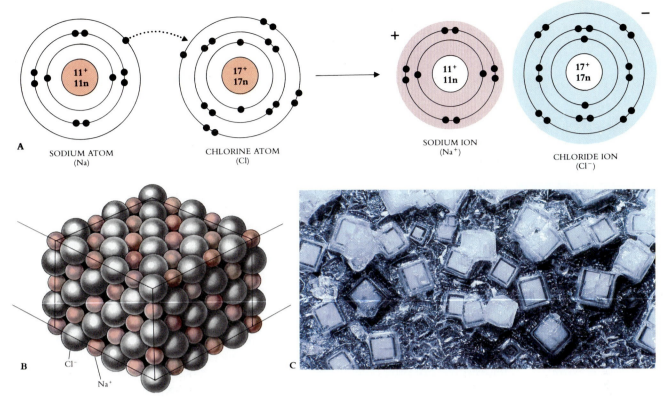

A   SODIUM ATOM
(Na)

CHLORINE ATOM
(Cl)

SODIUM ION
(Na⁺)

CHLORIDE ION
(Cl⁻)

B   Cl⁻

Na⁺

C

powerful chemical tendencies result: (1) the outer electron is unpaired ("free") and will seek to form a pair, and (2) the octet rule is not satisfied. A stable configuration is achieved, however, if the outer electron is lost. The loss of this electron results in the formation of a sodium ion (Na^+). This ion satisfies two of the three central chemical tendencies because it has a full outer octet and no free electrons. An Na^+ ion possesses a positive charge, however, and in this respect it violates the tendency toward electrical neutrality.

Here is where the chloride ion (Cl^-) comes in. The chlorine atom has 17 electrons, 2 at the inner K energy level, 8 at the next L energy level, and 7 at the outer M energy level. Just as in the sodium atom, there are two chemical tendencies that result from this electron distribution. These tendencies arise because (1) the outer energy level of the chlorine atom contains an unpaired electron, and (2) the octet rule is not satisfied. The addition of an electron to the outer energy level, however, satisfies both of these requirements and causes the formation of a negatively charged chloride ion.

When placed together, metallic sodium and gaseous chlorine react swiftly and explosively, the sodium atoms donating electrons to chlorine atoms. The result of this **chemical reaction** between the two kinds of atoms is the production of Na^+ and Cl^- ions. Because opposite charges attract, electrical neutrality (the third chemical tendency) can be achieved by the association of these ions with each other. The chemical bond between Na^+ and Cl^- ions resulting from this electrical attraction is not directional, however, and discrete sodium chloride molecules do not form. Instead, the ions aggregate, or come together, into a crystal matrix, which has a precise geometry. Such aggregations are what we know as crystals of salt. If a salt such as NaCl is placed in water, the electrical attraction of the water molecules, for reasons we will point out later in this chapter, disrupts the forces holding the ions in their crystal matrix, causing the salt to dissolve into a roughly equal mixture of free Na^+ and free Cl^- ions. Approximately 0.06% of the atoms in your body are free Na^+ or Cl^- ions—about the weight of your little finger.

An ionic bond is an attraction between ions of opposite charge. Such bonds are not directed between two specific ions but rather between a given ion and all of the oppositely charged ions in its immediate vicinity.

FIGURE 2-6

A *The formation of ionic bonds by sodium chloride. When a sodium atom donates an electron to a chloride atom, the sodium atom, lacking that electron, becomes a positively charged sodium ion; the chloride atom, having gained an extra electron, becomes a negatively charged chloride ion. These positive and negative ions cluster so that each charge is surrounded by ions of opposite charge.*
B *When water evaporates from solutions containing high concentrations of sodium chloride, a highly regular lattice of alternating Na^+ and Cl^- ions forms—a crystal. You are familiar with these crystals as table salt.*
C *Viewed as they form, such sodium chloride crystals can be very beautiful.*

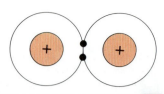

FIGURE 2-7

Hydrogen gas is a diatomic molecule composed of two hydrogen atoms each sharing its electron with the other—a molecular commune. Even more stable molecules are possible when the two hydrogen atoms share their electrons instead with an oxygen atom, forming water. The flash of fire that consumed the Hindenburg occurred when hydrogen gas that was used to inflate the balloon combined explosively with oxygen gas in the air to form water.

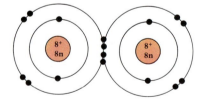

FIGURE 2-8

Oxygen gas is a diatomic molecule composed of two oxygen atoms each sharing two electrons with the other. Each of the two atoms thus obtains the two additional electrons it needs to satisfy the octet rule. The atmosphere of the early earth contained little oxygen gas; all that we now breathe has been formed by photosynthesis.

True molecules do not result from ionic bonds alone, however, for two reasons. First, the electrical attractive force is weak. Second, it is not directed between two specific ions with opposite charges; rather, it exists between any one ion and all neighboring ions of the opposite charge. True molecules are formed by much stronger bonds called covalent bonds, which involve far stronger attractive forces directed between specific atoms.

Covalent Bonds

Covalent bonds form when two atoms share one or more pairs of the electrons of their outer energy levels. Consider hydrogen (H) as an example. Each hydrogen atom has an unpaired electron and an unfilled *K* energy level; for these reasons the hydrogen atom is chemically unstable. When two hydrogen atoms are close enough to one another, however, each electron can orbit both nuclei. In effect, nuclei in proximity are able to share their electrons (Figure 2-7). The result is a diatomic molecule (one with two atoms) of hydrogen gas (H_2).

The diatomic hydrogen gas molecule that is formed as a result of this sharing of electrons is not charged, however, because it still contains two protons and two electrons. Each of the two hydrogen atoms can be considered to have two orbiting electrons at the *K* energy level. This relationship satisfies the octet rule because each shared *K* energy level electron orbits both nuclei and therefore is included in the outer energy level of *both* atoms. The relationship also results in the pairing of the two free electrons. For these reasons the two hydrogen atoms form a stable molecule. Note, however, that this stability is conferred by the electrons that orbit both nuclei; stability of this kind occurs only when the nuclei are very close together. For this reason the strong chemical forces tending to pair electrons and satisfy the octet rule will act to keep the two nuclei near one another. The bond between the two hydrogen atoms of diatomic hydrogen gas is an example of a covalent bond.

A covalent bond is a chemical bond formed by the sharing of one or more pairs of electrons. Covalent bonds are stronger than ionic bonds.

Covalent bonds can be very strong. The strength of the bond depends on the degree to which breaking the bond violates the three central tendencies toward electrical neutrality, paired electrons, and full outer energy levels. Thus covalent bonds that satisfy the octet rule by sharing *two* pairs of electrons, called **double bonds,** are stronger than covalent bonds that involve the sharing of only one electron pair, called **single bonds.** Covalent bonds are represented in chemical formulations as lines connecting atomic symbols. Each line between two bonded atoms represents the sharing of one pair of electrons. Hydrogen gas is thus symbolized H—H, and oxygen gas, O=O.

A double bond: oxygen gas

Oxygen gas (Figure 2-8) provides an example of the formation of a covalent double bond. An oxygen atom (atomic number, 8) contains eight protons and eight electrons. Two electrons are at the *K* energy level, and six electrons are at the *L* energy level. Thus, to satisfy the octet rule, an oxygen atom must gain two electrons. One way it can do this is by sharing two of the electrons of another oxygen atom, in which case it forms diatomic oxygen gas (O_2). In a diatomic oxygen gas molecule four of the electrons (two contributed by each atom) orbit both nuclei. This distribution of electrons results in the outer energy level of each of the oxygen atoms containing a full complement of eight electrons, thus satisfying the octet rule. A molecule such as O_2, in which two electron pairs are shared, forms for the following reasons: (1) the number of electrons (16) is the same as the number of protons in the two nuclei, and there is accordingly no net electrical charge; (2) there are no unpaired electrons; and, most important, (3) atoms lacking two electrons in their outer energy levels are now able to satisfy the octet rule. More chemical energy is required to disrupt a bond of this kind than a single bond.

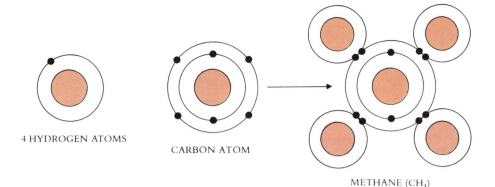

4 HYDROGEN ATOMS

CARBON ATOM

METHANE (CH₄)

FIGURE 2-9

Left *Methane is a gas composed of a carbon atom sharing a single electron with each of four hydrogen atoms. The hydrogen atoms are distributed symmetrically around the carbon atom, forming a pyramid-shaped tetrahedron in which each of the four hydrogen atoms is equally distant from the others. Because of this symmetry, methane is not a polar molecule; although each of the shared electrons is closer to the carbon atom than to the hydrogen, the resulting charge displacements cancel each other out.*

Below *Different methods of representing a methane molecule.*

Molecules with several covalent bonds

Molecules are often composed of more than two atoms. One reason that larger molecules may be formed is that a given atom is able to share electrons with more than one other atom. An atom that requires two, three, or four additional electrons to fill its outer energy level completely may acquire them by sharing its electrons with two or more other atoms.

For example, carbon (C) atoms (atomic number, 6) contain six electrons, two at the *K* energy level and the other four at the *L* energy level. To satisfy the octet rule, a carbon atom must gain access to four additional electrons; it must form the equivalent of four covalent bonds. Because there are many ways in which four covalent bonds may form, carbon atoms are able to participate in many different kinds of molecules.

One of the simplest arrangements in which carbon atoms participate involves four single covalent bonds with hydrogen atoms (Figure 2-9). This molecule is **methane** (CH_4), or marsh gas. Methane is a major component of natural gas, resulting from the breakdown of more complex carbon compounds. The structural formula for methane is:

$$
\begin{array}{c}
\text{H} \\
| \\
\text{H}-\text{C}-\text{H} \\
| \\
\text{H}
\end{array}
$$

Methane is said to be a highly "reduced" molecule. In general, **reduction** is the addition of electrons to a molecule, often accompanied by the gain of a hydrogen nucleus (a proton); the converse of reduction is **oxidation,** the removal of electrons from a molecule, often accompanied by the loss of a proton. A molecule that is deficient in electrons tends to gain electrons. Such an electron-poor molecule is said to be oxidized before it gains electrons; once it has gained the electrons, it is reduced. The carbon atom of methane, which has gained shared electrons from four hydrogen atoms, is thus fully reduced. In methane each of four hydrogen atoms forms a single covalent bond with the carbon atom. The single *K* energy level electron of the hydrogen atom orbits both the hydrogen nucleus and the carbon nucleus. Similarly, one of the four *L* energy level electrons of the carbon atom orbits both nuclei. All four *L* energy level electrons of the carbon atom are paired in this way with hydrogen atom electrons. As a result, eight *L* energy level electrons orbit the carbon nucleus, and two such electrons orbit each hydrogen nucleus. The octet rule is thus satisfied for all five atoms of the methane molecule, while at the same time the free electrons of all four hydrogen atoms are paired.

> **Oxidation is the loss of an electron by an atom, ion, or molecule, often accompanied by the loss of a hydrogen nucleus (a proton). The electron lost by one molecule is transferred with its proton to another, a process called reduction.**

MOLECULAR MODEL

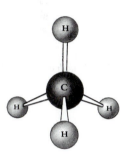

BALL AND STICK MODEL

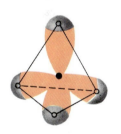

ELECTRON CLOUD MODEL

TABLE 2-2   THE MASS AND CHEMICAL CHARACTER OF THE LIGHTEST ELEMENTS

ELEMENT	SYMBOL	ATOMIC NUMBER	NUMBER OF ELECTRONS REQUIRED TO SATISFY OCTET RULE	APPROXIMATE PERCENT OF HUMAN BODY BY NUMBER OF ATOMS
Hydrogen	H	1	$+1\,K$	63.0
Helium	He	2	0 (inert)	—
Lithium	Li	3	$-1\,L$	—
Beryllium	Be	4	$-2\,L$	—
Boron	B	5	$-3\,L$	—
Carbon	C	6	$+4\,L$	9.5
Nitrogen	N	7	$+3\,L$	1.4
Oxygen	O	8	$+2\,L$	25.5
Fluorine	F	9	$+1\,L$	—
Neon	Ne	10	0 (inert)	—
Sodium	Na	11	$-1\,M$	0.03
Magnesium	Mg	12	$-2\,M$	0.01
Aluminum	Al	13	$-3\,M$	—
Silicon	Si	14	$+4\,M$	—
Phosphorus	P	15	$+3\,M$	0.22
Sulfur	S	16	$+2\,M$	0.05
Chlorine	Cl	17	$+1\,M$	0.03
Argon	Ar	18	0 (inert)	—
Potassium	K	19	$-1\,M$	0.06
Calcium	Ca	20	$-2\,M$	0.31
Scandium	Sc	21	$-3\,M$	—
Titanium	Ti	22	$+4\,M$	—
Vanadium	V	23	$+3\,M$	—

The above elements include all those which commonly occur in organisms. The elements that make up the bodies of organisms have a low atomic number and thus a low mass relative to the rest of the elements that are present on the earth's surface. They are also remarkable in that 99.7% of them require an *addition* of electrons to satisfy the octet rule and thus participate in covalent bonds.

ATOMS IN LIVING ORGANISMS

Of the 92 elements that formed the crust of the cooling earth, only 11 are common in living organisms. Table 2-2 lists the frequency with which various elements occur in the human body; the frequencies that occur in the bodies of other organisms are similar. Inspection of this table reveals that the distribution of elements in living systems is by no means accidental. As was evident in Table 2-1, the life elements—those which make up 0.01% (1 in 10,000) or more of the atoms of organisms—are not the elements that are most abundant in the earth's crust. Unlike the elements that occur most abundantly in this crust, all of the elements common in living organisms are light (Table 2-2), each having an atomic number less than 21 and thus a low mass.

The great majority of atoms in living things—99.4% of the atoms in the human body, for example—are either nitrogen, oxygen, carbon, or hydrogen. You can conveniently remember these elements by their first initials, NOCH. Why are these four elements the most abundant?

1. First, they all form gases, either alone or in combination with one another. Life is thought to have evolved from complex molecules formed from the interaction of these gases in the primitive earth's atmosphere. Many of these molecules are water soluble. As water vapor in the atmosphere cooled and fell as rain, it brought these dissolved molecules to the primitive oceans where life began.

2. The NOCH elements all require the *addition* of from one to four K or L electrons to satisfy the octet rule. This is evident in Table 2-2: the number of electrons

required to satisfy the octet rule is always positive. In different terms the NOCH atoms form molecules with one another by sharing electrons and so forming covalent bonds.

3. Molecules formed by the NOCH elements are characterized by instability, forming a variety of different chemical bonds that are not so strong that they cannot be broken.

4. Approximately 88.5% of the elements common in living organisms are hydrogen and oxygen, reflecting the predominant role of water (H_2O) in living systems.

THE CRADLE OF LIFE: WATER

On the primitive earth many of the atomic elements formed complex molecules with silicon and oxygen. These molecules bound to one another in strongly linked crystalline arrays, forming minerals, which in turn combined and formed rocks. In contrast to the others, the molecules of one compound that was released from these rocks as the primitive earth cooled did not form crystals: **water.** An oxide of hydrogen, water has the chemical formula H_2O. This seemingly simple molecule has many surprising properties (Figure 2-10). For example, of all the molecules that are common on earth, only water exists as a liquid at the relatively cool temperatures that prevail on the earth's surface. When life was originating, water provided a medium in which other molecules could move around and interact without being bound by strong covalent or ionic bonds. Life evolved as a result of these interactions.

FIGURE 2-10

Water takes many forms. As a liquid, it fills our rivers and runs down over the land to the sea, sometimes falling in great cascades, such as at Victoria Falls in Zimbabwe, Central Africa. The iceberg on which the penguins are holding their meeting was formed in Antarctica from huge blocks of ice breaking away into the ocean water. When water cools below 0° C, it forms beautiful crystals, familiar to us as snow and ice. However, water is not always plentiful. At Badwater, Death Valley, there is no hint of water except for the broken patterns of dried mud.

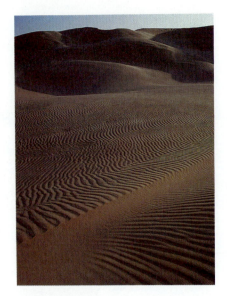

Life as it evolved on earth is thus inextricably tied to water. Three fourths of the earth's surface is covered by liquid water. You are composed of about two-thirds water, and you cannot exist long without it. All other organisms also require water. It is no accident that tropical rain forests are bursting with life, whereas dry deserts (Figure 2-11) are almost lifeless except when water becomes temporarily plentiful, such as after a rainstorm. Farming is possible only in those areas of the earth where rain is plentiful. No plant or animal can grow and reproduce in any but a water-rich environment.

The chemistry of life, then, is water chemistry. The way in which life evolved was determined in large part by the chemical properties of the liquid water in which that evolution occurred. The single most outstanding chemical property of water is its ability to form weak chemical associations with only 5% to 10% of the strength of covalent bonds. This one property of water, which as you will see derives directly from its structure, is responsible for much of the organization of living chemistry.

Water has a simple atomic structure. Just as a methane molecule results from the formation of four single covalent bonds between a carbon atom and four hydrogen atoms, thus satisfying the four vacancies in the outer energy level of the carbon atom, so water results from the formation of two single covalent bonds between an oxygen atom and two hydrogen atoms (Figure 2-12). The resulting molecule is stable: it satisfies the octet rule, has no unpaired electrons, and does not carry a net electrical charge.

The remarkable story of the properties of water does not end there, however. The oxygen atom has more protons than do the hydrogen atoms so that the electron-attracting power of oxygen (electronegativity) is much greater than that of hydrogen. As a result, the electron pair shared in each of the two single oxygen-hydrogen covalent bonds of a water molecule is more strongly attracted to the oxygen nucleus than to the hydrogen nuclei. Although electron orbitals encompass both nuclei, the negatively charged electrons are far more likely, at a given moment, to be found near the oxygen nucleus than near one of the hydrogen nuclei. This relationship has a profoundly important result: the oxygen atom acquires a partial negative charge. It is as if the electron cloud were denser in the neighborhood of the oxygen atom and less dense around the hydrogen atoms. This **charge separation** within the water molecule creates negative and positive electrical charges on the ends of the molecule. These partial charges are much less than the unit charges of ions.

What would the shape of such a molecule be? Consider water's two covalent bonds. Each will have a partial charge at each end, because of the charge separation resulting from the unequal attraction of electrons by hydrogen and oxygen atoms. Since like charges repel, the most stable arrangement of these charges is one in which the two negative and two positive charges are equidistant from one another. Such an arrangement is called a **tetrahedron.** In water the oxygen molecule occupies the center of such a tetrahedron, with hydrogen atoms at two of the apexes and the partial negative charges at the other two apexes. This arrangement results in a **bond angle** between the two covalent oxygen-hydrogen bonds of 104.5 degrees.

FIGURE 2-12

Water has a simple molecular structure. Each molecule is composed of one oxygen atom and two hydrogen atoms. The oxygen atom shares one electron with each hydrogen atom.

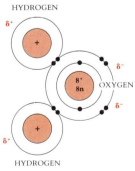

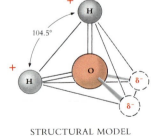

HYDROGEN

OXYGEN

HYDROGEN

BOHR MODEL

104.5°

STRUCTURAL MODEL

MOLECULAR MODEL

FIGURE 2-13

Some insects, such as the water strider, literally walk on water. In this photograph you can see the dimpling its feet make on the water as its weight bears down on the surface. Because the surface tension of the water is greater than the force that one foot brings to bear, the strider does not sink, but rather glides along.

The asymmetrical water molecule has distinct "ends," each with a partial charge, like the two poles of a magnet. Molecules, such as water, that exhibit charge separation are termed **polar** molecules because of these magnetlike poles. Water is one of the most polar molecules known. The polarity of water underlies its chemistry and thus the chemistry of life.

> **Much of the biologically important behavior of water results because the oxygen atom attracts electrons more strongly than do the hydrogen atoms and the water molecule has electron-rich (−) and electron-poor (+) regions, giving it magnetlike positive and negative poles.**

Polar molecules interact with one another. The partial negative charge at one end of a polar molecule is attracted to the partial positive charge of another polar molecule. This weak attraction is called a **hydrogen bond.** Water forms a lattice of such hydrogen bonds. Each of these hydrogen bonds is individually very weak and transient. A given bond lasts only $\frac{1}{100,000,000,000}$ of a second. However, even though each bond is transient, a very large number of such bonds can form, and the cumulative effects of very large numbers of these bonds can be enormous. The cumulative effects of very large numbers of hydrogen bonds are responsible for many of the important physical properties of water.

Water Clings to Itself

Water molecules, being very polar themselves, are attracted to other polar molecules. When the other polar molecule is another water molecule, the attraction is referred to as **cohesion.** When the other polar molecule is a different substance, the attraction is called **adhesion.** It is because water is cohesive, forming a lattice of hydrogen bonds with itself, that it is a liquid, and not a gas, at moderate temperatures.

The cohesion of liquid water is responsible for its **surface tension.** Insects, weighing little, can walk on water (Figure 2-13), even though they are denser than the water, because at the air-water interface all the hydrogen bonds face inward, causing the molecules of the water surface to cling together. Water is adhesive to any substance with which it can form hydrogen bonds. This is why things get "wet" when they are dipped in water—and why waxy substances do not (they are composed of nonpolar molecules).

The adhesion of water to substances with surface electrical charges is responsible for **capillary action:** if a glass tube with a narrow diameter is lowered into a beaker of water, the water will rise up in the tube above the level of the water in the beaker, held by adhesion of the water to the glass (Figure 2-14). The narrower the tube, the more the adhesive forces exceed gravity's pull on the smaller volume of water and the higher the water rises. Cap-

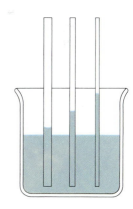

FIGURE 2-14

Capillary action will cause the water within a narrow tube to rise above the surrounding fluid; the adhesion of the water to the glass surface, drawing it upward, is stronger than the force of gravity, drawing it down. The narrower the tube, the greater the surface/volume ratio and the more the adhesion defies gravity.

THE ORIGIN OF MOLECULES

The most common element in the universe is the one with the simplest structure: hydrogen. Hydrogen involves the union of a single proton with a single electron. Of all the atoms in the universe, 92.8% are hydrogen atoms. Helium (He), with an atomic number of 2, comprises most of the remainder, 7.17%. Hydrogen and helium are thought to have been the first elements formed as the early universe began to cool. The gravitational attraction of these hydrogen and helium atoms to one another caused them to aggregate, forming stars. Within these primary stars, hydrogen and helium atoms began to collide with increasing frequency, their nuclei fusing to form new atoms. This process takes place with the release of nuclear energy and is essentially the same as that which occurs when a hydrogen bomb explodes. These hot balls of hydrogen and helium were the first stars. Within their interiors, all of the elements of the periodic table were formed, primarily by the progressive capture of helium nuclei. Every carbon atom in your body was once part of a star.

At some point the initial aggregation blew apart, scattering dust and debris through the universe in an explosion often referred to as the big bang. The stars that shine today, including our own star, the sun, formed about 10 billion years ago from the remnants of this explosion, from clouds of cold gas, dust particles, and debris. The matter of these clouds contained all of the elements created by the original solar fusion, although they were still predominantly hydrogen. Gravity caused the clouds to rotate, and matter slowly concentrated at the center, where the release of gravitational energy and the collisions of nuclei created great heat. In this way stars like our sun were formed. There are 250 billion such stars in our galaxy, the Milky Way, and some 10 billion galaxies.

Farther out on the revolving disk of gas, eddies formed and became centers of gravitational aggregation, eventually forming the planets. Our solar system began to form about 4.6 billion years ago, with the planets condensing about 4.5 billion years ago. The large outer planets—Jupiter, Saturn, Uranus, and Neptune—possessed gravitational forces strong enough to entrap all of the elements. In contrast, the smaller inner planets—Mars, Earth, Venus, and Mercury—did not have such strong

illary action plays an important role among living organisms. It is, for example, responsible for the uptake of water (imbibition) by swelling plant seeds; the water splits their hard coat and permits germination.

Water Stores Heat

The temperature of any substance is a measure of how rapidly its individual molecules are moving about. Because of the many hydrogen bonds that water molecules form with one another (Figure 2-15), a large input of thermal energy is required to disrupt the organization of water and raise its temperature. Water is said to have a high **specific heat.** The specific heat of water is twice that of most carbon compounds and nine times that of iron. Only ammonia, which forms very strong hydrogen bonds, has a higher specific heat than water.

FIGURE 2-15

When water cools below 0° C, it forms a regular crystal structure in which the four partial charges of each atom in the water molecule interact with opposite charges of atoms in other water molecules.

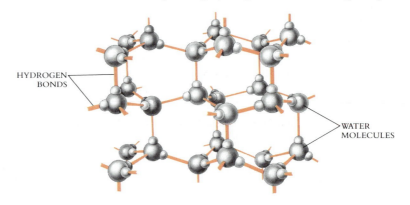

HYDROGEN BONDS

WATER MOLECULES

gravities. Most of the atoms of the lighter gases, such as hydrogen, helium, and neon, escaped the gravitational attraction of these smaller planets.

Earth is the third planet from our sun. When the earth was young and still a molten ball, high temperature and low gravity caused it to lose many of its gases and volatile constituents to outer space. Oxygen, however, trapped in molten silicon rock, was preserved. The common heavy elements, iron and nickel, tended to sink into the core of the planet. The remaining material formed a thick dense **mantle** of silicate, a complex ore containing iron, magnesium, silicon, and oxygen. After approximately half a billion years the surface of the earth cooled, forming a thin crust of lighter rock. Our continents are composed of this thin crust of light rock (Figure 2-A) and ride and drift on the denser mantle beneath. The crust of the earth is only about 100 kilometers thick; if you were to drill beneath it, the earth would still be hot, the result of continuing radioactive decay within the mantle. The oldest known rocks on earth are 3.9 billion years old, although older rocks are preserved on the moon.

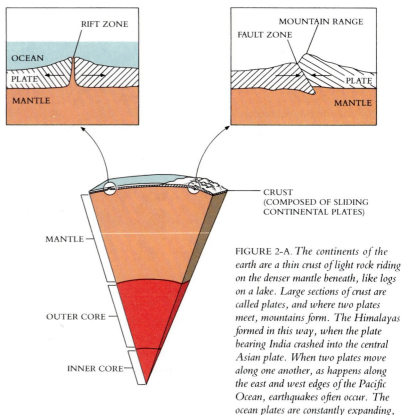

FIGURE 2-A. *The continents of the earth are a thin crust of light rock riding on the denser mantle beneath, like logs on a lake. Large sections of crust are called plates, and where two plates meet, mountains form. The Himalayas formed in this way, when the plate bearing India crashed into the central Asian plate. When two plates move along one another, as happens along the east and west edges of the Pacific Ocean, earthquakes often occur. The ocean plates are constantly expanding, new material welling up from the mantle in rift fissures such as have formed the Hawaiian Islands. The outer edges of the ocean plates dip back down into the mantle where they meet the continental plates.*

Because of its high specific heat, water heats up more slowly than almost any other substance and will hold its temperature longer when heat is lost. One consequence of organisms having evolved with a high water content is that even on land they are able to maintain a relatively constant internal temperature.

A considerable amount of heat energy (586 calories) is required to change 1 gram of liquid water into vapor. Water is said to have a high **heat of vaporization.** Every gram of water that evaporates from the human body removes 586 calories of heat from the body. This amount is equal to the energy released by lowering the temperature of 586 grams of water 1° Celsius (C). Thus the evaporation of water from a surface produces significant cooling. Many organisms dispose of excess heat by evaporative cooling. Humans, for example, sweat.

Water Is a Powerful Solvent

The attraction of water molecules to other charged or polar molecules may prevent these other molecules from associating with one another. Thus the reason that a crystal of table salt, sodium chloride, dissolves in water is that individual Na^+ and Cl^- ions form hydrogen bonds with water. Every time an ion dissociates from the salt crystal, water molecules orient around it in a cloud (Figure 2-16). This **hydration shell,** formed by the water molecules, prevents the ions of the salt from reassociating with one another, with the result that Na^+ and Cl^- exist in water as charged ions. All five of the ionic elements common among living things (sodium, potassium [K^+], chlorine, calcium [Ca^{++}], magnesium [Mg^{++}]) form salts that dissociate in water in this way.

FIGURE 2-16

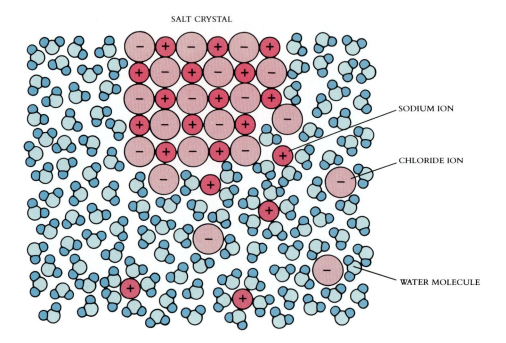

SALT CRYSTAL

SODIUM ION

CHLORIDE ION

WATER MOLECULE

When a crystal of table salt dissolves in water, individual Na⁺ and Cl⁻ ions break away from the salt lattice and become surrounded by water molecules. Water molecules orient around Cl⁻ ions so that their positive ends face in toward the negative chloride ion; water molecules surrounding Na⁺ ions orient in the opposite way, the negative ends facing the positive sodium ion. Surrounded by water in this way, Na⁺ and Cl⁻ ions never reenter the salt lattice.

Water will form a hydration shell around any molecule that exhibits an electrical charge, whether it carries a full charge (ion) or a charge separation (polar molecule). For example, sucrose (table sugar) is composed of molecules that contain slightly polar hydroxyl (OH^-) groups. A sugar crystal dissolves rapidly when it is placed in water because water molecules can form hydrogen bonds with the polar hydroxyl groups of the sucrose molecules. The resulting hydration shell around each sucrose molecule prevents it from associating with other sucrose molecules. Similarly, hydration shells form around all polar molecules. Polar molecules that dissolve in water in this manner are said to be **soluble** in water. Nonpolar molecules are not water soluble; oil is an example of a nonpolar molecule that is not soluble in water. Life originated in water not only because it was a liquid, but also because so many molecules are polar or ionized and therefore are soluble in water.

Water Organizes Nonpolar Molecules

Water molecules in solution always tend to form the maximum number of hydrogen bonds possible. When nonpolar molecules, which do not form hydrogen bonds, are placed in water, the water molecules act in such a way as to exclude them, instead preferentially forming hydrogen bonds with other water molecules. The nonpolar molecules are forced into association with one another, thus minimizing their disruption of the hydrogen bonding of water. The result is a sort of molecular slum, in which all of the nonpolar molecules are crowded together. It seems almost as if the nonpolar compounds shrink from contact with the water, and for this reason they are called **hydrophobic** (Greek, *hydros,* water + *phobos,* fearing—although "feared by water" might be a more apt description).

The tendency for nonpolar molecules to band together in water solution is called **hydrophobic bonding.** The hydrophobic forces induced by hydrogen bonding of water were important in the evolution of life, because some of the exterior portions of many of the molecules on which life came to be based are nonpolar. By forcing these hydrophobic portions of molecules into proximity to one another, water causes such molecules to assume particular shapes in solution. Different molecular shapes have evolved by alteration of the location and strength of nonpolar regions. As you will see, much of the evolution of life reflects changes in the shape of proteins, changes that are induced in just this way.

Water Ionizes

The covalent bonds of water sometimes break spontaneously. When this happens, one of the protons (hydrogen atom nuclei) dissociates from the molecule. Because the dissociated proton lacks the negatively charged electron that it had been sharing in the covalent bond with oxygen, its own positive charge is not counterbalanced. For these reasons, the positively charged ion H^+ is produced. The rest of the water molecule, which has retained the shared electron from the covalent bond, is negatively charged and forms the hydroxide ion (OH^-). This process of spontaneous ion formation is called **ionization:**

$$H_2O \rightarrow OH^- + H^+$$

To quantify the concentration of H^+ ions in solution, a scale based on the slight degree of spontaneous ionization of water has been constructed. In a liter of water roughly 1 water molecule out of each 550 million is ionized at any instant in time, corresponding to $\frac{1}{10,000,000}$ of a mole of H^+ ions. (A **mole** is defined as the weight in grams that corresponds to the summed atomic weights of all the atoms of a molecule, its molecular mass. In the case of H^+ the molecular mass equals 1, and a mole of H^+ ions would weigh 1 gram.) The molar concentration of hydrogen ions in pure water, $\frac{1}{10,000,000}$, can be written more easily by employing exponential notation. This is done by counting the number of decimal places after the digit "1" in the denominator:

$$[H^+] = \frac{1}{10,000,000}$$

Brackets are used to indicate the chemical concentration of H^+. Since there are 7 decimal places, the molar concentration is 10^{-7} moles per liter.

Any substance that dissociates to form H^+ ions when dissolved in water is called an **acid.** The exponents in the exponential notation of H^+ ion concentrations are used as a convenient indication of acid strength, called the **pH scale.** pH is normally expressed as a positive number and is determined by taking the negative of the exponent of the molar concentration. Thus pure water has a molar H^+ concentration of 10^{-7} and a pH of 7. The stronger an acid is, the more H^+ ions that it produces and the lower its pH. Hydrochloric acid (HCl), which is abundant in your stomach, ionizes completely, so the molar concentration of $[H^+]$ in water containing $\frac{1}{10}$ of a mole of hydrochloric acid (e.g., 0.1 mole per liter) is 0.1, or 10^{-1} moles per liter, corresponding to a pH of 1. Some acids, such as nitric acid, are even stronger than this, although such very strong acids are rarely found in living systems (Figure 2-17). The pH of champagne, which bubbles because of the carbonic acid dissolved in it, is about 2.

> pH refers to the relative concentration of H^+ ions in solution. The numerical value of the pH is the negative of the exponent of the molar concentration. Thus low pH values indicate high concentrations of H^+ ions (acids), and high pH values indicate low concentrations.

Positively charged hydrogen ions are not the only types of ions that are produced when water ionizes. Negatively charged OH^- ions are also produced in equal concentration. Any substance that combines with H^+ ions, as OH^- ions do, is said to be a **base.** In pure water the concentrations of H^+ and OH^- ions are both 10^{-7} moles per liter, reflecting the spontaneous rate of dissociation of water. Any increase in base concentration has the effect of lowering the H^+ ion concentration, because base and H^+ ions join spontaneously. Bases, therefore, have pH values above 7. Strong bases, such as sodium hydroxide (NaOH), have pH values of 12 or more.

FIGURE 2-17

pH scale, with some examples among familiar substances.

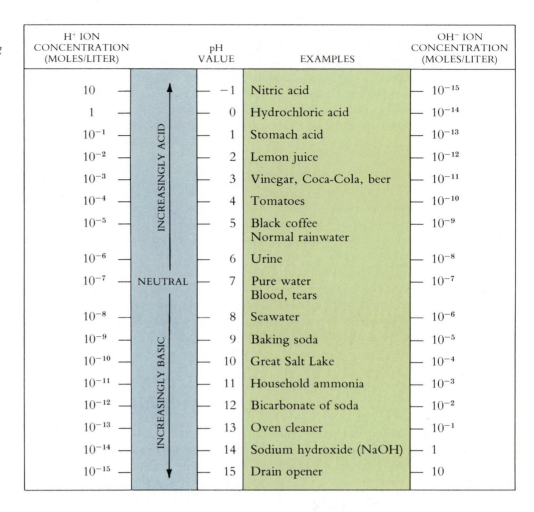

H⁺ ION CONCENTRATION (MOLES/LITER)		pH VALUE	EXAMPLES	OH⁻ ION CONCENTRATION (MOLES/LITER)
10	↑ INCREASINGLY ACID	-1	Nitric acid	10^{-15}
1		0	Hydrochloric acid	10^{-14}
10^{-1}		1	Stomach acid	10^{-13}
10^{-2}		2	Lemon juice	10^{-12}
10^{-3}		3	Vinegar, Coca-Cola, beer	10^{-11}
10^{-4}		4	Tomatoes	10^{-10}
10^{-5}		5	Black coffee / Normal rainwater	10^{-9}
10^{-6}		6	Urine	10^{-8}
10^{-7}	NEUTRAL	7	Pure water / Blood, tears	10^{-7}
10^{-8}		8	Seawater	10^{-6}
10^{-9}	INCREASINGLY BASIC	9	Baking soda	10^{-5}
10^{-10}		10	Great Salt Lake	10^{-4}
10^{-11}		11	Household ammonia	10^{-3}
10^{-12}		12	Bicarbonate of soda	10^{-2}
10^{-13}		13	Oven cleaner	10^{-1}
10^{-14}		14	Sodium hydroxide (NaOH)	1
10^{-15}	↓	15	Drain opener	10

THE CHEMICALS OF LIFE

The compounds formed by combinations of the 11 principal life elements constitute the chemical framework that makes life possible. Three such compounds are thought to have been common on the early earth: nitrogen gas (N_2), water (H_2O), and carbon dioxide (CO_2). It is thought that methane (CH_4), hydrogen sulfide (H_2S), and ammonia (NH_3) may have been present as well, although there is no general agreement on this point. Having considered water in detail, we will now briefly examine carbon dioxide, nitrogen gas, ammonia, and methane. By studying the chemical bonds within these important molecules, we can learn a great deal about their chemical properties.

Carbon Dioxide

Despite the fact that only about 0.033% of the air surrounding the earth at present is carbon dioxide, all of the carbon atoms in every living organism have come from the carbon atoms of this gas. In the CO_2 molecule each oxygen atom forms two covalent bonds with the carbon atom. In this way all three atoms achieve stable outer energy shells containing eight electrons. Carbon dioxide is soluble in water: it is responsible for the fizz in soda and champagne. Unlike water molecules, however, CO_2 molecules exhibit no significant overall charge separation, since the bonds of carbon are arrayed in a tetrahedron, with the result that the charge separations cancel each other out. As a result of this symmetry, the CO_2 molecule is not polar.

Carbon dioxide has played an important role in the evolution of life. When the earth first became rich in oxygen, carbon dioxide was the principal source of carbon atoms from which living organisms were formed. These organisms were in the sea, which does not absorb a lot of carbon dioxide, but a process occurs that concentrates carbon atoms in water: the CO_2 molecule chemically reacts with water. When carbon dioxide is dissolved in water, the CO_2 and H_2O molecules undergo a chemical reaction (Figure 2-18) in which the covalent bonds of the two molecules are rearranged, forming a single molecule called **carbonic acid** (H_2CO_3):

$$CO_2 \ + \ H_2O \ \leftrightarrow \ H_2CO_3$$

CARBON WATER CARBONIC
DIOXIDE   ACID

FIGURE 2-18

Carbon dioxide and water combine chemically to form carbonic acid (H_2CO_3), which dissociates in water, freeing H^+ ions. This reaction makes carbonated beverages acidic.

The reaction involves very little change in energy and is easily reversible.

The interaction of carbon dioxide and water has the important consequence that significant amounts of carbon enter into water solution in the form of carbonic acid. As we will discuss, biologists believe that life first evolved in the early oceans, which were rich in carbon because of this reaction.

Nitrogen Gas

Many molecules of which organisms are composed contain nitrogen atoms. Most nitrogen atoms on the surface of the earth today exist as diatomic nitrogen gas (N_2). Approximately 78% of the atmosphere is N_2. Diatomic nitrogen gas is a particularly stable molecule, since it contains a triple covalent bond ($N\equiv N$), which is very difficult to break. Most organisms cannot use N_2 gas for this reason, so that conversion of N_2 gas to other forms is essential to life as we know it. Because the N_2 triple bond is so difficult to break, all of the nitrogen atoms that occur within living organisms are obtained from one of only a few processes:

1. Lightning or intense heat occasionally causes atmospheric nitrogen to react with water, forming nitric acid:

$$N\equiv N \ + \ 6H_2O \ \rightarrow \ 2HNO_3 \ + \ 5H_2$$

NITROGEN WATER NITRIC ACID HYGROGEN
GAS   GAS

When this same bond is broken during the process of combustion in automobiles, atmospheric nitrogen is converted into nitrous oxides, which are the major components of air pollution.

2. This same reaction can be carried out within an organism, as long as no oxygen is present. Among all of the different kinds of organisms, however, only a few species of bacteria are able to carry out this reaction.

Ammonia

During the formation of the earth, hydrogen gas was much more common than it is now, and lightning was capable of splitting nitrogen gas by means of a different reaction:

$$N\equiv N \ + \ 3H_2 \ \rightarrow \ 2NH_3$$

NITROGEN HYDROGEN AMMONIA
GAS   GAS

Ammonia (Figure 2-19) is the simplest nitrogen compound that is soluble in water. It is thought to have been common during the early history of the earth and was probably the source of much of the nitrogen that was incorporated into organisms then, just as it is now. The nitrogen atom (atomic number, 7) has five electrons in its outer energy level and so attains a stable configuration of eight outer energy level electrons by forming three covalent bonds. In the case of ammonia these are formed with hydrogen atoms. As in the water molecule, the shared electrons are more strongly attracted to the larger atom, causing a

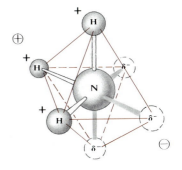

FIGURE 2-19

Ammonia is a very polar molecule. Each of the three shared electron pairs is more closely associated with the nitrogen, creating a charge displacement similar to that seen in water molecules. Although the individual displacements are somewhat less than the displacement in the O — H bonds of water, there are three of them, as opposed to only two in water molecules, with the result that ammonia is even more polar than water.

TABLE 2-3   COMPOUNDS OF HYDROGEN

	METHANE (CH_4)	AMMONIA (NH_3)	WATER (H_2O)
Number of hydrogen atoms in molecule	4	3	2
Polarity of individual bonds	Weak	Strong	Very strong
Charge distribution	Symmetrical	Asymmetrical	Asymmetrical
Chemical behavior	Nonpolar	Very polar	Polar

Three of the molecules that were critical to the origin of life are compounds of hydrogen. Water is an oxygen atom covalently bonded to two hydrogen atoms, ammonia is a nitrogen atom covalently bonded to three hydrogen atoms, and methane is a carbon atom covalently bonded to four hydrogen atoms. Although their chemical formulas seem similar, the properties of these three compounds are very different because of the different spatial distributions of the chemical bonds.

charge displacement. The larger nitrogen atom is more negative as a result of its greater attraction for the shared electrons. The charge displacement is not as great for the individual bonds of ammonia as it is for the individual bonds of water. Despite this smaller charge displacement, there are three such bonds in the ammonia molecule, a relationship that makes ammonia even more polar than water (Table 2-3).

Methane

In the hydrogen-rich atmosphere of the early earth many of the carbon atoms that existed were involved in complexes with hydrogen. The simplest carbon-hydrogen compound is methane, which exists as a gas at the temperatures that normally occur at the earth's surface. Methane is the simplest **hydrocarbon.** A hydrocarbon is a molecule consisting solely of carbon and hydrogen. Because carbon can form carbon-carbon covalent bonds, hydrocarbons with complex structures can be built up.

All organisms are composed primarily of a water solution of molecules that are chains composed of carbon atoms linked to one another. Consequently, we refer to any molecule containing carbon atoms as an *organic* molecule, excepting only CO_2 and the carbonic acids and carbonate salts that CO_2 forms. Carbon dioxide and molecules that do not contain carbon atoms are termed *inorganic* molecules. Thus methanol (CH_3^-OH), formic acid (HCOOH), and CH_4 are organic molecules, whereas CO_2 is not. Most of the inorganic molecules significant to the evolution of life are ions.

SUMMARY

1. The smallest stable particles of matter are protons, neutrons, and electrons, which associate to form atoms. The core, or nucleus, of an atom is composed of protons and neutrons; the electrons orbit around this core in a cloud. The farther out an electron orbits, the faster it goes and the more energy it possesses.

2. The chemical behavior of atoms is largely determined by the distribution of its electrons and in particular by the number of electrons in its outermost (highest) energy level. There is a strong tendency for atoms to have a fully populated outer level; electrons are lost or gained until this condition is reached.

3. A molecule is a stable collection of atoms. The forces holding the atoms together in a molecule are called chemical bonds. These bonds may involve the attraction of op-

positely charged ions or the sharing of electrons between the two bonded nuclei. This second class of bonds, covalent bonds, is by far the stronger and is responsible for the formation of most biologically important molecules.

4. Over 99% of the atoms in the human body are either nitrogen, oxygen, carbon, or hydrogen (NOCH), all of which form strong covalent bonds with one another.

5. The chemistry of life is the chemistry of water. The central oxygen atom in water tends to attract the electrons shared between its constituent oxygen and hydrogen atoms. As a result, the oxygen atom is electron rich (partial negative charge) and the hydrogen atoms are electron poor (partial positive charge). This charge separation is like that of a magnet with positive and negative poles, and water is termed a polar molecule.

6. A hydrogen bond is formed by the attraction of the partial positive charge of one hydrogen atom of a water molecule with the partial negative charge of the oxygen atom of another.

7. Water is cohesive, has a great capacity for storing heat, is a good solvent for other polar molecules, and tends to exclude nonpolar molecules.

8. Water ionizes spontaneously to a slight degree, perhaps 1 molecule in 550 million at any one time. The molar concentration of H^+ ions that results is $1/10,000,000$, or 1×10^{-7}. This concentration is usually signified as the negative of the exponent of the molar concentration, termed pH. The pH of water is thus 7, because 7 is the negative of the exponent of 1×10^{-7}.

9. Relatively few of the earth's carbon atoms are present in carbon dioxide gas, but this molecule represents the source of all the carbon atoms present in living organisms.

10. Ammonia, the source of much of the nitrogen present in living organisms, is formed from diatomic nitrogen gas. Ammonia is thought to have been common during the earth's early history.

11. The simplest carbon-hydrogen molecule is methane, or marsh gas, also thought to have been common on the primitive earth.

SELF-QUIZ

The following questions pinpoint important information in this chapter. You should be sure to know the answers to each of them.

1. Atoms that have the same number of protons but different numbers of neutrons are called _____ .

2. Atoms in which the number of protons and the number of electrons are not equal are called _____ .

3. Which of the following do not carry an electrical charge?
   (a) proton   (d) isotope
   (b) nucleus   (e) ion
   (c) electron

4. The maximum number of electrons within any one orbital of an atom is _____ .

5. The maximum number of electrons within an outer energy level of an atom is _____ .

6. Elements that possess eight electrons in their outer energy level are
   (a) very reactive
   (b) very unreactive

7. Some chemical bonds are more difficult to break than others. Arrange these bonds in the order of their strength, starting with the one most difficult to break.

(a) ionic bond
(b) single covalent bond
(c) hydrogen bond
(d) double covalent bond
(e) triple covalent bond

8. When a covalent bond forms between oxygen and hydrogen atoms, the number of electrons in the outer energy level of the oxygen atom is _____.

9. Which has a greater atomic number, carbon or nitrogen?

10. What four elements comprise over 99% of the body weight of living organisms?

11. What fraction of your body is water?

(a) ¼
(b) ⅓
(c) ½
(d) ⅔
(e) ¾

12. One mole of water would weigh how many grams?

THOUGHT QUESTIONS

1. Water has the chemical formula H_2O, which indicates that each water molecule contains one oxygen atom and two hydrogen atoms. Why is the shape of this molecule bent rather than straight? Why is the carbon dioxide molecule, CO_2, straight rather than bent?

2. Carbon (atomic number, 6) and silicon (atomic number, 14) both have four vacancies in their outer energy levels. Ammonia is even more polar than water. Why do you suppose life evolved in organisms composed of carbon chains in water solution rather than ones of silicon in ammonia?

3. Water and ammonia are polar molecules, but methane is not at all polar. Why not?

4. Carbon atoms have four vacancies in their outer energy level and form molecules by sharing four electron pairs. Of all the possible molecules that carbon can form, which would be the most stable?

5. Carbon atoms can share four electron pairs in forming molecules. Why do you suppose that carbon does not form a bimolecular gas, as hydrogen (one pair of shared electrons), oxygen (two pairs of shared electrons), and nitrogen (three pairs of shared electrons) do?

FOR FURTHER READING

FRIEDEN, E.: "The Chemical Elements of Life," *Scientific American,* January 1972, pages 52-64. A nice introduction to the diversity of atoms in living organisms, with emphasis on the diversity of trace elements.

HOFFMANN, B.: *The Strange Story of the Quantum,* 2nd ed., Dover Publications, New York, 1959. A classic recounting of the ideas and experiments that led to our modern view of atoms. A wonderful book, full of the excitement of science pushing its conceptual frontiers.

HOYLE, F.: *The Ten Faces of the Universe,* Freeman, Cooper, and Company, San Francisco, 1980. A readable account of how the universe is thought to have formed, written for the nonscientist.

MERTZ, W.: "The Essential Trace Elements," *Science,* vol. 213, pages 1332-1338, 1981. An account of the roles of rare elements in human metabolism and the effects of their absences.

MILLER, J.: *Chemistry: A Basic Introduction,* 2nd ed., Wadsworth Publishing Co., Inc., Belmont, Calif., 1981. A good basic text.

SAGAN, C.: *Cosmos,* Random House, Inc., New York, 1980. A popular description of modern astronomy and of the origin of the solar system.

SPEAKMAN, J.C.: *Molecules,* McGraw-Hill Book Company, New York, 1966. A short, lucid, and wonderfully engaging account of basic chemical principles, focusing on atomic orbitals and the nature of the chemical bond.

THE ORIGIN AND EARLY HISTORY OF LIFE

Chapter 3

Overview

Life originated on earth more than 3.5 billion years ago, within 1 billion years of our planet's formation. We do not know how life formed, although the evidence is consistent with the hypothesis that it evolved from nonliving materials by a process in which more stable associations of molecules persisted longer and gradually acquired the ability to reproduce themselves. Life is not a simple concept to define, and the term itself is often used subjectively. Of all the properties of "living" things, heredity is perhaps the most characteristic. The ability to preserve advantageous changes was, and is, an essential aspect of life.

For Review

Here are some important terms and concepts that you will encounter in this chapter. If you are not familiar with them, you should review them before proceeding.

- **Oxidation and reduction** (Chapter 2)
- **Formation of ammonia** (Chapter 2)
- **Covalent bonds** (Chapter 2)

The earth was formed about 4.5 billion years ago. We know nothing directly of these very early times, which are called Hadean times, since no rocks older than 3.9 billion years have been found on earth. The crust of the earth on which we live formed in this early period, solidifying over a hot and very thick mantle. You would not have recognized the early earth, a land of molten rock and violent volcanic activity. The oldest terrestrial rocks that have survived contain no definite traces of life, or at least none that can be recognized with our current level of technology. Today, however, the earth is teeming with life; except within a blast furnace, you would be hard pressed to think of a place anywhere on the surface of the earth where life does not exist in profusion.

The question of the origin of life is not simple. One cannot go back in time and watch how life originated; nor are there any witnesses. There is testimony, in the rocks of the earth, but it is not easily read, and often this record is silent on issues crying out for answers. Perhaps the most fundamental of these issues is the nature of the agency or force that led to the creation of life. There are, in principle, at least four possibilities:

1. *An unknowable agency.* The first life on earth may have possessed genetic systems very different from those characteristic of the organisms we know now, and they might even have lacked cellular organization. What we think of as

"life" may be a secondary stage of some earlier process, about whose origins we can learn nothing because no traces remain.

2. *Natural outside agency*. Life may not have originated on earth at all but instead may have been carried to it, perhaps as an extraterrestrial infection of spores originating on a planet of a distant star. How life came to exist on that planet is not a question we can soon hope to answer.

3. *Supernatural outside agency*. Life forms may have been put on earth by supernatural or divine forces. This viewpoint, common to most religions, is the oldest hypothesis and the most widely accepted.

4. *Evolution*. Life may have evolved from inanimate matter, associations of molecules achieving ever-greater degrees of complexity. In this view the force leading to life was selection; changes in molecules that increased their stability caused them to persist longer.

In this book we deal only with the fourth possibility, attempting to understand whether the forces of evolution could have led to the origin of life and, if so, how the process might have occurred. This is not to say that the fourth possibility is definitely the correct one. Any one of the four possibilities might be true. Nor does the fourth possibility preclude religion: a divine agency might have acted via evolution. Rather, we are limiting the scope of our inquiry to scientific matters. Of the four possibilities, only the fourth permits testable hypotheses to be constructed and so provides the only **scientific** explanation, that is, one which could potentially be disproven by experiment, by obtaining and analyzing actual information.

In our search for this understanding we must look back to the early times before life appeared, when the earth was just starting to cool. We must go back at least that far because there are fossils of simple living things, bacteria, in rocks about 3.5 billion years old, some of the oldest rocks that have persisted on earth. Because of the existence of these fossils, we know that life originated during the first billion years of the history of our planet. In attempting to determine how this process took place, we must first consider the mode of origin of organic molecules, the building blocks of organisms. Then we will consider how organic molecules might have become organized into living cells.

THE ORIGIN OF ORGANIC MOLECULES: CARBON POLYMERS

As the primitive earth cooled, many gases were released from its hot interior by volcanism. There is considerable dispute among geochemists about the exact composition of our original atmosphere, but it is thought to have been composed principally of nitrogen gas. It also contained significant amounts of carbon dioxide and water. It is probable that compounds containing hydrogen bonded to other light elements—sulfur, nitrogen, and carbon—were also present in the atmosphere of the early earth. These compounds would have been hydrogen sulfide (H_2S), ammonia (NH_3), and methane (CH_4). These gases formed a cloud around the earth, held as a part of the atmosphere by the earth's gravity.

The atmosphere of the early earth is also thought by most investigators to have been rich in hydrogen, although there is debate on this point. We refer to such an atmosphere as a reducing one, because of the ample availability of hydrogen atoms and associated electrons. There was little if any oxygen gas present. In such a reducing atmosphere it does not take as much energy to form the carbon-rich molecules from which life evolved. Later our atmosphere changed as living organisms began to harness the energy in sunlight to split water molecules and form complex carbon molecules, giving off gaseous oxygen molecules in the process. Our atmosphere is now approximately 21% oxygen. In the oxidizing atmosphere that exists today the spontaneous formation of complex carbon molecules cannot occur.

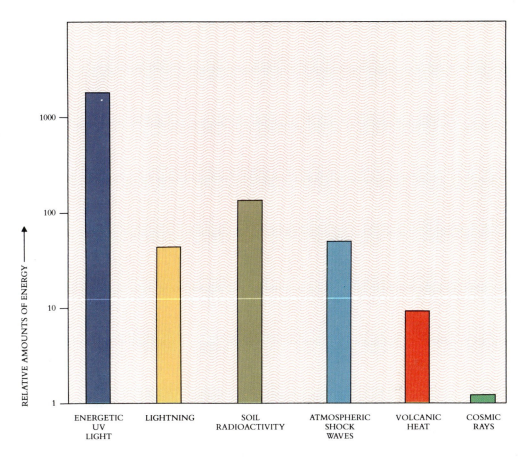

RELATIVE AMOUNTS OF ENERGY →

| ENERGETIC UV LIGHT | LIGHTNING | SOIL RADIOACTIVITY | ATMOSPHERIC SHOCK WAVES | VOLCANIC HEAT | COSMIC RAYS |

The first step in the evolution of life, therefore, probably occurred in a reducing atmosphere, devoid of gaseous oxygen and very different from the atmosphere that exists now. Those were violent times, and the earth was awash with geothermal energy (Figure 3-1): solar radiation, lightning from intense electrical storms, violent volcanic eruptions, and heat from radioactive decay. Living on earth today, shielded by a layer of ozone gas (O_3) in the upper atmosphere from the effects of solar ultraviolet radiation, it is particularly difficult to imagine the enormous flux of ultraviolet energy to which the early earth's surface must have been exposed. Subjected to ultraviolet energy and to the other sources of energy as well (Figure 3-2), the gases of the early earth's atmosphere underwent chemical reactions with one another, forming a complex assemblage of molecules. In the covalent bonds of these molecules some of the abundant energy present in the atmosphere was captured as chemical energy.

What kinds of molecules might have been produced? One way to answer this question is to repeat the process: (1) assemble an atmosphere similar to that thought to exist on the early earth; (2) place this atmosphere over liquid water, which was present on the surface of the cooling earth; (3) exclude gaseous oxygen from the atmosphere, since none was present in the atmosphere of the early earth; (4) maintain this mixture at a temperature somewhat below 100° C; and (5) bombard it with energy in the form of sparks. When Stanley L. Miller and Harold C. Urey carried out this experiment in 1953 (Figure 3-3), they found that within a week 15% of the carbon that was originally present as methane gas had been converted into simple polymers of carbon.

Among the first new substances that were produced in the Miller-Urey experiments (Figure 3-4) were molecules derived from the breakdown of methane, such as formaldehyde (CH_2O) and hydrogen cyanide (HCN). These then combined to form simple molecules, such as formic acid (HCOOH) and urea (NH_2CONH_2), and more complex mol-

FIGURE 3-1 (Left)

Before life evolved, the simple molecules in the earth's atmosphere combined to form more complex molecules. The energy that drove these chemical reactions came from lightning and other forms of geothermal energy.

FIGURE 3-2 (Above)

Sources of energy for the synthesis of complex molecules in the atmosphere of the primitive earth. Of the ultraviolet radiation, only the very short wavelengths (less than 200 nanometers) would have been effective in promoting chemical reactions. Electrical discharges are thought to have been more common on the primitive earth than they are now.

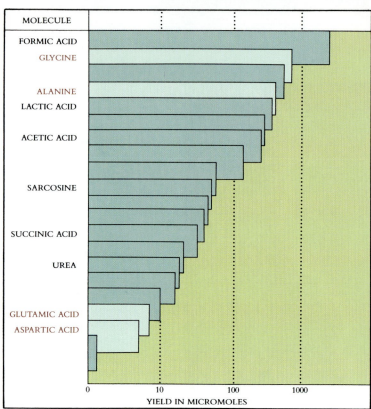

MOLECULE			
FORMIC ACID			
GLYCINE			
ALANINE			
LACTIC ACID			
ACETIC ACID			
SARCOSINE			
SUCCINIC ACID			
UREA			
GLUTAMIC ACID			
ASPARTIC ACID			

0　　　　10　　　　100　　　　1000
YIELD IN MICROMOLES

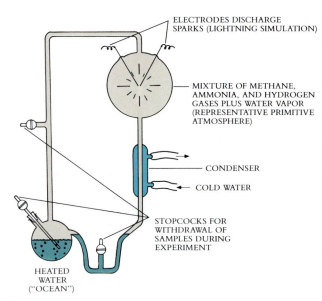

FIGURE 3-3

A *The Miller-Urey experiment. The apparatus consisted of a closed tube connecting two chambers. The upper chamber contained a mixture of gases thought to resemble the primitive earth's atmosphere. Through this mixture electrodes discharged sparks, simulating lightning. Condensers then cooled the gases, causing water droplets to form, which passed to the second heated chamber, the "ocean." Any complex molecules formed in the atmosphere chamber would be carried dissolved in these droplets to the ocean chamber, from which samples were withdrawn for analysis.*

B *Some of the 20 most common complex molecules detected in the original Miller-Urey experiments are indicated. Among these 20 molecules are 4 amino acids.*

ecules containing carbon-carbon bonds, including the amino acids glycine and alanine. **Amino acids** are molecules with the structure

$$H_2N-\overset{\displaystyle \overset{R}{|}}{\underset{\displaystyle \underset{H}{|}}{C}}-COOH$$

where R can be any of several possible chemical groups, referred to as side chains. In the amino acid glycine, for example, R is a hydrogen atom, H. In the amino acid alanine, on the other hand, R is the chemical group CH_3. Amino acids are important because they are the basic building blocks of proteins, which are one of the major kinds of molecules of which organisms are composed. About 50% of the dry weight of each cell in your body consists of amino acids, alone or linked together into protein chains.

In later experiments more than 30 different carbon polymers were identified, including the amino acids glycine, alanine, glutamic acid, valine, proline, and aspartic acid. The production of amino acids indicates that proteins could have formed under conditions similar to those thought to have existed on the early earth. Other biologically important molecules were also formed in these experiments: the presence of hydrogen cyanide was shown to lead to the production of a complex ring–shaped molecule called adenine. Adenine is an important component of **deoxyribonucleic acid** (DNA), the long molecule that organisms use to encode the information specifying which amino acids they place in a given protein and in what order. Thus the key molecules from which life evolved were created in the atmosphere of the early earth as a byproduct of its birth.

Among the molecules that form spontaneously under conditions thought to be similar to those of the primitive earth are those which form the building blocks of organisms.

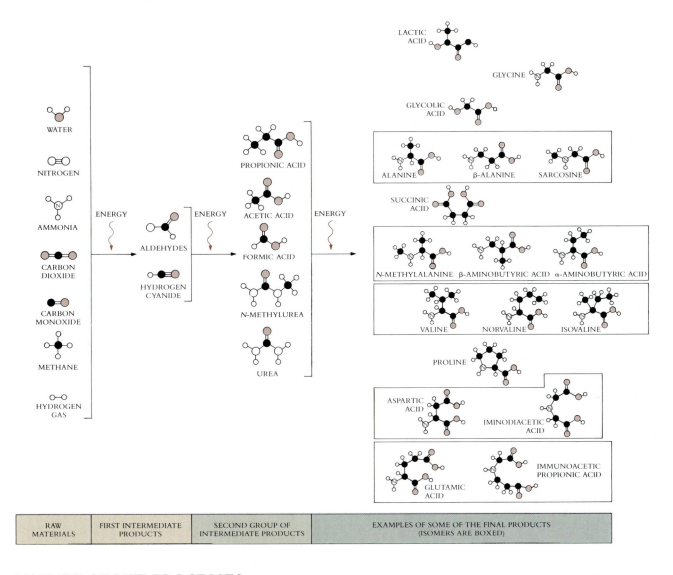

Labels within figure:

LACTIC ACID

GLYCINE

GLYCOLIC ACID

WATER

NITROGEN

AMMONIA

CARBON DIOXIDE

CARBON MONOXIDE

METHANE

HYDROGEN GAS

ENERGY

ALDEHYDES

HYDROGEN CYANIDE

ENERGY

PROPIONIC ACID

ACETIC ACID

FORMIC ACID

N-METHYLUREA

UREA

ENERGY

ALANINE   β-ALANINE   SARCOSINE

SUCCINIC ACID

N-METHYLALANINE   β-AMINOBUTYRIC ACID   α-AMINOBUTYRIC ACID

VALINE   NORVALINE   ISOVALINE

PROLINE

ASPARTIC ACID   IMINODIACETIC ACID

GLUTAMIC ACID   IMMUNOACETIC PROPIONIC ACID

| RAW MATERIALS | FIRST INTERMEDIATE PRODUCTS | SECOND GROUP OF INTERMEDIATE PRODUCTS | EXAMPLES OF SOME OF THE FINAL PRODUCTS (ISOMERS ARE BOXED) |

NATURE OF LIFE PROCESSES

As the earth cooled, much of the water vapor present in its atmosphere condensed into liquid water, which accumulated in the ever-expanding oceans. Judging from the results of the Miller-Urey experiments, we may presume that the water droplets carried the precursors of amino acids, nucleotides, and other compounds produced by chemical reactions in the atmosphere. The primitive oceans must not have been pleasant places. It is odd to think of life originating from a dilute, hot, smelly soup of ammonia, formaldehyde, formic acid, cyanide, methane, hydrogen sulfide, and organic hydrocarbons. Yet from such an ocean emerged the organisms from which all subsequent life forms are derived. The way in which this happened is a puzzle and may forever remain so. Nevertheless, one cannot escape a certain curiosity about the earliest steps that eventually led to the origin of all living things on earth, including ourselves. How did organisms evolve from complex molecules? What is the "origin of life"?

These are difficult questions to ask, largely because life itself is not a simple concept. If you try to write a simple definition of "life," you will find that it is not an easy task. The problem is not your ignorance, but rather the loose manner in which the concept of "life" is used. For example, imagine a situation in which two astronauts encounter a large amorphous blob on the surface of the moon. One might say to the other, "Is it alive?" What does this question mean? Observe what the astronauts do to find out. Probably they would first observe the blob to see if it moves.

FIGURE 3-4

Results of the Miller-Urey experiment. Seven simple molecules, all gases, were included in the original mixture; note that oxygen was not among them, the atmosphere instead being rich in hydrogen. At each stage of the experiment more complex molecules were formed: first aldehydes, then simple acids, then more complex acids. The molecules that are structural isomers of one another are grouped together in boxes. In most cases only one isomer of a compound is found in living systems today, although many may have been produced in the Miller-Urey experiment.

FIGURE 3-5

Movement. A graceful gull gliding through the air, a killer whale exploding from the sea—animals have evolved mechanisms that allow them to move about in any medium. Land-dwellers ourselves, we move on land. Whether it is the awkward galumphing of a giraffe, the laborious crawling of a weevil, or a human's first faltering steps, we have grown to expect some kind of movement from all land-dwelling animals. All living organisms, however, are not land-dwelling animals.

Movement. Most animals move about (Figure 3-5). Movement from one place to another is not in itself diagnostic of life, however. Some animals, and most plants, do not move about, whereas numerous nonliving objects, such as clouds, can be observed to move. The criterion of movement is thus neither *necessary*—possessed by all life—nor *sufficient*—possessed only by life.

The astronauts might prod the blob, to see if it responds, and thus test for another criterion, sensitivity.

Sensitivity. Almost all living things respond to stimuli (Figure 3-6). Plants grow toward light, and animals retreat from fire. Not all stimuli produce responses, however. Imagine kicking a redwood tree or singing to a hibernating bear. Thus this criterion, although superior to the first one, is still inadequate.

The astronauts might attempt to kill the blob.

Death. All living things die, whereas no inanimate objects do. Death is not easily discriminated from disorder, however; a car that breaks down does not die—it was never alive. Death is simply the absence of life. Unless one can detect life, death is a meaningless concept. This is a terribly inadequate criterion.

Finally, the astronauts might cut up the blob, to see if it is complexly organized.

Complexity. All living things are complex. Even the simplest bacteria contain a bewildering array of molecules, organized into many complex structures. Complexity is not diagnostic of life, however. A computer is also complex, but it is not alive. Complexity is a necessary criterion of life but not sufficient in itself to identify living things, since many complex things are not alive.

To determine whether the blob is alive, the astronauts must learn much more about it. Probably the best thing they could do would be to examine it more carefully and determine the ways in which it resembles living organisms. All organisms that we know about share certain general properties, ones that we think must ultimately have derived from the first organisms that evolved on earth. It is by these properties that we recognize other living things, and to a large degree these properties define what we mean by the process of life. Three fundamental properties shared by all organisms on earth are:

Cellular organization. All organisms are composed of one or more cells, complex organized assemblages of molecules enclosed within membranes (Figure 3-7).

Growth and metabolism. All living things assimilate energy and use it to grow. This process is called **metabolism.** Plants, algae, and some bacteria utilize sunlight to create carbon-carbon covalent bonds from CO_2 and H_2O by the process known as photosynthesis (Figure 3-8). Nearly all other organisms obtain their energy by metabolizing these bonds. The conversion of energy is essential to all life on earth. All living organisms are similar in another respect: in all organisms with which we are familiar, metabolic energy captured from carbon compounds is transferred from one molecule to another via special energy-storing phosphate bonds.

Reproduction. All living things reproduce, passing on traits from one generation to the next (Figure 3-9). Some organisms live for a very long time; for example, some of the bristlecone pines *(Pinus longaeva),* growing near timberline in the White Mountains of the western Great Basin, have been alive for nearly 5000 years. But no organisms live forever, as far as we know. Because all organisms die, life as an ongoing process is impossible without reproduction.

Are these properties adequate to define life? Is a membrane-enclosed entity that grows and reproduces alive? Not necessarily. Phospholipid molecules (long, linear carbon polymers with polar and nonpolar ends) in water solution spontaneously form hollow spheres, membranes that enclose a small volume of fluid (Figure 3-10). These spheres, called **coacervates,** may contain energy-processing molecules, and they may also grow and

FIGURE 3-6

Sensitivity. This father is responding to a stimulus—he has just been bitten in the rump by his cub. As far as we know, all organisms respond to stimuli, although not always to the same ones. Had the cub bitten a tree instead of its father, the response would not have been as dramatic.

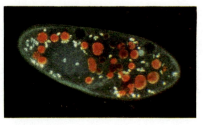

FIGURE 3-7

Cellular organization. This Paramecium, *a complex protozoan, has just ingested several yeast cells, stained red in this photograph. The* Paramecium *is a single cell. The yeasts that it ingested are enclosed within membrane vesicles called digestive vacuoles. A variety of other organelles are also visible.*

FIGURE 3-8

Metabolism. The energy that almost all organisms use to grow is obtained through the process of photosynthesis, which is carried out by bacteria and green plants. In this process the energy of light is captured by chlorophyll and used to build carbon-containing molecules from atmospheric carbon dioxide gas.

FIGURE 3-9

Reproduction. The reptile crawling out of this egg represents successful reproduction. All organisms reproduce, although all do not hatch from eggs. Some organisms reproduce several generations each hour; others, once in a thousand years.

subdivide. Despite these features, they are certainly not alive, any more than soap bubbles are alive. Therefore the three criteria just listed, although necessary for life, are not sufficient to define life. One ingredient is missing—heredity, a mechanism for the preservation of experience:

> *Heredity.* All organisms on earth possess a genetic system that is based on the replication of a complex linear molecule called DNA. The order of DNA subunits encodes the information that determines what that individual organism will be like.

To understand the role of heredity in our definition of life, we will return for a moment to coacervates. When examining an individual coacervate globule, we see it at that precise moment in time but learn nothing of its predecessors. It is likewise impossible to guess what future globules will be like. The globules are the passive prisoners of a changing

THE PUZZLE OF COMPLEX MOLECULES

It is not clear how in the dilute, nonliving (**prevital**) soup of the early earth the more complex molecules characteristic of life were formed. Among the many questions that arise, two are particularly significant:

1. Why do living things not contain all of the amino acids produced in the Miller-Urey experiments?

2. How can proteins aggregate spontaneously in water?

The first question is like a detective story in which some of the clues are missing. If the initial aggregation of amino acids into proteins was random, then it is difficult to understand why more than half of the kinds of amino acids produced are not present in living organisms. For example, the Miller-Urey experiments produce three **isomers** (alternative configurations) of the amino acid

$C_3H_7NO_2$, alanine, beta-alanine, and sarcosine, but only alanine is present in proteins of biological origin. The experiments produce seven isomers of the amino acid $C_4H_9NO_2$, none of which are present in organisms today. Perhaps early proteins contained many more kinds of amino acids than the 20 specified in the genetic code that exists today; after a series of false starts many may have been weeded out because they produced unsatisfactory proteins. If so, these early prototypes apparently have left no traces.

The second question is a puzzle because it seems to defy what we know of the laws of chemistry. Biological proteins form by the joining of subunits into chains. The addition of each element to the growing chain requires the input of energy and the removal of a water molecule. This addition is accomplished by the removal of an —OH group from the end of the chain and an —H from the

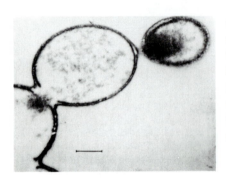

FIGURE 3-10

These coacervates, proteinlike microspheres formed from protein molecules, possess many of the characteristics of living cells. If you look carefully, for example, you can see that each is bounded by a bilayer membrane. The first cells are thought to have evolved from protein or lipid coacervates.

environment, and it is in this sense that they are not alive. The essence of being alive is the ability to encompass change and to reproduce the results of change permanently. Heredity, therefore, provides the basis for the great division between the living and the nonliving. Change does not become evolution unless it is remembered. A genetic system is the sufficient condition of life. When we look at any living organism, we are seeing its history, carried through its genes from the earliest time. Some changes are preserved because they improve chances of survival in a hostile world, whereas others are lost. Not only did life evolve, evolution is the very essence of life.

All living things on earth are characterized by cellular organization, growth, reproduction, and heredity. These characteristics serve to define the term *life*. **Other properties that are commonly exhibited by living organisms include movement and sensitivity to stimuli.**

ORIGIN OF THE FIRST CELLS

Many different kinds of molecules will aggregate with one another in water, much as people from the same foreign country tend to aggregate together in a large city. Sometimes the aggregation of molecules of one kind will form a cluster big enough to see, spherical coacervates 1 to 2 micrometers in diameter. Coacervates will form spontaneously from fat molecules when suspended in water. Simple coacervate microdrops that formed in the primeval soup by aggregation of fat molecules called lipids are thought to have been the first step in the evolution of cellular organization. Coacervates have several remarkably cell-like properties:

1. Coacervates form an outer boundary that resembles a biological membrane in that it has two layers.

2. Coacervates grow, accumulating more subunit lipid molecules from the surrounding medium.

incoming element. Because this reaction is chemically reversible, an excess of water should in principle drive the reaction in the direction of breakdown of the molecule rather than synthesis. The puzzle is that these reactions critical to the evolution of life are thought to have taken place within the oceans and therefore in a high excess of water. It is difficult to imagine that the spontaneous formation of proteins could have occurred under these conditions.

It has been suggested that the first proteins may have formed in the residue of evaporated pools. Dry mixtures of amino acids and phosphates at temperatures of 60° C will, within a few days, produce amino acid chains called proteinoids several hundred amino acid units long. It has also been suggested that the first formation may have taken place on the surfaces present within silicate clays. The interior of clays such as kaolinite is made up of thin layers, only 0.71 nanometer apart, separated by water. A nanometer is a billionth of a meter (10^{-9} m). Thus a cube of kaolinite 1 centimeter on a side has a surface area of 2800 square meters! Clay would thus have provided ample surfaces for protein formation to occur. In addition, there are many positive and negative charges on the surfaces of these layers, which would have facilitated the process. Supporting the hypothesis that clays may have played an important role in the origin of life, clays such as kaolinite have been shown to catalyze the formation of long polypeptide chains from amino acids when the amino acids are first joined to adenosine monophosphate (AMP), a molecule that produces the energy that makes the reaction possible. This same initial joining of amino acids to AMP is universally employed by all living systems in their synthesis of proteins.

3. Coacervates form budlike projections and divide by pinching in two, like bacteria.
4. Coacervates can contain amino acids and use them to facilitate the occurrence of a variety of acid-base reactions, including the decomposition of glucose.

It is not difficult to imagine a process of chemical evolution, involving such coacervate microdrops, that may have taken place before the origin of life. The early oceans must have contained untold numbers of these microdrops, billions in a spoonful, each one forming spontaneously, persisting for a while, and then dispersing. Some of the droplets would by chance have contained amino acids with side groups that were better able to catalyze growth-promoting reactions than the other droplets. These droplets would have survived longer than the others, because the persistence of both protein and lipid coacervates is greatly increased when they carry out metabolic reactions such as glucose degradation and when they are growing actively.

Over millions of years those complex microdrops which were better able to incorporate molecules and energy from the lifeless oceans of the early earth would have tended to persist more than the others. Also favored would have been those microdrops which could use these molecules to expand in size, growing large enough to break into daughter microdrops with features similar to those of their "parent" microdrop. The daughter microdrops would have been able to grow themselves by utilizing the same favorable combination of characteristics as their parents had. When a means occurred to facilitate this transfer of new ability from parent to offspring, heredity—and life—began.

HISTORY OF EARLY LIFE

Anyone who has been to a large museum has seen the fossilized remains of extinct animals—large dinosaurs and exquisite small trilobites (ancient, distant relatives of the horseshoe crabs). Creatures whose forms are preserved in rock, these fossils provide a record of our biological past (Figure 3-11). Looking at them, we can see what early life forms were

FIGURE 3-11

A fossil fish. The rocks of Colorado and Utah are rich with such fossils, since the western United States was submerged beneath an inland sea in prehistoric times.

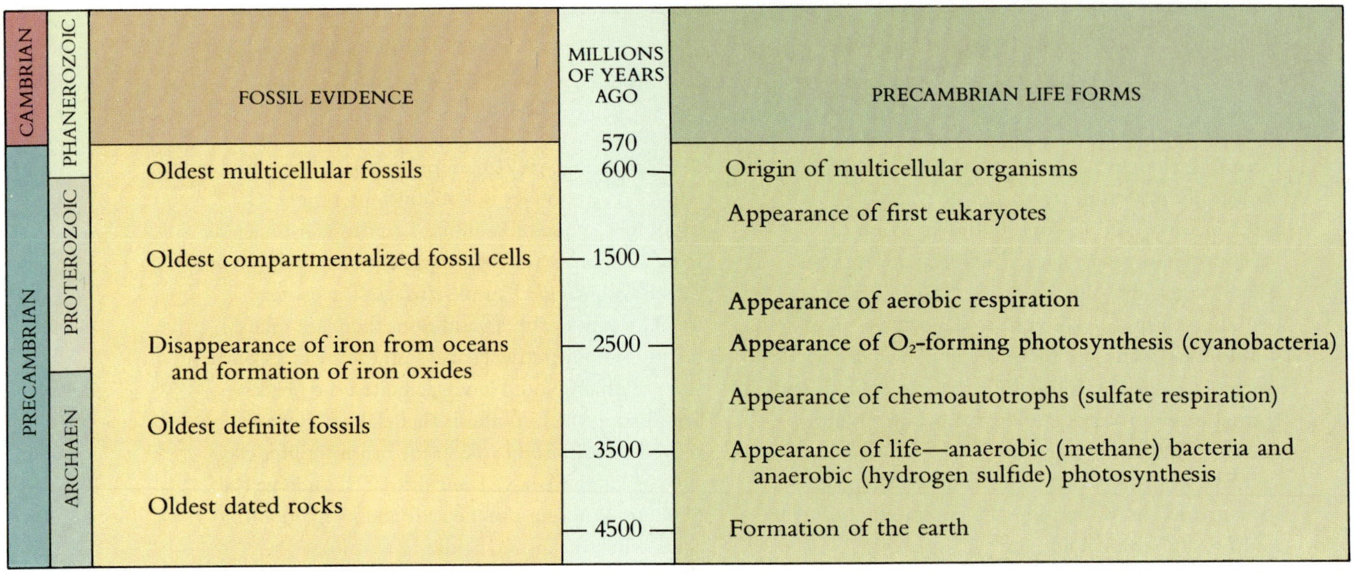

		FOSSIL EVIDENCE	MILLIONS OF YEARS AGO	PRECAMBRIAN LIFE FORMS
CAMBRIAN	PHANEROZOIC		570	
PRECAMBRIAN	PROTEROZOIC	Oldest multicellular fossils	600	Origin of multicellular organisms
				Appearance of first eukaryotes
		Oldest compartmentalized fossil cells	1500	
				Appearance of aerobic respiration
		Disappearance of iron from oceans and formation of iron oxides	2500	Appearance of O$_2$-forming photosynthesis (cyanobacteria)
	ARCHAEN			Appearance of chemoautotrophs (sulfate respiration)
		Oldest definite fossils	3500	Appearance of life—anaerobic (methane) bacteria and anaerobic (hydrogen sulfide) photosynthesis
		Oldest dated rocks		
			4500	Formation of the earth

FIGURE 3-12

Geological time scale. The periods refer to different stages in the evolution of life on earth. Certain fossils were found in the Archean Period, very different ones in the Proterozoic Period. The time scale is calibrated by looking at the rocks within which particular kinds of fossils are found; the fossils are dated by determining the degree of spontaneous decay of radioactive isotopes locked within rock when it was formed.

like. The record of fossil life is troubling, however, because it appears to begin abruptly and relatively recently. For a long time the earliest fossils known came from a geological period called the Cambrian Period, which began about 590 million years ago (Figure 3-12). In most Precambrian rock formations there are no fossils that can be seen with the naked eye. In the last few decades a few fossils of soft-bodied animals have been found in slightly older rocks, up to 630 million years old. From the evidence of these fossils, one might get the impression that life sprang forth fully formed about 630 million years ago.

The Puzzle of Precambrian Life

Charles Darwin, the great nineteenth-century English scientist who first suggested the mechanism that explains how evolution occurs, was greatly puzzled by the lack of Precambrian fossils. In his masterful synthesis *The Origin of Species,* Darwin wrote, "To the question why we do not find rich fossiliferous deposits belonging to periods prior to the Cambrian system, I can give no satisfactory answer. . . . The case at present must remain inexplicable: and may be truly urged as a valid argument against the views here entertained." Some of Darwin's critics seized on the apparent lack of Precambrian fossils as an argument against the progressive evolution of life on earth, and Darwin was unable to answer them.

We have only recently understood the nature of the Precambrian puzzle: the fossils, although actually abundant in certain rock strata, were simply too small to be seen with the naked eye (Figure 5-9, p. 89). We now have identified many such fossils, the earliest of them nearly as old as the oldest known rocks on earth, some 3.5 billion years. All fossils that are more than 630 million years old, however, are microscopic; they are, in fact, bacteria.

With the discovery of these Precambrian fossils over the course of the past 30 years, we have learned that the seeming lack of fossils from Precambrian times, which troubled Darwin so deeply, was apparent rather than real. Life is in fact abundantly represented in rocks over the entire past 3.5 billion years. The fossils that have been found represent an obvious progression from simple to complex organisms during that vast period of time, which begins no more than a billion years after the origin of the earth. Living things might have been present still closer to the time of the actual origin of the earth, but suitable rocks in which to look for their fossils are unknown.

What can be said about these early life forms? One lesson that we have learned from studying early microfossils is that for most of the history of life, organisms were very simple. Like the living bacteria, they were small (10 to 100 micrometers in diameter), single-celled creatures with little evidence of internal structure and no external appendages. More complex living forms did not appear until about 1.3 billion years ago, or possibly as early as 1.5 billion years ago. For at least 2 billion years, nearly half of the age of the earth, bacteria were the only organisms that existed on earth. Despite their apparent uniformity, however, major changes were taking place within the interiors of these bacteria. We know this by the effects they had on the world around them.

The First Autotrophs

The first evidence of major changes occurring within early bacteria concerns isotopes of carbon, the atomic building block of life. The carbon in the molecules of organisms is acquired from CO_2 molecules in the atmosphere. The carbon atoms are present in CO_2 as either of two isotopes: a light isotope of carbon, symbolized ^{12}C, and a heavier isotope, symbolized ^{13}C. In these symbols the numbers refer to the atomic mass of the atoms, ^{13}C possessing one more neutron than ^{12}C. The carbon atoms of atmospheric CO_2 are captured by organisms in the process of photosynthesis and incorporated into sugars, proteins, and other molecules. Organisms that carry out photosynthesis show a marked preference for the light isotope of carbon, ^{12}C, over the heavier ^{13}C isotope, with the result that the tissues of living organisms are enriched for the light isotope; that is, they contain a higher proportion of ^{12}C than does the atmosphere.

In rocks 3.4 billion years old investigators have found organic material that is enriched for the ^{12}C isotope to the same degree as are the tissues of contemporary organisms. This enrichment indicates that before these rocks were formed organisms were assimilating carbon into organic molecules from CH_4 and CO_2 in the atmosphere. From these "metabolic tracks" we are able to conclude that no less than 3.4 billion years ago, organisms had evolved that were not dependent on geochemically produced carbon molecules raining down into the oceans.

Starting about 3.4 billion years ago, rocks became enriched for light carbon isotopes, indicating the presence of organisms that were able to garner carbon atoms from atmospheric methane and carbon dioxide.

Enrichment for light carbon isotopes is thought to have first been the result of metabolism by organisms, **photosynthesizers,** that obtained carbon atoms from CO_2 dissolved in the water. This process, **anaerobic photosynthesis,** is a kind of photosynthesis that uses H_2S as a source of the hydrogen employed in the synthesis of organic molecules, generating elemental sulfur as a byproduct. It is called "anaerobic" because it does not generate gaseous oxygen (**O_2-forming photosynthesis**) as the byproduct, as modern plants do. This and other forms of photosynthesis are described in more detail in Chapter 12.

The carbon-gathering organisms whose presence is indicated by the ratios of carbon isotopes in the ancient rocks were the world's first **autotrophs.** An autotroph is an organism that is able to synthesize all the organic molecules it requires from simple inorganic substances and an energy source. These first autotrophs were able to synthesize all of the molecules of which their bodies were composed from simple inorganic molecules such as water and carbon dioxide and an energy source such as sunlight.

Other new organisms, called **methanogens,** reduced CO_2 to CH_4 using hydrogens derived from organic molecules produced by other organisms, just as the so-called methane bacteria do today.

Evidence of yet another biological process is first seen in a group of rocks about 2.7 billion years old, known as the Woman River iron formation. Organic material in these rocks is enriched for the light isotope of sulfur, ^{32}S, relative to the heavier isotope ^{34}S. No known geochemical process produces such enrichment, but biological sulfur reduction does. Apparently what these deposits reflect is the advent of **sulfate respiration,** in which organisms derive energy from the reduction of inorganic sulfates (SO_4) to H_2S, the hydrogen atoms being obtained from organics produced by other organisms. These organisms thus do the same thing that methanogens do but use SO_4 as the oxidizing agent in place of CO_2. Both methanogens and sulfate-reducing organisms are called **heterotrophs.** They derive their energy not from the sun but rather from the covalent bonds of organic molecules.

The Advent of Atmospheric Oxygen

Within the first billion years of life, a span known as the **Archean Period,** these autotrophs and heterotrophs (photosynthesizers, methanogens, and sulfate reducers) predominated. For all intents and purposes, they *were* life on earth. In rocks about 2.5 billion years old, however, we see evidence of something new. We know from examining early Archean sediments that the early oceans were rich in iron (Fe), dissolved as ferrous ions, whose compounds are water soluble. About 2.5 billion years ago, however, almost all of this iron was removed from the earth's oceans. Over a very short period of time, a few million years, the vast amounts of iron dissolved in the oceans were deposited in sediments as iron oxides. The iron had been oxidized—the oxidation of ferrous ion (Fe^{++}) produces ferric ions (Fe^{+++}), whose compounds are not water soluble and precipitate as iron oxides. The disappearance of the dissolved iron from the oceans marks the introduction of plentiful oxygen gas to the earth, and all of this oxygen appears to have been generated by living organisms carrying out O_2-forming photosynthesis.

From the evidence of these and other iron deposits we are able to conclude that at some point more than 2.5 billion years ago, a fourth kind of organism evolved. These organisms were able to obtain the hydrogen atoms needed to produce carbon compounds by breaking down water, a process in which oxygen gas was released as a byproduct. Contemporary photosynthetic bacteria, called cyanobacteria, are thought to be the direct descendants of these organisms, practically unchanged as far as we know for 3 billion years or more. The widespread release of oxygen brought about by the early cyanobacteria gradually saturated the oceans. Eventually, the gas escaped into the atmosphere and became a major component of it. This event marks the end of the Archean Period, usually placed at about 2.5 billion years ago, and the beginning of aerobic life.

> **A major change in the earth's atmosphere occurred about 2.5 billion years ago. At that time there appeared organisms that carried out a form of photosynthesis in which water is the source of hydrogen and oxygen gas a byproduct. The result was the enrichment of the earth's atmosphere for oxygen gas.**

For the next billion years, the first half of what is called the Proterozoic Era, the appearance of the microfossils that we see in rocks changed little. Although they were often linked together in filaments, their cells were still small and simple, with thin walls and little evidence of internal structure. Sometime during this period, cells arose that had lost the ability to carry out photosynthesis, obtaining their carbon atoms instead from other organisms. These organisms were heterotrophs, obtaining their energy by stripping hydrogens from organic food molecules and combining them with oxygen to form water. All animals that inhabit the earth today are their descendants.

> **The first organisms of which we have direct evidence were autotrophs, which obtained energy from the sun. These were followed by different types of heterotrophs, living off of organics synthesized by autotrophs.**

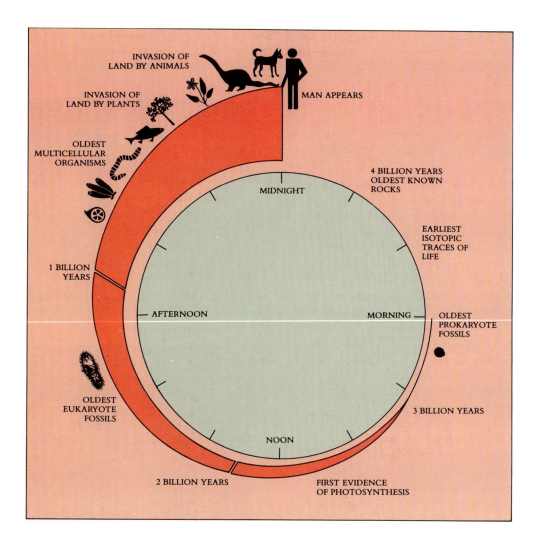

FIGURE 3-13

The clock of biological time. A billion seconds ago it was 1953, and most students using this text had not yet been born. A billion minutes ago Jesus was alive and walking in Galilee. A billion hours ago the first human had not been born. A billion days ago no biped walked on earth. A billion months ago the first dinosaurs had not yet been born. A billion years ago no creature had ever walked on the surface of the earth.

The Dawn of the Eukaryotes

We refer to the simple organisms we have been discussing, the bacteria, as **prokaryotes,** from the Greek words *pro,* before + *karyon,* kernel or nucleus. The name reflects their lack of a nucleus, a spherical organelle that evolved later in other kinds of organisms. In rocks more than a billion years old we begin to see for the first time microfossils noticeably different in appearance from the earlier simpler forms. These cells are much larger than those of bacteria, with thicker walls and internal membranes (Figure 5-10, p. 90). The oldest known cells of this kind are about 1.3 billion years old, although there are some fossils as much as 1.5 billion years old that seem similar. These new cells are called **eukaryotes** from the Greek words *eu,* good + *karyon,* kernel or nucleus, since they do possess a nucleus. Except for bacteria, all organisms are eukaryotes. Chapter 6 explores their structure in detail.

The early eukaryotes were all single-celled organisms. No multicellular organisms evolved until hundreds of millions of years later. When that important evolutionary event occurred, the detection of fossils became simple. The origin of multicellular organisms also marks the close of the Proterozoic Era and the start of the Phanerozoic Era, which extends to the present. The Cambrian Period, which used to mark the time of the first appearance of multicellular organisms, is pinpointed now at about 40 million years after the start of the Phanerozoic Era. Bacteria have survived, along with the eukaryotes, to the present; they are the oldest kind of organism. Humans are among the most recent (Figure 3-13). More details of the later evolution of life are presented in Chapter 23.

OTHER WORLDS WHERE LIFE MIGHT EXIST

There are many places other than earth where one can imagine life might be. Here are a few of the possibilities that have been considered.

In the nineteenth century it was commonly speculated that life might exist on the moon. In 1865 the French novelist Jules Verne described moonmen in *From Earth to the Moon*. We now know that life never evolved there. Life did, however, reach the moon in 1969 (Figure 3-A). The planet from which it came is in the background.

Ever since the erroneous reports of canals on Mars by nineteenth-century astronomers, it has been speculated that life in some form might exist there, or might have existed in the past. The speculation has not been limited to science fiction writers; in the mid-1970s the United States mounted a space probe to explore for life on Mars. Extensive tests were done on soil samples, but no clear evidence of life was obtained.

Venus was the last planet in our solar system where life might conceivably have been found. When space probes finally reached its surface, however, the temperatures proved far too hot to support life as we know it.

Currently the most likely candidate for life within the solar system is seen in this photograph of Europa taken by a passing space probe (Figure 3-B). Europa is one of the many moons of the large planet Jupiter. Whereas most of Jupiter's moons resemble ours, pockmarked by meteors and devoid of life, this moon is covered with ice. Perhaps below its frozen surface the pressures of gravity create enough heat to maintain water in a liquid form. In such an environment life may have evolved. The conditions on Europa now are far less hostile to life than those which existed in the oceans of the primitive earth. We have not yet traveled to Europa to investigate.

A

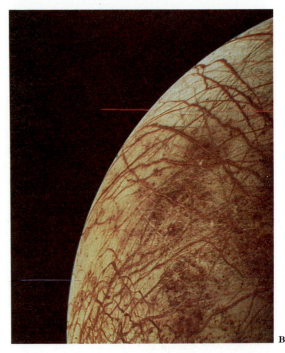

B

IS THERE LIFE ON OTHER WORLDS?

The life forms that evolved on earth closely reflect the nature of this planet and its history. If the earth were farther from the sun, it would be colder and chemical processes would be greatly slowed down. Water, for example, would be a solid, and many carbon compounds would be brittle. If the earth were closer to the sun, it would be warmer, chemical bonds would be less stable, and few carbon compounds would be stable and persist. The evolution of a carbon-based life form is probably possible only within the narrow range of temperatures that exist on earth, which is directly related to the distance from the sun.

Around the star Beta Pictoris (Figure 3-C) you can see a new solar system forming. Matter swirls in a broad ring around this star, much as it is thought to have done around our sun, and astronomers believe that within the swirl planets are forming. If a planet were to form at the right distance from this star, it might be like earth. Ten percent of all stars, like this one, are thought to have a solar system.

The nearest galaxy to ours is a spiral galaxy called Andromeda (Figure 3-D). It contains millions of stars, many of them resembling our own sun. It is difficult to believe that life has not evolved on some of them. If the evolution of life is a rare and unlikely occurrence, life would still be common in the Andromeda galaxy, since it contains so many stars.

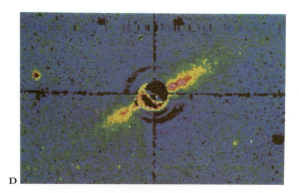

D

The universe contains over a billion galaxies. In this photograph of one galactic cluster (Figure 3-E), individual galaxies resemble grains of sand. The blue dot you see is a star. Every other speck is a galaxy. Each of these specks, like our own Milky Way, contains countless thousands of stars. On planets orbiting how many of these stars do students study photographs such as this and speculate on our existence?

C

E

The size of the earth has also played an important role because it has permitted a gaseous atmosphere. If the earth were smaller, it would not have a sufficient gravitational pull to hold an atmosphere. If it were larger, it might hold such a dense atmosphere that all solar radiation would be absorbed before it reached the surface of the earth.

Has life evolved on other worlds? In the universe there are undoubtedly many worlds with physical characteristics that resemble those of our planet. The universe contains some 10^{20} stars with physical characteristics that resemble those of our sun; at least 10% of these stars are thought to have planetary systems. If only 1 in 10,000 planets is the right size and at the right distance from its star to duplicate the conditions in which life originated on earth, the "life experiment" will have been repeated 10^{15} times (i.e., a million billion times). It does not seem likely that we are alone.

It seems very likely that life has evolved on other worlds in addition to our own.

Nor should we overlook the possibility that life processes might have evolved in different ways on other planets. A functional genetic system, capable of the accumulation and replication of changes and thus of adaptation and evolution, could theoretically evolve from molecules other than carbon, hydrogen, nitrogen, and oxygen in a different environment. Silicon, like carbon, needs four electrons to fill its outer energy level, and ammonia is even more polar than water. Perhaps under radically different temperatures and pressures, these elements might have formed molecules as diverse and flexible as those which carbon forms on earth.

The final question about life, of course, is whether it has an end. Does evolution tend toward any fixed form or state? It has been suggested that with 10^{15} worlds on which life might have emerged, we should have heard from someone by now—unless the life experiment does not work, with life always tending to destroy itself. We can only hope that this is not so and that peace and progress are perpetual possibilities.

SUMMARY

1. Of the many explanations of how life might have originated, only the theory of evolution provides a scientifically testable explanation.

2. The experimental re-creation of atmospheres similar to those thought to have existed on the primitive earth, with the energy sources and temperatures we think were prevalent, leads to the spontaneous formation of amino acids and other biologically significant molecules.

3. Life cannot be defined simply by movement or sensitivity. Its key characteristics are cellular organization, growth, reproduction, and especially heredity.

4. The first cells are thought to have arisen by a process of selecting for aggregations of molecules that were more stable and therefore persisted longer.

5. Before the Phanerozoic Era, which began about 630 million years ago, organisms were too small to be seen with the naked eye. Microscopic fossils—bacteria—are found continuously in the fossil record as far back as 3.5 billion years, in the oldest rocks suitable for the preservation of organisms.

6. Bacteria were the only life forms on earth for 2 billion years or more. At least four kinds of bacteria were present in ancient times: methane and sulfate utilizers, anaerobic photosynthesizers, and eventually O_2-forming photosynthesizers.

7. The first eukaryotes can be seen in the fossil record 1.3 to 1.5 billion years ago. All organisms other than bacteria are their descendants.

8. The number of stars in the universe similar to our sun exceeds 10^{20} (i.e., 100,000,000,000,000,000,000). It is almost certain that life has evolved on planets circling many of them.

SELF-QUIZ

The following questions pinpoint important information in this chapter. You should be sure to know the answers to each of them.

1. The oldest fossil that has ever been found is in a rock that is how old?
   - (a) 3 million years
   - (b) 4.5 million years
   - (c) 3.5 billion years
   - (d) 32 million years
   - (e) 3.9 million years

2. Which of the following gases was *not* present in the atmosphere of the primitive earth?
   - (a) nitrogen
   - (b) carbon dioxide
   - (c) oxygen
   - (d) water
   - (e) ammonia

3. An atmosphere rich in hydrogen is referred to as a _____ atmosphere.

4. Which of the following conditions is a sufficient criterion to define life?
   - (a) movement
   - (b) sensitivity
   - (c) growth
   - (d) reproduction
   - (e) none of the above

5. Small spheres of phospholipid that form spontaneously in water solution are called _____.

6. The earliest fossils that can be seen with the naked eye are how old?
   - (a) 3.5 billion years
   - (b) 590 million years
   - (c) 3.5 million years
   - (d) 3.9 billion years
   - (e) none of the above

7. For a very long period of time, bacteria were the only living creatures on earth. How long a period?

8. Organisms that carry out photosynthesis show a marked preference for one of the two isotopes of carbon in the CO_2 from which they obtain carbon atoms for biosynthesis. Which carbon isotope is preferred, ^{12}C or ^{13}C?

9. The advent of O_2-forming photosynthesis occurred some _____ billion years ago.

10. The most far-reaching consequence of the appearance of organisms carrying out O_2-forming photosynthesis was the introduction into the atmosphere of _____.

11. The descendants of the early O_2-forming photosynthesizers have changed very little. Their living representatives appear very similar to the fossil ones. The living forms are called ____.

12. The first eukaryotes that appear in the fossil record are about how old?
   - (a) 1.3 billion years
   - (b) 590 million years
   - (c) 3.5 billion years
   - (d) 32 million years
   - (e) 3 million years

THOUGHT QUESTIONS

1. In Fred Hoyle's science fiction novel *The Black Cloud,* the earth is approached by a large interstellar cloud of gas. The cloud orients around the sun. Scientists soon discover that the cloud is feeding, absorbing the sun's energy through the excitation of electrons in the outer energy levels of cloud molecules, a process similar to the photosynthesis that occurs on earth. Different portions of the cloud are isolated from one another by associations of ions created by this excitation. Electron currents pass between these sectors, much as they do on the surface

of the human brain, endowing the cloud with self-awareness, memory, and the ability to think. Using static electricity produced by static discharges, the cloud is able to communicate with human beings and describe its history, as well as to maintain a protective barrier around itself. It tells our scientists that it once was smaller, having originated as a small extrusion from an ancestral cloud, but has grown by the adsorption of molecules and energy from stars like our sun, on which it has been grazing. Soon the cloud moves off in search of other stars. Is it alive?

2. If 1 in 10 stars like our sun has planets, and 1 in 10,000 of these planets is capable of supporting life, and 1 in 1,000,000 life-supporting planets evolves an intelligent life form, how many planets in the universe support intelligent life? Can you think of any objections to this estimate?

FOR FURTHER READING

CAIRNS-SMITH, A.G.: "The First Organisms," *Scientific American,* June 1985, pages 90-101. Interesting article arguing that clay, not primordial soup, provided the fundamental materials from which life arose.

DICKERSON, R.: "Chemical Evolution and the Origin of Life," *Scientific American,* September 1978, pages 62-78. A lucid exposition of the chemical changes thought to have occurred during the evolution of life.

HARLAND, W.B., et al.: *A Geologic Time Scale,* Cambridge University Press, Cambridge, England, 1982. An up-to-date treatment of geological time.

HOROWITZ, N.: "The Search for Life on Mars," *Scientific American,* November 1977, pages 52-61. A fascinating account of what it means to search for alien life and the difficulty in knowing when you have found it.

MILLER, S.L.: "A Production of Amino Acids Under Possible Primitive Earth Conditions," *Science,* vol. 117, pages 528-529, 1953. The original paper describing the experiment which first demonstrated that biological molecules might have been generated by natural forces on the primitive earth.

OPARIN, A.: *The Origin of Life,* Dover Publishers, New York, 1938. The original presentation by a Russian biochemist of the idea that life arose spontaneously in the oceans of the primitive earth. A hallmark in scientific thinking.

SILK, J.: *The Big Bang: The Creation and Evolution of the Universe,* W.H. Freeman and Company, San Francisco, 1980. A description of current ideas about how the earth was formed, well written and broad in its coverage.

Overview

Living organisms are built from subassemblies, much as a house is. Some of these building blocks are long polymers, chains of similar units joined in a row. Among the polymers that make up the bodies of organisms are starches (molecules used to store chemical energy), proteins (molecules that facilitate specific chemical reactions), and nucleic acids (molecules in which hereditary information is stored). Other building blocks of organisms are composites, assemblies of very different elements. Among the important composites of organisms are the phospholipids, which form biological membranes, and the universal energy carrier of all organisms, adenosine triphosphate (ATP).

For Review

Here are some important terms and concepts that you will encounter in this chapter. If you are not familiar with them, you should review them before proceeding.

- **Amino acids** (Chapter 3)

- **Hydrophobic interactions** (Chapter 2)

- **Miller–Urey experiments** (Chapter 3)

- **Reduced carbon compounds** (Chapter 2)

All living organisms possess the same basic architecture, an arrangement that must have evolved early in the history of life. It is the architecture of a factory.

Like a factory, an organism contains *tools* that accomplish specific tasks. In an organism these tools act to facilitate certain chemical reactions that would not occur otherwise. The process of facilitating chemical reactions is **catalysis,** and the tools that accomplish it are **enzymes.** A different enzyme catalyzes each reaction within an organism. Enzymes are one example of a class of molecules, the proteins, that is very important to organisms.

In a factory *power* is required to operate the tools. In an organism the power is chemical power, which is used to aid the tools in catalyzing certain reactions. In a factory power is delivered to tools by cables. In an organism chemical power is packaged in a small portable unit, a molecule called **adenosine triphosphate** (ATP).

Like most factories, organisms store backup supplies of energy. Organisms store energy in the carbon-hydrogen bonds of carbon-rich molecules called carbohydrates. You are familiar with some of these carbohydrates as sugars and starches.

The tools that are present in any factory and the schedule in which they are used are governed by a *master plan*. In an organism the information of this master plan is contained within a long molecule called deoxyribonucleic acid (DNA).

When some of this information is to be used by the organism, a copy of the appropriate part of the master plan is made. This copy is called ribonucleic acid (RNA). It is used to direct the production of new enzymes, just as in a factory a blueprint copy is made of portions of the master plan, to specify the production of new tools.

The interior walls of a factory are usually flexible, permitting changes in arrangement. So are the interior walls of organisms, which are called membranes. They are composed of molecules called lipids, which are not soluble in water. Extending through the membranes are numerous doors made of clusters of protein molecules.

The outer walls of a factory are often, although not always, structurally reinforced with strong materials such as cinder block and steel. An organism is constructed in a similar manner. Some, but by no means all, of the cells of organisms possess structurally reinforced exteriors. The strength of these cell walls is achieved by networks of carbohydrate fibers.

This factory-like architecture is not the only way in which one could conceive of constructing an organism. It is, however, that evolutionary path taken by our earliest ancestors; we all share the same basic biological design.

In this chapter we will limit discussion to a description of the basic chemical building blocks of organisms, the mortar and bricks that are used to assemble a cell. In following chapters we will discuss cells, the basic units of organization that are characteristic of all organisms; how cells evolved; and what their internal structure is like.

THE BUILDING BLOCKS OF ORGANISMS

Most organic molecules are built up of groups of atoms with definite chemical properties. These groups of atoms are referred to as **functional groups.** For example, a hydrogen atom bonded to an oxygen atom, —OH, is a **hydroxyl** group. The most important functional groups are listed in Figure 4-1. Most of the chemical reactions that occur within organisms involve the transfer of a functional group from one molecule to another or the breaking of a carbon-carbon bond (Figure 4-2). The enzymes called kinases, for example, add or remove phosphate groups from molecules.

Some of the molecules in organisms are simple organic molecules, often with a single reactive functional group protruding from a carbon chain. Other molecules are much larger, particularly those which play a structural role in organisms or store information. Most of these large macromolecules are themselves composed of simpler elements, just as a wall is composed of bricks. Most of the macromolecules that occur in organisms fall into five classes:

Polymers	Composites
Storage carbohydrates	Lipids
Proteins	Nucleotides, such as ATP
DNA, RNA	

A **polymer** is a molecule built of a long chain of similar modules, like cars coupled together to form a railroad train. Many of the most important building blocks of organisms are polymers. Complex carbohydrates, for example, are polymers of simple ring mole-

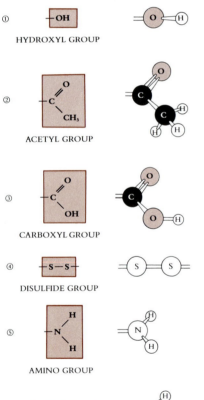

FIGURE 4-1

The principal functional chemical groups. These groups tend to act as units during chemical reactions and to confer specific chemical properties on the molecules that possess them. Hydroxyl groups, for example, make a molecule more basic, whereas carboxyl groups make a molecule more acidic.

① HYDROXYL GROUP

② ACETYL GROUP

③ CARBOXYL GROUP

④ DISULFIDE GROUP

⑤ AMINO GROUP

⑥ PHOSPHATE GROUP

FIGURE 4-2

SINGLE BONDS		DOUBLE BONDS	
BOND LENGTH (NANOMETERS)	BOND STRENGTH (KCAL/MOLE)	BOND LENGTH (NANOMETERS)	BOND STRENGTH (KCAL/MOLE)
N—H 0.102	93.4	C═O 0.123	171
—C—H 0.109	98.8	C═N— 0.127	147
—C—O— 0.143	84	C═C 0.133	147
—C—N 0.148	70	TRIPLE BOND	
—C—C— 0.154	83	—N≡N— 0.110	225

The length and strength of some covalent bonds that occur in biologically important molecules. The weakest bonds, those most easily broken, couple hydrogen atoms to carbon or nitrogen. The bonds binding more massive atoms, such as oxygen, are much stronger. For any of these atoms, double bonds linked to carbon are about twice as strong as single bonds. Triple bonds, such as the one formed in nitrogen gas, are even stronger.

cules called sugars. Enzymes, membrane proteins, and other proteins are polymers of amino acids. DNA and RNA are two versions of a long-chain molecule called a nucleic acid, a polymer that is composed of a long series of complex composite molecules called nucleotides.

A **composite** molecule is composed of several distinct elements that are joined, forming an operational whole. A mermaid is a mythical composite, assembled from the torso of a woman and the tail of a fish. Similarly, ATP is composed of three quite different elements: a sugar, a triphosphate group, and a carbon-nitrogen ring molecule called adenine. Many lipids, particularly those which form membranes, are also composites.

In discussing the polymers and composites that make up the bodies of organisms, we will begin with carbohydrates. Some carbohydrates are simple small molecules, whereas others form long polymers. Carbohydrates are important to organisms both as energy-storage vessels and as structural elements. After we have completed our discussion of carbohydrates, we will address lipids, amino acids and proteins, ATP, and nucleic acids.

ENERGY-STORING MOLECULES
Sugars

The **carbohydrates** are a loosely defined group of molecules that contain carbon, hydrogen, and oxygen. Because they contain many carbon-hydrogen (C—H) bonds, which release energy when they are broken, carbohydrates are well suited for energy storage. Most carbohydrates contain the three elements carbon, hydrogen, and oxygen in the molar ratio 1:2:1. A chemist would say that their empirical formula (a list of the atoms in a molecule with a subscript to indicate how many of each) is $(CH_2O)_n$, where n is the number of carbon atoms. Among the simplest of the carbohydrates are the simple sugars, or **monosaccharides** (from the Greek, *monos,* single + Latin, *saccharum,* sugar). Simple sugars may have as few as three carbon atoms, but those molecules which play the central role in energy storage have six. They have the empirical formula

$$C_6H_{12}O_6, \text{ or } (CH_2O)_6$$

FIGURE 4-3

Structure of the glucose molecule.
Glucose is a linear six-carbon
molecule that forms a ring in solution.
The structure of the ring can be
represented in many ways, of which
these are the most common.

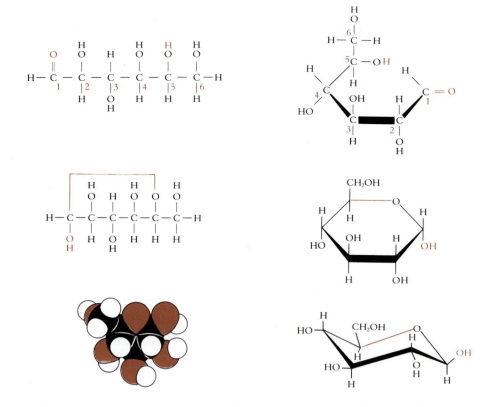

Sugars can exist in a straight-chain form, but in water solution they almost always form rings. The primary energy-storage molecule is **glucose** (Figure 4–3).

Among the most important energy-storage molecules in organisms are sugars. Simple sugars contain six carbon atoms and seven energy-storing C—H bonds.

Glucose is not the only sugar with the formula $C_6H_{12}O_6$. Among the other sugars that have this same formula are fructose and galactose (Figure 4–4). Since these molecules have the same empirical formula as glucose, they are isomers, or alternative forms, of glucose. Glucose and fructose are **structural isomers.** In fructose the double-bonded oxygen is attached to an internal carbon rather than to a terminal one. Your taste buds can tell the difference: fructose tastes much sweeter than glucose. This structural difference also has an important chemical consequence: the two sugars form very different polymers.

Unlike fructose, galactose has the same bond structure as glucose; the only difference between them is in the orientation of one hydroxyl (—OH) group. Because the two orientations are mirror images of one another, glucose and galactose are said to be **stereo-isomers.** Again, this seemingly slight difference has important consequences, since this hydroxyl group is one that is often involved when links are formed to create polymers.

Transport Disaccharides

Most organisms transport sugars within their bodies. In humans glucose circulates in the blood. In plants glucose is manufactured in the leaves as a result of photosynthesis and then transported to other portions of the plant. The glucose that circulates in our bloodstream does so as a simple monosaccharide. In many other organisms, however, particularly plants, glucose is converted to a *transport form* before it is moved from place to place within the organism. In such a form it is less readily metabolized while it is being transported from place to place. Transport forms of sugars are commonly formed by linking two monosaccharide sugar molecules together to form a **disaccharide** (Greek, *di-,* two). Disaccharides serve as very effective reservoirs of glucose, which cannot be metabolized by the normal

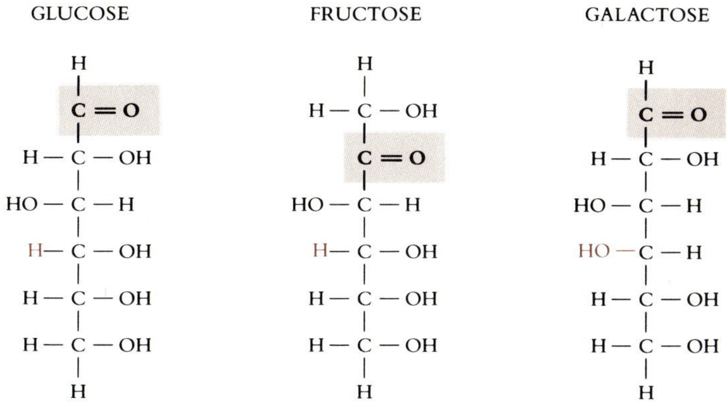

GLUCOSE

$C = O$
H
$|$
$H - C - OH$
$|$
$HO - C - H$
$|$
$H - C - OH$
$|$
$H - C - OH$
$|$
$H - C - OH$
$|$
H

FRUCTOSE

H
$|$
$H - C - OH$
$|$
$C = O$
$|$
$HO - C - H$
$|$
$H - C - OH$
$|$
$H - C - OH$
$|$
$H - C - OH$
$|$
H

GALACTOSE

H
$|$
$C = O$
$|$
$H - C - OH$
$|$
$HO - C - H$
$|$
$HO - C - H$
$|$
$H - C - OH$
$|$
$H - C - OH$
$|$
H

FIGURE 4-4

A structural isomer of glucose, such as fructose, has identical chemical groups bonded to different carbon atoms, whereas a stereoisomer of glucose, such as galactose, has identical chemical groups bonded to the same carbon atoms in a different orientation.

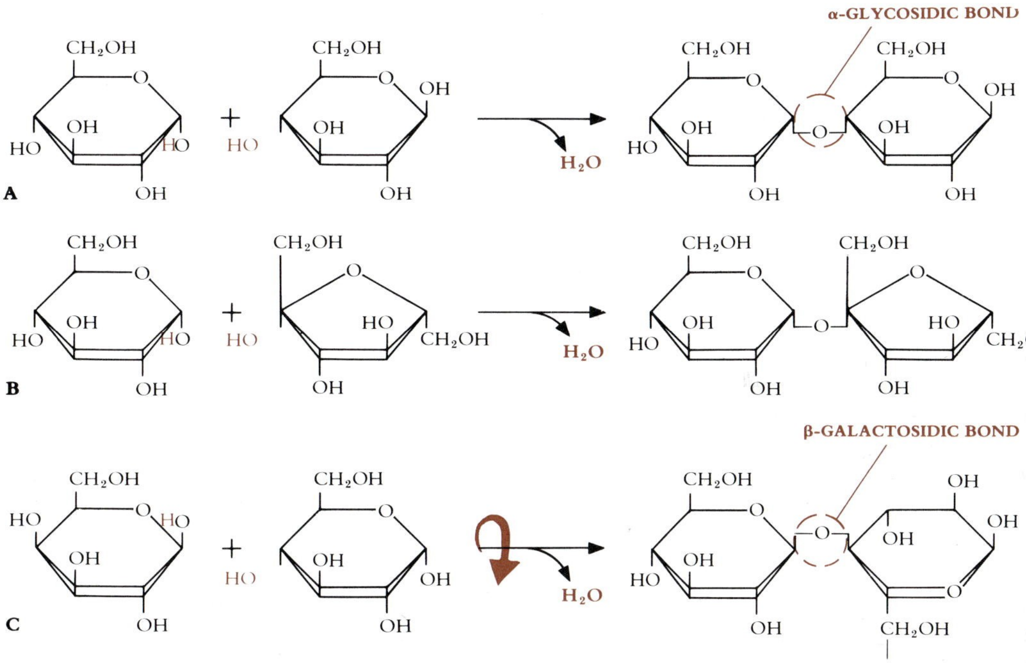

FIGURE 4-5

Disaccharides formed with glucose are often used to transport glucose from one part of an organism's body to another and to store it. Apples, for example, are rich in maltose (A), whereas sugar cane is rich in sucrose (B); both are disaccharides. Because many adult mammals lack an enzyme capable of splitting a rotated beta-linkage, they cannot metabolize glucose disaccharides containing these linkages. Children, by contrast, contain the appropriate enzyme and readily metabolize lactose (C). This simple strategy serves to reserve the food supply committed to lactose for the child.

α-GLYCOSIDIC BOND

β-GALACTOSIDIC BOND

A

B

C

glucose-utilizing enzymes of the organism to obtain energy until the bond linking the two monosaccharide subunits is broken by a specific enzyme. These enzymes are typically present only in the tissue where the glucose is utilized.

Different transport forms differ from one another in the identity of the monosaccharides linked in the disaccharide and in the nature of the chemical bond linking the two monosaccharides together. Glucose forms transport disaccharides with many other monosaccharides, including glucose itself, galactose, and fructose. When two glucose molecules form a bond between the front of one glucose molecule (carbon 1) and the back of the other (carbon 4), the resulting disaccharide is maltose (Figure 4-5). When glucose forms this same kind of linkage between two glucose molecules front to front (carbon 1 to carbon 1), the resulting disaccharide is trehalose, the form in which glucose is transported in many insects. When glucose forms a disaccharide with its structural isomer fructose by forming the same kind of front-to-front bond between glucose (carbon 1) and fructose (carbon 2), the resulting disaccharide is sucrose. Commonly called cane sugar, sucrose is the form in which most plants transport glucose and the sugar that most people (and other animals) eat (Figure 4-6).

A very different disaccharide results when glucose is linked to its stereoisomer galactose. The two hydroxyl groups involved in forming the link between them have opposite orientations, producing a beta-galactosidic bond. The resulting disaccharide is lactose. Many mammals supply energy to their babies in the form of lactose. Channeling food energy into lactose production has the effect of reserving the energy for the child, since

FIGURE 4-6

This hungry fly has its proboscis extended down into a sucrose solution. Sensors in the fly's footpads alert it to the presence of sugar and cause it to extend its proboscis. If the solution had been distilled water instead of sucrose, the reflex governing proboscis extension would not have been triggered.

many adults, including virtually all non-Caucasian humans, lack the beta-galactosidase enzyme required to cleave the disaccharide into its two monosaccharide components. Since they lack this enzyme, adults cannot metabolize lactose.

Starches

Organisms store the metabolic energy that is captured in glucose and transported in glucose-containing disaccharides by converting it into an insoluble form and depositing it in specific storage areas in their bodies. The disaccharides are rendered insoluble when they are joined together into long polymers called **polysaccharides** composed of monosaccharide sugar subunits. Polysaccharides formed from glucose are called **starches.**

The starch with the simplest structure is amylose. Amylose is composed of many hundreds of glucose molecules linked together in long unbranched chains. Each linkage is between the carbon 1 of one glucose molecule and the carbon 4 of another so that amylose is, in effect, a longer form of maltose. The long chains of amylose tend to coil up in water, a property that renders amylose insoluble. Potato starch is about 20% amylose. When amylose is digested by a sprouting potato plant (or by you), enzymes first break it into fragments of random length. These shorter fragments are soluble. Baking or boiling potatotes has the same effect. When the amylose has been split into such fragments, another special enzyme cuts them into segments that are two glucose molecules long—the disaccharide maltose. Finally the maltose is cleaved into glucose, which the cell is able to metabolize.

Most plant starch, including 80% of potato starch, is a somewhat more complicated variant of amylose called amylopectin (Figure 4-7). **Pectins** are branched polysaccharides, and amylopectin is a form of amylose with short, linear amylose branches consisting of 20 to 30 glucose subunits. In some plants these chains are cross-linked. These cross-links serve to create an insoluble mesh of glucose, which can be degraded only in the presence of yet another enzyme. The size of the mesh differs from one plant to another. In rice, for example, it is made up of about 100 amylose chains, each with one or two cross-links.

FIGURE 4-7

Storage polymers of glucose are called starches. The simplest starches are long chains of maltose called amylose. Most plants contain more complex starches called amylopectins, which contain branches. The nucleus of the liver cell in the micrograph inset is surrounded by dense granules of animal starch called glycogen, which is even more highly branched.

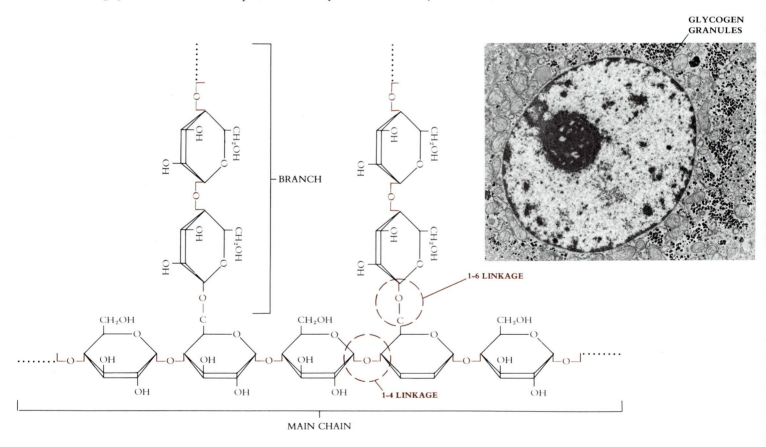

BRANCH

GLYCOGEN GRANULES

1-6 LINKAGE

1-4 LINKAGE

MAIN CHAIN

Animals also store glucose in branched amylose chains. In animals, however, the average chain length is much longer and there are more branches, producing a highly branched structure called glycogen (Figure 4-7, *inset*).

Starches are sugar polymers, which are often built from glucose subunits. Most starches are branched, and some are cross-linked. The branching and cross-linking serve both to render the polymer insoluble and to protect it from degradation.

Cellulose

All starches are composed of glucose. None is formed utilizing beta-linkages like those involved in the disaccharide lactose. The reason that none is formed becomes obvious when we consider the properties of a glucose polymer that is formed with beta-glucosidic links. When such a polymer is formed of glucose subunits linked in this way, the polysaccharide that results is **cellulose,** the chief constituent of plant cell walls. Cellulose is chemically similar to amylose, except for one very important difference (Figure 4-8): the starch-degrading enzymes that occur in most organisms cannot break a beta-glucosidic link. It is not that beta-links are stronger, but rather that their cleavage requires a different enzyme, one not usually present. Because cellulose cannot readily be broken down, it works well as a biological structural material and occurs widely as such in plants. The structural material that occurs in insects, many fungi, and certain other organisms is **chitin.** In effect, chitin is a modified form of cellulose in which a nitrogen group has been added to the glucose units (Figure 4-9). Chitin is a tough, resistant surface material.

Few organisms are able to digest cellulose or chitin. Those which can include some fungi, some protozoa, and a few kinds of insects, earthworms, wood-boring mollusks, and bacteria. Cows and other grass-eating mammals, termites, and a few other groups of plant-eaters digest cellulose by maintaining living colonies of the appropriate kinds of protozoa

FIGURE 4-8

The jumble of cellulose fibers below is from a yellow pine (Pinus ponderosa).
Each fiber is composed of microfibrils (below left), which are bundles of cellulose chains. Because each chain contains only beta-linkages, cellulose fibers can be very strong; they are quite resistant to metabolic breakdown, which is one reason that wood is such a good building material.*

CELLULOSE MICROFIBRILS

CELLULOSE CHAIN

FIGURE 4-9

Chitin, which might be considered to be a modified form of cellulose, is the principal structural element in the external skeletons of many invertebrates, such as this crab, and in the cell walls of fungi.

and bacteria in their digestive systems. Since humans do not possess colonies of such organisms in their digestive systems, they cannot digest cellulose.

Fats

When organisms store glucose molecules for long periods of time, they usually convert them to another kind of insoluble molecule containing more C—H bonds than carbohydrates have. These storage molecules are called **fats.** Whereas the hydrogen/oxygen ratio in carbohydrates is 2:1, in fat molecules it is much higher. Like starches, fats are insoluble and can therefore be deposited at specific storage locations within the organism. The insolubility of starches, as we have just described, results from the fact that they are long polymers. The insolubility of fats, on the other hand, occurs because they are nonpolar. Unlike the H—O bond of water, the C—H bond of carbohydrates is nonpolar and cannot form hydrogen bonds. Because of the large number of C—H bonds that they contain, fat molecules are hydrophobically excluded by water, since water molecules seek to form hydrogen bonds with other water molecules. The result is that fat molecules cluster together and are therefore insoluble in water.

Fats are classified as **lipids,** a loosely defined group of molecules that are insoluble in water but soluble in oil. Oils and waxes are also lipids. Fats are composite molecules, each comprising two different kinds of subunits:

1. *Glycerol.* A three-carbon alcohol, each of whose carbons bears a hydroxyl group. The three carbons form the backbone of the fat molecule, to which three fatty acids are attached.
2. *Fatty acids.* Long hydrocarbon chains ending in a carboxyl (—COOH) group. Three fatty acids are attached to each glycerol backbone.

The structure of an individual fat molecule, such as the one diagrammed in Figure 4–10, consists simply of a glycerol molecule with a fatty acid joined to each of its three carbon atoms:

$$
\begin{array}{c}
\text{H} \\
| \\
\text{H—C—Fatty acid} \\
| \\
\text{H—C—Fatty acid} \\
| \\
\text{H—C—Fatty acid} \\
| \\
\text{H}
\end{array}
$$

Because there are three fatty acids, the resulting fat molecule is called a **triglyceride.**

Fatty acids vary in length. The most common are even-numbered chains of 14 to 20 carbons. Fatty acids in which all of the internal carbon atoms possess hydrogen side groups are said to be **saturated** because they contain the maximum number of hydrogen atoms that are possible. Some fatty acids have double bonds between one or more pairs of successive carbon atoms (Figure 4–11). Fats that are composed of such fatty acids are said to be **unsaturated,** since the double bonds replace some of the hydrogen atoms and they therefore contain fewer than the maximum number of such atoms. If a given fat has more than one double bond, it is said to be **polyunsaturated.** Polyunsaturated fats have low melting points, because their chains bend at the double bonds so that the fat molecules cannot be aligned closely with one another. Consequently, the fat may be fluid. A liquid fat is called an **oil.** Many plant fatty acids, such as oleic acid (a vegetable oil) and linolenic acid (a linseed oil) are unsaturated. Animal fats, in contrast, are often saturated and occur as hard fats.

It is possible to convert an oil into a fat by adding hydrogen. The peanut butter that you buy in the store has usually been hydrogenated to convert the peanut fatty acids to hard fat and thus to prevent them from separating out as oils while the jar sits on the store shelf.

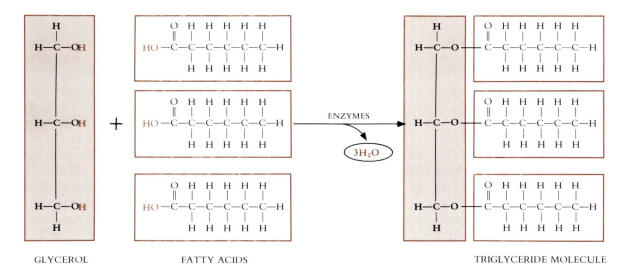

GLYCEROL   FATTY ACIDS   TRIGLYCERIDE MOLECULE

FIGURE 4-10

Triglycerides are composite molecules, made up of three fatty acid molecules coupled to a single glycerol backbone.

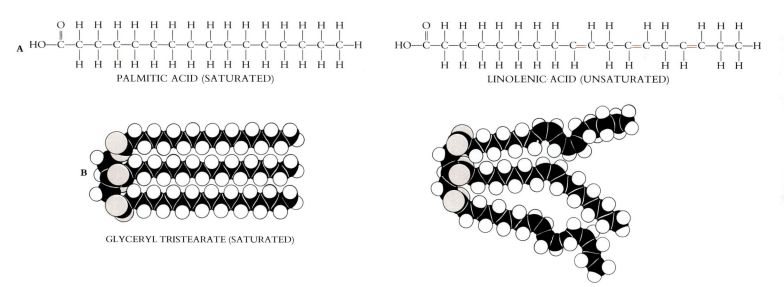

PALMITIC ACID (SATURATED)   LINOLENIC ACID (UNSATURATED)

GLYCERYL TRISTEARATE (SATURATED)

LINSEED OIL (UNSATURATED)

FIGURE 4-11

A *Palmitic acid, with no double bonds and thus a maximum number of hydrogen atoms bonded to the carbon chain, is a saturated fatty acid. Linolenic acid, with three double bonds and thus fewer than the maximum number of hydrogen atoms bonded to the carbon chain, is an unsaturated fatty acid.*
B *Many animal triglyceride fats are saturated. Because their fatty acid chains can fit closely together, these triglycerides form immobile arrays called hard fat. Plant fats, by contrast, are typically unsaturated, and the many kinks that the double bonds introduce into the fatty acid chains prevent close association of the triglycerides and produce oils such as linseed oil, which is obtained from flax (Linum) seed.*

Fats are very efficient energy-storage molecules because of their high concentration of C—H bonds. Most fats contain over 40 carbon atoms. The ratio of energy-storing C—H bonds to carbon atoms is more than twice that of carbohydrates, making fats much more efficient vehicles for storing chemical energy. On the average, fats yield about 9 kilocalories (kcal) of chemical energy per gram of fat, as compared with somewhat less than

4 kilocalories per gram obtainable from carbohydrates. As you might expect, the more highly saturated fats are richer in energy than less saturated ones. Animal fats contain more calories than do vegetable fats. Human diets that contain relatively large amounts of saturated fats appear to upset the normal balance of fatty acids in the body, a situation that may lead to heart disease.

The total carbohydrate that organisms consume or produce is allocated three different ways: (1) some is maintained as glucose and is available for immediate use; (2) a second portion is converted to transport disaccharides and can then be carried with minimum loss to other parts of the organism; and (3) some is converted to starches or fats and reserved for future use. The reason that many people gain weight as they grow older is that the amount of carbohydrate needed daily decreases with age, whereas the intake of food does not. A greater and greater portion of the ingested carbohydrate is available to be converted to fat.

PHOSPHOLIPIDS AND BIOLOGICAL MEMBRANES

All living organisms are enclosed within membranes, barriers that isolate them from the surrounding world. A membrane serves much the same function as the walls of a house: it prevents the dilution of the contents of the cell and permits selective control over which molecules enter and leave the cell.

Biological membranes have diverse structures, but they share several important characteristics:

1. Membranes are formed from sheets of molecules only a few molecules thick; most are 6 to 10 nanometers thick. It would take more than 10,000 membranes to equal the thickness of this sheet of paper.
2. The basic structure of a membrane is composed of lipids, which, as you will see, spontaneously form bimolecular layers that are impermeable to ions and polar molecules.
3. Proteins embedded within the lipid bilayer of membranes serve as gates, permitting and in some cases facilitating the passage of specific ions and polar molecules.
4. The forces holding membranes together are hydrophobic rather than covalent. Consequently, membranes are fluid structures, within which individual lipids are free to move from place to place.

The units of which most biological membranes are composed are **phospholipids,** lipids that are similar in structure to the triglycerides that you encountered as components of energy-storing fats. These membrane phospholipids (Figure 4-12, *A*) are composite molecules made up of four elements:

1. The simple three-carbon molecule called glycerol, which you encountered in the discussion of fats. Glycerol constitutes the backbone of the composite lipid molecule, just as it does for the fat molecule. The other three elements are attached to the glycerol molecule like ribs to a backbone, one element to each carbon.
2 & 3. Two fatty acid chains, which need not be the same as one another. Carbon chains with 16 carbon atoms (palmitic acid) and those with 18 carbon atoms (oleic acid) are most common, although fatty acid chains with as many as 24 carbons are also represented in different phospholipid molecules.
4. A phosphorylated alcohol. The third fatty acid chain found in triglycerides is absent in phospholipids. Instead, an alcohol is attached to the third glycerol carbon. The alcohol is usually serine, ethanolamine, choline, glycerol, or inositol. Since all of these alcohols possess polar groups, they render the alcohol end of the phospholipid polar, in contrast to the fatty acid end, which is markedly nonpolar.

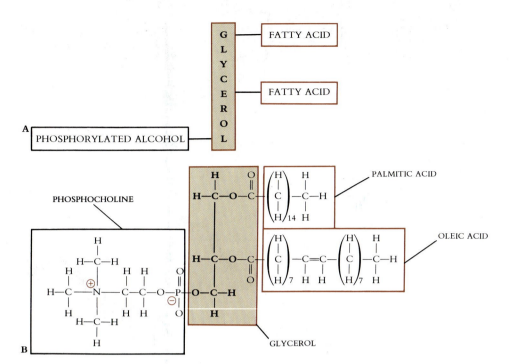

FIGURE 4-12

A *A phospholipid is a composite molecule similar to a triglyceride, except in this case only two fatty acids are bound to the glycerol backbone, the third position being occupied by a phosphorylated alcohol.*

B *One of the most common membrane phospholipids is lecithin.*

As an example, let us construct a phospholipid from these components:

Component	Molecule selected
Core backbone	Glycerol
Fatty acid 1	Palmitic acid (C = 16)
Fatty acid 2	Oleic acid (C = 18)
Phosphorylated alcohol	Choline

The resulting molecule is called phosphatidyl choline, or **lecithin** (Figure 4–12, *B*). It is a major component of the membranes of most higher organisms. All of the principal phospholipids are built up in this same simple way, from two fatty acids and a phosphorylated alcohol that are attached to glycerol.

In the phospholipids in Figure 4–12 you can see that each has one end that is hydrophilic (polar) and one end that is hydrophobic (nonpolar). Molecules that have a polar end and a nonpolar end are said to be **amphipathic.** The amphipathic nature of phospholipids can be seen clearly in a molecular model of phosphatidyl choline: the two nonpolar fatty acid chains extend in one direction, roughly parallel to each other, while the polar choline group points in the opposite direction. Because of this structure, phospholipids are often symbolized diagrammatically as a (polar) ball with two dangling (nonpolar) tails (Figure 4–13).

FIGURE 4-13

Phospholipids are usually oriented so that the phosphorylated alcohol, which is polar, extends from one end of the molecule, and the two fatty acid chains, which are nonpolar, extend from the other. For this reason phospholipids are often diagrammed as a polar ball with a nonpolar tail. Left, Diagrammatic representation. Right, Molecular model of lecithin.

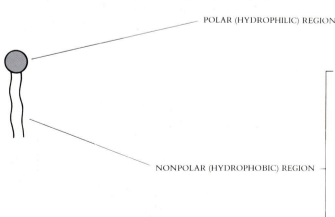

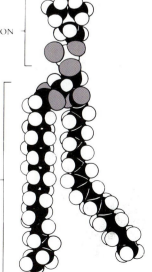

POLAR (HYDROPHILIC) REGION

NONPOLAR (HYDROPHOBIC) REGION

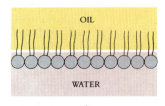

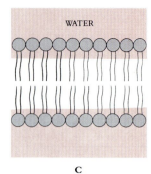

A | B | C

FIGURE 4-14

A *At an oil-water interface, phospholipid molecules will orient so that their polar heads are in the polar medium, water, and their nonpolar tails are in the nonpolar medium, oil.*
B *When the tails are short, micelles form. They are spheres with polar surfaces and nonpolar interiors.*
C *When tails are longer, they tend to adhere to one another, and a bilayer membrane forms instead.*

Why do phospholipids occur in all biological membranes? Their amphipathic nature ideally suits them to this role. Imagine what happens to phospholipids placed in aqueous medium. Their nonpolar ends, which tend to avoid water, aggregate together. When the nonpolar tails are short, small spherical aggregations called **micelles,** each about 10 to 20 nanometers in diameter, can form (Figure 4-14). When the nonpolar tails are long, however, and consist of chains with more than 16 carbons, the nonpolar tails tend to stick to one another. Because of this attraction, the phospholipid molecules that have long nonpolar tails orient themselves with their tails side-by-side. In doing this they spontaneously form a bimolecular sheet, one in which the polar groups are on the outside surface and the nonpolar tails are isolated in the interior. This **lipid bilayer** is the basic structure of cellular membranes. Its formation, which is driven by hydrophobic interactions, occurs spontaneously.

> **The basic structure of biological membranes forms spontaneously as a result of the aggregation of phospholipid molecules into a double layer. In such a layer the nonpolar heads of the phospholipid molecules point inward, forming a nonpolar zone in the interior of the bilayer.**

Although the lipids that occur in lipid bilayer membranes are not bonded to one another by covalent bonds, the membranes are quite stable for the following reasons: (1) the imprisonment of nonpolar tails within the interior of the bilayer is maintained by hydrophobic forces; (2) the hydrocarbon tails are attracted to one another by cohesive forces in areas where these tails are close to one another; (3) the polar groups form hydrogen bonds with water molecules; and (4) electrical charges on the polar groups interact with water molecules.

The net effect of these cooperating forces is to create a very stable structure. Interestingly, none of these forces depends on any given phospholipid molecule occupying a fixed position with respect to any other; only their orientation is important. As a result, the individual lipid molecules are free to move about within the membrane, but not to leave it. A lipid molecule may travel from one end of a bacterial cell to the other in about a second. The membrane is truly fluid.

> **Biological membranes are flexible because the lipid molecules that make them up move about freely within them. The degree of flexibility of a membrane depends on the ease with which this movement occurs.**

The fluidity of a particular membrane depends on the strength of the cohesive attraction between adjacent phospholipid tails, since the chains must be pulled apart for molecular movement to occur. When one chain of the phospholipid is more saturated than the other—contains fewer double C=C bonds and consequently more carbon-bonded hydrogens—the two chains do not align precisely with one another. The membranes that are formed from such molecules are more fluid than ones formed from phospholipids in which the two chains are more nearly equally saturated (Figure 4-15). You are soft rather than

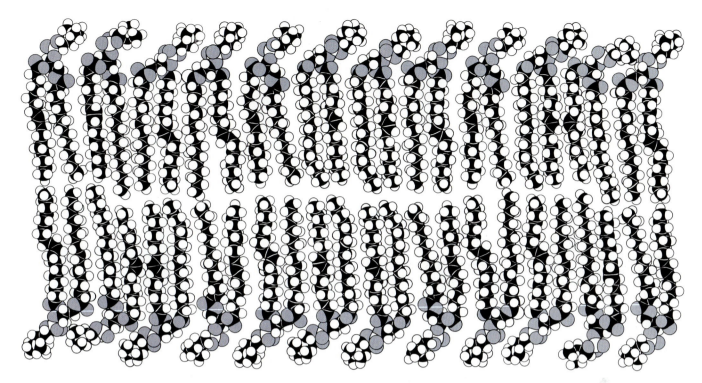

rigid, for example, because of the unequal degree of saturation that is characteristic of the two chains that make up the majority of your phospholipids.

> **Because their movement involves the disruption of cohesive attractions between the hydrocarbon tails, phospholipids whose tails line up least well are the most mobile. Tails with different degrees of saturation line up poorly.**

Transport Across Membranes

The most important property of lipid bilayer membranes is that they are impermeable to ions, as well as to most polar molecules. They act as firm barriers, preventing the movement of large molecules into and out of cells. If membranes were composed only of phospholipids, cells would be impermeable to many metabolites and life would not be possible. Impermeability such as this does not occur, however, because there are gates extending across the lipid bilayer through which specific ions and large polar molecules can pass.

These passages through the membrane are not simply holes. Holes would allow anything to pass and by doing so destroy the compartmentalization that is the essence of cellular organization. Rather, the passages are provided by large protein molecules embedded within the lipid bilayer (Figure 4-16). Proteins are able to change their shape when they interact with other molecules. Molecules pass into and out of the cell by binding to specific proteins that are embedded within the lipid bilayer membranes. Such binding induces a change in the shape of the protein and flips the transported molecule into the cells.

This kind of transport across a membrane sometimes requires metabolic energy. Such transport is often exquisitely specific. Membrane proteins can distinguish between Na^+ and K^+ ions, even though the difference in ionic diameters is much less than the thickness of the membrane itself. Each kind of ion or large polar molecule can pass through the membrane only if it has a specific port of entry which recognizes that individual kind of ion or molecule. In this way the cell controls the kinds of ions and large polar molecules that it allows in and out of itself. Only precise and specific binding to a particular kind of protein in the cell membrane permits a molecule to be transported into or out of that cell.

A

B

FIGURE 4-15

This human red blood cell **(A),** *like all the other cells of your body, is encased within a membrane bilayer, which is clearly visible in the electron micrograph* **(B).** *The diagram above illustrates how the long nonpolar tails of the phospholipids orient toward one another. Because some of the tails contain double bonds, which introduce kinks in their shape, the tails do not align perfectly and the membrane is "fluid"—individual phospholipid molecules can move from one place to another in the membrane.*

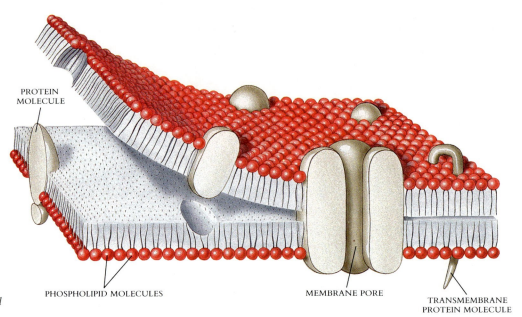

PROTEIN
MOLECULE

PHOSPHOLIPID MOLECULES

MEMBRANE PORE

TRANSMEMBRANE
PROTEIN MOLECULE

FIGURE 4-16

Cell membranes have embedded within them a variety of proteins, some of which extend all the way through the membrane and are called transmembrane proteins. In almost all membranes, aggregations of proteins form pores through which small polar molecules may pass.

PROTEINS

Proteins are the third major group of chemicals that make up the bodies of organisms. We have already encountered one class of proteins, the **enzymes,** biological catalysts that are able to facilitate specific chemical reactions. Because of this property, the appearance of enzymes was one of the most important events in the evolution of life. Other kinds of proteins also have important functions. The cartilage that binds your bone joints together is composed of a protein, collagen. Short proteins called **peptides** are used as chemical messengers within your brain and throughout your body. Despite their diverse functions, however, all proteins have the same basic structure: a long polymer chain of amino acid subunits linked end to end.

Amino Acids

Amino acids are thought to have been among the first molecules formed in the reducing atmosphere of the early earth. Recall that in the Miller–Urey experiments (Chapter 3) the passage of sparks through a mixture of methane, ammonia, hydrogen, and water produced the amino acids glycine, alanine, aspartic acid, and glutamic acid, along with many other organic molecules. Numerous other amino acids were produced in subsequent experiments, and it seems highly likely that the oceans which existed early in the history of the earth contained a wide variety of different amino acids.

An amino acid can be defined as a molecule containing an amino group ($-NH_2$), a carboxyl group ($-COOH$), and a hydrogen atom, all bonded to a central carbon atom:

$$H_2N-\underset{\underset{H}{|}}{\overset{\overset{R}{|}}{C}}-COOH$$

The identity and unique chemical properties of each amino acid are determined by the nature of the fourth covalent bond formed by the central, or alpha, carbon atom. This bond links the central carbon to any of a variety of different side groups, or R groups. Although many different amino acids occur in nature, only 20 are used in proteins. These 20 "common" amino acids and their side groups are illustrated in Figure 4-17. The different side groups of these 20 amino acids give each amino acid distinctive chemical properties. For example, when the side group is $-H$, the amino acid (glycine) is polar, whereas when

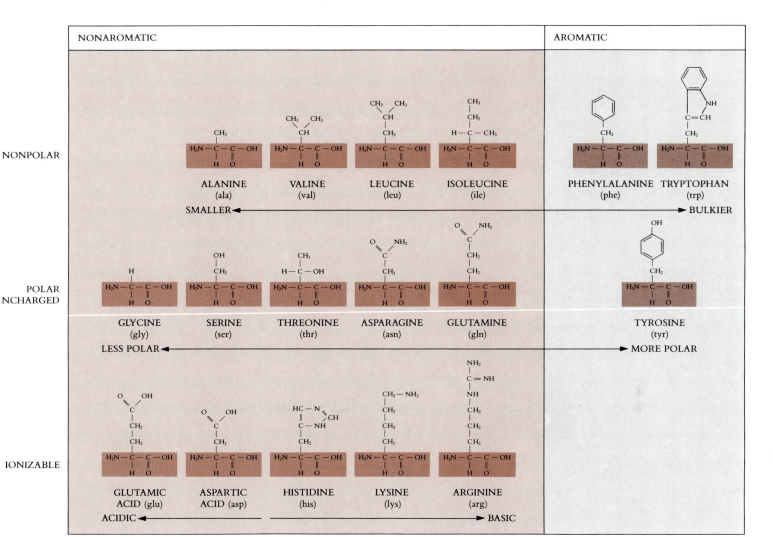

NONAROMATIC

AROMATIC

NONPOLAR

ALANINE (ala) VALINE (val) LEUCINE (leu) ISOLEUCINE (ile) PHENYLALANINE (phe) TRYPTOPHAN (trp)

SMALLER ◄——————————————————————————► BULKIER

POLAR NCHARGED

GLYCINE (gly) SERINE (ser) THREONINE (thr) ASPARAGINE (asn) GLUTAMINE (gln) TYROSINE (tyr)

LESS POLAR ◄——————————————————————————► MORE POLAR

IONIZABLE

GLUTAMIC ACID (glu) ASPARTIC ACID (asp) HISTIDINE (his) LYSINE (lys) ARGININE (arg)

ACIDIC ◄——————————————————————————► BASIC

SPECIAL STRUCTURAL PROPERTY

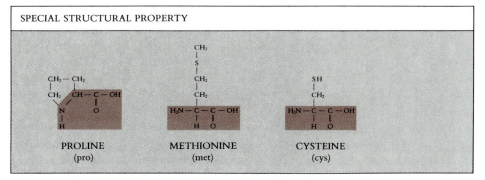

PROLINE (pro) METHIONINE (met) CYSTEINE (cys)

FIGURE 4-17

The 20 common amino acids. Each amino acid has the same chemical backbone, each differing from the others in the side or R group that it possesses. Six of the amino acid R groups are nonpolar, some more bulky than others (particularly the ones containing ring structures, which are called aromatic). *Another six are polar but uncharged; these differ from one another in how polar they are. Five are polar and capable of ionizing to a charged form; under typical cell conditions some of these are acids, others bases. The remaining three have special chemical properties that play important roles in forming links between protein chains or kinks in their shape.*

the side group is —CH_3, the amino acid (alanine) is nonpolar. The 20 amino acids that occur in proteins are commonly grouped into five chemical classes, based on their side groups:

1. *Nonpolar* amino acids, such as leucine
2. *Polar uncharged* amino acids, such as threonine
3. *Ionizable* amino acids, such as glutamic acid
4. *Aromatic* amino acids, such as phenylalanine, whose R groups contain an organic ring with alternating single and double bonds
5. *Special function* amino acids, such as methionine (which often initiates polypeptide chains), proline (which causes kinks in chains of amino acids), and cysteine (which links chains together)

The way in which each amino acid affects the shape of a protein depends on the chemical nature of its side group. Portions of a protein chain with numerous nonpolar amino acids, for example, tend to be shoved into the interior of the protein by hydrophobic interactions, with polar water molecules tending to exclude nonpolar amino acid side groups.

Proteins can contain up to 20 different kinds of amino acids. These amino acids fall into five chemical classes, which have properties that are quite different from one another. These differences determine what the polymers of amino acids, known as proteins, are like.

Note that in addition to its R group, each amino acid, when ionized, has a positive (amino, or NH_3^+) group at one end and a negative (carboxyl, or COO^-) group at the other end. These two groups can undergo a chemical reaction, losing a molecule of water and forming a covalent bond between two amino acids. A covalent bond that links two amino acids is called a **peptide bond.** As illustrated in Figure 4-18, the two amino acids linked by such a bond are not free to rotate around the N—C linkage, because the peptide bond has a partial double bond character. The stiffness of the peptide bond is one factor that leads chains of amino acids linked in this way to form spirals and other regular shapes (referred to as **secondary structures**).

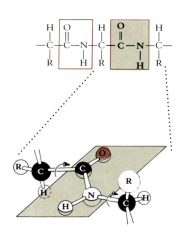

FIGURE 4-18

The peptide bond. Because of the partial double-bond nature of the C—N peptide bond, which forms when the —NH$_2$ end of one amino acid joins to the — COOH end of another, the resulting peptide chain cannot rotate freely around the peptide bond.

Polypeptides

As just mentioned, a protein is composed of one or more long chains of amino acids. These chains are called polypeptides. Within a polypeptide the amino acids are linked end to end by peptide bonds. The sequence of amino acids that make up a particular polypeptide chain is termed its **primary structure** (Figure 4-19). Because the R groups that distinguish the various amino acids play no role in the peptide backbone of proteins, a protein can be composed of any sequence of amino acids. A protein composed of 100 amino acids linked together in a chain might have any of 20^{100} different amino acid sequences. This is perhaps the most important property of proteins, permitting great diversity in the kinds of proteins that are possible.

Each of the amino acid R groups of a polypeptide interacts with the R groups of its neighbors, forming hydrogen bonds. Because of these near-neighbor interactions, polypeptide chains tend to fold spontaneously into sheets or wrap into coils (Figure 4-20). The form that a region of a polypeptide assumes is called its local *secondary structure*.

Amino acids in a protein chain also interact with water. Because the nonpolar side chains of a protein tend to aggregate together in such a way as to minimize disruption of the hydrogen bonding of water molecules, the secondary structure of polypeptide chains tends to further fold up in water into a complicated globular shape called the protein's **tertiary structure.** The shape, or conformation, of a protein's tertiary structure depends greatly on the order and nature of amino acids in the sequence. A change in the identity of a single amino acid can have very subtle, or profound, effects on protein shape. It is because proteins can assume so many different shapes that they are such effective biological catalysts.

When two protein chains associate to form a functional unit, the chains are referred to as subunits. Hemoglobin, for example, is a protein composed of four subunits. The subunits need not all be the same. Hemoglobin possesses two identical alpha-chain subunits and two beta-chain subunits. A protein that possesses two subunits is called a **dimer,** three subunits a **trimer,** and four subunits a **tetramer.** A protein's subunit structure is often referred to as its **quaternary** structure.

The shape that proteins assume is determined by the sequence of amino acids in the polymer. Because different amino acid R groups have very different chemical properties, the shape of a protein may be altered by a single amino acid change.

FIGURE 4-19

The amino acid sequence (primary structure) of the enzyme lysozyme. The enzyme contains four S-S bridges that link portions of the chain together into a particular form, which then folds into the enzyme's final shape.

FIGURE 4-20

A *The primary structure of a protein is its amino acid sequence.*
B *The secondary structure of a protein is the local form the polypeptide chain assumes as a result of hydrogen bonding between adjacent amino acids; this may be an alpha-helix, a side-by-side array of polypeptides, or any of several other arrangements.*
C *The tertiary structure of a protein is the overall shape the protein assumes as a result of interactions with water and between different portions of the polypeptide chain.*

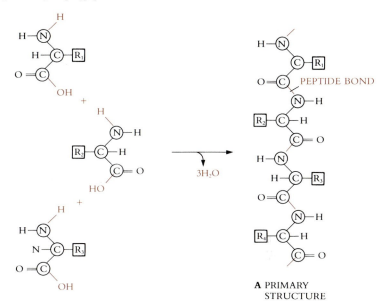

A PRIMARY STRUCTURE

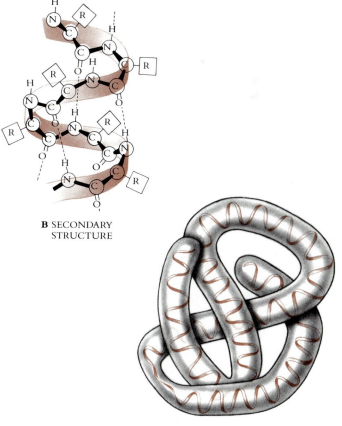

B SECONDARY STRUCTURE

C TERTIARY STRUCTURE

AMINO ACID SEQUENCE DETERMINES PROTEIN SHAPE

The great utility of proteins as biochemical catalysts lies in two of their properties: (1) any sequence of amino acids is possible in a polypeptide chain, and (2) different amino acid side groups have different chemical properties. Consequently, the side groups of a polypeptide chain interact with one another in different ways.

The shape that a protein assumes in water is determined to a large extent by the distribution of polar and nonpolar amino acids in its sequence. This relationship can be seen indirectly in a simple protein such as ribonuclease (Figure 4-A). This enzyme, which degrades the nucleic acid RNA, is locked into its correct shape by four disulfide bonds *(A)*. Disulfide bonds are sulfur-sulfur (—S—S—) bonds that link the R groups of two cysteine amino acids. When these bonds are broken by reducing agents to form —SH HS—, the protein maintains its correct shape in water *(B)*. However, if the uniform hydrogen bonding of water molecules with one another is disrupted by heat, or by the addition of a very polar molecule such as urea [H₂N—(CO)—NH₂], the reduced ribonuclease unfolds *(C)*.

Left in water, unfolded ribonuclease will slowly refold. This reconstituted ribonuclease has full catalytic activity and is in every respect a normal enzyme. The hydrophobic forces cause nonpolar amino acid side groups within the polypeptide chain to aggregate together in solution. At the same time they encourage polar amino acid side groups to form hydrogen bonds with water molecules. As a result of these forces, the ribonuclease molecule is forced into its proper shape. Its tertiary structure is determined by its primary structure and is imposed on it by the polar nature of water.

THE ENERGY CURRENCY OF ORGANISMS: ATP

Most of the activities of organisms that we associate with life, including growth, movement, communication, and the passing on of genetic information, require energy. In all organisms the bulk of this energy is supplied in the same way, carried to the places where it is used in the form of ATP. Because of the universal distribution of ATP among living organisms, it is clear that its role as a mechanism of energy transport evolved early in the history of life.

ATP is a composite molecule of a kind called a **nucleoside triphosphate** (Figure 4-21). Each ATP molecule is composed of three parts:

1. A five-carbon sugar called a **ribose,** which serves as the backbone of the composite to which the other two elements are bound

2. A triphosphate group (three phosphate groups linked in a chain)

3. Adenine, one of a class of carbon-nitrogen ring molecules called **organic bases**

In ATP the triphosphate group is bound to the nonring carbon of the ribose sugar, and the organic base adenine is bound to the opposite end of the sugar.

Where is the energy in ATP? It is in the high-energy bonds that link the phosphates to one another in the triphosphate group. These bonds, in which two phosphate groups are linked together by an oxygen atom, are called **phosphoanhydride bonds:**

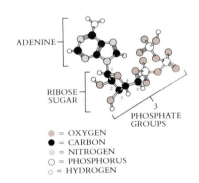

ADENINE

RIBOSE SUGAR

PHOSPHATE GROUPS

- = OXYGEN
- = CARBON
- = NITROGEN
- = PHOSPHORUS
- = HYDROGEN

FIGURE 4-21

ATP is a composite molecule. In ATP a chain of three phosphate groups is attached to one end of a five-carbon sugar, and the organic base adenine to the other end.

$$\text{Adenosine} - - - \begin{array}{ccc} \overset{\displaystyle O}{\underset{\displaystyle O}{|}} & \overset{\displaystyle O}{\underset{\displaystyle O}{|}} & \overset{\displaystyle O}{\underset{\displaystyle O}{|}} \\ P-O-P-O-P-O \end{array}$$

Phosphoanhydride bonds require considerable input of energy to form, are quite stable once formed, and release a large amount of chemical energy when they are broken. The prebiological formation of phosphoanhydride bonds utilized the energy of ultraviolet ra-

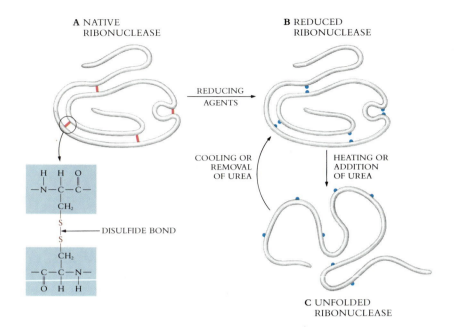

A NATIVE
RIBONUCLEASE

B REDUCED
RIBONUCLEASE

REDUCING
AGENTS

COOLING OR
REMOVAL
OF UREA

HEATING OR
ADDITION
OF UREA

DISULFIDE BOND

C UNFOLDED
RIBONUCLEASE

FIGURE 4-A

Ribonuclease reconstitution experiment.

---

diation and lightning to form long-chain polymers of phosphates linked by anhydride bonds. Such polymers are called **polyphosphates.** It is probable that cleavage of the phosphoanhydride bonds of these nonbiological polyphosphates provided the energy source for some of the earliest organisms.

> **ATP is the principal energy carrier of all organisms. In it, energy is stored in stable phosphoanhydride bonds. The energy is released only when these bonds are broken.**

NUCLEIC ACIDS

All organisms store the information specifying the structures of their proteins in **nucleic acids.** Nucleic acids are long polymers of repeating subunits called **nucleotides.** Each nucleotide, the basic repeating unit, is a composite molecule composed of three smaller building blocks, similar in this respect to ATP (Figure 4-22). In nucleic acids the building blocks are:

1. A five-carbon sugar
2. A phosphate group (PO_4)
3. An organic nitrogen-containing base

The nucleotide subunit is constructed from these building blocks in the same way that ATP was constructed, by binding the phosphate group to the nonring carbon of the

FIGURE 4-22

A *A deoxyribose sugar has a hydrogen atom (H) attached where a ribose sugar has a hydroxyl group (OH); other than that the two kinds of sugar are identical.*

B *The nucleotide bases of DNA and RNA have a composite structure similar to that of ATP, except that only one phosphate group is linked to the sugar rather than three.*

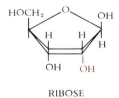

DEOXYRIBOSE

RIBOSE

A

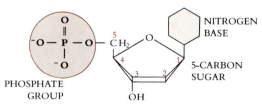

PHOSPHATE
GROUP

NITROGEN
BASE

5-CARBON
SUGAR

B

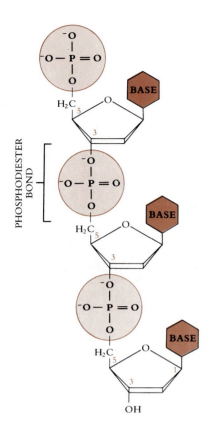

PHOSPHODIESTER
BOND

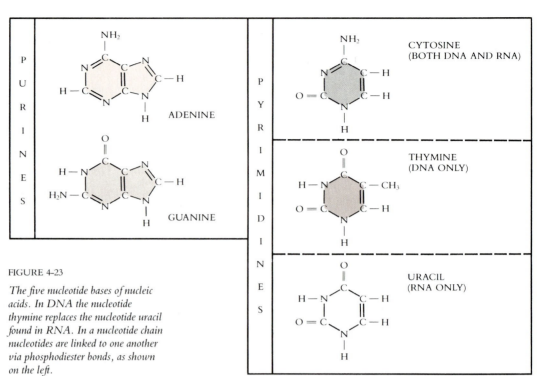

FIGURE 4-23

The five nucleotide bases of nucleic acids. In DNA the nucleotide thymine replaces the nucleotide uracil found in RNA. In a nucleotide chain nucleotides are linked to one another via phosphodiester bonds, as shown on the left.

ribose sugar (carbon 5 in Figure 4-22) and the nitrogen-containing base to the opposite end of the sugar (carbon 1 in Figure 4-22).

In the formation of a nucleic acid chain the individual nucleotide subunits are linked together in a line by the phosphate groups. Thus the phosphate group of one nucleotide binds to the hydroxyl group of another, forming an —O—P—O bond. This bond is called a **phosphodiester bond.** A nucleic acid, then, is simply a chain of ribose sugars linked together by phosphodiester bonds, with an organic base protruding from each sugar.

Organisms encode the information specifying the amino acid sequence of their proteins as sequences of nucleotides in DNA. This information is used in the everyday metabolism of the organism, as well as in storing and passing on the organism's hereditary traits. How does the structure of DNA permit it to store hereditary information? If DNA were a simple monotonously repeating polymer, it could not encode the message of life. Imagine trying to write a story using only the letter *E* and no spaces or punctuation. All you could ever say is "EEEEEEE. . . ." To write, you need more than one letter. We use 26 letters in the English alphabet, and the Chinese use thousands of individual characters to convey the same messages. You do not need so many individual symbols, of course, if the individual "letters" are grouped together into words. The Morse code, which is used to transmit messages by telegraph, employs only two elements (*dot* and *dash*), as do most modern computers (*0* and *1*). The reason nucleic acids can encode information is that they contain more than one kind of organic base. Each sugar link in a nucleic acid chain can have any one of four different organic bases attached to it. Just as in the English language, the sequence of letters encodes the information. In nucleic acids, however, there are not 26 letters, as in the English language, but only 4 letters, the four organic bases.

Information is stored in two forms of nucleic acid. One form, called deoxyribonucleic acid (DNA), provides the basic storage vehicle, the master plan. The other form, called ribonucleic acid (RNA), is similar in structure and is made as a template or copy of portions of the DNA. This copy passes out into the rest of the cell, where it provides a blue-

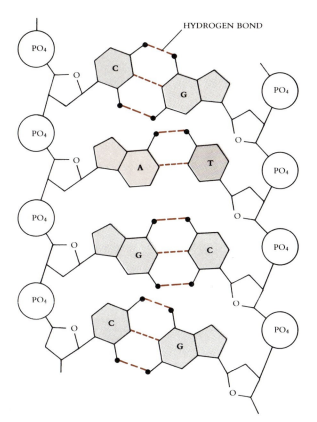

HYDROGEN BOND

FIGURE 4-24

Hydrogen bond formation between the organic bases, called base pairing, causes the two chains of a DNA duplex to bind to one another. The molecule is not a straight chain, but rather a graceful double helix, rising like a circular staircase.

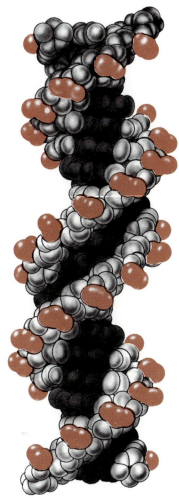

print specifying the amino acid sequence of proteins. RNA is thought to have evolved during the development of the hereditary system, with DNA evolving later because of the advantage of isolating the genetic information to protect its integrity and fidelity. We will consider DNA first and then address RNA.

> **Organisms store and use hereditary information by encoding the sequence of the amino acids of each of their proteins as a sequence of nucleotides in nucleic acids.**

DNA

Two of the four organic bases that make up DNA (Figure 4-23), adenine and guanine, are large, double-ring compounds called **purines.** You have already encountered adenine as the organic base in ATP. The other two organic bases in DNA, cytosine and thymine, are smaller, single-ring compounds called **pyrimidines.** In discussing the sequence of bases in DNA these organic bases are usually referred to by their first initials, *A, G, C,* and *T.*

The DNA chains in organisms exist not as single chains folded into complex shapes, as in proteins, but rather as double chains (Figure 4-24). Two of the polymers wind around each other like the outside and inside rails of a circular staircase. Such a winding shape is called a **helix,** and one composed of two molecules winding in concert like DNA is called a **double helix.** The steps of the helical staircase are hydrogen bonds between the bases in one polymer chain and those opposite them in the other chain. These hydrogen bonds hold the two chains together as a duplex. The chemical differences between the four organic bases have important influences on the hydrogen bonding that takes place, with two very important consequences:

1. *The two chains of DNA are complementary.* Because purines are much larger than pyrimidines, a nucleic acid chain has a rather jagged, irregular profile. If you line

up two different nucleic acid chains tooth to tooth, they will not match up at all and cannot get close to each other. If, however, one chain has a large purine at every position that the other has a small pyrimidine, and a small pyrimidine at every place where the other chain has a large purine, then the two chains do interdigitate. The bases at each position are then placed close to one another but do not overlap. Successful matching will only occur if in every place where one chain has a purine, the other has a pyrimidine, and vice versa.

2. *Purine-pyrimidine bases form hydrogen bonds.* These bonds stabilize the association of two nucleotide chains. Because of the different positions of the amino and carboxyl groups, however, only two of the four potential pairs can form:

Adenine *(A)* can pair with thymine *(T)*
Guanine *(G)* can pair with cytosine *(C)*

A cannot pair with *C,* nor *G* with *T,* since the side groups do not line up in such a way that hydrogen bonds, which are highly directional, can form.

Imagine now that you are going to form one or more nucleic acid chains complementary to an existing one. Every place in the existing chain that you encounter a purine, you place in your new chain a pyrimidine, and vice versa. Of the two possible choices that you can make at each position, you select for the new chain the base that is able to form hydrogen bonds with the opposite base of the existing chain. How many different nucleic acid sequences can you construct? Only one! Whereas the first chain can have any sequence or combination of the four "letters," the differences in structure between the four organic bases, both with respect to size and also with respect to their hydrogen-bonding potential, completely determine the sequence of the new chain. This is the way in which heredity works, by **complementarity.**

The hereditary information is reproduced faithfully by maintaining the polymer containing the information—DNA—in duplicate chains, one copy the mirror image of the other. For each molecule of DNA that is formed, a complementary image can be constructed from each of the two chains, yielding two identical duplexes.

RNA

RNA is similar in structure to DNA. There are only two differences:

1. RNA molecules contain ribose sugars in which carbon 2 has bound to it a hydroxyl group. In DNA molecules, in contrast, this hydroxyl group is replaced by a hydrogen atom.

2. One of the four organic bases is changed slightly. In RNA the base corresponding to thymine has the same structure as thymine except that one of the carbons lacks a CH_3 group (is not methylated); this new RNA base is called **uracil.** RNA molecules utilize uracil in place of thymine. When RNA pairs with DNA, the RNA base uracil pairs with adenine just as the DNA base thymine does.

Copying the DNA message onto a chemically different molecule such as RNA allows the cell to tell which is the original and which is the copy. Using two different molecules allows the separation of the role of DNA in storing hereditary information from the role of RNA in using this information to specify proteins.

RNA functions in several different ways to bring about the expression of the genetic message encoded within DNA. One form of RNA, which is called **messenger RNA** (mRNA), is synthesized as a complementary copy of the DNA sequence. This mRNA then passes into the cell cytoplasm, where it provides the information that specifies the production of specific proteins. To use the information stored within the base sequence of mRNA, the RNA must be single stranded. In a single-stranded condition, molecules are able to bind to it and so use its information to direct the synthesis of proteins.

This situation is dangerous, however, since an unprotected single-stranded RNA molecule is subject to random cleavage. The RNA strand would be protected, of course, if it were bound to its complementary strand and any cleavage that did occur could be repaired by using the complementary strand as a template. In some of the genetic systems that evolved first, RNA may actually have been double stranded. Such systems, however, have the great disadvantage that the complementary sequence must be stripped off every time the RNA is used in the process of replication. Although there is no reason that RNA cannot form double-stranded molecules, most cells lack enzymes able to catalyze the formation of RNA duplex molecules. With rare exceptions RNA is a single-stranded molecule.

An Evolutionary Perspective: DNA

DNA probably evolved from RNA as a means of preserving the genetic information, protecting it from the ongoing wear and tear that is associated with cellular activity. By storing the information in DNA and then using a complementary RNA sequence to direct protein synthesis, the information-encoding DNA chain is not exposed to the dangers of single-strand cleavage every time that the information is used. In all organisms the information encoded in the RNA base sequence is derived from a complementary "master" DNA sequence. (In some viruses, which are not organisms, other systems occur.) Cells have evolved enzymes, called DNA polymerases, that synthesize strands which are complementary to single-stranded portions of a DNA molecule. Other enzymes inspect the two strands, ensuring that the polymerase has made no mistakes and that no damage has occurred subsequently. As discussed in Chapter 16, cells employ a large battery of devices that ensure the accurate replication of the genetic message.

This genetic system, then, is the one that has come down to us from the very beginnings of life. The information necessary for the synthesis of all the protein molecules that are required by the cell is stored in a double-stranded DNA base sequence. The cell uses this information by making a working copy of it: RNA nucleotides are paired to complementary DNA ones at positions where the DNA is temporarily opened up into single strands. The resulting single-stranded, short-lived RNA copy is used to direct the synthesis of a protein that has a specified sequence of amino acids. In this way the information flows from DNA to RNA to protein. In prokaryotic cells this process, which has been termed the "central dogma" of heredity, can be visualized directly: RNA dangles from the DNA as it is synthesized, with protein chains already beginning their synthesis on each of the RNA strands while the RNAs are still being completed (Figure 4-25).

Proteins certainly evolved in synchrony with the evolution of the genetic system as a whole. They are in fact two aspects of a single system, neither part of which is functional without the other.

SUMMARY

1. Organisms store energy in carbon-hydrogen (C—H) bonds. Short-term storage usually occurs in carbohydrates. The most important of these is glucose, a six-carbon sugar.

2. Organisms usually transport sugars as disaccharides, two simple sugars that are linked together and cannot therefore be utilized while they are being transported.

3. Excess energy resources may be stored in complex sugar polymers called starches, especially in plants. Glycogen, which is a comparable storage polymer that is common in animals and fungi, is characterized by its complex branching.

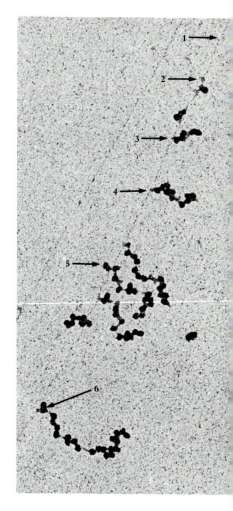

FIGURE 4-25

In bacteria the messenger RNA copy of the DNA begins to direct the production of its protein even before all of it is copied off of the DNA.
1 *DNA strand.*
2 *RNA polymerase molecule on the DNA.*
3 *A chain of ribosomes assembles on the mRNA molecule.*
4 *Each ribosome produces a protein as it moves along the mRNA.*
5 *Proteins produced by the ribosomes are too small to see.*
6 *This mRNA molecule is almost completed. The ribosomes at this end of it have moved all the way along the mRNA and have almost completed assembly of a protein.*

4. Fats are molecules that contain many more C—H bonds than carbohydrates and thus provide more efficient energy storage.

5. Biological membranes are composed of a phospholipid matrix with proteins embedded in it. Phospholipids are fatlike molecules in which one fatty acid is replaced by a polar alcohol, imparting to the molecule a polar and a nonpolar end.

6. The matrix of biological membranes assembles spontaneously as a result of the aggregation of the nonpolar tails of lipid molecules in water.

7. Proteins are linear polymers of amino acids. Because the 20 amino acids that occur in proteins have side groups with very different chemical properties, the function and shape of a protein are critically affected by its particular sequence of amino acids.

8. ATP is the principal energy carrier in all organisms.

9. Hereditary information is stored as a sequence of nucleotides in a linear nucleotide polymer called DNA. DNA is a double helix, one strand of which complements the other. When organisms reproduce, a complementary strand is assembled on each, yielding two duplexes.

10. The cell uses the information within DNA by transcribing a complementary single strand of RNA from the DNA duplex. Elsewhere in the cell, this single RNA strand directs the synthesis of a protein with an amino acid sequence corresponding to the nucleotide sequence of DNA from which it was transcribed.

SELF-QUIZ

The following questions pinpoint important information in this chapter. You should be sure to know the answers to each of them.

1. Which of the following kinds of atoms do not occur in carbohydrates?
   (a) carbon   (c) nitrogen
   (b) hydrogen   (d) oxygen

2. Stereoisomers differ from one another in one or more covalent bonds. True or false?

3. Fats yield more energy than carbohydrates when they are metabolized. About how much more?
   (a) a third again as much (30% more)
   (b) half again as much (50% more)
   (c) twice as much (100% more)
   (d) four times as much (300% more)
   (e) ten times as much (1000% more)

4. The difference between a phospholipid and a triglyceride is that one contains a glycerol backbone and the other does not. True or false?

5. Some lipids in water solution spontaneously form micelles, whereas others spontaneously form bilayer membranes. What determines whether a micelle or a membrane will form?
   (a) specific enzymes
   (b) random chance
   (c) the length of the nonpolar lipid chains
   (d) the degree of saturation of the phospholipids
   (e) the number of phosphorylated alcohols

6. Ignoring the side group, how many carbon atoms are present in an amino acid?

7. How many different sequences are possible for a three-amino-acid polypeptide containing the amino acids glycine, alanine, and methionine?

8. How many phosphoanhydride bonds are present in one ATP molecule?

9. How many phosphodiester bonds are present in a DNA molecule?
   (a) twice as many as the number of nucleotides
   (b) none
   (c) the same as the number of nucleotides
   (d) one less than the number of nucleotides

10. How many different three-base DNA sequences are possible?
    (a) 3   (d) 20
    (b) 9   (e) 64
    (c) 16

11. What is the nucleotide sequence of a messenger RNA molecule transcribed from a DNA molecule with the sequence ATGC?

12. Which is thought to have evolved first, DNA or RNA?

THOUGHT QUESTIONS

1. Both lactose and cellulose are composed of simple monosaccharide sugars joined to one another by beta-linkages. In lactose the linkage is between glucose and galactose, whereas in cellulose the linkage is between two glucose molecules. A human baby has no trouble digesting lactose, but no human baby can digest cellulose. Why not?

2. DNA exists in cells as a double-stranded duplex molecule, whereas messenger RNA, which is composed of very similar nucleotides linked together in the same way, does not form a double-stranded duplex within cells. If you were to place complementary single strands of DNA in a test tube, would they spontaneously form duplex molecules? If you were to place complementary single strands of mRNA in a test tube, would they spontaneously form duplex molecules?

3. Of all possible DNA nucleotide sequences, what sequence of base pairs would most easily dissociate into single strands upon gentle heating of the DNA duplex? Why did you choose this sequence?

4. Why do you suppose humans circulate free glucose in their blood, rather than employing a disaccharide such as sucrose as a transport sugar, like insects and plants?

FOR FURTHER READING

DICKERSON, R., and I. GEIS: *Chemistry, Matter, and the Universe,* R.K. Benjamin, Publisher, Menlo Park, Calif., 1976. A comprehensive overview of chemistry for the non–chemistry student, with a broad biological perspective.

EIGEN, M., and W. GARDINER: "The Origin of Genetic Information," *Scientific American,* April 1981, pages 88-120. The authors, one a Nobel prize winner, argue that in the beginning, genetic information was carried by RNA.

"The Molecules of Life," *Scientific American,* October 1985. An issue devoted entirely to presenting a comprehensive view of what is known about biologically important molecules, with individual articles on DNA, RNA, and proteins, among many others. Highly recommended.

SHARON, N.: "Carbohydrates," *Scientific American,* November 1980, pages 90-116. An overview of the structures of carbohydrates and the diverse roles they assume in organisms.

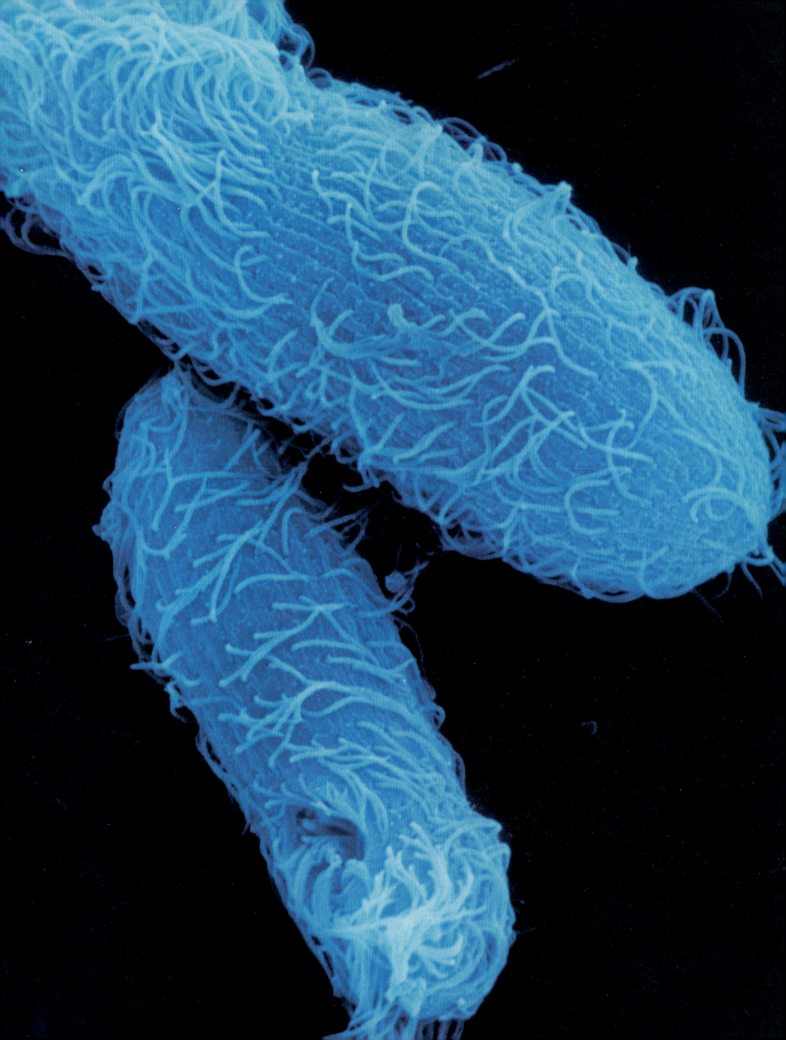

BIOLOGY OF THE CELL *Part Two*

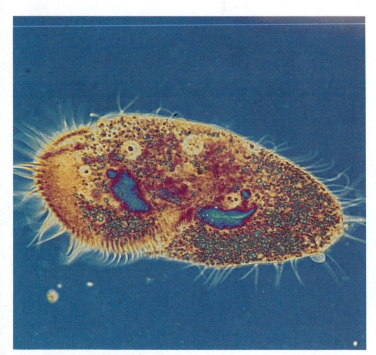

The diversity of life extends far beyond
the ability of the naked eye to see it.
These protists are vibrantly alive, their
numerous flagellae propelling individual
cells this way and that. Many bacteria,
by contrast, grow as members of
colonies. Underlying this diversity is a
great sameness—these and all other
organisms share one fundamental
architecture, that of the cell.

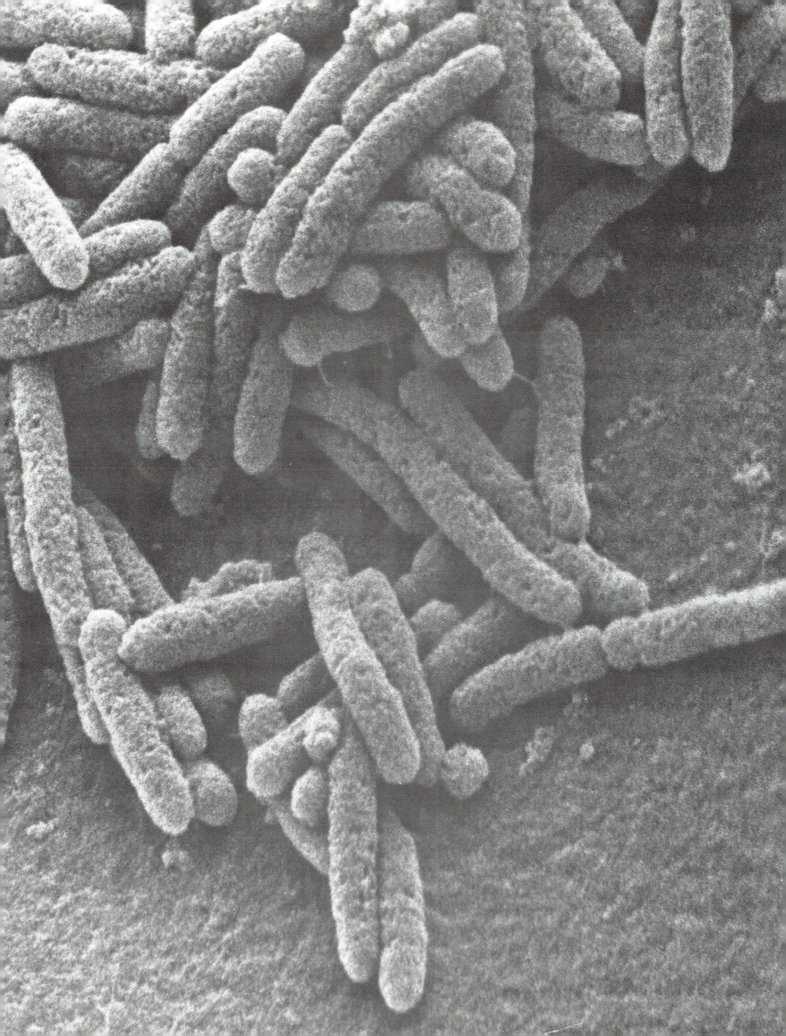

Overview

All organisms are composed of one or more cells, the basic units of life. All cells consist of a lipid bilayer membrane enclosing a cytoplasm that contains enzymes and DNA. Most prokaryotes resemble one another in form, having little internal organization and a strong cell wall encasing their exteriors. Eukaryotes, which evolved from prokaryotes, have a much more elaborate interior organization, in which organelles create separate compartments for functions such as photosynthesis, aerobic respiration, and the preservation and duplication of the cell's DNA. Among the eukaryotes, plants and fungi have evolved strong exterior cell walls, whereas animals have dispensed with cell walls and thus have gained mobility.

For Review

Here are some important terms and concepts that you will encounter in this chapter. If you are not familiar with them, you should review them before proceeding.

- **Lipid bilayer** (Chapter 4)
- **Membrane fluidity** (Chapter 4)
- **Autotrophs** (Chapter 3)
- **Evolution of eukaryotes** (Chapter 3)

FIGURE 5-1

Robert Hooke, a seventeenth-century English scientist, first described cells. This drawing of cork cells is reproduced from Hooke's Micrographia *(1665).*

Just as factories and other buildings come in a broad range of sizes and designs, so do organisms, from bacteria too small to be seen by the naked eye to elephants and redwood trees. Despite their apparent complexity, however, buildings and organisms exhibit a common organizing theme: both are composed of rooms. This fundamental attribute of organisms first became clear in 1665, when Robert Hooke, using a microscope that he had built, examined thin sections of plant tissue. Figure 5-1 is his drawing of what he saw. The slice of cork, like a honeycomb, is divided into myriad small compartments. Hooke called these compartments "cellulae," using the Latin word for a small room. His term has come down to us as "cells," the units of organization of living things.

The first living cells were observed a few years later by the Dutch naturalist Antonie van Leeuwenhoek. Just as the simplest buildings possess one room, so did the "animalicules" that were observed by van Leeuwenhoek. For almost 200 years, however, the general importance of cells was not appreciated by biologists, until in 1838 the German Matthias Schleiden, basing his conclusion on extensive observations of plant tissues, made the first statement of what is now called the cell theory. Schleiden stated that all plants "are aggregates of fully individualized, independent, separate beings, namely the cells themselves." The following year, Theodor Schwann reported that all animal tissues are also composed of individual cells. Twenty years later the German physiologist Rudolf Virchow deduced that not only were all organisms composed of cells, but also each cell can arise only from a preexisting cell.

The **cell theory,** in its modern form, states that all organisms are composed of one or more cells, within which the life processes of metabolism and heredity occur. Although life evolved spontaneously in the hydrogen-rich environment of the early earth, biologists have concluded that additional cells are not originating now. Life has evolved on earth only once, to our knowledge. Earth's organisms represent a continuous line of descent from those early cells.

> **All the organisms on earth are cells or aggregates of cells, and all humans are descendants from the first cells.**

CELLS

A cell, like a room, is a space that is defined both by what surrounds it and what is contained within it. All cells are surrounded by a lipid bilayer membrane, as described in Chapter 4. Every substance that enters or leaves the cell must pass through this membrane. A variety of proteins extend through the lipid matrix, acting as passages for certain molecules. The identity of the lipids that form the membrane matrix varies from one kind of organism to another. Animal membranes, for example, contain a complex lipid called cholesterol, which has four linked carbon rings. In contrast, bacterial membranes lack complex lipids. The general chemical nature of all biological membranes, however, is the same in all organisms.

Within every cell is a complex mixture of substances that are collectively called the **cytoplasm.** The cytoplasm contains many of the enzymes of the cell, as well as sugars, amino acids, and other small molecules, called metabolites, which participate in the metabolic life of the cell. All eukaryotic cells use ATP as an energy-carrying molecule, deriving the energy needed to make ATP from metabolizing food or from photosynthesis and expending ATP to drive chemical reactions.

Complex structures may sometimes be included within the cytoplasm of cells. Organized structures within the cytoplasm are called **organelles.** With only a very few exceptions, eukaryotic cells possess small self-contained organelles called mitochondria that have the appearance of bacteria living within the cell. The cells of plants and algae also possess bacteria-like organelles of a different kind—chloroplasts—in addition to their mitochondria. Other types of organelles also exist, each forming a separate compartment within the cell, like a closet within a room. The largest organelle that occurs in cells is the nucleus. Within the nucleus are most of the cell's DNA and the enzymes necessary to make copies of this DNA, as well as those enzymes that are necessary to produce RNA copies. All eukaryotic cells contain a nucleus, but bacteria do not: bacterial DNA is not set aside in an organelle.

> **A cell is a membrane-bound unit containing enzymes and other elements that render it capable of independent metabolism and reproduction. All cells possess hereditary machinery, in which DNA is used to store information; a cell membrane that is a lipid bilayer; and a metabolism based on ATP.**

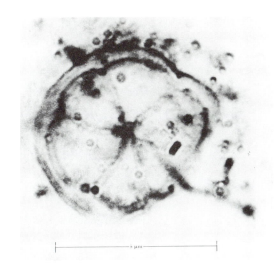

FIGURE 5-2

Masses of bacteria, some of them photosynthetic, discolor this hot spring in Yellowstone Park.

FIGURE 5-3

In 1964 S. Siegel of the University of Hawaii was testing the viability of soil microorganisms subjected to conditions similar to those thought to have existed in primitive atmospheres. When a soil sample from a stable in Wales was exposed to an atmosphere of 40% nitrogen, 25% ammonia, 25% methane, and 10% oxygen, several microorganisms in the soil survived and grew. Among them was the one you see here, Kakabekia umbellata. This prokaryote was thought to be extinct, but similar samples, nearly 2 billion years old, of this same microbe had been known in the Gunflint chert of southern Ontario. Kakabekia, a prokaryote, is anaerobic (it does not use oxygen) and lives by metabolizing ammonia.

PRIMITIVE PROKARYOTES

Most of the organisms living today resemble one another fundamentally, having the same kinds of membranes and hereditary systems and similar metabolisms. Not all living organisms are exactly the same in these respects, however. If one looks carefully in unusual environments, one occasionally encounters organisms that are quite unusual, differing in form and metabolism from most other living things. Sheltered from evolutionary alteration in unchanging habitats resembling those of earlier times, these organisms are living relics, the surviving representatives of the first ages of life on earth. In those ancient times biochemical diversity was the rule, and living things did not resemble one another as closely in their metabolic features as they do today. In places such as the oxygenless depths of the Black Sea or the boiling waters of hot springs (Figure 5-2), one can still find bacteria living without oxygen that display a bewildering array of metabolic strategies. Some of them have shapes similar to the fossils of bacteria that lived 2 to 3 billion years ago.

The search for primitive and unusual bacteria has led biologists to strange places, such as the ruins of an ancient stable near Harlech Castle in Wales. A stable existed on that site continuously for some 700 years, and, as you might expect, the soil beneath the stable is highly unusual. Seven centuries of urine and manure have rendered it rich in ammonia, raising the pH so high that little could be expected to live within it. There, however, researchers found *Kakabekia umbellata,* a curious prokaryote with a shape that resembles a miniature umbrella (Figure 5-3). This curious prokaryote proved to be alive and would grow quite well if it were provided with 5 to 10 molar ammonium hydroxide (the same solution that one uses to strip wax from a kitchen floor, only much stronger). *Kakabekia* was not, however, a novel life form. It was already well known to biologists—but as a fossil. Its distinctive form is common in the Gunflint chert, an outcrop of ancient rock in Canada that is about 2 billion years old; this rock teems with well-preserved fossil bacteria.

What were the early bacteria like? Perhaps the most primitive ones that still exist today are the **methane-producing bacteria.** These bacteria are mostly simple in form, and they are able to grow only in an oxygen-free environment. For this reason they are said to grow "without air," or anaerobically, and are poisoned by oxygen. The methane-producing bacteria convert CO_2 and H_2 into methane gas (CH_4). They resemble all other bacteria in that they possess hereditary machinery based on DNA, a cell membrane that is a lipid bilayer, an exterior cell wall, and a metabolism based on ATP. But there the resemblance ends.

When the details of membrane and cell wall structure of the methane-producing bacteria are examined, they prove to be different from those of all other bacteria. Similarly, there are major differences in some of the fundamental biochemical processes of metabolism, processes that are the same in other bacteria. Survivors from an earlier time, when there was considerable variation in the mechanisms of cell wall and membrane synthesis, in DNA translation, and in energy metabolism, the methane-producing bacteria appear to represent a road not taken, a side branch of evolution.

The early stages of the history of life on earth seem to have been rife with such evolutionary experimentation. Novelty abounded, and many biochemical possibilities seem to have been represented among the organisms alive at that time. From the array of different early living forms, representing a variety of different biochemical strategies, a very few forms became the ancestors of the great majority of organisms that are alive today. A few of the other "evolutionary experiments," such as *Kakabekia* and the methane-producing bacteria, have survived in very unusual habitats. Most of the others became extinct millions, or even billions, of years ago.

Most organisms now living are descendants of a few lines of early bacteria. Many other diverse forms have not survived.

The modern bacteria, for the most part, seem to have stemmed from a tough, simple little cell; its hallmark was adaptability. For at least 2 billion years bacteria—prokaryotes— were the only form of life on earth. All eukaryotes, including the animals, plants, fungi, algae, and unicellular organisms called protozoa, are their descendants.

STRUCTURE OF THE PROKARYOTIC CELL

About 2500 species of bacteria have been given names. These species are extremely diverse, ranging from anaerobic organisms to the ones that require oxygen to grow. Such oxygen-requiring organisms are called **aerobic.** Some bacteria obtain their energy from organic matter, either in the bodies of other organisms or in organic debris. Such organisms are called **heterotrophs,** "eaters of others." Other bacteria, such as the cyanobacteria, are photosynthesizers. Because they are able to harvest their own energy from the physical environment, they are called **autotrophs,** "self-feeders." In all of these bacteria the cellular organization is fundamentally the same (Figure 5-4): isolated small cells about 1 to 10 micrometers across (Figure 5-5), enclosed within a lipid bilayer membrane and encased within a rigid cell wall, with no interior membrane-bound organelles or other subcompartments.

Compared with the other kinds of cells that have evolved from them, prokaryotic cells—bacteria—are smaller and lack interior organization.

Exterior Cell Walls

Bacteria encase themselves within a strong cell wall. This wall is composed of a matrix of disaccharide sugars cross-linked by peptides several amino acids in length. Plant cell walls, in contrast, are made of cellulose and are not cross-linked by peptides. The cell walls of single-celled eukaryotes and of fungi vary widely in structure.

Bacterial cell walls stain purple when the cells are treated with **Gram stain,** a staining process developed by the Danish microbiologist Hans Christian Gram. This is why such cells are referred to as gram positive. By using this stain, Gram could detect the presence of certain disease-causing bacteria. Gram-positive bacteria possess a single, thick cell wall that retains the Gram stain within the cell, causing the stained cells to appear purple under the microscope. More complex cell walls have evolved in other groups of bacteria. In them the cell wall is thinner and it does not retain the Gram stain; such bacteria are called gram negative.

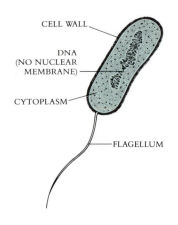

CELL WALL

DNA (NO NUCLEAR MEMBRANE)

CYTOPLASM

FLAGELLUM

FIGURE 5-4

Prokaryotic organisms—bacteria—all share a very simple body plan. They are unicellular, the single cell being encased within a strong cell wall. Within the cell there are no organelles or closed compartments, and the DNA is in direct contact with the cytoplasm. Many, although not all, bacteria possess one or more flagella, which they use to propel themselves through their liquid surroundings.

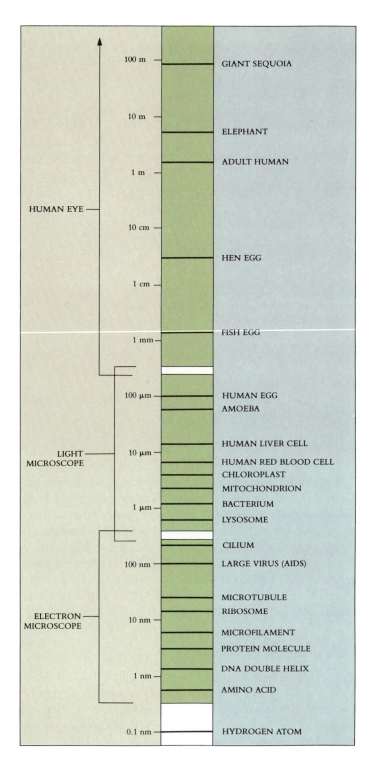

FIGURE 5-5

The size of things, from nanometers to meters.

The cell wall of a gram–positive bacterium (Figure 5-6) is relatively thick, about 25 nanometers, or four times the thickness of a lipid bilayer membrane. The entire cell wall consists of a single molecule. It is made up of simple repeating units, which are cross-linked together. The basic unit is a disaccharide to which is linked a T-shaped chain of amino acids. Long chains of individual disaccharide molecules are built up and then cross-linked by the amino acid chains, forming an enormous net around the outside of the cell. Antibiotics such as penicillin act on these bacteria by blocking the cross-linking reaction; cell walls without cross-links have no strength and rupture when the cell grows. Penicillin and similar kinds of antibiotics are not effective in treating diseases caused by gram–negative bacteria.

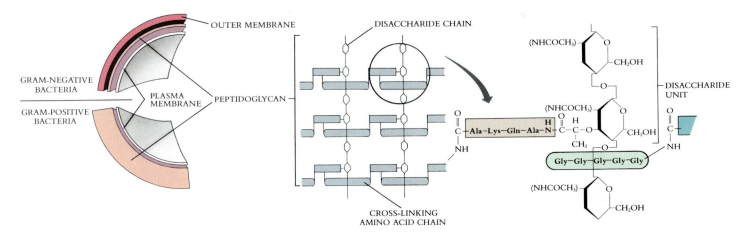

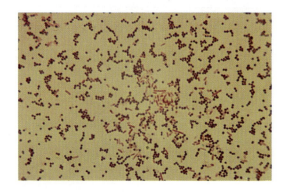

FIGURE 5-6

Bacterial cell walls are composed of an interlocking network of disaccharide molecules cross-linked by short amino acid chains. The complex, called peptidoglycan, forms what is in actuality a single molecule woven around the outside of the cell. The peptidoglycan layer is much thicker in gram-positive bacteria than in gram-negative ones, causing them to retain gentian violet dye and thus appear purple, as shown here, in a Gram-stained smear; gram-negative bacteria do not retain the violet dye and so exhibit the background red stain.

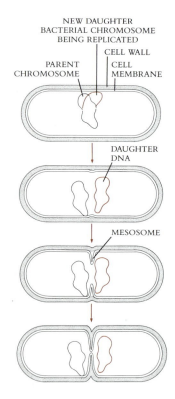

FIGURE 5-7

In a dividing bacterium the cell membrane invaginates inward at the point of DNA attachment, forming mesosomes. Because mesosomes form between the two replicas of a dividing cell's DNA, they serve to partition the replicas into the two daughter cells.

Bacterial cells are bounded within the cell wall by a single membrane. Although most molecules pass freely across the cell wall, which is very porous, the membrane acts as a barrier preventing free entry into the cell, permitting only specific molecules to be transported into and out of the interior cytoplasm. The bacterial cell membrane is composed of simple phospholipids (in contrast, the membranes of animals contain more complex ones). In aerobic bacteria special proteins necessary for aerobic metabolism are embedded within the membrane and bound to its inner surface. Similarly, the sites of photosynthesis in anaerobic photosynthetic bacteria, the so-called purple bacteria, are incorporated within the membrane.

Lack of Interior Organization

If you were to look at a cross section of a bacterial cell under a powerful microscope, its most obvious structural feature would be its simple interior organization. There are no internal compartments bounded by membranes in bacterial cells and no membrane-bound organelles. The entire cytoplasm of a bacterial cell is one continuous compartment, and there is no internal support structure. The entire strength of the cell comes from its rigid exterior wall.

The external membrane of bacterial cells often intrudes into the interior of the cell, where it may play an important role. In many dividing bacteria, for example, the cell membrane extends inward at the point of division, at a site where the replicated DNA has attached to it (Figure 5-7). These invaginations, or intrusions, are called **mesosomes**; they are thought to aid in the separation of DNA replicas during cell division. In the anaerobic photosynthetic bacteria the cell membrane is often extensively convoluted, with folds extending into the interior of the cell (Figure 5-8). These folded membranes bear the photosynthetic pigments. As a result of the folding, the surface area available for capturing light is greatly increased.

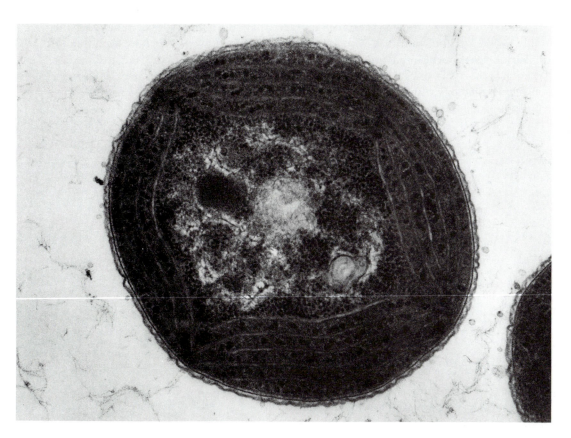

FIGURE 5-8

A cross section of a unicellular cyanobacterium exhibits extensive folding of the cell membrane into the interior of the cell, creating interior membranes that bear photosynthetic pigments and are specialized for photosynthesis.

In no case, however, does a bacterial membrane actually wall off any portion of the cell. Every area is equally accessible to the molecules that are present within the cell. The major macromolecules of prokaryotic cells exist free within the cytoplasm. Their genetic information, for example, is encoded within a single long circle of DNA, exposed directly to the cytoplasm where the information is used.

Bacterial cells are encased by an exterior wall composed of carbohydrates cross-linked by amino acid chains. They lack interior compartments.

ORIGIN OF EUKARYOTIC CELLS

All fossils more than 1.5 billion years old—and the oldest were formed 2 billion years earlier than that—are generally similar to one another. They are small simple cells (Figure 5-9), mostly 5 to 20 nanometers thick, with none more than about 60 nanometers. Starting about 1.5 billion years ago, larger and more complex cells appeared (Figure 5-10). Cells more than 100 nanometers thick rapidly increased in abundance. Some fossil cells 1.4 billion years old are as much as 600 nanometers thick. Others, 1.5 billion years old, contain what appear to be small membrane-bound structures. Many of these fossils have elaborate shapes, with some exhibiting highly branched filaments, tetrahedral configurations, or spines. These early fossil traces mark the first major transition in the evolutionary history of life on earth. A new kind of organism had appeared, the eukaryote.

For at least the first 2 billion years of life on earth, all organisms were prokaryotic. No less than 1.3 to 1.5 billion years ago, the first eukaryotes appeared.

We do not know the details of how the first eukaryotes evolved. The key events all took place in the interiors of cells, and we can learn little of them by studying fossils. No

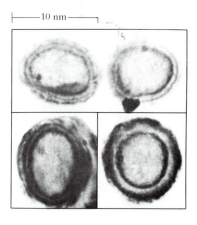

|—— 10 nm ——|

FIGURE 5-9

Fossil bacteria from the Bitter Springs Formation of Australia. These fossils, far too small to be visualized except with an electron microscope, are about 850 million years old. Similar fossil bacteria as much as 3.5 billion years old have been detected.

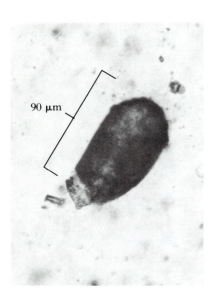

90 μm

Fossil unicellular eukaryote about 800 million years old from Spitsbergen. All life was unicellular until about the past 700 million years. This cell, however, is much larger and more complex than the cyanobacteria shown in Figure 5-9 and seems similar to the modern ciliates.

intermediate stages survive today. Just as was the case when the automobile supplanted the horse-drawn carriage, the eukaryotic cell appears to have burst into the evolutionary record fully formed and functional. We can say with confidence, however, that the evolution of eukaryotes was not a simple process and probably not a rapid one. In the course of the evolution of the eukaryotic cell four major kinds of evolutionary change occurred: (1) changes in cell morphology, (2) pooling of genetic resources, (3) development of genetic recombination, and (4) gene fragmentation.

Changes in Cell Morphology

Eukaryotic cells are far more complex than prokaryotic ones (Figure 5-11). Compared to their prokaryotic ancestors, eukaryotic cells exhibit great morphological improvements:

1. Their DNA is packaged tightly in chromosomes, which makes it easier to manipulate during cell division.
2. The interiors of eukaryotic cells are subdivided into membrane-bound compartments, which permits one biochemical process to proceed independently of others that may be going on at the same time. Complex metabolism is not possible without such subdivision.
3. Cell walls are thought to have been present in the common ancestor of the eukaryotes, but animals and many single-celled eukaryotes do not possess a cell wall.

The interiors of eukaryotic cells are subdivided by membranes in a complex way. The DNA of the cell is organized into segments called chromosomes. Many eukaryotes possess cell walls, but they are lacking in animals and some single-celled organisms.

Pooling of Genetic Resources

The total complement of hereditary information contained within a cell is called its **genome.** A major characteristic of eukaryotic cells is that their genomes appear to be composites containing contributions from several prokaryotic lineages. In other words, the evolution of the eukaryotic genome almost certainly involved the acquisition of new genetic material from other organisms. This acquisition is thought to have been accomplished by a process of **symbiosis**—the close association and mutual dependency of two different kinds of organisms—followed by **gene transfer.**

The mitochondria that occur in all eukaryotic cells are thought by most biologists to have originated as symbiotic, aerobic bacteria. The theory of the symbiotic origin of mitochondria has been controversial, and a few biologists still do not accept it as fully proven. The evidence supporting it is substantial, however, and in this text the symbiotic theory will be taken as established.

The theory proposes that sometime during the origin of eukaryotes the bacteria which became mitochondria were engulfed by cells that were the ancestors of eukaryotes. The cells that became the home of the ancestors of mitochondria were unable themselves to carry out the metabolic reactions necessary for living in an atmosphere that contained increasing amounts of oxygen. Reactions requiring oxygen are collectively called **aerobic metabolism.** The engulfed bacteria were able to carry out aerobic metabolism. Over time, the theory proposes, most of the genes contained in the genome of the aerobic bacterium were transferred to the chromosomes of the host cell nucleus, where they could more readily be replicated and controlled. The subcompartment provided by the engulfed cell was maintained, however, and became the mitochondrion. As you will see in later chapters, this organelle plays a critical role in the mechanism of aerobic metabolism. Because of its presence, all eukaryotes are capable of aerobic metabolism. A more detailed presentation of the evidence supporting the theory that mitochondria did indeed originate in this way is presented in Chapter 6.

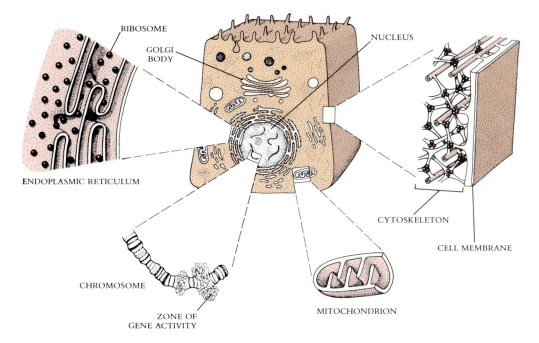

RIBOSOME

GOLGI BODY

NUCLEUS

ENDOPLASMIC RETICULUM

CYTOSKELETON

CELL MEMBRANE

CHROMOSOME

ZONE OF GENE ACTIVITY

MITOCHONDRION

FIGURE 5-11

Upper left *Diagram of a typical eukaryotic cell. The cell is encased within a simple bilayer membrane; this cell, which is from an animal, lacks a cell wall. The bulk of the cell's DNA is contained in chromosomes located within a double-membrane–bound compartment called the nucleus. Weaving through the cell's interior is a series of membranes called the endoplasmic reticulum, which serves to create separate compartments within the cell; the contents of these compartments are isolated by the membrane from other parts of the cell. Some of these compartments, the Golgi complex, are specialized for packaging certain proteins into vesicles for export from the cell. Within all eukaryotic cells are organelles called mitochondria, which closely resemble bacteria and are thought to be descended from bacterial guests. The entire cell structure is stiffened and is given its distinctive shape by a cytoskeleton composed of protein tubes.*

Lower left *Micrograph of a section through the center of a cell, showing the cell nucleus embedded within a mass of endoplasmic reticulum and mitochondria.*

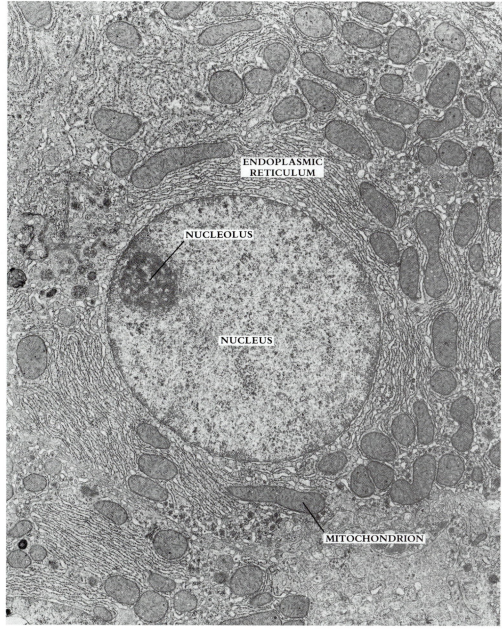

ENDOPLASMIC RETICULUM

NUCLEOLUS

NUCLEUS

MITOCHONDRION

Mitochondria occur in all but a few primitive species of eukaryotes and are characteristic of the group. Symbiotic events similar to those which are thought to have occurred during the origin of mitochondria evidently also have been involved in the origin of chloroplasts, which are characteristic of photosynthetic eukaryotes. Chloroplasts seem to represent the descendants of symbiotic, anaerobic, photosynthetic bacteria. They occur in the cells of most plants and algae, conferring on them the ability to perform photosynthesis. At least three groups of photosynthetic, anaerobic bacteria seem to have been involved in the origin of the chloroplasts in different groups of algae and plants.

Chloroplasts are rather similar to mitochondria structurally and, like them, are distinct subcellular organelles. As discussed in Chapter 12, the existence of such compartments is critical to the mechanism of photosynthesis.

The genomes of eukaryotes appear to be composites. Mitochondria are thought to have originated as symbiotic, aerobic bacteria. Chloroplasts seem to have been derived independently from symbiotic, anaerobic, photosynthetic bacteria at least three times, judged from the structural and metabolic diversity of the organelles.

Development of Genetic Recombination

The genetic information that specifies the synthesis of a particular protein occupies a segment of the DNA called a **gene.** The entire genetic complement of an organism, the sum of all its genes, is called its genome. The genomes of bacteria consist predominantly of a single molecule of DNA, which forms a circle. The genomes of most eukaryotes, in contrast, are partitioned among several to many chromosomes (Figure 5-12), each of which is independently passed to daughter cells during the process of cell division. By separating the DNA into several pieces in this way, organisms are able to exchange parts of their genomes with one another in a process that is called **genetic recombination.** The development of genetic recombination was an advance of great evolutionary significance, because such recombination leads to the rapid generation of many new combinations of genes and so speeds the rate of evolution. Recombination does occur in bacteria, but it is a much more limited process than that which is characteristic of eukaryotes.

Eukaryotes achieve recombination by fusing their cells periodically and joining their genomes in a process called *sexual* recombination. The subsequent separation of the two genomes in the formation of reproductive cells—eggs and sperm—occurs as the result of a special form of cell division, which is called **meiosis.** In this process the chromosomes are **reassorted,** shuffled like a deck of cards, with different combinations being dealt in each gamete's formation. In addition, portions of individual chromosomes are exchanged with similar duplicate chromosomes in a process called **crossing-over.** These two aspects of genetic recombination, reassortment and crossing-over, are the key to the variability, and consequently to the diversity, that underlies the evolution of eukaryotic organisms. Much of the rapid evolutionary rise of the eukaryotes probably reflects their increased capacity for genetic recombination.

The rapid evolution of eukaryotes probably reflects their much higher frequency of genetic recombination, as compared with that characteristic of the bacteria. Recombination in eukaryotes is made possible by a special form of cell division called meiosis.

Gene Fragmentation

In bacteria, individual genes encode the information specifying particular proteins in a straightforward way: a stretch of DNA corresponds to each protein. When the information encoded within a particular gene is used by a prokaryotic cell, a base-for-base RNA copy

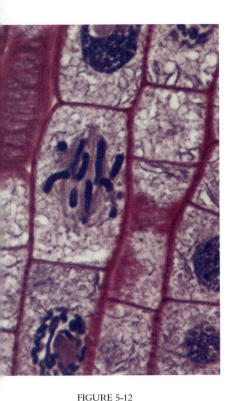

FIGURE 5-12

Unlike the simple chromosomes of bacteria, which are single closed circles, the complex chromosomes of eukaryotes, such as those in this onion root tip cell, are usually fragmented into several individual linear chromosomes, each of which carries only a portion of the genetic information.

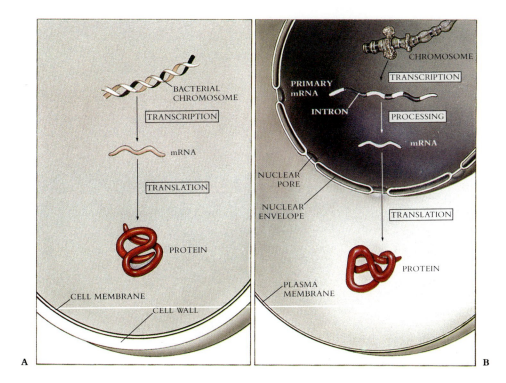

FIGURE 5-13

Genes are used very differently in prokaryotes and eukaryotes.
A *The genes on a bacterial chromosome are transcribed directly into protein, the sequence of DNA nucleotides corresponding to the sequence of amino acids in the encoded protein.*
B *The genes of a eukaryote are typically very different, containing long stretches of nucleotides called introns, which do not correspond to amino acids within the encoded protein but rather serve some other function. These introns are removed from the mRNA copy of the gene before the mRNA is used to direct the synthesis of the encoded protein.*

is made of the gene: the DNA duplex is unraveled in the vicinity of the gene, and each DNA nucleotide of one duplex strand pairs with a complementary RNA nucleotide; the result is a DNA-RNA duplex, from which the RNA strand is subsequently peeled off. This RNA copy of the gene is called a messenger RNA transcript, or mRNA. This mRNA is then used elsewhere in the cell to specify the amino acid sequence of a protein, the bases of the mRNA sequence being "read" from one end to the other in commanding the synthesis of a series of amino acids.

The organization of genes in eukaryotes is fundamentally different from that which is characteristic of the bacteria. The DNA encoding almost all eukaryotic genes is a composite, possessing two kinds of information rather than one. In addition to the nucleotide sequences that specify the amino acid sequence, eukaryotic genes also include additional sequences of DNA, which are not involved in this process. These extraneous bits of noncoding DNA are called intervening sequences, or **introns.** The intervening sequence is included within the RNA copy made from the DNA, but it is removed from the RNA copy by cytoplasmic enzymes before that copy is used in the manufacture of a protein (Figure 5-13). The amino acid sequence of the protein, therefore, does not reveal that additional information was originally spliced into the gene that encoded it. Surprisingly, more of a eukaryotic gene is normally devoted to introns than to the coding sequences themselves. With a few minor exceptions, bacteria do not have introns.

> **Segments of DNA called introns occur scattered about within eukaryotic genes. These introns are not part of the sequence of nucleotides encoding the amino acid sequence of a protein.**

EVOLUTIONARY DIVERGENCE OF THE EUKARYOTES

The evolution of eukaryotic cell organization is a benchmark in the evolutionary history of life. In the early eukaryotes a fundamental divergence of cell types soon occurred, by virtue of which the eukaryotes soon came to exceed by a wide margin their prokaryotic ancestors in cellular diversity. For perhaps half of their history on earth, a period of about

FIGURE 5-14

Major events in Precambrian evolution.

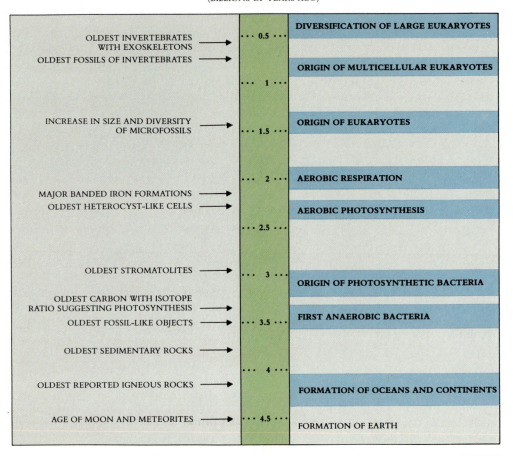

AGE OF EARTH
(BILLIONS OF YEARS AGO)

OLDEST INVERTEBRATES WITH EXOSKELETONS → ··· 0.5 ···
OLDEST FOSSILS OF INVERTEBRATES →

··· 1 ···

INCREASE IN SIZE AND DIVERSITY OF MICROFOSSILS → ··· 1.5 ···

··· 2 ···

MAJOR BANDED IRON FORMATIONS →
OLDEST HETEROCYST-LIKE CELLS →

··· 2.5 ···

OLDEST STROMATOLITES → ··· 3 ···

OLDEST CARBON WITH ISOTOPE RATIO SUGGESTING PHOTOSYNTHESIS →
OLDEST FOSSIL-LIKE OBJECTS → ··· 3.5 ···

OLDEST SEDIMENTARY ROCKS →

··· 4 ···

OLDEST REPORTED IGNEOUS ROCKS →

AGE OF MOON AND METEORITES → ··· 4.5 ···

DIVERSIFICATION OF LARGE EUKARYOTES

ORIGIN OF MULTICELLULAR EUKARYOTES

ORIGIN OF EUKARYOTES

AEROBIC RESPIRATION

AEROBIC PHOTOSYNTHESIS

ORIGIN OF PHOTOSYNTHETIC BACTERIA

FIRST ANAEROBIC BACTERIA

FORMATION OF OCEANS AND CONTINENTS

FORMATION OF EARTH

FIGURE 5-15

The evolutionary radiation of the eukaryotes. Of the five kingdoms of living organisms, four are eukaryotes. The most diverse of these, kingdom Protista, gave rise to the other three. The ancestral groups for animals and plants are well established, as we will discuss in detail in Chapter 23, but none of the many diverse groups of living protists seems to have the set of characteristics that would be expected in the ancestor of the fungi.

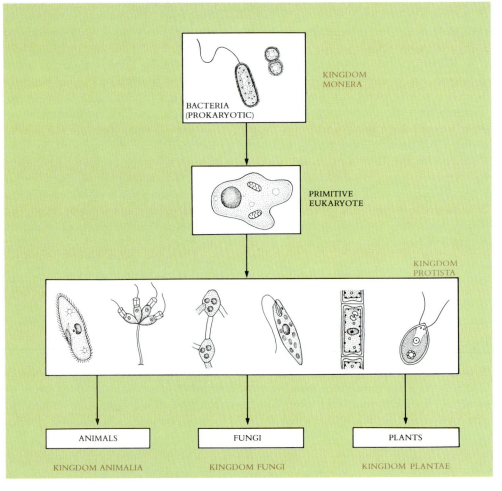

KINGDOM MONERA

BACTERIA (PROKARYOTIC)

PRIMITIVE EUKARYOTE

KINGDOM PROTISTA

ANIMALS   FUNGI   PLANTS

KINGDOM ANIMALIA   KINGDOM FUNGI   KINGDOM PLANTAE

700 million years, all eukaryotes were single celled (Figure 5-14). These single-celled organisms soon evolved many highly distinctive evolutionary lines. Some of them lived by engulfing other cells or particles of organic matter; others secreted enzymes and absorbed their food, becoming the ancestors of the fungi; and still others acquired symbiotic photosynthetic bacteria and thus gained the ability to manufacture their own food.

The single-celled eukaryotes, together with some of their later multicellular derivatives, are grouped together in the kingdom Protista. As discussed in Chapter 3, multicellular eukaryotes began to appear as fossils starting about 630 million years ago, at the beginning of the Phanerozoic Era. A number of different multicellular lines were derived independently from the single-celled eukaryotes (Figure 5-15). Among these are the major groups biologists now classify as kingdoms distinct from the Protista, namely the animals, plants, and fungi. Each of these three major, largely multicellular, evolutionary lines has become specialized in relation to its particular mode of nutrition.

Additional multicellular lines appeared among groups that are still considered to belong to the kingdom Protista because of their close relationship to unicellular organisms that are retained in that kingdom. These are the red, brown, and green algae. The multicellular green algae (Figure 5-16) ultimately gave rise to the plants. All three of these groups of algae, the plants, and some additional single-celled groups of Protista acquired symbiotic, photosynthetic bacteria and thus became photosynthetic themselves. Such events appear to have happened independently on many different occasions, judged from the evidence of the structure and biochemistry of existing chloroplasts and the ease with which similar relationships seem to be established today.

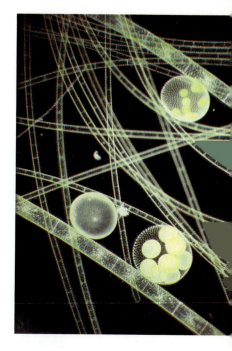

FIGURE 5-16

A multicellular green alga, a member of the kingdom Protista, gave rise to the plants. Shown here are some freshwater algae.

Differences Between Plant and Animal Cells

The cells of animals and plants differ in a number of fundamental respects, even though both are basically eukaryotic in structure.

Metabolic differences

Metabolism in animals and plants is fundamentally similar. The greatest and most obvious of the differences between plants and animals concerns the ways in which they obtain their energy. Plants are autotrophs, gaining their energy from photosynthesis, whereas animals are heterotrophs, gaining their energy from other organisms. Chloroplasts are present in plants and some Protista (algae) but are absent in animals, except in a few where they have been acquired as a result of recent symbiotic events. In other respects all eukaryotes are far more uniform metabolically than are the bacteria. Among the single-celled eukaryotes, kingdom Protista, there is also far more metabolic and even morphological diversity than exists in their multicellular descendants.

Structural differences

Two major structural differences are evident between plant and animal cells, differences that are related to one another: plant cells possess vacuoles and cell walls, whereas animal cells do not.

Vacuoles

Plant cells differ from those of animals in the way in which they use membrane-bound storage vesicles. In the cells of animals a series of internal membranes fuse, encapsulating highly concentrated digestive enzymes into mobile vesicles called **lysosomes.** These lysosomes roam about the cell absorbing debris and recycling mitochondria. The maintenance of their membranes keeps the powerful digestive enzymes that lysosomes contain separated from the other parts of the cells in which they occur. This maintenance demands the continual expenditure of energy, since the membranes of the lysosomes themselves are continually being digested by the enzymes within.

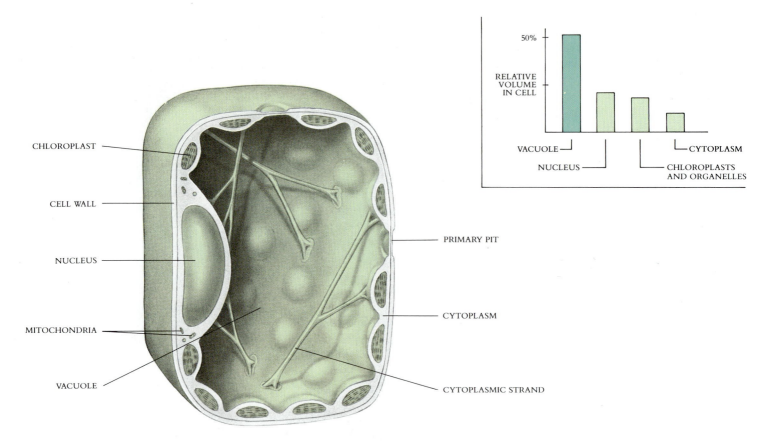

CHLOROPLAST

CELL WALL

NUCLEUS

MITOCHONDRIA

VACUOLE

PRIMARY PIT

CYTOPLASM

CYTOPLASMIC STRAND

RELATIVE VOLUME IN CELL

50%

VACUOLE
NUCLEUS
CHLOROPLASTS AND ORGANELLES
CYTOPLASM

FIGURE 5-17

Most mature plant cells contain large central vacuoles, which occupy a major portion of the internal volume of the cell. The cytoplasm occupies a thin layer between vacuole and cell membrane. Within it are all of the cell's mitochondria and other organelles.

In plants the waste disposal function is dealt with passively, in a way that does not involve the constant expenditure of energy that is required for the maintenance of lysosomes. Plants consolidate the vesicles; as a plant cell matures it contains fewer but larger vesicles, which are, when they become relatively large, called **vacuoles.** In a mature plant cell practically all of the vacuoles have characteristically fused to form one large central vacuole, which occupies a major part of the cell volume (Figure 5-17). The cytoplasm exists as a thin layer between the vacuole and cell membrane. Digestive and other enzymes are present in the central vacuole, which also stores a variety of carbohydrates and amino acids. In addition, the vacuole acts as a repository of cellular wastes. Because no part of the cytoplasm is far removed from the central vacuole, diffusion provides an effective means of exchange between it and the cytoplasm. The digestive enzymes are not present in high concentrations and so do not as readily attack the vacuole membrane as do those present in the lysosomes of animals. This fact may explain why plants live longer than animals do.

> **Plant cells contain large central reservoirs called vacuoles, in which metabolites, digestive enzymes, and waste products collect.**

Why Plants Have Cell Walls

Although a large central vacuole has many advantages, it also presents the plant with one serious problem. Because the vacuole contains a high concentration of metabolite molecules, water tends to rush into it from its surroundings. As a result, the vacuole swells, pushing outward against the cytoplasm and cell membrane.

Diffusion

Why does water rush into the vacuole? Consider a lipid bilayer membrane in aqueous solution, with a high concentration of metabolites on one side of the membrane. First imagine that the membrane has large holes in it—the internal metabolites would rush across until the concentrations of metabolites on both sides of the membrane were equal. Why? At any temperature above absolute zero, all molecules move about randomly. This random movement causes a net movement of molecules toward zones of lower concentration, a process called *diffusion*. Driven by this random movement, molecules always "explore" the space around them, diffusing out until they fill it uniformly. A simple experiment will demonstrate this. Take a jar, fill it with water, and add some ink to it. Now fill the jar to the brim, cap it, place it at the bottom of a full bathtub, and remove the cap. The ink molecules will slowly diffuse out until there is a uniform concentration in the tub and jar. A porous membrane does not interfere with this kind of diffusion.

The lipid bilayer surrounding a vacuole, however, does not have large holes in it. It is porous, but the pores are very small, slight imperfections in the matrix of lipid molecules. For this reason such a lipid bilayer is impermeable to internal metabolites such as sugars and amino acids, which are too large to pass through these pores. Unable to move across the membrane, the metabolites maintain a high concentration on one side. By contrast, small polar molecules such as water or urea move unimpeded through the small pores and so pass freely through lipid bilayers. The membrane is differentially permeable, because it retards the movement of some molecules and not others.

Osmosis

If the concentration of internal metabolites is high on one side of such a membrane, the metabolites cannot pass outward through the membrane and so dilute their concentration. Dilution can result, however, from the movement of water molecules inward. The concentration of *water* is lower on the high-solute (metabolite) side of the membrane, and so water streams across the membrane toward that side. The process continues as long as the concentration of water is higher on one side of the membrane than on the other and as long as it is not opposed by electrical charges or other forces, such as pressure. This phenomenon, in which water diffuses through a differentially permeable membrane from a region of lower metabolite concentration to one of higher metabolite concentration, is one example of a process called **osmosis.** Osmosis results in a net movement of water across the membrane (Figure 5-18). It is, in a sense, solvent diffusion, with the water molecules, rather than the solute molecules, moving to equalize concentrations. (See Chapter 9.)

Now recall that the membrane enclosing our hypothetical cell is a closed sphere. The movement of water into such a cell, driven by osmotic pressure, will cause the cell to swell and, if the swelling continues unabated, to burst. A bacterial cell from which the cell wall has been removed by chemical or other means will burst in water for just this reason. Plant cells would burst because of this, too, except for their rigid external cell walls. If such walls were not present, it would have been impossible for plants to have evolved the strategy of disposing of waste products by means of a large vacuole.

FIGURE 5-18

In this experiment a 3% salt solution is employed to demonstrate osmosis. Within cells, metabolites have the same effect.

A *The end of the tube containing the 3% salt solution is closed by stretching a differentially permeable membrane across its face that will pass water molecules but not salt molecules.*

B *When this tube is immersed in a beaker of distilled water, the salt cannot equalize its concentration in tube and beaker by passing into the beaker, since it cannot cross the membrane; water can achieve the same effect, however, by passing into the tube from the beaker. The added water will cause the salt solution to rise in the tube.*

C *Water will continue to enter the tube from the beaker until the weight of the column of water in the tube exerts a downward force equal to the force drawing water molecules upward into the tube. This force is referred to as the* **osmotic pressure** *generated by the 3% salt solution in distilled water.*

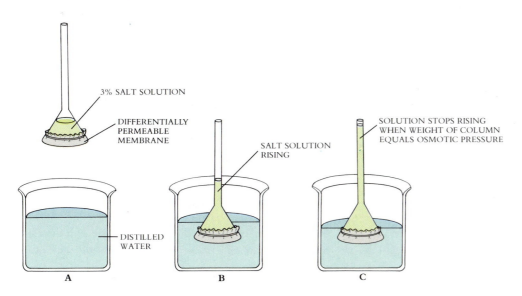

3% SALT SOLUTION

DIFFERENTIALLY PERMEABLE MEMBRANE

SALT SOLUTION RISING

SOLUTION STOPS RISING WHEN WEIGHT OF COLUMN EQUALS OSMOTIC PRESSURE

DISTILLED WATER

A

B

C

The rigid cell walls that are characteristic of all plants counteract the osmotic pressure that is created by the movement of water into the vacuole.

Plants, as well as some of the green algae from which they were derived, have rigid cellulose cell walls that envelop their membranes on the outside and preclude complex movements during the course of their development. Because of these rigid cell walls, multicellular plants are rigid as well and remain fixed at a single place throughout their lives. Although individual cells may swell up in response to osmotic changes (Figure 5-19), individual plants move only by growing. The heavy cell walls that encase the cells of plants and their absence in animals have in large measure determined the very different life-styles of these two most familiar kingdoms of organisms.

Plasma Membranes

The hallmark of an animal cell is that its flexible cell membrane, called the **plasma membrane,** is unconstrained by an exterior cell wall. Complex lipids, particularly cholesterol, are incorporated in the plasma membranes of animals but not those of plants. Changes in the degree of saturation of cholesterol within the membrane permit the fine tuning of its fluidity. The original advantage of flexible cell membranes in early eukaryotes probably had to do with the way in which fluid lipid bilayer membranes could be used to engulf other, smaller cells or bits of organic matter.

Other advantages are associated with the presence of flexible cell membranes and the absence of a cell wall. Perhaps the most important of these is the facilitation of cell-to-cell communication. By alterations in the permeability of shared membranes, two adjacent cells can achieve quite sophisticated communication with each other. The potential of such communication is realized to a very high degree among animals. Another advantage of flexible cell membranes is that they allow the movement of cells in relation to one another, as in the complex early development, or embryology, that is characteristic of animals. Finally, movement of the kind that is characteristic of animals, which actively go from place to place to search for food, to avoid being eaten themselves, or to seek a mate, depends directly on the flexible connections between their cells.

Most protists, both unicellular and multicellular, possess rigid cell walls. Animals seem to have lost such cell walls early in the course of their evolution. Although the differences between plant and animal cells are significant, all eukaryotic cells are basically similar to one another. Plants and animals, although superficially very different, share the same metabolism, the same membrane structure, and the same genetic apparatus. The differences between them have been added during the course of their divergent evolutionary paths.

FIGURE 5-19

The cells surrounding this opening in a Tradescantia *leaf are swollen by osmosis. The opening is called a stoma, and the swollen cells are called guard cells. When relaxed, these guard cells rest against one another, closing the opening. When swollen by osmosis, as they are here, the cells assume a rigid bowed shape, like two bananas placed end to end, creating an opening between them.*

SUMMARY

1. Biological cells were first described by Hooke in 1665. All organisms are composed of cells.

2. A cell is a volume of cytoplasm encased within a lipid bilayer membrane. Such a membrane serves to package together the enzymes and other materials that enable the cell to carry out metabolism and reproduce itself independently.

3. Primitive bacteria were quite diverse, and some of them had metabolic and genetic characteristics different from those of modern bacteria and eukaryotes. Some of these unusual bacteria have survived in odd habitats like those which existed on the early earth.

4. All bacteria and most single-celled eukaryotes possess exterior cell walls. In bacteria these cell walls are composed of carbohydrates cross-linked by proteins, whereas in eukaryotes other materials are employed. In plants, for example, the cell walls are composed mainly of cellulose. Cell walls seem to have evolved independently a number of times in eukaryotes.

5. The hallmark of the eukaryotic cell is interior organization, which the prokaryotic cell lacks. The membranes of eukaryotic cells form closed compartments called organelles within the cell. Included among these organelles is the nucleus, within which DNA is isolated. Eukaryotic cells also contain organelles—mitochondria and chloroplasts—that resulted from past symbiotic events.

6. Eukaryotic cells are genetically dynamic, their genomes actively exchanging versions of particular genes with one another. These processes of gene acquisition and genetic recombination are thought to be responsible to a large extent for the rapid evolution of the eukaryotes.

7. Most of the major groups of eukaryotes consist entirely of unicellular organisms, grouped as members of the kingdom Protista. Three of these, the green algae, brown algae, and red algae, are partly or wholly multicellular. In addition, three major multicellular lines—animals, plants, and fungi—have been derived from unicellular Protista and are recognized as distinct kingdoms.

8. The differences between animal and plant cells are striking, most being related to the presence of a central vacuole in plant cells. These cells possess a strong exterior cell wall, which compensates for the osmotic pressure that results from the functioning of the vacuole.

9. All eukaryotic cells are fundamentally similar. The difference between animal and plant cells, or between either one of these and fungal cells, is much less than the difference between eukaryotes and prokaryotes.

SELF-QUIZ

The following questions pinpoint important information in this chapter. You should be sure to know the answers to each of them.

1. Animal cells contain mitochondria. Plant cells contain chloroplasts. Do plant cells contain mitochondria?

2. Which of the following is *not* an organelle within cells?
   (a) nucleus
   (b) chloroplast
   (c) metabolite
   (d) mitochondrion
   (e) all the above are organelles

3. Is a photosynthetic bacterium a heterotroph or an autotroph?

4. The diameter of a typical bacterium is about
   (a) 10 micrometers   (b) 100 micrometers

5. Which type of bacterial cell has a thicker cell wall, gram positive or gram negative?

6. If you were to place a small sample of radioactive molecules into a bacterial cell and monitor where the radioactivity went, you would see that the radioactive molecules spread by diffusion everywhere within the bacterial cell; there would be no place that the radioactive molecules could not go simply by diffusing. True or false?

7. What is the largest subcompartment within the cytoplasm of a eukaryotic cell?

8. Mitochondria, which are present in almost all eukaryotic cells, are thought to represent the descendants of symbiotic prokaryotes called _____ _____ bacteria.

9. Do mitochondria contain organelles?

10. Are introns more likely to be found in a human cell or a bacterial cell?

11. There are five biological kingdoms of organisms. How many include only single-celled organisms?

12. Will an animal cell, placed in distilled water, swell or shrink?

THOUGHT QUESTIONS

1. Plants and fungi possess cell walls that are as rigid as those of bacteria. No bacterium is multicellular, however, whereas fungi and plants are. Why do you suppose that this is so?

2. All animals are heterotrophs, whereas plants are photosynthetic autotrophs. Why do you suppose no organism has evolved that is capable of both photosynthesis and heterotrophy—or have any organisms done so?

FOR FURTHER READING

DEWITT, W.: *Biology of the Cell: An Evolutionary Approach,* Saunders College/Holt, Rinehart and Winston, Philadelphia, 1977. A text with an evolutionary approach to cell biology.

GRAY, M.W.: "The Bacterial Ancestry of Plastids and Mitochondria," *BioScience,* vol. 33, pages 693-698, 1983. A clear and compelling presentation of the molecular evidence that mitochondria and chloroplasts are descended from the genomes of symbiotic bacteria, pitched at a level comprehensible to a general audience. The case in favor of the symbiont theory is overwhelming.

MARGULIS, L.: *Symbiosis in Cell Evolution,* W.H. Freeman and Company, San Francisco, 1980. A broad-ranging exposition of the theory that mitochondria, chloroplasts, and other organelles were acquired by eukaryotes from symbiotes, written by the chief proponent of the idea.

SCHOPF, J.W.: "The Evolution of the Earliest Cells," *Scientific American,* September 1978, pages 110-139. This article presents a wealth of data on the earliest cells and the evidence that they gave rise to the oxygen present in the earth's atmosphere today.

VIDAL, G.: "The Oldest Eukaryotic Cell," *Scientific American,* February 1984, pages 48-57. An intriguing glimpse of the oldest eukaryotes discovered to date.

WOESE, C.R.: "Archaebacteria," *Scientific American,* June 1981, pages 98-126. The author argues that these primitive prokaryotes should be classified not as bacteria or eukaryotes, but rather in a third kingdom. A very provocative paper.

THE STRUCTURE OF EUKARYOTIC CELLS

Chapter 6

Overview

Eukaryotic cells exhibit a considerable degree of internal organization, with a dynamic system of membranes forming internal compartments. Some of these compartments are relatively permanent, such as the nucleus that isolates the hereditary apparatus from the rest of the cell. Others are transient, such as the lysosomes that contain digestive enzymes. The partitioning of the cytoplasm into functional subcompartments is the most distinctive feature of the eukaryotic cell.

For Review　　*Here are some important terms and concepts that you will encounter in this chapter. If you are not familiar with them, you should review them before proceeding.*

- **Lipid membrane fluidity** (Chapter 4)

- **Origin of mitochondria** (Chapter 5)

- **Distinction between prokaryotic and eukaryotic cells** (Chapters 3 and 5)

The advent of the eukaryotic cell was one of the most profound changes that has occurred during the entire course of evolution on earth. The differences between eukaryotic cells and the prokaryotic cells from which they sprang affect all aspects of cell structure and function. Relatively little is known about the way in which the eukaryotic cell became stabilized in its modern form, a form that is now shared by all organisms except bacteria. Among the many differences between eukaryotes and the prokaryotes from which they evolved, two are of particular importance: (1) eukaryotic cells are extensively subdivided into compartments, within each of which a particular cellular activity is isolated; and (2) eukaryotic cells extract energy from organic foodstuffs by oxidizing them, a very efficient form of metabolism shared by some prokaryotes.

In this chapter we will describe the structure of eukaryotic cells. There is great diversity among eukaryotes in all of the features that we will discuss, and we cannot hope to address all of the variations here. Instead, we will highlight the structural elements that the many different kinds of eukaryotes have in common, the principles underlying the architecture of eukaryotic cells. In Chapter 10 we will discuss the evolution of eukaryotic metabolism.

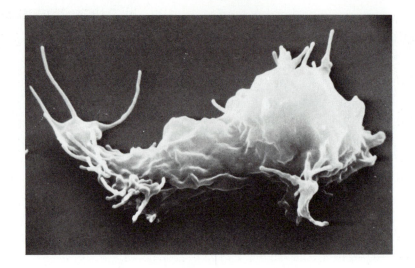

FIGURE 6-1

This eukaryotic cell, an amoeba, is moving toward you, its advancing edges extending projections outward. The moving edges have been said to resemble the ruffled edges of a woman's skirt.

THE CYTOSKELETON

Eukaryotic cells are highly variable in shape. In animal cells, which lack a cell wall, the shape continually changes. If you look at an animal cell under a microscope, you will find it alive with motion: the cytoplasm streams this way and that; mitochondria float in the moving cellular currrents; parts of the plasma membrane push inward, are pinched off as vesicles, and move into the cell's interior; on the exterior of the plasma membrane, projections shoot outward from the cell surface, then retract, only to shoot out elsewhere moments later; irregular edges form, swell, and change shape (Figure 6-1). In all of their diverse manifestations animal cells seem vibrantly alive.

Microfilaments

What is the source of the dynamic movements of animal cells? Let us try an experiment. It is known that mammalian skeletal muscles are inactivated by **cytochalasin B,** a chemical that alters the properties of the intracellular muscle protein actin. Specifically, cytochalasin B prevents the fibers of actin from interacting with fibers of the other muscle protein, myosin, and thus blocks the contraction of the muscle cell. What happens if you add cytochalasin B to a growing culture of eukaryotic cells? Movement within the cells ceases. You see no cytoplasmic streaming, no movement of internal vesicles or of surface protuberances. The cell is not dead, but it is paralyzed. The cytochalasin B has blocked the activity of very fine **microfilaments,** delicate threads only 3 to 6 nanometers thick that are composed of the two proteins actin and myosin. As we will discuss in Chapter 7, microfilaments such as these possess the property of contraction. They are able to shorten their length even when the shortening is opposed by considerable resistance. For example, microfilaments encircling a cell can, by contracting, divide the cell into two halves. The muscles of your body are composed of these same microfilaments; thus it is because of microfilament contraction that you are able to turn the pages of this book.

TUBULIN
DIMER

FIGURE 6-2

Microtubules are composed of a spiral array of tubulin subunits. The tubulin proteins first aggregate into dimers composed of two tubulin proteins, and these dimers then add onto the end of a growing spiral.

Microtubules

Microfilaments are not, however, the only fibers you will see when you examine the interior of a cell with a powerful microscope. There are other, much larger ones, called **microtubules.** As their name implies, microtubules are slender cylinders; their walls are constructed of a protein called **tubulin** (Figure 6-2). Each microtubule consists of a chain of tubulin subunits wound in a spiral like a bedspring. Microtubules are more rigid than microfilaments and about five times as thick: each microtubule is 15 to 34 nanometers in di-

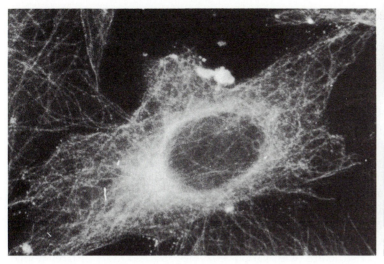
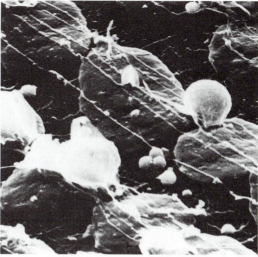

FIGURE 6-3

Glimpses of the cytomusculature of eukaryotic cells.
Left *Microtubules of baby hamster kidney cells growing in tissue culture.*
Above *Microtubules within plant cells.*

ameter. Like microfilaments, microtubules are able to contract, but the way in which they do so differs sharply.

The role of microtubules in cell architecture first became evident in studies involving the chemical **colchicine.** Colchicine blocks the movement of chromosomes during eukaryotic cell division, because it binds to tubulin and thus prevents formation of the microtubules that move the chromosomes. In studies of this effect another remarkably interesting observation was made: eukaryotic cells that lack a cell wall lose their shape if they are treated with colchicine. This was the first clue investigators had that eukaryotic cells are supported from within by a network of microtubules attached to the inner surface of the cell membrane. These microtubules radiate outward from a central region in the interior of the cell, much as the spokes of a bicycle wheel extend outward from a central ring, but in three dimensions. This network of supporting microtubules, called the **cytoskeleton** (Figure 6-3), lends support to the outer membrane of the cell, much as the bony skeleton supports your body. A microtubular cytoskeleton is much more dynamic than a bony skeleton, however. Microtubules are not stable components of the cell, except when they protrude from the cell as long waving fibers called flagella. Within the cell they are dynamic structures, constantly undergoing assembly and disassembly. Contraction or extension of portions of the cytoskeleton's network of microtubules can actually change the shape of an animal cell, because the network is attached to the cell membrane and supports it.

In Chapter 7 we will discuss microfilaments and microtubules in more depth. They provide the two basic mechanisms that enable eukaryotic cells to move chromosomes and other organelles to different locations within their interiors, as well as providing the mechanisms underlying the swimming of protists and the running of a cheetah. All animal motion depends on microfilaments and microtubules.

> **The cytoplasm of eukaryotic cells contains microfilaments and a cytoskeleton of microtubules. Changes in microfilaments permit the movement of organelles within the cell, and changes in the cytoskeleton alter the shape of the animal cell membranes.**

INTERNAL COMPARTMENTALIZATION: THE ENDOPLASMIC RETICULUM

Under a light microscope, both prokaryotic and eukaryotic cells look relatively homogeneous within. Both exhibit a relatively featureless matrix, within which various organelles are embedded. With the advent of electron microscopes, however, a very striking difference between the cells of these two groups of organisms became evident. The interiors of most prokaryotic cells still appear amorphous, even at the highest magnifications, but the interiors of eukaryotic cells are seen to be packed with membranes. So thin that they are not visible with the relatively low resolving power of light microscopes, these membranes fill the cell, dividing it into compartments, channeling the transport of molecules through the interior of the cell, and providing the surfaces on which enzymes act. Membranes are present within the cells of certain photosynthetic bacteria, but in eukaryotes they have become much more extensive and important in all aspects of cell functioning. In fact, the system of internal compartmentalization created by these membranes in eukaryotes constitutes the most fundamental distinction between eukaryotic and prokaryotic cells.

The extensive system of internal membranes that exists within the cells of eukaryotic organisms is called the **endoplasmic reticulum,** often abbreviated ER (Figure 6-4). The term *endoplasmic* means "within the cytoplasm," and the term *reticulum* comes from a Latin word that means "a little net." As in the membrane systems of some photosynthetic bacteria, the endoplasmic reticulum of eukaryotes is continuous with the cell membrane, which is usually called the **plasma membrane** in eukaryotes. The endoplasmic reticulum has a simple lipid bilayer structure, with various enzymes attached to its surface. Within the lipid bilayer, as usual, are embedded transport proteins. The endoplasmic reticulum, weaving in sheets through the interior of the cell, creates three kinds of compartments: a series of channels, another series of fluid-filled tubular or spherical spaces called **cisternae,** and a third series of interconnections between membranes that isolates some spaces as membrane-enclosed vesicles.

The endoplasmic reticulum is perhaps the most important structural feature of the eukaryotic cell. It carries out three different functions, each of which represents a significant evolutionary advance over the capabilities of unorganized prokaryotic cells: (1) it provides channels through the interior of the cell, (2) it provides a site for enzymes, and (3) it creates subcompartments within the cell.

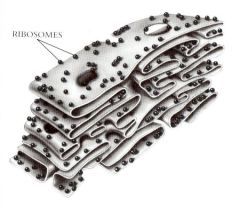

RIBOSOMES

FIGURE 6-4

Rough endoplasmic reticulum. In the drawing above you can see that the ribosomes (black dots) are associated with only one side of the rough ER; the other side bounds a separate compartment within the cell, into which the ribosomes extrude newly made proteins destined for secretion. The electron micrograph at right is of a rat liver cell, rich in ER-associated ribosomes. The rough ER is the large striated region in the center of the photograph.

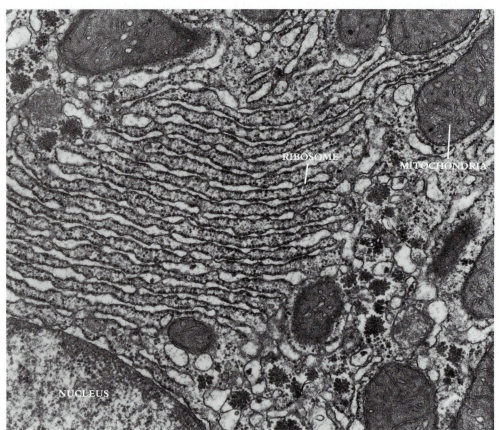

RIBOSOME

MITOCHONDRIA

NUCLEUS

Interior Channels

The spaces between folds of endoplasmic reticulum provide routes for the transport of molecules between various regions of the cytoplasm, much as a system of plumbing would do. Many eukaryotic cells, for example, synthesize proteins that are not used within the cell itself but are intended to be exported outside the cell. The genetic message specifying these particular proteins is recognized only by those ribosomes which are bound to the endoplasmic reticulum. **Ribosomes** are large molecular aggregates of protein and ribonucleic acid (RNA) that translate RNA copies of genes into protein. As new proteins are made on the surface of the endoplasmic reticulum, they are passed out across the endoplasmic reticulum membrane into the system of channels between adjacent layers of endoplasmic reticulum. They then travel through these channels to the inner surface of the cell, where they are released outside of the cell in which they were produced. From the time the protein is first synthesized on the ER-bound ribosome, it is, in a sense, already located outside the cell. The surfaces of those regions of the endoplasmic reticulum devoted to the synthesis of such transported proteins are heavily studded with ribosomes. Their membrane surfaces appear pebbly, like the surface of sandpaper, when viewed with an electron microscope. Because of this "rocky beach" appearance, the regions of endoplasmic reticulum that are rich in bound ribosomes are often termed **rough ER.** Regions of the endoplasmic reticulum in which bound ribosomes are relatively scarce are correspondingly called **smooth ER.**

Enzyme Sites

The complex folding of the endoplasmic reticulum provides an extensive surface, one on which enzymes carry out metabolic reactions. Many of the cell's enzymes cannot function when floating free in the cytoplasm; they are active only when they are associated with a membrane. Endoplasmic reticulum membranes contain embedded within them many such enzymes, which float within the fluid membrane like icebergs in the sea. Enzymes embedded within the endoplasmic reticulum, for example, catalyze the synthesis of a variety of carbohydrates and lipids. In cells that carry out extensive lipid synthesis, such as the cells of the testicles, smooth ER is particularly abundant. Enzymes that catalyze lipid biosynthesis are structural components of the smooth ER membrane and cannot be separated from it without losing their activity. Intestinal cells, which synthesize triglycerides, and brain cells, which synthesize steroid hormones, are also rich in smooth ER. In the liver, enzymes embedded within smooth ER are involved in the breakdown of stored glycogen to its constituent glucose molecules and in a variety of detoxification processes. Drugs such as amphetamines, morphine, codeine, and phenobarbital are detoxified in the liver by components of smooth ER.

Subcompartments

Because the structure of lipid bilayer membranes is fluid, two membrane surfaces that touch may fuse at the point of contact. The ability of two membranes to fuse (Chapter 4) underlies the remarkable ability of endoplasmic reticulum to form vesicles within the cell. Imagine a long, narrow balloon being pinched in the middle. In cells the pinched-in surfaces touch and fuse, resulting in the formation of two vesicles, each totally bound by ER membrane. It is in this way that a number of the major organelles of the eukaryotic cell are created, including several kinds of vesicles containing digestive enzymes and other vesicles that contain different, specialized enzymes. Many researchers believe that endoplasmic reticulum is also the source of the double membrane that forms around the chromosomes immediately after cell division. This double membrane, which is called the nuclear envelope, disappears once again at the beginning of the next cycle of cell division. Together with the chromosomes, it constitutes the nucleus. Large and centrally located, the nucleus is the most obvious interior compartment of eukaryotic cells.

The dynamic nature of endoplasmic reticulum lends great flexibility to the activities that can take place within eukaryotic cells. Transient vesicles are formed that, operating like cargo containers, are able to move molecules from one place in the cell to another. Membranes form, dissolve, and reform. Some membranes are more long lived than others, but few if any are permanent. Like the orange traffic cones used by highway workers during construction, the endoplasmic reticulum of the cell channels its metabolic traffic first one way and then another, to suit the needs of the moment.

> **The endoplasmic reticulum is an extensive system of membranes that divides the interior of eukaryotic cells into compartments and channels. The system of membranes is a dynamic structure, with membranes being organized and dissolved continuously.**

THE NUCLEUS

The largest and most easily seen organelle within a eukaryotic cell is the **nucleus,** which was first described by the English botanist Robert Brown in 1831. The word *nucleus* is derived from the Latin word for a kernel or nut. In fact, nuclei are more or less spherical in shape and do bear some resemblance to nuts. In animal cells the nucleus is typically located in the central region. In some cells the nucleus seems to be cradled in this position by a network of microfilaments. The nucleus is the repository of the genetic information that directs all of the activities of a living cell. Some kinds of cells, such as mature red blood cells, discard their nuclei during the course of development. The development of these cells terminates when their nuclei are lost. They lose all ability to grow, change, and divide, becoming merely passive vessels.

A nucleus can be considered to be composed of several elements (Figure 6-5), including:

1. An encircling system of membranes, the nuclear envelope
2. A fluid matrix, the nucleoplasm
3. DNA, complexed with protein and forming the chromosomes
4. One or more darkly staining bodies, the nucleoli

The Nuclear Envelope

The surface of the nucleus is bounded by *two* layers of membrane, the **nuclear envelope.** The nuclear envelope seems to be continuous with the endoplasmic reticulum and is conveniently thought of as its largest and most permanent compartment. Between the two membranes of the nuclear envelope is a cytoplasmic space whose precise function is unknown.

Scattered over the surface of the nuclear envelope are shallow depressions, like craters on the moon, called **nuclear pores.** These pores, 50 to 80 nanometers apart, form at locations where the two membrane layers of the nuclear envelope pinch together. The fluid nature of lipid bilayer membranes causes them to fuse together at these points, creating a pore.

The nuclear pores are not empty openings in the same sense that the hole in a doughnut is, with nothing filling the actual hole. Nuclear pores contain embedded within them many proteins that act as molecular channels, permitting passage into and out of the nucleus. Like gates, these protein channels are selective, permitting only some molecules to pass. Many proteins and smaller molecules that enter the cytoplasm of the cell through transport channels in the plasma membrane cannot pass through the nuclear envelope, since they are not transported through the protein channels in its pores. Passage through the pores, and thus passage into and out of the nucleus, is restricted primarily to two kinds of molecules: (1) proteins moving into the nucleus, where they will be incorporated into nuclear structures or serve to catalyze nuclear activities; and (2) RNA and protein-RNA complexes formed in the nucleus and subsequently exported to the cytoplasm.

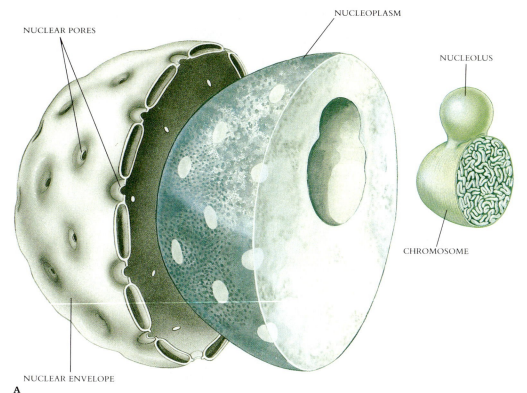

NUCLEAR PORES

NUCLEOPLASM

NUCLEOLUS

NUCLEAR ENVELOPE

CHROMOSOME

A

FIGURE 6-5

The nucleus is composed of a double membrane called a nuclear envelope that encloses a fluid-filled interior. Within the fluid, called nucleoplasm, chromosomes are suspended. When cells are not dividing, individual chromosomes cannot usually be seen with a microscope; here they are indicated as the end of a tube seen in circular cross section, blown up much larger than in a real nucleus to emphasize the tightly packed DNA strands. At one or more points on one of the chromosomes, one of which is shown in A, swelling occurs at a nucleolus, where particular genes are being copied into RNA intensively. The nuclear envelope is studded with pores, enlarged in the diagram for clarity. These pores are clearly visible as dimples in this freeze etch, B, of a nuclear envelope. In a cross section of the nuclear envelope, C, an individual pore (p) clearly extends through the two membrane layers of the envelope between the cytoplasm (c) and the nucleus (n); the dark material within the pore is protein, which acts to control access through the pore.

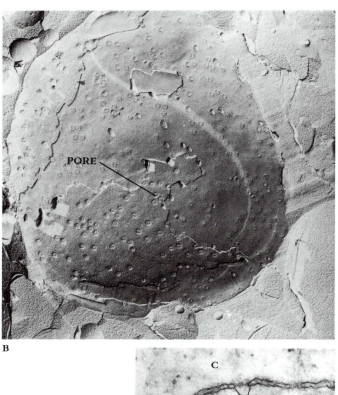

PORE

B

C

P

N

C

Like all of the derivatives of the endoplasmic reticulum, the nuclear envelope is a dynamic structure. The number of pores within it appears to increase when the cell is most active metabolically, such as immediately before cell division, and to decrease when the cell is relatively inactive, or quiescent. During cell division the nuclear envelope of most eukaryotes disintegrates, permitting the chromosomes in the nucleus to move outward to the poles of the cell. When cell division is completed, a new nuclear envelope re-forms around the chromosomes in the daughter cells.

> **The nucleus of eukaryotic cells is a semipermanent vesicle derived from the endoplasmic reticulum. The nucleus contains the cell's hereditary apparatus and isolates it from the rest of the cell. The nuclear envelope disintegrates when the cell divides and later re-forms around the two daughter nuclei.**

The Nucleoplasm

The **nucleoplasm,** the cell substance enclosed within the nuclear envelope, differs from the cytoplasm around it in that it contains no ribosomes. All of its proteins are made in the cytoplasm and imported into the nucleus through nuclear pores. Most of the DNA present in the cell is located within the nucleoplasm; none is present free in the cytoplasm.

Chromosomes

As in prokaryotes, all of the hereditary information that specifies cell structure and function in eukaryotes is encoded in DNA. Unlike prokaryote DNA, however, which exists as a single circular chromosome of naked DNA, the DNA of eukaryotic chromosomes is fragmented into several segments, each complexed with protein. Association with protein enables eukaryotic DNA to wind into a highly condensed form during cell division, which facilitates the moving of the daughter chromosomes to opposite sides of the dividing cell. Under a light microscope these condensed chromosomes are readily seen in dividing cells as densely staining rods. After cell division, eukaryotic chromosomes unravel and can no longer be distinguished individually with a light microscope. Unraveling the chromosomes into a more extended form permits the mRNA-producing enzymes of the cell to gain access to the DNA so that the hereditary information can be used to direct the synthesis of enzymes.

> **A distinctive feature of eukaryotes is the organization of their chromosomes, in which DNA is complexed with protein. Their chromosomes can be condensed into compact rods for ready movement when a eukaryotic cell divides, and later unraveled so that the information which the chromosomes contain may be used.**

The Nucleolus

A cell employs many ribosomes in its manufacture of proteins and enzymes and constantly requires replacements for the ones that are degraded or lost. This need is met by virtue of the fact that the portion of the DNA encoding the RNA components of ribosomes, called **ribosomal RNA (rRNA),** is actively copied throughout the life of the cell. In bacteria there is one copy of each rRNA gene. Eukaryotes, in contrast, have many copies of their rRNA genes; these copies are clustered together on one or more of the chromosomes. Because each member of a gene cluster may be copied simultaneously, it is possible to produce a great deal of rRNA within a short period. This amplification of rRNA production enables the cell to produce high levels of these rRNAs rapidly.

At any given moment many rRNA molecules dangle from the chromosome at the sites of rRNA gene clusters. The proteins that will later form part of the ribosome complex bind to the dangling rRNA molecules. These areas of intense ribosome synthesis are easily visible within the nucleus as one or more dark-staining regions, called **nucleoli** (sing. **nucleolus**). Nucleoli can be seen under a light microscope even when the chromosomes are extended, unlike the rest of the chromosomes, which are visible only when condensed. Because nucleoli are visible in nondividing cells, early scientists thought that nucleoli were distinct cellular structures. We now know that this is not true. Nucleoli are, in fact, those regions on the chromosomes where very active synthesis of rRNA is taking place.

THE GOLGI COMPLEX

The nucleus is not the only eukaryotic cellular organelle derived from the endoplasmic reticulum. At various locations in the cytoplasm the membranes of the endoplasmic reticulum coalesce, forming flattened stacks of membranes called **Golgi bodies** (Figure 6-6). These structures are named for Camillo Golgi, the nineteenth-century Italian physician who first called attention to them. Animal cells contain 10 to 20 Golgi bodies, whereas plant cells may contain several hundred. Collectively these Golgi bodies are referred to as the **Golgi complex.**

Golgi bodies are a functional extension of the endoplasmic reticulum, although they are typically not connected directly to it. Golgi bodies function in the collection, packaging, and distribution of molecules synthesized elsewhere in the cell. The proteins and lipids that are manufactured on the rough and smooth ER membranes are transported through the channels of the endoplasmic reticulum, or as vesicles budded off from it, into the Golgi bodies. Within the Golgi bodies, many of these molecules are complexed with polysaccharides. Almost all of the polysaccharides in the cell are manufactured within the Golgi bodies, and they are therefore immediately available there for complexing with the proteins and lipids synthesized on the endoplasmic reticulum membrane. A polysaccharide complexed to a protein is called a **glycoprotein,** whereas one complexed to a lipid is called a **glycolipid.** The newly formed glycoproteins and glycolipids collect at the ends of the membranous folds of the Golgi bodies; these ends are given the special name cisternae (Latin, collecting vessels). Intermittently in such regions, the membranes push together the sides of the cisternae, pinching off small membrane-bound vesicles containing glycoprotein and glycolipid molecules. These vesicles then move to other locations in the cell, distributing the newly synthesized molecules within them to their appropriate destinations.

> **The Golgi complex is the delivery system of the eukaryotic cell. It collects, packages, and distributes molecules that are synthesized at one location within the cell and used at another.**

Like all other derivatives of the endoplasmic reticulum, the Golgi complex is a dynamic structure, constantly budding off vesicles. Golgi bodies are most prominent in secretory cells, such as those lining your stomach, and change their shape as the level of secretory activity in these cells rises. Within the Golgi body the many catalytic proteins that are embedded in the membranes become much less common toward the bud-forming periphery. The membranes in the center of the Golgi body are more like those of the endoplasmic reticulum, whereas those toward the periphery are progressively more different. The tips of cisternae from which vesicles are pinched off do not include any catalytic proteins embedded within their membranes, and they resemble the plasma membrane in this respect. Thus one of the many results of Golgi complex activity is the conversion of complex endoplasmic reticulum membrane into simpler plasma membrane.

Vesicles formed in the Golgi complex carry out numerous important functions:
1. *They provide an exit from the cell.* Molecules synthesized within the cytoplasm but destined for use outside the cell are carried to the plasma membrane in small vesicles that fuse with the plasma membrane's inner surface and release their contents to the exterior (Figure 6-7). In this way glycolipids and glycoproteins, which are used as components of the outer surface of animal membranes, are

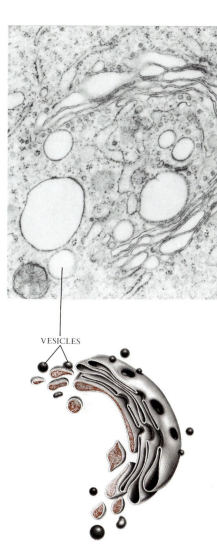

VESICLES

FIGURE 6-6

A Golgi body in cross section. Within the Golgi body the membrane is smooth and pinches together at its terminus to produce membrane-bound vesicles. These vesicles contain whatever enzymes or other substances have been transported from the endoplasmic reticulum to the Golgi body.

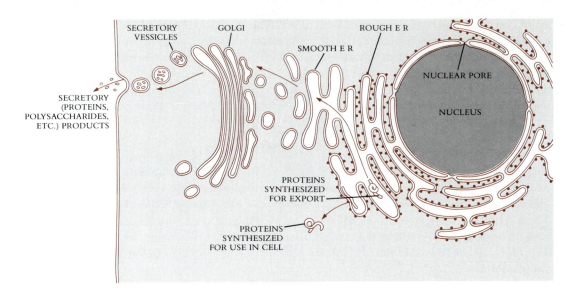

SECRETORY VESSICLES   GOLGI   ROUGH E R

SMOOTH E R

NUCLEAR PORE

NUCLEUS

SECRETORY
(PROTEINS,
POLYSACCHARIDES,
ETC.) PRODUCTS

PROTEINS
SYNTHESIZED
FOR EXPORT

PROTEINS
SYNTHESIZED
FOR USE IN CELL

FIGURE 6-7

How secretory proteins are exported from cells. Secretory proteins are assembled by ribosomes on rough ER, which extrude them across the endoplasmic reticulum into the channels that endoplasmic reticulum creates within the cell. The proteins move through these channels to a zone of smooth ER, where some of them are chemically modified by enzymes embedded within the smooth ER membrane. They then pass through more channels to a Golgi body, which is the end of the channel. There they are encapsulated within vesicles formed by the pinching together of membranes at the end of the Golgi body. These secretory vesicles pass through the cytoplasm to the interior surface of the plasma membrane, with which they fuse, releasing their contents to the exterior.

transported out across the plasma membrane. In plants this mode of cell secretion is a prominent feature of cell growth, with Golgi vesicles releasing complex polysaccharide carbohydrates to the exterior, where they are used in constructing the cell wall.

2. *They facilitate growth of the plasma membrane.* Golgi vesicles, after discharging their contents, are often incorporated into the plasma membrane, thereby increasing the surface area of the cell. This is the way that the plasma membrane expands during cell growth.

3. *They isolate certain cell enzymes within sacs.* Eukaryotic cells contain a variety of enzyme-bearing, membrane-bound vesicles derived from the Golgi complex. These vesicles are called **microbodies.**

MICROBODIES

The microbodies produced by the Golgi complex (Figure 6-8) include **lysosomes,** which carry digestive enzymes; **peroxisomes,** which carry oxidative enzymes; and **glyoxysomes,** which carry enzymes for converting fat to carbohydrate. Not all types of microbodies are present in all cells; for example, plant cells do not contain lysosomes, and animal cells do not contain glyoxysomes. The distribution of enzymes into microbodies is one of the principal ways in which eukaryotic cells organize their metabolism.

Lysosomes

Lysosomes provide an impressive example of the metabolic compartmentalization achieved by the activity of the Golgi complex. Lysosomes are vesicles that contain in a concentrated mix the digestive enzymes of the cell, enzymes that catalyze the rapid breakdown of proteins, nucleic acids, lipids, carbohydrates—in other words, the breakdown of essentially every major macromolecule in the cell. Lysosomes play many important roles in the life of a eukaryotic cell. Their role in digesting other cells is one example. When a cell such as an amoeba or a white blood cell encounters a bacterium, the outer membrane of the eukaryotic cell extends outward and engulfs the bacterium, a process known as phagocytosis. The bacterium is first encased by extending folds of the plasma membrane. This action produces, within the amoeba or white blood cell, a vacuole that surrounds the bacterial cell. When this process is complete, lysosomes fuse with the vacuole, their membranes coalescing around it and releasing into the vacuole lysosome-borne digestive enzymes. These enzymes proceed to digest the contents of the vacuole rapidly.

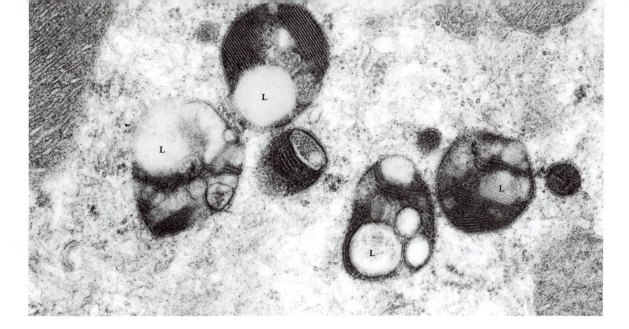

Lysosomes digest worn-out cellular components such as mitochondria and chloroplasts, making way for newly formed ones while recycling materials locked up in the old ones. This is perhaps their most important function in the cell. Cells can persist for a long time only if their component parts are constantly renewed. Otherwise the ravages of use and accident chip away at the metabolic capabilities of the cell, slowly degrading its ability to survive. Cells age for the same reason that people do, because of a failure to renew themselves. Lysosomes destroy the organelles of eukaryotic cells and recycle the organellar material at a fairly constant rate throughout the life of the cell. Mitochondria, for example, are replaced in some tissues every 10 days, with lysosomes digesting the old mitochondria as new ones are produced. In electron micrographs of lysosomes one can often observe fragments of partially digested mitochondrial carcasses.

In addition to their role in eliminating organelles and other structures within cells, lysosomes also eliminate whole cells. Selective cell death is one of the principal mechanisms used by multicellular organisms in their achievement of complex patterns of development. This directed cellular suicide is accomplished by the rupture of the lysosomes within the cells that are being eliminated. These lysosomes release their concentrated mix of digestive enzymes into the cytoplasm; once released, the enzymes proceed to digest the entire cell. Once this process of digestion has been initiated, it is irreversible.

Lysosomes are vesicles, formed by the Golgi complex, that contain digestive enzymes. The isolation of these enzymes in lysosomes protects the rest of the cell from their digestive activity.

If the contents of lysosomes are able to digest whole cells, it might occur to you to ask, "What prevents lysosomes from digesting *themselves?*" They do. Lysosomes are constantly being digested from within, their membrane being degraded by the digestive enzymes within. Lysosomes survive this digestion unruptured, however, because the cell in which the lysosomes reside constantly repairs the digestive damage to the membranes encasing the lysosomes. This repair process requires energy, of course, which is the reason why metabolically inactive eukaryotic cells die. Without the input of energy necessary to maintain lysosomal membranes, these membranes disintegrate and the digestive enzymes of the lysosomes pour out into the cytoplasm and destroy it. Bacteria, in contrast, do not possess lysosomes and do not die when they are metabolically inactive. They are simply able to remain quiescent until altered conditions restore their metabolic activity, a property that greatly heightens the ability of bacteria to persist under unfavorable environmental conditions. For you, and all eukaryotes, the very process that repairs your cells may also

FIGURE 6-8

Lysosomes (L) are digestive organelles within many eukaryotic cells. These lysosomes are within the cytoplasm of mouse kidney cells.

lead to their destruction. An absolute dependency on a constant supply of energy is the price that you pay for long life.

Peroxisomes

Among the other enzyme-containing microbodies formed by the Golgi complex are peroxisomes. Peroxisomes are vesicles containing **oxidative enzymes.** Such enzymes catalyze the reactions that remove electrons and associated hydrogen atoms. If these enzymes were not isolated within peroxisomes, they would tend to short-circuit the metabolism of the cytoplasm, which involves the addition of hydrogen atoms to oxygen.

Peroxisomes are present both in animal and plant cells. Their name refers to the chemical hydrogen peroxide (H_2O_2), which is produced as a byproduct of the activities of many of the oxidative enzymes within the peroxisomes. Hydrogen peroxide is a dangerous byproduct because it is violently reactive chemically. Peroxisomes cope with hydrogen peroxide by destroying it; they contain an enzyme, **catalase,** which breaks down the hydrogen peroxide into harmless constituents:

$$2H_2O_2 \quad \xrightarrow{\text{[CATALASE]}} \quad 2H_2O \;+\; O_2$$

HYDROGEN PEROXIDE $\qquad\qquad\qquad$ WATER $\quad$ OXYGEN GAS

Peroxisomes contain other enzymes that oxidize amino acids, an important capability for a cell attempting to persist on a protein diet. The peroxisomes of green plants also contain the enzymes that carry out **oxidative photorespiration.** This process, which will be discussed further in Chapter 12, is a kind of metabolic short-circuiting of photosynthesis.

Glyoxysomes

Some plants store metabolic food reserves in their seeds as fats or oils. On germination these fat reserves are mobilized for use by rapidly growing cells. The first step in this mobilization is the conversion of these fats to simple carbohydrates. This conversion is achieved by a battery of oxidative enzymes packaged by the Golgi complex into glyoxysomes. These enzymes are imprisoned within vesicles because the high enzyme concentrations necessary for rapid conversion of fat reserves must be prevented from being active before the appropriate time. Animal cells do not contain glyoxysomes. Instead they maintain low concentrations of fat-oxidizing enzymes free in the cytoplasm.

RELICT SYMBIONTS

Most of the membrane-bound structures within eukaryotic cells are derived from the endoplasmic reticulum, but there are important exceptions. As discussed in Chapter 5, eukaryotic cells contain complex cell-like organelles that most biologists believe were derived from ancient symbiotic bacteria. **Symbionts** perform specialized tasks for the eukaryotic cells in which they occur, tasks that would not otherwise be possible for these cells. Some symbionts are unique to particular organisms; these are referred to as *minor* symbionts. The *major* symbionts, as proposed by the symbiont theory discussed in Chapter 5, are the mitochondria, which occur in all eukaryotic cells, and the chloroplasts, which occur in all photosynthetic eukaryotes, the algae and plants.

Mitochondria

Within the cytoplasm of eukaryotic cells, mitochondria appear as long tubular inclusions 1 to 3 micrometers long (Figure 6-9). They are, therefore, about the same size as most bacteria. Although mitochondria are barely visible under the light microscope, they are readily studied with the greater resolution of the electron microscope. A variety of mitochondrial

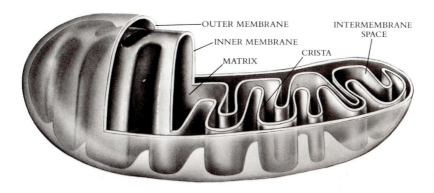

OUTER MEMBRANE
INNER MEMBRANE
INTERMEMBRANE SPACE
CRISTA
MATRIX

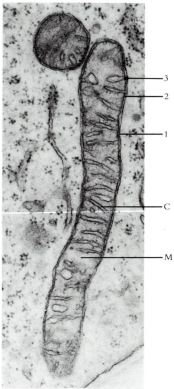

shapes are seen in electron micrographs, but the differences in shape are deceiving. They arise because examination with an electron microscope is carried out on a thin section cut from a block of solidified material. Mitochondria appear in these sections in a variety of shapes, from circular to tubular, because of variation in the angle with which the section cleaves the long tubular mitochondrial exterior.

Mitochondria appear much like one might expect bacteria to look after uptake by phagocytosis into a food vacuole, in that they are bounded by two membranes rather than one. These membranes are very different from one another: the outer mitochondrial membrane is smooth and differentially permeable, closely resembling the plasma membrane; the inner mitochondrial membrane is folded into numerous contiguous layers called **lamellae,** which resemble the membranes that occur within purple nonsulfur bacteria. The inner lamellae partition the mitochondrion into numerous compartments, called **cristae.** On the surfaces of the membrane lamellae and also submerged within them is the protein machinery that carries out aerobic respiration.

Eukaryotes derive most of their energy from metabolism that utilizes oxygen (aerobic respiration), and this aerobic respiration occurs within mitochondria. During the billion-years plus that mitochondria are thought to have been symbionts within eukaryotes, chromosomes of the eukaryotic nucleus have incorporated most of the original symbiont genes. The mitochondrion, however, still reflects its presumed history as an ancient, freeliving bacterium in that it maintains a circular DNA molecule encoding several elements that are functionally critical for its present role of aerobic respiration. All of these genes are copied into RNA within the mitochondrion and translated there into protein. In this process the mitochondria use small RNA molecules and ribosomal components that are also encoded within the mitochondrial DNA.

A eukaryotic cell does not assemble new mitochondria; all mitochondria within a eukaryotic cell are produced by the division of existing mitochondria (just as all bacteria are produced from existing bacteria by cell division). Mitochondria divide by simple fission, splitting into two just as bacterial cells do, and apparently replicate and partition their circular DNA molecule in much the same way as do bacteria. Mitochondrial reproduction is not autonomous (self-governed, independent), however, as is bacterial reproduction. Most of the components required to assemble a new mitochondrion are encoded as genes within the eukaryotic nucleus and translated into proteins by the cytoplasmic ribosomes of the cell itself. Mitochondrial replication is thus impossible without nuclear participation.

The rate of mitochondrial division is regulated by the nucleus, apparently in response to the metabolic needs of the cell. Metabolically active cells such as those of the liver may contain from 1000 to 3000 mitochondria, amounting to about a quarter of the volume of the entire cell; heart muscle cells contain even more. In contrast, metabolically less active epithelial (skin) cells or adipose (fat-storage) tissue cells contain far fewer mitochondria. The kind of eukaryotic cell that is involved even influences the way in which individual mitochondria are assembled. Thus the mitochondria of metabolically active cells, such as insect flight muscle, exhibit many more cristae than inactive cells, creating a far greater surface area for aerobic respiration.

FIGURE 6-9

Mitchondria in longitudinal and cross section. A mitochondrion possesses an outer membrane (1) with many of the characteristics of the plasma membrane, an inner matrix (M), and a highly convoluted inner membrane (2) whose inward projections (3) are called cristae (C). Mitochondria carry out the oxidative respiration of all eukaryotes. They are thought to have evolved from bacteria that took up residence within the ancestors of present-day eukaryotes long ago.

Chloroplasts

The photosynthetic symbionts that are thought to have given rise to chloroplasts were aerobic, photosynthetic bacteria belonging to at least three different groups. As in the case of mitochondria, the genes of these photosynthetic symbionts have apparently mostly been transferred to the nucleus of the eukaryotic cells during the course of the evolution of the particular groups. Because photosynthesis, like aerobic respiration, is a process that requires a closed compartment, the cell body of the presumed photosynthetic symbiont has been maintained as the body of the chloroplast. The chloroplast cytoplasm, or **stroma,** is bounded, like the mitochondrion, by two membranes (Figure 6-10). The outer membrane apparently has been derived from the endoplasmic reticulum of the host cell and the inner one from the original bacterial cell membrane.

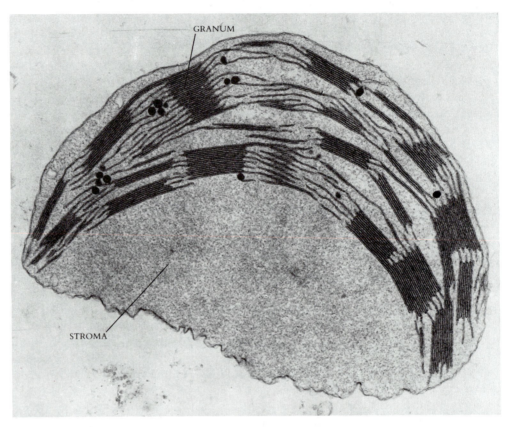

FIGURE 6-10

A chloroplast possesses interior and exterior membranes like mitochondria do, but in the case of chloroplasts these membranes, called lamellae, fuse to form stacks of closed vesicles called thylakoids. Within the thylakoids, photosynthesis takes place. Thylakoids are typically stacked one on top of the other in columns called grana.

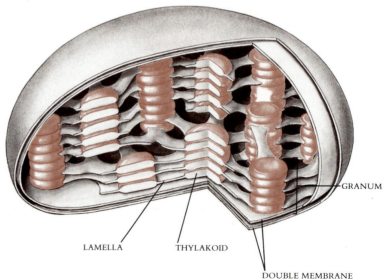

The nucleus of a plant cell contains most of the genes that encode chloroplast proteins. The chloroplast genome appears to have been assimilated by the plant nucleus in much the same manner as the genetic material of the original mitochondrial symbiont is thought to have been. Most genes specifying chloroplast components now reside in the nucleus, including all of the genes that encode the soluble enzymes that catalyze the CO_2 molecule–binding activities of photosynthesis. The chloroplast does, however, maintain a small circular DNA molecule, which encodes a few critical components of the photosynthetic apparatus, as well as the special RNA and protein components necessary to synthesize these components within the chloroplast.

Mitochondria and chloroplasts are similar in that both symbionts have lost the bulk of their genomes to the host chromosomes. They retain only the few genes necessary to maintain the specific structure of their relict cell bodies. The integrity of these structures is essential for the processes of aerobic respiration and photosynthesis.

Photosynthetic cells typically contain from one to several hundred chloroplasts, depending on the organism involved or, in the case of multicellular photosynthetic organisms, the kind of tissue. Although chloroplasts are much larger than mitochondria, they resemble mitochondria in that they possess a smooth outer membrane and a second invaginated internal membrane. In chloroplasts, however, the organization of the internal membranes is somewhat more complex than it is in mitochondria. Instead of forming an isolated compartment within the organelle, as do mitochondrial cristae, the invaginated lamellae of the internal chloroplast membrane lie in close association with one another. By fusing along their peripheries, two adjacent membranes form a disk-shaped closed compartment called a **thylakoid.** Chloroplasts contain stacks of such thylakoids, which, when viewed with a microscope, resemble stacks of coins. Each stack, called a **granum,** may contain from a few to several dozen thylakoids. A chloroplast may contain a hundred or more grana, each connected to another by part of the original invaginated lamellae.

On the surface of the thylakoids are the light-capturing photosynthetic pigments. In plants, algae, and some bacteria the photosynthetic pigment is a form of chlorophyll, a pigment that absorbs visible light very efficiently. The only wavelength that it does not absorb well, and so transmits, is the one corresponding to green light. That is why chloroplasts, and the organisms in which they occur, are green.

Organisms can synthesize chlorophyll only in the presence of light. In the dark the production of chlorophyll ceases and most of the lamellae are reabsorbed. When reabsorption occurs, the chloroplast, largely devoid of internal membrane invaginations, is called a **leucoplast.** In the root cells and various storage cells of plants, leucoplasts may serve as sites of storage for starch. A leucoplast that stores starch is sometimes termed an **amyloplast.** Many plant pigments other than chlorophyll likewise occur in chloroplasts. A collective term for these different kinds of organelles, all derived from chloroplasts, is **plastid.** Like mitochondria, all plastids come from the division of existing plastids.

Minor Relict Symbionts

Mitochondria and chloroplasts are not the only examples of symbionts among the eukaryotes. Whole unicellular algae live within some animal cells. Unlike mitochondria and chloroplasts, however, the algae can still reproduce outside of their host cells. Many other symbionts are true relicts, now totally dependent on their host cells to reproduce. In every case the symbiont seems to confer a useful ability on the host. In the protist *Paramecium,* for example, live small symbiotic bacteria called **Kappa particles.** These Kappa particles secrete a toxin called paramecin, which kills competing *Paramecium* individuals of different genetic makeup. The symbiosis of many of the other relict symbionts that occur in particular eukaryotes also depends on unique biochemical abilities (Figure 6-11).

A

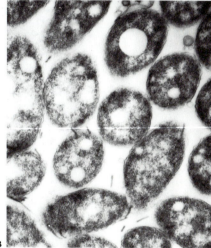
B

FIGURE 6-11

Symbiosis is a close relationship between dissimilar organisms that affects both members of the partnership in different ways. For example, the luminous spot below the eye of this deepsea fish, A, results from the presence of a colony of luminous bacteria, B, which grow on nutritious secretions that the fish provides.

AN OVERVIEW OF CELL STRUCTURE

The structure of a eukaryotic cell is much more complicated than that of a prokaryotic one (Figure 6-12; Table 6-1). The most distinctive difference is the extensive subdivision of the interior of eukaryotic cells by endoplasmic reticulum. The most visible of these endoplasmic reticulum–bound compartments, the nucleus, gives eukaryotes their name. There are no equivalent membrane-bound compartments within prokaryotic cells. Some photosynthetic prokaryotes possess extensive inward foldings of membrane, but these do not

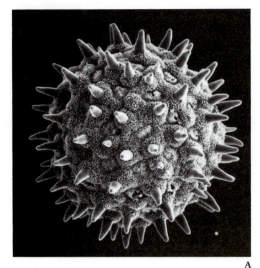

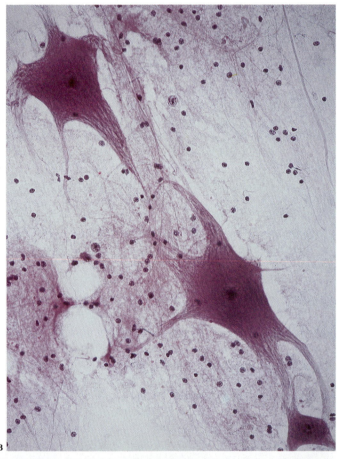

A

FIGURE 6-12

Cellular diversity. Eukaryotic cells take many forms.
A *A morning glory pollen grain starts as a single cell with one nucleus but has two internal cells when it is shed.*
B *A nerve cell of a human spinal cord.*
C *Striated muscle cells.*
D *Red and white blood cells.*

B

C

isolate one portion of the prokaryotic cell from another. A molecule can travel from any location in such a cell to any other location unimpeded.

In later chapters we will consider the consequences of these structural differences and how they influence the metabolism and biochemistry of eukaryotes. You will see that the metabolic processes that go on within eukaryotic cells differ from those of bacterial cells and that those differences, like the structural ones discussed in this chapter, are substantial.

TABLE 6-1   A COMPARISON OF BACTERIAL, ANIMAL, AND PLANT CELLS

	BACTERIA	ANIMAL	PLANT
EXTERIOR STRUCTURE			
Cell wall	Present (protein-polysaccharide)	Absent	Present (cellulose)
Cell membrane	Present	Present	Present
Flagella	Present (1 strand)	Often present (9 + 2)	Present in only a few (9 + 2)
INTERIOR STRUCTURE			
Endoplasmic reticulum	Absent	Usually present	Usually present
Microtubules	Absent	Present	Present
Centrioles	Absent	Present	Absent
Golgi bodies	Absent	Present	Present
ORGANELLES			
Nucleus	Absent	Present	Present
Mitochondria	Absent	Present	Present
Chloroplasts	Absent	Absent	Present
Chromosomes	A single circle of DNA	Multiple, with DNA complexed to protein	Multiple, with DNA complexed to protein
Ribosomes	Present	Present	Present
Lysosomes	Absent	Usually present	Absent
Vacuoles	Absent	Absent or small	Usually a large single vacuole in mature cell

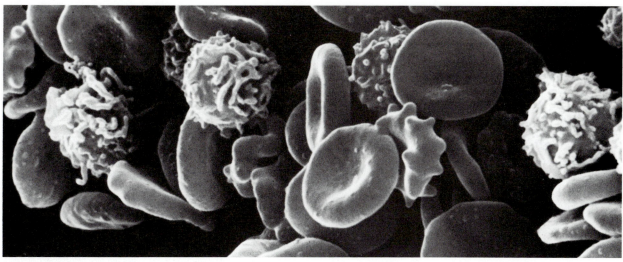

D

SUMMARY

1. The shape of animal cells is determined by a cytoskeleton of microtubules and microfilaments and may be altered by the contraction of these elements.

2. The endoplasmic reticulum is a dynamic series of membranes that subdivides the interior of eukaryotic cells into separate compartments. It is the most distinctive feature of eukaryotic cells and one of the most important, since it serves to compartmentalize the cell functionally.

3. The nucleus is the largest of the compartments created by the endoplasmic reticulum. It serves to isolate the cell's genomic DNA from the rest of the cytoplasm, which both protects the DNA from cytoplasmic elements and provides a means of controlling its exportation of information to the cytoplasm.

4. A second derivative of the endoplasmic reticulum is the Golgi complex, which serves as a cellular "express package service," packaging molecules within special membrane vesicles and transporting them to various locations in the cell.

5. One class of vesicles created by the Golgi complex consists of lysosomes, vesicles containing high concentrations of digestive enzymes. These enzyme-containing packages are delivered to digestive vacuoles or to the cell's exterior. A cell constantly expends energy to repair digestive damage to these vesicles, and when this repair activity ceases, the vesicles soon burst, digesting and killing the cell.

6. Not all of the eukaryotic cell's internal organelles are created by the endoplasmic reticulum. Others appear to have been derived from symbiotic bacteria. By far the most widespread of these organelles are mitochondria and chloroplasts.

7. Many, but not all, of the genes that originally were present in mitochondrial and chloroplast DNA seem to have been transferred to, or had their functions taken over by, the DNA in the chromosomes of the host cell. Both classes of organelles, however, have retained the genes necessary to create their particular distinctive structures. Without these structures, mitochondria would be unable to function in aerobic respiration, and chloroplasts in photosynthesis.

SELF-QUIZ

The following questions pinpoint important information in this chapter. You should be sure to know the answers to each of them.

1. What happens if you add cytochalasin B to a growing culture of eukaryotic cells?
   (a) the cells die
   (b) the microfilaments contract
   (c) the cells fuse with one another
   (d) the cells divide
   (e) the cell interiors cease to move

2. Which is thicker, a microtubule or a microfilament?

3. Microtubules are composed of protein subunits. The name of the protein that makes up the subunits is _____.

4. Cells that carry out extensive lipid synthesis are rich in
   (a) smooth ER
   (b) rough ER
   (c) both smooth and rough ER
   (d) neither smooth nor rough ER
   (e) either smooth or rough ER

5. The nuclear envelope of eukaryotes is composed of more than one lipid bilayer membrane. True or false?

6. Imagine a protein with 100 amino acids. In most cases the gene specifying that protein
   (a) would be longer in a prokaryote
   (b) would be the same length in a prokaryote as in a eukaryote
   (c) would be longer in a eukaryote

7. The regions of eukaryotic chromosomes devoted to ribosome synthesis are often visible in histological stains of cells as one or more darkly staining regions called _____.

8. A bacterial cell does not die when it is metabolically inactive. What prevents lysosomes within these cells from digesting their own membranes and then digesting and destroying the bacterial cells?
   (a) the membranes encasing bacterial lysosomes are resistant to oxidative enzymes
   (b) bacterial lysosomes do not use oxidative enzymes, since bacteria are anaerobic
   (c) bacterial lysosomes synthesize the ATP necessary to repair digestive damage
   (d) the metabolism of bacterial cells degrades cellular protein to obtain the ATP necessary to repair digestive damage
   (e) bacteria do not contain lysosomes

9. The enzyme catalase can be found in eukaryotic cells
   (a) within lysosomes
   (b) within peroxisomes
   (c) free in the cytoplasm
   (d) bound to the endoplasmic reticulum
   (e) within the nucleus

10. Oxidative photorespiration takes place within
    (a) the mitochondria
    (b) the chloroplasts
    (c) the peroxisomes
    (d) the cytoplasm
    (e) the nucleus

11. Which is bigger, a chloroplast or a mitochondrion?

12. From the center of a mitochondrion's interior, what is the minimum number of membranes that would have to be crossed to reach the exterior of the mitochondrion?

THOUGHT QUESTIONS

1. Mitochondria are thought to be the evolutionary descendants of living cells, probably of symbiotic purple nonsulfur bacteria. Are mitochondria alive?
2. How does the internal matrix of microtubules within eukaryotic cells manage to pass through the extensive system of endoplasmic reticulum?
3. How does the Golgi complex know where to send what vesicle?

FOR FURTHER READING

ALBERTS, B., et al.: *Molecular Biology of the Cell,* Garland Publishing, Inc., New York, 1983. An exceptionally good text on the molecular events that take place within cells and on how cells interact with one another.

HOLTZMAN, E., and A.B. NOVIKOFF: *Cells and Organelles,* 3rd ed., Saunders College Publishing, Philadelphia, 1984. A good overview of cell structure, with detailed descriptions of the chief organelles and how they function.

PORTER, K., and J. TUCKER: "The Ground Substance of the Living Cell," *Scientific American,* March 1981, pages 56–67. A lucid account of how the authors used high-resolution electron microscopy to detect the network of microfilaments within eukaryotic cells.

ROTHMAN, J.E.: "The Compartmental Organization of the Golgi Apparatus," *Scientific American,* September 1985, pages 74–89. A current view of how the Golgi complex operates.

SLOBODA, R.D.: "The Role of Microtubules in Cell Structure and Cell Division," *American Scientist,* vol. 68, pages 290-298, 1980. A description of how microtubules determine cell shape.

Chapter 7

Overview

Movement is a characteristic of almost all living things. Practically all movements of cells, of things within cells, and of organisms composed of cells are achieved by one of three mechanisms: flagellar rotation, microtubular movement, and microfilament contraction. Although the three mechanisms evolved independently and are very different, each has proved to be highly successful.

For Review   *Here are some important terms and concepts that you will encounter in this chapter. If you are not familiar with them, you should review them before proceeding.*

- **Characteristics of life** (Chapter 3)

- **ATP** (Chapter 4)

- **Plant vacuoles** (Chapter 5)

- **Microfilaments** (Chapter 6)

- **Microtubules** (Chapter 6)

Most living things share the same constellation of properties. All organisms have cellular organization, are capable of growth and metabolism, reproduce themselves, and possess the same hereditary machinery. In addition, almost all organisms are capable of movement. Active movement from one place to another is one of the most characteristic properties of animals. Even organisms that do not move from place to place have cell interiors that are alive with motion. In undergraduate biology texts the mechanisms by which organisms achieve movement are usually not discussed during consideration of cellular biology; this matter is deferred until vertebrate muscles are discussed in later chapters. In this text we will depart from that tradition and discuss motion as a fundamental property of all cells.

Because movement is a property of almost all cells, the mechanisms responsible for movement have played an important part in the evolution of life. The material in this chapter will provide an essential background for much of what is to come later in the text. To mention a few examples: The mechanism that bacteria use when swimming to power the rotation of their flagella, called a **proton pump,** has been adapted by the mitochondria and chloroplasts of eukaryotes to produce ATP during oxidative metabolism (Chapter 11) and photosynthesis (Chapter 12). The mechanism that protists use to wave their flagella, microtubular movement, is also employed to move chromosomes within eukaryotic cells during cell division (Chapter 8)—and to sense the movement of air within your ears even now as you hear yourself breathe (Chapter 47). The mechanism that a single cell uses to change its shape, microfilament contraction, is also the basic mechanism of muscle movement (Chapter 42). It provides the motive power for the movement of your eyes as you read this page, for your fingers as they turn the pages (Chapter 48)—indeed, for all animal movement. Although it is natural to think of "locomotion" in terms of muscles contracting, since that is how humans walk and move, the processes responsible for vertebrate motion are the same as those which occur within single cells that are changing their shape by contracting their cytoskeleton. A muscle is simply a collection of single cells contracting their microfilaments in concert.

Only a few different mechanisms underlie the great diversity of organismal movements. Practically all of the movement of cells, and of the matter within them, is attributable to one of three processes: **flagellar rotation, microtubular movement,** or **microfilament contraction.** Each of the three major motive processes provides an independent and very different solution to the challenge of providing a means of locomotion.

FLAGELLAR ROTATION

Of all living cells, bacteria are perhaps the ones most constantly in motion. Observed under a microscope in a droplet of water, most bacteria can be seen to dart around, constantly changing direction and rarely slowing down (Figure 7-1). Many individual bacteria can swim about 20 cell diameters per second. Imagine trying to run 20 body lengths per second!

Bacteria can swim through the water by rapidly rotating long protein fibers called **bacterial flagella,** which protrude out of their cells. *Flagellum* is the Latin word for "whip." Eukaryotic cells seen under a low-power microscope also possess fine filaments extending outward, which are used in swimming, and these too are called flagella. However, higher-power microscopes have shown that eukaryotic flagella have a complex internal structure, one that is totally unrelated to the structure of bacterial flagella. It is important not to confuse these two very different kinds of structures with one another, especially because eukaryotic flagella play such an important role in the biology of eukaryotes.

One or more bacterial flagella, each several times longer than the cell, trail behind a swimming bacterium. The waving movement of these flagella propels the organism through the water (Figure 7-2). The waving pattern of flagellar movement is not caused by a series of contractions within the bacterial flagellum, as, for example, is the undulating

PATH TAKEN
BY SWIMMING
E. COLI

FIGURE 7-1

The path taken by a swimming bacterium, indicated in color, is highly erratic—a "random walk." In the experiment illustrated here the position of a single individual of Escherichia coli *was monitored by serial photographs and its path projected in three dimensions.*

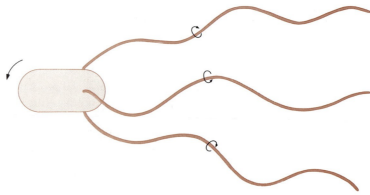

FIGURE 7-2

The bacterium Escherichia coli *possesses several flagella, which trail behind it. When these flagella all rotate in the same direction, they wind around one another and, whipping through the water, propel the cell forward. When the flagella rotate in different directions, they cause the cell to tumble rather than to move forward. Bacterial cells typically alternate between forward motion and tumbling. The periodic tumbling causes the direction of the bacterium's path to change randomly, which is an efficient way to explore an environment.*

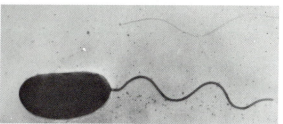

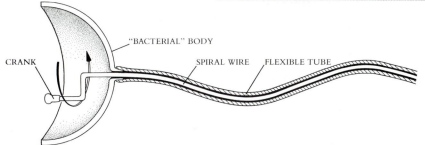

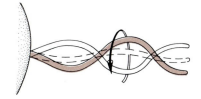

CRANK "BACTERIAL" BODY SPIRAL WIRE FLEXIBLE TUBE

of a crawling snake. You can demonstrate this by severing an individual flagellum at its base with a laser: it does not continue to wave, but instead lies still. The wave must originate from within the body of the bacterium. Nor is the wave generated by an up-and-down beat, like you get if you grasp the end of a long rope and shake it up and down, sending vertical waves down the rope. That the wave is not vertical can be demonstrated vividly by using a chemical called an antifilament antibody to glue the end of a flagellum or group of flagella to the surface of a microscope slide. When the flagellum is attached by its tip in this way, the tethered bacterium at the other end of the flagellum does not bob up and down, but instead rotates round and round. The flagellar fiber behaves not like an oar, but like a propeller (Figure 7-3). The rotation of bacterial flagella is carried out by a complex rotary "motor" embedded within the cell wall and membrane (Figure 7-4). This rotary motion is virtually unique to prokaryotes; only a very few eukaryotes have organs that truly rotate.

The movement of bacteria is achieved by the rotation of a long protein filament. Such rotating flagella are unknown in eukaryotes.

Since all bacteria have a basically similar flagellar apparatus, it is assumed that the structure of this apparatus has not changed much during the evolutionary history of the group. There is, however, some variation. Among the gram-negative bacteria, the flagellum has a second sleeve where the flagellar shaft passes through the outer membrane

FIGURE 7-3

Bacteria swim by rotating their flagella. The photograph is of Vibrio cholerae, *the microbe that causes the serious disease cholera. The unsheathed core visible at the top of the photograph is composed of a single crystal of the protein* **flagellin.** *In intact flagella this core is surrounded by a flexible sheath. Imagine that you are standing inside the* Vibrio *cell, turning the flagellum like a crank. You would create a spiral wave that travels down the flagellum, just as if you were turning a wire within a flexible tube. The bacterium employs this kind of rotary motion when it swims. Few other organisms practice truly rotary motion.*

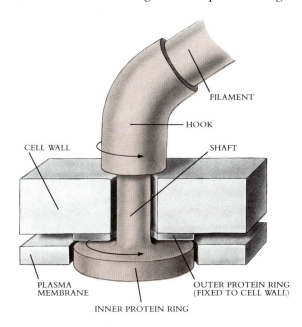

FILAMENT

HOOK

CELL WALL

SHAFT

PLASMA
MEMBRANE

OUTER PROTEIN RING
(FIXED TO CELL WALL)

INNER PROTEIN RING

FIGURE 7-4

A mechanical model of the flagellar motor of a gram-positive bacterium. The long flagellar filament, a single solid crystal of the protein flagellin, is attached to a curved hook whose shaft protrudes inward through a sleevelike base anchored within the cell wall. Fixed to the end of the shaft is an inner ring of proteins, like a ring of ball bearings. This inner ring is opposite a similar outer ring of proteins fixed to the interior surface of the sleeve through which the shaft protrudes. The shaft rotates when the inner protein ring attached to it turns with respect to the outer ring, which remains fixed. This outer ring cannot move, since the sleeve to which it is attached is firmly anchored in the rigid cell wall.

FIGURE 7-5

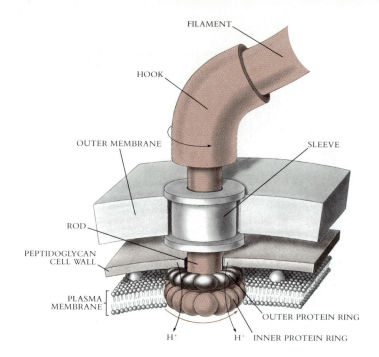

The flagellar motor of a gram-negative bacterium. The filament passes through a sleeve in the outer membrane and through a hole in the thin peptidoglycan cell wall to an inner protein ring. As in the gram-positive flagellum, this ring turns against an outer ring fixed to the interior surface of the peptidoglycan cell wall. The cell pumps H^+ ions out of it at other locations, and, being more concentrated outside as a result, the ions reenter the cell at ports that admit them through the cell membrane. The inner ring is one such port, and as it admits H^+ ions back into the cell, their passage causes it to rotate relative to the outer ring, spinning the flagellar shaft.

(Figure 7-5). In addition, the placement of the flagella varies from group to group. Some gram-negative bacteria, such as the common *Escherichia coli,* have four to eight different sites of flagellar attachment scattered over the cell surface. In contrast, most gram-positive bacteria have a bundle of flagella clustered together at one end of the cell.

MICROTUBULAR MOVEMENT

The much larger cells of eukaryotes could not move efficiently by the mechanisms that had evolved among bacteria. More elaborate mechanisms that greatly increased the mobility of the early eukaryotes soon evolved.

A eukaryotic cell moves by contracting a network of fibers embedded within the cytoplasm of the cell. As discussed in Chapter 6, this network, often referred to as the cytoskeleton of the cell, consists of large microtubules and very fine microfilaments. These microtubules and microfilaments, together with fibers of intermediate size, are interconnected in a cytoskeleton, or **microtrabecular lattice** (Figure 7-6), which fills the interior of

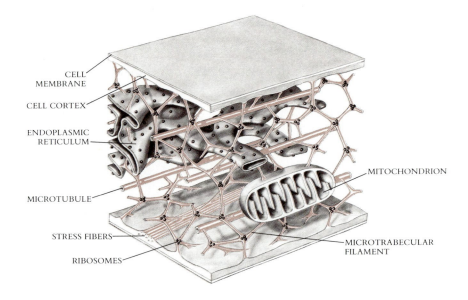

FIGURE 7-6

The microtrabecular lattice. In this diagrammatic cross section of a eukaryotic cell the mitochondria, ribosomes, and endoplasmic reticulum are all supported by a fine network of filaments, through which pass microtubules linking various portions of the cell together.

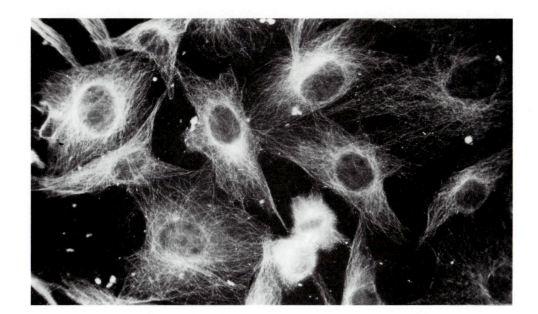

FIGURE 7-7

The microtubules in these baby hamster kidney cells have been rendered visible by treating them with a preparation of fluorescent proteins that specifically bind to tubulin.

the cell. The filaments and tubules of this lattice interact with one another, and all of the movements of eukaryotes are based on these interactions. In discussing the mechanism of movement in eukaryotes, we will first consider the microtubules.

The Structure of Microtubules

Microtubules are the largest and most transient fibers of the microtrabecular lattice. They are hollow tubes about 24 nanometers thick and up to 25 micrometers long that occur in all eukaryotic cells (Figure 7-7). Microtubules are long chains of protein subunits, assembled much like a brick smokestack. The basic building blocks of microtubules are two similar proteins called *a* and *b* **tubulin,** each containing about 500 amino acids. These two proteins aggregate to form ellipsoidal, double molecules usually loosely termed **tubulin subunits.**

The growth of microtubules in living cells is not spontaneous and occurs only in the presence of **microtubular organizing centers.** These centers often appear as dense bodies within the cytoplasm or associated with the cell membrane.

Microtubular organizing centers are located at a variety of points within the cell and play important roles in cell motility and division:

1. *Basal bodies.* The flagella by which motile eukaryotes and eukaryotic cells propel themselves through a liquid medium are each composed of a bundle of microtubules originating from an organizing center that is located within the cell immediately beneath the place from which the flagellum protrudes. These organizing centers are given a special name, **basal bodies,** to emphasize their association with flagella.

2. *Spindle fibers.* The spindle fibers by which all eukaryotes pull the replicated chromosomes to opposite poles during the process of cell division are composed of bundles of microtubules. In animal and many other eukaryotes (but not in fungi or plants) the spindle fibers originate from two organizing centers called **centrioles,** which are located perpendicular to each other near the nuclear envelope.

Microtubules probably evolved first as structural elements in eukaryotic cells. Because they are readily disassembled, microtubules provide the cell with a means of supporting temporary shapes and forms. Such adaptations can be seen dramatically in many

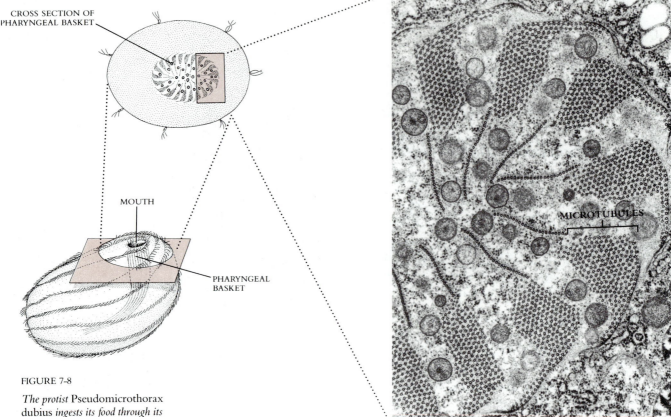

CROSS SECTION OF
PHARYNGEAL BASKET

MOUTH

PHARYNGEAL
BASKET

MICROTUBULES

FIGURE 7-8

The protist Pseudomicrothorax
dubius *ingests its food through its
mouth into a pharyngeal basket,
seen here in cross section. The
highly ordered arrays of
microtubules expand and contract
the walls of the basket, drawing
water and food inward.*

of the morphological structures of unicellular eukaryotes, members of the kingdom Protista (Figure 7-8). A particularly striking example is provided by the beautiful heliozoan protist *Actinosphaerium,* which radiates thin extensions called **axopods** outward in all directions (Figure 7-9). The core of each extension is a spiral of microtubules. When an axopod traps prey, the shortening of the microtubules at the cellular end of the spiral causes the axopod and its prey to be drawn into the cell.

> **Microtubules are a major structural component of the lattice of filaments
> and tubules that forms the cytoskeleton of eukaryotic cells. They are long,
> hollow tubes built from tubulin subunits.**

The Basis of Microtubular Movement

In addition to their structural role, however, microtubules are also used by single-celled eukaryotes to provide a mechanism of cell movement more effective than the complex rotary motor that drives bacterial flagella. Eukaryotic cells, using energy derived from ATP, have evolved mechanisms to move one microtubule past another. This microtubular movement pulls the ends of the sliding microtubules, and whatever is attached to them, together or apart, just as sliding your two index fingers past one another horizontally pulls your elbows together or apart.

A careful examination of the arrangement of microtubules reveals how cell locomotion works. Cell movement depends on *pairs* of microtubules lined up side by side. Each pair is cross-connected at numerous positions by molecules of a protein called **dynein.** In electron micrographs the dynein cross-links appear as projections or arms protruding from one microtubule toward the other. Occasionally, the two microtubules are also connected

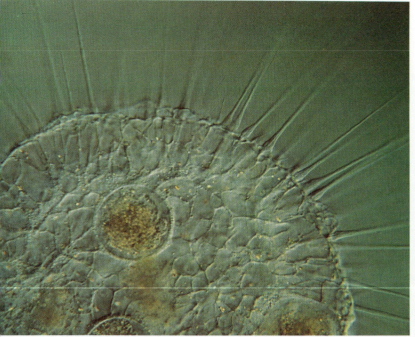
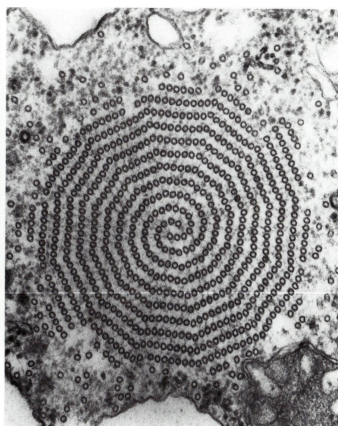

FIGURE 7-9

A *Axopods extend outward radially from* Actinosphaerium.
B *In cross section each axopod is composed of a complex array of microtubules.*

by elastic protein bands called **nexin links.** To investigate how the two microtubules interact, let us carry out a simple experiment. If we use a protein-digesting enzyme to remove the occasional nexin links from a preparation of microtubule pairs, and then add ATP, we obtain a dramatic result: the two members of the pair slide along and away from each other. The dynein cross-links are "walking" along the microtubules.

As a way of understanding how this system works, imagine a row of schoolgirls standing alongside a row of schoolboys. Each child holds firmly to the child in front with one hand and reaches across to a child in the opposite row with the other hand. Now further imagine that each boy releases the hand of the opposite girl, and instead reaches for the hand of the girl in front of her. The boys then step forward so that they are opposite their new partners. The result is that the two lines move past one another. With each wave of hand transfer, more movement would occur. Viewed from a distance, the two lines of schoolchildren would seem to slide past one another. This is how members of microtubular pairs move with respect to one another, sliding past each other by passing along a dynein handshake, each handshake movement using the energy from the cleavage of ATP to break the old dynein-microtubule cross-connection and form the new one.

The 9 + 2 Flagellum

Eukaryotic cells long ago used the dynein-driven sliding of microtubules past one another in a mechanism of cell propulsion so simple and effective that it has been maintained practically unaltered throughout the subsequent evolutionary history of the eukaryotes. This propulsion is based on a unique kind of microtubular cable, called the **9 + 2 flagellum** (Figure 7-10). This flagellum consists of a circle of nine microtubule pairs surrounding two central pairs. Completely different from bacterial flagella, this complex microtubular appa-

FIGURE 7-10

*A eukaryotic flagellum springs
directly from a basal body, is
composed of nine doublets, and has
two microtubes in its core connected
to the outer ring of doublets by
dynein arms.*

MICROTUBULES

PLASMA MEMBRANE

BASAL BODY

MICROTUBULE
PLASMA MEMBRANE
DYNEIN ARM
RADIAL SPOKE

CENTRAL SHEATH

MICROTUBULES

FIGURE 7-11

*The exterior surfaces of many protists
are covered with dense banks of cilia
or short, numerous flagella. The*
Tetrahymena *in this photograph has
a clearly visible mouth at its narrower
upper end, lined with microtubules.*

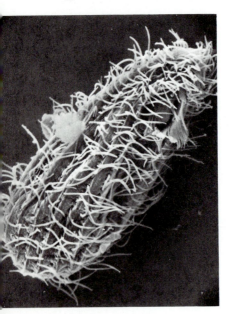

ratus evolved very early in the history of the eukaryotes. Although the cells of many
multicellular eukaryotes today no longer exhibit 9 + 2 flagella and are nonmotile, the 9 +
2 arrangement of microtubules can still be found within them, the apparatus having been
put to other uses. Like mitochondria, the 9 + 2 flagellum appears to be a fundamental component of the eukaryotic endowment.

Eukaryotic flagella are not difficult to see. If you were to examine a motile, single-
celled eukaryote under a microscope, you would immediately notice the flagella protruding from its cell surface (Figure 7-11). The constant beating of these flagella propels the cell
through its environment (Figure 7-12). When examined carefully, each flagellum proves
to be an outward projection of the interior of the cell, containing cytoplasm and enclosed
by the cell membrane. To emphasize that the flagellum contains cellular cytoplasm, the
individual flagellar protrusion is called an **axoneme.** This term is suggestive of the long cytoplasmic extensions of nerve cells, which are called axons.

An axoneme is basically an outward extension of the cell. Each axoneme contains
microtubules. These microtubules are derived from a basal body, situated just below the
point at which the axoneme protrudes from the surface of the cell membrane.

The swimming of single-celled eukaryotes results from a series of bending motions
along the axoneme (Figure 7-13). The sequential bends cause a wave to be propagated
down the axonemal flagellum. The beating motion that results propels the cell through the

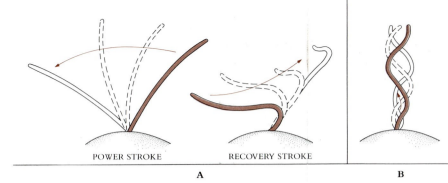

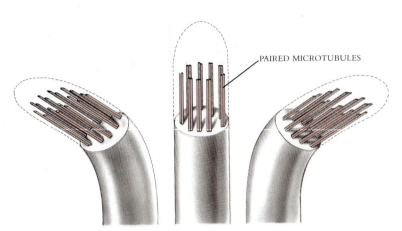

FIGURE 7-12

The movement of a eukaryotic flagellum, **A,** *or cilium, is not rotary like that of a bacterial flagellum,* **B,** *but rather a back and forward whip lash.*

POWER STROKE   RECOVERY STROKE

A

B

PAIRED MICROTUBULES

FIGURE 7-13

How a eukaryotic flagellum bends. Periodically the nine microtubule doublets are connected by nexin cross-links, not visible at the scale shown here. Because each end of a doublet segment is anchored in place by the nexin link, movement of the two members relative to each other causes a bend in the flagellum.

water. Continued swimming requires a constant input of ATP into the axoneme, but it does not require the active intercession of the basal body. Unlike bacterial flagella, which are inert if severed from their membrane-embedded motors, a eukaryotic axoneme that has been cut free from the cell by a laser will swim away—and will continue to do so as long as it can obtain ATP from the medium.

Different eukaryotic cells propel themselves by means of different numbers of axonemes. For example, some sperm cells have only one axoneme, which may be very long. Other cells may have dense banks of axonemes, which are then usually arranged in rows or rings. By convention, individual long axonemes are usually termed flagella, whereas shorter and more numerous ones are called **cilia.**

FIGURE 7-14

These highly ciliated epithelial cells line the palate of a frog. An exocrine gland can be seen emptying into the mouth cavity. The cilia in the mouth serve to move food material down into the digestive tract.

> **The 9 + 2 arrangement of microtubules is almost universal among eukaryotes. In eukaryotes that have flagella, flagellar waving is accomplished by the movement of pairs of microtubules past one another. The ends of these microtubules are anchored, their motion producing a wave that travels down the flagellum.**

The 9 + 2 flagella occur in nearly all kinds of eukaryotes and are thought to have been present in the cells from which eukaryotes evolved. Groups such as flowering plants and some insects lack flagella now but are closely related to other organisms that still have them, at least on their sperm cells. Thus these groups are considered to have lost 9 + 2 flagella during their evolution from ancestors that possessed them. In contrast, the red algae, the fungi, and the amoebas completely lack flagella and have no close relatives that possess them. It is likely that these groups had ancestors with 9 + 2 flagella, but this cannot be conclusively demonstrated with the evidence now available.

In many multicellular organisms, cilia carry out tasks far removed from their original function of propelling cells through water (Figure 7-14). In several kinds of human tissues, for example, the beating of rows of cilia moves water over the tissue surface. Interestingly, many of the sensory receptors of vertebrates appear to have evolved from cilia. The

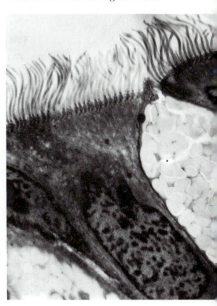

9 + 2 arrangement of microtubules occurs in the sensory cochlea hairs of the human ear, where the bending of these hairs by pressure constitutes the initial sensory input of hearing. The same 9 + 2 arrangement of microtubules is associated with the sense of smell, occurring in the olfactory fibers of the nasal epithelium. This arrangement is also involved in vision; it occurs in both the rods and cones of the retina of the eye.

MICROFILAMENT CONTRACTION

Single-celled organisms are not the only ones that possess cells that move about, nor are flagella the only means of propulsion. If you were to separate the cells of fibrous animal tissue from one another by digesting away the protein matrix between them with a protein-cleaving enzyme and then place the individual cells on a culture plate, the cells would start to explore the plate. Their movement would be imperceptible to your eye, for it is not rapid, but a slow-motion camera would reveal that the cells are in constant motion. As they move about, they constantly lift their edges—in appearance, the phenomenon has been compared to the slow ruffling of a skirt—and extend protrusions. If, like a Sherlock Holmes, you were to place powder on the surface of the culture plate to see if it had been disturbed—gold particles, taken up by cells that contact them, are used for this purpose—the tracks of moving cells would be seen readily (Figure 7-15).

Movements of individual cells play profoundly important roles in the development of animals. To a great extent, their evolution has depended on the ability of different kinds of cells to scramble around one another, of like cells to assemble together, and of periodic migrations of entire collections of cells from one location in a developing organism to another. Certain animal cells play specialized roles that depend particularly on their ability to move about the body. In your body, for example, white blood cells migrate from the vessels of the bloodstream into wounded areas, where they engulf invading microorganisms.

FIGURE 7-15

A human epithelial cell growing in tissue culture does not rest in one location, but rather slowly moves about the surface of the culture dish. Too slow a motion to see, it can be demonstrated indirectly. In this experiment the surface of the culture dish has been sprinkled with fragments of colloidal gold, which the cell takes up when it contacts them. You can see the irregular path it has taken, marked by the absence of gold particles, which the cell has removed like a vacuum cleaner.

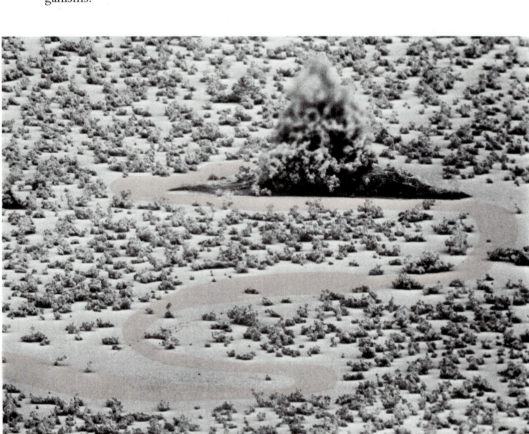

Most of the motile cells of animals lack flagella (sperm are an obvious exception) and do not swim free in liquid. Instead they crawl laboriously over one another. They move by expanding and contracting portions of their surfaces, looking much as you might if you were imprisoned within a sleeping bag, thrashing around to get out. This ability to alter surface relationships is the central mechanism underlying the movement of individual cells within multicellular animals. It arises, as you might expect, from dynamic changes in their cytoskeletons. In this case the key elements are not microtubules, but rather the fine microfilaments, which are only 60 nanometers thick.

The Structure of Microfilaments

Microfilaments are composed of five protein components. The major ones that we will be concerned with here are **actin** and **myosin.** In addition, there are three ancillary components, the proteins **tropomyosin, troponin,** and **alpha-actinin,** which we will discuss in more detail in Chapter 42.

1. *Actin.* Actin is a major constituent of microfilaments. The individual actin proteins are the size of a small enzyme and are sometimes called **G-actin** because of their globular shape. In many cells actin proteins do not exist free but rather are complexed with a stabilizing protein called **profilin,** which prevents their haphazard polymerization. After dissociating from profilin, actin molecules polymerize to form thin filaments called **F-actin** (Figure 7-16). The filaments consist of two strings of monomers wrapped around one another, like two strands of pearls loosely wound together. The result is a long, thin, helical filament with a diameter of about 70 nanometers.

2. *Myosin.* The other major constituent of microfilaments, myosin, is a protein molecule more than 10 times longer than an individual actin molecule. Myosin has an unusual shape: one end of the molecule consists of a very long rod (actually two identical polypeptides wound around one another); the other end consists of a double-headed globular region. In electron micrographs a myosin molecule looks like a two-headed snake. Like actin, myosin spontaneously forms filaments (Figure 7-17). These filaments have a very important property: their globular "heads" act like an ATP-splitting enzyme, cleaving ATP molecules into ADP + P$_i$ + energy. The energy made available by this reaction ultimately drives the shortening of the microfilament.

Microfilaments are composed of a highly ordered complex of actin and myosin. In solution, actin and myosin spontaneously form a random complex called **actomyosin,** a complex that is dissociated by the addition of ATP to the solution. When an ordered complex is formed, however, the addition of ATP causes the complex to contract rather than

FIGURE 7-16
An actin filament.

FIGURE 7-17

The structure of myosin.
A *Each myosin molecule is a coil of two chains wrapped around one another; at the end of each chain is a globular region referred to as the "head."*
B *Myosin molecules are usually combined into filaments, which are cables of myosin from which the heads protrude at regular intervals. The rod portions of the myosin molecules in the micrograph above are 150 mm long.*

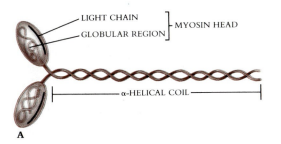

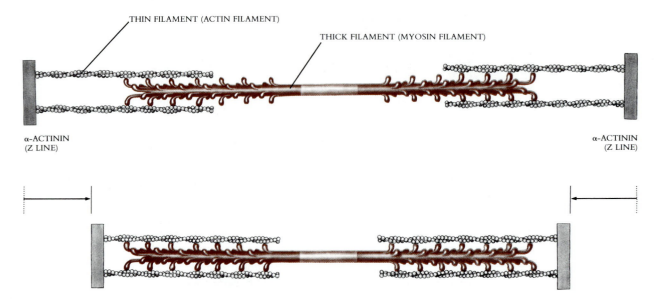

THIN FILAMENT (ACTIN FILAMENT)

THICK FILAMENT (MYOSIN FILAMENT)

α-ACTININ
(Z LINE)

α-ACTININ
(Z LINE)

FIGURE 7-18

The interaction of actin and myosin filaments in vertebrate muscle. The heads on the two ends of the myosin filament are oriented in opposite directions, so that as the right-hand end of the myosin molecules "walks" along the actin filaments going right and so pulling them and the attached alpha-actinin (a component that in muscles is called the Z line) leftward toward the center, the left-hand end of the same myosin molecule "walks" in a leftward direction, pulling its actin filaments and their attached Z line rightward toward the center. The result is that both Z lines move toward the center—contraction.

dissociate. The secret of contraction lies in the way in which the actin and myosin fibers are combined. When they interdigitate, the two fibers are able to contract with respect to each other. The ordered array as it occurs in vertebrate muscle is diagrammed in Figure 7-18, with a myosin filament interposed between two pairs of actin filaments, its globular heads (called **S-1 units**) jutting out toward the actin filaments on each side.

How Microfilaments Contract

The contraction of microfilaments occurs in a way that is reminiscent of the contraction of microtubules. In every microfilament that contracts, the myosin fiber slides past the actin filaments, using the cross-links provided by its S-1 heads to "walk" along the actin. The process is well understood, since it is the same one that occurs, on a grander scale, in vertebrate muscle. The contraction occurs in five stages (Figure 7-19):

Step 1: In a resting microfilament the S-1 heads of myosin do not make contact with actin. Instead they lie curled around the myosin fiber. Also bound to the myosin fiber are the chemicals ADP and P_i (phosphate ion), which, you will recall, are the breakdown products of ATP.

Step 2: When the microfilament is stimulated to contract, the S-1 heads of myosin move out from the myosin fiber, dragging along the central portion of the individual myosin molecules. The S-1 heads make contact with the actin fiber and become attached to it in a perpendicular orientation, each S-1 head sticking straight out from the actin fiber.

Step 3: Focus now on the perpendicular orientation of the S-1 head and the actin fiber. This orientation is about to change, and in this change lies the power stroke of microfilament contraction. After the S-1 end of the myosin molecule has made contact with actin, the other end of the myosin senses this, perhaps as a tension transmitted down the molecule. It responds by releasing the bound ADP and P_i molecules from the myosin fiber. This release changes the tension along the myosin molecule, sending a message back to the S-1 head that causes it to tilt back toward the central portion of the myosin molecule. Remember, though, that the S-1 head is still bound to the actin fiber. The tilting of the S-1 head does not pull it away from the actin, but it does cause the angle between them to change from perpendicular (90 degrees) to about 45 degrees. Consider now what has happened so far. The S-1 head has tilted its orientation and yet remains connected both to the rest of the myosin molecule and to actin. Something has to give. What

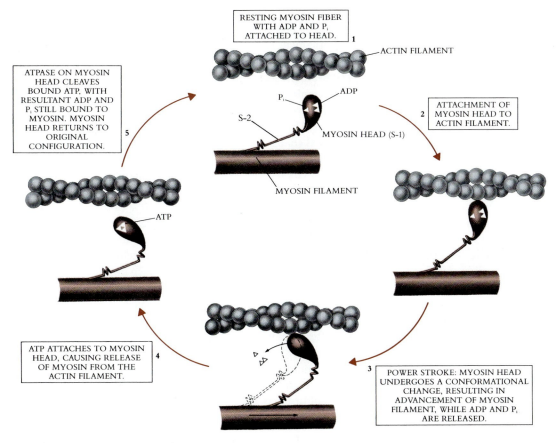

RESTING MYOSIN FIBER
WITH ADP AND P$_i$
ATTACHED TO HEAD. **1**

ACTIN FILAMENT

ATPASE ON MYOSIN
HEAD CLEAVES
BOUND ATP, WITH
RESULTANT ADP AND
P$_i$ STILL BOUND TO
MYOSIN. MYOSIN
HEAD RETURNS TO
ORIGINAL
CONFIGURATION. **5**

P$_i$ ADP

S-2

MYOSIN HEAD (S-1)

2 ATTACHMENT OF
MYOSIN HEAD TO
ACTIN FILAMENT.

MYOSIN FILAMENT

ATP

ATP ATTACHES TO MYOSIN
HEAD, CAUSING RELEASE
OF MYOSIN FROM THE
ACTIN FILAMENT. **4**

3 POWER STROKE: MYOSIN HEAD
UNDERGOES A CONFORMATIONAL
CHANGE, RESULTING IN
ADVANCEMENT OF MYOSIN
FILAMENT, WHILE ADP AND P$_i$
ARE RELEASED.

FIGURE 7-19

The mechanisms of microfilament contraction.

happens is that the myosin fiber is pulled about 75 nanometers relative to the actin fiber. As a result, the microfilament that is composed of the actin and myosin has contracted by that distance. This movement of the myosin fiber relative to the actin fiber, resulting from the tilting of the S-1 head, is the central event of microfilament contraction.

Step 4: At the end of the power stroke the S-1 head has a tilted orientation with respect to actin. This permits a molecule of ATP to bind to actin, which displaces the S-1 head. The S-1 head, no longer bound to actin, resumes its perpendicular orientation.

Step 5: The S-1 head, exercising its enzymatic activity as an ATP splitter, cleaves the actin-bound ATP. The resulting ADP and P$_i$ dissociate from actin and bind to the nearby myosin fiber, around which the S-1 heads coil.

Thus, at the end of a contraction cycle, everything is as it was in the beginning, except that the microfilament has contracted 75 nanometers and one molecule of ATP has been split. Repeated time and again, this cycle can lead to great changes in the length of microfilaments.

How does microfilament contraction lead to cell movement? Again, an analogy to our study of microtubules is helpful. Recall what happened when microtubular 9 + 2 flagellar arrangements were supplied with ATP. The microtubules immediately separated from one another, simply "walking" away, unless they were anchored by a rigid nexin cross-spoke. Microfilaments convert the sliding of fibers into motion in the same way, by anchoring one of the filaments. In this case it is the actin, the ends of which are bound to the anchoring protein alpha–actinin (Figure 7-18). Alpha–actinin is found widely distributed on the interior surfaces of the plasma membranes of eukaryotic cells. Because the actin is not free to move with respect to the alpha–actinin to which it is bound, the zones between alpha-actinin anchors shorten when the microfilament contracts. Because the microfilament may be attached at both of its ends to membrane, its overall shortening moves the membranes to which the microfilament is attached.

> **The contraction of vertebrate muscles and many other kinds of cell movement in eukaryotes result from the movements of microfilaments within cells. Microfilaments are composed of long parallel fibers of actin cross-connected by myosin. Their movement results from an ATP-driven conformational change in myosin.**

Unlike prokaryotic cells, all eukaryotic cells contain both actin and myosin. These proteins, therefore, probably evolved in the common ancestor of the eukaryotes.

Actin is a common protein within eukaryotic cells: about 10% of the protein in most kinds of eukaryotic cells is actin. In all eukaryotes the actin molecule has a similar amino acid sequence. For example, the actin protein molecules of rabbits and slime molds differ by only 17 of the 375 amino acids. The great similarity of these proteins among eukaryotes indicates that the amino acid sequence of the actin molecule has been highly conserved during the course of evolution of the eukaryotes.

There is usually much less myosin than actin in eukaryotic cells, and the myosin appears to have changed more than the actin in its amino acid sequence during the course of its evolution. There are many differences in amino acid sequence among the myosins found in eukaryotic cells. This is perhaps not surprising, since myosin is the active partner in the contraction process. Many different kinds of contractive mechanisms, requiring different speeds and sensitivities, have evolved. Your muscles contract in fractions of a second, whereas a skin cell growing in culture may take days to traverse a culture dish.

Muscles

Even though the cells of almost all eukaryotic organisms appear capable of shape changes mediated by microfilaments, many animals have evolved specialized cells devoted almost exclusively to this purpose. These cells contain far more myosin in their fibers than occurs in the fibers of single-celled eukaryotes, and these fibers are typically much thicker also. Such specialized animal cells are called **muscle cells.** In an animal the contraction of these cells brings about the movement of others attached to them. In this way the contraction of the arrays of microfilaments in the muscle cells bunched along the leg of a racehorse causes the leg to move with respect to the rest of the horse's body. The muscle cells of the horse have evolved a structure that permits very rapid and strong contraction. Other muscle cells have different functions that require less vigorous activity, such as the gentle rhythmic contractions of the vertebrate intestine. These cells have evolved a different arrangement of microfilaments, less highly ordered and compact. We will discuss these and other muscle tissues later in the text, when we consider the special biology of animals in greater detail.

Movements Within the Cytoplasm

Actin-myosin microfilaments are also responsible for a very important form of intracellular movement called **cytoplasmic streaming.** The contraction of microfilaments within the cell causes the progressive flowing of cytoplasm through the interior of the cell. The addition of actin inhibitors, such as the cytochalasin B we mentioned earlier, which blocks the binding of actin to cell membranes, results in the immediate cessation of cytoplasmic streaming. The mechanism that is responsible for generating the cytoplasmic flow is not clear, but it probably involves attachments of the contracting microfilaments to membranes of the endoplasmic reticulum, which would then sweep the cytoplasm along in swaying sheets. Cytoplasmic streaming is particularly evident in plant cells, where it has an important physiological function. The constant flow of cytoplasm past the central vac-

FIGURE 7-20

Intestinal microvilli.

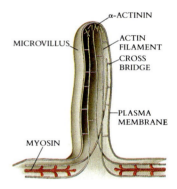

uole ensures ample opportunity for all portions of the cytoplasm to carry out the diffusion of molecules into and out of this storage and waste collection system.

Animal cells often produce fingerlike projections (Figure 7-20) called **microvilli.** You will encounter them later in a variety of guises, primarily as absorptive devices. In the microvilli of intestinal epithelial cells a bundle of actin filaments is anchored at the tip of the microvillus by alpha-actinin. This bundle extends downward through the microvillus into the interior of the cell, where it is complexed with myosin to form a microfilament. The contraction of the microfilament below the cell surface shortens the protruding microvillus as the actin fibers attached to its tip are drawn inward. Since such contractions of the microfilament base can be quite rapid, microvilli can change their length quickly. They often appear to pop up almost instantaneously and to disappear just as quickly.

MOVEMENT IS A BASIC PROPERTY OF CELLS

The fluttering of an eyelash, the flight of an eagle, the awkward gallop of a fleeing giraffe, the slow, almost imperceptible crawling of one cell over another all have in common three basic mechanisms. So do the movement of chromosomes to the opposite poles of cells during cell division, the streaming of plant cell cytoplasm past the central vacuole, and the explosive creation and disappearance of microvilli on the surfaces of intestinal cells. So do the darting movements of swimming bacteria, the closing of a protist's mouth around its prey, and the slow streaming movements of amoebas. Motion, a word that most often brings to mind a swiftly running cheetah or racing horse, is actually a property shared by all living things, single-celled as well as multicellular.

The basic mechanisms that make motion possible evolved very early in the history of life. First there evolved a mechanism that is shared by all prokaryotes, and then two more mechanisms evolved on which all eukaryotic motion is based. A *Paramecium* swimming in a cup of water uses the same microtubular sliding mechanism to wave its cilia and propel itself through the water as your body uses to separate your chromosomes when your cells divide; the band of microfilaments whose contraction divides a *Paramecium* into two daughter cells contracts using the same mechanism that drives the muscle of a running cheetah. Of the many elements that you study in your consideration of the biology of the cell, few are more fundamental than the cellular mechanisms that produce motion.

SUMMARY

1. Almost all locomotion achieved by organisms occurs by one of three mechanisms: flagellar rotation, microtubular movement, or microfilament contraction.

2. Flagellar rotation evolved among the prokaryotes, and it is still the characteristic means of locomotion of that group. Rotating flagella are not found in eukaryotes.

3. All eukaryotic cells contain microtubules, hollow tubes of protein that are a major component of the cell's cytoskeleton. Tubules are formed by centrioles within the cell.

4. Microtubules move within the cell by sliding past one another. Pairs of microtubules are connected by dynein cross-links, and these cross-links break and re-form, permitting one microtubule to "walk" along the other.

5. The 9 + 2 flagellum is almost universal among eukaryotes. A waving motion is achieved by the movement of parallel microtubules past one another. Because the ends of the microtubules are anchored, their relative movement creates a bend in the flagellum that can be propagated along its length.

6. A second mode of movement among eukaryotes uses another element of the cytoskeleton, the microfilaments. This is the movement responsible for muscle contraction and for the crawling of cells over one another.

7. Microfilaments are composed of the proteins actin and myosin, together with three other, less abundant proteins. The myosin is located between the actin filaments.

8. Changes in the shape of the end of the myosin molecule cause it to move along the actin, producing contraction.

SELF-QUIZ

The following questions pinpoint important information in this chapter. You should be sure to know the answers to each of them.

1. If you cut off a bacterial flagellum at its base, will it continue to rotate?

2. A eukaryotic flagellum is made up of how many sets of microtubule pairs?

3. The total number of microtubules in a flagellum is
   (a) 9   (c) 20   (e) 1
   (b) 11   (d) 3

4. If the nexin bands are digested away from a microtubule pair, and ATP is added,
   (a) the microtubules dissociate from one another
   (b) the microtubules slide past one another
   (c) the microtubules are unable to move past one another
   (d) the microtubules disaggregate into tubulin subunits

5. Do plants possess flagella?

6. Beating rows of cilia are common in human tissues. What is the difference between an individual cilium and a eukaryotic flagellum, such as that which propels a swimming sperm cell?
   (a) there is no difference
   (b) a cilium does not possess a 9 + 2 arrangement of microtubules
   (c) movement of a cilium is driven by contracting microfilaments rather than by microtubules
   (d) flagella are capable of rotary motion, whereas cilia are not
   (e) two of the above

7. Which is larger, actin or myosin?

8. In a resting microfilament the S-1 heads of myosin are not in physical contact with actin. True or false?

9. In microfilament contraction the cleavage of ATP is associated with
   (a) tilting of the S-1 head in the power stroke
   (b) establishment of contact between the S-1 heads and the actin fiber
   (c) displacement of the S-1 head from the actin fiber
   (d) relaxation of S-1 head

10. In a single microfilament contraction the tilting of the S-1 head causes the myosin molecule to move relative to the actin fiber. How far?

11. Cells are able to contract because of orderly arrangements of actin and myosin. Which of the two is anchored to the plasma membrane of cells, actin or myosin?

12. Do microvilli extend themselves by means of a bundle of microtubules or a bundle of microfilaments?

THOUGHT QUESTIONS

1. All eukaryotic cells appear to possess cytoskeletons of microfilaments and microtubules. No prokaryotic cells do so. What structural component of the prokaryotic cell plays the same functional role as the eukaryotic cytoskeleton?

2. Microfilaments can contract forcefully, pulling membranes attached to the two ends together. Microfilaments cannot expand, however, pushing membranes attached to the two ends of a microfilament apart from one another. Why is it that microfilaments can pull but not push?

FOR FURTHER READING

DE DUVE, C.: "Exploring Cells with a Centrifuge," *Science,* vol. 189, pages 186-194, 1975. A description of the studies that clearly defined the different organelles of eukaryotic cells, by an investigator who won a Nobel Prize for his work.

DUOTIN, P.: "Microtubules," *Scientific American,* August 1980, pages 66-80. Describes the ways in which microtubules are formed within cells and the many roles they play.

FAWCETT, D.: *The Cell,* 2nd ed., W.B. Saunders College Publishing, Philadelphia, 1981. A very good text on cell anatomy, with superb electron micrographs of subcellular structures.

SATIR, P.: "How Cilia Move," *Scientific American,* October 1980, pages 44-55. A fascinating account of how these hairlike appendages of living cells are molecular machines powered by ATP.

Chapter 8

THE CELL CYCLE

Overview

The essential problem in cell division is to achieve an equal partitioning of duplicate genomes among the daughter cells. Bacteria solve this problem by coupling the replication of their single DNA molecule with the fission of their cell body. Eukaryotes have a more complex genome structure and have evolved a correspondingly more complex form of cell division, called mitosis, in relation to it. In mitosis, microtubules attach to each replicated chromosome, pulling sister chromosomes to opposite poles of the cell. Once the redistribution of chromosomal material has been completed, the cell divides.

Review   *Here are some important terms and concepts that you will encounter in this chapter. If you are not familiar with them, you should review them before proceeding.*

- **Chromosome** (Chapter 6)

- **Nuclear envelope** (Chapter 6)

- **Microtubules** (Chapter 7)

- **Centriole** (Chapter 7)

All living organisms grow. From the smallest bacterium to the largest elephant, life involves the accumulation of material, the synthesis of new molecules. Simple as this observation seems, it has a profound consequence. Living organisms must divide or they would grow too large to function. The ways in which this division is achieved, and their biological consequences, have changed significantly during the course of the evolution of life on earth.

In bacteria, all of which are unicellular, division occurs by a simple splitting of the cell in two. The evolution of the eukaryotic cell required a different and far more complex process of division to accommodate the many morphological changes that had occurred. This process, which is called **mitosis**, achieves a precise segregation of the chromosomes during the course of cell division. In this chapter we will first discuss the simple cell division of bacteria and then examine in detail the more complex eukaryotic process of division.

CELL DIVISION IN BACTERIA

Among bacteria, the process of cell division is simple, and there is no reason to think that it has changed significantly during the history of the group. In bacteria the genetic information, or genome, exists as a single circle of double-stranded DNA, attached at one point to the interior surface of the cell membrane. Early in the life of a cell, a second copy of its DNA is synthesized. At a special site on the chromosome called the **replication origin,** a battery of more than 22 different enzymes goes to work and starts to make a complementary copy of the DNA. When these enzymes have proceeded all the way around the circle of DNA, the cell then possesses two copies of the genome, attached side by side to the interior cell membrane.

Growth of a bacterial cell to an appropriate size induces the onset of cell division, or **binary fission,** the division of a cell into two equal or nearly equal halves. First, new plasma membrane and cell wall materials are laid down in the zone between the attachment sites of the two "daughter" DNA genomes. As new material is added in the zone between the attachment sites, the growing plasma membrane pushes inward and the cell is progressively constricted in two (Figure 8-1). Initiating the constriction at a position between the membrane attachment sites of the two daughter DNA genomes provides a simple and effective mechanism for ensuring that each of the two new cells will contain one of the two identical genomes. Eventually the invaginating circle of membrane reaches all the way into the cell center, pinching the cell in two. A new cell wall forms around the new membrane, and what was originally one cell is now two.

> **Bacteria divide by binary fission, in which a cell pinches in two. The point where the constriction begins is located between where the two replicas of the chromosome are bound to the cell membrane, ensuring that one copy will end up in each daughter cell.**

The Cell Cycle

The process of cell division in prokaryotes can be described as a simple cycle. In such a description the various stages of division are designated by symbols. The growth phase can be designated G; the phase during which the new genome is synthesized, S; and the fission of the cytoplasm, C. The process is then

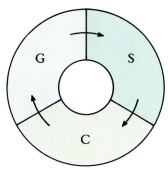

or, more simply,

$$G \rightarrow S \rightarrow C$$

This symbolic description of the life of an organism, in this case a bacterium, is called a **cell cycle diagram.** Such diagrams allow one to refer in a succinct way to the different elements of the life cycle of the organism. We will often refer to individual stages of cell cycles, since the progressive events that occur during cell division are discussed most conveniently as separate entities. It is important, however, not to lose sight of the fact that cell division is, in fact, continuous and that our separation of it into discrete elements is done merely for the sake of clarity.

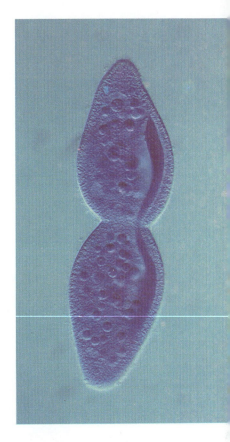

FIGURE 8-1

Bacteria and many protists, such as this Paramecium, *divide by a process of simple cell fission.*

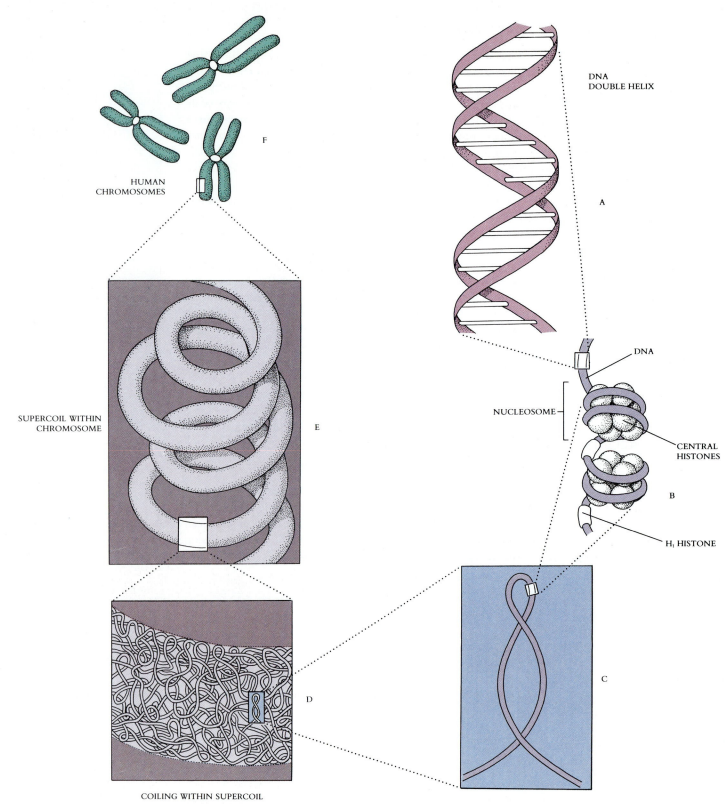

DNA
DOUBLE HELIX

A

DNA

NUCLEOSOME

CENTRAL
HISTONES

B

H₁ HISTONE

HUMAN
CHROMOSOMES

F

SUPERCOIL WITHIN
CHROMOSOME

E

COILING WITHIN SUPERCOIL

D

C

FIGURE 8-2

Levels of chromosomal organization. The DNA duplex, **A,** *is wound twice around aggregates of histone proteins to form nucleosomes,* **B,** *and the string of nucleosomes is further coiled in a series of condensation stages,* **C** *to* **E.** *During cell division chromosomes are fully condensed,* **F.** *In interphase (not shown here) much of the coiling is relaxed and chromosomes are not visible as discrete entities under the light microscope.*

THE CELL CYCLE OF EUKARYOTES

The evolution of the eukaryotes introduced several additional factors into the process of cell division. Eukaryotic cells are much larger than prokaryotic ones and contain genomes with much larger quantities of DNA. Unlike the naked DNA of prokaryotic chromosomes, eukaryotic DNA is complexed with histones and other proteins (Figure 8-2), which enables the DNA to be packaged very efficiently into tightly condensed coils during cell division (Figure 8-3). The genomes of most eukaryotes are partitioned into several different chromosomes, rather than being present on a single circular chromosome as prokaryotic genomes are. The evolutionary advantage to eukaryotes of possessing multiple chromosomes was great, since multiple chromosomes allowed the regular reshuffling of genetic information, called genetic recombination. Much of the rapid evolutionary advance of the eukaryotes undoubtedly reflected their ability to put genes together in novel combinations.

These changes in eukaryotic genome organization led in turn to radical changes in the way cells divide. Because the genomes of eukaryotes are large, have a complex organization, and are distributed over several chromosomes, new mechanisms for packaging and moving the genome were required to partition two replicas of the genome accurately, one into each of two daughter cells. These mechanisms did not evolve all at once; there are traces of very different, and possibly intermediate, mechanisms in some of the eukaryotes surviving today. We do not know whether these different kinds of cell division actually represent intermediate steps on the evolutionary journey to the form of mitosis that is characteristic of most eukaryotes today, or whether they are simply different ways of solving the same problem. There are no fossils in which we can see the interiors of dividing cells well enough to be able to trace the history of cell division.

The processes that occur during the division of eukaryotic cells can be diagrammed as a cell cycle, in much the same way that we diagrammed prokaryotic cell division. In this case, however, many more events occur. A generalized eukaryotic cell cycle can be diagrammed as follows:

$$G_1 \rightarrow S \rightarrow G_2 \rightarrow M \rightarrow C$$

G₁ phase The growth phase of the cell. For many organisms, this phase occupies the major portion of the cell's life span.

S phase The phase in which a replica of the genome is synthesized.

G₂ phase The stage in which preparations are made for genomic separation, including the replication of mitochondria and other organelles, chromosome condensation, and the synthesis and recruitment of microtubules.

M phase The phase in which the microtubular apparatus is assembled, binds to the chromosomes, and moves the sister chromosomes apart. This stage, called mitosis (Figure 8-4), is the essential step in separation of the two daughter genomes.

C phase The phase in which the cell itself divides, creating two daughter cells. Each daughter cell receives approximately half of the cytoplasmic contents as its patrimony, including one of the two replicated nuclei. This phase is called cytokinesis.

In fungi and some groups of protists the nuclear membrane does not dissolve and mitosis is confined to the nucleus. When mitosis is complete in these organisms, the nucleus divides into two daughter nuclei. One goes to each daughter cell during cytokinesis. This separate nuclear division phase of the cell cycle does not occur in most protists, or in plants and animals. The absence of nuclear division as a separate step is a simplification of the overall process, one that allows less room for error in chromosome separation and, by combining several steps into one, is less expensive in terms of energy.

Two phases of the eukaryotic cell cycle that have received close attention from biologists are those which accomplish the physical division of the cell and its contents: phase M (mitosis) and phase C (cytokinesis). The time devoted to these two phases often represents only a small portion of the cell cycle (Figure 8-5).

FIGURE 8-3

A single pair of human chromosomes, magnified 76,000 times. Each of the tiny projections is a chromatid, composed of DNA and protein.

FIGURE 8-4

This preparation of pollen cells of a spiderwort, Tradescantia, *was made by freezing the cells and then fracturing them. It shows several stages of mitosis.*

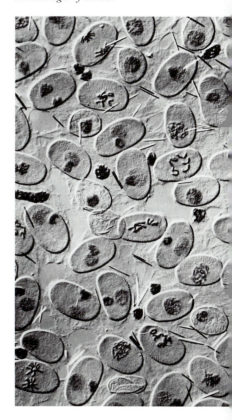

FIGURE 8-5

The eukaryotic cell cycle. This example of the cell cycle is of mouse fibroblasts growing in tissue culture. This cycle takes approximately 22 hours to complete.

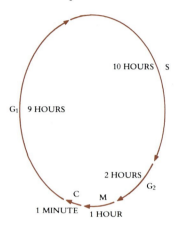

MITOSIS

The M phase of cell division, mitosis, has received more attention than any other aspect of the eukaryotic cell cycle. Biologists have long been fascinated by the intricate movements of the chromosomes as they separate. We will describe the process as it occurs in animals and plants. Among fungi and some protists, variant forms occur, but the process varies little in different animals and plants.

Traditionally mitosis is subdivided into four stages (Figure 8-6): prophase, metaphase, anaphase, and telophase. Such a subdivision is convenient, but the process is actually continuous, the stages flowing smoothly into one another.

Preparing the Scene: Interphase

Before the initiation of mitosis, four events have occurred in the preceding **interphase** (that is, the G_1, S, and G_2 phases), all of great importance for the successful completion of the mitotic process:

FIGURE 8-6

In this remarkable series of photographs the chromosomes of a rat kangaroo have been stained blue and the spindle fibers stained brown by binding to them antibodies directed against tubulin. The mechanical role of microtubules in the process of mitosis is clearly illustrated.

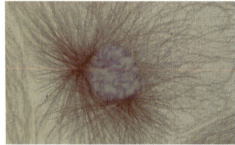

PROPHASE

PROMETAPHASE

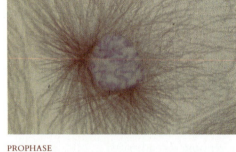

METAPHASE

EARLY ANAPHASE

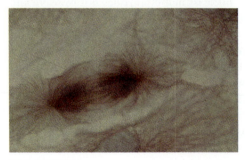

LATE ANAPHASE

TELOPHASE

1. Two centrioles of animal cells, each located near the nucleus, have replicated, creating two pairs of centrioles, each consisting of a normal-sized and a tiny member. Plant and fungal cells have no visible centrioles, nor do a number of groups of Protista; the members of all of these groups, however, undergo mitosis in ways that are generally similar to the process that occurs in animals.

2. The eukaryotic cell undertakes an intensive synthesis of tubulin, the protein of which microtubules are formed. New tubulin is made, and additional tubulin monomers are scavenged from other portions of the cell cytoskeleton. At this stage up to 10% of the cell's protein may be tubulin.

3. Although each chromosome has been replicated in the preceding S phase, the daughter replicates (called daughter chromosomes or sometimes **sister chromatids**) remain attached to one another at a point of constriction called the **centromere.** The centromere circles the two daughter chromosomes with a ring of protein. Each replicated chromosome has a single centromere that occurs at a characteristic site.

4. During S phase, when the chromosomes are replicated, they are fully extended and uncoiled and are not visible under the light microscope. In interphase the chromosomes begin the long process of **condensation,** coiling into more and more tightly compacted bodies.

Interphase is that portion of the cell cycle in which the condensed chromosomes are not visible under the light microscope. It includes the G_1, S, and G_2 phases. In the G_2 phase the cell mobilizes its resources for cell division.

Formation of the Mitotic Apparatus: Prophase

When the chromosome condensation initiated in G_2 phase reaches the point at which individual condensed chromosomes first become visible with the light microscope, the first stage of mitosis, **prophase,** is said to have begun. This condensation process continues throughout prophase so that chromosomes which start prophase as minute threads may appear quite bulky before its conclusion. Ribosomal RNA (rRNA) synthesis ceases when that portion of the chromosome bearing the rRNA genes is condensed, with the result that the nucleolus, which was previously conspicuous, disappears.

Prophase is the stage of mitosis characterized by the condensation of the chromosomes.

While the chromosomes are becoming condensed in prophase, another series of equally important events also occurs: the assembly of the microtubular apparatus, which will be employed to separate the daughter chromosomes. Early in prophase the two centriole pairs start to move apart, forming between them an axis of microtubules referred to as **spindle fibers.** The centrioles continue to move apart until they reach the opposite poles of the cell, with a bridge of microtubules extending between them. In plant cells a similar bridge of spindle fibers forms between opposite poles of the cell (Figure 8-7), although in this case the microtubular organizing center is not visible with the light microscope.

During the formation of the spindle apparatus the nuclear envelope breaks down and its components are reabsorbed into the endoplasmic reticulum. The microtubular spindle thus extends right across the cell from one pole to the other. Its position determines the plane in which the cell will subsequently divide—a plane that passes through the center of the nucleus at right angles to the spindle.

During prophase the nuclear envelope breaks down and a network of microtubules called the spindle forms between opposite poles of the cell. The position of the spindle determines the plane in which the cell will divide.

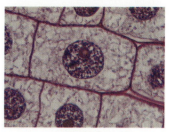

A INTERPHASE

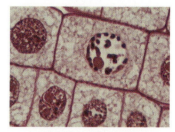

B EARLY PROPHASE

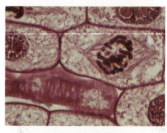

C LATE PROPHASE

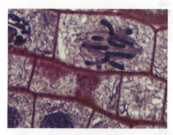

D METAPHASE

E ANAPHASE

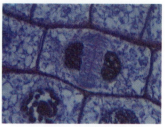

F TELOPHASE

FIGURE 8-7

The stages of mitosis in onion root cells.

FIGURE 8-8

A comparison of the stages of mitosis in an animal, **A,** and in a plant, **B.**

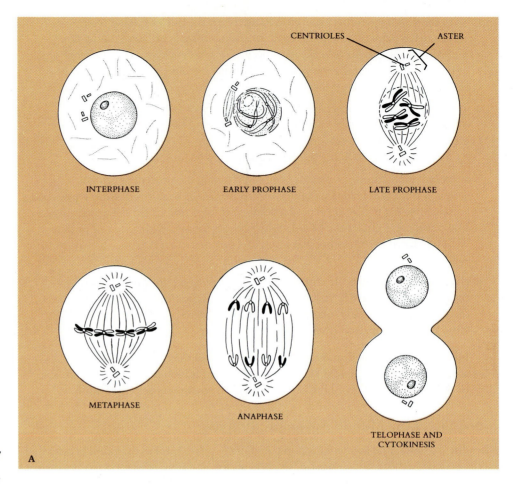

CENTRIOLES   ASTER

INTERPHASE   EARLY PROPHASE   LATE PROPHASE

METAPHASE   ANAPHASE   TELOPHASE AND CYTOKINESIS

A

FIGURE 8-9

A *A dividing sea urchin egg. Each of the two cells is dividing, with the spindle apparatus clearly visible.*
B *An isolated spindle apparatus from a sea urchin egg.*

A

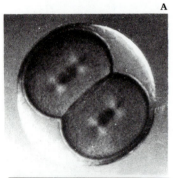

B

When the centrioles of dividing animal cells reach the poles of the cell, they extend a radial array of microtubules outward, thus bracing the centrioles against the cell membrane (Figure 8-8). This arrangement of microtubules is called an **aster.** The function of the aster is not well known, but is probably mechanical, acting to stiffen the point of microtubular attachment during the subsequent contraction of the spindle. Plant cells, which have rigid cell walls, lack centrioles and do not form asters.

As prophase continues, a second group of microtubules appears to grow out from the individual centromeres to the poles of the spindles. Two such microtubules extend from each chromosome (Figure 8-9), connecting opposite sides of the centromere to the two poles of the spindle. The two microtubules attached to a centromere continue to grow until both have made contact with one of the poles of the cell. This has the effect of drawing the chromosomes in toward the midline of the cell.

> **At the end of prophase the centromere joining each pair of sister chromatids is attached by microtubules to the poles of the spindle.**

Division of the Centromeres: Metaphase

The second phase of mitosis, **metaphase,** begins when the chromatid pairs align in the center of the cell. Viewed with a light microscope, the chromosomes appear to be lined up along the inner circumference of the cell in a circle perpendicular to the axis of the spindle. The three-dimensional configuration is that of a line inscribing a great circle on the periphery of a sphere, like the equator girdles the earth (Figure 8-10). An imaginary plane passing through this circle is called the **metaphase plate.** The metaphase plate is not a struc-

CELL WALL

INTERPHASE   EARLY PROPHASE   LATE PROPHASE

METAPHASE   ANAPHASE   TELOPHASE AND CYTOKINESIS   CELL PLATE

B

ture in the physical sense of the word but rather a statement about where the future axis of cell division will occur. The chromosomes do not all line up parallel to one another. Their long arms extend in all directions. Their centromeres form a linear array equidistant from the two poles of the cell. In mitosis the chromosomes follow their centromeres passively.

Metaphase is the stage of mitosis characterized by the alignment of the chromosomes in a ring along the inner circumference of the cell, each chromosome drawn to that position by the microtubules extending from it to the two poles of the spindle.

Each centromere has two faces, and a centromeric microtubule is attached to each face, extending to the opposite poles. This arrangement is absolutely critical to the process of mitosis. Any mistakes in this positioning of the microtubules are disastrous: the attachment of the two centromeric microtubules to the same pole, for example, leads to **nondisjunction,** with the sister chromatids failing to separate and ending up in the same daughter cell.

At the end of metaphase the centromeres divide. Each centromere splits in two, freeing the two sister chromatids from their attachment to one another. Centromere division is simultaneous for all the chromosomes; the mechanism that achieves this synchrony is not known.

At the end of metaphase the centromeres divide, freeing the sister chromatids to be drawn in the next phase to opposite poles of the spindle by the microtubules attached to opposite faces of the centromere.

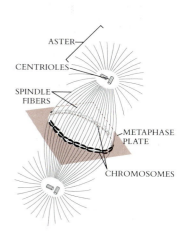

ASTER

CENTRIOLES

SPINDLE FIBERS

METAPHASE PLATE

CHROMOSOMES

FIGURE 8-10

The metaphase plate is a great circle perpendicular to the axis of the spindle, around the edge of which the chromosomes array themselves during metaphase.

THE EVOLUTION OF CELL DIVISION IN EUKARYOTES

The type of cell division that occurs in plants, animals, and most protists is simpler than the alternative ways in which it occurs in the fungi and in some other protists. These unusual types of mitosis may represent more primitive stages in the evolution of mitosis that have persisted to the present, although it is difficult to be sure.

One of the simplest, and perhaps most primitive, forms of cell division in a surviving eukaryotic organism occurs in the giant amoeba-like *Pelomyxa palustris* (Figure 27-4, p. 537). *Pelomyxa* is one of the very few eukaryotes that lacks mitosis. Within its nucleus, which is bounded by a nuclear envelope like the nuclei of all other eukaryotes, the chromosomes simply double in number and are reassorted to the daughter cells randomly. Each daughter nucleus still obtains the minimum amount of genetic material necessary to specify the organism because multiple copies of each chromosome are present within the nucleus. *Pelomyxa* has no centrioles, and there is no reason to believe that it has been derived from any more complex organism. In other words, *Pelomyxa* may be a eukaryote that traces its origin to ancestors more primitive than those in which mitosis evolved.

It also lacks mitochondria, which is another sign that it may be a very primitive eukaryote.

Another unusual form of mitosis occurs in a phylum of protists called the dinoflagellates. During the process of mitosis in dinoflagellates, the chromosomes become attached to microtubules that extend through the nucleus. When the nucleus divides, the two copies of a chromosome are separated from one another by the growth of the nuclear envelope in the zone between the places where the chromosomes are attached. Eventually the nucleus constricts into two nuclei by a process of binary fission, providing each daughter nucleus with one complement of chromosomes. The cell then divides, each daughter cell gaining one of the nuclei.

A form of mitosis that might be more advanced than that of the dinoflagellates is found among the diatoms (Figure 8-A) and among the fungi (Figure 8-B). In them, the microtubules are organized and the daughter chromosomes reassorted within the nuclear envelope. The cells divide after nuclear division is complete. In other eukaryotes, including animals and plants, the nuclear envelope disintegrates during the process of mitosis and re-forms after chromosomal division has been completed.

Separation of the Chromatids: Anaphase

Of all the stages of mitosis, **anaphase** is the most beautiful to watch and the shortest. Free of the conflicting demands of bipolar attachment, each sister chromatid now rapidly moves toward the pole to which its centromeric microtubule is attached. What forces move the chromosomes? You have encountered them before in Chapter 7, pushing cell membranes and moving mitochondria: the chromosomes are moved by microtubules.

Two forms of movement take place simultaneously, each driven by microtubules:

1. *The poles move apart.* The continuous microtubules slide past one another, which has the effect of moving the poles apart. Because the chromosomes are attached to these poles, they move apart too. In this process the cell, if it is bounded by a flexible membrane, becomes visibly elongate. The mechanism of microtubular sliding is probably not markedly different from the one that you encountered in studying flagellar motion in Chapter 7. The sliding is based on ATP-driven changes in the conformation of proteins bridging pairs of microtubules. Because the two members of each microtubular pair are physically anchored to opposite poles, their sliding past one another pushes the poles apart.

2. *The centromeres move toward the poles.* The centromeric microtubules shorten as tubulin subunits are continuously removed from their polar ends. This shortening process is not a contraction, since the centromeric microtubules do not get any

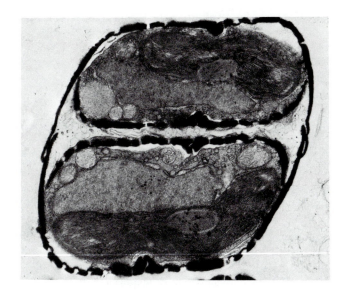

FIGURE 8-A *A dividing diatom.*

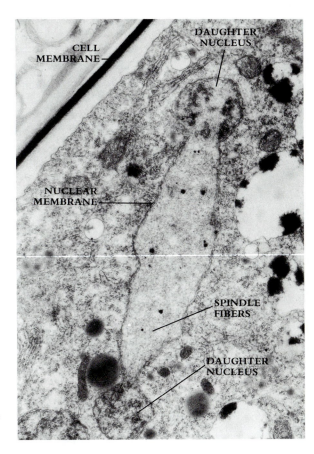

FIGURE 8-B

*The completion of mitosis within the closed nucleus of
a fungal cell.*

thicker. Instead, tubulin subunits are removed from the polar ends of the
centromeric microtubules by the organizing center. As more and more subunits
are removed, the progressive disassembly of the chromosome-bearing
microtubule renders it shorter and shorter, pulling the chromosome ever closer to
the pole of the cell.

Actin filaments are also present in spindles, just as they are in many other parts of
eukaryotic cells. Whether their presence is a chance event, or reflects the mobilization of
yet another dynamic force that aids in chromosome movement, is not known.

**Anaphase is the stage of mitosis characterized by the physical separation of
sister chromatids. The poles of the cell are pushed apart by microtubular
sliding, and the sister chromatids are drawn to opposite poles by the
shortening of the microtubules attached to them.**

Re-formation of Nuclei: Telophase

The chromatid separation achieved in anaphase completes the accurate partitioning of the
replicated genome, the essential element of mitosis. With the play complete, the only tasks
that remain in **telophase** are to dismantle the stage and remove the props. The spindle apparatus dissolves, the microtubules being disassembled into tubulin monomers ready for

reuse in constructing the cytoskeleton of the new cell. The nuclear envelope re-forms around each set of daughter chromatids, which are now chromosomes in their own right and begin to uncoil into the more extended form that permits gene expression. One of the early genes to regain expression is the rRNA gene, resulting in the reappearance of the nucleolus.

> **Telophase is the stage of mitosis during which the mitotic apparatus assembled during prophase is disassembled, the nuclear envelope is reestablished, and the normal utilization of the genes present in the chromosomes is reinitiated.**

CYTOKINESIS

At the end of telophase mitosis is complete. The eukaryotic cell has partitioned its replicated genome into two nuclei, which are positioned at opposite ends of the cell. The process of cell division is not yet complete, however. Indeed, the division of the cell proper has not yet begun. The cell contains many other elements that are necessary for the viability and well-being of its daughter cells. A successful process of cell division must ensure that all necessary cellular elements are as successfully partitioned between these daughter cells as the chromosomes are. For example, the daughter cells would not survive without mitochondria, the machinery of aerobic metabolism. Nor would cells be capable of photosynthesis without chloroplasts. These organelles arise only from the division of previous mitochondria or chloroplasts. Unlike the genome of the cell itself, however, these elements exist in multiple copies so that a simple cleavage of the cell is sufficient to ensure that both daughter cells receive all of the critical organelles. The stage of the cell cycle at which cell division actually occurs is called **cytokinesis.**

> **Cytokinesis is the physical division of the cytoplasm of a eukaryotic cell into two daughter cells.**

In eukaryotes, cytokinesis generally involves the cleavage of the cell into two roughly equal halves. Because of the intricate movements of the previous M phase (mitosis), this cellular cleavage may be brought about successfully by relatively simple mechanisms. Two mechanisms are widespread: one occurs in animals and other eukaryotes that lack cell walls, and another occurs in plants and the particular green algae from which they evolved.

Cytoplasmic Cleavage in Animal Cells

In animal cells, and in the cells of all other eukaryotes that lack cell walls, the dissolution of the spindle is rarely completed before the onset of cytoplasmic cleavage. The orientation of the spindle axis determines the formation of a belt of actin microfilaments around the circumference of the cell at its equator. Cytoplasmic cleavage is initiated by the contraction of these microfilaments (Figure 8-11). Drugs that destroy microfilament contraction, such as cytochalasin B, block this contraction. As the contraction proceeds, a **cleavage furrow** becomes evident around the circumference of the cell (Figure 8-12) where the cytoplasm is being progressively pinched inward by the decreasing diameter of the microfilament belt. As contraction proceeds, the furrow deepens until it eventually extends all the way into the residual spindle, and the cell is literally pinched into two.

Cytoplasmic Cleavage in Plants

Plant cells possess a rigid cell wall, one far too strong to be de-formed by microfilament contraction, so that the mechanism of cytokinesis employed by animal and other eukaryotic cells will not work for plants. Plants have evolved a different strategy of cell division. Plants redirect vesicles produced by the Golgi bodies that contain cell wall and membrane

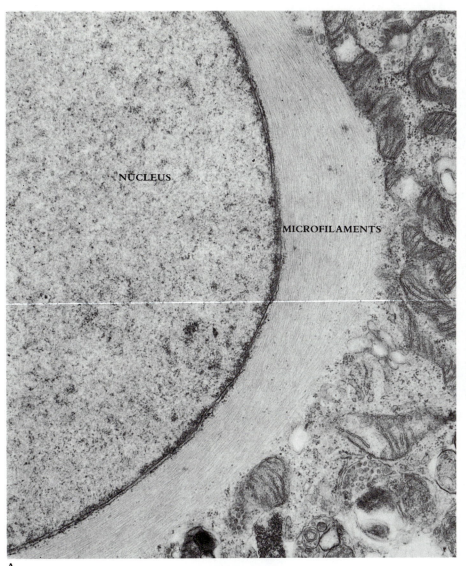

NUCLEUS

MICROFILAMENTS

A

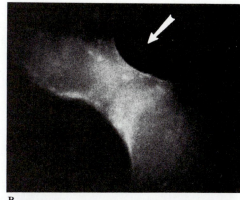

B

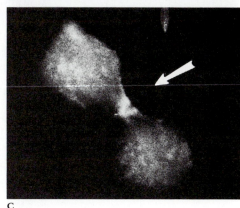

C

FIGURE 8-11

Animal cells, such as these rat kidney cells, divide in cytokinesis by the progressive constriction of a contracting ring, A, of microfilaments. First a constriction, B, appears (indicated by the arrows); it eventually tightens, C, to divide the cell in two.

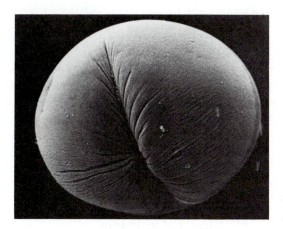

FIGURE 8-12

A cleavage furrow around a dividing frog egg.

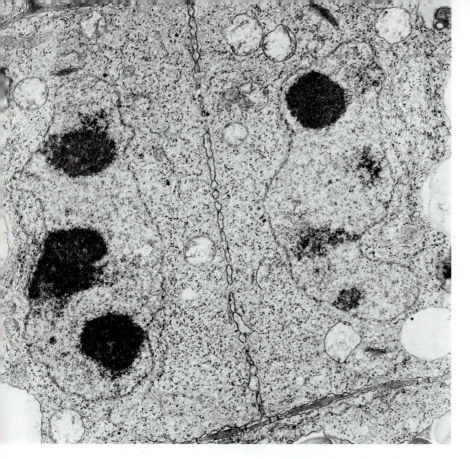

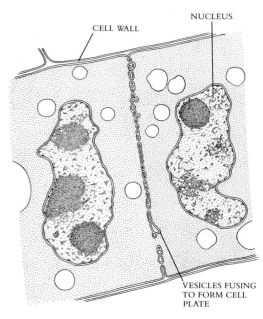

FIGURE 8-13

Plant cytokinesis involves the formation of a cell plate between the daughter cells.

components; instead of being exported to the cell exterior, the microtubules of the residual spindle apparatus direct these vesicles inward toward the spindle midline. There these vesicles accumulate and begin to fuse together, their contents aggregating into an enlarging double membrane (Figure 8-13). This expanding partition is called a **cell plate.** It continues to grow outward until it reaches the interior surface of the cell membrane and fuses with it, at which point it has effectively divided the cell into two. Cellulose is then laid down on the new membranes, creating two new cells. The space between the two new cells becomes impregnated with pectins and is called a **middle lamella.**

COMPARING CELL DIVISION IN EUKARYOTES AND PROKARYOTES

When we consider the process of cell division, it is impossible not to be struck by the sophisticated way in which eukaryotic cells partition their chromosomes among daughter cells, relative to the simple way in which this is achieved in prokaryotes. This complex mechanism of moving chromosomes to opposite ends of a dividing cell by attaching them to microtubules permits a rapid and accurate separation of sister chromatids into daughter cells. Separation of daughter chromosomes within a dividing bacterium, by contrast, is a slower process whose success depends on uninterrupted membrane growth. The process of binary fission that occurs in bacteria works well, but mitosis is a surer, more efficient, process.

Most eukaryotes possess much more DNA than do prokaryotes. Usually this DNA is fragmented into several separate chromosomes, which wind up tightly during cell division. Human genetic material, for example, is partitioned among 46 separate chromosomes. If human cells divided by binary fission, like those of bacteria, all 46 chromosomes would be stitched together in one enormous circle. The circle would be so large (about 3 meters in circumference) that it would be very difficult to partition it equally between the two daughter cells. Eukaryotes achieve this difficult feat by the process of mitosis.

SUMMARY

1. Bacterial cells divide by simple binary fission. A dividing bacterium ensures that each daughter cell contains one of the DNA replicas by attaching these replicas to its cell membrane and initiating fission between the points to which they are attached.

2. A cell cycle diagram is a symbolic representation of the life cycle of an organism. In bacteria the cycle is cell growth, synthesis of a new genome, and cytoplasmic division, symbolized $G \to S \to C$.

3. Eukaryotes contain much more DNA than do prokaryotes. Eukaryotic DNA is complexed with histones and divided among several to many chromosomes. These differences necessitate a more complex form of division than that which is found in the prokaryotes.

4. In plants, animals, and most protists, mitosis is streamlined, with the nuclear envelope dissolving before mitosis begins. In the fungi and some protists the nuclear envelope does not dissolve, and nuclear division occurs after mitosis. This latter state might reflect the original condition in the evolution of mitosis.

5. DNA replication is completed before mitosis begins. Immediately before the onset of mitosis, condensation of the chromosomes and synthesis of microtubules begin. This period preceding mitosis is called interphase.

6. The first stage of mitosis proper is prophase, during which the mitotic apparatus forms. At the end of prophase the nuclear envelope has dissolved and microtubules attach each pair of sister chromatids to the two poles of the cell.

7. The second stage of mitosis is metaphase, during which the chromosomes align along the periphery of a plane cutting through the center of the cell at right angles to the spindle axis. At the end of metaphase the centromeres joining each pair of sister chromatids divide, freeing each chromatid to be pulled to one of the poles of the cell by the microtubules attached to it.

8. The third stage of mitosis is anaphase, during which the chromatids physically separate, moving to opposite poles of the cell.

9. The fourth and final stage of mitosis is telophase, during which the mitotic apparatus is disassembled, the nuclear envelope re-forms, and the chromosomes uncoil.

10. Following mitosis, most cells undergo cytoplasmic cleavage, or cytokinesis. In cells without a cell wall the cell body is pinched in two by a belt of microtubules drawing inward around its midsection. In plant cells and other cells with a cell wall an expanding cell plate forms along the spindle midline.

SELF-QUIZ

The following questions pinpoint important information in this chapter. You should be sure to know the answers to each of them.

1. The constriction of a bacterial cell into two cells, each with a copy of the cellular DNA, is called

   _____ _____.

2. Which of the following is *not* a phase of cell division in human beings?
   (a) G_1—the growth phase
   (b) S—the DNA replication phase
   (c) G_2—preparations for cell division
   (d) M—the mitotic phase
   (e) All occur in human cell division

3. The two kingdoms of organisms in which at least some members carry out nuclear division before cell division are _____ and _____.

4. At the end of interphase, as prophase begins, each mitotically dividing human cell contains how many centrioles?
   (a) one
   (d) two pairs
   (b) two
   (e) many
   (c) one pair

5. Sister chromatids are attached to one another at points of constriction called _____.

6. Each replicated chromosome of a eukaryotic cell has how many centromeres?

7. Human cells are diploid, containing two copies of each of the 23 different chromosomes. At the end of interphase, as prophase begins, how many chromatids are present within each cell?

8. Do asters occur in animal cells or in plant cells?

9. What is the correct order of the phases of mitosis?
   (a) prophase-anaphase-metaphase-telophase
   (b) prophase-metaphase-anaphase-telophase

10. The phase of mitosis in which the nuclear envelope re-forms is called _____.

11. Which is the shortest phase of mitosis?

12. Mitosis, properly speaking, is nuclear division. Cell division usually begins after mitosis is complete. The stage of the cell cycle at which the cells themselves divide is called _____.

THOUGHT QUESTIONS

1. In fungi, mitosis takes place within a closed nucleus, whereas in most other eukaryotes the nuclear envelope dissolves during mitosis. Why do you suppose that fungi have not adopted the mode of mitosis used by so many other eukaryotes?

2. Continuous microtubules form the spindle of the dividing eukaryotic cell. These microtubules serve to move pairs of chromatids to opposite poles of the cell by sliding past one another. What other modes of mechanical motion might be possible for moving the chromosomes, and which of these possibilities are actually employed?

FOR FURTHER READING

MARGULIS, L.: *Symbiosis in Cell Evolution,* W.H. Freeman and Company, Publishers, San Francisco, 1981. An outstanding book for those who wish to learn more about the variations in the process of mitosis in different groups.

MAZIA, D.: "The Cell Cycle," *Scientific American,* January 1974, pages 54-64. An account of the cell cycle by one of those responsible for the development of the concept.

MITCHISON, J.M.: *The Biology of the Cell Cycle,* Cambridge University Press, New York, 1972. A solid basic text on cell division and the cell cycle.

SLOBODA, R.D.: "The Role of Microtubules in Cell Structure and Cell Division," *American Scientist,* vol. 68, pages 290-298, 1980. A good description of how the spindle apparatus works.

MEMBRANE TRANSPORT

Chapter 9

Overview

A cell membrane isolates the cell from most molecules in the exterior environment, although not from water. As a result, cells are faced with the task of preventing the excessive uptake of water while facilitating the uptake of nutrients. Cells meet these conflicting demands in a variety of ways, many of which involve specific protein channels through the membrane. Some of these protein channels pass molecules freely in either direction, others in one direction only; still others transport only certain molecules. It is largely the nature of the protein channels in the cell membrane that determines the pattern of movement of substances into and out of cells.

For Review

Here are some important terms and concepts that you will encounter in this chapter. If you are not familiar with them, you should review them before proceeding.

- **Phospholipid** (Chapter 4)
- **Bilayer membrane structure** (Chapter 4)
- **Transmembrane proteins** (Chapter 4)
- **Diffusion** (Chapter 5)
- **Osmosis** (Chapter 5)

Cell membranes are characteristic of all living things. From the smallest microbes to the cells in your brain that allow you to consider the words on this page, every living cell is enclosed within a membrane. This membrane isolates the cell so that it can accumulate nutrients from its environment and grow. No cell, however, lives in perfect isolation. Each communicates with its environment, taking in food, releasing wastes, and gathering information about its immediate surroundings. The overwhelming majority of these transactions involve the passage of molecules into and out of the cell across the cell membrane. The almost unbelievable rapidity with which a housefly dodges your swatting hand, the growth of a plant toward sunlight, a transient dream or perceptive thought—all derive from the movement of molecules into and out of cells.

In this chapter we will consider the various ways in which molecules pass into and out of cells. The mechanisms are for the most part ancient ones on an evolutionary time scale, shared by both primitive and modern organisms.

153

THE OSMOTIC DILEMMA: KEEPING WATER OUT OF CELLS

FIGURE 9-1

Cells contain a high concentration of solutes such as sugars, amino acids, and other biochemicals. In this representation each dot signifies a solute molecule. These solutes dilute the concentration of water within the cell, relative to the outside. Thus an equal number of molecules sampled from inside and outside a cell will not contain an equal number of water molecules, because more of the molecules taken from inside the cell will be solute molecules.

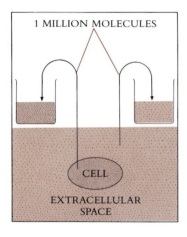

As you will recall from Chapter 4, biological membranes all have the same basic structure: they are lipid bilayers within which numerous proteins are embedded. All lipid bilayer membranes are freely permeable to water through small imperfections in the membrane surface. Because water molecules are free to cross lipid bilayer membranes, the two sides of any membrane tend to have the same concentration of water molecules. If one side of a membrane were to have more water molecules per cubic millimeter than the other, water molecules would simply pass across the membrane from the side of higher concentration to the side of lower concentration. This process is called diffusion and was discussed in Chapter 5.

> **Diffusion is the net movement of molecules down a concentration gradient as a result of random spontaneous molecular motions. Diffusion tends to distribute molecules uniformly.**

The free diffusion of water across a cell's membrane creates a potential problem. To understand this problem, we must concentrate for a moment on the molecules present in solution within a cell. A typical cell contains as part of its cytoplasm not only water, but also other molecules dissolved in water, such as sugars, amino acids, and ions. The mixture of these molecules and water is called a **solution;** water, the most common of the molecules in the mixture, is called the **solvent,** whereas the other molecules dissolved in water are called **solutes.** Because all of the molecules within a cell are not water molecules, the water within a cell can be thought of as being diluted by the solute molecules.

To illustrate this point, conduct a hypothetical experiment (Figure 9-1): imagine that you remove a small volume of liquid from the solution immediately outside a cell, just enough to contain exactly 1 million molecules; now remove a second volume of liquid, also containing 1 million molecules, from inside the cell. Which volume of liquid contains the most water molecules? The outside volume does because the second volume taken from within the cell contains many more solute molecules among its million molecules and thus fewer water molecules. It follows that the water within the cell is more dilute than the water outside the cell: it possesses fewer water molecules per million molecules. Now recall our earlier discussion of diffusion. Water, like any other molecule, will tend to move toward zones of lower concentration, driven by diffusion. Consequently, water will enter the cell.

A cell immersed in pure water is **hypertonic** (from the Greek prefix *hyper-,* more than) with respect to its surrounding solution, because it has a higher concentration of metabolites than the water does. The surrounding solution, which has a lower concentration of metabolites than the cell, is **hypotonic** (from the Greek prefix *hypo-,* less than) with respect to the cell.

FIGURE 9-2

Osmosis is the movement of water across a membrane impermeable to solutes in the direction of lower water concentration. When interior and exterior water concentrations are equal (isotonic), there is no net movement of water. When the interior of a cell contains more solute than does the surrounding solution, the outer solution is hypotonic with respect to the cell, and water will move into the cell. In the reverse situation, when the outer solution contains a higher concentration of solute than does the cell's interior, the surrounding solution is hypertonic with respect to the cell, and water will move out from the cell into the surrounding solution.

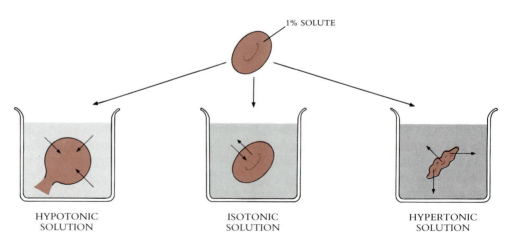

Because of diffusion, any molecule will exhibit net movement from a region where its concentration is greater to a region where its concentration is less. The free diffusion of water across cell membranes creates a serious problem for cells, because the cell membrane is differentially permeable: water can diffuse across it, but many solutes cannot. Metabolites and other solutes are imprisoned within the cell, unable to cross the membrane, whereas water molecules are free to diffuse inward. Water molecules stream into the cell across the membrane, as if they were attempting to dilute the high concentration of metabolites within the cell to match the lower concentration of the outside solution (Figure 9-2). This form of net water movement into a cell is called **osmosis.**

> **Osmosis is the diffusion of water across a membrane that permits the free passage of water but not the passage of one or more solutes.**

Intuitively you might think that as new water molecules diffuse inward, the internal hydrostatic pressure within a cell would build up; this is indeed what happens. As water molecules continue to diffuse inward, toward the area of lower *water* concentration, the hydrostatic water pressure within the cell increases. This pressure is called **osmotic pressure.** Because this water pressure opposes the movement of water inward, the inward diffusion of water will not continue indefinitely. The cell will eventually reach an equilibrium where the osmotic force driving water inward is exactly counterbalanced by the hydrostatic pressure driving water out. However, the hydrostatic pressure at equilibrium is typically so high that an unsupported cell membrane cannot withstand it. Such an unsupported cell, suspended in water, will burst like an overinflated balloon. Cells whose membranes are supported by cell walls, on the other hand, can withstand considerable internal hydrostatic pressures.

> **Within the closed volume of a cell that is hypertonic to its surroundings, the movement of water inward, which tends to lower the relative concentration difference of water, will at the same time increase the internal hydrostatic pressure. The net movement of water stops when an equilibrium condition is reached, or when the cell bursts.**

It is not possible to solve the osmotic dilemma by simply closing the passages through which water crosses the membrane, because the structure of lipid bilayers renders all membranes freely permeable to water. Cells deal with the problem of osmotic pressure in a variety of ways.

Bacterial Cell Walls

Bacteria have adopted a direct solution to the problem posed by osmosis. The hydrostatic pressure resulting from osmosis is resisted by a strong exterior cell wall. Composed of sugar polymers cross-linked by amino acids, the bacterial cell wall is one large molecule completely enveloping the cell. Because of the extensive cross-linking, this wall is very resistant to internal hydrostatic pressure.

Avoiding Osmotic Pressure

Early eukaryotes found other solutions to the osmotic dilemma. Some, living in the sea, adjusted their internal concentration of nonwater molecules (solutes) to be the same as that of the seawater in which they swam. When the concentration of solutes on the inside of the cell membrane is the same as the concentration on the outside, the cell is said to be **isotonic** (from the Greek prefix iso-, the same) with respect to its environment, and there is no tendency for a net flow of water into or out of the cell to occur. The cell is in osmotic balance with its environment. The problem is solved by avoiding it.

FIGURE 9-3

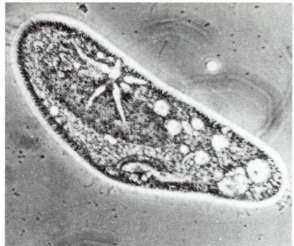
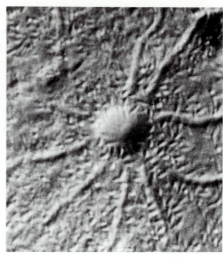

Protists such as this Paramecium *contain star-shaped contractile vacuoles. The long arms act as collecting ducts into which water diffuses by osmosis. The water is channeled to a central vacuole near the cell surface, which somewhat resembles a balloon. By contracting microfilaments surrounding this vacuole, the cell expels its water to the exterior.*

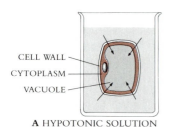

CELL WALL
CYTOPLASM
VACUOLE

A HYPOTONIC SOLUTION

B ISOTONIC SOLUTION

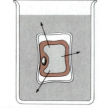

C HYPERTONIC SOLUTION

FIGURE 9-4

Osmotic pressure is a hydrostatic pressure that results from the inward diffusion of water molecules into cells.
A *In plant cells the large central vacuole contains a high concentration of solutes so that water tends to diffuse inward, causing the cells to swell outward against their rigid cell walls.*
B *If a plant cell is immersed in an isotonic solution, one with the same concentration of solutes as the vacuole, there is no net osmotic pressure.*
C *However, if a plant cell is immersed in a high-solute (hypertonic) solution, water will leave the cell, causing the cytoplasm to shrink and pull in from the cell wall.*

Many multicellular animals adopt similar solutions, circulating a fluid through their bodies that bathes cells in isotonic solution. By controlling the composition of its circulating body fluids, a multicellular organism can control the solute concentration of the fluid bathing its cells, adjusting it to match that of the cells' interiors. The blood in your body, for example, contains a high concentration of a protein called albumin, which serves to elevate the solute concentration of the blood to match that of your tissue so that osmosis does not occur.

Water Removal

When early eukaryotes colonized fresh water, they had to face the osmotic dilemma again. Eukaryotes that had adjusted their internal environment to match that of seawater found that in fresh water they were again hypertonic with respect to their environment. The solution adopted by many single-celled eukaryotes was **extrusion.** The water that moved into the cell was dumped back out by one means or another.

In *Paramecium,* for example, the cell contains one or more special organelles known as **contractile vacuoles.** In the photographs of *Paramecium* in Figure 9-3 the contractile vacuole resembles a spider, with many legs that collect osmotic water from various parts of the cell's interior and transport it to the central body, a vacuole near the cell surface. This vacuole is bounded by microfilaments and possesses a small pore that opens onto the outside of the cell. By rhythmic contractions, it pumps the accumulating osmotic water out through the pore. Because contraction of microfilaments involves the utilization of ATP, the use of a contractile vacuole to survive in a hypotonic environment entails the constant expenditure of energy.

Plant Cell Walls

Unlike animals, plants do not circulate an isotonic solution. Most plant cells are hypertonic with respect to their immediate environment, with a high concentration of solutes in the central vacuole. The resulting osmotic pressure presses the cytoplasm firmly against the interior of the plant cell wall (Figure 9-4), making the individual plant cells rigid. The internal pressure in these cells is referred to as **turgor pressure.** The shapes of plants that lack wood and of the newer, soft portions of trees and shrubs depend to a large extent on the turgor of their individual cells. Because plants depend on the cell rigidity imparted by turgor pressure to maintain their shape, they wilt when they lack sufficient water.

These indirect solutions are the best that evolution has achieved, each a compromise. They all satisfy one important criterion—they all work.

THE METABOLIC DILEMMA: LETTING METABOLITES IN

The lipid nature of biological membranes raises a second problem for growing cells, one that is in some respects the inverse of the osmotic one just discussed. The metabolites required by cells as food are for the most part polar molecules, which will not pass across the hydrophobic barrier interposed by a lipid bilayer. How then are organisms able to pass food molecules into the cell? Two fundamentally different approaches have been taken during the course of evolution.

Bulk Passage by Membrane Fusion

Particularly among single-celled eukaryotes, the dynamic cytoskeleton is employed to extend the cell membrane outward toward food particles, such as bacteria. The membrane encircles and engulfs a food particle; its edges surround the particle and eventually meet. Because of the fluid nature of the lipid bilayer, the membranes fuse together. This process is called **endocytosis** (Figure 9-5). Endocytosis involves the incorporation of a portion of the exterior medium into the cytoplasm of the cell as a vesicle.

If the material that is brought into the cell contains an organism (Figure 9-6) or some other fragment of organic matter, that particular kind of endocytosis is called **phagocytosis** (Greek, *phagein,* to eat + *cytos,* cell). When the liquid brought into the cell contains dissolved molecules, the endocytosis is referred to as **pinocytosis** (Greek, *pinein,* to drink). Pinocytosis is common among the cells of multicellular animals. Human egg cells, for example, are "nursed" by surrounding cells, which secrete nutrients that maturing egg cells take up by pinocytosis.

> **Phagocytosis is a process in which cells literally engulf organisms or fragments of organisms, enfolding them within vesicles. The interior membranes of these vesicles were once part of the exterior membrane of the cell.**

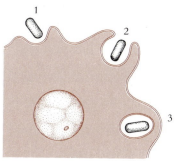

PHAGOCYTOSIS

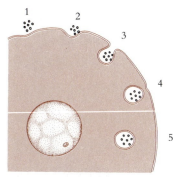

PINOCYTOSIS

FIGURE 9-5

Forms of endocytosis. Phagocytosis is the ingestion by cells of other cells or large fragments of cells. Pinocytosis is the ingestion by cells of dissolved molecules.

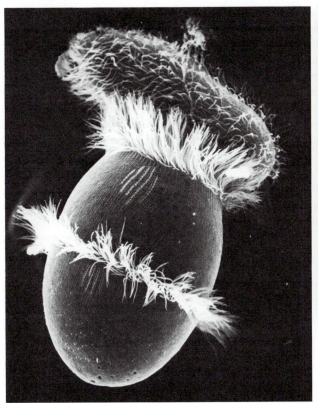

A

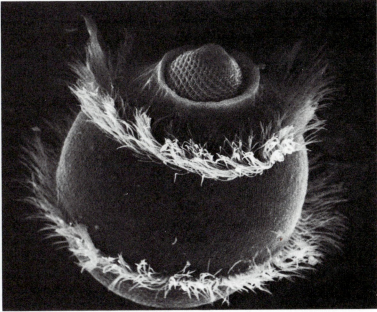

B

FIGURE 9-6

Phagocytosis in action. The large egg-shaped protist Didinium nasutum *is in the act of eating the smaller protist* Paramecium.
A *Its meal has just begun.*
B *The meal is practically over.*

FIGURE 9-7

Exocytosis. Proteins and other molecules are secreted from cells in small packets called vesicles, whose membranes fuse with the cell membrane, releasing their contents to the cell surface.

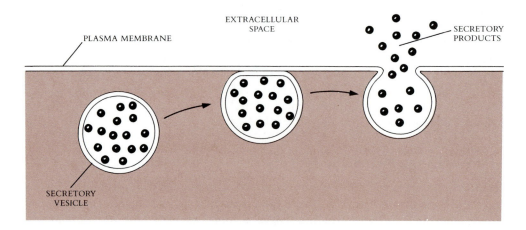

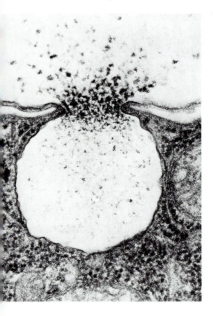

The antithesis of endocytosis is **exocytosis** (Figure 9-7), the extrusion of material from a cell by discharging it from vesicles at the cell surface. In Chapter 6 we considered the formation of such secretory vesicles by the Golgi apparatus. In plants these vesicles constitute a major means of exporting materials to be used in the construction of the cell wall through the plasma membrane. In animals many cells are specialized for secretion using the mechanism of exocytosis.

Channels Through the Membrane

Endocytosis is an awkward means of governing entrance to the interior of the cell. It is expensive from an energy standpoint, since a considerable amount of membrane movement is involved. More important, endocytosis is not selective. Particularly when the process is pinocytotic, it is difficult, if not impossible, for the cell to discriminate between different solutes, admitting some molecules into the cell and excluding others. The capacity to discriminate is achieved by cells in quite a different way. What they do is to take advantage of the protein mosaic embedded within biological membranes to render the membranes **selectively permeable,** that is, permeable to some molecules and not others.

The protein channels that traverse membranes do not contain lipids and therefore do not present a hydrophobic barrier to polar molecules. Some of the channels are like the holes in doughnuts, open to anything small enough to squeeze through; others are more like turnstiles at a subway station, permitting entry only when the proper ticket is presented. The pores in the nuclear envelope (Chapter 6) are of this second kind, passing nucleic acids through while retarding much smaller molecules. Much of the movement of molecules into and out of cells is determined by the nature of these **transmembrane channels,** and we will devote the remainder of this chapter to their properties.

> Selective permeability allows the passage across a membrane of some
> solutes but not others. Selective permeability of cell membranes is
> the result of specific protein-lined channels extending across the
> membrane through which some molecules can pass and others cannot.

THE IMPORTANCE OF SELECTIVE PERMEABILITY

The most essential property of any cell, the bottom line of cellular organization, is that the cell constitutes an isolated compartment within which certain molecules can be concentrated and associated together. This essential isolation is not possible if no molecule can enter the cell, or if any molecule can. Your home is private in the same sense—inoperable if no one, including you, can enter, but not private if everyone in the neighborhood is wandering through all the rooms. The solution adopted by cells is the same that most homeowners adopt: there are doors with keys, and only those possessing the proper keys can

TABLE 9-1   TYPES OF TRANSMEMBRANE CHANNELS

NONSELECTIVE CHANNELS	
Open channels	Some channels appear to be open to most molecules, although they may not be open all the time.
SELECTIVE CHANNELS	
Passive two-way channels	Molecules are free to pass in or out but may be restricted by the size or chemical character of the channel.
Passive one-way channels	Although the channel does not restrict one-way passage, it or other processes may act to prevent subsequent return.
Active channels	Passage involves the conformational change of a transport protein that is a component of the channel, a process driven by the cleavage of ATP molecules.
Coupled channels	Passage through a passive channel may be coupled to the activity of an active channel driven by ATP or excited electrons.
Secretory channels	Membrane-bound ribosomes may constitute a channel for the proteins that they synthesize.

enter or leave. The channels through cell membranes are the doors to cells. These doors are not open to any molecule that presents itself; only particular molecules can pass through a given kind of door. A cell is able to control the entry and exit of many kinds of molecules by possessing many different kinds of doors. These channels through its membrane are among the most important functional features of any cell.

MECHANISMS OF SELECTIVE TRANSPORT

The biology of cell membranes is the biology of selective permeability. Only certain molecules are free to enter or leave a cell. The precise discriminations that membranes carry out reflect the nature of the transmembrane channels. There are six general types of channels through the membranes of living cells (Table 9-1). Examples of all of them can be found in bacteria, as well as in eukaryotes. These channels, therefore, represent one of the oldest parts of our evolutionary heritage.

Open Channels

Although most channels through membranes are highly selective in determining which molecules may pass through them, some are not. Among bacteria, for example, the gram-negative ones have evolved a second lipid bilayer membrane that encloses a compartment between the cell wall and the interior plasma membrane. This compartment is called the **periplasmic space.** The periplasmic space contains many transport proteins that bind to sugars and other metabolites and that aid in their passage through the plasma membrane.

How do these metabolites cross the *outer* membrane? They do so through special open channels. The outer membrane is densely studded with open channels, each composed of a transmembrane protein called **porin.** Any molecule smaller than the size of a large monosaccharide sugar can pass unrestricted through these holes simply by diffusion. In this way small polar metabolites such as glucose easily traverse the outer membrane and enter the periplasmic space, from which transport systems can pass them through the plasma membrane and into the cell.

Small nonpolar molecules, on the other hand, have difficulty passing through porin channels. Although the channels are open, the porin subunits that line them possess charged and polar surface groups that repel nonpolar molecules. Nonpolar molecules, as well as larger polar molecules such as disaccharide sugars, must traverse other channels through the outer membrane. These channels will be described later in this chapter.

Open channels are protein-lined gaps in the membrane, which permit molecules that are small enough to pass freely in either direction.

Open channels are not limited to the special outer membranes of gram-negative bacteria. Similar open channels formed by porinlike molecules occur in the outer membranes of mitochondria and chloroplasts. Their presence is perhaps not surprising, since both of these organelles are thought to be relict bacteria that have become permanent residents in the eukaryotic cytoplasm.

Perhaps the most spectacular open channels occur between adjacent eukaryotic cells of multicellular organisms. In animals these open channels between adjacent plasma membranes are called **gap junctions** (Figure 9-8); they act as passageways between adjacent cells.

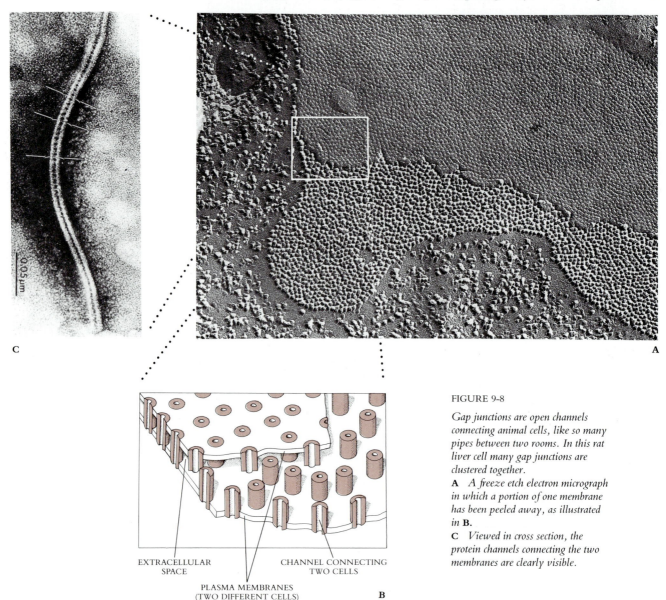

C

A

FIGURE 9-8

Gap junctions are open channels connecting animal cells, like so many pipes between two rooms. In this rat liver cell many gap junctions are clustered together.

A *A freeze etch electron micrograph in which a portion of one membrane has been peeled away, as illustrated in* **B.**

C *Viewed in cross section, the protein channels connecting the two membranes are clearly visible.*

EXTRACELLULAR SPACE

CHANNEL CONNECTING TWO CELLS

PLASMA MEMBRANES (TWO DIFFERENT CELLS)

B

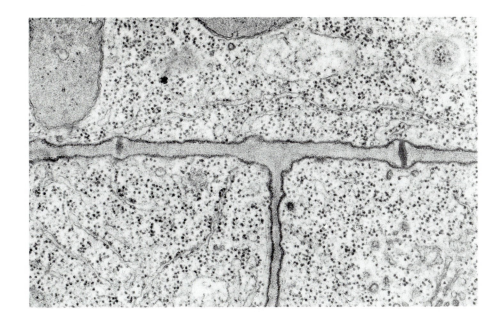

FIGURE 9-10

Plasmodesmata traversing primary walls of cells in the root tip of the common bean (Phaseolus vulgaris). *(×68,000.) The plasmodesmata extend through the primary cell walls, which are stained dark in this electron micrograph.*

These openings are about 20 nanometers in diameter and are created by transmembrane proteins that project across both membranes, as well as across the extracellular space between them. These junctions are like tunnels crossing the extracellular space between the two membranes. Such a large opening allows the free passage of most metabolites, such as sugars, amino acids, or nucleotides, while preventing the passage of macromolecules such as proteins or nucleic acids. These gap junctions play important physiological roles in tissues where the individual cells are not in contact with a circulating blood supply, since the gap junctions permit nutrients to be passed along from cell to cell. Gap junctions are to be clearly distingiushed from **desmosomes** (Figure 9-9), which are simple points of attachment between animal cells and provide no channels of transport.

Open channels also occur between adjacent plant cells. Such channels, called **plasmodesmata** (Figure 9-10), traverse not only the plasma membranes of both adjacent cells, but also the thick cell walls of the two cells.

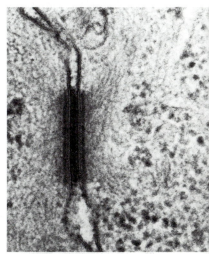

FIGURE 9-9

Desmosomes are simple points of attachment between animal cells. They do not contain channels connecting the interiors of the two attached cells.

Passive Two-Way Channels: Facilitated Diffusion

Some of the most important channels in the cell membrane are not open but rather are highly selective, facilitating the passage only of specific molecules—but in either direction. An example is provided by the **anion channel** of vertebrate red blood cell membranes. This channel plays a key role in the oxygen-transporting function of these cells. The anion channels of the red blood cell membrane readily pass chloride ion (Cl^-) or carbonate ion (HCO_3^-) across the red blood cell membrane. We can easily demonstrate that the ions are moving by *diffusion* through the channels in the red blood cell membrane: if there is more Cl^- ion within the cell, the net movement is outward, whereas if there is more Cl^- ion outside the cell, then the net movement is into the cell. Since the ion movement is always toward the direction of lower ion concentration, the transport process is one of diffusion.

Are these channels simply holes in the membrane, somehow specific for these anions? Not at all. If we repeat our hypothetical experiment, progressively increasing the concentration of Cl^- ion outside the cell, the rate of movement of Cl^- ion into the cell increases only up to a certain point, after which it levels off and will proceed no faster despite increases in the concentration of exterior Cl^- ion. The reason that the rate will increase no further is that the Cl^- ions are being transported across the membrane by a carrier, and all available carriers are in use: we have saturated the capacity of the carrier system. The transport of these anions, then, is a diffusion process facilitated by a carrier. Such transport processes are referred to as **facilitated diffusion** (Figure 9-11).

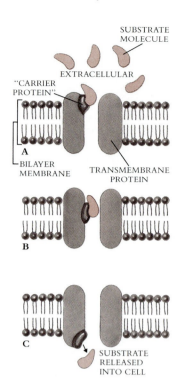

FIGURE 9-11

Facilitated diffusion is a carrier-mediated transport process.
A *The molecule to be transported from the cell's exterior, which can be a Cl^- ion, a sugar molecule, or any of a variety of other "substrate" molecules, binds to a specific carrier protein. The binding is selective, a particular carrier protein binding only one or a few types of substrate molecules.*
B *The carrier protein transports the substrate molecule through a protein channel to the other side of the membrane. The channel, diagrammed here in cross section, usually has a more complex shape.*
C *The carrier protein releases its passenger substrate molecule into the cell.*

Facilitated diffusion is the transport of molecules across a membrane by a carrier protein in the direction of lowest concentration.

Other molecules, particularly sugars, also pass into and out of cells by facilitated diffusion (Figure 9–12), and their passage also appears to be aided by channels created by transmembrane proteins like those of the anion channels.

Facilitated diffusion provides the cell with a ready means of preventing the buildup of unwanted molecules within the cell, or of gleaning from the external medium molecules that are present there in high concentration. Facilitated diffusion has two essential characteristics: (1) it is *specific,* with only certain molecules being able to traverse a given channel; and (2) it is *passive,* the direction of net movement being determined by the relative concentrations of the transported molecule inside and outside the membrane.

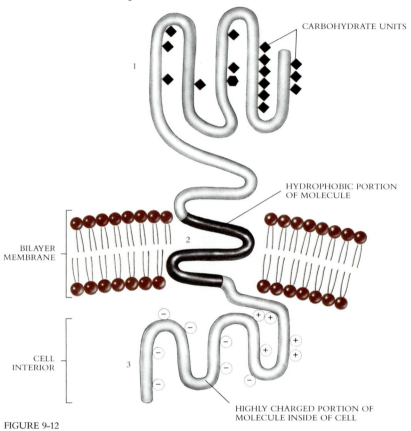

FIGURE 9-12

*One of the major sugar-transporting channels, the **glycophorin A** sugar transport channel, has recently been isolated and its structure determined in detail. This channel has three distinctly different domains, or zones. 1, The end of the protein protruding from the outside surface has carbohydrates bound to it; 2, the central zone possesses a sequence of hydrophobic amino acids, facilitating the burial of this portion of the protein within the lipid bilayer membrane; 3, the end of the protein protruding into the cell possesses a sequence of amino acids, many of which are polar and ionized, giving the interior of the channel a negative charge.*

Passive One-Way Channels: Group Translocation

In some cases it is advantageous for the cell to be able to retain specific molecules that it has acquired from the surrounding medium. Sugars such as glucose provide a clear example of such substances. Any glucose molecules gleaned from outside the cell provide it with an increment of metabolic fuel, and it is therefore clearly advantageous to accumulate these molecules within the cell. If facilitated diffusion were the only means of transporting glucose across the cell membrane, however, a cell in a medium in which glucose was rare might lose far more glucose than it could gain.

Many cells circumvent the difficulty of maintaining glucose concentrations higher than their surroundings by **phosphorylating** the sugar while transporting it into the cell. This process transfers a phosphate ($PO_4^=$) group from ATP to the sugar molecule and in this way prevents the subsequent exit of the sugar molecule from the cell. Exit is prevented because the newly acquired negative charge imparted by the added phosphate group prevents the sugar from approaching the inner surface of the membrane through which the channel protrudes: the membrane is lined with a lipid that carries a negative charge and so repels the negatively charged phosphate groups. This method of rendering a molecule that was originally able to pass through a membrane unable to do so subsequently is called **group translocation.** In this particular example of group translocation a phosphate group is translocated from ATP to the sugar. The sugar phosphate within the cell is a different molecule from the unmodified sugar without, so that high internal concentrations of the sugar phosphate present no osmotic barrier to the entry of scarce external sugar.

Active Channels

Sugars are not the only molecules that traverse the cell membrane against a concentration gradient. There are many other molecules that the cell admits across its membrane but that are maintained within the cell at a concentration different from that of the surrounding medium. In all such cases the cell must expend energy to maintain the concentration difference. Transport that requires the expenditure of energy is called **active transport.** Active transport may maintain molecules at a higher concentration within the cell than without by expending energy to pump more molecules in than would enter by diffusion; active transport may also maintain molecules at a lower concentration by expending energy to actively pump them out.

> **Active transport is the transport of a solute across a membrane to a region of higher concentration by the expenditure of chemical energy.**

Active transport is one of the most important functions of any cell. It is by active transport that a cell is able to concentrate metabolites. Without it, the cells of your body would be unable to harvest glucose molecules from the blood, since glucose concentration is often higher in cells than in blood. Imagine how difficult it would be to survive as a beggar if you could only obtain money from those who have less than you do! Active transport permits a cell, at the cost of ATP, to import molecules already present there in high concentration.

There are many molecules that a cell takes up or eliminates against a concentration gradient. These many different kinds of molecules possess a wide variety of selectively permeable transport channels, some permeable to one or a few sugars, others to a certain size of amino acid, and others to a specific ion or nucleotide. You might suspect that active transport occurs at each of these channels, but you would be wrong. There is *one* major active transport channel in cell membranes that transports sodium and potassium ions—all the others work by tying their activity to this all-important channel. The many metabolite- and ion-concentrating channels in the membrane are called **coupled channels.** We will first discuss the sodium-potassium channel and then these coupled channels.

The sodium-potassium pump

More than one third of all the ATP energy expended by a resting cell is used to actively transport sodium (Na^+) and potassium (K^+) ions. The remarkable channel by which these two ions are transported across the cell membrane is referred to as the **sodium-potassium pump (Na^+-K^+ pump).** Most animal cells have a low internal concentration of Na^+ ions relative to their surroundings and a high internal concentration of K^+ ions. They maintain these concentration differences by actively pumping Na^+ ions out of the cell and K^+ ions in. The transport of these ions is carried out by a highly specific transmembrane protein

CHLORIDE CHANNELS AND CYSTIC FIBROSIS

Cystic fibrosis is a fatal disease of humans in which affected individuals secrete thick mucus that clogs the airways of the lungs. These same secretions block the ducts of the pancreas and liver so that the few patients who do not die of lung disease die of liver failure. Cystic fibrosis is usually thought of as a children's disease, since few affected individuals live long enough to become adults. There is no known cure.

Cystic fibrosis is a genetic disease resulting from a defect in a single gene that is passed down from parent to child. It is the most common fatal genetic disease of Caucasians. One in 20 individuals possesses at least one copy of the defective gene.

Most carriers are not afflicted with the disease, however; only those children who inherit two copies of the defective gene, one from each parent, succumb to cystic fibrosis—about 1 in 1800 Caucasian children.

Cystic fibrosis has proven difficult to study. Many organs are affected, and until recently it was not possible to identify the nature of the defective gene responsible for the disease. In 1985 the first clear clue was obtained. An investigator named Paul Quinton seized on a commonly observed characteristic of cystic fibrosis patients, that their sweat is abnormally salty, and performed the following experiment. He isolated a sweat duct from a small piece of skin and

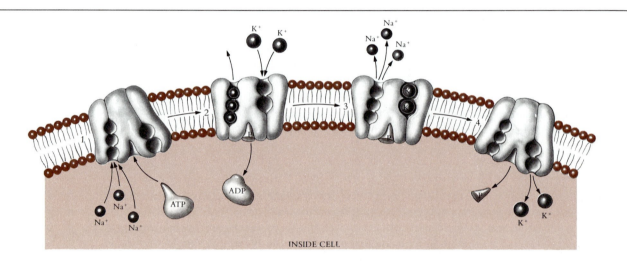

FIGURE 9-13

The sodium-potassium pump. This ATP-driven active transport channel plays a very important role in vertebrate biology. It is responsible for establishing the charge difference between nerve cell interior and exterior on which all nerve conduction depends. In addition, many other transport processes are driven by the pumping of Na^+ and K^+ ions by this channel. The pump works by a series of conformational changes in the transmembrane protein: Step 1. Three molecules of Na^+ bind to the interior ends of the two a subunits, producing a conformational change in the transmembrane protein complex. Step 2. The new shape of the complex binds a molecule of ATP and cleaves it into ADP + P_i. ADP is released, but the terminal phosphate group remains bound to the complex. Step 3. The binding of the phosphate group to the complex induces a second conformational change in the complex. This change results in the passage of the three Na^+ ions across the membrane, *where they are now positioned facing the exterior. In this new conformation the complex has a very low affinity for the Na^+ ions, which dissociate and diffuse away. The new conformation does, however, have a high affinity for K^+ ions and binds two of them as soon as it is free of the Na^+ ions. Step 4. The binding of the K^+ ions leads to another conformational change in the transmembrane complex, this time resulting in the dissociation of the bound phosphate group. Freed of the phosphate group, the complex reverts to its original conformation, with the two K^+ ions exposed on the interior side of the membrane. The new conformation has a low affinity for K^+ ions so that the two K^+ ions dissociate from it and diffuse away into the interior of the cell. This conformation, which is the one we started with, does have a high affinity for Na^+ ions. When these ions bind, they initiate another pump cycle.*

channel (Figure 9-13), passage through which entails conformational changes in the proteins within the channel. As in many other cellular processes, these conformational changes are driven by the hydrolysis of ATP.

The essential characteristic of the Na^+-K^+ pump is that it is an active transport process, transporting Na^+ and K^+ ions from areas of low concentration to areas of high concentration. This transport *up* a concentration gradient is just the opposite of what occurs

placed it in a solution of salt (NaCl) that was three times as concentrated as the NaCl inside the duct. He then monitored the movement of ions. Diffusion tends to drive both the Na^+ and Cl^- ions into the duct, because of the higher outer ion concentrations. In skin isolated from normal individuals Na^+ ions indeed enter the duct, transported by the Na^+-K^+ pump; Cl^- ions follow, passing through a passive channel. Both ions cross the membrane easily. In skin isolated from individuals with cystic fibrosis, the Na^+-K^+ pump transports Na^+ ions into the ducts, but no Cl^- ions enter. The passive chloride channels are not functioning in these individuals.

It appears that cystic fibrosis results from a defective channel within membranes, one that transports Cl^- ions across the membranes of normal individuals but not in affected persons. It is not known if the genetic defect is the result of an alteration in the transmembrane protein itself or in an enzyme that modifies that protein by attaching carbohydrate molecules to it. However, investigators are now actively attempting to isolate and study the gene that encodes the transmembrane protein. If this gene in cystic fibrosis patients differs from that in unaffected persons, then scientists will have finally identified the primary effect of the disease and can embark on the road to find a cure.

spontaneously in diffusion, and it is achieved only by the constant expenditure of metabolic energy in the form of ATP. Some membranes contain large numbers of Na^+-K^+ channels, whereas others have few. The conformational switching that goes on within an individual channel is very rapid. When working full tilt, each channel is capable of transporting as many as 300 Na^+ ions per second, utilizing 100 ATP molecules to do so.

Coupled Channels

The accumulation of many amino acids and sugars by cells is driven against a concentration gradient, the molecules being harvested from a surrounding medium in which their concentration is much lower than it is within the cell. However, the active transport of these molecules across the cell membrane does not involve passage through ATP-cleaving channels per se. Instead, molecules are **cotransported** by moving passively through a channel in concert with an Na^+ ion. For example, the active transport of glucose involves the simultaneous passage through a channel of a glucose molecule and an Na^+ ion, with no cleavage of ATP occurring at that channel site. Passage through such a coupled channel occurs by a process of facilitated diffusion, driven in the direction of uptake by the very low concentration of Na^+ ions within the cell (Figure 9-14).

Here is how coupled channels carry out active transport: the Na^+-K^+ pump keeps the intracellular concentration of Na^+ ions very low relative to the outside of the cell. The lower internal concentration of Na^+ ions (and the relatively more negative charge of the interior that results) acts to produce a strong inward diffusion gradient for Na^+ ions at the coupled channel. The transmembrane protein of the coupled channel, however, acts as a carrier of Na^+ only when the cotransported sugar or amino acid is also bound to the protein's exterior surface. As a result, the facilitated diffusion inward of the Na^+ ion also results in the importation into the cell of the sugar or amino acid.

The codiffusion inward of the two molecules at the coupled channel occurs only because the interior concentration of Na^+ is so low. The sugar or amino acid is literally dragged along osmotically because the inward concentration gradient of the ion is so much steeper than the outward gradient of the sugar or amino acid. Thus the transport of the sugar or amino acid inward to an area of its higher concentration occurs as a direct consequence of the activity of the Na^+-K^+ pump.

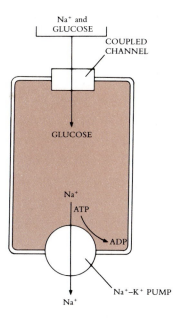

FIGURE 9-14

A coupled channel is a facilitated diffusion channel that transports two molecules in concert. Here glucose and Na^+ ions are transported together. Diffusion through the coupled channel occurs because there is a much lower Na^+ ion concentration within the cell because of the action of the Na^+-K^+ pump.

FIGURE 9-15

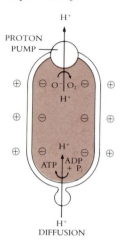

The proton pump

Transport channels coupled to the Na^+-K^+ pump are but one example of coupled channels. A second kind of coupled channel is of equal importance to eukaryotic cells. It is one that drives all of your oxidative metabolism. This channel couples the transport of protons to the production of ATP. It is responsible for the production of almost all the ATP you harvest from food that you eat—and for all of the ATP produced by photosynthesis. This coupled channel is a very ancient one, present in prokaryotes as well as eukaryotes. It appears to have first evolved as a mechanism to drive the rotation of bacterial flagella (Chapter 7). These same channels that couple proton transport to ATP production are seen in what we think are the descendants of bacteria, the mitochondria and chloroplasts, but here the channels carry out a very different function. At other sites in the membrane mitochondria and chloroplasts use electrons garnered from the oxidation of glucose or from light-excited pigments to actively pump protons (H^+ ions) outward. This pumping creates a **proton gradient,** in which the concentration of protons is higher outside the membrane than inside. As a result, diffusion acts as a force to drive protons back across the membrane toward a zone of lower proton concentration. This proton diffusion force, often referred to as the **proton motive force,** drives protons inward through the only channels available to them, those coupled to ATP production (Figure 9-15). The net result is the expenditure of electron excitation energy produced by metabolism or photosynthesis and the production of ATP.

Secretory Channels

Sugars, amino acids, and inorganic ions are not the only molecules that must pass across cell membranes in the course of a normal cell life. One obvious addition to the list is the passage of proteins. As you learned in Chapter 6, many of the secretory proteins of eukaryotic cells are synthesized on the surface of the endoplasmic reticulum by ribosomes. These proteins pass through this membrane during the course of their synthesis and subsequently are transported to the Golgi apparatus for encapsulation in vesicles.

Secretory proteins possess a polypeptide "tag" that marks them for export (Figure 9-16), a tag not present on proteins destined to remain within the cytoplasm. This export tag, called a **signal sequence,** is the first part of the protein to be assembled. The signal sequence of the still incomplete polypeptide binds to a specific secretory channel on the inner surface of the membrane, which passes it through. As synthesis of the protein by the ribosome continues, the emerging polypeptide is threaded through the membrane as well, the activity of the ribosome serving to push the growing amino acid chain through the channel. All during this process, the ribosome remains attached to the endoplasmic reticulum membrane, creating rough ER.

An essential characteristic of secretory channels is that they modify the molecule being transported in such a way as to prevent its subsequent passage back. When a newly made protein passes through a secretory channel, the signal sequence is cleaved off. Lacking this signal sequence, the protein can no longer pass through cellular membranes and thus is a prisoner of the intracellular space into which it has passed. When proteins are destined to pass through several membranes, as are those synthesized in the cytoplasm and destined for the interior of the mitochondria, the cleavage of the signal sequence is delayed or takes place in stages.

THE IMPORTANCE OF TRANSPORT CHANNELS

If it were not for the transport channels we have considered in this chapter, few molecules could enter or leave a cell. The lipid interior of the cell membrane blocks diffusion of polar molecules, which are insoluble in lipid, and the polar exterior of the membrane tends to repel nonpolar molecules. Transport channels provide a cell with practically all of its communication with the outside world. Different types of human cells possess different arrays

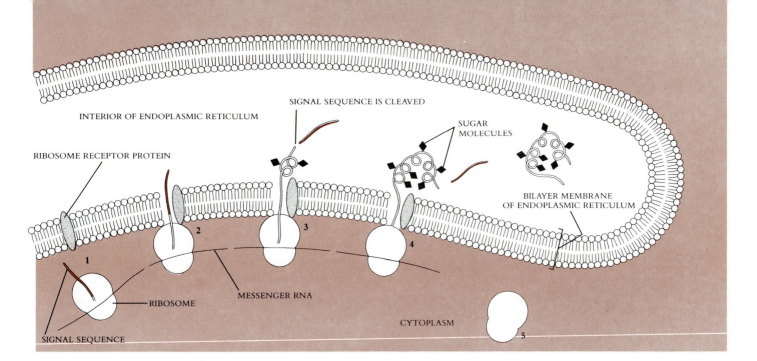

INTERIOR OF ENDOPLASMIC RETICULUM

SIGNAL SEQUENCE IS CLEAVED

RIBOSOME RECEPTOR PROTEIN

SUGAR MOLECULES

BILAYER MEMBRANE OF ENDOPLASMIC RETICULUM

MESSENGER RNA

RIBOSOME

CYTOPLASM

SIGNAL SEQUENCE

of transport channels, some cell types specializing in particular transport processes because of the physiological consequences of the process. Nerve cells, for example, have distributed over their membranes large numbers of Na^+-K^+–pumping channels, which play an important role in conducting nerve impulses. We will discuss these ion transport channels in greater depth in Chapter 46.

Perhaps the most important of all the cellular transport channels is the proton pump. Using the energy of light-excited electrons to drive proton-pumping channels, photosynthetic cells export protons from chloroplast interiors; the diffusion of protons back in through special channels is coupled to ATP synthesis. This is the basic mechanism whereby photosynthesis generates ATP. Proton-pumping channels play a similar role in the oxidative respiration carried out by mitochondria. Mitochondria possess a collection of enzymes that strip energetic electrons from food molecules. These electrons are delivered to proton-pumping channels within the mitochondrial membrane. The energy provided by the electrons is used by those channels to pump protons out of the mitochondrial interior; the return of these protons by diffusion occurs through special channels coupled to the synthesis of ATP.

Thus transport channels through a cell's membrane are of central importance. They have played an integral role in the evolution of cell metabolism—without them, photosynthesis and aerobic metabolism would be impossible—and are a major factor in the developmental processes that differentiate tissues of your body. The nature of a cell's transport channels is one of its most important attributes.

SUMMARY

1. Because cells contain significant concentrations of metabolites and other solutes, water tends to diffuse into them. As it does so, a hydrostatic pressure builds that will rupture cells lacking a wall or other means of support.

2. Cells solve the problem of osmosis in a variety of ways. Some pump excess water back out in order to become isotonic in relation to their surroundings.

3. Plasma membranes are selectively permeable and do not permit the diffusion of metabolites into the cell. Metabolite transport is achieved by a variety of transmembrane proteins embedded within the membrane, which act as transport channels.

FIGURE 9-16

Secretory channels transport specific proteins to the exterior of cells:

1 *As the ribosome reads along the messenger RNA molecule specifying the protein to be secreted, the first portion of the amino acid sequence that it assembles is a special signal sequence that recognizes a secretory transport channel on the membrane. These sequences are not long and predominantly contain amino acids with nonpolar side chains.*

2 *The nonpolar signal sequence binds to the secretory channel, a transmembrane protein called the ribosome receptor protein, and passes through the membrane like a thread through the eye of a needle. Behind it, the ribosome binds to the ribosome receptor protein.*

3 *As the ribosome continues to assemble the protein from the messenger RNA instructions, the newly assembled portions are pushed through the membrane. The signal sequence is cleaved off, and specific sugar molecules are added to the protruding end of the newly formed protein.*

4 *The ribosome continues to assemble the protein and pass it through the channel.*

5 *When assembly of the protein is complete, it passes across the channel, and the ribosome disengages from the ribosome receptor protein and the messenger RNA. The finished protein is released into channels of the endoplasmic reticulum and travels to the Golgi complex, which exports it from the cell.*

4. Some protein channels are open and permit free movement of molecules small enough to pass through them.

5. Other channels involve carriers that transport molecules across the membrane much like cars of a train carry passengers across a bridge.

6. Many ion and proton channels operate in only one direction, like a turnstile at the entrance to a store or amusement park. An ion or proton can enter or leave the cell by means of the appropriate channel but cannot pass back through that channel.

7. Some channels transport molecules against a concentration gradient by expending energy. The two most important ones are those which transport Na^+ ions and those which transport protons. Both of these channels are directed outward. They create very low concentrations of protons and Na^+ ions within the cell. These molecules can diffuse back in only through special coupled channels, where their passage is often coupled to transport of another molecule inward or to the synthesis of ATP.

8. Some channels recognize only specific amino acid sequences on the ends of proteins destined for secretion. They pass the appropriate proteins across the membrane of the endoplasmic reticulum as the proteins are being made.

SELF-QUIZ

The following questions pinpoint important information in this chapter. You should be sure to know the answers to each of them.

1. Lipid bilayer membranes are *not* freely permeable to which of the following molecules?
   (a) oxygen
   (b) water
   (c) amino acids
   (d) carbon dioxide
   (e) they are permeable to all of the above

2. When a molecule such as glucose, which cannot pass freely across a lipid bilayer membrane, is dissolved in water, and a membrane separates two such solutions of different concentrations, what happens?
   (a) nothing
   (b) glucose moves toward the solution of greater glucose concentration
   (c) glucose moves toward the solution of lesser glucose concentration
   (d) water moves toward the solution of greater glucose concentration
   (e) water moves toward the solution of lesser glucose concentration

3. When a cell is placed into a solution in which it is isotonic
   (a) there is no net movement of solutes in or out
   (b) there is a net movement of solutes inward
   (c) there is a net movement of solutes outward

4. The difference between pinocytosis and phagocytosis is that pinocytosis does not involve the ingestion of cellular fragments, whereas phagocytosis does. True or false?

5. Which of the following are *not* open channels through a cell membrane?
   (a) porin channels
   (b) gap junctions
   (c) plasmodesmata
   (d) anion channels
   (e) all of the above are open channels

6. Facilitated diffusion is capable of creating a net movement of a molecule toward an area of higher concentration. True or false?

7. How many Na^+ ions does a single Na^+-K^+ channel transport for every ATP it uses?

8. If a bacterial cell becomes depleted for ATP, its sodium-potassium pump ceases to transport ions. Why?
   (a) ATP is the carrier in the facilitated diffusion of the ions   (d) the cell dies
   (b) the energy of ATP is required to drive the pump   (e) none of the above
   (c) the cell swells and ruptures

9. A single Na^+-K^+ channel is capable of transporting how many Na^+ ions per second, running full tilt?
   (a) 30   (b) 300   (c) 3000   (d) 3 million   (e) 3 billion

10. Proton-pumping channels provide the force that drives the synthesis of ATP. True or false?

11. Which of the following are capable of producing a net movement of molecules across a membrane in the direction of higher concentration?
    (a) group translocation   (d) symports
    (b) facilitated diffusion   (e) the Na^+-K^+ pump
    (c) open channels

12. Which of the following directly utilize ATP in transporting molecules across membranes?
    (a) group translocation   (d) symports
    (b) facilitated diffusion   (e) the Na^+-K^+ pump
    (c) open channels

THOUGHT QUESTIONS

1. When a hypertonic cell is placed in water solution, water molecules move rapidly into the cell. However, if the introduced cell is hypotonic, water molecules leave the cell and enter the surrounding solution. How does an individual water molecule *know* what solutes are on the other side of the membrane?

2. Cells maintain many internal metabolites at high concentrations by coupling their transport into the cell with the transport of Na^+ and K^+ ions by the Na^+-K^+ pump. What happens to all of the K^+ ions that are constantly being pumped into the cell?

3. Why is a lipid bilayer membrane freely permeable to water, which is quite polar, and not freely permeable to ammonia, also polar and about the same size?

FOR FURTHER READING

SATIR, B.: "The Final Steps in Secretion," *Scientific American,* October 1975, pages 28-38. Substances that are secreted from a cell are first packaged into membrane-bound vesicles.

SLAYMAN, C.L.: "Proton Chemistry and the Ubiquity of Proton Pumps," *BioScience,* January 1985, pages 16-47. A collection of seven articles on proton pumps. Together they illustrate the central role that these transmembrane channels play in the biology of both eukaryotes and prokaryotes.

STEAHLIN, L.A., and B.E. HULL: "Junctions Between Living Cells," *Scientific American,* May 1978, pages 140-152. The many kinds of passages that exist between adjacent cells play important roles in the biology of organisms.

SWEADNER, K.J., and S.M. GOLDIN: "The Active Transport of Sodium and Potassium Ions: Mechanism, Function, and Regulation," *New England Journal of Medicine,* vol. 302, pages 777-783, 1980. An up-to-date discussion of what we know about the sodium-potassium pump.

UNWIN, N., and R. HENDERSON: "The Structure of Proteins in Biological Membranes," *Scientific American,* February 1984, pages 78-95. A well-illustrated account of how proteins embedded within membranes act to transport molecules.

WICKNER, W.: "Assembly of Proteins into Membranes," *Science,* vol. 210, pages 861-868, 1980. An excellent review.

Chapter 10 METABOLISM

Overview

The activities of organisms are fueled by chemical energy. By facilitating certain changes in chemical bonds, organisms can store energy or direct it to the synthesis of new molecules. The kinds of chemical activities that organisms carry out, particularly in the harvesting of chemical energy, have changed dramatically in the more than 3.5 billion years since life appeared, with each successive change building on the enzymatic machinery already present. As a result, the metabolism of modern organisms is a composite of primitive and more recently evolved processes.

For Review *Here are some important terms and concepts that you will encounter in this chapter. If you are not familiar with them, you should review them before proceeding.*

- **Nature of covalent bond** (Chapter 2)

- **Oxidation-reduction reactions** (Chapter 2)

- **Enzymes** (Chapter 4)

- **ATP** (Chapter 4)

- **Proton pumps** (Chapter 9)

Living things maintain their organization through the expenditure of energy. If you stopped eating, you would soon begin to lose weight as your body used up its stored energy. Eventually you would die of starvation if you were not supplied with some outside source of energy. The same is true of all other living things (Figure 10-1). Deprived of a source of energy, the active processes of life cease. Why does this happen? Once it has evolved into a highly organized state, why cannot life simply continue? The reason is that each of the significant properties by which we define life—growth, reproduction, and heredity—uses up energy.

Organisms acquire energy and use it in a complex series of chemical reactions that take place within cells. Which reactions occur at any given time depend on which enzymes a particular cell possesses. The chemistry of living things, the total of all the chemical reactions that an organism carries out, is called **metabolism** (Greek, *metabole,* change). In this chapter we will consider how energy is released by chemical reactions and how enzymes are able to speed up certain reactions, making them proceed thousands of times faster. We will then briefly review how the energy-obtaining set of metabolic reactions has evolved among organisms.

THE ENERGY OF CHEMICAL BONDS

All but a very few organisms use chemical bond energy to fuel life processes. Where does the energy in chemical bonds come from, and how is it used to promote living processes? To answer these questions, we need to consider for a moment the nature of energy. Instinctively we all know something about energy. It is "work," the force of a falling boulder, the pull of a locomotive, the swift dash of a horse; it is also "heat," the blast from an explosion, a warming fire. What do these images have to do with chemical bonds? The answer first became clear in 1847, when Hermann von Helmholtz wrote a scientific paper proposing that *work and heat are both forms of energy and that energy can be changed from one form to another, but never created or destroyed.* The paper was not accepted for publication (he later had it published privately), but von Helmholtz's proposition has become one of the cornerstones of modern chemistry, referred to as the **first law of thermodynamics.** The forces holding atoms together in molecules are one form of energy.

How much energy is there in a chemical bond? Recall the nature of covalent chemical bonds. They are created by the sharing of electrons between two atomic nuclei, a sharing that is promoted by the universal tendency of atoms to pair unpaired electrons, balance electrical charges, and satisfy the octet rule. To disrupt a covalent bond, these forces must be opposed, a process that requires energy. Indeed, the strength of a covalent bond is measured by the amount of energy required to break it. For example, it takes 98.8 kilocalories (kcal) of energy per mole to break carbon-hydrogen (C—H) bonds.

Enthalpy

When an alternative new bond forms, with the electrons rearranging themselves among the nuclei, the process is called a **chemical reaction** (Chapter 2). Not all bonds contain the same amount of energy. It is easier, for example, to break a C—H bond than a C—O bond. The total energy content of a molecule or other system at constant pressure is designated E. When a chemical reaction occurs that converts one molecule into another with less bond energy, the difference in energy (ΔE) is converted to heat or to a change in pressure (P) or volume (V). Under conditions where pressure and volume do not change, such as in most cells, the energy is released as heat. We call this release of energy a change in **enthalpy,** symbolized H. Enthalpy is just the energy of a system, corrected for any influence of pressure $(\Delta H = \Delta E + P\Delta V)$. If heat is emitted in a reaction, then the change in enthalpy, ΔH, is negative, and the reaction is said to be **exothermic.** If heat is absorbed, creating bonds with more chemical energy than the reactant, then ΔH is positive, and the reaction is **endothermic.** The magnitude of this energy is often expressed in thermal energy units, or calories, the amount of energy required to raise the temperature of 1 gram of water from $14.5°$ to $15.5°$ C.

> **A change in enthalpy is a measure of the amount of chemical bond energy converted to or from heat in a chemical reaction, under conditions of constant pressure.**

Entropy

Chemical reactions that release energy (such as ones involving a negative change in enthalpy at constant pressure) tend to occur spontaneously, producing more stable molecules. A second influence also acts on chemical reactions but in just the opposite way. This influence is the universal tendency toward disorder. The randomness or disorder of a system is given the name **entropy,** which is denoted by the symbol S. The **second law of ther-**

FIGURE 10-1

This lion is eating a giraffe. The tissue the lion is consuming will undergo many changes on its metabolic journey, some eventually becoming part of the lion and of the gases the lion exhales, as well as solid waste products.

modynamics states that *all objects in the universe tend to become progressively more disordered.* This is true of a child's room, of the desk you study or work at, of a waiting crowd of people—and of molecules. The random motion of molecules that occurs at any temperature above absolute zero tends to encourage disorder among molecular arrangements, a disordering effect that increases as higher temperatures induce the more rapid random movement of molecules.

> **Entropy is a measure of randomness or disorder. It occurs as a result of the random motion of molecules. Such random motion increases with a rise in temperature, which imparts energy of motion to molecules. Consequently, the disordering effects of entropy are greater at higher temperatures.**

The total amount of thermal energy present in a system may be expressed as *TS,* the product of entropy and of temperature measured on a scale of degrees above absolute zero. The number of centigrade degrees above absolute zero is referred to as the **temperature Kelvin** (for example, 0° C, the temperature at which water freezes, is 273° Kelvin, whereas 100° C, the temperature at which water boils, is 373° Kelvin).

Free Energy

Both stabilizing and disordering effects have a significant influence on molecules, the first tending to promote order and the second to reduce it. The net effect, the amount of energy actually available to form chemical bonds, is called the **free energy** of that molecule, denoted by the symbol G. Free energy in a more general sense is defined as the energy available in any system to do work. In a molecule this energy is equal to the energy contained in the chemical bonds *(H)* minus the total thermal energy at that temperature *(TS)* which is unavailable because of disorder:

$$G \quad = \quad H \quad - \quad TS$$

| FREE ENERGY | ORDERING INFLUENCES | DISORDERING INFLUENCES |

Because chemical reactions can involve changes in enthalpy, they can produce changes in free energy. Chemical reactions may also increase disorder, by promoting movements among molecules, for example. The change in the available energy that a chemical reaction produces is referred to as the **change in free energy,** symbolized ΔG. When a chemical reaction occurs at a constant temperature, the change in free energy is simply

$$\Delta G = \Delta H - T\Delta S$$

> **Free energy is the energy available to do work. In a chemical reaction carried out in a cell at constant pressure and volume, the change in free energy is the difference in bond energies between reactants and products, corrected for changes in the degree of disorder of the system.**

The change in free energy, ΔG, is the most fundamental property of any chemical reaction. Any reaction in which ΔG is negative will proceed spontaneously. Consider, for example, the combustion of paper. In the equation above, ΔG for the combustion of paper would be negative because the term ΔH has a negative value (lower bond energies) and also because the term ΔS has a positive value (higher disorder) and follows a negative sign. The chemical reactions that take place in all living things obey this fundamental rule: the balance between bond energies and disorder must be negative for a reaction to occur spontaneously.

> **Any reaction that produces products containing less free energy than the original reactants contained will tend to proceed spontaneously.**

CATALYSIS

If all chemical reactions that reduce free energy tend to occur spontaneously, it is fair to ask, "Why haven't all such reactions already occurred?" Clearly they have not. If you ignite gasoline, for example, the resulting chemical reaction proceeds with a net release of free energy. So why doesn't all the gasoline within all automobiles of the world, and beneath all filling stations, just burn up right now?

Activation Energy

To answer the question, consider once again the nature of the covalent bond. To form a new covalent bond in a chemical reaction that occurs spontaneously, the old bond must first be broken. In most chemical rections the atoms forming the bond are first pushed farther apart, making them less stable and thus more reactive. This intermediate stage in the chemical reaction is called the **transition state,** and the destabilizing of the chemical bonds that takes place to achieve the transition state is called **activation.** To push bonded atoms apart to reach a transition state, it is necessary to supply energy, called **energy of activation,** E_a (Figure 10-2). Only those substrate molecules which have acquired enough energy to pass through this transition state will proceed to form product with an overall release of free energy. The fact that a reaction has a negative ΔG—releases free energy—predicts only that the reaction will occur; it says nothing about how fast that outcome will be achieved.

Therefore the speed with which a chemical reaction proceeds, its **reaction rate,** depends not on ΔG, but rather on E_a, the amount of activation energy required. This rate will always be proportional to the fraction of substrate molecules that have an energy level equal to or greater than E_a. That is why reactions go faster when you heat the reacting molecules: heat imparts extra energy to molecules. This then is the answer to our question of why all the world's gasoline has not burned up yet: most chemical reactions, including the burning of gasoline, involve appreciable activation energies.

Activation energies are not fixed constants, however. Activation energies can be changed. In this regard activation energy is very different from free energy. The free energy change of a reaction is determined solely by the temperature and the chemical nature of the molecules involved in the reaction. The magnitude of the energy difference is a chemical fact and is not subject to change. No amount of experimenting, or of evolution, can change a reaction involving a positive free energy change into one with a negative free energy change. Activation energies, in contrast, can be changed, by altering the chemical nature of the transition state in such a way as to make the bonds of the reacting molecule easier to break. This process of lowering activation energies is called **catalysis.** Because reactions with lower activation energies are more likely to proceed, catalysis results in more rapid reaction rates.

Enzymes

The chemistry of living things, metabolism, is organized by controlling the points at which catalysis takes place. The agents of catalysis in living organisms are a class of proteins called enzymes (Chapter 4). Enzymes are able to bind to other small molecules in the cell because the three-dimensional shape of the enzyme fits the small molecule like a hand fits a glove. Molecules that bind closely in this way, and undergo a chemical reaction as a result, are called the **substrates** of that enzyme. The **specificity** of an enzyme refers to the range of molecules that will bind to it and subsequently be induced to undergo a chemical reaction.

Enzymes carry out catalysis by binding to substrate molecules and stressing certain substrate chemical bonds so that the free energy required to reach a particular transition state is lowered (Figure 10-3). The result is that the reaction occurs more rapidly. Most enzymes increase reaction rates by at least a million times (a factor of 10^6).

It is important to understand that a reduction in E_a accelerates both the forward and reverse reactions by exactly the same amount. For this reason enzymes cannot alter the ul-

FIGURE 10-2

A chemical reaction proceeds because the products of the reaction possess less free energy than the substrates so that there is a net release of energy.

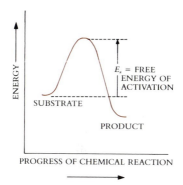

A *Before a reaction occurs, energy must be supplied to destabilize existing chemical bonds. This energy is called the energy of activation (E_a).*

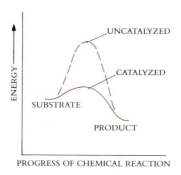

B *Enzymes are able to catalyze particular reactions because they lower the amount of activation energy required to initiate the reaction.*

FIGURE 10-3

Some enzymes bind their substrates with hydrogen bonds. The RNA-cleaving enzyme ribonuclease binds its substrate RNA by means of hydrogen bonds that form between the nitrogen bases of the RNA and the side chains of serine and threonine amino acids in the active site of ribonuclease.

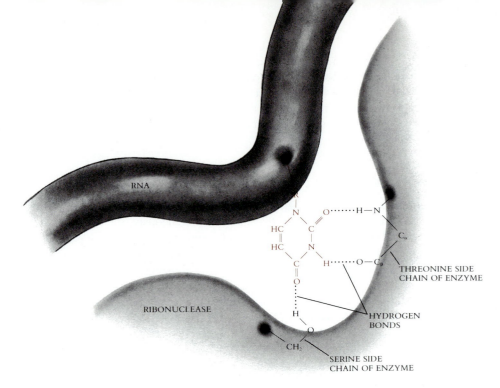

timate destination toward which a reaction will proceed, the final proportion of substrate converted to product. Enzymes only hasten the inevitable. An enzyme can never by itself make an energetically unfavorable reaction proceed; in fact it would simply accelerate the reverse process. Imagine lowering a fence between two identical pastures, one of which contains many more sheep than the other. There is no way to lower the barrier between the pastures in such a way that more sheep will jump from the sparsely populated pasture than from the full one; lowering the barrier simply makes movement in either direction easier. The lowering of the fence will promote a faster *net* movement of sheep into the empty pasture, no matter *how* it is lowered. The direction in which a chemical reaction proceeds is determined solely by differences in bond energies.

Enzymes alter the rates of reactions but not their final proportions.

THE MECHANISM OF ENZYME CATALYSIS
Why Enzymes Are Specific

Enzymes typically catalyze only one or a few different chemical reactions, because only one or a few different kinds of substrate molecules are able to bind to the surface of a given enzyme—each different kind of enzyme binds a different sort of substrate. To understand how enzymes are able to be so fastidious, so "picky," in their choice of substrate, consider for a moment what happens when two molecules come close to one another.

When atoms come very close to one another (0.3 to 0.4 nanometer apart, about twice the radius of a carbon atom), they begin to attract each other. The interactions that produce this attraction are called **van der Waals interactions.** Attraction occurs because at any given instant the distribution of electrons orbiting an atom may not be symmetrical. The transient charge separation, even though it is very weak, can serve to attract two atoms if they are sufficiently close to one another. However, they can only come so close, since their electron clouds repel one another when they come in contact.

The distance between two atoms at which maximum attraction occurs (a distance that is close enough but not too close) is approximately 0.1 nanometer (Figure 10-4). The strength of an individual attraction is only about 1 kcal/mole, which is not much more than the average thermal energy of molecules at room temperature (0.6 kcal/mole). However, this very weakness is the most important property of van der Waals interactions. Although no one of the bonds is significant by itself, a strong bonding may result when many atomic

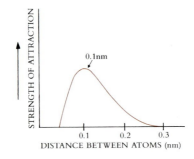

FIGURE 10-4

The strength of van der Waals interactions depends on distance, the interactions only being strong when molecules are about 0.1 nanometer apart. Nearer than that, or farther away, the interactions are weaker.

contacts are made at the same time. For this reason enzymes must precisely match the shape of their substrate like hand and glove in order to be able to bind substrate to them strongly. Only a substrate that has a shape that precisely matches the surface of an enzyme will be able to bind to it.

How Enzymes Bind Their Substrates

The **substrate-binding sites** of many enzymes have been studied in detail. They all share several fundamental characteristics:

1. Almost all binding sites are located within clefts or pockets on the surfaces of enzymes.

2. The clefts or pockets are shaped in such a way that they fit the appropriate substrate molecule. Protein shapes are not rigid, and in some enzymes the binding of substrate may induce in the protein a conformational adjustment (a slight alteration in enzyme shape) leading to a better **induced fit.**

3. The clefts or pockets within which binding sites are situated are often hydrophobic. Thus water is excluded, except when water molecules themselves take part in the reaction.

4. The **enzyme-substrate complex (ES complex),** which is formed when substrate binds to enzyme, is not strong in most reactions. Since the ES complex is weak, it readily dissociates. Only a fraction of the ES complexes that form are able to undergo catalysis before dissociation occurs. The rate at which an enzyme-catalyzed reaction proceeds is a function both of the speed of the catalytic process at the active site and of the persistence of the ES complex.

How Catalysis Occurs

Catalysis results from the interaction of an enzyme-bound substrate molecule with a nearby portion of the enzyme or, sometimes, with a metal ion. The precise chemical means that is employed by particular enzymes facilitating different biochemical reactions varies greatly from one reaction to another. Although enzymes employ many different catalytic mechanisms, these mechanisms are all variations on a central theme. Their mode of action consists of distorting substrate bonds. The distortion is induced by the binding of the substrate to the enzyme. This distortion is often enhanced by the chemical activity of enzyme R groups located close to where the substrate binds. These catalytically active R groups constitute the **active site** of the enzyme.

There are a very limited number of possible catalytic mechanisms. For example, only four ways are known in which the cleavage of a peptide bond can be catalyzed. Despite this limitation, the different kinds of enzymes are numerous because of the diverse ways that entry to an active site can be limited.

BIOCHEMICAL PATHWAYS

Your body contains over a thousand different kinds of enzymes. These enzymes catalyze a bewildering variety of reactions. Many of the reactions occur in sequences, called **biochemical pathways,** in which the product of one reaction becomes the substrate for another. For example, the amino acid proline is synthesized from atmospheric nitrogen (N_2) and a simple carboxylic acid, alpha-ketoglutarate, in a series of enzyme-catalyzed steps (Figure 10-5). Biochemical pathways are the organizational units of metabolism, the ele-

FIGURE 10-5

The synthesis of proline takes place in a series of enzyme-catalyzed steps. Shown here are the later steps in this series. This ordered sequence of reactions is an example of a biochemical pathway.

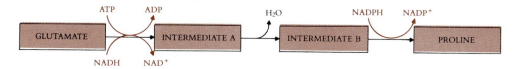

THE EVOLUTION OF SPECIFICITY

The specificity with which enzymes select their proper substrates is achieved by restricting access to the active site. Changes in specificity occur during evolution when one or more changes in the DNA lead to a protein with a different amino acid sequence. The change in the amino acid sequence leads to changes in the shape of the substrate-binding site, the place on the enzyme where the substrate binds to it. On most enzymes the substrate-binding site is not the same as the active site, although they are located close to each other on the surface of the enzyme. Changes in the shape of the substrate-binding site alter the range of substrates that can successfully gain access to the active site.

The evolution of different specificities can be seen clearly in the three peptide-cleaving proteases trypsin, chymotrypsin, and elastase. About 40% of the amino acid sequence of these three enzymes is identical, which indicates that they have evolved from a common ancestral protein and have changed gradually over time to achieve their present amino acid sequences and particular patterns of activity. Reflecting this similarity in amino acid sequence, all three of these enzymes have the same overall shape and identical active sites.

Despite their clear similarities, however, the three enzymes have evolved very different specificities, each recognizing substrates different from the others. Chymotrypsin cleaves polypeptides only at positions occupied by amino acids with large aromatic and nonpolar R groups; trypsin cleaves polypeptides only at amino acids with positively charged R

ments that an organism controls to achieve coherent metabolic activity. Just as the many metal parts of an automobile are organized into distinct subassemblies such as carburetor, transmission, and brakes, so the biochemical parts of an organism are organized into biochemical pathways.

How Biochemical Pathways Evolved

In primitive organisms the first biochemical processes were probably heterotrophic processes, in which energy-rich molecules were scavenged from the environment. (The evolution of heterotrophic organisms was discussed in Chapter 3.) Most of the molecules necessary for these processes are thought to have existed in the organic soup of the early oceans. The first catalyzed reactions were probably simple one-step processes that brought these energy-rich molecules together in various combinations. Eventually, of course, the original energy-rich molecules became depleted in the external environment, and only organisms that evolved some means of making the energy-rich molecule from other raw materials in the organic soup could survive. Thus a hypothetical process

$$F \longrightarrow H \quad \text{(with G entering)}$$

where two hypothetical energy-rich molecules *(G and F)* react to produce compound *H* and release energy, becomes more complex when the supply of *F* in the environment runs out. A new reaction is added in which the depleted molecule, *F,* is made from another molecule, *E,* present in the environment:

$$E \longrightarrow F \longrightarrow H \quad \text{(with G entering)}$$

When the supply of compound *E* in turn becomes depleted, organisms that are able to make it from some other available precursor, *D,* survive. When *D* becomes depleted, these or-

groups such as lysine or arginine; and elastase cleaves polypeptides only at amino acids with small uncharged R groups. Studies of the shapes of these enzymes have revealed the reason for the difference in choice of substrate: the different specificities of these enzymes result from small alterations in the shapes of their substrate-binding sites (Figure 10-A).

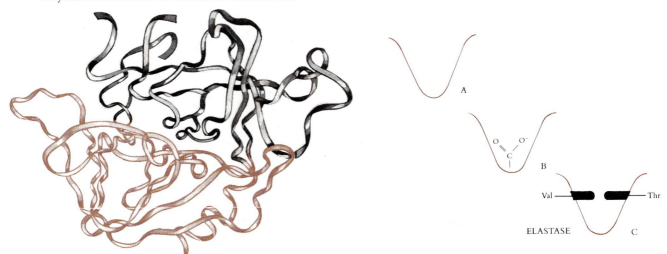

ganisms in turn are replaced by ones able to synthesize *D* from another molecule, *C*:

$$C \longrightarrow D \longrightarrow E \longrightarrow F \overset{G}{\underset{}{\longrightarrow}} H$$

Thus our hypothetical biochemical pathway slowly evolves through time, with the final reactions evolving first and earlier reactions evolving only later. Looking at this pathway now, we would say that the organism, starting with compound *C*, is able to synthesize *H* by means of a complicated series of steps. This is how the biochemical pathways within organisms are thought to have evolved—not all at once, but one step at a time, backward.

How Biochemical Pathways Are Regulated

Once organisms evolved the means to catalyze ordered sequences of biochemical pathways, it became necessary to develop ways of controlling the output of these pathways. Not only does it make little sense to synthesize compound *H* from compound *C* when there is already plenty of *H* present, but also energy and raw materials that could better be put to use elsewhere are used up. The problem, then, was to find a way to shut down biochemical pathways at times when their products were not needed.

Primitive organisms evolved an ingenious mechanism, the **feedback loop,** to solve the problem of shutting down biochemical pathways. To understand how the mechanism works, we need to briefly reconsider protein structure. Recall from Chapter 3 that proteins maintain their shape largely as a result of hydrophobic interactions and internal hydrogen bonds. Such bonding forces are weak, and relatively small changes in the surface character of an enzyme can lead to an alteration of its shape. Thus the binding of substrate to enzyme often induces a change in the shape of the enzyme, which enhances its catalytic activity.

The opposite action can also occur. The binding of a molecule to an enzyme can alter the shape of the enzyme in such a way as to make it less active. This effect underlies the simplest, and probably most primitive, means of regulating enzyme activity: enzymes evolved secondary binding sites to which nonsubstrate molecules could bind. These sites served as chemical on/off switches, because the binding of the nonsubstrate molecule

FIGURE 10-A

The enzyme chymotrypsin consists of two parts, which are mirror images of one another—like a left and right hand; the active site is formed by the cleft between the two halves of the molecule. Chymotrypsin is one of a family of different protein-degrading enzymes with very similar amino acid sequences that evolved from the same ancestral protein. They differ in which amino acid side groups they attack because of changes in some of the amino acids that line the border of the cleft.

A *The cleft is open in chymotrypsin, which attacks amino acids with large aromatic side groups, such as phenylalanine, that can fit into the big gap.*

B *The cleft possesses a negatively charged side group in trypsin, which attacks amino acids with a positive charge, such as arginine or lysine, that is attracted to the negative charge.*

C *The cleft is blocked by bulky valine and threonine side groups in elastase, which attacks amino acids with small side groups, such as alanine, that can fit into the small gap.*

changed the shape of the enzyme and thus its activity. If the nonsubstrate binding induced a change in enzyme shape that improved catalytic activity, then the binding of the nonsubstrate molecule acted as an "on" command; if the binding led to an altered enzyme shape that was not catalytically active, then the nonsubstrate molecule acted as an "off" signal. Enzymes controlled in this way are **allosteric** (Greek, *allos,* other + *steros,* form).

Enzymes can assume different forms, with different levels of catalytic activity, when bound by specific nonsubstrate molecules.

The regulation of simple biochemical pathways often takes place by means of an elegant feedback mechanism, in which the enzyme catalyzing the first step in a biochemical pathway possesses a second binding site, one which binds not the substrate of that enzyme, but rather the molecule that is the end product of the whole biochemical pathway. In our hypothetical example the enzyme catalyzing the reaction $C \rightarrow D$ would possess a secondary binding site for H, the end product of the pathway. As the pathway proceeds and the end product H becomes present in the cell in ever-greater quantities, it becomes increasingly likely that one of the H molecules will encounter the secondary binding site of the $C \rightarrow D$ enzyme and bind to it. When this occurs, the binding induces an allosteric change in the shape of the enzyme that causes it to stop catalyzing the reaction $C \rightarrow D$. Because shutting down this reaction stops the first reaction in the sequence, the effect is to shut down the whole pathway. Thus when the cell has made the end product H in sufficient quantity, it is able to stop making more. Such a mode of regulation is called **feedback inhibition.**

THE EVOLUTION OF METABOLISM

The way in which contemporary organisms harvest energy and employ it to drive chemical reactions is very different from the way in which the earliest organisms did. The metabolism of our world today is primarily an oxidative, or **aerobic,** metabolism, in which carbon atoms are cycled between organisms and the atmosphere. The portion of the cycle that acquires carbon atoms from atmospheric carbon dioxide and incorporates them into organisms is **photosynthesis,** carried out by green plants. The photosynthesis that is carried out by today's plants converts atmospheric CO_2 and the energy of sunlight into organic compounds with the release of oxygen gas (O_2). The other half of the cycle, in which organic compounds are burned, is **respiration,** carried out by all eukaryotes. Respiration converts carbon compounds and O_2 gas into CO_2 and chemical energy (ATP).

This kind of chemical machinery, which is so characteristic of biological systems, is nevertheless a relatively recent development in evolutionary history. For most of the period in which life has existed on earth, metabolic processes evolved in what was essentially an oxygen-free, or **anaerobic,** environment. Not only does anaerobic metabolism still exist, it remains the fundamental chemistry of all life processes. The more modern biochemical machinery of aerobic metabolism has simply been added on, like clothes over underwear.

The evolution of metabolism has proceeded something like the evolution of airplanes, in fits and starts. Gliders were the only airplanes until propeller-driven engines were introduced. Propeller-driven engines were subsequently modified and improved, until the advent some years later of a major innovation, the jet engine. Although small improvements are constantly being effected, the major steps in the evolution of the airplane have involved the introduction of a new capability. So it has been with the metabolism of organisms. We will review what are thought to have been the major steps in the evolution of metabolism.

Abiotic Synthesis

The first organic molecules were created by natural forces before life evolved. The creation of organic molecules by nonliving forces is called **abiotic synthesis.** These first organic molecules consisted of short chains of carbon atoms bonded to the other NOCH atoms, oxygen, nitrogen, and hydrogen. Among these molecules were amino acids, carboxylic acids, and hydrocarbons. Produced in an atmosphere rich in hydrogen gas and free of oxygen gas, these early organic molecules were highly reduced, with hydrogen atoms bonded to many of their carbon atoms. A considerable amount of energy is required to form C—H bonds, and much chemical energy is made available by their cleavage.

The input of energy required for the synthesis of these reduced molecules came directly from the sun in the form of intense ultraviolet radiation. There was no oxygen gas in the atmosphere to form ultraviolet light–absorbing ozone; for this reason the input of energy was uninterrupted and abiotic synthesis was continuous. Only with the advent of oxygen-generating photosynthetic organisms did gaseous oxygen become common, creating an atmosphere that blocked incoming ultraviolet radiation, and so brought abiotic organic synthesis to a halt. This was a point of no return in the evolution of life, because the generation of abundant oxygen in the atmosphere destroyed the conditions that had made the origin of life possible.

Degradation

The most primitive forms of life are thought to have obtained chemical energy by degrading, or breaking down, organic molecules that were abiotically produced. We know nothing of these organisms directly, but some bacteria, including the anaerobic methane-producing ones, carry out similar reactions today. These first simple reactions form the basic framework of all of the more complicated biochemical pathways that later evolved.

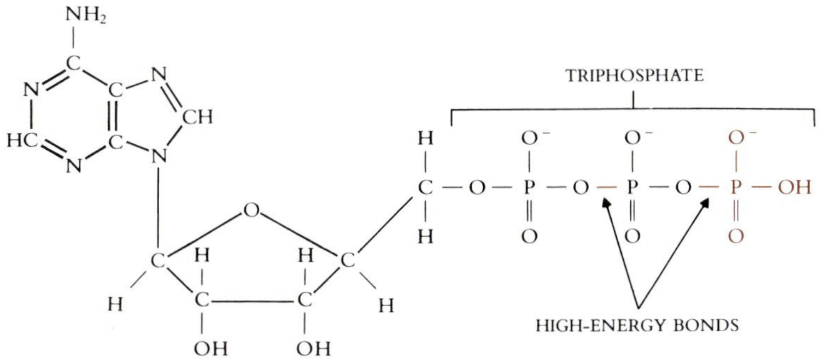

FIGURE 10-6

Adenosine triphosphate (ATP) contains two high-energy phosphate bonds. The terminal phosphate group is usually split off in energy transfers within cells, producing adenosine diphosphate (ADP) and a phosphate group.

For example, the universal presence in all organisms of ATP (Figure 10-6) as a carrier of chemical energy suggests that ATP acquired this role in metabolism at a very early stage.

> **The first major event in the evolution of metabolism was the origin of the ability to obtain energy from chemical bond energy. At an early stage organisms began to convert this energy into ATP, an energy carrier employed by all organisms today.**

Fermentation

As proteins evolved diverse catalytic functions, it became possible to capture a larger fraction of the chemical bond energy in abiotic organic molecules by breaking a series of chemical bonds in successive steps. For example, the progressive breakdown of the six-carbon sugar glucose into smaller two-carbon molecules was performed in a series of 10 steps that resulted in the net production of two ATP molecules. The energy for the synthesis of ATP was obtained by breaking a series of chemical bonds and forming new ones with less bond energy, the energy difference being channeled into ATP production. This biochemical pathway (Figure 10-7) is called **glycolysis** (Greek, *glyco,* sweet + *lysis,* to break; literally, the breakdown of sugar).

As you will learn in Chapter 11, one of the reactions of glycolysis is an oxidation-reduction reaction, in which an electron and an associated hydrogen atom are stripped from

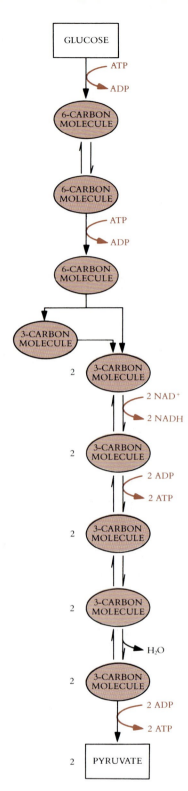

FIGURE 10-7

The glycolytic pathway is one of the oldest parts of our metabolic heritage. It consists of a series of 10 reactions and occurs in all living organisms. This pathway will be described in more detail in Chapter 11.

one of the glycolytic intermediates. The electron is used to generate ATP, and the hydrogen atom is donated to another organic molecule. Reactions that use organic molecules as final hydrogen acceptors are called **fermentations;** in this case the final hydrogen acceptor is a small carbon molecule, often alcohol. Thus when the sugar present within grapes is fermented by yeasts within vats that contain no oxygen, the result is the production of the alcohol ethanol. In making wine in this way, it is important that no bacteria be present. Bacteria carry out additional reactions, with the result that acetic acid (vinegar) is produced as the end product rather than alcohol. It is also important that no air be present, since in the presence of oxygen yeast metabolizes sugar all the way to CO_2 and H_2O, just as you do.

The glycolytic sequence of reactions undoubtedly evolved very early in the history of life on earth, since the biochemical pathway has been retained by all modern organisms. It is a chemical process that does not appear to have changed for well over 3 billion years.

The second major event in the evolution of metabolism was the development of organized sequences of catalyzed degradation reactions.

Anaerobic Photosynthesis

Early in the history of life, organisms evolved a different way of generating ATP, called photosynthesis. Instead of obtaining energy for ATP synthesis by reshuffling chemical bonds, as in glycolysis, organisms developed the ability to use light to pump protons out of their cells. Proton-pumping channels work like the sodium-potassium pump channels discussed in Chapter 9. Because pumping out protons creates a low concentration of protons within the cell, diffusion causes a flow of protons from outside the cell back in. To obtain entry, these protons pass through special proton-transporting channels within the cell membrane. The passage of protons through these channels induces a conformational change in the surrounding proteins that results in the synthesis of ATP from ADP and inorganic phosphate (P_i).

In the first stage of photosynthesis photons of light, striking chlorophyll pigments embedded within the cell membrane, excite electrons within the pigments. These excited electrons pass from the pigment to a series of proteins embedded within the membrane, eventually reaching and activating transmembrane proton-transport channels such as described in Chapter 9. This channel retains the electron and exports the proton across the membrane and out of the cell. The electron is then returned to the chlorophyll, completing its circular journey.

The light energy absorbed during photosynthesis is used to export protons out across the cell membrane.

In photosynthesis electrons do not travel naked, as free electrons. Rather, the single electron of a hydrogen atom (which as you will recall is composed of one electron orbiting around a nucleus composed of one proton) is boosted to a higher energy level. The proton of the excited hydrogen atom serves as a carrier of the very energetic electron, much as a school bus may carry quiet children to school or very lively and excited ones home. In early forms of photosynthesis the molecule from which the carrier hydrogen atom was extracted is thought to have been hydrogen sulfide (H_2S). Later versions of anaerobic photosynthesis used organic compounds as hydrogen donors.

Note that nowhere have we mentioned oxygen. The photosynthetic process just described evolved in its absence and works well without it. Dissolved H_2S, present in the oceans beneath an atmosphere free of oxygen gas, would serve as a ready source of hydrogen atoms for photosynthesis, since the S—H covalent bond (78 kcal/mole) is readily broken. The O—H bond of water (H_2O) requires 118 kcal/mole to break, or half again as much. When photosynthesis utilizes H_2S as a source of hydrogen atoms, free sulfur is produced as a byproduct. The atmosphere was little changed by this process, a process which has persisted to the present in a few groups of bacteria that exist in oxygen-free anaerobic environments.

The third major event in the evolution of metabolism was the advent of photosynthesis utilizing H_2S.

Nitrogen Fixation

Life evolved around four key biological polymers, all of which could be built up from subunits created abiotically in the primitive oceans: carbohydrates, lipids, proteins, and nucleic acids. Carbohydrates and lipids are chains of carbon atoms linked to oxygen and hydrogen atoms. As discussed in Chapter 4, they play important roles in energy storage and in formation of membranes. Both carbohydrates and lipids can be synthesized from the products of photosynthesis.

In contrast, proteins and nucleic acids cannot be synthesized from the products of photosynthesis, since both of these biologically critical molecules contain nitrogen. Life could not have persisted and expanded in the early oceans without a means of replacing organic nitrogen compounds. Chemically, this would not have been a simple task, since atmospheric nitrogen is thought to have been largely in the form of nitrogen (N_2) gas. Obtaining nitrogen atoms from N_2 gas, a process called **nitrogen fixation,** requires the breaking of an N≡N triple bond (bond energy $E = 225$ kcal/mole). This very strong bond is difficult to break. Nevertheless, all the nitrogen atoms in living organisms exist as a direct result of the breaking of this bond.

This important reaction, nitrogen fixation, is catalyzed in living systems by three proteins (Figure 10-8). The first, **ferredoxin,** obtains electrons from the breakdown of organic molecules or, among photosynthetic organisms, from the photosynthetic light reaction. Ferredoxin then carries the electrons to a second protein, **nitrogen reductase,** which channels them to the third protein, **nitrogenase.** With the transfer of six electrons and the hydrolysis of twelve ATP and four water molecules, nitrogenase converts nitrogen gas into two molecules of ammonia.

The nitrogen-fixation reactions evolved in the hydrogen-rich atmosphere of the early earth, an atmosphere in which no oxygen was present. Eventually, after this mechanism had evolved, our atmosphere began to contain increasingly high concentrations of gaseous oxygen as a result of the gradual buildup of oxygen produced by the photosynthetic splitting of water. This oxygen acts as a poison to nitrogen fixation, since the reaction mechanism will not work in its presence.

One consequence of the appearance of plentiful oxygen in the atmosphere was that the ability to fix nitrogen became restricted to organisms that lived in strictly anaerobic habitats and to the cyanobacteria. In the nitrogen-fixing cyanobacteria there are specialized inclusions called **heterocysts,** which contain the nitrogenase complex and are impermeable to oxygen. In the nodules on the roots of legume plants, other groups of nitrogen-fixing bacteria called *Rhizobium* live symbiotically. Any oxygen that diffuses into the nodule is combined with a respiratory protein produced by the plant, **leghemoglobin,** which removes it. *Rhizobium* bacteria, which when growing alone require oxygen for their metabolism, live within nodule cells and survive in this oxygen-free environment by utilizing the metabolites of these nodule cells. All biological nitrogen fixation today is strictly anaerobic, as it was when the process first evolved.

The fourth major event in the evolution of metabolism was the evolution of a mechanism for fixing atmospheric nitrogen.

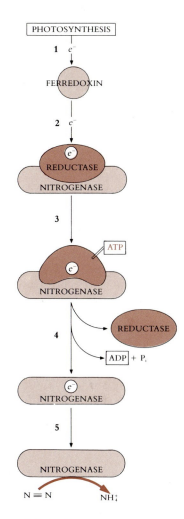

FIGURE 10-8

Nitrogen fixation is a strictly anaerobic process and is poisoned by even trace amounts of oxygen. The steps of the process are (1) photosynthesis generates high-energy electrons, donated to ferredoxin; (2) ferredoxin transfers its electrons to reductase; (3) ATP binds to reductase, changing the protein's shape and increasing its tendency to donate electrons; (4) electrons are transferred to nitrogenase; ATP is split and reductase dissociates from the complex; (5) nitrogen binds to nitrogenase and is reduced to ammonia ion ($NH_4{}^+$).

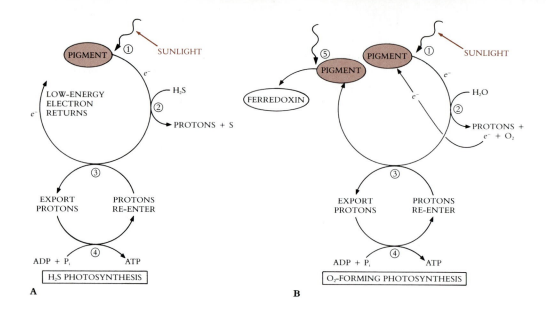

FIGURE 10-9

A simplified comparison of H₂S photosynthesis and oxygen-forming photosynthesis.

A *In H₂S photosynthesis light striking a pigment generates a high-energy electron (1), which in a series of steps causes protons derived from H₂S (2) to drive a proton transport channel, activates ATP-generating mechanisms (3 and 4), and returns to the original pigment.*

B *In oxygen-forming photosynthesis the process takes a similar flow, except that in step 2 H₂O provides both protons and an electron and in an additional step (5) the original electron is not returned to its source but to a different pigment. The end products of step 2 are also different: sulfur is generated in one process, oxygen gas in the second.*

Oxygen-Forming Photosynthesis

More than 2 billion years ago, over a comparatively brief period, practically all of the dissolved iron in the oceans became bound to oxygen. Only one process is known that could release oxygen at a rapid enough rate to account for this phenomenon: oxygen-forming photosynthesis.

Oxygen-forming photosynthesis employs H_2O rather than H_2S as a source of hydrogen atoms and their associated electrons (Figure 10-9). Because it garners its hydrogen atoms from reduced oxygen rather than from reduced sulfur, oxygen-forming photosynthesis results in the generation of oxygen gas rather than free sulfur. When the ability to capture protons from cleavage of the stronger O—H bond first evolved, those organisms in which it appeared gained a terrific competitive advantage. Not only was the process based on a far more plentiful resource, water, but the oxygen gas that it generated acted as a poison to H_2S photosynthesis.

As a consequence of this double advantage, small cells carrying out the new oxygen-forming photosynthesis, such as the cyanobacteria, became the dominant forms of life on earth. For several hundred million years the increasing amounts of oxygen gas that they produced were soaked up by the dissolved iron in the oceans. Eventually, however, the iron in the oceans was depleted, and oxygen gas began to disperse into the atmosphere. This was the beginning of the great transition that changed conditions on earth permanently. Our atmosphere is now 20.9% oxygen, every molecule of which is derived from an oxygen-forming photosynthetic reaction.

> **The fifth and pivotal event in the evolution of metabolism was the substitution of H₂O for H₂S in photosynthesis.**

Using Oxygen to Drive Redox Processes

To understand the profound consequences of the appearance of an oxygen atmosphere on earth after the evolution of photosynthesis, it is necessary to reconsider briefly the nature of oxidation-reduction reactions. An **oxidation-reduction reaction** (or **redox** reaction) is one in which an electron is transferred from one molecule to another. The molecule that loses the electron—the electron donor—is said to be oxidized, whereas the molecule that accepts the electron—the electron acceptor—is said to be reduced.

$$X[e^-] \quad + \quad Y \quad \rightarrow \quad X \quad + \quad Y[e^-]$$

ELECTRON DONOR	ELECTRON ACCEPTOR	OXIDIZED (LOST e^-)	REDUCED (GAINED e^-)

All redox reactions proceed in such a fashion that the energy (called **electrical potential**) of the transferred electron decreases, thus liberating energy. Although oxidation-reduction reactions can occur between a wide variety of molecules, the ones of paramount biological importance are those involving electrons associated with atoms of hydrogen and oxygen. Either atom may donate or accept an electron:

Reductions	*Oxidations*
Removal of oxygen: adds e^-	Addition of oxygen: removes e^-
Addition of hydrogen: adds e^-	Removal of hydrogen: removes e^-

Diagrammatically these possibilities can be represented as simple redox equations in which X stands for any molecule that donates electrons and Y for any molecule that receives them:

$$X + YO \rightarrow XO + Y$$
$$XH + Y \rightarrow X + YH$$

$$\underset{\substack{e^- \\ \text{DONOR}}}{} \quad \underset{\substack{e^- \\ \text{ACCEPTOR}}}{} \quad \underset{\text{OXIDIZED}}{} \quad \underset{\text{REDUCED}}{}$$

In an oxygen-containing atmosphere, the first of these two redox reactions becomes feasible, whereas it was not possible in the oxygen-free atmosphere of the primitive earth. In an oxygen-containing atmosphere a considerable amount of energy could be liberated by the addition of oxygen directly to reduced compounds. The lower redox reaction also becomes far more expeditious in an oxygen-containing atmosphere, since oxygen may serve as the ultimate acceptor of the proton and associated electron, thus forming water.

In the oxygen-rich atmosphere produced by oxygen-forming photosynthesis, both kinds of redox reactions were employed by organisms to derive usable energy:

1. *Oxygen addition.* Some bacteria obtained their energy by adding oxygen to inorganic compounds such as the elemental sulfur produced by a billion years of anaerobic photosynthesis or to the ammonia and other nitrogen compounds produced by photosynthetically driven nitrogen fixation. Among these are the chemoautotrophs described below.
2. *Hydrogen removal.* Other heterotrophic bacteria obtained their energy by the removal of hydrogen from organic compounds produced by photosynthesis, using oxygen as the ultimate acceptor of the electron and its associated hydrogen. This process is called **aerobic respiration.** Most of the organisms now living on earth are the descendants of these bacteria.

Chemoautotrophs

When oxygen-forming photosynthesis first began to enrich the oceans (and eventually the atmosphere) with oxygen, these oceans were already rich in elemental sulfur, the residue of eons of anaerobic photosynthesis. Organisms soon evolved that could combine this sulfur with oxygen and thus generate reducing power and energy from the reaction

$$2S + 2H_2O + 3O_2 \rightarrow 2SO_4^- + 4H^+$$

ELEMENTAL SULFUR	WATER	OXYGEN GAS	SULFATE ION	HYDROGEN ION
ELECTRON DONOR	ELECTRON ACCEPTORS		REDUCED	OXIDIZED

An organism that obtains its energy in this way is called a **chemoautotroph** (Greek, feeding itself by chemicals). Such an organism requires no additional source of energy other than the chemical to which it adds oxygen, and so it is, in a sense, "feeding" on inorganic compounds.

The oxidation of elemental sulfur offered a tempting evolutionary path, permitting the utilization of raw materials made plentiful in the oceans by the long history of anaerobic photosynthesis. Many organisms may have chosen this path. We do not know. Few, it is certain, survive today. The elemental sulfur used as raw material by the sulfur chemoau-

FIGURE 10-10

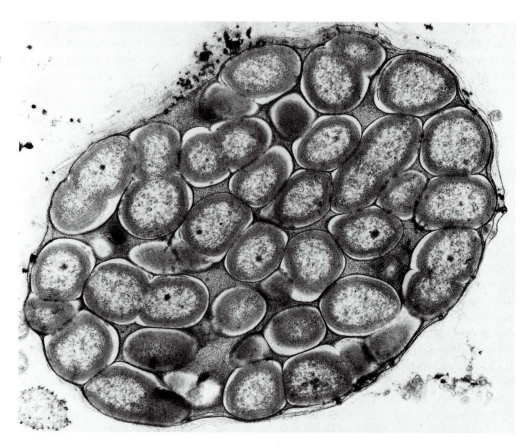

totrophs was the product of anaerobic photosynthesis, a process poisoned by oxygen. The oxygen-forming photosynthesis whose advent made this kind of energy harvesting possible also stopped the supply of its raw materials.

Other chemoautotrophic organisms have evolved that carry out the oxidation of other inorganic molecules. Because the various inorganic molecules were in most cases not plentiful, such evolutionary paths have proven to be of limited scope. Some of these organisms survive today in odd environments, and some play important roles in mineral cycling. *Nitrosomonas* bacteria (Fig. 10-10), for example, obtain metabolic energy by oxidizing ammonia derived from the decay of biological material, converting the ammonia to nitrite:

$$2NH_3 \;+\; 3O_2 \;\rightarrow\; 2NO_2^- \;+\; H_2O \;+\; 4H^+$$

| AMMONIA | OXYGEN GAS | NITRITE | WATER | HYDROGEN ION |

Other bacteria continue the oxidation process, converting nitrite to nitrate:

$$2NO_2^- \;+\; O_2 \;\rightarrow\; 2NO_3^-$$

| NITRITE | OXYGEN GAS | NITRATE |

These organisms thus carry out critical steps of the **nitrogen cycle.** The by-product of their metabolism, nitrate, is a form of nitrogen readily assimilated and utilized by plants in their synthesis of nitrogen-containing proteins and nucleic acids.

The mainstream of evolution has passed chemoautotrophs by, taking a different route for most of life. Nevertheless, we and all other organisms are still dependent on them. They are essential links in our chain of life. Without a means of mobilizing atmospheric nitrogen to convert it to ammonia, and transform it to nitrate, plants would have no means of obtaining the nitrogen necessary for the synthesis of proteins and nucleic acids—nor would we, who in turn live on plants.

Aerobic Respiration

Oxygen soon became plentiful in the oceans and atmosphere as a result of oxygen-forming photosynthesis. The oceans must have been teeming with life for the transition to an aerobic atmosphere to have been accomplished so quickly. This situation presented an evolutionary opportunity that has had profound consequences. In addition to the now absent abiotic synthesis of organic molecules and the photosynthetic fixation of carbon, organisms were presented with a third major source of chemical bond energy—the organic molecule of other living organisms.

It is not possible to harvest the energy of organic molecules in an efficient way simply by breaking them into smaller molecules. As you have already seen, glycolysis yields only two ATP molecules from the breakdown of a molecule of the sugar glucose. Two ATP molecules represent very little of the chemical energy that is present in sugar molecules. The efficient utilization of the organic molecules of other organisms as an energy source required a new approach, and a powerful one soon evolved. The new approach was based on the same fundamental process that had already evolved as the central mechanism of ATP generation during photosynthesis: the utilization of proton gradients across membranes to produce ATP. In the new approach reduced carbon-based molecules are removed from sugar molecules in a process called the citric acid cycle and are stripped of their hydrogen atoms and associated electrons. The hydrogen atoms are carried by special transport molecules to the inner surface of a membrane impermeable to protons. On the inner surface of this membrane the proton pumps strip the electrons from the arriving hydrogen atoms and export the nuclei (protons) out across the membrane. The return of protons back into the cell generates ATP in a process called **oxidative phosphorylation.**

The gleaning of energy from organic molecules in this manner is called aerobic respiration because oxygen gas molecules ultimately acquire the hydrogen atoms stripped from organic molecules, as well as their associated electrons, and form water molecules. In a very real sense aerobic respiration is the reverse of oxygen-forming photosynthesis, converting carbon polymers to carbon dioxide and oxygen gas to water with the net release of energy (as ATP). The photosynthetic process converts carbon dioxide to carbon polymers and water to oxygen gas, with a net input of energy from photons of light.

We think that the ability to carry out photosynthesis without H_2S first evolved among organisms that harvested electrons from organic molecules and used them to drive proton pumps and so generate ATP. Purple nonsulfur bacteria are the first organisms known to have evolved this ability. These bacteria carry out photosynthesis in the absence of H_2S, obtaining their hydrogens from organic compounds instead. They are able to harvest electrons from these organic molecules and drive ATP-generating proton pumps, even in the absence of light. Oxidative respiration is thought to have evolved in this fashion, as a modification of the photosynthetic machinery.

> **Aerobic respiration probably evolved from photosynthesis, as a modification of the basic photosynthetic machinery, and employs the same kinds of proton pumps. However, the hydrogens and their associated electrons are not obtained from H_2S or H_2O, as in photosynthesis, but rather from the breakdown of organic matter.**

The efficiency of this aerobic respiratory process is remarkable. The initial breakdown of glucose by glycolysis yields only two ATP molecules per molecule of glucose. In the process of aerobic respiration an additional 34 ATP molecules are formed. Overall, 38% of the chemical bond energy of the glucose that is broken down in aerobic respiration is recovered as high-energy ATP bonds.

It was perhaps inevitable that among the descendants of these respiring photosynthetic cells, some would eventually do without photosynthesis entirely, subsisting only on the energy and reducing power derived from the breakdown of organic molecules. The mitochondria within all eukaryotic cells are thought to be their descendants.

The sixth and final stage in the evolution of metabolism was the development of a mechanism, aerobic respiration, in which electrons were stripped from the glucose molecules manufactured by plants and employed to drive ATP-generating proton pumps.

THE METABOLIC JOURNEY

From our consideration of how metabolism has evolved, two fundamental lessons emerge. The first of these is that there is no one solution to the problem of obtaining energy (Figure 10-11). Organisms have employed a diverse collection of different approaches. The most primitive of these involves the breaking of chemical bonds; others involve inorganic chemical reactions; still others use light energy to fuel proton pumps. No one mechanism is the best in all environments.

FIGURE 10-11

The major ways in which a cell obtains energy from food. The three principal classes of macromolecules that provide most food energy are proteins, carbohydrates, and fats. These classes are broken down into amino acids, sugars, and fatty acids, which are used in glycolysis and the citric acid cycle to generate ATP.

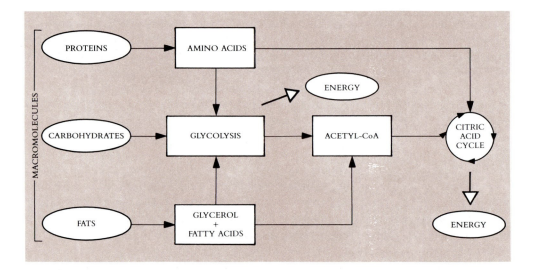

The most recent mechanism to evolve, the aerobic respiration that we employ to obtain energy, is a combination of a primitive mechanism, glycolysis, coupled to a mechanism for generating ATP modified from photosynthesis. The second fundamental lesson to emerge from our consideration of how metabolism evolved is that successive stages of metabolism have incorporated earlier processes as integral parts of new approaches, new houses built on old foundations. As we discuss metabolism in more detail in the next two chapters, the traces of our metabolic history will appear repeatedly.

SUMMARY

1. Two chemical forces act on all molecules, one tending toward minimum energy and the other toward maximum disorder. Because of these tendencies, reactions in which the bond energy decreases and the disorder increases occur spontaneously. Both factors, respectively called enthalpy and entropy, are important in determining whether a reaction will proceed.

2. Many reactions proceed only very slowly unless the energy barrier to the transition state is lowered. Enzymes act to lower these energy barriers and to thus increase the rate of reactions. The reactions continue to proceed in the same direction in which they would have gone anyway, but at a much faster rate.

3. Enzymes bind particular reactants, which are called substrates. Because the bonds are very weak, the three-dimensional fit between the enzyme and its substrate must be very close for successful binding between them to occur. Changes in the shape of an enzyme, such as those which are induced by its binding to a nonsubstrate molecule, can have a major influence on the ability of that enzyme to bind substrate molecules.

4. The earliest metabolic systems are thought to have obtained chemical energy from the breakdown of organic chemicals. Glycolysis, the progressive breakdown of six-carbon sugar molecules, is one of these primitive reaction sequences. It is retained by all living organisms.

5. The advent of transmembrane channels that could pump protons out of the cell made possible photosynthesis, in which light energy is used to activate electrons to drive the proton pumps. Rushing back into the cell, these protons promote the synthesis of ATP.

6. Nitrogen fixation uses excited electrons derived from photosynthesis, or from organic chemical bonds created with photosynthetic energy, to split atmospheric nitrogen gas (a triple bond) into two molecules of ammonia. This reaction evolved before the advent of significant amounts of atmospheric oxygen and is poisoned by even very small traces of oxygen gas.

7. Oxygen-forming photosynthesis employs water rather than hydrogen sulfide as a source of hydrogen atoms and their associated electrons. This process requires the capture of more light energy than does nitrogen fixation but has the benefit that water is far more plentiful than hydrogen sulfide. The consequence of this reaction, which is the production of gaseous oxygen rather than elemental sulfur, has forever altered the nature of the earth.

8. Aerobic respiration evolved as a modification of photosynthesis, in which electrons are obtained not from water or hydrogen sulfide but rather from organic molecules. Since these organic molecules are derived ultimately from photosynthesis, organisms such as ourselves that subsist on aerobic respiration can be thought of as parasites of photosynthesizers.

SELF-QUIZ

The following questions pinpoint important information in this chapter. You should be sure to know the answers to each of them.

1. What is enthalpy?
   (a) a measure of randomness
   (b) free energy
   (c) the energy required to break chemical bonds
   (d) the difference between ordering and disordering influences
   (e) none of the above

2. Will a reaction proceed spontaneously when the change in free energy is positive, or negative?

3. No enzyme can alter the final proportions of substrate and product to which a chemical reaction will ultimately proceed. True or false?

4. In a biochemical pathway such as A → B → C → D → E which reaction probably evolved last?
   (a) A → B   (d) D → E
   (b) B → C   (e) they all would have evolved at once
   (c) C → D

5. Does the fundamental chemistry of metabolism consist of enzyme-catalyzed biochemical pathways that evolved in an atmosphere that was aerobic, or anaerobic?

6. The correct order in which the following metabolic pathways first evolved is presented below, except that one process is in an incorrect position. Which one?
   (a) abiotic synthesis   (e) oxygen-forming photosynthesis
   (b) heterotrophic degradation   (f) chemoautotrophy
   (c) fermentation   (g) aerobic respiration
   (d) anaerobic photosynthesis

7. To obtain a nitrogen atom for biological molecules from a molecule of nitrogen gas, how many covalent bonds must be broken?

8. In nitrogen-fixing bacteria the nitrogenase complex is isolated within special cells called _____.

9. Does the addition of a hydrogen atom to a compound oxidize or reduce it?

10. Does the addition of an oxygen atom to a compound oxidize or reduce it?

11. After the advent of oxygen-forming photosynthesis created an oxygen-rich atmosphere, chemoautotrophs evolved that were able to obtain energy by metabolizing sulfur. In what chemical form was the sulfur?
    (a) SO_4 (b) SO_2 (c) SH   (d) S_2 (e) S

12. Oxygen-forming photosynthesis evolved before aerobic respiration. True or false?

THOUGHT QUESTIONS

1. After the advent of oxygen-forming photosynthesis, when the atmosphere of the earth became rich in oxygen, two very different forms of oxidative metabolism evolved, one based on adding oxygen atoms to elemental sulfur, the other on removing hydrogen atoms from organic compounds and adding them to oxygen. On the earth today one of these forms of oxidative metabolism has become very common, whereas the other is only rarely seen. What factors might have contributed to this?

2. Oxidation-reduction reactions occur between a wide variety of molecules. Why do you suppose those concerning hydrogen and oxygen are the ones of paramount importance in biological systems?

FOR FURTHER READING

DICKERSON, R.E.: "Chemical Evolution and the Origin of Life," *Scientific American,* September 1978, pages 70-86. A lucid account of how the major metabolic systems that we see today may have evolved.

DICKERSON, R.E.: "Cytochrome *c* and the Evolution of Energy Metabolism," *Scientific American,* March 1980, pages 137-153. An interesting presentation of the argument that the ancestors of respiring mitochondria were photosynthetic bacteria. The power of molecular data in resolving these issues is well illustrated.

GUTFREUND, H., (ed.): *Biochemical Evolution,* Cambridge University Press, New York, 1981. A collection of articles covering our current state of knowledge on the evolution of a variety of metabolic systems.

NEURATH, H.: "Evolution of Proteolytic Enzymes," *Science,* vol. 224, pages 350-357, 1984. A discussion of how a group of enzymes with relatively simple roles in primitive organisms has evolved into a complex and diverse collection of enzymes in higher organisms.

SCHOPF, J.W.: "The Evolution of the Earliest Cells," *Scientific American,* March 1978, pages 110-139. A presentation of the argument that the earliest cells gave rise to the oxygen in the atmosphere on which modern life depends.

SCHOPF, J.W., (ed.): *Earth's Earliest Biosphere: Its Origin and Evolution,* Princeton University Press, Princeton, N.J., 1983. A collection of articles by experts on the evolution of early forms of life and the geochemical evidence of what kinds of metabolism they conducted. Advanced, but fascinating.

WOESE, C.R.: "Archaebacteria," *Scientific American,* June 1981, pages 98-122. A challenging argument that the methanogenic bacteria and some related forms may represent a separate line of descent from the true bacteria and eukaryotes.

Chapter 11

THE METABOLIC LIFE OF A CELL

Overview

All organisms conduct their metabolism with energy from ATP. The simplest processes for generating ATP involve the rearrangement of chemical bonds, but organisms later evolved far more efficient means of carrying out this process. Plants, algae, and some bacteria obtain ATP by using energetic electrons to drive proton pumps, obtaining the electrons both by photosynthesis and by oxidizing sugars and fats. Other organisms, such as all animals and fungi, as well as most protists and bacteria, are heterotrophs; they obtain electrons only by the second process, which occurs within symbiotically derived mitochondria in eukaryotes.

For Review

Here are some important terms and concepts that you will encounter in this chapter. If you are not familiar with them, you should review them before proceeding.

- **Oxidation-reduction** (Chapter 2)

- **ATP** (Chapter 4)

- **The proton pump** (Chapter 9)

- **Free energy** (Chapter 10)

- **Chemiosmotic generation of ATP** (Chapters 9 and 10)

Life is driven by energy. All of the activities of living organisms—the swimming of bacteria, the purring of a cat, your reading of these words—use energy. The ways in which organisms acquire energy are many and varied, but all of life's energy ultimately has the same beginning: the sun. Plants, algae, and some bacteria harvest the energy of sunlight directly, converting it to chemical energy, whereas animals, fungi, and most protists and bacteria take advantage of the chemical energy that plants have trapped. Some bacteria use the oxygen produced by photosynthesis to derive energy from inorganic molecules. All living creatures—plants, bacteria, and you yourself—share that same dependency on the sun. We are all children of light.

In this chapter we will discuss the processes that cells use to derive energy from organic molecules. We will consider photosynthesis in detail in Chapter 12. We treat the harvesting of chemical energy first because all organisms, heterotrophic and photosynthetic alike, are capable of acquiring energy in this way, whereas only a small proportion of the kinds of organisms on earth—less than 1 in 10—is capable of carrying out photosynthesis. As you will see, though, the two processes have much in common.

USING CHEMICAL ENERGY TO DRIVE METABOLISM

The first organisms are thought to have acquired energy from the degradation of organic compounds created by geothermal processes and by solar radiation in the early oceans. Such processes, which obtain energy from the breakdown of organic molecules, are said to be **catabolic.** Conversely, those which expend energy to synthesize organic molecules are said to be **anabolic.** (The terms *catabolic* and *anabolic*, respectively, come from the Greek words meaning "throwing down" and "throwing up.")

There is nothing mysterious about obtaining energy by splitting up organic molecules. Recall from Chapter 2 that all molecules exist because of the tendency of atoms to satisfy certain fundamental conditions, such as a full outer shell of electrons and a balance of electrical charges. When a molecule can be changed so that these conditions are more readily met, a chemical reaction occurs that promotes the change. The tendency for the change to occur can be thought of as chemical energy driving the reaction.

In Chapter 10 we considered the evolution of the major catabolic systems of organisms and the biochemical pathways they use to harvest energy from organic compounds. The metabolic sequences we considered are driven by the release of free energy as the electrons rearrange themselves. A biochemical pathway is nothing more than a cascade of shuffling chemical bonds, each step facilitated by the release of free energy.

Those chemical reactions of a biochemical pathway that involve a release of free energy will occur spontaneously. Many of the reactions in biochemical pathways do not release free energy, however, but rather require it. *Reactions that require an input of free energy are not spontaneous and do not take place by themselves.* To drive such thermodynamically unfavorable reactions, all living organisms use the same mechanism: they *couple* the energy-requiring reaction to the energy-yielding splitting of ATP (Figure 11-1).

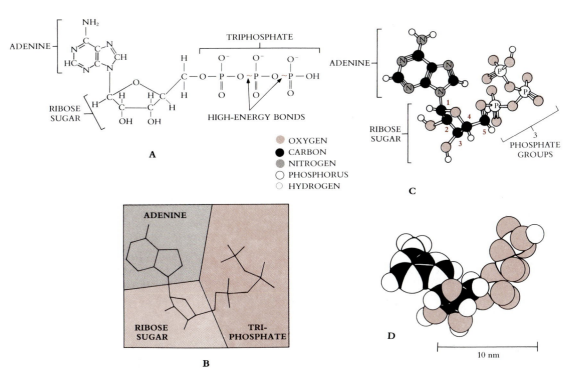

FIGURE 11-1

The structure of ATP is represented in a variety of ways.
A *Chemical formula.*
B *A "stick" diagram showing the covalent bonds to all the atoms but hydrogen.*
C *A "ball and stick" model, in which different kinds of atoms are represented by balls of different size and color.*
D *A three-dimensional or "space-filling" model, in which the approximate size and position of the atoms are represented.*

Coupled Reactions

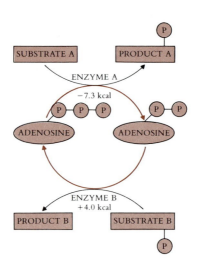

FIGURE 11-2

A coupled reaction is one that does not normally proceed spontaneously because it involves a net increase in free energy. It proceeds when the reaction is "coupled" to a reaction that involves a large decrease in free energy, such as the cleavage of the phosphodiester bond of ATP. Here ATP is signified as "adenosine-P-P-P."

To understand how coupled reactions work, recall for a moment the analogy of sheep jumping over a lowered gate between two pastures, which we employed in Chapter 10 to explain why enzymes cannot alter the direction of a reaction but only its speed. The net movement of sheep was always in the direction of the pasture with fewer sheep. Now imagine that the more vacant pasture also contains wolves and that sheep are afraid of wolves. Many of the sheep in the more vacant lot will now leap over the lowered gate into the more crowded pasture to escape the wolves, and more will hurry in that direction than will stray into the more vacant pasture. The net movement of sheep will now be into the more crowded pasture.

A coupled reaction works in just this way. The enzyme's catalysis of the reaction is equivalent to the lowering of the gate between the two pastures. The enzyme has substrate-binding sites that recognize both the metabolite *and* ATP, and catalysis in either direction is possible. On those occasions when the reaction occurs in the thermodynamically unfavorable direction (sheep moving into the more crowded pasture), a molecule of ATP is cleaved to ADP plus P_i, releasing the considerable free energy of the phosphoanhydride bond. This large release of free energy drives the coupled reaction, just as fear drove the movement of sheep in our analogy.

The tendency of any coupled reaction to occur is determined by the net change in free energy when both reactions are taken into account. In Figure 11-2, for example, the standard free energy change of the coupled reaction is the sum of the value of the unfavored reaction, +4 kcal/mole, and the value of the hydrolysis of the phosphoanhydride bond of ATP, −7.3 kcal/mole. Because the sum of +4 and −7.3 is negative, −3.3 kcal/mole, the reaction proceeds.

Organisms induce reactions to occur that require an input of free energy by coupling these reactions to the cleavage of ATP.

Because ATP can drive thermodynamically unfavored reactions, this molecule occupies a central position in the metabolism of organisms.

HARVESTING ENERGY FROM CHEMICAL BONDS: ATP

ATP is the energy currency of all living organisms. How do organisms harvest ATP from organic molecules? They do so in one of two ways:

1. *Substrate level phosphorylation.* The tendency for a particular organic bond to break can be so strong that this tendency can be used to drive the synthesis of ATP. The two reactions are said to be **coupled** in the same sense as described in the previous section. Because the formation of ATP from ADP + inorganic phosphate (P_i) requires an input of free energy, ATP formation does not occur spontaneously. When coupled to another reaction with a very strong tendency to occur, however, the synthesis of ATP from ADP + P_i does take place, because the release of free energy from the forward reaction is greater than the input of free energy required to drive the synthesis of ATP. The generation of ATP by coupling highly favored reactions to the synthesis of ATP from ADP + P_i is called **substrate level phosphorylation.**

2. *Chemiosmotic generation of ATP.* Many organisms possess transmembrane channels that function in pumping protons out of cells. We encountered these in Chapter 9. Proton-pumping channels use a flow of excited electrons to induce a conformational change in the transmembrane protein, which in turn causes protons to pass outward. As the proton concentration outside the membrane rises higher than that inside, the outer protons are driven inward by diffusion, passing *backward* through special proton channels that use their passage to induce the

formation of ATP from ADP + P$_i$. Because the chemical formation of ATP is driven by a diffusion force similar to osmosis, this process is referred to as a **chemiosmotic** one.

The harvesting of chemical energy can be considered to take place in one or more of three stages: (1) a reshuffling of chemical bonds to obtain ATP, (2) the harvesting of electrons and some ATP from the product of this bond shuffling, and (3) the transport of harvested electrons to a membrane where they drive a proton pump and so power the synthesis of ATP.

CATABOLIC METABOLISM

During the millions of years that preceded the development of aerobic photosynthesis, organisms obtained ATP by reshuffling and cleaving the bonds of organic molecules. There was a lot of organic material available, since organisms practicing anaerobic photosynthesis had already evolved and become common. Although organic materials were plentiful, however, much of the energy of their chemical bonds could not be applied directly to the synthesis of ATP, because only a few bonds of these organic molecules released enough free energy on splitting to drive the coupled synthesis of ATP. The basic approach of catabolic organisms was, and still is, to channel the degradation of carbon compounds toward these few very favored reactions.

The first step in catabolism is the degradation of complex molecules to simple ones. Cell walls and starches are degraded to simple sugar molecules such as glucose, proteins to amino acids, and nucleic acid chains to nucleotide monomers. These initial steps usually yield no usable energy, but they serve to marshal the resources from a diverse array of complex chemicals into a small number of simple molecules such as glucose.

GLYCOLYSIS

Among the many simple molecules that were available as a consequence of degradation, the metabolism of primitive organisms focused on the simple six-carbon sugar glucose, undoubtedly in part because glucose was a major constituent of the carbohydrates of the cell. Glucose molecules can be dismantled in many ways: primitive organisms evolved the ability to do it with reactions that drive the synthesis of ATP in coupled reactions. The process involves a sequence of ten reactions that convert glucose into two three-carbon molecules of pyruvate. For each molecule of glucose that passes through this transformation, the cell acquires two ATP molecules. The process is called **glycolysis.** In Chapter 10 we briefly mentioned glycolysis in our discussion of the evolution of metabolism.

An Overview of Glycolysis

Glycolysis consists of two very different kinds of processes, one wedded to the other:

1. Glucose is converted to two molecules of the three-carbon compound **glyceraldehyde-3-phosphate** (G3P), with the expenditure of ATP.
2. ATP is generated from G3P.

Perhaps the original catabolic process involved only the ATP-yielding breakdown of G3P; the generation of G3P from glucose may have evolved later, when alternative sources of G3P were depleted. Like many biochemical pathways, glycolysis may have evolved backward, the latest stages in the process being the most ancient.

As we understand it now, the ten reactions of glycolysis (Figure 11-3) proceed in four stages:

Stage a: Three reactions change glucose into a compound that can readily be cleaved into three-carbon phosphorylated units. Two of these reactions require

FIGURE 11-3

The four stages of the glycolytic pathway.

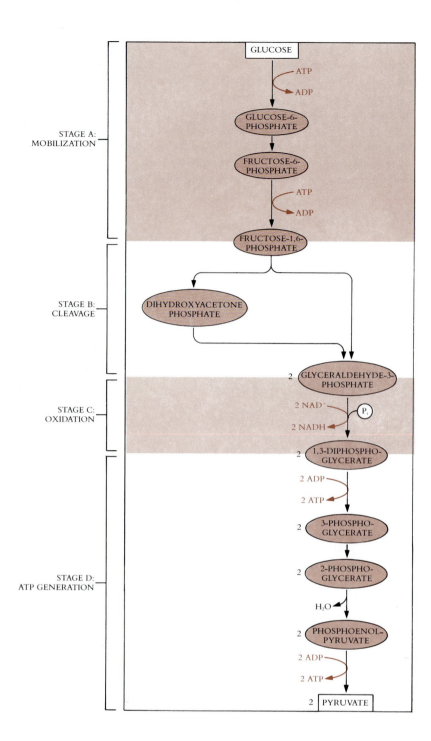

STAGE A: MOBILIZATION

GLUCOSE
ATP → ADP
GLUCOSE-6-PHOSPHATE
FRUCTOSE-6-PHOSPHATE
ATP → ADP
FRUCTOSE-1,6-PHOSPHATE

STAGE B: CLEAVAGE

DIHYDROXYACETONE PHOSPHATE
2 GLYCERALDEHYDE-3-PHOSPHATE

STAGE C: OXIDATION

2 NAD⁺ → 2 NADH, Pᵢ
2 1,3-DIPHOSPHO-GLYCERATE

STAGE D: ATP GENERATION

2 ADP → 2 ATP
2 3-PHOSPHO-GLYCERATE
2 2-PHOSPHO-GLYCERATE
H₂O
2 PHOSPHOENOL-PYRUVATE
2 ADP → 2 ATP
2 PYRUVATE

the cleavage of an ATP molecule, so this stage, **glucose mobilization,** requires the investment by the cell of two ATP molecules.

Stage b: The second stage is **cleavage,** in which the six-carbon product of the first stage is split into two three-carbon molecules. One is G3P, and the other is converted to G3P by another reaction.

Stage c: The third stage is **oxidation,** in which a hydrogen atom with a pair of electrons is removed from G3P and donated to a special **cofactor** molecule called **nicotinamide adenine dinucleotide (NAD⁺).** A cofactor is an accessory molecule that assists an enzyme in a metabolic process. Many cofactors are regenerated unchanged at the end of these processes. As an analogy, consider an armored car that transports money from a department store to a bank—the car is temporarily "rich" during the transaction, but the change in its wealth is only temporary. NAD^+ acts as an electron carrier in the cell, in this case accepting the proton and two electrons from G3P to form NADH. Note that NAD^+ is an ion and that *both* electrons in the new covalent bond come from G3P.

Stage d: The final stage, **ATP generation,** is composed of a series of four reactions that convert the product of G3P oxidation into another three-carbon molecule, pyruvate, and in the process generate two ATP molecules.

Because each glucose molecule is split into *two* G3P molecules, the overall net reaction sequence yields two ATP molecules:

$$\begin{array}{ll} -2\text{ATP} & \text{stage a} \\ \underline{2(+2\text{ATP})} & \text{stage d} \\ +2\text{ATP} & \end{array}$$

Two ATP molecules do not represent a great deal of energy. When you consider that the free energy of the total oxidation of glucose to CO_2 and H_2O is -680 kcal/mole and that ATP's high-energy bonds each have an energy content of -7.3 kcal/mole, then the efficiency with which glycolysis harvests the chemical energy of glucose is only 14.6/680, or 2.1%. Although far from ideal in terms of the amount of energy that it releases, glycolysis does generate ATP, and for more than a billion years during the long anaerobic prelude of the earth's history, this reaction sequence was the *only* way for heterotrophic organisms to generate ATP from organic molecules.

The glycolytic reaction sequence generates a small amount of ATP by reshuffling the bonds of glucose molecules. Glycolysis is a very inefficient process, capturing only about 2% of the available chemical energy of glucose. Most of the remaining energy is unrecovered in the molecules that glycolysis produces, particularly pyruvate.

The Universality of the Glycolytic Sequence

The glycolytic reaction sequence is thought to have been among the earliest of all biochemical processes to evolve. It uses no molecular oxygen and occurs readily in an anaerobic environment. All of its reactions occur free in the cytoplasm; none is associated with any organelle or membrane structure. Every living creature on earth is capable of carrying out the glycolytic sequence. Most present-day organisms, however, are able to extract considerably more energy from glucose molecules than did primitive prokaryotes limited to glycolysis. You, for example, obtain 36 ATP molecules from each glucose molecule that you metabolize, and only two of these are obtained by glycolysis. Why is glycolysis maintained even now, considering that its energy yield is comparatively so paltry?

This simple question has an important answer: evolution is an incremental process. Change occurs during evolution by improving on past successes. In catabolic metabolism, glycolysis satisfied the one essential evolutionary criterion: it was an improvement. Cells that could not carry out glycolysis were at a competitive disadvantage. By studying the metabolism of contemporary organisms, we can see that only those cells capable of glycolysis survived the early competition of life. Later improvements in catabolic metabolism built on this success. Glycolysis was not discarded but rather was used as the starting point for the further extraction of chemical energy. Nature did not, so to speak, go back to the drawing board and design a different and better metabolism from scratch. Rather, metabolism evolved as one layer of reactions was added to another. The metabolism that takes place in organisms living today is composed of a series of such additions, like successive layers of paint in an old apartment. We all carry glycolysis with us, a metabolic memory of our evolutionary past.

THE NEED TO CLOSE THE METABOLIC CIRCLE: FERMENTATION AND NAD$^+$

Inspect for a moment the net reaction of the glycolytic sequence:

$$Glucose + 2ATP + 2P_i + 2NAD^+ \rightarrow$$
$$2Pyruvate + 2ATP + 2NADH + 2H^+ + 2H_2O$$

For a catabolic organism growing in an anaerobic environment, this is an important balance sheet, since it describes the organism's only source of metabolic energy. What raw materials are needed to keep the reaction going? Well, glucose, of course, obtained by degrading organic material. And ATP, to prime the process. But all living cells contain ATP. Similarly, all cells are constantly generating $ADP + P_i$ when they utilize ATP. These four molecules, glucose, ATP, ADP, and P_i, are readily available in cells. However, a fifth molecule is also required. The overall glycolytic reaction sequence consumes a molecule of NAD^+, which was used to accept the electrons gleaned from G3P in the oxidation step (stage c, pp. 194–195).

Where does this NAD^+ in the cell come from? This is a critical question, because the glycolytic sequence does not itself produce any NAD^+. Without a separate NAD^+-generating process, therefore, continuous glycolysis is impossible.

In all anaerobic organisms the NAD^+ needed for glycolysis is obtained simply by reoxidizing the NADH that is produced by glycolysis back to NAD^+ and *contributing the hydrogen atom to some other molecule.* When the molecule to which the hydrogen atom is donated is an organic molecule, the process is called fermentation. Fermentation permits continuous glycolysis because the redox reaction of NAD^+ is cyclic, with no net production of NADH. Consequently, the process never runs out of NAD^+ and can continue as long as glucose is available.

Bacteria carry out many different kinds of fermentations, all employing some form of carbohydrate molecule to accept the hydrogen atom from NADH and thus reform NAD^+:

$$carbohydrate + NADH \rightarrow reduced\ carbohydrate + NAD^+$$

More than a dozen different fermentation processes evolved among the bacteria, each using a different carbohydrate as the hydrogen acceptor. Often the resulting reduced compound is an organic acid, such as acetic acid, butyric acid, propionic acid, or lactic acid. In other organisms, such as yeasts (Chapter 10), the reduced compound is an alcohol.

Of these various bacterial fermentations, only a few now occur among eukaryotes. In one fermentation process, which occurs in microorganisms such as yeast, the carbohydrate that accepts the hydrogen from NADH is the end product of the glycolytic process itself, pyruvate. Yeasts remove a terminal CO_2 group from pyruvate (a process called **decarboxylation**), producing a toxic two-carbon molecule, **acetaldehyde**, and CO_2. Because of this CO_2, bread with yeast rises and unleavened bread (without yeast) does not. Acetaldehyde then accepts the hydrogen from NADH, producing NAD^+ and ethyl alcohol, also called **ethanol** (Figure 11-4, *1*).

This particular type of fermentation has been of great interest to human beings, since it is the source of the ethyl alcohol in wine and beer (Figure 11-5). This alcohol is an undesirable byproduct from the yeast's standpoint, since it becomes toxic to the yeast when it reaches high levels. That is why natural wine contains only about 12% alcohol; 12% is approximately the amount of alcohol that it takes to kill the yeast whose fermentation of sugar produces the alcohol.

Most animals regenerate NAD^+ without decarboxylation, and the processes they use involve the production of byproducts that are less toxic than alcohol. You can see why it would be important not to accumulate toxic byproducts. Imagine you are running from a cloud of angry bees. The cells of your muscles will soon have used all available oxygen to produce ATP, and your blood supply cannot transport new oxygen rapidly enough to

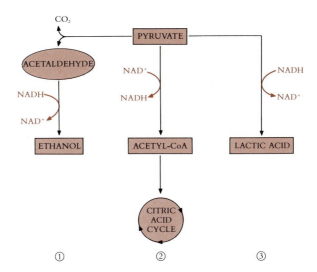

CO_2

PYRUVATE

ACETALDEHYDE

NAD⁺

NADH

NADH

NADH

NAD⁺

NADH

NAD⁺

ETHANOL

ACETYL-CoA

LACTIC ACID

CITRIC ACID CYCLE

① ② ③

FIGURE 11-4

Three alternative fates of pyruvate: 1, The conversion of pyruvate to ethanol, such as occurs during the production of wine and beer; 2, the conversion of pyruvate to acetyl-CoA, used in the citric acid cycle and elsewhere; 3, the conversion of pyruvate to lactic acid, such as occurs in your muscles. Fates 1 and 3 are types of fermentations.

provide sufficient ATP to continue rapid forceful muscle contraction. As a last resort, your muscle cells shift to glycolysis, which is an inefficient but available process, to obtain ATP. What is the cells' source of NAD^+? They use an enzyme called **lactate dehydrogenase** to add the hydrogen of the NADH produced during glycolysis back to the pyruvate that is the end product of glycolysis. This process converts pyruvate + NADH into lactic acid + NAD^+ (Figure 11-4, *3*).

This process closes the metabolic circle, allowing glycolysis to continue for as long as the glucose holds out. Blood circulation removes the lactic acid, a byproduct of this fermentation. When the lactic acid cannot be removed as fast as it is produced, your muscles cease to work well. Subjectively, they feel tired or leaden. Try raising and lowering your arm rapidly a hundred times and you will soon experience this sensation. A more efficient circulatory system will let you run longer before the accumulation of lactic acid becomes a problem—this is why people train before they run marathon races—but there is always a point past which lactate production exceeds lactate removal. It is essentially because of this limit that the world record for running a mile is just under 4 minutes, and not significantly less.

Thus there are many ways to close the glycolytic circle, all of which have one thing in common: they involve the accumulation of a reduced molecule. The key energetic step in the glycolytic sequence is the phosphorylation of G3P, a step driven by its oxidation. The hydrogen taken from G3P must have some final home, some molecule that accepts the hydrogen and is reduced. At first that molecule is NAD^+. Ultimately, some other molecule acquires the hydrogen gleaned from G3P; which molecule this is varies from organism to organism.

In fermentations, which are anaerobic processes, the electron generated in the glycolytic breakdown of glucose is donated to an oxidized organic molecule. In aerobic metabolism, in sharp contrast, such electrons are transferred to oxygen, generating ATP in the process.

No single evolutionary solution to the problem of disposing of this hydrogen during prolonged glycolysis has prevailed. The diversity of solutions probably reflects in large measure the very different role of prolonged glycolysis in the lives of different organisms. You run fast only occasionally, whereas a colony of yeast cells may exist in an oxygenless environment for many generations. The diversity of fermentations among organisms is one reflection of the diversity of life itself.

FIGURE 11-5

The conversion of pyruvate to ethanol takes place naturally in grapes left to ferment on vines, as well as in fermenting vats of crushed grapes. The process is carried out by yeasts, and when their conversion increases the ethanol concentration to about 12%, the toxic effects of the alcohol kill the yeasts. What is left is wine. Often wine will be aged in casks to allow time for more complex chemical reactions to occur that subtly alter its flavor. The casks must be airtight to prevent bacterial contamination, since bacteria will further ferment the ethanol to acetic acid (vinegar).

OXIDATIVE RESPIRATION

The mechanism that has evolved in organisms to extract the energy from photosynthetically produced organic molecules depends on a process central to photosynthesis itself, the chemiosmotic generation of ATP. Electrons (and associated hydrogens) are stripped from photosynthetically produced carbon compounds inside the cell of prokaryotes or mitochondrion of eukaryotes and carried to the inner surface of the surrounding membrane, where the electrons stimulate channels in the membrane to pump protons out across the membrane. The chemiosmotic synthesis of ATP is then driven by the diffusion of protons back in. The original electrons and their associated hydrogens remain in the cell, where they eventually combine with dissolved oxygen gas to form water. In considering the extraction of metabolically useful energy by aerobic respiration, also referred to as **oxidative respiration,** we will first focus on the chemical reactions used to harvest electrons and then trace the path of these electrons as they first generate ATP and then form water.

The Oxidation of Glucose: An Overview

In all aerobic organisms the oxidation of glucose, which began in stage c of glycolysis, continues where glycolysis leaves off—with pyruvate. The evolution of this new biochemical process was conservative, as is almost always the case in evolution: the new process is simply tacked onto the old one. The oxidation of pyruvate takes place in two stages: the decarboxylation of (removal of CO_2 from) pyruvate to form acetyl-CoA, and the subsequent oxidation of the acetyl-CoA.

The Decarboxylation of Pyruvate

In an oxidative reaction that yields electrons, one of the three carbons of pyruvate is cleaved off, departing as CO_2 and leaving behind two remnants (Figure 11-6): (1) a pair of electrons and their associated hydrogen, which reduce NAD^+ to NADH; and (2) a two-carbon fragment called an **acetyl group.** This fragment is then added to a cofactor, a carrier molecule called coenzyme A, forming a compound called **acetyl-CoA:**

$$\text{Pyruvate} + NAD^+ + \text{CoA} \rightarrow \text{Acetyl-CoA} + \text{NADH} + CO_2$$

This reaction is very complex, involving three intermediate stages, and is catalyzed by an assembly of enzymes, a **multienzyme complex.** Such multienzyme complexes serve to organize a series of reactions so that the chemical intermediates do not diffuse away or undergo other reactions. Within such a complex, component polypeptides pass the reacting substrate molecule from one enzyme to the next in line, without ever letting go of it. The complex of enzymes that removes CO_2 from pyruvate, called **pyruvate dehydrogenase** (Figure 11-7), is one of the largest enzymes known—it contains 48 polypeptide chains. This reaction contributes significantly to the harvesting of electrons, using them to reduce NAD^+ to NADH; the NADH molecule then carries these electrons to the membrane.

Of far greater significance than the reduction of NAD^+ to NADH, however, is the residual fragment, acetyl-CoA (Figure 11-8). Acetyl-CoA is important because it is the end product of many different metabolic processes. It is produced not only by the decarboxylation of the product of glycolysis as outlined above, but also by decarboxylation of the products of protein catabolism and by the breakdown of lipids in fat utilization. Acetyl-CoA thus provides a single focus for the many catabolic processes of the eukaryotic cell, all of whose resources are channeled into this single metabolite. Acetyl-CoA is the "currency" of oxidative respiration.

Although acetyl-CoA is formed by many catabolic processes in the cell, only a limited number of processes use acetyl-CoA, ones that expend this metabolic currency. Most acetyl-CoA is either directed toward energy storage—it is used in lipid synthesis—or it is oxidized to produce ATP. Which of these two processes occurs depends on the level of ATP in the cell. When ATP levels are high, the oxidative pathway is inhibited and acetyl-

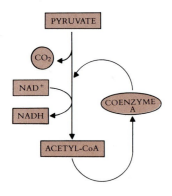

FIGURE 11-6

The decarboxylation of pyruvate. This complex reaction involves the reduction of NAD$^+$ to NADH and is thus a significant source of metabolic energy. Its product, acetyl-CoA, is the starting material for the citric acid cycle.

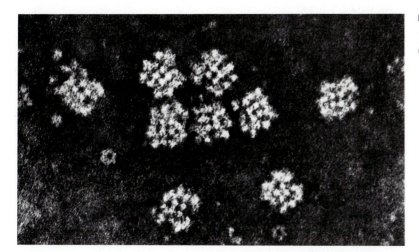

FIGURE 11-7

The pyruvate dehydrogenase complex. (×400,000.)

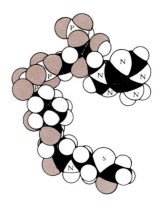

FIGURE 11-8

Acetyl-CoA. This molecule contains a high-energy bond similar to that of ATP.

CoA is channeled into fatty acid biosynthesis. This is why people become overweight when they eat too much. When ATP levels are low, the oxidative pathway is stimulated and acetyl-CoA flows into energy-producing oxidative respiration.

The Oxidation of Acetyl-CoA

Acetyl-CoA is oxidized in several stages: first, its acetyl group binds to a four-carbon carbohydrate; the resulting six-carbon molecule is then passed through a series of electron-yielding oxidation reactions, during which two CO_2 molecules are split off. A four-carbon carbohydrate remains, which is then free to bind another acetyl group. The process is thus a cycle. In each turn of this cycle a new acetyl group replaces the two CO_2 molecules lost, and more electrons are extracted. This cycle is usually referred to as the **citric acid cycle,** after the six-carbon citric acid molecule formed in its first step. However, it is also called the **tricarboxylic acid cycle** (citric acid has three carboxyl groups) or the **Krebs cycle** after the British biochemist Sir Hans Krebs, who discovered it.

The Reactions of the Citric Acid Cycle

The reactions of the citric acid cycle evolved in prokaryotes before the evolution of eukaryotes and still occur in the cytoplasm of many bacteria. These reactions do not occur in the cytoplasm of eukaryotic cells. Rather, the citric acid cycle in eukaryotic cells is carried out entirely within the mitochondria, in the same way as it is within the cytoplasm of bacterial cells. The fact that in eukaryotes these reactions are restricted to mitochondria may reflect the origin of mitochondria from bacteria possessing the citric acid cycle and the origin of eukaryotes from cells lacking it.

The citric acid cycle consists of nine reactions (Figure 11-9). There are three stages to the cycle:

Stage A: Three **preparation** reactions (reactions 1 to 3) set the scene. In the first reaction acetyl-CoA joins the cycle, and in the second two chemical groups are rearranged.

Stage B: In the second stage (reactions 4 to 7) **energy extraction** occurs. Three of the reactions are oxidations in which electrons are removed, and one generates an ATP equivalent directly by substrate level phosphorylation.

Stage C: The third stage (reactions 8 and 9) is one of **regeneration,** in which the original starting compound is re-formed.

The nine reactions together constitute a cycle, which begins and ends with oxaloacetate. At every turn of the cycle, acetyl-CoA enters and is oxidized to CO_2 and H_2O, the electrons being channeled off to drive proton pumps that generate ATP.

FIGURE 11-9

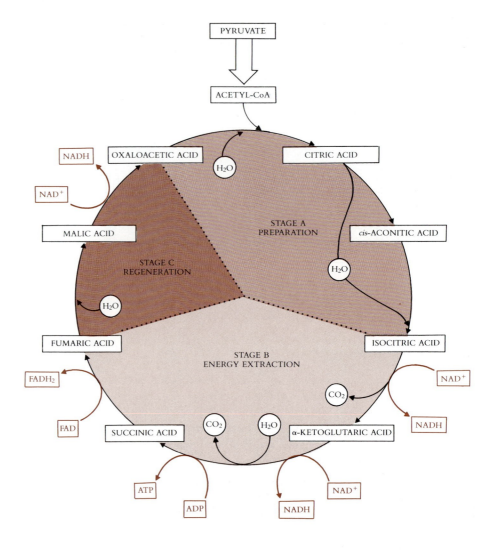

The Products of the Citric Acid Cycle

In the oxidation of glucose described in this chapter, we have generated many electrons and some ATP. The extracted electrons are temporarily stored on electron carrier molecules, NADH and **flavin adenine dinucleotide (FADH$_2$).** Table 11-1 enumerates the molecules of ATP and of electron carriers generated, starting from glucose.

In the process of aerobic respiration the glucose molecule has been consumed entirely. Its six carbons were first cleaved into three-carbon units during glycolysis. One of the carbons of each three-carbon unit was then lost as CO_2 in the conversion of pyruvate to acetyl-CoA, and the other two were lost during the oxidations of the citric acid cycle. All that is left to mark the passing of the glucose molecule is its energy (Figure 11-10), which is preserved in four ATP molecules and the reduced state of twelve electron carriers.

The systematic oxidation of the pyruvate left after glycolysis generates two ATP molecules, which is as many as glycolysis produced. More important, this process harvests many energized electrons, which can then be directed to the chemiosmotic synthesis of ATP.

TABLE 11-1   THE OUTPUT OF AEROBIC METABOLISM

	SUBSTRATE LEVEL PHOSPHORYLATION	OXIDATION	
Glycolysis	2ATP	2NADH	
Decarboxylation of pyruvate ($\times 2$)		2NADH	
Citric acid cycle ($\times 2$)	2ATP	6NADH	2FADH$_2$
Total	4ATP	10NADH	2FADH$_2$

The oxidative respiration of one molecule of glucose proceeds in three stages: glycolysis (which can be anaerobic), the oxidation of pyruvate, and the citric acid cycle. Both glycolysis and the citric acid cycle produce two ATP molecules by substrate level phosphorylation, coupling ATP production to a reaction that involves a large release of free energy. All three processes involve oxidation reactions, that is, reactions that produce electrons. The electrons are donated to a carrier molecule, NAD$^+$ (or in one instance FAD$^+$).

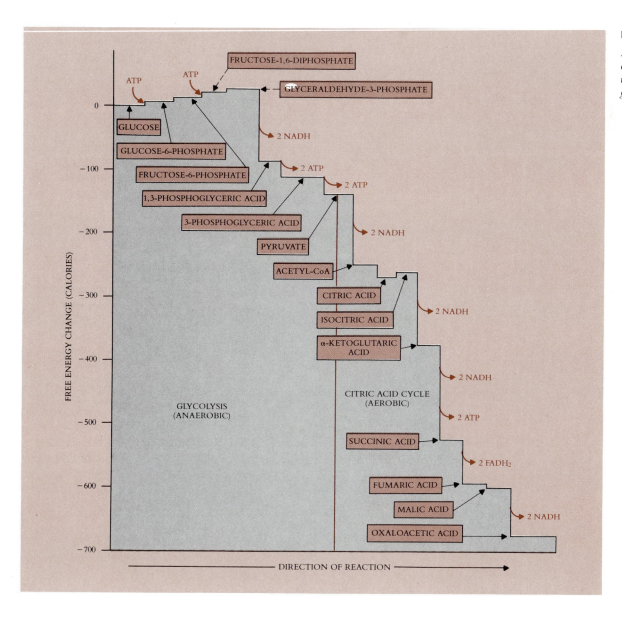

FIGURE 11-10

A summary of the energy released during the metabolism of glucose.

HOW THE CITRIC ACID CYCLE WORKS

The citric acid cycle consists of nine sequential reactions that the cell uses to extract energetic electrons to drive the synthesis of ATP. A two-carbon molecule enters the cycle at the beginning, and two CO_2 molecules and several electrons are given off during the cycle. The citric acid cycle represents one of the best examples of how a biochemical pathway can be organized to accomplish a sophisticated goal. It is clever, efficient, and, as chemistry, beautiful.

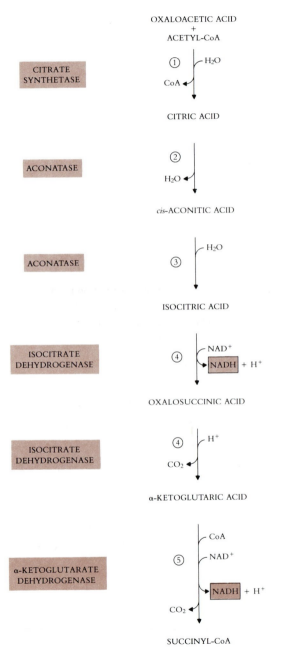

Reaction 1   Condensation

Acetyl-CoA is joined to the four-carbon molecule oxaloacetic acid to form the six-carbon molecule citric acid. This **condensation reaction** is irreversible, committing acetyl-CoA to the citric acid cycle. The reaction is inhibited by high ATP concentrations and stimulated by low ones so that when the cell possesses ample amounts of ATP, the cycle is shut off and acetyl-CoA is channeled instead into fat synthesis.

Reactions 2 and 3   Isomerization

Before the oxidation reactions begin, the hydroxyl group of citric acid must be repositioned. This is done in two steps: a water molecule is first removed, and then added back to a different carbon, so that an —H and an —OH change positions. The resulting molecule is an isomer of citric acid called **isocitric acid.**

Reaction 4   The First Oxidation

In the first energy-yielding step of the cycle, isocitric acid undergoes an **oxidative decarboxylation** reaction, such as you encountered in the conversion of pyruvate to acetyl-CoA. On the surface of the enzyme that catalyzes this reaction, isocitric acid is first oxidized, yielding a pair of electrons that reduce a molecule of NAD^+ to NADH. The reduced carbohydrate intermediate is then decarboxylated by splitting off the central carbon as a CO_2 molecule, yielding a five-carbon molecule called alpha-ketoglutaric acid.

Reaction 5   The Second Oxidation

Alpha-ketoglutaric acid is then oxidatively decarboxylated itself by a multienzyme complex much like pyruvate dehydrogenase. As in that reaction, the fragment left after removal of CO_2 is joined to a molecule of coenzyme A, in this case forming succinyl-CoA, and a pair of electrons is extracted that reduce a molecule of NAD^+ to NADH.

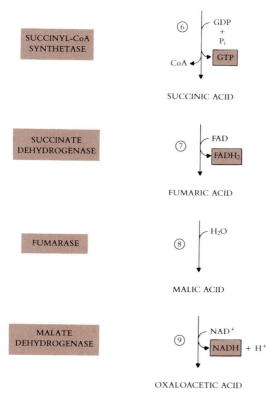

SUCCINYL-CoA
SYNTHETASE

SUCCINATE
DEHYDROGENASE

FUMARASE

MALATE
DEHYDROGENASE

Reaction 6   Substrate Level Phosphorylation

The bond linking the four-carbon succinyl group to the CoA carrier molecule is a high-energy ester linkage, similar in character to the high-energy phosphoanhydride linkage of ATP. Such a bond represents metabolic energy that is "ripe for the picking." In a coupled reaction similar to those encountered in glycolysis, this bond is cleaved, its energy driving the phosphorylation of guanosine diphosphate (GDP). The GTP that results is readily converted to ATP. The four-carbon fragment that remains at the completion of this reaction is called succinic acid.

Reaction 7   The Third Oxidation Reaction

In the third of the four oxidation-reduction reactions of the cycle, succinic acid is oxidized to fumaric acid. Because the free energy change in this reaction is not large enough to reduce NAD^+, a different electron acceptor, **flavin adenine dinucleotide (FAD$^+$),** is used. Unlike NAD^+, FAD^+ is not free to diffuse into the surroundings; rather, it is an integral part of the membrane. Its reduced form, $FADH_2$, contributes electrons directly to the electron transport network of the membrane.

Reactions 8 and 9   Regeneration of Oxaloacetic Acid

In the final two reactions of the cycle a water molecule is added to fumaric acid, and the resulting malic acid is oxidized. The oxidation yields a four-carbon molecule of oxaloacetic acid and two electrons that reduce a molecule of NAD^+ to NADH. Oxaloacetic acid, the molecule with which the cycle began, is now free to bind another molecule of acetyl-CoA and reinitiate the cycle.

USING THE ELECTRONS GENERATED BY THE CITRIC ACID CYCLE TO MAKE ATP

The NADH and FADH$_2$ molecules formed during glycolysis and the oxidation of pyruvate each contain a pair of electrons gained when NADH was formed from NAD$^+$ and FADH$_2$ was formed from FAD$^+$. The NADH molecules carry their electrons to the cell or mitochondrial membrane (the FADH$_2$ is already attached to it) and there transfer them to a complex membrane-embedded protein called **NADH dehydrogenase.** The electrons are then passed on to a series of cytochromes and other carrier molecules (Figure 11-11),

FIGURE 11-11

The structure of cytochrome c. Cytochromes are respiratory proteins that contain "heme" groups, complex iron-containing carbon rings often found in molecules that transport electrons.
A *In the molecular model of cytochrome c the heme group is indicated in color.*
B *Heme is an iron-containing pigment with a complex ring structure and many alternating single and double bonds. Such structures are often associated with the transport of electrons.*

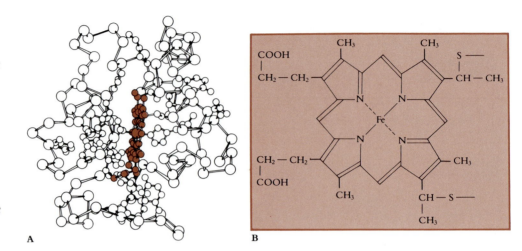

A   B

one after the other, losing much of their energy in the process by driving several transmembrane proton pumps. This series of membrane-associated electron carriers is collectively called the **electron transport chain** (Figure 11-12). At the terminal step of the electron transport chain the electrons are passed to the **cytochrome c oxidase complex,** which uses four of them to reduce a molecule of oxygen gas and form water:

$$O_2 + 4H^+ + 4e^- \rightarrow 2H_2O$$

> **The electron transport chain puts the electrons garnered from the oxidation of glucose to work driving proton-pumping channels.**
> **The ultimate acceptor of the electrons harvested from pyruvate is oxygen gas, which is reduced to form water.**

It is the availability of a plentiful electron acceptor (that is, an oxidized molecule) that makes oxidative respiration possible. In the absence of such a molecule, oxidative respiration was not possible in the early atmosphere. The electron transport chain used in aerobic respiration is similar to, and is thought to have evolved from, the one employed in aerobic photosynthesis.

CHEMIOSMOTIC SYNTHESIS OF ATP

Thus we see that the final product of oxidative respiration is water. This is not an impressive result in itself. Recall, however, what happened in the process of forming that water. The electrons contributed by NADH molecules passed down the electron transport chain, activating three proton-pumping channels. Similarly, the passage of electrons contributed by FADH$_2$, which entered the chain later, activated two of these channels. The electrons harvested from the citric acid cycle have in this way been used to pump a large number of protons out across the cell or mitochondrial membrane, and *that* is the payoff of oxidative respiration.

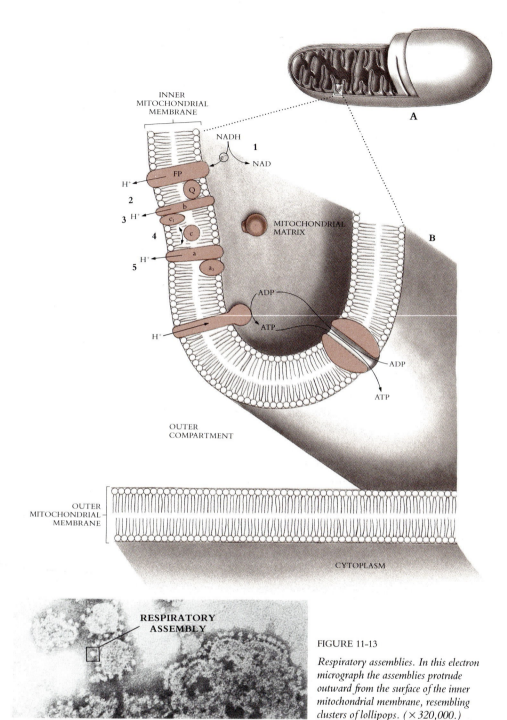

INNER MITOCHONDRIAL MEMBRANE

NADH

1

e⁻

NAD

FP

H⁺

Q

2

b

3 H⁺

c₁

4

c

H⁺

a

5

a₃

H⁺

MITOCHONDRIAL MATRIX

ADP

ATP

ADP

ATP

H⁺

OUTER COMPARTMENT

A

B

OUTER MITOCHONDRIAL MEMBRANE

CYTOPLASM

FIGURE 11-12

The internal structure of a mitochondrion.

A *Cross section through a mitochondrion, showing cristae.*

B *The electron transport chain, a series of electron carriers embedded within the inner mitochondrial membrane:*

1 *The first step in the chain is NADH dehydrogenase (FP), which accepts an electron from an NADH molecule within the mitochondrial matrix, where the reactions of the citric acid cycle take place. NADH dehydrogenase is a complex protein and is itself part of a proton-pumping channel. On accepting electrons, the channel changes conformation, pumps a proton out into the outer compartment, and passes the electrons along to a very nonpolar molecule called* **coenzyme Q** *(Q).*

2 *Being so nonpolar, coenzyme Q is free to diffuse within the hydrocarbon interior of the membrane bilayer and, in doing so, is able to transport the electrons from NADH dehydrogenase to a protein called* **cytochrome b** *(b). Coenzyme Q also picks up the electrons of the membrane-bound FADH₂ and transfers them to cytochrome b as well.*

3 *Cytochrome b serves as a second proton-pumping channel, driving a proton out of the cell on receiving the electron pair and passing the electrons along to the next link in the electron transport chain,* **cytochrome c₁** *(c₁). Cytochromes b and c₁ are both firmly embedded in the membrane.*

4 *When cytochrome c₁ receives the electrons, it passes them on to the next link in the chain, a small, loosely associated cytochrome,* **cytochrome c** *(c), which transports them to the last proton-pumping site.*

5 *Passing from cytochrome c, the electron finally arrives at* **cytochrome a** *(a), which pumps out a third proton. Cytochrome a then donates the electrons to* **cytochrome a₃** *(a₃), which uses four of them and four protons to reduce a molecule of oxygen to water.*

The electron transport chain thus serves to create a gradient in proton concentration, the outer compartment having many more protons than the inner matrix. The resulting diffusion force causes protons to reenter the matrix, passing through special channels. Their passage is coupled to the synthesis of ATP.

RESPIRATORY ASSEMBLY

FIGURE 11-13

Respiratory assemblies. In this electron micrograph the assemblies protrude outward from the surface of the inner mitochondrial membrane, resembling clusters of lollipops. (×320,000.)

Protons recross the membrane, driven by diffusion, through special channels called respiratory assemblies, which are visible in electron micrographs (Figure 11-13) as projections on the inner surface of the membrane. The proton channel proper is made up of polypeptide chains that traverse the membrane. At the inner boundary of the membrane the channel is linked by several other proteins (the "stalk") to a large protein complex. This protein complex synthesizes ATP when protons travel inward through the channel (Figure 11-14).

The electrons harvested from glucose and carried to the membrane by NADH drive protons out across membranes. The return of the protons by diffusion generates ATP.

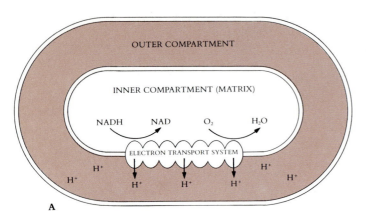

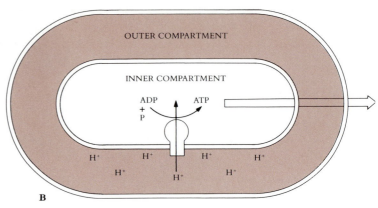

FIGURE 11-14

Proton movement in mitochondria.
A *The electron transport system utilizes energetic electrons from NADH to drive the pumping of electrons out of the inner matrix compartment. The result is a deficit of protons in the matrix and an excess in the outer compartment.*
B *Protons reenter the matrix, driven by diffusion, through special channels; their passage is coupled to the synthesis of ATP from ADP + phosphate (P$_i$). The ATP then passes out of the mitochondria by passive diffusion through special channels (large arrow).*

AN OVERVIEW OF GLUCOSE CATABOLISM: THE BALANCE SHEET

How much metabolic energy does the chemiosmotic synthesis of ATP produce? One ATP molecule is generated chemiosmotically for each activation of a proton pump by the electron transport chain. Thus each NADH molecule that the citric acid cycle produces ultimately causes the production of three ATP molecules, since its electrons activate three pumps. Each FADH$_2$ molecule, activating two of the pumps, leads to the production of two ATP molecules. Because eukaryotes carry out glycolysis in their cytoplasm and the citric acid cycle within their mitochondria, the electrons of the NADH created during glycolysis must be transported across the mitochondrial membrane, consuming one ATP molecule per NADH molecule. Thus glycolytic NADH produces only two ATP molecules in the final balance sheet, instead of three.

The overall reaction for the catabolic metabolism of glucose follows:

$$C_6H_{12}O_6 \text{ (glucose)} + 36NAD^+ + 36ADP + 36P_i + 36H^+ + 6O_2 \rightarrow$$
$$6CO_2 + 36ATP + 6H_2O + 36NADH$$

As you have seen in Table 11-1, 4 of the 36 ATP molecules result from substrate level phosphorylation; the remaining 32 ATP molecules result from chemiosmotic phosphorylation (Table 11-2).

The overall efficiency of glucose catabolism is very high: the aerobic oxidation (combustion) of glucose yields 36 ATP molecules ($-7.3 \times 36 = -263$ kcal/mole). The aerobic oxidation of glucose thus has an efficiency of 263/686, or 38%. Compared to the 2 ATP molecules generated by glycolysis, aerobic oxidation is 18 times more efficient, an enormous improvement (Figure 11-15).

> **Oxidative respiration is almost 20 times as efficient as glycolysis at converting the chemical energy of glucose into ATP. It produces 36 molecules of ATP from each glucose molecule consumed, compared to 2 produced by glycolysis.**

Oxidative respiration is so efficient at extracting energy from photosynthetically produced molecules that its development opened up an entirely new avenue of evolution. For the first time it became feasible for heterotrophic organisms to evolve that derived their metabolic energy exclusively from the oxidative breakdown of other organisms. As long as some organisms produced energy by photosynthesis, heterotrophs could exist solely by feeding on these autotrophs. The high efficiency of oxidative respiration was one of the key conditions that fostered the evolution of nonphotosynthetic heterotrophic organisms.

TABLE 11-2   THE OXIDATION OF GLUCOSE: A BALANCE SHEET

REACTION	NUMBER OF ATP MOLECULES	NUMBER OF NADH MOLECULES
GLYCOLYSIS (Glucose to Pyruvate)		
Phosphorylation of glucose	−1	
Phosphorylation of fructose-6-phosphate	−1	
Oxidation of two glyceraldehyde-3-phosphates		+2
Dephosphorylation of two diphosphoglycerates	+2	
Dephosphorylation of two phosphoenolpyruvates	+2	
PYRUVATE OXIDATION (Pyruvate to Acetyl-CoA)		
Decarboxylation of two pyruvates		+2
CITRIC ACID CYCLE (Acetyl-CoA to CO_2 + e^-)		
Cleavage of two succinyl-CoA	+2	
Oxidation of two isocitric acids		+2
Oxidation of two alpha-ketoglutaric acids		+2
Oxidation of two malic acids		+2
Oxidation of two succinic acids		[+2FADH$_2$]
CHEMIOSMOTIC PHOSPHORYLATION		
Two NADH from glycolysis	+4	
(transported across membrane)		
Two NADH from oxidation of pyruvate	+6	
Six NADH from citric acid cycle	+18	
Two FADH$_2$ from citric acid cycle	+4	
Total	+36	

ATP molecules are generated in two ways when glucose is oxidized during aerobic respiration. Six molecules of ATP are generated by three substrate level phosphorylation reactions, and two are expended preparing for them, for a net yield of 4 ATP molecules. An additional 32 ATP molecules are generated by oxidative phosphorylation: 6 ATP molecules are produced from NADH molecules generated during glycolysis, and 2 ATP molecules are expended transporting these NADH molecules across the membrane, for a net producton of 4 ATP molecules; 6 ATP molecules result from the oxidation of pyruvate, and another 22 ATP molecules result from the oxidative reactions of the citric acid cycle.

FIGURE 11-15

An overview of the ATP extracted from the oxidation of glucose.

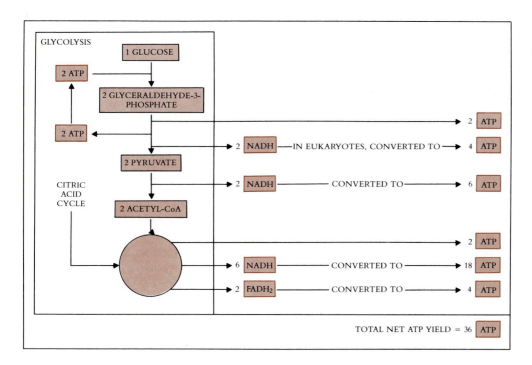

MITOCHONDRIA AND ATP GENERATION

As we discussed in Chapter 6, one of the key events in the evolution of eukaryotes is thought to have been the acquisition by a photosynthetic bacterium of a second symbiotic cell, a heterotrophic prokaryote that could carry out aerobic respiration. Although all of the genes encoding the enzymes of the citric acid cycle seem to have transferred to the nuclear chromosomes of the evolving eukaryotic cell, the shell of the symbiotic, respiring bacterium was maintained as the organelle now called the mitochondrion. Considering that the chemiosmotic production of ATP during aerobic respiration depends on an osmotic pressure driving protons inward across the mitochondrial membrane, you can understand why the outer covering of the mitochondrion would have been maintained intact. The mitochondrion provides a separate compartment from which protons can be pumped, creating the lower interior proton concentration that drives protons back in across ATP-generating channels. If there were a hole in the mitochondrion, then the protons could reenter without passing through the ATP-generating channels. The chemiosmotic generation of ATP could not take place without a closed compartment.

Mitochondria resemble bacteria in size and shape. They are oval and typically about 2 micrometers long. Their interior organization, however, differs from that of a bacterium in one important respect: their "cell" membrane is a double membrane (Figure 11-12). The inner of the two membranes extends into the interior of the mitochondrion (invaginates) in many places, forming highly folded layers called cristae. The presence of the many folds of the cristae has two consequences:

1. *Extended surface area.* The extensive folding of the cristae creates a very large surface area. Embedded within this membrane are protein complexes called **respiratory assemblies** (Figure 11-13), which are special channels that carry out electron transport, chemiosmotic pumping, and associated ATP synthesis. The greatly increased surface area created by the many folds of the inner membrane provides space for a large number of respiratory assemblies and thus greatly increases the oxidative capacity of the mitochondrion.

2. *Intracellular partition.* The invaginations of the inner membrane serve to divide the mitochondrial interior into two compartments. The true interior, within the inner membrane, is called the **matrix;** the space between the inner and outer membranes is called the **intermembrane space.**

The matrix contains within it the enzymes that carry out the reactions of the citric acid cycle. These enzymes are encoded in the cell nucleus and synthesized in the cytoplasm. They are then transported into the interior of the mitochondrion across special transmembrane channels. The synthesis of ATP occurs when the proton-pumping channels of the respiratory assembly actively pump protons from the matrix across the inner membrane into the intermembrane space. This pumping causes the intermembrane space to have a higher proton concentration than the matrix. The resulting diffusion of protons back into the matrix through special channels in the inner membrane drives ATP synthesis (Figure 11-14). Note that the actual synthesis of ATP occurs within respiratory assemblies that project from the inner membrane into the matrix.

The ATP that is synthesized by oxidative respiration is released into the mitochondrial matrix. Since the inner membrane of the mitochondrion is impermeable to ATP or ADP, it is fair to ask how the newly made ATP gets out of the mitochondrion and into the cytoplasm of the cell. This process occurs by facilitated diffusion. A special coupled channel within the inner membrane called **ATP-ADP translocase** binds ADP in the intermembrane space and transports it across the inner membrane into the matrix. There the channel exchanges the molecule of ADP for a molecule of ATP. Since respiration has built up a high concentration of ATP within the matrix, ATP molecules diffuse outward across this channel into the intermembrane space. From there they are free to move across the permeable outer mitochondrial membrane and into the cytoplasm.

AN OVERVIEW OF HETEROTROPHIC METABOLISM

The simplest heterotrophs are those anaerobic bacteria which first evolved glycolysis. These bacteria derive chemical energy by breaking carbon-hydrogen bonds, using the energy to promote ATP production via coupled reactions and disposing of the hydrogen atoms by fermentations. The inefficiency of the process, which extracts only a small percentage of the chemical energy available in organic compounds, probably placed an important constraint on early organisms: most heterotrophs must have lived by consuming photosynthetic organisms rather than other heterotrophs. If a heterotroph preserves only 2% of the energy of a photosynthetic organism that it consumes, then any other population of heterotrophs that consumes this kind of heterotroph has only 2% of the original energy available to it. From this 2% it can garner in turn by glycolysis only a portion, 2%. Thus only 2% of 2%, or 0.04%, of the original amount of available energy is available to it. A very large base of autotrophs would thus be needed to support a small number of secondary heterotrophs.

When organisms became able to extract energy from organic molecules by oxidative respiration (Figure 11-16), the constraint imposed by the low efficiency of glycolysis became far less severe, because oxidative processes are far more efficient. Since the efficiency of oxidative respiration is 38%, about two thirds of the available energy is lost at each trophic step. But a loss of 62% is a great improvement over the loss of 98% associated with anaerobic glycolysis. As a result of this greater efficiency, it became possible for heterotrophs to subsist entirely on other heterotrophs, and for still others to subsist entirely on the heterotroph-eaters. We refer to such a heterotrophic chain as a **trophic chain,** after the Greek word meaning "to eat." You will read more about trophic chains in Part Nine. The length of a trophic chain is dictated by the efficiency of oxidative respiration; most chains involve only three, and rarely four, organisms. Too much energy is lost at each transfer point to allow the chains to become much longer than that.

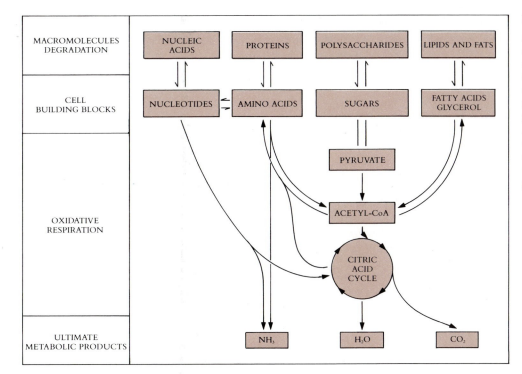

FIGURE 11-16

How cells extract chemical energy. All eukaryotes and many prokaryotes extract energy from organic molecules by oxidizing them. The first stage of this process, tearing down macromolecules into their constituent parts, yields little energy. The second stage, oxidative respiration, extracts energy, primarily in the form of high-energy electrons, and leaves simple inorganic molecules as final end products.

SUMMARY

1. Metabolism is driven by the release of free energy. Reactions that require a net input of free energy are coupled to the cleavage of ATP, which involves a large release of free energy.

2. Organisms acquire ATP from photosynthesis or by harvesting the chemical energy in the bonds of organic molecules.

3. Harvesting energy can be accomplished in two ways: bond rearrangement, in which some reactions involving a large decrease in free energy are coupled to ATP formation; or oxidation, in which electrons are used to drive proton pumps and thus the chemiosmotic synthesis of ATP.

4. Glycolysis harvests chemical energy via bond rearrangement by rearranging the chemical bonds of glucose to form 2 molecules of pyruvate and 2 molecules of ATP.

5. Organisms living in anaerobic environments require a mechanism to dispose of the electron and associated hydrogen produced in the oxidation step of glycolysis. They are donated to one of a number of carbohydrates in a process called fermentation.

6. Aerobic organisms direct electrons harvested from organic molecules to the membrane, where they drive proton pumps. Pyruvate is further oxidized, yielding many additional electrons, which are also channeled to the membrane to drive these pumps.

7. The excess of outside protons created by the proton pumps diffuses back across the membrane through special channels, and the passage of these protons drives the production of ATP.

8. In all, the oxidation of glucose results in the net production of 36 ATP molecules, all but 4 of them produced chemiosmotically.

9. Within all eukaryotic cells, oxidative respiration of pyruvate takes place within mitochondria, which act as closed osmotic compartments from which protons are pumped and into which protons pass by diffusion to create ATP. The ATP then leaves the mitochondrion by facilitated diffusion.

SELF-QUIZ

The following questions pinpoint important information in this chapter. You should be sure to know the answers to each of them.

1. In a coupled reaction in which the free energy of activation of the unfavored reaction is $+7.3$ kcal/mole and the free energy of activation of the favored reaction is -4.0 kcal/mole
   (a) the reaction proceeds in the favored direction
   (b) the reaction runs in the unfavored direction
   (c) the reaction does not proceed

2. The chemiosmotic generation of ATP is carried out by transmembrane proteins through which protons pass. In what direction are these protons moving?
   (a)   out of the cell   (b)   into the cell

3. For each molecule of glucose that passes through glycolysis, the cell acquires how many molecules of ATP?

4. Glycolysis involves an oxidation step, in which a hydrogen atom with a pair of electrons is removed. What happens to the hydrogen atom and its electrons?
   (a)   they diffuse from the cell   (c)   they combine with oxygen to form water
   (b)   they reduce NAD^+ (d)   they combine with carbon dioxide to form carbon compounds

5. In an anaerobic atmosphere prolonged glycolysis cannot proceed without fermentation. True or false?

6. What carbon compound is the end product of glycolysis?

7. In the passage of one glucose molecule through glycolysis and the oxidative cycle, what is the total number of electrons that are stripped away and passed to electron carriers?
(a) 10   (c) 24   (e) 22
(b) 36   (d) 12

8. In the passage of one molecule of glucose through glycolysis and the oxidative cycle, and the subsequent passage of the harvested electrons through the electron transport chain, how many ATP molecules are generated overall?

9. In the chemiosmotic synthesis of ATP, how many molecules of ATP are produced by the protons contributed by a single molecule of NADH?

10. In fermentation the ultimate acceptor of electrons stripped from glucose is an organic molecule. In oxidative respiration the ultimate acceptor of electrons stripped from glucose is _____.

11. Does glycolysis take place inside the mitochondria?

12. What fraction of the total energy available in glucose is extracted by oxidative respiration?
(a) 50%   (b) 38%   (c) 17%   (d) 2%   (e) 100%

THOUGHT QUESTIONS

1. If you poke a hole in a mitochondrion, can it still carry out oxidative respiration? What about fragments of mitochondria? Explain.

2. Why have eukaryotic cells not dispensed with mitochondria, instead placing the mitochondrial genes in the nucleus and incorporating the mitochondrial ATP-producing machinery and proton pumps into the cell's smooth ER?

FOR FURTHER READING

CAPALDI, R.: "The Structure of Mitochondrial Membranes." In Jamieson, J., and D. Robinson, (eds.): *Mammalian Cell Membranes,* Butterworth Publishers, Woburn, Mass., 1977. An advanced description of how the components of the electron transport chain and chemiosmotic ATP synthesis are incorporated into the mitochondrial membrane.

DICKERSON, R.: "Cytochrome *c* and the Evolution of Energy Metabolism," *Scientific American,* March 1980, pages 136-154. A superb description of how the metabolism of modern organisms evolved.

HINKLE, P., and R. MCCARTY: "How Cells Make ATP," *Scientific American,* March 1978, pages 104-125. A good summary of oxidative respiration, with a clear account of the events that happen at the mitochondrial membrane.

LEVINE, M., et al.: "Structure of Pyruvate Kinase and Similarities with Other Enzymes: Possible Implications for Protein Taxonomy and Evolution," *Nature,* vol. 271, pages 626-630, 1978. An advanced article, well worth the effort, that recounts how enzymes of oxidative metabolism may have evolved.

MITCHELL, P.: "Keilin's Respiratory Chain and Its Chemiosmotic Consequences," *Science,* vol. 206, pages 1148-1159, 1978. An account of the development of the chemiosmotic theory, by the man responsible for it.

STRYER, L.: *Biochemistry,* 2nd ed., W.H. Freeman and Company, San Francisco, 1981. Of all undergraduate biochemistry texts, this is the best. Its descriptions of the molecular mechanisms of metabolism are lucid and easy to understand, and the material is up-to-date without being difficult to comprehend.

Chapter 12

PHOTOSYNTHESIS

Overview

Photosynthesis is, along with glycolysis, one of the oldest and most fundamental life processes. It is ultimately responsible for the formation of all of the organic molecules that you, as heterotrophs, use to fuel your lives. Indeed, the aerobic respiration process, which is used to degrade organic molecules in order to obtain energy, appears to have evolved as a short-circuited form of photosynthesis. All of the major types of photosynthesis evolved among the bacteria. The earliest of these apparently split hydrogen sulfide and generated elemental sulfur as a byproduct. More advanced versions of the process added an additional stage that gave photosynthesis the power necessary to split water, generating oxygen as a byproduct and changing the atmosphere of the earth forever.

For Review

Here are some important terms and concepts that you will encounter in this chapter. If you are not familiar with them, you should review them before proceeding.

- **Advent of atmospheric oxygen** (Chapter 3)
- **Chloroplasts** (Chapter 6)
- **Evolution of photosynthesis** (Chapter 10)
- **Electron transport chain** (Chapter 11)
- **Chemiosmotic synthesis of ATP** (Chapter 11)
- **Glycolysis** (Chapter 11)

For all its size and diversity, our universe might never have spawned life, except for one characteristic of overriding importance: it is awash in energy. Everywhere in the universe, matter is continually being converted to energy by thermonuclear processes. The energy resulting from those processes streams from the stars, which include our sun, in all directions. Part of this energy reaches the earth as light, is trapped by living things, and is used by them to drive the synthesis of organic molecules from atmospheric carbon dioxide. This harnessing of the energy in starlight by living organisms is called photosynthesis. As you saw in Chapter 10, the energy that fuels all of the processes that occur in living things derives ultimately from photosynthesis.

In this chapter we will provide a detailed account of photosynthesis, both in an evolutionary and in a functional context. We will review the way in which photosynthesis is thought to have evolved among the photosynthetic bacteria, and we will then conduct a more detailed examination of the particular pattern of photosynthesis that is characteristic of the photosynthetic eukaryotes, which are the plants and algae.

THE BIOPHYSICS OF LIGHT

Where is the energy in light? What is there about sunlight that a plant can use to create chemical bonds? To answer these questions, we need to consider the physical nature of light itself. Perhaps the best place to start is with a curious experiment carried out in a laboratory in Germany in 1887. A young German physicist, Heinrich Hertz, was attempting to verify a highly mathematical theory that predicted the existence of electromagnetic waves. To see whether such waves existed, Hertz first constructed a spark generator in his laboratory—a machine consisting of two shiny metal spheres standing near one another on slender rods. When a very high static electric charge built up on one sphere, sparks would jump across to the other sphere.

After Hertz constructed this system, he set out to investigate whether the sparking would create invisible electromagnetic waves, as predicted by the mathematical theory. On the other side of the room, he placed a thin metal hoop on an insulating stand. This hoop was not a complete circle—it had a very narrow gap at one place where the two ends did not quite meet. When he turned on the spark generator across the room, tiny sparks could be seen to pass across the gap in the hoop. This was the first demonstration of radio waves. But Hertz noted a curious side effect. When light from a window shone on the ends of the hoop, the sparks came more readily. This unexpected facilitation, called the **photoelectric effect,** puzzled investigators for many years.

Especially perplexing was the fact that the strength of the photoelectric effect depended not only on the brightness of the light shining on the gap in the hoop, but also on its wavelength. Short wavelengths were much more effective than long ones in producing the photoelectric effect. This effect was finally explained by Albert Einstein as a natural consequence of the physical nature of light. Light was literally blasting electrons from the metal surface at the ends of the hoop, creating positive ions and thus facilitating the passage of the electronic spark induced by the radio waves. Light consists of units of energy called **photons,** and some of these photons were being absorbed by the metal atoms of the hoop. In this process some of the electrons of the metal atoms were being boosted into higher energy levels, and so ejected from the atoms.

Photons do not all possess the same amount of energy. Some possess a great deal of energy, others far less. The energy content of light is inversely proportional to the wavelength of the light: short-wavelength light contains photons of higher energy than those of long-wavelength light. For this reason the photoelectric effect was more pronounced with light of shorter wavelength: the metal atoms of the hoop were being bombarded with higher-energy photons. Sunlight contains photons of many different energy levels, only some of which our eyes perceive as visible light. The highest-energy photons, at the short-wavelength end of the **electromagnetic spectrum** (Figure 12-1), are gamma rays, with wavelengths of less than 1 nanometer; the lowest-energy photons, with wavelengths of thousands of meters, are radio waves.

As you saw in Chapter 2, electrons occupy discrete energy levels in their orbits around the atomic nucleus. Boosting an electron to a different energy level requires just the right amount of energy, no more and no less; similarly, if you are climbing a ladder, you must raise your foot just so much to climb a rung, not a centimeter more or less. Specific atoms can therefore absorb only certain photons of light—those which correspond to available energy levels. A given atom or molecule has a characteristic range, or **absorption spectrum,** of photons that it is capable of absorbing. The absorption spectrum of a given atom or molecule depends on the electron energy levels that are available in it.

FIGURE 12-1

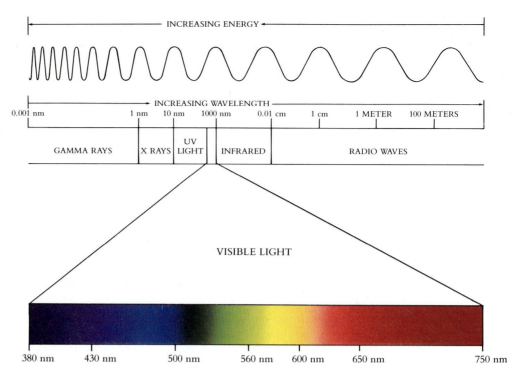

The electromagnetic spectrum. Light is a form of electromagnetic energy and is conveniently thought of as a wave. The shorter the wavelength of light, the greater the energy. Visible light represents only a small part of the electromagnetic spectrum, that between 380 and 750 nanometers.

Not all of the photons in sunlight actually reach the earth's surface. Most high-energy (short-wavelength) photons are absorbed by ozone and oxygen atoms in the upper atmosphere. Similarly most low-energy (long-wavelength) photons are absorbed by water vapor in the air. The light that reaches earth is composed of photons that have not been absorbed—the middle-energy photons in the range between ultraviolet light (wavelength greater than 100 nanometers) and infrared (wavelength less than 1 meter).

CAPTURING LIGHT ENERGY IN CHEMICAL BONDS

The energy of light is "captured" by a molecule that absorbs it, in that the photon of energy is used to boost an electron of the molecule to a higher energy level. Molecules that absorb light are called **pigments,** and organisms have evolved a variety of such pigments. There are two general kinds of pigments:

1. *Carotenoids.* Some pigments are carbon ring compounds linked to chains that possess alternating single and double bonds. The carbon ring, coupled to the network of alternating single and double bonds, can absorb photons of a broad range of energies, although not always with high efficiency. A typical carotenoid is **β-carotene** (Figure 12-2), in which two six-carbon rings are connected by a chain of 18 alternating single and double bonds. Cleavage of this chain into equal halves produces two molecules of **vitamin A.** When oxidized, vitamin A yields **retinal,** the pigment used in human vision.

   When retinal absorbs a photon of light, the resulting electron excitation causes a conformational change in the membrane-associated pigment, which triggers a nerve impulse. Retinal can absorb a broad range of middle-energy photons, from those with wavelengths of 380 nanometers (violet light) to those with wavelengths of 750 nanometers (red light). Photons within this range are called **visible light,** because the human eye can detect them by absorbing them. Other organisms may use different light-absorbing pigments for vision and thus "see" a different portion of the electromagnetic spectrum. Most insects, for example, have eye pigments that absorb at lower wavelengths than does retinal. As a result, bees perceive photons of ultraviolet wavelengths but are blind to those of red light; in the same way you are blind to ultraviolet light.

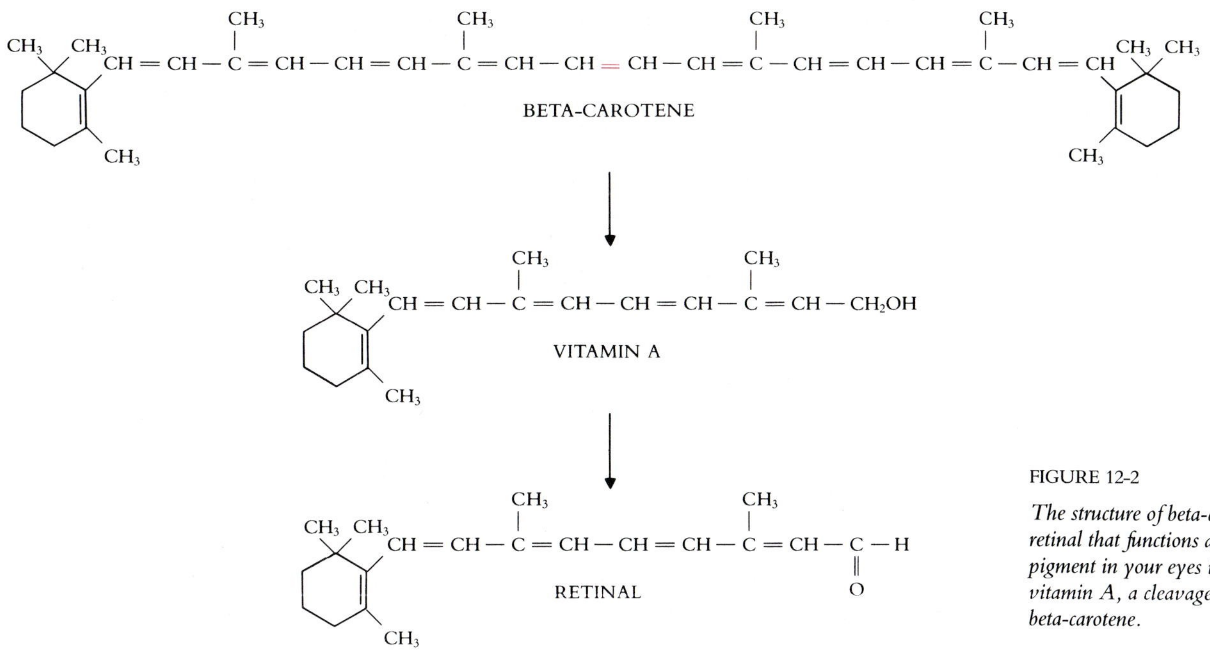

BETA-CAROTENE

VITAMIN A

RETINAL

FIGURE 12-2

The structure of beta-carotene. The retinal that functions as the key visual pigment in your eyes is produced from vitamin A, a cleavage product of beta-carotene.

2. *Chlorophylls.* Other biological pigments called **chlorophylls** absorb photons by carrying out an excitation process analogous to the photoelectric effect. These pigments utilize a metal atom (magnesium), which is ensnared within a network of alternating single and double carbon bonds. Photons absorbed by the pigment molecule excite electrons of the magnesium atom, which are then channeled away through the carbon bond system. In chlorophyll *a* the magnesium lies at the center of a complex ring structure of alternating single and double bonds, called a **porphyrin ring.** Outside the porphyrin ring, several small side groups are attached; these alter the absorption properties of the pigment. Substitution of an aldehyde (CHO) group for one of the CH_3 groups of chlorophyll *a,* for example, produces chlorophyll *b,* with an absorption spectrum shifted to higher wavelengths. The porphyrin ring is attached to a long hydrophobic chain, which probably serves to embed the pigment within the photosynthetic membrane.

Unlike retinal, chlorophylls do not absorb photons across a broad range of energies. As you can see from their absorption spectra (Figure 12-3), chlorophylls *a* and *b* absorb primarily violet and red light. The light between wavelengths of 500 and 600 nanometers is not absorbed by chlorophyll pigments and is reflected. The light reflected from a chlorophyll-containing plant has had all of its middle-energy photons absorbed by the chlorophyll except those in the range of 500 to 600 nanometers. When these photons are subsequently absorbed by the retinal in your eyes, you perceive them as green.

> **A pigment is a molecule that absorbs light. The wavelengths absorbed by a particular pigment depend on the energy levels available in that molecule to which light-excited electrons can be boosted.**

As you will see, at least one group of bacteria carries out photosynthesis using retinal-like pigments. However, all plants and algae, as well as the other photosynthetic bacteria, use chlorophylls as their primary light gatherers. It is reasonable to ask why these photosynthetic organisms do not use a pigment like retinal, which has a very broad absorption spectrum and can harvest light in the wavelength range of 500 to 600 nanometers, as well as at other wavelengths. The most likely hypothesis involves photoefficiency. Retinal absorbs a broad spectrum of light wavelengths but does so with relatively low efficiency;

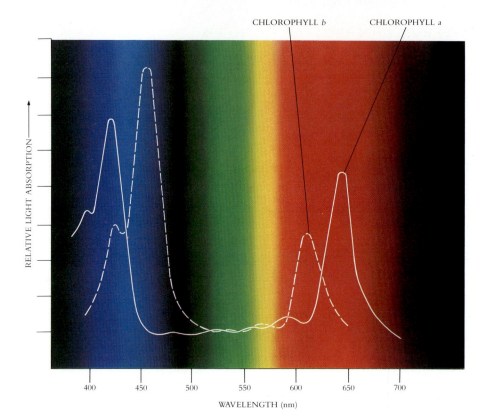

CHLOROPHYLL *b*   CHLOROPHYLL *a*

RELATIVE LIGHT ABSORPTION

WAVELENGTH (nm)

FIGURE 12-3

Chlorophylls absorb predominately violet and red light in two narrow bands of the spectrum.

A *Absorption spectra for chlorophyll a and chlorophyll b. The absorption of chlorophyll b is shifted to absorb more in the blue region than does chlorophyll a.*

B *The structures of chlorophyll a and chlorophyll b. The only difference between the two chlorophyll molecules is the substitution of a —CHO group in chlorophyll b for a —CH$_3$ group in chlorophyll a.*

PORPHYRIN HEAD

PHYTOL TAIL

CHLOROPHYLL *a*

PORPHYRIN HEAD

PHYTOL TAIL

CHLOROPHYLL *b*

chlorophyll, in contrast, absorbs in only two narrow bands, violet and red, but does so with very high efficiency. By using chlorophyll, therefore, plants and most other photosynthetic organisms achieve far higher overall photon capture rates than would be possible with a pigment that allows a broader but less efficient spectrum of absorption.

AN OVERVIEW OF PHOTOSYNTHESIS

Photosynthesis is a single term describing a complex series of events that involves three different kinds of chemical processes (Figure 12-4):

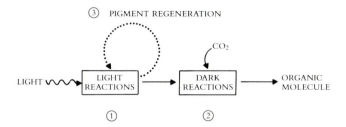

1. In the first process, sunlight is used to harvest electrons, which drive the chemiosmotic generation of ATP. The reactions involved in this process are called the **light reactions** of photosynthesis, since the photosynthetic synthesis of ATP takes place only in the presence of light.
2. The light reactions are followed by a series of enzyme-catalyzed reactions that use this newly generated ATP to drive the formation of organic molecules from atmospheric carbon dioxide (CO_2). These are called the **dark reactions** of photosynthesis, since they will occur as readily in the absence of light as in its presence, as long as ATP is available.
3. Finally, the pigment that absorbed the light in the first place is rejuvenated, ready to initiate another light reaction by absorbing another photon.

Absorbing Light Energy

The light reactions occur on membranes called **photosynthetic membranes.** In photosynthetic bacteria these membranes are part of the cell membrane itself; in plants and algae all photosynthesis is carried out by the evolutionary descendants of photosynthetic bacteria, the chloroplasts, and the photosynthetic membranes occur within the chloroplasts. The light reactions occur in three stages:

1. A photon of light is captured by a pigment. The result of this **primary photoevent** is the excitation of an electron within the pigment.
2. The excited electron is shuttled along a series of electron-carrier molecules embedded within the photosynthetic membrane to a transmembrane proton-pumping channel, where the arrival of the electron induces the transport of a proton out across the membrane.
3. The reentry of protons drives the chemiosmotic synthesis of ATP, just as it does in aerobic respiration (pp. 204–205).

Fixing Carbon

The chemical events that use the ATP generated by the light reactions take place readily in the absence of light, even though they are fueled by the ATP and reducing power produced by the light reactions. In the dark reactions atmospheric CO_2 is incorporated into carbon-containing molecules, a process that is often called **carbon fixation.**

Replenishing the Pigment

The third kind of chemical event that occurs during continuous photosynthesis involves the electron that was stripped from the chlorophyll at the beginning of the light reactions. This electron must be returned to the pigment, or another source of electrons must be used, to replenish the supply of electrons in the pigment. Otherwise, continual electron removal would cause the pigment to become deficient in electrons (bleached), and it could then no longer trap photon energy by electron excitation. As you will see, various organisms have evolved different approaches to this problem during their evolutionary history.

Photosynthesis involves three processes: the use of light-ejected electrons to drive the chemiosmotic synthesis of ATP, the use of the ATP to fix carbon, and the replenishment of the photosynthetic pigment.

THE CHEMICAL CONSEQUENCES OF PHOTOSYNTHESIS

Photosynthesis operates in somewhat different ways in different groups of organisms. Before considering these details, however, let us step back and view the overall process. There are three general chemical consequences of photosynthesis:

1. Carbon dioxide is consumed (fixed), and the carbon atom becomes part of a six-carbon sugar.
2. Protons are extracted from an oxidizable substance such as hydrogen sulfide (H_2S) or water (H_2O), generating elemental sulfur or oxygen gas.
3. Water is formed.

The overall process of photosynthesis may be summarized by a simple oxidation-reduction equation:

$$CO_2 \; + \; 2H_2x \; \xrightarrow{\text{light}} \; (CH_2O) \; + \; H_2O \; + \; 2x$$

CARBON DIOXIDE   REDUCED MOLECULE   NEW ORGANIC MOLECULE   WATER   OXIDIZED MOLECULE

where the x atom may be either sulfur (S), oxygen (O), or nothing at all if H_2 gas serves as the hydrogen donor.

The two most common hydrogen donors in photosynthetic processes are H_2S and H_2O. The former generates elemental sulfur in an anaerobic process, whereas the latter generates oxygen gas.

H_2S (anaerobic)

$$CO_2 + 2H_2S \rightarrow (CH_2O) + H_2O + 2S$$

H_2O (oxygen-forming)

$$CO_2 + 2H_2O \rightarrow (CH_2O) + H_2O + O_2$$

In oxygen-forming photosynthesis H_2O appears on both sides of the equation, but these are *not* the same water molecules. This can be demonstrated by carrying out photosynthesis with water vapor in which the oxygen atom of the water molecules is a heavy isotope. The majority of the heavy oxygen atoms (indicated below by the symbol ★O) end up in oxygen gas and not in water:

$$6CO_2 \; + \; 12H_2{\star}O \; \xrightarrow{\text{light}} \; C_6H_{12}O_6 + 6H_2O \; + \; 6{\star}O_2$$

ATMOSPHERIC CARBON DIOXIDE   WATER VAPOR   SUGAR   WATER   OXYGEN GAS FROM THE ORIGINAL WATER MOLECULES

The strength with which various oxidizable compounds retain their hydrogen atoms differs. The hydrogen atoms can be extracted relatively easily from hydrogen sulfide, whereas water bonds its two hydrogen atoms more strongly. Early photosynthetic processes apparently took the easier route, utilizing H_2S; this route still persists in some bacteria. Improvements in the photosynthetic mechanism of capturing photons developed later to permit the breaking of more powerful H_2O bonds.

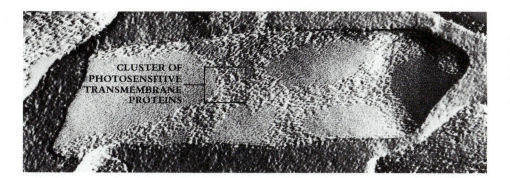

FIGURE 12-5

A freeze-etch preparation of the interior surface of Halobacterium. *In several places where the interior membrane has been stripped away, dense clusters of the photosensitive transmembrane protein can be seen.*

CLUSTER OF
PHOTOSENSITIVE
TRANSMEMBRANE
PROTEINS

HOW LIGHT DRIVES CHEMISTRY: THE LIGHT REACTIONS

Photosynthesis as it now occurs in plants, algae, and some bacteria is the result of a long evolutionary process. As this process took place, new reactions were added to older ones, thus making the process more complex. Much of the evolution has centered on the light reactions, which apparently have changed considerably since they first evolved.

A Primitive Form of the Light Reaction

An unusual genus of bacteria called *Halobacterium,* thought to be very primitive, has evolved a form of photosynthesis that may be the simplest version known. The halobacteria are salt-requiring, aerobic bacteria that live in saturated solutions of sodium chloride in the brine pools of saltworks all over the world. It is probable that the particular version of photosynthesis carried out by halobacteria evolved very early in the history of life on earth.

Halobacteria employ as their photosynthetic pigment the carotenoid retinal, the same pigment that is used in the vertebrate eye. In halobacteria the retinal is covalently attached to a small transmembrane protein (Figure 12-5). When the retinal is illuminated, a hydrogen atom is transported by the attached protein across the membrane and out of the cell. The transmembrane protein thus acts as a proton pump; the light-induced excitation of retinal induces its activity. In bright light the pigment-transmembrane protein complex, called **bacteriorhodopsin,** is capable of pumping several hundred protons per second out of the cell. The cell then synthesizes ATP chemiosmotically as these protons reenter the cell through independent ATP-synthesizing channels.

The two distinctive features of this primitive form of photosynthesis are that (1) it does not employ an electron transport chain, but instead links the photosynthetic pigment directly to the proton pump; and (2) it uses a pigment that absorbs photons of many wavelengths but at low photoefficiency.

Evolution of the Photocenter

The very simple photosynthetic system of the halobacteria represents an unusual evolutionary path. In the main line of evolution, in contrast, there developed among bacteria a photosynthetic system called a **photocenter,** in which the capture of light was achieved by a network of chlorophyll pigments working together. Each chlorophyll molecule within the photocenter was capable of efficiently capturing middle-energy photons. Pigment molecules are arrayed on a protein matrix to form the photocenter, and several of these photocenters are linked together to form a **photosynthetic unit.** The ordered arrangement of numerous pigment molecules permitted the channeling of excitation energy from anywhere in the array to a central point. The assembly of chlorophyll molecules could thus act collectively as a sensitive "antenna" to capture and focus photon energy. Except for the halobacteria, all photosynthetic processes in living organisms capture light in this way.

When light of the proper wavelength strikes any pigment molecule of the photosynthetic unit, the resulting excitation passes from one chlorophyll molecule to another. The excited electron does not transfer physically from one pigment molecule to the next. Instead, the pigment passes along the energy to an adjacent molecule of the photosynthetic unit, after which its electron returns to the low energy level it had before the photon was absorbed.

A crude analogy to this form of energy transfer exists in the initial "break" in a game of pool. If the cue ball squarely hits the point of the triangular array of 15 pool balls, the two balls at the far corners of the triangle fly off, and none of the central balls moves at all. The energy is transferred through these balls to the most distant ones. The protein matrix of the photocenters, in which the molecules of chlorophyll are embedded, serves as a kind of scaffold, holding individual pigment molecules in orientations that are optimum for energy transfer.

The photon-induced excitation energy might bounce around indefinitely within the photosynthetic unit, were it not for the fact that one of the pigment molecules is positioned adjacent to a membrane-bound molecule of ferredoxin. This proximity induces a change in the pigment molecule, increasing its absorption maximum to a wavelength of 700 nanometers. Because of this characteristic, this particular pigment molecule within the array is referred to as **P_{700}**. When P_{700} receives photon-induced excitation energy, it does not shuttle the excitation to yet another pigment molecule, but instead channels the excitation energy to the membrane-bound ferredoxin in the form of electrons.

> **A photocenter is an array of pigment molecules that acts as a light antenna, directing photon energy captured by any of its members toward a single pigment molecule and so amplifying the light-gathering powers of the individual pigment molecules.**

Photocenters represent a significant evolutionary advance over the directly coupled pigment-proton pump system of the halobacteria. Just as a magnifying glass, by focusing light, can generate enough heat energy at the point of focus to burn paper, so the photocenter, by focusing the excitation energy gathered by many pigments on one pigment molecule, can contribute to ferredoxin an electron of much higher energy than would otherwise be possible.

Cyclic Photophosphorylation

The absorption of a photon of light by the photosynthetic unit results in the transmission of an electron from P_{700} to ferredoxin. This process requires something to carry the electron. In biological systems the most commonly employed electron carriers are protons, the combination of proton and electron forming a hydrogen atom. Where does the proton come from in photosynthesis? In the first photosystems that evolved, it was extracted from hydrogen sulfide, H_2S, a process that produced elemental sulfur as a residual byproduct.

The ejection of an electron from P_{700} and its donation to ferredoxin leaves P_{700} one electron short. Before the photosynthetic unit can function again, another electron must be provided to replace it. Much of the subsequent evolution of photosynthesis has involved changes in how this electron replenishment is carried out.

Photosynthetic units are thought to have evolved more than 3 billion years ago in organisms similar to the green sulfur bacteria. These bacteria replenish the photosynthetic pigment by channeling the electron back to the pigment through an electron transport system similar to the one discussed in Chapter 11, in which the electron's passage drives a proton pump and thus promotes the chemiosmotic synthesis of ATP. Viewed overall (Figure 12-6), the path of the electron originally extracted from P_{700} is thus a circle, and the process is called **cyclic photophosphorylation.** Note carefully, however, that the process is not a true circle. The electron that left P_{700} was a high-energy-level electron, boosted to its high

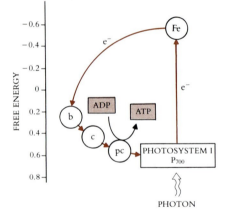

FIGURE 12-6

Cyclic photophosphorylation. Primitive forms of photosynthesis utilized a single photocenter, photosystem I. When an electron was ejected from this photocenter by a photon of light, it was passed to ferredoxin (Fe); from there it passed to a series of three cytochromes (b, c, pc) embedded within the photosynthetic membrane. Its passage to pc served to pump protons out across the membrane, creating a scarcity of protons within. From pc the electron passes back to photosystem I. Its passage is thus a circle, the electron returning to the photocenter from which it was initially ejected.

energy level by the absorption of a photon of energy; the electron that returns has only as much energy left as it had before the photon absorption. The difference in the energy of that electron is the photosynthetic payoff, the energy that drives the proton pump.

Cyclic photophosphorylation evolved in bacteria, which channel the light-excited electron past a proton pump and then return it to the photocenter where it originated.

The light reactions of all other photosynthetic systems are based on this simple cyclic photophosphorylation process. Just as glycolysis is retained as a fundamental component of the respiratory metabolism of all organisms, so cyclic photophosphorylation remains a fundamental component of photosynthesis in the chloroplasts of all plants, algae, and most bacteria.

LIGHT REACTIONS OF PLANTS

For more than a billion years, cyclic photophosphorylation was the only form of photosynthetic light reaction that organisms used. However, it has a fundamental limitation: it is geared only toward energy production, not toward biosynthesis. To understand this point, consider for a moment that the ultimate point of photosynthesis is *not* to generate ATP, but rather to fix carbon—to incorporate atmospheric carbon dioxide into new carbon compounds. Because the molecules produced during carbon fixation (sugars) are more reduced (have more hydrogen atoms) than their precursor (CO_2), it is necessary to provide a source of hydrogens ("reducing power"). Cyclic photophosphorylation does not do this, and bacteria that use this process must scavenge hydrogens from other sources, which is a very inefficient undertaking.

The Advent of Photosystem II

At some point after the appearance of the green sulfur bacteria, organisms evolved an improved version of the photocenter, one that solved the problem of reducing power in a neat and simple way. What they did was to graft a more powerful second photosystem onto the original one, using a new form of chlorophyll, chlorophyll *a*. This great evolutionary advance took place when the cyanobacteria and probably other groups of prokaryotes first evolved, no less than 2.8 billion years ago.

In this second photosystem, called **photosystem II,** molecules of chlorophyll *a* are arranged in the photocenter with a different configuration so that the energy gathered from absorbed photons boosts an electron to higher levels of excitation than in the more ancient bacterial photosystem (called **photosystem I** in plants). As in the bacterial photosystem, energy is transmitted from one pigment molecule to another within the photosynthetic unit until it encounters a particular pigment molecule, positioned near a strong membrane-bound electron acceptor. In photosystem II the absorption peak of this pigment molecule, called P_{680}, is not shifted as it is in the bacterial photosystem, but rather remains at 680 nanometers.

How Photosystems I and II Work Together

Plants use both photosystems—a two-stage photocenter (Figure 12-7). The new photosystem acts first. When a photon jolts an electron from photosystem II, the excited electron is donated to an electron transport chain that passes it along to photosystem I. In its journey to photosystem I the electron drives a proton pump and so generates ATP chemiosmotically (Figure 12-8).

When the electron reaches photosystem I, it has already expended its excitation energy in driving this proton pump and has remaining only the same amount of energy as

FIGURE 12-7

The two-stage photocenter: noncyclic photophosphorylation. Plants that carry out oxygen-forming photosynthesis employ two photocenters. First a photon of light ejects an electron from photosystem II. This ejected electron is replaced from water by a strongly electron-seeking protein called Z, in the process pumping a proton out across the photosynthetic membrane. The ejected electron passes to two cytochromes (Q and pQ) and then to the three cytochromes utilized in cyclic photophosphorylation, b, c, and pc. As in cyclic photophosphorylation, passage of the electron to pc drives a proton pump, causing protons to be passed out across the photosynthetic membrane. From pc the electron passes to photosystem I. When this photosystem absorbs a photon of light, it ejects an electron to ferredoxin (Fe), just as it does in cyclic photophosphorylation, but this time the electron is not passed to the cytochrome chain and returned to photosystem I (because photosystem II donates an electron to photosystem I, it has one to spare, so the ejected electron need not be returned); instead, it passes to a soluble form of ferredoxin, which utilizes it to drive the formation of NADPH.

A great advantage of the two-photocenter form of photosynthesis is that it generates both energy (ATP) and reducing power (NADPH), both of which are required to form organic molecules from carbon dioxide. In cyclic photophosphorylation, by contrast, NADPH must be generated indirectly and much less efficiently.

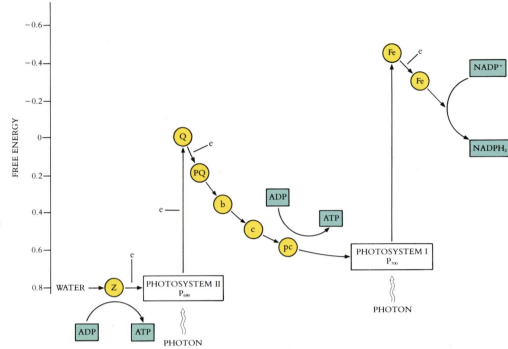

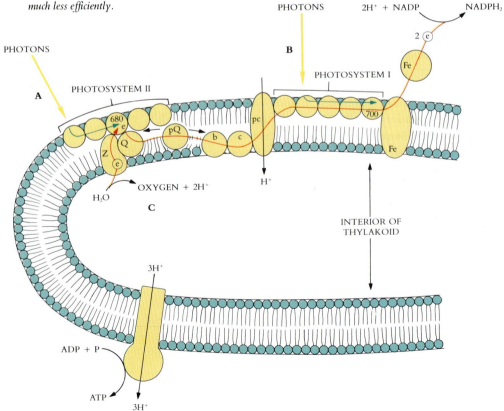

FIGURE 12-8

The photosynthetic electron transport system.
A *When a photon of light strikes a pigment molecule in photosystem II, the energy is channeled to P_{680}, which reacts by donating an electron to a molecule of membrane-bound plastoquinone (Q); the electron then passes to a mobile form of Q, and on to a series of electron carriers called cytochromes, first cytochrome b (b), then cytochrome c (c), and finally plastocyanin (pc), which acts as a proton pump, using the energy supplied by the electron to transport a proton across the membrane into the thylakoid. The resulting proton gradient drives the chemiosmotic synthesis of ATP (lower left).*
B *From pc the electron then passes to photosystem I. When photosystem I absorbs a photon of light, P_{700} passes a high-energy electron to membrane-bound ferredoxin (Fe), which passes it to free ferredoxin. Free ferredoxin donates this electron to drive nitrogen fixation and NADPH generation.*
C *The loss of an electron by P_{680} at the start of the entire sequence causes P_{680} to obtain one from the protein Z, which makes up the loss by capturing an electron from water. Water is split into a proton (H^+), an electron (e) (which is passed to P_{680} by Z), and an OH^- radical; the OH^- radicals are then reassembled into water and oxygen gas.*

the other electrons of this photosystem. Its arrival, however, does give the photosystem an electron that it can afford to lose. Photosystem I now absorbs a photon, boosting one of its pigment electrons to a high energy level. This electron is channeled to ferredoxin and used to generate reducing power in a way we will describe below.

Thus the energy produced in the first photoevent is spent in ATP synthesis; the energy of the second photoevent creates reducing power. These two processes together comprise the light reactions of eukaryotic photocenters.

> **Plants and algae use a two-stage photocenter. First a photon is absorbed by photosystem II, which passes an electron to photosystem I, using its photon-contributed energy to drive a proton pump and so generate ATP. Then another photon is absorbed, this time by photosystem I, which also passes on a photon-energized electron. This second electron is channeled away to provide reducing power.**

The Generation of Reducing Power

As soon as membrane-bound ferredoxin acquires an electron from P_{700}, that electron is passed along to a soluble ferredoxin, which is not embedded within the membrane. The primary outcome of the second photosynthetic event, which occurs in photosystem I, is thus to create a reduced molecule of free ferredoxin. This reduced ferredoxin has a strong tendency to donate its newly acquired electron to other molecules. In organisms that are capable of nitrogen fixation, ferredoxin can contribute a high-energy electron to the nitrogenase complex. In plants and algae ferredoxin contributes its electron to **nicotine adenine dinucleotide phosphate ($NADP^+$)**, generating NADPH.

Why employ a new electron carrier, $NADP^+$, instead of NAD? You will recall that in oxidative respiration, a catabolic process, hydrogens and their associated electrons are transmitted to the electron transport chain by NADH. It is not possible to use this same hydrogen carrier for anabolic (synthetic) reactions without creating metabolic confusion. If the same carrier were used for both catabolic and anabolic processes, electrons intended for synthesis might be channeled instead into ATP production, or vice versa. In the same sense you employ electrical energy both to cool a house with an air conditioner and to heat a house with a furnace, but you must use separate wires to air conditioner and heater lest the efforts of the two systems cancel each other. In photosynthesis the problem does not arise because the hydrogen carrier used in synthetic reactions is "tagged" with a phosphate group, NADP, which serves to distinguish it clearly from NAD: NADP, not NAD, is recognized by the enzymes that carry out synthetic reactions.

Note that this process removes an electron from P_{700}—the electron donated to ferredoxin. Before the photosynthetic unit can function again, another electron must be provided. The electron donated to ferredoxin need not be returned to P_{700}, however, since electrons are continually being fed to P_{700} from P_{680}. Therefore P_{700} does not need to retain the electrons passed to ferredoxin in order for the process to continue.

The Formation of Oxygen Gas

By now you should have begun to wonder about the fate of P_{680}, the pigment of photosystem II. It started the photosynthetic process by donating an electron. How does it make up for this loss if the electron is not returned but is instead expended to synthesize NADPH? As you might expect, an electron is obtained from another source. The loss of the excited electron from photosystem II converts P_{680} into a powerful oxidant (electron seeker), and it proceeds to obtain the required electron from a protein called **Z**. The removal of this electron from Z renders it a strong electron acceptor in turn. This change causes Z to seek electrons by splitting water molecules; Z then funnels these electrons back to P_{680}. Z catalyzes a complex series of reactions in which water is split into electrons (passed to P_{680}), H^+ ions, and OH radicals. The OH radicals are collected and reassembled as water and oxygen gas. The H^+ ions (protons) are exported across the membrane, augmenting the gradient in proton concentration established during the passage of electrons to photosystem I.

The electrons and associated protons that oxygen-forming photosynthesis employs to form reduced organic molecules are obtained from water, the residual oxygen atoms of the water molecules combining to form oxygen gas.

The use of a two-stage photocenter containing both photosystem I and photosystem II thus solves in a simple way the evolutionary problem of how to obtain reducing power for biosynthetic reactions. Anaerobic photosynthesis, even though it provides ATP by cyclic photophosphorylation, does not provide a ready means of generating NADPH. For this reason, organisms that use this form of photosynthesis must make NADPH in a roundabout way and expend a lot of ATP to do so.

Comparing Plant and Bacterial Light Reactions

It is useful to compare the two-stage P_{680}/P_{700} photocenter with the P_{700} photocenter from which it evolved. The removal of an electron from P_{700} yields enough energy to extract hydrogen from H_2S (78 kcal) but not from H_2O (118 kcal). The removal of an electron from P_{680}, by contrast, yields a considerably greater amount of energy, adequate to cleave water, producing gaseous oxygen as a byproduct. In cyanobacteria, algae, and plants, which use this double photocenter, there is no cyclical flow of electrons. Instead, electrons and associated hydrogen atoms are extracted continually from water, eventually being used to reduce $NADP^+$ to NADPH. As a result of stripping hydrogens from water, oxygen gas is continuously generated as a product of the reaction. This photosynthetic process generates all of the oxygen in the air that you breathe (Figure 12-9).

Every oxygen molecule in the air that you breathe was once split from a water molecule by an organism carrying out oxygen-forming photosynthesis.

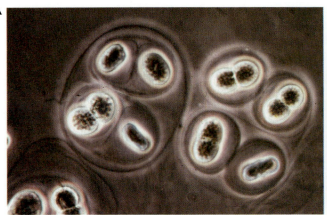

FIGURE 12-9

*All of the oxygen we breathe, indeed all the oxygen in earth's atmosphere, has been generated by photosynthesis. In the roughly 3 billion years between the advent of the cyanobacteria, **A**, and the cultivation of maize, **B**, enough oxygen has accumulated to account for 21% of the atmosphere.*

Some species of cyanobacteria are capable of both nitrogen fixation and oxygen-forming photosynthesis. They are also able to oxidize organic compounds. Such cyanobacteria are, in terms of energy, the most independent organisms on earth, requiring only sunlight, water, atmospheric nitrogen gas, carbon dioxide, and a few minerals to survive.

Accessory Pigments

As you saw earlier in this chapter, chlorophyll *b* differs from chlorophyll *a* only slightly. This slight difference has an important consequence, however. Chlorophyll *b* has an absorption spectrum shifted toward the green wavelengths. It acts as an **accessory pigment** within the photocenter of plants, one that is able to absorb photons that chlorophyll *a* cannot. By doing this, chlorophyll *b* greatly increases the proportion of the photons of sunlight that a plant can harvest.

THE DARK REACTIONS OF PHOTOSYNTHESIS: THE CALVIN CYCLE

The preceding section was concerned with the light reactions of photosynthesis. These reactions use light energy to produce metabolic energy in the form of ATP and reducing power in the form of NADPH. But this is only half of the story. Photosynthetic organisms employ the ATP and NADPH produced by the light reactions to build organic molecules from atmospheric carbon dioxide. This later phase of photosynthesis, which comprises the so-called dark reactions, is carried out by a series of enzymes that are present in the cells of many photosynthetic bacteria and in the chloroplasts of algae and plants. This series of reactions is not carried out by the green sulfur bacteria and so is presumed to have evolved at some point after the green sulfur bacteria did.

Carbon fixation depends on the presence of a molecule to which CO_2 can be attached. The cell produces such a molecule by reassembling the bonds of two of the intermediates of glycolysis, fructose-6-phosphate (F6P) and glyceraldehyde-3-phosphate (G3P), to form a five-carbon sugar, **ribulose 1,5 *bis*-phosphate (RuBP).** This sugar is also called ribulose di-phosphate, RuDP. The ability to form this particular molecule was the major evolutionary breakthrough that made carbon fixation possible.

The dark reactions of photosynthesis form a cycle, a circle of enzyme-catalyzed steps, just as the citric acid cycle is a circular biochemical pathway. This cycle of reactions is called the **Calvin cycle,** after its discoverer, Melvin Calvin. It is diagrammed in Figure 12-10. The cycle has three phases.

Phase 1: Carboxylation

The five-carbon molecule RuBP reacts with atmospheric CO_2 (a kind of reaction called **carboxylation**) to form a transient six-carbon intermediate; this intermediate is immediately cleaved, yielding two molecules of 3-phosphoglyceric acid (3PG, also referred to as PGA), a three-carbon molecule. This reaction is catalyzed by the enzyme **RuBP carboxylase.** RuBP carboxylase is a very large enzyme and comprises more than 16% of the total protein in the chloroplasts of plants and algae, making it the most abundant protein on earth.

Phase 2: Glycolytic Reversal

In a series of five reactions two molecules of 3PG join to form one molecule of F6P, which you will recognize from Chapter 11 as an early intermediate of glycolysis. Indeed, this *is* the glycolytic sequence, run in reverse. The sequence of five reactions, carried out in this direction, requires the input of two molecules of ATP and two molecules of NADPH. The light reactions of photosynthesis provide them.

FIGURE 12-10

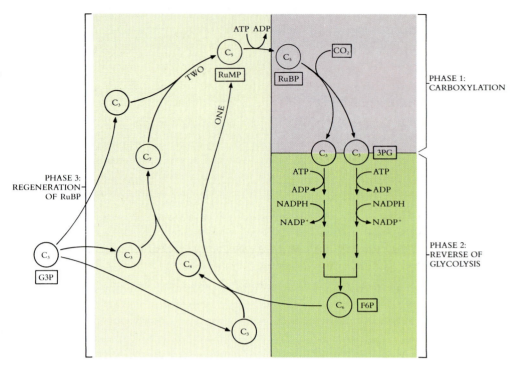

The Calvin cycle. Molecules are represented by the number of carbon atoms they contain. Thus RuBP, a five-carbon molecule, is represented as C_5; 3PG, as C_3; and so on. The cycle consists of three phases:

*In **phase 1** a molecule of carbon dioxide is joined to the five-carbon molecule ribulose 1,5 bis-phosphate (RuBP) to form two molecules of 3-phosphoglycerate (3PG).*

$6C_5$ (RuBP) + $6CO_2$ are converted to $12C_3$ (3PG)

*In **phase 2** the central reaction sequence of glycolysis is run in reverse, the two molecules of 3PG being utilized to form a single molecule of the six-carbon sugar fructose-6-phosphate (F6P).*

$6C_3$ are converted to $3C_6$ (F6P)

$6C_3$ remain as 3PG

*In **phase 3** some of these F6P molecules are combined with 3PG molecules to re-form RuBP, completing the cycle.*

Two of the three molecules of C_6 and all six unused molecules of C_3 are converted to six molecules of C_5 (RuBP), which is what the cycle started with.

One C_6 molecule of F6P remains left over.

Because a molecule of carbon dioxide (one carbon atom) is added for each round of the cycle, six rounds of the cycle result in the addition of six carbon atoms; in other words, there is an extra molecule of F6P generated every six turns of the cycle. This extra molecule, not needed to continue the cycle, is used in the synthesis of other organic molecules.

Phase 3: Regeneration of RuBP

In a complex series of reactions some of the six-carbon F6P molecules are converted back to RuBP. To understand how this is done, examine Figure 12-10 and watch where the carbons go over six rounds of the cycle. Every six turns of the cycle, six CO_2 molecules are incorporated into an extra molecule of F6P above and beyond what is needed to continue the cycle. This F6P molecule is the product of the Calvin cycle, the form in which fixed carbon enters the cell's metabolism. Glucose is readily formed from F6P by reversal of the initial steps of glycolysis.

> **In the dark reactions of photosynthesis RuBP is carboxylated, split, and the products run backward through the glycolytic sequence to form F6P molecules; some are used to reconstitute RuBP, and continue the cycle, while others are converted to glucose and used by the cell to build new molecules.**

To reverse glycolysis it is of course necessary to supply energy, which is provided in the form of two ATP molecules, and a source of hydrogen atoms, which comes in the form of two molecules of NADPH. The hydrogen atoms carried by NADPH provide reducing power. By using a different cofactor, NADP rather than NAD, the cell is able to keep the two processes of photosynthesis and carbohydrate metabolism separate. If this were not the case, one of these processes would short-circuit the other.

Once more we see the conservative nature of the evolutionary process. The dark reactions of photosynthesis evolved by building on glycolysis, a pathway that had evolved long before. Several glycolytic reactions, run in reverse, were coupled to new reactions catalyzing the synthesis of RuBP. Thus the process of carbon fixation represents an elaboration over evolutionary time of reactions that organisms were already capable of performing.

THE CHLOROPLAST AS A PHOTOSYNTHETIC MACHINE

In eukaryotes all photosynthesis takes place in chloroplasts (Figure 12-11). As you will recall from Chapter 6, the internal membranes of chloroplasts are organized into flattened sacs called thylakoids, and numerous thylakoids are stacked on top of one another in arrangements called grana (Figure 12-12). The photosynthetic pigments are bound to proteins embedded within the membrane of the thylakoids: a photosystem I complex consists of 14 chlorophyll *a* molecules bound to a large protein; a photosystem II complex consists of 3 chlorophyll *a* and 3 chlorophyll *b* molecules bound to a much smaller protein.

Each thylakoid is a closed compartment from which protons can be pumped. The thylakoid membrane is impermeable to protons and to most molecules, so transit across it occurs almost exclusively via transmembrane channels. As in mitochondria, the reentry of protons into the thylakoid interior, driven by diffusion, takes place at distinctive ATP-synthesizing proton channels. These channels protrude as knobs on the external surface of the thylakoid membrane, from which ATP is released into the surrounding fluid. The fluid inside the chloroplast, within which the thylakoids are embedded, is called the **stroma.** It contains the Calvin cycle enzymes that catalyze the dark reactions of carbon fixation, using the ATP and NADPH that the photosynthetic activity of the thylakoids produces (Figure 12-13).

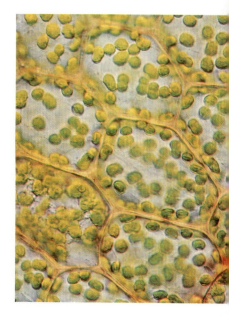

FIGURE 12-11

These moss cells are densely packed with bright green chloroplasts.

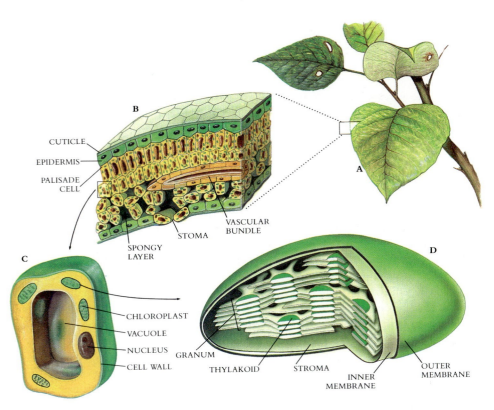

FIGURE 12-12

Journey into a leaf.
A *The leaf shown is from a plant that carries out the Calvin cycle.*
B *In cross section the leaf possesses a thick palisade layer whose cells,* **C,** *are rich in chloroplasts,* **D.**

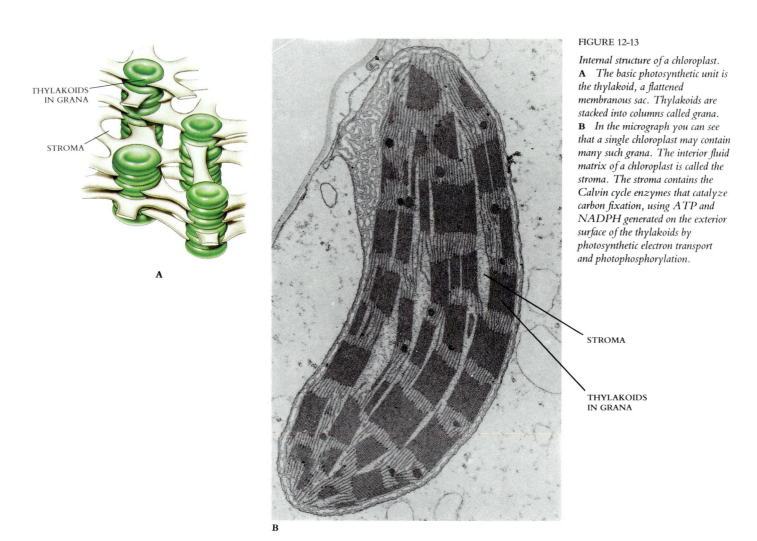

FIGURE 12-13

Internal structure of a chloroplast.
A The basic photosynthetic unit is the thylakoid, a flattened membranous sac. Thylakoids are stacked into columns called grana.
B In the micrograph you can see that a single chloroplast may contain many such grana. The interior fluid matrix of a chloroplast is called the stroma. The stroma contains the Calvin cycle enzymes that catalyze carbon fixation, using ATP and NADPH generated on the exterior surface of the thylakoids by photosynthetic electron transport and photophosphorylation.

THYLAKOIDS IN GRANA

STROMA

A

STROMA

THYLAKOIDS IN GRANA

B

THE LIMIT IMPOSED BY PHOTORESPIRATION

One of the ironies of evolution is that processes evolve to be only as good as they need to be, rather than as good as they potentially might be. Evolution favors not optimum solutions, but rather workable ones that can be derived from others that already exist. New reactions are often grafted onto old ones, as the citric acid cycle was grafted onto glycolysis. Photosynthesis is no exception. Several stages of the glycolytic pathway are used in the Calvin cycle. One of the carry-over enzymes, RuBP carboxylase, the enzyme that catalyzes the key carbon-fixing reaction of photosynthesis, provides a decidedly nonoptimum solution. This enzyme has another activity that interferes with the successful performance of the Calvin cycle: RuBP carboxylase catalyzes the conversion of RuBP to phosphoglyceric acid and the two-carbon molecule phosphoglycolate.

The reason this second activity of RuBP carboxylase is important concerns the fate of the phosphoglycolate that is produced: it is quickly split to form glycolate, which is oxidized by the cell's peroxisomes. In this process, called **photorespiration,** glycolate is oxidized and CO_2 is released without the production of ATP or NADPH. Such a release of CO_2 is called decarboxylation. Because this decarboxylation produces neither ATP nor NADPH, photorespiration acts to undo the work of photosynthesis.

The undesirable oxidative reaction catalyzed by RuBP carboxylase that initiates this chain of events is catalyzed at the same active site that also carries out the carbon-fixing carboxylation reaction so important to photosynthesis. When photosynthesis first evolved, there was little oxygen in the atmosphere, and, because the decarboxylation reaction requires oxygen, there was thus little or no photorespiration. Under these conditions the fact that the active site of RuBP carboxylase was capable of carrying out both reactions presented no problem. Only after millions of years of oxygen buildup in the atmosphere did the competition of CO_2 and O_2 for the same site lead to the problem that photorespiration now poses.

The loss of fixed carbon caused by photorespiration is not trivial. Plants that use the Calvin cycle to fix carbon lose between a fourth and a half of their photosynthetically fixed carbon in this way. The extent of carbon loss depends a great deal on temperature (Figure 12-14), since the oxidative activity of the RuBP carboxylase enzyme increases far more rapidly with temperature than does its carbon-fixing activity. In tropical climates, especially those in which the temperature is often above 28° C, the problem is a very severe one, and it has a major impact on tropical agriculture.

Plants adapted to these warmer environments have evolved a special way to deal with this problem. They expend a considerable amount of ATP to concentrate CO_2 within the cells that carry out the Calvin cycle. Because CO_2 binds to the same place within the RuBP carboxylase active site that O_2 does, high concentrations of CO_2 act to commandeer the available enzyme for the CO_2-fixing reaction.

Some of these plants concentrate CO_2 by carboxylating a three-carbon metabolite, phosphoenolpyruvate, within leaf cells. These leaf cells, called **mesophyll cells,** are freely permeable to CO_2. The resulting four-carbon molecule, oxaloacetic acid, is converted within the mesophyll cells to the citric acid cycle intermediate malic acid, which is transported to an adjacent **bundle-sheath cell.** The malic acid in the bundle-sheath cell is oxidatively decarboxylated to pyruvate, releasing a CO_2 molecule into the interior of the bundle-sheath cell. The bundle-sheath cell is impermeable to CO_2 and therefore holds CO_2 within it (Figure 12-15). Pyruvate returns to the mesophyll cell, where an ATP molecule is expended in changing the pyruvate back to phosphoenolpyruvate, thus completing the cycle (Figure 12-16).

This series of reactions is called the **C_4 pathway,** because the first product of the pathway (oxaloacetic acid) contains four carbons. The enzymes that carry out the Calvin cycle are located within the chloroplasts of bundle-sheath cells, and the increased CO_2 concentration inhibits photorespiration there. Because each CO_2 molecule released into the bundle-sheath cells requires the cleavage of two high-energy ATP bonds in the mesophyll cell,

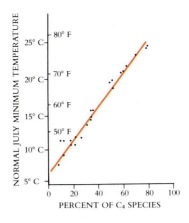

FIGURE 12-14

The occurrence of C_4 grasses as a function of climatic temperature.

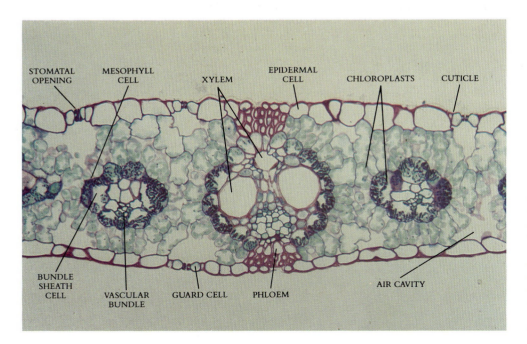

FIGURE 12-15

Structure of a C_4 plant leaf (corn maize). Carbon dioxide is taken up by the mesophyll cells, which incorporate it into C_4 compounds. These C_4 compounds then pass inward to the bundle sheath cells, where carbon dioxide is released. Because these cells are impermeable to carbon dioxide, this process can build up high carbon dioxide concentrations in the bundle sheath cells, which carry out the Calvin cycle. The high concentrations of carbon dioxide in these cells tends to counteract photorespiration, as carbon dioxide and oxygen molecules compete for the same sites on the enzyme RuBP carboxylase.

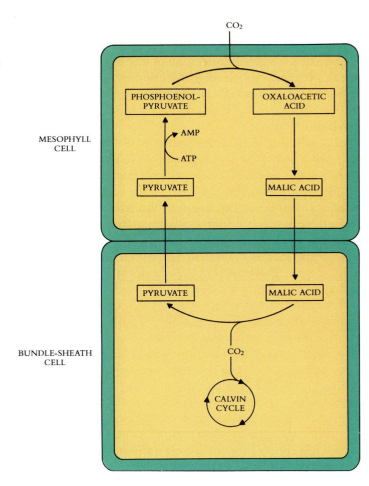

FIGURE 12-16

The path of carbon fixation in C_4 plants.

CO_2

MESOPHYLL
CELL

PHOSPHOENOL-
PYRUVATE

OXALOACETIC
ACID

AMP

ATP

PYRUVATE

MALIC ACID

BUNDLE-SHEATH
CELL

PYRUVATE

MALIC ACID

CO_2

CALVIN
CYCLE

and six such carbon atoms must be fixed to form one six–carbon molecule of glucose, it follows that, viewed overall, 12 additional molecules of ATP are required to form a molecule of glucose. These 12 molecules of ATP significantly increase the cost of forming glucose, from 18 to 30—almost double the cost. In a hot climate in which photorespiration would otherwise remove more than half of the carbon fixed, however, it is the best compromise available. Many of the plants that grow in hot climates are C_4 plants. North American plants are no exception (Figure 12-17).

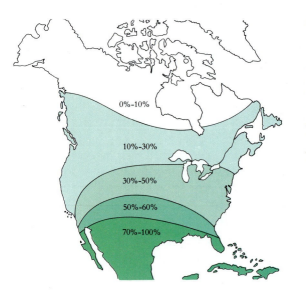

0%–10%

10%–30%

30%–50%

50%–60%

70%–100%

FIGURE 12-17

The distribution of C_4 grass species in North America. Many more C_4 species occur in southern locations, where the average temperatures are higher; at higher temperatures photorespiration wastes more of the products of photosynthesis, and the ability of C_4 plants to counteract photorespiration is more of an advantage.

A LOOK BACK

Perhaps more than any other phase in a cell's life, its metabolism betrays its evolutionary past. This is particularly true of photosynthesis. The two-stage photocenter of modern plants has as its second stage a photosystem that first evolved millions of years earlier, in anaerobic bacteria that utilized hydrogen sulfide rather than water as a source of reducing hydrogen. The Calvin cycle uses part of the ancient glycolytic pathway, run in reverse, to produce glucose. The principal chlorophyll pigments of plants are but simple modifications of bacterial chlorophylls employed for a billion years before by anaerobic photosynthetic bacteria. Thus, by looking at a plant today, we also see its history.

In Chapters 29 through 34 we will examine plants in detail; photosynthesis is but one aspect of plant biology, although an important one. We have treated photosynthesis here, in a section devoted to cell biology, because photosynthesis arose long before plants did. Many of its mechanisms may be seen in other metabolic processes such as aerobic respiration that are shared by all eukaryotes.

SUMMARY

1. Light consists of energy packets called photons; the shorter the wavelength of light, the more energy in its photons. Short and long wavelengths are absorbed in the earth's atmosphere; only middle-range wavelengths reach the earth's surface.

2. When light strikes a pigment, photons may be absorbed by boosting an electron to a higher energy level, an excited state. Most biological photopigments are carotenoids, such as the retinal of human eyes, or chlorophylls, such as those that make grass green.

3. In the primitive photosynthetic system of the halobacteria the photopigment retinal is attached directly to a transmembrane proton pump. Absorption of a photon leads to the pumping out of a proton, and the proton's reentry drives the chemiosmotic synthesis of ATP.

4. Green sulfur bacteria evolved a more efficient means of energy capture, using an array of chlorophyll molecules (a photosynthetic unit) to channel photon excitation energy to one pigment molecule, referred to as P_{700}. P_{700} then donates an electron to an electron transport chain, which drives a proton pump and returns the electron to P_{700}, in a process called cyclic photophosphorylation.

5. The descendants of these bacteria developed a two-stage photocenter in which a new photosystem, photosystem II, was grafted onto the old one. The new photosystem employed a new pigment, chlorophyll *a,* and was able to generate enough energy to utilize H_2O rather than H_2S as a hydrogen source.

6. In organisms with the two-stage photocenter, light is first absorbed by photosystem II, which jolts an electron out of one chlorophyll molecule, P_{680}. The loss of this electron has two effects: (1) the absence of the high-energy electron causes photosystem II to seek another one actively, which results in the eventual splitting of water to obtain it, with O_2 as the by-product; and (2) the high-energy electron is passed to photosystem I, driving a proton pump in the process and thus bringing about the chemiosmotic synthesis of ATP.

7. When the spent electron arrives at P_{700} from P_{680}, the P_{700} pigment absorbs a photon of light, boosting one of its electrons out. This electron is directed to ferredoxin, where it is used to drive the synthesis of NADPH from $NADP^+$, thus providing reducing power.

8. The ATP and reducing power produced by the light reactions are used to fix carbon in a series of dark reactions called the Calvin cycle. In this process RuBP is carboxylated, and the products run backward through the glycolytic sequence to form fructose-6-phosphate molecules, some of which are used to reconstitute RuBP. The remaining molecules enter the cell's metabolism as newly fixed carbon in glucose.

9. RuBP carboxylase, the enzyme that fixes carbon in the Calvin cycle, also carries out an oxidative reaction that burns up the products of photosynthesis, a wasteful process called photorespiration. Many tropical plants inhibit photorespiration by expending ATP to increase the intracellular concentration of CO_2. This process, called the C_4 pathway, nearly doubles the energetic cost of synthesizing glucose. In warm climates, though, the cost is less than the cost of the glucose molecules that would otherwise be converted back to carbon dioxide by photorespiration and lost.

SELF-QUIZ

The following questions pinpoint important information in this chapter. You should be sure to know the answers to each of them.

1. When Hertz turned on his static electricity generator, it generated radio waves that induced tiny sparks to cross a gap in a metal hoop across the room. If Hertz shined a light on the gap, the sparks came more readily. Which was more effective in generating the sparks, long-wavelength or short-wavelength light?

2. Which light contains photons of higher energy, violet light (wavelength = 380 nanometers) or red light (wavelength = 750 nanometers)?

3. A principal difference between carotenoids and chlorophylls is that
   (a) chlorophylls utilize a metal atom to snare electrons, whereas carotenoids do not
   (b) chlorophylls absorb light over a much broader range than do carotenoids
   (c) chlorophylls absorb light at much lower efficiency than do carotenoids
   (d) carotenoids contain carbon ring molecules with alternating single- and double-carbon bonds, whereas chlorophylls do not
   (e) all of the above

4. The immediate result of the primary photoevent is
   (a) the chemiosmotic generation of ATP
   (b) the fixation of one molecule of CO_2
   (c) the replenishment of one pigment electron
   (d) the excitation of one pigment electron

5. In photosynthesis, atmospheric CO_2 is attached to a five-carbon sugar to create a six-carbon molecule. The name of the five-carbon sugar is _____.

6. The dark reactions of photosynthesis require energy in the form of ATP and reducing power in the form of _____.

7. Oxygen-forming photosynthesis utilizes carbon dioxide and water as raw materials and generates organic molecules, water, and oxygen as products. In what molecule does the oxygen atom of the water utilized as raw material end up after photosynthesis?
   (a) CO_2 (d) O_2
   (b) H_2O (e) $C_6H_{12}O_6$
   (c) CH_2O

8. Which form of photosynthesis requires the harvesting of photons of more energy, the one extracting protons from H_2S or the one extracting protons from H_2O?

9. In early forms of photosynthesis, such as those carried out by bacteria, the electron extracted from the P_{700} pigment molecule finally ends up in which molecule?

(a) P_{680} (d) P_{700}

(b) CO_2 (e) H_2O

(c) NADPH

10. In later forms of photosynthesis, such as those carried out by plants, the electron extracted from the P_{700} pigment molecule finally ends up in which molecule?

(a) P_{680} (d) P_{700}

(b) CO_2 (e) H_2O

(c) NADPH

11. Which absorbs greener light, chlorophyll *a* or chlorophyll *b?*

12. Are the enzymes that catalyze the dark reactions of carbon fixation inside the thylakoids or outside the thylakoids?

THOUGHT QUESTIONS

1. What is the advantage of having many pigment molecules in each photocenter for every P_{700}? Why not couple *every* pigment molecule directly to an electron acceptor?

2. Why are plants that consume 30 ATP molecules to produce 1 molecule of glucose (rather than the usual 18 molecules of ATP per glucose) favored in hot climates but not in cold ones? What role does temperature play?

FOR FURTHER READING

BJÖRKMAN, O., and J. BERRY: "High Efficiency Photosynthesis," *Scientific American,* October 1973, pages 80-93. A description of C_4 photosynthesis and other strategies carried out by plants that live in Death Valley, where the problem posed by photorespiration is acute.

CLAYTON, R., and W. SISTROM: *The Photosynthetic Bacteria,* Plenum Publishing Corporation, New York, 1978. An advanced account of how the light reactions of plants and algae have evolved.

DANKS, S.M., E.H. EVANS, and P.A. WHITTAKER: *Photosynthetic Systems: Structure, Function, and Assembly,* John Wiley & Sons, Inc., New York, 1983. A concise, up-to-date review of the mechanism of photosynthesis in both bacteria and plants.

FOYER, C.H.: *Photosynthesis,* John Wiley & Sons, Inc., New York, 1984. A comprehensive treatment of the light and dark reactions of photosynthesis, emphasizing higher plants.

GOVINDJEE and R. GOVINDJEE: "The Absorption of Light in Photosynthesis," *Scientific American,* December 1974, pages 68-87. A good account of how a photocenter works.

MILLER, K.: "The Photosynthetic Membrane," *Scientific American,* October 1979, pages 100-113. An up-to-date description of the way in which the key components of the photosynthetic apparatus are embedded in membranes.

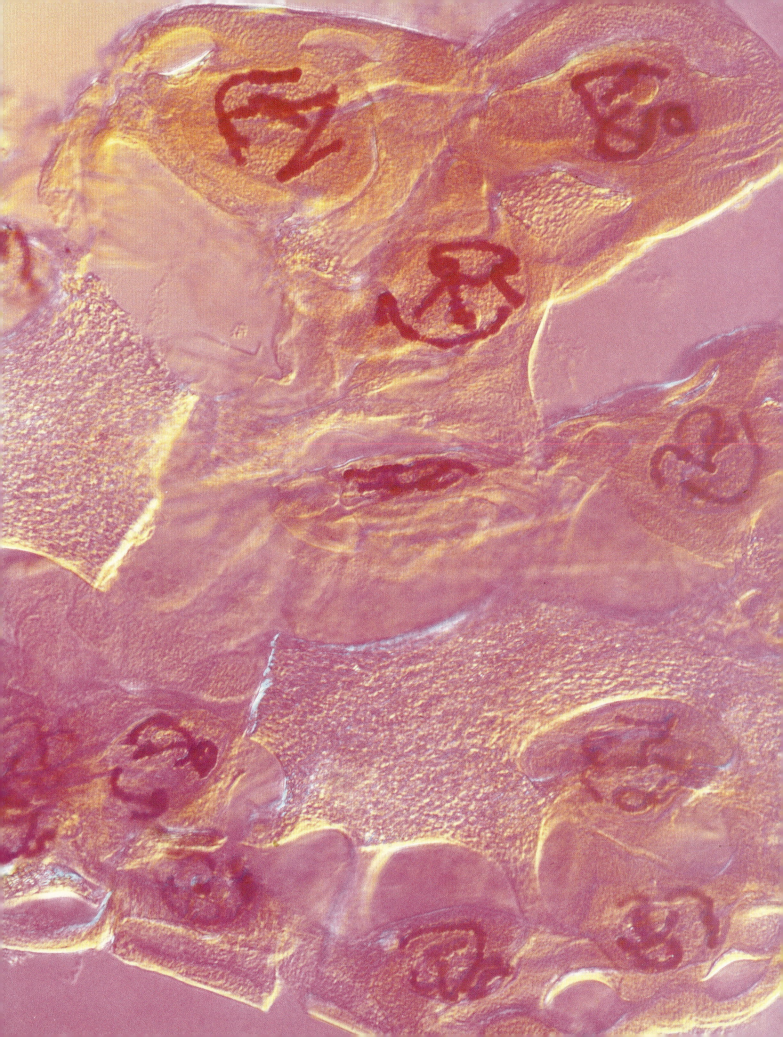

In the hundred years since Mendel investigated heredity in peas, we have learned a great deal about the basic mechanisms of heredity, and we are now beginning to be able deliberately to manipulate genes, such as those of these fruit flies. The fly on the right absorbs stain because investigators have added to its chromosomes a gene that its sister on the left lacks. The manipulation of genes is but one aspect of genetics, one of the most exciting areas of modern biology.

PATTERNS OF INHERITANCE

Chapter 13

Overview

Our understanding of how heredity works dates back to the studies of Gregor Mendel a century ago. Although Mendel was not the first person to observe patterns of inheritance, he was the first to quantify them. By doing so, he was able to perceive the underlying principle that inherited traits are specified by discrete factors that assort independently during gamete formation and then form new combinations during fertilization. We now know that traits such as those studied by Mendel are specified by genes, which are integral parts of chromosomes. Mendel proposed his model before the existence of chromosomes was known—one of the greatest intellectual accomplishments in the history of science.

For Review

Here are some important terms and concepts that you will encounter in this chapter. If you are not familiar with them, you should review them before proceeding.

- **DNA** (Chapter 4)

- **Chromosomes** (Chapters 5 and 8)

- **Mitosis** (Chapter 5)

Every creature now living is an end product of the long evolutionary history of life on earth. You, your dog or cat, the smallest bacterium—all of us share this history. Only humans, however, puzzle about the processes that led to their origin. Our attempts to understand where we come from have continued throughout the course of our history. We are still far from understanding everything about our origins, but we have learned a great deal. Like a partially completed jigsaw puzzle, the boundaries are now known, and much of the internal structure has become apparent. In this chapter we will discuss one piece of that puzzle, the enigma of heredity. Why do the members of a family tend to resemble one another more than they do members of other families?

THE FACES OF VARIATION

Some variation is evident to all of us. Groups of people from different parts of the world are often quite distinct in appearance. Within any of these groups, individuals of one family

FIGURE 13-1

often differ greatly from those of another. Look at your classmates. Rarely will any two of you resemble one another closely. Even your brothers and sisters do not resemble you exactly, unless of course you have an identical twin. Nor are we human beings unique in this respect. Great differences in appearance often exist within other species. For example, there is a bewildering array of varieties and breeds of dogs, of every size and form imaginable, and yet all are still dogs, able to breed with one another and produce puppies (Figure 13-1).

Variation is not surprising in itself. Differences in diet during development can have marked effects on adult appearance, as can variation in the environments that different individuals experience. Many arctic mammals, for example, develop white fur when they are exposed to the cold of winter and dark fur during the warm summer months. The remarkable property of some of the patterns of variation that we can observe—the property that has always fascinated and puzzled us—is that some of the differences that we observe between individuals are inherited, passed down from parent to child.

As far back as there is a written record, such patterns of resemblance among the members of particular families have been noted and commented on. Some of the features by which the members of families resemble one another are unusual, such as the protruding lower lip of the European royal family Hapsburg, which is evident in pictures and descriptions of family members from the thirteenth century onward. Other characteristics are more familiar, such as the common occurrence of red-headed children within families of red-headed parents. Such inherited features, the building blocks of evolution, will be our concern in this chapter.

EARLY IDEAS ABOUT HEREDITY: THE ROAD TO MENDEL

Like many great puzzles, the riddle of heredity seems simple now that it has been solved. The solution was not an easy one to find, however. Our present understanding is the result of a long history of thought, surmise, and investigation. At every stage we have learned more, and as we have done so, the models that we have used to describe the mechanisms of heredity have been changed to encompass new facts.

Two concepts provided the basis for most of the thinking about heredity before the twentieth century:

1. *Heredity occurs within species.* For a very long time people believed that it was possible to obtain bizarre hybrid animals by breeding (**crossing**) widely different species. The minotaur of Cretan mythology, a creature with the body of a bull and the torso and head of a man, is one example. The giraffe was thought to be another; its Latin name, *Giraffa camelopardalis,* suggests that it was believed to be the result of a cross between a camel and a leopard. From the Middle Ages onward, however, people discovered that such extreme crosses were not possible and that variation and heredity occur mainly within the boundaries of a particular species. Species were thought to have been maintained without significant change from the time of their creation.

2. *Traits are transmitted directly.* When variation is inherited by offspring from their parents, *what* is transmitted? The ancient Greeks suggested that parts of the bodies of parents were transmitted directly to their offspring. Reproductive material, which Hippocrates called *gonos,* meaning "seed," was thought to be contributed by all parts of the body; hence a characteristic such as a misshapen limb was transmitted directly to the offspring by elements that came from the misshapen limb of the parent. Information from each part of the body was thought to be passed along independently of the information from the other parts, and the new child was formed after hereditary material from all parts of the parents' bodies had come together.

     This idea was predominant until fairly recently. For example, in 1868 Charles Darwin proposed that all cells and tissues excrete microscopic granules, or "gemmules," that are passed along to offspring, guiding the growth of the corresponding part in the developing embryo. Most similar theories of the direct transmission of hereditary material assumed that the male and female contributions *blended* in the offspring. Thus parents with red and brown hair would be expected to produce children with reddish brown hair, and tall and short parents would produce children of intermediate height.

     Taken together, however, these two concepts lead to a paradox. If no variation enters a species from outside, and the variation within each species is blended in every generation, then all members within a species should soon resemble one another exactly. This does not happen, however. Individuals within most species differ widely from one another, and they differ in characteristics that are transmitted from generation to generation.

How can this paradox be resolved? Actually, the resolution had been provided long before Darwin, in the work of the German botanist Josef Koelreuter. In 1760 Koelreuter carried out the first successful hybridizations of plant species. He was able to cross different species of tobacco and obtain fertile offspring. The hybrids differed in appearance from both of their parent strains. When crosses were made within the hybrid generation, the offspring were highly variable. Some of these offspring resembled plants of the hybrid generation, and a few resembled not the hybrid generation, but rather the original parent strains (that is, the grandparents of these individuals).

Koelreuter's work provided an important clue about how heredity works: the traits that he was studying were capable of being masked in one generation, only to reappear in the next. This pattern is not predicted by the theory of direct transmission. How could a characteristic that is transmitted directly be latent and then reappear? Nor were Koelreuter's traits "blended." A contemporary account records that they reappeared in the next generation "fully restored to all their original powers and properties."

It is important to note that the offspring of Koelreuter's crosses were not all identical to one another. Some resembled the parents of the crosses, whereas others did not; the alternative forms of the traits Koelreuter was studying were distributing themselves among the offspring. A modern geneticist would say the alternative forms of a trait were **segregating** among the progeny of a single mating, meaning that some offspring exhibit one alternative form of a trait (for example, hairy leaves), whereas other offspring from the same mating exhibit a different alternative (smooth leaves). As you will see, the **segregation of alternative forms of a trait** provided the clue that led Mendel to his understanding of the nature of heredity.

Over the next hundred years, Koelreuter's work was elaborated on by other investigators. Prominent among them were English gentleman farmers who were trying to improve varieties of agricultural plants. In one such series of experiments, carried out in the 1790s, T.A. Knight crossed two **true-breeding** varieties of the garden pea, *Pisum sativum* (Figure 13-2), that is, varieties which were uniform from one generation to the next. One of these varieties had purple flowers; the other, white flowers. All of the progeny of the cross had purple flowers. Among the offspring of these hybrids, however, were some plants with purple flowers and others, less common, with white ones. Just as in Koelreuter's earlier studies, a character trait from one of the parents was hidden in one generation, only to reappear in the next.

FIGURE 13-2

The garden pea, Pisum sativum. Easy to cultivate and with many distinctive varieties, the garden pea had been a popular choice as an experimental subject in investigations of heredity as much as a century before Gregor Mendel's investigations.

Early geneticists demonstrated that (1) some forms of an inherited trait can be masked in some generations, but may subsequently reappear unchanged in future generations, (2) forms of a trait segregate among the offspring of a cross, and (3) some forms of a trait are more likely to be represented than are their alternatives.

In these deceptively simple results were the makings of a scientific revolution. Another century passed, however, before the process of segregation of genes was appreciated properly. Why did it take so long? One reason was that early workers did not quantify their results, and a numerical record of results proved to be crucial to the understanding of this process. Knight and later experimenters who carried out other crosses with pea plants noted that some traits had a "stronger tendency" to appear than others, but they did not record the numbers of the different classes of progeny. Science was young then, and it was not obvious that the numbers were important.

MENDEL AND THE GARDEN PEA

The first quantitative studies of inheritance were carried out by an Austrian monk, Gregor Mendel (Figure 13-3). Born in 1822 to peasant parents, Mendel was educated in a monastery and went on to study science and mathematics at the University of Vienna, where he failed his examinations for a teaching certificate. Returning to the monastery (where he spent the rest of his life, eventually becoming abbot), Mendel initiated a series of experiments on plant hybridization in its garden. The results of these experiments would ultimately change our views of heredity irrevocably.

For his experiments Mendel chose the garden pea, the same plant that Knight and many others had studied earlier. The choice was a good one for several reasons:

1. Because many earlier investigators had produced hybrid peas by crossing different varieties, Mendel know that he could expect to observe segregation among the progeny.

FIGURE 13-3

Gregor Johann Mendel. Cultivating his plants in the garden of his monastery in Czechoslovakia, Mendel studied how differences between varieties of peas were inherited when the varieties were crossed with one another. Such experiments with peas had been done often before, but Mendel was the first to appreciate the significance of the results of such crosses.

FIGURE 13-4

One of the differences among varieties of pea plants that Mendel studied affected the shape of the seed. In some varieties the seeds were round, whereas in others they were wrinkled. As you can see here, wrinkled seeds look like dried-out, shrunken versions of the round ones. Mendel established "lines" with round seeds and others with wrinkled seeds, checking the constancy of these characteristics by allowing self-fertilization to proceed for several generations. The line of peas with round seeds that Mendel established always produced only round seeds and never wrinkled ones, and the line that produced wrinkled seeds also did so without error, generation after generation.

2. A large number of true-breeding varieties of peas were available. Mendel initially examined 32. Then, for further study, he selected lines that differed with respect to seven easily distinguishable traits, such as smooth versus wrinkled seeds (Figure 13-4) and purple versus white flowers (a characteristic that Knight had studied 60 years earlier).

3. Pea plants are small, are easy to grow, and have a short generation time. Thus one can conduct experiments involving numerous plants, grow several generations in a single year, and obtain results relatively quickly.

4. The sexual organs of the pea are enclosed within the flower (Figure 13-5). The flowers of peas, like those of most flowering plants, contain both male and female sex organs. Furthermore, the gametes produced by the male and female parts of the same flower, unlike those of many flowering plants, can fuse to form a viable offspring. Fertilization takes place automatically within an individual flower if it is not disturbed. As a result of this process, the offspring of garden peas are the progeny of a single individual. Therefore one can either let self-fertilization take place within an individual flower or remove its male parts before fertilization, introduce pollen from a strain with alternative characteristics, and thus perform an experimental cross.

MENDEL'S EXPERIMENTAL DESIGN

Mendel usually conducted his experiments in three stages:

1. He first allowed pea plants of a given variety to produce progeny by self-fertilization for several generations. Mendel was thus able to assure himself that the forms of traits that he was studying were indeed constant, transmitted regularly from generation to generation. Pea plants with white flowers, for example, when crossed with each other, produced only offspring with white flowers, regardless of the number of generations for which the experiment was continued.

2. Mendel then conducted crosses between varieties exhibiting alternative forms of traits (Figure 13-6). For example, he removed the male parts from a flower of a

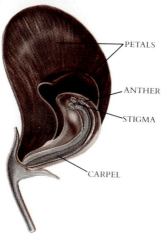

PETALS

ANTHER

STIGMA

CARPEL

FIGURE 13-5

In a pea plant flower the petals enclose the male and female parts, ensuring that self-fertilization will take place unless the flower is disturbed. (Longitudinal section shown.)

FIGURE 13-6

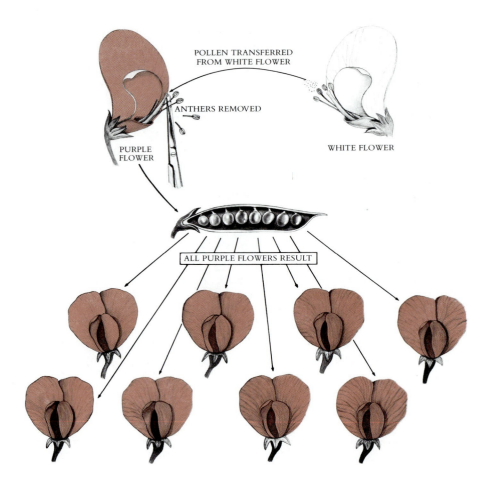

*How Mendel conducted his
experiments. In this cross of two
varieties with different colored flowers,
Mendel pushed aside the petals of a
purple flower and cut off the anthers,
the structures within which the male
gametes are produced, enclosed in
pollen grains—their units of dispersal.
He then collected pollen from a white
flower (he needed only to open the
petals and take out the anthers) and
placed that pollen onto the female parts
of the castrated purple flower, within
which cross-fertilization took place. All
of the seeds in the pod that resulted from
this pollination were hybrids, with a
purple-flowered female parent and a
white-flowered male parent. Planting
these seeds, Mendel observed what
kinds of plants they produced. In this
instance all of them had purple flowers.*

plant that produced white flowers and fertilized it with pollen from a purple-flowered plant. He also carried out the reciprocal cross, by reversing the procedure, using pollen from a white-flowered individual to fertilize a flower on a pea plant that produced purple flowers.

3. Finally, Mendel permitted the "hybrid" offspring produced by these crosses to self-pollinate for several generations. By doing so, he allowed the alternative forms of a trait to segregate among the progeny. This was the same experimental design that Knight and others had used much earlier. But Mendel added a new element: he counted the numbers of offspring of each type and in each succeeding generation. No one had ever done that before. The quantitative results that Mendel obtained proved to be of supreme importance in helping him (and us) understand the process of heredity.

WHAT MENDEL FOUND

When Mendel crossed two contrasting varieties, such as purple-flowered plants with white-flowered plants, the hybrid offspring that he obtained were not intermediate in flower color, as the theory of blending inheritance would have predicted. Instead, the hybrid offspring in every case resembled one of their parents. It is customary to refer to these hybrid offspring as the **first filial,** or **F$_1$ generation.** Thus in a cross of white-flowered with purple-flowered plants the F$_1$ offspring all had purple flowers, just as Knight and others had reported earlier. Mendel referred to the trait that was expressed in the F$_1$ plants as **dominant** and to the alternative form, which was not expressed in the F$_1$ plants, as **recessive.** For each of the seven pairs of contrasting forms of traits that Mendel examined, one of the pair proved to be dominant; the other, recessive.

After individual F_1 plants had been allowed to mature and self-pollinate, Mendel collected and planted the fertilized seed from each plant to see what the offspring in this **second filial, or F_2, generation** would look like. He found, just as Knight had earlier, that some of the F_2 plants exhibited the recessive form of the trait. Latent in the F_1 generation, the recessive alternative reappeared among some of the F_2 individuals.

At this stage Mendel instituted his radical change in experimental design. He counted the numbers of each type among the F_2 progeny. Mendel was investigating whether the proportions of the F_2 types would provide some clue about the mechanism of heredity. For example, he scored a total of 929 F_2 individuals in the cross between the purple-flowered F_1 plants described above. Of these F_2 plants, 705 had purple flowers and 224 had white flowers. Almost precisely one fourth of the F_2 individuals (24.1%) exhibited white flowers, the recessive trait.

Mendel examined seven traits with contrasting alternative forms (Figure 13-7), and the numerical result was always the same: three fourths of the F_2 individuals exhibited the dominant form of the trait, and one fourth displayed the recessive form of the trait. The dominant/recessive ratio among the F_2 plants was always $3:1$.

Mendel went on to examine how the F_2 plants behaved in subsequent generations. He found that the one fourth that were recessive were always true-breeding. In the

TRAIT	DOMINANT VS RECESSIVE	F_2 GENERATION RESULTS		RATIO
		DOMINANT FORM	RECESSIVE FORM	
FLOWER COLOR	 PURPLE × WHITE	705	224	3.15:1
SEED COLOR	 YELLOW × GREEN	6022	2001	3.01:1
SEED SHAPE	 ROUND × WRINKLED	5474	1850	2.96:1
POD COLOR	 GREEN × YELLOW	428	152	2.82:1
POD SHAPE	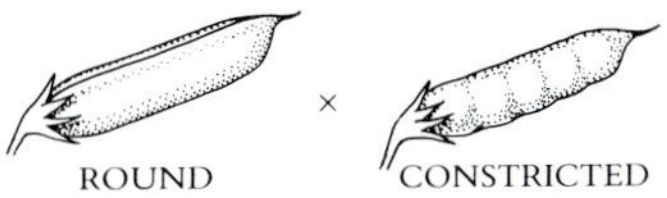 ROUND × CONSTRICTED	882	299	2.95:1
FLOWER POSITION	 AXIAL × TOP	651	207	3.14:1
PLANT HEIGHT	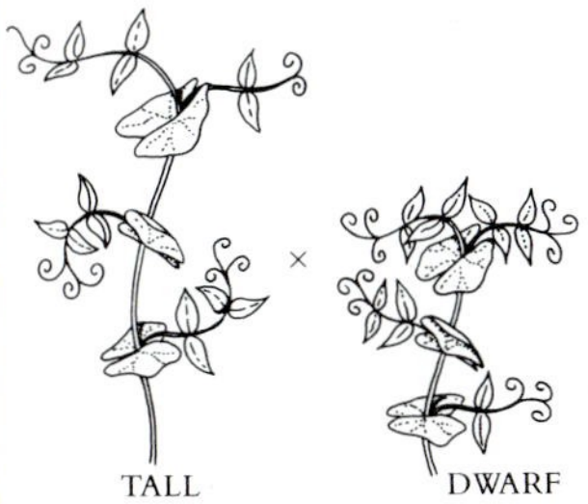 TALL × DWARF	787	277	2.84:1

FIGURE 13-7

The seven pairs of contrasting traits studied by Mendel in the garden pea. In each instance, one quarter of the F_2 generation individuals exhibited the recessive trait and three quarters did not. The ratio of individuals with contrasting traits in the F_2 generation, in this case 3:1, is known as a "Mendelian ratio." Every character that Mendel studied yielded proportions very close to a perfect 3:1 ratio.

cross of white-flowered with purple-flowered plants described above, for example, the white-flowered F_2 individuals reliably produced white-flowered offspring when they were allowed to self-fertilize. By contrast, only one third of the dominant purple-flowered F_2 individuals (one fourth of the total offspring) proved true-breeding, whereas two thirds were not. This last class of plants produced dominant and recessive F_3 individuals in a ratio of 3:1. This result suggested that, for the entire sample, the 3:1 ratio that Mendel observed in the F_2 generation was really a disguised 1:2:1 ratio: one fourth pure-breeding dominant individuals to one half not-pure-breeding dominant individuals to one fourth pure-breeding recessive individuals.

HOW MENDEL INTERPRETED HIS RESULTS

From these experiments Mendel was able to learn four things about the nature of heredity. First, plants exhibiting the traits that he studied did not produce progeny of intermediate appearance when crossed, as a theory of blending inheritance would have predicted. Instead, alternatives were inherited intact, as discrete characteristics that either were or were not seen in a particular generation. Second, for each pair of traits that Mendel examined, one alternative was not expressed in the F_1 hybrids, although it reappeared in some F_2 individuals. *The "invisible" trait must therefore have been latent (present but not expressed) in the F_1 individuals.* Third, the pairs of alternative forms of the traits that Mendel examined segregated among the progeny of a particular cross, some individuals exhibiting one form of a trait, some the other. Fourth, pairs of alternatives were expressed in the F_2 generation in the ratio of three fourths dominant to one fourth recessive. This characteristic 3:1 segregation is often referred to as the **Mendelian ratio.**

To explain these results, Mendel proposed a simple model. It has become one of the most famous models in the history of science, containing simple assumptions and making clear predictions. The model has five elements. For each, we will first state Mendel's assumption and then rephrase it in modern terms:

1. Parents do not transmit their physiological traits or form directly to their offspring. Rather, they transmit discrete information about the traits, what Mendel called **factors.** These factors later act in the offspring to produce the trait. In modern terms we would say that the forms of traits that an individual will express are *encoded* by the factors that it receives from its parents.

2. Each individual, with respect to each trait, contains two factors, which may code for the same form of the trait or which may code for two alternative forms of the trait. We now know that there are two factors for each trait present in each individual because these factors are carried on chromosomes, and each adult individual is *diploid* (from the Greek prefix *di-,* "two"), which means that there are two of each kind of chromosome present in the cells of the individual (therefore there are two factors for each trait). When the individual forms gametes (eggs or sperm), only one of each kind of chromosome is included in each gamete: it is *haploid* (from the Greek *haploos,* "one"). Therefore only one factor for each trait of the adult organism is included in the gamete. Which of the two factors for each trait is included in a particular gamete is random.

3. Not all copies of a factor are identical; the alternative forms of a factor, leading to alternative forms of a trait, are called **alleles.** When two haploid gametes containing exactly the same allele fuse during fertilization, the offspring that develops from that zygote is said to be **homozygous;** when the two haploid gametes contain different alleles, the individual offspring is **heterozygous.**

In modern terminology Mendel's factors are called **genes.** We now know that one of Mendel's "factors," a gene, is composed of a DNA nucleotide sequence. The position on a chromosome where a gene is located is often referred to as a **locus.** Most genes exist in alternative versions, or alleles, with differences at

TABLE 13-1   SOME DOMINANT AND RECESSIVE TRAITS IN HUMANS

TRAIT	PHENOTYPE	TRAIT	PHENOTYPE
RECESSIVE TRAITS		**DOMINANT TRAITS**	
Common baldness	M-shaped hairline receding with age	Mid-digital hair	Presence of hair on middle segment of fingers
Albinism	Lack of melanin pigmentation	Brachydactyly	Short fingers
Alkaptonuria	Inability to metabolize homogentisic acid	Huntington's disease	Degeneration of nervous system, starting in middle age
Red-green color blindnesses	Inability to detect red, or green, wavelengths of light	Phenylthiocarbamide (PTC) tasting	Ability to taste PTC as bitter
Cystic fibrosis	Abnormal gland secretion, leading to liver degeneration and lung failure	Camptodactyly	Inability to straighten the little finger
Duchenne muscular dystrophy	Wasting away of muscles during childhood	Hypercholesterolemia (the most common human Mendelian disorder—1:500)	Elevated levels of blood cholesterol and high risk of heart attack
Hemophilia	Inability of blood to clot	Polydactyly	Extra fingers and toes
Sickle cell anemia	Defective hemoglobin that collapses red blood cells		

one or more nucleotide positions in the DNA. Different alleles of a gene are usually recognized by the change in appearance or function that results from the nucleotide differences.

4. The two alleles, one each contributed by the male and female gametes, do not influence each other in any way. In the cells that develop within the new individual these alleles remain discrete (Mendel referred to them as "uncontaminated"). They neither blend with one another nor become altered in any other way. Thus when this individual matures and produces its own gametes, the alleles for each gene are segregated randomly into these gametes, just as described in point 2 on p. 244.

5. The presence of a particular element does not ensure that the form of the trait encoded by it will actually be expressed in the individual carrying that allele. In heterozygous individuals only one (dominant) allele achieves expression, the other (recessive) allele being present but unexpressed. In modern terms each element encodes the information that specifies an alternative form of a trait, rather than containing the trait itself. The presence of information does not guarantee its expression, as any undergraduate student taking an examination appreciates. To distinguish between the presence of an element and its expression, modern geneticists refer to the constellation of alleles that an individual contains as its **genotype** and to the physical appearance of an individual as its **phenotype.**

Mendel's results were clear because he was studying alternatives that exhibited complete dominance. Many traits in humans also exhibit dominant or recessive inheritance, in a manner similar to the traits Mendel studied in peas (although the majority do not). Albinism, for example, is a rare recessive condition that is characterized by the complete absence of the pigment melanin (Figure 13-8). Table 13-1 lists a few of the many human traits that are known to exhibit recessive or dominant alleles.

> **The phenotype of an individual is the observable outward manifestation of the genes that it carries. The phenotype is the end result of the functioning of the enzymes and proteins encoded by the genes of the individual, its genotype. The genotype is the blueprint; the phenotype, the realized outcome.**

These five elements, taken together, constitute Mendel's "model" of the hereditary process. Does Mendel's model predict the result that he actually obtained?

FIGURE 13-8

An African albino. This boy lacks all melanin pigment. His albinism is due to a recessive allele for which he is homozygous. One individual in every 130 carries a copy of this allele, but albino children are not common; it is unlikely that both parents of a family would be individuals that carry the allele. Even in the rare instances when such a marriage did take place, only a quarter of the children would be expected to be double-recessive albino individuals. About one child in 17,000 is born an albino.

FIGURE 13-9

1 *The parent from the white-flowered line is homozygous for the recessive allele w, and so produces only w gametes; similarly, the parent from the purple-flower line is homozygous for the dominant allele W and produces only W gametes. Thus the only possible offspring are Ww heterozygotes, purple in color. These individuals are known as the F₁ generation.*
2 *When two heterozygous F₁ individuals cross, each is capable of producing both W and w gametes. Three kinds of offspring are thus possible: WW homozygotes (purple), Ww heterozygotes (purple), which may form two ways, and ww homozygotes (white). Thus among these individuals, known as the F₂ generation, the ratio of dominant type to recessive type is 3:1.*

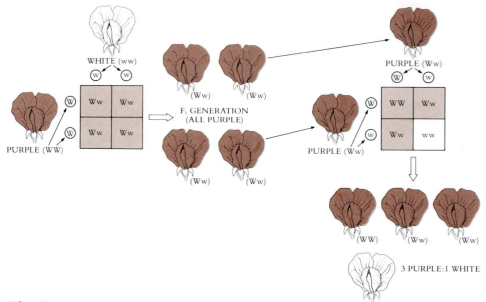

FIGURE 13-10

A Punnett square. If one arrays the different possible types of female gametes along one axis of a square and the different possible types of male gametes along the other, then each potentially different kind of zygote can be represented as the intersection of a vertical (female-gamete) and horizontal (male-gamete) line.

	(MALE) POLLEN	
	W	w
W	WW (PURPLE)	Ww (PURPLE)
(FEMALE) EGGS		
w	Ww (PURPLE)	ww (WHITE)

The F₁ Generation

Consider again Mendel's cross of purple-flowered with white-flowered plants. We will assign the symbol *w* to the recessive allele, associated with the production of white flowers, and the symbol *W* to the dominant allele, associated with the production of purple flowers. By convention, genetic traits are usually assigned a letter symbol referring to their less common state, in this case the letter "W" for **w**hite flower color. The recessive allele (white flower color) is written in lower case, as *w*; the alternative dominant allele (purple flower color) is assigned the same symbol in upper case, *W*.

In this system the genotype of an individual that is true-breeding for the recessive white-flowered trait would be designated *ww*. In such an individual both of the copies of the allele specify white flowers. Similarly, the genotype of a true-breeding purple-flowered individual would be labeled *WW*, and a heterozygote would be designated *Ww* (the dominant allele is usually written first). Using these conventions, and denoting a cross between two lines with ×, Mendel's original cross can be symbolized as *ww* × *WW* (Figure 13-9). Since the white-flowered parent can produce only *w* gametes and the purple-flowered parent can produce only *W* gametes, the union of an egg and a sperm from these parents can produce only heterozygous *Ww* offspring in the F₁ generation. Because the *W* allele is dominant, all of these F₁ individuals are expected to have purple flowers. The *w* allele is present in these heterozygous individuals, but it is not phenotypically expressed.

The F₂ Generation

When F₁ individuals are allowed to self-fertilize, the *W* and *w* alleles segregate at random during gamete formation. Their subsequent union at fertilization to form F₂ individuals is also random, not being influenced by which alternative alleles the individual gametes carry. What will the F₂ individuals look like? The possibilities may be visualized in a simple diagram called a **Punnett square,** named after its originator, the English geneticist Reginald Crundall Punnett (Figure 13-10). Mendel's model, analyzed in terms of a Punnett square, clearly predicts that in the F₂ generation one should observe three-fourths purple-flowered plants and one-fourth white-flowered plants, a phenotypic ratio of 3:1.

Further Generations

As you can see in Figure 13-10, there are really three kinds of F₂ individuals: one fourth are pure-breeding *ww* white-flowered individuals, one half are heterozygous *Ww* purple-flowered individuals, and one fourth are pure-breeding *WW* purple-flowered individuals. The 3:1 phenotypic ratio is really a disguised 1:2:1 genotypic ratio.

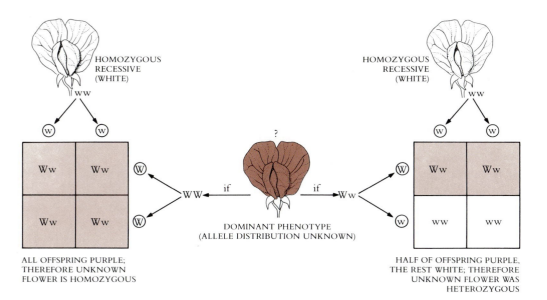

HOMOZYGOUS RECESSIVE (WHITE) ww

HOMOZYGOUS RECESSIVE (WHITE) ww

Ww	Ww
Ww	Ww

Ww	Ww
ww	ww

WW ← if

DOMINANT PHENOTYPE (ALLELE DISTRIBUTION UNKNOWN)

if → Ww

ALL OFFSPRING PURPLE; THEREFORE UNKNOWN FLOWER IS HOMOZYGOUS

HALF OF OFFSPRING PURPLE, THE REST WHITE; THEREFORE UNKNOWN FLOWER WAS HETEROZYGOUS

FIGURE 13-11

A testcross. To determine whether an individual exhibiting a dominant phenotype, such as purple flowers, is homozygous or heterozygous for the dominant allele (both such individuals would be expected to exhibit purple flowers), Mendel crossed the individual in question to a plant which he knew to be double recessive, in this case a plant with white flowers. If the purple-flowered individual being tested was homozygous dominant, then all of the progeny would have been heterozygotes with purple flowers; if the purple-flowered individual being tested was heterozygous for the gene controlling flower color, half of its progeny would have been homozygous for the recessive allele and therefore white-flowered.

THE TESTCROSS

To test his model further, Mendel devised a simple and powerful procedure called the **test-cross.** Consider a purple-flowered individual: is it homozygous or heterozygous? It is impossible to tell simply by looking at it. To learn its genotype, you must cross it with some other plant. With what kind of plant? If you cross it with a homozygous dominant individual, all of the progeny will show the dominant phenotype whether the test plant is homozygous or heterozygous. It is also difficult (but not impossible) to distinguish between the two possible test plant genotypes by crossing with a heterozygous individual. If you cross the test individual with a homozygous recessive individual, however, the two possible test plant genotypes give totally different results (Figure 13-11):

Alternative 1 **Test individual homozygous**

$WW \times ww$: all offspring have purple flowers *(Ww)*

Alternative 2 **Test individual heterozygous**

$Ww \times ww$: one half of offspring have white flowers *(ww)*

To test his model, Mendel crossed heterozygous F_1 individuals back to the parent homozygous for the recessive trait. He predicted that the dominant and recessive traits would appear in a 1:1 ratio:

Gametes of homozygous recessive parent

Gametes of F_1 individual

	w
W	Ww
w	ww

For each pair of alleles that he investigated, Mendel observed phenotypic backcross ratios very close to 1:1, just as his model predicted that he would.

Mendel's model thus accounted in a neat and satisfying way for the segregation ratios that he had observed. Its central assumption—that alternative alleles segregate independently of one another—has since been verified in countless other organisms, and it is commonly referred to as **Mendel's First Law of Heredity,** or the **Law of Segregation.** As you will see in Chapter 14, the segregational behavior of alternative alleles has a simple physical basis, one that was unknown to Mendel. It is a tribute to the intellectual power of Mendel's analysis that he arrived at the correct scheme with no knowledge of the cellular mechanisms of inheritance: neither chromosomes nor meiosis, the subjects of Chapter 14, had yet been described.

DID MENDEL CHEAT?

Mendel's results fit his prediction of a 3:1 ratio of dominant to recessive individuals very well—too well, some have argued. An analysis of Mendel's data by the father of modern statistics, Sir Ronald Fisher, suggested that there might indeed be a problem. Although it is possible to obtain precisely one-fourth recessive individuals in a given cross, just by good luck—just as one may flip precisely 50 heads in 100 flips of a coin—one would not expect perfection in every cross, such as Mendel reported. Try four coin-flipping experiments of 100 flips each; you will not get exactly the same proportion of heads each time.

Subsequent investigations of Mendel's experimental notebooks do not bear out the suggestion that he "cooked" his data. Instead, it appears that he simply kept counting pea plants until he obtained exact ratios. Statistics had not yet been invented, and it was not obvious that sample sizes should not be allowed to vary. Mendel, in looking for a clear result, violated what has since become a standard rule of experimental analysis: sample size should be independent of result.

Mendel's inadvertent error in no way alters the validity of his result. Other investigators who followed Mendel obtained 3:1 ratios almost as perfect as Mendel's (see table) but without violating statistical order.

Mendel's original paper describing his experiments, published in 1866, remains charming and interesting to read. His explanations are clear, and the logic of his arguments is presented in lucid detail. Unfortunately, Mendel failed to arouse much interest in his findings, which were published in the journal of the local natural history society. Only 115 copies of the journal were sent out, in addition to 40 reprints, which Mendel distributed himself. Only the German botanist Carl Naegeli was interested enough to correspond with Mendel about his findings. Naegeli believed that Mendel was wrong; he was convinced that the offspring of all hybrids must be variable. Although Mendel's results did not receive much notice during his lifetime, in 1900, 16 years after his death, three different investigators independently rediscovered his pioneering paper. They came across it while they were searching the literature in preparation for publishing their own findings (Table 13-2), which were similar to those Mendel had quietly presented more then three decades earlier.

> **Mendel's First Law states that (1) the alternative forms of a trait encoded by a gene are specified by alternative alleles of that gene and are discrete (do not blend in heterozygotes); (2) when gametes are formed in heterozygous diploid individuals, the two alternative alleles segregate from one another; and (3) each gamete has an equal probability of possessing either member of an allele pair.**

PROBABILITY AND ALLELE DISTRIBUTION

Many, although not all, alternative alleles produce discretely different phenotypes. Mendel's pea plants were either tall or dwarf, were green or white, and had wrinkled or smooth seeds. The eye color of a fruit fly may be red or white; the skin color of a human, pigmented or albino. When the number of alternative phenotypes is two, rather than three or some other number, the distribution of phenotypic types seen among the progeny of a cross is referred to as a **binomial distribution**.

A COMPARISON OF THE RESULTS OF EARLY GENETICISTS

GENETICIST	SEED SHAPE			SEED COLOR		
	NUMBER ROUND	NUMBER WRINKLED	PERCENT WRINKLED	NUMBER YELLOW	NUMBER GREEN	PERCENT GREEN
Mendel (1865)	5474	1850	25.2	6022	2001	24.9
Correns (1900)				1394	453	24.5
Tschermak (1900)	884	288	24.6	3580	1190	24.9
Hurst (1904)	1335	420	23.9	1310	445	25.4
Bateson (1905)	10,793	3542	24.8	11,903	3903	24.7
Lock (1909)	620	197	24.1	1438	514	26.2
Darbishire (1909)				109,060	36,186	24.9
Winge (1924)				19,195	6533	25.4

Two of the crosses carried out by Mendel were widely repeated by other students of heredity after Mendel's results were rediscovered in 1900. Their results were compatible with Mendel's prediction that the ratio of dominant to recessive traits in each cross would be 3:1—that the recessive traits, green and wrinkled, would appear in 25% of the progeny.

TABLE 13-2   THE SEXES OF CHILDREN IN HUMAN FAMILIES EXHIBIT A BINOMIAL DISTRIBUTION

COMPOSITION OF FAMILY	ORDER OF BIRTH	CALCULATION	PROBABILITY	
3 boys	♂ ♂ ♂	$p \times p \times p$	p^3	
2 boys and 1 girl	♂ ♂ ♀ ♂ ♀ ♂ ♀ ♂ ♂	$p \times p \times q$ $p \times q \times p$ $q \times p \times p$	p^2q p^2q p^2q	$3p^2q$
1 boy and 2 girls	♀ ♀ ♂ ♀ ♂ ♀ ♂ ♀ ♀	$q \times q \times p$ $q \times p \times q$ $p \times q \times q$	pq^2 pq^2 pq^2	$3pq^2$
3 girls	♀ ♀ ♀	$q \times q \times q$	q^3	

To illustrate the distribution of phenotypes that will occur in a cross as a result of the segregation of two alternative alleles, consider the distribution of sexes in human families that results from the segregation of particular sex-determining chromosomes. Imagine that you choose to have three children. Let the probability of having a boy at any given birth be symbolized p and the probability of having a girl be symbolized q. Table 13-2 describes all the possibilities for this family of three.

Because there are only eight ways in which a family of three can occur, the possibilities listed in Table 13-2 are all the possibilities that there are. The frequency with which any particular possibility occurs is referred to as its **probability** of occurrence, and the sum of the probabilities of the eight different possibilities must equal one:

$$p^3 + 3p^2q + 3pq^2 + q^3 = 1$$

To calculate the probability that the three children will be two boys and one girl, with $p = \frac{1}{2}$ and $q = \frac{1}{2}$, one calculates that $3p^2q = 3 \times [\frac{1}{2}] \times [\frac{1}{2}] \times [\frac{1}{2}] = \frac{3}{8}$. To test your understanding, try to estimate the probability that two parents heterozygous for the recessive allele producing albinism will have one albino child in a family of three:

Father's gametes

		A	a
Mother's	A	AA	Aa
gametes	a	Aa	aa

You can see that one fourth of the children are expected to be aa, albino. Thus for any given birth the probability of an albino child is one fourth. This probability can be symbolized as q. The probability of a nonalbino child is three fourths, symbolized as p. The probability that among the children there will be one albino child is $3p^2q = 27/64$, or 42%.

$p^3 + 3p^2q + 3pq^2 + q^3$ is what is known as a **binomial series;** it represents the result of raising ("expanding") the sum of two factors (a "binomial") to a power, n. In this case n is equal to 3, the number of children:

$$p^3 + 3p^2q + 3pq^2 + q^3 = (p + q)^3$$

In studying the distribution of two alternative phenotypes in families, n is the number of children in the family, and p and q are the probabilities of obtaining each of the phenotypes. Thus for a family of six children, the appropriate binomial is $(p + q)^6$ and the expanded series is $p^6 + 6p^5q + 15p^4q^2 + 20p^3q^3 + 15p^2q^4 + 6pq^5 + q^6$; from this you can locate the outcome that corresponds to two children of one sex and four of another. It is $15p^4q^2$. Expansion of binomials is awkward and laborious when the number of offspring, n, is large; fortunately, shorter approaches are available to determine the desired term of the expansion.

INDEPENDENT ASSORTMENT

After Mendel had demonstrated that different alleles of a given gene segregate independently of one another in crosses, it seemed logical to ask whether different genes also segregated independently of one another. For example, would the particular alleles of a gene for seed shape that a gamete possessed have any influence on which allele the gamete had of a gene affecting seed color?

Mendel set out to answer this question in a straightforward way. He first established a series of pure-breeding lines of peas that differed from one another with respect to two of the seven pairs of characteristics that he had studied. His second step was to cross contrasting pairs of the pure-breeding lines. In a cross involving different seed shape alleles (round, W, and wrinkled, w) and different seed color alleles (yellow, G, and green, g), all the F_1 individuals were identical, each being heterozygous for both seed shape (Ww) and seed color (Gg). The F_1 individuals of such a cross are **dihybrid** individuals. A dihybrid is an individual heterozygous for two genes.

The third step in Mendel's analysis was to allow the dihybrid individuals to self-fertilize. If the segregation of alleles affecting seed shape was independent of the segregation of those affecting seed color, then the probability that a particular pair of seed shape alleles would occur together with a particular pair of seed color alleles would be simply the product of the individual probabilities that each pair would occur separately. Thus the probability that an individual with wrinkled, green seeds would appear in the F_2 generation would be equal to the probability of observing an individual with wrinkled seeds (one fourth) times the probability of observing an individual with green seeds (one fourth), or one sixteenth.

Since the genes concerned with seed shape and those concerned with seed color are each represented by a pair of alternative alleles in the dihybrid individuals, four types of gametes are expected: WG, Wg, wG, wg. Thus in the F_2 generation there are 16 possible

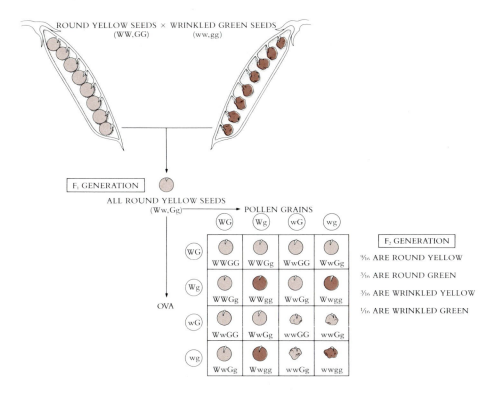

ROUND YELLOW SEEDS × WRINKLED GREEN SEEDS
(WW,GG)   (ww,gg)

F₁ GENERATION

ALL ROUND YELLOW SEEDS
(Ww,Gg) ──→ POLLEN GRAINS

	WG	Wg	wG	wg
WG	WWGG	WWGg	WwGG	WwGg
Wg	WWGg	WWgg	WwGg	Wwgg
wG	WwGG	WwGg	wwGG	wwGg
wg	WwGg	Wwgg	wwGg	wwgg

OVA

F₂ GENERATION

9/16 ARE ROUND YELLOW
3/16 ARE ROUND GREEN
3/16 ARE WRINKLED YELLOW
1/16 ARE WRINKLED GREEN

FIGURE 13-12

A dihybrid cross. The two pairs of contrasting traits employed by Mendel in this cross were round (W) versus wrinkled (w) seeds and yellow (G) versus green (g) seeds. In the F₁ generation, all seeds were round and yellow (Ww, Gg). Each such individual is able to make four different kinds of gametes, its W allele paired in a gamete with either G or g, and its w allele paired with either G or g. If the male parent contributes four kinds of gametes and the female parent contributes four kinds of gametes, then there are 4 × 4 = 16 different kinds of zygotes possible. As you can see in the Punnett square, nine of these possibilities produce round, yellow seeds (i.e., seeds that have at least one dominant allele for both traits and thus exhibit both dominant phenotypes). Three of the possible zygotes produce plants with round, green seeds (i.e., they have the dominant W allele but are homozygous gg), and three others produce plants with yellow, wrinkled seeds (i.e., they have at least one dominant G allele but are homozygous ww). Only 1 of the 16 possible combinations is a zygote that would give rise to a plant with wrinkled, green seeds (i.e., one that is both homozygous ww and gg). The ratio of the four possible combinations of phenotypes is thus expected to be 9:3:3:1, the ratio that Mendel found repeatedly in crosses of this nature.

combinations of alleles, each of them equally probable (Figure 13-12). Of the 16 combinations, 9 possess at least one dominant allele for each gene (usually signified *W–G–*, where the dash indicates the presence of either allele) and thus should have round, yellow seeds. Three possess at least one dominant *W* allele but are double-recessive for color *(W–gg)*, three others possess at least one dominant *G* allele but are double-recessive for shape *(wwG–)*, and one combination among the 16 is double-recessive for both genes *(wwgg)*. The hypothesis that color and shape genes assort independently thus predicts that the F₂ generation of this dihybrid cross will display a ratio of 9 individuals with round, yellow seeds to 3 individuals with round, green seeds to 3 individuals with wrinkled, yellow seeds to 1 with wrinkled, green seeds: a 9:3:3:1 ratio.

What did Mendel actually observe? He examined a total of 556 seeds from dihybrid plants that had been allowed to self-fertilize, and he obtained the following results:

315	Round, yellow	*W–G–*
108	Round, green	*W–gg*
101	Wrinkled, yellow	*wwG–*
32	Wrinkled, green	*wwgg*

This is very close to a perfect 9:3:3:1 ratio (313:104:104:35). Thus the two genes appeared to assort completely independently of one another. Note that this independent assortment of different genes in no way contradicts the independent segregation of alleles. Round and wrinkled seeds occur in the approximate ratio 3:1 (423:133), as do yellow and green seeds (416:140). Mendel obtained similar results for other pairs of traits.

Mendel's observation is often referred to as **Mendel's Second Law of Heredity,** or the **Law of Independent Assortment.** As you will see in Chapter 14, genes that assort independently of one another, as did Mendel's seven genes, often do so because the genes are located on different chromosomes, which segregate from one another during the process of gamete formation called meiosis. A modern restatement of Mendel's second law would be that *genes that are located on different chromosomes assort independently during meiosis.*

Mendel's Second Law of Heredity states that genes located on different chromosomes assort independently of one another.

THE GENE CONCEPT

It is important to keep in mind that the "gene" of Mendelian genetics is an abstract concept that is employed to refer to elements located on the chromosomes, elements that act in some unspecified way to produce differences among the progeny. We can see these differences and study their ratios. The molecular identity of the Mendelian gene is a matter that we will address in Chapter 14. Whatever the gene's physical basis, however, an investigator must realize that the relationship between the chromosomal gene and the phenotype that the investigator studies is not always a simple one; Mendel was lucky in his choice of traits. Often genes reveal more complex patterns, including the following:

Multiple alleles. Although a diploid individual may possess no more than two alleles at one time, this does not mean that only two allele alternatives are possible for a given gene. On the contrary, almost all genes that have been studied exhibit several different alleles.

Gene interaction. Few phenotypes are the result of the action of only one gene (Figure 13-13). Most traits reflect the action of many genes that act sequentially

FIGURE 13-13

In many pure-breeding lines of chickens, the individuals have distinctive combs. Two of these kinds of combs are "rose," a large irregular comb, and "pea," a more compact and solid one. When chickens with these two different kinds of combs are crossed, the resulting hybrid individuals have a new and very different kind of comb, rather than one characteristic of the parents. This new type of comb is called "walnut," which is what it looks like. When chickens with these walnut combs are crossed with one another, they produce some offspring with walnut combs, others with rose or pea combs, and others with still another comb type, called "single." When they are crossed to one another, chickens with a single comb produce pure lines of individuals with single combs, as in Leghorn chickens.

How can these results be explained? There is not one gene segregating here, but two. The walnut combs of the first F_1 generation occur in individuals that are heterozygous for both rose and pea genes, whereas individuals with single combs are homozygous recessive for both of these genes. In a dihybrid cross, as you might expect, the four comb types are produced in 9:3:3:1 proportions.

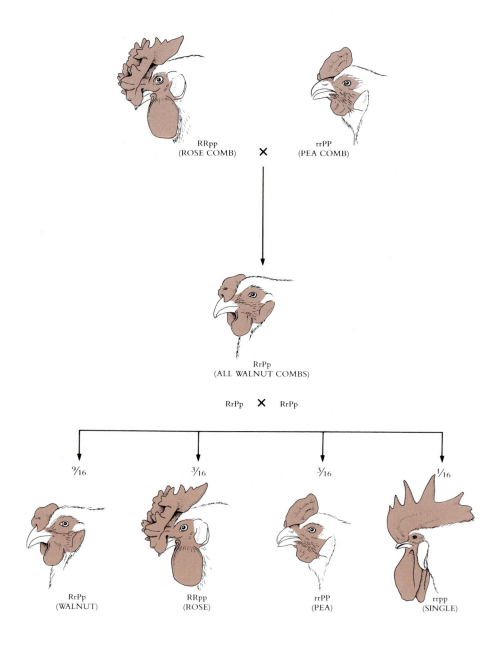

or jointly. When genes act sequentially, as in a biochemical pathway, an allele specifying a defective enzyme early in the pathway blocks the flow of material through the pathway and thus makes it impossible to judge whether the later steps of the pathway are functioning properly. Such interactions between genes are the basis of the phenomenon called **epistasis.** Epistasis is an interaction between the products of two genes in which one of them modifies the phenotypic expression of the other. Epistatic interactions between genes often make the interpretation of particular phenotypes very difficult.

Continuous variation. When multiple genes act jointly to determine a trait such as height or weight, the contribution caused by the segregation of one particular gene is difficult to monitor, just as it is difficult to follow the flight of one bee within a swarm. Because all of the participating genes play a role in determining the phenotype, and many are segregating independently of one another, one sees a gradation in degree of difference when many individuals are examined (Figure 13-14).

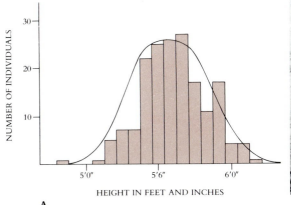

A

B

Pleiotropy. Often an individual genetic alteration will have more than one effect on the phenotype. Such an alteration is said to be **pleiotropic.** Thus, when the pioneering French geneticist Lucien Cuénot studied yellow fur in mice, a dominant trait, he was unable to obtain a true-breeding homozygous yellow strain by crossing individual yellow mice with one another—individuals that were homozygous for the yellow allele died. The yellow allele was pleiotropic: one effect was yellow color; another effect was a lethal developmental defect. Thus a pleiotropic gene alteration may be dominant with respect to one phenotypic consequence (yellow fur) and recessive with respect to another (lethal developmental defect). Pleiotropic relationships occur because in examining the characteristics of organisms, we are studying the consequences of the action of products made by genes, and these products often also perform other functions about which we are ignorant.

Incomplete dominance. Not all alternative alleles are fully dominant or recessive in heterozygotes. Sometimes heterozygous individuals do not resemble one parent precisely. Some pairs of alleles produce instead a heterozygous phenotype that (1) is intermediate between the parents (intermediate dominance), (2) resembles one allele closely but can be distinguished from it (partial dominance), or (3) is one in which both parental phenotypes can be distinguished in the heterozygote (lack of dominance).

Environmental effects. The degree to which many alleles are expressed depends on the environment. Some alleles encode an enzyme whose activity is more sensitive to conditions such as heat (Figure 13-15) or light than are other alleles.

FIGURE 13-15

An arctic fox in winter has a coat that is almost white, so it is difficult to see the fox against a snowy background. In summer the fox's fur darkens to a red brown, in which condition it resembles the color of the tundra over which it runs.

MODIFIED MENDELIAN RATIOS

When individuals heterozygous for two different genes mate (a dihybrid cross), four different phenotypes are possible among the progeny: the dominant phenotype of both genes is displayed, one of the dominant phenotypes is displayed, the other dominant phenotype is displayed, or neither dominant phenotype is displayed. Mendelian assortment predicts that these four possibilities will occur in the proportions $9:3:3:1$. Sometimes, however, it is not possible for an investigator to successfully identify each of the four possible phenotypic classes, because two or more of the classes look alike.

One example of such difficulty in identification is seen in analysis of particular varieties of corn, *Zea mays*. Some commercial varieties exhibit a purple pigment called anthocyanin in their seed coats, whereas others do not. When in 1918 the geneticist R.A. Emerson crossed two pure-breeding corn varieties, neither of which typically exhibits any anthocyanin pigment, he obtained a surprising result: all of the F_1 plants produced purple seeds. The two white varieties, which had never been observed to make pigment, would, when crossed, produce progeny that uniformly make the pigment.

When two of these pigment-producing F_1 plants were crossed to produce an F_2 generation, 56% were pigment producers and 44% were not. What was happening? Emerson correctly deduced that two genes were involved in the pigment-producing process and that the second cross had thus been a dihybrid cross such as described by Mendel.

Mendel predicted 16 possible genotypes in equal proportions ($9 + 3 + 3 + 1 = 16$) suggesting to Emerson that the total number of genotypes in his experiment was also 16. How many of these were in each of the two types Emerson obtained? He multiplied the fraction that were pigment producers ($.56) \times 16 = 9$, and multiplied the fraction that were not ($.44) \times 16 = 7$. Thus Emerson had in fact a **modified ratio** of $9:7$ instead of the usual $9:3:3:1$ ratio.

In this case the pigment anthocyanin is produced from a colorless molecule by two enzymes that work one after the other. In other words the pigment is the product of a two-step biochemical pathway:

For pigment to be produced, a plant must possess at least one good copy of each enzyme. The dominant alleles encoded functional enzymes; the recessive alleles, defective nonfunctional ones. Of the 16 genotypes predicted by random assortment, 9 contain at least one dominant allele of both genes—these are the purple progeny. The $9:7$ ratio that Emerson observed resulted from the pooling of the three phenotypic classes that lack dominant alleles at either or both loci ($3 + 3 + 1 = 7$) and so all looked the same, nonpigmented.

All of these potential complications can make the job of the Mendelian geneticist more difficult than it would be if they did not exist. Our study of genes need not be restricted to phenotypic phenomena, however, and more precise relationships can be expressed at the level of actual gene structure. In the following chapters we will reexamine genes as physical entities—discrete segments of DNA. What geneticists have learned about the molecular nature of genes, particularly within the last 10 years, is revolutionizing biology.

SUMMARY

1. Koelreuter noted the basic facts of heredity a century before Mendel. He found that alternative traits segregate in crosses and may mask each other's appearance. Mendel, however, was the first to quantify his data, counting the numbers of each alternative type among the progeny of crosses.

2. From counting progeny types, Mendel learned that the alternatives that were masked in hybrids appeared only 25% of the time when they subsequently segregated in the F_2 generation. This finding, which led directly to Mendel's model of heredity, is usually referred to as the Mendelian ratio of 3:1 dominant to recessive traits.

3. Mendel deduced from the 3:1 ratio that traits are specified by discrete "factors," which do not blend. We refer to Mendel's factors as genes and to alternative forms of his factors as alleles. Mendel deduced that pea plants contain two factors for each feature that he studied (we now know this is because they are diploid). When the two copies of a factor are not the same—when the plant is heterozygous—one factor, which Mendel described as dominant, determines the appearance, or phenotype, of the individual.

4. When two heterozygous individuals mate, an individual offspring has a 50% (that is, random) chance of obtaining the dominant allele from the father and a 50% chance of obtaining the dominant allele from the mother so that the probability of obtaining two dominant alleles—of being homozygous dominant—is $0.5 \times 0.5 = 0.25$, or 25%. Similarly, the probability of being homozygous recessive is 25%. The rest of the progeny, one half, are heterozygotes. Because the appearance of heterozygotes is specified by the dominant allele, the progeny thus appear as three fourths dominant and one fourth recessive, a ratio of 3:1 dominant to recessive.

5. When two genes are located on different chromosomes, the alleles included in an individual gamete are selected at random. The allele for one gene included in the gamete has no influence on which allele of the other gene is included in the gamete. Such genes are said to assort independently.

SELF-QUIZ

The following questions pinpoint important information in this chapter. You should be sure to know the answers to each of them.

1. Before Mendel, biologists believed that
   (a) traits were not inherited, but rather divinely dictated
   (b) traits were inherited indirectly as genes
   (c) traits were inherited directly, as parts of the body passed on to children

2. The first successful hybridizations between different plant species were carried out by a German named _____.

3. Of the four genetic properties listed below, which was *not* observed in these initial hybridizations?
   (a) segregation of alternative traits   (c) indirect transmission of traits
   (b) independent assortment   (d) inherited traits do not blend

4. What did Gregor Mendel do that earlier geneticists such as Knight did not do?
   (a) selected the garden pea as a suitable system to study heredity
   (b) carried out controlled crosses
   (c) employed true-breeding lines in his crosses
   (d) counted the numbers of individuals in each progeny class

5. When Mendel crossed two pea varieties, the offspring resembled one of the parental lines. He referred to the trait exhibited by the offspring as being _____, whereas the trait that was not apparent among the progeny was referred to as _____.

6. In a cross of purple-flowered and white-flowered plants the F_1 offspring observed by Mendel were all purple; a cross of two of these purple plants yielded F_2 purple-flowered and white-flowered plants in the ratio of _____.

7. If two of the F_2 white-flowered plants in the previous cross were to be crossed, the progeny
   (a) would all be white
   (b) would be one fourth purple
   (c) would be one half purple
   (d) would be all purple

8. If one of the purple-flowered F_2 plants in the progeny in question 7 were selected for more detailed examination, it might be either homozygous or heterozygous. To determine which, it is necessary to cross it with
   (a) another purple-flowered plant
   (b) a white-flowered plant
   (c) a purple-flowered plant that is known to be homozygous
   (d) a heterozygote

9. When two heterozygous individuals are crossed, the percentage of the progeny that exhibit the recessive trait is _____.

10. When two heterozygous individuals are crossed, the proportion of the progeny that are true-breeding for the dominant trait is _____.

11. The white color of a homozygous recessive plant is its
    (a) phenotype   (b) genotype

12. Of the following crosses, which is a testcross?
    (a) $WW \times WW$ (c) $Ww \times ww$
    (b) $WW \times Ww$ (d) $Ww \times Ww$

13. What is independent assortment?
    (a) different genes on the same chromosome segregate independently of one another in genetic crosses
    (b) different alleles of the same gene segregate independently in genetic crosses
    (c) different genes on different chromosomes segregate independently of one another in genetic crosses

14. How many alleles of a given gene are possible?
    (a) 1   (d) 4
    (b) 2   (e) any number
    (c) 3

15. For any two alleles of a gene, one is always dominant with respect to the other. True or false?

16. Often the genotype of one locus will prevent an investigator from learning how another locus is affecting the phenotype of an individual. This phenomenon is called _____.

17. Two individuals with different genotypes may have the same phenotype in one environment and a different phenotype in another. True or false?

18. In Sweden many more people possess blue eyes than all other colors combined. Thus, in Sweden, the allele that determines blue eyes is dominant to most other eye color alleles. True or false?

19. Under what circumstances does Mendel's second law (independent assortment) not hold?

THOUGHT QUESTIONS

1. Why did Mendel observe only two alleles of any given trait in the crosses that he carried out?

2. If you were to flip a coin six times, the most likely outcome would be three heads and three tails. What is the probability that you will *not* get three heads and three tails?

3. Figure 13-16 describes Mendel's cross of *wrinkled* and *round* seed characters. What is wrong with this diagram?

4. The annual plant *Haplopappus gracilis* has two pairs of chromosomes, which we may call *1* and *2.* In this species the probability that two traits, *a* and *b,* selected at random will prove to be on the same chromosome of *Haplopappus* is the probability that they will both be on chromosome 1 (½), times the probability that they will both be on chromosome 2 (also ½); ½ × ½ = ¼, or 25%. This is often symbolized

$$[½]^{\text{(number of pairs of chromosomes)}}$$

   Human beings have 23 pairs of chromosomes. What is the probability that any two human traits selected at random will prove to be on the same chromosome?

5. Among Hereford cattle, there is a dominant allele called *polled;* individuals that have this allele lack horns. After college you become a cattle baron and stock your spread entirely with polled cattle. You personally check each cow to be sure that none possess horns, and none do. Among the calves that year, however, some grow horns. Angrily you dispose of them and check to be sure that no horned adult has gotten into your pasture. None has. And yet the next year more horned calves are born. What is the source of your problem? What should you do to rectify it?

6. There is an inherited trait among humans in Norway that causes affected individuals to have very wavy hair, not unlike that of a sheep. The trait, called *woolly,* is very evident when it occurs in families: no child possesses woolly hair unless at least one parent does. Imagine you are a Norwegian judge, and that you have before you a woolly-haired man suing his normal-haired wife for divorce because although their first child has woolly hair, their second child does not, her hair being quite normal, long and blonde. The husband claims this constitutes evidence of infidelity on the part of his wife. Do you accept his claim? Justify your decision.

7. In humans, Down syndrome, a serious developmental abnormality, results from the presence of three copies of chromosome 21 rather than the usual two copies. If a female exhibiting Down syndrome mates with a normal male, what proportion of her offspring would be expected to be affected?

8. Many animals and plants bear recessive alleles for albinism, a condition in which homozygous individuals completely lack any pigments. An albino plant, for example, lacks chlorophyll and is white. An albino person lacks any melanin pigment. If two normally pigmented humans heterozygous for the same albinism allele marry, what proportion of their children would be expected to be albino?

9. Your uncle dies and leaves you his racehorse Dingle. To realize some money from your inheritance, you decide to put the horse out to stud. In looking over the stud book, however, you discover that Dingle's grandfather exhibited a rare clinical disorder that leads to brittle bones. This disorder is hereditary and results from homozygosity for a recessive allele. If Dingle is heterozygous for the allele, it will not be possible to use him for a stud, lest the genetic defect be passed on. How would you determine whether Dingle carries the allele?

10. In the fruitfly *Drosophila* the allele for dumpy wings (symbolized *d*) is recessive to the normal long-wing allele (symbolized *D*), and the allele for white eye (symbolized *w*) is recessive to the normal red-eye allele (symbolized *W*). In a cross of *DDWw* and *Ddww,* what proportion of the offspring are expected to be "normal" (long wing, red eye)? What proportion "dumpy, white"?

FIGURE 13-16

11. As a reward for being a good student, your instructor presents you with a fruit fly named Oscar. Oscar has red eyes, the same color that normal flies do. You add Oscar to your fly collection, which also contains Heidi and Siegfried, flies with white eyes, and Dominique and Ronald, which are from a long line of red-eyed flies. Your previous work has shown that the white-eye trait exhibited by Heidi and Siegfried is a result of their homozygosity for a recessive allele. How would you determine whether or not Oscar was heterozygous for this allele?

12. Sometimes in families children are born who exhibit recessive traits (and therefore must be homozygous for the recessive allele specifying the trait), even though one or both of the parents do not exhibit the trait. What can account for this?

FOR FURTHER READING

BLIXT, S.: "Why Didn't Gregor Mendel Find Linkage?" *Nature,* vol. 256, page 206, 1975. Modern information on the location on chromosomes of the genes that Mendel studied.

GOODENOUGH, U.: *Genetics,* 2nd ed., Saunders College Publishing, Philadelphia, 1984. A sound and very molecular treatment of genetics.

KLUG, W., and M. CUMMINGS: *Concepts of Genetics,* Charles Merrill Publishing Company, Columbus, Ohio, 1983. An excellent short text.

MENDEL, G.: "Experiments on Plant Hybridization," (1866). Translation, reprinted in C. Stern and E. Sherwood (eds.): *The Origins of Genetics: A Mendel Source Book,* W.H. Freeman and Company, San Francisco, 1966.

SUZUKI, D., A. GRIFFITHS, and R. LEWONTIN: *An Introduction to Genetic Analysis,* 2nd ed., W.H. Freeman and Company, San Francisco, 1981. A good undergraduate text.

VON TSCHERMAK-SEYSENEGG, E.: "The Rediscovery of Mendel's Work," *Journal of Heredity,* vol. 42, pages 163-171, 1951. An account of how Mendel's work, ignored for over 30 years, was rediscovered, given by one of the geneticists responsible.

MEIOSIS AND CHROMOSOMES

Chapter 14

Overview

Hereditary traits are specified by genes, which are integral parts of chromosomes. The patterns of heredity that Mendel observed result from the behavior of chromosomes during the sexual cycle. When diploid organisms produce haploid gametes, the particular chromosomal partner that is included in a particular gamete is chosen at random. This random assortment of chromosomes among gametes is the mechanism that leads to the random assortment of traits in Mendel's crosses. The process of gamete formation, in which these events occur, is one example of a special form of cell division called meiosis.

For Review

Here are some important terms and concepts that you will encounter in this chapter. If you are not familiar with them, you should review them before proceeding.

- **Crossing-over** (Chapter 5)

- **Microtubules** (Chapter 7)

- **Centromere** (Chapter 8)

- **Mitosis** (Chapter 8)

- **Mendelian segregation** (Chapter 13)

- **Independent assortment** (Chapter 13)

The hereditary elements that Mendel described are found on chromosomes—microscopic, threadlike bodies within the nuclei of eukaryotic cells (Figure 14-1). Chromosomes were first observed by the German embryologist Walther Fleming in 1882, while he was examining the rapidly dividing cells of salamander larvae. When Fleming looked at these cells through what we would now regard as a rather primitive light microscope, he saw minute threads within the nuclei of the cells. These threads appeared to be dividing lengthwise, and Fleming called their division mitosis, basing his term on the Greek word *mitos,* meaning "thread." We discussed mitosis in Chapter 8.

Since their initial discovery, chromosomes have proved to be present in the cells of all eukaryotes. Their number may vary enormously. A few kinds of plants and animals—such as *Haplopappus gracilis,* an annual relative of the sunflower that grows in North Amer-

FIGURE 14-1

A human chromosome as it appears immediately before nuclear division. Each of its DNA strands has already been replicated; the two sister copies are held together by the centromere. Human cells contain 23 different kinds of chromosomes. Each of these is present in two homologous versions, so that the total number of chromosomes is 46.

TABLE 14-1   CHROMOSOME NUMBER IN SELECTED EUKARYOTES

DIVISION	NUMBER OF CHROMOSOMES	DIVISION	NUMBER OF CHROMOSOMES
FUNGI		VERTEBRATES	
Penicillium	4	Opossum	22
Neurospora	7	Toad	22
Dictyostelium	7	Salamander	24
Saccharomyces (a yeast)	18	Frog	26
INSECTS		Vampire bat	28
Mosquito	6	Lungfish	38
Drosophila	8	Mouse	40
Housefly	12	Rat	42
Honeybee	32	Human	46
Silkworm	56	Chimpanzee	48
PLANTS		Orangutan	48
Haplopappus gracilis	2	Gorilla	48
Barley	14	Cow	60
Garden pea	14	Horse	64
Corn	20	Black bear	76
Bread wheat	42	Chicken	78
Tobacco	48	Dog	78
Sugarcane	80	Duck	80
Horsetail	216		
Adder's tongue fern	1262		

A wide range of organisms is presented to demonstrate the very different values that are possible. The number of chromosomes in the body cells of most eukaryotes falls between 10 and 50.

ican deserts, or the parasitic roundworm *Ascaris*—have as few as 2 pairs of chromosomes, whereas some ferns have more than 500 pairs (Table 14-1).

In humans and most other animals the number of copies of each chromosome present in body cells is the diploid number; the number of chromosomes in the gametes is the haploid number. In many plants more than two copies of the chromosomes are present in the body cells. Bread wheat, for example, is a hexaploid: it has six copies of each chromosome in each cell, except for the eggs and sperm, which have three.

The fusion of haploid gametes to form new diploid cells is called **syngamy.** As you can readily imagine, cell fusion is not a process that can continue to occur over and over. If it did, the number of chromosomes in each cell would become impossibly large. For example, in 10 generations the 46 chromosomes present in each of your cells would have increased to over 47,000 (46×2^{10}). It was clear even to early investigators that there must be some mechanism during the course of gamete formation to reduce the number of chromosomes to half the number that is characteristic of most cells of that species. If a special **reduction division** occurs, in which cells are formed with half the number of chromosomes characteristic of most cells of that species, then the subsequent fusion of these cells would make possible a stable chromosome number. The mature individuals of each successive generation would have the same number of chromosomes. The expected reduction division process was soon observed. It is called **meiosis.**

Only a year after Fleming's report, the Belgian cytologist Pierre-Joseph van Beneden was studying the chromosomes of *Ascaris* and found, to his surprise, that the number of chromosomes was different in different types of cells. Specifically, he observed that eggs and sperm each contain two chromosomes, whereas the cells of the young embryo contained four chromosomes, as do all the cells in the body of a mature individual of *Ascaris*. From his observations, van Beneden was able to outline the basic process of meiosis: the gametes—eggs and sperm—each contain a single basic complement of chromosomes (the egg and sperm are haploid), whereas the resulting zygote, and the adult that it becomes, contains two copies of each chromosome (the zygote is diploid).

Meiosis should be distinguished clearly from mitosis, the subject of Chapter 8. Both the **somatic** or body cells and the gamete-producing cells are usually diploid. When a somatic cell undergoes mitosis, it divides to form two diploid daughter cells exactly like it, whereas when a gamete-producing cell undergoes meiosis, it produces haploid cells with half the diploid number of chromosomes. In animals the cells that will eventually undergo meiosis to produce the gametes are set aside from the somatic cells early in the course of development.

> **Meiosis is a process of nuclear division in which the number of chromosomes in certain somatic cells is halved during gamete formation.**

THE STAGES OF MEIOSIS

Although meiosis is a continuous process, it is most easily studied when it is divided into arbitrary stages, just as is done in describing the process of mitosis in Chapter 8. Indeed, the two forms of nuclear division have much in common. Meiosis in a diploid organism consists, in a sense, of two rounds of mitosis, during the course of which two unique events occur:

1. In an early stage of the first of the two nuclear divisions, the two versions of each chromosome, called **homologous chromosomes**, or **homologues**, pair all along their length. While they are thus paired, genetic exchange occurs between them, physically joining them. The homologous chromosomes are then drawn together to the equatorial plane of the dividing cell; subsequently, each homologue is pulled by spindle fibers toward the opposite poles of the cell. When this process is complete, the group of chromosomes at each pole contains one of the two homologues of each chromosome. The clusters of chromosomes at each pole are thus haploid, since each chromosome within a cluster *has no homologue* in that cluster. Each pole contains half the number of chromosomes that were present in the original diploid cell.

2. The second meiotic division is identical to a normal mitotic division, except that *the chromosomes do not replicate between the two divisions*. Because of the crossing-over that occurred earlier, the sister chromatids that separate in this second division are not identical to one another.

> **Two important properties distinguish meiosis from mitosis: (1) In meiosis the homologous chromosomes pair lengthwise, and their chromatids exchange genetic material by crossing-over. (2) The sister chromatids, which are not identical following crossing-over, do not separate from one another in the first nuclear division, and the chromosomes do not replicate between the two nuclear divisions.**

The two stages of meiosis are traditionally called meiosis I and meiosis II, each stage being subdivided further into prophase, metaphase, anaphase, and telophase, just as in mitosis. In meiosis, however, prophase is more complex than it is in mitosis.

The First Meiotic Division
Prophase I

In prophase I individual chromosomes first become visible under the light microscope, as their DNA coils more and more tightly in a process called **condensation** and a matrix of very fine threads, or **chromonemata,** becomes apparent. Since the DNA has already replicated before the onset of meiosis, each of these threads actually consists of two sister chromosomes joined at their centromeres.

The ends of these chromatids attach to the nuclear envelope at specific sites. The membrane sites to which homologous chromosomes attach are near one another so that the members of each homologous pair of chromosomes are brought close together. They then line up side by side, a process called **synapsis.** A lattice of protein and RNA is laid down between the homologous chromosomes (Figure 14-2). This lattice holds the two chromonemata in precise register with one another, each gene located directly across from its corresponding sister on the homologous chromosome. The effect is much like that of zipping up a zipper. The resulting complex is called a **synaptonemal complex.** Within it the DNA duplexes unwind and single strands of DNA pair with their opposite number *from the other homologue.*

> **Synapsis is the close pairing of homologous chromosomes that takes place early in prophase I of meiosis. During synapsis, a molecular scaffold called the synaptonemal complex aligns the DNA molecules of the two homologous chromosomes side by side. As a result, a DNA strand of one homologue can pair with the corresponding DNA strand of the other.**

The process of synapsis initiates a complex series of events called crossing-over, in which DNA is exchanged between the two paired DNA strands. The molecular events that occur will be described in Chapter 17. Once the process of crossing-over is complete, the synaptonemal complex breaks down. At that point the homologous chromosomes are released from the nuclear envelope and the chromatids begin to move apart from one another. There is a total of four chromatids for each chromosome at this point: two homologous chromosomes, each replicated and so present twice (as sister chromatids). The four chromatids do not separate completely, however, because they are held together in two ways:

1. The two sister chromatids of each homologue, recently created by DNA replication, are held together by their common centromere.
2. The paired homologues are held together at the points where crossing-over occurred within the synaptonemal complex.

The points at which portions of chromosomes have been exchanged can often be seen under the light microscope as an **X**-shaped structure known as a **chiasma** (the Greek word for a cross; pl., **chiasmata**) (Figure 14-3). The presence of a chiasma indicates that two of the four chromatids of paired homologous chromosomes have exchanged parts, one participant from each homologue. The **X**-shaped chiasma figures soon begin to move out to the end of the respective chromosome arm as the chromosomes separate (imagine moving a small ring down two strands of rope).

> **In prophase I individual DNA strands of the two homologues pair with one another by complementary base pairing. Crossing-over occurs between the DNA paired strands, creating the chromosomal configurations known as chiasmata. These exchanges serve to lock the two homologues together, and they do not disengage readily.**

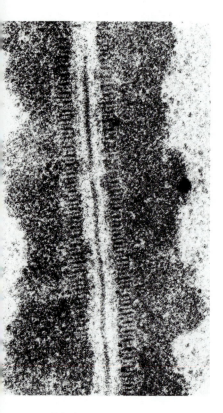

FIGURE 14-2

Electron micrograph of a portion of a synaptonemal complex of the ascomycete Neotiella rutilans, *a cup fungus. Similar complexes can be observed in a wide variety of eukaryotes during prophase I of meiosis.*

FIGURE 14-3

Chiasmata.

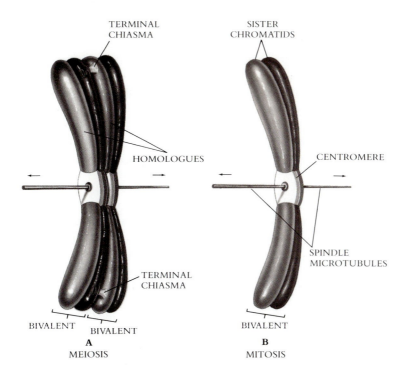

TERMINAL
CHIASMA

SISTER
CHROMATIDS

HOMOLOGUES

CENTROMERE

TERMINAL
CHIASMA

SPINDLE
MICROTUBULES

BIVALENT BIVALENT

BIVALENT

A
MEIOSIS

B
MITOSIS

FIGURE 14-4

A *A key mechanical element necessary for reduction division in meiosis is the presence of chiasmata. These chiasmata are created by crossing-over during prophase I, when the chromosome homologues are close to one another within the synaptonemal complex. The chiasmata hold the two bivalents together; consequently, the spindle microtubules are able to bind only one face of each centromere. Later, when these microtubules shorten, the terminal chiasmata break and the two bivalents are drawn to opposite poles—but there is no microtubular-driven separation of the individual bivalents.*

B *In mitosis, by contrast, microtubules attach to both faces of each centromere, and when they contract the bivalent is split, the two chromatids being drawn to opposite poles.*

Metaphase I

In the second phase of meiosis the nuclear envelope disperses and the microtubules form a spindle, just as in mitosis. There is, however, a crucial difference between this metaphase and that of mitosis: in meiosis one face of each centromere is inaccessible to spindle microtubules. To understand why, consider again the chiasmata produced by crossing-over in prophase. They continue their movement down the paired chromosomes from the original point of crossover, eventually reaching the ends of the chromosomes. At this point they are called **terminal chiasmata.** Their presence has an important consequence: a terminal chiasma holds the two homologous chromosomes together, like a couple dancing close, so that only one side of each centromere faces outward from the complex, just as only one side of each dancer's body does. The other face of each centromere is directed inward toward the homologue, like the belt buckles of the dancers. Microtubules from the spindle are able to attach only to the outside centromere faces (Figure 14-4). This one-sided attachment is in marked contrast to what occurs in mitosis, when *both* faces of a centromere are bound by microtubules.

As a result of microtubules binding only to the outside of each centromere, the centromere of one homologue becomes attached to microtubules originating from one pole, while that of the other homologue becomes attached to microtubules originating from the other pole. Each joined pair of homologues, called a **bivalent,** then lines up on the metaphase plate. For each bivalent the orientation on the spindle axis is random so that which homologue is oriented toward which pole is a matter of chance (Figure 14-5).

Anaphase I

After spindle attachment is complete, the microtubules of the spindle fibers begin to contract. As they do, they break the chiasmata apart and pull the centromeres toward the two poles, dragging the chromosomes along with them. Because the microtubules are attached to only one side of each centromere, the individual centromeres are not pulled apart to form two daughter centromeres, as they are in mitosis. Instead, the entire centromere proceeds to one pole, taking both sister chromatids with it. When the contraction of the spindle fibers is complete, each pole has a complete set of chromosomes, consisting of one member of each homologous pair. Because the orientation of each pair of homologous chromosomes

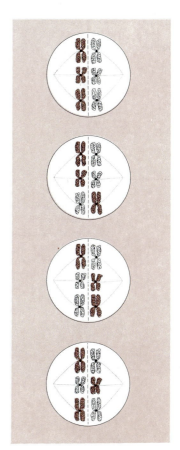

FIGURE 14-5

Independent assortment occurs because the orientation of chromosomes on the metaphase plate is random. Many combinations are possible—in fact, 2 raised to a power equal to the number of chromosomes. In this hypothetical three-chromosome cell, four of the eight possible orientations (2^3) are illustrated. Each orientation results in gametes with different combinations of parental chromosomes.

on the metaphase plate is random, the chromosome that a pole receives from each pair of homologues is random with respect to all other chromosome pairs.

> **The random orientation of homologous chromosome pairs on the metaphase plate and the subsequent separation of homologues from one another in anaphase are responsible for the independent assortment of traits located on different chromosomes.**

Telophase I

At the completion of anaphase I, each pole has a complete complement of chromosomes, one member of each pair of homologues. Each of these chromosomes replicated before meiosis began and thus contains two copies of itself, attached by a common centromere. These copies, however, are not identical, because of the crossing-over that occurred in prophase I. The stage at which the two complements of bivalents gather together at their respective poles to form two chromosome clusters is called telophase I. After an interval of variable length the second meiotic division, meiosis II, occurs.

> **The first meiotic division is traditionally divided into four stages:**
> **Prophase I—homologous chromosomes pair and exchange segments**
> **Metaphase I—homologous chromosomes align on a central plane**
> **Anaphase I—homologous chromosomes move toward opposite poles**
> **Telophase I—individual chromosomes gather together at the two poles**

The Second Meiotic Division

Meiosis II is simply a mitotic division, involving the products of meiosis I. At the completion of anaphase I each pole has a full haploid complement of chromosomes, each of which is composed of two sister chromatids attached by a common centromere. Because

FIGURE 14-6

The stages of meiosis in the Easter lily, Lilium longiflorum. *The first five stages (**A** to **E**) are stages of prophase I: in zygotene (**A**) homologous chromosomes begin to pair together; in pachytene (**B**) crossing-over occurs between homologues; in diplotene (**C**) the synaptonemal complex breaks down; in diakinesis (**D** and **E**) chiasmata move to the ends of the chromosomes. The next three stages (**F** to **H**) complete meiosis I. The final four stages (**I** to **L**) are those of meiosis II.*

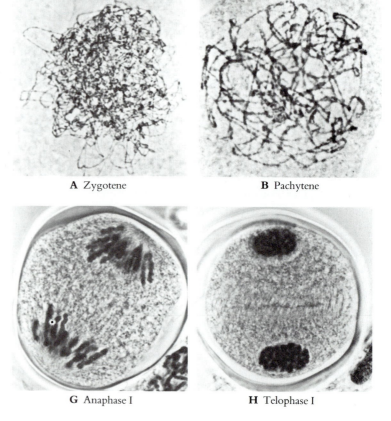

A Zygotene   **B** Pachytene

G Anaphase I   **H** Telophase I

of crossing-over in the first phase of meiosis, however, these sister chromatids are not identical to one another genetically. At both poles of the cell, these two complements of chromosomes now divide mitotically, spindle fibers binding each side of the centromeres, which divide and move to opposite poles. The end result of this mitotic division is four haploid complements of chromosomes. At this point the nuclei are reorganized, nuclear envelopes forming around each of the four haploid complements of chromosomes (Figure 14-6). The cells that contain these haploid nuclei may function directly as gametes, as they do in animals, or they may themselves divide mitotically, as they do in plants, fungi, and many protists.

Each of the four haploid products of meiosis contains a full complement of chromosomes. These haploid cells may function directly as gametes, as they do in animals, or may divide by mitosis, as they do in plants, fungi, and many protists, then eventually producing greater numbers of gametes.

CHROMOSOMES: THE VEHICLES OF MENDELIAN INHERITANCE

Chromosomes are not the only kinds of organelles that segregate regularly when eukaryotic cells divide. Centrioles also divide and segregate in a regular fashion, as do the mitochondria and chloroplasts in the cytoplasm. Thus in the early twentieth century it was by no means obvious that chromosomes were the vehicles for the information of heredity. A central role for them was first suggested in 1900 by the German geneticist Karl Correns, in one of the papers announcing the rediscovery of Mendel's work. Soon after, observations that similar chromosomes paired with one another in the process of meiosis led directly to the **chromosomal theory of inheritance,** first formulated by the American Walter Sutton in 1902. Sutton's argument was as follows:

1. Reproduction involves the initial union of only two cells, egg and sperm. If Mendel's model is correct, then these two gametes must make equal hereditary

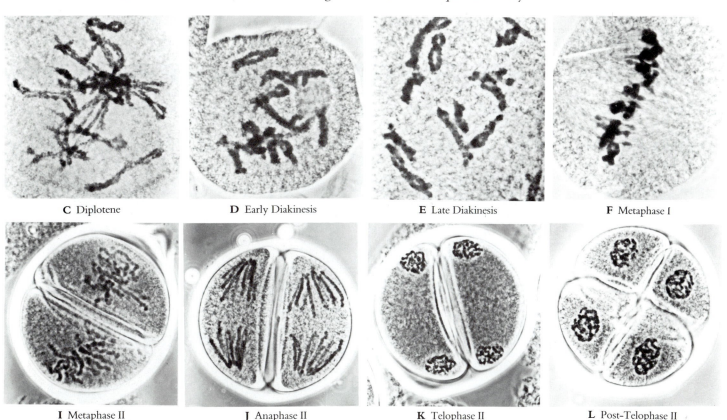

C Diplotene   D Early Diakinesis   E Late Diakinesis   F Metaphase I

I Metaphase II   J Anaphase II   K Telophase II   L Post-Telophase II

contributions. Sperm, however, contain little cytoplasm. Therefore the hereditary material must reside within the nuclei of the gametes.

2. Chromosomes segregate during meiosis in a manner similar to that exhibited by the elements of Mendel's model.

3. Gametes have one copy of each pair of homologous chromosomes; diploid individuals, two copies. In Mendel's model gametes have one copy of each element; diploid individuals, two copies.

4. During meiosis, each pair of homologous chromosomes orients on the metaphase plate independently of any other pair. As we noted above, this independent assortment of chromosomes is a process very reminiscent of the independent assortment of factors postulated by Mendel.

There was one problem with this theory, as many investigators soon pointed out. If Mendelian traits are determined by factors located on the chromosomes, and the independent assortment of Mendelian traits reflects the independent assortment of these chromosomes in meiosis, then why is it that the number of genes which assort independently of one another in a given kind of organism is often much greater than the number of chromosome pairs that the organism possesses? This seemed a fatal objection, and it led many early researchers to have serious reservations about Sutton's theory.

The essential correctness of the chromosomal theory of heredity was demonstrated long before this paradox was resolved. The proof was provided by a single small fly. In 1910 Thomas Hunt Morgan, studying the fly *Drosophila melanogaster,* detected a **mutant** fly, a male fly that differed strikingly from normal flies of the same species. In this fly the eyes were white (Figure 14-7), instead of the normal red.

Morgan immediately set out to determine if this new trait would be inherited in a Mendelian fashion. He first crossed the mutant male to a normal female to see if either red or white eyes were dominant. All F_1 progeny had red eyes, and Morgan therefore concluded that red eye color was dominant over white. Following the experimental procedure that Mendel had established long ago, Morgan then crossed flies from the F_1 generation with each other. Eye color did indeed segregate among the F_2 progeny, as predicted by Mendel's theory. Of 4252 F_2 progeny that Morgan examined, 782 had white eyes—an imperfect 3:1 ratio, but one that nevertheless provided clear evidence of segregation.

FIGURE 14-7

Red-eyed and white-eyed Drosophila. *The white-eyed defect in eye color is hereditary, the result of a mutation in a gene located on the sex-determining X chromosome. It was by studying this mutation that Morgan first demonstrated that genes are on chromosomes.*

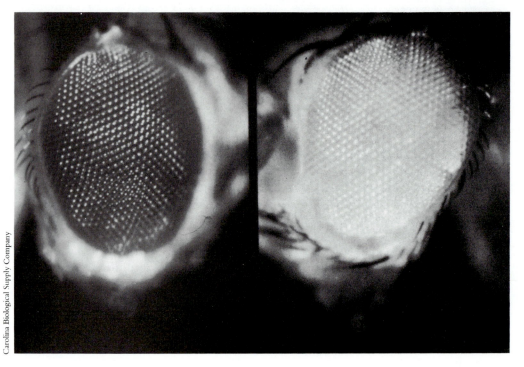

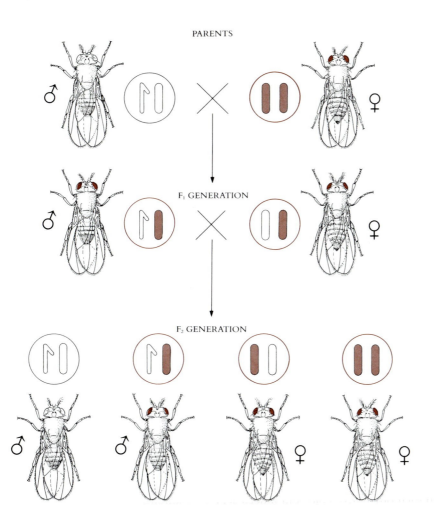

PARENTS

F₁ GENERATION

F₂ GENERATION

FIGURE 14-8

Morgan's experiment demonstrating the chromosomal basis of sex linkage in Drosophila. *The white-eyed mutant male fly was crossed to a normal female. The F₁ generation flies all exhibited red eyes, as expected for flies heterozygous for a recessive white-eye allele. In the F₂ generation every progeny fly obtains one sex chromosome from its father. If that chromosome is an X chromosome, then the progeny fly will be female; because the F₁ father's X chromosome carries the normal red-eye allele, which is dominant, all of these female progeny flies will be red eyed. If on the other hand the progeny fly obtains the Y chromosome of its father, it will be a male. Because the Y chromosome carries no genes, the X chromosome contributed by the F₁ mother will exhibit whatever allele it bears; since half of them bear the white allele (the F₁ mother is heterozygous), half of the male progeny will exhibit white eyes. This is exactly the result Morgan observed: all of the white-eyed F₂ generation flies were male.*

Something was very strange about Morgan's result, however, something that was totally unpredicted by Mendel's theory: all of the white-eyed F₂ flies were males!

How could this strange result be explained? Perhaps it was not possible to be a white-eyed female fly; such individuals might not be viable for some unknown reason. To test this idea, Morgan testcrossed the F₁ progeny back to the original white-eyed male and obtained white-eyed and red-eyed males and females in a 1:1:1:1 ratio, just as Mendelian theory predicted. So a female could have white eyes. Why then were there no white-eyed females among the progeny of the original cross?

The solution to this puzzle proved to involve sex. In *Drosophila* the sex of an individual is influenced by the number of copies of a particular chromosome, the **X chromosome,** that an individual possesses. An individual with two X chromosomes is a female, and an individual with only one X chromosome—which pairs in meiosis with a large, dissimilar partner called the **Y chromosome**—is a male. The female thus produces only X gametes, whereas the male produces both X and Y gametes. When fertilization involves an X sperm, the result is an XX zygote, which develops into a female; when fertilization involves a Y sperm, the result is an XY zygote, which develops into a male.

The solution to Morgan's puzzle lies in the fact that in *Drosophila* the white-eye trait resides on the X chromosome and is absent from the Y chromosome. (We now know that the Y chromosome carries almost no functional genes.) A trait that is determined by a factor on the X chromosome is said to be **sex linked.** Knowing the white-eye trait to be recessive to the red-eye trait, we can now see that Morgan's result was a natural consequence of the Mendelian assortment of chromosomes (Figure 14-8).

Morgan's experiment is one of the most important in the history of genetics, because it presented the first clear evidence that Sutton was right and that the factors determining Mendelian traits do indeed reside on the chromosomes. The segregation of the white-eye trait, evident in the eye color of the flies, has a one-to-one correspondence with the segregation of the X chromosome, evident from the sexes of the flies.

The white-eye trait behaves exactly as if it were located on an X chromosome, and this is indeed the case. The eye color gene, which specifies eye color in *Drosophila,* is carried through meiosis as part of an X chromosome. In other words, Mendelian traits such as eye color in *Drosophila* assort independently because chromosomes do. When Mendel observed the segregation of alternative traits in pea plants, he was observing a reflection of the meiotic segregation of chromosomes.

> **Mendelian traits assort independently because they are determined by genes located on chromosomes that assort independently in meiosis.**

CROSSING-OVER

Morgan's results led to the general acceptance of Sutton's chromosomal theory of inheritance. Scientists then attempted to resolve the paradox posed by the fact that there are more independently assorting Mendelian factors than there are chromosomes. In 1903 the Dutch geneticist Hugo de Vries had suggested that this paradox could be resolved only by assuming that homologous chromosomes exchange elements during meiosis. In 1909 the cytologist F. A. Janssens provided evidence for this suggestion. Investigating chiasmata produced during amphibian meiosis, Janssens noticed that of the four filaments involved in the X configuration, two crossed each other and two did not. He suggested that this crossing of chromatids reflected a switch in chromosomal arms between paternal and maternal chromatids, that the two homologues were exchanging arms with one another. His suggestion was not accepted widely, primarily because it was difficult to see how the two chromatids could break and rejoin at exactly the same position.

Janssens was right, however. Later experiments clearly established the correctness of his hypothesis, just as they had for the earlier hypotheses of Mendel and Sutton. One of these experiments, performed in 1931 by American geneticist Curt Stern, is described in Figure 14-9. Stern studied two sex-linked traits in strains of *Drosophila* whose X chromosomes were visibly abnormal at both ends. He first examined many flies and identified those in which an exchange had occurred with respect to the two eye traits. Stern then studied the chromosomes of these flies to see if their X chromosomes had exchanged arms. He found that all of the individuals that had exchanged eye traits also possessed chromosomes whose abnormal ends could be seen to have exchanged (Figure 14-10). The conclusion was inescapable: genetic exchanges of traits on a chromosome, such as eye color, involve crossing-over, the physical exchange of chromosome arms.

The chromosomal exchanges demonstrated by Stern provide the solution to our paradox. Crossing-over can occur between homologues anywhere along the length of the chromosome; where it actually occurs seems to be random. Thus if two different genes are located relatively far apart from one another on the chromosome, crossing-over is more likely to occur somewhere in the long interval between them than it is if they are located closer together and the interval between them is short. Two genes can be on the same chromosome and still show independent assortment if they are located so far apart on the chromosome that crossing-over occurs regularly between them (Figure 14-11).

Genetic Maps

Because crossing-over is more frequent between two genes that are relatively far apart than between another set of two genes that are relatively close to each other, the frequency with which crossing-over occurs can be used to map the relative positions of genes on chro-

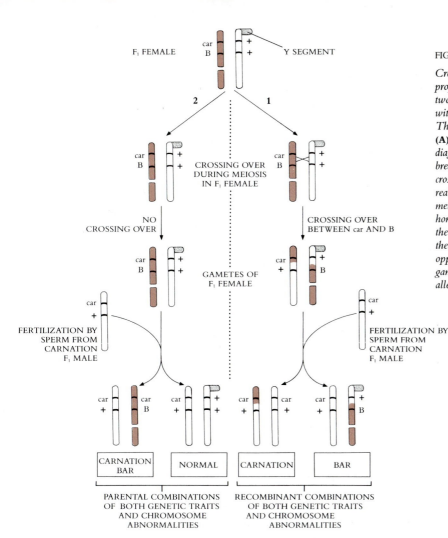

FERTILIZATION BY
SPERM FROM
CARNATION
F₁ MALE

FERTILIZATION BY
SPERM FROM
CARNATION
F₁ MALE

| CARNATION BAR | NORMAL | CARNATION | BAR |

PARENTAL COMBINATIONS
OF BOTH GENETIC TRAITS
AND CHROMOSOME
ABNORMALITIES

RECOMBINANT COMBINATIONS
OF BOTH GENETIC TRAITS
AND CHROMOSOME
ABNORMALITIES

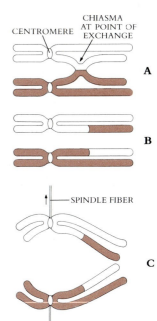

FIGURE 14-10

*Crossing-over occurs during prophase of meiosis I, when the two homologs are aligned together within the synaptonemal complex. The exchange creates a chiasma (**A**). For clarity, we have in this diagram allowed the chiasma to break apart immediately after crossing-over (**B**), although in reality many chiasmata persist into metaphase. As you can see, the two homologues have exchanged arms at the point of crossing-over. When the two homologs are drawn to opposite poles (**C**) they produce gametes with recombinant groups of alleles.*

FIGURE 14-9

Curt Stern's experiment demonstrating that a physical exchange of chromosomal arms occurs during crossing-over. Stern monitored crossing-over between two genes—recessive carnation eye color (car) and dominant Bar-shaped eye (B)—on chromosomes with physical peculiarities that he could see under a microscope. Stern separated out those F_2 generation flies in which crossing-over had occurred between car and B (1) from those in which it had not (2). Among the flies that experienced crossing-over, recombinant combinations of these alleles were seen (i.e., carnation-colored eyes of normal shape and Bar-shaped eyes of normal color). Stern then examined the chromosomes of these flies. All of them showed recombinant combinations of the chromosomal abnormalities as well. When he examined the F_2 flies with parental combinations of eye traits (2), they all proved to possess the parental combinations of chromosomal abnormalities. The result, then, is that whenever genes recombine, chromosomes do too; the recombination of genes reflects a physical exchange of chromosome arms.

FIGURE 14-11

The chromosomal locations of the seven genes studied by Mendel in the garden pea. In only two characteristics, seed shape and pod color, are the genes actually alone on a particular chromosome. The genes for flower color and seed color are both on chromosome 1 but are located so far apart that they recombine freely. Three different genes of this set are located on chromosome 4. Among these, the gene for flower position is so far from the other two that it freely recombines with them. In contrast, the genes for plant height and pod shape are very close to one another and do not recombine freely, tending instead to remain associated with one another, that is, to exhibit linkage. Pod shape and plant height were not among the pairs of traits that Mendel examined in dihybrid crosses. One wonders what Mendel would have made of the linkage he would surely have detected had he tested this pair of traits.

CHROMOSOME NUMBER	LOCI OF GENES	CHROMOSOME NUMBER	LOCI OF GENES
1	FLOWER COLOR SEED COLOR	5	POD COLOR
2		6	
3		7	SEED SHAPE
4	FLOWER POSITION PLANT HEIGHT POD SHAPE		

FIGURE 14-12

The first genetic map. This map of the X chromosome of Drosophila *was prepared in 1913 by A.H. Sturtevant, a student of Morgan. On it he located the relative positions of five recessive traits that exhibited sex linkage, by estimating their relative recombination frequencies in genetic crosses. Sturtevant arbitrarily chose the position of the* yellow *gene as zero on his map to provide a frame of reference.*

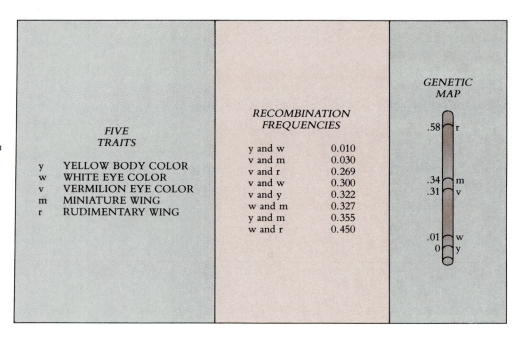

FIVE TRAITS		RECOMBINATION FREQUENCIES	
y	YELLOW BODY COLOR	y and w	0.010
w	WHITE EYE COLOR	v and m	0.030
v	VERMILION EYE COLOR	v and r	0.269
m	MINIATURE WING	v and w	0.300
r	RUDIMENTARY WING	v and y	0.322
		w and m	0.327
		y and m	0.355
		w and r	0.450

mosomes. In a cross the proportion of progeny derived from a gamete in which an exchange has occurred between two genes is a measure of the frequency with which crossover events occur between them, and thus of the distance separating them.

Thus the frequencies with which crossing-over occurs in crosses can be used to construct a **genetic map,** in which distance is measured in terms of the frequency of recombination. One "map unit" is defined as the distance within which a crossover event is expected to occur, on the average, in 1 out of 100 gametes. This unit, 1% recombination, is now called a **centimorgan,** after Thomas Hunt Morgan.

In constructing a genetic map, one simultaneously monitors recombination among three or more genes that are located on the same chromosome. By convention, the most common allele of a locus is often designated on the map with the symbol + and is referred to as **wild type,** whereas all other alleles are assigned specific symbols. Genes located on the same chromosome are said to be **syntenic.** When genes are close enough together on a chromosome that they do not assort independently, they are said to be **linked** to one another. A cross involving three linked genes, referred to as a **three-point cross,** is illustrated in the boxed essay on pp. 274–275 for three sex-linked traits that were studied by Morgan. These data were used by Morgan's student A.H. Sturtevant to draw the first genetic map (Figure 14-12).

THE IMPORTANCE OF MEIOTIC RECOMBINATION

The reassortment of genetic material that takes place during meiosis is the principal factor that has made possible the evolution of eukaryotic organisms, in all their bewildering diversity, over the past 1.5 billion years. Sexual reproduction represents an enormous advance in the ability of organisms to generate genetic variability. To understand why this is so, recall that most organisms have more than one chromosome. Human beings, for example, have 23 different pairs of homologous chromosomes. Each human gamete receives one of the two copies of each of the 23 different chromosomes, but which copy of a particular chromosome it receives is random. For example, the copy of chromosome number 14 that a particular human gamete receives in no way influences which copy of chromosome number 5 that it will receive. Each of the 23 pairs of chromosomes segregates independently of all the others, so there are 2^{23} (more than 8 million) different possibilities for the kinds of gametes that can be produced, no two of them alike. The subsequent fertilization of two gametes thus creates a unique individual, a new combination of the 23 chromosomes that probably has never occurred before and probably will never occur again.

Even this process does not fully explain the diversity of the gametes that results from meiosis, however. As you will recall, pairs of homologous chromosomes exchange physical segments during meiotic prophase I (if they did not, the pairs would not remain associated in metaphase I, and meiosis would not proceed normally). The exchange that occurs as a result of this crossing-over adds even more recombination to the random assortment of chromosomes that occurs subsequently in meiosis (Figure 14-10). Thus the number of possible genetic combinations that can occur among gametes is virtually unlimited.

Reassortment and crossing-over are two of the ways in which genes can reorganize during meiosis. These and other forms of genetic recombination will be discussed further in Chapter 17.

THE STRUCTURE OF EUKARYOTIC CHROMOSOMES

In the century since their discovery we have learned a great deal about chromosomes, their structure, and how they function. Eukaryotic chromosomes are composed of a complex of DNA, RNA, and protein called **chromatin.** Most chromosomes are about 60% protein and 40% DNA. There is also a significant amount of RNA associated with chromosomes, since that is where RNA is made. The DNA of a chromosome exists as one very long double-stranded fiber, a **duplex,** which extends unbroken through the entire length of the chromosome. A typical human chromosome contains about 5 billion (5×10^9) nucleotides in its DNA fiber. The amount of information it contains is estimated to be in excess of that on 600,000 printed pages of 500 words per page—a library of over 1000 books. If such a strand of human DNA were isolated and laid out in a straight line, it would be more than a meter long! This length is many thousands of times too long to fit into a cell. In the cell, however, the DNA is coiled (Figure 8–2), thus fitting into a much smaller space than would be possible if it were not.

How is the coiling of this long DNA fiber achieved? If you were to gently disrupt a eukaryotic nucleus and examine the DNA with an electron microscope, you would find that it resembles a string of beads (Figure 14–13). Every 200 nucleotides, the DNA duplex is coiled around a complex of **histone** proteins, forming an assembly called a **nucleosome** (Figure 14-14). The core of each nucleosome is composed of eight histone proteins. Each histone contains 100 to 200 amino acids, and of these, 20 to 40 are either lysine or arginine,

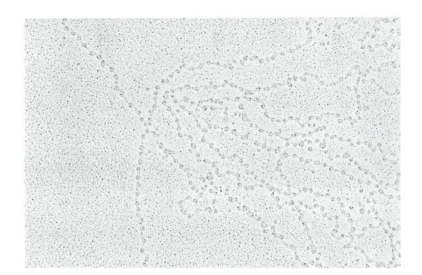

FIGURE 14-13

Nucleosomes are regions in which the DNA duplex is wound around an aggregation of histone proteins. In this electron micrograph of rat liver DNA, they exhibit the characteristic "beads on a string" structure that results.

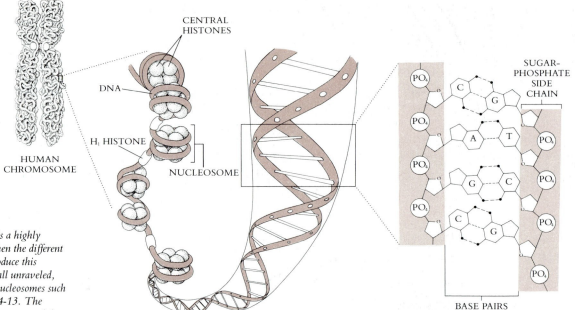

CENTRAL HISTONES

DNA

H_1 HISTONE

NUCLEOSOME

HUMAN CHROMOSOME

SUGAR-PHOSPHATE SIDE CHAIN

BASE PAIRS

FIGURE 14-14

A human chromosome is a highly condensed structure. When the different levels of folding that produce this condensed structure are all unraveled, the result is a string of nucleosomes such as that seen in Figure 14-13. The DNA molecule that wraps around the histones to produce nucleosomes is a double helix, two chains of nucleotides bound together by a series of hydrogen bonds, just as the rungs hold the two sides of a ladder together. From each nucleotide an organic base protrudes, and the hydrogen bonds form because two such bases located opposite each other are able to share the electrons of some of their hydrogen atoms.

amino acids with basic side groups. Because so many of their amino acids are basic, histone proteins are very positively charged. The DNA duplex, which is negatively charged, is strongly attracted to the histone proteins and wraps tightly around the histone core of each nucleosome. The core thus acts as a "form," which promotes and guides the coiling of the DNA. Further coiling of the DNA occurs when the string of nucleosomes wraps up into higher-order coils called **solenoids.** The solenoids themselves twist into supercoils in the continuing process of **chromatin condensation.**

When chromatin is condensed, it is called **heterochromatin.** Some portions of chromosomes are permanently condensed. The chromatin in these portions is called **constitutive heterochromatin,** or permanent heterochromatin; it appears primarily to perform a structural function. When a section of a chromosome is highly condensed, as it is in heterochromatin, its DNA is inaccessible to the enzymes that the cell uses to read the genes. For this reason the genes of the constitutive heterochromatin are never expressed. The remainder of the chromosome, in contrast, is not condensed into heterochromatin except during cell division, when the movement of the chromosomes is facilitated by the compact packaging that occurs at that stage. At all other times in the life of the cell, this DNA, called **euchromatin,** is dispersed in an "unwound" open configuration. While euchromatin is in this uncoiled form, it is readily accessible to the enzymes of the cell that use its information, although the chromosomes are invisible under a light microscope.

Only three elements are necessary for the proper separation of the DNA of chromosomes during cell division (Figure 14-15). First, there must be a point of initiation of chromosome replication so that the duplication of the chromosome can be complete before cell division begins. Second, there must be a **telomere** (Greek, *telos,* end + *meros,* part), or terminus. This structure is necessary for the recognition and binding of the chromosome to the attachment site on the nuclear envelope—the process that initiates the lengthwise pairing of homologous chromosomes. Third, there must be a centromere, the structure that is needed for the attachment of the microtubules to the chromosomes. The retraction of these microtubules is the mechanism that moves the chromosomes to the opposite poles of the cell.

During the initial stages of cell division, the euchromatic portions of chromosomes condense, becoming compact and easily visible under the light microscope. Condensation appears to proceed in stages. Initially, genes that are being used actively aggregate into tight coils, whereas the genes that are being used less actively remain uncoiled for a time. At this stage of condensation the chromosome resembles a beaded fiber when viewed with the light microscope. The chromosome's coiled regions, which are much larger than nucleosomes, are called **chromomeres.** Each chromomere contains enough DNA to code for from 1 to 10 genes.

The patterns of condensation are integral aspects of chromosome structure and function. In their fully extended forms chromosomes are "open for business," the genes in their euchromatin accessible to the enzymes that must read them in order for their information to be expressed. In their fully condensed form chromosomes are compact bodies that can move around the cell readily during cell division. In a fully condensed chromosome the DNA is about 1/8000 as long as it is in its extended form. By wrapping around nucleosomes, a process that occurs at the initial stage of condensation, the DNA is reduced sevenfold in length, and the subsequent formation of solenoid structures reduces the length another fivefold. Finally, the higher-order packaging mechanisms that operate during the process of chromosome condensation compact the DNA over 200-fold more.

Karyotypes

Chromosomes may differ widely from one another in appearance. They vary in such features as the location of their centromeres, the relative length of the two arms (regions on either side of the centromere), size, and the position of constricted regions along the arms. The particular array of chromosomes that an individual possesses, called its **karyotype,** may differ greatly between different species, or sometimes even between particular individuals.

Karyotypes of individuals are often examined to detect genetic abnormalities such as those arising from extra or lost chromosomes. Karyotypes are normally examined by photographing the chromosomes and then rearranging the parts of the photograph into a particular order, one that has been established as conventional for that kind of organism. In such reconstructions the chromosomes are conventionally arranged in descending order according to their size (Figure 14-16). In diploid cells the two homologous chromosomes

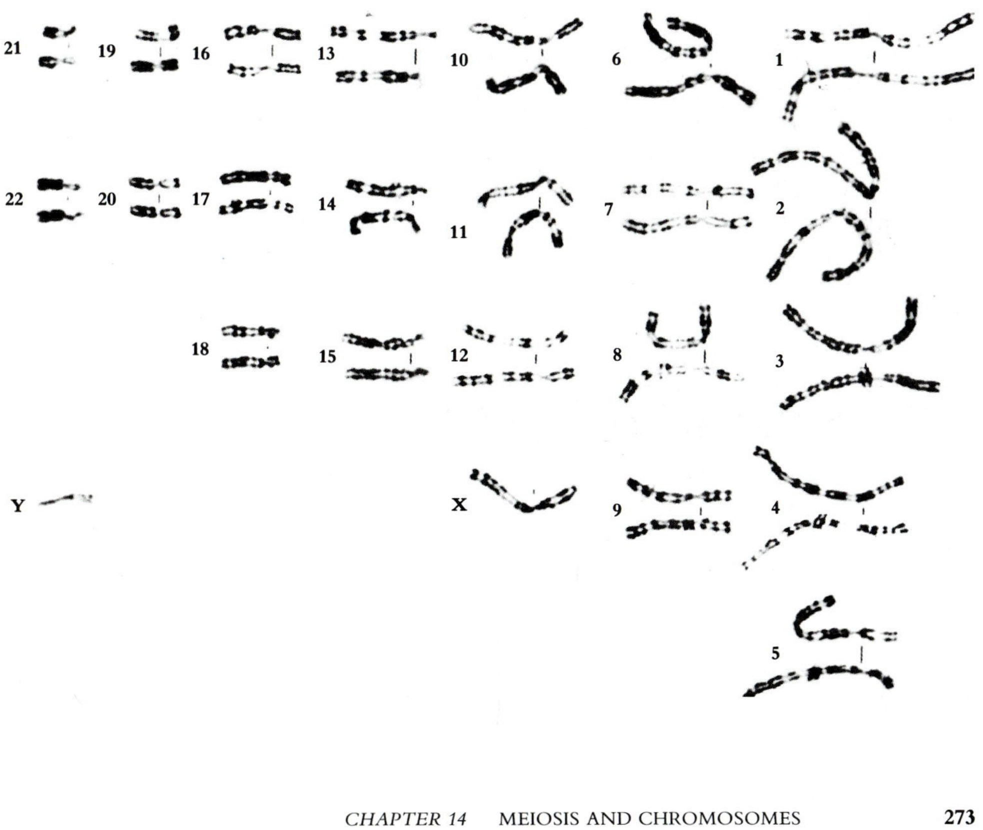

FIGURE 14-15

The three elements essential to the proper separation of a chromosome during cell division are (1) a point of initiation of chromosome replication, (2) a telomere, and (3) a centromere. It is possible to construct an artificial chromosome of random-sequence DNA containing these three elements, and when such a chromosome is injected into yeast cells it replicates and later separates normally during the process of cell division. Such artificial chromosomes are eventually lost, however, perhaps because they lack the efficient packaging of naturally occurring ones.

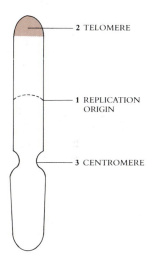

2 TELOMERE

1 REPLICATION ORIGIN

3 CENTROMERE

FIGURE 14-16

A human karyotype.

CONSTRUCTING A GENETIC MAP

The first genetic map was constructed by A.H. Sturtevant, a student of Morgan's, in 1913. He studied three traits of *Drosophila,* all of which exhibited sex linkage and thus were known to be encoded by alleles of genes residing on the same chromosome (the X chromosome). The three traits were:

y	yellow body color	(the normal body color is gray)
w	white eye color	(the normal eye color is red)
min	miniature wing	(the normal wing is 50% longer)

The mapping cross was carried out by crossing a female fly homozygous for the three recessive alleles with a male fly that was normal and carried none of them. All of the progeny were thus heterozygotes. Such a cross is conventionally represented by a diagram like the one presented below, in which the lines represent chromosomes and the symbols represent the locations of genes (+ indicates normal, usually referred to as wild type). Each fly participating in a cross possesses two homologous copies of the chromosome being mapped, and both are represented in the diagram. When crossing-over occurs in meiosis, it occurs between these two copies.

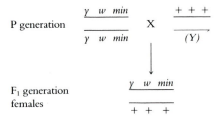

These heterozygous females, the F₁ generation, are the key to the mapping procedure. Because they are heterozygous, any crossing-over that occurs during meiosis will, if it occurs between where these genes are located, result in gametes with different combinations of alleles for these genes—recombinant chromosomes. Thus a crossover between the homologous X chromosomes of such a female in the interval between the *y* and *w* genes

$$\frac{y \qquad w}{\underset{+ \qquad +}{\overset{X}{\rule{2cm}{0.4pt}}}} \longrightarrow \frac{y \qquad +}{\underset{+ \qquad w}{\rule{2cm}{0.4pt}}}$$

will yield recombinant [y +] and [+ w] chromosomes, which are different combinations than we started with (in the parental chromosomes *w* is always linked with *y* and + linked with +).

To be able to see all the recombinant types that might be present among the gametes of these heterozygous flies, Sturtevant conducted a testcross. He crossed female heterozygous flies to males recessive for all three traits and examined the progeny. Since males contribute either a Y chromosome with no genes on it, or an X chromosome with recessive alleles at all three loci, the male contribution does not disguise the potentially recombinant female chromosomes.

Sturtevant obtained the results described in Table 14-2. Here is how he analyzed his data:

Consider the traits in pairs, and determine which classes involve a crossover event.

1. For the body trait *(y)* and the eye trait *(w),* the first two classes, [+ +] and [y w], involve no crossovers (they are parental combinations), so no progeny numbers are tabulated for these two classes on the "body-eye" column (enter a dash).

2. The next two classes have the same body-eye combination as the parents, [+ +] and [y w], so again no numbers are entered as recombinants.

3. The next two classes, [+ w] and [y +], do *not* have the same body-eye combinations as the parent chromosomes (which were [w y] and

are placed side by side in the reconstructions. In this way relatively small structural alterations in individual chromosomes can be detected. This method is now standard for detecting those kinds of human genetic abnormalities which are associated with structural changes in the chromosomes or with the duplication or absence of particular chromosomes. Down syndrome, for example, is associated with a duplication of chromosome 21 and can therefore be detected in karyotypes taken from embryonic or other cells.

Another important technique that is used in comparing individual chromosomes and

TABLE 14-2   STURTEVANT'S RESULTS

| | PHENOTYPES | | | NUMBER OF PROGENY | CROSSOVER TYPES | | |
	BODY	EYE	WING		BODY-EYE	EYE-WING	BODY-WING
Parental	+	+	+	758	—	—	—
	y	*w*	*min*	700	—	—	—
Single crossover	+	+	*min*	401	—	401	401
	y	*w*	+	317	—	317	317
	+	*w*	*min*	16	16	—	16
	y	+	+	12	12	—	12
Double crossover	+	*w*	+	1	1	1	—
	y	+	*min*	0	0	0	—
TOTAL				2205	29	719	746
Recombination frequency (%)					1.315	32.608	33.832

[+ +]), so one enters the observed numbers of progeny into the tabulation of crossover types, 16 and 12, respectively.

4. The last two classes also differ from parental chromosomes in their body-eye combination, so one again enters the observed numbers of each class into the tabulation, 1 and 0.

5. The sum of the numbers of observed progeny that are recombinant between body *(y)* and eye *(w)* is 16 + 12 + 1, or 29. Since the total number of progeny is 2205, this amount of crossing-over represents 29/2205, or 0.01315. The percentage of recombination between *y* and *w* is thus 1.315%, or 1.31 centimorgans.

To estimate the percentage of recombination between body *(w)* and wing *(min)*, one proceeds in the same manner, obtaining a value of 32.608%, or 32.61 centimorgans. Similarly, eye *(y)* and wing *(min)* are separated by a recombination distance of 33.832%, or 33.83 centimorgans.

This then is our genetic map. The biggest

distance, 33.83 centimorgans, separates the two outside genes, which are evidently *y* and *min*. The gene *w* is between them, near *y*.

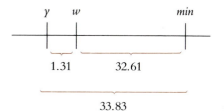

The two distances 1.31 and 32.61 do not add up to 33.83 but rather to 33.92. The difference, 0.09, represents chromosomes in which two crossovers occurred, one between *y* and *w* and another between *w* and *min*. These chromosomes are not counted in the tabulation of recombinants between *y* and *min*.

Genetic maps such as this are key tools in genetic analysis, permitting an investigator reliably to predict the way in which a newly discovered trait will recombine with many others, once it has been located on the chromosome map.

determining their karyotypes has to do with the functioning of their DNA. In different kinds of organisms particular genes are active at different times during the process of development. The chromomeres that condense in early cell division reflect the regions of gene activity, and their pattern is therefore characteristic of a particular kind of organism. By staining chromomeres chemically, it is possible to visualize this pattern as bands under the microscope (Figure 14–17). With suitable stains to reveal the condensed regions, the 23 different kinds of human chromosomes can be distinguished from one another.

THE SIGNIFICANCE OF MEIOSIS

When you first encountered mitosis in Chapter 8, it seemed a very complex process compared with the binary fission that divides the cells of bacteria. Relative to meiosis, however, mitosis is simple (Figure 14–18). Almost all of the complications occur during meiotic prophase, when homologous chromosomes are brought together for a brief time during which crossing-over occurs. Functionally, crossing-over holds homologous chromosomes together so that only one face of each centromere is available to bind to microtubules; if that were not the case, chromosome segregation would not be possible. It is not likely, however, that crossing-over first evolved as a means of binding homologues together during meiosis. Rather it seems probable that the pairing of homologous chromosomes together within the synaptonemal complex provided a mechanism for correcting damage to the DNA that affected both strands of a chromosome, such as a two-strand break of the helix. Such repair mechanisms and their relation to crossing-over are discussed in Chapter 17.

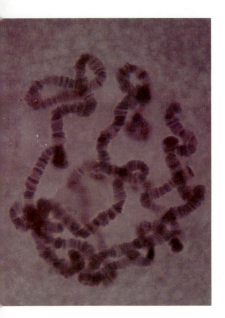

FIGURE 14-17

In Drosophila *the chromomeres on the salivary gland chromosomes undergo 11 cycles of duplication; as they do so, the daughter copies remain side by side, their chromomeres lined up with one another. The result of this process is a highly duplicated structure called a* **polytene chromosome.** *In cross section, a polytene chromosome is not unlike a telephone cable, with thousands of individual parallel strands. In each polytene chromosome, more than 1000 duplicates of each partially condensed chromomere are aligned with one another. Bands of various thicknesses reveal the position of one or more active genes. The pattern of bands is highly characteristic of species and even sometimes individuals of* Drosophila.

FIGURE 14-18

A comparison of meiosis with mitosis. Meiosis involves two serial nuclear divisions with no DNA replication between them, and so produces four daughter cells, each with half the original amount of DNA.

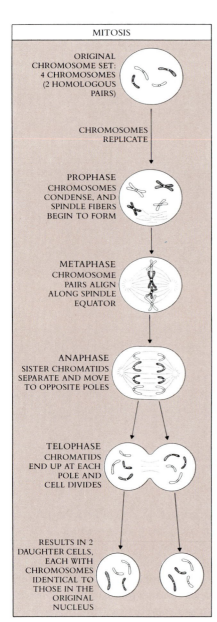

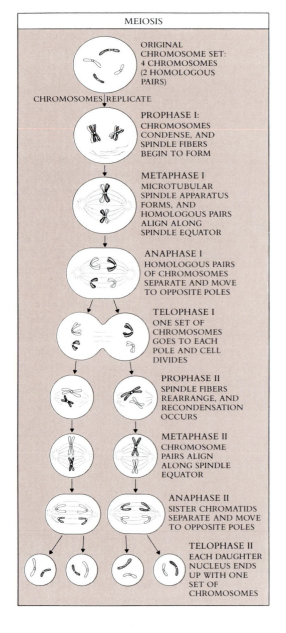

Most significant, however, are the genetic consequences of crossing-over—the recombination of segments of chromosomes. The continual generation of genetic variability by crossing-over during meiosis has provided eukaryotes with a wealth of genetic alternatives from which selection could choose better adapted types. The provision of these alternatives has been the critical factor in promoting rapid evolution in eukaryotes.

SUMMARY

1. Meiosis is a special form of nuclear division that occurs during gamete formation in most eukaryotes. In meiosis there is a single replication of the chromosomes and two chromosome separations.

2. Meiosis consists of a pair of serial nuclear divisions. Two unique events that occur during meiosis are synapsis—the intimate pairing of homologous chromosomes—and a lack of chromosome replication before the second nuclear division.

3. Crossing-over is an essential element of meiosis. The crossing-over between homologues that occurs during synapsis binds the two homologous chromosomes together, like dancers whose belts are linked together. As a result of this, the spindle fibers are able to bind to only one face of each homologous chromosome, since the other face is covered by the opposite homologue. For this reason they do not pull the sister chromatids apart. Rather, their contraction pulls the paired homologous chromosomes apart, ultimately breaking the terminal chiasmata that link them.

4. At the end of meiosis I one member of each bivalent is present at each of the two poles of the dividing nucleus. These chromosomes already consist of two chromatids, which differ from one another as a result of the crossing-over that occurred when the chromosomes were paired with their homologues. No further replication occurs before the next nuclear division, which is a normal mitosis that occurs at each of the two poles. The sister chromatids simply separate from one another. This process results in the formation of four clusters of chromosomes, each with a complement of half the number of chromosomes that were present initially. Because each daughter nucleus has one copy of each chromosome, any or all of the clusters can function as a gamete.

5. The first clear evidence that genes reside on chromosomes was provided by Thomas Hunt Morgan. Morgan demonstrated that the segregation of the white-eye trait in *Drosophila* was associated with the segregation of the X chromosome, the one responsible for sex determination.

6. The first evidence that crossing-over occurs between chromosomes was provided by Curt Stern. Stern showed that when two Mendelian traits exchange during a cross, so do visible abnormalities on the ends of the chromosomes bearing the traits.

7. The frequency of crossing-over between genes can be used to construct genetic maps, which are representations of the physical locations of genes on chromosomes, inferred from the degree of crossing-over between particular pairs of genes.

8. Eukaryotic chromosomes are composed of DNA complexed with protein, often associated with RNA molecules in the process of being made. Some of the DNA is permanently condensed into constitutive heterochromatin, while the rest is condensed only during cell division.

9. During early phases of cell division, condensation of euchromatin begins, active genes condensing first. These active regions are visible as chromomeres. Each chromosome exhibits a characteristic pattern of chromomeres.

10. The chromosomes of different individuals can be compared in shape, size, and other features of their overall morphology, as well as in the banding patterns revealed by different stains. The overall chromosomal constitution of an individual is called its karyotype.

SELF-QUIZ

The following questions pinpoint important information in this chapter. You should be sure to know the answers to each of them.

1. Chromosomes are tiny threadlike bodies that occur free in the cytoplasm of eukaryotic cells. True or false?

2. A cell that contains a single basic complement of chromosomes is said to be _____.

3. In some cell divisions the two daughter cells each contain half the number of chromosomes possessed by the parental cell. Such a division is called a _____ division and occurs in the eukaryotic cell in the process of _____.

4. Meiosis involves two sequential nuclear divisions. In this process how many times does the DNA replicate?

5. Human beings are diploid, because they have two copies of each of their chromosomes. These two copies are called _____.

6. Crossing-over occurs in what stage of meiosis?

7. In meiosis DNA replication occurs
   (a) in prophase
   (b) in metaphase
   (c) between the first and second meiotic divisions
   (d) during the first meiotic division
   (e) before prophase

8. When synapsis occurs, the structure that results is called a _____ complex.

9. The **X**-shaped chromosomal configuration that results from crossing-over is called a _____.

10. In what phase of meiosis does the nuclear envelope break down?

11. In what phase of meiosis do the chromosomes move to opposite poles?

12. How many haploid complements of chromosomes are present after the two nuclear divisions of meiosis?

13. When a male white-eye mutant fly mates with a normal red-eye female, all the progeny exhibit red eyes. Does this prove that the white-eye trait is recessive to the normal red-eye condition? Explain your answer.

14. When two of the red-eye progeny of the previous question are allowed to mate, one fourth of the progeny exhibit white eyes. Is this the result predicted for a simple recessive Mendelian trait?

15. The white-eye progeny in question 14 are all male. Why?
   (a) It is physiologically impossible for females to have white eyes.
   (b) The white-eye mutation originally occurred in a male.
   (c) The white-eye mutation occurred on an X chromosome.
   (d) A testcross was not carried out.

16. The closer together two genes are on a chromosome, the less likely it is that crossing-over will occur between them. True or false?

17. Chromosomes contain about how much protein?
    (a) none
    (b) half as much as DNA
    (c) twice as much as DNA
    (d) the same amount as DNA
    (e) trace amounts

18. Some portions of chromosomes are permanently condensed. These regions are referred to as _____.

19. Which is larger, a chromomere or a nucleosome?

20. The place on a chromosome where the spindle fibers attach during cell division can be seen as a constriction. It is called a _____.

THOUGHT QUESTIONS

1. Humans have 23 pairs of chromosomes, 22 pairs that play no role in sex determination and an XX (female) or XY (male) pair. Ignoring the effects of crossing-over, what proportion of your eggs or sperm contain all of the chromosomes you received from your mother?

2. You collect in your backyard two individuals of *Drosophila melanogaster,* one a young male and the other a young unmated female. Both are normal in appearance, with the vivid red eyes typical of *D. melanogaster.* You keep the two flies in the same vial, where they mate. Two weeks later there are hundreds of offspring flying around in the vial. They all have normal red eyes. From among them you select 100 individuals, some male and some female. You place each of these individuals in a vial and add a fly of the opposite sex that you know to be homozygous for a recessive allele called *sepia* (which when homozygous leads to black eyes). These flies thus have the genotype *se/se.* You thus conduct crosses in 100 vials. Examining the results of your 100 crosses, you observe that in about half of them, only normal red-eyed progeny flies are produced, whereas in the other half 50% have red eyes and 50% have black eyes. What must have been the genotypes of your original flies?

3. Hemophilia is a recessive sex-linked human blood disease that leads to failure of blood to clot normally. One form of hemophilia has been traced to the royal family of England, from which it spread throughout the royal families of Europe. It originated as a mutation either in Prince Albert or in his wife, Queen Victoria.
   a. Prince Albert did not himself have hemophilia. If the disease is a sex-linked recessive abnormality, how can it have originated in Prince Albert, who is a male and therefore is expected to exhibit recessive sex-linked traits, without Prince Albert suffering from hemophilia?
   b. Alexis, the son of Czar Nicholas II of Russia and Empress Alexandra (a granddaughter of Victoria) had the disease, while Anastasia, their daughter, did not. Anastasia died, a victim of the Russian revolution, before she bore any children. Can we assume that Anastasia would have been a carrier of the disease? What difference to your answer does it make if the disease originated in Nicholas II or in Alexandra?

4. There once was a lonely and rather sour buzzard named Clyde. It came as no surprise to any who knew him that Clyde had no offspring—no female buzzard would come anywhere near a buzzard with his personality. Clyde, however, nursed a secret desire to pass on his genes, and one day he hit upon a plan. He had heard from his boss, Professor Johnson, that the St. Louis Zoo practices birth control among its captive birds by the simple expedient of keeping only female birds. In dark of night he invaded the zoo and there wooed numerous female buzzards, none of whom knew the meanness of his nature. Soon buzzard eggs were hatching all over the zoo. Now a new issue arose to give Clyde pain. In reading his boss's genetics notes he made a dread discovery: genes recombine during meiosis. This meant that there was a

chance that his wonderfully horrible combination of characteristics might be diluted out by other more "normal" alternatives in subsequent generations. Clyde brooded on this for quite a while. But he finally decided that he need not worry, since he remembered his scrawny mother telling him on her knee that the two traits he most cared about, *small mind* and *hard heart,* were closely linked to one another; his mother was in fact homozygous for these traits, as well as for *scrawniness of frame.* Because Clyde shows all three traits, even though his father was normal and did not, all three traits are dominant.

Clyde asked his boss to look into this matter for him by examining the baby buzzards at the zoo. Dr. Johnson located 1000 baby buzzards apparently sired by Clyde in his nightly visits and checked out the smallness of their minds, the hardness of their hearts, and the scrawniness of their frames. Here is what he found:

NUMBER	MIND	FRAME	HEART
235	Normal	Normal	Normal
230	Small	Normal	Hard
226	Small	Scrawny	Hard
221	Normal	Scrawny	Normal
24	Small	Normal	Normal
23	Normal	Normal	Hard
21	Normal	Scrawny	Hard
20	Small	Scrawny	Normal

Dr. Johnson went home and told Clyde he had nothing to worry about, that hardness of heart and smallness of mind are very closely linked, relative to other Clyde traits such as scrawniness of frame. Was he right? Explain, backing up your argument with an appropriate genetic map.

FOR FURTHER READING

BAUER, W., F.H.C. CRICK, and J. WHITE: "Supercoiled DNA," *Scientific American,* July 1980, pages 118-134. The DNA double helix itself winds into a helix. This article describes current research into the mechanisms that cells employ to perform this feat.

GARDNER, E., and P. SNUSTAD: *Principles of Genetics,* 2nd ed., John Wiley & Sons, Inc., New York, 1984. A relatively simple introduction to genetics, written from a human perspective.

JOHN, B.: "Myths and Mechanisms of Meiosis," *Chromosoma,* vol. 54, pages 295-325, 1976. A fine description of meiosis as a physical process, with focus on mechanisms.

KORNBERG, R.D., and A. KLUG: "The Nucleosome," *Scientific American,* February 1981, pages 52-79. The DNA superhelix is wound on a series of protein spools.

MORGAN, T.H.: "Sex-Limited Inheritance in *Drosophila,*" *Science,* vol. 32, pages 120-122, 1910. Morgan's original account of his famous analysis of the inheritance of the white-eye trait.

PICKETT-HEAPS, J., D. TIPPIT, and K. PORTER: "Rethinking Mitosis," *Cell,* vol. 29, pages 729-744, 1982. An advanced but very rewarding assessment of the evidence concerning the evolution of mitosis and meiosis.

SNYDER, L., D. FREIFELDER, and D. HARTL: *General Genetics,* Jones and Bartlett, Publishers, Inc., Boston, 1985. An up-to-date text with a good discussion of meiosis.

STRICKBERGER, M.: *Genetics,* 3rd ed., Macmillan Publishing Co., New York, 1985. The best general undergraduate genetics text available, with a strong treatment of chromosomes, meiosis, crossing-over, and genetic mapping.

SUTTON, W.S.: "The Chromosomes of Heredity," *Biological Bulletin,* vol. 4, pages 213–251, 1903. The original statement of the chromosomal theory of heredity.

INVESTIGATING THE MECHANISM OF HEREDITY: AN EXPERIMENTAL JOURNEY

Chapter 15

Overview

This chapter describes the chain of experiments that have led us to our current understanding of the cellular mechanisms of heredity. The experiments are among the most elegant in science. Just as in a good detective story, each conclusion drawn has led to new questions. The intellectual path that the geneticists who finally unraveled these facts have taken in their pursuit of understanding has not always been a straight one, the best questions not always obvious. But however erratic and lurching the course of the experimental journey, our picture of heredity has become progressively clearer, the image more sharply defined.

For Review

Here are some important terms and concepts that you will encounter in this chapter. If you are not familiar with them, you should review them before proceeding.

- **Structure of DNA** (Chapter 4)

- **Amino acid sequence determines protein shape** (Chapter 4)

- **Ribosomes** (Chapter 6)

The realization, reached at the beginning of this century, that patterns of heredity can be explained by the segregation of chromosomes in meiosis was one of the most important advances in human thought. Not only did it lead directly to formulation of the science of genetics and thus to great progress in agriculture and medicine, but also it profoundly influenced the way in which we think about ourselves. By removing the mystery from the process of heredity, it made the biological nature of human beings seem much more approachable. It also raised the question that was to occupy biologists for more than half a century: what is the exact nature of the connection between hereditary traits and the chromosomes?

We are now able to answer this question. We understand in considerable detail the mechanism by which the information on the chromosomes is converted into organisms with eyes, arms, and inquiring minds. Our understanding was not acquired all at once—deduced in a single flash of insight—but rather was developed slowly over many years by a succession of investigators. How they did this and what they learned are the subjects of this chapter.

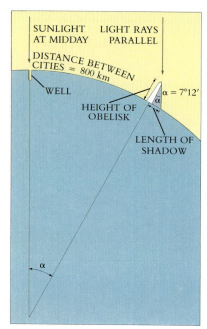

How Eratosthenes deduced the circumference of the earth.
1 *On the day when sunlight shone directly down into a deep well at Syene, near the present location of the Aswan Dam in Egypt, he measured the length of the shadow cast by a tall obelisk in the city of Alexandria.*
2 *The shadow's length and the obelisk's height formed two sides of a triangle. From principles of Euclidian geometry recently developed at that time, he was able to use the height and length to deduce the angle α. It proved to be 7 degrees 12 minutes, exactly one fiftieth of a circle (360 degrees).*
3 *If angle α = one fiftieth of a circle, then the distance between the obelisk (Alexandria) and the well (Aswan) must equal one fiftieth of the circumference of the earth.*
4 *Eratosthenes had heard that it was a 50-day camel trip to the vicinity of the well, and so, figuring that a camel travels about 18.5 kilometers per day, he estimated the distance between obelisk and well as 925 kilometers. (Eratosthenes used different units, of course.)*
5 *Eratosthenes thus estimated the circumference of the earth to be 50 × 925 = 46,250 kilometers. In reality the distance from well to obelisk is just over 800 kilometers, 15% less than Eratosthenes' crude estimate, so his estimate of the earth's circumference was 15% too large. Employing a value of 800 kilometers, Eratosthenes' estimate would have been 50 × 800 = 40,000 kilometers. The real value is 40,075.*

THE SCIENTIFIC METHOD

In Egypt more than 200 years before Christ was born, the Greek Eratosthenes correctly estimated the circumference of the earth. He waited until the longest day of the year, when at high noon the sun was directly above a deep well in the city of Syene, shining down on its deepest waters. On that midday he measured the length of the shadow of a tall obelisk in the city of Alexandria about 800 kilometers away. Because he knew the distance between the two cities, and the height of the obelisk, he was able to employ the principles of Euclidian geometry to correctly deduce the circumference of the earth (Figure 15-1). Reasoning in this way, from general principles to a specific case, is called **deductive reasoning.**

Almost two millenia later, in the early 1600s in Europe, Francis Bacon and other "new philosophers" (we would call them scientists) began to employ a different kind of reasoning to study nature, using knowledge gained from investigating specific cases to infer general principles. Reasoning in this way, from specific cases to general principles, is called **inductive reasoning.** The systematic application of inductive reasoning in attempts to answer questions about the world around us marked the beginning of the scientific revolution, which has changed our civilization greatly.

How does a scientist reason inductively? How do we learn which general principles are true, from among the many that might be? We do this by attempting systematically to demonstrate that certain proposals are *not* true, not consistent with what we learn from experimental observation. Those proposals which are not disproved by experiment are retained, for the moment, as working ideas. They are useful because they fit the facts as we know them but are subject to future rejection if, in the light of new information, they are found wanting. We call the proposals that scientists test hypotheses.

A hypothesis is a proposition that might be true. The test of a hypothesis is called an experiment. An experiment evaluates alternative hypotheses. "There is no light in this dark room because the light switch is turned off" is a hypothesis; an alternative hypothesis is "There is no light in the room because the bulb is blown"; another alternative might be "You are going blind." An experimental test works by eliminating one of the hypotheses. For example, you might test these alternatives by reversing the position of the light switch. Suppose that when you do this the light does not come on. The result of your experiment is thus to disprove the first of the hypotheses—something other than the setting of the light switch must be at fault. Note that a test such as this does not prove that any one alternative is true, but rather demonstrates that one of them is not. In this instance the fact that the light does not come on does not establish that the switch was in fact in the "on" position (it might have been either "on" or "off" if the bulb was burnt out), but rather that the switch setting alone is not the sole reason for the darkness. A successful experiment is one in which an alternative hypothesis is demonstrated to be inconsistent with experimental results and thus rejected. Scientific progress is made the same way a marble statue is, by chipping away unwanted bits.

Hypotheses that stand the test of time, never being shown false despite repeated tests, are called theories. Thus one speaks of the theory of gravity, the theory of relativity, and the theory of evolution. Theories are the solid ground of science, those things of which we are most certain.

It used to be fashionable to speak of the "scientific method" as consisting of an orderly sequence of logical "either/or" steps, each step rejecting one of two mutually incompatible alternatives, as if trial-and-error testing would inevitably lead one through the maze of uncertainty that always impedes scientific progress. If this were indeed so, a computer would make a good scientist. In fact, science is not done this way (p. 283). As the British philosopher Karl Popper has pointed out, if you ask successful scientists how they do their work, you would discover that without exception they design their experiments with a fair idea of how the experiments are going to come out—what Popper calls an "imaginative preconception" of what the truth might be. A hypothesis that a successful scientist tests is not just any hypothesis, but a "happy guess": the scientist integrates all that he or she knows, and also allows imagination full play, in attempting to get a sense of what *might*

THE NATURE OF SCIENTIFIC ACTIVITY

Scientists at work have the look of creatures following genetic instructions; they seem to be under the influence of a deeply placed human instinct. They are, despite their efforts at dignity, rather like young animals engaged in savage play. When they are near to an answer their hair stands on end, they sweat, they are awash in their own adrenalin. To grab the answer, and grab it first, is for them a more powerful drive than feeding or breeding or protecting themselves against the elements.

It sometimes looks like a solitary activity, but it is as much the opposite of solitary as human behavior can be. There is nothing so social, so communal, so interdependent. An active field of science is like an immense intellectual anthill; the individual almost vanishes into the mass of minds tumbling over each other, carrying information from place to place, passing it around at the speed of light.

There are special kinds of information that seem to be chemotactic. As soon as a trace is released, receptors at the back of the neck are caused to tremble, there is a massive convergence of motile minds flying upwind on a gradient of surprise, crowding around the source. It is an infiltration of intellects, an inflammation.

There is nothing to touch the spectacle. In the midst of what seems a collective derangement of minds in total disorder, with bits of information being scattered about, torn to shreds, disintegrated, reconstituted, engulfed, in a kind of activity that seems as random and agitated as that of bees in a disturbed part of the hive, there suddenly emerges, with the purity of a slow phrase of music, a single new piece of truth about nature. . . .

There is something like aggression in the activity, but it differs from other forms of aggressive behavior in having no sort of destruction as the objective. While it is going on, it looks and feels like aggression: get at it, uncover it, bring it out, grab it, halloo! It is like a primitive running hunt, but there is nothing at the end of it to be injured. More probably, the end is a sigh. But then, if the air is right and the science is going well, the sigh is immediately interrupted, there is a yawping new question, and the wild, tumbling activity begins once more, out of control all over again.

From THE LIVES OF A CELL by Lewis Thomas. Copyright © 1973 by Lewis Thomas. Originally appeared in *The New England Journal of Medicine*. Reprinted by permission of Viking Penguin, Inc.

---

be. It is because insight and imagination play such a large role in scientific progress that some scientists are so much better at science than others, for precisely the same reason that Beethoven and Mozart stand out above most other composers.

The large role played by creativity in formulating scientific hypotheses is also the reason that the schedules of governmental programs do not dictate the rate of scientific progress. "Go out and make a discovery by next Tuesday" is simply not a practical directive. What does succeed, time and again, is to turn a lot of creative scientists loose to do the best they can. This is how the riddle of heredity was solved. Several generations of biologists applied to the problem their creativity, coupled with rigorous inductive experimentation. In this chapter we will retrace their experimental journey. We will not dwell for long at any one stage, nor worry overly about technical details. Instead, we will concentrate on tracing the history of the major hypotheses in this field and the ways in which the results of specific experiments have changed these ideas. The unraveling of the mechanism of heredity constitutes one of our greatest intellectual journeys.

We will focus on six questions, a series that illustrates the incremental nature of scientific progress, a process in which each advance depends on and elaborates the results that precede it. As in a serialized mystery or a soap opera, what we have learned from answering each question has led naturally, inevitably, to the questions that follow:

1. Where do cells store the hereditary information?

2. Which component of the chromosomes contains the hereditary information?

3. How is the information in DNA reproduced so accurately?

4. What is the unit of hereditary information?

5. How are genes encoded?

6. How does DNA sequence dictate protein structure?

WHERE DO CELLS STORE HEREDITARY INFORMATION?

Perhaps the most basic question that one can ask about hereditary information is where it is stored in the cell. Of the many approaches that one might take to answer this question, let us start with a simple one: cut a cell into pieces and see which of the pieces are able to express hereditary information. For this experiment we will need a single-celled organism that is large enough to operate on conveniently and differentiated enough that the pieces can be told apart.

An elegant experiment of this kind was performed by the Danish biologist Joachim Hammerling in 1943. As an experimental subject, Hammerling chose the large unicellular green alga *Acetabularia* (Figure 15-2). Individuals of this genus have distinct foot, stalk, and cap regions, all of which are differentiated parts of a single cell. The nucleus of this cell is located in the foot. As a preliminary experiment, Hammerling tried amputating the caps or feet of individual cells. He found that when the cap is amputated, a new one regenerates from the remaining portions of the cell. When the foot is amputated and discarded, however, no new foot is regenerated. Hammerling concluded that the hereditary information resided within the foot, or basal portion, of *Acetabularia*.

To test this hypothesis, Hammerling selected individuals from two species of the genus in which the caps looked very different from one another: *Acetabularia mediterranea*, which has a disk-shaped cap, and *A. crenulata*, which has a branched, flowerlike cap. Hammerling cut the stalk and cap away from an individual of *A. mediterranea*; to the remaining foot he grafted a stalk cut from a cell of *A. crenulata* (Figure 15-3). The cap that formed looked something like the flower-shaped one characteristic of *A. crenulata*, although it was not exactly the same.

Hammerling then cut off this regenerated cap and found that a disk-shaped one exactly like that of *A. mediterranea* formed in the second regeneration and in every regeneration thereafter. This experiment strengthened Hammerling's earlier conclusion that the instructions which specify the kind of cap that is produced are stored in the foot of the cell—and probably therefore in the nucleus—and that these instructions must pass from the foot through the stalk to the cap. In his regeneration experiment the initial flower-shaped cap was formed as a result of the instructions that were already present in the transplanted stalk when it was excised from the original *A. crenulata* cell. In contrast, all subsequent caps used

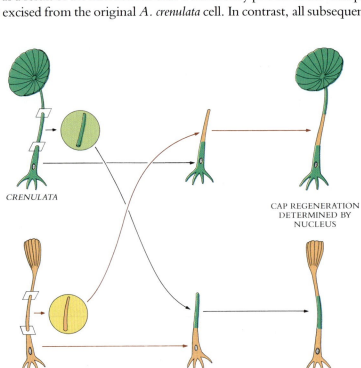

CRENULATA

CAP REGENERATION
DETERMINED BY
NUCLEUS

MEDITERRANEA

FIGURE 15-3

Hammerling's Acetabularia *reciprocal graft experiment. To the foot of each species he grafted a stalk of the other. In each case the appearance of the cap that eventually developed was dictated by the foot and not the stalk. Although the first cap to develop was sometimes intermediate in appearance, when it was cut away and another cap was allowed to form, that cap was always the one characteristic of the species from which the foot of the grafted cell was derived.*

new information, derived from the foot of the *A. mediterranea* cell onto which the stalk had been grafted. In some unknown way the original instructions that had been present in the stalk were eventually "used up."

Hammerling's experiments identified the nucleus as the likely repository of the hereditary information but did not prove definitely that this was the case. To do that, isolated nuclei had to be transplanted. Such an experiment was carried out in 1952 by American embryologists Robert Briggs and Thomas King. Using a glass pipette drawn to a fine tip and working with a microscope (Figure 15-4), Briggs and King removed the nucleus from a frog egg; without the nucleus, the egg would not develop. They then replaced the absent nucleus with one that they had isolated from a cell of a more advanced frog embryo (Figure 15-5). The implant of this nucleus ultimately caused an adult frog to develop from the egg. Clearly the nucleus was directing the frog's development.

Hereditary information is stored in the nucleus of eukaryotic cells.

Can each and every nucleus of an organism direct the development of an entire adult individual? The Briggs and King experiment did not answer this question definitively, since the nuclei that they took from more advanced frog embryos often caused the eggs to which they were transplanted to develop abnormally. But at Yale, John Gurdon, working with another amphibian, was able to transplant nuclei isolated from developed tadpole tissue into eggs from which the nuclei had been removed and obtain normal development.

Have the nuclei in the cells of adult animals lost their hereditary information? This question has proved very difficult to answer, since animal development is so complex. In plants, on the other hand, a simple experiment did yield a clear-cut answer. At Cornell University in 1958, plant physiologist F.C. Steward let fragments of fully developed carrot tissue (bits of conducting tissue called phloem) swirl around in a rotating flask containing liquid growth medium. Individual cells broke away and tumbled through the liquid. Steward observed that these cells often divided and differentiated into multicellular roots. If these roots were then immobilized by placing them in a gel, they would go on to develop into entire plants that could be transplanted to soil and develop normally into maturity. Figure

FIGURE 15-4

Working with a microscope, it is possible to pierce a cell with the fine tip of a glass micropipette without rupturing the cell.
A *The tip is just beginning to pierce the cell.*
B *The tip has entered the cell and is about to draw up the nucleus (arrow).*

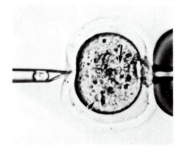

A

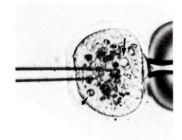

B

FIGURE 15-5

Briggs and King's nuclear transplant experiment. Two strains of frog were used that differed from each other in the number of nucleoli their cells possessed. The nucleus was removed from an egg of one strain, either by sucking the egg nucleus up into a micropipette or, more simply, by destroying it with ultraviolet light. Briggs and King then injected into this anucleate egg a nucleus obtained from a differentiated cell of the other species—in this case a cell isolated from the intestine of a tadpole of that species. The hybrid egg was then allowed to develop. One of three results was obtained in individual experiments: (1) no growth occurred, perhaps reflecting damage to the egg cell during the nuclear transplant operation; (2) normal growth and development occurred up to the blastula stage, but subsequent development was not normal and the embryo did not survive; or (3) normal growth and development occurred, eventually leading to the development of an adult frog. That frog was of the species that contributed the nucleus and not of the species that contributed the egg. Only a few experiments gave this third result, but they serve to clearly establish that the nucleus directs frog development.

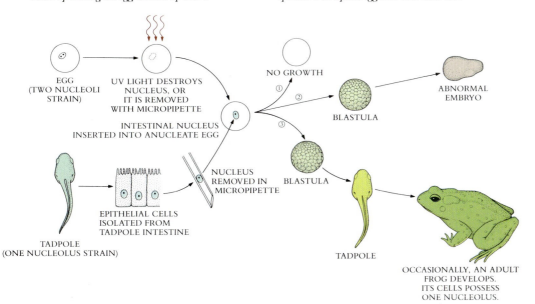

EGG (TWO NUCLEOLI STRAIN)

UV LIGHT DESTROYS NUCLEUS, OR IT IS REMOVED WITH MICROPIPETTE

INTESTINAL NUCLEUS INSERTED INTO ANUCLEATE EGG

NO GROWTH

BLASTULA

ABNORMAL EMBRYO

NUCLEUS REMOVED IN MICROPIPETTE

BLASTULA

EPITHELIAL CELLS ISOLATED FROM TADPOLE INTESTINE

TADPOLE (ONE NUCLEOLUS STRAIN)

TADPOLE

OCCASIONALLY, AN ADULT FROG DEVELOPS. ITS CELLS POSSESS ONE NUCLEOLUS.

32-3 presents a fuller description of this experiment. Steward's experiment makes clear that in plants at least some of the cells present in adult individuals do contain a full complement of hereditary information.

> With rare exceptions, the nuclei of all cells of multicellular eukaryotes contain a full complement of genetic information. In many of the tissues of adult animals, however, the expression of much of this information may be blocked.

WHICH COMPONENT OF THE CHROMOSOMES CONTAINS THE HEREDITARY INFORMATION?

The identification of the nucleus as the source of hereditary information focused attention on the chromosomes, which were already suspected to be the vehicles of Mendelian inheritance. Specifically, biologists wondered how the actual hereditary information was arranged in the chromosomes. It was known that chromosomes contain both protein and DNA. On which of these was the hereditary information written?

Over a period of about 30 years, starting in the late 1920s, a series of investigators addressed this issue, resolving it clearly. We will describe three very different kinds of experiments, each of which yields a clear answer in a simple and elegant manner.

The Griffith-Avery Experiments

As early as 1928, a British microbiologist, Frederick Griffith, made a series of unexpected observations while experimenting with pathogenic (disease-causing) bacteria. When Griffith infected mice with a virulent strain of *Pneumococcus* bacteria, the mice died of blood poisoning, but when he infected similar mice with a strain of *Pneumococcus* that lacked a polysaccharide coat like that possessed by the virulent strain, the mice showed no ill effects. The coat was apparently necessary for successful infection.

As a control, Griffith injected normal but heat-killed bacteria into the mice to see if the polysaccharide coat itself had a toxic effect. The mice remained perfectly healthy. As a final control, he blended his two ineffective preparations—living bacteria whose coats had been removed and dead bacteria with intact coats—and injected the mixture into healthy mice (Figure 15-6). Unexpectedly, the injected mice developed disease symptoms, and

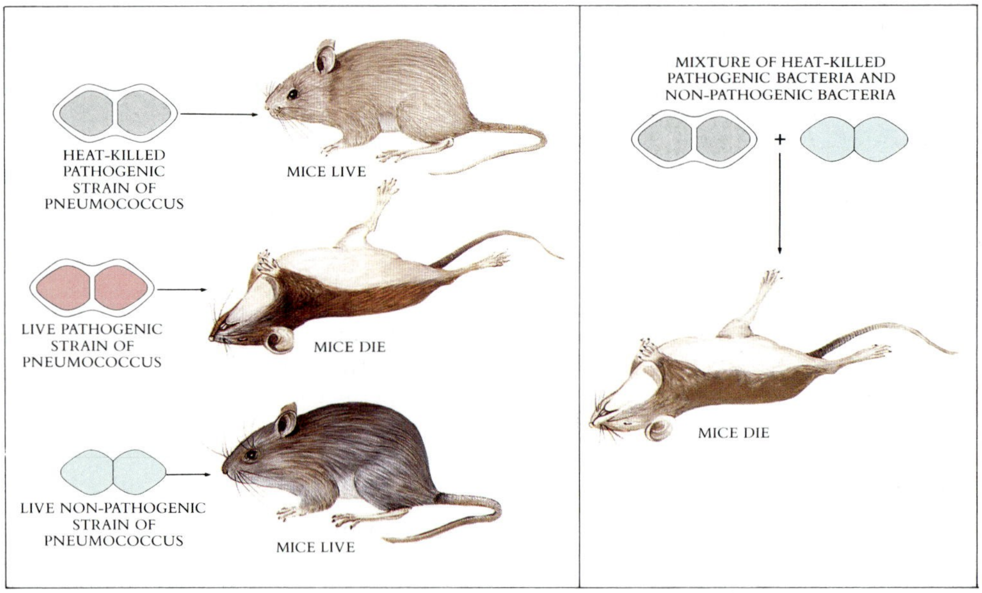

FIGURE 15-6

Griffith's discovery of transformation. The pathogenic bacterium Pneumococcus *will kill many of the mice into which it is injected, but only if the bacterial cells are covered with a polysaccharide coat. Living bacteria without such polysaccharide coats do not harm the mice, since the coat is necessary for successful infection. However, it is not the coat itself that is the agent of disease. When Griffith injected dead bacteria possessing polysaccharide coats, the mice were not harmed. However, if he injected a mixture of dead bacteria with polysaccharide coats (harmless) and live bacteria without such coats (harmless), many of the mice died. Griffith concluded that the live cells had been "transformed" by the dead ones—that the genetic information specifying the polysaccharide coat had passed from the dead cells to the living ones.*

HEAT-KILLED PATHOGENIC STRAIN OF PNEUMOCOCCUS — MICE LIVE

LIVE PATHOGENIC STRAIN OF PNEUMOCOCCUS — MICE DIE

LIVE NON-PATHOGENIC STRAIN OF PNEUMOCOCCUS — MICE LIVE

MIXTURE OF HEAT-KILLED PATHOGENIC BACTERIA AND NON-PATHOGENIC BACTERIA — MICE DIE

For the past two years, first with MacLeod and now with Dr. McCarty, I have
been trying to find out what is the chemical nature of the substance in the bacterial
extract which induces this specific change. The crude extract of Type III is full of
capsular polysaccharide, C (somatic) carbohydrate, nucleoproteins, free nucleic
acids of both the yeast and thymus type, lipids, and other cell constituents. Try to
find in the complex mixtures the active principle! Try to isolate and chemically
identify the particular substance that will by itself, when brought into contact with
the R cell derived from Type II, cause it to elaborate Type III capsular polysaccharide and to acquire all the aristocratic distinctions of the same specific type
of cells as that from which the extract was prepared! Some job, full of headaches
and heartbreaks. But at last *perhaps* we have it.

. . . if we prove to be right—and of course that is a big if—then it means that
both the chemical nature of the inducing stimulus is known and the chemical
structure of the substance produced is also known, the former being thymus
nucleic acid, the latter Type III polysaccharide, and both are thereafter reduplicated
in the daughter cells and after innumerable transfers without further addition of the
inducing agent and the same active and specific transforming substance can be
recovered far in excess of the amount originally used to induce the reaction. Sounds
like a virus—may be a gene. But with mechanisms I am not now concerned. One
step at a time and the first step is what is the chemical nature of the transforming
principle? Some one else can work out the rest. Of course the problem bristles with
implications. It touches the biochemistry of the thymus type of nucleic acids which
are known to constitute the major part of chromosomes but have been thought to
be alike regardless of origin and species. It touches genetics, enzyme chemistry,
cell metabolism and carbohydrate synthesis. But today it takes a lot of well
documented evidence to convince anyone that the sodium salt of deoxyribose
nucleic acid, protein free, could possibly be endowed with such biologically active
and specific properties and that is the evidence we are now trying to get. It is lots of
fun to blow bubbles but it is wiser to prick them yourself before someone else tries to.

Part of a letter from Oswald Avery to his brother Roy, written in May 1943. From R.D. Hotchkiss. In
Phage and the Origins of Molecular Biology, J. Cairns, G.S. Stent, and J.D. Watson, eds. (Cold Spring Harbor
Laboratory, 1966), pp. 185-186.

---

many of them died. The blood of the dead mice was found to contain high levels of normal
virulent *Pneumococcus* bacteria. Somehow the information specifying the polysaccharide
coat had passed from the dead bacteria to the live but coatless ones in the control mixture,
transforming them into normal virulent bacteria that infected and killed the mice.

Not until 1944 was the agent responsible for transforming *Pneumococcus* discovered.
In an elegant series of experiments Oswald Avery and his coworkers characterized what
they referred to as the "transforming principle" (see above). Its properties resembled those
of DNA rather than protein: the activity of the transforming principle was not affected by
protein-destroying enzymes but was lost completely in the presence of the DNA destroying enzyme DNAse.

When the transforming principle was purified, it indeed consisted predominantly of
DNA. Subsequently, it was shown that all but trace amounts of protein (0.02%) could be
removed without reducing the transforming activity. The conclusion was inescapable:
DNA is the hereditary material in bacteria. It has since proved possible to use purified DNA
to change the genetic characteristics of eukaryotic cells in tissue culture and even possible
to inject pure DNA into fertilized *Drosophila* eggs and thereby alter the genetic characteristics of the resulting adult.

The Hershey–Chase Experiment

Avery's results were not widely appreciated at first, many biologists preferring to believe
that proteins were the depository of the hereditary information. Another very convincing
experiment was soon performed, however, that was difficult to ignore. It was done in a

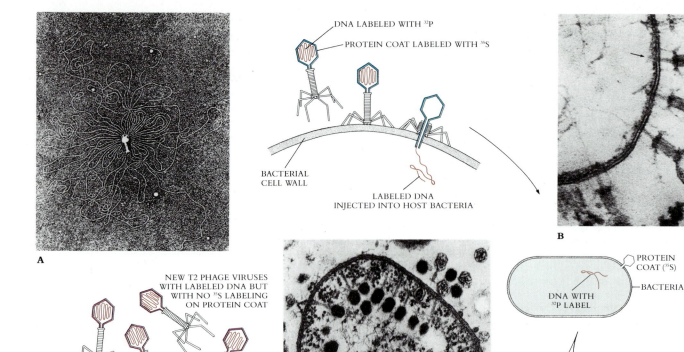

FIGURE 15-7

The Hershey-Chase double-label experiment. The T2 bacterial viruses that they employed have a simple structure: they are composed of a protein envelope within which DNA is packaged.

A *You can see the DNA released from a single virus particle. Hershey and Chase labeled the T2 virus particles with two radioisotopes: the protein bodies were labeled with* ^{35}S *(sulfur occurs in protein in the amino acids cysteine and methionine but does not occur in DNA), and the DNA molecules were labeled with* ^{32}P *(phosphorus occurs in the phosphate groups of DNA but does not occur in proteins). These T2 particles were then allowed to infect bacterial cells. Each virus body binds to the outside of the cell but, instead of entering, injects its DNA into the cell.*

B *Individual DNA strands are entering the cell from virus particles bound to its surface. Within the cell, the injected DNA commandeers the machinery of the cell and directs the synthesis of all the parts necessary to make new viruses.*

C *New virus particles are being assembled from parts within an infected cell. Eventually these new viruses will rupture the cell and be released into the surroundings.*

very simple system—viruses—so that a very direct experimental question could be asked. Viruses consist of either RNA or DNA with a protein coat; they are described in more detail in Chapter 25. These investigators focused on bacterial viruses, or bacteriophages, and carried out an experiment analogous to the transplant experiments described above. When a bacteriophage infects a bacterial cell, it first binds to the cell's outer surface and then injects its hereditary information into the cell. There the hereditary information directs the production of thousands of new virus particles within the cell. The host bacterial cell eventually falls apart, or lyses, releasing the newly made viruses.

In 1952 Alfred Hershey and Martha Chase set out to identify the material injected into the bacterial cell at the start of an infection. They used a strain of bacteriophage known as T2, which contains DNA rather than RNA, and designed an experiment to distinguish between the alternative hypotheses that the genetic material was DNA or that it was protein (Figure 15-7). Hershey and Chase labeled the DNA of these T2 bacteriophages with a radioactive isotope of phosphorus, ^{32}P, and, at the same time, labeled their protein coats radioactively with an isotope of sulfur, ^{35}S. Since the radioactive ^{32}P and ^{35}S isotopes emit particles of very different energies when they decay, they are easily told apart. The labeled viruses were permitted to infect bacteria. The bacterial cells were then agitated violently to shake the protein coats of the infecting viruses loose from the bacterial surfaces to which they were attached. Centrifuging the cells from solution, Hershey and Chase found that the ^{35}S label (and thus the virus protein) was now predominantly in solution with the dissociated virus particles, whereas the ^{32}P label (and thus the DNA) had transferred to the interior of the cells. The viruses subsequently released from the infected bacteria contained the ^{32}P label. The hereditary information injected into the bacteria that specified the new generation of virus particles was DNA and not protein.

The Fraenkel-Conrat Experiment

One objection might still be raised about accepting the hypothesis that DNA was the genetic material. Some viruses contain *no* DNA and yet manage to reproduce themselves quite satisfactorily. What is the genetic material in this case?

In 1957 Heinz Fraenkel-Conrat and coworkers isolated **tobacco mosaic virus (TMV)** from tobacco leaves. From ribgrass *(Plantago)*, a common weed, they isolated a second, rather similar kind of virus, **Holmes ribgrass virus (HRV).** In both TMV and HRV the viruses consist of protein and a single strand of RNA. When they had isolated these viruses, the scientists chemically dissociated each of them, separating their protein from their RNA. By putting the protein component of one virus with the RNA of another, they were able to reconstitute hybrid virus particles.

The payoff of the experiment was in its next step. To choose between the alternative hypotheses "the genetic material of viruses is protein" and "the genetic material of viruses is RNA," Fraenkel-Conrat now infected healthy tobacco plants with a hybrid virus composed of TMV protein capsules and HRV RNA, being careful not to include any non-hybrid virus particles (Figure 15-8). The tobacco leaves that were infected with the reconstituted hybrid virus particles developed the kinds of lesions that were characteristic of HRV and that normally formed on infected ribgrass. Clearly, the hereditary properties of the virus were determined by the nucleic acid in its core and not by the protein in its coat.

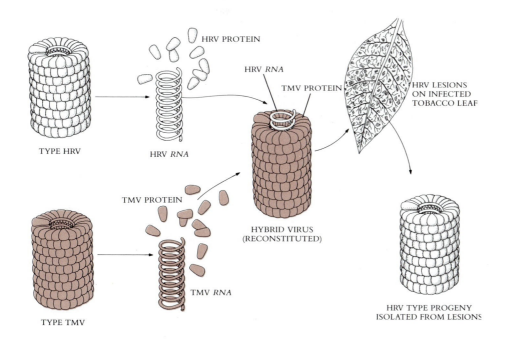

TYPE HRV — HRV PROTEIN — HRV *RNA* — HRV LESIONS ON INFECTED TOBACCO LEAF — TMV PROTEIN — HRV *RNA* — HYBRID VIRUS (RECONSTITUTED) — TMV PROTEIN — TMV *RNA* — TYPE TMV — HRV TYPE PROGENY ISOLATED FROM LESIONS

FIGURE 15-8

Fraenkel-Conrat's virus reconstitution experiment. Both TMV and HRV are plant RNA viruses that infect tobacco plants, causing lesions on the leaves. Because the two viruses produce different kinds of lesions, the source of any particular infection can be identified. In this experiment, TMV and HRV were both dissociated into protein and RNA, and the protein and RNA were separated from one another. Then hybrid virus particles were produced by mixing the HRV RNA and the TMV protein and allowing virus particles to form from these ingredients. When the reconstituted virus particles were painted onto tobacco leaves, lesions developed— of the HRV type. From these lesions, normal HRV virus particles could be isolated in great numbers; no TMV viruses could be isolated from the lesions. Thus the RNA (HRV) and not the protein (TMV) contains the information necessary to specify the production of the viruses.

Later studies have shown that many virus particles contain RNA rather than the DNA found universally in cellular organisms. When these viruses infect a cell, they make DNA copies of themselves, which can then be inserted into the cellular DNA as if they were cellular genes. The viruses responsible for human fever blisters are like this. Fever blisters are caused by an RNA virus called a herpesvirus, which lives within the cells of affected individuals as if it were a part of their genes; an outbreak of fever blisters results when the virus begins to reproduce RNA copies of itself.

DNA is the genetic material for all cellular organisms and most viruses, although some viruses use RNA.

HOW IS THE INFORMATION IN DNA REPRODUCED SO ACCURATELY?

As it became clear that the DNA molecule was the repository of the hereditary information, investigators began to puzzle over how such a seemingly simple molecule could carry out such a complex function. As we discussed in Chapter 4, a DNA molecule is simply a repeating chain of identical five-carbon sugars linked together head to tail. One of four ring-shaped organic bases—adenine (A), guanine (G), thymine (T), and cytosine (C)—protrudes from each of the sugars.

One reason why it was at first difficult to imagine that DNA plays such a complex role in heredity was that it was initially thought to be a simple repeating polymer, for example, one such as AGTC AGTC AGTC. By the late 1940s, however, careful chemical analyses by Erwin Chargaff and his colleagues at Columbia University had revealed that the base ratios of different DNAs varied widely; the sequences were not as simple as had originally been thought. Chargaff did observe an important underlying regularity: the amount of adenine present in all DNA molecules was equal to the amount of thymine, and the amount of guanine equaled the amount of cytosine. On the other hand, the amounts of adenine plus thymine often differed greatly from the amounts of guanine plus cytosine.

The significance of the regularities pointed out by Chargaff, although not immediately obvious, soon became clear. A British chemist, Rosalind Franklin, had carried out x-ray crystallographic analysis of fibers of DNA. In this process the DNA molecule is bombarded with an x-ray beam. When individual rays encounter atoms, their path is bent or defracted; the pattern created by the total of all these diffractions can be captured on a piece of photographic film. Such a pattern resembles the ripples created by tossing a rock into a smooth lake. By carefully analyzing the diffraction pattern, it is possible to develop a three-dimensional image of the molecule. The diffraction patterns that Franklin obtained suggested that the DNA molecule was a helical coil with repeating elements of 2 and 3.4 nanometers.

Learning informally of Franklin's results in 1953, James Watson and Francis Crick, two young investigators at Cambridge University, quickly worked out the probable structure of the DNA molecule (Figure 15-9). They analyzed the problem deductively: first they built models of the nucleotides, and then they tested how these could be assembled into a molecule that fit what they knew about the structure of DNA. They tried various possibilities, first assembling molecules with three strands of nucleotides wound around one another to stabilize the helical shape. None of these early efforts proved satisfactory. They finally hit on the idea that the molecule might be a simple double helix, one in which the bases of two strands pointed inward toward one another. By always pairing a purine, which is large, with a pyrimidine, which is small, the diameter of the duplex stays the same, 2 nanometers. Because hydrogen bonds can form between the two strands, the helical form is stabilized.

It immediately became apparent why Chargaff had obtained the results that he had (Figure 15-10)—because the purine adenine (A) will not form proper hydrogen bonds in this structure with cytosine (C) but will with thymine (T), every A is paired to a T. Similarly the purine guanine (G) will not form proper hydrogen bonds with thymine but will with cytosine, so that every G is paired with C.

The Watson-Crick model immediately suggested that the basis for copying the genetic information is **complementarity.** One chain of the DNA molecule may have any conceivable base sequence, but this sequence completely determines that of its partner in the duplex. If the sequence of one chain is ATTGCAT, the sequence of its partner in the duplex *must* be TAACGTA. Each chain in the duplex is a mirror image of the other. To copy the DNA molecule, one need only "unzip" it and construct a new complementary chain along each naked single strand.

The form of DNA replication suggested by the Watson-Crick model is called **semi-conservative,** because after one round of replication, the original duplex is not conserved;

FIGURE 15-9

Watson and Crick, delighted with their discovery, examine their first model of the DNA double helix.

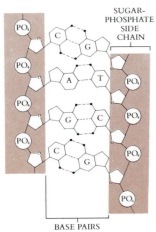

FIGURE 15-10

In a DNA duplex molecule only two base pairs are possible: adenine (A) with thymine (T), and guanine (G) with cytosine (C). A G-C base pair has three hydrogen bonds; an A-T base pair, only two.

FIGURE 15-11

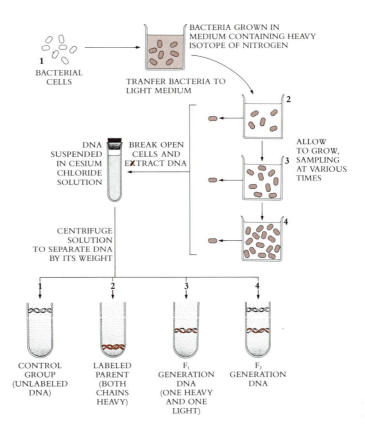

The Meselson and Stahl experiment. Bacterial cells were grown for several generations in a medium containing heavy nitrogen isotopes and then were transferred to a new medium containing only the normal lighter nitrogen isotope. At various times thereafter, samples were taken and the DNA was centrifuged in a cesium chloride solution. Because the cesium ion is so massive, the cesium chloride tends to settle in the rapidly spinning tube, establishing a gradient of cesium concentration. DNA molecules sink in the gradient until they reach a place where the cesium concentration has the same density as the density of the DNA; it then "floats" at that position. Because DNA built with heavy nitrogen isotopes is denser than normal DNA, it sinks to a lower position on the cesium gradient. What Meselson and Stahl found was that after one generation in "light" medium, a single band of intermediate density halfway between heavy and light was obtained (one strand of each duplex was labeled; the other was not). After a second cell division, two bands were obtained, one intermediate (one of the two strands was labeled) and one light (neither strand was labeled). Meselson and Stahl concluded that replication of the DNA duplex involves building new molecules by separating strands and assembling new partners on these templates.

instead, each strand of the duplex becomes part of another duplex. This prediction of the Watson-Crick model was tested in 1958 by Matthew Meselson and Frank Stahl of the California Institute of Technology (Figure 15-11). These two scientists grew bacteria for several generations in a medium containing the heavy isotope of nitrogen ^{15}N, so the DNA of their bacteria was eventually denser than normal. They then transferred the growing cells to a new medium containing the normal lighter isotope ^{14}N and harvested the DNA at various intervals.

At first the DNA that the bacteria manufactured was all heavy. But as the new DNA that was being formed incorporated the lighter nitrogen isotope, DNA density fell. After one round of DNA replication was complete, the density of the bacterial DNA had decreased to a value intermediate between all-light isotope and all-heavy isotope DNA. After another round of replication, two density classes were observed, one intermediate and the other light, corresponding to DNA that included none of the heavy isotope. These results indicate that after one round of replication, each daughter DNA duplex possessed one of the labeled strands of the parent molecule. When this hybrid duplex replicated, it contributed one heavy strand to form another hybrid duplex and one light strand to form a light duplex. Meselson and Stahl's experiment thus clearly confirmed the prediction of the Watson-Crick model that DNA replicates in a semiconservative manner.

The basis for the great accuracy of DNA replication is complementarity. A DNA molecule is a duplex, containing two strands that are mirror images of each other, so either one can be used to reconstruct the other.

WHAT IS THE UNIT OF HEREDITARY INFORMATION?

It is one thing to demonstrate that hereditary information resides within replicating molecules of DNA and quite another to understand what kind of information is stored there. Since the time of Mendel, geneticists had puzzled over this question. Mendelian traits in peas were colors and shapes, traits that were the end result of complex processes. What kind of change in the hereditary information would change a Mendelian trait?

The first answer to this question came soon after Mendel's work, but its significance was not readily appreciated. In 1902 a British physician, Archibald Garrod, who was working with one of the early Mendelian geneticists, his countryman William Bateson, noted that certain diseases among his patients were prevalent in particular families. Indeed, if one examined several generations within such families, some of these disorders seemed to behave as if they were controlled by simple recessive alleles. Garrod concluded that these disorders were Mendelian traits and that they had resulted from changes in the hereditary information that had occurred in the past to an ancestor of the affected families.

Garrod examined several of these disorders in detail. In one, **alkaptonuria,** the patients passed urine that rapidly turned black on exposure to air. Such urine contained homogentisic acid (alkapton), which air oxidized. In normal individuals homogentisic acid is broken down into simpler substances, but the affected patients were unable to carry out that breakdown. With considerable insight, Garrod concluded that the patients suffering from alkaptonuria lacked the enzyme necessary to catalyze this breakdown and, more generally, that many inherited disorders might reflect enzyme deficiencies.

From Garrod's finding it is but a short leap of intuition to surmise that the information encoded within the DNA of chromosomes is used to specify particular enzymes. This point was not actually established, however, until 1941, when a series of experiments by the Stanford University geneticists George Beadle and Edward Tatum finally provided definitive evidence on this point. Beadle and Tatum deliberately set out to create Mendelian mutations in the chromosomes; they then studied the effects of these mutations on the organism.

One of the reasons that Beadle and Tatum's experiments produced clear-cut results— and one of the characteristics of most successful laboratory experiments in biology—is that the researchers made an excellent choice of experimental organism. They chose the bread mold *Neurospora,* a fungus that can readily be grown in the laboratory on a **defined medium** (a medium that contains only known substances such as glucose and sodium chloride, rather than some uncharacterized cell extract such as ground-up yeasts). To increase the numbers of mutations, Beadle and Tatum took x rays of the spores of *Neurospora.* They then allowed the progeny to grow on **complete medium** (a medium that contained all necessary metabolites and would therefore supply them to the growing fungi, whether or not individual strains could manufacture the metabolites for themselves). In this way the investigators were able to preserve strains that, as a result of the earlier irradiation, had experienced damage to their DNA in a region encoding the ability to make one or more of the compounds that the fungus needed for normal growth. Change of this kind is called **mutation,** and the strains that have lost the ability to use one or more compounds are called **mutant strains.**

The next step was to test the progeny of the irradiated spores to see if any mutations leading to metabolic difficiency actually had been created by the x-ray treatment. Beadle and Tatum did this by attempting to grow subdivisions of individual fungal strains on **minimal medium,** which contained only sugar, ammonia, salts, a few vitamins, and water. A cell that had lost the ability to make a necessary metabolite would not grow on such a medium. Using this approach, Beadle and Tatum succeeded in identifying and isolating many deficient mutants.

To determine the nature of each deficiency, Beadle and Tatum tried adding various chemicals to the minimal medium in an attempt to find one that would make it possible for a given strain to grow (Figure 15-12). In this way they were able to pinpoint the nature of the biochemical problems that many of their mutants had. Many of the mutants proved unable to synthesize a particular vitamin or amino acid. The addition of arginine, for example, permitted the growth of a group of mutant strains, dubbed *arg* mutants. When the chromosomal position of each mutant *arg* gene was located, they were found to cluster in three areas (Figure 15-13).

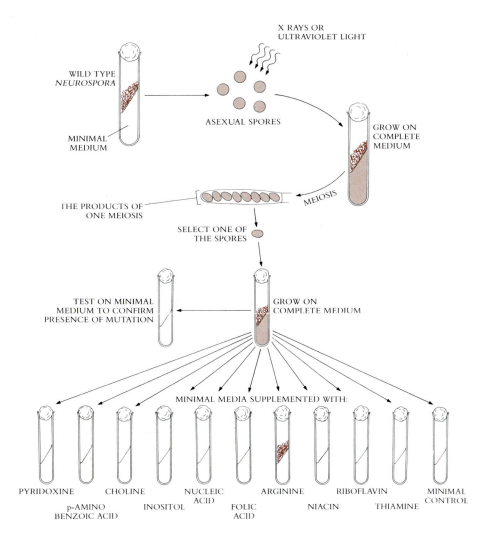

X RAYS OR
ULTRAVIOLET LIGHT

WILD TYPE
NEUROSPORA

MINIMAL
MEDIUM

ASEXUAL SPORES

GROW ON
COMPLETE
MEDIUM

THE PRODUCTS OF
ONE MEIOSIS

MEIOSIS

SELECT ONE OF
THE SPORES

TEST ON MINIMAL
MEDIUM TO CONFIRM
PRESENCE OF MUTATION

GROW ON
COMPLETE MEDIUM

MINIMAL MEDIA SUPPLEMENTED WITH:

PYRIDOXINE   CHOLINE   NUCLEIC   ARGININE   RIBOFLAVIN   MINIMAL
                         ACID   CONTROL
   p-AMINO   INOSITOL   FOLIC   NIACIN   THIAMINE
BENZOIC ACID   ACID

FIGURE 15-12

Beadle and Tatum's procedure for isolating nutritional mutations in Neurospora. This fungus grows easily on an artificial medium in test tubes. In this experiment spores were first irradiated to increase the frequency of mutation and then were placed on complete medium and allowed to grow. Any mutation that might have occurred in genes that were normally used by the fungus to produce its necessary amino acids or vitamins would not prevent growth, since all of these substances were present in the complete medium. Once the colonies were established, individual spores were taken and tested to see whether they would grow on minimal medium, which lacks the amino acids and vitamins that the fungus normally manufactures. Any strains that will not grow on minimal medium, but will grow on complete medium, contain one or more mutations in the genes that are necessary to produce one of the substances in the complete but not the minimal medium. To find out which one, the line is tested for its ability to grow on minimal medium supplemented with particular substances. The mutation illustrated here is an arginine mutant, a cell line that has lost the ability to produce arginine. It will not grow on minimal medium but will grow on minimal medium to which only arginine has been added.

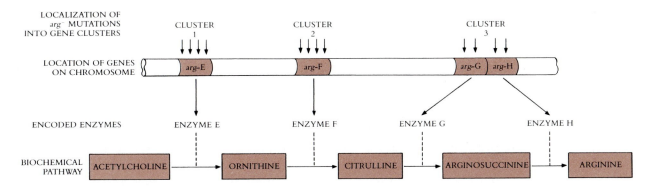

LOCALIZATION OF
arg⁻ MUTATIONS
INTO GENE CLUSTERS

CLUSTER 1   CLUSTER 2   CLUSTER 3

LOCATION OF GENES
ON CHROMOSOME

arg-E   arg-F   arg-G   arg-H

ENCODED ENZYMES   ENZYME E   ENZYME F   ENZYME G   ENZYME H

BIOCHEMICAL
PATHWAY   ACETYLCHOLINE → ORNITHINE → CITRULLINE → ARGINOSUCCININE → ARGININE

FIGURE 15-13

The chromosomal locations of the many arginine mutations isolated by Beadle and Tatum cluster around three locations, corresponding to the locations of the genes encoding the enzymes that carry out arginine biosynthesis.

For each enzyme in the arginine biosynthetic pathway, Beadle and Tatum were able to isolate a mutant strain with a defective form of that enzyme, and the mutation always proved to be located at *one* of a few specific chromosomal sites, a different site for each enzyme; that is, each of the mutants that Beadle and Tatum examined could be explained in terms of a defect in one (and only one) enzyme, which could be localized at a single site on one chromosome. The geneticists concluded that genes produce their effects by specifying the structure of enzymes. They called this relationship the **one gene–one enzyme hypothesis.**

> **Genetic traits are expressed largely as a result of the activities of enzymes. Organisms store hereditary information by encoding the structures of enzymes in the DNA of their chromosomes.**

What kind of information must a gene encode to specify a protein? For some time, the answer was not clear, since protein structure seemed to be impossibly complex. It was not evident, for example, whether or not a particular kind of protein had a consistent sequence of amino acids that would be the same in one individual molecule as it would be in another of that same kind of protein. The picture changed in 1953, the same year in which Watson and Crick unraveled the structure of DNA. The great English biochemist Frederick Sanger, after many years of work, announced the complete sequence of amino acids in the protein insulin. Sanger's achievement was extremely significant because it demonstrated for the first time that proteins consisted of definable sequences of amino acids. For any given form of insulin, each molecule has the same amino acid sequence as every other and this sequence can be learned and written down. All enzymes and other proteins are strings of amino acids arranged in a certain definite order. The information necessary to specify an enzyme, therefore, is an ordered list of amino acids.

In 1956 Sanger's pioneering work was followed by Vernon Ingram's analysis of the molecular basis of sickle cell anemia, a protein defect inherited as a Mendelian disorder. By analyzing the structure of normal and sickle cell hemoglobin (Figure 15-14), Ingram, working at Cambridge University, showed that sickle cell anemia was caused by the change of the amino acid at a single position in the protein from glutamate to valine. The alleles of the gene encoding hemoglobin differed only in their specification of this one amino acid in the hemoglobin amino acid chain.

These experiments, and other related ones, have finally brought us to a clear understanding of what the unit of heredity is. It is the information that encodes the amino acid sequence of an enzyme or other protein. The sequence of nucleotides that encodes this information is called a **gene.** Although most genes encode proteins, there are also genes devoted to the production of special forms of RNA, many of which play important roles in protein synthesis. These are discussed in Chapter 16.

> **The amino acid sequence of a particular protein is specified by a corresponding sequence of nucleotides in the DNA. This nucleotide sequence is called a gene.**

HOW ARE GENES ENCODED?

By the late 1950s the scientific pace was accelerating, and the results of the long experimental journey were beginning to come together and make coherent sense. It was now possible to phrase the old question of how hereditary information is stored in a new and very concrete way: how does the *order* of nucleotides in a DNA molecule encode the information that specifies the order of amino acids in a protein? What, in other words, is the nature of the DNA genetic code? The answer came in 1961, as the result of an experiment led by Francis Crick.

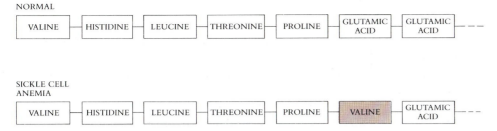

NORMAL

| VALINE | HISTIDINE | LEUCINE | THREONINE | PROLINE | GLUTAMIC ACID | GLUTAMIC ACID | _ _ _ |

SICKLE CELL
ANEMIA

| VALINE | HISTIDINE | LEUCINE | THREONINE | PROLINE | VALINE | GLUTAMIC ACID | _ _ _ |

FIRST SEVEN AMINO ACIDS IN NORMAL AND SICKLE CELL HEMOGLOBIN

VALINE

VALINE

STRUCTURE OF SICKLE-CELL HEMOGLOBIN

SICKLE FIBERS

NORMAL RED BLOOD CELL   FORMATION OF SICKLED RED BLOOD CELL   SICKLED RED BLOOD CELL

FIGURE 15-14

Sickle cell hemoglobin is produced by a recessive allele of the gene encoding the beta chain of hemoglobin. It represents a single amino acid change from glutamic acid to valine at the sixth position in the chain; in the folded beta-chain molecule this position contacts the alpha chain, and the amino acid change causes the hemoglobins to aggregate into long chains, altering the shape of the cell. Hemoglobin aggregated in this way within sickled cells does a very poor job of transporting oxygen with the result that individuals homozygous for the sickle cell hemoglobin allele, which can synthesize only this kind of hemoglobin, suffer from severe anemia and may die if they are not treated.

Crick and his colleagues reasoned that the "genetic code" most likely consisted of a series of blocks of information, each block corresponding to an amino acid in the encoded protein. They further hypothesized that the information within one block was probably a sequence of three nucleotides specifying a particular amino acid. They arrived at the number three because a two-nucleotide block would not yield enough different combinations to code for the 20 different kinds of amino acids that commonly occur in proteins. Imagine that you have an apple, an orange, a pear, and a grapefruit. How many different pairs of fruit can you construct? Only 4^2, or 16 different pairs. How many groups of three can you construct? A lot more—4^3, or 64 different combinations, more than enough to code for the 20 amino acids.

But are these three-nucleotide blocks read in simple sequence, one after the other? Or is there punctuation (a silent nucleotide) between each of the three-nucleotide blocks? The sentences of this book, for example, use spaces to punctuate between words. Without the spaces, this sentence would read "withoutthespacesthissentencewouldread." In principle, either way of reading will work. However, it is important to know which method is employed by cells, since the two ways of reading DNA imply different translating mechanisms. To choose between these two alternative hypotheses, Crick and his colleagues used a chemical that adds or deletes a single nucleotide to a DNA strand. Such an addition or deletion *changes the reading frame* of a genetic message, whether or not it is punctuated, but the two alternative hypotheses differ in what is required to restore the correct reading frame register (Figure 15-15).

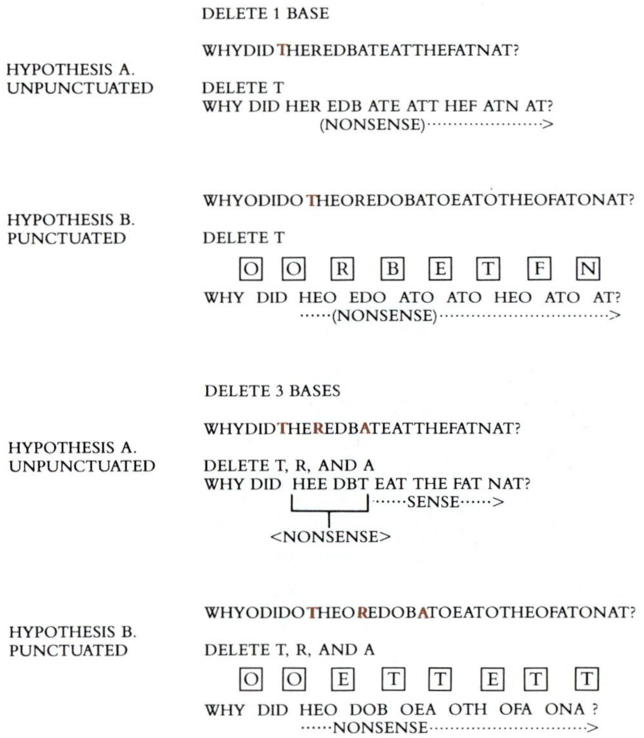

DELETE 1 BASE

WHYDID THEREDBATEATTHEFATNAT?

HYPOTHESIS A.
UNPUNCTUATED

DELETE T
WHY DID HER EDB ATE ATT HEF ATN AT?
           (NONSENSE)·················>

WHYODIDO THEOREDOBATOEATOTHEOFATONAT?

HYPOTHESIS B.
PUNCTUATED

DELETE T

| O | | O | | R | | B | | E | | T | | F | | N |

WHY DID HEO EDO ATO ATO HEO ATO AT?
    ·····(NONSENSE)·····························>

DELETE 3 BASES

WHYDID THEREDBATEATTHEFATNAT?

HYPOTHESIS A.
UNPUNCTUATED

DELETE T, R, AND A
WHY DID HEE DBT EAT THE FAT NAT?
      └─────┘······SENSE······>
               <NONSENSE>

WHYODIDO THEO REDOBATOEATOTHEOFATONAT?

HYPOTHESIS B.
PUNCTUATED

DELETE T, R, AND A

| O | | O | | E | | T | | T | | E | | T | | T |

WHY DID HEO DOB OEA OTH OFA ONA ?
    ·····NONSENSE····························>

FIGURE 15-15

Using the correction of frame-shift mutations to determine if the genetic code is punctuated. The hypothetical genetic message presented here is "Why did the red bat eat the fat nat?". Under hypothesis b, that the message is punctuated, the three-letter words are separated by nucleotides that are not read (here indicated by the letter "O").

From the examples in Figure 15-15, you can see that the deletion (or addition) of three nucleotides restores a three-digit unpunctuated code to the correct reading frame but does not restore a punctuated code. What Crick and his coworkers did was to use genetic recombination to put three deletions together near one another on a virus DNA and then look to see if the genes downstream were read correctly or as nonsense. When one or two deletions were placed at the beginning of the region, the downstream gene was translated as nonsense, but three deletions (or three additions) restored the correct reading frame so that sequences downstream were translated correctly. Thus the genetic code was read in increments consisting of three nucleotides, and reading occurs without punctuation between the three-nucleotide units.

Within genes that encode proteins, the nucleotide sequence of DNA is read in increments of three consecutive nucleotides, without punctuation between increments. Each block of three nucleotides codes for one amino acid.

Just what the code words are was soon worked out, the first results coming within the same year. Researchers had developed mixtures of RNA and protein isolated from ruptured cells ("cell-free systems") that would synthesize proteins in a test tube. To determine which three-nucleotide sequences specify which amino acids, researchers added artificial RNA molecules to these cell-free systems and then looked to see what proteins were made. For example, when Marshall Nierenberg, of the National Institutes of Health, added polyU (an RNA molecule consisting of a string of uracil nucleotides) to such a system, it proceeded to synthesize polyphenylalanine (a protein consisting of a string of phenylalanine amino acids). This result indicated that the three-nucleotide sequence (or **triplet**) specifying phenylalanine was UUU. In this and other ways all 64 possible triplets were examined, and the full genetic code was determined. It was now possible to say for the first time in exact detail how the information specifying a protein is encoded within the DNA.

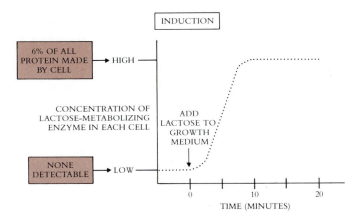

FIGURE 15-16

Bacterial cells growing in the absence of the sugar lactose do not contain measurable amounts of the enzyme beta-galactosidase, which splits lactose into glucose and galactose. When lactose is added to a growth medium containing bacteria, the cells begin to produce large amounts of the enzyme. Within 10 minutes after lactose is added, 6% of all the protein made by a typical cell will be the enzyme beta-galactosidase.

HOW DOES DNA SEQUENCE DICTATE PROTEIN STRUCTURE?

We have come a long way from *Acetabularia*. All that remains is to sort out how the genetic information encoded in DNA actually achieves expression. Of the many experiments that we might examine, we will consider two simple ones. First, let us repeat in a living organism the protein-synthesizing experiment described briefly above. We add to a growing eukaryotic cell culture a solution of the amino acid leucine that has been radioactively labeled with the carbon isotope ^{14}C. The cells are allowed to take up this labeled amino acid and use it briefly to synthesize protein. The cells are then fixed on a photographic plate. The plate is later examined with a microscope to determine the places in the cell where the radioactivity is localized. These are the sites where proteins, incorporating radioactive leucine, are being made.

Are these sites located in the nucleus, where the information is stored that specifies how to make the proteins? No. They are in the cytoplasm, associated with the small RNA-containing organelles called ribosomes. This means that the genetic information specifying proteins must pass from the nucleus to the cytoplasm before it can be used. We now know that the DNA never leaves the nucleus, and therefore some other molecule must act as the "genetic messenger" to carry the information to the cytoplasm. The messenger was demonstrated in 1961 to be an RNA molecule transcribed from the DNA, with a sequence complementary to that of the DNA. This class of molecule is dubbed messenger RNA (mRNA), because of its function. All protein-encoding genes are expressed via an mRNA intermediate.

Genes are not used directly; rather, information is copied from the DNA in the form of RNA gene copies called mRNA. The mRNA copies are transported out of the nucleus to the cytoplasm, where they direct the assembly of proteins on the ribosomes.

Our second experiment concerns the conditions under which bacteria produce a particular mRNA, the one encoding the enzyme beta-galactosidase. This enzyme cleaves the sugar lactose into digestible halves. Normal bacteria growing in a lactose-free environment do not produce this enzyme, apparently because the nucleotide region encoding this enzyme, the *lac* gene, is not normally transcribed into mRNA. If we add lactose to the medium, however, these same bacteria soon begin to transcribe the *lac* gene at a furious rate, producing vast quantities of the enzyme. The transcription of the enzyme is said to have been **induced** by the lactose.

The behavior of these bacteria reflects the fact that the *lac* gene is normally turned "off" in a cell that is not currently using lactose as a food source. Such a cell turns off the transcription of mRNA copies of this and other unnecessary genes simply by producing a special off-turning protein called a **repressor.** The repressor sits down on the DNA just in front of the gene to be turned off and blocks access to the gene by proteins needed to make mRNA. The repressor acts just like a bar across a door.

Induction results from bending the bar so it no longer fits across the door, leaving the door open. In the case of beta-galactosidase induction, lactose is capable of binding to the repressor protein. When this occurs, the binding changes the repressor protein's shape so that it can no longer bind to DNA; as a consequence, the segment of DNA containing the beta-galactosidase gene is once again free to be transcribed (Figure 15–16). Thus the induction of enzyme activity actually reflects a release from repression. The expression of most genes proves to be regulated by DNA-binding proteins of this kind.

The expression of genes is regulated by proteins that interact with DNA.

AN OVERVIEW OF THE HEREDITARY MECHANISM

This completes our retracing of the ideas that have led to our current understanding of the genetic machinery of the cell. In the following chapters we will treat much of this material in greater detail. In this chapter our goal has been to show the progression of ideas that have taken us, one step at a time, to our current vantage point. As we continue in the following chapters to survey the now-familiar molecular landscape of modern genetics, we must remember the long journey that has led to our understanding. It is an ongoing journey, every bit as exciting now as it has been in the past; the results are still often surprising, and the proper future experimental path is, as always, still undetermined. Science is at its heart a questioning enterprise, both intellectually challenging and enormous fun. Nowhere is this more evident than in the history of genetics that we have just reviewed.

SUMMARY

1. Eukaryotic cells store hereditary information within the nucleus. When one transplants the nucleus, one also transplants the hereditary specifications of an organism.

2. In viruses, bacteria, and eukaryotes, the hereditary information resides in nucleic acids. The transfer of pure nucleic acid can lead to the transfer of hereditary traits. In all cellular organisms the genetic material is DNA, although in some viruses the genetic material is RNA.

3. During cell division, the hereditary message is duplicated with great accuracy. The mechanism that achieves this high degree of accuracy is complementarity: the DNA molecule is organized as a double helix with two strands that are mirror images of one another. By adding complementary bases, either one of the strands can be used to re-form the helix.

4. Most hereditary traits reflect the actions of enzymes. The traits are hereditary because the information necessary to specify these enzymes is encoded within the DNA.

5. Enzymes are encoded within DNA by specifying their amino acid sequence with a sequence of nucleotides. Three consecutive nucleotides specify each amino acid.

6. Not all genes are expressed all the time. Many genes are silent because their transcription is blocked by gene-specified repressor proteins. Only when particular cell metabolites interact with these proteins are the otherwise silent genes transcribed.

SELF-QUIZ

The following questions pinpoint important information in this chapter. You should be sure to know the answers to each of them.

1. Match these investigators with the organism they worked with:
   - (a) *Acetabularia*
   - (b) bacteria
   - (c) carrot
   - (d) frog
   - (e) humans
   - (f) TMV virus
   - (g) *Neurospora*
   - (h) *Pneumococcus*
   - (i) T2 bacteriophage

   - (1) Beadle and Tatum
   - (2) Briggs and King
   - (3) Fraenkel-Conrat
   - (4) Garrod
   - (5) Griffith
   - (6) Hammerling
   - (7) Hershey and Chase
   - (8) Steward
   - (9) Meselson and Stahl

2. Match these investigators with the phenomenon they studied:
   - (a) Avery
   - (b) Chargaff
   - (c) Franklin
   - (d) Garrod
   - (e) Gurdon
   - (f) Ingram
   - (g) Meselson and Stahl
   - (h) Nierenberg
   - (i) Sanger
   - (j) Watson and Crick

   - (1) alkaptonuria
   - (2) base composition of DNA
   - (3) genetic code
   - (4) modeling DNA structure
   - (5) molecular basis of sickle cell anemia
   - (6) nuclear transplantation
   - (7) semiconservative replication of DNA
   - (8) sequence of insulin
   - (9) transforming principle
   - (10) x-ray analysis of DNA

3. When different combinations of stalk and foot of *Acetabularia* individuals are grafted to one another, the shape of the cap that eventually develops after several cap removals
   - (a) depends on the identity of the individual from whom the stalk was obtained
   - (b) depends on the identity of the individual from whom the foot was obtained
   - (c) represents the average of foot and stalk contributions

4. Are identical twins always of the same sex?

5. The radioactive isotopes ^{32}P and ^{35}S can be used to label a cell's proteins and its DNA. Which isotope has been used to label which substance?

6. When a healthy mouse is infected with *Pneumococcus*, it is likely to become ill and die if the *Pneumococcus* is
   - (a) heat-killed
   - (b) lacking a polysaccharide coat
   - (c) both heat-killed and lacking a polysaccharide coat
   - (d) a mixture of some heat-killed *Pneumococcus* and some *Pneumococcus* lacking a cell coat

7. In the DNA studied by Chargaff the amount of adenine in DNA varied widely from one organism to another, but in each of these DNAs the amount of adenine was always closely similar to the amount of which other base?

8. The most important result obtained by Rosalind Franklin in her probing of the structure of DNA with x rays was that the molecule has a spiral shape. A molecule with such a shape is called a _____.

9. In DNA, purines pair with pyrimidines. Adenine (A) pairs with thymine (T), for example. Why won't adenine pair with cytosine (C)?
   - (a) cytosine is a purine
   - (b) cytosine does not form hydrogen bonds with adenine
   - (c) cytosine is a pyrimidine
   - (d) cytosine is too large

10. If a DNA molecule replicates semiconservatively, the ratio of original to newly synthesized strands in the molecules after two rounds of replication is _____.

11. Citrulline and ornithine are intermediates in the arginine biosynthetic pathway. A *Neurospora arg* mutant strain that will not grow on minimal medium supplemented with ornithine but will grow on minimal medium supplemented with citrulline
    (a) possesses a mutation inactivating the enzyme catalyzing the reaction that converts citrulline to ornithine
    (b) possesses a mutation inactivating the enzyme catalyzing the reaction that converts ornithine to citrulline

12. If one attempts to make proteins in a test tube, and uses as mRNA the synthetic polynucleotide UGUGUGUG . . . , the result will be a protein composed of how many different amino acids?

THOUGHT QUESTIONS

1. In Hammerling's experiments with *Acetabularia* the grafting of a nucleus-containing basal portion of an *A. mediterranea* individual (with a disk-shaped cap) to the stalk of an *A. crenulata* individual (with a flower-shaped cap) results in the formation of a cap above the grafted stalk. The cap that forms is flower-shaped, like that of *A. crenulata*. If you remove that cap, another one forms, only this one is disk-shaped, like that of *A. mediterranea*. What is responsible for the change in morphology of these two regenerated caps?

2. From an extract of *Pneumococcus* cells you obtain a white fibrous substance. How would you distinguish whether it was DNA, RNA, or protein?

3. In analyzing your DNA, you are able to determine that 15% of the nucleotide bases which it contains is thymine. What percentage of the bases is cytosine?

4. You are presented with two test tubes, each containing purified DNA, and told that one of the tubes contains double-stranded human DNA; the other contains single-stranded virus DNA. You analyze the base composition of the two preparations, with the following results:

       Tube 1:   22.1% A : 27.9% C : 27.9% G : 22.1% T
       Tube 2:   31.3% A : 31.3% C : 18.7% G : 18.7% T

Which of the two tubes contains single-stranded virus DNA?

FOR FURTHER READING

CRICK, F.H.C.: "The Genetic Code," *Scientific American,* October 1962, pages 66-77. An account of the frame-shift experiments which established that the genetic code was read in contiguous units of three bases. Not always easy for a beginner to follow, this is one of the most elegant experiments in modern genetics.

CRICK, F.H.C.: "The Double Helix: A Personal View," *Nature,* vol. 248, page 766, 1974. Francis Crick's own recollections of the hectic days when he and James Watson deduced that the structure of DNA is a double helix.

JUDSON, H.F.: *The Eighth Day of Creation,* Simon and Schuster, New York, 1979. The definitive historical account of the experimental unraveling of the mechanisms of heredity, based on personal interviews with the participants. This book is full of the feel of how science is really conducted.

PLATT, J.R.: "On Strong Inference," *Science,* vol. 146, pages 347-352, 1964. A valuable discussion of why some fields of science, such as molecular genetics, progress faster than others.

WATSON, J.D.: *The Double Helix,* Atheneum Publishing Company, Inc., New York, 1968. A lively, often irreverent account of what it was like to discover the structure of DNA, recounted by someone in a position to know.

WATSON, J.D., and F.H.C. CRICK: "A Structure for Deoxyribose Nucleic Acid," *Nature,* vol. 171, page 737, 1953. The original report of the double helical structure of DNA. Only one page long, this paper marks the birth of molecular genetics.

Overview

The information encoded within the nucleotide sequence of a DNA molecule is used to direct the synthesis of a variety of proteins, some of them enzymes and others, such as muscle proteins, having different functions. An RNA copy of the DNA is first made in a process called transcription. Subsequently, in a process called translation, this RNA copy binds to ribosomes, which use the nucleotide sequence of the RNA to specify the identity of the amino acids that are added to a protein chain.

For Review   *Here are some important terms and concepts that you will encounter in this chapter. If you are not familiar with them, you should review them before proceeding.*

- **Structure of DNA and RNA** (Chapters 4 and 15)

- **Structure of eukaryotic chromosomes** (Chapter 14)

- **Messenger RNA** (Chapter 15)

- **Three-base nature of genetic code** (Chapter 15)

As you saw in Chapter 15, a long trail of investigation has led biologists to a broad overview of how heredity works: within the cells of living organisms, hereditary information is encoded in DNA as genes—sequences of nucleotides that specify the amino acid sequences of proteins. However, a profound mystery remained to be addressed. Although the evidence was clear that the sequence of nucleotides in a DNA molecule acts to specify the sequence of amino acids in a protein, the way in which this is done was not clear initially. Gene expression was like a black box into which we could not see, with DNA going in and proteins coming out. It was not possible to deduce by logic alone how the hereditary machine within the box worked. The structure of a nucleotide is nothing like that of an amino acid. Furthermore, as you learned in Chapter 15, a sequence of *three* nucleotides specifies each amino acid. How can bunches of one kind of molecule correspond to one single molecule of another kind? In this chapter we will answer this question and by doing so, open the black box and see the hereditary machine in operation.

AN OVERVIEW OF GENE EXPRESSION

As you have seen, the fundamental unit of hereditary information is the gene. When a gene is "expressed," a protein with an amino acid sequence that is specified by that gene is synthesized within the cell. In all living organisms gene expression occurs in two distinct phases: transcription and translation.

Transcription

In the first stage of gene expression an RNA copy is made of the gene. This process is called **transcription,** and the messenger RNA (mRNA) molecule is said to have been "transcribed" from the DNA. Transcription is initiated when a special enzyme, called an **RNA polymerase,** binds to a particular sequence of nucleotides on one of the DNA strands, a sequence located at the edge of a gene. Starting at that end of the gene, the RNA polymerase enzyme proceeds to assemble a single strand of mRNA with a nucleotide sequence complementary to that of the DNA strand it has bound (Figure 16-1). The enzyme moves along the strand into the gene, encounters each DNA nucleotide in turn, and selects a corresponding complementary RNA nucleotide for the growing mRNA strand. When the RNA polymerase arrives at a special "stop" signal at the far edge of the gene, it disengages from the DNA and releases the newly assembled mRNA chain. This new mRNA molecule is complementary in sequence to the DNA strand from which the polymerase assembled it; in effect, it is an **RNA transcript** (copy) of the DNA nucleotide sequence of the gene.

Translation

The second stage of gene expression is the synthesis of a protein, using the information contained on an mRNA molecule to direct the choice of amino acids. This process of mRNA-directed protein synthesis is called **translation,** because nucleotide sequence information is translated into amino acid sequence information. Translation begins when ribosomes, large aggregates of protein plus special RNA molecules called **ribosomal RNA (rRNA),** bind to one end of the mRNA transcript. Once a ribosome has bound to mRNA, it proceeds to assemble a protein. It does so by adding amino acids to the end of a lengthening chain, moving down the mRNA molecule in increments of three nucleotides. At each step it adds an amino acid to the growing protein chain. The ribosome continues to do this until it encounters a "stop" signal that indicates the end of the protein. It then disengages from the mRNA and releases the newly assembled protein.

> **The information encoded in genes is expressed in two stages: transcription, in which a polymerase enzyme assembles an mRNA molecule whose nucleotide sequence is complementary to that of the gene; and translation, in which a ribosome assembles a protein, using the mRNA to specify the amino acids.**

Regulating Gene Expression

It is important for an organism to be able to control which of its genes are transcribed and when. There is, for example, little point in producing an enzyme when its substrate, the target of its activity, is not present in the cell. Much energy can be saved if the enzyme is not produced until the appropriate substrate is encountered and the enzyme's activity will be of use to the cell. From a broader perspective, the growth and development of many multicellular organisms, including human beings, entails a long series of biochemical steps, each of which is delicately tuned to achieve a precise effect. Specific enzyme activities are called into play and used to effect a particular change. Once this change has occurred, those particular enzyme activities cease, lest they disrupt other activities that follow. Many genes

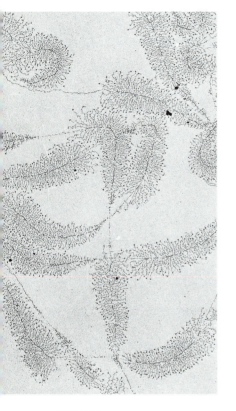

FIGURE 16-1

Transcription of the genes encoding ribosomal RNA in amphibian oocytes. Each of the strands of the Christmas tree–like formations is an rRNA molecule being transcribed from an rDNA gene. A line of polymerases reads these rDNA genes, the ones farthest along in transcription trailing the longest rRNAs.

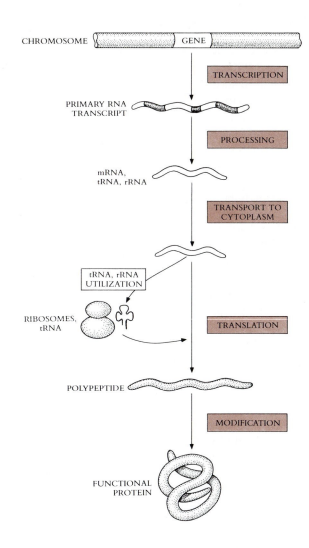

CHROMOSOME — GENE

TRANSCRIPTION

PRIMARY RNA TRANSCRIPT

PROCESSING

mRNA, tRNA, rRNA

TRANSPORT TO CYTOPLASM

tRNA, rRNA UTILIZATION

RIBOSOMES, tRNA

TRANSLATION

POLYPEPTIDE

MODIFICATION

FUNCTIONAL PROTEIN

FIGURE 16-2

Control of gene expression in a eukaryotic cell may occur at many stages in the expression of gene-encoded information:
1 *Most control occurs at the transcription stage, when an RNA copy, or transcript, is made of the gene.*
2 *Control may also be exercised at the processing stage, when the RNA transcript is modified and any extraneous nucleotide sequences removed.*
3 *A third stage at which control may occur is the translation of RNA transcripts into proteins by ribosomes.*
4 *Many proteins require modification after they are made but before they become active, and this post-translational modification process is also often under cellular control.*

are transcribed, in a carefully prescribed order, each gene for a specified period of time; the hereditary message is played like a piece of music on a grand organ, in which the genes are the notes and the developmental program is the score.

An organism controls the expression of its genes largely by controlling when the transcription of individual genes begins (Figure 16-2). Most genes possess special nucleotide sequences called **regulatory sites,** which act as points of control. These nucleotide sequences are recognized by specific proteins within the cell called **regulatory proteins,** which bind to the sites. In some cases this binding blocks transcription; in others the binding is necessary before the polymerase can initiate transcription.

In some cases the site at which the regulatory protein binds to the DNA is located between the site at which the polymerase binds and the beginning edge of the gene that the polymerase is to transcribe. Thus the presence of the regulatory protein bound to its site blocks the movement of the polymerase toward the gene. To understand this situation more clearly, imagine that you are shooting a cue ball at the eight ball on a pool table and that someone places a brick on the table between the cue ball and the eight ball. Functionally, this brick is like the regulatory protein that binds to the DNA: its placement blocks movement of the cue ball to the eight ball just as placement of the repressor protein between polymerase and gene blocks movement of the polymerase to the gene. This process of blocking transcription is called **repression.**

In other situations the binding of a regulatory protein to the DNA is necessary to initiate the transcription of genes. The "turning on" of the transcription of specific genes in this way is called **activation.** Activation can be achieved by a variety of mechanisms. In some cases the regulatory protein's binding promotes the unwinding of the DNA duplex. This unwinding facilitates the production of an mRNA copy of a gene, because the polymerase, although it can bind to a double-stranded DNA duplex, cannot produce an mRNA copy from such a duplex—mRNA is copied from a single strand of the duplex.

The cell can control the transcription of particular genes because it can influence the shape of the regulatory proteins. A regulatory protein not only recognizes a specific regulatory site on the DNA, but also recognizes and can bind certain small "signal" molecules within the cell. These signal molecules provide information about the function carried out by the gene that is being controlled. For example, the regulated gene might encode an enzyme catalyzing a biochemical reaction, and the protein which regulates that gene might be able to bind the enzyme's substrate molecules. When the regulatory protein encounters one of the substrate molecules, it binds that molecule to it. This binding causes the regulatory protein to undergo a change in shape. In some cases the protein in its new shape may no longer recognize the regulatory site on the gene. In other cases the recontoured regulatory protein may begin to recognize a regulatory site that it had previously ignored.

The presence of particular "signal" molecules within the cell thus can act to incapacitate particular regulatory proteins or to mobilize them for action. In turn, these regulatory proteins repress or activate the transcription of particular genes. Just as a computer uses keyboard strokes to set "on/off" switches on its microprocessor chip in certain configurations, so an organism uses the pattern of metabolites in its body to set "on/off" protein regulatory switches and, by doing so, to achieve a proper configuration of gene expression.

Organisms control the expression of their hereditary information by selectively inhibiting the transcription of some genes and facilitating the transcription of others. Control over transcription is exercised by modifying the shape of regulatory proteins and thus influencing their tendency to bind to sites on the DNA that influence the initiation of transcription.

THE COMPONENTS OF A GENE

With this simple overview of gene expression, we may now consider the structure of a gene in more detail. Imagine that you had to design a gene. What would be its functional requirements? What elements must it contain so that the patterns of gene expression just described can be carried out? A list of requirements would at a minimum contain the following (Figure 16-3):

1. A coding sequence
2. Translation signals
3. Ribosome recognition site
4. Transcription signals
5. Polymerase binding site
6. Regulatory sites

A Coding Sequence

The heart of every gene is a long string of nucleotides that, when read in groups of three, corresponds to the amino acid sequence of the encoded protein. Except for a few code words associated with the signal for "stop," the genetic code never varies; a given sequence of nucleotides directs the assembly of the same amino acid sequence in bacteria as it does in human beings or any other kind of organism. A typical protein contains 100 or so amino acids, so a typical gene's coding sequence is composed of a string of several hundred nucleotides.

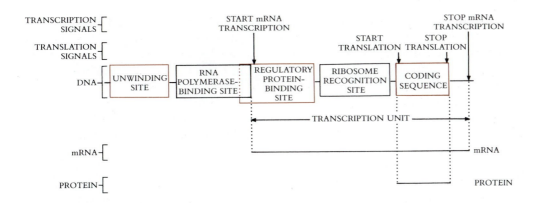

FIGURE 16-3

The components of a gene. Genes have a nested structure, with the coding sequence in the interior, bracketed by a series of sequences that define the boundaries of the protein to be made and of the mRNA which directs its assembly, and preceded by sequences which serve as recognition sites for polymerases, ribosomes, and regulatory proteins.

Translation Signals

To mark the boundaries of a gene, it is necessary that you position special nucleotide sequences, bracketing the coding sequence. These sequences act as signals to the ribosome, directing it when to start "paying attention" to the continuous chain of nucleotides and ensuring that it not read any more of the sequence than required to make the complete protein. A strand of DNA typically contains millions of nucleotide base pairs, and these translation "start" and "stop" signals serve to identify the proper run of a few hundred.

Ribosome Recognition Site

Translation is not possible unless a ribosome binds to the mRNA molecule. The simplest way to facilitate this binding is to place a sequence of nucleotides "upstream" from the coding sequence that is complementary to the nucleotide sequence of the ribosome's own RNA so that the ribosome will bind to the mRNA in front of the gene. (The term *upstream* refers to the direction of future movement of the ribosome, which moves *downstream* through the coding sequence, just as a log carried by a river's current moves downstream.)

Transcription Signals

The three elements just described—a ribosome recognition site upstream from a coding sequence bracketed by translation signals—are the components of a functional mRNA molecule, called a **transcription unit.** All of this transcription unit must be copied from the DNA by the RNA polymerase enzyme. It is thus necessary that you position additional signals at the beginning and end of the entire transcription unit, to guide the polymerase in its task of producing an RNA copy of the DNA.

Polymerase Binding Site

Transcription cannot occur unless the RNA polymerase molecule first binds to the DNA molecule. This binding requires a specific DNA nucleotide sequence upstream from the transcription unit, a sequence that the polymerase recognizes and to which it binds. These polymerase binding sites are called **promoters,** because they promote transcription. The promoters of different genes often have slightly different sequences; by changing the shape of polymerase molecules by binding different small molecules to them, a cell can alter which promoter the polymerase recognizes and so direct polymerases to different transcription units.

Regulatory Sites

To control when a transcription unit is transcribed, you must position regulatory sites near the promoter. If you locate a binding site for a regulatory protein between the promoter

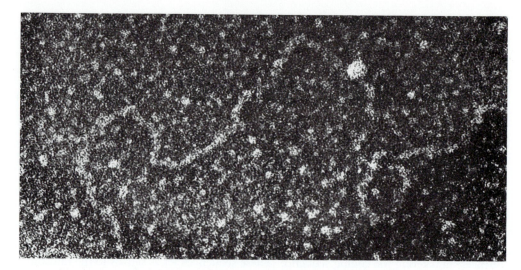

The regulatory protein controlling
transcription of the genes encoding
lactose-utilizing enzymes binds to
the DNA between the promoter
and the transcription unit. The
large whitish sphere bound to the
DNA strand is the regulatory
protein known as the lac repressor.

and the transcription unit, then the binding of the regulatory protein will act to block transcription (Figure 16-4). If, on the other hand, you locate your binding site upstream from the promoter, and the binding of regulatory protein to that site tends to unwind the DNA duplex there, then the binding of the regulatory protein would facilitate the binding of the polymerase to the promoter site, since it will not have to wait for spontaneous unraveling of the duplex to occur.

THE ARCHITECTURE OF A REAL GENE

In those cases in which it has been possible to examine the detailed structure of individual genes, researchers have found them by and large to be constructed along the lines of the hypothetical gene you designed above. Figure 16-5 presents the structure of the region of DNA in the bacterium *Escherichia coli* that encodes three genes involved in lactose utilization. This region exhibits each of the six classes of gene component present in our hypothetical gene:

1. There are three coding sequences.

2. Each of the three coding sequences is bracketed by translation signals, in this case three-base triplets that say "start" (AUG) and "stop" (UGA). The three coding sequences together constitute an operational unit called an **operon,** because all three are transcribed on the same piece of mRNA. This operon is called the *lac* operon, because all three genes are involved in lactose utilization. Such clustering of coding sequences onto single transcription units is common among bacteria but is not found in eukaryotes.

3. Upstream from the three coding sequences, a series of nucleotides is the binding site for the bacterial ribosome. This series of nucleotides is within an initial untranslated portion of the mRNA sometimes called a **leader region.**

4. Bracketing this entire constellation of sequences are signals that initiate transcription and end it; the position of these signals defines the transcription unit.

FIGURE 16-5

The lac region of the Escherichia coli chromosome. The left-most segment (P_i) is a promoter site where RNA polymerase binds. Immediately downstream of it (i.e., in the direction that the polymerase moves after it binds) is the gene i, which encodes a regulatory repressor protein. The RNA polymerase reads this sequence, producing the corresponding mRNA, and then dissociates from the DNA. To the right of the i gene is a second RNA polymerase binding site, called P_{lac}, and an operator site to which the repressor protein can bind. Just downstream of these two sites are three genes encoding three enzymes involved in the metabolism of lactose. Because they encode the primary structure (amino acid sequence) of enzymes, they are called structural genes.

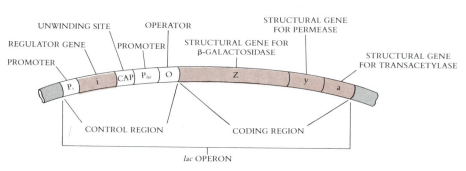

5. Upstream from the transcription unit is the promoter site, where the polymerase binds to the DNA.

6. Between the promoter site and the transcription unit is a regulatory site, where a repressor protein binds to block transcription. Upstream from the promoter is another regulatory site, where an activator protein binds, thus facilitating the unwinding of the DNA duplex and therefore polymerase binding to the nearby promoter.

Almost all genes share this same architecture, although, as you will see, eukaryotes have added a few new twists during the course of their evolution.

Genes have a nested architecture. The inner segment is the portion that is copied into mRNA. It is called a transcription unit and consists of the elements that are involved in the translation of the mRNA: the ribosome binding site, translation-start signal, coding sequence, and translation-stop signal. The outer segment, which surrounds the inner one, is composed of the elements involved in transcription of the inner segment from the DNA: the regulatory sites, polymerase binding site, and transcription-start and transcription-stop signals.

THE GENETIC CODE

The system that is used to "read" the genetic information contained within the nucleotide sequence of mRNA transcripts must have evolved very early in the history of life, since all living organisms share it. It is based on a collection of 20 enzymes called aminoacyl tRNA synthetases, or, more simply, **activating enzymes.** There is one activating enzyme for each of the 20 common amino acids. An activating enzyme binds the amino acid that it recognizes (Figure 16-6) to a small RNA molecule called a **transfer RNA (tRNA) molecule.** First, the activating enzyme binds its particular amino acid; then, from the many different kinds of tRNA molecules in the cell, the activating enzyme selects the one that possesses a particular three-nucleotide sequence at one end; finally, the activating enzyme adds the amino acid to the other end of that tRNA molecule.

If one considers the nucleotide sequence of mRNA to be a coded message, then the 20 activating enzymes are the code books of the cell—the instructions for decoding the message. The code word recognized by an activating enzyme is three nucleotides long. Since there are four different kinds of nucleotides in mRNA (cytosine, guanine, adenine, and uracil instead of thymine) there are 4^3, or 64, different three-letter code words possible. The three-nucleotide triplets of an mRNA coding sequence are called **codons.** The complementary three-base sequence on the tRNA molecule that binds to a codon is called an **anticodon.** Some of the activating enzymes recognize only one tRNA molecule, corresponding to one of these code words; others recognize two, three, four, or six different tRNA molecules, each containing a different anticodon.

There is reason to think that the primitive genetic code dealt with 16 or fewer amino acids and used only two nucleotides, with the third nucleotide acting as a "spacer." Subsequently, activating enzymes evolved that could discriminate among tRNA molecules which differed in the third nucleotide position of anticodons, an advance that allowed more kinds of amino acids to be encoded.

The complete **genetic code** is presented in Table 16-1. With a few minor exceptions, this genetic code is the same in all organisms, a given three-nucleotide sequence always corresponding to the same amino acid. Note that 3 of the 64 codons do not correspond to triplets that are recognized by any activating enzyme. These three codons, called **nonsense codons,** serve as "stop" signals in the mRNA message, marking the end of a protein. The "start" signal, which marks the beginning of a protein amino acid sequence within an mRNA message, is the triplet AUG, which also encodes the amino acid methionine. The

FIGURE 16-6

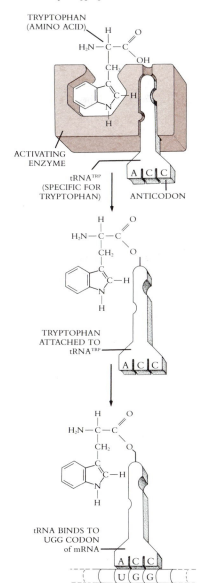

Activating enzymes are the elements within cells that "read" the genetic code. Each kind of activating enzyme recognizes and binds a specific amino acid such as tryptophan, on the one hand, and also recognizes and binds the tRNA molecules with anticodons specifying that amino acid, such as ACC for tryptophan. While both amino acid and tRNA are bound to the activating enzyme, they become covalently joined; the tRNA is said to be "charged" with its appropriate amino acid and will contribute that amino acid to a growing protein chain when the tRNA anticodon binds to a mRNA codon during protein synthesis. The codon is UGG in this case, the mRNA code word for tryptophan.

THE CONTROL OF TRANSCRIPTION

The regulatory sites associated with the *lac* operon of the bacterium *Escherichia coli* have been characterized in detail and provide a good example of how gene regulation is achieved. The line in Figure 16-3 represents a portion of the *E. coli* chromosome, the segment that contains the *lac* operon. It is transcribed from left to right in the diagram. The region that regulates the transcription of the *lac* genes is a continuous string of 122 nucleotides (Figure 16-A), almost all of which are involved in some phase of the regulatory process.

Beginning at the left (i.e., upstream), the first regulatory element is the gene labeled *i*, which encodes a repressor protein. At the left margin of the *i* gene you can see its promoter, the point where RNA polymerase first binds when transcribing the *i* gene. Exactly 17 nucleotides downstream from the *i* gene is a 14-nucleotide sequence that acts as a specific binding site for an activator protein called **CAP.** Immediately adjacent to that sequence is a stretch of 65 nucleotides that make up the RNA polymerase-binding site, or *lac* **promoter.** To the right of the *lac* promoter and actually overlapping it is a 24-nucleotide binding site recognized by the repressor protein that the *i* gene encodes. This site is referred to as the **operator.** Transcription starts just downstream from it. A 17-nucleotide leader region containing the ribosome recognition site is transcribed first, followed by the three genes encoding the enzymes involved in lactose utilization.

Here is how control of the *lac* operon works (Figure 16-B):

Stage 1: Activation

1. When the CAP activator protein binds the CAP site, it serves to destabilize the region of the duplex that is immediately adjacent to it, a region that contains the *lac* promoter site. The CAP protein acts as an activator of *lac* operon transcription because it facilitates the unwinding of the duplex, which is necessary for transcription.

2. The shape of the CAP protein usually prevents it from binding to the CAP site. Only when its shape has been changed by the binding of a small metabolite, **cyclic AMP (cAMP),** can the CAP activator protein bind to the CAP site on the DNA. Because cells usually contain little cAMP, the CAP protein is usually nonfunctional. However, when a cell's energy reserves become low, its level of cAMP goes up. Thus only in a cell low on energy is transcription of the *lac* operon activated by the CAP protein. As a result of this system, the enzymes needed for the metabolism of lactose are produced only when the cell requires the energy that lactose would provide.

Stage 2: Derepression

3. Overlapping the promoter region is the operator, the binding site recognized by the repressor protein. The repressor protein is usually bound very tightly to the operator site on the DNA. Can you see the consequences of this overlap? Because the binding site of the repressor protein overlaps the promoter region, the binding of the repressor prevents any subsequent binding of RNA polymerase (Figure 16-B, 1). So in most cells RNA polymerase is unable to bind the *lac* promoter.

triplet AUG is recognized both by an activating enzyme and by the ribosome, which uses the first AUG that it encounters in the mRNA message to signal the start of the translation of its specified amino acid sequence.

Most of the more than 20 different kinds of tRNA molecules have similar structures. Each is about 80 nucleotides long (Figure 16-7), and each contains two pairs of palindromic sequences—sequences that read the same backward and forward. Each tRNA molecule is folded back in such a way that the two pairs of palindromic sequences are opposite one another, forming two regions of stable duplex base pairing. The result of this pairing is a cloverleaf-shaped molecule. The unpaired nucleotides of the "leaflet" segments then form hydrogen bonds so that the final shape of the tRNA resembles a distorted L. The anticodon sequence that is recognized by the activating enzyme is located at one end, on the perimeter

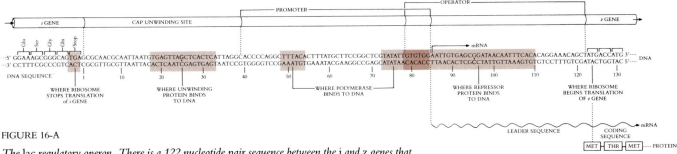

FIGURE 16-A

The lac *regulatory operon. There is a 122 nucleotide pair sequence between the* i *and* z *genes that is not translated into protein. Three different proteins bind to specific sites within this sequence: 1, An unwinding protein called CAP; 2, RNA polymerase; and 3, a regulatory protein encoded by the* i *gene, called the repressor protein. The colored boxes identify regions that are protected from enzymatic degradation when the proteins are bound to the DNA.*

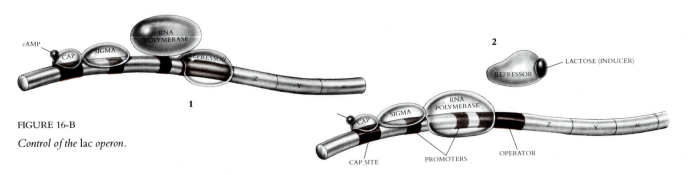

FIGURE 16-B

Control of the lac *operon.*

These cells are said to be *repressed* with respect to *lac* operon transcription.

4. Like the CAP protein, the *lac* repressor protein is capable of changes in shape. When lactose binds to the repressor protein, the protein assumes a different shape, one that does not recognize the operator sequence (Figure 16-B, *2*). In a cell with appreciable amounts of lactose the *lac* repressor proteins will be in an inactive shape, and transcription of the *lac* operon will not be repressed (will be derepressed). This system thus ensures that the *lac* operon is transcribed only in the presence of lactose.

5. When the CAP protein binds successfully and the repressor protein is unable to bind, the *lac* operon is poised for transcription by RNA polymerase (Figure 16-B, *2*). First, a subunit of the polymerase called **sigma** binds to one site of the promoter region; then the rest of the polymerase begins the transcription at a nearby site.

The *lac* operon is thus controlled at two levels: the lactose-utilizing enzymes are not produced unless the sugar lactose is available; even if lactose is available, these enzymes are not produced unless the cell has need of the energy.

of the central loop. The activating enzyme adds its amino acid to the other end of the tRNA molecule, at the end of the long stem.

All organisms possess a battery of 20 enzymes, called activating enzymes, each of which recognizes both a particular combination of three-base nucleotide sequences on tRNA and one of the 20 common amino acids. These 20 enzymes translate the genetic code by coupling particular amino acids to tRNAs with particular three-base nucleotide sequences. With a few minor exceptions, the code is the same in all organisms, a given three-base nucleotide sequence always corresponding to the same amino acid.

TABLE 16-1   THE GENETIC CODE

FIRST LETTER	SECOND LETTER				THIRD LETTER
	U	C	A	G	
U	Phenylalanine	Serine	Tyrosine	Cysteine	U
	Phenylalanine	Serine	Tyrosine	Cysteine	C
	Leucine	Serine	Stop	Stop	A
	Leucine	Serine	Stop	Tryptophan	G
C	Leucine	Proline	Histidine	Arginine	U
	Leucine	Proline	Histidine	Arginine	C
	Leucine	Proline	Glutamine	Arginine	A
	Leucine	Proline	Glutamine	Arginine	G
A	Isoleucine	Threonine	Asparagine	Serine	U
	Isoleucine	Threonine	Asparagine	Serine	C
	Isoleucine	Threonine	Lysine	Arginine	A
	(Start) Methionine	Threonine	Lysine	Arginine	G
G	Valine	Alanine	Aspartate	Glycine	U
	Valine	Alanine	Aspartate	Glycine	C
	Valine	Alanine	Glutamate	Glycine	A
	Valine	Alanine	Glutamate	Glycine	G

A codon consists of three nucleotides read in the sequence shown above. For example ACU codes threonine. The first letter, A, is read in the first column; the second letter, C, from the second letter columns; and the third letter, U, from the third letter column. Each of the codons is recognized by a corresponding anticodon sequence on a tRNA molecule. Some tRNA molecules recognize more than one codon sequence but always for the same amino acid. Most amino acids are encoded by more than one codon. For example, threonine is encoded by four codons (ACU, ACC, ACA, ACG), which differ from one another only in the third position.

FIGURE 16-7

Structure of a tRNA molecule.
A *The secondary structure. The TψC loop and the D loop function in binding to the ribosomes during protein synthesis. The activating enzyme adds an amino acid to the free single-stranded — OH end. The third loop contains the anticodon sequence.*
B *The tertiary structure. In water the tRNA molecule folds into an L shape, in which the anticodon forms the end of the long arm and the amino acid is carried by the end of the short arm.*

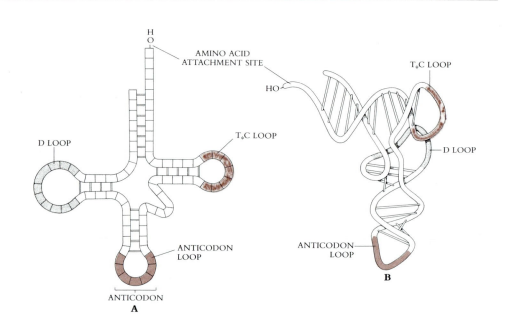

PROCESSING THE mRNA TRANSCRIPT

Most eukaryotic genes contain noncoding regions, sequences that are not translated into protein. These sequences, which intervene at various places within the gene, are called **introns.** The remaining segments of the gene—the nucleotide sequences that encode the amino acid sequence of the protein—are called **exons.**

Introns are common in vertebrate genes, essentially absent in prokaryotic ones. A typical eukaryotic gene is composed of several exons of roughly equal size scattered randomly within a much longer nontranslated sequence, like cars on a highway. The nontranslated portions of this highway that occupy the space between the scattered exons are introns. In a typical vertebrate gene the nontranslated (intron) portion of a gene is 10 to 30 times larger than the coding (exon) portion. For example, even though only 432 nucleotides are required to encode the 144 amino acids of hemoglobin, there are actually 1356 nucleotides in the primary mRNA transcript of the hemoglobin gene.

After the primary mRNA transcript is produced in the nucleus from the chromosomal DNA template, the introns are cut out; the exons are spliced together to form the mRNA that is actually translated in the cytoplasm. Thus the mRNA used by the ribosomes to produce a protein is a shortened, edited version of the primary transcript, the short version consisting only of the exons stitched together. This process—intron excision followed by exon splicing—is called **mRNA processing.**

Why are eukaryotic genes organized in pieces? Since the mid-1970s, when introns were discovered, this issue has been hotly debated. As the nucleotide sequences of more and more introns became known, a pattern has begun to emerge. It appears that each exon corresponds to a functional part of a protein. For example, one exon might encode that portion of a secretory protein that tags the protein for export; another exon might encode the transmembrane portion of a membrane channel protein; yet another exon might encode a portion of a polypeptide that forms an active site in an enzyme; and so on. Scientists now believe that enzymes have evolved by the shuffling of exons, achieved by recombination—different combinations of exons can quickly produce very different kinds of enzymes and other proteins. Typical of this process is a transmembrane channel protein that transports a lipid called cholesterol and its protein carrier across the cell membrane. The transmembrane channel protein has 18 exons. These exons are not unique to this gene, but rather have been recruited from several other genes. Thus it appears that the many kinds of proteins which now occur in organisms are mosaics that have evolved by reassembling simple elements in different ways. The recombination processes that move gene segments from one location on the chromosome to another will be described in Chapter 17.

TRANSLATING mRNA INTO PROTEIN

mRNA ultimately directs the synthesis of proteins within the cytoplasm. In bacteria the mRNA is translated soon after it is transcribed from the DNA (Figure 16-8). Indeed, the translation of mRNA has begun at one end before the other end has finished being transcribed from the DNA. In eukaryotes the situation is very different. The protein-synthesizing machinery is in the cytoplasm, and the genes are in the nucleus. The mRNA copied from the DNA (called the **primary transcript**) must first be transported across the nuclear membrane into the cytoplasm before it directs the synthesis of proteins.

The initial portion of the mRNA serves as a recognition region, guiding the binding of the message to the ribosomes, the cytoplasmic translating machinery. Ribosomes are among the more complex elements in living cells: they are composed of two major subunits (Figure 16-9) and contain more than 55 different proteins. The larger of the two subunits also contains two molecules of ribosomal RNA (rRNA), whereas the smaller subunit contains a third rRNA molecule. As far as we know, every ribosome that occurs within a cell's cytoplasm is identical.

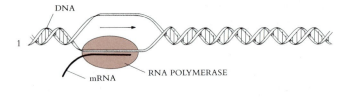

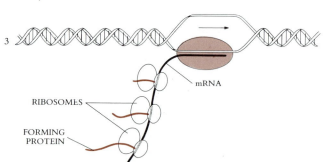

FIGURE 16-9

A ribosome is composed of two subunits. The smaller subunit fits into a depression on the surface of the larger one.

SMALL SUBUNIT

LARGE SUBUNIT

RIBOSOME

FIGURE 16-8

Bacteria have no nucleus and hence no membrane barrier between DNA and cytoplasm. As soon as the polymerase begins to copy an mRNA molecule from the DNA, a parade of ribosomes binds one after another to the mRNA's ribosome recognition site (the part of the mRNA copied from the DNA first), and each of these ribosomes begins in turn to translate the coding sequence into protein. From each mRNA molecule dangling from the DNA, a series of ribosomes is assembling proteins, one following the other down the mRNA.

FIGURE 16-11

These ribosomes are reading along an mRNA molecule of the fly Chironomus Tentans *from bottom to top, assembling proteins that dangle behind them like the tail of a tadpole. Clearly visible are the two subunits (arrows) of each ribosome translating the mRNA.*

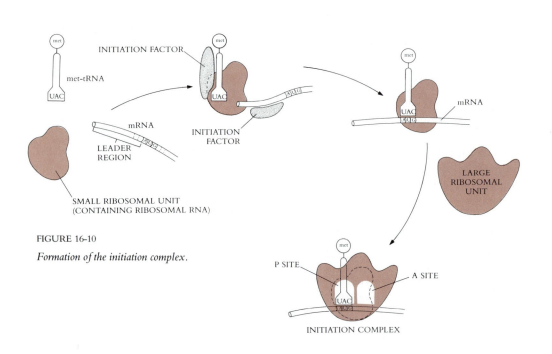

FIGURE 16-10

Formation of the initiation complex.

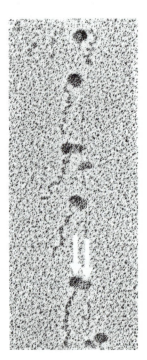

The leader region of an mRNA molecule is responsible for the first event in protein synthesis, the formation of an **initiation complex** (Figure 16-10). First a methionine-carrying tRNA (*met* tRNA) molecule binds to the small ribosomal subunit. Special proteins called **initiation factors** position the *met* tRNA on the small ribosomal subunit. The proper positioning of the initial amino acid that is carried by this tRNA is critical because it determines the reading frame (the groups of three) with which the nucleotide sequence will be translated into protein.

This complex, guided by another initiation factor, then binds to mRNA. Only after this binding is completed does the large ribosomal subunit become attached. In bacteria a portion of the roughly 100-nucleotide-long sequence of the mRNA leader region is homologous to an exposed region of one of the rRNA molecules on the larger subunit of the ribosome and binds to it. This step completes formation of the initiation complex.

After the initiation complex has been formed, the synthesis of the protein proceeds as follows (Figure 16-11):

1. The codon on the mRNA immediately adjacent to the initiating AUG codon interacts with the large ribosomal subunit, which positions the codon for interaction with another incoming tRNA molecule. The ribosome complex can thus be considered to have two tRNA binding sites: (a) the **P site,** or "polypeptide" site, to which the initial methionine-carrying tRNA binds and which carries the growing protein chain; and (b) the **A site,** or "amino acid" site, to which new incoming amino acid–bearing tRNA molecules bind. When a tRNA molecule with the *complementary* codon (an anticodon) appears, this new incoming tRNA binds to the mRNA molecule at the exposed codon position. Special proteins called **elongation factors** help to position the incoming tRNA. Binding the incoming tRNA to the large subunit in this manner places the amino acid at the other end of the incoming tRNA molecule directly adjacent to the initial methionine, which is dangling from the initiating tRNA molecule bound to the small subunit.

2. The two amino acids undergo a chemical reaction in which the initial methionine is released from its tRNA and is attached instead to the adjacent incoming amino acid by a peptide bond. The abandoned tRNA falls from its site on the small ribosomal subunit, leaving that site vacant.

3. In a process called **elongation** (Figure 16-12) the ribosome now moves along the mRNA molecule a distance corresponding to three nucleotides, guided by other elongation factors. This movement repositions the growing chain, at this point containing two amino acids, over the small ribosomal subunit and exposes the next codon of the mRNA over the large subunit. This same situation existed in step 1. When a tRNA molecule that recognizes this next codon appears, the anticodon of the incoming tRNA binds this codon, placing a new amino acid adjacent to the growing chain. The growing chain transfers to the incoming amino acid, as in step 2, and the elongation process continues.

4. When a chain-terminating nonsense codon is encountered (Figure 16-13) positioned over the large ribosomal subunit, no tRNA exists to bind to it. Instead, it is recognized by special **release factors,** enzymes that bind to nonsense codons and cleave the chemical bond attaching the adjacent polypeptide chain to the small ribosomal subunit. This cleavage releases the newly made protein.

Protein synthesis is carried out by ribosomes, which bind to sites at one end of the mRNA transcription unit and then move down the transcription unit in increments of three nucleotides. Each step of the ribosome's progress exposes a three-base sequence to binding by a tRNA molecule with the complementary nucleotide sequence. Ultimately the amino acid carried by that particular tRNA molecule is added to the end of the growing protein chain.

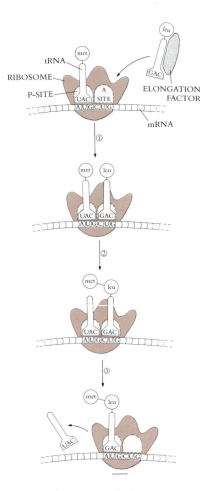

FIGURE 16-12

How protein synthesis proceeds. In the elongation process the A site is occupied by the tRNA with an anticodon complementary to the mRNA codon exposed in the A site (step 1). The growing polypeptide chain (in this illustration the first amino acid, methionine) is transferred to the incoming amino acid (in this illustration, leucine) (step 2); the tRNA to which the methionine was previously bound falls away, and the ribosome moves three nucleotides to the right (step 3) in preparation to receive the next incoming amino acid.

INTERVENING SEQUENCES: THE DISCOVERY OF INTRONS

Every nucleotide within the transcribed portion of a bacterial gene participates in an amino acid–specifying codon, and the order of amino acids in the protein is the same as the order of the codons in the gene. It was assumed for many years that all organisms would naturally behave in this logical way. In the late 1970s, however, biologists were amazed to discover that this relationship, one with which they had become completely familiar, did not in fact apply to eukaryotes. Instead, eukaryotic genes are encoded in segments that are excised from several locations along the transcribed mRNA and subsequently stitched together to form the mRNA that is eventually translated in the cytoplasm. With the benefit of hindsight, it is not difficult to design an experiment that reveals this unexpected mode of gene organization:

1. Isolate the mRNA corresponding to a particular gene. Much of the mRNA of red blood cells, for example, is hemoglobin and ovalbumin mRNA, making it easy to purify the mRNAs of these genes.

2. Using the enzyme reverse transcriptase, make a DNA version of the mRNA that has been isolated. This version of the gene is called **"copy" DNA (cDNA).**

3. Using genetic engineering techniques (Chapter 18), isolate from the nuclear DNA the portion that corresponds to the actual hemoglobin gene. This procedure is referred to as "cloning" the hemoglobin gene.

4. Mix single-strand forms of hemoglobin cDNA and nuclear DNA, and permit them to pair with each other ("hybridize") and form a duplex.

When this experiment was actually carried out, and the resulting duplex DNA molecules were examined with the electron microscope, the hybridized DNA did not appear as a simple duplex. Instead, unpaired loops were observed (Figure 16-C). At seven different sites within the ovalbumin gene, for example, the nuclear version contains long nucleotide sequences that are not present in the

FIGURE 16-13

Termination of protein synthesis.

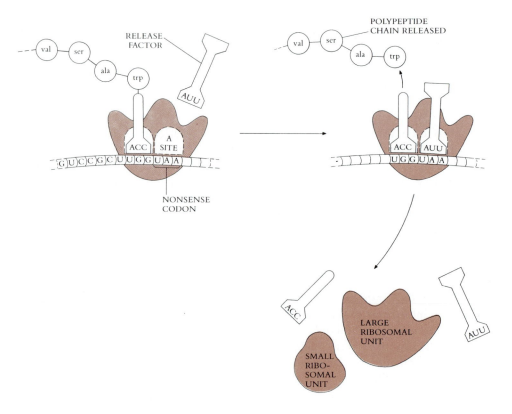

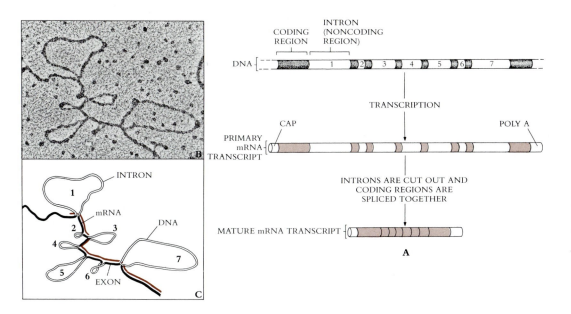

FIGURE 16-C

A *The ovalbumin gene and its primary transcript contain seven segments not present in the mRNA version, which the ribosomes use to direct the synthesis of protein. These intron segments are removed by enzymes that excise the introns and splice together the exons.*
B *The seven loops are the seven introns represented on the schematic drawing,*
C *of the DNA and primary mRNA transcript.*

cytoplasmic cDNA version. The conclusion is inescapable: nucleotide sequences are removed from within the gene transcript before the cytoplasmic mRNA is translated into protein. As we noted earlier, these internal noncoding sequences are called introns, and the coding segments are called exons. Because introns are removed from the mRNA transcript before it is translated, they do not affect the sequence of the protein that is encoded by the gene in which they occur.

HOW DNA REPLICATES

The complementary nature of the DNA duplex provides a ready means of duplicating the molecule. If one were to unzip the molecule, one would need only to assemble the appropriate complementary nucleotides on the exposed single strands in order to form two daughter duplexes of the same sequence. The density label experiments of Meselson and Stahl (Chapter 15) demonstrate that this is indeed what happens. When a DNA molecule replicates, the double-stranded DNA molecule separates at one end, forming a **replication fork** (Figure 16-14), and each separated strand serves as a template for the synthesis of a new complementary strand. Indeed, electron micrographs reveal Y-shaped DNA molecules at the point of replication, just as the model predicts.

To the surprise of those investigating the way in which DNA molecules replicate, it turned out that the two new daughter strands are synthesized on their templates in different ways. One strand is built by simply adding nucleotides to its end. This strand grows inward toward the junction of the Y as the duplex unzips. Because this strand ends with an —OH group attached to the third carbon of the ribose sugar (called the "3-prime," or 3′ carbon), the strand is said to grow from its 3′ end. The enzyme that catalyzes this process is called **DNA polymerase.**

However, when investigators searched for a corresponding enzyme that added nucleotides to the other strand (which ends with a —PO$_4$ group attached to the fifth carbon of the ribose sugar and is called the 5′ end), they were unable to find one. Nor has anyone ever found one. DNA polymerases add only to the 3′ ends of DNA strands.

NOT ALL ORGANISMS USE THE SAME GENETIC CODE

Perhaps the strongest evidence that all organisms share a common evolutionary heritage is that they all appear to use the same genetic code. For example, when a gene from a human cell is transferred by the techniques of molecular engineering (see Chapter 18) into a bacterial cell, that gene is translated by the decoding machinery of the bacterium into the same protein that would have been produced by a human cell. Bacteria use the same genetic code that humans do.

Starting in 1979, investigators began to determine the complete nucleotide sequences of mitochondrial genomes. The sequences for humans, cattle, and mice were all reported within a short period. It came as something of a shock when these investigators learned that the genetic code which these mammalian mitochondria were using was not quite the same as the "universal code" that had by then become so familiar to biologists. Most of the code words were the same—but not all. In mammalian mitochondria, what should have been a stop codon, UGA, was instead read as an amino acid, tryptophan; AUA was read as methionine rather than isoleucine; and AGA and AGG were read as "stop" rather than arginine. To make matters worse, yeast mitochondria also showed a few differences in these codons, but they were again different from those of the mammalian mitochondria.

FIGURE 16-14

A DNA replication fork. The colored upper left-hand strand is being copied from left to right by a DNA polymerase that is adding nucleotides to the growing 3' end. The colored lower left-hand strand is being copied from right to left by a DNA polymerase, which is adding nucleotides to the 3' end of a fragment; when the polymerase reaches the 5' end of the previous fragment, it binds this new fragment to it. It then goes back and starts a new fragment about 1000 bases ahead of where it started last time. Each fragment is started with a few bases of RNA, which are later removed.

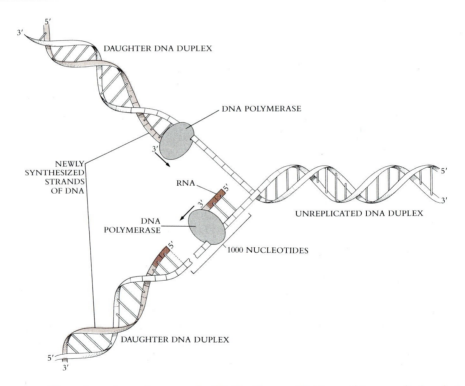

How does the polymerase build the 5' strand? Along this strand, the chain is also formed in the 3' direction, the polymerase jumping ahead and filling in backward. The DNA polymerase starts a burst of synthesis at the point of the replication fork and moves outward, adding nucleotides to the 3' end of a short new chain until this new segment fills in a gap of 1000 to 2000 nucleotides between the replication fork and the end of the growing chain to which the previous segment was added. The short new chain is then added to the growing chain and the polymerase jumps ahead again to fill in another gap. In effect, it copies the template strand in segments about 1000 nucleotides long and stitches each new fragment to the end of the growing chain. This mode of replication is referred to as **discontinuous synthesis.** If one looks carefully at electron micrographs showing DNA

When the first chloroplasts were sequenced, they too proved to contain minor differences from the universal code. These differences were associated with the same group of codons—and they were different from either of the mitochondrial versions of the code.

Nor does the story end with organelles. In 1985 the first gene sequences were reported for the ciliate *Tetrahymena*. Instead of reading the codon UAA as "stop," as in the universal code, *Tetrahymena* translates UAA as the amino acid glutamine. In another ciliate, *Stylonychia,* UAA also proved to be translated as glutamine. In yet a third ciliate, *Paramecium,* both UAA and another "stop" codon, UAG, were found to be translated as glutamine.

Thus it appears that the genetic code is not quite universal. Sometime, presumably after beginning their symbiotic existence, mitochondria and chloroplasts began to read the code differently, particularly that portion of the code associated with "stop" signals. Nor is this phenomenon limited to subcellular organelles. At some point very early in the evolution of ciliates, one of the phyla of the kingdom Protista, they too changed the way in which they read chain-terminating "stop" codons. Under most conditions, any genetic change involving termination signals would be expected to be lethal to an organism. How this change could have arisen in organelles, much less in eukaryotes such as ciliates, is a puzzle to which we as yet have no answer.

replication in progress, one of the daughter strands behind the polymerase appears single-stranded for about 1000 nucleotides.

A DNA molecule copies (replicates) itself by separating its two strands and using each as a template to assemble a new complementary strand, thus forming two daughter duplexes. The two strands are assembled in different directions.

How Bacteria Replicate Their DNA

The genetic material of bacteria is organized as a single, circular molecule of DNA. Such a structure cannot be compared directly with the complex chromosomes that are characteristic of eukaryotes. The replication of the DNA in bacteria is a complex process involving many enzymes; it is not, however, difficult to visualize the overall process. The entire genome can be replicated simply by nicking the duplex at one site and displacing the strand on one or both sides of the nick, creating one or two replication forks. These replication forks then proceed around the circle, creating a new daughter duplex as they go. In Figure 16-15 you can see a partially replicated circular DNA molecule, the displaced strand forming a loop. The mitochondria and chloroplasts of eukaryotic cells contain similar circular molecules of DNA, which replicate in the same way as they do in the bacteria from which they evolved.

ORIGIN

FIGURE 16-15

The circular chromosome of a bacterium initiates replication at a single site, moving out in both directions. When the two moving replication forks meet on the far side of the chromosome, the circular chromosome has been duplicated.

How a Eukaryote Replicates its Chromosome

The DNA within a eukaryotic chromosome is not circular, and, when examined under the electron microscope (Figure 16-16), it proves to have numerous replication forks spaced along the chromosome, rather than a single replication fork as prokaryotic chromosomes do. Each individual zone of the chromosome replicates as a discrete unit, called a **replication unit.** Replication units vary in length from 10,000 to 1 million base pairs; most are about 100,000 base pairs in length. They have been described for many different eukaryotes. Because each chromosome of a eukaryote possesses so much DNA (Figure 16-17), the orderly replication of DNA in eukaryotes undoubtedly requires sophisticated controls, which are as yet largely unknown.

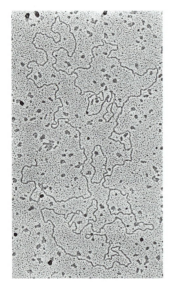

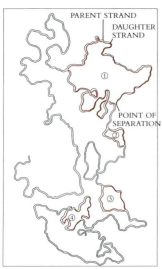

FIGURE 16-16

Eukaryotic chromosomes possess numerous replication forks spaced along their length. Four replication units are visible in this electron micrograph.

FIGURE 16-17

A human chromosome contains an enormous amount of DNA. The dark element at the bottom of the photograph is the protein matrix of a single chromosome. All the surrounding material is the DNA of that chromosome.

THE GENE MACHINE

For all of its seeming complexity, the apparatus that a cell uses to translate its hereditary information into protein, and to replicate that information from one generation to the next, is basically simple. A copy of the gene is made and sent to the ribosomes, where it directs the sequential assembly of a linear chain of amino acids. A major change occurred in this apparatus during the evolution of the eukaryotes, in which the copy of the gene came to undergo various modifications that we think play a role in controlling gene expression. Other than this single modification, however, the apparatus that organisms use to express and replicate genes appears to have persisted virtually unchanged since early in the history of life; it is shared by all living organisms. The same cannot be said about the genes themselves, which have undergone, and are still undergoing, major changes in organization. In the next two chapters we will focus on how these changes occur in nature, what kinds of changes seem to have occurred, and how we are learning to engineer our own changes in genes.

SUMMARY

1. The expression of hereditary information in all organisms takes place in two stages. First, an mRNA molecule with a nucleotide sequence that is complementary to a particular segment of the DNA is synthesized by an enzyme called RNA polymerase. Second, an amino acid chain is assembled by a ribosome, using the mRNA sequence to direct its choice of amino acids. The first process is called transcription; the second, translation.

2. Genes, the basic units of hereditary information, are composed of six elements: (1) the coding sequence, which directs the ribosome's choice of amino acids; (2) translation signals, which tell the ribosome where to start and stop; (3) a recognition sequence on the mRNA, to which the ribosome binds; (4) transcription signals, which tell the RNA polymerase when to start and stop the assembly of an mRNA molecule; (5) a recognition sequence on the DNA, to which the RNA polymerase binds; and (6) regulatory sites, to which other proteins can bind and influence the polymerase's activity.

3. The identity of the transcription units that are copied by a particular RNA polymerase molecule depends on which DNA "promoter" sites that particular polymerase is able to recognize and bind to. The recognition is carried out by a protein factor loosely bound to the polymerase. By substituting different factors, a cell may direct a polymerase to different genes.

4. The control of transcription is achieved largely by controlling the ability of RNA polymerase molecules to initiate transcription. In some instances a nucleotide sequence located between the promoter and the transcription unit acts as a binding site for a "repressor" protein. When bound to the site, this repressor protein prevents the read-through of the RNA polymerase.

5. In other instances nucleotide sequences located near the promoter site act as binding sites for "activator" proteins. The binding of the activator protein to the site facilitates the unwinding of the DNA duplex. Since RNA polymerase can translate only a single strand of DNA, this forced unwinding greatly facilitates the translation process.

6. The coding sequence of a gene is read in increments of three nucleotides from the mRNA by a ribosome. The ribosome positions the three-nucleotide segments of the message so that a tRNA molecule with the complementary base sequence can bind

to it. Attached to the other end of the tRNA molecule is an amino acid, which is added to the end of the growing protein chain.

7. The actual decoding of the genetic message is carried out by a family of 20 activating enzymes, each of which recognizes a few three-base mRNA nucleotide sequences and also recognizes one of the 20 common amino acids. These same 20 activating enzymes apparently function in all organisms in exactly the same way, always associating particular amino acids with particular three-base nucleotide sequences. The only known exceptions are a few code assignments often associated with termination.

8. In eukaryotes the mRNA transcript is transported from the nucleus to the cytoplasm before it is translated.

9. Most eukaryotic genes contain embedded within the coding sequences of their transcription units additional sequences called introns. These introns are removed from the transcript before it is translated.

10. DNA is replicated by a battery of enzymes, which include a DNA polymerase and a variety of proteins that unwind the DNA in front of the polymerase. DNA is replicated by unwinding the duplex and using each of the single strands as a template to assemble a complementary new strand. The two strands are assembled in different directions, one by the continuous addition of nucleotides to the growing end, the other in the opposite direction in 1000-nucleotide segments, which are then joined to the end of the growing strand.

SELF-QUIZ

The following questions pinpoint important information in this chapter. You should be sure to know the answers to each of them.

1. Which of the following entities are associated with transcription? Which with translation?
   (a) activating enzymes (f) promoter
   (b) initiation factors   (g) ribosomes
   (c) leader sequence   (h) RNA polymerase
   (d) nonsense codons   (i) transfer RNA
   (e) operon

2. Which of the following are *not* components of a transcription unit?
   (a) coding sequence
   (b) translation "start" and "stop" signals
   (c) ribosome recognition site
   (d) transcription "start" and "stop" signals
   (e) promoter sequence

3. The messenger RNA transcript of an operon will usually contain more than one transcription unit. True or false?

4. At one end of each mRNA transcript is a "leader region." To what does this region bind?

5. The order in which the elements of the *lac* genetic system are arrayed on the bacterial chromosome is
   (a) CAP binding site—*i* gene—promoter—operator—leader sequence—*lac* genes
   (b) *i* gene—CAP binding site—operator—promoter—leader sequence—*lac* genes
   (c) *i* gene—CAP binding site—promoter—operator—leader sequence—*lac* genes
   (d) CAP binding site—*i* gene—operator—promoter—leader sequence—*lac* genes

6. Lactose induces high levels of transcription of the *lac* genes by
   (a) blocking the binding of repressor protein to the operator site
   (b) destabilizing the DNA duplex so that it becomes locally single-stranded
   (c) binding to the operator site
   (d) binding to and thus changing the shape of RNA polymerase
   (e) Lactose does not induce high levels, but rather represses transcription

7. A particular activating enzyme binds with great selectivity a specific type of each of two different classes of molecules. What are these two classes of molecules?

8. How many different activating enzymes are there?

9. Can a given tRNA molecule be recognized by more than one activating enzyme?

10. How many different rRNA molecules are contained within one complete ribosome?

11. DNA replicates by unraveling the duplex, each single strand then acting as a template to guide the synthesis of a complementary strand on it. Are these two strands synthesized in the same direction, relative to the location of the replication fork?

12. In what direction does DNA polymerase proceed to assemble a new DNA strand, 3′ to 5′, or 5′ to 3′?

THOUGHT QUESTIONS

1. The DNA of human beings differs markedly from the DNA of bacteria in its base ratios. In humans the percentage of G + C is about 40%, whereas in bacteria it is about 60%. Does this necessarily imply that there are differences in the distribution of nucleotide bases in tRNA? rRNA? mRNA?

2. You are provided with a sample of aardvark DNA. As part of your investigation of this DNA, you transcribe mRNA from the DNA and purify it. You then separate the two strands of DNA and analyze the base composition of each strand, and of the mRNA. You obtain the following results:

	A	G	C	T	U
DNA strand 1	19.1	26.0	31.0	23.9	0
DNA strand 2	24.2	30.8	25.7	19.3	0
mRNA	19.0	25.9	30.8	0	24.3

Which strand of the DNA is the "sense" strand, serving as the template for mRNA synthesis?

3. Investigators first deciphered the genetic code by constructing artificial mRNA molecules and seeing what proteins were produced in test tube protein synthesis experiments. When a synthetic mRNA composed only of U (uracil) nucleotide bases was employed, the protein that resulted was a long chain of phenylalanine amino acids. Similarly, when the synthetic mRNA was composed only of A (adenine) nucleotide bases, a protein was obtained composed of a long chain of lysine amino acids. However, when a synthetic mRNA is employed that is composed of a random mixture of A and U, no long polypeptide is ever obtained. Why not?

FOR FURTHER READING

DARNELL, J.: "The Processing of RNA," *Scientific American*, October 1983, pages 90-102. An up-to-date description of how intron sequences are removed from mRNA transcripts before they are employed to direct the assembly of proteins.

DICKERSON, R.E.: "The DNA Helix and How It Is Read," *Scientific American*, December 1983, pages 94-112. X-ray analysis of DNA molecules of different sequence shows that sequence differences create subtle differences in shape that may influence DNA site recognition by proteins. A very clear exposition of a complex subject.

GRIVELL, L.A.: "Mitochondrial DNA," *Scientific American*, March 1983, pages 78-90. New research is revealing that the major cellular organelles of eukaryotes possess DNA that utilizes a slightly different genetic code than does the nucleus.

HOFFMAN, G.W.: "On the Origin of the Genetic Code and the Stability of the Translation Apparatus," *Journal of Molecular Biology*, vol. 86, pages 349-362, 1974. A thought-provoking discussion of how the gene machine evolved.

LAKE, J.: "The Ribosome," *Scientific American*, August 1981, pages 84-91. Recent studies have shown ribosomes to have unexpectedly complex shapes and provide a rather different view of protein synthesis than had previously been accepted.

LEWIN, B.: *Genes*, John Wiley & Sons, Inc., New York, 1983. A somewhat advanced but very readable account of how the genetic machinery works.

MILLER, O., B. HAMKALO, and C. THOMAS: "Visualization of Bacterial Genes in Action," *Science*, vol. 169, pages 392-395, 1970. Perhaps the best electron micrographs of coupled transcription and translation in bacteria.

RICH, A., and S. KIM: "The Three-Dimensional Structure of Transfer RNA," *Scientific American*, July 1978, pages 52-74. Transfer RNA molecules take the form of a cloverleaf folded into an L. This article discusses the functional consequences of that shape.

GENETIC CHANGE

Chapter 17

Overview

The evolution of life on earth depends critically on the existence of genetic variation in nature. Only when heritable differences exist can one life form supplant another. In this chapter we consider the mechanisms that are responsible for creating genetic change. Broadly speaking, there are two such factors, mutation and recombination, each occurring in many guises.

For Review   *Here are some important terms and concepts that you will encounter in this chapter. If you are not familiar with them, you should review them before proceeding.*

- **AIDS** (Chapter 1)

- **Independent assortment** (Chapters 13 and 14)

- **Crossing-over** (Chapter 14)

- **Triplet reading frames** (Chapters 15 and 16)

- **Repression of gene transcription** (Chapter 16)

There is an enormous amount of DNA within the cells of your body. This DNA represents a long series of DNA replications, starting with the DNA of a single cell, the fertilized egg. The replication of DNA that took place in producing the cells of your body is equivalent to producing a length of DNA nearly 97×10^9 kilometers long from an original 0.9-meter piece. Living cells have evolved many different kinds of mechanisms to avoid errors during DNA replication and to preserve the DNA from damage. These mechanisms ensure the accurate replication of DNA duplexes by proofreading the strands of each daughter cell against one another for accuracy and correcting any mistakes. The proofreading is not perfect, however. If it were, no mistakes would occur, no variation in gene sequence would result, and evolution would come to a halt.

In fact, living cells do make mistakes, and changes in the genetic message do occur, though only rarely. If changes were common, the genetic instructions encoded in DNA would soon degrade into meaningless noise. Typically, a particular gene is altered in only one out of a million gametes. Limited as it might seem, this steady trickle of change is the very stuff of evolution. Every single difference between the genetic message specifying you and the one specifying your cat, or the fleas on your cat, arose as the result of genetic change.

MUTATION

A change in the genetic message of a cell is referred to as a **mutation.** Some mutational changes affect the message itself, producing alterations in the sequence of DNA nucleotides. These alterations in the coding sequence are called **point mutations,** since they usually involve only one or a few nucleotides. Other classes of mutation involve changes in the way in which the genetic message is organized. In both bacteria and eukaryotes individual genes may move from one place on the chromosomes to another by a process called **transposition.** When a particular gene moves to a different location, there is often an alteration in its expression or in that of the neighboring genes. In eukaryotes large segments of chromosomes may change their relative location or undergo duplication. Such **chromosomal rearrangement** often has drastic effects on the expression of the genetic message. Sources and types of mutations are summarized in Table 17-1.

In the course of the differentiation of all multicellular organisms, there comes a point where the cells destined to form gametes are differentiated from those which form the rest of the body. In most animals this distinction occurs very early in development, when cells destined to form gametes (**germline** cells) are segregated from those destined to differentiate into other cells of the body (**somatic** cells). In other organisms such as plants and fungi, this developmental decision comes much later. Thus in plants a mutation of any cell can potentially be passed to the next generation, as any cell can potentially develop into an adult individual. In humans and most other animals, by contrast, only when a mutation occurs within a germline cell is the mutational change passed along to subsequent generations as part of the hereditary endowment contributed by the gametes derived from that cell. When a mutation occurs within a somatic cell, such as one of the cells lining your lung, the mutation is not passed along to subsequent generations of organisms, since the mutation does not affect the gametes. Mutation of a somatic cell (**somatic mutation**) may, however, have drastic effects on the individual organism in which it occurs, since it *is* passed on to all the cells of the organism that are descended from that cell. Thus if a mutant lung cell divides, the cells that are derived from it all carry the mutation. Somatic mutations of lung cells may lead to lung cancer.

Point mutations, involving only one or a few nucleotides, result either from chemical or physical damage to the DNA or from spontaneous pairing errors that occur during DNA replication. The first class of mutation is of particular practical importance, because modern industrial societies produce and release into the environment many chemicals capable of damaging DNA, chemicals called **mutagens.**

> **Point mutations are changes in the hereditary message of an organism. They may result from physical or chemical damage to the DNA or from spontaneous errors during DNA replication.**

DNA Damage

Although there are many different ways in which a DNA duplex can be damaged, three are of major importance: (1) ionizing radiation, (2) ultraviolet radiation, and (3) chemical mutagens.

Ionizing radiation

High-energy radiation such as x rays and gamma rays is highly mutagenic. When such radiation reaches a cell, it is absorbed by the atoms that it encounters, imparting energy to the electrons of their outer shells and causing these electrons to be ejected from the atoms. The ejected electrons leave behind ionized atoms with unpaired electrons, each called a **free radical.** Because most of a cell's atoms reside in water molecules and not in DNA, the great majority of free radicals created by **ionizing radiation** are produced from water molecules and not DNA.

FIGURE 17-1

Ultraviolet radiation is absorbed by pyrimidine bases, particularly thymine. When two thymine bases are adjacent to one another in a DNA strand, the absorption of UV radiation can cause the formation of a covalent bond between them—a thymine dimer, **A.** *Such a dimer introduces a "kink" into the double helix,* **B.** *The two linked thymines do not pair properly with the two adenines (AA) on the complementary DNA strand. This prevents replication of the duplex by DNA polymerase.*

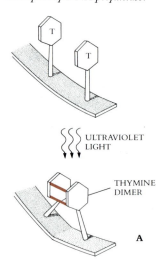

ULTRAVIOLET LIGHT

THYMINE DIMER

A

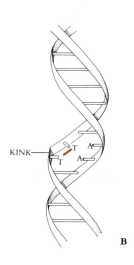

KINK

B

TABLE 17-1   SOURCES AND TYPES OF MUTATION

SOURCE	PRIMARY EFFECT	TYPE OF MUTATION
Ionizing radiation	Two-strand breaks in DNA	Deletions, translocations
Ultraviolet radiation	Pyrimidine dimers	Errors in nucleotide choice during repair
Chemical mutagens	Base analogue mispairing	Single nucleotide substitution
	Modification of a base leads to mispairing	Single nucleotide substitution
Spontaneous	Isomerization of a base	Single nucleotide substitution
	Slipped mispairing	Frameshift, short deletion
Transposition	Insertion of transposon into gene	Insertional inactivation
Mispairing of repeated sequences	Unequal crossing-over	Deletions, addition, inversions
Homologue pairing	Gene conversion	Single nucleotide substitutions

Most of the damage that the DNA suffers is thus indirect. It occurs because free radicals are highly reactive chemically, reacting violently with the other molecules of the cell, including DNA. The action of free radicals on a chromosome is like that of shrapnel from a grenade blast tearing into a human body.

The locations on the DNA where damage occurs are random, and the damage is often severe. Cells can heal some of this damage. Thus chemical changes can be repaired by excising altered nucleotides, and single ruptured bonds can be reformed. This **mutational repair** is not always accurate, however, and some of the mistakes are incorporated into the genetic message.

When a free radical breaks *both* phosphodiester bonds of a DNA helix—a **double-strand break**—the cell's usual mutational repair enzymes cannot fix the damage. To repair a double-strand break, the two fragments created by the break must be immobilized end to end so that the phosphodiester bond can be re-formed. Bacteria have no mechanism that can achieve this kind of alignment, and double-strand breaks are lethal to their descendants. Eukaryotes, almost all of which possess multiple copies of their chromosomes, are able to position the two fragments created by a double-strand break end to end by using the synaptonemal complex that is assembled in meiosis to pair the fragmented duplex with another copy of the chromosome. In fact, it is thought that meiosis may have initially evolved as a mechanism to repair double-strand breaks in DNA.

Ultraviolet radiation

Ultraviolet (UV) radiation, the component of sunlight that leads to suntan (and sunburn), is much lower in energy than are x rays. When molecules absorb UV radiation, the radiation does not impart enough energy to cause the molecules to eject electrons; consequently, free radicals are not formed. The only molecules that are capable of absorbing UV radiation, in fact, are certain organic ring compounds.

In DNA the principal absorption of UV radiation is by the pyrimidine bases thymine and cytosine. When these bases absorb UV energy, the electrons in their outer shells become reactive. If one of the nucleotides on either side of the absorbing pyrimidine is also a pyrimidine, a double covalent bond is formed between the two pyrimidines. The resulting cross-link between adjacent bases of the DNA strand is called a **pyrimidine dimer** (Figure 17-1). If such a cross-link is left unrepaired, it can potentially block DNA replication, in which case the damage would be lethal. What actually happens in most cases, however (Figure 17-2), is that cellular **UV repair systems** either (1) cleave the bond that links the adjacent pyrimidines or (2) excise the entire pyrimidine dimer from the strand and fill in the gap, using the other strand as a template. In those rare instances in which a pyrimidine

FIGURE 17-2

Thymine dimers are repaired by excising out the troublesome dimer, as well as a short run of nucleotides on either side of it, and then filling in the gap using the other strand as a template.

1

2

3

4

5

dimer does escape such cleavage or excision, the replicating polymerase simply fails to replicate the portion of the strand that includes the pyrimidine dimer, skipping ahead and leaving the problem area to be filled in later. This filling-in process is often error prone, however, and it may create mutational changes in the base sequence of the gap region.

Chemical mutagens

Many mutations result from the direct chemical modification of the DNA bases. The chemicals that act on DNA fall into three classes: (1) chemicals that look like DNA nucleotides but pair incorrectly when they are incorporated into DNA (Figure 17-3); (2)

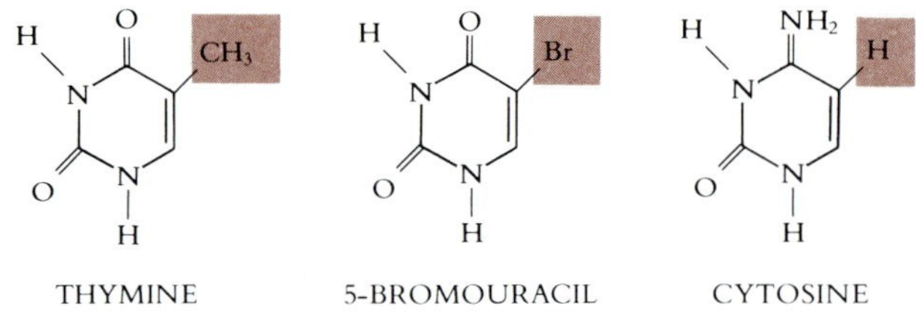

THYMINE   5-BROMOURACIL   CYTOSINE

FIGURE 17-3

Mutations may be caused by chemicals that look like DNA bases. For example, DNA polymerase cannot distinguish between thymine and 5-bromouracil, which are very similar in shape. Once incorporated into a DNA molecule, however, 5-bromouracil tends to rearrange to a form that resembles cytosine and pairs with guanine. When this happens, what was originally an AT base pair becomes a GC base pair.

chemicals that remove the amino group from adenine or cytosine, causing them to mispair; and (3) chemicals that add hydrocarbon groups to nucleotide bases, also causing them to mispair. This last group includes many particularly potent mutagens that are commonly used in the laboratory, as well as compounds that are sometimes released into the environment, such as mustard gas.

> **The three major sources of mutational damage to DNA are (1) high-energy radiation, such as x rays, which physically breaks the DNA strands; (2) low-energy radiation, such as UV light, which creates DNA cross-links whose removal often leads to errors in base selection; and (3) chemicals that modify DNA bases and thus alter their base-pairing behavior.**

Spontaneous Mispairing

Many point mutations occur spontaneously, without the intercession of mutagenic chemicals or radiation. Sometimes nucleotide bases spontaneously shift to alternative conformations, or **isomers.** These isomers form different kinds of hydrogen bonds than do the normal conformations. During the replication of a DNA strand the polymerase selects a different nucleotide to pair with an isomer than the one it would have otherwise selected (Figure 17-4). These spontaneous errors occur in fewer than one in a billion nucleotides per generation, but they are still an important source of mutation.

When chromosomes pair, sequences may sometimes misalign, looping out a portion of one strand (Figure 17-5). Usually such misaligned pairing, called **slipped mispairing,** is only transitory and the strands revert quickly to the normal arrangement. If the error-correcting system of the cell encounters such a mispairing before it reverts, however, it will attempt to "correct" it, usually by excising the loop. This results in a **deletion** of several hundred nucleotides from one of the strands. Many of the deletions created in this way start or stop in the middle of a codon. When they occur, therefore, they lead to the creation of a genetic message that is out of synchrony with the normal reading pattern, the three-base increments being displaced one or two positions. The deletion of the letter "F" from the sentence "THE FAT CAT ATE THE RAT" similarly shifts the reading frame of that mes-

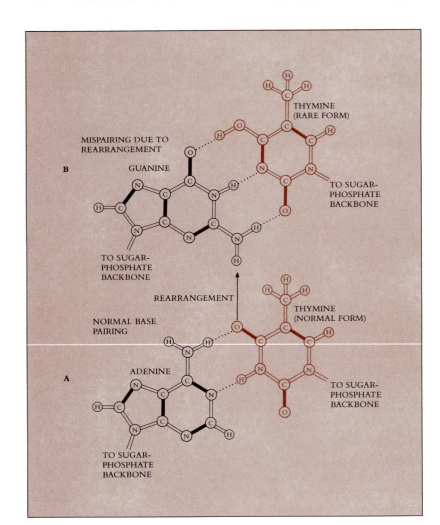

FIGURE 17-4

*Thymine, which normally pairs with adenine, **A**, occasionally shifts to an alternative isomeric conformation that pairs with guanine, **B**.*

FIGURE 17-5

Slipped mispairing occurs when a sequence is present in more than one copy on a chromosome and the copies on homologous chromosomes pair out of register. The loop that is produced by this mistake in pairing is sometimes excised by the cell's repair enzymes, producing a short deletion and often altering the reading frame. Any chemical that tends to stabilize the loops increases the chance that this will happen.

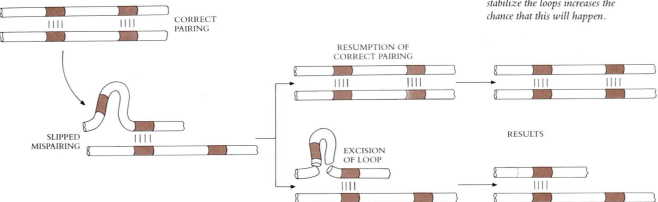

sage, producing the message "THE ATC ATA TET HER AT," which is meaningless. Such an event results in what is called a **frameshift mutation** and leads to a message that is read in the wrong three-base groupings. Some chemicals specifically promote the occurrence of frameshift mutations by stabilizing the loops that are produced by slipped mispairing, thus increasing the length of time during which the loops are vulnerable to excision.

Spontaneous errors in DNA replication occur very rarely. They result from transient changes in the conformation of nucleotides and also from the accidental mispairing of nonidentical but similar sequences.

AN OVERVIEW OF RECOMBINATION

If we define mutation as any change in an organism's genetic message, then you might think that mutation is the only source of genetic diversity and that all change must arise as a result of it. Change, however, can also result from moving existing elements of the genetic message around on the chromosomes. As an analogy, consider the pages of this book. A mutation would correspond to a change in one or more of the letters on the pages. For example, ". . . in one or more of the letters of the pages" is a mutation of the previous sentence, in which a letter "n" is changed to "f." A significant alteration is also achieved, however, by moving the position of words; ". . . in one or more of the pages on the letters," for example, alters the meaning of the sentence (and destroys it) by exchanging the position of the words "letters" and "pages." This second kind of genetic change, which is an alteration in the chromosomal position of a gene or a fragment of a gene, is a form of genetic recombination.

Viewed broadly, three kinds of genetic recombination occur in organisms: (1) **reciprocal recombination,** in which two chromosomes trade segments; (2) **gene transfer,** in which one chromosome donates a segment to another; (3) **chromosome assortment,** in which chromosomes assort during meiosis. The crossing-over that occurs between the homologous chromosomes of eukaryotes during meiosis is an example of reciprocal recombination, the acquisition of an AIDS-bearing virus by a human chromosome is an example of gene transfer, and the 9:3:3:1 dihybrid Mendelian ratio is an example of independent assortment. Gene transfer processes occur among both bacteria and eukaryotes; reciprocal recombination between chromosomes and independent assortment are strictly eukaryotic phenomena. For this reason, gene transfer is thought to be the more primitive process.

> **Genetic recombination is a change in the chromosomal association among genes. This often involves a change in the position of a gene or portion of a gene. Recombination of this sort may result from one-way gene transfer or from reciprocal gene exchange.**

GENE TRANSFER

Genes are not fixed in their locations on chromosomes, like words engraved in granite. They move around. Some genes move because they are part of small ancillary chromosomes called **plasmids,** which are able to enter and leave the main chromosome. Plasmids are able to do so, however, only at specific places on the chromosome where a nucleotide sequence occurs that is also present on the plasmid DNA. Other genes move as part of small fragments of the chromosome called **transposons,** which migrate from one chromosomal position to another at random, like fleas jumping along a string. Plasmids occur primarily in bacteria, in which the naked DNA can interact readily with other DNA fragments; transposons occur in both bacteria and eukaryotes. Both of these gene transfer processes, plasmid movement and transposon movement, probably represent the earliest form of genetic recombination to evolve, one that is still responsible for much of the genetic recombination that occurs today.

These two forms of gene movement were discovered within three years of one another, the first (plasmid movement) by Joshua Lederberg and Edward Tatum in 1947, the second (transposon movement) by Barbara McClintock in 1950. Because the second mode of gene movement is random and implies that gene position on chromosomes is not constant, its discovery was not readily accepted by researchers accustomed to viewing chromosomes as composed of genes in fixed positions, like beads on a string. Lederberg and Tatum received a Nobel prize for their discovery in 1958. McClintock was awarded a Nobel prize for hers 25 years later, in 1983.

Plasmids

Most, but not all, of the DNA of bacteria exists in the form of a single, circular, duplex molecule. About 5% of the total DNA, however, is found outside the main chromosome

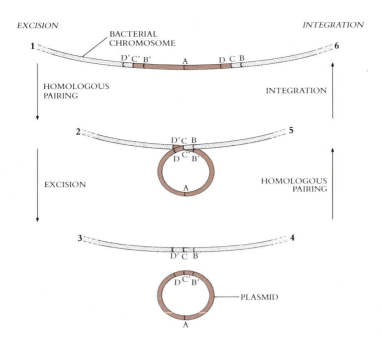

EXCISION   INTEGRATION

FIGURE 17-6

Integration and excision of a plasmid. Because the ends of the two sequences on the bacterial chromosome are the same— D', C', B', and D, C, B—it is possible for the two ends to pair. Steps 1 to 3 show the sequence of events if the strands exchange during the pairing. The result is excision of the loop and a free circle of DNA—a plasmid. Steps 4 to 6 show the sequence followed when a plasmid integrates itself into a chromosomal bacterium.

in small, circular DNA plasmids. Some plasmids are very small, containing only one or a few genes; others are quite complex and contain many genes.

To understand how plasmids arise, consider a hypothetical stretch of DNA along a bacterial chromosome that contains two copies of a repeated sequence. Because this sequence occurs twice on the chromosome, it is possible for the chromosome to form a transient "loop" in such a way that the two sequences base-pair with each other and create a transient double duplex. All cells have enzymes, called **recombination enzymes,** that can recognize such double duplexes. These enzymes can cause the two duplexes to exchange strands—to undergo a **reciprocal exchange.** The result of such an exchange, as you can see in Figure 17-6 (steps 1 to 3), is to free the loop from the rest of the chromosome. The resulting free loop is a plasmid. The DNA that makes up the plasmid corresponds to the genes, such as gene A between the two duplicated sequences in Figure 17-6. These genes are no longer present in the main chromosome and reside only on the plasmid.

Once a plasmid has been created by this kind of reciprocal exchange of DNA strands, it is a free element, able to move about within the cell. The plasmid may contain a **replication origin,** a site that the cell's DNA-replicating enzyme, DNA polymerase, recognizes. If it does, the plasmid will be replicated by the DNA polymerase, often without the controls that restrict replication of the chromosome to only once per cell division. Because their replication is independent from that of the cell, some plasmids may be represented by many copies and others by few copies in a given cell.

A plasmid that was created by recombination can reenter a chromosome the same way that it left. Sometimes the region of the plasmid DNA involved in the original exchange, called the **recognition site,** aligns with its mate on the main chromosome; if a second recombination event occurs anywhere in the common sequence while they are aligned, it will result in the reincorporation of the plasmid sequence into the chromosome. The plasmid is then said to have been **integrated** into the chromosome (Figure 17-6, steps 4 to 6). This integration can occur at any site at which the shared sequence (or any other shared sequence) exists so that plasmids sometimes integrate at positions different from those at which they arose. If they are reincorporated into the chromosome at a new position, they transport their genes with them to the new location.

The locations to which genes are transported by plasmids correspond to the positions of DNA sequences on the main chromosome that also occur in the plasmids.

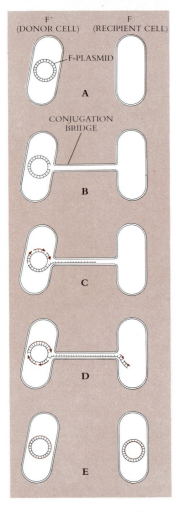

FIGURE 17-7

Gene transfer between bacteria. Donor cells contain an F plasmid that recipient cells lack (A). The long pilus connecting the two cells is called a conjugation bridge (B). Across it the F plasmid replicates a copy of its chromosome. One strand is passed across, the other serving as a template to build a replacement (C); when the single strand enters the recipient cell, it serves as a template to assemble a double-stranded duplex (D). When the process is complete, both cells contain a complete copy of the circular plasmid DNA molecule (E).

Gene Transfer Among Bacteria

A startling discovery that Lederberg and Tatum made in the 1950s was that bacteria can pass plasmids from one cell to another. The plasmid that was studied by Lederberg and Tatum was a fragment of the chromosome of *Escherichia coli*. It was given the name "F," for fertility factor, since only cells containing that plasmid could act as plasmid donors. The F plasmid contains a DNA replication origin and several special genes that promote its transfer to other cells. These genes encode protein subunits that assemble on the surface of the bacterial cell, forming a hollow tube called a **pilus.**

When the pilus of one cell makes contact with the surface of another cell that lacks pili, and therefore does not contain the F plasmid, the pilus forms a **conjugation bridge** between the two cells and initiates a series of changes within the first cell that mobilize the plasmid for transfer. First, the F plasmid binds to a site on the interior of the bacterial cell just beneath the pilus. The F plasmid then begins to replicate its DNA at that point of binding, passing the replicated copy out across the pilus and into the other cell (Figure 17-7). This **rolling-circle replication** eventually places an entire newly synthesized single-strand copy of the plasmid into the recipient cell, where a complementary strand is added, creating a new stable F plasmid.

The F plasmid does not have to be free in order to carry out transfers of this kind. It will transfer just as readily if it is integrated into the bacterial chromosome. The F region binds beneath the pilus and initiates synthesis across the bridge between the two cells as described above. In this case, however, the cell begins to replicate a copy of *the entire bacterial chromosome* across to the recipient cell. Transfer proceeds just as if the bacterial chromosome were simply part of the small plasmid. Such transfer of bacterial genes is called **conjugation.** Researchers have used knowledge of this process to locate the positions of different genes on bacterial chromosomes (Figure 17-8).

Plasmids such as the F plasmid transfer bacterial genes from one individual bacterium to another by integrating the entire circular bacterial chromosome into the small circular plasmid DNA molecule. When a copy of the plasmid DNA molecule is replicated across to another cell, the integrated bacterial DNA is replicated across as if it were part of the plasmid.

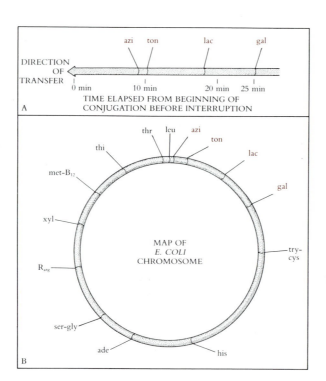

FIGURE 17-8

It takes over an hour for the entire bacterial chromosome to pass across the conjugation bridge from donor to recipient cell. Because this bridge is fragile, it is often broken before all the chromosome has passed across. It is possible for an investigator to break all the conjugation bridges in a cell suspension by agitating the suspension rapidly in a blender. By conducting parallel experiments in which a blender is turned on at different times after the start of conjugation, investigators have been able to locate the positions of various genes on the bacterial chromosome. The closer genes are to the origin of replication, the sooner one has to turn the blender on to block their transfer.
***A** The time elapsed from the beginning of conjugation before turning on the blender no longer blocks the genes indicated.*
***B** A map of the* Escherichia coli *chromosome developed using this method.*

Transposition

Transposition is a form of gene transfer that occurs both in bacteria and in eukaryotes. Transposing genes do not stay put on chromosomes. Every once in a while, after many generations in one location, a transposing gene will abruptly move to a new position on the chromosomes, the location of its new residence apparently chosen at random. Transposing genes move about the chromosomes like so many Mexican jumping beans. These nomadic genes behave in this unusual way because they are carried along as parts of randomly moving genetic elements called transposons, passengers in a car with no driver.

To understand the transposition process, it is useful to consider briefly **viral transfer,** using the example of the bacterial virus lambda. Like a plasmid, a lambda virus (Figure 17-9) may move from one chromosomal location to another. Rather than pairing a sequence shared in common with the chromosome, as plasmids do, lambda viruses possess a gene encoding an **integrase,** an enzyme that recognizes a specific sequence on the bacterial chromosome and inserts the lambda DNA into the bacterial chromosome at sites where these sequences occur. These sequences do not occur in the lambda DNA itself.

Transposons encode an enzyme similar to the lambda integrase, which is called a **transposase.** However, the transposase does not recognize any particular sequence on the host chromosome. Instead, it selects a site more or less randomly (Figure 17-10). Therefore the genetic locations of these elements change in a random manner.

> **Transposition is a one-way gene transfer to a random location on the chromosome. Genes move because they are associated with mobile genetic elements called transposons.**

Many transposons exist in multiple copies scattered about on the chromosomes. In *Drosophila,* for example, more than 30 different transposons are known, most of them present in numerous copies located at some 20 to 40 different sites throughout the genome. In all, the known transposons in a *Drosophila* cell account for perhaps 5% of its total DNA. Fewer different transposons are thought to occur in humans, but some of them are repeated many thousands of times.

FIGURE 17-9

Below *The bacterial virus lambda has the ability to introduce itself into a bacterial chromosome and also to move from one location on a chromosome to another.*
Left *The process a lambda phage uses to integrate itself into a bacterial chromosome or excise itself, after the virus has entered the bacterial cell.*

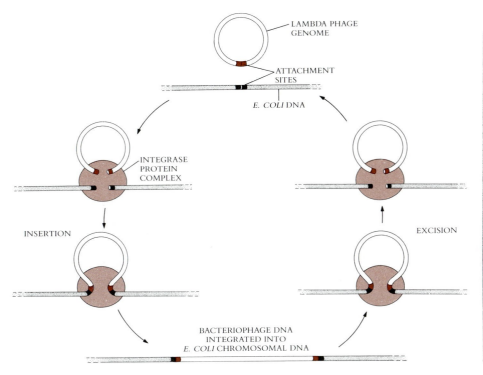

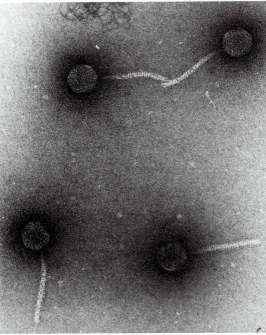

FIGURE 17-10

Transposition. The transposase does not recognize any particular DNA sequence, but rather selects one at random, with the result that the movement is to a random location. Some, but not all, transposons leave a copy of themselves behind when they move.

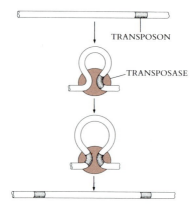

Retroviruses

The transposons that do occur in humans are of particular interest to us, because they are potential agents of disease. Both cancer and AIDS have been linked to a special class of transposons called **retroviruses.** Recall our discussion in Chapter 1 of the cat virus that acquired a portion of the cheetah HLA gene, thus becoming a virulent pathogen to cheetahs. That virus is a retrovirus. Leukemia in cats, a highly contagious killer, is caused by retroviruses, and other retroviruses also have been linked to human cancers. AIDS, the dangerous human disease that was also discussed in Chapter 1, is caused by a variant form of a retrovirus that is associated with leukemia. In all of these cases the disease results from the placing of a gene in an inappropriate location. Recombination does not always produce desirable results.

A retrovirus is an RNA transcript of a portion of eukaryotic DNA. Like other viruses, retroviruses pass from cell to cell enclosed within a protein sheath; once within a cell the retrovirus RNA can be replicated by the cell's machinery. Retroviruses can stitch themselves into cellular chromosomes: they encode a special enzyme, called **reverse transcriptase,** which makes a DNA transcript of the RNA strand that exists in the retroviruses. This process is transcription in reverse, which accounts for the name of the enzyme. The DNA copy can then be integrated into the eukaryotic chromosome at any random location by the retrovirus-encoded transposase enzyme.

The Impact of Transposition

The transposition of a given element is relatively rare, although an element is perhaps 10 times more likely to move than to undergo a random mutational change. The transposition of a particular mobile element occurs perhaps once in every 100,000 cell generations. There are many elements in most cells, however, and many generations to consider. Viewed over long periods of time, transposition has had enormous evolutionary impact. Some of the ways in which this has happened are as follows:

1. *Mutation.* The insertion of a mobile element within a gene often destroys the gene's function. Such a loss of gene function is termed **insertional inactivation.** It is thought that a significant fraction of the spontaneous mutations observed in nature has in fact resulted from this effect.

2. *Cancer.* The inclusion of eukaryotic genes within transposable retroviruses causes the transcription of these genes to be regulated by the genetic controls of the retrovirus. These controls may act very differently from those of the eukaryotic cell. Some genes are transcribed far more frequently when they have been freed from cellular control in this way. When the products of the overtranscribed eukaryotic genes are important growth-regulating substances, uncontrolled cell growth—cancer—may result.

3. *Gene mobilization.* Natural selection sometimes favors gene mobilization, the bringing together in one place of genes that are usually located at different positions on the chromosomes. In bacteria, for example, a number of different genes encode enzymes that make the bacteria resistant to one or more antibiotics, such as penicillin. Many of these genes are encoded within small plasmids, different genes on different plasmids. The administration of several antibiotic drugs simultaneously to sick people was a common medical practice some years ago. Unfortunately, this simultaneous exposure to many antibiotics favors the persistence of plasmids that have managed to acquire several resistance genes— and transposition can rapidly generate such composite plasmids (Figure 17-11), with the antibiotic resistance genes moving by transposition from one plasmid to another. Called **resistance transfer factors,** these composite plasmids each possess many antibiotic resistance genes. The bacteria in which these plasmids occur are able to resist or destroy a variety of antibiotic drugs and are thus immune to all of them simultaneously.

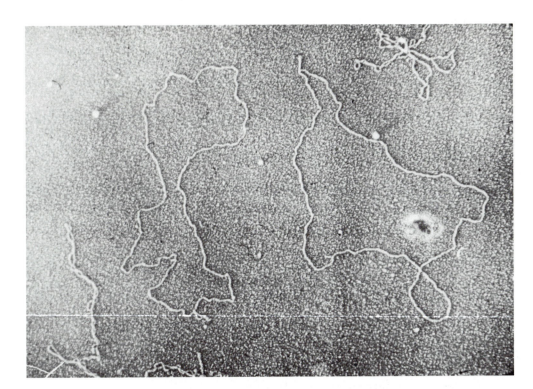

4. *Rapid recombination.* More and more examples are being discovered of novel functions made possible by the ability of organisms to use transposition to move genes to new locations. A particularly important instance is the gene mobilization practiced by African **trypanosomes,** the protozoa that are responsible for sleeping sickness, a serious disease of human beings and other animals (Chapter 27). The surface polysaccharides of these organisms determine their infective properties. Humans and other animals that are targets of trypanosome infection defend against these organisms by producing antibodies directed against the surface polysaccharides. Trypanosomes defeat these defenses with transposition. The trypanosome genome contains a cluster of several thousand different polysaccharide-encoding sequences. Because the cluster lacks a promoter, these sequences cannot be transcribed as they are. The necessary promoter is located at a second site, called the **transcription site.** This site is within a transposon that periodically jumps randomly from one position to another within the cluster. These sporadic changes in the cluster sequence transcribed result in the appearance of new surface polysaccharides before the immune system of the host has completely mobilized an attack against the old one.

CHROMOSOMAL REARRANGEMENT

Although transposable sequences move around within the genome, most genes appear to maintain relatively stable positions even over long evolutionary periods. Why are genes not scattered to different locations by recombination and transposition? For most genes, chromosome location plays an important role in the cell's control of gene transcription, of when particular genes are used and when they are not. For example, some genes are not capable of being transcribed if they are adjacent to a tightly coiled heterochromatic sequence, even though the same genes are transcribed normally in any nonheterochromatic location. The transcription of many chromosomal regions appears to be regulated in an analogous manner, with the binding of specific chromosomal proteins regulating the degree of coiling in local regions of the chromosome and so decreasing the accessibility of genes located within the region to RNA polymerase.

CHROMOSOMAL REARRANGEMENT: THE SOURCE OF ANTIBODY DIVERSITY

During mammalian development, the genes responsible for forming proteins called antibodies, which recognize foreign cells and molecules (antigens), achieve great diversity of expression by rearranging the position of individual genes. Antibody molecules contain several distinct regions, each encoded at a different site on the chromosome. These chromosomal sites are composed of a cluster of similar sequences, each sequence varying from the others in its cluster by small degrees. When an antibody is made (Figure 17-A), one sequence from each cluster is selected at random, and the DNA sequences from the various clusters are assembled together by DNA recombination to produce a segment of DNA containing the full gene sequence. The cell in which this happens then makes the antibody protein encoded by that sequence, and no other. Each of the many cells that produces antibody proteins undergoes this process independently during its maturation. Because the elements of the antibody sequence are selected from their clusters randomly, it is rare that two cells assemble the same antibody—several million different combinations are possible.

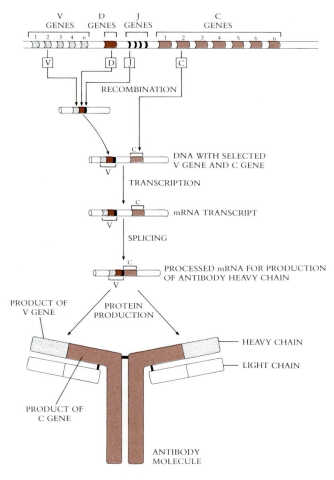

FIGURE 17-A

How an antibody is assembled. Every antibody molecule is composed of four polypeptides, two long ones called **heavy chains** *and two short ones called* **light chains.** *Both heavy and light chains are assembled from various materials. The assembly of a heavy chain is illustrated here. Every antibody-producing cell makes only one version of the heavy chain, but every cell's version is different from every other's. In a particular cell a randomly selected version of V is selected from among the many present on the chromosome and recombined in that cell near D; similarly, one randomly chosen copy of J is recombined to the vicinity of D. Then another recombination event places one of the many versions of C near the VDJ cluster. The space between C and the cluster is removed from the RNA transcript, producing a single VDJC sequence that encodes a single polypeptide. The reason that there are so many different antibody molecules is that there are many possible combinations of the four elements that are assembled into the heavy chain a particular cell produces.*

Chromosomes undergo several different kinds of gross physical alterations that have significant effects on the locations of their genes. The two most important are **translocations,** in which a segment of one chromosome becomes a part of another chromosome, and **inversions,** in which the orientation of a portion of a chromosome becomes reversed. Translocations, like transpositions, often have important effects on gene expression. Inversions, in contrast, usually do not alter gene expression but are important because of their effect on recombination. Recombination within a region of the chromosome that is inverted on one chromosome and not on the other (Figure 17-12) leads to serious problems during meiosis, because none of the gametes produced following such a crossover event will have a complete set of genes.

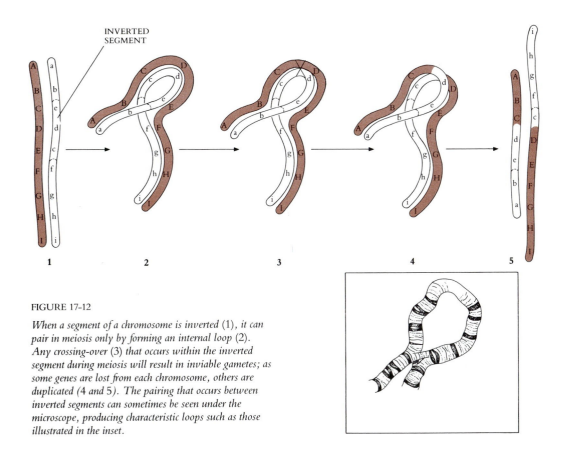

INVERTED
SEGMENT

1   2   3   4   5

FIGURE 17-12

When a segment of a chromosome is inverted (1), it can pair in meiosis only by forming an internal loop (2). Any crossing-over (3) that occurs within the inverted segment during meiosis will result in inviable gametes; as some genes are lost from each chromosome, others are duplicated (4 and 5). The pairing that occurs between inverted segments can sometimes be seen under the microscope, producing characteristic loops such as those illustrated in the inset.

Other chromosomal alterations change the number of gene copies that an individual possesses. Whole chromosomes may be gained or lost. An individual that has gained or lost a whole chromosome is said to be **aneuploid.** Whole diploid sets of chromosomes may be multiplied; an individual, tissue, or cell that has three or more sets of chromosomes is said to be **polyploid** (see Chapter 21). An individual chromosome segment may be deleted from the genome. Most deletions are very harmful; they halve the number of gene copies within a diploid genome and thus seriously affect the level of transcription. A portion of a chromosome may also be duplicated, a condition that also results in gene imbalance and thus is harmful in many cases.

RECIPROCAL RECOMBINATION

Eukaryotes typically possess two or more copies of their chromosomes at some stage of their life cycle. Human beings, for example, like most other animals, are diploid; they have two copies of each of their chromosomes in each of their cells throughout their entire life cycle, except when they form gametes. The possession of duplicate copies of chromosomes by eukaryotes has a profound consequence: because the two copies can be paired together, it is possible for them to exchange segments with each other. You encountered this process in Chapter 14 as crossing-over.

Crossing-Over

Crossing-over occurs in the first prophase of meiosis, when two homologous chromosomes are lined up side by side within the synaptonemal complex; an exchange of strands occurs between the two homologues at one or more locations. Such an exchange may re-

THE MOLECULAR MECHANISM OF CROSSING-OVER

Although many molecular mechanisms have been proposed to describe crossing-over, one seems most likely. It is called the **double-strand-break repair** model, proposed by Frank Stahl and his coworkers at the University of Oregon in 1983. If they are correct, and we think they are, then this is what happens during crossing-over (Figure 17-B):

1. The synaptonemal complex creates a double-strand break in one of the two homologous chromosomes *(ab)*, which are lined up side by side.

2. Cellular enzymes degrade away part of the broken duplex, like termites chewing the broken ends of a snapped pencil. The result is the creation of a *gap* at the site of the break.

3. One of the loose ends *(a,* third strand) then unwinds, and one of its strands aligns itself with the other undamaged duplex, pairing in a base-to-base manner with the strand that has a nucleotide sequence complementary to its own *(AB,* top strand). The displaced undamaged strand, which has no pairing partner, forms a single-strand loop *(AB,* second strand).

4. The strand that has invaded the undamaged duplex starts to grow. New nucleotides are added to its end, with the undamaged complementary strand being used as a template. As a result of this growth, the displaced loop in the unpaired undamaged strand *(AB,* second strand) gets larger and larger.

5. Eventually the growing strand fills in the entire gap created in stage 2, the bases selected to be complementary to those of the undamaged homologous strand. At the same time this is occurring, the damaged strand that did not invade *(b,* fourth strand) also fills in the gap, pairing with the displaced loop and using it as a template.

6. The broken ends are sealed together by enzymes so that the two duplexes are now both continuous, with no breaks. One strand of each, however, has been exchanged with the other in two positions.

Finally, the crossed-over strands break and re-form, thus eliminating the bridges between the two homologous chromosomes. As you can see in Figure 17-C, this can occur in either of two ways. In one case the original arrangement is re-formed as if a double-strand break had never occurred *(1).* The other way results in an exchange of both strands, in other words, a physical exchange of chromosomal arms *(2).* It is to this second kind of event that we are referring when we speak of crossing-over. We do not observe the first kind of event, because it does not lead to exchange of the chromosomal arms and therefore does not have genetic effects outside of the local area of these molecular events.

sult in the physical exchange of one of the paired chromosome strands at that point (see Figure 14-10).

Pairing of homologous chromosomes is thought to have evolved originally as a means of repairing double-strand breaks in DNA duplexes. The molecular events that take place during crossing-over are by and large the same events that are involved in the repair of double-strand breaks and use many of the same enzymes. Indeed, mutant forms of yeast that have been selected for their sensitivity to x rays (which produce double-strand breaks in DNA) prove unable to repair double-strand breaks in their DNA, and the same mutants are also incapable of meiotic crossing-over.

When a reciprocal crossover event has concluded, the participating chromosomes often have physically exchanged chromosome arms. If the two chromosomes contain dif-

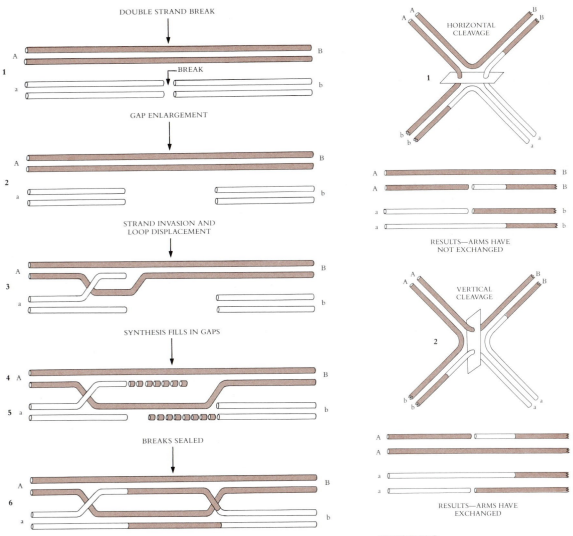

FIGURE 17-B

FIGURE 17-C

The second stage in crossing-over is the cleavage of the chiasmata produced by the strand exchange used to repair the double-strand break. Only one of the two possible cleavages results in the exchange of chromosomal arms, crossing-over.

ferent mutations on each side of the crossover event, the physical exchange of chromosomal arms produces chromosomes that have different combinations of mutations. When such chromosomes go on to segregate in meiosis, they form gametes that have new combinations of alleles.

Imagine, for example, that a giraffe has a gene encoding neck length on one of its chromosomes and a gene encoding leg length that is located elsewhere on that same chromosome. Now imagine further that a recessive mutation occurs within the giraffe population at the neck length locus, leading after several rounds of independent assortment to some individuals that are homozygous for a variant "long-neck" allele. Similarly, a recessive mutation occurs at the leg length locus, leading to individuals homozygous for a "long-leg" allele.

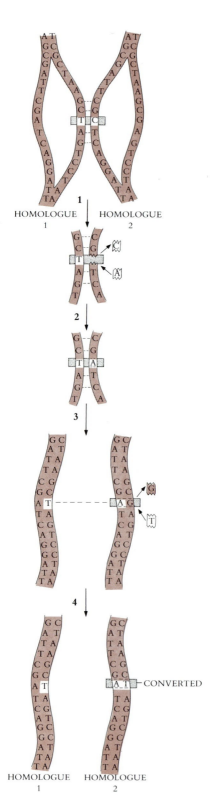

HOMOLOGUE 1   HOMOLOGUE 2

1

2

3

4

CONVERTED

HOMOLOGUE 1   HOMOLOGUE 2

FIGURE 17-13

It is very unlikely that the two independent mutations would occur at the same time in the same individual, since the probability of two independent events occurring together is the product of their individual probabilities. If the only way to produce a giraffe that had both a long neck and long legs would be to await the spontaneous occurrence of both of these mutations in a single individual, it would be extremely unlikely that such an individual would ever occur. By recombination, however, a crossover in the interval between the two genes could lead in one meiosis to the production of a chromosome bearing both variant alleles. It is because of this ability to reshuffle gene combinations rapidly that recombination is so important in the production of natural variation and in the process of evolution by natural selection in eukaryotes.

Gene Conversion

Because the two homologues that pair within a synaptonemal complex are not identical, there are places along the length of the paired chromosomes where one or more nucleotides of one homologue are not complementary to the strand with which it is paired. These occasional nonmatching pairs of nucleotides are called **mismatch pairs.**

As you might expect, the error-correcting machinery of the cell is able to detect mismatch pairing. Error correction is achieved by a search for mismatches between the strands of a duplex. This process occurs normally during DNA replication, when these same enzymes "proofread" the new DNA strand to ensure that no mistakes have been made in producing a sequence complementary to the template. If a mismatch is detected, it is "corrected" (Figure 17-13): one of the mismatched strands is excised and the gap is filled in in a way that is complementary to the other strand—producing two chromosomes with the same sequence. One of the two mismatched sequences has been lost. The conversion of one nucleotide sequence to another is called **gene conversion.**

> **Gene conversion is the alteration of one homologous chromosome by the cell's error detection and repair system to make it resemble the other.**

Unequal Crossing-Over

Recombination can occur between two chromosomes in any region where the sequences are similar enough to permit close pairing. Mistakes in pairing occasionally occur when a sequence exists at several different locations on a chromosome. In such cases one copy of a sequence may line up not with its homologous segment on the other chromosome, but rather with one of the other duplicate copies on that chromosome. Such misalignment is responsible for the slipped mispairing that leads to the small deletions and frameshift mutations that were discussed earlier in this chapter. If a crossover occurs in the region of pairing, this process is termed **unequal crossing-over,** because, as you can see in Figure 17-14, the two chromosomes exchange segments of unequal length.

In unequal crossing-over one chromosome gains extra copies of the multicopy sequences, and the other chromosome loses these copies. This process is capable of generating a chromosome that has hundreds of copies of a particular gene, lined up side by side along the chromosome in tandem array.

> **Unequal crossing-over is the occurrence of a crossover event between chromosomal regions that are similar in nucleotide sequence but are not homologous. Unequal crossing-over typically occurs when a sequence exists in many copies and the wrong copies pair with one another in meiosis.**

Gene conversion. When a cell's DNA repair enzymes detect an improper base pair (1), they set about "repairing" the problem base pair by excising one of the two bases, chosen at random, and replacing it with a base that pairs properly (2). When the two DNA strands that are pairing are not members of the same duplex, however, but rather homologues paired during meiosis, this repair process will tend to make one homologue a hybrid (3). This hybrid base, an AG pair in this example, will also be repaired, and half the time, on average, the G will be removed and replaced with a T (4). In these instances one strand (left) has actually been converted to the sequence of the other.

TABLE 17-2   CLASSES OF GENETIC RECOMBINATION

CLASS	OCCURRENCE
GENE TRANSFERS	
Plasmid transfer	Occurs predominantly, if not exclusively, in bacteria and is targeted to specific locations on their chromosomes.
Virus transfer	Common in bacteria; probably occurs in eukaryotes also. Like plasmid transfer, genes move to specific locations on chromosomes.
Transposition	Common in both bacteria and eukaryotes; genes move from one chromosomal location to another at random.
MEIOTIC RECOMBINATIONS	
Crossing-over	Occurs only in eukaryotes. Requires the pairing of homologous chromosomes, and may occur anywhere along their length.
Unequal crossing-over	The result of crossing-over between mismatched segments; leads to gene duplication.
Gene conversion	Occurs when homologous chromosomes pair and one is "corrected" to resemble the other.
Independent assortment	Haploid cells produced by meiosis contain only one member of each pair of homologous chromosomes, selected at random.

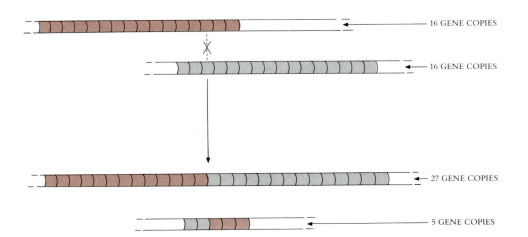

16 GENE COPIES

16 GENE COPIES

27 GENE COPIES

5 GENE COPIES

FIGURE 17-14

Unequal crossing-over. When a repeated sequence pairs out of register, a crossover within the region will produce one chromosome with fewer gene copies and one with more. It is thought that much of the gene duplication that has occurred during eukaryotic evolution is the result of unequal crossing-over.

Unequal crossing-over can occur between any two regions of the chromosome that are similar in sequence. Because the genomes of most eukaryotes possess multiple copies of transposons scattered throughout the chromosomes, unequal crossing-over between copies of transposons located at different positions on chromosomes has had a profound influence on gene organization in eukaryotes. Most of the genes of eukaryotes appear to have been subject to duplication at one or more times during the course of their evolution.

A summary of the different kinds of genetic recombination that we have discussed is presented in Table 17-2.

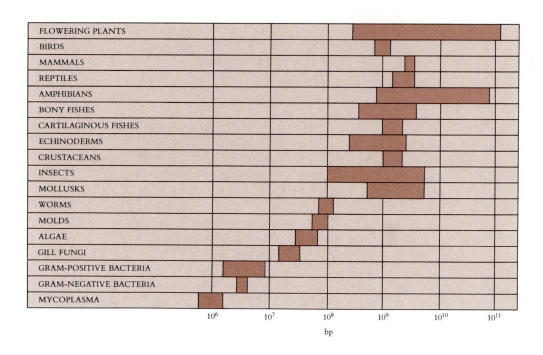

FIGURE 17-15

The DNA content of the haploid genomes of a variety of organisms, expressed as number of base pairs (bp). Genome size increases from the bacteria through the protists to other eukaryotes. Among animals and plants there is considerable variability within each phylum and no consistent differences between phyla. Note particularly the wide range among plant species.

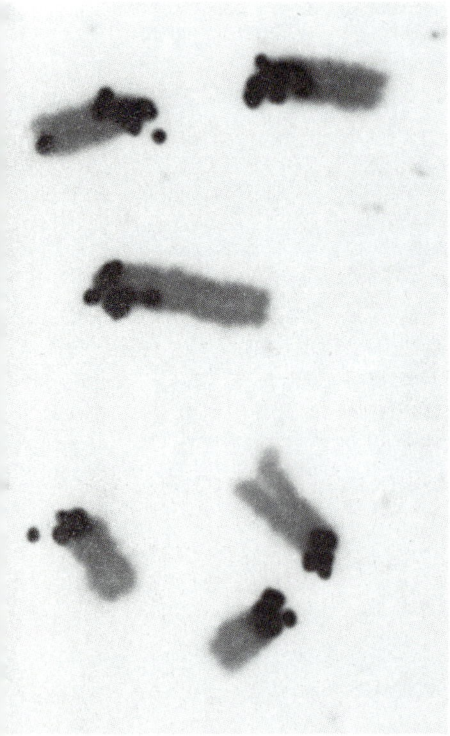

FIGURE 17-16

These mouse chromosomes have been exposed to a solution containing radioactive RNA complementary to mouse satellite DNA. The labeled RNA (dark spots) has bound to its complementary sequences on the DNA, showing that the satellite sequences are localized in centromeres.

THE EVOLUTION OF GENE ORGANIZATION

Bacterial genomes are relatively simple compared with those of eukaryotes; genes almost always occur in them as single copies. Unequal crossing-over between repeated transposition elements on a bacterial chromosome tends to delete material, fostering the maintenance of a minimum genome size. In eukaryotes, by contrast, the introduction of *pairs of homologous chromosomes* (presumably because of their importance in repairing breaks in two-strand DNA) has led to a radically different situation. *Unequal crossing-over between homologous chromosomes tends to promote the duplication of material rather than its reduction.* As a consequence, the eukaryotic genome has during the course of its evolution been in a constant state of flux, genes evolving multiple copies, some of which subsequently diverge in sequence to become new and different genes, which in their turn duplicate and diverge (Figure 17-15). Six different classes of eukaryotic DNA sequence are commonly recognized, based on copy number.

Satellite DNA

Some short nucleotide sequences are repeated several million times in the genomes of eukaryotes. About 4% of the human genome consists of repeated sequences of this kind. They are collectively called **satellite DNA.**

Why should such sequences be repeated so many times? Almost all of the copies of satellite DNA are clustered together around the centromere (Figure 17-16) or are located near the ends of the chromosomes. These regions of the chromosomes remain highly condensed, tightly coiled, and untranscribed throughout the cell cycle; the satellite DNA sequences thus seem to be serving structural functions.

Transposition Elements

Other sequences in eukaryotic genomes are repeated thousands of times, but are much longer than satellite sequences, and are scattered randomly throughout the genome. These segments are transposons. From one generation to the next, a few transposons in each cell jump sporadically to new, apparently random, locations. There appear to be fewer different transposons in mammals than in many other organisms, but the ones that are present in mammals are repeated more often. The family of human transposons called *ALU* elements, for example, typically occurs about 300,000 times in each cell. Transposons are transcribed but appear to play no functional role in the life of the cell that bears them.

Tandem Clusters

A third class of repeated sequences is transcribed and is clearly important to the cell: the repeated sequences that encode products the cell requires in large amounts. These sequences are repeated many times, one copy following another in tandem array. The cell is able to obtain large amounts of the encoded product rapidly by transcribing all of the numerous copies simultaneously. The genes encoding ribosomal RNA (rRNA), for example, are present in several hundred copies in most eukaryotes (Figure 17-17). Because these clusters are active sites of rRNA synthesis, they are readily visible in cytological preparations; they are called **nucleolar organizer** regions. When transcription of these rRNA gene clusters ceases during cell division, the nucleolar organizers "disappear" from view under the microscope as the localized expansion of the chromosome subsides, only to reappear when transcription is reinitiated.

All of the many transcribed genes that are present in a tandem cluster are similar in sequence. Not all of the copies are identical—some differ from others by one or a few nucleotides—but their overall similarity is great. Each of the genes in a cluster is separated from its neighboring copies by a short "spacer" sequence, which is not transcribed. Unlike the transcribed genes of a cluster, spacers are not similar to one another. They vary considerably within a cluster, both in sequence and in length.

FIGURE 17-17

A tandem cluster. The genes encoding rRNA are repeated several hundred times, each copy separated from the next by a spacer. **A** *In a region of active rRNA synthesis, progressively longer segments of rRNA primary transcript can be seen as more and more of the repeated gene copies are passed by the RNA polymerase.* **B** *A close-up view illustrates how the DNA duplex is opened in the zones corresponding to rRNA-encoding genes.* **C** *Once a full transcript of the repeat region is produced, it is processed in two stages: first it is cut into lengths corresponding to the repeat unit; then each unit is cut into segments corresponding to the three different rRNA molecules.*

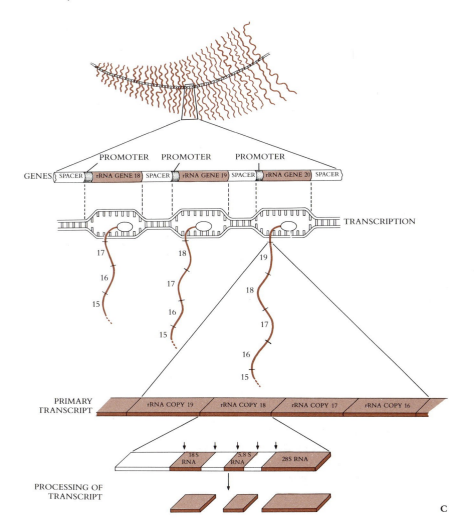

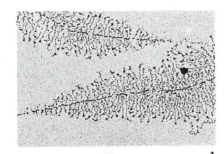

A

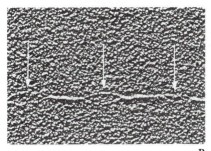

B

C

FIGURE 17-18

Evolution of the hemoglobin multigene family. The ancestral globin gene is at least 800 million years old. By the time that modern fishes had evolved, this ancestral gene had already duplicated, forming the α and β forms. Later, after the evolutionary divergence of amphibians and reptiles, these two forms moved apart on the chromosome (arrow); the mechanism is not known, but may have involved transposition. Each of these two sequences duplicated in mammals by the time their ancestors, and those of the birds, had differentiated as distinct lines of reptiles. In mammals two more waves of duplication occurred to produce the array of genes that occurs in humans. There are 11 globin genes in the human genome. Three of them (ψ) are silent, encoding nonfunctional proteins. Other forms are produced only during embryonic (ζ and ε) or fetal γ development. Of the 11 genes, only four (δ, β, α₁, α₂) encode the chains of adult human hemoglobin.

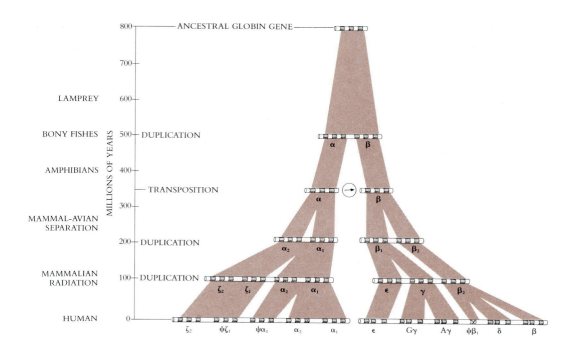

Multigene Families

Multigene families are groups of related but distinctly different genes that occur together in a cluster. They differ from tandem clusters in that they contain far fewer genes, and these genes are far more different from one another than those in the tandem clusters. Multigene families may contain as few as three or as many as several hundred genes. Although they do differ from one another, the members of a multigene family are clearly related in their sequences. All members of multigene families are thought to have arisen from a single ancestral sequence (Figure 17-18) through a series of unequal crossing-over events.

Dispersed Pseudogenes

Silent copies of a gene, which have been inactivated by mutation, are called **pseudogenes.** In some cases pseudogenes result from mutations in promoters; in other cases they result from frameshift mutations or small deletions. Pseudogenes may occur within a multigene family cluster; they may also occur at a distant site. We call distantly located pseudogenes **dispersed pseudogenes,** because they have somehow moved far from their original position within a multigene family gene cluster. The existence of such dispersed pseudogenes was not suspected as recently as a few years ago, but they are now thought to be of major evolutionary significance in eukaryotes.

Single-Copy Genes

Relatively few genes in eukaryotes exist as single unique copies in the genome. Over the long time since the appearance of the eukaryotes, processes such as unequal crossing-over between different copies of transposons have time and time again caused segments of the eukaryotic chromosome to duplicate; it appears that no portion of the genome has been immune to this process. Indeed, this process of duplication followed by the conversion of copies to pseudogenes has probably been the major source of "new" genes during evolution; pseudogenes accumulate mutational changes until a fortuitous combination of changes in one pseudogene results in an active gene encoding a protein with different properties. When the new gene encoding this useful protein first arises, it is a single-copy gene

TABLE 17-3.   THE CLASSES OF GENES FOUND IN EUKARYOTES

CLASS	COPY NUMBER
Satellite DNA	Short sequences present in millions of copies per genome
Transposition elements	Thousands of copies of transposons scattered around the genome
Tandem clusters	Clusters containing hundreds of nearly identical copies of a gene
Multigene families	Clusters of a few copies of related but distinctly different genes
Dispersed pseudogenes	Nonactive members of a multigene family located elsewhere
Single-copy genes	Genes that exist in only one copy

but it in its turn will undergo duplication. A single-copy gene is but one stage in the cycle of duplication and divergence that characterizes the evolution of the eukaryotic genome.

A summary of the classes of genes found in eukaryotes is presented in Table 17-3.

THE IMPORTANCE OF GENETIC CHANGE

All evolution begins with genetic change, with the processes described in this chapter: mutation, which creates new alleles; recombination, which shuffles and sorts the changes; transposition, which alters gene location; chromosomal rearrangement, which alters the arrangement of entire chromosomes. Some changes produce alterations in an organism that enable it to leave more offspring, and these changes are preserved as the genetic endowment of future generations. Other changes reduce the ability of an organism to leave offspring. These changes tend to be lost, the organisms that carry them contributing fewer members to future generations than those which lack them.

Evolution can be viewed as the selection of particular combinations of alleles from a pool of different alternatives. The rate of evolution is ultimately limited by the rate at which these different alternatives are generated. Genetic change provides the raw material for evolution.

SUMMARY

1. A mutation is any change in the hereditary message. Changes in one or a few nucleotides are called point mutations. They may arise as a result of physical damage from ionizing radiation or mistakes made during the correction of chemical damage caused by the absorption of ultraviolet light. Occasionally changes occur spontaneously as a result of errors in pairing that take place during the course of DNA replication.

2. Recombination is defined as the creation of new gene combinations. It occurs when the position on the chromosome of a gene or a fragment of a gene changes. Recombination encompasses the transfer of a gene to the same position on a homologous chromosome or to a different position on the same or a different chromosome. It also includes the reciprocal exchange of genes between two chromosomes and the exchange of entire chromosomes that occurs during meiosis.

3. Bacterial genes may be transferred by association with plasmids, which are small circles of DNA that regularly move from cell to cell by conjugation. Because the plasmid enters the bacterial genome by a reciprocal exchange of strands in a region of sequence similarity, it may carry genes only to certain locations on the bacterial chromosome. At these sites base sequences occur that are also represented on the plasmid.

4. Bacterial viruses may also transfer bacterial genes to different locations on the chromosome. As with plasmids, only certain chromosomal sites are potential targets. Unlike plasmids, however, these sites do not occur on the virus genome.

5. Transposition is similar to viral transfer, except that the transposing element may integrate itself (and any genes that are associated with it) anywhere on the chromosomes.

6. Certain transposition elements, or transposons, are potential disease vectors in humans. Called retroviruses, these elements are able to acquire random human genes and move them from one individual to another. Such gene acquisition by transposons is associated with cancer, AIDS, and other defective-gene disorders.

7. Crossing-over occurs between homologous chromosomes during the close pairing that occurs in meiosis. The strand switching results from a molecular process that we think originated as a means of repairing double-strand breaks in DNA.

8. Six different kinds of DNA sequences occur in eukaryotic chromosomes: satellite sequences, transposition elements, tandem clusters, multigene families, dispersed pseudogenes, and single-copy genes.

9. Satellite sequences are short sequences repeated millions of times. They are commonly associated with nontranscribed regions of chromosomes and probably play a structural role.

10. Transposition elements (transposons) are scattered throughout the genomes of prokaryotes and eukaryotes. In humans and other mammals there are only a few different sequences, but they are repeated hundreds of thousands of times. Transposons are transcribed, but they are not thought to play any regular functional role in the life of the cell.

11. Tandem clusters are genes that occur in thousands of copies clustered together at one or a few sites on the chromosome.

12. Single-copy genes are not common. Multigene families are genes that occur in tens of copies clustered at one site on the chromosome. Because these clusters are smaller than tandem clusters, less pairing occurs between members. Consequently, the copies in multigene families diverge in sequence more than do copies in tandem clusters. Occasionally, inactive members of multigene families move to other locations on the chromosome. Such "distant cousins" are called dispersed pseudogenes.

SELF-QUIZ

The following questions pinpoint important information in this chapter. You should be sure to know the answers to each of them.

1. The damage caused to DNA by ionizing radiation results from
   (a) excess energy imparted to electrons in the outer shell of nucleotides by radiation
   (b) ionized water molecules
   (c) breakage of phosphodiester bonds by high-energy radiation
   (d) ejection of nucleotides from the DNA

2. Of the four nucleotide bases, which two absorb ultraviolet radiation and form cross-links?

3. When a plasmid integrates into a bacterial chromosome, it does so at certain sites with a nucleotide sequence identical to one that occurs on the plasmid itself. The same is true of lambda virus, which integrates at special "lambda recognition sites." True or false?

4. When a crossover event occurs in a region of slipped mispairing, the process is termed _____.

5. A retrovirus is a transposon. True or false?

6. Mutations that result from the transposition of material into a gene are called _____.

7. A eukaryotic individual that has gained one extra copy of one of its chromosomes is said to be
(a) diploid   (c) aneuploid
(b) polyploid   (d) monoploid

8. Label each of the following "e" (eukaryotic) or "p" (prokaryotic):
(a) single circular chromosome
(b) extragenomic plasmids common
(c) most genes present in multiple copies
(d) genes contain introns
(e) crossing-over leads to deletions
(f) genome fragmented into several portions
(g) satellite DNA
(h) meiosis
(i) some of DNA in cytoplasmic organelles

9. Order the following in terms of number of copies per genome, from most to least:
(a) dispersed pseudogenes   (d) single-copy genes
(b) multigene families   (e) tandem clusters
(c) satellite DNA   (f) transposition elements

10. Of the following classes of gene element, which are transcribed?
(a) dispersed pseudogenes   (d) transposition elements
(b) multigene family   (e) tandem clusters
(c) satellite DNA

11. If one forms a heteroduplex DNA molecule by pairing a naturally occurring DNA strand with a DNA strand prepared from the mRNA transcribed off that strand (by using reverse transcriptase), and examines the result under the electron microscope, one of the two partner strands of the heteroduplex forms loops that jut out from the duplex, and the other does not. The loops are composed of the DNA transcribed from the mRNA. True or false?

12. For a given length of DNA, would bacteria be expected to contain as many pseudogenes as do humans?

THOUGHT QUESTIONS

1. In medical research, mice are often used as model systems in which to study the immune system and other physiological systems that are important to human health. Many medical centers maintain large colonies of mice for these studies. In one such colony under your supervision, a hairless mouse is born. What minimum evidence would you accept that this variant represents a genetic mutation?

2. An American couple both work in an atomic energy plant, and both are exposed daily to low-level background radiation. After several years the couple have a child, and this child proves to be affected by Duchenne muscular dystrophy, a sex-linked recessive genetic defect in which the mutant locus is located on the X chromosome. Both parents are normal, as are the grandparents. The couple sue the plant, claiming that the abnormality in their child was the direct result of radiation-induced mutation of their gametes, radiation against which the company should have protected them. Before reaching a decision, the judge insists on knowing the sex of the child. Which sex would be more likely to result in an award of damages, and why?

3. Most eukaryotic genes exist in multiple copies. Most bacterial genes, in contrast, exist in single copies. As a result, a newly arisen mutation usually alters the phenotype of the bacterium in which it occurs. In contrast, when a mutation arises in a eukaryotic gene, it alters only one member of a multigene family. Does this mean that mutations in eukaryotes do not usually alter the phenotype, and if so, how does evolution occur in eukaryotes?

4. The analysis of the structure of vertebrate dispersed pseudogenes suggests that RNA transcripts of genes, after being processed in the cytoplasm, may gain reentry into the genome as DNA copies of the transcripts. In light of the fact that many processes which go on within the cytoplasm may alter RNA molecules, does this passage of RNA back into the genome imply that Lamarck was right, and that experience may act to modify the genetic message?

FOR FURTHER READING

CHAMBON, P.: "Split Genes," *Scientific American,* May 1981, pages 60-71. One of the original reports that eukaryotic genes are in fact composites, containing translated and untranslated sequences scrambled together.

COHEN, S., and J. SHAPIRO: "Transposable Genetic Elements," *Scientific American,* February 1980, pages 40-49. Focusing on transposition in bacteria, this article provides a particularly lucid account of how the process may occur, stressing how experiments have led to our current understanding.

DARNELL, J.: "The Processing of RNA," *Scientific American,* October 1983, pages 90-102. An up-to-date description of how intron sequences are removed from messenger RNA transcripts before they are employed to direct the assembly of proteins.

DONELSON, J., and M. TURNER: "How the Trypanosome Changes Its Coat," *Scientific American,* February 1985, pages 44-51. An easily understood discussion of how transposition is used by the parasites responsible for sleeping sickness.

FEDEROFF, N.: "Transposable Genetic Elements in Maize," *Scientific American,* June 1984, pages 85-98. An excellent introduction to the studies for which Barbara McClintock was awarded the 1984 Nobel prize in physiology or medicine, rephrased in molecular terms that were not available when she made her important discoveries.

LEDER, P.: "The Genetics of Antibody Diversity," *Scientific American,* May 1982, pages 102-116. An account of how a few hundred genes are shuffled to make billions of antibody combinations, by the man who first worked it out.

NOVICK, R.P.: "Plasmids," *Scientific American,* December 1980, pages 103-127. A general review of bacterial plasmids and the few that occur in higher organisms.

SCHULL, W., M. OTAKE, and J. NEEL: "Genetic Effects of the Atomic Bombs: A Reappraisal," *Science,* vol. 213, pages 1220-1227, 1981. An important examination of the results of the Japan explosions of 1945, the only instance of mass human exposure to radiation.

SHEN, S., J. SLIGHTOM, and O. SMITHIES: "A History of the Human Fetal Globin Gene Duplication," *Cell,* vol. 26, pages 191-203, 1981. A concise summary of the evolution of perhaps the best understood multigene family.

GENE TECHNOLOGY

Chapter 18

Overview

Recent advances in molecular biology have given us powerful new tools with which to investigate genetics at the molecular level. Particularly important new techniques permit geneticists to isolate individual genes and to transfer them from one kind of organism to another. These techniques are revolutionizing technology by enabling the inexpensive commercial production of previously rare and expensive hormones and providing a general means of producing vaccines against diseases. Perhaps most importantly, these techniques have given us for the first time a clear view of the events leading to cancer. Tragically, these techniques have made it evident that many of the cancers that occur each year could be prevented.

For Review

Here are some important terms and concepts that you will encounter in this chapter. If you are not familiar with them, you should review them before proceeding.

- **Point mutation** (Chapter 17)

- **Transposons** (Chapter 17)

- **Retrovirus** (Chapter 17)

During the last decade, a revolution in genetics has taken place as new and powerful techniques have been developed to study and manipulate DNA. These techniques have allowed biologists for the first time to intervene directly in the genetic fate of organisms. In this chapter we will discuss one of these techniques, the "cloning" of individual genes, and consider its application to a specific problem of great practical importance, the riddle of cancer.

Many other new techniques have also generated exciting results during the past few years. One of the most important has been the development of procedures to determine rapidly the sequence of nucleotides in DNA. For the first time investigators have been able to view directly the changes that natural selection and evolution have brought about, a view unobscured by the complexities of gene expression.

Another important new technique depends on our newfound ability to fuse a single antibody-producing cell to a cancer cell, enabling investigators to produce "monoclonal antibodies," which react against only one specific target and therefore may be used safely

FIGURE 18-1

Douglas fir shoots growing in a sterile culture. These shoots are the result of genetic experimentation aimed at cloning trees that have commercially valuable traits. Similar techniques are being used increasingly to propagate organisms altered by genetic engineering.

as clinical agents to fight specific diseases. The study of novel genes, some of them created in the laboratory, is being facilitated by our ability to transfer specific genes into normal lines of *Drosophila* and other organisms by carrying them "piggyback" on transposons. We could also list many other new developments, all resulting from the application of new molecular approaches to biology. Any one of these new advances might have served equally well as the focus of this chapter—and we can be certain that there will be many more such advances in the near future as the field of molecular biology grows rapidly (Figure 18-1).

PLASMIDS AND THE NEW GENETICS

In 1980 geneticists succeeded for the first time in introducing a human gene, the one that encodes the protein interferon, into a bacterial cell. Interferon is a rare protein, difficult to purify in any appreciable amounts, that increases human resistance to viral infection. It may prove to be the basis of a useful therapy against cancer. This possibility has been difficult to explore, however, since the purification of the substantial amounts of interferon required for large-scale clinical testing would, until recently, have been prohibitively expensive. A cheap way to produce interferon was needed, and introducing the gene responsible for its production into a bacterial cell made this possible.

The bacterial cell that had acquired the human interferon gene proceeded to produce copious amounts of interferon and to grow and divide. Soon there were many millions of bacterial cells in the culture, all of them descendants of the original bacterial cell that had the human interferon gene, and all of them producing interferon. This procedure, called **cloning,** had succeeded in making every cell in the culture a miniature factory for the production of human interferon. In a similar way the successful cloning of insulin was commercially significant, providing the basis for producing large amounts of a clinically important drug relatively inexpensively; cloning also has considerable theoretical significance. Such molecular techniques can be used—and will be used increasingly—to manipulate genes and, by doing so, to learn more about them. This interferon experiment and others like it mark the beginning of a new genetics, the birth of **genetic engineering.**

Genetic engineering is based on the ability to cut up DNA into recognizable pieces and to rearrange these pieces in different ways. In the experiment just described the gene segment carrying the interferon gene was inserted into a plasmid, which brought the inserted gene in with it when it infected the bacterial cell. Most other genetic engineering approaches have used the same general strategy of carrying the gene of interest into the target cell by first incorporating it into an infective plasmid or virus.

The success of the initial step in a genetic engineering experiment is the key to the whole procedure. As you might expect, success depends on being able to cut up the source DNA (human DNA in the interferon experiment, for example) and the plasmid DNA in such a way that the desired fragment of source DNA can be spliced permanently into the plasmid genome. This cutting is performed by a special kind of enzyme called a **restriction endonuclease.** These restriction enzymes are able to recognize and cleave specific sequences of nucleotides in a DNA molecule. They are the basic tools of genetic engineering.

RESTRICTION ENZYMES

Scientific discoveries often have their origins in odd little crannies—seemingly unimportant areas that receive little attention by researchers before their general significance is appreciated. In the example we are considering now, the particular obscure topic was the warfare that takes place between bacteria and viruses.

Most organisms in nature eventually evolve means of defending themselves from predators, and bacteria are no exception to this rule. Among the natural enemies of bacteria are the viruses called bacteriophages, which infect bacteria, multiply within them, and eventually burst the bacterial cell, releasing their offspring. For a bacterium, a virus is a potentially lethal adversary. As a result of natural selection, those bacterial individuals which can somehow resist viral infection will be favored and leave more progeny, on the average, than those which lack such means. Some bacteria have powerful weapons against viruses: enzymes that chop up the foreign viral DNA as soon as it enters the bacterial cell without harming the bacterial DNA at all. When viruses insert their DNA into a bacterial cell that is protected in this way, the viral DNA is immediately attacked by these enzymes, called restriction endonucleases, and degraded. Why is the DNA of the bacteria not also degraded by the restriction enzymes? Because the bacterial cell has **modified** its own DNA in such a way that the restriction enzymes do not recognize it as DNA.

Restriction endonucleases recognize specific nucleotide sequences within a DNA strand, bind to DNA strands at sites where these sequences occur, and cleave the bound strand of DNA at a specific place in the recognition sequence. In this way they cut up DNA. Other bacterial enzymes called **methylases** recognize the same bacterial DNA sequences, bind to them, and add methyl (—CH₃) groups to the nucleotides. When the recognition sites of bacterial DNA have been modified with methyl groups in this way, they are no longer recognized by the restriction enzymes. Consequently the bacterial DNA is protected from being degraded. Viral DNA, on the other hand, is not protected because it has not been methylated.

The sequences that restriction enzymes recognize are typically four to six nucleotides long and are symmetrical. Their symmetry is of a special kind, called **twofold rotational symmetry** (Figure 18-2). The nucleotides at one end of the recognition sequence are complementary to those at the other end so that the two strands of the duplex have the same nucleotide sequence running in opposite directions for the length of the recognition sequence. This arrangement has two consequences. The first is of great importance to the bacteria; the second is of little significance in the bacterial systems, but of paramount importance to us:

1. Because the same recognition sequence occurs on both strands of the DNA duplex (running in opposite directions), the restriction enzyme is able to recognize and cleave both strands of the duplex, effectively cutting the DNA duplex in half. This ability to chop across both strands is almost certainly the reason that restriction enzymes have evolved in such a way that they specifically recognize nucleotide sequences with twofold rotational symmetry—it lets them use one sequence to bind *both* strands.

2. Because the position of the bond cleaved by a particular restriction enzyme is typically not in the center of the recognition sequence to which it binds and the sequence is running in opposite directions on the two strands, the sites at which the two strands of a duplex are cut are offset from one another. Take your two hands, one palm up and the other palm down, and fit the little fingers and ring fingers together—offset like that. After cleavage the two fragments of DNA duplex each possess a short single strand a few nucleotides long dangling from the end. Look carefully at Figure 18-3. *The two single-stranded tails are complementary to one another.*

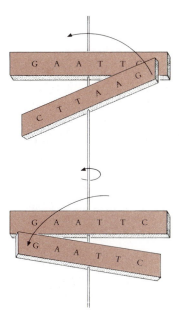

FIGURE 18-2

The nucleotide sequences recognized by restriction enzymes have twofold rotational symmetry. Thus if the sequence CTTAAG is rotated through the plane of the paper as illustrated here, it becomes its complementary sequence, GAATTC. The importance of twofold rotational symmetry is that because the two DNA strands are read in opposite directions both strands have the same sequence.

FIGURE 18-3

A restriction enzyme cleaves the sequence G★AATTC at the position indicated by ★. Because the same sequence occurs on both strands, both are cut. The position of ★ is not the same on the two strands, however, with the result that single-stranded tails are produced. Because of the twofold rotational symmetry of the sequence, the single-stranded tails are complementary to each other.

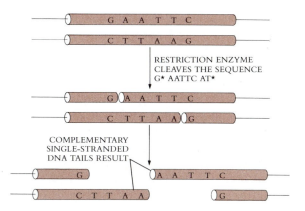

RESTRICTION ENZYME CLEAVES THE SEQUENCE G★ AATTC AT★

COMPLEMENTARY SINGLE-STRANDED DNA TAILS RESULT

Every cleavage by a given kind of restriction enzyme takes place at the same recognition sequence. By chance, this sequence will probably occur somewhere in any given sample of DNA so that a restriction endonuclease will cut DNA from any source into fragments. Each of these fragments will have the dangling sets of complementary nucleotides (sometimes called "sticky ends") characteristic of that endonuclease. Because the two single-stranded ends produced at a cleavage site are complementary, they can pair with each other. Once they have done so, the two strands can then be joined back together with the aid of a sealing enzyme called a **ligase,** which re-forms the phosphodiester bonds. This latter property makes restriction endonucleases the invaluable tools of the genetic engineer: *any* two fragments produced by the same restriction enzyme can be joined together. Fragments of elephant and ostrich DNA cleaved by the same bacterial restriction enzyme can be joined to one another just as readily as can two bacterial fragments because they have the same complementary sequences at their ends.

> **A restriction enzyme cleaves DNA at specific sites, generating in each case two fragments whose ends have one strand of the duplex longer than the other. Because the tailing strands of the two cleavage fragments are complementary in nucleotide sequence, *any* pair of fragments produced by the same enzyme, from any DNA source, can be joined together.**

There are hundreds of different bacterial restriction enzymes, recognizing a wide variety of four- to six-nucleotide sequences. How extensively any particular restriction enzyme degrades a DNA molecule depends both on the length of the recognition sequence for that enzyme and on the base composition of the target DNA. The probability that a particular restriction sequence will occur by chance in a stretch of DNA is simply the probability of selecting by chance a specific one of the four nucleotides (i.e., ¼) in position one, times the probability of selecting by chance a specific nucleotide in position two, and so on, for as many positions as there are in the recognition sequence.

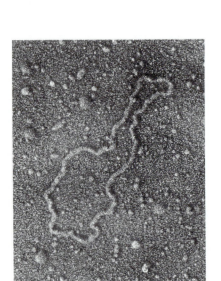

FIGURE 18-4

The first plasmid used successfully to clone a vertebrate gene, pSC101. Its name refers to it being the hundred-and-first plasmid isolated by Stanley Cohen.

NUMBER OF NUCLEOTIDES IN RECOGNITION SEQUENCE	EXPECTED FREQUENCY
Four nucleotides	$(¼) \times (¼) \times (¼) \times (¼) = \frac{1}{256}$
Five nucleotides	$(¼) \times (¼) \times (¼) \times (¼) \times (¼) = \frac{1}{1024}$
Six nucleotides	$(¼) \times (¼) \times (¼) \times (¼) \times (¼) \times (¼) = \frac{1}{4096}$

Thus a five-nucleotide restriction sequence will occur in a DNA fragment by chance alone at a frequency averaging about 1 per every 1000 nucleotides of the fragment.

CONSTRUCTING CHIMERIC GENOMES

A chimera is a mythical creature with the head of a lion, the body of a goat, and the tail of a serpent. No chimera ever existed in nature. Human beings, however, have made them—not the lion-goat-snake variety, but chimeras of a more modest kind.

The first actual chimera was a bacterial plasmid that American geneticists Stanley Cohen and Herbert Boyer made in 1973. Cohen and Boyer used a restriction endonuclease to cut up a large bacterial plasmid called a resistance transfer factor. From the resulting fragments, they isolated one fragment 9000 nucleotides long, which contained both the sequence necessary for replicating the plasmid—the **replication origin**—and a gene that conferred resistance to an antibiotic, tetracycline.

Because both ends of this fragment were cut by the same restriction enzyme (called *Escherichia coli* restriction endonuclease 1, or **Eco R1**), they could join together to form a circle, a small plasmid that Cohen dubbed pSC101 (Figure 18-4). Cohen and Boyer used the same restriction enzyme, Eco R1, to cut up DNA that they had isolated from an adult amphibian, the African clawed toad, *Xenopus laevis*. They then mixed the toad DNA fragments with opened-circle molecules of pSC101, allowed bacterial cells to take up DNA from the mixture, and selected for bacterial cells that had become resistant to tetracycline (Figure 18-5). From among these pSC101-containing cells they were able to isolate ones containing the toad ribosomal RNA gene. These versions of pSC101 had the toad gene spliced in at the Eco R1 site. Instead of joining to one another, the two ends of the pSC101 plasmid had joined to the two ends of the toad DNA fragment that contained the ribosomal RNA gene.

The pSC101 containing the toad ribosomal RNA gene is a true chimera. It is an entirely new creature that never existed in nature and would never have evolved there. It is a form of **recombinant DNA,** a DNA molecule created in the laboratory by molecular geneticists who joined together bits of several genomes into a novel combination.

The first recombinant genome produced by human genetic engineering was a bacterial plasmid into which an amphibian ribosomal RNA gene was inserted in 1973.

The insertion of fragments of foreign DNA into bacterial cells by carrying them into the cells piggyback in plasmids or viruses has become common in molecular genetics. Newer-model plasmids with exotic names, such as pBR322, can be induced to make many hundreds of copies of themselves and thus of the foreign genes that are included in them within such bacterial cells. Even easier entry into bacterial cells can be achieved by inserting

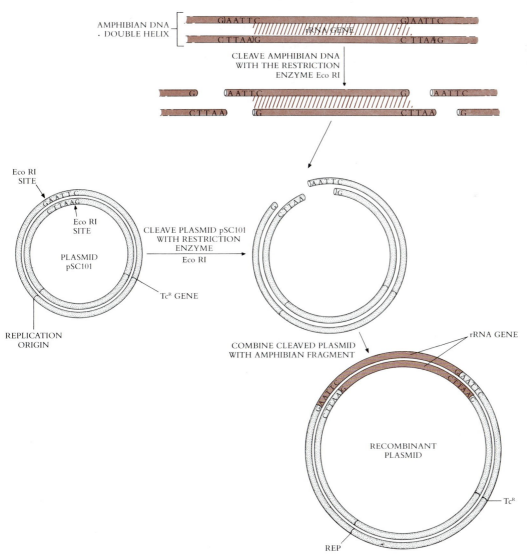

FIGURE 18-5

Cohen and Boyer's insertion of an amphibian gene encoding rRNA into a bacterial plasmid; pSC101 contains a single site cleaved by the restriction enzyme Eco RI. The rRNA-encoding region was inserted into pSC101 at that site by cleaving the rRNA region with Eco RI and allowing the complementary sequences to pair.

the foreign DNA fragment into the genome of a bacterial virus, such as lambda virus, instead of into a plasmid. The infective genome that harbors the foreign DNA and carries it into the target cell is called a **vector.** Not all vectors have bacterial targets. Animal viruses, for example, have been used as vectors to carry bacterial genes into monkey cells. Animal genes have even been carried into plant cells by using methods of this kind.

There has been considerable discussion about the potential danger of inadvertently creating an undesirable life form in the course of a recombinant DNA experiment. What if one fragmented the DNA of a cancer cell and then incorporated the fragments at random into viruses that are propagated within bacterial cells? Might there not be a danger that among the resulting bacteria there could be one capable of constituting an infective form of cancer?

Even though most recombinant DNA experiments are not dangerous, such concerns are real and need to be taken seriously. Both scientists and individual governments monitor these experiments to detect and forestall any hazard of this sort. Experimenters have gone to considerable lengths to establish appropriate experimental safeguards. The bacteria used in many recombinant DNA experiments, for example, are unable to live outside of laboratory conditions; many of them are obligate anaerobes, poisoned by oxygen. Decidedly dangerous experiments such as the cancer cell shotgun experiment described above are prohibited.

GENETIC ENGINEERING

The 1980s have seen an explosion of interest in applying genetic engineering techniques to practical human problems. Perhaps the most obvious commercial application and the one that was seized on first was to introduce human genes that encode clinically important proteins into bacteria. Because they can be grown cheaply in bulk, bacteria that incorporate human genes can produce the human proteins that they specify in large amounts. This method has been used to produce several forms of human interferon, to place rat growth hormone genes into mice (Figure 18-6), and to manufacture many commercially valuable nonhuman enzymes.

Perhaps the greatest difficulty with this general approach is the difficulty of separating the desired protein from all the other proteins that the bacteria make. The purification of proteins from such complex mixtures is both time-consuming and expensive. Recently, however, it has proved possible to produce RNA transcripts of cloned genes and to use such transcripts to produce the protein directly in a test tube containing only the transcribed RNA, ribosomes, cofactors, amino acids, tRNA, and ATP.

A second major area of genetic engineering activity is the manipulation of the genes of key crop plants. In plants the primary experimental difficulty has been identifying a suitable vector for introducing genes into target organisms. Plants do not possess the many plasmids that bacteria do, so the choice of potential vectors is therefore limited.

The most successful results obtained thus far with plant systems have involved a bacterial plasmid called T_i that infects plants and causes crown-gall tumors to develop. Part of this T_i plasmid integrates into the plant DNA; it has proved possible to attach other genes to this portion of the plasmid. Attempts are currently under way to use such techniques to introduce genes that will improve the resistance of crop plants to disease, frost, and other forms of stress; improve their nutritional balance and protein content; and confer herbicide resistance on them. Recently, for example, a bacterial gene endowing resistance to the herbicide Roundup has been made to work in plants (Figure 18-7). This advance is of great interest to farmers, since Roundup is a particularly powerful and broad-spectrum herbicide that will kill any green plant, and a crop resistant to Roundup would never have to be weeded if the field were simply treated with the herbicide. A more long-range goal is to increase yield and plant size in crop plants; however, we do not yet know, for the most part, which genes are responsible for these complex characteristics. It has also proved difficult to introduce the nitrogen-fixing genes from bacteria into plants, because these genes do not

FIGURE 18-6

These two mice are genetically identical, except that the large one has one extra gene—the gene encoding a potent rat growth hormone not normally present in mice. The gene was added to the mouse genome by human genetic engineers and is now a stable part of the mouse's genetic endowment.

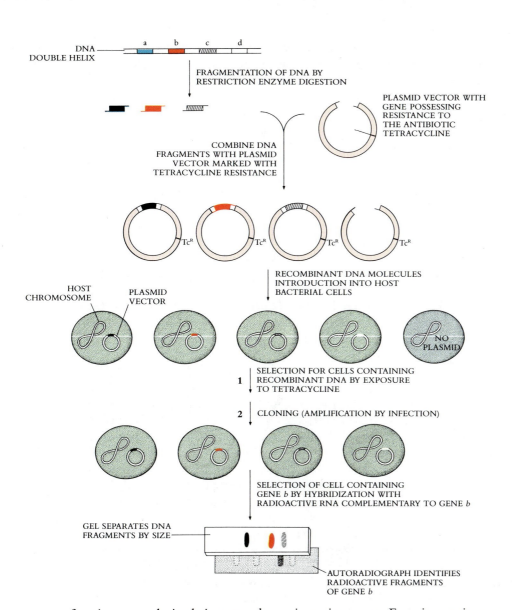

DNA DOUBLE HELIX

a b c d

FRAGMENTATION OF DNA BY
RESTRICTION ENZYME DIGESTION

PLASMID VECTOR WITH
GENE POSSESSING
RESISTANCE TO
THE ANTIBIOTIC
TETRACYCLINE

COMBINE DNA
FRAGMENTS WITH PLASMID
VECTOR MARKED WITH
TETRACYCLINE RESISTANCE

Tc^R Tc^R Tc^R Tc^R

RECOMBINANT DNA MOLECULES
INTRODUCTION INTO HOST
BACTERIAL CELLS

HOST
CHROMOSOME

PLASMID
VECTOR

NO
PLASMID

1   SELECTION FOR CELLS CONTAINING
RECOMBINANT DNA BY EXPOSURE
TO TETRACYCLINE

2   CLONING (AMPLIFICATION BY INFECTION)

SELECTION OF CELL CONTAINING
GENE b BY HYBRIDIZATION WITH
RADIOACTIVE RNA COMPLEMENTARY TO GENE b

GEL SEPARATES DNA
FRAGMENTS BY SIZE

AUTORADIOGRAPH IDENTIFIES
RADIOACTIVE FRAGMENTS
OF GENE b

FIGURE 18-7

A generalized scheme for cloning a gene. Imagine that you wish to clone the gene conferring Roundup resistance from the bacterial strain in which it was discovered. You would first fragment the bacterial DNA with a restriction enzyme. Then you would incorporate the fragments into a plasmid containing a resistance gene, and introduce the plasmid population into bacterial cells, adjusting concentrations so that on the average one plasmid enters each cell. You would then expose the cells to the antibiotic tetracycline, killing any cell that does not now harbor a plasmid. The surviving cells are plated onto a solid growth medium so that each cell forms a separate colony. A portion of each colony is then checked for the presence of the Roundup-resistant gene. One way in which such checks are carried out is to check each colony for the presence of a DNA fragment that will pair to a radioactive "probe," a fragment of mRNA or DNA known to contain the sequence.

seem to function properly in their new eukaryotic environment. Experiments in many similar areas are also being pursued actively.

A third area of potential significance involves the use of genetic engineering techniques to help produce important vaccines. For example, vaccines against herpesvirus and hepatitis viruses have recently been manufactured by using gene-splicing techniques. Genes specifying the protein-polysaccharide coat of the herpes simplex virus or hepatitis B virus were spliced into a fragment of the DNA genome of the cowpox, or vaccinia, virus—the same virus that British physician Edward Jenner used almost 200 years ago in his pioneering vaccinations against smallpox.

Live vaccinia virus is introduced into a mammalian cell culture, along with spliced fragments of vaccinia virus DNA, into which have been incorporated genes encoding the protein-polysaccharide coat of the herpes simplex virus genome (or, in other experiments, of hepatitis B virus). Under such circumstances the spliced herpes simplex gene is able to recombine into the vaccinia virus genome, producing a recombinant vaccinia virus. When this recombinant virus is injected into a mouse or rabbit, it dictates the production not only of the proteins that the vaccinia virus genome specifies, but also of the protein-polysaccharide coat specified by the herpes genes that it carries. The viruses that it produces are thus somewhat like sheep wearing wolf's clothing—they have the interior of a benign vaccinia virus and the exterior surface of a herpes virus. Infected individuals contract cowpox and not herpes, and cowpox is relatively harmless. At the same time infected individuals make antibodies against herpes and so become immune to it. The antibody-producing cells of the infected animal react to the *outer surface* of the recombinant virus particles and so cause the animal to develop a high level of antibodies directed against the cell surface of the herpes virus. Vaccines produced in this way are harmless, since only a small fragment of the DNA of the disease-associated virus is introduced via the recombinant virus.

The great attraction of this approach is that it does not depend on the nature of the virus-caused disease. In the future similar recombinant viruses injected into humans may prove able to confer resistance to a wide variety of viral diseases. The approach is also currently being used to develop a vaccine against malaria, a protozoan disease for which there is currently no effective protection. By incorporating genes encoding factors specifying the cell surface of the protozoan that causes this disease into the vaccinia virus, it should be possible to develop such a vaccine. Malaria affected 250 million people in 1984, and about 2 million of these people died.

> **In principle, the surface-specifying genes of any disease-causing virus or organism might be introduced into a benign virus, such as vaccinia virus, and the recombinant virus could be infected safely into humans, thereby producing antibodies directed against the disease-causing factor.**

THE RIDDLE OF CANCER

Cancer is a growth disorder of cells. It starts when an apparently normal cell begins to grow in an uncontrolled and invasive way (Figure 18-8). The result is a ball of cells, called a **tumor,** that constantly expands in size. When this ball remains a hard mass, it is called a **sarcoma** (if connective tissue, such as muscle, is involved) or a **carcinoma** (if epithelial tissue, such as skin, is involved). If its cells leave the mass and spread throughout the body (Figure 18-9), forming new tumors at distant sites, the spreading cells are called **metastases.** Cancer is perhaps the most pernicious human disease. Of the children born in 1985, one third will contract cancer; one fourth of the male children and fully one third of the female children will someday die of cancer. Each of us has had family or friends affected by the disease. Many of us will die of cancer. Not surprisingly, a great deal of effort has been expended to learn the cause of this disease. Using the new techniques of molecular biology, a great deal of progress has been made, and the rough outlines of understanding are now emerging.

The search for a cause of cancer has focused in part on those environmental factors which were considered potential agents in causing the disease (Figure 18-10). Many were found, including ionizing radiation, such as x rays, and a variety of chemicals. Many of

FIGURE 18-8

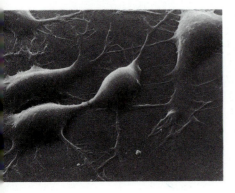

Cancer cells. These mouse nerve cells, growing over one another on a culture dish, were isolated from a tumor.

FIGURE 18-9

Portrait of a cancer. The ball of cells is a carcinoma, developing from epithelial cells lining the interior surface of a human lung. As the mass of cells grows, it invades surrounding tissues, eventually penetrating into lymphatic vessels and blood vessels, both of which are plentiful within the lung. These vessels carry metastatic cancer cells throughout the body, where they lodge and grow, forming new foci of cancerous tissue.

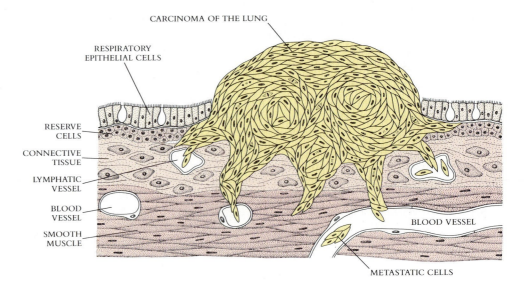

CARCINOMA OF THE LUNG

RESPIRATORY EPITHELIAL CELLS

RESERVE CELLS

CONNECTIVE TISSUE

LYMPHATIC VESSEL

BLOOD VESSEL

SMOOTH MUSCLE

BLOOD VESSEL

METASTATIC CELLS

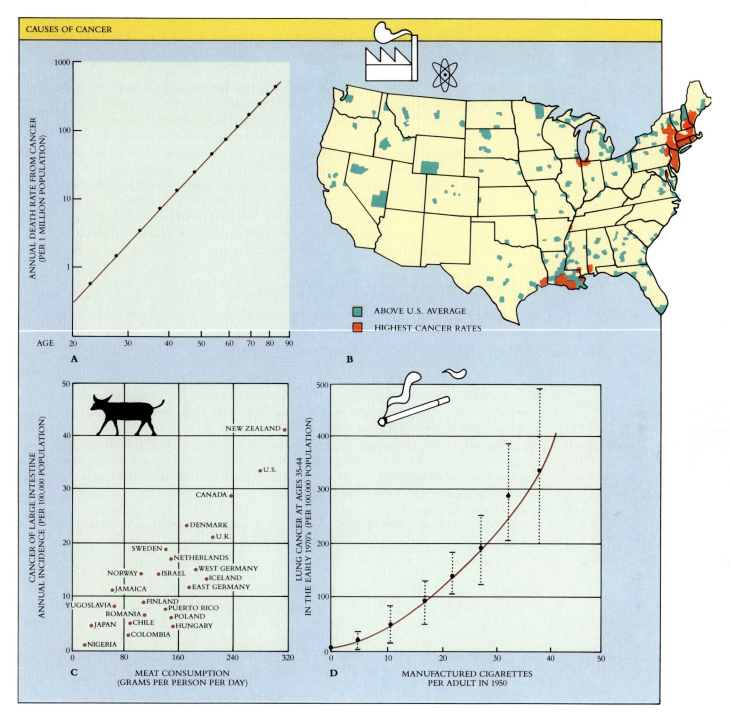

FIGURE 18-10

Potential cancer-causing agents.

A *The annual death rate from cancer is a function of age; when the logarithm of each is plotted, a straight line is obtained. This exponential rise suggests that several independent events are required to give rise to cancer.*

B *The incidence of cancer per thousand people is not uniform throughout the United States. Rather, it is highest in cities and in the Mississippi delta. This suggests that pollution and pesticide runoff may contribute to cancer.*

C *One of the most common deadly cancers in the United States, cancer of the large intestine, is not at all common in many other countries, such as Japan. Its incidence appears to be related to the amount of meat an average person consumes; a diet high in meat intake relative to fiber intake slows passage of food through the intestine, prolonging exposure of the intestinal wall to the oxidation by-products of digestion.*

D *The biggest killer among cancers is lung cancer, and the most important environmental agent producing lung cancer is cigarettes. It takes 20 years for the full effect to be seen. When levels of lung cancer in many countries are compared, the incidence of lung cancer among adult men between 40 and 50 years of age is strongly correlated with the cigarette consumption in that country 20 years earlier. Each dot represents a different country.*

these cancer-causing agents, or **carcinogens,** had in common the property of being potent mutagens. This observation led to the suspicion that cancer might be caused, at least in part, by the creation of mutations.

Mutagens were not the only cancer-causing agents that were found, however. Some tumors seemed almost certainly to have resulted from viral infection. As early as 1910, American medical researcher Peyton Rous demonstrated that a virus was associated with a cancerous sarcoma in chickens. Over the next 50 years a number of experiments demonstrated that viruses could be isolated from certain tumors and that these viruses would cause virus-containing tumors to develop in other individuals.

What do mutagens and viruses have in common, that they can both induce cancer? At first it seemed that they had little in common: mutagens alter genes directly, whereas viruses introduce foreign genes into cells. However, as you shall see below, this seemingly fundamental difference is less significant than it might first appear, and the two causes of cancer have in fact proved to be one and the same.

THE STORY OF CHICKEN SARCOMA

As early as 1920, Peyton Rous reported the presence of a virus, subsequently named Rous avian sarcoma virus (RSV), that was associated with chicken sarcomas; he was awarded the 1966 Nobel prize in physiology or medicine for his discovery. The RSV virus proved to be an RNA virus, one of the retroviruses, a group that we discussed in Chapter 17. As you will recall, retroviruses have the unusual ability to insert DNA transcripts of their RNA genomes into animal genomes. RSV virus was able to initiate cancer in chicken fibroblast (connective tissue) cells growing in culture when they were infected with the virus; from these **transformed** cells, more virus could be isolated.

How does RSV act to initiate cancer? When RSV was compared with a closely related virus, RAV-O, which was not able to transform chicken cells into cancer cells, the two viruses proved to be identical, except for one gene that was present in the RSV virus but absent from the RAV-O virus. The cancer-causing gene was called the *src* gene (for sarcoma) (Figure 18-11).

These and other results led to the hypothesis that cancer results from the action of a specific tumor-inducing *onc* gene, a hypothesis called the **oncogene theory.** The term **on-**

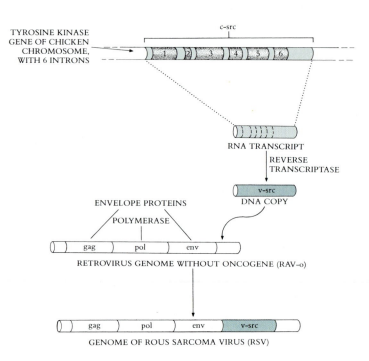

FIGURE 18-11

Structure of the RSV retrovirus. The virus contains only a few virus genes, encoding the virus protein envelope (env and gag) and reverse transcriptase, which produces a DNA copy of its RNA genome (pol). It also contains a gene, src, that it picked up from chicken DNA during some previous infection. The RAV-O virus lacks this src gene but is identical to the RSV retrovirus in all other respects. RSV causes cancer in chickens; RAV-O does not.

cogene was derived from the Greek word *oncos,* which means "cancer." This theory soon proved to be true for a broad range of tumor types. The RSV *src* gene is but one example of such *onc* genes. Seemingly spontaneous leukemia in mice, for example, is inherited in a Mendelian manner and can be mapped to specific sites on the mouse chromosomes. When investigated in detail, these sites proved to be sites of integrated virus genomes.

What might be the nature of a virus gene that causes cancer? An essential clue came in 1970, when RSV mutants were isolated that were temperature-sensitive. These mutants would transform tissue culture cells into cancer cells at 35° C, but not at 41° C. Temperature sensitivity of this kind is almost always associated with proteins. It seemed very likely, therefore, that the *src* gene was actively transcribed by the cell, rather than serving as a recognition site for some unknown regulatory protein. This was a very exciting result, because it suggested that the protein specified by this *onc* gene could be isolated and its properties studied.

The *src* protein was first isolated by Ericson and coworkers in 1977. It proved to be an enzyme of moderate size that acts to phosphorylate (add a phosphate group to) the tyrosine amino acids of proteins; such an enzyme is called a **tyrosine kinase.** Tyrosine kinases are not common in animal cells. Among the few that are known is **epidermal growth factor,** a protein that signals the initiation of cell division by binding to a special receptor site on the plasma membrane and phosphorylating key protein components. This fact raised the exciting possibility that RSV perhaps causes cancer by introducing into cells an active form of a normally quiescent enzyme. As you will see below, this indeed proved to be the case.

Does the *src* gene actually integrate into the host chromosome with the RSV virus? One way of investigating this question is to prepare a radioactive version of the *src* gene. We can then permit this radioactive *src* DNA to bind to complementary sequences on the chicken genome and examine where the chicken chromosomes become radioactive. Sites of radioactivity are sites where a sequence complementary to *src* occurs. As expected, radioactive *src* DNA binds to the site where RSV is inserted into the chicken genome—but unexpectedly it also binds to a second site, where there is no RSV.

From such experiments investigators learned that the *src* gene is not a virus gene at all, but rather a growth-promoting gene that evolved in and occurs normally in chickens. This chicken gene is the second site where *src* binds to chicken DNA. Somehow a copy of the normal chicken gene was picked up by an ancestor of the RSV virus in some past infection. Now part of the virus, the gene is active in a pernicious new way, escaping the normal regulatory controls of the chicken genome; its transcription is governed instead by virus promoters, which are actively transcribed during infection.

Thus the picture of cancer that emerges from the study of RSV is of a malady that results from the inappropriate activity of growth-promoting cellular genes that are normally less active or completely inactive.

> **Sarcoma in chickens results from the inappropriate activity of a normal chicken gene. Taken up by the RSV virus, it escapes the genetic regulation that is imposed by the chicken genome and induces cancer by being active when it should not be.**

THE MOLECULAR BASIS OF CANCER

Not all cancer is associated with viruses. Other kinds of tumors have been studied by using techniques different from those which we have just been exploring. The most important of these techniques, called **transfection,** consists of (1) the isolation of the nuclear DNA from human cells that have been isolated from tumors, (2) its cleavage into random fragments by using restriction endonucleases, and (3) the testing of the fragments individually for the ability of any particular fragment to induce cancer in the cells that assimilate it.

By using transfection techniques, researchers found that a single gene isolated from a cancer cell is all that is needed to transform normally dividing cells in tissue culture to cancerous ones. The cancerous cells differed from the normal ones only with respect to this one gene. In some cases the cancer-inducing gene identified by transfection proved to be the same as one of the *onc* genes that had previously been identified as having been included in a cancer-causing virus.

Onc genes are normal genes gone wrong. By identifying *onc* genes by transfection and then isolating them, investigators have been able to compare them to their normal counterparts. In this way researchers have been able to study why normal genes were converted into cancer-causing ones. The analysis of a number of *onc* genes associated with cancers of various tissues has led to the following conclusions:

1. The induction of many cancers involves changes in cellular activities that occur at the inner surface of the plasma membrane. In a normal cell these activities are associated with the initiation of cell division (Figure 18-12): a cellular regulatory protein, epidermal growth factor (EGF), binds to a specific receptor on the inner plasma membrane, acts as a tyrosine kinase, and triggers cell division. In this process a protein encoded by a gene called *ras* is associated with the membrane EGF receptor site and acts to determine what cellular levels of EGF are adequate to initiate cell division. In several forms of human cancer the cancerous, or *onc*, version of the *ras*-encoded protein activates the receptor site in response to much lower levels of EGF than does the normal version of the protein.

2. The difference between a normal gene encoding the proteins that carry out cell division and a cancer-inducing *onc* version need only be a single-point mutation in the DNA. For example, in a human bladder carcinoma induced by *ras* a single nucleotide alteration from G to T converts a glycine in the normal *ras* protein into a valine in the cancer-causing protein. There is no other difference between the normal and cancer-inducing forms of the *ras* gene.

FIGURE 18-12

Mutations that lead to cancer often involve proteins of the plasma membrane. The proteins depicted here, which have been implicated in cancers, are associated with cell division. In a normal cell, cell division is triggered by a protein called epidermal growth factor (EGF), which binds to an EGF receptor protein on the exterior surface of the cell and phosphorylates one or more of the EGF's tyrosine amino acids. This phosphorylation alters the shape of the portion of the receptor protruding into the cell, initiating a signal that passes to the cell nucleus and initiates cell division. Shown here is the phosphorylation of a cellular protein, which then passes to the nucleus. The level of EGF necessary to start this process is affected by another protein, ras.

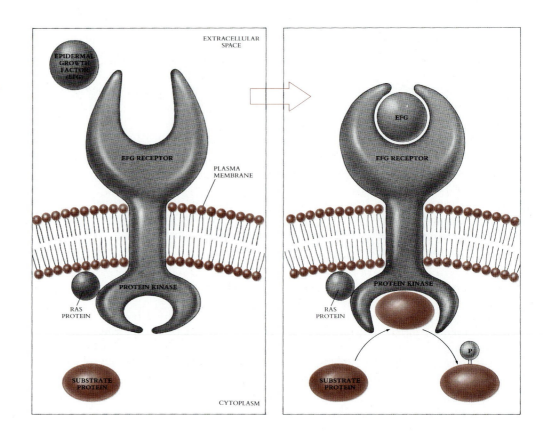

3. The mutation of a gene such as *ras* to an *onc* form in different tissues can lead to different forms of cancer. There are probably no more than a few dozen different genes whose mutation can lead to cancer, and there may be far fewer than that.

4. The induction of many cancers involves the action of two or more different *onc* genes. The initiation of cancer may require changes at both the plasma membrane and in the nucleus. This may be the reason why most cancers occur in people over 40 years old. It is as if human cells accumulate mutational changes, and time is required for several such mutations to occur in the same cells.

The emerging picture of cancer is one that involves aborted regulation of the genes that normally signal the onset of cell proliferation (Figure 18-13). Cancer seems to occur when several of the controls that cells normally impose on their own growth and division become inoperative. Among the examples that have been studied in detail, the specific means whereby these controls are evaded vary, but many of them involve one of two general causes. In the first case cancer may result from the mutation of a cellular gene with a regulatory function, as in the case of the *ras* gene, whose mutant *onc* form induces human bladder carcinoma. In the second case cancer may result from the introduction of a normal cellular gene with a regulatory function into a cell as part of a virus, in which the gene escapes the cell's regulation. In experiments in which a normal counterpart to an *onc* gene is placed by an active viral promoter and is not changed in any other way, it acts as an *onc* gene and successfully induces cancer; the increased level of transcription is the only difference. Again, the cancer is associated with an increase in the level of activity of a normal protein.

> **Cancer is a growth disease of cells, in which the controls that normally restrict cell proliferation do not operate. The critical factor in inducing cancer seems to be the inappropriate activation of one or more proteins that regulate cell division, transforming cells to a state of cancerous growth.**

SMOKING AND CANCER

How can we prevent cancer? The most obvious strategy is to minimize mutational insult. Anything that we do to increase our exposure to mutagens will result in an increased incidence of cancer, for the unavoidable reason that such exposure increases the probability of mutating a potential *onc* gene. It is no accident that the most reliable tests for carcinogenic substances are those which measure their mutagenic ability.

Of all the environmental mutagens to which we are exposed, perhaps the most tragic are cigarettes. They are tragic because the cancers they cause are preventable. About a third of all cases of cancer in the United States can be attributed directly to cigarette smoking. The association is particularly striking for lung cancer. The dose-response curve obtained for male smokers (Figure 18-14) shows a highly positive correlation, with the risk of lung cancer increasing with increasing amounts of smoking. For those smoking two or more packs a day, the risk of contracting lung cancer is 40 times or more greater than it is for nonsmokers. Note that the curve extrapolates back approximately to zero. Clearly, an effective way to avoid lung cancer is to not smoke. Life insurance companies have computed that on a statistical basis smoking a single cigarette lowers your life expectancy 10.7 minutes. (That is more than the time is takes to smoke the cigarette!) Every pack of 20 cigarettes bears an unwritten label:

> "The price of smoking this pack of cigarettes is 3½ hours of your life."

Do we have any idea how to cure cancer, once it has started? Not yet. The solution lies in a better understanding of the thwarted cellular controls that are responsible for cancerous growth so that therapies can be devised to reinstitute appropriate cellular regulation within cancerous tissue. As you can imagine, this is an area of intense research activity.

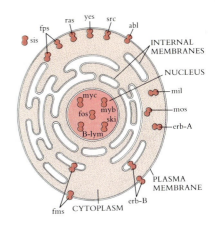

FIGURE 18-13

A representation of a eukaryotic cell, showing sites of action of the proteins encoded by 16 of the 30 known cancer-causing genes. The activities of some genes are associated with plasma membranes; others are associated with internal membranes, the nucleus, or the cytoplasm of cells.

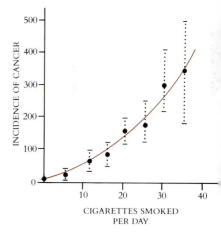

FIGURE 18-14

The annual incidence of lung cancer per 100,000 men is clearly related to the number of cigarettes smoked per day. In addition, smoking is a major contributor to heart disease. Of every 100 American college students who smoke cigarettes regularly, 1 will be murdered, about 2 will be killed on the roads, and about 25 will be killed by tobacco.

HOW TO CATCH CANCER

Of the 1.7 million people who died in the United States last year, one fourth died of cancer and a third of these—140,000 people—died of lung cancer. About 160,000 cases of lung cancer will be diagnosed each year in the 1980s, and 90% of these persons will die within 3 years. Of those who die, 96% will be cigarette smokers.

Smoking is a popular pastime in the United States. One third of the U.S. population smokes. American smokers consumed 443 billion cigarettes in 1983. These cigarettes emit in their tobacco smoke some 3000 chemical components, among them vinyl chloride, benzo(a)pyrenes, and nitroso-nor-nicotine, all of them potent mutagens. Smoking introduces these mutagens to the tissues of your lungs. Figure 18-A, a photograph of lung cancer in an adult human, illustrates the result. The bottom half of the lung is normal; the top half has been taken over completely by a cancerous growth. As you might imagine, a lung in this condition does not function well, but difficulty in breathing is rarely the cause of death from lung cancer. As the cancer grows within the lung, its cells invade the surrounding tissues, as diagrammed in Figure 18-9, and eventually break through into the lymph and blood vessels. Once the cancer cells have done this, they spread rapidly through the body, lodging and growing at many locations, particularly in the brain. Death soon follows.

Particularly among cigarette manufacturers, it has been popular to argue that the causal connection

FIGURE 18-A

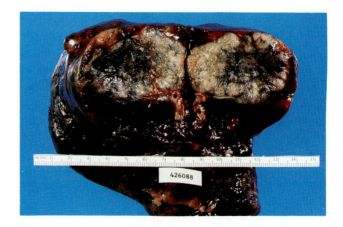

between smoking and cigarettes has not been proved, that somehow the relationship is coincidental. Look carefully at the data presented in Figure 18-B and see if you agree. The upper graph presents data collected for American males, including the incidence of smoking from the turn of the century until now and the incidence of lung cancer over the same period. Note that as late as 1920, lung cancer was a rare disease. With a lag of some 20 years behind the increase in smoking, it became progressively more common.

Now look at the lower graph, which presents data on American females. Because of social mores, significant numbers of American females did not

SUMMARY

1. Genetic engineering involves the isolation of specific genes and their transfer to new genomes.

2. The key to genetic engineering technology is a special class of enzymes called restriction endonucleases, which cleave DNA molecules into fragments. These fragments have short one-strand tails that are complementary in nucleotide sequence to each other.

3. Because the sequences of the tails of the two cleavage products are complementary, they can pair with one another and rejoin together, whatever their source. For this reason DNA fragments from very different genomes can be combined.

4. The first combination of fragments from different genomes (recombinant DNA) was achieved by Stanley Cohen and Herbert Boyer in 1973. They inserted an amphibian ribosomal RNA gene into a bacterial plasmid.

smoke until after World War II, when many social conventions changed. As late as 1963, when lung cancer among males was near current levels, this disease was still rare in females. In the United States that year, only 6588 females died of lung cancer. But as their smoking increased, so did their incidence of lung cancer, with the same inexorable lag of about 20 years. American females today have achieved equality with their male counterparts in the numbers of cigarettes that they smoke—and their lung cancer death rates are now no different than those for males. In 1983 more than 36,000 females died of lung cancer in the United States.

Among smokers, the current rate of deaths resulting from lung cancer is 180 per 100,000, or about 2 of each 1000 smokers *each year*. Smoking is very much like going into a totally dark room, standing still, and then calling in someone with a gun and closing the door. The person with the gun cannot see you, does not know where you are, and so just shoots once in a random direction and leaves the room. Every time an individual smokes a cigarette, he or she is "shooting" mutagens at their genes. Just as in the dark room, a hit is unlikely, and most shots will miss potential *onc* genes. As one keeps shooting, however, the odds of eventually scoring a hit get shorter and shorter. Nor do statistics protect any one individual: nothing says the first shot will not hit. Older people are not the only ones to die of lung cancer.

Except for eating powerful radioisotopes, there is probably no more certain way to catch cancer than to smoke.

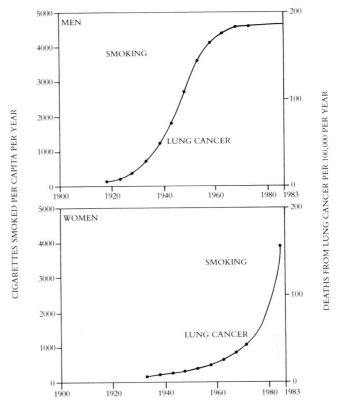

FIGURE 18-B

5. Gene splicing holds great promise as a clinical tool, particularly in the prevention of disease. Several vaccines are being prepared by stitching the coat-specifying genes of disease-causing viruses or cells into the genomes of harmless viruses and using the recombinant form—which is harmless and has the outside coat of the disease form—as a vaccine to induce antibody formation against the disease.

6. Cancer is a growth disease of cells in which the regulatory controls that usually restrain cell division do not work.

7. Sarcoma in chickens is caused by the presence in a retrovirus of an otherwise normal chicken gene whose activity mimics that of epidermal growth factor, a regulator of cell division. Because it is located in the virus genome, the gene is transcribed far more often than it otherwise would be, triggering cell division.

8. The testing of DNA fragments from tumor cells to identify those fragments capable of inducing cancer has led to the isolation of a variety of cancer-causing genes. In

every case the cancer-causing gene is a normal gene whose product has a role in cell proliferation but that becomes active when it should not be.

9. In some cases cancer induction results from a single nucleotide mutation in a normal gene. In other cases a normal gene is acquired by a transposable retrovirus and is transcribed at the high rate characteristic of the virus.

10. The best way to avoid cancer is to avoid those things which cause mutation, notably smoking.

SELF-QUIZ

The following questions pinpoint important information in this chapter. You should be sure to know the answers to each of them.

1. The first step in a genetic engineering experiment is to cut up the source DNA with a special kind of enzyme called a _____.

2. These DNA-cutting enzymes are isolated from bacteria. The reason that these enzymes do not cut up the DNA of these bacterial cells is that
   (a) they are not in contact with the bacterial DNA, which is in the nucleus
   (b) the enzymes attack only viral DNA
   (c) the DNA in these cells has been modified so that the enzymes do not recognize it
   (d) the enzymes must be modified before they work

3. The DNA-cutting enzyme Hind III recognizes a six-base-pair sequence on DNA. The first three nucleotides on one strand are AAG__ __ __. Because of the twofold rotational symmetry of sequences such as this, we know that the other three nucleotides of the six- base-pair sequence must be _____.

4. If one cuts up elephant DNA with an enzyme such as Hind III and uses the same enzyme to cut up DNA from other organisms
   (a) the elephant DNA fragments will rejoin with each other more readily than with fragments from other organisms
   (b) the elephant DNA fragments will rejoin with human fragments more readily than with bacterial fragments
   (c) the elephant fragments will rejoin with bacterial fragments, but not with human ones
   (d) none of the fragments will rejoin with one another
   (e) all of the fragments will rejoin with one another at random

5. The DNA-cutting enzyme Eca I recognizes the seven-base-pair sequence GGTAACC. With what frequency is this sequence expected to occur in human DNA?
   (a) 1 per 1024 nucleotides   (d) 1 per 65,536 nucleotides
   (b) 1 per 4096 nucleotides   (e) 1 per 262,144 nucleotides
   (c) 1 per 16,384 nucleotides

6. A genetically engineered vaccine against malaria is currently being developed. Geneticists are inserting into benign vaccinia virus a few of the genes of the malaria-causing protozoan. The genes that they are inserting
   (a) are the ones that cause malaria
   (b) are the ones specifying the protein exterior of the parasite
   (c) encode interferon
   (d) encode the outer surface of a herpesvirus
   (e) none of the above

7. A tumor that remains a hard mass of cells and does not spread throughout the body is called a _____.

8. Some cancer cells spread throughout the body, forming tumors at distant sites. These cells are called _____.

9. Of the people born in 1985, what percentage will contract cancer?

10. One third of the deaths from cancer each year are caused by lung cancers. What habit do over 96% of those who will die in this way this year share?

11. Avian sarcoma in chickens is caused by a retrovirus. Only one gene of the retrovirus is necessary to initiate the cancer. The gene encodes a protein with enzyme activity. On what kind of molecule does it act?

12. Smoking two packs a day increases the risk of lung cancer about _____ times.

THOUGHT QUESTIONS

1. The evidence associating lung cancer with smoking is overwhelming, and, as you have learned in this chapter, we now know in considerable detail the mechanism whereby smoking induces cancer. It introduces powerful mutagens into the lungs, which cause mutations to occur; when a growth-regulating gene is mutated by chance, cancer results. In light of this, why do you think that cigarette smoking is not illegal?

2. In a study of the segregation of leukemia in several generations of laboratory mice, a line selected for high incidence of leukemia was crossed with a line that never develops leukemia (it presumably lacks a leukemia-inducing oncogene). All of the F_1 progeny exhibited a high incidence of leukemia. Among the progeny resulting from a testcross of these F_1 individuals to the leukemia-free parent line, 25% were healthy and leukemia-free; the other 75% were leukemic. The result is not a 1:1 ratio, but rather a 3:1 ratio. How would you interpret this?

FOR FURTHER READING

BISHOP, J.M.: "Oncogenes," *Scientific American*, March 1982, pages 82-92. The story of *src* and how it causes cancer, by the man who first identified the protein product of the *src* gene. This article is particularly good at showing the chain of reasoning that underlies a major scientific advance.

CHILTON, M. "A Vector for Introducing New Genes into Plants," *Scientific American*, June 1983, pages 50-60. A description of the only successful plant genetic engineering vector developed to date, important because of its agricultural implications.

COHEN, S.: "The Manipulation of Genes," *Scientific American*, July 1975, pages 24-34. A lucid description of how genes can be cut into fragments and then incorporated into plasmids that can infect organisms.

DOLL, R.: "An Epidemiological Perspective of the Biology of Cancer," *Cancer Research*, vol. 38, pages 3573-3583, 1978. An advanced article that presents data on who contracts cancer and at what age. The data constitute a damning indictment of smoking as the leading cause of cancer.

GILBERT, W., and L. VILLA-KAMAROFF: "Useful Proteins from Recombinant Bacteria," *Scientific American*, April 1980, pages 74-94. An important by-product of the new gene technology is the movement of human genes into bacteria, which produce their products in large amounts. This article, by a Nobel prize winner, provides a clear description of how this is done.

GODSON, G.: "Molecular Approaches to Malaria Vaccines," *Scientific American*, May 1985, pages 52-59. A revealing look at how the new techniques of molecular engineering are being used to defeat an ancient scourge.

GROBSTEIN, C.: "The Recombinant DNA Debate," *Scientific American*, July 1977, pages 22-34. An illuminating account of the discussions among scientists and between them and society concerning the potential benefits and dangers of gene technology.

WEINBERG, R.: "The Molecular Basis for Cancer," *Scientific American*, November 1983, pages 126-144. This important paper by a developer of the transfection approach describes how transfection experiments yielded our first look at a cancer-causing gene.

EVOLUTION

Part Four

The incredible diversity of life on earth has evolved over the course of at least 3.5 billion years. The process through which this occurred was outlined by Charles Darwin in the mid-nineteenth century.

DEVELOPMENT OF EVOLUTIONARY THEORY

Chapter 19

Overview

Charles Darwin provided the first satisfactory explanation for evolution in the middle of the nineteenth century, but his work was based on centuries of observation and thought. Realizing that any organism has a tendency to produce more offspring than it needs to re-place itself, Darwin concluded that only some of the progeny would survive. He based this conclusion on the observation that the numbers of individual plants and animals in natural populations did not increase year after year but remained relatively constant. Furthermore, the progeny that survived would tend to be those which had the characteristics best fitted to survival. These individuals, in turn, would leave the largest numbers of progeny, and their characteristics would dominate more and more in succeeding generations. Darwin's proposal of natural selection as the mechanism of evolution was the centerpiece of one of the great intellectual revolutions of human history, a revolution that has completely changed our perception of the world and of our own nature.

For Review *Here are some important terms and concepts that you will encounter in this chapter. If you are not familiar with them, you should review them before proceeding.*

- **The origin of life** (Chapter 3)
- **The nature of genes** (Chapter 13)
- **The hereditary mechanism** (Chapter 15)
- **Mutation and recombination** (Chapter 17)

FIGURE 19-1

When European explorers first came to Australia, they found bizarre, unexpected kinds of plants and animals, such as this koala and Banksia coccinea, *one of the hundreds of species of the protea family that occur on that continent. Such new discoveries greatly stimulated their desire to catalogue and understand the diversity of life on earth.*

For as far back as we can trace human thought, people have wondered about the diversity of the plants and animals that they saw around them and, increasingly, about those which were being discovered in foreign lands (Figure 19-1). Why were there so many kinds of organisms? Why did they display such marvelous adaptations to their particular ways of life, such intricate structures, and such remarkable physiological characteristics? Finally, why did organisms tend to fall together in groups, resembling one another in their features and often occurring together in continuous geographical areas, rather than differing from one another in random and unpredictable ways?

The search for explanations for this diversity, these adaptations, and these patterns has proved to be challenging. The answers that we have found all point in one direction—to evolution, the progressive change of living things through time. The theory of evolution is the cornerstone of modern biology.

FROM SPECIAL CREATION TO EVOLUTION

The idea of evolution—the notion that living things change gradually from one form into another over the course of time—was not, contrary to the popular view, originated by Charles Darwin in the nineteenth century. From the very beginnings of organized thought about the nature of life, some philosophers noted that many groups of organisms exhibit continuous variation from the very simple to the extraordinarily complex. The traditional explanation for these patterns was that the various kinds of organisms and their individual structures were the result of direct action by the Creator. The fact that these organisms could be arranged in an almost continuous series, called the chain of being, or *scala naturae*, was considered to be a coincidence.

Most scientists believed fully that each kind of organism, with all of its individual adaptations, resulted from the direct intervention of God. They studied these organisms, their structures, and the relationships between them as a way of learning more about the Creator. In the seventeenth century, for example, the English scientist-clergyman John Ray (1627-1705) explicitly declared his belief that each kind of animal and plant had remained unchanged from the day it was created. His major work, which appeared in 1691, was entitled *The Wisdom of God as Manifested in the Works of the Creation.* Other early biologists agreed with Ray's view. Carolus Linnaeus (1708-1778) was a Swedish biologist who, as you will see, established the system of naming organisms that is still in use. Like John Ray, he believed that every feature of the organisms he studied resulted from direct divine action.

By the first part of the eighteenth century, many more kinds of organisms were being discovered than previously. In addition, a great deal of information was beginning to accumulate about fossil plants and animals. These discoveries gradually began to revive discussions of evolution, which was seen as a possible explanation of the patterns that were being observed and charted in increasing detail. The great French biologists Georges-Louis Leclerc, Comte de Buffon (1707-1788), and Joseph Pitton de Tournefort (1656-1708) spoke explicitly about a sort of natural affinity between kinds of organisms. A century before Darwin, Buffon wrote of nature, "She sends the bat flying among the birds, while she imprisons the armadillo under the armor of a crustacean; she molded the whale in the shape of a quadruped, and merely truncated the quadruped in the walrus and the seal who, born on the land, plunge into the billows and join the whale, as if to demonstrate the universal kinship of all generations born from a common mother." He could see no explanation for the common features of all mammals except their evolution from a common ancestor.

Thoughts similar to those which Buffon had expressed led, within 50 years, to those of Jean Baptiste A.P.M. de Lamarck (1744-1829). Lamarck explicitly proposed evolution as a theory to account for the patterns observed in nature. In 1801 he suggested that all species, including human beings, were descended from other species. Lamarck thought of life as having evolved progressively from simple to more complex and was the first to propose a coherent theory of evolution.

Lamarck's theory was based on the inheritance of acquired characteristics. We now know that this assumption was incorrect, although it was consistent with the knowledge available to Lamarck at the time. He thought that certain organs and structures in animals and plants would become stronger through use. When they did so, they would be passed on, improved, to subsequent generations (Figure 19-2). Even though the mechanisms that Lamarck proposed to account for evolution were incorrect, he called wide attention to the process itself and, by doing so, set the stage for the acceptance of the correct, and much simpler, explanation that was developed by Darwin half a century later.

FIGURE 19-2

A giraffe browsing on the Serengeti Plains of Tanzania, stretching its neck to reach the uppermost leaves on the tree on which it is feeding. In Lamarck's view the giraffe would be likely to acquire a longer neck by browsing in this way and to be able to pass on its longer neck to its offspring. We now know that this is an incorrect theory of inheritance.

Darwin's intellectual predecessors, men like Buffon, Tournefort, and Lamarck, accepted evolutionary relationships between organisms as a fact but could not provide a satisfactory explanation for them.

Both Lamarck and Darwin thought that there was a natural progression from simple to more complex or "perfect" organisms. In fact, they both went much further in this direction than modern theory would accept. The major difference between them was that Darwin provided an understandable, simple, and correct explanation for the process of evolutionary change. He did this by proposing the mechanism of natural selection.

CHARLES DARWIN AND MODERN EVOLUTIONARY THOUGHT

Charles Robert Darwin (1809-1882) was an English naturalist who, at the age of 50, after 30 years of study and observation, wrote one of the most famous and influential books of all time. The full title of this book, *On the Origin of Species by Means of Natural Selection, or the Preservation of Favoured Races in the Struggle for Life,* expressed both the nature of its subject and the way in which Darwin treated it. The book created a sensation when it was published in 1859, and the ideas expressed in it have played a central role in the development of human thought ever since.

Darwin doubted neither the existence of a Divine Creator nor the fact that this Creator was personally responsible for all things. He simply believed that this Creator expressed Himself through the operation of natural laws that could be studied and observed, rather than simply by creating things individually and leaving them forever as they were. These views put Darwin at odds with most people, who believed in a literal interpretation of the Bible and accepted the idea of a fixed and constant world. In essence his theory was a revolutionary one that troubled many of his contemporaries, as well as Darwin himself, very deeply.

One of Darwin's professors at Cambridge University, from which he had recently graduated, recommended him, as a young man of 22, to accompany a 5-year voyage to chart the southern coasts of South America. This was the now-famous voyage of H.M.S. *Beagle* (1831-1836) (Figure 19-3). Darwin went on the voyage both as a gentleman companion to the ship's captain and as naturalist. Naturalists had routinely been posted on all British voyages to distant lands since the 1770s so that each voyage could add to the store of human knowledge. During this long voyage, Darwin had the chance to study plants and animals widely on continents and islands and in far-flung seas. He was able to experience firsthand the biological richness of the tropical forests, the extraordinary fossils of huge extinct mammals in Patagonia (Figure 19-4), and the remarkable series of related

FIGURE 19-3

Replica of H.M.S. Beagle. Although small by modern standards, the Beagle was a sturdy ship. Charles Darwin sailed on this ship for 5 years among the continents, islands, and oceans of much of the world, but mainly around South America.

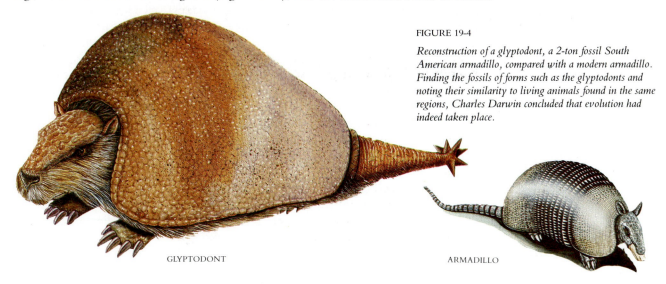

FIGURE 19-4

Reconstruction of a glyptodont, a 2-ton fossil South American armadillo, compared with a modern armadillo. Finding the fossils of forms such as the glyptodonts and noting their similarity to living animals found in the same regions, Charles Darwin concluded that evolution had indeed taken place.

GLYPTODONT

ARMADILLO

but distinct forms of life on the Galapagos archipelago, off the west coast of South America. Such an opportunity clearly played an important role in the development of his thoughts about the nature of life on earth.

When Darwin returned from the voyage at the age of 27, he began a long life of study and contemplation (Figure 19-5). During the next 10 years, he published important books on several different subjects, including coral reefs and the formation of atolls, volcanic islands, and the geology of South America. He also devoted 8 years of study to barnacles, writing a four-volume work on their classification and natural history. In 1842 Darwin and his family moved out of London to a country home at Downe, in the county of Kent. In these pleasant surroundings he lived, studied, and wrote for the next 40 years. During this period, he formulated for the first time a consistent theory of the process of evolution, an explanation that provided a mechanism for evolution. He presented his ideas in such convincing detail that they could logically be accepted as explaining the diversity of life on earth, the intricate adaptations of those living things, and the ways in which they are related to one another.

DARWIN'S EVIDENCE

So much information had accumulated by 1859 that the acceptance of the theory of evolution now seems, in retrospect, to have been inevitable. Darwin was successful where many others had failed because he rejected supernatural explanations for the phenomena that he was studying. He and his scientific contemporaries had come to realize that only by studying the facts that were available to them, facts about which they could make additional observations and experiments to test their hypotheses, could they advance knowledge. They certainly did not believe that they were in contradiction with the personal belief in the existence of a Divine Being that most of them held. Rather, they felt that they were affirming the fact that the works of this Divine Being could be studied and understood by human beings.

One of the obstacles that blocked the acceptance of any theory of evolution was the idea that the earth was relatively young. In the seventeenth century, for example, James Ussher, Archbishop of Armagh, in what is now Northern Ireland, had calculated that the earth was created precisely in the year 4004 BC; he based his conclusion on his study of the Bible. As more geological knowledge accumulated, however, Archbishop Ussher's conclusion came to seem less and less likely. The thick layers of rocks, evidences of extensive and prolonged erosion, and increasing numbers of diverse and unfamiliar fossils that were being discovered began to convince people that the earth must be more than 6000 years old.

Notable among those who began to claim that the earth was far older than Archbishop Ussher had calculated was a new school of geologists that had arisen by the time that Darwin was starting to think about evolution. Charles Lyell (1797-1895) was the most important of these geologists. Only by assuming a greater age of the earth could these geologists account for its present condition and appearance. In addition, Lyell, whose works Darwin read eagerly while he was sailing on the *Beagle,* outlined for the first time the story of an ancient world of plants and animals in flux. In this world, some species were constantly becoming extinct, while others were emerging. In relation to Lyell's important theory of **uniformitarianism,** it was logical for Darwin to assume that these processes were still continuing. Although Darwin was not able to convince Lyell until 1862 that natural selection was the mechanism of evolution, Lyell's own views had profound importance in shaping those of his younger contemporary.

What Darwin Saw

When the *Beagle* set sail, Darwin was fully convinced that species were unchanging and immutable. Indeed, he wrote that he did not consider seriously the possibility that they could change until 2 or 3 years after his return. Nevertheless, during his 5 years on the ship,

A

B

FIGURE 19-6

Darwin observed a number of phenomena that were of central importance to him in reaching his ultimate conclusion. For example, in rich beds of fossils in southern South America he observed extinct armadillos that were related directly to the armadillos that still lived in the same area (Figure 19-4). Why would there be such unusual living and fossil organisms, directly related to one another, in the same area, unless one had given rise to the other?

Repeatedly, Darwin saw that the characteristics of different species varied from place to place. These patterns suggested to him that organisms change gradually as they migrate from one area into another. On the Galapagos Islands, off the coast of Ecuador, Darwin encountered giant land tortoises. Surprisingly, these tortoises were not all identical. Indeed, local residents and the sailors who captured the tortoises for food could tell which island a particular animal had come from just by looking at it (Figure 19-6). This pattern of variation suggested that all of the tortoises were related but had changed slightly in appearance after they had become isolated on the different islands.

In a more general sense Darwin was struck by the fact that on these relatively young volcanic islands there was a profusion of living things, but that these plants and animals resembled those of the nearby coast of South America. If each one of these plants and animals had been created independently and simply placed on the Galapagos Islands, why did they not resemble the plants and animals of faraway Africa, for example? Why did they resemble those of the adjacent South American coast instead?

If Charles Darwin had known, as we do now, that the Galapagos Islands arise from ocean floor rocks that are no more than 10 million years old, and that the islands themselves arose from the sea starting only about 3.3 million years ago, he would have been even more impressed with his evidence. As it was, he found ample evidence to stimulate his active mind to fruitful deductions about the process of evolution.

The patterns of distribution and relationship of organisms that Darwin observed on the voyage of the *Beagle* made him certain that a process of evolution had been responsible.

Darwin and Malthus

Of key importance to the development of Darwin's thoughts was his study of Thomas Malthus' *Essay on the Principles of Population*. Darwin read this book 2 years after returning to England. Coincidentally, Malthus had been stimulated to write the book, which was published in 1798, by the writings of Darwin's grandfather, the philosopher Erasmus Darwin. The elder Darwin's views of evolution, which were somewhat comparable to the contemporary views of Lamarck in France, had been both extensive and influential.

*Galapagos tortoises. Tortoises with large, domed shells, **A**, are found in relatively moist habitats, where the competition for food may not be as intense as where it is drier. The lower, saddleback-type shells, in which the front of the shell is bent up, exposing the head and part of the neck, **B**, are found among the tortoises of arid habitats. Tortoises of this kind are likewise less tolerant of the presence of other individuals and more antagonistic. In addition to these relationships, larger tortoises occur in moister places than do smaller ones. All of these differences are genetic, and in combination they make it possible to identify the races of tortoises that inhabit the different islands of the group. Different kinds of tortoises also occur on some of the large individual islands.*

Charles Darwin wrote, in The Origin of Species, *"Believing that it is always best to study some special group, I have, after deliberation, taken up domestic pigeons. . . . The fantail has thirty or even forty tail feathers, instead of twelve or fourteen, the normal number in all members of the great pigeon family; and these feathers are kept expanded, and are carried so erect that in good birds the head and tail touch. . . . Although at least a score of pigeons might be chosen, which, if shown to an ornithologist, and he were told that they were wild birds, would certainly, I think, be ranked by him as well-defined species."* The differences that have been obtained by artificial selection of birds similar to the wild European rock pigeon, **A,** and such domestic races as the red fantail, **B,** and the fairy swallow, **C,** with its fantastic tufts of feathers around its feet, are indeed so great that the birds probably would, if wild, be classified in different genera. In a way similar to that in which these races were derived, widely different species have originated in nature by means of natural selection.

A

B

C

In his book Malthus presented the observation that populations of plants and animals tend to increase geometrically. A **geometric progression** is one in which the elements progress by a constant factor, such as 2, 6, 18, 54, and so forth. An **arithmetic progression,** in contrast, is one in which the elements increase by a constant difference, such as 2, 6, 10, 14, and so forth. Malthus pointed out that populations of plants and animals, including human beings, tend to increase geometrically, while, in the case of people, our ability to increase food supply grows only arithmetically.

Virtually any kind of animal or plant, if it could reproduce unchecked, would cover the entire surface of the world within a surprisingly short period of time. In fact, this does not occur; instead, populations of species remain more or less constant year after year, because death intervenes and limits population numbers. Malthus was especially concerned with the effects of this geometric increase on the human population, which he saw as growing unchecked and, therefore, ultimately reaching a condition of widespread misery. For Darwin, however, Malthus provided the key ingredient in the development of his theory of evolution by natural selection.

Natural Selection

Every organism has the potential to produce more offspring than are able to survive, but only a limited number actually survive and produce offspring. Combining this observation with what he had seen on the voyage of the *Beagle,* as well as with his own experiences in breeding domestic animals, Darwin made the key association: *those individuals which possess superior physical, behavioral, or other attributes are more likely to survive than those which are not so well endowed.* In surviving, they have the opportunity to pass on their favorable characteristics to their offspring. Since these characteristics will increase in the population, the nature of the population as a whole will gradually change. Darwin called this process **natural selection,** and called the driving force that he had identified the *survival of the fittest.*

Darwin was thoroughly familiar with variation in domesticated animals, and he began his *Origin of Species* with a detailed discussion of pigeon breeding. He knew that varieties of pigeons and other animals, such as dogs, could be selected to exhibit certain characteristics. Once this selection had been carried out, the animals would breed true for the characteristics that had been concentrated in them. Darwin had also observed that the differences that could be developed between domesticated races or breeds in this way were often greater than those which separated wild species. The breeds of the domestic pigeon shown in Figure 19-7, for example, are much more different from one another in various ways than are all of the hundreds of wild species of pigeons found throughout the world. Such relationships suggested to Darwin that evolutionary change could occur very rapidly under the appropriate circumstances. Artificial selection provided Darwin with useful information for understanding natural selection.

> **Malthus had observed the tendency of all organisms to increase geometrically in number. Realizing that they did not in fact do so, Darwin reasoned that those with characteristics that enabled them to survive and produce more offspring would take over successive populations. Their characteristics would then come to be shared by most or all of the individuals in the population. The nature of the population would gradually change as more and more individuals with those traits appeared. This process is known as *natural selection*.**

PUBLICATION OF DARWIN'S THEORY

Darwin read Malthus' book in 1838. He drafted the overall argument for evolution by natural selection in 1842, continuing to enlarge and refine his argument for many years. He discussed his ideas with a very few friends and colleagues, who urged him to perfect and publish his theory. They feared that eventually someone else might reach the same conclusions and that Darwin might be deprived of the credit that he so richly deserved.

The stimulus that finally brought Darwin's work into print, although in a much more concise form than he had hoped, was an essay that he received on June 18, 1858, 20 years after he had first conceived the idea of evolution by natural selection. A young English naturalist named Alfred Russel Wallace (1823-1913) (Figure 19-8) had sent the essay from Ternate, one of the Molucca Islands, in the Malayan archipelago. It concisely set forth the theory of evolution by means of natural selection! Like Darwin, Wallace had been greatly influenced in his development of this theory by reading Malthus' 1798 essay.

Once he had received Wallace's communication, Darwin was deeply concerned that he might treat his young colleague unfairly in some way. He consulted his friends Charles Lyell and William Dalton Hooker, who was the first director of the Royal Botanic Gardens at Kew, outside London, about this problem. In response, they organized a joint presentation of Darwin's and Wallace's work at the Linnaean Society, an important scholarly society in London, on July 1, 1858. The papers were presented by Lyell and Hooker, and neither Wallace nor Darwin attended. Little attention was paid to the papers at the time they were read. The fact of their presentation, however, stimulated Darwin to complete his book *On the Origin of Species*. Since he wanted to finish it soon, he wrote only an abbreviated version of the lengthy essay on which he had been working for some 16 years. The shorter book appeared in November, 1859, and caused an immediate sensation. Although he lived for 23 more years, Darwin never completed the longer explanation of his theories.

Many people were deeply disturbed by the implications of Darwin's book, especially by the implication that human beings were, so to speak, descended from the apes. The fact that human beings were clearly members of the group of mammals called Primates and closely similar to the apes in all of their characteristics had been well established for more than a century. Nevertheless, the possibility that there might be a direct evolutionary relationship between apes and human beings was unacceptable to many people, especially in the decade following the publication of Darwin's book. Darwin's arguments were compelling, however, and his supporters were numerous, active, and influential. With their help, the theory of evolution by natural selection prevailed almost completely in scientific and learned circles after the 1860s.

Two serious concerns continued to occupy Darwin's attention: (1) the age of the earth, about which there was much dispute; and (2) the mechanism of heredity—the way in which desirable characteristics were transmitted to offspring. In the first area Darwin was bothered by the great difficulties that he and his contemporaries met in attempting to estimate the age of the earth. He was personally convinced that all living things were descended from some single original ancestor, but such a process would have required a great deal of time. In Darwin's day the arguments of some physicists, which we now know to have been flawed, seemed to indicate that sufficient time was not available.

Darwin also remained extremely interested in the subject of heredity, and he collected a considerable amount of observational evidence that provided further documentation for the fact that some characteristics are indeed inherited. In fact, in the progeny of one of his crosses with plants, Darwin obtained results similar to Mendel's, with segregation in the F_2 generation in roughly 3:1 proportions. Unlike Mendel, however, Darwin failed to see the significance of his results, that unchanging hereditary units were being segregated in a predictable fashion. Consequently, he was never able to provide a clear explanation of heredity. Darwin and Mendel were contemporaries, but they never met and never commented on each other's work, if they were aware of it.

Scientists did not adopt a satisfactory explanation of the process of heredity until two decades after Darwin's death. The lack of understanding of this area during the nineteenth century was a serious obstacle to the further development of evolutionary theory.

Although problems remained, Darwin's works firmly established that evolution accounts for the diversity of life on earth. Huge masses of facts, otherwise seemingly without logic, began to make sense when they were viewed in light of his theory. These facts in-

FIGURE 19-8

Alfred Russel Wallace in his later years.

clude the common features of the members of different groups or phyla, the ways in which embryos resemble and differ from one another, and the increasing complexity that is observed in the fossil record through time. Evolution is the main unifying theme of the biological sciences. It provides one of the most important insights that human beings have achieved into their own nature and that of the earth on which they have evolved.

FIGURE 19-9

Reconstructions of fossils, illustrating the evolution of horses. Animals called hyracotheres, which included the earliest member of the evolutionary line that is illustrated, Hyracotherium, *gave rise to several groups of mammals, including tapirs and rhinoceroses, in addition to the horses. As grasslands have spread on a continental scale during the past 30 million years, horses have become more abundant, and more different kinds have appeared. Horses are an excellent example of the way in which abundant fossil evidence has allowed the reconstruction of the evolution of a particular vertebrate group.*

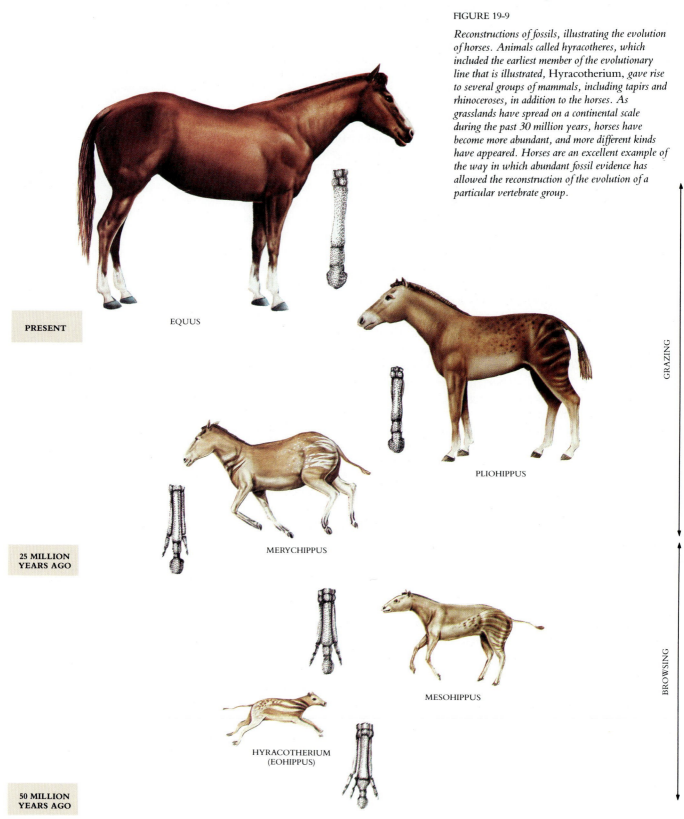

PRESENT

25 MILLION YEARS AGO

50 MILLION YEARS AGO

EQUUS

PLIOHIPPUS

MERYCHIPPUS

MESOHIPPUS

HYRACOTHERIUM (EOHIPPUS)

GRAZING

BROWSING

EVOLUTION AFTER DARWIN

More than a century has elapsed since Charles Darwin's death in 1882. During this period, there have been many significant advances in our understanding of the process of evolution. In general, these have not altered our basic understanding of the process, but we have learned much more about the mechanisms by which it has occurred. Some of the significant ways in which other fields of inquiry have affected progress in the field of evolution are discussed in the following pages.

Paleontology

Darwin explicitly postulated that the fossil record should yield intermediate links between the great groups of organisms. Paleontologists, those who study fossils or organisms, provide concrete evidence of the history of life on earth—evidence that cannot be obtained in any other way. For example, one of the kinds of evolutionary intermediate forms that Darwin had predicted, the early bird-reptile *Archaeopteryx* (pp. 457-458), was discovered in 1861, within 2 years of the publication of Darwin's book. Somewhat surprisingly, however, Darwin did not emphasize the importance of this fossil.

The fossil record is now known to a degree that would have been unthinkable in the nineteenth century. Recent discoveries of microscopic fossils have extended back the known history of life on earth to more than 3.5 billion years. The discovery of other fossils has shed light on the ways in which organisms have, over the course of this enormous time span, evolved from the simple to the complex. Most important, the evolution of the major groups of vertebrates and their relationships to one another have become reasonably well understood. The bones and teeth of vertebrates are often encountered in the fossil record, and we therefore know more about the evolution of this group of organisms than we do about most others. The evolutionary lineages of certain vertebrate groups are now understood in reasonable detail. Among these, the evolution of the horses provides an especially good example (Figure 19-9). Invertebrates that have only soft body parts, on the other hand, are not nearly as well represented as fossils as are the vertebrates, and their history is not nearly as well understood.

In interpreting the fossil record, or even specifying relationships between contemporary groups of organisms, one of the most important concepts that must be employed is that of homology. **Homology** is defined as similarity that occurs because of a common evolutionary origin. A familiar example has to do with the homology between the wing of a bat, the flipper of a porpoise, and a human arm (Figure 19-10). Scientists have become better and better able to establish homologies between organs, both in morphological and anatomical features, as well in biochemical ones. As a result, we have come to understand more and more clearly how and when particular groups have evolved, and we can also classify organisms more appropriately, because we understand their relationships better. We will present many examples of homologies throughout this book.

Genetics

Genetics was established as a science at the start of the twentieth century. Since then, its principles have been applied to improve our understanding of evolutionary processes. For the past 50 years, genetics has been harmonized completely with descriptive evolutionary studies, and the concepts of both fields have been sharpened as a result. Until the laws of inheritance were understood, it was impossible to account for the origin of new variations or for their persistence. In fact, it was in this area that Darwin received some of his sharpest criticism. Without the concept of genes, which determine the features of every organism and do not blend with one another, it is not possible to explain completely how evolution occurs.

FIGURE 19-10

Homologies between the forelimbs of four mammals and a frog, showing the ways in which the proportions of the bones have changed in relation to the particular way of life of the organism.

MAN

FROG

BAT

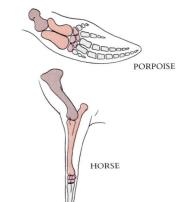

PORPOISE

HORSE

GEL ELECTROPHORESIS

Gel electrophoresis has provided a lot of extremely valuable information about genetic variation in natural populations in a way that would have seemed impossible just a short time ago. In applying this technique, a tissue sample is prepared from each of a series of individual organisms. Each sample is then applied to a uniform, porous gel, often one made with starch, and an electric potential is applied across the gel. Within the gel, the many proteins from the tissue sample migrate along the electric field for different distances, depending on their individual electric charges. By the use of gel electrophoresis, the variability of the loci that code for proteins can be examined.

Gel electrophoresis enables investigators to detect many of the mutations that are responsible for the appearance of allozymes. If the amino acid constitution of one allozyme differs from that of another with respect to an amino acid with a charged side group, the differing amino acid gives the allozymes different net charges, and the two allozymes will migrate at different rates through the electrophoresis gel. The position of an allozyme on a gel is located by flooding the gel with a substrate

Biochemistry

Biochemical tools are now of major importance in our efforts to reach a better understanding of evolution. The sequences of amino acids in samples of the same kind of protein derived from different organisms, for example, can now be determined directly. When the sequences that are characteristic of different individuals, populations, species, and larger groups are compared, their degree of relationship can be specified precisely. In addition, inferences can often be made about such additional factors as the rate of evolution in the different groups.

After a sufficient number of comparisons of this kind had been made, students of evolutionary biology began to use them to make inferences about the amounts of time required for different degrees of biochemical change. Within the last few years, evolutionists also have begun to determine the nucleotide sequences of particular genes. By this method, the base substitutions and rearrangements that are responsible for differences among different kinds of organisms can be studied directly and comparatively. Biochemical advances of the kinds we have just outlined are helping to make possible an improved understanding of both the specifics and generalities of evolutionary change.

Ecology

The synthesis of ecology with evolution is proving to be an exciting field for investigation as the twentieth century draws to a close. In the last sections of this book we will explore some of the principles involved in this synthesis. Darwin had a wide knowledge of organisms as they occur in the field, and he understood a great deal about the ways in which they interact. He pointed out clearly that the ecological role that an organism played would greatly influence the course of its evolution. For example, he concluded that organisms had possibilities for evolutionary change on the Galapagos Islands that were not available to their relatives on the adjacent mainland.

Issues similar to the ones that Darwin thought about 150 years ago are still of great interest to contemporary studies of evolution. Central among these issues is the nature of selection and change in natural populations, the degree to which selection is local, and the ways in which ecological interactions between organisms affect the ways in which they evolve. The development of more precise solutions to these problems will bring further progress in the field of evolution.

with which that enzyme reacts, together with a suitable stain. The accumulation of that particular enzyme at one or more positions in the gel shows up as a band of enzyme activity, as illustrated in the photograph. In such stained gels, different homozygotes for particular allozymes can be discriminated from one another. In addition, heterozygotes often produce additional intermediate bands in the gels, and in such cases can be identified by this method also.

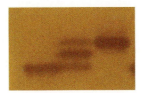

FIGURE 19-A

The results of gel electrophoresis, carried out on three individuals of the common killifish, Fundulus heteroclitus. *This gel was stained to detect activity of the enzyme glucose phosphate isomerase. Proteins have migrated from the bottom of the gel (as shown here) towards the top, with the sample from each fish filling a separate lane. The fish from which the sample in the left lane was obtained is homozygous for one allele of this enzyme, and the fish on the right is homozygous for a second, alternative allele. The fish represented by the middle sample is a heterozygote individual that has a copy of both of these alleles.*

CONTEMPORARY ISSUES IN EVOLUTION

To illustrate what some of these future directions may be, we have selected for discussion three topics among those which most interest contemporary students of evolution: (1) levels of genetic variation in natural populations, (2) evolutionary rates, and (3) the steadiness of the pace of evolution. These are some of the modern equivalents of the problems that occupied Darwin. By discussing them, we hope to provide some of the flavor of this active and exciting field as it exists now.

Levels of Variation and Their Significance

Mutation constantly creates new alleles within populations. When more than one allele of a given gene is present in a population, the population is said to be **polymorphic** for that gene. Over the course of the past 20 years, we have developed techniques that make it a relatively simple matter to determine the degree to which individual genes are polymorphic. If these same techniques are used to measure the polymorphism that is present at a number of different gene loci in the population, we can estimate the proportion of the total genes that are polymorphic in that population. Such results are of great interest to students of evolution, because a population that has a higher degree of genetic polymorphism—genetic variability—can respond more rapidly to selection than one that is less polymorphic. If the alternative genes are not present in the population, they cannot be selected for. If the degree of variability of the population overall is high, then it can respond relatively rapidly to diverse environmental challenges by changing its characteristics.

One very efficient way in which the number of different forms of a gene that are present in a population can be determined is by the study of **allozymes.** Allozymes are forms of particular enzymes that differ from one another by one or a few amino acids; they are alternative alleles at the locus encoding the enzyme. Allozymes have been found to occur frequently in natural populations of nearly all species that have been investigated. The reason that they are of such interest is because of the key role that enzymes play in all biological processes; the significance of variation in enzymes is therefore great.

First used in 1966, the technique of gel electrophoresis has subsequently been widely applied to the study of natural populations. By using it, scientists have discovered a truly remarkable degree of polymorphism in these populations. In one of the earliest studies, for example, Richard Lewontin and John Hubby found that the *Drosophila* populations they were investigating were polymorphic at about a third of their loci. In other words, about a third of the genes they investigated by this method had more than one allele present in the populations they were studying. More remarkably, these polymorphic loci had up to six alleles each. An individual fly in the species that they examined, on the average, was heterozygous—had more than one allele present—at about 12% of its loci. Since there are approximately 10,000 loci in *Drosophila*, an average fly may be heterozygous for about 1200 of these, a tremendous and totally unsuspected degree of genetic variation. The flies are not nearly so variable in their external features, and so the geneticists and evolutionists who had studied them earlier had not realized just how variable they were and, consequently, how readily they were able to change their overall characteristics in relation to selective forces.

But subsequent studies have shown that these early estimates of heterozygosity, surprising as they were, actually presented much too conservative a picture. For example, many allozymes differ by amino acid substitutions that are not revealed by the techniques employed in gel electrophoresis. The actual degree of polymorphism in natural populations is therefore even greater than indicated by the figures given in the preceding paragraph. In fact, refined methods of detection have shown that in *Drosophila* and other insects, more than *half* of the loci are usually polymorphic in each population. On average, plants that regularly outcross are heterozygous at about 18% of their loci, invertebrates at about 13%, and vertebrates at about 6%.

In insects and plants more than half of the loci in a given population are usually polymorphic. An individual insect or outcrossing plant may be heterozygous at about 15% of its loci. The corresponding value for vertebrates is about 6%.

Finding a satisfactory explanation for the thousands of polymorphisms that are present in almost every population has been one of the major tasks of modern evolutionary genetics. So much variation is present in virtually every investigated population that the traditional concept of a "normal" or wild-type individual must certainly be abandoned. The variability of populations enables them to change more readily in the course of their adaptation to different environments than if they were less variable, as we will discuss in Chapter 20.

Evolutionary Rates

On the basis of a relatively complete fossil record, it has been estimated that most living species of mammals go back at least 100,000 years, almost none a million years. A good average value for the life of a species of mammal might be about 200,000 years. Mammal genera, on the other hand, seem to persist several million years on the average. The American paleontologist George Gaylord Simpson has pointed out that certain other groups, such as the lungfishes (Figure 19-11), are apparently evolving much more slowly than the mammals. In fact, Simpson estimated that there has been little evolutionary change among the lungfishes during the past 150 million years. Even slower evolutionary rates are known. For example, a marine animal of the genus *Lingula* (Figure 19-12) appears to have changed little during the past 400 million years—approximately half of the entire history of multicellular life on earth.

Rates of evolution, therefore, differ greatly from group to group, and they are very difficult to compare directly. One question that interests evolutionists is whether groups regularly have more and less active phases in their evolution. New genera and species may be produced at a rapid rate during the active phases and then may remain essentially constant during the less active ones. The fossil record provides evidence for such variability in evolutionary rate, and evolutionists are anxious to understand the differences that govern such fluctuations. The kinds of environments in which the individual organisms live cer-

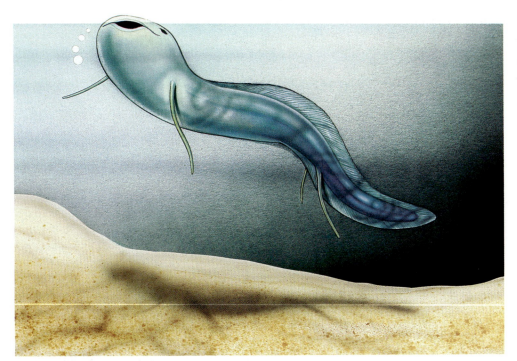

FIGURE 19-11

A lungfish, a member of a group of vertebrates that has changed very little during the past 150 million years. The surviving members of this group occur in South America, Africa, and Australia, continents that were connected with one another early in the history of the lungfishes.

tainly appear to play a role in determining their evolutionary rates. For example, harsh environments, in which survival for many organisms is difficult, seem to be important sites of evolutionary innovation, whereas constant environments are evidently much less so.

Does Evolution Occur in Spurts?

One of the areas that is under the most active contemporary discussion concerns the degree to which the rate of evolution is periodically accelerated and then reduced. In 1972 the paleontologists Niles Eldredge of the American Museum of Natural History in New York and Stephen Jay Gould of Harvard University proposed that it was the norm of evolution to proceed in spurts. They claimed that the process of evolution proceeds by a series of **punctuated equilibria.** Evolutionary innovations would occur and give rise to new lines; then these lines might persist unchanged for a very long time. Eventually there would be a new spurt of evolution, when one or more lines might suddenly disappear, creating a "punctuation," or break, in the fossil record. Eldredge and Gould contrasted their theory of punctuated equilibria with that of **gradualism,** or gradual evolutionary change, which they claimed was what Darwin and most earlier students of evolution had considered normal. Whether they did so or not is debatable.

FIGURE 19-12

Lingula, one of the ancient group of lampshells or brachiopods. This animal lives on the bottom of the oceans, mostly in shallow water, and appears little changed from fossils dating back hundreds of millions of years.

> **The punctuated equilibrium model assumes that evolution occurs in spurts, between which there are long periods in which there is little evolutionary change. The gradualism model assumes that evolution proceeds gradually, with progressive change in a given evolutionary line.**

All models of evolution ascribe an important role to the origin of species, since their origin, which will be discussed in Chapter 21, ultimately provides the basis for the origin of the major groups of organisms. If a particular species has important new adaptive features that give it an unusual advantage in what Darwin called "the struggle for life," it may become the progenitor of such a group. The rules that govern whether a given species will or will not do so, rules to which we will return in Chapter 23, are obviously of special interest.

SCIENTIFIC CREATIONISM

Based on their evaluation of a very large body of evidence, there are virtually no differences of opinion among modern biologists, and indeed nearly all scientists, on the following major points: (1) the earth has had a history of approximately 4.5 billion years; (2) organisms have inhabited it for the greater part of that time; and (3) all living things, including human beings, have arisen from earlier, simpler living things. The antiquity of the earth and the role of the process of evolution in the production of all organisms, living and extinct, have been accepted by those who have examined with an open mind the evidence that has accumulated steadily since the 1860s.

This evidence is bolstered by new discoveries, which are being made constantly. There is no scientific evidence to support the hypothesis that the earth is much younger and none that indicates that every species of organisms was created separately. These conclusions can be reached only on the basis of arbitrary faith; they are untestable, and, as such, they lie outside of the realm of science. Even assuming that they were true, one would also have to assume that the Creator had made the world so that everything in it would be consistent with its appearing to be about 4.5 billion years old and with the hypothesis that all living things had evolved from simpler forms.

In the face of all this evidence it seems surprising that a number of individuals, mainly in the United States and starting largely in the 1970s, have titled a collection of observations in this area **scientific creationism** and put forward the view that the theory of evolution represents some sort of bizarre, "secular humanistic" plot. Scientific creationists believe that the biblical account of the origin of the earth is literally true, that the earth is therefore much younger than all of the available evidence would indicate, and that all species of organisms were individually created looking just like they do today.

The central problem with "scientific creationism" is that it implicitly denies the whole intellectual basis for the set of facts that human beings have assembled over the centuries about the nature of life on earth. It implies that a Creator has made a world in which all phenomena are deceptively consistent with a great age for the earth and an evolutionary relationship between organisms. Evolutionary theory, on the other hand, provides a coherent scientific explanation for the nature of the world in which we live and enables us to make predictions about that world. Certainly there is ample controversy among serious students as to the details of how evolution has occurred, just as there is controversy in every active scientific field. There is no controversy, however, about the facts presented at the start of this section or about Darwin's basic finding that natural selection has played and is continuing to play the central role in the process of evolution.

The future of the human race depends largely on our collective ability to deal with the science of biology and all the phenomena that it comprises. We need the information that we have gained to deal with the problems, challenges, and uncertainties of the world in an appropriate way. We cannot afford to discard the advantages that this knowledge gives us because some of us wish to do so as an act of what we construe as religious faith. Instead, we must use all of the knowledge that we are able to gain for our common benefit. With its help, we can come to understand ourselves and our potentialities better. In no way should such rational behavior be taken as denial of the existence of a Supreme Being; it should rather be considered by those who do have religious faith as a sign that they are using their God-given gifts to reason and to understand.

SUMMARY

1. The relationships among organisms and their obvious progression from simple to complex forms have been well known for centuries. By the end of the eighteenth century, the problem of how this pattern might be explained occupied many of the best minds.

2. Charles Darwin satisfactorily explained evolution by his theory of natural selection. Populations of organisms theoretically tend to increase without limit but in fact remain relatively constant. The individuals that survive are those which, on the average, possess traits that contribute directly to their survival and transmit these traits to their offspring. By this process, the nature of populations gradually changes and new species originate.

3. By the 1860s, naural selection was widely accepted as the correct explanation for the process of evolution. The field of evolution did not progress much further, however, because of the lack of a suitable explanation of how hereditary traits are transmitted.

4. By the techniques of gel electrophoresis, it has been possible to determine that most populations of insects and plants are polymorphic at more than half of their loci, whereas vertebrates are somewhat less so. An individual insect or outcrossing plant may actually be heterozygous at about 15% of its loci; the corresponding value for vertebrates is about 6%. These levels of genetic variations are much higher than any imagined before methods suitable to detect them were introduced in 1966.

5. Evolutionary rates vary greatly among groups of organisms. Species of mammals last, on the average, about 200,000 years; genera of mammals, several million years. In other groups of organisms, evolutionary rates appear to be much slower or much more rapid.

6. The theory of punctuated equilibrium suggests that evolution proceeds in spurts, interrupted by long, quiet periods when there is little change. Gradualism is a school of evolutionary thought that holds that organisms changed slowly, almost imperceptibly, from one form to the next. Both patterns are in accordance with some analyses of the fossil record.

SELF-QUIZ

The following questions pinpoint important information in this chapter. You should be sure to know the answer to each of them.

1. Who first proposed a coherent theory in the field of evolution?
   - (a) Charles Darwin
   - (b) Jean Baptiste de Lamarck
   - (c) Ernst Mayr
   - (d) Alfred Russel Wallace
   - (e) Julian Huxley

2. Charles Darwin proposed that _____ is the mechanism by which evolution takes place.

3. Lamarck's theory of evolution was based on
   - (a) the will of a Divine Creator
   - (b) the arrival in our universe of superior beings
   - (c) the genetic code based on RNA and DNA
   - (d) the inheritance of acquired characteristics

4. Aboard the H.M.S. *Beagle,* Darwin was influenced by
   (a) populations of species varying from place to place
   (b) a profusion of living things on relatively young volcanic islands that resembled species on neighboring coasts
   (c) the fossil evidence he found
   (d) all of the above

5. Thomas Malthus' theory on population sizes stated that
   (a) current populations of animals and plants are decreasing in number
   (b) populations remain more or less constant each year
   (c) populations increase geometrically
   (d) there is no way to predict the size of populations

6. Darwin's theory of evolution stated that each individual
   (a) has the potential to produce more offspring than are able to survive
   (b) has the potential to produce only limited numbers of offspring that will survive
   (c) has the potential to produce only one offspring to replace itself

7. Which of the points below caused Darwin serious concern about his theory of evolution?
   (a) the age of the earth
   (b) the mechanism of heredity
   (c) the way in which favored characteristics were lost
   (d) the appearance of series of fossils with no gaps
   (e) none; Darwin had no doubts about his theory

8. _____ is defined as similarity that occurs because of a common evolutionary origin.

9. Allozymes are
   (a) enzymes involved with increasing the mutation rate of organisms
   (b) different forms of a particular enzyme
   (c) temperature-sensitive enzymes that catalyze the reaction for red blood cell production
   (d) enzymes that have been condensed through evolution

10. A wild-type individual is one that is considered
    (a) hyperactive and uncontrollable   (c) to have a social disease
    (b) to have psychiatric disorders   (d) to be normal

11. A theory of evolution that assumes that the rate of evolution remains constant through time is
    (a) punctuated equilibrium   (c) scientific creationism
    (b) gradualism   (d) Lamarckian

12. The approximate age of the earth is
    (a) 3.6×10^6 years old
    (b) 4.5×10^6 years old
    (c) 4.5×10^9 years old
    (d) 3.6×10^9 years old

THOUGHT QUESTIONS

1. What kind of evidence would you use in evaluating whether the theory of punctuated equilibrium provides a correct explanation of the process of evolution in most groups of organisms?

2. Darwin believed that genetic factors were inherited by a kind of "blending" process and that the characteristics of the organisms themselves somehow flowed into, and were transmitted by, their genes. Why would these theories have made a complete understanding of the process of evolution more difficult than it would have been otherwise?

3. What are the essential differences between natural selection and artificial selection?

FOR FURTHER READING

APPLEMEN, P., (ed.): *Darwin,* 2nd ed., W.W. Norton & Company, New York, 1979. A collection of critical essays, contemporary opinion, and original correspondence. Highly recommended to students interested in Darwin and his times.

DARWIN, C.R.: *On the Origin of Species by Means of Natural Selection, or the Preservation of Favoured Races in the Struggle for Life,* Cambridge University Press, New York, 1975. One of the most important books of all time, Darwin's long essay is still comprehensible and interesting to modern readers, even though it was first published in 1859.

FUTUYMA, D.: *Science on Trial: The Case for Evolution,* Pantheon Books, New York, 1983. An excellent exposition of the basic reasons that the creationist argument is flawed by serious errors.

IRVINE, W.: *Apes, Angels, and Victorians,* McGraw-Hill Book Company, New York, 1954. The story of Darwin, Huxley, and the early years of evolutionary theory.

LA FOLETTE, M.: *Creation Science and the Law: The Arkansas Case,* MIT Press, Cambridge, Mass., 1983. A collection of the legal arguments and other documents and essays associated with a famous court case testing the constitutionality of a law mandating the teaching of "scientific creationism" in public schools.

MOORE, J.A.: "Science as a Way of Knowing—Evolutionary Biology," *American Zoologist,* vol. 23, pages 1-68, 1983. An outstanding exposition of the whole field of evolution.

STEBBINS, G.L., and F. AYALA: "The Evolution of Darwinism," *Scientific American,* July 1985, pages 72-82. A lucid review of how modern molecular approaches are contributing to our understanding of evolution, this article makes clear the dynamic nature of evolutionary theory.

Chapter 20

ADAPTATION

Overview

Organisms survive and diversify as a result of progressive change in relation to their environments. Such change is defined as adaptation, a microevolutionary process. In contrast, the grand outline of the history of life on earth, a process in which the origin of species plays a central role, is called macroevolution. Organisms that are more fit produce higher numbers of surviving offspring than others, and their characteristics consequently become more and more common in their populations. In this way the characteristics of populations of organisms change, and they may do so rapidly under the right conditions. The reasons that different alleles, or combinations of alleles, are maintained in natural populations at such high levels are not obvious.

For Review

Here are some important terms and concepts that you will encounter in this chapter. If you are not familiar with them, you should review them before proceeding.

- **Alleles** (Chapter 13)
- **Heterozygosity and homozygosity** (Chapter 13)
- **The hereditary mechanism** (Chapter 14)
- **Natural selection** (Chapter 19)
- **Allozymes** (Chapter 19)

When the word *evolution* is mentioned, it is difficult not to conjure up images of dinosaurs, of woolly mammoths frozen in blocks of ice, or of Darwin confronting a monkey. Traces of ancient forms of life, now extinct, survive as fossils that help us piece together the evolutionary story. With such a background, we usually think of evolution as meaning changes in the kinds of species on earth, changes that take place over long periods of time, with new forms replacing old ones. This kind of evolution is called **macroevolution.**

Much of the evidence for Darwin's theory, however, was concerned not with the way in which new species are formed from old ones, but rather with the way that changes occur within species. Natural selection, the explanation that Darwin proposed for evolutionary change, is the process whereby some individuals in a population—those with favored characteristics—produce more surviving offspring than others that lack these char-

FIGURE 20-1

Two peppered moths, Biston betularia, *resting on a piece of bark. The dark moth is more easily seen on this light bark than is the light moth. This chapter deals with the ways in which differences such as the colors in these two moths arise, are maintained, and change in frequency.*

acteristics. As a result, the population gradually will include more and more individuals with the favored characteristics, generation after generation, assuming that the variation is heritable. The process by which such genetic change occurs is called **adaptation.** Change of this kind within populations provides an example of a **microevolutionary** process.

The essence of Darwin's explanation of evolution is that progressive adaptation and change by natural selection are responsible for evolutionary change. This process eventually leads to the formation of new species, which is responsible for the appearance of organic diversity. Thus adaptation, a microevolutionary process, leads to species formation, a macroevolutionary process. Genetic change involves both the generation of novelty by mutation and the assembly of particular combinations of mutations by recombination. Similarly, evolution involves the creation of novelty by adaptation and the partitioning of novel changes into groups by species formation. In this chapter we will consider adaptation, and in Chapter 21, the origin of species.

> **Microevolution is a term that describes the process of change in populations, a process that eventually leads to the formation of new species. The appearance of new species, in turn, leads to the major outlines of life on earth, a pattern that is described as macroevolution.**

THE NATURE OF ADAPTATION

In general terms adaptation can be defined as an increase in the frequency of a heritable trait in a population that renders the individuals possessing the trait better able to survive and reproduce (Figure 20-1). Adaptation is the central theme of the process of evolution.

When a pigeon breeder (as Darwin was) imposes selection on his flock, let us say by allowing only those individuals which have longer wings to breed, any alleles that increase wing length are adaptive. In other words, these alleles enable the individuals that carry them to escape the trip to the kitchen and instead to live and to reproduce more pigeons like themselves. In this instance the pigeon breeder imposes the environmental demand—longer wings—whereas in nature the environmental demands are imposed by the requirements of survival. The kind of selection that the pigeon breeder is carrying out is called artificial selection; conversely, the kind of selection that occurs in nature is called natural selection. If an organism lives in a cold climate, those traits which insulate it better are adaptive because they increase the probability of survival. As a result of natural selection, such traits will become more common in the population.

> **Selection is the differential reproduction of genotypes. Artificial selection arises from conscious choice on the part of the breeder. Natural selection occurs in nature and results from the interaction between the inheritable phenotypic characteristics of populations and the environmental conditions in which they occur.**

Adaptation is measured in terms of a successful response to the demands of the environment. Individuals that produce the most reproducing offspring are said to be the most fit. **Fitness** is defined as the tendency to leave behind more offspring that reproduce than competing individuals do. In this sense Darwinian evolution is indeed the "survival of the fittest." In all cases the environment sets the conditions under which survival occurs.

> **Fitness is defined as a genetically determined tendency to leave behind more reproducing offspring than do competing individuals.**

TWO KINDS OF ADAPTATION

To study evolution, then, is to study adaptation. To learn how the process of adaptation occurs, we will begin by considering the genetic variability that occurs in all populations of organisms. We might proceed in one of two ways. We could consider simple characteristics in the hope that they would respond to fewer variables and thus be understood more readily (Figure 20-2). This approach is a good one when the function of the trait being considered is clear, but it is difficult to apply if the function of the trait is not understood. For example, the allele in human beings that leads to albinism clearly leads to a lower degree of fitness than its alternative. On the other hand, it is not possible at present to determine the significance of different esterase enzyme alleles in a population, because we do not know the functions of these allozymes.

The second possible approach involves the consideration of complex characteristics. The adaptive significance is often clear, even if the genetic basis is not. For instance, are the more efficient gills of a fish adaptive? Certainly, because they increase the rate of oxygen capture that is possible for a fish and thus improve its ability to survive and prosper under water. Which genes are responsible for the efficiency of gills? That question is much more difficult to answer. Nevertheless, by studying the physiology of gills, we might be able to obtain clues as to which genetic factors are important. Both approaches, however, are valuable for the study of adaptation.

In the following discussion, we will focus mainly on the approach offered by analyzing simple genetic traits. Moving on then to a discussion of a simple model that has been used widely in connection with the analysis of population genetics, the Hardy-Weinberg equilibrium, we will conclude the chapter with a discussion of several examples of changes in natural populations. In these, more complex combinations of characteristics appear to be involved, and the observed changes conform to principles that are more in line with the analysis of complex characteristics, the second approach just outlined.

GENE FREQUENCIES IN NATURE
Human Blood Groups

Blood transfusions, which offer great medical advantages, were first attempted nearly a century ago, but many of them led to serious complications. The physician Karl Landsteiner (1868-1943) was one of the first to try to understand why they did so. He undertook

FIGURE 20-2

Hurler's syndrome, a genetic disease associated with a recessive gene, is the result of an allele that leads to lowered fitness. Homozygous individuals, such as this young girl, lack the ability to break down mucopolysaccharides, which then accumulate in almost all body tissues, resulting in facial deformities, dwarfism, mental retardation, and heart damage. People with Hurler's syndrome rarely live past their teens.

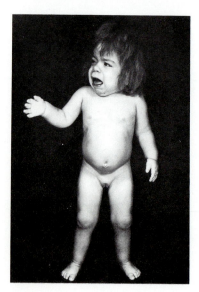

a number of experiments and published his results in 1900. When Landsteiner mixed samples of blood from different people, he found that sometimes the red blood cells in these samples would agglutinate, or clump together, and sometimes they would not (Figure 20-3). Those transfusions between persons whose red cells would agglutinate when the blood samples were mixed were unsafe.

By using the same samples in various combinations, Landsteiner was able to detect the existence of four major blood groups, which he called A, B, AB, and O. People with AB blood can safely receive any of the other types; people with O blood can receive only O blood; people with A blood can receive either A or O blood; and people with B blood can receive either B or O blood.

We now know that highly specialized protein molecules in the blood plasma called **antibodies** recognize and bind to a particular site, called the **antigenic determinant,** on the surface of the foreign red blood cells, which are called **antigens.** In a more general sense an antibody is one of millions of blood proteins that specifically recognizes a foreign substance and initiates its removal from the body. An antigen is any substance that stimulates the production of an antibody. An antigenic determinant is the specific binding site on an antigen to which an antibody binds.

In the ABO system people with type A blood have antibodies to B antigens, so if type A blood plasma comes into contact with red blood cells from type B blood, clumping will occur. Similarly, people with type B blood have antibodies to A antigens; red cells from type A blood will cause clumping if they come into contact with type B blood. People with type AB blood have antibodies to neither A and B antigens, so that they can receive blood safely from people who have either type A or type B blood; such people are called *universal recipients.* Finally, people with type O blood have neither antigen but both antibodies. For these reasons, they are called *universal donors;* they can donate blood to type A, B, AB, or O recipients. The red cells in their blood do not react with the antibodies present in any kind of blood. These relationships are summarized in Figure 20-3.

In making blood transfusions, it is always best to use blood from a donor who has the same blood type as the recipient. If such a donor is not available, however, blood of another type may be used, provided that the patient's plasma and the donor's red blood cells are compatible. The reason for doing so is because unless the transfusion is an unusually massive one, the donor's plasma is usually so diluted during the process that little agglutination of the recipient's red blood cells will occur.

In the decades following Landsteiner's important discovery the nature of the A, B, AB, and O blood types was studied genetically. It was discovered that these blood types were determined by the polymorphism of alleles at a single locus, of the same kind that we discussed in Chapter 19. AB individuals are heterozygous for A and B alleles; the genotype of A individuals is either AA or AO; that of B individuals is either BB or BO; and that of O individuals is OO. These three alleles are found in primates other than humans and thus appear to have evolved in our ancestors. All three alleles occur in nearly all human populations that have been studied, although in different proportions. Here are a few examples of actual allele frequencies:

	A	B	O
Russians (Minsk)	23.8%	17.2%	59.0%
U.S. whites (Boston)	24.6%	7.3%	68.2%
Navahos	16.7%	0.1%	83.2%

Note carefully that these are frequencies of alleles in the population and not frequencies of phenotypes. You can calculate the phenotypic frequencies by using the materials that will be presented later in this chapter.

Why Is the Human ABO System Maintained?

Do the differences that we see in ABO blood group polymorphisms represent adaptive changes, or do they reflect some other process, not important to the fitness of populations in which they occur? Some scientists argue that these polymorphisms have nothing to do

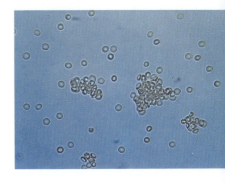

BLOOD TYPES AND AGGLUTINATION

RECIPIENT (SERUM TYPE)

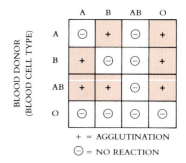

+ = AGGLUTINATION

⊖ = NO REACTION

FIGURE 20-3

Type A red blood cells agglutinating in type O serum. The unclumped red blood cells are type O cells. The diagram shows the combinations in which agglutination will occur.

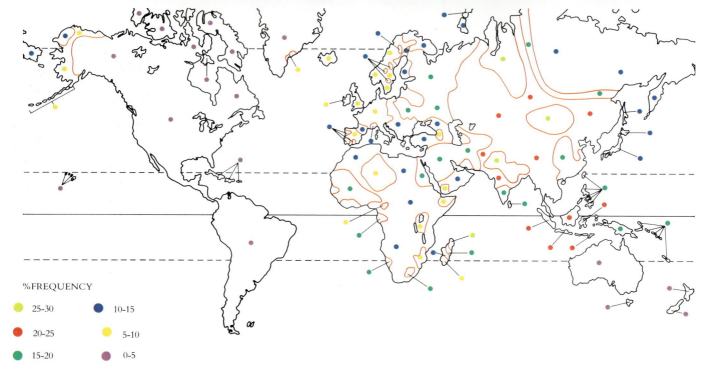

%FREQUENCY

● 25-30   ● 10-15

● 20-25   ● 5-10

● 15-20   ‖ 0-5

FIGURE 20-4

Frequency distribution of the B allele throughout the world. By means of such distributions, students of human populations can often trace past migrations.

with the characteristics of the alleles themselves. Two important lines of evidence, however, indicate that the different alleles may indeed play an adaptive role.

First, there are major trends in the geographical frequency of these alleles. Most American Indians are type O, whereas the B allele was totally absent both in America and in Australia before the advent of Europeans. In Europe the frequency of the B allele increases from the west toward Central Asia (Figure 20-4). Even in relatively small areas significant differences in the frequency of these alleles may occur from place to place. These differences could be explained if the individual alleles have different adaptive values, but they could also have originated by random processes and have been maintained simply because there was nothing causing selection against them. We will consider the second possibility in more detail later.

A second line of evidence, however, suggests more strongly that these alleles do in fact have different adaptive values for the people who possess them. For white males the life expectancy is greatest for those with type O blood and least for those with B type blood; for white females the opposite is true. People with type O blood are more likely than others to contract Asian flu and are more subject to duodenal ulcers (ulcers that occur in the first part of the small intestine leading from the stomach). People with type A blood are more likely than others to develop stomach cancer or pernicious anemia, a serious disease in which the level of red cells in the blood becomes progressively lower. Relationships such as these certainly indicate factors that are important to survival and thus have to do with the maintenance of the polymorphism. Different environmental factors occur in different areas and may have led to the selection of different proportions of blood types in them.

Genetic Polymorphism

The ABO blood group polymorphism is representative of a broad class of genetic variation found not only in human beings but also in all other organisms that have been examined for such variation. Electrophoretic analysis has revealed the existence of more than 30 different blood group loci, in addition to the ABO locus, in humans. At least a third of these loci are routinely found to be polymorphic in human populations. In addition to these, more than 45 additional polymorphic genes are also known that encode proteins in human blood cells and plasma, but which are not considered to define blood groups. Thus there are more than 75 polymorphic genes in this one system alone; for most of them, we do not know the function.

Earlier we indicated that there were two ways of studying adaptation: using simple or complex characteristics. The kinds of data for the distribution of type B blood in human populations (Figure 20-4) are just the kind of information about a simple system that is suitable for the study of adaptation. These data show the distribution of individual alleles rather than that of complexes of characters. Similar changes in the frequency of particular alleles along geographical gradients have been described for many of the other human blood types and for a number of other polymorphic loci. Local populations of humans are often quite distinct from one another in their allele frequencies, and gradients in allele frequencies can sometimes be observed in stable human populations over distances as short as a few kilometers.

The geographical distribution of different alleles often suggests that at least some of them contribute to individual fitness.

Such differences in allele frequencies can be used to study the process of adaptation. To evaluate the differences, however, we first need to establish baseline criteria for the behavior of genes in populations so that we will be able to compare different populations in this respect. The development of the field of population genetics has provided a tool with which to establish such baseline criteria.

POPULATION GENETICS

In the early part of this century the science of genetics contributed relatively little to our understanding of the process of evolution. At first, geneticists were involved primarily with understanding the action of individual genes. Evolutionists in turn could not understand how such observations would have a bearing on the evolution of a complex structure, such as an eye. The extent of the genetic polymorphism that actually exists in nature was not even imagined, either by the geneticists or by the evolutionists.

The gap between the geneticists and the evolutionists finally started to close with the development of the field of **population genetics** from the 1920s onward. Brilliant mathematicians, including R.A. Fisher (1890-1962), J.B.S. Haldane (1892-1964) (Figure 20-5), and Sewall Wright (1889-   ), began to formulate a comprehensive theory of how alleles behave in populations and of the ways in which changes in their frequencies lead to evolutionary change within populations. These scientists saw local populations as the units of adaptation and regarded changes in allele frequencies in these populations as providing the basis for evolutionary change. Successful adaptation was defined, very simply, as an increase in fitness, and increased fitness was evidenced by an increase in the number of surviving offspring that an individual with a particular genotype left in the next generation. Collectively the study of these phenomena makes up the field of population genetics.

This simplified way of viewing the functioning of populations in nature does not assign a value to the important role of the environment in evolution. It does, however, allow the construction of models on which explicit predictions can be based. These predictions, in turn, can be tested and evaluated in real populations. In such a view the overall process of adaptive evolution is a summation of individual changes in fitness. These individual changes in fitness collectively alter the genetic constitution of the population as a whole over time. The most fundamental model in the field of population genetics, the Hardy-Weinberg equilibrium, was developed in the early years of this century; all other aspects of this synthetic field can be viewed in relation to it.

The Hardy-Weinberg Equilibrium

Polymorphism was a puzzle to Darwin and his contemporaries. Indeed, they thought that it provided some of the best evidence against the theory of evolution. They knew of no mechanism by which rare characteristics would be maintained from one generation to the

A

B

FIGURE 20-5

*R.A. Fisher, **A**, and J.B.S. Haldane, **B**, had significant impacts on the field of population genetics through their development of statistical and mathematical models that helped explain the dynamics of population changes.*

FIGURE 20-6

H.H. Hardy, above, and G. Weinberg discovered independently the principle that underlies the frequency of alleles in populations. This principle is now called the Hardy-Weinberg equilibrium.

next. The early geneticists were still concerned with the problem of why a dominant allele would not eventually drive a recessive one, especially a rare recessive one, out of the population. They found this problem all the more puzzling, since a recessive allele was so rarely expressed and, therefore, could not be maintained in the population by selection.

The solution to this problem was developed independently and published almost simultaneously in 1908 by G.H. Hardy, an English mathematician, and G. Weinberg, a German physician (Figure 20-6). They pointed out that in a large population in which there is random mating, and in the absence of forces that change the proportions of the alleles at a given locus (which will be discussed below), the original proportions of the alleles will not be disturbed by Mendelian segregation, and will remain constant from generation to generation. This relationship is now called the **Hardy-Weinberg equilibrium.**

> **In a large population, in which there is random mating, and in the absence of forces that would change their proportions, the proportion of alleles at a given locus will remain constant. This relationship is called the Hardy-Weinberg equilibrium.**

To demonstrate this relationship, consider a single gene with two alleles, A and a. Let the symbol p represent the frequency of the A allele and the symbol q represent the frequency of the a allele. Since these are the only two alleles present at this particular locus, $p + q = 1$. For purposes of discussion, assume that p, the frequency of the A allele, is 0.8 and that q, the frequency of the a allele, is 0.2. Further assume that, as is usually the case, the frequencies of these alleles are the same in males and females and that they mate at random with respect to whether they have the A or the a allele. Male gametes with the A allele will come together with female gametes also carrying the A allele at a frequency p^2, or 0.64; male gametes with the a allele will combine with female gametes with a at the frequency q^2, or 0.04. The chance of A gametes coming together with a gametes, taking into account the possibility that either allele could come from either the male or female parent, is $2pq$, or $(2)(0.8)(0.2) = 0.32$.

Now ask what the proportion of allele A will be in the *next* generation. This can be demonstrated as follows:

Every allele in an AA homozygote is an A allele $\longrightarrow p^2$

½ of the alleles in an Aa heterozygote are A alleles $\longrightarrow ½ \cdot 2pq = pq$

None of the alleles in an aa homozygote is an A $\longrightarrow 0$

By simple algebra, this factors to $p (p + q)$

But $p + q = 1$

Therefore the total contribution of A in the next generation equals p, just as it was in the first generation.

For the population as a whole, therefore, the proportion of AA genotypes will be 0.64; that of aa genotypes, 0.04; and that of Aa genotypes, 0.32. Neither these proportions nor those of the alleles will change, regardless of how many generations of mating may occur, under the ideal baseline conditions outlined for the Hardy-Weinberg equilibrium above: a large population, random mating, and an absence of other factors that would change the proportion of the alleles.

In algebraic terms the Hardy-Weinberg equilibrium is simply expressed by the binomial theorem

$$(p + q)^2 = p^2 + 2pq + q^2$$

If additional alleles are considered, they are simply taken into account by the expansion of the algebraic expression

$$(p + q + r \ . \ . \ .)^2$$

in which r and additional terms represent the frequencies of additional alleles.

REASONS FOR CHANGE IN POPULATIONS

We established above the fact that the proportions of alleles in a population, as well as those of the genotypes and phenotypes associated with them, will remain constant under the conditions we have been discussing. We will now consider the nature of the factors that lead to a departure from this ideal state. The model provided by the Hardy-Weinberg equilibrium is a convenient baseline against which to measure such change. More important, it allows us to understand the nature of the forces that actually bring change about. Against this background, we can better understand the Darwinian process of natural selection.

Five principal groups of factors are responsible for deviations from the Hardy-Weinberg equilibrium: mutation, migration, sampling error (random loss of alleles, which is more likely in small populations), nonrandom mating, and selection. We have already mentioned the importance of a large population and of random mating; now we will examine the details of all five factors.

Five factors can bring about a deviation from the Hardy-Weinberg equilibrium: mutation, migration, sampling error (random loss of alleles), nonrandom mating, and selection.

Mutation

Mutation from one allele to another can obviously change the proportions of particular alleles in the population, unless there is a corresponding rate of mutation in the opposite direction. Mutation rates are generally so low, however, that they could be expected to affect only very slowly the kind of balance reflected by the Hardy-Weinberg equilibrium. Since most environments are constantly changing, populations that are stable enough to reflect in a detectable way the influence of differential mutation rates in their allelic frequencies are very rare.

If organisms have short generations, they may accumulate new mutations more rapidly than if they have longer ones. Mutation is obviously of greater importance in changing the frequencies of rare alleles than of common ones, because the common ones are already so well represented that mutations will not affect their proportions much. For the kinds of polymorphisms that we have been discussing in this chapter and in Chapter 19, many alleles are very rare. It is difficult to study the effects that such alleles may have in contributing to the fitness of the population as a whole because they are so rare.

Migration

Migration is defined in genetic terms as the movement of individuals from one population into another. It can be a very powerful force in upsetting the genetic stability of natural populations. Migration can be obvious, as in the movement of animals from one place to another. If the characteristics of the newly arrived individuals differ from those of the existing ones, the genetic composition of the receiving population will be altered.

Many kinds of migration that are of great evolutionary importance are not obvious, however. These subtler movements include, for example, the drifting of gametes or immature stages of marine animals or plants from one place to another (Figure 20-7). The male gametes of flowering plants are often carried great distances by insects and other animals that visit their flowers. Their seeds may also be blown in the wind or carried by animals or other agents to new populations far from their place of origin.

The movement of plants or animals from one area to another may be common or uncommon, depending on the distances involved and the nature of the organism. Population geneticists have defined the movement of alleles that accompanies such migration as **gene flow.** They visualize the alleles as "flowing" from one place to another. If they do so, the variability of the receiving population may be increased. If there is a great deal of migration between a series of populations, they may all remain (or become) similar to one

FIGURE 20-7

The cloud around these individuals of Monterey pine, Pinus radiata, *in California, is pollen, dispersed by the wind. The male gametes within the pollen reach the egg cells of the pine passively in this way. In genetic terms such dispersal is a form of migration.*

another. Thus migration may counteract the effects of selection, which will be discussed on p. 394, in a way set by the characteristics of the organisms involved.

Sampling Error

In small populations individual rare alleles may be represented in very few individuals. Under such circumstances these alleles may be lost from the population by chance alone. The same kind of chance is involved that occurs in rolling dice or in flipping a coin. Viewing the generations as a series of samples taken from the total assemblage of alleles present in the population, the changes that occur from generation to generation are also considered to result from **sampling error,** which is a statistical term that reflects the fact that successive samples taken from a mixed population will differ from one another by chance alone.

Even in large populations, rare alleles have a good chance of being lost. The individuals that have these rare alleles may simply die without any progeny. If a particular allele makes an extraordinarily important contribution to the fitness of an individual in which it occurs, of course, it may persist; if such an allele is very rare, however, it is still in greater danger of being lost from the population.

In small populations even the frequencies of common alleles may be changed, sometimes drastically, by chance alone. So few individuals may be involved that many of the alleles originally present in the population may be lost from it. Since this process appears to occur randomly, as if it were drifting, it is known as **genetic drift.** A series of small populations that are isolated from one another may come to differ strongly as a result of genetic drift.

> **Genetic drift is the random loss of alleles from loci as a result of sampling error. It is of special significance in small populations.**

If a population decreases drastically in size, for whatever reason, it becomes subject to drift. Such a reduction in size also tends to lead to close inbreeding among the relatively small number of remaining individuals. Under such circumstances those alleles which are still

present will tend to express themselves phenotypically, even if they are recessive (see Figure 20-2). Plants and animals that occur in climatic zones, or particular habitats, in which the size of their populations is frequently decreased drastically, have a tendency to evolve into distinctive local populations. Such populations are especially prone to extinction because of their limited variability; a particular example of this phenomenon was provided by our discussion of cheetahs in Chapter 1. The possible role of this process in the production of new species will be discussed in Chapter 21. The limited variability that is characteristic of small populations also has a role in promoting species extinction when it is coupled with the loss of habitat and dwindling population sizes that characterize more and more plant and animal species.

Sometimes one or a few individuals are dispersed and become the founders of a new, isolated population at some distance from their place of origin. When this occurs, the alleles that they carry are of special significance. Not only may they be ones that are rare in the source population, but also they will participate in their new area in combinations unusual for the species as a whole. This effect is called the **founder principle.** It is particularly important for the evolution of organisms on distant oceanic islands, such as the Hawaiian Islands. Most of the kinds of organisms that occur in such areas have probably been derived from one or a few initial "founders." Because of the ways in which combinations of alleles interact with one another, the populations that these "founders" establish may soon be very different from the ones from which they came (Figure 20-8). In a similar way isolated human populations are often dominated by the genetic features that were characteristic of their founders, if only a few individuals were involved initially.

Nonrandom Mating

Individuals with certain genotypes sometimes mate with one another more commonly than would be expected on a random basis. If they do so, the representation of genotypes in the population may deviate from the expected values. Criteria involved in the selection of mates and the availability of mates with certain characteristics may also play a role in nonrandom mating.

FIGURE 20-8

The tarweed Adenothamnus validus (below) *is a primitive member of a group of plants of the sunflower family,* Asteraceae, *which includes more than 100 species found in western North America. In the Hawaiian Islands a group of unusual species, including the silversword,* Argyroxiphium sandwicense *subspecies* macrocephalum (above), *have evolved from ancestors that reached Hawaii by chance dispersal over long distances. The particular genetic constitution of the founding individuals played key roles in the evolution of Hawaiian plants and animals.*

Self-fertilization, which is a type of nonrandom mating that is characteristic of many groups of organisms, will also cause the frequencies of particular genotypes to differ greatly from expectations. In such situations the frequencies of the alleles themselves may not change. Thus, if the population consists primarily of homozygous individuals that fertilize themselves, heterozygotes may be present at frequencies much lower than would be expected if the organism were **outcrossing**—mainly or exclusively interbreeding with individuals different from themselves. The repeated self-fertilization of heterozygotes also greatly decreases the proportion of such heterozygotes in the population.

Selection

The major factor that causes deviations from Hardy-Weinberg equilibrium is selection, and natural selection is, of course, the cornerstone of Darwin's theory of evolution. Selection is the nonrandom reproduction of genotypes. As Darwin pointed out, some individuals leave behind more progeny than others, and the rate at which they do so is affected by their inherited characteristics. We describe the results of this process as selection and speak both of artificial selection and of natural selection. In the former, the breeder selects for the desired characteristics; in the latter, conditions in nature differentiate the fitness of the individuals involved.

Selection is a complex process because of the interactions of alleles at different loci. Because of these interactions, selection for a single set of characteristics is apt to reach limits, associated with the factors that we are about to discuss, before many generations have gone by. The characteristics that are most desirable (in artificial selection) or most important to the production of large numbers of viable progeny (in natural selection) are likely to be complex themselves. They often involve many genes, particular alleles of which may have contradictory effects on the chance of success of the individual in which they occur. In addition, these genes are linked on the chromosomes with other genes, and, again, this situation may give rise to combinations of alleles that may have drastically different effects on the organism from those which are being selected. Finally, the integration of alleles at all of these loci—an individual *Drosophila* has about 10,000 genes—to produce a single individual adds a further level of biochemical and developmental complexity in the determination of the phenotype.

In other words, strong selection is apt to result in rapid change initially, but the change is apt to come to a halt soon because of the kinds of interactions just outlined. Moreover, the features of an organism for which selection is taking place must combine with all of the other features to produce an organism that can still function in its environment. For this reason, we do not have gigantic cattle that yield twice as much meat as our leading strains, chickens that lay twice as many eggs as the best layers do now, or corn with an ear at the base of every leaf. Interactions within the whole genotype limit the extent to which the alleles associated with the desired characteristics can express themselves.

Only those characteristics which are expressed in the phenotype of an organism can affect the ability of that organism to produce progeny. For this reason, selection does not operate efficiently on rare recessive alleles, simply because they do not often come together as homozygotes—there is no way of selecting them unless they do so. For example, at an allelic frequency q for a recessive allele a of 0.1, 10% of the alleles for that particular gene will be a, but only one out of a hundred individuals will display the phenotype associated with this allele; if $q = 0.01$, the frequency of homozygotes will be only 1 in 10,000.

> **Selection, the differential reproduction of genotypes, is the major reason for deviation from Hardy-Weinberg equilibrium and the principal process that leads to evolutionary change.**

What this means is that selection against undesirable genetic traits in humans or domesticated animals is very difficult unless the heterozygotes can also be detected. For example, if, for a particular undesirable recessive allele *r*, $q = 0.01$, and none of the homozygotes for this allele were allowed to breed, it would take 1000 generations, or about 25,000 years in humans, to lower the allelic frequency by half to 0.005. At this point, after 25,000 years of work, the frequency of homozygotes would still be one in 40,000, or 25% of what it was initially. This is the basic reason that few geneticists advocate **eugenics,** efforts to change the genetic characteristics of the human race by selection. Aside from the moral implications, such efforts are essentially doomed to failure by the sheer difficulty of producing the desired results.

The differential reproduction of genotypes, or selection, is the primary force that shapes the pattern of life on earth. Highly polymorphic individuals in populations are acted on simultaneously by different kinds of selective forces, many of which affect the numbers and kinds of progeny that the individual will leave. These forces change from place to place and also from time to time, since the environment is anything but static. Because of the central role that selection plays, many scientists are attempting to understand the process better. In our effort to do so, let us return to the question that was posed in Chapter 19, concerning the maintenance of high levels of polymorphism in natural populations.

MAINTAINING POLYMORPHISMS

In the broadest sense, sexual reproduction and meiosis are devices for maintaining and recombining diversity. In haploid organisms, including bacteria, every allele expresses itself in the phenotype and must immediately be favored or eliminated. Among the organisms that have sexual recombination, including most eukaryotes, a more complex response to variable and changing environments is possible. Sexual recombination is reinforced in both plants and animals through the reinforcement of outcrossing because of its importance in the adaptation of organisms to their environment (Figure 20-9).

FIGURE 20-9

This wild bee (Ptilothrix), *loading the pollen baskets on its hind legs with pollen from the desert "poppy"* (Kalstroemia) *in eastern Arizona, illustrates the way that flowering plants have coopted animals to the task of dispersing their pollen precisely from plant to plant. The pollen that the bee has gathered will be used to provision the cell in which its larva will reach maturity; other pollen grains, adhering to the hairy body of the bee, will have a good chance of reaching the female parts of other flowers of this same species. Outcrossing is as important for plants as it is for all eukaryotic organisms, a major factor in making possible their adaptation to the environment.*

Genetic polymorphism can be maintained in populations because the individual alleles make different contributions to the fitness of the organisms in which they occur; because individual alleles are maintained in different frequencies in different parts of a subdivided population; or simply because they are in Hardy-Weinberg equilibrium. In the last instance the variability would be **neutral.** In general it is difficult to tell whether individual allozymes, such as those mentioned in Chapter 19, are being maintained because they have a selective advantage or not. Distinctive patterns of distribution of different alleles have often been demonstrated, however, both in space and in time. Such relationships suggest that many of the allozymes may have selective advantages under these circumstances.

> **Allozymes may be maintained in populations because they make different contributions to individual fitness, or simply because of the structure and dynamics of the population.**

CHANGES IN NATURAL POPULATIONS

As you have seen thus far, our attempt to study adaptation by examining simple systems has met with only limited success. We will now consider adaptation from a different perspective, looking at complex situations, at nature "in the rough." Of particular interest are situations in which allele frequencies are actively changing at the time they are being studied. At such times we should be able to evaluate more readily the adaptive significance of the individual alleles present in the populations.

We have already considered, at some length, one example of polymorphism in a natural system: that concerned with the ABO blood types of humans. In this section we will consider some additional examples of polymorphisms, analyze them to the extent possible in relation to their maintenance and role in natural change, and then attempt to synthesize what is now known about the course of evolution in nature.

Sickle Cell Anemia

Sickle cell anemia is a heritable disease in which afflicted individuals have defective hemoglobin molecules. When oxygen is scarce, these defective molecules become insoluble and combine with one other, forming stiff, rodlike structures. Surprisingly, there is only one difference between the hemoglobin that occurs in normal and defective red blood cells: in defective hemoglobin, as compared with normal hemoglobin, out of a total of about 300 amino acid molecules, 1 molecule of glutamic acid is replaced by 1 of valine. Red blood cells that contain large proportions of such molecules become sickle shaped and stiff; normal red blood cells are disk shaped and much more flexible (Figure 20-10). As a result of their stiffness and irregular shape, the sickle-shaped red blood cells are able to move through the smallest blood vessels only with great difficulty. They also tend to accumulate in the blood vessels, forming clots. As a result, people who have large proportions of sickle-shaped red blood cells tend to have intermittent illness and a shortened life span.

Individuals homozygous for the sickle cell allele show the characteristics just mentioned; those who are heterozygous for the allele are generally indistinguishable from normal persons. In the blood of people who are heterozygous for this trait, however, some of the red cells show the sickling characteristic when they are exposed to low levels of oxygen. The sickle cell allele is particularly common among blacks. In the United States, for example, about 9% of blacks are heterozygous for it, and about 0.2% are homozygous, showing the symptoms of the disease. In some African groups up to 45% of the people are heterozygous for this allele.

People who are homozygous for the sickle cell allele almost never reproduce, because they usually die too young to do so. This fact raises the question as to why the sickle cell allele is not eliminated from all populations, rather than being maintained at high levels in some. The answer to this question has proved much clearer than has that for the distribution of the ABO blood groups. People who are heterozygous for the sickle cell allele are

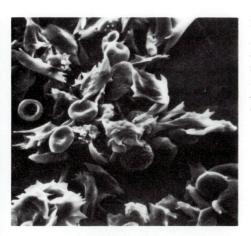

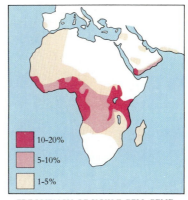

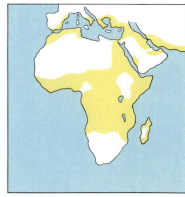

FREQUENCY OF SICKLE CELL GENE   DISTRIBUTION OF *FALCIPARUM* MALARIA

FIGURE 20-10

Scanning electron micrograph of normal and sickle-shaped red blood cells. The irregularly shaped cells are the sickle cells; the circular cells are the normal blood cells.

10-20%

5-10%

1-5%

FIGURE 20-11

Frequency of sickle cell allele, left, and distribution of falciparum malaria, right. Falciparum malaria is one of the most devastating kinds of the disease; its distribution in Africa is closely correlated with that of the allele for the sickle cell character.

much less susceptible to malaria, which is one of the leading causes of illness and death, especially among young children, in the areas where the allele is common. In addition, for reasons that are not understood, women who are heterozygous for this allele are more fertile than are those who lack it. Consequently, even though most people who are homozygous recessive die before they have children, the sickle cell allele is maintained at high levels in these populations because of its role in resistance to malaria in heterozygotes and its association with increased fertility in female heterozygotes. For the people living in areas where malaria is frequent, therefore, maintaining a certain level of the sickle cell allele in the populations has adaptive value (Figure 20-11).

Peppered Moths and Industrial Melanism

The peppered moth, *Biston betularia,* is a European moth that rests on the trunks of trees during the day (see Figure 20-1). Until the mid–nineteenth century, almost every individual of this species that was captured had light-colored wings. Since then, individuals with dark-colored wings increased in frequency in the moth populations near industrialized centers until they comprised almost 100% of these populations. The black individuals have a dominant allele that was present in populations before 1850, although very rare then. In industrialized regions the tree trunks were very dark, and the dark moths were much less conspicuous resting on them than were the light moths. Why was it an advantage for the moths to be less conspicuous?

Although initially there was no evidence, the ecologist H.B.D. Kettlewell hypothesized that the moths were eaten by birds while they were resting on the trunks of the trees during the day. He tested the hypothesis by rearing populations of peppered moths in which dark and light individuals were evenly mixed. Kettlewell then released these populations into two sets of woods: one, near Birmingham, was heavily polluted; the other, in Dorset, was unpolluted. Kettlewell set up rings of traps around the woods to see how many of both kinds of moths survived. In order to be able to evaluate his results, he had marked the moths that he had released with a dot of paint on the underside of their wings, where it could not be seen by the birds.

In the polluted area near Birmingham, Kettlewell trapped 19% of the light moths, but 40% of the dark ones. This indicated that the dark moths had had a far better chance of surviving in these polluted woods, where the tree trunks were dark. In the relatively unpolluted Dorset woods, Kettlewell recovered 12.5% of the light moths but only 6% of the dark ones. These results indicated that where the trunks of the trees were still light colored, the light moths had a much better chance of survival than the dark ones. Kettlewell later solidified his argument by placing hidden blinds in the woods and actually filming birds eating the moths. The birds that Kettlewell observed sometimes actually passed right over or next to a moth that was of the "correct" color and thus well concealed.

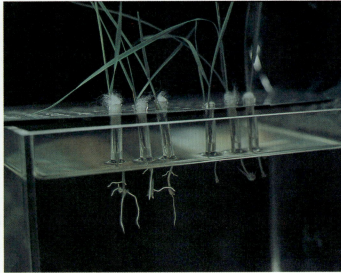

A

B

FIGURE 20-12

A *Mine at Drws-y-Coed, Wales, showing soil contaminated by lead-rich tailings.*

B *Bent grass,* Agrostis tenuis, *tolerant and nontolerant strains growing in 0.5 millimolar copper, one of the metals that contaminates the mine tailings.*

Industrial melanism is a phrase used to describe the evolutionary process whereby initially light-colored organisms become dark, as a result of natural selection. The process, which is common among moths that rest on tree trunks, takes place because the dark organisms are better concealed from their predators in habitats that have been darkened by soot and other forms of industrial pollution.

Dozens of other species of moths have changed in the same way as the peppered moth in England, with dark forms becoming more common from the mid–nineteenth century onward as industrialization spread. In the second half of the twentieth century, with the widespread implementation of pollution controls, the trends are being reversed, not only for the peppered moth in many areas in England but also for many other species of moths throughout Eurasia and North America. Such examples provide some of the best-documented instances of changes in allelic frequencies of natural populations in response to selective forces. The increase in drug-resistant bacteria that occurred following the widespread introduction of antibiotics after World War II has resulted from the responses of the bacteria to the antibiotics in a way similar to that in which the moths responded to the industrial pollution.

Welsh Grasses

A.D. Bradshaw and other investigators have studied the grasses that grow on the tailings, or refuse, around lead mines in Wales (Figure 20-12). These tailings, areas in which the soil is rich in lead, are almost bare of plants, but a few kinds, including bent grass, *Agrostis tenuis,* are found there. Bradshaw compared the growth patterns of plants of *Agrostis* taken from nearby pastures and areas where lead is not abundant in the soil with those of plants from the mine tailings. He grew plants taken from the two different kinds of soils side by side in samples of both soil types.

In normal pasture soil the plants from the mine soil were smaller and grew more slowly than the ordinary pasture plants, but they did survive. In the lead-rich mine soil the plants that Bradshaw had collected there grew well. In complete contrast the pasture plants were, with a few exceptions, unable to grow in the mine soil, most of them dying within a few months. The exceptions, however, were significant: in one sample of 60 plants from a pasture, 3 showed some ability to grow in the lead-rich soil. Such plants were undoubtedly the kind from which those which could grow in the mine soil had been selected originally. In this way a race of bent grass that was able to grow well in lead-rich mine soil evolved.

Since races of plants able to grow on lead-rich soil are found in association with mines less than a century old, they clearly are able to evolve quickly under the appropriate circumstances. In general, populations or organisms may change their characteristics rapidly when there is strong selection for such change.

THE SHIFTING-BALANCE THEORY OF EVOLUTION

In 1932 the American geneticist and evolutionist Sewall Wright (Figure 20-13) proposed an integrated theory of adaptation. This theory combined for the first time both the random effects of genetic drift and the environmentally directed effects of selection into an overall synthesis. It provided a comprehensive view of how evolution proceeds.

Wright's view of the process of evolution provides an appropriate summary for this chapter. He brought to bear on the problem of evolution the conceptual advances associated with genetics. Starting with the essential elements of the Darwin-Wallace view of evolution—that heritable phenotypic polymorphism is widespread in natural populations and that this polymorphism affects individual fitness—Wright added two further observations about natural populations:

1. Many genes interact in specifying most physiologically important phenotypes. As a result of these interactions, the effects of polymorphisms on fitness extend to many genes, not just one or a few. The appropriate balance of alleles at many different loci determines individual fitness.

2. In nature most populations are subdivided into relatively small units. As a result, most populations experience continual genetic drift. Different combinations of alleles will eventually come to predominate in different subpopulations as a result of random events.

In Wright's view the essential elements that make adaptive shifts in populations possible are coordinated changes in combinations of alleles that contribute significantly to fitness. The extensive polymorphism that is now known to occur in natural populations, which goes far beyond anything that Wright imagined, tends to make the array of possible combinations of alleles almost unbelievably complex.

Wright defined populations as those groups of organisms in nature between which interbreeding took place regularly. He was one of the first scientists to realize that the actual size of populations in nature, as defined in this way, was very small. Wright pointed out that small units of this kind, which he called *effective breeding populations,* were the critical units of evolutionary change. Even when we see a population of trees, birds, or some other obvious organism, which looks large and continuous, the actual units between which interbreeding regularly takes place are often very local (see Figure 21-5). Each of these units is to some extent independent of all others, and it evolves independently by adjusting to its local environment.

Each small, local population may experience a change in its own particular representation of alleles to another combination. The new combination may produce a higher proportion of phenotypes that are highly fit under the particular local conditions in which it occurs. Such changes in effective breeding populations may occur as a result of sampling error (genetic drift) or selection. Large populations and species are in fact mosaics of small populations that are separated more or less sharply from all other elements of that large population or species. Ultimately the changes in these small populations from one adaptive combination of alleles to another determine the direction of evolution as a whole.

Adding to the classical nineteenth-century view of evolution by natural selection the two new elements of (1) phenotypes determined by the balance of alleles at many loci and (2) small effective population size, Wright developed a theory called the **shifting-balance theory of evolution.** He proposed that the balance of alleles determining phenotypes was constantly shifting in natural populations. Wright attributed these shifts largely to genetic drift, operating generation after generation in small, local populations. Once a certain par-

FIGURE 20-13

Sewall Wright, an important figure in the field of population genetics, made a major contribution to our understanding of adaptation with his shifting-balance theory of evolution, first proposed in 1932.

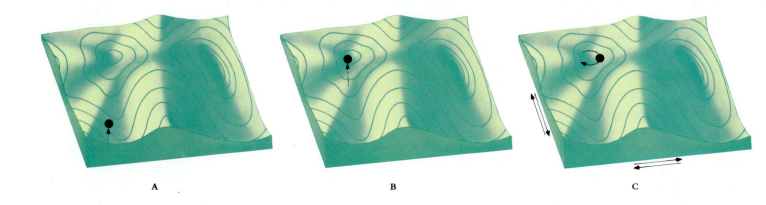

A   B   C

FIGURE 20-14

Sewall Wright's adaptive landscape, inverted.

ticularly suitable combination of alleles had arisen by genetic drift, selection would act to increase the frequency of that particular combination.

In his shifting-balance theory of evolution Sewall Wright proposed that the balance of alleles that determined particular phenotypes was constantly shifting within natural populations. Within small effective breeding populations, particular combinations of alleles that improve the fitness of their members arise constantly and are reinforced by natural selection.

In Wright's shifting-balance theory random effects such as genetic drift are one of the principal driving forces in evolution. Genetic drift, by generating random combinations of alleles, randomly tests all kinds of allele combinations. When certain favored combinations occur, they are so highly favored that they quickly predominate.

Wright visualized the shifting balance of alleles within populations as a kind of a topographical map, a surface in which the vertical dimension represented adaptive success. Favorable allele combinations projected up from the surface of this evolutionary landscape as **adaptive peaks.**

Another, more convenient, way to visualize the shifting-balance theory is to imagine a marble rolling around on a gently vibrating, dented tabletop. Think of the table's surface as representing an adaptive landscape, with every position on the tabletop corresponding to a different combination of alleles (Figure 20-14). The vibration reflects the random perturbations of genetic drift and very local selection within these populations. The marble represents the allele frequencies within a population. It moves randomly from place to place on the surface of the vibrating table. Occasionally it encounters a depression and rolls into it. This movement corresponds to the action of selection reinforcing the frequency of a favorable combination of alleles once that combination has appeared. The deeper the depression, figuratively, the more advantageous the particular combination of alleles. If the combination were particularly favorable, it would be difficult for the marble to be jostled up out of the depression by the vibrations of drift, since in effect the depression would be so "deep."

In a sense this analogy also illustrates why evolution has direction. The marble will tend to remain longer in progressively deeper holes, since they are more difficult to get out of than are the shallower ones. Those deeper holes represent the most favorable combinations of alleles. Adaptation is actually much more complicated than this simple analogy suggests, but it does provide a simplified means of visualizing the process of adaptive change in natural populations. In general, Wright's conception of a large series of small populations in which the allele frequencies are constantly shifting is an accurate description of the way in which adaptation seems to occur in nature.

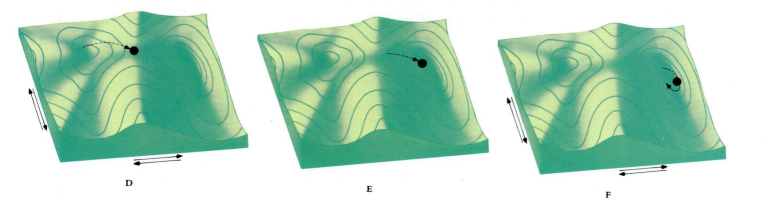

D   E   F

SUMMARY

1. Macroevolution describes the outlines of evolution, and is based on the evolution of species. Microevolution, also called adaptation, refers to the evolutionary process itself. Adaptation leads to species formation and thus ultimately to macroevolution.

2. Fitness is the tendency of some organisms to leave more offspring than do competing ones. The genetic traits possessed by the fit individuals will be progressively better represented in succeeding generations. This process is described as selection.

3. Some genetic characteristics have a relatively simple basis. The one that we considered in detail was that of ABO blood groups in human beings. These blood groups can be defined by their antigen-antibody reactions, which, in some combinations, lead to blood clotting. The alleles at the locus responsible for these blood groups differ greatly in frequency among different groups of people, but the reasons for these differences are poorly understood, although they seem to contribute somewhat to individual fitness.

4. Population genetics is concerned with the behavior of genes in populations. Understanding population genetics is essential to understanding evolution.

5. The Hardy-Weinberg equilibrium provides the baseline for all population genetic theory. This relationship states that in a large population in which there is random mating, and in the absence of forces that would change their proportions, the proportion of alleles at a given locus will remain constant. Five factors can bring about a deviation from the Hardy-Weinberg equilibrium: mutation, migration, sampling error (random loss of alleles), nonrandom mating, and selection.

6. Allozymes, forms of a given enzyme that differ slightly, are abundant in natural populations and provide clear examples of simple genetic differences. Population geneticists are very interested in learning whether they are maintained by selection or are adaptively neutral and simply maintained by the operation of the genetic systems in the populations where they occur.

7. Other characteristics that occur in organisms have a more complex genetic basis, although their adaptive significance may be clear. For example, the disease called sickle cell anemia occurs when an altered hemoglobin molecule, associated with a particular allele, is well represented in the red blood cells. If this allele is present in homozygous form, it is almost invariably lethal; if it is present in heterozygous form, it not only does not produce anemia, but also it confers resistance to malaria and increased female fertility on heterozygous persons. For these reasons the allele has reached high frequencies in certain African populations in areas where malaria occurs frequently.

8. The peppered moth, *Biston betularia,* and the grass *Agrostis tenuis* provide examples of rapid changes in natural populations. Changes in both kinds of organisms have resulted from adaptation to conditions associated with human activities.

9. The shifting-balance theory of evolution was developed by the American geneticist Sewall Wright. In it, Wright proposed that the balance of alleles that determines particular phenotypes is constantly shifting in natural populations as a result of genetic drift. The new combinations often are enhanced subsequently by selection. Collectively these changes result in evolution.

SELF-QUIZ

The following questions pinpoint information in this chapter. You should be sure to know the answers to each of them.

1. The major outlines of life on earth arise as a result of the process known as _____.

2. Darwin's explanation of the way in which evolution occurs is that
   (a) God determines which species should evolve
   (b) progressive adaptions enable one species to leave more offspring
   (c) certain species have "built-in" plans of evolution
   (d) those traits used most often persist longer

3. _____ is defined as the propensity to leave more offspring behind than do competing individuals.

4. _____ sets the conditions under which survival occurs.
   (a) the environment   (c) the strength of the individual
   (b) sexual attractiveness   (d) random chance

5. Successful adaptation is defined simply as
   (a) an increase in fitness   (c) producing offspring
   (b) moving to a new location   (d) evolving new traits

6. Hardy and Weinberg stated in their equilibrium model that in a large population in which there is random mating, and in the absence of forces that would change their proportions, the proportion of alleles at a given locus will _____.
   (a) increase   (c) remain the same
   (b) decrease

7. Mutation rates are generally _____ in most organisms.
   (a) high   (c) low
   (b) medium

8. _____ is the random loss of alleles from loci as a result of sampling error.

9. _____ populations are more subject to sampling error.
   (a) very large   (c) small
   (b) large

10. The repeated self-fertilization of heterozygotes will _____ the proportion of heterozygotes in the population.
    (a) maintain unchanged   (c) increase
    (b) decrease

11. The major factor causing deviations from Hardy-Weinberg equilibrium is
    (a) random mating   (d) nonrandom mating
    (b) mutation   (e) selection
    (c) migration

12. Only selection can maintain a high degree of heterozygosity in natural populations. True or false?

THOUGHT QUESTIONS

1. Why will repeated self-fertilization lead to a decrease in the number of heterozygotes in a population?

2. If individuals heterozygous for an allele were more fit than either of the homozygous classes, how would the representation of the two alleles in the population be affected? Assume that the allele exists in only two states, *A* and *a*.

3. Will a recessive allele that is lethal in the homozygous condition ever be removed from a large population as a result of natural selection?

FOR FURTHER READING

BENDALL, D. (ed.): *Evolution from Molecules to Men,* Cambridge University Press, New York, 1983. A broad overview of contemporary evolutionary biology.

HARTL, D.: *A Primer of Population Biology,* Sinauer Associates, Inc., Sunderland, Mass., 1981. The best short introduction to what can sometimes seem a bewilderingly mathematical field.

STANLEY, S.M.: *Macroevolution: Pattern and Process,* W.H. Freeman and Company, San Francisco, 1979. A thorough discussion of adaptation, species formation, and all aspects of evolution, which expands on many of the concepts presented in this chapter.

Chapter 21

THE ORIGIN OF SPECIES

Overview

As a result of adaptive change, populations that are originally identical eventually become more and more dissimilar. No two populations exist under exactly the same circumstances, and none has exactly the same genetic material combined in the same way. When selective forces are strong, for example, when conditions are changing rapidly or organisms are moving into a new habitat, the characteristics of the population either change rapidly or the population becomes extinct. When conditions are stable, these characteristics may remain constant for long periods even if the populations are widely separated. As a result of the process of adaptive change, some populations eventually become different enough that they are recognized as species. On islands, or in places where there are many available habitats and little competition from organisms that are already established, large numbers of distinct species may originate rapidly. For this reason, Darwin learned a lot about the process of evolution during his stay on the Galapagos Islands.

For Review   *Here are some important terms and concepts that you will encounter in this chapter. If you are not familiar with them, you should review them before proceeding.*

- **Meiosis** (Chapter 14)

- **Effective breeding populations** (Chapter 20)

- **Adaptive peaks** (Chapter 20)

- **Fitness** (Chapter 20)

- **Change in populations** (Chapter 20)

ADAPTIVE CHANGE

The kinds of changes that have occurred over the last few hundred years in populations of peppered moths and bent grass (Chapter 20) occur continuously in nature. Not surprisingly, many of the changes in natural populations that have been documented best have occurred in response to alterations in the environment caused by people. Human activities have dominated habitats all over the world for centuries, and populations of moths, grasses, bacteria, and other organisms have had to react to them or have become extinct.

FIGURE 21-1

Thrush "anvil," with shells of Cepaea. Thrushes hunt and eat snails of this genus. By studying the accumulations of broken shells found around their anvils—the stones on which they break the snails open—scientists were able to determine that the thrushes found higher proportions of lighter-colored snails in dark habitats and higher proportions of darker-colored snails in light habitats. The birds thus exert a strong selective pressure on the snails.

Changes similar to those which have occurred in peppered moths and bent grass also occur in populations of organisms undisturbed by human activities, however. These changes allow the organisms in which they occur to live in diverse habitats, by changing their characteristics so that they can live successfully in the new areas (Figure 21-1). All such changes come about as a result of changes in allele frequencies.

Sickle cell anemia in human populations provides an illustration of the way in which such a process of adaptation works. As people migrated for the first time into regions where malaria was frequent, the frequency of the sickle cell allele would have increased. In populations of people who lived in other regions the allele would have remained rare (see Figure 20-11).

Many natural changes are occurring that have effects on organisms just as profound, or more profound, than those which result from the impact of human activities. As successive periods of expansion and contraction of continental glaciers have occurred over the past few million years, for example, climates have gotten warmer and cooler, wetter and drier. In response to these changes whole communities of organisms have had to adjust in their characteristics, primarily by selective shifts in the frequencies of those alleles which are related to survival under the new circumstances. Species of plants and animals that could not adjust have become extinct, at least locally, while others have taken their places.

THE NATURE OF SPECIES

How do such adaptive changes in natural populations lead to the origin of species? Darwin was extremely interested in this question because he considered species to be the most important evolutionary units. As we begin to think about the problem, we must first examine what the concept **species** means and how this concept has changed through the years.

John Ray, the English clergyman and scientist discussed in Chapter 20, was one of the first to propose a definition of the category species. In about 1700 he pointed out how a species could be recognized: all of the individuals that belonged to it could breed with one another and produce progeny that were still of that species. Even if two different-looking individuals appeared among the progeny of a single mating, they were still considered to belong to the same species. All dogs were one species, all pigeons, and so on; carp, however, were not the same species as goldfish, nor mallards the same species as teal, and so forth (Figure 21-2).

In an informal way, of course, people had always recognized species. Indeed, the word "species" is simply Latin for "kind." With Ray, however, the species began to be regarded as an important biological unit that could be catalogued and understood. Along with other scientists of his time, Ray believed that species did not change, a view that held until it was shattered by Darwin in 1859. Darwin was interested in those situations in which

FIGURE 21-2

The mule, a sterile hybrid between a female horse (mare) and a male donkey. By any standard, the horse and donkey are distinct species. The combination of their characteristics in the mule makes it a highly desirable work animal. Mules, however, being sterile, cannot interbreed with one another; they are always obtained by repeating the same interspecific cross.

A

B

C

FIGURE 21-3

Distinct species in one genus of animals and one of plants. This page: Butterflies of the genus Vanessa.
A *American painted lady, V. virginiensis;* **B,** *western painted lady, V. annabela;* **C,** *red admiral, V. atalanta.*
Next page: *Annual plants of the genus* Clarkia, *all from California.* **D** *C. cocinna;* **E,** *C. speciosa;* **F,** *C. rubicunda. This plant genus, which has about 40 species in and near California and 1 in areas of Chile and Argentina with a similar climate, was discovered on the Lewis and Clark Expedition and named in honor of George Rogers Clark.*

it was not clear whether he was dealing with distinct species or not. He considered the fact that species intergraded to be important evidence in support of his theory of evolution. Darwin explained the relative constancy of species by saying that each had its own distinctive role in nature, a role that we would in modern terms call a **niche.**

From the 1920s onward, with the emergence of population genetics, there was a desire to define the category species more precisely. The definition that began to emerge was stated by the American evolutionist Ernst Mayr as follows: species were "groups of actually or potentially interbreeding natural populations which are reproductively isolated from other such groups." In other words, hybrids between species occur rarely in nature. On the other hand, individuals that belong to the same species are able to interbreed freely. In practice, however, scientists usually have little or no experimental evidence on which to base their decisions about what constitutes a species in a particular group. They recognize species primarily because they differ from one another in their features (Figure 21-3). In fact, there are essentially no barriers to hybridization between the species in some groups of organisms, and strong barriers to hybridization between the species of other groups.

Within the units that are classified as species, the populations that occur in different places may be more or less distinct from one another (Figure 21-4) but still intergrade when

FIGURE 21-4

Subspecies of the seaside sparrow, Ammodramus maritimus. *Seaside sparrows are fairly common throughout their range in grassy tidal marshes but are rarely seen inland. The subspecies are quite local in distribution, and some of them are in immediate danger of extinction, largely as the result of alteration of their habitats. The widespread subspecies* Ammodramus maritimus maritimus (*adult,* 1, *and juvenile,* 2) *is the most common and is grayish-olive;* A. m. fisheri (3) *occurs along the Gulf Coast and has a buffier breast. The Cape sable seaside sparrow,* A. m. mirabilis (4), *is greener; it occurs in a small area of southwestern Florida. Of special interest is the dusky seaside sparrow,* A. m. nigrescens (5), *which occurs only near Titusville, Florida, and is near extinction, as discussed in the boxed essay on pp. 412-413.*

D   E   F

they occur together. In areas where these different-looking races, which may be classified taxonomically as **subspecies** or **varieties,** approach one another, many individuals may occur in which the distinctive features characteristic of each of the intergrading races are combined. Many of these individuals may not match either race in their characteristics. In contrast, when *species* occur together, they usually do not intergrade, although they may hybridize occasionally.

In some groups of organisms even local races are not capable of interbreeding with one another. This pattern occurs, for example, in many annual plants. In contrast, species of trees, some groups of mammals, and fishes generally *are* able to form fertile hybrids with one another, even though they may not do so in nature. For still other kinds of plants and animals, we do not know whether the species can form hybrids. We define a species, therefore, as a group of organisms that is unlike other such groups of organisms and does not intergrade extensively with them in nature.

> **Species are groups of organisms that differ in one or more characteristics and do not intergrade extensively if they occur together in nature.**

THE DIVERGENCE OF POPULATIONS

Local populations are usually more or less separated geographically, and the conditions in which they occur are dissimilar. Effective breeding populations are often extremely small (Figure 21-5), and there may be very limited exchange of individuals, and thus of genetic material, between such populations. For these reasons local populations often are able to adjust individually and effectively to the demands of their particular environment. As they do so, their characteristics change. The rate at which they change will depend primarily on the selective forces to which they are responding. If these forces are strong, the populations will change rapidly; populations certainly do not need to be isolated on islands in order to be able to become distinct species.

> **The characteristics of populations tend to diverge more or less rapidly depending on the features of their particular environment. Since there is only a limited exchange of genetic material between populations, even if they are geographically close, all populations tend to become increasingly divergent from one another in their characteristics over time.**

Many genes interact in the production of most phenotypic characteristics. Furthermore, all of the developmental processes that occur within a single organism are closely integrated with one another. In addition, every population begins with a different endowment of alleles, as we saw in our discussion of the founder effect in Chapter 20. Because of these three factors, the response of a given population of organisms to a combination of selective factors tends to be unpredictable. Even if two populations respond to similar selective forces, and even though they may be geographically close to one another, they still tend to change and to differ more and more from one another over time. In Sewall Wright's

A

B

FIGURE 21-5

The butterfly Euphydryas editha bayensis
(A) *occurs on Jasper Ridge* **(B),** *a biological
preserve in the foothills above Stanford
University, just south of San Francisco,
California (see also Figure 53-3). The
butterflies apparently constitute one continuous
population throughout the open grassland,
which occurs as an "island" surrounded by oak
forest and chaparral. Extensive studies by Paul
Ehrlich and his colleagues over many years,
involving marking the butterflies with dots of
ink on their wings and recapturing them, have
revealed that the butterfly actually exists as a
series of discontinuous populations between
which there is very little movement of unmated
individuals. Each of these local populations,
separated by dashed lines on the maps* **(C),** *is
free to respond to the selective forces that are
characteristic of its particular part of the ridge.
Changes in these populations, which have
remained essentially constant in overall
distribution for more than 25 years, are shown
in the maps. Illustrated (arrows) are the
locations of first capture of individual butterflies
in 1961, 1962, and 1963.*

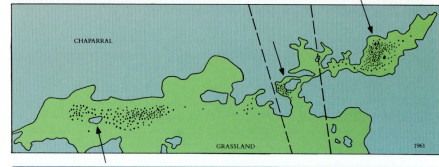

C

terminology, they eventually occupy different adaptive peaks; if their environments are dissimilar enough or change rapidly enough, the populations may diverge rapidly and strikingly, and soon come to look very different from one another.

BARRIERS TO HYBRIDIZATION

As isolated populations become more different from one another in their overall characteristics, they may eventually occupy different niches; in other words, they may exploit different resources in different ways. If such differentiated populations do ever migrate back into contact with one another, they may still remain distinct in their characteristics. The populations may occur in different habitats, have different feeding habits, or otherwise be separated by differences that arose in them while they were apart. For these or other reasons the individuals of the differentiated populations may not hybridize with each other; if they do, functional hybrid individuals may not be formed, or the hybrids that are formed may be sterile. For animals that choose their mates, the differences between the populations that have originated in isolation may be so great that they may choose mates of their own kind, rather than those of the other, formerly isolated population. In other words, and for a variety of possible reasons, the populations may have become species.

> **Populations of organisms tend to become increasingly different from one another in all of their characteristics. If the process continues long enough, or the selective forces are strong enough, the populations may become so different that they are considered distinct species.**

Once species have formed, how do they keep their identity? The reasons that they do retain their identity may be grouped into two categories: **prezygotic mechanisms,** those preventing the formation of zygotes; and **postzygotic mechanisms,** those preventing the proper functioning of zygotes after they are formed. Some of these mechanisms also occur, although often in a less marked form, within species. When they do, they illustrate stages in the evolution of new species. In the following sections we will discuss various mechanisms in these two categories and offer examples that illustrate how the mechanisms operate to help species retain their identity.

Prezygotic Mechanisms
Geographical isolation

Most species simply do not exist together in the same places. Species are generally adapted to different kinds of climates, or to different habitats. If this is the case, there is no possibility of natural hybridization between them. They may, however, hybridize if they are brought together in zoos, parks, or botanical gardens.

As an example of geographical isolation, the English oak, *Quercus robur,* occurs throughout those areas of Europe which have a relatively mild, oceanic climate. In its characteristics it is quite similar to the valley oak, *Quercus lobata,* of California, and quite different from the scrub oak, *Quercus dumosa,* also of California and adjacent Baja California (Figure 21-6). All of these species can hybridize with one another and form fertile hybrids. The English oak does not hybridize with the others in nature, however, simply because its geographical range does not overlap with theirs. Similarly, although lions, *Panthera leo,* and tigers, *P. tigris,* do not now occur together in nature, they do mate and produce hybrids in zoos. The hybrids in which the tiger is the father, called "tiglons," are viable and fertile; less is known about "ligers," hybrids in which the lion is the father.

Ecological isolation

Even if two species occur in the same area, they may occur in different habitats and thus may not hybridize with one another. If, on the other hand, they do hybridize with one an-

FIGURE 21-6

Leaf and acorn characteristics of three species of oaks, Quercus robur, Q. lobata, *and* Q. dumosa. *The range of* Q. robur, *which is in Europe, does not overlap with those of the other two species, which are in North America. Although* Q. robur *is interfertile with the North American species, hybridization cannot occur under natural circumstances.*

A

B

FIGURE 21-7

A Quercus lobata *in Yosemite Valley.*
B Q. dumosa, *a shrub, in the Coast Ranges south of San Francisco. The ranges of these two very distinct oak species, which are nevertheless interfertile, are shown in Figure 21-6, along with drawings illustrating details of their characteristics.*

other, the hybrids may not be well represented in the overall population, since they may not be as fit in the habitat of either of their parents. In the latter case one would speak of a postzygotic isolating mechanism of the kind that will be discussed in the next section.

For example, in India the ranges of lions and tigers overlapped until about 150 years ago. Even when they did, however, there were no records of natural hybrids. Lions stayed mainly in the open grassland and hunted in groups called *prides;* tigers tended to be solitary creatures of the forest. Because of the ecological differences between them, lions and tigers rarely came into direct contact with one another, even though their ranges overlapped over thousands of square kilometers.

Similar situations occur among plants. We have already mentioned two species of oaks that occur in California: the valley oak, *Quercus lobata,* and the scrub oak, *Q. dumosa* (Figures 21-6 and 21-7). Valley oak, a graceful deciduous tree, which can be as much as 35 meters tall, occurs in the fertile soils of open grassland on gentle slopes and valley floors in central and southern California. In contrast, scrub oak, *Quercus dumosa,* is an evergreen shrub, usually only 1 to 3 meters tall, which often forms the kind of dense scrub known as chaparral. The scrub oak is found on steep slopes, in less fertile soils. Hybrids between these very different oaks do occur, and they are fully fertile, but they are rare. The sharply distinct habitats of their parents limit their occurrence together, and there is no intermediate habitat where the hybrids might flourish.

Seasonal isolation

Lactuca graminifolia and *L. canadensis,* two species of wild lettuce, grow together along roadsides throughout the southeastern United States. Hybrids between these two species can easily be made experimentally and are completely fertile. Such hybrids are rare in nature, however, because *L. graminifolia* flowers in early spring and *L. canadensis* flowers in summer. When their blooming periods overlap, as they do occasionally, the two species do form hybrids, which may even be abundant locally.

Many species of birds and amphibians that are closely related have different breeding seasons. Differences of this kind prevent hybridization between such species. For example, five species of frogs of the genus *Rana* occur together in most of the eastern United States. The peak time of breeding is different for each of them; because of this difference, hybrids are rare. In insects such as termites and ants, mating occurs when winged, reproductive individuals swarm from the nest. Species often differ in their swarming times, which eliminates the possibility of hybridization between them.

A

B

C

D

Behavioral isolation

In Chapter 49 we will consider the often elaborate courtship and mating rituals of some groups of animals. Related species of organisms such as birds often differ in their mating rituals, which tend to keep these species distinct in nature even if they do occur in the same places. Indeed, much animal communication is related to the selection of mates.

In the Hawaiian Islands there are more than 500 species of flies of the genus *Drosophila*. This is one of the most remarkable concentrations of species in a single animal genus found anywhere. Many of these flies differ greatly from other species of *Drosophila*, exhibiting characteristics that can only be described as bizarre (Figure 21-8). The genus occurs throughout the world, but nowhere are the flies more diverse in their external appearance or behavior than in Hawaii.

The Hawaiian species of *Drosophila* are long-lived and often very large as compared with their relatives on the mainland. The females are more uniform than the males, which are often bizarre and highly distinctive. The males display complex territorial behavior and elaborate courtship rituals. Some of these are shown in Figure 21-8, *A* to *C*; another kind of courtship behavior is exhibited by *D. claviseta* (Figure 21-8, *D*).

FIGURE 21-8

A Drosophila silvestris. *After approaching the female from the rear, the male has lunged forward while vibrating his wings with his head under the wings of the female, and raised his forelegs up and over the female's abdomen. Specialized hairs on the dorsal surface of one of the leg segments are then "drummed" over the dorsal surface of the female's abdomen.*
B *and* **C** D. heteroneura. *The males of this species have heads that are greatly expanded laterally.* **B** *shows a male with extended wings approaching and displaying toward a female in typical courtship posture. In* **C,** *two males have locked antennae as part of the aggressive behavior involved in the defense of their territories. Groups of males, called* leks, *form in defended mating grounds.*
D *In Drosophila clavisetae, the males spray a chemical signal over the female before mating.*

FACING EXTINCTION

The patterns of mating behavior among the Hawaiian species of *Drosophila* are of great importance in maintaining the distinctiveness of the individual species. Despite the great differences between them, which are so evident in Figure 21-8, for example, *D. heteroneura* and *D. silvestris* are very closely related. Hybrids between them are fully fertile. They occur together over a wide area on the island of Hawaii, and yet hybridization has been observed at only one locality. The very different and complex behavioral characteristics of these flies obviously play the major role in maintaining their distinctiveness.

Mechanical isolation

Structural differences between some related species of animals prevent mating. Aside from such obvious features as size, the structure of the male and female copulatory organs may be so incompatible that mating cannot occur. In many insect and other arthropod groups the sexual organs, particularly those of the male, are so diverse that they are used as a primary basis for classification. This diversity in structure is generally presumed to have some importance in maintaining differences between species.

Similarly, the flowers of related species of plants often differ significantly in their proportions and structures. Some of these differences limit the transfer of pollen from one plant species to another and thus decrease the frequency of hybridization between them.

Prevention of gamete fusion

In animals that simply shed their gametes into water the eggs and sperm derived from different species may not attract one another. Many hybrid combinations involving land animals are not realized because the sperm of one species may function so poorly within the reproductive tract of another that fertilization never takes place. The growth of the pollen tubes may be impeded in hybrids between different species of plants. In both plants and animals the operation of such mechanisms may prevent the union of gametes even following successful mating. The prevention of fusion between gametes is the last kind of prezygotic isolating mechanism possible before hybrids are formed.

Prezygotic mechanisms lead to reproductive isolation by preventing the formation of hybrid zygotes. The principal mechanisms are geographical, ecological, seasonal, behavioral, and mechanical isolation and the prevention of gamete fusion.

backcrossed with females of the Gulf Coast subspecies of the seaside sparrow. The three males are thought to be 8 to 11 years old, very near the end of their natural life spans. Charles Cook, curator of Discovery Island, reported in August, 1985, that there were one 25% dusky seaside sparrow, one 50% dusky, and two 88% duskys as a result of captive breeding at Walt Disney World. The fate of the dusky seaside sparrow depends on the breeding efforts that are now under way, with each season critical for the survival of this fascinating small bird, now very near extinction.

FIGURE 21-A

One of the three remaining male dusky seaside sparrows. All of them are now kept in captivity at Discovery Island. (© Walt Disney Productions.)

Postzygotic Mechanisms

All of the factors that we have discussed up to this point tend to prevent hybridization. If hybridization does occur, however, and zygotes are produced, there are still many factors that may prevent those zygotes from developing into normal, functional, fertile F_1 individuals. Development in any species is a complex process. In hybrids the genetic complements of two species may be so different that they cannot function together normally in embryonic development. For example, hybridization between sheep and goats usually produces embryos that die in the earliest developmental stages, although this barrier has recently been overcome (Figure 21-9).

The leopard frogs (*Rana pipiens* complex) of the eastern United States are a series of very similar species between some of which it is difficult or impossible to produce viable hybrids. Before the nature of these species was elucidated, it was assumed that there were simply difficulties in producing hybrids between these frogs because they came from very different regions. The species are very similar, and hybrids between them are usually rare (Figure 21-10).

Many examples of this kind, in which similar species initially have been distinguished only as a result of hybridization experiments, are known in plants. Sometimes the hybrid embryos can be removed at an early stage and grown in an appropriate medium. When these hybrids are supplied with extra nutrients or other growth requirements that compensate for their weakness or inviability, they may complete their development normally.

Postzygotic mechanisms are those in which gametes fail to fuse, hybrid zygotes develop abnormally, or hybrids cannot become established in nature.

Even if the hybrids survive the embryo stage, they may not develop normally. If the hybrids are weaker than their parents, they will almost certainly be eliminated in nature. Even if they are vigorous and strong, as in the case of the mule—a hybrid of the horse and the donkey (Figure 21-2)—they may still be sterile and thus incapable of contributing to succeeding generations. The development of sex organs in hybrids may be abnormal, the chromosomes derived from the respective parents may not pair properly, or their fertility may simply be lower than normal for other reasons.

FIGURE 21-9

Artificial sheep-goat hybrid, reported in 1984. This hybrid is actually a chimera, a mosaic of sheep-goat characteristics; different parts of the animal reflect the characteristics of one or the other of its parents. Thus its coat has a mixture of hairy and woolly areas; its horns are goatlike but twisted like those of a sheep; and its blood contains both sheep and goat red blood cells. The chimera is about 1 year old and was produced by mixing one sheep embryo at the 8-cell stage of development with three goat embryos at the same stage. The resulting composite embryo was transferred to the uterus of a goat, where it completed its development. It has not yet been possible to obtain hybrids, in the conventional sense, between these two species.

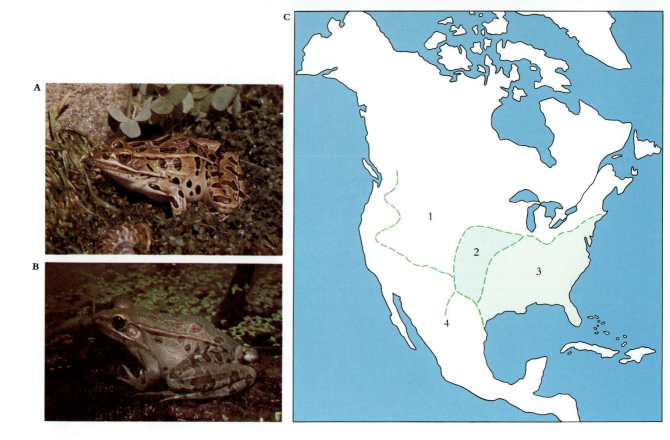

FIGURE 21-10

A   *The leopard frog,* Rana pipiens, *in Wisconsin.*
B   *The southern leopard frog,* Rana berlandieri, *in California.*
C   *Numbers indicate the following species in the geographic ranges shown: 1,* Rana pipiens; *2,* Rana blairi; *3,* Rana sphenocephala; *4,* Rana berlandieri. *These four species resemble one another closely in their external features. Their existence was first suspected when hybrids between them produced defective embryos in some combinations. At first it was thought that this occurrence indicated differentiation within a species along a geographical gradient, but when it was demonstrated that the mating calls of these four species differed in their structure, their distinctiveness was recognized. These species usually do not hybridize where they occur together, although hybrids do occur locally and are even abundant in certain areas. Such closely related species, often distinguished primarily by behavioral or other nonevident characteristics, are called sibling species.*

Reproductive Isolation: A Summary

All of the kinds of reproductive isolation that we have discussed can arise in populations that are diverging from one other. These kinds of reproductive isolation all occur within certain species, as would be expected from the way in which such differences evolve. Some populations of particular species, therefore, cannot hybridize with one another, just as if they were distinct species as judged by this criterion.

The formation of species is a continuous process, one that we can understand because of the existence of intermediate stages at all levels of differentiation. If populations that are partly differentiated come into contact with one another, they may still be able to interbreed freely, and the differences between them may then disappear over the course of time. If their hybrids are partly sterile, or not as well adapted to the existing habitats as their parents, these hybrids will be at a disadvantage. As a result, there may be selection for factors that limit the ability of the differentiated populations to hybridize. If hybrids are sterile or not as successful as their parents, individual plants or animals that do not hybridize may be more fit, in the sense explained in Chapter 20, than those which do so.

Most species are separated by combinations of the factors that we have discussed. For example, two related species may occur in different habitats, produce their gametes at different times of the year, have different behavioral patterns, and produce inviable embryos even if hybridization does take place. Such patterns, in which more than one factor functions in limiting the frequency of hybrids between two species, presumably arise for two reasons.

1. The factors that limit hybridization arise primarily as by-products of adaptive change in populations. Consequently, several different kinds of factors that limit hybridization often emerge simultaneously and may characterize the differentiated populations.

2. If differentiated populations do come into contact with one another, natural selection may strengthen the isolating mechanisms that are already present. If, for example, the hybrids do not complete their development beyond the embryo stage, then any factor that limits the hybridization that produces them (prezygotic mechanisms) would be an advantage. Individuals that form hybrids that do not function well in nature waste reproductive energy by doing so and are less fit than individuals that do not form such hybrids.

Both kinds of forces that limit hybridization are built up as a part of the overall process of change in populations that are isolated from one another, although they may sometimes be strengthened when and if the differentiated populations migrate into contact with one another.

As a result of the way in which they originate, species are often separated by more than one kind of factor. Some of these factors prevent the species from hybridizing at all; others limit the success of the interspecific hybrids once they are formed.

ECOLOGICAL RACES

What kinds of patterns can be expected as a result of the differentiation of populations? One consequence is that the individuals of a species that occur in one part of its range often look different from those which occur elsewhere (Figure 21-4). Such groups of distinctive individuals are informally called races, and they may, as we have mentioned, be classified taxonomically as subspecies or varieties. The existence of such races proved fascinating to Darwin, because he considered them to be an intermediate stage in the evolution of species. The kinds of ecologically defined races that we will discuss first may change over time to the clusters of species that we will consider next. Both provide important examples of the evolution of populations in nature.

Ecological races were first studied in detail in plants. As every gardener knows, the same plants may differ greatly in appearance depending on the places where they are grown (Figure 21-11). This is true even for genetically identical divisions of the same plant—

A

B

FIGURE 21-11

Two forms of Bishop pine, Pinus muricata, *in Marin County, just north of San Francisco, California.*
A *This tree is growing in a flat field behind Inverness Ridge, where it is protected from the ocean winds.*
B *This tree, growing in an exposed place, has been dwarfed and shaped by the winds. Without experimental study, it would not be possible to determine whether these two individuals differ genetically or not.*

EXPERIMENTAL STUDIES OF PLANT ECOTYPES IN CALIFORNIA

The important studies that were initiated by Turesson were continued in California by a group of scientists from the Department of Plant Biology of the Carnegie Institution of Washington, located on the campus of Stanford University. California is a state of varied environments, and the team of Jens Clausen, David Keck, and William Hiesey were able to put these environmental contrasts to good use in designing their experiments.

Clausen, Keck, and Hiesey studied plant species that occurred at a wide range of different elevations, from the moist, foggy coast all the way up to and across the Sierra Nevada. They improved on Turesson's methods (discussed below) by growing clones of the same individuals at a series of transplant stations. At each of these stations they established cleared garden plots. They kept these plots free of weeds but did not water their experimental plants once they were established.

The Carnegie Institution botanists used three major transplant stations for their experiments:

1. Stanford, near sea level on the inner side of the Coast Ranges, in a summer-dry climate with mild winters.

2. Mather, at about 1400 meters elevation near Yosemite Valley, on the west side of the Sierra Nevada. This is a region of snowy winters and relatively long, hot, sunny, dry summers.

3. Timberline, near the crest of the Sierra at Tioga Pass above Tuolumne Meadows, again in the Yosemite region, at about 3050 meters elevation. Here there were long, hard, snowy winters and short, relatively cool and dry summers.

As a result of these studies, Clausen, Keck, and Hiesey were able to demonstrate the existence of major races, or ecotypes, in a number of plant species (Figure 21-B). These ecotypes were distinct from one another in morphological and physiological characteristics that were determined genetically. Many of the individual populations that were grouped into these races also differed in characteristics that were genetically determined.

These ecotypes were dissimilar in their physiological, as well as their morphological, features. For example, plants of ecotypes from the Coast Ranges usually grew actively both in the winter and in the summer at Stanford. At Mather, the mid-elevation station, plants of such coastal ecotypes become dormant during the 5-month winter. Despite this, such plants were often able to manufacture and store enough food during the warm part of the year to survive. At Timberline, plants from the low elevations of the Coast Ranges almost invariably died during the first year. Near the summit of the Sierra Nevada, these plants simply could not accumulate enough food in the short summers to survive through the long winters.

---

clones. The part of an individual plant that is usually in the sun often produces leaves that are unlike those which the same plant produces in the shade. For example, shade leaves are usually thinner and broader and have more internal air spaces than do sun leaves. Thus it seemed to many botanists in the nineteenth century that environmental factors, rather than genetic differences, might account for many of the differences between races and even between species of plants.

Ecotypes in Plants

In the 1920s and 1930s the Swedish botanist Göte Turesson performed a series of experiments that were designed to test whether differences between the races of plants were largely genetically determined or mainly caused by environmental factors. Turesson observed that many plants had distinctive races that grew in different habitats. These races differed from one another in characteristics such as height, leaf size and shape, degree of hairiness, flowering time, and branching pattern. Turesson dug up individuals representing these races and cultivated them together in his experimental garden at Lund, Sweden. In nearly every case he found that the unique features of individual races were maintained when the plants were grown in a common environment. Most of the characteristics that

he observed, therefore, had a genetic basis; a few of them were environmental. Turesson called the ecological races that he studied and showed to have a genetic basis **ecotypes.**

Ultimately the kinds of studies first put on a scientific basis by Turesson and carried on by others led to the conclusion that most of the differences between individuals, populations, races, and species of plants are genetically determined. Despite the fact that plants change their characteristics in relation to the environments in which they grow, the differences between them are usually fixed genetically in the course of their evolution.

Ecological Races in Animals

Similar patterns of variation are, of course, also found in animals (Figure 21-4). The differences may be morphological or physiological, and they have the same basis as do ecotypes in plants. The differences between subspecies may be striking. For example, the larger races of some species of birds may often consist of individuals that weigh three or four times as much as individuals of the smaller races. Races may differ from one another in their tolerance to different temperatures, in the speed of their larval development, in their behavioral characteristics—in short, in virtually any feature that can be measured and studied. Their features are almost always genetically determined.

CLUSTERS OF SPECIES

One of the most visible manifestations of evolution is the existence of groups of closely related species in various genera. These species often have evolved relatively recently from a common ancestor. Such clusters are often particularly impressive on groups of islands, in series of lakes, or in other sharply discontinuous habitats. We will discuss two examples from islands to illustrate these species clusters.

Darwin's Finches

Thirteen species of Darwin's finches occur on the Galapagos Islands, and one lives on Cocos Island, which lies about 1000 kilometers to the north. This group of birds provides one of the most striking and best-studied examples of evolution on islands. Although Darwin was, as we have seen, most impressed by the differences in the land tortoises from the different islands, these finches have excited subsequent students of evolution.

Just as Darwin found them on his voyage, the Galapagos Islands are a particularly striking natural laboratory of evolution. The islands are all relatively young in geological terms (several million years), and they have never been connected with the adjacent mainland of South America or with any other source area. The lowlands of the Galapagos Islands are covered with thorn scrub; at higher elevations, which are attained only on the larger islands, there are moist, dense forests. All of the organisms that occur on these islands have reached them by crossing the sea, as a result of chance dispersal in the water, by wind, or by transport via another organism.

On islands there is often a disproportionate representation of certain groups of organisms. For example, in addition to the 13 species of Darwin's finches on the Galapagos, there are only 7 other species of land birds. Presumably the ancestor of Darwin's finches reached these islands earlier than the ancestors of the other birds. All of the kinds of habitats where birds occur on the mainland were unoccupied, and the ancestor of Darwin's finches was able to take advantage of them all. As the new arrivals moved into these vacant niches, adopting new life-styles in doing so, they were subjected to diverse sets of selective pressures. Under these circumstances, the ancestral finches rapidly split into a series of diverse populations, and some of these became species.

In addition, since the Galapagos are a group of islands, small flocks of birds could occasionally reach new islands where no other land birds were present. On these new islands the immigrants would become adjusted to the local conditions through the process of natural selection. In doing so, they would have become progressively more different from the original populations.

The descendants of the original finches that reached the Galapagos Islands now occupy many different kinds of habitats on the islands (Figure 21-12). These habitats encompass a variety of niches comparable to those occupied by several distinct groups of birds on the mainland (Figure 21-13). Among the 13 species of Darwin's finches that inhabit the Galapagos, there are three main groups. Some scientists consider these birds to belong to several different genera, but most now assign all of them to the genus *Geospiza*. The three groups are:

1. *Ground finches.* There are six species of ground finches (Figure 21-12, *A*). Most of the ground finches feed on seeds of different sizes. The size of their bills is related to the size of the seeds on which the birds feed. One of the ground finches is the cactus ground finch, *Geospiza conirostris*. It feeds primarily on cactus flowers and fruits and has a longer, larger, more pointed bill than the others.
2. *Tree finches.* There are six species of tree finches, belonging to several diverse groups. These species also differ from one another primarily in bill size and shape, which reflect adaptations to their food. Four species have bills that are suitable for feeding on insects: the ones with larger bills feed on larger insects, and those with smaller bills feed on smaller insects. One of the tree finches has a parrotlike beak; it feeds on buds and fruit in trees. The sixth species of tree finch, the woodpecker

FIGURE 21-12

Darwin's finches.
A *Medium ground finch,* Geospiza fortis, *on Daphne Island. Note the large, cone-shaped bill, adapted for eating seeds and presumably similar to that of the common ancestor of the group.*
B *Warbler finch,* Certhidea olivacea, *Santa Cruz Island. The most distinctive of Darwin's finches.*

A

B

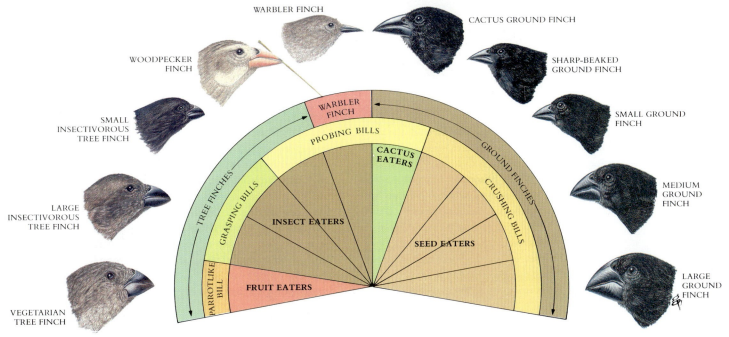

FIGURE 21-13

Ten species of Darwin's finches from Indefatigable Island, one of the Galapagos Islands, showing differences in bills and feeding habits. The bills of several of these species resemble those of different distinct families of birds on the mainland, a condition that presumably arose when the finches evolved new species in habitats where other kinds of birds normally occur on the mainland and which are not available to finches there. The woodpecker finch uses cactus spines to probe in crevices of bark and rotten wood for food. All of these birds are thought to have been derived from a single common ancestor, a finch like that shown in Figure 21-14.

finch, has a chisel-like beak. This unusual bird carries around a twig or cactus spine, which it uses to probe for insects in deep crevices. It is an extraordinary example of a bird that uses a tool.

3. *Warbler finch.* This unusual bird differs from the ground finches and the tree finches in its habits and morphological adaptations. The warbler finch plays the same ecological role in the Galapagos woods that warblers play on the mainland, searching continually over the leaves and branches for insects. It has a slender, warbler-like beak (Figure 21-12, *B*).

The warbler finch is the most distinctive of all of Darwin's finches, and the question has been raised as to whether it is directly related to the other species. Detailed studies, however, have shown that it is derived from the same common ancestor as the others: presumably a pair of windblown finches, or a pregnant female, that reached the Galapagos Islands perhaps 100,000 or more years ago.

David Steadman, of San Diego State University, believes that this ancestor was the blue-black grassquit, *Volatinia jacarina* (Figure 21-14). This species of bird is abundant throughout the lowlands of Latin America and occurs from Mexico to Chile along the Pacific Coast. Steadman bases his conclusion on the characteristics of this widespread bird, including its short, stout, cone-shaped bill, adapted for seed crushing. He believes that the blue-black grassquit colonized the Galapagos Islands and Cocos Island independently and gave rise to similar derivatives in both areas. On the basis of his morphological and behavioral studies, Steadman places all of Darwin's finches together with the blue-black grassquit in the genus *Geospiza*.

The evolution of Darwin's finches on the Galapagos Islands and Cocos Island provides one of the classic examples of species formation. Although spectacular, it illustrates the same kinds of processes by which species are originating continuously in all groups of organisms. Isolated populations subjected to unique combinations of selective pressures diverge from one another and may ultimately become so different that they are distinct species.

FIGURE 21-14

The blue-black grassquit, Volatinia jacarina, *the most likely common ancestor of Darwin's finches.*

HUMAN RACES

Human beings, like all other species, have differentiated as they have spread through the world. The local populations in one area often look impressively different from those that live elsewhere. For example, northern Europeans often have blond hair, fair skin, and blue eyes, whereas Africans often have black hair, dark skin, and brown eyes. In addition, other features, such as the blood group alleles we considered in Chapter 21, also differ in proportion from area to area. These traits may play an adaptive role; thus the blood types may be associated with immunity to diseases characteristic of certain geographical areas, and dark skin may shield the body from the damaging effects of ultraviolet radiation, which is much stronger in the tropics than in temperate regions.

All human beings are completely interfertile. The reasons that they do or do not choose to associate with one another are purely psychological and behavioral (cultural). The number of races into which the human species might logically be divided, and which ought to be given names, has long been a point of contention. Some contemporary anthropologists divide people into as many as 30

Hawaiian *Drosophila*

Our second example of a cluster of species is provided by the fly genus *Drosophila* on the Hawaiian Islands, which we mentioned earlier in this chapter as an example of behavioral isolation as a mechanism for keeping species distinct. There are about 1250 species of this genus throughout the world, and more than a third of them are found only in the Hawaiian Islands, a tiny area on a world scale. More species of *Drosophila* are still being discovered in Hawaii, although the rapid destruction of the native vegetation is making the search more difficult. Aside from their sheer number of species, these flies are unusual in the morphological and behavioral traits discussed earlier (Figures 21-8 and 21-9). No comparable species of *Drosophila* are found anywhere else in the world.

A second, closely related genus of flies, *Scaptomyza,* is also found in Hawaii, where it may be represented by as many as 300 species. A few species of *Scaptomyza* are found outside of Hawaii, but the genus is better represented there than elsewhere. In addition, species intermediate between *Scaptomyza* and *Drosophila* exist in Hawaii, but nowhere else. The genera are so closely related that scientists have suggested that all of the estimated 800 species of these two genera that occur in Hawaii may have been derived from a single common ancestor.

The native Hawaiian flies of this group are closely associated with the remarkable native plants of these islands and are often abundant in the native vegetation. Evidently, when their ancestors first reached these islands, they encountered many "empty" habitats, ones in which other kinds of flies had not become established earlier. The evolutionary opportunities that the ancestral *Drosophila* flies found were similar to those encountered by the ancestors of Darwin's finches in the Galapagos Islands, and both groups evolved in a similar way. Many of the Hawaiian *Drosophila* species are highly selective in their choice of host plants for their larvae and in the part of the plant that they use. The larvae of various species live in rotting stems, fruits, bark, leaves, or roots or feed on sap.

New islands have continually arisen from the sea in the region of the Hawaiian Islands. As they have done so, they appear to have been invaded successively by the various groups of *Drosophila* that were present on the older islands. New species have evolved as new islands have been colonized. The Hawaiian species of *Drosophila* have had even greater evolutionary opportunities than have Darwin's finches, because of their restricted ecological niches and the variable ages of the islands. They clearly present one of the most unusual evolutionary stories found anywhere in the world.

races, others as few as three: Caucasoid, Negroid, and Oriental. American Indians, Bushmen, and Aborigines are particularly distinctive units that are sometimes given racial status.

The problem with classifying people, or other organisms, into races is that the characteristics that we use to define the races are usually not well correlated with one another. The way in which we determine the race to which a given population should be assigned is, therefore, always somewhat arbitrary. In humans, and in many other species, it is simply not possible to delimit clearly defined races that can be recognized by a particular combination of characteristics. Geographical variation certainly does occur, but the products of that variation are often difficult to assign in a meaningful way to arbitrary units of classification. For people specifically, we know that the different groups have constantly intermingled and interbred with one another during the entire course of their history. Today the differences between human "races" are tending to break down rapidly as large numbers of individuals are constantly moving over the face of the globe, mixing with one another and recombining their characteristics.

As a result of the geographical position of the Hawaiian Islands, which emerged from the sea in isolation, the individuals of *Drosophila* that reached them first were able to move into many adaptive zones that were occupied elsewhere by other kinds of insects and other animals. The extraordinary adaptive radiation that followed resulted in the evolution of at least 800 species of this group in a relatively small area and provides an excellent model for similar processes, usually taking place on a smaller scale, that have resulted in the evolution of species generally.

Rapid Evolution

As you saw in the discussion above, the evolution of organisms on islands differs from that which occurs in mainland areas only in the degree of opportunity afforded for *rapid* evolution. Because evolution on islands is often rapid, and therefore recent, Darwin and other investigators have found the study of island plants and animals especially informative.

In continental areas, too, the process of evolution sometimes results in the production of clusters of species. These clusters are more frequent in certain regions. For example, more than a quarter of the total number of plant species in the United States and Canada, about 6000 out of perhaps 22,000 species, occur in California, and nearly half of these are found nowhere else. In the dry, semiarid, and variable habitats of that state, there are opportunities for evolution comparable to those found on islands throughout the world. Climatic changes in relatively recent times, together with a diversity of habitats, have provided conditions in California that are especially suitable for the rapid evolution of new species.

In general, rapid evolution takes place in areas where the habitats are varied and located near one another. The pace of evolution is accelerated further by rapid climatic change.

THE ROLE OF HYBRIDIZATION IN EVOLUTION

Among perennial plants, especially trees and shrubs, there are often few postzygotic barriers to hybridization among the species of a given genus. For example, all white oaks are capable of forming fertile hybrids with one another. So are all of the members of the major groups of *Eucalyptus,* many of the cottonwoods and aspens, and so forth. Such hybrids

FIGURE 21-15

A parthenogenetic species of the desert whiptail lizard, Cnemidophorus. *All individuals of this particular species are females, and all of their descendants are genetically identical to their mother and to one another because of the way in which they reproduce.*

recombine the characteristics of their parents in unusual ways, and these genetic combinations may be particularly suited to some habitats, where they may flourish to a greater degree than either of their parents. If so, the hybrids may give rise to distinctive populations that may replace their parental species in the appropriate habitats. The opportunity to recombine alleles from different species may be important in helping to bring about rapid changes in the overall populations. If it is, hybridization can be considered an important force in the production of new species in some groups, especially among plants.

Even if such hybrids are sterile, they may have a future. Sometimes hybrid plants simply reproduce vegetatively; for example, broken-off pieces of stems or roots may become established away from the original plant. If they are able to do so, and if such a clone has a genotype that is particularly well adapted to a certain habitat, the sterile hybrid individuals may become quite common. Among some groups of animals, the corresponding phenomenon of **parthenogenesis** is of importance. In it, egg cells may give rise directly to new individuals or other somatic cells may function to produce embryos (Figure 21-15). Depending on the way in which they are produced, these individuals may either be diploid or haploid. Most races of dandelions, including the ones you see in lawns, reproduce in this same way; their seeds contain "embryos" that are produced asexually and that have the same genetic constitution as the parent plant.

Fertile individuals may also arise from sterile ones by the process of **polyploidy,** by which the chromosome number of the original sterile hybrid individual is doubled. A **polyploid** cell, tissue, or individual has more than two sets of chromosomes. Polyploid cells and tissues derived from them occur spontaneously and reasonably often in all organisms, although in many they are soon eliminated. A hybrid may be sterile simply because the sets of chromosomes derived from its male and female parents do not pair with one another, having come from different species. If the chromosome number of such a hybrid doubles, the hybrid, as a result of the doubling, will have a duplicate of each chromosome. In that case the chromosomes will pair, and the fertility of the polyploid hybrid individual may be restored (Figure 21-16). It is estimated that about half of the approximately 260,000 species of plants are of polyploid origin, including many that are of great commercial importance, such as bread wheat, cotton, tobacco, sugarcane, bananas, and potatoes.

FIGURE 21-16

Bread wheat, Triticum aestivum. *Known only in cultivation, bread wheat has 42 chromosomes in its somatic cells, 21 chromosomes in its eggs and sperm. It appeared at least 8000 years ago, probably in central Europe, following spontaneous hybridization between a species of cultivated wheat with 28 chromosomes and a wild relative with 14 chromosomes. The chromosomes in such a hybrid doubled by chance, thus giving rise to one of our most important crop plants.*

In this chapter we have traced the way in which adaptive changes in populations—changes that are occurring at all times and in all populations—lead to the origin of diverse races of organisms. Factors of the kind that make it difficult or impossible for species to hybridize arise as a consequence of such adaptive change but are sometimes strengthened if the unlike populations come back into contact with one another. Even when differentiation is relatively complete, however, hybridization between species can often have important evolutionary consequences, particularly among plants. Processes of the kinds described in this chapter have ultimately been responsible for the origin of at least several million species of plants, animals, and microorganisms. In Chapter 22 we will discuss the ways in which we deal with these species conceptually by classifying them.

SUMMARY

1. Species are kinds of organisms that differ from one another in one or more characteristics and that do not normally hybridize freely with other species when they come into contact in nature. They often cannot hybridize with one another at all. Individuals within a given species, on the other hand, usually are able to interbreed with one another freely.

2. Populations change as they adjust to the demands of the environments where they occur. Even populations of a given species that are close to one another geographically are normally effectively isolated. Such populations are free to diverge in ways that are responsive to the needs of their particular environment.

3. Because of the complex interactions of genetic and developmental factors, as well as initially different genetic endowments, populations inevitably change in unpredictable ways even when they respond to the same selective pressures.

4. In general, if the selective forces bearing on populations differ greatly, the populations will diverge rapidly; if these selective forces are similar, the populations will diverge slowly.

5. Among the factors that separate populations, and species, from one another are geographical, ecological, seasonal, behavioral, and mechanical isolation, as well as factors that inhibit the fusion of gametes or the normal development of the hybrid organisms. Reproductive isolation between species arises as a normal byproduct of the progressive differentiation of populations.

6. Ecological races and subspecies differentiate within species but often still intergrade with one another. The differences between them, in both plants and animals, are mostly genetically fixed.

7. Clusters of species arise when the differentiation of a series of populations proceeds further. On islands such differentiation often is rapid, because of the numerous open habitats that are available. In many continental areas, differentiation is not as rapid, but there are local situations, such as those where many different kinds of habitats are developing close to one another, where differentiation may be rapid.

8. Hybridization between differentiated species, especially in trees and shrubs, affords a source of new genotypes on which selection can act. If the chromosomes of the parents cannot pair in a hybrid, spontaneous doubling may produce a polyploid individual, one with multiple complements of chromosomes, in which each chromosome has a partner with which it can pair.

SELF-QUIZ

The following questions pinpoint important information in this chapter. You should be sure to know the answer to each of them.

1. Each species has a distinctive role in nature, which is known as its _____.

2. Populations must be separated by great distances to become distinct from one another. True or false?

3. Are most phenotypic characteristics exhibited by organisms in natural populations controlled by one gene, or by many?

4. _____ is a mechanism of prezygotic isolation that does not allow hybridization because the species simply do not exist in the same area.

5. Two species that may not form viable hybrids because they become sexually active at different times in the year are said to be _____ isolated.

6. Morphological differences in breeding structures that prevent mating between related species of animals are known as _____ isolating mechanisms.

7. "_____ mechanism" is a collective term for the failure of development of a hybrid zygote.

8. Genetically identical organisms are known as _____.

9. Ecotypes are ecological races. True or false?

10. A polyploid individual has how many sets of chromosomes?
    (a) one   (c) more than two
    (b) two   (d) none

11. Among perennial plants, such as trees and shrubs, there are many postzygotic barriers. True or false?

12. When natural populations that have diverged sufficiently to form new species are brought back into contact with each other
    (a) their descendants will hybridize
    (b) their descendants will not hybridize
    (c) their descendants may or may not hybridize

THOUGHT QUESTIONS

1. Under what circumstances is hybridization between species selected against? When it is selected for?

2. There are polyploid animals in some groups, but they are far less frequent than they are among plants, particularly perennial plants. Why do they think this might be so?

3. Do species usually originate within a single population, or in another way? Describe the way in which the process normally occurs.

FOR FURTHER READING

AYALA, F.J., and J.W. VALENTINE: *Evolving: The Theory and Processes of Organic Evolution,* Benjamin Publishing Co., Menlo Park, Calif., 1979. Excellent summary of modern views of evolution.

Biological Journal of the Linnaean Society, vol. 21, pages 1-258, 1984. Special issue on Charles Darwin and the Galapagos Islands. A fine collection of papers, including historical accounts and details of modern scientific work. They indicate that the Galapagos Islands are still of great interest as a natural laboratory today, a century and a half after Darwin's 6-week visit in 1835.

COLE, C.J.: "Unisexual Lizards," *Scientific American,* January 1984, pages 94-100. An extraordinary account of the populations of whiptail lizards that consist of females only and reproduce themselves by parthenogenesis.

FUTUYMA, D.: *Evolutionary Biology,* Sinauer Associates, Inc., Sunderland, Mass., 1979. This text presents a good discussion of the development of evolutionary thought, combined with a process-oriented treatment of the whole field.

GOULD, S.J.: *Ever Since Darwin,* W.W. Norton & Company, New York, 1977. A delightfully entertaining collection of insightful essays on evolution and Darwinism.

MINKOFF, E.C.: *Evolutionary Biology,* Addison-Wesley Publishing Company, Inc., Reading, Mass., 1983. A well-illustrated account, simpler than that of Futuyma.

Chapter 22   CLASSIFYING THE PRODUCTS OF EVOLUTION

Overview

People have always classified and given names to organisms in order to be able to communicate about them. Starting in the late seventeenth century, these systems began to be standardized, and attempts were made to use them to describe all kinds of organisms everywhere. The basic unit of classification was the genus, subdivided into species that were at first designated by polynomial names—ones that were made up of many words—and later, after the work of Linnaeus in the mid–eighteenth century, by the binomial (two-part) names that we still use today. Species are kinds of organisms, characterized in part by their genetic features. As additional kinds of organisms were discovered and studied in more detail, genera were grouped into more inclusive categories, including families, orders, classes, and phyla (or divisions). The original two kingdoms of living organisms, plants and animals, were broken up and redefined. Today the prokaryotes are placed in the kingdom Monera. The eukaryotes are placed in four kingdoms: a catch-all kingdom, Protista, which consists mostly of unicellular organisms, and three almost exclusively multicellular kingdoms derived from Protista. These latter three kingdoms are the plants, animals, and fungi. Viruses, which are nonliving, are portions of the genomes of bacteria and eukaryotes that have broken loose and taken up an independent existence of their own.

For Review   *Here are some important terms and concepts that you will encounter in this chapter. If you are not familiar with them, you should review them before proceeding.*

- **Prokaryotic versus eukaryotic cells** (Chapter 3)

- **Development of evolutionary thought** (Chapter 19)

- **Definition of species** (Chapter 21)

- **Origin of species** (Chapter 21)

People everywhere have always invented names for all kinds of objects, including organisms, in order to be able to communicate with one another. Initially, of course, these names were only spoken. By Greek and Roman times, however, students of natural history prepared written descriptions of familiar organisms and recorded their names. Some of the writings of these scholars, both herbals (books about plants) and bestiaries (books about animals), have survived in manuscript form.

In effect, these scholars were recording the set of names that was used in their region to communicate about plants and animals. Even today, people who have no written language still deal with plants and animals by means of limited systems of classification similar to those used by the Greeks and Romans. In these systems, several hundred kinds of animals and several hundred kinds of plants are usually recognized; this seems to be roughly the number of names that speakers of a given language can remember and with which they can deal.

THE HISTORY OF CLASSIFICATION OF ORGANISMS

Before Johannes Gutenberg set up printing presses with movable type in Europe in the mid–fifteenth century, there was not much effort to make classification systems consistent or to relate them to one another. What people wanted was simply to communicate accurately about the organisms in their own area. They were not concerned with telling others who lived in different regions about the plants and animals they knew. It did not even matter much whether names used in different regions duplicated one another or were very similar. It was not until the time of John Ray, about 200 years after the introduction of movable type, that the effort to produce systems of classification that were generally applicable began in earnest. The specialists who produced these systems became known as **taxonomists** or **systematists**, and their subject is called **taxonomy** or **systematics.**

In sixteenth-century Europe, printed books were distributed widely for the first time. Scientists began to think about producing comprehensive works in which all kinds of plants and animals could be named and described for use by everyone. By the end of the seventeenth century, scientists were producing encyclopedias in which all knowledge was collected and presented in a convenient form. From that time until today—a period of about three centuries—people have agreed that systems of classification should be applicable generally, and they have labored to improve such systems.

> **The introduction of movable type—an efficient method of printing—in Europe in the mid–fifteenth century first made it possible for scholars to contemplate a universal system of classification for organisms. The earlier systems had all been only local in application.**

Since Latin was the traditional language of scholarship in the Middle Ages, botanists and zoologists naturally used Latin to name organisms (Figure 22-1). About 300 years ago the concept of **genus** (pl. **genera**) became important as the initial word used in the names of all organisms. Genera are the names for the basic kinds of organisms, such as oaks, cats, or horses. All languages, written and unwritten, have names for units of this sort. The names that were adopted for these units from the seventeenth century onward were often the Latin names by which they had been known since ancient times. Thus oaks were assigned to the genus *Quercus,* cats to *Felis,* and horses to *Equus*—names that the Romans applied to these groups of organisms. For genera that were not known in antiquity, of course, new names had to be invented.

> **The names of genera were the basic point of reference in taxonomic systems; since the names of organisms were given in Latin, the names of genera were often simply the Latin names for those kinds of organisms.**

The Polynomial System

Before the 1750s, scholars usually added a series of additional descriptive terms to the name of the genus when they wanted to designate a particular species. These phrases, starting with the name of the genus, made up what came to be known as **polynomials,** strings of Latin words and phrases consisting of up to 15 words or even more. Not only were such

FIGURE 22-1

This page from a text of the early 1700s illustrates the continued use of Latin as the scholarly language, as well as polynomial names for identifying organisms, in this case various poppy species (genus Papaver).

polynomials cumbersome, but they also could be altered at will by subsequent authors. Thus a reader could not count on seeing the same name used for the same organism in different books.

The polynomials in a given book were written in such a way that they contrasted the characteristics of a given species with those of its relatives. Therefore the choice of the best polynomial for a particular species depended on which additional organisms were described in the same book. As a result, a given organism really did not have a single name that was its alone. Instead, its names were a series of different descriptive phrases, used in different books, that could be related to one another only by scholars. This was a burdensome and imprecise system of naming. It was certainly not one to which information could easily be added about the rapidly growing numbers of organisms that were being discovered in this age of world exploration.

The Binomial System

The simplified system of naming plants, animals, and microorganisms that has been standard for more than two centuries stems from the works of the Swedish biologist Carolus Linnaeus (1707-1778) (Figure 22-2). Linnaeus' ambition, like that of many of his predecessors, was to catalogue all of the known kinds of organisms and minerals. In the 1750s he produced several major works that, like his earlier books, employed the polynomial system, which had been well established for nearly a century. In attempting to catalogue every kind of object, living and nonliving, Linnaeus worked particularly hard to perfect his polynomials. He considered them to be the basis of his system. Consequently, he tried to make them directly comparable with one another while at the same time forming them in such a way that they concisely showed the distinctive features of the organisms that he was discussing. Linnaeus studied the polynomials that had been used earlier, either by himself or by others. If they fulfilled his criteria, he used them again; if they did not, he invented new ones.

The way that Linnaeus' system worked may be illustrated with an example from his *Species Plantarum* ("The Kinds of Plants"), a two-volume book that appeared in 1753. All of the oaks were grouped in the genus *Quercus,* as they had been since Roman times. The willow oak of the southeastern United States (Figure 22-3, *A*) had by that time been described by a number of travelers and noted by several scientists. Linnaeus took the name that he adopted for this particular kind of oak, *Quercus foliis lanceolatis integerrimis glabris* ("oak with spear-shaped, smooth leaves with absolutely no teeth along the edges"), from an account of the plants of Virginia published from 1739 to 1743 by his friend, the Dutch botanist Jan Fredrik Gronovius. For the common red oak of eastern temperate North America (Figure 22-3, *B*) Linnaeus devised a new name. That name was *Quercus foliis obtuse-sinuatis setaceo-mucronatis* ("oak with leaves with deep blunt lobes bearing hairlike bristles"). Linnaeus needed to invent a new name that fit in better than the earlier ones with the names for the other 13 kinds of oaks he included, according to the principles mentioned above.

In his *Species Plantarum,* Linnaeus introduced a novel practice that soon gave rise to the names that have been used ever since. In the margin opposite the "true" (polynomial) name for each species, he included a single word that he considered apt for making a brief reference to that particular kind of plant. Thus, opposite the polynomial name for the willow oak, Linnaeus entered the word "phellos," and opposite the name for the red oak, "rubra." For each kind of oak that he included, he wrote a distinctive word in the margin of the page. Soon these single words began to be combined by Linnaeus and others with the generic name. Written in this form, they were **binomials,** two-part names that were different for each species. The names for the two kinds of oaks that we have been discussing then became, respectively, *Quercus phellos* and *Quercus rubra.* These have remained their official names since 1753, even though Linnaeus did not intend this when he first used them in his book. He considered the polynomials the true names of the species.

FIGURE 22-2

Carolus Linnaeus (1707-1778), The Swedish biologist who devised the system of naming organisms that is still in use today.

QUERCUS PHELLOS
(WILLOW OAK)

A

QUERCUS RUBRA
(RED OAK)

B

FIGURE 22-3

A *Willow oak,* Quercus phellos.

B *Red oak,* Quercus rubra.

Advantages of the binomial system

The advantage that such binomial names have over the earlier, much longer polynomials is that they are specific and simple to remember. Perhaps an even greater advantage is that biologists have agreed on the principle of priority, by which the earliest name applied to a particular kind of animal, plant, or microorganism remains its official name, regardless of how many other different names are applied to it subsequently. The second part of the binomial name of a species, the **specific epithet,** always remains associated with that species, even if it is decided later that that particular species was originally misclassified in the wrong genus. If that particular specific epithet has already been used in the new genus, then the next specific epithet that has been used for the same species is used, or a new one is invented by the scientist who reclassifies the species, if no earlier one is available.

A fictitious example will illustrate the principle of correcting a misclassification. Suppose that it were decided that the red oak was really a species of beech, *Fagus,* and that it had been misplaced in *Quercus.* If that were the case, the plant would then take the name *Fagus rubra.* Since Linnaeus was the first person to employ the binomial system, scientists have agreed to use his 1753 book as the first publication in which names of plants could be published validly; all earlier names, and all subsequent names that do not use the binomial system, are regarded as invalid. They also have agreed to use the tenth edition of his comparable book about animals, *Systema Naturae,* written in 1758, in the same way for the names of animals. The earliest name for any species, starting in 1753 (for plants) or 1758 (for animals), is the correct name for that species.

> **In Linnaeus' binomial system the name of a species consists of two parts, the first of which is the generic name. The specific epithet—the second word in the name of the species—has priority over other epithets for that species, even if the species is transferred to another genus.**

The Type Method

A further improvement in the system of naming organisms came about with the introduction of the **type** method, starting near the end of the eighteenth century for certain groups of animals. Under this method, an individual specimen is designated by the scientist who names a given species as the type of specimen of that species (Figure 22-4). This specimen, preserved and recognized as a type, becomes the standard against which additional specimens of that species can be compared. If the original author did not designate a type, the type is either the specimen he had—if there is only one—or it is designated formally by

FIGURE 22-4

Type specimen of bristlecone pine, Pinus aristata. *This specimen, preserved in the herbarium of the Missouri Botanical Garden, was collected by C.C. Parry in August, 1862, at timberline in the Rocky Mountains of Colorado near Pike's Peak. It was sent for study to George Engelmann of St. Louis, the leading student of conifers (cone-bearing plants) of his day. Engelmann gave this species the name by which it is still known, and this particular specimen serves as a point of reference for that name.*

a later author. The type method is a useful device for helping to ensure that the names of organisms will be applied consistently to the same species. Before its introduction, taxonomists were free to change the definition of a species if it suited their purposes so that a given name could be used in a sense that differed from that intended by the original author. This led to much confusion about certain names.

> **A type specimen is one that is designated as the official point of reference for the name of a species, either by the original author or by a later one. It is a standard of comparison for applying that name subsequently.**

The scientific rules that govern the names of plants, animals, and microorganisms are outlined in successive editions of three publications: the *International Code of Botanical Nomenclature*, the *International Rules of Zoological Nomenclature*, and the *International Bacteriological Code of Nomenclature*. Legally these documents are completely separate from one another, but for the most part they differ only in details. It is possible, but fortunately rare, for a plant and an animal to have the same name. The scientific names of organisms are the same throughout the world and thus provide a standard precise way of communicating about organisms, whether the language of a particular scholar is Chinese, Arabic, Russian, or English. The system established by Carolus Linnaeus for naming plants, animals, and microorganisms has served the science of biology well for more than 200 years.

WHAT IS A SPECIES?

In Chapter 21 we reviewed the nature of species and saw that there are no absolute criteria that can be applied to the definition of this category. Individuals that belong to a given species, for example, dogs (Figure 22-5), may look very unlike one another. Nevertheless, they are generally capable of hybridizing with one another, and the different forms can appear in the progeny of a single mated pair. On the other hand, the members of a given species often cannot hybridize with those of a second species. For example, dogs are not capable of interbreeding with foxes, which, although they are generally similar to dogs, are members of another, completely distinct, group of mammals. In contrast, dogs can and do form fully or partly fertile hybrids with related species such as wolves and coyotes, which are also members of the genus *Canis*. The transfer of characteristics between these species has, in some areas, changed the characteristics of both of the interbreeding units.

About the only points that can be made about species generally are that they differ from one another in at least one characteristic and that they generally do not interbreed freely where their ranges overlap in nature. In some groups of organisms, including bacteria and many eukaryotes, asexual reproduction predominates and classification systems clearly do not have a genetic basis. Biologists agree, in general, on the kinds of units that they classify as species, but these units share no biological characteristics uniformly.

> **Species differ from one another in at least one characteristic and generally do not interbreed freely with one another where their ranges overlap in nature.**

HOW MANY SPECIES ARE THERE?

As we have seen, the basic unit of biological classification is the species. Since the time of Linnaeus, about 1.5 million species have been named. This is a far greater number of organisms than Linnaeus suspected to exist when he was developing his system of classification in the eighteenth century. The actual number of species in the world, however, is undoubtedly much greater, as illustrated by the following reasoning.

About two thirds of all kinds of organisms that have been named occur in temperate regions, amounting to perhaps 1 million species. However, there may be as many as 1.5

B

A

C

FIGURE 22-5

A *Dogs all belong to one species,* Canis familiaris, *and are capable of interbreeding with one another within the limits of what is possible physically, even if they are as different as the Great Dane and long-haired dachshund shown here.*
B *Coyotes,* Canis latrans, *occur widely in North America.*
C *The gray wolf,* Canis lupus, *extends all over the Northern Hemisphere. Coyotes and wolves belong to the same genus as dogs and produce fertile hybrids with them where their ranges come together. Together with foxes and a few other kinds of carnivores that belong to other genera, dogs, coyotes, and wolves comprise the family Canidae.*

FIGURE 22-6

Scientists attempting to determine the number of species in the canopy of moist tropical forest, the richest biological community on earth. In these experiments the insects and other animals living in the canopy are killed with insecticide and then sampled as they drop onto the sheets below. By such methods, Terry L. Erwin, Jr., of the Smithsonian Institution, has estimated that there may be as many as 30 million kinds of organisms in the world. Taxonomists have thus far recognized only about 1.5 million of these.

million species in temperate regions, judged from the numbers of new species that are discovered when many groups of organisms are studied in detail. For some well-known groups of organisms such as birds, butterflies, and mammals, there are usually at least twice as many species in the tropics as in temperate regions. This implies that there are no fewer than 3 million kinds of organisms in the tropics, because the proportion of numbers of temperate to tropical species is presumably similar in other groups of organisms that are not so well known (Figure 22-6). Since only about 500,000 tropical species of plants, animals, and microorganisms have been named, we may conclude that at least five sixths of all tropical organisms are still unknown scientifically. From these relationships, it appears reasonable to assume that there are at least three times as many species of organisms in the world as have yet been named—a total of at least 4 to 5 million species.

> **There are at least 4 to 5 million kinds of organisms in the world, at least two thirds of them in the tropics. Only about two thirds of the organisms in temperate regions and no more than a sixth of those in the tropics and subtropics have been described and named.**

Our binomial system of nomenclature is adequate to deal with the problem of naming all of these organisms. We are losing kinds of organisms much faster than we are naming them, however, largely because of the destruction of habitats in the tropics. This loss is of great importance because species make up communities and ecosystems that support all life on earth, including our own. To lose them so rapidly, while knowing so little about most of them, is to threaten our survival and our chances of obtaining many unknown or poorly known organisms that might be of great importance to us as crops; sources of medicine, fuel, or fiber, or for other purposes.

THE TAXONOMIC HIERARCHY

When relatively few kinds of plants and animals were known, there was no need to group genera into more inclusive units to reflect their common characteristics. In the taxonomies used when there was no written language, all of the genera in the whole system were known to everyone who spoke that particular language and lived in that particular place. Grouping the genera would not provide additional information about them to anyone and would merely add useless additional names to the language. People might group series of genera together, but they usually did not give names to these more inclusive groups. In a universal system of classification, however, one that deals with the classification of literally millions of species, no one can be familiar with more than a very few genera out of the total recognized. Grouping genera into larger, more inclusive categories is therefore an effective way to remember and study their characteristics. Groups of organisms such as "finches," "legumes," and "mammals" include many genera and have characteristics of their own. If one knows that a genus belongs to such a group, one will be familiar with many of its features.

Linnaeus was aware of the desirability of grouping genera into larger, inclusive categories, but he did not succeed in creating them. As the years passed, other scientists began to group genera into **families.** Thus, for example, the oaks *(Quercus),* beeches *(Fagus),* and chestnuts *(Castanea)* were grouped, along with other genera, into the beech family, Fagaceae, because of the many features they had in common. Similarly, the tree squirrels *(Sciurus),* Siberian and western North American chipmunks *(Eutamias),* and marmots *(Marmota)* were grouped with other related mammals in the family Sciuridae (Figure 22-7).

Eventually, even more inclusive units than families were recognized in the taxonomic system. These units function in the same way as do the names of families, making possible more efficient communication about organisms. Families were eventually grouped into **orders,** orders into **classes,** and classes into **phyla** (sing. **phylum**), which are called **divisions** among plants, fungi, and some microorganisms. Such a system, in which there are more and more inclusive categories at higher and higher levels, is called a **hierarchical** system. The derivation of the word hierarchical is interesting: it comes from the Greek words *hieros,* "sacred," and *archos,* "leader" or "ruler." Eventually, by an extension of the meaning of "chief of a sacred order," the term came to imply levels of rule, or order. Biological classification, and indeed most classification, is hierarchical.

A

B

C

TABLE 22-1   SAMPLE CLASSIFICATIONS OF THREE REPRESENTATIVE ORGANISMS

CATEGORY	HUMAN BEING	HONEYBEE	RED OAK
Kingdom	Animalia	Animalia	Plantae
Phylum (division)	Chordata	Arthropoda	Anthophyta
Class	Mammalia	Insecta	Dicotyledones
Order	Primates	Hymenoptera	Fagales
Family	Hominidae	Apidae	Fagaceae
Genus	*Homo*	*Apis*	*Quercus*
Species	*Homo sapiens*	*Apis mellifera*	*Quercus rubra*

By convention, species are grouped into genera, genera into families, families into orders, orders into classes, and classes into phyla (or divisions). Phyla are the basic units within kingdoms. Such a system is hierarchical.

Table 22-1 shows how the human species, the honeybee, and the red oak are placed in taxonomic categories at different hierarchical levels. The categories at the different levels may include many, a few, or only one item, depending on the nature of the relationships in the particular groups involved. Thus, for example, there is only one living genus of the family Hominidae, but, as we have mentioned, there are several living genera of Fagaceae. Each of the categories at every level implies to someone familiar with the system, or with access to the appropriate books, both a series of characteristics that pertain to that group and also a series of organisms that belong to it.

The Form of Scientific Names

The scientific names of genera and species of organisms are written, by convention, in italics or other distinctive print, or are underlined. The scientific names of the other taxonomic units are not printed this way or underlined. The names of all families of animals end in -idae, and those of the families of plants and microorganisms end in -aceae. Among the plants are a handful of optional exceptions to this rule, such as using the name "Legumino-

FIGURE 22-7

Three genera of the squirrel family, Sciuridae, which differ greatly in their characteristics and are adapted for different modes of life.
A *The red squirrel,* Tamiasciurus hudsonicus, *an agile dweller in the trees.*
B *Townsend's chipmunk,* Eutamias townsendii, *a member of a genus of diurnal, brightly colored, very active ground-dwelling squirrels.*
C *Yellow-bellied marmot,* Marmota flaviventris, *a large squirrel that lives in burrows. Yellow-bellied marmots are social, tolerant, and playful; they live in harems that consist of a single territorial male together with numerous females and their offspring. The flying squirrel shown in Figure 22-9 represents a fourth genus of the family Sciuridae, one that is nocturnal; the three species shown here are all diurnal.*

FIGURE 22-8

Animals or plants may resemble one another in their overall features but differ greatly in many other respects. Such a pattern indicates that the obvious ways in which they resemble one another have been acquired relatively recently, and do not reflect a close common origin. For example, these thick-trunked trees efficiently conserve water in arid habitats, and they have evolved independently in widely separated desert areas.

A Elephant tree, Pachycormus discolor, in the deserts of Baja California, Mexico; this plant is a member of the family Anacardiaceae, to which the very different-looking poison ivy belongs. Elephant trees and poison ivy are relatively close relatives, sharing many features; elephant trees are very remotely related to the superficially similar plants shown in **B** and **C**.

B A treelike member of the grape family (Vitaceae), Cissus, in Namibia.

C Baobab, Adansonia muelleri, in Western Australia. Most species of baobabs, which are members of the silk-cotton family (Bombacaceae) and thus relatively closely related to the mallows, occur on Madagascar, with one species each in continental Africa and Australia. The process whereby such similar forms are produced from different ancestors is called **convergent evolution.**

A

sae" in place of "Fabaceae" for the legumes. Because of their conventional endings, the names of families and certain other taxonomic categories can easily be recognized as such. The names of genera may be abbreviated when they are part of a species name and it is clear from the context what is intended (for example, Q. rubra for Quercus rubra). On the other hand, specific epithets, the second words in the names of species, are never used by themselves to designate the species; such words often occur in many different genera. For example, the sandhill crane, Grus canadensis; the red-breasted nuthatch, Sitta canadensis; and a common northern species of shadbush or serviceberry, Amelanchier canadensis, share the same specific epithet but of course have different names. Just as your first name will not by itself fully identify you, so the specific epithet means nothing by itself.

CLASSIFICATION AND EVOLUTION

Today scientists consider their system of classification to be an evolutionary one, and they therefore believe that, ideally, each of the named entities in the taxonomic hierarchy represents a group of microorganisms, animals, or plants that had a common ancestor. Since the time of Darwin, taxonomists have attempted to construct their systems of classification in such a way that these systems would reflect as far as possible the course of evolution in the particular group. Such systems are called **phylogenetic** ones: they are supposed to reflect the **phylogeny,** or pattern of descent, of the group. Each group is believed to have evolved from a single common ancestor, and the subunits used in the classification of the group are supposed to reflect the evolutionary branches that diverged from one another during the history of the group.

The way that scientists deduce which groups were actually derived from a common ancestor, however, is precisely the same way that systems of classification were constructed long before Darwin's time. Similar kinds of organisms were grouped together, and dissimilar ones were separated from them. A system of classification that has been constructed according to this principle will generally provide the best estimate of the course of evolution in the group also. Whether overall resemblance or the evolutionary history of the group is stressed, the way that it is classified probably will not differ greatly. The relationships of the giant panda and the red panda, as interpreted in light of modern evidence, are presented on pp. 838-839 as an example of the way such reasoning is developed.

The underlying reason for this observation is that the characteristics of organisms have in fact resulted from the process of evolution. Groups of organisms can be recognized

B

C

as a result of their descent from common ancestors because the members of a group share characteristics that were present in these common ancestors. It is the correlation of these characteristics that is used in the construction of classifications. This same correlation makes possible the construction of useful hierarchical systems of classification. However, classifications based on one or a few features may be highly misleading and are said to be **artificial.** They often group unrelated organisms that may not have much in common beyond the few features on which the classification was based. If those features are of overwhelming importance, the artificial systems of classification may be useful, but for more general purposes they tend to be highly misleading. In contrast to artificial systems of classification, those based on many features are said to be **natural** ones (Figure 22-8).

Natural systems of classification are more useful than artificial ones. They are more likely to reflect the phylogeny of the group being classified and therefore are more likely to be useful in predicting other features of the organisms that are classified. Many characteristics of organisms change during the course of evolution, and these characteristics are correlated with one another; thus, if we know the evolutionary history of a group, we can construct a classification that will convey more information about the organisms being classified than if we do not. **Numerical taxonomy** is a field in which the attempt is made to correlate numerous characteristics and use them in the classification of organisms.

> **Artificial systems of classification are those based on one or a few features of the organisms being classified; natural systems, in contrast, are based on many features. Natural systems are more likely than artificial ones to be phylogenetic, that is, to reflect the lines of descent in the group being classified.**

If the kinds of organisms that exist had been created individually, one might expect their characteristics to be distributed randomly. In such a case neither comprehensive classification schemes nor deductions about the probable course of evolution would be possible. For example, if we were to produce a scheme of classification that put together all flying organisms in one category, it would place bats, butterflies, and birds together and put mice, spiders, and reptiles in some other category. Regardless of evolutionary theory, however, bats clearly have much more in common with mice than they do with butterflies; any careful study of their overall characteristics immediately reveals this. Grouping the two kinds of mammals together on the basis of multiple characteristics, therefore, produces a much more useful classification than one based on the single attribute of flight.

FIGURE 22-9

For obvious reasons, people have always considered plants and animals to be very distinct from one another. The southern flying squirrel, Glaucomys volans, top, shown feeding on berries, appears completely unlike the ladyslipper orchid, Cypripedium calceolus, bottom. Animals ingest their food, and most move about from place to place; plants manufacture their food through the process of photosynthesis and are stationary. These and other major differences have traditionally caused plants and animals to be regarded as distinct kingdoms, with their similarities being appreciated only in the past hundred years or so.

CLADISTICS

An important advance in the methodology of taxonomy was made recently, based on principles that were intuitively clear earlier. This advance has been based on the principles of using many features to construct systems of classification and on ordering of these characteristics in such a way as to illuminate the course of evolution in the group being classified.

In practice only a limited number of the features of organisms are usually taken into consideration in the construction of classification systems. This tendency is almost unavoidable, given the enormous numbers of organisms that exist and the small numbers of people studying them. Taxonomists, therefore, have tended to construct schemes of classification based on only a few features. This is especially true for groups of organisms that are not well known. These features are generally the ones that can be observed relatively easily or that have been emphasized in earlier systems of classification for the group. If the particular characteristics are well chosen and highly correlated with many other features, the schemes of classification will be more or less natural; if the characteristics are not well selected, they will be artificial. We cannot predict the characteristics of organisms efficiently from their position in artificial systems of classification; in natural systems we can presume that the kinds of organisms grouped together will have a number of features in common, and thus we can predict unknown features with a high degree of accuracy.

An improved understanding of the nature of the characteristics used in classification was made possible in the 1940s by the German entomologist (student of insects) Willi Hennig, who first distinguished clearly the significance of different kinds of characteristics. Some features have evolved relatively recently, and these **advanced** or **specialized** features therefore mark groups of organisms that have had a common ancestor. Other **generalized** features evolved earlier in the history of the particular group. These generalized features may in a sense be evolutionary "holdovers" that have been retained in different members of the overall group. Such features therefore do not indicate a direct relationship among the organisms in which they occur in the same way that the advanced features do.

A group of systematic biologists called **cladists** attempt to differentiate between specialized and generalized features and by doing so to construct classification systems that reflect the evolutionary history of a group. Cladists emphasize the more telling specialized features in setting up appropriate schemes of classification and evolution for particular groups of organisms.

> **Cladistics attempts to determine which characteristics of the organisms are specialized, derived ones that truly reflect common descent, and it emphasizes such features in classification. The goal of cladistics is to produce phylogenetic systems of classification.**

THE MAJOR GROUPS OF ORGANISMS

To Linnaeus and his contemporaries, the most important distinction in the living world was that between plants and animals. Animals moved about, ate things, actively sought their mates, and preyed on other animals or on plants. Plants, in contrast, were stationary, did not seem to eat anything, and had obscure mating systems that were just beginning to be understood. Following well-established traditions (Figure 22-9), Linnaeus considered animals and plants to belong to separate **kingdoms** and minerals to constitute a third.

Problems with Traditional Groupings

Generally these traditional distinctions persisted into the twentieth century, even though some additional kinds of organisms had been discovered that were troublesome to the established definitions. Some of these microscopic organisms swam about but manufactured their own food like plants while they were doing it. Others remained in one place and not only manufactured food but also seized prey. By analogy, however, it was still possible to regard the first of these groups as plants, the second as animals. This merely avoided the question, however; the traditional criteria for recognizing kingdoms were no longer applicable. The application of increasingly sophisticated biochemical techniques and the development of the electron microscope began to reveal a more detailed picture of the similarities and differences between the major groups of organisms. These techniques made especially clear the tremendous diversity among the many forms of life that had traditionally been grouped as "plants" and which are still treated as plants in many books.

The Five Kingdoms of Organisms

During the past 50 years, biologists have recognized that the most fundamental distinction among groups of organisms is not that between animals and plants, but that between prokaryotes and eukaryotes. The prokaryotes, or bacteria (Figure 22-10), differ from all other organisms in their lack of membrane-bound organelles and microtubules, as well as in their possession of simple flagella. Their DNA is not associated with proteins and is not located in a nucleus. Bacteria do not undergo sexual recombination in the same sense that eukaryotic organisms do, although forms of genetic recombination occur occasionally in some bacteria. The cell walls of most bacteria contain muramic acid, which is only one of the many biochemical peculiarities by which they differ from all eukaryotes. Bacteria existed for at least 2 billion years before the appearance of eukaryotes. In recognition of their highly distinctive features, the bacteria have been assigned to a kingdom of their own, **Monera,** which will be treated in detail in Chapter 26.

The features of eukaryotes contrast sharply with those of prokaryotes, as we stressed in Chapter 5 and will now review. All eukaryotes have microtubules in their cytoplasm, and their cells are characterized by classes of membrane-bound organelles. They have a clearly defined nucleus that is bounded by a double membrane, the nuclear envelope. Their chromosomes are complex structures in which the DNA is characteristically associated with proteins; the chromosomes divide and are distributed in a regular manner to the daughter cells as a result of the process of mitosis. When flagella and cilia are present in eukaryotes, they have a characteristic 9 + 2 structure that consists of microtubules. Mitochondria are among the complex organelles that occur in the cells of all eukaryotes. In addition, eukaryotes exhibit integrated multicellularity of a kind that is not found in bacteria, as well as true sexual reproduction.

Despite their often bewildering diversity of form, eukaryotes are much less diverse metabolically than bacteria. All eukaryotes are believed to have evolved from a common ancestor. Eukaryotes consist basically of a number of groups of organisms that are exclusively or primarily single-celled, together with several multicellular groups that were derived from single-celled ancestral eukaryotes (Figure 22-11).

The classification of organisms, especially eukaryotes, into kingdoms is clearly somewhat arbitrary, and no scheme has been universally accepted. Here we have adopted what is called the **five-kingdom system,** in which four of the kingdoms are eukaryotic organisms. The fifth kingdom, Monera, contains the bacteria, as we mentioned above. In the five-kingdom system of classification the array of diverse eukaryotic phyla consisting predominantly of single-celled organims is assigned to the kingdom **Protista,** which is discussed in Chapter 27. Three major multicellular lines that have originated from this kingdom, each with a great many species, are assigned to separate kingdoms: **Animalia, Plantae,** and **Fungi** (Figure 22-12). Although the fungi include a few single-celled forms, the yeasts, these have been derived from multicellular ancestral forms. Each of the three multicellular kingdoms—plants, animals, and fungi—has evolved from a different ancestor among Protista.

The members of the three multicellular kingdoms derived from Protista differ in their modes of nutrition: animals ingest their food, plants manufacture it, and fungi digest

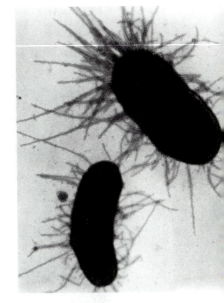

FIGURE 22-10

Bacteria, such as these individuals of Escherichia coli, *are the only members of the kingdom Monera. They are prokaryotes, a term that describes a pattern of cellular organization completely different from that of all other organisms. The slender projections extending from the surfaces of these bacteria are called pili.*

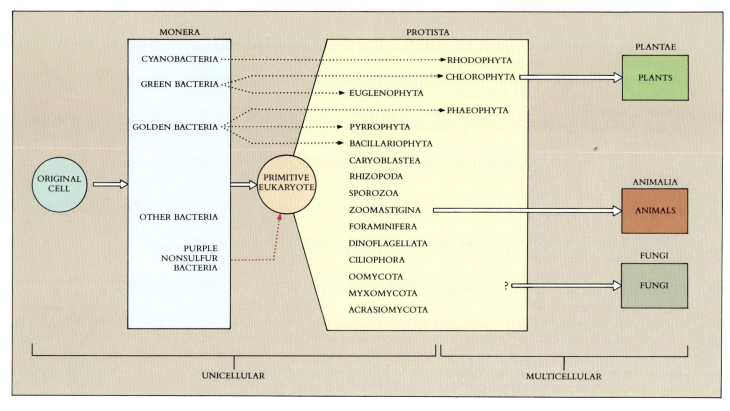

FIGURE 22-11

The evolutionary relationships of the phyla of organisms.

A   B   C

FIGURE 22-12

Animals, fungi, and plants are the three multicellular kingdoms derived from members of the kingdom Protista.
A *Tortoise beetle, family Chrysomelidae, in Costa Rica.*
B *Part of the complex leaf of a palm.*
C *Mushroom, Mill Valley, California.*

it by means of enzymes and absorb it. Fungi are filamentous and have no motile cells; plants are mainly stationary, but some of them have motile sperm; and many animals are highly motile.

Protists are extraordinarily diverse both in their form and in their biochemistry. Three of the phyla of protists—the red, brown, and green algae—have attained multicellularity independently of the animals, plants, and fungi (Figure 22-14). These three groups are considered plants in some schemes of classification because they are photosynthetic and multicellular. Furthermore, multicellular green algae were almost certainly the direct ancestors of the plants, as we will discuss in more detail in Chapters 27 and 29. Since the fungi, plants, and animals are reasonably clearly defined, however, and are major evolutionary groups that presumably have each had a single common ancestor different from the others, we will treat them as separate kingdoms and leave all remaining eukaryotic organisms as members of the diverse kingdom Protista.

Viruses present a special classification problem. They are nonliving, but capable of replication. They are basically fragments of nucleic acids derived from the genome of a eukaryote or that of a prokaryote. The viruses have the capacity to organize protein coats around themselves (Figure 22-13). They are able to direct the machinery of their host cells to manufacture more virus material, but they cannot exist on their own. Viruses cannot logically be placed in any of the kingdoms, since they are not organisms. They are discussed in detail in Chapter 25.

> **We recognize five kingdoms of organisms: Monera, Protista, Fungi, Animalia, and Plantae. Among these, the Monera, also called bacteria, are prokaryotes, and the members of the other four kingdoms are eukaryotes. Protista are predominantly unicellular but have several wholly or partly multicellular lines. Three of the major multicellular lines—plants, animals, and fungi—are recognized as distinct kingdoms. Viruses are nonliving and therefore are not included in any of the kingdoms of organisms.**

In the remaining chapters of this section we will first outline the major features of the evolution of life on earth and then discuss the evolution of our own species, *Homo sapiens*, in detail. The following section will present a detailed discussion of the major groups of organisms, stressing their evolutionary relationships with one another.

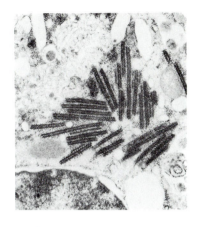

FIGURE 22-13

A rod-shaped virus called GD7 replicating itself in a human white blood cell. Viruses are segments of DNA or RNA, ultimately derived from the genomes of either bacteria or eukaryotes, which have the ability to take over the protein-synthesizing machinery of their host cells. They are nonliving but derived from organisms. Viruses manufacture protein coats around their nucleic acid cores.

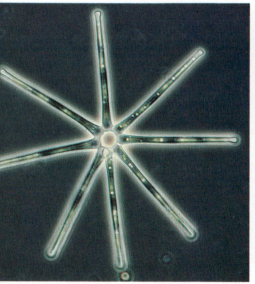
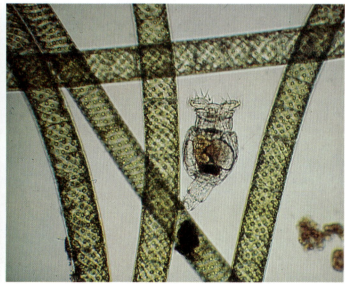
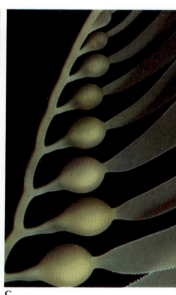

| A | B | C |

FIGURE 22-14

The members of the kingdom Protista are very diverse, as suggested by these photographs.
A *Freshwater diatom, a member of a mostly photosynthetic, unicellular group of organisms, characterized by silica shells, that occurs widely in fresh and salt water.*
B *A green alga, Spirogyra, an organism with a filamentous growth form and a single, ribbon-shaped spiral chloroplast, by means of which it manufactures its own food, with a rotifer, a small, motile animal that captures and ingests its food.*

C *A brown alga, or kelp, of a completely multicellular member of the kingdom Protista, the individuals of which may be many meters in length. Brown, red, and green algae are divisions (phyla) of Protista not classified as plants, even though they are all photosynthetic, because they were derived independently from different ancestors among the Protista. Many green algae and a few red algae are unicellular.*

SUMMARY

1. Systems of classification have been a part of all languages; they are necessary for communication.

2. The introduction of printing into Europe about 500 years ago made possible the widespread distribution of books, and the subsequent era of exploration brought thousands of new kinds of plants and animals to the attention of scientists. As a result, people found that they needed encyclopedic works and universal systems of classification.

3. Carolus Linnaeus, an eighteenth-century Swedish scientist, developed the system of naming organisms that is still used today. In it, every species of organism has a binomial name consisting of two words, the first of which is the name of the genus and the second of which, when combined with the name of the genus, designates the species. Other, more inclusive categories have been added from the late eighteenth century onward and provide additional means of communicating about organisms and their properties.

4. There are at least 4 million species of plants, animals, and microorganisms in the world. We have named only about a third of them so far, but we are driving them to extinction faster than we are naming them.

5. Evolutionary or phylogenetic schemes of classification attempt to arrange organisms according to their patterns of descent. In practice, these schemes are based on overall correlations of characteristics, which may be ordered by the methods of cladistics. Artificial systems of classification are based on only a few characteristics of the organisms being classified. They may still be useful if the characteristics used are well chosen. Artificial systems are often the only ones available for groups of organisms that are not well known.

6. The categories that are used in the taxonomic hierarchy of organisms, from the largest and most inclusive to the smallest, are kingdoms, phyla (divisions), classes, orders, families, genera, and species. Species are sometimes divided into subspecies or varieties.

7. The fundamental distinction among living organisms is that which separates the prokaryotes (kingdom Monera) from the eukaryotes. Among the latter are a large number of diverse phyla. Each of the most prominent multicellular evolutionary lines— the animals, plants, and fungi—is recognized as a separate kingdom. The remaining eukaryotes, which are mainly unicellular phyla, are grouped together in the diverse kingdom Protista. In all, we classify organisms into five kingdoms. Viruses are not organisms and are not included in the classification of organisms.

SELF-QUIZ

The following questions pinpoint important information in this chapter. You should be sure to know the answers to each of them.

1. In a polynomial system for naming animals and plants, the author constructs the names by
   (a) naming the new organisms after the founder
   (b) giving the organisms a number
   (c) putting together various descriptive words after the name of the genus

2. In the binomial system of naming organisms, the name of a species consists of two words, the first of which is the
(a) genus
(d) family
(b) specific epithet
(e) class
(c) phylum

3. In general, the specific epithet remains associated with a particular species even if the species is moved to a new genus. True or false?

4. The _____ of a species is the standard specimen with which others are compared to determine whether they belong to the same species.

5. A general characteristic of species is that
(a) they are all the same color and shape
(b) they all live in the same location
(c) they cannot produce hybrids with other species
(d) all of the above
(e) none of the above

6. Asexually reproducing species can be recognized by
(a) hybridization to other organisms
(d) two of the above
(b) morphological characteristics
(e) all of the above
(c) physiological characteristics

7. The scientific name of a species is written in which style?
(a) Quercus rubra
(c) *Quercus rubra*
(b) Quercus Rubra
(d) *Quercus Rubra*

8. Related genera are grouped into a taxonomic category called the _____.

9. Which of the following lists presents a correct order of taxonomic categories?
(a) kingdom, class, phylum, order
(d) class, order, family, genus
(b) class, family, order, genus
(e) none of the above
(c) phylum, order, family, species

10. The _____ of a group, its pattern of evolutionary descent, is often reflected in the taxonomic hierarchy.

11. Classifications based on one or a few features of the organisms being classified are not as useful as those based on many features. Those based on one or a few features are said to be _____.

12. _____ features mark groups of organisms that have had a common ancestor, whereas _____ features evolved earlier in the history of the organisms.

THOUGHT QUESTIONS

1. What are some of the advantages of the binomial system of naming organisms?

2. How do you think the computer might affect the system of classification of organisms, both as an information retrieval device and as a way of using the information? Comment on the statement that the introduction of the computer in the mid–twentieth century has caused an information revolution comparable to that which accompanied the introduction of movable type and the resulting wide distribution of printed books in Europe.

3. Do you think a better system of classification would be to group all photosynthetic eukaryotes, regardless of whether they are single-celled, as plants, and to group all nonphotosynthetic ones as animals? Present arguments in favor of doing so and other arguments in favor of the system of classification used in this book.

FOR FURTHER READING

ELDREDGE, N., and J. CRACRAFT: *Phylogenetic Patterns and the Evolutionary Process,* Columbia University Press, New York, 1980. Good introduction to modern systematic theory.

MARGULIS, L., and K.V. SCHWARTZ: *Five Kingdoms. An Illustrated Guide to the Phyla of Life on Earth,* W.H. Freeman and Company, San Francisco, 1982. A marvelous account of the diversity of organisms, beautifully illustrated and presented in a logical fashion.

MAYR, E.: "Biological Classification: Toward a Synthesis of Opposing Methodologies," *Science,* vol. 214, pages 510-516, 1981. A thoughtful consideration of alternative methods of classifying, with emphasis on genera and species.

ROSS, H.H.: *Biological Systematics,* Addison-Wesley Publishing Co., Inc., Reading, Mass., 1974. A nicely balanced overview of taxonomy.

Overview

No more than a billion years after the condensation of the earth, which was complete by about 4.5 billion years ago, bacteria were abundant. After another 2 billion years, eukaryotic cells appeared—about 1.5 billion years ago. Half that long ago, the eukaryotic cells began to aggregate as multicellular organisms that were more efficient than their unicellular ancestors, thus extending the mastery of living things over their environment. Just over 400 million years ago the ancestors of the groups that are so successful on land today, the insects and the plants, appeared for the first time; 50 million years later they were followed by the first terrestrial vertebrates. More kinds of plants, animals, and microorganisms inhabit the earth now than have ever existed before, and the great majority of them live on the land.

For Review

Here are some important terms and concepts that you will encounter in this chapter. If you are not familiar with them, you should review them before proceeding.

- **Isotopes** (Chapter 2)
- **The origin of life** (Chapter 3)
- **Punctuated equilibrium** (Chapter 19)
- **The origin of species** (Chapter 21)
- **Taxonomic categories** (Chapter 22)

Individuals that belong to distinct major groups of organisms differ so greatly from one another superficially that it is difficult at first to identify the features they have in common. Nevertheless, the biochemical processes that occur within an oak tree, a bear, and a bacterium in your gut resemble one another so closely that we are justified in concluding that all of these kinds of organisms are related by descent. The purpose of this chapter is to provide an overview of the major features of the evolution of life on earth, relating this overview both to time and to geography. Further details will be given in Chapters 24 to 39, in the course of our discussion of individual groups.

THE FOSSIL RECORD

Over the course of time, it has become clear to students of evolution that we can interpret the origins of the major groups of organisms by applying the same principles that underlie the origin of species. If we thoroughly understand the homologies of the structures in different major groups of organisms, we will be able to outline in detail the relationships between them. An additional very important line of evidence about the major outlines of evolution, however, is provided by the fossil record, which is studied and interpreted by paleontologists (Figure 23-1).

Only fossils provide definite evidence about what extinct kinds of organisms looked like, including the ancestors of those that are living now. On the basis of the characteristics of living organisms, we can make deductions about their ancestors, but fossils provide the means whereby these deductions can be checked. Fossils provide an actual record of organisms that once lived, an accurate understanding of where and when they lived, and some appreciation of the environment in which they lived. Fossils also enable us to trace lines of evolutionary progression among individual groups of organisms, provided that enough fossils of various ages are available. For many reasons, however, the fossil record is not as complete as we might wish, and the samples that we have been able to obtain are not as extensive as would be ideal. Both mechanical and theoretical problems have contributed to this lack of completeness.

Mechanical Problems in Preservation

Only a minute fraction of the organisms living at any one time are preserved as fossils. Most of the fossils that we do have are preserved in **sedimentary** rocks. Sedimentary rocks are formed of particles of other rocks that are weathered off, deposited at the bottom of bodies of water or accumulated by wind, and then hardened into strata of the sort that Darwin found so filled with intriguing fossils in South America. During the formation of sedimentary rocks, dead organisms are sometimes washed down along with inorganic debris, such as mud or sand, and eventually reach the bottom of some pond, lake, or the ocean itself. Occasionally, dead organisms are covered by debris accumulated by wind. In exceptional circumstances fossils may be preserved in an organic substance, such as tar: the La Brea Tar Pits in Los Angeles (Figure 23-2) illustrate this phenomenon. In other instances

FIGURE 23-1

A reconstruction of a community of marine organisms in the Cambrian Period, 500 to 550 million years ago. The swimming animals, which have jointed legs and large compound eyes, are trilobites, early arthropods (phylum Arthropoda); their features have been reconstructed from fossils. Trilobites disappeared about 245 million years ago. On the sea floor is a colony of tube sponges (phylum Porifera).

organisms themselves may form the whole deposit, as in the formation of coal or oil deposits. However, many organisms do not live in places where their remains are likely to be preserved in sedimentary rocks; for this reason they are unlikely to be preserved as fossils.

Sedimentation proceeds steadily in the sea, but the seabed is constantly being renewed, as you will see in the next section. Most of the existing seabed is less than 100 million years old, but despite that limitation, we do find a more complete record of life in the sea than anywhere else. For the older record to reach us, however, the seafloor must be located in strata that are eventually uplifted and become land.

For specific organisms to become preserved as fossils, they must usually be buried before their decay is complete. In addition, the process of decay, for at least some portions of the organism, must stop after they are buried. Since the soft parts usually decay rapidly, normally only structures such as bones, teeth, shells, scales, leaves, and wood are available in the fossil record. If fossils are exposed, they often disintegrate quickly; most of those we find, therefore, were exposed recently. When fossils are formed, the actual parts of the organisms are almost always replaced by minerals; a fossil rarely contains any of the material that made up the body of the organism originally, but rather is a mineralized replica of that body part.

In interpreting the structure of fossil organisms, we must normally use the hard parts to try to interpret what the soft parts looked like. For organisms such as worms, which have no hard parts, fossils are understandably rare. Even though soft-bodied animals undoubtedly evolved before their hard-bodied counterparts, we have little evidence of their history in the fossil record. Sometimes soft-bodied animals are preserved in exceptionally fine-grained muds, in conditions under which the supply of oxygen was poor while the muds were being deposited and deterioration was therefore slowed (Figure 23-3).

Theoretical Problems in Interpretation

The fossil record is relatively complete over the past 650 million years, but even from that point to the present it contains substantial gaps. These gaps may represent periods when, for one reason or another, sedimentary rocks were not laid down. On the other hand, as we discussed in Chapter 19, many of the apparent gaps may be genuine and reflect "spurts" of evolution, with the rapid loss of intermediate forms. The fossil record provides the pri-

FIGURE 23-2

Life in what is now downtown Los Angeles about 15,000 years ago. The skeletons of these animals, which were similar in size and abundance to those which have survived from the Pleistocene Epoch on the plains of East Africa, were preserved in large pools of tar (asphalt) into which they fell.

mary evidence for deciding between gradualism and punctuated equilibrium as major patterns of evolution, but it is rarely complete enough to allow us to decide about the particular pattern of evolution for an individual group.

> **Fossils provide the concrete means whereby we can judge our deductions about the history of particular groups of organisms. Fossils are found mainly in sedimentary rocks, and those organisms which live in areas where sedimentary rocks are being formed are the most likely to be preserved. The fossil record is rarely complete enough to allow whole evolutionary sequences to be studied for a particular group.**

Dating Fossils
Correlations of strata

One of the ways of determining how old particular fossils are is to compare the sequences in which they appear in different strata. Such sequences demonstrate the evolutionary progression of life on earth. By comparing the assemblages of fossils that occur in strata at different places, paleontologists can estimate their relative dates of occurrence and thereby determine regional patterns of evolution. By making such correlations, they were able to establish the broad patterns of progression of life on earth and the relative ages of different sequences of fossils well before Darwin completed his formation of the theory of evolution.

Many of the geological periods were recognized and given names before methods were available that would allow their actual dates to be established. An outline of the eras, periods, and epochs, including their dates (which were assigned later) and major features, is presented on the inside cover of this book. The past periods were usually named for the areas where they were especially well represented or first studied, or for other outstanding characteristics. For example, rocks from the Cambrian Period are abundant in Wales, and the period is named for "Cambria," the Roman name for Wales. In contrast, the name for the Carboniferous Period comes from the Latin words for "coal-bearing," since many of the major coal deposits that were first studied dated from this period. The relative sequence of these distinctive strata was established well over a century ago, but their absolute dates are still under active study.

> **The geological eras, periods, and epochs, summarized on the inside cover of this book, provide a means for organizing data about the history of life on earth. They were named for the characteristic strata associated with them long before their actual dates had even been estimated.**

It is usually assumed that strata which contain similar fossils were deposited at about the same time. One obvious limitation of this method, however, results from the fact that a particular kind of organism often originates in one area and then migrates to other areas over a long period of time. When whole groups of fossils are considered together, this limitation is lessened. However, because of migration and the spread of organisms over the earth, it cannot be assumed automatically that the occurrences of a particular kind of organism in different places date from the same period of time.

Direct methods of age determination

Direct methods of dating rocks and fossils first became available in the late 1940s. Naturally occurring radioactive isotopes of certain elements are employed in this process. Such isotopes are unstable and decay over the course of time at a steady rate, producing other isotopes. One of the most widely employed methods of dating, the **carbon-14** (^{14}C) method, employs estimates of the different isotopes present in samples of carbon.

Most carbon atoms have an atomic weight of 12; the symbol of this particular isotope of carbon is ^{12}C. A fixed proportion of the atoms in a given sample of carbon, however,

FIGURE 23-3

A *Paradoxides carolinaensis, a Precambrian fossil from the Piedmont of south-central North Carolina. This fossil, perhaps about 630 million years old, represents a segmented animal about 9 centimeters long and about 2.5 centimeters wide; it had 26 segments. Its relationships, and those of most other animals that were contemporary with it, are unknown. It is the best-preserved multicellular fossil of its age that has yet been reported from the United States.*

B *Mawsonites spriggi, a fossil jellyfish (phylum Cnidaria) from the Ediacara formation in South Australia, illustrating the remarkable preservation of soft-bodied fossils in these fine-grained sedimentary rocks. This jellyfish was originally conical, with distinct lobes on its upper surface, and about 11 centimeters across. During Precambrian time, nearly all of the major phyla and most of the classes and orders of animals (other than chordates) first appeared.*

consists of carbon with an atomic weight of 14 (^{14}C), an isotope that has two more neutrons than ^{12}C. ^{14}C is produced from ^{12}C as a result of bombardment by particles from space. The carbon that is incorporated into the bodies of living organisms consists of the same fixed proportion of ^{14}C and ^{12}C that occurs generally. After an organism dies, however, and is no longer incorporating carbon, the ^{14}C in it gradually decays over time, by the loss of neutrons, to ^{12}C. It takes 5730 years for half of the ^{14}C present in a sample to be converted to ^{12}C by this process; this length of time is called the **half-life** of the ^{14}C isotope.

It is possible, therefore, to date accurately fossil material that contains carbon, as all organic material does, provided that the fossil material is less than about 50,000 years old. This can be done by estimating as accurately as possible the proportion of ^{14}C that is still present in the carbon. For older fossils, the amount of ^{14}C remaining is so small that it is not possible to measure it precisely enough to provide accurate estimates of age.

To determine the ages of older samples, one can sometimes study the decay of other radioactive isotopes. For example, ^{40}K (potassium–40) decays very slowly into ^{40}Ar (argon–40) and can be used to date much older rocks. The half-life of ^{40}K is 1.3 million years; other isotopes have even longer half-lives. With the use of such **radiometric** dating, our knowledge of the ages of various rocks has become more precise. Unfortunately, such dating methods can establish only the ages of the materials in sedimentary rocks and not the time when they were incorporated into those rocks. They do, however, allow us to estimate the ages of all of the rocks on earth, regardless of how old they may be.

The ages of fossils may be estimated by correlating the strata in which they occur. A more accurate method is to determine the proportions of different isotopes of carbon in the organic material preserved in them, or the proportions of different isotopes in the rocks where they are found. Radioactive isotopes change from one form to another over time, so that the proportions in a given sample will provide an absolute date for the age of that sample.

THE FORMATION AND MOVEMENT OF CONTINENTS

As you saw in Chapter 2, the planets of our solar system, including the earth, began to coalesce approximately 4.6 billion years ago. The accretion of the early earth seems to have been completed about 4.5 billion years ago. After several hundred million years, the lighter silica (SiO_2)-rich rocks that make up the continents had formed. The oldest known rocks on earth are 3.8 billion years old; by the time they were consolidated, some portions of the earth's crust had begun to move in relation to the others. As these sections moved, they became thicker.

About 200 million years ago, in the early Jurassic Period, the major continents were all together in one great supercontinent (Figure 23-4). Alfred Wegener, the German scientist who in 1915 first proposed the idea of continental movement in his book, *The Origin of Continents and Oceans,* called this giant land mass *Pangaea.* Because the specific mechanism Wegener proposed to account for continental movement was not feasible, his theory fell into discredit with the great majority of geologists and biologists. Indeed, although it is now generally accepted, Wegener's theory was denied by most scientists for nearly half a century.

In the early 1960s, new evidence developed that provided a mechanism for continental movement. The theory that emerged visualized heavy, basaltic, ocean-floor rocks moving like a conveyor belt away from the mid–ocean ridges where they were formed. In the process, they carried the lighter continental rocks along with them. The mid–Atlantic Ridge, for example, is an enormous zone of upwelling basaltic lava. It generates basaltic, ocean-floor rocks that move out from the Ridge both toward Europe and Africa and toward North and South America. As these rocks continue to be formed, both pairs of continents ride farther and farther apart from one another (see Figure 2-A).

The earth's crust and associated upper mantle, which together are about 100 to 150

A 200 MILLION YEARS AGO

B 135 MILLION YEARS AGO

C 65 MILLION YEARS AGO

D PRESENT

FIGURE 23-4

The positions of the continents at 200, 135, and 65 million years ago, and at present.

FIGURE 23-5

The Himalayas have been uplifted to the highest elevations on earth as a result of the collision of the Indian subcontinent with the main body of Asia, a process that has been going on for tens of millions of years and continues today.

kilometers thick, are divided into plates. There are seven enormous ones and a series of smaller ones that lie between them. Because of the existence of these plates, the theory that explains the movement of continents in this way is called **plate tectonics.** Earthquakes, in general, are caused by the relative movement of these plates. Thus an earthquake along the San Andreas Fault in California, such as the one that destroyed San Francisco in 1906, results from the relative movement of two gigantic segments of the earth's crust and mantle. Most mountains are thrust up by plate movements; for example, the Himalayas have been thrust up to the highest elevations on earth as a result of the grinding, prolonged collision of the Indian subcontinent with Asia (Figure 23-5).

> **Basaltic lavas, flowing out of fissures in the earth's crust like the mid–oceanic ridges, form great plates of relatively heavy rocks 100 to 150 kilometers thick. These rocks move, carrying the lighter continents along with them. There are seven huge plates, all moving in relation to one another, and a number of smaller ones.**

History of Continental Movements

The continents have gradually moved apart from their positions as parts of Pangaea 200 million years ago. For example, the opening of the South Atlantic Ocean began about 125 to 130 million years ago, with Africa and South America connected directly earlier. Subsequently South America has moved slowly toward North America, with which it eventually became connected by a land bridge, the Isthmus of Panama, between 3.1 and 3.6 million years ago. When the bridge was complete, many South American plants and animals, such as the opossum (Figure 23-6) and the armadillo (see Figure 19-4), migrated overland into North America. At the same time, numerous North American plants and animals such as oaks, deer, and bears moved into South America for the first time. Changes on a greater or lesser scale occurred in the positions of all the continents, playing a major role in the patterns of distribution of organisms that we see today. For example, the ratite birds—ostriches (see Figure 39-14, *A*), emus, rheas, cassowaries, and kiwis—now thought to have diverged very early in the course of their evolution from all other living groups of birds, undoubtedly migrated between southern continents and islands, now widely separated from one another, at a time when direct overland migration was possible: the Mesozoic Era.

Australia provides a striking example of the way in which continental movements have affected the nature and distribution of organisms. Until about 53 million years ago, Australia was joined with a much warmer Antarctica, as was South America. Marsupials,

FIGURE 23-6

The North American opossum, Didelphis virginiana. *Marsupials, which once lived in Europe, Africa, and Antarctica, are now best represented in Australia and South America. They formerly occurred in North America but became extinct there about 40 million years ago. The opossum, like the armadillo, is a mammal that migrated into North America from South America once the Isthmus of Panama had been uplifted above the sea, an event that occurred several million years ago.*

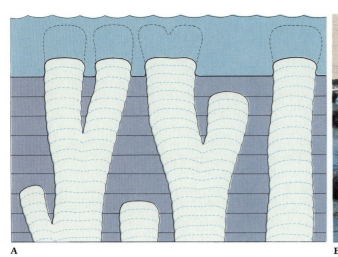

A

B

which are best represented in Australia and South America at present, seem clearly to have moved overland between these continents via Antarctica when this was still possible (see Figure 39-18). As Australia moved northward toward the tropical islands fringing Southeast Asia, its peculiar plants and animals (see Figure 19-1) evolved in isolation.

As Australia moved northward, the Antarctic Ocean opened, and the cold Circumantarctic Current began to flow around the Southern Hemisphere. A land connection of Australia with Antarctica, through the present position of Tasmania, seems to have persisted until about 38 million years ago. On the other side of the world, the connection of southern South America with West Antarctica lasted until about 23 million years ago. With the warmer climates of the early Cenozoic Era, more or less direct overland migration between Australia and South America, via Antarctica, was possible until about 40 million years ago.

One important effect of Australia and South America separating from Antarctica was the development of the cold Circumantarctic Current. Its formation led directly to the development of the Antarctic ice sheet, which reached its full size by about 10 million years ago. Ultimately, the cold temperatures associated with the formation of that enormous mass of ice triggered the onset of widespread continental glaciation in the Northern Hemisphere during the past few million years.

> **The movements of continents over the past 200 million years have profoundly affected the distribution of organisms on earth. Some of the major events include the linkage of South America with North America about 3.1 to 3.6 million years ago, and the separation of Australia and South America from Antarctica, which triggered the formation of the southern and ultimately the northern ice sheets.**

THE EARLY HISTORY OF LIFE ON EARTH

The main divisions of earth history, as outlined by contemporary authors, are Hadean, Archean, Proterozoic, and Phanerozoic. From Hadean time, before 3.8 billion years ago, we have no extant rocks on earth. Archean time, during which the oldest recognizable rocks were formed, extended until an oxygen-containing atmosphere was stabilized, an event that marked the start of Proterozoic time about 2.5 billion years ago. The oldest known fossils, all bacteria, are from Archean time, about 3.5 billion years ago.

Massive limestone deposits called **stromatolites** (Figure 23-7) became frequent in the fossil record by about 2.8 billion years ago. Produced by cyanobacteria, stromatolites were abundant in virtually all freshwater and marine communities until about 1.6 billion years

FIGURE 23-7

A This diagram shows how aggregations of calcium carbonate build up around massive colonies of cyanobacteria. Such deposits are known as stromatolites.
B Stromatolites in the intertidal zone at Shark Bay, Western Australia. The largest structures are about 1.5 meters across. These stromatolites formed during a time of slightly higher sea level perhaps 1000 to 2000 years ago; they are now exposed to the atmosphere and cemented by the form of calcium carbonate ($CaCO_3$) known as aragonite. In other words, they are already rocks.

HOMOLOGY AND THE EVOLUTION OF THE EYE

In considering the evolution of major groups of organisms, we sometimes find that the differences between them are so great that it is difficult to trace the lines of descent. Especially complex features, such as the eye, appear so unlike any possible earlier structure from which they may have been derived that the question of their origin greatly puzzled early students of evolution.

When we consider eyes in more detail, however, we find that they have in fact evolved many times, as shown by the fundamental differences between the eyes found in various groups of organisms. Some eyes consist of a single photoreceptor cell, and others are complex, like those of the vertebrates,

with focusing lenses and color sensitivity. Zoologists, analyzing these differences, have calculated that eyes have evolved independently, and through intermediate stages, at least 38 different times in various groups of animals (Figure 23-A).

In other words, all of the structures we call eyes are not homologous with one another. By analyzing and understanding their similarities and differences, biologists can more precisely interpret the relationships among the groups in which they occur and more accurately understand the outlines of their evolution. Thus the difficult problem of how eyes evolved, when solved, assists us in interpreting broad evolutionary problems of great interest.

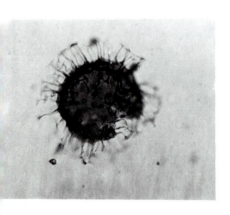

FIGURE 23-8

An acritarch from the early Cambrian Period of Spitsbergen, Norway. This fossil is about 700 million years old, but acritarchs are the earliest fossils that are considered probable eukaryotes; some fossils of this group are as much as 1.5 billion years old.

ago. Today stromatolites are still being formed, but only under conditions of high salinity, aridity, and high light intensities. About 2 billion years ago many different kinds of bacteria existed, including single, rounded cells; filaments apparently divided by cross-walls; tubular structures; branching filaments; and several unusual forms that do not fit well into any of these categories.

During Proterozoic time, 2.5 billion to 630 million years ago, oxygen was stabilized as a component of the atmosphere, and large continents coalesced from the light crustal rocks. During all of Archean and Proterozoic time, the only organisms in existence were bacteria and protists. Specialized microfossils called **acritarchs** (Figure 23-8) first appeared in the fossil record about 1.5 billion years ago; they may represent the earliest eukaryotes, judged from their relatively large size. The available fossils do not allow us to determine the internal structure of the acritarchs, however. Fossils that definitely seem to represent eukaryotes, namely unicellular protists, appeared about 1.3 billion years ago.

The oldest probable eukaryotic fossils, the acritarchs, appeared about 1.5 billion years ago. Within 200 million years of that date, other fossils, definitely representing eukaryotes, appeared in the fossil record.

The onset of Phanerozoic time is marked by the appearance of easily visible, multicellular fossils. The earliest such fossils are found in rocks at least 630 million years old, in the strata of the Ediacara series of southern Australia. Of the roughly 250,000 different kinds of fossils that have been identified, described, and named, only a few dozen are more than 630 million years old.

Within Phanerozoic time, we begin to speak in terms of geological eras, periods, epochs, and ages, as summarized inside the front cover of this book. For many years the oldest known fossils were those from the Cambrian Period (590 to 505 million years ago), the first period of the Paleozoic Era. The geological eras, with their dates in millions of years before the present, are as follows:

1. Paleozoic Era, 590 to 248 million years ago. The name of this era is derived from the Greek words *paleos,* "old," and *zoos,* "life." Until the discoveries outlined above were made, this was the oldest period from which fossils were known.

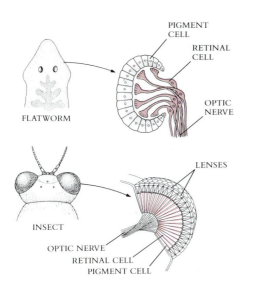
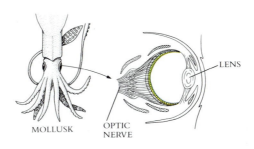
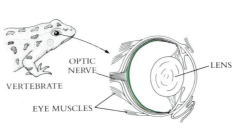

2. Mesozoic Era, 248 to 65 million years ago. The name is from the Greek *mesos,* "middle."

3. Cenozoic Era, 65 million years ago to the present. This name is from the Greek *coenos,* "recent."

All of the strata older than the Cambrian Period, and thus older than the Paleozoic Era, are classified as Precambrian. To earlier scientists, there seemed to be no fossils at all in Precambrian rocks, and their absence was regarded as a great mystery. Now, however, we know that fossil organisms were not detected simply because they were unicellular and therefore so small that they were difficult to observe. The evolution of skeletons dates the start of the Cambrian Period: even those multicellular animals which lived earlier were not often well preserved. Skeletons seem to have evolved about 570 million years ago, first mainly of calcium phosphate and later mostly of calcium carbonate. The presence of atmospheric oxygen was necessary before skeletons could evolve.

Some Precambrian fossils had simply been missed. The Ediacaran Period, comprising the last 40 million years of Precambrian time, from 630 to 590 million years ago, is a period from which we now have abundant and diverse fossils of multicellular animals (see Figure 23-3). Members of the phylum Cnidaria (Chapter 35) were abundant and diverse worldwide during this period. Among the contemporary members of this phylum are corals, sea anemones, and hydroids. Other animal fossils from the Ediacaran Period include some that resemble annelids, as well as possible early arthropods and some that cannot be assigned clearly to any phylum that we know. Annelida (Chapter 36) is a large phylum of worms, and Arthropoda (Chapter 37) is an enormous phylum that includes insects, crustaceans, and spiders.

Even older than the Ediacaran fossils are some traces of soft-bodied, multicellular animals. These include what are apparently worm burrows that have been found in Australia, China, and the USSR, suggesting that multicellular organisms actually may have evolved about 700 million years ago. Such fossils are preserved only under unusually favorable conditions, such as when they were buried in fine mud.

With the evolution of hard external skeletons, shells, and other structures that were easily preserved, the fossil record changed dramatically. The evolution of nearly all major

MARPOLIA

VAUXIA

ELDONIA

OPABINIA

HALLUCIGENIA

FIGURE 23-9

FIGURE 23-9

Reconstruction of a shallow seafloor during the Cambrian Period, about 530 million years ago. Fossils of the organisms shown here, together with well over 100 others, are found together in the Burgess Shale, a formation that has been uplifted to high elevations in the Rocky Mountains of British Columbia, Canada. Opabinia, which was about 12 to 15 centimeters long, had a jointed grasping organ unlike anything seen in living animals. Hallucigenia is a very peculiar organism that also has no known relatives among living organisms. Eldonia is a cnidarian, Vauxia a sponge, and Marpolia a multicellular alga. These three organisms represent living phyla. These organisms are beautifully preserved in the Burgess Shale, which was formed from a kind of fine-grained mud.

groups of organisms of which we are aware took place within a few tens of millions of years after the appearance of skeletons in their ancestors about 570 million years ago. The ability of organisms to manufacture such skeletons seems to have been a major evolutionary advance that made the further evolutionary radiation of these groups possible.

Multicellular fossils appear in Ediacaran time, 630 to 590 million years ago, and include members of the phyla Cnidaria, Annelida, and Arthropoda. Older traces of multicellular life, including what are apparently worm burrows, go back to about 700 million years ago.

THE PALEOZOIC ERA

Except for plants, all of the main phyla and divisions of organisms that exist today had evolved by the end of the first period of the Paleozoic Era, the Cambrian. The Cambrian Period was a time of intense evolutionary radiation, the greatest period of diversification for multicellular animals that has occurred during the entire history of life on earth. These animals were entering new **adaptive zones,** which might be thought of as major niches that differ greatly from one another, and were changing their characteristics in relation to those of their new ways of life.

The six periods of the Paleozoic Era, together with the approximate times that they began and ended, expressed in millions of years before the present, follow:

	START	END
Cambrian	590	505
Ordovician	505	438
Silurian	438	408
Devonian	408	360
Carboniferous	360	286
Permian	286	248

We present these periods at this point so that they may provide a point of reference for the subsequent discussion.

A

FIGURE 23-10

Cambrian Period fossils from the Burgess Shale, British Columbia, Canada, about 530 million years old.
A *and* **B** Sidneyia inexpectans, *an ancient arthropod; shown are a specimen and a model.*
C Hallucigenia sparsa, *shown in Figure 23-9. The relationships of this bizarre animal are obscure; this specimen is 12.5 millimeters long. The animal seems to have been supported on seven pairs of elongate spines; its trunk bore seven long tentacles and an additional group of tentacles near the rear end.* Hallucigenia *may have been a scavenger on the bottom of the sea.*

B

C

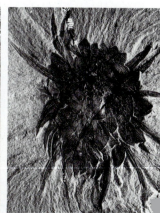

D

E

Origins of Phyla and Divisions

The Paleozoic Era was the time of origin or early diversification for most of the phyla, divisions, and classes that survive at the present time, except for the plants. As far as we know, all animal phyla, living and extinct, arose in the sea. Plants, on the other hand, originated on land, although they were derived from aquatic ancestors, the green algae. The basic record of the diversification of animal life on earth, then, is a marine record, and the record for the Paleozoic Era is exclusively marine.

Many kinds of multicellular organisms that were present in the early Paleozoic Era have no living relatives. In studying their remains in the rocks, one senses that this was a period of "experimentation" with different body forms and ways of life (Figures 23-9 and 23-10). Multicellular organisms were stronger and more mobile than their unicellular ancestors. Consequently, they could obtain their food and satisfy other requirements in more diverse, active ways. These possibilities led them to diversify along new evolutionary pathways, some of which ultimately led to the contemporary divisions and phyla of animals (Figure 23-10). As another example, the trilobites (see Figure 23-1) appear to be the ancestor of at least one living group of arthropods, the horseshoe crabs (Chapter 38).

The early Paleozoic Era was a time of extensive diversification for marine animals. Many new kinds of animals appeared, and some of them have persisted to the present.

Never again would there be a period of opportunity for the evolution of completely novel kinds of organisms such as that which existed at the start of the Paleozoic Era and earlier, about 600 to 680 million years ago. During this period, nearly all of the different kinds of body plans originated that occur on earth now or that lived earlier and have become extinct. Particularly during the second half of the Cambrian Period, there was a very rapid expansion of the kinds of living organisms. Because so many new kinds appeared in such a short time, paleontologists speak of a Cambrian "explosion" of living forms.

Before the close of the Cambrian Period, all of the animals in the sea fed on unicellular

FIGURE 23-10, cont'd

D Wiwaxia corrugata. *The body of this animal, also of unknown relationships, was largely covered with scales and also bore elongate spines; this specimen is 30.5 millimeters across, excluding the spines. It may have been a distant relative of the mollusks.*
E Burgessochaeta setigera. *A polychaete worm, a member of the phylum Annelida. The photograph shows the anterior end of the worm, which had a pair of tentacles. Parapodia, footlike structures, are borne in pairs along the sides of the body. This specimen is 16.5 millimeters long.*

organisms that floated freely in the water. During the subsequent period, the Ordovician, predators that fed on multicellular animals appeared, and the diversity of animals became even more extensive. The first corals also originated during the Ordovician Period and began to change the structure of marine communities permanently. By the end of this period, practically every mode of life that has ever existed had already evolved: bottom feeders, scavengers, carnivores, colonies, drifting multicellular forms, and so forth. With all of the adaptive zones in which these animals occur filled, subsequent opportunities for the additional evolution of novel forms have been much more limited.

The Invasion of the Land

The first organisms that colonized the land were the plants, and they did so near the close of the Silurian Period of the Paleozoic Era, about 410 million years ago. The features of plants evolved in relation to their colonization of the land and were those that made them best suited for such colonization. The ancestors of plants were specialized green algae (division Chlorophyta; Chapter 27). Although most of the members of this division are aquatic, the immediate ancestors of plants might themselves have been semiterrestrial. It was, however, with the plants themselves that the occupation of the land truly began (Figure 23-11).

Many major groups of animals also originated in connection with their invasion of the land. Among the arthropods, the insects originated on land from an evolutionary line derived from annelid worms (Chapter 36). This occurred at about the same time as the evolution of the plants, about 410 million years ago. The body plan of arthropods has proved so well adapted to life on land that insects and other classes of this phylum have radiated extensively, producing a probable majority of all species of organisms. During the subsequent history of the group, only a few kinds of insects have moved back into freshwater habitats, and a handful have returned to marine habitats. It seems certain that the plants colonized the land before animals did, since they would have been essential sources of food and shelter for the first arthropods that emerged from the sea.

Among the vertebrates, the amphibians originated, as did the plants, from some of the first forms that colonized the land. The earliest amphibians known are from the Devonian Period, appearing about 360 million years ago (Figure 23-12). Among their descendants on land are the reptiles. Different groups of reptiles, in turn, ultimately became the ancestors of the birds and the mammals, as we will discuss further in Chapter 39.

The fact that all three of these major groups of organisms—plants, arthropods, and vertebrates—colonized the land within a few tens of millions of years of one another is probably related to the development of suitable environmental conditions, for example, the formation of a layer of ozone in the atmosphere, which blocked ultraviolet radiation. These conditions have allowed the existence of multicellular organisms in terrestrial habitats for somewhat more than 400 million years, about a tenth of the age of the earth.

Only a few phyla of organisms have invaded the land successfully, most remaining exclusively marine. Another group that has done so, in addition to the three that have been mentioned thus far, is the fungi, which constitute a distinct kingdom of organisms. The success of the fungi on land, where they probably have been present for as long as any group of organisms, might be related to the structure of their cell walls, which are rich in chitin, a drought-resistant substance that also characterizes the outer skeletons of the arthropods. The members of one division of fungi, the Zygomycetes, were associated with the roots of the first terrestrial plants. We will discuss this relationship in more detail in Chapter 28.

Plants and arthropods colonized the land about 410 million years ago; amphibians did so about 50 million years later. Fungi may also have colonized the land at about the same time as plants. Earlier, there were no terrestrial, multicellular organisms.

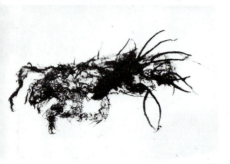

FIGURE 23-11

A *This fragment of an unknown plant from the Silurian Period, about 423 million years ago, shows the earliest evidence of an apparently well-organized conducting strand, consisting of banded, elongate conducting cells surrounded by unbanded, smooth-walled cells. It cannot be determined from the material found so far whether this organism was terrestrial or aquatic, or whether it was actually a plant, a large alga, or a transitional form.*
B *A mite (phylum Arthropoda) from the Devonian Period, about 376 to 379 million years ago, in what is now New York State. This mite is a member of a family that includes living members. Land animals appeared in the period before the Devonian, in the uppermost Silurian Period (about 415 million years ago), but they were very scarce then. The first known terrestrial arthropods were millipede-like.*

FIGURE 23-12

Reconstruction of Ichthyostegia, *one of the first amphibians with efficient limbs for crawling on land, an improved olfactory sense associated with a lengthened snout, and a relatively advanced ear structure for picking up airborne sounds. Despite these features,* Ichthyostegia, *which lived about 350 million years ago, was still quite fishlike in overall appearance and represents a very early amphibian.*

Mass Extinctions

One of the most prominent features of the history of life on earth has been the periodic occurrence of major episodes of extinction. During the course of geological time there have been five such events, in each of which a large proportion of the organisms on earth at that time became extinct. Four of these events occurred during the Paleozoic Era, the first of them near the end of the Cambrian Period (about 505 million years ago). At that time most of the existing families of trilobites, a well-known group of marine arthropods (Figures 23-1 and 23-13), became extinct. A second extinction event marked the close of the Ordovician Period, about 438 million years ago, and a third marked the close of the Devonian Period, about 360 million years ago.

The fourth and most drastic extinction event in the history of life on earth happened during the last 10 million years of the Permian Period, which ended the Paleozoic Era, about 248 to 238 million years ago. It is estimated that approximately 96% of all species of marine animals that were living at that time may have become extinct! All of the trilobites, and many other groups of organisms as well, disappeared forever. The fifth major extinction event occurred at the close of the Mesozoic Era, 65 million years ago.

Major episodes of extinction produce conditions similar to those which existed at the start of Phanerozoic time, or to those which exist on an island that has been swept free of vegetation by volcanic eruption or has just arisen from the sea. For those relatively few plants, animals, and microorganisms which survive the extinction episode, there may be great opportunities for further expansion. Many paleontologists believe that the great episodes of extinction that have occurred periodically have had much to do with the pattern of evolution of life on earth. In each case, however, only a few of the kinds of organisms that were in existence before the extinction episode have radiated, giving rise to the dominant life forms of the subsequent period. In contrast to these later situations, there seem to have been almost completely unlimited opportunities for evolutionary radiation and innovation among multicellular organisms before and during the Cambrian Period.

Large-scale extinction events other than the five major ones summarized above also have occurred. One surprising correlation that was reported in late 1983 by J. John Sepkowski and David M. Raup of the University of Chicago is that such events appear to occur regularly every 26 to 28 million years. The most recent such event occurred about 11 million years ago. The search for the sort of cosmic event, such as periodic comet showers, that could cause major events of extinction to occur with such regularity is now under way, as other scientists attempt to test the validity of the correlation made by Sepkowski and Raup. It has been hypothesized that the major extinction event that ended the Mesozoic Era is correlated with the impact of a large meteorite, as we will discuss below. Numerous hypotheses have been advanced to explain the other major extinction events, but there is no general agreement concerning the ones that are most likely.

FIGURE 23-13

A fossil trilobite.

CRETACEOUS   JURASSIC

FIGURE 23-14

Some of the remarkable diversity of dinosaurs, as shown in a reconstruction from the Peabody Museum, Yale University. This painting covers a span of approximately 320 million years (reading right to left) during the later Paleozoic and Mesozoic Eras, ending 65 million years ago at the end of the Cretaceous Period (far left). Throughout this vast period of time, the remarkable increase in structural complexity and overall diversity of the dinosaurs can be seen—until they abruptly became extinct, giving way to the dominant mammals of the Cenozoic Era. Flowering plants can be seen for the first time at the left-hand side of the illustration. The names of the geological periods are given along the bottom of the painting.

THE MESOZOIC ERA

The Mesozoic Era, which began about 248 million years ago and ended about 65 million years ago, was a time very different from the present and one of intensive evolution of terrestrial plants and animals. The major evolutionary lines on land had been established during the mid–Paleozoic Era, but the evolutionary radiation of these lines—a radiation that led to the establishment of the major groups of organisms living today—took place in the Mesozoic Era. In tracing the evolution of these lines, we need to refer to events in the Permian Period (286 to 248 million years ago), a period of drought and extensive glaciation that concluded the Paleozoic Era. Many of the evolutionary events that occurred then had important consequences during the Mesozoic Era, influences that have continued to the present.

The Mesozoic Era has traditionally been divided into three periods:

	START (MILLIONS OF YEARS AGO)	END (MILLIONS OF YEARS AGO)
Triassic	248	213
Jurassic	213	144
Cretaceous	144	65

The Triassic Period marked a time of abrupt change from the preceding Permian Period. In the sea, only about 4% of the species that were present earlier in the Permian Period also occurred in the Triassic Period. The Mesozoic and Paleozoic eras were initially recognized as distinct from each other because of the effects of the major extinction event at the end of the Permian Period: the marine animals of the Paleozoic Era can be recognized instantly as different from those of the Mesozoic Era. The few kinds of marine organisms that survived into the Mesozoic Era—including gastropod and bivalve mollusks, crustaceans, fishes, and echinoderms, among others—began to evolve into many new species, genera, and families. In addition, a number of new life forms evolved in the sea during the Mesozoic Era. For example, the first efficient burrowers appeared among the echinoderms (Chapter 38).

Both on land and in the sea, the number of species of almost all groups of organisms has been climbing steadily for the past 250 million years and is now at an all-time high. Even though the evolutionary radiation of marine organisms during this period has been spectacular, the story of the evolution of life on land during the Mesozoic Era is of even greater interest for us, who are products of that history.

The History of the Vertebrates

We trace the history of vertebrates here because so many of its significant events took place during the Mesozoic Era. Although we think of the vertebrates as advanced organisms, the phylum that includes them, Chordata, is nearly as old as any other. We will consider this phylum in more detail in Chapters 38 and 39. The first chordates for which we have a fossil record, the lancelets and lampreys, appeared about 540 to 550 million years ago, in the middle Cambrian Period. Over the next 100 million years, during the early Paleozoic Era, fishes were becoming abundant and diverse: the Devonian Period, 408 to 360 million years ago, has been called the age of fishes. Amphibians, the descendants of fishes, appear in the fossil record by the end of this period. During the following period, the Carboniferous, amphibians became abundant in the great swamps that formed at that time.

The amphibians gave rise to the first reptiles during the Carboniferous Period. The reptiles, being better suited for a terrestrial existence than their amphibian ancestors, replaced them as dominants over the next 50 million years, becoming abundant during the Permian Period. During the course of the Mesozoic Era, there were extensive evolutionary radiations of flying reptiles, the **pterosaurs,** and swimming reptiles, the **plesiosaurs,** in addition to the well-known radiation of terrestrial ones.

Dinosaurs, the major group of reptiles that was dominant throughout the Mesozoic Era, first appeared in the late Triassic Period, perhaps 220 million years ago, and persisted to the end of the Cretaceous Period—in other words, more than 150 million years. The dinosaurs became more diverse as the Mesozoic Era proceeded (Figure 23-14), and they finally became extinct, for reasons that are still much debated. The fossil record indicates that this occurred in the last few million years of the Cretaceous Period. The tuatara (*Sphenodon;* see Figure 39-12) is a reptilian survivor from the Mesozoic Era.

Two evolutionary lines that were derived from the dinosaurs did not, however, become extinct: the crocodiles and the birds. Crocodiles are not directly related to any other group of living reptiles; and birds, despite the fact that we often link them with mammals, are very probably more closely related to extinct reptiles than they are to mammals. The earliest known bird, *Archaeopteryx,* from the late Jurassic Period (about 150 million years ago) is structurally little more than a bipedal dinosaur with feathers (Figure 23-15; see also Figure 39-16). *Archaeopteryx* had feathers, but apparently not enough of them to fly very effectively. Its ancestor must have been a dinosaur that ran on its hind legs and used its front legs for grasping. Such bipedal dinosaurs were swifter and more active than their quadripedal ancestors, and their bones became lighter too. The feathers that evolved in some of these bipedal dinosaurs, including *Archaeopteryx,* would have been valuable, since they

FIGURE 23-16

A multituberculate, a member of one of the most abundant groups of mammals that lived during the Mesozoic Era and survived well into the Cenozoic Era. The multituberculates were seed and fruit eaters that probably had squirrel-like habits and therefore may have played an important role in the dispersal of seeds. This multituberculate, Ptilodus kummae, *lived in the late Paleocene Period in western North America, about 60 million years ago. Its prehensile tail and several other features indicate that it was an agile climber. The modern egg-laying mammals, the monotremes (echidna and platypus; see Chapter 39), may have evolved from the multituberculates.*

would have served to insulate the animals that had them. In addition, feathers may have appeared first in animals that lived in trees, helping them to glide from tree to tree. For whatever reason feathers first evolved, they made possible the subsequent evolution of birds that *could* fly efficiently. The ability of birds to fly, and thus to exploit an unoccupied adaptive zone, made possible the evolution of the entire class Aves, the birds, a group that comprises about 9000 living species.

An interesting point about this story is that if *Archaeopteryx* had left no descendants, it would be regarded merely as an unusual bipedal dinosaur. Because it, or forms similar to it, give rise to the birds, it is classified with them, and this large group of organisms is regarded as a class of vertebrates, on the same level as mammals, reptiles, and fishes. This example clearly illustrates the way in which the evolution of any major group, up to the level of a kingdom, starts with the evolution of a single species; the fate of that species cannot be predicted in advance.

Mammals originated during the Triassic Period from a different group of reptiles and represent a line parallel to the dinosaurs (Figure 23-16). Mammals became dominant during the Cenozoic Era, replacing the dinosaurs as ecological dominants, and it might therefore be assumed that they originated later. In fact, mammal fossils are known from about the same time as the fossils of the first dinosaurs. Therefore the times of origin of these two great groups seem to have been about the same. Mammals never became abundant and diverse, however, until the Cenozoic Era, after the dinosaurs had become extinct. They seem to present an example of evolutionary radiation into an adaptive zone, or rather a series of adaptive zones, left vacant by the extinction of the dinosaurs. For example, many of the dinosaurs weighed thousands of kilograms, but no land animal weighing more than 20 kilograms survived into the Cenozoic Era. The ancestral small mammals clearly had ample opportunity to take up ways of life similar to those of the extinct larger dinosaurs, occupying vacant adaptive zones in doing so.

During the Mesozoic Era the reptiles, which had evolved from amphibians during the Carboniferous Period of the preceding Paleozoic Era, became dominant and in turn gave rise to the mammals (about 200 million years ago) and the birds (at least 150 million years ago).

The History of Plants

The earliest known fossil plants are from the late Silurian Period, about 410 million years ago. By the close of the Paleozoic Era, plants had become abundant and diverse. Shrubs and then trees evolved and came to form forests, which dominated many terrestrial habitats by the Carboniferous Period (360 to 286 million years ago). These Carboniferous forests in turn formed many of the great coal deposits that we are consuming now. Much of the land was low and swampy during the Carboniferous Period, providing excellent conditions for the preservation of plant remains. In these coal deposits, we have a relatively complete record of the horsetails, ferns, and primitive seed-bearing plants that made up these ancient forests (Figure 23-17).

The Permian Period (286 to 248 million years ago) was, as we have seen, cool and dry. The swamps of the preceding Carboniferous Period, which existed at a time of worldwide moist and warm climates, largely disappeared. The Permian Period seems to have been one of ecological stress and evolutionary innovation; for example, the conifers—a group of seed-bearing plants that is represented today by pines, spruces, firs, and similar trees and shrubs—originated then. The efficient water-conducting systems of the conifers and other vascular plants and their waxy cuticles (protective covering layers) helped to make them successful in the Permian Period and subsequently. Today the descendants of these conifers are dominant in relatively cool and dry habitats at high latitudes and in the mountains. Seed-bearing plants with featherlike leaves, similar to the living cycads, were abundant in the Mesozoic Era and helped to give that period its nickname, "the age of dinosaurs and cycads." The conifers, which first appeared during the Permian Period, and later the flowering plants, have gradually come to dominate terrestrial habitats.

The oldest fossils definitely known to be flowering plants are from the early Cretaceous Period, about 127 million years ago. It seems likely that the group actually originated somewhat earlier, but no one is certain how much earlier. Like the mammals, the flowering plants, originating much earlier, ultimately became dominant and diverse during the Cenozoic Era. Unlike the mammals, however, they became dominant before the close of the Cretaceous Period. The flowering plants have been more abundant than any other group of plants for about 100 million years.

The evolution of the flowering plants, which began in the Mesozoic Era, has continued strongly to the present. Today there are about 240,000 species of this large group,

FIGURE 23-17

Reconstruction of a Carboniferous Period forest, in which early vascular plants were dominant. This reconstruction was made at the Field Museum of Natural History in Chicago.

which greatly outnumbers all other kinds of plants. As the flowering plants became more diverse, so did the insects, which had feeding habits that were closely linked with the characteristics of the flowering plants (Figure 23-18). Insects and flowering plants have coevolved with one another; indeed, all groups of terrestrial organisms, including mammals, birds, and fungi, have evolved their characteristics largely in relation to those of the flowering plants. These groups now dominate life on the land, literally making our world look the way it does. Although all of them existed during the Mesozoic Era, none was as dominant then as it has become since.

The earliest fossil plants, about 410 million years ago, evolved into others that formed extensive forests within 50 million years of their appearance. Conifers evolved during the Permian Period (286 to 248 million years ago), the closing period of the Paleozoic Era. Together with the flowering plants, which appeared in the fossil record about 127 million years ago, conifers have come to dominate the modern landscape.

The Extinction of the Dinosaurs

Everyone is generally familiar with the disappearance of the dinosaurs, an event of global importance that took place at the end of the Cretaceous Period, the closing period of the Mesozoic Era. Less discussed, but actually of more fundamental importance, was the disappearance of many other kinds of organisms that took place at about the same time. Recently, important strides have been made in understanding the nature of this, the most recent of the major extinction events. We will discuss some of these findings here.

The Mesozoic Era has traditionally been separated from the Cenozoic Era by abrupt shifts in the kinds of marine organisms that occur in strata exposed in Europe. Among the plankton—free-drifting protists and members of other groups of organisms that are still abundant in the sea—many of the larger (but still microscopic) forms suddenly disappeared about 65 million years ago and a much lower number of smaller ones took their place. The same rapid changes occurred in at least some nonplanktonic marine animal groups, such as the bivalve mollusks (clams and their relatives). The ammonites, a large and diverse group of relatives of octopuses with shells (see Figure 36-13), abruptly disappeared. In 1980 a group of distinguished scientists headed by physicist Luis W. Alvarez of the University of California, Berkeley, presented a hypothesis about the reasons for these drastic changes.

Alvarez and his associates discovered that the usually rare element iridium was abundant in a thin layer that marked the end of the Cretaceous Period, not only in the strata where the period had first been defined, in Italy, but also in many other parts of the world at the same time (Figure 23-19). Iridium is rare in the earth's outer crust but common in meteorites. Alvarez and his colleagues proposed that if a large meteorite, or asteroid, had struck the surface of the earth, a dense cloud would have been thrown up which would for a time have darkened the earth. The cloud would have been rich in iridium, and as its particles settled, the iridium would have been incorporated in the layers of sedimentary rock that were being deposited at that time. By darkening the world, the cloud would have greatly slowed or temporarily halted photosynthesis and driven many kinds of organisms to extinction.

Subsequent calculations have shown that the cloud produced by a meteorite about 20 kilometers in diameter would certainly have caused these effects to occur. Daytime conditions would have resembled those on a moonless night, with less than the amount of light required for photosynthesis, for several months. In biological communities such as the marine plankton, which are based directly on the continuous production of food by photosynthesis, such darkness would have had seriously disruptive effects and could have produced the sudden change seen in the fossil record. Disruption of photosynthesis may also have been responsible for the extinction of certain other kinds of organisms, but is it reasonable to assume that it would have done away with the dinosaurs?

A

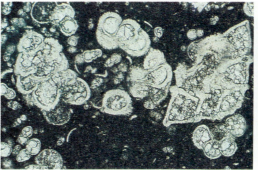

B

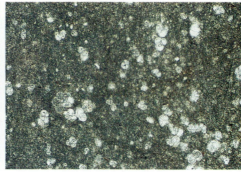

C

FIGURE 23-19

Evidence of the abrupt change that occurred at the end of the Cretaceous Period, 65 million years ago, in the vicinity of Gubbio, northern Italy.

A   The white limestone below was deposited under the sea in the closing years of the Cretaceous Period, and the red limestone above was deposited during the first years of the Tertiary Period. They are separated by a layer of clay about 1 centimeter thick in which iridium is abundant.

B   Large, ornamented, diverse planktonic forams in sediments laid down during the late Cretaceous Period. Some of these species are 600 micrometers or more in diameter.

C   Small, relatively unornamented, much less diverse planktonic forams deposited in the limestone formed during the early years of the Tertiary Period. These forams are about 100 micrometers in diameter. It took several million years for the population of forams to regain the diversity that it had near the end of the Cretaceous Period.

As we mentioned earlier, correlations are often difficult between different sedimentary layers; because of these difficulties, it is not clear whether the dinosaurs became extinct suddenly and thus whether their extinction is related to the meteorite collision postulated by Alvarez and his associates. There does not, however, appear to have been a major change in the representation of particular groups of mammals or fishes, for example, before and after the end of the Cretaceous Period. The same is apparently true of the flowering plants. Furthermore, Charles Officer and Charles Drake of Dartmouth College reported in early 1985 that the iridium layer actually seems to have been deposited over a period of from 10,000 to 100,000 years. Their results are consistent with its deposition as a result of volcanic activity, and not as a result of the impact of an asteroid, in which case the layer would have been deposited over a period of a few years. The relationships involved are under active investigation.

Summing up the evidence, it appears that worldwide, iridium-rich clouds did cause widespread darkening near the end of the Cretaceous Period. This darkening apparently had extensive effects on some groups of organisms, such as the marine plankton, but it does not appear sufficient, based on present evidence, to account for those changes in the kinds of terrestrial plants and animals that occurred on land before and after the end of the Cretaceous Period, including the extinction of the dinosaurs. Whether the clouds were caused by a meteorite colliding with the earth, extensive volcanic activity, or some other source is unclear at this time. Scientists today are still actively pursuing the answer to this riddle.

A worldwide layer rich in iridium, an element that is much more common in meteorites than it is in general on earth, marks the end of the Cretaceous Period and therefore of the Mesozoic Era. It appears likely that this iridium settled out of an enormous, worldwide cloud cover. Such a cloud cover would have darkened the earth greatly and seems to have been responsible for at least some of the extinction events that occurred at this time.

THE CENOZOIC ERA: THE WORLD WE KNOW

We conclude this chapter with a brief account of some of the major evolutionary changes that have occurred during the past 65 million years, changes that have resulted in the conditions we now experience. The relatively warm and moist climates of the early Cenozoic Era have gradually given way to today's climates. As we saw earlier, the final separation of South America and Antarctica about 27 million years ago set the stage for (1) the establishment of the Circumantarctic Current; (2) glaciation in Antarctica, fully established by about 13 million years ago; and (3) glaciation in the north over the past several million years. The ice mass that has been formed as a result of this glaciation has made the climate cooler near the poles, warmer near the equator, and drier in the middle latitudes than ever before.

In general, forests covered most of the land area of the continents, except for Antarctica, until about 15 million years ago, when they began to recede rapidly. During the time when these forests were receding and modern plant communities were appearing, some of the continents were approaching one another again after having been widely separated during most of the Cenozoic Era. The organisms in Australia and South America, particularly, evolved in isolation from all those in the rest of the world, mainly during the Cenozoic Era. Evolution in isolated regions of this kind has been responsible for the distinctive characteristics of the groups of plants and animals found in different regions of the world. During the past several million years, the formation of extensive deserts in northern Africa, the Middle East, and India has made migration very difficult for the organisms of tropical forests between Africa and Asia. This formation of desert barriers, in turn, has provided further opportunities for evolution in isolation. In general, the overall character of the Cenozoic Era has been marked by a deteriorating climate, sharp differences in habitats even within small areas, and the regional evolution of distinct groups of plants and animals.

Throughout the 65 million years of the Cenozoic Era, the world climate has deteriorated steadily and the distributions of organisms have become more and more regional in character.

THE EVOLUTION OF LIFE ON EARTH

The structure of life in the modern world depends largely on the existence of multicellular organisms. These evolved starting about 700 million years ago, after more than 2.5 billion years of earlier evolution on earth. For the first 300 million years of their evolution, up to the end of the Cambrian Period, many novel life forms appeared and diversified in relation to one another. After the extinction event that occurred about 500 million years ago, most of the marine phyla and divisions of multicellular organisms were established. Subsequent evolution seems to have resulted largely from the reassortment and diversification of these evolutionary lines.

Somewhat more than 400 million years ago, the ancestors of plants and terrestrial arthropods came onto the land; fungi presumably also did so at about this time. After another 50 million years or so, the first amphibians (the first terrestrial vertebrates) appeared on land. The diversification of and interplay between these major groups of organisms has been responsible for the world as we see it now. At first the climate on land was relatively mild, but the Permian Period, which ended about 248 million years ago, was a time of aridity and glaciation. Evolutionary lines that were established at that time have been especially successful during the past 65 million years, the Cenozoic Era, when world climates have progressively deteriorated and become more diverse. As they have done so, the insects, flowering plants, fungi, and vertebrates have evolved millions of new species on land, their evolution accelerated by the intense interactions among the members of these groups. The products of this evolution dominate life on earth today.

SUMMARY

1. The evolutionary processes that result in the evolution of genera, families, orders, and classes are basically the same as those that result in the evolution of species. In other words, particular species may give rise to major evolutionary lines when the circumstances are suitable for this to happen.

2. The fossil record is relatively complete for marine organisms, since there is a steady rate of sedimentation in the sea. For the fossil-bearing rocks of marine origin to be preserved, however, they must be uplifted onto the land, since the sea floor is constantly disappearing and being renewed.

3. Gaps in the fossil record may represent either the operation of punctuated equilibria or simply times and places when appropriate conditions for preservation did not occur.

4. The earth originated about 4.6 billion years ago, and distinct crustal plates had been organized by about 3.8 billion years ago. By 200 million years ago, the continents were all clustered together, and they have been separating from one another since, resulting in realignments in their positions.

5. The main divisions of earth history are Hadean, Archean, Proterozoic, and Phanerozoic time. Our oldest rocks on earth date the start of Archean time and are about 3.8 billion years old. The oldest bacterial fossils are about 3.5 billion years old. Stromatolites, massive deposits of limestome formed by cyanobacteria, appear in the fossil record starting about 2.8 billion years ago.

6. During Proterozoic time, 2.5 billion to 630 million years ago, oxygen was stabilized as a component of the atmosphere. The first unicellular eukaryotes appeared about 1.3 to 1.5 billion years ago; all earlier forms of life were bacteria. Although multicellularity appeared about 700 million years ago, the first multicellular animals were soft-bodied and are poorly represented in the fossil record.

7. Phanerozoic time, which began about 630 million years ago and extends to the present, is a period when fossils of multicellular organisms are frequent. It is divided into three eras: the Paleozoic Era, 590 to 248 million years ago; the Mesozoic Era, 248 to 65 million years ago; and the Cenozoic Era, 65 million years ago to the present.

8. All of the phyla and divisions, as well as most of the classes, of organisms that we know about evolved during the Cambrian Period (590 to 505 million years ago), with the exception of the plants. The evolution of hard skeletons, including shells and similar structures, about 570 million years ago seems to have been a fundamental evolutionary advance that made this evolutionary radiation possible.

9. During the Paleozoic Era, plants and terrestrial arthropods appeared about 410 million years ago and terrestrial vertebrates (amphibians) appeared about 360 million years ago. These groups of organisms, together with the fungi, have dominated life on the land since then.

10. The Paleozoic Era closed with the Permian Period (286 to 248 million years ago), a cold and dry period during which conifers and reptiles first became evident.

11. The Mesozoic Era, the "age of dinosaurs and cycads," was a time when the outlines of life on earth as we know it were established. Flowering plants were dominant by the end of the era, and insects, mammals, birds, and other groups had begun to evolve in relation to the diversity of these plants.

12. There have been five major episodes of mass extinction during the history of life on earth. The most drastic was at the end of the Permian Period, about 248 million years ago, when about 96% of marine animals became extinct. At the end of the Cretaceous Period, 65 million years ago, the dinosaurs and many other kinds of organisms disappeared.

13. The diversity of living things, measured in terms of species number, has been increasing steadily for the past 248 million years and is now at an all-time high. The wide separation of the continents and their arrangement into novel configurations probably has played a role in this trend.

SELF-QUIZ

The following questions pinpoint important information in this chapter. You should be sure to know the answer to each of them.

1. Extinction _____ the magnitude of differences between different groups of organisms.
   (a) decreases   (c) does not affect
   (b) increases

2. _____ with $^{14}C/^{12}C$ and $^{40}K/^{40}Ar$ can be used to date materials in sedimentary rocks.

3. The oldest known fossils, which are all _____, occur during the Archean time, 3.5 billion years ago.
   (a) viruses   (d) dinosaurs
   (b) *Archaeopteryx*   (e) plants
   (c) bacteria

4. The oldest probable eukaryotic fossils, the _____, appeared about 1.5 billion years ago.

5. The oldest known multicellular fossils came from the _____ period, at which time all of the phyla and divisions that exist now, except for plants, were already in existence.

6. Plants evolved on land. True or false?

7. Which of the following major groups of organisms did not originate in the sea?
   (a) plants   (c) vertebrates
   (b) arthropods   (d) worms

8. During the course of geological time there have been _____ major episodes of mass extinction.
   (a) three   (d) six
   (b) four   (e) more than a hundred
   (c) five

9. The giant land mass that was formed when all of the continents united is called _____.

10. The _____ era is called "the age of dinosaurs and cycads."
    (a) Mesozoic   (c) Tessierozoic
    (b) Paleozoic   (d) Cenozoic

11. _____ became common during the Cenozoic Era, replacing the dinosaurs as ecological dominants.
    (a) plesiosaurs   (c) pterosaurs
    (b) mammals   (d) amphibians

12. There are more flowering plants than any other type of plant. True or false?

THOUGHT QUESTIONS

1. How can you tell if a new species will give rise to a new genus? Under what circumstances is it most likely to do so?

2. Why do you think that at least 2 billion years of evolution preceded the origin of multicellularity, and more than 3 billion years preceded the invasion of the land? What kinds of information would you need to test your ideas?

3. There seems to be no iridium layer preserved in the rocks from four of the five major periods when mass extinction occurred. How could one account for these four events?

FOR FURTHER READING

ATTENBOROUGH, D.: *Life on Earth,* Little, Brown & Co., Boston, 1979. A readable and well-presented outline of the evolution of life on earth, based on the materials collected for a popular BBC television series.

CLOUD, P., and M.F. GLAESSNER: "The Ediacaran Period and System: Metazoa Inherit the Earth," *Science,* vol. 218, pages 783-792, 1982. A comprehensive review of what we know about the first multicellular animals.

DOBZHANSKY, T., et al.: *Evolution,* W.H. Freeman & Co., Publishers, San Francisco, 1977. A balanced account of the history of life on earth, with an accurate description of the process of evolution.

DOTT, R., and R.L. BATTEN: *The Evolution of the Earth,* 3rd ed., McGraw-Hill Book Co., New York, 1980. An outstanding introduction to the physical development of the earth.

HORNER, J.R.: "The Nesting Behavior of Dinosaurs," *Scientific American,* April 1984, pages 130-137. The members of a group of dinosaurs named the hadrosaurs that lived in Montana 80 million years ago appear to have built nests and to have exhibited social behavior. An interesting example of paleontological detective work.

LEWIN, R.: *Thread of Life: The Smithsonian Looks at Evolution,* Smithsonian Books, Washington, D.C., 1982. A beautifully illustrated chronicle of the history of life on earth.

SCHOPF, J.W. (ed.): *Earth's Earliest Biosphere: Its Origin and Evolution,* Princeton University Press, Princeton, N.J., 1983. An outstanding and authoritative compendium of articles containing the latest information on the earliest traces of the evolution of life on earth.

VIDAL, G.: "The Oldest Eukaryotic Cells," *Scientific American,* February 1984, pages 48-57. An up-to-date review of the evidence for the origin of eukaryotic cells; the author concludes that this event occurred about 1.5 billion years ago.

WHITTINGTON, H.B.: *The Burgess Shale,* Yale University Press, New Haven, Conn., 1985. A comprehensive account of what is known about multicellular life during one of its earliest and most active periods of evolution.

Chapter 24

EVOLUTION OF HUMAN BEINGS

Overview

Human beings are biologically similar to the apes—chimpanzees, gorillas, orangutans, and gibbons—and, together with them, comprise an evolutionary line called the hominoids. The hominoids are primates, an order of mammals that includes the lemurs and lorises (prosimians) in addition to the monkeys, apes, and ourselves. The story of our evolution as mammals began more than 210 million years ago and continued with the appearance of primates about 65 million years ago and that of early anthropoid primates about 36 million years ago. *Australopithecus,* a genus that includes the direct ancestors of humans, appeared in Africa perhaps 5 million years ago, having separated earlier from the evolutionary line leading to chimpanzees and gorillas. Ultimately *Australopithecus* gave rise to our genus, *Homo,* also in Africa, through a series of stages involving an increase in brain size from the average 350 to 450 cubic centimeters of *Australopithecus* to the 1400 to 1450 cubic centimeters that is characteristic of our brains. The genus *Homo* has existed for about 2 million years, *Homo sapiens*—our species—for perhaps 500,000 years.

For Review

Here are some important terms and concepts that you will encounter in this chapter. If you are not familiar with them, you should review them before proceeding.

- **DNA hybridization** (Chapter 18)

- **Evolutionary radiation and the concept of adaptive zones** (Chapters 21 and 22)

- **Taxonomic categories** (Chapter 22)

- **Evolution of the vertebrates** (Chapter 23)

We conclude our treatment of evolution with a detailed account of human evolution, doing so not only because of our great personal interest in the story of our own origins, but also because the evolution of our species, *Homo sapiens,* is better understood and documented than the evolution of any other species. We know as much as we do because of the enormous effort that has been made to unravel the details of this particular evolutionary story—our own.

EARLIER VIEWS OF HUMAN ORIGINS

Through the seventeenth century, philosophers such as Descartes emphasized the differences between human beings and animals, citing the souls of humans as a basis for this distinction. By the mid–eighteenth century, however, Linnaeus, a master of classification, had placed the human species squarely among the primates, using a wealth of evidence that had accumulated over the years. In taking this position, Linnaeus was acting in accordance with the views prevailing in his century. Scientists had increasingly been impressed by the detailed similarities between humans and other animals, especially the primates.

In formulating his theory of evolution, Charles Darwin had at his disposal much more information than had been available to Linnaeus a century earlier. In the light of this added information, Darwin was able to conclude that humans, at least insofar as their physical and mental properties were concerned, had evolved, just as had all of the other mammals. Although he avoided the controversial question of human origins in his 1859 book *On The Origin of Species,* Darwin in 1871 stated his views on this issue explicitly in another book, *The Descent of Man.* Many people attacked Darwin's position, feeling that to regard people as related to other living things somehow removed their dignity and special status. However, others realized that the special features of human beings could be explained and understood best in the light of the evolutionary process. Although much new information about the evolution of human beings and their detailed relationships to the other primates has emerged since Darwin's time (Figure 24-1), the essential outlines of the story remain as he presented them.

THE EVOLUTION OF MAMMALS

The mammal-like group of reptiles known as therapsids originated in the late Permian Period, about 250 million years ago, and continued to flourish for another 40 million years. Toward the end of that time, or about when ancestral dinosaurs first appeared—in the early Jurassic Period, about 210 million years ago—the first mammals appeared. For the remainder of the Mesozoic Era, a period of approximately 150 years, mammals existed as a relatively inconspicuous group of vertebrates. They lived as minor components in communities dominated by the often large and diverse reptiles that were so abundant at that time.

During the Mesozoic Era, mammals had not yet occupied most of the adaptive zones into which their descendants later evolved. These habitats were already occupied by the abundant, large, and well-adapted reptiles. Then, at the end of the Cretaceous Period, some 65 million years ago, all of the larger terrestrial animals became extinct. This event set the stage for the movement of the ancestral mammals and their descendants into the vacant adaptive zones that opened up with the disappearance of most of the dominant reptiles.

Three major groups of mammals existed during the Mesozoic Era. Most of the members of all three groups were about the size of mice, although some were as large as cats. One of the groups was the multituberculates, which were illustrated in Chapter 23 (see Figure 23-16); although the echidnas and platypus of Australia and its neighboring lands may be descendants of the multituberculates, the group became extinct tens of millions of years ago. The multituberculates were weed and fruit eaters, but the other two major groups of Mesozoic Era mammals fed mostly on insects and other small animals. There were two evolutionary lines: (1) the **marsupials** and (2) the common ancestors of the **insectivores** and **primates,** two orders of mammals that became important during the course of the Cenozoic Era.

Marsupials bear their young very small and keep them within a pouch. Today they are especially well represented in South America and in Australia, where kangaroos (see Figure 39-18), wombats, and koalas (see Figure 19-1) are among the familiar examples. Fossil marsupials are known from Europe, North America, Africa, and Antarctica, and the members of this order clearly spread among the continents when the continents were con-

FIGURE 24-1

The pygmy chimpanzee, or bonobo, Pan pygmaeus. *This rare animal, a close relative of the common chimpanzee, survives in the wild only in one small region of Zaire, in a bend of the Zaire (Congo) River. It has been proposed that the pygmy chimpanzee is the closest living equivalent of the common ancestor of human beings and the African apes. It walks on its two lower limbs more often than does its better-known relative and resembles in many details the first fossil humans. The pygmy chimpanzee is the only great ape not yet protected in any park or reserve.*

nected with one another. As we mentioned in Chapter 23, the opossum (see Figure 23-6) crossed from South America into North America after the emergence of the Isthmus of Panama from the sea. Marsupials, like multituberculates, have no direct connection to the evolutionary line leading to human beings. The other "insectivores," however, as we will discuss in the next section, were the common ancestors of both the modern Insectivora and the Primates. They constitute the roots of the human evolutionary line.

> **Three of the major groups of mammals that existed in the Mesozoic Era were (1) multituberculates, (2) marsupials, and (3) the ancestors of the modern insectivores and primates.**

THE EVOLUTION OF PRIMATES

The small Mesozoic mammals that ultimately gave rise to the insectivores and the primates were less numerous than the marsupials for much of their history (Figure 24-2). They were rare during the Paleocene Epoch (Table 24-1), roughly the first 10 million years of the Cenozoic Era (65 to 55 million years ago), when other kinds of mammals appeared and diversified very rapidly. The earliest insectivores lived on insects and other small animals, probably supplementing their diets with plant material when the animals were rare. Some of their descendants became specialized, feeding only on insects or only on plants, as we can tell from the structure of their teeth.

All of these small, primarily insectivorous mammals that lived in the Mesozoic Era and the first 10 million years of the Cenozoic Era were probably nocturnal, judged by their large eyes and small size. Although it depends on how some of the fossil forms are classified, scientists generally believe that no primates of modern aspect existed more than 55 million years ago. Some groups of mammals did exist, however, that exhibited generalized characteristics of primates, much as do the contemporary tree shrews (Figure 24-3). At some point, probably late in the Cretaceous Period, the order Primates first appeared, little differentiated from the insectivores, order Insectivora, that existed at the same time.

Early History of the Primates

Although the primates may have originated in the late Cretaceous Period, more than 65 million years ago, they were not well represented in the fossil record at first. By the Eocene Epoch (55 to 38 million years ago), however, an archaic group of mammals called the **plesiadapoids** was quite common (Figure 24-4). The earliest fossils that resemble any living primates belonged to the group known as **prosimians.** They are similar in structure to the earlier plesiadapoids, but their exact origins are obscure. Plesiadapoids occupy a position halfway between primates and the common ancestor of the primates and insectivores; their assignment to a particular order of mammals depends on the way in which these orders are defined.

By the end of the Eocene Epoch, prosimians were abundant in North America and Eurasia and were probably also present in Africa. Their descendants now live only in tropical Asia and adjacent islands, in Africa, and especially on the island of Madagascar, an island that is about twice the size of the state of Arizona and lies in the Indian Ocean about 325 kilometers off the east coast of Africa (Figure 24-5). All of the surviving lemurs, one of the groups of prosimians, are restricted to Madagascar; there are about 24 species. In contrast, there are only a few surviving species of the other prosimians in Asia and on the mainland of Africa. The kinds of tropical climates that are characteristic of the areas in which prosimians now occur were much more widespread during the Eocene Epoch. Modern primates—monkeys and apes—had not yet evolved at that time.

The prosimians of the Paleocene and early Eocene epochs seem to have had few competitors. They became diverse in their feeding habits as well as in their morphological and behavioral specializations. In the Eocene prosimians, the large eyes are located at the front

FIGURE 24-2

Reconstruction of an insectivore from the early Cenozoic Era, a mouse-sized mammal that probably resembled the common ancestor of the primates and insectivores, two contemporary orders of mammals.

FIGURE 24-3

A tree shrew, representative of a group of living mammals that combine the characteristics of Insectivora with those of Primates and cannot be assigned clearly to either class.

TABLE 24-1   SUBDIVISIONS OF THE CENOZOIC ERA

	START (MILLIONS OF YEARS AGO)	END (MILLIONS OF YEARS AGO)
Tertiary Period	65	2
Paleocene Epoch	65	55
Eocene Epoch	55	38
Oligocene Epoch	38	25
Miocene Epoch	25	5
Pliocene Epoch	5	2
Quaternary Period	2	Present
Pleistocene Epoch	2	0.01
Holocene Epoch	0.01	Present

FIGURE 24-4

Reconstruction of a plesiadapoid from the Eocene Epoch. The actual fur color of this mammal is unknown.

A

B

C

FIGURE 24-5

Living prosimians.
A *Ringtail lemur,* Lemur catta.
B *Indri,* Indri indri, *the largest living lemur and the only one without a tail. All living lemurs are restricted to the island of Madagascar, an island about twice the size of the state of Arizona that lies off the east coast of Africa.*
C *Tarsier,* Tarsius syrichta, *tropical Asia. Note the binocular vision and large eyes of the tarsier, adaptations to nocturnal living in the trees. The aye-aye, a very unusual nocturnal lemur, is mentioned and shown in the boxed essay on p. 477.*

A

B

C

D

FIGURE 24-6

Old and New World monkeys. Old World (catarrhine) monkeys:
A *Rhesus macaque,* Macaca mulatta, *from the Indian subcontinent, an animal that was formerly captured extensively for use in vaccine production and in medical research (the* Rh[*esus*] *factor is named for this monkey).*
B *Chacma baboon,* Papio ursinus, *southern Africa.*
C *Red monkey,* Erythrocebus patas, *a ground-dwelling monkey of the tropical African savannas.*
New World (platyrrhine) monkeys:
D *Uakari,* Cacajao rubicundus, *South America.*
E *Owl monkey,* Aotus trivirgatus, *South America, a nocturnal monkey (note the large eyes).*
F *White-faced marmoset,* Callithrix geoffroyi, *Atlantic forests of Brazil, representative of a group of small monkeys, the marmosets, with very active, squirrel-like behavior.*

E

F

of a shorter face, thus allowing both eyes to focus on the same object at one time. The resulting binocular vision allows a good perception of a three-dimensional habitat and excellent discrimination during night feeding. Judged from their structure, these prosimians appear to have fed on fruits, leaves, and flowers among the slender branches of trees, often at night. Unlike their ancestors, they probably did not depend to any great extent on their sense of smell, judged from the anatomical modifications just discussed. For animals living in the trees, the increased visual acuity made possible by their large eyes and binocular vision would have been important, as would their grasping hands. These appear to have been among the elements that made possible the early success of the prosimians.

Monkeys, Apes, and Human Beings

The ancestors of the monkeys, apes, and humans appear for the first time in the fossil record in the early Oligocene Epoch, about 36 million years ago. These **anthropoid** primates—monkeys (Figure 24-6), apes, and **hominids** (human beings and their direct ancestors)—seem to have replaced most of the prosimians, which were abundant earlier, rather rapidly. In the warm climates that prevailed during the Oligocene, Miocene, and Pliocene Epochs, both the surviving prosimians and the anthropoids ranged much farther north and south of the equator than do the members of either group now.

Monkeys and apes are almost all diurnal, or active during the day. They are agile animals that feed mainly on fruits and leaves. Most of them live in trees. Monkeys and apes differ markedly in these respects from the smaller, mostly nocturnal (night-active) prosimians and the mostly smaller, diurnal birds that feed mainly on fruits and insects. With these different habits and characteristics, the monkeys and apes entered a new adaptive zone, one in which they had relatively little competition and in which they were very successful. Their well-developed binocular vision evolved in relation to a reduction in their snouts; this evolutionary trend resulted in the flat faces that are characteristic of the group, although this feature has been reversed in some members, such as the baboons.

Like most mammals, the anthropoids have five digits; in the members of this group, the thumb stands out at an angle to the other digits and can be bent back against them, as when the animal grasps an object. Some of the advanced anthropoids have developed an extraordinary ability to manipulate objects by using this opposable thumb.

The anthropoid primates are more adaptable in their behavior than their prosimian ancestors. They have larger brains, with expansion especially of the upper part of the cortex. Because of their larger brains, the brain case of the anthropoids forms a larger portion of the head than it does in the prosimians and other mammals. Monkeys and apes, like the relatively few diurnal prosimians, live in groups in which complex social interaction occurs. In addition, anthropoids tend to care for their young for prolonged periods of time. Such a pattern of care makes possible the prolonged period of learning that is characteristic of these animals (Figure 24-7). The maternal care given the young of anthropoids seems to be related in some way to their large brains and the long periods of time that are necessary for their development.

> **Contemporary primates are characterized by the expansion and elaboration of their brains; the shortening of their snouts, together with well-developed binocular vision; the free mobility of the fingers and toes, with the forefinger and thumb opposable; and an increasing complexity and quantity of social behavior.**

Monkeys

Different groups of monkeys occur in the Old World (Eurasia and Africa) and in the New World (North and South America). The New World monkeys are called **platyrrhine** monkeys. They have flat, spreading noses and differ in other anatomical features from their Old World relatives. The Old World monkeys are the **catarrhine** monkeys. Their nostrils are close together, and their nose points downward. Some New World monkeys have the ability to hang by their tails, but no Old World monkeys can do so. All of the New World monkeys live in trees, but a number of the Old World monkeys, such as macaques, baboons, and, among the apes, gorillas, live primarily on the ground.

Where the monkeys and apes originated is not entirely clear, but the initial stages of their evolution probably took place in Africa. The New World, or platyrrhine, monkeys are probably descended from ancestors that came from the Old World more than 30 million years ago. In both hemispheres, primates occurred much farther north and south when the climate was warmer. Thus the Old World monkeys have ranged through the warmer parts of Eurasia, including Europe, and Africa. The New World monkeys have never occurred outside of Central and South America.

Because the Old World and New World monkeys are really quite different from one another, one might assume that they have evolved independently from different groups of prosimians. However, by comparing the sequences of amino acids in some of their proteins, and by other comparative biochemical methods, researchers have concluded that these two groups of monkeys share a common ancestor and are much more similar to one another than either is to any of the living prosimians. The platyrrhine and catarrhine monkeys therefore did have a common ancestor, which evolved from the prosimians and almost certainly lived during the Oligocene Epoch.

FIGURE 24-7

Vervet monkey, Cercopithecus aethiops, *nursing its young. Primates exhibit more sustained and intense parental care than other mammals. Vervet monkeys are widespread in the drier forests of tropical and subtropical Africa.*

The Old World, or catarrhine, monkeys and the New World, or platyrrhine, monkeys are derived from a common ancestor that lived in the early Oligocene Epoch. The New World monkeys have flat, spreading noses and differ in other anatomical features from the Old World monkeys, which have noses that point downward and closely spaced nostrils.

Apes

The apes, together with the hominids (human beings and their direct ancestors), comprise a group called the hominoids. The earliest known fossils of hominoids are from the early Miocene Epoch, 20 to 24 million years ago, and occur in the Old World. The living apes are usually classified as members of two families, the Pongidae (chimpanzees, gorillas, and orangutans) and Hylobatidae (gibbons). Pongidae include the single species of the genus *Gorilla* (Figure 24-8), with two or three subspecies; the two species of chimpanzee, genus *Pan* (Figures 24-1 and 24-9), both of central and western Africa; and the orangutan, genus *Pongo* (Figure 24-10), now confined to the large islands of Sumatra and Borneo in tropical Asia. Hylobatidae comprise several species of gibbons, genera *Hylobates* (Figure 24-11) and *Symphalangus* (the simiang), of tropical Asia. The ranges of the living genera of apes are shown in Figure 24-12 and illustrate the fact that the group is now entirely tropical in distribution.

FIGURE 24-8

Adult male lowland gorilla. The other subspecies, the mountain gorilla, is shown in the boxed essay on p. 477.

FIGURE 24-9

Common chimpanzee, Pan troglodytus.
A *Adult and young.*

B *Adult chimpanzee using a straw as a tool to pull termites, which seize the straw with their jaws, out of their nest to eat.*

FIGURE 24-10

Orangutans.
A *Adult male.*
B *Baby.*

A

PART FOUR   EVOLUTION   B

Apes have larger brains than monkeys. They exhibit the most adaptable behavior of all mammals except for human beings, which are assigned to a separate but closely related family, the Hominidae, the hominids (not hominoids!). All apes lack tails, as do some of the Old World monkeys. With the exception of the gibbons, which are relatively small, all apes are larger than any surviving monkey.

All apes are capable of some form of suspensory locomotion, moving along by hanging from branches with their arms. In gibbons, an extreme form of suspensory locomotion called **brachiation** is the primary mode of moving from place to place; gibbons swing rapidly through the branches, arm over arm, with their bodies erect. Suspensory locomotion has played an important role in the evolution of body form and way of life in some of the apes. It is not, however, characteristic of human beings, although the similarities between the human trunk and forelimbs and those of chimpanzees have suggested to some investigators that our ancestors might have engaged in locomotion of this sort at some time in their history.

The four genera of living apes differ in their habits and modes of social organization. Among them, gorillas and chimpanzees exhibit more complex patterns of social organization than do orangutans and gibbons. Both gorillas and chimpanzees live essentially on the ground, although they frequently climb trees. Gorillas, for example, may sleep in nests

FIGURE 24-12

Ranges of the living apes. All are now confined to tropical regions; they occur exclusively in the Old World.

C *Adult male chimpanzee.*

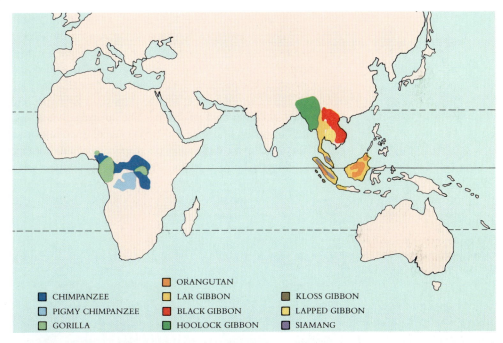

ORANGUTAN
CHIMPANZEE   LAR GIBBON   KLOSS GIBBON
PIGMY CHIMPANZEE BLACK GIBBON   LAPPED GIBBON
GORILLA   HOOLOCK GIBBON   SIAMANG

B C

FIGURE 24-11

A *and* B *Lar gibbons* (Hylobates lar).
C *Mueller gibbon* (Hylobates muelleri). *Gibbons move from place to place by brachiating, swinging rapidly and continuously from the branches of trees. Sometimes, however, they walk on the ground, as may be seen in* B. *The gibbons apparently diverged from the other living apes about 10 million years ago, judged from molecular evidence.*

among the lower branches. Gibbons and orangutans, in contrast, live mostly in the trees. Both gorillas and chimpanzees move around in groups in which males are dominant. Gorillas browse steadily all day, feeding mainly on the leaves of plants. Chimpanzees eat mainly fruit and occasionally prey on insects and small vertebrates. Gibbons are the only apes known to be consistently monogamous; the others are polygamous.

There are four groups of living apes, of which gorillas and chimpanzees occur in Africa, gibbons and the orangutan in tropical Asia. Apes exhibit the most adaptable behavior of all mammals except for humans. The African apes live mainly on the ground, whereas the Asian ones live primarily in trees.

Human beings

Humans and their direct ancestors—hominids (family Hominidae) and apes (families Pongidae and Hylobatidae)—are anatomically very similar to one another, which is why the members of these families are grouped as hominoids. There are, however, some significant differences between them. It is important to understand these differences in order to appreciate the evolutionary progression that ultimately led to the origin of modern human beings.

The normal posture of humans, while walking, is fully erect. In this respect we differ from all other mammals. Because we constantly hold our trunk erect, related structural modifications have occurred in our bodies. Most notably, the bones of our pelvis have become bowl shaped. Such a pelvis provides an effective attachment point for the muscles connecting the pelvis to the legs and the spine (Figure 24-13), and so is important in upright walking. The spine itself has developed an S-shaped curve, which can sometimes cause strains and other back problems. The human pelvis is so characteristic and so closely linked with our upright posture that it has been of great importance in the interpretation of fossils. The characteristic differences between the ape and human pelvis are shown in Figure 24-13.

In apes the arms are longer than the legs, whereas in people the opposite is true (Figure

FIGURE 24-13

Comparison between the skeleton of an ape and that of a human being, showing details of the pelvis and other important structures.

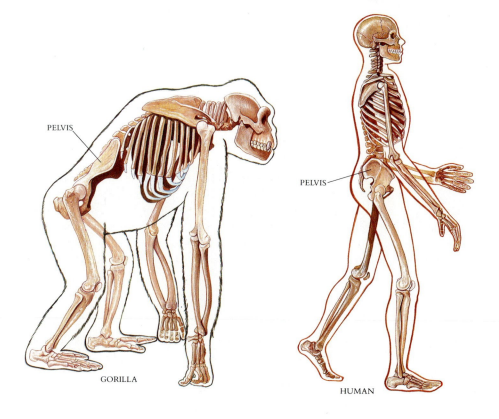

PELVIS

PELVIS

GORILLA

HUMAN

24-13). During the course of evolution both the arms and the upper trunk of human beings have become much less muscular than those of the apes. Human arms have become specialized in such a way that they are efficient in handling various objects, such as tools, in a precise manner. Modern human beings, in fact, can be defined almost completely by their total dependency on elaborate tools. Few other animals use tools (e.g., Figure 24-9); certainly none do so with the precision or to the extent that we do.

Our ability to make and use tools and to transmit our cultural advances by means of a spoken language has made us the dominant ecological force on earth. Human brains are, on the average, three or more times larger than those of the apes or those of any other primates, mostly ranging from 1400 to 1450 cubic centimeters in volume. Although some other very large mammals, such as whales and elephants, have larger brains than we do, their brains are smaller in proportion to their body mass. In addition, the neocortex of the human brain—the gray matter on the surface of the cerebral hemispheres (see Chapter 46)—is proportionately greatly enlarged in primates as compared with other mammals, and proportionately even more enlarged in humans than in other primates. This may be related directly to our use of a spoken language.

As a result of our huge brains and large heads, childbirth in human beings is difficult; the child's head can barely squeeze through its mother's pelvis. As a consequence, human babies are born at an earlier stage of development than the infants of many other species of mammals and are helpless at birth. They must be nurtured and guided for many years before they are completely independent. Our culture is transmitted during this prolonged period of training, which is therefore essential for our physical and social development.

Human beings differ from apes and other mammals in several significant ways: in being fully erect, with proportionately very long legs and somewhat short arms; the possession of very large brains with a proportionately enlarged neocortex; the prolonged period of nurturing and training given to our infants; the possession of a spoken language; and the ability to make and use elaborate tools and to transmit our culture.

THE EVOLUTION OF HOMINOIDS
Fossil Evidence

The evolution of the hominoids—apes and hominids—apparently begins with *Proconsul* and related genera, which were abundant in East Africa from 21 to about 10 million years ago. In the middle Miocene Epoch, about 15 million years ago, additional evolutionary lines of apes appeared. One of these certainly gave rise to the hominids, but we have not yet identified that group. More fossils are being discovered each year, and eventually we will probably be able to identify the line that includes our ancestors.

Among the better-known Miocene apes were the closely related genera *Ramapithecus* and *Sivapithecus,* both of which lived in Africa and Asia. These apes fell within the size range of modern chimpanzees and had robust jaws and thickly enameled teeth, presumably features that were connected with grinding seeds. *Ramapithecus* and *Sivapithecus* were once thought to be members of the hominid line, but now that they are better understood, this appears not to be the case. Certainly they are closely related to the orangutan and probably ancestral to it.

Biochemical Evidence

Another way of approaching the question of hominoid evolution and ultimately that of human ancestry involves biochemistry. For example, a comparison of the sequences of amino acids in selected proteins provides a measure of the similarities between organisms. Mutations altering the identity of one or more amino acids take place constantly in the

THE THREAT TO PRIMATES

Not surprisingly, we are intensely interested in the other primates, our closest relatives, because of what they can teach us about ourselves. The comparative study of their behavior, for example, may illuminate many aspects of the evolution of our own behavioral patterns. In addition, primates play a key role in medical research.

There are about 190 species of living primates other than human beings. Approximately 175 of them live in the tropical forests of Asia, Africa, and tropical North and South America. As the human population of the tropics grows rapidly, these forests are rapidly being cut, cleared, and destroyed. If current trends continue, all of the forests will have been destroyed or severely damaged within the next 50 years. As a result, many species of primates will also be destroyed in the wild, although some of them will survive in zoos.

Aside from habitat destruction, there are several other major reasons that so many primates are in immediate danger of extinction. In the Amazon region of South America and in West and Central Africa, primates are a major source of food for people, and thousands are killed each year. Some primate species are hunted for their pelts or because they are considered pests of crops. Many are captured as pets, for zoos, or for medical research. In these cases the mother is usually shot and the babies are taken from her dead body.

The net effect of all these forces is that a great many primates are in danger of extinction now. Approximately 67 species, or about a third of the total, are officially classified by the International Union for the Conservation of Nature and Natural Resources (IUCN) as threatened or endangered. Of these, 26 species are so severely threatened that they could become extinct before the end of the century. Some of these endangered primates are:

Golden lion tamarin (*Leontopithecus rosalia*); southeastern Brazil. Fewer than 200 individuals exist in the wild. Golden lion tamarins have bred well in captivity, however, and some have been reintroduced into natural populations.

Muriqui (*Brachyteles arachnoides*); southeastern Brazil. The largest and most apelike of the New World monkeys. Fewer than 300 individuals are living, and they have never been bred successfully in captivity.

genes encoding the proteins, as discussed in Chapter 17. At the point of divergence for different evolutionary lines, the proteins of the organisms involved would have been identical. By accumulating amino acid substitutions, these proteins would have become increasingly different from one another. Therefore, the degree of difference provides some indication of how long ago, and in what sequence, the ancestors of these organisms became distinct from one another.

When relationships determined by a comparison of proteins are compared with the fossil record, they can likewise provide an indication of the *rate* at which the respective pro-

Aye-aye *(Daubentonia madagascariensis)*. The aye-aye is a unique nocturnal lemur, which probes for insects in bark with its long digits. There are probably only a few surviving individuals. Like the other lemurs, the aye-aye occurs wild only on the island of Madagascar. More than 90% of the native vegetation of Madagascar has been severely disturbed or destroyed already.

Mountain gorilla *(Gorilla gorilla beringeri)*. Severely threatened by poachers, the highly endangered mountain gorilla survives in a few scattered bands in misty forests on the high volcanoes of Central Africa. There are fewer than 300 surviving individuals in nature.

Drill *(Papio leucophaeus)*, a handsome baboon that lives in troops of up to 200 animals in the coastal rain forests of Cameroon in West Africa. A very popular source of food, the drill is actively hunted with dogs and is protected only in Kourup National Park. It is rapidly becoming extinct throughout its range, which was less than 200 kilometers across at its largest extent. This is a subadult male.

Lion-tailed macaque *(Macaca silenus)*. An endangered relative of the more abundant rhesus and bonnet macaques, the lion-tailed macaque is confined to the monsoon forests of southwestern India, with fewer than 1000 individuals known to be living in the wild.

teins changed. When these data are tabulated and cross-checked for many groups, the rate of amino acid substitution in particular proteins can be used as a kind of "molecular clock." The more different two proteins are, the longer ago they diverged or, speaking figuratively, the longer the "clock" has run. For this reason, the clock can be used to estimate the time of divergence of organisms from one another, even in the absence of a fossil record. Several different proteins can be studied and compared independently in this respect, and if calculations based on them are pooled, a more precise estimate can be obtained of the time of divergence for the groups of organisms in which the proteins occur.

Since mutations altering the identity of one or more amino acids occur constantly, the proteins in two separated populations of organisms become more different with the passage of time. Consequently, this process can be used to assign relative dates to the times of separation of different evolutionary lines, especially when more than one protein has been compared. Tied in with the fossil record, it becomes what has been called a "molecular clock" of evolution.

Let us now consider some examples. The hemoglobin molecule of humans differs from that of gorillas to a degree that could be accounted for by one base substitution in the DNA coding for the alpha chain and one in the DNA coding for the beta chain. The hemoglobin molecules of humans and chimpanzees, on the other hand, are identical. The alpha chain in human hemoglobin differs from that of macaques, one of the groups of Old World monkeys, in a way that could be accounted for by 5 base substitutions, whereas the differences in the beta chain would require 10 base substitutions. For humans versus rabbits, the respective substitutions required are 28 and 16. Even greater degrees of substitution are necessary to account for the differences between the hemoglobin of humans and those of other groups of vertebrates.

A second method that can be used to analyze the relationship of different evolutionary lines and the sequences in which they diverged from one another is that of DNA hybridization. Duplexes are formed when strands of DNA are brought together. The more similar the two samples of DNA, the more extensive the duplex formation. Within a species, the formation of **homoduplexes**—the binding of identical DNA strands—is complete. When a sample of DNA from another organism is introduced, forming **heteroduplexes,** the binding is incomplete. If the comparison is with human DNA, for example, the DNA from the other organism "competes" with the binding of human–human DNA samples. In experiments of this kind, the inhibition of human–human DNA binding by chimpanzee DNA samples is 98%; that by macaque DNA, 93%; and that by lemur and mouse DNA, probably lower than 50%. These and other methods of evaluating the degree of difference between samples of DNA have been extended to many other kinds of organisms and compared with the known fossil record. As a result of their application, the time necessary for the attainment of different degrees of divergence has been calculated.

Molecular studies have demonstrated that gorillas and chimpanzees are very closely related to one another and also to humans, that orangutans are about twice as different from this line, and that gibbons are the most distinct of all the living apes. Using the molecular clock method with the protein albumin and assuming a divergence between the New World and Old World anthropoids 35 million years ago, some investigators have calculated that the evolutionary line leading to gibbons diverged from that leading to the other apes about 10 million years ago; that leading to orangutans, about 8 million years ago; and the split between hominids and the line leading to chimpanzees and gorillas, about 4 million years ago. The time of divergence between chimpanzees and gorillas appears to have been a little less than 3.2 million years ago, as determined by the same methods. It is important to bear in mind, while considering these postulated dates, that they are not certain and not yet wholly compatible with the fossil evidence. Much further study is needed before the dates of divergence of the different lines of hominoids are fully established and generally accepted (Figure 24-14).

Molecular evidence suggests that chimpanzees and gorillas diverged from one another a little less than 3.2 million years ago, hominids from that line about 4 million years ago, orangutans from all of them about 8 million years ago, and gibbons from the rest about 10 million years ago. The absolute timing of these events depends on the interpretation of the fossil record and on the relationship of molecular patterns to time.

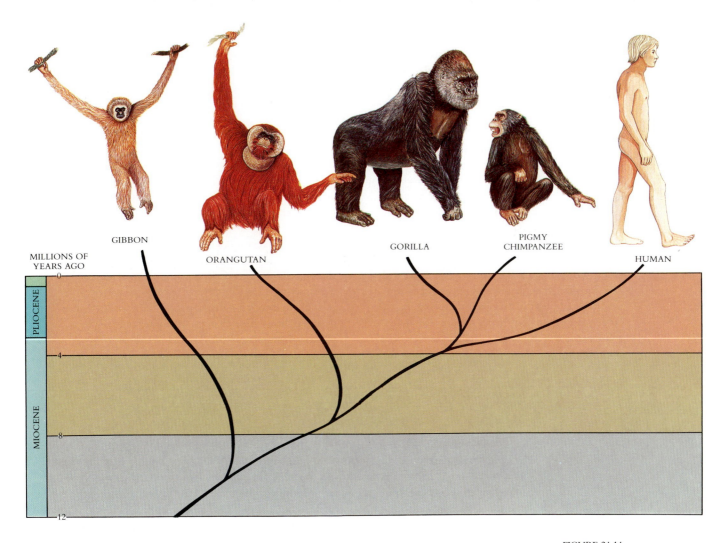

GIBBON

ORANGUTAN

GORILLA

PIGMY
CHIMPANZEE

HUMAN

MILLIONS OF
YEARS AGO

PLIOCENE

MIOCENE

0

4

8

12

FIGURE 24-14

The evolution of the living apes and of human beings.

When divergence dates like these were suggested by Allan Wilson and Vincent Sarich of the University of California, Berkeley, in 1967, they created a sensation because they were greatly at variance with the interpretation of the fossil record that was generally accepted at that time. However, much additional information has accumulated subsequently, which suggests that the dates Wilson and Sarich proposed are both reasonable and consistent with an improved interpretation of the fossil record. The relationship between chimpanzees, gorillas, and human beings, as measured biochemically, is so close that if they were members of any other group of organisms, they would almost certainly be classified as members of the same genus, despite their substantial morphological differences.

The Pattern of Evolution

We are now ready to consider in greater detail the evolution of hominoids, the apes and hominids. As we have just seen, chimpanzees, gorillas, and hominids are closely related to one another, and it has often been assumed that the ancestor of hominids would have had characteristics similar to those of gorillas and chimpanzees. For example, chimpanzees and gorillas often walk on their knuckles (Figure 24–15), and one might suppose that the ancestral hominids did so too. Some believe that chimpanzees and gorillas shared a common ancestor distinct from hominids, however, and if they did, it would be just as logical to assume that knuckle-walking, an unusual trait, evolved in their common ancestor *after* it split from the hominid line. Even though hominids certainly evolved in Africa, there is no reason to regard the traits of the other living African apes as necessarily being ancestral ones.

FIGURE 24-15

Knuckle-walking in a gorilla, Gorilla gorilla.

ALLOMETRY IN HUMAN EVOLUTION

By Alan R. Templeton, Washington University

When judging the evolutionary relationships of humans to apes or to monkeys, it is important to keep in mind that all apes and humans are very large primates. The large size of humans and apes is important because of a phenomenon known as **allometry.** For many organisms, when size changes, so does shape (e.g., relative limb proportions, head-to-body proportions, and so forth). Often these shape changes associated with size changes are not haphazard but rather follow well-defined rules. In other words, as size increases, some parts of the body grow more or less than the other parts, thereby causing changes in their relative proportions (i.e., shape). Allometry refers to such a relative growth relationship, in which size changes induce predictable shape changes.

Allometric relationships can be found both within and between species. For example, a newborn human baby has arms that are relatively larger with respect to the legs than is the case in adults. Similarly, smaller monkeys (as adults) tend to have forelimbs that are relatively larger with respect to their hindlimbs than do larger monkeys (as adults). By studying a large number of species, it is often possible to discover a mathematical relationship that describes how shape changes with size, and then to use this mathematical relationship to make predictions about the expected shapes of larger or

Well-dated fossils suggest an origin for hominids more than 5 million years ago, somewhat earlier than suggested by the biochemical evidence that we have just reviewed but not in fundamental contradiction with it. Chimpanzees and gorillas seem not to have diverged from one another until about 3.2 million years ago. These time relationships tend to support the suggestion that the common ancestor of humans, chimpanzees, and gorillas was at least partly bipedal and not a knuckle-walker. If knuckle-walking originated after the separation of the hominid line from that leading to chimpanzees and gorillas, the proportionately long arms of gorillas and chimpanzees probably did so also. In that case the proportionately shorter and smaller arms of human beings might have been the condition in the common ancestor of humans, chimpanzees, and gorillas.

In this connection, however, it is intriguing that the pygmy chimpanzee (see Figure 24-1) is often bipedal and has shorter upper than lower limbs and a more slender trunk than the common chimpanzee. In these respects, it resembles the hypothetical common ancestor of the African apes and humans more closely than does the common chimpanzee or the gorilla.

Summarizing, we may present the evolution of hominids and African apes as follows. Approximately 5 million years ago a species of relatively small apes, well adapted to climbing trees but sometimes bipedal and sometimes quadripedal when on the ground, diverged into the hominid and African ape lineages. We do not know how much earlier the ancestors of this line may have occasionally walked from place to place on two limbs. In the African ape lineage, specialization for long arms and of the hand for knuckle-walking, coupled with a great increase in size, rapidly altered the appearance of the animals. In the hominid line, bipedalism was well established as the primary mode of locomotion at least 3.7 million years ago, which led in turn to the kinds of structural modifications reviewed on pp. 474–475.

Some features of both the African ape line and the hominid line rapidly became specialized. In the African ape line, we have stressed the proportionately long arms, for example; the cheek teeth and jaws are also obviously specialized. In the hominid line, on the other hand, the front teeth, skull, and foot changed more rapidly than they did in the African ape line. The retention of more generalized features in the hominid line, however,

smaller animals. For example, studies on many different monkey species have been used to predict the body and limb proportions that one would expect to find in ape- or human-sized primates that have retained the monkey developmental plan.

Such allometric predictions are very important in understanding human evolution, because all evolutionary biologists agree that apes and humans evolved from monkeylike ancestors. However, apes and humans differ in many ways from monkeys and from each other. For example, apes and humans both have brains that tend to be larger than monkey brains, but human brains are much larger than ape brains. However, when one takes the brain/body size allometric relationship of monkeys and extrapolates it to primates of the size of humans or apes, one discovers that the size of an ape brain is just about what one would expect for a monkey of that size, but the human brain is much larger than is expected for a comparably sized monkey. In other words, the larger brain size of apes can be explained simply by their large body size, but the large brain size of humans requires a drastic alteration from the basic developmental plan seen in monkeys.

On the other hand, apes tend to have forelimbs that are longer than their hindlimbs, whereas humans have just the opposite. Using the forelimb/hindlimb allometric relationship found in monkeys, the extrapolation to large-sized primates indicates that apes have limb proportions that represent severe deviations from the basic monkey developmental plan, whereas humans do not. Hence, for brain size, humans are evolutionarily derived and apes are primitive; but for overall body plan, humans have retained the primitive monkey allometries, whereas the apes are highly derived.

such as the proportionately short arms and unspecialized jaws, resulted in what amounts to a relatively limited alteration of the basic body proportions found in most primates, as compared with the more extensive alterations of these body proportions that are characteristic of the African apes. About 2 million years ago the rapid evolution of large brains in humans began to alter the features of the human skull. In addition, these rapidly developing brains made possible the evolution of the cultural properties that characterize human beings.

FIGURE 24-16

Reconstruction of an adult female Australopithecus, *a member of the genus that gave rise to our own,* Homo.

Ancestral hominids appear to have been derived from a common ancestor with chimpanzees and gorillas about 5 million years ago. The features of this common ancestor may be approached most closely among the living apes by those of the pygmy chimpanzee.

THE APPEARANCE OF HOMINIDS
The Australopithecines

The two critical steps in the evolution of human beings were the evolution of bipedalism, which was well established in our ancestors more than 3.7 million years ago, and the enlargement of the brain, which began about 2 million years ago. The earliest definite hominids, which share many of the characteristics that we regard as distinctively human, are assigned to the genus *Australopithecus* (Figure 24-16). Among their human features were bipedal locomotion and rounded jaws. The australopithecines lived on the ground in the open savannas of eastern and southern Africa, weighed upwards of 20 kilograms and were up to 1.3 meters tall. Their brains were about the size of those of gorillas, about 350 to 450 cubic centimeters in volume, as compared with an average volume of about 1450 cubic centimeters in modern human beings. Individuals of *Australopithecus* appear to have eaten many different kinds of plant and animal food, judged from the characteristics of their teeth and jaws.

The oldest known fossils that may represent *Australopithecus* are 5 to 5.5 million years old; they thus date from close to the postulated time of divergence of the hominid and chimpanzee-gorilla evolutionary lines. Other fossils of *Australopithecus* are commonly found from 4 million years ago onward at a number of localities in East and South Africa (Figure 24-17). Among them, the fossil discoveries made by the anthropologists Louis and Mary Leakey, and by their son Richard, have been of special importance in the development of the story of human evolution (Figure 24-18). At Laetoli in Kenya the Leakeys found fossil bones and also footprints 3.7 million years old (see photograph on p. 2).

Two species of *Australopithecus*, *A. robustus* and *A. boisei*, consist of robust individuals with blunt, grinding teeth and are therefore thought to have been vegetarians to a greater extent than other members of the genus. They lived in eastern and southern Africa from about 2.3 to 1.3 million years ago. Other species of *Australopithecus* are more slender and seem to have been at least partly carnivorous, in view of their dental characteristics. Although the classification of this genus is a matter of intense interest, there is no general agreement about how many species existed in total; many scientists believe that there were four: the two just mentioned and two others that had a more slender build.

The first definite hominids, *Australopithecus*, appeared about 5 million years ago in Africa. They lived on the ground, weighed less than 20 kilograms, and were less than 1.3 meters tall. They disappeared about 1.3 million years ago.

The Use of Tools: *Homo habilis*

Starting about 2 million years ago, tools appear in the sedimentary beds of East Africa. Such tools have been associated with the fossils of the first known human beings, the earliest members of the genus *Homo*. The fossils and tools were discovered in the Olduvai Gorge, Tanzania, which has now become famous as a fossil site. Because these people used tools, the Leakeys named them *Homo habilis* (Latin *habilis*, "able"). Fossils of this species have since been discovered in Kenya, Ethiopia, and South Africa, in addition to Tanzania, so *H. habilis* seems to have been widespread. It appears in the fossil record about 2 million years ago and persists for about half a million years.

Judged from its physical characteristics, *Homo habilis* was certainly derived from *Australopithecus*. As the use of tools was developed in its earliest populations, the characteristics of the genus *Homo* may have begun to emerge. *Homo habilis* had a brain somewhat larger than that of *Australopithecus*, probably related to its larger size overall. Judged from the structure of its hands, it still regularly climbed trees, like *Australopithecus*, although it was mainly bipedal and ground dwelling. The interpretation of the sites where many tools and bones are found together varies; some scientists believe that they were campsites; others, like Richard Potts of Yale University, believe that they were merely areas where stone implements were stored and to which carcasses were carried. In its characteristics, *Homo habilis* was apparently about midway between the australopithecines and the other members of the genus *Homo*. The transition between these distinct hominids, and therefore the existence of *Homo habilis*, appears to have occupied a relatively short period of time and ultimately to have resulted in the emergence of those features that we consider distinctively human.

ted>*Homo habilis* was the first member of our own genus and the first primate to use tools consistently. This species appeared in East Africa about 2 million years ago and survived for about 500,000 years. In this species, which certainly evolved from *Australopithecus*, fully developed human characteristics first emerged.

Homo erectus

All of the early evolution of the genus *Homo* seems to have taken place in Africa. There, fossils belonging to the second and only other extinct species of *Homo*, *Homo erectus*, are

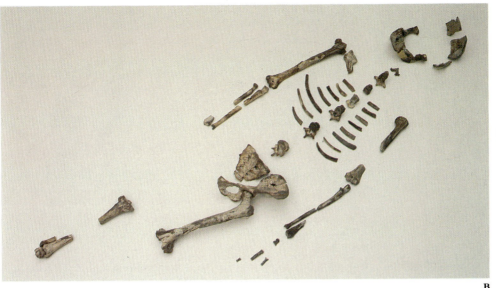

A

FIGURE 24-17

A *Skull of* Australopithecus robustus *from South Africa, about 1.7 million years old. This individual lived towards the end of the history of the genus.*
B *"Lucy," from Ethiopia, the most complete skeleton of* Australopithecus *discovered so far. Many more such complete skeletons are needed for a better understanding of our ancestors.*

A

B

FIGURE 24-18

A *From localities like this, near Lake Rudolph in northern Kenya, have come many of the most important fossils of human beings.*
B *Richard Leakey, Director of the National Museums of Kenya, patiently cleaning a cranium of* Australopithecus *at this locality.*

FIGURE 24-19

Reconstruction of the skulls of Homo erectus *and* Homo sapiens, *showing comparative features.*

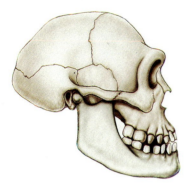

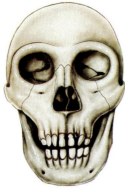

HOMO ERECTUS

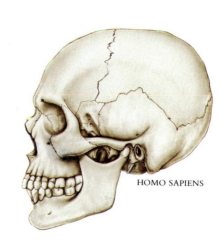

HOMO SAPIENS

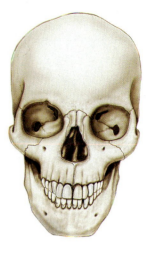

FIGURE 24-20

Tools made by Homo erectus, *from St. Acheul, France. Left, a hand axe; above right, a flake; below right, a pointed flake scraper. These tools were much more advanced than those of* Homo habilis. *They are widespread throughout Africa, India, Europe, and western Asia.*

widespread and abundant from 1.7 million until about 500,000 years ago. By 1 million years ago, *Homo erectus* had migrated into Asia and Europe. These large humans, who probably did not differ in size from *Homo sapiens,* displayed the essential human characteristics. Their brains were from 700 to 1300 cubic centimeters in volume, roughly twice as large as the brains of *Australopithecus* or *Homo habilis.* The skulls of *Homo erectus* were larger and heavier than those of *Homo sapiens,* with prominent brow ridges, rounded jaws, and massive teeth (Figure 24–19). Both "Java man" and "Peking man," two of the most famous early fossils of the genus *Homo,* belong to the species *Homo erectus.* They are only about 250,000 years old, which indicates that the species survived much longer in Asia than it did in Africa.

The earliest campsites associated with *Homo erectus* are about 1.5 million years old. Most of these seem to have been temporary camps beside streams and lakes and in other places where other animals would have been plentiful and where the people could obtain water. Some of the members of this species also lived in caves, and they were apparently the first people to have done so.

Groups of *Homo erectus* probably subsisted mainly by gathering food, scavenging dead animals, and sometimes hunting, at least locally. They may have stampeded game off of cliffs occasionally, but the evidence for such a practice is limited. The stone tools made by *H. erectus* were much more complex than those associated with *H. habilis.* Among these tools were carefully fashioned hand axes and scrapers (Figure 24–20), which appeared at about the same time as the earliest campsites of *Homo erectus,* about 1.5 million years ago. These tools, which spread rapidly throughout Africa, were used for such purposes as food preparation, including cleaning and cutting meat; scraping hides; digging; and cutting wood. None of these tools appears especially suitable for killing game, although *H. erectus* may have used hand axes as projectiles.

The wide distribution of the characteristic tools made by *Homo erectus* suggests that their widely dispersed bands may have been in communication with one another. The first evidence of the use of fire by humans occurs at the campsites of this species in the Rift Valley of Kenya at least 1.4 million years ago, and was characteristically associated with populations of *Homo erectus* subsequently. As we mentioned above, the species was replaced by our own, *Homo sapiens,* in Africa by about 500,000 years ago, and in Asia about 250,000 years ago.

Homo erectus, the second species of our genus to evolve, appeared in Africa about 1.7 million years ago and migrated throughout Eurasia by 1 million years ago. It was replaced by our species, *Homo sapiens*, in Africa about 500,000 years ago and in Asia about 250,000 years ago. *Homo erectus* used fire, starting at least 1.4 million years ago, and made characteristic stone tools.

Modern Human Beings: *Homo sapiens*

We recognize the first appearance of members of our own species in the fossil record because of their larger brains, reduced teeth, and enlargement of the rear of their skulls as compared with those of *Homo erectus*. Although the skulls of contemporary *Homo sapiens* are much less massive than those of *Homo erectus,* the skulls of the earliest representatives of *Homo sapiens* were comparable to those of the earlier species. The earliest fossils that have such features are about half a million years old and appear first in Europe. In other words, these characteristics appeared just when *Homo erectus* was becoming rarer. It appears likely that *Homo sapiens* represents a derivative of *Homo erectus* with certain improved, "modern" features that we will now review. These features probably were associated with the ability of early populations of *Homo sapiens* to replace their ancestor. At any rate, *Homo erectus* clearly persisted much longer in Asia than in Europe, where populations of *Homo sapiens* appeared first.

Neanderthals

Populations of *Homo sapiens* of the kind that are called Neanderthals were abundant in Europe and western Asia between about 70,000 and about 32,000 years ago, judged from the fossil record. Such fossils were first discovered in the valley of the Neander River in Germany in 1856, and the name that is applied to humans is derived from that of the valley.

Compared with contemporary people, Neanderthals were powerfully built, short, and stocky. Their skulls were massive, with protruding faces, projecting noses, and rather heavy bony ridges over the brows (Figure 24-21). Their brains were even larger than those of modern human beings, a fact that may have been related to their heavy, large bodies. The distinctive physical features associated with the Neanderthals first became apparent in Europe about 200,000 years ago and can be traced back even farther. Similar populations of *Homo sapiens* were also widespread in Africa at that time.

The Neanderthals made diverse tools, including scrapers, borers, spearpoints, and hand axes. Some of the Neanderthal tools were used for scraping hides, which they used for clothing. They lived in hutlike structures or in caves. They took care of their injured and sick and commonly buried their dead, often placing food and weapons, and perhaps even flowers, with the bodies. Such attention to the dead suggests strongly that they believed in a life after death. The first indications of thought processes that are characteristic of modern *Homo sapiens,* including symbolic thought, are evident in these acts.

> **Our species, *Homo sapiens,* may be as much as 500,000 years old; it has been numerous for the past 150,000 years. Neanderthals were abundant in Europe and western Asia from about 70,000 to about 32,000 years ago.**

Cro-Magnons

About 34,000 years ago the European Neanderthals were abruptly replaced by people of essentially modern character, the so-called Cro-Magnons. We can only speculate about why this sudden replacement occurred, but it was complete all over Europe in a short period of time. There is some evidence that the Cro-Magnons came from Africa, where fossils of essentially modern aspect, but as much as 100,000 years old, have been found in southern Africa and in Ethiopia. They seem to have replaced the Neanderthals completely in southwest Asia by 40,000 years ago and then spread across Europe during the next few thousand years, coexisting and possibly even interbreeding with the Neanderthals for several thousand years.

The Cro-Magnons used sophisticated stone tools that rapidly became more diverse and thus more useful for many purposes (Figure 24-22). They also made a wide variety of tools out of bone, ivory, and antler, materials that had not been used earlier. They were hunters who killed game with complex tools. In addition, it is believed on the basis of their evident social organization that they may have been the first people to have fully modern

FIGURE 24-21

A Neanderthal man. Despite their heavy brows and jaws, Neanderthals appear not to have been dramatically different from us in mental capacity or any other way, except for the knowledge that we have accumulated over the course of our history.

FIGURE 24-22

Some of the tools made by Cro-Magnon people. (A) blade; (B) pointed stone tool; (C) antler harpoon point; (D) bone needle. Other tools included knives, awls, and chisels, some of which were used to make other tools and carvings out of bone and ivory, including spearheads, fishhooks, needles like the one shown here, and, ultimately, arrow tips.

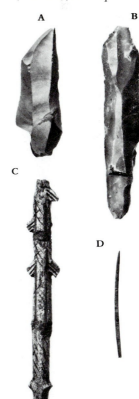

language capabilities. Soon after Cro-Magnon people appeared, they began to make what were apparently ritual paintings in caves (Figure 24-23). The culture associated with these impressive and mysterious works of art existed during a period that was cooler than the present. Such conditions prevailed at the time of the last great expansion of continental ice. During that period grasslands, tundra, and coniferous forest, inhabited by large herds of grazing animals, occurred across Europe. The animals that occurred in such habitats are often depicted in the cave paintings.

Human beings of modern appearance eventually spread across Siberia to the New World, which they reached at least 12,000 to 13,000 years ago, after the ice had begun to retreat at the end of the last glacial period. As people spread throughout the world, they were apparently responsible for the extinction of many populations of animals and plants, a process that has been accelerated recently. Human remains are often associated with the bones of the large animals they hunted, and these animals often disappeared promptly from different parts of the world shortly after human beings arrived for the first time. By the end of the Pleistocene Epoch, about 10,000 years ago, there were only about 10 million people throughout the entire world, as compared with nearly 5 billion now, so the reasons for the increased human pressure on world ecosystems are clear.

THE DOMESTICATION OF ANIMALS AND PLANTS

As the glaciers retreated at the end of the most recent expansion of continental ice, starting about 18,000 years ago, forests spread northward in Eurasia and North America. With the disappearance of the herds of large grazing mammals that occurred during the last period of glacial expansion, people who lived in the areas that were formerly open grassland probably turned increasingly to hunting smaller game. Along the seashores, where food was plentiful, they appear to have adopted a more sedentary way of life. In the drier regions to the south and southeast, in areas that are now parts of Egypt, Turkey, Iran, and Iraq, we find traces of early agriculture starting about 11,000 years ago.

People first began to gather, and then to plant for future harvest, the seeds of cereals such as barley and wheat (Figure 24-24, *A*). Other plants, such as the broad bean, lentil, and pea among the legumes, as well as grapes and olives, were also brought into cultivation in the Near East. At about the same time these same people brought into captivity and then into domestication many different kinds of animals, including dogs, cattle, horses, chickens, cats, hogs, goats, and sheep (Figure 24-24, *B*). The combination of herds of domestic animals with cultivated plants made possible the formation of towns and cities in the Near East and thus led to a rapid acceleration of both human knowledge and the size of the human population.

From the Near East, agriculture spread westward and northward into Europe, with the addition along the way of crops such as rye, oats, cabbage, lettuce, hops, carrots, and beets. Agricultural communities reached eastern Europe by about 6000 BC, then central Europe a thousand years later and Britain a thousand years after that—about 4000 BC.

Agriculture was developed independently in other parts of the world. For example, the water buffalo, elephant, and camel were domesticated long ago in tropical and arid portions of Asia. The various kinds of citrus, as well as the mango, were cultivated for the first time in southern Asia, whereas soybeans and perhaps rice, one of the most important of all cereals, were first cultivated somewhat farther north. Additional crops such as okra, sorghum, various kinds of millets, and perhaps cotton were domesticated in Africa or in the drier parts of western or southwestern Asia.

A parallel development of agriculture took place in the New World. Dogs were brought to the region by migrating people, but they are the only domesticated plant or animal that seems to have been carried there. Otherwise, the domestic plants and animals of the New World were entirely different from those of the Old World and were unknown in the Old World before the voyages of Columbus (Figure 24-25). In America, corn was the staple cereal; kidney beans, lima beans, and peanuts were the legumes; and tomatoes,

A

B

FIGURE 24-24

A *Berber people harvesting wheat in the Atlas Mountains of Morocco, North Africa. Such simple methods of cultivation probably resemble those used during the period when agriculture was first being developed.*

B *Herds of sheep or goats, like those moving with this nomad caravan in Afghanistan, were domesticated about 11,000 years ago in southwestern Asia. Rapidly increasing in numbers, they soon had a devastating effect on the grasslands of the region.*

pumpkins, squashes, sunflowers, potatoes, chili peppers, and tobacco were among the other crops. Subtract all of these and you can imagine what European cooking was like five centuries ago.

> **Plants and animals were domesticated in the Near East, temperate and tropical Asia, and Latin America starting about 11,000 years ago. Different crops and domestic animals were involved in each center, with full interchange occurring only in the last few centuries.**

In the past few hundred years, there has been a human population explosion on an unprecedented scale. The adoption of a more sedentary way of life, made possible by the domestication of plants and animals, has provided the opportunity for the rapid growth of human populations. In turn, it has made necessary the greater productivity of the agricultural systems that were associated with these populations. The course of this population growth, and its realized and potential effect on human beings and other aspects of the global ecosystem, will be examined in Chapter 54.

SUMMARY

1. Reptiles called therapsids gave rise to mammals about 210 million years ago, near the start of the Jurassic Period. By the end of the Mesozoic Era, 65 million years ago, seed- or fruit-eating mammals called multituberculates, largely insectivorous marsupials, and the common ancestors of the insectivores and primates were all frequent.

2. Primates first appear in the Eocene Epoch, 55 to 38 million years ago, and the first anthropoid primates—a group that now includes monkeys, apes, and human beings—appeared in the following Oligocene Epoch, about 30 million years ago. Primates have proportionately large brains, binocular vision, and five digits, including an opposable thumb, and they exhibit complex social interactions.

3. The Old World monkeys, also called the catarrhine monkeys, and the New World, or platyrrhine, monkeys diverged from a common ancestor more than 35 million years ago. The ancestor of the New World monkeys presumably dispersed from the Old World to the New World at about that time.

4. In catarrhine monkeys the nostrils are close together and the nose points downward. Some catarrhine monkeys dwell on the ground. In platyrrhine monkeys the nostrils are widely spaced and the nose is flat and spreading. All platyrrhine monkeys dwell in trees, and some are able to hang by their tails.

5. The apes, which appeared 20 to 25 million years ago, gave rise in a way not well understood to the gibbons and orangutans, and then to the African apes (chimpanzees and gorillas) and hominids (humans and their direct ancestors), starting more than 10 million years ago. Apes and hominids collectively are termed hominoids.

FIGURE 24-25

This attractive drawing of corn, Zea mays, is from a herbal published by the Bavarian botanist Leonard Fuchs in 1542, just 50 years after the first voyage of Christopher Columbus. By that time, corn was being grown in the warmer parts of Europe and the Near East. Fuchs did not know that it had come from America. Corn is one of the many plants that were brought back to Europe from the New World.

6. The earliest hominids, family Hominidae, are assigned to the genus *Australopithecus*. The members of this genus were the direct ancestors of human beings. They appeared in Africa about 5 million years ago and were small, up to about 1.3 meters tall and up to about 20 kilograms in weight, with brains about 450 cubic centimeters in volume.

7. About 2 million years ago, with an enlargement of brain size that was perhaps associated with the increased use of tools, the genus *Australopithecus* gave rise to human beings, belonging to the genus *Homo*. Early fossils from Africa about 2 million years old that seem to have been associated with the use of tools have been called *Homo habilis*. They were evidently derived from populations of relatively slender individuals of *Australopithecus* and became extinct about 1.5 million years ago.

8. The second extinct species of *Homo*, *Homo erectus*, appeared in Africa at least 1.7 million years ago. *Homo erectus* migrated from Africa to Eurasia about 1 million years ago; lived by gathering, scavenging, and sometimes hunting; and included the first humans to use fire (starting at least 1.4 million years ago) and to live at times in caves. They were replaced by our species, *Homo sapiens*, about 500,000 years ago in Africa and about 250,000 years ago in Asia.

9. Modern *Homo sapiens* formed societies that increased greatly in complexity, especially since the development of agriculture at a number of scattered places during the past 10,000 years or so.

SELF-QUIZ

The following questions pinpoint important information in this chapter. You should be sure to know the answers to each of them.

1. Which of the following was *not* a major type of mammal in existence by the end of the Mesozoic Era?
   (a) multituberculates
   (b) marsupials
   (c) insectivores
   (d) primates
   (e) none of the above; all were present

2. The earliest fossils that can unequivocally be considered primates were of the group known as _____.

3. Anthropoid primates are often larger than the prosimians and more adaptable in their behavior. True or false?

4. Complex social interactions are characteristic of prosimians. True or false?

5. Anthropoid primates have which of the following characteristics?
   (a) expanded and elaborated brain compared with prosimians
   (b) a nonopposable thumb and forefinger
   (c) lengthened gestation period
   (d) free mobility of fingers and toes
   (e) all of the above

6. Monkeys and apes are primarily carnivores. True or false?

7. Some New World monkeys are the only monkeys that
   (a) live in trees
   (b) eat meat as a normal constituent of their diet
   (c) have the ability to hang by their tails
   (d) are still alive today

8. _____ is a mode of transportation used by apes by which they swing from one arm and then the other with their bodies upright.

9. Hominids differ from all other mammals because their normal posture while walking is erect. True or false?

10. The two critical steps in the evolution of human beings were the evolution of bipedalism and the enlargement of the brain. True or false?

11. The first hominids appeared about 5 million years ago in Africa. They belong to the genus _____.

12. How old is our species?

THOUGHT QUESTIONS

1. Which human characteristics result in our being classified among the mammals? Among the primates? Among the hominoids? What are some of the differences among *Australopithecus, Homo erectus,* and *Homo sapiens?*

2. How did Neanderthal people differ from modern humans? Develop a hypothesis that you consider plausible for their extinction, and list the reasons for and against your hypothesis.

FOR FURTHER READING

ALTMANN, J.: *Baboon Mothers and Infants,* Harvard University Press, Cambridge, Mass., 1980. A beautifully written account of the results of a 9-year study of baboons in Kenya. Provides clear insights into the way primate research is conducted.

FOSSEY, D.: *Gorillas in the Mist,* Houghton Mifflin Co., Boston, 1983. Dian Fossey has devoted many years to a study of mountain gorillas, and this book summarizes her findings.

HAY, R.L., and M.D. LEAKEY: "The Fossil Footprints of Laetoli," *Scientific American,* September 1982, pages 50-57. An account of this fascinating discovery, which we discussed in Chapter 1.

JOHANSON, D., and M. EDEY: *Lucy: The Beginnings of Humankind,* Simon & Schuster, Inc., New York, 1981. The account of an expediton to the Afar region of Ethiopia and the significance of the findings of early fossils of *Australopithecus* to an understanding of early human evolution.

KUMMER, H.: *Primate Societies: Group Techniques of Ecological Adaptations,* Davidson, Harlan, Inc., Arlington Heights, Ill., 1971. An outstanding book on primate behavior and ecology, easily read and understood.

LEAKEY, R.E.: *The Making of Mankind,* E.P. Dutton, Inc., New York, 1981. A beautifully illustrated and well-written book on human evolution during the past several million years, based on much personal research and reflection.

MOLNAR, S.: *Races, Types, and Ethnic Groups,* 2nd ed., Prentice Hall, Inc., New York, 1983. An excellent introduction to human biology, including a discussion of the adaptive significance of variation in human beings.

PILBEAM, D.: "The Descent of Hominoids and Hominids," *Scientific American,* March 1984, pages 84-96. Good digest of contemporary research on human evolution.

SMITH, F.H., and F. SPENCER (eds.): *The Origin of Modern Humans: A World Survey of Fossil Evidence,* Alan R. Liss, Inc., New York, 1984. An outstanding summary of the available fossil evidence for the origin of modern human beings.

TRINKAUS, E., and W.W. HOWELLS: "The Neanderthals," *Scientific American,* December 1979, pages 54-66. Contemporary research on an earlier form of *Homo sapiens.*

VAN LAWICK-GOODALL, J.: *In the Shadow of Man,* Houghton Mifflin Co., Boston, 1971. An account of the first 11 years of a detailed study of the social organization of a group of chimpanzees in Tanzania, with many observations on the significance of this work in relation to the field in general.

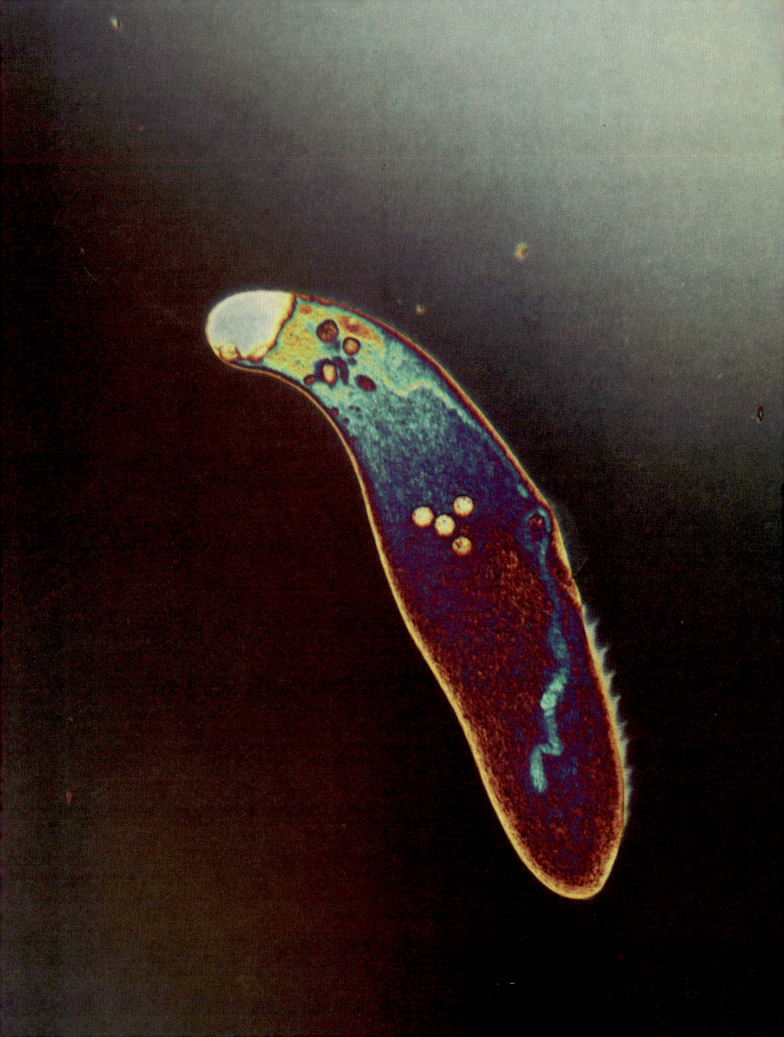

BIOLOGY OF SIMPLE ORGANISMS   *Part Five*

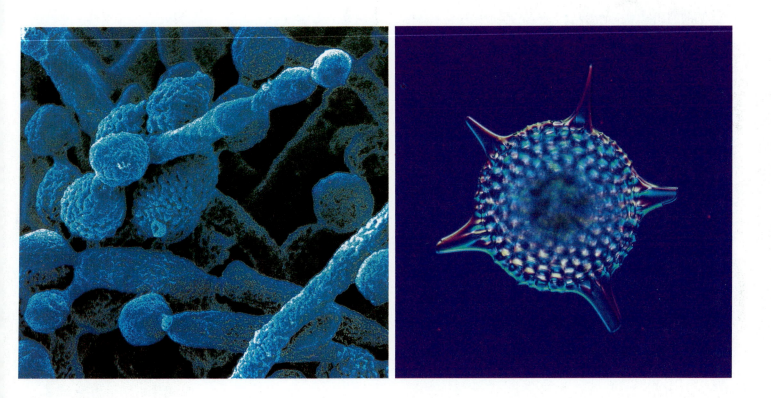

In the evolution of their incredible
morphological and physiological
diversity, the protists set the stage for
the character of the three dominant
kingdoms of multicellular organisms:
plants, fungi, and animals.

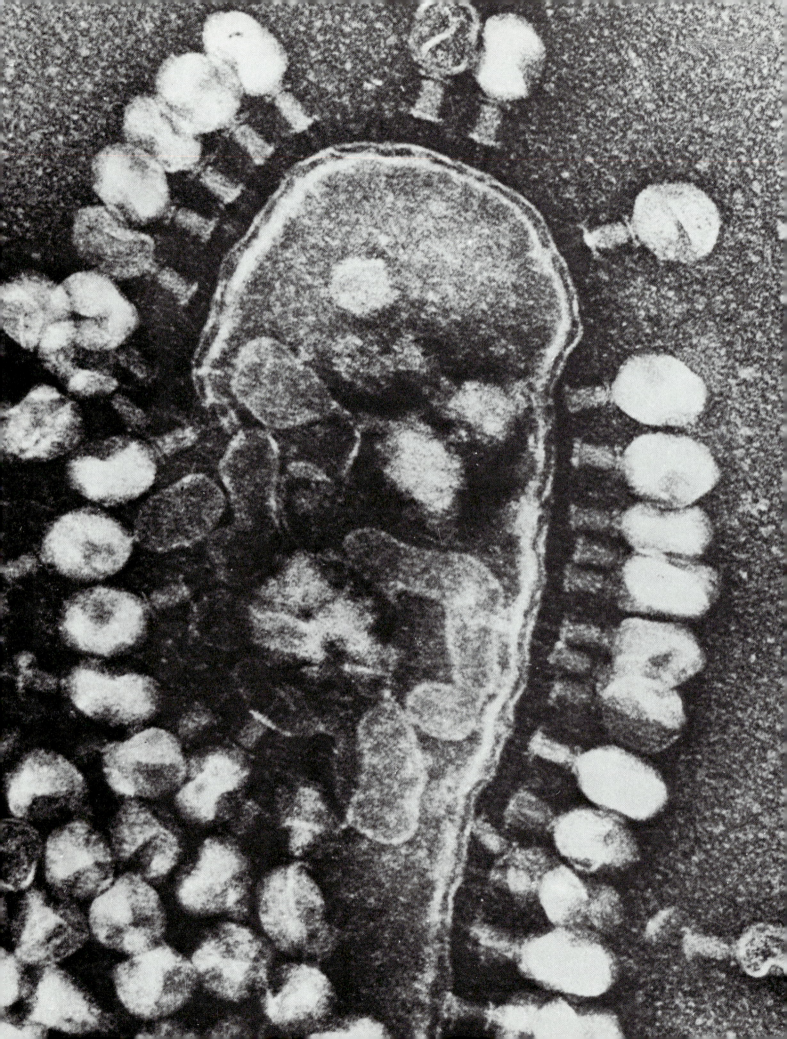

Overview

A virus is a fragment of another genome, often a small fragment. It is able to replicate, if it gains entry to a cell, by using that cell's machinery. Many viruses have evolved complex sets of genes that enable them to infect host cells and multiply within them, but viruses may also produce as side effects serious diseases in their host organisms. Viruses are not alive and are not organisms, but because of their disease-producing potential, they are very important biological entities. Much of what we know about molecular biology has been learned from studies of viruses. It is thought that all organisms possess viruses and that new viruses are "escaping" from bacterial and eukaryotic genomes and evolving even now.

For Review

Here are some important terms and concepts that you will encounter in this chapter. If you are not familiar with them, you should review them before proceeding.

- **Definition of organism** (Chapter 3)

- **TMV and T2 viruses** (Chapter 15)

- **Lambda virus integration** (Chapter 16)

- **Retroviruses and cancer** (Chapter 17)

- **Vaccinia virus and vaccination** (Chapter 18)

The simplest organisms living on earth today are bacteria, and we think that they closely resemble the first living organisms to evolve on earth. Even simpler than the bacteria, however, are the viruses. **Viruses** are strands of nucleic acid that are encased within a protein coat. No viruses have the ability to grow or replicate on their own. Viruses reproduce only when they enter cells and utilize the cellular machinery of their hosts (p. 288). They are able to reproduce themselves in this way because they carry genes that are translated into proteins by the cell's genetic machinery, leading to the production of more viruses. In view of their characteristics, earlier theories that viruses represent a kind of "halfway house" between life and nonlife have largely been abandoned. Instead, viruses are now viewed as fragments of the genomes of organisms; they could not have existed independently of preexisting organisms.

Any fragment of DNA that includes a recognition site for DNA polymerase can replicate itself autonomously within a cell. Such fragments have become detached on numerous occasions from the genomes of both bacteria and eukaryotes. Subsequently they have acquired the ability to move about on their own, co-opting the genetic machinery of the host's cells or the cells of another organism to reproduce themselves. Sometimes the original genomic fragments have been altered considerably in the further course of evolution of a particular virus, and most viruses have acquired the ability to manufacture protective protein sheaths and envelopes during the course of their evolution. Some viruses have RNA in place of DNA, but no virus has both kinds of nucleic acid; viruses differ in this respect from all organisms.

As an analogy to a virus, consider a computer whose operation is directed by a set of instructions in a program, just as a cell is directed by DNA-encoded instructions. A new program can be introduced into the computer that will cause the computer to cease what it is doing and instead devote all of its energies to making copies of the introduced program. The new program is not itself a computer, however, and cannot make copies of itself when outside the computer, lying on the desk. The introduced program, like a virus, is simply a set of instructions.

Viruses are fragments of DNA or RNA that have become detached from the genomes of bacteria or eukaryotes and have the ability to replicate themselves within cells. Viruses are nonliving and are not organisms. Their genetic material consists of DNA or RNA, but not both.

THE DISCOVERY OF VIRUSES

Outside of its host cell, a virus is simply a fragment of nucleic acid encased in protein, a helpless portion of a cellular genome. Depending on which genes a virus carries, it can often seriously disrupt the normal functioning of the cells that it infects. Such effects occur in

FIGURE 25-1

Edward Jenner inoculating patients with cowpox, and thus protecting them from smallpox, in the 1790s. The underlying principles were not understood until more than a century later.

humans and other animals, plants, protists, and bacteria; in fact, cells that are completely free of viruses may be rare.

Diseases caused by viruses have been known and feared for thousands of years. Among these are smallpox, chickenpox, measles, German measles (rubella), mumps, influenza, colds, infectious hepatitis, yellow fever, polio, rabies, and AIDS, as well as many other diseases not as well known. Viral diseases were studied centuries before it was understood that viruses caused them. In the 1790s, Edward Jenner, an English country doctor, observed that milkmaids often caught a mild form of "the pox," presumably from cows, and that milkmaids who had caught cowpox rarely contracted smallpox, a very serious disease. It was as if the cowpox were protecting them from the smallpox. Using this observation, Jenner developed the process of **vaccination.** He deliberately infected people with a relatively harmless cowpox virus, causing the recipient's body to develop antibodies against cowpox, antibodies that also were effective against smallpox (Figure 25-1). Jenner learned how to carry out this process despite total ignorance of the nature of antibodies and viruses. The explanation of how vaccination works was not established until more than a hundred years later.

The earliest indirect observations of viruses, other than simple observations of their effects, were made near the end of the nineteenth century. At that time several groups of European scientists, working independently, concluded that the infectious agents associated with a plant disease known as tobacco mosaic and those associated with hoof-and-mouth disease in cattle were not bacteria. They reached this conclusion because these infectious units were not filtered out of solutions by the kinds of fine-pored porcelain filters that were routinely used to remove bacteria from various media.

Building on these observations and on the properties of the filtered material, Martinus Beijerinck in the Netherlands, at about the same time as Friedrich Loeffler and Hans Frosch in Germany, concluded that viruses not only were much smaller than any known bacteria but also were different in nature. Viruses could reproduce themselves only in the living cells of their hosts and therefore lacked some of the critical machinery by which living cells are able to reproduce themselves.

The next and most important advance in the understanding of viruses was made possible in 1933 when Wendell Stanley of the Rockefeller Institute prepared an extract of tobacco mosaic virus and purified it. Surprisingly, the purified virus precipitated in the form of crystals (Figure 25-2). Stanley was able to show by this method that viruses can better be regarded as chemical matter than as living organisms, at least in any normal sense of the word "living." The purified crystals still retained the ability to reinfect healthy tobacco plants and so clearly were the virus itself, not merely a chemical derived from it. With Stanley's experiments, scientists began to understand the nature of viruses for the first time.

Within a few years other scientists were able to follow up on Stanley's discovery and demonstrate that tobacco mosaic virus consisted of a protein in combination with a nucleic acid. They discovered that the particles that constituted this virus were rods about 300 nanometers long (Figure 25-3). Subsequently it was shown that tobacco mosaic virus consists of RNA surrounded by a protein coat. Many plant viruses have a similar composition, but most other viruses have DNA in place of RNA. Nearly all viruses form a protein sheath, or **capsid,** around their nucleic acid core. In addition, many viruses form an **envelope,** rich in proteins, lipids, and glycoprotein molecules, around the capsid.

> **Most viruses form a protein sheath, or capsid, around their nucleic acid core, and a lipid-rich protein envelope around the capsid.**

The simple structure of viruses, the large numbers that are produced in an infection of a cell, and the fact that their genes are related to those of their host, have led many scientists to study viruses in attempts to unravel the nature of genes and how they work. For more than three decades, the history of virology has been thoroughly intertwined with those of genetics and molecular biology.

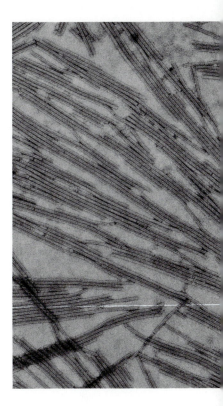

FIGURE 25-2

Tobacco mosaic virus, showing virus particles taken from infected tobacco leaves in a crystalline array.

RNA COIL

PROTEIN COAT

FIGURE 25-3

Diagram of tobacco mosaic virus structure.

THE NATURE OF VIRUSES

Viruses occur in virtually every kind of organism that has been investigated for their presence. Under normal circumstances, the vast majority of these viruses produce no perceptible outward sign of their presence—no disease, not even any change in the appearance of the organism in which they occur. Viruses are almost always highly specific in the hosts they infect and do not reproduce anywhere else. This means that there should be nearly as many kinds of viruses as there are kinds of organisms—perhaps millions of them. Since there often is more than one kind of virus in a given organism, the actual number of kinds of viruses may even be much greater.

Most viruses are able to transmit their nucleic acid component from one host individual to another. Having entered the host cell, some viruses "take over" the genetic machinery of the host cell and use it to produce more viruses. The cell usually **lyses** as a result of such an infection, rupturing and releasing the newly made virus particles. Viruses that cause lysis to occur in their host cells are called **virulent viruses.** Other viruses become established within the host cell as stable parts of its genome; they are called **temperate viruses.**

The fact that individual kinds of viruses contain only a single type of nucleic acid, either DNA or RNA, is one of the major reasons that they can reproduce only within living cells. All true organisms contain both DNA and RNA, and both are essential components of their genetic machinery. Viruses also lack ribosomes, as well as all of the enzymes necessary for protein synthesis and energy production. In general, viruses contain only those enzymes which are necessary for them to replicate their own nucleic acids, capsids, and envelopes, and thus make possible the invasion of new host cells.

Viruses are so completely integrated into the normal metabolism of their host that they are very difficult to control by chemical means, although immunization is effective. Once the viruses have become established, anything that inhibits them also inhibits the proper development of their host. Antibiotics, which selectively kill bacteria but not their hosts, are useless against viruses. On the other hand, viruses themselves are being used as agents of biological control against insects and are considered increasingly as control agents for some kinds of bacteria.

Viral diseases are difficult to treat because a virus becomes an integral part of the cell it infects and has no metabolic traits of its own, only those of the host cell.

THE STRUCTURE OF VIRUSES

The smallest viruses are only about 17 nanometers in diameter, the largest ones up to 1000 nanometers (1 micrometer) in their greatest dimension. The largest viruses, therefore, are barely visible with a light microscope. Most viruses can be detected only by using the higher resolution of an electron microscope. Viruses are directly comparable to molecules in size, a hydrogen atom being about 0.1 nanometer in diameter and a large protein molecule being several hundred nanometers in its greatest dimension.

The simplest viruses consist of a single molecule of a nucleic acid surrounded by a capsid, which is made up of one to a few different protein molecules, repeated many times. In more complex viruses, there may be several different kinds of molecules of either DNA or RNA in each virus particle and many different kinds of proteins. Most viruses have an overall structure that is usually either **helical** or **isometric.** Helical viruses, such as the tobacco mosaic virus, have a rodlike or threadlike appearance; isometric ones have a roughly spherical shape.

The only structural pattern that has been found among the isometric viruses is the **icosahedron,** a figure with 20 equilateral triangular facets, characteristically subdivided in viruses into 60 or more subunits (Figures 25-4 and 25-5). The icosahedron is the basic design of the geodesic dome; it is the most efficient symmetrical arrangement that subunits can take to form an external shell with maximum internal capacity.

FIGURE 25-4

An icosahedral virus of the reovirus group, showing the basic appearance of many viruses as viewed with an electron microscope. In the reoviruses, the genetic material consists of double-stranded RNA.

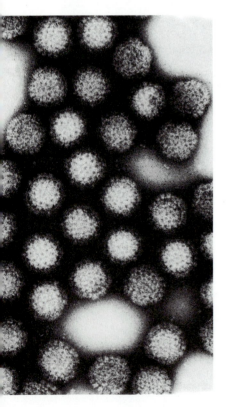

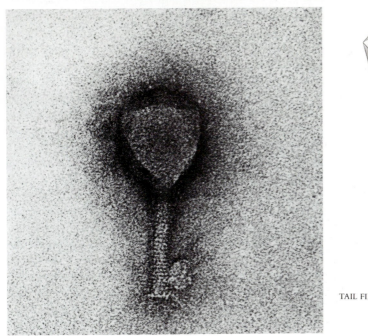

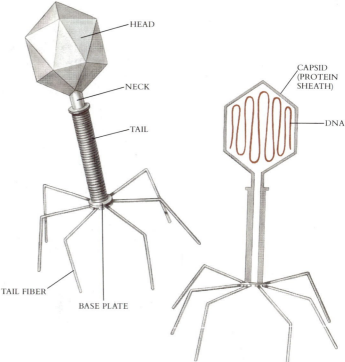

HEAD

NECK

TAIL

TAIL FIBER

BASE PLATE

CAPSID (PROTEIN SHEATH)

DNA

Most viruses are icosahedral in basic structure; an icosahedron has 20 equilateral triangular faces, each of which is characteristically subdivided in viruses. Many other viruses are helical, such as tobacco mosaic virus. Some viruses have more elaborate forms.

Some bacterial viruses, or **bacteriophages,** are among the most complex viruses (Figure 25-5). Each of them is made up of at least five separate proteins; these make up the hexagonal head, the tail core, the molecules of the capsid, the base plate of the tail, and the tail fibers. A long DNA molecule is coiled within the hexagonal head.

VIRUS REPLICATION

Before any virus can infect an animal cell, it must first bind to a specific receptor molecule in the cell membrane, probably a glycoprotein. As mentioned earlier, many, but not all, viruses have a lipid-rich envelope around the capsid. From the envelope of many viruses protrude spikes that may contain glycoproteins and lipids (Figure 25-6). The properties of the molecules that make up the outer covering of the virus have much to do with its adhesion to various substrates. If an envelope is not present, the properties of the capsid determine the adhesive qualities of the virus. Mammals protect themselves from viral infections by producing antibodies to envelope and capsid proteins, enabling their immune system to identify and remove the virus particles.

Changes can occur in the base sequence in the genetic material of a virus that affect the detailed structure of its protein or glycoprotein capsid, envelope, and spikes; these changes may make that virus infectious to host organisms that had earlier been immune (p. 501). The relationships may be very complex: in one kind of virus, it has been shown that one of the proteins in the capsid binds to surface cell receptors; another determines the capacity for growth of the virus on mucosal surfaces, like those that line your nasal cavities; and a third is responsible for inhibiting the synthesis of proteins by the host cell. The lipids that are present in the envelopes of many viruses are determined by the genetic machinery of their hosts.

Among the most important characteristics of a virus is the nature of the proteins, especially the glycoproteins, that make up its capsid, envelope, and spikes. These proteins determine the infective properties of the virus.

FIGURE 25-5

Electron micrograph and diagram of the structure of T4 bacteriophage. Viruses of this group are discussed on p. 506.

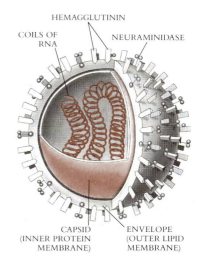

HEMAGGLUTININ

COILS OF RNA

NEURAMINIDASE

CAPSID (INNER PROTEIN MEMBRANE)

ENVELOPE (OUTER LIPID MEMBRANE)

FIGURE 25-6

Diagram of a flu virus. In this diagram, the coiled RNA has been revealed by cutting through the outer, lipid-rich envelope, with its two kinds of projecting spikes, and the inner protein capsid. The infective qualities of the flu virus are determined largely by the composition of the proteins that are present in the two different kinds of spikes.

After a virus has become bound to the surface of an animal cell, it proceeds to enter. Animal viruses appear to enter by endocytosis, with the cell membrane folding inward to form a deep cavity around the virus. Once they have been replicated within the cell, some animal viruses—for example, those responsible for influenza or rabies—may be released from the cell by a budding process that is, in effect, the opposite of endocytosis. Plant viruses normally enter the cells of their hosts at points of injury, whereas bacteriophages shed their coats outside their host cell, injecting their nucleic acids through the host cell wall. When a bacteriophage attacks a bacterial cell, it first becomes attached to the cell by its tail and then injects its DNA; some of the bacteriophages seen in the electron micrograph in Figure 15-7, *B*, have already done so, as revealed by their empty heads. The complete cycle of virus infection takes about 20 minutes. At the end of that time, the bacterial cell lyses and several hundred new bacteriophage particles are released from it.

Once within the host cell, those viruses which have retained their coats—plant and animal viruses—shed them and begin the process of replication, just as do bacteriophages, which have left their coats outside. Often hundreds, even thousands, of new virus particles are produced more or less simultaneously within a host cell. Ultimately the newly formed virus particles escape from the original host cell, and some find their way to a new host cell.

The simplest viruses use the enzymes of the host cell both for protein synthesis and for gene replication. For example, some of the parvoviruses, which are single-stranded DNA viruses mainly known from rodents, may contain only a single gene. This gene codes for a protein molecule that subsequently splits, forming the three proteins needed to make up the parvovirus coat. The enzymes of the infected cell copy the gene of the parvovirus into mRNA, after which the host enzymes translate the mRNA into viral protein. The host enzymes also copy the viral DNA, which is then packaged, forming new virus particles. Parvoviruses normally replicate only in cells that are actively dividing, presumably because they depend on the dividing cell to provide them with the enzymes necessary for their own replication. Their simple mode of replication illustrates how completely viral replication is integrated into the machinery of a normal cell.

At the other end of the scale, some large viruses contain more than 200 genes, which are used to form many structural proteins and enzymes that are peculiar to the virus. No matter how complex a particular virus may be, however, there are always some stages of its replication at which the virus depends on the synthetic machinery of its host cell.

In many DNA viruses the viral genetic material is integrated into the DNA of the infected cell's chromosomes. All of us probably have some viral DNA of this kind integrated into our chromosomes. Once a virus has entered a chromosome, its activity is usually suppressed and the virus is said to be **latent.** A latent bacteriophage, which consists only of a fragment of DNA, is called a **prophage,** and a cell in which a prophage exists is termed a **lysogenic** cell, because it may be lysed in the future. Those bacteriophage strains that have the ability to exist as prophages are, of course, temperate viruses.

Temperate bacteriophages carry one or more bacterial genes. Once they have gained entry into a bacterial cell, such bacteriophages can integrate themselves into the bacterial genome by crossing-over directly with the DNA of the host bacterium. The first gene of a prophage that is read by the host genetic apparatus stops the further transcription of the remaining viral DNA.

Some viruses have the ability to incorporate themselves into the chromosomes of their hosts; in this form they are said to be latent. A latent bacteriophage is called a prophage, and the cell in which a prophage occurs is termed a lysogenic cell. Bacteriophage strains that have the ability to become prophages are called temperate bacteriophages.

In animals, latent viruses may be released from the host chromosome by certain external stimuli, such as ultraviolet radiation and some chemicals. This is precisely what happens when a fever blister develops on your lip because of the activation of a latent her-

TABLE 25-1   CHARACTERISTICS OF SOME GROUPS OF VIRUSES

NAME OF GROUP	DNA OR RNA	NUMBER OF STRANDS	ENVELOPE PRESENT	EXAMPLES OF HOSTS	EXAMPLES OF DISEASES
Unenveloped plus-strand RNA	RNA	One	No	Plants, bacteria, animals	Many colds; poliomyelitis
Enveloped plus-strand RNA	RNA	One	Yes	Arthropods, vertebrates	Some cancers, AIDS
Minus-strand RNA	RNA	One	Yes	Plants, animals	Flu, mumps, rabies
Viroids	RNA	One	Naked	Plants only	
Double-stranded RNA	RNA	Double	Yes	Plants and animals	Colorado tick fever
Small-genome DNA	DNA	One or Double	Yes	Mainly animals	Viral hepatitis; warts
Medium- and large-genome DNA	DNA	Double	Yes	Animals	Herpes; shingles; some cancers; poxes
Bacteriophages	DNA	Double	Yes	Bacteria	

pesvirus, or when a lung cell bearing a latent retrovirus is suddenly converted into a rapidly dividing cancerous cell by a smoke-borne carcinogen, a process described in Chapter 18.

The interactions between viruses and their host cells are often very complex. For example, some bacteria, including *Corynebacterium diphtheriae,* the causative agent of diphtheria, and *Streptococcus pyogenes,* the causative agent of scarlet fever, can produce their characteristic toxins only if they are infected by the appropriate bacteriophages. In the absence of these bacteriophages, the respective bacteria are harmless.

DIVERSITY AMONG THE VIRUSES

The diversity of the viruses is great and is almost certainly related to their modes of origin. To provide a systematic idea of some of this diversity, we will discuss the viruses under eight main headings: (1) unenveloped plus-strand RNA viruses, (2) enveloped plus-strand RNA viruses, (3) minus-strand RNA viruses, (4) viroids and similar forms, (5) double-stranded RNA viruses, (6) small-genome DNA viruses, (7) medium- and large-genome DNA viruses, and (8) bacteriophages. The main characteristics of these groups are summarized in Table 25-1.

Unenveloped Plus-Strand RNA Viruses

Unenveloped plus-strand RNA viruses are called plus strands because they act directly as mRNA on infection of a host cell, attaching to the host's ribosomes and being translated. As indicated by their name, these viruses lack envelopes and consist only of a nucleic acid core surrounded by a protein capsid. They all have similar shapes and dimensions. Some are bacteriophages; others occur in plants. In both of these groups the protein coats are made up of multiple copies of a single molecule. Still others occur in the cells of animals and have coats made up of as many as four different kinds of proteins. The divided-genome plant viruses are unusual members of this group; they are called **divided-genome viruses** because in them, several different kinds of RNA are packaged separately, but collectively determine the characteristics of a particular virus.

> **In plus-strand RNA viruses the RNA acts directly as messenger RNA when it has entered the host cell.**

We have already discussed one of the best-known plant viruses in this group, tobacco mosaic virus (Figures 25-2 and 25-3). Other viruses classified here cause various diseases

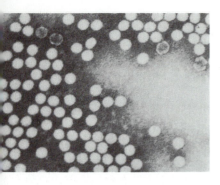

FIGURE 25-7

Polio virus. Soon after he took office in 1933, President Franklin Delano Roosevelt, himself a victim of polio, proclaimed a "war on polio." This effort, largely funded by private contributions, was won with the development of vaccines against polio approximately 20 years later.

of plants called **mosaics** or **stunt diseases,** such as alfalfa mosaic or tomato bushy stunt. The plant viruses in this group are much more numerous in kinds and diverse in structure than are either bacteriophages or animal viruses with similar characteristics. The plus-strand RNA viruses that cause plant diseases vary greatly in size; some are icosahedral, but many are elongate like tobacco mosaic virus.

Several of the animal viruses classified here cause important diseases. Poliomyelitis was a widespread, crippling disease through the first half of the twentieth century (Figure 25-7). The disease has now largely been controlled in the developed world through vaccination, although it remains a serious and frequent disease in the tropics and elsewhere in the less developed parts of the world. Hoof-and-mouth disease, which plays a major role in determining the geographical limits within which cattle can be raised successfully, is also caused by an unenveloped plus-strand RNA virus.

Colds are viral infections of the upper respiratory tract. About a third of all colds are caused by the **rhinoviruses,** which are unenveloped plus-strand RNA viruses. The colds caused by other kinds of viruses differ in their symptoms and in the seasons at which they occur most frequently. There are dozens of different strains of cold-causing rhinoviruses alone, each of them with different properties and none conferring cross-immunity to the others. More than 200 strains of viruses that cause colds have been identified, which makes the development of appropriate inoculation methods very difficult, if not impossible. For about 40% of all colds, the responsible agents have not even been identified.

Enveloped Plus-Strand RNA Viruses

The enveloped plus-strand RNA viruses, all of which parasitize animals, are distinguished from the members of the preceding group by their lipid-rich envelopes. Members of this group cause many widespread diseases, such as many kinds of encephalitis, dengue, and yellow fever. They are classified as **arboviruses** (arthropod-borne viruses), together with many unrelated viruses, because they are transmitted by insects and other arthropods.

One of the most fascinating stories in the annals of disease control is that relating to the unraveling of the mystery of yellow fever by Walter Reed and his colleagues during the period when the Panama Canal was being built. Yellow fever is maintained in native populations of monkeys and is spread to humans by mosquitoes of the genus *Aedes* (Figure 25-8). The disease is now largely controlled through a combination of the eradication of mosquito populations and vaccination of people who live in or visit areas where it occurs. Rubella, or German measles, is also caused by a virus of this group; it too is largely controlled in many parts of the world through vaccination.

The retroviruses discussed in Chapter 18 in connection with cancer in humans and other animals are enveloped plus-strand RNA viruses. They are called retroviruses because they have the enzyme reverse transcriptase, which catalyzes the synthesis of a double strand

FIGURE 25-8

A Aedes *mosquito.*
B *The togavirus that causes yellow fever. The building of the Panama Canal was possible only after the eradication of this mosquito, which has been a serious scourge of humankind for as long as we have records. The reservoir of the virus is in monkeys, from which the mosquitoes spread it to human beings.*

A

B

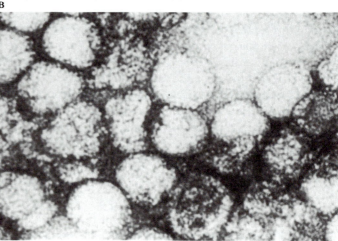

of DNA complementary to the RNA genome of the virus. This double-stranded DNA is integrated into the chromosomes of the host cell, from which it is copied into both mRNA (used to produce viral proteins) and the RNA copies of the viral genetic material itself. A virus of this group, Human T Cell Lymphotropic Virus, type III, or HTLV III, was discovered in 1985 to be the cause of acquired immune deficiency syndrome (AIDS), a fatal disease discussed in Chapter 1.

Minus-Strand RNA Viruses

Minus-strand RNA viruses are distinguished from plus-strand RNA viruses because they carry the RNA strand complementary to the mRNA that carries the genetic information of the virus. After a minus-strand RNA virus enters a host cell, it directs the formation of the appropriate mRNA, which then functions in the cell.

Among the members of this group are the rhabdoviruses, which are rod-shaped or bullet-shaped viruses about 180 nanometers long and about 70 nanometers thick, with a helical capsid and a lipid-rich envelope. Different rhabdoviruses infect humans and other vertebrates, arthropods (some rhabdoviruses are arboviruses), and plants. Rabies, which was the subject of the path-breaking discoveries of Louis Pasteur in the nineteenth century, is caused by a rhabdovirus. A second group of minus-strand RNA viruses, the paramyxoviruses, includes those that cause measles and mumps in humans, Newcastle disease in poultry, and distemper in various other animals.

Rabies is occasionally spread by small mammals, such as skunks, raccoons, and foxes; in 1984 more than 25,000 people in the United States had to get shots after being bitten by animals suspected of having rabies. Outbreaks in the vicinity of Washington, D.C., and in other areas are a matter of great local concern. Worldwide, it is estimated that some 70,000 people die of rabies each year. Fortunately it is a simple matter to vaccinate pets against rabies, and human vaccines are also being developed. The rabies virus is spread by the saliva of the host animal, often after bites, but it can also be contracted from handling a dead infected animal. An animal infected with rabies may go into a mad frenzy, often running great distances in its confusion.

An important third group of minus-strand RNA animal viruses consists of the influenza, or flu, viruses. Type A flu viruses are responsible for most of the serious flu epidemics in humans; they occur in mammals and birds also. Type B and also probably type C viruses are restricted to humans. Type B viruses cause relatively small outbreaks of flu, whereas type C viruses rarely cause serious health problems.

An individual flu virus resembles a ball studded with spikes (Figures 25-6 and 25-9); these spikes are composed of two kinds of protein. One of these proteins, hemagglutinin (HA), is the substance that allows the virus access to the cell interior, and the other, neuraminidase (NA), permits the daughter viruses to break free of the host cell once the viral replication has been completed. The "types" of flu virus (A, B, C, and so forth) are defined by their serological reactions; their capsid proteins, which are associated with these reactions, are similar within each type and differ between the types. In contrast, the proteins that make up the spikes and the envelope are highly variable and specific, and they determine the infective characteristics of the individual strains. For example, the A-type flu viruses can be classified into 13 distinct HA subtypes and nine distinct NA subtypes. The virus that caused the Hong Kong flu epidemic of 1968 is characterized by one specific type of HA together with one specific type of NA.

Flu viruses are able to change their genetic constitution so rapidly that vaccines against them soon become obsolete. The structures of both the HA and NA molecules are now known in detail. The HA molecule, for example, is a trimer, which, in effect, stands on the surface of the virus somewhat like a tripod and has clublike projections on top. Each of its three legs has four "hot spots," or segments of amino acids that display unusual tendencies to change readily as a result of changes in the viral RNA. These segments function as antigens against which the body's antibodies are directed. Each time the segments change, new antibodies are required to lock them up and thus neutralize the virus.

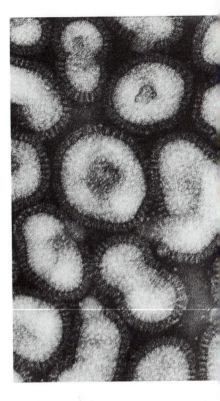

FIGURE 25-9

These densely packed objects are the more-or-less icosahedral particles characteristic of a flu virus.

PRIONS

Stanley Prusiner, of the University of California, San Francisco, has described an infectious particle that consists only of protein. This particle, which he has called the **prion**, exists in extracts prepared from the brains of diseased animals. The name is derived from the first letters of *proteinaceous* (pr) and *infectious* (i). The particles can only be visualized with an electron microscope. Prions apparently cause scrapie, a neurological disorder of sheep and goats, and Creutzfeldt-Jakob disease, a rare neurological disorder of humans. They also seem to be the causative agent for several other rare human diseases. There is circumstantial evidence linking them with such human diseases as multiple sclerosis, Parkinson's disease, Alzheimer's disease, and Lou Gehrig's disease; this evidence is under active study. So far, it has not been possible to infect animals with any of these last-mentioned diseases, which makes it difficult to investigate their causes.

The prions isolated from diseased brains retain their infectiveness even when bombarded with agents such as heat, radiation, and chemicals, which

---

A change in one of the amino acids of the spike protein occurs as a result of point mutation in 1 of 100,000 viruses during the course of each generation. Such changes rarely alter the shape of the spike protein enough for it to escape the surveillance of the human immune system. The problem in combating the flu viruses arises through recombination. Viral genes are readily reassorted and recombined, even between fundamentally different viruses. Such reassortment can produce major shifts in the amino acids that make up the spikes. When such shifts occur, worldwide epidemics may result, since a large change in the shape of the spike protein makes it unrecognizable to human antibodies specific for the old spike configuration. Viral recombination of this kind seems to have been responsible for the three major flu epidemics that have occurred in this century: the "killer flu" of 1918, the Asian flu of 1957, and the Hong Kong flu of 1968. The flu epidemic of 1918-1919 resulted in 20 million deaths; yet the characteristics of that particular virus are not known, nor do we know why that particular strain was so virulent.

The infectivity of flu viruses is determined by the properties of the two proteins that make up their envelopes and spikes: hemagglutinin (HA) and neuraminidase (NA).

Somewhat surprisingly, it appears that wild ducks may be the major reservoirs for flu viruses. Viruses are abundant in ducks, infecting about 1 of 30 adult individuals and most juveniles. They live in the cells that line the intestinal tract, cause no disease in the ducks, and are spread readily in the feces. All of the known types of HA and NA molecules occur in ducks, so that recombination can occur readily. Since ducks fly long distances, they spread newly originated mutant flu viruses to all parts of the globe.

Flu virus researcher Robert G. Webster, of St. Jude Children's Research Hospital in Memphis, Tennessee, believes that Hong Kong flu originated when the genetic material of a human flu virus recombined with another virus from a duck. This unfortunate event probably occurred in China, where someone infected with Asian flu was working with an infected domestic duck and contracted a second viral infection from the duck. As a result of this postulated event, 50 million people in the United States alone contracted flu, 70,000 died, and approximately $4 billion was lost in medical expenses and absences from work.

Viroids

Viroids are infectious circular molecules of RNA that include from 250 to 400 nucleotides. Viroids lack capsids and have no proteins associated with them. They are functionally related to the RNA viruses that we have been discussing but are less than one tenth the size

would destroy nucleic acids but not protein. Careful chemical analysis indicates that prions contain only protein—even though prions are able to reproduce themselves within mammalian cells, they contain no DNA or RNA! How do they do this? It appears that the prion protein binds to the cell's DNA, initiating transcription of a normally quiescent gene encoding the amino acid sequence of the prion protein. The prion is a product of one of the host's genes.

Because the prion is a normal product of the host genome, the diseases caused by prions never trigger the production of antibodies and thus are unaffected by the body's immune system, a highly unusual property. Prions seem to cause their characteristic symptoms very slowly: when an extract from the brains of sheep infected by scrapie is injected into healthy sheep, they do not become ill for about 3 years. It had been thought that the diseases mentioned above might be caused by "slow viruses," which somehow become incorporated into the body and exhibit their symptoms much later.

of the next smallest known agents of infectious disease. The first viroid to be identified was the causative agent of potato spindle tuber disease (Figure 25-10). All viroids, as far as is known, infect plants: different viroids have been associated with about a dozen plant diseases.

Viroids are naked, infectious, circular molecules of RNA that cause diseases in certain plants. They have from 250 to 400 nucleotides.

FIGURE 25-10

Symptoms of potato spindle disease, with an electron micrograph of viroids. The viroids (labeled PSTV for potato spindle tuber viroid) are much smaller than a typical virus such as the T₇ virus DNA that here looks like a rope by comparison.

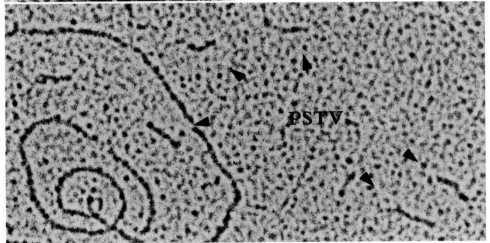

Because viroids are circular, their RNA has no end groups. They do not seem to encode any polypeptides, so all components necessary to their replication—which involves direct RNA-to-RNA copying—must be present in the cells of the host plant. They probably produce their disease symptoms by the proliferation of viroid RNA in a normal synthetic pathway. Viroids may have originated from the introns of their hosts, which they resemble closely. Other kinds of RNA particles, which are less well understood than viroids, have also been isolated.

Double-Stranded RNA Viruses

The reoviruses, which are double-stranded, icosahedral RNA viruses, occur in mammals, arthropods, and plants. Some are pathogenic; Colorado tick fever is an example. Double-stranded RNA was first discovered as a stable natural product in the reoviruses, which were then studied intensively. Reoviruses (Figure 25-4) have a double envelope about 80 nanometers in diameter, with three kinds of proteins in the outer shell and four in the inner one. From 10 to 13 molecules of double-stranded RNA are present in each reovirus, the exact number depending on the kind of reovirus. Each virus contains a molecule of RNA-dependent RNA polymerase.

Small-Genome DNA Viruses

Many DNA viruses have small genomes; some of these viruses have single-stranded DNA, others double-stranded DNA. Among them are the parvoviruses, mentioned earlier, which infect animals. The parvoviruses, which are icosahedral and about 20 nanometers in diameter, include some of the smallest and simplest viruses. They contain a single molecule of single-stranded DNA and a coat that consists of either three or four different proteins. A second group of small-genome DNA viruses, the papilloma viruses, causes warts in animals, including human beings. Among the 12 different kinds of papilloma virus that have been identified in human beings, some lead to cancer; cervical cancer may be associated with viruses of this group.

The most intensively studied of the small-genome DNA viruses are those that cause viral hepatitis. These viruses have been difficult to study because it has not been possible to infect animals other than human beings or to grow the viruses in cell culture. Not all kinds of hepatitis are caused by viruses of this group; for example, hepatitis A is caused by an RNA virus that is poorly understood, but hepatitis B (serum hepatitis) is caused by a very unusual DNA virus. The causative agent of hepatitis B contains a small, circular molecule of partly, but not completely, double-stranded DNA. The viral genome encodes two kinds of proteins, a core protein and a surface protein, as well as DNA polymerase. The amount of genetic information in the hepatitis B viral genome—359 nucleotides—is less than that of any other pathogen, except for the viroids.

> **The virus that causes hepatitis B has the smallest amount of genetic information in any infectious particle known except for the viroids. It has a partly double-stranded circular molecule of DNA consisting of 359 nucleotides.**

The hepatitis B virus poses a serious public health problem, particularly among Asians, Africans, and male homosexuals. It often persists in carriers without causing any symptoms but may still be highly infectious. People infected early in life often become carriers, and it is estimated that there are about 200 million such carriers worldwide. Since most of these people are not recognized as carriers, there is a real possibility of the frequent transmission of hepatitis during immunization, blood transfusion, and similar medical procedures. Although the proportion of carriers in Europe and the United States appears to be less than 0.1%, it may be as high as 20% in the tropics. Inoculations, medical procedures, and any kind of skin contact may lead to transmission of the virus. Not only is the hepatitis itself serious, but the virus also appears in some cases to play a role in causing human liver cancer, even among carriers who show no other symptoms. Developing a vaccine against hepatitis B is of great importance, especially for those people who require frequent blood transfusions, as well as for homosexuals, prostitutes, and others who run a severe and continuing risk of infection.

Medium-Genome and Large-Genome DNA Viruses

Herpesviruses

The herpesviruses, one of the major groups of medium- and large-genome, double-stranded DNA viruses, are of considerable medical importance. Herpesviruses are icosahedral and about 200 nanometers in diameter (Figure 25-11). They occur frequently in most vertebrates and are involved in cold sores, shingles, venereal disease, mononucleosis, birth defects, and probably several kinds of cancer in humans (Figure 25-12). They have a very complex set of proteins that includes more than 30 different kinds.

The five major categories of human herpesviruses and their associated diseases are:

Herpesvirus 1 (herpes simplex 1): cold sores and fever blisters

Herpesvirus 2 (herpes simplex 2): the genital virus, which causes the venereal disease

Herpesvirus 3: chickenpox and, if it becomes persistent, shingles, an acute inflammatory disease of the ganglia

Herpesvirus 4: Epstein-Barr virus, which causes mononucleosis and certain human cancers

Herpesvirus 5: cytomegaloviruses, which may cause fatal systemic infections of newborn babies

All of these different kinds of human herpesviruses are related to forms that occur widely among other mammals and birds, and there is some evidence for cross-infectivity between humans and other vertebrates. Herpesviruses of the first three groups replicate rapidly and are present in many hosts and tissues. They cause persistent infections in which the outward symptoms may be visible only during periodic outbreaks, perhaps half a dozen or so times a year. Herpesvirus 2, which is spread primarily by sexual contact, afflicts 10 to 20 million people in the United States alone. It attacks the lining of the womb in females, sometimes damaging or destroying an embryo. A baby that passes through the birth canal of a mother with active herpes can be infected and sometimes blinded or killed; this risk can be eliminated by the performance of a cesarean section. In addition, evidence is accumulating that herpesvirus 2 infections may be responsible for about a third of all miscarriages. Women with genital herpes are five to eight times as likely to develop cervical cancer as those who do not have the disease. Group 4 herpesviruses are generally present only in chronic infections, particularly those associated with the lymphatic system. Group 5 herpesviruses replicate slowly, mainly in fibroblasts, and tend to be host specific.

The herpesviruses belong to five groups and are responsible for cold sores, shingles, venereal diseases, chickenpox, and certain cancers. They are large-genome, double-stranded DNA viruses.

Epstein-Barr virus (EBV) is a widespread member of herpesvirus group 4. Many people are infected with EBV but display no signs of the infection. Others may develop the signs of infectious mononucleosis, also called glandular fever; and in some, infection with EBV leads to cancer—either Burkitt's lymphoma, a cancer of the lymph nodes below the jaw, or nasopharyngeal carcinoma, a tumor that develops in the space behind the nose. EBV is often isolated from the milk of women whose families have a history of breast cancer. It primarily infects lymphocytes, a type of white blood cell, but apparently replicates in skin cells. Once the lymphocytes have been infected, the virus will remain in them, and

FIGURE 25-11

Electron micrograph of herpes simplex virus. Different viruses of this group are responsible for cold sores and genital diseases of the kinds shown in Figure 25-12.

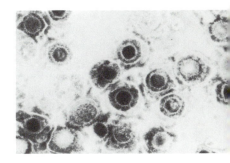

FIGURE 25-12

The symptoms of different herpesviruses include cold sores (A) and genital diseases (B), as well as shingles (C) and other diseases.

A

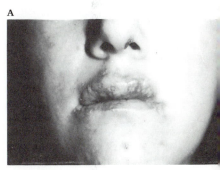

B

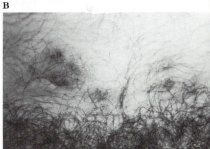

C

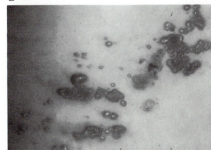

THE ERADICATION OF SMALLPOX

One of the greatest triumphs of modern medicine has been the eradication of smallpox everywhere in the world. This dread disease has killed millions of people, and when first introduced to new populations, as in the Spanish conquest of Mexico, it has sometimes eliminated half or more of the total population. Fortunately there are no known hosts of the smallpox virus other than human beings, so when all of the people that were susceptible in all of the areas where smallpox still occurred were inoculated in the late 1970s, the disease was eliminated completely.

Officials of the World Health Organization, established in 1948 as a special agency of the United Nations, noted that vaccination had already eliminated smallpox in North America and Europe. By 1959 the disease had been eliminated throughout the Western Hemisphere, except for five South American countries, and an intensive, worldwide campaign was initiated. As late as 1967, smallpox was still endemic in 33 countries; the campaign appeared to be faltering. Countries throughout the world, including the Soviet Union and the United States, manufactured and donated large quantities of

---

thus resident in the body, permanently. Recently, EBV has also been linked with a chronic flulike condition that sometimes persists for a year or more and is accompanied by several low-grade symptoms, among them fatigue, headache, depression, sore throat, fever, aches, and pains.

Poxviruses

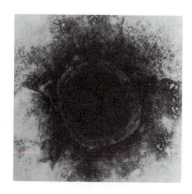

FIGURE 25-13

Cowpox virus, one of the poxviruses. Large viruses of this kind were the basis of Jenner's inoculations in the eighteenth century; they are being used now, in genetic engineering, as the basis for vaccines against several different diseases simultaneously.

The poxviruses are the largest and most complex animal viruses. Often about 400 nanometers in the largest dimension, poxviruses can be seen with a light microscope. About 90% of a poxvirus is protein. Living cells contain about the same percentage; all other viruses contain much less. The viruses are somewhat brick-shaped (Figure 25-13). One of the poxviruses, smallpox virus, was the first virus actually to be seen. Another well-known poxvirus is the vaccinia virus, which causes cowpox and which Jenner used many years ago in successful inoculations against smallpox. As we discussed in Chapter 18, scientists can use this virus in the production of herpes and hepatitis vaccines by splicing the genes specifying the envelopes of these viruses into a fragment of the vaccinia virus genome and inoculating mammals with the engineered vaccinia virus. Vaccinia virus is especially suitable for such procedures because of its large size.

Bacteriophages

Although double-stranded DNA bacteriophages are logically grouped with the other medium- and large-genome DNA viruses, they are so important in relation to molecular biology that we will discuss them separately here. As we mentioned earlier, many of these bacteriophages are large and complex, with relatively large amounts of DNA and proteins. The bacteriophages are very diverse both structurally and functionally, united by their occurrence in bacterial hosts rather than by other common features. Some of them have been named as members of a "T" series (T1, T2, and so forth); others have been given different kinds of names. To illustrate the diversity of these viruses, T3 and T7 bacteriophages are icosahedral and have short tails. In contrast, the so-called T-even bacteriophages (T2, T4, and T6) have an icosahedral head; a capsid that consists primarily of three proteins; a long tail; a connecting neck with a collar and long "whiskers;" and a complex base plate (Figure 25-5).

During the process of bacterial infection by T4 phage, at least one of the tail fibers of the phage—they are normally held near the phage head by the "whiskers"—contacts the lipoproteins of the host bacterial cell wall. The other tail fibers set the phage perpendicular

vaccine, and improved methods of vaccination were developed.

Although reporting was poor, there were probably 10 to 15 million cases of smallpox worldwide in 1967. Attention was focused on areas in which cases had actually occurred. The last case of smallpox on the Indian subcontinent was contracted by a 3-year-old girl on October 16, 1975. By 1978, no more cases were reported in Somalia, the last country of the world in which the scourge of smallpox persisted, and none have been reported since, anywhere in the world. Edward Jenner's great work had been completed.

FIGURE 25-A

Ali Maow Maalin, of Merka, Somalia, contracted the last known case of smallpox reported anywhere in the world in 1977. At the time, he was 23 years old.

to the surface of the bacterium and bring the base plate into contact with the cell surface. The tail contracts, and the tail tube passes through an opening that appears in the base plate, piercing the bacterial cell wall. The contents of the head, mostly DNA, are then injected into the host cytoplasm.

The T-series bacteriophages are all virulent, invariably multiplying within infected cells and eventually lysing them. Among the temperate bacteriophages is the lambda (λ) phage of *Escherichia coli,* discussed in Chapter 17. We know as much about this bacterio-phage as we do about virtually any other biological particle; the complete sequence of its 48,502 bases has been determined. At least 23 proteins have been identified with the development and maturation of lambda phage, and many enzymes are involved in the integration of these viruses into the host genome.

Bacteriophages are a diverse group of viruses that attack bacteria. Some of them have a very complex structure.

VIRUSES: PARTICLES OF GENOMES

Viruses are best thought of as particles of genomes. Because viruses originate as fragments of bacterial and eukaryotic genomes, their great diversity is understandable. We have learned a great deal about the structure of viruses, but there is doubtless much more to be discovered, probably including the existence of kinds of particles that are unsuspected at present. As parts of the genetic machinery of cells before they became independent, viruses are still intimately associated with these cells. The nature of viruses suggests that new forms are evolving constantly. While this process continues, we clearly have a great deal to learn about the existing viruses.

SUMMARY

1. Viruses are fragments of bacterial or eukaryotic genomes that are able to replicate within cells by using the genetic machinery of those cells. In most cases, the fragments subsequently acquire the ability to synthesize protein coats, and sometimes envelopes, around the DNA or RNA that originally split from the genome from which they were derived. They are not alive and are not organisms.

2. Viruses are either virulent, destroying the cells in which they occur, or temperate, becoming integrated into their genomes and remaining quiescent there for long periods of time.

3. Viruses are basically either helical or isometric. Most isometric viruses are icosahedral. The adhesion properties of viruses are determined by those of the proteins that make up their coats and envelopes.

4. The simplest viruses use the enzymes of the host cell for both protein synthesis and gene replication; the more complex ones contain up to 200 genes and are capable of synthesizing many structural proteins and enzymes.

5. Among the major groups of viruses are the unenveloped plus-strand RNA viruses, which act directly as messengers upon infection of the host cell; the enveloped plus-strand RNA viruses, which are similar but have an envelope; the minus-strand RNA viruses, which must synthesize mRNA in the cells they infect; viroids and similar forms, which consist of RNA alone with no associated proteins; double-stranded RNA viruses; small-genome DNA viruses, which have either single-stranded or double-stranded DNA; medium- and large-genome DNA viruses; and bacteriophages.

6. The flu viruses, minus-strand RNA viruses, resemble balls studded with spikes. The spikes are composed of two kinds of protein, hemagglutinin and neuraminidase, both of which play a role in determining the infective qualities of the virus. Recombination of the genetic material of the flu viruses appears to play the critical role in causing worldwide flu epidemics.

SELF-QUIZ

The following questions pinpoint important information in this chapter. You should be sure to know the answers to each of them.

1. Wendell Stanley was able to purify tobacco mosaic virus in the form of
   (a) chemicals that subsequently were unable to infect tobacco
   (b) cellular debris, including membranes and organelles
   (c) crystals that could still infect tobacco
   (d) an organism closely resembling bacteria

2. Viruses contain
   (a) only DNA
   (b) only RNA
   (c) either DNA or RNA, but not both simultaneously
   (d) both DNA and RNA together
   (e) neither DNA nor RNA

3. Virus particles contain enzymes for
   (a) protein synthesis and sexual reproduction   (c) replication of their own nucleic acids
   (b) energy production   (d) invasion of the host cell

4. Viral diseases are difficult to treat because
   (a) the dead viruses are just as dangerous as the "living" ones
   (b) chemicals tend to have little effect on them
   (c) chemicals used to kill them will also harm the host cell
   (d) they are too scarce
   (e) their nucleic acids cannot be broken

5. Which of the following is *not* a structural element of T4 bacteriophages?
   (a) hexagonal head
   (b) tail core
   (c) capsid
   (d) base plate, tail, and tail fibers
   (e) polymerase enzyme

6. Before infecting an animal, a virus must
   (a) replicate its DNA (or RNA) within its capsid
   (b) shed its tail fibers
   (c) shed its protein coat
   (d) bind to a specific receptor on the animal cell surface
   (e) translate its DNA (or RNA)

7. Unenveloped plus–strand RNA viruses are called "plus" because
   (a) they lack envelopes
   (b) they possess ribosomes and mitochondria
   (c) they interact directly with the nuclear genome of the host cell
   (d) they function directly as messengers upon infection, attaching to the ribosomes of the host and becoming translated
   (e) only their plus strand functions to direct mRNA synthesis

8. _____ developed the process of vaccination.

9. Flu viruses are able to change the nature of their protein coats. Does this change involve a change in nucleotide sequence of the DNA, or does it involve a rearrangement of unchanged genes?

10. _____ are the smallest known agents of infectious diseases.

11. _____ are the largest of all animal viruses.

12. Changes in genes of flu viruses that alter the shape of the spike proteins enough for them to be unrecognizable to human antibodies specific for the old spike configuration are caused by
    (a) point mutations of the DNA
    (b) recombination and reassortment of the DNA
    (c) point mutations of the RNA
    (d) recombination and reassortment of the RNA

THOUGHT QUESTIONS

1. Why are viruses not considered organisms?

2. What are the advantages to a virus of having a protein coat or envelope? The disadvantages?

3. In what ways might the early, self-replicating particles that gave rise to the first organisms have resembled, or differed from, viruses?

4. Did the different groups of viruses arise from one another? Describe the process by which they probably evolved.

FOR FURTHER READING

COOPER, J.I., and F.O. MACCALLUM: *Viruses and the Environment,* Chapman & Hall, London, 1984. A concise and well-written book, presenting an up-to-date account of all aspects of virus biology.

DIENER, T.O.: "Viroids: Minimal Biological Systems," *BioScience,* vol. 32, pages 38-44, 1982. An account of the studies that established the existence of viroids.

FRAENKEL-CONRAT, H., and P.C. KIMBALL: *Virology,* Prentice-Hall, Inc., Englewood Cliffs, N.J., 1982. An excellent and well-balanced account of all aspects of the field, thoroughly modern and up-to-date.

HENDERSON, D.A.: "The Eradication of Smallpox," *Scientific American,* October 1976, pages 25-33. A description of the campaign that ultimately led to the worldwide elimination of this deadly disease.

PRUSINER, S.B.: "Prions," *Scientific American,* September 1984, pages 50-59. A fascinating introduction to prions, agents of disease that contain no DNA or RNA and yet are able to reproduce within cells.

SIMONS, K., H. GAROFF, and A. HELENIUS: "How an Animal Virus Gets Into and Out of Its Host Cell," *Scientific American,* February 1982, pages 58-70. The process is followed in illuminating detail in this excellent article.

Overview

Bacteria are the most ancient, the morphologically simplest, and the most abundant of organisms, and the only ones characterized by a prokaryotic organization of their cells. Because they are so distinct, they are classified as the only members of the kingdom Monera. Bacteria have been mentioned repeatedly in this book because their key position in the evolution of life places them at the beginning of many evolutionary developments. Life on earth would not exist without bacteria, which make possible many of the essential functions of ecosystems. This chapter presents an overview of the bacteria: their structure, life history, ecology, and diversity.

For Review   *Here are some important terms and concepts that you will encounter in this chapter. If you are not familiar with them, you should review them before proceeding.*

- **Prokaryotic cell structure** (Chapter 5)

- **Bacterial flagella** (Chapter 7)

- **Bacterial cell division** (Chapter 8)

- **Bacterial metabolism and chemoautotrophy** (Chapter 10)

- **Bacterial photosynthesis** (Chapter 12)

- **Transduction and bacterial conjugation** (Chapter 17)

Bacteria are the oldest, the structurally simplest, and the most abundant forms of life on earth; they are also the only organisms with prokaryotic cellular organization. Represented in the oldest rocks from which fossils have been obtained—rocks about 3.5 billion years old—bacteria were abundant for well over 2 billion years before eukaryotes appeared in the world (Figures 3-13 and 5-9). Bacteria are the only members of the kingdom Monera.

About 2500 different kinds of bacteria are currently recognized, but there are doubtless many thousands more awaiting proper description (Figure 26-1). The morphological differences among various bacteria, even as viewed with an electron microscope, often are not great, and species are recognized largely by their metabolic characteristics. Bacteria can be characterized properly only when they are grown on a defined medium (see Chapter

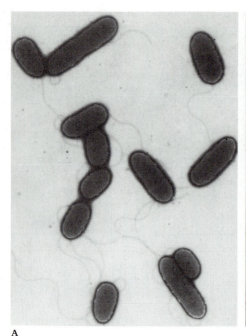

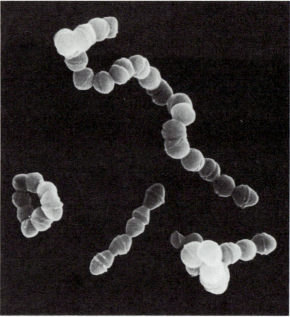

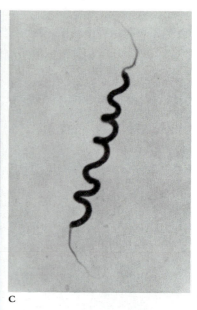

A

B

C

FIGURE 26-1

The diversity of bacteria.

A Pseudomonas aeruginosa, *a rod-shaped, flagellated bacterium (bacillus).* Pseudomonas *includes the bacteria that cause many of the most serious plant diseases.*

B Streptococcus. *The spherical individual bacteria (cocci) adhere in chains in the members of this genus.*

C Spirillum volutans, *one of the spirilla. This large bacterium, which occurs in stagnant fresh water, has a tuft of flagella at both ends.*

D Stigmatella aurantiaca, *one of the gliding bacteria. The rod-shaped individuals move together, forming the composite spore-bearing structures shown here; millions of spores, basically individual bacteria, eventually are released from these structures.*

E Calothrix, *a cyanobacterium in which the individuals adhere within a gelatinous capsule in groups of four.*

15), because the characteristics of these organisms often change depending on their growth conditions and the substances with which they are in contact.

Bacteria were largely responsible for creating the properties of the atmosphere and the soil during the more than 2 billion years in which they were the only kind of life on earth. They are metabolically much more diverse than are the eukaryotes, which is why they are able to exist in such an exceptionally wide range of habitats and are so important ecologically in so many of them. Many bacteria are autotrophic—both photosynthetic and chemoautotrophic—and make major contributions to the world carbon balance in terrestrial, freshwater, and marine habitats. Others are heterotrophic and play a key role in world ecology in breaking down organic compounds. Some of these heterotrophic bacteria cause major diseases of plants and animals, including human beings. One of the most important roles of bacteria in the world ecosystem relates to the fact that only a few genera of bacteria—and no other organisms—have the ability to fix atmospheric nitrogen and thus make it available for use by organisms.

Bacteria are the oldest and most abundant organisms on earth. The maintenance of life depends on them, since they play a vital role both in productivity and in cycling the substances essential to all other life forms. The only organisms capable of fixing atmospheric nitrogen are bacteria.

Bacteria are very important, and becoming more so, in industrial processes. Bacteria are used now, for example, in the production of acetic acid and vinegar, various amino acids, and enzymes, and especially in the fermentation of lactose into lactic acid, which coagulates milk proteins and is used in the production of almost all cheeses, yogurt, and similar products. In the production of bread and other foods, the addition of strains of bacteria with appropriate characteristics can lead to the enrichment of the final product with respect to its mix of amino acids, a key factor in its nutritive value. Many products that have traditionally been manufactured by using yeasts, such as ethanol, can also be made by using bacteria; the comparative economics of these processes will determine which group of organisms is used in the future. Also, many of the most widely used antibiotics, including streptomycin, aureomycin, erythromycin, and chloromycetin, are derived from bacteria.

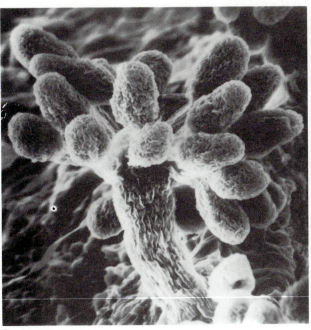

D

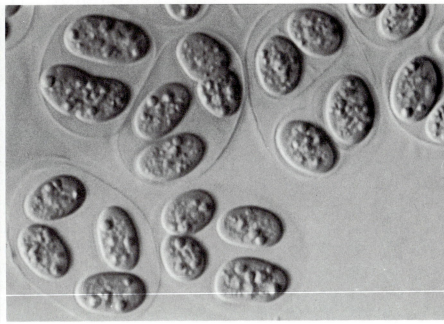

E

The application of genetic engineering methods to the production of improved strains of bacteria for commercial use has begun (Figure 26-2), and it holds enormous promise for the future. For example, bacteria are being investigated increasingly as non-polluting methods of insect control; *Bacillus thuringiensis,* which attacks insects naturally, is particularly useful in this regard. Improved, highly specific strains of B. *thuringiensis* greatly increase its usefulness as a biological control agent. Considering the enormous metabolic diversity of the bacteria, it seems logical to assume that we have barely begun to understand and use them fully.

FIGURE 26-2

Recombinant DNA technologies will be used increasingly to create strains of bacteria that produce substances of commercial importance. One of the first major achievements of this sort, by Eli Lilly & Company, was the production of human insulin for diabetics by genetically engineered bacteria (Escherichia coli) that had been produced for Lilly by Genentech, a California company, in 1978. Eli Lilly began marketing bacterially synthesized human insulin (under the trade name Humulin) in July, 1983.

This photo shows hollow fiber ultrafiltration columns used in the purification of human insulin produced by bacteria. Earlier, insulin had been obtained from pork and beef pancreatic glands, which were collected as by-products from the meat-packing industry. The natural supply is inadequate, however.

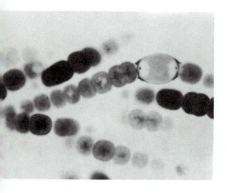

FIGURE 26-3

The cyanobacterium Anabaena, *in which the individual cells adhere in filaments. The large cell is a heterocyst, a specialized cell in which nitrogen fixation occurs. Heterocysts are essentially like the ordinary cells, but they form an outer bilayered envelope consisting of a polysaccharide and an inner layer of glycolipids. Within them, the membranes of the cyanobacterium are reorganized into a concentric or netlike pattern. It has recently been learned that some of the nitrogen fixation genes are actually rearranged during heterocyst formation in cyanobacteria. These organisms exhibit one of the closest approaches to multicellularity among the bacteria.*

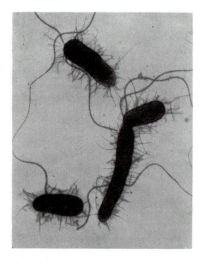

FIGURE 26-4

Pili in the common colon bacterium, Escherichia coli. *The long strands are flagella. A greater proportion of the genetic information is known for this much-studied bacterium than for any other organism. Conjugation pili in it, which are longer and connect the individuals between which a plasmid is being transferred, are shown in Figure 17-7.*

PROKARYOTES VERSUS EUKARYOTES

Prokaryotes, or bacteria, differ from eukaryotes in numerous features of fundamental importance. These differences have been discussed at many places in this book, since a consideration of the simpler features of bacteria illustrates the evolutionary origins of many of the structures and capabilities of eukaryotes. We will review the differences briefly, with cross-reference to the earlier discussion.

1. *Multicellularity.* All bacteria are fundamentally single celled. Even though in some of them the cells may adhere within a matrix and form filaments (Figures 26-1, *E,* and 26-3), especially in some cyanobacteria, their cytoplasm is not directly interconnected as often is the case in multicellular eukaryotes. The activities of a bacterial colony are less integrated and coordinated than in multicellular eukaryotes. A primitive form of colonial organization occurs in the gliding bacteria, which move together and form spore-bearing structures (Figure 26-1, *D*). Such approaches to multicellularity, however, are rare among the bacteria.

2. *Cell size.* Most bacteria have cells only about 1 micrometer thick (Chapter 5), whereas most eukaryotic cells are well over 10 times that size.

3. *Chromosomes.* Eukaryotic cells have a membrane-bound nucleus containing chromosomes that include both nucleic acids and proteins. Bacteria do not have a nucleus, nor do they have chromosomes of the kind present in eukaryotes, in which DNA is structurally complexed with proteins.

4. *Cell division and genetic recombination.* Cell division in eukaryotes takes place by mitosis and involves spindles made up of microtubules; cell division in bacteria takes place mainly by binary fission (Chapter 8). True sexual reproduction is present only in eukaryotes and involves syngamy and meiosis, with an alternation of diploid and haploid forms. Despite their lack of sexual reproduction, bacteria do have mechanisms that lead to the transfer of genetic material. These mechanisms are far less regular than those of eukaryotes and do not involve the equal participation of the individuals between which the genetic material is transferred.

5. *Internal compartmentalization.* In eukaryotes the enzymes for the oxidation of 3-carbon organic acids are packaged in mitochondria; in bacteria the corresponding enzymes are not packaged separately but are bound to the cell membranes (see Figure 5-8; Chapter 10). The cytoplasm of bacteria, unlike that of the eukaryotes, contains no internal compartments or organelles.

6. *Flagella.* Bacterial flagella (Figure 26-1, *A* and *C* and Figure 26-4; Chapter 7) are simple, composed of a single fiber of the protein flagellin; the flagella and cilia of eukaryotes are complex and have a 9 + 2 structure of microtubules (Figure 7-10).

7. *Autotrophic diversity.* In photosynthetic eukaryotes the enzymes for photosynthesis are packaged in membrane-bound organelles, the plastids. Only one kind of photosynthesis is found in eukaryotes, and it involves the release of oxygen. In photosynthetic bacteria the enzymes for photosynthesis are bound to the cell membrane. These bacteria have several different patterns of anaerobic and aerobic photosynthesis (Chapter 12), involving the formation of end products such as sulfur, sulfate, and also oxygen. This photosynthetic diversity again underscores the metabolic diversity of the bacteria. **Chemosynthesis** is the process whereby certain bacteria obtain their energy from the oxidation of inorganic compounds and obtain their carbon from carbon dioxide; chemoautotrophs are discussed in Chapter 10. In addition, only bacteria have the ability to fix atmospheric nitrogen.

The profound differences between bacteria and eukaryotes began to be understood properly only about 50 years ago. Scientists began to realize that these differences were far greater than the traditional ones used to separate plants and animals. Indeed, they represent the most fundamental differences that separate any two groups of organisms.

BACTERIAL STRUCTURE

Bacterial cell walls consist of a network of polysaccharide molecules connected by polypeptide cross-links (Figure 5-6). Most species of bacteria are gram positive; in them, this network makes up the basic structure of the bacterial cell wall, which is about 15 to 80 nanometers thick. In the less frequent gram-negative bacteria, large molecules of **lipopolysaccharide**—a polysaccharide chain with lipids attached to it—are deposited over this layer. Gram-negative bacteria are resistant to many antibiotics to which gram-positive ones, lacking the lipopolysaccharide layer, are susceptible. Beyond these layers a gelatinous layer, the **capsule**, often surrounds bacterial cells.

Bacteria are mostly simple in form, varying mainly from straight and rod shaped **(bacilli)** or spherical **(cocci)** to long and spirally coiled **(spirilla)**. Some bacteria change into stalked structures, grow long, branched filaments, or form erect structures that release **spores,** single-celled bodies that grow into new bacteria (Figure 26-1, *D*). Among the rod-shaped bacteria, some adhere end-to-end after they have divided; if they do, they form a chain.

The members of one of the major photosynthetic groups, the cyanobacteria (formerly misleadingly called "blue-green algae"), often form filaments (Figure 26-3) that may occur in large masses up to 1 meter or more in length. In these filaments, the cells are connected to one another by their outer walls or by gelatinous sheaths, and there is a degree of coordination between them. Some of these filamentous bacteria are capable of gliding motion, often combined with rotation around a longitudinal axis. The mechanism by which they move is uncertain.

> **Bacterial cells have a very simple structure. The cytoplasm contains no internal compartments or organelles and is bounded by a membrane encased within a cell wall composed of one or more layers of polysaccharide.**

Many kinds of bacteria have very slender, rigid, helical flagella composed of flagellin (see Figure 7-3). These flagella range from 3 to 12 micrometers in length and are very thin—only 10 to 20 nanometers thick. Flagella may be distributed all over the cells of the bacteria in which they occur, or they may be confined to one or both ends of the cell.

Pili (sing. **pilus**) are other kinds of rodlike outgrowths that occur on the cells of some bacteria (Figure 26-4). They are shorter than bacterial flagella, up to several micrometers long, and about 7.5 to 10 nanometers thick. Pili evidently function in the attachment of bacterial cells to appropriate substrates and also may function in bacterial conjugation.

Some bacteria have the ability to form thick-walled **endospores** around their chromosome and a small portion of the surrounding cytoplasm when they are exposed to drying conditions and high temperatures. These endospores (Figure 26-5) are highly resistant to environmental stress; they may germinate and form new bacterial individuals after decades or even centuries. The formation of endospores is the reason that the bacterium responsible for botulism, *Clostridium botulinum,* sometimes persists in improperly sterilized cans and bottles—ones that have not been kept at high enough temperatures for a long enough time—and then is able to multiply within them.

BACTERIAL REPRODUCTION

Bacteria, like all other living cells, divide. In bacteria, the mode of division is binary fission. In this process, an individual cell simply increases in size and divides in two. The cell membrane and cell wall grow inward and eventually divide the cell by the formation of a new wall. This new wall ultimately splits from the outside toward the center of the old cell, and the two new cells become distinct entities (Figure 5-7). If the new cell wall splits incompletely, or if the derivative cells are held together by another sheath, chains of cells may be

FIGURE 26-5

The round circle at the bottom is an endospore forming within a cell of the botulism bacterium, Clostridium botulinum. *These resistant endospores allow the bacterium to survive in improperly sterilized canned and bottled goods.*

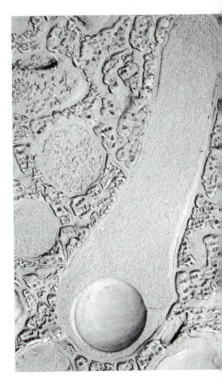

BACTERIAL MAGNETIC NAVIGATION

A few species of aquatic bacteria synthesize and then carry within themselves crystals of magnetite (Fe_3O_4). The magnetic properties of this mineral, also known as lodestone, have been recognized for about 2500 years. With the aid of these crystals, these bacteria are able to orient themselves to the earth's magnetic field. They swim steadily in one direction, instead of tumbling about like the chemotactic bacteria described in Chapter 7.

One of these bacteria, *Aquaspirillum magnetotacticum,* is shown in the accompanying photograph (Figure 26-A). It has a flagellum at either end and can swim either forward or backward. About 20 crystals of magnetite form a chain parallel to the long axis of the cell. The bacteria form such

crystals only when they are supplied with a sufficient quantity of iron. In the Northern Hemisphere, bacteria of this and other species that are able to orient themselves to the magnetic field swim consistently to the north. Since they are following the magnetic field, which also slopes downward toward the earth's surface, the bacteria also swim downward at about 70 degrees and thus keep among the sediments at the bottom of the water in the ponds where they live. In the Southern Hemisphere the corresponding bacteria are predominantly south-seeking. At the geomagnetic equator, in Brazil, samples of the bacteria were equally divided among north-seeking and south-seeking strains.

Life evolved in a magnetic field, and so it is

---

formed and persist. While the cell is dividing, its circular DNA molecule replicates, forming two identical circular DNA molecules. These new molecules are attached at specific places on the cell membrane. As the cell divides, one of the DNA molecules is carried to each of the two cells formed from the original dividing one (Figure 8-1).

BACTERIAL VARIATION
Mutation

In a bacterium such as *Escherichia coli* there are about 5000 genes, judged from the amount of DNA that an individual cell contains. In considering any one of these genes, we may conservatively expect one mutation to have occurred by chance among every million copies. Since 200 individual bacteria contain a total of a million genes (of 5000 kinds), we can expect one of every 200 bacteria to have a mutant characteristic (Figure 26-6). A spoonful of soil typically contains over a billion bacteria and therefore should contain something on the order of 5 million mutant individuals!

The variation created within bacterial populations by random mutation and their rapid growth rate means that populations of bacteria are capable of changing their characteristics rapidly. For example, when adequate food and nutrients are available, a pop-

FIGURE 26-6

*Mutations in bacteria can be detected by the technique of replica plating. The bacterial colonies, growing on a semisolid agar medium, are transferred from **A** to **B** and **C** by using a sterile velveteen disk pressed on the plate. Plates **A** and **B** have a medium that includes all of the factors necessary for the growth of the bacterium, whereas **C** has a medium that lacks some of the essential growth factors. The colonies marked with arrows in **B** were not able to grow in the deficient medium in **C**; they were mutant colonies, already present but undetected in **A**. This experiment demonstrates that these mutants were not caused by the new environment, but were selected by it, just like the mutants in Welsh bent grass discussed in Chapter 20.*

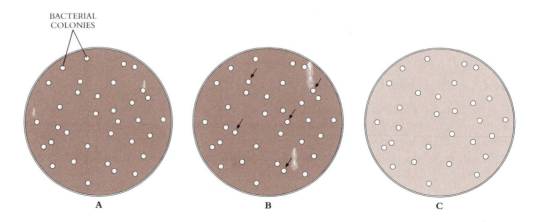

BACTERIAL COLONIES

A   B   C

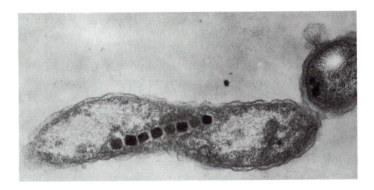

FIGURE 26-A

The magnetotactic bacterium Aquaspirillum magnetotacticum. *The particles that form a chain of objects that are opaque to the electron beam are magnetite. Individuals of this species from the Northern Hemisphere uniformly swim north.*

perhaps not a surprise that the ability to orient to this field has been found recently in a species of *Chlamydomonas* (a green alga) or that crystals of magnetite have been found in dolphins, fishes, homing pigeons, and turtles, especially around their nasal cavities. Such crystals have also been found in cancerous tumors in mice and in human beings, where their function, if any, is unknown. Further discoveries will undoubtedly be made in this fascinating new field of investigation.

ulation of *Escherichia coli* growing under optimal conditions will double every 12.5 minutes, so that a new favorable mutation can quickly become represented in large numbers of individuals and so contribute significantly to the ability of the bacterial population to adjust to new environmental circumstances.

Mutation is a powerful generator of genetic diversity in bacteria because of the large numbers of individuals and short generation times in bacterial populations.

The ability of bacteria to change very rapidly in response to new challenges often has adverse effects on human beings. For example, in the past decade or so, a number of strains of *Staphylococcus aureus* associated with serious infections have appeared, some of them with alarming frequency, in hospitals. Unfortunately, many of these strains have acquired resistance to penicillin; other strains have acquired the ability to produce an enzyme that breaks down the molecules of penicillin. *Staphylococcus* infections provide an excellent example of the way in which mutation and intensive selection can bring about rapid change in bacterial populations. Such changes occur in all bacteria and have serious medical implications, as mentioned in Chapter 17, particularly when strains of bacteria emerge that are resistant to many different kinds of antibiotics. In the mid-1980s, the routine administration of low doses of antibiotics to livestock to improve their weight gains came into question because of the danger to humans posed by the antibiotic-resistant strains of bacteria that inevitably result from this practice.

Genetic Recombination

Another source of genetic variation in populations of bacteria is recombination, which occurs by the transfer of genes from one cell to another as portions of viruses, plasmids, or other DNA fragments (see Chapters 14 and 16). The rapid transfer of newly produced, antibiotic-resistant genes by plasmids has been an important factor in the appearance of the resistant strains of *Staphylococcus aureus* discussed above. An even more important example in terms of human health involves the Enterobacteriaceae, the family of bacteria to which the common colon bacterium, *Escherichia coli*, belongs. In this family, there are many im-

TRAVELER'S DIARRHEA

One of the most common diseases of travelers in foreign countries is diarrhea, with its accompanying fever and chills, nausea, and vomiting. People who travel from areas where infectious diarrhea is rare into areas where it is common are most often affected, and often a third or more of such visitors will become ill. It may surprise you to learn that the most frequent cause of the disease is the common gut bacterium, *Escherichia coli,* although other bacteria, as well as the protozoan *Giardia,* are also involved in certain areas. Although *Escherichia coli* constitutes about a quarter of the mass of normal feces, the strains that are normally present do not cause diarrhea.

The infectious strains of *Escherichia coli* produce toxins that indirectly cause the movement of fluid into the gut and inhibit the absorption of sodium. When this occurs, diarrhea follows. It is best treated by replacing the lost fluid, and especially by drinking weak salt solutions, with sugars, to replace what has been lost. Antibiotics and drugs that control the contraction of the intestines should be used sparingly and normally only on the advice of a physician. In areas where it occurs frequently, the disease can be avoided to some extent by avoiding tap water, uncooked vegetables and fruits, and incompletely cooked food—that is, by following normal sanitary procedures. Diarrhea is of great interest from a world health standpoint, since it plays a major role in the death of malnourished infants in the less developed countries of the world.

portant pathogenic bacteria, including the causative organisms of dysentery, typhoid, and other major diseases. At times, some of their genetic material is exchanged with or transferred to *E. coli,* often with very serious health consequences.

In a sense, the rapid generation time of bacteria may be seen as an evolutionary strategy alternative to the complex patterns of recombination and segregation that occur in eukaryotes. Eukaryotes have a much more complex cell structure and a much slower generation time than do bacteria, and they may be simply unable to reproduce as fast. In both bacteria and eukaryotes, ample variation is present to form the raw material of evolution, but the variation is organized and presented in very different ways.

BACTERIAL ECOLOGY AND METABOLIC DIVERSITY

Bacteria are the most abundant organisms in most environments, whether measured by numbers of individuals or by absolute weight. For example, in a single gram of fertile agricultural soil, there may be 2.5 billion bacteria, in addition to 400,000 fungi, 50,000 algae (photosynthetic eukaryotes other than plants), and 30,000 individual protozoa (heterotrophic protists). Another calculation has shown that in a hectare (2.47 acres) of wheat land in England, the weight of the microorganisms in the soil is approximately equal to that of 100 sheep! Bacteria may be equally abundant in the sea, where a major proportion of the productivity depends on the activities of photosynthetic bacteria.

Bacteria occur in the widest possible range of habitats and play key ecological roles in virtually all of them. Some thrive in hot springs, for example, where the usual temperatures may range as high as 78° C (see Figure 5-2); others have been recovered living beneath 430 meters of ice in Antarctica. Bacteria are abundant in groundwater, where they were once thought to be absent. Still other bacteria, capable of dividing only under high pressures, exist around deep-sea thermal vents, where the water is at temperatures as high as 360° C. These bacteria supply energy to all the other organisms that live around the vents, obtaining it by converting hydrogen sulfide to elemental sulfur (see boxed essay on pp. 762-763).

The metabolic diversity of bacteria is what allows them to play such varied ecological roles. Among the members of this group are strict anaerobes, **facultative anaerobes** (organisms that may function either as anaerobes or as aerobes), and aerobes. In contrast, almost all eukaryotes are aerobes—they need oxygen at high concentrations to live—and the very few anaerobic eukaryotes clearly have been derived from ancestors that were aerobes. Different kinds of bacteria vary extensively in almost all aspects of their metabolism, in marked contrast to eukaryotes, which are relatively uniform. All eukaryotes, for example, follow the same general patterns of respiration, glucose breakdown, photosynthesis, and synthesis of nucleic acids and proteins. Bacteria are far more diverse in their ways of carrying out each of these vital functions. In view of their abundance and metabolic versatility, it is not surprising that bacteria play a key ecological role in virtually all habitats on earth.

Autotrophic Bacteria

The life modes of bacteria are as diverse as their metabolisms; different groups exhibit very different ecological patterns, or relationships with their environments. Most bacteria are heterotrophs, obtaining their energy from organic material formed by other organisms. Less common than their heterotrophic relatives, but exceedingly important in the world ecosystem, are the autotrophic bacteria, which obtain their energy from nonorganic sources. There are two major groups of bacterial autotrophs: photosynthetic bacteria and chemoautotrophic bacteria.

Photosynthetic bacteria

Like plants, the photosynthetic bacteria contain chlorophyll, but it is not held within plastids. One of these groups of photosynthetic bacteria, the cyanobacteria, was discussed extensively in Chapters 3 and 23 in connection with the evolution of life on earth. The other major groups of photosynthetic bacteria are the green sulfur bacteria, the purple sulfur bacteria, and the purple nonsulfur bacteria. The different colors of these organisms are contributed by their photosynthetic accessory pigments. The photosynthetic processes carried out by these different bacteria are diverse and reflect the long evolutionary path to green plant photosynthesis. Such recently discovered organisms as *Prochloron* (Figure 26-11) and *Heliobacterium* (boxed essay, p. 526) are thought to represent steps in the evolution of chloroplasts in different groups of eukaryotes.

Chemoautotrophic bacteria

In contrast to the photosynthetic bacteria, the chemoautotrophic ones derive the energy they use for biosynthesis from the oxidation of inorganic molecules such as nitrogen, sulfur, and iron compounds, or from the oxidation of gaseous hydrogen. The metabolic chemistry of these remarkable organisms was described in Chapter 10, and some groups are discussed further on pp. 531-532.

Heterotrophic Bacteria

Most bacteria are heterotrophs: they cannot make organic compounds from simple inorganic substances but must obtain them from other organisms. Together with the fungi, bacteria play the leading role in breaking down the organic molecules that have been formed by biological processes. By doing so, bacteria and fungi make the nutrients in these molecules available once more for recycling. Decomposition is just as indispensable as photosynthesis to the continuation of life on earth.

Among the heterotrophic bacteria, the best represented are **saprobes,** those bacteria that obtain their nourishment from dead organic material. Many of the characteristic odors associated with soil come from substances produced by saprobic bacteria. Mutant bacteria have appeared in recent decades that are able to break down synthetic products such as nylon, herbicides, and pesticides, which are released by humans into the air, water, and soil in large amounts. The techniques of genetic engineering will be used increasingly in the future to produce bacteria capable of disposing of such waste materials. On the other hand, bacteria break down herbicides and pesticides so rapidly in some situations that these chemicals are much less effective than they would be otherwise.

Other activities of the heterotrophic bacteria may be destructive, threatening, or beneficial to human beings. Some by-products of the metabolism of heterotrophic bacteria, such as antibiotics, are commercially valuable. Others are harmful; for example, one of the species of the bacterial genus *Clostridium, C. botulinum* (Figure 26-5), is responsible for the type of food poisoning known as botulism, in which it produces one of the most powerful toxins known: a gram is enough to kill about 14 million people. As mentioned earlier, the endospores of this species are relatively resistant to sterilization. Botulism is very rare in the United States, but another kind of food poisoning, associated with *Staphylococcus aureus* and fortunately much milder, is common. The bacterium secretes a toxin into food that causes severe reactions, including nausea with diarrhea and vomiting, within a few hours of the ingestion of the food. If foods are kept refrigerated, this type of food poisoning does not occur, since *Staphylococcus* is unable to grow at low temperatures. Bacteria of the genus *Salmonella* that are ingested with food may cause gastrointestinal disease, which often develops several days after contaminated food is eaten. Infections of this kind are especially frequent in the summer months.

Nitrogen-Fixing Bacteria

Both autotrophic and heterotrophic bacteria play an important role in the nitrogen cycle (see Figure 34-13). Although nitrogen gas makes up 78% of the atmosphere, most organisms cannot use this vast reservoir to make amino acids and other nitrogen-containing compounds; they are dependent on the nitrogen that is present in the soil. Such nitrogen, however, is often present in short supply, and its availability is one of the chief factors limiting plant growth—and therefore, indirectly, animal growth—locally throughout the world. Two of the more interesting and ecologically critical genera of free-living, heterotrophic bacteria, *Azotobacter* and *Clostridium* (Figure 26-5), both common in soils, play a key role in this cycle: they are able to fix atmospheric nitrogen, as are many of the cyanobacteria, actinomycetes, and some of the symbiotic bacteria that we will discuss later. Bacteria also participate in the nitrogen cycle by releasing fixed nitrogen, which they do by breaking down proteins. Some microorganisms break down the original protein molecules into their constituent peptides, and many are able to break these peptides down into their constituent amino acids, with the consequent release of ammonium ions.

> One of the most essential links in the chain of life, the cycling of nitrogen, is carried out exclusively by the bacteria. Both heterotrophic and autotrophic genera are involved, the latter only in the fixation of atmospheric nitrogen.

BACTERIA AS PATHOGENS

Many costly diseases of plants are associated with particular bacteria; almost all kinds of plants are susceptible to one or more kinds of bacterial disease. The symptoms of these plant diseases vary, but they are commonly manifested as spots of various sizes on the stems, leaves, flowers, or fruits. Other common and destructive diseases of plants, including blights, soft rots, and wilts, also are associated with bacteria. Fire blight, which destroys pears, apples, and related plants, is a well-known example of bacterial disease (Figure 26-7). Most bacteria that cause plant diseases are members of the group of rod-shaped bacteria known as pseudomonads (Figure 26-1, *A*).

An outbreak in Florida of a bacterial disease known as citrus canker, first detected in August, 1984, led to the destruction of more than 4 million citrus seedlings in four months in an effort to halt the spread of the disease. Citrus canker, which had not been detected in Florida for more than 40 years, probably was introduced on fruits that were brought from abroad by a traveler and escaped detection at a customs station. The disease is caused by one of more than 100 distinct varieties of the pseudomonad *Xanthomonas campestris;* other

A B

FIGURE 26-7

Fire blight, a bacterial disease that is very destructive in apples, pears, cotoneaster, and allied plants, is caused by the bacterium Erwinia amylovora. *Members of this genus are the most frequent and serious plant pathogens.*
A *Canker on the branch of an apple tree.*
B *Dead branch of a pear, killed by fire blight.*

varieties of this bacterium cause diseases with similar symptoms in beans, cabbages, peaches, and other plants. Citrus canker now threatens the continued existence of the Florida citrus industry, a $2.5 billion business that was weakened earlier by the harsh winter of 1983-1984.

Bacteria also cause many human diseases, including cholera, leprosy, tetanus, bacterial pneumonia, whooping cough, and diphtheria. Enormous sums are spent annually in the effort to reduce the likelihood of these infections and in limiting the other destructive activities of bacteria.

Several genera of pathogenic bacteria are of particular importance for humans. For example, members of the genus *Streptococcus* (Figure 26-1, *B*) are associated with scarlet fever, rheumatic fever, and other infections. As mentioned in Chapter 25, the scarlet fever bacterium produces its characteristic and deadly toxin only if it is infected with the appropriate bacteriophage. Tuberculosis, another bacterial disease, is still a leading cause of death in humans. These diseases are mostly spread through the air, as are some of those caused by the pathogenic genus *Staphylococcus,* which causes widespread hospital infections.

Toxic shock syndrome, which is characterized by fever, a skin rash that appears first on the palms of the hands and soles of the feet before spreading to other parts of the body, and a serious drop in blood pressure that can lead to shock, is also caused by certain strains of *Staphylococcus aureus.* These strains produce toxic shock syndrome toxin 1 (TSST-1), a bacterial toxin that causes the rare but sometimes fatal disease. About 85% of the cases of toxic shock syndrome reported in the United States have occurred in menstruating women who are using tampons at the time the disease occurs, but both men and women can contract the disease. In 1985, Dr. Edward Kass and associates of the Harvard Medical School reported that two fibers contained in some so-called superabsorbent tampons, polyester foam and polyacrylate rayon, absorb strong concentrations of magnesium from the surrounding blood and tissue. These high concentrations enhanced the production of TSST-1 by the bacteria, thus explaining the connection of tampon use with toxic shock syndrome, a correlation that had been under active investigation for several years.

About 95% of all people develop antibodies to TSST-1 by age 20, and although tampon use is very common in the United States, only 5 to 15 cases of toxic shock are reported per 100,000 menstruating women per year. Once recognized, toxic shock syndrome usually responds well to antibiotic therapy and supportive care, but if untreated, it can lead to kidney and liver failure, coma, and eventually death. Tampons containing the highly absorbent fibers have now been taken off the market in the United States, and only 186 cases of toxic shock syndrome were reported in 1984, as compared with a high of 890 in 1980.

Many bacterial diseases are dispersed primarily in food or water, including typhoid, paratyphoid, and bacillary dysentery. Typhus is spread among rodents and humans by insect vectors. The bacterium *Brucella abortus,* which causes the disease called brucellosis in animals, may also cause undulant fever in humans who drink milk from an infected cow. The bacteria are destroyed by pasteurization of the milk, however, and as a result of widespread pasteurization practices, undulant fever is becoming rare throughout the world.

FIGURE 26-8

One of the bacteria that causes legionellosis, a species of the bacterial genus Legionella.

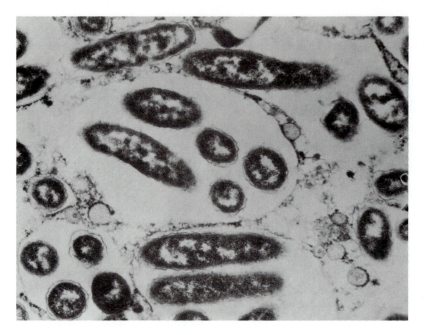

One of the more recently detected bacterial diseases of humans is legionellosis (Legionnaires' disease), which is believed to affect about 125,000 people in the United States annually. The disease develops into a severe form of pneumonia, which proves fatal to about 15% to 20% of its victims if it is untreated. Discovered in 1976, when a mysterious lung ailment proved fatal to 35 American Legion members attending a convention in Philadelphia, legionellosis is caused by small, flagellated, rod-shaped bacteria with pointed ends that have been given the name *Legionella* (Figure 26-8). These bacteria are common in water, preferring warm water at about 40° to 50° C. In the human body they attack the monocytes, a kind of white blood cell that normally plays a major defensive role against most microorganisms. The gram-negative bacteria can be destroyed with erythromycin treatment.

A number of important bacterial diseases are sexually transmitted; they are called **venereal diseases.** Among the most common of these diseases are gonorrhea, caused by the bacterium *Neisseria gonorrhoeae,* and syphilis, caused by *Treponema pallidum,* a spirochete. Both of these diseases are easily controlled with penicillin. Gonorrhea is much more common and less serious than syphilis, which can be fatal; gonorrhea has infected about 500 people per 100,000 in the United States each year during the 1980s, syphilis fewer than 20. More common than either of these are chlamydial infections caused by the bacterium *Chlamydia trachomatis.* These usually are relatively mild infections and are controllable with the antibiotic tetracycline.

One human disease—dental caries, which causes cavities—that we do not usually consider bacterial in origin arises in the film on our teeth, the **dental plaque;** this film consists largely of bacterial cells surrounded by a polysaccharide matrix. Most of the bacteria are filamentous cells, classified as *Leptotrichia buccalis,* which extend out perpendicular to the surface of the tooth, but many other bacterial species are present in plaque also. Tooth decay is caused by the bacteria present in the plaque, which persists especially in places that are difficult to reach with a toothbrush. Diets that are high in sugars are especially harmful to the teeth, because lactic acid bacteria (especially *Streptococcus sanguis* and *Streptococcus mitis*) ferment the sugars to lactic acid, a substance that causes the local loss of calcium from the teeth. Once the hard tissue has started to break down, the lysis of proteins in the matrix of the tooth enamel starts, and tooth decay begins in earnest. Fluoride makes the teeth more resistant to decay because it retards the loss of calcium. Since tooth decay is an infectious disease caused by bacteria, it theoretically can be controlled by antibiotics; a number of studies on this problem are now under way.

TABLE 26-1   CHARACTERISTICS OF SOME PHYLA OF BACTERIA

NAME OF GROUP	FORM[a]	MOTILITY[b]	METABOLISM[c]	ECOLOGICAL ROLE
Methanogens	R,S,C	N,F	C,P	Some digest cellulose; some evolve methane; some reduce sulfur
Omnibacteria	R	N,F	H	Saprobes, pathogens, decomposers
Cyanobacteria	R,C,M	G,N	P	Carbon and nitrogen fixation
Chloroxybacteria	C	N	P	Symbiotic in sea squirts
Mycoplasmas, spiroplasmas	No wall[d]	N	H	Pathogens of plants and animals
Spirochetes	S	F[e]	H	Decomposers and pathogens
Pseudomonads	R	F	H,C	Decomposers and plant pathogens
Actinomycetes	M,R	N	H	Soil, plants, decomposers, nitrogen fixers
Myxobacteria	R[f]	G	H	Soil, some in animals
Nitrogen-fixing aerobes	R	N,F	H	Free-living and in nodules on plant roots
Chemoautotrophs	R,C	N,F	C	Stages in nitrogen cycle; oxidize sulfur compounds; oxidize methane or methanol

[a] R, Rods (bacilli); C, cocci; S, spirilla; M, regularly forming chains or aggregations.
[b] F, Flagellated; N, nonmotile; G, gliding.
[c] H, Heterotrophic; C, chemosynthetic; P, photosynthetic.
[d] Either more or less spherical or elongate and twisted.
[e] Flagella inserted below the outer lipoprotein membrane of cell wall.
[f] The bacteria move together and form complex spore-bearing structures at certain stages in their life cycle.

Bacteria are important disease-causing organisms in plants and animals, including humans. Among the human diseases that they cause are scarlet fever, rheumatic fever, tuberculosis, typhoid, and legionellosis.

BACTERIAL DIVERSITY

Bacteria are not easily classified according to their forms, and only in recent years has enough been learned about their metabolic characteristics to begin to develop a satisfactory overall classification comparable to that used for other organisms. Recently Lynn Margulis and Karlene Schwartz proposed a useful classification system that divides bacteria into 16 phyla according to some of their most significant features. We have organized the following discussion, which deals with 10 of these phyla, according to this system. Table 26-1 outlines some of the major features of the phyla that we describe.

The Methane-Producing Bacteria

One phylum of chemoautotrophic bacteria that has attracted great attention during the last decade is the methane-producing bacteria, or methanogens (Figure 26-9). These bacteria, which we discussed in Chapter 3 in connection with the evolution of life on earth, are strictly anaerobic. They produce methane (CH_4) from carbon dioxide and hydrogen and obtain their energy from this process. They are the source of marsh gas, which is produced in swamps and sewage-treatment plants, and of a major portion of our natural gas reserves as well. Recently it has been shown that the methanogens are able to reduce elemental sulfur to form hydrogen sulfide, just as do other sulfur-reducing bacteria not of this phylum. By reducing elemental sulfur, these bacteria, in effect, produce their own anaerobic environments, which is important for them because they cannot function when oxygen is present.

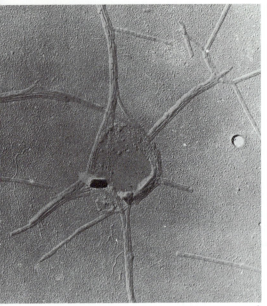

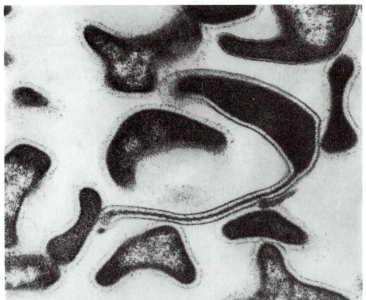

A

B

C

FIGURE 26-9

Two examples of the remarkable diversity of Archaebacteria.

A Pyrodictium occultum, *a sulfur-reducing bacterium that grows on sulfur vents on the sea floor off Vulcano, Italy. (×45,000.) The optimum temperature for growth by this bacterium, which forms "cobwebs," shown here, on the surface of the sulfur, is 105° C; growth does not take place below 80° C. Pyrite was precipitated where* Pyrodictium occultum *was growing actively; this suggests that the bacterium plays a role in the formation of ores.*

B *Individuals of* Pyrodictium occultum. *(×75,000.)*

C *The "square" bacterium, a species of* Halobacterium *from a salt pond near the shore of the Red Sea, swims by means of a posterior flagellum (not visible here). In most organisms, such a shape would have been precluded by the osmotically generated internal hydrostatic pressure, but these bacteria, which live in ponds saturated with brine, have little or no cell turgor pressure. These thin, nearly transparent bacteria derive buoyancy from their gas vacuoles and float at the surface of the brine. Square organisms are very uncommon in nature.*

A methanogen capable of fixing atmospheric nitrogen was discovered in 1984, thus adding an unexpected dimension to the fundamental metabolic diversity of the phylum.

Methane-producing bacteria play a key role in the rumen stomachs of cows, elephants, and other herbivores that are able to utilize cellulose. No vertebrate is able to digest cellulose directly; in those vertebrates that do utilize this extremely abundant substance as a source of food, it is digested by the bacteria that live inside the animals and then becomes available as food. Recently, David B. Wilson, a biochemist from Cornell University, has been able to transfer genes that code for the enzyme cellulase, which breaks down cellulose, from the methane-producing bacterium *Thermomonospora* to *Escherichia coli*. The altered individuals of *E. coli* then make stable and active cellulase. Their ability to do so may make them important agents in disposing of the vast amounts of industrial cellulose waste produced throughout the world by pulp factories and similar installations.

Carl Woese and his colleagues at the University of Illinois have shown that although the methane-producing bacteria are structurally diverse, they resemble one another and several other seemingly dissimilar groups of bacteria in a number of fundamental characteristics and differ from other bacteria in these same respects. For example, the base sequences in portions of the ribosomal RNA (rRNA) in different methane-producing bacteria are virtually identical. These sequences differ sharply from the corresponding ones found in other bacteria, as well as from those found in eukaryotes. The cell walls of all methane-producing bacteria and the related groups mentioned below lack muramic acid, which is a component of the cell walls of all other bacteria.

On the basis of their rRNA sequences, cell wall composition, and other characteristics, the methane-producing bacteria, together with *Halobacterium* and a few other groups, have been grouped by Woese and his colleagues as the Archaebacteria. Because the Archaebacteria are so distinct from other bacteria, the scientists have concluded that they may have diverged from the main stem of life before the evolution of other living bacteria; they actually treat them as a separate kingdom, but this conclusion has not yet been accepted generally. The divergence of Archaebacteria from other bacteria, however, may have taken place more than 3 billion years ago, when there was an anaerobic atmosphere rich in CO_2 and H_2. After the divergence, the Archaebacteria persisted in certain habitats where conditions continued like those that existed throughout the world when these bacteria first evolved.

The Omnibacteria

Bacteria similar to *Escherichia coli* (Figure 26-4) constitute the phylum Omnibacteria, also called the Eubacteria, a phylum of rigid, rod-shaped, heterotrophic, gram-negative bacteria that includes many important pathogens. Most of these bacteria have flagella and none of them produce spores. The members of this very large and diverse phylum are usually aerobic, but in the absence of oxygen they can usually continue to function, using compounds such as nitrate (NO_3^-) as the terminal electron acceptor in respiration.

In addition to the gram-negative rods similar to *Escherichia coli*, the Omnibacteria also include a series of comma-shaped bacteria, most of which have a single terminal flagellum. Such bacteria of this form are called **vibrios.** Members of the genus *Vibrio* itself (see Figure 7-3) cause cholera, a serious and widespread disease. At least three genera of Omnibacteria are bioluminescent, including the genus *Photobacterium,* which is associated with deep-sea fishes, where the bacteria occur in special light organs (see Figure 6-11). Another group of very small Omnibacteria, the rickettsias, are **obligate parasites** (organisms that can live only as parasites) within the cells of vertebrates and arthropods. Rocky Mountain spotted fever, which is spread by ticks, is an example of a human disease caused by rickettsias.

Cyanobacteria

Cyanobacteria are among the most prominent of the photosynthetic bacteria (Figures 26-3 and 26-10). We have already discussed the critical role that the members of this ancient phylum have played for at least 3 billion years in the history of the earth and its atmosphere (Chapters 3 and 23). The activities of cyanobacteria appear to have been decisive in bringing about the increase of free oxygen in the earth's atmosphere from far less than 1% to about 20%. This change, in turn, appears to have been crucial for the evolution of eukaryotic life based on aerobic respiration. Cyanobacteria were responsible for the accumulation of the massive limestone deposits known as stromatolites (Figure 23-7), which became abundant about 2.8 billion years ago but are still being formed where appropriate conditions exist. The few hundred living species of cyanobacteria are important relics of the earth's history, still of great interest and ecological importance today.

Cyanobacteria contain chlorophyll *a,* as do all photosynthetic eukaryotes. The process of photosynthesis is identical in both groups. Accessory pigments in the cyanobacteria consist of both carotenoids and blue and red, water-soluble pigments known as **phycobilins.** Carotenoids are widespread in photosynthetic organisms, both prokaryotes and eu-

FIGURE 26-10

Spirulina, *a cyanobacterium that is being grown increasingly as a source of protein.*
A *Individuals of* Spirulina, *each about 250 micrometers long.*
B *A mat of* Spirulina *on a conveyor belt in a processing plant in Mexico.*
C *Ponds for the production of* Spirulina *in Thailand.*

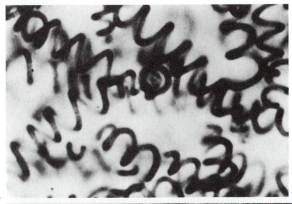

A

B

C

A LIVING RELICT?

In late 1983, Indiana University scientists Howard Gest and Jeffrey Favinger reported the discovery of a novel photosynthetic bacterium, which they named *Heliobacterium chlorum.* This bacterium is strictly anaerobic, brownish, and nitrogen-fixing. It was found growing in the bottom of a laboratory flask discarded from an undergraduate student's experiment; it had come from soil in front of the biology building on campus. The bacterium is unique, as far as is known, in its combination of photosynthesis, strict anaerobic nature, and nitrogen fixation. It would have been able to function in an atmosphere that lacked oxygen—in other words, the kind of atmosphere in which photosynthetic bacteria first evolved.

There are at least three biochemically distinct classes of chloroplasts. These occur in (1) euglenoids, green algae, and plants, (2) red algae, and (3) diatoms and brown algae. In its structure and its photosynthetic pigments, *Heliobacterium* resembles the golden-brown chloroplasts of the third group more closely than does any other bacterium, suggesting that it may be related to, or the same as, the organism that originally gave rise to them after symbiosis. Similarly, the chloroplasts of green algae, euglenoids, and plants resemble *Prochloron,* and those of red algae resemble the cyanobacteria. With the discovery of *Heliobacterium,* a possible source of the remaining class of chloroplasts is suggested (Figure 26-B).

---

karyotes. Phycobilins, on the other hand, are known to occur only in cyanobacteria, in the phylum of protists known as red algae (p. 555), and in one other small phylum of eukaryotic organisms, the cryptomonads. Because the accessory pigments shared by the cyanobacteria and the chloroplasts of red algae are unique and also resemble each other in additional biochemical and structural characteristics, it seems probable that the chloroplasts of red algae have been derived directly from symbiotic cyanobacteria.

Cyanobacteria usually have a mucilaginous sheath, which is often deeply pigmented and can be yellow, brown, green, blue, red, violet, or blue-black. Colorful "blooms" may occur in polluted water as a result of the rampant growth of cyanobacteria under conditions in which these organisms flourish temporarily, such as an excess of phosphorus; the colors of such "blooms" usually result from the photosynthetic pigments of the cyanobacteria involved. The cyanobacteria form masses near the water surface because of the formation of gas vacuoles within their cells. When such population explosions occur, the results may become traumatic for the bodies of water in which they happen. Certain cyanobacteria can, under appropriate circumstances, produce toxic substances that may poison other living things that occur with them.

Recently, scientists have noted that "blooms" of *Spirulina* are used as a source of human food in Africa, Mexico, and elsewhere (Figure 26-10). This cyanobacterium thrives in very alkaline water (up to pH 11) in which few other kinds of organisms will grow. Dried cakes of *Spirulina* have a higher protein content than soybeans, and the ponds in which they are grown are about ten times more productive, on an area basis, than wheat fields. *Spirulina* requires only carbon dioxide, water, inorganic salts, and light to grow and is being cultivated experimentally in Mexico, Italy, France, Thailand, and Japan to see whether it can be produced commercially as a food source.

Many cyanobacteria can fix atmospheric nitrogen, and for this reason they are especially important in rice fields, where they augment the supply of available nitrogen to a considerable extent (see Figure 34-15). As mentioned in Chapter 10, nitrogen fixation among the cyanobacteria occurs only in specialized cells called heterocysts (Figure 26-3)—enlarged cells that occur in many of the filamentous members of this phylum. These cells begin to form when available nitrogen falls below a certain threshold; when nitrogen is abundant, their formation is inhibited.

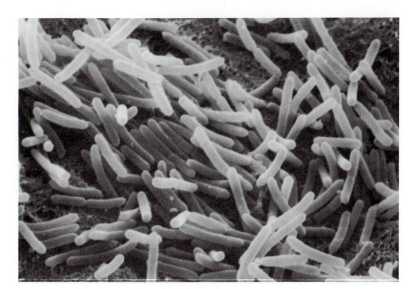

FIGURE 26-B

Heliobacterium chlorum, a newly discovered bacterium, has features that link it with the postulated ancestor of the plastids of the golden algae and the brown algae.

Chloroxybacteria

Chloroxybacteria comprise a single genus of photosynthetic bacteria, *Prochloron*. This organism is especially interesting because it contains the same photosynthetic pigments, chlorophylls *a* and *b*, that the eukaryotic green algae and the plants do. *Prochloron* was discovered in 1975 by Ralph A. Lewin of the Scripps Institution of Oceanography. It lives mainly in association with colonial ascidians (sea squirts or tunicates, phylum Chordata; Figure 38-17) along tropical and subtropical seashores (Figure 26-11). *Prochloron* is of great importance

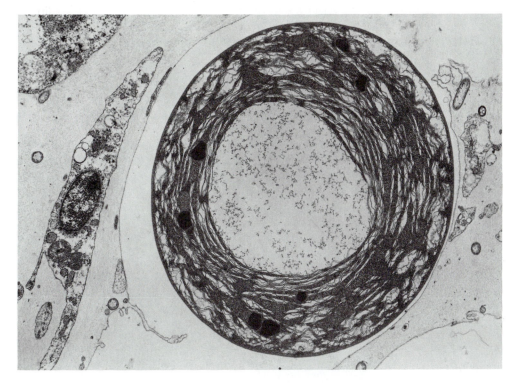

FIGURE 26-11

Electron micrograph of an individual of Prochloron *from the cloacal cavity of a tunicate.*

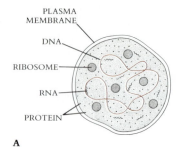

A

FIGURE 26-12

A *Diagram of a mycoplasma, one of the smallest bacteria known.*
B *The mycoplasma that causes pneumonia in humans.*

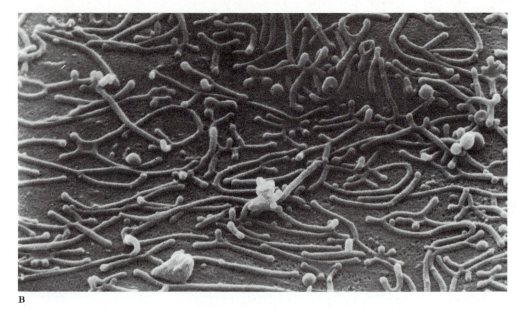

B

to students of evolution because of its biochemical characteristics, including its possession of chlorophyll *b*. These characteristics are compatible with those postulated for the prokaryote that gave rise, by symbiosis, to the chloroplasts of green algae and thus, indirectly, to those of plants.

Mycoplasmas and Spiroplasmas

The two groups of small bacteria known as mycoplasmas and spiroplasmas have been placed by Margulis and Schwartz into a separate phylum, which they call Aphragmabacteria. These organisms differ from all other bacteria in that they lack cell walls. Their cells are bounded by a single triple-layered membrane composed of lipids (Figure 26-12). Be-

FIGURE 26-13

A *A dying coconut palm in south Florida, one of more than 1.5 million individuals that have been killed by lethal yellowing, a spiroplasma-caused disease, since 1971.*
B *Spiroplasmas in plant tissue.*

B

A

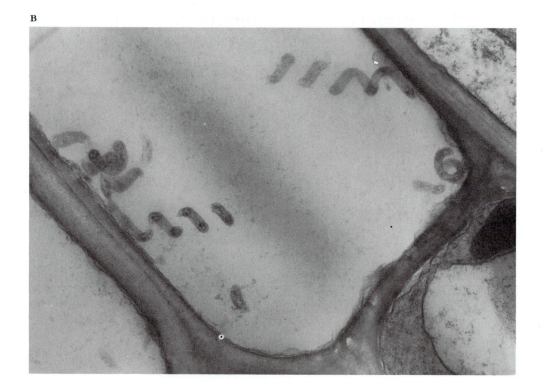

cause they lack cell walls, mycoplasmas and spiroplasmas are resistant to penicillin and other antibiotics that work by inhibiting cell wall growth. Some members of this phylum are less than 0.2 micrometer in diameter—tiny irregular blobs that cannot be distinguished with a light microscope.

Some mycoplasmas cause diseases in mammals and birds, being especially important as the causative agents of certain types of pneumonia in humans and some domestic mammals. Some have recently been associated with newly detected kinds of sexually transmitted diseases in humans.

Spiroplasmas and other members of this phylum cause significant plant diseases, having been found in more than 200 species thus far (Figure 26-13). Among the most severe diseases are aster yellows, the stubborn and little-leaf diseases of citrus, and the lethal yellowing disease of coconuts and at least 30 other species of palms. Since it was first observed in 1971, the latter disease has killed an estimated 1.5 million coconut palms in south Florida, some 80% of the total that were originally cultivated in this area; these are now being replaced with resistant strains. The disease is spread from tree to tree by a leafhopper, a kind of insect. It can be controlled by injecting an antibiotic, such as tetracycline, into the trees. In general, spiroplasmas are abundant in insects, which are the principal agents that spread them from plant to plant.

Spirochetes

The spirochetes are long spirilla in which the flagella are inserted beneath the outer lipoprotein membrane of their gram-negative outer cell wall. Spirochetes may have from 2 to more than 100 flagella in this space. They can move through thick, viscous liquids with great speed and through liquids of thinner consistency with more complex motions. Each flagellum originates near one end of the cell and extends about two thirds of the body length. Thus the flagella that originate at the different ends of the cell often overlap. As discussed in Chapter 7, the internal rotation of these flagella is thought to be responsible for the mobility of the spirochetes.

Spirochetes of the genus *Treponema* are the causative agents of syphilis and of yaws, a disfiguring eye disease that is frequent in some tropical areas. They are also apparently responsible for Lyme disease, an inflammatory ailment of humans that is named for the Connecticut village of Old Lyme. Lyme disease, which is spreading in the eastern United States, is apparently caused by spirochetes that live in wild mammals and may be transmitted to human beings by ticks. An episode of Lyme disease begins with fevers, chills, muscle aches, and pains, and may lead to chronic arthritis.

Pseudomonads

Pseudomonads are straight or curved gram-negative rods that have one or several flagella at one end. Members of this phylum are found nearly everywhere in soil and water and have a very broad ability to break down organic compounds of many kinds. Some pseudomonads are autotrophic, and many, especially those of the genera *Pseudomonas* and *Xanthomonas,* are serious plant pathogens. Pseudomonads of the genus *Bdellovibrio* occur in salt water and fresh water, and prey on other bacteria.

Actinomycetes

The actinomycetes are a distinctive phylum of bacteria that are often mistaken for fungi because of their filamentous growth form. Members of this phylum produce spores by the division of their terminal, erect branchlets into chains of small segments that are tough and very resistant to drought and other unfavorable environmental conditions (Figure 26-14). Actinomycetes of the genus *Frankia* form nitrogen-fixing nodules on the roots of a wide range of flowering plants. *Mycobacterium tuberculosis,* the causative agent of tuberculosis, and *Mycobacterium leprae,* which causes leprosy, are actinomycetes.

FIGURE 26-14

An actinomycete, Streptomyces ambofaciens, *showing the characteristic filamentous growth habit of members of this phylum of bacteria.*

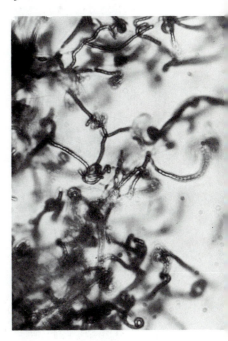

The genus *Streptomyces,* which is abundant in soil, is the source of streptomycin and other antibiotics. In addition, many other common antibiotics, including tetracycline, chloramphenicol, aureomycin, erythromycin, and neomycin, were derived originally from actinomycetes, although some of them are now synthesized commercially. Ivermectin is a potent new antiparasitic agent that has been derived from actinomycetes and is used widely in the treatment of diseases that previously have been very difficult to cure. Of the roughly 3000 antibiotics that had been discovered by 1974, about 300 were from bacteria other than actinomycetes, approximately 600 were from fungi, and all of the remainder—roughly 2000—were from actinomycetes. Some 150 new antibiotics from actinomycetes are being discovered each year.

Myxobacteria

Myxobacteria are gliding bacteria, some of which form upright spore-bearing structures of unusual complexity for bacteria (Figure 26-1, *D*). The members of this phylum are unicellular rods usually less than 1.5 micrometers thick but up to 5 micrometers long. The cells are frequently embedded in a slimy mass of polysaccharides that they excrete. These bacteria often aggregate into gliding masses, and they may form compound, spore-filled cysts that open when wet, releasing large numbers of individual gliding bacteria. All myxobacteria are obligate aerobes; some lyse (break down) other bacteria or protists enzymatically and then digest the contents of their cells. A few others live by oxidizing sulfides to sulfates. Most myxobacteria occur in soils.

FIGURE 26-15

Bacteria of the genus Rhizobium form nodules (A) *on the roots of many legumes. The bacteria* (B) *move into the roots of the legumes through cellulose tubes, which the plant forms when the bacteria are present. In the nodules, the bacteria are very abundant in the individual plant cells* (C).

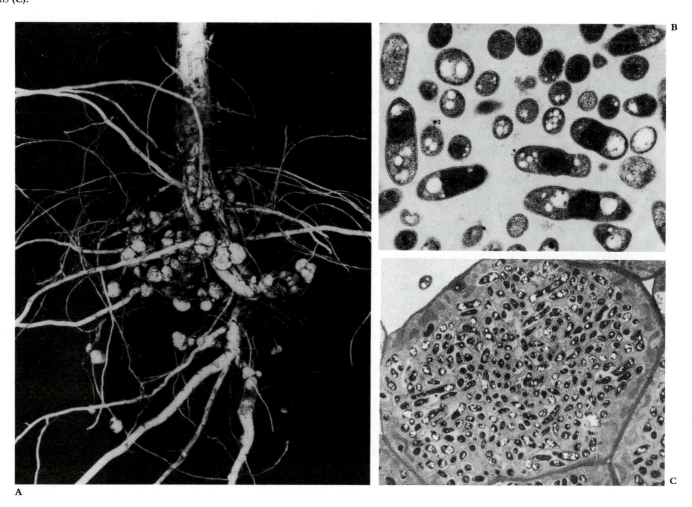

BACTERIA AND FROST DAMAGE IN PLANTS

Frost inflicts some $6 billion in damage on plants each year. Farmers try to protect their plants with smudge pots and gas heaters, but these solutions are very expensive and short term. Recently it was discovered that bacteria of the genus *Pseudomonas* (Figure 26-1, *A*), which live on plant surfaces, form the nuclei around which ice crystals form. If the crystals do not form, the plants apparently can withstand much colder temperatures without damage. This fact is highly significant, since 90% of all frost damage occurs within 5 to 10 degrees of freezing. One promising line of attack on frost damage, therefore, is to eliminate these bacteria.

Natural mutants of the bacteria that do not form nuclei for the formation of ice crystals occur in natural bacterial populations, differing from normal bacteria only at a single locus on a chromosome. Bacteria with these same properties have also been produced by excising the gene that leads to ice-crystal nucleation, using the techniques of genetic engineering. If such bacteria replace the natural populations of ice-forming ones on the plants—and they seem to do so—then the ice crystals will not form and the plants will remain healthy at much lower temperatures than normal. Alternative approaches to the same problem are to select for bacteriophages that will attack the ice-forming bacteria and to select for strains of the crops to which the natural populations of bacteria will not adhere. One or more of these methods will probably succeed and thus result in great savings for our agricultural crops in the future.

Nitrogen-Fixing Aerobic Bacteria

The nitrogen-fixing aerobic bacteria are a phylum of great economic importance. All members of this group are gram negative and most are flagellated. *Azotobacter,* mentioned earlier, is a free-living member of this phylum; with three other recognized genera, it is common in soil and water everywhere, playing a key role in converting the vast reservoir of atmospheric nitrogen into a form in which it can be used by itself and other organisms. *Clostridium,* also a free-living, nitrogen-fixing bacterium, is a member of another phylum, the fermenting bacteria, which is not treated separately in this book. It is an obligate anaerobe—its growth is inhibited, or it is killed, by oxygen. The cyanobacteria are also important in the global nitrogen cycle (Figure 34-13; Chapter 50), sharing with the other bacteria just mentioned the ability to fix atmospheric nitrogen.

One of the most important genera of nitrogen-fixing aerobic bacteria is *Rhizobium,* a group of motile bacteria that are symbiotic in nodules that they form on the roots of legumes and a few other groups of plants (Figure 26-15; p. 181). Bacteria of this genus make possible the growth of many legumes, a huge family of plants with about 18,000 species, in soils that are poor in nitrogen. Nitrogen fixation by *Rhizobium* often exceeds 100 kilograms of nitrogen per hectare per year provided directly to the legumes. All of the nitrogen fixed in this way is taken up directly by the legume, as compared with about half of the nitrogen applied as fertilizer, so the beneficial effect is even greater than the comparison implies. The individuals of *Rhizobium* enter the root hairs of their host legumes when the plants are still seedlings and move into the outer layers of the root tissue. In these tissues they stimulate repeated division of the outer cells of the root to form tumorlike nodules. Specific strains of *Rhizobium* that will form nodules on the roots of different crop legumes, such as soybeans, alfalfa, and clover, are now available commercially.

Chemoautotrophic Bacteria

Only bacteria are capable of chemoautotrophy, a form of metabolism that depends on chemical sources of energy such as the reduced gases ammonia (NH_3), methane (CH_4), or hydrogen sulfide (H_2S). Chemoautotrophs do not require sunlight; in the presence of one of the chemicals just mentioned, and with nitrogenous salts, oxygen, and carbon dioxide, they can manufacture all of their own nucleic acids and proteins. The chemoautotrophs appear to be a very ancient and primitive phylum of bacteria.

One group in this phylum consists of bacteria that use nitrogen compounds to gain energy. Among these organisms is the genus *Nitrosomonas* (see Figure 10-10), which oxidizes ammonium ions (NH_4^+) to nitrite ions (NO_2^-). The nitrite ions themselves may subsequently be oxidized to nitrate ions (NO_3^+) by other chemoautotrophic bacteria of the genus *Nitrobacter*. The process of converting ammonium ions to nitrites and nitrates is called **nitrification.** This process releases energy, which is used to reduce carbon dioxide to carbohydrates. Other bacteria are capable of converting the nitrates to nitrogen gas or nitrous oxide; by this process, known as **denitrification,** the nitrogen is lost from the soil.

Other members of this phylum oxidize inorganic sulfur compounds to gain energy. These bacteria live in habitats in which there are high concentrations of hydrogen sulfide or other reduced sulfur compounds; their cells contain sulfur granules.

The members of a third group of chemoautotrophic bacteria use methane or methanol (CH_3OH) in the same way that the other members of the phylum use nitrogen or sulfur compounds.

SIMPLE BUT VERSATILE ORGANISMS

Bacteria are the simplest organisms, but in the diversity of their metabolism they far surpass all the rest. Although many bacteria are decomposers of one kind or another, many are photosynthetic or chemosynthetic; in addition, only bacteria have the ability to convert atmospheric nitrogen into a form in which it can be used by organisms. For at least two billion years—more than half of the history of life on earth—bacteria were the only organisms in existence. They have survived to the present in rich variety by exploiting an amazingly diverse set of habitats, some of them unchanged since the beginnings of the world as we know it. Indirectly their metabolic activities are of fundamental importance; but bacteria also contribute directly to the functioning of all but a few eukaryotes in their role as mitochondria. Considering the high probability that all chloroplasts are symbiotic bacteria, it may be said that even photosynthesis, like chemosynthesis and nitrogen fixation, is exclusively a property of the bacteria.

The ways in which symbiotic relationships of eukaryotes with bacteria have contributed to the formation of evolutionary lines among the eukaryotes will be an important theme in the next chapter, as we begin to trace the evolution of eukaryotic organisms.

SUMMARY

1. The kingdom Monera comprises the prokaryotes, or bacteria, with about 2500 species named so far. Bacteria are the oldest and simplest organisms, but they are metabolically much more diverse than all of the other forms of life on earth combined.

2. Bacteria have cell walls consisting of a network of polysaccharide molecules connected by polypeptide cross-links. In the gram-negative bacteria, large molecules of lipopolysaccharide are deposited over this layer.

3. Bacteria are rod shaped (bacilli), spherical (cocci), or spiral (spirilla) in form. Bacilli or cocci may adhere in small groups or chains.

4. Bacteria and fungi break down organic compounds and thus recycle them for use by other living organisms. Different groups of bacteria fix the energy from the sun in different ways, and some bacterial groups use chemical processes as a source of energy also.

5. Only bacteria are capable of fixing atmospheric nitrogen, thus making it available for their own metabolic activities and those of other organisms. Some of these bacteria live in nodules on the roots of plants, whereas others are free-living, mainly in the soil. The best-known nitrogen-fixing bacterium is *Rhizobium*, which forms nodules on the roots of legumes.

6. Bacteria reproduce asexually by binary fission. Mutation is the most important source of variability in bacteria, but they also exhibit genetic recombination mediated by bacteriophages or plasmids.

7. The methane-producing bacteria and a few related groups differ markedly from the other bacteria and from the eukaryotes in their ribosomal sequences and in other respects. Because they are so different from other bacteria, some investigators have suggested that they should be considered a separate kingdom, Archaebacteria. This suggestion is not yet generally accepted, however.

8. Cyanobacteria, formerly and misleadingly called "blue-green algae," are photosynthetic bacteria in which chlorophyll *a* occurs together with characteristic accessory pigments known as phycobilins. The plastids of the eukaryotic red algae are almost certainly derived from cyanobacterial ancestors. On the other hand, the plastids of the green algae, and therefore those of the plants, are more similar to the newly discovered genus *Prochloron*.

9. The actinomycetes are bacteria that occur in chains and are the source of a majority—about 2000 kinds—of the antibiotics discovered thus far, including erythromycin, neomycin, and tetracycline. Some actinomycetes fix nitrogen in nodules on the roots of nonleguminous plants.

SELF-QUIZ

The following questions pinpoint important information in this chapter. You should be sure to know the answers to each of them.

1. Eukaryotes were abundant for well over 2 billion years before bacteria appeared in the world. True or false?

2. The kingdom Monera contains
   (a) eukaryotes   (d) two of the above
   (b) prokaryotes   (e) none of the above
   (c) viruses

3. Bacteria are much less diverse metabolically than are the eukaryotes and hence can survive only in a limited range of habitats, under very special circumstances. True or false?

4. Bacteria are
   (a) autotrophic and make major contributions to the world carbon balance
   (b) heterotrophic and, with the fungi, play the leading role in breaking down organic molecules from biological processes
   (c) able to break down synthetic products such as nylon and herbicides
   (d) capable of "fixing" nitrogen to a form that is usable for life processes
   (e) all of the above

5. Bacteria are always single celled. True or false?

6. Many functions that are performed by organelles in eukaryotes are carried out by membrane-bound enzymes in prokaryotes. True or false?

7. All bacteria reproduce asexually by binary fission. True or false?

8. Is the major source of genetic variation in bacteria mutation or recombination?

9. Certain bacteria are capable of living in an environment containing H_2 rather than O_2. True or false?

10. Cyanobacteria contain chlorophyll *a,* as do all photosynthetic eukaryotes, and have a kind of photosynthesis that is identical to that of plants. True or false?

11. *Prochloron* possesses all of the characteristics postulated for the bacteria that gave rise by symbiosis to plant chloroplasts. True or false?

12. Chemoautotrophs require all but one of the following. Which is not required?
    (a) sunlight
    (b) nitrogenous salts
    (c) oxygen
    (d) carbon dioxide
    (e) none of the above

THOUGHT QUESTIONS

1. Why do we say that bacteria are alive and viruses are not?

2. What are the three groups of bacteria that may have been involved in the origins of plastids in different groups of eukaryotic organisms? Outline the evidence for this assertion.

3. What do you think the functions of antibiotics are in the bacteria that produce them?

FOR FURTHER READING

BOYD, R.: *General Microbiology,* The C.V. Mosby Co., St. Louis, 1984. Excellent review of the bacteria, viruses, and some of the unicellular protists.

BROCK, T.D., D.W. SMITH, and M.T. MADIGAN: *Biology of Microorganisms,* 4th ed., Prentice-Hall, Inc., Englewood Cliffs, N.J., 1984. A comprehensive outline of the bacteria, algae, protozoa, and fungi.

FERRIS, F.G., and T.J. BEVERIDGE: "Functions of Bacterial Cell Surface Structures," *BioScience,* vol. 35, pages 172-176, 1985. An interesting review of the significance of bacterial cell wall features.

FOX, G.E., et al.: "The Phylogeny of Prokaryotes," *Science,* vol. 209, pages 457-463, 1980. Interesting account, particularly of the relationships of the Archaebacteria.

LAPPÉ, M.: *Germs That Won't Die,* Anchor Press, Garden City, N.Y., 1982. An interesting account of the author's theories about the overuse and misuse of antibiotics and his suggestions for making them more effective.

MARGULIS, L.: *Early Life,* Science Books International, Inc., Boston, 1982. An outstanding short treatise on the evolution of life on earth.

MARGULIS, L., and K.V. SCHWARTZ: *Five Kingdoms: An Illustrated Guide to the Phyla of Life on Earth,* W.H. Freeman & Co., San Francisco, 1982. An excellent outline of the diversity of life, concisely but thoroughly presented, and especially valuable for its treatment of the bacteria.

PALLERONI, N.J.: "The Taxonomy of Bacteria," *BioScience,* vol. 33, pages 370-377, 1983. An interesting essay about the methods by which bacteria are classified.

SONEA, S., and M. PANISETT: *A New Bacteriology,* Jones & Bartlett Publishers, Inc., Boston, 1980. Marvelous short account of all aspects of the biology of bacteria; strongly recommended for anyone with an interest in these fascinating organisms.

WOESE, C.R.: "Archaebacteria," *Scientific American,* June 1981, pages 98-122. Presents arguments that the Archaebacteria represent an independent kingdom of organisms.

Overview

Of the five kingdoms of organisms, the protists, a eukaryotic group, are far and away the most diverse structurally and in terms of their life cycles. Two important characteristics—multicellularity and sexuality—evolved within this group, although the great majority of protists are unicellular. The kingdom Protista contains the ancestors of each of the three kingdoms of more advanced multicellular organisms—the fungi, the plants, and the animals. Many other protists are unique single-celled forms that have not given rise to multicellular groups.

For Review

Here are some important terms and concepts that you will encounter in this chapter. If you are not familiar with them, you should review them before proceeding.

- **Symbiotic origin of mitochondria and chloroplasts** (Chapter 6)
- **Flagella and cellular motility** (Chapter 7)
- **Evolution of mitosis** (Chapter 8)
- **Cyanobacteria and** *Prochloron* (Chapter 26)

FIGURE 27-1

The kingdom Protista predominantly includes single-celled organisms, such as Paramecium *(phylum Ciliophora)* **(A),** *but multicellularity has evolved in many lines. One of these, representative of the phylum Phaeophyta, is a gigantic kelp* **(B;** *note the rowboat in this aerial photograph). Still others have more unusual forms of multicellular construction, such as the multinucleate plasmodium of a plasmodial slime mold (phylum* Myxomycota) **(C).** *Individual cells cannot be distinguished in such a plasmodium, but they retain their distinctness in other modes of multicellular construction.*

Between the bacteria and the eukaryotes there is a gap in the evolutionary record, a dark curtain through which we cannot yet peer. The evolution of the first eukaryotes could not have happened at once, quickly, because they differ from bacteria in so many complex ways (Chapter 5). No organisms that are transitional between bacteria and eukaryotes survive, however. The following chapters deal with the eukaryotes. For all their incredible diversity (Figure 27-1), the various groups of eukaryotes are not nearly as different from one another as are the eukaryotes from the bacteria from which they must have evolved.

EVOLUTIONARY RELATIONSHIPS OF PROTISTS

Of the four kingdoms of eukaryotes, the most diverse kingdom is Protista, the protists. It contains the simplest of the eukaryotes as well as many groups that consist of very complex organisms. Few protists are large; most are microscopic and single celled. The multicellular animals and plants were derived independently from different groups of protists, as were

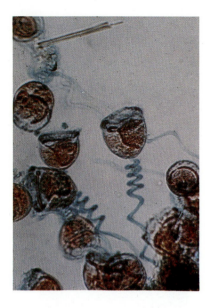

FIGURE 27-2

Complex unicellular organisms like Vorticella *(phylum Ciliophora)— which is heterotrophic, feeds largely on bacteria, and has a retractable stalk—do not really fit the definition of either animals or plants.*

the primarily multicellular fungi. Animals, plants, and fungi are considered distinct kingdoms; the remaining, extremely diverse groups of eukaryotic organisms are assigned to the kingdom Protista.

Among the protists are the ancestors of the primarily multicellular fungi and those of the entirely multicellular animals and plants. We no longer consider protists themselves as belonging to any one of these groups, however.

The flagella and cilia of all eukaryotes, when they are present, have the complex 9 + 2 arrangement of microtubules discussed in Chapter 7. Such flagella and all of the other characteristics of eukaryotes are common to the protists. Protists are distinguished from plants, animals, and fungi primarily because of their lack of the specialized features characteristic of those groups. For example, protists do not form embryos, and they do not develop the kinds of complex, multicellular sexual organs that characterize plants. Some protists have chloroplasts and manufacture their food, like the plants; others ingest their food, like the animals; and still others absorb their food, like the fungi.

In the older systems of classification, the green, photosynthetic protists were considered plants, together with all protists that absorbed their food, like the fungi. Historically, the fungi have been included among the plants as well. Those protists that ingested their food like the animals were considered small, very simple animals (Figure 27-2). These schemes of classification do not reflect the close and complex relationships between the different groups of protists, however, and thus they generally have been replaced by the classification scheme used in this book.

Although those phyla of protists that do have chloroplasts and carry out photosynthesis are classified according to the botanical rules of nomenclature, and therefore should technically be called *divisions,* we will use the term *phyla* for all such groups for clarity here. Protists that have chloroplasts are informally called **algae;** the prokaryotic cyanobacteria are even loosely included among the algae. The scientists who study algae are called **phycologists,** whereas those who study heterotrophic protists are called **protozoologists,** since these organisms collectively are called **Protozoa,** a term similar to *algae* in its general, non-technical meaning.

Two phyla of algae are primarily multicellular: the brown algae, phylum Phaeophyta, and the red algae, phylum Rhodophyta. A third phylum of algae, the green algae, phylum Chlorophyta, includes many kinds of multicellular organisms and even larger

FIGURE 27-3

Three phyla of algae have become multicellular independently, and differ greatly in their photosynthetic pigments and structure.
A *The brown algae (phylum Phaeophyta) are large, complex algae, which often dominate north-temperate seacoasts.*
B Ulva *is a green alga (phylum Chlorophyta), a group that includes many unicellular organisms as well as a number that are multicellular.*
C *Some red algae (phylum Rhodophyta), shown here on sponges, occur at greater depths in the sea than do the members of either of the other phyla. Different groups of bacteria appear to have given rise to the chloroplasts of each of these phyla of algae as a result of ancient symbiotic events.*

A

B

C

numbers of unicellular ones (Figure 27-3). Multicellularity has been achieved separately in each of these three phyla, which have direct relationships to different unicellular protists. The green algae include the ancestors of the plants, but we will retain them in the kingdom Protista here, since plants and green algae are so sharply distinct from each other.

> **Three phyla of photosynthetic protists have become wholly or partly multicellular: green algae, brown algae, and red algae. These phyla are assigned to the kingdom Protista, even though the green algae include the ancestors of the plants.**

Grouping all of these predominantly unicellular, eukaryotic organisms together allows us to compare them directly and to emphasize the similarities and differences among them. It should be kept in mind, however, that the kingdom Protista is a much more diverse group than any of the other three kingdoms of eukaryotic organisms.

SYMBIOSIS AND THE ORIGIN OF EUKARYOTES

Much of the confusion about the classification of the phyla of the kingdom Protista and their relationships to one another stems from an important factor in the evolution of the eukaryotic cell. This factor concerns the origin of some of the organelles of the eukaryotic cell by a process involving the symbiosis of formerly free-living prokaryotic organisms. As discussed in Chapter 6, two of the principal organelles of eukaryotic cells, mitochondria and chloroplasts, share a number of unusual characteristics with each other and with the bacteria from which they apparently were derived.

Mitochondria

The first eukaryotes appeared about 1.5 billion years ago, and their descendants soon became abundant and diverse. With very few exceptions, all modern eukaryotes possess mitochondria, and it is therefore clear that these organelles were acquired early in the history of the group. Mitochondria are most similar to nonsulfur purple bacteria, the group that probably gave rise to them originally (see Chapter 6).

Among the eukaryotes, mitochondria are lacking only in many of the zoomastigotes (p. 547) and in *Pelomyxa palustris* (p. 146; Figure 27-4), an unusual amoeba-like organism that occurs on the muddy bottoms of freshwater ponds. *Pelomyxa* also lacks mitosis; its nuclei divide somewhat like those of bacteria do, by simply pinching apart into two nuclei, with new membranes forming around the derivatives. The cells of *Pelomyxa* are much larger than bacteria and are visible to the naked eye. Although *Pelomyxa* lacks mitochondria, its cells do have two kinds of bacterial symbionts within them, and these may play the same role that mitochondria do in all other eukaryotes. Possibly along with the zoomastigotes, *Pelomyxa* seems to represent an early stage in the evolution of eukaryotic cells, one before they had acquired mitochondria.

> **Mitochondria, which probably originated from symbiotic nonsulfur purple bacteria, are absent in two unusual groups of protists. Of these, *Pelomyxa* may represent a stage of evolution before ancestral eukaryotes acquired these organelles.**

Chloroplasts

In contrast to mitochondria, which are relatively uniform from group to group, chloroplasts fall into three classes that are distinct in their biochemistry. They are:

1. The chloroplasts of red algae, with chlorophyll *a,* carotenoids, and phycobilins. These chloroplasts were almost certainly derived from symbiotic cyanobacteria.

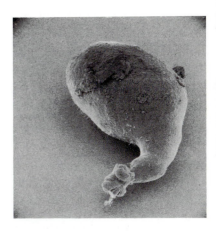

FIGURE 27-4

Pelomyxa palustris, *a unique, amoeba-like protist, which lacks mitochondria and mitosis.* Pelomyxa *may represent a very early stage in the evolution of eukaryotic cells. This species is the only member of the phylum Caryoblastea.*

2. The chloroplasts of brown algae, diatoms, and dinoflagellates, with chlorophylls *a* and *c,* carotenoids, and distinctive yellowish-brown pigments. The newly discovered bacterial genus *Heliobacterium* resembles these chloroplasts most closely and may resemble the bacteria that gave rise to them originally.

3. The green chloroplasts of plants, green algae, and another, very distinctive phylum, the euglenoids (p. 548). These chloroplasts contain chlorophylls *a* and *b,* in addition to carotenoids. *Prochloron* has essentially the same biochemical features and apparently resembles the bacteria that gave rise to the chloroplasts of the phyla mentioned above.

Even today, so many bacteria and unicellular protists are symbiotic that the incorporation of smaller organisms with desirable features into cells appears not to be a difficult process. In view of the great differences among the phyla with each of the three kinds of chloroplasts just outlined, it appears likely that multiple symbiotic events were involved in the origins of chloroplasts in the different groups. Thus, for example, the same kind of bacterium might have become symbiotic and given rise to the chloroplasts of green algae and euglenoids independently of each other.

Eukaryotic cells acquired chloroplasts by symbiosis not once but at least several times during the evolution of different groups of protists.

Flagella and Centrioles

Lynn Margulis of Boston University believes that symbiosis has contributed even more to eukaryotic cells than mitochondria and chloroplasts. She has proposed that the mitotic apparatus is also a result of symbiotic events. She hypothesizes that amoeba-like organisms somewhat similar to *Pelomyxa* acquired spirochete-like, symbiotic organisms that lived on the surface of their cells. Margulis proposes that these spirochetes, conferring motility on their host cells, eventually became the 9 + 2 flagella that are so characteristic of eukaryotes. She believes that the subsequent specialization of these spirochetes eventually led to the evolution of centrioles and of chromosomal centromeres during the course of evolution of mitosis. As you saw in Chapter 8, there are important differences in the process of mitosis as it occurs in different groups. These differences suggest the complexity of the evolutionary steps that may have led to the full development of the process of mitosis as it occurs in most contemporary organisms. A bizarre modern organism that is able to move largely because of the bacteria that adhere to its surface is shown in Figure 27-18.

EVOLUTION OF MULTICELLULARITY AND SEXUALITY
Multicellularity

True **multicellularity,** a condition in which the activities of the individual cells are coordinated and the cells themselves are in contact, is a property of eukaryotes alone and one of their major characteristics. The cell walls of bacteria occasionally adhere to one another, and bacterial cells may also be held together within a common sheath (Chapter 26). Consequently, some bacteria form filaments, sheets, or three-dimensional bodies, but the degree of integration of individual bacteria within these groups of cells is much more limited than is the case in multicellular eukaryotes.

The evolution of multicellularity allowed organisms to deal with their environments in novel ways (Figure 27-5). Distinct types of cells, tissues, and organs can be differentiated within the complex bodies of multicellular organisms. With such functional division within its body, a multicellular organism can protect itself, move about, seek mates and prey, and carry out other activities on a scale and with a complexity that would have been impossible for its unicellular ancestors. With all of these advantages, it is not surprising that multicellularity has arisen independently so many times.

FIGURE 27-5

Volvox (phylum Chlorophyta). Individual biflagellated, motile, unicellular green algae comparable to Chlamydomonas *are united in* Volvox *as a hollow ball of cells, which moves by means of the beating of the flagella of its individual cells. Some of these cells are specialized morphologically and functionally: a few, near the posterior end of the colony, are reproductive cells, but most are relatively undifferentiated. Some species of* Volvox *have cytoplasmic connections between the cells, which function in the overall coordination of the activities of the colony.*

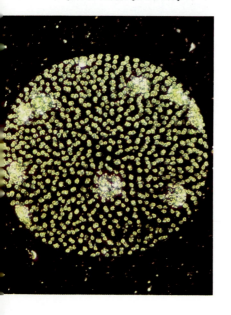

Multicellularity evolved numerous times among the protists. All of the complex differentiation that we associate with advanced life forms depends on multicellularity, which must have been highly advantageous to have evolved independently so often.

Sexuality

The second major characteristic of eukaryotic organisms as a group is sexuality. Although some interchange of genetic material does occur in bacteria (Chapters 8 and 26), it is certainly not a regular, predictable mechanism in the same sense that sex is in eukaryotes. The regular alternation between syngamy—the union of male and female gametes—and meiosis constitutes the sexual cycle that is characteristic of eukaryotes. It differs sharply from anything found in bacteria.

Many eukaryotes are haploid, but adult animals and plants are diploid, with few exceptions. In the cells of animals and plants there are two sets of chromosomes, derived from their male and female parents. These chromosomes segregate regularly by the process of meiosis; since meiosis involves crossing-over, no two products of a single meiotic event are ever identical. As a result, the offspring of sexual, eukaryotic organisms vary widely, thus providing the raw material for evolution.

Sexual organisms are able to change over generations in relation to the demands of their environment because they produce variable progeny. Sexual reproduction, involving the regular alternation between syngamy and meiosis, is the process that makes their evolutionary adjustment possible.

In many of the unicellular phyla of protists, sexual reproduction occurs only during times of stress. Meiosis may have originally evolved as a means of producing new, well-adapted forms that would increase the chances of survival during such times.

Judged by their ancestry and the abundance of such forms now, it is all but certain that the first eukaryotes were haploid. Diploidy evolved among them by the incorporation of two sets of chromosomes in one organism. The evolution of diploidy undoubtedly occurred on a number of separate occasions. It probably took place when two haploid cells occasionally combined to form a diploid cell, which then was essentially a zygote. Such a diploid cell may in turn have divided by mitosis and thus multiplied. Eventually meiosis evolved in such cells; initially it was probably a way of restoring the haploid complement of chromosomes in the gametes. Diploid cells probably also originated many times by chromosome division within haploid cells. Since both sets of chromosomes in such cells would have been identical, however, it is unlikely that they would have possessed any selective advantage over their haploid ancestors.

Eukaryotes are characterized by three major types of life cycles. (1) In the simplest of these, the zygote is the only diploid cell. Such a life cycle is said to be characterized by **zygotic meiosis,** since the zygote immediately undergoes meiosis. (2) In most animals the gametes are the only haploid cells. They exhibit **gametic meiosis;** meiosis produces the gametes, which fuse, giving rise to a zygote. (3) In plants, there is a regular alternation between a multicellular haploid phase and a multicellular diploid phase. The diploid phase produces spores that give rise to the haploid phase, and the haploid phase produces gametes that fuse to form the zygote. The zygote is the first cell of the multicellular diploid phase. This kind of life cycle is characterized by **alternation of generations** and has **sporic meiosis.** These three major types of eukaryotic life cycles are diagrammed in Figure 27-6.

In a life cycle characterized by zygotic meiosis, the zygote is the only diploid cell; in one characterized by gametic meiosis, the gametes are the only haploid cells. In plants, there is an alternation of generations, in which both diploid and haploid cells divide by mitosis; their type of life cycle is characterized by sporic meiosis.

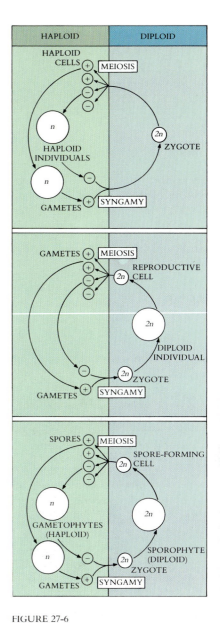

FIGURE 27-6

Diagrams of three major kinds of life cycles in eukaryotes. Top to bottom: zygotic meiosis, gametic meiosis, and sporic meiosis.

TABLE 27-1   FEATURES OF 15 PHYLA OF KINGDOM PROTISTA

NOCTILUCA

PTYCHODISCUS

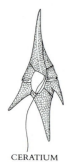

CERATIUM

GONYAULAX

FIGURE 27-7

Dinoflagellates, including Noctiluca, Ptychodiscus, Ceratium, *and* Gonyaulax. Noctiluca, *which lacks the heavy cellulosic armor characteristic of most dinoflagellates, is one of the bioluminescent organisms that causes the waves to sparkle in warm seas at certain times of year. In the other three genera, the shorter, encircling flagellum may be seen in its groove, with the longer one projecting away from the body of the dinoflagellate.*

NAME*	NUMBER OF SPECIES	CHLOROPHYLLS	WALL/SHELL	FLAGELLA
Acrasiomycota	65	None	None	0
Bacillariophyta	11,500	*a* + *c*	Opaline silica	0
Caryoblastea	1	None	None	0
Chlorophyta	7000	*a* + *b*	Cellulose†	2
Ciliophora	8000	None	Flexible	Many
Dinoflagellata	1000	*a* + *c*	Mostly cellulose plates	2; Un‡
Euglenophyta	800	*a* + *b* or none	Flexible pellicle	2; Un‡
Foraminifera	Hundreds	None	Shells	0 (podia)
Myxomycota	450	None	None	0 or 2
Oomycota	475	None	Cellulose†	2
Phaeophyta	1500	*a* + *c*	Cellulose†	2
Rhizopoda	Hundreds	None	Naked	0
Rhodophyta	4000	*a*	Cellulose†	0
Sporozoa	3900	None	Flexible	Variable
Zoomastigina	Thousands	None	Flexible	Many

*Phyla whose names are printed in italics consist of multicellular organisms; Chlorophyta also includes many unicellular ones, and Rhodophyta includes a few unicellular ones. Myxomycota move about in a mass, called a plasmodium, in which there are many nuclei.
†These phyla have cellulose as the primary constituent in the cell walls of many or all members, but other substances are present, sometimes exclusively so in some members of these phyla.
‡Un, Unequal in length.

Now that we have set the stage for our general consideration of the diversity of eukaryotes by our discussion of the overall features of the group, we will proceed to consider some of the more important protist phyla.

MAJOR GROUPS OF PROTISTS

We consider 15 phyla of protists in this book. Because the kingdom is so diverse, we can discuss its features effectively only by dealing individually with some of its members.

One of the organisms discussed earlier, *Pelomyxa palustris* (p. 537; Figure 27-4), is assigned to a phylum of its own, Caryoblastea. Additionally, we will present some of the characteristics of the following phyla of protists: (1) dinoflagellates (Dinoflagellata), (2) amoebas (Rhizopoda), (3) forams (Foraminifera), (4) sporozoans (Sporozoa), (5) cellular slime molds (Acrasiomycota), (6) plasmodial slime molds (Myxomycota), (7) zoomastigotes (Zoomastigina), (8) euglenoids (Euglenophyta), (9) brown algae (Phaeophyta), (10) diatoms (Bacillariophyta), (11) green algae (Chlorophyta), (12) ciliates (Ciliophora), (13) water molds and their relatives (Oomycota), and (14) red algae (Rhodophyta). Table 27-1 provides a summary of some of the highly diverse characteristics of these phyla.

Dinoflagellates

Dinoflagellates (phylum Dinoflagellata) consist of about 1000 known species of unicellular, photosynthetic organisms, most of which have two flagella. A majority of the dinoflagellates are marine, but some occur in fresh water. These organisms are often abundant as members of the marine **plankton** (organisms that float freely and passively in the water). Plankton is much more abundant in the upper, relatively well-lighted layers of the water than elsewhere, because photosynthesis and therefore food production can take place only

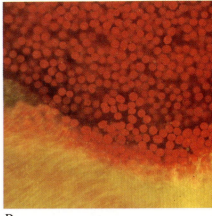

A   B   C   D

near the surface of the water, where there is enough light. Some of the planktonic dinoflagellates are luminous and contribute to the twinkling or flashing effects that sometimes are seen in the sea at night, especially in the tropics.

The flagella, protective coats, and biochemistry of the dinoflagellates are distinctive. Their flagella beat in two grooves, one encircling the body like a belt, and the other perpendicular to it. By beating in their respective grooves, these flagella cause the dinoflagellate to rotate like a top as it moves. Many dinoflagellates are clad in stiff cellulose plates, often encrusted with silica, which give them a very unusual appearance (Figure 27-7). Most have chlorophylls *a* and *c,* in addition to carotenoids, so that in the biochemistry of their chloroplasts, they resemble the brown algae and the diatoms. The members of each of these groups probably acquired their chloroplasts as a result of independent symbiotic events.

Some dinoflagellates are capable of ingesting other cells, and a number are colorless and heterotrophic. Other dinoflagellates occur as symbionts in many other groups of organisms, including jellyfish, sea anemones, mollusks (see Figure 36-3), and, notably, corals (Figure 27-8, *A*). When dinoflagellates grow as symbionts within other cells, they lack their characteristic cellulose plates and their flagella, appearing as spherical, golden-brown globules in their host cells. In such a state they are called **zooxanthellae** (Figure 27-8, *B* to *D*). Zooxanthellae are the primary factor responsible for the productivity of corals and their ability to grow in tropical waters, which are often extremely low in nutrients. Most of the carbon that the zooxanthellae fix is translocated to the host corals, which can obtain all of the fixed carbon that they need from zooxanthellae if they are growing in well-lighted places.

> **Zooxanthellae, the symbiotic form of dinoflagellates, are primarily responsible for the productivity of coral reefs, which usually occur in waters that are poor in nutrients. The zooxanthellae live within the bodies of the coral animals.**

The poisonous and destructive "red tides" that occur frequently in coastal areas are often associated with great population explosions, or "blooms," of dinoflagellates (Figure 27-9). The pigments in the individual, microscopic cells of the dinoflagellates, or, in some cases, other organisms, are responsible for the color of the water. Such red tides have a profound, detrimental effect on the fishing industry in the United States. Approximately 20 species of dinoflagellates are known to produce toxins under red-tide conditions. These powerful toxins, mostly poorly known chemically, inhibit the diaphragm and cause respiratory failure in many vertebrates, although they are not toxic to the shellfish that feed on the dinoflagellates by straining them out of the water. When they consume enough of the dinoflagellates, however, the shellfish become highly poisonous themselves. Unfortunately, relatively little can be done to reduce the effects of such outbreaks of poisonous dinoflagellates other than to keep humans from eating the contaminated seafood.

FIGURE 27-8

A *A coral reef—a complex, highly productive community of tropical and subtropical waters—fringes each of these small islands off the coast of Java.*
B *In their symbiotic form, dinoflagellates, then known as zooxanthellae, contribute most of the productivity to the well-illuminated portions of coral reefs. Each of the tentacles of this coral animal is packed with golden-brown zooxanthellae.*
C *Individuals of the sea anemone Anthopleura elegantissima, freshly collected from the intertidal zone on the Pacific Coast of the United States. The brown-green pigmentation of these animals is due almost entirely to pigments in the zooxanthellae; animals of the same species that lack zooxanthellae are milky pink.*
D *Densely packed zooxanthellae in tissue of the anemone shown in* **C**. *This photo was taken with filters that make the zooxanthellae fluoresce red, while the tissue of the anemone fluoresces yellowish or bluish-green.*

FIGURE 27-9

The effects of a red tide on the beaches of Tampa Bay, Florida, in 1963; these workers are cleaning up large numbers of dead fishes. This red tide was caused by the dinoflagellate Ptychodiscus brevis. When dinoflagellates and other toxin-producing organisms become abnormally abundant, they are consumed by marine animals. The toxins often kill animals such as fish in great numbers and poison human beings and other animals that consume the poisoned fish.

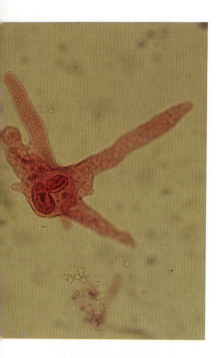

FIGURE 27-10

Amoeba proteus, *a relatively large amoeba that is commonly used in teaching and for research in cell biology. The projections are pseudopods; an amoeba moves simply by flowing into them. The nucleus of the amoeba is plainly visible.*

FIGURE 27-11

Entamoeba histolytica, *the cause of amoebic dysentery. At left are two amoebas within resistant cysts.*

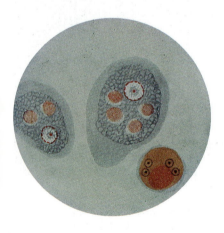

Dinoflagellates reproduce primarily by longitudinal cell division, but sexual reproduction has also been shown in more than 10 genera. Their form of mitosis is unique (see boxed essay, p. 146): the chromosomes remain within the nucleus but are distributed along the sides of channels containing bundles of microtubules that run through the nucleus. The unique chromosomes of the dinoflagellates consist of a framework of nucleic acids, largely DNA, within a protein matrix. Histones are absent, and the chromosomes remain condensed permanently. The chromosomes and mitotic cycle of the dinoflagellates seem to have evolved independently of those of the other eukaryotes, judged from the great differences between them.

Mitosis in dinoflagellates is unique and takes place within the nucleus. The dinoflagellates do not appear to have any close relatives.

Amoebas

Amoebas (phylum Rhizopoda) are found throughout the world in both fresh and salt waters (Figure 27-10). They are also abundant in soil. Many kinds of amoebas are parasites of animals. Reproduction in amoebas occurs by fission—the direct division into two cells of equal volume. Amoebas of the phylum Rhizopoda lack cell walls, flagella, meiosis, and any form of sexuality. They do undergo mitosis and have a spindle apparatus that resembles that of most other eukaryotes. Heliozoa (see Figure 7-9) are unusual amoebas.

Amoebas move from place to place by means of their **pseudopods,** from the Greek words for "false" and "foot" (Figure 6-1). The pseudopods are flowing projections of the cytoplasm that extend and pull the amoeba forward or engulf food particles. An amoeba puts a pseudopod forward and then flows into it. Actin and myosin proteins similar to those found in muscles are associated with these movements.

Some kinds of amoebas form resistant cysts. In parasitic species such as *Entamoeba histolytica,* which causes amoebic dysentery, the cysts enable the amoebas to resist digestion by their animal hosts, which include humans (Figure 27-11). Mitotic division takes place within the cysts, which may then ultimately break and release four, eight, or even more amoebas within the digestive tracts of their host animals. The primary infection takes place in the intestine, but it often moves into the liver and sometimes into other parts of the body. The cysts are dispersed in feces and may be transmitted from person to person in infected food and water, by flies, and by direct contact. It is estimated that up to 10 million people in the United States have infections of parasitic amoebas, and some 2 million show symptoms of the disease, ranging from abdominal discomfort with slight diarrhea to much more serious conditions. In some tropical areas, more than half of the people may be infected. The spread of amoebic dysentery can be limited by proper sanitation and hygiene.

Amoebas are unicellular, heterotrophic protists that lack cell walls, flagella, meiosis, and sexuality; they move from place to place by means of extensions called pseudopods.

Forams

The forams (phylum Foraminifera) are heterotrophic, marine protists. They range in diameter from about 20 micrometers to several centimeters. Characteristic of the group are pore-studded shells (called **tests**) composed of organic materials usually reinforced with grains of inorganic matter (Figure 27-12). These grains may be calcium carbonate or sand or even plates from the shells of echinoderms or **spicules** (minute needles of calcium carbonate) from sponge skeletons. Depending on the building materials they use, forams may have shells of very different appearance. Some of them are brilliantly colored—red, salmon, or yellow-brown.

Most forams live in sand or are attached to other organisms, but two families consist of free-swimming, planktonic organisms. The tests of forams may be single chambered but more often are multichambered. These tests tend to have a spiral shape resembling that of a tiny snail. Thin cytoplasmic projections called **podia** emerge through openings in the tests (Figure 27-13); the podia are used for swimming, gathering materials for the tests, and feeding. Forams eat a wide variety of small organisms.

The life cycles of forams are extremely complex, involving an alternation between haploid and diploid generations like that described on p. 549 for the brown algae. Forams are among the very few heterotrophic protists that have an alternation of generations. In this alternation, an individual characteristically has two or more different kinds of nuclei that differ greatly in size; some of these nuclei may engage in meiosis, others may fuse with one another, and still others may simply disintegrate at certain stages.

Forams have contributed massive accumulations of their tests to the fossil record ever since the Triassic Period, about 230 million years ago. Because of the excellent preservation of their tests and the often striking differences among them, forams are very important as geological markers (see Figure 23-19). The pattern of occurrence of different forams, for example, is often used as a guide in searching for oil-bearing strata. Forams are excellent markers for locating geological strata of equivalent age in different areas. Limestones all over the world are often rich in forams, and the White Cliffs of Dover, the famous landmark of southern England, are made up almost entirely of their tests (Figure 1-6).

> **The pore-studded tests, or shells, of the forams are characteristic of this group of unicellular protists. Their life cycles involve a complex alternation of generations.**

Sporozoans

All sporozoans (phylum Sporozoa) are nonmotile, spore-forming parasites of animals. Their spores are small, infective bodies that are transmitted from host to host. These organisms are distinguished by a unique arrangement of fibrils, microtubules, vacuoles, and other cell organelles at one end of the cell. There are about 3900 described species of this phylum; best known among them is the malarial parasite, *Plasmodium.*

Sporozoans have complex life cycles that involve both asexual and sexual phases. Sexual reproduction involves an alternation of haploid and diploid generations. Both haploid and diploid individuals can also undergo a process known as **schizogony,** a rapid series of mitotic events that results in the production of a large number of small infective individuals. Sexual reproduction involves the fertilization of a large female gamete by a small, flagellated male gamete. The zygote that results soon becomes a thick-walled cyst called an **oocyst,** which is highly resistant to drying out and other unfavorable environmental factors. Within the oocyst, meiotic divisions produce infective haploid spores.

An alternation between different hosts often occurs in the life cycles of the sporozoans. The sporozoans of the genus *Plasmodium,* for example, are spread from person to person by mosquitoes of the genus *Anopheles* (Figure 1-5). When an *Anopheles* mosquito bites a human, it injects saliva mixed with an anticoagulant. If the mosquito is infected with *Plasmodium,* it will also inject elongated *Plasmodium* cells known as **sporozoites**—a stage of its life cycle—into the bloodstream of its victim. The parasite makes its way through the bloodstream to the liver, where it rapidly divides asexually. After this division phase, **merozoites,** which are the next stage of the life cycle, are formed, either reinvading other liver cells or entering the host's bloodstream. In the bloodstream, they invade the red blood cells, dividing rapidly within them and causing them to become enlarged and ultimately to rupture. This event releases toxic substances throughout the body of the host, bringing about the well-known cycle of fever and chills that is characteristic of malaria. The cycle repeats itself regularly every 48 hours, 72 hours, or longer, depending on the species of *Plasmodium* involved.

Plasmodium enters a sexual phase when some merozoites develop into **gametocytes,** cells capable of producing gametes. There are two types of gametocytes, male and female.

FIGURE 27-12

Forams (phylum Foraminifera).
A *Forams lying on an ocean bed.*
B *Scanning electron micrograph of some representative tests.*

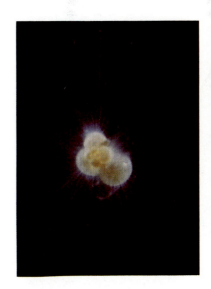

FIGURE 27-13

A living foram, Globigerina, showing the podia, thin cytoplasmic projections that extend through pores in the calcareous test, or shell, of the organism.

MALARIA

Malaria is one of the most serious diseases in the world. About 250 million people are affected by it at any one time, and 2 to 4 million of them die each year. The symptoms, familiar throughout the tropics, include paroxysmal chills, fever, and sweating, an enlarged and tender spleen, confusion, and great thirst. Ultimately, a victim of malaria may die of anemia, kidney failure, or brain damage, or the disease may be brought under control by the person's immune system or by drugs. People do develop immunity to malaria, which kills most children under 5 years old who contract it; in areas where malaria is prevalent, most survivors more than 5 or 6 years old do not become seriously ill again from malaria infections.

Efforts to eradicate malaria have focused on (1) the elimination of the mosquito vectors, (2) the development of drugs to poison the parasites once they have entered the human body, and (3) the development of vaccines. Regarding the first method, the wide-scale application of DDT from the 1940s to the 1960s led to the elimination of the mosquito vectors in the United States, Italy, Greece, and certain areas of Latin America. For a time, the elimination of malaria worldwide appeared possible, but this hope was soon rendered impossible by the development of DDT-resistant strains of the mosquitoes in many different regions; no less than

64 resistant strains were identified in a 1980 survey. Even though the worldwide use of DDT, long banned in the United States, nearly doubled from its 1974 level to more than 30,000 metric tons in 1984, its effectiveness in controlling mosquitoes is dropping; furthermore, there are serious environmental concerns about the continued use of this long-lasting chemical anywhere in the world. In addition, strains of *Plasmodium* have appeared that are resistant to all of the drugs that have historically been used to kill them.

As a result of these problems, the number of new cases of malaria per year roughly doubled from the mid-1970s to the mid-1980s, largely as a result of the spread of resistant strains of the mosquito and the parasite. In many regions of the tropics, malaria is blocking permanent settlement. Scientists have therefore redoubled their efforts to produce an effective vaccine against the disease. Antigens to the parasites can be isolated and produced by monoclonal antibody techniques (Chapter 18), which are starting to produce promising results. The three different stages of the life cycle of *Plasmodium* (Figure 27-A) each produce different antigens, however, and are sensitive to different antibodies.

The gene encoding the sporozoite antigen was cloned by scientists at the New York University Medical Center in 1984, permitting the antigen to be

Gametocytes are incapable of producing gametes within their human hosts, and do so only when they are extracted from an infected human by a mosquito. Within the gut of the mosquito, the male and female gametocytes form sperm and eggs, respectively. Zygotes develop within the mosquito's intestinal walls and ultimately differentiate into oocysts. Within the oocysts, repeated mitotic divisions take place, producing large numbers of sporozoites. These sporozoites migrate to the salivary glands of the mosquito, from which they may be injected by the mosquito into the bloodstream of a human, thus starting the life cycle of the parasite again.

In the life cycle of *Plasmodium*, a mosquito injects sporozoites into the bloodstream of its victim. These pass to the liver, where they divide rapidly by mitosis, eventually liberating large numbers of merozoites, which infect the red blood cells. Sexual reproduction may then occur.

Cellular Slime Molds

There are about 65 species of cellular slime molds (phylum Acrasiomycota), a phylum with extraordinarily interesting features (Figure 27-14). "Mold" is a general term for funguslike organisms; the members of this group were once thought to be related to fungi. "Slime molds" are those "molds" that move along like amoebas or look like slime. In fact,

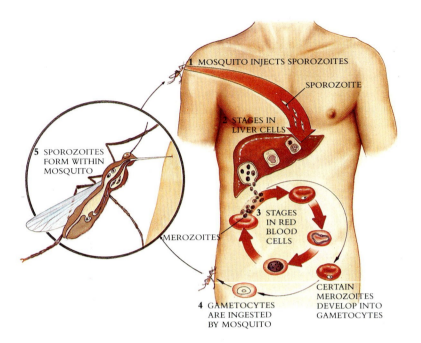

1 MOSQUITO INJECTS SPOROZOITES

SPOROZOITE

5 SPOROZOITES FORM WITHIN MOSQUITO

2 STAGES IN LIVER CELLS

MEROZOITES

3 STAGES IN RED BLOOD CELLS

CERTAIN MEROZOITES DEVELOP INTO GAMETOCYTES

4 GAMETOCYTES ARE INGESTED BY MOSQUITO

FIGURE 27-A

Life cycle of Plasmodium, *the zoomastigote that causes malaria.*

mass produced by genetic engineering techniques. It is not certain, however, how effective a program of vaccination against sporozoites alone might be. When a mosquito bites a person, it injects a thousand or so sporozoites. They travel to the liver within a few minutes, where they are no longer exposed to the antibodies circulating in the blood. If even one sporozoite reaches the liver, it will multiply rapidly there and still cause malaria. The number of

malaria parasites increases roughly eightfold every 24 hours after they enter the body of the host. A compound vaccination against sporozoites, merozoites, and gametocytes would probably be the most effective preventive measure, and such a vaccine is now under development. No greater achievement for molecular biology could be imagined than the control of malaria, a disease that causes so much misery throughout the world.

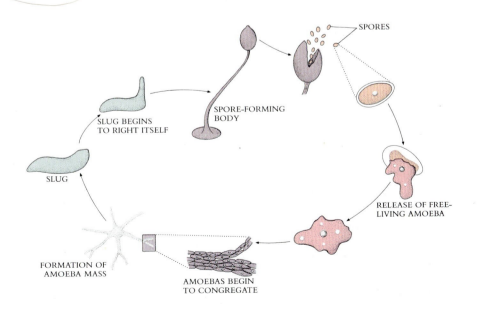

SPORES

SPORE-FORMING BODY

SLUG BEGINS TO RIGHT ITSELF

SLUG

FORMATION OF AMOEBA MASS

AMOEBAS BEGIN TO CONGREGATE

RELEASE OF FREE-LIVING AMOEBA

FIGURE 27-14

Cellular slime mold life cycle.

the cellular slime molds are probably more closely related to the amoebas, phylum Rhizopoda, than to any other group, but they have many special features that mark them as distinct. Cellular slime molds are common in fresh water, in damp soil, and on rotting vegetation, especially on fallen logs. They have become one of the most important groups of organisms for studies of differentiation, because of their relatively simple developmental systems and the ease of analyzing them.

The individual organisms of this group behave as separate amoebas, moving through the soil or other substrate and ingesting bacteria and other smaller organisms. At a certain phase of their life cycle, the individual organisms aggregate and form a moving mass that eventually transforms itself into a spore-forming body, or **sporangium,** within which the amoebas persist as individual cysts. Meiosis occurs within these cysts, leading to the formation of gametes; the gametes may fuse, starting the life cycle again.

The development of *Dictyostelium discoideum,* a cellular slime mold, has been studied extensively because of the implications of its unusual life cycle for understanding the developmental process in general. In it, apparently identical amoebas inhabiting a given area come together to form a unified mass that displays some of the features that characterize a multicellular organism. Later, this mass differentiates into several different kinds of cells. How these processes are accomplished is still not entirely clear.

When the individual amoebas of this species exhaust the supply of bacteria in a given area and are near starvation, they aggregate and form a compound, motile mass. This mass then migrates to a new habitat, where it develops and releases spores. The aggregation of the individual amoebas is induced by pulses of cyclic adenosine monophosphate (cAMP), which the cells begin to secrete when they are starving. In the new habitat the colony transforms itself into a basal portion, a stalk, and a terminal swollen portion, within which spores differentiate. Each of these spores, if it falls into a suitably moist habitat, releases a new amoeba, which begins to feed, and the cycle is started again.

Individual amoebas of *Dictyostelium,* one of the cellular slime molds, aggregate in response to pulses of cyclic adenosine monophosphate when they are starving and migrate as a mass to a new area. There they form a multicellular sporangium, within which spores differentiate.

Plasmodial Slime Molds

The plasmodial slime molds (phylum Myxomycota) are a group of about 450 species. These bizarre organisms stream along in a **plasmodium,** which resembles a moving mass of slime and in which individual cells are indistinguishable (Figure 27-1, *C*). The plasmodia may be orange, yellow, or another color. Plasmodia show a back-and-forth pulsating of protoplasm, a flow that is very conspicuous, especially under a microscope. They are able to pass through the mesh in cloth, or simply to flow around or through other obstacles. As they move, they engulf and digest bacteria, yeasts, and other small particles of organic matter. Plasmodia contain many nuclei, but these are not separated by cell walls; each nucleus undergoes mitosis independently, with the nuclear envelope breaking down.

When either food or moisture is in short supply, the plasmodium migrates relatively rapidly to a new area. Here it stops moving and either forms a mass in which spores differentiate or divides into a large number of small mounds, each of which produces a single, mature sporangium, the structure in which spores are produced. These sporangia are often extremely complex in form and beautiful (Figure 27-15). In many species, crystals of calcium carbonate are deposited in the sporangium or in the walls of the spores that are formed within these sporangia. These spores are diploid, like the nuclei within the plasmodium. After the spores shed, meiosis occurs within them, leading to the production of four haploid nuclei each. Three of the four nuclei in each spore disintegrate, leaving each spore with a single haploid nucleus.

FIGURE 27-15

Sporangia of three genera of plasmodial slime molds.

A Arcyria.
B Fuligo.
C Tubifera.

A

B

C

The spores are highly resistant to unfavorable environmental influences and may last for years if dry. When conditions are favorable, they split open and release their **protoplast,** the protoplasm of the individual spore; the protoplast may be amoeboid or bear two flagella. These two stages appear to be interchangeable, and conversions in either direction occur readily. Later, after the fusion of haploid protoplasts (gametes), a usually diploid plasmodium may be reconstituted by repeated mitotic divisions.

> **The plasmodial slime molds consist of a moving mass of slime called a plasmodium, within which individual cells are not distinguishable; a plasmodium can flow through a silk mesh and rejoin. If the plasmodium begins to dry out or is starving, it forms often elaborate sporangia. Meiosis occurs in the spores once they have been shed.**

Zoomastigotes

The zoomastigote protozoa (phylum Zoomastigina) are unicellular, heterotrophic organisms that are highly variable in form (Figure 27-16). Each has at least one flagellum, and some species have thousands. Both free-living and parasitic organisms are included in this large phylum. Many zoomastigotes apparently reproduce only asexually, but sexual reproduction by gametes is known in some. Members of one of the included groups, the amoebaflagellates, alternate between an amoeboid stage and a flagellated one, depending on environmental conditions. The members of another group, the trypanosomes, include the genera *Trypanosoma* (Figures 27-16, *C,* and 27-17, *A*) and *Crithidia,* which are important pathogens of human beings and domestic animals.

Among the diseases for which the trypanosomes are responsible are trypanosomiasis (sometimes called "sleeping sickness"), East Coast fever, and Chagas' disease, all of great importance in tropical areas. The trypanosomes that cause these diseases are spread by biting insects, including tsetse flies (Figure 27-17, *B*) and assassin bugs. A serious effort is now under way to produce a vaccine for trypanosome-caused diseases, which deprive a large portion of Africa of meat and milk from domestic cattle and thus pose a serious obstacle to the alleviation of hunger there with high-quality food. With their complex life cycles and ability to change their surface proteins, these organisms are especially difficult to control. In the guts of the flies that spread them, trypanosomes are noninfective; when they are prepared for transfer to a mammalian skin or bloodstream, they acquire a thick coat of glycoprotein antigens that protect them from the host's antibodies. Trypanosomes have the

A

CODOSIGA

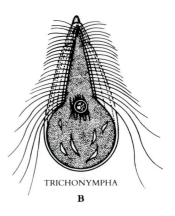

TRICHONYMPHA

B

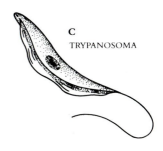

C

TRYPANOSOMA

A 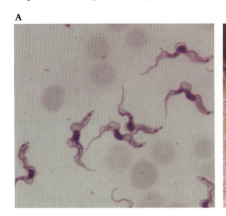 B

FIGURE 27-17

A Trypanosoma, *the zoomastigote that causes sleeping sickness, among red blood cells. The nuclei (dark-staining bodies); anterior flagella; and undulating, changeable shape of the trypanosomes may be seen in this photograph.*

B *Tsetse fly, shown here sucking blood from a human arm, in Tanzania. Sleeping sickness blocks the establishment of productive herds of many breeds of domestic cattle over large areas of Africa.*

FIGURE 27-16

Three genera of zoomastigotes (phylum Zoomastigina), a highly diverse group.

A Codosiga, *a colonial choanoflagellate that remains attached to its substrate; other colonial choanoflagellates swim around as a colony, like* Volvox *(Figure 27-5).*

B Trichonympha, *one of the zoomastigotes that inhabit the guts of termites and the wood-feeding cockroach* Cryptocerus *and digest cellulose there. Trichonympha has rows of flagella in its anterior regions.*

C Trypanosoma, *which causes sleeping sickness, an important tropical disease. It has a single, anterior flagellum.*

FIGURE 27-18

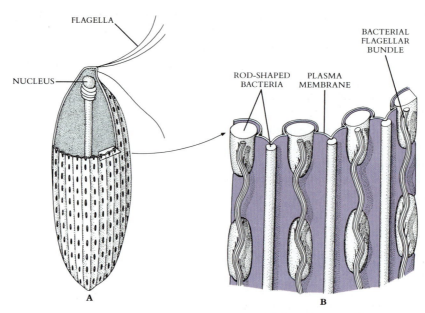

A devescovinid protozoan (A) *from the gut of a Florida termite,* Cryptotermes cavifrons. *Similar protozoa are responsible for the digestion of cellulose by termites and cockroaches throughout the world. This protozoan has four whiplike flagella at its anterior end, three of which lash to and fro while the fourth trails backward like a rudder. The organism largely moves, however, not because of these flagella but because of the activities of roughly 2000 or 3000 rod-shaped bacteria, 2 to 3 micrometers long, which are attached to its surface* (B). *The bacteria lie end-to-end in more than 100 rows, embedded in grooves on the outer surface of the protozoan. Each of the bacteria has a dozen flagella on its outer side, and these beat in an aligned and synchronized way. These activities result in the rapid gliding of the protozoan, but only when it is in contact with other cells or with a substrate. The "nose cone" of this protozoan rotates continuously, a feature unique in the eukaryotes.*

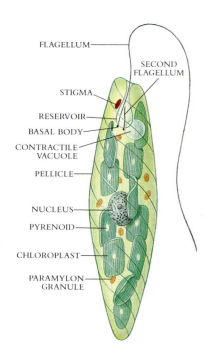

FIGURE 27-19

Diagram of an individual of the genus Euglena *(phylum Euglenophyta).*

ability to alter this coat more rapidly than the immune systems of cattle and humans can respond; as a result, the trypanosomes maintain their ability to infect their hosts successfully (Chapter 18). When they are taken up by a fly, the trypanosomes again shed their coats. Trypanosomes have now been cultured successfully in both their infective and noninfective stages, which makes it much easier to work with them for the development of vaccines. Field tests of some of the prototype vaccines are now under way.

Other zoomastigotes (Figures 27-16, *B,* and 27-18) occur in the guts of termites and other wood-eating insects; they possess the enzymes that allow them to digest the wood and thus make it available to their hosts. The relationship is similar to that between certain bacteria and protozoa that function in the rumens of cattle and related mammals (see Chapters 26 and 43).

Another group of zoomastigotes, the choanoflagellates (Figure 27-16, *C*), are certainly the group from which the sponges (phylum Porifera), and probably all other animals, were derived (see Chapter 35). Choanoflagellates have a single emergent flagellum surrounded by a funnel-shaped, contractile collar composed of closely placed filaments, a unique structure that is exactly matched in the sponges. These protists feed mostly on bacteria, which are pulled in a column of water into the collar and strained out of the water there.

Euglenoids

Most of the approximately 800 known species of euglenoids (phylum Euglenophyta) live in fresh water. The members of this phylum clearly illustrate the impossibility of distinguishing "plants" from "animals" among the protists. About a third of the approximately 40 genera of euglenoids have chloroplasts and are fully autotrophic; the others lack chloroplasts, ingest their food, and are heterotrophic. These organisms are not significantly different from some groups of zoomastigotes, and many biologists believe that the two phyla should be merged into one.

Some of the euglenoids that have chloroplasts may also lose them and thus become heterotrophic if the organisms are kept in the dark. If they are put back in the light, they may become green within a few hours. In one species of euglenoid, the chloroplasts may be removed permanently by treatment with antibiotics or ultraviolet light. Even without such drastic treatments, the normally photosynthetic euglenoids may sometimes feed on dissolved or particulate food.

Individual euglenoids range from only 10 to 500 micrometers long and are highly variable in form. Interlocking proteinaceous strips arranged in a helical pattern form a flexible structure called the **pellicle,** which lies within the cell membrane of the euglenoids. Since its pellicle is flexible, a euglenoid is able to change its shape. Reproduction in this phylum occurs by mitotic cell division. The nuclear envelope remains intact throughout the process of mitosis. No sexual reproduction is known to occur in this group.

In *Euglena* (Figure 27-19), the genus for which the phylum is named, two flagella are attached at the base of a flask-shaped opening called the **reservoir,** which is located at the anterior end of the cell. Contractile vacuoles (p. 156) collect excess water from all parts of the organism and empty it into the reservoir, which apparently functions in regulating the osmotic pressures within the organism. One of the flagella is long and has a row of very fine, short, hairlike projections along one side. A second, shorter flagellum is located within the reservoir but does not emerge from it.

Cells of *Euglena* contain numerous small chloroplasts. These chloroplasts, like those of the green algae and plants, contain chlorophylls *a* and *b,* together with carotenoids. Although the chloroplasts of the euglenoids differ somewhat in structure from those of the green algae, they probably had a common origin, as discussed on p. 538.

> **Euglenoids, phylum Euglenophyta, consist of about 40 genera, about a third of which have chloroplasts similar biochemically to those of green algae and plants. Euglenoids probably acquired these chloroplasts independently, because their structure closely resembles that of certain zoomastigotes and is very different from green algae and plants.**

Brown Algae

The brown algae (phylum Phaeophyta), consisting of about 1500 species of multicellular protists, occur almost exclusively in the sea. They are the most conspicuous seaweeds in many northern regions, dominating rocky shores almost everywhere in temperate North America and Eurasia. In the habitats where the larger brown algae, and in particular a group known as the kelps (order Laminariales), occur abundantly, they are responsible for most of the food production through photosynthesis. Many kelps are conspicuously differentiated into flattened blades, stalks, and grasping basal portions that anchor them to the rocks (Figures 27-1, *B,* and 27-3, *A*). The organic matter that these kelps produce supports the fishes and invertebrates, as well as some marine mammals and birds, that live among these seaweeds and feed on them or on other animals that do so (Figure 27-20).

Among the larger brown algae are genera such as *Macrocystis,* in which some individuals may reach 100 meters in length (Figure 27-1, *B*). The flattened blades of this kelp float out on the surface of the water, while the base is anchored tens of meters below the surface. Another ecologically important member of this phylum is sargasso weed, *Sargassum,* which forms huge floating masses that dominate the vast Sargasso Sea, an area of the Atlantic Ocean northeast of the Caribbean. The stalks of the larger brown algae often exhibit a complex internal differentiation of conducting tissues analogous to that of plants. The chloroplasts of brown algae resemble those of diatoms and dinoflagellates in having chlorophylls *a* and *c.* All of these chloroplasts were probably derived from a symbiotic bacterium that resembled *Heliobacterium.*

The life cycle of the brown algae is marked by alternation of generations between a diploid phase, or **sporophyte,** and a haploid phase, or **gametophyte.** The large individuals that we recognize, for example, as the kelps, are the sporophytes. The gametophytes—the individuals of the haploid phase—are often much smaller, filamentous individuals, perhaps a few centimeters across. Sporangia, in which haploid, swimming spores are produced after meiosis, are formed on the sporophytes. These spores divide by mitosis, giving rise to individual gametophytes. There are two kinds of gametophytes in the kelps; one produces sperm and the other produces eggs. If sperm and eggs fuse, the resulting zygotes grow into the mature kelp sporophytes, provided that they reach a favorable site.

FIGURE 27-20

Brown algae (phylum Phaeophyta). The massive "groves" of giant kelp that occur in relatively shallow water along the coasts of the world provide food and shelter for many different kinds of organisms.

The life cycle of brown algae is marked by an alternation of generations between the diploid phase, or sporophyte, and the haploid phase, or gametophyte, in which the individuals are much smaller than the sporophytes in many species. The sporophytes produce spores after meiosis, and the spores develop into gametophytes by mitosis. Gametes—eggs and sperm—are borne on two different kinds of gametophytes. They fuse, producing a zygote, the first cell of the sporophyte generation.

Some of the larger kelps form extensive beds that are harvested commercially for sodium and potassium salts, iodine, and alginates (carbohydrates that are used in the formation of gels). The possibility of using kelp, which grows continuously and produces large amounts of organic material, as an inexpensive source of fuel is also being investigated.

Diatoms

The diatoms (phylum Bacillariophyta) are photosynthetic, unicellular organisms with unique double shells, made of opaline silica, which are often strikingly and characteristically marked (Figure 27-21). The shells of diatoms are like small boxes with lids, one half of the shell fitting inside the other. The chloroplasts of these organisms contain chlorophylls *a* and *c*, as well as carotenoids. In many respects, these chloroplasts resemble the bacterium *Heliobacterium chlorum* (p. 526), as do the chloroplasts of the brown algae and those of the dinoflagellates. The chloroplasts of diatoms, brown algae, and dinoflagellates were probably derived from an ancestor similar to *Heliobacterium chlorum*. The symbiotic event that led to the origin of their chloroplasts probably occurred at least once independently in each of the three phyla involved, judged by the great differences among the three eukaryotic phyla. In addition to having chloroplasts, some diatoms are facultative heterotrophs; they are able to absorb carbon-containing molecules.

There are more than 11,500 living species of diatoms, with many more known in the fossil record. The shells of fossil diatoms often form very thick deposits, which are sometimes mined commercially. The resulting "diatomaceous earth" is used as an abrasive or

FIGURE 27-21

Diatoms (phylum Bacillariophyta).
A *Scanning electron micrograph of a diatom, showing most of the inner shell and the edge of the outer one.*
B *A pennate (bilaterally symmetrical) diatom, with a colonial member of the same phylum.*
C *Several different kinds of diatoms.*

A

B

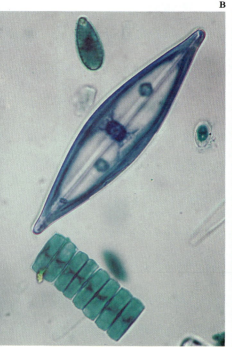

C

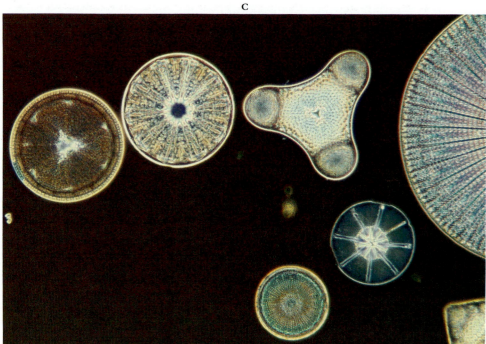

to add the sparkling quality to the paint used on roads, among other purposes. Living diatoms are often abundant both in the sea and in fresh water, where they are important food producers. Diatoms occur both in the plankton and attached to submerged objects in relatively shallow water. Many species are able to move by means of a secretion that is produced from a fine groove along each shell. The diatoms exude and perhaps also retract this secretion as they move.

There are two major groups of diatoms, one with radial symmetry (like a wheel) and the other with bilateral (two-sided) symmetry. Diatom shells are rigid, and the organisms reproduce asexually by dividing the two halves of the shell, each half then regenerating another half shell within it. Because of this mode of reproduction, there is a tendency for the shells and consequently the individual diatoms to get smaller and smaller with a given sequence of asexual reproduction. When the resulting individuals have diminished to about 30% of their original size, one may slip out of its shells, grow to full size, and regenerate a full-sized pair of new shells. Cell division in a diatom is shown on p. 147.

Individual diatoms are diploid. Meiosis occurs more frequently under conditions of starvation. Among marine diatoms, some individuals then produce numerous sperm and others produce a single egg. If fusion occurs, the resulting zygote regenerates a full-sized individual. In some freshwater diatoms, the gametes are amoeboid, and all of them appear similar.

Green Algae

Green algae (phylum Chlorophyta) are of special interest because they are believed to be the ancestors of the plants. Their chloroplasts are biochemically similar to those of the plants; they contain chlorophylls *a* and *b,* as well as carotenoids, and are similar to the bacterium *Prochloron* in these respects. It seems probable that a bacterium with characteristics similar to those of *Prochloron* gave rise to the chloroplasts of the originally heterotrophic ancestor of the green algae (see pp. 527 and 538).

Green algae are an extremely varied group of more than 7000 species, mostly aquatic but also semiterrestrial in moist places, such as on tree trunks or in soil. Many of these algae are microscopic and unicellular, but some, such as sea lettuce, *Ulva* (Figure 27-3, *B*), are tens of centimeters across and easily visible on rocks and pilings around the coasts.

Among the unicellular green algae, *Chlamydomonas* (Figure 27-22) is a well-known genus. Individuals are microscopic—usually less than 25 micrometers long—green, and rounded, and have two flagella at the anterior end. They move rapidly in water as a result of the beating of their flagella in opposite directions. Each individual has an eyespot, which contains about 100,000 molecules of rhodopsin, the same pigment that is used in the vision of vertebrates. Light received by this eyespot is used by the alga to help direct its swimming. Most individuals of *Chlamydomonas* are haploid. *Chlamydomonas* reproduces both asexually, by cell division, and sexually, by the functioning of some of the products of cell division as gametes, which fuse to form a four-flagellated zygote that ultimately enters a resting phase in which the flagella disappear. Meiosis occurs at the end of this resting period and results in the production of four haploid cells.

From organisms like *Chlamydomonas* there have been several lines of evolutionary specialization. The first is the evolution of nonmotile unicellular green algae. Even *Chlamydomonas* itself is capable of retracting its flagella and settling down as an immobile, unicellular organism if the ponds in which it lives dry out. Some common algae of soil and bark, such as *Chlorella,* are essentially like *Chlamydomonas* in this condition but do not have the ability to form flagella. *Chlorella* is widespread both in fresh and salt water and in soil and is only known to reproduce asexually. Recently, it has been widely investigated as a possible food source for humans and other animals, and pilot farms have been established in Israel, the United States, Germany, and Japan.

The second major line of specialization from cells like *Chlamydomonas* concerns the formation of motile, colonial organisms. In these green algae, the individual *Chlamydo-*

FIGURE 27-22

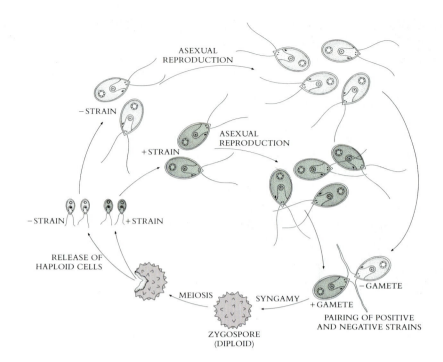

Life cycle of Chlamydomonas *(phylum Chlorophyta). Individual cells of this microscopic, biflagellated alga, which are haploid, divide asexually, producing identical copies of themselves. At times, such haploid cells act as gametes— fusing, as shown at the lower right-hand side of the diagram, to produce a zygote. The zygote develops a thick, resistant wall, becoming a zygospore. Within this diploid zygospore, meiosis takes place, ultimately resulting in the release of four haploid individuals. Because of the segregation during meiosis, two of these individuals are what are called the + strain, the other two the − strain. Only + and − individuals are capable of mating with one another when syngamy does take place, although both may divide asexually and reproduce themselves in that way also.*

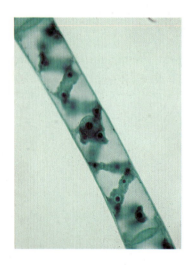

FIGURE 27-23

Spirogyra, a green alga with a filamentous growth form, consisting of strands of interconnected cells. The ribbonlike structure is the chloroplast; dotlike, starch-forming bodies called pyrenoids can be seen at intervals along its length. The nucleus is visible in the center of the cell.

monas-like cells retain some of their individuality. The most highly elaborated of these organisms is *Volvox* (Figure 27-5), a hollow sphere made up of a single layer of 500 to 60,000 individual cells, each provided with two flagella. Only a small number of the cells are reproductive. The flagella of all of the cells beat in such a way as to rotate the colony, which has a definite anterior and posterior end, in a clockwise direction as it moves forward through the water. The reproductive cells of *Volvox* are located mainly at the posterior end of the colony. Some may divide asexually, bulge inward, and give rise to new colonies that initially are held within the parent colony. Others produce gametes. In some species of *Volvox* there is a true "division of labor" among the different types of cells, which are specialized in relation to their ultimate function throughout the development of the organism; in these species, therefore, true multicellularity exists.

From green algae like *Chlamydomonas*—a biflagellated, unicellular organism—have been derived nonmotile, unicellular algae and multicellular, flagellated colonies.

In addition to these two lines of specialization from *Chlamydomonas*-like cells, there are many other kinds of green algae of less certain derivation. Many filamentous genera, such as *Spirogyra,* with its ribbonlike chloroplasts (Figure 27-23), differ substantially from the remainder of the green algae in their modes of cell division and reproduction. Some of these genera have even been placed in separate phyla. The study of the green algae, involving modern methods of electron microscopy and biochemistry, is beginning to reveal unexpected new relationships with this phylum.

Ulva, or sea lettuce (Figure 27-3, *B*), is a genus of marine green algae that is extremely widespread. The glistening individuals of this genus, often about a decimeter or more across, consist of undulating sheets only two cells thick. Sea lettuce attaches by protuberances of the basal cells to rocks or other substrates on which it occurs. *Ulva* has an alternation of generations in which the gametophytes and sporophytes resemble one another closely.

The stoneworts, a group of about 250 living species of green algae, many of them in the genera *Chara* and *Nitella,* have complex structures (Figure 27-24). Whorls of short

branches arise regularly at their nodes, and the **gametangia** (sing. **gametangium**)—gamete-producing structures—are complex and multicellular. Stoneworts are often abundant in fresh to brackish water and are common as fossils.

Ciliates

The ciliates (phylum Ciliophora) are a group of about 8000 species of unicellular, heterotrophic protists that range from about 10 to 3000 micrometers long (Figures 7-8, 7-11, 27-1, *A,* 27-2, and 27-25). The members of this phylum, which are bewildering in their morphological complexity, characteristically have large numbers of cilia, mostly arranged either in longitudinal rows or in spirals around the body of the organism (Figure 27-25, *B*). Some ciliates, such as *Paramecium* (Figures 9-3 and 27-1, *A*), are extremely well known.

FIGURE 27-25

Ciliates (phylum Ciliophora). **A** Blepharisma japonica, *showing micronuclei (small round objects) and a macronucleus (elongate structure within cell).* **B** Stentor, *a funnel-shaped ciliate, showing spirally arranged cilia.*

A

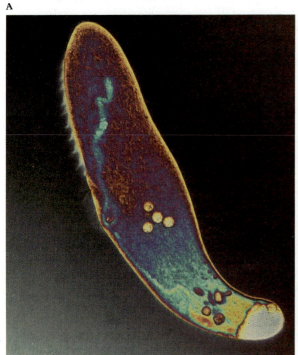

B

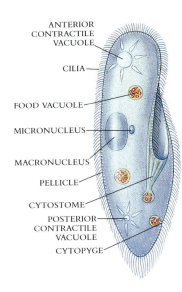

ANTERIOR
CONTRACTILE
VACUOLE

CILIA

FOOD VACUOLE

MICRONUCLEUS

MACRONUCLEUS

PELLICLE

CYTOSTOME

POSTERIOR
CONTRACTILE
VACUOLE

CYTOPYGE

FIGURE 27-26

Diagram of Paramecium.

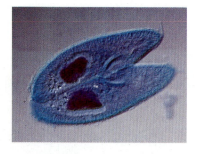

FIGURE 27-27

Paramecium *in conjugation.*

The cilia of this phylum are embedded in an outer proteinaceous layer of the cell. Each cilium ends in a basal body called a **kinetosome.** The kinetosomes are connected by complex, fibrous strands, and the movement of the whole organism is coordinated by means of this system. In some ciliates the cilia have been modified for different specialized locomotory and feeding functions, becoming fused into sheets, which may then function as mouths, paddles, teeth, or feet. The body wall of ciliates is a tough but flexible outer pellicle that enables the organism to squeeze through or move around many kinds of obstacles. The pellicle consists of an outer membrane with numerous fluid-filled cavities beneath it (Figure 27-26).

Ciliates form vacuoles for ingesting food and regulating their water balance. Food first enters the **cytostome** (gullet), which in *Paramecium* is lined with cilia fused into a membrane. After entering the cytostome the food passes into food vacuoles, where enzymes and hydrochloric acid aid in its digestion. After absorption of the digested material has been completed, the vacuole empties its waste contents in a special region of the pellicle known as the **cytopyge.** The cytopyge appears periodically when solid particles are ready to be expelled. The contractile vacuoles, which function in the regulation of water balance, periodically expand and contract as they empty their contents to the outside of the organism.

Most ciliates that have been studied have two very different types of nuclei within their cells, small micronuclei and larger macronuclei. The micronuclei, which contain apparently normal chromosomes, divide by mitosis. They play the same role as the nuclei in the cells of most other eukaryotes. The macronuclei are derived from certain micronuclei by a complex series of steps. Within the macronuclei, in place of chromosomes, are a large number of small chromatin bodies, each containing hundreds or even thousands of copies of only one or two genes. Macronuclei divide by elongating and constricting and play an essential role in routine cellular functions, such as the production of mRNA to direct protein synthesis.

Ciliates usually reproduce asexually by transverse fission of the parent cell across its short axis, thus forming two equal offspring (see Figure 8-1). In this process of cell division, the mitosis of the micronuclei proceeds normally and the macronuclei divide as we just described.

Most ciliates likewise have a sexual process of reproduction called conjugation (Figure 27-27). Although it occurs infrequently, the process has been of great interest to geneticists. In *Paramecium,* which has been the most carefully studied ciliate, there are up to 20 different mating types, and the cells that are able to conjugate with one another must represent two different types. The conjugating cells remain attached to each other for as long as several hours. Meiosis in the micronuclei of each individual produces several haploid micronuclei, and a pair of these are then exchanged through a cytoplasmic bridge that appears between the two partners.

In each conjugating individual, the new micronucleus fuses with one of the micronuclei that was already present in that individual, resulting in the production of a new diploid micronucleus in each individual. After conjugation, the macronucleus in each cell disintegrates, while the new diploid micronucleus undergoes mitosis, thus giving rise to two new identical diploid micronuclei within each individual. One of these micronuclei becomes the precursor of the future micronuclei of that cell, while the other micronucleus undergoes multiple DNA replication, becoming the new macronucleus. This kind of complete segregation of the genetic material is a unique feature of the ciliates and makes them ideal organisms for the study of certain aspects of genetics.

Conjugating ciliates exchange a pair of haploid micronuclei, which fuse in both cells to form new diploid micronuclei, while the macronucleus in each disintegrates. After their fusion, the new micronuclei divide by mitosis. One of the products of this division gives rise to more micronuclei, while the other undergoes multiple DNA replication and becomes the new macronucleus.

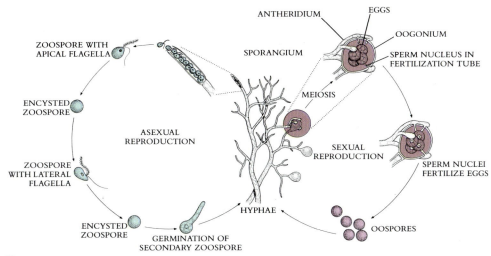

FIGURE 27-28

Life cycle of Saprolegnia, *an oomycete. Asexual reproduction by means of flagellated zoospores is shown at left, sexual reproduction at right. In both cases the resulting forms are tubular filamentous growths called hyphae.*

Oomycetes

The oomycetes (phylum Oomycota) comprise about 475 species. Among them are the water molds, white rusts, and downy mildews. All of the members of the group are either parasites or saprobes, organisms that live by feeding on dead organic matter. The cell walls of the oomycetes are composed of cellulose or polymers that resemble cellulose, and these walls therefore differ remarkably from the chitin cell walls of the fungi, with which the oomycetes have at times been grouped. Most oomycetes live in fresh water or soil, but some are parasites that depend on the wind to spread their spores.

Oomycetes are distinguished from other protists by their motile spores, or **zoospores,** which bear two unequal flagella, one of which is directed forward, the other backward. Such zoospores are produced asexually in a sporangium. Sexual reproduction in the group involves gametangia of two kinds. The female gametangium is called an **oogonium,** and the male gametangium is called an **antheridium.** The antheridia contain numerous male nuclei, whereas the oogonia contain only one or a few eggs. The fusion of the contents of an antheridium with those of an oogonium is followed by the thickening of the cell wall around the resulting zygote, producing a special kind of thick-walled cell called an **oospore,** the structure that gives the phylum its name. Details from the life cycle of one of the oomycetes, *Saprolegnia,* are shown in Figure 27-28.

The aquatic oomycetes, or water molds, are common and easily cultured. Some water molds cause fish diseases, which may be seen as a kind of white fuzz on aquarium fishes. Among their terrestrial relatives are some oomycetes of great importance as plant pathogens, including *Plasmopara viticola,* which causes downy mildew of grapes, and *Phytophthora infestans,* which causes the late blight of potatoes. This oomycete was responsible for the Irish potato famine of 1845 and 1847, during which about 400,000 people starved to death or died of diseases complicated by starvation. Millions of Irish people emigrated to the United States and elsewhere as a result of this disaster.

Red Algae

The characteristic colors of the red algae (phylum Rhodophyta) result from their possession of the same pigments, phycobilins, that are responsible for the colors of the cyanobacteria. Chlorophyll *a* also occurs with the phycobilins, just as it does in cyanobacteria. Undoubtedly cyanobacteria became symbiotic in the cells of the heterotrophic ancestors of the red algae and gave rise to their chloroplasts.

Almost all red algae are multicellular. The great majority of the estimated 4000 species occur in the sea, where they are common (Figures 27-3, *C,* and 27-29). In warm waters, they are more common than brown algae. Phycobilins are especially efficient in absorbing the green, violet, and blue light that penetrates into the deepest waters. For this reason red

FIGURE 27-29

A coralline alga (phylum Rhodophyta). In the coralline algae, the cellulose cell walls are heavily impregnated with calcium carbonate. Other species of coralline algae contribute greatly to overall food production in coral reefs, but members of the phylum, such as the one shown here, also occur widely elsewhere.

algae are able to grow at greater depths than are brown algae or green algae. In 1985, a red alga was reported growing attached to rocks 268 meters below the surface of the sea in the Bahamas, almost 100 meters below the limits to which sunlight penetrates. This record depth is also more than 100 meters below the greatest depths at which any other algae have been reported.

Red algae have complex bodies made up of interwoven filaments of cells. Some, the coralline algae, deposit calcium carbonate in their cell walls, which otherwise are mostly cellulose. In the cell walls of other red algae there is a mucilaginous outer component usually composed of sulfated polysaccharides such as agar and carrageenan, which make these algae important economically. Agar is used to make gelatin capsules, as a material for making dental impressions, and as a base for cosmetics. It is also the basis of the laboratory media on which bacteria, fungi, and other organisms are often grown. In addition, agar is used to prevent baked goods from drying out, for rapid-setting jellies, and as a temporary preservative for meat and fish in warm regions. Carrageenan is used mainly for the stabilization of emulsions such as paints, cosmetics, and dairy products such as ice cream. In addition to these uses, red algae such as *Porphyra,* called "nori," are eaten and, in Japan, are even cultivated as a food crop for human consumption (Figure 27-30).

The life cycles of red algae are complex but usually involve an alternation of generations like that of the brown algae. None of the red algae have flagella or cilia at any stage in their life cycle, and they may have descended directly from ancestors that never had them, especially since the red algae also lack centrioles. Together with the fungi, which also lack flagella and centrioles, red algae may be one of the most ancient groups of eukaryotes.

THE MOST DIVERSE KINGDOM OF EUKARYOTES

By now, you are probably convinced of the diversity of the protists. The 15 phyla we have considered in this chapter consist mostly of microscopic, unicellular organisms, but there are many multicellular ones that have been derived from them. In their ways of life, the habitats in which they occur, and the details of their life cycles, the protists are extraordinarily diverse.

Among the phyla that make up this great group, we are especially interested in those that apparently gave rise to the other exclusively or predominantly multicellular evolutionary lines that we call animals, plants, and fungi. The choanoflagellates, one of the groups of zoomastigotes, appear certainly to be the ancestors of the sponges and probably of all animals; the fundamental similarity between their unique structure and that of the collar cells of sponges, which we will describe in Chapter 35, is simply too great to be explained in any other way. Green algae, which share with plants their photosynthetic pigments, cell wall structure, and chief storage product (starch), doubtless include the ancestors of plants, a kingdom that achieved its distinctive features in connection with its invasion of the land. With regard to the third kingdom of predominantly multicellular organisms, the fungi, we will see in Chapter 28 that their ancestry is not clear.

Other groups of protists did not give rise to other kingdoms of organisms, but several of them, notably the brown algae and red algae, have achieved multicellularity. Taken as a whole, the protists are extremely diverse and, as we learn more about the details of their structures and their lives, endlessly fascinating.

FIGURE 27-30

Commercial harvesting of a red alga cultivated on bamboo rafts submerged in deep seawater, off the coast of Japan.

SUMMARY

1. The kingdom Protista consists of the exclusively or predominantly unicellular phyla of eukaryotes, together with three phyla that include large numbers of multicellular organisms: the red algae, brown algae, and green algae. The phyla of protists are highly diverse.

2. Chloroplasts originated a number of times among the protists, and the process seems to have involved the members of at least three different groups of bacteria: cyanobacteria (in the red algae); *Prochloron*-like organisms (in the green algae and euglenoids); and *Heliobacterium chlorum*–like organisms (in the diatoms, dinoflagellates, and brown algae).

3. The three major multicellular groups of eukaryotes—plants, animals, and fungi—all originated from protists, but they are so large and important that they are considered distinct kingdoms. They are not related directly to one another. The plants originated from green algae; at least the sponges, and probably all animals, originated from the choanoflagellates, one of the groups of zoomastigotes. The ancestors of the fungi are unknown.

4. True multicellularity and sexuality are both exclusively properties of the eukaryotes. Multicellularity confers a degree of protection from the environment and the ability to carry out a wider range of activities than is available to unicellular organisms. Sexuality permits extensive, orderly, genetic reproduction.

5. *Pelomyxa palustris,* an amoeba-like organism, lacks both mitosis and mitochondria and apparently represents a very early stage in the evolution of the eukaryotes. Many zoomastigotes likewise lack mitochondria.

6. Dinoflagellates (phylum Dinoflagellata) are a major phylum of unicellular organisms that have unique chromosomes and a very unusual form of mitosis that takes place entirely within the nucleus. They are responsible for many of the poisonous red tides. In the symbiotic form known as zooxanthellae, dinoflagellates are widespread and are responsible for most of the productivity of coral reefs and of many marine organisms.

7. The malarial parasite, *Plasmodium,* is a member of the phylum Sporozoa. Carried by mosquitoes, it multiplies rapidly in the liver of humans and other primates and brings about the cyclical fevers characteristic of malaria by releasing toxins into the bloodstream of its host.

8. Plasmodial slime molds move about as a plasmodium, containing numerous nuclei, within which individual cells cannot be distinguished. When drying out or starving, plasmodial slime molds form sporangia within which spores that undergo meiosis are formed.

9. Euglenoids have chloroplasts that share the biochemical features of those found in green algae and plants; independent symbiotic events probably resulted in these similarities. With their flexible proteinaceous pellicles, euglenoids closely resemble certain zoomastigotes, a group of heterotrophic, mostly unicellular protists that includes the organism responsible for sleeping sickness.

10. Brown algae are multicellular, marine protists, some reaching 100 meters in length. The kelps—larger brown algae—contribute greatly to the productivity of the sea, especially along the coasts in relatively shallow areas. Kelps have an alternation of generations in which the gametophytes are very small and filamentous and the sporophytes are large and evident.

11. Green algae are a highly diverse group of organisms that are abundant in the sea, fresh water, and damp terrestrial habitats, such as on tree trunks and in soil. Plants were derived from a multicellular green alga.

12. Ciliates are a very large phylum of protists, with about 8000 described species. They have a very complex morphology with numerous cilia. They also have a life cycle that involves micronuclei, which function like the nuclei of other organisms, and macronuclei, which contain a large number of small chromatin bodies.

13. Red algae, which have chlorophyll *a* and phycobilins, have more species than the other two phyla of seaweeds—about 4000—and occur at greater depths, down to more than 250 meters. They have very complex life cycles involving an alternation of generations. Their chloroplasts were almost certainly derived from symbiotic cyanobacteria. Red algae lack flagellated cells and centrioles and may be descendants of one of the most ancient lines of eukaryotes.

SELF-QUIZ

The following questions pinpoint important information in this chapter. You should be sure to know the answers to each of them.

1. Protista is the kingdom of eukaryotes that do not have the special characteristics of plants, animals, or fungi. True or false?

2. Protists are considered the ancestors of
   (a) fungi   (c) plants   (e) all of the above
   (b) animals   (d) two of the above

3. The groups of photosynthetic bacteria that are thought to have become chloroplasts after entering into symbiotic relationships with different groups of eukaryotes are _____, _____, and _____.

4. The organism *Pelomyxa* and some of the zoomastigotes do not contain mitochondria. Are they eukaryotes?

5. Is true multicellularity a property of eukaryotes alone?

6. Individual protist life cycles may contain
   (a) zygotic meiosis   (c) sporic meiosis   (e) all of the above
   (b) gametic meiosis   (d) alternation of generations

7. The existence of one type of prokaryote is the primary factor responsible for the ability of tropical coral reefs to grow in size. True or false?

8. _____ are multicellular structures in which haploid cells are produced as a result of meiosis.

9. Fusion of an egg and a sperm creates a _____, which is a diploid cell.

10. Sporozoa, which are nonmotile, spore-forming parasites of animals, can undergo a process known as _____, a rapid series of mitotic events that results in the production of a large number of small, infective individuals.

11. Micronuclei are found in most ciliates and
    (a) contain apparently normal chromosomes   (d) divide by elongation
    (b) contain mutated chromosomes that are nonfunctional   and constriction
    (c) contain little bodies of chromatin, each containing hundreds   (e) none of the above
        or thousands of copies of only one or two genes

12. Alternation of generations means that one generation is male, the next generation is female, and so forth. True or false?

THOUGHT QUESTIONS

1. If plants were derived from green algae, why don't we classify green algae as plants in this book?

2. Different phyla of protists consist of organisms that have different numbers and kinds of flagella. Enumerate some of these different conditions and suggest how one may have given rise to another.

3. If mitochondria and chloroplasts originated as symbiotic bacteria, what would you suggest were the characteristics of the organism in which they became symbiotic?

4. List the evidence for and against the hypothesis that multicellularity arose only once.

FOR FURTHER READING

ALEXOPOULOS, C.J., and C.W. MIMS: *Introductory Mycology,* 3rd ed., John Wiley & Sons, Inc., New York, 1979. A standard account of the fungi, including treatments of some phyla treated here as protists: Acrasiomycota, Myxomycota, and Oomycota.

DAWES, C.J.: *Marine Botany,* John Wiley & Sons, Inc., New York, 1981. A comprehensive discussion of the features of photosynthetic marine protists, ecological in orientation.

DONELSON, J.E., and M.J. TURNER: "How the Trypanosome Changes Its Coat," *Scientific American,* February 1985, pages 44-51. Trypanosomes survive in the bloodstream by evading the immune system; they do this by switching on new genes encoding new surface antigens.

GODSON, G.N.: "Molecular Approaches to Malaria Vaccines," *Scientific American,* May 1985, pages 52-59. This article chronicles recent advances in the important quest for an effective malaria vaccine.

HICKMAN, C.P., Jr., L.S. ROBERTS, and F.M. HICKMAN: *Integrated Principles of Zoology,* 7th ed., Times Mirror/Mosby College Publishing, St. Louis, 1984. The outstanding treatment of general zoology, including the heterotrophic phyla discussed in this chapter.

KOSIKOWSKI, F.V.: "Cheese," *Scientific American,* May 1985, pages 88-99. A fascinating account of the ways in which more than 2000 varieties of cheese are made, and of the participation of bacteria and fungi in the process.

MARGULIS, L.: *Symbiosis in Cell Evolution,* W.H. Freeman & Co., San Francisco, 1980. Outstanding treatment of the origin of eukaryotic cells by serial symbiosis.

MARGULIS, L., and K.V. SCHWARTZ: *Five Kingdoms: An Illustrated Guide to the Phyla of Life on Earth,* W.H. Freeman & Co., San Francisco, 1982. Excellent synoptic treatment of all phyla of protists, in the context of other living organisms.

RAPER, K.B.: *The Dictyostelids,* Princeton University Press, Princeton, N.J., 1984. An up-to-date account of these fascinating protists, written from a developmental point of view.

SCAGEL, R.F., R.J. BANDONI, J.R. MAZE, G.E. ROUSE, W.B. SCHOFIELD, and J.R. STEIN: *Nonvascular Plants: An Evolutionary Survey,* Wadsworth Publishing Co., Belmont, Calif., 1982. A thorough and well-illustrated treatment of the diversity of photosynthetic protists, plants, and fungi.

SLEIGH, M.: *The Biology of Protozoa,* University Park Press, Baltimore, Md., 1975. A general book on the protists, including most of the phyla discussed in this chapter.

Overview

The filaments of fungi grow through soil, wood, and other substrates, which are included, in a sense, within the fungi. By secreting enzymes, fungi digest organic matter and absorb the products of this external digestion. Together with the bacteria, fungi are the major decomposers of the biosphere, breaking down organic molecules and making the material in them available for recycling. This process is as necessary for the continuation of life on earth as photosynthesis is. Cytoplasm, with included nuclei and other organelles, streams through the bodies of fungi; their filaments either lack divisions completely, except when reproductive structures are formed, or have only perforated ones. Thus fungi are not multicellular in the same sense that plants and animals are. They are a unique kingdom of organisms that have no evident affinities with any other group.

For Review   *Here are some important terms and concepts that you will encounter in this chapter. If you are not familiar with them, you should review them before proceeding.*

- **Classification** (Chapter 22)
- **Major features of evolutionary history** (Chapter 23)
- **Diversity of protists** (Chapter 27)

The fungi are a distinct kingdom of organisms, comprising about 100,000 named species, but **mycologists,** scientists who study them, believe that there may be several times more actually in existence. Although fungi have traditionally been included in the plant kingdom, they lack chlorophyll and resemble plants only in lacking mobility and growing from the ends of somewhat linear bodies (Figure 28-1). Even these similarities prove to be misleading when they are examined closely, however. Some plants have motile sperm with flagella; no fungi do. Fungi are basically filamentous in their growth form (they consist of slender filaments), whereas plants are three dimensional. Unlike plants, fungi obtain their food by secreting enzymes into their substrate and absorbing the materials that these enzymes make available. Fungi are an ancient group of organisms, no less than 400 million years old, and possibly much older.

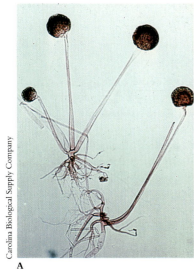

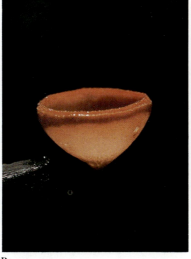

A

B

C

Carolina Biological Supply Company

FIGURE 28-1

Representatives of the three divisions of fungi.
A Rhizopus, *a zygomycete that grows on moist bread and other similar substrates. The individuals are about a centimeter tall.*
B *A cup fungus, a kind of ascomycete, in the rain forest of the Amazon Basin.*
C Amanita muscaria, *the fly agaric, a toxic basidiomycete. In the cup fungi, the spore-producing structures line the cup; in basidiomycetes that form mushrooms, like* Amanita, *they line the gills beneath the cap of the mushroom. All visible structures of fleshy fungi like the ones shown here arise from an extensive network of filamentous hyphae that penetrates and is interwoven with the substrate on which they grow.*

Many fungi are harmful, because they decay, rot, and spoil many different materials as they obtain food and cause serious diseases of plants and animals, including human beings. Others, however, are extremely useful. The manufacture of both bread and beer depends on the biochemical activities of yeasts, single-celled fungi that produce abundant quantities of ethanol and of carbon dioxide (Figure 28-2). Both cheese and wine achieve their delicate flavors because of the metabolic processes of certain fungi, and others make possible the manufacture of such Oriental delicacies as soy sauce and tofu. Vast industries depend on the biochemical manufacture of organic substances such as citric acid by fungi in culture, and yeasts are now being used on a large scale to produce protein for the enrichment of animal food. Many antibiotics, including the first one that was used on a wide scale, penicillin, are derived from fungi. Other fungi are used on a major scale to convert complex organic molecules into others, such as in the synthesis of many commercially important steroids. Fungi are of great importance to us in both harmful and beneficial ways.

> **Fungi absorb their food after digesting it by secreting enzymes. This unique mode of nutrition, combined with a filamentous growth form, makes the members of this kingdom highly distinctive.**

FUNGAL ECOLOGY

Fungi, together with bacteria, play a role of enormous importance as decomposers in the biosphere. They break down organic materials and return the substances locked in those molecules to circulation in the ecosystem. In this way, critical biological building blocks—such as compounds of carbon, nitrogen, and phosphorus—that have been incorporated into the bodies of living organisms are released and made available for other organisms.

In breaking down organic matter, fungi may make no distinction, for example, between a tree that is growing in the forest and one that has fallen, or a dead animal and one that is living. Fungi are thus often important as disease-causing organisms for both plants and animals; they are responsible for billions of dollars in agricultural losses every year. Not only are fungi the most harmful pests of living plants, but they also have important effects on the storage life of food products after the food has been harvested. In addition, fungi often secrete substances into the foods that they are attacking that make these foods unpalatable or even poisonous.

Two kinds of associations between fungi and autotrophic organisms are of critical importance ecologically. **Lichens,** which are prominent nearly everywhere in the world, especially in unusually harsh habitats such as bare rocks, are symbiotic associations between fungi and either algae or cyanobacteria. **Mycorrhizae,** specialized symbiotic associations

FIGURE 28-2

Baking bread involves the metabolic activities of the yeasts, single-celled ascomycetes. Winemaking also uses the metabolic properties of yeasts, and the same species, Saccharomyces cerevisiae, *is usually employed for both processes. In baking bread, it is the generation of carbon dioxide that is important; in making wine, it is the production of alcohol.*

CYCLOSPORINE—A MODERN WONDER DRUG

Fungi manufacture a remarkable arsenal of chemicals, which they employ in their interactions with the environment. The members of this kingdom display an intimate relationship with their environment, obtaining the nutrients they require and protecting themselves solely through the materials that they secrete. Furthermore, in their metabolism the fungi are exceedingly diverse. In view of these properties, it is not surprising that so many drugs and other useful substances are obtained from the fungi or that fungi play such an important role in various industrial processes.

In 1970 microbiologists working at the Sandoz Corporation in Basel, Switzerland, isolated two new strains of Fungi Imperfecti from soil samples that they had obtained from Norway and Wisconsin. Both of these strains produced a substance that was later named cyclosporine, which was first studied because it controlled the growth of some other fungi and was relatively nontoxic for such a metabolically active substance. One of these two fungi, *Tolypocladium inflatum,* was selected as the commercial source of cyclosporine because it grows in submerged culture, an important property for industrial production. After these original

discoveries, cyclosporine was found to be a cyclic molecule consisting of 11 amino acids, one of which is not known to occur anywhere else in nature.

Soon it was discovered that cyclosporine suppresses immunity of the kind that causes organ transplants to be rejected, and that it does not have the harmful side effects associated with the standard drugs used to suppress immune reactions. Cyclosporine causes these effects by inactivating lymphocytes (see Chapter 45), and, best of all, the inactivation ceases when the drug is no longer being administered, so that the body can return to normal. The other drugs used to suppress immune reactions kill bone marrow cells, which are the source of all blood cells; loss of bone marrow cells causes intolerable side effects centering on the greatly increased susceptibility of the patient to disease and infection. Cyclosporine does not damage bone marrow cells.

Before cyclosporine became available in 1979, about one in three patients who had received liver grafts survived for a full year. With the new drug, the percentage of survival has doubled. Up to 90% of kidney transplants now succeeed, as opposed to no more than half just a few years ago. Heart

---

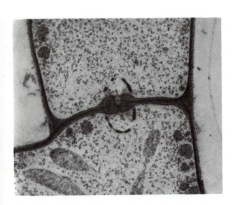

FIGURE 28-3

Transmission electron micrograph of a section through a hypha of the basidiomycete Laetisaria arvalis, *showing the kind of septum characteristic of secondary mycelia (those mycelia formed by the fusion of hyphae of two different mating types) in basidiomycetes. Such septa are lined with thick, barrel-shaped structures, as seen here in transection. The septa of ascomycetes and those of the primary mycelia of basidiomycetes do not have a barrel-shaped structure associated with them.*

between the roots of plants and fungi, are characteristic of about 80% of all plants. Both of these associations will be discussed further in this chapter. In each of them the photosynthetic organisms fix atmospheric carbon dioxide and thus make available a source of organic material to the fungi, and the metabolic activities of the fungi in turn enhance the overall ability of the symbiotic association to exist in a particular habitat. In the case of mycorrhizae, the fungal partner expedites the absorption of essential nutrients such as phosphorus by the plant.

FUNGAL STRUCTURE

Fungi exist mainly in the form of slender filaments, barely visible with the naked eye, which are called **hyphae** (sing. *hypha*). These hyphae may be divided into cells by cross walls called **septa** (sing. *septum*). These septa rarely form a complete barrier, however, except for those separating the reproductive cells. Cytoplasm characteristically flows or streams freely throughout the hyphae, passing right through the major pores in the septa (Figure 28-3). Because of this streaming, proteins, which are synthesized throughout the hyphae, may be carried to their actively growing tips. As a result, the growth of fungal hyphae may be very rapid when food and water are abundant and the temperature is high enough.

A mass of hyphae is called a **mycelium** (pl. *mycelia*). This word and the term *mycologist* are both derived from the Greek word for fungus, *myketos*. The mycelia of fungi (Figure 28-4) constitute a system that may be many kilometers long. This system grows through

transplants, which had essentially been abandoned before cyclosporine was available, are being performed again, and there have been no instances of rejection in those cases in which the drug has been used. The drug is also extremely promising for application in diseases that involve autoimmunity. These diseases, which are known or suspected to include multiple sclerosis, juvenile diabetes, and lupus, have been extremely difficult to treat in the past. Cyclosporine also shows great promise in treating certain diseases caused by parasites, including schistosomiasis (it kills the worms themselves), malaria, and trichinosis. Since the antiparasitic effects of cyclosporine are not dependent on its ability to bring about suppression of immune systems, scientists are searching for derivative molecules that would be able to kill the parasites without causing immune system suppression, which generally is not desirable.

Although research on cyclosporine continues and new applications are being reported almost monthly, scientists are wondering how many additional wonder drugs may be represented among the remarkable chemical arsenal of the fungi. It seems that these organisms have a virtually limitless potential both for harming and for helping human beings.

FIGURE 28-A

1 *Synthetic crystal of cyclosporine.*
2 *Conidia-bearing branches of* Tolypocladium inflatum. *(×4100.)*

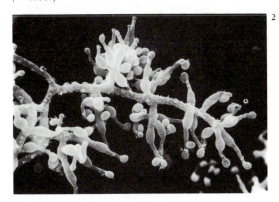

and penetrates the substrate of the fungus, resulting in a unique relationship between a fungus and its environment. All parts of a fungus are metabolically active, continually interacting with the soil, wood, or other material in which the mycelia are growing.

Fungi exist primarily in the form of filamentous hyphae, which are completely divided by septa only when reproductive organs are formed. These hyphae surround and penetrate the substrate within which the fungi are growing.

In two of the three divisions of fungi, reproductive structures formed of interwoven hyphae—such as mushrooms, puffballs, and morels—are produced at certain stages of the life cycle. These structures expand rapidly because of rapid growth of the hyphae. For this reason, mushrooms can appear suddenly in your lawn.

The cell walls of fungi are formed of polysaccharides and chitin, not cellulose like those of plants and many groups of protists. Chitin is the same material that makes up the major portion of the hard shells, or exoskeletons, of arthropods, a group of animals that includes insects and crustaceans (see Chapter 37). It is far more resistant to microbial degradation than is cellulose.

Mitosis in fungi differs from that found in other organisms: the nuclear envelope does not break down and re-form, and the spindle apparatus is formed within it (Figure 8-B). Centrioles are lacking in all fungi, which instead regulate the formation of microtubules

FIGURE 28-4

Fungal mycelium growing through leaves on the forest floor in Maryland.

during mitosis with small, relatively amorphous structures called **spindle plaques.** This unique combination of features strongly suggests that fungi originated from some unknown group of single-celled eukaryotes with these characteristics. There is no indication of a direct evolutionary relationship between fungi and any other group of living organisms.

FUNGAL REPRODUCTION

All fungal nuclei except for the zygote are haploid, and there are many such haploid nuclei in the common cytoplasm of each individual fungus, which, as described earlier, consists of an extensive system of threadlike hyphae. In the sexual reproduction of fungi, hyphae of two genetically different mating types come together and fuse. In two of the three divisions of fungi, the genetically different nuclei that are associated in a common cytoplasm after such fusion do not fuse immediately. Instead, the two types of nuclei simply coexist for most of the life of the fungus. A fungal hypha that has nuclei within it derived from two genetically distinct individuals is called a **heterokaryotic** hypha; if all of the nuclei are genetically similar to one another, the hypha is said to be **homokaryotic.** If there are two genetically distinct nuclei within each cell of the hyphae, they are **dikaryotic;** if each cell has only a single nucleus, it is **monokaryotic.** A dikaryotic hypha will always be heterokaryotic; a monokaryotic one will always be homokaryotic. As you will see, these distinctions are important in understanding the life cycles of the individual groups.

> **Fungal hyphae in which two genetically distinct kinds of nuclei occur together are said to be heterokaryotic; those in which all of the nuclei are of the same type are homokaryotic.**

The cytoplasm in fungal hyphae normally flows right through perforated septa or moves freely in their absence, but we have already mentioned one important exception to this general pattern. When reproductive structures are formed, they are cut off by septa that either lack perforations or have perforations that soon become blocked. Two kinds of reproductive structures occur in fungi: (1) **sporangia,** which are involved in the formation of spores (asexual bodies), and (2) **gametangia,** structures within which gametes form.

Spores, always nonmotile, are a common means of reproduction among the fungi. They may be formed as a result of either asexual or sexual processes, as we will see in our discussion of the individual divisions. When the spores land in a suitable place, they germinate, giving rise to a new fungal hypha. Since the spores are very small, they may remain suspended in the air for long periods of time. Because of this property, fungal spores may be blown great distances from their place of origin, a factor in the extremely wide distributions of many kinds of fungi. Unfortunately, many of the fungi that cause diseases of plants and animals are spread rapidly and widely by such means.

FUNGAL DIVISIONS

There are three divisions of fungi: Zygomycota, the zygomycetes; Ascomycota, the ascomycetes; and Basidiomycota, the basidiomycetes. They are called divisions instead of phyla because fungi are named according to the same rules used for plants. Several other groups that historically have been associated with fungi, such as the slime molds and water molds (phylum Oomycota; see Chapter 27), now are considered to be protists, not fungi. Oomycetes are sharply distinct from fungi in (1) their motile spores, (2) their cellulose-rich cell walls, and (3) their normal mitosis.

The three divisions of fungi are distinguished primarily by their sexual reproductive structures. In the zygomycetes, the fusion of hyphae leads directly to the formation of a zygote, which divides by meiosis when it germinates. In the other two divisions, an extensive growth of dikaryotic hyphae usually leads to the formation of massive structures

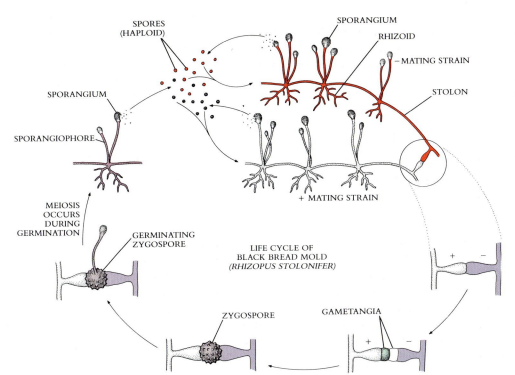

FIGURE 28-5

Life cycle of Rhizopus, *a zygomycete. The hyphae grow over the surface of the bread or other material on which the fungus feeds, producing erect sporangium-bearing stalks in clumps. If both + and − strains are present in a colony, they may grow together and their nuclei may fuse, producing a zygote. This zygote, which is the only diploid cell of the life cycle, acquires a thick, black coat and is then called a zygospore. Meiosis occurs during its germination, and normal, haploid hyphae grow from the haploid cells that result from this process.*

of interwoven hyphae within which is formed the distinctive kind of reproductive cell characteristic of that particular division. Syngamy, followed immediately by meiosis, occurs within these cells, and haploid spores are formed. Upon release they are dispersed, some of them giving rise to new hyphae.

Zygomycetes

The zygomycetes (division Zygomycota) lack septa in their hyphae except when they form sporangia or gametangia. Zygomycetes are by far the smallest of the three divisions of fungi, with only about 600 named species. Included among them are some of the more frequent bread molds (Figure 28-1, *A*), as well as a variety of other microscopic fungi found on decaying organic material. The group is named after a characteristic feature of the life cycle of its members, the production of temporarily dormant structures called **zygospores.**

In the life cycle of the zygomycetes (Figure 28-5), sexual reproduction occurs by the fusion of equal-sized gametangia, which contain numerous nuclei. The gametangia are cut off from the hyphae by complete septa. These gametangia may be formed on hyphae of different mating types or on a single hypha. If different mating types are involved, fusion between the pairs of nuclei occurs immediately. Once the nuclei have fused, forming diploid nuclei—zygotes—the area in which the fusion has taken place develops into an often massive and elaborate zygospore. This zygospore characteristically forms between the two gametangia that fused to form the zygote. Except for the zygote, all nuclei of the zygomycetes are haploid. Meiosis occurs during the germination of the zygospore.

> **Zygomycetes form characteristic resting structures, called zygospores, around the cells in which their zygotes are formed. The hyphae of zygomycetes are multinucleate, with septa only where gametangia or sporangia are cut off.**

Asexual reproduction occurs much more frequently than sexual reproduction in the zygomycetes. In asexual reproduction, haploid spores are produced in more or less specialized sporangia formed on the hyphae. These sporangia are cut off by septa, and are usually formed at the tips of erect hyphae in zygomycetes. Their spores are thus shed above the substrate, in a position where they may be picked up by the wind and blown about.

Ascomycetes

The second division of fungi, the ascomycetes (division Ascomycota), is a very large group of about 30,000 named species, with many more being discovered each year. Among the ascomycetes are such familiar and economically important fungi as yeasts, common molds, morels, and truffles (Figures 28-1, *B*, and 28-6). Also included in this division are many of the most serious plant pathogens, including the chestnut blight, *Endothia parasitica*, and Dutch elm disease, *Ceratocystis ulmi* (Figure 28-7).

The ascomycetes are named for their characteristic reproductive structure, the microscopic, club-shaped **ascus** (pl. **asci**). The zygote, which is the only diploid nucleus of the ascomycete life cycle (Figure 28-8), is formed within the ascus. The asci are differentiated within a structure that is made up of densely interwoven hyphae, corresponding to the visible portions of a morel or cup fungus, called the **ascocarp** (Figures 28-1, *B*, and 28-6).

Asexual reproduction is very common in the ascomycetes, as it is among the zygomycetes. It takes place by means of spores called **conidia** (sing. **conidium**), which are cut off by septa at the ends of modified hyphae called **conidiophores** (Figures 28-14 and 28-15; see also Figure 28-A, *2* in boxed essay on p. 563). Most conidia are multinucleate. The hyphae of ascomycetes are divided by septa, but the septa are perforated and the cytoplasm flows along the length of each hypha. The septa that cut off the asci and conidia are initially perforated like all other septa, but later they often become blocked.

The hyphae of ascomycetes may be either homokaryotic or heterokaryotic. The cells of these hyphae usually contain from several to many nuclei, as do the gametangia. The female gametangia, which are called **ascogonia,** each have a beaklike outgrowth called a **trichogyne.** When the **antheridium,** or male gametangium, forms, it fuses with this trichogyne. Both kinds of gametangia contain a number of nuclei. After the fusion of their contents, the nuclei derived from the two different mating types pair with one another. Heterokaryotic hyphae then arise from the area of the fusion. Throughout such hyphae,

FIGURE 28-6

A morel, Morchella esculenta, *a delicious edible ascomycete that appears in early spring in the north-temperate woods, especially under oaks. Morels are gathered in large quantities in nature, but mushroom growers began to learn how to cultivate them commercially only in the 1980s.*

FIGURE 28-7

An elm, dying, like millions of others in North America and Europe, of Dutch elm disease. The disease is caused by an ascomycete, Ceratocystis ulmi, *the spores of which are carried from tree to tree by a bark beetle. The spread of this devastating disease can be controlled by a combination of sanitation (removing dead and dying trees promptly), killing the beetles in traps baited with chemicals, and treating the trees in advance with fungicides and bacteria that inhibit the growth of the fungi.*

FIGURE 28-8

Life cycle of an ascomycete.

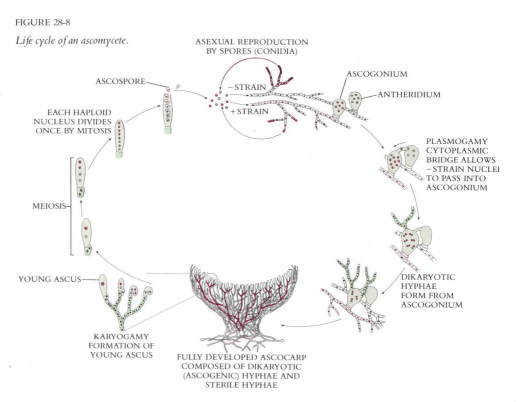

ASEXUAL REPRODUCTION BY SPORES (CONIDIA)

ASCOSPORE

ASCOGONIUM

ANTHERIDIUM

– STRAIN

+ STRAIN

EACH HAPLOID NUCLEUS DIVIDES ONCE BY MITOSIS

PLASMOGAMY CYTOPLASMIC BRIDGE ALLOWS – STRAIN NUCLEI TO PASS INTO ASCOGONIUM

MEIOSIS

DIKARYOTIC HYPHAE FORM FROM ASCOGONIUM

YOUNG ASCUS

KARYOGAMY FORMATION OF YOUNG ASCUS

FULLY DEVELOPED ASCOCARP COMPOSED OF DIKARYOTIC (ASCOGENIC) HYPHAE AND STERILE HYPHAE

nuclei that represent the two different original mating types occur. Several nuclei, some derived from each of the parents, occur within each cell of the hyphae.

The asci are cut off by the formation of septa at the tips of the heterokaryotic hyphae. There are two haploid nuclei within each ascus, one of each of the two mating types represented in the dikaryotic hypha. Fusion of these two nuclei occurs within each ascus, forming a zygote. Each zygote divides immediately by meiosis and then usually divides again by mitosis, giving rise to eight haploid **ascospores** (Figure 28-9). In most ascomycetes the ascus becomes highly turgid at maturity and ultimately bursts. When this occurs, the ascospores may be thrown as far as 30 centimeters, an amazing distance considering that most ascospores are only about 10 micrometers long. Aside from differences in scale, this would be equivalent to throwing a baseball (diameter, 7.5 centimeters) 1.25 kilometers—about 10 times the length of a home run!

Ascocarps (Figures 28-1, *B,* and 28-6) are made up of tightly interwoven monokaryotic and dikaryotic hyphae. Within an ascocarp, on special fertile layers of dikaryotic hyphae, the asci are formed. The ascocarps of the cup fungi and the morels are open, with the asci lining the open cups. Other ascocarps, such as those of *Neurospora,* an important organism in genetic research, are closed.

> **Ascomycetes form their zygotes within a characteristic club-shaped structure, the ascus. Meiosis follows zygote formation immediately and results in the production of ascospores. Asexual reproduction by conidia, which are multinucleate, is common among the members of this class of fungi.**

Yeasts

The yeasts, which are unicellular, are one of the most interesting and important groups of microscopic ascomycetes (Figure 28-10). There are about 40 genera of yeasts, with about 350 species. Most of their reproduction is asexual and takes place by means of cell fission or by budding (the formation of a smaller cell from a larger one). Sometimes, however, whole yeast cells may fuse. One of these cells, containing two nuclei, may then function as an ascus, with the resulting ascospores functioning directly as new yeast cells.

Because they are single-celled, yeasts might be considered primitive fungi; it appears certain, however, that instead they are reduced in structure and were originally derived from multicellular ancestors, most of which were ascomycetes. The word *yeast,* in fact, actually relates to the fact that these fungi are single celled, and to that fact alone; some yeasts have been derived from each of the three classes of fungi, although ascomycetes are best represented. Even the yeasts that were derived from filamentous ascomycetes are not necessarily directly related to one another, but instead seem to have been derived from different groups of ascomycetes.

The ability of yeasts to ferment carbohydrates, breaking down glucose to produce ethanol and carbon dioxide in the process, is of fundamental importance in the production of bread, beer, and wine. Through the millennia, many different strains of yeast have been domesticated and selected for these processes. Wild yeasts—ones that occur naturally in the areas where wine is made—often played an important role in winemaking historically, but domesticated yeasts are normally used now. The most important yeast, and often the only one, used in all of these processes is *Saccharomyces cerevisiae.* This particular yeast has been used by human beings throughout recorded history. Other yeasts are important pathogens and cause diseases such as thrush and cryptococcosis.

Over the past few decades yeasts have become very important organisms in genetic research. They were the first eukaryotes to be extensively manipulated by the techniques of genetic engineering, and they still play the leading role as models for research in eukaryotic cells. In 1983 investigators synthesized a functional artificial chromosome in *S. cerevisiae* by chemically assembling the appropriate DNA molecule; this has not yet been possible in any other eukaryote. With their rapid generation time and rapidly increasing pool of genetic and biochemical information, the yeasts in general and *S. cerevisiae* in par-

FIGURE 28-9

Asci from the lining of the cup of Peziza, *a cup fungus like the one shown in Figure 28-1,* B. *The eight spores in each ascus result from meiosis, followed by a mitotic division of each spore. The fact that the spores remain lined up in order during the process makes it possible to use* Neurospora, *another ascomycete, to study crossing-over and the inheritance of metabolic traits. The ascospores of* Neurospora *can be dissected individually out of the asci and grown on media of selected composition.*

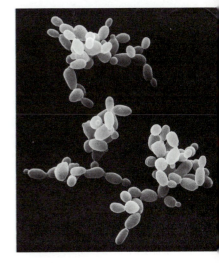

FIGURE 28-10

Scanning electron micrograph of a yeast, showing the characteristic method of cell division by budding. The cells in this colony are tending to hang together in chains, a feature that calls to mind the derivation of the yeasts, which are single celled, from multicellular ancestors.

ticular are becoming the eukaryotic cells of choice for many types of experiments in molecular and cellular biology. Yeasts have become, in this respect, comparable to *Escherichia coli* among the bacteria and are continuing to provide significant new insights into the functioning of eukaryotic systems.

Yeasts are unicellular fungi, mainly ascomycetes, which have evolved from hypha-forming ancestors; not all yeasts are directly related to one another. Long useful for baking, brewing, and winemaking, yeasts are now becoming very important in genetic research; they are more easily manipulated than any other eukaryote.

Other ascomycetes

Ascomycetes are by far the most frequent fungal partners in lichens (p. 572). Most species of fungi that normally lack sexual reproduction likewise are ascomycetes, as judged by the features of their hyphae and conidia. Such fungi are classified among the Fungi Imperfecti (p. 571), since they cannot be placed in the proper group of ascomycetes unless the details of their sexual structures, such as their asci and ascocarps, are known.

Basidiomycetes

The last division of fungi, the basidiomycetes (division Basidiomycota), has about 25,000 named species. More is known about the members of this group than about the other fungi. Among the basidiomycetes are not only the mushrooms, toadstools, puffballs, jelly fungi, and shelf fungi, but also many important plant pathogens among the groups called rusts and smuts (Figure 28-11). Many mushrooms are used as food, but others are deadly poisonous. Still other species are poisonous to some people and harmless to others.

FIGURE 28-11

Representative basidiomycetes.
A *A shelf fungus,* Coriolus versicolor, *growing on a tree trunk. Basidia line the underside of shelf fungi.*
B *A puffball. Basidia form within puffballs, which have no external openings; they release their basidiospores by rupturing when mature.*
C *A jelly fungus, growing in the Amazonian rain forest of Brazil. Although many familiar basidiomycetes form their basidia within mushroomlike basidiocarps, there are a number of other kinds, as these photographs illustrate.*

A

B

C

Basidiomycetes are named for their characteristic sexual reproductive structure, the **basidium.** A basidium is club shaped, like an ascus. Syngamy occurs within the basidium, giving rise to the zygote, the only diploid cell of the life cycle (Figure 28-12). As in all fungi, meiosis occurs immediately after the formation of the zygote. In the basidiomycetes, the four haploid products of meiosis are called **basidiospores,** and in most basidiomycetes, they are borne at the end of the basidium on slender projections called **sterigmata** (sing. **sterigma**). In this way the structure of a basidium differs from that of an ascus, although functionally the two are identical. The septum that cuts off the young basidium is initially perforated but often becomes blocked, as it does in the ascomycetes.

The life cycle of a basidiomycete starts with the production of homokaryotic hyphae after spore germination. These hyphae lack septa at first. Eventually, however, septa are formed between the nuclei of the monokaryotic hyphae. A basidiomycete mycelium made up of monokaryotic hyphae is called a **primary mycelium.** Ultimately, different mating types of monokaryotic hyphae may fuse, forming a dikaryotic or **secondary mycelium.** Such a mycelium is heterokaryotic, with two nuclei, representing the two different mating types, between each pair of septa. Most of the mycelium of basidiomycetes is dikaryotic.

The dikaryotic hyphae of basidiomycetes have a unique pattern of cell division. They grow by the simultaneous division of the nuclei in the cell at the tip of each hypha and the progressive formation of new, perforated septa. Other cells within the mycelium that will form lateral branches divide in the same way. The basidiocarps, or mushrooms, are formed entirely of secondary (dikaryotic) mycelium.

Basidia line the gills—radiating pleated areas like the folds of an accordion, which occur on the undersurface of the cap of a mushroom. On these gills, the basidia form dense layers from which vast numbers of minute spores may be released at maturity. It has been estimated, for example, that a mushroom with a cap that is 7.5 centimeters across produces

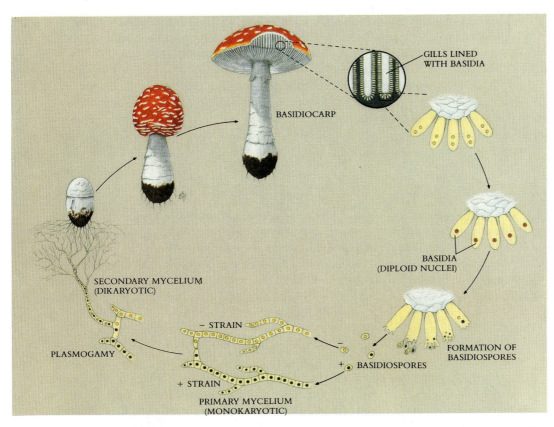

FIGURE 28-12

Life cycle of a basidiomycete.

GILLS LINED WITH BASIDIA

BASIDIOCARP

BASIDIA (DIPLOID NUCLEI)

SECONDARY MYCELIUM (DIKARYOTIC)

− STRAIN

PLASMOGAMY

+ STRAIN

FORMATION OF BASIDIOSPORES

BASIDIOSPORES

PRIMARY MYCELIUM (MONOKARYOTIC)

FIGURE 28-13

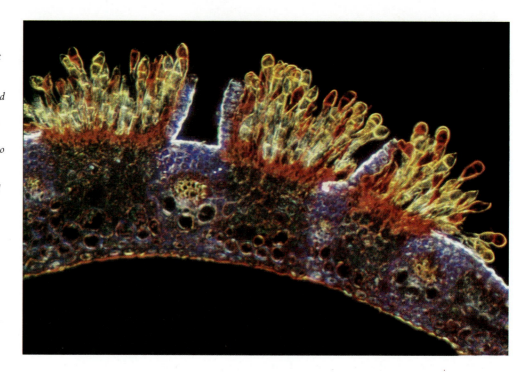

Wheat rust, Puccinia graminis, *one of about 7000 species of rusts, all plant pathogens. This species causes enormous economic losses to wheat wherever it is grown, and it is combated largely by breeding resistant wheat varieties. Mutation and recombination in wheat rust constantly produce new virulent strains and make it necessary to replace the existing wheat varieties constantly. Wheat rust alternates between two different hosts, wheat and barberries, and needs both to complete its sexual cycle, although it may continue to reproduce indefinitely asexually on wheat and other grasses. The sexual stages of wheat rust take place only on barberries, and eradicating them helps to control this disease. As shown in the photograph, rusts and smuts do not form basidiocarps; the erect structures are basidia.*

as many as 40 million spores per hour! In contrast to their effective sexual reproduction, asexual reproduction is rare among the basidiomycetes.

> **Most of the hyphae of basidiomycetes are dikaryotic, with two nuclei, one of each mating type, within each cell. These divide in a unique manner and ultimately mass to form basidiocarps, within which basidia line the gills or pores. Meiosis immediately follows syngamy in these basidia.**

Although we are more familiar with the mushrooms, another kind of basidiomycete is represented by the rusts and smuts (Figure 28-13), which are important plant pathogens. In these groups, basidiocarps are not formed. Instead, the basidia, which differ in structure from those of the mushrooms and other large, fleshy fungi, arise from hyphae at the surface of the host plant.

A COMPARISON OF THE FUNGAL DIVISIONS

The three divisions of fungi have a number of features in common, but they are separated by many important differences, some of which are summarized in Table 28-1. The zygomycetes have undivided hyphae; the ascomycetes and basidiomycetes have hyphae that are divided by perforated septa. The septa that cut off the reproductive structures of the ascomycetes and basidiomycetes (asci or basidia, respectively) may become blocked and thus isolate the respective structures, but they are still perforated initially. The sexual reproductive organs of the zygomycetes are relatively unspecialized and are simply called gametangia; in contrast, the ascogonia of ascomycetes have a distinctive structure, with a beak, through which the contents of the antheridium empty. Basidiomycetes have no distinct sexual reproductive organs.

In zygomycetes the fusion of gametangia results in the production of a zygote, around which a specialized, resistant structure called a zygospore forms. Sexual fusion in the ascomycetes occurs between ascogonia and antheridia; in the basidiomycetes it occurs between normal-looking hyphae. In the ascomycetes, sexual fusion is followed by the formation of heterokaryotic hyphae in which the individual cells normally are multinucleate.

TABLE 28-1   CHARACTERISTICS OF THE DIVISIONS OF FUNGI

	DIVISION		
	ZYGOMYCOTA	ASCOMYCOTA	BASIDIOMYCOTA
Number of species	600	30,000	25,000
Asexual reproduction	Sporangia	Conidia	Fragments
Sexual reproduction	Zygospores	Asci	Basidia
Septa★	Absent	Present, perforated	Present, perforated, specialized

★The reproductive structures are cut off by complete septa in all fungi; these descriptions refer to the septa in the ordinary, nonreproductive hyphae.

FIGURE 28-14

Scanning electron micrograph of the whorled conidiophores of alfalfa verticillium wilt, Verticillium algoatrum, *an important pathogen of alfalfa. The single-celled conidia of this member of the Fungi Imperfecti are borne at the ends of the conidiophores.*

In the basidiomycetes, sexual fusion is followed by the formation of dikaryotic hyphae, in which each cell has two genetically distinct nuclei.

The sexual spores of the ascomycetes are produced in complex structures called ascocarps, formed of both homokaryotic and heterokaryotic mycelia. Those of basidiomycetes are produced in similar structures, called basidiocarps, which are formed exclusively of dikaryotic, or secondary, mycelia. The characteristic asci and basidia of these divisions occur within or on the surfaces of their ascocarps or basidiocarps. In them, syngamy, meiosis, and the production of haploid spores occur. Asexual reproduction is very common in zygomycetes, in which the spores are produced in sporangia, and in ascomycetes, in which multinucleate spores called conidia are formed. Asexual reproduction is rare among the basidiomycetes.

It is generally assumed that the zygomycetes are the most primitive fungi. They lack septa except for those that are formed when reproductive structures are cut off from their hyphae. The perforated hyphae of the ascomycetes and basidiomycetes clearly seem to be specializations, and these groups appear to be more advanced than the zygomycetes. Both zygomycetes and ascomycetes have specialized and sometimes quite elaborate asexual spores. Basidiomycetes, which probably have lost asexual reproduction in the course of their evolution and have a very elaborate mode of cell division, evidently represent a different line of evolution from ascomycetes; both groups may have been derived from zygomycete ancestors.

FIGURE 28-15

A *Characteristic conidiophores of* Penicillium, *as viewed with the scanning electron microscope.*

FUNGI IMPERFECTI

Most of the so-called Fungi Imperfecti, a group that is also called Deuteromycetes, are ascomycetes that have lost the ability to reproduce sexually. The fungi that are classified in this group, however, are simply those in which the sexual reproductive stages have not been observed, whether they are believed to be ascomycetes or not. The division of fungi from which a particular nonsexual strain has been derived usually can be determined by the features of its hyphae and asexual reproduction. It cannot, however, be classified by the standards of that group because the classification systems are based on the features related to sexual reproduction. Especially in view of the great economic importance of many of the Fungi Imperfecti, it is necessary to classify them separately, and on the basis of their asexual structures, so that the individual species and genera can be identified (Figures 28-14 and 28-15; see also figure in boxed essay, p. 563).

There are some 25,000 described species of Fungi Imperfecti. Although the great majority of them are ascomycetes, there are a few basidiomycetes and several genera that have features suggesting they are zygomycetes. Even though sexual reproduction is absent among Fungi Imperfecti, there is a certain amount of genetic recombination. This becomes possible when hyphae of different genetic types fuse, as sometimes happens spontaneously.

Within the heterokaryotic hyphae that arise from such fusion, genetic recombination of a special kind called **parasexuality** may occur. In parasexuality, the exchange of portions of chromosomes between the genetically distinct nuclei within a common hypha takes place. Recombination of this sort also occurs in other groups of fungi and seems to be responsible for some of the production of new pathogenic strains of wheat rust (Figure 28-13), for example.

Among the economically important genera of Fungi Imperfecti are *Penicillium* (Figure 28-14) and *Aspergillus* (Figure 28-15). Not only are some species of *Penicillium* the sources of the well-known antibiotic penicillin, but other species of the genus give the characteristic flavors and aromas to such cheeses as Roquefort and Camembert. Species of *Aspergillus* are used for fermenting soy sauce and soy paste, processes in which certain bacteria and yeasts also play important roles. Citric acid is produced commercially with members of this genus under highly acid conditions. In addition, the enrichment of livestock feed by the products of fermentation of other species is being investigated. Most of the fungi that cause skin diseases in human beings, including athlete's foot and ringworm, are also Fungi Imperfecti.

> **Fungi Imperfecti are fungi that have lost the capacity for sexual reproduction and cannot be classified by the standards applied to, or placed among the members of, the three divisions of fungi. Instead, they are classified by their asexual reproductive structures. The great majority of Fungi Imperfecti are clearly ascomycetes.**

LICHENS

A lichen is a symbiotic association between a fungus and a photosynthetic partner. Ascomycetes are the fungal partners in all but about 20 of the 25,000 described species of lichens (Figure 28-16); the exceptions, mostly tropical, are basidiomycetes. Most of the visible body of a lichen consists of its fungus, but within the tissues of that fungus are found either cyanobacteria or green algae (Figure 28-17). Specialized fungal hyphae penetrate the photosynthetic cells within them and transfer nutrients directly to the fungal partner. Biochemical "signals" given by the fungus apparently direct its cyanobacterial or green algal component to produce certain metabolic substances that it does not produce when growing

A

B

C

CULTIVATING MUSHROOMS

Mushrooms have long been considered a delicacy as human food. Two species, *Agaricus campestris* and *Lentinus edodes,* are commonly cultivated. The button mushroom, *Agaricus campestris,* was first domesticated around Paris in about 1650. By the early 1980s, it was being grown in over 70 countries, with an annual crop of over 940,000 metric tons and a value of more than $14 billion. The Shiitake mushroom, *Lentinus edodes,* has been cultivated in the Orient for more than 800 years. The annual crop now amounts to about 192,000 metric tons, most of which is dried, with an annual value of more than $500 million.

An effort is being made to increase the cultivation of additional species of mushrooms, especially in Asia. Mushrooms can be grown on waste products and may be able to make a significant contribution to the world food supply, especially in view of their high protein content. If they can be grown cheaply enough, this promise may become a reality. In addition, many varieties of mushrooms are highly desirable because of their delicate flavors, and ways are being sought to bring them into cultivation to ensure a dependable supply.

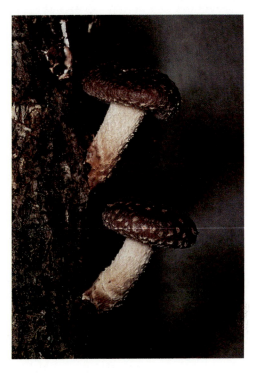

FIGURE 28-B

Lentinus edodes, *one of the most commonly cultivated mushrooms.*

---

independently of the fungus. The photosynthetic member of the association normally is held between thick layers of interwoven fungal hyphae and is not directly exposed to the light. Enough light penetrates the translucent layers of fungal hyphae to make photosynthesis possible, however. The fungi found in lichens are unable to grow normally without their photosynthetic partner. Overall, this particular symbiotic relationship might be characterized as one of controlled parasitism of the photosynthetic organism by the fungus.

The reproduction of lichens may take place by a combination of normal sexual processes: the formation of ascospores in the fungal component and the reproduction, usually asexual, of the photosynthetic component. These two components may come together under favorable circumstances and produce a lichen body. Alternatively, lichens can simply break up into pieces, each of which contains both the fungal and photosynthetic components and can give rise to a new individual.

The durable construction of the fungus, combined with the photosynthetic properties of its partner, has enabled lichens to invade the harshest of habitats at the tops of mountains, in the farthest north and south latitudes, and on dry, bare rock faces in the desert. In such harsh, exposed areas, lichens are often the first colonists, breaking down the rocks and setting the stage for the invasion of other organisms. Those lichens in which the photosynthetic partner is a cyanobacterium are able to fix atmospheric nitrogen, which is leached to the environment, where it can be used by other pioneering organisms.

Lichens are often strikingly colored because of the presence of pigments that probably play a role in protecting the photosynthetic partner from the destructive action of the sun's rays. These same pigments may be extracted from the lichens and used by people as natural

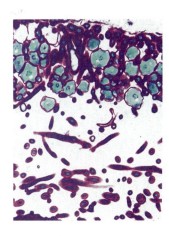

FIGURE 28-17

Transection of a lichen, showing the fungal hyphae more densely packed into a protective layer on the top, and, especially, the bottom layer of the lichen. The green cells near the upper surface of the lichen are those of a green alga. Penetrated by fungal hyphae, these cells supply carbohydrate to the fungus.

FUNGI AS CARNIVORES

It might surprise you to know that some fungi are predatory. For example, the mycelium of the oyster fungus, *Pleurotus ostreatus,* a well-known edible mushroom, excretes a substance that anesthetizes the tiny worms known as nematodes (Chapter 36). When the worms become sluggish and inactive, the fungal hyphae envelop and penetrate their bodies and absorb their nutritious contents. Oyster fungi usually grow on dead logs or old stumps, periodically producing tiers of large, fleshy mushrooms. Their mycelium obtains the bulk of its glucose through the enzymatic digestion of cellulose from the wood, so that the nematodes it consumes apparently serve mainly as a source of nitrogen, which, it seems, is almost always in short supply in biological systems. Other fungi are even more active predators than *Pleurotus,* snaring, trapping, or firing projectiles into the nematodes, rotifers, and other small animals on which they prey.

FIGURE 28-C

The oyster mushroom, Pleurotus ostreatus, *immobilizes nematodes, which the fungus uses as a source of food.*

FIGURE 28-18

In the seemingly lifeless "dry valley" region of Antarctica (A), *lichens* (B) *live just beneath the exposed surface of the sandstone. This habitat has been explored by microbiologists E.I. Friedmann and his wife Roseli Ocampo of Florida State University. In the photograph of the fractured rock in* B, *each colored band is a distinct biological zone, involving several kinds of heterotrophic bacteria as well as the lichens. The black and white zones are formed by a lichen (probably* Buellia pallida) *that lives only in situations of this kind within rocks; the lower, fainter green zone is produced by a free-living green alga,* Hemichloris antarctica. *There is a steep light gradient in the rock between the lichen zone and the lower, algal zone;* Hemichloris *is therefore adapted to exist at extremely low light intensities. Temperatures in these valleys rise almost to freezing in the summer but probably fall as low as −60° C in the winter; there is almost no precipitation, and blasts of cold, dry wind scour the valleys constantly.*

dyes, as, for example, in the traditional method of manufacturing Harris tweed—now, however, colored with synthetic dyes.

Lichens are symbiotic associations between a fungus—an ascomycete in all but a very few instances—and a photosynthetic partner, which may be either a green alga or a cyanobacterium.

Lichens are able to survive in inhospitable habitats partly by being able to dry or freeze to a condition that we might call suspended animation (Figure 28-18). Once the drought or cold has passed, the lichens recover quickly and resume their normal metabolic activities, including photosynthesis. The growth of lichens may be extremely slow in harsh environments: many relatively small ones actually appear to be thousands of years old, and therefore among the oldest living things on earth.

A

B

Lichens and Pollution

Paradoxically, lichens are extremely sensitive to pollutants in the atmosphere, and thus they can be used as bioindicators of air quality. Their sensitivity is the result of their ability to absorb substances dissolved in rain and dew readily, a property on which their existence depends, but one that can also prove fatal to them in a polluted environment. The distribution of lichens around cities is affected by the presence of sulfur dioxide, ozone, and toxic metals. These pollutants are absorbed by lichens and cause the destruction of chlorophyll, decreases in photosynthesis, and alterations in membrane permeability, among other problems. Similarly, the physiological balance between the fungus and the alga or cyanobacterium is upset and degradation or destruction of the lichen results.

Lichens vary in their response to pollutants, and the best bioindicators are the moderately sensitive species that in addition to distributional differences show visible effects related to pollution levels. For example, *Hypogymnia enteromorpha,* found in the mountains surrounding Los Angeles, is extremely bleached and reduced in size in severely ozone-impacted areas (Figure 28-19, *A*) as compared with samples from relatively unpolluted areas (Figure 28-19, *D*). Samples from moderately polluted areas show less severe damage (Figure 28-19, *B* and *C*). Lichens also are being used as a means of assessing the pollution by radionuclides that occurs in the vicinity of uranium mines, crashed satellites, and other similar sites, just as they were used in the days of atmospheric nuclear weapon testing to estimate the dispersal of radioactive fallout from the explosions. Other methods used to describe the response of lichens to air pollutants include (1) mapping the variety and abundance of species around pollution sources, (2) chemical analysis of lichens to determine pollutant accumulation in relation to distance from pollutant sources, and (3) transplant studies in which healthy lichens are placed along a pollution gradient and monitored over time for signs of stress.

Observations of the effects of air pollutants on lichens date back to the mid-nineteenth century. These past reports and the results of modern field studies are important for the documentation of baseline data for future reference. However, the real challenge in **lichenology** (the study of lichens) today is to determine the role that lichens have in the structure and function of natural ecosystems and the potential effects of their continued loss due to air pollution.

MYCORRHIZAE

The roots of about 80% of all kinds of plants normally are involved in symbiotic relationships with certain specific kinds of fungi; it has been estimated that these fungi probably amount to 15% of the total weight of the world's plant roots. Associations of this kind are termed **mycorrhizae** (from the Greek words for "fungus" and "roots"). To a certain extent, the fungi involved in mycorrhizal associations replace and play the same function as the fine projections from the epidermis, or outermost cell layer, of the terminal portions of the roots—root hairs. When mycorrhizae are present, they aid in the direct transfer of phosphorus, zinc, copper, and probably other nutrients from the soil roots. The plant, on the other hand, supplies organic carbon to the symbiotic fungus (Figure 28-20).

There are two principal types of mycorrhizae: **endomycorrhizae,** in which the fungal hyphae penetrate the outer cells of the plant root, forming coils, swellings, and minute branches, and also extend out into the surrounding soil; and **ectomycorrhizae,** in which the hyphae surround but do not penetrate the roots. In both kinds of mycorrhizae, the mycelium extends far out into the soil.

Endomycorrhizae (Figure 28-21) are by far the more common of these two types. The fungal component in them is a zygomycete. Only about 30 species of zygomycetes are involved in such relationships throughout the world. These few species of zygomycetes become associated with many species of plants, perhaps more than 200,000. Endomycorrhizal fungi are being studied intensively because their applications to cultivated fields—much like a fertilizer—are potentially capable of leading to increased yields with lower

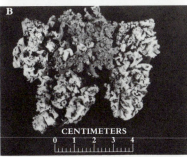

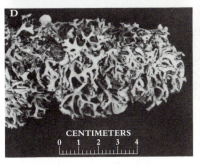

FIGURE 28-19

Samples of the lichen Hypogymnia enteromorpha *from the mountains near Los Angeles.*
A *Bleached, reduced individual from a severely ozone-impacted area.*
B *and* **C** *Samples from less polluted areas.*
D *Sample from a relatively unpolluted area.*

FIGURE 28-20

Soybean plants without endomycorrhizae (left) *and with different strains of endomycorrhizae* (center and right).

A

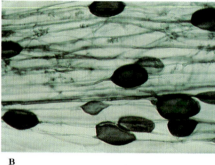

B

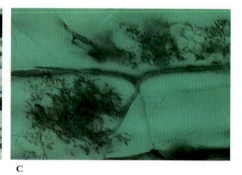

C

FIGURE 28-21

Endomycorrhizae. The zygomycete Glomus versiforme *is shown here growing in the roots of leeks,* Allium porrum.
A *General view of the leek root, showing vesicles, photographed with a dissecting microscope. The root has been squashed in fluid. (×16.)*
B *Vesicles in leek root. (×63.)*
C *Arbuscules in leek root. (×63.) Arbuscules are characteristic of young infections, with vesicles predominating later.*

phosphate and energy inputs. When the fields are fumigated—for example, to control nematodes—the populations of endomycorrhizal zygomycetes may be suppressed and need to be restored for full productivity.

Ectomycorrhizae (Figure 28-22) involve far fewer kinds of plants than do endomycorrhizae, perhaps only a few thousand. They are characteristic of certain groups of trees and shrubs, particularly those of temperate regions, including pines, firs, oaks, beeches, and willows. The fungal components in most ectomycorrhizae are basidiomycetes, but some are ascomycetes. At least 5000 species of fungi are involved in ectomycorrhizal relationships, and most of them are restricted to a single species of plant.

Mycorrhizae are symbiotic associations between plants and fungi. There are two main types: (1) endomycorrhizae, which occur in about 80% of all plants and in which the fungal partner is a zygomycete, and (2) ectomycorrhizae, which form in certain kinds of trees and shrubs and in which the fungal partner usually is a basidiomycete.

The kinds of plants involved in ectomycorrhizal associations evidently are more resistant to drought, cold, and other harsh environmental conditions than are the plants that lack such fungal partners. This may be why they are so well represented among trees and shrubs that grow at timberline in the mountains and in the far north and south. Recently, it has been found that some kinds of ectomycorrhizal fungi provide better protection to the trees with which they are associated from the effects of acid precipitation than do others. In some way, the fungi often seem to protect the plants with which they are associated from the high accumulations of heavy metals, such as copper and zinc, which are absorbed by the plants under acid conditions.

Another type of mycorrhizal relationship involves orchids. The seeds of orchids, which are so minute as to appear dustlike, germinate in nature only in the presence of suitable fungi, members of a particular group of basidiomycetes. Apparently the minute seeds of the orchids do not have enough stored food material to allow them to grow without the fungus; the fungi involved in this relationship actually supply them with carbon. Orchid growers achieve the same results with many species of orchids by growing them on agar in which a carbon source is included. Some of these, however, may germinate in a matter of days rather than months in association with the appropriate fungus. Perhaps 100 species of basidiomycetes are associated with orchids in this way.

The earliest fossil plants often have endomycorrhizal roots. Such associations, which were common during the initial period of invasion of the land by plants, may have played an important role in allowing this invasion to take place. The soils that were available at such times would have been sterile and completely lacking in organic matter. Since plants that form mycorrhizal associations are particularly successful in infertile soils now, and considering the fossil evidence, the suggestion that mycorrhizal associations were characteristic of the earliest plants seems reasonable.

SUMMARY

1. The fungi are a distinct kingdom of eukaryotic organisms characterized by their filamentous growth form, lack of chlorophyll and motile cells, chitin-rich cell walls, and external digestion of food by the secretion of enzymes. Together with the bacteria, they are the decomposers of the biosphere.

2. Fungal filaments, called hyphae, collectively make up a mass called the mycelium. A hypha may contain genetically uniform nuclei, and thus be homokaryotic, or it may contain two or more genetically different kinds of hyphae and be heterokaryotic. Mitosis in fungi occurs within the nuclear envelope.

3. Meiosis occurs immediately after the formation of the zygote in all fungi; the zygote therefore is the only diploid nucleus of the entire life cycle in these organisms.

4. There are three divisions of fungi: Zygomycota, the zygomycetes; Ascomycota, the ascomycetes; and Basidiomycota, the basidiomycetes. Zygomycetes form cell walls only when gametangia or sporangia are cut off at the ends of their hyphae; otherwise they are multinucleate. Most hyphae of ascomycetes and basidiomycetes have perforated septa through which the cytoplasm, but not necessarily the nuclei, flows freely.

5. Cells within the heterokaryotic hyphae of the ascomycetes are multinucleate; those within the heterokaryotic hyphae of the basidiomycetes are dikaryotic. Zygotes in ascomycetes form within club-shaped structures known as asci, and those in basidiomycetes form within structures known as basidia, which initially are somewhat similar. Asci and basidia are formed inside or on the surfaces of often large structures like mushrooms, which are formed of interwoven hyphae.

6. Asexual reproduction in zygomycetes takes place by means of multinucleate sporangia; that in ascomycetes takes place by means of multinucleate conidia. Asexual

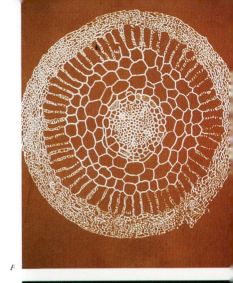

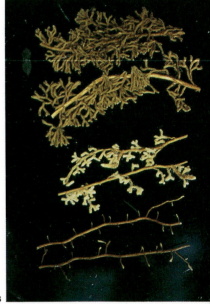

FIGURE 28-22

Ectomycorrhizae.
A *Transection of a pine root, showing thick mantle of ectomycorrhizae.*
B *Ectomycorrhizae on roots of pines. At the top are the yellow-brown mycorrhizae formed by Pisolithus tinctorius; in the center are the white mycorrhizae formed by Rhizopogon; at the bottom are pine roots not associated with a fungus. The different basidiomycetes that participate in the formation of ectomycorrhizae are usually highly specific, and, as seen here, several kinds may form mycorrhizal associations with one plant species. Different combinations have different effects on the physiological characteristics of the plant and its ability to survive under different environmental conditions.*

reproduction in basidiomycetes is rare; when it does occur, it usually involves simple fragmentation of the hyphae.

7. The yeasts are a group of unicellular fungi, mostly ascomycetes. They are of great commercial importance because of their roles in baking and fermentation.

8. The Fungi Imperfecti are a large artificial group of fungi composed largely of ascomycetes in which sexual reproduction does not occur or is not known. They are classified according to the characteristics of their spore-forming structures and spores. Many are of great commercial importance.

9. Symbiotic systems involving fungi include lichens and mycorrhizae. The fungal partners in lichens are almost entirely ascomycetes, which derive their nutrients from green algae or cyanobacteria.

10. Mycorrhizae are symbiotic associations between plants and fungi that are characteristic of the great majority of plants. Endomycorrhizae, which are more common, have zygomycetes as the fungal partner, whereas ectomycorrhizae have mainly basidiomycetes.

SELF-QUIZ

The following questions pinpoint important information in this chapter. You should be sure to know the answers to each of them.

1. Every fungus possesses flagella at some part of its life cycle. True or false?

2. Fungi obtain their food by
   (a) engulfing unicellular protists and bacteria
   (b) secreting digestive enzymes onto organic matter and absorbing the resulting nutrients
   (c) photosynthesis

3. The cell walls of fungi are made of
   (a) cellulose   (d) chitin
   (b) glucose   (e) nothing; there is no cell wall
   (c) cellulase

4. A dikaryotic fungus is one in which
   (a) each cell is homokaryotic
   (b) each cell is heterokaryotic
   (c) all of the nuclei are genetically similar
   (d) the two nuclei of each cell are genetically distinct
   (e) b and d

5. All fungal nuclei except for the zygote are
   (a) haploid   (b) diploid   (c) polyploid

6. Fungi reproduce only sexually. True or false?

7. What event immediately follows the formation of the zygote in all fungi? _____.

8. Zygomycetes form characteristic resting structures, called _____, around the cells in which their zygotes are formed.

9. An ascomycete zygote is formed within a structure called a _____.

10. Yeasts possess the ability to break down glucose to produce ethanol and carbon dioxide. True or false?

11. Fungi Imperfecti are so named because of defective nuclei. True or false?

12. Individual fungal hyphae may contain a set of
    (a) genetically identical nuclei
    (b) two or more different kinds of nuclei
    (c) segregating chromosomes
    (d) a and b

THOUGHT QUESTIONS

1. Why do we say there are three divisions of fungi when we have mentioned five major groups: Ascomycota, Zygomycota, Basidiomycota, Fungi Imperfecti, and lichens? What are the major distinguishing characteristics of these groups? Why are only three considered divisions?

2. If you had a sample of fungal hyphae, without characteristic reproductive structures, could you determine the major group to which the organism belonged? How?

3. What is there about the way fungi live in nature that helps to make them particularly valuable in industrial processes?

FOR FURTHER READING

AHMADJIAN, V.: "The Nature of Lichens," *Natural History*, March 1982, pages 30-37. Experimental studies of the way lichens are formed.

CHANG, S.T., and P.G. MILES: "A New Look at Cultivated Mushrooms," *BioScience*, vol. 34, pages 358-362, 1984. A fascinating account of this rapidly growing industry.

COURTENAY, B., and H.H. BURDSALL, JR.: *A Field Guide to Mushrooms and Their Relatives*, Van Nostrand Reinhold Co., Inc., New York, 1982. Excellent identification manual, illustrating and describing more than 350 species of ascomycetes and basidiomycetes.

DEMAIN, A.L.: "Industrial Microbiology," *Science*, vol. 214, pages 987-995, 1981. An elegant and readable review of the potential for further contributions by fungi and bacteria to industrial processes.

HALE, M.E.: *The Biology of Lichens*, 3rd ed., University Park Press, Baltimore, Md., 1983. A concise account of all aspects of the biology of lichens.

LAWREY, J.D.: *Biology of Lichenized Fungi*, Praeger Publishers, New York, 1984. An excellent account of all aspects of the biology of lichens, emphasizing their structure, chemistry, physiology, and ecology.

MICROBES FOR HIRE. *Science 85*, July-August 1985, pages 28-46. A series of articles outlining the challenges and opportunities posed by the industrial cultivation of fungi and bacteria, a field of critical importance for the future.

ROSS, I.K.: *Biology of the Fungi: Their Development, Regulation, and Associations*, McGraw-Hill Book Co., New York, 1979. An excellent overview of the fungi from many different points of view; probably the best overall reference on the group.

STROBEL, G.A., and G.N. LANIER: "Dutch Elm Disease," *Scientific American*, August 1981, pages 56-66. This deadly fungal infection of elms may be brought under control with the aid of fungicides, beetle traps, and bacteria that inhibit its growth.

STRUHL, K.: "The New Yeast Genetics," *Nature*, vol. 305, pages 391-397, 1983. This review article outlines the growing importance of yeast as a tool for understanding genetics and development in eukaryotes.

WEBB, A.D.: "The Science of Making Wine," *American Scientist*, vol. 72, pages 360-367, 1984. Traces the ways in which an ancient practical art has become a modern science.

BIOLOGY OF PLANTS

Part Six

More than a quarter of a million species of plants dominate the land, capturing the energy of the sun, building the soil, and determining the structure of all communities on earth.

DIVERSITY OF PLANTS

Chapter 29

Overview

Plants evolved more than 430 million years ago from a multicellular green alga. They became diverse on land over a relatively short time, creating the structure for the great forests and other plant communities in which the vertebrates, insects, and fungi became diverse. Roughly 266,000 species of plants are now living. The two major groups of plants are the vascular plants, which consist of nine living divisions, and the bryophytes. Bryophytes and ferns still require free water so that sperm can swim between the male and female sex organs, but most plants are fully terrestrial. Vascular plants have elaborate water- and food-conducting strands of cells, cuticles, and stomata; many of them are much larger than any bryophyte. Seeds evolved among the vascular plants and provide a means to protect young individuals. Flowers, which occur in only one division of living vascular plants, guide the activities of insects and other pollinators so that they will carry pollen rapidly and precisely from one flower to another of the same species.

For Review *Here are some important terms and concepts that you will encounter in this chapter. If you are not familiar with them, you should review them before proceeding.*

- **Plant classification** (Chapter 22)
- **Evolutionary history of plants** (Chapter 23)
- **Alternation of generations** (Chapter 27)
- **Mycorrhizae** (Chapter 28)

Of the five kingdoms of living organisms, the three we have discussed so far consist of organisms that are all small relative to us. Bacteria and most protists are unicellular, and fungi, though multicellular, are rarely as large as your fist. In the plant kingdom, however, we encounter really large individuals. It is not at all unusual for a tree to be more than 30 meters tall, with a mass of many metric tons. Plants are the dominant organisms of the terrestrial landscape, and we think that they were the first organisms to invade the land successfully from the sea, where life first evolved and flourished (Figure 29-1).

Despite all the diversity of life on earth, only three groups are responsible for virtually all of the fixation of carbon that occurs and thus for providing the substance of all living

FIGURE 29-1

From the air, one can see how completely plants dominate the land, as in this scene of tropical forest in northern South America.

organisms. The algae and photosynthetic bacteria carry out most marine photosynthesis, and plants are the dominant photosynthetic organisms on land. Plants are multicellular eukaryotic organisms that have cellulose-rich cell walls, chloroplasts that contain chlorophylls *a* and *b* together with carotenoids, and, as their primary carbohydrate food reserve, starch. In this chapter, we will discuss the characteristics of the major plant groups, focusing on differences in their life cycles. In Chapters 30 to 34, we will treat specific aspects of the biology of plants in more detail.

THE EVOLUTIONARY ORIGINS OF PLANTS

Plants are almost certainly derived from an organism that, if it existed today, would be classified as a multicellular green alga (division Chlorophyta). The great biochemical and morphological similarities between green algae and plants make this conclusion inescapable. Recall, for example, the patterns of cytokinesis discussed in Chapter 8: after mitosis is complete, animal cells pinch inward toward the center, whereas plant cells form an interior cell plate, which grows outward between the two daughter cells. Aside from plants, the only organisms that form a cell plate are a few groups of green algae.

By tradition, the major groups of plants are called *divisions,* rather than phyla, although the two units of classification are identical in rank and significance. There are two main groups of plants, the bryophytes and the vascular plants. The living vascular plants are grouped, in the system of classification used in this book, into nine divisions; several additional divisions are known only as fossils. The bryophytes, all of which fall into the division Bryophyta, include three major groups of plants: mosses, liverworts, and hornworts. All other plants, including ferns, conifers, flowering plants, and related groups, are vascular plants. Examples of both groups are shown in Figure 29-2.

A

FIGURE 29-2

Representatives of four divisions of plants.
A *A moss gametophyte; sporophytes are not visible.*
B *A branch of Norway spruce,* Picea nigra, *in the Alps. Seeds are produced in the large cones, and pollen is produced in the smaller ones.*
C *Maidenhair fern,* Adiantum pedatum.
D *Only flowering plants, such as this silk-cotton tree in the Peruvian Andes, add color to the landscape. Their flowers are specialized to attract insects and other animals. In all vascular plants, including the plants here (except for the moss), the sporophyte is the conspicuous generation.*

The bryophytes and vascular plants diverged a long time ago. The similarities between them are so great, however, that we believe they had a single common ancestor among the green algae. This ancestor was probably semiterrestrial. Fossils of simple vascular plants are known from the Silurian Period, some 430 million years ago; fossils of bryophytes are known from the Devonian Period, more than 360 million years ago. The common ancestor of these groups, therefore, must have invaded the land more than 430 million years ago. It was presumably intermediate between the green algae and the simplest known living plants.

> **Plants evolved from a multicellular green alga that possessed cell walls composed predominantly of cellulose and accomplished cell division by means of a cell plate. Both the bryophytes and the vascular plants were derived from such an organism, which may already have been semiterrestrial in its habits.**

THE GREEN INVASION OF THE LAND

Plants, fungi, and insects are the only major groups of organisms that occur almost exclusively on land; several other phyla, including chordates (see Chapter 38) and mollusks (see Chapter 36), are very well represented on land. The groups that occur almost exclusively in terrestrial habitats, however—the plants, fungi, and insects—all probably evolved there, whereas the chordates and mollusks certainly originated in the water. Of the groups that did evolve on land, the ancestors of the plants were almost certainly the first to become terrestrial. As we have seen in the case of fungi, and will see in the next section when we discuss insects, a major evolutionary challenge in the transition from a marine to a terrestrial

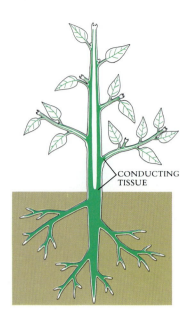

FIGURE 29-3

The conducting tissues of a vascular plant.

FIGURE 29-4

A stoma. The guard cells flanking the stoma, unlike the other epidermal cells, contain chloroplasts. Water passes out through the stomata, and carbon dioxide enters by the same portals.

habitat is desiccation—the tendency of organisms to lose water to the air. A key factor that doubtless helped to make possible the invasion of the land by the first plants was the possession by their green algal ancestors of relatively tough and durable cell walls that resisted drying out. Cell walls of this kind also made possible the evolution of reproductive structures that could pass through dry air without damage.

As in the case of the insects, efficient conducting systems helped the earliest vascular plants to achieve their ecological dominance of the land. Vascular plants are so named because they, unlike the bryophytes, have **vascular tissue.** The word *vascular* comes from the Latin *vasculum,* meaning a vessel or duct, and refers, in the case of plants, to their conducting systems. Vascular tissue consists of specialized strands of elongated cells that run from near the tips of a plant's roots through its stems and into its leaves. The vascular tissue conducts both water with dissolved minerals, which comes in through the roots, and carbohydrates, which are manufactured in the green, illuminated portions of the plant (Figure 29-3). Therefore, water and nutrients reach all parts of the plant, as do the carbohydrates that provide energy for the synthesis of its different structures. The bryophytes do not have such specialized conducting systems, but it is not clear whether they never had them or whether they lost them. In either case, bryophytes, being smaller than most vascular plants, do not have as great a need for transport within the plant.

Most plants also are well protected from desiccation in air by their **cuticle,** an outer covering formed from a waxy substance called **cutin.** The cuticles that cover the exposed surfaces of plants are impermeable to water and so provide a key barrier to water loss. However, there must be some way for water to leave a plant, since water flows in constantly through the roots; it cannot accumulate within the plant indefinitely, even though a constant supply is necessary for the maintenance of any plant. This problem is solved by passages through the cuticle—specialized pores called *stomata* (sing. *stoma;* Figure 29-4) in the leaves and sometimes in the green portions of the stems. Stomata allow carbon dioxide to pass into the plant bodies and water to pass out of them; the cells that border stomata expand and contract, thus controlling the movement of water and carbon dioxide. Cuticle apparently is absent in many bryophytes, and stomata occur only rarely among them, but a cuticle and stomata are found in virtually all vascular plants.

Vascular plants are defined primarily by their possession of a specialized conducting system, the vascular system, which involves strands of elongated, specialized cells that transport water and dissolved nutrients, and other strands that transport carbohydrate molecules. The possession of a cuticle and stomata is also more characteristic of vascular plants than of bryophytes.

In addition to their structural features, plants developed a special kind of relationship with fungi that evidently has been of key importance in their successful occupation of terrestrial habitats. Endomycorrhizae are characteristic of some 80% of all plants and are frequently seen in fossils of the earliest plants (see Figure 28-20). We presume that they probably also occurred in the common ancestor of plants. These symbiotic associations were probably critical to the ancestral plants as they adjusted to life on land. Certainly, mycorrhizae play an important role among today's plants in the assimilation of phosphorus and probably other ions, and they undoubtedly did so in the raw, unaltered soils that existed before the plants changed them. Just as lichens (see Chapter 28), which are symbiotic associations of fungi and algae or cyanobacteria, grow in many places where neither of their component organisms could survive alone, so the early plants, with their mycorrhizae, probably fared better on terrestrial soils than would have been possible otherwise.

Many other features developed gradually and aided the evolutionary success of plants on land. For example, in the first plants there was no fundamental difference between the aboveground and underground parts. Later, roots and shoots with specialized structures evolved, each better suited to its particular environment than the other (see Chapter 31).

Leaves—expanded areas of photosynthetically active tissue—evolved and diversified in relationship to the varied habitats that existed on land. Specializations in key reproductive features improved the methods by which plants protected their young and were dispersed from place to place.

The newly evolved and more specialized roots, stems, leaves, and reproductive features that evolved in plants all were important factors in the rise and overwhelming success of this group on land. There are some 250,000 species of vascular plants and about 16,000 additional species of bryophytes now in existence. Today plants dominate every part of the terrestrial landscape, except for the extreme polar regions and the tops of the highest mountains.

THE PLANT LIFE CYCLE
Alternation of Generations

Understanding the different kinds of life cycles that occur among plants provides an important key to understanding their evolutionary relationships. All plants exhibit alternation of generations, which also occurs in the brown, red, and green algae (see Chapter 26). In all of them, the diploid generation, or sporophyte, alternates with the haploid generation, or gametophyte. *Sporophyte* literally means "spore-plant," and *gametophyte* means "gamete-plant"; these terms indicate the kinds of reproductive structures that the respective generations produce (Figure 29-5).

Most adult animals are diploid, and in this respect they resemble the sporophyte generation of a plant. Such animals, however, produce eggs and sperm, which fuse directly to form a zygote. In contrast, the sporophyte generation of a plant does not produce gametes as a result of meiosis. Instead, meiosis takes place in specialized cells, called **spore mother cells,** and results in the production of haploid **spores,** the first cells of the gameto-

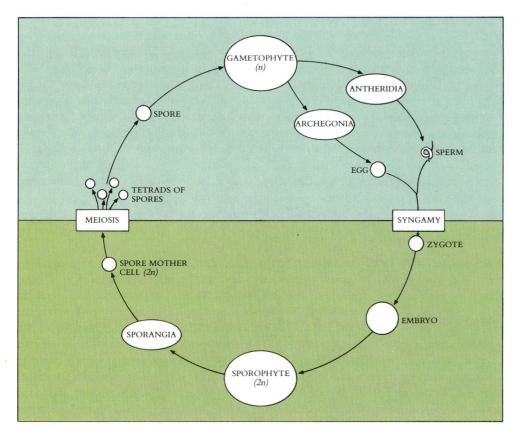

FIGURE 29-5

The life cycle of a vascular plant.

FIGURE 29-6

The fern gametophyte at right is about 1 centimeter across, whereas the pine gametophytes (pollen grains) below are just large enough to be visible with the naked eye.

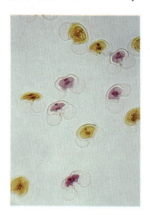

phyte generation. Spores do not fuse with one another, as gametes do; instead, they divide by mitosis, producing a multicellular haploid individual, the gametophyte (Figure 29-6).

In turn, the gametes—eggs and sperm—eventually are produced by the gametophyte, as a result of mitosis. They are haploid, like the gametophyte that produces them. When they fuse to form a zygote, the first cell of the next sporophyte generation has come into existence. The zygote grows into a sporophyte, in which meiosis ultimately occurs.

> **Plant life cycles are marked by an alternation of generations of diploid sporophytes with haploid gametophytes. Sporophytes, as a result of meiosis, produce spores, which grow into gametophytes. Gametophytes produce gametes, as a result of mitosis.**

The basic cycle involved in alternation of generations, then, is zygote-sporophyte-meiosis-spores-gametophyte-gametes (eggs and sperm)-syngamy-zygote (Figure 29-5). Such a cycle is characteristic of all plants. In some, including bryophytes and ferns, the gametophyte is green and free living; in others it is not green and is nutritionally dependent on the sporophyte. When you look at a moss or liverwort (bryophytes), what you see is largely gametophyte tissue; the sporophytes are usually smaller brownish or yellowish structures attached to or enclosed within the tissues of the gametophyte. In vascular plants, the gametophytes are always much smaller than the sporophytes, and, in most groups, are nutritionally dependent on them and enclosed within their tissues. What you see when you look at a vascular plant, with rare exceptions, is a sporophyte.

The Specialization of Gametophytes

The gametes of the first plants differentiated within specialized organs called **gametangia** (sing. **gametangium**), in which the eggs and sperm were surrounded by a jacket of cells. Complex multicellular gametangia of this sort are still found in all of the less specialized members of the plant kingdom, including the bryophytes and several divisions of vascular plants. Eggs and sperm are formed within different kinds of gametangia. Those in which eggs are formed are called **archegonia** (sing. **archegonium**), and those in which sperm are formed are called **antheridia** (sing. **antheridium**). An archegonium produces only one egg; an antheridium produces many sperm. These structures usually look very different from one another. In some bryophytes and less advanced vascular plants, including the ferns, antheridia and archegonia exist together on the same gametophyte; in other members of these groups, and commonly in bryophytes, the two kinds of gametangia are borne on

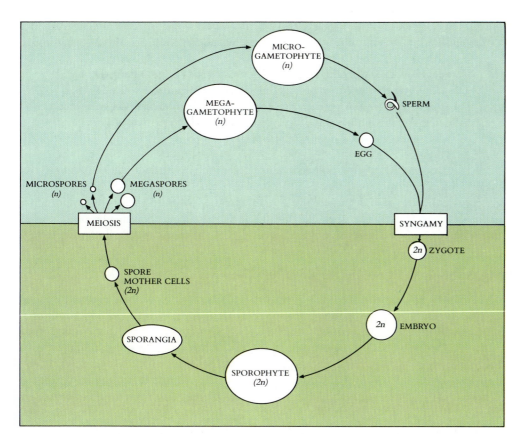

FIGURE 29-7

Diagram of the life cycle of a heterosporous vascular plant. Although for convenience both megaspores and microspores are shown here as arising from a single meiotic event, each such event would actually result in the production of all megaspores or all microspores.

separate gametophytes. In the more specialized vascular plants, including all of the vascular plants that form seeds, the gametangia have been lost during the course of evolution, and the eggs or sperm differentiate from individual cells of the respective gametophyte. Some of the gametophytes bear only eggs, and others bear only sperm.

Gametophytes may produce two different kinds of gametangia: (1) antheridia, which produce sperm, and (2) archegonia, each of which produces a single egg. In bryophytes these gametangia are multicellular, but these complex structures have been lost during the evolution of the more advanced vascular plants.

When one kind of gametophyte bears antheridia and another kind bears archegonia, the two kinds of gametophytes may look different from one another. If they do, the gametophytes that form antheridia are called **microgametophytes,** and those that form archegonia are called **megagametophytes.** These two formidable-looking terms literally mean no more than "small gametophytes" and "large gametophytes." The terms were used originally because the two kinds of gametophytes differ greatly in size in many kinds of plants.

In nearly all plants that do have two different kinds of gametophytes, the gametophytes arise from two different kinds of spores, **microspores** and **megaspores.** Plants that produce two morphologically different kinds of spores are called **heterosporous;** those that produce only one kind of spore are **homosporous** (Figure 29-7). Under this terminology, all bryophytes are homosporous, even though in many of them morphologically identical spores may give rise to either male or female gametophytes.

Spores are formed as a result of meiosis in the sporophyte generation. Their differentiation occurs within specialized, multicellular structures called **sporangia** (sing. **sporan-**

gium). If a plant forms both megaspores and microspores, each of these will be formed in a different kind of sporangium, called, respectively, **microsporangia** and **megasporangia**.

> **Gametophytes may produce either one or two kinds of gametangia. The gametophytes that produce only archegonia—or in more advanced forms, only eggs—are often larger than those that produce only antheridia, or sperm. In such cases, the female gametophytes are called megagametophytes; the male ones are called microgametophytes.**

Having now completed our overview of plant life cycles, we will consider the two major groups of plants, bryophytes and vascular plants. As we do so, we will see a progressive reduction of the gametophyte from group to group, a loss of multicellular gametangia, and increasing specialization for life on the land, culminating with the remarkable structural adaptations of the flowering plants, the dominant plant group today.

MOSSES AND OTHER BRYOPHYTES

The bryophytes consist of mosses, liverworts, and hornworts (Figure 29-8). Most bryophytes are small, and few exceed 2 centimeters in length. They are especially common in relatively moist places both in the tropics and in temperate regions. In the Arctic and the Antarctic, bryophytes are the most abundant plants. They boast not only the highest number of individuals in these harsh regions, but also the highest number of species. Many bryophytes, especially some of the mosses, are able to withstand prolonged periods of drought, contributing to their success in regions where the temperatures remain below freezing for much of each year. Most bryophytes, however, like lichens, are remarkably sensitive to air pollution, and they are rarely found in abundance in or near cities or other sources of industrial pollution. Also, regardless of how well or how long they can resist drought, the bryophytes require free water to reproduce sexually, as well as for their growth and development; they resemble the amphibians among the vertebrates in this respect. The places where they grow must be wet, at least during certain seasons.

The gametophytes of bryophytes are green and manufacture their own food; they are relatively large and evident (Figures 29-2, *A*, and 29-8) as compared with the sporophytes, which obtain their food directly from the gametophytes. Some sporophytes in the bryophytes are completely enclosed within gametophyte tissue; others, which are not enclosed, turn brownish or straw-colored at maturity.

FIGURE 29-8

Bryophytes.
A *A hair-cup moss,*
Polytrichum. *The leaves below belong to the gametophyte. Each of the yellowish-brown stalks, with the capsule at its summit, is a sporophyte. Although moss sporophytes may be green and carry out a limited amount of photosynthesis when they are immature, they are soon completely dependent, in a nutritional sense, on the gametophyte.*
B *A liverwort,* Marchantia. *The sporophytes are borne within the tissues of the umbrella-shaped structures that arise from the surface of the flat, green, creeping gametophyte. These particular structures develop archegonia within their tissues; another kind of similar structure, borne on different plants of* Marchantia, *produces the antheridia.*

A

B

The two major features that distinguish bryophytes from vascular plants are:

1. The general lack of specialized vascular tissues in bryophytes. Many moss sporophytes do have a central strand of somewhat specialized water-conducting tissue in their stems, a condition analogous to that which occurs in vascular plant sporophytes, but food-conducting tissue has been identified in only a few genera of bryophytes. Even if such tissues are present, their structures are much less complex than those found in the vascular plants.

2. The fact that the sporophytes of bryophytes are almost always smaller than, and always derive their food from, the gametophytes.

Rhizoids, which are slender, usually colorless projections that consist of one or a few cells, anchor most bryophytes to their substrate. Unlike the roots of vascular plants, rhizoids do not appear to play a major role in taking up water and minerals, which often enter a bryophyte directly through its stems or leaves.

Bryophytes are relatively simple plants. Taking all of their features into account, we may view them either as primitive plants or as plants that have become simplified during the course of their evolution. Which of these two possibilities accurately reflects the status of the bryophytes is not known.

Bryophytes lack the specialized vascular tissues that the vascular plants have. Their sporophytes photosynthesize only to a limited extent, if at all, and draw their food from the gametophytes on which they are produced.

The Classes of Bryophytes

The three classes of bryophytes are liverworts, with about 6500 species; hornworts, with about 100; and mosses, with nearly 10,000. The differences among these three classes are great, and some scientists have concluded that they are better considered separate divisions of plants, in which the common features do not reflect an immediate common ancestry.

Liverworts

Liverworts were given their name in medieval times, when the common belief in a "Doctrine of Signatures" held that a plant that resembled a particular part of the body was probably good for treating a disease of that organ; some common liverworts have an outline that resembles a liver, and thus they were thought to be useful in treating liver ailments. The ending -*wort* simply means "herb"—i.e., not a tree or shrub—and is found in many plant names of Anglo-Saxon derivation. In some liverworts, the gametophytes (Figure 29-9, *A*) grow flat along the ground and are relatively undifferentiated; they have a growing point at one end, where cell division takes place continuously, adding to the length of the liverwort. Other liverwort gametophytes have simple leaves, stems, and rhizoids, resembling mosses in their complexity. The sporophytes of liverworts often are more or less spherical and usually are held within the gametophyte tissue until they shed their spores.

Hornworts

Hornwort gametophytes resemble those of *Marchantia* (Figure 29-9, *B*) and similar genera among the liverworts. The sporophytes of hornworts, however, differ remarkably from those of liverworts: they are elongated capsules that stand up from the surface of the creeping gametophytes like horns, thus giving the English name to this class of plants.

Mosses

The largest class of bryophytes, and probably the one most familiar to you, is the mosses. Many small tufted plants are mistakenly called "mosses"—for example, "Spanish moss" is actually a flowering plant, a relative of the pineapple. Even eliminating these impostors,

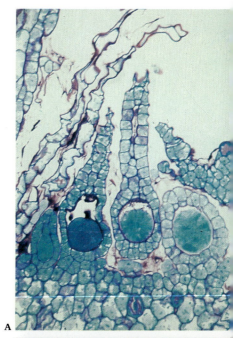

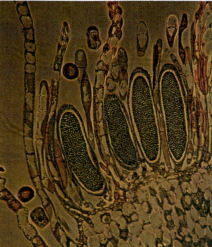

FIGURE 29-9

Gametangia of bryophytes.
A *Transection through archegonia of the liverwort* Marchantia. *Such archegonia are borne on the umbrella-shaped structures shown in Figure 29-8,* B. *A single egg differentiates after meiosis within the lower, swollen portion of each archegonium.*
B *Transections through a group of moss antheridia. The small cells in each of the elongate structures will undergo meiosis, each then giving rise to four biflagellated sperm. The sperm, when liberated by the rupturing of the antheridium, will swim through free water to the mouth of the archegonium.*

THE VERSATILITY OF MOSSES

Although mosses may seem to be a dull group of plants and unimportant in the global scheme of things, they are, like most things, surprisingly interesting when examined more closely. Here are three examples of why they are such interesting plants.

Fissidens, a common moss worldwide, often forms large masses in wet places or actually under the water. Shown here (Figure 29-A, *1*) is *Fissidens* growing on limestone along a small stream 12 kilometers west of Yalta, Crimea, USSR. Mosses are often ecological dominants where they occur, making important contributions to the fixation of carbon dioxide, the breakdown of rocks, and the stabilization of the soil. Probably the most important moss in terms of its ecological role, however, is *Sphagnum* (Figure 29-11).

Mosses also grow in the most extreme habitats where plants occur, as illustrated here. Mount Melbourne (Figure 29-A, *2*) is located at about 75 degrees south latitude in Antarctica; at about 3000 meters elevation on this mountain, the daily temperatures in summer mainly range between −10° and −30° C! In the small open patches of soil visible in Figure 29-A, *3*, however, is a moss of the genus *Campylopus,* growing in volcanically heated soil where the ground temperatures may reach as

high as 30° C. *Campylopus,* discovered for the first time in Antarctica at this site in late 1984, is mainly a tropical genus, but it has about six species in Australia and New Zealand.

You may not even think that the plants in these two photographs could be mosses. They are members of the genus *Splachnum,* which grow on mammal dung and are frequent in the cooler parts of the world. In these unusual mosses a portion of the stalk of the sporophyte, just below the capsule, is expanded up to 2 centimeters across and is purple, yellow, red, or otherwise colored. The spores of these mosses are sticky, unlike those of other mosses, and they are dispersed to other pieces of dung by flies, which visit the flowerlike structures shown here, attracted by the bright colors and sometimes also by the special odors that the mosses produce. Shown are two species that grow on moose dung in central Alberta, Canada: the purple *Splachnum rubrum* (Figure 29-A, *4*) and the yellow *Splachnum luteum* (Figure 29-A, *5*).

These three examples illustrate why the study of the world's approximately 10,000 species of mosses is so fascinating. There are doubtless many other stories just as interesting as these that remain to be discovered.

---

however, there are still some 10,000 species of true mosses, and they are found almost everywhere on earth. The gametophytes of mosses are almost always leafy, with small, simple leaves, and they may be tufted (Figure 29-8, *A*) or creeping (Figure 29-2, *A*). Moss sporophytes are often yellowish or brownish at maturity, bear a sporangium or capsule near their tip, and are borne individually on the gametophytes.

The gametophytes of liverworts may be either leafy or somewhat strap shaped and creeping; those of hornworts are always strap shaped; and those of mosses are almost always leafy. Liverwort sporophytes are unstalked; those of mosses are stalked and often very elaborate; and those of hornworts are elongated capsules.

Bryophyte Life Cycles

The flask-shaped archegonia of bryophytes are found among the leaves of those that have leafy gametophytes or near the growing ends of the liverworts that have strap-shaped gametophytes (Figure 29-9, *A*). A single egg is produced in the swollen basal portion of each archegonium. When this egg is mature, the central cells in the neck of the archegonium disintegrate, leaving an opening at the top and a column of fluid through which the sperm, which meanwhile have been released from the nearby antheridia, swim to the egg.

FIGURE 29-A, *4* Splachnum rubrum   FIGURE 29-A, *5* Splachnum luteum

Bryophyte antheridia commonly are stalked (Figure 29-9, *B*). In them, an outer layer of cells surrounds the specialized cells that become the sperm, which are twisted and bear two flagella each. These sperm swim through drops of water from rain, dew, or other sources to the neck of the archegonium and then into it to the egg. Fertilization takes place within the swollen basal portion of the archegonium. Inside the archegonium, the zygote develops into a young sporophyte, which grows out of the archegonium and, in mosses, differentiates into a slender, basal stalk, or **seta,** which has a swollen capsule, or **sporangium,** at its apex. In most liverworts, the sporangium remains within the tissues of the gametophyte.

Haploid spores are produced within the sporangium after meiosis. In mosses, the capsule often is covered, at least at first, with a cap formed from the swollen archegonium. Moss spores are shed from the capsule, usually in dry weather, after a specialized lid drops off, usually revealing a complex set of interlocking toothlike projections below it. The moss sporophyte, with its stalk and capsule, usually contains chlorophyll and may produce some of its own food, especially when young. By the time the sporophyte is mature, however, the chlorophyll usually has disintegrated; mature moss sporophytes draw their food from the photosynthetic gametophytes, remaining attached to them.

When the spores of mosses germinate, they give rise to threadlike filaments called **protonemata** (sing. **protonema**), which resemble green algae and grow over moist surfaces. The characteristic leafy gametophytes of mosses arise from buds that form on the

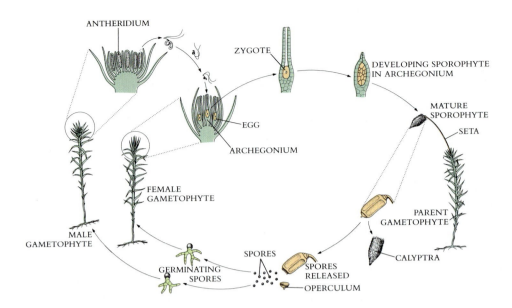

FIGURE 29-10

Moss life cycle.

ANTHERIDIUM

ZYGOTE

DEVELOPING SPOROPHYTE IN ARCHEGONIUM

MATURE SPOROPHYTE

SETA

EGG

ARCHEGONIUM

FEMALE GAMETOPHYTE

PARENT GAMETOPHYTE

MALE GAMETOPHYTE

CALYPTRA

GERMINATING SPORES

SPORES

SPORES RELEASED

OPERCULUM

FIGURE 29-11

Peat mosses, Sphagnum.
A *The glistening black sporangia emerge from the green, leafy gametophytes of* Sphagnum *on a stalk of gametophytic tissue. Unlike the sporangia of other mosses, those of* Sphagnum *have a lid that blows off explosively, releasing the spores.*
B *A peat bog being drained by the construction of a ditch. In many parts of the world, accumulations of peat, which may be many meters thick, are used as fuel.*

protonemata. Some moss gametophytes are cushiony, with erect stems; others are feathery, with creeping stems. In the mosses with erect stems, the leaves appear to be spirally arranged around the stems; in the feathery ones, they are crowded into two rows on opposite sides of the stem. Since they mostly lack the specialized and efficient conducting tissues and cuticle of the vascular plants, bryophytes cannot achieve large size, and thus they have never become as conspicuous or dominant as the vascular plants, although bryophytes are important in many kinds of terrestrial communities. A moss life cycle is summarized in Figure 29-10.

One of the kinds of mosses that is most important economically is *Sphagnum,* the peat mosses (Figure 29-11). These mosses, a highly distinctive group with more than 290 species, grow in boggy places. Here they form dense and deep masses that often are dried and used as fuel for domestic as well as industrial purposes. Peat is also used extensively in gardening and lawn care.

A

B

VASCULAR PLANTS

The first vascular plants appeared no later than the early Silurian Period, some 430 million years ago. The first group for which we have a relatively complete record belonged to the extinct division Rhyniophyta, which lived some 410 million years ago. One of these plants, *Rhynia,* was little more than a simple branching axis with sporangia at the tips of the branches (Figure 29-12). Other ancient vascular plants evolved leaves, which first appeared as simple emergences from the stems, and more complex arrangements of sporangia.

The vascular plants are distinguished from the bryophytes by their large, dominant, and nutritionally independent sporophytes; efficient conducting tissues (Figure 29-3); specialized leaves, stems, and roots; cuticles and stomata (Figure 29-4); and the fact that seeds evolved in the group (Figure 29-15).

Heterospory, which occurs only in the vascular plants, has evolved several times independently in different groups. Recall that heterospory is a condition in which a kind of plant regularly forms two different kinds of spores, one of which gives rise to egg-producing gametophytes, and the other of which gives rise to sperm-producing gametophytes (Figure 29-7). Seeds, which occur only in heterosporous plants, have evolved independently in a number of different divisions and classes of vascular plants. They represent the culmination of a major trend of evolution among the life cycles of vascular plants in which a progressive reduction in size of the gametophyte and its complete nutritional subordination, in most vascular plants, to the sporophyte, have taken place. Before illustrating this trend with concrete examples, however, we will discuss specialization in the conducting systems of the vascular plants, a feature that has been critical to the success of the group.

FIGURE 29-12

Drawing of structure of Rhynia.

> **Seeds occur only in heterosporous plants, and heterospory occurs only in vascular plants, not in bryophytes. The vascular plants are also characterized by their efficient conducting tissues; specialized stems, leaves, and roots; cuticles; and stomata.**

Conducting Systems of the Vascular Plants

Even in the very first vascular plants whose fossil record we know, the sporophytes were complex, with specialized conducting tissues. Just as in contemporary plants, they were characterized by **primary growth,** growth that results from cell division at the tips of the stems and roots (see Chapter 31). Within the stems formed as a result of primary growth were well-marked vascular bundles, which played the same conducting roles that they do in contemporary vascular plants. In the earliest vascular plants, however, there was no differentiation of the plant body into shoots and roots. The sharp separation of these two kinds of organs in a number of anatomical details is a property of nearly all modern plants. It reflects increasing specialization in relation to the demands of a terrestrial existence.

Secondary growth was an important early development in the evolution of vascular plants. It is basically an elaboration of the types of tissues that make up the primary vascular tissues. In secondary growth, cell division takes place actively in regions around the plant's periphery (see Chapter 31). In it, the conducting elements of the primary tissues are multiplied many times over as a result of the repeated divisions that occur in a cylindrical zone of dividing cells. Secondary growth makes it possible for a plant to increase in diameter. Only after the evolution of secondary growth could vascular plants become thick trunked and therefore tall. This evolutionary advance made possible the development of forests and, consequently, the domination of the land by plants. Judged from the fossil record, secondary growth had evolved independently in several different groups of vascular plants by the middle of the Devonian Period, some 380 million years ago (see Figure 23-14).

> **Primary growth results from cell division at the tips of stems and roots. Secondary growth results from the division of a cylinder of cells around the plant's periphery.**

There were two types of conducting elements in the earliest plants, elements that have become characteristic of the vascular plants as a group. **Sieve elements** are soft-walled cells that conduct carbohydrates away from the areas where they are manufactured. **Tracheary elements** are hard-walled cells that transport water and dissolved minerals up from the roots. Both kinds of cells are elongated, and both occur in strands. Sieve elements are the characteristic cell types of a kind of tissue called **phloem;** tracheary elements are the characteristic cell types of a kind of tissue called **xylem.** In primary tissues, which result from primary growth, these two types of tissue often are associated with one another in the same vascular strands.

> **Water and nutrients are carried in the xylem, which consists primarily of hard-walled cells called tracheary elements. Carbohydrates, in contrast, are carried in the phloem, which consists of soft-walled cells called sieve elements.**

The Divisions of Living Vascular Plants

The nine divisions of living vascular plants may be summarized briefly as follows:

Psilophyta. Whisk ferns. Homosporous. Motile sperm. No differentiation between root and shoot. Two genera and several species.

Lycophyta. Lycopods (including club mosses and quillworts). Homosporous or heterosporous. Motile sperm. Four genera and about 1000 species.

Sphenophyta. Horsetails. Homosporous. Motile sperm. One genus, 15 species.

Pterophyta. Ferns. Homosporous; a very few heterosporous. Motile sperm. About 12,000 species.

Coniferophyta. Conifers (including pines, spruces, firs, yews, redwoods, and others). Heterosporous, seed-forming. Sperm lack flagella. About 50 genera, with about 550 species.

Cycadophyta. Cycads ("sago palms"). Heterosporous, seed-forming. Sperm flagellated and motile but carried to the vicinity of the egg by a pollen tube. Palmlike plants with sluggish secondary growth compared with that of the conifers. Ten genera, about 100 species.

FIGURE 29-13

Representatives of four of the divisions of vascular plants.
A *The club moss* Lycopodium lucidulum *(division Lycophyta). Although superficially similar to the gametophytes of mosses, club moss plants are sporophytes. Meiosis occurs in the sporangia of the cones, ultimately resulting in the shedding of haploid spores. These spores grow into gametophytes that, in* Lycopodium, *bear both antheridia and archegonia.*
B *A horsetail,* Equisetum, *the only living genus of division Sphenophyta. The spores produced by meiosis in the cones give rise to a single kind of tiny, green, nutritionally independent gametophytes.*
C *An African cycad,* Encephalartos kosiensis. *The cycads are seed plants with fernlike leaves and sperm that swim within the pollen tubes.*
D *Maidenhair tree,* Ginkgo biloba, *the only living representative of the division Ginkgophyta, a group of plants that was abundant 200 million years ago. Among living seed plants, only the cycads and* Ginkgo *have swimming sperm.*

A

B

Ginkgophyta. Ginkgo. Heterosporous, seed-forming. Sperm flagellated and motile but carried to the vicinity of the egg by a pollen tube. Deciduous tree with fanlike leaves and seeds resembling a small plum, with fleshy, ill-scented outer covering (see p. 11). One species.

Gnetophyta. Mormon tea (*Ephedra*), *Gnetum, Welwitschia* (see Figure 50-B, *6,* p. 1099). Heterosporous, seed-forming. Sperm not motile. Shrubs, vines, three very diverse genera, about 70 species.

Anthophyta. Flowering plants, or **angiosperms.** Heterosporous, seed-forming. Sperm not motile. Range from very diverse, tiny plants to some of the largest trees known. About 235,000 species.

Representatives of four of these divisions are shown in Figure 29-13, and photographs of ferns, conifers, and flowering plants are included elsewhere in this chapter. Among living plants, the flowering plants (division Anthophyta) are by far the largest and most dominant group. Even though the 550 species of conifers (division Coniferophyta) are of great importance ecologically and are dominant over vast areas—particularly in harsh environments such as the far north and south, or at high elevations (Figure 29-14)—they are by no means as abundant or dominant on a global scale as are the flowering plants.

Seeds

Seeds are a characteristic feature of some groups of vascular plants but not of others. A seed is a developing sporophyte individual, surrounded by a tough protective coat, whose embryonic development has been temporarily arrested. Seeds are units by means of which plants, being rooted in the ground, are dispersed to new places; many seeds have devices, like the wings on the seeds of pines or maples, or the plumes on the seeds of a dandelion, that help them to travel very efficiently. From a morphological point of view, the outer layers of a seed consist of tissues derived from the parent sporophyte on which it occurs; inside are the derivatives of the megasporangium of that sporophyte, enclosing the derivatives of the gametophyte, and, within them, the young sporophyte of the next generation, the embryo (Figure 29-15).

To understand how seeds are produced, we need to return to our consideration of the plant life cycle. Microspores, which are produced within the microsporangia as a result

FIGURE 29-14

This forest in Rocky Mountain National Park in Colorado, like many temperate forests, is dominated by conifers. Conifers are able to outcompete the angiosperms at relatively high altitudes and latitudes, probably owing in part to their ectomycorrhizal fungi.

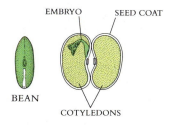

FIGURE 29-15

Diagram of a seed, with tissues in it and their sources identified.

of meiosis, ultimately differentiate into microgametophytes. The microgametophytes of seed plants are called **pollen grains** when they are ready to be shed by their parent plant. Some of these pollen grains are transported to the vicinity of the megagametophyte, which remains held within the protective coverings of the parent sporophyte.

Inside the pollen grain, the cells divide further, leading ultimately to the production of a mature microgametophyte. Among the products of this division are the cells that become the sperm. Within the megagametophyte, cell division also takes place, ultimately giving rise to an egg. After a pollen grain has reached the vicinity of the megagametophyte, it germinates, and a slender **pollen tube** grows toward the egg. When this tube bursts, the sperm and the egg come into contact, and syngamy takes place, all within the protective coverings surrounding the original megasporangium. The resulting zygote begins to divide and eventually grows into an embryo, a multicellular individual that actually is already a young plant belonging to the succeeding sporophyte generation.

After a certain point, the embryo stops growing and dividing. The megagametophyte remains enclosed within the megasporangium at all times, disintegrating in some plant divisions and becoming enlarged in others in which it is specialized for food storage. Around the megagametophyte and embryo, the seed coat becomes mature and more or less hard or tough, depending on the particular kind of plant. When the seed is mature, it is shed by its parent plant. Within the seed, the young embryo is enclosed in a protective envelope formed from the tissues of the parent sporophyte. Germination occurs when there is a sufficient supply of water and the other conditions are appropriate. The seed coat splits or cracks open, while the embryo resumes the process of cell division and grows, eventually becoming an adult sporophytic plant.

> **A seed is a special kind of dispersal unit that occurs in several groups of vascular plants. In a seed, the megasporangium is retained within tissues derived from the parent sporophyte. Food is stored within the seed, and the young sporophyte of the next generation, called the embryo, remains dormant once it reaches a certain size.**

The seed is a crucial adaptation to life on land because it protects the embryonic plant from drying out or being eaten when it is at its most vulnerable stage. Most kinds of seeds have abundant food stored in them, either inside the embryo or in specialized storage tissue. This food is used as a ready source of energy by the rapidly growing young plant, playing the same role as the yolk of an egg. The evolution of the seed clearly was a critical step in the domination of the land by plants.

Evolution of the Seed Plants

Seeds have evolved independently a number of times since their first appearance in the fossil record in the Devonian Period more than 360 million years ago. In seeds, all of the products of the mature megagametophyte, together with the young plant of the next generation, are included together in a neat, drought-resistant package. Both of the gametophytes are greatly reduced, and a basically new kind of life cycle has been established.

Of the nine divisions of vascular plants that were listed on pp. 596–597, the living representatives of the Psilophyta, Lycophyta, Sphenophyta, and Pterophyta do not have seeds, and the other vascular plants do. Some extinct groups of lycopods and horsetails did have seeds, however. The living seedless vascular plants form antheridia and archegonia, and produce free-swimming sperm that require the presence of free water for fertilization. In contrast, most seed plants have nonflagellated sperm; none form antheridia, and few form archegonia.

The living seed plants are remarkably diverse. For example, the whisk ferns, some lycopods, horsetails, and ferns are homosporous, whereas all other vascular plants, including all seed plants, are heterosporous. Microsporangia may be produced on the same

FIGURE 29-16

Scanning electron micrographs of two very different kinds of angiosperm pollen grains.
A *False aloe, Agave virginica. The pollen tube that develops in pollen of this sort grows through a single furrow, which is on the other side of the pollen grain.*
B *A plant of the sunflower family, Hyoseris longiloba. There are three pores hidden among the ornamentation of this pollen grain; the pollen tube may grow out through any one of them.*

A

B

individual plants that bear megasporangia, or on other plants. The microspores develop into microgametophytes; these microgametophytes, at the stage when they are shed from the plant that formed them, are called pollen grains (Figures 29-6, *B*, and 29-16), and they complete their development, as you saw earlier, after reaching their destination. The structures within which the megagametophytes of seed plants are held are called **ovules** (Figure 29-17); pollen grains reach the vicinity of the ovules as a result of a process called **pollination,** which may take place passively, the pollen blowing in the wind, or as a result of the activities of insects or other animals (see Chapter 30).

Gymnosperms

The four divisions of living seed plants in which the ovules are not completely enclosed by the tissues of the sporophytic individual on which they are borne at the time of pollination—conifers, cycads, ginkgo, and gnetophytes—are called **gymnosperms.** This name combines the Greek root *sperma,* or "seed," with *gymnos,* or "naked": in other words, naked-seeded plants. In fact, the seeds of gymnosperms often become enclosed by the tissues of their parents by the time they are mature, but their ovules are indeed naked at the time of pollination.

Botanists cannot agree whether the four divisions of gymnosperms had an immediate common ancestor or not. Although they resemble one another in the basic features of their leaves, they are diverse in other respects. For example, the cycads and ginkgo still have motile sperm, as do the vascular plants that do not form seeds. Some of the gymnosperms have archegonia, and others have lost them in the course of evolution, as the flowering plants, or angiosperms, have done. Archegonia are found in the bryophytes, as we have seen, as well as in all of the vascular plants that do not have seeds; their progressive reduction in size and eventual loss constitutes a clear evolutionary trend.

> **Because the five divisions of living seed plants are so diverse, the concept of "gymnosperms"—all seed plants that are not angiosperms (flowering plants)—is now recognized as an artificial one.**

Angiosperms

The flowering plants (division Anthophyta) differ from other seed plants in that their ovules are enclosed within the tissue of the parent sporophyte in structures called **carpels** (Figure 29-18). The pollen grains of the flowering plants reach a specialized portion of the carpel and then germinate; thus pollination in the plants of this division is indirect. Pollen tubes that emerge from the pollen grains grow down through the tissue of the carpel, reaching the ovule and ultimately the egg. This contrasts with the gymnosperms, in which the ovules are directly exposed to the air, at least at the time of pollination. The pollen reaches them directly or falls in their vicinity. Because of their enclosed ovules and seeds, the flowering plants are called **angiosperms,** a name derived from the Greek words *angion,* or "vessel," and *sperma.* The "vessel," or carpel, ultimately matures into a **fruit** that encloses the mature ovules, or seeds. The fruit itself has become an important unit of dispersal in the flowering plants, as we will consider further in the next chapter.

> **The angiosperms are characterized primarily by features of their reproductive system. A unique structure known as the carpel encloses the ovules and matures into the fruit. Since the ovules are enclosed, pollination is indirect.**

The angiosperms, although they too have seeds, could not have evolved directly from any of the living divisions of gymnosperms; they differ from each other in too many important ways. However, we would classify the ancestor of the angiosperms as a gym-

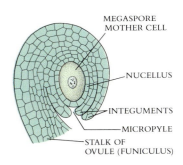

FIGURE 29-17

Diagram of an ovule. The outer covering, or integument, will mature to form the seed coat.

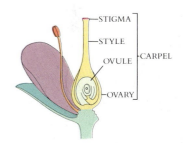

FIGURE 29-18

Diagram of a carpel.

nosperm if we knew what it was—it must have formed seeds and had naked ovules. The form of enclosure of the seeds is what defines an angiosperm, and seeds must originally have been present—and naked—for this form of enclosure to have evolved.

VASCULAR PLANT LIFE CYCLES

To illustrate the diversity among vascular plants, we will now outline briefly and comparatively the major features of the life cycles of a fern (division Pterophyta), a pine (division Coniferophyta), and a flowering plant (division Anthophyta). These three life cycles progress, in a sense, from simple to complex; but it should not be assumed that one of these groups has given rise to another. Nevertheless, the kind of progression that these life cycles represent must be similar to that which occurred during the evolution of these groups. For example, the ancestors of the flowering plants must have had complex gametophytes.

The Fern Life Cycle

Fern gametophytes are green, photosynthetic plants that live in relatively moist places and resemble the gametophytes of certain liverworts. They are flat, thin, and heart shaped, usually no more than 1 centimeter in length and somewhat less than that in width (Figure 29-6, *A*). Rhizoids project from their lower surface. Archegonia form on the lower surface of the gametophytes near the **apical notch,** the region of the most rapid cell division. The antheridia are formed somewhat farther back, but also on the lower surface. When the sperm (Figures 29-19 and 29-20) are mature, they are released from the antheridia. They require the presence of droplets of free water to make their way to the mouth of the archegonium. The sperm probably are attracted by chemicals released from the archegonium. When the sperm reach the mouth of the archegonium, they swim into the neck and eventually reach the egg in the swollen lower portion of the archegonium. When they do, a sperm fuses with the single egg, producing a zygote.

The zygote begins to divide within the archegonium of the fern. Soon the growing sporophyte becomes much larger than the gametophyte and also nutritionally independent from it. When this occurs, the sporophyte anchors itself in the substrate and grows into a plant like the familiar fern plants that we see in the woods. Most ferns have more or less horizontal stems that creep along below the ground; subterranean stems of this kind are called **rhizomes.** True roots arise from these rhizomes and absorb water and nutrients from the soil in which the fern is growing. The leaves are borne on the rhizomes, just as leaves are borne on the stems of vascular plants in general, but they stand vertically in ferns and are more or less tufted, depending on the nature of the rhizome. On these leaves, which are called **fronds,** sporangia ultimately differentiate. Meiosis takes place within these sporangia, followed by differentiation of the haploid spores, which ultimately are released from the sporangia. Only a few genera of ferns are heterosporous; one of them, incidentally, is the water fern *Azolla,* cultivated in rice paddies to encourage nitrogen fixation by its symbiotic cyanobacterium *Anabaena* (see Figure 26-3). Fern spores germinate if they fall on a suitable substrate; they then start a new gametophyte generation.

The fern life cycle differs from that of a moss primarily in the much greater development, independence, and dominance of the fern's sporophyte. In addition, the fern's sporophyte is much more complex than that of the moss, having vascular tissue and well-differentiated roots, stems, and leaves.

> For the most part, fern life cycles resemble those of mosses. Fern sporophytes, however, are green and become independent nutritionally. The fern plants that we see are sporophytes; fern gametophytes are only a centimeter or two across. Most ferns are homosporous.

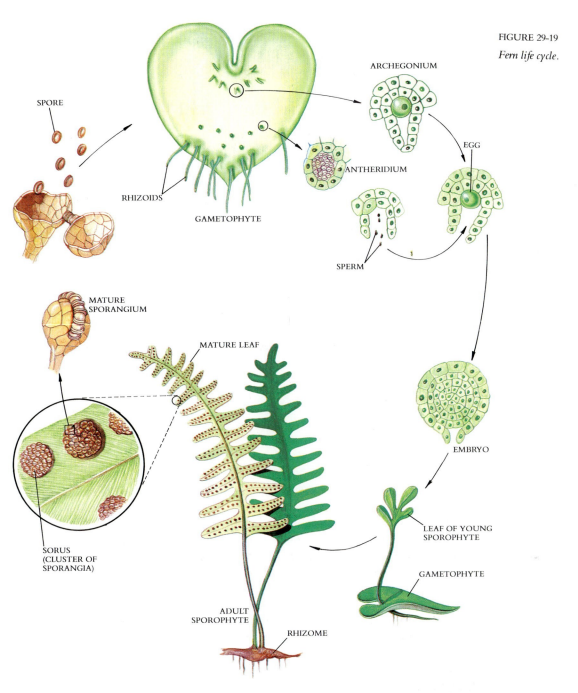

FIGURE 29-19

Fern life cycle.

SPORE

ARCHEGONIUM

RHIZOIDS

GAMETOPHYTE

ANTHERIDIUM

EGG

SPERM

MATURE
SPORANGIUM

MATURE LEAF

EMBRYO

SORUS
(CLUSTER OF
SPORANGIA)

LEAF OF YOUNG
SPOROPHYTE

GAMETOPHYTE

ADULT
SPOROPHYTE

RHIZOME

FIGURE 29–20

*The sporangia of most ferns
develop on the underside of the
leaves, and they are usually
clustered, as shown here in a fern
from the rain forests of Sumatra.*

FIGURE 29-21

Pine life cycle.

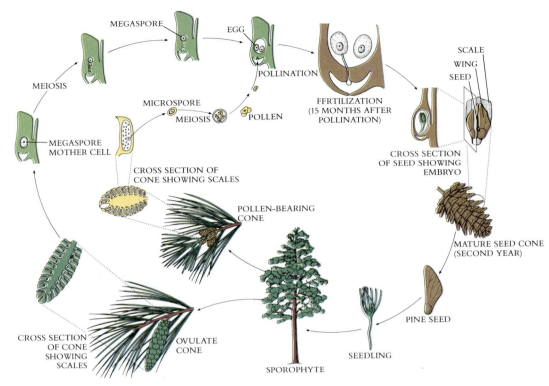

FIGURE 29-22

A *Sugar pine,* Pinus lambertiana, *in the Sierra Nevada of California. One of the largest and most impressive of all the pines.*

B *Ovulate cone and a group of microsporangiate cones of lodgepole pine,* Pinus contorta, *in Montana.*

The Pine Life Cycle

In the pine the megaspores are held within ovules, as they are in all other seed plants (Figure 29-21). The megagametophytes differentiate within the protective tissue of the parent sporophyte. In the pine and other gymnosperms, the megagametophyte continues to grow and becomes the place in which food is concentrated in the seed. The seeds of pines, by virtue of this stored food, not only protect but also nourish the young embryos that they contain.

Pine microsporangia are borne in pairs on the surface of the thin scales of the relatively delicate pollen-bearing cones (shown in the closely related genus *Picea* in Figure 29-2, *B*). When the young microgametophytes, or pollen grains, are shed, they consist of four cells. The pollen grains of pines are shed in huge quantities (see Figure 20-7), often appearing as a yellow scum on the surface of ponds, lakes, and even windshields.

The familiar seed-bearing cones of pines (Figure 29-22, *B*) are much heavier and more substantial structures than the pollen-bearing cones. Two ovules, and ultimately two seeds, are borne on the upper surface of each scale. In spring, when these cones are small and young, their scales are slightly separated. Drops of sticky fluid, to which the airborne pollen grains adhere, form between these scales. After pollination, the cone scales grow together, better protecting the developing ovules. The pollen tube grows slowly through the tissue of the ovule and ultimately, by two further cell divisions, becomes six celled and mature. Two of these six cells then become sperm. These sperm lack flagella and are carried passively to the egg, with which one of them ultimately fuses, producing a zygote. The whole process takes place so slowly that this fusion occurs some 15 months after pollination.

The growth and differentiation of the embryo take place within the developing seed, where the embryo is nourished by the food-rich tissue of the expanded megagametophyte and protected by the developing seed coat. The seed coat and embryo are diploid, whereas the nutritious tissue of the megagametophyte is haploid. This food-rich tissue, therefore, represents the gametophytic generation that intervenes between the two sporophytic ones. Dormant within the seed, the embryo consists of a shoot-root axis and several (often eight) seed leaves. Pine seeds usually are shed from the cone during the autumn of the year that follows the cone's first appearance. These seeds are winged and flutter to the ground, germinating if conditions are favorable and then giving rise to a new tree.

> **In seed plants, a group that includes gymnosperms and angiosperms, the megagametophytes are completely enclosed within sporophytic tissue. In the seeds of gymnosperms, stored food is provided for the embryo within tissue derived from the megagametophyte.**

The Flowering Plant Life Cycle

In angiosperms, the life cycle is generally similar to that of pines and other gymnosperms, but there are some major differences. To illustrate these differences, it is necessary to describe the basic structure of the flower. A complete flower consists of four **whorls** (Figure 29-23), circles of similar parts arising at about the same point on an axis. Our discussion will go from the inside parts out.

The innermost whorl contains the carpels. The carpels are folded, often leaflike structures that enclose the ovules (Figure 29-18). The receptive area of the carpel, the **stigma,** is often separated from the body of the carpel by a stalk called the **style.** Pollination occurs in angiosperms when pollen reaches the stigma. Subsequently, the pollen tube must grow through the stigma and style into the carpel to reach the ovule.

Because individual carpels, with their thick lower portion, slender stalk, and terminal stigma, were once thought to resemble the pestles used to grind powders in mortars, they have traditionally been called **pistils.** They are often fused together; when that is the case, they may collectively resemble a pestle even more. The term *carpel* is morphologically more precise, however, since it refers to one member of this whole whether it is fused with others or not; we will therefore use *carpel* in this book. Many angiosperms may have only one carpel in each flower, but the whorl in which the carpels occur collectively is still called the **gynoecium,** a term derived from the Greek words *gynos,* which means "female," and *oikos,* or "house." Just as the mature ovules become the seeds, so the mature carpel, with its enclosed seeds and sometimes other adhering flower parts, becomes the fruit. Fruits are just as characteristic of the angiosperms as are the flowers from which they develop.

The next whorl of the flower below the gynoecium on the floral axis is the **androecium** (Greek *andros,* "male"). The individual members of the androecium are the **stamens.** In most angiosperms, the stamens consist of a slender, threadlike portion, the **filament,** and a terminal, thicker, two-lobed **anther.** The anther contains four microsporangia, or **pollen sacs** (Figure 29-23), within which the microspores are formed. The products of differentiation within these microspores are shed as pollen grains at either a two- or three-celled stage. When mature, the microgametophytes have three cells; two are sperm, and the third may play a role in pollen tube growth. The sperm of angiosperms lack flagella. If the pollen

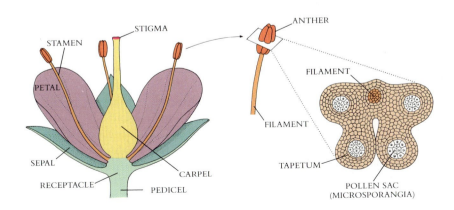

FIGURE 29-23

Diagram of an angiosperm flower.

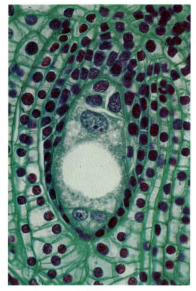

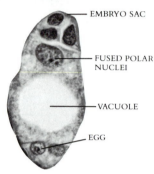

EMBRYO SAC

FUSED POLAR
NUCLEI

VACUOLE

EGG

FIGURE 29-24

The mature embryo sac of a lily
(Lilium), *showing six of the eight*
nuclei present at maturity.

FIGURE 29-25

Endosperm-rich (corn) and
endosperm-poor (bean) seeds.

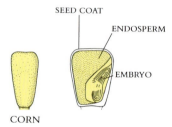

SEED COAT

ENDOSPERM

EMBRYO

CORN

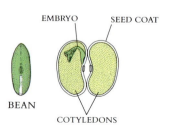

EMBRYO   SEED COAT

BEAN

COTYLEDONS

grains are shed at the two-celled stage, one of the cells divides after the germination of the pollen tube, while it is growing through the carpel toward the ovule. In such plants, the microgametophyte reaches its mature, three-celled state within the tissue of the carpel.

A mature angiosperm microgametophyte has three cells. As pollen, it may be shed at either the two- or the three-celled stage. If it is shed when it has only two cells, one further mitotic division occurs afterward.

In many angiosperms, the pollen grains are carried from flower to flower by insects and other animals that visit the flowers for food or other rewards, or are deceived into doing so because the characteristics of the flower suggest that they may offer such rewards. The ways in which pollination occurs, and some of the floral characteristics associated with different modes of pollination, will be discussed in the next chapter. In other angiosperms, the pollen is blown about by the wind, as it is in most gymnosperms, and reaches the stigmas passively. Although in a few instances insects have been observed feeding on the pollen of gymnosperms, such visits apparently are rare. In still other angiosperms, the pollen does not reach other individuals at all; instead, it is shed directly onto the stigma of the same flower. This results in self-pollination and inbreeding.

The mature megagametophyte of an angiosperm develops from the functional megaspore derived from a single meiotic event, which is the same way this process occurs among the gymnosperms. Meiosis occurs within the **nucellus,** a term used to describe the specific kind of megasporangium found in angiosperms. The nucellus is enfolded by specialized tissues, called the **integuments,** which are derived from the parent sporophyte. Together, these structures make up the ovule (Figure 29-17), the structure that ultimately matures to form the seed.

A mature angiosperm megagametophyte is called an **embryo sac.** An embryo sac is produced by the division of the functional megaspore, which is haploid, within the tissues of the nucellus. In the kind of embryo sac that is characteristic of about 70% of the angiosperms, three successive mitotic divisions, coupled with the differentiation of cell walls around the resulting haploid nuclei, result in an embryo sac with eight nuclei and seven cells (Figure 29-24). Within the embryo sac are the following kinds of cells: the egg, flanked by two cells, at one end; three cells at the far end; and a large cell at the approximate center of the embryo sac that contains two **polar nuclei.** The embryo sacs of the other 30% of angiosperm species also have an egg at one end and one or more polar nuclei in their central regions, but they differ in other respects.

In angiosperms, the mature megagametophyte is called an embryo sac. The most frequent type of embryo sac has eight nuclei, enclosed within seven cells. At one end is the egg, flanked by two cells; at the other end are three other cells; and in the center are two nuclei, the polar nuclei, in a single, large cell.

Fertilization in angiosperms is a unique process that is highly characteristic of this division of plants. Both of the sperm in the mature microgametophyte are functional. The first fuses with the egg, as in all organisms, forming the zygote. The second sperm, however, fuses with the polar nuclei, forming the **primary endosperm nucleus,** which divides more rapidly than the zygote and gives rise to a nutritious tissue called the **endosperm** (Figure 29-25). If two polar nuclei fuse with the second sperm, then the primary endosperm nucleus, and consequently the endosperm, is **triploid,** since triploid cells have three sets of chromosomes. The process whereby both the egg and the polar nuclei are fertilized, forming the zygote and the primary endosperm nucleus, is called **double fertilization.**

Endosperm, which only angiosperms possess, is the primary nutritional tissue on which the developing embryos of these plants depend. In some angiosperms, such as the common bean or pea, the endosperm is fully used up by the time the seed is mature. In such

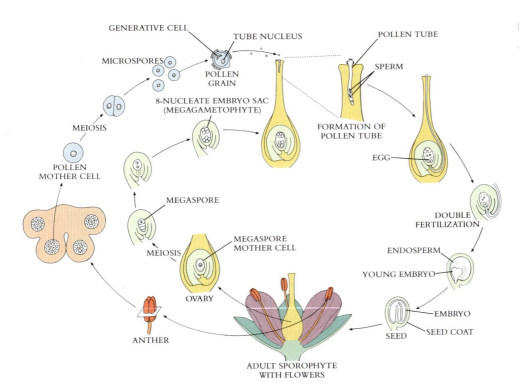

FIGURE 29-26

Angiosperm life cycle.

GENERATIVE CELL
TUBE NUCLEUS
POLLEN TUBE
MICROSPORES
POLLEN GRAIN
SPERM
8-NUCLEATE EMBRYO SAC (MEGAGAMETOPHYTE)
MEIOSIS
FORMATION OF POLLEN TUBE
POLLEN MOTHER CELL
EGG
MEGASPORE
DOUBLE FERTILIZATION
MEIOSIS
MEGASPORE MOTHER CELL
ENDOSPERM
YOUNG EMBRYO
OVARY
ANTHER
EMBRYO
SEED COAT
SEED
ADULT SPOROPHYTE WITH FLOWERS

plants, the seedling leaves often become swollen and fleshy and contain most of the stored food reserves. In other plants, such as corn, the mature seed contains abundant endosperm, which the embryo draws on as it germinates and starts to grow. In any case, endosperm plays the same role in nourishing angiosperm embryos that the haploid tissue of the megagametophyte does in gymnosperms.

Double fertilization, a process that is unique to the angiosperms, occurs when one sperm nucleus fertilizes the egg and the second one fuses with the polar nuclei. These two events result in the formation of, respectively, the zygote and the primary endosperm nucleus. The latter divides to produce the endosperm, the nutritive tissue that occurs in the seeds of angiosperms.

Apart from double fertilization, the development of angiosperm seeds is roughly the same as that of gymnosperm seeds. When the seed coats are mature and tough, the seed is shed. If it germinates in favorable conditions, a sporophyte is established. The angiosperm life cycle is summarized in Figure 29-26.

Continuing our consideration of the angiosperm flower (Figure 29-23), let us now turn to the two outer whorls, which together are called the **perianth.** Although not directly connected with the production of gametes, they also help to establish the distinctive character of the angiosperms. Below the androecium is the **corolla,** the members of which are called **petals.** Petals are often brightly colored and may perform an important role in attracting insects and other pollinators. The fourth and outermost whorl of the flower is the **calyx,** made up of **sepals.** The calyx, like the petals, may be colored and may play a role in luring pollinators, but it is more often greenish and not particularly attractive. It forms the outer layer of the bud before the flower opens and appears to serve mainly a protective function.

In the course of evolution, the members of any whorl of the flower may have become fused with one another or with those of another whorl. Tubular corollas made up of fused petals are frequent, for example. Another common floral plan is one in which the calyx, corolla, and stamens are fused around the gynoecium, whose members are also fused together. The arrangement of the parts in such a flower gives the impression that all the flower parts arise at the top of the **ovary,** which is a term used for the portion of the flower that becomes the fruit—it includes the gynoecium. Examples of several different kinds of angiosperm flowers may be seen in the next chapter.

A Very Successful Group

The vascular plants, which originated in connection with their invasion of the land, have come to dominate terrestrial habitats everywhere, except for the highest mountains and the polar regions. All plants, including bryophytes, have an alternation of generations, but this alternation has become modified by the reduction of the gametophyte during the evolution of the various divisions of vascular plants.

The reduction in size and complexity of the gametophyte generation in vascular plants can be traced easily. In ferns, as in bryophytes, the gametophyte is a green, photosynthetic, independent plant that forms multicellular archegonia and antheridia. Pines and flowering plants, in contrast, have two kinds of gametophytes, the egg-producing megagametophytes and the sperm-producing microgametophytes. In the pine, the megagametophyte is a complex, multicellular structure that bears archegonia within which eggs are formed. In flowering plants, the megagametophyte is an embryo sac, most often containing eight nuclei and seven cells at maturity.

Accompanying this reduction in the size and complexity of the gametophytes has been the appearance of the seed. Seeds are highly resistant structures that are well suited to protect a plant embryo from drought and predators. In addition, most seeds contain a supply of food for the young plant. Seeds are dispersed in many different ways, some involving animals. Flowers, which evolved among the angiosperms, are primarily a device whereby animals are induced to spread the pollen from the flowers of one individual plant to those of another. They are, in effect, a mechanism whereby plants can overcome their rooted, immobile nature and secure the benefits of wide outcrossing in promoting genetic diversity.

SUMMARY

1. Plants are derived from an aquatic ancestor, but the evolution of their conducting tissues, cuticle, stomata, and seeds has made them progressively less dependent on water. The oldest plant fossils date from the Silurian Period, some 430 million years ago.

2. The common ancestor of the plants was a green alga. The similarity of the members of these two groups can be demonstrated by their photosynthetic pigments (chlorophylls *a* and *b,* carotenoids); chief storage product (starch); cellulose-rich cell walls (in some green algae only); and cell division by means of a cell plate (in certain green algae only).

3. The major groups of vascular plants are the bryophytes (division Bryophyta)—mosses, liverworts, and hornworts—and the vascular plants, which make up nine other divisions. Vascular plants have well-defined conducting strands; some of these strands are specialized to conduct water and dissolved minerals, and others are specialized to conduct the food molecules the plants manufacture.

4. All plants have an alternation of generations, in which haploid gametophytes alternate with diploid sporophytes. The spores that sporophytes form as a result of meiosis grow into gametophytes, which produce gametes—sperm and eggs—as a result of mitosis.

5. The gametophytes of bryophytes and of ferns are green and nutritionally independent. The sporophytes of bryophytes are mostly brown or straw-colored at maturity and nutritionally dependent, at least in part, on the gametophytes. In ferns, sporophytes and gametophytes usually are both green and nutritionally independent. Among the gymnosperms and angiosperms, the gametophytes are nutritionally dependent on the sporophytes.

6. In all seed plants—gymnosperms and angiosperms—and in certain lycopods and a few ferns, the gametophytes are either female (megagametophytes) or male (microgametophytes). Megagametophytes produce only eggs; microgametophytes produce only sperm. These are produced, respectively, from megaspores, which are formed as a result of meiosis within megasporangia, and microspores, which are formed in a similar fashion within microsporangia.

7. In gymnosperms, the ovules are exposed directly to pollen at the time of pollination; in angiosperms, the ovules are enclosed within a carpel, and a pollen tube grows through the carpel to the ovule.

8. The nutritive tissue in gymnosperm seeds is derived from the expanded, food-rich gametophyte. In angiosperm seeds, the nutritive tissue is unique and is formed from a cell that results from the fusion of the polar nuclei of the embryo sac with a sperm cell.

9. The pollen of gymnosperms is usually blown about by the wind; that of angiosperms is carried from flower to flower by various insects and other animals. The ripened carpels of angiosperms grow into fruits, structures that are as characteristic of members of the division as flowers are.

SELF-QUIZ

The following questions pinpoint important information in this chapter. You should be sure to know the answers to each of them.

1. All plants exhibit an alternation of generations in their life cycles. True or false?

2. In which of the following groups of plants is the sporophyte the dominant or more conspicuous generation?
   (a) conifers   (d) ferns
   (b) mosses   (e) angiosperms
   (c) liverworts

3. The basic pattern of the life cycle in the alternation of generations is meiosis-_____-gametophyte-_____-syngamy-_____-sporophyte.

4. Which of the following is not one of the whorls that make up a flower?
   (a) calyx   (d) androecium
   (b) corolla   (e) gynoecium
   (c) antheridium

5. Tissue that carries water and minerals up in plants is called _____; carbohydrates are carried throughout the plant by _____ .

6. Plants in the divisions Bryophyta, Psilophyta, Lycophyta, Sphenophyta, and Pterophyta have one reproductive characteristic in common, namely, _____ .

7. Among the living plants, the largest group is
   - (a) Bryophyta
   - (b) Coniferophyta
   - (c) Gnetophyta
   - (d) Anthophyta
   - (e) Cycadophyta

8. Fruits contain parental sporophytic tissue. True or false?

9. Seed plants differ in how they store the reserve food for the young developing sporophyte in the seed. In pines the reserve is stored in
   - (a) the integuments
   - (b) the megagametophyte
   - (c) the young sporophyte itself
   - (d) the seed coat

10. Ferns produce pollen grains. True or false?

11. The embryo sac of angiosperms normally contains eight nuclei and seven cells. True or false?

12. Which two whorls in the flower are not directly related to the production of gametes?

THOUGHT QUESTIONS

1. Compare and contrast the adaptations of fungi and of plants to a terrestrial existence. Which group of organisms has been more successful, and why?

2. Why is the ability to form seeds always associated with a separation of the gametophyte generation into two distinct kinds, megagametophytes and microgametophytes? How has this relationship changed during the course of evolution?

3. Why do mosses and ferns both require free water to complete their life cycles? At what stage of the life cycle is the water required? Do angiosperms also require free water to complete their life cycles? What are the reasons for the difference, if any?

FOR FURTHER READING

BELL, P.R., and C.L.F. WOODCOCK: *The Diversity of Green Plants,* 3rd ed., Edward Arnold (Publishers), Ltd., London, 1983. A brief account of all plants, fungi, algae, and cyanobacteria, clearly and well presented.

CONRAD, H.S., and P.L. REDFERN, JR.: *How to Know the Mosses and Liverworts,* 2nd ed., William C. Brown Co., Dubuque, Iowa, 1979. Well-illustrated key to the bryophytes of the United States.

FOSTER, A.S., and E.M. GIFFORD: *Comparative Morphology of Vascular Plants,* 2nd ed., W.H. Freeman & Co., San Francisco, 1974. A comprehensive introduction to the vascular plants.

RICHARDSON, D.H.S.: *The Biology of Mosses,* John Wiley & Sons, Inc., New York, 1981. Excellent, concise account of the mosses.

TAYLOR, T.N.: *Paleobotany: An Introduction to Fossil Plant Biology,* McGraw-Hill Book Co., New York, 1981. A survey of fossil plants.

FLOWERING PLANTS

Chapter 30

Overview

The flowering plants dominate every spot on earth except for the polar regions, the high mountains, and the driest deserts. Despite their overwhelming success, they are a group of relatively recent origin. The oldest definite angiosperm fossils are about 123 million years old, and the group achieved world dominance only about 80 or 90 million years ago. Among the features that have contributed to the success of angiosperms are their unique reproductive features, which include the flower and the fruit. Flowers bring about the precise transfer of pollen by insects and other animals, a critical adaptation enabling an organism that is rooted in one place to achieve outcrossing. Fruits play a role of extraordinary importance in the dispersal of angiosperms from place to place. Both structures were not only important in making possible the early success of the flowering plants, but they are also chiefly responsible for the remarkable diversity of the group.

For Review   *Here are some important terms and concepts that you will encounter in this chapter. If you are not familiar with them, you should review them before proceeding.*

- **Hybridization and the origin of species** (Chapter 22)

- **Evolution of the seed** (Chapter 29)

- **Gymnosperms** (Chapter 29)

- **Flowering plant life cycle** (Chapter 29)

Flowering plants, division Anthophyta, are the dominant photosynthetic organisms nearly everywhere on land. Included in this great group of some 235,000 species are our familiar trees (except conifers), shrubs, and herbs; grasses, vegetables, and grains—in short, nearly all of the plants that we see on a daily basis (Figure 30-1). Virtually all of our food is derived, directly or indirectly, from the flowering plants; in fact, more than 85% of it comes from just 20 species! In a sense, the remarkable evolutionary success of the angiosperms is the culmination of the plant line of evolution, and, as such, this division deserves special attention.

B

C

A

FIGURE 30-1

Some of the remarkable diversity of angiosperms is shown in these photographs.

A *Fragrant water lily, Nymphaea odorata, a flower with numerous, free, spirally arranged parts that intergrade with one another.*

B *Wild geranium, Geranium, a perennial that occurs in woodlands. It illustrates a reduction in the number of flower parts compared with less specialized angiosperm flowers.*

C *A gigantic cactus, Opuntia, one of the plants that Darwin saw in the Galapagos Islands.*

D *An angiosperm that lacks chlorophyll, Indian pipe, Monotropa uniflora. This plant obtains its food from other organisms.*

E *Tiger lily, Lilium canadense, with six free colored perianth parts (not differentiated into petals and sepals) and six free anthers. The three carpels are fused, forming a compound gynoecium, or pistil.*

F *Scarlet gilia, Ipomopsis aggregata; in the flowers of this species, the sepals and petals are fused with one another, each forming a tube. The greenish calyx, formed of the fused sepals, can be seen at the base of the flower. There are five fused petals, and the stamens are fused to the inside walls of the corolla tube.*

HISTORY OF THE FLOWERING PLANTS

The flowers and fruits that are characteristic of angiosperms, as well as many of their other structural features, are unique; consequently, it has been difficult to determine the nature of the relationships between this division and any other group of plants. In view of the ecological dominance and overwhelming importance of the angiosperms, however, the problem of their relationships to other groups has continued to interest biologists greatly. We will now consider the problem of the origin and evolution of the angiosperms.

Ancestors

As we mentioned in the last chapter, the common ancestor of the angiosperms would clearly be classified as a gymnosperm if we knew what it was. Calling it a gymnosperm, however, means only that it was a seed-bearing plant that did not possess a carpel. Once the carpel had evolved, this hypothetical ancestor of the flowering plants would itself have been classified as a flowering plant. The other features of the flower would have evolved later, early in the history of the group.

Botanists used to think of the gymnosperms as a closely knit group that was derived from an immediate common ancestor. The divisions Coniferophyta, Ginkgophyta, Cycadophyta, and Gnetophyta were viewed as the major subdivisions of this group and were assumed to be directly related. Viewed in this light, the gymnosperms were treated as a single division of plants that probably had evolved from fernlike ancestors, and had in turn given rise to the angiosperms.

A great deal of evidence that has accumulated more recently contradicts this view, however. The four divisions of gymnosperms are very distinct from one another with respect to a whole series of features, some of which were mentioned in the last chapter. Furthermore, we now know that some extinct plants in the division Lycophyta and others in the division Sphenophyta evolved seeds independently from the gymnosperms. These unrelated plants, therefore, would be regarded in a technical sense as "gymnosperms" too, if the group is defined simply by the lack of a carpel surrounding their seeds.

Being a gymnosperm, in this sense, merely involves having seeds but not being an angiosperm. This definition represents a particular stage of evolutionary development, not

D   E

F

a closely knit group of organisms. And since seeds have evolved many times during the history of vascular plants, the relationships among the divisions of living seed-bearing plants must be judged on grounds other than their common possession of seeds.

> **Vascular plants with seeds are called gymnosperms, unless they have the unique features that would mark them as angiosperms (especially flowers and fruits). Seeds have evolved a number of different times in the past, and they therefore cannot be taken to define a group that has a common ancestor. The gymnosperms represent a level of evolutionary development, not a unified group of organisms.**

Age

Except for the angiosperms, all of the divisions of plants appear in the fossil record well before the start of the Cretaceous Period, 144 million years ago. Our earliest record of flowering plants, however, is from about 123 million years ago, in the first part of the Cretaceous. The division Anthophyta, therefore, is by far the youngest of all the divisions of plants, despite its world dominance for most of the last hundred million years.

We recognize the earliest flowering plants in the fossil record from their distinctive pollen. This ancient fossil pollen so closely resembles the pollen of some living angiosperms in some of its key features that there can be no doubt that it was produced by angiosperms as well (Figure 30-2). Pollen is often characteristic of particular groups of plants, and since pollen and spores are quite resistant to destruction by chemicals or other agents, they often are preserved as fossils, providing valuable evidence about plants that existed in the past.

Unfortunately, some primitive angiosperms produced pollen that cannot be distinguished clearly from that of other groups of seed-bearing plants. It is almost certain, for this reason, that the angiosperms originated more than 123 million years ago and that we actually have fossils of their pollen, but we cannot recognize this pollen as having come from angiosperms.

We would like to understand the nature of the earliest angiosperms in order to be able to determine the nature of their direct ancestors. Such knowledge would enable us to interpret the flower and the other characteristics of the angiosperms more exactly than is pos-

sible now, and in an evolutionary context. Unfortunately, flowers are not often well preserved in the fossil record, compared with pollen or leaves. In addition, early fruits and fruitlike structures are difficult to interpret, especially if no similar structures are known in living plants. It is highly probable, however, that the living angiosperms were derived from a single common ancestor. Their distinctive flowers and fruits resemble one another so closely that they seem unlikely to have evolved more than once. In addition, all of the characteristic and sometimes unique features of their embryology that were outlined in the last chapter, including double fertilization and the presence of endosperm, seem even less likely to have evolved more than once.

All living angiosperms are so specialized in one or more of their features that we can be sure that no living member of the division would resemble its common ancestor closely. Nevertheless, some real progress is being made in understanding what the features of this ancestor may have been like, as increasingly precise interpretations are being made of the structures of fossil plants. In addition, more and more fossil flowers are being found and studied in detail (Figure 30-3). It appears probable that these studies will reveal the nature of the ancestor of the flowering plants in the relatively near future.

FIGURE 30-2

Drimys winteri, of the family Winteraceae, is one of the few dicots in which the pollen has a single groove, a condition that is characteristic of the monocots. The highly distinctive pollen of this archaic family, which is shed as tetrads—groups of four pollen grains—is among the oldest angiosperm pollen known. Drimys is also one of the very few angiosperms in which vessels are lacking; all of the conducting elements in its wood are tracheids, as shown in Figure 30-6. As shown here, the perianth parts of Drimys are numerous and spirally arranged, as are the stamens; the green carpels form a single whorl.

B

C
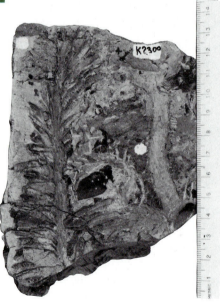

A

FIGURE 30-3

A *Archaeanthus linnenbergeri, an angiosperm discovered in the Cretaceous Period clays of Kansas by David Dilcher of Indiana University. This plant, which lived about 100 million years ago in forests dominated by ferns and gymnosperms, has no close living relatives; it is the only known member of an extinct family of angiosperms. Its flower parts are free and mostly spirally arranged, as in a magnolia or water lily flower, although the petals are reduced in number and are attached at one level on the axis, forming a whorl. A fossil of one of the unusual, deeply lobed leaves of Archaeanthus is shown in* **B**; **C** *shows a fossil of the fruiting axis, which becomes elongate, like that of Magnolia today. Flowers of this sort may well have been visited and pollinated by beetles, like the contemporary flowers they resemble. The reconstruction was drawn by Megan Rohn.*

Monocots and Dicots

There are two classes in the angiosperms, division Anthophyta: the Monocotyledones, or **monocots** (about 65,000 species), and the Dicotyledones, or **dicots** (about 170,000 species). Among the monocots are the lilies, grasses, cattails, palms, agaves, yuccas, pondweeds, orchids, and irises; the dicots include the great majority of familiar angiosperms of all kinds.

Monocots and dicots differ from one another in a number of features. For example, the venation in the leaves of monocots usually consists of parallel veins, and that in the leaves of dicots usually consists of netlike (reticulate) veins. In the flowers of dicots, the members of a given whorl are generally in fours or fives; in monocots, they are commonly in threes (Figure 30-4). The embryos of dicots, as the name of the class implies, generally have two seedling leaves, or **cotyledons;** those of monocots have one cotyledon. The condition in monocots has been derived from that in dicots by the suppression of one of the cotyledons. Similarly, monocots share with the most primitive dicots a very similar kind of single-pored pollen (see Figure 29-16, *A*).

> **The two classes of angiosperms are monocotyledons (monocots) and dicotyledons (dicots). Monocots have one cotyledon, or seedling leaf, and usually have parallel venation in their leaves; and their flower parts are often in threes. Dicots have two cotyledons, usually net (reticulate) venation, and flower parts in fours or fives. There are a number of other anatomical differences between the members of these classes.**

Aside from the differences just enumerated, monocots and dicots differ fundamentally in many other ways. Here are a few examples. About a sixth of the species of dicots, but very few monocots, are annuals (plants that complete their entire growth cycle within a year). Monocots have underground swollen storage organs, such as bulbs, much more frequently than do dicots. In mature monocot seeds, endosperm is usually present; it is often absent in the mature seeds of dicots. We will mention more differences between monocots and dicots in the rest of this chapter and in the next one, in relation to particular features.

Monocotyledons and dicotyledons are known to have been distinct from one another from nearly the time of the first appearance of angiosperms. Dicots are the more primitive of these two classes, and monocots probably were derived from early dicots before the middle of the Cretaceous Period.

Why Were the Angiosperms Successful?

When flowering plants first appeared in the fossil record, some 123 million years ago, Africa and South America were still connected to one another, to Antarctica and India, and, via Antarctica, to Australia–New Zealand (see Figure 23-4). These land masses formed the great continent known as Gondwanaland. In the north, Eurasia and North America also were united, forming another supercontinent called Laurasia. The huge land mass formed by the union of South America and Africa spanned the equator and probably had a climate characterized by extreme temperatures and aridity in its interior. Similar climates occur in the interiors of the major continents at present. In the patches of drier and less favorable habitat found in the interior of Gondwanaland, much of the early evolution of angiosperms probably took place. (Many of the features of the flowering plants seem to be correlated with successful growth under such arid and semiarid conditions.)

Flowers effectively ensure the transfer of gametes over substantial distances and therefore promote outcrossing, for reasons that we will explore later in this chapter (Figure 30-5). The ability to transfer pollen between widely separated individuals may have been important in linking the small populations of early angiosperms that we can envision in the vast interior of South America–Africa. Fruits that were likely to have been carried from place to place by animals would have been especially effective in assisting the spread of some

A

B

FIGURE 30-4

The bright red flowers of fire-pink, Silene virginica (A), with their flower parts in fives, are typical of the dicots; those of Trillium ozarkanum var. pusillum (B), with flower parts in threes, are typical of the monocots. Fire-pink and trillium occur side by side in the understory of the woods throughout the temperate parts of eastern and central Canada and the United States. Fire-pink is pollinated primarily by hummingbirds. The flowers of this trillium, like those of many angiosperms, change their color after pollination, thus signaling the pollinator which flowers are still unvisited and have a supply of food.

FIGURE 30-5

A soldier beetle, Chauliognathus pennsylvanicus, *visiting a daisy flower (family Asteraceae). Beetles were abundant and diverse at the time the angiosperms first evolved; they still predominate among the visitors to the flowers of many relatively primitive angiosperms— and some very advanced ones, as seen here. Beetles very probably played a key role in the early evolution of flowers.*

of these early flowering plants from one patch of favorable habitat to another. The tough, often leathery leaves of angiosperms, their efficient cuticles and stomata, and their specialized conducting elements all would have been important in survival and growth under such conditions, just as they are in stressful conditions today. The many natural insecticides produced by these plants, which we will discuss in Chapter 53, also would have been important for their survival. As the early angiosperms evolved, all of these features that contributed to their success became further elaborated and developed, and the group continued to evolve more and more rapidly.

In summary, the flowering plants were successful because they influenced the actions of insects and other animals through the features of their flowers, thus ensuring wide out-crossing and genetic diversity; because their fruits evolved numerous means of dispersal; and because their chemical diversity, and therefore their ability to cope with different diseases and pests, was high. In addition, many angiosperms evolved effective strategies for drought resistance, which have become increasingly important as the world climate has deteriorated throughout the history of this division.

The Rise to Dominance

The known history of the angiosperms, as we mentioned above, began about 123 million years ago. For the first 25 million years of this period, none of the angiosperms closely resembled any of the plants living today (Figure 30-3). Instead, they appear to have belonged to a series of forms that have either become extinct or changed over time and ultimately given rise to the families that now have living representatives. Over most of the world, angiosperms gradually increased in number and diversity. They eventually became more frequent than other plant groups about 80 to 90 million years ago, during the second half of the Cretaceous Period (see Figure 23-14). We can document the relative abundance of the different groups of plants by studying groups of fossils that occur together at the same time and place. More than 80 million years ago, the remains of various other divisions of plants, including lycopods, horsetails, ferns, and gymnosperms, are more numerous than those of the angiosperms in the strata in which they occur; afterwards, and with local exceptions, angiosperms predominate.

The angiosperms became more abundant than other groups of plants throughout the world during the 35 to 45 million years after their first known occurrence, 123 million years ago. By 80 to 90 million years ago, they were dominant throughout the world.

At about the time that angiosperms first became the most abundant group of plants in the fossil record, there also began to appear in this record the pollen, leaves, flowers, and fruits of some of the families that still survive. Such families as the magnolias, beeches, and legumes were in existence before the end of the Cretaceous Period (65 million years ago). Among the remains of these modern families, we can even recognize a few genera that are still living. A number of the orders of insects that are particularly associated with the flowers of the angiosperms also appeared at about this time. Other, more ancient, orders became more abundant and diverse; these trends have continued up to the present.

As new groups of insects appeared, angiosperm flowers doubtless became more complex and varied; we will discuss these evolutionary trends in the next section. Angiosperms apparently show no episode of extinction like that of the dinosaurs or many groups of marine plankton (see Figure 23-19) at the close of the Cretaceous Period. As far as we can tell from their fossil record, angiosperms simply continued to evolve progressively through this period, traumatic as it apparently was for many other groups of organisms. Since it appears very probable that a worldwide cloud, perhaps caused by the impact of an asteroid or extensive volcanic activity, led to dark conditions at that time, one can only assume that the darkness did not last long enough to cause widespread extinction among the flowering plants.

EVOLUTION OF THE FLOWER

As we saw in the last chapter, pollination in the angiosperms is indirect; the pollen reaches the stigma, germinates, and grows down to transport the sperm nuclei to the ovule. There, double fertilization occurs, the endosperm and embryo both grow, the seed ripens within the fruit, and the life cycle begins again. Pollen matures within the anthers and is transported, often by insects or other animals that visit the flowers for food, to another flower. The flower as a whole is attractive mainly because of the petals, and it is protected by the sepals in bud, which are themselves colored and attractive in some angiosperms.

Successful pollination in many angiosperms depends on the plants attracting insects and other animals regularly enough that these animals will carry pollen from one flower of that particular species to another (Figure 30-5). Flowers that are visited regularly by animals normally provide some kind of food reward, often in the form of a liquid called **nectar.** Nectar is rich not only in sugars but also in amino acids and other substances. Other forms of reward that animals may find in the flowers of the angiosperms will be discussed later in this chapter, in relation to the particular groups of visitors.

The relationship between such animals, which are known as **pollinators,** and the flowering plants, has been one of the central features of the evolution of both groups. By virtue of the evolution of such pollen transfer systems, the flowering plants are able to disperse their gametes on a regular and more or less controlled basis, even though they, like all other plants, are anchored to their substrate. In this way, and in a certain sense, the animals perform the same functions for the flowering plants that they do for themselves when they actively search out mates.

Analyzing Relationships Among the Angiosperms

How do we go about determining which sorts of flowers led to the early success of the angiosperms and ultimately gave rise to all of the kinds of flowers in the division? It is important to solve this problem to appreciate fully the nature of the adaptive radiation seen in this division, one of the most successful of all groups of organisms.

One way that we can learn about trends of specialization in flowers is to correlate—at least in direction if not in rate—such trends with those that affect other parts of the plant. To give an example of how such correlations are useful, most vascular plants, but only a very few genera of angiosperms, have exclusively a relatively unspecialized type of conducting cell, known as a **tracheid,** in their water-conducting tissue. Most angiosperms, and a few other vascular plants, have—in addition to or in place of tracheids—a more specialized type of conducting cell known as a **vessel element.** The few angiosperms that lack vessel elements also have certain floral characteristics which suggest that these particular families are relatively primitive (Figure 30-6).

A similar correlation can be made between relatively primitive floral features, wood that lacks vessel elements, and the single-pored pollen that is characteristic of the monocots and a few dicot groups (see Figure 29-16, A). Those dicot groups that have single-pored, instead of three-pored or many-pored pollen (Figures 29-16, B, and 30-13, B) are also primitive, as judged from their other characteristics.

> **Single-pored pollen and a lack of vessel elements in the wood both correlate with primitive flowers in angiosperms. Three-pored pollen and vessel elements, both of which are characteristic of most angiosperms today, clearly evolved long after the division Anthophyta originated.**

By making correlations between primitive and advanced features in living forms, and with a limited amount of information about fossils, botanists have deduced that the primitive flower probably had numerous, spirally arranged sepals, petals, stamens, and carpels (Figures 30-1, A, and 30-7). The differences between the petals and sepals of such a flower may not have been great, and the members of both of these floral whorls may have had

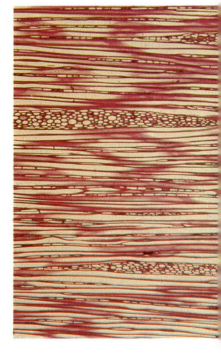

FIGURE 30-6

Section of wood of Drimys winteri *(Winteraceae), illustrating the absence of vessels. All of the uniform conducting cells in this photograph are tracheids. Winteraceae are one of the oldest families of angiosperms that still survive. They are best represented in New Caledonia and other areas in and near Australia where many relict plants and animals are found.*

FIGURE 30-7

Species of Magnolia.
A *A flower, illustrating the numerous, free, spirally arranged carpels and stamens and the whorled petals, an arrangement precisely like that of the 100-million-year-old* Archaeanthus *shown in Figure 30-3.*
B *Fruiting axis, with numerous, spirally arranged carpels. The bright red seeds of this species emerge from the individual fruits and hang by slender threads. These seeds are attractive to and dispersed by birds, which eat their fleshy outer layers.*

similar coloration and form, grading into one another along a continuous spiral. The members of all whorls in primitive flowers tended to be free from—not fused with—both the other members of the same whorl and the members of other floral whorls.

The first angiosperms had numerous, free, spirally arranged flower parts.

General Characteristics of Floral Evolution

Flowers are distinguished from the ordinary leafy shoots of the same plant by being **determinate** in growth. This means that their **apical meristem**—the terminal group of dividing cells from which all shoots and roots in plants, including flowers, arise—does not continue to divide after the flower is formed during the course of its development in the young bud. In contrast, the leafy shoots are **indeterminate** and continue to grow, progressively differentiating leaves along the way. In primitive flowers, the flower parts tend to be organized in one continuous spiral, more or less like the arrangement of leaves on a shoot; in the flowers of more advanced groups of plants, the flower parts are arranged in definite whorls (see Figure 29-23).

> **Flowers are determinate shoots; their apical meristem does not continue to grow after the flower is formed. Vegetative shoots are indeterminate, growing continuously.**

During the largely unknown early evolution of the angiosperms, the flower evolved as a unit, with increasing trends of specialization in its various parts. Most modern angiosperms have flowers that closely resemble one another, although some of the flowers borne by the more primitive groups of angiosperms include sepals, petals, stamens, carpels, or combinations of these that are archaic in structure and appearance. These primitive flowers provide valuable clues to the early evolution of flower structure and thus of the angiosperms as a whole. They represent, however, only a small selection from the wide range of different kinds of "experimental" flowers and flower parts that must have occurred early in the evolution of the group. Some of the early, extinct types of angiosperms are now being reconstructed from the fossil record (Figure 30-3).

Calyx

The calyx is the outer whorl of a complete flower and consists of the sepals. The similarities between sepals and leaves are extensive and include their pattern of veins, as well as coloration and form. In addition, many genes are known that affect both sepals and leaves, but not petals. For these reasons, it is reasonable to conclude that there is a close relationship, probably an evolutionary one, between sepals and leaves. Sepals, which function largely to protect the flower when it is a bud, may be thought of in effect as modified leaves, specialized for this purpose. But whether this is an accurate description of their evolutionary history is difficult to establish for certain.

Corolla

The evolution of petals, which collectively make up the corolla, is more complex and difficult to interpret than that of sepals. Plant morphologists, who study the form and function of the parts of plants in an evolutionary context, have demonstrated convincingly that the petals of many angiosperms are, in terms of their origin, flattened stamens that have lost their microsporangia and become specialized in relation to their function of attracting insects and other animals. This possible origin has been deduced partly from structural similarities between petals and stamens comparable to those between sepals and leaves, and partly because of the presence of structures intermediate between petals and stamens in certain kinds of flowering plants. In addition, certain genes are known that affect both stamens and petals, but not the other whorls of the flower, a relationship that indicates homology between these two structures.

The difficulty of clarifying the situation is compounded by the fact that in other kinds of flowers, such as water lilies (Figure 30-1, *A*), there are obvious transitions between sepals and petals. In these plants, the petals are derived from specialized sepals that have become larger and colored during the course of evolution and that attract insects and other animal visitors rather than protecting the bud.

Petals, then, appear to have had two different evolutionary origins, one—in most flowering plants—from stamens that have become flattened and leaflike in the course of evolution, and the other—in a minority of flowering plants—from sepals that have undergone the same kind of evolutionary specialization. Although it is possible that sepals should actually be regarded as leaves that have become specialized in the course of evolution, most petals and, as we will see, all stamens and carpels, are not directly comparable to leaves.

The petals of most plants evolved from stamens. During the course of their evolution, the stamens that gave rise to petals became flattened and lost their sporangia. Petals became specialized in this way because of their value in attracting visitors to flowers, an important factor in bringing about cross-pollination. In some plants, such as water lilies, petals evolved from sepals and thus perhaps indirectly from leaves.

Androecium

As we have seen, stamens, which collectively make up the androecium, are specialized organs that bear the microsporangia of angiosperms. In the pollen cones of gymnosperms, there are similar structures on which the microsporangia are borne. In both cases, the structures, in terms of their evolutionary history, probably evolved from systems of small branches among which the microsporangia were borne. These branch systems have progressively become reduced and modified over the course of evolutionary history. The great majority of living angiosperms have stamens whose filaments are slender—often threadlike—and whose four microsporangia are evident in a swollen portion, the anther, at the apex.

In contrast to this general pattern, a few primitive angiosperms have flattened, leaflike stamens whose microsporangia are either embedded in or protrude from the upper or lower surface (Figure 30-8). The existence of such stamens has sometimes been cited as evidence that the first stamens were derived from leaves. Perhaps more likely is the suggestion that stamens, like the microsporangium-bearing structures in other seed plants, are modified branch systems that have evolved in a way similar to leaves. Viewed in this light, it appears logical that stamens should share certain features with leaves without actually having been derived from them.

A

FIGURE 30-8

A *Stamen of* Magnolia, *viewed from above. The stamens of primitive angiosperms tend to be somewhat leaflike, with the sporangia occupying proportionately little of the bulk of the stamen. In most angiosperms, the leaflike structure of these primitive stamens has been modified during the course of evolution so that the sporangia appear as a swollen body, which is called the anther (here they are yellow), at the end of a slender stalk, which is called the filament.* **B** *A flower of strawberry (*Fragaria ananassa*), showing the many separate stamens around the central fleshy portion of the flower.* **C** *Central column from a flower of* Hibiscus. *In* Hibiscus *and other plants of the mallow family, the stamens are also fused with the style and appear to arise from it. The stigmas are bright red in this flower.*

B

C

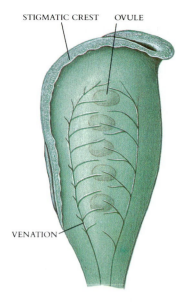

VENATION

FIGURE 30-9

Outside view of a primitive carpel, open along the side, with a stigmatic crest along the top. In more advanced carpels, found in most angiosperms, the swollen basal ovary portion is well differentiated from a slender stalk, the style, which has a receptive stigma at the end. Carpels are often fused together in various ways in different groups of angiosperms; if the other floral whorls are fused around them, the ovary is said to be inferior.

Gynoecium

Carpels, collectively the gynoecium, are highly specialized organs that are unique in the flowering plants. Some of the more primitive flowering plants, as judged by their other features, have rather leaflike carpels; in some cases, the margins of the carpel barely fit together and the pollen may then be deposited along a suture (Figure 30-9). In other primitive angiosperms, the fertile area is organized toward one end of the carpel in what is called a **stigmatic crest.** The evolution of the carpel appears to have been analogous to that of the stamen.

TRENDS OF FLORAL SPECIALIZATION

The evolution of the wide array of modern flowers can be understood largely in terms of trends involving the aggregation—grouping together—and reduction, or loss, of parts. The central axis of many flowers has been shortened in the course of their evolution. In such flowers, the whorls are close to one another. The spiral patterns that characterize the attachment of all floral parts in primitive angiosperms give way in the course of evolution to a single whorl at each level (Figure 30-10). In more advanced angiosperms, the number of parts in each of these whorls has often been reduced from many to few. The members of each whorl have tended to become **connate,** or adherent to one another and sometimes actually joined into a tube, and also **adnate,** or joined to the members of another whorl (Figure 30-11). Whole whorls may even be lost from the flower, which may lack sepals, petals, stamens, carpels, or various combinations of these.

In flowers in which the floral parts are highly fused together, the sepals, petals, and stamens appear to arise at the summit of the ovary, which is then said to be **inferior.** In a flower in which fusion has not occurred, the members of these whorls appear to arise at the base of the ovary, which is then said to be **superior.** Recall that the ovary (see Chapter 29) is the enlarged basal portion of a carpel or of a gynoecium composed of fused carpels; when mature, the ovary becomes the fruit.

A

B

C

FIGURE 30-10

Three kinds of structural modifications that have happened during the evolution of angiosperms.

A *Baobab,* Adansonia digitata. *In the hanging flowers of this species, the numerous stamens are united into a tube around the style.*

B *An orchid,* Brassavola digbyana, *from Costa Rica. In orchids, the three carpels are fused and the flower parts appear to be attached at the top of the ovary, which is therefore called an inferior ovary. In most orchids, there is only one stamen, which is fused together with the style and stigma into a single complex structure, the column.*

C *The large, brown, ill-scented flowers of this African milkweed,* Stapelia, *are mistaken for rotten meat by blowflies.*

Factors Promoting Outcrossing

Outcrossing, as we have stressed repeatedly, is of critical importance for the adaptation and evolution of all eukaryotic organisms. One of the ways in which outcrossing is enhanced in certain plant species is by the loss of the androecium or gynoecium. If the androecium has been lost during the evolution of a particular flower, that flower produces only ovules, not pollen. Such a flower is called **pistillate** (from the term *pistil*, discussed in Chapter 29); functionally it is female. If the gynoecium is absent, that flower produces only pollen and is functionally male; it is called **staminate.**

Staminate and pistillate flowers may occur on separate individuals in a given plant species, as for example in willows. Such plants, whose sporophytes produce either only ovules or only pollen, are called **dioecious,** from the Greek words for "two houses." In other kinds of plants—oaks, birches (Figure 30-12), and ragweed (Figure 30-13) are familiar examples—the two types of flowers may be produced on the same plant; such plants are called **monoecious**—"one house." (The Greek root *oikos,* which appears in these words, is the same one that has given rise to the words *ecology* and *economy,* where it is used in the sense of a functioning household.) The separation of ovule- and pollen-producing flowers that occurs in dioecious and monoecious plants makes outcrossing even more likely than it would otherwise be. Most angiosperms are neither dioecious nor monoecious, however; each of their flowers includes both pollen-producing and ovule-producing structures—that is, stamens and carpels.

Even if functional stamens and carpels are both present in each flower of a particular plant species, as is usually the case, these organs may reach maturity at different times. Plant species in which this occurs are said to be **dichogamous.** If the stamens become mature first, shedding their pollen before the stigmas are receptive, the flower is effectively staminate at that time. Once the stamens have completed shedding pollen, the stigma or stigmas may then become receptive, and the flower may then be pistillate. The flower may thus achieve the same effect as if it completely lacked either functional stamens or functional carpels, and may thereby significantly increase its outcrossing rate.

FIGURE 30-11

In the composites (family Asteraceae), the individual flowers are grouped together in a head. They are usually spirally arranged and open sequentially, as in this head of Melanthera nivea, *photographed in the Amazon Basin of Brazil. In some composites, the marginal flowers are bilaterally symmetrical and expanded into rays, which increase the attractiveness of the whole head, as shown in Figure 30-5. A single seed results from the pollination of an individual composite flower, but the head itself may function as an attractive unit for many days.*

FIGURE 30-12

The wind-pollinated flowers of a birch, Betula. *Birches are monoecious; their staminate flowers hang down in long, yellowish tassles, whereas their pistillate flowers, which mature into characteristic conelike structures, occur in small, reddish-brown clusters.*

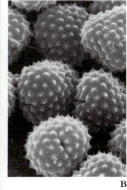

A

B

FIGURE 30-13

A *Flowers of great ragweed,* Ambrosia trifida.
B *Pollen of western ragweed,* Ambrosia psilostachya. *Ragweed pollen causes allergic rhinitis—more commonly known as hay fever—in millions of sufferers each year.*

FIGURE 30-14

Flowers of snapdragon,
Antirrhinum majus. *These flowers are bilaterally symmetrical, with both petals and sepals fused into a tube and the anthers fused with the inside of the tube. Flowers of this sort can be pollinated effectively only by large bees, which force their way between the lips and come into a precise contact with the anthers and the stigma while doing so.*

An even simpler way in which many flowers promote outcrossing is through the physical separation of the anthers and the stigmas. If the flower is constructed in such a way that these organs do not come into contact with one another, there will naturally be a tendency for the pollen to be transferred to the stigma of another flower, rather than to the stigma of its own flower.

Another device that occurs widely in flowering plants and increases outcrossing is **genetic self-incompatibility.** In self-incompatible plants, the pollen from a given individual does not function on the stigmas of that individual, or the embryos resulting from self-fertilization do not function. Self-incompatibility, which is widespread in flowering plants, is a mechanism that increases outcrossing even though the flowers of such plants may produce both pollen and ovules, and their stamens and stigmas may be mature at the same time.

> **Plants may promote outcrossing through self-incompatibility, dioecism, monoecism, or by the separation of the different parts of the flower in space (within the flower) or of their maturation in time.**

Trends in Floral Symmetry

In addition to those we have discussed, other evolutionary trends in floral evolution have affected the symmetry of the flower. Primitive flowers are radially symmetrical; those of many advanced groups are bilaterally symmetrical. That is, they are equally divisible into two parts along a single plane. The snapdragon (Figure 30-14) is an example of a tubular, bilaterally symmetrical flower whose sepals, petals, and carpels are connate but whose anthers, although adnate to the corolla (fused petals) and apparently therefore arising from it, are separate from one another. Such tubular, bilaterally symmetrical flowers are especially frequent in advanced groups of flowering plants, in which they are connected with advanced and often highly precise pollination systems.

POLLINATION IN FLOWERING PLANTS
Pollination by Wind

Early seed plants were pollinated passively, by the action of the wind. They simply shed great quantities of pollen, which was blown about and occasionally reached the vicinity of the ovules of the same species. For such a system to operate efficiently, the individuals of a given plant species must grow relatively close together. Otherwise, the chance that any pollen will arrive at the appropriate destination is very small, because the average distance that pollen is carried by the wind in large quantities is actually quite short, compared with the long distances to which pollen is routinely carried very accurately by insects and other animals that visit flowers.

Keeping this fact in mind, one can begin to understand why many of the uniform stands of trees that dominate huge areas in North America and in Eurasia, including gymnosperms such as spruces, pines (see Figure 20-7), and hemlocks, and angiosperms such as birches, alders, and ragweed, are wind pollinated (Figures 30-12 and 30-13). As we mentioned earlier, there is little evidence for any system except wind pollination among the living gymnosperms. Such angiosperms as the oaks and hickories that dominate the temperate forests and the grasses of our pastures are also wind pollinated. However, in tropical forests, which have many more species of plants than do temperate forests, with many different kinds growing close together, and where the distance to the next individual of the same kind is often great, wind pollination is almost completely absent.

Pollination by Animals

The rapid rise of the angiosperms to worldwide dominance was doubtless fostered in part by their development of a more precise system of pollen transfer than the wind. In a world where wind pollination was the only system that was available, those plants that acquired a system involving animal-mediated, and therefore more precise, transfer of pollen would have been at a considerable selective advantage. During the course of evolution of the flowering plants, interactions with their pollinators have brought about increasing specialization in many groups (Figure 30-5). Profound changes in the morphology, color, shape, odor, and other characteristics of individual flowers have occurred. These changes increase the attractiveness of the individual kinds of flowers to particular pollinators. By doing so, they increase the frequency with which these animals will visit the flowers of that individual species. Provided that they find sufficient food reward at those flowers and that the characteristics of the flowers are such that the pollinators will return to them consistently, increasing specialization may result. These interactions may lead to a gradual evolutionary modification both of the plant and of its pollinators, depending on the constancy of the visits.

Because a number of angiosperms are wind pollinated, like the living gymnosperms, and because angiosperms clearly were derived from gymnosperms originally, some scientists have conjectured that wind pollination might be the original condition among the angiosperms. The majority view, however, is that insect pollination is the original condition and that flowers somewhat like that of the magnolia (Figure 30-7) reflect the condition in the first angiosperm flowers. Such ancestral flowers probably were smaller than most of the familiar magnolias, however.

> **The pollen of the early seed plants was spread about by the wind, a very inefficient system. When pollination by insects and other animals arose, it was highly advantageous to the plants that had it, being much less wasteful and more precise.**

Bees and Flowers

Among the insect-pollinated angiosperms, the most numerous are those pollinated by bees (Figure 30-15). Like most insects, bees locate sources of food largely by odor at first, but they then orient themselves on the flower or group of flowers by its shape, color, and texture. Many flowers that are characteristically visited by bees are blue or yellow. Many have characteristic lines of dots or stripes that indicate the position in the respective flowers of the **nectaries,** the glands where the nectar is produced, which often are located within the throats of specialized flowers. Some bees visit flowers for nectar, primarily as a source of food for the adults and sometimes also for the larvae; most of the approximately 20,000 species of bees, however, visit flowers to obtain pollen. They use this pollen to provision the cells in which their larvae complete their development (Figure 30-16).

FIGURE 30-15

A bumblebee, Bombus, *covered with pollen while visiting flowers. This bee will transfer large quantities of pollen to the next flower that it visits.*

FIGURE 30-16

An underground nest of the solitary bee Nomia melanderi. *The female bee that dug this burrow has provisioned it with a large ball of alfalfa pollen, on which the larval bee can be seen feeding. In order to be able to set seed, alfalfa requires visits by bees to trip its floral mechanism. Since alfalfa seed is an important crop, this species of* Nomia *and other bees are often introduced in large numbers into the neighborhood of alfalfa fields.*

A

B

FIGURE 30-17

Flower visitation by a butterfly and a moth.
A *The copper butterfly* Lycaena gorgon, *visiting the flowers of a composite. The proboscis of the butterfly, which coils like a watchspring when at rest, is being thrust into one flower after another to extract the nectar.*
B *A hawkmoth, one of a family of moths that hover like hummingbirds, visiting flowers at night in Brazil. The moths thrust their proboscis into the flowers while hovering in front of them.*

Only a few hundred species of bees are social or semisocial in their nesting habits. These occur in colonies like the familiar honeybee, *Apis mellifera* (see Figure 37-26, *C*), or the bumblebees, *Bombus* (Figure 30-15). Such bees produce several generations during a year and must shift their attentions to different kinds of flowers as the season progresses. Because of the size of their colonies, they also must often use the flowers of more than one kind of plant as a food source at any given time.

Except for these social and semisocial bees, and about 1000 species that are parasitic in the nests of other bees, the great majority of bees—at least 18,000 species—are solitary. Solitary bees characteristically have only a single generation in the course of a year in temperate regions. Often they are active as adults for as little as a few weeks a year.

Solitary bees often use the flowers of a given group of plants exclusively, or nearly so, as sources of their larval food. The highly constant relationships of such bees with these flowers may lead to modifications, over time, in both the flowers and the bees. For example, the time of day when the flowers open may be correlated with the time when the bees appear; the mouthparts of the bees may become elongated in relation to tubular flowers; or the bees' pollen-collecting apparatus may reflect in its characteristics those of the pollen of the plants that they normally visit. When such relationships are established, they provide both an efficient mechanism of pollination for the flowers and a constant source of food for the bees that specialize on them.

Bees are the most frequent and characteristic pollinators of flowers. About 90% of the roughly 20,000 species of bees are solitary; morphological and physiological specializations often link these solitary bees with particular kinds of plants.

Bees first appeared perhaps 75 million years ago. They later became abundant and diverse as the world climate became drier. As the abundance and diversity of bee species increased, their activities clearly exerted a powerful selective influence on the pattern of angiosperm diversification. In addition, the bees themselves have become even more diverse as their relationships with flowers have grown in complexity and specificity. Because of the great precision of these relationships, which are the most specific of any between flowering plants and their insect visitors, you might assume that bees were responsible for much of the evolutionary diversification of angiosperms. There is some truth to this statement, but there are only about 20,000 species of bees in the world, and about 235,000 species of angiosperms. Clearly, then, bees have not been responsible for the overall diversification of angiosperms, even though they have had an important influence on the evolution of many individual species.

Coevolutionary relationships of a similar nature can be documented for the other major groups of animals that regularly visit flowers. Most of these animal groups appeared and became diverse during the same time frame as the bees—the last 75 million years. Even though most other animals are less specific than bees in their visits to flowers, they also have had a role in stimulating the evolutionary diversification of the flowering plants. We will now consider some of these other kinds of animals that regularly visit flowers.

Insects Other Than Bees

Among these additional groups of flower-visiting animals, a few are especially prominent. Flowers that are visited regularly by butterflies, for example phlox, often have flat "landing platforms" on which the butterflies perch. They also tend to have long, slender floral tubes filled with nectar that is accessible to the long, coiled proboscis characteristic of the Lepidoptera, the order of insects that includes butterflies and moths (Figure 30-17). Other flowers visited regularly by moths are often pale in color, white or yellow, like jimsonweed or evening primrose; heavily scented, thus serving to make the flowers easy to locate at night; and long tubed, like the flowers visited regularly by butterflies.

Pollination by Birds

A particularly interesting group of plants are those regularly visited and pollinated by birds, especially by hummingbirds in North and South America (Figure 30-18). Such plants must produce large amounts of nectar. If they do not, the birds, which consume a great deal of energy, will not be able to find enough food to maintain themselves or to continue visiting the flowers of that plant, for energy reasons alone. If these flowers do produce such large amounts of nectar, however, it is not advantageous for them to be visited by insects, because the insects could supply their energy requirements at perhaps a single flower and therefore would not move to another flower. Although the nectar of that flower might be very adequate for the insect, the plant would not be cross-pollinated as a result of the insect's visit. How do flowers that "specialize" on hummingbirds balance these different selective forces?

To birds, red is a very conspicuous color, just as it is to us. To insects, however, ultraviolet is a highly visible color, but red is not. Carotenoids, yellow pigments frequently found in plants, are responsible for the yellow colors of many flowers, such as sunflowers and mustard. Carotenoids reflect both in the yellow range and in the ultraviolet range, the mixture resulting in a distinctive color called "bee's purple." Such yellow flowers may then be marked in distinctive ways, normally invisible to us but highly visible to bees and other insects (Figure 30-19).

Red, in contrast, does not stand out as a distinct color to most insects. To most insects, the red upper leaves of poinsettias look just like the other leaves of the plant. Even though the red upper leaves of poinsettias are obvious to us, they do not attract insects to the small, yellowish-green flowers of the same plants, even though these flowers produce abundant supplies of nectar. Consequently, insects are not apt to visit these flowers, but hummingbirds are. Thus red is a perfect color both to signal birds of the presence of abundant nectar and, at the same time, to make that nectar as inconspicuous as possible to insects.

As hummingbirds migrate northward through North America every summer, they encounter a progression of red flowers that provide an abundant source of food for them along the way. Even if they are unfamiliar with these flowers, the red color of the flowers signals the birds that they are suitable sources of food.

Hummingbird-visited flowers are also specialized in other, analogous ways. Such flowers generally are odorless, because birds normally do not have well-developed olfactory senses and do not orient to odor. Insects, as we have mentioned, often do, and odors strongly signal the presence of food to them. The nectar in hummingbird-visited flowers is often held within strong tubes or otherwise protected where it can be obtained only by the action of the beak of the bird and not, in general, by insects.

FIGURE 30-18

Hummingbirds and flowers.
A *Ruby-throated hummingbird extracting nectar from the flowers of a trumpet-flowered creeper.*
B *Poinsettia flowers. The individual flowers, shown here, each have a large nectary at one side. The clusters of yellowish flowers are made more attractive to birds because of the large red leaves that surround them.*

FIGURE 30-19

The yellow flower of Ludwigia peruviana *(Onagraceae) photographed in normal light* **(A)** *and with a filter that selectively transmits ultraviolet light* **(B)**. *The outer sections of the petals reflect both yellow and ultraviolet, a mixture of colors called "bee's purple"; the inner portions of the same petals reflect yellow only and therefore appear dark in the photograph that shows ultraviolet reflections. To a bee, this makes the flower appear as if it has a conspicuous central bull's-eye.*

FIGURE 30-20

Flowers of a grass in the prairies of eastern Oklahoma. The modified sepals and petals of grasses are greenish. At this stage, the large yellow anthers, dangling on very slender filaments, are hanging out, about to shed their pollen to the wind; later these flowers will become pistillate, with long, feathery stigmas—well suited for trapping wind-blown pollen—sticking far out of them. Many grasses, like this one, are therefore dichogamous.

Insects often are attracted by the odors of flowers and do not perceive red as a distinct color. Birds are not attracted by odors and do perceive red. Bird-pollinated flowers, which must produce large quantities of nectar to support the birds that visit them, and to keep them visiting, are characteristically odorless and red, with the nectar well protected. Insects are not especially attracted to such flowers, and, in any case, often cannot gain access to the nectar. If they did, they might simply use it up, without being stimulated to move on to another flower.

Wind-Pollinated Angiosperms

Among the wind-pollinated angiosperms are such familiar plants as oaks, birches, cottonwoods, grasses, sedges, and nettles. The flowers of these plants are small, greenish, and odorless, and their corollas are reduced or absent (Figures 30-12, 30-13, and 30-20). Such flowers often are grouped together in fairly large numbers and may hang down in tassels that wave about in the wind and shed pollen freely. Many wind-pollinated plants are dioecious or monoecious, with the staminate and pistillate flowers separated on separate individuals or on a single individual. If the pollen-producing and ovule-bearing flowers are separated, it is certain that pollen released to the wind will fertilize a flower other than the one that sheds it, a strategy that greatly promotes outcrossing.

Self-Pollination

All of the modes of pollination that we have considered thus far tend to lead to outcrossing, which is as highly advantageous for plants as it is for eukaryotic organisms generally. Notwithstanding this, self-pollination is also very frequent among angiosperms. In fact, probably more than half of the angiosperms that occur in temperate regions self-pollinate regularly. Most of these have small, relatively inconspicuous flowers in which the pollen is shed directly onto the stigma, sometimes even before the bud opens (Figure 30-21). You might logically ask why there are so many self-pollinated plant species, if outcrossing is just as important genetically for plants as it is for animals. There are two basic reasons for the very frequent occurrence of self-pollinated angiosperms. They are as follows:

1. Self-pollination obviously is ecologically advantageous under certain circumstances, because the plants in which it occurs do not need to be visited by animals to produce seed. As a result, self-pollinated plants can grow in areas where the kinds of insects or other animals that might visit them are absent or very scarce—as in the Arctic, or at high elevations.

2. In genetic terms, self-pollination produces progenies that are more uniform than those that result from outcrossing. Such progenies may contain high proportions of individuals well adapted to particular habitats. Self-pollination initially tends to produce large numbers of ill-adapted individuals, many of which may be incapable of reaching reproductive size. Functionally, however, this kind of bottleneck in fitness may be overcome within a few generations, with no further obstacle to continued self-pollination. If the habitat to which the uniform progenies produced by self-pollination are well adapted continues to exist, it may be advantageous for the plant to continue self-pollinating indefinitely. This is the main reason that self-pollinating plant species are well represented as weeds—the habitat of weeds has been made uniform and spread all over the world by human beings.

Self-pollinated angiosperms are frequent where there is a strong selective pressure to produce large numbers of genetically uniform individuals adapted to particular relatively uniform habitats.

A

B

C

FIGURE 30-21

In the genus Epilobium, *most species are self-pollinating, but a few are outcrossing.* Epilobium angustifolium *is strongly outcrossing and was one of the first plants in which the process of pollination was studied (in the 1790s). In it, the anthers first shed pollen, then the style swings up into a similar position in the flower and its four lobes open; these flowers are dichogamous, functionally staminate first and pistillate about 2 days later. The flowers open progressively up the stem, so that the lowest ones are visited first. Working up the stem, the bees encounter pollen-shedding, staminate-phase flowers, become covered with pollen, and carry it passively to the lower, functionally pistillate flowers of another plant. Shown here are flowers in* **(A)** *the staminate phase and* **(B)** *the pistillate phase.* **C** Epilobium ciliatum. *In this species, which illustrates the situation that is more common in the genus, the anthers shed their pollen directly onto the stigma of the same flower, which is mature at the same time, thus bringing about self-pollination.*

THE EVOLUTION OF FRUITS

Paralleling the evolution of flowers of the angiosperms, and nearly as spectacular, has been the evolution of their fruits. Aside from the many ways in which fruits can be formed, they exhibit a wide array of modes of specialization in their dispersal.

Fruits that have fleshy coverings, often shiny black or bright blue or red, normally are dispersed by birds and other vertebrates (Figure 30-22). Like the red flowers that we discussed in relation to pollination by birds, red fruits signal an abundant food supply. By feeding on these fruits, birds and other animals may carry seeds from place to place and thus transfer the plants from one suitable habitat to another.

Other kinds of specialized fruit dispersal have evolved many different times in the flowering plants. Fruits with hooked spines, like snakeroot, beggar-ticks, and burdock (Figure 30-22, *A*), are particularly characteristic of several genera of plants that occur in the northern deciduous woods. Such fruits are often spread from place to place by mammals, now including humans! In addition, mammals like squirrels disperse and bury seeds, as the multituberculates may have done during the period of early evolution of the angiosperms (see Figure 23-16). Other fruits or seeds have wings and are blown about by the wind; the

A

B

C

FIGURE 30-22

Animal-dispersed fruits.
A *Fruits of burdock thistle,* Arctium lappa, *on the fur of a dog.*
B *The bright red berries of dogwood,* Cornus florida, *are highly attractive to birds, just as are red flowers. The birds carry the seeds they contain for great distances.*
C *Figs,* Ficus carica, *are whole clusters of flowers turned inside out. They are pollinated by tiny wasps, which enter through the hole in the end of the fig. When mature, figs are consumed and the seeds of the individual tiny flowers are scattered about by birds and mammals.*

A   B   C

FIGURE 30-23

Wind-dispersed fruits.
A *Dandelion,* Taraxacum officinale. *The "parachutes" disperse the fruits widely in the wind, much to the gardener's despair.*
B *The double fruits of maples,* Acer, *when mature, are blown considerable distances from their parent trees.*
C *Tumbleweed,* Salsola, *in which the whole dead plant becomes a light, windblown structure that rolls about scattering seeds.*

A   B

FIGURE 30-24

Water-dispersed fruits.
A *Mangrove,* Rhizophora mangle. *The embryos begin to grow on the parent plants and soon become established when they wash away to another silty shore.*
B *Coconuts,* Cocos nucifera, *sprouting on a sandy beach. One of the most useful plants for humans in the tropics, coconuts have become established even on the most distant islands by drifting in the waves.*

wings on the seeds of pines (see Figure 29-21) and those on the fruits of ashes or maples (Figure 30-23, *B*) play identical ecological roles. The dandelion (Figure 30-23, *A*) provides a familiar example of a kind of fruit that is dispersed by the wind, and the dispersal of the seeds of plants such as milkweeds, willows, and cottonwoods is similar.

Still other fruits, like the coconut and those of certain other plants that characteristically occur on or near beaches, are regularly spread from place to place by water (Figure 30-24). Dispersal of this sort is especially important in the colonization of distant island groups, such as the Hawaiian Islands. It has been calculated that the seeds of about 175 original plant species must have reached Hawaii to have evolved into the roughly 1200 species found there today (see Figure 20-6 for an example). Some of these seeds blew through the air, others were transported on the feathers or in the guts of birds, and still others drifted across the Pacific to Hawaii, nearly a third of them from North America. Although the distances are rarely as great as that between Hawaii and the mainland, dispersal is just as important for those mainland plant species in which the habitats are discontinuous, such as mountaintops, marshes, or north-facing cliffs. All in all, dispersal is one of the most fascinating aspects of plant biology.

The examples given and illustrated above provide some idea of the kinds of evolutionary changes that have been involved in the evolution of dispersal mechanisms in fruits and seeds. You may be interested in studying the fruits and seeds produced by the plants in your area and in trying to discover the mechanisms by which they are dispersed. For some plants it is advantageous for the seeds to be widely dispersed, and for others it is advantageous to keep them near the parent plant. You may find it of interest to consider the factors that are important in determining which of these conditions applies to a particular plant species.

SUMMARY

1. Since the ancestor of the angiosperms clearly had seeds, it would be classified as a gymnosperm if it were known. Seeds, however, evolved many times in the history of the vascular plants, and no living seed-bearing plant that is not an angiosperm could have been the direct ancestor of the group. All such plants are too specialized in other ways to fit this role.

2. The oldest known fossil evidence of angiosperms is pollen from the early Cretaceous Period, some 123 million years ago. By 80 to 90 million years ago, angiosperms were more common worldwide than other plant groups. They became abundant and diverse as drier habitats became widespread during the last 30 million years or so.

3. Among the reasons that angiosperms have been successful are their relatively drought-resistant vegetative features, including their vascular systems, cuticles, and stomata. Most important, however, are their flowers and fruits. Flowers make possible the precise transfer of pollen and therefore outcrossing, even when the stationary individual plants are widely separated. Fruits, with their complex adaptations, facilitate the wide dispersal of angiosperms.

4. The primitive flower had numerous, separate, spirally arranged flower parts, as we know from the correlation of flowers of this kind with primitive pollen, wood, and other features. Sepals probably are specialized leaves; petals, with some exceptions, are sterile stamens, modified to play a role in attracting pollinators; and stamens and carpels probably are modified branch systems whose spore-producing organs were incorporated into the flower during the course of evolution.

5. Outcrossing in different angiosperms is promoted by the separation of the pollen- and ovule-producing structures into different flowers, or even onto different individuals; by genetic self-incompatibility; and by the separation of the pollen and the stigmas within a given flower with respect to space or time of maturation.

6. Bees are the most frequent and constant visitors of flowers. The solitary bees, which constitute about 90% of the estimated 20,000 bee species, are the most specialized and specific visitors. They often have morphological and physiological adaptations related to their specialization in visiting the flowers of particular plant species.

7. Flowers visited regularly by birds must produce abundant nectar to provide the birds with enough energy so that they will continue to be attracted to them. If insects, most of which need less energy than birds, visit such flowers, they may not be stimulated to move from flower to flower. Many bird-visited flowers, therefore, are red, a color not attractive to most insects, and are odorless; insects often orient by odor, but birds usually do not. The nectar of bird-visited plants tends to be well protected by the structure of the flowers.

8. Fruits are highly diverse in terms of their dispersal, often displaying wings, barbs, or other structures that aid their dispersal. Means of fruit dispersal are especially important in the colonization of islands or other distant patches of suitable habitat.

SELF-QUIZ

The following questions pinpoint important information in this chapter. You should be sure to know the answers to each of them.

1. Fossil angiosperms first appeared in the fossil record approximately
   - (a) 10 million years ago
   - (b) 125 million years ago
   - (c) 200 million years ago
   - (d) 450 million years ago
   - (e) 2 billion years ago

2. Three-pored pollen and the presence of vessel elements are considered advanced features in angiosperms. True or false?

3. The sepals of an angiosperm flower probably evolved from _____, whereas the petals evolved from flattened _____ or from _____.

4. If a flower contains no stamens it is called
   - (a) pistillate
   - (b) staminate
   - (c) connate
   - (d) vericate

5. All flowering plants are either dioecious or monoecious. True or false?

6. Which of the following is *not* a characteristic of an *advanced* flower
   - (a) adnation of floral parts
   - (b) connation of floral parts
   - (c) radial symmetry
   - (d) fused carpels
   - (e) inferior ovary

7. A flower is described to you by a friend as being a long white tube containing a sickly-sweet smelling liquid. You would expect that
   - (a) the flower is very primitive in nature
   - (b) the flower is hummingbird pollinated
   - (c) the flower is butterfly pollinated or moth pollinated
   - (d) the flower is wind pollinated

8. You get the sniffles and watery eyes every fall. You visit your allergist, who skin tests you and says that you are allergic to pollen of a plant named *Ambrosia*. You guess that *Ambrosia* is
   - (a) a very big plant
   - (b) wind pollinated
   - (c) bee pollinated
   - (d) a plant with red flowers

9. Cross-pollination is always more advantageous than self-pollination. True or false?

10. A heterogenous environment promotes _____ pollination.

11. The fossil record proves conclusively that the earliest flowering plants were wind pollinated. True or false?

12. Many angiosperms are self-pollinated. True or false?

THOUGHT QUESTIONS

1. Why haven't we found the "missing link" between the flowering plants and their ancestors? What would you suggest we do to try to solve this riddle?

2. What role did bees play in the origin of angiosperms and in their subsequent diversification?

3. Angiosperms usually produce both pollen and ovules within a single flower, or at least on a single plant. Why don't all angiosperms simply self-pollinate?

FOR FURTHER READING

BARTH, F.G.: *Insects and Flowers: The Biology of a Partnership,* Princeton University Press, Princeton, N.J., 1985. An outstanding account of the biology of the insect-flower relationship; takes many modern findings into account.

BATRA, S.W.T.: "Solitary Bees," *Scientific American,* February 1984, pages 120-127. Excellent article on these diverse and fascinating pollinators, some of which are of great commercial importance.

FAEGRI, K., and L. VAN DER PIJL: *The Principles of Pollination Ecology,* 3rd ed., Pergamon Press, Inc., Elmsford, N.Y., 1979. A thorough and comprehensive review of pollination ecology.

HEYWOOD, V.H. (ed.): *Flowering Plants of the World,* Mayflower Books, Inc., New York, 1978. An outstanding guide to the families of flowering plants.

PETTITT, J., S. DUCKER, and B. KNOX: "Submarine Pollination," *Scientific American,* March 1981, pages 134-145. Sea grasses—marine angiosperms—flower underwater, shedding pollen that is borne from plant to plant by the waves.

PROCTOR, M., and P. YEO: *The Pollination of Flowering Plants,* Taplinger Publishing Co., Inc., New York, 1973. An excellent introduction to pollination biology, clearly and interestingly presented.

RADFORD, A.E., et al.: *Vascular Plant Systematics,* Harper & Row, Publishers, Inc., New York, 1974. Excellent source book on all aspects of the field.

TYRRELL, E.Q., and R.A. TYRRELL: *Hummingbirds: Their Life and Behavior. A Photographic Study of the North American Species,* Crown Publishers, Inc., New York, 1985. A magnificently illustrated account of the most highly specialized birds that visit flowers.

Chapter 31   VASCULAR PLANT STRUCTURE

Overview

In vascular plants, roots and shoots have very different structures, although growth at the apices is characteristic of both. All parts of a plant have an outer covering, called dermal tissue, and ground tissue, within which is embedded vascular tissue. One component of the vascular tissue, the xylem, conducts water and its dissolved nutrients, and the other component, phloem, distributes carbohydrates, which are manufactured in the green parts of the plant. Plants are immobile, but they adjust to their environment by growing and changing their form. Root branches arise deep within the tissue, ultimately bursting through to the surface; in stems, branches arise on the surface, in the axils of the leaves. The leaves themselves are flattened organs, specialized for photosynthesis. When flowers form on an axis, the stem stops growing permanently along that axis.

For Review   *Here are some important terms and concepts that you will encounter in this chapter. If you are not familiar with them, you should review them before proceeding.*

- **Cell structure** (Chapter 6)
- **How cells divide** (Chapter 8)
- **Major groups of plants** (Chapter 29)
- **Determinate and indeterminate growth** (Chapter 30)
- **Monocots and dicots** (Chapter 30)

A plant never becomes an adult in the way in which an animal does. Rather, plants simply keep growing, adding new cells, tissues, and organs at the ends of their shoots and their roots, becoming ever larger. Some large patches of prairie grasses are thought to be single individuals that have been growing in one place since the glaciers receded, more than 10,000 years ago. Trees attain great ages, and potatoes and many other crops are simply propagated over and over again as parts of a single cloned plant, producing generation after generation of genetically identical individuals.

The ways in which vascular plants grow and develop into mature individuals will be our major focus in this chapter. We will discuss the fundamental differences between the underground portions of the plant—the roots—and the aboveground portions—the

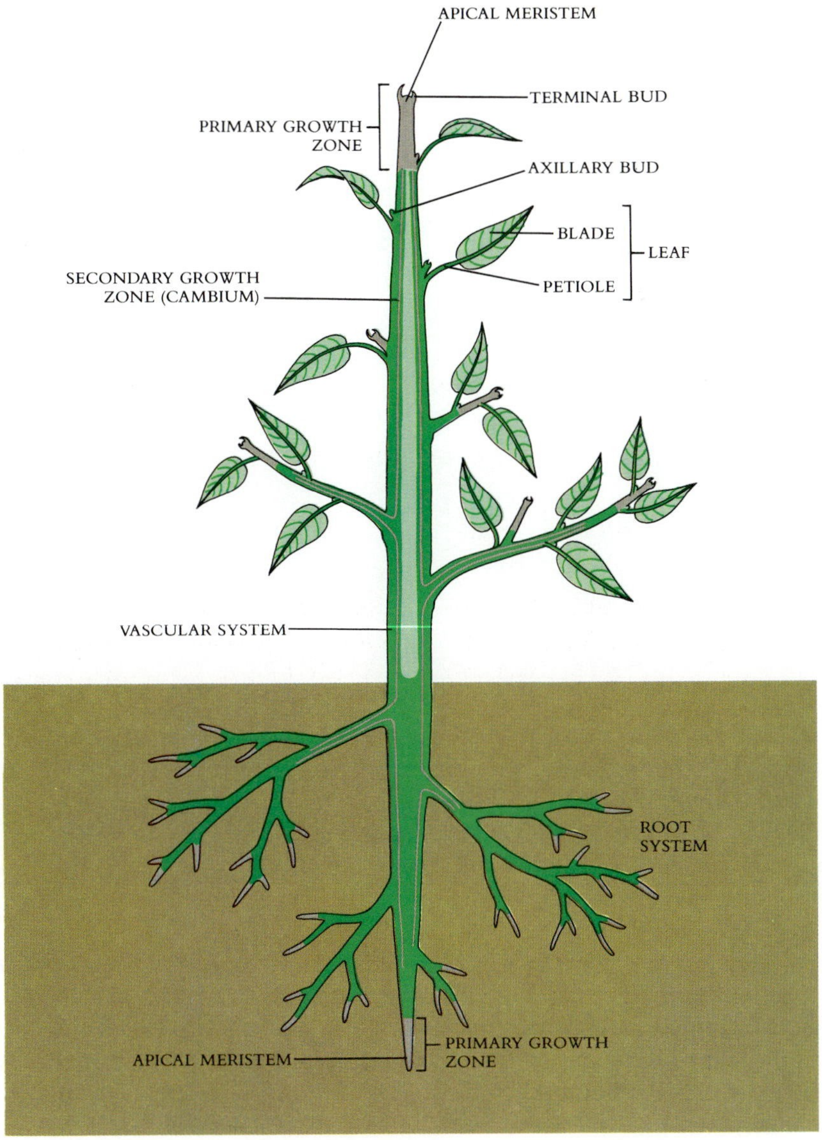

APICAL MERISTEM

TERMINAL BUD

PRIMARY GROWTH ZONE

AXILLARY BUD

BLADE — LEAF

PETIOLE

SECONDARY GROWTH ZONE (CAMBIUM)

VASCULAR SYSTEM

ROOT SYSTEM

APICAL MERISTEM — PRIMARY GROWTH ZONE

FIGURE 31-1

Diagram of a plant body, showing the parts.

shoots—as well as the functional and structural relationships between them. First, though, we will analyze the way in which these organs, together with their specialized cell and tissue types and their appendages, form the plant body.

Although the similarities between a cactus, an orchid, and a pine tree might not be obvious at first sight, plants have a fundamental unity of structure (Figure 31-1). This unity is reflected in the construction plan of their respective plant bodies; in the way that they grow, produce, and transport their food; and in the means by which they regulate their development.

EARLY DEVELOPMENT

The patterns of growth and development that are characteristic of plants begin to emerge in the early stages of embryo development, which take place within the growing seed. We began to consider this process in the last chapter, and we will now describe the early development of the embryo, as an introduction to our consideration of plant structure.

A mature angiosperm embryo consists of an axis with either one or two **cotyledons,** or embryonic leaves (Figure 31-2). In general, monocot embryos have only one cotyledon and dicot embryos have two. In dicots, the food stored initially in the endosperm may either be absorbed into the cotyledons, which may then become quite thick and fleshy as in

A

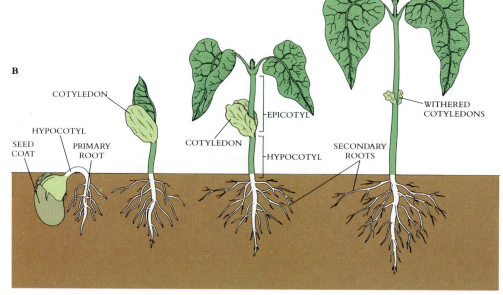

B

COTYLEDON

HYPOCOTYL

SEED COAT

PRIMARY ROOT

EPICOTYL

COTYLEDON

HYPOCOTYL

WITHERED COTYLEDONS

SECONDARY ROOTS

FIGURE 31-2

Seeds and stages of germination in a dicot, the common bean, Phaseolus vulgaris **(A** *and* **B),** *and a monocot, corn,* Zea mays **(C** *and* **D).**

FIGURE 31-3

A germinating seedling of radish, Raphanus sativus, *showing the abundant fine root hairs that form back of the root apex.*

peas and beans, or it may remain in the seed at maturity. In monocots, the single cotyledon functions mainly as a food-absorbing organ, transfering food from the endosperm to the young embryo during the germination of the seed. In either case, the food that is initially concentrated in the endosperm during the early development of the embryo is used during the germination and early establishment of the young plant. Starches, fats, and oils are converted into sugars and used to nourish and sustain the young plant before its photosynthesis begins.

In the embryo, the **apical meristems**—regions of active cell division that occur relatively close to the tips of roots and stems—differentiate early. These meristems continue dividing actively throughout the entire life of the plant and constitute one of the important ways in which the pattern of growth and development in plants differs from that which is characteristic of animals. The apical meristem of the shoot in a plant embryo is located at the tip of the **epicotyl,** the portion of the axis that extends above the cotyledons. The epicotyl may be short and relatively undifferentiated, or it may be longer and even include one or more embryonic leaves. The epicotyl, together with its young leaves, is called a **plumule.** The axis of the embryo below the cotyledons is called the **hypocotyl.** A distinct embryonic root at the lower end of the hypocotyl is called the **radicle.** Often, however, a radicle cannot be distinguished in the embryo.

> **An apical meristem is a region of active cell division that occurs relatively close to the tips of the roots and shoots of plants.**

The ways in which the embryonic root and shoot emerge from the seed during germination vary widely from species to species. In many, the root emerges first and anchors the plant in the soil before the shoot appears (Figure 31-3). The cotyledons may be held either below ground or above it, and they may or may not become green and then contribute to the nutrition of the seedling while it is becoming established. The period from the germination of the seed to the establishment of the young plant is a very critical one for the plant's survival. Before the plant is fully independent nutritionally, it is extremely susceptible to attack by bacterial and fungal diseases and by insects and other pests. If water is in short supply, the survival of the seedling may be threatened. After its establishment, however, the seedling grows on to produce all of the types of cells, tissues, and organs that are characteristic of the mature plant and are the subject of this chapter.

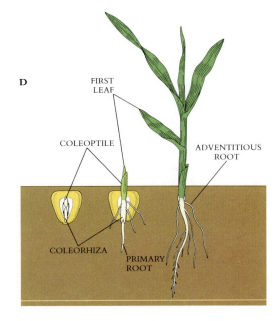

ORGANIZATION OF THE PLANT BODY

A vascular plant is basically an axis consisting of root and shoot. The **root** penetrates the soil and absorbs water and various ions, which are crucial for plant nutrition, and it also anchors the plant. The **shoot** consists of stem and leaves. The **stem** serves as a framework for the positioning of the **leaves,** the principal places where photosynthesis takes place. The arrangement, size, and other characteristics of the leaves are of critical importance in the plant's production of food. Flowers, other reproductive organs, and ultimately fruits and seeds are formed on the shoot as well.

Tissue Types in Plants

There are three major types of tissue in plants. One is the conducting tissue, or **vascular tissue.** There are two kinds of vascular tissue: (1) **xylem,** which conducts water and dissolved minerals, and (2) **phloem,** which conducts carbohydrates, mainly sucrose, which the plant uses for food. These two kinds of vascular tissue differ not only in function but also in structure.

   The other two major tissue types in plants are **ground tissue,** in which the vascular tissue is embedded, and **dermal tissue,** the outer protective covering of the plant. The dermal tissue often is covered with a waxy substance known as cuticle. Each of the three major tissue types—vascular tissue, ground tissue, and dermal tissue—consists of its own distinctive cell types, which are related to the functions of the tissues in which they occur. Some of the characteristic cell types will be discussed later in this chapter.

> **The three major types of tissues in plants are (1) dermal tissue, which covers the outside of the plant, (2) ground tissue, which fills its interior, and (3) vascular tissue, which conducts water, minerals, carbohydrates, and other substances throughout the plant and is embedded in ground tissue.**

Types of Meristems

Primary growth in plants is initiated by the apical meristems. These are regions of active cell division that occur relatively close to the tips of roots and stems. The growth of these meristems results primarily in the extension of the plant body. As it elongates, it forms

THE FLOWERING OF BAMBOOS AND THE STARVATION OF GIANT PANDAS

In many Asian bamboo species, all the individuals of the species flower and set seed simultaneously. These cycles tend to occur at very long intervals, ranging from 3 years upward. Most of the cycles are between 15 and 60 years long. The most extreme example is that of the Chinese species *Phyllostachys bambusoides,* which seeded massively and throughout its range in 919, 1114, between 1716 and 1735, in 1833-1847, and in the late 1960s. The last event involved cultivated plants in widely separated areas, including England, Russia, and Alabama. *Phyllostachys bambusoides,* therefore, has a flowering cycle of about 120 years, set by an internal clock that runs more or less independently of environmental circumstances!

The bamboos that have cycles of this kind spread mainly by the production of rhizomes. When they do flower, nearly all individuals flower at once, set large quantities of seed, and die. Huge numbers of animals, including rats, pigs, and pheasants, often migrate into areas where bamboos are fruiting to feed on the seeds; humans also gather them for food. Apparently the plants put all of their energy into producing numbers of seeds huge enough for large numbers of seedlings to survive, even though most of the seeds are eaten.

A particular conservation problem that has attracted widespread notice in the 1980s is the relationship between the flowering of bamboos (primarily of the genus *Fargesia*) and the survival of the giant panda, a spectacular animal that still survives in a few mountain ranges in southwestern China. Humans have occupied so much of the former range of the panda that the wild population now consists of only a few thousand individuals, which live only in several widely separated areas. There, only one or a few species of bamboo provide virtually all of their food (Figure 31-A). The mass flowering of some of these bamboo species in 1984, which led to the death of these plants over large areas (Figure 31-B), drove many of the pandas to the brink of starvation. Only a massive international effort made it possible to rescue many of the animals. The predictable episodes of mass flowering in bamboos will need to be considered carefully in planning for the pandas' survival.

---

what is known as the primary **plant body,** which is made up of **primary tissues.** The primary plant body comprises the young, soft shoots and roots of a tree or shrub, or the entire plant in some short-lived plants.

Secondary growth involves the activity of the **lateral meristems.** Lateral meristems are cylinders of meristematic tissue; the continued division of their cells results primarily in the thickening of the plant body. There are two kinds of lateral meristems: the **vascular cambium,** which gives rise to ultimately thick accumulations of secondary xylem and phloem, and the **cork cambium,** from which arise the outer layers of the bark in both roots and shoots. The tissues formed from the lateral meristems, comprising most of the bulk of trees and shrubs, are known collectively as the **secondary plant body,** and its tissues are known as **secondary tissues.** Figure 31-1 compares primary and secondary growth zones.

> **The primary plant body, which includes the young, soft shoots and roots, arises from the apical meristems. Once the lateral meristems begin to function, they produce the secondary plant body, which is characterized by thick accumulations of conducting tissue and the other cell types associated with it.**

The earliest plants had only primary growth. Secondary growth first appeared in the middle of the Devonian Period, about 360 million years ago (Figure 23-17 shows an early forest). It apparently originated independently in several unrelated groups of vascular plants, including some groups that had seeds and others that did not. Secondary growth

FIGURE 31-A

A giant panda in a large enclosure among shoots of the bamboo Fargesia spathacea. *Bamboos of the genus* Fargesia *are the most important panda food in the Min Mountains of northern Sichuan Province, China, where this photograph was taken.*

FIGURE 31-B

A Chinese scientist examining a stand of the bamboo Fargesia nitida *that has flowered and dropped its leaves; the whole plant will soon be dead. This bamboo covers large areas of the Min Mountains and other areas of the mountains of southwestern China where giant pandas occur. It flowered extensively in the Min Mountains between 1974 and 1976, and again in 1982 and 1983; as a result, many giant pandas needed to be rescued and fed in captivity.*

made possible the evolution of trees, because plant bodies became strong enough to achieve great size; because of this, it literally changed the face of the globe as forests appeared for the first time.

HOW LONG DO INDIVIDUAL PLANTS LIVE?

The ways in which the meristems function in the production of a mature plant depend on the kinds of cells and tissues that they produce. In turn, the overall pattern of organization is affected greatly by the habit—the general appearance—of the individual kind of plants, and by its cycle of growth through the year. Some plants, for example, are **woody,** whereas others are **herbaceous** (Figure 31-4). Woody plants are those in which secondary growth has been extensive; herbaceous plants are those in which it is limited or absent. Herbaceous plants send up new stems above the ground every year, producing them from woody underground structures, or germinate and grow to flowering just once. Depending on the length of their life cycles, herbaceous plants may be annual, biennial, or perennial in duration. Shorter-lived plants rarely become very woody, since there is not enough time in their limited life spans for the massive accumulation of secondary tissues.

Annual plants grow, flower, and form fruits within a period of less than a year and die when the process is complete. Many crop plants, for example, are annuals, including corn, wheat, and soybeans. Annuals generally grow rapidly under favorable conditions and increase greatly in size in proportion to the availability of additional water or fertilizer. Some annuals, like sunflowers or giant ragweed, do form wood as a result of secondary growth, if they live long enough. Many, however, are entirely herbaceous.

FIGURE 31-4

Plants live for very different lengths of time, desert annuals like those shown in **A** *completing their entire life span in a few weeks, and trees, such as the giant redwood,* Sequoiadendron giganteum **(B),** *which occurs in scattered groves along the west slope of the Sierra Nevada in California, living 2000 years or more.*

Biennial plants, of which there are many fewer kinds than there are annuals, have life cycles that take 2 years to complete. During the first year, biennials generally form a tuft of leaves, or **rosette.** In the second year of these plants' growth, when they flower, the energy stored in the rosette and in the underground parts of the plant is used to produce flowering stems (see Figure 33-9). Many crop plants, including carrots, cabbage (see Figure 33-9, *B*), and beets, are biennials, but these plants generally are harvested for food during their first season, before they flower. They are grown for their leaves or roots, not for their fruits or seeds.

Perennial plants grow on from year to year, and may be herbaceous, such as many woodland and prairie wildflowers, or woody, as are trees and shrubs. The great majority of vascular plant species are perennials. Herbaceous perennials rarely have any secondary growth in their stems; the stems die back promptly each year after a period of relatively rapid growth and food accumulation. The food is often stored in their roots, which may then become quite large in relation to their delicate aboveground stems.

> **Annual plants complete their whole growth cycle, including flowering and fruit production, within a single year. Biennial plants flower only once, after two seasons of growth. Perennials, which may be either herbaceous or woody, flower repeatedly once they begin to do so, and live for an indefinite period of time.**

Perennials that do have woody stems above ground are either trees or shrubs. Trees are large woody plants, usually with a single stem, whereas shrubs are smaller and often have several stems. There is, however, no real dividing line between trees and shrubs, and some plants may grow in either form depending on the local conditions. Trees and shrubs may be either **deciduous,** with all the leaves falling at one particular time of year and the plants then remaining bare for a period, or **evergreen,** with the leaves dropping more or less throughout the year and the plants never appearing completely bare (Figure 31-5). In north temperate regions, conifers are the most familiar evergreens; but in tropical and subtropical regions, most angiosperms are evergreen, except for those that characteristically occur where there is severe drought at a particular season of the year. Many of the kinds

FIGURE 31-5

*The contrast between the deciduous quaking aspens (*Populus tremuloides*) and the dark green, evergreen Engelmann spruces (*Picea engelmannii*) is evident in this autumn scene near Durango, Colorado.*

of tropical and subtropical angiosperms that do grow in regions where there is a drought may be deciduous, losing their leaves during the period of drought and thus conserving water.

Plants that flower only once and then die, like annuals and biennials, are termed **monocarpic** ("once-fruiting"); perennials, which flower repeatedly, are called **polycarpic** ("many-fruiting"). Century plants (*Agave* species) are well-known examples of monocarpic plants; they grow for very long periods of time and then produce massive flowering stalks. After they have set fruit, the entire plant dies. Many plants that are considered biennials actually do not flower for 3 years or more.

PLANT CELL TYPES
Meristems

Meristems, both apical and lateral, consist of small, usually unspecialized cells, some of which divide repeatedly (Figure 31-6). When one of these cells divides, one of its two derivative cells remains in the meristem and the other becomes part of the plant body. The cells that ultimately become part of the plant body usually divide further before they begin to differentiate and assume the characteristics of the type of cell that they will become.

The apical meristem gives rise to three types of **primary meristems.** These are partly differentiated tissues; in them, cell division continues to take place as they develop into the tissues of the plant body (Figure 31-7). The three primary meristems are the **protoderm,** which differentiates further into epidermis; the **procambium,** which differentiates further into primary vascular strands; and the **ground meristem,** which differentiates further into ground tissue. The basic tissue patterns are established by early activity in the region of the apical meristem, when the primary meristems begin to differentiate.

The primary meristems, which differentiate from the apical meristems, are (1) the protoderm, which becomes the epidermis, (2) the procambium, which becomes the primary vascular strands, and (3) the ground meristem, which becomes the ground tissue.

FIGURE 31-7

Diagram of primary meristems and their relation to apical meristem.

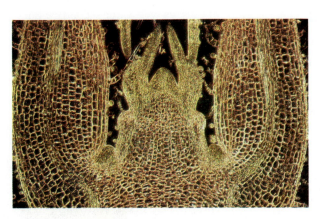

FIGURE 31-6

Transection of a shoot apex in Coleus.

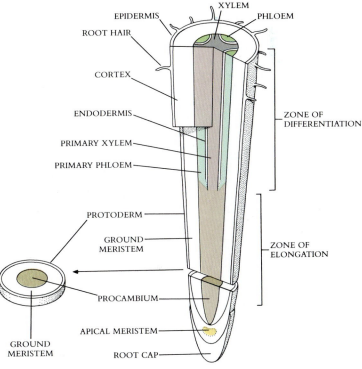

EPIDERMIS
ROOT HAIR
XYLEM
PHLOEM
CORTEX
ENDODERMIS
PRIMARY XYLEM
PRIMARY PHLOEM
ZONE OF DIFFERENTIATION
PROTODERM
GROUND MERISTEM
ZONE OF ELONGATION
GROUND MERISTEM
PROCAMBIUM
APICAL MERISTEM
ROOT CAP

Parenchyma and Collenchyma

Parenchyma cells, which often are somewhat spherical, are the least specialized and the most common of all plant cell types (Figures 31-9 and 31-17, for example, show parenchyma cells). They form masses in leaves, stems, and roots in which secondary growth has not taken place, and also form masses in secondary tissues. Parenchyma cells, unlike some of the other cell types that we will consider below, are characteristically alive at maturity, with fully functional protoplasts and a nucleus. They are capable, therefore, of further division. Most parenchyma cells have only **primary walls,** which are cell walls that are laid down while the cells are still growing. **Secondary cell walls,** in contrast, are deposited inside the primary walls of fully expanded cells. In secondary tissue, which consists mainly of densely packed, specialized conducting cells, parenchyma cells exist mainly as **rays,** horizontal strands that extend radially from the center of the stem or root. The rays conduct water laterally in a mature, woody root or stem.

 Collenchyma cells, which also are living at maturity, form strands or continuous cylinders beneath the epidermis of stems or leaf stalks and along veins in leaves. They are usually elongated, with unevenly thickened primary walls (Figure 31-8), which are their distinguishing feature. Strands of collenchyma provide much of the support for plant organs in which secondary growth has not taken place. The "strings" of celery leaves—the part of celery that we eat is the stalk of the leaf—consist mainly of collenchyma.

> **Parenchyma cells, which usually are living at maturity, are the most common type of cells in the primary plant body. They often are somewhat spherical, and, for the most part, lack secondary cell walls. Collenchyma cells, which also are living at maturity, are elongate cells that are characterized by their unevenly thickened walls. They provide much of the support to plant organs in which secondary growth has not taken place.**

Sclerenchyma

Sclerenchyma cells have tough, thick secondary walls; they usually do not contain living protoplasts when they are mature. Their secondary cell walls are often impregnated with **lignin,** a complex polymer; cell walls in which lignin is present are said to be **lignified.** Lignin is deposited in the primary walls of some kinds of cells, as well as the secondary ones. It is a rigid substance and makes the walls in which it is deposited more rigid. Lignin is common in the walls of plant cells that have a supporting or mechanical function.

 There are two types of sclerenchyma: **fibers,** which are long, slender cells that usually form strands, and **sclereids,** which are variable in shape but often branched. Both of these tough, thick-walled cell types serve to strengthen the tissues in which they occur. For example, linen is woven from strands of fibers that occur mainly in association with the phloem of flax. Sclereids, on the other hand, may occur singly or in groups. The gritty texture of pears is caused by the groups of sclereids that occur among the soft flesh of the fruit (Figure 31-9).

> **Sclerenchyma cells have thick, often lignified secondary walls and are nonliving at maturity. The two types of sclerenchyma are fibers, which are elongated, and sclereids, which are not.**

Xylem

Xylem is the principal water-conducting tissue of plants. It forms a continuous system running throughout the plant body. Within this system, water passes from the roots up through the shoot in an unbroken stream. Dissolved minerals also are taken into plants through their roots, as a part of this stream of water. When the water reaches the leaves, it passes into the air as water vapor, mainly through the stomata. **Primary xylem** is derived

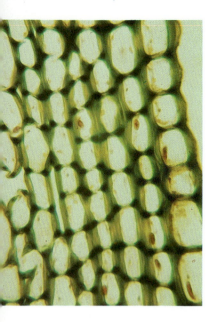

FIGURE 31-8

Collenchyma cells, with thickened side walls, from a young branch of elderberry (Sambucus). *In other kinds of collenchyma cells, the thickened areas may occur at the corners of the cells, or in other kinds of strips.*

FIGURE 31-9

Clusters of sclereids ("stone cells") in the pulp of a pear; the surrounding, thin-walled cells are parenchyma. Such clusters of sclereids give pears their gritty texture.

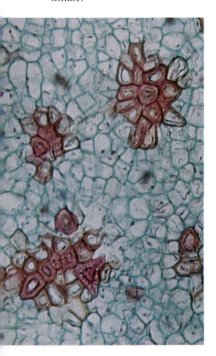

from procambium, which in turn comes from the apical meristem of that respective root or shoot. **Secondary xylem,** on the other hand, is formed by the vascular cambium, a lateral meristem that develops later.

The two principal types of conducting elements in the xylem are **tracheids** and **vessel elements** (Figure 31-10), both of which resemble fibers in having thick, lignified secondary walls, in being elongated, and in having no living protoplast at maturity. The continuous stream of water in a plant flows from tracheid to tracheid through pits in their secondary walls. In contrast, vessel elements have not only pits, but also definite perforations in their end walls by which they are linked together, and through which water flows even more efficiently. A linked row of vessel elements forms a **vessel.**

Vessels apparently conduct water much more efficiently than strands of tracheids do. We conclude this partly because vessel elements have evolved from tracheids independently in several groups of plants, indicating that they have been favored strongly by natural selection. It is probably also the case that at least some kinds of fibers have evolved from tracheids (Figure 31-10, *A*), becoming specialized for a strengthening, rather than a conducting, function. As we saw in the last chapter, primitive angiosperms have only tracheids (see Figure 30-6), but the great majority of angiosperms have vessel elements. In addition, it seems evident that the perforated end walls of vessel elements are better suited to the passage of water than are the pits that connect tracheids.

In addition to the conducting cells, xylem likewise includes fibers and parenchyma cells. The arrangements of these and other kinds of cells in the xylem make possible the identification of various plant species from their wood alone.

> **The major types of conducting cells of the xylem are tracheids and vessel elements. Incorporated in the xylem, however, are also fibers and parenchyma cells.**

Phloem

Phloem (Figure 31-11) is the principal food-conducting tissue in the vascular plants. Different kinds of plants have one of the two different kinds of elongate, slender conducting cells that occur in phloem: **sieve cells** and **sieve-tube members;** clusters of pores known as

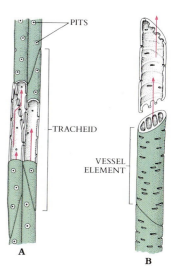

FIGURE 31-10

Comparison between vessel elements and tracheids. In tracheids **(A)** *the water passes from cell to cell by means of pits, whereas in vessel elements* **(B)** *it moves by way of pores, which may be simple or interrupted by bars, as in the example shown. In gymnosperm wood, tracheids both conduct water and provide support; in most kinds of angiosperms, vessels conduct the water and fibers provide the support.*

FIGURE 31-11

Phloem of squash (Cucurbita)
A *A sieve-tube member, connected with the cells above and below to form a sieve tube. Note the thickened end walls, which are at right angles to the sieve tube. The narrow cell with the nucleus at the right of the sieve-tube member is a companion cell.*
B *Transection of phloem, showing callose deposited around the sieve areas in the end walls of the sieve tube elements. Callose is believed mostly to be deposited when a plant is injured, and not to be present in normally functioning phloem.*

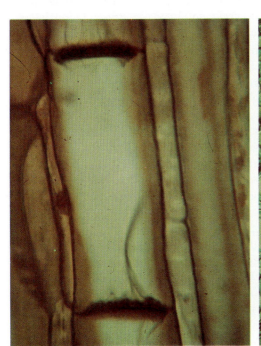

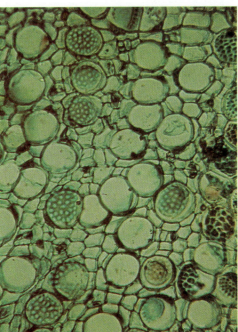

A   B

sieve areas occur on both types of cells. Such clusters are more abundant on the overlapping ends of the cells; they connect the protoplasts of adjoining sieve cells and sieve-tube members. Both of these types of cells are living, but they lack a nucleus.

In sieve-tube members, the pores in some of the sieve areas are larger than those in the others; such sieve areas are called **sieve plates.** Sieve-tube members occur end to end, forming longitudinal series called **sieve tubes.** Sieve cells are less specialized than sieve-tube members; the pores in all of their sieve areas are roughly the same diameter. They are the only type of food-conducting cell that most vascular plants have, except for the angiosperms, which have only sieve-tube members. In an evolutionary sense, sieve-tube members clearly are advanced over sieve cells; they are more specialized and presumably more efficient.

Specialized parenchyma cells known as **companion cells** occur regularly in association with sieve-tube members (Figure 31-11, *A*). Companion cells apparently carry out some of the metabolic functions that are needed to maintain the individual sieve-tube members with which they are associated. They have all of the components of the normal parenchyma cells of the plants in which they occur, including nuclei, and their numerous pores connect their cytoplasm with that of the conducting cells.

> **The principal cell types of phloem are sieve cells or sieve-tube members. If sieve-tube members are present, they are associated with specialized parenchyma cells called companion cells. Sieve cells and sieve-tube members lack a nucleus at maturity, whereas companion cells are living.**

In prepared microscope slides, a substance known as **callose** is very prominent in and around the pores of both sieve cells and sieve-tube members (Figure 31-11, *B*). Callose is a polysaccharide composed of spirally wound chains of glucose residues. It apparently forms around these pores as a wound response—a response to the killing and sectioning of the material that takes place in the preparation of the slides, for example—and either does not occur or is much less abundant in undisturbed sieve cells and sieve-tube members. Nevertheless, callose is so conspicuous in prepared material that it has long been considered an important feature of sieve cells and sieve-tube members.

Epidermis

Flattened epidermal cells, which often are covered by a thick layer of cuticle, cover all parts of the primary plant body. A number of types of specialized cells occur in the epidermis, including guard cells, trichomes, and root hairs.

Guard cells are paired; with the opening that lies between them, they make up the **stomata.** Stomata (Figure 31-12) occur frequently in the epidermis of leaves, and they sometimes occur on other parts of the shoot, such as stems or fruits. The passage of oxygen and carbon dioxide into and out of the leaves, as well as the loss of water from them, takes place almost exclusively through the stomata. In most kinds of plants, stomata are more numerous on the lower surface of the leaf than on its upper surface, and in many plants they are entirely confined to the lower surface. The stomata open and shut in relation to external factors such as the supply of moisture. During periods of active photosynthesis, the stomata are open, allowing the free passage of carbon dioxide into the leaf. We will consider the mechanism that governs such movements in Chapter 34.

Trichomes are outgrowths of the epidermis that vary greatly in form in different kinds of plants. They also occur frequently on stems, leaves, and reproductive organs (Figure 31-13). A "fuzzy" or "woolly" leaf is covered with trichomes, which can be seen clearly with a microscope under low magnification. Trichomes play an important role in regulating the heat and water balance of the leaf.

Near the tips of roots occur extensions of the epidermis called **root hairs.** The root hairs occur in masses just behind the very ends of the roots. They keep the root in intimate

FIGURE 31-12

Stomata are frequently among the epidermal cells of this member of the aralia family (Araliaceae). Around the guard cells are specialized epidermal cells, found in association with the guard cells, called subsidiary cells.

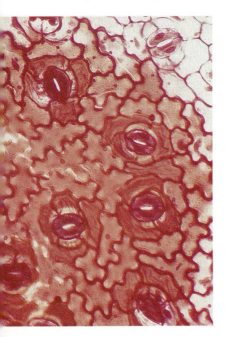

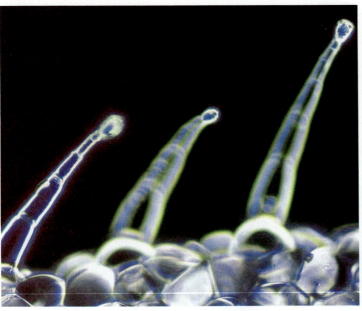

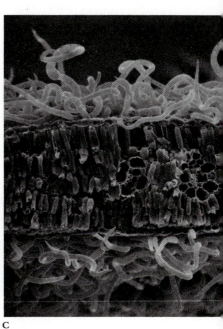

A

B

C

contact with the particles of soil, and are soon worn off as the root continues to grow. Virtually all of the absorption of water and minerals occurs by way of the root hairs in herbaceous plants (Figure 31-3). In mature woody plants, water is also absorbed in substantial quantities directly through the roots and other tissues.

LEAVES

Leaves, which arise as outgrowths of the shoot apex, are the most important light-capturing organs in the great majority of plants. The only exceptions to this generality are found in some plants, such as cacti, in which the stems are green and have largely taken over the function of photosynthesis for the plant. Since leaves are so crucial to a plant, such features as their arrangement, shape, size, and other aspects are highly significant. These factors differ greatly in plants that grow in different environments.

The apical meristems of stems and roots are capable of growing indefinitely under the appropriate circumstances. Leaves, on the other hand, grow by means of **marginal meristems,** which flank their thick central portions. These marginal meristems grow outward and ultimately form the blade of the leaf, while the central portion becomes the midrib. Once a leaf is fully expanded, its marginal meristems cease to function. The growth of roots and stems is indeterminate, but that of leaves, like that of flowers, is determinate.

General Features

Most leaves have a flattened portion, the **blade,** and a slender stalk, the **petiole** (Figure 31-14). In addition, there may be two leaflike organs, the **stipules,** that flank the base of the petiole, where it joins the stem. These stipules may be microscopic or relatively large, leaflike, and conspicuous. Veins, usually consisting of both xylem and phloem, run through the leaves. In many monocots, the veins are **parallel;** in most dicots, the pattern is **net** or **reticulate venation** (Figure 31-15). What are called **simple leaves** are undivided, although they may be deeply lobed. In contrast, **compound leaves** consist of clearly separated **leaflets.**

In their placement on the stem, leaves follow definite patterns (Figure 31-16). In most plants they are spirally arranged, a condition that is often called **alternate.** In many others the leaves are **opposite,** occurring in pairs, and in a few they are **whorled,** with more than two leaves attached at one level on the stem.

FIGURE 31-13

Trichomes.
A *Three of the multicellular trichomes that cover the leaves of African violets,* Saintpaulia. *A covering of trichomes creates a layer of more humid air near the surface of a leaf, enabling the plant to conserve available supplies of water.*
B *A complex multicellular hair from the leaves of the sundew,* Drosera. *Such hairs secrete the enzymes that the plant uses to digest the bodies of its insect prey.*
C *The dense mats of trichomes on both sides of the silvery leaves of the desert composite* Encelia farinosa, *seen in this scanning electron micrograph, cause the leaf absorption of visible sunlight to drop from about 85%, as in a green leaf, to about 30%; as a consequence, the leaf temperatures are about 8° to 10° C cooler than if they were green. This lower leaf temperature reduces water loss by some 30% to 50%.*

FIGURE 31-14

In both the simple leaves of a tulip tree, Liriodendron tulipifera **(A),** *and the compound leaves of the sensitive plant* Mimosa pudica **(B),** *the expanded blades of the leaves are joined to the stem by a slender stalk, the petiole.*

A

B

A

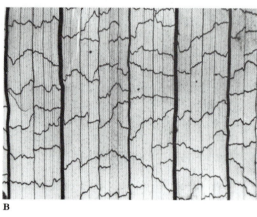

B

FIGURE 31-15

The leaves of dicots, such as this African violet relative from Sri Lanka **(A),** *have net, or reticulate, venation; those of monocots, like this Latin American palm* **(B),** *have parallel venation.*

FIGURE 31-16

The three common types of leaf arrangements.

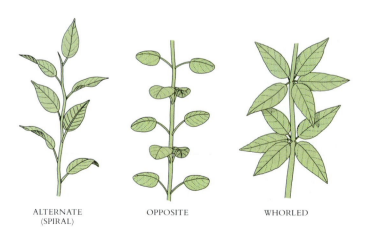

ALTERNATE (SPIRAL)   OPPOSITE   WHORLED

Structure and Organization

A typical leaf contains masses of parenchyma, through which the vascular bundles, or veins, run. The masses of parenchyma that occur in leaves are called **mesophyll** ("middle-leaf"). Beneath the upper epidermis of a leaf, there are one or more layers of columnar parenchyma cells, which are called **palisade parenchyma** (Figure 31-17). Certain kinds of plants, including some species of *Eucalyptus,* have leaves that hang down, rather than being extended horizontally; their palisade parenchyma is found on both sides of the leaf and there is, in effect, no upper side. But regardless of whether palisade parenchyma is found on one or both sides of a leaf, the rest of the interior, except for the veins, consists of a tissue called **spongy parenchyma.** Spongy parenchyma is composed of parenchyma cells that are more or less rounded in outline, or **isodiametric.** Between these cells are located large intercellular spaces that function in gas exchange and the passage of water vapor from the leaves. These intercellular spaces are connected, directly or indirectly, with the stomata.

The cells of the mesophyll, especially those near the leaf surface, are packed with chloroplasts. These cells constitute the plant's primary site of photosynthesis. Water and minerals are brought from the roots to the leaves in the xylem strands of the veins. Once the water reaches the ends of the veins in the leaves, it passes into the intercellular spaces of the mesophyll and gradually evaporates from the surfaces of the cells that border these spaces. The intercellular spaces have within them a saturated atmosphere that slows down the evaporation of additional water into them from the surfaces of the cells. At the same time that water is passing out through the stomata, the products of photosynthesis are transported from the leaves to all other parts of the plant via the phloem strands of the same veins.

> **The mesophyll in a leaf consists of two types of parenchyma cells, both packed with chloroplasts, especially in the cells that are near the surfaces of the leaves. Palisade parenchyma cells are columnar and closely packed together, whereas spongy parenchyma cells are loosely packed, mostly isodiametric, and separated by large intercellular spaces.**

Modified Leaves

Because of their great significance in the life of the plant, leaves have become modified in various ways during the course of their evolution (Figure 31-18). The huge, soft, highly dissected leaves of a tree fern lose water much more rapidly than do the tough, scalelike leaves of a juniper. Furthermore, the arrangement of leaves on a stem, the distance between the leaves, and the height and branching pattern of the stem are closely related to the way that a particular kind of plant functions in its environment.

As mentioned above, the stems have taken over the function of photosynthesis in some plants. For example, the large, succulent stems of cacti are protected from predators by their spines, which are highly modified leaves that do not function in photosynthesis and contain no living tissue at maturity. A few genera of cacti still have leaves of normal appearance, but in most they have been lost in the course of evolution.

Tendrils are organs that assist plants in climbing to relatively well-lighted places. They are sometimes modified stems (as in grapes or ivy) and sometimes modified leaves (as in the garden pea; Figure 31-18, *A*). Another example of modified leaves is provided by the layers of an onion, which are actually fleshy leaves, modified for food storage and clustered around an underground stem (Figure 31-18, *C*). The scales around a bud are modified leaves that fall off as the bud unfolds and the shoot within begins to grow. Perhaps the most spectacular modified leaves, however, are those of the carnivorous plants (see pp. 712-713).

Succulent leaves, like those of jade plants (Figure 31-18, *B*), hen and chickens, and ice plant, function in water storage and thus help plants to withstand drought. Such plants often have a modified form of photosynthesis called **CAM** (Crassulacean acid metabolism)

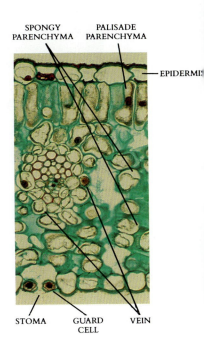

FIGURE 31-17

Transection of a lily leaf, showing palisade and spongy parenchyma; a vascular bundle, or vein; and the epidermis, with paired guard cells flanking the stoma that is visible on the lower surface of the leaf, and the substomatal space.

A   **B**   **C**

FIGURE 31-18

Modified leaves.
A *A tendril in the garden pea,* Pisum sativum.
B Jade plant, Crassula, *with succulent, opposite leaves and CAM photosynthesis.*
C *An onion,* Allium sativum, *in which the fleshy organs that make up the bulb are colorless, underground leaves modified for starch storage.*

photosynthesis, which is a variant of the C_4 pathway that we discussed in Chapter 12. In plants that have CAM photosynthesis, the stomata are closed during the day and C_4 compounds are accumulated at night, entering the Calvin cycle during the day. Such a process allows the fixation of much more CO_2 during a 24-hour cycle than would be possible otherwise, and is highly advantageous in arid regions when it is coupled with a succulent habit.

SHOOTS
Primary Growth

Leaves first appear as **leaf primordia** (sing. **primordium**), or rudimentary young leaves, which cluster around the apical meristem, unfolding and growing as the stem itself elongates (Figures 31-7 and 31-19). The places on the stem at which leaves are attached are called **nodes,** and the portions of the stem between these points of attachment are called the **internodes.** As the leaves expand to maturity, a **bud,** which is a tiny, undeveloped side shoot, develops in the axis of each (Figure 31-7). These buds, which have their own leaves, may elongate and form lateral branches, or they may remain small and dormant. As we will discuss in detail in Chapter 33, a hormone diffusing downward from the terminal bud of the shoot continuously suppresses the expansion of the lateral buds. These buds begin to

FIGURE 31-19

Scanning electron micrograph of shoot apex of sugar maple, Acer saccharinum, *showing a developing shoot in mid-August. The apical meristem, leaf primordia, and trichomes are plainly visible at this stage.*

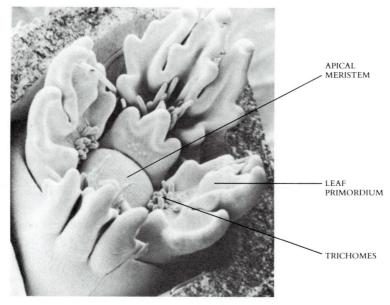

APICAL MERISTEM

LEAF PRIMORDIUM

TRICHOMES

expand when the terminal bud is removed. Therefore, gardeners who wish to produce bushy plants or dense hedges crop off the tops of the plants, removing their terminal buds.

Within the soft, young stems, the stands of procambium either occur as a cylinder in the outer portion of the ground meristem, as is common in dicots, or are scattered throughout it, an arrangement that is common in monocots (Figure 31-20). At the stage when only primary growth has occurred, the inner portion of the ground tissue of a stem is called the **pith,** and the outer portion is the **cortex.** The outermost parenchyma cells in the cortex frequently are packed with chloroplasts, in which case the stem is green and photosynthetically active.

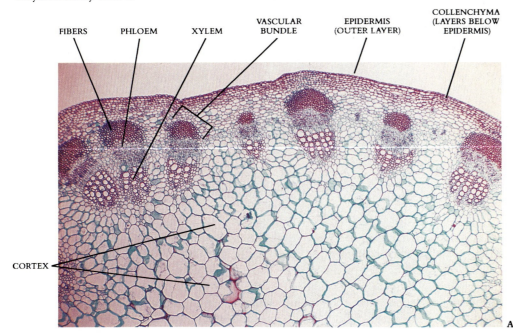

FIGURE 31-20

Transections of a young stem in **(A)** *a dicot, the common sunflower,* Helianthus annuus, *in which the vascular bundles are arranged around the outside of the stem; and* **(B)** *a monocot, asparagus,* Asparagus officinalis, *with the scattered vascular bundles characteristic of the class*

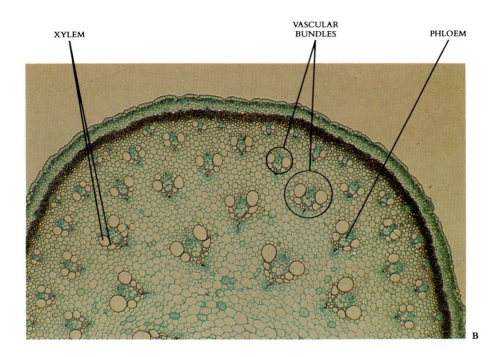

FIGURE 31-21

Transection of vascular bundles from a buttercup, Ranunculus acris, *showing the xylem and phloem.*

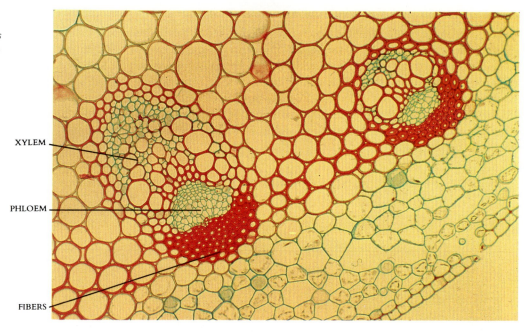

XYLEM

PHLOEM

FIBERS

As the stem matures, the strands of procambium within it differentiate into vascular bundles, which contain both primary xylem and primary phloem (Figure 31-21). The procambial strands of the stem grow upward into the developing leaf primordia by the differentiation of additional procambial cells as the primordia expand. As a consequence of this pattern of development, one or more vascular bundles diverge from the group of strands in the stem at each node and enter the leaf or leaves attached at that point. The pattern of vascular strands in the stem of a particular kind of plant, therefore, directly reflects the arrangement of the leaves.

Secondary Growth

Secondary growth is initiated by the differentiation of the **vascular cambium,** which consists of a thin cylinder of cells that in mature woody plants is located near the area where the bark and the main stem come together. The vascular cambium differentiates from parenchyma cells within the vascular bundles of the stem, between the primary xylem and the primary phloem. The cylindrical form of the vascular cambium is completed by the differentiation of some of the parenchyma cells that lie between the bundles (Figure 31-22). The pattern just described is the predominant pattern of formation of the vascular cambium in dicots, but other patterns also occur, especially in monocots, where true secondary growth is rare.

> **The vascular cambium differentiates, in most dicots at least, as a cylinder of dividing cells. Portions of this cylinder are derived from the parenchyma of the ground tissue, or cortex, and other portions are derived from the region within the primary vascular bundles that lies between the xylem and the phloem.**

The vascular cambium consists of elongated, somewhat flattened cells with large vacuoles. Such cells are very different in appearance from the small, isodiametric ones that occur in the apical meristems, which have small vacuoles. Cell division in the vascular cambium takes place both outwardly, in which case secondary phloem is produced, and

FIGURE 31-22

Early stage in differentiation of vascular cambium in an elderberry (Sambucus canadensis) stem.

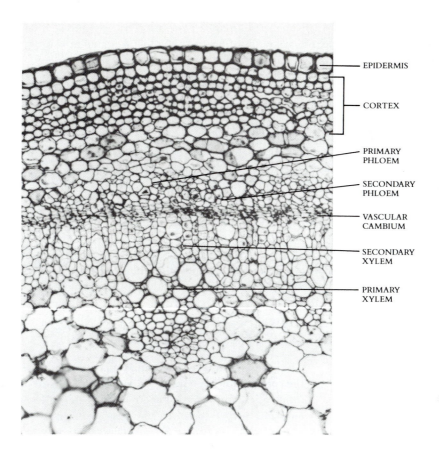

EPIDERMIS

CORTEX

PRIMARY PHLOEM

SECONDARY PHLOEM

VASCULAR CAMBIUM

SECONDARY XYLEM

PRIMARY XYLEM

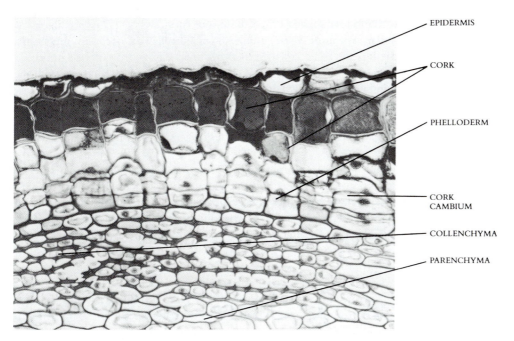

EPIDERMIS

CORK

PHELLODERM

CORK CAMBIUM

COLLENCHYMA

PARENCHYMA

FIGURE 31-23

An early stage in the development of periderm in elderberry (Sambucus canadensis).

inwardly, in which case secondary xylem is the product. The cells of the vascular cambium also divide laterally, allowing the cambium to increase in diameter as the stem thickens. The elongated cambial cells that divide to produce the conducting elements of the xylem and phloem, and other cambial cells similar to themselves, are called **fusiform initials.**

FIGURE 31-24

Lenticels, the numerous small indentations seen here, in the bark of bigtooth aspen, Populus grandidentata. *Oxygen reaches the layers of living tissue beneath the bark by way of such lenticels.*

Much smaller, more nearly isodiametric **ray initials** divide, giving rise to the **rays,** which are radial strands of parenchyma that function in water movement through the stem.

While the vascular cambium is first becoming established, a second kind of lateral cambium, the **cork cambium,** normally also develops in the outer layers of the stem. The cork cambium usually consists of plates of dividing cells that, in effect, move deeper and deeper into the stem as they divide. The cells that the divisions of the cork cambium produce outwardly are mainly radial rows of densely packed **cork** cells. Their inner layers contain large amounts of a fatty substance, **suberin,** which makes the layers of cork nearly impermeable to water. Cork cells are dead at maturity. The **phelloderm,** a dense tissue composed primarily of parenchyma, may be differentiated inwardly from the cork cambium.

Together, the cork, cork cambium, and phelloderm make up the **periderm** (Figure 31-23). The periderm constitutes the outer protective covering of the mature plant's stem or root; it is formed only after a considerable amount of secondary growth has taken place. The periderm is renewed continuously by the cork cambium, which is why cork for commercial use can be harvested from the bark of the cork oak *(Quercus suber)* without killing the trees. Gas exchange through the periderm is necessary for the metabolic activities of the living cells of the phelloderm and vascular cambium beneath. This exchange takes place through **lenticels,** areas of loosely organized cells, which often are easily identifiable on the outer surface of bark (Figure 31-24).

> **The periderm, a layer differentiated from the lateral meristem known as the cork cambium, consists of cork to the outside and phelloderm to the inside of a stem or root. Penetrated by areas of looser tissue known as lenticels, the periderm retards water loss from the secondary plant body.**

Cork, which covers the surfaces of mature stems or roots, takes the place of the epidermis, which performs a similar function in the younger portions of the plant. **Bark** is a term used to refer to all of the tissues of a mature stem or root outside of the vascular cambium. Since the vascular cambium has the thinnest-walled cells that occur anywhere in a secondary plant body, it is the layer at which the bark breaks away from the accumulated secondary xylem. The inner layers of the bark are primarily secondary phloem, with the remains of the primary phloem crushed among them. Its outer layers consist of the periderm, and the very outermost ones are cork (Figure 31-25).

Wood

Wood is one of the most useful, economically important, and beautiful products that we obtain from plants. From an anatomical point of view, wood is accumulated secondary xylem. As the secondary xylem ages, its cells become infiltrated with gums and resins, and the wood becomes darker. For this reason, the wood located nearer the central regions of a given trunk, called **heartwood,** is often darker in color and denser than the wood nearer the vascular cambium. That wood, called **sapwood,** is still actively involved in transport within the plant. The proportion of sapwood to heartwood, and even whether they are distinct or not, differs widely from one kind of tree to another (Figure 31-26).

Because of the way in which it is accumulated, wood often displays rings. In temperate regions, these rings are **annual rings.** They reflect the fact that the vascular cambium divides actively when water is plentiful and temperatures are suitable for growth and ceases to divide when water is scarce and the weather is cold. In most temperate regions, growth rings form during the spring and summer. The structural basis for their appearance is the difference in density between the wood that forms early in the growing season and the wood that forms later. The abrupt discontinuity between the layers of larger cells, with proportionately thinner walls, that form early in the growing season and those that form later is often very evident. Growth rings generally are thicker when formed in years of plen-

FIGURE 31-25

A *Bark of valley oak,* Quercus lobata, *in California. Bark is highly characteristic of individual trees and shrubs, which often can be recognized by the bark's characteristics even when the trees are leafless.*
B *Galleries constructed by elm bark beetles, which spread Dutch elm disease from tree to tree, in the cambium of an elm tree. Such beetles bore through the thin layer of living cells that separates bark from the trunk of a tree, thus gaining access to the carbohydrates passing through the phloem.*

A   B

FIGURE 31-26

*Heartwood and sapwood in section of a white poplar (*Liriodendron tulipifera*) trunk.*

tiful rainfall and thinner when formed in dry years. For this reason, the annual rings in a tree trunk can be used not only to calculate the age of the tree, but also to learn about past climates.

Commercially, wood is divided into hardwoods and softwoods. **Hardwoods** are the woods of dicots, regardless of how hard or soft they actually may be; **softwoods** are the woods of conifers. Many hardwoods used commercially come from the tropics, whereas almost all softwoods come from the great forests of the north temperate zone. When they are used as building materials or for other commercial purposes, woods often are given names that are not related directly to the names of the trees from which they come. Thus, the wood called "Oregon pine" does not come from a pine at all, but from Douglas fir *(Pseudotsuga menziesii).*

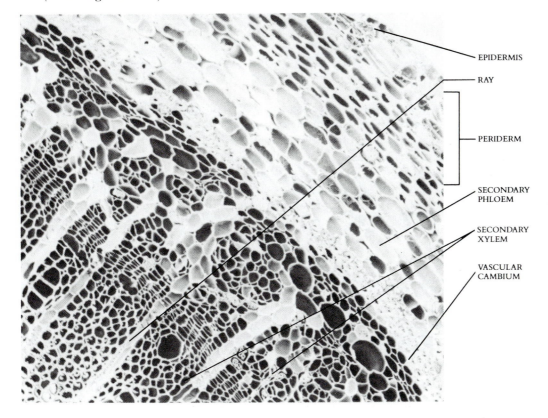

EPIDERMIS

RAY

PERIDERM

SECONDARY PHLOEM

SECONDARY XYLEM

VASCULAR CAMBIUM

FIGURE 31-27

Scanning electron micrograph of the outer portion of a trunk of sugar maple, Acer saccharinum, *showing the tracheids, vessels, and rays in the secondary xylem; the vascular cambium; phloem; and periderm.*

Individual woods differ widely in their microscopic characteristics, including such features as the width and height of the rays, type of pits in the conducting elements, abundance of fibers, and nature of the conducting elements and other cells (Figure 31-27). They can be identified readily by experts and often provide valuable clues to the evolutionary position of the plants that form them.

Wood is accumulated secondary xylem. Wood that is formed in regions with a seasonal climate is marked by annual rings. The woods of dicots are called hardwoods; those of conifers are called softwoods.

Modified Stems

Stems have been modified to perform many different functions during the course of plant evolution. One example that we mentioned earlier concerns plants such as grapes, Virginia creeper, and ivy, whose tendrils are modified stems (Figure 31-28). The tendrils of other plants, including those of garden peas, are modified leaves.

Stems running underground, which are called **rhizomes,** may be important in the vegetative propagation of plants; they often give rise to new individuals, sometimes quite far from the parent plant. Rhizomes are primarily responsible for the rapid spread of some of our most noxious weeds, such as nutgrass, and they also occur in many other plants, including irises and violets. As we saw in the last chapter, fern fronds arise from rhizomes also. Similar kinds of stems that run horizontally aboveground are called **stolons;** they are characteristic of plants such as strawberries (Figure 31-29) and Bermuda grass. The clumps of bamboo discussed in the boxed essay (pp. 634–635) also are connected by rhizomes.

Tubers are underground stems modified to store food; the most familiar tuber is that of the potato. The "eyes" of the potato are buds, each arising in the axis of a scale leaf and capable of giving rise to a new potato plant. **Corms** are thick, fleshy underground stems that are modified for food storage, as in cyclamens and gladiolus. Finally, **bulbs** are short underground stems that bear thickened, fleshy scale leaves. Onions have bulbs whose scales are modified leaves that enfold the stem (Figure 31-18, *C*); lilies have bulbs whose scales are small, very fleshy, and easily detached. When the scales of lily bulbs do become detached, they can give rise to new plants.

The Formation of Flowers

When a flower or group of flowers forms at a vegetative shoot apex, the meristematic activity of that apex ceases, usually permanently. In other words, the shoot changes from indeterminate to determinate growth. A number of internal and external factors that we will discuss in the following chapters are involved in the switch from vegetative growth to the production of a flower. As the change occurs, the shoot apex often becomes broad and domelike. The organs of a flower develop from primordia that resemble leaf primordia; they are bumps or shelves of tissue that project from the shoot apex. During the development of these organs, they may fuse with members of the same or another whorl, thus giving rise to the complex structures that occur in mature flowers.

THE ROOT

Roots have simpler patterns of organization and development than do stems (Figure 31-30). Although different patterns exist, we will describe a kind of root that is found in many dicots. There is no pith in the center of the vascular tissue of the root in most dicots. Instead, these roots have a central column of primary xylem with radiating arms. Between these arms are located strands of primary phloem (Figure 31-31). Around the column of vascular tissue, and forming its outer boundary, is a cylinder of cells one or more cell layers thick called the **pericycle.**

FIGURE 31-28

The tendrils of this wild grape, photographed in Kenya, are modified stems.

STOLON

FIGURE 31-29

Strawberry plants, Fragaria ananassa, *propagated here from a single individual by means of stolons, which are looping stems aboveground that root where they touch the soil.*

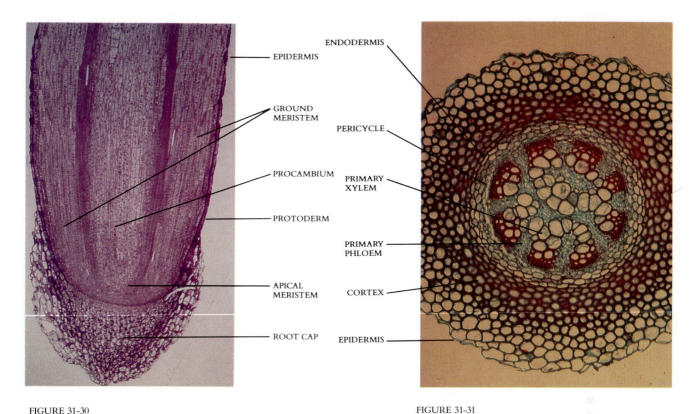

EPIDERMIS

GROUND
MERISTEM

PROCAMBIUM

PROTODERM

APICAL
MERISTEM

ROOT CAP

ENDODERMIS

PERICYCLE

PRIMARY
XYLEM

PRIMARY
PHLOEM

CORTEX

EPIDERMIS

FIGURE 31-30

*Median transection of a root tip in
corn,* Zea mays, *showing the root
cap, apical meristem, protoderm,
epidermis, and ground tissue.*

FIGURE 31-31

*Cross section through a root of a
club moss,* Lycopodium, *which
has an anatomical structure like that
of a dicot. The central xylem and
phloem, cortex, and epidermis are
visible.*

The vascular tissue of a root, surrounded by the pericycle, is located in the center of a mass of parenchyma called the cortex. The innermost layer of the cortex is a cylinder of specialized cells one cell layer thick called the **endodermis.** The cells of the endodermis have a thickened waxy band, the **Casparian strip,** extending around them through their lateral walls. The Casparian strip regulates the flow of water between the vascular tissues and the outer portion of the root. The outer layer of the root, as in the shoot, is the epidermis.

> **Most dicot roots have a central column of primary xylem with radiating
> arms, and strands of primary phloem between these arms. Around the
> column of vascular tissue is a layer one or more cells thick called the
> pericycle. The innermost layer of the cortex, or endodermis, consists of
> cells surrounded by a thickened waxy band called the Casparian strip.**

The apical meristem of the root divides and produces cells both inwardly—back toward the body of the plant—and outwardly. Outward cell division results in the formation of a thimblelike mass of relatively unorganized cells, the **root cap,** which covers and protects the root's apical meristem as it grows through the soil.

The root elongates relatively rapidly just back of its tip. Above that zone are formed abundant root hairs, which, as noted earlier, are slender projections of the epidermis (Figure 31-3). Virtually all of the absorption of water and minerals from the soil takes place through the root hairs. These root hairs greatly increase the surface area and therefore the absorptive powers of the root. In plants that have ectomycorrhizae (see Chapter 28), the root hairs often are greatly reduced in number; the fungal filaments take their place in promoting absorption.

Branching in Roots

Roots have a simpler structure than stems do, primarily because roots have no appendages comparable to leaves. Branching in stems occurs by means of buds, which form in leaf axils; the formation of the lateral branches of stems is superficial and does not involve the deeper layers of the stem. Branching in roots, on the other hand, is initiated well back of the root apex and deep within the tissues of the root. Branch roots in both angiosperms and gymnosperms are initiated by cell divisions in the pericycle.

Deep within the cortex, the lateral root primordia grow out toward the surface of the root (Figure 31-32). As they do so, they develop the characteristics of the main root apex, including a root cap. Eventually, they break through the surface of the root and become established as fully formed lateral roots.

Branch root primordia form well beneath the surface of the root and grow out through the cortex. There are no externally formed organs in roots comparable to leaves or lateral buds.

Secondary Growth

Secondary growth in roots is quite similar to that in shoots. The vascular cambium is initiated from undifferentiated procambial cells between the primary xylem and the primary phloem. The areas where this cell division is initiated are connected by corresponding areas of cell division in the pericycle. Ultimately, these areas of cell division fuse, and the vascular cambium then forms a cylinder that completely surrounds the primary xylem. The differentiation of cork cambium and periderm occurs in the pericycle.

Adventitious Roots and Shoots

Many dicots have a single **taproot,** which is sometimes the only major underground structure (Figure 31-33, *A*). In some, like carrots and radishes, the taproot is fleshy and modified for food storage. The taproot that develops in monocots, on the other hand, often dies during the early growth of the plant, and new roots differentiate from the tissues of the stem's lower part (Figure 31-33, *B*). These **adventitious roots**—which arise from kinds of tissue other than root tissue—take over the function of the taproots and branching root systems that are characteristic of dicots.

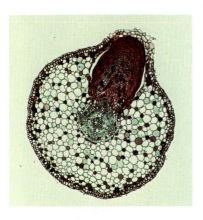

FIGURE 31-32

A lateral root growing out through the cortex of black willow, Salix nigra. *The origin of lateral roots occurs beneath the surface of the main root, whereas the origin of lateral stems is superficial.*

FIGURE 31-33

A *Taproots in dandelion,* Taraxacum officinale. *Even a small portion of such a taproot can regenerate a new plant.*
B *Prop roots in corn,* Zea mays. *Such roots, which are adventitious—they arise from stem tissue—take over the function of the main root, which is soon lost, in many monocots.*

A

B

FIGURE 31-34

Fireweed, Epilobium angustifolium, *in a burned area in Wyoming. These flowering stems have arisen mainly from adventitious shoots that have been produced on the widely spreading root system of this aggressive species.*

Similarly, **adventitious shoots** may arise from roots, often at some distance from the parent plant. Such shoots may then give rise to new plants that eventually may become fully independent of the parent plant. This method of propagation is characteristic of fireweed *(Epilobium angustifolium),* for example, and is primarily responsible for the abundance of this plant in burned or cleared areas (Figure 31-34). The beautiful and conspicuous clumps of quaking aspen *(Populus tremuloides)* shown in Figure 31-5 also are clones (and therefore genetically identical) that originated from the roots of a single individual. Most plants that spread underground, however, do so by means of rhizomes, which, as we have seen, are subterranean stems directly connected with other stem tissue.

PLANT GROWTH AND DEVELOPMENT

As we have shown in this chapter, plants grow by an open system—one that keeps functioning throughout their lives and producing similar structures over and over. The apical meristems of the shoot differentiate leaves and the primary tissues of the shoot, whereas the apical meristems of the root, protected by a root cap, simply produce the primary tissues of the root. No structure in roots is comparable to leaves; if such structures did exist, they would tend to retard the growth of the root through the soil. A major difference in the patterns of growth of roots and shoots is that each branch shoot develops from buds, whereas branch roots develop as a result of the functioning of areas of rapid cell division that lie deep within the root cortex. There is a bud at the place where each leaf joins the stem, but most of these buds simply remain as buds rather than growing into lateral branches.

The control of this seemingly simple and repetitive growth pattern by hormones, which the plant produces internally, and by external factors is complex and only partly understood. The three remaining chapters of this section will be devoted to a review of our knowledge concerning the way in which external and internal factors interact to produce a mature, functioning plant.

SUMMARY

1. An angiosperm embryo consists of an axis with either one or two cotyledons, or seedling leaves. In the embryo, the epicotyl will become the shoot, and the hypocotyl will become the root. Food for the developing seedling may be stored either in the endosperm at maturity or in the embryo itself.

2. A plant body is basically an axis that includes two parts, root and shoot. Within the plant are three principal tissue types: vascular tissue, ground tissue, and dermal tissue. Dermal tissue covers the outside plant; vascular tissue conducts substances through it; and ground tissue is the matrix in which the vascular tissue is embedded.

3. Plants grow by means of their apical meristems, zones of active cell division at the ends of the roots and the shoots. The apical meristem gives rise to three types of primary meristems, partly differentiated tissues in which active cell division continues to take place. These are the protoderm, which gives rise to the epidermis; the procambium, which gives rise to the vascular tissues; and the ground meristem, which becomes the ground tissue.

4. Plants that complete their entire life cycle within a year are called annuals; those that require 2 years to reach maturity and then flower just once are called biennials; and those that flower year after year once they have reached maturity are known as perennials. Plants that flower only once are called monocarpic; perennials are called polycarpic. Perennial plants may be soft and herbaceous or else woody, such as trees and shrubs.

5. Parenchyma cells are the most abundant cell type in the primary plant body, the portion of the plant that differentiates from the primary meristems. They usually have only primary walls, which are laid down while the cells are still growing. Parenchyma cells are living at maturity.

6. Collenchyma cells often form strands or continuous cylinders in the primary plant body, for which they provide the chief source of strength. They may be recognized by the uneven thickenings of their primary cell walls.

7. Sclerenchyma cells have tough, thick walls with secondary thickening—thickening that is laid down after the cells have reached full size. Fibers, which are elongate strengthening cells, are sclerenchyma cells. Tracheids and vessel elements, the principal conducting elements of the xylem, are similar to sclerenchyma cells in having thick cell walls. They lack protoplasts at maturity.

8. Carbohydrates are conducted through the plant primarily in the phloem, whose elongated conducting cells are living but lack a nucleus. These are called sieve cells and sieve-tube members. Associated with sieve-tube members, which have larger pores in their end wall than sieve cells do, are specialized parenchyma cells called companion cells.

9. The growth of leaves is determinate, like that of flowers; the growth of stems and roots is indeterminate.

10. Water reaches the leaves of a plant after entering it through the roots and passing upward via the xylem. Water vapor passes out of leaves by entering the intercellular spaces, evaporating, and then moving out through the stomata.

11. Stems branch by means of buds that form externally at the point where the leaves join the stem; roots branch by forming centers of cell division within their cortex. Young roots grow out through the cortex, eventually breaking through the surface of the root.

12. Secondary growth in both stems and roots takes place after the formation of lateral meristems known as vascular cambia. These cylinders of dividing cells form xylem internally and phloem externally. As a result of their activity, the girth of a plant increases.

13. The cork cambium forms in both roots and stems during the initial stages of establishment of the vascular cambium. It produces cork externally and phelloderm, a dense tissue composed mainly of parenchyma, internally. The cork, cork cambium, and phelloderm collectively are called the periderm. The periderm, the outermost layer of bark, is perforated, in the stem, by areas of loosely organized tissue called lenticels.

SELF-QUIZ

The following questions pinpoint important information in this chapter. You should be sure to know the answers to each of them.

1. Primary growth takes place from _____ _____.

2. How many types of cambium are there in plants?

3. Primary phloem is derived from
   (a) procambium
   (b) cork cambium
   (c) vascular cambium
   (d) primary meristems
   (e) protoderm

4. Which of the following cell types is dead at maturity?
   (a) parenchyma
   (b) collenchyma
   (c) sclerenchyma
   (d) cork cambium
   (e) phloem

5. You and your friend are digging in your garden and find something that is either a rhizome or a root. You section it and view the sections on a microscope. Which of the following structures, if visible, would convince you that it was a rhizome?
   (a) xylem
   (b) phloem
   (c) Casparian strip
   (d) endodermis
   (e) none of the above

6. Periderm consists of _____, which is mainly parenchyma, _____ _____, a type of lateral cambium, and _____, the outermost layer.

7. Growth at the vascular cambium increases stem _____, not stem _____.

8. In most monocots, the veins in the leaves are _____.

9. Branches of roots arise in the _____.

10. Which of the following is an adaptation to reduce water loss?
    (a) cork
    (b) cuticle
    (c) guard cells
    (d) Casparian strip
    (e) all of the above

11. Rhizomes, tubers, and bulbs are all stems. True or false?

12. If a plant has no secondary growth it will remain herbaceous. True or false?

THOUGHT QUESTIONS

1. If you hammer a nail into the trunk of a tree 2 meters above the ground when the tree is 6 meters tall, how far above the ground will the nail be when the tree is 12 meters tall?

2. What is the difference between primary and secondary growth?

3. Can you suggest a reason why branches are not formed the same way in roots that they are in stems?

FOR FURTHER READING

CUTLER, D.F.; *Applied Plant Anatomy,* Longman, Inc., New York, 1978. A concise treatment of the fundamentals of plant anatomy.

ESAU, K., *Anatomy of Seed Plants,* 2nd ed., John Wiley & Sons, Inc., New York, 1977. Short but outstanding textbook on plant anatomy.

GALSTON, A.W., P.J. DAVIS, and R.L. SATTER: *The Life of the Green Plant,* 3rd ed., Prentice-Hall, Inc., Englewood Cliffs, N.J., 1980. Very well-written, physiologically oriented treatment of the structure and functioning of plants.

JANZEN, D.H.: "Why Bamboos Wait So Long to Flower," *Annual Reviews of Ecology and Systematics,* vol. 7, pages 347-391, 1976. A fascinating essay about the natural history of bamboos.

NORTHINGTON, D.K., and J.R. GOODIN: *The Botanical World,* Times Mirror/Mosby College Publishing, St. Louis, 1984. A comprehensive review of all aspects of botany, with a humanistic orientation.

RAVEN, P.H., R.F. EVERT, and H. CURTIS: *Biology of Plants,* 4th ed., Worth Publishers, Inc., New York, 1986. A comprehensive treatment of general botany, emphasizing structural botany.

RAY, P.M., T.A. STEEVES, and S.A. FULTZ: *Botany,* Saunders College, Philadelphia, 1983. Well-integrated discussion of the structure and function of plants.

SCHALLER, G.B., H. JINCHU, P. WENSHI, and Z. JING: *The Giant Pandas of Wolong,* University of Chicago Press, Chicago, Ill., 1985. A fascinating account of the mutual adaptations of pandas and bamboo, this book offers the first glimpse of the life of pandas in their remote mountain home in western China.

ZIMMERMANN, M.H.: *Xylem Structure and the Ascent of Sap,* Springer-Verlag New York, Inc., New York, 1983. An outstanding monograph on the way xylem is put together and functions.

PLANT DEVELOPMENT

Chapter 32

Overview

Plant development and animal development differ in many important respects. Unlike animals, plants are always undergoing development. Patterns of plant growth are not under as strong an internal control as are those of animals. The cells of plants, unlike those of animals, do not move in relation to one another during their development. Furthermore, all nucleated, living plant cells that have not developed a secondary wall are capable of developing into other cells; when individual cells taken from mature plants are cultured under appropriate conditions, they are capable of growing into whole plants. Plant embryos are shed from their parents enclosed in seeds, enabling young plants to delay further development until appropriate conditions occur and to increase greatly the likelihood that they will be dispersed safely and efficiently.

For Review

Here are some important terms and concepts that you will encounter in this chapter. If you are not familiar with them, you should review them before proceeding.

- **Eukaryotic cell structure** (Chapter 6)

- **Mechanism of gene action** (Chapter 16)

- **Reproduction in plants** (Chapter 30)

- **Plant structure** (Chapter 31)

In this chapter, we begin to consider the dynamic aspects of plant growth. We will deal first with the development patterns of individual plant cells and their integration into tissues; then we will consider the integration of these tissues into individuals. More detail about hormones and ways in which plants regulate their growth will be presented in the next chapter. We will merely introduce them here in relation to their role in plant development. The movement in plants of water and solutes, including both minerals and carbohydrates, will be discussed in Chapter 34. Taken together, these three chapters present an overall picture of the way in which plants function as organized biological units, an area of study that is of great interest to many scientists throughout the world today.

CONTINUOUS DEVELOPMENT:
A CHARACTERISTIC OF PLANTS

A tree, like a human being, consists of an intricate array of tissues. Each of these tissues differs from the others, and each stands in precise physiological and morphological relation to the others. Similarly, the control of the developmental processes that affect the organization of complex tissues is often exceedingly specific. In animals, as you will see in Chapter 41, these developmental controls are largely internalized. They are exercised via cytoplasmic commands expressed as early as the initial formation of the female gamete, the egg. The responses to such controls in animals involve both the movement of cells and the irreversible commitment of cells to particular developmental paths.

Animal development proceeds as if it were unfolding according to a faithfully followed blueprint. At each stage, particular and highly specific cell movements and changes in gene expression occur—just as when the notes in a musical score are read, both the order and the timing are critical to the overall result. Such a program permits a high degree of tissue-specific specialization and physiological interaction. The final, complex arrangement of the cells and tissues in an adult animal normally is not influenced much by outside environmental changes during development, and no environmental cues are used during the process. Instead, all of the instructions that pertain to it are internal. Indeed, the very essence of animal development is environmental **homeostasis,** or stability, which provides freedom from the influences of unprogrammed disturbances that might upset the delicate balance required to produce complex organized tissues. We can upset that balance with unusual stimuli—for example, in human beings, cigarette smoking or consuming alcohol during pregnancy is apt to be harmful to the development of the embryo—but for the most part, animal development is tightly controlled.

Plants, in contrast, are always undergoing development. They develop, like all other organisms, according to a genetic blueprint—but the way in which their particular blueprint is expressed is greatly influenced by external factors. The differentiation of specific tissues in plants is carried out under the direction of **hormones**—chemical substances produced usually in minute quantities in one part of an organism and transported to other parts on which they have specific effects. The particular expression of these hormones is mediated not only by genes, but also by external environmental factors (Figure 32-1). Much of our discussion of plant growth and development in this chapter and in the next will be concerned with the patterns of production and functioning of these hormones. We also will address the ways in which plant growth responds to external environmental signals.

> **Both animals and plants develop according to a predetermined blueprint, but in animals development reaches a conclusion—the adult individual— whereas in plants, some cells undergo development continuously. In plants, the course of individual development is greatly influenced by signals received from the external environment, mediated by hormones. In animals, external influences exert much weaker influences on the system.**

Plants cannot move and therefore must continue to exist in the particular microhabitat where they develop. Much of what a plant does (photosynthesis, flowering, seed dispersal, growth) is sensitive to the exact nature of local conditions. In view of this close and important relationship, it is highly adaptive for a plant to respond to, and to be modified by, the conditions in which it will continue to grow and develop. Animals, of course, respond to their environments too—think of the thick coats that many mammals grow in the winter—but the range of such responses is only a tiny fraction of what it is in plants. Animals can meet the demands of their environment, or move to a new environment, by means of their complex patterns of behavior; plants do not have such opportunities.

Because they must adapt to highly individualized microhabitats, plants have more flexibility in their developmental patterns than do animals. If animal development is a com-

FIGURE 32-1

The development of an individual plant is strongly affected by its environment. This Jeffrey pine (Pinus jeffreyi) is growing on the granitic slopes of Sentinel Dome in Yosemite National Park, California.

plex symphony, always conducted so as to produce a specific musical effect (Beethoven's Fifth Symphony always sounds like Beethoven's Fifth Symphony), plant development is a jam session with no fixed scores. The "music" of plant development, unlike that of animal development, reflects the mood and ambience of the environmental audience. There are still rules for the jam session—the genetic factors within the plant—but, to a much greater extent than in animals, these rules provide only general guidelines for development.

DIFFERENTIATION IN PLANTS: EXPERIMENTAL EVIDENCE

Perhaps related to the lack of a fixed developmental blueprint in plants is the fact that almost all of their differentiation is fully reversible. In living plant cells—those that retain their protoplasts at maturity—gene expression can be reactivated. When it occurs, such reactivation may lead to alternative modes of differentiation, or even to a complete plant.

The process of differentiation in plants has been studied in detail throughout the twentieth century. Most of the impetus for this study has come from a single hypothesis: in 1902 the German botanist Gottlieb Haberlandt (Figure 32-2) proposed that all living plant cells are **totipotent,** suggesting that each of them possesses the full genetic potential of the organism. Haberlandt explained the differentiation of tissues by theorizing that only part of this potential is normally realized in any given differentiated tissue. He went on to suggest that one ought to be able to develop a whole mature plant from isolated plant cells if one could devise the correct medium to support growth. He tried repeatedly and unsuccessfully to do this himself. Indeed, for more than 50 years no one was able to do it.

> **Haberlandt suggested that all living plant cells are totipotent, each containing the ability to express the full genetic potential of that organism. His hypothesis was not confirmed for more than 50 years, however.**

Cell Culture

Before scientists could determine whether plant cells were indeed totipotent, they needed to learn how to grow them in culture. This proved to be a difficult technical problem. It was not difficult to isolate a single plant cell, but for differentiation to occur, that cell would have to divide repeatedly. The conditions promoting such rapid cell division had not been identified. It has since turned out that the substances needed for cell division diffuse out of individual cells when they are isolated in an agitated medium, and the cells then will not

FIGURE 32-2

Gottlieb Haberlandt, famous student of plant development, who hypothesized in 1902 that all living plant cells were totipotent.

divide further. This problem was finally solved in several different ways in the 1950s. One solution was to place an isolated single cell within a tiny drop of culture medium onto a piece of filter paper floating upon an established cell culture. In this way the single cell was isolated from other cells by the filter paper, but at the same time it could be influenced by them. Under these conditions, the isolated cell could obtain from the medium that it shared with the established cell culture any substances necessary to promote cell growth and division, and the isolated cell grew and divided rapidly, establishing the kind of mass of dividing, undifferentiated cells that is called **callus.** This callus could then be grown indefinitely in culture.

Using such methods, it has been possible to determine the nutritional requirements of some plants and to grow them in defined media. For other plants, this has not been possible, and it is necessary to add coconut milk, the liquid endosperm that forms within coconuts, to the medium. This substance, which we will discuss in more detail in the next section, clearly contains additional factors necessary for the proper functioning of these plant cells.

> **Cell cultures of some plants will divide and produce whole individuals if they are kept in culture in such a way that they can be influenced by other cells. Some cell cultures will develop in this way only if coconut milk, the liquid endosperm of coconuts, is added to the medium.**

Tissue Culture

A second solution to the problem of culturing plant cells was devised by Cornell University plant physiologist F.C. Steward and his coworkers in 1958. His methods also depended on supplying differentiated cells with substances obtained from dividing cells. Steward isolated small bits of secondary phloem tissue from carrot root and placed them in liquid growth medium in a flask (Figure 32-3). Such growth media contained sucrose and the minerals that were essential for plant growth, as well as certain vitamins—organic molecules that the particular plant parts cannot manufacture for themselves.

In such a medium, many of the new cell clumps, each of which had originated from a single phloem cell, differentiated roots (Figure 32-4, *A*). Placed on agar, the clumps also developed shoots (Figure 32-4, *B*) and eventually grew into whole plants, thus confirming Haberlandt's hypothesis (Figure 32-4, *C*). Steward's results clearly demonstrated that the

FIGURE 32-3

The technique of isolating phloem tissue from carrots (Daucus carota), *as carried out in the laboratory of F.C. Steward at Cornell University. The disks of tissue were grown in a flask in which the medium was constantly agitated so as to bring a fresh supply of nutrients to the masses of callus that soon formed.*

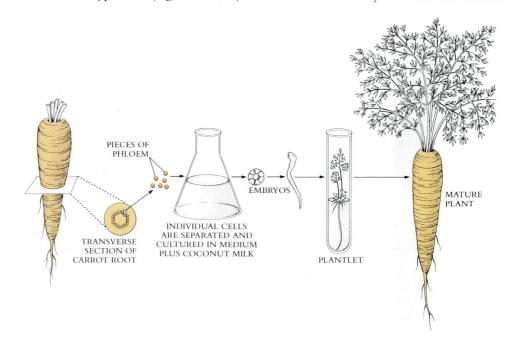

PIECES OF PHLOEM

TRANSVERSE SECTION OF CARROT ROOT

INDIVIDUAL CELLS ARE SEPARATED AND CULTURED IN MEDIUM PLUS COCONUT MILK

EMBRYOS

PLANTLET

MATURE PLANT

A   **B**   **C**

original differentiated phloem tissue still contained all of the genetic potential needed for the differentiation of whole plants. Indeed, the embryonic growth of isolated cells resembled closely that of normal zygotes. In their initial stages of development, the dividing cells differentiated into masses that resembled embryos closely. These "embryoids" had shoot and root apices from which further development proceeded, as well as distinct tissue systems that ordinary embryos have at comparable stages of development.

> **Steward experimentally confirmed Haberlandt's hypothesis: under appropriate circumstances, and provided that they still have a living protoplast and nucleus, individual plant cells can resume growth and differentiation.**

Many plants will differentiate "embryoids" and ultimately whole plants only if they form masses of callus first. In others, such as recently demonstrated in orchard grass, *Dactylis glomerata* (Figure 32-5), it has been possible to induce the formation of embryos directly from plant tissue, an impressive demonstration of the totipotency of plant cells.

Our general understanding of the factors that lead to the production of roots and shoots in masses of callus tissue is still limited, and the results obtained with a given treatment often vary widely from species to species. Despite these difficulties, however, more plant species are added each year to the list for which regeneration is possible. Researchers are successful mainly by "tinkering" with tissue cultures derived from individual kinds of

FIGURE 32-4

Tomato (Lycopersicon esculentum) *plants regenerated from single parenchyma cells.*
A *Roots differentiating from masses of callus tissue in culture.*
B *A mass of callus tissue anchored on agar, a firm medium, with a shoot emerging.*
C *Regenerated tomato plant, which will mature, producing flowers and fruits, if it is transferred to soil. F.C. Steward obtained similar results in his classical experiments with carrots, which confirmed Haberlandt's hypothesis.*

FIGURE 32-5

Direct embryogenesis from leaf mesophyll cells in orchard grass, Dactylis glomerata. *Grass leaves grow largely from a meristem at their base, which is why, when you mow your lawn, it remains green and healthy. In these experiments, carried out by scientists from the University of Tennessee, sections of such meristems were brought into an agar culture containing the necessary nutrients, to which the chemical 3,6-dichloro-o-anisic acid (dicambra) had been added. Meristematic regions farther from the base of the leaf, and therefore older, produced masses of callus, but the younger ones produced typical embryos directly, as shown in the photographs.*
A *Scanning electron micrograph of a well-developed embryo arising directly from a grass segment.*
B *A very young embryo protruding from the leaf surface. Each of these embryos is connected to the leaf surface by a suspensor.*

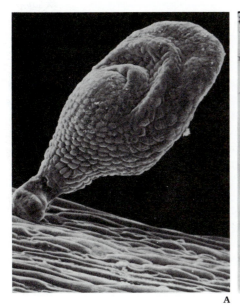
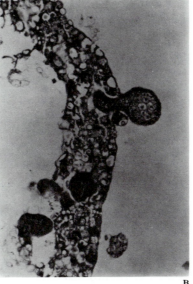

A   **B**

CLONING IN COCONUTS

Palms (family Arecaceae, or Palmae) are probably second only to grasses (family Poaceae, or Gramineae) as the most valuable plant family to human beings. In tropical and subtropical areas, the roughly 2600 species of this family provide building materials, fuel, food, and drink to millions of people. In addition, palms, with their beautiful divided leaves, provide one of the most characteristic aspects of the tropical landscape. We often grow them in our homes and offices for their beauty, and they are highly resistant to both overheating and air conditioning!

One of the most commercially important palms is the coconut palm. Coconut palms traditionally have been grown from seeds, and, of course, such progenies are highly variable with respect to traits such as resistance to disease (see Chapter 26 for a discussion of a major new disease of coconuts and other palms) and yield. In date palms, another leading member of the family, a grower can take shoots from the base of a tree and propagate them to produce individuals that have the same characteristics as their parent—for example, high yields or especially tasty fruits. Coconut palms, however, cannot be propagated in this way because the trees have only one apical meristem each.

At Wye College, University of London, research scientists successfully produced coconut plants in 1983 from cultures of coconut parenchyma tissue taken from flowers. They found that high concentrations of the synthetic hormone auxin 2,4-D, normally a herbicide, would promote the formation of callus tissue, but that activated charcoal also must be added to the medium to prevent the chemical from killing the cells. The high levels of auxin that encouraged the formation of callus, however, discouraged the formation of the embryoids from it. Normally, a young coconut plant growing inside a coconut develops not only a root and shoot, but also a large cotyledon, or

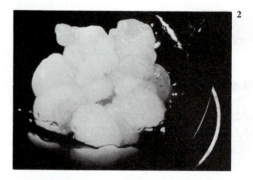

plants; experiments now being carried out by commercial firms involved in the application of biotechnology to agriculture are making a great contribution to the available stock of information. Cotton, tomatoes, and black cherries are among the plants that recently have been regenerated in this way, assuring a supply of genetically identical individuals growing under specified experimental conditions. For woody plants, the techniques are particularly important, since they greatly shorten the life cycle of the plant and thus allow the selection for improved characteristics to proceed much more rapidly than would otherwise be possible. This area of biotechnology is a very promising one that will continue to afford satisfying careers to many scientists in the future; Figure 18-1 illustrates its application to the breeding of an important forest tree of western North America, the Douglas fir (*Pseudotsuga menziesii*).

An important conclusion that follows from these experimental results is that the differentiated plant tissue is indeed capable of *expressing* its hidden genetic complement when suitable environmental signals are provided. What is the mechanism that halts the expres-

haustorium, which digests the material in the coconut and transfers it to the embryo. If the auxin was withdrawn too early, in an attempt to promote the formation of embryoids, the haustorium in the experimental plants continued to enlarge, but the shoot failed to develop.

Ultimately this problem was solved by lowering the concentration of auxin in the medium at exactly the right rate so that shoots and roots developed first. By this method, small coconut plants can be produced in about 9 months. It is now possible, in principle, to provide growers with large quantities of uniform, high-quality coconut plants. There are additional technical difficulties to overcome, but it soon should be possible to use these techniques to increase yields of coconuts at least fivefold—an advance of great importance to hundreds of millions of people throughout the tropics. It also should be possible to use genetic engineering methods to improve coconuts still more in culture and to be able to solve the problems of their diseases.

FIGURE 32-A

1 *Small pieces of flower tissue on the nutrient medium. The hormones in this medium will stimulate cell division, and a lump of callus eventually will be produced.*
2 *Callus produced from the small pieces of flower tissue. The nodules of callus will develop into embryoids and then into plantlets.*
3 *Germination of an embryoid, producing a shoot.*
4 *A coconut plantlet produced by the use of tissue culture.*

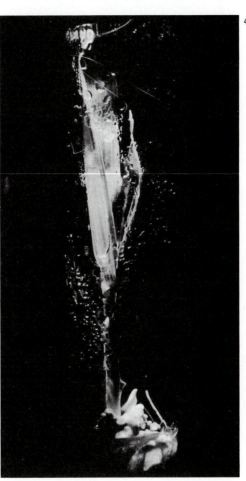

sion of genetic potential when the same kinds of cells are incorporated into normal, growing plants? As we will discuss in this chapter and the next, the expression of these genes is largely controlled by plant hormones.

Embryo Culture

Methods analogous to those discussed above have been used widely in recent years to learn about the development of plant embryos. Early efforts in this field involved dissecting out whole ovules and placing them in a suitable nutrient solution. If this is done carefully, the zygote within usually will develop into a mature embryo. When an ovule is taken from an ovary and placed in a nutrient solution by itself, it exhibits complex nutrient requirements. Once the embryo is mature, it is autotrophic; from that point on, it will grow and develop normally if it is simply kept at a favorable temperature and provided with water and oxygen. Earlier, however, when it is still heterotrophic, growth and development will not occur, even if the embryo is supplied with sucrose and the necessary mineral nutrients. In addition, young embryos will grow to maturity only if they are supplied with coconut milk, which provides a number of substances that the embryo needs to develop properly. The older an isolated embryo is, the better its chances for normal growth and survival will be. Embryo culture techniques are important in the development of new kinds of crops, and they allow the development of a much wider range of genotypes than would be possible if they remained attached to the plants on which they were formed.

Pollen and Anther Culture

In recent years, whole plants have been regenerated from pollen grains, under certain circumstances. These plants, of course, have the same number of chromosomes as the pollen grains that give rise to them. If the sporophyte plant that produced them is diploid, then its pollen grains and the plants regenerated from them will be haploid (Figure 32-6). Such haploid sporophytes may be of great importance in experimental plant breeding, because of the direct expression of the genes that are present in that particular haploid set of chromosomes.

> **Haploid individuals can be obtained by culturing the pollen of some kinds of plants. Such haploid plants have a great potential importance for plant breeding, since all of their genes are expressed directly.**

These regeneration techniques have been applied successfully to only a few kinds of plants since they were first developed in the mid-1960s. The stage of pollen development in the particular anther is critical in determining whether the pollen grain can be induced to develop into a new plant, but the importance of this factor seems to vary widely among species and even perhaps within a single species. The actual culture methods used, however, are similar to those used for embryo growth and for the growth of isolated plant cells or masses of tissue. The further refinement of these methods appears to hold great promise for plant breeding in the future.

FACTORS IN THE GROWTH MEDIUM
Experimental Studies

One way to learn more about the details of plant growth is to investigate the growth medium in which development takes place. Experimental results obtained throughout the twentieth century, some of which we have just reviewed, have made it abundantly clear that there are special chemical factors present in various growth media that play an important role in controlling plant development. The success of coconut milk in the regeneration of certain kinds of tissue caused it to come under investigation early. Eventually it was discovered that coconut milk is rich in reduced nitrogen compounds, such as amino acids, which are necessary for growth. In addition, coconut milk was found to contain growth hormones known as **cytokinins,** which are necessary for the differentiation of plant tissues.

Additional studies have demonstrated, however, that other types of differentiated plant cells exhibit the ability to redifferentiate even when they are not grown in a medium containing coconut milk. No one set of conditions is broadly effective, and to some extent each plant cell type is a special case. For example, Steward failed to regenerate whole potato plants, even when he used the apparently undifferentiated cells of potato tubers and exposed them to the same treatment as the carrot cells with which he had been successful earlier.

These results make it clear that plant development is not entirely molded by the environment in which the cells exist: genetically specified controls also play an important role. Control of the highly ordered progression of cell differentiation that occurs to produce a

mature plant tissue is vested in the genes of the plant cells. The environment determines the ultimate outcome by triggering the expression of one particular developmental pattern from among several genetically determined possibilities.

> **Plant development occurs as a result of the interactions between genetic factors and the external environment. The environment determines the ultimate expression of the genes in producing the phenotype.**

Regeneration in Nature

If the cues that trigger plant development come from the external environment, then they ought to be familiar stimuli that we see around us. Are everyday plants regenerating whole individuals from differentiated tissue out in the real world? Yes, indeed. This is particularly evident after injury or removal of tissue from plants. The common practice of using cuttings to produce mature plants reflects this property. At the base of a severed stem, adventitious roots may arise from mature pericycle tissue, which has divided to form a root meristem.

For reasons that are not well understood, adventitious roots seem to form easily in plants, adventitious shoots much less so. Cuttings from plants such as beans or garden geraniums (*Pelargonium*) will form roots if they are simply left with their lower ends in water or in sand that is watered (Figure 32-7). In some plants, however, root formation occurs only with much difficulty, if at all. Stems that have leaves on them will form roots much more readily than those from which the leaves have been removed, apparently because the buds produce auxin, a plant hormone that stimulates root production in such situations.

In some kinds of plants, for example succulent ones like jade plants (see Figure 31-18, *B*), whole plants may be regenerated from bits of differentiated leaf tissue. The tiny plantlets that differentiate along the edge of the leaves of the familiar houseplant *Kalanchoë*—aptly called "mother-of-thousands"—provide an example of the way in which such vegetative propagation happens frequently in nature. As you saw in the last chapter, many plants may likewise differentiate whole new individuals from rhizomes, stolons, or horizontal roots (see Figure 31-34). For example, colonies of aspen, *Populus tremuloides*, often consist of a single individual that has given rise to a whole genetically identical colony by producing adventitious buds on its roots (see Figure 31-5). Sometimes plants propagate by modification of their normal reproductive structures. In some century plants (species of *Agave*, for example), plantlets may form among the flowers. Seeds can also form that con-

FIGURE 32-7

A leaf of African violet, Saintpaulia, *forming adventitious roots and new plantlets at the base of the petiole.*

tain unfertilized embryos, a process that is genetically equivalent to the other kinds of asexual reproduction discussed here.

Plant stems regularly produce adventitious roots under the appropriate circumstances; roots produce adventitious stems less frequently.

PLANT EMBRYONIC DEVELOPMENT

We will now consider the factors that control the normal development of a plant embryo, comparing its development with that of the sort of callus tissue differentiation that we have been discussing. The first stage in the development of a plant zygote is active cell division. The zygote divides repeatedly to form an organized mass of cells, the embryo. Within such an embryo, one can observe the initial differentiation of the organs and tissues that make up the mature plant. Meristems are established early at the shoot and root apices.

In angiosperms, the differentiation of cell types in the embryo begins almost immediately after fertilization (Figure 32-8). The zygote first divides twice in one plane. After this division, the **basal cell,** which is the cell nearest the **micropyle**—the opening in the ovule through which the pollen tube entered—undergoes a series of further lateral divisions. As a result, it forms a narrow column of cells, the **suspensor.**

The other three cells into which the zygote first divided develop differently. After several days, they come to form a mass of about 40 ordered cells arranged in layers, like stacked bricks. By about day 5 of cell division, the principal tissue systems of the developing plant can be detected within this mass, and differentiation has begun. Soon thereafter, perhaps 6 days after the initiation of cell division, the root and shoot apical meristems can be detected as a zone of small, densely staining cells at the pole of the embryo opposite the suspensor. The shoot meristem grows upward and differentiates leaves and lateral branches; the root meristem grows downward, differentiating the various structures of the root. Both meristems continue to function throughout the life of the plant, so the shoot and root systems have a potentially unlimited pattern of growth. In addition, the lateral meristems—the vascular cambium and the cork cambium—make possible the plant's potentially unlimited increase in girth by secondary growth.

A developing angiosperm embryo consists of a suspensor and a mass of layered cells. When about 40 of these cells are present, the axes of the shoot and root begin to become evident, and differentiation begins.

In gymnosperms (Figure 32-9) the zygote nucleus divides repeatedly after fertilization, but in many gymnosperms, no cell walls form initially between the daughter nuclei. After eight rounds of cell division, a single large embryonic cell contains about 256 nuclei. Cell walls then form, producing a mass of 256 cells of equal size. Differentiation then commences. The cells farthest from the micropyle divide faster than those nearer to it. This soon results in a marked gradient in cell size, since the rapidly dividing cells are smaller than the others. The larger cells near the micropyle develop into a suspensor. At the same time, the smaller cells on the opposite pole of the embryo, which constitute about a third of the total cell mass, give rise first to the apical meristem of the root and then to that of the shoot.

Thus, although the details are different, the same basic pattern can be seen in the embryogenesis of angiosperms and gymnosperms. A particularly important feature, and one that clearly separates plants from animals, is that cell movement does not occur in plants during the course of embryonic development: plant cells differentiate where they are formed. Their position in relation to other cells is important in determining how they differentiate; their specific course of development probably is determined in part by chemical gradients, although this has not been demonstrated fully.

The pattern of development of an animal zygote, as we will see in Chapter 41, is strongly affected by specific chemical signals that are present in the egg. In plants, however,

FIGURE 32-8

Stages of development in an angiosperm embryo.

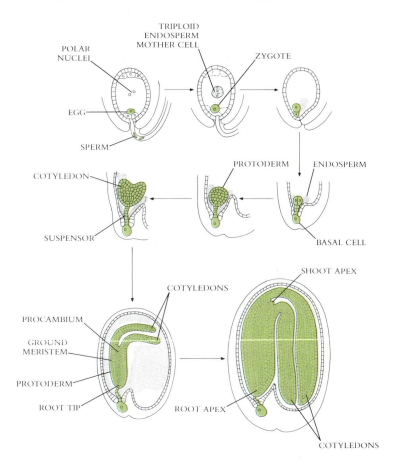

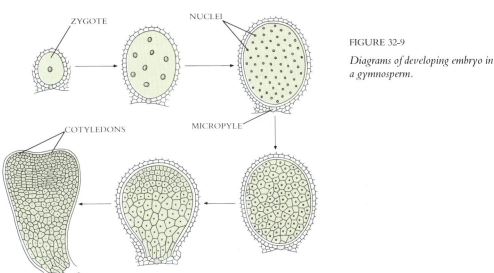

FIGURE 32-9

Diagrams of developing embryo in a gymnosperm.

this is not the case; instead, the pattern of embryo development in plants has been shown experimentally to be an expression of the genotype of the zygote itself. A plant embryo will develop normally even if it is removed from its ovule; this observation makes it clear that no chemical signals other than those the plant embryo received initially play a role in determining its subsequent pattern of development. An animal embryo, on the other hand, will not develop properly without such stimuli. In plants, the function of the tissue in the egg and in the zygote that is formed when the egg is fertilized appears to be primarily a nutrient one. The developmental pattern of an individual plant is also specified in detail by the environment, which alters the concentrations and distributions of the hormones that direct the patterns of differentiation.

Microchemical gradients within the egg of an animal establish developmental signals that persist in the zygote and in the cells formed by its division. In plants, the pattern of development is a direct result of gene expression, mediated by the environment.

GERMINATION IN PLANTS
The Role of Seed Dormancy

Early in the development of an angiosperm embryo, a profoundly significant event occurs: the embryo simply stops developing! In many plants, the development of the embryo is arrested soon after apical meristems and the first leaves, or cotyledons, are differentiated. The integuments—the coats surrounding the embryo—develop into a relatively impermeable seed coat, which encloses the quiescent embryo, together with a source of stored food, within the seed. Seeds are important adaptively in at least three respects:

1. They permit plants to postpone development when conditions are unfavorable and to remain dormant until more advantageous conditions arise. Under marginal conditions, a plant can "afford" to have some seeds germinate, because others remain dormant.

2. By tying the reinitiation of development to environmental factors, seeds permit the course of embryo development to be synchronized with critical aspects of the plant's habitat.

3. Perhaps most important, the dispersal of seeds facilitates the migration and dispersal of genotypes into new habitats and also offers maximum protection to the young plant at its most vulnerable stage of development.

Once a seed coat is formed around the embryo, most of the embryo's metabolic activities cease; a mature seed contains only about 10% water. Under these conditions, the seed and the young plant within it are very stable. Germination (Figure 32-10) cannot take place until water and oxygen reach the embryo, a process that sometimes involves cracking

FIGURE 32-10

The germination of a seed, such as this garden pea (Pisum sativum), *involves the fracture of the seed coat, from which the epicotyl and hypocotyl emerge.*

FIGURE 32-11

A seedling grown from seeds of lotus, Nelumbo nucifera, *recovered from the mud of a dry lake bed in Manchuria, northern China. The radiocarbon age of this seed indicates that it was formed in about 1515; another seed that was germinated was estimated to be at least a century older.*

the seed. Seeds of some plants have been known to remain viable for hundreds of years (Figure 32-11).

Specific adaptations often help to ensure that the plant will germinate only under appropriate conditions. Sometimes the seeds are held within tough fruits that themselves will not crack until, for example, they are exposed to the heat of a fire, a strategy that clearly results in the germination of a plant in an open, fire-cleared habitat; nutrients will also be relatively abundant there, having been released from the plants burned up in the fire. The seeds of other plants will germinate only when inhibitory chemicals have been leached from their seed coats, thus guaranteeing their germination when sufficient water is available. Still other plants will germinate only after they pass through the intestines of birds or mammals or are regurgitated by them, which both weakens the seed coats and ensures the dispersal of the plants involved.

In terms of gene expression, the events underlying embryo dormancy are not complex. The encapsulation of the embryo within the seed coat denies it water and oxygen and simply causes its metabolism to run down. To reactivate metabolism, the embryo has only to be provided with water, oxygen, and a source of metabolic energy. The final release of arrested development, germination, is then cued to specific signals from the environment.

Seed dormancy is an important evolutionary factor in plants, ensuring their survival in unfavorable conditions and allowing them to germinate when the chances of survival for the young plants are the greatest.

Germination

The first step in the germination of a seed occurs when it imbibes water. Because the seed is so dry at the start of germination, it takes up water with great force. Once this has occurred, metabolism within the seed resumes. Initially, metabolism may be anaerobic, but when the seed coat ruptures, oxidative metabolism takes over. At this point, it is important that oxygen be available to the developing embryo, because plants, which drown for the same reason people do, require oxygen for active growth. Few plants produce seeds that germinate successfully underwater, although some, like rice, have evolved a tolerance of anaerobic conditions (see boxed essay, pages 704-705).

A dormant seed, although it may have imbibed a full supply of water and may be respiring, synthesizing proteins and RNA, and apparently carrying on perfectly normal metabolism, may nonetheless fail to germinate without an additional signal from the environment. This signal may be light of the correct wavelengths and intensity (see Chapter 33), a series of cold days, or simply the passage of time at the appropriate temperatures for germination. The seeds of many plants will not germinate unless they have been **stratified**—held for a period of time at low temperatures. Stratification, which is also called *after-ripening,* prevents the seeds of many plants that occur in cold areas from germinating until they have passed the winter, thus protecting their seedlings from cold conditions.

Germination can occur over a wide temperature range (5° to 30° C), although it is generally optimum over a relatively narrow one, 25° to 30° C. Even under the best conditions, not all seeds will germinate. In some species, a significant fraction of seeds remain dormant, providing a genetic reservoir, or **seed pool,** of great evolutionary significance to the future of the plant population. One of the most interesting questions that has puzzled students of plant population biology is whether the genotypes of the seeds that germinate readily differ from those of the seeds that germinate later. This is a difficult question to investigate, but a highly significant one.

Because oxidative metabolism usually takes over soon after a plant embryo starts to grow, most seeds require oxygen for germination. Many seeds in a population may not germinate immediately, perhaps doing so when appropriate conditions occur later.

The Mobilization of Reserves

Germination occurs when all internal and external requirements are met. Germination and early seedling growth require the mobilization of metabolic reserves stored in the starch grains of **amyloplasts** (chloroplasts that are specialized to store starch) and protein bodies. Fats and oils also are important food reserves in some kinds of seeds. They can readily be digested during germination to produce glycerol and fatty acids, which yield energy through oxidative respiration and can also be converted to glucose. Any of these reserves may be stored in the embryo itself or in the endosperm, depending on the kind of plant.

In the cereal grains, the cotyledon is modified into an organ called the **scutellum,** a term that comes from the Latin word meaning "shield." The mobilization of the endosperm in these plants is not required during germination and does not occur then. The abundant food stored in the scutellum is mostly used up first. Later, while the seedling is becoming established, the scutellum absorbs the additional food that is stored in the endosperm, doing so by means of a two-stage process:

1. The initial mobilization of the starch in the endosperm is accomplished by hydrolases, which are secreted by the epithelial layer of the scutellum.

2. The later and more extensive mobilization of the starches in the endosperm is achieved by the secretion of amylase and other hydrolytic enzymes from the **aleurone layer,** a layer of specialized endosperm cells that lies just inside the seed coat. The synthesis and secretion of these aleurone hydrolases are controlled by a class of hormones called **gibberellins,** which are synthesized in the embryo (Figure 32-12).

How are the genes that transcribe the enzymes involved in the mobilization of food resources activated? Experimental studies have shown that, in the endosperm of the cereal grains at least, this occurs when the gibberellins initiate a burst of messenger RNA (mRNA) and protein synthesis. It is not known whether the gibberellins act directly on the DNA or via chemical intermediates in the cytoplasm. DNA synthesis apparently does not occur during the early stages of seed germination, but becomes important when the radicle, or embryonic root, has grown out of the seed coats.

> **During early germination and seed establishment, the vital mobilization of the food reserves stored in the embryo or the endosperm is mediated by hormones, which in at least some cases are gibberellins.**

THE ROLE OF THE APICAL MERISTEMS

Once a seed has germinated, the plant's further development depends on the activities of the meristematic tissues, which themselves interact with the environment. As described earlier, the shoot and root apical meristems give rise to all of the other cells of the adult plant. It is interesting to study the way in which a plant's genetic programming interacts with environmental influences to determine its structure.

A series of simple experiments sheds light on this matter. First, one may ask whether the meristems are influenced by other parts of the plant or whether they function independently. To test this most directly, one can isolate the apical meristem so that it is free of influences from the other parts of the plant. In one experiment on the wood fern, *Dryopteris,* four deep incisions were cut into the tip of the stem, separating its apical meristem from the surrounding tissue (Figure 32-13). The isolated stem tip continued to function as a normal meristem, continuing to act by itself as an autonomous source of differentiated tissue.

The results of this experiment suggest that shoot apical meristems cut from plants ought to be able to develop normally in a medium containing the necessary nutrients, vitamins, and a source of energy. Indeed, this has proved to be the case. The apical meristem produces only shoots and leaves, however, and never roots. Complete, mature plants

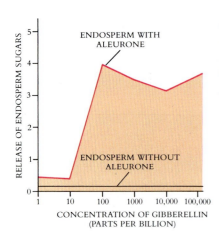

FIGURE 32-12

The release of sugar from endosperm is induced by gibberellin treatment. These data show that sugars are produced only when the aleurone layer is present. It is, in fact, the aleurone layer that is the source of the enzyme, α-amylase, which digests the starches stored in the endosperm.

never develop, regardless of how long the experiment proceeds. Eventually the excised apical meristem will even cease producing additional leaves and stems.

Even transferring the apical meristem to fresh medium will not help. The apical meristems of plants produce a hormone called **auxin** that continually diffuses back from the shoot apex and suppresses the active growth of the lateral branches. Auxins act largely by promoting cell elongation, apparently by affecting the plasticity of the cell walls. For them to work properly in normal cell division, however, another group of hormones, the cytokinins, also must be involved. Cytokinins are also necessary to cause masses of plant tissue being grown in a culture to differentiate and produce roots. When shoots are grown experimentally in a medium, they produce auxin themselves, but not cytokinins, which must be added for normal growth and development to take place (Figure 32-14). When this is done, complete little plants develop.

You will encounter auxins repeatedly in your study of plant development. They have simple structures, similar to the amino acid tryptophan. The naturally occurring auxin, **indoleacetic acid** (IAA), is of critical importance in determining the patterns of plant growth. Some of the synthetic auxins, which act in a similar way, have wide commercial use, as you will see in the next chapter.

The Autonomy of Apical Meristems

The experiments described above, taken as a whole, offer convincing evidence that apical meristems can function autonomously. Their activities are influenced by the other parts of the plant only in generalized ways, such as by supplying nutrients and hormones. The apical meristem itself is maintained as a coherent unit because its cells somehow act to inhibit the differentiation of the surrounding meristematic cells. As a result, new meristems normally do not grow out of an existing one. This relationship must reflect the presence of an inhibitor, which itself is presumably another plant hormone.

We have already noted the phenomenon of **apical dominance,** in which auxin, diffused backward from the apical meristem of the stem, inhibits the active growth of the lateral buds. The auxin may do so by stimulating production of the hormone **ethylene**

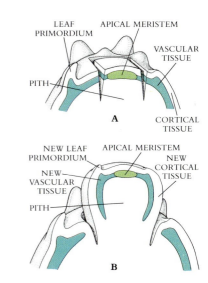

FIGURE 32-13

Surgical isolation of the apical meristem of the wood fern Dryopteris dilatata.
A *A view of the shoot tip, showing the position of the four incisions that isolated the terminal meristem from the surrounding leaf primordia and differentiating tissues. (×30).*
B *Longitudinal section of a surgically isolated meristem, showing the tissues that have been severed by the incisions, and the intact pith connection below the meristem. (×70.)*

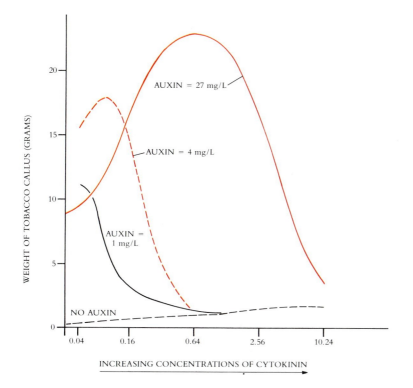

FIGURE 32-14

Auxin-cytokinin interactions. Cytokinin alone has little effect on the growth of tobacco callus in tissue culture. Auxin alone, regardless of the concentration used, causes the culture to grow to a weight of about 10 grams. When both hormones are present, growth is greatly increased. Notice, however, that when optimum concentrations are exceeded, the growth responses cease.

($H_2C{=}CH_2$). Ethylene apparently is produced around the nodes of the stem, and acts to inhibit differentiation of the surrounding cells, thus preventing lateral bud formation. It might also act in a similar way to maintain the integrity of the apical meristem of the stem. Cytokinins may likewise be involved in this complex system. By promoting cell division, they allow the development of lateral branches in some circumstances, even in the presence of auxin.

Leaf Primordia: Irreversibly Determined?

Finally, let us explore the degree to which the leaf primordia and buds that the apical meristem produces are irreversibly determined to be leaf primordia and buds at the time they are produced. Initially, these structures appear as simple cellular outgrowths at the base of the shoot meristem. How are the major differences in leaf form among, say, a fern, a pine, and an oak determined? Are they under genetic control, or are they set by environmental demands? Since the primordia of these different kinds of leaves look similar at first, the question may be restated another way: can structures in plants be determined before they differentiate?

This has been tested directly in ferns, which tend to have large leaves with complex patterns of division (see Figures 29-2, C, and 29-19). The leaf primordia of ferns have been isolated in culture medium at various stages of development, and the ability of these different isolates to produce mature leaves has been evaluated. The results of this experiment have been unequivocal: when the proper nutrients, vitamins, and hormones have been present in the medium, normal fern leaves have differentiated regardless of the stage at which the leaf primordia were isolated (Figure 32-15). No additional specific factor or factors from the original plant are necessary for the primordia to develop properly.

The leaf primordium therefore is an independent developmental subsystem. By the time it is differentiated, its ultimate "fate" has already been determined by the meristem on which it was produced. When it is cut off from that meristem, it proceeds along its own predetermined developmental path. The executive decision is made by the meristem ("What tissue will this cell be?"), and the operational decisions are left to be expressed by the tissue itself ("When will it differentiate, and into what kind of leaf?").

THE VASCULAR CAMBIUM: A DIFFERENTIATED MERISTEM

The apical meristems of plants retain the ability to proliferate indefinitely. As they do so, they produce cells that are differentiated in various ways and to various degrees. Some of these cells give rise to the vascular cambium, a permanently meristematic tissue. The ways in which the vascular cambium produces different kinds of cells inwardly and outwardly apparently involve both (1) the identity of the cells that are immediately adjacent, and (2) differences in physical factors, such as pressure, to which the derivatives of the vascular cambium are subjected. In this sense, the vascular cambium is partly differentiated, although permanently meristematic.

The vascular cambium is a differentiated, permanently meristematic tissue.

In this way, the vascular cambium represents an intermediate stage of differentiation. A developmental commitment to produce vascular tissue, which comprises a variety of different cell types, has already been made. At the same time, the proliferation of the meristematic tissue itself continues; the subsequent commitment of individual derivative cells to xylem, phloem, or ray cells has yet to be made. It is convenient to visualize a series of discrete stages of differentiation (Figure 32-16), each of which represents an increased level of developmental commitment, with each stage functioning more or less autonomously.

The ways in which the processes we have discussed in this chapter are integrated during the course of development in the whole plant involve several kinds of hormones, some

FIGURE 32-15

Development of a frond of the cinnamon fern, Osmunda cinnamomea, *excised from the plant at the end of the third growing season and grown in sterile culture to maturity. The normal course of frond development, from its inception to the end of the fourth growing season: (1) soon after inception, (2) end of the first season, (3) end of the second season, (4) end of the third season; (5-7) developmental stages during the fourth season. The coiled form of the frond, which resembles a crozier, develops entirely during the fourth season.*

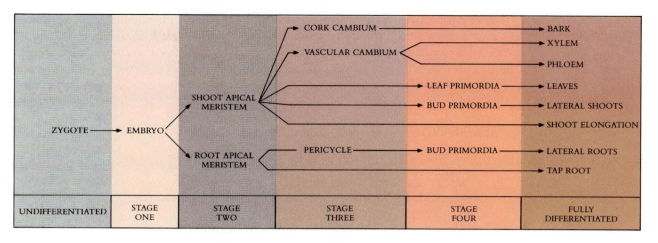

UNDIFFERENTIATED	STAGE ONE	STAGE TWO	STAGE THREE	STAGE FOUR	FULLY DIFFERENTIATED

FIGURE 32-16

Stages in plant differentiation.

of which we have already mentioned. These hormones strongly influence one another as they affect plant growth, and the form of the plant that results from their action is also strongly influenced by the environment in which that plant is growing. These interactions, and the ways in which plant growth is regulated, will form the subject of the next chapter.

SUMMARY

1. Animals undergo development according to a fixed blueprint that is followed faithfully until they are mature. Plants, in contrast, develop continuously. The course of their development is mediated by hormones, which are produced as a result of interactions with the external environment.

2. Almost all of the differentiation in plants is fully reversible; plant cells are totipotent as long as they retain a living protoplast and a nucleus.

3. Whole plants can be regenerated from cultures of single cells, provided that these cells are kept in fairly close contact with others. If they are not, they usually tend to lose the factors that make it possible for them to divide.

4. Embryo development in animals involves extensive movements of cells in relation to one another, but the same process in plants consists of an orderly production of cells, rigidly bound by their cellulose-rich cell wall. By the time about 40 cells have been produced in an angiosperm embryo, differentiation begins; the meristematic shoot and root apices are evident.

5. In most gymnosperms, cell walls do not form in the first part of the developmental process. After about 256 nuclei have been produced by successive mitotic divisions within a common cytoplasm, cell walls are formed and differentiation begins.

6. Seeds are held within a rigid, relatively impermeable seed coat, which may crack only when it receives appropriate environmental cues. These may include the quality and periodicity of light, cold temperatures, "ripening" (the passage of time), or mechanical or chemical cracking. All of these cues tend to concentrate seed germination at the times when the young plants are most likely to become established.

7. In the germination of seeds, the mobilization of the food reserves stored in the cotyledons and in the endosperm is critical. In the cereal grains, this process is mediated by hormones of the kind known as gibberellins, which appear to activate transcription of the loci involved in the production of amylase and other hydrolase enzymes.

8. Apical meristems will develop normally into whole plants when they are excised, provided that cytokinins, which stimulate root production and are not produced in the apical meristems or in the shoot, are added. Leaf primordia also are capable of development independent of the plant on which they are formed, starting with the point at which they are first visible as independent structures.

SELF-QUIZ

The following questions pinpoint important information in this chapter. You should be sure to know the answers to each of them.

1. Plants that are generated from tissue cultures of pollen grains are _____.

2. The formation of embryos in gymnosperms does not follow the same pattern as that of angiosperms, in that early nuclear replication is not followed by the formation of _____ _____.

3. In the development of plant embryos, cells migrate from the suspensor to an area near the shoot apical meristem. True or false?

4. Hormones are chemicals that "work at a distance." Plants have only one class of hormones, called gibberellins. True or false?

5. Many plants exhibit apical dominance, a phenomenon in which the shoot apex actively suppresses the formation of lateral branches. Apical dominance is caused by
   (a) the action of auxin
   (b) the action of cytokinins
   (c) the formation of gibberellins

6. One of the major problems in getting individual plant cells to differentiate is getting them to _____.

7. Coconut milk is in fact liquid _____.

8. A mass of dividing, undifferentiated plant cells is called
   (a) a primordium   (d) an ovule
   (b) callus   (e) a zygote
   (c) an embryoid

9. _____, also called after-ripening, refers to moist, low-temperature storage of seeds, usually for 4 to 6 weeks, to break dormancy.

10. DNA synthesis is not required for seed germination. True or false?

11. Ginseng, a perennial plant, produces only a few leaves every year. In fact, in August the _____ _____ have already formed, and the number of leaves the plant will have the next spring can easily be determined.

12. Which of the following is required for the differentiation of plant tissues?
   (a) IAA   (c) coconut milk
   (b) gibberellins   (d) cytokinins

THOUGHT QUESTIONS

1. Why is plant development so much more closely linked with environmental cues than animal development is?

2. If you take parenchyma cells from the stem of a plant and put them in a culture medium, what will that medium have to contain in order for the cells to develop into whole plants?

3. What experiments would you suggest for studying the differentiation of individual plant cells? How could you begin to understand the factors involved in their assuming their mature forms?

FOR FURTHER READING

COOK, R.E.: "Clonal Plant Populations," *American Scientist,* vol. 71, pages 244-253, 1983. Excellent discussion on the role of asexual reproduction among plants in natural populations.

DALE, J.E., and F.L. MILTHORPE (eds.): *The Growth and Functioning of Leaves,* Cambridge University Press, New York, 1983. An excellent series of articles on all aspects of leaf development and functioning.

GALSTON, A.W., P.J. DAVIES, and R.L. SATTER: *The Life of the Green Plant,* 3rd ed., Prentice-Hall, Inc., Englewood Cliffs, N.J., 1980. Excellent introduction to plant physiology, stressing development.

KOLLER, D.: "Germination," *Scientific American,* April 1959, pages 75-84. An interesting review of the different mechanisms of seed dormancy.

MAYER, A., and A. POLJAKOFF: *The Germination of Seeds,* 2nd ed., Pergamon Press, Inc., Elmsford, N.Y., 1975. Excellent and clearly written book on seed biology.

SHEPARD, J.F.: "The Regeneration of Potato Plants from Leaf-Cell Protoplasts," *Scientific American,* May 1982, pages 112-121. Provides valuable insight into the way in which experiments in this area are designed.

WAREING, P.F., and I.D.J. PHILLIPS: *Growth and Differentiation in Plants,* 3rd ed., Pergamon Press, Ltd., Oxford, England, 1981. Outstanding, up-to-date review of plant development and differentiation.

Chapter 33   REGULATION OF PLANT GROWTH

Overview

The major classes of plant hormones—auxins, cytokinins, gibberellins, ethylene, and abscisic acid—interact in complex ways to produce a mature, growing plant. Unlike the highly specific hormones of animals, plant hormones are not produced in definite organs, nor do they have definite target areas. They function rather as generalized stimulators or inhibitors of growth. By integrating the complex signals that reach the plant from its environment, they allow the plant to respond efficiently to the demands of its environment. Gibberellins promote the elongation of stems in most plants, whereas auxins do the same in young grass seedlings and some other herbs. Gibberellins also play a key role in breaking dormancy in buds, seeds, and the rosettes of biennials. Cytokinins promote mitosis and cell division. Plants respond, in complex ways that we are just starting to understand, to stimuli such as touch, light, gravity, and day length, by producing growth movements, flowers, or other responses appropriate to their survival in a particular habitat.

For Review

Here are some important terms and concepts that you will encounter in this chapter. If you are not familiar with them, you should review them before proceeding.

- **Turgor pressure** (Chapter 9)
- **Flowering plant life cycle** (Chapter 29)
- **Seed germination** (Chapter 31)
- **Auxin and plant growth** (Chapter 32)
- **Cytokinins and differentiation** (Chapter 32)

As we emphasized in the last chapter, plants respond to their environment by growing. Unlike animals, plants cannot move from place to place to seek more favorable circumstances, except by the very slow process of growth. We also introduced plant hormones and how they are stimulated by external factors and interact with one another during plant growth. In this chapter we will consider each of the major classes of plant hormones in greater detail. We will then explore some of the ways in which external factors affect plant growth and thus the appearance of individual plants.

PLANT HORMONES

Hormones are chemical substances that are produced in small, often minute, quantities, in one part of an organism and then are transported to another part of the organism, where they bring about physiological responses. The activity of hormones results from their ability to stimulate certain physiological processes and to inhibit others. How they act in a particular instance is influenced both by what the hormones themselves are and by how they affect the particular tissue that receives their message.

In animals, hormones usually are produced at definite sites, normally in organs that are solely concerned with hormone production. In plants, on the other hand, hormones are produced in tissues that are not specialized for that purpose but that carry out other, usually more obvious, functions as well. There are at least five major kinds of hormones in plants: auxin, cytokinins, gibberellins, ethylene, and abscisic acid. Other kinds of plant hormones certainly exist, but they are less well understood. The study of plant hormones, especially our attempts to understand how they produce their effects, is an active and important field of current research.

> **Five major kinds of plant hormones are reasonably well understood: auxins, cytokinins, gibberellins, ethylene, and abscisic acid.**

Auxins

In the last chapter, we mentioned auxin with respect to its ability to control the growth of lateral buds on the stem. Not only was auxin the first plant hormone to be discovered, but related compounds that have similar modes of action and are classified as auxins are also the most important commercially. Although the rate of stem elongation in mature flowering plants is regulated largely by gibberellins, auxin regulates stem elongation in young grass seedlings and in herbs generally. Because the effects of auxin on young grass seedlings are so evident, they were studied more than a century ago.

Discovery of auxin

In his later years, the great evolutionist Charles Darwin became increasingly devoted to the study of plants. In 1881 he and his son Francis published a book called *The Power of Movement in Plants*. In this book, the Darwins reported their systematic experiments concerning the way in which growing plants bend toward light, a phenomenon known as **phototropism.** They made many observations in this field and also conducted experiments using young grass seedlings.

Charles and Francis Darwin found that the seedlings they were studying normally bent strongly toward a source of light, if the light came primarily from one side. However, if they covered the upper part of a seedling with a cylinder of metal foil, so that no light reached its tip, the shoot would not bend (Figure 33-1). The Darwins obtained this result even though the region where the bending normally occurred was still exposed. Light reached this part of the seedling directly, but bending did not occur. However, if they covered the end of the shoot with a gelatin cap, which transmitted light, the shoot would bend as if it were not covered at all.

In explaining this unexpected finding, the Darwins hypothesized that when the shoots were illuminated from one side, an "influence" arose in the uppermost part of the shoot, was transmitted downward, and caused the shoot to bend. For some 30 years, the Darwins' perceptive experiments remained the sole source of information about this interesting phenomenon. Then a series of experiments was performed by several other botanists. Some of the most significant were those conducted independently by the Danish plant physiologist Peter Boysen-Jensen and the Hungarian plant physiologist Arpad Paal, who demonstrated that the substance that caused the shoots to bend was a chemical. They showed that if one cut off the tip of a grass seedling and then replaced it, but separated it from the rest of the seedling by a block of agar, the seedling would react as if there had been

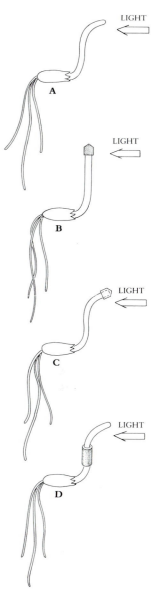

FIGURE 33-1

The Darwins' experiment. Young grass seedlings normally bend toward the light **(A)**. *When the tip of a seedling was covered by a lightproof collar* **(B)** *(but not when it was covered by a transparent one* **[C]***), this bending did not occur. When the collar was placed below the tip* **(D)**, *the characteristic light response took place. From these experiments, the Darwins concluded that, in response to light, an "influence" that causes bending was transmitted from the tip of the seedling to the area below the tip, where bending normally occurs.*

FIGURE 33-2

Frits Went's experiments.
A *Went removed the tips of the grass seedlings and put them on agar.*
B *Blocks of the agar were then put on one side of the ends of other grass seedlings from which the tips had been removed.*
C *The seedlings bend away from the side on which the agar block was placed. Went concluded that the substance that he named auxin promoted the elongation of the cells, and that it accumulated on the side of a grass seedling away from the light.*

FIGURE 33-3

Phototropism and auxin, diagrams of experiments originally performed by Winslow Briggs. Experiments A and B, performed in the dark, showed that splitting the tip of the seedling leaf and inserting a barrier did not significantly affect the total amount of auxin that was diffused from its tip. The amount of curvature produced, which is related to the amount of auxin produced, is shown by the numbers below the agar block. The other experiments (C to F) were performed with light coming from the right side, as indicated by the shading. A comparison of C and D with A and B shows that auxin production is not dependent on light. The slight differences in curvature shown are not significant. If a barrier was inserted in the agar block (E), light caused the displacement of the auxin away from the light. Finally, experiment F showed that it was displacement that had occurred, and not different rates of auxin production on the dark and light sides, because when displacement was prevented with a barrier, auxin production was not significantly different in the two sides.

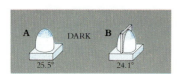

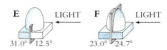

no change. Something evidently was passing from the tip of the seedling through the agar into the region where the bending occurred. On the basis of these observations, Paal suggested that an unknown substance, under conditions of uniform illumination or of darkness, continually moves down the grass seedlings from their tips and promotes growth on all sides. Such a pattern would not, of course, cause the shoot to bend.

The next telling results in this area were obtained by Frits Went, a Dutch plant physiologist, and published in 1926. Went obtained these results in the course of studies for his doctoral dissertation. Carrying Paal's experiments an important step further, Went cut off the tips of grass seedlings that had been illuminated normally and set these tips on agar. He then took grass seedlings that had been grown in the dark and cut off their tips in a similar way. Finally, Went cut tiny blocks from the agar on which the tips of the light-grown seedlings had been placed and put them on the tops of the decapitated dark-grown seedlings, but set off to one side (Figure 33-2). Even though these seedlings had not been exposed to the light themselves, they bent *away* from the side on which the agar blocks were placed.

Went then observed the seedlings on which untreated agar blocks had been placed. He put blocks of pure agar on the decapitated stem tips and noted either no effect or a slight bending *toward* the side where the agar blocks were placed. Finally, Went cut sections out of the lower portions of the light-grown seedlings to see whether the active principle was present in them. He placed these sections on the tips of decapitated, dark-grown grass seedlings, and again observed no effect.

As a result of his experiments, Went was able to show that the substance that had flowed into the agar from the tips of the light-grown grass seedlings could make seedlings curve when they otherwise would have remained straight. In other words, it enhanced rather than retarded cell elongation. Went also showed that this chemical messenger caused the tissues on the side of the seedling into which it flowed to grow more than those on the opposite side. He named the substance that he had discovered **auxin,** from the Greek word *auxein,* which means "to increase."

Went's experiments provided a basis for understanding the responses that the Darwins had obtained some 45 years earlier. The grass seedlings bent toward the light because the auxin contents on the two sides of the shoot differed. The side of the shoot that was in the shade had more auxin, and its cells therefore elongated more than those on the lighted side, bending the plant toward the light.

Thus auxin acts to fit the form of the plant to its environment in a highly advantageous way. It is a "releaser" of growth and elongation, and the means by which the plant is able to respond to its environment; signals received from the environment influence the distribution of auxin in the plant directly. How does the environment exert this influence? It might theoretically destroy the auxin, or decrease the cell's sensitivity to auxin, or cause the auxin molecules to migrate away from the light into the shaded portion of the shoot. This last possibility has since proved to be the case.

In a simple but effective experiment, Winslow Briggs, of the Department of Plant Biology of the Carnegie Institution of Washington, inserted a thin sheet of glass vertically

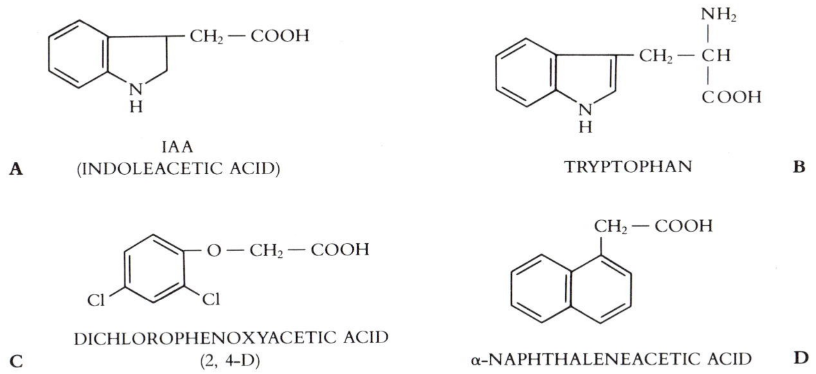

CH₂—COOH structures...

IAA
(INDOLEACETIC ACID)

A

NH₂

CH₂—CH

COOH

TRYPTOPHAN **B**

O—CH₂—COOH

Cl   Cl

DICHLOROPHENOXYACETIC ACID
(2, 4-D)

C

CH₂—COOH

α-NAPHTHALENEACETIC ACID **D**

FIGURE 33-4

A *Indoleacetic acid (IAA), the only known naturally occurring auxin.*
B *Tryptophan, the amino acid from which plants synthesize IAA.*
C *Dichlorophenoxyacetic acid (2,4-D), a synthetic auxin, is a widely used herbicide.*
D *Napthalanacetic acid, another synthetic auxin, is commonly employed to induce the formation of adventitious roots in cuttings and to reduce fruit drop in commercial crops. The synthetic auxins, unlike IAA, are not broken down readily by the enzymes that occur in plants and microbes, and so are well suited for commercial purposes.*

between the half of the shoot oriented toward the light and the half oriented away from it (Figure 33-3). He found that lateral light does not cause a plant into which such a glass barrier has been inserted to bend. When Briggs examined the illuminated plant itself, he found that the auxin levels in the light and dark sides of the barrier were equal. He concluded that auxin in normal plants migrates from the light side to the dark side in response to light. The presence of his glass barrier, by preventing the migration, blocked the light response. The exact nature of the auxin migration process is not known, but it is thought to involve a light-sensitive pigment that perhaps alters membrane permeability to auxin. Only light of wavelengths less than 500 nanometers can promote the lateral migration of auxin.

Subsequent experiments have revealed the chemical structure of what seems to be the only naturally occurring auxin (Figure 33-4, *A*), indoleacetic acid (IAA). These experiments have also yielded a great deal of information about how auxins act in plants. The term *auxin* is now used to refer both to the naturally occurring substance and to those related synthetic molecules that produce similar effects. As we saw in the last chapter, indoleacetic acid is produced at the shoot apex, in the region of the apical meristem, and diffuses continuously downward, suppressing the growth of lateral buds. It resembles the amino acid tryptophan (Figure 33-4, *B*) in structure, and it is in fact synthesized from tryptophan by plants.

> **The only known naturally occurring auxin, indoleacetic acid (IAA), is synthesized from the amino acid tryptophan. It is produced in the apical meristems of shoots and diffuses downward, suppressing the growth of lateral buds. In young grass seedlings and other herbs, it plays a major role in stem elongation, migrating from the illuminated portions of the stem to the dark portions and thus causing the stems to grow toward the light.**

Auxin and plant growth

Auxin increases the plasticity of the plant cell wall. A more plastic wall will stretch more during active cell growth, while its protoplast is swelling. Since very low concentrations of auxin promote cell wall plasticity, the hormone must be broken down rapidly to prevent its accumulation. Plants do this by means of the enzyme **indoleacetic acid oxidase.** By controlling the levels of both IAA and IAA oxidase, plants can regulate their growth very precisely. Using methods involving monoclonal antibodies (see Chapter 18), scientists have recently been able to locate the transport sites involved in the movement of auxin downward from the shoot apex through the plant. These sites are in the plasma membranes near the basal ends of the cells in which the auxins are promoting elongation. The further application of such methods will certainly help to clarify other properties of the auxin-mediated process of cell elongation.

The speed with which auxin increases cell plasticity—grass seedlings may begin to bend within 3 minutes of its application—has made it difficult to determine the chemical basis of these reactions. It seems highly unlikely that these rapid-response effects could be caused by changes in the rates of transcription or translation of genes. To act so rapidly, auxin must affect some system that already exists. Extensive changes have been reported in the polysaccharides of plant cell walls, involving the formation and breaking of covalent bonds, in response to treatment with auxin. An increase in the concentration of H^+ ions in the wall is also very likely to occur. In addition, auxin appears to mediate the stimulation of mRNA transcription, which would lead to long-term growth changes.

Auxin controls various different plant responses in addition to those involved in phototropism. One of these is the suppression of lateral bud growth, an effect we have mentioned earlier. How can auxin, a growth promoter, also inhibit growth? Apparently the cells around lateral buds, in the nodal regions of the stem, produce ethylene under the influence of auxin. The ethylene, in turn, inhibits the growth of the lateral buds. When the terminal bud on a stem is removed and the lateral buds grow, bushy plants are produced and the number of flowers on an individual plant is increased. Similar effects may be achieved chemically or by selectively breeding individuals that have such characteristics.

Auxin also promotes the growth of vascular tissue in stems and the growth of the vascular cambium itself. It likewise increases fruit growth, an effect that has important commercial implications. Fruits normally will not develop if fertilization has not occurred and seeds are not present, but they often will do so if auxins are applied. Apparently auxins, which are present in pollen in large quantities, play a key role in making the development of a mature fruit possible. Synthetic auxins are used commercially for the same purpose.

Auxin increases the plasticity of plant cell walls and thus the elongation of cells. By interacting with other hormones, it likewise produces a number of other effects, including the suppression of lateral buds and the promotion of vascular tissue growth.

Synthetic auxins

Synthetic auxins have many uses in agriculture and horticulture. One of their most important uses is based on their prevention of **abscission,** the process by which a leaf or other organ falls from a plant. Before abscission occurs, a **separation layer** or **abscission layer** forms at the base of the organ. Such a layer consists of a thin layer of often small cells oriented at right angles to the stalk of the particular structure. The walls of the cells in the separation layer become soft and gelatinous, so that the leaf, flower, or fruit readily breaks off at that point and falls from the plant when the wind moves it.

Before a leaf, flower, or fruit is ready to fall, it becomes **senescent;** its cells begin to die, and the organ often turns yellowish or brownish. In this process, the slowdown or cessation in the production of auxin by the dying leaf is one of the critical factors. Leaf fall often can be retarded, therefore, by applying auxin to the leaf blade. Synthetic auxins are also used to prevent fruit drop in apples before they are ripe, to hold the berries on holly that is being prepared for shipping, to promote flowering and fruiting in pineapples, and to induce the formation of roots in cuttings.

Furthermore, synthetic auxins are routinely used to control weeds. When they are used as herbicides, they are applied in higher concentrations than those at which IAA normally occurs in plants. One of the most important of the synthetic auxins used in this way is 2,4-dichlorophenoxyacetic acid, usually known as 2,4-D (Figure 33-4, *C*), which effectively kills weeds in lawns, selectively eliminating broad-leaved dicots. The weeds literally grow to death, their stems elongating from a few centimeters to several decimeters within days.

A molecule closely related to 2,4-D is the herbicide 2,4,5-trichlorophenoxyacetic acid, better known as 2,4,5-T, which has long been widely used to kill woody seedlings and weeds. Notorious as the "Agent Orange" of the Vietnam war, 2,4,5-T is easily contaminated with a by-product of its manufacture, 2,3,7,8-tetrachlorodibenzo-*para*-dioxin. This substance, known as **dioxin** for short, is extremely toxic to people; it is the subject of great environmental concern in the United States and elsewhere (Figure 33-5).

FIGURE 33-5

Personnel from the Missouri Department of Natural Resources and the U.S. Environmental Protection Agency taking earth samples in Times Beach, Missouri. These samples are then tested for dioxin. The entire community of Times Beach was abandoned in 1983, after it was found to be heavily contaminated with dioxin that had been an impurity in oil spread on the roads.

Cytokinins

Another group of naturally occurring growth hormones in plants is the **cytokinins.** Cytokinins were discovered because of their role in promoting the differentiation of organs in masses of plant tissue growing in culture. Studies by Haberlandt at about the turn of the century demonstrated the existence of a chemical that would induce the parenchyma tissue of potatoes to become meristematic. Subsequent studies have focused on the role of cytokinins in bringing about the differentiation of callus tissue.

A cytokinin is a plant hormone that, in combination with auxin, stimulates cell division in plants and determines the course of differentiation. Substances that have these properties are widespread both in bacteria and in other eukaryotes. In vascular plants, most cytokinins seem to be produced in the roots and then transported from there throughout the rest of the plant. Developing fruits apparently are also important sites of cytokinin synthesis. In mosses, cytokinins cause the formation of vegetative buds on the protonemata. In all plants, cytokinins seem to have been incorporated into the regulation of growth patterns, working together with other hormones.

The naturally occurring cytokinins all appear to be derivatives of the purine base adenine (Figure 33-6). Other molecules, not known to occur naturally, have similar effects but are very diverse chemically. In contrast to auxins, cytokinins *promote* the growth of lateral branches. Similarly, auxins promote the formation of lateral roots, whereas cytokinins inhibit their formation. As a consequence of these relationships, the balance between cytokinins and auxin determines, among other factors, the appearance of a mature plant. In addition, the application of cytokinins prevents the yellowing of leaves that are detached from the plant.

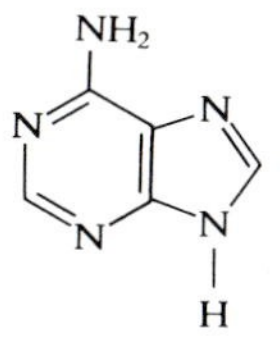

> **Cytokinins are plant hormones that, in combination with auxin, stimulate cell division and determine the course of differentiation. The naturally occurring cytokinins are derivatives of the purine base adenine.**

The action of cytokinins, like that of other hormones, has been studied in terms of its effects on the growth and differentiation of masses of tissue growing in defined media. Cytokinins seem to be necessary for mitosis and cell division to take place. They apparently work by influencing the synthesis or activation of proteins that are specifically required for mitosis.

Gibberellins

Gibberellins are named for the fungus genus *Gibberella,* which causes a disease of rice in which the plants grow to be abnormally tall. This "foolish seedling" disease of rice was investigated in the 1920s by Ewiti Kurosawa and other Japanese scientists. They found that if they grew the fungus in culture, they could obtain a chemical, completely free of the fungus itself, that would affect the rice plants in a way similar to the fungus. This substance was isolated in 1939 and chemically characterized in 1954. Although it was first thought to be only a curiosity, it has since turned out to belong to a large class of naturally occurring plant hormones called the **gibberellins** (Figure 33-7).

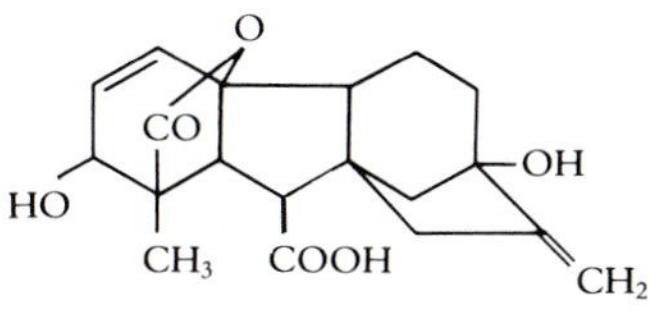

ADENINE

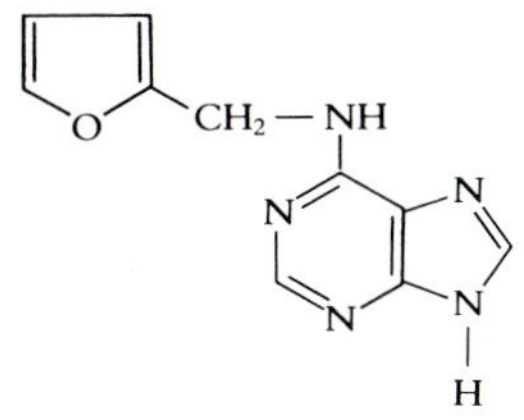

KINETIN

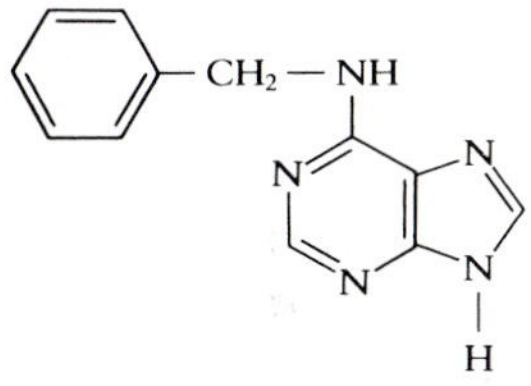

6-BENZYLAMINO PURINE
(BAP)

FIGURE 33-6

Two commonly used synthetic cytokinins: kinetin and 6-benzylamino purine. Note their resemblance to the purine base adenine, also shown.

FIGURE 33-7

Although more than 60 gibberellins have been isolated from natural sources, apparently only GA_1 is active in shoot elongation. All of the other gibberellins have a similar structure. Shown here are three such gibberellins. The arrows in GA_7 and GA_4 indicate points of structural difference among the three.

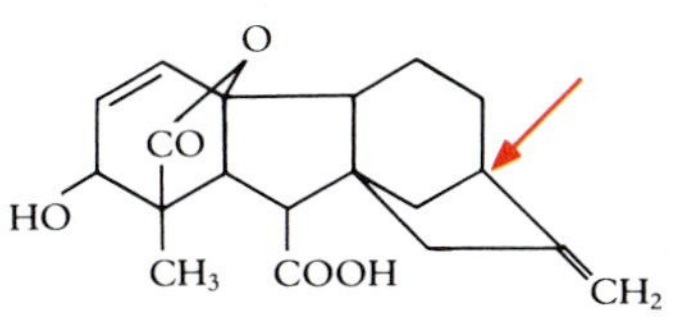

GIBBERELLIC ACID
(GA_3)

GA_7

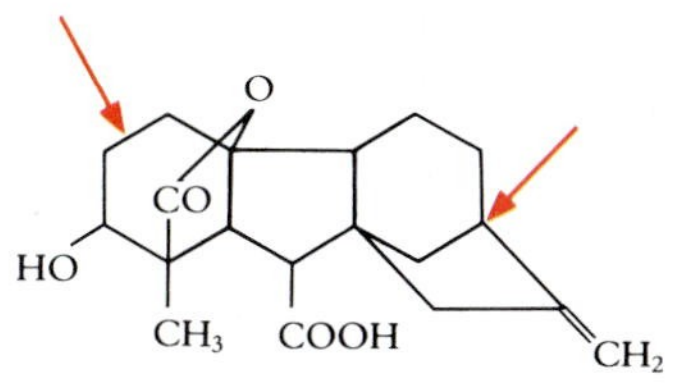

GA_4

FIGURE 33-8

A genetically dwarf variety of the garden pea (Pisum sativum). When treated with gibberellin, the plants at right attained the normal stature.

Synthesized in the apical portions of both stems and roots, gibberellins have important effects on stem elongation in plants and play the leading role in controlling this process in mature trees and shrubs. In these plants, the application of gibberellins characteristically promotes internode elongation, and this effect is enhanced if auxins are present also. Many dwarf mutants of plants are known in which normal growth and development can be restored if gibberellins are applied (Figure 33-8). It is supposed that such mutants lack naturally occurring gibberellins, at least in sufficient quantities to promote normal growth.

Gibberellins are also involved with many other aspects of plant growth. In the last chapter, for example, we examined their role in stimulating production of amylase and other hydrolytic enzymes that are mobilized during the germination and establishment of cereal seedlings. Biennial plants often can be induced, by the application of gibberellins, to grow out of the rosette stage and to flower (Figure 33-9). These hormones also hasten seed germination, apparently because they can substitute for the effects of cold or sometimes light requirements in this process.

Gibberellins are an important class of plant hormones, which are produced in the apical regions of shoots and roots. They play the major role in controlling stem elongation for most plants, acting in concert with auxin and other hormones.

According to recent analyses carried out by B.O. Phinney of the University of California, Los Angeles, probably only one kind of gibberellin, GA_1, occurs naturally in the flowering plants. This hormone is active in the control of shoot elongation. The natural functions of the other gibberellins that occur widely both in bacteria and in eukaryotes other than flowering plants are not known, although a number of them function like GA_1 if they are applied to flowering plants. As far as we know, therefore, the properties of the gibberellins by which we recognize them as plant hormones originated in the early angiosperms.

Like all other plant hormones, the gibberellins produce their effects largely on the basis of their interactions with other hormones. We still have a great deal to learn about these interactions. Gibberellins are enjoying an increased commercial use and doubtless will become even more important in the future. They are particularly significant in increasing fruit size and set and cluster size in grapes. They are used to delay the ripening of citrus fruits on the trees and to speed up the flowering of strawberries. The most important commercial use of gibberellins is to stimulate the partial digestion of starches in germinating barley during the process of brewing beer.

Ethylene

Long before its role as a plant hormone was appreciated, the simple hydrocarbon ethylene ($H_2C\!=\!\!CH_2$) was known to defoliate plants when it leaked from gaslights in street lamps. Ethylene is, however, a natural product of the metabolism of plants that interacts with other plant hormones in minute amounts. We have already mentioned the way in which auxin, diffusing down from the apical meristem of the stem, may stimulate the production of ethylene in the tissues around the lateral buds and thus retard their growth. Ethylene also suppresses stem and root elongation, probably for similar reasons.

However, ethylene plays other important roles in plants as well. For example, ethylene appears to be the main factor in the formation of the separation layer during abscission. Both pollinated flowers and fruits that are developing properly produce large amounts of auxin; if such amounts of auxin are present, or if, as we described earlier, auxin is applied to the flowers or fruits, abscission will not occur. Ethylene, however, which is being produced constantly in all parts of the plant, counteracts the effects of the auxin and promotes the formation of the separation layer.

Ethylene is also produced in large quantities during a certain phase of the ripening of

FIGURE 33-9

Effects of gibberellins.
A *Head lettuce* (Lactuca sativa) *will "bolt"— flower—when it is treated with gibberellins.*
B *Cabbage* (Brassica oleracea), *a biennial that is native to the seacoasts of Europe, will also bolt when treated with gibberellin. Most biennials exhibit similar behavior. In nature, they usually flower during their second year of growth.*

fruits, when their respiration is proceeding at its most rapid rate. At this phase, which is called the **climacteric,** complex carbohydrates are broken down into simple sugars, chlorophylls are also broken down, cell walls become soft, and the volatile compounds associated with flavor and scent in the ripe fruits are produced. When ethylene is applied to fruits, it hastens their ripening. One of the first lines of evidence that led to the recognition of ethylene as a plant hormone was the observation that gases which came from oranges caused premature ripening in bananas. Such relationships have led to major commercial uses. For example, tomatoes are often picked green and then artificially ripened later by the application of ethylene. Ethylene is widely used to speed the ripening of lemons and oranges as well. Carbon dioxide produces effects in fruits opposite to those of ethylene, and fruits that are being shipped are often kept in an atmosphere of carbon dioxide if they are not intended to ripen yet.

> **Ethylene, a simple gaseous hydrocarbon, is a naturally occurring plant hormone. It plays the key role in controlling the abscission of leaves, flowers, and fruits from the plants on which they form; their abscission is counteracted by auxin. Auxin probably retards the development of lateral buds by stimulating the production of ethylene around them.**

Ethylene can also be used to hasten fruit drop, when the fruits are ready to be harvested, and leaf fall, when this is desirable. It is used to induce flowering in pineapples and other members of the same plant family, Bromeliaceae, a number of which are strikingly beautiful ornamentals. Ethylene also stimulates the germination of seeds and is used commercially in weed control because of this property: the weed seedlings can be killed easily by the application of a herbicide, but their seeds, being dormant, are much more difficult to kill. A problem in the commercial use of ethylene, as you might imagine, is that it is a

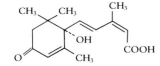

FIGURE 33-10

Structure of abscisic acid.

gas. Plant scientists are seeking chemicals that produce a steady supply of ethylene once they have been applied to a plant but that remain liquid themselves.

Abscisic Acid

Abscisic acid (Figure 33-10), a naturally occurring plant hormone, appears to be synthesized mainly in mature green leaves, in fruits, and in root caps. The hormone was given its name because applications of it stimulate leaf senescence and abscission, but there is little evidence that it plays an important natural role in this process. When it is spotted on a green leaf, the spots turn yellow. Thus abscisic acid has the exact opposite effect on a leaf from that of the cytokinins: a yellowing leaf will remain green in an area where cytokinins are spotted.

Abscisic acid may induce the formation of winter buds by converting leaf primordia into bud scales, and, like ethylene, it may suppress the growth of dormant lateral buds. From what we know, it appears likely that abscisic acid suppresses the growth and elongation of buds and promotes senescence, also counteracting some of the effects of the gibberellins (which stimulate the growth and elongation of buds) and auxin (which tends to retard senescence). Abscisic acid is also important in controlling the opening and closing of stomata, a subject that we will consider further in the next chapter.

> **Abscisic acid, which is produced chiefly in mature green leaves and in fruits, suppresses the growth of buds and plays an important role in controlling the opening and closing of stomata. It also promotes leaf senescence.**

Abscisic acid occurs in all groups of plants and apparently has been functioning as a growth-regulating substance since early in the evolution of the plant kingdom. Relatively little is known about the exact nature of its physiological and biochemical effects. These effects are so rapid, however—often taking place within a minute or two—that they must be at least partly independent of protein synthesis. Some longer-term effects of abscisic acid do involve the suppression of protein synthesis, but the way in which this occurs is poorly understood. Recently, the binding sites for abscisic acid have been demonstrated to be proteins located on the outer surface of the plasma membrane and not involved with the transport of the hormone into the cells. Like other plant hormones, abscisic acid probably will prove to have valuable commercial applications when its mode of action is better understood.

The Combined Action of Plant Hormones

Hormones are relatively difficult to study because they are produced in such small quantities; also, the actions of plant hormones are so thoroughly integrated with one another that they have sometimes been difficult to separate. Indeed, the most outstanding characteristic of plant hormones may be the way in which they interact to produce their effects. For example, auxins promote cell elongation, but only in certain kinds of stems and under certain conditions; cytokinins promote cell division; ethylene inhibits cell division; and gibberellins promote both cell division and cell elongation. An example of the interaction between auxin and cytokinin was shown in Figure 32-14, and other evidence of hormonal interaction has been mentioned in this chapter.

All of these hormones are produced at varying rates, and all are also broken down by enzymes produced within the plant body. When it comes to interpreting a particular plant growth response, therefore, it is usually the reinforcing or contradictory actions of two or more hormones that prove to be critical. As a result, it likely will be many years before the interactions of hormones in determining the form of a mature plant will be well understood.

TROPISMS

Tropisms, or responses to external stimuli, control the growth patterns of plants and thus their appearance. Plants adjust to the conditions of their environment by growth responses. Here we will consider three major classes of plant tropisms: phototropism, gravitropism, and thigmotropism.

Phototropism

We have already introduced **phototropism,** the bending of plants toward unidirectional sources of light, in our discussion of the action of auxin (pp. 677–680). In general, stems are positively phototropic, growing toward the light, whereas roots are negatively phototropic, growing away from it. The phototropic reactions of stems clearly are of adaptive value because they allow plants to capture greater amounts of light than would otherwise be possible. They are also of very great importance in determining how the organs of plants develop, and consequently in the appearance of the plant. Individual leaves display phototropic responses, as do whole stems, and the position of these leaves is of great importance to the photosynthetic efficiency of the plant. On the other hand, the phototropic responses of roots are also of adaptive value; they cause roots to grow downward, away from light, and so to reach a supply of water and nutrients. Auxin is certainly involved in most, if not all, of the phototropic growth responses of plants.

> **Phototropisms are growth responses of plants to a unidirectional source of light. They are mostly, if not entirely, mediated by auxin and are of great importance in determining the form of a plant.**

Gravitropism

Another familiar plant response is **gravitropism,** formerly known as **geotropism.** This kind of tropism causes stems to tend to grow upward and roots downward (Figure 33-11); both of these responses clearly are of adaptive significance. Stems that grow upward are apt to receive more light than those that do not; roots that grow downward are more apt to encounter a favorable environment than those that do not. The phenomenon is now called gravitropism because it is clearly a response to gravity and not to the earth (prefix "geo") as such.

In shoots that are placed horizontally, differences in concentrations soon develop between the upper and lower sides for both gibberellins and auxin. These differences cause the growth responses that are responsible for the shoots, positive gravitropism, or growth upward against the force of gravity. Abscisic acid also appears to be involved in this process. In roots, such gradients in hormone concentration have not been as well documented. Nevertheless, the upper sides of the roots that are growing somewhat to the side grow more rapidly than their lower sides.

Accompanying the gradients in hormone concentration are gradients in the concentration of calcium. These gradients are highly significant in regulating plant growth also, because calcium strengthens cell walls and inhibits their growth by making them stiff and rigid. Higher concentrations of calcium occur on the lower sides of roots and the upper sides of shoots. This causes the cells on these sides to elongate less than those on the other sides of the respective organs and is responsible in part for their characteristic growth patterns.

It seems likely that **amyloplasts,** starch-containing chloroplasts, play an important role in the perception of gravity by plants. The amyloplasts are heavy capsules that contain large amounts of calcium and starch. In roots, the cells in which amyloplasts occur are apparently located in the central cells of the root cap; removing the root cap stops the root's responses to gravity in most cases. In shoots, on the other hand, gravity is clearly sensed

FIGURE 33-11

The stems of this fallen tree are growing straight up because they are negatively gravitropic and also because they are positively phototropic.

along the whole length of the stem, probably by the functioning of similar amyloplasts in certain cells. Gravity causes the amyloplasts to fall to the lower side of a given cell. There the amyloplasts may release calcium and set in motion a series of reactions that eventually causes the shoots and roots to bend. The amyloplasts reach the lower side of their cells within a minute, and the bending of the root or shoot may occur within as little as 10 minutes. How the calcium and hormone gradients are set up to direct these relatively rapid movements is poorly understood, although electrical currents may play a role in the redistribution of the Ca^{++} ions.

> **Gravitropism, the response of a plant to gravity, generally causes shoots to grow up and roots to grow down. The force of gravity apparently is sensed in special cells with amyloplasts, starch-containing chloroplasts. In the root, these cells occur in the central zone of the root cap; in the stem, they are more widely distributed.**

Thigmotropism

Still another commonly observed plant response is **thigmotropism,** a name derived from the Greek root *thigma,* meaning "touch." Thigmotropism is defined as the response of plants to touch. The responses by which tendrils curl around and cling to stems or other objects are surprisingly rapid and clear (Figure 33-12); the ways in which twining plants, such as bindweed, coil around objects are analogous. This behavior is the result of rapid growth responses to touch. Specialized groups of cells in the epidermis appear to be concerned with thigmotropic reactions, but again, their exact mode of action is not well understood.

> **Thigmotropisms are growth responses of plants to touch. The way in which such stimuli are perceived is poorly understood.**

FIGURE 33-12

These tendrils coil around the stem because of their positive thigmotropism.

The tendrils of many kinds of plants will coil around an object in the direction from which the tendril is touched, but the tendrils of other plants always coil in the same direction, regardless of which side of the tendril is touched first. Both auxin and ethylene appear to be involved in the movements of tendrils; they can induce coiling in the absence of any contact stimulus.

TURGOR MOVEMENTS

Some kinds of plant movements are based on reversible changes in the turgor pressure of specific cells, rather than on differential growth or patterns of cell enlargement or division. One of the most familiar of these reversible changes has to do with the changing leaf positions that certain plants exhibit at night and in the day (Figure 33-13). For example, the attractively spotted leaves of the prayer plant (*Maranta*) spread horizontally during the day but become more or less vertical at night.

Many other kinds of plants exhibit comparable leaf movements, associated with the bending or straightening of a jointed, multicellular structure at the base of the leaf called the **pulvinus.** Enlarged cells called **motor cells** flank the pulvinus on both sides; when those on one or the other side imbibe water differentially, the leaf moves away from the side on which the most water has been imbibed. Using similar mechanisms, the leaves of certain plants track the sun. Although the ways in which they direct their orientation is not understood, such leaves often can move rapidly—as much as 15 degrees an hour—as a result of the changes in turgor pressure of the pulvini.

The ways in which leaves orient in relation to the sun depend on the kind of plant and its habitat. For example, in annual plants that grow in the desert and must complete their life cycles while water is available and before temperatures become too high, many

A   B

FIGURE 33-13

Wood sorrel (Oxalis) *in the day with leaflets held horizontally* (**A**), *and at night with leaflets held down* (**B**).

kinds of plants orient their leaves at right angles to the sun, which makes rapid growth possible. Plants that are able to orient their leaves in this way are from one and a third to nearly two times as efficient in assimilating carbon as those that cannot do so. In contrast, plants that grow in the hot sun and photosynthesize in the summer, such as wild lettuce, may hold their leaves parallel to the sun. Doing so keeps down the temperatures in the leaves of these plants and thus retards their loss of water but not their rates of photosynthesis. Plants that grow in habitats where it freezes every night, as on mountains in the tropics, may fold their leaves up around their shoot apical meristem and thus protect it from freezing.

> **Turgor movements of plants are reversible and involve changes in the turgor pressure of specific cells. They allow plants to orient their leaves and flowers in ways that allow them to photosynthesize more successfully.**

Especially rapid are the movements of the leaves and leaflets of the sensitive plant *(Mimosa pudica),* which fall into a folded position when they are touched (Figure 33-14). The leaves are able to do so because the motor cells on one side of their pulvini become permeable to potassium and other ions when they are touched or when they are stimulated by the electric currents that result from being touched. When this happens, ions rapidly flow out of the cells, thus equalizing their concentrations on both sides of the plasma membrane. The cells then rapidly lose much of their water content and therefore their turgor, which results in the rapid collapse of the leaf.

Even more spectacular are the rapid movements of the leaves of the carnivorous Venus flytrap *(Dionaea muscipula)* when capturing their animal prey (see Figure 34-16, *A* and *B*). These movements, however, are not caused by changes in turgor pressure; there are no pulvini in these leaves. In the lobes of the leaves are located trigger hairs, under which there are motor cells. When an insect or any other object touches a trigger hair, the leaf snaps shut. The process involves irreversible cell enlargement, which is initiated by a drop in the pH of the cell walls. The leaves close most rapidly at a pH of 3 to 4, a level at which the walls of the motor cells are very flexible. When the leaves are snapping shut, the motor cells expand their own walls rapidly by releasing H^+ ions, expending energy in the form of ATP. During the 1 to 3 seconds required for the leaves to close, about 29% of the ATP in the cells is lost! The ATP apparently is used in the very fast transport of H^+ ions from the motor cells to the outer leaf cells, where the ions acidify the cell walls. This rapid transfer of H^+ ions is, in effect, an electric signal. During closure, the outer surface of the leaf expands rapidly; as the leaf slowly opens, over a period of about 10 hours, its inner surface expands.

The movement of flowers, such as sunflowers, that "follow" the sun is controlled by structures similar to pulvini. By tracking the sun, the sunflower keeps the inside of the flower head warm, creating a favorable environment for its insect visitors, even when the

FIGURE 33-14

Sensitive plant (Mimosa pudica). *The leaves of* Mimosa *are divided into numerous segments, or leaflets. An undisturbed leaf is shown in* **A**; **B** *shows the effects of striking the leaf partway down from its tip.*

A

B

temperature of the surrounding air is relatively cool. For this reason, it is possible for insects to visit the flower heads over a wider range of conditions than would otherwise be possible. The flowers of many species of angiosperms close at night, and their closing is controlled by pulvini also.

PHOTOPERIODISM

Essentially all eukaryotic organisms are affected by the cycle of night and day, and many features of plant growth and development are keyed to the changes in the proportions of light and dark in the daily 24-hour cycle. Such responses constitute **photoperiodism,** a mechanism by which organisms measure seasonal changes in relative day and night length. One of the most obvious of these photoperiodic reactions concerns the production of flowers by angiosperms.

Flowering Responses

Day length changes with the seasons; the farther from the equator one is, the greater the variation. The flowering responses of plants fall into two basic categories in relation to day length. **Short-day plants** begin to form flowers when the days become shorter than a critical length (Figure 33-15). **Long-day plants,** on the other hand, initiate flowers when the days become longer than a certain length.

The critical day lengths for both long-day and short-day plants tend to fall in the 12- to 14-hour range. At middle latitudes, therefore, short-day plants bloom in late summer and autumn when the days have become short enough, whereas long-day plants bloom in spring and early summer when the days have become long enough. Many fall flowers, including chrysanthemums, goldenrods, and poinsettias; important crops, such as soybeans; and weeds, like ragweed, are short-day plants. Conversely, many spring and early-summer flowers, including clover, irises, and hollyhocks, are long-day plants. Commercial plant growers use these responses to day length to bring plants into bloom when they are wanted for sale. Supplementary light is used to "force" (as the process is termed) long-day plants, and artificial protection from light is used to bring short-day plants into bloom.

In addition to the long-day and short-day plants, a number of plants are described as **day neutral.** These plants produce flowers whenever environmental conditions are suitable, without reference to day length. Day-neutral plants include roses, snapdragons, and tomatoes.

> **Short-day plants start to form flowers when the days become shorter than a certain critical length; long-day plants form flowers when the days become longer than a certain length. Day-neutral plants do not have specific day-length requirements for flowering.**

Flowering responses to day length may be important in controlling the distribution of certain plants. For example, in the tropics, where the days may never be long enough to satisfy the requirements of certain long-day plants or short enough to satisfy those of some short-day plants, these plants may simply be unable to flower, and therefore to reproduce.

The Chemical Basis of the Photoperiodic Response

Students of flowering responses made an early discovery that was to prove extremely valuable in their later studies: short-day plants require a certain amount of darkness in each 24-hour cycle to initiate flowering; however, if the period of darkness is interrupted by a brief period of light, often less than a minute, the plants will not flower. Once this discovery had been made, scientists proceeded to determine the particular wavelengths of light that were most effective in inhibiting flowering. This was found to be red light with a wavelength

FIGURE 33-15

Chrysanthemum, a short-day plant. Millions of chrysanthemums are induced to flower by darkening them for part of the 24-hour cycle during times of year when the days are not short enough for them to flower naturally. This plant is being exhibited in Japan.

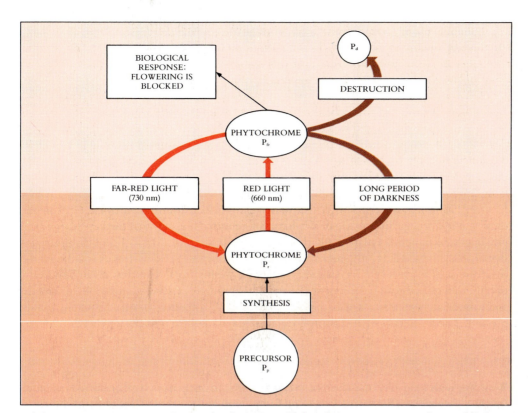

FIGURE 33-16

Phytochrome is synthesized in the P_r form from amino acids, designated P_p for precursor. When exposed to red light, P_r changes to P_{fr}, which is the active form that elicits a response in plants. P_{fr} is converted to P_r when exposed to far-red light, and also in darkness, where it may also be destroyed. The destruction product is designated P_d.

of about 660 nanometers. Curiously, however, if the plants were exposed to red light at 660 nanometers and then to far-red (longer-wavelength red) light at a wavelength of about 730 nanometers, the effect was canceled. What is the chemical basis for this strange observation?

Plants contain a pigment, **phytochrome,** which exists in two interconvertible forms, P_r and P_{fr}. In the first form, phytochrome absorbs red light; in the second, it absorbs far-red light. When a molecule of P_r absorbs a photon of red light, it is instantly converted into a molecule of P_{fr} and when a molecule of P_{fr} absorbs a photon of far-red light, it is quickly converted into P_r. P_{fr} is biologically active; P_r is biologically inactive. In other words, when P_{fr} is present, a given biological reaction that is affected by phytochrome will occur. When most of the P_{fr} has been replaced by P_r, the reaction will not occur.

Phytochrome is a receptor of light and does not itself act directly to bring about the reactions to light. In short-day plants, however, the presence of P_{fr} leads to the suppression of the flowering reaction. The amount of P_{fr} steadily declines in darkness, the molecules being converted to P_r—and when the period of darkness has been long enough, the flowering response is triggered. A single flash of red light at a wavelength of about 660 nanometers, however, will convert most of the molecules of P_r to P_{fr}, and the flowering reaction will be blocked (Figure 33-16). Since in darkness most of the P_{fr} is converted to P_r within 3 to 4 hours, however, the conversion of P_r to P_{fr} cannot be the full explanation for the flowering responses of short-day plants; other factors, still not understood, must also be involved.

The existence of phytochrome was conclusively demonstrated in 1959 by Harry A. Borthwick and his collaborators at the U.S. Department of Agriculture Research Center at Beltsville, Maryland. It has since been shown that the molecule consists of two parts: a smaller one that is sensitive to light and a larger portion that is a protein. The phytochrome pigment is blue, and its light-sensitive portion is similar in structure to the phycobilins that occur in cyanobacteria and red algae. Phytochrome is present in all groups of plants and in a few genera of green algae, but not in bacteria, fungi, or protists other than those few green algae. Therefore it is provisionally assumed that phytochrome systems for measuring light evolved among some of the green algae and were present in the common ancestor of the plants.

Two forms of phytochrome, a large molecule that includes a protein, are interconverted by exposure to red light of different wavelengths. Phytochrome plays an important role in controlling the expression of flowering in short-day-sensitive plants.

Phytochrome is also involved in many other plant growth responses. For example, seed germination is inhibited by far-red light and stimulated by red light in many kinds of plants. Another example is the elongation of the shoot of an **etiolated** seedling, a very slender and colorless one that has been kept in the dark. Such plants become normal when exposed to light, especially red light; but the effects of such exposure are canceled by far-red light, indicating a relationship similar to that observed in seed germination (Figure 33-17).

In 1983 Dina Mandoli and Winslow Briggs, working at the Department of Plant Biology at the Carnegie Institution of Washington, reported findings that further enlarged the possible ways in which a plant could receive light and react to it. Mandoli and Briggs found that an etiolated seedling, in essence, acts like a bundle of fiber optic strands, guiding light over distances as great as 4.5 centimeters through the interiors of dozens, even hundreds, of cells, as well as through the junctions between them. Such a system enables light directly and immediately to affect and begin to coordinate the responses of plant portions below ground. These findings open a promising avenue for research into plant growth.

The Flowering Hormone: Does it Exist?

Working with long-day and short-day plants, some investigators have gathered evidence for the existence of a flowering hormone. It has been shown that plants will not flower in response to day-length stimuli if their leaves have been removed before exposure to the light. However, the presence of a single leaf or the exposure of a single leaf to the appropriate stimuli will, in general, bring about flowering. If the leaf is removed immediately after exposure, the plant will not produce flowers; but if it is left on the plant for a few hours and then removed, flowering will occur normally. These results indicate that a substance passes from the leaves to the apices of the plant and induces flowering. Other experiments have shown that, unlike auxin, the substance cannot be transmitted through agar, but actually requires a connection through living plant parts.

Scientists have searched for a flowering hormone for more than 50 years, but their quest has not yet been successful. A considerable amount of evidence demonstrates the existence of substances that promote flowering and others that inhibit it. These poorly understood substances appear to interact in complex ways. It is probably the complexity of their interactions as well as the fact that multiple chemical messengers are evidently involved that has made this scientifically and commercially interesting matter such a difficult one to resolve.

DORMANCY

Plants respond to their external environment largely by changes in growth rate. As you might imagine, the ability to cease growing altogether when conditions are not favorable is a critical factor in their survival.

In temperate regions, we generally associate dormancy with winter, when low temperatures and the unavailability of water because of freezing make it impossible for plants to grow. During this season, the buds of deciduous trees and shrubs remain dormant, and the apical meristems remain well protected inside enfolding scales. Perennial herbs spend the winter underground as stout stems or roots packed with stored food. Many other kinds of plants, including most annuals, pass the winter as seeds.

In climates that are seasonally dry, dormancy will occur primarily during the dry season, whenever in the year it falls. In dry conditions, plants stay in a dormant condition by using strategies similar to those that the plants of temperate areas rely on in winter.

FIGURE 33-17

Lima bean (Phaseolus lunatus) *seedlings grown in light and darkness, showing the characteristics of an etiolated plant. Seedlings are naturally etiolated when they are growing through the ground, and etiolation may be considered a strategy to concentrate the energy of the plant in such a way that it will reach the surface, and light, as rapidly as possible. When it does so, normal growth commences.*

FIGURE 33-18

Palo verde (Cercidium floridum), *a desert tree with tough seeds that germinate only after they are cracked.*

Annual plants occur frequently only in areas of seasonal drought. Seeds are ideal for allowing annual plants to bypass the dry season when there is insufficient water for growth. When it rains, they can germinate and the plants can grow rapidly to take advantage of the relatively short periods when water is available. In the last chapter, we considered some of the mechanisms involved in breaking seed dormancy and allowing germination under favorable circumstances. These include leaching out from the seed coats chemicals that inhibit germination, or cracking the seed coats mechanically, a procedure that is particularly suitable for promoting growth in seasonally dry areas. Rains will leach out the chemicals in the seed coats whenever they occur, and the hard coats of other seeds may be cracked when they are being washed down along arroyos in temporary floods (Figure 33-18).

Seeds may remain dormant for surprisingly long periods of time. Many legumes (plants of the pea and bean family, Fabaceae—also known as Leguminosae), for example, have tough seeds that are virtually impermeable to water and oxygen and often last decades, possibly even longer, without special care. They will germinate eventually, after their seed coats have been cracked and when water is available. Seeds of the oriental lotus *(Nelumbo nucifera)* have germinated after a minimum of 400 years (see Figure 32-11). A few samples of seeds that are thought to be even older have been successfully germinated, but it is difficult to be certain of their exact age.

A period of cold is necessary before some kinds of seeds will germinate, as we mentioned in the last chapter. Similarly, a period of cold is necessary before the buds of some trees and shrubs will break dormancy and develop normally. For this reason, many of the plants that normally grow in temperate regions do not do well in the tropics—even at high elevations where it may be relatively cool all year—because it does not get cold enough and because the day-length relationships are different from those that occur in temperate regions.

In this chapter, we have considered the ways in which plant hormones are produced by the plant, are influenced by the external environment, and interact with one another to produce the form of a mature plant by controlling its growth responses. The ways in which such a plant functions, drawing water and nutrients from the soil and circulating the products of photosynthesis, will be the subject of the next chapter.

SUMMARY

1. Hormones are chemical substances produced in small quantities in one part of an organism and transported to another part of the organism, where they bring about physiological responses. The tissues in which plant hormones are produced are not specialized particularly for that purpose, nor are there usually clearly defined receptor tissues or organs.

2. There are five major classes of naturally occurring plant hormones: auxins, cytokinins, gibberellins, ethylene, and abscisic acid. They often interact with one another in bringing about growth responses.

3. Auxins are produced at the tips of shoots and diffuse downward, suppressing the growth of lateral buds. In young grass seedlings and other herbs, they play a major role in promoting stem elongation. In such stems, the auxin migrates away from the light and makes the cells on the side to which they have migrated more plastic. By expanding more than the less plastic cells on the other side of the stem, the elongating cells cause the stem to bend away from themselves.

4. Cytokinins are necessary for mitosis and cell division in plants. They promote the growth of lateral buds and inhibit the formation of lateral roots.

5. Gibberellins play the major role in stem elongation in most plants. They also tend to hasten seed germination, to break dormancy in buds, and to cause the stems in the rosettes of biennials to elongate and the plants ultimately to flower.

6. Ethylene is a gas that functions as a plant hormone. Auxin may retard the growth of lateral buds because it stimulates the production of ethylene near them. Ethylene is widely used to hasten fruit ripening.

7. Abscisic acid promotes the abscission of plant parts. It also plays a key role in the opening and closing of stomata.

8. Tropisms in plants are growth responses to external stimuli. A phototropism is a response to light, gravitropism is a response to gravity, and thigmotropism is a response to touch.

9. Turgor movements are reversible but important elements in the adaptation of plants to their environments. By means of turgor movements, leaves, flowers, and other structures of plants track light and take full advantage of it.

10. The flowering responses of plants fall into two basic categories in relation to day length. Short-day plants begin to form flowers when the days become shorter than a given critical length; long-day plants do so when the days become longer than a certain length. In temperate regions, the progression of the seasons stimulates plants of both classes to flower—the short-day plants in the autumn, and the long-day plants in the spring.

11. A blue pigment known as phytochrome is interconverted between two forms by red light of different wavelengths. It plays a role in determining the flowering response and in mediating certain other plant responses.

12. Dormancy is a necessary part of plant adaptation that allows a plant to bypass unfavorable seasons, such as winter, when the water may be frozen, or periods of drought. Dormancy also allows plants to survive in many areas where they would be unable to grow otherwise.

SELF-QUIZ

The following questions pinpoint important information in this chapter. You should be sure to know the answers to each of them.

1. If a grass seedling tip is differentially illuminated with light, which side has a higher concentration of auxin: the illuminated side or the darkened side?

2. Auxin affects cell wall
   - (a) plasticity
   - (b) porosity
   - (c) thickness
   - (d) deposition
   - (e) density

3. Cytokinins, which are derivatives of _____, actively promote the formation of lateral _____.

4. Which of the following plant hormones can cause some dwarf mutants to regain their normal stature?
   - (a) IAA
   - (b) IAB
   - (c) cytokinin
   - (d) auxin
   - (e) ethylene
   - (f) gibberellins

5. Which of the following plant hormones is primarily responsible for senescence?
   - (a) ethylene
   - (b) gibberellins
   - (c) cytokinins
   - (d) auxin
   - (e) abscisic acid

6. Gravitropism in plant roots is sensed by starch-containing chloroplasts called _____, which are located in the central cells of the _____ _____.

7. There is a fungus that grows on plant leaves, called the green island fungus. It is easy to spot in autumn when the leaves are turning yellow, because it causes circular green spots on the leaves it has infected. You might expect this fungus to be producing
   - (a) abscisic acid
   - (b) gibberellins
   - (c) IAA
   - (d) ethylene
   - (e) cytokinins

8. The closing of the leaves of a Venus flytrap
   - (a) requires no energy expenditure
   - (b) is triggered by contact with the motor cells
   - (c) is caused by changes in turgor pressure
   - (d) requires energy in the form of ATP
   - (e) is caused by the action of motor cells on the pulvini

9. Flowering in short-day and day-neutral plants is controlled by the pigment phytochrome. True or false?

10. You find some potatoes sprouting in the dark recesses of your root cellar and expose them to 10 minutes of far-red light. This will cause the sprouts to lose their etiolated appearance. True or false?

11. Pears are picked green and stored in a gaseous atmosphere of nitrogen, _____, and _____, with the concentrations adjusted so that the ripening of the pears is controlled precisely.

12. Phytochrome is converted to the far-red form upon exposure to far-red light. True or false?

THOUGHT QUESTIONS

1. If the length of the days in a particular place, and at a particular time of year, were 10 hours, which would produce flowers: a short-day plant, a long-day plant, both, or neither? Why? Do you think there are any short-day plants in the tropics? Why?

2. How can we be certain that it is the length of the dark period, rather than the length of the light period, that actually determines whether a short-day or a long-day plant will flower? Describe experiments that have led scientists to this conclusion.

3. When poinsettias are kept inside a house after the holiday season, they rarely bloom again. Why do you think this might be, and what might you do to get them to produce flowers a second time?

FOR FURTHER READING

ADDICOTT, F.T.: *Abscission,* University of California Press, Berkeley, 1982. Excellent treatment of the process.

ADDICOTT, F.T. (ed.): *Abscisic Acid,* Praeger Publishers, New York, 1983. Monograph on this important plant hormone.

GALSTON, A.W., P.J. DAVIES, and R.L. SATTER: *The Life of the Green Plant,* 3rd ed., Prentice-Hall, Inc., Englewood Cliffs, N.J., 1980. Good discussion of the functioning and development of the green plant, tied in well with the underlying chemical processes.

MANDOLI, D.F., and W.R. BRIGGS: "Fiber Optics in Plants," *American Scientist,* vol. 251, pages 90-98, 1984. A good summary of the subject.

RAY, P.M.: *The Living Plant,* 2nd ed., Holt, Rinehart & Winston, Inc., New York, 1972. Outstanding short treatment of all aspects of plant growth and development.

SALISBURY, F.B., and C.W. ROSS: *Plant Physiology,* 3rd ed., Wadsworth Publishing Co., Belmont, Calif., 1985. A fine account of the entire subject.

SISLER, E.C., and S.F. YANG: "Ethylene, The Gaseous Plant Hormone," *BioScience,* vol. 33, pages 233-238, 1984. An up-to-date review of this important plant hormone.

TREWAVAS, A.: "How Do Plant Growth Substances Work?" *Plant, Cell, and Environment,* vol. 4, pages 203-228, 1981. An excellent, concise introduction to plant hormones.

WAREING, P.F., and I.D.J. PHILLIPS: *Growth and Differentiation in Plants,* 3rd ed., Pergamon Press, Inc., Elmsford, N.Y., 1981. Excellent treatment of the subject.

Overview

The body of a plant is basically a tube embedded in the ground and extending up into the light, where expanded surfaces—the leaves—capture the sun's energy and participate in gas exchange. The warming of the leaves by sunlight increases evaporation from them, creating a suction that draws water into the plant through the roots and up the plant through the xylem to the leaves. Transport from the leaves and other photosynthetically active structures to the rest of the plant occurs through the phloem. This transport is driven by osmotic pressure; the phloem actively picks up sugars near the places where they are produced and unloads them where they are used. Most of the nutrients critical to plant metabolism are accumulated by the roots, which expend ATP to transport these nutrients actively throughout the plant.

For Review   *Here are some important terms and concepts that you will encounter in this chapter. If you are not familiar with them, you should review them before proceeding.*

- **Adhesion and cohesion** (Chapter 4)

- **Capillary movement** (Chapter 4)

- **Osmosis** (Chapter 9)

- **Transport of ions across membranes** (Chapter 9)

- **Nitrogen chemoautotrophs** (Chapters 10 and 26)

- **Pores in xylem elements** (Chapter 31)

Structurally, a plant is essentially a tube embedded in the ground. At the base of the tube are roots, and at its top are leaves. For a plant to function, two kinds of transport processes must occur. First, the carbohydrate molecules that are produced in the leaves by photosynthesis must be carried to all of the other living cells of the plant. To accomplish this, liquid, with these carbohydrate molecules dissolved in it, must move both up and down the tube. Second, nutrients and water in the ground must be taken up by the roots and ferried to the leaves and other cells of the plant. In this process, liquid must move up the tube.

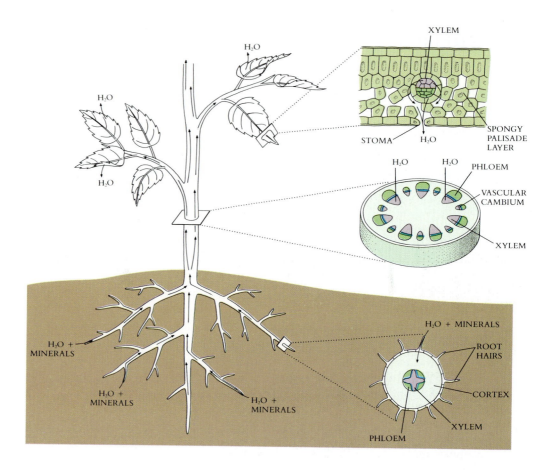

As you saw in Chapter 31, plants accomplish these two processes by using chains of specialized cells; those of the phloem transport photosynthetically produced carbohydrates up and down the tube, and those of the xylem carry water and minerals upward (Figure 34-1). Perhaps you are wondering why plants use the narrow channels in xylem and phloem to transport liquids, instead of large-diameter pipes like our blood vessels, in which the rapid movement of water would be possible. The answer is that we actively pump our blood, whereas plants rely on passive forces to drive the movement of liquids through their bodies. As you will see, these forces depend heavily on the existence of very narrow transport tubes.

In this chapter we will deal with the forces that move water and solutes in plants, with the nutritional requirements of plants, with the ways in which plants conserve water, and with some of the ecological consequence of these relationships. We will begin with some remarks about the soil, the substrate in which nearly all plants grow, and then consider how the water and minerals derived from the soil move into and through the plant.

THE SOIL

Soils are produced by the weathering of rocks in the earth's crust and vary according to the composition of those rocks. The crust includes about 92 naturally occurring elements; the abundance of some of them in the crust is listed in Table 2-1. Most elements are combined into inorganic compounds called **minerals;** most rocks consist of several different minerals. There are three groups of rocks:

1. **Igneous rocks.** Igneous rocks are derived directly from molten material, which reaches the surface only by way of volcanoes, vents, or other passageways that extend from the earth's interior (Figure 34-2). Examples are granite and basalt.

FIGURE 34-2

Volcanoes, like this one in El Salvador, bring up supplies of nutrient-rich materials, which form soils that are highly favorable for plant growth.

FIGURE 34-3

Soil organisms. About 5 metric tons of carbon are tied up in the organisms that are present in the soil under a hectare of wheat land in England, an amount that approximately equals the weight of 100 sheep!

2. **Sedimentary rocks.** Sedimentary rocks consist of broken elements from preexisting rocks that have been reworked and redeposited by water, wind, glaciers, or other forces. Shale, sandstone, and limestone are examples.
3. **Metamorphic rocks.** Metamorphic rocks are igneous or sedimentary rocks that have been transformed by the extreme heat and pressure deep within the earth. Examples are quartzite, which is formed from sandstone, and marble, which is formed from limestone.

Millions of years of weathering of rocks have produced the particles from which soils are formed. All soils on earth show unmistakable signs of the biological processes that have been taking place here for more than 3.5 billion years. These signs are both direct (shown in the structure of soils and their load of organic materials) and indirect. For example, the atmosphere itself, which plays the major role in weathering rocks on earth, has been changed profoundly over time by the effects of biological evolution. Biological effects are obvious not only in the topsoil, where organisms and organic debris are abundant and diverse (Figure 34-3), but also lower down. The ways in which this organic debris is broken down and recycled, mainly as a result of the activities of organisms living in the soil, have a great deal to do with the fertility of the soil and play a key role in controlling the cycling of substances through it.

> **Rocks, which may be igneous, sedimentary, or metamorphic, are made up of one or more kinds of minerals. They weather to give rise to soils, which differ according to the composition of their parent rocks. A large and active organic component in soils greatly affects their fertility and other properties.**

Soil is made up of particles of varying size, from coarse sand, with particles that are 200 to 2000 micrometers in diameter, to clay, with particles that are less than 2 micrometers in diameter. The particles in fine sand and silt are intermediate in size between these two extremes, and most soils are a mixture of particles of different sizes.

About half of the total soil volume is occupied by empty space, which may be filled with air or water depending on moisture conditions. Not all of the water in soil, however, is available to plants, because of the nature of water itself.

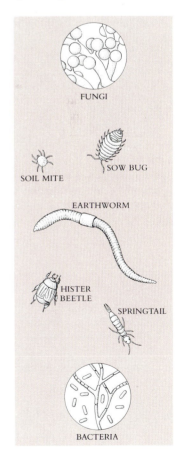

FUNGI

SOIL MITE

SOW BUG

EARTHWORM

HISTER BEETLE

SPRINGTAIL

BACTERIA

You will recall from Chapter 4 that the chemical and physical behavior of water is dictated largely by its tendency to form hydrogen bonds with itself and other materials. Two aspects of the behavior of water are particularly important here. First, water is tightly bound to objects that bear hydrostatic changes and to those which are able to form hydrogen bonds. Many soils contain minerals that have exactly these properties. Clays, for example, are often negatively charged, and water is tightly bound to clay particles by hydrogen bonds.

Second, the greater the surface area of the soil particles, the more water will adhere to them. Far more water will adhere to soils with very small particles than to soils with large ones. Thus water drains rapidly through sandy soils, which consist of relatively coarse particles. Clay soils may hold a great deal more water, but much of this water may be held so tightly that plants cannot extract it from the soil. For these reasons, as well as for those concerned with the availability of nutrients, a soil composed of a balanced mixture of coarse and fine particles—of sand, silt, and clay—is ideal for plant growth. Such soils are known as **loams.**

Water is tightly bound to objects that bear hydrostatic changes and to those which form hydrogen bonds. It also adheres more tightly to objects that have relatively large surface areas. For these reasons, water runs through sandy soils and is so tightly bound to clay soils that it is often unavailable to plants.

Some of the water that reaches a given soil will drain through it immediately, because of the force of gravity. Another fraction of the water is held in the smaller soil pores, which are generally less than about 50 micrometers in diameter. Such water remains in the soil and is readily available to plants; it is called **capillary water.** The amount of water held in a given soil after gravity has removed the excess is defined as the **field capacity** of that soil. When evaporation and removal by plants have eliminated the capillary water from the soil, leaving only water that is not available to plants, that soil is said to have reached its **permanent wilting point.** The permanent wilting point is defined as the moisture content of a given soil at which plants will wilt permanently unless additional water is added.

The permanent wilting point of a soil is the moisture content of that soil at which plants will wilt permanently unless additional water is added.

WATER MOVEMENT

It is not unusual for many of the leaves of a large tree to be more than 10 stories off the ground. Did you ever wonder how a tree manages to raise water so high? To understand how this happens, imagine filling a long, hollow tube with water, closing it at one end, and placing the tube, open end down, in a full bucket of water. Gravity affects the water in the tube in two ways. First, the weight of the water tends to move it down the tube. Second, gravity acts on the column of air over the bucket, and the air's weight (at sea level) exerts an amount of pressure that is defined as 1 atmosphere downward on the water in the bucket and thus up into the tube.

These two effects counteract each other, and there is a limit to how high the tube can be and still be filled with water. That limit is set by the pressure exerted by the atmosphere. At sea level, this pressure is sufficient to raise the water in a tube to a height of about 10.4 meters. The weight of the water in a tube any taller than about 10.4 meters pulls itself down, leaving a vacuum at the upper, closed end of the tube that fills with water vapor. The water is said to **cavitate** (Figure 34-4).

Could adhesion of water to the sides of small conducting vessels such as those in the xylem or phloem create capillary movement of water up the plant? No. The capillary forces are not strong enough. Water is lifted less than 1 meter in a glass tube the diameter of a xylem element. How then does water get to the top of the tree?

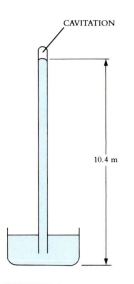

CAVITATION

10.4 m

FIGURE 34-4

Closed tube in a bucket, with water rising 10.4 meters, and cavitation at the upper end of the tube.

The answer was suggested by Otto Renner in Germany in 1911. To understand his suggestion, we will return for a moment to our analogy of a plant as an open tube embedded in soil. Imagine that water fills the tube to a certain height, and, for simplicity, further assume that the water is held at that height by capillary action—although we will later identify the true cause as something else. Now blow across the top of the tube. The stream of relatively dry air will cause water molecules to evaporate from the surface of the water in the tube. Does the level of water in the tube fall? No. As water molecules are drawn from the top, they are replenished by new ones that are taken in from the bottom. This, in essence, is what Renner proposed actually happens in plants. The passage of air across leaf surfaces results in the loss of water by evaporation, creating a suction force at the open upper end of the "tube," while water is pushed up from below by atmospheric pressure.

Water rises in a plant beyond the point at which it would be supported by atmospheric pressure (10.4 meters at sea level) because evaporation from its leaves produces a suction that pulls up on the entire water column all the way down to the roots.

But what of cavitation? Why does a column of water in a tree more than 10.4 meters tall not cavitate simply because of its weight? The answer is that water has an inherent tensile strength that arises from the cohesion of its molecules, their tendency to form hydrogen bonds with one other. The tensile strength of a column of water varies inversely with the diameter of the column; that is, the smaller the diameter of the column, the greater the tensile strength. Therefore plants must have very narrow transporting vessels in order to take advantage of the tensile strength of water.

Cavitation can occur locally within individual conducting cells in the xylem because of deformation or freezing. If it does, it will result in the formation of small bubbles of air within the cell. You will recall from Chapter 31 that individual vessel elements and tracheids are connected by one or more pores at their end walls, rather than simply being open and directly connected, like the segments of a sewer pipe. The bubbles formed by cavitation are larger than the pores, so they cannot pass through them. Furthermore, the cohesive force of water is so great that the bubbles are forced into rigid spheres. These spheres are unable to change their shape and thereby deform and squeeze through the pores of the vessel element. Because vessel elements and tracheids are connected by these fine-diameter pores, any cavitation that does occur is limited to the cells where it begins, and water may continue to rise in parallel elements.

When cavitation occurs within a column of water in a plant, the bubbles formed are held as rigid spheres by the cohesive force of the water and are unable to pass through the pores between the vessel elements or tracheids. For this reason they do not interrupt the passage of water upward in the plant.

Plant biologists often discuss the forces that act on water within a plant in terms of the water's potential energy. Water high in the plant has more potential energy than water at ground level, the difference being that more energy is required to oppose gravity and hold the water up there. In general, water in a plant may possess potential energy for two reasons: (1) **pressure potential,** or pressure exerted by the atmosphere, as we have discussed above; and (2) water movements driven by diffusion forces. Thus water tends to move by osmosis into a root because the root cells contain many metabolites and other solutes. For this reason the root is said to have a greater **solute potential** than the fluid outside: solutes are more concentrated in it. The algebraic sum of the pressure potential and the solute potential represents the total potential energy of the water in the plant and is called its **water potential.** In these terms water rises in a plant because it is driven by a water potential that is created by the positive pressure of the atmosphere and the negative pressure (pull) that is caused by the evaporation of water from the leaves.

Water rises in a plant because it is driven by a water potential that is created by the positive pressure of the atmosphere and the negative pressure caused by the evaporation of water from the leaves.

The water potential in a plant, and thus the movement of water through the plant, depends on its osmotic absorption by the roots, the positive pressures driving its movement through the xylem, and the negative pressures created by its transpiration from the leaves and other plant surfaces. These three processes are closely linked together, and, as you will see, the negative pressure generated by transpiration is largely responsible for the operation of the other two processes.

Transpiration

More than 90% of the water that is taken in by the roots of a plant is ultimately lost to the atmosphere, almost all of it from the leaves. It passes out primarily through the stomata in the form of water vapor. The process by which water leaves the plant is known as **transpiration.** First, the water passes into the pockets of air within the leaf from the walls of the spongy mesophyll cells that line the intercellular spaces (Figure 31-17). As you saw in Chapter 31, these intercellular spaces open to the outside of the leaf by way of the stomata. The water that evaporates from the surfaces of the spongy mesophyll cells that line the intercellular spaces is continuously replenished from the tips of the veinlets in the leaves. Since the strands of xylem conduct water within the plant in an unbroken stream all the way from the roots to the leaves, when a portion of the water vapor in the intercellular spaces passes out through the stomata, the supply of water vapor in these spaces is continually renewed.

Because they are constantly losing so much water to the atmosphere, and because the presence of this water is essential to their metabolic activities, growing plants depend on the continuous stream of water entering and leaving their bodies at all times. Water must always be available to their roots. Such structural features as the stomata, the cuticle, and the substomatal spaces have evolved in response to two contradictory requirements: to minimize the loss of water to the atmosphere, on the one hand, and to admit carbon dioxide into the plant, on the other. Before considering how plants resolve this problem, however, we must consider the absorption of water by the roots.

The Absorption of Water by Roots

Most of the water absorbed by the plant comes in through the root hairs, which collectively have an enormous surface area (Figure 31-3). Water enters the root hairs, which are always turgid, because of their greater solute potential. Once it is inside the roots, water passes inward to the conducting elements of the xylem, either directly through the cells themselves by way of the plasmodesmata (see Chapter 6) or through and between the cell walls. On its journey inward, the water eventually reaches the endodermis. When it finally passes through this layer, it must, because of the presence of the waxy Casparian strips, which we discussed in Chapter 31, go directly through the protoplasts of the endodermal cells.

Not only water enters the roots by passing into the cells of the root hairs. The membranes of these cells contain a variety of ion transport channels, which actively pump specific ions into them even against large concentration gradients. These ions, many of which are plant nutrients, are then transported throughout the plant as a component of the water flowing through the xylem. At night, when the relative humidity may approach 100%, there may be no transpiration from the leaves. Under these circumstances, the negative pressure component of the water potential—suction caused by evaporation—becomes very small or nonexistent. At such times, water does not travel upward in the xylem.

Active transport of ions into the roots, however, continues to take place under these circumstances. It results in an increasingly high ion concentration within the cells, which causes water to be drawn into the root hair cells by osmosis. In terms of water potential, we say that active transport increases the solute potential of the roots. The result is the movement of water into the plant and up the xylem columns, caused by the phenomenon called **root pressure**. In reality, root pressure is an osmotic phenomenon that results from the lack of transpiration-driven water movement from the roots. Thus root pressure is primarily a nighttime activity that disappears when transpiration resumes the next morning.

> **Root pressure, which is active primarily at night, is caused by the continued, active accumulation of ions by the roots of a plant at times when transpiration from the leaves is very low or absent.**

Under certain circumstances, root pressure is so strong that water will ooze out of a cut plant stem for hours or even days or will be forced up into a glass tube attached to the end of the stump. When root pressure is very high, it may force water up to the leaves, from which the water may be lost in a liquid form through a process known as **guttation** (Figure 34-5). Guttation takes place not through the stomata, but through special groups of cells that are located near the ends of small veins that function only in this process. Root pressure is never sufficient to push water up great distances, and guttation thus takes place only in relatively short plants.

Water Movement in Plants

Because of the interactions of the forces that we have just considered, water is transported to the tops of tall trees and to the uppermost branches of vines. The tension that arises within the xylem of trees during periods of active transpiration may create such a strong pull on the conducting cells that their walls may be pulled closer together, and the diameter of the tree trunk as a whole may be measurably less during the day than it is at night. Negative water potential occurs in columns of dead vessel elements or tracheids, as the newer and still living xylem elements draw water laterally from them. This negative water potential created in the column of dead xylem cells becomes better established as the season progresses from spring to autumn.

Each morning, the sun illuminates and warms the leaves and small branches of a tree, and this warming leads to the increased evaporation of water from them. This increased evaporation creates a pull at the upper end of the water column, in the veinlets of the leaves; this pull is transferred to the water column in the small branches and through them to that in the main trunk, ultimately creating a force that pulls water into the roots. As the sun sets and transpiration from the leaves decreases, water begins to accumulate first in the upper portions of the plant. In the movement of water through the plant, therefore, the sun is the ultimate source of potential energy. The water potential that is responsible for water movement is the product largely of the negative pressure generated by transpiration, which is driven by the warming effects of sunlight.

The Regulation of Transpiration Rate

The only way that plants can control water loss on a short-term basis is to close their stomata. Many plants can do this when they are subjected to water stress. The stomata must be open at least part of the time, however, so that CO_2, which is necessary for photosynthesis, can enter the plant. In its pattern of opening or closing its stomata, a plant must respond both to the need to conserve water and the need to admit CO_2.

When CO_2 has entered the intercellular spaces, it must dissolve before it can enter the plant's cells. The gas dissolves mainly on the walls of the intercellular spaces below the stomata. These walls are kept moist by the continuous stream of water that reaches the leaves from the roots.

The stomata open and close because of changes in the water pressure of their guard cells (Figure 34-6). The guard cells are the only epidermal cells with chloroplasts, and they stand out for this reason and because of their distinctive shape—thicker on the side next to the stomata opening and thinner on their other sides and ends. When the guard cells are

FIGURE 34-5

Guttation, which occurs only in herbaceous plants. The water passes through specialized groups of cells at the edges of the leaves; it is visible here as small droplets around the edge of the leaf of this strawberry plant (Fragaria ananassa).

turgid, or plump and swollen with water, they become bowed in shape, as do the thick inner walls, thus opening the stomata as wide as possible.

The guard cells use ATP-powered ion transport channels through their membranes to concentrate ions actively. This concentration creates a solute potential within the guard cells that causes water to enter osmotically. As a result, these cells accumulate water and become turgid, opening the stomata. The guard cells remain turgid only so long as the active transport channels pump ions into the cells and so maintain the higher solute concentration there. Thus keeping the stomata open requires a constant expenditure of ATP. When the active transport of ions into the guard cells ceases, the higher concentration of ions within the guard cells causes ions to move by diffusion out into the surrounding cells. This diffusion reduces the solute potential; water leaves the guard cells, which become flaccid; and the stomata between them close.

Stomata open when their guard cells become turgid. Their inner surfaces are thickest, and they bow inwardly when the pressure within the cells is high. Keeping the guard cells turgid requires a constant expenditure of ATP.

The ion most important in controlling the turgidity of guard cells is the potassium ion, large amounts of which are held in the cells surrounding the guard cells. The gradient of K^+ between the guard cells and these surrounding cells is capable of changing rapidly in certain circumstances, K^+ moving from one to the other. When such a change occurs, the stomata may open or close rapidly. Apparently photosynthesis in the guard cells provides an immediate source of ATP, which drives the active transport of K^+ ions into and out of the guard cells. In some species Cl^- ions accompany the K^+ ions into and out of the guard cells, thus maintaining electrical neutrality. In many other species, H^+ ions move in the direction opposite to the K^+ ions.

Potassium ions, which are abundant in the cells surrounding the guard cells, pass into the guard cells in large quantities when these cells become turgid and pass out when they become flaccid. In this way the opening and closing of the stomata is controlled.

When water is so scarce that the whole plant wilts, the guard cells may become limp and the stomata may close as a result. The guard cells of many plant species, however, regularly become turgid in the morning, when photosynthesis is possible, and flaccid in the evening, regardless of the availability of water. When they are turgid, the stomata open, and CO_2 enters freely; when they are flaccid, CO_2 is largely excluded, but water loss is retarded also.

Abscisic acid plays a primary role in allowing the K^+ ions to pass rapidly out of the guard cells and therefore in causing the stomata to close. This hormone is produced by leaf tissue under water stress, and it brings about a direct response from the guard cells, binding to specific receptor sites in their plasma membranes. Perhaps abscisic acid will prove useful in the conservation of water in crops grown in dry regions, provided that enough CO_2 would still be able to reach the mesophyll of the plants whose stomata had been closed by applying the hormone.

Factors such as CO_2 concentration, light, and temperature can also affect stomatal opening. When CO_2 concentrations are high, the guard cells of many plant species become flaccid, and their stomata close. At such times the plant has no need to bring in additional CO_2 and conserves water by closing its guard cells. When the temperature exceeds 30° to 34° C, the stomata also close. In the dark the stomata will open at low concentrations of CO_2. In Chapter 31 we mentioned CAM photosynthesis, which occurs in some kinds of succulent plants such as cacti. In this process CO_2 is taken in at night and fixed during the day. CAM photosynthesis conserves water in dry environments where succulent plants grow.

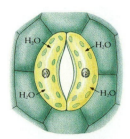

A GUARD CELLS OPEN

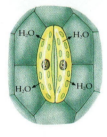

B GUARD CELLS CLOSED

FIGURE 34-6

A *The thick inner side of each of the two guard cells that make up a stoma bow outwardly, opening the stoma, when solute pressure is high (water pressure is low) within the guard cells.*
B *When the solute pressure is low (water pressure is high) within the guard cells, they become flaccid and close the stoma.*

Many mechanisms have evolved in plants by which they regulate their rate of water loss. One strategy involves dormancy at times of the year when water is in short supply. The deciduous habit, which is so common in plants that grow in areas that experience a severe drought at some season of the year, is perhaps the most familiar of these; one way in which a plant can avoid water loss is to lose its leaves. When water is locked up in ice and snow, plants experience drought and thus are often deciduous in regions with severe winters. Annual plants conserve water, in a sense, by simply not being present, except as seeds, when conditions are unfavorable.

Very thick, hard leaves, often with relatively few stomata—and frequently with stomata on the lower side of the leaf only—lost water much more slowly than large, soft leaves with abundant stomata. Leaves covered with masses of woolly-looking trichomes retain a more humid layer of air near their surfaces than average; an even more important factor in slowing down the loss of water vapor to the air is that the trichome layers hold down leaf temperatures greatly (Figure 31-13, C). Similarly, plants that live in arid or semiarid habitats often have their stomata in crypts or pits in the leaf surface. Within these depressions the water content of the air may be high, therefore reducing the rate of water loss.

NUTRIENT MOVEMENT

Apparently most of the movement of ions into a plant takes place through the protoplasts of the cells rather than through their walls. Ion passage through cell membranes seems to be active and carrier mediated, although the details are far from well understood. We do know, however, that the initial movement of nutrients into the roots is an active process that requires energy and that, as a result, specific ions can be maintained within the plant at very different concentrations from those in which they exist in the soil. When roots are deprived of oxygen, they lose their ability to absorb ions, a definite indication that they require energy for this process to occur successfully. A starving plant—one from which light has been excluded—will eventually exhaust its nutrient supply and be unable to replace it.

Once the ions reach the xylem, they are distributed rapidly throughout the plant, eventually reaching all metabolically active parts and being used in diverse ways. Ultimately the ions are removed from the roots and relocated to other parts of the plant, their passage taking place in the xylem, where phosphorus, potassium, nitrogen, and sometimes iron may be abundant in certain seasons. In this way a plant is often able to retain the ions that it has accumulated instead of losing them along with the leaves and twigs that it is constantly shedding as it grows. In contrast to many such ions, however, calcium is not removed once it has been deposited for the first time.

> **The accumulation of ions by plants is an active process that usually takes place against a concentration gradient and requires the expenditure of energy.**

THE MOVEMENT OF CARBOHYDRATES

Most of the carbohydrates that are manufactured in the leaves and other green parts of the plant are moved through the phloem to other parts of the plant. This process, known as **translocation,** is responsible for the availability of suitable carbohydrate building blocks at the actively growing regions of the plant. The carbohydrates that are concentrated in storage organs such as tubers, often in the form of starch, are also converted into transportable molecules, such as sucrose, and moved through the phloem. The pathway that sugars and other substances travel within the plant has been demonstrated precisely by using radioactive tracers, and the route is fairly well understood, despite the fact that living phloem is delicate and the process of transport within it is easily disturbed.

Aphids, a group of insects that suck the sap of plants, have been valuable tools in understanding translocation. Aphids thrust their piercing mouthparts into the phloem cells

PLANT RESPONSES TO FLOODING

The continued existence and growth of plants depends on an adequate supply of water. Plants lack closed circulation systems, and only the continuous stream of water that flows through them keeps them healthy.

Even plants, however, can receive too much water. This occurs when the soil is flooded, a condition that can arise when rivers or streams overflow their banks or when rainfall is heavy, irrigation is excessive, or drainage is poor. Flooding rapidly depletes the available oxygen in the soil and blocks the normal reactions that take place in roots and make possible the transport of minerals and carbohydrates. Abnormal growth patterns may result, and the plants may ultimately "drown." Hormone levels change in flooded plants—ethylene, for example, often increases, while gibberellins and cytokinins usually decrease—and these changes may also contribute to the abnormal growth patterns. Flooding involving moving water, which brings in new supplies of oxygen, is much less harmful than flooding involving standing water, which does not; flooding that occurs when a plant is dormant is much less harmful than flooding when it is growing actively.

Physical changes that may occur in the roots as a result of oxygen deprivation may halt the flow of water through the plant, paradoxically drying out the leaves—even though the roots of the same plant may be standing in water. Because of such stresses,

1

the stomata of flooded plants often close. In some plants the closing of the stomata maintains the turgor of the leaves.

Many plants, of course, grow in places that are often flooded naturally; they have adapted to these conditions during the course of their evolution (Figure 34-A, *1*). One of the most frequent adaptations among such plants is the formation of

---

FIGURE 34-7

Aphids like this individual of Schizaphis graminarum, *shown here on the edge of a wheat leaf, feed on the food-rich contents of the phloem, which they suck out with their piercing mouthparts.*

of leaves and stems to obtain the abundant sugars there (Figure 34-7). When the aphids are decapitated, the liquid continues to flow from the detached mouthparts and is thus available in pure form for analysis. The liquid in the phloem contains 10% to 25% dry matter, almost all of which is sucrose. Using aphids to obtain the critical samples and radioactive tracers to mark them, it has been possible to demonstrate that the movement of substances in the phloem can be remarkably fast; rates of 50 to 100 centimeters per hour have been measured.

The materials transported in the phloem move as a result of hydrostatic pressure, which develops as a result of osmosis (Figure 34-8). First, sucrose, which is produced as a result of photosynthesis, is actively **loaded** into the phloem tubes of the veinlets. This process increases the solute potential of the sieve tubes, so water passes into them by osmosis. An area where the sucrose is made is called a **source;** an area where it is being taken from the sieve tubes is called a **sink.** Sinks include the roots and other regions where the sucrose is being **unloaded.** There, the solute potential of the sieve tubes is decreased as the sucrose is removed. As a result of these processes, water moves in the sieve tubes from the areas where sucrose is being taken in to those areas where it is being withdrawn, and the sucrose moves passively with it.

aerenchyma, loose parenchymal tissue with large air spaces in it. Aerenchyma is very prominent in water lilies and many other aquatic plants. Oxygen may be transported from the parts of plants above the water to the those below by way of passages in the aerenchyma. This supply of oxygen allows oxidative respiration to take place even in the submerged portions of the plant. Some plants normally form aerenchyma, whereas others, subject to periodic flooding, can form it when necessary. In corn, ethylene, which becomes abundant under the anaerobic conditions of flooding, induces aerenchyma formation. Plants also respond to flooded conditions by forming larger lenticels, which facilitate gas exchange, and additional adventitious roots.

Plants such as mangroves (Figure 34-A, *2*, and 34-A, *3*), which are normally flooded with salt water, must not only provide a supply of oxygen for their submerged parts, but also control their salt balance. The salt must be excluded, actively secreted, or diluted as it enters. The arching stilt roots of mangroves are connected to long, spongy, air-filled roots that emerge above the mud. These spongy air roots have large lenticels on their abovewater portions through which oxygen enters; it is then transported to the submerged roots. In addition, the succulent leaves of mangroves contain large quantities of water, which dilutes the salt that reaches them. Many plants that grow in such conditions secrete large quantities of salt through more or less specialized glands.

FIGURE 34-A

Adaptations of plants to flooded conditions.
1 *The "knees" of the bald cypress (*Taxodium*) are formed wherever it grows in wet conditions.*
2 *White mangrove (*Rhizophora*), with its stilt roots, is a familiar sight along shores throughout the world's tropics and subtropics; this scene is in Honduras.*
3 *The air roots of black mangrove (*Avicennia nitida*), like those of white mangrove, bring oxygen to the roots of the plants. Although they have evolved a similar appearance, white mangroves and black mangroves are totally unrelated plants—an example of convergent evolution in a common habitat.*

The transport of water, sucrose, and other substances through the sieve tubes does not require energy, but the loading and unloading of these substances into and from the sieve tubes does. This energy is supplied in the form of ATP by the companion cells, if they are present, or other parenchyma cells, if they are not, through plasmodesmata into the sieve tubes. As you will recall from Chapter 31, sieve cells and sieve-tube elements lack nuclei at maturity and are not directly involved in providing this energy. However, the manufacture of sugar in some regions of the plant and its use in others make possible the mass flow of solutes through the phloem at such rapid rates.

Water moves through the phloem as a result of decreased water potential in areas of active photosynthesis, where sucrose is actively being loaded into the sieve tubes, and increased water potential in those areas where sucrose is being unloaded. Energy for the loading and unloading of the sucrose and other molecules is supplied by companion cells or other parenchyma cells. However, the movement of water and dissolved nutrients within the sieve tubes is a passive process that does not require the expenditure of energy.

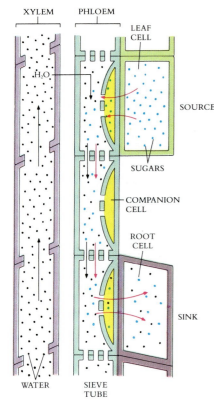

XYLEM PHLOEM

LEAF CELL

H₂O

SOURCE

SUGARS

COMPANION CELL

ROOT CELL

SINK

WATER

SIEVE TUBE

FIGURE 34-8

Diagram of mass flow. In this diagram, sucrose molecules are indicated by colored dots, water molecules by black ones. Moving from the parenchyma cells of a leaf or another part of the plant into the conducting cells of the phloem, the sucrose molecules are then transported to other parts of the plant by mass flow and unloaded where they are required.

PLANT NUTRIENTS

Plants require a number of inorganic nutrients. Some of these are **macronutrients,** which the plants need in relatively large amounts, and others are **micronutrients,** those required in trace amounts. There are nine macronutrients: carbon, hydrogen, and oxygen—the three elements found in all organic compounds—as well as nitrogen, potassium, calcium, phosphorus, magnesium, and sulfur. Each of these nutrients approaches or exceeds 1% of the dry weight of a healthy plant. The seven micronutrient elements that constitute from less than one to several hundred parts per million by dry weight in most plants are iron, chlorine, copper, manganese, zinc, molybdenum, and boron.

> **The macronutrients—substances required by plants in relatively large amounts—are carbon, hydrogen, oxygen, nitrogen, potassium, calcium, phosphorus, magnesium, and sulfur. Each of these approaches or exceeds 1% of a plant's dry weight. Micronutrients, which plants require in only trace amounts of one to several hundred parts per million, are iron, chlorine, copper, manganese, zinc, molybdenum, and boron.**

The macronutrients were, in general, discovered in the last century, but the micronutrients were detected only more recently because plants need them in such small quantities. For studying their nutritional requirements, plants are usually grown in **hydroponic culture;** their roots are suspended in aerated water containing nutrients. Most plants grow satisfactorily in hydroponic culture, and the method is occasionally practical for commercial purposes, although expensive.

The solutions in which test plants are grown contain all of the necessary nutrients in the right proportions but with certain known or suspected nutrients left out. The plants are then allowed to grow and are studied for the presence of abnormal symptoms that might indicate a need for the missing element (Figure 34-9). However, the water or vessels used often contained enough micronutrients to allow the plants to grow normally, even though these substances were not added deliberately to the solutions. To give an idea what small

FIGURE 34-9

Mineral deficiencies in plants, as demonstrated in individuals of Marglobe tomatoes (Lycopersicon esculentum) grown in hydroponic solution—water to which nutrients have been added.
A *Control (healthy) plant, grown in a complete nutrient solution (one that contains all of the essential minerals), with a plant showing the symptoms of calcium deficiency. Calcium was not added to the nutrient solution of the plant at right, resulting in its unhealthy, dwarfed appearance.*
B *Leaf of a healthy control plant.*
C *Chlorine-deficient plant, with curled and necrotic leaves (leaves with patches of dead tissue).*
D *Copper-deficient plant, with blue-green, curled leaves. Copper is required only in very small amounts for normal plant growth.*
E *Zinc-deficient plant, with small, necrotic leaves.*
F *Manganese-deficient plant, with chlorosis (yellowing) between the veins. This chlorosis is caused by a drastic reduction in chlorophyll production. The agricultural implications of these deficiencies are obvious; a trained observer can determine the nutrient deficiencies that are affecting a plant simply by inspecting it.*

A

B

C

quantities of micronutrients plants may require, the standard dose of molybdenum added to seriously deficient soils in Australia amounts to about 34 grams (about one handful) per hectare, once every 10 years!

There are many ways in which the six macronutrients other than carbon, hydrogen, and oxygen and the seven micronutrients are involved in plant metabolism. As we have just discussed, potassium ions regulate the turgor pressure of guard cells and therefore the rate at which the plant loses water and takes in carbon dioxide. Calcium is an essential component of the middle lamellae that are laid down between plant cell walls, and it also helps to maintain the physical integrity of membranes. Magnesium is a part of the chlorophyll molecule. The presence of phosphorus in many key biological molecules, such as nucleic acids and ATP, has been explored in detail in the earlier chapters of this book. Nitrogen is an essential part of amino acids and proteins, of chlorophyll, and of the nucleotides that make up nucleic acids.

A number of ions are components of enzyme systems and serve as cofactors in essential biochemical reactions. Potassium, for example, which reaches higher concentrations in plants than any other elements except carbon and oxygen, affects the conformation of many proteins and probably affects at least 60 different enzymes while they are functioning. Zinc appears to play a similar role in the synthesis of auxin: plants with an inadequate supply of zinc display symptoms that derive mainly from a lack of cell elongation, apparently reflecting a shortage of auxin.

Some kinds of plants have specific nutritional requirements that are not shared by others. Silica, for example, is essential for the growth of many grasses—it helps to retard their complete destruction by herbivores (Figure 34–10)—but not for plants in general. Cobalt is necessary for the normal growth of the nitrogen-fixing bacteria associated with the nodules of legumes and is therefore an essential element for the normal growth of these plants. Nickel seems to be essential for soybeans, and its role in the nutrition of other plants should be investigated further.

In general, the elements that animals require reach them through plants, which therefore form an indispensable link between animals and the reservoirs of chemicals in nature. Some of the elements that animals require, such as iodine, come by way of plants but are not required by the plants. Iodine is very rare in soils; a shortage of iodine in the human diet can lead to the condition known as goiter. Selenium, which might be an essential element for at least some plants, can sometimes be toxic to animals. Plants also may concentrate minerals of economic interest, such as gold, and they are sometimes used as an assay for the presence of these minerals in a particular region.

FIGURE 34–10

These zebras graze selectively on the grass species that make up this East African savanna, depending in part on the degree to which their leaves are reinforced with silica. In general, the more silica in the leaf cells, the less likely zebras and other grazing animals are to eat that particular kind of grass. Silicon, an essential nutrient for plants of the grass family, is used in constructing their leaf cells.

NUTRIENT CYCLES

Although plants obtain carbon primarily from the atmosphere in the form of carbon dioxide, all of the other plant nutrients come in through the roots with water. Nitrogen is fixed and becomes available to the plant in the soil in the form of ammonium ion (NH_4^+) or nitrite

D   E   F

(NO_2^-). In other words, the reservoirs of all of the elements that a plant uses except for carbon, hydrogen, nitrogen, and oxygen are minerals present in the soil. Some of these, such as phosphorus, are often present in short supply. The cations (positive ions) in the soil, including Ca^{++}, K^+, and Na^+, are usually bound to clay particles, which, as we discussed at the beginning of this chapter, are very small. If these cations are abundant enough in the parent rock from which the soil has weathered, they are usually plentiful and easily available for plant growth. They steadily leach off the clay particles and are then absorbed by the plant through its root hairs.

The principal anions (negative ions) in the soil, NO_3^-, $SO_4^=$, HCO_3^-, and OH^-, are leached out from the soil much more rapidly than cations are because they do not become attached to soil particles (which are themselves negatively charged). For this reason, in particular, the supply of nitrogen is often so limited that healthy plant growth is not possible. The phosphate ion, $PO_4^=$, unlike the anions mentioned above, forms insoluble precipitates and is held tightly by certain compounds containing iron, calcium, and aluminum that are abundant in some soils. Phosphorus is often present in very short supply.

Plants obtain carbon from carbon dioxide in the atmosphere, oxygen and hydrogen primarily from the water they take in through their roots, nitrogen from mainly biologically fixed ions present in the soil, and their other nutrients from the soil.

The Phosphorus Cycle

Phosphorus is, more often than any of the other required plant nutrients except nitrogen, apt to be so scarce that it limits plant growth. It is particularly rare in old, weathered soils—those of much of Australia, for example. Its scarcity often determines the limits of distribution of certain kinds of plants in nature. It is also so tightly bound by the usually positively charged soils characteristic of the tropics that it often limits normal plant growth even if an adequate supply is actually present in the soils. As phosphorus weathers out of soils, it is transported by rivers and streams to the oceans, where it is precipitated. It is naturally brought up again only by the uplift of lands, such as along the Pacific coast of North and South America, or by sea birds, which deposit enormous amounts of guano (feces) rich in phosphorus along certain coasts (these deposits have traditionally been used for fertilizer) (Figure 34-11). Crushed phosphate-rich rocks, found in certain regions, are also used in this way, but the seas are the only inexhaustible source of phosphorus, one of the reasons that deep-seabed mining now looks so commercially attractive.

Every year, millions of tons of phosphate are added to agricultural lands in the belief that it becomes fixed to, and enriches, the soil. In general, four times as much phosphate as a crop requires is added each year, usually in the form of **superphosphate,** which is soluble calcium dihydrogen phosphate, $Ca(H_2PO_4)_2$, derived by treating bones or apatite, the mineral form of calcium phosphate, with sulfuric acid. But the enormous quantities of phosphates that are being added annually to the world's agricultural lands are not leading to proportionate gains in crops; plants can apparently use only so much of the phosphorus that is added to the soil. Agricultural scientists are actively seeking new ways to approach the problem of phosphate supply.

As you saw in Chapter 28, mycorrhizal fungi play a major role in absorbing phosphorus from the soil. They, or root hairs in plants that lack mycorrhizae, transfer soluble phosphorus compounds back into the plants when they are released by the decay of parts of other plants or of animals that have eaten the plants. Plants are especially likely to form mycorrhizal associations when they are short of phosphate. The application of mycorrhizal fungi to agricultural soils and the fostering of conditions suitable to their growth would certainly aid greatly in making supplies of phosphorus available to many plants more efficiently than the application of superphosphate. The recycling of animal and human wastes to recover phosphates would also make a very important contribution to the availability of phosphorus in developed countries. The world's rivers carry 17 million metric tons of

FIGURE 34-11

Pelicans roosting on a small island in the Gulf of California, Mexico. Sea birds bring up phosphorus from the deeper layers of the sea by eating fish and other marine animals and depositing their remains as guano on the rookeries.

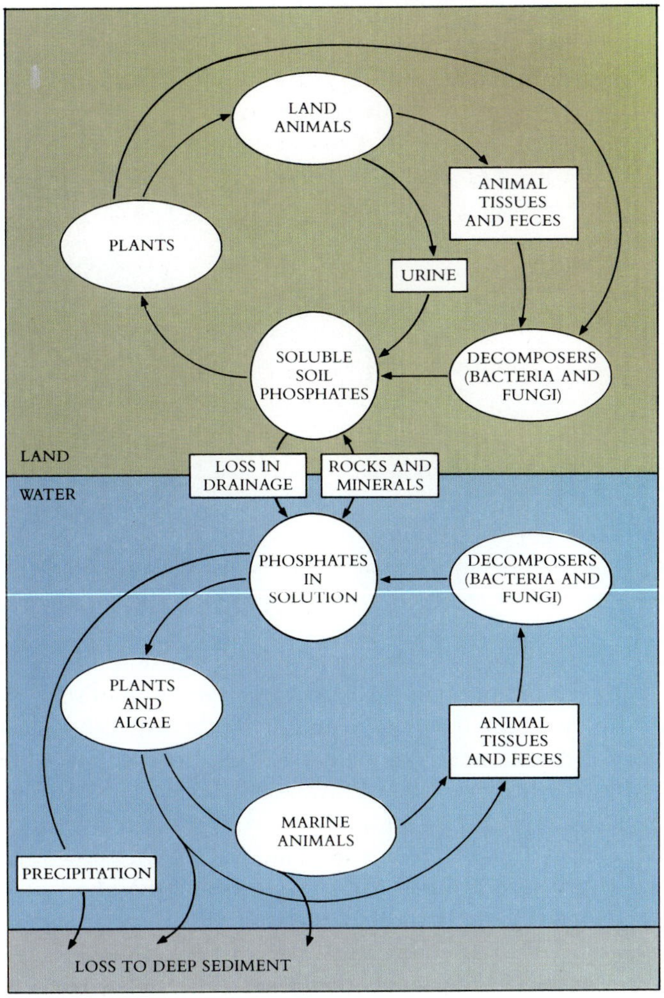

FIGURE 34-12

The phosphorus cycle.

phosphorus to the sea every year, half from natural erosion and half from domestic and industrial use and from sewage. Human sewage, in fact, represents a potential source of about 1 kilogram of phosphorus per hectare of cultivated land worldwide per year. The phosphorus cycle is summarized in Figure 34-12.

The Nitrogen Cycle

Even though the earth's atmosphere is 78% nitrogen, very few kinds of organisms—all of them bacteria—can convert it into a form that can be used for biological processes. As you saw in Chapter 10, the triple bond that links together the two atoms that make up diatomic atmospheric nitrogen (N_2) makes it a very stable molecule. In living systems the cleavage of atmospheric nitrogen is catalyzed by three proteins—ferredoxin, nitrogen reductase, and nitrogenase (p. 181). The process uses ATP as a source of energy, electrons derived from photosynthesis or respiration, and a powerful reducing agent. The overall reaction can be written this way:

$$N_2 + 3H_2 \rightarrow 2NH_3$$

All living organisms depend on nitrogen fixation. Without it, they would be unable to synthesize proteins, nucleic acids, and other necessary nitrogen-containing compounds.

Although dozens of genera of bacteria, most of them free-living, have the ability to fix atmospheric nitrogen, only those which form symbiotic relationships with plants usu-

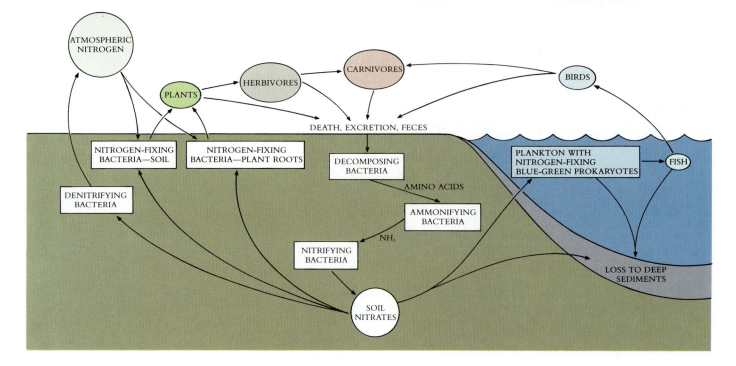

FIGURE 34-13

The nitrogen cycle.

ally fix enough nitrogen to be of major significance in plant production. Plants that have formed symbiotic associations with nitrogen-fixing bacteria can grow in soils that have such low amounts of available nitrogen that they are unsuitable for most other plants. Like phosphorus, nitrogen often limits the ability of many kinds of plants to grow in the areas where it is scarce. The cycle by which it is made available and renewed is shown in Figure 34-13.

FIGURE 34-14

The alder trees, which form the dark band in the foreground, are among the very first trees to colonize these relatively infertile, glaciated soils in Alaska. They can grow better than most other plants in these circumstances because actinomycetes that occur in nodules on their roots are able to fix nitrogen. The alders are preceded by willows and eventually will give way to poplars and then spruces.

Although nitrogen gas constitutes about 78% of the earth's atmosphere, it becomes available to organisms only through the activities of a few genera of bacteria, some of which are free-living and others of which live symbiotically on the roots of legumes and other plants. The molecules into which the nitrogen is incorporated are readily broken down by other kinds of bacteria.

As you saw in Chapter 26, bacteria of the genus *Rhizobium*, which inhabit nodules that they form on the roots of legumes (Figure 26-15), fix greater amounts of atmospheric nitrogen than do any other organisms. Because of the presence of *Rhizobium*, legumes not only fix enough nitrogen for their own use, but they also release excess fixed nitrogen into the soil, where it can be used by other plants. When legumes are grown to enrich the soil, they are often plowed under so that all of the nitrogen they have fixed is available for future crops.

The roots of a few other kinds of plants form associations with nitrogen-fixing actinomycetes of the genus *Frankia*, as mentioned in Chapter 26. Some of the plants involved are alders *(Alnus)* (Figure 34-14), sweet gale *(Myrica),* and mountain lilac *(Ceanothus).*

Nitrogen-fixing cyanobacteria of the genus *Anabaena* contribute large amounts of nitrogen to the rice paddies of China and Southeast Asia (Figure 34-15). The introduction of a foreign, cold-tolerant species of *Azolla, A. filiculoides,* into China in 1977 has allowed the expansion of *Azolla* cultivation to enrich rice and to feed to animals over a much wider area than previously. Such cultivation is, however, extremely labor intensive, so much so that it cannot be used everywhere.

It is estimated that the activities of all free-living bacteria may add, on the average, about 7 kilograms of fixed nitrogen per hectare of soil per year—about the same amount of fixed nitrogen that is added with the rain, which dissolves ammonia that reaches the atmosphere from various sources. In contrast, a crop of the legume alfalfa that is plowed back

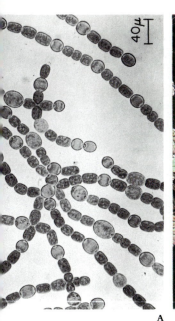

A   B   C

FIGURE 34-15

The nitrogen-fixing cyanobacterium Anabaena azollae (**A**) *lives in cavities between the leaves of the floating water fern* Azolla (**B**), *which is deliberately introduced into the rice paddies of the warmer parts of Asia. Rice* (**C**), *here cultivated in Sri Lanka, is the major food for well over one fourth of the human race.*

into the soil may add as much as 300 to 350 kilograms of nitrogen per hectare per year, and a healthy crop of *Azolla* growing in a rice paddy as much as 50 kilograms. In recent experiments at the International Rice Research Institute in the Philippines, yields of as much as 450 kilograms of nitrogen per hectare have been obtained with *Azolla* growing by itself under ideal conditions; the fixed nitrogen obtained in this way can then be applied to the crop. These figures clearly indicate the overwhelming importance of symbiotic associations in nitrogen fixation.

Ammonia, in the presence of water, is in equilibrium with the ammonium ion (NH_4^+). The chemoautotrophic nitrifying bacterium *Nitrosomonas* is of primary importance for the oxidation of NH_4^+, which is called **nitrification.** This reaction releases energy:

$$2NH_3 + 3O_2 \longrightarrow 2NO_2^- + 2H^- + 2H_2O$$

Another genus of bacteria, *Nitrobacter,* in turn oxidizes the nitrite ion (NO_2), which is toxic to higher plants, to nitrate (NO_3) in which form it is absorbed by plants and is directly available to them:

$$2NO_2^- + O_2 \longrightarrow 2NO_3^-$$

This reaction, too, yields energy; *Nitrobacter,* like *Nitrosomonas,* is chemoautotrophic.

Inside the plant cells, the nitrate ions are reduced back to ammonium ions, an energy-requiring process, and then are transferred to amino acids and other nitrogen-containing molecules. Almost all of the conversion of nitrates to ammonium ions and their incorporation into organic compounds takes place in the young, actively growing portions of the roots. Some amino acids are formed directly; others are formed by the transfer of the amino group (—NH_2) from one amino acid to another suitable molecule. Plants can produce all of the 20 amino acids they require. Most animals, however, can produce only 8; they obtain the other 12 from the plants they consume either directly (herbivores) or indirectly (carnivores).

All of the nitrogen that heterotrophic protists and animals use results from the nitrogen-fixing activities of certain bacteria and reaches them mostly by way of algae or plants. Most animals obtain 12 of the 20 amino acids they require from plants.

FERTILIZER

In natural communities, available nutrients are recycled and made available to organisms on a continuous basis. When these communities are replaced by cultivated crops, the situation changes drastically. First, the soil is much more exposed to erosion and to the loss of nutrients, which may simply wash away. Second, large amounts of nutrients may be removed with the crops themselves or with the animals to which they are fed. For this reason, cultivated crops and garden plants must usually be supplied with additional mineral nutrients. In the tropics the soils are often especially poor in nutrients, and fertilizers would significantly increase their yield.

The most important mineral nutrients that need to be added to soils are usually nitrogen (N), phosphorus (P), and potassium (K). All of these elements are needed in large quantities, and they are the ones most apt to become deficient in the soil. Commercial fertilizers are normally given a "grade," which reflects the percentages of N, P, and K by dry weight that they contain. The suitable proportions are best determined in relation to the tested fertility of the soil and the requirements of the particular crop. The nitrogen used in fertilizers is produced through the **Haber process,** in which nitrogen gas is recombined with hydrogen gas at high temperatures and pressures to yield ammonia. Nitrogen is often

A

B

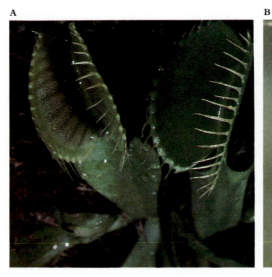

FIGURE 34-16

Carnivorous plants.
A *Venus flytrap,* Dionea muscipula, *which inhabits low boggy ground in North and South Carolina.*
B *A Venus flytrap leaf has snapped together, imprisoning a fly.*
C *Sundew,* Drosera. *A small fly has been trapped by the glandular hairs.*
D *A tropical Asian pitcher plant,* Nepenthes. *Insects enter the pitchers, which are modified leaves, seeking nectar, and are trapped and digested.*
E *Yellow pitcher plant,* Sarracenia flava, *which grows in bogs in the southeastern United States. Its pitchers, with their yellow borders, resemble flowers and secrete a sweet-smelling nectar that aids the plants in trapping insects. The flowers of this species, with their hanging petals, are also evident in this photograph.*
F *In the boggy uplands of the Cerro de la Neblina, the "mountain of the mists" discussed in Chapter 1, the old, weathered, boggy soils are very poor in nitrogen. Several carnivorous plants grow here, including the South American pitcher plant,* Heliamphora, *and the only partly carnivorous members of the pineapple family,* Bromeliaceae.

Nitrogen-containing compounds, such as proteins, are rapidly decomposed by certain bacteria and fungi. These bacteria and fungi use the amino acids they obtain in this way to synthesize their own proteins and to release excess nitrogen in the form of ammonium ions (NH_4^+), a process known as **ammonification.** These ammonium ions can be converted to nitrites and nitrates by certain kinds of organisms and then be absorbed by plants.

A certain proportion of the fixed nitrogen in the soil is steadily lost. Under anaerobic conditions, nitrate is often converted to nitrogen gas (N_2) and nitrous oxide (NO_2), both of which return to the atmosphere. This process, which several genera of bacteria carry out, is called **denitrification.**

CARNIVOROUS PLANTS

Some plants are able to use other organisms directly as sources of nitrogen and some required minerals, just as animals normally do. These are the carnivorous plants. Carnivorous plants often grow in acidic soils such as bogs—habitats that are not favorable for the growth of most legumes or of nitrifying bacteria. By capturing and digesting small animals

supplied to crops in very large amounts, since it seems to be especially important in affecting crop yield. Nutrients other than these three may also be scarce in an individual soil, but such situations are rather rare and must be dealt with individually.

Organic fertilizers were used long before chemical fertilizers were understood and applied widely. Such substances as manure or remains of dead animals have traditionally been applied to crops, and plants are often plowed into the soil to increase its fertility as well. There is no basis for believing that organic fertilizers supply any element to plants that inorganic fertilizers cannot also provide. However, organic fertilizers do build up the humus content of the soil,

which often enhances its water- and nutrient-retaining properties. For this reason the availability of nutrients to plants at different times of year may be improved, under certain circumstances, with organic fertilizers. Compost also serves to build up the humus content of the soil, although it may or may not be rich in the nutrients that the plants require, depending on the original source of the compost.

C

D

E

F

directly, such plants obtain adequate supplies of nitrogen and thus are able to grow in these seemingly unfavorable environments.

The carnivorous plants have adaptations that are used to lure and trap insects and other small animals. The plants digest their prey with enzymes secreted from various kinds of glands. In the sundews (Figure 34–16, *C*), the glandular trichomes (Figure 31-13, *B*) secrete both sticky mucilage, which traps small animals, and digestive enzymes. Pitcher plants (Figure 34–16, *D* to *F*) attract insects by the bright, flowerlike colors within their pitcher-shaped leaves and perhaps also by sugar-rich secretions. Once inside the pitchers, the insects may slide down into the cavity of the leaf, which is filled with water, digestive enzymes, and half-digested prey. Bladderworts, *Utricularia,* are aquatic; they sweep small animals into their bladderlike leaves by the rapid action of a springlike trapdoor and then digest these animals.

Carnivorous plants characteristically grow in nitrogen-poor soils. They supplement their supply of nitrogen by capturing and digesting small animals.

A TROPICAL ANT PLANT

In *Hydnophytum,* a tropical epiphyte—a plant that grows on the branches of other plants—the base of the stem is swollen and bulblike (Figure 34-B, *1*). All of the approximately 60 species of *Hydnophytum* share this peculiarity, which is rare in other members of the huge plant family Rubiaceae, the coffee family. Within the swollen base of *Hydnophytum* is an extensive, hollow honeycomb system that specific kinds of ants inhabit (Figures 34-B, *2* and *3*). Foraging throughout the area, these ants gather food—for example, other insects—and bring it back to their nest in the plant. There, they feed it to their larvae and to the ants that remain at home for defensive and other purposes. Pieces of this food that are not eaten, as well as the ants' excrement, are rich in nitrogen. In a sense, then, the ants play a role similar to that of an enormous, constantly moving system of roots, by means of which the plant can obtain nitrogen from throughout the entire area. Since *Hydnophytum* is a epiphyte, it has no direct access to the soil and therefore to nutrients other than nitrogen. These nutrients also are provided by ants' activities. *Hydnophytum* is maintained in its unpromising habitat by means of this complex symbiotic relationship. These photographs were taken on the island of Borneo, in Indonesia.

FIGURE 34-B

1 Hydnophytum *plants growing as epiphytes on the branches of a gymnosperm of the genus* Dacrydium *on infertile, white sand in Borneo.*
2 *Plants of* Hydnophytum *cut open to reveal the galleries inhabited by ants in their swollen bases.*
3 *Ants swarming out of a recently picked plant of* Hydnophytum.

The Venus flytrap *(Dionaea muscipula)* (Figure 34-16, *A* and *B*), which grows in the bogs of coastal North and South Carolina, has three sensitive hairs on each side of each leaf, which, when touched, trigger the two halves of the leaf to snap together (see p. 687). Once enfolded by a leaf, the prey of a Venus flytrap is digested by enzymes secreted from the leaf surfaces.

In their strange adaptations the carnivorous plants provide a somewhat bizarre conclusion to our treatment of plants. They use animals as a supplementary source of nitrogen and other essential nutrients and resemble animals in this respect. Overall, plants are very different from animals. Their rigid cell walls lead to rigid bodies, by virtue of which they are confined to a fixed, or sessile, mode of existence. Manufacturing their own food, plants bathe themselves in a continuous stream of water that they take in through their roots and lose through their leaves. They have evolved a whole series of adaptations that fit them as well to the demands of their primarily terrestrial existence as the animals, but they do so in an entirely different way. Animals have flexible cells and often move freely, hunting food, seeking mates, and fleeing from predators. We will now consider the ways in which animals have solved the problems of existence in their varied habitats.

SUMMARY

1. Soils are formed by the weathering of rocks; soils have been deeply modified, in most cases, by the action of biological systems. Clay soils, which are composed of very fine particles, hold a great deal of water but do so very tightly, so the water is mostly unavailable for plant growth.

2. Water flows through plants in a continuous column, driven mainly by transpiration through the stomata. The plant can control water loss primarily by closing its stomata. The cohesion of water molecules and their adhesion to the walls of the very narrow cell columns through which they pass are additional important factors in maintaining the flow of water to the tops of plants.

3. Root pressure develops at night, when there is high osmotic concentration in the root's cells and the stomata are closed. In most circumstances root pressure makes only minor contribution to the flow of water through plants.

4. Stomata open when their guard cells are turgid and bow out, thus causing the thickened inner walls of these cells to bow away from the opening. Abscisic acid, which is produced in the leaf under conditions of water stress, causes water to flow out of the guard cells and the stomata to close.

5. Nutrients move primarily through plants as solutes in the water column in the xylem. Their selective admission into the plant and their subsequent movement between living cells require the expenditure of energy.

6. The movement of water, with its dissolved sucrose and other substances, in the phloem does not require energy. Sucrose is loaded into the phloem near sites of synthesis, or sources, using energy supplied by the companion cells or other nearby parenchyma cells. The sucrose is unloaded in sinks, at the places where it is required. The water pressure is lowered where the sucrose is loaded into the sieve tube and raised where it is unloaded.

7. The nine macronutrients, substances that are each present at concentrations of 1% or more of a plant's dry weight, are carbon, hydrogen, oxygen, nitrogen, potassium, calcium, phosphorus, magnesium, and sulfur. The seven micronutrients, each present at concentrations of from one to several hundred parts per million of dry weight, are iron, chlorine, copper, manganese, zinc, molybdenum, and boron. These elements play different roles in the plant's metabolism.

8. Phosphorus is a key component of many biological molecules; it weathers out of soils and is transported to the world's oceans, where it tends to be lost. Phosphorus is relatively scarce in rocks; this scarcity often limits or excludes the growth of certain kinds of plants.

9. Atmospheric nitrogen is converted to ammonia by several genera of symbiotic and free-living bacteria. The ammonia, in turn, is converted to nitrites and they to nitrates by other bacteria. Nitrates are incorporated into the bodies of plants and there converted back into ammonium ions, which are used in the manufacture of many kinds of molecules in the bodies of living organisms. The breakdown of these molecules either converts them to recyclable forms or results in the release of atmospheric nitrogen.

10. Carnivorous plants often grow in nitrogen-poor habitats. They obtain their nitrogen from the insects and other small animals that they consume.

SELF-QUIZ

The following questions pinpoint important information in this chapter. You should be sure to know the answers to each of them.

1. If you made a barometer out of water instead of mercury, the height of the column at one atmosphere of pressure would be
   (a) 760 millimeters   (c) 10.4 meters
   (b) 1 meter   (d) 50 meters

2. The _____ of water to the sides of conducting vessels and the _____ of water molecules to other water molecules are the two kinds of interactions that allow water to travel upward in plants.

3. The _____ potential of a plant is the sum of its _____ potential and its _____ potential.

4. The process by which water leaves the plant is known as
   (a) transportation   (d) cavitation
   (b) transpiration   (e) translocation
   (c) differentation

5. Unlike water uptake, the uptake of nutrient ions requires _____.

6. When guard cells are turgid the stomata are open. True or false?

7. The hormone that controls the outflow of potassium from guard cells is
   (a) indoleacetic acid   (c) gibberellin
   (b) abscisic acid   (d) ethylene

8. The best agricultural soils, in general, are
   (a) clay   (c) loam
   (b) sand   (d) mineral soils

9. The accumulation of ions by roots at times when transpiration is low results in _____.

10. Carbohydrates move in the phloem by a process known as _____ from the _____, usually from the leaves to the _____.

11. The movement of carbohydrates within the phloem does not require energy. True or false?

12. The two macronutrients that usually play the most important role in limiting plant growth globally are _____ and _____.

THOUGHT QUESTIONS

1. Why do gardeners often remove many of a plant's leaves after transplanting it?

2. Where do potassium ions go when they flow from guard cells? What effect does their movement have on the stomata?

3. Are any legumes carnivorous? Why do you think so? If you did find one, what else would you expect to be true about it?

4. Are organic fertilizers better for plants than inorganic ones? Why?

5. If you grew a plant that initially weighed 200 grams, but eventually weighed 50 kilograms, in a pot, would you expect the soil in the pot to change weight? If so, how much, and why?

FOR FURTHER READING

BRADY, N.C.: *The Nature and Properties of Soils,* Macmillan Publishing Company, Inc., New York, 1974. A fine elementary text on soil science.

CRAFTS, A.S., and C.E. CRISP: *Phloem Transport in Plants,* W.H. Freeman and Company, San Francisco, 1971. Excellent, experimentally oriented treatment of the topic.

EPSTEIN, E.: *Mineral Nutrition of Plants,* John Wiley and Sons, New York, 1972. Good overview of the field.

HESLOP-HARRISON, Y.: "Carnivorous Plants," *Scientific American,* February 1978, pages 104–115. Provides experimental evidence about the way that carnivorous plants function.

KOZLOWSKI, T.T.: "Plant Responses to Flooding of Soil," *BioScience,* vol. 34, pages 162–167, 1984. A review of the metabolic, structural, and evolutionary responses of plants to excess water.

POSTGATE, J.R.: *The Fundamentals of Nitrogen Fixation,* Cambridge University Press, New York, 1982. Concise, authoritative account of biological nitrogen fixation.

SALISBURY, F.B., and C.W. ROSS: *Plant Physiology,* 3rd ed., Wadsworth Publishing Co., Inc., Belmont, Calif., 1985. An outstanding textbook of plant physiology.

ZIMMERMAN, M.H.: *Xylem Structure and the Ascent of Sap,* Springer-Verlag, New York, 1983. A comprehensive treatment of the structure of xylem and the way in which water moves through it.

BIOLOGY OF INVERTEBRATE ANIMALS

Part Seven

Invertebrate animals, constituting
millions of species, are the most
abundant living things. Found in every
conceivable habitat, they bewilder us
with their diversity.

Overview

In this chapter we begin to trace the long evolutionary history of the animals; in it we encounter the simplest members of this kingdom—sponges, jellyfish, and several kinds of worms. Although these animals are not as complex in their structures or behavioral patterns as are those we will discuss later, they are important ecologically. They are also significant because they illustrate the advent of the major characteristics that are important in the more advanced animal phyla. These characteristics include the development of tissues and organs, the use of internal digestion, the appearance of radial and then bilateral body organization, and the appearance of internal body cavities. The evolution of the primitive invertebrates accomplished the first major organization of the animal body, an organization that was maintained and elaborated by all later forms.

For Review

Here are some important terms and concepts that you will encounter in this chapter. If you are not familiar with them, you should review them before proceeding.

- **Evolutionary theory** (Chapter 19)

- **Classification** (Chapter 22)

- **Major features of evolution** (Chapter 23)

- **Choanoflagellates** (Chapter 27)

The two most conspicuous kingdoms of organisms are the plants, which we have just considered, and the animals, which we begin to consider here. Animals are a very distinct group of organisms and seem always to have been recognized as such. Animals are the eaters of the earth, and all of them are heterotrophs. All animals depend directly or indirectly for their nourishment on plants, photosynthetic protists, or autotrophic bacteria. Many animals are able to move from place to place in search of their food, which they ingest. In most of them, ingestion of food is followed by digestion in an internal cavity.

All animals are multicellular. Several decades ago a biologist would not have made that statement. In recent years the generally accepted scientific definition of the kingdom **Animalia,** which comprises the animals, has been changed in one respect. The unicellular, heterotrophic organisms called "Protozoa," which had in the past been regarded as very

 A

 B

 C

FIGURE 35-1

Diversity of lower invertebrates.
A *Yellow tube sponge (phylum Porifera). The cells of sponges are not organized into tissues.*
B *Sea anemone (phylum Cnidaria). The cnidarians are radially symmetrical animals in which the tissues are not grouped into organs.*
C *Free-living flatworm (phylum Platyhelminthes)—an acoelomate, a member of one of the groups of animal phyla in which the members are bilaterally symmetrical.*

simple animals, are now considered to be members of the kingdom Protista, the large and diverse group that we discussed in Chapter 27.

There are many kinds of animals—at least 4 million living species, and perhaps many times more (see Figure 22-6). Of these, only about one out of a hundred has a dorsal backbone. Those animals are the **vertebrates;** they will be discussed in Chapters 39 to 49. The other 99% of animal species lack such a backbone and are collectively referred to as the **invertebrates.** In this chapter we will consider the simplest of these (Figure 35-1), and in Chapters 36, 37, and 38, the more complex ones. But first we will step back a bit, and consider animals as a group. Despite their great diversity, they have much in common.

THE ANIMAL KINGDOM

Animals are extraordinarily diverse in form. They range in size from a few that are smaller than many protists to others like the truly enormous whales and giant squids. The animal kingdom, referred to as Animalia, includes about 35 phyla. Most of these phyla are found in the sea, with fewer in fresh water, and fewer still on the land. Only two phyla, Arthropoda—a gigantic group that includes crustaceans, spiders, and insects—and Chordata—our own phylum, which includes the vertebrates—have among their members organisms that are truly land-dwellers. Even though other kinds of animals do occur in terrestrial habitats, they require a constant supply of moisture—for example, the terrestrial snails and slugs (phylum Mollusca), which are frequent in many natural and manmade communities.

The cells of animals are exceedingly diverse in structure and function. Lacking rigid walls, these cells are flexible for the most part. Animal cells, except for those of sponges, are organized into **tissues,** groups of cells organized into a structural and functional unit. In most animals the tissues are organized into **organs,** complex structures that are made up by the composition of two or more kinds of tissues. Especially complex in animals are the junctions between cells, which play key roles in development, function, and locomotion. The structure of these junctions is becoming known through studies with the electron microscope. This knowledge, coupled with the results of exacting biochemical studies, is making it possible to understand the role of these junctions more completely.

In more complex animals, such as vertebrates, specialized cells compose many kinds of distinctive tissues, including many that are not found in any other kingdom. These include the incredibly complex tissues that are concerned with the senses and with locomotion. The ability of animals to move—and to move more rapidly and in more complex ways than the members of other kingdoms—is perhaps their most striking characteristic, one that is directly related to the flexibility of their cells. Animals usually move by means of muscle cells, contractile cells that contain the proteins actin and myosin. The most remarkable form of animal movement, perhaps, is flying, an ability that is well developed among both vertebrates (birds, bats, and some extinct reptiles) and insects.

> **The kingdom Animalia consists of about 35 phyla, most of which occur only in the sea. The diverse cells of animals, except for sponges, are organized into structurally and functionally distinctive groups called tissues, and the tissues of most animals in turn are organized into complex structures called organs, which are made up of two or more kinds of tissues.**

Most animals reproduce sexually. Their eggs, which are nonmotile, are much larger than their small, usually flagellated sperm. In animals the gametes do not divide by mitosis as they do in plants and some protists but rather fuse directly with one another. Consequently there is no counterpart among animals of the alternation of gametophytic (haploid) and sporophytic (diploid) generations, which is characteristic of plants. With few exceptions, some of which we will discuss, animals are diploid: the gametes are the only haploid cells in their life cycles.

Many animal phyla include species that reproduce either consistently or occasionally by **parthenogenesis,** a form of asexual reproduction in which an unfertilized egg develops directly into a mature individual. In parthenogenetic animals the egg may be diploid, as are the other cells of the animal that produces it, or it may be haploid and thus give rise to a haploid individual.

The complex form of a given animal develops from a zygote by a characteristic process of embryonic development (Figures 41-6, 41-8, and 41-9). The zygote first undergoes a series of mitotic divisions and becomes a hollow ball of cells, the **blastula,** a developmental stage that occurs in all animals. In most, the blastula folds inward at one point to form a hollow sac with an opening at one end called the **blastopore.** An embryo with a blastopore is called a **gastrula.** The subsequent growth and movement of the cells of the gastrula produce the digestive system, also called the **gut** or **intestine.** The details of embryonic development differ widely from one phylum of animals to another, as we will see. These details often provide important clues to the evolutionary relationships among the phyla.

THE LOWER INVERTEBRATES

The so-called "lower" or "primitive" invertebrates, which have a less complex organization of their tissues than the "higher" invertebrates, make up about 14 phyla. In this chapter we will focus on four of these that have been the most successful, as measured by their size, distribution, and important ecological roles. The four phyla of lower invertebrates that we will consider in detail include the sponges (phylum Porifera), which lack any tissue organization; the radially symmetrical jellyfish, hydroids, and sea anemones, and corals (phylum Cnidaria); the bilaterally symmetrical flatworms (phylum Platyhelminthes); and the nematodes (phylum Nematoda), a phylum that includes various free-living and parasitic worms.

THE SPONGES—ANIMALS WITHOUT TISSUES

Two subkingdoms are generally recognized within the kingdom Animalia: (1) **Parazoa**—animals that for the most part lack a definite symmetry, and possess neither tissues nor organs, and (2) **Eumetazoa**—animals that have a definite shape and symmetry and, in most cases, tissues, which are organized into organs and organ systems. The subkingdom Parazoa consists of only two phyla: Placozoa, which includes only one species, a very simple marine animal (Figure 35-2), and Porifera, the sponges. All of the other animals, composing some 33 phyla, belong to the subkingdom Eumetazoa. An outline of the animal phyla treated in this book follows:

OUTLINE OF ANIMAL CLASSIFICATION: KINGDOM ANIMALIA

I. Subkingdom Parazoa: phyla Placozoa and Porifera

II. Subkingdom Eumetazoa
   A. Radially symmetrical animals: phyla Cnidaria and Ctenophora; probably unrelated to one another (see pp. 727 and 735)
   B. Bilaterally symmetrical animals
      (1) Acoelomates (animals that lack a body cavity): phyla Mesozoa, Platyhelminthes, Rhynchocoela
      (2) Pseudocoelomates (animals with a pseudocoel): phyla Nematoda, Nematomorpha, Rotifera, Loricifera
      (3) Coelomates (animals with a coelom)
         (a) Protostomes (coelomates in which the mouth develops from or near the blastopore during the course of embryonic development): phyla Mollusca, Annelida, Pogonophora, Onychophora, and Arthropoda
         (b) Deuterostomes (coelomates in which the anus forms from or near the blastopore): phyla Echinodermata, Chaetognatha, Hemichordata, and Chordata

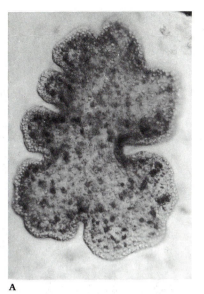

A

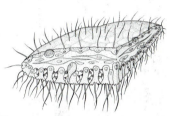

B

FIGURE 35-2

Photograph of a placozoan (A) and diagram of a cross section (B). This animal, Trichoplax adhaerens, is the only member of the phylum Placozoa. Like the sponges, the only other members of the subkingdom Parazoa, it lacks tissues and organs. Each individual is composed of a few thousand cells; most of those near the surface bear cilia. Barely visible to the naked eye, Trichoplax resembles an amoeba and changes shape constantly. The upper surface bears scattered cilia, and the lower surface is evenly ciliated; there is no differentiation between the anterior and posterior ends of the animal, nor between its two sides. Trichoplax is widespread in seawater but is rarely observed. It is grouped with the sponges because of its very simple organization, and the presumption that it has been derived from very simple ancestors.

FIGURE 35-3

The phylogeny of the major groups of animals.

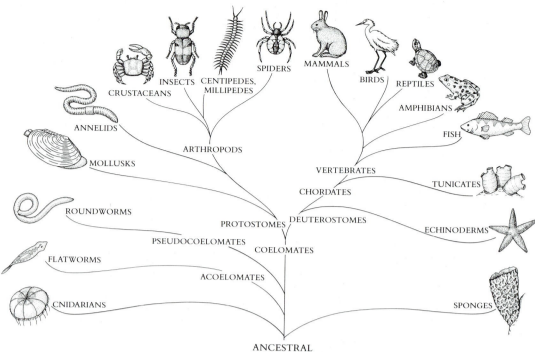

A

B

FIGURE 35-4

Diversity in sponges (phylum Porifera).
A *Barrel sponge, a large sponge in which the form is somewhat organized.*
B *Red boring sponge (Cliona delitrix), a crustose (encrusting) sponge.*

The animal kingdom is traditionally divided into two subkingdoms. In the simpler subkingdom, Parazoa, the animals lack symmetry and possess neither tissues nor organs. In the more advanced, Eumetazoa, the animals are symmetrical, and nearly all possess tissues.

It was once considered likely that the sponges and Eumetazoa had different unicellular ancestors. Most scientists now believe that Parazoa and Eumetazoa both evolved from a choanoflagellate ancestor (phylum Zoomastigina; see Figures 27-16 and 35-3). At any rate, the members of these subkingdoms have so many features in common that it is convenient to group them in a single kingdom.

There are about 10,000 species of marine sponges (Figures 35-1, *A,* and 35-4), and about 150 additional species that live in fresh water. In the sea, sponges are abundant at all depths. A few of the smaller ones are radially symmetrical, but most of the members of this phylum lack symmetry completely. Sponges lack specialized tissues and organs. Although some sponges are tiny, no more than a few millimeters across, some, like the loggerhead sponges, may reach 2 meters or more in diameter. Many sponges are colonial. Some are low and encrusting, while others may be erect and lobed, sometimes in complex patterns. Although larval sponges are free-swimming, the adults are **sessile,** or anchored in place. Individual adults remain attached to submerged objects once they have settled down.

There is relatively little coordination among the cells of sponges. Like a plasmodial slime mold (Chapter 27), a sponge can pass through a fine silk mesh and then reaggregate on the other side. In sponges, reaggregation takes a few weeks. No matter how large or elaborate they may be, the bodies of sponges consist of little more than masses of cells embedded in a gelatinous matrix.

Sponges have a unique form of feeding that involves the flow of water through a system of tiny pores and canals. The water is forced through these passageways by the beating of the flagella with which they are lined. The name of the phylum, *Porifera,* refers to this system of pores.

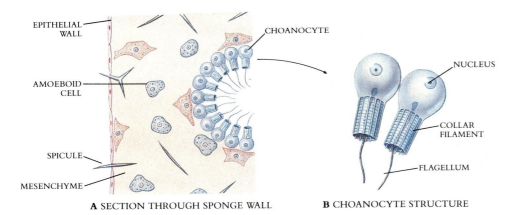

EPITHELIAL WALL

AMOEBOID CELL

SPICULE

MESENCHYME

CHOANOCYTE

NUCLEUS

COLLAR FILAMENT

FLAGELLUM

A SECTION THROUGH SPONGE WALL

B CHOANOCYTE STRUCTURE

FIGURE 35-5

A *The microstructure of a sponge, as seen in a transection through the animal; note the choanocytes lining the pore.*
B *Details of two choanocytes.*

The Body Plan of Sponges

The basic structure of a sponge can best be understood by examining the form of a young individual. When it attaches to a substrate, a young sponge grows into a vaselike shape. The walls of such a vase have three functional layers:

1. Facing into the internal cavity are specialized, flagellated cells called **choanocytes,** or collar cells (Figure 35-5). These cells line either the entire body cavity or, in many large and more complex sponges, specialized chambers.

2. The bodies of sponges are bounded by an outer **epithelial layer,** consisting of flattened cells somewhat like those that make up the epithelia, or outer layers, of other animal phyla. Some portions of this layer contract when touched or exposed to the appropriate chemical stimuli, and this contraction may cause some of the pores to close.

3. Between these two layers, sponges consist mainly of a gelatinous, protein-rich matrix called the **mesenchyme,** within which various types of amoeboid cells occur. In addition, many kinds of sponges have minute needles of calcium carbonate or silica known as **spicules** (Figure 35-6, *A*), fibers of a tough protein called **spongin** (Figure 35-6, *B*), or both, within this matrix. Spicules and spongin strengthen the bodies of the sponges in which they occur. A spongin skeleton is the model for the bathtub sponge, once the skeleton of a real animal, but now largely known from its cellulose and plastic mimics!

Water is drawn into a sponge through very numerous small pores and is circulated internally through a system of channels. Eventually the water is forced out through an **osculum,** a specialized, larger pore.

The Choanocyte

Each choanocyte closely resembles a protist with a single flagellum, a similarity that reflects its evolutionary derivation. The beating of the flagella of the many choanocytes that line the body cavity draws the water in through the pores and drives it through the sponge, thus providing the means by which the sponge acquires food and oxygen and expels wastes. Each choanocyte beats independently, and the pressure they create collectively in the cavity forces water out of the osculum. In some sponges the inner wall of the body cavity is highly convoluted, increasing the surface area exposed to water and therefore the number of flagella that can drive the water. In such a sponge, one cubic centimeter of tissue can propel more than 20 liters of water a day!

Sponges are unique in the animal kingdom in possessing choanocytes, special flagellated cells whose beating drives water through the body cavity.

FIGURE 35-6

Strengthening materials in sponges.
A *Spicules, as viewed with polarized light.*
B *The spongin skeleton of a glass sponge (Euplectella aspergillum).*

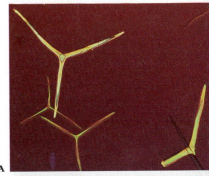

A

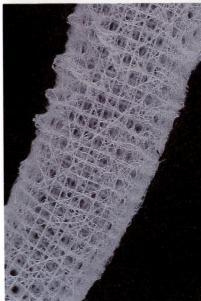

B

The microscopic examination of individual choanocytes (Figure 35-5, *B*) reveals an important substructure: the base of each flagellum is surrounded by a collar of small, hair-like projections, analogous to a picket fence. The strands that make up the collar of each choanocyte are connected to one another by delicate microfibrils. The beating flagellum of the choanocyte draws water, and any food particles in it, through the openings between these microfibrils. The flagellum then forces this water out through the open top of the collar. Any food particles that enter the collar are trapped in the mucus at the base of the flagellum. The food of a sponge is about four fifths particulate organic matter, one fifth small organisms. Once trapped at the base of the flagellum, this food is digested either by the choanocyte itself or by a neighboring amoeboid cell.

Symbiosis in Sponges

Many kinds of animals, such as crabs, fishes, and various mollusks, actually live inside sponges. These animals move in and out through the system of pores and may remain closely associated with the sponges throughout their lives; they may or may not actually feed on the sponges. The spicules that occur in certain kinds of sponges apparently make these sponges distasteful to most predators.

Another kind of symbiotic relationship is common in most kinds of sponges that occur in relatively shallow water, where there is enough light for photosynthesis to take place. Symbiotic algae or cyanobacteria usually live in the bodies of such sponges. These photosynthetic organisms provide food and possibly also oxygen for their hosts and may also help to screen them from sunlight in the shallow waters to which they are confined. Large populations of sponges with symbiotic cyanobacteria have been studied on the Great Barrier Reef, off the east coast of Australia, where they contribute significantly to the productivity of the reefs. Other symbionts of sponges are green and red algae and dinoflagellates. Because of these algae and cyanobacteria, as well as the natural pigments of the sponges themselves, mature sponges may be white, red, orange, yellow, blue, green, purple, or brown.

Reproduction in Sponges

As you might suspect from their ability to re-form themselves once they have passed through a silk mesh, sponges frequently reproduce themselves by simply breaking into fragments. If a sponge breaks up, the resulting fragments usually are able to reconstitute whole new individals. In addition, many kinds of sponges have more regular processes of asexual reproduction. Some of them form external buds that may eventually become detached and float away from their parent, regenerating whole new animals when they come to rest. If the buds remain attached to the parent, they may give rise to colonies. Some sponges reproduce asexually by means of **gemmules,** specialized aggregations of amoeboid cells within a hard outer layer that protects them.

Sexual reproduction in sponges is complex. In a few kinds of sponges, the eggs and sperm are shed into the water, and fertilization is external. In most sponges, however, the process is internal. The gametes form within specialized amoeboid cells in the mesenchyme or within certain choanocytes. Once formed, the sperm are expelled in large, milky clouds through the osculum. The sperm may then be pulled into the pores of another individual, where, like food particles, they may be captured by the choanocytes. Subsequently the sperm are transferred by amoeboid cells to the ripe eggs. Most kinds of sponges can form both male and female gametes in the same individual, but others consist of individuals of separate sexes.

Larval sponges, which may undergo their initial stages of development within their parent, have numerous external, flagellated choanocytes and are free-swimming. After a short planktonic stage, they settle down on a suitable substrate, where they begin their transformation into adults by turning inside out so that their choanocytes are internal.

The Evolution of Sponges

There is an exact morphological similarity between the choanocytes of sponges and a group of unicellular protists that belong to the phylum Zoomastigina (Chapter 27). The similarity is so striking that these zoomastigotes are called choanoflagellates. It seems almost certain that sponges evolved from choanoflagellates through a process that involved the multiplication of cells, resulting in multicellularity, and ultimately cellular specialization. As mentioned earlier in this chapter, many scientists have concluded that choanoflagellates may have been the ancestors not only of the sponges but of all animals.

The first fossils that we can definitely identify as sponges occur in the fossil record about 550 million years ago (Figure 23-9). Sponges are unique in their structure, and no other phylum has been derived from them during the course of their long history.

CNIDARIANS: THE RADIALLY SYMMETRICAL ANIMALS

The subkingdom Eumetazoa, animals that have a definite shape and symmetry and nearly always distinct tissues, is divided into two major groups. The first, which includes only two phyla, consists of radially symmetrical organisms. **Radial symmetry** is a condition in which the parts are arranged around a central axis in such a way that any plane passing through the central axis divides the organism into halves that are approximate mirror images. These two phyla are Cnidaria, or cnidarians—hydroids, jellyfish, sea anemones, and corals (Figures 35-1, *B,* and 35-7)—and Ctenophora, the comb jellies or ctenophores. The bodies of all other eumetazoans are marked by a fundamental bilateral symmetry. The radially symmetrical eumetazoans have often been considered primitive; the fact that they are the only eumetazoans in which the tissues are not organized into organs appears to be in agreement with this conclusion. Cnidarians were widespread, abundant, and diverse in Precambrian times, as much as 630 million years ago (see Figure 23-9 for a reconstruction of a Cambrian Period animal of this group). Whether ctenophores are actually related to them will be considered further when we discuss that phylum.

The most primitive eumetazoans belong to the phylum Cnidaria, which consists of radially symmetrical animals. Their tissues are not organized into organs.

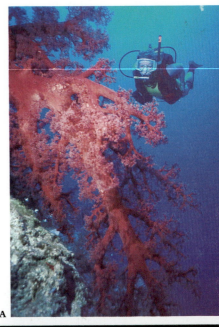

A

B

C

FIGURE 35-7

Representatives of the three classes of cnidarians (phylum Cnidaria).
A *Soft coral (class Anthozoa).*
B *Hydra (class Hydrozoa).*
C *Jellyfish, Aurelia aurita (class Scyphozoa).*

Embryonic Layers of Eumetazoans

Three distinct embryonic layers form in the embryos of all eumetazoans: an outer **ectoderm,** an inner **endoderm,** and an intermediate **mesoderm.** These layers ultimately differentiate into the tissues of the adult animal. In general, the nervous system and outer covering layers, or **integuments,** develop from the ectoderm; the muscles and skeletal elements from the mesoderm; and the intestine and digestive organs from the endoderm.

Because they possess three such embryonic layers, eumetazoans are said to be **triploblastic.** In some cnidarians, a gelatinous layer called the **mesoglea,** in which cells are usually absent, lies between the derivatives of the ectoderm and those of the endoderm. Since these animals appear to have only two distinct cellular layers, they were once considered to be **diploblastic,** but it is now generally accepted that there is no fundamental difference between their mesoglea and the products of the mesoderm in other kinds of animals. The interpretation of the mesoglea and of the presence of mesoderm in cnidarians is not accepted by all scientists, however; the matter clearly requires further study for full resolution.

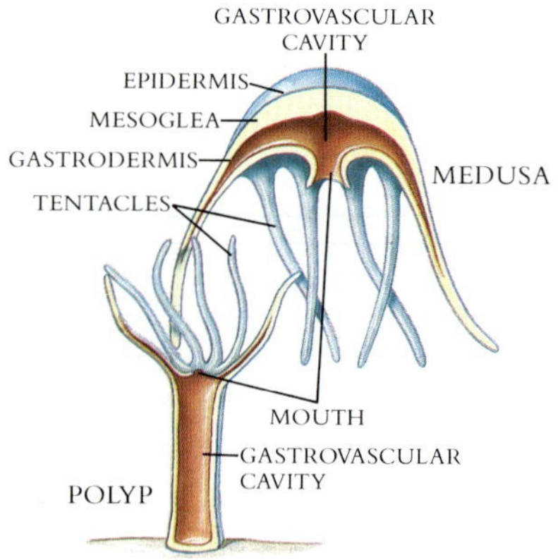

All eumetazoans are triploblastic, possessing three embryonic tissues: ectoderm, mesoderm, and endoderm.

The Body Plan of Cnidarians

Cnidarians are nearly all marine, although a few live in fresh water. These fascinating and simply constructed animals are often abundant, especially in shallow, warm-temperature or subtropical waters. They are basically gelatinous in composition. Like all eumetazoans, they differ markedly from the sponges in their organization: their bodies are made up of distinct tissues; as we just mentioned, however, the tissues of cnidarians are not grouped into organs. All of the members of this phylum are carnivores. For the most part, they do not move from place to place, but rather capture their prey (which include fishes, crustaceans, and many other kinds of animals) with the tentacles that ring their mouth.

There are two basically different body forms among the cnidarians, **polyps** and **medusae** (sing. **medusa**) (Figure 35-8). Polyps are cylindrical animals, usually found attached to a firm substrate, where they may be solitary or colonial. In a polyp the mouth faces away from the substrate on which the animal is growing, and therefore often upward. Many polyps build up a chitinous or calcareous (made up of calcium carbonate) external or internal skeleton, or both. Some polyps are free-floating, but most are anchored to the substrate. In contrast, most medusae are free floating and are often umbrella shaped. Their mouths usually point downward, and the tentacles hang down around them. Medusae, particularly those of the class Scyphozoa, are commonly known as jellyfish because of their thick, gelatinous mesoglea.

Many cnidarians occur only as polyps, while others exist only as medusae; still others alternate between these two phases during the course of their life cycles. Polyps may reproduce asexually by budding; if they do, they may produce either new polyps or medusae. The consequences of these two forms of reproduction are as follows:

1. If new polyps arise from an existing one, they may all remain together; the animal thereby becomes a colony of polyps. Alternately, the polyps may separate and subsequently remain isolated from one another.

2. If the polyps give rise to medusae, these are produced in specialized buds. Such cnidarians alternate between polyp and medusa forms. The medusae that are produced in this way often reproduce sexually, while the polyps may continue to reproduce asexually.

In most cnidarians, fertilized eggs give rise to free-swimming, ciliated larvae, known as **planulae** (Figure 35-9). Planulae are common in the plankton at times, and may be dispersed widely in the currents.

FIGURE 35-8

The two kinds of cnidarians, the medusa (above) and the polyp (below). These two phases alternate in the life cycles of many cnidarians, but a number—including the corals and sea anemones, for example—exist only as polyps.

Individual cnidarian species may either be medusae—floating, bell-shaped animals with the mouth directed downward—or polyps—anchored animals with the mouth directed upward. In some cnidarians, these two forms alternate during the life cycle of the organism.

Cnidarians have a digestive cavity with only one opening (Figure 35-8), a feature that they share with the ctenophores. This digestive cavity, usually called the **gastrovascular cavity,** has sometimes been called the **coelenteron.** Based on this word, a phylum was once recognized that included both Cnidaria and Ctenophora and was named "Coelenterata." The term seemed especially appropriate, since some of the hydroids—members of one of the characteristic groups of Cnidaria—apparently consist of little more than a gut. When the phylum was divided, however, its two parts were given different names, and the more inclusive group Coelenterata is no longer used.

The cnidarian gut is lined by endoderm and is much more complex in structure and function than the corresponding system of channels and pores in sponges. The mesoglea—the thick, gelatinous layer that lies between the derivatives of the endoderm and those of the ectoderm—is a loosely organized, gelatinous mass in some cnidarians. In others, such as the anthozoans (sea anemones), it is complex, cellular, and well organized.

The muscles of a cnidarian consist of specialized cells. Many of these cells are derived from the ectoderm and thus are part of the epithelial layer, whereas others are derived from the endoderm and thus line the gastrovascular cavity. Nerve nets, also derived in part from each of these layers, coordinate the animal's activities. In most cnidarians, the nerve net apparently has little central control. However, in some cnidarians, such as the jellyfish, there is a degree of centralization of the nerve net elements. Despite this limited degree of centralization, cnidarians cannot be said to have a central nervous system, an evolutionary advance that is characteristic of bilaterally symmetrical animals. Some of the nerve junctions of cnidarians transmit impulses in both directions, whereas others, like those of more advanced animals, transmit impulses in one direction only. A stimulus received in a nerve net simply spreads through the system away from its point of origin; the production of such a stimulus results in relatively limited coordination of behavior. Cnidarians have no blood vessels, no respiratory system, and no specialized internal cavity, such as those that occur in more advanced animals.

Nematocysts

The tentacles of the cnidarians bear stinging cells, called **cnidocytes.** The name of the phylum Cnidaria refers to these cells, which are highly distinctive and occur in no other group of organisms. Within each cnidocyte there is a **nematocyst,** a capsule with walls that are made of a substance resembling chitin (Figure 35-10).

A nematocyst is best thought of as a small but very powerful harpoon. Each one features a coiled, threadlike tube, which is essentially an extension of the capsule. Lining the inner wall of the tube is a series of barbed spines. Cnidarians use the threadlike tube to spear their prey and then draw the harpooned prey back with the tentacle containing the cnidocyte. To propel the harpoon, the cnidocyte uses water pressure. Before firing, the cnidocyte contains a very high concentration of ions, and its membrane is not permeable to water. Within the undischarged nematocyst, the osmotic pressure reaches about 140 atmospheres!

If a flagellum-like trigger on the cnidocyte is touched (and correct chemical stimuli are also present), the nematocyst is stimulated to discharge. Its walls become permeable to water, which rushes inside and violently pushes out the barbed filament. As the filament flies outward, it turns inside out, exposing the barbs. The entire process takes place in about 3 milliseconds, with a maximum velocity of 2 meters per second. Such a velocity corresponds to an acceleration of 40,000 times the force of gravity (40,000 Gs)! Nematocyst discharge is one of the fastest cellular processes in nature.

FIGURE 35-9

Planula larva of Aurelia, *a jellyfish.*

FIGURE 35-10

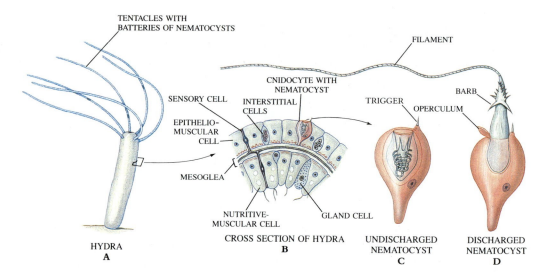

A Hydra.
B *Details of the structure of Hydra, showing the position of the cnidocytes and the structure of the nematocysts they contain. An undischarged* (**C**) *and a discharged* (**D**) *nematocyst from Hydra.*

Because the projection of the filament from the nematocyst takes place so forcefully, the barb can penetrate even the hard shell of crustaceans. When penetration occurs, chemicals sting and often stun the animal with which the nematocyst has made contact. Then the tentacle from which the nematocyst has been discharged, and sometimes additional tentacles as well, grasps the prey and pulls it into the cnidarian's mouth. The stinging sensation associated with the discharge of the nematocysts results from the injection of a toxic protein. Cnidocytes can discharge their contents only once. When they have done so, the cnidocytes are reabsorbed into the body of the animal. New cnidocytes are being formed continually to replace those that have been discharged.

Cnidarians characteristically possess a specialized kind of cell called a cnidocyte. Each cnidocyte contains a nematocyst, a harpoon used to attack prey. Nematocysts are found in no phylum other than Cnidaria.

The powerful **neurotoxins** (nerve poisons) that are secreted by the nematocysts of the Portuguese man-of-war, a large, floating, colonial, marine cnidarian (Figure 35-11), are dangerous and occasionally even fatal to humans. A number of species of jellyfish also have stings that are exceedingly painful. If the jellyfish are abundant, they can drive swimmers away from beaches.

Some nudibranchs—marine slugs, members of the phylum Mollusca (Chapter 36)—and flatworms are able to eat cnidarians, tentacles and all, without causing the nematocysts to discharge. These predators have the ability to retain the nematocysts within their own bodies and use them to defend themselves. Similar symbiotic systems, in which certain animals may obtain poisonous chemicals from the plants that they eat and use them in their own defense, are described in Chapter 53. Another kind of symbiotic relationship exists between many kinds of cnidarians and either green algae or dinoflagellates, members of the phyla Chlorophyta or Dinoflagellata (see Figure 27-8). The protists live within the cells of the cnidarians and provide some if not all of their food.

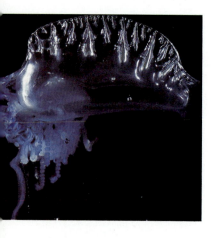

FIGURE 35-11

Portuguese man-of-war, Physalia utriculus. *The Portuguese man-of-war is a colonial hydrozoan that has adopted the way of life characteristic of the jellyfish. The float is probably a modified medusa, and the tentacles are modified polyps. This highly integrated colonial organism can ensnare good-sized fishes by utilizing its painful stings and tentacles, which are sometimes over 15 meters long.*

Extracellular Digestion

A major evolutionary innovation in cnidarians, compared with sponges, is the extracellular digestion of food—digestion within a gut cavity, rather than within individual cells. This evolutionary modification has been retained and elaborated in all more advanced groups of animals. In cnidarians, digestive enzymes (primarily proteases) are released from cells lining the walls of the cavity; these partially break down food. Unlike the process in more advanced invertebrates, however, digestion is not completely extracellular; rather, food is fragmented extracellularly into small bits that are subsequently engulfed by the cells lining the gut—the process of phagocytosis (Chapter 9).

The extracellular fragmentation that precedes phagocytosis and intracellular digestion allows cnidarians to digest animals larger than individual cells, an important improvement over the strictly intracellular digestion of sponges. Any food particles that cannot be digested are released out the same opening through which they were pulled into the animal originally.

Cnidarians are the most primitive animals that exhibit extracellular digestion.

Classes of Cnidarians

There are three classes of cnidarians: Hydrozoa (hydroids); Scyphozoa (jellyfish); and Anthozoa (corals and sea anemones). Hydroid medusae are generally considered to represent the most primitive form of any cnidarian. The Scyphozoa, which are complex in overall structure and in musculature, are medusae, alternating with inconspicuous polyps during the course of their life cycles. They are generally thought to have evolved from hydroid medusae. Increasing emphasis on the polyp generation, which has developed in relation to a particular, sedentary way of life, is pronounced among many hydroids. This evolutionary trend reaches full development in the Anthozoa, all of which are polyps. Like the Scyphozoa, the Anthozoa are believed to have evolved from the Hydrozoa.

Hydroids

Most of the approximately 2700 species of hydroids (class Hydrozoa) have both polyp and medusa stages in their life cycle. Most of these animals are marine and colonial; an example we have already mentioned is the Portuguese man-of-war (Figure 35-11). Some of the marine hydroids are bioluminescent.

A well-known hydroid is the abundant freshwater genus *Hydra,* which is exceptional in that it has no medusa stage and exists as a solitary polyp (Figure 35-10, *A*). Each polyp sits on a basal disk, by means of which it can glide around, aided by mucous secretions. It can also move like an inchworm by bending over, attaching itself to the substrate by its tentacles, and then looping over to a new location. If the polyp detaches itself from the substrate, it can float to the surface.

The individual *Hydra* polyps have an outer epithelial layer, which contains cnidocytes, and an inner endoderm. The mouth is surrounded by 6 to 10 hollow tentacles and opens directly into the gastrovascular cavity, which is also connected to hollow spaces within the tentacles. Contractile fibrils at the base of the cells of the epithelial layer run lengthwise through the animal, while the contractile fibrils of the endoderm run around the inside of the animal in a circular direction. When the *Hydra* is hungry, it extends itself by contracting these fibrils and especially by pulling in water, because the tentacles beat water actively into the gastrovascular cavity. When an individual small animal brushes against the tentacles, it is immediately harpooned by the threads discharged from the nematocysts and thrust into the hydra's mouth, which expands around the prey and engulfs it.

In some individuals of *Hydra,* buds project from the sides of an individual polyp, each bud with its own mouth and tentacles. These buds may either remain attached to the parent or become detached. If they detach, they grow into separate animals, a process of asexual reproduction. Testes or ovaries, when present, project from the sides of the individual polyp like buds. The eggs usually remain in the ovary, while the sperm are shed into the water. When a zygote is formed, it develops into a blastula, which subsequently settles down and gives rise to a new polyp.

An interesting relative of *Hydra* is the genus *Chlorhydra.* Symbiotic green algae color the polyps of *Chlorhydra,* and provide supplementary food for these animals.

Among the colonial hydroids, such as the genus *Obelia* (Figure 35-12), the polyps reproduce asexually by budding. The individuals are generally connected by branches, called **stolons,** that grow along near the substrate and include elements derived from all

FIGURE 35-12

The life cycle of Obelia, *a marine colonial hydroid.*

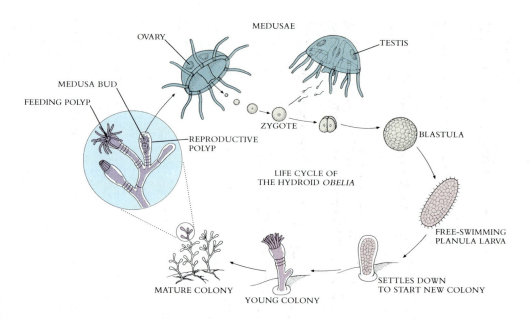

three embryonic layers. The polyps generally form an outer skeleton, or **exoskeleton,** which is usually chitinous. In addition to their asexual reproduction, the polyps may also give rise by the formation of specialized buds to medusae, which may swim off or remain attached to the polyps. In either case, the medusae ultimately give rise to gametes. The zygotes that are produced by the fusion of these gametes develop into planulae, which swim about for a time and eventually settle down, developing into a minute polyp. This polyp forms asexual buds, thus producing a new colony. The medusae of hydroids are small— usually between a few millimeters and several centimeters in diameter—translucent, and far less conspicuous than are the polyps.

Jellyfish

The approximately 200 species of jellyfish (class Scyphozoa) are transparent or translucent marine organisms (Figures 35-7, *C,* and 35-13), some of a striking orange, blue, or pink color. These animals spend most of their time floating horizontally near the surface of the sea. In all of them, the medusa stage is dominant. The polyp stage, which alternates with the medusa stage in most jellyfish, is small, inconspicuous, and simple in structure.

The medusa stages of jellyfish are very much larger and more structurally complex than are the polyps. The medusae are firm, gelatinous, flat or domed bells. The tentacles are borne around the margins of the bell and hang down from it. Cnidocytes occur not only on the tentacles but also around the edge of the mouth. Within the bell the gut may be divided into four or more sections by partitions. Jellyfish range from small animals about 2 centimeters across to large ones as much as 4 meters in diameter with trailing tentacles up to 10 meters long! Symbiotic algae are frequent in jellyfish, usually occurring in the mesoglea.

The outer layer, or epithelium, of jellyfish has a number of specialized **epithelio- muscular cells.** Each of these cells has contractile fibrils at its base and thus can contract individually. Together the cells form a muscular ring around the margin of the bell that pulses rhythmically and propels the animal through the water. The jellyfish nervous system consists of (1) a net near the surface of the bell, which controls the rhythmic pulsations; and (2) a second, more diffuse net, by means of which the jellyfish controls more local movements and reactions, such as feeding.

Jellyfish have separate male and female individuals. After fertilization, planulae are formed, which develop into the polyps. The polyps reproduce themselves asexually, in addition to budding off medusae. In some jellyfish that live in the open ocean, the polyp stage is suppressed and the planulae develop directly into medusae.

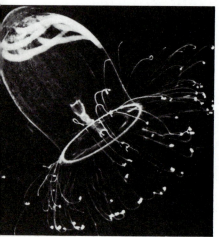

FIGURE 35-13

Aglantha digitale, *a jellyfish in which the bell is about 1 centimeter long, photographed in the waters near Friday Harbor, Washington. The individuals of* Aglantha, *like those of many other jellyfish, actually spend most of their time floating and sinking passively through the water.*

FIGURE 35-14

*Some of the diversity of
anthozoans.*
A *The polyps of a sea fan.*
B *The polyps of a reef-forming
coral, open at night.*
C *Fishes, such as this individual
of* Amphiprion perideraion
*in Guam, often form symbiotic
associations with sea anemones,
gaining protection by remaining
among their tentacles and gleaning
scraps from their food.*

Sea anemones and corals

By far the largest class of cnidarians is the class Anthozoa, the "flower animals" (from the Greek word *anthos,* meaning "flower"). The approximately 6200 species of this group are solitary or colonial marine animals (Figures 35-1, *B;* 35-7, *A;* 35-14; and 37-18). They include the stonelike corals (Figure 35-15), the soft-bodied sea anemones, and other groups known by such fanciful names as sea pens, sea pansies, sea fans, and sea whips. All of these names reflect a plantlike body topped by a tuft or crown of hollow tentacles, built on a plan involving multiples of six. Like other cnidarians, the anthozoans use these tentacles in feeding. Nearly all members of this class that live in shallow waters harbor symbiotic algae, which supplement the nutrition of their hosts through photosynthesis (see Figure 27-8, *B*).

The fertilized eggs of anthozoans usually develop into planulae that settle and develop into polyps; no medusae are formed by the members of this class. Each polyp is generally a hollow cylinder divided by longitudinal partitions, with a crown of tentacles at the top. Asexual reproduction, which takes place by budding or fragmentation, is common. When the members of this class reproduce sexually, their gametes are usually disseminated into the sea. Many corals, however, apparently hold their eggs in cavities, where they are fertilized and where the development of their planulae subsequently takes place.

Sea anemones are a large group of soft-bodied anthozoans. They are found in coastal waters all over the world and are especially abundant in the tropics. When they are touched, most sea anemones retract their tentacles into the body cavity and fold up. Sea anemones are highly muscular and relatively complex organisms, with greatly divided internal cavities. These animals range from a few millimeters to about 10 centimeters in diameter and are perhaps twice that high, but some of them are considerably larger. Many sea anemones are quite colorful.

FIGURE 35-15

*Corals, such as this yellow cup
coral, are inhabited by
zooxanthellae, the symbiotic form
of dinoflagellates.*

The corals are another major group of anthozoans. Many of them secrete tough outer skeletons, or exoskeletons, and are thus stony in texture. Others, including the large group called gorgonians, or soft corals, do not secrete such exoskeletons. Some of the hard corals participate in the formation of coral reefs, which are shallow-water limestone ridges that occur in warm seas. Coral reefs are formed as a result of the accumulation of carbonates over long periods. Although the waters where these reefs develop are often nutrient poor, the coral animals themselves, particularly because of their abundant symbiotic algae, are able to grow actively. Coralline red algae usually grow intermixed with the corals and contribute to the food supply of the reef; sponges with symbiotic cyanobacteria may also be abundant. The reef-forming coral animals themselves are colonial; they divide by asexual reproduction.

Coral reefs provide food and shelter for a diverse array of other organisms, including members of all major phyla. Reefs may form to a depth of about 60 meters, which is about as far as enough light can penetrate for photosynthesis to take place efficiently. Coral animals also live in deeper waters, however, but they do not form reefs there; many species of corals are solitary in any case, regardless of where they grow.

Coral reefs are built up in warm seas by colonial coral animals, which harbor symbiotic algae and are thus able to manufacture food to support themselves. Coralline algae (red algae) likewise make an important contribution to the food supply of coral reefs.

CTENOPHORA: THE COMB JELLIES

Traditionally the ctenophores (phylum Ctenophora) have been considered to be radially symmetrical and closely related to the cnidarians. The members of this small phylum are known as comb jellies or sea walnuts. Ctenophores are structurally more complex than cnidarians; for example, they have anal pores, so that water and other substances pass completely through the animal. The comb jellies are so soft-bodied that the first fossil—an animal from West German rocks about 400 million years old—was not discovered until 1983. We now know that the basic comb jelly body plan has not changed during at least these 400 million years. The phylum includes about 90 living species. Comb jellies themselves are transparent and usually a few centimeters long; they are often abundant in the open ocean (Figure 35-16).

Ctenophores may be either more or less spherical or ribbonlike. In at least some stage of their life cycles they propel themselves through the water by means of eight comblike plates of fused cilia that beat in a coordinated fashion. Ctenophores are the largest animals that use cilia for locomotion. The name of the phylum is derived from the Greek word *ctenos,* which means "comb," and refers to these characteristic plates. In addition, a few species of ctenophores use the muscular movements of their body walls for swimming. Many of the members of this phylum are luminescent, giving off bright flashes of light that are particularly evident in certain areas of the open ocean at night.

The members of one large group of ctenophores are animals that have two long, retractable tentacles. These tentacles may be as much as 20 times as long as the body and are used to capture food. They include complex cells, the **colloblasts,** which have sticky heads and are used to ensnare other animals. The prey of ctenophores includes larval fishes, small crustaceans, and other small animals. Once captured, they are enveloped with mucus and carried into the comb jelly's digestive cavity by ciliary action. Many comb jellies have symbiotic algae within their bodies, which may color the animals greenish, yellowish, reddish, or various other colors.

In the ctenophores, the zygotes usually develop into free-swimming larvae. Each individual is hermaphroditic and can form both male and female sex organs. Self-fertilization is common, but asexual reproduction is unknown.

FIGURE 35-16

A comb jelly (phylum Ctenophora). Note the comblike plates and two tentacles.

Ctenophores, phylum Ctenophora, are radially symmetrical animals that propel themselves through the water by means of eight comblike plates of fused cilia. The members of one group have two long, retractable tentacles.

Recently the longstanding assumption that cnidarians and ctenophores are closely related has been seriously questioned. Part of the reason is that the comb jellies are never wholly radially symmetrical; for example, the tentacles are always paired (when they are present), and other structures are paired in different members of the phylum. In fact, the apparent radial symmetry of the ctenophores is confined to the eight rows of combs and the gastrovascular canals that supply them. On closer examination, the features linking the ctenophores and cnidarians have proved to be mostly general ones, which certainly do not point to a close relationship. It may even be that the ctenophores are more closely related to another group, such as the phylum Platyhelminthes (p. 737), for example, than they are to the cnidarians. In any case, they are an ancient and fascinating small group of animals.

THE EVOLUTION OF BILATERAL SYMMETRY

Unlike a radially symmetrical animal, a bilaterally symmetrical one has a right and a left half that are mirror images of one another; there is a top and a bottom, better known respectively as the **dorsal** and **ventral** portions of the animal. There is also a front, or **anterior** end, and a back, or **posterior,** end, and therefore right and left sides. Even though adult echinoderms (Chapter 38) are radially symmetrical, their larvae are bilaterally symmetrical, and this pattern is also reflected in the arrangement of their internal organs. The echinoderms, therefore, have certainly evolved from bilaterally symmetrical ancestors; in contrast, we assume that the ancestors of cnidarians and ctenophores were never bilaterally symmetrical.

The bilaterally symmetrical animals constitute a major evolutionary line; a bilaterally symmetrical body plan is fundamental to all of the structural advances that have occurred during the evolution of the remaining animal phyla. This unique form of organization allowed the differential adaptation of the various parts of the body. Bilaterally symmetrical animals move from place to place more efficiently than radially symmetrical ones, which, in general, lead a sedentary existence. Bilaterally symmetrical animals are, therefore, efficient in seeking food and mates and avoiding predators.

During the early evolution of bilaterally symmetrical animals, those structures that were important to the organism in monitoring its environment, and thereby capturing prey or avoiding enemies, came to be grouped at the anterior end. Other functions tended to be located farther back in the body. The number and complexity of sense organs are much greater in bilaterally symmetrical animals than they are in radially symmetrical ones.

Much of the nervous system in bilaterally symmetrical animals is concentrated into major longitudinal nerve cords, which constitute the animal's **central nervous system.** In a very early evolutionary advance, nerve cells also become grouped around the anterior end of such animals. These nerve cells probably first functioned mainly to transmit impulses from the anterior sense organs to the rest of the nervous system. This trend ultimately led to the evolution of the head and the brain, as well as to the increasing dominance and specialization of these organs in the more advanced animal phyla.

> **Among bilaterally symmetrical animals different parts of the body are specialized in relation to different functions. All of the higher animals are bilaterally symmetrical, although in some of them the pattern is evident only in the embryos.**

Body Plans of Bilaterally Symmetrical Animals

Three basic body plans occur in bilaterally symmetrical animals (Figure 35-17).

1. In the more advanced phyla, the tissues of the mesoderm open during development, leading to the formation of a particular kind of body cavity called the **coelom.** The digestive, reproductive, and other internal organs develop within or around the margins of this coelom, suspended within it by double layers of mesoderm known as **mesenteries.** Animals in which a coelom develops are called **coelomates.**

2. In another group of phyla, the body cavity develops between the mesoderm and the endoderm, rather than within the mesoderm. Because of the way in which it originates, the body cavity of these animals lacks the characteristic epithelial lining derived from mesoderm that is found in a true coelom. This kind of cavity is called a **pseudocoel,** and the animals in which it occurs are called **pseudocoelomates.**

3. Some bilaterally symmetrical animals have no body cavity at all, other than the digestive system. Animals that have this kind of a body plan are called **acoelomates.**

FIGURE 35-17

Three body plans for bilaterally symmetrical animals, representing these patterns:
A *Acoelomate*
B *Pseudocoelomate*
C *Coelomate*

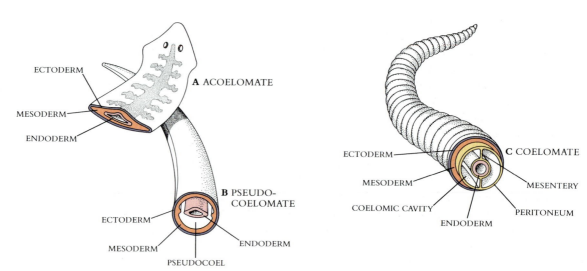

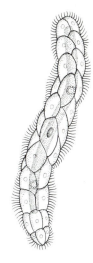

The body architecture of bilaterally symmetrical animals follows one of three patterns:

1. **Acoelomate, possessing no body cavity**
2. **Pseudocoelomate, possessing a cavity between endoderm and mesoderm**
3. **Coelomate, possessing a cavity bounded by mesoderm**

Each of these three fundamentally different body plans is characteristic of a series of animal phyla that belong to a distinct evolutionary line. The acoelomates and pseudocoelomates are all worms, and each of these lines comprises several phyla. Coelomates, however, are the most numerous and the most diverse of the three groups. All vertebrates and more advanced invertebrates are coelomates.

SOLID WORMS: THE ACOELOMATE PHYLA

Among bilaterally symmetrical animals, those with the simplest body plan are the acoelomates; they lack any internal cavity other than the digestive tract (Figure 35-17, *A*). We will discuss three acoelomate phyla: the mesozoans, the flatworms, and the ribbonworms.

FIGURE 35-18

A mesozoan, illustrating the very simple body plan that is characteristic of the members of this phylum.

Mesozoans

There are several phyla of acoelomate animals, but most have fewer than 100 species. One of these, the phylum Mesozoa, deserves special mention because its body plan is extraordinarily simple, even for an acoelomate (Figure 35-18). The mesozoans are minute worm-like animals. Their bodies consist of a layer of 20 to 30 outer ciliated cells—the exact number depends on the species—surrounding a long, slender, axial cell. Within this axial cell may be formed one or more reproductive cells. Mesozoans have no skeletal, muscular, nervous, digestive, respiratory, or excretory organs. All of them live as parasites within the bodies of marine invertebrates, such as cephalopod mollusks and annelids. Some mesozoans are hermaphroditic, but others have separate sexes. Because of their extremely simple structure, the mesozoans have at times been considered to be transitional between the kingdoms Protista and Animalia; in fact, the name of the phylum comes from the Greek prefix *meso-*, meaning "middle." The phylum Mesozoa may represent either a very simple evolutionary stage or a degenerate one; we do not yet know which of these alternatives is correct.

A

Flatworms

By far the largest of the phyla of acoelomates, with about 15,000 species, is that which includes the flatworms (phylum Platyhelminthes; Figures 35-1, *C*, and 35-19). These ribbon-shaped, soft-bodied animals are so called because they are flattened **dorsoventrally,** from top to bottom. They are the simplest in structure of all the bilaterally symmetrical animals, except for the mesozoans. As in other acoelomates, the bodies of flatworms are solid; the only internal space consists of the digestive cavity.

Flatworms range in size from a millimeter or less to many meters long, as in some of the tapeworms. Most species of flatworms are parasitic; the members of this phylum occur within the bodies of members of almost every other animal phylum.

Many flatworms, however, are free living, occurring in a wide variety of sea and freshwater habitats, as well as moist places on land. Free-living flatworms are carnivores and scavengers; they eat various small animals and bits of organic debris. They move from place to place by means of their ciliated epithelial cells, which are particularly concentrated on their lower surfaces. Some of the ventral cells secrete mucus, which provides traction for the cilia and allows the worms to move about efficiently. In the larger flatworms, these cilia are used to move water and other substances along the surface of the animal rather than

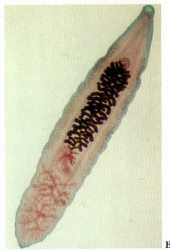

B

FIGURE 35-19

Flatworms (phylum Platyhelminthes).
A *A marine, free-living flatworm.*
B *The human liver fluke,* Clonorchis sinensis.

for locomotion. Cilia are absent in those stages of the parasitic flatworms that live in the bodies of other animals.

Flatworms exhibit traces of many of the evolutionary trends that are so highly developed among the members of more advanced phyla. The members of this phylum are the simplest animals in which organs occur. The cnidarians and ctenophores have tissues, but these tissues do not form organs. Flatworms have distinct bilateral symmetry and a definite head at the anterior end.

The acoelomates, typified by the flatworms, are the most primitive bilaterally symmetrical animals, and the simplest animals in which true organs composed of two or more primary tissues occur.

Organ systems of flatworms

Many flatworms have a gut with only one opening (Figure 35-20). Since they cannot feed, digest, and eliminate undigested particles of food simultaneously, flatworms cannot feed continuously, as more advanced animals can. In this respect, flatworms resemble the cnidarians, which also have only one opening to their digestive cavity. Muscular contractions in the upper end of the gut of flatworms cause a strong sucking force by which the flatworms ingest their food and tear it into small bits. The gut is branched and extends throughout the body, functioning both in digestion and transport of food. The cells that line the gut engulf most of the food particles by phagocytosis and digest them, but, as in the cnidarians, some of these particles are partly digested extracellularly. Tapeworms, which are parasitic flatworms, lack digestive systems; they absorb their food directly through their body walls.

Unlike the cnidarians, flatworms do have an excretory system, which consists of a network of fine tubules (little tubes) that runs throughout the body. In the hollow centers of the bulblike **flame cells,** located on the side branches of the tubules, cilia beat, moving water with the substances to be excreted into a system of tubules and then to exit pores located between the epidermal cells. Flame cells were named because of the flickering movements of the tuft of cilia within them. They primarily regulate the water balance of the organism. The excretory function of the flame cells appears to be a secondary one, since

FIGURE 35-20

Diagram of flatworm anatomy. The organism shown is Dugesia, *the familiar freshwater "planaria" of many biology laboratories.*

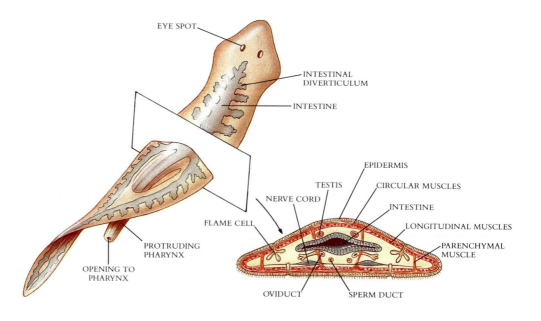

a large proportion of the metabolic wastes excreted by flatworms probably diffuses directly into the gut and is eliminated through the mouth.

Flatworms lack circulatory systems, which transport oxygen and food molecules in higher organisms. Consequently all of the cells of flatworms must be within diffusion distance of oxygen and food. Flatworms have thin bodies and highly branched digestive cavities, which make such a relationship possible.

The nervous system of flatworms is very simple compared to that of the other bilaterally symmetrical animals. Some primitive flatworms have only a nerve net, like that of the cnidarians. In most of the members of this phylum, however, there are definite longitudinal nerve cords that constitute a simple central nervous system. The tiny swellings at the anterior end of these cords are, in essence, primitive brains. Although they are fairly well developed in some of the free-living flatworms, these brains exert only a limited dominance over the activities of the animals in which they occur. Between the cords there are cross-connections, so that the flatworm nervous system resembles a ladder.

Flatworms lack circulatory systems, and most of them have a gut with only one opening. They excrete wastes directly from the gut, but also by means of a network of fine tubules that has ciliated flame cells on the side branches. Their nervous systems are simple.

Free-living flatworms use sensory pits or tentacles along the sides of their heads to detect food, chemicals, or movements of the fluid in which they are moving. Once they have sensed food, they move directly toward it and begin to feed. The free-living members of this phylum also have eyespots on their heads. These are inverted pigmented cups containing light-sensitive cells, connected with the nervous system. These eyespots enable the worms to distinguish light from dark; they move away from strong light.

Flatworms are far more active than are cnidarians or ctenophores. Such activity is characteristic of bilaterally symmetrical animals; in flatworms, it seems to be related to the greater concentration of sensory organs and, to a degree, of the nervous system elements in the heads of these animals. This concentration marks the beginning of an evolutionary trend of critical importance for the kingdom Animalia.

The reproductive systems of flatworms are complex. Most flatworms are hermaphroditic; in many of them, fertilization is internal. When they mate, each partner deposits sperm in the **copulatory sac** of the other. The sperm travel along special tubes to reach the egg. In free-living flatworms, the fertilized eggs are laid in cocoons strung in ribbons and hatch into miniature adults; in some parasitic flatworms, there is a complex succession of distinct larval forms.

Flatworms are capable of asexual regeneration to a much greater degree than are the members of most more advanced animal phyla. In some genera, when a single individual is divided into two or more parts, each part can regenerate an entire new flatworm.

Diversity in the flatworms

The free-living flatworms, class Turbellaria. Turbellarians constitute a large class of organisms, with 12 orders and some 121 families. The animals themselves are mostly small, inconspicuous, and difficult to identify. Most species are only about a millimeter long, but the more familiar ones are much larger. One of the most familiar is the freshwater genus *Dugesia,* the common planaria (Figure 35-20), which is used worldwide in biology laboratory exercises. Other members of this class are widespread and often abundant in lakes, ponds, and in the sea. Some of them also occur in moist places on land.

The free-living flatworms move about by means of their ciliated epidermis. These animals exhibit some very unusual features. For example, their sperm cells have two flagella—not one, as is usual in most animals. Furthermore, these flagella have a 9 + 1 rather than a 9 + 2 microtubular structure; there is only one strand of microtubules, instead of

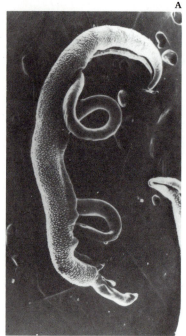

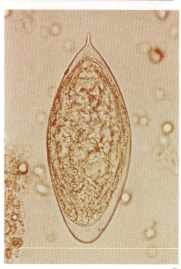

FIGURE 35-21

Schistosoma.

A *Scanning electron micrograph of copulating male and female individuals. The thicker male has a canal that runs the length of its body, within which the female is held during insemination and egg laying. The worms mate in the human bloodstream.*

B *An egg containing a mature larva.*

two, in the center of each flagellum. The evolutionary interpretation of both of these features is disputed.

The flukes, class Trematoda. There are two classes of parasitic flatworms: flukes (class Trematoda) and tapeworms (class Cestoda). In both groups the worms have epithelial layers and linings of the gut that are resistant to the digestive enzymes produced by their hosts—an important feature in relation to their parasitic way of life. However, they lack certain features of the free-living flatworms, such as cilia in the adult stage, eyespots, and other sensory organs, all of which lack adaptive significance for an organism that lives within the body of another animal.

Flukes take in food through their mouth, just like their free-living relatives. There are more than 8100 named species, ranging in length from less than 1 millimeter to more than 8 centimeters. Their attachment organs may consist of suckers, anchors, or hooks; their outer layer is often spiny. Although the digestive system is usually well developed, most flukes have no anus. The eggs of flukes are produced in capsules that are tough and resistant to the digestive juices of the host.

Some of the flukes have a life cycle that involves only one host—usually a fish; in some flukes, the host may be a mollusk, amphibian, or reptile. Flukes with a life cycle involving only a single host belong to the subclass Monogenea. Most flukes, however, have indirect life cycles involving two or more hosts. These flukes belong to the subclass Digenea. Their larvae almost always occur in snails, and there may be secondary intermediate hosts. The final host of the *digenetic* flukes, however, is almost always a vertebrate. Among the intermediate hosts—those that may occur in the life cycle between the snail and the vertebrate—are a wide range of animals in various phyla.

One of the most important flukes to human beings is the human liver fluke, *Clonorchis sinensis* (Figure 35-19, *B*), which lives in the bile passages of the liver of humans, cats, dogs, and pigs. It is especially common in the Orient. The worms are 1 to 2 centimeters long. Although they are hermaphroditic, with each individual containing a full complement of male and female sexual structures, cross-fertilization between different individuals is usual. The eggs, each of which contains a complete, ciliated first-stage larva, or **miracidium,** are passed out in the feces. If they reach water, they may be ingested by a snail, within which they transform into a **sporocyst**—a baglike structure with embryonic germ cells. Within the sporocysts are produced **rediae,** which are elongate, nonciliated larvae. These larvae continue growing within the snail, giving rise there to several individuals of the tadpolelike next larval stage, the **cercariae.**

The cercariae, which are produced within a snail, escape into the water, where they swim about. If they encounter a fish of the family Cyprinidae—the family that includes carp and goldfish—they bore into the muscles or under the scales, lose their tails, and transform into **metacercariae** within cysts in the muscle tissue. If a human being or other mammal eats raw infected fish, the cysts dissolve in the intestine and the young flukes migrate to the bile duct, where they mature. An individual fluke may live for 15 to 30 years in the liver. In humans a heavy infestation of liver flukes may cause cirrhosis of the liver and death.

Other very important flukes are the blood flukes of genus *Schistosoma* (Figure 35-21), which afflict about 1 in 20 of the world's population, some 200 to 300 million people throughout tropical Asia, Africa, Latin America, and the Middle East. Three species of *Schistosoma* cause the disease called schistosomiasis.

The tiny yellow eggs of these worms (Figure 35-21, *B*) leave the human body through the urine and feces. If they reach water, they hatch into miracidia. The miracidia, which are completely formed within the eggs, resemble ciliated protozoa. They must reach their freshwater snail hosts within a few hours to survive; if they do, they develop into sporocysts. These sporocysts divide asexually to give rise to daughter sporocysts, and the daughter sporocysts in turn give rise to cercariae like those of *Clonorchis.* A single infected snail may release 100,000 cercariae during its lifetime of several months; it starts to release them 25 to 40 days after it is infected. Instead of attacking fishes, the cercariae of *Schistosoma* burrow into human skin. They ultimately reach the bloodstream and are swept along to the lungs, where they remain for about 10 days. Subsequently they reenter the circulatory system and migrate to the hepatic and portal veins, which supply the liver. The worms become sexually mature and pair (Figure 35-21, *A*) on their journey to this destination, where they mate, and then migrate joined together to the veins that supply the upper intestine, the lower intestine, or the bladder, depending on the species of *Schistosoma* involved. In these sites, the adult worms live from several to more than 30 years, copulating and shedding from 300 to more than 3000 eggs per day continuously, but not reproducing themselves within the human body.

The very large numbers of eggs that these worms produce are the primary cause of the inflammation, extreme discomfort, and sometimes death associated with schistosomiasis. They cause widespread ulceration and bleeding of the intestinal and bladder walls and may also cause cirrhosis of the liver by blocking the small capillaries in that organ. Proper sanitation is important in controlling this disease, since it can continue only if urine or feces reach bodies of fresh water in which snails are living. It can also be controlled by eliminating the snail populations or reducing them to very low levels, since the miracidia can live only a few hours before they must reach such a host. The widespread introduction of irrigation in the tropics has contributed greatly to the spread of this serious disease, by spreading habitats for the snails that carry the worms.

Recently there has been a great deal of effort to control schistosomiasis. The worms protect themselves in part from the body's immune system by coating themselves with the host's own antigens, but even young worms that lack such coatings are also largely immune to attack. Despite these difficulties, the search for a vaccine is being actively pursued. The disease can be cured with drugs, and people who have been infected previously are often immune, an observation that offers some hope for the eventual development of a vaccine for this important and widespread disease.

The tapeworms, class Cestoda. Tapeworms, in contrast to flukes, simply hang on to the inner walls of their hosts by means of specialized terminal attachment organs and absorb food through their skins. Tapeworms lack digestive cavities as well as digestive enzymes. Thus they are extremely specialized in relation to their parasitic way of life (Figure 35-22). Most species of tapeworms occur in the intestines of vertebrates.

The long, flat bodies of tapeworms are divided into three zones: the **scolex,** or attachment organ; the unsegmented neck; and a series of repetitive segments, the **proglottids.** The scolex usually bears several suckers, and it may also have hooks. Each proglottid is a complete hermaphroditic unit, containing both male and female reproductive organs. Proglottids are formed continuously in an actively growing zone at the base of the neck, with the maturing ones moving farther back as new ones are formed in front of them. Ul-

FIGURE 35-22

Structure of the beef tapeworm, Taenia saginata.

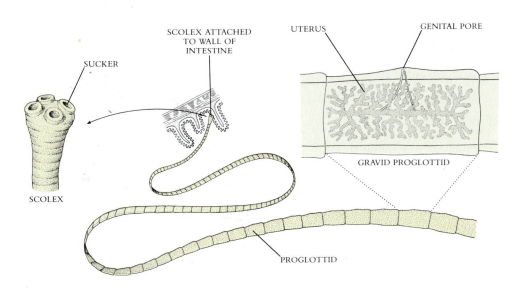

SCOLEX ATTACHED TO WALL OF INTESTINE

UTERUS

GENITAL PORE

SUCKER

SCOLEX

GRAVID PROGLOTTID

PROGLOTTID

timately the proglottids near the end of the body form mature eggs. As these eggs are fertilized, the zygotes in the very last segments begin to differentiate, and these segments become filled with embryos. These embryos, each surrounded by a shell, emerge from the proglottid either through a pore or the ruptured body wall. Their emergence may occur either while the proglottid is still attached to the tapeworm or after it breaks off, as the posterior segments eventually do. Normally, they leave their host with the feces and are deposited on leaves, in water, or other places from which they may be picked up by another animal. The production of such large numbers of embryos at one time—a phenomenon called **polyembryony**—is a clear adaptation to a parasitic way of life, since the likelihood of any one embryo reaching a suitable host is small. The numbers of potential offspring from a single tapeworm are enormous.

In all but a few species of tapeworms, the embryos are ingested by an intermediate host. They may develop into a larval form in this intermediate host and then into another stage that passes to the ultimate host, which may be infected by eating the intermediate host. Most tapeworms grow very rapidly, usually beginning to shed shelled embryos within a few days of infecting their ultimate host. Adults of some species of these parasites live for only a few days, whereas others may live 20 years or more. About a dozen kinds of tapeworms regularly occur as parasites in human beings.

As an example, the beef tapeworm, *Taenia saginata* (Figure 35-22), occurs as a juvenile primarily in the intermuscular tissue of cattle but as an adult in the intestines of human beings. A mature adult may reach a length of 10 meters or more. These worms attach themselves to the intestinal wall of their host by a scolex with four suckers. The segments that are shed off the end of the worm pass out of the human in the feces and may crawl onto vegetation; they ultimately rupture and scatter the embryos. The embryos may remain viable for up to 5 months. If they are ingested by cattle, they burrow through the wall of the intestine and ultimately reach the muscle tissues by way of the blood or lymph vessels. About 1% of the cattle in the United States are infected, and some 20% of the beef consumed is not federally inspected; when such beef is eaten rare, infection of human beings by these tapeworms is likely to result. As a result, the beef tapeworm is quite frequent as a human parasite.

Evolution of flatworms

Similarities between the planula stage of the cnidarians and the larvae of the flatworms have led some scientists to speculate that there may be a direct evolutionary relationship between the two groups. It is not likely that one phylum has directly given rise to the other, but rather that both derived from a common multicellular ancestor that resembled a planula larva. Varying degrees of similarity between the larvae of flatworms and those of other phyla, such as annelids and mollusks, suggest additional patterns of relationship, whose exact interpretation is controversial.

Within the phylum, it seems clear that the free-living flatworms are the most primitive of the three classes. In view of the great differences between them, it appears likely that the two classes of parasitic Platyhelminthes evolved independently from free-living ancestors.

Ribbon Worms

The members of phylum Rhynchocoela, the ribbon worms (Figure 35-23), share many characteristics with free-living flatworms but differ from them in several significant ways. These acoelomate aquatic worms, the great majority of which are marine, may be thread-shaped or ribbon-shaped. There are about 650 species. They are characterized by their **proboscis**, a long muscular tube that can be thrust out quickly from a sheath to capture prey. Ribbon worms are large, often 10 to 20 centimeters and sometimes many meters in length.

Among the smaller phyla of acoelomates, the ribbon worms are distinguished not only by their remarkable proboscis, but also by their complete digestive system. Although the larvae have no anus, the adults do; the ribbon worms, therefore, are the most primitive

FIGURE 35-23

A ribbon worm, Lineus *(phylum Rhynchocoela).*

animals in which this significant evolutionary advance occurs. They are also the simplest animals with a circulatory system in which the blood flows in vessels. In these ways, the ribbon worms provide early indications of important evolutionary trends that become fully developed in more advanced animals.

THE EVOLUTION OF A BODY CAVITY

The body organization of the other bilaterally symmetrical animals differs from the solid worms in an important respect: all the other bilaterally symmetrical animals possess an internal body cavity. Among them, seven phyla are characterized by their possession of a pseudocoelomate body plan (Figure 35-17, *B*). Only one of the seven, however, the phylum Nematoda, includes a large number of species. In all pseudocoelomates, the pseudocoel serves as a **hydrostatic skeleton**—one that gains its rigidity from being filled with fluid under pressure. The animal's muscles can work against this "skeleton," thus making the movements of pseudocoelomates far more efficient than those of the acoelomates.

> The pseudocoelomates were the first animals to possess an internal body cavity. Among many other advantages, this cavity makes the animal's body rigid, permitting resistance to muscle contraction and thus opening the way to muscle-driven body movement.

The pseudocoelomates lack a defined circulatory system; its role is performed by the fluids that move within the pseudocoel. All pseudocoelomates have a complete, one-way digestive tract, as do all coelomates. This digestive tract acts like an assembly line, with the food being acted on in different ways in each section. First it is broken down, then absorbed, then the wastes are treated and stored, and so on. Considerable specialization has evolved in the digestive tracts of the pseudocoelomate and coelomate phyla, as we will see.

FIGURE 35-24

Many nematodes are serious pests of plants.
A *Nematode damage to a field of onions caused by lesion nematodes (Pratylenchus penetrans) in Indiana. The area at right was treated with a nematicide to kill most of the nematodes before the onions were planted; the area at left was untreated.*
B *Cysts of the golden nematode (Heterodera rostochiensis) on roots of potato. Although most nematode species remain wormlike throughout their life, the mature females of some species that feed on plants become round cysts filled with eggs, as shown here.*
C *Swellings on crown and roots of African violet (Saintpaulia), caused by root knot nematode (Meloidogyne).*

Nematodes

The nematodes, eelworms, and roundworms comprise a large phylum, Nematoda, with some 12,000 recognized species. The members of this phylum are ubiquitous. Nematodes are abundant and diverse in marine and freshwater habitats, and many members of this phylum are parasites of vertebrates, invertebrates, and plants (Figure 35-24). Many nematodes are microscopic animals that live in soil; it has been estimated that a spadeful of fertile soil may contain, on the average, a million nematodes (Figure 35-25).

Almost every species of plant and animal that has been studied has been found to have at least one parasitic species of nematode living in it. Many of these worms live in environments that are very hot, dry, cold, or salty; when these conditions occur seasonally, the nematodes may be able to avoid them by entering a dormant state.

Relatively few specialists are concerned with the identification and classification of the members of this phylum. Those who are usually confine their attentions primarily to the nematodes that are animal or plant parasites; the numbers and kinds of free-living nematodes that must be present are, therefore, virtually unknown in most parts of the world. So many new ones are found when sampling different environments that some scientists think there actually may be 500,000 or more species of this phylum, the great majority of which have never been collected or studied.

A

B

C

FIGURE 35-25

One square meter of ordinary garden, lawn, or forest soil teems with 2 to 4 million nematodes. Although most are similar in form, they range from about 0.2 millimeter to about 6 millimeters long. A trained nematologist (student of nematodes) must examine slide mounts with a compound microscope to determine which species are present.

Nematodes (Figure 35-26) are bilaterally symmetrical, cylindrical, unsegmented worms. They are covered by a flexible, thick cuticle, which is molted as they grow. Their muscles constitute a layer beneath the epidermis and extend along the length of the worm, rather than encircling its body. These longitudinal muscles push both against the cuticle and against the pseudocoel, which forms a hydrostatic skeleton. When nematodes move, their bodies whip about from side to side.

Near the mouth of a nematode, at its anterior end, there are usually 16 raised, often hairlike, sensory organs (Figure 35-26, C). The mouth is often equipped with piercing organs called **stylets.** In the complete digestive tract of nematodes, the food first passes through the mouth as a result of the sucking action of a muscular chamber called the **pharynx.** After passing through a short corridor into the pharynx, food continues through the other portions of the digestive tract, where it is broken down and then digested. Some of the water with which the food has been mixed is reabsorbed near the end of the digestive tract, and the material that has not been digested is excreted through the anus.

Nematodes completely lack flagella or cilia, even on the sperm cells. Their excretory system consists of cells that function as glands, or systems of canals, and does not depend on flagella or cilia for its functioning, thus differing from the excretory systems of many animals. Reproduction in nematodes is sexual, with the sexes usually separate (Figure 35-26, A and B).

FIGURE 35-26

Diversity in nematodes.
A *Male and female of* Lindseyus costatus, *a free-living nematode that is about 5 millimeters long. Because nematodes are transparent, scientists can easily use both internal and external characteristics for identification.*
B Tetradonema plicans, *a parasite that lives in the larvae of mycetophilid flies. After fertilization by one or more males (m), the female nematode becomes an encapsulated egg sac. The tiny males continue to cling to the encapsulated female until they die.*
C *A predaceous nematode, called a "nematode monster" by early United States nematologist N.A. Cobb. The three jaws, characteristic of nematodes, are well developed in this species, which may be marine.*

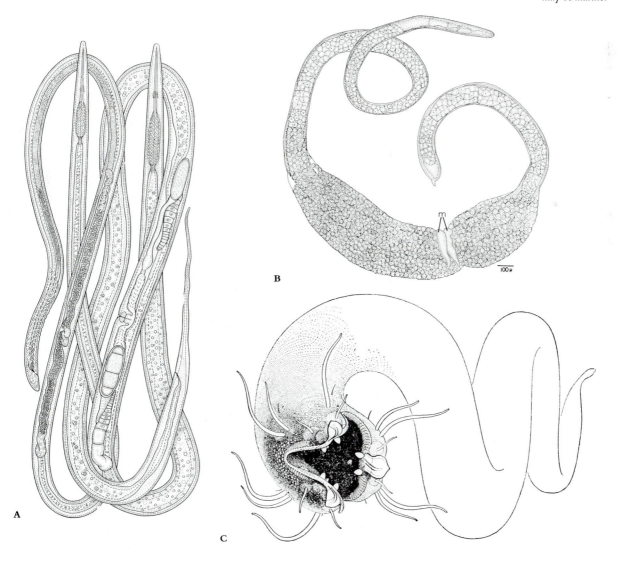

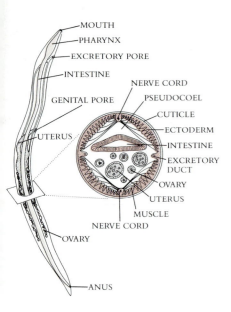

FIGURE 35-27

Structure of the large roundworm of humans, Ascaris lumbricoides, *illustrating the structure of a pseudoceolomate animal.*

Labels (top to bottom, left to right):
MOUTH
PHARYNX
EXCRETORY PORE
INTESTINE
NERVE CORD
GENITAL PORE
PSEUDOCOEL
CUTICLE
ECTODERM
INTESTINE
EXCRETORY DUCT
UTERUS
OVARY
UTERUS
MUSCLE
NERVE CORD
OVARY
ANUS

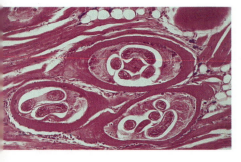

FIGURE 35-28

Cysts of Trichinella *in pork. These worms cause the disease trichinosis in human beings.*

Nematodes as parasites

About 50 species of nematodes, including several that are rather common in the United States, regularly parasitize human beings. Hookworms, mostly of the genus *Necator,* can be common in certain warm regions. The larvae occur in soil and burrow into skin that comes into contact with them. Their anterior end curves dorsally, suggesting a hook, which is why they are so named. Inside the host they attach to the wall of the small intestine and feed on blood. Because the worms suck much more blood from the intestinal wall than they use, their activities often result in anemia if they are abundant enough.

Pinworms, *Enterobius,* are abundant throughout the United States, where it is estimated they infect about 30% of all children and about 16% of adults. Fortunately the symptoms they cause are not severe, and the worms can easily be controlled by drugs. Adult pinworms live in the human large intestine and blind gut, laying fertilized eggs in the anal region of their hosts at night, which causes itching. The adult female worms are up to about 12 millimeters in length. The eggs, which become infective within 6 hours, are spread by scratching the anal region. Pinworms can be irritating but are not regarded as a particularly serious parasite.

The intestinal roundworm, *Ascaris,* infects approximately one of six people worldwide but is rare in most areas with modern plumbing. Like pinworms, intestinal roundworms occur only in human beings, and there is no intermediate host. They live in the intestine, and their fertilized eggs are spread in feces; these eggs can remain viable for years in the soil. In relatively unsanitary conditions, the eggs may be ingested easily because of improper washing of the hands. The young worms hatch in the small intestine, bore through the wall of the gut, pass through the heart and lungs, and eventually out of the breathing passage. They are then swallowed and reach the intestine again. The adult females, which are up to 30 centimeters long (Figure 35-27), contain up to 30 million eggs and can lay up to 200,000 of them each day! Obviously very few of these eggs ever reach another host.

The most serious common nematode-caused disease in temperate regions is trichinosis, caused by worms of the genus *Trichinella.* These worms live in the small intestine of pigs, where the fertilized females burrow into the intestinal wall. Once it has penetrated these tissues, each female produces about 1500 live young. The young enter the lymph channels and thereby reach muscle tissue throughout the body, within which they mature and form highly resistant, calcified cysts (Figure 35-28). Infection in human beings or other animals arises from eating undercooked or raw pork in which the cysts of *Trichinella* are present. If the worms are abundant, a fatal infection can result, but such infections are rare. It is thought that about 2.4% of the people in the United States are infected with trichinosis, but only about 20 deaths have been attributed to this disease during the past decade. The same parasite occurs in bears, and when humans eat improperly cooked bear meat, there seems to be an even greater chance of infection than with pork.

Other nematode-caused diseases are extremely serious in the tropics. *Filaria* and other genera cause filariasis, which infects at least 250 million people worldwide. These worms can be up to 10 centimeters long and live in the lymphatic system, which they may seriously obstruct, causing potentially severe inflammation and swelling. The female worms release the larvae into the blood and lymph, from which they may be picked up by female mosquitoes along with their blood meal. When an infected mosquito bites another animal, the larval worms may penetrate into the wound and thus reach a new host. The condition known as elephantiasis, a grotesque swelling of the legs or other extremities, may result from extreme filariasis. In certain tropical regions, filariasis-causing nematodes are the most serious health problem, effectively blocking settlement in areas where they are abundant.

Other Pseudocoelomates

There are several remaining phyla of pseudocoelomates, of which we will briefly discuss two: the horsehair worms (phylum Nematomorpha) and the rotifers (phylum Rotifera).

Horsehair worms

There are about 230 species of horsehair worms. The larvae are parasites of arthropods, including many different kinds of aquatic and terrestrial insects. The adults, which reproduce in water, are free living (Figure 35-29). The animals of this phylum were named "horsehair worms" because it was once believed that they originated from the long, coarse hairs in the tails of horses and suddenly came to life in water. Like nematodes, they are unsegmented and have bundles of longitudinal muscles beneath their epidermis. Adult horsehair worms range from 10 to more than 100 centimeters long, and from less than 1 to more than 3 millimeters thick. They have separate sexes, and the males differ morphologically from the larger females. The worms completely lack functional digestive tracts, and it is therefore presumed that the larvae absorb their food through their body walls. The short-lived reproductive adults probably live entirely on stored food, acquired during the larval stage. All but one genus of horsehair worms occur in fresh water, mainly along the edges of streams and ponds; the other genus is marine.

Rotifers

Rotifers (Figure 35-30) are common, small, basically aquatic animals that have a crown of cilia at their heads. There are more than 1500 species of this phylum. Some of them live in the soil, or in the capillary water in cushions of mosses; most occur in fresh water and are common everywhere. A very few rotifers are marine. Most rotifers are between 100 and 500 micrometers in length. They depend on their cilia for both feeding and locomotion, sweeping their prey into their mouths and swimming from place to place. Rotifers are often called "wheel animals" because the cilia of an individual rotifer, when they are beating together, resemble spokes radiating from a wheel.

Free-living rotifers feed on bacteria, protists, and small animals. Rotifers have a muscular pharynx and complex jaws inside it. They also have flame cells like those of the flatworms, and they regulate their osmotic pressure by excreting water from these cells. Some rotifers are sessile, while others creep, float, or swim. The sexes of rotifers are separate, although parthenogenesis is common and males do not occur in many species. Each female rotifer lays between 8 and 20 large eggs during the course of its life.

EVOLUTIONARY EXPERIMENTS IN BODY PLAN

In this chapter we have examined representatives of most of the kinds of body plans that occur in animals. We began with the sponges, very simple animals in which there are no tissues differentiated, and continued with the radially symmetrical phyla, animals that lack organs. These radially symmetrical animals, the cnidarians and ctenophores, apparently were derived originally from radially symmetrical ancestors. Even though the echinoderms, which we will meet in Chapter 38, are often radially symmetrical as adults, their larvae are bilaterally symmetrical. From this fact we are justified in concluding that the radial symmetry of adult echinoderms developed secondarily during the course of their evolution.

Once bilateral symmetry had developed among the animals, the stage was set for extensive specialization of the different parts of the body. Among the bilaterally symmetrical animals, there are three fundamentally different kinds of body plans; we call these the acoelomate, pseudocoelomate, and coelomate patterns. All of the representatives of the first two groups are worms, although some of them are very elaborate. It is among the coelomates, which we will begin to consider in the next chapter, that animal diversity reaches its full extent.

FIGURE 35-29

A horsehair worm (phylum Nematomorpha).

FIGURE 35-30

Rotifers (phylum Rotifera), common small aquatic animals.

A NEWLY DISCOVERED PHYLUM

In 1983 the scientific world was surprised by the announcement of the discovery of a new phylum of pseudocoelomates, Loricifera. The members of this phylum are tiny animals whose larvae are less than 195 micrometers long and adults are about 230 micrometers. They are members of the **meiofauna,** a group of animals adapted to living in the spaces between grains of sand in the ocean. Previously members of five or six other phyla had been known to live in this habitat.

The new phylum was christened by its discoverer, Reinhardt Kristensen of the University of Copenhagen, as Loricifera, from Greek words meaning "girdle-bearers." The most complete samples were obtained from gravel or sand taken from the ocean floor off France, Greenland, and Florida, at depths of 25 to 30 meters. The animals cling tenaciously to the grains of sand or gravel by means of clawlike and club-shaped spines on their head. They can be shocked into releasing their grip on the grains by flooding the samples with fresh water. They grow by molting and have unusual mouthparts, consisting of a unique flexible tube that can be telescopically retracted into the animal. The larvae swim by means of a pair of very unusual appendages, which can propel them through the water in any direction; these appendages are attached to a kind of ball-and-socket joint. In contrast, the adults lack such appendages and may be sedentary. More and more Loricifera are turning up in samples that have been screened in the proper way, and the animals may actually be fairly common but undetected because of their unusual habits. Larvae and adults of *Nanaloricus mysticus,* a species collected off the coast of France, are shown at left.

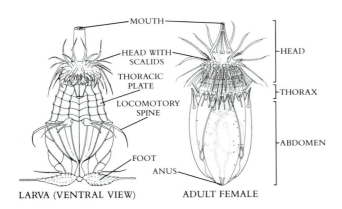

LARVA (VENTRAL VIEW)   ADULT FEMALE

SUMMARY

1. The animals, kingdom Animalia, comprise some 35 phyla and at least 4 million species—well over 80% of all living species. Animals are heterotrophic, multicellular organisms that ingest their food.

2. A mature animal develops from a zygote by a characteristic embryological process involving blastula and gastrula stages and complex cell movements.

3. The sponges (phylum Porifera) are characterized by specialized, flagellated cells called choanocytes, by a lack of symmetry in their body organization in most species, and by a lack of tissues and organs.

4. The remaining animals have radially or bilaterally symmetrical body plans and distinct tissues. The cnidarians (phylum Cnidaria) are predominantly marine, radially symmetrical animals with unique stinging cells called cnidocytes, each of which contains a specialized harpoon apparatus, or nematocyst. The comb jellies (phylum Ctenophora) are also radially symmetrical, but probably unrelated.

5. Bilaterally symmetrical animals include three major evolutionary lines: acoelomates, which lack a body cavity; pseudocoelomates, which develop a body cavity (pseudocoel) between the mesoderm and the endoderm; and coelomates, which develop a body cavity (coelom) within the mesoderm.

6. Acoelomates are the most primitive bilaterally symmetrical animals. They lack an internal cavity, except for the digestive system, and are the simplest animals that have organs—structures made up of two or more tissues. The most prominent phylum of acoelomates, Platyhelminthes, includes the free-living flatworms and the parasitic flukes and tapeworms.

7. Pseudocoelomates, exemplified by the nematodes (phylum Nematoda), have a body cavity that develops between the mesoderm and the endoderm. This pseudocoel provides a place to which the muscles can attach, giving these worms enhanced powers of movement.

SELF-QUIZ

The following questions pinpoint important information in this chapter. You should be sure to know the answers to each of them.

1. All animals are heterotrophic. True or false?

2. Larval sponges, after settling on a suitable substance, begin their transformation into adults by
   (a) turning inside out
   (b) entering into a protective covering and going through a metamorphosis
   (c) shedding their skin
   (d) going through a meiotic division

3. Of all the animals, only the _____ have no distinct tissues.

4. Cnidarians obtain their prey by "harpooning" it with structures called _____.

5. In an animal possessing _____ symmetry, the right and left halves of the body are mirror images of one another.

6. The nervous system develops from the endoderm. True or false?

7. Animals that develop a body cavity between mesoderm and endoderm are called

   _____.

8. Flatworms
   (a) lack a circulatory system, and therefore all of their cells must be within diffusion distance of oxygen and food
   (b) have a closed circulatory system with a heart
   (c) have an open circulatory system
   (d) have no need for a circulatory system because they have no organs

9. In a nematode's complete digestive tract, the food first passes through the mouth because of the sucking action of a muscular chamber called the _____.

10. Young sponges are free-swimming, while adult sponges are sessile—anchored in place. True or false?

11. The number and complexity of sense organs are much greater in the radially symmetrical animals than in bilaterally symmetrical ones. True or false?

12. Is a nematode a solid worm or a worm that has a body cavity?

THOUGHT QUESTIONS

1. Outline some of the advantages of bilateral symmetry.

2. Parasites, especially those that require two or more hosts to complete their life cycles, often produce very large numbers of offspring. What advantage would this present to them?

3. How might you investigate whether sponges shared a common ancestor with the other animals or not?

FOR FURTHER READING

BURNETT, A.L.: *Biology of Hydra,* Academic Press, Inc., New York, 1973. An overall account of this fascinating cnidarian.

CROLL, N.A.: *The Organization of Nematodes,* Academic Press, Inc., New York, 1976. These simple animals are being used increasingly for studies of experimental embryology.

DONNER, J.: *Rotifers,* Frederick Warne & Co., Inc., New York, 1966. A good overall account of this phylum.

GOSNER, K.L.: *Guide to Identification of Marine and Estuarine Invertebrates: Cape Hatteras to the Bay of Fundy,* Interscience Publishers, New York, 1971. An excellent field guide.

HICKMAN, C.P.: *Biology of the Invertebrates,* 2nd ed., The C.V. Mosby Co., St. Louis, 1973. A fine general text on invertebrates, very clearly written.

HICKMAN, C.P., JR., L.S. ROBERTS, and F.M. HICKMAN: *Integrated Principles of Zoology,* 7th ed., Times Mirror/Mosby College Publishing, St. Louis, 1984. An outstanding general zoology text, with thorough coverage of all groups; the standard in its field.

PENNAK, R.W.: *Freshwater Invertebrates of the United States,* The Ronald Press Co., New York, 1953. The standard identification manual.

RICKETTS, E.F., J. CALVIN, and J.W. HEDGPETH: *Between Pacific Tides,* 5th ed., Stanford University Press, Stanford, Calif., 1985. An outstanding, ecologically oriented treatment of the marine organisms of the Pacific Coast of the United States.

SCHMIDT, G.D., and L.S. ROBERTS: *Foundations of Parasitology,* 3rd ed., Times Mirror/Mosby College Publishing, St. Louis, 1985. General treatment of parasitology, with much information about parasitic groups of animals.

Overview

The evolution of a coelom—an internal body cavity surrounded by mesoderm—was a major advance for the animal kingdom that facilitated the development of complex internal organs. Perhaps the most primitive major coelomate phylum is that of the mollusks (the snails, bivalves, and octopuses), phylum Mollusca. Most mollusks are unsegmented, but it is not certain whether their ancestors were segmented or not. In the segmented worms, or annelids (marine worms, earthworms, and leeches), phylum Annelida, the body consists of a few to many dozen similar segments, lined one behind the other like a string of beads. Mollusks share with some annelids (the polychaetes) a very similar trochophore larva, and the two phyla probably share a relatively close common ancestor.

For Review *Here are some important terms and concepts that you will encounter in this chapter. If you are not familiar with them, you should review them before proceeding.*

- **Adaptation** (Chapter 20)

- **Classification** (Chapter 22)

- **Major features of evolutionary history** (Chapter 23)

- **Body plans of bilaterally symmetrical animals** (Chapter 35)

There are three types of bilaterally symmetrical animals. Two were introduced in Chapter 35: the acoelomates, represented by the flatworms, and the pseudocoelomates, worms with a body cavity that develops between their mesoderm and endoderm. Although the body plans of both of these types have proven very successful, a third way of organizing the body has been adopted by most of the animal kingdom, the "higher" invertebrates and vertebrates. It involves the development of a coelom, a body cavity that originates within the mesoderm.

We will begin our discussion of the coelomate animals with the mollusks and annelids, two major phyla that include such familiar animals as clams, snails, slugs, octopuses, earthworms, and clamworms (Figure 36-1). In these phyla we can observe all of the major evolutionary advances that are associated with the evolution of the coelom. Other groups of coelomate animals will be discussed later in this section and in the following one.

THE ADVENT OF THE COELOM

The evolution of an internal body cavity made possible a significant advance in animal architecture. Consider for a moment the limitations of a solid body: a solid worm has no internal circulatory or digestive systems, and all of its internal organs are pressed on by muscles and deformed by muscular activity. Although flatworms do have digestive systems, they are subject to problems of this kind.

An internal body cavity circumvents these limitations; its development was an important step in animal evolution. Perhaps the most important advantage of an internal body cavity is that the body's organs are located within a fluid-filled enclosure in which they can function without having to resist pressures from the surrounding muscles. In addition, the fluid that fills the cavity may act as a circulatory system, freely transporting food, water, waste, and gases throughout the body. Without the free circulation made possible by such a system, every cell of an animal must be within a short distance of oxygen, water, and all of the other substances that it requires.

The digestive system of an animal, like its circulatory system, functions much more efficiently within an open internal body cavity such as a coelom than it can when embedded in other tissues. Food can pass through the gut freely, and at a rate controlled by the animal, because the opening and closing of the gut does not depend on the movements of the animal. When the rate of passage of food is controlled, it can be digested much more efficiently than would be possible otherwise. Waste removal is also carried out much more efficiently under such circumstances.

In addition, an internal body cavity provides space within which the gonads (ovaries and testes) can expand, allowing the accumulation of large numbers of eggs and sperm. Such accumulation helps to make possible all of the diverse modifications of breeding strategy that characterize the more advanced phyla of animals. Furthermore, large numbers of gametes can be released when the conditions are as favorable as possible for the survival of the young animals.

Both pseudocoelomate and coelomate animals possess a fluid-filled body cavity, a great improvement in body design compared with the less advanced solid worms. So what is the functional difference between a pseudocoel and a coelom, and why has the latter kind of body cavity been so much more overwhelmingly successful in evolutionary terms? The digestive tract of a coelomate animal is composed of an endoderm lining and a mesoderm outer portion, suspended within the coelom cavity. In pseudocoelomates, the mesoderm outer portion of the digestive tract is missing, which evidently makes the differentiation of the gut into regions more difficult.

The evolutionary success of coelomates seems to reflect the repositioning of the body cavity. In coelomates, the cavity develops not between endoderm and mesoderm, but entirely within the mesoderm, and all of the internal organs and the circulatory system can now be assembled entirely by a single system of embryological controls—the mesoderm's; these controls direct their development, making it easier for complex organ systems to develop. The evolutionary specialization of the internal organs of the coelomates has far exceeded that of the pseudocoelomates. Very few pseudocoelomates, for example, possess a true circulatory system (Rhynchocoela are an exception). Many coelomates, however, have developed a system of blood vessels derived from mesoderm, by means of which they circulate blood throughout the body.

The evolution of the coelom was a major improvement in animal body architecture. It permitted the development of a closed circulatory system, provided a fluid environment within which digestive, sexual, and other organs could be suspended, and facilitated muscle-driven body movement.

The gut is positioned, along with other organ systems of the animal, suspended in the coelom; the coelom, in turn, is surrounded by a layer of cells, the **epithelium,** which is entirely derived from the mesoderm. The portion of the epithelium that lines the outer

FIGURE 36-1

A *A mollusk, the terrestrial snail* Allogona townsendiana.
B *An annelid, the Christmas-tree worm,* Spirobranchus giganteus. *This annelid is a polychaete, a member of the largest of the three annelid classes.*

A

B

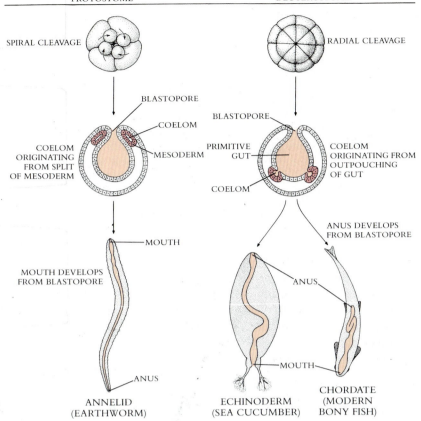

Within the figure:

PROTOSTOME   DEUTEROSTOME

SPIRAL CLEAVAGE   RADIAL CLEAVAGE

BLASTOPORE

COELOM

COELOM
ORIGINATING
FROM SPLIT
OF MESODERM   MESODERM

BLASTOPORE

PRIMITIVE
GUT   COELOM
ORIGINATING FROM
OUTPOUCHING
OF GUT

COELOM

ANUS DEVELOPS
FROM BLASTOPORE

MOUTH

MOUTH DEVELOPS
FROM BLASTOPORE

ANUS

ANUS

MOUTH

ANNELID
(EARTHWORM)   ECHINODERM
(SEA CUCUMBER)   CHORDATE
(MODERN
BONY FISH)

wall of the coelom is called the **parietal peritoneum,** and the portion that lines the internal organs suspended within the cavity is called the **visceral peritoneum.**

In the smaller coelomates, the coelom serves as a hydrostatic system, one based on the pressure of enclosed fluids, like the brake systems in most automobiles. It then plays a role similar to that of the pseudocoel, as described in Chapter 35. Such a hydrostatic system greatly increases the efficiency of the movements of animals that possess it, compared with those of the acoelomates.

AN EMBRYONIC REVOLUTION

There are two major branches of coelomate animals, representing two distinct evolutionary lines. In the first, which includes the mollusks, annelids, and arthropods, as well as some smaller phyla, the mouth (stoma) develops from or near the blastopore. This pattern of embryonic development also occurs in all noncoelomate animals. Animals whose mouth develops in this way are called **protostomes.** If such an animal has a distinct anus or anal pore, it develops later in another region of the embryo. The fact that this kind of developmental pattern is so widespread makes it virtually certain that it is the original one for animals as a whole and that it was characteristic of the common ancestor of all eumetazoan animals.

A second, very distinct pattern of embryological development occurs in the echinoderms, the chordates, and a few other small related phyla. In these animals, the anus forms from or near the blastopore, and the mouth forms subsequently on another part of the blastula (Figure 36-2). This group of phyla consists of animals that are called the **deuterostomes.** They are clearly related to one another by their shared pattern of embryonic development, which differs radically from the protostome pattern and was undoubtedly derived from it.

In addition to the pattern of blastopore formation, deuterostomes differ from protostomes in a number of other fundamental embryological features:

1. The progressive division of cells during embryonic growth is called **cleavage.** The pattern of cell cleavage relative to the embryo's polar axis determines the way in which the cells are arrayed. In nearly all protostomes, each new cell buds off at an angle oblique to the polar axis. As a result, a new cell nestles into the space between the adjacent older ones, resulting in a closely packed array. This kind of pattern is called **spiral cleavage,** because a line drawn through a sequence of dividing cells spirals outward from the polar axis (Figure 36-2). In deuterostomes, however, the cells divide parallel to and at right angles to the polar axis. As a result, the pairs of cells that result from each division are positioned directly above and below one another; this process gives rise to a loosely packed array of cells. Such a pattern is called **radial cleavage,** because a line drawn through a sequence of dividing cells describes a radius outward from the polar axis.

2. In protostomes, the developmental fate of each cell in the embryo is fixed when that cell first appears. Even at the four-celled stage, each cell is different, and no one of them, if separated from the others, can develop into a complete animal, because the chemicals that act as developmental signals have already been localized in different parts of the egg. Consequently the cleavage divisions that occur after fertilization separate different signals into different daughter cells. In deuterostomes, on the other hand, the first cleavage divisions of the fertilized embryo result in identical daughter cells, any one of which can, if separated, develop into a complete organism. The commitment to prescribed developmental pathways occurs later.

3. In all coelomates, the coelom originates from mesoderm. In protostomes, this occurs simply and directly: the cells simply move away from one another as the coelomic cavity expands within the mesoderm. In deuterostomes, however, whole groups of cells usually move around to form new tissue associations. The coelom is normally produced by an invagination of the gut cavity lining, forming a depression that is closed over, which becomes the hollow cavity of the coelom.

As we discussed in Chapter 23, the first abundant and well-preserved animal fossils are about 630 million years old; they occur in the Ediacara series of Australia and similar formations elsewhere. Among these fossils, many represent groups of animals that no longer exist (see Figure 23-3, *A*). In addition, however, these ancient rocks bear evidence of the two major evolutionary lines of coelomates, the protostomes and the deuterostomes. The coelomates are certainly the most advanced evolutionary line of animals, and it is remarkable that their two major subdivisions were differentiated so early. It is clear that deuterostomes are derived from protostomes, since the protostome pattern of development is characteristic of all animals except for the four deuterostome phyla. The event, however, occurred very long ago and presumably did not involve groups of organisms that closely resemble any that are living now.

> **The deuterostomes—echinoderms, chordates, and two minor phyla—are an evolutionary line descended from a common ancestor in which the embryo develops in a new and different way from the protostomes. All other coelomates are protostomes. The orientation of the embryo, the pattern of its cell cleavage, the timing of the developmental commitment of its cells, and the degree of cell movement during embryo development—all of these features distinguish deuterostomes from protostomes.**

In this chapter we will consider two of the three major phyla of protostomes, the mollusks and the annelids. The third major phylum, the arthropods, will be discussed in the

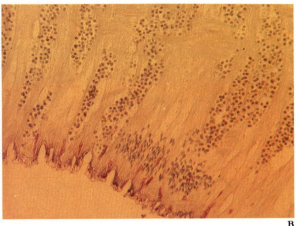

A   B

next chapter. The deuterostomes, the two largest phyla of which are the echinoderms and the chordates, will be covered in Chapter 38.

MOLLUSKS

The mollusks (phylum Mollusca) include the snails, clams, scallops, oysters, cuttlefish, octopuses, slugs, and many other familiar animals. The durable shells of mollusks are often beautiful and elegant; they have long been favorite objects for professional scientists and amateurs alike to collect, preserve, and study. The mollusks are one of the most successful of all phyla. They are the largest animal phylum, except for the arthropods, in terms of named species; there are at least 110,000, and probably at least that many more still to be discovered. Mollusks are widespread and often abundant in marine, freshwater, and terrestrial habitats.

A number of mollusks have also invaded the land, including the snails and slugs that may live in your garden. Terrestrial mollusks are often abundant in places that are at least seasonally moist. Some of these places, such as the crevices of desert rocks, may appear very dry, but even these habitats have at least a temporary supply of water at certain times of the year. There are so many terrestrial mollusks, in fact, that only the arthropods have more species adapted to a terrestrial way of life. With about 35,000 species, the terrestrial mollusks far outnumber the roughly 20,000 species of terrestrial vertebrates.

As a group, mollusks are an important source of food for humans. Oysters, clams, scallops, mussels, octopus, and squid are among the culinary delicacies that belong to this large phylum. The mollusks are also of economic significance to us in many other ways—for example, the production of pearls and of the shell material that is used as mother-of-pearl in jewelry and other decorative objects. Mollusks are not wholly beneficial to human beings, however. For example, certain mollusks are major destructive agents of timbers that are submerged in the sea including those used to build boats, docks, and pilings. Slugs and terrestrial snails often cause extensive damage to garden flowers, vegetables, and crops. Other mollusks serve as hosts to the intermediate stages for many serious diseases, including several carried by nematodes and flatworms, which we discussed in Chapter 35.

Most mollusks measure several centimeters in their largest dimension, and a number are minute. Some, however, reach formidable sizes. The giant squid, which is occasionally cast ashore but has rarely been seen alive, may be up to 21 meters long! Weighing up to at least 250 kilograms, the giant squid is the largest and heaviest invertebrate. There are thought to be millions of giant squid in the ocean, even though they are seldom caught, and they are probably increasing in number as whales, their major competitors as predators, are decreasing. Among the bivalves, one species of giant clam, *Tridacna maxima,* may have a pair of shells as much as 1.5 meters long and may weigh as much as 270 kilograms (Figure 36-3).

FIGURE 36-3

A *A giant clam,* Tridacna maxima, *on the bottom of the sea at Rose Atoll, American Samoa. The chocolate-brown color of the mantle is caused by the presence of symbiotic dinoflagellates (zooxanthellae), which probably contribute most of the food supply of the clam, although it remains a filter feeder like most bivalves. Some individual giant clams may be nearly 1.5 meters long and weigh up to 225 kilograms.*
B *Zooxanthellae in the modified hemal (blood-containing) sinuses of a giant clam. Like the corals, which also harbor zooxanthellae, the giant clams grow in nutrient-poor water and, despite that, attain very large sizes.*

FIGURE 36-4

*Body plans among the mollusks,
showing the evolution of four
of the major classes: gastropods,
cephalopods, bivalves, and chitons.*

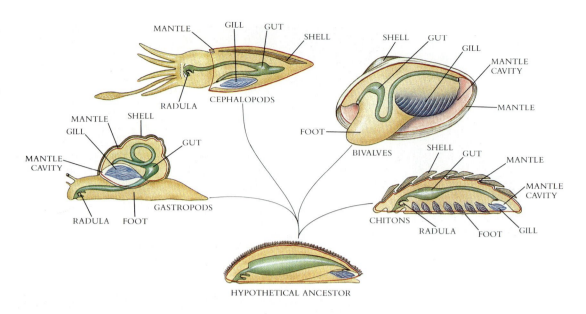

Body Plan of the Mollusks

In their basic body plan, mollusks have distinct bilateral symmetry (Figure 36-4). They have a **visceral mass** covered with a soft epithelium and a muscular **foot** that is used in locomotion. There may also be a differentiated **head** at the anterior end of the body. Within the visceral mass are found the organs of digestion, excretion, and reproduction. Folds, often two in number, arise from the dorsal body wall and enclose a cavity between themselves and the visceral mass; these folds comprise the **mantle.** Within the mantle cavity are the mollusks' gills or lungs. **Gills** are specialized portions of the mantle that usually consist of a system of filamentous projections, rich in blood vessels. These projections greatly increase the surface area available for gas exchange and therefore the animal's overall respiratory potential. Oxygen moves inward, carbon dioxide outward. Mollusk gills are very efficient, and many gilled mollusks extract 50% or more of the dissolved oxygen from the water that passes through the mantle cavity. Finally, in most members of this phylum, the outer surface of the mantle also secretes a protective **shell.**

A mollusk shell consists of a horny outer layer, rich in protein, which protects the underlying calcium-rich layers from erosion. A middle layer consists of densely packed crystals of calcium carbonate. The inner layer is pearly and increases in thickness throughout the animal's life. Pearls are occasionally produced inside the bodies of **bivalve** (two-shelled) **mollusks,** including clams and oysters, as a result of secretions of shell material around foreign bodies.

Many mollusks can withdraw for protection into their mantle cavity, which lies within the shell. In aquatic mollusks, a continuous stream of water is passed into and out of this cavity by the cilia on the gills. This water brings in oxygen and, in the case of the bivalves, food; it also carries out waste materials. When the gametes are being produced, they are carried out in the same stream. In the squids and octopuses, the mantle cavity has been modified to create the jet-propulsion system that enables the animals to move rapidly through the water.

The foot of a mollusk is muscular and may be adapted for locomotion, for attachment, for food capture (in squids and octopuses), or for various combinations of these functions. Some mollusks often secrete mucus, forming a path that they glide along on their foot. In the cephalopods—squids and octopuses—the foot is divided into arms, also called tentacles. In some pelagic forms—mollusks that are perpetually free-swimming—the foot is modified into winglike projections or thin, mobile fins.

One of the most characteristic features of mollusks is the **radula,** a rasping, tonguelike

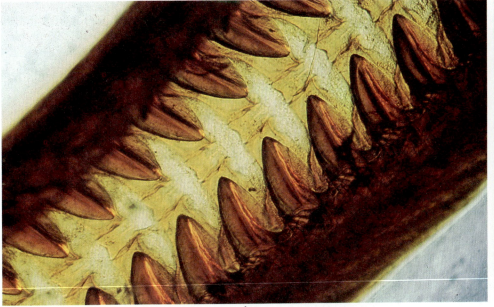

A

B

FIGURE 36-5

A *The radula, or rasping tongue, of a snail.*
B *Enlargement of the rasping teeth on a radula.*

organ that all members of the phylum have—except bivalves, in which it has almost certainly been lost; the bivalves obtain their food by filter-feeding. The radula consists primarily of chitin, can be protruded or moved by the animal, and is covered with rows of pointed, backward-curving teeth (Figure 36-5). It is used by some of the snails and their relatives, which are members of a group known as the gastropods, to scrape algae and other food materials off their substrates and then to convey this food to the digestive tract. Other gastropods are active predators; they use their radula to puncture their prey and extract food from it.

> **Mollusks are the second-largest phylum of animals in terms of named species. They characteristically have bodies with three distinct sections: head, visceral mass, and foot. All mollusks except the bivalves also possess a unique rasping tongue called a radula.**

The circulatory system of mollusks consists of a heart and, in most cases, an open system in which the blood circulates freely. The mollusk heart usually has three chambers, two of which collect aerated blood from the gills, while the third pumps it to the other body tissues. In the cephalopods, a closed system of vessels carries blood to and from the heart. These dexterous and rapidly moving animals also have auxiliary hearts, which increase the efficiency of their overall pumping system. In mollusks, the coelom is primarily represented by a small area around the heart.

In most coelomate animals, most of which have a closed circulatory system, the blood vessels are intimately associated with the excretory organs, making possible the direct exchange of materials between these two systems. For this reason, the excretory systems of these coelomates are much more efficient than the flame cells of the acoelomates, which pick up substances only from the body fluids. Nitrogen-rich wastes are removed from the mollusk by one or two tubular structures called **nephridia.** Mollusks were one of the earliest evolutionary lines to have developed an efficient excretory system.

A typical nephridium has an open funnel, the **nephrostome,** which is lined with cilia. From the nephrostome, a coiled tubule runs into an enlarged bladder, which in turn is connected to an excretory pore. Wastes are gathered by the nephridia from the coelom, which is located around the heart only, and discharged into the mantle cavity, from which they are expelled by the continuous pumping of the gills. As the wastes leave the mollusk, sug-

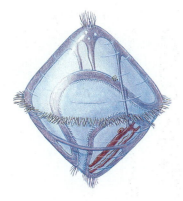

FIGURE 36-6

The trochophore larva of a mollusk. Similar larvae, as you will see, are characteristic of the annelid worms.

FIGURE 36-7

Veliger stage of a mollusk.

FIGURE 36-8

A chiton, Tonicella lineata, *class Polyplacophora. The shells of chitons, unlike those of other mollusks, consist of overlapping calcareous plates.*

ars, salts, water, and other materials are absorbed by the walls of the nephridia and returned into the animal's body as needed to achieve an appropriate osmotic balance. In animals with a closed circulatory system, such as annelids, some mollusks, and the vertebrates, the coiled tubule of a nephridium is surrounded by a network of capillaries. Wastes are extracted from the circulatory system through these capillaries, and transferred into the nephridium, from which they are subsequently discharged, while salts, water, and other associated materials may also be reabsorbed from the tubule of the nephridium back into the capillaries. All coelomates except for arthropods and chordates have basically similar excretory systems.

> **Mollusks were among the earliest animals to evolve an efficient excretory system, based on nephridia, small tubes that gather wastes by diffusion from the coelomic fluid.**

Reproduction in Mollusks

Most mollusks have distinct male and female individuals, although a few bivalves and many gastropods are hermaphroditic, with both sexes occurring in a single individual. Even in hermaphroditic mollusks, however, cross-fertilization, rather than self-fertilization, is most frequent. Remarkably, some sea slugs and oysters are able to change from one sex to the other several times during a single season.

Many marine mollusks have free-swimming larvae called **trochophores** (Figure 36-6). Trochophores are very distinctive in structure; they are propelled through the water by a row of cilia that encircles the middle of their body. In most marine snails and in the bivalves, there is a second free-swimming stage that follows the trochophore stage. In this **veliger** stage the beginnings of a foot, shell, and mantle can be seen (Figure 36-7). Mollusks are dispersed from place to place largely as trochophores and veligers. The trochophores and veligers drift widely in the currents to new areas.

The Classes of Mollusks

There are seven classes of mollusks. Some of the smaller classes have greatly helped us to understand the limits of variation in the phylum and the nature of its probable ancestor. By comparative study of these animals, some scientists have concluded that the ancestral mollusk was probably a dorsiventrally flattened, unsegmented, wormlike animal that glided on its ventral surface. These scientists believe that this hypothetical animal had a moderate amount of chitinous cuticle and overlapping calcareous scales. Alternative hypotheses are certainly possible, but this view of ancestral mollusks provides one plausible way of viewing the evolution of the phylum.

A contemporary class of mollusks in which many of these characteristics still exist is the class Polyplacophora, the chitons. These marine mollusks have oval bodies with eight overlapping calcareous plates (Figure 36-8). Underneath the plates is a broad, flat foot that chitons use to creep along, surrounded by a groove or mantle cavity in which the gills are arranged. Most of the chitons are grazing herbivores that live in shallow marine habitats, but some have been found in depths of more than 7000 meters.

We will treat three classes of mollusks in some detail as representatives of the phylum: (1) Gastropoda—snails, slugs, limpets, and their relatives; (2) Bivalvia—clams, oysters, scallops, and their relatives; and (3) Cephalopoda—squids, octopuses, cuttlefishes, and nautilus.

Snails and slugs

About 80,000 named species of snails and slugs compose the gastropods (class Gastropoda; see Figures 36-1, *A*, and 36-9). This class is primarily a marine group that is also abundant in both freshwater and terrestrial habitats. Gastropods have a single shell, which has been

A

C

B

D

FIGURE 36-9

Gastropod mollusks.
A *The textile cone,* Conus textilis. *The radulas of the cone shells, which are active hunters, are modified into harpoons.*
B *Hawaiian limpet,* Helconiscus exaratus, *viewed from below.*
C *A slug,* Arion ater, *one of the approximately 500 species of terrestrial mollusks that lack shells.*
D *A nudibranch,* Hermissenda crassicornis.

lost in some groups, and a body that is generally divisible into a head, foot, and visceral mass. These animals generally creep along on their foot, which may be modified for swimming. Slugs and nudibranchs (marine slugs) are clearly descended from ancestors that had a shell. The shell of most marine gastropods is closed by a plate, the **operculum,** that the animal can pull into place. Most land gastropods lack an operculum as adults.

On their head, most gastropods have a pair of tentacles, on which the eyes may be located. The tentacles have been lost, however, in some of the more advanced forms of the class. The mouth opening in gastropods may be simple, or it may be modified as a proboscis. Within the mouth cavity of many members of this class are horny jaws and a radula.

The visceral mass of the gastropods is twisted and has become asymmetrical during the course of evolution. This twisting, or **torsion,** occurs during the embryological development of the gastropod and results in a thorough rearrangement of the body. It takes place separately from the twisting of the shell and results from the fact that one side of the larva grows much more rapidly than the other. This evolutionary process, which is considerably advanced in certain members of this class, is sometimes associated with other changes. For example, many gastropods have lost their right gill and sometimes also their right nephridium.

> **Gastropods typically, but not always, possess a hard shell, in which they live, and a "door" called an operculum. During gastropod development, one side of the embryo grows more rapidly than the other, producing a characteristic twisting of the visceral mass.**

A   **B**

FIGURE 36-10

Bivalves.
A *Scallop,* Chlamys hericia.
Note the blue eyes around the
margin of the body.
B *File shell,* Lima scabra,
swimming.

Gastropods display extremely varied feeding habits. Some are predaceous, others scrape algae off rocks (or aquarium glass), and others are scavengers. Many are herbivores, and some of the terrestrial ones, as we have mentioned, can be serious garden and agricultural pests. The radula of whelks is stalked and is used to bore holes in the shells of other mollusks, through which the contents are then sucked out. In cone shells (Figure 36-9, *A*), the stalked radula has been modified into a kind of harpoon, which is shot suddenly into the prey with an injection of poison. Nudibranchs (Figure 36-9, *D*) are active predators; a few species of nudibranchs protect themselves with nematocysts they obtain from the polyps they eat.

In terrestrial gastropods, an area under the mantle that is extremely rich in blood vessels serves, in effect, as a lung. This lung evolved in animals living in environments with plentiful oxygen. It absorbs oxygen more efficiently than a gill does for an animal living on land and breathing air. The lung is essentially an empty mantle cavity, in the space that gills occupy in the aquatic ancestors of terrestrial mollusks. Some aquatic snails were probably derived from terrestrial forms that have returned to the water, as evidenced by the fact that they still come to the surface and breathe by means of lungs.

Bivalves

Members of the class Bivalvia—which includes clams, scallops, mussels, and oysters—have two shells hinged together dorsally and a wedge-shaped foot (see Figures 36-3 and 36-10). A ligament hinges the shells together and causes them to gape open. Pulling against this ligament are one or two large **adductor muscles** that can draw the shells together. The mantle secretes the shells and ligament and envelops the internal organs within the pair of shells. Additionally the mantle is drawn out to form two siphons, one for the incoming and one for the outgoing stream of water. There is a pair of gills on each side of the visceral

A

B

FIGURE 36-11

Pearly freshwater mussels, members of the family Unionidae, from the United States. More than 50 of the estimated 500 species in North America may already have become extinct, and many of the remainder are threatened with extinction by the pollution of the waters in which they occur.
A Epioblasma obliquata perobliqua, *from the St. Joseph River, Indiana. This species was formerly widespread in the Wabash and upper Maumee River systems in Indiana. It is almost certainly extinct in the former drainage, and there are no records at all for more than 20 years, despite intensive searching in the areas where it formerly occurred.*
B Quadrila sparsa, *officially listed by the U.S. government as "endangered," once occurred widely in the Clinch and Holston rivers of Tennessee and in the Cumberland River, Kentucky. Reduced greatly in extent by dam construction, pollution, gravel dredging, and strip-mining, this species now occurs only in the Powell River, Tennessee, where it is endangered by silting, which occurs primarily as a result of strip-mining.*

mass, each pair covered by a fold of the mantle. These gills consist of pairs or sheets of filaments that contain many blood vessels. Within the gills is a rather elaborate pattern of water circulation.

Bivalves do not have distinct heads or radulas, differing from gastropods in this respect (see Figure 36-4). They also lack tentacles, which are characteristic of cephalopods. However, their foot is large and muscular. It may be adapted, in different species, for creeping about, burrowing, cleansing the animal, or anchoring it in place in its burrow. Some species of clams can dig into sand or mud very rapidly by means of muscular contractions of their foot.

Most bivalves are sessile filter-feeders. They extract small organisms from the water that they filter through their mantle cavity by the ciliary action of their gills, squirting it out through the outgoing siphon at the other end of the animal. The food particles that bivalves ingest are entangled by masses of mucus secreted by glands, mainly located on the gills. On either side of the mouth is usually a pair of organs, the **palps,** which aid in handling particles of food.

Bivalves disperse from place to place largely as larvae. Most of the adults are fixed and relatively immobile, or even confined to a burrowing way of life. Some adult bivalves do, however, move about readily. Many genera of scallops, for example, can move swiftly through the water by using their large adductor muscles, which are what we usually eat as "scallops," to clap their shells together. (Small circular plugs of muscle cut from skates and rays, cartilaginous fishes, are sometimes also served as "scallops," however.) One of a scallop's shells is larger than the other, and the edge of their body is lined with tentacle-like projections. Their complex eyes can differentiate light and darkness and therefore, presumably, the shadows that potential predators, such as starfish, cast. Scallops can also detect predators by means of chemical signals, which prompts them to flee.

There are about 10,000 species of bivalves. Most of these are marine, although many species also live in fresh water. One of the more interesting freshwater groups is the pearly freshwater mussels, or naiads. There are nearly 1200 species of this group, which has a worldwide distribution. One of its families, Unionidae, includes more than 500 species that occur in the rivers and lakes of North America (Figure 36-11). The larvae of this family of mollusks parasitize fishes; they are brooded in a special pouch before they are released, thus exhibiting a very unusual life cycle for a mollusk.

Octopuses, squids, and nautilus

The more than 600 species of the class Cephalopoda—the octopuses, squids, and nautilus—are the most intelligent of the invertebrates. They are active marine predators that swim, often swiftly, and compete successfully with fish. The giant squid, mentioned earlier, plays an ecological role like that of the large marine mammals, such as the killer whales, and the large, predaceous fishes. Cephalopods feed primarily on fishes, other mollusks, crustaceans, and worms. The foot has evolved into a series of tentacles equipped with suction cups, adhesive structures, or hooks that seize prey efficiently. Squids have 10 tentacles; octopuses (as indicated by their name) 8; and the nautilus about 80 to 90 (Figure 36-12). Once the tentacles have snared the prey, it is bitten with strong, beaklike paired jaws and pulled into the mouth by the tonguelike action of the radula.

Cephalopods have highly developed nervous systems, and their brains are unique among the mollusks. Their rapid responses are made possible by a bundle of giant nerve fibers attached to the muscles of the mantle. Their eyes are very elaborate, and the retina has a structure very much like that of vertebrate eyes, although there are fundamental differences (see p. 451). Squid eyes can also be very large; in fact, the eyes of a giant squid that was washed up on a beach in New Zealand in 1933 were 40 centimeters across, the largest eyes known in any animal! Many cephalopods exhibit complex patterns of behavior and a high level of intelligence. For example, octopuses can be trained easily to distinguish among classes of objects. Most members of this class have closed circulatory systems; they are the only mollusks that do have such systems.

A

B

C

FIGURE 36-12

Cephalopod diversity.
A *An octopus. Octopuses generally move slowly along the bottom of the sea.*
B *A squid. Squids are active predators, competing effectively with fish for prey.*
C *Pearly nautilus,* Nautilus pompilius, *from the Philippines.*

GIANT TUBE WORMS

There are about 100 known species of the phylum Pogonophora, the beardworms. These worms are protostome coelomates, related to the annelids and mollusks. Some of these sedentary worms live in cool waters a few hundred meters deep at relatively high latitudes; others occur at great depths—up to 3000 meters or more—nearer the equator. Most of the beardworms are less than a millimeter in diameter, but they range from 10 centimeters to 3 meters or more long. The worms live within tubes that they construct of chitin and rigid protein. They are the only nonparasitic animals that have no trace of a mouth, a gut, or an anus. They apparently absorb their nutrients, including amino acids, glucose, and fatty acids, through their more than 200,000 slender tentacles, which protrude from their tubes and give their anterior regions a feathery appearance.

The largest and most unusual members of this curious phylum were discovered in 1977 in extraordinary communities that occur along warm-water vents in fissures deep in the Pacific Ocean, which scientists have been exploring with submersible vehicles. Water jets from these fissures at a scalding 350° C but is soon cooled to the 2° C temperature of the surrounding water. The abundant hydrogen sulfide emerging from the vents makes possible the existence of a rich community of bacteria, both inside and outside the bodies of the vent organisms. Some of these bacteria have been shown to be capable of growing at temperatures of at least 250° C, much higher than previously suspected for any living organism. Of course, the atmospheric pressure must be high enough to maintain the existence of liquid water at that temperature—no problem at the bottom of the sea.

The productivity of the chemoautotrophic bacteria in turn makes possible the existence of a diverse animal community. An incredible concentration of animals—including dense clusters of clams, shrimps, limpets, crabs, fishes, and, most spectacular of all, these giant worms—lives around these vents. One of the body segments of each beardworm is packed full of chemoautotrophic

---

The octopuses and other cephalopods are efficient and often large predators. They possess well-developed brains and are the most intelligent of the invertebrates.

Although they evolved from shelled ancestors, living cephalopods, except for the few species of nautilus, lack an external shell. The squids and cuttlefish retain an internal remnant of their ancestral shells, which serve as internal stiffening supports, but the octopuses have no trace of a shell. Cuttlefish "bones" are used in canary cages to provide calcium for the birds.

Like other mollusks, cephalopods take water into the mantle cavity and expel it through a siphon. The cephalopods, however, have modified this system into a means of jet propulsion. When threatened, they eject water violently and shoot themselves through the water. Squids, which have streamlined bodies and fins, have a particularly efficient jet-propulsion system; they also have lateral undulating fins running along their sides, which help to propel them through the water. Octopuses can also swim, but they spend most of their time climbing over rocks and other objects on the ocean floor. The members of both groups can swim in various directions by turning their siphon. Both can also release from special sacs a dark fluid, once used for ink, that clouds the water and thus helps to disguise the direction of their escape.

The sexes of cephalopods are separate. The sperm are stored in spermatophores in a sac that opens into the mantle cavity of the male. During copulation the male uses a specialized arm, or tentacle, to transmit a spermatophore from its own mantle cavity into the female's. The female then fertilizes the eggs as they leave the oviduct, attaching them to stones and other objects.

bacteria, with up to 4 billion bacteria per gram of contents. These bacteria oxidize hydrogen sulfide, which is produced abundantly along the vents, and provide the worms with organic molecules. Pure sulfur is deposited within the bodies of the worms as a result of this process. The worms and other animals that live around the vents must bind the hydrogen sulfide to avoid being poisoned by it. A protein in the worm's blood, apparently a modified hemoglobin, forms a complex with oxygen and also binds the hydrogen sulfide, transporting it to the symbiotic bacteria in the worm's body. The hydrogen sulfide is immediately metabolized by the bacteria when it is released. Since the vents are active for only a few years, these whole communities of animals must reach maturity rapidly and disperse efficiently to new localities while their vents are still active.

Similar communities have recently been found in relatively cold water where there is an abundant, energy-rich substrate like hydrogen sulfide. They have also been discovered in the fossil record in rocks nearly 100 million years in age. These ecosystems have proved to be of extraordinary interest, since they are among the most important on earth that do not depend in any way on photosynthesis or on energy from the sun.

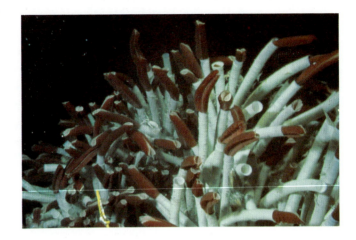

FIGURE 36-A

A colony of giant beardworms at great depth near a hot sulfur vent along the Galapagos Trench, eastern Pacific Ocean.

The pearly nautilus, or chambered nautilus (Figure 36-12, C), comprises a single genus of three to five species, which live in the western Pacific Ocean and perhaps in parts of the Indian Ocean as well. The kinds of chambered shells that are characteristic of the living pearly nautilus are frequent in the fossil record for the past 200 million years. For example, the ammonites are a related group that became extinct at the end of the Cretaceous Period, 65 million years ago (Figure 36-13). Fossils as much as 75 million years old are very similar to the living species. The shells of these living species are about 10 to 27 centimeters across their flattened plane. The exterior of these shells resembles porcelain, while the interior is pearly, hence one of the names of these animals. The animal occupies the outermost chamber of the shell. A tube leads from the inner chambers, by means of which the animal can regulate the relative proportions of gases and fluids in these chambers, and thus its density.

The 80 to 90 tentacles of a pearly nautilus are adhesive, but they lack suction cups. The outermost tentacles have thickened sheaths and, when held together, form a kind of hood that helps to protect the animal when it is retracted. In the center of the circle of tentacles is a strong, horny beak. The eyes are small and apparently function like a pinhole camera. The nautilus uses a specialized funnel leading from its mantle cavity to propel itself through the water. The animals are nocturnal, remaining hidden during the day; they feed on crustaceans and dead animals on and just above the ocean floor. They grow slowly, living a considerable length of time after they reach sexual maturity at about 10 to 15 years of age. Reproduction occurs year-round, with a few large eggs laid at any one time. In contrast, the octopuses and squids typically grow very quickly, reproduce once, and die. The nautilus is usually most abundant at depths of 200 to 500 meters but is sometimes found floating near the surface.

FIGURE 36-13

An ammonite from the Cretaceous Period. The ammonites, shelled cephalopods, were abundant during the Mesozoic Era, but abruptly went extinct at its end.

THE RISE OF SEGMENTATION

Of the major phyla of protostomes, the annelids and arthropods have segmented bodies, while most living mollusks do not. Just as it is efficient for workers to construct a tunnel from a series of identical prefabricated parts, so these advanced protostome coelomates are "assembled" from a serial succession of identical segments. During the animal's early development, these segments become most obvious in the mesoderm but later are reflected in the ectoderm and endoderm as well. Two advantages result from early embryonic segmentation:

1. Each segment of mesoderm may go on to develop a more or less complete set of the several organ systems. Damage to any one segment need not be fatal to the individual, since the other segments duplicate that segment's functions.
2. Locomotion is far more effective when individual segments can move independently, because the animal as a whole has more flexibility of movement.

Segmentation underlies the organization of all advanced animals. In some adult arthropods, the segments are fused, making it difficult to perceive the underlying segmentation, but it is usually apparent in embryological development. Even the backbone and muscular areas of the vertebrates are segmented.

Segmentation is a feature of the advanced coelomate phyla, notably the annelids and the arthropods, although it is not obvious in some of them.

ANNELIDS

The annelids (phylum Annelida), one of the major animal phyla, are segmented worms. They can be recognized by their strongly marked segmentation and are abundant in marine, freshwater, and terrestrial habitats throughout the world. Their segments are visible externally as a series of ringlike structures running the length of the body. Internally the segments are divided from one another by partitions called **septa.** In each of the cylindrical segments of these animals, the digestive and excretory organs are repeated in tandem.

The basic body plan of the annelids (Figure 36-14) is a tube within a tube, with the internal digestive tract suspended within the coelom—a tube running from mouth to anus. The anterior segments have become modified according to the habits of the particular kind of annelid, and there is a well-developed **cerebral ganglion,** or **brain,** in one of these segments. The diverse sensory organs are mainly concentrated near the anterior end of the worm's body. Some of these are sensitive to light, and elaborate eyes with lenses and retinas

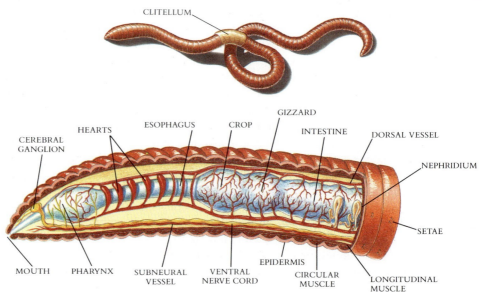

FIGURE 36-14

An earthworm, in external view **(A),** *with a cross section showing the anatomy in simplified form* **(B).**

have evolved in certain members of the phylum. Separate nerve centers, or ganglia, are located in each segment but are interconnected by nerve cords. These nerve cords are responsible for the coordination of the worm's activities.

Annelids use their muscles to crawl, burrow, and swim. In each segment, the muscles play against the coelomic fluid, which creates a hydrostatic skeleton that gives the segment rigidity. Because the septa isolate each segment into an individual hydrostatic skeletal unit, each is able to contract or expand autonomously. Therefore, a long body can move in ways that are often quite complex. When an earthworm crawls on a flat surface, for example, it lengthens some parts of its body while shortening others (Figure 36-15).

Each segment of an annelid typically possesses **setae,** bristles of chitin, that help to anchor the worms during locomotion or when they are in their burrows. In a crawling worm, for example, the setae are retracted (withdrawn into the animal) in the expanding segments, and extended to make contact with the surface in the contracted segments, anchoring them. Because of these setae, annelids are sometimes called "bristle worms." The setae may also aid some of the small marine annelids in swimming. They are absent in the leeches, except for a single species.

The members of one class of annelids, the polychaetes, develop from free-swimming trochophore larvae very similar to those of the mollusks (see Figure 36-6). The existence of such larvae in both phyla makes it extremely probable that they share a common ancestor, which may have resembled a flatworm. A primitive coelom would then have developed in this common ancestor. Clear segmentation is found in one small class of mollusks, the chitons, but their other features suggest that chitons are not particularly primitive and that the common ancestor of the mollusks was unsegmented. This indicates that the common ancestor of the annelids and mollusks was also unsegmented and that segmentation evolved within the annelids. Certainly annelids are an ancient phylum; Figure 23-10, *C,* shows a polychaete from the Cambrian Period, 530 million years ago. Segmentation clearly arose very early in the evolution of coelomate animals.

In any case, the elaborate and obvious segmentation of the annelid body clearly represents an evolutionary specialization. Many scientists believe that segmentation is an adaptation for burrowing; it makes possible the production of strong peristaltic waves along the length of the worm, thus enhancing vigorous digging. Segmentation itself is, however, also characteristic of the arthropods, a phylum that may share a common ancestor with the annelids. The segmentation that is characteristic of the vertebrates, on the other hand, very likely evolved independently, judging from their overall features.

Like all coelomates except for the arthropods and most mollusks, the annelids have a closed circulatory system. Blood moves through a closed circulatory system faster and more efficiently than it does through an open system. The annelids exchange oxygen and carbon dioxide with the environment through their body surfaces; they lack gills, lungs, and similar organs. Much of their oxygen supply reaches the different parts of their bodies by way of their blood vessels, however. Some of these vessels are enlarged and heavily muscular, serving as hearts in the pumping of blood. Earthworms, for example, have five pulsating blood vessels on each side that serve as hearts, helping to pump blood from the main dorsal vessel, which is their major pumping structure, to the main ventral one (Figure 36-14). The blood of the larger annelids has respiratory pigments, such as hemoglobin, dissolved in it.

The excretory system of the annelids consists of ciliated, funnel-shaped nephridia generally similar to those of the mollusks. These nephridia—each segment has a pair—collect waste products and transport them out of the body through the coelom by way of specialized excretory tubes. In mollusks, these waste products pass first into the mantle cavity, which is basically exterior to the body.

The annelids are characterized by serial segmentation. The body is composed of numerous similar segments, each with its own circulatory, excretory, and neural elements and each with its own array of setae.

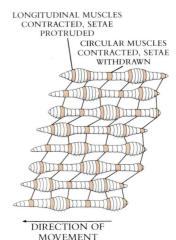

LONGITUDINAL MUSCLES CONTRACTED, SETAE PROTRUDED

CIRCULAR MUSCLES CONTRACTED, SETAE WITHDRAWN

←DIRECTION OF MOVEMENT

FIGURE 36-15

Locomotion in an earthworm. As the worm moves, some segments become long and thin while others become thick and fat.

A

B

FIGURE 36-16

Polychaete annelids.
A *Fan worm,* Sabella melanostigma.
B *Tube worms,* Eudistylia polymorpha, *withdrawn into their tubes.*
C (Opposite page) *Shiny bristleworm,* Oenone fulgida.

Classes of Annelids

The roughly 12,000 described species of annelids occur in many different habitats. They range in length from as little as 0.5 millimeter to more than 3 meters in some giant Australian earthworms and some polychaetes. There are three classes of annelids: (1) the Polychaeta, which are free-living, almost entirely marine bristle worms, comprising some 8000 species; (2) the Oligochaeta, the terrestrial earthworms and related marine and freshwater worms, with some 3100 species; and (3) the Hirundinea, the leeches, mainly freshwater predators or bloodsuckers, with about 500 species. The annelids are believed to have evolved in the sea, with the polychaetes being the primitive class. Oligochaetes seem to have evolved from polychaetes, perhaps by way of brackish water to freshwater estuaries and then to streams. Leeches share with oligochaetes an organ called a **clitellum,** which secretes a cocoon specialized to receive the eggs. It is generally agreed that leeches evolved from oligochaetes by specialization in relation to their bloodsucking habits and way of life as external parasites.

Polychaetes

The polychaetes (class Polychaeta) include the clamworms, plumed worms, scaleworms, lugworms, twin-fan worms, sea mice, peacock worms, and many others. These worms are often surprisingly beautiful, with unusual forms and sometimes iridescent colors (see Figures 36-1, *B,* and 36-16). However, other members of this large class may be unusually grotesque. Polychaetes live in such places as burrows, under rocks, in tubes of hardened mucus that they manufacture, and inside shells. They are often a crucial part of marine food chains, being extremely abundant in certain habitats. A number of polychaetes are **commensal;** they live inside sponges, in the shells of mollusks, within echinoderms or crustaceans, and in other animals, eating the food particles left over by these organisms. A very few of these worms are parasites; some are active predators.

Polychaetes have a well-developed head with specialized sense organs; they differ from other annelids in this respect. Their bodies are often highly organized into distinct regions formed by groups of segments related in function and structure. Their sense organs include eyes, which range from simple eyespots to quite large and conspicuous stalked eyes; all of these kinds of eyes primarily serve to concentrate light and indicate the direction from which it is coming.

Another distinctive characteristic of the polychaetes is the paired, fleshy, paddlelike flaps, called **parapodia,** on most of their segments. These parapodia, located along the often numerous setae, are used in swimming, burrowing, or crawling. They also play an important role in gas exchange, because they greatly increase the surface area of the body. In some polychaetes that live in burrows or tubes, the parapodia may feature hooks that help to anchor the worm. Slow crawling is carried out by means of the parapodia, while rapid crawling is aided by undulating motions of the body. In addition, the polychaete epidermis often includes ciliated cells, which the worms use to set up currents of water, thus aiding in respiration and food procurement.

The sexes of polychaetes are usually separate, and fertilization is often external, occurring in the water away from both parents. Also unlike other annelids, polychaetes usually lack permanent male gonads, the sex organs in which sperm are produced. These polychaetes produce their sperm directly from cells in the lining of the coelom or in their septa. The eggs of polychaetes may be laid in gelatinous masses and attached to the substrate or incubated in a tube or brood chamber in or on the female's body. Fertilization results in the production of a ciliated, mobile trochophore larva similar to the larva of a mollusk. The trochophores develop for long periods in the plankton before beginning to add segments and thus changing to a juvenile form that more and more closely resembles the adults.

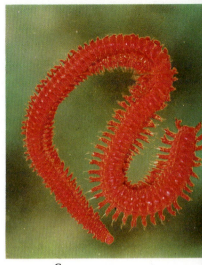

C

Earthworms

The earthworms (class Oligochaeta) literally eat their way through the soil, anchoring themselves by means of their setae. Earthworms suck in organic and other material by contracting their strong pharynx; everything that they ingest passes through their long, straight digestive tracts. In one region of this tract, the gizzard, muscles are concentrated; here the worm grinds up the organic material with the help of the soil particles that it also takes in. The body of an earthworm (Figure 36-14) consists of 100 to 175 similar segments, with a mouth on the first of these and an anus on the last.

The material that passes through an earthworm is deposited outside of its burrow in the form of castings, familiar objects that look as if they had been extruded from a toothpaste tube. In this way, earthworms aerate and enrich the soil, a subject that fascinated Charles Darwin, particularly in his later years. In his book *The Formation of Vegetable Mould through the Action of Worms,* Darwin pointed out that a worm could eat its own weight in soil every day and that the equivalent of 22 to 40 metric tons of soil per hectare pass through their intestines every year. This indicates a population of well over 16,000 worms per hectare!

In view of the underground life-style that earthworms have evolved, it is not surprising that they have no eyes. However, earthworms do have light-sensitive and touch-sensitive organs. These are not as elaborate as are the corresponding organs in polychaetes. The light-sensitive organs of the oligochaetes are concentrated in those segments near each end of the body—those regions that are most likely to be exposed to light. Earthworms also have sensitive moisture-detecting cells, an important feature since they, like all annelids, carry out their gas exchange, a highly moisture-sensitive process, through their skins. Earthworms have fewer setae than do the polychaetes, and no parapodia. In addition, earthworms lack the distinct head regions that characterize polychaetes.

Earthworms are hermaphroditic, another way in which they differ from most polychaetes. When they mate (Figure 36-17), their anterior ends point in opposite directions, and their ventral surfaces touch. The clitellum is obvious as a thickened band on an earthworm's body. The mucus it secretes holds the worms together during copulation. Sperm cells are released from pores in specialized segments of one partner into the sperm receptacles of the other, the process going in both directions simultaneously.

Two or three days after the worms separate, the clitellum of each worm secretes a mucus cocoon, surrounded by a protective layer of chitin. As this sheath passes over the female pores of the body—a process that takes place as the worms move—it receives eggs. As it subsequently passes along the body, it incorporates the sperm that were deposited during copulation. When the mucous sheath finally passes over the end of the worm, its ends pinch together. The sheath then encloses the fertilized eggs in a cocoon from which the young worms ultimately hatch.

FIGURE 36-17

Earthworms mating.

FIGURE 36-18

*A freshwater leech, with
young leeches visible inside
its body.*

Leeches

Leeches (class Hirundinea) occur mostly in fresh water, although a few are marine and
some tropical leeches occur in terrestrial habitats. Most leeches are 2 to 6 centimeters long,
but one tropical species reaches up to 30 centimeters. Leeches are usually flattened dorsiventrally, like flatworms (Figure 36-18). They are hermaphroditic, like the oligochaetes,
and develop a clitellum during the breeding season; cross-fertilization is obligatory. Leech
eggs are enclosed in cocoons and laid on damp earth, attached to objects in the water, or
occasionally attached to the host or to the parent leech itself.

The coelom of leeches is reduced and continuous throughout the body. It is not divided into individual segments like those of the polychaetes and oligochaetes. Leeches have
evolved suckers at either one or both ends of the body. Those that have a sucker at both
ends move by attaching first one and then the other end to the substrate and looping along.
Except for one species, leeches have no setae.

Most leeches are predators or scavengers, but some have evolved the habit of sucking
blood from mammals, including humans, and other vertebrates; a few suck blood from
crustaceans. Many freshwater leeches live as external parasites, remaining on their hosts for
long periods and sucking their blood from time to time. Some terrestrial leeches even climb
trees to seek mammals and birds.

The best-known leech is the medical leech, *Hirudo medicinalis,* which was used for
centuries to remove what was thought to be "excess" blood responsible for certain ill-
nesses. Individuals of *Hirudo* are 10 to 12 centimeters long and can suck a great deal of blood
from a person relatively rapidly. They were gathered in large numbers from the ponds of
Europe for this purpose until the nineteenth century. Their mouth has chitinous teeth that
rasp through the skin of the victim, and the leech secretes an anticoagulant into the wound
to prevent clotting of the blood as it flows out. The powerful sucking muscles of the leech
pump the blood out quickly once the hole has been opened. Bloodsucking leeches gorge
themselves when they feed, and the animals can survive for months between meals.

The medical leech is now used as a source of anticoagulant, which is used in research
into bloodclotting. The animal is still widely collected by European pharmaceutical companies for this purpose and is becoming rare in certain areas as a result. Attempts are now
being made to transfer the genes associated with the desired compounds to bacterial plasmids and thus to be able to produce them commercially—an application of molecular biology to a relationship that has been familiar to, and used by, human beings for millennia.

Mollusks and annelids are the two great phyla of protostome coelomate animals,
both abundantly represented in the sea, but with large numbers of species also in terrestrial
and freshwater habitats. The phyla originated early in the course of evolution of multicellular animals, and have proliferated extensively over hundreds of millions of years subsequently. In the next chapter, we will consider the third great group of protostomes, the
arthropods, a phylum that in the number of its species exceeds all others.

SUMMARY

1. The coelomates, or higher animals, are characterized by their internal body cavity, or coelom, which develops entirely within the mesoderm.

2. An internal body cavity, whether a pseudocoel or a coelom, not only makes an animal rigid but also provides an area where circulatory, digestive, sexual, and other systems can function free of the pressure of surrounding muscles.

3. A pseudocoel—which develops between the endoderm and the mesoderm—and a coelom are functionally equivalent, but the developmental processes that lead to the formation of organ systems within a coelom are much simpler, because the stimuli originate within a single type of embryonic tissue.

4. The two major evolutionary lines of coelomate animals—the protostomes and the deuterostomes—were both represented among the oldest known fossils of multicellular animals, dating back some 630 million years.

5. In the protostomes, the mouth develops from or near the blastopore, and the early divisions of the embryo are spiral. At early stages of development, the fate of the individual cells is already determined, and they cannot develop individually into a whole animal.

6. In the deuterostomes, the anus develops from or near the blastopore, and the mouth forms subsequently on another part of the blastula. The early divisions of the embryo are radial. At early stages of development, each cell of the embryo can differentiate into a whole animal.

7. The major phyla of protostomes are Mollusca, Annelida, and Arthropoda; the major phyla of deuterostomes are Echinodermata and Chordata.

8. The mollusks constitute the second largest phylum of animals in terms of named species. Their body plan consists of distinct parts, a visceral mass, a foot, a head, and a mantle, all of which have been modified greatly in the course of evolution of the different classes—including the gastropods (snails and slugs), bivalves, and cephalopods (octopuses, squids, and nautilus).

9. Annelids are segmented worms, consisting of the largely marine polychaetes; the largely terrestrial earthworms; and the largely freshwater leeches. These animals have a highly repetitive body plan.

SELF-QUIZ

The following questions pinpoint important information in this chapter. You should be sure to know the answers to each of them.

1. In coelomates, the internal cavity is created entirely within a single tissue. Which tissue?

2. Fluid-filled body cavities are characteristic of
   - (a) pseudocoelomates
   - (b) coelomates
   - (c) both
   - (d) neither

3. Which of the following is not an advantage of an internal body cavity?
   (a) a space in which the gonads can expand, allowing the accumulation of large numbers of eggs and sperm
   (b) a space in which the ribs could expand to protect the developing heart
   (c) a space for a more efficient digestive tract
   (d) a space for a more efficient circulatory system
   (e) a space in which the body's organs can carry out their activities without pressure from surrounding muscles

4. Animals in which the mouth develops from or near the blastopore are known as _____, and differ from _____, in which the anus forms from or near the blastopore while the mouth forms on another part of the embryo.

5. Mollusks have _____ symmetry.

6. Within the mantle cavity of mollusks are located
   (a) the gills or lungs   (c) the stomach, pancreas, and liver
   (b) the reproductive organs   (d) the heart

7. Mollusks usually have a closed circulatory system. True or false?

8. The coelom in mollusks is mainly an area for
   (a) digestion   (c) reproduction
   (b) respiration   (d) the heart

9. Nitrogen-rich wastes in mollusks are removed from the body by structures called _____.

10. In the annelids, the partitions that separate the segments internally are known as septa. True or false?

11. Segmentation is an evolutionary specialization for
    (a) burrowing   (c) crawling
    (b) swimming   (d) all of the above

12. Leeches have an organ, the _____, which secretes a cocoon specialized for the reception of their eggs.

THOUGHT QUESTIONS

1. Would you expect the pattern of cell division in the early embryos of the acoelomate and pseudocoelomate animals to be spiral? Why?

2. The ancestral mollusk is thought to have had a very limited shell, consisting mainly of calcareous plates; many contemporary mollusks now seem to be in the process of losing their shells. Most mollusks, however, still have well-developed shells. What is the evolutionary advantage of having a shell? Of not having one?

3. In what ways is an earthworm more complex than a flatworm?

4. Why do we believe that the ancestor of the deuterostomes is a protostome?

FOR FURTHER READING

DALES, R.P.: *Annelids,* 2nd ed., The Hutchinson Publishing Group, Ltd., London, 1967. A good overall account of this important phylum.

GOSLINE, J.M., and M.E. DEMONT: "Jet-Propelled Swimming in Squids," *Scientific American,* January 1985, pages 96–103. Describes how a squid jets through the water as rapidly as a fish for short distances by contracting radial and circular muscles in its boneless mantle wall.

HICKMAN, C.P., JR., L.S. ROBERTS, and F.M. HICKMAN: *Integrated Principles of Zoology,* 7th ed., Times Mirror/Mosby College Publishing, St. Louis, 1984. An outstanding general zoology text, with thorough coverage of all groups; the standard in its field.

JONES, D.S.: "Sclerochronology: Reading the Record of the Molluscan Shell," *American Scientist,* vol. 71, pages 384–391, 1983. The study of annual growth increments in the shells of bivalves has revealed a great deal about climatic changes.

MORTON, J.E.: *Mollusks,* 5th ed., The Hutchinson Publishing Group Ltd., London, 1979. A comprehensive account of the biology of mollusks.

RICKETTS, E.F., J. CALVIN, and J.W. HEDGPETH: *Between Pacific Tides,* 5th ed., Stanford University Press, Stanford, Calif., 1985. An outstanding, ecologically oriented treatment of the marine organisms of the Pacific Coast of the United States.

SATCHELL, J.E.: *Earthworm Ecology,* Chapman & Hall, London, 1984. A collection of interesting papers on these fascinating animals.

THOMPSON, T.E.: *Nudibranchs,* T.F.H. Publications, Neptune, N.J., 1976. Beautiful color photographs of some of the members of the fascinating seaslugs, with an interesting discussion of their diversity.

WILLOWS, H.O.D.: "Giant Brain Cells in Mollusks," *Scientific American,* February 1971, pages 69–75. An account of the nervous system of the certain mollusks that have become of key importance as experimental systems.

YONGE, C.M.: "Giant Clams," *Scientific American,* April 1975, pages 96–105. These enormous undersea bivalves, although they are filter-feeders, probably owe their large size to the fact that photosynthetic dinoflagellates live within their tissues and nourish them.

Chapter 37

ARTHROPODS

Overview

Arthropods are the most successful of all living organisms in number of individuals, number of species, total mass, and complete occupation of terrestrial habitats. There are more arthropods living on earth than all other kinds of animals put together. Arthropods all have an exoskeleton that is rich in chitin and often hard. The major evolutionary advance that led to their success was the development of jointed appendages, a development that may have occurred at least three times independently. The insects and their relatives, Uniramia, were clearly derived from the annelids, but the origins of the other two major lines of arthropods are not clear. The nature of their respiratory, circulatory, excretory, and skeletal systems limits the size that arthropods can attain. This in turn may be responsible for the failure of intelligence to evolve in this otherwise uniquely successful group; their maximum potential size may not be sufficient to accommodate a large enough brain.

For Review   *Here are some important terms and concepts that you will encounter in this chapter. If you are not familiar with them, you should review them before proceeding.*

- **Water relationships** (Chapter 9)

- **The origin of species** (Chapter 20)

- **Evolutionary history** (Chapter 23)

- **Protostomes–deuterostomes** (Chapter 36)

With the evolution of the first annelids, many of the major innovations of animal structure had already appeared: the division of tissues into three primary types (endoderm, mesoderm, and ectoderm), bilateral symmetry, coelomic body architecture, and segmentation. An innovation yet remained, however, whose appearance marks the origin of the body plan that is characteristic of the most successful of all animal groups, the arthropods. This innovation was the development of jointed appendages. Such legs seem to have evolved after the ancestors of the arthropods acquired a relatively rigid, protective exoskeleton. For legs of this kind to be able to move, joints were virtually a necessity; therefore, jointed appendages should not be taken as strong evidence for a common ancestry.

Arthropods are by far the most successful of all life forms. Approximately a million species—about two thirds of all the named species on earth—are members of this gigantic phylum, and it is estimated that there may be a *minimum* of 2 million more still awaiting discovery. One scientist has recently estimated, based on the number and diversity of insects in tropical forests, that there might be as many as 30 million species of this one class alone (see Figure 22-6). If he is correct, there would be 10 times as many species of insects as of all other kinds of living animals put together! At the very least, there are several times as many species of arthropods as there are of all other plants, animals, and microorganisms. A hectare of lowland tropical forest is estimated to be inhabited by as many as 41,000 species of insects, on the average; many suburban gardens may have 1500 or more species of this gigantic class of organisms. The insects and other arthropods are abundant in all of the habitats on this planet, but they especially dominate the land, where, along with the flowering plants and the vertebrates, they determine the very structure of life. In terms of individuals, it has been estimated that approximately a billion billion insects are alive at any one time!

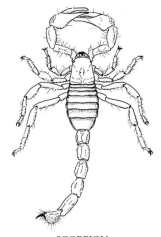

SCORPION

GENERAL CHARACTERISTICS OF ARTHROPODS

The name "arthropod" comes from two Greek words, *arthros,* jointed, and *podes,* feet. We recognize the members of this phylum especially because of their jointed appendages. The numbers of these appendages are progressively reduced in the more advanced members of the phylum, and their nature differs greatly in different subgroups. Thus individual appendages may be modified into antennae, mouthparts of various kinds, or legs. Still others—for example, the wings of certain insects—are not homologous to the other appendages of arthropods; insect wings evolved separately.

Arthropod bodies are segmented like those of annelids, a phylum to which at least some of the arthropods are clearly related. The members of some classes of arthropods have many body segments. In others, the segments have become fused together into functional groups, or **tagmata** (sing. **tagma**), such as the head or thorax of an insect (Figure 37-1), by a process known as **tagmatization,** which is of central importance in the evolution of the arthropods. But even in these arthropods, the original segments can be distinguished during the development of the larvae. To a certain extent, the segments can also be reconstructed on the basis of careful morphological comparisons of the adult structures. All arthropods have a distinct head, sometimes fused with the thorax. This is another distinguishing feature of the phylum.

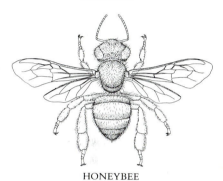

HONEYBEE

FIGURE 37-1

Evolution from many to few body segments among the arthropods; the scorpion and the honeybee are examples of arthropods with different numbers of body segments.

> **Arthropods, like annelid worms, are segmented, but in many arthropods the individual segments are fused into functional assemblies called tagmata. All arthropods have a distinct head, sometimes fused into a single unit with the thorax.**

The arthropods have a rigid external skeleton, or **exoskeleton.** As an individual outgrows an exoskeleton, that exoskeleton splits open and is shed, allowing the animal to increase in size. The eggs of arthropods develop into immature forms that may bear little or no resemblance to the adults of the same species; most members of this phylum change their characteristics as they develop from stage to stage, a process called **metamorphosis** (see Figure 37-31).

The great majority of arthropod species consist of small animals—mostly about a millimeter in length—but members of the phylum range in size as adults from some parasitic mites that are only about 80 micrometers long to a gigantic crab found in the sea off Japan, which is up to 3.6 meters across. Some lobsters are as much as 60 centimeters in length. The largest living insects are about 33 centimeters long, but the giant dragonflies that lived during the Carboniferous Period, about 300 million years ago, had wingspans of as much as 60 centimeters!

ONYCHOPHORA

The Onychophora, although closely related to certain arthropods, are different enough that they are placed in a separate phylum. They are wormlike animals with a chitinous exoskeleton; they show no evidence of segmentation except for their antennae. Onychophora range from about 1.5 to more than 15 centimeters long. They crawl over and through leaf litter in tropical and subtropical regions (Figure 37-A), and they also extend into some temperate regions of the Southern Hemisphere. The members of this group emerge and crawl about only on humid nights. Their fossil record indicates that they have changed little for more than 500 million years, and they undoubtedly attained their present distribution long before the existing continents separated from one another. These are very unusual animals, relics of the distant past that have survived in favorable habitats.

Onychophora have tracheae, like the Uniramia and some of the chelicerates; these are the only three groups in which such structures occur. The details of their embryology are identical to those of the Uniramia and the earthworms (phylum Annelida). Onychophora have ciliated nephridia, as do the annelids; Uniramia, like other arthropods, lack cilia completely. The open circulatory systems of

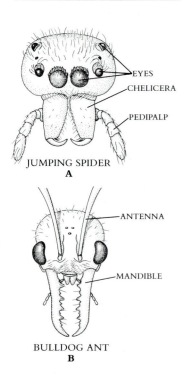

FIGURE 37-2

In the chelicerates, such as the jumping spider (A), the chelicerae are the foremost appendages of the body. Mandibulates, in contrast, such as the bulldog ant (B), have the antennae at the anterior end of the body.

Arthropods and especially the largest class of arthropods, the insects, are of enormous economic importance and affect all aspects of human life. They compete with humans for food of every kind and cause billions of dollars of damage to crops, both before and after harvest. They are by far the most important herbivores in all terrestrial ecosystems; virtually every kind of plant is eaten by one or many species of insect. Insects also prey on, parasitize, or otherwise obtain food from most kinds of animals. The diseases that they spread cause enormous financial damage each year and strike every kind of domesticated animal and plant, as well as human beings.

On the positive side, the pollination of certain crops by insects is of key significance, as is their role in controlling other insects and weeds. A number of products, such as silk and honey, are produced directly by insects. The crustaceans, a class of primarily marine and freshwater arthropods that includes crabs, lobsters, shrimps, and related animals, are important as sources of human food. Insects and other arthropods are important elements in recycling organic matter within the soil and elsewhere, and herbivores play an especially important role in this process. At normal population levels, insects and other arthropods in the ecosystem speed the cycling of nutrients, have little effect on the amount of living plant material, and strongly influence the stock of dead and decaying organic matter. These are only a very few of the ways in which arthropods are involved in human welfare.

MAJOR GROUPS OF ARTHROPODS

Many arthropods, traditionally called the **mandibulates,** have jaws, or mandibles, formed by the modification of one of the pairs of anterior appendages, but not the one nearest the anterior end. The mandibulates include the crustaceans, insects, centipedes, millipedes, and a few other small groups. The remaining arthropods, which include the spiders, mites, scorpions, and a few other groups, lack mandibles. These animals are called the **chelicerates.** Their mouthparts, called **chelicerae** (Figure 37-2), evolved from the appendages nearest the anterior end of the animal. They often take the form of pincers or fangs.

In the mandibulates, the appendages nearest the anterior end are one or more pairs of sensory **antennae,** and the next are the mandibles. The mandibles of the crustaceans, sometimes called the "aquatic mandibulates," are fundamentally similar to their other appendages. All crustacean appendages are fundamentally **biramous,** or two branched (Figure 37-3), although some of these appendages have become single branched by reduction in the course of their evolution. The mandibulates other than crustaceans have traditionally

Onychophora are similar to those of the Uniramia.

Onychophora are clearly members of the evolutionary line that connects the annelids with the Uniramia. Detailed study of their features has made it clear that they did not give rise to any of the classes of Uniramia directly. Rather, they represent an independent evolutionary line in which characteristics more primitive than those of the Uniramia have been retained for hundreds of millions of years. The way in which the Onychophora bridge the morphological gap between the annelids and the Uniramia makes it difficult to argue that there is a direct relationship between the Uniramia and either the crustaceans or the chelicerates, which differ from annelids, Onychophora, and Uniramia in many other features.

FIGURE 37-A

One of the Onychophora, Peripatodes, *in the moist forests of Queensland, Australia.*

been called the "terrestrial mandibulates." They include the insects, centipedes, millipedes, and two other small classes. Another phylum, the Onychophora, also seems to be related to this group of arthropod classes. The mandibles of the "terrestrial mandibulates" are fundamentally similar to their other appendages, which are **uniramous,** or single branched. It is clear that mandibles evolved independently in the crustaceans and the "terrestrial mandibulates"; the ancestors of both groups evidently lacked mandibles.

Among the chelicerates, the second pair of appendages is usually pincerlike or feelerlike, and the remaining pairs of appendages are legs. Chelicerae resemble the other appendages of the chelicerates more closely than mandibles resemble the other appendages of the mandibulates. The fundamental differences in the derivation and structure of their mouthparts indicate that the chelicerates and the mandibulates represent different evolutionary lines among the arthropods and that neither group gave rise to the other.

Evolutionary Relationships of the Major Groups

The late English zoologist S.M. Manton argued convincingly that the arthropods consist of three groups that evolved independently; she called these groups the subphyla Chelicerata, Biramia, and Uniramia. Manton concluded that the exoskeleton, which is such a characteristic feature of arthropods, evolved independently in each of these lines. Once it had done so, segmentation of both the body and its appendages became necessary. Otherwise the animals would have been unable to move!

Although all scientists do not agree that Manton was correct about the independent derivation of these three lines, strong support for her hypothesis is provided by embryology. The terrestrial mandibulates, which Manton called the Uniramia, and the crustaceans both have a spiral pattern of cell division in early embryology, like most protostomes, but their subsequent development is unique. The embryological development of the Uniramia closely resembles that of certain annelids, but with the elimination of the trochophore stage, whereas crustaceans exhibit a very different pattern that results in a special kind of swimming larva called a **nauplius** (Figure 37-4). It seems clear that the Uniramia were derived from the annelids; the ancestors of the crustaceans are unknown.

The embryology of the chelicerates differs greatly from that of both the crustaceans and the Uniramia. They do not have spiral cleavage but instead feature a unique pattern of embryological development similar to radial cleavage. It is not certain that the chelicerates

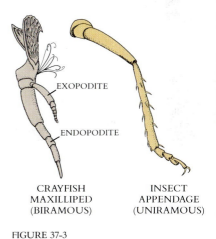

EXOPODITE

ENDOPODITE

CRAYFISH MAXILLIPED (BIRAMOUS)

INSECT APPENDAGE (UNIRAMOUS)

FIGURE 37-3

Diagrams of a biramous leg in crustacean (crayfish) and of a uniramous leg in an insect.

FIGURE 37-4

The nauplius larva of a crustacean is an important unifying feature found in all members of this very diverse group.

are even derived from ancestors that had spiral cleavage. They could possibly have originated independently of the annelids and other arthropods. Their limbs are somewhat similar to those of the extinct trilobites (see Figure 23-1), but not to those of the crustaceans or Uniramia. In view of their specialized features, it is clear that the chelicerates did not give rise to any other group of organisms.

Although these three evolutionary lines are almost certainly only distantly related to one another, they do have a number of features in common, especially those related to their exoskeletons and their jointed bodies and appendages. But these may well have originated as a result of **convergent evolution**—evolution that reaches a similar end product from different starting points by adopting similar ways of life. We will deal with all of these groups comparatively in this book. Because of their distinctiveness, we will accept Manton's conclusions about the Chelicerata, Uniramia, and Crustacea, and treat them as subphyla.

> **The phylum Arthropoda is divided into three subphyla: Chelicerata, Crustacea, and Uniramia. The chelicerates are characterized by chelicerae, mouthparts that often take the form of pincers or fangs, which evolved from the anterior appendages. Members of the other two subphyla have mandibles, originally biting jaws that also evolved from appendages, but from the second or third pair back from the anterior end. All appendages in crustaceans are fundamentally biramous, or two branched, whereas those in Uniramia are uniramous, or single branched.**

MAJOR GROUPS OF PHYLUM ARTHROPODA

Subphylum **Chelicerata**: chelicerates
    Class Arachnida: arachnids
        Order Scorpiones: scorpions
        Order Araneae: spiders
        Order Acari: mites (including ticks and chiggers)
        Order Opiliones: harvestmen or daddy longlegs
    Class Merostomata: horseshoe crabs
    Class Pycnogonida: sea spiders

Subphylum **Crustacea**: crustaceans
    Class Crustacea: crustaceans
        Order Cladocera: water fleas
        Order Anostraca: fairy shrimps
        Order Ostracoda: ostracods
        Order Copepoda: copepods
        Order Cirripedia: barnacles
        Order Isopoda: isopods
        Order Amphipoda: amphipods
        Order Decapoda: crabs, lobsters, and shrimps

Subphylum **Uniramia**: insects, centipedes, and millipedes
    Class Insecta: insects (many orders are mentioned in text)
    Class Chilopoda: centipedes
    Class Diplopoda: millipedes

EXTERNAL FEATURES

Aside from such features as the segmentation and appendages of arthropods, two characteristic features of the phylum deserve special discussion: the exoskeleton and the compound eye.

FIGURE 37-5

Some arthropods have a very tough exoskeleton, like this South American scarab beetle, Dilobderus abderus *(order Coleoptera)* **(A)**; *others have a fragile exoskeleton, like the green darner dragonfly,* Anax junius *(order Odonata)* **(B)**.

A

B

Exoskeleton

The bodies of all arthropods are covered by a hardened exoskeleton or cuticle (Figure 37-5). This tough outer covering is secreted by the epidermis and fused with it. The exoskeleton varies greatly in toughness and thickness in different arthropods; its versatility is doubtless one of the keys to the amazing success of the phylum. In most crustaceans, the exoskeleton is impregnated with calcium carbonate and is thus relatively inflexible. In other arthropods, the exoskeleton may be fairly soft and flexible, although in some large insects, horsehoe crabs, and other groups, it can also be thick and very hard.

Despite these differences and the fact that it may have evolved independently in each of the three subphyla, the exoskeleton is fundamentally similar in all arthropods. Its evolution was a key adaptation that facilitated the invasion of the land by arthropods, especially Uniramia. This protective outer shell is composed of an outer waxy or calcareous layer, a hardened middle layer, and a flexible inner one. The middle and inner layers are composed primarily of chitin and proteins. In addition to covering an arthropod's exterior, the exoskeleton lines the anterior and posterior portions of its digestive tract. In the terrestrial mandibulates, the exoskeleton also lines the tracheae, the tubes by which air enters the body. Arthropods are able to move because the exoskeleton is thin and flexible at the joints, regardless of how tough and thick it may be elsewhere. Internal projections of the exoskeleton serve as points of attachment for the muscles.

Chitin is also a major component in the external walls of most fungi, another group of organisms that has become very successful on land. It probably evolved independently in both groups because of its water-retaining characteristics. The fact that chitin is digested by so few organisms, which require a special enzyme called *chitinase* to do so, may be another reason that it has evolved as an outer covering in so many successful groups of organisms.

Arthropods periodically undergo **ecdysis:** when they outgrow their exoskeletons, they form a new one underneath the old one. This process is controlled by hormones. When the new exoskeleton is complete, it becomes separated from the old one by fluid. This fluid dissolves the chitin and, if it is present, calcium carbonate, from the old exoskeleton, and the fluid increases in volume. Finally, the original exoskeleton cracks open, usually along the back, and is shed. The arthropod emerges, clothed in a new, pale, and still somewhat soft exoskeleton. As a result of the blood circulation to all parts of the body, the arthropod ultimately expands to full size; many insects and spiders take in air to assist them in this expansion. Subsequently the exoskeleton hardens because of its exposure to the air or water in which the animal lives. The exoskeleton protects arthropods from water loss and helps to protect them from predators, parasites, and injury. While the exoskeleton is soft, however, the animal is especially vulnerable. At this stage arthropods often hide under stones, leaves, branches, or other safe places.

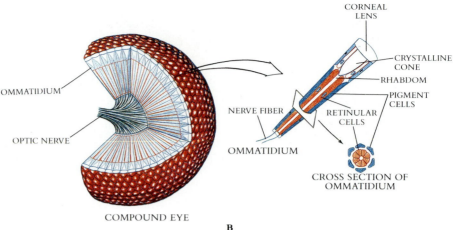
CORNEAL
LENS
OMMATIDIUM
OPTIC NERVE
COMPOUND EYE
B
NERVE FIBER
OMMATIDIUM
RETINULAR
CELLS
CRYSTALLINE
CONE
RHABDOM
PIGMENT
CELLS
CROSS SECTION OF
OMMATIDIUM

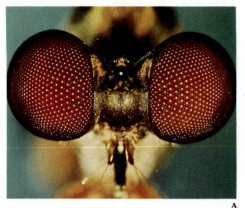

A

FIGURE 37-6

A *Three ocelli can be seen between the compound eyes of the robberfly (order Diptera).*
B *The compound eyes found in insects are complex structures.*

All arthropods have a rigid, chitinous exoskeleton that provides places for muscle attachment, protects the animal from predators and injury, and, most important, impedes water loss.

The need to shed the exoskeleton periodically is a definite liability, especially in view of the metabolic loss that this process involves. Nevertheless, this rigid external skeleton efficiently supports the basic structure of the body, a structure against which the muscles work.

Compound Eye

An important structure of many arthropods is the compound eye (Figure 37-6, *A*). Compound eyes are composed of many independent visual units, often thousands of them, called **ommatidia.** Each ommatidium is covered with a lens and is linked to a complex of eight **retinula** cells and a light-sensitive central core, or **rhabdom.** There are two main types of compound eyes among the insects: apposition eyes and superposition eyes. **Apposition eyes** are like those of bees, in which each ommatidium acts in isolation, collecting light from one sector of the external world and throwing an inverted image of that scene on the rhabdom of that ommatidium. In such an eye, the individual ommatidia are surrounded by **pigment cells,** which keep the light that reaches each ommatidium separate. The individual images are combined in the insect's brain to form its visual image of the external world. In the second kind of compound eye, **superposition eyes,** found, for example, in moths, the images from a series of ommatidia are combined on a cornea that lies at the back of the compound eye, and there are no screening pigment cells. Such an image is right-side-up.

Compound eyes occur only in horseshoe crabs among the chelicerates, but they are characteristic of both the crustaceans and the Uniramia. An evaluation of the overall characteristics of these groups suggests that compound eyes evolved independently in the trilobites (probable ancestors of horseshoe crabs), in the crustaceans, and in the Uniramia.

Simple eyes, or **ocelli,** with single lenses are found in the other groups and sometimes occur, as is often the case in insects, together with compound eyes (Figure 37-6, *B*). Ocelli function in distinguishing light and darkness; the ocelli of some flying insects, namely locusts and dragonflies, function as horizon detectors and are involved in the visual stabilization of their course in flight. It has recently been shown in a walking insect, the desert ant *Cataglyphis,* that the ocelli can read "compass" information from the blue sky, using the pattern of polarized light in the sky as a compass cue; remarkably, the ants find their way back to their nest by using this cue.

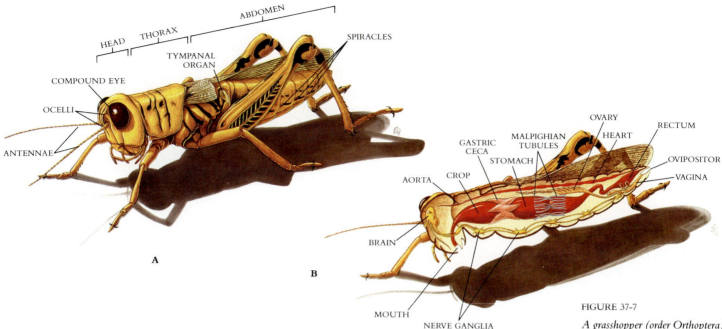

FIGURE 37-7

A grasshopper (order Orthoptera) illustrates the major structural features of the insects, the most numerous group of arthropods.
A *External anatomy.*
B *Internal anatomy.*

INTERNAL FEATURES

In the course of arthropod evolution, the coelom has become greatly reduced. Most scientists, in fact, view the arthropod coelom as consisting only of the cavities that house the reproductive organs and some glands. Arthropods completely lack cilia, both on the external surfaces of the body and on the internal organs. Like the annelids, the arthropods have a tubular gut that extends from the mouth to the anus. In the following paragraphs we will discuss the circulatory, respiratory, excretory, and nervous systems of the arthropods (Figure 37-7).

Circulatory System

The circulatory system of arthropods is open; their blood flows through the cavities between the internal organs and not through closed vessels (Figure 37-7, *B*). The principal component of an insect's circulatory system is a longitudinal vessel, which is called the heart. This vessel runs near the dorsal surface of the thorax and abdomen. When it contracts, blood flows from its anterior end into the head region of the insect. From there, the blood circulates gradually through all of the internal spaces of the body. Blood flows most rapidly when the insect is running, flying, or otherwise active. At such times the blood efficiently delivers nutrients to the tissues and removes wastes from them.

When an insect's heart relaxes, blood returns to it through a series of valves. These valves are located in the posterior region of the heart and allow the blood to flow inward only. Thus blood from the head and other anterior portions of the insect gradually flows through the spaces between the tissues toward the posterior end and then back through the one-way valves into the heart.

Respiratory System

Insects and other Uniramia, which are fundamentally terrestrial, depend on their respiratory system rather than on their circulatory system to carry oxygen to their tissues. In vertebrates, the blood moves within a closed circulatory system to all parts of the body and carries the oxygen with it. This is a much more efficient arrangement than exists in the arthropods, in which all parts of the body need to be near an air passage in order to obtain oxygen. As a result, the bodies of arthropods are much more limited in size than those of the vertebrates.

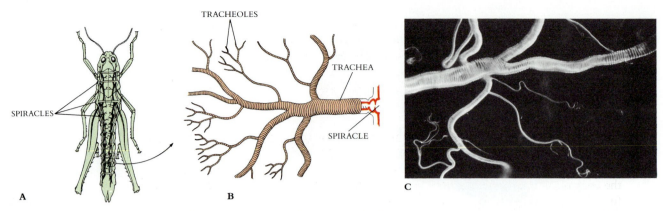

A   B   C

FIGURE 37-8

*Tracheae and tracheoles, connected to the exterior by specialized openings called spiracles, carry oxygen to all parts of an insect's body. **A** and **B** show the tracheal system of a grasshopper. **C** shows a portion of the tracheal system of a cockroach.*

Unlike most animals, the arthropods have no single major respiratory organ. The respiratory system of the terrestrial arthropods consists of small, branched, cuticle-lined air ducts called **tracheae** (Figure 37-8). These tracheae, which ultimately branch into very small **tracheoles,** are a series of tubes that transmit oxygen throughout the body. The tracheoles are in direct contact with the individual cells, and oxygen diffuses from them to other cells directly across the cell membranes. Air passes into the trachea by way of specialized openings called **spiracles,** which can, in most insects, be closed and opened by valves. The ability to prevent water loss by closing their spiracles was a key adaptation that facilitated the invasion of the land by arthropods. In many insects, especially the larger ones, the contraction of the muscles helps to increase the flow of gases into and out from the trachea. In other terrestrial arthropods, the flow of gases is essentially a passive process.

> **Arthropods have no cilia and only a limited coelom. Their circulatory system is open. Instead, a network of tubes called tracheae transmits oxygen from the outside to the organs; external tracheal openings are controlled by opening and closing spiracles.**

Many spiders and some other chelicerates have a unique system of respiration that involves **book lungs,** a series of leaflike plates within a chamber into which air is drawn and from which it is expelled by muscular contraction. Such book lungs may exist alongside tracheae, or they may function instead of tracheae, in particular groups of organisms. One small class of marine chelicerates, the horseshoe crabs, have **book gills,** which are analogous to book lungs but function in water. Trachea, book lungs, and book gills are all structures that are found only in arthropods and in the phylum Onychophora (see boxed essay, pp. 774-775), which have tracheae and are related to the insects. The crustaceans lack such structures and have gills.

Excretory System

Although there are several different kinds of excretory systems in different groups of arthropods, we will focus here on the unique excretory system that evolved in terrestrial arthropods in relation to their open circulatory system. The principal structural element in such an excretory system is the **Malpighian tubules,** which are slender projections, or **diverticula,** from the digestive tract. These are attached at the junction of the midgut and hindgut (Figure 37-7, *B*). Fluid is absorbed through the walls of the Malpighian tubules from the blood in which the tubules are bathed. As this fluid passes through the tubules toward the hindgut, the nitrogenous wastes in it are precipitated as concentrated uric acid or guanine. These substances are then emptied into the hindgut and thus eliminated. Most of the water and salts in the fluid are reabsorbed from it by the Malpighian tubules and returned to the arthropod's body. Malpighian tubules constitute a very efficient mechanism for water conservation and were another key adaptation facilitating invasion of the land by arthropods.

Arthropods eliminate metabolic waste by a unique system of Malpighian tubules that extend from the digestive tract into the blood. Fluid enters the tubes, waste is precipitated, and the fluid is reabsorbed, passing out through the tube walls back into the body cavity.

Nervous System

The central feature of the arthropod nervous system is a double chain of segmented ganglia running along the animal's ventral surface (Figure 37-7, *B*). At the anterior end of the animal are three fused pairs of dorsal ganglia, which constitute the brain. However, much of the control of an arthropod's activities is relegated to ventral ganglia. Therefore, many functions, including eating, movement, and copulation, can be carried out even if the brain has been removed. Indeed, the brain of arthropods seems to be a control point, or inhibitor, for various actions, rather than a stimulator, as it is in vertebrates. On the other hand, the degree of coordinated response by many arthropods is high, and their movements may be very rapid.

EVOLUTIONARY HISTORY OF THE ARTHROPODS

Fossils whose features resemble those of the chelicerates occur along with the earliest well-preserved multicellular animals; this group of arthropods, therefore, goes back at least 630 million years. The terrestrial chelicerates evolved from marine ancestors that had thick exoskeletons and clearly marked segmentation.

The trilobites were another very important group of early arthropods, but they have been extinct for 250 million years (see Figure 23-1). Trilobites, which lived in the early seas, were the first animals whose eyes were capable of a high degree of resolution. These eyes were very unusual in structure; the eye of a trilobite contained as many as 15,000 individual elements! Each of these elements provided a separate image, which the animal could integrate into a highly detailed composite. Although these eyes were compound, they were clearly not equivalent to those of the crustaceans or Uniramia, and they evolved independently of the eyes of members of those groups.

The fossil record of chelicerates goes back as far as that of any multicellular animals, about 630 million years. A major group of extinct arthropods, the trilobites, was also abundant then.

Fossil crustaceans more than 500 million years old are known, but they are so similar to modern forms that they do not help us to understand the derivation of the group. The crustaceans are very diverse, but many have a fundamentally similar nauplius larva (Figure 37-4) and all were clearly derived from a common ancestor. Aside from a generalized relationship with the annelids and the Uniramia, there is little that can be said about the evolutionary relationships of these animals. Their distinctive embryology definitely separates them from the chelicerates and the Uniramia, however.

Fossils similar to Onychophora go back some 500 million years. Because of their close relationship to Uniramia, we may assume that the evolutionary line leading to Onychophora and Uniramia had originated, probably from ancestors similar to oligochaetes, before that time. Among the Uniramia are the most diverse and abundant of all arthropods, the insects. The earliest known insect fossils are more than 300 million years old. Some groups of insects that go back that far in the fossil record still have living representatives—cockroaches and dragonflies, for example. One of the early and important evolutionary innovations among the insects was the ability to fly. Flying insects have been in existence for more than 300 million years and have achieved remarkable success (Figure 37-9). For more than 100 million years, insects were the only flying organisms in existence, and even today, among all the living groups of organisms, only birds, bats, and insects truly fly.

FIGURE 37-9

This African blister beetle (order Coleoptera) has spread its two tough, leathery forewings, exposing its delicate flying wings, which are normally concealed beneath them, and taken off. More than 350,000 species of beetles—a fifth of all animal species—have been described.

FIGURE 37-10

The scorpion Uroctonus mordas, *showing the characteristic pincers and segmented abdomen, ending in a sting, raised over the animal's back. White young scorpions cluster on this individual's back.*

Despite their great age and enormous evolutionary success, the arthropods have not evolved intelligent forms. Large brains, a prerequisite for intelligence, can develop only within large bodies. The external skeletons of the arthropods are more massive in relation the support that they provide than the internal ones of the vertebrates. Because of this large mass to support ratio, exoskeletons do not provide the structural basis for the evolution of large bodies. In addition, tracheae, because of their small diameters and the need for direct connections to the outside, are not as efficient as lungs in distributing oxygen throughout a large body; it would be impossible for tracheae to function in a human-sized organism. Furthermore, a closed circulatory system seems to be necessary for a large animal, judging from the fact that all large animals have systems of this sort. Nevertheless, the drought-resistant cuticles and Malpighian tubules of the arthropods are excellent adaptations for a terrestrial existence. Their versatile, highly manipulable appendages are another important element in their remarkable success.

THE CHELICERATES

Chelicerates (subphylum Chelicerata) are a distinct evolutionary line of arthropods, one in which the appendages nearest the anterior end of the body have been modified into chelicerae. These chelicerae, which often function as fangs or pincers, had a different evolutionary origin from the mandibles that originated separately in the crustaceans and in the Uniramia, the other two subphyla of arthropods.

Arachnids

By far the largest of the three classes of chelicerates is the largely terrestrial class Arachnida, with some 57,000 named species—the spiders, ticks, mites, scorpions, and daddy longlegs. Arachnids such as spiders have a pair of chelicerae, a pair of **pedipalps,** and four pairs of walking legs. The chelicerae are the foremost appendages; they consist of a stout basal portion and a movable fang, which is connected with a poison gland by a duct. The pedipalps, the next pair of appendages, may resemble the legs, but they have one less segment. In male spiders, the pedipalps are specialized copulatory organs; in other groups of arthropods, they may be specialized in other ways. Despite their appearance, the pedipalps of carnivorous arachnids, such as spiders, are often used for catching and handling prey—spiders also chew with the basal portions of their pedipalps—rarely for locomotion.

Although arachnids are sometimes confused with insects, they are actually very different in many structural and other features. Most arachnids are carnivorous, although the mites are largely herbivorous. Arachnids can ingest only preliquefied food, which they often digest externally by secreting enzymes into their prey and then suck up by means of their muscular, pumping pharynx. Most arachnids occur in terrestrial habitats, where they have evolved habits that lead to the direct transfer of sperm—thus protecting it from drying out—including the complex mating rituals of many species of spiders. In other arachnids, such as the scorpions, sperm transfer is indirect, but the sperm is held in packets and thus protected from drying out. Not all arachnids live on land; some 4000 known species of mites and one species of spider live in fresh water, and a few mites live in the sea. Arachnids breathe by means of tracheae, book lungs, or both.

There are 11 orders of arachnids that include living species. Of these, we will briefly discuss four: Scorpiones, the scorpions; Araneae, the spiders; Acari, the mites and ticks; and Opiliones, the harvestmen or daddy longlegs. These are the most familiar of the orders of arachnids, and our discussion of them will serve as an introduction to the group.

Scorpions

The scorpions (order Scorpiones) are a familiar group of arachnids whose pedipalps are modified into pincers. Scorpions use these pincers to handle their food and tear it apart (see Figures 37-1 and 37-10). The venomous stings of scorpions are used mainly to stun their

A

B

C

FIGURE 37-11

Spiders.

A *Tarantulas are large hunting spiders, powerful enough to prey on small lizards, mammals, and birds in addition to insects. They do not spin webs.*
B *An orb-weaving spider, the arrowhead spider* Micranthena, *in Peru.*
C *A wolf spider,* Phidippus audax, *in Maryland. Wolf spiders, like the tarantulas, do not spin webs, instead actively hunting their prey. Most spiders have eight pairs of simple eyes; these differ in size, as seen here.*

prey, less commonly in self-defense. The sting is located in the terminal segment of the body, which is slender towards the end. A scorpion holds its abdomen folded forward over its body when it is moving about. The elongated, jointed abdomens of scorpions are distinctive; in most chelicerates, the abdominal segments are more or less fused together and appear as a single unit.

Scorpions are probably the most ancient group of terrestrial arthropods; they are known from as early as the Silurian Period, some 425 million years ago. The adults of this order of arachnids range in size from 1 to 18 centimeters. Respiration in scorpions occurs by means of book lungs. There are some 1200 species of scorpions, all terrestrial, which occur throughout the world, although they are more common in tropical, subtropical, and desert regions. The courtship of scorpions is elaborate, with the spermatophores being fixed to a substrate by the male and then picked up subsequently by the female. The young are born alive, with 1 to 95 of them in a given litter.

Spiders

There are about 35,000 named species of spiders (order Araneae; Figure 37-11). These animals play a major role in virtually all terrestrial ecosystems, where they are particularly important as predators on insects and other small animals. Spiders hunt their prey or catch it in webs that they construct. The silk of the webs is formed from a fluid protein that is forced out of **spinnerets,** modified appendages on the posterior portion of the spider's abdomen (Figure 37-12). There may be up to six pairs of these silk glands, with which different kinds of spiders produce many adaptive modifications of webs; some spiders, for example, can spin balloons by means of which they may float away in the breeze to a new site. The webs and habits of individual kinds of spiders are often distinctive.

FIGURE 37-12

Spinnerets of a Peruvian orb-weaving spider in action producing fine strands of silk.

SPIDERS AND THEIR WEBS

Spider webs can be found everywhere—in gardens, woods, and fields, and even in the corners of your house or apartment. Spiders are the only organisms that make silk to catch their prey; they extrude it from paired glands in their abdomen. Each individual spider has several different kinds of these glands, each producing a different kind of silk. For example, one kind of gland may produce the strong, sticky silk used to attach the web to branches or other surfaces; another may produce the silk used to wrap the prey; and a third may make the silk used for the web scaffolding. Spiders also use this silk to construct nests and trapdoors, to balloon about from place to place, to build special webs for their sperm and cases for their eggs, and for many other purposes.

Liquid silk flows from the glands where it is produced through tubes to flexible, external nozzles called spinnerets, located on the underside of the abdomen near its end. The spinnerets can be used singly, or they may be used in different combinations to blend different kinds of silk. The silk solidifies as the spider attaches it to a substrate and pulls away, or draws it out with its foot. The faster it is drawn out, the thinner the thread becomes.

The ability of different species of spiders to produce different web designs—including orbs, sheets, funnels, domes, and other forms—is genetically programmed. Orb webs, perhaps the most familiar kind, always have a densely meshed hub, with strong radial threads extending from the hub to the outer frame. A more or less circular spiral of sticky silk is then spun across the radii. Most orb weavers build their web in a single night, working mainly just before dawn and replacing the torn and

Many kinds of spiders, like the familiar wolf spiders (Figure 37-11, *C*), do not spin webs but instead hunt their prey actively. Others, the trap-door spiders, construct silk-lined burrows with lids from which they seize their prey as it passes by. One species of spider, *Argyroneta aquatica,* lives in fresh water, spending most of its time below the surface. Its body is surrounded by a bubble of air, while its legs, which are used both for underwater walking and for swimming, are not. Several other kinds of spiders walk about freely on the surface of water.

Spiders have poison glands leading through their chelicerae, which are pointed and used to bite and paralyze prey. Some members of this order, such as the black widow or

broken webs used the preceding day. Some spiders put up their webs at night and take them down before the next dawn; others catch their prey mainly during the day, and spin more durable webs. The spiders that do have webs that function during the day often spin loosely constructed bands or patches of white silk on them, easily detectable by a vertebrate eye. These patches (Figure 37-B, *2*), or bands, called stabilimenta, apparently make the webs more visible to birds, which then avoid them.

About 2500 species of orb weavers are known. Most have poor vision and find their prey by sensing vibration and tension. They detect vibration by holding strands of the web with their front legs. The nature of the vibration often tells the spider what kind of animal has hit the web. When spiders locate prey, they immobilize it with venom and swathe it in silk, turning it over and over in the process. It may then either be hung up for later use or consumed immediately. The webs, which are spun by female spiders, also advise the males as to which species has spun that web.

An extraordinary behavioral pattern is exhibited by nocturnal spiders of the genera *Kaira* and *Mastophora*, which release substances that attract male moths like the sex pheromones released by the female moths. Spiders of the genus *Kaira* hang upside down from a trapeze line with their legs outstretched, seizing male moths when they come close enough. In contrast, spiders of the genus *Mastophora* suspend a short line of silk with a drop of glue at one end from one of their legs. These spiders face downwind; as a male moth approaches, the spider flicks its line at it, snares the moth, pulls it in, and immobilizes it by biting it and injecting venom. Because of these bizarre habits, one of the species of *Mastophora* has been given the appropriate name *Mastophora dizzydeani* (after the baseball pitcher)!

FIGURE 37-B

1 *Orb web, Sri Lanka.*
2 *A stabilimentum in the web of the orb weaver* Agriope aurantia, *in Florida.*
3 *A funnel web made by a spider of the genus* Maymena *in the lowland forest of Costa Rica. Insects flying upward become trapped within the funnel.*
4 *Web constructed by the spider* Eidmanella pallida *in Tennessee.*
5 Mastophora *sp., awaiting male moths, with a sticky drop of glue suspended at the end of its line.*

brown recluse (Figure 37-13), have bites that are poisonous to humans and other large mammals.

Male spiders make a sperm web from special silk glands in the anterior portion of their abdomen. A drop of sperm deposited on the sperm web is picked up by the spider with the pedipalps. Many spiders have an elaborate courtship, following which the pedipalps fit into a special plate on the lower side of the female's abdomen, permitting the sperm to enter special receptacles. In many species, the female eats the male once fertilization has been completed. The eggs are enclosed in a silken egg sac, which is abandoned, guarded, or carried about, depending on the species of spider. Young spiders resemble adults and become mature after 3 to 15 molts, depending on the species.

A

B

Mites

The order Acari, the mites, is the largest, in terms of number of species, and most diverse of the arachnids (Figure 37-14). Although only about 30,000 species of mites have been named, students of the group estimate that there may actually be a million or more members of this order in existence. In many regions of the world, the members of this group are poorly known; moreover, relatively few scientists study them.

Most mites are small, less than 1 millimeter long, but the adult length of different species ranges from 100 millimicrons to 2 centimeters. In most mites, the segments corresponding to the **cephalothorax**—a segment consisting of the head and thorax—and abdomen are fused into an unsegmented ovoid body. Respiration occurs either by means of tracheae or directly through the exoskeleton. Many mites pass through several distinct stages during their life cycle. In most, an inactive eight-legged larva hatches from the egg and gives rise to an active six-legged larva, which in turn produces a succession of three eight-legged stages and finally the adult males and females. In a number of different mites, however, various juvenile stages have become reproductive, and the development of those particular species then stops at that stage. The evolutionary process whereby juvenile stages become reproductive is known as **neoteny.**

Mites are very diverse, not only in their structure but also in their habits. They are found in virtually every terrestrial, freshwater, and shallow marine habitat known, feeding on fungi, plants, and animals, as well as acting as predators and as internal and external parasites of both invertebrates and vertebrates.

Many mites are well known to human beings because of their irritating bites and the diseases that they transmit. Follicle mites live in the hair follicles and wax glands of the human forehead, nose, and chin, but usually cause no symptoms. Other mites, however, cause mange in dogs and cats, often with severe consequences. One mite, *Dermatophagoides pteronyssinus,* is an important cause of house-dust allergy. Chiggers, which belong to several related genera of mites, are also very annoying. They are frequent throughout the warmer parts of the world. Adult chiggers are predaceous on invertebrates; the larvae suck blood from vertebrates and produce irritating pink to bright-red welts in many people.

Ticks, which are also members of this order, are blood-feeding **ectoparasites**—parasites that occur on the surface of their host—of vertebrates (Figure 37-14, *C*). They are larger than most other mites and cause discomfort directly by sucking the blood of human beings and other animals. Some of them also inject toxins into their hosts. A few ticks of this kind are known to cause paralysis in people they have bitten. Ticks are also carriers of many diseases, including some caused by viruses, bacteria, and even protozoa. The spotted fevers, of which Rocky Mountain spotted fever is a familiar example, are caused by members of the group of very small omnibacteria known as rickettsiae, whereas Colorado tick

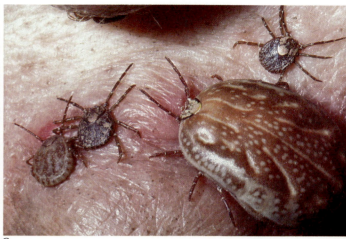

A   B   C

fever is caused by a virus. Lyme fever is apparently caused by spirochetes transmitted by ticks; tularemia is another example of a bacterial disease spread, in part, by ticks. Red-water fever or Texas fever is the best-known and most important tick-borne protozoan disease of cattle, horses, sheep, and dogs.

In addition to the diseases they cause in humans and other animals, mites cause extensive and often severe damage to plants. The group known as spider mites, or red spiders, are often the most serious pests of house plants. Mites of this group also damage many crops. On the positive side, in recent years mites have been used as agents of biological control: certain mites attack harmful insects or other mites and are introduced to control the numbers and therefore the harmful effects of their prey.

Daddy longlegs

Another familiar group of arachnids consists of the daddy longlegs or harvestmen (order Opiliones). The members of this order are easily recognized by their oval, compact bodies and extremely long, slender legs (Figure 37-15). They respire by means of a single pair of tracheae and are very unusual among the arachnids in that they engage in direct copulation. The males have a penis, and the females an **ovipositor,** or egg-laying organ, by means of which they deposit their eggs in cracks and crevices. Most daddy longlegs are predators of insects, other arachnids, snails, and worms, but some live on plant juices and many scavenge dead animal matter. The order includes about 5000 species. Although it occurs throughout the world, it is best represented in the tropics of Asia and South America.

FIGURE 37-14

Mites.
A *The tiny red flecks on this flower of creamcups,* Platystemon californicus, *are mites, which have eaten all of the pollen.*
B *A water mite,* Hydracarina. *Note the feathery legs, modified for swimming.*
C *Ticks on the hide of a tapir in Peru. Many ticks spread diseases in human beings and other vertebrates.*

FIGURE 37-15

A harvestman, or daddy longlegs, representative of a large order of arachnids that is better represented in the tropics but also common in temperate regions.

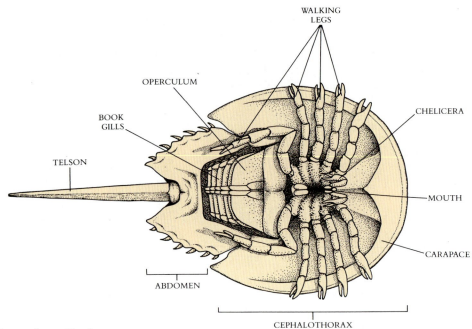

FIGURE 37-17

A pair of horseshoe crabs, Limulus, *emerging from the sea to mate at the edge of Delaware Bay, New Jersey.*

FIGURE 37-18

The sea spider Pycnogonum littorale *crawling over a sea anemone.*

Horseshoe Crabs

A second class of chelicerates is that of the horseshoe crabs (class Merostomata). There are three genera of horseshoe crabs. One, *Limulus* (Figure 37-16), is very common along the east coast of North America. The other two genera of horseshoe crabs live in the seas surrounding tropical Asia and its neighboring islands. The horseshoe crabs are an ancient group; fossils virtually identical to *Limulus* date back 220 million years, to the Triassic Period. Other members of the class, the eurypterans, are known from as much as 400 million years ago. Horseshoe crabs may have been derived from the trilobites, a relationship that is suggested by the appearance of their larvae.

Individuals of *Limulus* are up to 30 or even 60 centimeters long. They mature in 9 to 12 years, and have a life span of 14 to 19 years. Horseshoe crabs live in deep water, but they migrate to shallow coastal waters every spring, emerging from the sea to mate on moonlit nights when the tide is high (Figure 37-17). While they are mating, the male grasps the back of the female and holds on. Later the females lay their eggs in the sand. The young horseshoe crabs that emerge from these eggs swim actively. They lack the tail spine that is characteristic of the adults and resemble trilobites.

Horseshoe crabs feed at night, primarily on mollusks and annelids. They swim on their backs by moving their abdominal plates. They can also walk on their five pairs of legs, which are protected, along with their chelicerae, by their shell. Each of the five pairs of book gills of the horseshoe crab is located under a pair of covers, or opercula, posterior to the legs.

Sea Spiders

The third class of chelicerates is the sea spiders (class Pycnogonida). Sea spiders are relatively common, especially in coastal waters, and there are more than 1000 species in the class. These animals are not observed often, however, because they are small—usually only about 1 to 3 centimeters long—and rather inconspicuous (Figure 37-18). They are found in oceans throughout the world but are most abundant in the far north and far south. Adult sea spiders are mostly external parasites or predators on other animals, including sea anemones.

Sea spiders have a sucking proboscis, with the mouth located at its end. Their abdomen is much reduced, and their body appears to consist almost entirely of the cephalothorax, with no well-defined head. There are usually four, or less commonly five or six, pairs of legs. Male sea spiders carry the eggs on their legs until they hatch, thus providing a measure of parental care for them. Sea spiders completely lack excretory and respiratory systems. They appear to carry out these functions by the direct diffusion of waste products outward through the cells and of oxygen inward through them. Sea spiders are not closely related to either of the other two classes of arachnids.

CRUSTACEANS

The crustaceans (subphylum Crustacea, with a single class) are a large group of primarily aquatic organisms, consisting of some 35,000 species of crabs, shrimps, lobsters, crayfish, barnacles, water fleas, pillbugs, and related groups. Often incredibly abundant in marine and freshwater habitats and playing a role of critical importance in virtually all aquatic ecosystems, crustaceans have been called "the insects of the water." Most crustaceans have two pairs of antennae, three pairs of chewing appendages, and various numbers of pairs of legs. All of the appendages of crustaceans, with the possible exception of the first pair of antennae, are basically biramous. One of these branches has been lost during the course of evolutionary specialization in many crustaceans, however, so that the appendages then have only a single branch. The nauplius larva stage that crustaceans pass through (Figure 37-4) indicates that all of the members of this very diverse group are descended from a common ancestor. In many groups, however, this nauplius stage has been lost during the course of evolution, and development to the adult form is direct. The nauplius hatches with only three pairs of appendages but metamorphoses through several other stages before reaching maturity.

Crustaceans differ from the insects, but resemble the centipedes and millipedes, in that they have legs on their abdomen as well as on their thorax. They are the only arthropods with two pairs of antennae. Their mandibles are thought to have originated from a pair of limbs that during the course of evolution took on a chewing function, a process that apparently occurred independently in the common ancestor of the terrestrial mandibulates. Many crustaceans have compound eyes. In addition, they have delicate tactile hairs that project from the cuticle all over the body. The larger crustaceans have feathery gills near the bases of their legs. In the smaller members of this class, gas exchange takes place directly through the thinner areas of the cuticle or the entire body.

The excretion of nitrogenous wastes in crustaceans also takes place mostly by diffusion across thin areas of cuticle, such as the gills. However, there is also a pair of glands, located in the ventral region of the head, that consist of a bladder, a tubule, and a spongy mass, called the **labyrinth,** nearest the point of excretion. Salts, amino acids, and some water is extracted as the waste products pass along the tubule, eventually to be excreted as urine through the labyrinth. The primary role of these glands is to regulate the osmotic and ionic composition of the crustacean's blood, rather than excretion.

Large, primarily marine crustaceans such as shrimps, lobsters, and crabs, along with their freshwater relatives the crayfish, are collectively called the decapod crustaceans (Figure 37-19). The term *decapod* means "ten-footed." In these animals, the exoskeleton is usually reinforced with calcium carbonate. Most of the segments of their bodies are fused into a cephalothorax, which is covered on the dorsal side by a shield, or **carapace,** which arises from the head. The crushing pincers common in many decapod crustaceans are used to obtain food—for example, by crushing mollusk shells.

In lobsters and crayfish, appendages called **swimmerets,** which are used in reproduction and also for swimming, occur in lines along the sides of the abdomen (Figure 37-20). In addition, flattened appendages known as **uropods** form a kind of compound "paddle" at the end of the abdomen. By snapping their abdomen, the animals propel themselves through the water rapidly and forcefully. Crabs (Figure 37-19, *C*) differ from lobsters and crayfish in proportion; their carapace is much larger and broader, and the abdomen is tucked up under it. Shrimps (Figure 37-19, *B*) and their relatives also have different proportions; their carapace is proportionately smaller than that of lobsters or crabs.

A

B

C

FIGURE 37-19

Decapod crustaceans.
A *A freshwater crayfish,* Procambarus.
B *A colorful shrimp from the Red Sea.*
C *Sally lightfoot crab,* Grapsus grapsus, *in the Galapagos Islands. This crab species is semiterrestrial, ranging widely overland in search of food. Many of the decapod crustaceans are important sources of food for humans.*

FIGURE 37-20

The principal external features of a lobster, Homarus americanus.

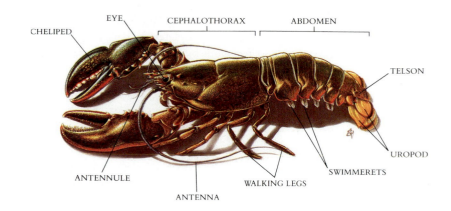

FIGURE 37-21

Diversity in crustaceans.
A *Sowbugs,* Porcellio scaber, *representatives of the terrestrial isopods (order Isopoda).*
B *A marine amphipod,* Eurius, *(order Amphipoda) thriving in water at − 1.9° C in Antarctica. Many amphipods, known as beach fleas or sand fleas, are terrestrial or semiterrestrial in their habits.*
C *A copepod, member of an abundant group of marine and freshwater crustaceans (order Copepoda), most of which are a few millimeters long. Copepods are important components of the plankton.*

Although most crustaceans are marine, many occur in fresh water and a few have become terrestrial. These include some crabs, which may play an important role in some tropical and subtropical areas, and the pillbugs and sowbugs (Figure 37-21, *A*), which are the terrestrial members of a large, predominantly marine order of crustaceans known as the **isopods** (order Isopoda). Terrestrial isopods live primarily in places that are moist, at least seasonally; there are about 1000 terrestrial species of this group. The sand fleas, or beach fleas (order Amphipoda) are another familiar group that includes many terrestrial crustaceans (Figure 37-21, *B*).

Minute crustaceans, including the larvae of larger species, are abundant in the **plankton,** the assemblage of small organisms that is suspended in the upper layers of the ocean. Especially significant among them are the tiny copepods (order Copepoda; Figure 37-21, *C*), which are among the most abundant multicellular organisms on earth. Other orders of minute aquatic crustaceans include the water fleas (Cladocera; Figure 37-22, *A*), ostracods (Ostracoda), and fairy shrimps (Anostraca). One of the fairy shrimps, the brine shrimp, is a standard food of aquarium fish (Figure 37-22, *B*).

Barnacles (order Cirripedia; Figure 37-23) are a group of crustaceans that are sessile as adults. Barnacles have free-swimming larvae. These ultimately attach themselves to a piling, rock, or other submerged object by their head and then stir food into their mouth with their feathery legs. Calcareous plates protect their body, and these plates are usually attached directly and solidly to the substrate.

Most crustaceans have separate sexes. Barnacles are hermaphroditic, but they generally cross-fertilize. Many different kinds of specialized copulation occur among the crustaceans, and the members of some orders carry their eggs with them, either singly or in egg pouches, until they hatch.

UNIRAMIA

The subphylum Uniramia is a great group that includes millipedes, centipedes, and insects, three distinct but clearly related classes. The wormlike Onychophora (see the boxed essay, pp. 774-775) are clearly related to the Uniramia. The Uniramia were certainly derived from annelids, probably ones similar to the oligochaetes, which they resemble in their embryology. All Uniramia respire by means of tracheae and excrete their waste products by means of Malpighian tubules.

Millipedes and Centipedes

The millipedes and centipedes both have bodies that consist of a head region, followed by numerous segments, all more or less similar and nearly all bearing paired appendages (Figure 37-24). Although the name *centipede* would imply an animal with a hundred legs and

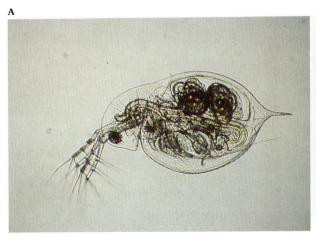

the name *millipede* one with a thousand, centipedes actually have 30 or more legs, millipedes 60 or more. Centipedes have one pair of legs on each body segment, millipedes two. Actually, each segment of a millipede is a tagma that originated during the group's evolution by the fusion of two ancestral segments. This fact explains the difference of a factor of two in the numbers of legs per segment between centipedes and millipedes.

In both centipedes and millipedes, fertilization is internal, and takes place by direct transfer of sperm. The sexes are separate and all species lay eggs. Young millipedes usually hatch with three pairs of legs; they go through a number of growth stages, adding segments and legs as they mature, but with no change in their general appearance.

The centipedes, of which some 2500 species are known, are all carnivorous and feed mainly on insects. The appendages of the first trunk segment are modified into a pair of poison fangs. The poison is often quite toxic to human beings, and many centipede bites are extremely painful, sometimes even dangerous.

In contrast, most millipedes are herbivores, feeding mainly on decaying vegetation; a few millipedes are carnivorous, like the centipedes. Many millipedes can roll their bodies into a flat coil or sphere, because the dorsal area of each of their body segments is much longer than the ventral one. More than 10,000 species of millipedes have been named, but this is estimated to be no more than one sixth of the actual number of species that exists. Most millipedes have a pair of complex glands in each segment of their body, from which

FIGURE 37-23

Gooseneck barnacles, Lepas anatifera, *feeding in Puerto Rico.*

A

B

FIGURE 37-24

A *Centipede,* Scolopendra.
B *Millipede,* Sigmoria, *in North Carolina. Centipedes are active predators, whereas millipedes are sedentary herbivores.*

the animals are able to exude an ill-smelling fluid for defensive purposes through openings along the sides of the body. The chemistry of the secretions of different millipedes has become a subject of considerable interest because of the diversity of the compounds involved and their effectiveness in protecting millipedes from attack. Millipedes live primarily in damp, protected places, such as under leaf litter, in rotting logs, under bark or stones, or in the soil.

C

A

B

Insects

The insects, class Insecta, are by far the largest group of organisms on earth, whether measured in terms of numbers of species or numbers of individuals. Insects live in every conceivable habitat on land and in fresh water, and a few have even invaded the sea. One scientist has calculated that there are probably about 200 million insects alive at any one time for each person on earth! More than 70% of all the named animal species are insects, and the actual proportion is doubtless much higher, because millions of additional forms await detection, classification, and naming. There are about 90,000 described species in the United States and Canada, and the actual number of species in this area probably approaches 125,000. In the tropics, the estimated numbers are truly amazing (see Figure 22-6). A glimpse at the enormous diversity of insects is presented in Figures 37-25 and 37-26.

External features

Insects are primarily a terrestrial group, and most, if not all, of the aquatic insects probably had terrestrial ancestors. Most insects are relatively small, ranging in size from 0.1 millimeter to about 30 centimeters in length or wingspan. Insects have three body sections, the head, thorax, and abdomen; three pairs of legs, all attached to the thorax; and one pair of antennae (see Figure 37-7). In addition, they may have one or two pairs of wings. Most insects have compound eyes, and many have ocelli as well.

D

E

F

FIGURE 37-25

Insect diversity.
A *Cockroach,* Supella longipalpa *(order Blattoidea), with an egg mass, in California.*
B *Termite,* Macrotermes bellicosus *(order Isoptera), in Ghana. The large, sausage-shaped individual is a queen, specialized for laying eggs; the smaller individuals around it are nonreproductive workers.*
C *Mayfly,* Hexagenia limbata *(order Ephemeroptera), immediately after molting to reach the adult form. Adult mayflies, the reproductive stage, do not feed and live for only a few hours or a few days.*
D *Caddisfly,* Grammotaulius betteni *(order Trichoptera), in California.*
E *Copulating grasshoppers (order Orthoptera).*
F *A bizarre leaf-footed bug (order Hemiptera), in the Amazon Basin of Peru.*

A

B

C

D

E

F

FIGURE 37-26

More examples of insect diversity.
A *Luna moth,* Actias luna, *in Virginia. Luna moths and their relatives are among the most spectacular insects.*
B *A soldier fly,* Ptecticus trivittatus *(order Diptera), in Virginia.*
C *Honeybee,* Apis mellifera *(order Hymenoptera), widely domesticated and an efficient pollinator of flowering plants.*
D *A leaf beetle (order Coleoptera) in Arizona.*

E *The human flea,* Pulex irritans *(order Siphonaptera), in California. Fleas are flattened laterally, slipping easily through hair.*
F *A sucking louse,* Echinophthirus horridus *(order Anoplura), from a Lake Baikal seal. Flattened dorsoventrally, sucking lice have habits similar to those of fleas. Both fleas and lice have evolved from winged ancestors.*

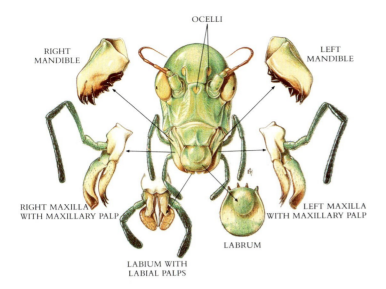

FIGURE 37-27

The complex chewing mouthparts of a grasshopper, showing the various modified appendages.

FIGURE 37-28

Modified mouthparts in three kinds of insects.
A *Mosquito,* Culex; *the mouthparts have been modified for piercing.*
B *Housefly,* Musca domestica; *mouthparts modified for sopping up liquids.*
C *Alfalfa butterfly,* Colias; *mouthparts modified for sucking nectar from flowers.*

A

MOSQUITO

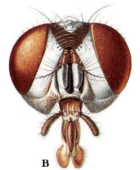

B

HOUSEFLY

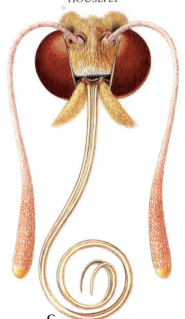

C

BUTTERFLY

The mouthparts of insects are elaborate (Figure 37-27). They usually consist of the jaws, or **mandibles,** which are tough and unsegmented; a secondary pair of mouthparts, the **maxillae,** which are segmented; and the lower lip, or **labium,** which probably evolved from the fusion of another pair of maxilla-like structures. The upper lip, or **labrum,** is of less certain origin. The **hypopharynx** is a short tonguelike organ (in chewing insects) that lies between the maxillae and above the labium; the salivary glands usually open on or near the hypopharynx.

Within this basic structural framework, the mouthparts vary widely among groups of insects, mainly in relation to their feeding habits. Many orders of insects—such as Coleoptera, the beetles; Hymenoptera, the bees, wasps, and ants; Isoptera, the termites; and Orthoptera, the grasshoppers, crickets, and their relatives (Figure 37-27)—have chewing, or mandibulate, mouthparts. In other orders, the mouthparts may be elongated or styletlike (Figure 37-28). For example, in some flies, order Diptera, such as mosquitoes, blackflies, and horseflies, there are six piercing stylets: the labrum, the mandibles, the maxillae, and the hypopharynx; the labium sheaths the stylets (Figure 37-28, *A*). In more advanced flies, the labium may be the principal piercing organ, or may be expanded into large, soft lobes through which liquid food is absorbed (Figure 37-28, *B*). In the order Lepidoptera, the moths and butterflies, a long, coiled, sucking proboscis has evolved from the maxillae (see Figures 30-17 and 37-28, *C*), and the mandibles are absent or highly reduced in all but a few families. The proboscis of the Lepidoptera uncoils due to blood pressure and coils because of its own elasticity.

The insect thorax consists of three segments, each of which has a pair of legs. Occasionally one or more of these pairs of legs is absent. Legs are completely absent in the larvae of certain groups—for example, in most members of the order Hymenoptera, the bees (see Figures 30-15 and 30-16), wasps, and ants—and also among the flies, order Diptera (Figures 37-29 and 37-31, *B*). If two pairs of wings are present, they are attached to the middle and posterior segments of the thorax; if only one pair of wings is present, it is usually attached to the middle segment. The thorax is almost entirely filled with muscles that operate the legs and wings.

The wings of insects arise as saclike outgrowths of the body wall; in adult insects, they are solid, except for the veins. Insect wings, therefore, are not homologous to the other appendages. Basically, insects have two pairs of wings (Figures 37-9 and 37-26, *A*), but one

INSECT PHEROMONES AND PEST CONTROL

Insecticides present numerous environmental problems, as our experience since World War II has taught us. When they are used in large amounts, these chemicals may poison many kinds of organisms in addition to those for which they are intended. Furthermore, resistant strains of the target organisms soon appear, rendering further use of the insecticide ineffective. The insects' normal predators may have virtually disappeared when the pest populations were low and thus may be unable to control these insects when their numbers return to higher levels. For such reasons, scientists have increasingly been examining and applying the principles of integrated pest management, including selective use of pesticides, development of resistant strains of crops and domestic animals, biocontrol through the introduction or manipulation of predator populations that attack the agricultural pests, and other biological methods, alone or in combination.

One interesting approach to integrated pest control involves insect pheromones. These molecules or mixtures of molecules have potent effects at low levels. They are used to trap insects for various reasons, to monitor and survey their populations, to infect them with a disease or toxic substance that they may spread, to disrupt their ability to locate mates, or simply to destroy large quantities of them. All such strategies depend, of course, on a thorough knowledge of the behavior and biology of the pest species.

Approximately one third of all the insecticides produced worldwide are applied to cotton crops. By 1958 it had become obvious that the boll weevil, *Anthonomus grandis*, an important cotton pest, included so many resistant strains in the United States that further use of pesticides would be of doubtful value. A decade of research led to the development of methods in which the boll weevil pheromone, actually a mixture of four compounds,

---

FIGURE 37-29

Larvae ("wigglers") of a mosquito, Culex pipiens. *Most immature stages of flies resemble the maggots (larvae) shown in Figure 37-31, B, but the aquatic larvae of mosquitoes are quite active. They breathe through tubes from the surface of the water, as shown here. Covering the water with a thin film of oil, a method used widely in mosquito control, causes them to drown.*

of these has been lost during the course of evolution in some groups—for example, in the flies (Figure 37-26, *B*). Most insects can fold their wings over their abdomen when they are at rest, but a few, such as the dragonflies and damselflies, order Odonata, cannot do this and keep their wings erect or outstretched at all times (Figure 37-5, *B*).

Insect forewings may be tough and hard; if they are, they form a cover for the hindwings and usually open during flight. The beetles (Figures 37-5, *A* and 37-26, *D*) are an example in which the forewings remain folded over the back except during flight (Figure 37-9). The tough forewings also serve a protective function in the order Orthoptera, which includes grasshoppers (Figures 37-7 and 37-25, *E*) and crickets. Insect wings are made of sheets of chitin; their strengthening veins are tubules of chitin. The moths and butterflies have wings that are covered with detachable scales, which provide most of the basis for their bright colors (Figure 37-30). In some wingless insects, such as the springtails or silverfish, wings never evolved; many others, such as the fleas (Figure 37-26, *E*) and lice (Figure 37-26, *F*), are derived from ancestral groups of insects that did have wings.

All insects possess three body segments: the head, the thorax, and the abdomen. The three pairs of legs are attached to the thorax. Fused and specialized mouthparts, ultimately derived from appendages, as well as the antennae, are located on the head. Most insects have compound eyes as well as simple eyes, or ocelli, and many have one or two pairs of wings.

Internal organization

The internal features of insects resemble those of the other arthropods in many ways (Figure 37-7). The digestive tract is a tube, usually somewhat coiled. It is often about the same length as the body, but in the order Homoptera, which consists of the leafhoppers, cicadas,

was used to monitor and disrupt populations that existed at low enough levels. These applications of pheromone were combined with the introduction of sterile males, which mated with the females and thus prevented them from mating normally, and as a result limited the number of boll weevil offspring. Similarly, pheromones have been used successfully to disrupt mating in populations of the pink bollworm, *Pectinophora gossypiella,* another important cotton pest. This is done by permeating the air with the male sex attractant.

Pheromones that attract the European elm bark beetle, *Scotylus multistriatus,* have been used to catch these pests before they enter lightly infected or uninfected elm groves. Massive trapping of the spruce bark beetle, *Ips typographicus,* was achieved by the use of pheromones in a joint Swedish-Norwegian project in 1979, resulting in the elimination of about 2.9 billion beetles. In general the most important uses of pheromones to date have been based on their incredible ability to attract pest insects even when they are very rare, and thus provide an early warning of their presence in a given area.

FIGURE 37-C

The boll weevil, Anthonomus grandis *(order Coleoptera), shown here on a cotton boll (immature fruit), causes billions of dollars of damage to cotton crops throughout the world. Its pheromone is a mixture of four compounds, which are synergistic and serve to aggregate males and females in the field.*

and related groups, and in many flies, order Diptera, it may be greatly coiled and several times longer than the body. Such long digestive tracts are generally found in insects that feed on juices, rather than on protein-rich solid foods, because they offer a greater opportunity to absorb fluids and their dissolved nutrients. The digestive enzymes of the insect, also, are much less effective in a highly liquid medium than in a more solid one. Longer digestive tracts give these enzymes more time to work while food is passing through the insect's body. As a result, long, coiled digestive tracts often occur in those insects that have sucking mouthparts.

The anterior and posterior regions of an insect's digestive tract are lined with cuticle. Digestion takes place primarily in the stomach, or midgut, and excretion takes place through the Malpighian tubules, which arise from the junction between the midgut and hindgut and empty into the anterior end of the hindgut. The digestive enzymes are mainly secreted by cells that line the midgut, although some are contributed by the salivary glands, located closer to the mouth.

The tracheae of insects extend throughout the body and permeate its different tissues (Figure 37-8). In many winged insects, the tracheae are dilated in various parts of the body, forming air sacs. Such air sacs are surrounded by muscles and form a kind of bellows system to force air deep into the tracheal system. The spiracles, a maximum of 10 on each side of the insect, are paired and located on or between the segments along the sides of the thorax and abdomen. In most insects the spiracles can be opened by muscular action. In some parasitic and aquatic groups of insects, however, the spiracles are permanently closed. The tracheae run just below the surface of the insect, and gas exchange takes place by diffusion. Closing the spiracles at times may be important in retarding water loss. Longitudinal tubes connect the tracheae and make the outer part of the system an important route for flowing air.

FIGURE 37-30

Scales on the wing of Parnassius imperator, *a butterfly from China. Scales of this sort account for most of the colored patterns seen on the wings of butterflies and moths.*

FIGURE 37-31

*Simple metamorphosis (this page),
in which the young stages resemble
the adult fairly closely, in a chinch
bug (order Hemiptera), contrasted
with complete metamorphosis
(opposite page), in which the
younger stages differ greatly in
appearance from the adult, in a
housefly, Musca domestica. In
complete metamorphosis, there is a
resting stage, called the pupa or
chrysalis stage, before the emergence
of the adult.*

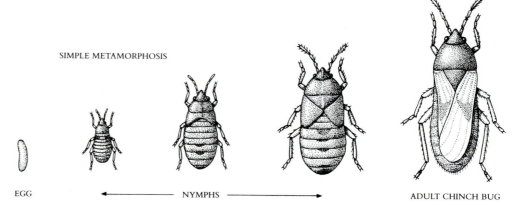

SIMPLE METAMORPHOSIS

EGG   ◄————— NYMPHS —————►   ADULT CHINCH BUG

The **fat body** of insects is a group of cells located in the body cavity. This structure may be quite large in relation to the size of the insect, and it serves as a food-storage reservoir. It is often more prominent in immature insects than in adults, and it may be completely depleted when metamorphosis is finished. However, insects that do not feed as adults retain their fat bodies and live on the food stored in them throughout their adult life.

Sense receptors

In addition to their eyes, insects have several characteristic kinds of sense receptors. These include sensory hairs, usually widely distributed over their bodies. The sensory hairs are linked to nerve cells and are sensitive to touch. They are particularly abundant on the antennae and legs—those parts of the insect most likely to come into contact with other objects. Similar organs are sensitive to chemicals or changes in position of the different parts of the insect. Taste organs are mostly located on the mouthparts; organs of smell, primarily on the antennae.

Sound, which is of vital importance to insects, is detected in groups such as grasshoppers and crickets, cicadas, and some moths, by **tympanal organs,** paired structures composed of a thin membrane, the **tympanum,** associated with the tracheal air sacs (see Figure 37-7). In many other groups of insects, the sound waves are detected by sensory hairs similar to those discussed above. Male mosquitoes, for example, detect the sounds made by the vibrating wings of female mosquitoes by means of thousands of sensory hairs on their antennae.

The detection of sound in insects is important not only for protection but also for communication. Many insects communicate by making sounds, most of which are very soft, very high-pitched, or both and thus inaudible to human beings. Only a few groups of insects, especially the grasshoppers, crickets, and cicadas, make sounds that people can hear. Male crickets and longhorned grasshoppers produce sounds by rubbing their two front wings together; shorthorned grasshoppers do so by rubbing their hind legs over specialized areas on their wings. Male cicadas make their sounds by vibrating the membranes of air sacs located on the lower side of the most anterior abdominal segment. Other insects communicate by tapping some part of their body against an external object.

In addition to sound, nearly all insects communicate by means of chemicals or synergistic mixtures of chemicals known as **pheromones.** These compounds, which are extremely diverse in their chemical structure, are sent forth into the environment, where they are active in very small amounts and convey a variety of messages to other individuals. Such messages may include not only the attraction and recognition of members of the same species for mating, for example, but also the marking of trails for members of the same species, as in the ants.

Insects possess sophisticated means of sensing their environment, including sensory hairs to detect touch, tympanal organs to detect sound, and chemoreceptors to detect chemical signals called pheromones.

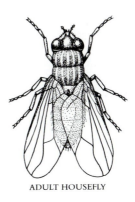

| EGG | EARLY LARVA | FULL-SIZED LARVA | PUPA | ADULT HOUSEFLY |

Life histories

Most young insects hatch from fertilized eggs laid outside their mother's body. In a few insects, the eggs hatch within the mother's body. The zygote develops within the egg into a young insect, which escapes by chewing its way out or by bursting the shell in various ways. Some immature insects have specialized projections, often on the dorsal side of the head, that assist in this process.

During the course of their development into adults, young insects undergo ecdysis a number of times before they become adults and stop molting permanently. Most insects molt 4 to 8 times during the course of their development, but some may molt as many as 30 times. The stages between the molts are defined as **instars.** When an insect first emerges following ecdysis, it is usually pale, soft, and especially susceptible to predators. Its exoskeleton generally hardens in an hour or two, and it must grow to its new size, usually by taking in air or water, during this brief period (Figure 37-25, C). The wings are expanded by forcing blood into their veins.

There are two principal kinds of metamorphosis in insects: **simple** and **complete** (Figure 37-31). In simple metamorphosis, the wings, if present, develop externally during the juvenile stages, and there is ordinarily no "resting" stage before the last molt. In complete metamorphosis, the wings develop internally during the juvenile stages and appear externally only during the resting stage that immediately precedes the final molt (Figure 37-32). During this stage, the insect is called a **pupa** or **chrysalis,** depending on the group to which it belongs. A pupa or chrysalis does not normally move around much, although the pupae of mosquitoes do move around freely and thus provide one well-known exception to this general rule (Figure 37-29). A very large amount of internal reorganization of the insect's body takes place while it is a pupa or chrysalis.

In insects with simple metamorphosis, the immature stages are often called **nymphs.** They are usually very similar to the adults, differing mainly in their smaller size, less well-developed wings, and sometimes different colors. The most primitive orders of insects, such as the springtails and silverfish, are descended from ancestors that never had wings during their evolutionary history. In them, development is **direct:** there is no particular change between the appearance of the nymphs and adults, except in size. In some other orders of insects with simple metamorphosis, such as the mayflies and dragonflies, the nymphs are aquatic and extract oxygen from the water by means of gills, which have evolved in these groups. The adult stages are terrestrial and look very different from the nymphs. In still other groups, such as the grasshoppers and their relatives, the nymphs and adults live in the same habitat. Such insects usually change gradually during their life cycle with respect to wing development, changes in body proportions, the appearance of ocelli, and other features.

More than 90% of the insects, however, including the members of all of the largest and most successful orders, display complete metamorphosis, in which the juvenile stages and adults often live in distinct habitats, have different habits, and are usually extremely different in form. In these insects, development is **indirect. Larvae** in insects are immature stages, often wormlike, which differ greatly in appearance from the adults of the same spe-

FIGURE 37-32

The life cycle of the monarch butterfly, Danaus plexippus *(order Lepidoptera), illustrates complete metamorphosis.*

A *Egg.*

B *Larva (caterpillar) feeding on a leaf. The pseudopods, or false legs, that occur on the abdomen of larval butterflies and moths are not related to the true legs that occur in the adults.*

C *Larva preparing to shed its skin and become a chrysalis.*

D *Chrysalis.*

E *The colors of the wings of the adult butterfly can be seen through the thin outer skin of the chrysalis; the adult is nearly ready to emerge.*

F *to* **H** *Adult butterfly emerging.*

I *Adult monarch butterfly. Monarch butterflies feed on poisonous plants of the milkweed family (Asclepiadaceae), and they are protected from their predators, primarily birds, by the chemicals they obtain from these plants. Their conspicuous warning coloration advertises their toxic nature.*

cies. Larvae do not have compound eyes, and they may be legless or have legs and sometimes also leglike appendages on the abdomen (Figures 30-16, 37-29, 37-31, and 37-32). They usually have chewing mouthparts, even in those orders in which the adults have sucking mouthparts. Chewing mouthparts are therefore clearly the primitive condition in all of these groups, and the sucking mouthparts of the adults have evolved from ancestral chewing mouthparts. When larvae and adults play different ecological roles, they do not compete directly for the same resources, a fact that is clearly an advantage to the species.

Pupae do not feed and are usually, but not always, relatively inactive. As pupae, insects are extremely vulnerable to predators and parasites, but they are often covered by a cocoon or some other protective structure. Groups of insects with complete metamorphosis include the moths and butterflies; beetles; bees, wasps, and ants; flies; and fleas. Fleas, order Siphonaptera, are among those orders of wingless insects that are descended

from winged ancestors; their winglessness is an advantage in their specialized way of life.

Both ecdysis and metamorphosis are controlled by hormones. Molting hormone, or **ecdysone,** is released from a gland in the thorax when that gland has been stimulated by brain hormone, which in turn is produced by neurosecretory cells in the brain and released into the blood. The effects of ecdysone are held in check by juvenile hormone, which is present during the immature stages, but declines in quantity as the insect passes through successive molts. When the level of juvenile hormone is low enough, the ecdysone stimulates the formation of the pupa and then the final development of the adult (Figure 37-33).

In this chapter we have presented an overview of the arthropods, the largest and most successful group of organisms on earth. Bewildering in their diversity, extraordinary in their numbers and ecological dominance in every habitat, the arthropods represent the culmination of one evolutionary line. We will now move on to consider the parallel line that led to the chordates, and ultimately to ourselves.

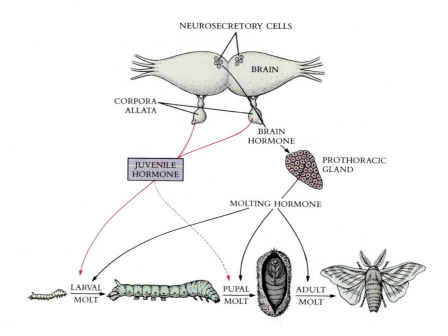

FIGURE 37-33

The hormonal control of metamorphosis in the silkworm moth, Bombyx mori. *Neurosecretory cells on the surface of the brain secrete brain hormone, which in turn stimulates the prothoracic gland to produce molting hormone. For the caterpillars to molt, both ecdysone and juvenile hormone, produced by bodies near the brain called the* corpora allata, *must be present, but juvenile hormone counteracts molting hormone when the caterpillar sheds its skin to become a pupa, and when the pupa becomes an adult. At these stages, therefore, it is important that the corpora allata not produce juvenile hormone.*

SUMMARY

1. Arthropods are the most successful of all animals in terms of numbers of individuals and numbers of species, as well as in terms of ecological diversification. Like the annelids, arthropods have segmented bodies; but in arthropods, some of the segments have become fused into tagmata during the course of evolution.

2. All arthropods possess a rigid external skeleton, or exoskeleton, which provides a surface for muscular attachment, a shield against predators and diseases, and a means of conserving water.

3. Arthropods have an open circulatory system, which is not as important in conducting oxygen to the tissues as the tracheae, a system of tubes that open to the outside by way of spiracles.

4. Arthropods consist of three subphyla: chelicerates (subphylum Chelicerata); crustaceans (subphylum Crustacea); and insects, centipedes, and millipedes (subphylum Uniramia). These subphyla probably evolved from different ancestors, as suggested by the structure and derivation of their mouthparts and other appendages and by the important embryological differences among them.

5. The ancestry of the chelicerates and crustaceans is not clear, although they are generally related to the annelids and to the Uniramia. Crustaceans, like Uniramia, have spiral cleavage in the embryo, but chelicerates do not.

6. Chelicerates consist of three classes, of which the Arachnida are the largest. Spiders, the best known arachnids, have a pair of chelicerae, a pair of pedipalps, and four pairs of walking legs. Spiders secrete digestive enzymes into their prey by means of the fangs on their chelicerae and then suck the contents out.

7. Crustaceans comprise some 35,000 species of crabs, shrimps, lobsters, barnacles, sowbugs, beach fleas, and many other groups. Their appendages are basically biramous, and their embryology is distinctive.

8. Uniramia, comprising the insects, centipedes, millipedes, and two other small classes, share a unique pattern of embryological development with the oligochaete annelids and leeches. Thus it seems clear that they evolved from ancestors similar to oligochaete annelids. The small phylum Onychophora resembles the Uniramia closely, and may share a common ancestor with it.

9. Insects and other Uniramia eliminate wastes by a unique system of Malpighian tubules, which extend from the digestive tract into the blood. This system is a very efficient means of water conservation, and, along with the exoskeleton and the spiracles, was a key factor in the invasion of land by the ancestors of insects.

10. Insects are the largest class of organisms, constituting at least 2 million and perhaps as many as 30 million species. They exhibit either simple metamorphosis, in which a succession of forms similar to the adult progressively matures, or complete metamorphosis, in which an often wormlike larva becomes a pupa, usually relatively sedentary, and then an adult; each of these stages is quite distinct from the others.

SELF-QUIZ

The following questions pinpoint important information in this chapter. You should be sure to know the answers to each of them.

1. The characteristic that separates the arthropods from the annelids is
   (a) division of tissue into three primary types
   (b) coelomic body architecture
   (c) segmentation
   (d) bilateral symmetry
   (e) jointed legs

2. _____ are fused segments that become functional groups, such as the head or thorax of an insect.

3. The process by which an animal sheds its exoskeleton, enabling it to grow, is known as _____.

4. Chelicerates gave rise to mandibulates. True or false?

5. The phylum Arthropoda actually consists of three independent evolutionary lines. They do have a number of features in common. Name one feature that they do *not* have in common.
   (a) segmented appendages
   (b) hard exoskeleton
   (c) radial embryonic cleavage
   (d) segmented body

6. Arthropods have
   (a) cilia
   (b) closed circulatory systems
   (c) only simple eyes
   (d) all of the above
   (e) none of the above

7. _____ are slender projections extending from the digestive tract into the blood that provide a means for arthropods to eliminate metabolic wastes.

8. The evolutionary process whereby juvenile stages become sexually mature is known as _____.

9. Insects have _____ pairs of legs.

10. Nearly all insects communicate by means of chemicals or synergistic mixtures of chemicals known as _____.

11. Insects do not have legs on their abdomen. Do crustaceans?

12. Pupae are a developmental stage in insects when the developing individuals
    (a) feed and are very active   (b) do not feed and are relatively inactive

THOUGHT QUESTIONS

1. What are the main factors that limit the size of terrestrial arthropods? What effect has this had on the evolution of intelligence in the group?

2. What is the evidence that the annelids and the arthropods had a common ancestor? Which arthropods were involved? When would this evolutionary transition have taken place?

3. What is the evidence that the arthropods consist of groups that evolved from different ancestors at different times in the past? Why have the arthropods been treated as a single phylum if this evidence is so convincing?

FOR FURTHER READING

ADKISSON, P.L., G.A. NILES, J.K. WALKER, L.S. BIRD, and H.B. SCOTT: "Controlling Cotton's Insect Pests: A New System," *Science,* vol. 216, pages 19-22, 1982. The development of integrated insect control programs is especially important for cotton, which is more heavily treated with insecticides than any other crop in the United States.

BORROR, D.J., D.M. DELONG, and C.A. TRIPLEHORN: *An Introduction to the Study of Insects,* 5th ed., Holt, Rinehart & Winston, New York, 1980. The standard text in the field, readable and well presented; strongly oriented to systematics and morphology.

CAMERON, J.N.: "Molting in the Blue Crab," *Scientific American,* May 1985, pages 102-109. An interesting discussion of the way in which the blue crab molts; soft-shelled crabs are individuals caught right after molting.

EVANS, H.E.: *Insect Biology: A Textbook of Entomology,* Addison-Wesley Publishing Co., New York, 1984. An outstanding modern textbook of entomology, less oriented to morphology and systematics than the text by Borror et al.

FOELIX, R.F. *Biology of Spiders,* Harvard University Press, Cambridge, Mass., 1982. A readable account of all aspects of the life of spiders, ecologically oriented and well illustrated.

GREANY, P.D., S.B. VINSON, and W.J. LEWIS: "Insect Parasitoids: Finding New Opportunities for Biological Control," *BioScience,* vol. 34, pages 690-695, 1984.

Parasitoids—insects that live within the bodies of other insects—may be mass produced and even genetically engineered to improve their value in controlling insect pests.

HORRIDGE, G.A.: "The Compound Eye of Insects," *Scientific American,* July 1977, pages 108-120. Fascinating account of the way in which compound eyes function in different kinds of insects.

KINGSOLVER, J.G.: "Butterfly Engineering," *Science,* August 1985, pages 106-113. Discusses the strategies by which butterflies regulate their temperatures.

MANTON, S.M.: *The Arthropoda: Habits, Functional Morphology, and Evolution,* Clarendon Press, Oxford, England, 1977. A brilliant exposition of the features of this great group; highly recommended.

MICHAELSEN, A.: "Insect Ears as Mechanical Systems," *American Scientist,* vol. 67, pages 696-706, 1979. The ways in which insect ears function.

PRESTWICH, G.D.: "The Chemical Defenses of Termites," *Scientific American,* August 1983, pages 78-98. Termites, which are soft bodied and blind, attack intruders with an array of sophisticated irritants, toxins, anticoagulants, and glues.

SILVERSTEIN, R.M.: "Pheromones: Background and Potential for Use in Insect Pest Control," *Science,* vol. 213, pages 1326-1332, 1981. Increasing attention is being paid to the ways in which pheromones can be applied to insect control.

ECHINODERMS AND CHORDATES

Chapter 38

Overview

More than 630 million years ago, a revolution occurred within the animal kingdom, leading to the evolution of a novel pattern of embryonic development. A new line of animals appeared, a line that we call the deuterostomes. Two superficially dissimilar animal phyla comprise most of this group: the echinoderms (including sea stars, sea urchins, and their relatives) and the chordates (including vertebrates). The group also includes two other small phyla of marine animals. Although the echinoderms and chordates are very different in appearance, they are fundamentally similar in a number of features, which indicates that they were derived from a common ancestor.

For Review

Here are some important terms and concepts that you will encounter in this chapter. If you are not familiar with them, you should review them before proceeding.

- **Evolutionary theory** (Chapter 21)

- **Classification** (Chapter 22)

- **Evolutionary history** (Chapter 23)

- **Evolution of human beings** (Chapter 24)

- **Protostomes and deuterostomes** (Chapter 36)

Among the animals, there are two great evolutionary branches, corresponding to fundamental differences in the architecture of the developing embryo. In Chapters 35 to 37, we have been traveling along one of these branches. Most invertebrates, including the annelids, mollusks, and arthropods, share the same pattern of embryological organization and are collectively called the protostomes. We will now consider the second major evolutionary branch, the deuterostomes.

DEUTEROSTOMES

Two outwardly dissimilar large phyla, Echinodermata and Chordata, together with two smaller phyla have a series of key embryological features that are very different from those of the other animal phyla (see Figure 36-2). Because it is extremely unlikely that these fea-

A

B

C

FIGURE 38-1

Diversity in echinoderms (phylum Echinodermata).

A *Sea star,* Oreaster occidentalis *(class Asteroidea), in the Gulf of California, Mexico.*

B *Sea cucumber,* Stichopus *(class Holothuroidea), Philippines.*

C *Feather star (class Crinoidea), on the Great Barrier Reef in Australia.*

D *Brittle star,* Ophiothrix *(class Ophiuroidea).*

E *Giant red sea urchin,* Strongylocentrotus franciscanus *(class Echinoidea).*

F *Sand dollar,* Echinarachnius parma *(class Echinoidea), New England.*

tures evolved more than once, it is believed that these four phyla share a common ancestry. They are the members of a group called the deuterostomes, which we introduced in Chapter 36. As we saw there, deuterostomes diverged from protostome ancestors, which would not have resembled any living animals, more than 630 million years ago.

Deuterostomes, like protostomes, are coelomates. They differ fundamentally from protostomes, however, in the way in which the embryo grows. The blastopore of a deuterostome becomes the animal's anus, and the mouth develops at the other end; they have radial, rather than spiral, cleavage; daughter cells are identical for a brief period of development—each of which, if separated from the others at an early enough stage, can develop into a complete organism; and whole groups of cells move around during the course of embryonic development to form new tissue associations. The coelom is normally produced by an evagination of the **archenteron**—the main cavity within the gastrula, which is lined with endoderm and opens to the outside by way of the blastopore. The archenteron closes over, forming the hollow cavity of the coelom, and later becomes the gut cavity.

In this chapter we will discuss the echinoderms, the two minor phyla of deuterostomes, and the simpler groups of chordates, thus encompassing the diversity of the class. The more advanced members of the phylum Chordata, the vertebrates, will be reviewed in the next chapter. The following section, which starts with Chapter 39, considers the features of the vertebrates in detail.

ECHINODERMS

Echinoderms (phylum Echinodermata) are an ancient group of marine animals, very well represented in the fossil record. The term *echinoderm* means "spine-skin," an apt description of many members of this phylum, which consists of about 6000 living species (Figure 38-1). Many of the most familiar animals seen along the seashore—the sea stars (starfish), brittle stars, sea urchins, sand dollars, and sea cucumbers—are echinoderms. All are radially symmetrical as adults. In the sea cucumbers and some other echinoderms, however, the animal's axis lies horizontally, and radial symmetry is not as obvious as it is in most other

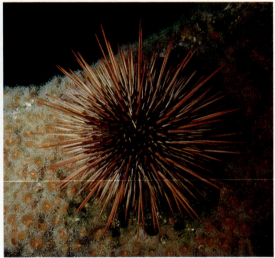

D   E   F

members of the phylum. Echinoderms are well represented not only in the shallow waters of the sea but also in its abyssal depths. All of them are bottom–dwellers except for a few swimming sea cucumbers. The adults range from a few millimeters to more than a meter in diameter (for one species of sea star) or in length (for a species of sea cucumber).

Basic Features of Echinoderms

Echinoderms have a delicate epidermis, containing thousands of neurosensory cells, stretched over an endoskeleton made up of either movable or fixed calcium-rich (calcite) plates called **ossicles.** The animals grow more or less continuously, but their growth slows down with age. When the plates first form, they are enclosed in living tissue. In most cases, these plates bear spines—hence the name of the phylum. The plates in certain portions of the body of most echinoderms are perforated. Through these perforations extend the **tube feet,** a part of the **water vascular system,** which is a unique feature of this phylum.

Echinoderms have a five-part body plan corresponding to the arms of a sea star or the design on the "shell" of a sand dollar. As adults, these animals have no head or brain. Their nervous systems consist of central nerve rings from which branches arise. The animals are capable of complex response patterns, but there is no centralization of function. Apparently, the centralization of the nervous system is not feasible in animals with radial symmetry.

The water vascular system of an echinoderm radiates from a ring canal that encircles the animal's esophagus. Five radial canals, the position of which are determined early in the development of the embryo, extend into each of the five parts of the body and determine its basic symmetry (Figure 38-2). Water enters the water vascular system through a **madreporite,** a sievelike plate on the animal's surface, from which it flows to the ring canal through a tube, or **stone canal,** so named because of the rings of calcium carbonate that surround it. The five radial canals in turn extend out of the animal through short side branches into the hollow tube feet (Figure 38-3). In some echinoderms, each tube foot has a sucker at its end; in others, suckers are absent. At the base of each tube foot is a muscular sac, the

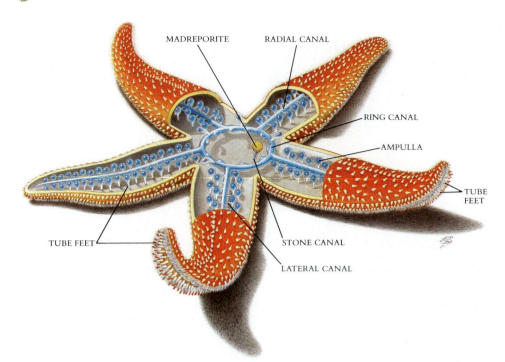

FIGURE 38-2

The water vascular system of a sea star.

MADREPORITE   RADIAL CANAL

RING CANAL

AMPULLA

TUBE FEET

TUBE FEET

STONE CANAL

LATERAL CANAL

FIGURE 38-3

The extended tube feet of a sea star, Ludia magnifica, *in Hawaii.*

FIGURE 38-4

A slow-moving sea urchin, Calocidaris micans, *perched on a rock at about 300 meters' depth in the Bahamas. This species grazes on small organisms living on the rock surface as it moves over the rocks.*

ampulla, which contains fluid. When the ampulla contracts, the fluid is prevented from entering the radial canal by a one-way valve and is forced into the tube foot, thus extending it. When extended, the foot attaches itself to the substrate. Longitudinal muscles in the tube foot's wall contract, shortening it, and the animal is pulled forward to the new position as the water in the tube foot is forced back into the ampulla. By such action, repeated in many tube feet, sea stars, sand dollars, and sea urchins slowly move along (Figure 38-4). In a sea star, any one of the five arms may go first.

Echinoderms are radially symmetrical animals whose bodies typically have a pattern that repeats itself five times. They have characteristic calcium-rich plates called ossicles and a unique water vascular system.

In sea cucumbers (Figure 38-1, *B*), which are shaped like their vegetable namesakes, there are usually five rows of tube feet on the body surface that are used in locomotion. Sea cucumbers also have modified tube feet around their mouth cavity; these are used in feeding. In sea lilies, which most often remain attached at a particular place in the sea, the tube feet arise from the branches of the arms, which extend from the margins of an upward-directed cup. With these tube feet, the animals take food from the surrounding water (Figure 38-1, *C*). In brittle stars (Figure 38-1, *D*), the tube feet are pointed and specialized for feeding.

In echinoderms, the coelom, which is proportionately large, connects with a complicated system of tubes and helps to provide for circulation and respiration. In many echinoderms, respiration takes place by means of skin gills, which are small, fingerlike projections that occur near the spines. Waste removal also takes place through these skin gills. The digestive system is simple but usually complete, consisting of a mouth, gut, and anus.

Many echinoderms are able to regenerate lost parts, and some, especially sea stars and brittle stars, drop various parts when under attack. In a few echinoderms, asexual reproduction takes place by splitting, and the broken parts of sea stars, for example, can sometimes regenerate whole animals. Some of the smaller brittle stars, especially tropical ones, regularly reproduce by breaking into two equal parts; each half then regenerates a whole animal. Sea cucumbers, when irritated, sometimes eject a portion of their intestines by a strong muscular contraction that may send the intestinal fragments through the anus or even rupture the body wall. This is apparently a defensive mechanism.

Despite the ability of many echinoderms to break into parts and regenerate new animals from them, most reproduction in the phylum is sexual. The sexes in most echinoderms are separate, although there is usually little external difference between them. The fertilized eggs of echinoderms usually develop into free-swimming, bilaterally symmetrical larvae (Figure 38-5), which are very different from the larvae of trochophores of mollusks and annelids. These larvae form a part of the plankton until they metamorphose through a series of stages into the more sedentary adults.

Diversity of Echinoderms

There are more than 20 extinct classes of echinoderms and an additional five with living members: (1) Crinoidea, the sea lilies and feather stars; (2) Asteroidea, the sea stars or starfish; (3) Ophiuroidea, the brittle stars; (4) Echinoidea, the sea urchins and sand dollars; and (5) Holothuroidea, the sea cucumbers. It is almost certain, based on the existence of their bilaterally symmetrical, free-swimming, ciliated larvae, that adult echinoderms were originally free swimming and bilaterally symmetrical as well. During the course of their evolution, they seem to have settled down to a more sedentary life. After doing so, they developed radial symmetry, the water vascular system, and the other distinctive characteristics of the phylum as we now know it.

Sea lilies and feather stars

The sea lilies and feather stars, or crinoids (class Crinoidea; see Figures 38-1, C, 38-6, and 38-7) differ from all other living echinoderms in that the mouth and anus are located on their upper surface in an open disk. They are connected by a simple gut. These animals have simple excretory and reproductive systems and a very extensive water vascular system. The arms, which are the food-gathering structures of crinoids, are located around the margins of the disk. Different species of crinoids may have from 5 to more than 200 arms extending upward from their bodies, with smaller structures called pinnules branching from the arms. In all crinoids, the number of arms is initially small; species with more than 10 arms add additional arms progressively during growth. Crinoids are filter feeders, capturing the microscopic organisms on which they feed by means of the mucus that coats their tube feet, which are abundant on the animal's pinnules. Consequently, the position of an individual animal is very critical to its success in obtaining food (Figure 38-7).

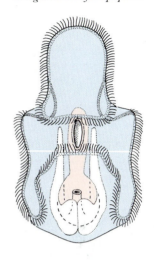

FIGURE 38-5

The free-swimming larva of an echinoderm; the bands of cilia by which the larva moves are prominent in this drawing. Such bilaterally symmetrical larvae demonstrate convincingly that the ancestors of the echinoderms were not radially symmetrical like the living members of the phylum.

FIGURE 38-6

A view of the Great Barrier Reef at Lizard Island, off the northeast coast of Australia, shows several different kinds of feather stars coexisting on a single coral head. The clawlike cirri by which these animals grip the substrate are plainly visible at the base of the large crinoid in the center. Its arms, which are arranged in a semicircular pattern for filter-feeding, measure about 25 centimeters long when extended. Over 30 species of feather stars were found in a transect along 400 meters of the reef, and up to 14 species were observed in a single area about 3 meters across.

FOSSIL ECHINODERMS—A PARADE OF UNSUCCESSFUL EXPERIMENTS

By David L. Pawson, National Museum of Natural History, Smithsonian Institution

Echinoderms are common fossils; many ancient limestones are composed almost exclusively of their remains. The earliest echinoderm fossils are about 600 million years old, from Precambrian times. Over the next 160 million years, during the Cambrian and Ordovician Eras, a dramatic evolutionary radiation occurred in this group. Fifteen classes of echinoderms are known from the Cambrian Era, and 19 classes from the Ordovician. Only one class of echinoderms has arisen during the subsequent 440 million years. It can therefore be said that virtually all echinoderm evolution on a major scale was over by the end of the Ordovician Era.

It is evident from this early fossil record that the echinoderms "experimented" with a great variety of bizarre body types before settling on the pattern of five-part symmetry that is characteristic of the living members of the phylum. Several extinct classes consisted of animals that were completely asymmetrical; the only reason we can tell that they were echinoderms

FIGURE 38-7

Sea lilies, Cenocrinus asterius. *Two specimens showing a typical parabola of arms, forming a "feeding net." The water current is flowing from right to left, carrying small organisms to the stalked crinoid's arms. Prey, when captured, are passed down the arms to the central mouth. This photograph was taken at a depth of about 400 meters in the Bahamas from the Johnson-Sea-Link Submersible of the Harbor Branch Foundation, Inc.*

There are two basic crinoid body plans. In the sea lilies, the animal is attached to its substrate by a stalk that may be as much as a meter long in some living forms and up to 20 meters long in some of the extinct ones. If they are removed from their points of attachment to a substrate, some sea lilies can move slowly by means of their featherlike arms. There are about 80 living species of sea lilies, all of them found below a depth of 100 meters in the ocean. In the second group of crinoids, the feather stars, which comprise about 520 species, the disk detaches from the stalk at an early stage of development. Adult feather stars are usually anchored directly to their substrate by clawlike structures. However, some feather stars are able to swim for short distances, and many of them can move along the substrate. Feather stars range into shallower water than do sea lilies, and only a few species of either group are found at depths greater than 500 meters. Along with the sea cucumbers, crinoids are the most abundant and conspicuous large invertebrates in the warm waters and among the coral reefs of the western Pacific Ocean. They have separate sexes, with the sex organs simply being masses of cells in special cavities of the arms and pinnules. Fertilization is usually external, with the male and female gametes shed into the water, but brooding occurs occasionally.

Students of echinoderms believe that when the common ancestors of this phylum first settled down to the substrate, they gave rise to sessile, sedentary, radially symmetrical animals that resembled the crinoids. Such animals were abundant in the ancient seas, and crinoids were present when the Burgess Shale (see Figure 23-9) was deposited about 530 million years ago. Some fossil crinoids were as much as 25 meters in length, although the living ones are rarely more than a hundredth of that size. More than 6000 fossil species of this class are known, in comparison with the approximately 600 living species. The sea lilies are the only living echinoderms that are fully sessile.

Sea stars

The sea stars, or starfish (see Figures 38-1, *A,* 38-2, and 38-8), class Asteroidea, are perhaps the most familiar echinoderms. These attractive animals are among the most important predators in many marine ecosystems. Their arms are more or less prominent and sharply set off from the disk. Although many sea stars have five arms, conforming to the basic symmetry of the phylum, the members of some families have many more. The mouth of a sea star is located in the center of its lower surface. There is striking variation in color in this group, and many species of sea stars are remarkably beautiful. They are abundant in the

is by their calcite (a hexagonal crystalline form of calcium carbonate) skeletons, a characteristic of the group. The members of at least one group had a three-part symmetry, possibly an intermediate evolutionary step toward five-part symmetry.

Most early echinoderms were sessile, or sedentary, attaching themselves in various ways to hard or soft substrates and feeding on passing small organisms or on organic detritus. The free-living habit was not common among the members of the group until the middle Ordovician Era, about 470 million years ago. From then onward, most echinoderms were mobile, with the exception of the stalked crinoids.

After the Ordovician Era—during the past 440 million years—echinoderms remained fairly conservative in structure. Only one new class arose, and many gradually became extinct. There was a rapid increase in the number of genera that rose to a peak near the start of the Triassic Era, about 250 million years ago, followed by a dramatic "crash" during the next 30 million years. Since the early part of the Jurassic Era (about 210 million years ago), the number of genera of echinoderms has steadily increased, until today there is a relatively high number of living forms in this ancient and fascinating phylum.

FIGURE 38-8

Crown-of-thorns sea star, Acanthaster planci. *This echinoderm was first reported in large numbers in 1957 in the Ryukyu Islands, between Japan and Taiwan, and rapidly spread over the western Pacific Ocean, destroying coral reefs as it went. The animals are able to move along the reef at a rate of up to 20 meters per hour. After the 1960s, the crown-of-thorns became much less abundant, for unknown reasons, and the reefs began to recover. This individual, photographed on Lodestone Reef, part of the Great Barrier Reef near Townsville, Australia, has just finished feeding on the table coral* (Acropora hyacinthus) *at the top left; the sea star is an adult, about 30 centimeters in diameter.*

intertidal zone (the zone between the high and low tide marks), but they also occur at depths as great as 10,000 meters. The roughly 1500 species of sea stars occur throughout the world.

Some sea stars have an extraordinary way of feeding on bivalve mollusks. They grasp both of a bivalve's shells with their tube feet (Figure 38-9), extrude their stomach through its mouth, and push the stomach into the bivalve. A sea star can push its stomach into a bivalve through an opening as little as 0.1 millimeter wide, corresponding to the natural openings in many irregular shells! Within the mollusk, the sea star secretes its digestive enzymes and digests the soft tissues of its prey, retracting its stomach when the process is complete. Most sea stars do not feed in this way, however.

Most sea stars have separate sexes, with a pair of gonads lying in the space between each pair of arms. The eggs and sperm are shed into the water so that fertilization is external. In some species, the fertilized eggs are brooded in special cavities or simply under the animal. They mature into larvae that swim by means of conspicuous bands of cilia.

FIGURE 38-9

A sea star attacking a clam.

Brittle stars

The brittle stars (see Figure 38-1, *D*), class Ophiuroidea, have very slender branched arms. These arms are more sharply set off from their central disk than in many sea stars. Brittle stars move by "rowing" along the substrate—moving their arms, often in pairs or groups, from side to side. In a number of species the arms are covered with spines, which also assist in their movements. Some of the brittle stars use their arms to swim, a very unusual habit among the echinoderms.

The brittle stars feed by capturing suspended microplankton with their tube feet, long arm spines, or branching arms or by climbing over the objects on the ocean floor in search of small animals. In addition, the tube feet are important sensory organs and assist in directing food into the mouth once the animal has captured it. As implied by their English name, the arms of brittle stars detach easily, a characteristic that helps to protect the brittle stars from their predators. There are about 2000 species. In shallow water, they occur largely on hard substrates and are secretive, but in the deep ocean, they are one of the most abundant organisms. More closely related to the sea stars than to the other classes of the phylum, they are sometimes grouped with them in a single class.

Brittle stars usually have separate sexes, with the male and female gametes in most species being released into the water and fusing there. Some brood their young in special cavities and release the larvae, which are free-swimming by means of their conspicuous bands of cilia, when they are mature enough.

Sea urchins and sand dollars

The members of the class Echinoidea, the sea urchins and sand dollars (see Figures 38-1, *E* and *F,* and 38-4), lack distinct arms but have the same five-part body plan as all other echinoderms. Five rows of tube feet protrude through the plates of the calcareous skeleton, and there are also openings for the mouth and anus. These different openings can be seen in the globular skeletons of sea urchins and in the flat skeletons of sand dollars (Figure 38-10). Both types of skeleton, which are sometimes common along the seashore, consist of fused calcareous plates. There are about 950 living species of the class Echinoidea.

Echinoids walk either by means of their tube feet or, in the case of sea urchins, their movable spines, which are hinged to the skeleton by a joint that makes free rotation possible. Sea urchins and sand dollars move along the sea bottom, feeding on algae and small fragments of organic material (see Figure 38-4). They scrape these off the substrate with the large, triangular teeth that ring their mouth. The gonads of sea urchins, often eaten raw, are considered a great delicacy by people in different parts of the world, especially along the Mediterranean and in Japan. Sea urchins and sand dollars, because of their calcareous plates, are well preserved in the fossil record, from which more than 5000 additional species have been described.

As with most other echinoderms, the sexes of sea urchins and sand dollars are separate. The eggs and sperm are shed separately into the water, where they fuse. Some brood their young, and others have free-swimming larvae, with bands of cilia extending onto their long, graceful arms.

Sea cucumbers

Sea cucumbers (see Figure 38-1, *B*), class Holothuroidea, differ from the preceding classes in that they are soft, sluglike organisms, often with a tough, leathery outside skin. The class consists of about 1500 species found worldwide. Except for a few forms that swim, sea cucumbers lie on their sides at the bottom of the ocean. Their mouth is located at one end and is surrounded by 8 to 30 modified tube feet called tentacles; the anus is at the other end. The tentacles around the mouth may secrete a mucous net that is used to capture the small planktonic organisms on which the animals feed. The tentacle is periodically wiped off within the esophagus and then brought out again, covered with a new supply of mucus. Some sea cucumbers ingest bottom sediment, as earthworms do, and extract the organic matter from it. When they do, they leave castings of excreted material similar to those deposited by earthworms.

Sea cucumbers are soft because their calcareous skeleton is reduced to widely separated microscopic plates. These interesting animals have extensive internal branching systems called **respiratory trees,** which arise from the **cloaca,** or anal cavity. Water is pulled into and expelled from the respiratory tree by contractions of the cloaca, and gas exchange to the various parts of the body takes place as this process occurs. Most kinds of sea cucumbers have tube feet on the body in addition to tentacles. If they do, these additional tube feet, which might be restricted to five radial grooves or scattered over the surface of the body, may enable the animals to move about slowly. On the other hand, sea cucumbers may simply wriggle along whether or not they have additional tube feet. Most sea cucumbers are quite sluggish, but some, especially among the deep-sea forms, swim actively. Sea cucumbers are considered a great delicacy in many parts of the world.

The sexes of sea cucumbers are separate, but some of them are hermaphroditic—that is, both male and female. Their sex organ is in the form of a single gonad consisting of one or two clusters of tubules that join together at a duct. The eggs and sperm are shed into the water and give rise to free-swimming larvae.

TWO MINOR PHYLA: ARROW WORMS AND ACORN WORMS

Arrow Worms

There are roughly 70 species of arrow worms (phylum Chaetognatha; Figure 38-11). Given their English name because of their shape, arrow worms are abundant predators in the marine plankton. Arrow worms occur all over the world, but they are most abundant in warm, shallow seas. In most marine planktonic environments, their numbers are exceeded only by the copepod crustaceans. They are thus the most abundant primary carnivores in the marine food web. They can dart rapidly forward or backward through the water to capture their prey, which includes mainly copepods but also cnidarian medusae, fish larvae, and other small animals. Arrow worms that feed on fish larvae may be harmful predators in areas where commercially important fishes breed.

Arrow worms, which range from 0.6 to 7 centimeters long as adults, are translucent and bilaterally symmetrical coelomates. Transverse septa divide arrow worms into head,

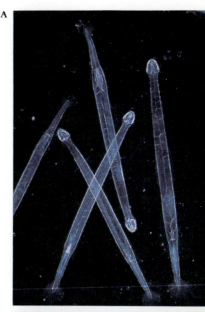

A

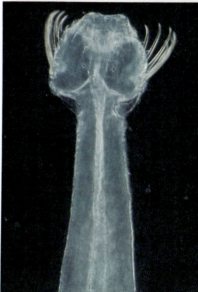

B

FIGURE 38-11

Arrow worms (phylum Chaetognatha).
A *Several individuals of* Sagitta elegans.
B *The head of an arrow worm,* Sagitta setosa, *showing the eyes and the hooks ringing the jaws. The individual shown here is dead, having been captured by a jellyfish,* Obelia.

trunk, and tail segments, foreshadowing the kind of specialization that occurs in chordates. The arrow worms have large eyes, powerful jaws, and a head ringed with numerous sharp movable hooks (Figure 38-11, *B*), all key factors in their predatory role. These animals, which are linked to the other deuterostomes only in the details of their embryonic development, seem to lack close relatives. Arrow worms date back in the fossil record at least 500 million years and appear to have remained essentially unchanged throughout this long history.

Acorn Worms

The exclusively marine acorn worms consist of about 90 species comprising the phylum Hemichordata, meaning "half chordates" (Figure 38-12). In addition to those features that mark them as deuterostomes, the acorn worms share a number of features with both the echinoderms and the chordates; their ciliated larvae closely resemble those of sea stars, for example. A dorsal nerve cord is present in hemichordates, in addition to a ventral one; and in some hemichordates, part of of the dorsal nerve cord is hollow, a feature otherwise found only among the chordates. Also, only in acorn worms and in chordates is the pharynx, or throat region, perforated with holes known as pharyngeal slits; these structures are found in no other animals.

The acorn worms are soft-bodied animals that live in burrows in sand or mud in the sea. Most range between about 2.5 centimeters and 2.5 meters in length. Their bodies are fleshy and contractile and consist of a proboscis, collar, and trunk. A second group of animals included in this phylum, the class Pterobranchia, consists of sessile and colonial animals less than 7 millimeters long. The members of this class have the same basic body regions and structure as the acorn worms, even though they differ so remarkably in size and appearance. Fossils similar to pterobranchs, which if confirmed as members of this class would be the oldest fossil hemichordates, occur in rocks 450 million years old. This evidence suggests that the phylum is an ancient one.

CHORDATES

The chordates (phylum Chordata) are the best known and most familiar group of animals. There are some 45,000 species of chordates, a phylum that includes the birds, reptiles, amphibians, fishes, and mammals (and therefore our own species).

Characteristic Features

Three principal features characterize the chordates (Figure 38-13):

1. A single, hollow **nerve cord,** which runs just beneath the dorsal surface of the animal.
2. A flexible rod, the **notochord,** which forms on the dorsal side of the primitive gut in the early embryo and is present at some stage of the life cycle in all chordates. The nerve cord is located just above the notochord.
3. **Pharyngeal slits.**

Each of these three distinguishing features of the chordates has played an important role in the evolution of the phylum. In the more advanced vertebrates, as we will see in Chapter 39, the dorsal nerve cord becomes more and more differentiated into the brain and spinal cord. The notochord, which persists throughout the life cycle of some of the invertebrate chordates, becomes surrounded and then replaced during the embryological development of vertebrates by the vertebral column. Pharyngeal slits are present in the embryos of all vertebrates but are lost later in the development of terrestrial vertebrates. However, the presence of these structures in all vertebrate embryos provides a clue to the aquatic ancestry of the group.

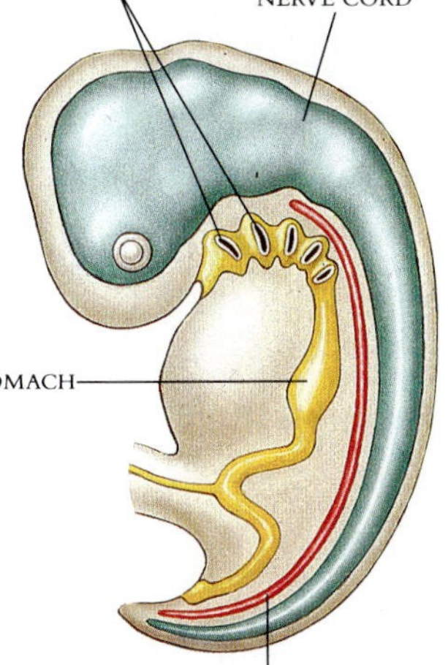

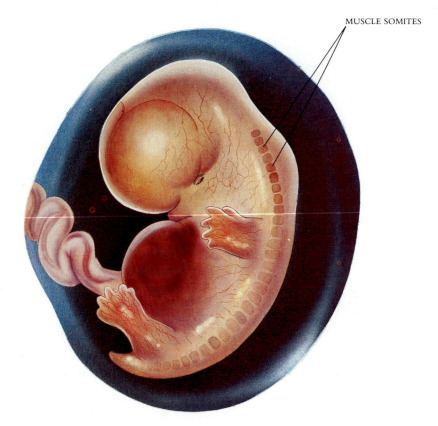

PHARYNGEAL
SLITS

HOLLOW, DORSAL
NERVE CORD

STOMACH

NOTOCHORD

MUSCLE SOMITES

FIGURE 38-13

Some of the principal features of the chordates.

FIGURE 38-14

A human embryo. The muscle is evidently divided into segments called somites at this stage, reflecting the fundamentally segmented nature of all chordates.

Chordates are characterized by a single, hollow dorsal nerve cord. At some point in the embryonic development of all chordates, the notochord, a flexible rod, forms dorsal to the gut, and slits are present in the pharynx. The notochord persists into the adult stage in the less advanced chordates.

In addition to these three principal features, a number of other characteristics also tend to distinguish the chordates. In their body plan, chordates are more or less segmented, and distinct blocks of muscles can be seen clearly in many less specialized forms (Figure 38-14). Chordates have an internal skeleton against which the muscles work; either this skeleton or the notochord makes possible the extraordinary powers of locomotion that characterize the members of this group. Finally, chordates have a tail that extends beyond the anus, at least during their embryonic development; nearly all other animals have a terminal anus. In all of these respects, chordates differ fundamentally from other animals.

Chordate Subphyla

There are three subphyla of chordates. Two of them, Urochordata, the tunicates, and Cephalochordata, the lancelets, are **acraniates;** they lack a distinct skull. The **craniate** chordates are the vertebrates, subphylum Vertebrata. Most of the vertebrates have a bony skeleton, although two classes of fishes—Agnatha, the lampreys and hagfish, and Chondrichthyes, the sharks and rays—have a cartilaginous one.

FIGURE 38-15

*Tunicates (phylum Chordata,
subphylum Tunicata).*
A *The sea peach,* Halocynthia
aurantium.
B *A beautiful blue-and-gold tunicate.*
C *Diagram of the structure
of a tunicate.*

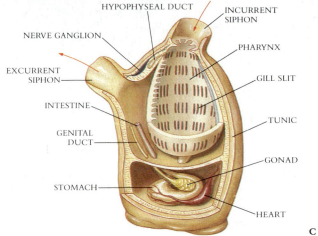

HYPOPHYSEAL DUCT
INCURRENT SIPHON
NERVE GANGLION
PHARYNX
EXCURRENT SIPHON
GILL SLIT
INTESTINE
TUNIC
GENITAL DUCT
GONAD
STOMACH
HEART

C

Tunicates

The tunicates (subphylum Urochordata) are a group of about 1250 species of marine animals. Most of them are sessile as adults (Figure 38-15), and in these, only the larva has a
notochord and a nerve cord. As adults, these animals also lack visible signs of segmentation
or of a body cavity. Most species occur in shallow waters, but some are found at great
depths. In some tunicates, the adults are colonial, living in masses on the ocean floor. Regardless of whether they are solitary or colonial, the tunicates obtain their food by ciliary
action; their pharynx is lined with numerous cilia. These cilia beat, drawing a stream of
water into the pharynx, where the microscopic particles that are eaten are removed from
the current of water by mucous secretions.

FIGURE 38-16

Diagram of the structure of a larval tunicate, showing the characteristic tadpolelike form. Larval tunicates resemble the postulated common ancestor of the chordates.

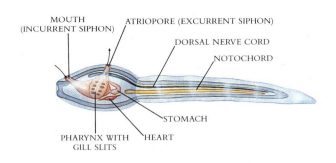

MOUTH (INCURRENT SIPHON)
ATRIOPORE (EXCURRENT SIPHON)
DORSAL NERVE CORD
NOTOCHORD
STOMACH
PHARYNX WITH GILL SLITS
HEART

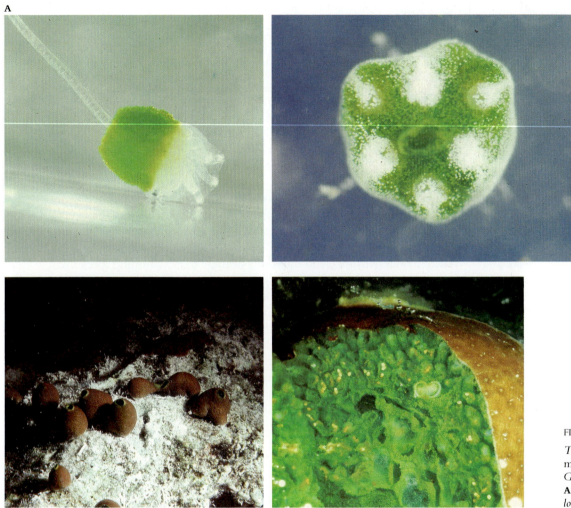

A

B

C

D

FIGURE 38-17

The colonial tunicate Didemnum molle, *at Lizard Island, on the Great Barrier Reef, Australia.*
A *A larva, about 2.5 millimeters long; the green material inside the larva is the symbiotic bacterium* Prochloron.
B *A juvenile individual, which, having settled, has divided into six parts and is forming a colony.*
C *Colonies of the tunicate at about 2 meters' depth. One of the colonies is in the process of dividing.*
D *A colony broken open, showing the masses of* Prochloron *within. The larvae are released at midday, swim for approximately 1 to 6 minutes, and settle preferentially on dark or shaded substrates; full sunlight is lethal to them.*

The tadpolelike larvae of tunicates plainly exhibit all of the basic characteristics of chordates and mark the tunicates as having the most primitive combination of features found in any chordate (Figures 38-16 and 38-17, *A*). They do not feed and have a poorly developed gut. The larvae remain free swimming for no more than a few days; then they settle to the bottom and attach themselves to a suitable substrate by means of a sucker.

Tunicates change so much as they mature and adjust developmentally to a sessile, filter-feeding existence that it would be very difficult to discern their evolutionary relationships by examining an adult of this subphylum. Many adult tunicates secrete a tunic, which is a tough sac composed mainly of cellulose, a substance that is frequent in the cell walls of plants and algae but is very rare in animals. The tunic surrounds the animal and

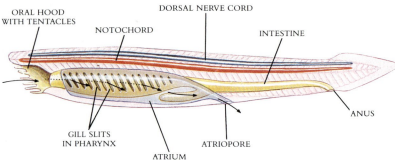

ORAL HOOD
WITH TENTACLES

NOTOCHORD

DORSAL NERVE CORD

INTESTINE

ANUS

GILL SLITS
IN PHARYNX

ATRIUM

ATRIOPORE

B

FIGURE 38-18

A *Two lancelets,* Branchiostoma lanceolatum *(phylum Chordata, subphylum Cephalochordata), partly buried in shell gravel, with their anterior ends protruding. The muscle segments are clearly visible in this photograph; the square, pale yellow objects along the side of the body are gonads, indicating that these are male lancelets.*

B *The structure of a lancelet, showing the path by which the water is pulled through the animal by its cilia.*

A

gives the subphylum its name. In colonial tunicates, there may be a common sac and a common opening to the outside. Many tunicates are inhabited by photosynthetic bacteria (Figure 38-17), shielding them from the light by means of their tough, pigmented exteriors.

In some tunicates that occur in the warmer regions of the world, the tadpole-shaped larvae never change to another form, and the adults of these free-swimming species may resemble the larvae of other groups. Some of these tunicates are gelatinous animals that have several unique adaptations to their planktonic way of life, including high feeding rates and much shorter generation times than other marine planktonic herbivores.

Because it is the larval tunicate rather than the highly modified adult that shows close similarities to other chordates, it has been proposed that larval tunicates became sexually mature at some stage in their evolution and then developed directly into the ancestors of other groups. The process of neoteny, or **paedomorphosis,** is one by which larvae become sexually mature and reproduce; the development of the organism then ceases at that stage. We saw this process in connection with the evolution of mites (p. 786), and it has occurred during the evolution of a number of other groups of organisms, as well.

Lancelets

The lancelets (subphylum Cephalochordata; Figure 38-18) are scaleless, fishlike marine chordates a few centimeters long; they occur widely in shallow water throughout the oceans of the world. There are about 23 species of this subphylum. Lancelets were given their English name because of their similarity in appearance to a lancet—a small, two-edged surgical knife. Most of them belong to the genus *Branchiostoma,* formerly and incorrectly called *Amphioxus,* a name that is still used widely for them. In the lancelets, the notochord persists through the animal's life and runs the entire length of the dorsal nerve cord.

Lancelets spend most of their time partly buried in sandy or muddy substrates, with only their anterior ends protruding. They can, however, swim efficiently, although they seem to do so only rarely. Their muscles can easily be seen as a series of discrete blocks. Lancelets have many more pharyngeal gill slits than do fishes, which they resemble in overall shape. They lack pigment in their skin, which has only a single layer unlike the multi-layered skin of all vertebrates. The lancelet body is pointed at both ends. There is no distinguishable head and no separate eyes, nose, or ears, although there are pigmented light receptors.

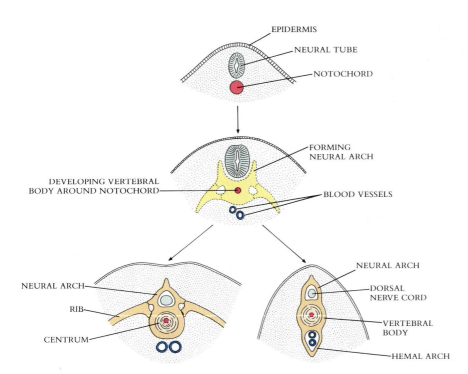

EPIDERMIS
NEURAL TUBE
NOTOCHORD

FORMING NEURAL ARCH

DEVELOPING VERTEBRAL BODY AROUND NOTOCHORD

BLOOD VESSELS

NEURAL ARCH

RIB

CENTRUM

NEURAL ARCH

DORSAL NERVE CORD

VERTEBRAL BODY

HEMAL ARCH

FIGURE 38-19

Embryonic development of a vertebra. During the course of evolution, or of development, the flexible notochord is surrounded and eventually replaced by a cartilaginous or bony covering, the centrum. The neural tube is protected by an arch above the centrum, and the vertebra may also have a hemal arch, which protects the major blood vessels below the centrum. The vertebral column is a strong, flexible rod against which the muscles pull when the animal swims or moves.

Lancelets feed on microscopic members of the plankton, filtering them through a current that they create by means of cilia that line the anterior end of their alimentary canal. They have an oral hood that projects beyond the mouth and bears sensory tentacles, which also ring the mouth itself. The males and females are separate, but there are no obvious external differences between them.

Biologists are not sure whether lancelets are primitive chordates that have survived with a number of ancestral features or are actually degenerate fishes. In the latter case, they would be vertebrates whose structural features have been reduced and simplified during the course of evolution so that they resemble their ancestors. The facts that lancelets feed by means of cilia and have a single-layered skin, coupled with distinctive features of their excretory systems, seem to make it unlikely that they are simply degenerate fishes. The recent discovery of fossil forms, similar to living lancelets, in rocks some 550 million years old—well before the appearance of any organisms that could reasonably be called fishes—also argues for the antiquity of this group and its importance in understanding the evolution of the vertebrates.

Vertebrates

Vertebrates (subphylum Vertebrata) differ from other chordates in that they usually possess a vertebral column, which replaces the notochord to a greater or lesser extent in different members of the subphylum. In addition, the vertebrates, or craniate chordates, have a distinct head with a skull and brain. The hollow dorsal nerve cord of most vertebrates is protected within a U-shaped groove formed by paired projections from the vertebral column (Figure 38-19).

Among the internal organs of the vertebrates, their livers, kidneys, and endocrine organs are characteristic. The endocrine organs are ductless glands that secrete hormones, which play a critical role in controlling the functions of the vertebrate body. All vertebrates have a heart and closed blood vessels. In both their circulatory and their excretory functions, vertebrates differ markedly from other animals.

There are seven classes of living vertebrates, three of them fishes and four of them terrestrial **tetrapods.** (The name *tetrapod* comes from two Greek words meaning "four-footed.") The classes of fishes are Agnatha, the lampreys and hagfish; Chondrichthyes, the cartilaginous fishes, sharks, skates, and rays; and Osteichthyes, the bony fishes that are the dominant group of fishes today. The four classes of tetrapods are Amphibia, the amphibians, including salamanders, frogs, and toads; Reptilia, the reptiles; Aves, the birds; and Mammalia, the mammals.

The members of the class Agnatha differ from all other vertebrates in that they lack jaws. Partly because of this fact, they are usually taken as representative of the very earliest stages of vertebrate evolution. This relationship has in the past sometimes been recognized by designating the agnathans as a distinct group of vertebrates, more different from the other classes than the other classes are from one another. They, together with the members of the class Chondrichthyes, which we will also discuss in the next chapter, have a cartilaginous skeleton at maturity. In five other classes of vertebrates, one of fishes and four of tetrapods, the mature skeletons are bony.

SUMMARY

1. Deuterostomes differ from protostomes in several significant embryological features: their mouth, not their anus, is formed at or near the blastopore; they have radial, rather than spiral, cleavage; the fate of their embryonic cells is settled at a later stage of development; and their coelom is produced by evagination, involving extensive cell movement. The major phyla of deuterostomes are the echinoderms (phylum Echinodermata) and the chordates (phylum Chordata).

2. Echinoderms are exclusively marine deuterostomes that are radially symmetrical, at least as adults. They have a system of separate or fused calcium-rich plates and a unique water vascular system that includes tube feet, by means of which some echinoderms move.

3. The arrow worms, phylum Chaetognatha, and acorn worms, phylum Hemichordata, are two small, unique phyla of marine deuterostomes, not closely related to the echinoderms or to the chordates.

4. The chordates are characterized by their single, hollow dorsal nerve cord; by the presence, at least early in their development, of a dorsal, cartilaginous rod, the notochord; and by pharyngeal slits, at least during their embryological development.

5. The approximately 45,000 species of chordates are classified into two small, exclusively marine subphyla, the tunicates and the lancelets, which seem to represent ancient evolutionary offshoots from the group, and one large subphylum, the vertebrates, whose members include the fishes as well as four classes that have radiated extensively in terrestrial habitats.

SELF-QUIZ

The following questions pinpoint important information in this chapter. You should be sure to know the answer to each of them.

1. Echinoderms have a
   (a) water vascular system
   (b) brain
   (c) delicate epidermis containing thousands of neurosensory cells
   (d) all of the above

2. The water vascular system of the echinoderms is employed by some of them in locomotion. True or false?

3. The _____ differ from all other living echinoderms because their mouth and anus are on their upper surface in an open disk.

4. Do echinoderms have brains?

5. Which phylum includes the vertebrates?

6. Which of the following features do chordates lack?
   (a) a hollow dorsal nerve cord
   (b) a notochord
   (c) pharyngeal slits
   (d) muscles and bone
   (e) none of the above

7. Lancelets have fewer layers of skin than vertebrates. True or false?

8. Larval tunicates exhibit all the characteristics of other chordates, while adult tunicates exhibit few, if any, of them. True or false?

9. Many adult tunicates secrete a _____, which is a tough sac composed mainly of cellulose.

10. A tunicate is a vertebrate. True or false?

11. How many classes of living vertebrates are there?
    (a) five   (c) seven
    (b) six   (d) eight

12. The members of the class Agnatha differ from all other vertebrates in that they lack _____.

THOUGHT QUESTIONS

1. How are the characteristics of the three subphyla of chordates related to the way of life for each group?

2. Why is it believed that echinoderms and chordates, which are so dissimilar, are members of the same evolutionary line?

FOR FURTHER READING

ALLDREDGE, A.L., and L.P. MADIN: "Pelagic Tunicates: Unique Herbivores in the Marine Plankton," *BioScience,* vol. 32, pages 655–663, 1982. Pelagic tunicates have many unique adaptations that fit them for a motile way of life.

BARRINGTON, E.J.W.: *The Biology of Hemichordata and Protochordata,* Oliver & Boyd, London, 1965. A readable account of the structure, physiology, and behavior of the primitive chordates.

HOUSE, M.R. (ed.): *The Origin of Major Invertebrate Groups,* Academic Press, Inc., New York, 1979. An outstanding, scholarly discussion of the origins of invertebrate groups, including echinoderms and chordates.

MACURDA, D.B., JR., and D.L. MEYER: "Sea Lilies and Feather Stars," *American Scientist,* vol. 71, pages 354–365, 1983. By observing living crinoids, scientists have learned a great deal about the earliest echinoderms.

BIOLOGY OF VERTEBRATE ANIMALS   *Part Eight*

The key to understanding our own structure and function is found in the long evolutionary history of the vertebrates—a history that has led to the elaboration of the most complex forms of life on earth.

Overview

The vertebrates originated in the sea as jawless fishes. The bony fishes, which evolved from them, are the most plentiful vertebrates today. The first vertebrate land dwellers were amphibians, but they are not truly terrestrial because they still require frequent access to water. The first true terrestrial vertebrates were the reptiles, which gave rise independently to the birds and to the mammals, including humans.

For Review

Here are some important terms and concepts that you will encounter in this chapter. If you are not familiar with them, you should review them before proceeding.

- **Major features of evolutionary history** (Chapter 23)

- **Mammalian and human evolution** (Chapter 24)

- **Evolution of chordates** (Chapter 38)

The approximately 45,000 species of chordates, including birds, reptiles, amphibians, fishes, and mammals, are distinguished by three principal features (Figure 38-13): (1) a single, dorsal, hollow nerve cord, (2) a rod-shaped notochord, which forms between the nerve cord and the primitive gut in the early embryo, and (3) pharyngeal slits. With the exception of the tunicates and lancelets, which we considered in the last chapter, all chordates are vertebrates.

Vertebrates differ from other chordates in usually having a vertebral column—a series of bones called vertebrae that encloses and protects the dorsal nerve cord and sometimes the vertebral arteries as well (see Figure 38-19). Vertebrates derive their name from this column of bones, a novel structure that replaces the notochord to a greater or lesser extent in the different members of the subphylum. In addition, vertebrates (except for the agnathans, as we will show) have a distinct and well-differentiated head; as a result, they are sometimes called the craniate chordates. Most vertebrates have a bony skeleton, although the living members of two of the classes of fishes—Agnatha, the lampreys and hagfishes, and Chondrichthyes, the sharks and rays—have a cartilaginous one.

The vertebrates are a diverse group, containing members adapted to life in the sea, on land, and in the air. There are seven principal classes of living vertebrates. Three of the classes are fishes that live in the water, and four are land-dwelling tetrapods. As we men-

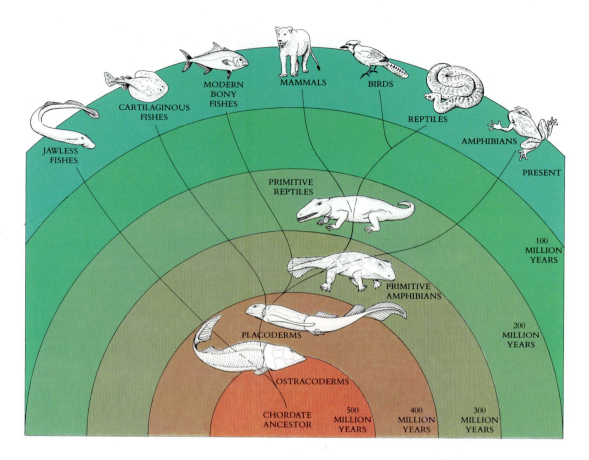

CARTILAGINOUS FISHES

MODERN BONY FISHES

MAMMALS

BIRDS

REPTILES

JAWLESS FISHES

AMPHIBIANS

PRIMITIVE REPTILES

PRESENT

100 MILLION YEARS

PRIMITIVE AMPHIBIANS

200 MILLION YEARS

PLACODERMS

OSTRACODERMS

CHORDATE ANCESTOR

500 MILLION YEARS

400 MILLION YEARS

300 MILLION YEARS

FIGURE 39–1

Diagram of the evolutionary relationships of the seven classes of vertebrates that include living representatives, and some of their extinct relatives.

tioned in the last chapter, the classes of fishes are Agnatha, the lampreys and hagfishes; Chondrichthyes, the cartilaginous fishes, sharks, skates, and rays; and Osteichthyes, the bony fishes, the dominant group of vertebrates in the world today. The four classes of tetrapods are Amphibia, the amphibians, including salamanders, frogs, and toads; Reptilia, the reptiles; Aves, the birds; and Mammalia, the mammals. The evolutionary relationships between these classes and some of their extinct relatives are shown in Figure 39–1.

Vertebrates are a subphylum of chordates characterized by a vertebral column surrounding a dorsal nerve cord. Most species of vertebrates are bony fishes.

In this chapter, we will first review the jawless fishes, the first vertebrates, and then discuss the evolution of the Chondrichthyes and bony fishes, Osteichthyes, from them. Later, we will consider the invasion of the land by the vertebrates, led by the amphibians, starting about 350 million years ago. The reptiles were probably derived from the amphibians, but they may have originated directly among the kinds of fishes that also gave rise to the amphibians. From the reptiles, in turn, were derived separately the two most successful terrestrial groups of vertebrates, the mammals and the birds. With the evolution of these last two classes, the conquest of the land was complete.

JAWLESS FISHES

The members of the class Agnatha differ from all other vertebrates in that they lack jaws. They are representative of the first stages in the evolution of the vertebrates; for more than 100 million years, they were the only vertebrates. Many of the major groups of agnathans are extinct, and the living ones—the lampreys and hagfishes—belong to a single, fairly uniform group. Both are naked, eel-like fishes.

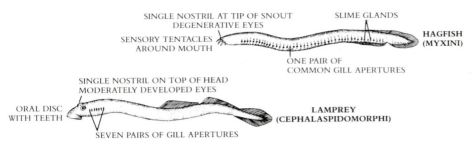

SINGLE NOSTRIL AT TIP OF SNOUT
DEGENERATIVE EYES
SENSORY TENTACLES
AROUND MOUTH
SLIME GLANDS
HAGFISH
(MYXINI)
ONE PAIR OF
COMMON GILL APERTURES

SINGLE NOSTRIL ON TOP OF HEAD
MODERATELY DEVELOPED EYES
ORAL DISC
WITH TEETH
LAMPREY
(CEPHALASPIDOMORPHI)
SEVEN PAIRS OF GILL APERTURES

FIGURE 39-2

Comparison of hagfish and lamprey, representative agnathans.

In the agnathans, the notochord persists throughout the life of the animal. Within the body of the living agnathans are portions of cartilaginous skeleton similar to the skeletons of the Chondrichthyes. We know from their fossils, however, that the ancestral agnathans had bony skeletons comparable to those of the Osteichthyes. In contrast to the naked living lampreys and hagfishes, many of these extinct agnathans had elaborate and well-developed bony scales.

The feeding habits of the living agnathans are specialized; they are parasites on other fish or are scavengers. Indeed, their success in exploiting these particular ecological niches, or ways of life, may be the reason that they alone have survived into the modern world as representatives of what was once the dominant group of vertebrates. There are about 31 living species of hagfishes and about 32 living species of lampreys (Figure 39-2). All of the other major groups of this class have been extinct for hundreds of millions of years.

One of the two groups of living agnathans, the lampreys, have round mouths that function like suction cups (Figure 39-3). Using these specialized mouths, lampreys attach themselves to bony fishes. Once attached, they rasp through the skin of the fish with their tongues, which are covered with sharp spines, sucking out the blood of the fish through the hole that they make. Sometimes lampreys can be so abundant that they constitute a serious threat to commercial fisheries.

Spawning in lampreys takes place in the spring and is accompanied by the building of depressions in the stony bottoms of freshwater streams. The tiny transparent larvae grow into opaque, eel-like fishes, which may be up to more than 15 centimeters long. They live buried in the mud, feeding on plankton, and apparently move only to find richer sources of food. Eventually, these larvae mature and metamorphose into the parasitic adults. The adults occur in the sea or in brackish water, are free swimming, and feed on other fishes in the manner just described. In contrast to the lampreys, the hagfishes, which are externally similar, are scavengers, often feeding on the insides of dead fishes or large invertebrates.

Agnathans were the first vertebrates; for over 100 million years they were the only vertebrates. Fossils indicate that agnathans originally had bony skeletons, but all present-day agnathans have only cartilaginous skeletons and lack jaws. Most agnathans are now extinct; the survivors are the parasitic lampreys and hagfishes.

Larval lampreys and hagfishes resemble lancelets and, like them, feed on plankton. Lancelets move a stream of water through their pharynx by means of ciliary action, whereas larval lampreys, lacking cilia, do so by muscular contraction. The muscular action of the larval lampreys is much more efficient than that of lancelet cilia in forcing a stream of water through their gills; consequently, larval lampreys can extract much more food from a stream of water in a given period of time than can the lancelets. Because they obtain much more food, larval lampreys grow much more rapidly than lancelets when they are living under identical conditions. Primitive chordates doubtless fed like the lancelets, by means of cilia; and the evolution of the ability to use muscular contractions to force water through the gills represents an important evolutionary advance. The oldest fossils of the lampreys' relatives date from about 540 million years ago and are approximately as old as the oldest fossil lancelets yet found.

FIGURE 39-3

Lampreys attach themselves to the fishes on which they prey by means of their suckerlike mouths. When they have done so, they bore a hole in the fish with their teeth and feed on its blood.

THE APPEARANCE OF JAWED FISHES
The Evolution of Jaws

Jaws first developed among vertebrates that lived 410 million years ago, toward the close of the Silurian Period, or about 130 million years after the first appearance of the vertebrates. For a very long period of time, therefore, agnathans were the only vertebrates.

The jaws that evolved in ancient fishes during the later Silurian Period allowed them to become proficient predators, to defend themselves, and to collect food from places and in ways that had not been possible earlier. They evolved more and more efficient fins and then were able to propel themselves through the seas rapidly and accurately to hunt, to seek mates, and to avoid predators. Fishes were able to bite and chew their food instead of sucking it or filtering it in like all of the more primitive chordates did. The early jawed fishes outcompeted their jawless ancestors, and agnathans have survived to the present only as parasites or scavengers, two specialized ways of life in which they have relatively few competitors. Over the same period of time, fishes have become dominant throughout the waters of the world.

The biting jaws of the vertebrates were produced by the evolutionary modification of one or more of the gill arches, originally the areas between the gill slits. In this process the gill arches moved forward in relation to the animal's body and changed in form (Figure 39-4). The progressive modification of the arches just behind the mouth produced the kind of jaw that is characteristic of modern **gnathostome** ("tooth-mouth") vertebrates—all vertebrates except the agnathans. The teeth of vertebrates originated through the evolutionary modification of the skin. After jaws had first evolved in the earliest gnathostomes, their descendants developed many different kinds of jaws, which vary greatly in different vertebrate classes.

The Jawed Fishes

Of the approximately 42,500 species of living chordates, about half are fishes. Fishes are also by far the most poorly known of all groups of vertebrates; there are undoubtedly many more awaiting discovery, especially in the fresh waters of South America and Africa, which are exceptionally rich in members of this group. Fishes have efficiently and completely occupied the waters of the globe ever since their origin more than 410 million years ago. In addition, jawed fishes, through a series of important evolutionary advances, have given rise to all of the other vertebrates.

Chondrichthyes

The first jawed fishes to evolve were members of the class Chondrichthyes, the sharks, dogfish sharks, skates, and rays (Figure 39-5). Many hundreds of extinct species of class Chondrichthyes are known from the fossil record, but there are only about 850 living ones. These fishes are mostly scavengers and carnivores, and the internal skeleton of the living forms is composed of cartilage—a soft, light, and elastic material. Such skeletons are lighter and more buoyant than those of the early agnathans that the Chondrichthyes replaced; modern agnathans, in contrast, have evolved cartilaginous skeletons much like those of Chondrichthyes. In addition to the evolutionary advance represented by such a skeleton, the Chondrichthyes were the first vertebrates to develop efficient fins. With these strong advantages, they soon largely replaced agnathans throughout the world.

> **The sharks and other Chondrichthyes were the first jawed vertebrates and the first to develop efficient fins. They evolved from agnathans that had bony skeletons, but they soon developed cartilaginous ones.**

The skin of Chondrichthyes is covered everywhere with small, pointed **denticles**, similar in structure to the teeth of other vertebrates. As a result, the skin of these fishes has

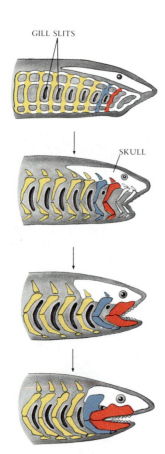

FIGURE 39-4

Jaws evolved from the anterior gill arches of ancient, jawless fishes.

GILL SLITS

SKULL

A

B

C

a rough, sandpapery texture. Their teeth, which are abundant in the fossil record, are enlarged versions of these denticles.

Sharks drive themselves through the water by sinuous motions of the whole body and by their thrashing tails. Such motion tends to drive the sharks downward, but this tendency is corrected by their two spreading pectoral fins. In the skates and rays, these pectoral fins have become enlarged and are undulated when these fishes move, which gives these animals a very characteristic appearance. Their tail, which is not the principal means of locomotion that it is in the sharks, is thin and whiplike, and it is sometimes armed with a poisonous spine. Skates and rays spend most of their time at or near the ocean floor, feeding mainly on invertebrates. Feeding along the ocean floor is facilitated because their mouths are located on their lower surfaces.

For their supply of oxygen, many sharks (and those bony fishes that swim constantly) depend on a constant stream of water that is forced past the gills and brings dissolved oxygen with it; others are able to pump water through their gills while they are stationary. Fishes that obtain their oxygen by moving constantly can literally drown if they are prevented from swimming (Figure 39-6). Drowned sharks are often found, for example, trapped in the nets used to protect Australian beaches. Bony fishes that do not swim constantly have pumping muscles in their gills, which they use to draw water through their gills. The action of these muscles makes the extraction of oxygen from the water possible even if the fish is stationary. The muscles that are used in this process are lacking in sharks and other fishes that swim continuously. In skates and rays, which feed on or near the ocean floor and have mouths on their lower surfaces, the openings through which they draw water are located on the upper surface of their heads.

BONY FISHES
Origin and Evolution

Of the two classes of fishes, Chondrichthyes includes only about 850 living species. The vast majority of the more than 18,000 known species of fishes belong to the class Osteichthyes, which has become the dominant class of fishes in the modern world. The first Chondrichthyes apparently had bony skeletons like those of Osteichthyes and early agnathans. Such skeletons, however, evidently were modified soon to become the cartilaginous ones that are characteristic of living Chondrichthyes and agnathans.

Although bones are heavier than cartilaginous skeletons, bony fishes are still buoyant because they possess a **swim bladder,** a gas-filled sac that allows them to regulate their buoyant density and so remain suspended at any depth in the water with great accuracy (Figure 39-7). Swim bladders apparently evolved from the lungs that originated as outpocketings of the pharynx, specialized for respiration, in Devonian Period fishes. The use

FIGURE 39-5

Members of the class Chondrichthyes, which are mainly predators or scavengers, are constantly and gracefully in motion. As they move, they create a flow of water, from which they extract oxygen, past their gills. Three representatives of this class are shown here.
A *Blue shark.*
B *Diamond sting ray.*
C *Manta ray.*

FIGURE 39-6

A few sharks feed by straining plankton out of the water. One of them is this extraordinary fish, dubbed the megamouth ("big-mouth") shark, Megaschisma pelagios. It is the only species of a unique family. Although this shark was known from fossils, the first living individual was caught off Hawaii in November, 1976; the only other one that has been captured is shown here. Entangled in a gill net on a commercial fishing boat in deep water off Santa Catalina Island in southern California in 1984, this individual is a gray-black mature male measuring nearly 5 meters long and weighing approximately a ton.

THE DISCOVERY OF A LIVING COELACANTH

In 1938 scientists were stunned by the announcement that a trawler fishing in the Indian Ocean off the coast of South Africa had landed a large, very strange fish. The specimen, which was about 2 meters long, became rotten before it was seen by an ichthyologist (a scientist who specializes in the study of fishes) and was skinned so that some of it could be saved. Nevertheless, as soon as J.L.B. Smith of the University of Grahamstown in South Africa saw it, he recognized it as a member of an ancient group of lobe-finned fishes that had been described from fossils more than a century earlier, and which had been thought to have been extinct for about 70 million years! The members of this group were called coelacanths, and the newly discovered fish was dubbed *Latimeria chalumnae*.

Coelacanths are well represented in the fossil record for more than 300 million years, from about 390 million years to about 70 million years ago. Until *Latimeria* was discovered alive, it was assumed that the group, which was related to the lungfishes and therefore similar to the ancestors of the terrestrial vertebrates, had become extinct in the late Cretaceous Period. Here, however, was indisputable evidence that one of their descendants was still alive, swimming in the warm waters of the western Indian Ocean.

Because the first specimen had been skinned, scientists had no opportunity at first to learn about its internal parts, features that were of great interest in terms of its relationship to terrestrial vertebrates and other fishes. Leaflets and posters were

FIGURE 39-7

Diagram of a swim bladder. This structure, which evolved as an outpocketing of the pharynx, is used by the bony fishes to control their buoyancy in water. Eventually, the swim bladders of some of the early bony fishes gave rise to lungs.

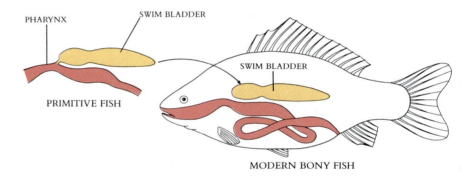

of such lungs to regulate buoyancy among the early bony fishes seems to have evolved secondarily. Lacking such bladders, the sharks swim perpetually to keep from sinking. This single strategic difference—the ability to swim or not to swim depending on the circumstances—has apparently had much to do with the overwhelming success of the bony fishes.

Osteichthyes comprise about half of the species of living vertebrates (see Figures 35-14, *C,* and 39-8). Chondrichthyes may originally have evolved in the sea, but bony fishes, which originated as an offshoot of Chondrichthyes, clearly evolved in fresh water, judged by the places where early fossils of the class are found and the characteristics of these ancient fishes. Bony fishes apparently entered the sea only after the origin of a number of distinctive evolutionary lines. They are abundant both in the modern seas and in fresh water, and many kinds of fishes spend a portion of their lives in fresh water and another portion in the sea.

From the early Chondrichthyes were derived the bony fishes (class Osteichthyes). The bony fishes possess swim bladders, which enable them to increase their buoyancy and thus to offset the greater weight of a bony skeleton. They are the most successful of the vertebrates, accounting for more than half of all living vertebrate species.

distributed among the fishing communities of southern and eastern Africa, offering rewards for another specimen of this remarkable fish. It was not until 1952, however, that a second coelacanth was brought to the attention of scientists. It was landed in the Comoro Islands, about 3000 kilometers northeast of the place where the first specimen was found. Coelacanths were landed occasionally in the Comoros and were well known to local fishermen.

Living mostly at depths of 150 to 300 meters in the sea, *Latimeria* has a slender, fat-filled swim bladder. Some individuals that have been captured are nearly 3 meters long. The eggs are large, about 9 centimeters in diameter, and few in number; they hatch inside the mother, who produces live young. Although *Latimeria* is a very strange animal, its features mark it as a member of the evolutionary line that gave rise to the terrestrial tetrapods.

FIGURE 39-A

The living coelacanth, Latimeria.

By studying the dozens of specimens that have been landed since 1952, scientists have been able to shed additional light on the nature of this ancient and archaic group of vertebrates.

FIGURE 39-8

Only a little of the incredible diversity of the bony fishes, class Osteichthyes, is shown here.
A *Koran angelfish,* Pomacanthus semicircularis, *in Fiji; one of the many striking fishes that live around coral reefs in tropical seas.*
B *Puffers, which are also marine fishes, avoid being eaten by inflating themselves when attacked, thus projecting their spines outward.*
C *A sea horse, an example of the bizarre forms of some fishes. Sea horses move slowly and are difficult to see among the marine algae and sea grasses where they live.*
D *Moray eel, Rangiroa, French Polynesia.*

Bony fishes have a number of features that are peculiar to this class alone. Their scales are a covering of thin, overlapping bony plates, sometimes spiny along the edges, which provide some protection for the animal. These fishes have a highly developed **lateral line system,** which consists of organs in sunken pits connected with one another by a canal below the surface, and with the surface by the openings of the pits. This lateral line system enables fishes to detect changes of pressure in the water and thus the movements of predators, prey, and other objects in the water. Lateral line systems are found in the Chondrichthyes, but they are not as elaborate or as well developed as those in the Osteichthyes. Bony fishes detect sound, which travels better in water than in air, especially well and respond to it. In many kinds of fishes the swim bladders are linked to the inner ear by bony connections, thus providing a very efficient sound-detecting organ (Figure 47-5).

Subclasses of Osteichthyes

The class Osteichthyes is divided into two subclasses, the lobe-finned fishes (subclass Sarcopterygii) and the ray-finned fishes (subclass Actinopterygii). There are only four living genera of lobe-finned fishes, but the group is known from fossils that go back to the very beginnings of the class about 390 million years ago, in the Devonian Period. The paired fins of the lobe-finned fishes have a characteristic scaly, lobed form, for which the group is named. These lobed fins, which differ so much in appearance from the fins of most living fishes, are thought to have given rise to the limbs of the tetrapods.

The living lobe-finned fishes include three genera of lungfishes, one each in Australia, Africa, and South America. The African genus of lungfishes (see Figure 19-11) has four species, whereas the others have one each. In the lungfishes the swim bladder is paired and serves as a functional lung, which allows these remarkable organisms to breathe air and to survive in conditions of drought, dormant in burrows at the bottom of ponds that dry up seasonally. The lungfishes are specialized, and they represent a distinct evolutionary line within the lobe-finned fishes. The remaining genus of lobe-finned fishes contains one species, the coelacanth, a remarkable fish that occurs in the seas off southeastern Africa. Scientists do not agree whether it is the lungfishes or the coelacanths that are most closely related to the terrestrial tetrapods, and therefore to humans.

Some kinds of ray-finned fishes, such as the mudskipper (Figure 39-9), spend a great deal of their time out of the water. These unusual fishes have evolved lungs similar to those of the lungfishes, which allow them to gulp air directly. The evolution of such lungs from their swim bladders seems to have been an adaptation to freshwater conditions, in which individual ponds can become stagnant and thus very poor in oxygen. Some of the modern fishes that have lungs, including the mudskippers, have the habit of climbing onto the land occasionally, using their fins. In doing so, they remind us of the sort of event that must have occurred when the vertebrates first invaded the land, in the latter part of the Carboniferous Period.

THE INVASION OF THE LAND

About 350 million years ago, the lobe-finned fishes that were to become the ancestors of all tetrapods were starting to explore a terrestrial existence at the margins of freshwater ponds or swamps. In these animals, which ultimately became the earliest amphibians, the lungs gradually developed into the kinds of efficient air-breathing organs seen in modern tetrapods. These lungs originally had been the kinds of supplementary organs that are present in modern lungfishes (see Figure 19-11). Efficient locomotion on land was gradually made possible by the evolution of strong thoracic (body) skeletal supports. Such supports provided a more rigid base for the limbs, which were derived from the kinds of fins found in the lobe-finned fishes. These two evolutionary advances, which led to the ability to use gaseous oxygen for respiration and to move from one place to another on land, became the basis of the success and subsequent evolutionary radiation of vertebrates on land.

FIGURE 39-9

A male mudskipper (Boleophthalmus pectinirostris), *breathing air on the mudflats near Nagasaki, Japan, raising his dorsal fins to signal to females of his species and thus to begin a complex mating ritual. Mudskippers are widespread along ocean shores from Japan throughout Southeast Asia to Africa. One species actually climbs up onto plants at the edge of the water to feed on insects and other small animals, using their lungs when they are doing so. Although they are ray-finned fishes, and not directly related to the ancestors of the tetrapods, mudskippers illustrate the kinds of habits that probably were characteristic of the ancestor of the amphibians 350 million years ago.*

AMPHIBIANS

The first land vertebrates were amphibians, members of a class that first became abundant about 300 million years ago during the Carboniferous Period. The early amphibians had rather fishlike bodies, short stubby legs, and lungs (see Figure 23-12). Ultimately these amphibians probably gave rise to all of the other tetrapods, although there is a possibility that the reptiles evolved directly from the fishes; in that case the amphibians would be an evolutionary "dead-end." Even so, the amphibians constitute the group of tetrapods most like the ones that originally evolved from the fishes. As a group, amphibians, unlike reptiles, birds, and mammals, depend on the availability of water during their early stages of development. Many amphibians live in moist places even when they are mature. There are about 3150 species. Members of this class were the dominant vertebrates on land for about a hundred million years, until they were gradually replaced in their dominant role by the reptiles.

The two most familiar orders of amphibians are those that possess tails—the salamanders, mud puppies, and newts, which comprise the order Caudata, with about 350 species; and those that do not have tails—the frogs and toads, which comprise the order Anura, with about 2800 species (Figure 39-10). Amphibians possess the first true lungs, but these organs are relatively inefficient. Respiration also takes place through their thin, moist, glandular skins, which lack scales in both the orders Caudata and Anura, and through the lining of their mouth. The constant loss of water through the skins of amphibians is one of the reasons that these animals must remain, for the most part, in moist habitats. Amphibian larvae and the adults of those species that remain permanently in the water—certain salamanders—respire by means of gills.

FIGURE 39-10

Some of the types of amphibians, representing the diversity of the class Amphibia.
Frogs and toads (order Anura):
A *Giant toad,* Bufo marinus.
B *Red-eyed tree frog,* Agalychnis callidryas.
C *A tadpole, the larval form of the pickerel frog,* Rana palustris, *during metamorphosis to the adult form. The hind legs develop at this stage, and the animal will soon become terrestrial.*
Salamanders and newts (order Caudata):
D *Tiger salamander,* Ambystoma tigrinum.
E *Tennessee cave salamander,* Gyrinophilus palleucus. *This and other neotenic salamanders remain aquatic for their entire lives; permanently larval in form, they become sexually mature and breed without ever changing.*

A

B

FIGURE 39-11

Representatives of the major groups of reptiles (class Reptilia).
A *River crocodile, Crocodilus acutus. The crocodiles and their relatives resemble birds and mammals in having a four-chambered heart; all other living reptiles have a two-chambered one. Crocodiles, like birds, are related to dinosaurs, rather than to any of the other living reptiles.*
B *Red-bellied turtles, Pseudemys rubriventris. This attractive turtle occurs frequently in the northeastern United States.*
C *An Australian skink, Sphenomorphus. Some burrowing lizards lack legs, and the snakes evolved from one line of legless lizards.*
D *Smooth greensnake, Opheodrys vernalis.*

FIGURE 39-12

The tuatara (Sphenodon punctatum) is the sole living survivor of the order Rhynchocephalia, a group that has characteristics more primitive than those of the lizards—which probably were derived from Rhynchocephalia. Surviving virtually unchanged for more than 100 million years, the tuatara exhibits one of the slowest rates of evolution known.

Amphibians lay their eggs, which lack water-retaining external membranes and shells and dry out rapidly, directly in water or in moist places. Anuran larvae are tadpoles, which usually live in the water, where they feed on minute algae; their adults, which are highly specialized for jumping and very different from the larvae in appearance, are carnivorous. In contrast to the anurans, young salamanders are carnivorous like the adults and look like small versions of them. Many salamanders swim efficiently and return to water to breed. Although some salamanders remain in the water even as adults, most of them live in moist places, such as under stones or logs, or among the leaves of tropical bromeliads (relatives of the pineapple).

Amphibians are terrestrial, but they still depend on a moist environment; their eggs are laid in water and the development of their larvae takes place there.

Although amphibians appear primitive, they are in fact members of a successful group—one that has survived for over 300 million years. The amphibians evolved long before the dinosaurs and have, thus far, outlasted them by 65 million years.

REPTILES

Reptiles (class Reptilia) have a dry skin, covered with scales, that efficiently retards water loss. As a result, the members of this large class, consisting of nearly 6000 living species, are independent of free water, unlike their ancestors. Nonetheless, of the four orders of reptiles that have living representatives (Figure 39-11), the Crocodilia—the crocodiles, alligators, and caimans—live in water, as do most of the members of the order Chelonia—the turtles and tortoises. Members of the third, and by far the largest, order, Squamata—the lizards, snakes, and other reptiles—are almost entirely terrestrial. The scales of crocodiles and turtles are replaced as they are worn away, whereas those of lizards and snakes are shed several times a year when the animals molt. Snakes evolved from lizards through the loss of their limbs and other evolutionary modifications; some lizards have also lost their limbs in relation to specialization for burrowing. There are about 3300 living species of lizards, about 2300 species of snakes, about 220 species of turtles and tortoises, and 22 species of crocodiles and alligators.

The fourth order of living reptiles, Rhynchocephalia, is represented only by the tuatara, *Sphenodon punctatum* (Figure 39-12), which is the sole survivor of a group of primitive reptiles that otherwise became extinct about 100 million years ago, and which was abundant about 200 million years ago. The tuatara is a long-lived (up to nearly a century), lizardlike animal, up to about 0.7 meter long, that lives only on about a dozen small islands off the coast of New Zealand. It retains a primitive skull structure similar to that of the ancestors of the modern lizards.

C

D

Scientists do not agree about whether the reptiles evolved from amphibians or from lunged fishes directly, like the amphibians themselves. Regardless of the answer to this question, the general direction of their ancestry and subsequent evolution is well established both by a rather complete fossil record and by the relationships of the living forms. The first reptiles appeared during the age of dominance of the amphibians, the Carboniferous Period, about 300 million years ago. Reptiles became abundant over the next 50 million years, gradually replacing amphibians. Immediately following the Carboniferous Period, the Permian Period (286 to 248 million years ago) was a period of widespread glaciation and drought. The special adaptations of reptiles, including their water-resistant skin and more efficient lungs, probably gave them an advantage as arid climates became widespread. The dinosaurs, by far the best-known group of fossil organisms, appeared near the start of the Mesozoic Era, at least 225 million years ago. The reptiles dominated the land for about 180 million years, until the start of the Tertiary Period (see Figure 23-14).

Reptiles were the first truly terrestrial vertebrates. They have more efficient lungs than amphibians do and so do not need to respire through their skin, which is dry and efficiently retards water loss.

One of the most critical adaptations of reptiles in relation to their life on land is the evolution of the **amniotic egg,** which protects the embryo from drying out, nourishes it, and enables it to develop outside of water (Figure 39-13). Amniotic eggs, characteristic of

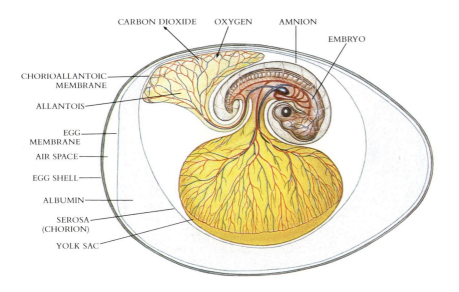

CARBON DIOXIDE   OXYGEN   AMNION

EMBRYO

CHORIOALLANTOIC MEMBRANE

ALLANTOIS

EGG MEMBRANE

AIR SPACE

EGG SHELL

ALBUMIN

SEROSA (CHORION)

YOLK SAC

FIGURE 39-13

The amniotic egg is perhaps the most important feature that allows reptiles to live in a wide variety of terrestrial habitats.

reptiles and birds (and of the very few egg-laying mammals), retain their own water. They contain a large yolk, the primary food supply for the embryo, and abundant **albumin,** or egg white, which provides additional nutrients and water. The yolk is enclosed in a yolk sac, which is attached to the embryo. The embryo's nitrogenous wastes are excreted into the **allantois,** a sac that grows out of the embryonic gut. These wastes are stored in it until the egg hatches, at which time they are simply left behind. Nitrogen waste in an insoluble form—uric acid—does not add to the osmotic environment of the shelled egg. Blood vessels grow out of the embryo through the membranes of the yolk sac and allantois to the surface of the egg, where they take in oxygen and release carbon dioxide. The egg is more easily permeable to these gases than to water. The egg membrane, called the **amnion,** surrounds the developing embryo, enclosing a liquid-filled space within which the embryos pass their gilled stage. Around the embryo, amnion, yolk sac, and allantois is a membrane called the **chorion,** which controls the permeability of the egg and lies just within the shell. In most reptiles, the egg shell is leathery.

> **The key to successful invasion of the land by vertebrates, as it was by arthropods, was the elaboration of ways to avoid desiccation in dry air. In addition to a dry skin that retained water, a second pivotal innovation of reptiles was the watertight amniotic egg, which contains embryonic nutrients and is permeable to gases but not to water.**

TEMPERATURE CONTROL IN LAND ANIMALS

Reptiles, like amphibians and fishes but unlike birds and mammals, are **ectothermic** (from the Greek words *ectos,* outside + *thermos,* heat), regulating their body temperature by taking in heat from the environment. They have evolved a wide array of behavioral mechanisms that enable them to control their temperatures remarkably precisely. Even though ectothermic animals are often called "cold-blooded," they often may actually be able to

A

B

C

D

maintain bodily temperatures much warmer than their surroundings, as a result of their behavior. Several scientists have postulated that at least some dinosaurs, like the birds and mammals, were **endothermic** (Greek *endos,* inside), capable of regulating their body temperature internally. Although endothermic animals are sometimes called "warm-blooded," the temperatures of their bodies may often be cooler than that of their surroundings, a condition that they maintain by the expenditure of energy and by appropriate behavior, as when a dog pants. Endothermic animals are also called **homeotherms** (Greek *homos,* similar), and ectothermic ones are also called **poikilotherms** (Greek *poikilos,* many colored—i.e., variable), and we will use both sets of terms interchangeably in this book, since both are used commonly.

Birds (class Aves) and mammals (class Mammalia) are unique among living organisms in being endothermic. Consequently, the members of these two classes are able to remain active at night, even if temperatures are cool. They are also able to live at higher elevations and farther north and south than most reptiles and amphibians. Endotherms can maintain their internal organs within a very narrow temperature range more or less independently of external conditions, an adaptation that makes possible the more efficient maintenance and functioning of these animals as compared with the ectotherms. About 80% of the calories in the food of endotherms, however, goes to maintain their body temperature; a reptile the same size as a mammal therefore can subsist on about a tenth of the amount of food. For this reason, reptiles are able to live in places, such as some extreme deserts, where food may be too scarce to support abundant bird and mammal fauna.

BIRDS

There are approximately 9000 species of living birds (class Aves; Figures 39-14 and 30-18, *A*). In birds, the wings are forearms, modified in the course of evolution. The scaly skin of birds is provided with feathers, flexible, strong organs that can be replaced and that make up an excellent airfoil—a surface designed to obtain resistance on its surfaces from the air

FIGURE 39-14

The birds (class Aves) are a large and very successful group of about 9000 species, more than any other class of vertebrates except the bony fishes. These photographs show some of their diversity.

A *Ostrich,* Struthio camelus. *The group of wingless birds that includes the ostrich (Africa), rhea (South America), emu (Australia), and kiwi (New Zealand)—the ratite birds—seems to represent an evolutionary line distinct from all other birds. They migrated between their now widely separated southern lands before these drifted apart.*

B *A pair of wood ducks,* Aix sponsa, *showing the marked difference between males and females.*

C *Tufted puffin,* Fratercula cirrhata, *one of many unique groups of sea birds.*

D *A woodpecker,* Dryocopus lineatus, *in Venezuela.*

E *A bee-eater,* Merops, *in Tanzania. The bee-eaters nest communally, with many adult individuals participating in the care of the young.*

F *Purple finch,* Carpodacus purpureus, *one of the very large group of birds called passerines. Darwin's finches and their relatives, shown in Figures 21-12 to 21-14, are passerines, as are sparrows, warblers, chickadees, and many other familiar birds.*

G *Northern saw-whet owl,* Aegolius acadius. *Owls, hawks, eagles, and similar predatory birds are called raptors.*

E

F

G

WHAT ARE THE RELATIVES OF THE GIANT PANDA?

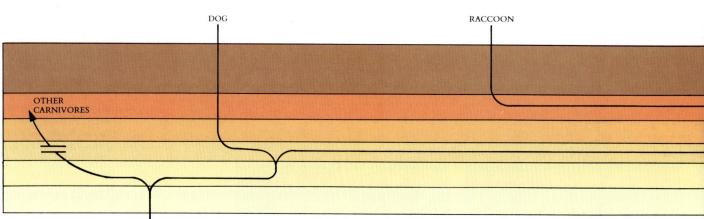

DOG

RACCOON

OTHER
CARNIVORES

Ever since the giant panda, a unique mammal of the mountain forests of southwestern China, was brought to the attention of European scientists, scientists worldwide have speculated about its true relationships. Although it was widespread throughout most of central China as recently as 150 to 200 years ago, the giant panda is now confined to six widely separated areas amounting in total to 29,500 square kilometers. These areas are mostly in central Sichuan Province, with an outlier in the Qinling Mountains of Shaanxi Province, to the northeast. The total wild population of giant pandas at present probably numbers as few as 1150 to 1200 individuals. In addition, 13 giant pandas are still alive in zoos outside of China, and 60 to 70 are in zoos within China.

Although Pere David, who provided the first scientific description of the giant panda for European scientists (in 1869), considered it to be a bear, others soon pointed out that its bones and teeth resembled those of the raccoons, a family of mammals from the New World. The argument has raged ever since and although there is general acceptance of the fact that pandas resemble both bears and raccoons, it has not been possible to reach general agreement about which group of mammals they resemble more closely.

The lesser panda, or red panda *(Ailurus fulgens),* which is another fascinating Asian mammal, seems similar to giant pandas in many respects, yet it is clearly similar to the raccoons. Partly because of these relationships, the red panda has often been grouped with the giant panda in a subfamily of the raccoon family. The red panda lacks the remarkable false thumb of the giant panda, an "extra" structure with which it grasps bamboo stalks while eating them; but otherwise the two are similar in behavior; both feed on bamboos, and they occur in the same general area.

By matching reactions between blood serum samples from these animals, a comparison that was

A MOLECULAR SOLUTION

LESSER PANDA   GIANT PANDA   BEAR

PRESENT

MILLIONS OF YEARS AGO

20

30

40

50

60

first made in the 1950s, scientists soon came to the conclusion that the giant panda was indeed related to the bears. By the same methods, the red panda showed up as a separate evolutionary line, independent from, but related to, both bears and raccoons. Zoologist George Schaller, of the New York Zoological Society's Wildlife Conservation International Division, who has probably spent more time studying the behavior of pandas in the wild than anyone else, has concluded on the basis of their behavioral traits that red pandas and giant pandas are closely related to one another, and that they are both more closely related to bears than to raccoons.

Recently, Stephen O'Brien of the U.S. National Cancer Institute and his coworkers at the U.S. National Zoological Park have neatly solved this problem with molecular methods of analysis—DNA hybridization, isozyme similarities, immunological comparisons, and the study of chromosomal morphology after the application of special staining techniques that reveal satellite DNA and other differentiated regions. Their conclusions are illustrated in the diagram (Figure 39–B). They clearly indicated that the red panda represents an evolutionary line that diverged from the raccoons fairly soon after they had separated from the bears, and that the giant panda diverged from the bears much more recently. As to timing, the original split between raccoons and bears probably occurred 30 to 50 million years ago; that between the red panda and the raccoons about 10 million years later; and that between the giant panda and the bears about 10 million years later than that, or about 10 to 20 million years ago. Although the red panda and the giant panda share a number of structural and behavioral features, they do not share a common ancestor closer than the common ancestor of the bears and the raccoons. In these studies, modern molecular information has led to the solution of a longstanding evolutionary problem.

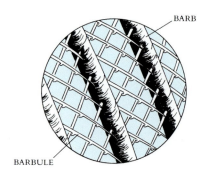

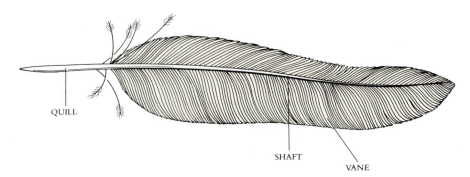

FIGURE 39-15

A feather. The enlargement above shows the way in which the vanes—secondary branches—are linked together by the microscopic barbules.

through which it moves—for flying (Figure 39-15). The membranes that make up the flying surface of a bat, or of one of the extinct groups of flying reptiles, can be severely damaged by a single rip. In contrast, the feathers of birds are individually replaceable. Even after the loss of many individual feathers, a bird can keep flying. Several hundred microscopic hooks along the sides of the individual barbules of a feather attach the barbules to one another. These hooks unite the feather so that in a marvelous way, unique to birds, a feather is perfectly adapted for flight. Overall, flight in birds is made possible by light, hollow bones, by the replacement of scales with feathers, and by the development of highly efficient lungs that supply the large amounts of oxygen necessary to sustain muscle contraction during prolonged flight.

The earliest known fossil bird is *Archaeopteryx* (see Figures 23-15 and 39-16), known from rocks about 150 million years old, in the middle of the age of the dinosaurs, the Mesozoic Era. The discovery of *Archaeopteryx* in the rocks of Bavaria (now a state of West Germany) in 1860 created a sensation, coming as it did on the heels of Charles Darwin's presentation of the theory of evolution by natural selection. *Archaeopteryx* seemed, and still seems, to provide an actual link between the reptiles and the birds and to provide concrete evidence for the evolution of the birds from the reptiles. It had not only feathers, but also a fused collarbone ("wishbone"), an additional distinctive feature of the birds, although it also had archaic features such as teeth and a long, jointed tail (see pp. 457-458).

Birds evolved from reptiles and, with the bats, are one of the two living vertebrate groups that have achieved a full mastery of the air.

MAMMALS

The first known mammals are from the early Mesozoic Era, about 200 million years ago. The first mammals were clearly derived from reptiles; in fact, a number of the early fossil organisms are difficult to classify satisfactorily either as reptiles or as mammals. The first mammals probably were small; fed on insects, as suggested by their teeth; and, judged from their large eyes, may have been nocturnal. The mammal from the early Tertiary Period shown in Figure 24-2 probably represents this general sort, even though it occurred more than 130 million years after the origin of the class. Mammals remained secondary to the reptiles, both in absolute numbers and in numbers of species, until about 65 million years ago, after the start of the Tertiary Period. Following the extinction of the dinosaurs at that time (see p. 458), mammals began to become abundant.

FIGURE 39-16

Fossil of the earliest known bird, Archaeopteryx. *The outstanding preservation of the few known fossils of this genus has aided greatly in their interpretation. A reconstruction of* Archaeopteryx *is shown in Figure 23-15.*

Characteristics of Mammals

There are about 4500 species of living mammals (class Mammalia), including human beings. Like birds, mammals are homeotherms. Their skin is covered with hair during at least some stage of their life cycle. Mammals, also like birds and crocodiles, have a four-chambered heart with complete double circulation, that is, separate systems for circulating oxygen-rich and oxygen-poor blood. Mammals nourish their young with milk, a nutritious substance produced in the mammary glands of the mother. Their locomotion is advanced over that of the reptiles, which in turn is advanced over that of the amphibians; the legs of mammals are positioned much farther under the body than those of reptiles and are suspended from limb girdles, which permit greater leg mobility.

Mammals nourish their young with milk and are covered with hair rather than scales or feathers.

The evolution of specialized teeth in mammals represents a major evolutionary advance. In fishes, amphibians, and reptiles, all of the teeth are essentially the same size and shape; in mammals there has been evolutionary specialization of these teeth into incisors, chisel-like teeth used for cutting; canines, for gripping and tearing; and molars, for crushing and breaking (Figure 43-4). These evolutionary changes, and the subsequent specializations that have taken place in different groups of mammals, have constituted one of the major reasons for their evolutionary success.

Diversity of Mammals
Monotremes

In all but a few mammals, the young are not enclosed in eggs when they are born. The **monotremes,** or egg-laying mammals, consisting of the duckbilled platypus (Figure 39-17, *A*) and the echidna, or spiny anteater (Figure 39-17, *B*), are the only exceptions to this rule. These animals occur only in Australia and New Guinea, and they are not known from the fossil record. The eggs of monotremes are similar to those of reptiles. In the echidna, they are transferred by the mother to a special marsupial pouch, where they hatch. The young in both living genera of monotremes enter the pouch, and they are fed by milk there, just as in other mammals; the milk is produced within the pouch from specialized sweat glands. The ducts of these sweat glands are not united to open on nipples. Monotremes have a cloaca, a common channel for digestive, excretory, and reproductive products; a horny beak or bill (they lack teeth); and a pouch like those of marsupials. The ability of monotremes to regulate their own temperature is less efficient than that of the other mammals. Monotremes have no clear relationships, but they may be descended from the multituberculates (see Figure 23-16), an order of mammals that flourished from the Mesozoic Era well into the Tertiary Period.

FIGURE 39-17

Monotremes (class Mammalia).
A *Duckbilled platypus,* Ornithorhynchus anatinus, *at the edge of a stream in Australia. The "duck bill" and webbed feet of this unique mammal can readily be seen in this photograph.*
B *Spiny echidna,* Tachyglossus aculeatus, *a monotreme.*

B

A

FIGURE 39-18

Kangaroo with young in its pouch; a marsupial.

Marsupials

No mammals other than monotremes lay eggs. The **marsupials** are mammals in which the young are born early in their development—sometimes as little as 8 days after fertilization—and are retained in a pouch, or **marsupium.** Marsupials have a cloaca, as do the monotremes. Living marsupials are found only in Australia, where they are abundant and diverse (see Figures 19-1 and 39-18), and in North and South America. Immediately before the Isthmus of Panama rose above the sea between 3.1 and 3.6 million years ago, marsupials were found only in Australia and South America. Subsequently, some of them, including the opossum (see Figure 23-6), migrated into North America. Ancient fossil marsupials, 40 to 100 million years old, are found in North America, where they became extinct until their geologically recent reintroduction, in South America, and in Europe. Recently a fossil marsupial was found in Antarctica, thus indicating the pathway by which marsupials passed between South America and Australia, presumably before the end of the Mesozoic Era. Surprisingly, ancient fossil marsupials have not yet been found in Australia, although they may be expected there.

Placental mammals

Most modern mammals are **placental mammals.** The first organ to form during the course of their embryonic development is the placenta. The placenta is a specialized organ, held within the womb in the mother, across which she supplies the offspring with food, water, and oxygen and through which she removes wastes (Figure 39-19). Both fetal and maternal blood vessels are abundant in the placenta, and substances thus can be exchanged efficiently between the bloodstreams of the mother and her offspring. The fetal placenta forms from

FIGURE 39-20

Placental mammals (class Mammalia).
A *Snow leopard,* Uncia, *a cat (order Carnivora).*
B *Starnosed mole,* Condylura cristata, *a burrowing insectivore (order Insectivora).*
C *A pangolin,* Manis, *order Pholidota, in Yunnan, southwestern China. The bodies of pangolins are covered with large, horny epidermal scales, interspersed with hairs. Pangolins have no teeth and feed on ants.*
D *White-tailed deer,* Odocoileus virginianus, *abundant in eastern and central temperate North America (order Artiodactyla).*
E *Orca,* Orcinus orca, *a carnivorous whale (order Cetacea).*

B

A

C

the membranes of the chorion and allantois. The maternal side of the placenta is part of the wall of the uterus, the organ in female mammals in which the young develop. In placental mammals, unlike marsupials, the young undergo a considerable period of development before they are born.

Some mammals (monotremes) lay eggs; others (marsupials) give birth to embryos that continue their development in pouches (as do the monotremes); and still others (placental mammals) nourish their developing embryos within the body of the mother, by means of a placenta, until development is almost complete.

Placental mammals are extraordinarily diverse, as can be indicated merely by mentioning some of the constituent orders. Among these are the insectivores (order Insectivora), including shrews, moles (Figure 39-20, B), and hedgehogs; the Primates, including monkeys, apes, and human beings (many illustrations in Chapter 24); the bats (order Chiroptera); the rodents (order Rodentia), including squirrels, rats, mice, woodchucks, and marmots (see Figure 22-7); the meat eaters (order Carnivora), including bears, wolves, foxes, cats (Figure 39-20, A), dogs, minks, weasels (Figure 39-21), and raccoons (see Figure 49-14); elephants (order Proboscidea; Figures 39-22, A, and 49-10); and the cetaceans (order Cetacea), including whales (Figures 39-20, E, and 39-22, B) and dolphins. One evolutionary line of mammals, the bats, has joined insects and birds in the only group of living animals that truly flies; another, the Cetacea, together with the Sirenia, or manatees, has reverted to an aquatic habitat, like that from which the ancestors of mammals came hundreds of millions of years ago.

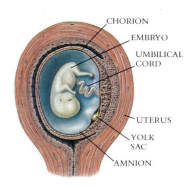

FIGURE 39-19

The placenta is characteristic of the largest group of mammals, which are called placental mammals. It evolved from the amniote egg. The umbilical cord evolved from the allantois. The chorion, or outermost member of the amniote egg, forms most of the placenta itself. The placenta serves as the provisional lung, intestine, and kidney of the embryo, without ever allowing the maternal and fetal blood to mix with one another.

FIGURE 39-21

The black-footed ferret, Mustela nigripes, *is one of the most attractive weasels of North America. This animal preys on prairie dogs, colonial rodents that have been exterminated by poisoning over much of their former large area. As a result, black-footed ferrets became so rare that for a time they were feared to be extinct. One small population, in which these photographs were taken, was discovered by ranchers in September, 1981, near Meeteetse, Wyoming; the minimum size of this population was established to include about 129 individuals in the summer of 1984, but an outbreak of canine distemper reduced the population to about 30 individuals in 1985.*

D

E

BATS

1

With nearly 1000 species, bats make up a fourth of all the kinds of living mammals. In the warmer parts of the world, bats are more numerous, both in numbers of species and in numbers of individuals, than any other order of mammals, although they are outnumbered by the rodents on a worldwide basis. In the tropics, bats constitute nearly half of the mammal species. Bats range in size from a few that are little larger than a bumblebee to some of the "flying foxes" with a wingspan of nearly a meter.

Only one species of bat, which occurs in Central and South America, actually sucks blood; nevertheless, the idea of vampires and a general fear of bats have contributed to the widespread misunderstanding of this fascinating order of mammals. The notion that bats spread rabies is widespread, yet only 10 people in the United States and Canada are known to have contracted rabies from them during the past 30 years; the likelihood of such an event occurring, therefore, is exceedingly small. Nonetheless, for these and other reasons, bats are persecuted worldwide, and a number of species are threatened with extinction.

Bats are mostly nocturnal, and relatively difficult to study for that reason. As these photographs, taken by Merlin Tuttle of the Milwaukee Public Museum and Bat Conservation International, illustrate, bats play key ecological roles in eliminating insect pests, pollinating flowers, and dispersing seeds. They are the only consistent predators on nocturnal flying insects.

Bats are often exceedingly abundant locally, and play a major role in controlling insect pests. For example, every night, Mexican free-tailed bats (*Tadarida brasiliensis*) emerge by the millions from Bracken Cave in west Texas (Figure 39-C, *1*). Huge columns of bats can be seen for kilometers and may climb more than 3000 meters above the ground. The estimated 20 million bats that inhabit this one cave cover thousands of square kilometers nightly and will eat more than 100,000 kilograms of insects before their return to the cave at dawn. In addition to helping control night-flying insects, these bats produce some of the world's best natural fertilizer. Some bat species form enormous colonies like these, whereas others roost alone. All bats hang upside down by their hind feet while at rest.

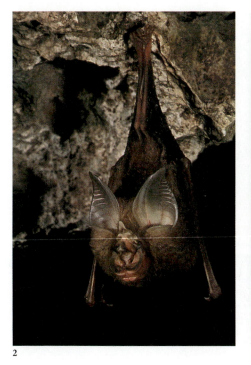

2

3

4

Bats often have elaborate nose leaves, flaps, and other strange-looking facial characteristics, which are critically important to the bat's use of echolocation and which vary greatly between different species of bats. Dobson's horseshoe bat *(Rhinolophus yunanensis),* perched on a rock in Thailand, provides an example (Figure 39-C, *2*). The highly sophisticated sonar of bats, which is based on pulses of sound ranging from 20 to more than 500 per second, has been the subject of a great deal of scientific study. Using this sonar, some bats can detect and avoid obstacles no thicker than a human hair.

Many tropical plants depend on bats for pollination, and some that occur within the borders of the United States, like the agaves of the Southwest, are frequently visited by bats during their annual migration northward. Figure 39-C, *3*, shows a greater short-nosed bat *(Cynopterus sphinx)* pollinating a wild banana flower in Thailand; note the bat's pollen-covered face. Bananas depend on bats for pollination and are visited by bats in the Western Hemisphere, where bananas were introduced as cultivated plants, not wild ones.

When bananas are grown in plantations, they are propagated from buds at the base, not from seed; therefore it is a matter of indifference to the grower whether they are pollinated or not. Pollination by bats, however, supports and preserves their wild relatives, which might ultimately prove to be sources of commercially valuable traits.

In the tropics, where bats are most abundant, many species also play important roles in dispersing plant seeds. Dwarf epauletted bats *(Micropteropus pusillus),* for example, eat ripe figs (Figure 39-C, *4*) and other kinds of fruits in West Africa. This is one of several bat species that play a critical role in the regeneration of disturbed areas in this region, spreading the seeds from forested areas into those that have been cut over. Through their seed dispersal and pollination activities, bats such as these play a critical role in the survival of rain forests worldwide. Products from plants that originally depended on bats include avocados, dates, figs, cashews, cloves, carob, and even tequila (a member of the genus *Agave*); when these plants are cultivated, they are often pollinated by hand, are self-pollinating, or produce fruits without seeds.

FIGURE 39-22

A *Parental care is a feature of the life of mammals, as shown in these photographs, taken immediately after the birth of an African elephant (Loxodonta africana); the baby elephant, stroked by the mother, rapidly gets to its feet and seeks the protection of its mother.*
B *It is also seen in this underwater photograph of a mother and calf humpbacked whale (Megaptera) taken in the waters off Hawaii.*

A

B

In Chapters 40 to 49, we will examine in detail the biology of the vertebrates, treating in some detail the way their bodies function, the way they reproduce, and the ways in which they behave. Underlying all of these functions is the marvelous diversity we have reviewed in this chapter, a diversity that has unfolded during the evolutionary course of this dominant group of animals.

SUMMARY

1. Chordates are characterized by a single, dorsal, hollow nerve cord; a flexible rod, the notochord, which forms on the dorsal side of the primitive gut in the early embryo; and pharyngeal slits.

2. Vertebrates are a subclass of the phylum Chordata. The members of this subclass differ from other chordates in that they usually possess a vertebral column, a distinct and well-differentiated head, and a bony skeleton.

3. Members of the class Agnatha differ from the other vertebrates in that they lack jaws. Once abundant and diverse, they are represented among the living mammals only by the lampreys and hagfishes, which are parasitic on other fishes.

4. The two classes of fishes other than Agnatha consist of animals that have jaws, as do the members of the other four classes of vertebrates. Jawed fishes constitute more than half of the estimated 45,000 species of vertebrates and are dominant in fresh and salt waters everywhere. Of the two classes of jawed fishes, the Chondrichthyes, or cartilaginous fishes, consist of about 850 species of sharks, rays, and skates; the Osteichthyes, or bony fishes, comprise about 25,000 species.

5. The first land vertebrates were the amphibians, one of the four classes of tetrapods (basically, four-footed vertebrates). Amphibians are quite dependent on water and lay their eggs in moist places. In many species the larvae live in water and in some species the adults do too. The amphibians probably gave rise to the reptiles, which otherwise would have been derived directly from the fishes.

6. The reptiles were the first vertebrates that were fully adapted to terrestrial habitats. Amniotic eggs, which evolved in this group but are also characteristic of the birds and the very few egg-laying mammals, represent a significant adaptation to the dry conditions that are widespread on land.

7. The birds and mammals were derived from the reptiles and are now the dominant groups of animals on land. The members of these two classes have independently become homeothermic (endothermic), capable of regulating their own body temperatures; all other living animals are poikilothermic (ectothermic), their body temperatures set by external influences.

8. The living mammals are divided into three major groups: (1) the monotremes, or egg-laying mammals, consisting only of the echidnas and the duckbilled platypus; (2) the marsupials, in which the young are born at a very early stage of development and complete their early development in a pouch, or marsupium (as do the young of monotremes); and (3) the placental mammals, which lack pouches and suckle their young.

SELF-QUIZ

The following questions pinpoint important information in this chapter. You should be sure to know the answers to each of them.

1. Vertebrates differ from other chordates in having
   (a) a rod-shaped notochord
   (b) pharyngeal slits
   (c) a single, hollow dorsal nerve cord
   (d) a series of stacked bones that protect and enclose the dorsal nerve cord

2. The biting jaws of the vertebrates were produced, in an evolutionary sense, by the
   (a) modification of the amniotic arches
   (b) modification of the silurian bone
   (c) modification of the gill arches
   (d) modification of the gnathostome

3. Although bones are heavier than cartilaginous skeletons, bony fishes became buoyant during the course of their evolution by employing _____.

4. Which evolutionary advance became the basis of the success and subsequent evolutionary radiation of vertebrates on land?
   (a) the development of efficient means of locomotion on land
   (b) the development of a specialized digestive system
   (c) the development of the ability to use gaseous oxygen for respiration
   (d) a and b
   (e) a and c

5. One of the pivotal innovations by reptiles is their watertight egg, which contains embryonic nutrients and is permeable to gases but impermeable to water. It is called a(n) _____ egg.

6. Reptiles regulate their body temperature by taking in heat from the environment. This type of temperature regulation is characteristic of
   (a) ectotherms   (d) a and b
   (b) endotherms   (e) a and c
   (c) poikilotherms

7. Birds evolved from reptiles, and with the bats they are one of the two living vertebrate groups that have achieved a full mastery of the air. Which of the following was an important modification in the birds' transition to flight?
   (a) evolution of light, hollow bones   (d) a and b
   (b) evolution of highly efficient lungs   (e) b and c
   (c) evolution of bipedal ability

8. Which of the following is considered to have been a major evolutionary advance for mammals?
   (a) specialized eyes   (d) endothermy
   (b) specialized teeth   (e) locomotion on four limbs
   (c) specialized taste

9. The first mammals were derived from
   (a) reptiles   (c) amphibians
   (b) birds   (d) fishes

10. No mammals lay eggs. True or false?

11. Which of the following do not contain members able to fly, and have never in the past contained any members capable of flight?
    (a) mammals   (c) birds
    (b) reptiles   (d) amphibians

12. Marsupials are restricted to Australia. True or false?

THOUGHT QUESTIONS

1. List some of the advantages that the early birds, in which flight was not nearly as efficient as it is in most of their modern descendants, might have had as a result of their feathered wings.

2. What limits the ability of amphibians to occupy the full range of terrestrial habitats, and allows other terrestrial vertebrates to occur in them successfully?

3. Describe the structure of the amniotic egg, and explain why it is ecologically significant.

FOR FURTHER READING

BELLAIRS, A., and R. CARRINGTON: *The Life of Reptiles,* Universe Books, New York, 1970. A good overall introduction to this class.

COLBERT, E.H.: *Evolution of the Vertebrates,* 2nd ed., John Wiley & Sons, Inc., New York, 1969. Evolutionary account of the vertebrates.

DESMOND, A.J.: *The Hot-Blooded Dinosaurs: A Revolution in Paleontology,* The Dial Press, Inc., New York, 1976. A very readable account of the development of evolutionary history, focused on the dinosaurs, and concerning the controversy about whether dinosaurs could control their own temperatures internally.

EISENBERG, J.F.: *The Mammalian Radiations: An Analysis of Trends in Evolution, Adaptation, and Behavior,* University of Chicago Press, Chicago, Ill., 1981. A marvelous, biologically oriented account of the mammals.

ELDREDGE, N., and S.M. STANLEY (eds.): *Living Fossils,* Springer-Verlag New York, Inc., New York, 1984. An excellent account of many archaic animals, including coelacanths, lungfishes, crocodilians, and several other vertebrates.

FEDDUCIA, A.: *The Age of Birds,* Harvard University Press, Cambridge, Mass., 1980. The evolution of birds from their origin to the present; very well illustrated.

LITTLE, C.: *The Colonisation of Land: Origins and Adaptations of Terrestrial Animals,* Cambridge University Press, New York, 1983. An outstanding book that contains an excellent account of the conquest of the land by vertebrates.

MOYLE, P.B., and J.J. CECH, JR.: *An Introduction to Ichthyology,* Prentice-Hall, Inc., Englewood Cliffs, N.J., 1982. This excellent book, which treats all aspects of the biology and diversity of fishes, is highly recommended.

OSTROM, J.: "Bird Flight: How Did It Begin?" *American Scientist,* vol. 67, pages 46-56, 1979. Fascinating account of theories of the origin of feathers and flight.

WEBB, P.W.: "Form and Function in Fish Swimming," *Scientific American,* July 1984, pages 72-82. The correlation between form and function in swimming reveals a great deal about how the members of different groups of fishes live.

Chapter 40

SEX AND REPRODUCTION

Overview

We begin our consideration of the biology of vertebrates with sexual reproduction, the process that initiates their development. Although the evolutionary origins of sex are not clear, almost all vertebrates reproduce sexually. Sex evolved in the sea, long before the vertebrates, and its modification for organisms living on land has entailed evolutionary innovations to avoid desiccation. The solutions adopted by vertebrates to the problem of water loss during reproduction include the amniotic egg, which first evolved among the reptiles. Amniotic eggs occur in most terrestrial vertebrates, including the birds and a few primitive mammals. The placental mammals have adopted a different solution, nourishing their developing young within the mother's body. Sex in human beings plays an important role in pair bonding as well as in reproduction. A variety of procedures have been developed that permit people to engage in sex while blocking reproduction.

For Review

Here are some important terms and concepts that you will encounter in this chapter. If you are not familiar with them, you should review them before proceeding.

- **Meiosis** (Chapter 14)

- **Adaptation** (Chapter 20)

- **Amniotic egg** (Chapter 39)

- **Mammals** (Chapter 39)

On any dark night you can hear evolution happening. The cry of a cat in heat, insects chirping outside the window, frogs croaking in swamps, wolves howling in a frozen northern scene—all of these are the sounds of evolution's essential act, reproduction. Nor are we human beings immune. Few subjects pervade our everyday thinking more than sex; few urges are more insistent. They are no accident, these strong feelings, no trap laid by the Devil to ensnare the innocent in sin. The frog in its swamp knows no sin, only an urgent desire to reproduce itself, a desire that has been patterned within it by a long history of evolution. It is a pattern that we share. The occasional sad perversion of these deeply ingrained feelings only serves to point out how very fundamental they are. For almost all of us, the reproduction of our families spontaneously elicits a sense of rightness and fulfillment. It is

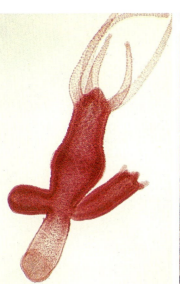

A

B

C

difficult not to return the smile of a new infant, not to feel warmed by it and by the look of wonder and delight on the faces of the parents.

This chapter deals with sex and reproduction among the vertebrates, of which humans are one kind. It seems fitting that we begin our detailed consideration of vertebrate animals here, where the reproduction of each successive generation begins.

WHY SEX?

Sexual reproduction evolved long before the advent of the vertebrates. However, it is not the only way that reproduction can occur among animals (Figure 40-1). Consider, for example, the sponges discussed in Chapter 35. They, as well as many cnidarians and some echinoderms, can reproduce by simply breaking up their bodies. In such a process of asexual reproduction a small portion of the animal divides from it and gives rise to a new organism. In asexual reproduction there is no alternation of haploid and diploid cells, no meiosis, and no gametes. Instead, differentiated diploid tissue simply undergoes a new cycle of developmental events.

Even when meiosis and the production of gametes occur, reproduction may still occur without sex. The development of an adult from an unfertilized egg (called **parthenogenesis**) is a common form of reproduction in arthropods. Among bees, for example, the development of eggs into adults does not require fertilization. The fertilized eggs develop into diploid females, but the unfertilized eggs also undergo development, becoming haploid males. Parthenogenesis even occurs among the vertebrates (see Figure 21-15). Some lizards, fishes, and amphibians are capable of reproducing in this way, their unfertilized eggs undergoing a mitotic division without cell cleavage to produce a diploid cell, which then undergoes development as if it were a diploid zygote.

If reproduction can occur without sex, it is fair to ask why sex occurs at all. There has been considerable discussion of this question, particularly among evolutionary biologists. Sex is of great evolutionary advantage for populations or species, which benefit from the variability that is generated by meiotic recombination and gene segregation. However, evolution occurs because of changes that occur at the level of *individual* fitness, rather than the fitness of populations, and it is not immediately obvious what advantage accrues to the progeny of an individual that engages in sexual reproduction. The segregation of chromosomes, which occurs in meiosis, tends to disrupt advantageous combinations of genes more often than it creates new, better-adapted ones; as a result, some of the diverse progeny that are produced as a result of any sexual mating will be less well adapted than their

FIGURE 40-1

Reproduction without sex.
A *This curiously shaped invertebrate is a reproducing* Hydra. *The main body of the animal has budded several times, and one of the buds has begun to develop tentacles.*
B *The marked queen honeybee (bottom right center of photo) is laying eggs in the cells constructed by the workers, seen here surrounding her on the hive. In almost all species of the order Hymenoptera, an order that includes bees, wasps, and ants, fertilized eggs produce females, and unfertilized eggs produce males—which therefore are haploid, asexual derivatives of the females.*
C *These ants were trapped in resin that oozed from pine trees that grew along the shores of the Baltic Sea about 40 million years ago, when the climate of the area was warm-temperate to subtropical. Although the mechanism by which honeybees produce workers is unique, the phenomenon of male haploidy probably goes back to near the origin of the order Hymenoptera—in other words, for more than 100 million years.*

DIFFERENT APPROACHES TO SEX

Human beings, like most vertebrates, have two sexes and reproduce exclusively by the union of the male's sperm with the female's egg. A few vertebrates, however, violate this rule.

One of the first cases of unusual modes of reproduction among vertebrates was reported by the Russian biologist Ilya Darevsky in 1958. He observed that some populations of small lizards of the genus *Lacerta* were exclusively female, and suggested that lizards could lay eggs that were viable even if they were not fertilized—virgin birth, or parthenogenesis, reproduction in the absence of sperm. Further work has shown parthenogenesis to occur among populations of other lizard genera, one of which is illustrated in Figure 21-15. Such reproduction is asexual, with offspring being clones—genetically identical replicas of their mothers.

Numerous fish genera contain species in which individuals can change their sex. Among coral reef fishes, for example, both **protogyny** (a change from female to male) and **protandry** (a change from male to female) occur. In fishes that practice protogyny (Figure 40-A, *1*), the sex change appears to be under social control. These fishes commonly live in large groups, or schools, where successful reproduction is typically limited to one or a few large dominant males. If these large males are removed, the largest female rapidly changes sex, becoming a dominant male.

Deepsea fishes commonly practice yet another reproductive variant that is unusual among the parents were. It is therefore a puzzle to know what a well-adapted individual gains in fitness from participating in sexual reproduction, since *all* of its progeny could maintain its successful gene combinations if that individual were simply to reproduce asexually.

Recombination is both a destructive and a constructive process in evolution. The more complex the adaptation of an individual organism, the less likely it is that recombination will improve it, and the more likely it is that recombination will disrupt it. It is no accident that asexually reproducing plants are more common in harsh habitats, such as the Arctic, where there is a premium on large numbers of genetically uniform, well-adapted individuals.

If recombination is often detrimental to an individual's progeny, where, then, is the benefit from sex that promoted the evolution of sexual reproduction? The answer to this puzzle is not yet known. Meiotic recombination is often absent among the protists, which typically undergo sexual reproduction only occasionally. The flagellates, for example, are predominantly haploid and asexual. Many other protists, such as the green algae, exhibit a true sexual cycle, but the diploid phase of such a cycle may be transient. In many protists the great majority of the cells are haploid almost all of the time, and such cells reproduce themselves asexually. Often the fusion of two haploid cells occurs only under stress, creating a diploid zygote, and the diploid stage that results from such a fusion event may not persist.

To understand how sex evolved, let us consider the protists more carefully. Why do some protists form a diploid cell in response to stress? We think this occurs because only in a diploid cell can certain kinds of chromosome damage be repaired effectively, particularly double-strand breaks in DNA. Such breaks are induced both by radiation and by chemical events within cells. As organisms became larger and longer lived, it must have become increasingly important for them to be able to repair such damage. The synaptonemal complex, which in early stages of meiosis precisely aligns pairs of homologous chromosomes (see Chapter 14), may well have evolved originally as a mechanism for repairing double-strand damage to DNA by utilizing the homologous chromosome as a template. A transient diploid phase would have provided an opportunity for such repair.

The origins of crossing-over in meiosis appear to have been in the gene repair processes. The current view of the molecular mechanism of crossing-over, described in the

vertebrates: they are **hermaphrodites**—both male and female at the same time. Among these fishes, two individuals fertilize each other's eggs (Figure 40-A, *2*)!

FIGURE 40-A

1 *The bluehead wrasse,* Thalassoma bifasciatium, *is protogynous—females sometimes turn into males. Here a large male or sex-changed female is pair mating with a female, typically much smaller.*
2 *The hamlet bass (genus* Hypoplectrus) *is a deepsea fish that is simultaneously male and female—a hermaphrodite. In the course of a single pair mating, one fish may alternate sexual roles as many as four times, by turns offering eggs to be fertilized and fertilizing its partner's eggs. Here the fish acting as male curves around the motionless partner, fertilizing the upward-floating eggs.*

boxed essay on pp. 336–337, is that it evolved from a preexisting synaptonemal mechanism for repairing accidental double-strand breaks in DNA. In yeast, mutations that inactivate the repair system for double-stranded breaks of the chromosomes also act to prevent crossing-over, suggesting a common mechanism.

> **Sexual reproduction and the close association between homologous chromosomes that occurs during meiosis probably evolved as mechanisms to repair chromosomal damage by using the homologous chromosome as a template. This repair mechanism then provided a ready means for generating genetic recombination by crossing-over.**

THE EVOLUTIONARY CONSEQUENCES OF SEX

Whatever the forces that led to sexual reproduction, its evolutionary consequences have been profound. No genetic process generates diversity more quickly, and, as you have seen, genetic diversity is the raw material of evolution, the fuel that drives it and determines its potential directions. In many cases the pace of evolution appears to be geared to the level of genetic diversity: the greater the genetic diversity, the greater the evolutionary pace. Programs for selecting larger stature in domesticated animals such as cattle and sheep, for example, proceed rapidly at first, but then slow as all of the existing genetic combinations are exhausted; further progress must then await the generation of new gene combinations.

> **The genetic recombination associated with sexual reproduction has had an enormous evolutionary impact because of the extensive variability that it can rapidly generate within the genome.**

Paradoxically, the evolutionary process is thus both revolutionary and conservative. It is revolutionary in that the pace of evolutionary change is quickened by genetic recombination, much of which results from sexual reproduction. It is conservative in that evolutionary change is not always favored by selection, which may instead act to preserve existing combinations of genes. These conservative pressures appear to be greatest in some asexually reproducing organisms that do not move around freely and live in habitats that are especially demanding. In vertebrates, on the other hand, the evolutionary premium appears to have been on versatility, and sexual reproduction is, by an overwhelming margin, the predominant mode of reproduction.

SEX EVOLVED IN THE SEA

Sexual reproduction first evolved among marine organisms, long before the advent of the vertebrates. The eggs of most marine fishes are produced in batches by the females; when they are ripe, the eggs are simply released into the water. Seawater itself is not a hostile environment for gametes or for young organisms. Fertilization is achieved by the male's release of sperm into the water containing the eggs.

The effective union of free gametes in the sea by **external fertilization** poses a significant problem for all marine organisms. Eggs and sperm become diluted rapidly in seawater, so their release by females and males must be almost simultaneous if successful fertilization is to occur. Because of this necessity, most marine vertebrates restrict the release of their eggs and sperm to a few brief and well-defined periods. There are few seasonal cues in the ocean that organisms can use as signals, but one that is all-pervasive is the cycle of the moon. Each month, the moon approaches closer to the earth at one time; when it does so, its increased gravitational attraction causes tides. Many different kinds of marine organisms sense the changes in water pressure that accompany the tides, and much of the reproduction that takes place in the sea is entrained to the lunar cycle. Some marine vertebrates reproduce once a year, and others once every month, timing the development, production, and joint release of their male and female gametes by the lunar clock.

The invasion of the land by creatures from the sea meant that these organisms faced for the first time the danger of desiccation, or drying up. This problem was all the more severe for their small and vulnerable reproductive cells. Obviously the gametes could not simply be released near one another on land; they would soon dry up and perish. There are many possible solutions to this problem. We have already considered the adaptations of the plants, the fungi, and the arthropods to an existence on dry land, including the ways in which the transmittal of their gametes has been modified for successful union under terrestrial conditions. We now consider the vertebrates.

VERTEBRATE SEX AND REPRODUCTION: FOUR STRATEGIES

The five major groups of vertebrates have evolved reproductive strategies that are quite different, and that range from external fertilization to bearing of live young. In large measure these differences reflect different approaches to protecting gametes from desiccation, since four of the five major groups of the vertebrate subphylum are terrestrial.

Fishes

Some vertebrates, of which the bony fishes are the most abundant, have remained aquatic. Although some fishes have acquired lungs that enable them to gulp oxygen when it is advantageous to do so, all fishes continue to reproduce in the water and in essentially the same way as all other aquatic animals. Fertilization in most species is external, the eggs containing only enough yolk material to sustain the developing zygote for a short time. After the initial dowry of yolk has been exhausted, the growing individual must seek its food from the waters around it. Many thousands of eggs are fertilized in an individual mating, but few of the resulting zygotes survive the rigors of their aquatic environment and grow to maturity. Some of them succumb to microbial infection, many others to predation. The development of the fertilized eggs is speedy, and the young that survive achieve maturity rapidly.

Although most fish fertilize their eggs externally, fertilization is internal in a few

FIGURE 40-2

Viviparous fishes carry live, mobile young within their bodies while these young complete their development. They are then released as small but competent adults. Here a lemon shark has just given birth to a young shark, which is still attached by the umbilical cord.

groups of fishes, the male injecting sperm into the female. Biologists distinguish three kinds of internal fertilization:

1. When the fertilized eggs are laid outside the mother's body to complete their development, the practice is called **ovipary.** Most fishes are oviparous.

2. When the eggs are retained within the mother and complete their development there, with fully developed young eventually hatching from the eggs and being released, the practice is called **ovovivipary.** Mollies and guppies are ovoviviparous, as are mosquito fish.

3. Some fishes are even **viviparous,** the undeveloped young hatching within their mother and obtaining further nourishment from her as their development proceeds (Figure 40-2). Young are released fully developed at birth. Hammerhead and blue sharks are among the truly viviparous fishes.

The usual pattern in fishes, however, is external fertilization and development of the eggs.

Amphibians

A second group of vertebrates, the amphibians (exemplified by the frogs and toads), have invaded the land without fully adapting to the terrestrial environment. The life cycle of the amphibians is still inextricably tied to the presence of free water. Among most amphibians, fertilization is still external, just as it is among the fishes and other aquatic animals. Many female amphibians lay their eggs in a puddle or a pond of water. Among the frogs and toads, the male grasps the female and discharges fluid containing the sperm onto the eggs as they are produced (Figure 40-3).

In most amphibians, as in most fishes, large numbers of eggs become mature, are released from the ovary, and are fertilized together. It was only with the advent of reptiles, which enclose fertilized eggs within an amniotic membrane and devote considerable metabolic resources to each egg, that fewer eggs are produced at one time—typically tens of eggs rather than hundreds.

The development time of the amphibians is much longer than that of the fishes, but amphibian eggs do not include a significantly greater amount of yolk. Instead, the process of development consists of two distinct components, a larval and an adult stage, in a way reminiscent of some of the life cycles found among the insects.

The development of the aquatic larval stage of the amphibians is rapid, utilizing yolk supplied from the egg. The larvae then function, often for a considerable period of time, as independent food-gathering machines. They scavenge nutrients from their environ-

FIGURE 40-3

When frogs mate, as these two are doing, the clasp of the male induces the female to release a large mass of mature eggs, over which the male projects his sperm.

ments and often grow rapidly. Tadpoles (Figure 39-10, *C*), which are the larvae of frogs and toads, grow in a matter of days from creatures no bigger than the tip of a pencil into individuals as big as goldfish. When an individual larva has grown to a sufficient size, it undergoes a developmental transition, or **metamorphosis,** into the terrestrial adult form.

> **The invasion of land by vertebrates was initially tentative, with the amphibians maintaining the external fertilization and oviparous reproduction that are characteristic of most of the fishes, the group from which they evolved.**

Reptiles and Birds

The reptiles were the first group of vertebrates to abandon the marine habitat completely. Their eggs are fertilized internally within the mother before they are laid, the male introducing his sperm directly into her body (Figure 40-4). By this means, fertilization still occurs in a nondesiccating environment, even though the adult animals are fully terrestrial. Most vertebrates that fertilize internally utilize a tube, the **penis,** to inject sperm into the female. Encased in erectile tissue, the penis can become quite rigid and penetrate far into the female reproductive tract.

Many reptiles are oviparous, the eggs being laid and then abandoned to their developmental destiny; others are viviparous or ovoviviparous, forming eggs that hatch within the body of the mother. The young are then born alive from the mother's body.

Most birds (swans are an exception) lack a penis and achieve internal fertilization by the simple expedient of the male slapping sperm against the reproductive opening of the female. This kind of mating occurs more quickly than that of most reptiles. All birds are oviparous, the young emerging from their eggs outside of the body of the mother. Birds encase their eggs in a harder shell than do their reptilian ancestors. They also hasten the development of the embryo within each egg by warming the eggs with their bodies. The young that hatch from the eggs of most bird species are not able to survive unaided, since their development is still incomplete. The young birds, therefore, are fed and nurtured by their parents, and they grow to maturity gradually.

FIGURE 40-4

The introduction of sperm by the male into the female's body is called copulation. Reptiles such as these turtles were the first vertebrates to develop this form of reproduction, which is particularly suited to a terrestrial environment.

The shelled eggs of reptiles and birds constitute one of their most important adaptations to life on land, since these eggs can be laid in dry places. Such eggs, as we discussed in Chapter 39, develop a system of internal cell layers, which enclose the space where the embryo will develop in a fluid-filled amniotic cavity. The outermost layer, the chorion, lines the shell membrane of the egg. The inner cellular layer, the amnion, forms a sac around the embryo. Each egg is provided with a large amount of yolk and is encased within a leathery cover. The zygote develops within the egg, eventually achieving the form of a miniature adult before it completely depletes its supply of yolk and leaves the egg to face its fate as an adult.

The first major evolutionary change in reproductive biology among land vertebrates was that of the reptiles and their descendants, the birds. Although both groups are still oviparous like fishes, they practice internal fertilization, with the zygote encased within a watertight egg.

Mammals

The most primitive mammals, the monotremes, are oviparous like the reptiles from which they evolved. The living monotremes (Figure 39-17) consist solely of the duckbilled platypus and the echidna. No other mammals lay eggs. All other members of this class are viviparous.

The young of viviparous mammals are nourished and protected by their mother, an outstanding characteristic of the members of this class. The mammals other than monotremes have approached the problem of nourishing their young in two ways:

1. Marsupials give birth to embryos at a very early stage of development. The tiny animals continue their development in pouches on the mother's body, eventually emerging when they are able to function sufficiently well.

2. The placental mammals retain their young for a much longer period within the body of the mother. To nourish them, they have evolved a specialized, massive network of blood vessels called a placenta, through which nutrients are channeled to the embryo from the blood of the mother.

The second and third major evolutionary changes in reproductive biology among land vertebrates were those of the marsupials and placental mammals. These groups nourish their young within a pouch or inside the body until they have reached a fairly advanced stage of development.

THE EVOLUTION OF LIFE HISTORY TRAITS

The way in which an animal is born, how and when it reproduces, and how long it lives are referred to as its **life history.** The very different life histories that we see among the vertebrates are evidence that the reproductive life of organisms is subject to great changes during the course of evolution. A population geneticist would say that life history traits are sensitive targets of selection. As discussed in Chapters 19 to 21, evolution occurs when certain genotypes gain greater representation in subsequent generations than they had initially. Obviously, life history factors bear heavily on reproductive success. What forces are at play here? Does evolution always favor those reproductive alternatives which result in the production of the most offspring? No. Evolution does not maximize the number of young, the **reproductive rate,** but rather the number of young that results in the most survivors. A mother that bears more young than she can care for may produce fewer surviving adults than a mother that bears fewer young. Evolution selects for those reproductive traits which optimize participation in the next generation.

FIGURE 40-5

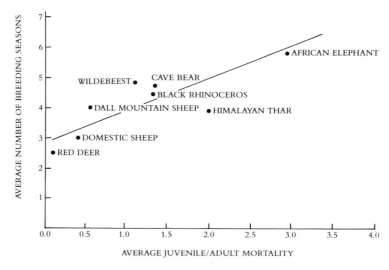

*Among mammals, the average
number of breeding seasons per
lifetime depends on the proportion
of deaths that occur among
juveniles. The more mortality is
focused on the young, the greater
the number of broods typical of
an individual lifetime.*

Evolutionists generally agree that there are five aspects of an organism's life history that are of major importance in its reproductive success:

1. **Age at first reproduction.** A population that bears its young at an earlier age than others grows more rapidly, since the young that are produced have their own broods earlier in turn, and so on, like compound interest.

2. **Litter size.** The more young in an individual brood, the greater the reproductive rate. In many organisms, however, reproduction is dangerous. Mothers are vulnerable to predation and are subject to severe metabolic demands—and a mother who does not survive cannot reproduce the next year. In general, organisms seem to optimize the size of their broods (often called **clutch size**) by reducing the number of young as far as possible without reducing the number of offspring that survive birth and rearing. Birds, for example, do not produce as many eggs as they are physiologically capable of doing, but only as many as they can successfully raise while surviving to reproduce again.

3. **Number of litters per lifetime.** The number of broods that an individual will produce in its lifetime depends on the point in that life cycle at which mortality is most likely to occur. The greater the proportion of immature individuals (**juveniles**) among those who die, the greater the number of broods per lifetime (Figure 40-5). When the survivorship of young is low relative to that of adults (that is, most deaths occur early in life), an organism must have either very big litters or lots of them to replenish the pool of young. Mammals, which nourish their developing young internally during their early growth, are physically constrained from having overly large litters. A human mother, for example, cannot safely bear eight young at one time. Mammal species in which juvenile mortality is relatively high tend to replenish the pool of juveniles by having more litters per lifetime. Birds and other oviparous animals are not constrained physically in their litter size, since their young develop outside the mother's body, and in these organisms the proportion of juvenile mortality is related to the clutch size rather than to the number of litters per lifetime.

4. **The interbrood interval.** Some organisms devote only a fraction of their life span to reproduction, whereas others reproduce continuously after reaching adulthood. The number of litters per lifetime depends both on the length of the interval between broods and the fraction of the life span that is devoted to reproduction.

5. **The reproductive cost per young.** The optimum investment for a mother or father to commit to each of their offspring is a balance between cost and benefit. The larger each offspring is at birth, the greater the cost and the danger to the parent, and the greater the chance of survival of the offspring.

All five aspects of an organism's life history are important, and all tend to evolve in concert when an organism's reproductive approach changes. The particular collection of life history traits exhibited by a group of organisms is conventionally referred to as the **reproductive strategy** of the group. Many combinations of traits are possible. There are, however, two extreme strategies:

1. **r-selection:** early age of first reproduction
   large brood size
   single breeding season
   small, numerous offspring
   no parental care
   short generation time
   large reproductive effort per brood

2. **K-selection:** delayed reproduction
   small broods
   multiple breeding seasons
   few, large offspring
   parental care
   long generation time
   smaller reproductive effort per brood

r-Selection reproductive strategies are characteristic of unstable and unpredictable environments, where mortality is often inflicted by the environment. In such habitats, for example, extremes of weather may wipe out entire populations. In these situations the optimum strategy is to fill a vacant habitat as rapidly as possible, while conditions remain favorable. The accent is on numbers of offspring rather than their quality, since mortality tends to be a random event.

K-Selection reproductive strategies are characteristic of stable environments; in such environments mortality often results from unsuccessful competition for limiting resources. For example, a limited number of nest sites or of available food resources places a premium on biological competition. In these situations the optimum strategy is "efficiency": producing a smaller number of offspring, each better able to survive in a highly competitive world. The accent is on quality of offspring rather than numbers of them, since mortality depends on individual competitive abilities.

Life history traits have evolved to maximize reproductive success. The specific constellation of traits that is characteristic of a particular species is critically influenced by the way in which mortality affects its age distribution. Species in which mortality is focused on reproductively immature individuals tend to have more and larger broods and to initiate reproduction earlier than those in which mortality is higher among adults.

During the evolution of the vertebrates there has been a general tendency toward a more K-selected reproductive strategy, although great variation exists within individual groups. Nor do groups adopt K strategies deliberately, reviewing all the options and making the rational choice—the general tendency is rather the result of many trial-and-error evolutionary responses involving whatever reproductive possibilities were available to a particular group at the time.

SEXUAL CYCLES

For efficient reproduction, mating or the release of sperm must occur as soon as the mature egg or eggs become available. Otherwise, few of the sperm will survive long enough to achieve successful fertilization. Among those vertebrates that practice internal fertilization, the females typically signal the successful development and release of an egg, **ovulation,** by the release of chemical signals called **pheromones.** Female dogs signal their reproductive readiness in this manner, and the male is able to detect their pheromones in the air, even at very low concentrations. In many mammals the female does not actually release a mature egg until mating has occurred. For them, the physical stimulus of copulation causes the pituitary gland to release a signal that triggers ovulation.

When a female vertebrate does not possess a mature egg, she will often reject the sexual advances of males. Most female mammals are sexually receptive, or "in heat," for only a few short periods each year. The period in which the animal is "in heat" is called **estrus;** the periods of estrus correspond to ovulation events during a periodic cycle, the **estrous cycle.** In general, small mammals have many estrous cycles in rapid succession, whereas larger ones have fewer cycles spaced farther apart.

Human beings and some apes provide the sole exception to the rule of cycles of receptivity. Human females are sexually receptive throughout the reproductive cycle. The reproductive cycle of egg production and release takes on average about 28 days. One cycle follows another, continuously, and a human female may mate at any time during a cycle. Successful fertilization, however, is possible only during a period of 3 to 4 days starting the day before ovulation.

HORMONES

A lot of signaling goes on during sex. Not only is it necessary for a viviparous female to signal the male when a mature egg has been released at the beginning of an estrous cycle, it is also necessary that many different processes within the female and within the male be coordinated during the process of gametogenesis and reproduction. The delayed sexual development that is common in mammals, for example, entails a nonsexual juvenile period, after which changes occur that produce sexual maturity. These changes occur in many parts of the body, a process that requires the simultaneous coordination of further development in many different kinds of tissues. The production of gametes is another closely orchestrated process, involving a series of carefully timed developmental events. Successful fertilization initiates yet another developmental "program," in which the female body prepares itself for the many changes of pregnancy.

All of this signaling is carried out by a portion of the brain called the **hypothalamus.** The hypothalamus is specialized to process information about the body's internal functions. It regulates many of the basic activities of the body, dictating the responses to pain and emotion, coordinating the digestion of food, maintaining a constant body temperature in mammals and birds, as well as coordinating reproductive events. The control of body functions by the hypothalamus involves regulating organs and glands that are situated at distant locations in the body. To accomplish this control, command signals must pass from the hypothalamus to the organs, and feedback signals must pass from that organ back to the hypothalamus.

The signals by which the hypothalamus regulates reproduction are not carried by nerves. Although a nerve-conducted signal would be very fast and specific, many of the desired changes are long-term ones, which would require a continuous nerve signal to run for days or months. The reproductive signals are thus of a slower, longer-lasting variety. They are **hormones** produced by the brain, which are carried by the bloodstream to the various organs of the body. Many reproductive hormones are complex carbon-ring lipids

FIGURE 40-6

The steroid sex hormones estrogen and progesterone. Progesterone prepares and maintains the uterine lining for pregnancy; estrogen stimulates thickening of the uterine lining during pregnancy.

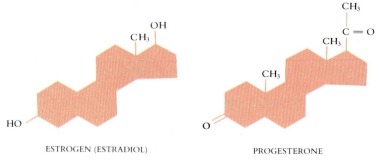

ESTROGEN (ESTRADIOL)   PROGESTERONE

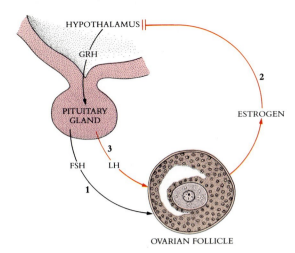

FIGURE 40-7

Mammalian egg maturation is under hormonal control. 1, The releasing hormone GRH, produced by the hypothalamus, causes the pituitary to produce the hormone FSH and release it into the blood circulation; when FSH reaches the ovaries, it initiates final egg development; 2, FSH also causes the ovaries to produce the hormone estrogen; rising levels of estrogen in the blood cause the pituitary to shut down its production of FSH and instead produce LH; 3, when this LH circulates back to the ovaries, it inhibits estrogen production and initiates ovulation.

called **steroids** (Figure 40-6); others are peptides. The target organs of the reproductive hormones possess on their cell surfaces, or in the case of steroid hormones within their cell cytoplasm, specific receptors to which a particular hormone binds. Once bound to a receptor, a hormone molecule is transported to the nucleus, where it binds to specific locations on the chromosomes and alters the pattern of gene transcription and by doing so initiates the physiological changes associated with that hormone.

> **The body uses hormones as signals to control various body functions in the reproductive cycle. Many reproductive hormones are complex carbon-ring molecules called steroids. Steroid hormones bind to specific receptor sites on target cells, enter the cells, and alter the pattern of gene expression in the target cells.**

Reproduction is only one of many functions that the vertebrate body controls with such chemical signals from the brain. Vertebrate hormones and their production will be discussed in more detail in Chapter 48. Here we need only note that like any complex organization, the vertebrate brain is hierarchical. The hypothalamus initiates a reproductive change by issuing an order to a subordinate, a hormone-producing region of the brain called the **pituitary gland,** with which it is in intimate contact. These orders are themselves hormones, although they do not travel far.

When the pituitary gland receives a "releasing hormone" from the hypothalamus, it initiates the production of the particular pituitary hormone that is called for by the releasing hormone. This pituitary hormone is released into the bloodstream, where it acts at a distance to bring about the change commanded by the hypothalamus. For example, the maturation of eggs, which begins the estrous cycle, is initiated when the hypothalamus produces a releasing hormone called **gonadotropin-releasing hormone (GRH)** (Figure 40-7). This hormone passes to the nearby pituitary, where it initiates the production of **follicle-stimulating hormone (FSH)** by the pituitary. FSH passes from the pituitary into the bloodstream, eventually reaching the ovaries. There, FSH is bound by receptors on the surface of the **follicles,** capsules in the ovary within which the eggs are developing. The binding of FSH initiates the final development and maturation of the eggs. Other tissues of the body also possess FSH receptors and respond to the hormone by producing pheromones and undergoing other physiological changes associated with the initiation of the estrous cycle.

THE REPRODUCTIVE CYCLE OF MAMMALS

The reproductive cycle of female mammals, including that of human beings, is composed of two distinct phases, the **follicular phase** and the **luteal phase.** The first, or follicular, phase of the reproductive cycle is marked by the hormonally controlled development of eggs within the ovary. As we have described, the pituitary initiates the cycle by secreting FSH, which binds to receptors on the surface of the follicles, initiating the final development and maturation of the egg. Normally, only a few eggs at any one time have developed far enough to respond immediately to FSH. FSH levels are reduced before other eggs reach maturity so that in every cycle only a few eggs ripen.

The reduction of the levels of FSH is achieved by a feedback command to the pituitary. In addition to initiating final egg development, FSH also triggers the production of the female sex hormone **estrogen** by the ovary. Rising estrogen levels in the bloodstream feed back to the pituitary and cut off the further production of FSH. In this way, only the few eggs that are already developed far enough for their maturation to be initiated by FSH

FIGURE 40-8

The human menstrual cycle. A single cycle occurs every 28 days. The growth and thickening of the uterine (endometrial) lining is governed by levels of the hormone progesterone; menstruation, the sloughing off of this blood-rich tissue, is initiated by lower levels of progesterone. Egg maturation and ovulation (egg release) are governed by the hormones FSH and LH, whose production by the pituitary is governed by the hypothalamus of the brain. When a fertilized egg implants within the endometrium, progesterone levels do not fall and pregnancy ensues.

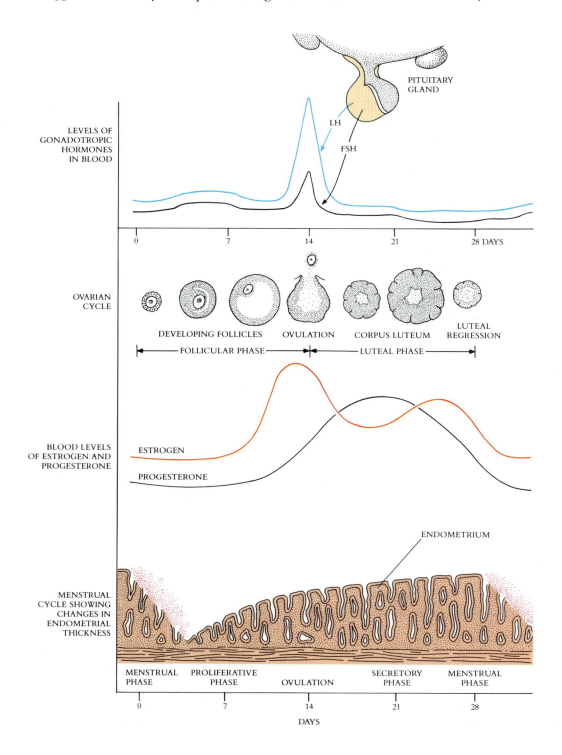

are taken into the final stage of development. The rise in estrogen level and the maturation of one or more eggs complete the follicular phase of the estrous cycle.

The second, or luteal, phase of the cycle follows smoothly from the first. The pituitary responds to estrogen by secreting a second hormone, called **luteinizing hormone (LH),** which is carried in the bloodstream to the developing follicle. LH inhibits estrogen production and causes the wall of the mature follicle to burst. The egg within the follicle is released into one of the fallopian tubes, which extend from the ovary to the uterus. This process is called ovulation. Meanwhile, the ruptured follicle repairs itself, filling in and becoming yellowish. In this condition it is called the **corpus luteum,** which is simply the Latin phrase for "yellow body." The corpus luteum soon begins to secrete a hormone, **progesterone,** which initiates the many physiological changes associated with pregnancy. The body is preparing itself for fertilization. The corpus luteum continues its production of progesterone after fertilization and throughout pregnancy. If, however, fertilization does not occur soon after ovulation, then production of progesterone slows and eventually ceases, marking the end of the luteal phase.

> **The estrous cycle of mammals is composed of two phases, which alternate with one another. During the follicular phase some of the eggs within the ovary complete their development. During the following luteal phase the mature eggs are released into the fallopian tubes, a process called ovulation. If fertilization does not occur, ovulation is followed by a new follicular phase, the start of another cycle.**

In the absence of estrogen and progesterone the pituitary can again initiate production of FSH, thus starting another estrous cycle. In human beings the next cycle follows immediately after the end of the preceding one. A cycle occurs every 28 days, or a little more frequently than once a month (Figure 40-8). The Latin word for month is *mens,* which is why the estrous cycle in humans is called the **menstrual cycle,** or monthly cycle. In human beings and some other primates the hormone progesterone has among its many effects a thickening of the walls of the uterus in preparation for the implantation of the developing embryo. When fertilization does not occur, the lowering levels of progesterone cause this thickened layer of blood-rich tissue to be sloughed off, a process that results in the bleeding associated with **menstruation.** Menstruation, or "having a period," usually occurs about midway between successive ovulations, or roughly once a month.

THE HUMAN REPRODUCTIVE SYSTEM

Human reproductive cells, like those of all vertebrates, consist of egg cells, which are richly endowed with nutrients, and sperm cells, which are practically bereft of them. As we will describe in more detail in the next chapter, the egg is fertilized within the female, and the zygote develops into a mature fetus there. Most reproductive activity, then, is centered on the female. The male contributes little but his sperm.

Males

The human male gamete, or sperm, is a very simple cell, highly specialized for its role as a carrier of genetic information. Unlike the other cells of the body, sperm do not complete their development successfully at 37° C (98.6° F), the normal human body temperature. The sperm-producing organs, or **testes,** move during the course of fetal development out of the body proper and into a sac called the **scrotum.** The scrotum, which hangs between the legs of the male, maintains the testes at a temperature about 3° C cooler than the rest of the body (Figure 40-9).

The testes (Figure 40-10) are composed of several hundred compartments, each of which is packed with large numbers of tightly coiled tubes called **seminiferous tubules.** In-

FIGURE 40-9

The human male reproductive system.

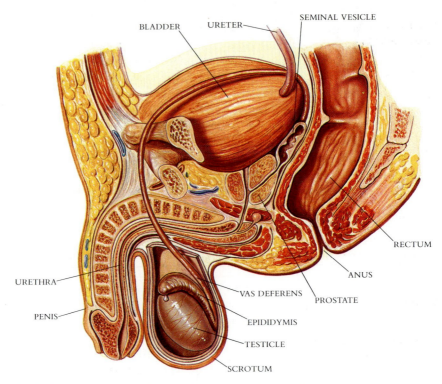

FIGURE 40-10

Human testes. The testicle is the darker sphere in the center of the photograph; within it sperm are formed. Cupped beneath the testicle is the epididymis, a highly coiled passageway within which sperm complete their maturation. Extending away from the epididymis is a long tube, the vas deferens, in which mature sperm are stored.

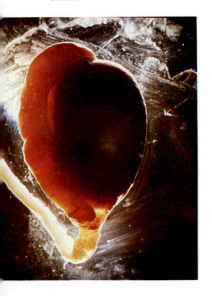

terstitial cells between the tubules produce the male sexual hormone **testosterone.** The tubes themselves are the sites of sperm cell synthesis, or **spermatogenesis** (Figure 40-11), a process that occurs in two phases:

1. **Meiosis.** Packed against the basement membrane of the seminiferous tubules are diploid cells called **spermatogonia,** which are constantly dividing mitotically. Some of the daughter cells move inward toward the interior of the tubule, the **lumen,** and begin the path of differentiation that will eventually lead to the production of a sperm cell. These cells are **primary spermatocytes.** They undergo meiosis, producing four haploid gametes, which are called **spermatids.** Each of these spermatids contains one member of each chromosome pair (an X or a Y sex chromosome and 22 autosomes).

2. **Development.** The undifferentiated spermatids then undergo a process of development, producing mature sperm cells, or **spermatozoa.** Spermatozoa are relatively simple cells, containing only a cell membrane, a compact nucleus, and a vesicle, called an **acrosome,** which is derived from the Golgi body. The acrosome is located at the leading tip of the sperm cell. It contains enzymes that aid in the penetration of the protective layers surrounding the egg. Each spermatozoan also has a propulsive mechanism, consisting of a flagellum, which propels the cell; a centriole, which acts as a basal body for the flagellum; and mitochondria, which generate the necessary metabolic power.

The full process of sperm development, from spermatogonia to spermatozoa, takes about 2 months. The number of sperm that are produced is truly incredible. A typical adult male produces several hundred million sperm each day of his life. Those which are not ejaculated from the body are reabsorbed, in a continual cycle of renewal.

After the sperm cells complete their differentiation within the testes, they are delivered to a long coiled tube called the **epididymis,** where they mature further. The sperm cells are not motile when they arrive in the epididymis, and they must remain there for at least 18 hours before their motility develops. From the epididymis, the sperm are delivered to another long tube, the **vas deferens,** where they are stored. When they are delivered during intercourse, the sperm travel through a tube from the vas deferens to the **urethra,** where the reproductive and urinary tracts join, emptying through the penis.

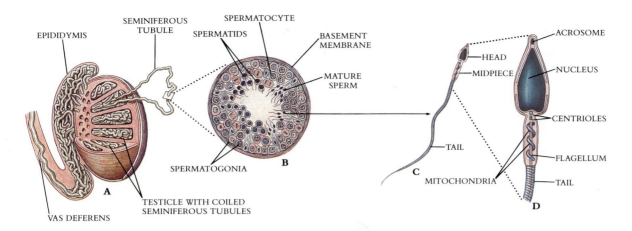

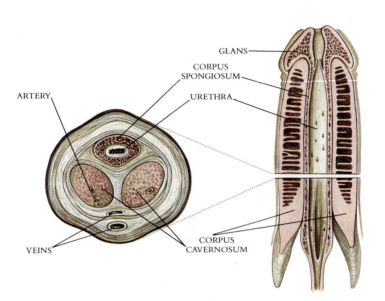

FIGURE 40-11

The interior of the testes, site of spermatogenesis. Within the seminiferous tubules of the testes (A), cells called spermatogonia (B) develop into sperm, passing through the spermatocyte and spermatid stages. Each sperm (C) possesses a long tail coupled to a head (D), which contains a haploid nucleus.

FIGURE 40-12

A penis in cross section.

The penis is an external tube composed of three cylinders of spongy tissue (Figure 40-12). In cross section the arteries and veins can be seen to run along the dorsal surface, beneath which two of the cylinders sit side by side. Below the pair of cylinders is a third cylinder, which contains in its center the urethra, through which both sperm (during intercourse) and urine (during liquid elimination) pass. The spongy tissue that makes up the three cylinders is riddled with small spaces between its cells. When nerve impulses from the central nervous system cause the dilation of the arteries leading into this tissue, blood collects within the spaces. This causes the tissue to become distended and the penis to become erect and rigid. Continued stimulation by the central nervous system is required for this erection to be maintained.

Erection can be achieved without any physical stimulation of the penis at all. Fantasy by the central nervous system is a common initiating factor. Regardless of how it is achieved, the physical stimulation of the penis is required for any delivery of sperm to take place. Prolonged stimulation of the penis, as by repeated thrusts into the vagina of a female, leads first to the mobilization of the sperm. In this process muscles encircling the vas deferens contract, moving the sperm into the urethra. Eventually the prolonged stimulation leads to the violent contraction of the muscles at the base of the penis. The result is **ejaculation,** the ejection of about 5 milliliters of **semen** out of the penis. Semen is a mixture of sperm and other liquids collectively called **seminal fluid.** Within this small volume are several hundred million sperm. The odds against any one of them successfully completing the long journey to the egg and fertilizing it are extraordinarily high. Successful fertilization requires a high sperm count (Figure 40-13); males with less than 20 million sperm per milliliter are generally considered sterile.

An adult male produces sperm continuously, several hundred million each day of his life. The sperm are stored and then are delivered during sexual intercourse.

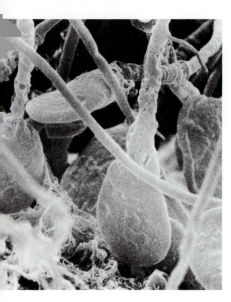

FIGURE 40-13

Human sperm. Only the heads and a portion of the long slender tails of these sperm are shown in this scanning electron micrograph.

Females

Fertilization requires more than insemination. There must be a mature egg to fertilize. Eggs are produced within the **ovaries** of females (Figure 40-14). The ovaries are compact masses of cells, 2 to 3 centimeters long, located within the abdominal cavity. Eggs develop from cells called **oocytes,** which are located in follicles in the outer layer of the ovary. Unlike males, whose gamete-producing spermatogonia are constantly dividing, females have at birth all of the oocytes that they will ever produce. At each cycle of ovulation one or a few of these oocytes initiate development; the others remain in a developmental holding pattern. This long maintenance period is one reason that developmental abnormalities crop up with increasing frequency in pregnancies of women who are over 35 years old. The oocytes are continually exposed to chromosomal mutation throughout life, and after 35 years the odds of a harmful mutation having occurred become appreciable.

At birth a female's ovaries contain some 2 million oocytes, all of which have initiated the first meiotic division. At this stage they are called **primary oocytes.** Meiosis is arrested, however, in prophase of the first meiotic division. The development of very few oocytes ever proceeds further. With the onset of puberty, the female matures sexually. At this time the release of FSH initiates the resumption of the first meiotic division in a few oocytes, but a single one soon becomes dominant, the others regressing (Figure 40-15). Every 28 days thereafter, another oocyte matures. It is rare for more than about 400 out of the approximately 2 million oocytes with which a female was born to mature during her lifetime. When they do mature, the egg cells are called **ova,** the Latin word for eggs.

Unlike male gametogenesis, the process of meiosis in the oocytes does not result in the production of four haploid gametes. Instead, a single haploid ovum is produced, the other meiotic products being discarded as **polar bodies.** The process of meiosis is stop-and-go rather than continuous (Figure 40-16):

1. **Developmental arrest.** The first phase of meiosis in female gametogenesis is arrested in prophase of meiosis I, leaving the gamete-forming cell as a primary oocyte.

2. **Ovulation.** The second phase of meiosis completes the first meiotic division. It is triggered by hormonal changes associated with ovulation. At the end of meiosis I the nuclear envelope disintegrates and the chromosomes move to the surface of the cell. There a bulge in the cell surface forms a pocket into which one set of chromosomes settles. This pocket is pinched off, forming a polar body and leaving the other set of chromosomes within the egg.

3. **Fertilization.** The second meiotic division does not occur until after fertilization. It results in the production of an ovum, which is haploid, and another polar body. The polar bodies eventually disintegrate, all reproductive investment being devoted to the single ovum. It is a very large cell, over 100 micrometers in diameter, and is rich in metabolites and biosynthetic machinery.

When the ovum is released from the follicle at ovulation, it is swept by the beating of cilia into the **oviducts,** or **fallopian tubes,** which lead away from the ovaries (Figure 40-17). Smooth muscles lining the oviduct tubes contract rhythmically; these rhythmic contractions, called **peristaltic contractions,** move the egg down the tube to the **uterus.** The journey is a slow one, taking about 3 days to complete. If the egg is unfertilized, it is dead

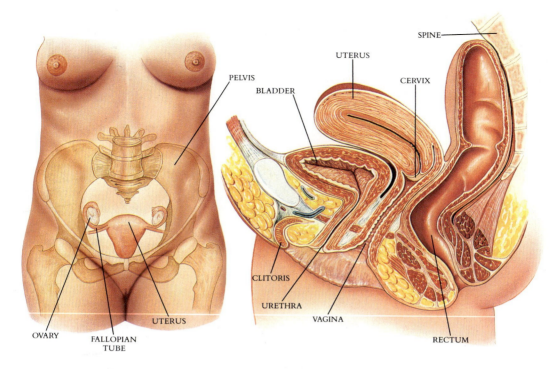

FIGURE 40-14

The human female reproductive system.

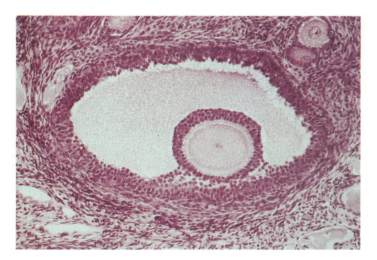

FIGURE 40-15

A mature egg within an ovarian follicle of a cat.

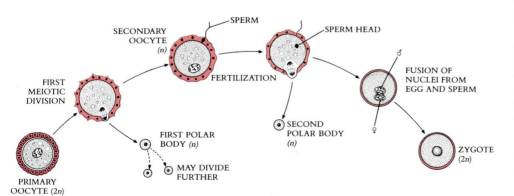

FIGURE 40-16

The meiotic events of oogenesis. A primary oocyte is diploid. In its maturation the first meiotic division is completed, and one division product is eliminated as a polar body. The other product, the secondary oocyte, is released during ovulation. The second meiotic division does not occur until after fertilization and results in production of a second polar body and a single haploid egg nucleus. Fusion of the haploid egg nucleus with a haploid sperm nucleus produces a diploid zygote, from which an embryo subsequently forms.

FIGURE 40-17

The journey of an egg. Produced within a follicle and released at ovulation, an egg is swept up into a fallopian tube and carried down by waves of contraction of the tube walls. Fertilization occurs within the tube, by sperm journeying upward. Several mitotic divisions occur while the fertilized egg continues its journey down the fallopian tube, so that by the time it enters the uterus, it is a hollow sphere of cells, a blastula-stage embryo. The blastula implants itself within the wall of the uterus, where it continues its development.

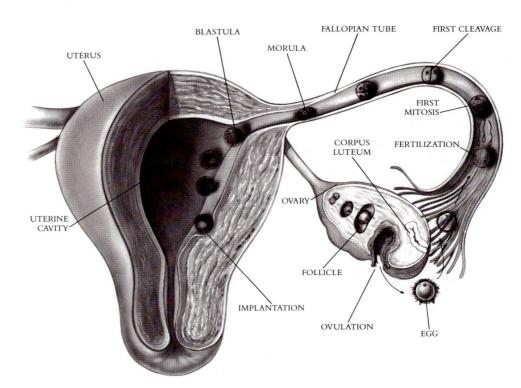

when it reaches the uterus. It can live only about 24 hours unless it has been fertilized. For this reason the sperm cannot simply lie in wait within the uterus. Any sperm that is to fertilize an egg successfully must make its way up the fallopian tubes, a long passage that few survive.

Sperm are deposited within the **vagina,** a muscular tube about 7 centimeters long, which leads to the mouth of the uterus. This opening is bounded by a muscular sphincter called the **cervix.** The uterus is a hollow pear-shaped organ about the size of a small fist. Its inner wall, the **endometrium,** has two layers. The outer of these layers is shed during menstruation, while the one beneath it generates the next layer. Sperm entering the uterus are carried upward by waves of motion in the walls of the uterus, enter the oviduct, and swim upward against the current generated by peristaltic contractions, which is carrying the ovum downward toward the uterus.

> **All of the eggs that a woman will produce during her life develop from cells that are already present at birth. Their development is halted at prophase I of meiosis, and one or a few of these cells resume meiosis each 28 days to produce a mature egg. On maturation, eggs travel to the uterus. Fertilization by a sperm, if it occurs, happens en route.**

Chapter 41 recounts the fate of a successfully fertilized egg. It does not die within the oviduct, as do its unfertilized sisters; rather, it proceeds to undergo mitosis and continues its journey. When it reaches the uterus, the new embryo soon attaches itself to the endometrial lining and thus starts the long developmental journey that eventually leads to the birth of a child.

THE PHYSIOLOGY OF HUMAN INTERCOURSE

Few physical activities are more pleasurable to humans than sexual intercourse. The sex drive is one of the strongest drives directing human behavior, and, as such, it is circumscribed by many rules and customs. Sexual intercourse acts as a channel for the strongest of human emotions: love, tenderness, and personal commitment on the one hand, rage and aggression on the other. Few subjects are at the same time more private and of more general

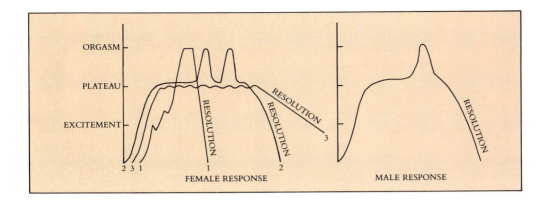

FIGURE 40-18

The human orgasmic response. Among females, the response is highly variable. It may be typified by one of the three patterns illustrated here. 1, One intense peak of response, with a rapid resolution phase; 2, several peaks of intense response, with a somewhat longer resolution phase; 3, a large number of response peaks, of less intensity, and a long resolution phase. Among males, the response does not vary as much as it does among females, with a single intense response peak being followed quickly by a resolution phase.

interest. Here we will limit ourselves to a very narrow aspect of sexual behavior, its immediate physiological effects. The emotional consequences are no less real, but they lie beyond what we can hope to cover in this book.

Until relatively recently, the physiology of human sexual activity was largely unknown. Perhaps because of the prevalence of strong social taboos against the open discussion of sexual matters, research on the subject was not being carried out, and detailed information was lacking. Each of us learned from anecdote, from what our parents or friends told us, and eventually from experience. Largely through the pioneering efforts of William Masters and Virginia Johnson in the last 25 years and an army of researchers who have followed them, this gap in the generally available information about the biological nature of our sexual lives has now largely been filled.

The sexual act is referred to by a variety of names, including intercourse, copulation, and coitus, as well as a host of more informal ones. It is common to partition the physiological events that accompany intercourse into four periods, although the division is somewhat arbitrary, with no clear divisions between the four phases. The four periods are **excitement, plateau, orgasm,** and **resolution** (Figure 40-18).

Excitement

The sexual response is initiated by a part of the nervous system devoted to passing commands from the brain to the body's organs. In both males and females, commands from the brain passing along this nerve network increase the heartbeat, blood pressure, and rate of breathing. These changes are very similar to the ones that the brain induces in response to alarm. Other changes increase the diameter of blood vessels, leading to increased circulation. In some women these changes may produce a reddening of the skin around the face, breasts, and genitals (the **sex flush**). The nipples commonly harden and become more sensitive. In the genital area this increased circulation leads to the vasocongestion in the penis of the male that produces erection. Similar swelling occurs in the **clitoris** of the female, a small knob of tissue composed of a shaft and glans much like the male penis but without the urethra running through it. The female experiences additional changes that prepare the vagina for sexual intercourse. The increased circulation leads to swelling and parting of the lips of tissue, or **labia,** which cover the opening to the vagina; the vaginal walls become moist; and the muscles encasing the vagina relax.

Plateau

The penetration of the vagina by the thrusting penis results in the repeated stimulation of nerve endings both in the tip of the penis and in the clitoris. The clitoris, which is now swollen, becomes very sensitive and withdraws up into a sheath or "hood." Once it has withdrawn, the stimulation of the clitoris is indirect, with the thrusting movements of the penis rubbing the clitoral hood against the clitoris. The nervous stimulation that is produced by the repeated movements of the penis within the vagina elicits a continuous sympathetic nervous system response, greatly intensifying the physiological changes that were initiated in the excitement phase. In the female, pelvic thrusts may begin, while in the male the penis reaches its greatest length and rigidity.

Orgasm

The **climax** of intercourse is reached when the stimulation of intercourse is sufficient to initiate a series of reflexive muscular contractions. The nerve signals producing these contractions are associated with other nervous activity within the central nervous system, activity that we experience as intense pleasure. In females the contractions are initiated by impulses in the hypothalamus, which causes the pituitary to release large amounts of the hormone **oxytocin.** This hormone in turn causes the muscles in the uterus and around the vaginal opening to contract and the cervix to be pulled upward. Contractions occur at intervals of about a second. There may be one intense peak of contractions (an "orgasm") or several, or the peaks may be more numerous but less intense (Figure 40-18).

Analogous contractions occur in the male, initiated by nerve signals from the brain. These signals first cause **emission,** in which the rhythmic peristaltic contraction of the vas deferens and of the **prostate gland,** where the vas deferens and urethra join, causes the sperm and seminal fluid to move to a collecting zone of the urethra. This collecting zone, which is located at the base of the penis, is called the **bulbourethra.** Shortly after the sperm move into the bulbourethra, nerve signals from the brain induce violent contractions of the muscles at the base of the penis, resulting in the ejaculation of the collected semen out through the penis. As in the female orgasm, the contractions are spaced about a second apart, although in the male they continue for a few seconds only. The orgasmic contractions in the male, however, are almost invariably restricted to a single intense wave of contractions.

Resolution

After ejaculation, males rapidly lose their erection and enter a **refractory period** lasting 20 minutes or longer, in which sexual arousal is difficult to achieve and ejaculation is almost impossible. By contrast, many women can be aroused again almost immediately. After intercourse, the bodies of both men and women return slowly, over a period of several minutes, to their normal physiological state.

CONTRACEPTION AND BIRTH CONTROL

In most vertebrates, sexual intercourse is associated solely with reproduction. Reflexive behavior that is deeply ingrained in the female limits sexual receptivity to those periods of the sexual cycle when she is fertile. In human beings sexual behavior serves a second important function, the reinforcement of pair bonding, the emotional relationship between two individuals living together. The evolution of strong pair bonding is certainly not unique to human beings, but it was probably a necessary precondition for the subsequent evolution of our increased mental capacity. The associative activities that make up human "thinking" are largely based on learning, and learning takes time. Human children are very vulnerable during the extended period of learning that follows their birth, and they require parental nurturing. It is perhaps for this reason that human pair bonding is a continuous process, not restricted to short periods coinciding with ovulation. Among all of the vertebrates, human females and a few species of apes are the only ones in which the characteristic of sexual receptivity throughout their reproductive cycle has evolved and come to play a role in pair bonding.

Not all human couples want to initiate a pregnancy every time they have sexual intercourse, yet sexual intercourse may be a necessary and important part of their emotional lives together. Among some religious groups, this problem does not arise, or is not recognized, since members of these groups believe that sexual intercourse has only a reproductive function and thus should be limited to situations in which pregnancy is acceptable—among married couples wishing to have children. Most couples, however, do not limit sexual relations to procreation, and among them unwanted pregnancy presents a real problem. The solution to this dilemma is to find a way to avoid reproduction without avoiding sexual intercourse, an approach that is commonly called **birth control.**

At least seven different approaches are commonly taken to achieve birth control. These methods differ from one another in their effectiveness and in their acceptability to different couples. Some of them are shown in Figure 40-19.

FIGURE 40-19

Shown here are six of the common means used to achieve birth control.
A *Condom.*
B *Spermicidal jelly.*
C *Diaphragm.*
D *IUD.*
E *Oral contraceptives.*
F *Vaginal sponge.*

Abstinence

The simplest and most reliable way to avoid pregnancy is not to have sex at all. Of all methods of birth control, this is the most certain—and the most limiting, since it denies a couple the emotional support of a sexual relationship.

A variant of this approach is to avoid sexual relations only on the two days preceding and following ovulation, since this is the only period during which successful fertilization is likely to occur. The rest of the sexual cycle is relatively "safe" for intercourse. This approach, called the **rhythm method,** is satisfactory in principle but difficult in application, since ovulation is not easy to predict and may occur unexpectedly. The effectiveness of the rhythm method is low; the failure rate is estimated to be 13% to 21% (13 to 21 pregnancies per 100 women practicing the rhythm method per year).

Another variant of this approach is to have only incomplete sex—withdraw the penis before ejaculation, a procedure known as **coitus interruptus.** This requires considerable willpower, often destroys the emotional bonding of intercourse, and is not as reliable as it might seem. Prematurely released sperm can be secreted by the penis within its lubricating fluid, and a second sexual act may transfer sperm ejaculated earlier. The failure rate of this approach is estimated at 9% to 25%, which is no better than that of the rhythm method.

Sperm Blockage

If sperm are not delivered to the uterus, fertilization cannot occur. One way to prevent the delivery of sperm is to encase the penis within a thin rubber bag, or **condom.** Many males do not favor the use of condoms, since they tend to decrease the sensory pleasure of the male. In principle this method is easy to apply and foolproof, but in practice it proves to be less effective than one might expect, with a failure rate of from 3% to 15%. Nevertheless, it is the most commonly employed form of birth control in the United States, with over a billion condoms sold annually.

A second way to prevent the entry of sperm into the uterus is to place a cover over the cervix. The cover may be a relatively tight-fitting **cervical cap,** which is worn for days at a time, or a rubber dome called a **diaphragm,** which is inserted immediately before intercourse. Because the dimensions of individual cervices vary, a cervical cap or diaphragm must be fitted by a physician. Failure rates average from 5% to 20% for diaphragms, perhaps because of the propensity to insert them carelessly when in a hurry. Failure rates for cervical caps are somewhat lower, about 8%.

Sperm Destruction

A third general approach to birth control is to remove or destroy the sperm after ejaculation. This can in principle be achieved by washing out the vagina immediately after intercourse, before the sperm have a chance to travel up into the uterus. Such a procedure is called by the French name for "wash," **douche.** This method turns out to be difficult to apply well, since it involves a rapid dash to the bathroom immediately after ejaculation and a very thorough washing. The failure rate has been estimated as high as 40%.

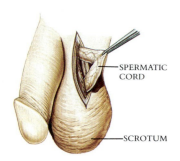

- SPERMATIC CORD
- SCROTUM

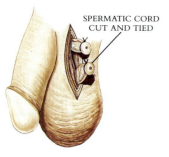

SPERMATIC CORD CUT AND TIED

A

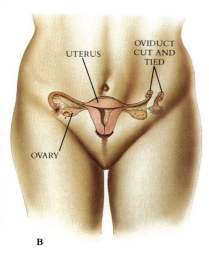

- UTERUS
- OVIDUCT CUT AND TIED
- OVARY

B

FIGURE 40-20

Surgical means of birth control.
A *Vasectomy.*
B *Tubal ligation.*

Sperm delivered to the vagina can be destroyed there with spermicidal jellies, sponges, or foams. These generally require application immediately before intercourse. The failure rate varies widely, from 3% to 22%.

Prevention of Egg Maturation

Since about 1960 a widespread form of birth control in the United States has been the daily ingestion of hormones, or **birth control pills.** These pills contain estrogen and progesterone, either taken together in the same pill or in separate pills taken sequentially. In the normal sexual cycle of a female, these hormones act to shut down the production of the pituitary hormones FSH and LH. The artificial maintenance of high levels of estrogen and progesterone in a woman's bloodstream fools the body into acting as if ovulation had already occurred, when in fact it has not: the ovarian follicles do not ripen in the absence of FSH, and ovulation does not occur in the absence of LH. For these reasons, birth control pills provide a very effective means of birth control, with a failure rate of 0% to 10%. A small number of women using birth control pills experience undesirable side effects, such as blood clotting and nausea. The long-term consequences of the prolonged use of these pills are not yet known, since they have been in widespread use for only 25 years. To date, however, there has been no conclusive evidence of any serious side effects for the great majority of women.

Surgical Intervention

A completely effective means of birth control, although it is permanent, is the surgical removal of a portion of the tube through which gametes are delivered to the reproductive organs (Figure 40-20). After such an operation no gametes are delivered during intercourse, which is unaffected in all other respects. The failure rate of such surgical approaches is 0. The great disadvantage of surgical intervention is that it generally renders the person permanently sterile.

In males, such an operation involves the removal of a portion of the vas deferens, the tube through which sperm travel to the penis. This procedure is called a **vasectomy.** The operation is simple and can be carried out in a physician's office. Reversing a vasectomy, however, is difficult and sometimes impossible, requiring general anesthesia and long microsurgery (using the magnification of a surgical microscope to guide the reconnection of the many microscopic tubules of the vas deferens individually).

In females the comparable operation involves the removal of a section of each of the two fallopian tubes through which the egg travels to the uterus. Since these tubes are located within the abdomen, the operation, called a **tubal ligation,** is more difficult than a vasectomy and is even more difficult to reverse. It is usually carried out through a small incision made in the vaginal wall just below the cervix.

The surgical removal of the entire uterus, an operation that is not uncommon but is usually performed for medical reasons other than birth control, is called a **hysterectomy.**

Prevention of Embryo Implantation

The insertion of a coil or other irregularly shaped object into the uterus is an effective means of birth control, since the irritation in the uterus prevents the implantation of the descending embryo within the uterine wall. Such **intrauterine devices** (IUDs) have a failure rate of only 1% to 3%. Their high degree of effectiveness probably reflects their convenience; once they are inserted, they can be forgotten. The great disadvantage of this method is that almost a third of the women attempting to use IUDs cannot; the devices cause them cramps, pain, and sometimes bleeding.

Another effective way to prevent embryo implantation is the use of the "morning after" pill, which contains 50 times the dose of estrogen present in birth control pills. The failure rate is 1% to 10%, but many women are uneasy about taking such high hormone doses.

Abortion

Reproduction can be prevented after fertilization if the embryo is removed before its development and birth. During the first trimester this can be accomplished by **vacuum suction** or by **dilation and curettage,** in which the cervix is dilated and the uterine wall is scraped with a spoon-shaped surgical knife called a curette. In the second trimester the embryo can be removed by injecting a 20% saline solution into the uterus, which induces labor and delivery of the fetus. In general, the more advanced the pregnancy, the more difficult and dangerous the abortion is to the mother.

As a method of birth control, abortion takes a great emotional toll, both on the woman undergoing the abortion and often on others who know and care for her. Abortion also presents serious moral problems. Many people feel that the fetus is a living person from the time of conception and that abortion is simply murder. In many countries the law defines abortion as a crime. In the United States a fetus is not legally considered a person until birth, and abortions are not prohibited by law during the first two trimesters. They are illegal in the third trimester, however, except when the mother's life is endangered. The Supreme Court's ruling legalizing abortion in the first two trimesters is a relatively recent one, and it is still the subject of intense controversy. The abortion rate in the United States (and in most other developed countries) exceeds the birthrate.

Among methods of birth control, the rhythm method, coitus interruptus, and douches are relatively ineffective. Condoms and diaphragms are effective when used correctly, but mistakes are not uncommon. Birth control pills and IUDs are very effective. Vasectomies and tubal ligations are completely effective, although permanent.

AN OVERVIEW OF VERTEBRATE REPRODUCTION

Vertebrates carry out reproduction in many different ways, reflecting the evolutionary course of their invasion of the land. These differences are reflected in whether fertilization is external or internal, in whether zygotes begin life as eggs or develop internally, in the size of broods and the frequency with which they are produced. Nor does the story end here. The course of development of the zygote is influenced in many ways by the reproductive strategy of the organism. A zygote nourished by the yolk of an egg develops differently from one nourished by its mother's blood supply. In the next chapter you will consider these differences in developmental processes in detail.

Much of the diversity seen among the vertebrates in reproduction and development, however, reflects the transition to the land. As you will see in the chapters that follow, the biology of all vertebrates is similar in most other respects. You employ a different mechanism than a fish does to obtain the oxygen necessary for metabolism, because extracting oxygen from air presents different problems than extracting it from water. However, having obtained oxygen, you circulate it to your tissues in much the same manner that a fish does. The similarities among vertebrates are more striking than the differences between them.

SUMMARY

1. Sexual reproduction appears to have evolved first as a mechanism to permit the repair of double-strand breaks in chromosomal DNA. The repair mechanism involved pairing of homologous chromosomes, and the process of crossing-over is thought to have evolved from it. The genetic recombination that is effected by crossing-over is perhaps the most important source of genomic variation in natural populations.

2. Sexual reproduction evolved in the sea. Among most fishes and the amphibians, fertilization is external, with gametes being released into water. Amphibians and the great majority of fishes are oviparous, the young being nurtured by the egg rather than by the mother.

3. Successful invasion of the land by vertebrates involved major changes in reproductive strategy. The first change was the watertight egg of the reptiles, which could survive in dry places. Internal fertilization, which is also characteristic of the reptiles, is another important survival strategy on land. Birds, as well as a primitive group of mammals called monotremes, have the same kinds of eggs as reptiles do.

4. The second important change in reproductive adaptation to life on land was that of the marsupials. In them, fertilization is internal. The young are typically born within a few weeks of fertilization, but then they are nourished and protected during the course of their further development within a pouch.

5. A third change was that of the placental mammals. Their young are nourished within the mother's body by means of a large, complex structure, the placenta, which exchanges material from her bloodstream with her offspring via a common circulation of blood.

6. The life history traits seen in any particular group of vertebrates reflect the nature of the mortality typically experienced by the group. Many species that live in stable habitats tend to exhibit reproductive characteristics that maximize the competitive abilities of the young; such species are said to be K-selected. In contrast, species that live in unstable environments typically exhibit reproductive characteristics that maximize the numbers of young produced and minimize the parental investment of any individual offspring; they are said to be r-selected.

7. Reproduction in mammals is regulated by hormones, typically produced by the pituitary on commands from the hypothalamus region of the brain. These chemical signals circulate in the bloodstream. When a hormone molecule reaches its target tissue, it binds to specific receptors of the target cell.

8. The reproductive cycle of mammals is called the estrous cycle. It is composed of two phases: (1) a follicular phase, in which some eggs in the ovary are hormonally signaled to complete their development; and (2) a luteal phase, in which one or more mature eggs are released into the fallopian tube, a process called ovulation. A complete estrous cycle in a human female takes about 28 days.

9. The male gametes, or sperm, of mammals are produced within the testes. In human males hundreds of millions of sperm are produced each day. Sperm mature and become motile in the epididymis, and they are stored in the vas deferens. Stimulation of the penis causes it to become distended and erect and causes the sperm to be delivered from the vas deferens to the urethra at the base of the penis. Further stimulation causes violent muscle contractions, which ejaculate the sperm from the penis.

10. At birth, female mammals contain all of the gametes, or oocytes, that they will ever have. A human female has approximately 2 million oocytes. All but a very few of these are arrested in meiotic prophase. At each ovulation the first meiotic division of

one or a few eggs is completed. The second meiotic division does not occur until after fertilization.

11. Fertilization occurs within the fallopian tubes. The journey of an egg to the uterus takes 3 days, and it is viable for only 24 hours unless it is fertilized. Consequently, only those eggs which have been reached within 24 hours by sperm swimming up the fallopian tubes from the uterus can be fertilized successfully. Fertilized eggs continue their journey down the fallopian tubes and attach to the lining of the uterus, where their development proceeds.

12. Human intercourse is marked by four physiological periods: excitement, plateau, orgasm, and resolution. Orgasm in women is highly variable and may be prolonged. Orgasm in men is uniformly abrupt; it coincides with the ejaculation of sperm.

13. A variety of birth control procedures is practiced by humans, of which condoms used by men and birth control pills used by women are perhaps the most common. Intrauterine devices (IUDs) and surgical procedures that block the delivery of gametes are becoming increasingly common. Birth control is not always effective, and some couples terminate unwanted pregnancies. In the United States, as in other developed countries, the abortion rate exceeds the birthrate.

SELF-QUIZ

The following questions pinpoint important information in this chapter. You should be sure to know the answers to each of them.

1. A common form of reproduction among arthropods involves the development of an adult from an unfertilized egg, a process called
   (a) parthenogenesis
   (b) crossing-over
   (c) ovipary
   (d) K-selection
   (e) meiotic recombination

2. A fish that lays an egg in the water is oviparous. True or false?

3. Do reptiles have larvae?

4. Which of the following is not an aspect of an organism's history that is of major importance in its reproductive success?
   (a) age at death
   (b) number of litters per lifetime
   (c) the interbreed interval
   (d) litter size
   (e) the reproductive cost per young

5. Which of the following would characterize r-selection?
   (a) large brood size
   (b) multiple breeding seasons
   (c) parental care
   (d) long generation time

6. The name for the sexually receptive period, sometimes referred to as being "in heat," is
   (a) the menstrual cycle
   (b) the ovulation cycle
   (c) estrus
   (d) the amnion cycle
   (e) the chorion cycle

7. The part of the brain that coordinates reproductive events is the _____.

8. The estrous cycle of female mammals is composed of two distinct phases, the _____ and the _____ phases.

9. The _____ is located at the leading tip of the sperm cell and contains enzymes that aid in the penetration of the protective layers surrounding the egg.

10. The process of meiosis in the oocyte results in a single haploid ovum and other meiotic products, which are discarded as
    (a) oocyte fragments
    (b) meiotic spindles
    (c) ova bodies
    (d) secondary oocytes
    (e) polar bodies

11. Birth control pills contain the hormone
    (a) estrogen
    (b) progesterone
    (c) testosterone
    (d) a and b
    (e) a and c

12. The surgical removal of the vas deferens, the tube through which sperm travel to the penis, is a
    (a) tubal ligation
    (b) hysterectomy
    (c) vasectomy
    (d) vacuum curettage

THOUGHT QUESTIONS

1. Perhaps the most controversial issue to arise in our consideration of sex and reproduction is abortion. List arguments for and against abortion. Under what conditions would you permit abortions? Forbid them? Do you think the disadvantages of abortion are in any way counterbalanced by the advantages it provides to underdeveloped countries with very high birthrates and swelling populations? Should the United States promote or oppose dissemination of information about abortion in underdeveloped countries?

2. Some fishes and many reptiles are viviparous, retaining fertilized eggs within a mother's body to protect them. Birds, however, even though they evolved from reptiles, never employ this means of protecting their eggs. Can you think of a reason why?

3. Relatively few kinds of animals have both male and female sex organs in the same animal, whereas most plants do. Propose an explanation for this.

4. Explain why animals that develop by means of parthenogenesis are usually female.

FOR FURTHER READING

BELL, G.: *The Masterpiece of Nature: The Evolution and Genetics of Sexuality,* University of California Press, Berkeley, 1982. A somewhat advanced but very rewarding look at the various theories of why sexual reproduction evolved the way it did.

DALY M., and M. WILSON: *Sex, Evolution, and Behavior: Adaptations for Reproduction,* Wadsworth Publishing Co., Belmont, Calif., 1978. A general presentation of the various strategies of sexual reproduction among vertebrates and arthropods.

DJERASSI, C.: *The Politics of Contraception,* W.H. Freeman & Co., San Francisco, 1981. The best of the many books about contraception.

HAPGOD, F.: *Why Males Exist: An Inquiry into the Evolution of Sex,* William Morrow & Co., New York, 1979. An accurate but informal treatment that serves as a good introduction to the issues involved in this area.

LEIN, A.: *The Cycling Female,* W.H. Freeman & Co., San Francisco, 1979. A short, informal description of the human menstrual cycle and its physical and emotional effects on women.

MAYNARD-SMITH, J.: *The Evolution of Sex,* Cambridge University Press, New York, 1978. An important viewpoint on the origin of sex, which outlines clearly the issues currently being argued by researchers.

PENGELLEY, E.T.: *Sex and Human Life,* Addison-Wesley Publishing Co., Reading, Mass., 1974. A general and relaxed treatment of sexuality in human beings, discussing many physiological and social issues.

SEGAL, S.J.: "The Physiology of Human Reproduction," *Scientific American,* September 1975, pages 52-62. An overview of the way in which hormones and the nervous system interact to regulate the sexual cycle.

STEARNS, S.: "Life History Tactics: A Review of the Ideas," *Quarterly Review of Biology,* vol. 51, pages 3-47, 1976. A discussion of the factors that influence the evolution of life history characteristics.

WILLIAMS, G.C.: *Sex and Evolution,* Princeton University Press, Princeton, N.J., 1975. Presents some important ideas on the evolution of sex.

PATHS OF EMBRYONIC DEVELOPMENT

Overview

Vertebrate embryonic development may be divided artificially into several stages, although in reality these stages are parts of a continuous, dynamic process. Development starts with the union of egg and sperm to form a diploid zygote, followed by a series of rapid "cleavage" divisions, the pattern of which is influenced by the presence of yolk. The ball of cells that results from cleavage then differentiates into a structure that resembles an indented tennis ball and possesses the three primary cell types. Although the details differ, this process occurs in all deuterostomes. The development of the specific tissues of the body follows. In chordates the notochord and dorsal nerve tube are among the first tissues to develop. Vertebrates go on to elaborate the neural crest, whose cells migrate to other positions in the developing embryo and ultimately give rise to most of the distinctive features of vertebrates. The developmental process takes longer in mammals than in other vertebrates; it lasts about 266 days in human beings.

For Review

Here are some important terms and concepts that you will encounter in this chapter. If you are not familiar with them, you should review them before proceeding.

- **Deuterostomes** (Chapters 36 and 38)
- **Chordates** (Chapter 38)
- **The amniotic egg** (Chapter 39)
- **Terrestrial reproductive strategies** (Chapter 40)

Sex is almost universal among the vertebrates. Reproduction in all but a few members of this subphylum involves two haploid gametes, which unite to form a single diploid cell called a zygote. This zygote grows by a process of cell division and differentiation into a complex multicellular animal, composed of many different tissues and organs. The process of development comprises the events that occur after the union of the two haploid gametes. Although some of its details differ from group to group, the process of development is fundamentally the same in all vertebrates.

In vertebrates, development occurs in six stages:

1. Fertilization	The male and female gametes form a zygote.
2. Cleavage	The zygote rapidly divides into many cells, with no overall increase in size. These divisions set the stage for development, since different cells receive different portions of the egg cytoplasm and hence different regulatory signals.
3. Gastrulation	The cells of the zygote move, forming three cell layers. These layers are the primary cell types: ectoderm, mesoderm, and endoderm.
4. Neurulation	In all chordates the first organ to form is the notochord, followed by formation of the dorsal nerve cord.
5. Neural crest formation	The first uniquely vertebrate event is the formation of the neural crest. From it develop many of the uniquely vertebrate structures.
6. Organogenesis	The three primary cell types then proceed to combine in various ways to produce the organs of the body.

In this chapter we will discuss each of these stages in turn, then consider the mechanisms governing developmental changes, and conclude with a detailed description of the events that occur during the course of human development.

INITIAL STAGE OF REPRODUCTION: FERTILIZATION

In vertebrates, as in all sexual animals, the first step in reproduction is the union of male and female gametes, a process called **fertilization.** Fertilization consists of three stages: (1) penetration, (2) activation, and (3) fusion. The male gametes of vertebrates, like those of other animals, are small, motile sperm. Each sperm is shaped like a comet, with a head containing a haploid nucleus and a long tail. Sperm are among the smallest cells in the body. The female gametes, called eggs or **oocytes,** are large cells. In many vertebrates the eggs contain significant amounts of yolk.

Penetration

In fishes and amphibians, fertilization is typically external, whereas in all other vertebrates it occurs internally. Internal fertilization is achieved by the release of a mature egg into a body cavity into which sperm can be introduced. There the egg can be fertilized by one of the many sperm that are introduced into the female reproductive tract during mating. The actively swimming sperm migrate up the oviduct until they encounter a mature egg.

Like a traveling princess, the mammalian egg is surrounded by a great deal of baggage (Figure 41-1). The egg cell itself is encased within an outer membrane called the **zona pellucida,** which is in turn surrounded by a protective layer of follicle cells. The first sperm to worm its way through this barrier adheres to the egg membrane by the tip of the sperm cell head, the acrosome. From its acrosome, the sperm releases enzymes that cause the plasma membranes of the sperm and egg cell to fuse. Egg cytoplasm bulges out at this point, engulfing the head of the sperm and permitting the sperm nucleus to enter the cytoplasm of the egg (Figure 41-2).

Activation

Entry of the sperm nucleus into the egg has three effects:

1. The series of events initiated by sperm penetration is collectively called **egg activation.** In frogs, reptiles, and birds, many sperm may gain entry to the egg cell, but only the first one that enters is successful in fertilizing it. In mammals, by contrast, the penetration of the first sperm initiates changes in the egg cell membrane that prevent the entry of other sperm.

A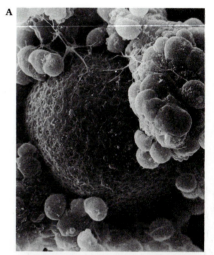

B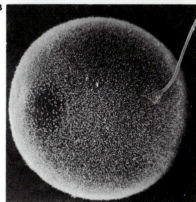

FIGURE 41-1

Eggs such as this human ovum **(A)** (the large cell in the center of the photograph) *are often surrounded by numerous nutritional cells. Often many sperm crowd around such a cell, whereas in other vertebrates a solitary sperm may effect penetration, as the hamster sperm has done in* **B.**

2. In the development of a mammalian egg, a polar body, one of the two products of the first meiotic division, is extruded from the egg cell as soon as that division has been completed. The second meiotic division does not occur until the first sperm has penetrated into the egg cell. Upon sperm penetration, the chromosomes in the egg nucleus complete meiosis, producing two egg nuclei. One of these two newly formed nuclei is extruded from the egg cell as a second polar body, leaving a single haploid egg nucleus within the egg.

3. A third effect of sperm penetration in vertebrates is the rearrangement of the egg cytoplasm. Around the point of sperm entry a series of cytoplasmic movements is initiated within the egg. These movements ultimately establish the bilateral symmetry of the developing organism. In frogs, for example, sperm penetration causes an outer pigmented cap of egg cytoplasm to rotate toward the point of entry, uncovering a **gray crescent** of interior cytoplasm opposite the point of penetration (Figure 41-3). The position of the gray crescent determines the orientation of initial cell division. A line drawn between the point of sperm entry and the gray crescent would bisect the right and left halves of the future adult.

In some vertebrates it is possible to activate an egg without the entry of a sperm, simply by pricking the egg membrane. If the development of any egg is stimulated in this way, the egg may go on to develop parthenogenetically. A few kinds of amphibians, fishes, and reptiles rely entirely on parthenogenetic reproduction in nature, as we mentioned in Chapter 40.

FIGURE 41-2

Sperm penetration of a sea urchin egg.
A *The stages of penetration.*
B *An electron micrograph of penetration. (×50,000.)*

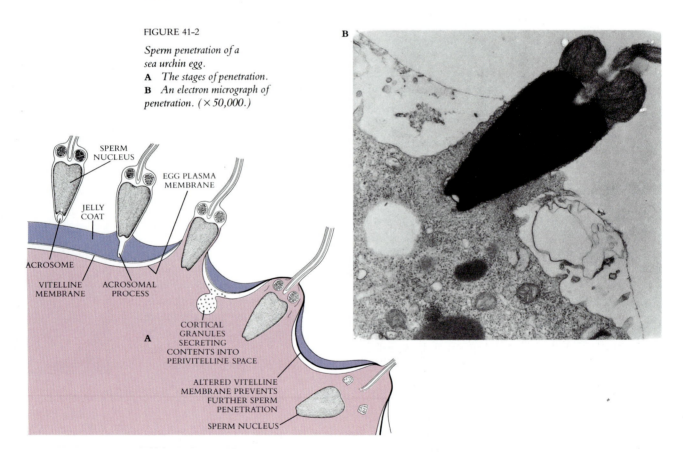

SPERM NUCLEUS

EGG PLASMA MEMBRANE

JELLY COAT

ACROSOME

VITELLINE MEMBRANE

ACROSOMAL PROCESS

CORTICAL GRANULES SECRETING CONTENTS INTO PERIVITELLINE SPACE

ALTERED VITELLINE MEMBRANE PREVENTS FURTHER SPERM PENETRATION

SPERM NUCLEUS

Fusion

The third stage of fertilization is the fusion of the entering sperm nucleus with the haploid nucleus of the egg to form a diploid zygotic nucleus. This fusion is triggered by the activation of the egg. If a sperm nucleus is introduced by microinjection without activation of the egg, fusion of the two nuclei will not take place. The nature of the signals that are exchanged between the two nuclei, or sent from one to the other, is not known.

> **The three stages of fertilization are penetration, activation, and fusion. Penetration initiates a complex series of developmental events, including major movements of cytoplasm, which eventually lead to the fusion of the egg and sperm nuclei.**

SETTING THE STAGE FOR DEVELOPMENT: CELL CLEAVAGE

The second major event in vertebrate reproduction is the rapid division of the zygote into a larger and larger number of smaller and smaller cells. This period of division, called **cleavage,** is not accompanied by any increase in the overall size of the embryo. The resulting tightly packed mass of about 32 cells is called a **morula,** and each individual cell in the morula is referred to as a **blastomere.** Blastomeres are by no means equivalent to one another. Different blastomeres contain different components of the egg cytoplasm, and these components dictate different developmental fates for the cells in which they are present. The cells of the morula continue to divide without an overall increase in size, each cell secreting a fluid into the center of the cell mass. Eventually a hollow ball of 500 to 2000 cells is formed, called a **blastula.** This hollow ball of cells surrounds a fluid-filled cavity called the **blastocoel.**

Cell Cleavage Patterns

The pattern of cleavage division is greatly influenced by the yolk (Figure 41-4). As we discussed in the last chapter, vertebrates have embraced a variety of reproductive strategies that involve different patterns of yolk utilization.

Primitive aquatic vertebrates

When eggs contain little or no yolk, cleavage occurs throughout the whole egg (Figure 41-5). This pattern is called **holoblastic cleavage.** Holoblastic cleavage was characteristic of the ancestors of the vertebrates and is still seen in groups such as the lancelets and agnathans. It results in the formation of a symmetrical blastula, composed of cells of approximately equal size.

Amphibians and advanced fishes

The eggs of bony fishes and frogs contain much more yolk in one hemisphere than in the other. Because yolk-rich cells divide much more slowly than those which are poor in yolk, holoblastic cleavage of these eggs results in a very asymmetrical blastula (Figure 41-6), with large cells containing a lot of yolk at one pole and a concentrated mass of small cells containing very little yolk at the other.

Reptiles and birds

Some eggs are composed almost entirely of yolk, with a small amount of cytoplasm concentrated at one pole. Such eggs are typical of reptiles, birds, and some fishes. In such eggs, cleavage occurs only in the tiny disk of polar cytoplasm, called the **blastodisc,** which lies astride the large ball of yolk material (Figure 41-7). Such a pattern of cleavage is called **meroblastic cleavage.** The resulting blastoderm is not spherical, but rather has the form of a hollow cap perched on the yolk.

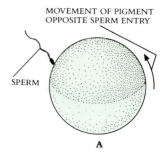

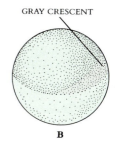

FIGURE 41-3

Gray crescent formation in frogs.
A *Sperm penetration.*
B *Appearance of the gray crescent opposite the point of penetration.*

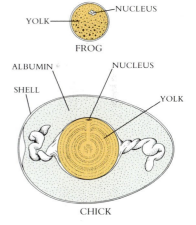

FIGURE 41-4

Three kinds of eggs. In the primitive vertebrate Amphioxus the organization of the egg is simple, with a central nucleus surrounded by yolk. In a frog egg there is much more yolk, and the nucleus is displaced toward one pole. Bird eggs are complexly organized, with the nucleus astride the surface of a large central yolk like a spot painted on a balloon.

FIGURE 41-5

Holoblastic cleavage is symmetrical, dividing an egg into equal portions. The egg dividing here is that of a mouse. Holoblastic cleavage was typical of the ancestors of the vertebrates.

Carolina Biological Supply Company

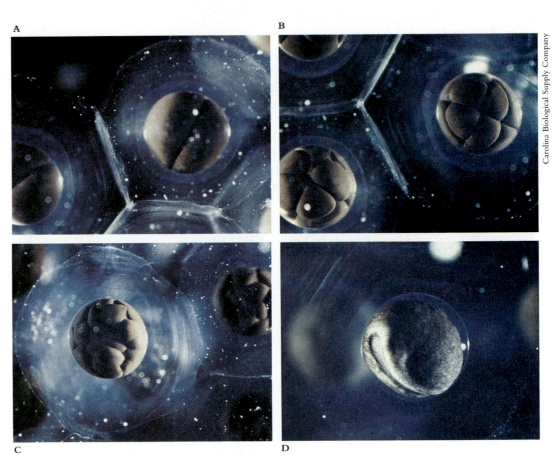

A **B**

C **D**

FIGURE 41-6

Cleavage divisions producing a frog embryo. The initial divisions (A) are, in this case, on the side of the zygote facing you, producing a cluster of cells on this side of the embryo (B), which soon expands to become a compact mass of cells (C). This mass eventually invaginates into the interior of the embryo (D), forming a gastrula-stage embryo.

FIGURE 41-7

Meroblastic cleavage is asymmetrical, with only a portion of a fertilized egg actively dividing to form a cell mass. The dividing cells of this fish embryo are magnified 400 times.

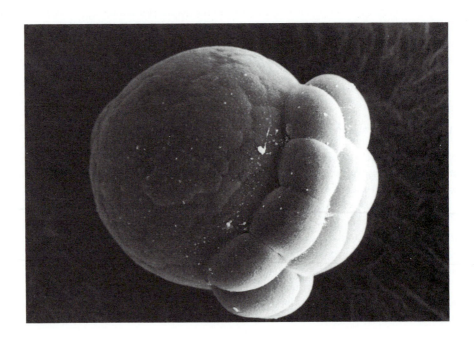

Mammals

Mammalian eggs are in many ways similar to the reptilian eggs from which they evolved, except that they contain very little yolk. Because there is no mass of yolk to impede cleavage in mammalian eggs, the cleavage of the developing zygote is holoblastic. Such cleavage forms a ball of cells surrounding a blastocoel. In mammalian eggs, however, there is an inner cell mass concentrated at one pole (Figure 41-8). This interior plate of cells is analogous to the blastodisc of reptiles. It goes on to form the developing embryo. The outer sphere of cells is called a **trophoblast.** It is analogous to the cells that form the membrane which functions as a watertight covering and conserves water within the reptilian egg. These cells have changed during the course of mammalian evolution to carry out a very different function. The trophoblast develops into a complex series of membranes, the placenta, which connects the developing embryo to the blood supply of the mother.

> **Development is initiated in the zygote by a series of rapid cell divisions called cleavage divisions, producing a ball of cells called a blastula. The evolution of the amniotic egg in reptiles caused an alteration in the pattern of cleavage, related to the presence of a yolk. This kind of cleavage pattern is carried on by mammals, a reflection of their ancestry.**

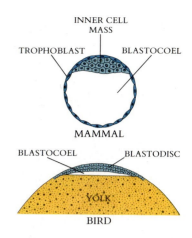

FIGURE 41-8

A mammalian blastula is composed of a sphere of cells, the trophoblast, surrounding a cavity, the blastocoel, and an inner cell mass. An avian (bird) blastula is a disk rather than a sphere, a blastodisc resting astride a large yolk mass; in bird eggs the blastocoel is a cavity between the blastodisc and the yolk.

The Blastula

Viewed from the outside, the blastula looks like a simple ball of cells that all resemble one another. The apparent close similarity of these cells is misleading, however. In fact, they differ from one another in three essential respects:

1. Each cell contains a different portion of cytoplasm derived from the egg.

2. Some cells are larger than others, containing more yolk and dividing more slowly.

3. Each cell is in contact with a different set of neighboring cells.

Egg cells contain many substances that act as genetic signals during the early stages of zygote development. These signal substances are not distributed uniformly within the egg cytoplasm, however. Instead, they are clustered at specific sites within the egg. The location of each site is genetically determined by information encoded on the mother's chromosomes. When the egg is activated during fertilization, its cytoplasm reorients itself with respect to the site of sperm entry. During the cleavage divisions that follow, the signal substances within this cytoplasm are partitioned into different daughter cells. The signals endow the different daughter cells with distinct developmental instructions. The egg is therefore **prepatterned,** since the pattern of its cytoplasm determines the future orientation of the different embryonic cells.

THE ONSET OF DEVELOPMENTAL CHANGE: GASTRULATION

The first visible results of prepatterning and of the cell orientation within the blastula can be seen immediately after the completion of the cleavage divisions. Certain groups of cells *move* inward from the surface of the sphere in a carefully orchestrated migration called **gastrulation.** How can cells move within a cell mass? Recall from Chapter 7 that cell shape can readily be changed by microfilament contraction. Apparently the migrating cells creep over the stationary ones by means of a series of microfilament contractions. The migrating cells move as a single mass because they adhere to one another. How do cells "know" to which other cell to adhere? Within an adhering cell, genes have been expressed that bring about the synthesis of particular polysaccharides on the cell surface which adhere to similar polysaccharides on the surfaces of the other adhering cells.

Alternative Patterns of Gastrulation

During gastrulation about half of the cells of the blastula move into the interior of the hollow ball of cells. By doing so, they form a structure that looks something like an indented tennis ball. Just as the pattern of cleavage divisions in different groups of vertebrates depends heavily on the amount and distribution of yolk in the egg, so the pattern of gastrulation varies among the vertebrates, depending on the shape of the blastulas produced by the earlier cleavage divisions.

Aquatic vertebrates

In fishes and other aquatic vertebrates with asymmetrical yolk distribution in their eggs, the blastula produced by the cleavage divisions has two distinct poles, one more rich in yolk than the other. The hemisphere of the blastula that comprises cells rich in yolk is called the **vegetal pole;** the opposite hemisphere, comprising cells that are relatively poor in yolk, is called the **animal pole.** In primitive chordates such as lancelets, the animal hemisphere bulges inward, invaginating into the blastocoel cavity. Eventually the inward-moving wall of cells pushes up against the opposite side of the blastula, and then it ceases to move. The resulting two-layered, cup-shaped embryo is the **gastrula.** The hollow crater resulting from the invagination is called the **archenteron,** and it becomes the progenitor of the gut. The opening of the archenteron, the future anus of the lancelets, is the **blastopore.**

Gastrulation in the lancelets produces an embryo with two cell layers, an outer ectoderm and an inner endoderm. A third cell layer, the mesoderm, forms soon afterward between these two layers from pouches pinched off of the endoderm. The formation of these three primary cell types sets the stage for all subsequent tissue and organ differentiation, since the descendants of each cell type are destined to have very different developmental fates:

Ectoderm:	Skin, central nervous system, sense organs, neural crest
Mesoderm:	Skeleton, muscles, blood vessels, heart, gonads
Endoderm:	Digestive tract, lungs, many glands

Amphibians

In the blastula of amphibians the yolk-laden cells of the vegetal pole are fewer and far larger than the yolk-free cells of the animal pole. Because of this distribution of cells, it is mechanically not feasible to invaginate the blastula at the vegetal pole. Instead, a layer of cells from the animal pole folds down over the yolk-rich cells and then invaginates inward (Figure 41-9). The place where the invagination begins is called the **dorsal lip.** As in the lancelets, the invaginating cell layer eventually eliminates the blastocoel cavity, its cell pressing against the inner surface of the opposite side of the embryo. In both fishes and amphibians, the cavity produced by the invagination is called the archenteron, and its opening is called the blastopore. In this case the blastopore is filled with yolk-rich cells, the **yolk plug.** The outer layer of cells in the gastrula, which is formed as a result of these cell movements, is the ectoderm, and the inner layer is the endoderm. Cells from the dorsal lip migrate between these two cell layers, forming the mesoderm layer.

Reptiles, birds, and mammals

In the blastodisc of a chick the developing embryo is not shaped like a sphere. Instead, it is a hollow cap of cells situated over the animal pole of the large yolk mass. Despite this seemingly great difference between the developing embryos of birds, reptiles, and mammals on the one hand, and amphibians on the other hand, the pattern of establishment of the three primary cell layers is basically similar in all of these groups.

There is no yolk separating the two sides of the blastodisc in reptiles, birds, and mammals (Figure 41-10). Consequently, the lower cell layer is able to differentiate without cell movement into endoderm, the upper layer into ectoderm. Just after this differentiation, the mesoderm layer arises by the invagination of cells from the upper layer inward, along the

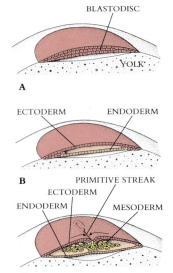

FIGURE 41-10

A *and* **B** *Gastrulation of the chick blastodisc. The upper layer of the blastodisc differentiates into ectoderm, the lower layer into endoderm.*

C *Among the cells that migrate into the interior through the dorsal primitive streak are future mesodermal cells.*

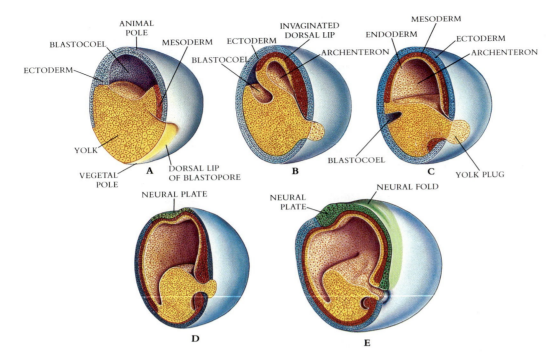

FIGURE 41-9

Frog gastrulation.
A *A layer of cells from the animal pole folds down over the yolk cells, forming the dorsal lip.*
B *The dorsal lip zone then invaginates into the hollow interior, or blastocoel, eventually pressing against the far wall. The three principal tissues (ectoderm, mesoderm, endoderm) become distinguished here.*
C *The inward movement of the dorsal lip creates a new internal cavity, the archenteron, which opens to the outside through the plug of yolk remaining at the point of invagination.*
D *The neural plate later forms from ectoderm and folds down over the surface of the embryo,* **(E)** *before moving to the interior.*

edges of a furrow that appears at the longitudinal midline of the embryo. The site of this invagination, which is analogous to an elongate blastopore, appears as a slit on the surface of the gastrula. Because of its appearance, it is called the **primitive streak.** Gastrulation occurs at the site of formation of a primitive streak in the reptiles and their descendants, the birds and mammals (Figure 41-11).

> **The many cells of the blastula gain unequal portions of egg cytoplasm during cleavage. This asymmetry results in the activation of different genes and a repositioning of cells with respect to one another, which establishes the three primary cell types: ectoderm, mesoderm, and endoderm.**

The events of gastrulation determine the basic developmental pattern of the vertebrate embryo. By the end of gastrulation, the distribution of cells into the three primary cell types has been completed. Although the position of the yolk mass dictates changes in the details of gastrulation, the end result of the process is fundamentally the same in all deuterostomes: the ectoderm is destined to form the epidermis and neural tissue; the mesoderm to form the connective tissue, muscle, and vascular elements; and the endoderm the lining of the gut and its derivatives.

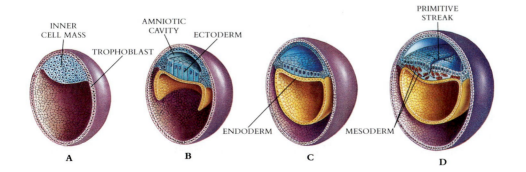

FIGURE 41-11

Mammalian gastrulation. The amniotic cavity forms within the inner cell mass **(A),** *and at its base, layers of ectoderm and endoderm differentiate* **(B** *and* **C),** *as in the chick blastodisc. A primitive streak develops, through which cells destined to become mesoderm migrate into the interior* **(D),** *again reminiscent of gastrulation in the chick.*

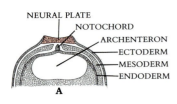

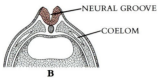

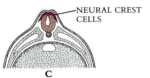

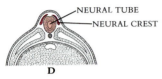

FIGURE 41-12

Neural tube formation. The neural tube forms above the notochord (A) when cells of the neural plate fold together to form the neural groove (B), which eventually closes (C) to form a hollow tube (D). As this is happening, some of the cells from the dorsal margin of the neural tube differentiate into the neural crest, which is characteristic of vertebrates.

THE DETERMINATION OF BODY ARCHITECTURE: NEURULATION

In the next step in vertebrate development the three primary cell types begin their development into the tissues and organs of the body. In all chordates the process of tissue differentiation begins with the formation of two characteristic morphological features, the notochord and the hollow dorsal nerve cord. This stage in development, called **neurulation,** occurs only in the chordates.

The first of these two structures to form is the notochord (see Figure 38-19). It is first visible soon after gastrulation is complete, forming from mesoderm tissue along the midline of the embryo, below its dorsal surface.

After the notochord has been laid down, the dorsal nerve cord forms. The region of the ectoderm that is located above the notochord later differentiates into the spinal cord and brain. The process is illustrated in Figure 41-12. First, a layer of ectodermal cells situated above the notochord invaginates inward, forming a long groove called the **neural groove** along the long axis of the embryo. The edges of this groove then move toward each other and fuse, creating a long hollow tube, the **neural tube,** which runs beneath the surface of the embryo's back.

> The key developmental event that marks the evolution of the chordates is neurulation, the elaboration of a notochord and a dorsal nerve cord.

While the neural tube is forming from ectoderm, the rest of the basic architecture of the body is being determined rapidly by changes in the mesoderm. On either side of the developing notochord, segmented blocks of tissue form (see Figure 38-14). Ultimately, these blocks, or **somites,** give rise to the muscles, the vertebrae, and the connective tissue. As the process of development continues, more somites are formed progressively. Many of the significant glands of the body, including the kidneys, adrenal glands, and gonads, develop within another strip of mesoderm that runs alongside the somites. The remainder of the mesoderm layer moves out and around the inner endoderm layer of cells and eventually surrounds it entirely. As a result of this movement, the mesoderm forms a hollow tube within the ectoderm. The space within this tube is the coelom; it contains the endoderm layers that ultimately form the lining of the stomach and gut.

EVOLUTIONARY ORIGIN OF THE VERTEBRATES: THE NEURAL CREST

Neurulation occurs in all chordates; the process is much the same in a lancelet as it is in a human being. The next stage in the development of a vertebrate, however, is unique to that group and is largely responsible for the characteristic body architecture of its members. Just before the neural groove closes over to form the neural tube, the edges of the neural groove develop a special strip of cells, the **neural crest,** that becomes incorporated into the roof of the neural tube (see Figure 41-12, *C* and *D*). Subsequently, the cells of the neural crest shift laterally to the sides of the developing embryo. The appearance of the neural crest was a key event in the evolution of the vertebrates, because neural crest cells, migrating to different parts of the embryo, ultimately develop into the structures that are characteristic of the vertebrate body.

The various cells of the neural crest strip develop very differently, depending on their location. At the anterior end of the embryo they merge with the anterior portion of the brain, the forebrain. Nearby clusters of ectoderm cells associated with the neural crest cells thicken into the **placodes,** structures that subsequently develop into parts of the sense organs of the head. The neural crest and associated placodes consist of two lateral strips, a pattern responsible for the fact that the sense organs of vertebrates come in pairs.

The remaining cells of the neural crest, located behind the anterior ones, have a very different developmental fate. These cells migrate away from the nerve tube to other locations in the head and trunk where they form connections between the nerve tube and the surrounding tissues. At these new locations they dictate the development of a series of different structures, discussed below, that are particularly characteristic of the vertebrates. This migration of neural crest cells is a unique process, in that it is not simply a change in the relative position of the cells, such as is seen in gastrulation. Instead, the migrating neural crest cells actually pass through other tissues.

Structures Derived from the Neural Crest

Many of the differences between vertebrates and the more primitive chordates from which they evolved are related to the products of the neural crest. The characteristic structures that are derived from neural crest cells or from placodes induced by the anterior neural crest include the following body elements.

The gill chamber

Primitive chordates such as lancelets were filter-feeders, using the rapid beating of cilia to draw water into their bodies through slits in their pharynx. These pharyngeal slits have become elaborated greatly in the course of vertebrate evolution, forming the **gill chamber,** a structure that provides a greatly improved means of respiration (Figure 41-13). The evolution of the gill chamber was certainly a key event in making possible the transition from filter-feeding to active predation.

In the development of the gill chamber, some of the neural crest cells form cartilaginous bars between the embryonic pharyngeal slits. Other neural crest cells induce portions of the mesoderm to form muscles along the cartilage. Still others form nerves between the nerve cord and these muscles. A major blood vessel, called an **aortic arch,** passes through each of the bars. Lined by still more neural crest cells, these bars, with their internal blood supply, become highly branched and form the gills.

Because the stiff bars of the gill chamber can be bent inward by powerful muscles controlled by nerves, the whole structure is a very efficient pump, which serves to drive water past the gills. The gills themselves act as highly efficient oxygen exchangers, greatly increasing the respiratory capacity of the organism. The structure and operation of the gill chamber are described in more detail in Chapter 44.

 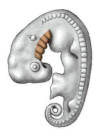 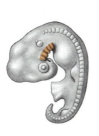 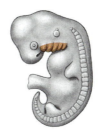

FISH   REPTILE   BIRD   HUMAN

FIGURE 41-13

The stiff bars of the gill chamber (in color) are readily visible in a fish embryo—and in the embryos of the major vertebrate groups that evolved from fishes.

Elaboration of the nervous system

Some neural crest cells migrate downward toward the notochord and form sensory ganglia. Others become specialized as Schwann cells, which insulate nerve fibers and thus permit the rapid conduction of nerve impulses. Most vertebrate motor neurons are derived from neural crest cells. Other neural crest cells form the **adrenal medulla,** the key element of the sympathetic nervous system. The adrenal medulla produces adrenaline at times of stress or danger, allowing the animal to respond. Together, these structures make possible

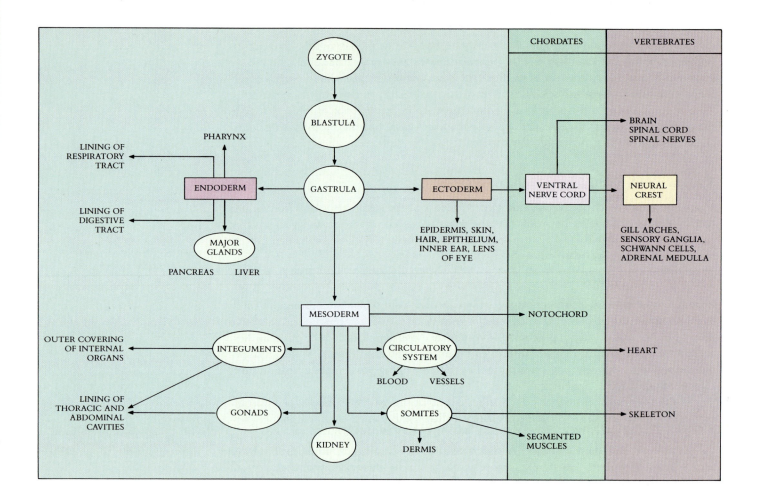

FIGURE 41-14

Derivation of the major tissue types. Many of these tissues are discussed in more detail in Chapter 42. The key role of the neural crest is evident from the many characteristically vertebrate features that derive from it.

a great improvement in the ability of the nervous system to respond to sensory information precisely and quickly.

Skull and sensory organs

A variety of sense organs develop from the placodes, which are formed from the cells of the anterior neural crest. Included among them are the olfactory (smell) and lateral line (primitive hearing) organs, which are discussed in Chapter 46. The teeth develop from neural crest cells, as do the cranial bones that protect the brain, which is the enlarged anterior end of the nerve cord.

The Role of the Neural Crest in Vertebrate Evolution

The adaptations just described, and still others that are associated with neural crest cells, are thought to have played a key role in the evolution of the vertebrates (Figure 41-14). The primitive chordates were initially slow-moving, filter-feeding animals with relatively low metabolic rates. Adaptations derived from neural crest allowed the vertebrates to assume a very different ecological role. The vertebrates became fast-swimming predators with much higher metabolic rates. This metabolic pattern allowed a much higher level of activity than was possible for the vertebrate ancestors. Other evolutionary changes associated with the derivatives of the neural crest made possible improved detection of prey, a better ability to orient spatially during the capture of prey, and a greatly improved ability to respond quickly to sensory information. The evolution of the neural crest cells and of the structures derived from them was therefore a crucial step in the evolution of the vertebrates.

> The appearance of the neural crest in the developing embryo marked the beginning of the first truly vertebrate phase of development, since many of the structures that are characteristic of vertebrates are derived directly or indirectly from neural crest cells.

HOW CELLS COMMUNICATE DURING DEVELOPMENT

In the process of vertebrate development (Figure 41-15), the relative position of particular cell layers determines, to a large extent, the organs that develop from them. By now, you may have wondered how these cell layers know where they are. For example, when cells of the ectoderm that are situated above the developing notochord give rise to the neural groove, how do these cells know they are above the notochord?

The solution to this puzzle is one of the outstanding accomplishments of **experimental embryology,** the study of how embryos form. It was worked out by the great German biologist Hans Spemann and his student Hilde Mangold early in this century. In their investigations of this question Spemann and Mangold removed cells from the dorsal lip of an amphibian blastula (Figure 41-16) and transplanted them to a different location on another blastula. The dorsal lip region of amphibian blastulas develops from the gray crescent zone and is the site of origin of those mesoderm cells which later produce the notochord. The new location corresponded to that of the future belly of the animal. What happened?

FIGURE 41-15

Development in the chick.
A *After 24 hours development of the neural groove is well advanced.*
B *After 36 hours the embryo has grown much larger.*
C *After 72 hours (3 days) the eyes have already formed. In this photograph the shell has been removed to show the extensive blood circulation around the developing embryo.*
D *After 7 days most of the body's internal organs are present.*
E *After 21 days the chick pecks its way out of the egg. Note that it hasn't much room to maneuver.*
F *This pensive chick is 1 day old. Although not yet an adult, it is fully able to live on its own.*

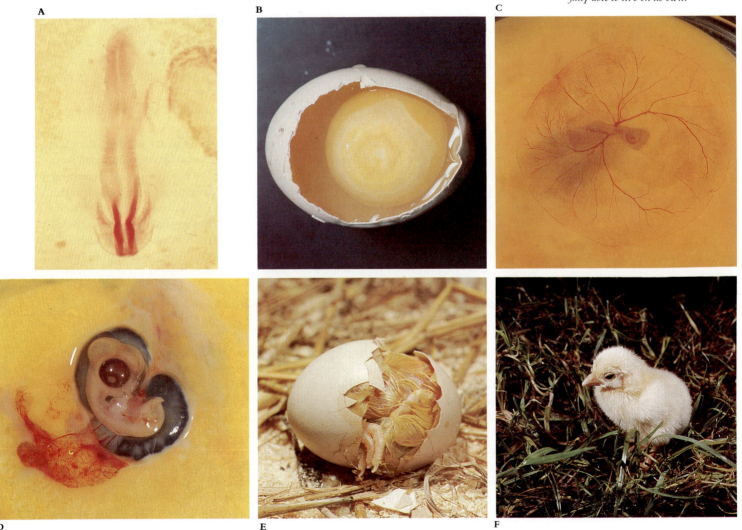

A B C

D E F

FIGURE 41-16

Spemann and Mangold's dorsal lip transplant experiment.

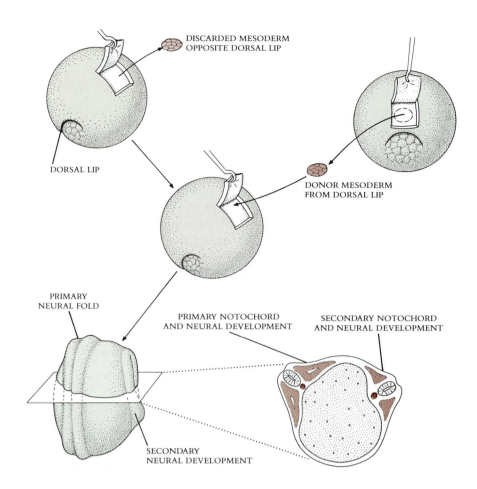

The embryo developed *two* notochords, one normal dorsal one and a second one along its belly!

By using genetically different donor and host blastulas, Spemann and Mangold were able to show that the notochord produced by transplanting dorsal lip cells contained host cells as well as transplanted ones. The transplanted dorsal lip cells had acted as **organizers** of notochord development. As such, these cells stimulated a developmental program in the belly cells of embryos to which they were transplanted: the development of a notochord. The belly cells clearly contained this developmental program but would not have expressed it in the normal course of their development. The transplantation of the dorsal lip cells caused them to do so. These cells had *induced* the ectoderm cells of the belly to form a notochord. This phenomenon as a whole is known as **induction.**

> **Induction is the determination of the course of development of one tissue by another tissue.**

The process of induction that Spemann discovered appears to be the basic mode of development in vertebrates. Inductions between the three primary tissue types—ectoderm, mesoderm, and endoderm—are referred to as **primary inductions.** Inductions between tissues that have already been differentiated are called **secondary inductions.** The differentiation of the central nervous system during neurulation by the interaction of dorsal ectoderm and dorsal mesoderm to form the neural tube is an example of primary induction. In contrast, the differentiation of the lens of the vertebrate eye from ectoderm by interaction with tissue from the central nervous system is an example of secondary induction.

The eye develops as an extension of the forebrain, a stalk that grows outward until it comes in contact with the epidermis (Figure 41-17). At a point directly above the growing stalk, a layer of the epidermis pinches off, forming a transparent lens. When the optic stalks

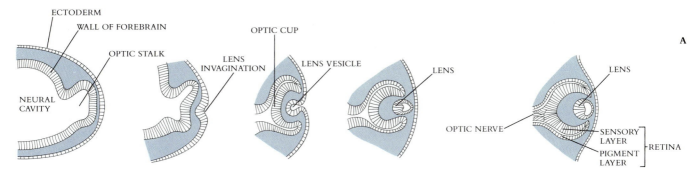

ECTODERM
WALL OF FOREBRAIN
OPTIC STALK
OPTIC CUP
LENS INVAGINATION
LENS VESICLE
LENS
NEURAL CAVITY
LENS
OPTIC NERVE
SENSORY LAYER
PIGMENT LAYER
RETINA

A

FIGURE 41-17

Development of the vertebrate eye proceeds by induction.
A *The eye develops as an extension of the forebrain called the optic stalk that grows out until it contacts the ectoderm. This contact induces the formation of a lens from the ectodermal tissue.*
B *Scanning electron micrograph of a developing vertebrate eye in cross section.*

of the two eyes have just started to project from the brain and the lenses have not yet formed, one can remove one of the budding stalks and transplant it to a region underneath a different epidermis, such as that of the belly. When this critical experiment was performed by Spemann, a lens still formed, this time from belly epidermis cells in the region above where the budding stalk had been transplanted.

The chemical nature of the induction process is not known in detail. If one interposes a nonporous barrier, such as a layer of cellophane, between the inducer and the target tissue, no induction takes place. A porous filter, in contrast, does permit induction to occur. It is thought that the inducer cells produce a protein factor that binds to the cells of the target tissue, stimulating mitosis in them and initiating changes in gene expression.

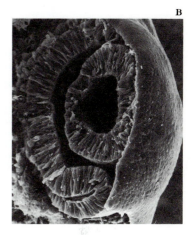

B

THE NATURE OF DEVELOPMENTAL DECISIONS

All of the cells of the body, with the exception of a few specialized ones that have lost their nuclei, have a full complement of genetic information. Despite the fact that all of its cells are genetically identical, an adult vertebrate contains hundreds of different cell types, each expressing different aspects of the total genetic information. What factors determine which genes are expressed in a particular cell and which are not? In a liver cell, what mechanism keeps the genetic information that specifies nerve cell characteristics turned off? Does the differentiation of that particular cell into a liver cell entail the physical loss of the information specifying other cell types? No. But cells progressively lose the capacity to *express* ever-larger portions of their genomes. Development is a process of progressive restriction of gene expression.

Some cells become **determined** very early. For example, all of the egg cells of a human female are set aside very early in the life of the embryo, yet some of these cells will not achieve differentiation into functional oocytes for more than 40 years. To a large degree, the fate of a particular cell is determined by the place at which it is located in the developing embryo. By changing a cell's location, an experimenter can alter its developmental destiny. However, this is only true up to a certain point in a cell's development. At some stage, the ultimate fate of every cell becomes fixed and irreversible, a process referred to as **commitment**.

> When a cell is "determined," it is possible to predict its developmental fate; when a cell is "committed," that developmental fate cannot be altered. Determination often occurs very early in development, commitment somewhat later.

THE EVOLUTION OF DEVELOPMENT

The patterns of development that occur in the vertebrate groups that evolved most recently reflect in many ways the simpler patterns that occur among earlier forms. Thus mammalian development and bird development are elaborations of reptile development, which is an elaboration of amphibian development, and so forth. During the development of a mammalian embryo, traces can be seen of appendages and organs that are apparently relicts

DEVELOPMENT GONE WRONG: DOWN SYNDROME

Vertebrate development is one of the most complex of biological processes. Successful development is the product of a long series of carefully orchestrated gene-directed processes, and expression of even one gene at an inappropriate time or to too great a degree can have severe effects. One of the most important developmental defects to occur in vertebrates is **Down syndrome** (formerly called "Down's syndrome"), named for J. Langdon Down, who first described the symptoms in 1866. In human children with Down syndrome, maturation of the skeletal system is delayed, leading to short stature and poor muscle tone. Mental development is also affected—children with Down syndrome invariably are mentally retarded.

Down syndrome occurs frequently in all human racial groups (about 1 in every 750 children), and very similar conditions occur in chimpanzees and other primates. In humans the defect is associated with a particular small portion of chromosome 21. When this chromosomal segment is present in three copies instead of two, Down syndrome results. In 97% of the human cases examined, all of chromosome 21 is present in three copies; in the other 3% a small portion containing the critical segment has been added to another chromosome, in addition to the two chromosome 21 copies (Figure 41-A, *1*).

The developmental role of the genes whose duplication produces Down syndrome is not known in detail, although clues are beginning to emerge from current research. The gene or genes that produce Down syndrome appear to be closely related to, if not the same as, genes associated with cancer. When human cancer-causing genes, identified by transfection techniques as described in Chapter 18, were localized on the human chromosomes in 1985, one of them turned out to be located on chromosome 21, at precisely the location of the segment associated with Down syndrome. Is cancer more common in children with Down syndrome? Yes. The incidence of leukemia, for example, is 11 times higher in children with Down syndrome than in age-matched children who do not have it.

---

of more primitive chordates. For example, there is a time when embryonic human beings possess pharyngeal slits, which occur in all chordates and are homologous to the gill slits of fishes. At a later stage in the course of their development human embryos also exhibit a tail (see Figure 38-14).

In a sense the patterns of development in given chordate groups were built up over the evolutionary history of those groups in incremental steps. Speaking teleologically (that is, as if "nature" were a person), one might even say that nature first learned how to build a fish, then an amphibian, then a reptile, and then a mammal. The developmental instructions for each new form seem to have been layered on top of the previous instructions, becoming additional steps in the developmental journey. Such a hypothesis, first proposed in the nineteenth century by Ernst Haeckel, is referred to as the "biogenic law." Haeckel's law is usually stated as an aphorism, that *ontogeny recapitulates phylogeny:* that is, the course of vertebrate development (ontogeny) involves the same progression of changes that have occurred during the course of vertebrate evolution (phylogeny). As stated, the phrase is not literally true. Embryonic stages are *not* reflections of adult ancestors. Embryonic stages of particular vertebrates do, however, often reflect the embryonic stages of those vertebrates' ancestors. The pharyngeal slits of mammalian embryos are not like the pharyngeal gill slits their ancestors had when they were adults, but rather like the pharyngeal slits they had when they were embryos.

> **Vertebrates seem to have evolved largely by the addition of new instructions to the developmental program. Development of a mammal thus proceeds through a series of stages, the earlier stages unchanged from those which occur in the development of more primitive vertebrates.**

Among Down syndrome children, 97% possess three copies of chromosome 21. The extra chromosome appears to result from primary nondisjunction (failure of chromosomes to separate in meiosis) in the female gamete, the egg. The cause of these primary nondisjunctions is not known, but like cancer, the incidence increases with age (Figure 41-A, *2*). In mothers under 20 years of age, the incidence of Down syndrome is only about 1 per 1700 births; in mothers 20 to 30 years old, the risk is only slightly greater, about 1 per 1400. In mothers 30 to 35 years old, however, the risk doubles, to 1 per 750. In mothers over 45 years of age the risk is as high as 1 in 16 births.

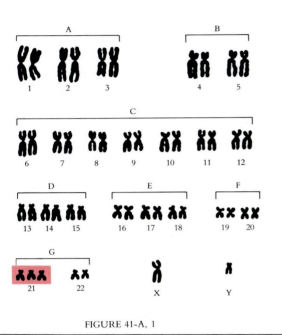

FIGURE 41-A, 1

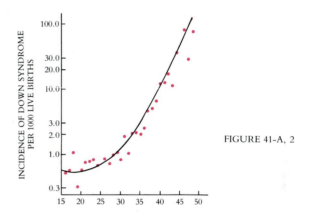

FIGURE 41-A, 2

THE COURSE OF HUMAN DEVELOPMENT

The development of the human embryo shows its evolutionary origins. If we did not have an evolutionary perspective, we would be unable to account for the fact that human development proceeds in much the same way as development in the chick. In both embryonic chickens and embryonic human beings, the blastodisc is flattened. In a chick egg, for example, the blastodisc is pressed against a yolk mass; in a human embryo the blastodisc is similarly flat despite the absence of a yolk mass. In human blastodiscs a primitive streak forms and gives rise to the three primary cell types, just as it does in the chick blastodisc.

Human development takes much longer than chicken development, an average of 266 days from fertilization to birth, the familiar "9 months of pregnancy." What may not be so readily apparent, however, is how very early the critical stages of development outlined in this chapter occur during the course of human pregnancy.

First Trimester
The first month

In the first week after fertilization occurs, the fertilized egg undergoes cleavage divisions. The first of these divisions occurs about 30 hours after the fusion of the egg and the sperm, and the second 30 hours later. Cell divisions continue until a blastodisc forms within a ball of cells, the trophoblast. During this period the embryo continues the journey down the mother's oviduct, a journey that the egg initiated. On about the sixth day the embryo reaches the uterus, attaches to the uterine lining, or **endometrium,** and penetrates into the tissue of the lining. The trophoblast begins to grow rapidly, initiating the formation of

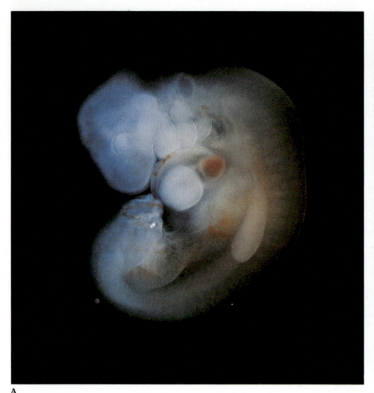

A

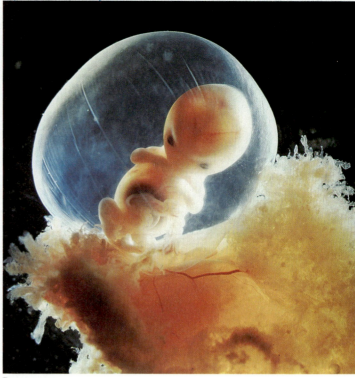

B

FIGURE 41-18

The developing human.
A *4 weeks.*
B *7 weeks.*
C *3 months.*
D *4 months.*

membranes. One of these membranes, the **amnion,** will enclose the developing embryo, while another, the **chorion,** will interact with uterine tissue to form the placenta that will nourish the growing embryo.

In the second week after fertilization, gastrulation takes place. The primitive streak can be seen on the surface of the embryo, and the three primary cell types are differentiated. Around the developing embryo the placenta starts to form from the chorion.

In the third week neurulation occurs. It is marked by the formation of the neural tube along the dorsal axis of the embryo, as well as by the appearance of the first somites, from which the muscles, vertebrae, and connective tissue develop. By the end of the week, over a dozen somites are evident, and the blood vessels and gut have begun to develop. At this point the embryo is about 2 millimeters long.

In the fourth week **organogenesis** (the formation of body organs) occurs (Figure 41-18, *A*). The eyes form, and the tubular heart begins to pulsate, develops four chambers, and begins a rhythmic beating that stops only with death. At 70 beats per minute, the little heart is destined to beat more than 2.5 billion times during a lifetime of 70 years, before it ceases. Over 30 pairs of somites are visible by the end of the fourth week, and the arm and leg buds have begun to form. The embryo more than doubles in length during this week, to about 5 millimeters.

All of the major organs of the body have begun their formation by the end of the fourth week of development. Although the developmental scenario is now far advanced, many women are not aware that they are pregnant at this stage.

Early pregnancy is a very critical time in development, since the proper course of events can be interrupted easily. In the 1960s, for example, many pregnant women took the tranquilizer **thalidomide** to minimize discomforts associated with early pregnancy. Unfortunately, this drug had not been adequately tested. It interferes with fetus limb bud development, and its widespread use resulted in many deformed babies. Also during the first and second months of pregnancy, contraction of rubella (German measles) by the mother can upset organogenesis in the developing embryo. Most spontaneous abortions occur in this period.

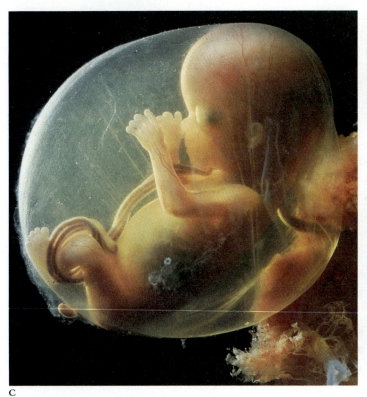

C

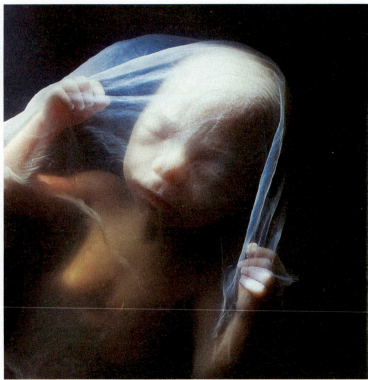

D

The second month

Morphogenesis (the formation of shape) takes place during the second month (Figure 41-18, *B*). The miniature limbs of the embryo assume their adult shapes. The arms, legs, knees, elbows, fingers, and toes can all be seen—as well as a short bony tail (see Figure 38-14)! The bones of the embryonic tail, an evolutionary reminder of our past, later fuse to form the **coccyx.** Within the body cavity, the major organs, including the liver, pancreas, and gallbladder, become evident. By the end of the second month, the embryo has grown to about 25 millimeters in length, weighs perhaps a gram, and begins to look distinctly human.

The third month

The nervous system and sense organs develop during the third month (Figure 41-18, *C*). By the end of the month, the arms and legs begin to move. The embryo begins to show facial expressions and carries out primitive reflexes such as the startle reflex and sucking. By the end of the third month, all of the major organs of the body have been established. Development of the embryo is essentially complete. From this point on, the developing human being is referred to as a **fetus** rather than an embryo. What remains is essentially growth.

Second Trimester

In the fourth and fifth months of pregnancy, the fetus grows to about 175 millimeters in length, with a body weight of about 225 grams. Bone enlargement occurs actively during the fourth month. During the fifth month the head and body become covered with fine hair. This downy body hair, called **lanugo,** is another evolutionary relict and is lost later in development. By the end of the fourth month, the mother can feel the baby kicking. By the end of the fifth month, she can hear its rapid heartbeat with a stethoscope. In the sixth month growth begins in earnest. By the end of that month, the baby weighs 0.6 kilogram (about 1½ pounds) and is over 0.3 meter (1 foot) long; but most of its prebirth growth is still to come. The baby cannot yet survive outside the uterus without special medical intervention.

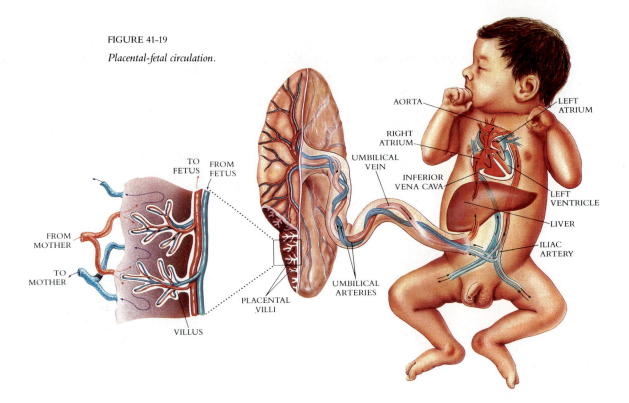

FIGURE 41-19

Placental-fetal circulation.

Third Trimester

The third trimester is predominantly a period of growth, rather than one of development. In the seventh, eighth, and ninth months of pregnancy the weight of the fetus doubles several times. This increase in bulk is not the only kind of growth that occurs, however. Most of the major nerve tracks are formed within the brain during this period, as are new brain cells. All of this growth is fueled by nutrients provided by the mother's bloodstream. Within the placenta these nutrients pass into the fetal blood supply (Figure 41-19). The undernourishment of the fetus by a malnourished mother can adversely affect this growth and result in severe retardation of the infant. Retardation resulting from fetal malnourishment is a severe problem in many underdeveloped countries where poverty is common—a tragedy discussed in Chapter 54.

By the end of the third trimester, the neurological growth of the fetus is far from complete, and, in fact, it continues long after birth. By this time, however, the fetus is able to exist on its own. Why doesn't development continue within the uterus until neurological development is complete? Because physical growth would continue as well, and the fetus is probably as large as it can get and still be delivered through the pelvis (Figure 41-20) without damage to mother or child. Birth takes place as soon as the probability of survival is high. For better or worse, the infant is then on its own, a person.

The critical stages of human development take place quite early. All the major organs of the body have been established by the end of the third month. The following 6 months are essentially a period of growth.

POSTNATAL DEVELOPMENT

Growth continues rapidly after birth. Babies typically double their birth weight within 2 months. Different organs grow at different rates, however. The reason that adult body proportions are different from infant ones is that different parts of the body grow at different rates or stop growing at different times (see boxed essay, pp. 480-481). The head, for example, which is disproportionately large in infants, grows more slowly in infants

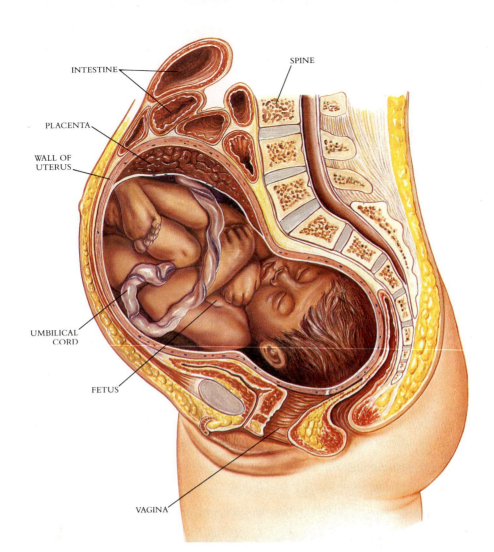

INTESTINE

SPINE

PLACENTA

WALL OF
UTERUS

UMBILICAL
CORD

FETUS

VAGINA

FIGURE 41-20

*A developing fetus is a major
addition to a woman's anatomy.
The stomach and intestines are
pushed far up, and there is often
considerable discomfort from
pressure on the lower back. In a
natural delivery the fetus exits
through the vagina, which must
dilate (expand) considerably to
permit passage.*

than does the rest of the body. Such a pattern of growth, in which different components grow at different rates, is referred to as allometric growth.

In most mammals, brain growth is entirely a fetal phenomenon. In chimpanzees, for example, the growth of the brain and that of the cerebral portion of the skull rapidly decelerate after birth, while the bones of the jaw continue to grow. The skull of an adult chimpanzee, therefore, looks very different from that of a fetal chimpanzee. In human beings, on the other hand, the brain and cerebral skull continue to grow at the same rate after birth as before. During gestation and after birth the developing human brain generates neurons (nerve cells) at an average rate estimated at more than 250,000 per minute; it is not until about 6 months after birth that this astonishing production of new neurons ceases permanently. Because both brain and jaw continue to grow, the jaw-skull proportions do not change after birth, and the skull of an adult human being looks very similar to that of a human fetus. It is primarily for this reason that a young human fetus seems so incredibly adult-like.

The bulk of the next eight chapters is devoted to describing what an adult vertebrate body is like and how it functions. We will describe the various tissues of the vertebrate body and then consider how vertebrates eat, breathe, and sense the world around them. All the complexity you will encounter, the beauty and cleverness of design, are established during the developmental process we have briefly considered in this chapter. Development is not simply the beginning of life—it is the process that determines life's form and substance. The following eight chapters describe how adult vertebrates play out the roles that development has written for them.

SUMMARY

1. Fertilization is the union of an egg and a sperm to form a zygote. Fertilization is external in fishes and amphibians, internal in all more advanced vertebrates. The three stages of fertilization are (1) penetration, in which the sperm cell moves past the cells surrounding the egg and penetrates the egg membrane; (2) activation, in which a series of cytoplasmic movements are initiated by penetration; and (3) fusion, in which the sperm and egg nuclei fuse.

2. Cleavage is the rapid division of the newly formed zygote into a mass of perhaps a thousand cells, without any increase in overall size. Because the egg is structured with respect to the location of developmentally important regulating signals, the future embryo becomes structured by the cleavage divisions. These divisions, in effect, partition the egg cytoplasm into small portions that contain different regulatory elements.

3. Gastrulation is the mechanical movement of portions of the blastosphere, forming the three basic cell types: ectoderm, endoderm, and mesoderm. In eggs that lack a yolk the movement is one of simple invagination. When a yolk is present, the movement of the cells is affected by it. In amphibians the cell layers move down and around the yolk. In reptiles, birds, and mammals, cells establish the three primary cell types as an upper layer (ectoderm), a lower layer (endoderm), and a layer that invaginates inward from the upper layer (mesoderm).

4. Neurulation in chordates is the formation of the first tissues, particularly the notochord and the dorsal nerve cord, from the primary cell types.

5. The formation of the neural crest is the first development event that is unique to vertebrates. Most of the distinctive structures that are associated with vertebrates are derived from cells of the neural crest.

6. Cells influence one another during development by a process of induction. In this process, substances that exist on the surface of one cell induce other cells to divide.

7. At some point during animal development the ultimate developmental fate of cells becomes fixed and unalterable. The cells are then said to be committed, even though they may not exhibit any of the characteristics they will eventually assume.

8. Most of the critical events in the development of a human occur in the first month. Cleavage occurs during the first week, gastrulation during the second week, neurulation during the third week, and organ formation during the fourth week.

9. The second and third months of the first trimester are devoted to morphogenesis and to the elaboration of the nervous system and sensory organs. By the end of this period, the development of the embryo is essentially complete.

10. The last 6 months of human pregnancy are essentially a period of growth, devoted to increase in size and to formation of nerve tracks within the brain. Most of the weight of a fetus is added in the final 3 months of pregnancy.

SELF-QUIZ

The following questions pinpoint important information in this chapter. You should be sure to know the answer to each of them.

1. Which of the following does not occur when the sperm penetrates the egg?
   (a) The membrane of the egg changes, preventing more sperm from entering.
   (b) The egg chromosomes divide, producing a polar body.
   (c) Rearrangement of the egg cytoplasm occurs.
   (d) The egg constructs an acrosome, necessary for further division.

2. The second major event in vertebrate reproduction is the rapid division of the zygote into a larger number of smaller cells, a process called _____.

3. When cleavage occurs through the whole egg as it does in amphibians and advanced fishes, the egg is undergoing _____ cleavage.
   (a) holoblastic   (b) meroblastic

4. Which of the following develops into the placenta in mammals?
   (a) blastodisc   (c) trophoblast
   (b) blastocoel   (d) chorion

5. The hemisphere of the blastoderm that comprises the cells that contain more yolk is called the animal pole. True or false?

6. Which cell layer eventually evolves into the skeleton, muscles, blood vessels, heart, and gonads?
   (a) ectoderm   (c) endoderm
   (b) mesoderm

7. The notochord and the hollow dorsal nerve cord form during the stage of development known as
   (a) gastrulation   (d) organogenesis
   (b) neurulation   (e) syngamy
   (c) neural crest formation

8. The cells that become incorporated into the roof of the neural tube are collectively known as the _____.

9. A variety of sense organs develop from the placodes, which are formed from the cells of the anterior neural crest. True or false?

10. When a cell directs other cells to form a structure (as is the case with the notochord), the process is referred to as _____.

11. When a cell is differentiated into a liver cell, it physically loses the information required to specify other types of cells. True or false?

12. For each of the following five events of human development, indicate the time period during which it occurs.
    (1) development of lanugo   (a) first trimester
    (2) neurulation   (b) second trimester
    (3) allometric growth   (c) third trimester
    (4) bone formation   (d) postnatal
    (5) organogenesis

THOUGHT QUESTIONS

1. In reptiles and birds the fetus is basically masculine, and fetal estrogen hormones are necessary to induce the development of female characteristics. In mammals the reverse is true, the fetus being basically female, with fetal hormones acting to induce the development of male characteristics. Can you suggest a reason why the pattern that occurs in reptiles and birds would not work in mammals?

2. Female armadillos always give birth to four offspring of the same sex. Can you suggest a mechanism that would account for this?

FOR FURTHER READING

BROWDER, L.: *Developmental Biology,* 2nd ed., Saunders College Publishing, Philadelphia, 1984. An exceptionally well-illustrated undergraduate text, with many striking photographs.

CALOW, P.: *Life Cycles: An Evolutionary Approach to the Physiology of Reproduction, Development, and Aging,* Chapman & Hall, London, 1978. An interesting treatment that emphasizes ecological issues.

DALE, B.: *Fertilization in Animals,* Edward Arnold, Publishers, London, 1983. A brief text devoted entirely to the process of animal fertilization, with good illustrations and up-to-date discussions of physical mechanisms.

DRYDEN, R.: *Before Birth,* Heinemann Educational Books, Inc., London, 1978. A detailed description of human development.

GANS, C., and R.G. NORTHCUTT: "Neural Crest and the Origin of Vertebrates: A New Head," *Science,* vol. 220, pages 268-274, 1983.

GOULD, S.J.: "Dr. Down's Syndrome," *Natural History,* vol. 89, pages 142-148, 1980. An account of the history of "mongolism," a relatively common chromosomal abnormality that results in severe mental retardation.

NILSSON, L., and J. LINDBERG: *Behold Man,* Little, Brown & Co., Boston, 1974. A wonderful collection of color photographs of the developing human fetus.

RUGH, R., and L.B. SHETTLES: *From Conception to Birth: The Drama of Life's Beginnings,* Harper & Row, Publishers, New York, 1971. A detailed treatment of human development, from fertilization to birth, with excellent photographs.

SAUNDERS, J.: *Developmental Biology,* Macmillan Publishing Co., New York, 1982. A very clear undergraduate text, easily understood by students with little background. A particularly strong point of this text is its emphasis on how experiments have established the basic information.

TRINKAUS, J.: *Cells into Organs: The Forces that Shape the Embryo,* 2nd ed., Prentice-Hall, Inc., Englewood Cliffs, N.J., 1984. An advanced but easily understood discussion of the physical mechanisms underlying developmental change.

WESSELLS, N.K.: *Tissue Interactions and Development,* The Benjamin-Cummings Publishing Co., Menlo Park, Calif., 1977. A brief and lucid account of the experiments that led to our current understanding of the mechanisms of development.

ORGANIZATION OF THE VERTEBRATE BODY

Chapter 42

Overview

The three primary cell types determined early during the development of vertebrates—ectoderm, mesoderm, and endoderm—differentiate during the course of later developmental stages into over 200 kinds of cells. These cells make up the four major tissue types: epithelium, connective tissue, muscle, and nerve. Each of these tissue types may be recognized by its structure, function, and origin. The organs of the body often are composed of several different tissue types.

For Review

Here are some important terms and concepts that you will encounter in this chapter. If you are not familiar with them, you should review them before proceeding.

- **Microfilament contraction** (Chapter 7)

- **The sodium-potassium pump** (Chapter 9)

- **Variability of immunoglobulin genes** (Chapter 17)

- **Pseudocoel** (Chapter 35)

At the end of its long developmental journey, a newborn vertebrate confronts the world as a very sophisticated apparatus, designed and refined by many years of evolution to survive and reproduce in a variety of environments. By birth its three primary cell types have differentiated into many different kinds of cells with a bewildering variety of form and function.

The bodies of adult vertebrates contain between 50 and several hundred different kinds of cells, depending upon which vertebrate is being considered and how finely one differentiates between cell types. These diverse kinds of cells are traditionally grouped into four types of tissues: epithelium, connective tissue, muscle, and nerve (Figure 42-1). Tissues, as we saw in Chapter 36, are groups of similar cells organized into structural and functional units.

FIGURE 42-1

Vertebrate tissue types.

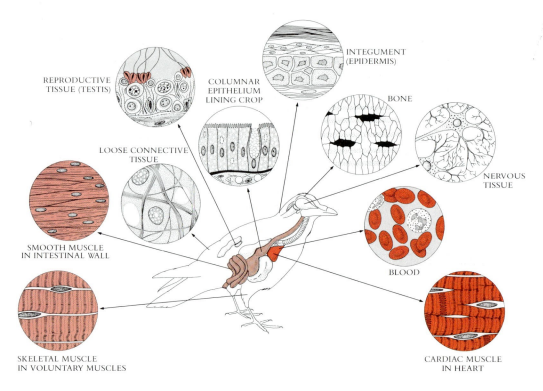

REPRODUCTIVE
TISSUE (TESTIS)

COLUMNAR
EPITHELIUM
LINING CROP

INTEGUMENT
(EPIDERMIS)

BONE

NERVOUS
TISSUE

LOOSE CONNECTIVE
TISSUE

SMOOTH MUSCLE
IN INTESTINAL WALL

BLOOD

SKELETAL MUSCLE
IN VOLUNTARY MUSCLES

CARDIAC MUSCLE
IN HEART

EPITHELIUM

Epithelial cells are the guards and protectors of the body. They cover its surface and determine which substances enter it and which do not. The organization of the vertebrate body is fundamentally tubular, with ectodermal cells covering the outside (skin), endodermal cells lining the hollow inner core (alimentary canal and gut), and mesodermal cells lining the body cavity (coelom). All of these kinds of **epidermal** or "skin" cells are broadly similar in form and function, and they are collectively called the **epithelium.** The epithelial layers of the body function in four different ways:

1. They protect the tissues beneath them from dehydration and mechanical damage, a particularly important function in land-dwelling vertebrates.
2. They provide a selectively permeable barrier that can facilitate or impede the passage of materials into the tissue beneath. Because epithelium encases all of the body's surfaces, every substance that enters or leaves the body must cross an epithelial layer.
3. They provide sensory surfaces. Many sensory nerves end in epithelial cell layers.
4. They secrete materials. Most secretory glands are derived from invaginations of layers of epithelial cells that occur during development.

Layers of epithelial tissue are usually only one or a few cell layers thick. Individual epithelial cells possess a small amount of cytoplasm and have a relatively low metabolic rate. Few blood vessels pass through the epithelium; instead, the circulation of nutrients, gases, and wastes in epithelial tissue occurs via diffusion from the capillaries of neighboring tissue.

Epithelial tissues possess remarkable regenerative powers. The cells of epithelial layers are constantly being replaced throughout the life of the organism. The liver, for example, which is a gland formed of epithelial tissue, can readily regenerate substantial portions of tissue that have been surgically removed from it.

There are three general classes of epithelial tissue (Figure 42-2): simple epithelium, stratified epithelium, and glands.

Simple Epithelium

Simple epithelial tissues are a single cell layer thick. The three kinds of simple epithelium come in two shapes, which occur in very different kinds of membranes. **Squamous** epithelial cells have an irregular flattened shape with tapered edges. The membranes that line the lungs and the major cavities of the body (enclosing the heart, lungs, and intestines) are composed of squamous cells. These cells, like most other epithelial cells, are relatively inactive metabolically; their presence is generally associated with the passive movement of water and electrolytes into and out of the body. **Cuboidal** and **columnar** epithelial cells have a fuller, less flat appearance, possess many mitochondria and extensive endoplasmic reticulum, and are much more active metabolically than the squamous epithelial cells. The respiratory tract and the ducts of the testes are lined with epithelium composed of cuboidal and columnar cells. Individual cells within these epithelial layers often bear cilia or have secretory functions.

Stratified Epithelium

Stratified epithelial tissues are several cell layers thick. Typically the upper layer is squamous, the middle one is cuboidal, and the bottom one is columnar. The most abundant and prominent kind of stratified epithelial tissue is the epidermis, or skin. The cornea of the eye is composed of stratified epithelium in which the individual cells have become highly specialized. A characteristic property of the cells of stratified epithelium is that they are able to lay down **keratin,** a very strong fibrous protein. The deposition of keratin usually occurs as a response to stress. The callouses that occur on the palms of the hands of manual laborers are composed of keratin. Hair, which is derived from skin, is also composed of keratin.

> **Epithelium makes up the skin of the body and the lining of the respiratory and digestive tracts. Much of the internal lining of respiratory and digestive tracts is simple epithelium, only one cell thick, whereas external skin is composed of stratified epithelium that is many cell layers thick.**

Glands

The vertebrate body contains two major gland systems, both of which are derived from epithelial layers. These body glands produce sweat, milk, saliva, digestive enzymes, hormones, and a diverse catalogue of other substances. In one gland system, the **exocrine glands,** the connection of the gland to the epithelial sheet from which it invaginated is maintained as a duct. The liver and pancreas are exocrine glands. The other major gland system, which comprises the **endocrine glands,** or ductless glands, also arises during development as a tubular invagination of a sheet of epithelial cells; however, the connection with the epithelium is lost during the course of development. The endocrine glands, therefore, are not connected to the epithelium in the adult animal. Many of the hormone-producing glands are endocrine glands.

CONNECTIVE TISSUE

The cells of connective tissue provide the body with its structural building blocks and also with its most potent defenses. Unlike epidermal cells, the cells that make up loose connective tissue are not stacked tightly; instead, they are spaced well apart from one another (Figure 42-3). A connective tissue is a supporting tissue containing a large amount of extracellular material called a matrix. Connective tissue cells are derived from the mesoderm and fall into three categories: defensive, structural, and sequestering connective tissues. Cells of defensive connective tissue, those which are involved in the defense of the body, have a matrix of blood plasma and roam the circulatory system as mobile hunters of in-

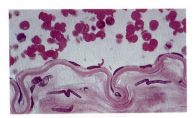
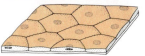

A SIMPLE SQUAMOUS

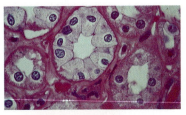
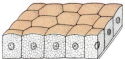

B SIMPLE CUBOIDAL

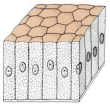

C COLUMNAR

FIGURE 42-2

Types of epithelial tissue.
A *Squamous epithelium lines the artery seen here. The nuclei are characteristically flattened. The round cells above the epithelium are blood cells in the lumen, or hollow interior, of the artery.*
B *Cuboidal epithelium forms the walls of these kidney tubules, seen in cross section.*
C *Columnar epithelium forms the outer cell layer of this human ileum. Interspersed among the epithelial cells are goblet cells, which secrete mucus. The striated appearance of the border results from large numbers of microvilli.*

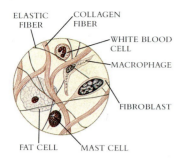

ELASTIC FIBER COLLAGEN FIBER
WHITE BLOOD CELL
MACROPHAGE
FIBROBLAST
FAT CELL MAST CELL

FIGURE 42-3

Loose connective tissue. Macrophages, lymphocytes, and mast cells are involved in the body's defenses against infection by microorganisms. Fibroblasts play a structural role, secreting collagen fibers. Collagen is important in wound healing as well as in bone formation. Other connective tissues such as fat cells serve to store and transport particular molecules.

FIGURE 42-4

Lymphocytes and macrophages. The lymphocytes are small and spherical; the macrophages are larger and more irregular in form.

vading bacteria and foreign substances. Structural connective tissue cells are static, secreting proteins into the empty spaces between individual cells and thus creating a loose fibrous matrix. Sequestering connective tissue cells act as storehouses, accumulating and sequestering specific substances such as fat, the skin pigment melanin, or hemoglobin.

Connective tissues are derived from mesoderm. Some are mobile hunters of invading bacteria, others secrete a protein matrix to provide structural support, and still others act as sites of storage for fat or hemoglobin.

Defensive Connective Tissues

The three principal types of defensive connective tissues are macrophages, lymphocytes, and mast cells. All of these tissues are dispersed; they are composed of many relatively small, rounded cells that move individually within the circulating fluids of the body (Figure 42-4).

1. **Macrophages.** Macrophages are abundant in the bloodstream. They usually are mobile but sometimes are attached to fibers. The key property of these cells is that they are phagocytic: they are able to engulf and digest cellular debris and invading bacteria. Macrophages provide the key phagocytic element in the body's two defenses against invasive elements. As elements of the **reticuloendothelial system,** they patrol the capillaries of the circulatory system, engulfing and digesting any bacteria that they encounter and disposing of any damaged tissue. As elements of the **immune system,** macrophages phagocytize any particles that are coated with antibodies that they encounter in the capillaries. Antibodies are special proteins that bind to molecules that are not normally present in the body. For this reason,

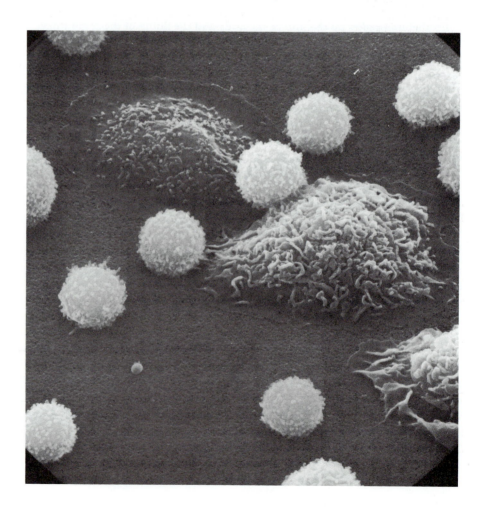

the binding of an antibody to a particle acts as a flag that alerts the macrophages to the foreign nature of the tagged molecule.

2. **Lymphocytes.** Lymphocytes normally are not numerous except when an organism is sick. For example, patients suffering from leukemia have high lymphocyte levels. The key property of these cells is their ability to synthesize antibody proteins. When they are stimulated by a foreign substance called an **antigen,** some of the lymphocytes enlarge and begin to divide and synthesize an antibody protein capable of binding to that particular antigen. Each mature lymphocyte cell, called a **plasma cell,** possesses within its nucleus a unique version of the antibody-encoding gene, different from the version of almost any other lymphocyte. As described in Chapter 17, the gene encoding a particular antibody is assembled during the lymphocyte's development by the splicing together of segments selected at random from several clusters of nucleotide sequences. Because each cluster contains many different alternative versions of a segment, and only one of them is selected at random from each cluster as a particular lymphocyte matures, many combinations of gene segments are possible. Imagine constructing a sentence by selecting at random one word from each of the previous 10 pages of this book—very many different sentences would be possible. Because of the random way in which its component segments are selected, each lymphocyte is different from every other one and encodes a different antibody protein.

   The protein encoded by this antibody-encoding gene is present in many copies on the surface of the lymphocyte cell. Here it is able to interact with other proteins that the lymphocyte encounters circulating in the bloodstream. The lymphocyte ignores most of these proteins, but when it encounters one to which it can bind (an antigen), a signal goes through the lymphocyte cytoplasm to its nucleus. This signal commands proliferation and the production of more of the antibody protein, which is excreted from the cell. The excreted antibody binds to any other molecules of the inducing antigen that are present, marking the bound antigen molecule for eventual consumption by macrophages. Because there are so many different kinds of lymphocytes, each with a different antibody studding its surface, any foreign molecule inevitably encounters a lymphocyte cell that is sensitive to it and stimulates that cell to production. The body's own molecules are not normally bound by antibodies, since all of the lymphocytes encoding an antibody that recognizes any of the body's molecules are removed from circulation early in development. The T cells described in Chapter 1 are lymphocytes.

3. **Mast cells.** Mast cells are not numerous. Their key property is that they synthesize the molecules that are involved in the body's response to physical injury, or **trauma.** The cytoplasm of mast cells is filled with many vesicles. These vesicles contain **histamine** and **serotonin,** molecules that produce the inflammation response, as well as **heparin,** a molecule that prevents blood clotting. The mast cells release the contents of their vesicles into the bloodstream in response to mechanical or chemical injury. The molecules that they release act to enlarge the capillaries and speed the healing process. Interestingly, mast cells also disgorge their contents in response to very high levels of individual antibodies. Antigens that the body has encountered frequently in the past thus evoke this response, which is called **hypersensitivity.** Most allergies are the result of hypersensitivity.

The vertebrate body uses connective tissues to defend against foreign infection. Lymphocytes produce proteins called antibodies, which recognize and bind to foreign bacteria and other elements, after which macrophages engulf and digest the invader. Mast cells enlarge the blood vessels in response to trauma, speeding the healing process.

THE MANY ROLES OF SKIN

The skin of a vertebrate is far more than simply a container of epithelial cells encasing the body's muscles, blood, and bones. Unlike the "skin" of a sausage, the skin of a vertebrate body is a dynamic organ with many functions:

1. Skin is a protective barrier. It keeps out microorganisms that otherwise would infect the body. Because skin is waterproof, it keeps the fluids of the body in and other fluids out; when you soak in a bathtub, your body absorbs little or no water. Skin cells contain a pigment called **melanin,** which absorbs potentially damaging radiation from the sun.

2. Skin provides a sensory surface. Sensory nerve endings in skin act as your body's pressure gauge, telling you how gently to caress a loved one or how hard to hold a pencil. Other sensors embedded in skin detect pain, heat, and cold. Skin is the body's point of contact with the outside world.

3. Skin compensates for body movement. Skin stretches when you reach for something and contracts quickly when you stop reaching. It expands when you grow and shrinks when you lose weight.

4. Skin controls the body's internal temperature. When it is cold, the capillary blood vessels in the skin contract so that less of the body's heat is lost to the surrounding air. When it is hot, these capillaries expand, and glands in the skin release sweat, the evaporation of which cools the body surface.

Skin is the largest organ of the vertebrate body. In an adult human, 15% of the total weight is skin. Much of the multifunctional role of skin reflects the fact that many other specialized cells are crammed in among skin cells. One square centimeter of human skin contains 200 nerve endings, 10 hairs and muscles, 100 sweat glands, 15 oil glands, 3 blood vessels, 12 heat-sensing organs, 2 cold-sensing organs, and 25 pressure-sensing organs.

Vertebrate skin is composed of three layers (Figure 42-A): an outer **epidermis,** a lower **dermis,** and an underlying layer of **subcutaneous tissue.** The epidermis of skin is from 10 to 30 cells thick— about as thick as this page. The outer cell layer, called the **stratum corneum,** is the one you see when you look at your arm or face. Cells from this layer are continuously subject to damage. They are abraded, injured, and worn by friction and stress during the body's many activities. They also lose moisture and dry out. The body deals with this damage not by repairing cells but by replacing them. Cells from the stratum corneum are shed continuously. They are replaced by new cells produced deep within the epidermis. The cells of the innermost layers of the epidermis, called the **basal layer,** are among the most actively dividing cells of the vertebrate body. New cells formed there migrate upward, and as they move they form the protein keratin, which makes the skin tough. Each cell eventually arrives at the outer surface; like a new gladiator in the arena, it takes its turn in the stratum corneum, ready to suffer an uncertain fate and eventual replacement by a newer cell. A cell normally lives in the stratum corneum for about a

Structural Connective Tissues

There are three principal structural connective tissues: fibroblasts, cartilage, and bone. The three are distinguished principally by the nature of the matrix laid down between their individual cells.

Fibroblasts

Of all connective tissue cells, fibroblasts are the most common within vertebrate bodies. They are flat, irregular, branching cells that secrete structurally strong proteins into the matrix between cells. The most commonly secreted protein is **collagen** (Figure 42-5), which is the most abundant protein in the bodies of vertebrates. One quarter of all animal protein is collagen. Collagen is an unusual protein in that the amino acid glycine constitutes a third

month. **Psoriasis,** familiar to some 4 million Americans as persistent dandruff, is a chronic skin disorder in which new cells reach the epidermal surface every 3 or 4 days—about seven times faster than normal.

The dermis of skin is from 15 to 40 times thicker than the epidermis. The thick dermis provides structural support for the epidermis and a matrix for the many nerve endings, muscles, and specialized cells residing within skin. A fine network of blood vessels passes through it. The leather of a belt is animal dermis.

The layer of subcutaneous tissue below the dermis is composed primarily of fat-rich cells. They act as shock absorbers and provide insulation, which conserves body heat. This tissue varies greatly in thickness in different parts of the body. The eyelids have none of it, whereas the buttocks and thighs may have a lot. The subcutaneous tissue of the skin on the soles of your feet may be a half a centimeter thick, or even more.

Embedded within the epidermis are millions of specialized cells called **melanocytes,** which produce a brownish pigment called melanin. People of all races (see boxed essay, Chapter 21) have about the same number of melanocytes in their skin, but they differ in the amount of melanin that their individual melanocytes produce. The result is a range of human skin tones; the more melanin the melanocytes produce, the darker the skin. Melanin production is influenced by a variety of different genes, so that it exhibits "blending" inheritance: offspring usually exhibit a shade of skin color that is intermediate between those of their parents.

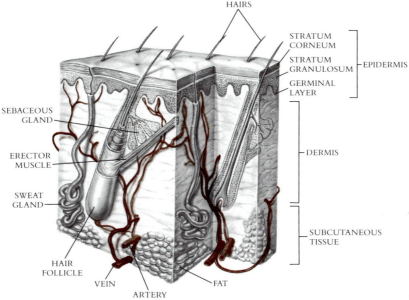

FIGURE 42-A

of its total amino acids. Most proteins contain only about 5% glycine. Collagen also contains many more proline amino acids than do normal proteins; these prolines form attachment sites for sugar side chains.

Because glycine is a very small amino acid, chains of collagen are able to wind intimately around one another, forming a triple helix. The side chains of the proline amino acids form cross-links that lock the three strands together. Collagen fibers are made up of many individual collagen strands arrayed side by side and connected with one another by additional cross-links. The strength of the fiber is determined by the nature of its cross-links. Fibroblasts are active in wound healing. They multiply rapidly in wound tissue, forming granulated fibrous scar tissue that possesses a collagen matrix.

The way in which collagen fibers are laid down is of particular interest. The loose fibrous matrix of the vertebrate body, tying different tissues together, is composed of fi-

FIGURE 42-5

Collagen fibers. Each fiber is composed of many individual collagen strands, and can be very strong.

broblast-secreted collagen. The deformation of preexisting collagen fibers by stress exposes charged lysine side chains that otherwise would be buried in the fibers. As a result of the exposure of these lysine side chains, a very minute electric current, called a **piezoelectric current,** is generated along the deformed fiber. Fibroblasts are able to sense these minute currents, which polarize the surfaces of the cells and thus determine their lines of movement. As a result, the new collagen fibers are laid down along the lines of stress, helping to strengthen the area and heal breaks.

Collagen is not the only fibrous tissue that is produced by the fibroblasts. **Reticulin** is a fine-branching fiber that forms the framework of many glands, such as the spleen and the lymph nodes, and also makes up the junctions between many kinds of tissues. **Elastin** is fibrous tissue that is the principal component of the lung. Elastin cross-links are longer than those of collagen, giving great elasticity to fibers of elastin. When elastin fibers are stretched, they return promptly to their former length after being released.

Cartilage

Cartilage is a specialized connective tissue in which the collagen matrix between cells is formed at positions of mechanical stress. In cartilage the fibers are laid down along the lines of stress in long parallel arrays. The result of this process is a firm and flexible tissue, far tougher than the defensive or sequestering connective tissues, that has great tensile strength. Cartilage binds together the bones that meet in joints, such as the knee, ankle, and elbow. It also makes up the entire skeletal system of the modern agnathans and Chondrichthyes (see Chapter 39), having replaced the body skeletons that were characteristic of the ancestors of these vertebrate groups.

Bone

Bone is a special form of cartilage in which the collagen fibers are coated with a calcium phosphate salt. The great advantage of bone over the chitin of invertebrates as a structural material is that bone is strong without being brittle. To understand the properties of bone, consider those of fiberglass. Fiberglass is composed of glass fibers embedded in epoxy glue. The individual glass fibers are rigid, giving great strength, but they are brittle, like the cuticle of arthropods. The epoxy component of fiberglass, on the other hand, is flexible but weak. The composite, fiberglass, is both rigid and strong. When a glass fiber breaks because of stress and a crack starts to form, the crack runs into glue before it reaches another fiber. The glue distorts and reduces the concentration of the stress, and the adjacent fibers consequently are not exposed to the same high stress. In effect, the glue acts to spread the stress over many fibers.

Bone is constructed in a similar manner. Small needle-shaped crystals of a calcium-containing mineral, hydroxyapatite, surround and impregnate the collagen fibrils of bone. The long axis of the crystals is parallel to the axis of the fibril. Within bone, the collagen is laid down along the lines of stress, the bond-forming cells following the piezoelectric currents that are set up because of pressure. Bone is formed along long stress lines down the shanks of long bones and along curved stress lines at the curved ends of bones in joints. No crack can penetrate far into bone without encountering myriad hydroxyapatite crystals in a collagenous matrix. Bone is more rigid than collagen, just as fiberglass is more rigid than epoxy. Conversely, bone is more flexible and resistant to fracture than hydroxyapatite—or chitin.

The bones of the vertebrate skeleton (Figure 42-6) are composed of two kinds of tissue. The ends and interiors of long bones are composed of an open lattice of bone called **spongy bone tissue,** or **marrow.** Within this lattice framework, most of the body's red blood cells are formed. Surrounding the spongy bone tissue at the core of bones are concentric layers of **compact bone tissue** in which the collagen fibrils are laid down in a pattern that is far denser than the marrow. The compact bone tissue gives the bone the strength to withstand mechanical stress.

New bone is formed by cells called **osteoblasts,** which secrete collagen fibers on which calcium is subsequently deposited. Bone is laid down in thin concentric layers called **lamellae,** like so many layers of paint on an old pipe. The lamellae are laid down as a series of tubes around narrow channels called **Haversian canals,** which run parallel to the length of the bone (Figure 42-7). The Haversian canals are interconnected and carry blood vessels and nerve cells. The blood vessels provide a lifeline to living bone-forming cells, and the nerves control the diameter of the blood vessels and thus the flow through them. When bone is first formed in the embryo, osteoblasts use the cartilage skeleton as a template for bone formation. Later, new bone is formed along lines of stress.

Bone is formed in two stages: first, collagen is laid down in a matrix of fibrils along lines of stress, and then calcium minerals impregnate the fibrils. The calcium salt provides rigidity, and the collagen provides flexibility.

FIGURE 42-7

The Haversian canals are evident in this cross section of compact bone.

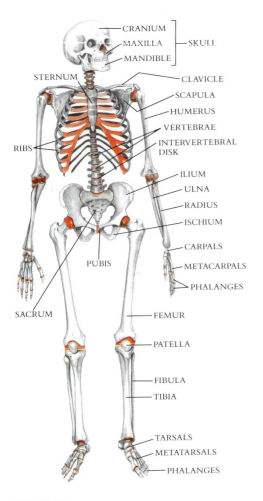

FIGURE 42-6

Human skeleton. The human skeleton contains 206 different bones, indicated here in white. It also contains numerous deposits of cartilage, indicated here in color.

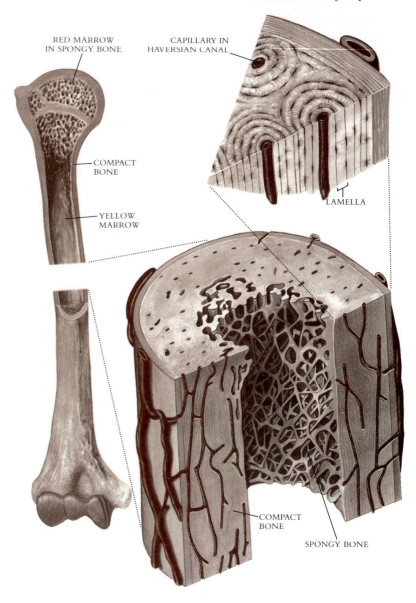

Sequestering Connective Tissues

The third general class of connective tissue is composed of cells that are specialized for the accumulation and transport of particular molecules. Sequestering connective tissues include the fat cells of adipose tissue, as well as pigment-containing cells. The most important tissues of this class are red blood cells.

Blood cells are classified, according to their appearance (Figure 42-8), as **erythrocytes** (red blood cells) or **leukocytes** (white blood cells). We have described white blood cells before; they include the macrophages and lymphocytes which defend the body. The role of red blood cells is very different. They act as mobile transport units, picking up and delivering gases. Blood also contains fragments of large bone-marrow cells called **platelets.**

Erythrocytes are the most common blood cells; there are about 5 million in every milliliter of blood. They act as the transporters of oxygen in the vertebrate body. The primary cellular components of blood, erythrocytes are also the "ghosts" of mammalian circulatory systems. During their maturation in mammals they lose their nucleus and mitochondria, and their endoplasmic reticulum dissolves. As a result of these processes, mammalian erythrocytes are relatively inactive metabolically, but they are not empty. Large amounts of the iron-containing protein hemoglobin are produced within the erythrocytes and remain in the mature cell. Hemoglobin is the principal carrier of oxygen and carbon dioxide in vertebrates, as it is in many other groups of animals. Each erythrocyte contains about 300 million molecules of hemoglobin.

The fluid intracellular matrix, or **plasma,** in which the erythrocytes move is both the banquet table and the refuse heap of the vertebrate body. Practically every substance used by cells is dissolved in plasma. These include the sugars, lipids, and amino acids, which are the fuel of the body, as well as the products of metabolism. The plasma also contains inorganic salts, such as the calcium used to form bone; fibrinogen from the liver; albumin, which gives the blood viscosity; and antibody proteins produced by lymphocytes. Every substance secreted or discarded by cells, such as urea, is also present in the plasma.

> **Many connective tissues are specialized for accumulating particular classes of molecules, such as fat or pigment. Blood cells called erythrocytes accumulate the oxygen-carrying protein hemoglobin.**

MUSCLE TISSUE

Muscle cells, formed from the mesoderm-derived somites laid down early in development, are the workhorses of the vertebrate body. The distinguishing characteristic of muscle cells, the one that makes them unique, is the relative abundance of actin and myosin microfilaments within them. These microfilaments are present as a fine network in all eukaryotic cells, but they are far more common in muscle cells. In muscle cells, the microfilaments are bunched together into many thousands of strands called **myofibrils.** As in all eukaryotic cells, the shortening of these microfilaments can cause the cell to change shape. Because there are so many filaments all aligned in parallel in muscle cells, a considerable force is generated when all of the microfilaments contract at the same time. Vertebrates possess three different kinds of muscle cells (Figure 42-9): smooth muscle, striated muscle, and cardiac muscle.

Smooth Muscle

Smooth muscle was the earliest form of muscle to evolve, and it is found throughout the animal kingdom. Smooth muscle cells are long and spindle shaped, each cell containing a single nucleus. The interiors of smooth muscle cells are packed with actin-myosin myofibrils, but the individual myofibrils of a cell are not aligned with respect to one another into organized arrays. Smooth muscle tissue is organized into sheets of cells. In some tissues the muscle cells contract only when they are stimulated by a nerve or hormone, and then all of the cells contract together as a unit. Examples of this are the muscles found lining the

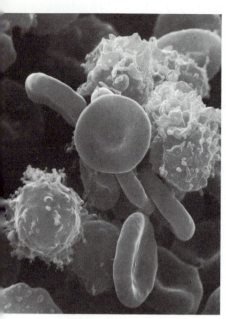

FIGURE 42-8

Blood cells. White blood cells, or leukocytes, are roughly spherical and have irregular surfaces with numerous extending pili. Red blood cells, or erythrocytes, are flattened spheres, typically with a depressed center, not unlike doughnuts with incomplete holes.

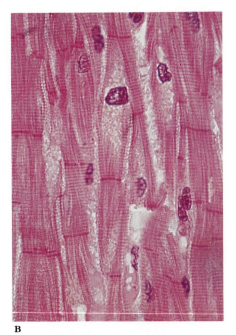

A

B

C

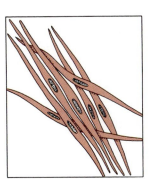

SMOOTH MUSCLE

CARDIAC MUSCLE

SKELETAL MUSCLE

walls of many vertebrate blood vessels and those that make up the iris of the vertebrate eye. In other smooth muscle tissue, such as that found in the wall of the gut, the individual cells contract spontaneously, leading to a slow, steady contraction of the tissue.

Striated Muscle

Striated muscles are the muscles of the skeleton (Figure 42-10). They are voluntary muscles, contracting only in response to stimulation by nerves. Each striated muscle is a tissue made up of numerous individual muscle cells acting in concert. These striated muscle cells represent a distinct improvement in muscle cell organization, as compared with that found in the smooth muscle cells. Imagine a large raft being towed upstream by many small canoes, each canoe bound to the raft by its own towline. This is analogous to the contraction of smooth muscle, each smooth muscle cell participating individually in the contraction of the muscle. Now imagine placing all the rowers in one galley so that they are able to row in concert, pulling the raft far more effectively. This is analogous to the contraction of striated muscle, in which numerous muscle cells pool their resources.

Striated muscle cells are produced during development by the fusion of several cells, end to end, to form a very long fiber (Figure 42-11). A single fiber will typically run the entire length of a vertebrate muscle. Each cell, or **muscle fiber,** still contains all of the original nuclei, pushed out to the periphery of the cytoplasm by a central cable made up of 4 to 20 myofibrils. The cytoplasm in striated muscle is given a special name, the **sarcoplasm.**

FIGURE 42-9

Types of muscle.
A *Smooth muscle cells are long and spindle shaped, with a single nucleus. These smooth muscle cells are from the intestinal lining of a monkey.*
B *Cardiac muscle cells such as these from a human heart also contain a single nucleus, and are organized into long branching chains that interconnect, forming a lattice.*
C *Skeletal or striated muscle cells are formed by the fusion of several muscle cells end to end to form a long fiber with many nuclei.*

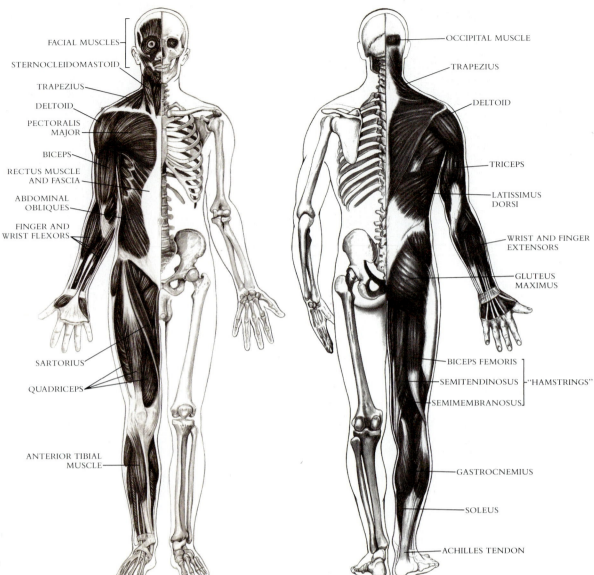

FACIAL MUSCLES
STERNOCLEIDOMASTOID
TRAPEZIUS
DELTOID
PECTORALIS MAJOR
BICEPS
RECTUS MUSCLE AND FASCIA
ABDOMINAL OBLIQUES
FINGER AND WRIST FLEXORS
SARTORIUS
QUADRICEPS
ANTERIOR TIBIAL MUSCLE

OCCIPITAL MUSCLE
TRAPEZIUS
DELTOID
TRICEPS
LATISSIMUS DORSI
WRIST AND FINGER EXTENSORS
GLUTEUS MAXIMUS
BICEPS FEMORIS
SEMITENDINOSUS ⎱ "HAMSTRINGS"
SEMIMEMBRANOSUS ⎰
GASTROCNEMIUS
SOLEUS
ACHILLES TENDON

FIGURE 42-10

Human musculature.
A *Anterior view.*
B *Posterior view.*

The myofibrils that run down the center of a muscle cell are highly organized to promote simultaneous contraction. Contraction occurs in the following way:

1. A myofibril is made up of a long chain of contracting units called **sarcomeres,** lined up like the cars on a train.

2. Each sarcomere is composed of interdigitating filaments of actin and myosin. One collection of actin filaments is attached to the front and another to the back of the sarcomere, to plates referred to as **Z lines.** These front and back assemblies of actin filaments are not long enough to reach each other in the center of the sarcomere. They are joined to one another by interdigitating myosin filaments.

3. As you may recall from Chapter 7 (Figure 7-19), the sarcomere contracts when the heads of the myosin filaments change their shape. Since these heads are in contact with the actin filaments, the effect of their contraction is to pull the myosin along the actin. When the myosin filament has slid along the actin about 75 nanometers, the strain is released. The cleavage of ATP then provides the energy for the myosin head to regain its original orientation, positioning it for another pulling movement.

FIGURE 42-11

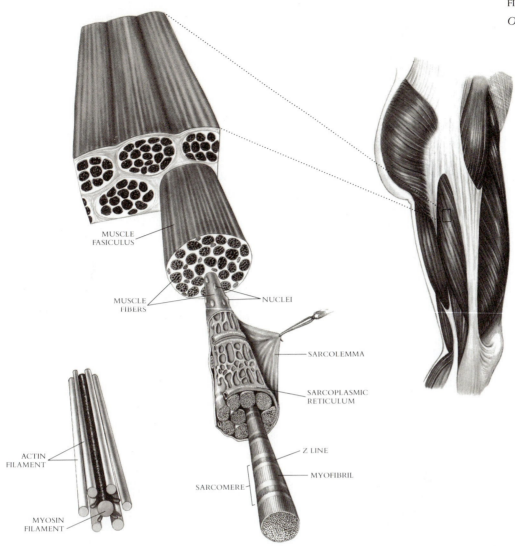

MUSCLE
FASICULUS

MUSCLE
FIBERS

NUCLEI

SARCOLEMMA

SARCOPLASMIC
RETICULUM

ACTIN
FILAMENT

Z LINE

MYOFIBRIL

SARCOMERE

MYOSIN
FILAMENT

4. The orientation of myosin is such that it moves along the actin towards the Z line. Because both ends of the myosin filaments move in this manner simultaneously, the effect is to pull the two Z lines together, contracting the sarcomere.

5. Simultaneous contraction of all of the sarcomeres of a myofibril results in an abrupt and forceful shortening of the myofibril. All of the myofibrils of a muscle fiber usually contract at the same time, producing a very strong contraction in the length of the muscle fiber cell.

The sarcomeres are in register all along the length of the stacked myofibrils. This gives striated muscle, as viewed with a light microscope, a distinctive pattern of bands, or **striations.** It is this pattern that gives striated muscle its name.

Cardiac Muscle

The hearts of vertebrates are composed of striated muscle fibers arranged very differently from the fibers of skeletal muscle. Instead of very long multinucleate cells running the length of the muscle, heart muscle is composed of chains of single cells, each with its own nucleus. These chains of cells are organized into fibers that branch and interconnect, forming a latticework. This lattice structure is critical to how heart muscle functions. As we will describe in Chapter 46, heart contraction is initiated by the opening of transmembrane channels that admit ions into muscle cells, altering the charge of their membranes. This change in membrane charge is called **depolarization.** When two cardiac muscle fibers touch one another, their membranes make an electrical junction. As a result, the electrical depolarization of any one fiber initiates a wave of contraction throughout the heart, with the wave of depolarization rapidly passing from one fiber to another across these junctions. Thus a mass of heart muscle tends to contract all at once rather than a bit at a time.

Muscle cells are rich in actin and myosin, which form microfilaments that are capable of contraction. Many muscle cells contracting in concert can exert considerable force.

NERVE TISSUE

The fourth major class of vertebrate tissue is nervous tissue. It is composed of two kinds of cells: (1) **neurons,** which are specialized for the transmission of nerve impulses; and (2) **supporting cells,** which are specialized to assist in the propagation of the nerve impulse and to provide nutrients to the neuron.

Neurons (Figure 42-12) are cells specialized to conduct an electric current. Their membranes are rich in ion-transporting channels, which pump ions out of the cell and so maintain a charge difference between the interior and exterior of the cell. A current is initiated when channels in a local area of the membrane open, permitting ions to reenter from the exterior and eliminate the charge difference (this is the process of depolarization mentioned in our discussion of cardiac muscle). This depolarization in a local area of the membrane tends to open nearby channels of the neuron membrane, starting a wave that prop-

FIGURE 42-12

A human neuron. The cell body is at the upper left, with the axon extending up out of view. The branching network of fibers extending down from the cell body is made up of dendrites, which carry signals to the cell body.

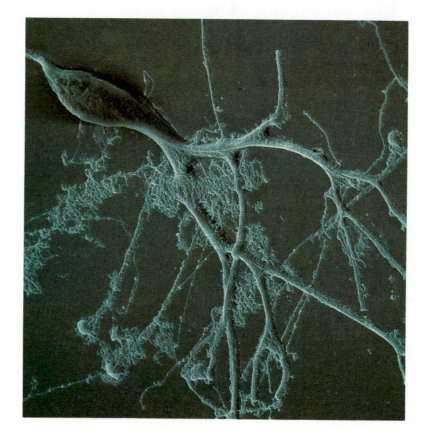

agates down the nerve as a nerve impulse. The nature of nerve impulses, described in more detail in Chapter 46, can vary, but all nerve impulses propagate along nerves as waves of membrane depolarization.

The cell body of a neuron (Figure 42-13) contains the cell nucleus. From the cell body project two kinds of cytoplasmic elements that carry out the transmission functions of the neuron. The first group of elements consists of the **dendrites,** which are threadlike protrusions. The dendrites act as antennas for the reception of nerve impulses from other cells or sensory systems. The second kind of cytoplasmic element that projects from the cell body of a neuron is the **axon.** An axon is a long tubular extension of the cell that provides a channel for the transmission of the nerve impulse away from the cell body. Because axons can be quite long, some nerve cells can be very long indeed. A single neuron that innervates the muscles in your thumb, for example, may have its cell body in the spinal cord and possess an axon that extends all the way across your shoulder and down your arm to your thumb. Single neuron cells over a meter in length are common.

The nerves of the vertebrate body, which appear as fine white threads when they are viewed with the naked eye, are actually composed of clusters of axons and dendrites. Like a telephone trunk cable, they include large numbers of independent communications channels—bundles of hundreds of axons and dendrites, each connecting a different nerve cell with a different muscle fiber. In addition, the nerve contains numerous supporting cells bunched around the axons. In the brain and spinal cord, which together make up the **central nervous system,** these supporting cells are called **glial cells.** The supporting cells associated with projecting axons and all other nerve cells, which make up the **peripheral nervous system,** are derived from neural crest cells and are called **Schwann cells.**

> **Neurons are cells that are specialized to conduct electrical currents. Their membranes are rich in ion-transporting transmembrane channels that establish an electrical charge on the membrane surface. Opening some of these channels removes the charge, an effect that tends to open nearby channels and so to propagate the charge reversal along the surface of the nerve as a nerve impulse.**

How Nerves Cause Muscles to Contract

In the striated skeletal muscle of vertebrates, contraction is initiated by a nerve impulse. The nerve impulse arrives as a wave of depolarization along the nerve membrane. The nerve fiber is embedded in the surface of the muscle fiber, forming a **neuromuscular junction** (Figure 42-14). When such a wave of depolarization reaches the end of a neuron, at the point where the neuron attaches to a muscle, its arrival causes that membrane to release the chemical **acetylcholine** from the nerve cell into the junction. The acetylcholine passes across to the muscle membrane and opens the ion channels of that membrane, depolarizing it. Acetylcholine is called a **neurotransmitter** because it transmits the nerve impulse across the junction.

How does the depolarization of the nerve fiber membrane cause the contraction of the muscle fibers? The endoplasmic reticulum of a striated muscle cell, which is called the **sarcoplasmic reticulum,** wraps around each myofibril like a sleeve. As a result, the entire length of every myofibril is very close to the intracellular space bounded by the sarcoplasmic reticulum membrane system. Within the sarcoplasmic reticulum are embedded numerous calcium ion transport channels. In resting muscle, calcium ions are actively pumped through these channels, concentrating all of the calcium ions of that cell within the spaces of the sarcoplasmic reticulum. The depolarization of the muscle fiber membrane opens ion channels in the sarcoplasmic reticular membrane and so causes the release of this concentrated calcium across the reticular membrane into the cytoplasm (sarcoplasm), where it is free to interact with myofibrils. This calcium acts as a trigger to release contraction of the myofibril. It does so in the following way:

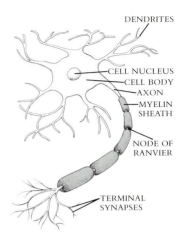

FIGURE 42-13

Idealized structure of a vertebrate neuron. Many dendrites lead to the cell body, from which a single long axon extends. In some neurons specialized for rapid signal conduction, the axons are encased at intervals within myelin sheaths. At its far end an axon may terminate at one cell, or branch to several cells, or connect to several locations on one cell.

FIGURE 42-14

Neuron axons typically branch as they connect with skeletal muscle as shown here. Each connection at a neuromuscular junction ends in a motor end plate.

1. In resting muscle, myosin filaments are not free to interact with actin, because the sites on actin where the myosin heads must make contact are not available. The heads are covered by a long thread of **tropomyosin,** which winds around the actin.

2. At regular intervals along the tropomyosin thread are spaced globular proteins, **troponin** molecules, which hold the tropomyosin in position on the actin.

3. Troponin molecules are able to bind calcium ions, and when they do, they change their shape. As a result of this change in shape, the tropomyosin thread is repositioned to a new location where it does not interfere with myosin interaction. Only when this repositioning has occurred can contraction take place.

Thus the release of calcium by the stimulation of the sarcoplasmic reticulum releases the tropomyosin trigger and results in the contraction of the myofibril.

> **In a resting state, skeletal muscles are not contracted. Troponin molecules are bound to the actin, preventing actin from moving relative to myosin. When a nerve impulse depolarizes the membrane of a muscle, the depolarization causes the calcium-transporting transmembrane channels in the muscle membrane to open, resulting in an influx of calcium ions that release troponin from actin. This release permits the actin to move relative to the myosin and so causes the muscle to contract.**

THE MAJOR ORGAN SYSTEMS OF VERTEBRATES

The four classes of vertebrate tissues that we have discussed in this chapter are the building blocks of the vertebrate body. Each organ of the body is a structure that carries out a specific function and is composed of these tissues, arrayed and assembled in various ways. A muscle, for example, is composed of muscle tissue wrapped in connective tissue and connected with nervous tissue. Different combinations of tissues are found in different organs.

The many organs that carry out the principal activities of the body are grouped together and are referred to as **organ systems.** For example, the digestive organ system is composed of individual organs concerned with breaking up of food (teeth), passage of food to the stomach (esophagus), digestion of food (stomach), absorption of food (intestine), and expulsion of solid residue (anus). The vertebrate body includes 10 organ systems (Table 42-1).

Much of the biology of vertebrates is concerned with the functioning of organ systems. We have already discussed the reproductive system; the immune system, which the body uses to defend itself from infection; muscles; and skin. In the next seven chapters, we will discuss the other major organ systems.

TABLE 42-1   THE TEN MAJOR VERTEBRATE ORGAN SYSTEMS

Musculoskeletal	The bony framework of the body and the muscles that move it
Integumentary	The skin of the body
Respiratory	The lungs and the tubes involved in oxygen capture and gas exchange
Circulatory	The heart and blood vessels, which transport cells and materials throughout the body
Digestive	The teeth, esophagus, stomach, and intestines, which capture soluble nutrients from ingested food
Excretory	The kidneys and the passages that remove metabolic wastes from the blood
Endocrine	The glands and hormones, which regulate the activities of the body
Immune	The lymphocytes, macrophages, and antibodies, which remove foreign bodies from the bloodstream
Nervous	The brain, nerves, and sensory organs, which direct the body
Reproductive	The gonads and reproductive organs

SUMMARY

1. The four basic types of tissue are epithelium, connective tissue, muscle, and nerve.

2. Epithelium covers the surface of the body. The lining of the major body cavities is composed of simple epithelium, whereas the exterior skin is composed of stratified epithelium. The major gland systems are also derived from epithelium.

3. Connective tissue supports the body mechanically and defensively. The structural connective tissues are fibroblasts, cartilage, and bone. The defensive connective tissues are macrophages, which engulf foreign bacteria and antibody-coated cells or particles; lymphocytes, each of which produces a single unique antibody; and mast cells, which release healing chemicals at sites of trauma. The sequestering connective tissues are pigment-producing cells, fat cells, and blood cells.

4. Muscle contraction provides the force for mechanical movement of the body. There are three kinds of muscle: smooth muscle, which is organized into sheets and contracts spontaneously; striated muscle, which is organized into trunks of long fibers and contracts when stimulated by a nerve; and cardiac muscle, in which the fibers are interconnected and may initiate contraction spontaneously.

5. Nerve cells provide the body with a means of rapid communication. There are two kinds of nerve cells: neurons and supporting cells. Neurons are specialized for the conduction of electrical impulses.

SELF-QUIZ

The following questions pinpoint important information in this chapter. You should be sure to know the answers to each of them.

1. Which is not one of the four basic types of tissue?
   (a) epithelial   (d) muscle
   (b) connective   (e) nerve
   (c) digestive

2. Which type of epithelial cell has an irregular flattened shape with a tapered edge, such as the cells that line the lungs?
   (a) squamous   (c) columnar
   (b) cuboidal   (d) striated

3. Sarcomeres of striated muscle tissue are composed of _____ and _____, protein filaments important in muscle contraction.

4. Which is larger, a sarcomere or a myofibril?

5. Which type of cell produces antibodies?
   (a) macrophages   (d) fibroblasts
   (b) lymphocytes   (e) leukocytes
   (c) mast cells

6. Which type of cell secretes collagen?
   (a) cartilage cells   (d) leukocytes
   (b) fibroblasts   (e) erythrocytes
   (c) bone cells

7. At the core of long bones are dense layers of bone tissue in which the collagen fibrils are laid down in a highly condensed pattern. True or false?

8. Which type of muscle would be found lining the digestive tract?

9. The contractile unit of striated muscle is called the
   (a) myofibril
   (b) sarcomere
   (c) filament
   (d) myosin
   (e) muscle fiber

10. A conducting nerve cell, such as one of the neurons that innervate leg muscles, possesses a long cytoplasmic tube extending from the body of the nerve cell. The cytoplasmic tube is called _____.

11. Cells whose role is to metabolically support the nerve cells of the central nervous system are
    (a) Schwann cells
    (b) interneurons
    (c) glial cells
    (d) osteoblasts
    (e) two of the above

12. Place the following events describing muscle contraction in the order in which they occur.
    (a) depolarization of muscle fiber membrane opens ion channels
    (b) tropomyosin repositions to a new location, freeing myosin to interact with troponin
    (c) nerve impulse causes membrane to release acetylcholine, which transmits the nerve impulse across to the muscle membrane
    (d) troponin molecules bind calcium ions, thus changing shape
    (e) calcium is released into the sarcoplasm, where it interacts with myofibrils
    (f) contraction can then take place

THOUGHT QUESTIONS

1. Among long-distance runners and committed joggers, the long bones of the legs often develop "stress fractures," numerous fine cracks running parallel to one another along the lines of stress. In most instances, stress fractures occur when runners push themselves much farther than they are accustomed to running. Runners who train by gradually increasing the distance they run develop stress fractures rarely, if ever. What protects this second kind of runner?

2. Land was successfully invaded four times—by plants, fungi, arthropods, and vertebrates. Each of these four groups evolved a characteristic hard substance to lend mechanical support, since bodies are far less buoyant in air than in water. Describe and compare these four substances, discussing their advantages and disadvantages. Can you imagine any other substance that would have been superior to any of these? What about plastic?

3. Your body contains 206 bones. As you grow, each of these bones must increase in size, all 206 of them maintaining proper proportions with one another. How do your bones manage to coordinate their growth?

FOR FURTHER READING

CURREY, J.: *The Mechanical Adaptations of Bones,* Princeton University Press, Princeton, N.J., 1984. A functional analysis of why different bones are structured the way they are, with an unusually well-integrated evolutionary perspective.

EYRE, D.R.: "Collagen: Molecular Diversity in the Body's Protein Scaffold," *Science,* vol. 207, pages 1315-1322, 1980. A good description of how collagen is formed and how it is utilized by the vertebrate body.

KENT, G.: *Comparative Anatomy of the Vertebrates,* 5th ed., The C.V. Mosby Co., St. Louis, 1983. A very readable introduction to the comparative anatomy of the vertebrates.

KROMMENHOCH, W., J. SEBUS, and G. VAN ESCH: *Biological Structures,* University Park Press, Baltimore, 1979. A striking collection of color photographs of mammalian cells and tissues.

NILSSON, L., and J. LINDBERG: *Behold Man,* Little, Brown & Co., Inc., Boston, 1974. A visually exciting collection of photographs of the tissues of the human body.

ROMER, A.S., and T.S. PARSONS: *The Vertebrate Body,* W.B. Saunders Co., Philadelphia, 1977. Perhaps the best-known and most-quoted comparative anatomy text, with a focus on body tissues and organs.

FUELING BODY ACTIVITIES: DIGESTION *Chapter 43*

Overview

Digestion is the conversion of parts of organisms into their constituent sugars, lipids, and amino acids. Although in many simple animals digestion is intracellular, in vertebrates it is extracellular, food being digested as it passes through a long, one-way digestive tract. Food is first pulverized in the mouth and then degraded into molecular fragments in the stomach. The digestion of fragments to simple molecules is completed in the small intestine. The small molecules that are the product of digestion are then absorbed into the body through the walls of the small intestine, and the residual solids are concentrated in the large intestine and eliminated.

For Review

Here are some important terms and concepts that you will encounter in this chapter. If you are not familiar with them, you should review them before proceeding.

- **Acid** (Chapter 2)

- **Lysosome** (Chapter 6)

- **Phagocytosis** (Chapter 9)

- **Oxidative respiration** (Chapter 11)

The vertebrate body is a complex colony of many cells. Like a city, it contains many individuals that carry out specialized functions. It has its own police (macrophages), its own construction workers (fibroblasts), and its own telephone company (the nervous system). The many individual cells of the vertebrate body, like the people in a city, need to be fed with food that is trucked in from elsewhere. Among the cells of the vertebrate body, there are no farmers. No vertebrate contains photosynthetic cells. All of the cells of the body are nourished with food that the body obtains outside itself and transports to the individual cells. Many of the major organ systems of the vertebrate body are involved in this acquisition of energy. The digestive system acquires organic foodstuffs; the respiratory system acquires the oxygen necessary to metabolize the food; the circulatory system transports both food and oxygen to the individual cells of the body; and the excretory system rids the body of wastes produced by metabolism. In this chapter we will consider the first of these activities, digestion.

THE NATURE OF DIGESTION

All heterotrophs obtain the metabolic energy needed for growth and activity by degrading the chemical bonds of organic molecules. In Chapters 10 and 11 we considered these degradation processes in detail. They include the breakdown of sugar molecules in glycolysis and the oxidation of pyruvate in the citric acid cycle. What these processes have in common is that they act on simple molecules: on amino acids, lipids, sugars, and fragments of these molecules. Heterotrophs, then, live by degrading simple organic molecules. Therein lies a major evolutionary challenge: few organisms contain significant concentrations of free sugars and amino acids. Instead, the simple molecules are incorporated into long chains, into starches, fats, and proteins. Thus, eating another organism does not in itself provide a source of metabolic energy to a heterotroph. It is first necessary to degrade that organism's macromolecules into the simple compounds from which they were built. This process is called **digestion.**

Protein Digestion

The chemical bonds that bind amino acids together into proteins are peptide bonds, and they are cleaved by protein-digesting enzymes called **proteases** (Figure 43-1, *A*). Some proteases cleave long protein chains into shorter segments by attacking specific amino acids. Others then chew away at the ends of these short segments. All of the protein-digesting enzymes of animals use similar reaction mechanisms, ones in which an electron-seeking component of the enzyme's active site draws an electron from the peptide bond. There are

FIGURE 43-1

The fundamental processes of digestion involve breaking the chemical bonds that bind together complex molecules with repeating units, such as protein (A) and starch (B).

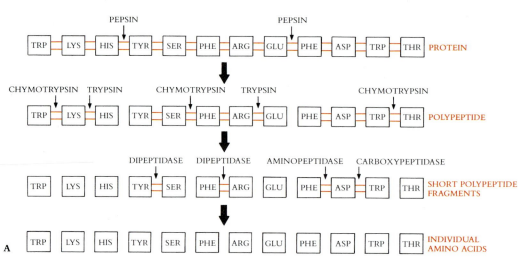

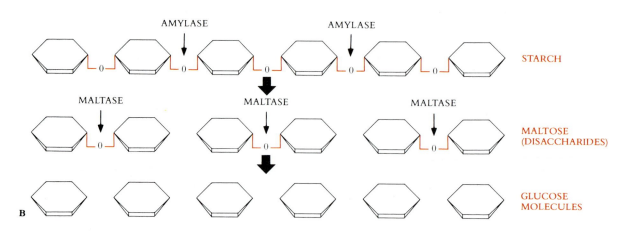

three classes of proteases, differing from one another in the way that the electron is attracted: (1) zinc proteases, such as carboxypeptidase, in which a zinc atom bound to the active site attracts the peptide bond electron; (2) serine proteases, such as trypsin, chymotrypsin, and elastase, in which a serine amino acid side chain within the enzyme's active site draws the electron; and (3) acid peptidases, such as pepsin, in which an aspartate amino acid side chain is the active element.

These differences in the reaction mechanism have a major influence on the ways in which the proteases are utilized. Zinc proteases and serine proteases operate at near-neutral pH values, because at these pH values zinc and serine are electron seeking, whereas at extremes of pH they are not. In mammals these enzymes are synthesized by special cells called **acinar cells** in the pancreas, from which they are secreted into a duct leading to the upper part of the small intestine—the **duodenum.** The pH of the duodenum is near neutral. Acid proteases, in contrast, are inactive at neutral pH values. Their active sites contain two aspartate residues and cannot catalyze protease activity unless one of these aspartate residues is ionized and the other is un-ionized; this occurs only at very low pH values, between 2 and 3. In mammals the principal acid peptidase is **pepsin,** which is synthesized by glands in the walls of the stomach and secreted into the gastric juice, which has a very low pH.

Just as guns possess safety catches to prevent their users from inadvertently discharging them, so proteases also possess a safety mechanism to prevent their digesting the cells that synthesize and transport them. In every case, proteases are first synthesized as **zymogens,** with one or more extra amino acid sequences added on to the basic sequence of the protein. As long as these zymogens possess the extra peptide segments, they are unable to assume an enzymatically active shape. Thus pepsin is synthesized as pepsinogen, with an extra segment of polypeptide added to one end of the protein that must be cleaved off before the protein can act as an enzyme. Similarly, trypsin is synthesized as trypsinogen, chymotrypsin as chymotrypsinogen, and so on. Such proteolytic enzymes as these are activated in the duodenum or stomach by the enzyme-catalyzed removal of these activity-blocking segments.

Starch and Fat Digestion

Starches, polymers of glucose, are digested by enzymes called **amylases** (Figure 43-1, *B*). The name of this group of enzymes is based on the Latin term for starch, *amylum*. Amylases cleave starches into short disaccharide fragments (maltose units), which are then further digested into glucose by other enzymes. Amylases are active at neutral pH values and are secreted by mammals into the saliva by special salivary glands lining the mouth, as well as into the small intestine by the pancreas. The pancreatic amylase carries out most of the starch digestion.

Fats are digested by enzymes called **lipases,** another class of enzyme that is active at neutral pH values, synthesized in the mammalian pancreas, and used in the small intestine. Lipases break up fat molecules into their constituents, glycerol and fatty acids, small molecules that can be broken down directly by the metabolic machinery of the body's cells. Unlike many other digestive enzymes, lipases are not very good at their job, only partially digesting fats. Much of the fat that is taken in by an animal remains undigested. This is not the problem it might at first seem, however, since fats and partly digested fats are lipid soluble and so, unlike partly digested proteins and starches, they can be absorbed across cell membranes. Once absorption has occurred, fats can be carried by the bloodstream to individual cells, in which their digestion can be completed by cellular enzymes.

> **In the stomach, proteins are hydrolyzed into short fragments by enzymes
> called proteases; the fragments are then cleaved into individual amino acids
> by a variety of peptidases in the small intestine. Also in the small intestine,
> starches are digested by amylases and fats are digested by lipases.**

THE EVOLUTION OF DIGESTIVE SYSTEMS
Phagocytosis

The first problem confronting an organism that attempts to digest part or all of another organism is where to do it. Fungi solve this problem by secreting their digestive enzymes onto food and then absorbing the products of enzyme digestion back into their hyphae. The members of most of the phyla of protists have taken different approaches. The majority of them are motile, not sitting for protracted periods in one spot digesting, as fungi do. In general, protists absorb food particles into their bodies and digest them there.

Protists avoid digesting themselves with their own digestive enzymes by enclosing their food particles within a membrane-bound vesicle called a **food vacuole** and digesting them there. This food vacuole is the "stomach" of the protists. The digestive enzymes of protists are compartmentalized in lysosomes when they are first synthesized. During digestion these lysosomes fuse with the membrane of a food vacuole, and the digestive enzymes pass into the food vacuole. This system serves to expose the food particle to the enzymes while protecting the cell from them. The sugars, amino acids, and other metabolites that are produced in the food vacuole by the action of the digestive enzymes are absorbed across the vacuole membrane into the cytoplasm of the *Paramecium* by a variety of transport channels. Fats, which are soluble in other lipids, pass freely across the membrane.

A Digestive Cavity

The simple feeding strategy of phagocytosis was greatly improved on by various groups of multicellular animals (Figure 43-2). In the sponges, the metazoans with the simplest structure, each cell resembles an individual zoomastigote, and the process of digestion is what would be expected—phagocytosis by each cell. All other metazoan animals, however, have a digestive cavity that is specialized and efficient.

In the cnidarians, the digestive cavity is a blind sac with only one opening—an opening that serves as both mouth and anus. Within this sac the first stage of digestion occurs. Large food particles are broken down into smaller ones, which are then phagocytized by the individual cells of the cnidarian. Further digestion takes place within the food vacuoles that are formed in the individual cells. The advent of a digestive cavity was a great modification in heterotrophy: the animals that possessed such a sac were able, for the first time, to digest other organisms larger than an individual cell.

> **The first great evolutionary change in digestion was the advent of a digestive cavity, which for the first time permitted animals to digest particles larger than a cell.**

During their evolution, the ancestors of the flatworms (phylum Platyhelminthes) developed a pharynx, by means of which they control the entry of external materials into the gut. The digestive cavity of the flatworms branches far more than that of the cnidarians,

FIGURE 43-2

Types of digestion. The most ancient animals used intracellular digestion (type I), which limited the size of the food particle to a fraction of the animal's size. Among the cnidarians we first see a digestive cavity, in which food particles are broken down extracellularly, outside of cells (type II); digestion within the cavity is only preliminary, however; final digestion is carried out within the cells that line the cavity. The round-worms are the oldest organisms that practice true extracellular digestion (type III), in which food passes through a "digestive tube" in one direction; digestive products are absorbed into the organisms during passage, and the residual matter is excreted from the far end of the tube.

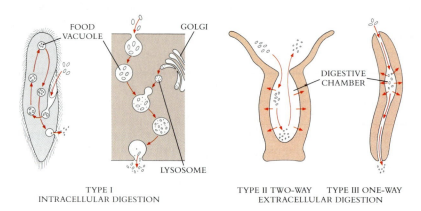

FOOD VACUOLE   GOLGI   DIGESTIVE CHAMBER   LYSOSOME

TYPE I
INTRACELLULAR DIGESTION

TYPE II TWO-WAY   TYPE III ONE-WAY
EXTRACELLULAR DIGESTION

although functionally it is similar. Again, digestion within the cavity is only preliminary; once the food particles have been broken into fragments, they are engulfed by cells in the lining of the cavity and then digested intracellularly.

Extracellular Digestion

The first true extracellular digestion occurred with the evolution of the roundworms, or nematodes (phylum Nematoda). The members of this group have a tubular gut composed of endoderm; the gut runs from their mouth to their anus. Food moves through it on a one-way journey, being digested and absorbed along the way. Although the details have been modified in many ways, this same general strategy is employed by all of the more complex animal phyla.

Much of the evolution of extracellular digestion has been concerned with how food is processed during its one-way journey through the gut. The principal elements evolved early; most of the elements found in the digestive system of vertebrates can be seen in the members of older phyla, such as the annelids. In earthworms, which are representative annelids, digestion proceeds in four stages:

1. **Acquisition.** Annelids have a muscularized pharynx, the contraction of which forces food into and through the nonmuscular gut.

2. **Storage.** From the pharynx, food passes into and eventually through a storage sac called a **crop.**

3. **Fragmentation.** From the crop, the food passes into a thickly muscled chamber, the **gizzard.** Here it is pulverized by churning, often together with grit or small stones that the animal has eaten.

4. **Extracellular digestion.** From the gizzard, food passes into a long **intestine.** Glands that secrete hydrolytic enzymes into the intestinal fluid are located in the epithelial lining of the intestine. These enzymes digest the food particle as it passes through the intestine. The sugars, amino acids, and other end products of digestion are absorbed across the intestinal wall.

In other groups of advanced invertebrates, such as mollusks and arthropods, the tubular gut is partitioned into three functional zones. The first zone combines the several functions of the anterior portion of the earthworm digestive tract, carrying out acquisition, storage, and fragmentation of food. The second zone is devoted exclusively to extracellular digestion. The third zone is devoted to the aborption of the end products of digestion. Among arthropods, for example, the mouth and foregut act to mix and prepare the food. The midgut in arthropods is the primary digestive organ; it functions in the absorption of the nutrients. Later, water is reabsorbed in the hindgut. This segmentation of the digestive tract into three functional zones is a pattern that is characteristic of all of the more advanced groups of animals.

> The more advanced invertebrates exhibit two fundamental advances in digestion, as compared with groups such as the cnidarians and flatworms: (1) extracellular digestion—the secretion by cells of digestive enzymes, coupled to the absorption by cells of the products of digestion; and (2) the one-way flow of food particles through the zone of digestion.

Among nonvertebrate chordates such as the lancelets, which obtain both food and oxygen from a stream of water passing through the mouth, the pharynx acts as a filtering device. It passes food particles through an esophagus into the digestive tract, while diverting the water into another channel. A pouch that is derived from a section of midgut in the lancelets has become specialized for the secretion of digestive enzymes. This pouch is the forerunner of the pancreas of vertebrates, a more specialized organ that has exactly the same function.

FIGURE 43-3

The human digestive system. Food passes from the mouth through the pharynx and esophagus to the stomach, where acids break up proteins and other molecules. From the stomach, food passes to the small intestine, where nutrients are absorbed into the body's bloodstream, and then to the large intestine or colon, where water is resorbed and the residual material is compacted. Exit is through the rectum and out the anus. Above the stomach is the liver, which plays many important roles in digestion, and nested below the stomach is the pancreas, which secretes many of the digestive enzymes.

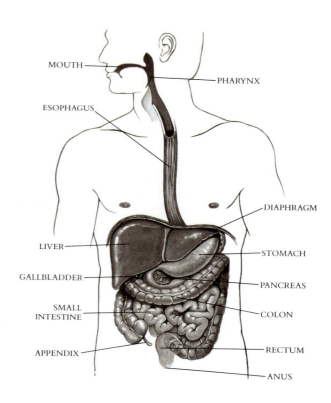

TABLE 43-1

PROCESS	CHAMBER
Ingestion	Mouth
Mixing	Pharynx
Macerating	Gizzard
Storage	Crop
Digestion	Stomach, intestine
Absorption	Intestine
Elimination	Colon

Organization of Vertebrate Digestive Systems

The general organization of the digestive tract (Figure 43-3) is the same in all vertebrates, although different elements are emphasized in different groups (Table 43-1).

Specializations among the digestive systems of different kinds of vertebrates reflect differences in the way these animals live. Fishes have a large pharynx with gill slits, not unlike that of the lancelets, whereas air-breathing vertebrates have a greatly reduced pharynx. Adult amphibians, which are carnivores, have a short gut; the food they ingest is readily digested. Many birds, in contrast, subsist on plant material; they have long, convoluted small intestines by means of which they prolong the process of digestion and aid absorption of digestion products. Birds, which lack teeth to masticate their food, accomplish this process instead in their stomachs, which have two chambers. In one of these chambers—the gizzard—small pebbles that are ingested by the bird are churned together with the food by muscular action; this churning serves to grind up the seeds and other hard plant material into smaller chunks before their digestion in the second chamber.

WHERE IT ALL BEGINS: THE MOUTH

The food of every vertebrate is taken in through its mouth, which in all groups except for the birds usually contains teeth (Figure 43-4). Vertebrates capture their food in many different ways, from biting it off of other animals to grazing on plants. As their different modes of feeding suggest, the teeth of different kinds of vertebrates are specialized in many different ways. The teeth of carnivorous mammals, for example, are pointed and lack flat grinding surfaces. Such pointed teeth are adaptations to cutting and shearing, and characterize meat eaters. Carnivores must capture and kill their prey but have little need to chew it, since digestive enzymes can act directly on animal cells. Recall how a cat or dog gulps

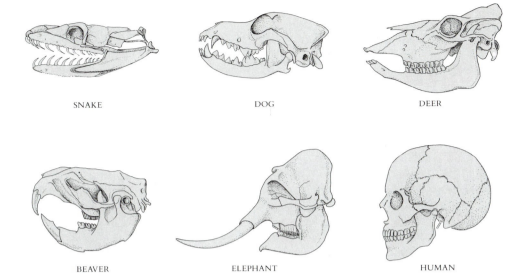

SNAKE DOG DEER

BEAVER ELEPHANT HUMAN

FIGURE 43-4

The arrangement of teeth in a variety of vertebrates. Teeth are in each case specialized for particular tasks. In the snake the teeth slope backward, to aid in retention of prey during swallowing. In carnivores such as the dog, "canine" teeth specialized for ripping food predominate. In herbivores such as deer, grinding teeth predominate. In the beaver the foreteeth are specialized as incisors—chisels. In the elephant, two of the upper front teeth are specialized as weapons. Humans as omnivores have a relatively broad-function mouth containing canine, chisel, and grinding teeth.

down food. By contrast, grass-eating herbivores, such as cows and horses, have very large and flat teeth, with complex ridges that are well suited to grinding. Broad, flat teeth characterize vegetarians, which must pulverize the cellulose cell wall of plant tissue before digesting it.

Human beings are **omnivores,** eating both plant and animal food regularly. As a result, our teeth are structurally intermediate between those of carnivores and those of herbivores. Viewed simply, humans are carnivores in the front of the mouth and herbivores in the back (Figure 43-5). The four front teeth in the upper and lower jaws are **incisors**—sharp, chisel shaped, and used for biting. On each side of the incisors, on both jaws, are sharp, pointed teeth called **canines,** which are used in tearing food. Behind each canine, on each side of the mouth and on both top and bottom jaws, are two **premolars** and three **molars,** both of which have flattened, ridged surfaces for grinding and crushing food. Human beings have only 20 teeth when they are babies. These "baby teeth" are lost later in childhood, being replaced then by the 32 adult teeth.

Within the mouth, the tongue of vertebrates mixes the food with a mucous solution, the **saliva.** In humans, saliva is secreted into the mouth by three pairs of salivary glands, which are located in the mucosal lining of the mouth. The saliva moistens and lubricates the food so that it is swallowed more readily and does not abrade the tissue it passes on its way down through the esophagus. The saliva also contains the hydrolytic enzyme amylase, which initiates the breakdown of starch and other complex polysaccharides into smaller fragments. This action by amylase is the first of the many digestive processes that occur as the food passes through the digestive tract.

The secretions of the salivary glands are controlled by the nerves, which maintain a constant secretion of about half a milliliter a minute in humans. This process goes on every minute, all of the time, and always keeps the mouth moist. The presence of food in the mouth acts as a stimulus to increase the rate of secretion by the salivary glands. Chemoreceptors in the mouth send a signal to the brainstem, which responds by stimulating the salivary glands. The most potent stimuli are acid solutions: lemon juice can increase the rate of salivation eightfold. Actually, even thinking about food acts as an effective stimulus in some vertebrates. The sight, sound, or smell of food is an effective stimulus to salivation in dogs, although it has only a slight effect in humans.

Vertebrate teeth serve to shred animal tissue and to grind plant material. Saliva secreted into the mouth moistens the food, aiding its journey into the digestive system, and begins the enzyme-catalyzed process of degradation.

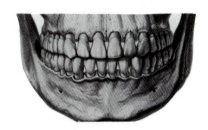

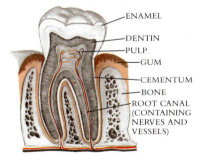

ENAMEL
DENTIN
PULP
GUM
CEMENTUM
BONE
ROOT CANAL (CONTAINING NERVES AND VESSELS)

FIGURE 43-5

Humans are carnivores in front of their mouths and herbivores in the back.

A *The front six teeth on the upper and lower jaws are canines and incisors. The remaining teeth, running along the sides of the mouth, are grinders, called premolars and molars.*

B *Each tooth is alive, with a central pulp containing both nerves and blood vessels. The actual chewing surface is hard enamel, layered over the softer dentin that forms the body of the tooth.*

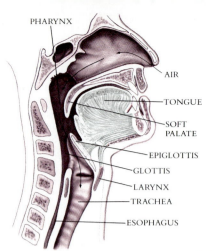

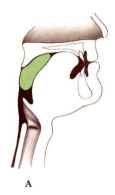
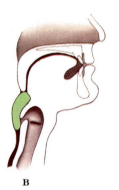
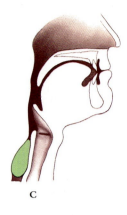

A   B   C

FIGURE 43-6

How humans swallow. As food passes back past the rear of the mouth (A), it presses the soft palate against the back wall of the pharynx, sealing off the nasal passage; as the food passes on down, a flap of tissue called the epiglottis folds down (B), sealing the respiratory passage; after the food enters the esophagus, the soft palate relaxes and the epiglottis is raised (C), opening the respiratory passage between the nasal cavity and the trachea.

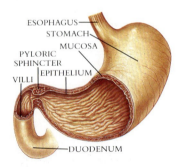

FIGURE 43-7

The human upper gastrointestinal area. Food enters the stomach from the esophagus. The epithelial walls of the stomach are dotted with gastric pits, which contain glands that secrete acid and digestive enzymes. The entrance to the small intestine—the duodenum—is controlled by a band of muscle called the pyloric sphincter. The surface of the duodenum is covered with microscopic villi, which greatly increase its surface area and so aid in absorption of nutrients.

THE JOURNEY OF FOOD TO THE STOMACH

When food passes beyond the teeth to the back of the mouth in a mammal, three things happen (Figure 43-6):

1. The **palate** (which is soft in mammals) elevates, pushing against the back wall of the pharynx. This seals off the nasal cavity and prevents any food from entering it.
2. Pressure against the pharynx stimulates receptors within its walls to send nerve signals to the swallowing center in the brainstem.
3. The swallowing center sends out nerve signals that keep food from getting into the respiratory tract, which branches off just below the pharynx. These signals both inhibit respiration and seal the windpipe **(trachea)** by raising the larynx and thus closing the passage between the larynx and the pharynx, called the **glottis.** A flap of tissue, called the **epiglottis,** folds back over the opening, providing a seal.

After passing the tracheal opening, the food encounters a muscular constriction, or **sphincter,** which opens in response to the pressure exerted by the food. Passing through this sphincter, the food enters a tube called the **esophagus,** which connects the pharynx to the stomach. No further digestion takes place in the esophagus. Its role is that of an escalator, moving food down toward the stomach. In adult human beings the esophagus is about 25 centimeters long, and its lower end opens into the stomach proper. The upper portion of the esophagus is enveloped in skeletal muscle, and the lower two thirds is enveloped in smooth muscle. Successive waves of contraction of these muscles, which are stimulated by the swallowing center, move food down through the esophagus to the stomach. These rhythmic sequences of waves of muscular contraction in the walls of a tube are called **peristalsis.** Recall that the movement of human eggs through the fallopian tubes, described on page 866, also occurs by peristalsis. Because the movement of food through the esophagus is primarily caused by these peristaltic contractions, humans can swallow even if they are upside down.

The exit of food from the esophagus to the stomach is controlled by another sphincter. When this second sphincter is contracted, it prevents the food in the stomach from moving back up the esophagus.

PRELIMINARY DIGESTION: THE STOMACH

The stomach (Figure 43-7) is a saclike portion of the digestive tract. In it, the digestive process is organized; the stomach collects ingested food, hydrolyzes it into small molecular fragments, and feeds it in a controlled fashion into the primary digestive organ, the small intestine. The interior of the stomach, like that of the rest of the digestive tract, is continuous with the outside of the body. The epithelium overlays a deep layer of connective tissue (Figure 43-8), called **mucosa,** below which are located a complex array of muscles, blood vessels, and nerves.

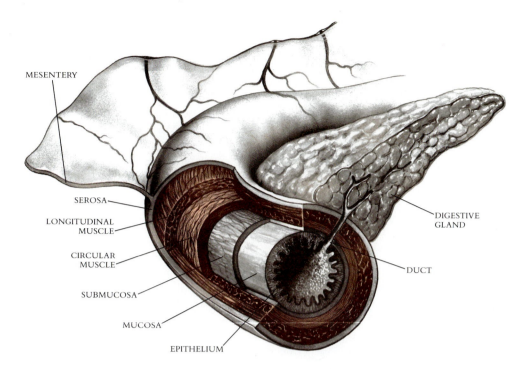

MESENTERY

SEROSA

LONGITUDINAL MUSCLE

CIRCULAR MUSCLE

SUBMUCOSA

MUCOSA

EPITHELIUM

DIGESTIVE GLAND

DUCT

FIGURE 43-8

Organization of the digestive tract. Both the stomach and the intestinal tract have the same general organization, being tubes encased within successive layers of different tissues. The innermost layer of tissue is a thin cylinder of epithelium. This is encased by a thick layer of mucosa, a connective tissue rich in blood vessels, glands, and nerves. Surrounding the mucosa are layers of muscle tissue and an envelope of tough connective tissue called serosa, often connected to body structures by extensions called mesentery.

The epithelium of the stomach is the source of the "digestive juices," which actually are a complex mixture of substances that continue the breakdown of food that begins in the saliva. The upper epithelial surface of the stomach is dotted with deep depressions called **gastric pits** (Figure 43-9). In them the epithelial membrane is invaginated inward, forming exocrine glands within the mucosa. These exocrine glands contain two kinds of secreting cells, **parietal cells,** which secrete hydrochloric acid (HCl), and **chief cells,** which secrete the protein pepsinogen. Within the gland, HCl cleaves a terminal fragment from the secreted pepsinogen, converting it into the protein-hydrolyzing enzyme pepsin.

Many of the epithelial cells that line the stomach are specialized for the secretion of mucus. This mucus, which is produced in large quantities, lubricates the stomach wall and facilitates the movement of food within the stomach. It also protects the cells of the stomach wall from abrasion by the food and, perhaps most important, protects the walls of the stomach from its own digestive juices, the **gastric fluid.**

Food within the stomach is attacked by both acid and enzymes. The human stomach secretes about 2 liters of HCl every day, creating a very concentrated acid solution. This solution is actually about 150 millimolar, and thus 3 million times more acidic than the blood. The hydrochloric acid breaks up connective tissue. It does this because the very low pH values, between 1.5 and 2.5, that are created by the HCl change the ionization of carboxyl and amino side groups of proteins. This process causes the folded proteins of connective tissue to open out and disrupts their associations with one another.

The action of the acid disintegrates the food into molecular fragments, and thus is an essential prelude to digestion. The acid itself does not carry out any further digestive activity. It has a very limited ability to break proteins down into amino acids, or carbohydrates into their constituent sugars, and it does not attack fats at all. The further digestion of protein is accomplished by the enzyme pepsin, which cleaves proteins into short polypeptides. Because the fragments produced by pepsin are large and frequently charged, they cannot pass across the epithelial membrane. Consequently, no absorption takes place through the stomach wall.

It is important that a stomach not produce *too* much acid. If it did, it would be impossible for the body to neutralize the acid later in the small intestine, a step that is essential for the terminal stages of digestion. The stomach controls the production of acid by means of a hormone produced by endocrine cells that are scattered throughout its epithelial layer (Figure 43-10). This hormone, called **gastrin,** regulates the synthesis of hydrochloric acid

FIGURE 43-9

Gastric pits are deep invaginations of the stomach epithelium down into the underlying mucosa, at the base of which parietal and chief cells secrete hydrochloric acid and pepsinogen into the lumen of the pit.

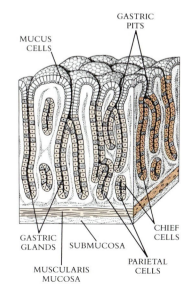

MUCUS CELLS

GASTRIC PITS

GASTRIC GLANDS

SUBMUCOSA

MUSCULARIS MUCOSA

CHIEF CELLS

PARIETAL CELLS

BEAKS

Most vertebrates have teeth. A notable exception are the birds, which instead possess **beaks,** hard, horny extensions of the mouth that serve the same purpose. Just as teeth have evolved many specializations suiting them to particular tasks, so too have beaks. Here are five specializations of many:

Chiseling. This woodpecker (Figure 43-A, *1*) is extracting a grub from a tree. It captured it by driving its sharp beak into the bark. Woodpecker "holes" in trees are commonly seen.

Tearing. The prairie falcon is a raptor. Like other hawks and eagles, it tears its prey apart with its sharp beak and eats the flesh. Raptors are the carnivorous birds, and among them this little falcon is a leopard.

Crushing. The cardinal feeds upon seeds, which it crushes in its strong beak. Many of the most common finches, sparrows, and other songbirds also eat primarily seeds, and have similar beaks.

Digging. The roseate spoonbill shown here has a long shovel-shaped bill, which it uses to spoon along the muddy bottom under the surface of the water for food.

Sucking. Hummingbirds suck nectar from flowers and live upon the sugar that the nectar contains. Hummingbirds do this while hovering in the air, their long slender beaks protruding down into the flower.

FIGURE 43-A

1 *Woodpecker,*
a beak specialized
for chiseling.

2 *Prairie falcon, a beak*
specialized for tearing.

by the parietal cells of the gastric pits, permitting such synthesis to occur only when the pH of the stomach contents is higher than about 1.5.

Some stomachs greatly overproduce gastrin, which results in excessive acid production. The excessive acid may attack the walls of the small intestine, burning holes through the wall. These holes are called **duodenal ulcers.** The contents of the small intestine are not normally acidic, and this organ is much less able to withstand the disruptive actions of stomach acids than is the wall of the stomach. For this reason, over 90% of all ulcers are duodenal, although other ulcers do sometimes occur in the stomach.

The ability of the acid solution within the stomach to break down cells is also a very effective means of killing bacteria that are ingested with the food. Most bacteria are simply

3

5

4

3 *Cardinal, a beak specialized for crushing.*

5 *Hummingbird, a beak specialized for sucking.*

4 *Roseate spoonbill, a beak specialized for digging.*

torn apart by the strong acid. The few that survive their harrowing journey through the stomach and make it to the intestine intact are able to grow and multiply there, particularly in the large intestine. There are enough survivors so that most vertebrates harbor thriving colonies of bacteria within their intestines, with the result that bacteria are a major component of feces. As we will discuss later, bacteria that live within the digestive tract of cows and other ruminants play a key role in the ability of these mammals to digest cellulose.

In the stomach, concentrated acid breaks up connective tissue and protein into molecular fragments, which are further digested by pepsin into short polypeptides. Carbohydrates and fats are not digested in the stomach.

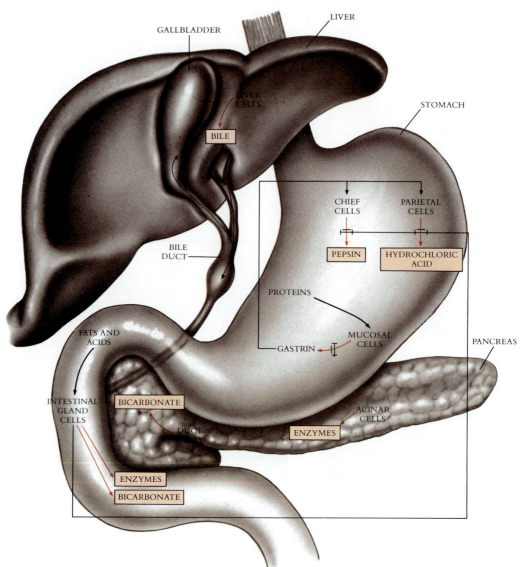

GALLBLADDER

LIVER

LIVER
CELLS

BILE

STOMACH

BILE
DUCT

CHIEF
CELLS

PARIETAL
CELLS

PEPSIN

HYDROCHLORIC
ACID

PROTEINS

FATS AND
ACIDS

MUCOSAL
CELLS

GASTRIN

PANCREAS

INTESTINAL
GLAND
CELLS

BICARBONATE

DUCT

ACINAR
CELLS

ENZYMES

ENZYMES

BICARBONATE

FIGURE 43-10

Regulation of digestive enzyme production. The production of pepsin and HCl by the endothelial glands of the stomach, and of bile by the liver, is regulated by the hormone gastrin. The production of gastrin, and of pancreatic digestive enzymes, is regulated by intestinal gland cells, which also directly regulate production of pepsin, HCl, and bile. In this way, levels of HCl production in the stomach are increased when stomach pH rises or duodenum pH falls below critical values.

The inner surface of the stomach is highly convoluted. For this reason it can fold up when empty and open out like an expanding balloon as it fills with food. There is, of course, a limit to how much food a stomach can hold. The human stomach has a volume of about 50 milliliters when empty; when full, it may have a volume 50 times larger, from 2 to 4 liters. Carnivores, for which sporadic gorging is an important survival strategy, possess stomachs that are able to distend much more than our stomachs, or those of most other mammals, can.

The digestive tract exits from the stomach at a third muscular constriction, the **pyloric sphincter** (Figure 43-7). The pyloric sphincter is the gate to the small intestine, the organ within which the final stages of digestion occur. The pyloric sphincter, therefore, is the traffic light of the digestive system. The capacity of the small intestine is limited, and its digestive processes take time. Consequently, efficient digestion requires that only relatively small portions of food be introduced from the stomach into the small intestine at any one time. When a small volume of food, which is by now highly acidic, passes into the small intestine, the acid introduced with the food acts as a signal, prompting the closing of the pyloric sphincter. As time passes, the food is digested and the acid that entered the small intestine with it is neutralized. At a certain point in the process, the pH of the small intestine reaches a level that signals the pyloric sphincter to open once again. Another small portion of food is introduced from the stomach into the small intestine, and the processs continues.

TERMINAL DIGESTION AND ABSORPTION: THE SMALL INTESTINE

The small intestine is the true digestive vat of the vertebrate body. Within it, carbohydrates, proteins, and fats are broken down into sugars, amino acids, and fatty acids. Once these small molecules have been produced, they all pass across the epithelial wall of the small intestine into the bloodstream. Some of the enzymes necessary for these digestive processes are secreted by the cells of the intestinal wall. Most, however, are introduced into a short initial segment of the small intestine, the duodenum, through a duct from a gland called the pancreas.

The **pancreas** secretes a number of different enzymes that break down carbohydrates, proteins, and fats in a variety of ways. This important gland also possesses cells similar to the parietal cells of the gastric pits. However, instead of secreting an acid, HCl, the specialized cells of the small intestine secrete a base, **bicarbonate.** The alkaline bicarbonate that is secreted by the pancreas is critical to successful digestion, since most of the enzymes secreted by the pancreas will not work in acid solution. The introduction of bicarbonate into the duodenum neutralizes the acid derived from the stomach and thus permits the digestive enzymes to function.

Because fats are insoluble in water, they tend to enter the small intestine as small globules that are not attacked readily by the enzymes secreted by the pancreas. Before fats can be digested, they must be made soluble. This process is carried out by a collection of detergent molecules secreted by a second gland, the **liver.** The liver is the body's principal metabolic factory, turning foodstuffs arriving from the digestive tract in the bloodstream into substances that are utilized by the different cells of the body. It is the largest internal organ of the body. In an adult human being the liver weighs about three pounds and is the size of a football.

The liver carries out a wide variety of metabolic functions, many of which we will discuss in later chapters. It supplies quick energy, metabolizes alcohol, makes proteins, stores vitamins and minerals, regulates blood clotting, regulates the production of cholesterol, and detoxifies poisons. It also produces and secretes through a duct into the duodenum the detergent molecules mentioned above, known as **bile salts.** These bile salts act as superdetergents, rendering fats within the digestive tract soluble. In addition, the liver also secretes cholesterol and the phospholipid lecithin, both of which also play a role in rendering fat soluble. Human beings store excess secreted bile in the **gall bladder.**

Of the approximately 6-meter length of the human small intestine (Figure 43-11), only the duodenum region (the first 25 centimeters) is actively involved in digestion. The rest of the small intestine is highly specialized to aid in the absorption of the products of digestion by the bloodstream. The epithelial wall of the small intestine is covered with fine fingerlike projections called **villi,** which are so small that it takes a microscope to see them (Figure 43-12). In turn, each of the epithelial cells covering the villi is covered on its outer surface by a field of cytoplasmic projections called **microvilli.** Both kinds of projections greatly increase the absorptive surface of the small intestine's epithelium lining. The average surface area of the small intestine in an adult human being is about 300 square meters, a greater area than the surface covered by many houses. The membranes of the epithelial cells contain carrier systems that actively transport sugars and amino acids across the membrane; fatty acids cross passively by diffusion.

> **Most digestion occurs in the first 4% of the length of the small intestine, a zone called the duodenum. The rest of the small intestine is devoted to the absorption of the products of digestion.**

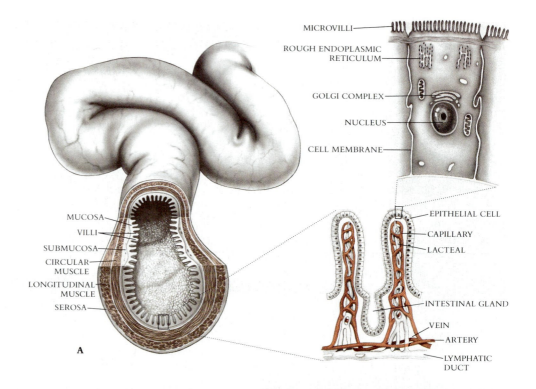

MICROVILLI

ROUGH ENDOPLASMIC
RETICULUM

GOLGI COMPLEX

NUCLEUS

CELL MEMBRANE

MUCOSA
VILLI
SUBMUCOSA
CIRCULAR
MUSCLE
LONGITUDINAL
MUSCLE
SEROSA

A

EPITHELIAL CELL
CAPILLARY
LACTEAL

INTESTINAL GLAND

VEIN
ARTERY
LYMPHATIC
DUCT

FIGURE 43-11

A *Cross section of the small intestine.*

B *Villi, shown in a scanning electron micrograph, are very densely clustered, giving the small intestine an enormous surface area, which is very important in efficient absorption of the digestion products.*

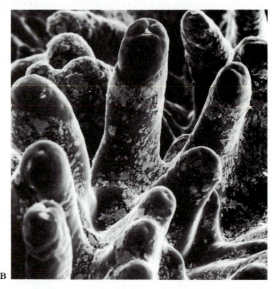

B

The amount of material passing through the small intestine is startlingly large. An average human consumes about 800 grams of solid food and 1200 milliliters of water each day, for a total volume of about 2 liters. To this amount is added about 1.5 liters of fluid from the salivary glands, 2 liters from the gastric secretions of the stomach, 1.5 liters from the pancreas, 0.5 liter from the liver, and 1.5 liters of intestinal secretions. The total is a remarkable 9 liters. However, although the flux is great, the *net* passage is small. Almost all of these fluids and solids are reabsorbed during their passage through the intestines, with about 8.5 liters passing across the walls of the small intestine and an additional 350 milliliters through the wall of the large intestine. Of the 800 grams of solid and 9 liters of liquid that enter the digestive tract, only about 50 grams of solid and 100 milliliters of liquid leave the body as feces. The normal fluid absorption efficiency of the digestive tract thus approaches 99%, which is very high indeed.

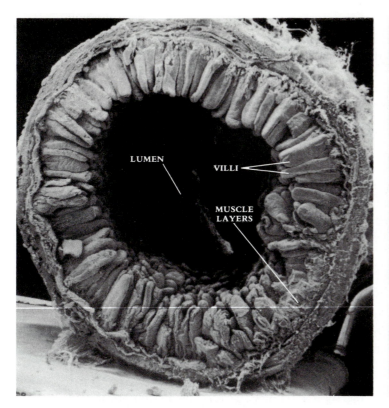

LUMEN

VILLI

MUSCLE
LAYERS

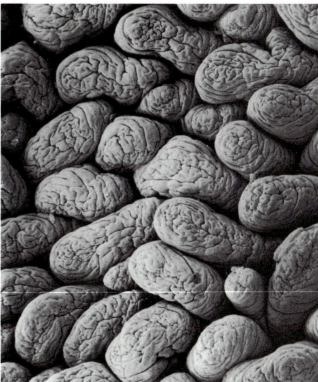

FIGURE 43-12

The surface of the duodenum is covered with small villi, fingerlike projections that greatly increase its surface area.

When this efficient reabsorption of the body's fluids is interrupted, the body rapidly dehydrates. The consumption of excessive amounts of alcohol produces such a temporary dehydrating effect by inhibiting reabsorption of water through the small intestine and the kidney (described in Chapter 48). When significant interference with reabsorption occurs, severe diarrhea may result, a condition that quickly leads to serious dehydration. The high mortality associated with diseases such as typhus reflects this kind of dehydration. Dehydration due to failure of intestinal reabsorption is one of the most important causes of death among infants in the less developed parts of the world. Some of the causes of traveler's diarrhea are discussed in the boxed essay on p. 518.

CONCENTRATION OF SOLIDS: THE LARGE INTESTINE

The large intestine, or **colon,** is much shorter than the small intestine, occupying approximately the last meter of the intestinal tract. No digestion takes place within the large intestine, and only about 4% of the absorption of fluids by the body occurs there. The large intestine is not convoluted, lying instead in three relatively straight segments, and its inner surface does not possess villi. Consequently, the large intestine has less than one thirtieth the absorptive surface area of the small intestine. Although sodium, vitamin K, and some other products of bacterial metabolism are absorbed across its wall, the primary function of the large intestine is to act as a refuse dump. Within it, undigested material, primarily bacterial fragments and cellulose, is compacted and stored. Many bacteria live and actively divide within the large intestine; the excess bacteria are incorporated into the refuse material. Bacterial fermentation produces gas within the colon at a rate of about 500 milliliters per day. This rate increases greatly after the consumption of beans or other vegetable matter because the passage of undigested plant material into the large intestine provides material for fermentation.

The large intestine serves primarily to compact the solid refuse remaining after digestion of food, thus facilitating its elimination.

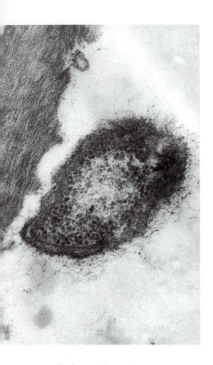

FIGURE 43-13

A bacterium, Bacteroides succinogenes, *digesting a cellulose fiber. Its approach is not unlike that employed by the fungi: the cell adheres to the fiber by means of a loose carbohydrate mesh, and secretes bacterial enzymes such as cellulase and xylulase onto the fiber. These enzymes then digest the fiber in that immediate vicinity.*

Like all good things, digestion eventually comes to an end. In this case the end is a short extension of the large intestine called the **rectum.** Compacted solids within the colon, called **feces,** pass into the rectum as a result of the peristaltic contractions of the muscles encasing the large intestine. From the rectum, the solid material passes out the anus through two anal sphincters. The first of these is composed of smooth muscle; it opens involuntarily in response to a pressure-generated nerve signal from the rectum. The second sphincter, in contrast, is composed of striated muscle. It is subject to voluntary control from the brain, thus permitting a conscious decision to delay defecation.

In all vertebrates except placental mammals, the reproductive and urinary tracts also exit into the anal cavity. This common cavity is called the **cloaca.** In placental mammals (but not in monotremes and marsupials), the rectum is a nonabsorptive cavity that acts solely to control the exit of material from the large intestine. The reproductive and urinary tracts of placental mammals do not join the digestive tract in the rectum, instead making their own separate exits from the body.

SYMBIOSIS WITHIN THE DIGESTIVE SYSTEMS OF VERTEBRATES

Vertebrates lack the enzymes that are necessary to digest some kinds of potentially useful foodstuffs. In certain kinds of vertebrates, bacteria and protozoans living within the digestive tract possess the enzymes necessary to convert these materials into substances that the host vertebrate can digest. The relationship is mutually beneficial and provides an excellent example of symbiosis.

Although bacterial digestion within the gut plays a relatively small role in human metabolism, it is an essential element in the metabolism of many other vertebrates. Herbivorous animals are one example. No vertebrate produces the enzyme cellulase, and therefore no vertebrate can digest cellulose, the chief carbohydrate components of plants, without outside help. Herbivorous animals achieve the digestion of cellulose by using bacteria that live within their digestive tracts to produce this necessary enzyme (Figure 43-13). Termites, cockroaches, and a few other groups of insects operate in the same way, utilizing protozoans rather than bacteria. Silverfish, which are members of a primitive order of insects, seem to have the ability, extremely unusual among animals, to produce cellulase themselves.

In horses, rodents, and lagomorphs (an order that includes mammals such as rabbits and hares), the digestion of cellulose by bacteria takes place in a large pouch of the intestine called the **cecum.** In cows and related mammals, the digesting pouch is called the **rumen;** it is located in front of the stomach instead of in the intestinal region. This forward location proves to be an important difference, because it allows a cow to regurgitate and rechew the contents of the rumen ("chewing the cud") (Figure 43-14). This process leads to the far more efficient digestion of cellulose in mammals such as cows, which have a rumen, than in those which lack such an organ, such as horses.

Rabbits have evolved a bizarre but effective way to digest cellulose, a way that achieves a degree of efficiency similar to that of ruminate digestion, despite the fact that a rabbit's cecum is positioned behind the stomach, which precludes regurgitation and redigestion in it. Rabbits do this by eating their feces, thus passing their food through the digestive tract for a second time. The second passage provides the rabbit with many of the important products of bacterial metabolism; rabbits cannot remain healthy if they are prevented from eating their feces and thus gaining the opportunity of digesting more of the cellulose in them.

Cellulose is not the only indigestible plant product that can be used by vertebrates as a food source because of the digestive activities of intestinal bacteria. Wax, a substance that is indigestible by almost all animals, is digested by symbiotic bacteria that live within the gut of the honey guides, African birds that eat the wax in bee nests.

Another example of the way in which intestinal bacteria function in the metabolism of their host animals is provided by the synthesis of vitamin K. All mammals rely on intestinal bacteria to synthesize this vitamin, which is necessary for the coagulation of blood. Birds, which lack the necessary bacteria, must consume the necessary quantities of vitamin K in their food. In humans, prolonged treatment with antibiotics greatly reduces the populations of bacteria in the intestine; under such circumstances, it may be necessary to provide supplementary vitamin K.

Much of the food value of plants is tied up in cellulose, which no vertebrate is able to digest unaided. Many vertebrates, however, harbor colonies of microorganisms in their digestive tract that do have the ability to digest cellulose and thus make it available as a source of food for their hosts. Intestinal microorganisms often also excrete molecules important to the well-being of their vertebrate hosts.

NUTRITION

The ingestion of food by vertebrates serves two ends: it provides a source of energy, and it also provides raw materials that an animal is not able to manufacture for itself (Figure 43-15). Energy is derived from sugars, fatty acids, and amino acids, largely through the oxidative respiration of glucose. Such oxidative metabolism is carried out in muscle cells and in many other cells that receive glucose from the bloodstream. Fats and amino acids are converted to glucose in the liver. As you will recall from Chapter 10, fats are made from glucose, with the investment of energy. Their breakdown provides twice as many calories as does the breakdown of glucose or other simple sugars and amino acids.

Fats represent a very efficient way to store energy. The liver maintains a very constant level of glucose in the blood, and stores several hours' reserve of glucose in the form of glycogen. Any intake of food in excess of that required to maintain the glycogen reserve

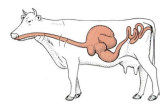

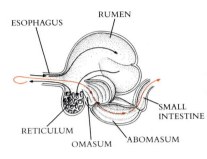

FIGURE 43-14

Digestive tract of a ruminant. In a cow, the grass and other plants that an individual eats pass first into the rumen, the first of four chambers, where they are partially digested; after passing through a second chamber, the reticulum, food may be regurgitated for rechewing before passing through the rear chambers.

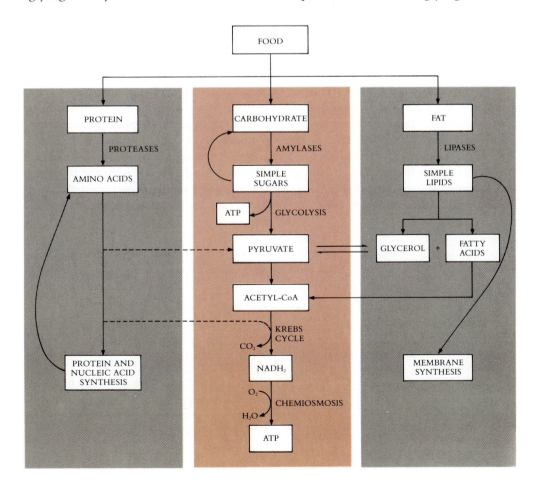

FIGURE 43-15

The fate of food. Most food can be broadly considered to consist of carbohydrate, fat, or protein. During digestion, protein is broken down into amino acids, carbohydrate is broken down into simple sugars, and fats are absorbed often without degradation. The resulting sugars and lipids may be considered members of a common pool of metabolites, as each may be converted to the other. Amino acids form a distinct metabolic pool— although there is some interconversion of amino acids and sugars, the amount is relatively small. In general, amino acids are channeled into synthesis of nitrogen-containing molecules such as proteins, hormones, and nucleotides, and the carbohydrates and fats are used to store energy and assemble structural components of the cell. (Modified from A. Vander et al., Human Physiology, *third edition, McGraw-Hill.)*

OBESITY STARTS HERE*			
*FOR ADULTS OF MEDIUM BUILD BETWEEN THE AGES OF 25 AND 59, INCLUDING CLOTHES AND ALLOWING FOR 1″ HEELS. (BASED ON THE 1983 METROPOLITAN LIFE TABLES.)			
MEN		WOMEN	
HEIGHT	WEIGHT	HEIGHT	WEIGHT
5′6″	174	5′2″	150
5′7″	178	5′3″	154
5′8″	181	5′4″	157
5′9″	185	5′5″	161
5′10″	188	5′6″	164
5′11″	192	5′7″	168
6′0′	196	5′8″	172
6′1″	200	5′9″	175
6′2″	205	5′10″	179

FIGURE 43-16

Obesity is usually characterized as the state of being more than 20% heavier than the average person of the same sex and height. For a variety of heights, these are the weight values at which obesity begins in Americans, using average 1985 weights.

results in one of two consequences. Either the excess glucose is metabolized by the muscles and other cells of the body, or it is converted to fat and stored within fat cells. Thus we have the simple equation

$$FOOD - EXERCISE = FAT$$

In wealthy countries such as those of North America and Europe, the obesity that results from chronic overeating and from imbalanced diets high in carbohydrates is a significant human health problem. In the United States, about 30% of middle-aged women and 15% of middle-aged men are classified as overweight. In other words, they weigh at least 20% more than the average weight for their height (Figure 43-16). Being overweight is strongly correlated with coronary heart disease and many other disorders.

Over the course of their evolution, many vertebrates have lost the ability to synthesize different substances that nevertheless continue to play critical roles in their metabolism. They simply no longer have the genes that encode the enzymes necessary to produce these substances. Perhaps substances that these vertebrates required were dependably present in the food they ate, so that mutations rendering these enzymes nonfunctional were not disadvantageous, and came to predominate.

Substances that an animal cannot manufacture for itself but which are necessary for its health must be obtained in other ways—in its diet. Such essential organic substances utilized by organisms in trace amounts are called **vitamins.** Human beings, apes, monkeys, and guinea pigs, for example, have lost the ability to synthesize ascorbic acid (vitamin C) and will develop the disease scurvy if vitamin C is not supplied in sufficient quantities in their diets. All other mammals, as far as is known, are able to synthesize ascorbic acid. Vitamin K, as we just mentioned, is obtained by mammals from their symbiotic intestinal bacteria but must be consumed by birds with their food. Humans require at least 13 different vitamins.

> **A vitamin is an organic substance that is required in minute quantities by an organism for growth and activity, but which the organism cannot synthesize.**

Some of the substances that vertebrates are not able to synthesize are required in more than trace amounts. Many vertebrates, for example, require one or more of the 20 amino acids that are necessary to synthesize proteins. Human beings are unable to synthesize 8 of these 20 amino acids: lysine, tryptophan, threonine, methionine, phenylalanine, leucine, isoleucine, and valine. These amino acids must be obtained by humans from proteins in the food they eat (Figure 43-17). All vertebrates have lost the ability to synthesize certain polyunsaturated fats that provide backbones for fatty acid synthesis. Some essential substances that vertebrates do synthesize for themselves cannot be manufactured by the mem-

FIGURE 43-17

The protein content of a variety of common foods. The dashed and solid horizontal lines represent the minimum daily requirements for children and adults, respectively.

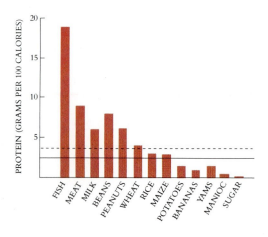

bers of other animal groups. Insects, for example, are unable to synthesize cholesterol, which is required for fatty acid synthesis; they must therefore obtain cholesterol in their diet.

Besides supplying energy and those organic compounds that cannot be synthesized (termed **essential amino acids** and **essential vitamins**), the food that an animal consumes must also supply essential minerals such as calcium and phosphorus. In addition, it must include a wide variety of **trace elements,** which are minerals that are required in very small amounts. Among the trace elements are iodine (a component of thyroid hormone), cobalt (a component of vitamin B_{12}), zinc and molybdenum (components of enzymes), manganese, and selenium. All of these, with the possible exception of selenium, are essential for plant growth also (see Chapter 34); they are obtained by the animals that require them either directly from plants or from animals that have eaten the plants.

Interestingly, one essential characteristic of food is simply bulk, its content of undigested fiber. The large intestine of humans, for example, has evolved as an organ adapted to process food that has a relatively high fiber content. Diets that are low in fiber, which are common in the United States, result in a slower passage of food through the colon than is desirable. This low dietary fiber content is thought to be associated with the levels of colon cancer in the United States, levels that are among the highest in the world.

SUMMARY

1. Digestion is the rendering of parts of organisms into amino acids and sugars, which can be metabolized by heterotrophic organisms.

2. Primitive digestion was carried out within cells, in food vacuoles that isolated the digestive process from the rest of the cell. The more advanced invertebrates evolved extracellular digestion, in which digestive enzymes are secreted into a digestive cavity and the products of digestion are absorbed from that cavity.

3. The digestive tract of vertebrates is one way. The initial portion leads from a mouth and pharynx through an esophagus to a stomach. In some vertebrates, food is further macerated in a gizzard, and in some birds it is stored in a crop. In most mammals, food is stored in the stomach, and its digestion begins there.

4. The stomach juices are concentrated acid, in which the protein-digesting enzyme pepsin is active.

5. Food passes from the stomach to the small intestine, where the pH is neutralized and a variety of enzymes synthesized in the pancreas act to complete digestion. Most digestion occurs in the first 25 centimeters of the small intestine, in a zone called the duodenum.

6. The products of digestion are absorbed across the walls of the small intestine, which possess numerous villi and so achieve a very great surface area. Amino acids and sugars are transported by specific transmembrane channels, whereas fats, which are lipid soluble, pass readily across the membranes of the villi.

7. The large intestine has little digestive or absorptive activity; it functions principally to compact the refuse that is left over from digestion for easier elimination.

8. Vertebrates lack the enzymes necessary to digest cellulose. Ruminants, such as cows and sheep, are able to digest grass by maintaining colonies of cellulase-excreting bacteria in their gastrointestinal tracts, much as termites maintain symbiotic protozoa in their guts that are able to digest the cellulose in wood.

9. Vertebrates also lack the essential organic substances required to synthesize many necessary compounds and must obtain these organic substances, called vitamins, from their diet. A number of trace elements also must be present in the diet.

SELF-QUIZ

The following questions pinpoint important information in this chapter. You should be sure to know the answers to each of them.

1. Protein-digesting enzymes are called _____.

2. Human beings are
   (a) herbivores   (c) omnivores
   (b) carnivores   (d) anivores

3. Match the corresponding pairs.
   (a) gastric pits   (1) production of hydrochloric acid
   (b) parietal cells   (2) production of mucus
   (c) chief cells   (3) production of pepsinogen

4. Many enzymes, such as those that digest proteins, are initially synthesized in inactive forms, called _____.

5. When a lipase digests fat, the digestion is
   (a) complete   (b) partial

6. The stomach regulates hydrochloric acid secretion by use of the hormone
   (a) pepsin   (c) lipase
   (b) gastrin   (d) trypsin

7. The pancreas secretes
   (a) bicarbonate   (d) a and b
   (b) hydrochloric acid   (e) a and c
   (c) pepsinogen

8. The liver produces detergent substances known as _____.

9. Most food digestion and absorption by the human body occurs in the
   (a) duodenum   (d) stomach
   (b) liver   (e) large intestine
   (c) pancreas

10. In cows, digestion of cellulose is accomplished by bacteria in the pouch in front of the stomach called the
    (a) rumen   (b) gall bladder   (c) colon   (d) cecum

11. Which material is the most efficient form for humans to store excess energy?
    (a) fats   (d) glycogen
    (b) proteins   (e) ATP
    (c) cellulose

12. _____ are organic substances required in minute quantities by an organism for growth and activity, but which the organism cannot synthesize.

THOUGHT QUESTIONS

1. One of the great evolutionary changes in the process of digestion was the advent of extracellular digestion. Why was this such a beneficial change? What advantage does it confer over intracellular digestion?

2. Many birds possess crops, although few mammals do. Suggest a reason for this difference.

3. Human beings obtain vitamin K from symbiotic bacteria living in their gastrointestinal tract. Many bacteria also produce ascorbic acid, vitamin C. Can you suggest a reason why people have not evolved a symbiotic relationship with bacteria that would result in their obtaining bacterial vitamin C?

4. Many proteolytic enzymes are synthesized in the pancreas and released into the duodenum. Since this is the case, why do mammalian digestive systems go to the trouble of producing pepsin in the stomach? What does pepsin do that the others do not?

FOR FURTHER READING

DAVENPORT, H.W.: "Why the Stomach Does not Digest Itself," *Scientific American,* January 1972, pages 86-93. An interesting description of how the structure of the stomach is suited to its peculiar task.

JENNINGS, J.B.: *Feeding, Digestion, and Assimilation in Animals,* St. Martin's Press, Inc., New York, 1973. A comparative approach to vertebrate digestion, with a good account of feeding mechanisms.

MADGE, D.S.: *The Mammalian Alimentary System: A Functional Approach,* Edward Arnold Publishers, Ltd., London, 1975. A good elementary text.

MOOG, F.: "The Lining of the Small Intestine," *Scientific American,* November 1981, pages 154-176. A clear description of the most important absorptive surface in the human body.

SCRIMSHAW, N.S., and V.R. YOUNG: "The Requirements of Human Nutrition," *Scientific American,* September 1976, pages 50-64. If you want to know what you should eat and what perhaps you should not, this article will point the way. A particularly good treatment of the important roles of trace elements in the human diet.

Chapter 44

THE CAPTURE OF OXYGEN: RESPIRATION

Overview

Oxygen enters the bodies of animals by diffusion into water. The evolution of respiratory mechanisms among the vertebrates has favored changes that maximize the rate of this diffusion. The most efficient aquatic mechanism to evolve is the gill of bony fishes; the most efficient aerial respiration mechanism is the two-cycle lung of birds. Both of these structures achieve high efficiency by countercurrent flow. The respiratory mechanisms of other terrestrial vertebrates have evolved in various ways, which although less efficient than those of the fishes and birds, adapt them well to their terrestrial habitat. In the vertebrates, hemoglobin plays a critical role in the transport of oxygen and carbon dioxide in the respiration process.

For Review

Here are some important terms and concepts that you will encounter in this chapter. If you are not familiar with them, you should review them before proceeding.

- **Chemistry of carbon dioxide** (Chapter 2)
- **Diffusion** (Chapters 5 and 9)
- **Oxidative respiration** (Chapter 11)
- **Structure of hemoglobin** (Chapter 15)
- **Adaptation of vertebrates to terrestrial living** (Chapter 39)
- **Red blood cells** (Chapter 42)

No vertebrate is capable of photosynthesis. Vertebrates are heterotrophs, obtaining carbon compounds by consuming other organisms and then burning these carbon compounds to obtain energy. The biochemical mechanism of this **oxidative metabolism** was discussed in Chapter 11. Basically, heterotrophs obtain their energy by oxidizing carbon compounds. They first remove electrons from them and then channel these electrons through a series of proton-pumping ports in the membrane to generate ATP. Finally, they donate the electrons to oxygen gas (O_2), forming water. In the process of stripping electrons from organic

TABLE 44-1	WATER BALANCE IN KANGAROO RATS AND IN HUMANS	
	KANGAROO RATS*	ADULT HUMANS†
WATER GAIN (in milliliters)		
Ingested as solids	6.0	1200
Ingested as liquid	0	1000
Produced metabolically	54.0	350
Total	60.0	2550
WATER LOSS (in milliliters)		
Urine	13.5	1500
Feces	2.6	100
Evaporation	43.9	950
Total	60.0	2550

*Data from K. Schmidt-Nielsen, Animal Physiology, 1975.
†Data from A. Vander et al., Human Physiology, 1980.

carbon compounds, individual carbon atoms are cleaved from the organic molecules and released as carbon dioxide (CO_2). The process thus uses up oxygen and generates carbon dioxide and water.

The water given off by this process is called **metabolic water** to emphasize its source. In some desert vertebrates the generation of metabolic water provides most of the water that they require to live (Table 44-1). In considering the oxidative metabolism of most vertebrates, however, metabolic water is usually ignored, because most vertebrates have plentiful supplies of water available to them. In these vertebrates the metabolic water produced is simply diluted into the much larger volume of the body's internal water.

For most vertebrates, then, metabolic water can be ignored and oxidative metabolism may be viewed as a process that utilizes oxygen and produces carbon dioxide. This final balance sheet determines one of the principal physiological challenges facing all animals—how to obtain oxygen and dispose of carbon dioxide. The uptake of oxygen and the release of carbon dioxide together are called **respiration.**

HOW GAS ENTERS A CELL
The Composition of Air

The oxygen gas molecules that are the raw material of respiration have—each one of them—been produced by photosynthesis. As we saw in Chapters 10 and 12, photosynthetic organisms released oxygen gas molecules into the oceans, where they dissolved in the water. Oxygen gas molecules diffused from the oceans into the atmosphere as the oxygen concentration of the oceans increased. The present atmosphere, which we call **air,** is rich in oxygen.

Dry air has a very constant composition: 78.09% nitrogen, 20.95% oxygen, 0.93% argon and the other noble gases, and 0.03% carbon dioxide. By convention, physiologists group the noble gases present in air (they constitute less than 1% of the total volume) with nitrogen, because like nitrogen they are inert in most physiological processes. Convection currents cause air to have a constant composition to an altitude of at least 100 kilometers, although there is much less air present at high altitudes (Figure 44-1).

The amount of air present at a given altitude is usually expressed in terms that depend on its weight. Imagine a column of air standing on the ground and extending up into the sky as far as the atmosphere goes. This column has a lot of gas molecules in it, and they all experience the force of gravity. For this reason the column, which you might consider

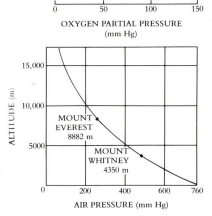

FIGURE 44-1

The relationship between air pressure and altitude above sea level. At the high altitudes characteristic of mountaintops, air pressure is much less than at sea level. At the top of Mount Everest, the world's highest mountain, there is only one third the air present at sea level. (Modified from Schmidt-Nielsen, Animal Physiology, 1975.)

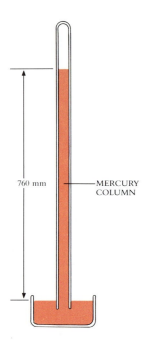

FIGURE 44-2

A simple mercury barometer. The weight of air pressing down on the surface of the mercury in the open dish pushes the mercury down into the dish and up the tube. The greater the air pressure pushing down on the mercury surface, the farther up the tube the mercury will be forced. At sea level, air pressure will cause a standard column of mercury to rise 760 millimeters.

"as light as air," actually weighs a lot. People and other terrestrial animals are not usually aware of this weight, since land-dwelling animals evolved under its influence and possess bodies that are structurally able to withstand the pressure that is exerted by the weight of the air.

How much does all this air weigh? It weighs enough to push down on one end of a U-shaped column of mercury sufficiently to raise the other end of the column 760 millimeters at sea level under a set of specified, standard conditions (Figure 44-2). An apparatus that measures air pressure in this way is called a barometer; 760 mm Hg (millimeters of mercury) is therefore the **barometric pressure** of the air at sea level, on the average. This pressure is also defined as **one atmosphere** of pressure.

Within the column of air, each separate gas exerts a pressure as if it existed alone. The total air pressure is simply made up of the component **partial pressures** of each of the individual gases. Thus at sea level the total of 760 millimeters air pressure is composed of

$$760 \times 79.02\% = 600.6 \text{ millimeters of nitrogen}$$
$$760 \times 20.95\% = 159.2 \text{ millimeters of oxygen}$$
$$760 \times 0.03\% = 0.2 \text{ millimeter of carbon dioxide}$$

At altitudes above 6000 meters human beings do not survive long. The air still contains 20.95% oxygen, but there is much less air and thus less total oxygen available. The atmospheric pressure at such a height is about 380 millimeters (Figure 44-1), so that the partial pressure of oxygen is only

$$380 \times 20.95 = 80 \text{ millimeters of oxygen}$$

This figure is only half of the amount of oxygen that is available at sea level.

The Diffusion of Oxygen into Water

Molecules of oxygen and other gases are constantly diffusing into the earth's atmosphere from the oceans and diffusing back into the oceans from the air in a dynamic exchange. As in any diffusion process, the net movement is in the direction of least concentration. Thus the number of oxygen gas molecules that will diffuse from the air into a volume of water is directly proportional to the amount of oxygen that is present in that air, and inversely proportional to the amount of oxygen already present in the water. Oxygen will diffuse into the volume of water until an equilibrium is reached, an event that occurs when the partial pressure of the oxygen in the water is the same as the partial pressure of the oxygen in the air.

The Passage of Oxygen into Cells

Life evolved in an aqueous environment, and in a real sense terrestrial organisms are still prisoners of water. The cell membrane of a terrestrial organism presents no barrier to the entry of oxygen gas molecules directly from air. Like water, oxygen and carbon dioxide molecules can diffuse freely across the phospholipid layer of the cell membrane, passing through small imperfections in it. The membrane of a cell, however, cannot maintain its integrity without the presence of water surrounding it, because it is the hydrophobic reaction to water that causes the lipid bilayer to form. For this reason no terrestrial organism obtains gases across a dry cell membrane. Instead organisms import and export gases as dissolved components within the water that passes freely across cell membranes.

Like passengers on a train, oxygen and carbon dioxide gas molecules are transported into and out of cells by water. How many molecules enter or leave depends on how many are carried there by water, in which they are dissolved—in other words, on the concentrations of these gases in the water. Any factor that affects these concentrations will thus be an important one in respiration. How many molecules of a particular gas will be present in water depends on four factors:

1. *Composition of the air.* Molecules from the air enter and leave water by diffusion. Recall that diffusion is the net movement of molecules to areas of lower concentration; it occurs as a result of random molecular motion. The rate of diffusion depends on molecular concentration. The more oxygen that is in air, for example, the more oxygen molecules that are available to enter the water.
2. *Solubility.* Some gases are more soluble in water than others.
3. *Temperature.* Random molecular motion increases with increasing temperature, with the result that diffusion proceeds more rapidly at higher temperatures. Because molecules diffuse out of liquids as well as into them, higher temperatures decrease the solubility of gases in liquids. Oxygen is only half as soluble in water at 30° C as it is at 0° C.
4. *Solute concentration.* Salts reduce the solubility of gases. Thus the solubility of oxygen in seawater at 15° C is only 82% of what it is in fresh water.

Air enters cells by diffusion into water. Air contains 21% oxygen, but the rate of diffusion of oxygen from air into water, and thus into cells, depends on the concentration of air, which is less at higher altitudes.

Some gases are more soluble in water than others, even when they are present at the same partial pressures. In dry air at sea level, at 15° C, and a partial pressure of one atmosphere (760 millimeters), 16.9 milliliters of nitrogen, 34.1 milliliters of oxygen, and 1019 milliliters of carbon dioxide will dissolve in a liter of water. Carbon dioxide is 30 times as soluble as oxygen.

THE EVOLUTION OF RESPIRATION

The capture of oxygen and the discharge of carbon dioxide by organisms depend on the diffusion of these gases into water. In vertebrates the gases diffuse into the aqueous layer covering the epithelial cells that line the respiratory system of the animal. The diffusion process is in all cases passive, driven only by the difference in oxygen concentration between the interior of the organism and the exterior environment. The process is described by a relationship known as **Fick's Law of Diffusion:**

$$R = D \times A \times \frac{\Delta p}{d}$$

where

R = the rate of diffusion. In this case, the speed of oxygen or carbon dioxide movement is the rate of diffusion.

D = the diffusion constant, whose value depends on the material through which the diffusion is occurring.

A = the area across which diffusion takes place.

Δp = the difference in partial pressures, or concentration difference. In this case, the difference is the partial pressure of oxygen or carbon dioxide in the environment minus that in the organism.

d = the distance a molecule must travel to get to an area of lower concentration.

Major changes in the mechanism of respiration have occurred during the evolution of animals (Figure 44-3). In general these changes have tended to optimize the rate of diffusion, R. By inspecting Fick's Law you can see that natural selection can optimize the rate of diffusion by favoring changes that (1) increase A (surface area); (2) decrease d (distance); (3) increase Δp (concentration difference). The evolution of respiration systems has involved changes of all three sorts.

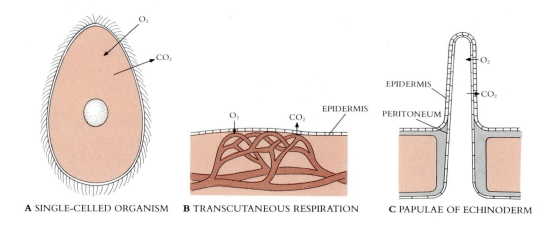

A SINGLE-CELLED ORGANISM   **B** TRANSCUTANEOUS RESPIRATION   **C** PAPULAE OF ECHINODERM

FIGURE 44-3

Gas exchange in animals may take place in a variety of ways.
A   Gases diffuse directly into single-celled organisms.
B   Amphibians and many other multicellular organisms respire across their skin.
C   Echinoderms have protruding papulae, which provide an increased respiratory surface.
D   Insects respire through spiracles, openings in their cuticle.
E   The gills of fish provide a very large respiratory surface and countercurrent exchange.
F   Mammalian lungs provide a large respiratory suface, but do not permit countercurrent exchange.

Simple Diffusion

Oxygen diffuses slowly. The levels of oxygen required by oxidative metabolism in most organisms cannot be obtained by diffusion alone over distances greater than about 0.5 millimeter. This factor severely limits the size of organisms that obtain their oxygen entirely by diffusion from the environment to the metabolizing cytoplasm. Protozoa are small enough that diffusion distance presents no problem, but as the size of an organism increases, the problem soon becomes significant. An organism's mass increases as the cube of the diameter, whereas its surface area increases only as the square. In other words, as the size of an organism increases, an increasingly greater proportion of the volume of that organism is farther and farther away from the surface. Unless metabolism is greatly slowed down, as it is, for example, in the dilute cytoplasm of jellyfish, an increase in organism size must be accompanied by changes that facilitate the diffusion of gases into that organism.

Creating a Water Current

Most of the more primitive metazoan phyla possess no special respiratory organs. The sponges (phylum Porifera), cnidarians (phylum Cnidaria), many flatworms (phylum Platyhelminthes) and roundworms (phylum Nematoda), and some annelids (phylum Annelida) all obtain their oxygen by diffusion directly from surrounding water. How do they overcome the limits imposed by diffusion? They increase the term Δp, the difference in concentration, in the Fick equation. In a number of different ways, many of which involve beating cilia, these organisms create a water current, by means of which they continuously replace the water over the diffusion surface. Because of this continuous replenishment with water containing fresh oxygen, Δp *does not decrease as diffusion proceeds*. Although each oxygen gas molecule that passes into the organism has been removed from the surrounding volume of water, the exterior oxygen concentration does not fall, as a new volume of water is constantly replacing the depleted one. The result is a higher realized value of R, the rate of diffusion.

Increasing the Diffusion Surface

All of the more advanced invertebrates (mollusks, arthropods, echinoderms), as well as the vertebrates, possess special respiratory organs that increase the surface area available for diffusion and provide intimate contact with the internal fluid, which is usually circulated throughout the body. These special organs thus increase the rate of diffusion in the Fick equation by increasing A and decreasing d. These organs are of two kinds: (1) those that facilitate exchange with water, and (2) those that facilitate exchange with air. As a rough

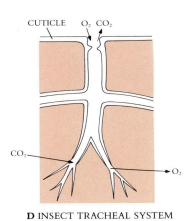

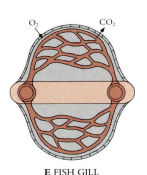

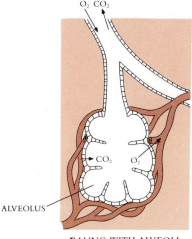

D INSECT TRACHEAL SYSTEM   **E** FISH GILL   **F** LUNG WITH ALVEOLI

rule of thumb, aquatic respiratory organs increase the diffusion surface by extensions of tissue, called gills, that project from the body out into the water. Atmospheric respiratory organs, on the other hand, involve invaginations into the body.

Among the simplest of the water-exchange respiratory organs is the **external gill,** a highly convoluted outfolding membrane system with a very large surface area exposed to water. The great increase in diffusion surface provided by gills enables organisms living in water to extract far more oxygen from it than would be possible from their body surface alone. Nevertheless, frogs, worms, and some aquatic animals without external gills obtain significant amounts of oxygen by simple diffusion across the cells of their wet skin.

Perhaps the simplest of the air-exchange organs are the tracheae of arthropods (Chapter 37). Tracheae are extensive series of passages connecting the surface of the animal to all portions of its body. Oxygen diffuses from these passages directly to the cells, without the intervention of a circulation system. Piping air directly to the cells in this manner works very well in such organisms as insects, which have small bodies relative to those of vertebrates. Air must move only a relatively short distance within their bodies, but this relationship severely limits the potential body size of such organisms.

Enclosing the Respiratory Organ

External gills provide a greatly increased diffusion surface *(A),* but they do have one great disadvantage. It is very difficult to circulate water constantly past the diffusion surface to maintain a high Δp. Many fish larvae and amphibian larvae, as well as developmentally arrested **(neotenic)** amphibian larvae that remain permanently aquatic, such as the axolotl (Figure 44-4), circulate water past the respiratory surface of external gills by physically moving the gill through the water. This is not a very effective solution to the problem, especially since the highly branched gills offer significant resistance to the movement.

Among other organisms, specialized **branchial chambers** evolved, which provide a means of pumping water past the gills. Mollusks, for example, have an internal cavity called the **mantle cavity,** which opens to the outside and contains the gill. The contraction of the muscular walls of the mantle cavity causes the chamber to draw water in and then to expel it. In crustaceans the branchial chamber lies between the bulk of the body and the hard outer shell of the animal. This chamber contains gills and opens to the surface underneath a limb. Movement of the limb serves to bail out the branchial chamber, drawing water through it and thus creating currents over the gills.

> **Aquatic animals have optimized the rate of diffusion of oxygen by increasing the diffusion surface with gills and by actively moving water past the diffusion surface.**

FIGURE 44-4

The gills of the axolotl are external. The axolotl sweeps them through the water in order to move water over the diffusion surface.

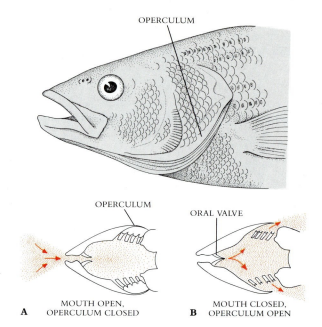

FIGURE 44-5

How a fish breathes. The gills are suspended between the mouth cavity and the outside, under a hard cover called the operculum, which when pressed down seals off the opening. Breathing occurs in two stages.

A *When the oral valve of the mouth is open, closing the operculum increases the volume of the mouth cavity so that water is drawn in.*

B *When the oral valve is closed, opening the operculum decreases the volume of the mouth cavity, forcing water out past the gills to the outside.*

OPERCULUM

OPERCULUM

ORAL VALVE

A MOUTH OPEN, OPERCULUM CLOSED

B MOUTH CLOSED, OPERCULUM OPEN

THE GILL AS AN AQUEOUS RESPIRATORY MACHINE

By far the most successful branchial chamber evolved among the bony fishes. In the members of this group, water passes through the mouth into two **opercular cavities,** which are situated on each side of the head behind the mouth. From these cavities the water passes out of the body. The gills are situated between the mouth and the entrance to each of the opercular cavities, separating the cavities from the mouth just as curtains would do.

Many fishes that swim continuously, such as tuna, have practically immobile gill covers over the opercular cavities. These fishes swim with their mouths partly open, forcing water constantly over the gills, a process that amounts to a form of **ram ventilation.** Most bony fishes, however, have flexible gill covers that permit a pumping action (Figure 44-5). When the fish swallows water, the water is forced past the gills during the process of inhalation. By expanding the sides of the opercular cavities during exhalation, a negative pressure (suction) is created, which draws water from the mouth past the gills and through the opercular cavities. Thus there is an uninterrupted one-way flow of water past the gills all through the inhalation cycle. This continuous movement of water over the gills maintains Δp at a high value, since the layer of water immediately over the gills is constantly being replaced as oxygen is taken from it.

FIGURE 44-6

Structure of a fish gill. Several rows of gills underlie the operculum on each side of a fish's head (A). A gill is composed of two rows of filaments (B), each of which bears rows of thin, disk-like lamellae (C). Water passes from the gill arch out over the filaments (from left to right in the diagram). Water always passes the lamellae in the same direction—which is opposite to the direction the blood circulates across the lamellae. The success of the gill's operation critically depends on this opposite orientation of blood and water flow.

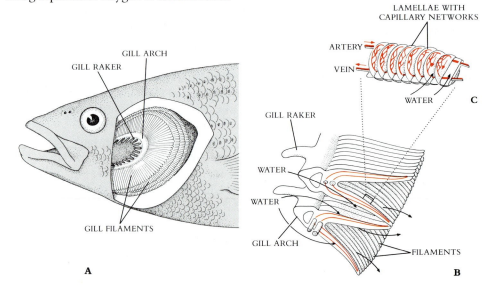

LAMELLAE WITH CAPILLARY NETWORKS

ARTERY

VEIN

WATER

C

GILL ARCH

GILL RAKER

GILL FILAMENTS

A

GILL RAKER

WATER

WATER

GILL ARCH

FILAMENTS

B

In addition to maintaining a high value of Δp by a continuous flow of water, the gills of fishes are constructed in such a way that they actually increase the value of Δp. Each gill is composed of two rows of **gill filaments,** thin membranous plates stacked one on top of the other and projecting out into the flow of water (Figure 44-6). Each filament, in turn, carries rows of thin disklike lamellae arrayed parallel to the direction of water movement. Water flows past these lamellae from front to back. Within each lamella the blood circulation is arranged so that blood is carried in the direction opposite the movement of the water, from the back of the lamella to the front.

Because the water flowing over the lamellae and the blood flowing within lamellae run in opposite directions (Figure 44-7), Δp is maximized. At the back of the gill the least oxygenated blood meets the least oxygenated water and is able to remove oxygen from it.

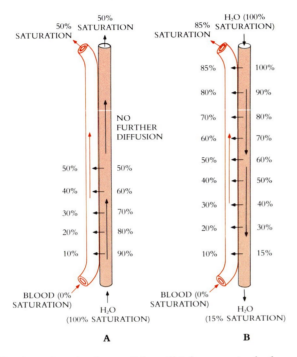

A

B

FIGURE 44-7

Countercurrent exchange.

A *When blood and water flow in the same direction, dissolved oxygen gas can diffuse from the water into the blood rapidly at first, because of the large concentration difference (0% in blood versus 100% in water), but the difference decreases as more oxygen diffuses from water into blood, until finally the concentrations of oxygen in water and blood are equal. At this point there is no concentration difference to drive any further diffusion. In this example, blood can obtain no more than 50% dissolved oxygen in this fashion.*

B *When blood and water flow in opposite directions, the initial concentration difference between water and blood is not as great (0% in blood versus 15% in water) but is sufficient for diffusion to occur from water to blood. As more oxygen diffuses into the blood, raising its oxygen concentration, it encounters water with higher and higher oxygen concentrations; at every point, the oxygen concentration is higher in the water, so that diffusion continues. In this example, blood obtains 85% dissolved oxygen.*

By the time the blood reaches the front of the gill it has acquired a lot of oxygen, but it is able to acquire still more oxygen by diffusion from the water entering the gill. The reason that it can do this is because the new water is richer in oxygen than the water that has already flowed past the gills and lost some of its oxygen. This kind of **countercurrent flow** ensures a continuous gradient of concentration, and diffusion continues to occur all along the gill.

If the flow of water and blood had been in the same direction, Δp would have been high initially, as the oxygen-free blood met the new water that was entering. The concentration difference would have fallen rapidly, however, as the water lost oxygen to the blood. Since the blood oxygen concentration rises as the water oxygen falls, much of the oxygen in the water would remain there when the blood and water concentrations become equal, and diffusion would cease before maximizing diffusion of oxygen from the water into the blood. In countercurrent flow, in contrast, the blood oxygen level encountered by the water becomes lower and lower as the level of oxygen in the water falls. The result is that in countercurrent flow the blood can attain oxygen concentrations as high as those which exist in the water entering the gills. Fish gills are the best Fick machines that occur among organisms. They are able to maximize the rate of diffusion in an oxygen-poor medium, obtaining as much as 85% of the available oxygen.

> **The gill is the most efficient of all respiratory organs. This great efficiency derives from the countercurrent flow of water past the blood vessels of the gills.**

FROM AQUATIC TO ATMOSPHERIC BREATHING: THE LUNG

Water is relatively poor in dissolved oxygen; it contains only 5 to 10 milliliters of oxygen per liter. Air, in contrast, is rich in oxygen, containing about 210 milliliters of oxygen per liter. Not surprisingly, many members of otherwise aquatic groups utilize atmospheric air as a source of oxygen; these include many mollusks, crustaceans, and fishes. As discussed in Chapter 39, the first lungs seem to have evolved from the swim bladders of fishes, organs that functioned originally to set the buoyancy of the fishes in water.

When organisms first became fully terrestrial, the air became the source of their oxygen. An entirely new respiratory apparatus evolved, one that was based on internal passages rather than on gills. Why were gills not maintained in terrestrial organisms, in view of the fact that they are such superb oxygen-capturing mechanisms? Gills were lost for two principal reasons:

1. Air is less buoyant than water. Because the fine membranous lamellae of gills lack structural strength, they must be supported by water to avoid collapsing on one another. A fish out of water, although awash in oxygen, soon suffocates because its gills collapse into a mass of tissue. This collapse greatly reduces the diffusion surface of the fish. Unlike gills, internal air passages can remain open, the body itself providing the necessary structural support.

2. Water diffuses into air through the process of **evaporation.** Atmospheric air is rarely saturated with water vapor, except immediately after a rainstorm. Consequently, terrestrial organisms that live surrounded by air are constantly losing water to the atmosphere. Gills would have provided an enormous surface for water loss.

Two main systems of internal respiratory exchange evolved among terrestrial organisms. One was the tracheae of insects, mentioned earlier, and the other was the lung. Both systems sacrifice respiratory efficiency in order to maximize water retention. Insects prevent excessive water loss by closing the external openings of the tracheae whenever possible. They do this whenever body carbon dioxide levels are below a certain point.

Lungs, in contrast, minimize the effects of desiccation by eliminating the one-way flow of air that was such an effective means of increasing Δp in aquatic respiratory systems. In organisms that respire by means of lungs, the air moves into the lung through a tubular passage and then back out again via the same passage. When each breath is completed, the lung still contains a volume of air, the so-called "dead volume." In adult human beings this volume is about 150 milliliters. Each inhalation adds from 500 milliliters (resting) to 3000 milliliters (exercising) of additional air. Each exhalation removes approximately the same volume as inhalation added, reducing the air volume in the lung once more to about 150 milliliters. Because the diffusion surfaces of the lungs are not exposed to fully oxygenated air, but rather to a mixture of fresh and partly depleted air, Δp is far from maximal, and the respiratory efficiency of lungs is much less than that of gills. Oxygen capture is lessened by this two-way flow of air—but so is water loss.

Amphibians

There is so much more oxygen in air than in water that low respiratory efficiency does not appear to have presented a critical problem to early land-dwellers; the amphibian lung is hardly more than a sac with a convoluted internal membrane (Figure 44-8, A). This internal sac is connected by a windpipe or trachea to the rear of the mouth chamber or oral cavity, and the opening is controlled by a valve, the **glottis.** A series of passages called **sinuses** connect the oral cavity to the nose, which opens to the outside, enabling the animal to breathe with its mouth closed. Because the inner membrane surface of the lung is convoluted, the surface area of amphibian lungs available for diffusion is much greater than that of a fish swim bladder, although it is still not large. Amphibians obtain much of their oxygen through their moist skin.

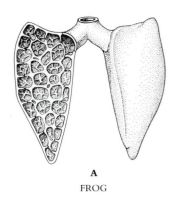
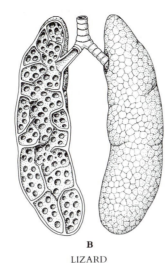
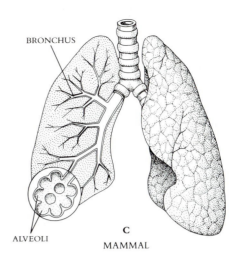

A
FROG

B
LIZARD

BRONCHUS

ALVEOLI

C
MAMMAL

FIGURE 44-8

Evolution of the vertebrate lung.
A *The amphibian lung.*
B *The reptile lung.*
C *The mammalian lung.*

Reptiles

Reptiles are far more active than amphibians, and they have significantly greater metabolic demands for oxygen. The early reptiles could not rely on their skins for respiration; living fully on land, they are "watertight," avoiding desiccation by means of a dry scaly skin. Again, as in aquatic organisms, the respiratory apparatus has changed in ways that tend to optimize respiratory efficiency. The lungs of reptiles possess many small chambers within their surface called **alveoli,** which are clustered together like grapes (Figure 44-8, *B*). Each cluster of alveoli is connected to the main air sac in the lung by a short passageway called a **bronchiole.** Air within the lung enters the alveoli, where all gas exchange with the circulatory system takes place. The alveoli greatly increase the diffusion surface of the lung.

Mammals

The metabolic demands for oxygen became even greater with the evolution of mammals, which, unlike reptiles and amphibians, maintain a constant body temperature by heating their bodies metabolically. In the lungs of mammals the bronchioles branch, each branch connecting to many clusters of alveoli (Figure 44-8, *C*). Human beings have about 300 million alveoli in their two lungs. The branching of the bronchioles and the increase in number of alveoli combine to increase yet again the total diffusion surface of the lung. In human beings the total surface devoted to diffusion can be as much as 80 square meters, an area about 42 times the surface area of the body.

Some mammals are far more active than others, but the more active ones do not have proportionally larger lungs. Lung mass is proportional to body mass in all mammals. In active mammals, however, the individual alveoli are smaller and more numerous, increasing the diffusion surface (*A* in the equation on p. 943). In addition, the epithelial layer of cells separating the alveoli from the bloodstream in such active mammals is thinner, a factor reducing the diffusion distance (*d* in the equation on p. 943).

> **When amphibians evolved lungs, one-way flow through the respiratory organ was abandoned in favor of a saclike lung. Increases in efficiency among the reptiles and mammals have been achieved by increases in the lung's internal surface area.**

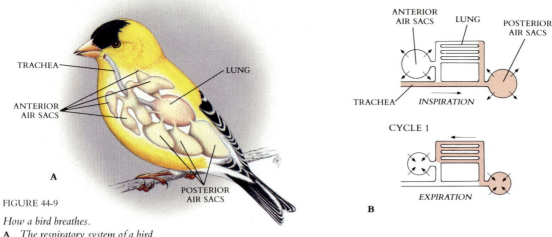

FIGURE 44-9

How a bird breathes.

A *The respiratory system of a bird is composed of anterior air sacs, lungs, and posterior air sacs.*

B *Breathing occurs in two cycles. Cycle 1: Air is drawn from the trachea into the posterior air sacs and then is exhaled through the lungs. Cycle 2: Air is drawn from the lungs into the anterior air sacs and then is exhaled through the trachea. Passage of air through the lungs is always in the same direction, from posterior to anterior (right to left in this diagram). Because blood circulates in the lung from anterior to posterior, the lung achieves countercurrent flow and thus is very efficient at picking up oxygen from the air.*

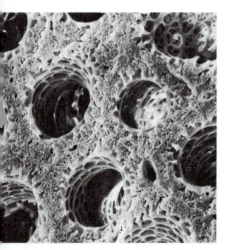

FIGURE 44-10

Cross section of a lung of a domestic chicken, magnified 75 times. Air travels through the tunnels, called parabronchi, while blood circulates in the opposite direction within the fine lattice.

Birds

There is a limit to the improvements that can be realized by increases in the diffusion surface of the lung, a limit that is probably approached by the more active mammals. With the advent of birds, flying introduced respiratory demands that exceed the capacity of a saclike lung. Unlike bats, whose flight involves considerable gliding, many birds rapidly beat their wings for prolonged periods during flight. Such rapid wing beating uses up a lot of energy quickly, because it depends on the frequent contraction of wing muscles. Flying birds must carry out intensive oxidative respiration within their cells to replenish the ATP expended by contracting flight muscles—and thus require a great deal of oxygen, more oxygen than a saclike lung, even one with a large surface area such as a mammalian lung, is capable of delivering. The lungs of birds cope with the demands of flight by employing a new respiratory mechanism, one that produces a significant improvement in respiratory efficiency.

An avian lung works like a two-cycle pump (Figure 44-9). When a bird inhales air, the air passes directly to a nondiffusing chamber called the **posterior air sac.** When the bird exhales, the air flows into the lung. On the following inhalation, the air passes from the lung to a second air sac, the **anterior air sac.** Finally, on the second exhalation, the air flows from the anterior air sac out of the body. What is the advantage of this complicated passage? It creates a unidirectional flow of air through the lungs!

Air flows through the lung system of birds in one direction only, from posterior to anterior. There is no dead volume, as there is in the mammalian lung. For this reason, the air passing across the diffusion surfaces of the avian lung is always fully oxygenated. Unidirectional flow also permits a second major improvement in respiratory efficiency. Just as in the gills of fish, the flow of blood past the avian lung runs in the direction opposite that of air flow in the lung (Figure 44-10). The oxygenated blood leaving the lung can thus contain more oxygen than exhaled air, a capacity not achievable by mammalian lungs.

Because of the high efficiency of countercurrent flow in gathering oxygen from the air into the blood, a sparrow has no trouble breathing at an altitude of 6000 meters, whereas a mouse, which has approximately the same body mass, blood with the same affinity for oxygen, and a similar high metabolic rate, cannot respire successfully at such an elevation. In terms of the Fick equation, one would say that in the kind of lungs that birds possess, Δp is greatly increased. Just as fish gills are the most efficient aquatic respiratory machines, so avian lungs are the most efficient atmospheric ones. Both achieve high efficiency by utilizing countercurrent flow.

The most efficient atmospheric respiratory organs are the lungs of birds, which are arranged to permit a one-way flow of air without significant water loss. This arrangement permits the establishment of a countercurrent flow of blood, the key to high efficiency.

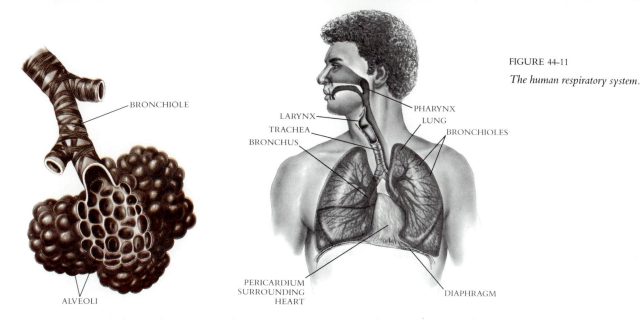

FIGURE 44-11

The human respiratory system.

BRONCHIOLE

LARYNX

TRACHEA

BRONCHUS

PHARYNX

LUNG

BRONCHIOLES

ALVEOLI

PERICARDIUM
SURROUNDING
HEART

DIAPHRAGM

THE MECHANICS OF HUMAN BREATHING

Thus we see that humans are by no means the most efficient breathers among terrestrial vertebrates. Understanding how humans breathe is, however, of considerable practical interest, and we will now examine the mechanics of human breathing in more detail.

Humans possess a pair of lungs located in the chest, or **thoracic,** cavity. The two lungs hang free within the cavity, being connected to the rest of the body only at the one position where the lung's blood vessels and air tube enter (Figure 44-11). This air tube is called a **bronchus.** It connects each lung to a long tube, the **trachea,** which passes upward in the body, past the voice box, or **larynx,** and opens into the rear of the mouth (Figure 44-12). Air normally enters through the nostrils, which are lined with hairs that filter out dust and other particles. As the air passes through the nasal cavity, an extensive array of cilia on its epithelial lining further filters the air and moistens it (Figure 44-13). The air then passes through the back of the mouth, crossing the path of food as it enters first the larynx and then the trachea. From there it passes down through the bronchus and the bronchioles, and into the alveoli of the lungs.

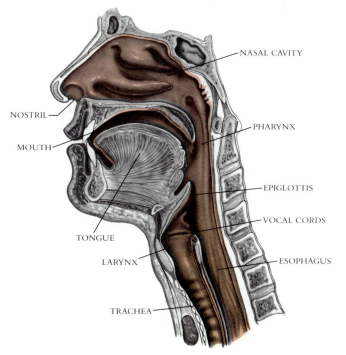

NASAL CAVITY

NOSTRIL

MOUTH

PHARYNX

EPIGLOTTIS

VOCAL CORDS

TONGUE

LARYNX

ESOPHAGUS

TRACHEA

FIGURE 44-12

Side view of the human upper respiratory tract.

FIGURE 44-13

Respiratory cilia such as these line the trachea. They moisten the air and aid its passage.

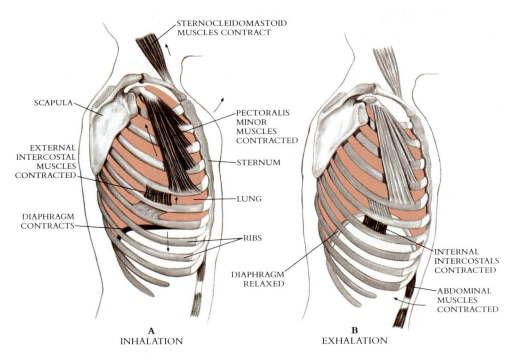

STERNOCLEIDOMASTOID
MUSCLES CONTRACT

SCAPULA

PECTORALIS
MINOR
MUSCLES
CONTRACTED

EXTERNAL
INTERCOSTAL
MUSCLES
CONTRACTED

STERNUM

LUNG

DIAPHRAGM
CONTRACTS

RIBS

DIAPHRAGM
RELAXED

INTERNAL
INTERCOSTALS
CONTRACTED

ABDOMINAL
MUSCLES
CONTRACTED

A
INHALATION

B
EXHALATION

FIGURE 44-14

How a human breathes.
A *Inhalation. The diaphragm and walls of the chest cavity expand, increasing its volume. As a result of the larger volume, air is sucked in through the trachea.*
B *Exhalation. The diaphragm and chest walls return to their normal positions, reducing the volume of the chest cavity and forcing air outward through the trachea.*

The human respiratory apparatus is simple in structure, functioning as a one-cycle pump. The thoracic cavity is bounded on its sides by ribs, which are capable of flexing, and on the bottom by a thin layer of muscle, the **diaphragm,** which separates the thoracic cavity from the abdominal cavity. Each lung is covered by a very thin, smooth membrane called the **pleural membrane.** A second pleural membrane marks the interior boundary of the thoracic cavity, into which the lungs hang. Within the cavity the weight of the lungs is supported by water, the **intrapleural fluid.**

The intrapleural fluid not only supports the lungs, but also plays another important role by permitting an even application of pressure to all parts of the lung. You can visualize the pleural membranes as a system of two balloons of different sizes, one nested inside the other, with the space between them completely filled with water and the inner balloon opening out to the atmosphere. Two forces act on this inner balloon: air pressure pushing it outward and water pressure pushing it inward.

During **inhalation** (Figure 44-14), the walls of the chest cavity expand. The rib cage moves outward, and the diaphragm moves downward by stretching taut. In effect, we have enlarged the outer balloon by pulling it in all directions. This expansion of the fluid space causes the fluid pressure to decrease to a level less than that of the internal air pressure within the lung (the inner balloon). As a result, the wall of the lung is pushed out. As the lung expands, its internal air pressure decreases, and air moves in from the atmosphere. During **exhalation,** the ribs and diaphragm return to their original resting position. In doing so, they exert pressure on the fluid. This pressure is transmitted uniformly by the fluid over the entire surface of the lungs, forcing air from the inner cavity back out to the atmosphere.

Breathing—the cycle of inhalation and exhalation—thus depends on the regular contraction of the muscles surrounding the thoracic cavity. The rate of breathing is regulated by a respiratory center in the brainstem. Chemoreceptors in the arterial walls detect changes in carbon dioxide levels and transmit nerve impulses to the respiratory center, which sends appropriate signals to the muscles of the diaphragm and rib cage.

The active pumping of air in and out through the lungs is called breathing. The lungs are pumped by suction, unlike the heart, which pushes the blood. The expansion of the chest cavity draws air into the lungs, and the return of the ribs to their resting position drives air from the lungs.

HOW RESPIRATION WORKS: GAS TRANSPORT AND EXCHANGE

When oxygen has diffused from the air into the moist cells lining the inner surface of the lung, its journey has just begun. Passing from these cells into the bloodstream, the oxygen is carried throughout the body by the circulation system to be described in the next chapter. It has been estimated that it would take a molecule of oxygen 3 years to be transported from your lung to your toe if the transport depended only on diffusion, unassisted by a circulatory system.

Oxygen moves within the circulatory system on carrier proteins that bind dissolved molecules of oxygen. This binding occurs in the capillaries surrounding the alveoli of the lungs. The carrier proteins subsequently release their oxygen molecules to metabolizing cells at distant locations in the body.

Hemoglobin and Gas Transport

The carrier protein that is used by all vertebrates is hemoglobin. Hemoglobin is a protein composed of four polypeptide subunits; each of the four polypeptides is combined with an iron ion in such a way that oxygen can be bound reversibly to the ion. Hemoglobin is synthesized by erythrocytes, or red blood cells, and remains within these cells, which circulate in the bloodstream like ships bearing cargo.

Hemoglobin is an ancient protein, one that is also utilized as a carrier of oxygen in the circulatory systems of annelids, mollusks, and even some protozoans. Many invertebrates, however, also employ a second carrier protein, **hemocyanin.** Hemocyanin utilizes copper to bind oxygen in the same way that hemoglobin utilizes iron. Hemocyanin never occurs within blood cells as hemoglobin does but exists free in the circulating fluid (called **hemolymph**) of invertebrates. When hemocyanin combines with oxygen, it gives the hemolymph a blue color; when hemoglobin combines with oxygen, it gives blood a bright red color.

The higher the partial pressure of oxygen in the air within the lungs, the more of the hemoglobin in the blood that will combine with oxygen. When data expressing this relationship are plotted, an S-shaped curve, called an **oxygen-hemoglobin dissociation curve,** is obtained (Figure 44-15). Hemoglobin is an efficient oxygen carrier because of the value of its oxygen association/dissociation constant. Different species of vertebrates have different dissociation curves, depending on the particular amino acid sequence of their hemoglobin. At the oxygen partial pressures encountered in the blood supply of the lung (110 millimeters of mercury), most hemoglobin molecules are saturated with oxygen. However, the curve can be shifted if the hemoglobin assumes a different shape. In the presence of carbon dioxide, for example, the curve is shifted to the right, a phenomenon called the **Bohr effect.** A higher carbon dioxide concentration causes hemoglobin to give up more

FIGURE 44-15

Oxygen-hemoglobin dissociation curve for human hemoglobin. In tissue such as exercising muscle, the concentration of oxygen, measured as pressure in millimeters of mercury, is low, about 30 millimeters. At such low pressures, oxygen tends to dissociate from hemoglobin, with the hemoglobin molecules giving up about 60% of the oxygen molecules they carry. In the lungs, where oxygen concentrations are much higher (typically 110 millimeters), oxygen molecules do not tend to dissociate from hemoglobin; hemoglobin molecules are fully saturated with all the oxygen they can carry.

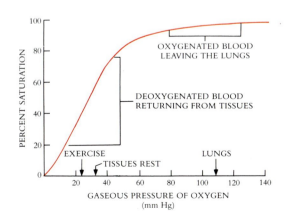

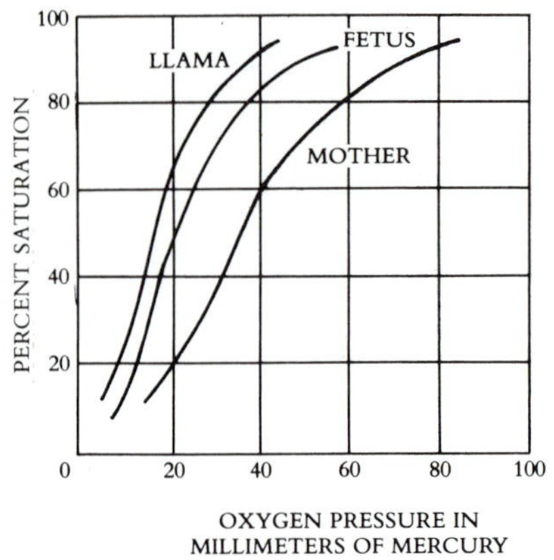
100
80
60
40
20
0
PERCENT SATURATION
LLAMA
FETUS
MOTHER
20
40
60
80
100
OXYGEN PRESSURE IN
MILLIMETERS OF MERCURY

FIGURE 44-16

Oxygen-hemoglobin dissociation curves are shifted in organisms adapted to environments with low oxygen concentrations. Llamas living at high altitudes in the Andes exhibit such shifts, as does the hemoglobin of a human fetus, which must garner oxygen secondhand from the maternal circulatory system.

oxygen at any given oxygen concentration than it otherwise would have done. This effect is of real importance to the regulation of vertebrate respiration, since carbon dioxide is produced at the site of cell metabolism by the tissues. Because of the facilitation induced by the Bohr effect, the blood unloads oxygen more readily to those tissues undergoing metabolism and generating carbon dioxide.

Other molecules also bind to hemoglobin, influencing the ease of oxygen unloading. One of these is **2,3-diphosphoglycerate (DPG),** which shifts the dissociation curve even further to the right than carbon dioxide, greatly facilitating oxygen unloading. DPG binds to hemoglobin in many rapidly metabolizing tissues. The levels of DPG increase within a few days in the bodies of human beings who stay at high altitudes (Figure 44-16). Because the partial pressure of oxygen is lower at high altitudes, it is necessary that the dissociation curve be shifted to the right for tissues to receive adequate amounts of oxygen.

Oxygen and carbon dioxide are not the only gases that bind to hemoglobin. Carbon monoxide (CO) binds to hemoglobin 300 times more readily than does oxygen. Because it binds so strongly, it does not readily dissociate and so prevents the CO-bound hemoglobin from acting as an oxygen carrier. As a result, even a small amount of carbon monoxide in the air can lead to respiratory failure; as little as 0.1% of carbon monoxide in the air is dangerous.

The Exchange Process

When oxygen-rich air enters the alveoli of the lungs, it encounters oxygen-depleted blood in the capillaries surrounding the alveoli. The oxygen diffuses down the concentration gradient into the blood plasma. Within the plasma, it diffuses into the red blood cells, where it is bound by hemoglobin, one oxygen molecule to each of the four subunits of each hemoglobin protein. The binding of oxygen molecules to hemoglobin effectively removes them from solution, so that the concentration of free oxygen in the red blood cells remains low, facilitating the additional diffusion inward of more oxygen from the plasma and lungs. This soaking up of oxygen by hemoglobin enables whole blood to carry 50 times as much oxygen as the plasma alone could absorb by diffusion.

When the oxygen-loaded red blood cell reaches a capillary in tissue where the carbon dioxide concentration is high as a result of oxidative respiration carried out by the tissue cells, the hemoglobin molecules react to the higher carbon dioxide concentration by unloading their oxygen—the Bohr effect. This dissociation greatly elevates the free oxygen concentration within the plasma of the capillary, and the oxygen rapidly diffuses outward into the surrounding tissue, where the concentration of oxygen is much lower.

At the same time that the red blood cells are unloading oxygen, they are also absorbing carbon dioxide from the tissue. Perhaps one fifth of the carbon dioxide that the blood absorbs is bound to hemoglobin. The remaining 80% of the carbon dioxide diffuses from the plasma into the cytoplasm of the red blood cells. There an enzyme, **carbon anhydrase,** catalyzes the combination of carbon dioxide with water to form carbonic acid, which dissociates into bicarbonate and hydrogen ions. This process removes large amounts of carbon dioxide from the plasma, facilitating the diffusion of more carbon dioxide inward from the surrounding tissue. The facilitation is critical to carbon dioxide removal, since the difference in carbon dioxide concentration between blood and tissue is not large (only 5%).

The red blood cells carry their cargo of carbonate ions back to the lungs. Here the lower carbon dioxide concentration in the air causes the carbonic anhydrase reaction to proceed in the reverse direction, releasing gaseous carbon dioxide, which diffuses outward into the alveoli. With the next exhalation, this carbon dioxide leaves the body. Hemoglobin has a greater affinity for oxygen than for carbon dioxide at low carbon dioxide concentrations. For this reason, the diffusion of carbon dioxide outward from the red blood cells causes the hemoglobin within these cells to release its bound carbon dioxide and take up oxygen instead. The red blood cells, with their bound oxygen, then start the next respiratory journey immediately.

In the next chapter we will consider in more detail the journey of the red blood cell. Traveling from the lungs, where they acquire oxygen and release carbon dioxide, to respiring tissues, where they release oxygen and acquire carbon dioxide, the body's red blood cells traverse a complex highway that passes to all parts of the body. Nor are red blood cells the only traffic on this highway. Just as the roads and sidewalks transport all the commerce of a city (although not all the information, as much as this passes over phone lines), so the circulatory system of the vertebrate body transports all the material that moves from one part of the body to another (although not all information, because much of this moves by way of the nerves).

SUMMARY

1. All animals obtain oxygen by its diffusion into water. The rate of this diffusion depends on temperature and on atmospheric pressure. Air at high altitudes has the same percentage of oxygen as does air at sea level, but because there is less air at high elevations, the rate of diffusion is slower.

2. The evolution of respiratory mechanisms among animals has tended to favor changes that improve the rate of diffusion. These include changes that decrease the length of the path over which diffusion occurs, changes that increase the surface area over which diffusion occurs, and changes that maximize the difference in oxygen concentration between environment and tissue.

3. The most efficient aquatic respiratory organ is the gill of bony fishes. Fishes take water in through their mouths, move it past their gills, and pass it out of their bodies. This one-way flow is the secret to high respiratory efficiency, since it permits fishes to establish a countercurrent flow of blood: blood vessels are located within the gills in such a way that blood flows in a direction opposite that of water.

4. Gills will not work in air, since air is not buoyant enough to support their fine latticework of passages. That is why fishes drown in air, even though there is much more oxygen available in air than there is in the water in which they normally live.

5. The first successful atmospheric respiratory organ was the lung of amphibians, which was a simple sac with two-way flow in and out. It was not efficient, and gaseous diffusion across its small internal surface area has been supplemented, over the course of evolution, by diffusion across the moist skin.

6. The further evolution of the lung in reptiles and mammals has involved no fundamental changes. Instead there has been a progressive increase in the internal surface area of the lung, achieved by partitioning the inner surface into increasingly numerous chambers called alveoli. A mammalian lung possesses about 300 million alveoli, with a combined surface area 40 times the body surface.

7. A fundamental change in the atmospheric lung is characteristic of the birds, which utilize a series of air chambers and two-cycle breathing to effect a one-way flow of air. Just as in the gill of bony fishes, the establishment of a one-way flow of the diffusing medium permits countercurrent flow, which is far and away the most efficient diffusion mechanism.

8. Terrestrial vertebrates breathe by expanding and contracting the cavity within which the lungs hang. These actions expand the lungs, sucking air inward, and compress the air, forcing it out of the lungs.

9. In the respiration of terrestrial vertebrates, hemoglobin within the red blood cells binds the oxygen, which diffuses across the lung capillaries into the blood cells to the respiring tissues of the body.

10. In respiring tissues, the partial pressure of oxygen is much lower than it is in the blood and the partial pressure of carbon dioxide is higher as a result of the consumption of oxygen and generation of carbon dioxide by respiring cells. Hemoglobin responds to the higher carbon dioxide concentration by unloading its oxygen, which diffuses out of the red blood cells into the tissue. The carbon dioxide is absorbed into the red blood cells and carried to the lungs, where it is discharged.

SELF-QUIZ

The following questions pinpoint important information in this chapter. You should be sure to know the answers to each of them.

1. The number of molecules of a particular atmospheric gas that will be present in water depends on four things. Which of the following is not a determining factor?
   (a) composition of the air
   (b) partial pressure of the gas in air
   (c) temperature
   (d) partial pressure of the gas in water
   (e) partial pressure of other gases in the water

2. Air contains _____ percent oxygen.
   (a) 6   (d) 65
   (b) 21   (e) 95
   (c) 48

3. Increasing volume in proportion to surface area is a favorable change for animals that rely on diffusion to obtain oxygen. True or false?

4. Aquatic respiratory organs increase the diffusion surface by extensions of tissues that project from the body out into the water. These projections are known as _____.

5. This type of chamber provides a means of pumping water past the gills:
   (a) opercular cavities   (d) a and b
   (b) branchial chamber   (e) a and c
   (c) bronchial chamber

6. In the gills blood runs in the direction opposite to the water flow, thus maximizing efficiency of the diffusion by a process known as _____ _____.

7. Lungs are much more efficient respirators than gills. True or false?

8. The opening of the trachea, or windpipe, in amphibians is regulated by a valve called the _____.

9. The thin layer of muscle that separates the thoracic cavity from the abdominal cavity in mammals is known as the _____.

10. During inhalation the pressure of intrapleural fluid
    (a) decreases   (b) increases

11. The oxygen carrier protein used by all vertebrates is _____.

12. In the presence of carbon dioxide, the oxygen-hemoglobin dissociation curve is shifted
    (a) up   (c) left   (e) remains the same
    (b) down   (d) right

THOUGHT QUESTIONS

1. Often people who appear to have drowned can be revived, in some cases after being under water for as long as half an hour. In every case of full recovery after extended submergence, however, the person had been submerged in very cold water. Is this observation consistent with the fact that oxygen is twice as soluble in water at 0° C as it is at 30° C?

2. Can you think of a reason why a respiratory system has not evolved in which oxygen is actively transported across respiratory membranes, in place of the passive process of diffusion across these membranes that is universally employed?

3. If by accident your pleural membrane were punctured, would you be able to breathe?

FOR FURTHER READING

BAKER, P.: "Human Adaptation to High Altitude," *Science,* vol. 163, pages 1149-1156, 1969. A study of the physiological changes observable in people who live in very high places.

DICKERSON, R.E., and I. GEIS: *Hemoglobin: Structure, Function, Evolution, and Pathology,* Benjamin-Cummings Publishing Co., Menlo Park, Calif., 1983. A comprehensive treatment of all aspects of hemoglobin, from the genes that encode it to the way in which the protein functions in the vertebrate respiratory system.

NEGUS, V.: *The Biology of Respiration,* Livingstone, Edinburgh, 1965. A comprehensive treatment of gas exchange in animals, viewed from an evolutionary and comparative perspective.

PERUTZ, M.F.: "Hemoglobin Structure and Respiratory Transport," *Scientific American,* December 1978, pages 92-125. An account of how hemoglobin changes its shape to facilitate oxygen binding and unloading, by the man who won a Nobel prize for unraveling the structure of hemoglobin.

RANDALL, D.J., W.W. BURGREN, A.P. FARRELL, and M.S. HASWELL: *The Evolution of Air Breathing in Vertebrates,* Cambridge University Press, Cambridge, England, 1981. An account of the physiological changes in respiration that occurred during the evolution of the terrestrial vertebrates.

SCHMIDT-NIELSEN, K.: "How Birds Breathe," *Scientific American,* December 1971, pages 73-79. A fascinating account of the discovery of unidirectional flow in avian lungs, by a great comparative physiologist.

Chapter 45

TRANSPORT WITHIN THE BODY: CIRCULATION

Overview

In vertebrates, circulatory systems are like highways, over which red blood cells carry oxygen to the tissues and remove carbon dioxide. Blood fluid also transports glucose and amino acids to the cells and carries away nitrogenous wastes. Blood composition is kept constant by the liver, which monitors and adjusts metabolite levels in the blood. The key to circulation in vertebrates is the organ that pumps the blood through the system, the heart. The heart has evolved in concert with the respiratory system of vertebrates. The double-pump architecture of the heart in mammals and birds plays an important role in their ability to be warm-blooded (endothermic).

For Review

Here are some important terms and concepts that you will encounter in this chapter. If you are not familiar with them, you should review them before proceeding.

- **Glycogen** (Chapter 4)

- **Erythrocytes** (Chapter 42)

- **Depolarization** (Chapter 42)

- **How hemoglobin carries oxygen** (Chapter 44)

Of the many tissues of the vertebrate body, few have the emotive impact of blood. In movies and in real life the sight of blood connotes violence, injury, and bodily harm. In literature, blood symbolizes the life force, which is not a bad analogy in real life either. This chapter is about blood, its circulation through the body, and the many ways in which it affects the body's functions. Although vertebrates possess many other organ systems that are necessary for life, it is the activities of the blood that bind them together into a functioning whole.

THE EVOLUTION OF CIRCULATORY SYSTEMS

The capture of nutrients and gases from the environment is one of the essential tasks that all living organisms must carry out. When a bacterium has transported nutrients across its

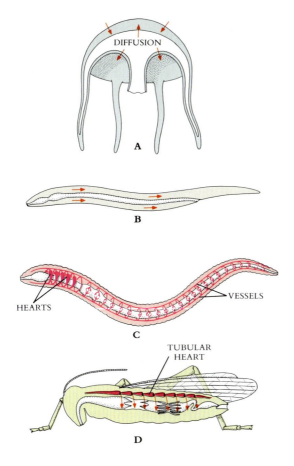

DIFFUSION

A

B

HEARTS   VESSELS

C

TUBULAR
HEART

D

cell membrane into its cytoplasm, its job is essentially over. The molecules it has brought in are able to move throughout the entire bacterium, a single cell, by diffusion. Among the single-celled protists and some of the simpler metazoans, the movement of nutrients occurs in a similar way. Some portion of every cell of the organism is exposed to the external liquid environment (Figure 45-1, *A*), enabling the individual cells to capture materials from that environment and to release materials directly into it.

As metazoans have become larger, however, with layers of cells stacked on one another, their more interior cells have experienced greater difficulty in exchanging materials with the environment by simple diffusion. This dilemma was solved with the advent of the body cavity (Figure 45-1, *B*), which can be seen in one of its first manifestations among the roundworms, phylum Nematoda (see Chapter 35). Fluid within this body cavity constitutes a primitive kind of circulatory system, one that permits materials to pass from one cell to another without leaving the organism. In organisms with a body cavity, one cell is able to take up material from the environment, and another distant cell can receive and use that material. This transport of material from one place to another within an organism by passage through an internal fluid is called **circulation.**

A circulatory system may be either open or closed. In a **closed system** the fluid in the circulatory system is separated from the rest of the body's fluids and does not mix freely with them. Materials pass into and out of the circulating fluid by diffusion. In an **open system,** by contrast, there is no distinction between circulating fluid and body fluid generally—the circulating fluid *is* the body fluid.

Either an open or a closed circulatory system will work, and both function successfully in different phyla of animals. Annelids (see Chapter 36), for example, have a closed circulatory system (Figure 45-1, *C*) in which two major tubes or vessels extend the length of the animal, one on top (dorsal) and the other on the bottom (ventral), with branches extending from each tube out to the muscles, skin, and digestive system. Movement of

FIGURE 45-1

Evolution of circulatory systems.
A *Simple diffusion; a hydra. Without any movement of the organism, diffusion reaches all parts of the body.*
B *Circulation in a body cavity; a nematode. As the worm moves, its contracting muscles push fluid back and forth through the body cavity.*
C *A closed circulatory system; an annelid. The series of hearts in each segment push blood out through a system of vessels to the body's tissues, then back via a second vessel system.*
D *An open circulatory system; an insect. The tubular heart of insects pumps blood out to the tissues of the body, from which it perfuses back rather than traveling in vessels.*

fluid within this network of vessels is promoted by the contraction of the muscles surrounding the main dorsal vessels and by five conduits connecting the dorsal and ventral tubes. All vertebrates have closed circulatory systems.

Arthropods (see Chapter 37), by contrast, evolved an open circulatory system (Figure 45-1, *D;* see also Figure 37-7). In this system, a muscled tube within the central body cavity forces the cavity fluid out into the body through a network of interior channels and spaces. The fluid then flows back into the central cavity.

A great advantage of closed circulatory systems is that they permit regulation of fluid flow by means of muscle-driven changes in the diameters of the vessels. In other words, with a closed circulatory system the different parts of the body can maintain different circulation rates.

THE FUNCTION OF CIRCULATORY SYSTEMS IN VERTEBRATES

The closed circulatory system of vertebrates is like a roadway connecting the various muscles and organs of the body with one another. It serves four principal functions, each of which may be thought of as a circuit.

Nutrient and Waste Transport

The food molecules that fuel cell metabolism are transported to the cells of the body by the circulatory system. The products of digestion pass into the bloodstream largely through the wall of the small intestine, diffusing into a fine net of blood vessels below the mucosa. Metabolites, principally sugars, amino acids, and fatty acids, are first carried from the intestines to the liver. In the liver some of these metabolites are converted to glucose, which is released into the bloodstream. Others, such as the essential amino acids, fatty acids, and vitamins, pass through the liver unchanged. Still others—excessive amounts of molecules that are sources of energy—are converted to storage compounds and accumulated for later use. These conversion processes are discussed in more detail later in this chapter.

From the liver, blood carries the dissolved glucose and other metabolites to all body cells. In these cells glucose is used as an energy source, and the other metabolites become building blocks. These metabolizing cells release into the bloodstream the wastes that are produced during the course of their metabolism. The blood carries these wastes to a cleansing organ, the kidney, which captures them and concentrates them for excretion in the urine (Chapter 48). The cleansed blood then passes back to the heart, completing the metabolic circuit.

Oxygen and Carbon Dioxide Transport

Most of the cells of the vertebrate body carry out oxidative respiration and therefore require oxygen. This necessary oxygen is transported to the cells of the body by the circulatory system. Within lungs or gills, oxygen molecules diffuse into the circulating blood through the walls of very fine blood vessels. This oxygen soon passes into cells suspended in the blood, the erythrocytes, which circulate in very large numbers (Figure 45-2). The erythrocytes contain within them large amounts of the protein hemoglobin, each molecule of which binds several of the oxygen molecules. Oxygen is thus well packaged: hemoglobin is the primary carrier; hemoglobin is in turn carried by erythrocytes; and erythrocytes are carried within the circulating blood. From the lungs, the blood carries the carrier-bound oxygen to all of the metabolizing cells of the body, where the oxygen is released from the blood and is used to carry out oxidative respiration. The metabolizing cells release the end product of this metabolism, carbon dioxide, into the bloodstream. There some of it is bound by the hemoglobin carrier molecules. The blood then returns to the lungs, where the release of carbon dioxide and the capture of new oxygen complete the respiratory circuit.

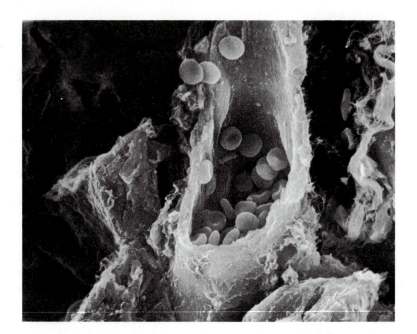

FIGURE 45-2

This ruptured tube is a blood vessel, an element of the human circulatory system. It is full of red blood cells, which move through these blood vessels transporting oxygen and carbon dioxide from one place to another in the body.

Temperature Maintenance

As discussed in Chapter 39, mammals and birds maintain a constant body temperature by expending metabolic energy. They differ in this respect from most other organisms, which typically do not use metabolism to maintain a constant internal temperature. In the vertebrates, regardless of how a given body temperature is attained, the circulating blood distributes the heat more or less uniformly throughout the body. This heat circulation is accomplished first by means of a network of fine blood vessels, which passes immediately beneath the external surfaces of the body. These surfaces are in contact with the environment. The blood that circulates within this network absorbs heat from a hot environment and releases heat to a cold one. The further circulation of this heated or cooled blood to the interior of the animal's body tends to adjust the internal temperature of that animal in the direction of its external environment. At the same time the circulation of blood back out from the body's interior to its extremities completes the thermal circuit. Many vertebrates conserve heat by short-circuiting this pattern of circulation. They do so by constricting the blood vessels beneath the surface extremities of their body so that little blood flows through them.

Hormone Circulation

The metabolic and other activities of the body's many organs are coordinated both by the nervous system and by a family of small molecules called **hormones.** The hormones are signals that initiate or modulate various body functions. They are produced by glands situated at various locations within the body. Once the hormones have reached the blood, they are transported throughout the body. They soon reach the target tissues that are capable of responding to them. Most hormones persist in the bloodstream for only a short time, since they are continuously destroyed by the enzymes present in the body. As a result of this turnover, the appearance of hormones within the bloodstream is a dependable signal for the initiation of whatever process the hormones affect.

> **The closed circulatory system in vertebrates transports oxygen, nutrients, and metabolites to cells, transports carbon dioxide and metabolic wastes away from metabolizing cells, helps to minimize temperature differences, and transmits hormonal regulatory signals to the cells of the body.**

THE CARDIOVASCULAR SYSTEM

The closed circulatory system of vertebrates is composed of three elements: (1) the **heart,** a muscular pump that we will consider later in this chapter; (2) the blood vessels (Figure 45-3), a network of tubes that passes through the body; and (3) the blood, which circulates within these vessels. The plumbing of the closed circuit, the heart and vessels, is known collectively as the **cardiovascular system.** Blood moves within this cardiovascular system, leaving the heart through vessels known as **arteries.** From the arteries the blood passes into a larger network of **arterioles,** or smaller arteries. From these it is eventually forced through the **capillaries,** a fine latticework of very narrow tubes, which get their name from the Latin word *capillus,* "a hair." It is while passing through these capillaries that the blood exchanges gases and metabolites with the cells of the body. After traversing the capillaries, the blood passes into a third kind of vessel, the **venules** or small **veins.** The venules lead to a network of larger veins, which collect the circulated blood and carry it back to the heart.

Because the vertebrate cardiovascular system is closed, it has certain structural requirements that open circulatory systems do not. To illustrate this relationship, we will consider the properties of one very familiar kind of open "vascular" system, a garden hose. If you turn the water on, a pulse of water passes down the hose, emerging a few moments later out the end as a stream of water. If you turn the water on harder, the water just moves through the hose faster. Now close the nozzle at the end of the hose. You have now converted the hose into a closed vascular system. If you turn the water on even harder, will more water move into the hose? Where is it going to go? The answer is that no more water will move into the hose, because the walls of the hose cannot expand to accommodate a larger volume. They are strong, but they are not elastic.

The heart of a vertebrate faces the same plumbing problem as the closed garden hose: it must push a pulse of fluid through a closed system of vessels that meets a resistance at one end, a network of small capillaries. The capillaries have a much smaller diameter than the other blood vessels of the body. To understand why reduction in tube diameter leads to increased resistance to flow, we must recall a little physics. When a fluid, such as blood, flows through a horizontal tube, it meets a frictional resistance that is inversely proportional to the fourth power of the radius of the tube (Figure 45-4). For example, when the radius of a tube is reduced by half, the resistance to flow is 16 times greater. Now you can see why flow through the capillaries creates resistance. Blood leaves the human heart through the **aorta,** a tube that has a radius of about 1 centimeter, but when it reaches the capillaries, it passes through vessels with an average radius of only 4 micrometers, a reduction in radius of some 2500 times!

The capillary network's resistance to flow has an important biological consequence: the vessels of the cardiovascular system must have elastic walls. An adult human heart pumps out approximately 70 milliliters of blood with every beat. Consequently, the vessels leading out from the heart must be able to expand enough to accommodate this added volume.

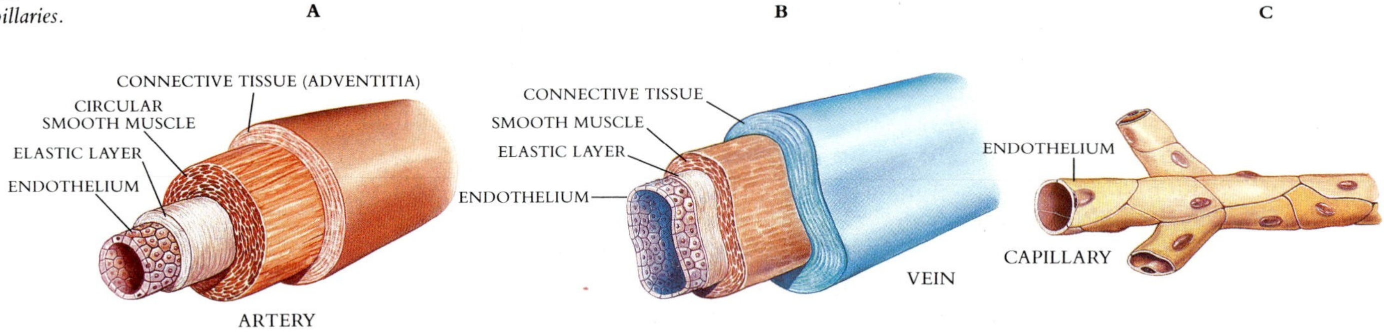

FIGURE 45-3

The structure of blood vessels.
A *Arteries.*
B *Veins.*
C *Capillaries.*

A

CONNECTIVE TISSUE (ADVENTITIA)
CIRCULAR SMOOTH MUSCLE
ELASTIC LAYER
ENDOTHELIUM

ARTERY

B

CONNECTIVE TISSUE
SMOOTH MUSCLE
ELASTIC LAYER
ENDOTHELIUM

VEIN

C

ENDOTHELIUM

CAPILLARY

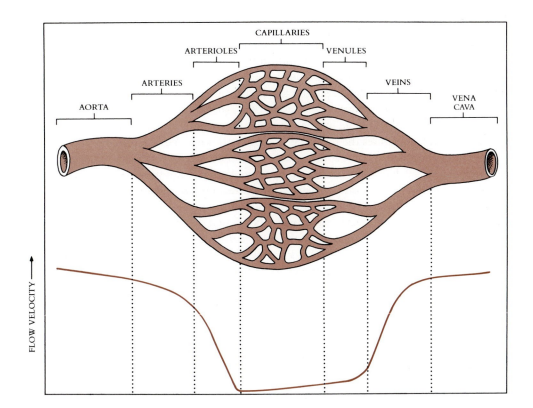

FIGURE 45-4

The effect of vessel diameter on resistance and flow velocity. The narrower a tube, the greater the resistance it presents to liquid flowing through it. The resistance is inversely proportional to the fourth power of the radius of the tube.

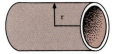

FLOW RESISTANCE = R

FLOW RESISTANCE = 16R

Arteries

Artery walls are made up of four layers of tissue. The innermost one is composed of a thin layer of cells. Surrounding these cells is a thick layer of smooth muscle and another of elastic fibers, which in turn are encased within an envelope of connective tissue. Because this sheath is elastic, the artery is able to expand its volume considerably in response to a pulse of hydrostatic pressure, much as a tubular balloon might. The steady contraction of the muscle layer strengthens the wall of the vessel against overexpansion.

Arterioles

Arterioles differ from arteries simply in that they are smaller in diameter. The muscle layer that surrounds them can be relaxed under the influence of hormones and metabolites. When this happens the blood flow can be increased, an advantage during times of high metabolic activity. Conversely, most arterioles are in contact with many nerve fibers. When stimulated, these nerves cause the muscular lining of the arteriole to contract and thus constrict the diameter of the arteriole. Such contraction limits the flow of blood to the extremities during periods of low temperature or stress. You turn pale when you are scared because contraction of this kind constricts the arterioles in your skin. When you blush or flush from embarrassment or because you are overheated, the opposite effect occurs: the nerve fibers connected to muscles surrounding the arterioles are inhibited, relaxing the smooth muscle and causing the arterioles in the skin to dilate.

> **The major vessels of the circulatory system are tubes of cells encased within three sheaths: (1) a layer of elastic fibers, which renders the diameter of the vessel elastic to accommodate pulses of blood pumped from the heart; (2) a layer of muscle serviced by nerves, which permits the body to control the diameter of the vessel and strengthens the wall against overexpansion; and (3) a layer of connective tissue, which protects the vessel.**

Capillaries

Capillaries have the simplest structure of any element in the cardiovascular system. They are little more than tubes one cell thick and on the average about 1 millimeter long; they connect the arterioles with the veins. The internal diameter of the capillaries is, on the average, about 8 micrometers. Surprisingly, this is little more than the diameter of a red blood cell (5 to 7 micrometers). However, red blood cells squeeze through these fine tubes without difficulty (Figure 45-5). The intimate contact between the walls of the capillaries and the red blood cells facilitates the diffusion of gases and metabolites across the walls of the red blood cell and those of the capillary, as well as those of the surrounding cells. *No cell of the body is more than 100 micrometers from a capillary.* At any one moment, about 5% of your blood is in your capillaries. Some capillaries, called **through-flow channels,** connect arteries and veins directly (Figure 45-6). From these channels, loops of true capillaries leave and return. It is through these loops that almost all exchange between the blood and the cells of the remainder of the body occurs. The entry to each loop is guarded by a ring of muscle called a **precapillary sphincter,** which, when closed, blocks flow through the capillary. Such restriction of entry to the capillaries in surface tissue is another powerful means of limiting heat loss from an animal's body during periods of cold.

The entire body is permeated with a fine mesh of these capillaries, a network that amounts to several thousand kilometers in overall length. If all of the capillaries in your body were laid end to end, they would extend across the United States. While individual capillaries have high resistance to flow because of their small diameters, the cross-sectional area of the extensive capillary network is greater than that of the arteries leading to it, so that the blood pressure is actually lower in the capillaries.

Veins

Veins do not have to accommodate the pulsing pressures that arteries do, because much of the force of the heartbeat is attenuated by the high resistance and great cross-sectional area of the capillary network. The walls of veins, although similar in structure to those of the arteries, have much thinner layers of muscle and elastic fiber (Figure 45-7). An empty artery is still a hollow tube, like a pipe, but when a vein is empty, its walls collapse like an empty balloon.

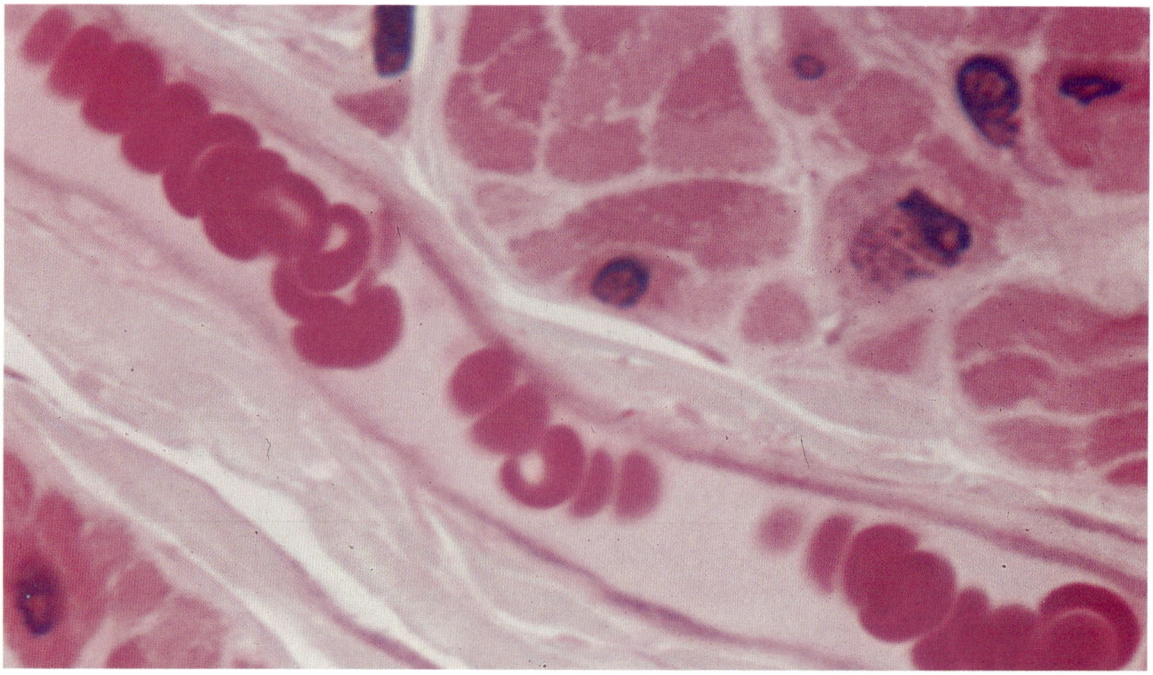

FIGURE 45-5

The red blood cells in this capillary are passing along in single file. Many capillaries are even narrower than those shown here, in the bladder of a monkey. However, red blood cells will even pass through capillaries narrower than their own diameter, pushed along by the pressure generated by a pumping heart.

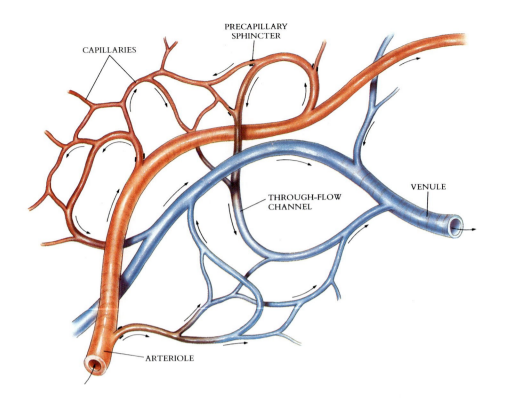

CAPILLARIES

PRECAPILLARY SPHINCTER

THROUGH-FLOW CHANNEL

VENULE

ARTERIOLE

FIGURE 45-6

The capillary network connects arteries with veins. The most direct connection is via through-flow channels that connect arterioles directly to venules. Branching from these through-flow channels is a network of finer channels, the capillary network. Most of the exchange between body and red blood cells occurs while they are in this capillary network. Entrance to the capillary network is controlled by bands of muscle called precapillary sphincters at the entrance to each capillary. When a sphincter is contracted, it closes off the capillary. By contracting these sphincters, the body can limit the amount of blood in the capillary network of a particular tissue, and thus control the rate of exchange in that tissue.

The internal passageway of veins is often quite large. The diameter of the largest vein in the human body, the **vena cava,** which leads into the heart, is fully 3 centimeters. The reason that veins are so much larger than arteries is related to the lower pressure of blood flowing within veins back toward the heart—it is advantageous to minimize any further resistance to its flow. Recall, as we noted above, that resistance to flow varies as the inverse of the fourth power of the radius of a tube—a larger tube presents much less resistance to flow. Veins are large because larger veins present less resistance to the flow of blood back to the heart.

FIGURE 45-7

The vein (left) has the same general structure as an artery (right), but much thinner layers of muscle and elastic fiber. An artery will retain its shape when empty, but a vein will collapse.

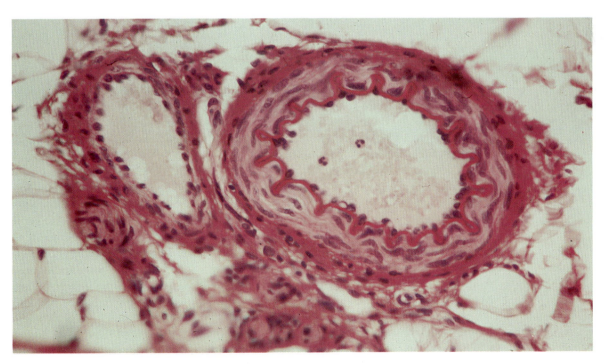

FIGURE 45-8

The human lymphatic system. The lower portion of the body and the left side drain into a single lymphatic vessel, the thoracic duct, which empties into the left subclavian vein through a one-way valve. The upper right side of the body drains into a second collecting vessel, the right lymphatic duct, which drains into the right subclavian vein.

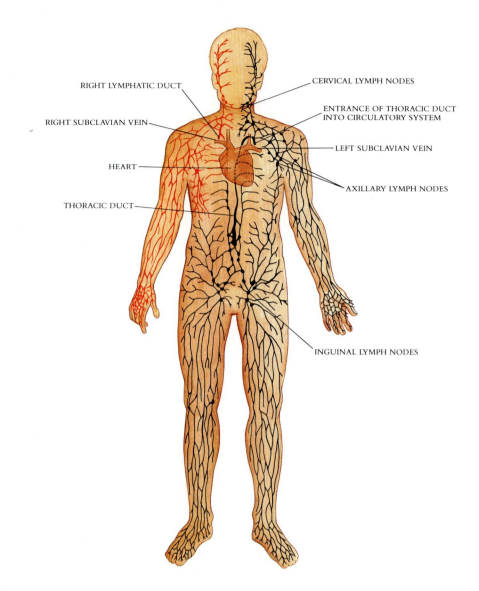

RIGHT LYMPHATIC DUCT

CERVICAL LYMPH NODES

ENTRANCE OF THORACIC DUCT INTO CIRCULATORY SYSTEM

RIGHT SUBCLAVIAN VEIN

LEFT SUBCLAVIAN VEIN

HEART

AXILLARY LYMPH NODES

THORACIC DUCT

INGUINAL LYMPH NODES

The Lymphatic System

The cardiovascular system is considered a closed system because all of its tubes are connected with one another and none are simply open ended. In another sense, however, the system is open—to diffusion through the walls of the capillaries. This process is a necessary part of the functioning of the circulatory system, but it poses difficulties in maintaining the integrity of that system.

Difficulties arise because diffusion from the capillaries is accompanied by the loss of large quantities of liquid from the cardiovascular system. When blood passes through the capillaries, it loses more water to the body than it reabsorbs from it. In a human being about 3 liters of fluid leave the cardiovascular system in this way each day, a quantity amounting to more than half the body's total supply of about 5.6 liters of blood. To counteract the effects of this process, the body uses a second *open* circulatory system called the **lymphatic system.** The elements of the lymphatic system gather liquid from the body and return it to cardiovascular circulation (Figure 45-8). Open-ended lymph capillaries gather up fluids by diffusion and carry them through a series of progressively larger vessels to two large lymphatic vessels, which resemble veins. These two lymphatic vessels drain into veins through one-way valves. No heart pumps fluid through the lymphatic system; instead, fluid is driven through it when its vessels are squeezed by the movements of the body's

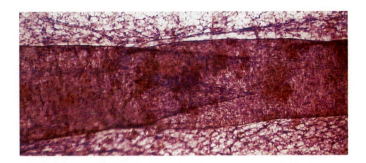

FIGURE 45-9

A lymphatic vessel valve, magnified 25 times. Flow from left to right is not retarded since such flow tends to force open the inner cone; flow from right to left is prevented because such flow tends to force the inner cone closed.

muscles. The lymphatic vessels contain a series of one-way valves (Figure 45-9), which permit movement in only one direction.

Much of the water within blood plasma diffuses out during passage through the capillaries. This water is collected by an open circulatory system, the lymphatic system, and returned to the bloodstream.

THE CONTENTS OF VERTEBRATE CIRCULATORY SYSTEMS

About 8% of the body mass of most vertebrates is taken up by the blood circulating through their bodies. This blood is composed of a fluid plasma and several different kinds of cells that circulate within that fluid.

Blood Plasma

Blood plasma is a complex solution of three very different components:

1. *Metabolites and wastes*. If the circulatory system is thought of as the "highway" of the vertebrate body, the blood contains the "traffic" passing on that highway. Dissolved within the plasma are all of the metabolites, vitamins, hormones, and wastes that circulate among the cells of the body.

2. *Salts and ions*. Like the water of the seas in which life arose, plasma is a dilute salt solution. The chief plasma ions are sodium, chloride, and bicarbonate ions. In addition, there are trace amounts of other salts, such as calcium and magnesium, as well as of metallic ions, including copper, potassium, and zinc. The composition of the plasma, therefore, is not unlike that of seawater.

3. *Proteins*. Blood is 90% water. Passing by all the cells of the body, blood would soon lose most of its water to them by osmosis if it did not contain as high a concentration of proteins as the cells that it passed. The water would move from the blood vessels into the surrounding cells, diluting their contents. Blood plasma does contain the antibody and globulin proteins that are active in the immune system, as well as a small amount of **fibrinogen,** a protein that plays a role in blood clotting. Taken together, however, these proteins make up less than half of the amount of protein that is necessary to balance the protein content of the other cells of the body. The rest consists of a protein called **serum albumin,** which circulates in the blood as an osmotic counterforce. Human blood contains 46 grams of serum albumin per liter. Protein deficiency diseases such as **kwashiorkor** produce swelling of the body because the body's cells take up water from the albumin-deficient blood.

Blood contains metabolites, wastes, and a variety of ions and salts. Blood also contains high concentrations of the protein serum albumin, which functions to keep the blood plasma in osmotic equilibrium with the cells of the body.

HOW BLOOD CLOTS

When you cut your finger, the reason that you do not bleed to death is that the body seals the breach in your circulatory system. The blood in the immediate area of the cut is converted to a solid gel that seals the cut, in much the same way that a tubeless tire seals a puncture. A blood clot forms from the polymerization of protein fibers circulating in solution, in exactly the same way that Jell-O hardens. In blood the protein that polymerizes and thus controls the bleeding is **fibrinogen.** Fibrinogen is present in significant concentrations in the blood plasma.

Blood clotting is a "cascade" process in which one or a few molecules initiate a wave of events, much as a nuclear reaction starts with a few events but soon involves many atoms. Blood clotting starts when blood plasma encounters a rough surface, such as torn tissue. When such an encounter occurs, a protein factor in the blood initiates a progressive series of enzyme activations. As each enzyme in the series is activated, it then activates many molecules of the next enzyme in the series in a spreading wave of activations. Phospholipids that occur on blood platelets are used as a catalytic surface for many of these enzyme activations.

The final result of this cascade of reactions is the activation of an enzyme called **thrombin.** This enzyme is normally present in blood plasma in an inactive form called **prothrombin.** Thrombin cleaves small polypeptides from fibrinogen, creating a very sticky polypeptide fragment called **fibrin.** Fibrin fragments bind to one another, forming a tight mesh across the surface of the wounds (Figure 45-A, *1*). This mesh later contracts, drawing the sides of the wound together.

It was once thought that exposure to air played a role in the clotting process. If you place some blood in a test tube, it will form a clot, even though no wound is present. The mechanism by which this clotting occurs turns out to be exactly the same as that which happens at the site of a wound. In the test tube the initiating surface is the roughness of the surface of the glass. Proper clotting requires fibrinogen, platelets, and all of the many factors in the cascade of reactions leading to the production of thrombin from prothrombin. A mutation causing the loss of activity of any of these factors leads to a form of **hemophilia,** a condition in which the blood is slow to clot or does not clot at all.

Hemophilias are hereditary diseases, because they

FIGURE 45-A, *2*

Queen Victoria of England in 1894, surrounded by some of her descendants. Of Victoria's four daughters who lived to bear children, two, Alice and Beatrice, were carriers of Royal hemophilia. Two of Alice's daughters are standing behind Victoria (wearing feathered boas): to Victoria's right is Princess Irene of Prussia; to her left, Alexandra, who would soon become Tsarina of Russia. Both Irene and Alexandra were also carriers of hemophilia.

entail mutations in genes encoding one of the clotting factors. Hemophilia is a recessive condition, expressed only when an individual does not possess at least one copy of the gene that is normal and so cannot produce one of the clotting factors. Thus individuals homozygous for a clotting factor mutation cannot clot blood. Most clotting factor genes are on nonsex chromosomes (autosomes), but two (designated VIII and IX) are known to be located on the X chromosome. In these two instances, any male who inherits a mutant form will express hemophilia, because his other chromosome is the inactive Y and thus he has no good copy of the factor-encoding gene. Such a mutation in factor IX occurred in one of the parents of Queen Victoria of England (1819-1901) (Figure 45-A, *2*), and for this reason factor IX mutations are often referred to as **Royal hemophilia.** In the five generations since Queen Victoria (Figure 45-A, *3*), 10 of her male descendants have had the disease.

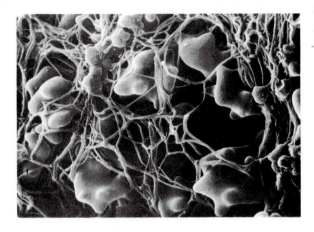

FIGURE 45-A, *1*

A mesh of fibrin fibers forming a clot.

FIGURE 45-A, *3*

The Royal hemophilia pedigree. From Queen Victoria's daughter Alice, the disorder was introduced into the Russian and Austrian royal houses, and from her daughter Beatrice it was introduced into the Spanish royal house. Victoria's son Leopold, himself a victim, also transmitted the disease in a third line of descent.

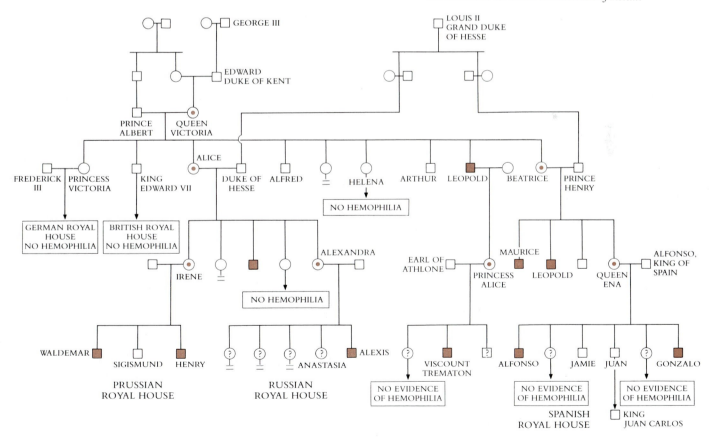

FIGURE 45-10

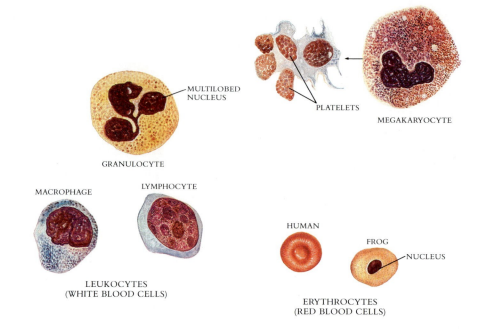

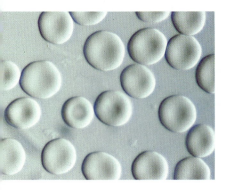

FIGURE 45-11

Human erythrocytes, magnified 1000 times. Human erythrocytes lack nuclei, which gives them a characteristic collapsed appearance, a bit like a pillow on which someone has sat.

Types of Blood Cells

The fraction of the total volume of the blood that is occupied by cells is referred to as the blood's **hematocrit.** In human beings, cells typically occupy about 45% of the blood's volume. There are three principal types of cells in the blood (Figure 45-10).

Erythrocytes

Each milliliter of blood contains about 5 million erythrocytes, or red blood cells. Each erythrocyte is a flat disk with a central depression (Figure 45-11), something like a doughnut with a hole that does not go all the way through. Attached to the outer membranes of the erythrocytes is a collection of polysaccharides, which determines an individual's blood group (see Figure 20-3). Almost the entire interior of each cell is packed with hemoglobin.

Erythrocytes in mammals lose their nuclei and protein-synthesizing machinery when they mature. Because they lack a nucleus, these cells are unable to repair themselves, and they therefore have a rather short life; any one erythrocyte lives for only about 4 months. As erythrocytes age, they are removed from the bloodstream by macrophages, which ingest them and break them down. Balancing this loss, new erythrocytes are constantly being synthesized and released into the blood by cells within the soft interior marrow of bones. Each erythrocyte develops from a bone-marrow cell, accumulating hemoglobin and gradually losing its nucleus, a process referred to as **erythropoiesis.** When the oxygen levels in the blood fall below normal values, increased amounts of the hormone **erythropoietin** are secreted into the blood by the kidney and other organs. This hormone stimulates the bone marrow to produce more erythrocytes, thus increasing the oxygen-carrying capacity of the blood.

Leukocytes

Perhaps 1% of human blood cells are leukocytes, or white blood cells. There are several kinds of leukocytes, which have different functions. All of these functions, however, are related to the body's defense against invading bacteria and other foreign substances. Granular leukocytes, which contain many cytoplasmic granules, develop from cells within the bone marrow. They travel within the blood to sites of injury or trauma, where they leave the blood and enter the tissue. In this injured tissue the granular leukocytes ingest injured cells and release histamines. Other leukocytes produced by bone marrow cells develop into macrophages; still others develop into lymphocytes, the cells that are responsible for antibody production (see page 905).

Platelets

Certain large cells within the bone marrow, called **megakaryocytes,** regularly pinch off bits of their cytoplasm. These cell fragments, which contain no nuclei, enter the bloodstream, where they play an important role in controlling blood clotting. The cell fragments are called **platelets.**

HOW THE LIVER REGULATES BLOOD GLUCOSE LEVELS

It is important that the blood plasma of vertebrates maintain a relatively constant composition, since the different tissues of the body have evolved complex specializations that depend in important ways on the composition of the blood, and particularly on its metabolites. Brain cells, for example, can store very little glucose and lack the enzymes to convert fat or amino acids into glucose. Despite these factors, brain cells are very sensitive to the level of available glucose. They are totally dependent on blood plasma for glucose and cease to function if the level of glucose in the blood falls much below normal values.

Maintaining a constant level of metabolites in blood plasma requires active control by the body's organs. A moment's reflection shows why. Most vertebrates eat sporadically. In the United States, for example, most people eat three meals a day. Food enters the digestive system at intervals separated by long periods of fasting. Much of the food is digested relatively quickly, with the metabolites, such as glucose and amino acids, passing through the lining of the stomach and small intestine into the bloodstream. Without active control of the levels of these and other metabolites, they would suddenly become much more abundant in the blood right after a meal and then fall rapidly during a period of starvation, when the metabolites are being removed from the bloodstream by metabolizing cells and not being replenished.

Control over metabolite levels in blood plasma is achieved in a very logical way: by establishing a reservoir, or metabolic bank, in the liver. In Chapter 43 we discussed the liver as one of the body's principal synthetic factories. Of all its many functions, one of the most important is its regulation of the blood's metabolite levels. This regulation is achieved by means of a special shunt in the circulatory system. Blood returning from the stomach and small intestine flows into a special conduit called the **portal vein,** which carries it not to the vena cava and the heart but rather to the liver (Figure 45-12). Within the liver, blood passes through a network of fine passages called **sinuses.** Only after perfusing through the liver tissue is the blood collected into the **hepatic vein** and delivered to the vena cava and the heart.

When excessive amounts of glucose are present in blood passing through the liver, a situation that occurs soon after a meal, the liver converts the excess glucose into the starch-like glucose polymer, glycogen. The liver then stores this glycogen. When blood glucose levels fall low, as they do in a period of fasting or between meals, the glucose deficit in the blood plasma is made up from the glycogen reservoir. The human liver, for example, stores enough glycogen to supply glucose to the bloodstream for about 24 hours of fasting. If fasting continues, the liver begins to convert other molecules, such as amino acids, into glucose to maintain the level of glucose in the blood. The liver, the body's metabolic reservoir, thus acts much like a bank, making deposits and withdrawals in the "currency" of glucose molecules.

In addition to its many other roles, the liver acts to regulate the level of blood glucose, maintaining it within narrow bounds.

FIGURE 45-12

Hepatic-portal circulation. Blood from the stomach and small intestine, rich with the metabolites of digestion, is collected into the portal vein, which carries it to the liver. After flowing through the sinuses of the liver, the liver-processed blood is re-collected into the hepatic vein, which carries it back toward the heart.

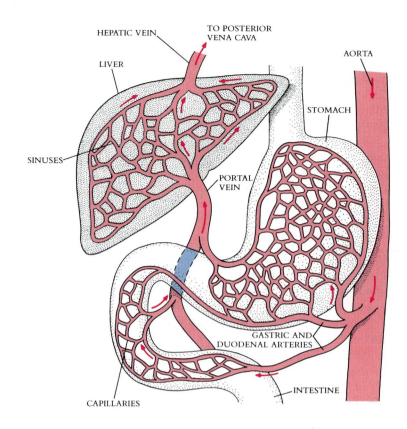

FIGURE 45-13

Nitrogenous wastes.
A *In many freshwater fishes, the primary nitrogenous waste is ammonia.*
B *Sharks maintain high internal concentrations of urea, which many other organisms excrete.*
C *Birds and land reptiles convert much of their urea to uric acid in the liver, and excrete it in urine.*

Also like a bank, the liver exchanges currencies, converting other molecules, such as amino acids and fats, to glucose for storage. The liver cannot store significant concentrations of amino acids. Instead, excess amino acids that may be present in the blood are converted to glucose by liver enzymes, and the glucose is stored as glycogen. The first step in this conversion is the removal of the amino group (NH_4^+) from the amino acid, a process called **deamination.** Unlike plants, animals cannot reuse the nitrogen from these amino groups, and must excrete it as nitrogenous waste. The product of amino acid deamination, ammonia (NH_3), complexed with carbon dioxide to form urea, or further complexed to form uric acid (Figure 45-13), is released by the liver into the bloodstream. Subsequently the kidneys remove urea or uric acid from the bloodstream.

The liver has a finite storage capacity. When its glycogen reservoir is full it continues to remove excess glucose molecules from the blood by converting them to fat, which is stored elsewhere in the body. In human beings, for example, long periods of overeating

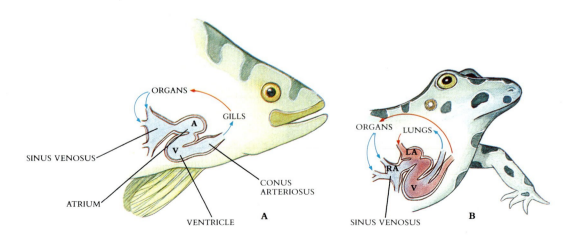

and the resulting chronic oversupply of glucose frequently result in the deposition of fat around the stomach or on the hips.

THE EVOLUTION OF THE VERTEBRATE HEART

Any closed circulatory system requires both a system of passageways through which fluid can circulate and a pump to force the fluid through them. In the circulation of blood the pump is the heart. The evolution of the heart among the vertebrates reflects two great transitions in their history. The first of these was the shift from filter feeding to active prey capture, an event that occurred at the dawn of the vertebrates. The second was the invasion of the land. Active predation by the first fishes entailed a greatly improved respiration system, the gill (see Chapter 44). The transition to land involved the evolution of a different type of breathing apparatus (the lung), a significant lessening of pressure from sea to air, and the development of intrinsic control of body temperature. All of these changes had major influences on the evolution of the heart.

The Early Chordate Heart: A Peristaltic Pump

The chordates that were ancestral to the vertebrates are thought to have had simple tubular hearts, not unlike those now seen in lancelets. The heart was little more than a specialized zone of the ventral artery, more heavily muscled than the rest of the arteries, which beat in simple peristaltic waves. Peristaltic pumps such as these hearts are not very efficient—blood is pushed in *both* directions as the heart contracts peristaltically. The pumping action results because the blood moves more readily in the direction of the wave of peristaltic contractions, because the as–yet–uncontracted portion of the blood vessel has a larger diameter than the contracted portion and thus less resistance to flow.

The Fish Heart: A One-Cycle Chamber Pump

The development of gills by fishes necessitated a more efficient pump to force blood through the fine capillary network, and in fishes we see the evolution of a true "chamber-pump" heart. The fish heart can be considered a tube that has four chambers arrayed one after the other (Figure 45-14). The first two chambers (**sinus venosus** and **atrium**) are collection chambers; the second two (**ventricle** and **conus arteriosus**) are pumping chambers.

As might be expected from the nature of the hearts of the early chordates from which the fishes evolved, the sequence of the heartbeat is the peristaltic sequence, starting at the rear and moving to the front. The first of the four chambers to contract is the sinus venosus, then the atrium, then the ventricle, and finally the conus arteriosus. Despite shifts in the relative positions of the chambers in the vertebrates that evolved later, this heartbeat sequence is maintained and unchanged in all vertebrates.

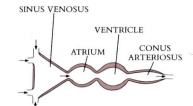

FIGURE 45-14

Schematic diagram of the fish heart. Sinus venosus. *Because water pressure pushing in on the fish's body tends to collapse veins and thus to oppose the movement of blood through the veins back to the heart, efficient operation requires that the fish heart contribute the least possible resistance to venous flow, which is minimized by a large collection chamber.* Atrium. *To deliver blood to the pump quickly in increments of suitable volume, a vestibule about the size of the ventricle is required to receive blood from the sinus venosus.* Ventricle. *To provide the energy to move the blood through the gills, a thick-walled pumping chamber is required.* Conus arteriosus. *To smooth the pulsations and add still more thrust, blood leaving the ventricle passes through a second elongated chamber.*

FIGURE 45-15

Evolution of the vertebrate heart.
A *Fishes.*
B *Amphibians.*
C *Reptiles.*
D *Birds and mammals.*
LA, *Left atrium.*
RA, *Right atrium.*
LV, *Left ventricle.*
RV, *Right ventricle.*

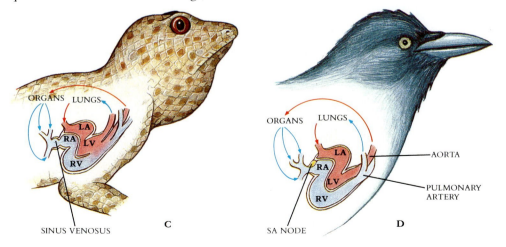

FIGURE 45-16

Comparison of fish and mammalian circulatory systems. The blood pressure in the fish circulatory system drops precipitously after passage of the blood through the gills, because of the great resistance their narrow passages present to flow. Because the blood flows from the gills directly to the rest of the body, circulation is sluggish at the low pressure that remains. In mammals, the blood is repumped after leaving the lungs; thus the resistance of the capillaries in the lungs does not affect the force of the heart's contraction, and circulation to the rest of the body remains vigorous.

The fish heart is admirably suited to the gill respiratory apparatus and represents one of the major evolutionary innovations of the vertebrates. Perhaps its greatest advantage is that the blood it delivers to the tissues of the body is fully aerated (oxygenated). Thus blood is pumped directly to and through the gills, where it becomes fully aerated; from the gills it flows through a network of arteries to the rest of the body and then returns to the heart through veins (Figure 45-15, *A*). This arrangement has one great limitation, however. After passing through the fine network of capillaries in the gills, the flow of blood has lost much of the force contributed by the contraction of the heart (Figure 45-16), so the circulation from the gills to the rest of the body is sluggish. This means that oxygen cannot be delivered to body muscles at a high rate. Fishes have never evolved a means to overcome this limitation.

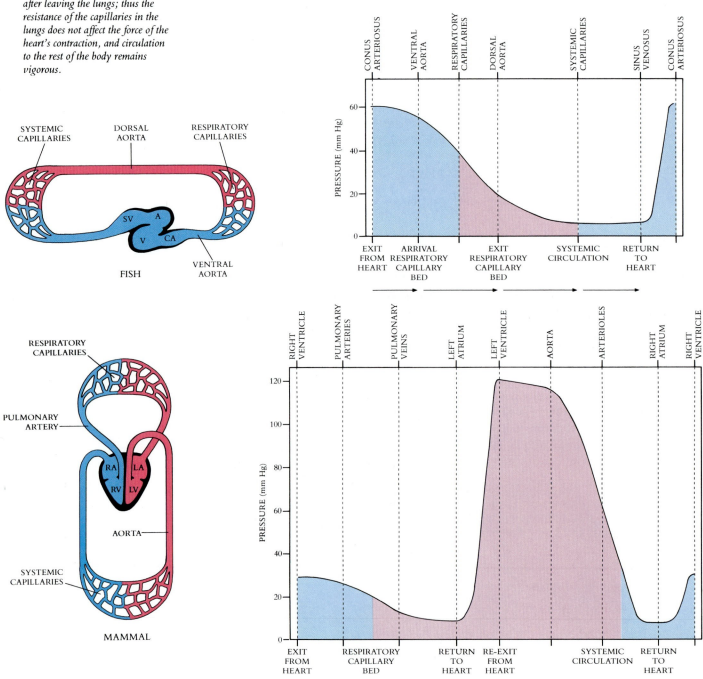

Amphibian and Reptile Hearts: The Advent of Pulmonary Circulation

The advent of gaseous respiration in the lungs involved a major change in the pattern of circulation. Ultimately this evolutionary change enabled vertebrates to overcome the inherent limitations of the circulatory system that evolved in fishes. The change consisted of the development of an additional pair of major veins. After blood is pumped through the fine network of capillaries in the lungs, it is not dispersed to the tissues of the body but is instead returned to the heart through large veins called **pulmonary veins** for repumping. The development of these new veins led to a great improvement in the performance of the circulatory system, since the blood being pumped to other tissues of the body could be pumped at a much higher pressure than if it were not returned to the heart at this stage. The disadvantage of pulmonary veins is that the aerated blood from the lungs is mixed in the heart with the nonaerated blood that is constantly being returned to the heart from the rest of the body. Consequently the heart pumps out a mixture of aerated and nonaerated blood rather than fully aerated blood. Many of the subsequent changes in the evolution of the heart in birds and mammals have been modifications that minimize this disadvantage.

The amphibian heart is altered from the arrangement seen in fishes in two ways (Figure 45-15, *B*) that tend to lessen the effect of the mixing of aerated and nonaerated blood:

1. The atrium is divided into two chambers—a right atrium, which receives nonaerated blood from the sinus venosus for circulation to the lungs, and a left atrium, which receives aerated blood from the lungs through the pulmonary vein for circulation through the body.

2. The conus arteriosus is partially separated by a septum, or dividing wall, which directs aerated blood into the aorta and directs nonaerated blood into the pulmonary arteries, which lead to the lungs. The aorta leads to the body's network of arteries.

These two changes tend to separate the blood circulation of the body into two separate paths: (1) the **pulmonary circulation,** in which blood travels from the heart to and from the lungs; and (2) the **systemic circulation,** in which blood travels from the heart to and from the rest of the body. However, the separation of the pulmonary and systemic circulations is imperfect. A significant amount of mixing of aerated and nonaerated blood still occurs in the heart. It is no accident that amphibians are usually sluggish.

Among reptiles, further modifications occurred that reduced the amount of mixing in the heart (Figure 45-15, *C*):

1. A septum partially subdivides the pumping chamber of the heart, the ventricle. The presence of this septum results in a far more effective separation of aerated and nonaerated blood within the heart. In one order of reptiles, the crocodiles, the separation is complete.

2. The conus arteriosus is absent. It has become fully subdivided, forming the trunks of the large arteries leaving the heart, which thus become directly connected to the ventricle.

Because the reptilian heart achieves a better separation of aerated and nonaerated blood, it represents a distinct increase in efficiency over the amphibian heart. The separation within the ventricle is not complete, however, a factor that still limits the overall efficiency of the circulatory system, some of the nonaerated blood still mixing in the ventricle with aerated blood.

Mammal and Bird Hearts: A True Two-Cycle Pump

Only a relatively slight alteration is seen in the hearts of mammals and of birds (Figure 45-15, *D*), although the change, which occurred independently in these two lines of evolution, has great consequences: the septum within the ventricle is closed, dividing the pumping chamber into two parts.

The closure of the ventricular septum in mammals and birds created for the first time the double circulatory system toward which evolutionary pressures had been modifying the heart since the advent of the pulmonary vein. These hearts have the advantage that is conferred by repumping the blood after its passage through the lungs, without the disadvantage of mixing aerated and nonaerated blood. The blood that is pumped by the heart of a mammal or bird into the systemic arterial system is fully aerated.

In mammals and birds the four-chambered heart acts as a double pump, the left side of the heart pumping aerated blood to the general body circulation and the right side pumping nonaerated venous blood to the lungs. Note that the four chambers that occur in a bird or mammalian heart evolved from only two of the chambers of the four-chambered fish heart: the atrium and the ventricle. The great increase in efficiency that is realized by the double circulatory system in mammals and birds is thought to have been important in the evolution of endothermy in them. More efficient circulation is necessary to support the great increase in metabolic rate that is required to generate body heat internally. Also, blood is the carrier of heat within the body, and an efficient circulatory system is required to distribute heat evenly throughout the body.

The separation of the blood flow into two circuits also has a second favorable result. Because the overall circulatory system is closed, in each full passage through the system the same volume of blood has to move through the smaller lung circulation path as through the much more extensive body circulation path. This means that the blood must move through the lungs very much faster than through the rest of the body. This more rapid circulation is not accomplished by higher pressure in the pulmonary circuit. Instead the blood vessels in the lung are larger in diameter than those in the rest of the body and offer less resistance to flow. The favorable result is that the rapid flow of blood through the lungs greatly increases the efficiency with which oxygen is captured by the bloodstream.

The Pacemaker: Preservation of the Sinus Venosus

The sinus venosus, a major chamber in the fish heart, is reduced in size in amphibians and is further reduced in reptiles. In mammals and birds the sinus venosus is no longer evident as a separate chamber. The disappearance of the sinus venosus is not really complete, however. Although the chamber is gone, some tissue remains, and it has a very important function. Throughout the evolutionary history of the vertebrate heart, the sinus venosus has had two functions, one as a collection chamber and the other as the site of origin of the heartbeat. This second function is indispensable, and mammals have retained the excitatory tissue of the sinus venosus in the wall of the right atrium—where it was previously located in the fish heart, near the point where the veins now empty directly into the atrium. This tissue is called the **sinoatrial node (SA node).** In mammals it is the point of origin of each heartbeat—the pacemaker.

> **The hearts of mammals and birds evolved independently from the central two chambers of the four-chambered fish heart, each central chamber becoming divided into two chambers separated by a septum. A vestige of the initial chamber of the fish heart remains as the sinoatrial node, or pacemaker.**

Increase in Pumping Capacity

During the course of vertebrate evolution, hearts have gotten bigger, and thus are able to pump harder. Within any one class of vertebrate, the size of the heart is a constant fraction of the body mass. In mammals, for example, the heart is very close to 0.6% of the total body mass, whether that body mass is 0.01 kilogram or 100 kilograms (Figure 45-17). However, the transition to more complex hearts during the evolution of the vertebrates has involved an increase in the size of the heart relative to that of the rest of the body in different classes, as shown in the following table:

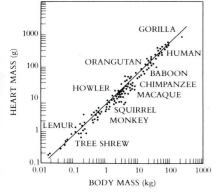

FIGURE 45-17

The heart is about 0.6% of the total body mass of any mammal, regardless of its size. This is true of a mouse and equally true of an elephant.

CLASS	$\dfrac{\text{HEART MASS (kg)}}{\text{BODY MASS (kg)}}$
Fishes	0.20%
Amphibians	0.46%
Reptiles	0.51%
Mammals	0.60%
Birds	0.82%

THE HUMAN HEART

The human heart, like that of all mammals and birds, is really two separate pumping systems operating together within a single sac. One of these pumps the circulation to the lungs, while the other pumps the circulation to the rest of the body. A cross section through the heart shows its organization clearly (Figure 45-18). The left side has two connected chambers, and so does the right. However, even though the right and left sides of the heart pump in concert with one another, they are not connected.

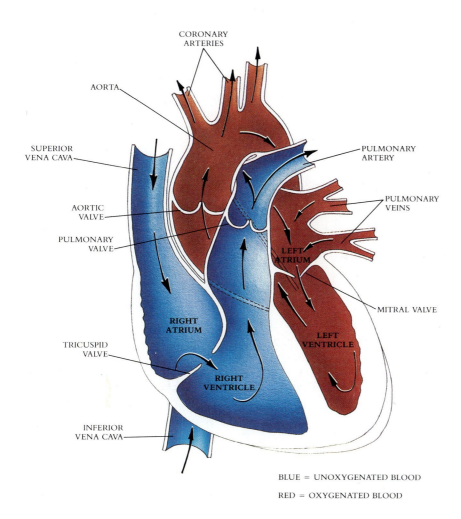

FIGURE 45-18

The path of blood through the human heart.

BLUE = UNOXYGENATED BLOOD

RED = OXYGENATED BLOOD

Circulation Through the Heart

Let us follow the journey of blood through the human heart, starting with the entry into the heart of oxygenated blood from the lungs. Oxygenated blood from the lungs enters the left side of the heart, emptying into the **left atrium** through the **pulmonary veins,** which open directly into the atrium. From the atrium, blood flows through an opening into the connecting chamber, the **left ventricle.** Most of this flow, roughly 80%, occurs while the heart is relaxed. When the heart starts to contract, the atrium contracts first, pushing the remaining 20% of its blood into the ventricle.

After a slight delay, the ventricle contracts. The walls of the ventricle are far more muscular than those of the atrium, and as a result this contraction is much stronger. It forces most of the blood out of the ventricle in a single strong pulse. The blood is prevented from going back into the atrium by a large one-way valve, the **mitral valve,** whose flaps are pushed shut as the ventricle contracts. Strong fibers that prevent the flaps from moving too far when closing are attached to their edges. If the flaps did move too far, they would project out into the atrium. The fibers that prevent this operate in much the same way as a rope might if tied from the steering wheel of a car to the driver's door handle, with a bit of slack—the door can be opened only as far as the slack in the rope permits.

FIGURE 45-19

The human circulatory system.

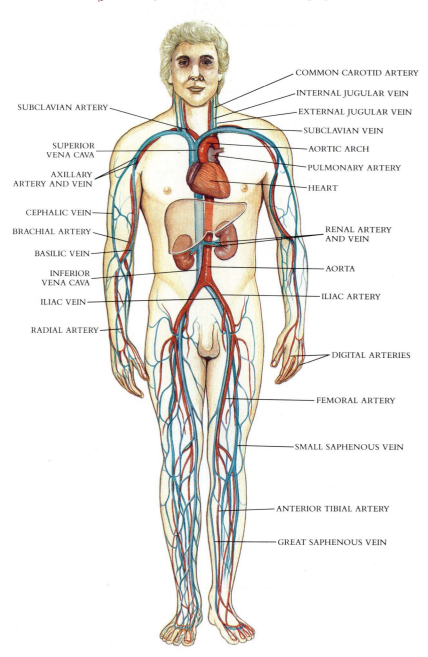

COMMON CAROTID ARTERY
INTERNAL JUGULAR VEIN
EXTERNAL JUGULAR VEIN
SUBCLAVIAN VEIN
AORTIC ARCH
PULMONARY ARTERY
HEART
RENAL ARTERY AND VEIN
AORTA
ILIAC ARTERY
DIGITAL ARTERIES
FEMORAL ARTERY
SMALL SAPHENOUS VEIN
ANTERIOR TIBIAL ARTERY
GREAT SAPHENOUS VEIN

SUBCLAVIAN ARTERY
SUPERIOR VENA CAVA
AXILLARY ARTERY AND VEIN
CEPHALIC VEIN
BRACHIAL ARTERY
BASILIC VEIN
INFERIOR VENA CAVA
ILIAC VEIN
RADIAL ARTERY

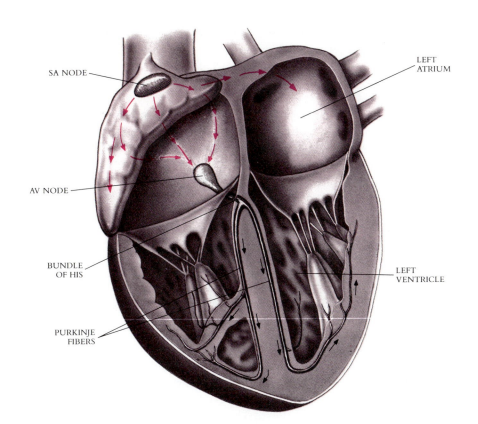

SA NODE

AV NODE

BUNDLE
OF HIS

PURKINJE
FIBERS

LEFT
ATRIUM

LEFT
VENTRICLE

FIGURE 45-20

Contraction of the human heart is initiated by a wave of depolarization that begins at the SA node, the evolutionary vestige of the sinus venosus. After passing over the right and left atria and causing their contraction, the wave of depolarization reaches the AV node, from which it passes to the ventricles. The depolarization is conducted rapidly over the surface of the ventricles by a set of fibers called Purkinje fibers, which make up the bundle of His and branch throughout the ventricle.

Prevented from reentering the atrium, the blood within the ventricle takes the only other passage out of the contracting left ventricle. It moves through a second opening that leads into a large vessel called the **aorta.** The aorta is also bounded by a one-way valve, the **aortic valve.** Unlike the mitral valve, the aortic valve is oriented to permit the *outward* flow of the blood. Once this outward flow has occurred, the aortic valve closes, thus preventing the reentry of blood from the aorta into the heart.

The aorta and all of the other blood vessels that carry blood away from the heart are arteries (Figure 45-19). Many of these arteries branch from the aorta, carrying oxygen-rich blood to all parts of the body. The first to branch are the coronary arteries, which carry fresh oxygenated blood to the heart itself; the muscles of the heart do not obtain their supply of blood from within the heart.

The blood that flows into the arterial system, after delivering its cargo of oxygen to the cells of the body, eventually returns to the heart. In doing so, it passes through a series of veins, eventually entering the right side of the heart. Two large veins collect blood from the systemic circulation. The **superior vena cava** drains the upper body, the **inferior vena cava** the lower body. These veins empty oxygen-depleted blood into the right atrium. The right side of the heart is similar in organization to the left side. Blood passes from the right atrium into the right ventricle through a one-way valve, the **tricuspid valve.** It passes out of the contracting right ventricle through a second valve, the **pulmonary valve,** into the **pulmonary artery,** which carries the oxygen-depleted blood to the lungs. The blood then returns from the lungs to the left side of the heart with a new cargo of oxygen, which is pumped to the rest of the body.

How the Heart Contracts

The contraction of the heart consists of a carefully orchestrated series of muscle contractions. Contraction is initiated by the sinoatrial node (Figure 45-20), the small cluster of excitatory cardiac muscle cells derived from the sinus venosus that is embedded in the upper

DISEASES OF THE HEART AND BLOOD VESSELS

Cardiovascular diseases are the leading cause of death in the United States. More than 42 million people in this country (about one person in five) have some form of cardiovascular disease. **Heart attacks** (myocardial infarctions) are the main cause of cardiovascular deaths in the United States, accounting for about a fifth of all deaths. They result from an insufficient supply of blood reaching an area of heart muscle. Heart attacks may be caused by a blood clot forming somewhere in the blood vessels and blocking the passage of blood through those vessels (Figure 45-B, *1*). They may also result if a vessel is blocked sufficiently by deposits of materials. Recovery from a heart attack is possible if the segment of the heart tissue damaged was small enough that the other blood vessels in the heart can enlarge their capacity and resupply the damaged tissues. **Angina pectoris,** which literally means "chest pain," occurs for reasons similar to those which cause heart attacks, but it is not as severe. The pain may occur in the heart, and often also in the left arm and shoulder.

The amount of heart damage that occurs from a small heart attack may be relatively slight and thus difficult to detect. It is important that such damage be detected, however, so that the overall condition of the heart can be evaluated properly. Electrocardiograms are very useful for this purpose, since they reveal abnormalities in the timing of heart contractions—abnormalities that are associated with

the presence of damaged heart tissue. Damage to the AV node, for example, may delay as well as reduce the second (ventricular) pulse. Unusual conduction routes may lead to continuous, disorganized contractions called **fibrillations.** In many fatal heart attacks ventricular fibrillation is the immediate cause of death.

Strokes are caused by an interference with the blood supply to the brain. They often occur when a blood vessel bursts in the brain. Strokes may be associated with a **thrombus,** or coagulation (clotting) of blood cells or other elements in one of the vessels. Such a thrombus may be caused by cancer or other diseases. The effects of strokes depend on how severe the damage is and where the stroke occurs in the brain.

Atherosclerosis is a buildup of debris and deposits within the arteries (Figure 45-B, *2*). Atherosclerosis contributes to heart attacks and strokes. The accumulation within the arteries of fatty materials, abnormal amounts of smooth muscle cells, deposits of cholesterol or fibrin, or cellular debris of various kinds can all impair the arteries' proper functioning. When this condition is severe, the arteries can no longer expand and contract properly, and the blood moves through them with difficulty. The accumulation of cholesterol is thought to be the prime contributor to atherosclerosis, and diets low in cholesterol are now prescribed to help to prevent this condition.

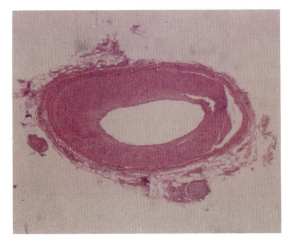 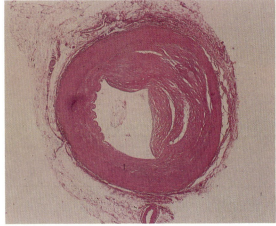

Arteriosclerosis, or hardening of the arteries, occurs when calcium is deposited in arterial walls. It tends to occur when atherosclerosis is severe. Not only is flow through such arteries restricted, but they also lack the ability to expand as normal arteries do to accommodate the volume of blood pumped out by the heart; the heart is forced to work harder.

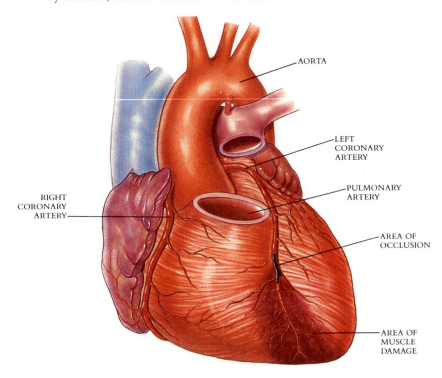

AORTA

LEFT CORONARY ARTERY

PULMONARY ARTERY

RIGHT CORONARY ARTERY

AREA OF OCCLUSION

AREA OF MUSCLE DAMAGE

FIGURE 45-B, *1*

A heart attack. This heart attack resulted from occlusion (blockage) of the left coronary artery. The heart muscle serviced by the portion of the coronary artery beyond the point of blockage is damaged.

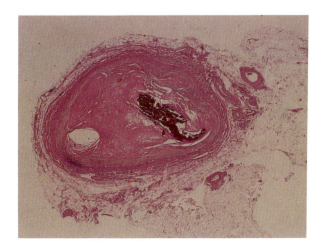

FIGURE 45-B, *2*

Atherosclerosis is a buildup in an artery that blocks the passage of blood. The coronary artery on the left shows only minor blockage. The artery in the center exhibits severe atherosclerosis—much of the passage is blocked by buildup on the interior walls of the artery. The coronary artery on the right is essentially completely blocked.

wall of the right atrium. The cells of the SA node act as a pacemaker for the rest of the heart, their membranes spontaneously depolarizing with a regular rhythm that determines the rhythm of the heart's beating. Each depolarization initiated within this pacemaker region passes quickly from one cardiac muscle cell to another in a wave that envelops both the left and right atria almost instantaneously.

> **The contraction of the heart is initiated by the periodic, spontaneous depolarization of cells of the SA node; the resulting wave of depolarization passes over both the left and the right atria and causes their cells to contract.**

The wave of depolarization does not immediately spread to the ventricles, however. Almost 0.1 second passes before the lower half of the heart starts to contract. The reason for the delay is that the atria of the heart are separated from the ventricles by connective tissue, and connective tissue cannot propagate depolarization. The depolarization would not pass to the ventricles at all except for a slender connection of cardiac muscle cells known as the **atrioventricular node (AV node).** These cells have a small diameter and thus propagate the depolarization slowly, causing the delay just noted. This delay permits the atria to finish emptying their contents into the corresponding ventricles before those ventricles start to contract.

From the AV node, the wave of depolarization is conducted rapidly over both ventricles by a special network of fibers called the **bundle of His.** The passage of this wave triggers the almost simultaneous contraction of all the cells of the right and left ventricles.

> **The passage of the wave of depolarization over the narrow AV node delays the depolarization of the ventricles for a fraction of a second. When the ventricles receive the wave of depolarization, they conduct it so rapidly over their surface that all of the cells of the two ventricles contract almost simultaneously.**

Monitoring the Heart's Performance

As you can see, the heartbeat is not simply a squeeze-release, squeeze-release cycle but rather is like a little play in which a series of events occur in a predictable order. You may watch the play in several ways, depending on the events that you are observing. The simplest way to monitor heartbeat is to listen to the heart at work. The first sound you hear, a low pitched *lub,* is the closing of the mitral and tricuspid valves at the start of ventricular contraction. A little later you hear a higher pitched *dub,* the closing of the pulmonary and aortic valves at the end of ventricular contraction. If the valves are not closing fully, or if they open too narrowly, turbulence is created within the heart. This turbulence can be heard as a **heart murmur.** It often sounds like sloshing.

A second way to examine the events of the heartbeat is to monitor blood pressure. During the first part of the heartbeat the atria are filling and contracting. At this time the pressure in the arteries leading from the left side of the heart out to the tissues of the body decreases slightly as the blood moves out f the arteries, through the vascular system, and into the atria. This period is referred to as the **diastolic period.** During the contraction of the left ventricle a pulse of blood is forced into the systemic arterial system, immediately raising the blood pressure within these vessels. This pushing period, which ends with the closing of the aortic valve, is referred to as the **systolic period.** Blood pressure values are measured in millimeters of mercury (mm Hg); like atmospheric pressure, they reflect the height to which a column of mercury would be raised in a tube by an equivalent pressure. Normal blood pressure values are 70 to 90 mm Hg diastolic and 110 to 130 systolic. When the inner walls of the arteries accumulate fats, as they do in the condition known as atherosclerosis, the diameters of the passageways are narrowed. If this occurs, the systolic blood pressure is elevated.

FIGURE 45-21

An electrocardiogram.

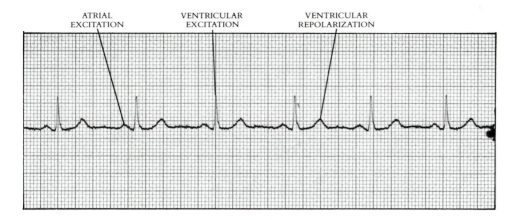

ATRIAL EXCITATION   VENTRICULAR EXCITATION   VENTRICULAR REPOLARIZATION

A third way to monitor the progress of events during a heartbeat is to measure the waves of depolarization. Because the human body consists primarily of water, it conducts electrical currents rather well. A wave of membrane depolarization passing over the surface of the heart generates an electric current that passes in a wave throughout the body. The magnitude of this electrical pulse is tiny, but it can be detected with sensors placed on the skin. A recording made of these impulses (Figure 45-21) is called an **electrocardiogram.**

In a normal heartbeat three successive electrical pulses are recorded. First, there is an atrial excitation, caused by the depolarization that is associated with atrial contraction. A tenth of a second later there is a much stronger ventricular excitation, reflecting both the depolarization of the ventricles and the relaxation of the atria. Finally, perhaps 0.2 second later, there is a third pulse, caused by the relaxation of the ventricles.

Cardiac Output

Cardiac output is the rate at which the heart beats multiplied by the volume of blood delivered with each heartbeat. In a normal resting person, cardiac output will be 72 beats per minute $\times$ 0.07 liter per beat = 5.0 liters per minute.

When a person exercises heavily, the oxygen in the bloodstream is used more quickly and becomes depleted. When the central nervous system detects such depletion in the blood circulating to itself, it sends out signals through the nerves of the sympathetic nervous system to the SA node, stimulating these cells to depolarize at a more rapid rate. The hormone adrenaline may also be released into the bloodstream in response to nerve signals to the adrenal glands that produce it. Adrenaline also increases the rate of the heartbeat. As a result of these two stimuli, the heart beats faster—as much as twice as fast as normal. Its ventricles are squeezed much more tightly than normal. Since the ventricles never empty completely, a stronger contraction will deliver more blood. The combined result of these two changes is a greatly increased cardiac output: 140 beats per minute $\times$ 0.21 liter per beat = 30 liters per minute.

THE CENTRAL IMPORTANCE OF CIRCULATION

The evolution of multicellular organisms has depended critically on the ability of organisms to circulate nutrients and other materials to the various cells of the body, and to carry metabolic wastes away from them. The digestive processes described in Chapter 43, by which vertebrates obtain metabolizable foodstuffs, and the respiratory processes described in Chapter 44, by which vertebrates obtain the oxygen necessary for aerobic metabolism, both depend critically on the transport of food and oxygen to cells and the removal of the end products of their metabolism. Vertebrates carefully regulate the operation of their circulatory systems; by doing so, they are able to integrate their body activities. This regulation is carried out by the nervous system, the subject of the next three chapters.

SUMMARY

1. The circulatory system of vertebrates is closed, permitting control of circulation rates.

2. The circulatory fluid is blood, which contains cells rich in hemoglobin that transport oxygen to metabolizing cells and transport carbon dioxide away from them. The plasma of the circulating blood contains the proteins and ions necessary to maintain the blood's osmotic equilibrium with the surrounding tissues.

3. Glucose and other metabolic products of digestion do not enter the general circulation directly but rather flow to the liver. The liver removes and stores any excess metabolic products and maintains blood glucose levels within narrow bounds.

4. The general flow of blood circulation through the body is a circuit starting from the heart, which pumps blood out via muscled arteries to the capillary networks that permeate the tissues of the body; the blood returns to the heart from these capillaries via the veins. The water that is lost by diffusion during transit is recovered by the lymphatic system.

5. The heart of mammals and birds is a double pump, pushing both pulmonary (lung) circulation and systemic (general body) circulation. Because the blood is repumped after it passes through the narrow passages of the lung, the systemic circulation of mammals and birds is much more vigorous than that of the fishes. Because the two circulations are kept separate within the heart, the systemic circulation receives only fully aerated blood, a great improvement in efficiency over the heart of amphibians and reptiles. Mammals and birds can be warm-blooded largely because of the kind of improved heart that they evolved independently.

6. The four-chambered mammalian or bird heart evolved from two of the chambers of the four-chambered fish heart, by the creation of septa (dividing walls) within the two central chambers. The other two chambers of the fish heart were gradually lost, although the pacemaker cells of the sinus venosus have been retained in their original location.

7. The contraction of the heart is initiated at the SA node, or pacemaker, as a periodic spontaneous depolarization of these cells. The wave of depolarization spreads across the surface of the two atrial chambers, causing all of these cells to contract.

8. The passage of the wave of depolarization to the ventricles is briefly delayed by insulating the two segments of the heart from one another. Only a narrow channel of cardiac muscle cells connects the atria and the ventricles. The delay in the passage of the wave of depolarization permits the atria to empty completely into the ventricles before ventricular contraction occurs.

SELF-QUIZ

The following questions pinpoint important information in this chapter. You should be sure to know the answers to each of them.

1. Which of these cells contain large amounts of the oxygen carrier protein hemoglobin?
   (a) leukocytes   (d) erythrophages
   (b) macrophages   (e) erythrocytes
   (c) macrocytes

2. Blood leaves the heart through vessels known as
   (a) capillaries   (c) arteries
   (b) arterioles   (d) veins

3. Arterioles differ from arteries simply in being smaller in diameter. True or false?

4. Which of the following is not involved in the vessels of the circulatory system?
   (a) a layer of elastic fibers
   (b) a layer of metabolically supportive glia cells
   (c) a layer of muscle serviced by nerves
   (d) a layer of connective tissue
   (e) all are involved

5. Much of the water that diffuses out of the capillaries is collected and returned to the blood-stream by an open circulatory system known as the _____ system.

6. The primary protein that serves to keep the blood plasma in osmotic equilibrium with the cells of the body is
   (a) albumin   (d) myoglobin
   (b) glycogen   (e) trypsin
   (c) hemoglobin

7. Cell fragments that contain no nuclei and enter the bloodstream, where they play an important role in blood clotting, are called _____.

8. The point of origin of each heartbeat in mammals is the _____.

9. Which class has the highest heart mass relative to body mass?
   (a) fishes   (d) reptiles
   (b) birds   (e) amphibians
   (c) mammals

10. From the atrioventricular (AV) node, the wave of depolarization spreads over both ventricles of the heart by a special network of fibers called the bundle of _____ .

11. Contraction of the left ventricle along with increased blood pressure is referred to as the _____ period.
    (a) diastolic
    (b) systolic

12. How many of the chambers of the fish heart are represented as chambers in the mammalian heart?
    (a) none   (d) two
    (b) all   (e) three
    (c) one

THOUGHT QUESTIONS

1. Why have mammals not improved the efficiency of their circulation by evolving hearts larger than 0.6% of their body mass?

2. Instead of evolving an entire second open circulatory system, the lymphatic system, to collect water lost from the blood plasma during passage through the capillaries, why haven't vertebrates simply increased the level of serum albumin in their blood?

3. Starving animals often exhibit swollen bodies rather than emaciated ones, in early stages of their deprivation. Why?

4. The hearts of the more advanced vertebrates pump blood entirely by pushing action. Why do you suppose hearts have not evolved that act like suction pumps, drawing blood into the heart as it expands, rather than pushing it out as the heart contracts?

FOR FURTHER READING

BROOKS, S.M.: *Basic Facts of Body Water and Ions,* 3rd ed., Springer-Verlag, Inc., New York, 1973. A clear account of the principles of fluid and electrolyte balance.

DOOLITTLE, R.F.: "Fibrinogen and Fibrin," *Scientific American,* December 1981, pages 126-135. How the structure of the fibrin proteins determines the course of clot formation.

MAYERSON, H.S.: "The Lymphatic System," *Scientific American,* June 1963, pages 80-90. A description of how the lymphatic system recovers water lost from the circulatory system.

SMITH, H.M.: *Evolution of Chordate Structure,* Holt, Rinehart & Winston, New York, 1960. A comprehensive and thoughtful treatise on the comparative anatomy of vertebrates, with a strong evolutionary perspective. A classic.

WIGGERS, C.J.: "The Heart," *Scientific American,* May 1957, pages 74-87. A general account of the structure and functioning of the human heart.

ZUCKER, M.B.: "The Functioning of the Blood Platelets," *Scientific American,* June 1980, pages 86-103. A description of the many roles of platelets in human health, with emphasis on their role in blood clotting.

REGULATING BODILY ACTIVITIES: THE NERVOUS SYSTEM

Chapter 46

Overview

The nervous system controls the operation of the vertebrate body by transmitting information from sensors located in muscles, organs, and special sensory receptors to an associative center, the brain, and from the brain to the various muscles and glands. A neuron carries an electrical signal as a wave of disturbance (depolarization), abolishing a difference in ion concentration across the cell membrane. All such depolarization signals are equal in strength, but their frequency and destinations may differ. Neurons integrate neural information by the combination of stimulatory and inhibitory junctions (synapses) that impinge on the neuron, the summed effect of which determines whether a further signal will be initiated. The major center of integration is the brain, in which complex zones of activity have evolved.

For Review

Here are some important terms and concepts that you will encounter in this chapter. If you are not familiar with them, you should review them before proceeding.

- **Sodium/potassium pump** (Chapter 9)

- **Passive ion channels** (Chapter 9)

- **Depolarization** (Chapter 42)

- **Neuron** (Chapter 42)

There is no such thing as a simple vertebrate. Just as a town is a complex and integrated unit, with different people performing different jobs on which everyone depends, so vertebrates such as ourselves are composed of many different organs and tissues, each specialized in a different way to accomplish growth, survival, and reproduction. The essential condition of specialization is stability. No town or city can long survive in chaos, without economic, social, and legal order, and the different tissues of a complex organism likewise need to maintain stable relationships with one another if the organism is to function as a coherent whole. The internal order, an essential property of all organisms, is given a special name, **homeostasis** (Greek, *homoios,* same + *stasis,* standing). Just as order in a city is impossible without communication (tax bills, traffic tickets, phone calls, and the like), so the maintenance of homeostasis within an organism requires communication between tissues. In this chapter we will consider that communication system. In the following three chapters, we will consider the ways in which the communication system is used to achieve homeostasis under a variety of conditions and to mediate interactions with the external environment.

TRANSMITTING INFORMATION WITHIN THE BODY: AN OVERVIEW

There are many ways in which one cell can communicate with another. One simple way is by direct contact, with an open channel between the two cells that permits the passage of ions and small molecules. An example of this kind of communication is provided by gap junctions, discussed in Chapter 9. They provide a ready means of communication between adjacent cells but ordinarily are not able to provide communication between distant tissues. In a similar manner, depending on face-to-face interactions between adjacent people would be too slow and uncertain a way to run a city.

It would be better, in terms of distant communication, if the city manager sent a letter instructing various persons what to do. The body organizes various tissues in just this way, sending chemical instructions to these tissues. The instructions are in the form of **hormones,** small chemicals that act as messengers within the body. The hormone "letters" are produced by one of several different **endocrine glands,** secreted into the bloodstream, and then carried around through the body by the circulatory system. Like a letter, each hormone has an address, a chemical shape that only the target tissue will recognize and act on. However, this kind of command system can be too slow, just as the mail may sometimes be when you are waiting for an important letter.

If the message to be delivered to the leg muscles of your body is "Contract quickly, we are being pursued by a leopard," a quicker means of communication than hormones is desirable. In a city, a person in an emergency does not mail a letter, but rather uses the telephone to shout for help, dialing 911 and requesting assistance. That, in effect, is just what the vertebrate body does. All higher animals possess specialized **neuron** cells that maintain an electrical charge on their outer surface by actively pumping certain ions out across their membranes. An electrical signal can pass down the length of a neuron just as electrical impulses pass down a phone line.

The command center of higher organisms is the **brain,** a precisely ordered but complicated maze of interconnected neurons, a large biological "computer." The brain is connected by a network of neurons both to the hormone-producing glands and to the individual muscles and other tissues. This dual channel of command permits great flexibility: the signals can be slow and persistent (hormones), fast and transient (nerve signals), or any combination of the two.

Two forms of communication integrate body functions in vertebrates. They are neurons, rapid electrical signals that report information or initiate a quick response in a specific tissue; and hormones, slower chemical signals that initiate a widespread prolonged response, often in a variety of tissues.

ORGANIZATION OF THE VERTEBRATE NERVOUS SYSTEM

All nervous systems can be said to have one underlying mechanism and three basic elements. The underlying mechanism is the nerve impulse; the three basic elements are (1) a central processing region or brain, (2) nerves that bring information to the brain, and (3) nerves that transmit commands from the brain. This chapter will focus on the vertebrate nervous system, but the basic architecture of nervous systems is similar throughout the animal kingdom. Before treating the mechanism of the nerve impulse in detail, it is useful to consider briefly how the basic elements of the vertebrate nervous system are organized.

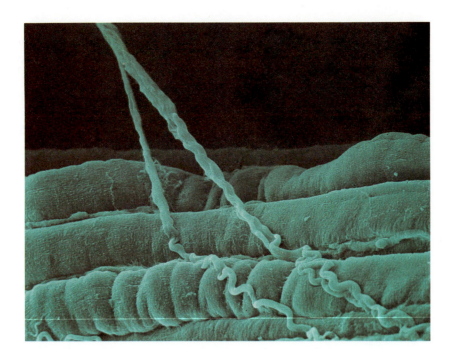

FIGURE 46-1

Motor nerves transmit commands from the central nervous system to individual muscles and glands. The slender, twisted threads are motor nerves, and the long, thick strands at the bottom are muscle fibers. Often a nerve will carry many independent fibers, which branch off to establish contact with different target muscles.

The vertebrate nervous system is traditionally subdivided in binary fashion, each level of its organization partitioned into two contrasting functional groups. Thus the nervous system as a whole is partitioned into the **central nervous system** and the **peripheral nervous system.** The central nervous system is composed of the brain and the spinal cord, and is the site of information processing within the nervous system. The peripheral nervous system includes all the nerve pathways of the body. They are also commonly partitioned into two groups: the **sensory** pathways, which transmit information to the central nervous system; and the **motor** pathways, which transmit commands from the central nervous system (Figure 46-1). The motor pathways are in turn partitioned into **somatic** ("voluntary") paths, which relay commands to skeletal muscles, and **autonomic** ("involuntary") paths, which stimulate the glands and other muscles of the body. The autonomic system is itself further partitioned into **sympathetic** and **parasympathetic** nerves, one of which stimulates and the other of which inhibits a given target tissue.

Both the central nervous system and the peripheral nervous system connected to it are composed of neurons, which (as you will recall from Chapter 42) are nerve cells specialized for signal conduction. A neuron typically is composed of (1) a cell body, (2) numerous dendrites capable of transmitting electrical signals to the cell body, and (3) a single long axon, which carries signals away from the cell body. The pathways of the nervous system are composed of large numbers of axons and long dendrites bundled together like the strands of a telephone cable. Within the central nervous system these bundles of nerve fibers are called **tracts;** in the peripheral nervous system they are called **nerves.** The cell bodies from which pathways extend are often clustered into groups, called **nuclei** if they are within the central nervous system and **ganglia** if they are outside it.

Neurons carry out only a minimal metabolism on their own; they subsist on metabolites provided by other supporting nerve cells, known as **neuroglia.** In the peripheral nervous system, specialized neuroglia cells called **Schwann cells** play an important role in nerve signal conduction, as you will see below.

In this chapter we will first consider the neuron in detail, since it is the basic functional unit of the nervous system. We will then examine the central nervous system, focusing on its functional organization and on the way in which that organization has evolved among the vertebrates. The peripheral nervous system will be the subject of the next two chapters, with Chapter 47 devoted to the sensory nervous system and much of Chapter 48 devoted to the autonomic nervous system.

FIGURE 46-2

*Vertebrate neurons. Vertebrate
neurons differ widely in structure.
Neurons within the brain often
possess extensive highly branched
dendrites, whereas sensory neurons
(which carry signals from sense
organs to the brain) typically have
dendrites only in specific receptor
cells. The axons of many motor
neurons (which carry commands
from the brain to the muscles and
glands) are encased at intervals by
Schwann cells, as are some of the
axons of sensory neurons.*

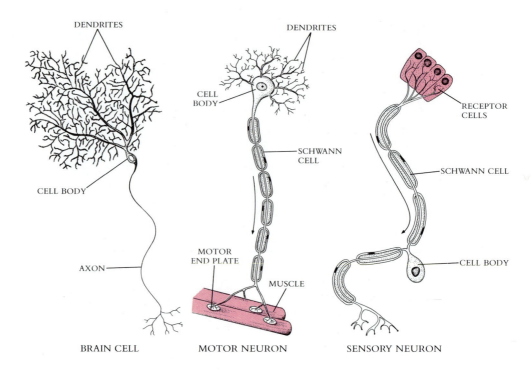

BRAIN CELL   MOTOR NEURON   SENSORY NEURON

THE NEURON: FUNCTIONAL UNIT OF THE NERVOUS SYSTEM

The nerve cells of vertebrates are diverse in appearance. Typical neurons are illustrated in Figure 46-2. The most obvious characteristics shared by these cells are the long projections, or **processes,** extending out from the cell body. In some neurons the major processes are dendrites, whereas in others a single long axon predominates. The processes of some neurons are only a few microns in length; others are as long as several meters.

The Functional Architecture of Neurons

Despite their varied appearances, all neurons have the same functional architecture. Each neuron receives, processes, transmits, and transfers information; each of these activities occurs in a different region of the nerve cell.

Information reception. Extending from the body of all but the simplest nerve cells are one or more cytoplasmic extensions called **dendrites** (Figure 46-3). The name "dendrite" comes from the Greek word *dendron,* "tree," because of the branching that is characteristic of these neuronal extensions. Most nerve cells possess a profusion of dendrites, any one of which may be branched, so that the body of the cell can receive inputs from many different sources simultaneously. Some of these inputs are excitatory, others inhibitory.

Information processing. The surface of the cell body acts to integrate the information arriving from the many different dendrites, summing the excitatory and inhibitory inputs. Often the result is an alteration in the level of membrane polarization.

Information transmission. The membrane excitation that results from arriving dendrite signals, if it is large enough, travels outward from the cell body as an electrical **nerve impulse,** along a long cytoplasmic projection called an **axon** (Figure 46-3). Most nerve cells possess a single axon, which may be quite long. The axons controlling motor activity in your legs are more than a meter long, and even longer ones occur in larger mammals. In a giraffe a single axon travels from the toe all the way up to the back of the skull, a distance of from 4 to 5 meters.

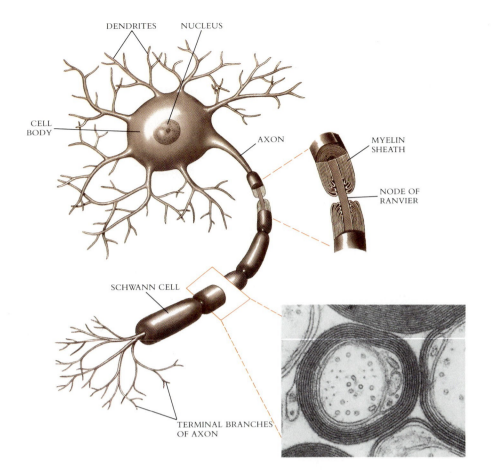

DENDRITES   NUCLEUS

CELL
BODY

AXON   MYELIN
SHEATH

NODE OF
RANVIER

SCHWANN CELL

TERMINAL BRANCHES
OF AXON

FIGURE 46-3

Structure of a typical neuron. Many dendrites project out from the cell body; these serve to carry information to it. Outward from the cell body extends a single long axon that carries information away from the cell body. This axon often ends in branches that connect to several muscles or to other nerves. In neurons that transmit information over long distances, the axons are encased at intervals by Schwann cells, whose many membrane layers, called a myelin sheath, electrically insulate the axon and so facilitate conduction of electrical impulses.

Information transfer. At the far end of most axons, the axonal membrane contains packets of specialized chemicals called **transmitters.** When a nerve impulse arrives at the axon terminus, these chemicals are released to interact with the cell to which the end of the nerve is in contact. The tip of the nerve axon may extend to a muscle cell, or a cell of an endrocrine gland, or some other nerve cell. All of these nerve-cell junctions are called **synapses.** In some instances, chemical transmitters are not used to transfer a nerve impulse from one neuron to another. Instead, the neurons may establish direct cell-to-cell contact so that the electrical signal passes directly from one cell to another.

Cell maintenance. A fifth activity shared by most neurons is interaction with support cells. Most neurons are unable to survive alone for long; they require the nutritional support that is provided by companion neuroglia cells. More than half the volume of vertebrate nervous systems is composed of supporting neuroglia cells. In many neurons, including the motor neurons that extend from brain to muscles, the transmission of impulses along the very long axons is facilitated by neuroglial Schwann cells, which envelop the axon at intervals (Figure 46-3) and act as electrical insulators. Despite their importance to the survival and functioning of neurons, supporting nerve cells are not well understood. An axon and its associated Schwann cells form a **myelinated fiber.** Bundles of such fibers collected in a nerve look in cross section (Figure 46-4) somewhat like telephone cables.

Nerve cells specialized for electrical signal transmission are called neurons. Typically a signal is received by a dendrite branch and passes to the cell body, where its influence is merged with that of other incoming signals; the resulting signal is passed outward along a single long axon. When the signal reaches the end of the axon, it usually is transmitted chemically to a target cell.

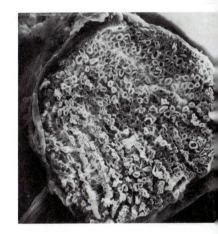

FIGURE 46-4

A nerve is a bundle of axons and dendrites bound together by connective tissue. In this cross section of a bullfrog nerve magnified 1600 times, many myelinated neuron fibers are visible, each looking in cross section something like a Cheerio.

Excitable Membranes

Although long processes are characteristic of neurons, long processes alone do not account for the unique specialization of these cells as information carriers. For example, the other kinds of cells in the vertebrate body that possess long processes do not conduct information. The single structural feature of a neuron that makes it unique is that the neuron cell possesses an **excitable membrane.** An excitable membrane is one that is capable of transmitting an electrical signal. How does the membrane of a neuron acquire this ability? It does so by selectively and actively transporting ions across its membrane.

The resting potential

Two modes of ion transport, active and passive, occur across the neuron cell membrane. Active transport of sodium (Na^+) and potassium (K^+) ions across the neuronal membrane is carried out by an array of transmembrane proteins called sodium-potassium pumps (described in detail in Chapter 9). Ions also pass across the neuron membrane through passive ion-specific channels. Each of these passive transmembrane channels may be open or closed; when a specific channel is open, those ions able to pass through it diffuse across the membrane in the direction of their lowest concentration. In a resting cell, K^+ ions diffuse out of the cell through open passive K^+ ion channels; by contrast, Na^+ ions cannot diffuse through passive Na^+ ion channels in such a cell, because these channels are closed in resting cells.

It is the interaction of these two kinds of channels—the active sodium-potassium ion pumps on the one hand and the passive channels on the other—that creates an "excitable" membrane. The sodium-potassium pumps are oriented so that the K^+ movement is directed inward, maintaining a concentration of potassium ions within the cell that is far higher than that outside. The sodium-potassium pumps transport Na^+ in the opposite direction—outward. Because these pumps actively transport sodium ions out of the cell, the concentration of sodium ions inside the cell is far less than that of the fluid surrounding it.

Now consider the electrical effects of these differences in ion concentrations. Na^+ ions cannot move into the cell, despite the low internal Na^+ ion concentration, because there are no open inward channels through which the Na^+ ions can pass. K^+ ions, on the other hand, can diffuse outward toward lower concentration, and do so. This flow of positively charged K^+ ions out of the cell gives the interior of the cell a more negative charge than the outside, creating an electrical potential. This kind of ion-generated electrochemical gradient is called a **resting potential.**

> An excitable membrane is one in which differently permeable ion channels produce an electrochemical gradient. Because of the activity of the Na^+ ion and K^+ ion transmembrane pumps and the impermeability of the cell to sodium but not to potassium, the outside surface of the neuron carries a positive charge relative to its interior.

The Nerve Impulse

The transmission of a signal by a nerve occurs in four phases, which we will consider in order (Figure 46-5): initiation of the impulse, its transmission along a nerve fiber, its transfer to a target muscle or nerve, and its effect on the target tissue.

Initiating a nerve impulse

Depolarization. A nerve is stimulated when pressure, chemical activity, or some other stimulus is applied at some site on the excitable membrane. Such events alter the conformation of the transmembrane proteins that comprise the passive Na^+ ion channels. The result is a transient flood of Na^+ ions diffusing into the cell from the outside, where there are many more of them. This movement of positive ions into the cell wipes out the local

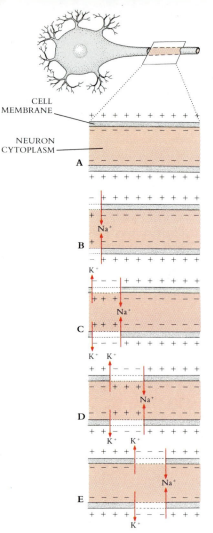

CELL MEMBRANE

NEURON CYTOPLASM

FIGURE 46-5

Transmission of a nerve impulse.
A *The interior of a resting neuron is more highly negatively charged than its exterior, due to the active transport of Na^+ ions outward.*
B *An impulse is initiated when the membrane becomes locally depolarized, Na^+ ions flooding inward across the membrane so that its interior becomes positively charged.*
C *The depolarization spreads, because the interior positive charge opens nearby transmembrane channels, thus permitting more Na^+ ions to enter; meanwhile, the initial site of depolarization gradually regains its original internal negative charge as a result of active transport of Na^+ ions out of the cell.*
D *and* **E** *As this process continues, a wave of depolarization moves down the neuron membrane, depolarizing in front and recovering behind. It is this self-propagating wave of depolarization that we call a nerve impulse.*

electrochemical gradient and therefore is called a **depolarization.** Ions rush into the cell in response to the large concentration difference built up by the Na^+ pump, in such numbers that the interior of the cell actually develops a positive charge relative to the outside.

Recovery. Depolarization causes the passive Na^+ ion channels in the membrane to close, so that no more Na^+ ions enter the cell. At the same time, it opens additional passive K^+ ion channels, so that K^+ ions are able to move out of the cell by diffusion. Active transport of Na^+ and K^+ ions by the sodium and potassium pumps then reestablishes the concentration differences that existed before the stimulus occurred. The whole process of stimulation and recovery takes about 5 milliseconds.

Why transmission occurs

During the few milliseconds when the site of stimulation is depolarized by the inward movement of sodium ions and before depolarization closes the passive Na^+ ion channels, the site of stimulation has less charge than the membrane surface surrounding it. This potential difference establishes a small, very localized current in the immediate vicinity, which is called a **gating current.** Many of the passive Na^+ ion channels are composed of transmembrane proteins with charged amino acid side groups, and these proteins respond to the gating current by changing their conformation. These channels are called **voltage-sensitive channels.** These voltage-sensitive passive channels open in response to the gating current, permitting Na^+ ions to enter the cell—and depolarizing these sites as well.

The depolarization thus spreads, depolarization at one site producing a local gating current, which induces nearby voltage-sensitive channels to open and so to depolarize the nearby site. In this way the initial depolarization passes outward over the membrane, spreading out in all directions from the site of stimulation. Like a burning fuse, the signal is usually initiated at the end and travels in one direction, but it would travel out from both directions if it were "lit" in the middle.

The time required for the Na^+ and K^+ pumps to restore the original ion concentrations is called the **refractory period.** The nerve impulse moves in the outward direction only, because during the refractory period the membrane is not sensitive to depolarization. By the time the recovery process has been completed, the signal has moved too far away for the site to be influenced by it.

> **A nerve impulse arises because of a disturbance of the ion distribution of an excitable membrane. The transient disturbance has electrical consequences to which nearby transmembrane proteins respond, spreading the disturbance.**

Some invertebrate neurons, specialized for long-range message conduction, have a very large cross-sectional diameter, which lowers the electrical resistance and so speeds the wave of depolarization down the neuron. Electrical resistance is inversely proportional to diameter, and in general the wider the diameter of a neuron, the less the threshold potential difference required to generate a self-propagating nerve impulse.

The action potential

The self-propagating wave of depolarization, the nerve impulse, is an all-or-nothing affair. The threshold value of stimulation required to open enough passive sodium channels is called the **action potential.** An action potential is formally defined as the electrochemical gradient that results from the entry of a sufficient number of sodium ions to initiate a nerve impulse. Because different neurons possess different densities of passive Na^+ ion channels, different neurons may exhibit different action potentials. For any one neuron, however, the action potential is always the same. Any stimulus that opens enough Na^+ ion channels to exceed the action potential of that particular neuron will propagate an impulse outward as a wave of depolarization with a constant amplitude (the height or strength of the wave) corresponding to wide-open sodium ion channels. The stimulus is said to have "fired" the

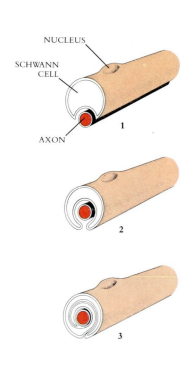

NUCLEUS

SCHWANN
CELL

AXON

1

2

3

4

FIGURE 46-6

Development of the myelin sheath that surrounds many neuron axons involves the envelopment of the axon by a Schwann cell. It is the progressive growth of the Schwann cell membrane around the axon (see steps 1 to 4) that contributes the many membrane layers characteristic of myelin sheaths. These layers act as an excellent electrical insulator.

nerve. Every nerve impulse traversing that nerve has the same amplitude, signals differing from one another only in frequency of these impulses.

> **A nerve impulse is transmitted when the local current induced by depolarization is strong enough to induce the depolarization of the neighboring portion of the membrane. The electrochemical gradient established by the spreading wave of depolarization is called an action potential.**

Saltatory conduction

Many vertebrate neurons possess axons sheathed at intervals by neuroglial Schwann cells. These cells envelop the axon (Figure 46-6), wrapping their cell membrane around it many times to produce a baklava-like series of layers. This lipid-rich envelope of membrane layers is called a **myelin sheath.** The myelin sheath prevents the transport of ions across the neuron membrane beneath it and thus acts as an electrical insulator, creating a region of high electrical resistance on the axon.

Schwann cells are spaced along such an axon one after the other, with a short interval separating each Schwann cell from the next (Figure 46-7). This interval, referred to as the **node of Ranvier,** is critical to the propagation of the nerve impulse in these cells. Within the small gap represented by each node, the surface of the axon is exposed to the fluid surrounding the nerve. The pumps and channels that move ions across the neuron membrane are concentrated in this zone. The direct fluid contact permits ion transport to occur through the pumps and an action potential to be generated. The action potential is not propagated by a wave of membrane depolarization traveling down the axon, since this is prevented by the insulating Schwann cells. Instead, the action potential jumps as an electrical current from one node to the next. When the current reaches a node, it acts as a gating current that opens voltage-sensitive Na^+ ion channels; in so doing, it generates a potential difference large enough to create a current that reaches the next node. The arrival of the current at that node opens its voltage-sensitive sodium channels, creating another current that passes on to the next node, and so on. This very fast form of nerve impulse conduction is known as **saltatory conduction** (from the Latin *saltare,* to jump). An impulse conducted in this fashion moves very fast, up to 120 meters per second for large-diameter neurons. Saltatory conduction is also very cheap metabolically for the cell, since there is far less membrane depolarization for the ion pumps to deal with when only the nodes are undergoing depolarization than when the entire nerve surface is.

> **Not all nerve impulses propagate as a wave of depolarization spreading along the neuronal membrane. On some vertebrate nerves, impulses travel very much faster by jumping along the membrane, leaping electrically over insulated portions.**

Transferring Information from Nerve to Tissue

An action potential passing down an axon eventually reaches the end of the axon. That end is often branched; it may be associated either with several dendrites of other nerve cells or with sites on muscle or secretory cells. These associations of nerves with other cells are the synapses mentioned earlier.

The structure of the synapse

When the tip of a vertebrate axon is examined carefully, it becomes apparent that it does not actually make contact with the target cell it approaches. There is a narrow intercellular gap, 10 to 20 nanometers across, separating the axon tip and the target cell (Figure 46-8). This gap is called a **synaptic cleft.**

Synaptic clefts do not occur in some of the nerve junctions of many of the less ad-

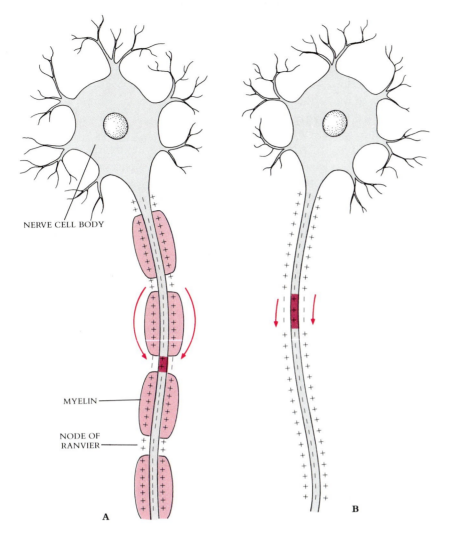

NERVE CELL BODY

MYELIN

NODE OF
RANVIER

A

B

FIGURE 46-7

The insulating properties of myelin sheaths have a major impact on the way in which nerve impulses are propagated along axons.

A In a myelinated fiber, the wave of depolarization jumps electrically from node to node without ever traversing the insulated membrane segment between nodes. This is called saltatory conduction. Just as running is much faster than walking because there is more distance between each step, so saltatory conduction is much faster than direct conduction, because there is more distance between each gating-current-induced depolarization.

B In an unmyelinated fiber, the wave of depolarization traverses the entire length of the axon, each portion of the membrane becoming depolarized in turn, like a row of falling dominoes.

vanced invertebrates; in such cases there is direct electrical contact between the neurons. Such clefts are, however, characteristic of all vertebrate nerve junctions, except for some of the synapses in the brain. When a nerve signal arrives at a synaptic cleft, it passes across the gap chemically. The membrane on the axonal side of the synaptic cleft is called the **presynaptic membrane.** When a wave of depolarization reaches the presynaptic membrane, it stimulates the release of transmitter chemicals into the cleft. These chemicals rapidly pass to the other side of the gap. Once there, they combine with receptor molecules in the membrane of the target cell, which is called the **postsynaptic membrane.** By doing so, they cause ion channels to open.

> **A synapse is a junction between an axon tip and another neuron, almost always including a narrow gap that the impulse cannot bridge. Passage of the impulse across the gap is by chemical signal from the axon.**

The great advantage of a chemical junction such as this, compared with a direct electrical contact, is that the nature of either the chemical transmitter or the receptor can be different in different junctions, permitting different kinds of responses. Over 60 different chemicals have been identified that act as specific neurotransmitters or modify the activity of neurotransmitters. The events that occur within the synaptic cleft when a nerve signal arrives depend very much on the identity of the particular neurotransmitter chemical that is released into the cleft, and on the nature of the receptor to which it binds. To understand what happens, we will first look at the junction between a nerve and a muscle cell, where the situation is a simple one, and then consider nerve-nerve junctions, where the situation is more complex.

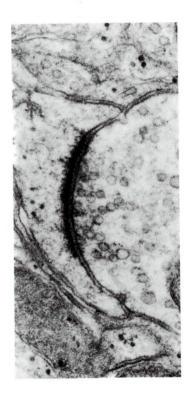

FIGURE 46-8

A synaptic cleft between two nerves. The larger of the two processes is rich in presynaptic vesicles; the dark zone on the left side of the synaptic cleft is the postsynaptic membrane.

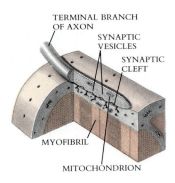

TERMINAL BRANCH
OF AXON
SYNAPTIC
VESICLES
SYNAPTIC
CLEFT

MYOFIBRIL

MITOCHONDRION

FIGURE 46-9

The neuromuscular junction. The terminal branch of the axon does not actually touch the muscle; the intervening space is the synaptic cleft. The axonal tip is typically rich in neurotransmitter-containing vesicles. In most neuromuscular junctions, stimulation of the nerve occurs by fusion of these vesicles with the depolarized axonal membrane, releasing the neurotransmitter into the synaptic cleft. In other synapses, the vesicles appear to serve a storage function, the depolarization of the axonal membrane serving to open transmembrane channels specific to the neurotransmitter and so permitting cytoplasmic molecules of the neurotransmitter to stream out into the cleft.

FIGURE 46-10

A neuromuscular junction, magnified 100 times. The axon has branched, different branches extending to different muscle fibers. Each junction of axon tip and muscle fiber forms an individual neuromuscular junction; such a junction is called a motor end plate.

How a nerve-muscle synapse works

When an axon makes connection with a muscle cell (Figure 46-9), the synapse is called a **neuromuscular junction.** Within the tip of the axon are numerous **synaptic vesicles,** small membrane-enclosed microbodies that store molecules of the neurotransmitter acetylcholine. Additional acetylcholine is present in the cell's cytoplasm. In neuromuscular junctions (Figure 46-10), the synaptic vesicles fuse with the axon membrane when the membrane becomes depolarized by an arriving nerve impulse, emptying their acetylcholine by exocytosis directly into the cleft. In some synapses, current evidence suggests that the release of acetylcholine occurs differently: when a wave of depolarization reaches the axon tip, transmembrane channels in the membrane open and acetylcholine passes from the axon cytoplasm out into the synaptic cleft.

The amount of acetylcholine released into the synaptic cleft depends on the degree to which the presynaptic side of the cleft, or **end plate,** is depolarized by the incoming nerve impulse. The arrival of a full action potential causes the release of about a million molecules in less than a millisecond (Figure 46-11). Passing across the gap, the acetylcholine molecules bind to receptors in the postsynaptic membrane, opening a kind of ion channel that admits both Na^+ and K^+ ions. During the millisecond that the channels are open, some 10^4 ions flow inward. This ion flow depolarizes the postsynaptic muscle cell membrane, which initiates a wave of depolarization that passes down the muscle. This wave of depolarization permits the entry of calcium, which in turn triggers muscle contraction.

> **At a neuromuscular junction, acetylcholine released from an axon tip depolarizes the muscle cell membrane, permitting the entry of calcium ions, which trigger muscle contraction.**

For a neuromuscular synapse to transmit more than one impulse, it is necessary to destroy the residual neurotransmitter remaining in the synaptic cleft after the last impulse. If this does not occur, the postsynaptic membrane simply remains depolarized. This removal of leftover acetylcholine is accomplished by an enzyme, **acetylcholinesterase,** which is present in the synaptic cleft. Acetylcholinesterase is one of the fastest-acting enzymes in the vertebrate body, cleaving one acetylcholine molecule every 40 microseconds. The rapid removal of neurotransmitter by acetylcholinesterase permits as many as 1000 impulses per second to be transmitted across the neuromuscular junction. Many organic phosphate compounds, such as the nerve gases tabun and sarin and the agricultural insecticide parathion, are potent inhibitors of acetylcholinesterase. Because they produce continuous neuromuscular transmission, such compounds can be lethal to vertebrates. Breathing, for example, requires muscular contraction, as does blood circulation.

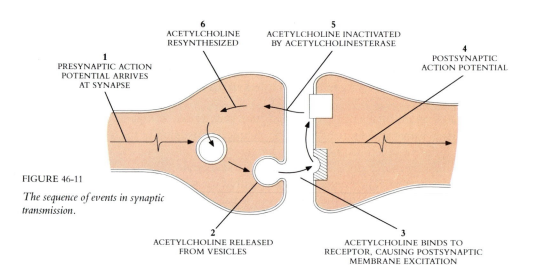

6
ACETYLCHOLINE
RESYNTHESIZED

5
ACETYLCHOLINE INACTIVATED
BY ACETYLCHOLINESTERASE

4
POSTSYNAPTIC
ACTION POTENTIAL

1
PRESYNAPTIC ACTION
POTENTIAL ARRIVES
AT SYNAPSE

FIGURE 46-11

The sequence of events in synaptic transmission.

2
ACETYLCHOLINE RELEASED
FROM VESICLES

3
ACETYLCHOLINE BINDS TO
RECEPTOR, CAUSING POSTSYNAPTIC
MEMBRANE EXCITATION

Transferring Information from Nerve to Nerve

When the axon connection is with another nerve cell rather than with a muscle, the outcome of a synaptic event is far less predictable. Vertebrate nervous systems utilize a variety of neurotransmitters and receptors. Some of these depolarize the postsynaptic membrane, in which case the synapse is an **excitatory synapse.** Other synapses, called **inhibitory synapses,** have the reverse effect, stabilizing the postsynaptic membrane to depolarization.

The integration of nerve impulses

An individual nerve cell can possess both kinds of synaptic connections to other nerve cells. When signals arrive from both excitatory and inhibitory synapses, the depolarizing effects, which cause less internal negative charge, and the stabilizing ones, which cause more internal negative charge, interact with one another as the input from the dendrites reaches the body of the neuron (Figure 46-12). The result is a process of *integration* in which the various excitatory and inhibitory electrical effects tend to cancel or reinforce each other.

In an *excitatory synapse,* neurotransmitters (Figure 46-13, *A*) such as acetylcholine, norepinephrine, and dopamine open the joint Na^+ and K^+ channel in the postsynaptic membrane. Opening these channels increases the membrane's permeability to both sodium and potassium ions simultaneously, which partially depolarizes the membrane. When the contributions of the excitatory synapses are summed at the cell body, the total depolarization that they induce may create a gating current strong enough to generate an action potential and so initiate a nerve impulse that will proceed down the axon of the cell.

In an *inhibitory synapse,* on the other hand, a neurotransmitter (Figure 46-13, *B*) such as γ-aminobutyric acid (GABA) acts to increase the permeability of the postsynaptic membrane to chloride ion (Cl^-) or to K^+ ion, but not to Na^+ ion. Either of these increases in permeability has the effect of stabilizing the membrane against depolarization, because increased permeability to negatively charged ions (which are in higher concentration outside) renders the interior of the membrane *more* negative with respect to the exterior, rather than

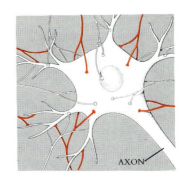

FIGURE 46-12

The cell body of a neuron serves to integrate the information arriving from many sources. In this representation, the dendrites and cell body of a neuron are contacted in many places by axonal projections from other nerves. The synapses made by some of these axons are inhibitory, tending to counteract depolarization of the cell body membrane; these are indicated in color. The synapses made by other axons are stimulatory, tending to depolarize the cell body membrane. The summed influences of all these inputs determine whether the axonal membrane will be sufficiently depolarized to initiate a propagating nerve impulse.

ACETYLCHOLINE

EPINEPHRINE

TABUN

SARIN

NOREPINEPHRINE

SEROTONIN

A

PARATHION

B

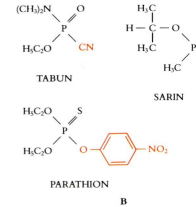

FIGURE 46-13

*The chemical structure of representative excitatory (**A**) and inhibitory (**B**) neurotransmitters.*

less so. When the net effect of the synaptic influences reaching the cell body is inhibitory, the neuron may be rendered less likely to respond to a subsequent incoming wave of depolarization. In such a condition, that neuron may inhibit the passage of nerve impulses.

> **The integration of neural information occurs within individual neurons as the result of the summed effect on a neuron's membrane of the electrical effects of different synapses. Some of these synapses facilitate depolarization; others inhibit it.**

Integration in the central nervous system

The processing of information in the central nervous system of vertebrates occurs by synaptic integration. The sensory neurons comprise the chain of nerves that extends between the sensory receptors and the central nervous system. At every synapse in this chain, information derived from many sensory nerve signals is integrated. By this means, sensory information is processed at each stage of its transmission. Within the brain, connecting neurons, or **interneurons,** join each nerve cell to many others, forming a complex web of signal integration that is responsible for associative activity, in which one neuron signal influences the transmission of other associated neurons. Thought and consciousness are the result of associative activity.

An Overview of Neuron Function

As we have shown, the essence of the way in which neurons are specialized for carrying information is the arrangement of transmembrane channels in their membranes. These channels perform a variety of functions critical to nerve conduction:

1. The sodium-potassium pumps are arranged so that they pump K^+ in and Na^+ out. Passive K^+ channels are open and passive Na^+ channels are closed, resulting in a net negative charge within the cell relative to the exterior—the resting potential. All nerve stimulation and conduction is based on the disturbance of this charge difference.

2. The normally closed passive Na^+ channels of the neuron membrane open in response to pressure or other stimulation, abolishing the resting potential and creating a localized gating current. In this way, nerves are able to receive information.

3. Some of the passive Na^+ channels are voltage sensitive and will open in response to a localized gating current. In this way, a nerve impulse moves down a nerve.

4. The tips of some axons respond to a nerve impulse by transporting acetylcholine out of the neuron into the synaptic cleft. In this way, muscles receive commands from the nervous system.

5. Acetylcholine is not the only neurotransmitter that may occur in axon tips. When two neurons synapse, the postsynaptic membrane may be excited or inhibited, depending on the identity of the neurotransmitter that is released from the presynaptic membrane in response to the arrival of a nerve impulse and upon the receptor that receives it. The integration of excitatory and inhibitory inputs that takes place on the neuron membrane is the basis of all associative activity, including thought.

Some nervous systems contain very few neurons. Others, like those of vertebrates, contain many millions. To understand the vertebrate nervous system, it is necessary to comprehend not only how one neuron functions, but also how the many neurons of the nervous system are organized into a functioning whole. There is a great deal that we do not yet know about the organization and functioning of vertebrate nervous systems, but a lot has been learned. The rest of this chapter will serve as an introduction to this very active field of research.

THE EVOLUTION OF NERVOUS SYSTEMS

Sponges are the only major phylum of multicellular animals that lack nerves. If you prick a sponge, the nearby surface contracts very slowly. The protoplasm of each individual cell conducts the impulse, which fades within a few millimeters. No messages dart from one part of the sponge body to others, as they do in all other multicellular animals.

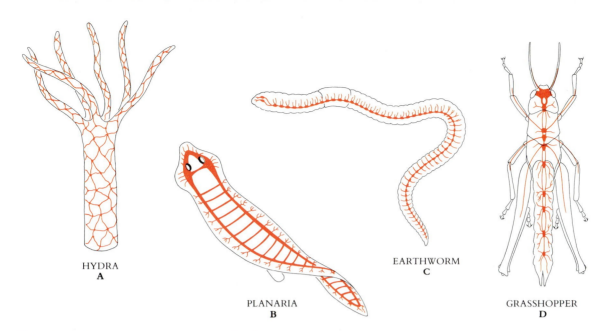

HYDRA
A

PLANARIA
B

EARTHWORM
C

GRASSHOPPER
D

The Simplest Nervous Systems: Reflex Arcs

The simplest nervous systems occur among the cnidarians (Figure 46-14, *A*), which were discussed in Chapter 35. In these animals, all neurons are similar, each having fibers of approximately equal length. Cnidarian neurons are linked to one another in a web, or **nerve net,** which is dispersed throughout the epithelium of the body. The conduction of signals through the nerve net is slow, but a stimulus anywhere can eventually spread throughout the whole net.

A nerve impulse in the cnidarians passes through the nerve net, eventually reaching the body muscles and causing them to contract. The motion that results is called a **reflex,** because it is an automatic consequence of the nerve stimulation. The simple chain of events is called a **reflex arc,** in this case sensory cell → neuron → muscle. There is no associative activity, no control of complex actions, and little or no coordination. The nerve net of the cnidarians possesses only the barest essentials of nervous reaction.

In the simplest reflex arcs, a sensory neuron synapses directly with a motor neuron. In the human body, only the knee jerk and a few other reflexes are so simple. Reflex arcs in complex animals more typically involve a relay of information from a sensory neuron through one or more interneurons to a motor neuron. The functioning of a reflex arc may be modulated by the interneurons.

> **A reflex is a simple behavior produced by a neural information-response circuit called a reflex arc.**

The Advent of More Complex Nervous Systems: Associative Activities

Reflex arcs are not associative. The passage of a signal along an arc is not influenced by the level of activity of other neurons, but rather depends only on whether or not the arc is stimulated. The first associative activity in nervous systems is seen in the free-living flatworms, phylum Platyhelminthes (Figure 46-14, *B*). Running down the body of these flatworms are two nerve cords; peripheral nerves extend outward from the nerve cords to the muscles of the body. The two nerve cords converge at the front end of the body, forming an enlarged mass of nervous tissue that also contains associative neurons with synapses connecting neurons to one another in various ways. This primitive "brain" is a rudimentary central nervous system. It permits a far more complex control of muscular responses than was possible in the cnidarians, since sensory input may induce any of a variety of responses, depending on the associative activity that occurs in the central nervous system.

FIGURE 46-14

The evolution of nervous systems in invertebrates has involved a progressive elaboration of organized nerve cords and the centralization of complex responses in the front end of the nerve cord. This can be seen in these four invertebrates.

The Evolutionary Path to the Vertebrates

All of the subsequent evolutionary changes in nervous systems can be viewed as a series of elaborations of the characteristics that are already present in the flatworms. Five principal evolutionary trends can be identified, each becoming progressively more pronounced as nervous systems evolved greater degrees of complexity. Vertebrate nervous systems, which are the most complex to have evolved, represent the farthest extension of each of these five trends.

1. *The elaboration of more sophisticated sensory mechanisms.* As such mechanisms evolved, they provided the nervous system with better information. By far the most complex sensory systems are found among the vertebrates; they are described in the next chapter.

2. *The differentiation of the nerve network into central and peripheral systems.* As nervous systems evolved, most cell bodies of neurons came to be concentrated in one or a few nerve cords (Figure 46-14, *C*), or in masses of nerve cell bodies that are part of, or located near, a nerve cord. Such a central nervous system is connected to all other parts of the body by peripheral nerves.

3. *The differentiation of afferent and efferent nerve fibers.* Individual neurons carry impulses in one direction only. As nervous systems increased in complexity, neurons operating in particular directions became specialized. Those carrying impulses toward the central nervous system are called **afferent** nerves, whereas those conducting impulses away from the central nervous system are called **efferent** nerves. In general, the sensory neurons are afferent, whereas the nerves to glands or muscles are efferent.

4. *The increased complexity of association.* Within the central nervous system, the afferent and efferent neurons are linked by associative interneurons. The path between sensory receptor and motor effector becomes more complex, the axon of interneurons often being branched, with processes leading to several alternative cells. The particular branch that carries the impulse is determined by the degree of membrane polarization in that branch, which in turn is influenced by synapses with many other nerves, some of them excitatory and some inhibitory. Thus, as central nervous systems with more numerous interneurons evolved, more different kinds of activities within the central nervous system were able to influence the destination of a particular impulse.

5. *The elaboration of the brain.* The central coordination of complex responses came to be increasingly localized in the front end of the nerve cord (Figure 46-14, *D*). As this region evolved, it came to contain a progressively larger number of associative interneurons, and to develop tracts, which are highways within the brain that connect associative elements.

As nervous systems became more complex, there was a progressive increase in associative activity. It is important to note that no matter how complex the associations are that form within a brain, all nerve impulses are identical because of the "all-or-nothing" nature of nerve transmission. The difference between a strong signal and a weak one is solely in terms of the frequency of impulses and the number of neurons carrying the signal. All the information that a brain has to work with is the source, frequency, and number of signals that it receives.

EVOLUTION OF THE VERTEBRATE BRAIN

The brains in primitive chordates are little more than swellings at the end of the nerve cord. In the ancestors of the vertebrates, such primitive brains were present, serving primarily as sensory centers, receiving messages from eyes and other sensory receptors. They also received information from sensors embedded within muscles and used this information to modulate neuromuscular commands. Far more complex brains and associated neuron-rich

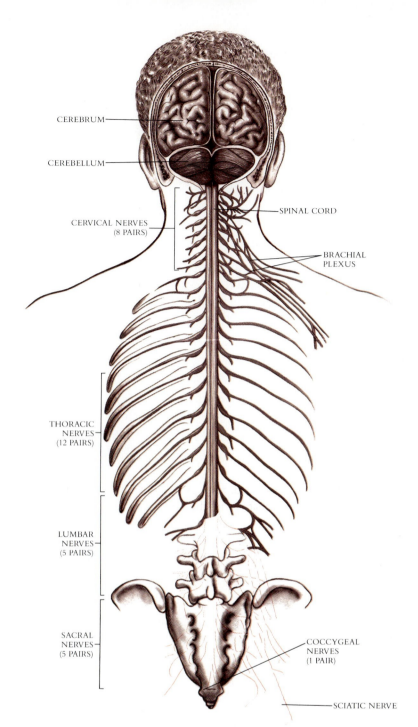

FIGURE 46-15

The human central nervous system is composed of the brain and spinal cord, from which 31 pairs of spinal nerves extend that direct the various activities of the body.

CEREBRUM

CEREBELLUM

SPINAL CORD

CERVICAL NERVES
(8 PAIRS)

BRACHIAL
PLEXUS

THORACIC
NERVES
(12 PAIRS)

LUMBAR
NERVES
(5 PAIRS)

SACRAL
NERVES
(5 PAIRS)

COCCYGEAL
NERVES
(1 PAIR)

SCIATIC NERVE

nerve trunks, collectively called central nervous systems, developed both in vertebrates (Figure 46-15) and, independently, in cephalopods and arthropods.

Basic Organization of the Vertebrate Brain

The earliest vertebrates had far more complex brains than their ancestors. Casts of the interior braincases of fossil agnathans, fishes that swam 500 million years ago, have revealed a lot about the early evolutionary stages of the vertebrate brain. Although they were very small, these brains already had the three principal divisions that characterize the brains of all contemporary vertebrates: (1) the hindbrain, or **rhombencephalon;** (2) the midbrain, or **mesencephalon;** and (3) the forebrain, or **prosencephalon.**

The hindbrain is the principal component of these early brains (Figure 46-16), as it

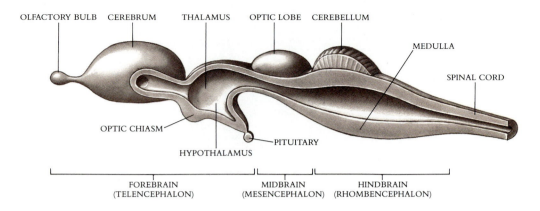

OLFACTORY BULB   CEREBRUM   THALAMUS   OPTIC LOBE   CEREBELLUM   MEDULLA   SPINAL CORD   OPTIC CHIASM   PITUITARY   HYPOTHALAMUS

FOREBRAIN (TELENCEPHALON)   MIDBRAIN (MESENCEPHALON)   HINDBRAIN (RHOMBENCEPHALON)

FIGURE 46-16

The basic organization of the vertebrate brain can be seen in the brains of primitive fishes. These brains are divided into the same regions that exist in different proportions in all vertebrate brains: the hindbrain, which is the largest portion of the brain in fishes; the midbrain, which in fishes is a small zone devoted to processing visual information; and the forebrain, which in fishes is devoted primarily to processing olfactory (smell) information. In the brains of terrestrial vertebrates, the forebrain plays a far more dominant role than it does in fishes.

still is in fishes today. Composed of the **pons** and the **medulla oblongata,** the hindbrain may be considered an extension of the spinal cord devoted primarily to coordinating motor reflexes. Tracts, cables composed of large numbers of nerve fibers, run up and down the spinal cord to the hindbrain. The hindbrain in turn integrates the many afferent signals from the muscles and determines the pattern of efferent response.

Much of this coordination is carried on within a small extension of the hindbrain called the **cerebellum** ("little cerebrum"). In the advanced vertebrates, the cerebellum plays an increasingly important role as a coordinating center and is correspondingly larger than it is in the fishes. In all vertebrates, the cerebellum processes data on the current position and movement of each limb, the state of relaxation or contraction of the muscles involved, and the general position of the body and its relation to the outside world. These data are gathered in the cerebellum and synthesized, and the resulting orders are issued to efferent pathways.

In fishes the remainder of the brain is devoted to the reception and processing of sensory information. The second major division of the brain, the midbrain, is composed primarily of the **optic lobes,** which receive and process visual information. The third major division, the forebrain, is devoted to processing olfactory (smell) information.

The brains of fishes continue growing throughout their lives. This is in marked contrast to the brains of the more advanced classes of vertebrates, which have completed their development by infancy and, with a few minor exceptions, form no new neurons thereafter.

In early vertebrates the principal component of the brain was the hindbrain, which is devoted largely to coordinating motor reflexes.

The Advent of a Dominant Forebrain

Starting with the amphibians, and much more prominently in the reptiles, a pattern arises that becomes a dominant evolutionary trend in the further development of the vertebrate brain (Figure 46-17): the processing of sensory information becomes increasingly centered in the forebrain.

The forebrain in reptiles, amphibians, birds, and mammals is composed of two elements that have distinct functions. The **diencephalon** (Greek, *dia,* between) is devoted to the integration of sensory information. The **telencephalon,** or "end brain" (Greek *telos,* end) is located at the front and is devoted largely to associative activity.

The diencephalon, as its name indicates, is a "between brain," located between the telencephalon and the midbrain. It has two components, situated one atop the other above the midbrain: the **thalamus** and the **hypothalamus.**

Sensory integration

The thalamus is the primary site of sensory integration in the brain. Auditory, optical, and

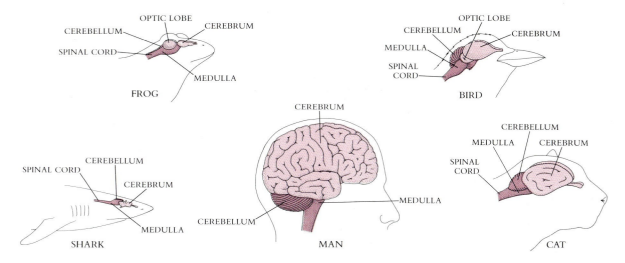

(Labels in figure: CEREBELLUM, OPTIC LOBE, CEREBRUM, SPINAL CORD, MEDULLA — FROG; OPTIC LOBE, CEREBELLUM, MEDULLA, SPINAL CORD, CEREBRUM — BIRD; SPINAL CORD, CEREBELLUM, CEREBRUM, MEDULLA — SHARK; CEREBRUM, CEREBELLUM, MEDULLA — MAN; CEREBELLUM, MEDULLA, CEREBRUM, SPINAL CORD — CAT)

other information is relayed to the thalamus, passing from there to the outer surface of the brain. For example, information about posture, derived from the muscles, and information about orientation, derived from sensors within the ear, passes from the cerebellum of the hindbrain to the thalamus; the thalamus then processes the information and channels it to the appropriate motor center on the outer surface of the brain.

Arousal and attention

A complex interwoven network of neurons called the **reticular system** weaves through the brainstem and midbrain and is connected to the thalamus. All of the sensory systems have fibers that feed into this system, which serves to "wiretap" all of the incoming and outgoing communications channels of the brain. In doing so, the reticular system monitors information concerning the incoming stimuli and identifies important ones. When the reticular system has been stimulated to arousal, it increases the activity level of many parts of the brain.

The reticular system is periodically repressed by the release of the neurotransmitter **serotonin** within the brain. Serotonin causes the level of brain activity to fall, bringing on sleep. It is easier to sleep in a dark room than in a lighted one because there are fewer incoming stimuli to trigger the reticular system. Drugs such as barbiturates facilitate sleep by blocking the reticular system; overdoses, however, can irreversibly repress the system, resulting in permanent coma.

Integrating the body's responses

The hypothalamus integrates the visceral activities. It controls body temperature, respiration, and heartbeat; it also directs the secretions of the brain's master hormone-producing gland, the **pituitary** gland. The hypothalamus is linked by a network of neurons to the cortex. This network, together with the hypothalamus, is the so-called **limbic system.** The operations of the limbic system are responsible for many of the most deep-seated drives and emotions of vertebrates, including pain, anger, sex, hunger, thirst, and pleasure.

Stereotyped behavior

The telencephalon of reptiles and birds consists largely of a layer of nervous tissue called the **corpus striatum,** which controls complicated stereotyped behavior. There is nothing like the corpus striatum in any other group. Much of what we think of as "behavior" in birds and reptiles is, in reality, composed of completely determined neural reactions. For example, the complex mating rituals of some birds are often choreographed by the structure of the brain; their pattern is fixed by nerve paths within the corpus striatum.

In land-dwelling vertebrates the processing of information is increasingly centered in the forebrain.

FIGURE 46-17

The evolution of the vertebrate brain has involved pronounced changes in the relative sizes of different regions of the brain. In fishes, the hindbrain (indicated in dark color) is predominant, the rest of the brain serving primarily to process sensory information. In amphibians and reptiles, the forebrain is far larger, and a large cerebrum devoted to associative activity can be seen. In birds, which evolved from reptiles, the cerebrum is more pronounced than in reptiles. In mammals, which evolved from reptiles independently of birds, the cerebrum is the largest portion of the brain. The dominance of the cerebrum is greater in humans, where it envelops much of the rest of the brain, than in any other animal.

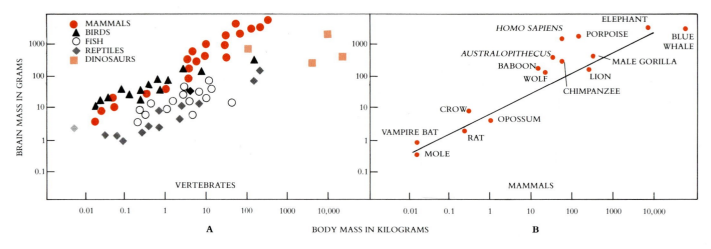

FIGURE 46-18

Among most vertebrates, brain weight is a relatively constant proportion of body weight, so that a plot of brain mass versus body mass gives a straight line.

A *However, the proportion of brain mass to body mass is much greater in birds than in reptiles, and even greater in mammals.*

B *Among mammals, humans have the greatest brain mass per unit of body mass (i.e., the farthest perpendicular distance from the plotted line). In second place are not monkeys but dolphins!*

The Recent Expansion of the Cerebrum

In examining the ratio of brain mass to body mass among the vertebrates (Figure 46-18), one sees a remarkable discontinuity between fishes and reptiles, on the one hand, and birds and mammals, on the other. Mammals have brains that are particularly large relative to their body mass; this is especially true of human beings and porpoises. The increase in brain size in the mammals largely reflects a great enlargement of the **cerebrum,** the dominant part of the mammalian brain.

The cerebrum is the center for correlation, association, and learning in the mammalian brain. It receives sensory data from the thalamus. From the cerebrum, motor fibers extend to the motor columns of the spinal cord, which contain groups of neurons servicing different muscles. Efferent motor neurons extend from these motor columns directly to the muscles. These motor columns pass straight through the brain to the voluntary motor regions of the brainstem, and can be seen as a cable of fibers called the **pyramidal tract,** a structure that dominates the structure of primate brains particularly.

> The brains of mammals and birds are unusually large relative to their body size. This largely reflects great enlargement of the cerebrum, which is the center for correlation, association, and learning.

ANATOMY AND FUNCTION OF THE HUMAN BRAIN

The cerebrum, which is at the very front of the human brain, is so large relative to the rest of the brain that it appears to envelop it (Figure 46-19). In the brains of humans and other primates, the cerebrum is split into two halves, or hemispheres, which are connected only by a nerve tract called the **corpus callosum.** Each hemisphere is divided further by two deep grooves into four lobes, designated the **frontal, parietal, temporal,** and **occipital** lobes of the brain (Figure 46-20).

The Cerebral Cortex

Much of the neural activity of the cerebrum occurs within a thin layer only a few millimeters thick on its outer surface. This layer, called the **cerebral cortex,** is densely packed with nerve cells. The human cerebral cortex contains over 10 billion nerve cells, amounting to roughly 10% of all the neurons in the brain. The surface of the cerebral cortex is highly convoluted, particularly in human brains, a property that increases its surface area threefold.

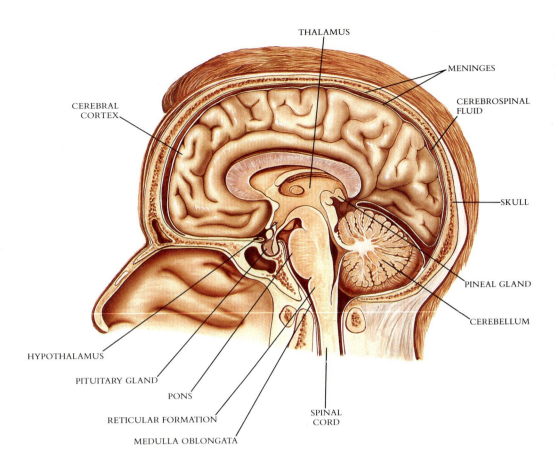

FIGURE 46-19

A cross section of the human brain demonstrates how the cerebrum envelops the rest of the brain.

THALAMUS

MENINGES

CEREBROSPINAL FLUID

CEREBRAL CORTEX

SKULL

PINEAL GLAND

CEREBELLUM

HYPOTHALAMUS

PITUITARY GLAND

PONS

RETICULAR FORMATION

SPINAL CORD

MEDULLA OBLONGATA

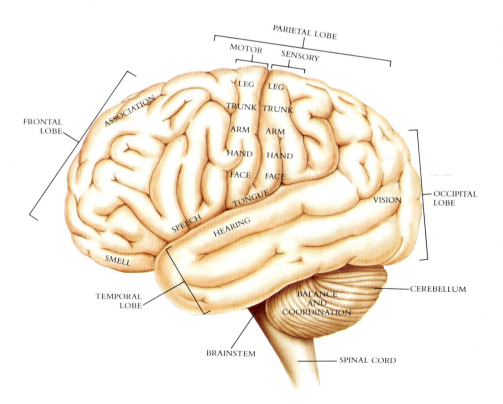

FIGURE 46-20

One hemisphere of the human brain, showing the four lobes and the major functional regions.

PARIETAL LOBE

MOTOR

SENSORY

LEG

LEG

TRUNK

TRUNK

ARM

ARM

HAND

HAND

FACE

FACE

TONGUE

ASSOCIATION

FRONTAL LOBE

VISION

OCCIPITAL LOBE

SPEECH

HEARING

SMELL

CEREBELLUM

TEMPORAL LOBE

BALANCE AND COORDINATION

BRAINSTEM

SPINAL CORD

By examining the effect of injuries to particular sites on the cerebrum, it has been possible to plot roughly the location of its various associative activities. There are four major regions of such activity on the cerebral cortex, each of them referred to as a specialized cortex: the motor, sensory, auditory, and visual cortexes (Figure 46-20).

The motor cortex straddles the rearmost portion of the frontal lobe. Each point on its surface is associated with the movement of a different part of the body. Right behind the motor cortex, on the leading edge of the parietal lobe, lies the sensory cortex. Each point on the surface of the sensory cortex represents sensory receptors from a different part of the body, such as the pressure sensors of the fingertips or the taste receptors of the tongue. The auditory cortex lies within the temporal lobe; different surface regions of this cortex correspond to different sound frequencies. The visual cortex lies on the occipital lobe, with different sites corresponding to different positions on the retina.

Associative Organization of the Cerebral Cortex

Only a small portion of the total surface of the cerebral cortex is occupied by the motor and sensory cortexes. The remainder of the cerebral cortex is referred to as **associative cortex.** This appears to be the site of higher mental activities, such as planning and contemplation. The associative cortex represents a far greater portion of the total cortex in primates than it does in any other mammals and reaches its greatest extent in human beings. In a mouse, for example, 95% of the surface of the cerebral cortex is occupied by motor and sensory areas; in humans, only 5% of the surface is devoted to motor and sensory functions; the remainder is associative cortex.

The two hemispheres of the cerebrum constitute two associative brains in miniature. All four of the specialized cortex regions occur on the surface of each hemisphere, the ones on the right hemisphere related to the left side of the body and vice versa. The switch occurs because the nerve tracts descending from the hemispheres cross over in the pyramidal tract via the corpus callosum, where the two sides of the brain are connected.

Although the two hemispheres of the brain each contain the four cortexes, the two hemispheres are responsible for different associative activities. Injury to the left hemisphere of the cerebrum, for example, often results in the partial or full loss of speech, whereas a similar injury to the right side does not. There are several **speech centers** that control speech. They are almost always located on the left hemisphere of right-handed people; in about one third of left-handed people they are located on the right hemisphere. An injury to one speech center produces halting but correct speech; injury to another speech center produces fluent, grammatical, but meaningless speech; injury to a third center abolishes speech altogether.

Injuries to other sites on the surface of the brain's left hemisphere result in impairment of the ability to read, write, or do arithmetic. Comparable injuries to the right hemisphere have very different effects, resulting in the impairment of three-dimensional vision, musical ability, or the ability to recognize patterns and solve inductive problems.

It has been a popular activity, particularly among journalists, to speculate that the human brain is composed of a "rational" left hemisphere and an "intuitive" right hemisphere, particularly since many of the more imaginative associative activities seem to be carried out by the right hemisphere. Some people even argue that the brain actually possesses two consciousnesses, one dominant over the other, or that the right brain hemisphere is the more ancient, with language and reasoning having evolved much later. Actually, the significance of the clustering of associative activities in different areas of the brain is not at all clear—but it remains a subject of much interest.

Memory and Learning

One of the great mysteries of the brain is the basis of memory and learning. If portions of the brain are removed, particularly the temporal lobes, memory is impaired but not lost;

there is no one part of the brain in which memory appears to reside. Investigators who have tried to probe the physical mechanisms underlying memory often have felt that they were grasping at a shadow. An understanding of these mechanisms has continued to elude them.

However, researchers have learned a little about the physical basis of memory. The first stage, **short-term memory,** is transient, lasting only a few moments. Such memories can readily be removed from the brain by application of an electrical shock. When this is done, short-term memories are wiped from the circuits, but the longer-term memories are preserved. This result suggests that short-term memories are stored electrically, in the form of short-term neural excitation. **Long-term memory,** in contrast, appears to involve structural changes in the neural connections within the brain. Very little in known about how such changes might occur. Perhaps differences in electrical stimulation alter the patterns of synapse formation, long-term memory corresponding to a pattern of nerve impulse integration.

THE IMPORTANCE OF THE PERIPHERAL NERVOUS SYSTEM

In this chapter, after considering how individual neurons function, we have focused on the central nervous system—the brain and spinal cord. It is within the central nervous system that all associative and most integrative activities occur, and it is natural that our attention should first be directed there. The central nervous system does not function in isolation, however. It functions as an information-processing center and central command post, a manager of the body's functions; like any manager it requires both information on which to act and a means of carrying out commands. The central nervous system obtains information from the afferent nerves of the sensory nervous system, the subject of the next chapter, and issues commands via the motor nerves of the somatic and autonomic nervous systems, described in Chapter 48. These other elements comprise the peripheral nervous system. The sensory, integrative, and command functions of nervous systems evolved in concert, and the system functions as a unified whole.

SUMMARY

1. A neuron is a nerve cell with an excitable membrane. It is specialized for the transmission of waves of depolarization along the membranes of processes extending out from the cell body.

2. An excitable membrane is created by the active pumping by the sodium-potassium pump of sodium ions out across the membrane and of potassium ions in. Because the membrane is not permeable to sodium ions but is permeable to potassium ions, potassium streams back out while sodium does not flow back in. The result is a net negative charge within the cell.

3. A nerve impulse is initiated by depolarization of an excitable membrane, that is, by the opening of sodium ion channels. Such an opening equalizes the ion concentrations and thus removes the charge difference.

4. A nerve impulse propagates because the opening of some sodium ion channels facilitates the opening of adjacent channels, causing a wave of depolarization to travel down the membrane of the nerve cell. All signals traversing a given neuron have the same amplitude, differing only in the frequency of impulses.

5. When a nerve impulse reaches the far end of a nerve cell, the axon tip, it depolarizes the membrane at the tip, causing the release of chemicals from the tip. These chemicals pass across a synaptic gap and interact with channels in the membrane of another

neuron or of muscle or gland cells. They either open these channels by depolarizing them—an "excitatory" response—or stabilize them in a closed conformation—an "inhibitory" response.

6. The integration of nerve signals occurs on the cell body membranes of individual neurons, which receive both depolarizing waves from dendrites associated with excitatory synapses and stabilizing waves from dendrites associated with inhibitory synapses. These waves tend to cancel each other out, the final amount of depolarization depending on the mix of the signals received.

7. The evolution of the nervous system has involved the extensive elaboration of networks of associative interneurons in the brain.

8. The hindbrain is the principal component of the brain in fishes.

9. In the more advanced vertebrates, associative activity is increasingly centered in the forebrain. The midbrain serves as a conduit, linking the lower brain to the forebrain. In fishes and reptiles, the corpus striatum of the frontal lobes governs complicated stereotyped behavior. In mammals this tissue is largely replaced by the cerebrum, which is the center for associative activity and learning.

SELF-QUIZ

The following questions pinpoint important concepts in this chapter. You should be sure to know the answers to each of them.

1. In a nerve, excitation travels outward from the cell body as a nerve impulse, along a long cytoplasmic projection called a _____.

2. The specialized chemicals that carry out communication across synaptic junctions are called
   (a) synaptonemal complexes   (d) neurotransmitters
   (b) enzymes   (e) myelin
   (c) neuroglia

3. Most neurons are unable to survive alone for long; they require nutritional support, which is provided by companion cells called _____.

4. The ion-generated electrochemical gradient, called a *resting potential,* generates a charge inside the cell. Is it positive or negative?

5. A nerve causes a depolarization that propagates down a nerve fiber. This impulse has a constant amplitude for any given nerve. True or false?

6. Acetylcholinesterase is an enzyme that occurs in neurons. Which of the following statements concerning acetylcholinesterases is not true?
   (a) they destroy acetylcholine
   (b) they operate only inside of neuron cells
   (c) they are one of the fastest-acting enzymes in the body
   (d) they are required during breathing

7. Integration is a process that occurs on which part of a neuron?
   (a) dendrite   (d) two of the above
   (b) cell body   (e) all of the above
   (c) axon

8. Much of the coordination of the motor reflexes is carried out by a small extension of the hindbrain called the _____.

9. The brain's master hormone-producing gland works closely with the hypothalamus. It is called the
   (a) limbic gland   (c) reticular gland
   (b) corpus gland   (d) pituitary gland

10. Which is not one of the four lobes of the cerebrum?
    (a) pyramidal   (d) frontal
    (b) parietal   (e) temporal
    (c) occipital

11. The corpus callosum is a nerve tract that connects the two hemispheres of the cerebrum. True or false?

12. The associative cortex is highly developed in humans. It is
    (a) proportionately larger in non-primates than in primates
    (b) responsible for motor coordination
    (c) the means of control of heart rate and other involuntary functions
    (d) the site of higher mental activities

THOUGHT QUESTIONS

1. When the brain is starved for oxygen even briefly, it dies. When the body is starved for energy, it begins to metabolize its own tissues, channeling the products preferentially to the brain. This behavior points out the importance to the brain of ongoing oxidative respiratory metabolism. Why is active oxidative respiration so very important to the continued well-being of the brain's nerve cells?

2. Why do most synapses contain gaps across which an electrical impulse cannot pass, when a direct physical connection would enable the uninhibited passage of the impulse?

3. Why do excitable membranes utilize K^+ channels? Why aren't Na^+ ions simply pumped out of the neuron to achieve the internal negative charge, and passive Na^+ ion channels opened to depolarize the membrane?

FOR FURTHER READING

AXELROD, J.: "Neurotransmitters," *Scientific American,* June 1974, pages 59-71. An account of the chemicals that lead to the different synaptic responses underlying neuronal integration.

BULLOCK, T.H., R. ORKAND, and A. GRINNELL: *Introduction to Nervous Systems,* W.H. Freeman & Co., San Francisco, 1977. A very good, comprehensive, comparative treatment.

DUNANT, Y., and M. ISRAEL: "The Release of Acetylcholine," *Scientific American,* April 1985, pages 58-66. A challenge to the accepted theory that acetylcholine is emitted by synaptic vesicles.

KATZ, B.: *Nerve, Muscle, and Synapse,* McGraw-Hill Book Co., New York, 1966. A classic description of how a nerve impulse arises and is propagated; concise and unusually lucid.

KEYNES, R.D.: "Ion Channels in the Nerve-Cell Membrane," *Scientific American,* March 1979, pages 126-135. An up-to-date account of the structure of nerve-cell membranes.

KUFFLER, S.W., and J.G. NICHOLLS: *From Neuron to Brain: A Cellular Approach to the Function of the Nervous System,* 2nd ed., Sinauer Associates, Inc., Sunderland, Mass., 1984. A superb overview of the mechanisms of nerve excitation and transmission.

NAUTA, W., and M. FREITAG: "The Organization of the Brain," *Scientific American,* September 1979, pages 88-111. An overview of the complex circuitry of the human brain.

STEVENS, C.F.: "The Neuron," *Scientific American,* September 1979, pages 54-65. A description of the structure and functioning of a typical nerve cell.

Chapter 47

SENSORY SYSTEMS

Overview

All sensory information is acquired through the depolarization of sensory nerve endings. From a knowledge of which neurons are sending signals and how often they are doing so, the brain builds a picture. Various specialized sensory receptors cause the depolarization of sensory nerve endings and the firing of neurons in response to particular aspects of the body's internal or external environment—chemical stimuli, mechanical deformation, or electromagnetic stimulation.

For Review

Here are some important terms and concepts that you will encounter in this chapter. If you are not familiar with them, you should review them before proceeding.

- **Carotene** (Chapter 12)

- **Evolution of the eye** (Chapter 23)

- **Sensory neuron** (Chapter 46)

- **Depolarization** (Chapter 46)

Imagine floating on a still surface of water, your eyes closed, in a quiet place where no air is stirring. It would be hard not to sleep. The reticular system of your brain, experiencing little or no stimulation, would not arouse your brain to consciousness. You could not long tolerate this peace, however. Without any stimulation, you would become disoriented and awaken; if you did not find a frame of reference for yourself, you would eventually go mad. The human brain cannot endure sensory deprivation for long—its proper functioning depends on continuous sensory input.

All of the input from sensory neurons to the central nervous system arrives in the same form, as nerve impulses carried on afferent sensory neurons. Every arriving nerve impulse is identical to every other one. The information that the brain derives from sensory input is based on the frequency with which these impulses arrive and on the neuron that transmits the input. A sunset, a symphony, searing pain—to the brain they are all the same, differing only in the source of the impulse and its frequency. Thus if the auditory nerve is artificially stimulated, the central nervous system perceives the stimulation as a noise. If the optic nerve is artificially stimulated in exactly the same manner and degree, the stimulation is perceived as a flash of light. To understand sensory input to the nervous system, then, we must examine the sources of sensory signals, as well as the factors that influence the frequency with which these sources send nerve impulses to the brain.

THE NATURE OF NEUROSENSORY COMMUNICATION

The path of sensory information to the central nervous system is a simple and ancient one, composed of three elements:

1. **Stimulation.** A physical stimulus impinges on a neuron or an accessory structure, called a **sensory receptor.**
2. **Transduction.** The sensory receptor initiates the depolarization of a sensory neuron.
3. **Transmission.** The sensory neuron conducts an action potential along an afferent pathway to the central nervous system.

All sensory receptors are able to initiate nerve impulses by opening ion channels within sensory neuron membranes and thus depolarizing them. They differ from one another with respect to the nature of the environmental input that triggers this event. Many sorts of receptors have evolved among the vertebrates, with each receptor sensitive to a different aspect of the environment. Broadly speaking, we can recognize three classes of environmental stimuli:

MECHANICAL FORCES	CHEMICALS	ELECTROMAGNETIC ENERGY
pressure	taste	light
gravity	smell	heat
inertia	humidity	electricity
sound		magnetism
touch		
vibration		

Receptors capable of responding to these stimuli, singly or in concert, constitute the sensory repertory of vertebrates.

SENSING INTERNAL INFORMATION

Traditionally, the sensing of information that relates to the body itself—its internal condition and its position—is known as **interoception,** or inner perception (Figure 47-1, *A*). In contrast, the sensing of the exterior is called **exteroception** (Figure 47-1, *B*). Many of the neurons and receptors that monitor body functions are simpler than those that monitor the external environment. These receptors are perhaps more similar to what primitive sensory receptors were like than are any other present-day receptors.

Nerve Endings as Sensory Receptors

How would you design a sensory receptor? Recall for a moment the nature of nerve synapses. Within the nervous system, a nerve impulse is initiated when the postsynaptic membrane of a synapse is depolarized. The simplest sensory receptors work in the same way. They are simply nerve endings that depolarize in response to direct physical stimulation— to temperature, to chemicals like oxygen diffusing into the nerve cell, or to a bending or stretching of the neuron cell membrane.

> **The simplest sensory receptors are nerve endings that depolarize in response to direct physical stimuli such as pressure, change in temperature, or concentrations of chemicals.**

FIGURE 47-1

By tradition, the senses are grouped into two classes.
A Dancers are able to maintain their proper stance by using their internal sense of balance. Sensing of this sort is interoception.
B The other class senses the external environment. This leaf frog (Phyllomedusa tarsius), photographed in the Amazon rain forest of Ecuador, learns about the world around it by using eyes to perceive patterns of light, just as you are doing in reading this page. Sensing of this sort is exteroception.

The vertebrate body uses a variety of simple interoceptors to obtain information about its internal condition. Among the simplest are those that report on changes in temperature and blood chemistry.

Temperature change receptors

There are two populations of nerve endings in the skin that are sensitive to changes in temperature. One set is stimulated by a lowering of temperature (the cold receptors), the other by a raising of temperature (the heat receptors). Depolarization of temperature-change receptors is thought to arise from temperature-induced conformational changes in the proteins that constitute the Na^+ channels of the neuronal membrane. Temperature-sensitive neurons occur within the hypothalamus, where they monitor the temperature of the circulating blood and thus provide the central nervous system with reliable information on changes in the body's internal temperature.

Blood chemistry receptors

Embedded within the walls of arteries, at several locations in the circulatory system, are receptors called **carotid bodies.** The carotid bodies provide the sensory input that the body uses to regulate its rate of respiration. When oxygen levels fall below normal limits (hypoxia), this information is conveyed to the central nervous system, which reacts by increasing the respiration rate. These receptors, composed of the nerve endings of afferent fibers, are bathed by the blood that flows through the arteries. The nerve endings sense the level of oxygen in the arterial blood by changes in the rate of oxygen diffusion into the neuron. CO_2 concentration is sensed, in a similar way, by the diffusion into the neuron of hydrogen ions, which are associated with the carbonic acid formed when CO_2 is dissolved in the bloodstream.

Simple Mechanical Receptors

Physical twisting, bending, or stretching of a nerve ending results in depolarization of its membrane. A variety of interoceptors consist of nerve endings that contain specific ion-transporting channels sensitive to mechanical force applied to the membrane; these channels open in response to lateral stretching or distortion, initiating depolarization and causing the nerve to fire. These **mechanical receptors** differ from one another primarily in their locations and in the way that the neuron is oriented with respect to the mechanical stress. The different ways in which the central nervous system uses interoceptor location and orientation to sense particular stimuli are illustrated by the following examples.

Pain

A stimulus that causes or is about to cause tissue damage is perceived as pain. Such a stimulus elicits an array of responses from the central nervous system, including reflexive withdrawal of a body segment from a source of stimulus, and changes in heartbeat and blood pressure. The receptors that produce these effects are called **nociceptors.** They consist of the ends of small, sparsely myelinated nerve fibers located within tissue, usually near the surfaces where damage is most likely to occur. Physical deformation of the ends of the nerves initiates depolarization. Different nociceptors have different thresholds. Some respond only to actual tissue damage, while others fire when subjected to pressure or temperature that has not yet damaged tissue. All nociceptors fire more often as the strength of the stimulus increases, with the result that the frequency of afferent nerve impulses corresponds to the strength of the stimulus.

Muscle contraction

Buried deep within the muscles of all vertebrates, except the bony fishes, are specialized muscle fibers called **spindle fibers.** Wrapped around each of these fibers is the end of an afferent neuron, called a **stretch receptor** (Figure 47-2). When a muscle is stretched, the spin-

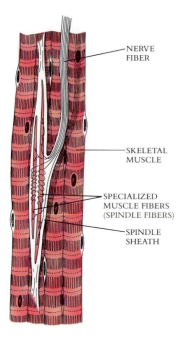

FIGURE 47-2

A stretch receptor embedded within skeletal muscle. Stretching of the muscle elongates the spindle fiber, which deforms the nerve endings, causing them to fire and send a nerve impulse out along the nerve fiber.

NERVE FIBER

SKELETAL MUSCLE

SPECIALIZED MUSCLE FIBERS (SPINDLE FIBERS)

SPINDLE SHEATH

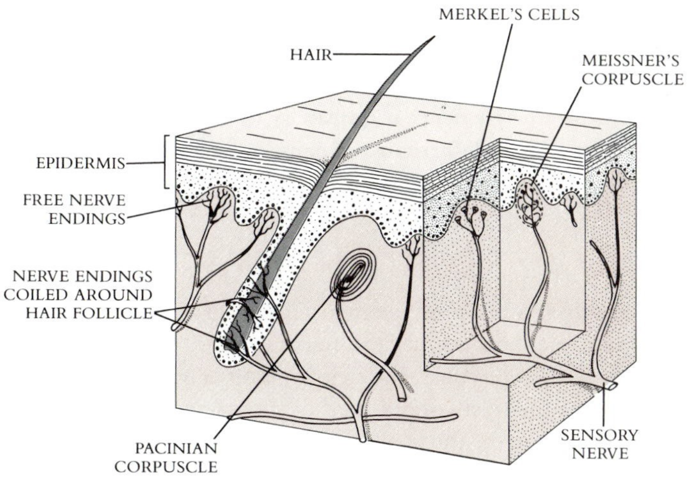

FIGURE 47-3

Sensory receptors in human skin.

dle fiber elongates, stretching the spiral nerve ending of the stretch receptor and repeatedly stimulating it to fire. When the muscle contracts, the tension on the fiber lessens and the stretch receptor ceases to fire. Thus the frequency of stretch receptor discharges along the afferent nerve fiber indicates the degree of muscle contraction at any given moment. The central nervous system uses this information to control those movements that involve the combined action of several muscles, such as those that are involved in breathing or walking.

Blood pressure

The brain is supplied with blood by two major arteries, the carotids, which branch high up in the neck at a cleft called the **carotid sinus.** In this cleft, the wall of the artery is thinner than usual. Within this wall is a highly branched network of nerve endings called a **baroreceptor,** or pressure receptor.

These baroreceptor neurons fire at a low rate at all times, sending a steady stream of afferent impulses to the cardiovascular center of the brain stem. When the blood pressure increases, the arterial wall balloons out where it is thinnest, in the region of the baroreceptor, increasing the rate of firing of the afferent neuron. A fall in blood pressure causes the wall to move inward and lowers the rate of neuron firing. Thus the frequency of impulses arriving from these afferent fibers provides the central nervous system with a continuous measure of blood pressure. The greater the blood pressure, the higher the frequency of impulses. Many of the other large arteries and veins possess similar baroreceptors.

Vibration

Deep below the skin of vertebrates lie pressure receptors of yet another kind (Figure 47-3). These pressure receptors are not sensitive to pressure itself but rather to *changes* in pressure. Receptors of this type, called **Pacinian corpuscles,** are the principal receptors of the sexual organs. Each receptor is an afferent nerve terminal, surrounded by a capsule made up of alternating layers of cells and extracellular fluid. When sustained pressure is applied to the corpuscle, the elastic capsule absorbs much of it; only the rapid onset of a pressure pulse causes the nerve ending to fire. The frequency of impulses along the nerve fiber thus provides information about the *rate* of pressure application rather than its intensity.

Touch

Other types of pressure receptors are similar in general design to Pacinian corpuscles, but their nerve endings are less protected from deformation and are located closer to the surface of the skin. These receptors are responsible for the sense of touch. The two principal kinds of receptors in this class, called **Meissner's corpuscles** and **Merkel cells,** are responsible for the great sensitivity of your fingertips. Meissner's corpuscles fire in response to rapid changes in pressure, while Merkel cells measure the duration of pressure and the extent to which it is applied. These cells are present on many of the body surfaces that do not contain hair—the fingers, palms, and nipples, for example.

Mechanical stimuli initiate nerve impulses by deforming the sensory neuron membrane. Such deformation opens ion channels in the membrane and initiates depolarization.

SENSING THE BODY'S POSITION IN SPACE

A second class of receptors, more complex than simple nerve endings, are also mechanical, relying on the physical deformation of the nerve ending to initiate depolarization and neuron firing. In this class of receptor, the deformation is not direct, however; instead, it employs a lever. Perhaps the least complicated of these lever-operated receptors are hair-associated touch receptors such as those on the back of your hand; an afferent nerve ending is wrapped around the base of each hair so that when a hair moves, its nerve fires. The hairs are used as mechanical amplifiers that deform the neuron and initiate depolarization—you cannot feel a mosquito alight on your hand until it brushes against a hair.

Most lever-operated receptors are more complicated than this. To illustrate how the more complex mechanical receptors work, imagine a pencil standing in a glass. No matter which way you tip the glass, the pencil will roll along the rim, applying pressure to the lip of the glass. If you wish to know the direction in which the glass is tipped, you need only inquire where on the rim the pressure is being applied. The body uses receptors of this sort, called **proprioceptors** (Greek, *proprios,* self) to sense its position in space. Gravity serves as a reference point, with changes in the pressure applied to a nerve ending by a heavy object within the receptor providing the stimulus. There are many types of lever-operated receptors, all built along this same general design and differing primarily in the disposition of the lever. Three examples will illustrate that this type of receptor is capable of much diversity of response.

Gravity

All vertebrates possess gravity receptors known as **statocysts.** Statocysts function in much the same way as the pencil-in-a-glass model described above. In a series of hollow chambers within the inner ear, for example, are located the receptors that obtain the information our brain uses to perceive balance. The outermost chambers of the inner ear are called the sacculus and the utriculus. Within each of these chambers, a small pebble of calcium carbonate called a **statolith** rests on a bed of cilia that is connected with sensory receptor cells. Those cilia located directly beneath the statolith are bent by its weight; each bent cilium exerts pressure on the membrane of the sensory cell from which it arises just as a pencil exerts pressure on the lip of a glass. Any shift in a statolith's position results in different cilia being bent by its weight and a different collection of ciliary sensory cells sending discharges along afferent nerves to the central nervous system. In this way, the brain always knows the orientation of the receptor relative to "up."

Angular Acceleration

Vertebrates sense motion in a way similar to the way in which they detect vertical position, by employing a receptor in which fluid deflects cilia in a direction opposite to that of the motion. Within the inner ear, past the sacculus and utriculus, are three fluid-filled **semicircular canals** (Figure 47-4), each oriented in a different plane at right angles to the other two so that motion in any direction can be detected. Protruding into the canals are sensory cells, which are connected to afferent nerves. From the ends of these sensory cells extend a series of short cilia, with one long cilium located at one side of each cilia bundle.

The pressure of the long cilium against the shorter ones initiates the depolarization

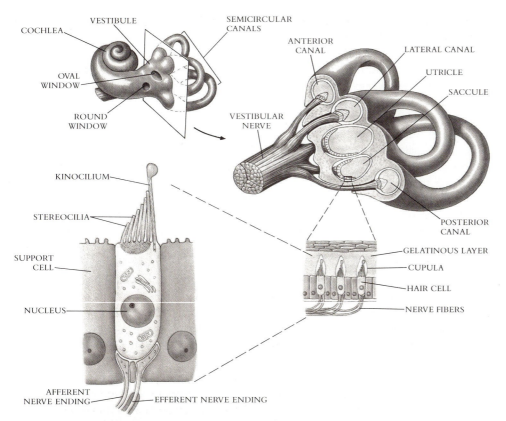

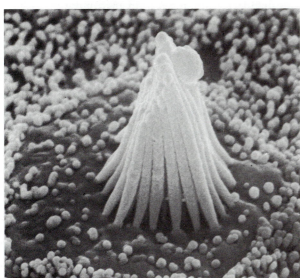

FIGURE 47-4

Vertebrates sense movement as pressure of a long cilium, the kinocilium, against a battery of surrounding cilia, the stereocilia. As you can see in the electron micrograph, these cilia (magnified 20,000 times) form tentlike assemblies called cupulae that project out into passages within the semicircular canals of the ear's vestibular apparatus. Movement causes fluid in the canal to press against the cupula. Because there are three semicircular canals at right angles to one another, movement in any plane can be detected.

of the cell membrane, which triggers a nerve impulse. The long cilia from many cells are commonly grouped together in a bundle that is covered with a jellylike coating. Such a bundle is called a **cupula,** the Latin word for a little cup. When the cupula is bent by the movement of liquid within the semicircular canal, the long fibers bend in the direction of the fluid movement. Because of inertia, the direction of their bending is opposite to the direction in which the body is moving. When the long cilia bend, they either (1) depress the shorter cilia further and so increase the frequency of nerve firing, or (2) relieve the pressure on the shorter cilia and thus decrease the rate of firing of the afferent nerve. Because the three canals are oriented in all three planes, movement in any plane is sensed by at least one of them. Complex movements can be analyzed by comparing the sensory input from each canal.

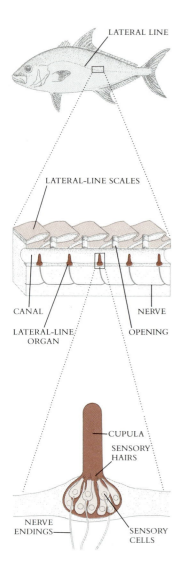

Lateral Line

Fishes perceive their movement relative to the water surrounding them in the same way that you perceive motion when you shake your head, with fishes having cupulae receptors similar to those in your semicircular canals. In fishes, however, such sensory cells are located on the surface of the body. The cupulae of fishes extend out into grooves that run laterally along the head and sides of the body and make up the lateral line discussed in Chapter 39 (Figure 47-5).

Within the lateral line of a fish, the long cilia of adjacent cells are oriented in opposite directions so that some are stimulated by movement of water in one direction and others by movement in the opposite direction. Nerve impulses from the lateral-line receptors permit the fish to assess its rate of movement through the water. They also enable a fish to detect motionless objects at a distance by the movement of water reflected off the objects, and to sense movement of the water around the fish's body as pressure waves against its lateral line. This is how a trout orients with its head upstream.

In a very real sense, this latter form of lateral-line perception is the fish's equivalent of hearing. It does not differ in its mechanism from what happens within your ears when you listen to a symphony, a pattern of pressure waves in the air around you. To hear, terrestrial vertebrates use receptors within the ear that are thought to have evolved through the modification of lateral-line receptors. The other structures of the vertebrate ear function primarily as stimulus amplifiers that convert weak pressure waves in air into stronger pressure waves like those that occur in the water.

> **Complex mechanical receptors can respond to pressure, to gravity, or to angular acceleration. In each case, the receptors employ mechanical devices such as weights or levers to convert the information to a mechanical stimulus, which then initiates the depolarization of a sensory membrane by deformation.**

SENSING THE EXTERNAL ENVIRONMENT

So far, most of the sensory receptors that we have considered have been directed at the internal environment of the body or at determining its position in space. Sensing the nature of other objects in the external environment presents quite a different problem. Imagine, for example, the problem of learning about an object standing some distance away. The amount of information that any sensory receptor can obtain about that object is limited both by the nature of the stimulus sensed by that receptor and by the medium, either air or water, through which the stimulus must move to reach the receptor. There are three levels of information that different sensory systems can provide:

Attention. Some sensory systems provide only enough information to determine that the object is present, saying little or nothing about where it is located.

Location. Other sensory systems provide information about the direction of the object. They may therefore permit the organism to locate the object by moving toward it.

Imaging. Still other sensory systems provide information concerning distance, as well as direction. By doing so, they enable the central nervous system to construct a three-dimensional image of the object and its surroundings.

All three of these levels of sensory information are used by the sensory receptors of vertebrates. Almost all of the exterior senses of vertebrates evolved in the sea before vertebrates invaded the land. Consequently, the senses of terrestrial vertebrates tend to emphasize those stimuli that travel well in water. Some, like vision, use stimuli that travel as well as or better in air than they do in water; such senses use receptors that have been carried over from the sea to the air virtually unchanged. Others, like hearing, convert an airborne stimulus to a waterborne one, as we have just seen, and use receptors that are very similar

FIGURE 47-5

The lateral line is a system of motion-sensing receptors very similar in structure to the cupulae of the semicircular canals that humans use to sense motion. A series of these receptors project into a canal beneath the surface of the skin. The canal is open to the exterior; it runs the length of the fish's body. Movement of water past the fish forces water through the canal, deflecting the cupulae and initiating nerve impulses.

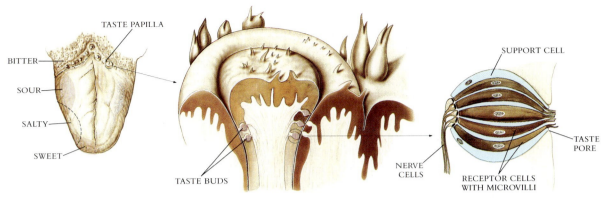

A

to the ones that originally evolved in the water. A few vertebrate sensory systems that function well in the water, such as the electric organs of fish, cannot function in the air and are not found among terrestrial vertebrates. On the other hand, some land dwellers have sensory systems, such as infrared vision, that could not function in the sea.

The four primary senses that detect objects at a distance use different classes of receptors. Taste and smell use chemical receptors; hearing uses mechanical receptors; and vision uses electromagnetic receptors. We will consider each of these four in turn, and then go on to examine other sensory systems employed by vertebrates.

Taste and Smell

The simplest exteroceptors, just like the simplest interoceptors, are chemical ones. Embedded within the membranes of afferent nerve endings or of sensory cells associated with afferent neurons are specific chemical receptors that induce depolarization when they are bound by particular molecules. There are two types of chemical sensory systems that utilize different receptors and process information at different locations in the brain: **taste,** in which the receptors are specialized sensory cells, and **smell,** in which the receptors are neurons. Despite the differences in how stimuli are received and the information processed, there is little difference in the chemical stimuli to which the two chemical senses respond.

Taste

The taste and olfactory (smell) receptors of fishes are the most sensitive vertebrate chemoreceptors known. The taste receptors, or **taste buds,** are not located in the mouth like those of terrestrial vertebrates but rather are scattered over the surface of the fish's body. These taste buds are exquisitely sensitive to amino acids. A catfish, for example, can distinguish between two different amino acids at a concentration of less than 100 micrograms (0.000001 gram) per liter of water. The ability to taste the surrounding water in this way is very important to bottom-feeding fishes, enabling them to sense the presence of food in an often murky environment.

One group of chemoreceptors in human beings are also composed of special sensory cells and are also called taste buds, although a human's taste buds are not nearly as sensitive as those of a fish. In all terrestrial vertebrates, the taste buds are located in the mouth (Figure 47-6). Each taste bud is intimately associated with an afferent neuron. Their function is limited to screening food chemically to determine whether it is suitable for eating.

Human beings have four kinds of taste buds, each of which responds to a broad range of chemicals. We refer to the four classes of stimuli to which the different kinds of taste buds respond as salt, sweet, sour, and bitter. Our complex perception of taste is composed of different combinations of impulses from these four types of chemoreceptors. The chemoreceptors are concentrated on different parts of the tongue, with sweet and salt on the front, sour on the sides, and bitter at the back. These chemoreceptors are capable of a rich and diverse array of combined sensory input, but this input is difficult for us to describe in words.

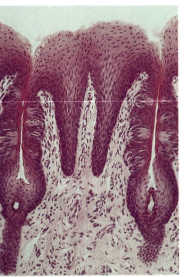

B

FIGURE 47-6

A Human beings have four kinds of taste buds (bitter, sour, salty, sweet), located on different regions of the tongue. Groups of taste buds are typically organized in sensory projections called papillae.
B Individual taste buds are bulb-shaped collections of chemical receptor cells that open out into the mouth through a pore.

FIGURE 47-7

Humans smell by using olfactory receptor cells located in the lining of the nasal passage. The receptor cells are neurons. Axons from these sensory neurons project back through the olfactory nerve directly to the brain.

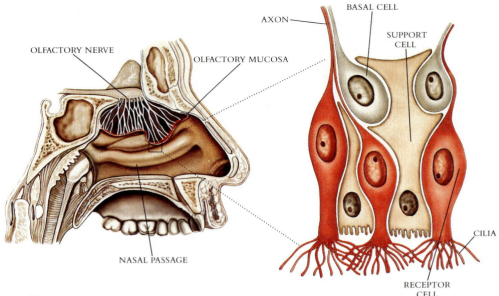

Smell

The chemoreceptors that human beings use to smell are located in the upper portion of the nasal passage (Figure 47-7). These olfactory chemoreceptors are neurons whose cell bodies are embedded in the nasal epithelium; from the cell bodies, dendrites extend to the surface of the epithelium and project sensory cilia into the surface mucus layer. A terrestrial vertebrate uses the sense of smell in much the same way that a fish uses the sense of taste—to sense the chemical environment around it. Because terrestrial vertebrates live surrounded by air rather than water, their sense of smell has become specialized to detect airborne particles.

> **External chemical stimuli are sensed by specific receptors that are components of sensory neuron membranes (smell) or of the membranes of associated receptor cells (taste). These receptors depolarize in response to the binding of specific chemicals.**

The information that chemoreceptors provide to the central nervous system is sketchy at best. The stimulus arrives slowly, and the medium through which the chemical must diffuse is itself moving constantly. Chemoreception functions primarily to alert the central nervous system to the presence of a particular chemical, but it conveys little information about the location of the source of that chemical. Some specific chemical cues, the pheromones that we discussed in Chapter 37, are detectable by chemoreceptors in trace amounts. They are employed by animals of several different phyla to attract potential mates or other individuals.

Hearing

Just as fishes detect vibration in water by means of their lateral line receptors, so terrestrial vertebrates detect vibration in air by means of mechanical receptors within the ear. These receptors, as we noted earlier, evolved from lateral-line organs. When the stimulus that these receptors detect passes through air, we call the sense "hearing." Hearing actually works less well in air than in water because water transmits pressure waves better. Despite this, hearing is a sense widely used by terrestrial vertebrates to monitor their environments and particularly to detect possible sources of danger. Auditory stimuli travel farther and more quickly than chemical ones, and auditory receptors provide better directional information than do chemoreceptors. Auditory stimuli alone, however, provide little information about distance.

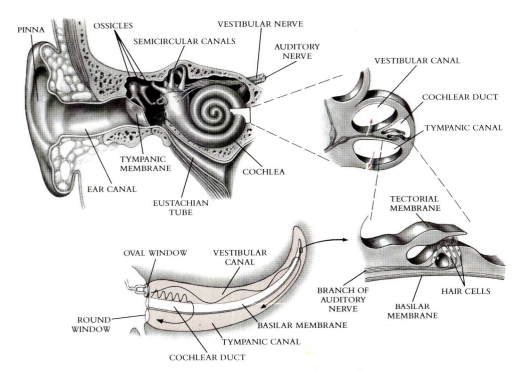

FIGURE 47-8

Structure of the human ear. Sound waves passing through the ear canal beat on the tympanic membrane, pushing a set of three small bones, or ossicles (hammer, anvil, and stirrup) against an inner membrane called the oval window. This sets up a wave motion in the fluid filling the cochlea that travels through the vestibular and tympanic canals. Where the sound wave beats against the sides of the canals, the tectorial membrane is pushed against the basilar membrane, bending hair cells and firing associated neurons.

The evolutionary modification of the lateral line receptors of fishes into a terrestrial vertebrate hearing organ that functions in air—that is, an **ear**—has involved ingenious solutions to a serious mechanical problem: sound waves are much weaker in air than in water, so that to utilize the same receptor, it is necessary to amplify the sound. The ears of terrestrial vertebrates achieve this amplification in two ways.

Structure of the ear

In the ears of mammals, sound waves beat against a large membrane of the outer ear called the **tympanic membrane** (Figure 47-8), causing corresponding vibrations in three small bones, or ossicles: the **hammer, anvil,** and **stirrup.** These bones act in concert as a lever system, increasing the force of the vibration. The third in line of these levers, the stirrup, pushes against another membrane, the **oval window.** Because the oval window is smaller than the tympanic membrane, vibration against it produces more force per unit area of membrane. The chamber in which all of these events occur is called the **middle ear.** It is connected to the pharynx by the Eustachian tube in such a way that there is no difference in air pressure between the middle ear and the outer ear. The familiar "ear popping" that is associated with landing in an aircraft or with the rapid descent of an elevator in a tall building is a result of pressure equalization between these two sides of the eardrum.

The membrane covering the oval window is the door to the **inner ear,** where hearing actually takes place. The chamber of the inner ear is shaped like a tightly coiled snail shell and is called the **cochlea,** from the Latin name for "snail." To understand the cochlea's structure and function, imagine a tube that is sealed at each end by a membrane. Further imagine that the tube has another long membrane running down its center, bisecting it into an upper and a lower channel. This long membrane begins at the outer end of the tube but does not extend all the way to its inner end, so that the two chambers are not separated there. Now pound on the part of the membrane covering the top half of the tube at its outer end, where the tube is completely divided into two halves lengthwise. Every inward movement that you create in the membrane moves in the form of a pressure wave through the liquid in the upper chamber. When it reaches the far (inner) end of the tube, beyond the point at which the dividing membrane stops, the pressure wave crosses into the lower chamber and moves back outward, eventually reaching the outer end of the tube again but on the lower half. That is how sound moves through the inner ear.

Now imagine that the bisecting lengthwise membrane of our hypothetical tube is stiffer at the beginning of the tube than at the far end and that you have rolled the tube into a coil. In the coiled tube, the pressure waves will beat against the sides of the tube, thus creating a corresponding vibration in the bisecting membrane. Because the bisecting membrane is stiffer at one end, however, it does not vibrate uniformly along its entire length. Sound waves of different lengths (frequencies) will cause different parts of the middle membrane to vibrate in their passage down the tube. This is the way in which the ear differentiates between sounds of different frequencies.

The ears of terrestrial vertebrates mechanically amplify sound waves and then direct them at a fluid-filled chamber. Within this chamber, receptors respond to the mechanical deformation that is produced by the motion of the sound waves traveling through the fluid.

The auditory receptors

The auditory receptors, the same hair cells that we previously encountered as lateral line receptors in fishes, are located on the middle **basilar membrane,** which bisects the cochlea. These hair cells do not project out into the fluid filling the cochlea. Instead, they are covered with another membrane called the **tectorial membrane.** The bending of the basilar membrane as it vibrates causes the hairs of the receptor cells pressed against the tectoral membrane to bend, firing their associated afferent neurons. This arrangement, in which the receptor is sandwiched between two membranes, is sometimes called the **organ of Corti.** Sounds of different frequencies cause different portions of the basilar membrane to vibrate and thus to fire different afferent neurons. The central nervous system perceives sound intensity in terms of the frequency of discharge; it perceives sound frequency in terms of the pattern of neurons on the basilar membrane that fire afferent impulses.

Our ability to hear depends on the flexibility of the basilar membrane, a flexibility that changes as we grow older. We are not able to hear low-pitched sounds, below 20 vibrations or cycles per second, although some other vertebrates can. As children, human beings can hear high-pitched sounds, up to 20,000 cycles per second, but this ability decays progressively throughout middle age. Other vertebrates can hear sounds at far higher frequencies than these. Dogs, for example, readily detect sounds of 40,000 cycles per second. Thus dogs can hear a high-pitched dog whistle when it seems silent to a human observer.

The information provided by hearing can be used by the central nervous system to determine direction with some precision. Sound sources vary in strength, however, and sounds are attenuated (weakened) to various degrees by the presence of objects in the environment. For these reasons, auditory sensors do not provide a reliable measure of distance, and auditory input alone does not permit the human central nervous system to construct a three-dimensional image of objects.

Sonar

A few groups of mammals that live and obtain their food in dark environments have circumvented the limitations of darkness. Under such circumstances, light is not available as an alternative cue. A bat flying in a completely dark room easily avoids objects that are placed in its path—even a wire less than a millimeter in diameter (Figure 47-9). Shrews use a similar form of "lightless vision" beneath the ground, as do whales and dolphins beneath the sea.

All of these mammals perceive distance and depth by means of **sonar.** They emit sounds and then determine the time that it takes these sounds to reach an object and return to the animal. A bat, for example, emits clicks that last from 2 to 3 milliseconds and are repeated several hundred times per second. The three-dimensional imaging achieved with such an auditory sonar system is quite sophisticated. While humans do not use a similar sensing system, you can gain some appreciation of its capabilities in viewing the ultrasound

FIGURE 47-9

This bat (Antrozous pallidus) *is emitting high-frequency "chirps" as it flies. The bat listens for the sound's reflection against distant objects, and by timing how long it takes for a sound to return, it can effectively navigate in the dark.*

imaging of a growing fetus within a pregnant woman; utilizing the same principle, this technology yields quite a detailed view, often permitting visual sex determination months before birth. Because bats have solved the problem of being active and efficient in the dark, they are one of the most numerous and successful of all orders of mammals.

Vision

Sonar sensory systems are not common among vertebrates because most terrestrial vertebrates live in environments where light is present and vision provides a better picture of the environment than does sonar. Because the visual stimulus, electromagnetic energy, travels in a straight line and arrives virtually instantaneously, visual information can be used to determine both the direction and the distance of an object. No other stimulus provides as much detailed information.

The problem of perceiving an electromagnetic stimulus is one that we have not encountered before in our consideration of sensory systems. All of the receptors that we have described up to this point have been either chemical or mechanical ones. None of them are able to respond to the electromagnetic energy provided by photons of light. Vision, the perception of light, is carried out in vertebrates by a specialized sensory apparatus called an **eye.** Eyes have evolved many times independently in different groups of animals (see the boxed essay on pp. 450–451).

Eyes contain sensory receptors that detect photons of light. These receptors are located in the back of the eye, which is organized something like a camera; light that falls on the eye is focused by a lens on the receptors in the rear, just as the lens of a camera focuses light on film. We will discuss the detailed structure of the vertebrate eye shortly, but first we will discuss the nature of the sensory receptors themselves. How does a photon-detecting sensory receptor work? We have encountered photon receptors once before, in our consideration of photosynthesis in Chapter 12. Just as in photosynthesis, the primary photoevent of vision is the absorption of a photon of light by a pigment. The visual pigment is called *cis*-retinal (Figure 47-10), a substance that we encountered as the cleavage product of carotene, a photosynthetic pigment in plants. In vertebrates, the *cis*-retinal pigment is coupled to a transmembrane protein called **opsin** to form **rhodopsin.**

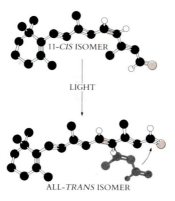

11-*CIS* ISOMER

LIGHT

ALL-*TRANS* ISOMER

FIGURE 47-10

*When light is absorbed by the 11-cis isomer of the visual pigment retinal, the pigment undergoes a change in shape: the linear end of the molecule (right end) rotates about a double bond, indicated here in color. The new isomer is referred to as all-*trans* retinal. This change in the pigment's shape induces in turn a change in the shape of the protein opsin to which the pigment is bound, initiating a chain of events that leads to the generation of a nerve impulse.*

FIGURE 47-11

A *The broad tubular cell diagrammed on the right is a rod, the shorter tapered cell next to it a cone.*

B *Although not obvious from the electron micrograph, the pigment-containing outer segments of these cells are separated from the rest of the cells by a partition through which there is only a narrow passage (seen in the connecting cilium in **A**).*

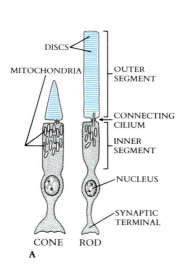

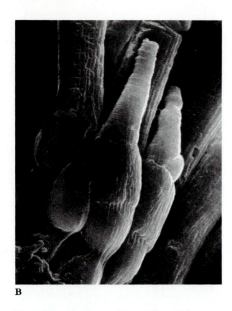

In vertebrate eyes, the visual pigment is located within the tips of specialized sensory cells in the eye called **rod** and **cone cells** (Figure 47-11). Rod cells are responsible for black-and-white vision, and cone cells function in color vision. We will first consider rod cells.

Rod cells

Rod cells are composed of four elements:

1. An outer segment packed with rhodopsin. Pigment molecules are densely packed onto the surface of about 1000 flattened disks, stacked one on top of the other within the outer segment of a rod cell.

2. An inner segment connected to the outer one only by a narrow passageway. This inner segment is rich in mitochondria.

3. A cell nucleus.

4. A synaptic connection to a neuron.

The outer segment of a rod cell has many open sodium channels. Because there is more sodium outside the cell than within it, sodium ions flow into this outer segment and across the narrow passageway into the inner segment. The inner segment of a rod cell possesses a very active sodium pump, which pumps the sodium ions back out again and maintains a negative charge within the rod cell. The reception of a single photon of light by one of the rhodopsin molecules in the outer segment causes the sodium channels of the outer segment to close. At the same time, the sodium pump of the inner segment continues to function. The result is **hyperpolarization**—the interior of the rod becomes even more neg-

FIGURE 47-12

A single photon of light initiates a nerve impulse by blocking the entry into the outer segment of a rod or cone of more than a million sodium ions. It does this by initiating a cascade of chain reactions. When the retinal within a rhodopsin molecule absorbs a photon of light, it changes shape, causing the opsin portion of the molecule to change as well. In its new shape, the rhodopsin molecule acts like an enzyme, altering hundreds of molecules of a protein called transducin. The alteration makes transducin an enzyme in turn, and each altered transducin molecule modifies and activates hundreds of phosphodiesterase protein molecules. Each activated phosphodiesterase enzyme molecule rapidly cleaves large numbers of guanosine triphosphate (GTP, a molecule similar to ATP but with guanine rather than adenine as the nucleotide base) molecules. The GMP molecules that result close sodium channels. In this way more than a million GMP molecules are released as a result of the absorption of one photon.

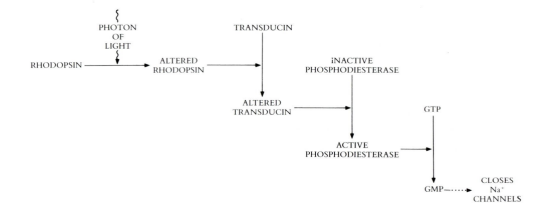

atively charged than before. This hyperpolarization is transmitted to the synapse, where it initiates a nerve impulse in the postsynaptic membrane of a neuron.

Thus the stimulation of a rod cell, unlike that of any other sensory cell, results in hyperpolarization rather than depolarization. It is easier to close channels than to open them, so this arrangement makes the eye a uniquely sensitive instrument.

The rod cell uses a single photon of light to block the entry of more than a million sodium ions. How does the pigment accomplish this prodigious feat? Each rhodopsin pigment/protein complex that absorbs a photon of light is changed in shape as a result and so is transformed into an active enzyme, which proceeds in turn to activate several hundred molecules of a protein called **transducin** (Figure 47-12). Each of these activated transducin molecules in turn activates several hundred molecules of another enzyme called **phosphodiesterase,** and each phosphodiesterase molecule when activated generates **GMP** (guanine monophosphate) molecules that close sodium channels at a rate of about 1000 GMP molecules per second.

The photoreceptors of rod cells thus greatly amplify the original sensory stimulus by employing a cascade of interactions. Each photon that is absorbed by rhodopsin leads to the hydrolysis of more than 10^5 molecules of cyclic GMP!

> **The rod cells of vertebrates are able to detect low levels of light because even one photon can stimulate a rod cell. This incredible sensitivity results from a multistage cascade arrangement, at each level of which every activated molecule activates numerous molecules of the next stage in an ever-expanding chain reaction.**

Cone cells

Color vision is achieved by cone cells in a way that is very similar to the action of rod cells. Cone cells have the same general internal architecture as rod cells and are stimulated by light in the same way. Cone cells are shorter than rod cells (see Figure 47-11) and, as their name suggests, are shaped like inverted ice cream cones. There are three kinds of cone cells, each of which possesses an opsin molecule with a distinctive amino acid sequence and thus a different shape. These differences shift the absorption maximum of *cis*-retinal from the 500 nanometers characteristic of rod cells to 455 nanometers (blue-absorbing), 530 nanometers (green-absorbing), or 625 nanometers (red-absorbing), as shown in Figure 47-13. A battery of cone cells containing these three kinds of pigment can distinguish wavelengths of from 313 to 760 nanometers.

The lens of the vertebrate eye is constructed to filter out low-wavelength light. This solves a difficult optical problem: any uniform lens refracts short wavelengths more than it does longer ones, a phenomenon known as **chromatic aberration.** Consequently, these short wavelengths cannot be brought into focus simultaneously with longer wavelengths. Unable to focus the short wavelengths, the vertebrate eye eliminates them. Insects, whose

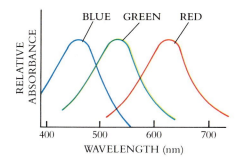

FIGURE 47-13

The absorption spectrum of cis-retinal is shifted in cone cells from the 500 nanometers characteristic of rod cells. The amount of the shift determines what color the cone absorbs: a shift down to 455 nanometers yields blue absorption, a shift up to 530 nanometers yields green absorption, and a shift farther up to 625 nanometers yields red absorption. There is no reason in principle why additional kinds of cones could not be generated by other alterations in cis-retinal that absorb light of other wavelengths, such as yellow, but they are not needed. By comparing the relative intensities of the signals from the three cones, the brain can calculate the intensity of yellow light.

A   B   C

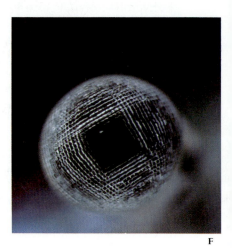

D   E   F

FIGURE 47-14

Variation in eyes.
A *A mollusk, the sea robin.*
B *A bird, the great horned owl.*
C *An amphibian, the frog.*
D *A mammal, the goat.*
E *An arthropod, the deer fly.*
F *A marine invertebrate, the shrimp.*

eyes do not focus light, perceive these lower, ultraviolet wavelengths quite well and often use them to locate food or mates; recall our discussion of this in Chapter 37 (see also Figure 30-19).

The evolution of vision

The evolution of photoreceptors like those involved in the human visual system is only one step in the development of a true visual sense. Less advanced animals, whose photoreceptors are clustered together in an eye spot, can perceive light but cannot "see." The eye spot, however, can be used to perceive the direction from which light is arriving. True image-forming eyes probably evolved from such comparatively simple structures.

As we discussed in Chapter 23, eyes may have evolved independently numerous times among different groups of animals (Figure 47-14). The members of four phyla—annelids, mollusks, arthropods, and vertebrates—have each evolved well-developed image-forming eyes. Not surprisingly, the eyes found in these various phyla differ in structural design from one another; they certainly evolved separately. Interestingly, all of them use the same visual pigment, *cis*-retinal, suggesting that not many alternative pigments are able to play this role. In photosynthesis, a pigment needs only to be able to absorb light and to donate an electron. In vision, on the other hand, the pigment must undergo a conformational change large enough to alter the shape of the protein (opsin) associated with it. The conformational change that *cis*-retinal undergoes when it absorbs a proton is unusually large.

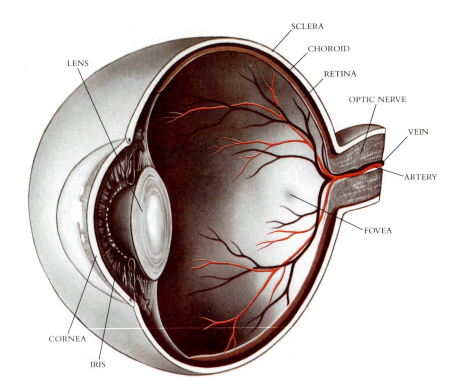

SCLERA
CHOROID
RETINA
OPTIC NERVE
VEIN
ARTERY
FOVEA
LENS
CORNEA
IRIS

FIGURE 47-15

Structure of the human eye. Light passes through the transparent cornea and is focused by the lens on the rear surface of the eye, the retina, at a particular location called the fovea. The retina and particularly the fovea are rich in rods and cones.

Mechanical Design of the Vertebrate Eye

Only two basic designs have evolved for image-forming eyes. In a **multifaceted eye,** the receptors are arrayed on the outside of the eye, each receptor facet at a slight angle to another so that their ends form a hemisphere (see p. 778). By means of such an eye, information is obtained about the entire horizon. In a **lens-focused** eye, which works like a camera, images from different distances are focused to different positions on an inverted hemisphere of receptor cells within the interior of the eye.

Vertebrate eyes are lens-focused eyes (Figure 47-15). In them, light first passes through a transparent layer, the **cornea,** which begins to focus the light onto the rear of the eye. Light then passes through the **lens,** a structure that completes the focusing. The lens is a fat disk filled with transparent jelly, somewhat resembling a flattened balloon. In mammals, the lens is attached by suspending ligaments to **ciliary muscles.** When these muscles contract, they change the shape of the lens (Figure 47-16) and thus the point of focus on the

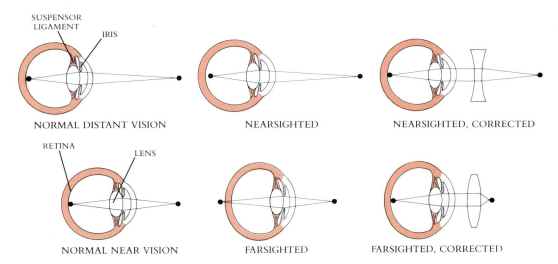

SUSPENSOR LIGAMENT
IRIS

NORMAL DISTANT VISION

NEARSIGHTED

NEARSIGHTED, CORRECTED

RETINA
LENS

NORMAL NEAR VISION

FARSIGHTED

FARSIGHTED, CORRECTED

FIGURE 47-16

Focusing the human eye: contraction of ciliary muscles pulls on suspensor ligaments and changes the shape of the lens, which alters its point of focus forward or backward. In many people the ciliary muscles place the point of focus in front of the fovea rather than on it. Such people are said to be nearsighted. The problem can be corrected with glasses or contact lenses, which extend the focal point back to where it should be. In others, the ciliary muscles make the opposite error, placing the point of focus behind the retina; such farsighted individuals can correct their vision with lenses that shorten the focal point.

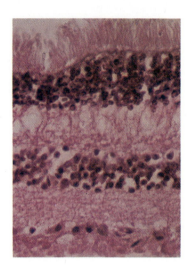

FIGURE 47-17

Structure of the retina. Note that the rods and cones are at the rear of the retina, not the front. Light passes through several layers of ganglion and bipolar cells before it reaches the rods and cones.

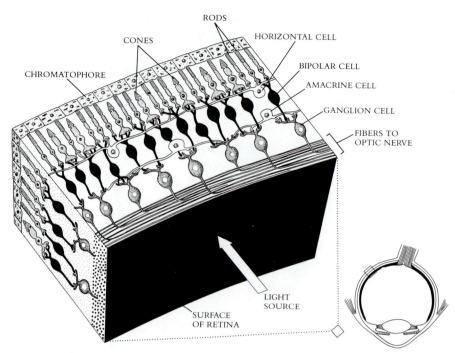

rear of the eye. In amphibians and fishes, the lens does not change shape. These animals instead focus their images by moving the lens in and out, thus operating in exactly the same way that a camera does. In all vertebrates, the amount of light entering the eye is controlled by a shutter, called the **iris,** between the cornea and the lens. The iris reduces the size of the transparent zone, or **pupil,** of the eye through which the light passes.

The field of receptor cells—rods and cones—that lines the back of the eye is the **retina** (Figure 47-17). The retina contains about 3 million cones, most of them located in the central region of the retina called the **fovea,** and approximately 1 billion rods. A simple camera lens has many inherent limitations, such as the chromatic aberration mentioned above. Because of these limitations, the eye forms a sharp image only in the central fovea region of the retina, a region composed almost entirely of cone cells.

Each foveal cone cell makes a one-to-one connection with a special kind of neuron called a **bipolar cell.** Each of the bipolar cells is connected in turn to an individual visual ganglion cell, whose axon is part of the **optic nerve.** The bipolar cells receive the hyperpolarization stimulus from the cone cell and transmit a depolarization stimulus to a nearby ganglion neuron. The optic nerve transmits visual impulses directly to the brain. The frequency of pulses transmitted by any one receptor provides information about light intensity; the pattern of firing among the different foveal axons provides a point-to-point image; and the different cone cells provide information about the color of the image.

The relationship of receptors to bipolar cells to ganglion cells is 1:1:1 within the fovea. Outside of the fovea, however, the output of many receptor cells is channeled to single bipolar cells, many of which converge in turn on individual ganglion cells. In peripheral (outer edge) portions of the retina, more than 125 receptor cells feed stimuli to each ganglion axon in the optic nerve. In this peripheral region, many additional neurons crossconnect the ganglion cells with one another and carry out extensive processing of visual information. This portion of the retina does not transmit a point-to-point image, as does the fovea, but transmits instead a processed version of the visual input. It is with this portion of the eye, for example, that we detect movement and boundaries. It has been said that we use the periphery of the eye as a detector and the fovea as an inspector.

Only the central portion of the eye, the fovea, transmits a point-to-point image. The periphery of the retina transmits a processed version of the image in the form of information about boundaries and movement.

Most vertebrates have two eyes, one located on each side of the head. When both of them are trained on the same object, the image that each sees is slightly different because each eye views the object from a different angle. Because we view objects simultaneously with our two eyes, we do not perceive a "blind spot" in the back of either eye, representing the point at which the optic ganglia axons pass out of the eye; one eye supplies the image for that area that the other eye misses. In the camera-lens eyes of annelids and mollusks, there is no analogous break in the retinal field, no blind spot—and no bilateral arrangement of paired eyes.

The bilateral arrangement of eyes, with its slight displacement of images—an effect called **parallax**—also permits sensitive depth perception. By comparing the differences between the images provided by the two eyes with the physical distance to particular objects, vertebrates learn to interpret different degrees of disparity between the two images as representing different distances—this is called stereoscopic vision. We are not born with the ability to perceive distance; we learn it. Stereoscopic vision develops in babies only over a period of months.

Other Environmental Senses in Vertebrates

Vision is the primary sense used by all vertebrates that live in a light-filled environment, but the wavelength and intensity of visible light are by no means the only stimuli that are available to vertebrates for assessing this environment. We will now consider some examples of other ways in which vertebrates function in this area of sensory perception.

Polarized light

In addition to the ability of vertebrate eyes to detect the wavelength and intensity of light, eyes with suitable filters can also detect when light waves are all vibrating in the same plane—**polarized** light. Because sunlight becomes polarized as it moves through the air, an animal can tell where the sun is by being able to detect how light is polarized, an important aid in navigation. Most insect eyes detect polarized light (see p. 778), as do the eyes of the octopus, a mollusk. There is some evidence that fishes are sensitive to polarized light as well. Human beings, however, are not.

Heat

Electromagnetic radiation with wavelengths longer than those of visible light is not detected by visual receptors because such radiation has low quantum energy. Radiation from this **far red** portion of the spectrum is what we normally think of as radiant heat. Heat is an extremely poor environmental stimulus in water because the water has a high thermal capacity and readily absorbs heat, causing any thermal stimulus that is introduced into a body of water to decay very rapidly. Air, in contrast to water, has a low thermal capacity, and heat in air is therefore a potentially useful environmental stimulus.

The potential of using heat as a source of information about the environment, however, has not been tapped by most groups of terrestrial vertebrates. Snakes are an exception. The pit vipers, for example, possess a pair of heat-detecting **pit organs**, one located on each side of the head between the eye and the nostril (Figure 47-18). A snake that has such organs

FIGURE 47-18

The depression between the nostril and the eye of this timber rattlesnake opens into the pit organ. In the cutaway portion of the diagram you can see that the organ is composed of two chambers separated by a membrane. The organ is thought to function by constantly comparing the temperatures in the two chambers.

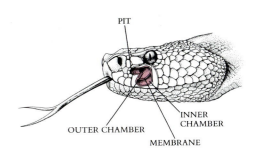

PIT

INNER CHAMBER

OUTER CHAMBER

MEMBRANE

can locate and strike a motionless warm animal in total darkness by perceiving the thermal radiation emanating from the body of the prey. A pit organ has an outer and an inner chamber, separated by a membrane. The organ apparently operates by comparing the temperatures of the two chambers. The nature of the pit organ's thermal receptor is not known; it probably consists of temperature-sensitive receptors in the neurons innervating the two chambers. The two pit organs appear to provide stereoscopic information, in much the same way that two eyes do. Indeed, the information that is transmitted from the pit organs is processed by the visual center of the snake brain.

Electricity

Air does not readily conduct an electric current, and electricity is thus a poor environmental stimulus for terrestrial animals. Water, on the other hand, conducts electricity readily. A number of different groups of fishes apparently use weak electrical discharges to locate their prey, doing so by constructing a three-dimensional image of their environment even in murky water. Although electricity is inferior to light as a stimulus—it has a limited range—it can provide these fishes with very detailed information.

The electric discharges in these fishes are produced by special organs that consist of modified muscle. We can get some idea of how these organs work by studying an unusual adaptation: in certain fishes, such as electric eels and some rays, electric discharges are used to create a shock. In these animals, each electrical organ is composed of numerous columns of disk-shaped cells stacked one on top of the other. Such a column may contain from 5000 to 10,000 of these **electroplates,** and there may be more than 50 columns. One surface of each disk has nerve endings; the other does not. At rest, the individual disks carry a positive charge on both their upper and lower surfaces, actively transporting positive ions out from the cell. The electrical discharge is initiated by the neurons, which depolarize the surface with which they synapse. As a result of this discharge, there is a transient voltage difference across each disk amounting to about 150 millivolts. Because the electroplates are arranged in series, the amount of voltage adds up. An electric eel can produce 500 volts in a single discharge!

Most electric fishes use far weaker discharges to survey their surroundings. The electric catfish (Figure 47-19), for example, discharges electricity continuously—some 300 discharges per second—producing an electric current in the water around it. The presence of any object with higher or lower conductivity distorts the field by changing the lines of flow of the current. The receptors that detect the changes in current, the **ampullae of Lorenzini,** are very sensitive, being able to respond to extremely weak currents. These receptors are located in chambers that lie beneath the surface of the skin, which are connected to the surrounding water by long canals.

Magnetism

Eels, sharks, and many birds appear to navigate along the magnetic fields of the earth. Even some bacteria use such forces to orient themselves (see the boxed essay on pp. 516-517). Birds that are kept in blind cages, with no visual cues to guide them, will peck and attempt to move in the direction in which they would normally migrate. They will not do so, however, if the cage is shielded from magnetic fields by steel. Indeed, if the magnetic field of a blind cage is deflected 120 degrees clockwise by an artificial magnet, a bird that normally orients to the north will orient east-southeast. There have been many speculations about the nature of the magnetic receptor in these vertebrates, but it is still very poorly understood.

AN OVERVIEW OF SENSORY SYSTEMS

The sensory nervous systems of vertebrates inform the brain about the condition of the body and also provide it with a detailed picture of its surroundings. As we have seen, sensory systems have evolved that utilize a broad variety of cues. Most of the sensory systems of vertebrates evolved early, while vertebrates were still confined to the sea. Some, such

FIGURE 47-19

The electric catfish, Malapterus electricus, *surveys its surroundings by producing a continuous electric current in the water around it and monitoring the lines of current flow in the water. It can detect objects that disturb the lines of current flow, and so "see" in very murky water.*

as sensors that detect electric fields, are not used by land vertebrates, presumably because air is a very poor conductor of electric currents. Others, such as sensors that respond to liquid pressure waves on the lateral line, are used by land vertebrates to detect pressure waves in air—that is, to hear. Because the terrestrial environment is so physically different from the sea, it has presented some new sensory opportunities. Air, for example, transmits heat much better than water does, and some terrestrial vertebrates have evolved heat sensors that provide quite detailed three-dimensional images.

No one vertebrate has a perfect repertoire of sensory systems, able to finely discriminate all potential cues. Rather, vertebrates have evolved sensory nervous systems that meet particular evolutionary challenges. Thus some hunting mammals such as dogs have evolved highly capable olfactory systems, but primates have not. It is of some interest to note, for example, the existence of detailed sensory information in our own environment that we could potentially utilize if we possessed the proper sensory receptors. We cannot utilize thermal radiation, as snakes do, or ultrasound, as bats do, to construct a three-dimensional image. Only by using a computer to convert these sensory cues to differences in light emission are we able to utilize them to construct a visual analog.

SUMMARY

1. Sensory receptors respond to three classes of stimuli: mechanical, chemical, and electromagnetic. No matter what type of stimulus is involved, a response usually takes the same form—the depolarization of a sensory neuron membrane.

2. Many of the body's internal receptors are simple receptors in which a nerve ending becomes depolarized as a response to a chemically induced or temperature-induced opening of ion channels in the nerve membrane.

3. Muscle contraction, blood pressure, and touch are sensed by simple mechanical receptors, in which deformation of a nerve membrane by stretching or distortion opens ion channels and thus initiates depolarization.

4. Gravity, angular acceleration, and lateral-line hearing are sensed by complex mechanical receptors, which use a mechanical device (a stone or a lever) to convert the sensory information into a mechanical stimulus.

5. The receptors that vertebrates use to sense their external environments all evolved in the sea, with the exception of the heat vision that pit vipers possess.

6. The most sensitive chemical receptors are the taste and smell receptors of fishes, which can discriminate the presence of a few molecules of specific substances. Human taste and smell receptors are much less acute.

7. Hearing in terrestrial vertebrates uses an evolutionary modification of the fish lateral-line receptor, in which the airborne sound waves are amplified and directed at a water-containing chamber within the ear. Mechanical receptors in the chamber are then deformed by the waterborne sound waves.

8. Vision, like photosynthesis, uses a pigment as a primary photoreceptor. The reception of a single photon is greatly amplified by a cascade arrangement of sequential reactions that results in a spreading chain reaction of effects.

9. The vertebrate eye is designed like a lens-focused camera. The fovea at the center of the retina transmits a point-to-point image to the brain; the edges of the retina transmit information on movement and boundaries rather than a full picture.

10. Other stimuli sensed by vertebrate receptors include polarized light, heat, electricity, and magnetism.

SELF-QUIZ

The following questions pinpoint important information in this chapter. You should be sure to know the answers to each of them.

1. The receptors that produce the response to pain are known as _____.

2. Receptors responsible for the detection of vibration are
   (a) Meissner's corpuscles
   (b) Merkel corpuscles
   (c) carotid corpuscles
   (d) Pacinian corpuscles
   (e) baro corpuscles

3. Gravity receptors are known as _____.

4. Humans detect angular acceleration in their semicircular canals. Are these canals filled with air or fluid?

5. A change in temperature induces a response in a temperature-change receptor neuron by
   (a) conformational changes in the neuron's sodium channels
   (b) physical deformation of the nerve ending
   (c) stretching or distortion of the neuron's membrane
   (d) diffusion into the neuron of hydrogen ions
   (e) increasing the activity of enzymes in the neuron

6. The sensory hair cells of the ear are located on the _____ membrane.

7. The sensing of reflected sound waves as a means of depth perception is known as _____.

8. The human eye can detect a single photon of light. True or false?

9. In the vertebrate eye the light first passes through the
   (a) cornea
   (b) iris
   (c) pupil
   (d) retina
   (e) fovea

10. The central region of the retina that is composed mainly of cone cells is called the _____.

11. The "shutter" of the eye is the
    (a) retina
    (b) pupil
    (c) iris
    (d) fovea
    (e) cornea

12. Certain snakes can perceive thermal radiation with special organs known as _____ organs.

THOUGHT QUESTIONS

1. Most of us have sensed at one time or another an oncoming storm by detecting the increase in humidity in the air. What sort of receptors detect humidity? Why do you suppose that hot days seem so much hotter when it is humid?

2. The heat-detecting pit receptor of snakes is a very effective means of "seeing" at night. It is the same sort of sensory system employed by soldiers in "snooperscopes" and in heat-seeking missiles. Why do you suppose other night-active vertebrates, such as bats, have not evolved this sort of sensory system?

3. In zero gravity, how would you expect a statocyst to behave? What would you expect the subjective impression of motion to be? Would the semicircular canals detect angular acceleration equally well at zero gravity?

FOR FURTHER READING

BLACKMORE, C.: *The Mechanics of the Mind,* Oxford University Press, New York, 1977. A very good explanation of how the mind integrates sensory information.

BURTON, M.: *The Sixth Sense of Animals,* Taplinger Publishing Co., New York, 1972. A semipopular account of animal senses, taking a comparative approach.

HUDSPETH, A.J.: "The Hair Cells of the Inner Ear," *Scientific American,* January 1983, pages 54-64. An explanation of the mechanism of hearing.

LYTHGOE, J.N.: *The Ecology of Vision,* Oxford University Press, Oxford, England, 1979. A comprehensive comparative treatment of animal vision.

MILNE, L., and M. MILNE: *The Senses of Animals,* Atheneum Publishing Co., New York, 1972. A general and entertaining introduction to the animal senses; recommended.

SCHMIDT-NIELSEN, K.: *Animal Physiology: Adaptation and Environment,* 2nd ed., Cambridge University Press, New York, 1984. A brilliant comparative treatment of vertebrate physiology, with an excellent section on sensory systems.

Chapter 48

INTEGRATION AND HOMEOSTASIS

Overview

In vertebrates, the central nervous system coordinates and regulates the diverse activities of the body. It does this by using a network of motor neurons to direct the voluntary muscles, a second network of motor neurons to control cardiac and smooth muscles, and chemical commands to effect long-term changes in physiological activities. These controls maintain physiological conditions within narrow bounds, and most of them function similarly in all vertebrates. However, the control of some functions, such as water balance, has changed considerably as vertebrates have adapted to different environments. Both in water and on land, vertebrates control water balance by means of kidneys, which use blood pressure to drive blood plasma through filters that retain blood cells but pass chemical wastes; the water and salts of the plasma are then largely reabsorbed, and the wastes are excreted.

For Review

Here are some important terms and concepts that you will encounter in this chapter. If you are not familiar with them, you should review them before proceeding.

- **Mechanism of muscle contraction** (Chapters 7 and 42)

- **FSH and LH hormones** (Chapter 40)

- **Countercurrent exchange** (Chapter 44)

- **Control of blood glucose levels by the liver** (Chapter 45)

- **Nitrogenous wastes** (Chapter 45)

The tissues and organs of an adult mammal orchestrate a symphony of activities, all of them regulated so as to avoid conflict and maximize interaction. This is why we think of ourselves as an *organism,* rather than as a smoothly functioning collection of organs.

Integration of the vertebrate body's many activities is the primary function of the central nervous system. All control of the bodily functions, voluntary and involuntary, is vested in this system. It directs voluntary and involuntary functions in different ways. **Voluntary** functions are movements of skeletal (striated) muscle that are directed by somatic motor pathways from the brain and spinal cord. These are the pathways that coordinate your fingers when you grasp a pencil, spin your body when you dance, and put one foot

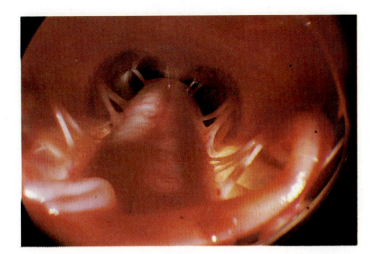

FIGURE 48-1

A view down the human spinal cord. Pairs of spinal nerves can be seen extending out from it. It is via these nerves that the central nervous system communicates with the body.

ahead of the other when you walk. The direction of these muscular movements by the central nervous system is in large measure subject to conscious control by the associative cortex.

Many of the functions directed by the central nervous system are not subject to conscious control, however; these are referred to as **involuntary** or **autonomic** (Greek, *auto,* self + *nomos,* law—self-controlling). The motor pathways that carry commands from the central nervous system to regulate the glands and nonskeletal muscles of the body are collectively called the **autonomic nervous system.** The autonomic nervous system takes your temperature, monitors your blood pressure, and sees to it that your food is properly digested. The body's internal physiological condition is regulated by the autonomic nervous system. Most physiological conditions are maintained within relatively narrow bounds, the condition of homeostasis discussed in Chapter 46.

Both the voluntary and the involuntary nervous systems are motor pathways. In both cases, sensory input from receptors such as those described in the previous chapter provides the central nervous system with information. The central nervous system processes this information and reaches decisions about the appropriate responses. The central nervous system then issues a series of commands to bring these responses about. These commands travel to the body's muscles and organs (Figure 48-1) along the motor pathways of the voluntary and involuntary nervous systems.

HOW THE CENTRAL NERVOUS SYSTEM REGULATES BODY ACTIVITIES

The central nervous system harmoniously controls the body's many functions by employing two very effective management procedures:

1. **Feedback loops.** The organs that receive central nervous system commands report back to the central nervous system on the degree to which these commands have evoked a response. In this way the central nervous system can decide when to cease issuing a command and when to become more insistent. Like a boss with a hidden microphone in every office, the body's management always knows exactly how well its orders are being carried out.

2. **Antagonistic controls.** All of the principal bodily functions are controlled by both positive and negative commands that have opposite effects. The commands are said to be **antagonistic** because they are mutually opposed to one another. The control center never turns a control channel on or off; instead, it modulates the relative strengths of two opposing channels. For this reason, it is possible to eliminate great swings in the function of the target organ, permitting instead a finely graded response.

FIGURE 48-2

The central nervous system issues commands via three different systems: (1) The voluntary nervous system is a network of often-myelinated nerves that extend to skeletal muscle. These are the nerves that carry the commands to arm muscles when you lift your arm. (2) The autonomic nervous system is a network of nerves that extend to cardiac and smooth muscles. These are the nerves that carry the commands to the smooth muscles encasing your intestines, causing these muscles to undergo regular peristaltic contractions and moving food through your digestive tract. (3) The neuroendocrine system is a network of endocrine glands whose hormone production is controlled by commands from the central nervous system. These are the hormones that regulate sexual cycles.

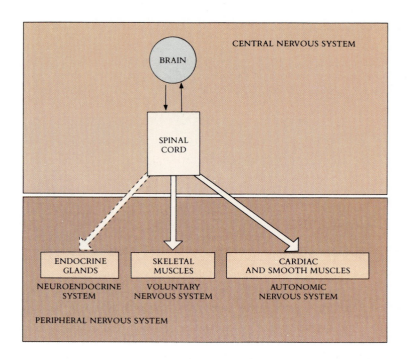

Both of these management strategies are applied to almost all regulatory activities of the central nervous system. The only exceptions are certain bodily functions such as digestion that have their own built-in feedback mechanisms.

The central nervous system employs three separate motor command systems (Figure 48-2). These systems differ in their speed of expression, duration of response, and narrowness of application. They are (1) the **neuromuscular control** of striated muscles by the motor neurons of the voluntary nervous system, (2) the **neurovisceral control** of cardiac and smooth muscles by the motor neurons of the autonomic nervous system, and (3) the **neuroendocrine control** of hormone-producing glands by the hypothalamus.

In this chapter we will first consider how these three command systems function. We will then examine in detail how they act together to maintain constant internal physiological conditions. We will focus on one organ—the kidney—as an example, and consider how this organ meets the challenge of maintaining a constant water and salt balance in the body's bloodstream.

NEUROMUSCULAR CONTROL

The muscles of the body are directed by the voluntary nervous system, a network of motor nerves connecting the central nervous system with the individual striated muscles. An electrical impulse travels from the cell body of a motor neuron within the spinal cord out along an axon that extends all the way to the muscle; the point at which the axon reaches the muscle is called an end plate. The arriving impulse transmits acetylcholine across the end-plate synapse, initiating the depolarization of the muscle membrane, or sarcolemma. This depolarization stimulates the release of calcium ions from the sarcoplasmic reticulum. As described in Chapter 42, these ions attach to troponin, shifting the tropomyosin on the actin filaments to expose the myosin binding sites and thus initiate muscle contraction.

Feedback Loops

The voluntary nervous system functions by means of both feedback loops and antagonistic controls. Feedback loops are established by stretch receptors within the muscles. When a muscle extends or contracts, the receptors are depolarized, initiating a nerve impulse that travels along afferent nerve fibers to a cell body in a ganglion outside the spinal cord. From that cell body the impulse passes into the spinal cord, where the nerve cell synapses. Two sorts of synapses occur, corresponding to two different kinds of feedback loop: mono-synaptic and interneuron-mediated.

Monosynaptic reflex arcs

The simplest feedback loop is a **monosynaptic reflex arc,** in which the afferent nerve cell synapses with a motor neuron in the spinal cord whose axon travels back to the muscle. Note that the nerve impulse from the stretch receptor of the muscle passes over only one synapse, directly stimulating the motor nerve that causes the muscle to contract. A mono-synaptic reflex arc thus is not subject to control by the central nervous system. All the vol-untary muscles of your body possess such monosynaptic reflex arcs, although usually in conjunction with other more complex feedback arcs involving interneurons. It is through these more complex paths, to be discussed below, that voluntary control is established.

In the few cases in which the monosynaptic reflex arc is the only feedback loop pres-ent, its function can be clearly seen. The well-known "knee jerk" is one such reflex. If the patellar ligament just below the kneecap is struck lightly by the edge of your hand or by a doctor's rubber hammer, the resulting sudden pull stretches the muscles of the upper leg, which are attached to the ligament. Stretch receptors in these muscles immediately send an impulse along afferent nerve fibers to the spinal cord, where these fibers synapse directly with motor neurons that extend back to upper leg muscles, stimulating them to contract and the leg to jerk upward. Such reflexes play an important role in maintaining posture.

Interneuron-mediated reflex arcs

Most muscles also possess a more complicated feedback loop, in which the afferent nerve cell synapses in the spinal cord with an interneuron (Figure 48-3). This interneuron in turn synapses with an efferent motor neuron, whose axon extends back to the muscle that was contracting. Depending on the nature of the interneuron and on the other synapses that influence the interneuron, the arriving impulse may stimulate the motor neuron. If it does so, it will be channeled back to the contracting muscle, continuing the contraction. If, how-ever, the arriving impulse inhibits the interneuron, the interneuron will decrease the fre-quency of signals passing out over the motor neuron, causing the muscle to relax. In this way, the central nervous system modulates feedback information to control muscular con-traction.

FIGURE 48-3

An interneuron-mediated reflex arc.

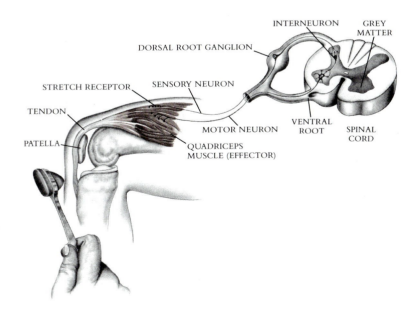

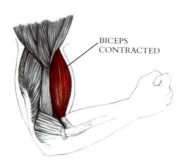

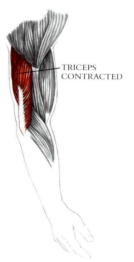

FIGURE 48-4

To raise your arm, one set of muscles, the flexors, contracts and another set, the extensors, relaxes; to lower your arm, the extensors contract and the flexors relax.

Antagonistic Control of Skeletal Muscles

The movements of limbs and organs are sensitively controlled through the use of two opposing sets of muscles to move every limb and organ (Figure 48-4). One set of muscles is called **flexors,** the other **extensors.** Muscles, you will recall, can only pull (when they contract); they cannot push. To move a limb in one direction, one set of muscles pulls it by contracting while the other set of muscles relaxes. To reverse the direction, the first set of muscles relaxes while the second contracts, pulling the limb back. All voluntary motor movements of the vertebrate body, and all of its reflex movements, follow this model.

Feedback information is used to coordinate the flexor-extensor muscle pair. The stretching of one muscle, in response to an increase in load on it, stimulates stretch receptors, sending afferent signals to the spinal cord. In the spinal cord, an interneuron inhibits the motor neuron that extends to the antagonistic (opposing) muscle, causing it to relax, while it excites the motor neuron that travels to the contracting muscle, causing it to contract. If the stretching continues, more motor neurons leading to the muscle are excited, increasing contraction until the load is balanced. In this way, the muscle is maintained at a constant length despite variations in the load it bears. This is why we are able to hold a heavy weight in our hands.

> **The voluntary muscles of the vertebrate body are organized in antagonistic pairs of flexors and extensors. Stretch receptors within them report to the central nervous system on their state of contraction.**

Some weights are too heavy to lift for long, as any aspiring weight lifter can tell you. If a contracted muscle is stretched very forcefully, members of another set of sensory receptors called **tendon organs,** embedded in the protein fibers (tendons) that link muscle to bone, are deformed. The tendon organs initiate impulses that travel to interneurons within the spinal cord that inhibit the motor neurons driving contraction of the muscle. The muscle relaxes and ceases to resist the load. This reflex protects the muscle from rupture by excessive loads.

NEUROVISCERAL CONTROL

The glands, smooth muscle, and cardiac muscle of the body (the "viscera") respond to another network of motor neurons, the autonomic nervous system. This network is the other part of the central nervous system's motor command network. It plays a major role in fine-tuning the body's internal environment. Like the voluntary nervous system, the autonomic nervous system achieves its control by both feedback loops and antagonistic controls.

The Autonomic Nervous System

The nature of feedback loops in the autonomic nervous system varies for different target tissues. The feedback loop from smooth muscle is the same as in the neuromuscular control of striated muscle, consisting of afferent neurons from stretch receptors embedded within the muscle. The feedback loop in the glands and heart is frequently chemical. We will discuss these chemical feedback loops later in this chapter.

Antagonistic control of the autonomic nervous system is achieved by two complete antagonistic nervous systems, each acting to oppose the other (Figure 48-5):

1. The **parasympathetic nervous system** consists of a network of long efferent central nervous system axons that synapse with organ-associated ganglia in the immediate vicinity of an organ, and of short efferent neurons extending from the ganglia to the organ.

2. The **sympathetic nervous system** is composed of a net of short efferent central nervous system axons that extend to ganglia located near the spine, and of long efferent neurons extending from the ganglia directly to each target organ.

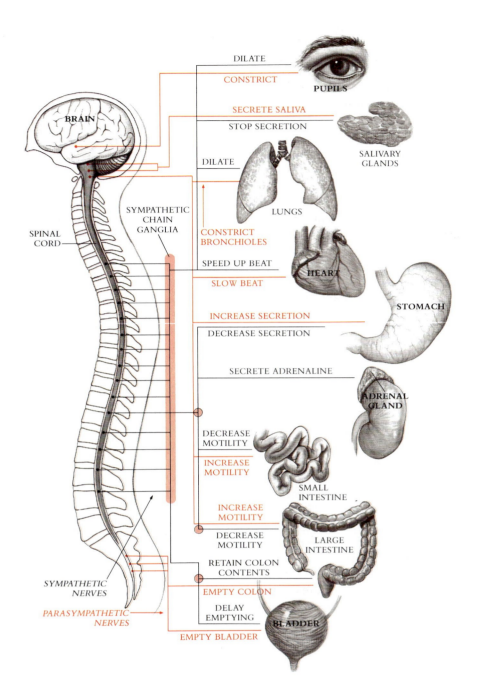

DILATE

CONSTRICT

PUPILS

SECRETE SALIVA

STOP SECRETION

SALIVARY
GLANDS

BRAIN

DILATE

SYMPATHETIC
CHAIN
GANGLIA

LUNGS

SPINAL
CORD

CONSTRICT
BRONCHIOLES

SPEED UP BEAT

SLOW BEAT

HEART

INCREASE SECRETION

STOMACH

DECREASE SECRETION

SECRETE ADRENALINE

**ADRENAL
GLAND**

DECREASE
MOTILITY

INCREASE
MOTILITY

SMALL
INTESTINE

INCREASE
MOTILITY

DECREASE
MOTILITY

LARGE
INTESTINE

RETAIN COLON
CONTENTS

*SYMPATHETIC
NERVES*

EMPTY COLON

DELAY
EMPTYING

BLADDER

*PARASYMPATHETIC
NERVES*

EMPTY BLADDER

FIGURE 48-5

*The sympathetic and
parasympathetic nervous systems.
The ganglia of sympathetic nerves
are located near the spine, and the
ganglia of parasympathetic nerves
are located far from the spine, near
the organs they innervate.*

Antagonistic Control of the Autonomic Nervous System

An autonomic nerve signal crosses two synapses in traveling from the central nervous system out to its target organ, whether it travels out along sympathetic or parasympathetic nerves. The first synapse is in the ganglion, between the axon of a neuron extending from the central nervous system and the dendrites of the autonomic neuron's cell body; the second synapse is between the autonomic neuron's axon and the target organ. The neurotransmitter within the ganglion is acetylcholine for both sympathetic and parasympathetic nerves. However, the neurotransmitter between the terminal autonomic neuron axon and the target organ is different in the two antagonistic autonomic nerve systems. In the parasympathetic system, the neurotransmitter at the terminal synapse is acetylcholine, just as it is in the ganglion. In the sympathetic system, the neurotransmitter at the terminal synapse is either adrenaline or noradrenaline, both of which have an effect *opposite* to that of acetylcholine. Thus, depending on which of the two nerve paths is selected by the central nervous system, an arriving signal will either stimulate or inhibit the organ.

Each gland, smooth muscle, and cardiac muscle constantly receives stimulatory signals via one nerve and inhibitory signals via the other nerve. The central nervous system controls activity in each case by varying the ratio of the two signals.

The glands and involuntary muscles of the body are innervated by two antagonistic sets of efferent nerves.

Thus an organ receiving nerves from both visceral nervous systems will be subjected to the effects of two opposing neurotransmitters. If the sympathetic nerve ending excites a particular organ, the parasympathetic synapse usually inhibits it. For example, the sympathetic system speeds up the heart and slows down digestion, whereas the parasympathetic system slows down the heart and speeds up digestion. In general, the two opposing systems are organized so that the parasympathetic system stimulates the activity of normal body functions—for example, the churning of the stomach, the contractions of the intestine, and the secretions of the salivary glands. The sympathetic system, on the other hand, generally mobilizes the body for greater activity, as in increased respiration or a faster heartbeat.

Heartbeat: How Neurovisceral Control Works

To maintain reasonably constant internal conditions, the vertebrate body employs its visceral nervous system to modulate the activities of a wide variety of organs. The way in which the body controls the rate at which the heart beats, and thus the rate of flow through the circulatory system, serves as an example of how these neurovisceral controls work. As you will recall from Chapter 45, each heartbeat is initiated by a spontaneous discharge of excitatory tissue within the pacemaker, the SA node. The rate at which the heart beats is simply the rate at which these discharges occur. Although these discharges will occur spontaneously, their occurrence may be facilitated or inhibited by autonomic nerve fibers that synapse at the SA node. The sympathetic fibers excite the SA node; the parasympathetic fibers inhibit it. The balance of their opposing effects determines the rate at which the heart beats.

When you exercise, your right atrium becomes stretched by an unusual amount of blood. At such times, stretch receptors within this chamber initiate impulses that travel along afferent nerves to a complex of interneurons called a **coordinating center** located in the brainstem or medulla oblongata. This coordinating center responds by increasing the rate at which impulses are transmitted back to the SA node via the sympathetic nerve fibers. This increases the excitatory contribution of the sympathetic synapses relative to the inhibitory contribution of the parasympathetic synapses, and so increases the heart rate.

Similarly, an increased level of carbon dioxide in the blood is detected by receptors in the neck arteries. These receptors initiate impulses that travel to the coordinating center and also cause it to transmit more stimulatory impulses along sympathetic fibers to the SA node, increasing the heart rate.

As heart rate quickens, blood pressure increases. Pressure receptors in the aorta detect this increased hydrostatic pressure and send impulses to a second coordinating center in the medulla, which in this case responds by increasing the rate of transmission of impulses back to the SA node via parasympathetic nerves. The increase in these impulses inhibits the SA node, thus acting to slow down the heart rate.

At any one moment the SA node is receiving impulses from both sympathetic and parasympathetic nerves. It is the balance between these two inputs that determines the rate at which the heart beats.

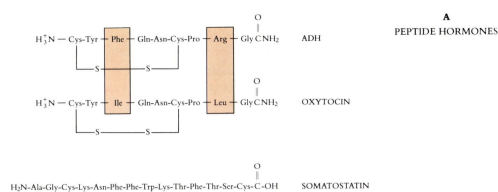

A
PEPTIDE HORMONES

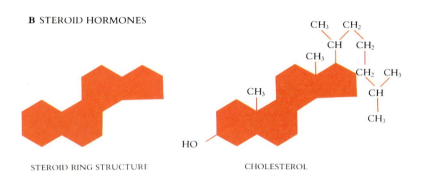

B STEROID HORMONES

STEROID RING STRUCTURE   CHOLESTEROL

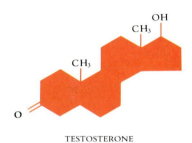

TESTOSTERONE

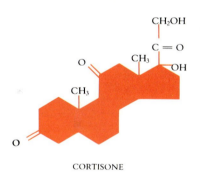

CORTISONE

FIGURE 48-6

Chemical structure of some important hormones.
A *Peptide hormones. Antidiuretic hormone (ADH) regulates water loss by the kidneys; oxytocin, which is very similar in structure to ADH, acts on the mammary glands to stimulate milk ejection. Somatostatin inhibits the secretion of growth hormone.*
B *Steroid hormones. Many steroid hormones are similar in structure to the blood lipid cholesterol. Testosterone stimulates development of the genital tract; cortisone promotes the breakdown of muscle proteins in metabolism. See Figure 48-6 for the structure of estrogen and progesterone.*

NEUROENDOCRINE CONTROL

The effective regulation of many of the body's functions requires not only the constant modulation and integration of its internal activities, but also the ability to induce longer-term changes in levels of activity, such as initiating the production of milk by the mammary glands of nursing mothers, or carrying out the sexual maturation that occurs when a young boy or girl goes through puberty. In the vertebrate body, these changes are achieved by chemical signals rather than by nervous ones. These signals are hormones that the circulatory system delivers to target organs (Figure 48-6). As we have noted earlier, hormones are regulating chemicals that are made at one place in the body and exert their influence at another. They are not themselves enzymes, but they act by regulating preexisting processes. When the hormones reach their destination, they are recognized by specific receptors that only their target glands or tissues possess. Because most hormones circulating in the bloodstream are produced by endocrine glands whose activity is under the direct control of the nervous system, *these hormones constitute a chemical extension of the visceral nervous system.* They permit the central nervous system to issue long-term commands that may influence the level of activity of numerous organs.

Neuroendocrine control is vested in the hypothalamus. As you will recall from Chapter 46, the hypothalamus is part of the diencephalon, located at the base of the brain just above the pituitary gland, which is also called the **hypophysis.** Within the hypothalamus, interoceptive information about the body's many internal functions and regulatory commands is processed. These commands involve matters such as the regulation of body temperature, the intake of food and water, reproductive behavior, and response to pain and emotion. The commands are issued to the pituitary gland, which in turn sends chemical

FIGURE 48-7

The release of thyroid hormone (TH) from the thyroid gland is the end result of a long series of events mediated by the central nervous system. First, the hypothalamus produces a specific thyroid releasing hormone, TRH, which circulates through special very short blood vessels to the anterior region of the pituitary. Second, the arrival of TRH stimulates the pituitary to produce the pituitary hormone thyroid stimulating hormone, TSH, which it releases into the general blood circulation, which carries it to the thyroid gland. Third, the arrival of TSH stimulates the thyroid gland to produce thyroid hormone, TH, which is released to circulate throughout the body. Levels of both TSH and TH in the bloodstream influence the rate at which the central nervous system produces TRH.

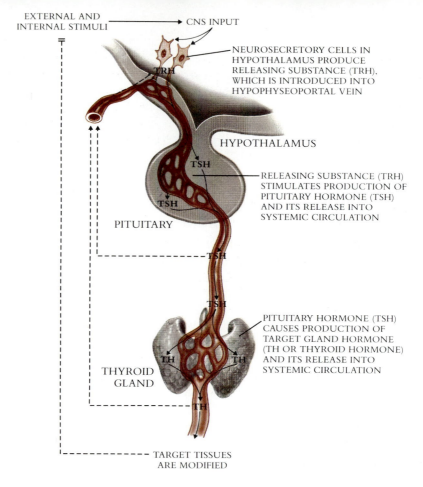

EXTERNAL AND INTERNAL STIMULI → CNS INPUT

NEUROSECRETORY CELLS IN HYPOTHALAMUS PRODUCE RELEASING SUBSTANCE (TRH), WHICH IS INTRODUCED INTO HYPOPHYSEOPORTAL VEIN

HYPOTHALAMUS

RELEASING SUBSTANCE (TRH) STIMULATES PRODUCTION OF PITUITARY HORMONE (TSH) AND ITS RELEASE INTO SYSTEMIC CIRCULATION

PITUITARY

PITUITARY HORMONE (TSH) CAUSES PRODUCTION OF TARGET GLAND HORMONE (TH OR THYROID HORMONE) AND ITS RELEASE INTO SYSTEMIC CIRCULATION

THYROID GLAND

TARGET TISSUES ARE MODIFIED

signals to the various hormone-producing glands of the body. The hypothalamus thus regulates the body through a chain of command, in the same way that a general gives his orders to a chief of staff, who relays them to lower-ranking commanders. Figure 48-7 uses the thyroid gland to illustrate this type of control mechanism.

Hormones Released by the Posterior Pituitary

The pituitary gland really consists of two glands, which are very different from each other. The posterior region of the pituitary, or **neurohypophysis,** is linked directly to the hypothalamus by neural connections and secretes hormones into the general blood circulation. One of these hormones is **antidiuretic hormone** (ADH), also called vasopressin. ADH is involved in regulating the rate of water reabsorption in the kidneys and intestines. Alcohol suppresses ADH release, which is why excessive drinking leads to the production of excessive quantities of urine and to dehydration.

Another hormone secreted by the neurohypophysis is **oxytocin.** Oxytocin aids in initiating milk release by causing the contraction of the muscles around the ducts into which mammary glands secrete milk. Oxytocin also causes the uterus to contract during the labor associated with childbirth. This is why the uterus of a nursing mother returns to normal size after its extension during pregnancy more quickly than does one of a mother who does not nurse her baby.

Both vasopressin and oxytocin are actually formed within the hypothalamus. From there, they are transported within nerve axons to the nerve endings of the neurohypophysis, which releases them into the blood.

Hormones Released by the Anterior Pituitary

The anterior region of the pituitary, called the **adenohypophysis,** is connected to the hypothalamus by special blood vessels only a few millimeters long. Through these vessels pass a group of regulating hormones, which are produced in the hypothalamus. These "releasing hormones" command the adenohypophysis to initiate the production of specific hormones in distant endocrine glands (Figure 48-8). When these glands have manufactured their hormones, they release them into the general circulation.

For each releasing hormone secreted by the hypothalamus, there is a corresponding hormone synthesized by the adenohypophysis. When the pituitary receives a releasing hormone from the hypothalamus, the adenohypophysis responds by secreting the corresponding "pituitary hormone." The seven principal pituitary hormones are:

Thyroid-stimulating hormone, or TSH. This hormone stimulates the thyroid gland to produce thyroxin, which in turn stimulates oxidative respiration and regulates the level of calcium in the blood.

Luteinizing hormone, or LH. Luteinizing hormone plays an important role in the female menstrual cycle, as described in Chapter 40. It also stimulates the male gonads to produce testosterone, which initiates and maintains the development of male secondary sexual characteristics, those external male sexual features not involved in reproduction.

Follicle-stimulating hormone, or FSH. This hormone is also significant in the female menstrual cycle, as described in Chapter 40. In males, it stimulates certain cells in the testes to produce a hormone that regulates the development of sperm.

Adrenocorticotropic hormone, or ACTH. ACTH stimulates the adrenal cortex to produce corticosteroid hormones. Some of these hormones regulate the production of glucose from fat; others regulate the balance of sodium and potassium ions in the blood; and still others contribute to the development of the male secondary sexual characteristics.

Somatotropin, or growth hormone, GH. This hormone stimulates the growth of muscle and bone throughout the body.

Prolactin, or P. Prolactin stimulates the breasts to produce milk, and at high levels reduces fertility.

Melanocyte-stimulating hormone, or MSH. In reptiles and amphibians, MSH stimulates color changes in the epidermis. This hormone has no known function in mammals.

Aside from their endocrine functions, these and many other hormones have been shown to be associated with particular populations of cells within the central nervous system. This is an area of very active research; biologists are attempting to uncover the role of these hormones in the central nervous system. Whatever their function, it is clear that the same hormone may play different roles in different parts of the body. Hormones are signals, used in different tissues for different reasons, just as raising your hand is a signal that has different meanings in different contexts: in class, it indicates you have a question; in a football game it signals a "fair catch"; when a policeman does it on a street it means "stop." The evolving use of hormones by vertebrates has been a conservative process. Rather than produce a new hormone for every use, vertebrates have often adopted for a new use a hormone already at hand, where its new use does not cause confusion.

Control over the production of these hormones is exercised in two ways, which correspond to the two management procedures that the other elements of the nervous system also use.

Antagonistic controls

The production of GH, P, and MSH hormones is controlled by both releasing and inhibitory signals produced by the hypothalamus. It is the brain, rather than the levels of these three hormones in the blood, that determines the rate of their production. Consider the growth-promoting hormone somatotropin (GH) (Figure 48-9): the releasing signal for GH is the growth-hormone–releasing hormone, GHRH, produced by the hypothalamus; GHRH stimulates the adenohypophysis to produce somatotropin. The inhibiting signal,

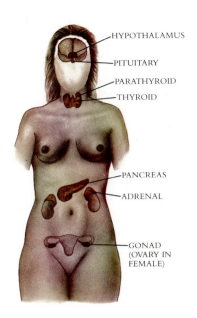

FIGURE 48-8

Locations of some of the major human endocrine glands.

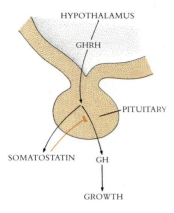

HYPOTHALAMUS

GHRH

PITUITARY

SOMATOSTATIN   GH

GROWTH

FIGURE 48-9

Levels of the growth-promoting hormone GH are regulated by both stimulatory and inhibitory signals. The production of GH by the pituitary is released by a hypothalamic releasing hormone, GHRH, and is inhibited by another pituitary hormone, somatostatin. If levels of GHRH become too high, the excess somatostatin that is induced in the pituitary shuts down some of the production of GH (this effect is indicated by the colored line and double bar), so that effective levels of GH circulating in the blood do not fluctuate.

which is also produced at the same time by the hypothalamus, is somatostatin. Somatostatin inhibits the adenohypophysis from producing somatotropin. The hypothalamus thus regulates growth by mediating the relative rates of production of GH. In a similar way, the hypothalamus regulates the production of P with an inhibiting hormone, prolactin-release–inhibiting hormone (PIH). Isolated in 1985, PIH has been shown to be synthesized as part of the same molecule as the releasing hormone that stimulates LH and FSH production, called gonadotropin-releasing hormone (GnRH). The two hormones PIH and GnRH are made as a single precursor molecule that is later split into the two active components.

Feedback controls

The levels of all other pituitary hormones produced by the adenohypophysis are controlled by negative feedback from the target glands. For example, when LH stimulates the gonads to release testosterone into the bloodstream, that testosterone in turn inhibits the hypothalamus. The hypothalamus then ceases to transmit LH-releasing hormone to the adenohypophysis.

The regulation of pituitary hormone production is achieved by (1) pairs of hypothalamic hormones with opposite effects; and (2) feedback loops, in which the hypothalamus is sensitive to hormone levels in the blood.

Non-Pituitary Hormones

The production of a small number of endrocrine hormones is not controlled by the hypothalamus at all. These hormones are of two sorts:

1. *Antagonistic pairs of hormones concerned with basal metabolism.* Among these hormones, **parathyroid hormone** increases the levels of calcium ion in blood plasma; an antagonistic hormone, **calcitonin,** lowers the concentrations of calcium ion. The final levels of calcium in the blood are determined by the balance between the two systems. The islets of Langerhans in the pancreas (see below) produce another pair of opposing hormones, **insulin** and **glucagon,** which together regulate the level of blood glucose.

2. *Noradrenaline and adrenaline.* These hormones are produced by the **adrenal medulla** (the inner portion of the adrenal gland, located just above the kidneys in the center of the body). They are used in two ways: (1) as neurotransmitters in the sympathetic nervous system, and (2) as hormones released into the bloodstream. Circulating in the bloodstream, noradrenaline and adrenaline produce an "alarm" response throughout the body that is identical to the individual effects achieved by the sympathetic nervous system, but is longer lasting. Among the effects of these hormones is an accelerated heartbeat, increased blood pressure, higher levels of blood sugar, dilated blood vessels, and increased blood flow to the heart muscles and lungs. These adrenal hormones act as extensions of the sympathetic nervous system.

How Neuroendocrine Control Works

The body's many hormones fall into two general chemical categories. The members of each category operate by very different mechanisms:

Peptide hormones. Some hormones, such as adrenaline and thyroxin, are small molecules derived from the amino acid tyrosine; others are short polypeptide chains. Most of the hormones that circulate within the brain belong to this second class. They include small peptides called **enkephalins,** which are only 5 amino acid units long. The enkephalins appear to play a role in integrating afferent impulses from the pain receptors. A third type of peptide hormone is larger. The class of brain hormones called **endorphins** are polypeptides 32 amino acid units long.

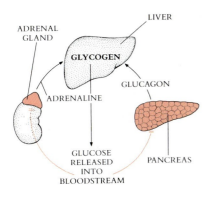

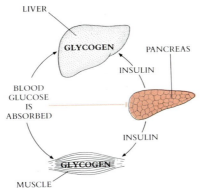

A LOW BLOOD SUGAR

B HIGH BLOOD SUGAR

FIGURE 48-10

Hormonal control of blood glucose levels.
A *When blood sugar levels are low, cells within the pancreas release the hormone glucagon into the bloodstream, and other cells within the adrenal gland, situated on top of the kidneys, release the hormone adrenaline into the bloodstream. These two hormones, when they reach the liver, both act to increase the liver's breakdown of glycogen to glucose.*
B *When blood glucose levels are high, other cells within the pancreas produce the hormone insulin, which stimulates the liver and muscles to convert blood glucose into glycogen. The glucose level in the blood determines the levels of insulin and glycogen in the blood via feedback loops (noted by the colored lines with the double bars in **A** and **B**) to the pancreas and adrenal gland.*

Endorphins appear to regulate emotional responses in the brain. Morphine has such a potent analgesic (pain-relieving) effect on the central nervous system because it mimics the effects of the endorphins. Still other hormones are proteins, which consist of even longer polypeptide chains; insulin is an example. Peptide hormones do not enter their target cells. Instead, they interact with a receptor on the cell surface and thus initiate a chain of events within it.

Steroid hormones. All steroid hormones are derived from cholesterol, a complex molecule composed of three six-membered carbon rings and one five-membered carbon ring—it looks something like a fragment of chicken wire. The hormones that promote the development of the secondary sexual characteristics are steroids. They include cortisone and testosterone, as well as the hormones estrogen and progesterone used in birth control pills. Steroid hormones act by altering the pattern of gene expression. They enter a target cell and penetrate its nucleus, where they initiate the transcription of some genes while repressing the transcription of others.

Control systems

As an example of how a neuroendrocrine control system works, consider how the body regulates the level of glucose in the blood (Figures 48-10 and 48-11). When blood glucose levels fall below normal, a small group of cells in the pancreas, the **islets of Langerhans** (Figure 48-12), respond by releasing the hormone glucagon into the bloodstream. Simultaneously, the cells of the adrenal medulla also detect the lowered blood glucose and secrete adrenaline into the bloodstream. Both of these hormones act within the liver to increase the rate of conversion of glycogen to glucose and thereby raise the level of glucose in the blood.

When blood glucose levels become excessive, other cells within the islets of Langerhans produce a different hormone, insulin. Insulin mediates the uptake of glucose by cells from the blood and stimulates the formation of glycogen from blood glucose, causing the level of glucose in the blood to fall. The amounts of insulin and glucagon that are pro-

FIGURE 48-11

The mechanism of stimulation of blood glucose levels by adrenaline. (1) The hormone adrenaline is secreted by the adrenal gland; (2) adrenaline binds to receptors on the surface of liver cells; (3) this binding causes a conformational change in the transmembrane receptor protein, promoting the synthesis of cyclic AMP (cAMP) from ATP; (4) cAMP activates an enzyme (1), which in turn activates many molecules of a second enzyme (2); each molecule of enzyme 2 in turn activates many molecules of enzyme 3; (5) each molecule of activated enzyme 3 breaks down glycogen granules into glucose, which passes from the cell into the bloodstream. The mechanism thus is a cascade, one molecule of adrenaline ultimately resulting in the release of millions of molecules of glucose.

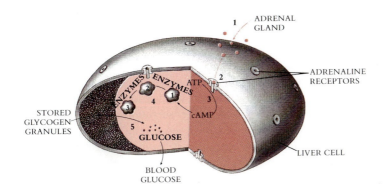

FIGURE 48-12

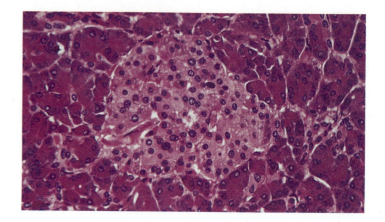

duced are determined by the level of glucose in the blood—a feedback loop. The overall level of glucose is determined by the balance between the two hormones.

Certain individuals are genetically incapable of producing sufficient amounts of insulin. These people have the hereditary disorder **diabetes.** Lacking insulin, affected individuals are unable to take up glucose from the blood, even though glucose levels in the blood become very high in the absence of insulin. These individuals lose weight, and if untreated may eventually suffer brain damage and death. The disorder may be treated by administering insulin to the affected individual daily. Because the body's digestive enzymes degrade insulin, the hormone cannot be taken orally—it must be administered by injection directly into veins. Active research on the possibility of transplanting islets of Langerhans holds much promise of a lasting treatment for diabetes.

THE REGULATION OF PHYSIOLOGICAL FUNCTIONS: WATER BALANCE

Vertebrates maintain relatively constant physiological conditions within their bodies. Neurovisceral and neuroendrocrine controls maintain temperatures, rates of respiration, blood pressures, blood glucose levels, and plasma ion concentrations within narrow bounds. These controls all act homeostatically; they tend to restore to normal levels any process that has been disturbed. These homeostatic controls for the most part function similarly in all vertebrates. There are two notable exceptions to this generalization: (1) alone among the vertebrates, mammals and birds are homeotherms, maintaining a constant body temperature physiologically; and (2) during the evolution of vertebrates, mechanisms that maintain a proper balance of water and ions within the blood plasma have changed greatly. As an example of diversity in homeostatic control, we will focus on the second of these.

Osmoconformers

The first vertebrates evolved in water, and the physiology of all members of the subphylum still reflects this origin. Approximately two thirds of every vertebrate's body is water. If the amount of water in the body of a vertebrate falls much lower than this, the animal will die.

Plasma membranes, as we have seen, are freely permeable to water but impermeable to salts and ions. This property of differential permeability forms the basis for many life processes, including the nerve conduction that we discussed in the last chapters. If the concentration of salts and ions dissolved in the water surrounding a vertebrate's body were the same as that within the body, the differential permeability of the cells would present no problem; there would be no tendency for water to leave or enter the body. In other words, the osmotic pressure of the body fluids would be the same as that of the surroundings. Most marine invertebrates are **osmoconformers.** They maintain the osmotic concentration of

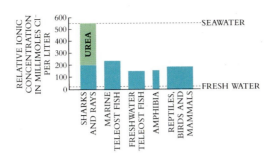

FIGURE 48-13

The concentration of ions is roughly similar in the bodies of different classes of vertebrates. Sharks become isotonic or even slightly hypertonic with respect to their environment by adding urea to their bloodstream. Marine fishes contain a far lower ion concentration than their environment, seawater—they are hypotonic with respect to their environment, tend to lose water by osmosis, and must struggle to retain as much water as possible. Freshwater fishes have the opposite problem: they contain a far higher ion concentration than their environment, are hypertonic, tend to gain water by osmosis, and must lose as much of this excess water as possible. Terrestrial vertebrates have ion concentrations not unlike those of the fishes from which they evolved.

their body fluids at the same level as that of the medium in which they are living, and they change the osmotic concentrations of their body fluids when the osmotic concentration of the medium changes.

The Problems Faced by Osmoregulators

Among the vertebrates, by contrast, only sharks are osmoconformers, and they are only imperfect ones. All other vertebrates are **osmoregulators** (Figure 48-13). Osmoregulators maintain an internal solute concentration that does not vary, regardless of the environment in which the vertebrate lives. The maintenance of a constant internal solute concentration has permitted the vertebrates to evolve complex patterns of internal metabolism. Maintaining this concentration, however, does require constant regulation of the animal's internal water level.

Freshwater vertebrates must maintain much higher salt concentrations in their bodies than those in the water surrounding them. In other words, they are **hyperosmotic** relative to their environment, and water tends to enter their bodies. They must, therefore, exclude water to prevent self-dilution.

Marine vertebrates have only about one-third the osmotic concentration of the surrounding seawater in their bodies. They are therefore said to be **hypoosmotic** relative to their environments; water tends to leave their body. These animals must retain water to prevent dehydration.

On land, surrounded by air, the bodies of vertebrates have a higher concentration of water than does the air surrounding them. They therefore tend to lose water to the air by evaporation. This situation is faced to some degree by the amphibians, which live on land only part of the time. It is also faced by all reptiles, birds, and mammals, which must conserve water to prevent dehydration.

How Osmoregulation Is Achieved

Animals have evolved a variety of mechanisms to cope with these problems of water balance, all of them based in one way or another on the animal's excretory systems. In many animals the removal of water or salts is coupled to the removal of metabolic wastes from the body. Simple organisms, such as many protists and sponges, employ contractile vacuoles for this purpose. Many freshwater invertebrates employ nephrid organs, in which water and waste pass from the body across a membrane into a collecting organ, from which they are ultimately expelled to the outside through a pore. The membrane acts as a filter, retaining proteins and sugars within the body while permitting water and dissolved waste products to leave.

Insects use a similar filtration system, with a significant improvement that helps them guard against water loss. The excretory organs in insects are the Malpighian tubules, which we discussed in Chapter 37 (see Figure 37-7, *B*). Malpighian tubules are tubular extensions of the digestive tract that branch off before the hindgut. Potassium ion is secreted into the tubules, causing body water and organic wastes to flow into them from the body's circulatory system because of the osmotic gradient. Blood cells and protein molecules are too large to pass across the membrane into the Malpighian tubules, so that the circulatory system is not short-circuited. Because the system of tubules empties into the hindgut, however, the water and potassium can be reabsorbed by the hindgut, and only small molecules and waste products are excreted. Malpighian tubules provide a very efficient means of water conservation.

Like the insects, the vertebrates use a strategy that couples water balance and salt concentration with waste excretion. Instead of relying on the secretion of salts into the excretory organ to establish an osmotic gradient, however, vertebrates rely on pressure-driven filtration. Whereas insects use an osmotic gradient to *pull* the blood through the filter, the vertebrates *push* blood through the filter, using the higher blood pressure that a closed circulatory system makes possible. Fluids are forced through a membrane that retains proteins and large molecules within the body but passes the small molecules out. Water is then reabsorbed from the filtrate as it passes through a long tube.

Terrestrial animals require efficient means of water conservation. In eliminating metabolic wastes, both insects and vertebrates filter the blood to retain blood cells and protein molecules, then collect the water from the filtrate by passing it through a long tube across whose walls it is reabsorbed.

Like the osmotically collected waste fluid of insects, the fluid that is passed through the filter of vertebrates contains many small molecules that are of value to the organism, such as glucose, amino acids, and vitamins. Vertebrates have evolved a means of selectively reabsorbing these valuable small molecules without absorbing the waste molecules that are also dissolved in the filtered waste fluid, **urine.** Selective reabsorption gives the vertebrates great flexibility, since the membranes of different groups of animals can evolve different active transport channels and thus the ability to reabsorb different molecules. This flexibility is a key factor underlying the ability of different vertebrates to function in many diverse environments. They can reabsorb small molecules that are especially valuable in their particular habitat, and not absorb wastes. Among the vertebrates, the apparatus that carries out the process of filtration and reabsorption is the **kidney.** It can function, with modifications, in fresh water, in the sea, and on land.

THE EVOLUTION OF KIDNEYS AMONG THE VERTEBRATES

The evolution of the kidney among the vertebrates has been greatly influenced by the transition to the land, as were the respiratory and circulatory systems. The flexibility with which evolution has molded and adapted the kidney to suit very different circumstances is truly remarkable.

Freshwater Fishes

Vertebrate kidneys are thought to have evolved first among freshwater fishes. As we noted above, the osmotic problem faced by animals living in fresh water is that their bodies tend to absorb water. In freshwater fishes, the kidney functions to remove much of this water. The water balance of a freshwater fish is achieved by two segments of its kidney, which carry out different functions:

1. *Filtration.* Blood is passed through a filter that retains blood cells and proteins but passes water and small molecules
2. *Reabsorption.* Desirable ions and metabolites are recaptured from the filtrate, leaving metabolic wastes and water behind for later elimination

Filtration

The filtration device of the freshwater fish kidney consists of a large number of individual tubular filtration-reabsorption devices called **nephrons.** At the front end of each nephron

FIGURE 48-14

Malpighian corpuscles.

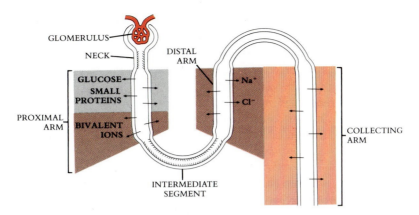

FIGURE 48-15

The basic organization of the vertebrate nephron is seen in the nephron tube of the freshwater fish, a basic design that has been retained in the kidneys of marine fishes and terrestrial vertebrates, which evolved later. Sugars, small proteins, and bivalent ions are recovered in the proximal arm; monovalent ions are recovered in the distal arm; and water is recovered in the collecting arm.

tube is a filtration apparatus called a **Malpighian corpuscle** (Figure 48-14). In each Malpighian corpuscle, a small artery enters and splits into a fine network of capillaries called a glomerulus. It is the walls of these capillaries that act as the filtration device. Blood pressure forces fluid through the differentially permeable capillary walls. These walls withhold the proteins and other large molecules while passing water and small molecules such as glucose, ions, and waste products such as ammonia.

Reabsorption

At the back end of each nephron tube is a reabsorption device that operates like the mammalian small intestine discussed in Chapter 43. The fluid that passes out of the capillaries of each glomerulus, called the **glomerular filtrate,** enters the tube portion of that nephron, within which reabsorption takes place. The tube is divided into several zones (Figure 48-15):

1. The first section of the tube encountered by the filtrate is a narrow **neck** in which numerous cilia vigorously beat, speeding the filtrate's flow. This serves to increase the amount of water that can pass through the kidney, a significant advantage to a fish that must get rid of a lot of water.

2. The tube then opens into two consecutive zones called the **proximal segments.** The membranes of the first zone, or **first proximal segment,** contain many transmembrane channels devoted to the active transport of glucose and small proteins. The channels are directed outward from the tube, so that as filtrate passes through this zone of the tube, glucose and small proteins are reabsorbed by the body from the filtrate. In a similar fashion, the **second proximal segment** is devoted to the active reabsorption of divalent ions such as Mg^{++}, Ca^{++}, and $SO_4^=$.

3. Next comes an **intermediate segment** of the tube, which contains many more cilia. This segment serves as a secondary pump, reinforcing the velocity of the filtrate through the tube.

4. The **distal segment** (far end) of the tube possesses within its membranes a variety of channels devoted to the reabsorption by active transport of monovalent ions such as Na^+, K^+, and Cl^-.

5. The ends of the many nephron tubes of the kidney join to form **ducts,** conducting the residue filtrate, or urine, to the urinary bladder. From this bladder, urine is periodically excreted from the body.

This basic design has been retained in all vertebrate kidneys, although, as you will see, there have been some changes.

The nephron tube of the kidney in freshwater fishes is a sophisticated reabsorption device, with several zones devoted to different transport channels and to cilia that drive the filtrate through the tube.

A freshwater fish drinks little and produces large amounts of urine. Because a freshwater fish is hypertonic to the water, water is not reabsorbed in its nephrons. The excess water that enters its body passes instead through the nephron tubes to the bladder, from which it is eliminated as urine. Within the urine is not only the excess water, but also all of the small molecules that were not reabsorbed while passing through the nephron tubes. Notable among these molecules is ammonia, the principal nitrogen waste product of metabolism. Ammonia in higher concentrations is toxic, but because the urine contains so much water, the concentration of ammonia is low enough not to harm the fish.

Marine Fishes

Marine bony fishes probably evolved from freshwater ancestors, as we mentioned in Chapter 39. In making the transition to the sea, marine fishes faced a significant problem of water balance. In the sea, fishes are hypotonic with respect to the water that surrounds them. For this reason, water tends to leave their bodies, which are less concentrated osmotically than is seawater. To compensate, marine fishes drink a lot of water, excrete salts instead of reabsorbing them, and reabsorb water. This places radically different demands on their kidneys, turning the tables on an organ that had originally evolved to eliminate water and reabsorb salts! As a result, the kidneys of marine fishes have evolved important differences from those of their freshwater relatives:

1. The kidneys of saltwater fishes lack the two cilia-rich pumping zones that the freshwater fishes have. A fast flow of filtrate through the kidney is no longer advantageous in a marine environment; a slower flow permits the more efficient reabsorption of water.

2. The first proximal segment of the nephron in marine fishes, like that in freshwater fishes, is devoted to the reabsorption of glucose. The second proximal segment, however, possesses active divalent ion transport channels whose direction is reversed compared with those of the freshwater fishes—ions are actively pumped *out* of the body *into* the tube. This is particularly important in a marine environment, where the water that the fish drinks is rich in divalent ions such as Ca^{++}, Mg^{++}, $SO_4^=$, and $PO_4^=$, all of which must be subsequently excreted.

3. The distal segment of the tube acts as a water-collecting vessel. The absorption of water by the kidneys of marine fishes is not efficient, however. Most of the water that passes through the glomerulus filter is lost from the body. Marine fishes compensate for this water loss by drinking a lot of seawater, using their kidneys to remove the large concentrations of divalent ions that such water contains.

Except for one species of shark found in Lake Nicaragua, all sharks, rays, and their relatives live in the sea (Figure 48-16). Some of these elasmobranchs (members of the class Chondrichthyes) have solved the osmotic problem posed by their environment in a different way than have the bony fishes. Instead of actively pumping ions out of their bodies through their kidneys, these fishes use their kidneys to reabsorb the metabolic waste prod-

FIGURE 48-16

This tiger shark is swimming in the ocean in French Polynesia. Although the seawater contains a far higher concentration of ions than the shark's body does, the shark avoids losing water osmotically by maintaining such a high concentration of urea in its body that it is osmotically isotonic or even hypertonic with the sea around it.

uct urea, creating and maintaining urea concentrations in their blood 100 times as high as those that occur among the mammals. As a result, the sharks and their relatives become isotonic or even slightly hypertonic with the surrounding sea. They have evolved enzymes and tissues that tolerate these high concentrations of urea. Because they are isotonic with the water in which they swim, they avoid the problem of water loss that other marine fishes face. Since the elasmobranchs do not need to drink large amounts of seawater, their kidneys do not have to remove large amounts of divalent ions from their bodies.

Amphibians and Reptiles

The first terrestrial vertebrates were the amphibians; the amphibian kidney is identical to that of the freshwater fishes. This is perhaps not surprising, since amphibians spend a significant portion of their time in fresh water, and when on land they generally stay in wet places.

Reptiles, on the other hand, live in diverse habitats, many of them very dry. The reptiles that live mainly in fresh water, like some of the crocodilians, occupy a habitat similar to that of the freshwater fishes and amphibians and have similar kidneys. Marine reptiles, which consist of some crocodilians, some turtles, and a few lizards, possess kidneys similar to those of their freshwater relatives. They eliminate excess salts not by kidney excretion but rather by means of salt glands located near the nose or eye.

The terrestrial reptiles reabsorb much of the water in kidney filtrate before it leaves the kidneys, and so excrete a concentrated urine. This urine, however, cannot become any more concentrated than the blood plasma; otherwise, the body water would simply flow into the urine while it was in the kidneys. In the relatively concentrated urine of the land reptiles, nitrogenous waste is no longer excreted in the form of ammonia. If it were, the urine would be toxic to the animals that excrete it. In place of the ammonia, freshwater and marine reptiles excrete urea, and the terrestrial ones excrete uric acid. These metabolic conversions take place in the liver, and were described in Chapter 45.

Mammals and Birds

Your body possesses two kidneys, each about the size of a small fist, located in the lower back region. These kidneys represent a great increase in efficiency compared with those of reptiles. Mammalian, and to a lesser extent avian, kidneys can remove far more water from the glomerular filtrate than can the kidneys of reptiles and amphibians. Human urine is 4.2 times as concentrated as blood plasma. Some desert mammals achieve even greater efficiency: a camel's urine is eight times as concentrated as its plasma, a gerbil's 14 times as concentrated, and the pocket mouse (*Perognathus*, Figure 1-7), which occurs in deserts, has urine 22 times as concentrated as its blood plasma (Figure 48-17). Mammals and birds achieve this remarkable degree of water conservation by using a simple but superbly designed mechanism: they greatly increase the local salt concentration in the tissue through which the nephron tube passes, and then use this osmotic gradient to draw the water out of the tube.

FIGURE 48-17

Kangaroo rats (shown here is Dipodomys panamintensis) *have very efficient kidneys that can concentrate urine to a high degree by reabsorbing water, and so avoid losing any more moisture than necessary. They resemble pocket-mice in this respect but are less efficient.*

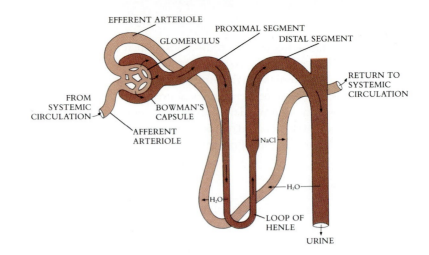

FIGURE 48-18

Organization of a mammalian renal tubule. The glomerulus is enclosed within a filtration device called Bowman's capsule. Blood pressure forces liquid through the glomerulus and into the proximal segment of the tubule, where glucose and small proteins are reabsorbed from the filtrate. The filtrate then passes through a double loop arrangement consisting of the loop of Henle and the collecting duct, which act to remove water from the filtrate. The water is then collected by blood vessels and transported out of the kidney to the systemic (body) circulation.

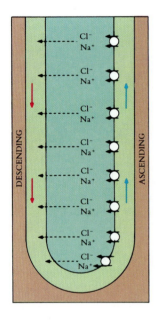

FIGURE 48-19

How an active countercurrent multiplier would work. The filtrate passes down the descending arm and turns the corner of the loop. As it begins to rise up the ascending arm, active transport passes some of the ions in the filtrate out into the surrounding tissue—the farther up the ascending arm the filtrate rises, the fewer ions remain in the filtrate. The exported ions have not abandoned the game, however. They pass into the descending arm by diffusion and reenter the cycle. The farther down the descending arm you go, the more ions have reentered and the higher the ion concentration in the filtrate.

HOW THE MAMMALIAN KIDNEY WORKS

Remarkably, mammals and birds have brought about this major improvement in efficiency by a very simple change—they bend the nephron tube. A single mammalian kidney contains about a million nephrons. Each of these is composed of a glomerulus, which is connected to a nephron tube, called a **renal tubule** in mammals. Between its proximal and distal segments, each renal tubule is folded into a hairpin loop called the **loop of Henle** (Figure 48-18).

The Countercurrent Multiplier Hypothesis

How does a bend in the renal tubule lead to greatly improved efficiency? Before we explain, it is instructive to consider briefly an obsolete hypothesis, once widely held (you will still find it in many freshman biology texts). In the 1960s, the idea was advanced that the bend allowed the renal tubule to work as a countercurrent flow device, much as the fish's gill and bird's lung, described in Chapter 44. Filtrate from the glomerulus passes down the beginning or descending arm of the loop and then back up the far arm of the loop, the ascending arm. The **countercurrent multiplier hypothesis** proposed that as the filtrate rose out of the loop through the ascending arm, salt (NaCl) was pumped out of the ascending arm into the surrounding tissue by active transport (Figure 48-19). Because the direction of flow in the two arms is opposite ("counterflow"), salt ions that have been pumped out of the ascending arm can reenter the flow at an earlier position by diffusing passively into the nearby descending arm. Removing salt from the ascending arm and adding it back to the descending arm results in a greater salt flow downward into the loop than upward out of it, so that salt becomes more concentrated at the bottom of the loop. The longer the length of the ascending arm after the tube bends, the greater the multiplication effect and the greater the salt concentration that is achieved at the bottom of the loop. Such an arrangement is called a **countercurrent multiplier.**

Note particularly the high salt concentration in the tissue surrounding the bottom of the loop. This is the key to the mechanism—the renal tubule empties into a collecting duct, which turns around and passes through this zone of high salt concentration again, but the collecting duct, unlike the ascending arm of the loop, is permeable to water, which streams out of the collected filtrate and into the high-salt tissue. The water that diffuses into the salt-rich region is carried away to the other parts of the body by the circulation of the blood, whose vessels are permeable to water but not to salt.

The problem with this countercurrent multiplier hypothesis is that it is predicated on the assumption that Na^+ and Cl^- ions are actively pumped out of the ascending arm—but when the cells composing the walls of the long lower portion of the ascending arm are examined carefully, their membranes do not contain active ion transport channels! Nor does the descending arm prove to be permeable to salt. However appealing it may have seemed, the countercurrent multiplier hypothesis is clearly wrong.

The Two-Solute Model

The currently accepted hypothesis of how the bend in the renal tubule achieves high efficiency, the **two-solute model** proposed in 1972, also postulates a mechanism driven by active transport, but one that employs a passive countercurrent mechanism. As its name implies, this hypothesis involves *two* solutes: salt (NaCl) and urea, the waste product of nitrogen metabolism. It has long been known that animals fed high-protein diets, yielding large amounts of urea as waste products, can concentrate their urine better than animals excreting lower amounts of urea. This suggests that urea plays a role in kidney function. In the two-solute model, urea's role is pivotal (Figure 48-20):

1. Filtrate from the glomerulus passes down the descending arm of the loop. The walls of this portion of the tubule are impermeable to either salt or urea, but are freely permeable to water. Because (for reasons we will come to) the surrounding tissue has a high osmotic concentration of urea, water passes out of the descending arm by osmosis, leaving behind a more concentrated filtrate.

2. At the turn of the loop, the walls of the tubule become permeable to salt, but much less permeable to water. As the concentrated filtrate passes up the ascending arm, salt passes out into the surrounding tissue by diffusion (the surrounding tissue, though it has lots of urea, does not contain as much salt as the concentrated filtrate). This makes salt more concentrated at the bottom of the loop.

3. Higher in the ascending arm, the walls of the tubule contain active-transport channels that pump out even more salt. This active removal of salt from the ascending loop encourages even more water to diffuse outward from the filtrate.

FIGURE 48-20

The two-solute model of kidney function.

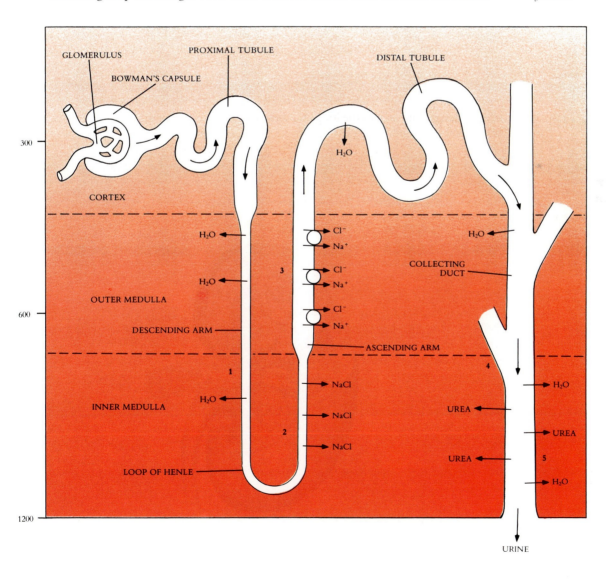

Left behind in the filtrate is the urea that initially passed through the glomerulus as nitrogenous waste; eventually the urea concentration becomes very high in the tubule.

4. Finally, the tubule empties into a collecting duct that passes back through the tissue—and unlike the tubule, the lower portions of the collecting duct are permeable to urea. During this final passage, the concentrated urea in the filtrate diffuses out into the surrounding tissue, which has a lower urea concentration. A high urea concentration in the tissue results, which is what caused water to move out of the filtrate by osmosis when it first passed down the descending arm.

5. As the filtrate passes down the collecting duct, even more water passes outward by osmosis, since the osmotic concentration of the surrounding tissue reflects both the urea diffusing out from the duct *and* the salt diffusing out from the ascending arm. This sum is greater than the osmotic concentration of urea in the filtrate (the salt has already been removed).

In effect, the interior of the kidney is divided into two zones (Figure 48-21). The outer portion of the kidney, called the **outer medulla,** contains the upper portion of the loop, including the upper ascending arm where reabsorption of salt from the filtrate by active transport occurs. The inner portion of the kidney, or **inner medulla,** contains the lower portion of the loop and also contains the bottom of the collecting duct, which is permeable to urea.

The active reabsorption of salt in the outer medulla of the kidney drives the process. This reabsorption of salt from the filtrate in one arm of the loop establishes a gradient of salt concentration, with salt concentrations higher in the inner medulla at the bottom of the loop. It is this high salt concentration that raises the total tissue osmotic concentration so high that water passes by osmosis out of the collecting duct. Just as importantly, the active reabsorption of salt in the outer medulla concentrates the renal filtrate with respect to urea.

FIGURE 48-21

Structure of the human kidney. The outer cortex of the kidney is isosmotic with the rest of the body. The outer medulla has a somewhat higher osmotic concentration, primarily salt. The osmotic concentration within the inner medulla becomes progressively higher at greater depths; this part of the kidney maintains substantial concentrations of urea, which is a major component of its high osmotic concentration.

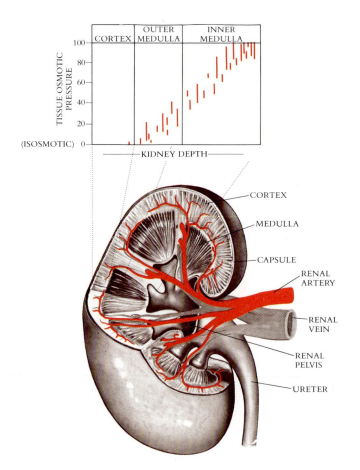

This high concentration of urea in the filtrate causes urea to diffuse outward into surrounding tissue in the only zone where it is able to do so, the lower collecting duct, creating a high urea concentration in the inner medulla. It is this high urea concentration in the inner medulla that causes water to diffuse out from the filtrate in the initial descending arm. The water is then collected by blood vessels in the kidney, which carry it into the systemic circulation of the body.

> **The mammalian kidney achieves a high degree of water reabsorption by using the salts and urea in the glomerular filtrate to increase the osmotic concentration of the kidney tissue. This facilitates the movement of the water from the filtrate out into the surrounding tissue, where it is collected by blood vessels impermeable to the high urea concentration but permeable to the water.**

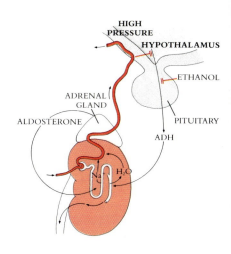

INTEGRATION AND HOMEOSTASIS

The kidney is one example of an organ whose principal function is homeostasis, the maintenance of constant physiological conditions within the body. The kidney is concerned with water and ion balance. Other organs have other homeostatic functions. In almost all cases these homeostatic functions are integrated by the central nervous system, which uses the voluntary, autonomic, and hormonal controls we have discussed in this chapter.

The kidney provides numerous examples of how integration by the central nervous system maintains homeostasis. Consider water balance (Figure 48-22). It is not always desirable for you to retain the same amount of water. If you have consumed an unusually large amount of water, the maintenance of homeostasis requires that you retain less of it than you would otherwise. When the uptake of water is excessive, the hypothalamus detects the resulting increase in blood pressure and decreases its output of antidiuretic hormone (ADH). ADH controls the permeability to water of the collecting ducts of the renal tubules. When ADH is present, the collecting ducts are freely permeable to water, so that water passes out into the surrounding tissue, which has a higher osmotic concentration of solutes. A decrease in levels of ADH renders the collecting ducts of the renal tubules less permeable to water and, in that way, inhibits the reabsorption of water and increases urine volume.

Another example of integration by the central nervous system leading to homeostasis in the kidney is provided by the salt balance in your body (Figure 48-22). The amount of salt in your diet can vary considerably, and yet it is important to many physiological processes that salt levels in your blood not vary widely. When levels of sodium ion in the blood fall, the adrenal gland increases its production of the hormone **aldosterone,** a steroid hormone that stimulates active sodium ion reabsorption across the walls of the ascending arms of your kidneys' renal tubules. In this way, it decreases the amount of sodium that is lost in the urine. In the total absence of this hormone, human beings may excrete up to 25 grams of salt a day.

FIGURE 48-22

Control of water and salt balance within the kidney is centered in the hypothalamus. The hypothalamus produces antidiuretic hormone (ADH), which renders the collecting ducts of the kidneys freely permeable to water and maximizes water retention. If too much water retention leads to high blood pressure, baroreceptors in the hypothalamus detect this and cause the production of ADH to be shut down. If levels of sodium in the blood fall, the adrenal gland initiates production of the hormone aldosterone, which stimulates salt reabsorption by the renal tubules of the kidney.

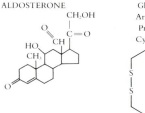

SUMMARY

1. The somatic muscles are directed by the motor neurons of the central nervous system. The activity of the central nervous system in directing the somatic muscles is modulated by stretch receptors, which are embedded in the muscles linked to afferent fibers returning to the central nervous system. The motor and afferent fibers constitute a control loop.

2. Most somatic striated muscles are organized in antagonistic pairs, one pulling in one direction, the other in the opposite direction; muscles do not push.

3. Smooth muscles, cardiac muscles, and glands are directed by antagonistic command nerve pairs of the autonomic nervous system, one of which stimulates while the other inhibits. In general, the parasympathetic nerves stimulate the activity of normal internal body functions and inhibit alarm responses, and the sympathetic nerves do the reverse.

4. The brain maintains long-term control over physiological processes by synthesizing releasing hormones in the hypothalamus. These hormones direct the synthesis of specific circulating hormones by the pituitary gland. Pituitary hormones travel out into the body and initiate the synthesis of particular hormones in target tissues.

5. Most brain hormones are peptides that interact with receptors on the target cell surface, thus activating enzymes within. Many of the hormones produced by endocrine glands such as the pituitary are steroids, which enter cells and alter their transcription patterns.

6. Water retention was one of the most significant problems posed by terrestrial life. Both insects and vertebrates evolved efficient means of filtering their blood to eliminate metabolic wastes without at the same time eliminating much water. The members of both of these phyla achieved water conservation by passing the waste-containing filtrate through a long tube across whose walls water could be reabsorbed.

7. The kidneys of freshwater fishes are adapted for water loss; the transition to salt water made by bony fishes required the reversal of the direction of many pumping channels within the wall of the absorption tube, and the slowing down of the filtrate's flow by loss of the cilia.

8. The kidneys of terrestrial vertebrates are organized similarly to those of the freshwater fishes, but with a small change in geometry that has far-reaching effects: the nephron tube is bent, forming the loop of Henle. The mammalian kidney reabsorbs water efficiently by using the salts and urea in the glomerular filtrate to increase the osmotic concentration of the kidney tissue. This facilitates the movement of the water from the filtrate out into the surrounding tissue, where it is collected by blood vessels impermeable to the high urea concentration but permeable to the water.

SELF-QUIZ

The following questions pinpoint important information in this chapter. You should be sure to know the answers to each of them.

1. All striated muscles possess monosynaptic reflex arcs. True or false?

2. The hormone primarily involved in regulating the rate of water reabsorption within the kidneys and intestines is
   (a) vasopressin (ADH)   (c) somatotropin
   (b) oxytocin   (d) aldosterone

3. Noradrenaline and adrenaline act primarily to increase the levels of sodium ion in the blood plasma. True or false?

4. The neurotransmitters of the terminal synapses of the sympathetic nervous system are _____.

5. The bodies of marine vertebrates relative to their surrounding environment are said to be
   (a) hyperosmotic   (b) hypoosmotic

6. The filtration device of the freshwater-fish kidney is composed of a large number of individual, tubular, filtration-reabsorption devices called _____.

7. At the front of each fish nephron tube is a filtration apparatus known as a _____.

8. A freshwater fish drinks little and produces large amounts of urine. True or false?

9. Marine fishes lose a lot of water to their environment; to compensate for the loss they drink a lot of water and
   (a) absorb salts   (b) excrete salts

10. Reptiles produce concentrated urine; however, reptile urine can never become more concentrated than the blood plasma. True or false?

11. Between the proximal and distal segment of each mammalian renal tubule is a hairpin loop known as _____.

12. Which of the following hormones acts primarily to stimulate the reabsorption of sodium?
    (a) oxytocin   (c) vasopressin
    (b) somatotropin   (d) aldosterone

THOUGHT QUESTIONS

1. In the mammalian kidney, water is reabsorbed from the filtrate across the wall of the collecting duct into the salty tissue near the bottom of the loops of Henle, and the water is then taken away by blood vessels. Why doesn't the blood in these vessels become very salty?

2. If you are lost in the desert with a case of liquor and are desperately thirsty, should you drink the liquor? Explain your answer.

3. Why do you suppose the brain goes to the trouble of synthesizing releasing hormones, rather than simply directing the production of the pituitary hormones immediately?

FOR FURTHER READING

BEEUWKES, R.: "Renal Countercurrent Mechanisms, or How to Get Something for (Almost) Nothing." In C.R. Taylor et al., editors: *A Companion to Animal Physiology,* Cambridge University Press, New York, 1982. A clearly presented summary of current ideas about how the kidney works, with an account of the evidence that led to the rejection of the countercurrent multiplier hypothesis.

BENTLEY, P.G.: *Comparative Vertebrate Endocrinology,* Cambridge University Press, New York, 1976. A good comparative account of endrocrine function, stressing ecological and evolutionary factors.

BLOOM, F.E.: "Neuropeptides," *Scientific American,* October 1981, pages 148-168. An account of recent advances in the study of endorphins and other brain hormones.

DAVIS, J.: *Endorphins: New Waves in Brain Chemistry,* Doubleday & Co., Inc., New York, 1984. A popular account of current research on brain hormones.

O'MALLEY, B.W., and W.T. SCHRADER: "The Receptors of Steroid Hormones," *Scientific American,* February 1976, pages 32-43. Steroid hormones recognize and bind to specific sites on the membranes of target cells.

SCHALLY, A., A. KASTAN, and A. ARIMURA: "Hypothalamic Hormones: The Link between Brain and Body," *American Scientist,* vol. 65, pages 712-719, 1977. A description of how the synthesis of the pituitary hormones is regulated by the hypothalamus.

SMITH, H.W.: *From Fish to Philosopher,* Little, Brown & Co., Boston, 1953. A broad and well-written account of the evolution of the vertebrate kidney.

VANDER, A., J. SHERMAN, and D. LUCIANO: *Human Physiology: The Mechanisms of Body Function,* 4th ed., McGraw-Hill Book Co., New York, 1985. This basic physiology text contains a very good chapter on human body fluids and the physiology of the kidney.

Overview

The vertebrates' evolution of a complex nervous system has led not only to a physiologically well-controlled body, capable of systematic patterns of behavior, but also to an increased ability to learn and modify new behaviors. This ability is not unique to the vertebrates. However, as the size of the associative areas of the vertebrate brain has increased, the associative "learning" abilities of the vertebrates have also increased, to a level not found in any other animals. The relative contributions of learning and of innate, genetically determined traits to complex human behaviors are difficult to assess.

For Review   *Here are some important terms and concepts that you will encounter in this chapter. If you are not familiar with them, you should review them before proceeding.*

- **Adaptation** (Chapter 20)
- **Interneuron** (Chapter 46)
- **Memory and learning** (Chapter 46)

The life of a bacterium is a simple one. Cells metabolize, grow and divide, and die. Some swim a little. Occasionally cells encounter one another and may then exchange cytoplasm or DNA. This is neither a complex life nor a long one. The life of a protist is a little more exciting. An individual *Paramecium* may engulf debris or even other cells, may swim around and around a bubble of carbon dioxide, and may engage in complex genetic exchange. These single-celled organisms often respond to their environment—moving toward a chemical or avoiding light, for example—but their responses are invariably simple. Only in multicellular animals do complex behavioral responses to external stimuli occur, because only multicellular organisms possess the mechanisms that are necessary to process complex environmental information. Only with the development of a nervous system do the complex responses that we call **behavior** begin. A sensory system, an associative center, and a network of command fibers are the essential ingredients of an organism that exhibits true behavior.

Arthropods, and particularly insects, have evolved very complex patterns of behavior. The hallmark of insect behavior, however, is fixed, unvarying patterns of activity. Although arthropods are capable of modifying some behavior patterns based on experience, many of their complex behaviors cannot be altered. Once elicited by environmental or other stimuli, the pattern is always the same and never varies. Such insect behaviors are said to be fully **determined,** the behaviors reflecting neural circuits within the nervous system specified by the genes and not subject to change.

Much vertebrate behavior is also of this kind, the pattern of a behavior predetermined before birth. For example, the sucking response of a newborn baby is fully determined. So is a chick's pecking to free itself from its egg at the time of hatching. Such innate patterns of behavior are called **instincts.** Vertebrates, however, are also capable of **associative responses,** behavioral responses that are adopted by trial and error. Your ability to walk upright is learning of this kind, as is writing with a pen or driving a car. These responses are no more complex than instinctive ones, but they are more flexible. Associative responses involve interneurons within the brain. Sensory input to interneurons may alter the functioning of circuits in complex ways, because the interneurons are integral parts of the circuit. With different feedback from the environment, the pattern of interneuron stimulation changes and a different behavioral response may result. This process of accumulating associative responses is called **learning.**

> **Behaviors are typically composed of two elements, that which is innate and that which is learned. Both elements involve associative activity among interneurons within the brain, but the learned components of behaviors are subject to modification.**

Vertebrate behavior is a mixture of instinctive and learned responses to environmental cues (Figure 49-1). At the beginning of this century there was considerable disagreement among biologists about how much of behavior was instinctive and how much was learned. Some researchers believed, for example, that very little human behavior was really instinctive, others that almost all of our behavioral repertory is genetically determined. This "nature-nurture" controversy has largely died down now; most biologists agree that behavior contains important elements of both types of behavior: heredity determines the limits within which a behavior can be modified, and learning determines what that behavior will be like within those limits. The limits, or range of flexibility, depend largely on the associative power of the central nervous system. Among insects these limits are narrow; among vertebrates they are much broader; and among human beings they are the broadest of all.

FIGURE 49-1

Learning is one of the most important activities of young vertebrates. Much of this learning takes place during play. These lion cubs are not trying to hurt each other, but are only exploring their capabilities, testing what it feels like to be fierce.

Traditionally there has been a wide gap between the approaches of biologists studying instinctive behavior and those studying learning. The former have tended to concentrate on behavior as it occurs in natural situations, an approach called **ethology,** while the latter have emphasized laboratory experimentation, founding a field called **experimental psychology.** In this chapter we will first consider instinctive behaviors, which are modified little by learning, in an attempt to see how the central nervous system organizes a complex motor response. We will then consider how learning relies on sensory feedback from the environment to modify these responses.

INNATE PATTERNS OF BEHAVIOR
Recurring Patterns in Behavior

In considering innate patterns of behavior, an ethologist must above all be a careful observer. What patterns recur in different instances of a behavior? What stimuli elicit these patterns? How predictable are the patterns and responses? A half century of ethological investigation, pioneered by Konrad Lorenz and later by Niko Tinbergen, has enabled us to sketch a general picture of instinctive behavior. There are five elements to any instinctive response, discussed in the following sections.

Motivation state

Within the brain, the hypothalamus associates the many sensory inputs of the body, which influence certain patterns of neural activity. Hunger, for example, is the way in which we perceive the response of the hypothalamus to the sensory input that results from lowered blood glucose levels. As a result of these neural interactions, we are "motivated" to eat. Many hormones also have strong feedback effects on the hypothalamus, altering mood, sexual receptivity, or activity level.

In some cases a motivational state reflects an innate rhythm, rather than a response to sensory input. For example, the motivation to be active is commonly a response of the reticular system to an internal clock. The cycle of sleeping and waking in humans is driven by a clock of this kind, called a **natural rhythm.** If the natural rhythms of different individuals are tested by isolating them in a dark room for long periods, most people show a natural 24-hour rhythm of sleeping and waking—but some people show a 48-hour rhythm. They sleep once each 48 hours instead of every 24 hours and sleep for about 16 hours instead of 8. The urge to migrate in birds and mammals is another motivational state (Figure 49-2). The "clock" is run by an innate natural rhythm, its timing set by the seasons.

FIGURE 49-2

Annual migration is a behavior common to many vertebrates.
A Among birds such as these snow geese, migration may cover a significant fraction of the globe's circumference.
B In caribou the seasonal migration across northern Canada may cover 1500 kilometers, although hunting and development are disrupting the migration routes that the herds have long used.

A

B

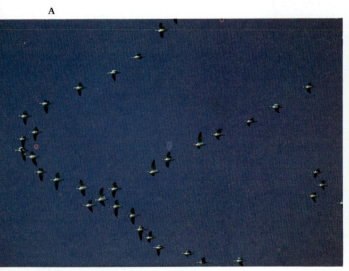

FIGURE 49-3

This chameleon is catching an insect with its tongue. The complex series of muscular contractions of tongue and mouth required to complete this capture are always the same and never vary once they have been initiated for the first time.

Response threshold

The neural response to a motivational state is not automatic. Rather, a behavior is released when a certain threshold is exceeded. Below the threshold value of motivation, no response occurs. Thus hunger increases as levels of blood glucose fall lower and lower, until a point is reached at which behaviors associated with foraging for food are released.

Behavior releasers

Many behavioral thresholds are lowered by specific stimuli, often referred to as **sign stimuli.** These stimuli are often quite specific. A male European robin, for instance, will attack a bunch of red feathers as if it were another male robin threatening its territory: the redness alone is sufficient to elicit the behavorial response. A sign stimulus may constitute only a part of the available sensory information. The distress calls of newborn chicks are sign stimuli of this kind. The hen will respond quickly to these calls if she can hear them—but she will ignore a chick tethered under a glass dome, even though she can see its agitated movements, because she does not hear its cry. The sound, not the visual image, is the behavioral releaser in this case.

A sign stimulus does not automatically guarantee a response; it merely lowers the motivational threshold and thus makes the response more likely. Thus the intensity of the sign stimulus that is required to release a particular pattern of behavior is inversely related to the level of motivation: the stronger the motivation, the more easily a sign stimulus will elicit the behavior. In some cases the behavior may occur without any releaser stimulus. Lorenz, for example, once noticed one of his tame starlings darting about the room collecting insects—although there were no insects in the room. Conversely, when motivation is low, a normally effective sign stimulus may fail to elicit any response.

Innate releasing mechanisms

The neural mechanisms that are responsible for behavioral thresholds, and the ways in which sign stimuli lower them, are not known in detail. The brain probably perceives a set of stimuli that correspond to a releaser pattern, stimulating certain neural centers, the **innate releasing mechanisms,** which trigger a chain of precisely orchestrated neuromuscular events.

Fixed action patterns

The chain of neuromuscular events initiated by a releaser stimulus never varies; it is fully determined by the neurological circuits of the innate releasing center. The tongue flick that a chameleon uses to catch a fly (Figure 49-3), for example, is the same every time in every

FIGURE 49-4

A male (right) and female (left) stickleback in courtship.

chameleon. Once initiated, the rapid sequence of tongue and mouth motions cannot be altered or stopped, and it will continue whether the fly moves or not. Initiation of a fixed action pattern is often coordinated with environmental information. The chameleon, for example, will orient itself toward the fly, in a sense aiming the tongue flick. The pattern of the response itself, however, does not vary and is not altered in any way by the orientation. Many fixed action patterns mature only after birth and are refined by feedback learning. A baby mouse, for example, will scratch vigorously and with perfectly coordinated movements—and touch only air. Only later does it learn the proper orientation.

Innate patterns of behavior are thought to involve motivational states in the hypothalamus that induce a fixed neural response, called an action pattern, whenever stimulated. A sign stimulus, for example, activates an innate releasing mechanism. The threshold stimulus required to trigger the response can vary.

Hierarchical Behaviors

Many innate patterns of behavior are composed of a complex series of fixed action patterns, produced by a hierarchy of innate releasing mechanisms, each responding to an increasingly specific stimulus. The sequence of courtship behaviors described by Tinbergen in a small freshwater fish, the stickleback, provides an example of how such complex patterns of behavior are organized (Figure 49-4). The motivation for reproduction in the male stickleback is triggered by hormones and keyed by an internal clock to the season of the year. This is the most general input, and it will activate one of four different innate releasing mechanisms, depending on the subsequent visual input. The sight of another male will release aggression; the sight of a female will release mating; the sight of a suitable territory may elicit nest building; and the sight of eggs will elicit the care of offspring.

Any of these four innate releasing mechanisms, in turn, offers a number of choices, depending on the specific characteristics of the visual stimulus. If the intruding male attacks, the defending male will bite; if the intruding male remains motionless, the defending male will threaten; and if the intruding male flees, the defending male will chase him. Thus the particular behaviors that a male fish selects from among the complex repertory available to him depend on a series of increasingly specific signals from the environment.

Latent Behaviors

Some behaviors may remain **latent,** or inactive, until they are stimulated. For example, the imprinting of song in the white-crowned sparrow is triggered by hearing the appropriate song within a critical window of from 10 to 50 days of age. During this period, the connections of the circuit, stimulated by the sound of the appropriate song, become stronger and stronger. The young birds themselves do not sing until much later in their lives, when they are more than 100 days old, but in order for them ever to be able to do so, the correct neural circuits must be established during the earlier period.

The songs of most birds are composed of only a few notes, six or seven in the swamp sparrow, for example. Each note in these songs corresponds to a fixed action pattern of muscular contractions. Establishing the proper song during the period of imprinting is apparently a matter of selecting the appropriate sequence of these few fixed action patterns (Figure 49-5). A white-crowned sparrow cannot learn the song of another bird, even during its imprinting period. A number of other birds, however, apparently possess more flexible circuits; they will adopt whatever song they hear during the imprinting period. If a white-crowned sparrow hears no songs, or only the songs of other species of birds, during its imprinting period, it simply sings the notes that it would normally have sung, but it arranges these notes into a very simple tune unlike that of other white-crowned sparrows.

FIGURE 49-5

The singing of a sparrow, which to us sounds so spontaneous and free, is actually a carefully orchestrated performance that the bird learns while it is immature.

LEARNING

Learning is the alteration of behavior by experience. It is formally defined as the creation of long-term changes in behavior that arise as a result of experience, rather than as a result of maturation. There are generally considered to be two broad categories of learning. The simplest learned behaviors are **nonassociative,** ones that do not require the animal to form an association between two stimuli or between a stimulus and a response. Learned behaviors that do require associative activity within the central nervous system are **associative.**

Nonassociative Learning

Habituation

Habituation can be regarded as learning *not* to respond to a stimulus. Learning to ignore unimportant stimuli is a critical ability to an animal confronting a barrage of stimuli in a complex environment. In many cases the stimulus evokes a strong response when it is first encountered, but then the magnitude of response gradually declines after repeated exposure. When you sit down in a chair, you feel your own weight at first, but soon you are not conscious of it. Your brain "tunes out" this information.

Sensitization

Sensitization is learning to be hypersensitive to a stimulus. After encountering an intense stimulus, such as an electric shock, an animal may often react vigorously to a mild stimulus that it would previously have ignored.

> **Associative learning is the alteration of behavior by experience leading to the formation of an association between two stimuli or between a stimulus and a response. Nonassociative learning involves no such associations.**

Associative Learning

Among the many different ways in which associative learning occurs, we can distinguish four major classes, discussed in the following sections.

Imprinting

Strong associations between stimuli and motivational states are made early in the life of many vertebrates. For example, the young of hooved animals and of birds such as geese, ducks, partridges, and chickens will follow the first moving object they see (see Figure 1-10). They will form a lasting bond with it, particularly if they can hear it make a sound. This kind of learning probably evolved among animals that move about soon after birth, to help them recognize their parents. Normally, the parent is the first moving noisy object a newborn individual encounters. The initial association does not depend on any sensory characteristic other than motion and sound. Once it has been formed, other information intensifies the identification.

However, mistakes in imprinting can occur. If a newborn goose is presented with a moving box containing a clock, it will adopt the box as its parent. Imprinting is usually possible only within a relatively narrow time span when the nervous system is developing.

Classical conditioning

The repeated presentation of a stimulus in association with a response can cause the brain to form an association between them, even if the stimulus and the response have never been associated before. If you present meat (stimulus) to a dog, it will salivate (response). In the classic study of this kind of **conditioning** (Figure 49-6) the Russian psychologist Ivan Pavlov also presented a second unrelated stimulus. He shined a light on a dog at the same time that meat powder was blown into its mouth. As expected, the dog salivated. After repeated

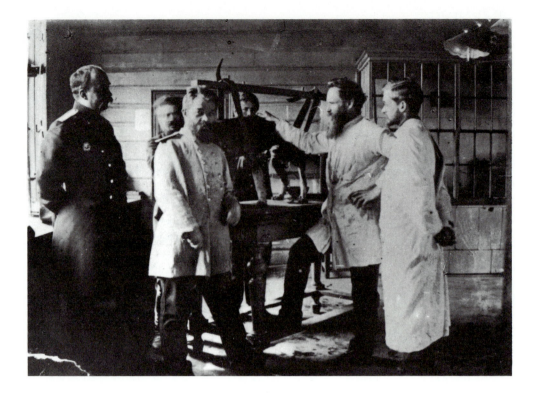

trials, the dog eventually would salivate in response to the light alone. The dog had learned to associate the unrelated light stimulus with the meat stimulus.

Operant conditioning

In classical conditioning, association by the animal does not influence whether or not the reinforcing stimulus is received. In **operant conditioning,** by contrast, the reward follows only after the correct behavioral response, so the animal must make the proper association before it receives the reinforcing stimulus. The American psychologist B.F. Skinner studied such conditioning in rats by placing them in a box of a kind that came to be called a "Skinner box" (Figure 49-7). Once inside, the rat would explore the box feverishly, running this way and that. Occasionally, it would accidentally press a lever, and a pellet of food would appear. At first a rat would ignore the lever and continue to move about, but soon it learned to press the lever to obtain food. When it was hungry, it would spend all of its time pushing the bar. This kind of trial-and-error learning is of major importance to most vertebrates.

Inductive reasoning

Anthropoid primates, and especially human beings, are able to associate certain outcomes not only with stimuli but also with abstract concepts (Figure 49-8). In this way they can learn to respond correctly to a situation the first time that it is encountered. For example, if a chimpanzee is placed in a room full of boxes with a bunch of bananas dangling from the ceiling on a string out of reach, it is likely to stack the boxes on top of one another, climb up, and seize the bananas. Many forms of abstract reasoning, such as symbolism, concept formation, and logic, can be distinguished.

> **Associative learning may involve relatively permanent associations
> between stimuli and motivational states (imprinting), less permanent
> associations between a stimulus and other stimuli (classical conditioning)
> or between a stimulus and a response (operant conditioning), or transient
> associations among abstract concepts (reasoning).**

FIGURE 49-7

*A rat in a Skinner box. The rat
rapidly learns that pressing the
lever results in the appearance of a
food pellet. This kind of learning,
trial and error with a reward for
success, is called operant
conditioning.*

FIGURE 49-8

This chimpanzee is fashioning a tool. It is stripping the leaves from a twig, which it will then use to dig into a termite nest. (See Figure 24-9, B.) Advance preparations like this strongly suggest that the chimpanzee is consciously planning ahead, with full knowledge of what it intends to do.

THE NATURE OF SOCIAL BEHAVIOR

Most behavior is a mixture of innate and learned components. Nowhere is this relationship more obvious than in social behavior, the way in which one animal interacts with other members of its own species. When one animal meets another of its species, they rarely ignore each other; each usually regards the other either as a competitor or as an ally. We call the belligerent behavior directed toward potential competitors **aggression** and the cooperative behavior directed toward allies **sociality.**

In insects the distinction between ally and competitor is very simple. If two individuals are part of one family, they are allies; if not, they are competitors. Many species of insects, particularly bees and ants, are organized into societies composed of highly integrated groups of individuals, each of which performs a special task (Figure 49-9). Their specialization is so extreme and their organization so rigid that these insect societies as a whole exhibit many of the properties of an individual organism. A human body such as yours, for example, relies on millions of individual cells that are specialized to perform many different tasks. Similarly, a beehive or ants' nest is a coherently organized group of individuals, in which certain individuals perform specialized tasks on which the survival of the entire group depends (see Figure 40-1, *B*). Only one component of a human body, the gonads, is responsible for reproduction. In a similar way, only one component of the beehive or ants' nest, the queen, is involved in the reproduction of that colony. All the cells of a human body are related to one another by descent from one fertilized zygote. Similarly, all of the members of a nest or hive are descended from an individual queen.

Some groups of vertebrates are also highly organized, and related individuals, or **families,** may play an important role in organizing groups of individuals. Vertebrate societies are not as rigidly constructed as are insect ones, however, and they are rarely as large. Their membership is usually not restricted to individuals that are closely related, and specialization is not as prevalent, or as pronounced in degree, as it is in bees, termites, or ants. Vertebrate societies share with insect ones, however, the central characteristic of any society—the ability to direct competitive behavior (aggression) outward, toward individuals that are not members of the group.

FIGURE 49-9

This ant, a member of the genus Polyergus, is capturing a slave from the nest of an ant of the genus Formica. It is carrying off a pupa, which it will take to its own nest, where the emerged adult will live as a nonreproductive worker slave.

APLYSIA AND THE NEUROLOGICAL BASIS OF LEARNING

We do not know a great deal about the neural basis of learning, although intriguing glimpses of what the mechanisms may be are emerging from studies of simple systems. Some of the most interesting results have been obtained by Eric Kandel and coworkers with a marine slug, *Aplysia* (Figure 49-A). *Aplysia* possesses only a few large neurons, a property that makes its reactions simple enough to study in detail. Like many mollusks, this animal has a single large gill, which is protected by a sheet of tissue called the mantle. The mantle ends in a spout called the siphon. A gentle tap on the mantle or siphon of *Aplysia* will cause the animal's gill to withdraw into its central cavity. This response involves only two neurons, a sensory neuron and a motor neuron with which it synapses directly.

Aplysia can learn to modify this withdrawal response, and its learning exhibits habituation, sensitization, and classical conditioning.

Habituation. If the mantle is gently prodded several times, *Aplysia* will learn to ignore it.

Sensitization. An electric shock applied to its tail will render *Aplysia* more sensitive to prodding; when the mantle is prodded after a shock very gently, so gently that the prod would not previously have elicited a response, *Aplysia* withdraws its gill abruptly.

Classical conditioning. When the tail of *Aplysia* is shocked a short time after a gentle tap on the siphon, the gill is withdrawn rapidly. After about 15 exposures to this pair of stimuli, a gentle tap alone elicits the strong withdrawal.

The habituation of gill withdrawal in *Aplysia* results from the overloading of the Ca^{++} ion channels in the mantle's sensory neuron synapses. When a wave of depolarization reaches the presynaptic membrane, it opens Ca^{++} channels. The inrushing calcium causes the synaptic vesicles to release neurotransmitter into the synaptic cleft. The repeated stimulation of the same nerve uses up all available neurotransmitter quicker than it can be supplied. Thus repeated stimulation leads to the gradual loss of function of these calcium channels, with the result that less neurotransmitter is released.

The sensitization of *Aplysia* gill withdrawal involves an interneuron. Stimulation of the tail sensory nerve cell stimulates this interneuron, which synapses with the mantle's sensory neuron, to release the inhibitory neurotransmitter serotonin. The serotonin blocks the K^+ channels of the mantle's sensory neuron. When a weak action potential arrives along this sensory neuron, the blockage of the K^+ channels prolongs the action potential and thus allows the calcium channels of its presynaptic membrane to stay open longer, releasing more neurotransmitter into the synaptic cleft and thereby facilitating gill withdrawal.

Conditioning of gill withdrawal in *Aplysia* occurs by a form of enhanced sensitization. If a sensory neuron has just "fired," its synapse is more efficient. Apparently the residual Ca^{++} from the preceding release of neurotransmitter lowers the level of Ca^{++} that is required to achieve a new release. In *Aplysia* conditioning the greater activity of the mantle's sensory nerve-motor nerve synapse (sensitization)

Aggression

The founder of ethology, Konrad Lorenz, proposed that aggressive behavior is innate in all vertebrates, including human beings. He considered aggression to be an inherent motivational state that is released by certain sign stimuli and whose threshold is conditioned by learning. Lorenz's idea has not been popular among biologists and social scientists, many of whom believe that human beings at least are born "with a clean slate." Nor has it sat well with Western religious and political traditions. Christianity holds that human behavior is conditioned by learning and modified by "free will"; Marxism holds that human behavior is profoundly conditioned and modified by society. Both Marxists and Christians believe that because human behavior is modifiable, people are responsible for it, either individually (according to Christians) or collectively (according to Marxists).

Despite these widespread beliefs, and although the behavioral evidence is not extensive, the weight of available evidence favors Lorenz's assertion. In many nonhuman ver-

FIGURE 49-A

Aplysia.

that is elicited by the repeated shocks facilitates the passage of subsequent impulses across the synapse, even in the absence of the shocks. Conditioning in this case appears to be only a special form of sensitization.

We do not know whether all learning processes depend on the same simple repertory of habituation and sensitization that is responsible for learning in *Aplysia*. One idea that is rapidly gaining acceptance is that the circuits of the vertebrate brain are fully wired at birth or soon thereafter and that no new connections are added by the growing brain. Instead, sensory experience modifies the strength of the various synapses. Fixed action patterns would not be subject to much modification, whereas behaviors capable of conditioning or other learning could be modified. Such modification might occur by means of the selective facilitation of the kind we have examined in *Aplysia*.

Thus *Aplysia* teaches us that learning need not involve the laying down of new circuits in the brain. In simple animals, learning involves modification of the strength of preexisting synapses, strengthening some (sensitization) and weakening others (habituation). In human beings, however, the associative areas of the brain undergo constant neuron turnover, suggesting that some new circuits are established.

tebrates, aggression does appear to be an innate drive (Figure 49-10). Among certain species of cichlid fishes, for example, the males must fight before they can mate. When they have been reproductively stimulated by the sight of a female during their mating period, the male fishes will dart about furiously, searching for someone to fight. If they cannot find another male, they will frequently attack the female, often killing her. Such behavior seems to be innate; indeed, mating behavior in these fishes cannot be released except by this kind of aggressive behavior. Many other animals exhibit aggressive behavior when they are approached by another member of the same species.

Innate behavioral aggression is responsible for **territoriality** (Figure 49-11), a form of behavior in which individual members of a species will mark off a fixed area or some other limiting resource, such as a foraging ground or a group of females, and then will defend these reserved resources against invasion or use by any other member of the same species. Territoriality is very common both in birds and in mammals. You can see it when

FIGURE 49-10

It is important not to confuse aggression with physical contact. These two male elephants (Loxodonta africana) are greeting one another, and aggression is the farthest thing from their minds.

FIGURE 49-11

These male hippopotami are not saying hello. They are contesting for territory and are in deadly earnest. They will fight until one is too injured or too discouraged to continue. The winner will then dominate that particular part of the lake.

birds sit on a tree branch, spaced evenly so that no bird gets too close to its neighbor. You can see it in your own behavior, when a stranger intrudes into your "personal space"; most people back away when someone starts to talk to them from a distance of less than about 60 centimeters, although the exact distance varies among members of different human societies.

Like any innate behavior, however, aggression in vertebrates is subject to modification by other innate behaviors, by learning, and in some animals by conscious choice.

Aggression appears to be an innate drive in most if not all animals, although its expression is conditioned by many other behaviors.

Patterns of Aggressive Behavior

Because aggression in vertebrates is subject to modification by learning, patterns of aggressive behavior would be expected to be subject to evolutionary modification, more successful patterns being favored by selection. Particularly where members of a species are in active competition with one another, evolution will be expected to favor those combinations of modifying influences that result in the most successful competitions. One can imagine at least five strategies by which this might occur:

1. **Chicken.** An individual can avoid aggressive behavior entirely, fleeing from any encounter. This strategy always rewards the aggressor and thus would be expected to favor the evolution of more aggressive behaviors, rather than less aggressive ones.

2. **Killer.** An individual can always attack a potential competitor. This is how rats behave when a new individual is introduced to a cage containing an established group. The new rat is tolerated for a while, but eventually it is attacked repeatedly until it is killed. Many predatory animals behave in the same way, attacking and attempting to kill intruding members of their own species. This is not an optimum evolutionary strategy, however, since it also exposes the attacker to the danger of being killed. Most animal species avoid fighting to the death in aggressive encounters.

3. **Bluffer.** An individual can always pretend that it is going to fight seriously, without ever doing so. The problem with this approach is that any opponent that really does fight, even by accident, always wins. For this reason, selection is always expected to favor the nonbluffing opponent over the consistent bluffer.

4. **Retaliator.** An individual that bluffs with ritual behavior unless it is seriously attacked, but then responds by attacking in return, has been referred to as a "retaliator." This is a very successful strategy. In computer war games it gives near-optimum results. This strategy appears to have been adopted by many animal species.

5. **Law abider.** An individual that behaves in conflict situations according to rules that all parties understand and acknowledge can avoid conflict; the outcome is determined by the rules themselves. This approach has been adopted by many complex vertebrate societies, including most human ones.

FIGURE 49-12

These caribou deer are contesting a female. They are not really intent on hurting one another—their hooves are far more potent weapons—but rather are engaged in a ritual test of strength and determination.

Examples of all five of these approaches to conflict can be found among vertebrates, but the last two are by far the most common. The retaliator strategy is particularly common, especially in conflicts involving competing males. Most attacks are not to the death, and they usually involve little harm to the opponents. Although deer may use their sharp hooves as effective weapons against wolves or other predators, they do not use them in fighting each other. Instead, they use their antlers, whose primary function is sexual display (Figure 49-12). The prongs of the antlers are usually formed in such a way that they can inflict only limited damage on an opponent. Most animals that employ ritual will fight back, however, if an opponent seriously attacks them.

Vertebrates employing ritual often adopt highly stylized courtship displays. In many of these displays the animals appear to be trying to respond with one display to two or more conflicting drives. Such displays are called **agonistic displays.** An agonistic display, for example, may suggest both attack and escape motivations. The conflicting emotions often modify the behavior that one emotion alone would elicit. As a result, the animal's ritual may include acts that are out of context (displacement activity), directed at the wrong object (redirected activity), or not completed (intention movements). A bird on a perch shocked by a loud noise may experience such conflict between escape and attack that it suddenly starts to preen, to eat, or even to go to sleep. This is displacement activity. The bird might peck vigorously at its perch, the right response, but one that is directed at the wrong object—a form of redirected activity. It might hunch down as if it were going to fly, signaling its intention to do so, and then not fly; the hunching down is an intention movement. In a ritual display these behaviors are usually highly exaggerated, and this exaggeration increases their effectiveness as signals of the performer's motivational state.

The law abider strategy is common in large and relatively permanent social groups (Figure 49-13). Thus a newly formed flock of hens will quickly establish dominant-subordinate relationships among its members. These relationships constitute a dominance hierarchy, or "pecking order," which is established by the initial aggressive encounters among the hens. Such encounters are not repeated; the ground rules are laid down by the outcome of the first ones, which are acknowledged by all. The most dominant hen is free to peck any of the other birds without being pecked back, the next most dominant hen may peck all but the most dominant one, and so on. Conflict is avoided by a set of rules. The elimination of conflict benefits not only the dominant individual but all members of the group. The birds of a stable flock are bigger and lay more eggs than birds of flocks that are frequently disrupted by additions and removals of birds. Dominance hierarchies are common among vertebrate societies. They are found in wolf packs, in hyena groups—and in primates. Conflict within primate and human groups is usually lessened by such adherence to commonly recognized rules, and a sophisticated body of rules, or **laws,** is commonly held to be a hallmark of human civilization. Reduction of conflict among primates is not as complete, however, as among some other vertebrates; war is not known among any vertebrate group except the primates.

FIGURE 49-13

Baboons form cohesive social groups with a dominant male and levels of subordinate relationships. Where an individual sleeps in a tree reflects his or her position in the social group.

NAKED MOLE RATS—A RIGIDLY ORGANIZED VERTEBRATE SOCIETY

One exception to the general rule that vertebrate societies are not highly organized is the naked mole rat, a tiny, naked rodent that lives in East Africa. Adult naked mole rats (Figure 49-B) are about 8 to 13 centimeters long and weigh up to about 60 grams—about the size of a sausage. Although they are mammals, they have virtually no hair. Unlike other kinds of mole rats, which live alone or in small family groups, naked mole rats congregate in large underground colonies with a far-ranging system of tunnels and a central nesting area. It is not unusual for a colony to contain 80 or more animals.

Naked mole rats live by eating large, underground tubers, which they locate by tunneling. Naked mole rats tunnel in teams. Each mole rat has protruding front teeth, which make it look something like a pocket-sized walrus; it uses these teeth to chisel away the earth from the blind face at the end of a tunnel. When the leading mole rat has loosened a pile of earth, it pushes the pile between its

feet and then scuttles backward through the tunnel, moving the pile of earth with its legs. When this animal finally reaches the opening, it gives the pile to another animal, which kicks the dirt out of the tunnel. Then, free of its pile of dirt, the tunneler returns to the end of the tunnel to dig again, crawling on tiptoe over the backs of a long train of other tunnelers that are moving backward with their own dirt piles.

Naked mole rat colonies are unusual vertebrate societies not only because they are large and well organized but also because they have a breeding structure that one might normally associate with bees, termites, or other social insects. All of the breeding is done by a single "queen," who has one or two male consorts. The worker caste, composed of both sexes, keeps the tunnels clear and forages for food. As long as the queen is healthy, there is no fighting among the workers. This is very unusual— few mammals surrender breeding rights without

Altruistic Behavior

Vertebrate societies present a puzzle to evolutionists. They are often organized in such a way that the activities of certain individuals benefit the group at the expense of those individuals themselves. A flock of crows might maintain a lookout, whose call of alarm warns the flock of an approaching hawk or hunter. In doing so, however, the lookout draws attention to itself and thus exposes itself to greater danger than if it were not a sentry. Behaviors such as this which are not in an individual's self interest are referred to as **altruistic.** What forces could promote the evolution of altruistic behavior? The difficulty in answering this question stems from the fact that evolution acts on *individuals,* not on populations—selection favors the genes borne by those individuals which leave the most offspring. How can it be beneficial in an evolutionary sense—that is, lead to the production of more offspring bearing your genes—to sacrifice your potential offspring for the good of the group? However noble it seems to lay down one's life for a friend, the evolution of innate altruistic behaviors at first seems contrary to everything we know about how evolution proceeds.

Inclusive fitness

A possible way out of this quandary was suggested in 1932 by a passing remark made by the great population geneticist J.B.S. Haldane. He expressed his suggestion by saying that he would willingly lay down his life for two brothers or four first cousins. Do you see what he was getting at? Each brother and Haldane share half of their genes in common—Haldane's brothers each had a 50% chance of receiving a given allele that Haldane obtained, and so forth for all other alleles. Consequently, it is statistically true that two of his brothers would carry as many of Haldane's particular combination of alleles to the next generation as Haldane himself would. Similarly, Haldane and a first cousin would share one fourth of their alleles, and four first cousins would therefore pass on as many of these genes to the

FIGURE 49-B

A naked mole rat.

contest or challenge. If the queen is removed from the colony, however, havoc breaks out among the workers; one individual attacks another, and all of them compete to become part of the new power structure. When another female becomes dominant and starts breeding, things settle down once again and discord disappears. Biologists studying naked mole rats speculate that the dominant female secretes a chemical that prevents other individuals from maturing sexually.

next generation as Haldane himself would. The point is that evolution will favor any strategy that increases the net flow of a combination of genes to the next generation. It makes "evolutionary sense" to sacrifice one's own individual fitness if doing so increases one's **inclusive fitness,** the total number of one's genes that are passed to the next generation. Thus altruistic behavior can evolve in societies whose members are related.

The reproductive sacrifice seen in the rigid caste system of insect societies makes sense in this context. The members of the different caste systems are closely related to one another. Among many of the social insects, for example, the males are haploid and females diploid so that diploid workers share three fourths of their alleles with the queen. In insect societies such as these, sterile workers can achieve representation in the next generation by aidir ⸱e queen.

FIGURE 49-14

This raccoon is apparently washing its dinner before eating, a characteristic behavior for the species. To what degree these behaviors are learned—passed down from parent to offspring—and to what degree they are hereditary—reflecting the action of specific genes—is difficult to determine.

Reciprocal altruism

There has been considerable controversy among biologists about the degree to which the concept of heritable behaviors applies to animals whose behaviors are less fixed. In a seminal text, *Sociobiology*, E.O. Wilson has argued that these concepts do apply and that among the vertebrates complex behaviors, including altruistic ones, are often heritable, the product of specific behavioral genes (Figure 49-14). Many other biologists, though, have not agreed with Wilson and the sociobiologists. Much of the argument has centered on altruistic behavior. The problem is that in vertebrate groups, inclusive fitness is not in itself a powerful enough genetic incentive to have fostered the evolution of altruistic behavior, since the members of vertebrate societies are not as closely related to one another as the members of insect societies are. It has been suggested by Wilson and others that vertebrates practice a kind of "I'll-scratch-your-back-if-you-scratch-mine" strategy, known as **reciprocal altruism.** The concept of reciprocal altruism implies that individuals perform altruistic behavior in the expectation of receiving similar treatment.

The difficulty with this, of course, is that the optimum strategy for transmitting *your*

FIGURE 49-15

This sea otter is having dinner while swimming on its back. It is using the rock as a tool to break open a clam, bashing the clam on the rock "anvil." Often a sea otter will keep a favorite rock for a long time, suggesting that it has a clear idea of what it is going to use the rock for. The sea otter may learn this pattern of eating behavior from others while young, but the capacity to use tools, and consciously to foresee their future use, certainly depends on inherited abilities.

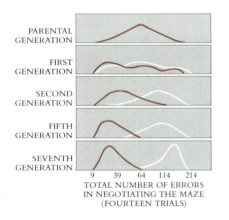

FIGURE 49-16

Tryon was able to select among rats for ability to negotiate a maze, demonstrating that this ability is directly influenced by genes. What he did was to test a large group of rats, select the few that ran the maze in the shortest time, and let them breed with one another; he then tested their progenies and again selected those with the quickest maze-running times for breeding. By seven generations he succeeded in halving the average time a naive rat required to negotiate the maze. Parallel selection for slow running time was also successful—it more than doubled the average running time.

genes is to cheat! If someone saves you from drowning, and then you observe that person drowning, you will leave the most offspring if you do not endanger your life by attempting to return the favor and rescue him. The evolutionary basis of altruistic behavior remains very much a mystery.

THE EVOLUTION OF BEHAVIOR

Both aggressive and altruistic behaviors are well documented in natural populations, and most biologists agree that at least a significant portion of these behaviors reflect the contributions of two kinds of behavior-specific genes: some totally specify a behavior, and others bestow the capacity to learn the behavior (Figure 49-15). The assumption here is that specific genes can act to determine components of specific behaviors. In discussions of specific human behaviors, this point has proven very controversial, usually because the behaviors often considered—such as criminality, homosexuality, and intelligence—are very complex ones, in which the roles of genes and environment are inextricably intertwined.

Unlike complex human behaviors, simpler behaviors of other vertebrates are often clearly influenced by genes specific to the behavior. In a famous experiment carried out in the 1940s, Robert Tryon studied the ability of rats to find their way through a maze with many blind alleys and only one exit, where a reward of food awaited (Figure 49-16). It took awhile, as false avenues were tried and rejected, but eventually some individuals learned to zip right through the maze to the food, making no false turns. Other rats never seemed to learn the correct path. Tryon bred the "maze-bright" rats to one another, establishing a colony from the fast learners, and similarly established a second "maze-dumb" colony by breeding the slowest-learning rats to each other. He then tested the offspring in each colony to see how quickly they learned the maze. The offspring of maze-bright rats learned even more quickly than their parents had, and the offspring of maze-dumb parents were even poorer students. After repeating this artificial selection procedure over several generations, Tryon was able to produce two behaviorally distinct types of rats with very different maze-learning abilities. Clearly the ability to learn the maze was hereditary, governed by genes that were being passed from parent to offspring. The genes are specific to this behavior, rather than being general ones that influence many behaviors—the abilities of these two groups of rats to perform other behavioral tasks did not differ.

Among vertebrates, some associative behaviors can be shown to have a significant genetic component, although environmental influences are also important.

FIGURE 49-17

The emperor penguins in Antarctica nest in rookery sites. There are only a limited number of such sites available, which may reflect the limited amount of resources available to the breeding penguins.

Natural Selection and Behavior

Experiments such as Tryon's demonstrate that learning ability is influenced by genes and suggest that other behaviors may be similarly influenced. If this is true, then we would expect natural selection to favor those gene combinations which adapt animals more completely to particular habitats. This relationship is exactly what ethologists have observed in natural populations. Examples of heritable behaviors that better adapt individuals to particular habitats are discussed in the following sections.

Territoriality

When resources such as nesting sites or food are limited (Figure 49-17), more birds will be produced each generation if the birds that are breeding ("breeding pairs") obtain sufficient resources, even though other birds not engaged in breeding may have to do without. Without a mechanism to ensure this kind of "rationing," competition for the limited resources might result in the resource being spread too thinly, with no one pair obtaining adequate resources—and the entire population might be in danger of unsuccessful breeding. One mechanism that has evolved to deal with this problem is **territorial behavior.** Resources within a territory are reserved for a particular breeding pair or group, other individuals being left to fend for themselves. In many mammalian societies, territoriality often plays an important role in the establishment of new populations, with the males defeated by the dominant male leaving to found their own populations. Individuals surviving in the surrounding marginal habitats can repopulate any territories that become vacant and can occupy any new habitats that they encounter.

Migratory memory

Many birds breed in temperate or arctic habitats and spend their winters at other locations far away, nearer to the equator where winters are less severe. In many cases the birds must cross thousands of miles, sometimes over open ocean. This migratory behavior is genetically determined, although it is facilitated by learning when young individuals fly for the first time with migratory flocks. When colonies of bobolinks became established in the western United States, far from their normal range in the Midwest and East, these birds did not migrate directly to their winter range in South America; rather they migrated east to their ancestral range and then south along the old flyway (Figure 49-18). The old pattern was not changed, but rather added to.

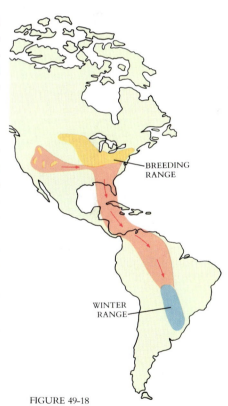

BREEDING RANGE

WINTER RANGE

FIGURE 49-18

The migratory path of California bobolinks. These birds came recently to the Far West from their more established range in the Midwest. When they migrate to South America in the winter, they do not fly directly, but rather fly to the Midwest first and then use the ancestral flyway.

THE LONG JOURNEY OF THE MONARCH BUTTERFLY

The monarch butterfly *(Danaus plexippus)* is a handsome orange and black insect that occurs over most of temperate North America. Aggregations of monarchs form every winter in certain groves of trees along the Pacific Coast from San Francisco to San Diego, and possibly farther south, from which they spread out in the spring to localities where milkweeds (species of the plant genus *Asclepias*) grow. The monarchs lay their eggs on these plants; the caterpillars feed on them; and the butterflies soon emerge, migrating in summer all over the western United States, except for the most extreme deserts, and the southern part of western Canada. The adults return each winter to the same overwintering sites along the Pacific Coast. A second series of monarch populations does not migrate, remaining in the lowlands of Mexico and in other parts of Latin America (Figure 49-C, *1*).

Most of the monarchs that are such a familiar sight over the central and eastern United States and southeastern Canada in summer migrate to several small and geographically isolated areas of coniferous forests at about 3000 to 4000 meters' elevation in the mountains of central Mexico, where they overwinter. These localities are so remote that the first one was not discovered until 1976, by University of Toronto biologist Fred A. Urquhart. Nevertheless, most of the monarchs in North America overwinter at these few limited sites, where Mexican conservationists are attempting to counter agricultural and logging operations to protect them. Smaller populations of monarchs also overwinter at scattered localities along the Gulf Coast of Florida.

The monarch is the only insect that regularly migrates, like many birds, between its summer and winter habitats. Each August, the butterflies begin a flight southward to their overwintering sites. In early November, wave after wave of monarchs arrive at their destination in the Oyamel fir forests in Michoacán, Mexico (Figure 49-C, *2*). By December, millions of monarchs have settled down on the fir branches in sheltered valleys, where they can avoid the impact of winter storms (Figure 49-C, *3*). After a storm, the thousands of butterflies that were knocked down are too cold to fly back into the trees and are forced to climb up the tree trunks, where they form fantastic clusters (Figure 49-C, *4*). As the dry season wears on, the monarchs fly out of their

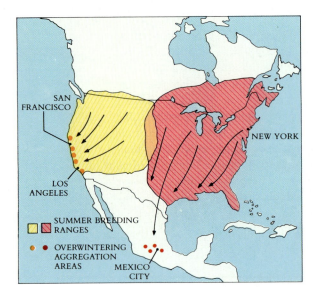

FIGURE 49-C, 1

Monarch butterflies from western North America overwinter in areas of mild climate, such as Pacific Grove, California, along the Pacific Coast; those from central and eastern North America overwinter in the mountains of central Mexico.

C, 2

roosting trees to nearby streams, where they drink copious amounts of water before returning to their clusters. Black-headed grosbeaks (Figure 49-C, *5*, center) and black-headed orioles (top) are able to eat the butterflies, despite the poisons they contain. During the 1978-1979 overwintering season, these birds consumed an average of 15,000 butterflies each day for a total of more than 2 million butterflies. Some of the monarchs that overwinter in Mexico reach southern Canada, thus completing a flight of some 3000 kilometers. What is amazing about the migration of the monarch, however, is that two to five generations may be produced as the butterflies

fly north. The butterflies that migrate in the autumn to the precisely located overwintering grounds in Mexico have never been there before. Thus the autumn migration occurs with no opportunity for the transfer of information from experienced individuals, as it does in all migrating birds. For reasons that we do not know, the monarchs are able to keep migrating in a south-westerly direction until they reach their overwintering sites. Genetic programming must be involved, at least in part, in the ability of the butterflies to achieve this unique feat, a striking example of behavioral diversity.

C, 3

C, 5

C, 4

FIGURE 49-19

A dominant male lion with a lioness.

Dominance hierarchies

In vertebrates living in groups, it is common for a few individuals—often but not always males—to dominate the group and, frequently, for only these individuals to mate with the opposite sex (Figure 49-19). This form of dominance behavior, as you saw above, limits conflict and competition within the group, since all members of the group recognize the rules and abide by them. The patterned behaviors that identify rank within such groups are often highly exaggerated.

Reproductive behaviors

Very early in his studies of behavior, Tinbergen noted that certain birds remove the shell fragments of broken eggs from their nests and that this behavior is innate. Any shell fragment that Tinbergen added was removed quickly. It turned out that the odor from shell fragments attracts predators, and their prompt removal was very smart housekeeping. Many other reproductive behaviors seem to have evolved to aid in the protection and rearing of young. Male parental care, for example, is rare among vertebrates, except for birds and primates, both of whose young go through long periods of learning before they can survive on their own (Figure 49-20). Male parental care in these groups greatly increases the probability of raising the young successfully and may have evolved for that reason.

There is little doubt that selection for specific behaviors has played an important role in vertebrate evolution, and it probably was significant in human evolution as well. The degree to which human behavior today reflects our genetic heritage is a controversial question, with no clear answer.

AN OVERVIEW OF VERTEBRATE BEHAVIOR

Complex behaviors are exhibited by all animals with nervous systems. It should come as no surprise that associative behaviors are more common among the vertebrates than among insects or other animals, since the portion of the brain devoted to associative activity is so much larger in vertebrates. The associative cortex is particularly prominent in mammals, humans being the most extreme case, and these are the animals with the most flexible behaviors, those most subject to modification by learning.

There can be little doubt that much of the behavioral capacity of vertebrates is genetically determined, the neural circuits specified by genes. It is not really clear to what degree these circuits can be overridden by the associative activity of the brain, but some degree of influence is very likely. In humans, where so much of the brain is devoted to associative cortex, most biologists would say that although some simple physical behaviors are innate and invariant, few complex social behaviors are fully genetically determined. The consensus rather is that we carry with us as evolutionary baggage the aggression and territoriality of our ancestors, modified by selection, and that in any individual these behaviors are subject to change by learning.

FIGURE 49-20

The young of primates and birds go through long periods of learning before they can survive on their own. This young lady, not yet 2 years old, still has much to learn before she can wear Daddy's hat.

Because the consequences of aggressive behavior among humans are so much more severe than among other vertebrates, the degree to which human aggressive behavior can be tempered by learning is an issue of very real importance. As you will see in the following chapters, human activities are rapidly altering the complex web of interactions on which life on earth depends. Organized human conflict, common throughout recorded human history, is now able to devastate the earth, and perhaps to end the history of life there. The knowledge that enables humans to act thus is itself a result of the great enlargement of the associative cortex. The evolutionary success of this modification of the vertebrate brain will depend critically on the ability of the greatly enlarged human associative cortex to modify behaviors such as aggression.

SUMMARY

1. Some patterns of behavior are innate (instincts), whereas others are associative responses acquired by trial and error (learned behaviors). Vertebrate behaviors combine both kinds of behavioral elements.

2. Heredity determines the limits within which a behavior can be modified; within those limits, learning determines what that behavior will be like. These limits are narrow or nonexistent for instincts, but they may be very broad for learned behaviors.

3. Innate patterns of behavior are thought to involve motivational states in the hypothalamus, which induce a fixed action pattern when a sign stimulus activates an innate releasing mechanism. The threshold stimulus required for this activation can vary.

4. Many innate behaviors consist of hierarchies of fixed action patterns; a variety of different responses are possible, depending on the environmental cues.

5. The simplest forms of learning involve sensitization and habituation. More complex associative learning may also occur in this way, by the strengthening and weakening of existing synapses, although learning may also involve the formation of entirely new synapses.

6. Behaviors directed toward other members of one's own species are usually classified as either aggressive (competitive) or social (cooperative).

7. Members of vertebrate groups often exhibit altruistic behavior, a form of social behavior in which an animal aids other members of its group, at possible risk to itself. The way in which such altruistic behaviors evolve is not clear.

8. Human behavior has both instinctive and learned components, and it is difficult if not impossible to sort out their relative contributions to particular behavioral patterns.

SELF-QUIZ

The following questions pinpoint important information in this chapter. You should be sure to know the answers to each of them.

1. Which is not one of the five elements involved in any instinctive response?
   (a) motivation state   (d) innate releasing mechanisms
   (b) response threshold   (e) fixed action patterns
   (c) behavior inhibitors

2. In what portion of the brain do the circuits reside which are responsible for motivational states such as hunger?

3. Can a fixed action pattern be modified by learning?

4. The simplest type of learning is
   (a) imprinting   (c) habituation   (e) reasoning
   (b) classical conditioning   (d) operant conditioning

5. Sensitization is _____ learning.
   (a) nonassociative
   (b) associative

6. The repeated presentation of a stimulus in association with a response to another stimulus can cause the brain to form an association between them. This is referred to as _____ conditioning.

7. Which pattern of aggressive behavior is most common among vertebrates?
   (a) chicken   (c) killer   (e) law abider
   (b) retaliator   (d) bluffer

8. One ritual display responding to two or more conflicting drives is called a(n) _____ display.

9. Would you increase your inclusive fitness if you laid down your life to save the lives of one brother and three first cousins?

10. We speak of the belligerent behavior directed toward potential competitors as _____ and of the cooperative behavior directed toward allies as _____.

11. Are vertebrate societies more or less rigidly constructed than insect societies?

12. In _____ conditioning a reward is obtained only if the proper response is associated with the stimulus.

THOUGHT QUESTIONS

1. War is so common among humans that it must be considered a basic behavior of our species. It appears to be absent in all other animal groups (with the possible exception of some other primates). Do you think this behavior has a genetic basis? If so, why might its evolution have been favored by natural selection?

2. It can be argued that our current nuclear standoff is a form of "retaliator" behavior, in which each side postures but neither attacks in earnest unless attacked first. Does this proposition conform with what you know of how animals deal with aggression? Explain your answer.

3. Swallows often hunt in groups, whereas hawks and other predatory birds usually are solitary hunters. Can you suggest a plausible explanation for this difference?

4. Can you suggest an evolutionary reason why many vertebrate reproductive groups are composed of one male and numerous females, rather than the reverse?

FOR FURTHER READING

ALCOCK, J.: *Animal Behavior: An Evolutionary Approach,* 3rd ed., Sinauer Associates, Inc., Sunderland, Mass., 1984. An evolutionary view of animal behavior, with a strong genetic perspective.

ALEXANDER, R., and D. TINKLE (eds.): *Natural Selection and Social Behavior,* Chiron Press, New York, 1981. A collection of essays by individual behaviorists that reveals the current state of research.

BLOOM, F., A. LAZERSON, and L. HOFSTADTER: *Brain, Mind, and Behavior,* W.H. Freeman & Co., San Francisco, 1985. A general introduction to how the brain is organized and to how its different parts form "alliances" among themselves; written for the layman but authorative enough to be useful to any student.

DAWKINS, R.: *The Selfish Gene,* Oxford University Press, New York, 1976. An entertaining account of the sociobiologist's view of behavior.

GRIER, J.: *Biology of Animal Behavior,* Times Mirror/Mosby College Publishing, St. Louis, 1984. A very clear and unusually comprehensive undergraduate text.

GRIFFIN, D.: "Animal Thinking," *American Scientist,* vol. 72, pages 456-463, 1984. An exciting discussion of the possibility that animals have conscious awareness, this article describes many examples of what appears to be conscious thinking by animals.

JOLLY, A.: "The Evolution of Primate Behavior," *American Scientist,* vol. 73, pages 230-239, 1985. A fascinating survey of behavior among primates that indicates a progressive development of intelligence, rather than a sudden full-blown appearance when humans evolved.

SHASHOUA, V.: "The Role of Extracellular Proteins in Learning and Memory," *American Scientist,* vol. 73, pages 364-370, 1985. An exciting speculation that the synthesis of proteins contributes to synapse formation during learning. This article is a good example of the intellectual ferment in this very active field.

WILLIAMS, G.: *Adaptation and Natural Selection,* Princeton University Press, Princeton, N.J., 1966. The first of two books that led to the current revolution in the way that we view behavior, with emphasis on explaining its evolution in terms of selection that acts on individual members of populations.

WILSON, E.O.: *Sociobiology: The New Synthesis,* Belknap Press, Cambridge, Mass., 1975. The second important book in the revolution of our understanding of the biological basis of behavior, with a controversial last chapter on the sociobiology of human beings.

ECOLOGY

Part Nine

The global ecosystem, which consists
of millions of species of organisms, has
evolved to capture the sun's energy and
control the cycling of nutrients. Record
numbers of human beings, growing at
an unprecedented rate, are now
challenging the ecosystem's ability to
function.

Overview

There are great differences in climate across the face of the globe that over billions of years have resulted in the evolution of diverse terrestrial biomes and comparable associations of organisms in the sea. These are major assemblages of plants, animals, and microorganisms that occur over wide areas and have distinctive characteristics that separate them from others. The major circulation patterns of the atmosphere and the oceans, driven by the unequal distribution of heat from the sun, amplify the differences. Among these biomes, the tropical forests are by far the richest biologically; they contain at least two thirds of all species of organisms.

For Review

Here are some important terms and concepts that you will encounter in this chapter. If you are not familiar with them, you should review them before proceeding.

- **Major features of evolution** (Chapter 23)
- **Nutrition and water transport in plants** (Chapter 34)
- **Vertebrate diversity** (Chapter 39)

Ecology, the study of the relationships of organisms with one another and with their environment, is a complex but fascinating area of biology that has many important implications for each of us. Every principle of ecology would hold true even if there were no human beings in the world, but the operation of these principles is now being affected profoundly by human activities. A human population that is climbing rapidly, now at an unprecedented level of about 5 billion people, is severely straining the earth's capacity to sustain us all. In the face of this situation the principles of ecology may enable us to chart a sound future.

Ecology, however, is not intrinsically an action-oriented field; it is an area of scientific research that is concerned with the most complex level of biological integration. Ecology attempts to tell us why particular kinds of organisms can be found living in one place and not another—the physical and biological variables that govern their distribution; the factors that control the numbers of particular kinds of organisms and maintain them at certain levels; and the principles that may allow us to predict the future behavior of assemblages of organisms (Figure 50-1).

FIGURE 50-1

Macaws, Ara macao, beautiful parrots of the Amazon Basin, obtaining salt from a bank in Manu National Park, Peru. The existence of these and all other organisms is determined by their position in the local community.

FIGURE 50-2

A diver explores a rich undersea community, with schools of fishes, soft corals (cnidarians), and many other phyla of invertebrates. The same ecological principles apply to the organization of communities in the sea as on land, but the details are very different.

Groups of different organisms may be considered at three progressively more inclusive levels of organization. In ecological terms, populations of different organisms that live together in a particular place are called **communities.** A community, together with the nonliving factors with which it interacts, is called an **ecosystem.** An ecosystem regulates the flow of energy, ultimately derived from the sun, and the cycling of the essential elements on which the lives of its constituent plants, animals, and other organisms depend. **Biomes** are major terrestrial assemblages of plants, animals, and microorganisms that occur over wide geographical areas and have definite characteristics that identify them as distinct from other such assemblages. They include deserts, tropical forests, and grasslands. Similar major groupings can be distinguished in marine and freshwater habitats.

> **A community is the interacting set of different kinds of organisms that occur together at a particular place. An ecosystem is that set of organisms, together with the nonliving factors with which it interacts. A biome is an assemblage of organisms that has a characteristic appearance and that occurs over a wide geographical area on land.**

In our exploration of ecological principles we will first consider in this chapter the properties of biomes and similar major aggregations in the seas (Figure 50-2) and in fresh water; move on to the dynamics of ecosystems in Chapter 51; population dynamics in Chapter 52; and community ecology in Chapter 53. Chapter 54 will deal with the future of the biosphere, a future that will depend to an ever-increasing extent on the ecological principles presented here.

The major biomes are easily recognized on the basis of their overall appearance and the characteristic climates associated with them. Biomes could be classified in a number of ways; those which we use here are chosen merely as convenient means for discussing the properties of life on earth from an ecological perspective.

The distribution of biomes results from the interaction of the actual features of the earth, such as different soil types or the occurrence of mountains and valleys, with two key physical factors: (1) the amounts of heat from the sun that reach different parts of the earth and the seasonal variations in that heat; and (2) global atmospheric circulation and the resulting patterns of oceanic circulation. Together these factors determine the local climate, including the amounts and distribution of precipitation.

THE GENERAL CIRCULATION OF THE ATMOSPHERE
The Sun and Atmospheric Circulation

The earth receives heat from the sun in the form of short-wave radiation, and it radiates an equal amount of heat back to space in the form of long-wave radiation. About 10^{24} calories arrive at the upper surface of the earth's atmosphere each year, or about 1.94 calories per square centimeter per minute, a quantity known as the **solar constant**. About half of this energy reaches the earth's surface. The wavelengths that reach the earth's surface are not identical to those which reach the outer atmosphere; for example, most of the ultraviolet radiation is absorbed by the oxygen and ozone in the atmosphere.

At a particular place and at a particular time, the amount of heat gained only rarely equals the amount of heat radiated back into space. The regions near the equator absorb more heat than they lose, whereas those near the poles radiate more than they absorb. However, the equatorial regions do not become increasingly warmer and the polar regions increasingly cooler: heat flows continuously from warm to cold regions, thus maintaining the actual temperatures. This transfer is brought about by the movement of the atmosphere and of the oceans, and it constitutes the reason for their general patterns of circulation (Figure 50-3).

The higher latitudes of the earth receive little short-wave radiation during the winter months, since much of this radiation is reflected back into space from clouds, ice, and snow. As a result, the heat differences between polar and equatorial regions are strengthened.

The earth's annual orbit around the sun and its daily rotation around its own axis are both important in determining world climate (Figure 50-4). Because of the daily cycle, the climate at a given latitude is relatively constant; there is a constant mixing of climates and temperatures at that latitude. Because of the annual cycle, and the inclination of the earth's axis at approximately 23.5 degrees from its plane of revolution around the sun, there is a progression of seasons in all parts of the earth away from the equator. One of the poles is closer to the sun than the other at all times, because the angle and direction of the earth's inclination are maintained as it rotates around the sun.

Major Circulation Patterns

Near the equator, warm air rises and flows toward the poles (see Figure 50-3). As it rises, it loses most of its moisture, so that rainfall is abundant near the equator. This region of rising air is one of low pressure, the doldrums, which draws air from both north and south of the equator. When the air masses that have risen reach about 30 degrees north and south latitude, the air, now cooler, sinks and becomes reheated, producing a zone of decreased precipitation. The air is still warmer than it is in the polar regions, and it continues to flow toward the poles. It rises again at about 60 degrees north and south latitude and flows back

FIGURE 50-3

General patterns of atmospheric circulation.
A *The pattern of air movement out from and back to the earth's surface.*
B *The major wind currents across the face of the earth.*

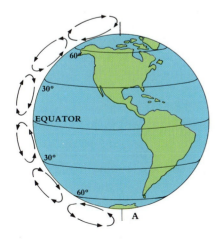

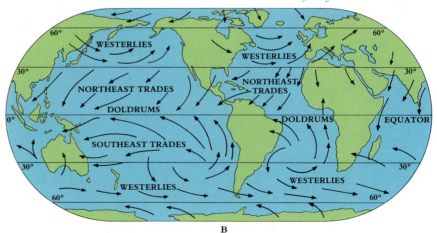

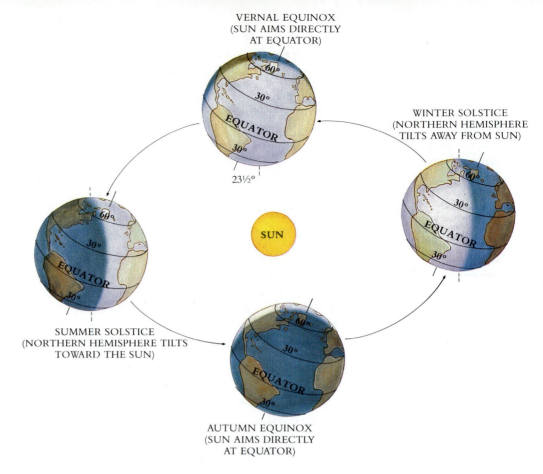

VERNAL EQUINOX
(SUN AIMS DIRECTLY
AT EQUATOR)

WINTER SOLSTICE
(NORTHERN HEMISPHERE
TILTS AWAY FROM SUN)

SUN

SUMMER SOLSTICE
(NORTHERN HEMISPHERE TILTS
TOWARD THE SUN)

AUTUMN EQUINOX
(SUN AIMS DIRECTLY
AT EQUATOR)

toward the equator. At this latitude is located another low-pressure area, the polar front. The air that rises here descends near the poles, producing a zone of very low precipitation.

Related to these bands of north-south circulation are three major air currents, generated mainly by the interaction of the earth's rotation with the patterns of worldwide heat gain. Between about 30 degrees north latitude and 30 degrees south latitude, the **trade winds** blow, from the east-southeast in the Southern Hemisphere and from the east-northeast in the Northern Hemisphere. The trade winds blow all year long and are the steadiest winds found anywhere on earth; they are stronger in winter and weaker in summer. Between 30 and 60 degrees north and south latitude, the **prevailing westerlies,** often very strong winds, blow from west to east and tend to dominate climatic patterns in these latitudes, particularly in lands that lie along the western edges of the continents. Zones of weaker winds, blowing from east to west, occur farther north and south in their respective hemispheres.

> **Warm air rises near the equator, descends at about 30 degrees north and south latitude, flows toward the poles, then rises again at about 60 degrees north and south latitude and moves back toward the equator; part of this air, however, moves toward the poles, producing a zone of very low precipitation.**

Atmospheric Circulation, Precipitation, and Climate

Since the moisture-holding capacity of air increases when it is warmed and decreases when it is cooled, precipitation is generally low near 30 degrees north and south latitude, where air is falling and being warmed, and relatively high near 60 degrees north and south latitude, where it is rising and being cooled. Partly as a result of these factors, all of the great deserts of the world lie near 30 degrees north or 30 degrees south latitude (see boxed essay, pp. 1098-1099), and some of the great temperate forests are near 60 degrees north and south

FIGURE 50-5

Moisture-laden winds that blow from the ocean to the southwest over India first stir up great dust storms and then bring rain. This is a typical scene in southern India in April at the start of the monsoon season.

latitude. Other major deserts are formed in the interiors of the large continents; these areas have limited precipitation because of their distance from the sea, and sometimes because mountain ranges intercept the moisture-laden winds from the sea also.

Four relatively small areas, each located on a different continent, share a climate that resembles that of the Mediterranean region of southern Europe, North Africa, and portions of the Near East. Such a climate is found in portions of California, southwestern Oregon, and northwestern Baja California, Mexico; in central Chile; in southwestern and part of South Australia; and in the Cape region of South Africa. In all of these areas the prevailing westerlies blow during the summer from a cool ocean onto warm land. As a result, the air's moisture-holding capacity is increased and precipitation is completely blocked or limited during the summer. Such climates are very unusual on a world scale; in the regions where they occur, many unusual kinds of plants and animals, often local in distribution, have evolved (see Figure 53-5). Because of the prevailing westerlies, the great deserts of the world (other than those in the interior of continents) and the areas of mediterranean climate lie on the western sides of the continents.

The great deserts of the world lie along the western sides of the continents at about 30 degrees north and south latitude. Other major deserts occur in the interiors of the large continents.

The monsoon climatic conditions that are characteristic of India and southern Asia occur during the summer months (Figure 50-5). During the winter the trade winds blow from the east-northeast off the cool land onto the warm sea. From June to October, though, when the land is heated, the direction of the air flow is reversed, and the winds blowing from the east-southeast south of India veer around to blow onto the Indian sub-continent and adjacent areas from the southwest. The duration and strength of the monsoon winds spell the difference between food sufficiency and starvation for hundreds of millions of people in this region each year. Under normal conditions, the monsoons bring heavy rains to the entire area, and the year's crop succeeeds; if the rains do not come or are inadequate, the crop fails.

EL NIÑO

The west coast of South America is normally a highly productive area, cooled by the Humboldt Current, which sweeps up from the south and even makes it possible for penguins, which are cold-water birds, to live on the Galapagos Islands near the equator. The productivity of the surface waters depends on this current and on the cool, nutrient-rich waters that well up from the depths to replace the warm, exhausted ones at the surface. In a normal January, however, warm water flows down the coast from the tropics to southern Peru and northern Chile. The local fishermen named this current El Niño ("The Christ Child") because it occurs near Christmas and is normally benevolent. But scientists now reserve this name for a catastrophic version of the same phenomenon, one that is felt not only locally but on a global scale.

Such events have been recorded as far back as 1726, and they occur about every 4 years. The most recent occurrence began in the summer of 1982. At that time the trade winds failed and warm waters spread north and south along the coast starting in September. Between August and October, the temperature of the coastal ocean had risen drastically, from about 16.5° C to nearly 22° C. By early 1983, weather patterns had changed along the west coast of both North and South America. Commercial fish stocks disappeared from the waters of Peru and northern Chile, and the plankton dropped to a twentieth of its normal abundance. Violent winter storms lashed the coast of California, and there was flooding in areas where it had not occurred for years.

On islands throughout the Central Pacific, many birds starved to death, and a number of huge bird colonies simply gave up breeding. On Christmas Island, an important breeding site about 2000 kilometers south of Honolulu, 95% of the 14 million birds that normally nest on the island, representing 18 species, simply left, abandoning their eggs, young, and nests; by June 1983, there were only 150,000 birds on the island. Even on the Farallon Islands, which lie about 45 kilometers off San Francisco, many birds starved, and the large breeding ground was tragically disrupted. Many eggs failed to hatch, and thousands of chicks starved.

Off the coast of Panama, more than two thirds of the coral reefs died; apparently not only the warm

Patterns of Circulation in the Ocean

In the ocean, patterns of circulation are ultimately determined by the patterns of atmospheric circulation but are modified substantially by the location of the land masses around which and against which the ocean currents must flow. Salinity—and therefore the density of the water—also significantly affects the patterns of oceanic circulation.

In general, oceanic circulation is dominated by huge surface **gyrals** (Figure 50-6) that move around the subtropical zones of high pressure between approximately 30 degrees north and 30 degrees south latitude. These gyrals move clockwise in the Northern Hemisphere and counterclockwise in the Southern Hemisphere. They profoundly affect life not only in the oceans but also on coastal lands by the ways in which they redistribute heat. For example, the Gulf Stream, in the North Atlantic, swings away from North America near Cape Hatteras, North Carolina, and reaches Europe near the southern British Isles. Therefore western Europe is much warmer and thus more temperate than is eastern North America at similar latitudes. There is a similar relationship in the North Pacific. Because western Eurasia and North America are more temperate than the eastern portions of the same continents, the distribution of biomes and the organisms that occur in them tends to rise in latitude from east to west; the same biomes occur farther north on the western sides of these continents than on their eastern sides.

In South America the Humboldt Current carries cold water northward up the west coast and helps to make possible an abundance of marine life that supports the fisheries of Peru and northern Chile. Marine birds, which feed on these organisms, are responsible for

temperatures but also herbicide runoff from the mainland contributed to this problem. Fisheries can be altered permanently by such conditions. After the 1972 El Niño, for example, overfishing in Peru so reduced the anchovy populations there that the anchovies have been unable to recover. At the same time, abundant rain and humidity in naturally dry areas like the Galapagos caused the populations of some species to explode. Fortunately, by July 1983, cooler waters and productive conditions had largely returned to the eastern tropical Pacific.

FIGURE 50-A

1 *Daphne Crater, Galapagos Islands, containing about 800 pairs of nesting blue-footed boobies in 1975, a normal year.*

2 *The same crater in 1983, an El Niño year, ringed with abundant vegetation but with no nesting boobies.*

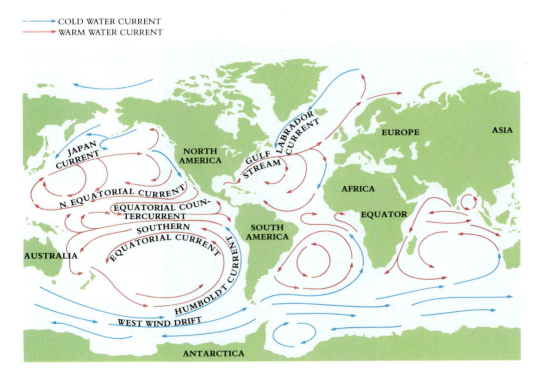

COLD WATER CURRENT
WARM WATER CURRENT

FIGURE 50-6

The circulation in the oceans moves in great surface gyrals and profoundly affects the climate on adjacent lands.

the commercially important guano deposits of these countries (see Figure 34-11). These deposits are rich in phosphorus, which is brought up from the ocean depths by the up-welling of cool water that occurs as a result of the frequent offshore wind from the generally mountainous slopes that border the Pacific. When the coastal waters are not as cool as usual, a devastating phenomenon called "El Niño" occurs (see boxed essay, pp. 1086-1087), which influences climatic patterns globally.

In the ocean, huge surface gyrals move around the subtropical zones of high pressure between approximately 30 degrees north and south latitude, clockwise in the Northern Hemisphere and counterclockwise in the Southern Hemisphere. As a result, the western sides of continents in the temperate zones of the Northern Hemisphere are warmer than their eastern sides; the opposite is true of the Southern Hemisphere.

THE OCEANS

More than three fourths of the earth's surface is covered by ocean. The seas have an *average* depth of more than 3 kilometers, and they are, for the most part, cold and dark. Hetero-trophic organisms are found at the greatest ocean depths, which reach nearly 11 kilometers in the Marianas Trench of the western Pacific Ocean, but photosynthetic organisms are confined to the upper few hundred meters of water (Figure 50-7). Organisms that live be-low this level obtain almost all of their food indirectly, as a result of photosynthetic activi-ties that occur above; these activities result in organic debris that drifts downward.

FIGURE 50-7

Fishes and many other kinds of animals find food and shelter among the kelp beds that occur in the coastal waters of temperate regions.

Because water is much denser than air, the minerals and gases dissolved in it diffuse much more slowly. The supply of oxygen can often be critical in the ocean, since it is present in some areas only in limited quantities. In addition, the warmer the water, the less the solubility of oxygen; the amount of available oxygen is therefore extremely important in limiting the occurrence of organisms in the warmer marine regions of the globe. Carbon dioxide, in contrast, is always plentiful in the oceans, and the distribution of minerals in the ocean is much more uniform than it is on land, where individual soils reflect faithfully the composition of the parent rocks from which they have weathered.

Despite the many new forms of animals, protists, and bacteria still being discovered in the sea and the huge biomass that occurs there, many fewer species live in the sea than on land. Probably more than 90% of all species of organisms occur on land, including the great majority of members of a few large groups—especially insects, mites, nematodes, fungi, and plants. Each of these groups has marine representatives, but these comprise only a very small fraction of the total number of species. On land the barriers between habitats are sharper, and variations in elevation, parent rock, degree of exposure, and other factors have all been crucial to the evolution of the millions of species of terrestrial organisms. This pattern of radiation into sharply defined habitats seems to account for the much larger numbers of species that live on land than in the sea. In terms of overall diversity, the pattern is very different: every phylum occurs in the sea, whereas relatively few phyla are found on land or in freshwater habitats. Most phyla originated in the sea, but only a few of them have been successful on land, although some of these have given rise to extraordinarily large numbers of species.

Although representatives of almost every phylum occur in the sea, an estimated 90% of living species of organisms are terrestrial. This occurs because of the enormous evolutionary success of a few phyla on land, where the boundaries between different habitats are sharper than they are in the sea.

The marine environment consists of three major kinds of habitats: (1) the **neritic** zone, the zone of shallow waters along the coasts of the continents; (2) the **surface layers** of the open sea; and (3) the **abyssal** zone, the deep-water areas of the oceans.

The Neritic Zone

The neritic zone is small in area, but it is inhabited by very large numbers of species as compared with other parts of the ocean (Figure 50-8). The intense and sometimes violent interaction between sea and land in this zone gives a selective advantage to well-secured organisms that can withstand being washed away by the continual beating of the waves. Part of this zone, the **intertidal**, or **littoral**, region, is exposed to the air whenever the tides recede.

Because of the way in which it affords access to the land, the intertidal zone must have been home for the ancestors of those organisms that originally colonized terrestrial habitats. Perhaps the greater complexity needed to anchor and fasten the animals and plants that dwell in this zone constituted a kind of preadaptation to life on land, where the environmental stresses are even more extensive. The organisms that live successfully in habitats that are regularly exposed to the air must also have some kind of waterproof covering or habits that protect them from the drying action of the air when the tide is out (Figure 50-9); such adaptations are of central importance for terrestrial organisms.

The world's great fisheries occur in shallow waters over continental shelves, either near the continents themselves or in the open ocean, when huge banks come near the surface. The preservation of these fisheries, which are a source of high-quality protein exploited throughout the world, is a matter of growing concern. In areas such as Chesapeake Bay (Figure 50-10), where complex systems of rivers enter the ocean from heavily pop-

FIGURE 50-8

Tidepools—depressions in coastal rock shelves in which water remains after the tide goes out—are often occupied by rich and varied communities of organisms. They are subject to exposure and drying out, depending on how far the tide goes out.

FIGURE 50-9

Drought-resistant intertidal organisms, such as these periwinkles, gastropod mollusks of the genus Littorina, often remain above the limits of the highest tides, obtaining moisture only from the sea spray.

ulated areas, the environmental stresses on highly productive environments have become so serious that they not only threaten the continued existence of formerly highly productive fisheries, but also diminish the quality of human life in these regions. For example, the increased amounts of runoff from farms and other activities and sewage effluent in areas like Chesapeake Bay provide nutrients for greater numbers of marine organisms, which in turn use up more and more of the oxygen in the water and thus may disturb the established populations of organisms such as oysters. Such effects may be enhanced by climatic shifts, and large numbers of marine animals may die suddenly as a result.

In tropical waters, where the water temperature remains at about 21° C, coral reefs occur. These are highly productive ecosystems that can successfully concentrate nutrients, even from the relatively nutrient-poor waters that are characteristic of the tropics (see Figure 27-8 and Chapter 36). The coral animals, which provide most of the structure of the reefs, together with the photosynthetic coralline algae and the many other kinds of organisms that live in and around the reefs, make up one of the most complex and fascinating living systems on earth. The dinoflagellates that live symbiotically within the coral animals contribute the greatest proportion of the reef's productivity.

The Surface Zone

Drifting freely in the upper, better-illuminated waters of the ocean, a diverse biological community exists, primarily consisting of microscopic organisms called **plankton.** Fish and other larger organisms that swim in these same waters constitute the **nekton,** whose members feed mainly on plankton. Together, the organisms that make up the plankton and the nekton provide all of the food for those which live below. Some of the members of the plankton, including the algae and some bacteria, are photosynthetic. Collectively, these organisms account for about 40% of all the photosynthesis that takes place on earth, even more by some calculations. Most of the plankton occurs in the top 100 meters of the sea, the zone into which light from the surface penetrates freely.

Many heterotrophic protists and animals also live in the plankton and feed directly

A

B

FIGURE 50-10

Chesapeake Bay, which has more than 11,300 kilometers of shoreline, drains more than 166,000 square kilometers in one of the most densely populated and heavily industrialized areas in North America. The body of open water is about 320 kilometers long and, at some points, nearly 50 kilometers wide.

A *Hampton Roads and Baltimore make the Bay one of the busiest natural harbors in the world.*

B *One of the most biologically productive bodies of water in the world, the Bay yielded an annual average of about 275,000 kilograms of fish in the 1960s but only a tenth as much in the 1980s. The human population of the area grew 50% during the same period.*

C *More than 290 oil spills were reported in the Bay in 1983, with oil transport and commercial shipping expected to double by the year 2020. This diving duck is coated from an oil spill off the mouth of the Potomac River.*

D *Uncontrolled erosion from certain agricultural practices, pesticides, and increases in nutrients block the light needed for photosynthesis and upset the delicate ecological balance on which the productivity of the Bay depends. The states that border the Bay are cooperating, with the assistance of the Environmental Protection Agency, to try to bring back its former productivity.*

C

D

on the photosynthetic organisms, as well as on one another. Gelatinous animals, especially jellyfish and ctenophores, which may consist of as much as 95% water, are abundant in the plankton but are relatively poorly known because their fragility makes them difficult to collect and study. The whales are the largest animals that graze on the plankton and nekton; indeed, they are the largest that have ever existed on earth—heavier than the largest dinosaurs known. A number of heterotrophic, free-swimming organisms move up into the plankton at times—in some instances on a regular daily or other cycle—to feed on the other organisms there at these times.

In recent years the existence of large numbers of minute photosynthetic organisms in the plankton, mostly 2 to 10 micrometers long, has been recognized. These organisms, which include photosynthetic bacteria (largely cyanobacteria) and some of the smallest algae, make up the **picoplankton.** They apparently contribute half or more of the total productivity in many marine ecosystems. The cyanobacteria are mostly only 0.2 to 1 micrometer long. Most of them are probably eaten by heterotrophic, flagellated protists 2 to 10 micrometers long, themselves a group of organisms that was scarcely known until recent years, even though they may consume a tenth or more of the total organic matter produced by all of the world's biological activity.

The populations of organisms that make up the plankton are able to increase so rapidly, and the turnover of nutrients in the sea is so great, that the amount of productivity in these systems has been seriously underestimated in the past. Even though nitrogen and phosphorus are often present in only small amounts and organisms may be relatively scarce, this productivity reflects rapid use and recycling rather than the abundance of these nutrients. The smallest organisms turn over the phosphorus much more rapidly than do the larger ones, and their role in these complex and productive ecosystems is just starting to be understood properly.

> **About 40% of the world's photosynthetic productivity is estimated to occur in the oceans. Of this, perhaps half is carried out by organisms less than 10 micrometers long. The turnover of nutrients in the plankton is much more rapid than in most other ecosystems, and the total amounts of nutrients are very low.**

The Abyssal Zone

In the deep waters of the sea, below the top 300 meters, occur some of the most bizarre organisms found on earth (see Figure 39-6). Many of these animals have some form of bioluminescence (see Figure 6-11, *A*), by means of which they communicate with one another or attract their prey. In the mud of the ocean floor, or along rifts from which warm water issues, live similar assemblages of peculiar creatures (see the boxed essay in Chapter 36). Bacteria are apparently rather common in the deeper layers of the sea and are as important in their role as decomposers in this zone as they are on land and in freshwater habitats.

FRESH WATER

Freshwater habitats are distinct from both marine and terrestrial ones, but they are very limited in area. Inland lakes cover about 1.8% of the earth's surface, and running water covers about 0.3%. Freshwater habitats are strikingly discontinuous; although the dispersal of organisms can occur through stream or river systems, ponds and lakes are often separated by relatively great distances, which organisms must cross to get from one body of water to another.

In two senses, freshwater habitats are strongly connected with terrestrial ones. First, marshes and swamps are intermediate habitats between bodies of fresh water and the land that surrounds them. Second, the organic and inorganic material that enters freshwater habitats is often derived on land—such as the terrestrial plant and animal matter that falls continually into a stream and is consumed there (Figure 50-11). Organisms must be able

FIGURE 50-11

Abundant organic material falls into a stream, which thus obtains much of its biological productivity from outside itself.

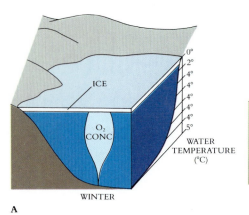
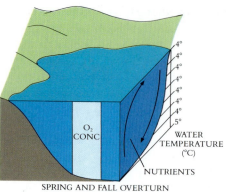
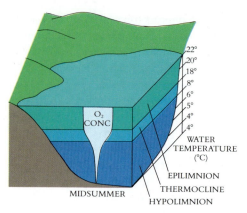

0°
2°
4°
4°
4°
4°
4°
5°

ICE

O₂ CONC

WATER TEMPERATURE (°C)

WINTER

4°
4°
4°
4°
4°
4°
5°

O₂ CONC

WATER TEMPERATURE (°C)

NUTRIENTS

SPRING AND FALL OVERTURN

22°
20°
18°
8°
6°
5°
4°
4°

O₂ CONC

WATER TEMPERATURE (°C)

EPILIMNION
THERMOCLINE
HYPOLIMNION

MIDSUMMER

A

FIGURE 50-12

The pattern of stratification in a large pond or lake (B) in temperate regions is upset in the spring and fall overturns (A). Conc, concentration.

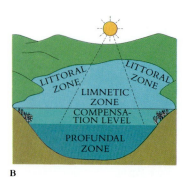

LITTORAL ZONE
LITTORAL ZONE
LIMNETIC ZONE
COMPENSATION LEVEL
PROFUNDAL ZONE

B

to attach themselves in such a way as to resist or avoid the effects of current in a stream or river, or risk being swept away; in bodies of standing water such matters are of much less importance.

Ponds and lakes, like the ocean, have three zones in which organisms occur: a **littoral** zone; a **limnetic** zone, inhabited by plankton and other organisms that live in open water; and a **profundal** zone, below the limits of effective light penetration (Figure 50-12, *A*). **Thermal stratification** is characteristic of the larger lakes in temperate regions (Figure 50-12, *B*). Since water is densest at about 4° C, water at that temperature sinks beneath water that is either warmer or cooler. In winter, water at 4° C sinks beneath cooler water, and that cooler water at the surface freezes at 0° C. Below the ice, the water remains between 0° and 4° C, and plants and animals survive there. In spring, as the ice melts, the surface water is warmed to 4° C and sinks below the cooler water, bringing that water to the top with nutrients from the lower regions of the lake. This process is known as the **spring overturn.**

In summer, warmer water forms a layer, known as the **epilimnion,** over the cooler waters (about 4° C) that lie below and that are called the **hypolimnion.** There is an abrupt change in temperature, the **thermocline,** between these two layers. Depending on the climate of the particular area, the epilimnion may become as much as 20 meters thick during the summer. In the autumn the temperature of the epilimnion drops until it is the same as that of the hypolimnion, 4° C. When this occurs, the epilimnion and hypolimnion mix— a process called the **fall overturn.** Colder waters reach the surfaces of lakes, therefore, in the spring and fall, bringing up near the surface fresh supplies of dissolved nutrients that have accumulated in them.

FIGURE 50-13

Eutrophic farm pond, with evident bloom of green algae, rapidly filling in with vegetation.

FIGURE 50-14

Lake Tahoe, an oligotrophic lake that lies high in the Sierra Nevada on the border between California and Nevada. The drainage of fertilizers applied to the plantings around residences, business concerns, and recreational facilities bordering the lake poses an ever-present threat to the maintenance of the deep blue color of its water.

In temperate lakes and ponds, there is an upper layer, or epilimnion; a lower layer, or hypolimnion; and a thermocline, which is a zone of abrupt temperature change, between them. The epilimnion is heated in summer but cooled in autumn to 4° C, the point at which water is densest. At that point, when their temperatures are equal, the shallower and deeper waters of the lake mix, with oxygen being carried to the depths and nutrients being brought to the surface.

Lakes can be divided into two categories, based on their production of organic matter. In **eutrophic** lakes there is an abundant supply of minerals and organic matter (Figure 50-13). Oxygen depletion occurs below the thermocline in the summer, because of the abundant organic material and the high rate at which aerobic decomposers in the hypolimnion use oxygen. These stagnant waters again reach the surface after the fall overturn. In **oligotrophic** lakes, on the other hand, organic matter and nutrients are relatively scarce; such lakes are often deeper than eutrophic ones (Figure 50-14), and their deep waters are always rich in oxygen.

BIOMES

As we have discussed, biomes are climatically delineated assemblages of organisms that have a characteristic appearance and that are distributed over a wide area on land. Biomes are classified in several ways, but, for our purposes, we will group them into seven categories: (1) tropical rain forests, (2) savannas, (3) deserts, (4) grasslands, (5) temperate deciduous forests, (6) taiga, and (7) tundra. They differ remarkably from one another because they have evolved in regions with very different climates. We will now examine some of their distinctive features. These and some of the related communities are shown in Figure 50-15. We will now examine some of the distinctive features of these seven biomes.

FIGURE 50-15

The distribution of biomes throughout the world.

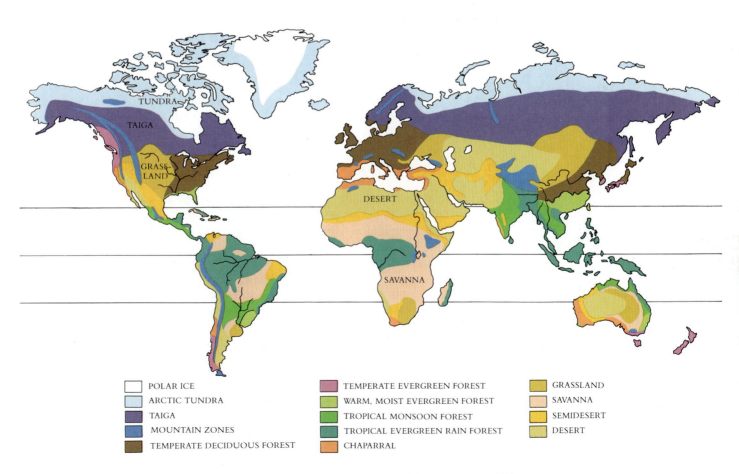

POLAR ICE	TEMPERATE EVERGREEN FOREST	GRASSLAND
ARCTIC TUNDRA	WARM, MOIST EVERGREEN FOREST	SAVANNA
TAIGA	TROPICAL MONSOON FOREST	SEMIDESERT
MOUNTAIN ZONES	TROPICAL EVERGREEN RAIN FOREST	DESERT
TEMPERATE DECIDUOUS FOREST	CHAPARRAL	

Tropical Rain Forests

The tropical rain forests (Figure 50-16, *A*) are the richest biome in terms of number of species, probably containing at least half of the species of terrestrial organisms—more than 2 million species. In such rain forests neither water nor temperature is a limiting factor, and an evergreen forest of giant trees supports a rich and diverse assemblage of plants and animals on its branches, high above the forest floor (Figure 50-16, *B* and *C*). As you saw in Chapter 31, evergreen plants retain their leaves all year and do not shed them in a particular season; for example, most conifers—a group that is very poorly represented in the tropics—are evergreen. Plants that grow on the branches of other plants are called **epiphytes** (see boxed essay, Chapter 34), and, along with the numerous vines that occur in tropical forests, they reach the light high up in the forest canopy. The communities that make up tropical forests are rich in species and diverse, so that each kind of plant, animal, or microorganism is often represented in a given area by very few individuals. There are seldom fewer than 40 species of tree per hectare, four or five times as many as are typical in temperate forests. The ways of life of tropical organisms are often specialized and highly unusual (Figure 50-16, *D*); we will discuss some further examples in Chapter 53.

The rainfall in areas where tropical rain forests occur is generally 200 to 450 centimeters per year, with little difference in its distribution from season to season. About two thirds of the soils of the tropics are acidic and deficient in phosphorus, potassium, calcium, magnesium, and other nutrients. In addition, the phosphorus in them tends to combine with iron or aluminum to form insoluble compounds that are not available to plants. Such soils also tend to have toxic levels of aluminum. The roots of the trees spread out in a thin layer of soil, often no more than a few centimeters thick, and their roots, virtually all involved in mycorrhizal associations (see Chapter 28), transfer the nutrients from the leaves and other fallen organic debris quickly and efficiently back to the trees themselves. Below the thin layer of topsoil, there is virtually no organic matter, and nutrients are scarce; most of the nutrients that exist in the system are concentrated in the trees themselves.

Tropical rain forests are highly productive, even though they exist mainly on infertile soils. Most of the nutrients are held within the plants themselves and are rapidly recycled when the plants die or when parts, such as leaves, are lost.

Tropical rain forests are widespread in South America, particularly in and around the Amazon Basin; in Africa, particularly in Central and West Africa; and in Southeast Asia. Their ecological properties are so unusual that we simply do not know, in most cases, how to cultivate them in such a way as to maintain their agricultural productivity year after year. Even so, these forests are being cut down at an ever-increasing rate, mainly by people living at the edge of starvation. The fields that these people clear can generally be cultivated for only a few years, after which they are worthless, unless they are given a period of many decades to recover. When wide areas are cleared, their forests will probably never recover. The thin soils erode rapidly, and their minerals are carried away with the trees and the crops that are harvested from the cultivated plants.

The human populations of the tropics and subtropics now constitute more than half of the world total, and their numbers are growing rapidly. Extremely poor people constitute more than a third of the population of most of these countries, and these people are cutting the forests very rapidly. We know little about how to cultivate tropical soils on a sustainable basis when their forest cover has been removed. It is estimated that, as a result of the interaction of these factors, few tropical forests will be left in an undisturbed condition anywhere in the world by the first part of the next century.

The destruction and disturbance of all tropical rain forests will be accompanied by the extinction of a major proportion of all plant, animal, and microorganism species on earth—perhaps a quarter—during the lifetime of many of us. Studying the plants and animals of these forests—which are the most poorly known, as well as the most numerous

A

B

C

D

FIGURE 50-16

A *Huge trees, many with buttresses like the ones prominent in this photograph, dominate tropical rain forests.*
B *More plants and animals occur in the canopy layers of a tropical rain forest, high above the ground, than anywhere else in this richest of all biological communities. An epiphytic orchid growing on a tree trunk.*
C *A leaflike katydid from the forest canopy in Brazil. The markings on its wings resemble holes bitten in a genuine leaf by other insects, making the camouflage more nearly perfect.*
D *Rafflesia, a leafless angiosperm that is parasitic on the roots of a wild relative of grapes in the forest of Southeast Asia, has the largest flowers of any angiosperm; these flowers sometimes are nearly 1 meter across. They are ill-scented and are pollinated by carrion flies, which mistake them for rotten meat and lay their eggs in them, spreading the plant's pollen in the course of these activities.*

in terms of numbers of species on earth at present—is a matter of pressing importance from both a scientific point of view and from the point of view of improving the conditions of life for human beings. Such study would undoubtedly uncover many species of organisms of great scientific interest, and of potential importance in terms of contributions to the quality of human life.

Savannas

In areas of reduced annual precipitation, or prolonged annual dry seasons, open tropical and subtropical deciduous forests give way to the kind of open grassland called **savannas** (Figure 50-17). The savanna biome, on a global scale, is transitional between tropical rain forest, which is evergreen, and desert. Generally 90 to 150 centimeters of rain fall each year in savannas. There is a wider fluctuation in temperature here during the year than in the tropical rain forests, and there is seasonal drought. These factors have led to the evolution of an open landscape, often with widely spaced trees, in which large grazing mammals are sometimes characteristic, as in Africa.

FIGURE 50-17

The herds of grazing mammals that inhabit the African savannas are one of the world's greatest sights.

Some tens of thousands of years ago, toward the close of the Pleistocene Epoch, vegetation similar in appearance to that of today's African savannas was widespread in North America (see Figure 23-2). It disappeared as the climate became more and more like it is now—with even greater extremes of temperature and longer periods of seasonal drought, especially in the West. In many areas, human beings seem to have slaughtered the remaining large animals that made up the extraordinary Pleistocene herds, and the vegetation evolved into the modern communities that we know today.

In savannas, the trees are usually deciduous and lose their leaves in the dry season. Savannas often gradually give way on their drier borders to thorn forest, plant communities dominated by thorny trees that are seasonally deciduous. In Southeast Asia, a similar plant community called monsoon forest occurs in such dry regions. Under and among the scattered trees of these communities, perennial grasses and other plants with food stored in roots or underground stems are common.

Desert

Less than 25 centimeters of annual precipitation usually falls in the world's desert areas—so low an amount that water is the predominant controlling variable for most biological processes and is also highly variable in quantity both during a given year and between years. In desert regions the vegetation is characteristically sparse (see Figures 2-10 and 2-11). Such regions occur around 20 to 30 degrees north and south latitude, where the warm air that rises near the equator falls and precipitation is limited. Deserts are most extensive in the interiors of continents, especially in Africa (the Sahara Desert), Eurasia, and Australia. Less than 5% of North America is desert.

As air currents pass over mountains, the air is cooled. When this occurs, its moisture-holding capacity decreases, resulting in increased precipitation on the windward side of the mountains—the side from which the wind is blowing. As the air descends the other side

A

B

FIGURE 50-18

A *Some of the heaviest snowfalls on earth have been recorded on the western slope of the Sierra Nevada in California, where Yosemite Valley cuts through a rich coniferous forest.*
B *A few miles away as the crow flies are the sagebrush flats of the Great Basin desert, which lie east of the range. Moisture-laden air from the Pacific Ocean is cooled as it ascends the Sierra Nevada, resulting in abundant precipitation. As the drier air descends the east side of the range, it is heated, gaining in moisture-holding capacity and decreasing the likelihood of additional precipitation east of the mountains.*

of the mountains, the lee side, it is warmed, and its moisture-holding capacity increases, tending to block precipitation. As a result, the two sides of a north-south mountain range often look quite different if the range intersects a strong east-west or west-east air current (Figure 50-18).

Because the vegetation is sparse and the skies are usually clear, deserts radiate heat rapidly at night. This leads to large daily changes in temperature, sometimes exceeding 30° C, between day and night. Summer daytime temperatures in deserts are extremely hot, frequently exceeding 40° C. Indeed, atmospheric temperatures of 58° C have been recorded both in Libya and in San Luis Potosí, Mexico—the highest temperatures that have been recorded on earth.

In deserts the cover of trees, shrubs, and succulents, such as cacti, is often sparse; a thicker cover indicates greater average annual precipitation than a sparse one, or underground sources of water. Annual plants often are abundant in deserts (see Figure 31-5, *A*) and simply bypass the unfavorable dry season in the form of seeds. After the rainfall, they germinate and grow rapidly, sometimes forming spectacular natural displays. Exhibiting a similar seasonal rhythm, animals such as fairy shrimps (see Figure 37-22, *B*) often appear and breed rapidly in temporary ponds.

The trees and shrubs that live in deserts often have deep roots that reach sources of water far below the surface of the ground. Thus trees may grow, even in regions that essentially lack precipitation, such as in the Atacama Desert of northern Chile. The woody plants that grow in deserts may be either deciduous, losing their leaves during the hot, dry seasons of the year, or evergreen, with hard, reduced leaves: the creosote bush *(Larrea)* of the deserts of North and South America is an example of an evergreen desert shrub. In some desert plants the leaves may be densely covered with trichomes, which tend to slow down transpiration and cool the leaves (see Figure 31-13, *C*). Near the coasts, in areas where there is cold water offshore, deserts may be foggy, and the water that the plants obtain from the fog may allow them to grow quite luxuriantly.

Succulent plants are more common in desert regions than elsewhere. They often have CAM photosynthesis, in which CO_2 is taken up largely at night and fixed into C_4 intermediates, largely malic acid. The malic acid and similar C_4 compounds are decarboxylated, and the products are then fixed by the Calvin cycle during the day. Thus in CAM photosynthesis there is a temporal separation of the C_4 pathway and the Calvin cycle, which takes the place of the spatial separation of these pathways that occurs in ordinary C_4 plants. The stomata of plants in which CAM photosynthesis occurs open at night and close during the day, thus conserving water. Succulent plants that have this system are thus able to avoid the effects of the intense heat of the desert days and still photosynthesize efficiently.

Desert animals, too, have fascinating adaptations that enable them to cope with the limited water and high temperatures of the deserts. They often limit their activity to a relatively short period of the year when water is available, or even plentiful: they resemble annual plants in this respect (Figure 50-19). Many desert vertebrates live in deep, cool, and sometimes even somewhat moist burrows, and some of them that are active over a greater portion of the year emerge from these burrows only at night, when temperatures are rel-

FIGURE 50-19

Spadefoot toad, Scaphiophus. Spadefoot toads, which live in the deserts of North America, can burrow nearly a meter below the surface and remain there for as much as 9 months of each year. Under such circumstances, their metabolic rate is greatly reduced, and they depend largely on their fat reserves. When moist, cool conditions return to the desert, they emerge and breed rapidly. The young toads mature rapidly and burrow back underground, using the horny projections on their feet, which give these unusual desert toads their name.

THE DUNES OF THE NAMIB

The vast dunes of the Namib Desert stretch for more than 2000 kilometers along the southwestern coast of Africa. These dunes, some of which are up to 300 meters high, feature a wide variety of habitats, including open sand, gravel plains, rocky hills, and dry watercourses. These diverse habitats are home to one of the most remarkable arrays of animals found anywhere on earth.

Near the coast, the annual precipitation is less than 5 *millimeters;* the moisture derived from fog is the main source of water for plants and animals. The rainfall gradually increases to about 100 millimeters about 100 kilometers from the coast; inland, fog is rare. On foggy nights near the coast, certain kinds of lizards and insects emerge from the dunes to collect moisture, which condenses on their bodies. Some beetles construct trenches, which serve the same purpose, and then return later to collect water from them. Organic debris tends to accumulate on the sands that form on the leeward crests of the dunes, and a whole community of insects has evolved that uses this debris as a source of energy. Insects and gerbils rely on this material as their main source of food, and predators—including the golden mole,

FIGURE 50-B

1 *The dunes of the Namib Desert.*
2 *An individual of the beetle* Onymacris unguicularis *collecting fog water by holding up its abdomen at the crest of a dune as shown here, with the water condensing on its body.*
3 *Dr. William Hamilton, of the University of California, Davis, a leading student of the ecology of the Namib Dunes, watching an individual of* Onymacris unguicularis. *When they are warm, they will not tolerate such close approach.*
4 Onymacris bicolor. *The ambient temperature at the level of the beetle, about 2 centimeters above the sand, is 40° C; the body temperature of beetles in comparable situations was measured at 41° C. The beetle, which is very active at this temperature, is alternately lifting its legs from the sand, which has a surface temperature of 54° C.*
5 *Trenches made by the beetle* Lepidochora kahani. *These trenches, which are perpendicular to the fog-laden winds, concentrate moisture during fogs; the beetles then return along the ridges of the trenches, collecting water from them.*
6 Welwitschia mirabilis, *a gymnosperm, one of the most bizarre plants in the world. The two large, strap-shaped leaves are continuously regenerated from the base, break into segments, and fray off at the end.*

1

2

atively cool (see Figure 1-7). Some, like camels, can drink large quantities of water when it is available and can then safely withstand the loss of much of it. Many animals simply migrate to or through the desert, where they exploit food that may be abundant seasonally; when the food disappears, the animals move on to more favorable areas.

Many arthropods combine several of these strategies with a tough exoskeleton and colors suited to the absorption or reflection of heat, depending on their habits. The cuticles

which is a burrowing mammal; small reptiles; and trap-door spiders—feed on the herbivores. Farther inland, plants grow rapidly during the infrequent periods of rainfall, and animals breed at such times.

The nights in the Namib Desert are cool, but the days warm rapidly. Temperatures on the surface of the sand can reach 66° C in the afternoon, so animals must carry out their activities in the mornings and late afternoons, when temperatures are more moderate and favorable. Many of the animals are largely subterranean, like the golden mole and a number of the insects, and others burrow beneath the sand to insulate themselves from the searing surface temperatures.

Welwitschia, one of the most bizarre plants anywhere, is found only in the region of the Namib Desert. It has two large, leathery, strap-shaped leaves that keep growing from a meristem at their bases and become ragged at their ends. Stalks with the pollen-bearing and seed-bearing cones are also produced by this meristem, on different individuals. The carrotlike roots of *Welwitschia* reach deep into the gravelly rivercourses of the Namib, allowing the plants to survive in these extremely dry areas. Once, as shown by the record of fossil pollen, *Welwitschia* was common over large areas on several continents; it now exists only in or near the Namib Desert.

and exoskeletons of all arthropods constitute an excellent preadaptation to desert conditions and an important reason why the members of this phylum are so successful in deserts.

Desert survival depends on water conservation, achieved by structural, behavioral, or physiological adaptations. Plants and animals may restrict their activity to favorable times of the year, when water is present; they must also avoid high temperatures.

One of the means by which desert animals avoid seasonal extremes in heat and dryness in the desert is **estivation.** Estivation is a prolonged state of torpor that occurs under hot, dry conditions; **hibernation** describes the more profound and prolonged condition of dormancy that certain animals undergo during cold winters. Ground squirrels (*Citellus*) hibernate in the very cold regions of North America, whereas the species of the same genus that occur on the deserts estivate, each thereby avoiding the extremes of the most unfavorable seasons. Estivation occurs in a number of desert rodents and some other animals, including a few birds such as the poorwill. It also occurs, as you saw in Chapter 39, in the lungfishes, when the ponds and lakes in which they live dry up.

Grasslands

Temperate grasslands cover much of the interior of North America, and they are widespread in Eurasia and South America as well. Such grasslands are often highly productive when they are converted to agriculture, and many of the rich agricultural lands in the United States and southern Canada were originally occupied by prairies, another name for temperate grasslands. The roots of perennial grasses characteristically penetrate far into the soil, and grassland soils tend to be deep and fertile. Temperate grasslands differ from savannas in that they occur in regions with relatively long, cold winters, whereas savannas occur where there is a relatively cool dry season and a hot, rainy one.

Where precipitation is relatively abundant, as in the eastern portion of the North American prairies, a kind of vegetation known as tall-grass prairie once existed (Figure 50-20). Where there is even more rainfall, as occurs in North America east of the prairie zone, forests occur. The line between tall-grass prairie and forest in the United States lies roughly along the border between Illinois (prairie) and Indiana (forest), but this boundary is not sharp. The forests of the eastern United States contain numerous local patches of prairie, sometimes called **glades** (Figure 50-21), in which prairie plants and animals occur, often far from their main ranges. Similarly, patches of woods may occur in prairies, and often are found along streams.

FIGURE 50-20

Tall-grass prairie stretched over thousands of square kilometers in the interior of North America when Europeans first came to the area. This scene, in which a native sunflower (Helianthus grosseserratus) *is prominent, is in western Illinois. Over much of its former area, tall-grass prairie has been replaced by cultivated fields.*

FIGURE 50-21

Glades, rocky openings in temperate deciduous forest, are inhabited mostly by prairie plants and animals. In this glade in the Ozarks near St. Louis, coneflower (Echinacea purpurea) dominates the view in June.

Farther west in the United States and Canada, another kind of grassland, the short-grass prairie, occurs (Figure 50-22). Short-grass prairie receives less rain than tall-grass prairie and is exceptionally sensitive to disturbance through overuse. The infamous "Dust Bowl" conditions that occurred in the United States in the 1930s stemmed from the mismanagement of short-grass prairies, linked with a succession of dry years that was not abnormal for such areas.

Temperate grasslands are often populated by herds of grazing mammals, among which the North American bison is a familiar example. In this respect, temperate grasslands resemble tropical savannas. Both biomes are characterized by large quantities of perennial grasses that are highly productive when properly managed and that can support such herds.

FIGURE 50-22

Short-grass prairie with grazing bison.

Temperate Deciduous Forests

In areas of the Northern Hemisphere with relatively warm summers, relatively cold winters, and sufficient precipitation, temperate deciduous forest occurs (Figure 50-23). Temperate deciduous forest and related vegetation types cover very large areas, including much

FIGURE 50-23

The leaves of the trees in the temperate deciduous forest often change color before they fall.

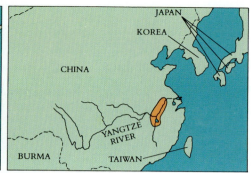

FIGURE 50-24

Distribution of alligators in North America and China. Above is an American alligator; the Chinese species has become very rare and is regarded as endangered. These are the only two species of the genus Alligator.

FIGURE 50-25

Many species of perennial herbs that grow on the floor of temperate deciduous forest, such as this false rue anemone (Isopyrum biternatum), flower and are visited by insects in the well-lighted conditions that exist before the trees have put out their leaves in spring.

of the eastern United States and Canada and an extensive region in Eurasia. Similar vegetation types are, in contrast, essentially unknown in the Southern Hemisphere, where climates are generally less extreme and deciduous trees are infrequent. All parts of the Southern Hemisphere are relatively near the oceans, which moderate the climates on the continents greatly.

The existing deciduous forests are remnants of a more continuous, richer forest that stretched across the Northern Hemisphere before the Pleistocene Epoch but that has become more restricted in distribution and in the array of species it houses as the world climate has deteriorated, especially during the past several million years. For this reason the plants and animals of the rich forests of southwestern and western China resemble those of eastern North America (Figure 50-24). Many fewer species are found in the similar forests of Europe, which were so disturbed by the successive advances of the ice over the past several million years. Apparently north of the Alps, the European forests were simply unable to move to more favorable areas, and many species of plants and animals, now common to Asia and North America, became extinct in Europe at this time.

Annual precipitation in areas of temperate deciduous forest is generally from about 75 to about 250 centimeters. Precipitation is well distributed throughout the year, but water is generally unavailable during the winter because it is frozen. Where there is less precipitation, temperate deciduous forests are replaced by grassland, as in the prairies of North America or the steppes of Eurasia. Where conditions are more limiting—where restricted, for example, by intense cold—these forests may be replaced by others that consist of fewer species, as in the taiga (see below) or the predominantly coniferous forests of western North America (Figure 50-18, *A*). Many of the same kinds of deciduous trees that now occur in eastern North American and eastern Asia, but which are absent in western North America, existed in that region a million years or more in the past; they disappeared largely because regular summer precipitation has stopped. The same is also true for the regions occupied by taiga and even tundra, where rich mixed forests occurred in the later Cretaceous and early Tertiary periods.

Many perennial herbs live in areas of temperate deciduous forest (Figure 50-25). They characteristically grow rapidly and flower early in the year, before the trees have begun to unfold their leaves. The herbs are able to grow so rapidly because of their underground storage organs, swollen roots or shoots in which the photosynthetic products of earlier seasons have been stored.

The **chaparral** of California—as well as related evergreen scrub associations that occur in the five areas of the world with a mediterranean, or summer-dry, climate—is historically derived from forests like the richest temperate deciduous forests that now grow in eastern Asia and eastern North America (Figure 50-26). Because of the climatic patterns

FIGURE 50-26

Frequently burned slopes with chaparral above Clear Lake in the North Coast Ranges of California.

we reviewed earlier in this chapter, the regions where this kind of evergreen scrub grows are dry in summer and moist in the winter growing season. In the Mediterranean area itself, in central Chile, in the Cape region of South Africa (see Figure 53-5), and perhaps to a lesser extent in southern and southwestern Australia and in California, human activities have also contributed significantly to the formation of such plant communities.

Taiga

Taiga is the northern coniferous forest, primarily spruce, hemlock, and fir, that extends across vast areas of Eurasia and North America (Figure 50-27). It is characterized by long, cold winters, in which the cold air contains little moisture; most of the precipitation falls in the summers. Because of the latitude where taiga occurs, the days are short in winter (as

FIGURE 50-27

A *Taiga, here a forest dominated by spruces and alders, in Alaska.*
B *Moose are one of the characteristic mammals of the taiga.*

A

B

little as 6 hours) and correspondingly long in summer. During the summer, plants may grow rapidly, and crops often attain a large size in a surprisingly short time. Marshes, lakes, and ponds are common here, and they are often fringed by willows or birches. Most of the trees in the taiga form ectotrophic mycorrhizae (see Chapter 28), and they tend to occur in dense stands of one or a few species. Alders, which are common, harbor nitrogen-fixing actinomycetes in nodules on their roots; partly for this reason, they are able to colonize raw, infertile soils such as those left behind recently by the retreat of glacial or other ice (see Figure 34-14).

Many large mammals live in the taiga, including elk, moose, deer, and carnivores such as wolves, bear, lynx, and wolverines. Traditionally, much fur trapping has gone on in this region, which is also an important lumber-producing region. To the south, taiga grades into forests or grasslands, depending on the amount of precipitation. Northward, it gradually gives way to open tundra. Coniferous forests also occur in the mountains to the south, but these are often richer and more diverse in species than those of the taiga. In the early Tertiary Period, 40 to 60 million years ago, all of these conifers grew in richer, mixed forests that extended far to the north, above the Arctic Circle. As more severe winters developed over much of the region and sufficient summer rainfall disappeared from the ranges of western North America, ecosystems formed that were poorer in their representation of species, such as taiga.

FIGURE 50-28

A *Tundra in Mt. McKinley National Park, Alaska.*
B *Caribou live in large herds that migrate across the tundra. Sometimes called reindeer, they have been domesticated by the Lapps in northern Scandinavia as a source of meat, milk, and hides.*
C *Arctic ground squirrels live in burrows and consume plant material actively during the short Arctic summers.*

A

B

C

Tundra

Farthest north in Eurasia, North America, and their associated islands, between the taiga and the permanent ice, occurs the open, often boggy, grassland community known as the **tundra** (Figure 50-28). This is an enormous biome, extremely uniform in appearance, that covers a fifth of the earth's land surface. Trees are small and are mostly confined to the margins of streams and lakes; in general the tundra's appearance is like that of some parts of the prairies.

Annual precipitation in the tundra is very low, usually less than 25 centimeters, and the water is unavailable for most of the year because it is frozen. During the brief Arctic summers, water sits on frozen ground, and the surface of the tundra is often extremely boggy then. **Permafrost,** or permanent ice, usually exists within a meter of the surface.

The tundra receives little precipitation, usually less than 25 centimeters per year, but the water is often trapped near the surface by the widespread permafrost. For that reason, the tundra is often boggy.

As in the taiga, the herbs of the tundra are perennials that grow rapidly during the brief summers, using food stored underground. Large grazing mammals, including musk-oxen, caribou, reindeer, and carnivores such as wolves, foxes, and lynx, live in the tundra, which teems with life in the short summer. Lemmings, a genus of small rodents, are animals of the tundra, and their populations rise rapidly and then crash on a long-term cycle, with important effects on the populations of the animals that prey on them.

SUMMARY

1. Populations of different organisms that live together in a particular place are called communities. A community, together with the nonliving components of its environment, is called an ecosystem.

2. Biomes are major terrestrial assemblages of plants, animals, and microorganisms that occur together over wide geographical areas and have definite characteristics that distinguish them from other such assemblages.

3. Warm air rises near the equator and flows toward the poles, descending at about 30 degrees north and south latitude. Since the air is falling in these regions, it is warmed and its moisture-holding capacity is therefore increased. The great deserts of the world are formed in these latitudes.

4. Ocean circulation is dominated by huge surface gyrals that move around the subtropical high-pressure zones between approximately 30 degrees north and 30 degrees south latitude. They move clockwise in the Northern Hemisphere and counterclockwise in the Southern Hemisphere.

5. The ocean comprises three major environments: the neritic zone, the surface layers, and the abyssal zone. The neritic zone, which lies along the coasts, is small in area but very productive and rich in species. The surface layers are the habitat of plankton—drifting organisms—and nekton—actively swimming ones. The productivity of this zone has been underestimated because of the very small size (less than 10 micrometers) of many of its key organisms, and because of its rapid turnover of nutrients.

6. Freshwater habitats comprise only about 2.1% of the earth's surface; most of them are ponds and lakes. These possess a littoral zone, a limnetic zone, and a profundal zone. As autumn changes to winter, cooler water forms and sinks, since water is most dense at 4° C, and the water of the lake is mixed by this means. When winter changes to spring, warmer water (at 4° C) sinks, producing a similar mixing.

7. We recognize seven major biomes, or terrestrial communities, in this book: (1) tropical forests, (2) savannas, (3) deserts, (4) grasslands, (5) temperate deciduous forests, (6) taiga, and (7) tundra. Tropical forests contain at least two thirds of the species of plants, animals, and microorganisms. These are the most poorly known organisms on earth and the most directly threatened by extinction in the near future.

8. Savannas, which were much more widespread during the Pleistocene Epoch, are highly productive ecosystems. They are often inhabited by large herds of grazing mammals and their predators, including human beings.

9. Deserts are the hottest and driest habitats on earth. They are of great biological interest because of the extreme behavioral, morphological, and physiological adaptations of the plants and animals that live in them.

10. Temperate grasslands cover a large area in the interior of North America; they are also widespread in Eurasia and South America. When they receive sufficient precipitation, they are often well suited to agriculture.

11. Temperate deciduous forests once dominated huge areas of the Northern Hemisphere. Today, though, they are confined largely to eastern North America and parts of southwestern China. Areas with mediterranean (summer-dry) climates feature distinctive life forms that are historically derived from those of the temperate deciduous forests.

12. Taiga is a vast coniferous forest that stretches across Eurasia and North America. Its winters are long and cold, but plants may grow rapidly here during the long summer days. Even farther north is the tundra, which covers about 20% of the earth's land surface and consists largely of open grassland, often boggy in summer, which lies over a layer of permafrost.

SELF-QUIZ

The following questions pinpoint important information in this chapter. You should be sure to know the answers to each of them.

1. _____ are terrestrial assemblages of organisms that have a characteristic appearance and that occur over a wide geographical area.

2. What percentage of solar energy is allowed through the earth's atmosphere?
   (a) 1   (d) 80
   (b) 20   (e) 100
   (c) 50

3. The moisture-holding capacity of air _____ when it is warmed and _____ when it is cooled.
   (a) decreases/increases
   (b) increases/decreases

4. More than three fourths of the surface of the earth is covered by oceans. True or false?

5. The great fisheries of the world occur in
   (a) shallow waters over continental shelves
   (b) shallow waters not over continental shelves
   (c) deep water not over continental shelves
   (e) two of the above

6. The abrupt change in temperature between the epilimnion and the hypolimnion is known as the _____.

7. In oligotrophic lakes, organic matter and nutrients are relatively abundant. True or false?

8. Plants that grow on the branches of other plants are called
   (a) eutrophytes   (d) endophytes
   (b) photophytes   (e) autophytes
   (c) epiphytes

9. The biome that is transitional between tropical rain forests and deserts is
   (a) grasslands   (d) temperate deciduous forests
   (b) taiga   (e) savannas
   (c) tundra

10. What leads to the large daily changes in temperature between day and night in the deserts?
    (a) rapid radiation of heat   (d) prevailing westerly winds
    (b) constant wind   (e) a and c
    (c) sparse vegetation

11. CAM photosynthesis
    (a) conserves water   (d) is more common in cold climates
    (b) conserves CO_2 (e) does not involve the Calvin
    (c) is another name for   cycle.
        C_4 photosynthesis

12. The prolonged state of torpor similar to hibernation, which occurs in some endothermic animals living in regions that at times are very hot and dry, is called _____.

THOUGHT QUESTIONS

1. How do you think the decomposition rates of organic matter would differ in a tropical rain forest, a temperate deciduous forest, a desert, and the tundra? How would these differences affect human existence in each area?

2. What kinds of biological communities would you expect to find on the windward and leeward sides of a mountain range in an area where the annual precipitation ranged between 20 and 100 centimeters per year and was distributed mainly in one rainy season, and there were no extremes of climate? How would the height of the mountain range affect the situation?

3. Near the coast in southern California there are two major plant communities. Right along the ocean occurs the coastal sage community, which is dominated by low shrubs that often wither or lose their leaves in the summer. Higher up, on the ridges, occurs the chaparral, a community dominated by tall evergreen shrubs. Which of these communities would you think grows in the area that receives the most rainfall? Which grows on the better soils? Which is the more productive on an annual basis? Where would you expect to find the most annual plants?

FOR FURTHER READING

ATTENBOROUGH, D.: *The Living Planet: A Portrait of the Earth,* William Collins Sons & Co., Ltd., and British Broadcasting Corporation, London, 1984. A beautifully written and illustrated account of the biomes.

BARBER, R.T., and F.P. CHAVEZ: "Biological Consequences of El Niño," *Science,* vol. 222, pages 1203-1210, 1983. This and the two articles that precede it in the magazine, dealing with the oceanic and atmospheric aspects of the 1982-1983 event, provide a clear picture of what we do and don't know about this impressive phenomenon.

DUFFEY, E.: *The Forest World: The Ecology of the Temperate Woodlands,* A & W Publishers, Inc., New York, 1980. A well-illustrated account of the temperate forests of the world, with an emphasis on Europe.

GLEASON, H.A., and A. CRONQUIST: *The Natural Geography of Plants,* Columbia University Press, New York, 1964. A very readable survey of the principles of plant geography.

GOLLEY, F.B. (ed.): *Tropical Rain Forest Ecosystems: Structure and Function,* Ecosystems of the World, vol. 14A. Elsevier Scientific Publishing Company, New York, 1983. An up-to-date collection of reviews of various aspects of tropical forest ecology by leading experts.

LAWS, R.: "Antarctica: A Convergence of Life," *New Scientist,* vol. 99, pages 608–616, 1983. A dynamic and beautifully illustrated view of one of the most productive oceans on earth.

LOUW, G.N., and M.K. SEELY: *Ecology of Desert Organisms,* Longman, Inc., New York, 1982. A short but excellent book on deserts and the mechanisms by which plants and animals cope with the high temperatures, limited moisture, and unpredictability of this harsh environment.

McNAUGHTON, S.J.: "Grazing Lawns: Animals in Herds, Plant Form, and Coevolution," *American Naturalist,* vol. 124, pages 863-886, 1984. The fascinating story of the ways in which differences in grasses and in the populations of animals that depend on them help to determine the structure of the savanna biome.

MYERS, N.: *The Primary Source: Tropical Forests and Our Future,* W.W. Norton, New York, 1984. An excellent overall account of the characteristics of tropical forests and their importance for us all.

PERRY, D.R.: "The Canopy of the Tropical Rain Forest,"*Scientific American,* November 1984, pages 138-147. This article describes efforts to explore the richest and least understood biological community on earth.

RASMUSSON, E.M.: "El Niño and Variations in Climate," *American Scientist,* vol. 73, pages 168-177, 1985. An analysis of the way in which large-scale interactions between the ocean and the atmosphere over the tropical Pacific Ocean can dramatically affect weather patterns around the world.

WHITTAKER, R.H.: *Communities and Ecosystems,* 2nd ed., Macmillan Publishing Co., Inc., New York, 1975. A good introductory text that clearly outlines the characteristics of the biomes.

Overview

Ecosystems are complex associations of plants, animals, and microorganisms that interact with their nonliving environment in such a way as to regulate the flow of energy through them and the cycling of nutrients within them. Carbon dioxide, nitrogen gas, and water are the reservoirs of the carbon, nitrogen, and hydrogen that are used in biological processes. All of the other elements that organisms incorporate into their bodies come from the earth's rocks. Through photosynthesis, plants growing under favorable circumstances capture and lock up about 1% of the sun's energy that falls on their green parts. They may then be eaten by herbivores (primary consumers), which in turn may be eaten by secondary consumers (carnivores or decomposers), and so on. This type of sequence constitutes a food chain, and the different links in it, called trophic levels, include organisms that can transfer about 10% of the energy that exists at each level to the next level. Communities become more complex and stable through succession, but human beings are threatening this stability everywhere because of activities of their very large and rapidly growing population.

For Review   *Here are some important terms and concepts that you will encounter in this chapter. If you are not familiar with them, you should review them before proceeding.*

- **Respiration** (Chapter 10)
- **Nitrogen cycle** (Chapters 26 and 34)
- **Phosphorus cycle** (Chapter 34)
- **Biomes** (Chapter 50)

Ecosystems are the most complex level of biological organization. They are systems in which there is a regulated transfer of energy and an orderly, controlled cycling of nutrients. The individual organisms and populations of organisms in an ecosystem act as parts of an integrated whole, adjust over time to their role in the ecosystem, and relate to one another in complex ways that we only partly understand. Despite their differences, all ecosystems regulate the flow of energy—ultimately derived from the sun—and the cycling of nutrients, and all are governed by the same principles and restricted by the same limitations.

The earth is a closed system with respect to the chemicals, but an open one in terms of energy. Ecosystems function to regulate the capture and expenditure of that energy and the cycling of those chemicals. As we will see in this chapter, all organisms, including human beings, depend on the ability of a few other organisms—plants, algae, and some bacteria in the case of carbon, and certain bacteria in the case of nitrogen, for example—for the basic components of life.

As distinct functional units, different kinds of ecosystems have more or less clearly recognizable boundaries, but they also intergrade into one another, sometimes almost imperceptibly; the task of recognizing their boundaries then becomes arbitrary (Figure 51-1). Ecosystems also change over time and slowly become modified into new ecosystems, whose characteristics come to differ increasingly from those that preceded them. Thus the complex ecosystems of the tropical rain forests have changed gradually in adapting to the particular conditions of temperature, seasonality, and soil that typify these places. The ecosystems of the tundra have developed in a similar way, but in relation to the very different environmental conditions of the far north. As the climate changes in a given place, the ecosystems that are present there change along with it, as do the individual populations within these ecosystems. By this process, the overall characteristics of the populations gradually adjust to the new conditions. Not all ecosystems are natural; we may also speak of an ecosystem in an aquarium or in a cultivated field.

All ecosystems are now being strained in an unprecedented way by the rapidly growing human population, which is now about 5 billion worldwide. Our numbers have doubled since 1950, when the total human population was about 2.5 billion people, so that the continued productive capacity of many ecosystems, and even their ability to function, are being pushed to near the breaking point. In terms of the way we interact with natural eco-

FIGURE 51-1

In the Coast Ranges of California, the boundary between the evergreen shrub association known as chaparral and grassland is often sharp, as shown in this photograph taken along the western edge of the Santa Clara Valley near Morgan Hill. The plant associations shown here are distinct ecosystems with different characteristics. The oak trees represent the edge of an oak woodland community that occurs in the same region.

systems, we do know how to modify some of these ecosystems—for example, those that occur on the rich prairie soils of the midwestern United States or the fertile plains of France—in such a way that they will produce food abundantly (Figure 51-2). It has not yet been possible, however, to modify many other kinds of natural ecosystems, including most tropical forests, in such a way as to produce food or other desired substances on a sustained basis. The ways in which we are using such ecosystems are simply destroying them, and their future productive capacity, as the human population climbs ever higher (Figure 51-3). The explosive growth of human populations and their expansion into new areas throughout the world have made them an ecological force that is changing the nature of ecosystems worldwide on an unprecedented scale. This change is proceeding so rapidly and in such an uncontrolled manner that we cannot yet estimate how many people the earth is capable of supporting on a long-term basis. It is proceeding so rapidly that the adjustment of the ecosystems by the evolution of their constituents is impossible, and many of the plants, animals, and microorganisms that make them up are becoming extinct.

BIOGEOCHEMICAL CYCLES

All of the substances that occur in organisms, including water, carbon, nitrogen, and oxygen, as well as a number of others that are ultimately derived from the weathering of rocks, cycle through ecosystems. These cycles, geological ones that involve the biologically controlled cycling of chemicals, are called **biogeochemical cycles.** Among the elements that are cycled are some that come from rocks, including phosphorus, potassium, sulfur, magnesium, calcium, sodium, iron, cobalt, and all the other elements that are essential for plant growth (see Chapter 34). All organisms require carbon, hydrogen, oxygen, nitrogen, phosphorus, and sulfur in relatively large quantities; the other elements are required in smaller amounts. We outlined the nitrogen and phosphorus cycles in Chapter 34, but we will review them here from an ecological point of view.

We speak of the **cycling** of materials in ecosystems because they first are incorporated from the atmosphere or from weathered rock into the bodies of organisms; they then sometimes pass from these organisms into the bodies of other organisms that feed on these primary ones, and ultimately are returned to the nonliving world. When this occurs, the nutrients may possibly be incorporated again into the bodies of other organisms. Some examples will help to clarify the ways in which different cycles function.

FIGURE 51-2

One of the most productive crops in the world is rice, the staple grain in Asia. This scene was photographed near Kunming, Yunnan Province, China.

FIGURE 51-3

The hungry people of the tropics, who constitute between one fourth and one half of the populations of most countries, often move to the cities looking for new opportunities. This photograph was taken along the banks of the Amazon at Iquitos, Peru.

FIGURE 51-4

The water cycle.

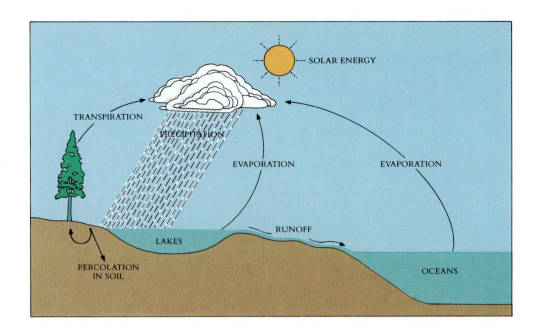

The Water Cycle

All life directly depends on the presence of water, since the bodies of most organisms consist mainly of this substance. Water is the source of the hydrogen ions whose movements generate ATP in organisms, and for that reason alone it is indispensable to their functioning. Thus the water cycle (Figure 51-4) is the most familiar of all biogeochemical cycles. The oceans cover three fourths of the earth's surface. From their surface, water evaporates into the atmosphere, a process that is powered by energy from the sun. This water eventually precipitates back to earth and passes into surface and subsurface bodies of fresh water. Most of it falls directly into the oceans, but some falls onto the earth. Only about 2% of all the water on earth, however, is fixed in any form—frozen, held in the soil, or incorporated into the bodies of organisms. All of the rest is free water, circulating between the atmosphere and the earth. Regardless of where this water is held temporarily, however, it eventually returns to the atmosphere and the oceans.

Organisms live or die based on their ability to capture some of this water and incorporate it into their bodies. Plants take up water from the earth in a continuous stream; crop plants require about 1000 kilograms of water to produce one kilogram of food, and the relationships in natural communities are similar. Animals obtain water directly or from the plants or other animals they eat. The amount of free water present at a particular place largely determines the nature and abundance of the living organisms present there.

Much less obvious than the surface waters, which we see in streams, lakes, and ponds, is the groundwater, which occurs in **aquifers**—permeable, saturated, underground layers of rock, sand, and gravel. In many areas, groundwater is the most important reservoir of water; for example, it amounts to more than 96% of all fresh water in the United States. It flows much more slowly than surface water, anywhere from a few millimeters to as much as a meter or so per day. In the United States, groundwater provides about a quarter of the water used for all purposes and provides about half of the population with drinking water. Three fourths of this country's cities and most of its rural areas rely at least in part on groundwater reserves. The use of groundwater throughout the world is growing much more rapidly than is the use of surface water.

Some 96% of the fresh water in the United States consists of groundwater. This groundwater, which already provides a quarter of all the water used in this country, will be used even more extensively in the future.

FIGURE 51-5

The carbon cycle.

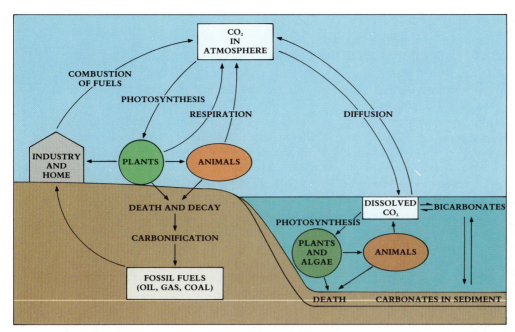

Because of the greater rate at which groundwater is being used, and also because it flows so slowly, the increasing chemical pollution of groundwater is a very serious problem. It is estimated that about 2% of the groundwater in the United States is already polluted, and the situation is worsening. One of the most important sources of groundwater pollution consists of the roughly 200,000 surface pits, ponds, and lagoons that are actively used for the disposal of chemical wastes. Even more numerous are underground storage tanks, from which vast quantities of waste material, such as gasoline, are seeping. The Environmental Protection Agency (EPA) is addressing this problem, which is by no means confined to the United States, as an urgent national priority. It is seeking ways to minimize the contamination of our groundwater and to clean up the supplies that are polluted.

The Carbon Cycle

The carbon cycle is based on carbon dioxide, which makes up only about 0.03% of the atmosphere. Perhaps 3 billion years ago, certain bacteria evolved the ability to fix the sun's energy through photosynthesis, using these relatively scarce molecules of carbon dioxide as the basic building block for the carbon compounds that make up all living things on earth. Some of the descendants of these early photosynthetic bacteria, as we saw in Chapters 12 and 26 particularly, still exist as free-living organisms. Others became incorporated as chloroplasts into the cells of different groups of eukaryotes, which in turn gave rise to the algae and ultimately to plants.

The worldwide synthesis of organic compounds from carbon dioxide and water results in the fixation of about a tenth of the roughly 700 billion metric tons of carbon dioxide in the atmosphere each year (Figure 51-5). This enormous amount of biological activity takes place as a result of the combined activities of photosynthetic bacteria, algae, and plants. All heterotrophic organisms—including the nonphotosynthetic bacteria and protists, the fungi, the animals, and a relatively few plants, such as dodder, that have lost the ability to photosynthesize—obtain their carbon indirectly, from the organisms that fix it. As a result of oxidative respiration and the ultimate decomposition of their bodies, organisms release carbon dioxide to the atmosphere again. Once there, it can be reincorporated into the bodies of other organisms.

About a tenth of the estimated 700 billion metric tons of carbon dioxide in the atmosphere is fixed annually by the process of photosynthesis.

CARBON DIOXIDE AND RISING WORLD TEMPERATURES

Before widespread industrialization, the concentration of carbon dioxide in the atmosphere was approximately 260 to 280 parts per million (ppm). During the 25-year period starting in 1958, this concentration increased from 315 ppm to more than 340 ppm and is continuing to rise rapidly. Even though human beings can tolerate concentrations of 1000 ppm of carbon dioxide without apparent harm, the levels of the gas in the atmosphere profoundly affect the way in which the earth is heated by the sun, primarily because carbon dioxide traps the longer wavelengths of infrared light, or heat, and prevents them from radiating into space. By doing so, it creates what is known as a **greenhouse effect.**

The current increase in carbon dioxide concentration can be attributed mainly to the burning of coal, oil, and other fuels. There is roughly seven times as much carbon dioxide locked up in these fossil fuels—roughly 5 trillion metric tons—as exists in the atmosphere today. Deforestation and changes in land-use patterns over the last century have also made a major contribution to rising carbon dioxide levels, but the proportion of this contribution has not been accurately determined.

What will happen to global atmospheric levels of carbon dioxide in the future is difficult to predict with certainty. The rate of consumption of the fossil fuels will be tied to economic developments that cannot be forecast precisely; it will be closely tied to the overall global demand for energy and to the attractiveness of fossil fuels in comparison with other sources of energy. In a recent study, however, the U.S. National Research Council estimated that the concentration of carbon dioxide in the atmosphere would pass 600 ppm, roughly double the current level, by the third quarter of the twenty-first century, and that this level might be reached as soon as 2035.

Such concentrations of carbon dioxide, if maintained indefinitely, would lead to a global surface-air warming of between 1.5° and 4.5° C. The actual increase, however, could be even greater than this, because a number of gases that are present in the atmosphere in trace amounts have effects similar to those of carbon dioxide, and most of them are, like carbon dioxide, increasing rapidly as a result of human activities. These gases include nitrous oxide, methane, ozone, and chlorofluorocarbons. Although they are much less concentrated in the atmosphere than carbon dioxide is, they all absorb the infrared wavelengths of light much more efficiently. Some scientists have estimated that increases in these trace

Most of the organic compounds that are formed as a result of carbon dioxide fixation in the bodies of photosynthetic organisms are ultimately broken down and released back into the atmosphere or water in the form of bicarbonate ion, HCO_3^-. Certain carbon-containing compounds, such as cellulose, are more resistant to breakdown than others; in Chapter 43 (see Figure 43-14), we discussed the breakdown of cellulose by bacteria in the rumens of cattle and related grazing mammals, and by rabbits and a few other kinds of animals as well. Some cellulose, however, accumulates as undecomposed organic matter, such as peat, or as insoluble carbonates in aquatic systems. This cellulose may eventually form fossil fuels, such as oil or coal, or minerals such as limestone, which is formed from the calcium carbonate that makes up the shells of mollusks and other animals.

In addition to the roughly 700 billion metric tons of carbon dioxide in the atmosphere, approximately 1000 billion metric tons are dissolved in the ocean; more than half of this quantity is in the upper layers, where photosynthesis takes place. The fossil fuels, primarily oil and coal, contain more than 5000 billion additional metric tons of carbon, and between 600 and 1000 billion metric tons are locked up in living organisms at any one time. The release of this carbon as carbon dioxide, a process that is proceeding rapidly as a result of human activities, currently appears to be changing global climates, and may do so even more rapidly in the future.

gases could produce an effect on global climate equal to that of the increase in carbon dioxide.

Although the increased concentration of carbon dioxide would have a beneficial effect on plant growth, the increased temperatures would create major regional and global climatic problems. In the American Southwest, for example, increases in temperature of the magnitude predicted would probably lead to a 40% to 75% reduction in the amount of water in the rivers. Consequently, agriculture might have to be abandoned in states such as California, Texas, and Arizona, unless sufficient water could be imported from elsewhere. On the other hand, stream runoff might increase markedly in certain regions because of decreased transpiration from plants—their stomata tend to close when carbon dioxide is abundant. Increased levels of damage by herbivores have likewise been demonstrated in plants grown at high concentrations of carbon dioxide, probably because of a dilution of the relative amounts of nitrogen—which the insects require—in the plants.

Most troubling of all, a global warming of 3° to 4° C over the next 75 years or so would cause the worldwide sea level to rise about 70 centimeters, on top of a rise of about 15 centimeters during the past century. At a certain point, the West Antarctic ice sheet and the ice mass in the Arctic Ocean would start to melt off. An additional rise in sea level of about a meter per century could follow, even without additional increases in the atmospheric concentrations of carbon dioxide. Ultimately, a rise in the sea level of some 5 to 6 meters could take place, with serious consequences for all coastal areas. To give an idea of the magnitude of the temperature changes involved in this process, the total rise in global temperatures since the end of the last glacial maximum ("ice age"), which occurred about 11,500 years ago, has been only about 8° C!

Both the rising sea levels and the regional climatic shifts, with their profound effects on agriculture in a world that is increasingly hungry, are potentially very critical issues. The carbon dioxide problem must continue to be monitored with great care, even though it does not yet seem possible to ameliorate it by any available means. Meanwhile, greater priority should be given to the more rapid development of long-term energy options that do not involve the use of fossil fuels. Processes that contribute to increased concentrations of the trace gases in our atmosphere should be studied and evaluated with special care so that the increase may be limited as much as possible.

The Nitrogen Cycle

Nitrogen gas constitutes nearly 80% of the earth's atmosphere by volume, but the total amount of fixed nitrogen in the soil, oceans, and the bodies of organisms is only about 0.03% of that figure. Nitrogen fixation, as you saw in Chapter 34, is a complex process that is crucial to the existence of life on earth. The nitrogen cycle was illustrated in Figure 34-13 and discussed on pages 709-712.

The activities of bacteria of the genus *Rhizobium,* which form nodules on the roots of legumes, are the most important factors in the nitrogen-fixing process (see Figure 26-15). If the nitrogen-fixing genes could be transferred from *Rhizobium* to other organisms, especially to the major crops themselves, in an energy-efficient system, the potential gains in crop productivity would be huge. Because insufficient nitrogen so often limits plant growth, and because the element reaches animals only by way of plants, the process of nitrogen fixation is receiving a great deal of worldwide attention from scientists and all those concerned with agriculture, forestry, and allied fields.

Bacteria form nitrites from ammonia, which is the first product of nitrogen fixation. Other bacteria may then convert these nitrites to nitrates. Because nitrites and nitrates are highly soluble and are widely used as fertilizers, they have become serious contaminants in the groundwater in many regions, especially where agriculture is important. Nitrates and nitrites also wash constantly into the sea, where they contribute to the growth and development of marine organisms. Ultimately, they may be lost to deep ocean sediments, or brought back to the surface waters by upwellings of cold water from the depths. The nitrogen compounds that do reach these deep sediments return to the surface layers only very slowly, a process that occurs because of the activities of fish and other animals that move from layer to layer in the sea. Ultimately, they may be brought back to the land by birds that feed on the fish.

The Oxygen Cycle

Earth is the only place in the solar system where free oxygen exists in significant quantities. This free oxygen, which constitutes nearly a fifth of our atmosphere by volume, is a product of photosynthesis. Free oxygen reacts rapidly with reduced organic matter everywhere in the biosphere in the processes of respiration and decay. Such processes would occur even in the absence of living organisms, but they are now catalyzed by organisms for the most part. The processes therefore proceed much more actively, resulting in a residence time of oxygen in the atmosphere of only about 4500 years between the time it is produced by photosynthesis and the time it is removed by respiration or decay.

If photosynthesis were to stop, and respiration and decay continued to consume oxygen and organic matter containing oxygen, all such matter would be exhausted in about 50 years. The completion of this process, however, would reduce the amount of atmospheric oxygen by only about 1%. It would require about 2.5 million years for the reservoir of free oxygen in the atmosphere to disappear solely as a result of chemical activities by nonliving oxidizing substances in the surface rocks. At such a point, the atmosphere would have returned to its condition early in the history of the earth, before the net rate of oxygen production by photosynthesis exceeded the net release rate of reduced gases by volcanoes and of reduced ions and compounds in the oceans. Such a condition last existed some 2.5 billion years ago.

If photosynthesis were to cease, it would take about 2.5 million years for all of the oxygen in the atmosphere to be consumed.

The Phosphorus Cycle

In all biogeochemical cycles other than those involving water, carbon, oxygen, and nitrogen, the reservoir of the nutrient exists in mineral form, rather than in the atmosphere. The phosphorus cycle (see Figure 34-12) is presented as a representative example of all other mineral cycles because of the critical role phosphorus plays in plant nutrition worldwide.

Phosphates—phosphorus anions—exist in the soil only in small amounts, because they are relatively insoluble and are present only in certain kinds of rocks. If the phosphates are not lost to the sea via rivers and streams, they may be absorbed by plants and incorporated into compounds such as ATP, nucleic acids, and membrane proteins. Mycorrhizae facilitate the transfer of phosphates from the soil to plants, as we discussed in Chapter 28. Other plants that grow on very poor soils, such as the South African and Australian Proteaceae, do not have mycorrhizae but instead have specialized clusters of very fine roots that perform a similar function.

When animals or plants die, slough off parts, or, in the case of animals, excrete waste products, the phosphorus may be returned to the soil or water and then cycle again through other organisms. When phosphates are lost to the deep sea, they may be recycled through the activities of sea birds that eat fishes and other animals that feed in deep waters (see Figure 34-11).

Phosphates are relatively insoluble and are present in most soils only in small amounts. They often are so scarce that they limit plant growth.

Recycling in a Forested Ecosystem

The overall recycling pattern of some nutrients has been revealed in impressive detail by an ongoing series of studies conducted at the Hubbard Brook Experimental Forest in New Hampshire. The way in which this ecosystem functions, and especially the way in which nutrients cycle within it, has been the subject of study since 1963 by Herbert Bormann, of the Yale School of Forestry and Environmental Studies; Gene Likens, of The New York Botanical Garden (Figure 51-6); and their colleagues. These studies have yielded much of the information that we now have about the cycling of nutrients in forest ecosystems. They have also provided the basis for the development of much of the experimental methodology that is being applied successfully to the study of other ecosystems.

Hubbard Brook is the central stream of a large watershed that drains a region of temperate deciduous forest. For measurement of the flow of water and nutrients within the Hubbard Brook ecosystem, concrete weirs with V-shaped notches were built across six tributary streams that were selected for study (Figure 51-7). All of the water that flowed out of those valleys had to pass through the notch, since the weirs were anchored in bedrock. The precipitation that fell in the six valleys was measured, and the amounts of nutrients that were present in the water flowing in the six streams were also determined. By these methods, it was demonstrated that the undisturbed forests in this area were very efficient at retaining nutrients. The small amounts of nutrients that precipitated from the atmosphere with the rain and snow were approximately equal to the amounts of nutrients that ran out of the valleys. These quantities were very low in relation to the total amount of nutrients in the system. There was a small net loss of calcium—about 0.3% of the total calcium in the system per year—and small net gains of nitrogen and potassium.

In 1965-1966, the investigators felled all of the trees and shrubs in one of the six watersheds and then prevented their regrowth by spraying the area with herbicides. The effects of these activities were dramatic. The amount of water running out of that valley was increased by some 40%, indicating that water that normally would have evaporated into the atmosphere from the leaves of the trees and shrubs was now running off. For the 4-month period of June to September 1966, the runoff was actually four times higher than it had been in comparable periods during the preceding years. The amounts of nutrients running out of the system also increased greatly. For example, the loss of calcium was 10 times higher than it had been previously. Phosphorus, on the other hand, did not increase

FIGURE 51-6

Gene E. Likens, one of the principal investigators at Hubbard Brook, collecting a sample of water from a precipitation collector at Hubbard Brook.

FIGURE 51-7

Experimental weir at Hubbard Brook. All of the water is forced over the concrete, and samples of it are representative of the flow from the valley where the stream is located.

The tropical rain forest has a large biomass and is highly productive.

in the stream water; it apparently was locked up in the soil. A great deal of the available phosphorus may have reached deeper levels in the soil and thus become less available for plant growth.

The change in the status of nitrogen in the disturbed valley was especially striking. The undisturbed ecosystem in this valley had been accumulating nitrogen at a rate of about 2 kilograms per hectare per year, but the cut-down ecosystem *lost* it at a rate of about 120 kilograms per hectare per year! The nitrate level of the water rapidly increased to a level exceeding that judged safe for human consumption, and the stream that drained the area generated massive blooms of cyanobacteria and algae. In other words, the fertility of this logged-over valley decreased rapidly, while at the same time the danger of flooding greatly increased.

When the trees and shrubs in one of the valleys in the Hubbard Brook watershed were cut down and the area was sprayed with herbicide, the loss of nutrients such as calcium from that valley became much greater than it had been previously. Nitrogen, which had been accumulating at a rate of about 2 kilograms per hectare per year, was lost at a rate of 120 kilograms per hectare per year.

THE FLOW OF ENERGY

An ecosystem includes two different kinds of living elements, autotrophic and heterotrophic ones. The autotrophic elements, consisting of plants, algae, and some bacteria, are able to capture light energy and manufacture their own food. To support themselves, the heterotrophic ones, including animals, fungi, most protists and bacteria, and nongreen plants, must obtain organic molecules that have been synthesized by autotrophs.

Once energy enters an ecosystem, mainly after it is captured as a result of photosynthesis, it is slowly released as metabolic processes proceed. The autotrophs that first acquire this energy provide all of the energy that heterotrophs use. Ecosystems, as well as the organisms that make them up, can in one sense be viewed as systems that have become adapted through time to delay the release of the energy obtained from the sun back into space.

Looking at an ecosystem as a whole, we can speak of its **net primary productivity,** which we define as the total amount of energy that is converted to organic compounds in a given area per unit of time. The **net productivity** of the ecosystem is the total amount of energy fixed per unit of time, minus that which is expended by the metabolic activities of the organisms in the community. The total weight of all of the organisms living in the ecosystem, its **biomass,** increases as a result of its net production. Some ecosystems—for example, a cornfield or a cattail swamp—have a very high net primary productivity. Others, such as tropical rain forests (Figure 51-8), also have a relatively high net primary productivity, but a rain forest has a much larger biomass than a cornfield, so the net primary productivity of rain forest is much lower in relation to its total biomass.

In tropical forests and in marshlands, between 1500 and 3000 grams of organic material is normally produced per square meter per year. Corresponding figures for other communities are: temperate forests, 1100 to 1500 grams; dry deserts, 200 grams. For such highly productive communities as estuaries, coral reefs, and sugarcane fields, the figures may range from 10 to 25 grams *per day,* for comparable annual yields of from 3600 to 9100 grams.

In such communities as sugarcane fields, coral reefs, and estuaries, the net primary productivity per square meter per year may range from roughly 3500 to 9000 grams. The productivity of marshlands and tropical forests is somewhat less, and that of deserts is about 200 grams.

A
B
C

Trophic Levels

Green plants, the primary producers of an ecosystem, generally capture about 1% of the energy that falls on their leaves, converting it to food energy. In especially productive systems, this percentage may be a little higher. When these green plants are consumed by other organisms, usually only about 10% of the plant's accumulated energy is actually converted into the bodies of the organisms that consume them.

Among these consumers, several levels may be recognized. The **primary consumers,** or herbivores, feed directly on the green plants. **Secondary consumers,** carnivores and the parasites of animals, feed in turn on the herbivores. **Decomposers** break down the organic matter accumulated in the bodies of other organisms. Another, more general, term that includes decomposers is **detritivores.** Detritivores are organisms that live on the refuse of an ecosystem—not only on dead organisms but also on the cast-off parts of organisms. They include large scavengers, such as crabs, vultures, and jackals, as well as decomposers (Figure 51-9).

All of these levels, and probably additional ones, are represented in any fairly complicated ecosystem. They are called **trophic levels,** from the Greek word *trophos,* which means "feeder." Organisms from each of these levels, feeding on one another, make up a series called a **food chain.** The length and complexity of food chains vary greatly. In real life, it is rather rare for a given kind of organism to feed on only one other kind of organism; usually, each will feed on two or more other kinds, and in turn will be fed on by several other kinds of organisms. When diagramed, the relationship appears as a series of branching lines, rather than as one straight line; it is called a **food web** (Figure 51-10).

FIGURE 51-9

A *Elephants browsing in a savanna in Tsavo National Park, Kenya, East Africa. The widely spaced trees and dense cover of perennial grasses are characteristic.*
B *This crab,* Gecarcinus quadratus, *photographed on the beach at Mazatlán, Mexico, is a detritivore, playing the same role that vultures and similar animals do in other ecosystems.*
C *Fungi, such as the basidiomycete whose mycelium is shown here growing through the soil in Costa Rica, are, together with the bacteria, the primary decomposers of terrestrial ecosystems.*

FIGURE 51-10

The food web in a salt marsh, showing the complex interrelationships between organisms.

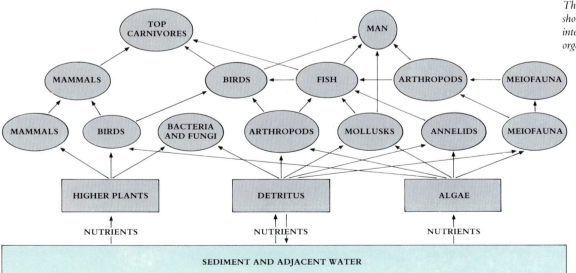

The food web diagram shows nodes: TOP CARNIVORES, MAN, MAMMALS, BIRDS, FISH, ARTHROPODS, MEIOFAUNA, MAMMALS, BIRDS, BACTERIA AND FUNGI, ARTHROPODS, MOLLUSKS, ANNELIDS, MEIOFAUNA, HIGHER PLANTS, DETRITUS, ALGAE, NUTRIENTS, NUTRIENTS, NUTRIENTS, SEDIMENT AND ADJACENT WATER

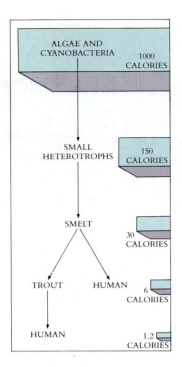

In Cayuga Lake, one of New York's Finger Lakes, autotrophic plankton fix the energy of the sun, heterotrophic plankton feed on them, and both are consumed by smelt. The smelt are eaten by trout, with about a tenfold loss in fixed energy; for humans, the amount of biomass available in smelt is at least 10 times greater than that available in trout, which they prefer to eat.

A certain amount of the energy that is ingested and retained by the organisms at a given trophic level goes toward heat production. A great deal of the energy is used for digestion and work, and usually 40% or less goes toward growth and reproduction. An invertebrate typically uses about a quarter of this 40% for growth; in other words, about 10% of the food that an invertebrate eats is turned into its own body, and thus into potential food for its predators. Although the comparable figure varies from approximately 5% in carnivores to nearly 20% for herbivores generally, 10% is a good average value for the amount of organic matter that is present at each step in a food chain, or each successive trophic level, and the amount that reaches the next level.

A plant fixes about 1% of the sun's energy that falls on its green parts. The successive members of a food chain, in turn, process into their own bodies about 10% of the energy available in the organisms on which they feed.

Lamont Cole, of Cornell University, studied the flow of energy in a freshwater ecosystem in Lake Cayuga in upstate New York. He calculated that about 150 calories of each 1000 calories of potential energy that are fixed by algae and cyanobacteria are transferred into the bodies of small heterotrophs (Figure 51-11). Of these, about 30 calories are incorporated into the bodies of smelt, the principal secondary consumers of the system. If human beings eat the smelt, they gain about 6 calories from each 1000 calories that originally entered the system. If, on the other hand, trout eat the smelt and we eat the trout, we gain only about 1.2 calories from each original 1000.

Relationships of this kind make it clear that organisms, including people, that subsist on an all-plant diet have more food available to them than do carnivores. Such considerations will become increasingly important in the future, not only for the efficient management of fisheries, but in general in the effort to maximize the yield of food for a hungry and increasingly overcrowded world.

Pyramids of numbers, biomass, and energy. In the aquatic community, biomass is greater in the zooplankton (heterotrophic plankton) than in the phytoplankton (autotrophic plankton) because of rapid turnover among the organisms engaged in primary production. This is an unusual situation, and in many aquatic ecosystems the phytoplankton has a greater biomass than the zooplankton.

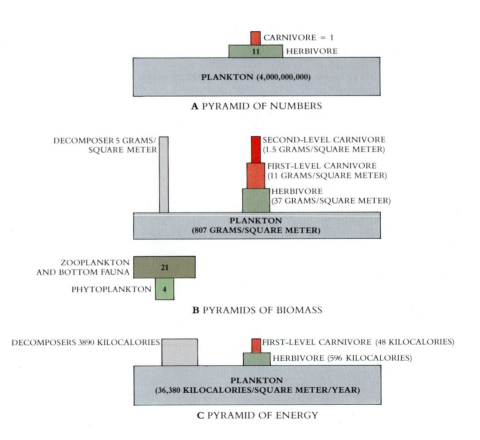

A PYRAMID OF NUMBERS

B PYRAMIDS OF BIOMASS

C PYRAMID OF ENERGY

Food chains generally consist of only three or four steps. The loss of energy at each step is so great that very little of the original energy remains in the system as usable energy after it has been incorporated successively into the bodies of organisms at four trophic levels. There are generally far more individuals at the lower trophic levels of any ecosystem than at the higher ones. Similarly, the biomass of the primary producers present in a given ecosystem is greater than that of the primary consumers, with successive trophic levels having a lower and lower biomass and correspondingly less potential energy. Larger animals characteristically are members of the higher levels; to some extent, they *must* be larger in order to capture enough prey to support themselves.

These relationships, if shown diagramatically, appear as pyramids (Figure 51-12). We may therefore speak of "pyramids of biomass," "pyramids of energy," "pyramids of number," and so forth, as characteristic of ecosystems. Occasionally, the pyramids are inverted. For example, in a planktonic ecosystem, as we saw in the last chapter, the turnover at the lower levels is rapid and the autotrophic organisms may reproduce rapidly, thus supporting a population of heterotrophs that is larger in biomass and more numerous than themselves. Pyramids of energy, on the other hand, cannot be inverted because of the necessary loss of energy at each step.

ECOLOGICAL SUCCESSION

Even when the climate of a given area remains stable year after year, ecosystems have a tendency to change from simple to complex in a process known as **succession.** This process is familiar to anyone who has seen a vacant lot or cleared woods slowly but surely become occupied with larger and larger plants and more and more different kinds of them, or a pond become filled with vegetation that encroaches from the sides and gradually turns it into dry land (Figure 51-13).

FIGURE 51-13

A pond near Tallahassee, Florida, rapidly filling with aquatic vegetation. As the process of succession proceeds, the pond will eventually be filled, and then gradually the area where it once existed will become indistinguishable from the surrounding vegetation.

LIFE DEVELOPS ON SURTSEY

The most spectacular instances of primary succession occur when a new lake is formed or an island first rises above the surface of the sea. The island of Surtsey, on the mid–Atlantic Ridge off the south coast of Iceland, rose above the sea on November 14, 1963, as a result of a volcanic eruption that began earlier on the ocean floor 120 meters below the surface. Even before the volcanic activity that created Surtsey began to subside, sea gulls were observed nesting on the warm cinders. The island continued to grow rapidly to the south, as a result of the continued accumulation of lava, until 1967. Its area was then about 280 hectares, and its maximum elevation was about 140 meters. Since then, Surtsey has been changed mainly by erosion, which has removed about 7.5 hectares annually.

The rocks on Surtsey are extremely low in phosphates, and nitrogen has only slowly been accumulating, as a result of biological activity. After the island formed, bacteria and fungi soon appeared, their spores blown in from other areas. Among them were the free-living, nitrogen-fixing bacteria *Nitrosomonas* and *Nitrobacter;* their activities led rapidly to the establishment of a nitrogen cycle on the island. Nitrogen-fixing and other cyanobacteria likewise became established very early, as did some of the green algae. The widespread weedy moss *Funaria hygrometrica* appeared in 1967, forming a patch by a sand bank. In 1970 the first lichens were found, and a number of additional species of mosses and lichens have since been discovered.

The first vascular plant to be found on Surtsey

A   B

FIGURE 51-14

Mount St. Helens in the state of Washington erupted violently on May 18, 1980; the lateral blast devastated more than 600 square kilometers of forest and recreation lands within 15 minutes, as seen in (A), which shows an area near Clearwater Creek 4 months after the eruption. Five years later, succession was under way at the same spot (B), with shrubs such as blackberries, blueberries, and dogwoods following the first herbaceous plants that became established immediately after the blast.

Succession is continuous and worldwide in scope. If a wooded area is cleared and the clearing is left alone, plants will slowly reclaim the area. Eventually, the traces of the clearing will disappear and the whole area will again be woods. This kind of succession, which occurs in areas that have been disturbed and that were originally occupied by living organisms, is called **secondary succession.** Human beings are often responsible for initiating secondary succession throughout portions of the world that they inhabit. Secondary succession may also take place after fire has burned off an area, for example, or after the eruption of a volcano (Figure 51-14).

Primary succession, in contrast to secondary succession, occurs on some bare, lifeless substrate, such as rocks or open water, where organisms gradually occupy the area and change its nature. Primary succession that occurs on places such as dry rocks is called **xerarch** succession, to distinguish it from the **hydrarch** primary succession that occurs in open water. On bare rocks, lichens may grow first, forming small pockets of soil and breaking down the stone (see Figure 28-16). Acidic secretions from the lichens and from the plants that grow on the rocks later may help to break down the substrate and add to the accu-

was a sea rocket, *Cakile edentula,* which appeared in 1965, only two years after the island emerged from the sea. This plant grows in open, sandy habitats; its fruits contain a large amount of corky tissue and float readily in seawater. Over the succeeding 20 years, an additional 19 species of vascular plants have taken hold on the island, and some of them have become quite abundant. Apparently all of these plants either drifted in from the sea or were imported inadvertently by birds.

Animals appeared on the island almost as soon as it was formed; the first fly was found in 1964. By 1976, nearly 200 species of insects and arachnids had become established. The introduction of these animals took place in several ways—by flying, being blown by the wind, drifting on flotsam in the sea, or, despite extensive precautions, as a result of the visits of scientists and other people to the island. More than 60 species of birds have been recorded, and some of them have been observed carrying seeds to Surtsey. Five of the bird species now nest on the island, further enriching the soil and increasing the complexity of the food webs.

Although the climate of Surtsey is cool, windy, and rainy, the many species of plants, animals, and microorganisms that became established there in only two decades provide striking testimony to their powers of dispersal and establishment. This small island in the North Atlantic has been colonized in the same way as the early Hawaiian Islands or the Galapagos, and it has taught us a great deal about the ways in which an extensive biota can develop rapidly under such circumstances.

---

mulation of soil. Mosses may then colonize these pockets of soil, eventually followed by ferns and the seedlings of flowering plants. Over many thousands of years, or even longer, the rocks may be completely broken down, and the vegetation over an area where there was once a rock outcrop may be just like that of the surrounding grassland or forest.

In a similar example involving hydrarch succession, an oligotrophic lake may gradually, by the accumulation of organic matter, become eutrophic (Figure 51-13). Plants standing along the edges of the lake, such as cattails and rushes, and those growing submerged, such as pondweeds, together with other organisms, may contribute to the formation of a rich organic soil. As this process continues, the pond may increasingly be filled in with terrestrial vegetation. Eventually, the area where the pond once stood, like the rock outcrop we just described, may become an indistinguishable part of the surrounding vegetation.

FIGURE 51-15

A stage in succession toward a climax community on the west side of the San Francisco Peaks in northern Arizona. Here ponderosa pine (Pinus ponderosa) *is replacing aspen* (Populus tremuloides).

> **Primary succession takes place in areas that initially are open, such as dry rock faces or oligotrophic ponds. Secondary succession, in contrast, takes place in areas that have been disturbed after having been occupied by living things earlier.**

Oligotrophic ponds and bare rocks in a given region may eventually feature the same kind of vegetation as one another—the vegetation that is characteristic of the region as a whole. This relationship led the American ecologist F. E. Clements, at about the turn of the century, to propose the concept of **climax vegetation,** a term referring to the fact that this characteristic vegetation type is thought to be controlled by the *climate* of that region (Figure 51-15). However, with an increasing realization that (1) the climate keeps changing, (2) the process of succession is often very slow, and (3) the nature of a region's vegetation is being determined to a greater and greater extent by human activities, ecologists no longer consider the term "climax vegetation" to be as useful as they once did.

One of the most interesting features of succession is that the organisms involved in its early stages often are participants in symbiotic systems. Lichens, so important in the early colonization of rocks, are symbiotic associations of cyanobacteria or algae and fungi.

FIGURE 51-16

*Intensively cultivated areas are kept
permanently at an early successional
stage, as shown here in this recently
cut hay field.*

This association appears to enable the lichens to withstand harsh conditions of moisture and temperature. Legumes and other plants that harbor nitrogen-fixing bacteria in nodules on their roots are often among the first colonists of relatively infertile soils. Trees with ectotrophic mycorrhizae are apparently more resistant to conditions of moisture stress and perhaps other environmental extremes than are those trees that have endotrophic mycorrhizae or no mycorrhizae at all. It is believed that the first land plants had endotrophic mycorrhizae and that this association was highly beneficial to them in the sterile, open habitats that they encountered (see Chapters 28 and 29).

**Symbiotic associations, such as lichens, legumes with their associated
nitrogen-fixing bacteria, and plants with mycorrhizal fungi, appear to play
an unusually important role in early successional communities.**

The characteristics that we will now outline are general ones that appear to hold for succession of all sorts. As ecosystems mature, there is an increase in total biomass but a decrease in net productivity. The earlier successional stages are more productive than the later ones. Agricultural systems are examples of early successional stages in which the process is intentionally not allowed to go to completion and the net productivity is high (Figure 51-16). There are more kinds of species in mature ecosystems than in immature ones, and the number of heterotrophic species increases even more rapidly than the number of autotrophic species. This progression is related to the decreasing net productivity of increasingly mature ecosystems and to the fact that mature ecosystems have a greater ability to regulate the cycling of nutrients than do disturbed and immature ones. It appears that the plants and animals that appear in the later stages of succession may be more specialized, in general, than those that exist in the earlier stages. The late-successional species appear to fit together into more complex communities and to have much narrower ecological requirements, or niches, as we will discuss in the next chapter.

**Communities at early successional stages have a lower total biomass, higher
net productivity, fewer species, many fewer heterotrophic species, and less
capacity to regulate the cycling of nutrients than do communities at later
successional stages.**

Some species are **fugitive** species, which occur at the earlier successional stages and disappear from the area as succession proceeds. Such species often have high reproductive rates and efficient means of dispersal. Foxglove *(Digitalis purpurea),* for example, is a fugitive species that temporarily becomes abundant in forest clearings and then disappears as the canopy cover of the trees becomes more complete. Other opportunities for fugitive species are afforded by fires; certain species may only be seen growing on burned areas. Certain fugitive species may occur only in newly formed ponds or islands and then apparently disappear from the area. Many weeds, also, are fugitive species, disappearing soon from the communities where they appear.

FIGURE 51-17

Smog, associated with a temperature inversion, in the Los Angeles Basin.

POLLUTION

The activities of human beings are now so extensive that our waste products are threatening the continued productivity, and in some cases even the continued existence, of not only many natural communities but also some of the communities that we have created for our own benefit. The by-products of our industry and the chemicals that we use in agriculture pass into various kinds of ecosystems and cause various kinds of disturbances. These disruptions depend on the nature of these ecosystems and the ways in which the chemicals circulate through them.

Air pollution is one of the most familiar forms of pollution with which we must contend. Much of it arises from the use of fossil fuels such as coal and oil without adequate steps being taken to remove the by-products that result from their combustion. Automobiles, which run on gasoline, an oil derivative, contribute heavily to the smog that hangs over cities such as Los Angeles (Figure 51-17) and Tokyo and makes living in such places relatively unhealthy for anyone who suffers from lung or heart problems. Physicians therefore have encouraged many patients to move away from such centers if possible. Many plants, too, are highly susceptible to damage from air pollution, which causes significant losses in natural vegetation, around homes and factories, and in agricultural areas.

Many related kinds of pollution have arisen as a result of human industry. For example, the polymers known as plastics are produced in abundance but are essentially impervious to breakdown by known natural forces. Even though scientists are trying to develop strains of bacteria that can decompose plastics, their efforts have been largely unsuccessful. Consequently, virtually all of the polymers in the plastic items that have been produced since the 1950s, made almost entirely from fossil fuels, are still with us. Collectively, these plastics constitute a new form of pollution of the world ecosystem for which there is, as yet, no solution.

Water pollution is another very serious problem that exists on a global scale (Figure 51-18). In connection with the water cycle, which we considered earlier in this chapter, we have already considered some of the ways in which the groundwater can be polluted. Furthermore, there is simply not enough water available to dispose of the diverse substances that today's enormous human population produces continuously. Despite the implementation of ever-improved methods of sewage treatment throughout the world, our lakes, streams, and groundwater are becoming increasingly polluted. For example, household detergents, which contain phosphates, may flow into oligotrophic lakes and lead to the production of algal blooms. These blooms may in turn change the overall pattern of ecological relationships of a lake and the quality of its water.

FIGURE 51-18

This "contaminated" sign near San Diego is symbolic of a condition that is rapidly spreading worldwide—the progressive pollution of clean water.

All of these problems are starting to be better understood, but at the same time, an increasingly industrialized and rapidly growing human population is putting greater and greater demands upon the available supplies of water. These problems can only increase; not only is the human population expected to double in the next 40 years or so, but people everywhere expect the quality of their lives to improve. The ways in which we manage our water supplies, therefore, will have an important impact on our future.

Widespread agriculture, carried out increasingly by modern methods, also causes very large amounts of many new kinds of chemicals to be introduced into the global eco-

ACID RAIN

An ecological problem that has particularly perplexed scientists and public officials in the United States and Europe since the late 1970s is acid rain, better called acid precipitation. "Acid rain" is a blanket term for the processes whereby pollutants such as nitric and sulfuric acids are spread over wide areas by the prevailing winds and then fall to earth with the precipitation. The pH of unpolluted, natural precipitation often lies between 5.0 and 5.6. At present, however, the pH in some areas of the United States often approaches 4.0—perhaps 10 times lower than normal (recall that pH is a log scale measurement). In rare cases, the pH of precipitation or fog may even approach 2.0! Many studies have shown that the industrial emission of sulfur dioxide is the major factor in the acidification of the precipitation and that nitrogen oxides, ozone, and other atmospheric contaminants also play a role. Acid precipitation is certainly adversely affecting lakes, streams, crops, forests, and buildings in much of the Northern Hemisphere.

The effects of acid rain have been especially dramatic in Europe and eastern North America, where tens of thousands of lakes are dying biologically (Figure 51-A, *1*). At pH's below 5.0, many fish species and other aquatic animals will die; and even at pH's below 6.0, food chains may be thrown off balance. Accompanying the increased acidity of the waters has been a substantial increase in dissolved aluminum, which is toxic to fishes. In southern Sweden, groundwater now regularly is found to have a pH between 4.0 and 6.0, its acidity resulting from the acid precipitation that is slowly filtering down into the underground reservoirs. As with all problems involving groundwater, this warning probably signals only the start of a process that will become worse with the passage of time, even if corrective measures are taken soon.

The effects of acid precipitation in terrestrial habitats are particularly complex. Key nutrients, such as calcium and potassium, may be leached from the soil; microorganisms of critical importance in the cycling of nutrients may be killed; and toxic elements such as magnesium and aluminum may be dissolved out of the substrate and brought into the water supply. The growing presence of aluminum in water is being viewed with increasing alarm as a possible human health hazard; for example, aluminum is concentrated in the nerve tangles that form in the brains of patients suffering from Alzheimer's disease. Such diseases also appear to be associated with a deficiency of calcium, which likewise is associated with acid precipitation.

Although the connection has not been demonstrated clearly, many scientists also believe that the widespread decline in forests around the Northern Hemisphere that began between 1955 and the early 1960s may have been caused by these nutrient relationships with acid precipitation. A third of the forests in Germany are dying, and growth has slowed in trees all over the eastern United States. An analysis of the tree rings in core samples taken from many thousands of trees has confirmed that their growth has slowed remarkably during this period. Atmospheric pollution of some kind, if not specifically acid precipitation, seems to be to blame.

Nitrogen oxides (NOx), which are generated primarily by automobile exhaust and industrial sources, are a major part of the acid precipitation problem. Many researchers believe that these oxides, which combine with water to form nitric acid, are the most important factor in the decline of forests in Europe and parts of North America (Figure 51-A, *2*). Ozone (O_3), which is also a by-product of air pollution, may also play an important role in damaging trees.

Coal-burning power plants and factories are responsible for about 86% of the sulfur dioxide in the atmosphere. This sulfur dioxide, in turn, is the most important cause of acid precipitation. At

system. These include pesticides, herbicides, and fertilizers. Developed countries like the United States now attempt to monitor the side effects of these chemicals carefully. Unfortunately, however, large quantities of many toxic chemicals that were manufactured in the past still circulate in the ecosystems of developed nations.

For example, the chlorinated hydrocarbons, a class of compounds that includes

Hubbard Brook, where pollutants in the precipitation have been measured precisely since 1965, these pollutant levels exactly mirror known changes in industrial emissions. What is not clear about the process, however, is how and where the sulfur dioxide travels through the atmosphere once it leaves the plants. It is certain that the problems in one state or country are often caused by industrial activities that take place in another. For example, some—perhaps a great deal—of the acid precipitation that is damaging the ecosystems of the northeastern United States and eastern Canada is caused by industrial activity in the American Midwest. The fact that three to five times as much sulfur dioxide reaches Canada from the United States as travels in the other direction is severely straining relations between the two countries.

Because the distribution patterns of the pollution have not yet been demonstrated precisely, and because the solutions often do not seem relevant locally, governments have been slow to act on this problem. Environmental scientists generally agree with a report by the U.S. National Research Council, which in June 1983 concluded that the reduction in sulfur dioxide emissions from smokestacks would result in a proportionate reduction in acid rain. However, they are not precisely sure where this reduction would take place.

In June 1984, representatives of 31 Western and Soviet bloc nations met in Munich to consider the problem of acid precipitation. They agreed to take steps to implement a 30% reduction in the 1980 level of sulfur dioxide emissions by 1993. The United States and Britain refused to join in this agreement, however, citing the need for further study. It has been estimated that it would cost 5 to 8 billion dollars annually for the United States to achieve such a reduction in sulfur dioxide emissions. An expenditure of that level would, if evenly distributed among the 31 eastern and midwestern states that appear to be most affected, add approximately 4.6% to utility rates.

FIGURE 51-A, 1

Twin Pond in the Adirondacks of upstate New York, one of the many lakes in the region in which the levels of acidity in the lake have killed the fishes, amphibians, and most of the other organisms that once lived there.

FIGURE 51-A, 2

Dying trees in the eastern United States, victims of a complex of environmental factors that probably include acid precipitation.

DDT, chlordane, lindane, and dieldrin, have all been banned for normal use in the United States, where they were once used very widely. These molecules break down very slowly and accumulate in animal fat. Furthermore, as they pass through a food chain, they are increasingly concentrated. DDT, for example, caused serious problems by leading to the production of very thin, fragile eggshells in many bird species in the United States and elsewhere until the late 1960s, when it was banned in time to save the birds from extinction. Chlorinated hydrocarbons have many other undesirable side effects, several of which are still poorly understood.

Developing countries still use large quantities of toxic chemicals that were banned years ago in the United States and most of Europe; in countries where the use of these chemicals has long been banned, a number of them are still being manufactured for export. In the developing countries, some of these chemicals are creating immediate problems, and resistant races of all the important disease-causing organisms and pests of crops are appearing widely. The accumulation of these chemicals in third world countries actually may pose serious obstacles to development in the future, since the chemicals break down very slowly. The chemicals also ultimately find their way back to the United States, Europe, and Japan as a result of the global circulation of the atmosphere and the oceans. Such problems will become much more acute in both the developed and the developing countries, as the demands on the productive capacity of agricultural lands everywhere rise sharply.

Obviously, a "back-to-nature" approach—one that ignores the important contributions made to our standard of living by the intelligent use of chemicals—will not allow us to care adequately for the needs of the current world population. Even less would such a backward approach allow us to feed the additional billions of people who will join us during the next few decades. On the other hand, it is essential that we use our technology as intelligently as possible and with due regard for the protection of the productive capacity of all parts of the earth, on which we all depend.

SUMMARY

1. Ecosystems are assemblages of organisms, along with the nonliving factors of their environment, which regulate the flow of energy, ultimately derived from the sun, and the cycling of nutrients.

2. Only 2% of the water on earth is fixed in any way; the rest is free. In the United States, 96% of the fresh water is groundwater. The contamination of our limited supplies of water is a serious problem.

3. About 10% of the roughly 700 billion metric tons of free carbon dioxide in the atmosphere is fixed each year through photosynthesis. An additional trillion metric tons of carbon dioxide is dissolved in the ocean, and five times that amount is locked up as coal, oil, and gas. About as much carbon exists in living organisms at any one time as there is in the atmosphere.

4. Carbon, nitrogen, and oxygen have gaseous or liquid reservoirs, as does water. All of the other nutrients, such as phosphorus, have solid reservoirs.

5. The Hubbard Brook experiments vividly illustrate the role of a forested ecosystem in regulating the cycling of nutrients. When a particular tributary valley was deforested, 4 times as much water and 10 times as much calcium ran off as had been the case previously. Nitrogen, which had been accumulating in the area at a rate of about 2 kilograms per hectare per year, was now lost at a rate of about 120 kilograms per hectare per year.

6. Plants convert about 1% of the light energy that falls on their leaves to food energy. The herbivores that eat the plants, and the other animals that eat the herbivores, constitute a set of trophic levels. At each of these levels, only about 10% of the energy fixed in the food is fixed in the body of the animal that eats that food. For this reason, food chains are always relatively short.

7. Primary succession takes place in areas that are originally bare, like rocks or open water. It is said to be either xerarch, if it occurs in dry places, or hydrarch, if it occurs in water or wet places. Secondary succession takes place in areas where the communities of organisms that existed initially have been disturbed.

8. Both types of succession lead ultimately to the formation of climax communities, whose nature is controlled primarily by the climate of the area concerned, although the human influence on many of these communities is increasing. Such communities have more total biomass, less net productivity, more species, many more heterotrophic species, and a higher capability of regulating the cycling of nutrients within them than do the earlier successional stages.

9. The pollution of the air we breathe and the water we drink with the by-products of our activities has become an increasingly serious problem as the size of the human population continues to increase. All over the world, ecosystems are now being severely stressed by a rapidly growing human population that has rising expectations.

SELF-QUIZ

The following questions pinpoint important information in this chapter. You should be sure to know the answers to each of them.

1. The total human population, now about 5 billion people, has doubled in size since approximately
   (a) 1970   (c) 1900
   (b) 1950   (d) 1600

2. Groundwater accounts for approximately what percentage of the fresh water in the United States?
   (a) 5   (d) 75
   (b) 25   (e) 95
   (c) 50

3. _____ constitutes nearly four fifths of the earth's atmosphere.

4. Earth is the only place in the solar system where free oxygen is present in significant quantities. True or false?

5. In which of the following cycles does the reservoir of the nutrient exist in a mineral form?
   (a) carbon cycle   (d) water cycle
   (b) phosphorus cycle   (e) oxygen cycle
   (c) nitrogen cycle

6. Experiments conducted at Hubbard Brook showed that
   (a) microorganisms play a large part in nitrogen fixation
   (b) nitrogen is lost from the environment in the absence of vegetation
   (c) trees lose water to the atmosphere
   (d) two of the above
   (e) all of the above

7. The total weight of all the organisms living in an ecosystem is called the _____ of that ecosystem.

8. Herbivores are
   (a) primary consumers   (d) detritivores
   (b) secondary consumers   (e) parasites
   (c) decomposers

9. A plant fixes about _____ of the sun's energy that falls on its green parts.
   (a) 1%   (c) 50%   (e) 99%
   (b) 25%   (d) 75%

10. Communities at early successional stages have _____, relative to the more advanced communities.
    (a) higher biomass   (d) better capacity to regulate cycling of nutrients
    (b) lower productivity   (e) higher productivity
    (c) more species

11. Household detergents contain _____, which may alter the overall ecology of an environment.

12. DDT belongs to the class of chemicals called
    (a) polyunsaturated hydrocarbons   (d) chlorinated hydrocarbons
    (b) oxygenated hydrocarbons   (e) carbonated hydrocarbons
    (c) nitrogenated hydrocarbons

THOUGHT QUESTIONS

1. What are the similarities between a field of wheat and an early stage of succession in a natural ecosystem? The differences? What are the effects of the differences?

2. Would you expect secondary succession to proceed more rapidly on the slopes of a volcano that had just erupted or around a smelter that had killed all the vegetation in an extensive area and then had been abandoned. Why?

3. How could you increase the net primary productivity of a desert?

4. Why does the productivity of an ecosystem increase as it becomes more mature?

FOR FURTHER READING

BORMANN, F.H. and G.E. LIKENS: *Pattern and Process in a Forested Ecosystem,* Springer-Verlag, New York, 1979. A fascinating and well-written description of the Hubbard Brook experiments.

DRURY, W.H., and I.C.T. NISBET: "Succession," *Journal of the Arnold Arboretum,* vol. 54, pages 331-368, 1973. An outstanding digest of contemporary thinking about, and evidence for, the ways in which communities replace one another through time.

MAY, R.M.: "The Evolution of Ecological Systems," *Scientific American,* September 1978, pages 160-175. An important essay on the interplay between ecology and evolution, by one of the masters of both fields.

NATIONAL RESEARCH COUNCIL BOARD ON ATMOSPHERIC SCIENCES AND CLIMATE: *Changing Climate,* National Academy Press, Washington, D.C., 1983. A thorough assessment of the carbon dioxide problem worldwide, including its potential impacts on agriculture and on sea level.

PYE, V.I., R. PATRICK, and J. QUARLES: *Groundwater Contamination in the United States,* University of Pennsylvania Press, Philadelphia, 1983. An excellent overall view of this important subject.

RICKLEFS, R.E.: *The Economy of Nature,* 2nd ed., Chiron Press, Newton, Mass., 1983. This well-illustrated text is an excellent introduction to all aspects of ecology.

SMITH, R.L.: *Ecology and Field Biology,* 2nd ed., Harper & Row, Publishers, Inc., New York, 1974. A fine, field-oriented introduction to ecology and evolution.

VALIELA, I.: *Marine Ecological Processes,* Springer-Verlag, New York, 1984. An outstanding introduction to life in the sea, with an emphasis on the flow of energy and the cycling of nutrients.

Overview

Natural populations are affected by the ways in which they interact with others in biological communities. Each population grows in size until it eventually reaches the limits of its habitat to support it. Some of the limits to the growth of a population are related to the density of that population, but others are not. Populations of different organisms differ greatly; some grow much more rapidly than others, and the actual patterns of growth differ as well. Competition between two species limits their coexistence; if they are competing for the same scarce resources, one will win and drive the other one to extinction where they occur together. Predation and other forms of interaction among populations at different trophic levels also play an important role in limiting population size.

For Review

Here are some important terms and concepts that you will encounter in this chapter. If you are not familiar with them, you should review them before proceeding.

- **Adaptation** (Chapter 20)
- ***r*-Selection and *K*-selection** (Chapter 40)
- **Trophic levels** (Chapter 51)

A population consists of the individuals of a given species that occur together at one place and at one time. This is a flexible definition, which allows us to speak in similar terms of the world's human population, the population of protozoa in the gut of an individual termite, or the deer that inhabit a wood.

All of these specific populations have characteristic features, such as size, density, dispersion, and demography. Each occupies a particular place and plays a particular role in its ecosystem; that role, about which we will have more to say later, is defined as the **niche** of that population. Although the properties of populations are ultimately determined by the genetic properties of the individuals that they include, a population is fundamentally an operating system with its own distinct properties. Understanding these properties is crucial for understanding the nature of life on earth.

FIGURE 52-1

A

These animals exist in very small populations and therefore are in danger of extinction unless the activities that are threatening them are brought under control.

A *The Sumatran rhinoceros,* Didermoceros sumatrensis, *reduced to a few hundred individuals on the island of Sumatra and on the adjacent Asian mainland. Poaching to obtain the horns, which are in great demand for medicinal uses, and destruction of habitat are responsible for the decline of this species.*

B *Mediterranean Monk seal,* Monachus monachus. *Fewer than 500 individuals of this species are believed to exist; they are confined to remote cliffbound coasts and islands in the Mediterranean. The seals have a low birth and survival rate, and are killed by fishermen.*

C *The kagu,* Rhynochetus jubatus, *the only member of a family of birds that is restricted to the island of New Caledonia, in the southwestern Pacific Ocean. It has been brought to the brink of extinction by feral dogs and habitat destruction.*

POPULATION SIZE AND DISPERSION

One of the critical features of any population is its size. Population size has a direct bearing on the ability of a given population to survive in a given area. Very small populations—including those in which the numbers have been reduced by past events, or ones that are just becoming established in a new area—are the most likely to become extinct. If a species consists of only one or a few small populations, that species is likely to become extinct, especially if it occurs in areas that have been or are being changed radically (Figure 52-1).

The size of a population, however, is meaningful only in terms of the **density** of the population in the area under consideration. If the individuals are very widely spaced, they may rarely if ever encounter one another, and the reproductive capabilities—and therefore the future—of the population may be very limited, even if the absolute numbers of individuals are relatively large. A related measure is **dispersion,** the way in which the individuals of a population are arranged. In practice, they may be **randomly spaced, evenly spaced,** or **clumped.** Each of these patterns has a different impact on the way in which a population functions in a given ecosystem, and, very importantly, on its evolutionary potentialities.

Individuals may be evenly spaced or clumped regardless of how abundant they are in a given habitat, and the patterns may not be at all obvious. For example, creosote bush (*Larrea divaricata*) is often dominant over wide areas in the deserts of Mexico and the southwestern United States. The individuals are well spaced and evenly dispersed, probably because the shrubs secrete chemicals that retard the establishment of other individuals near established ones. On the other hand, in tropical rain forests, any given plant species may be represented by only a few individuals per hectare; when measurements are taken that reveal the overall pattern, however, these scattered individuals may be shown to be evenly spaced, clumped, or randomly distributed.

Clumped distributions are by far the most frequent in nature, because of the relationship between animals, plants, and microorganisms and their habitats. Specific environmental conditions—for example, soil types, ponds, and certain kinds of host trees—are not themselves randomly distributed in general, nor are they evenly dispersed over wide areas. As a result, the organisms associated with these environments are themselves clumped in relation to them. Furthermore, animals often congregate together for a variety of different reasons, such as hunting, mating, or caring for their young. In addition, the young of a given organism are more apt to be found near its parents than far away from them, regardless of the mode of reproduction.

B

C

Although the individuals in a population may be evenly spaced or randomly dispersed, clumped distribution patterns are the most frequent in nature.

POPULATION GROWTH

A second key characteristic of any population is its capacity to grow. From observing nature in ways that range from the informal to the highly sophisticated, we know that most populations tend to remain relatively constant in number, regardless of how many offspring the individuals produce. As you saw in Chapter 19, Darwin partly based his theory of natural selection on this seeming contradiction. Many calculations have been done, showing, for example, that houseflies, bacteria, or even elephants would soon cover the world if their reproduction were unchecked. Under certain circumstances, however, populations do grow very rapidly in size for a time (Figure 52-2). We must consider, therefore, what these circumstances are and what factors operate in nature to limit population growth.

Biotic Potential

When discussing these matters it is useful to consider the innate capacity for increase, or **biotic potential,** of a population, a term that refers to the rate at which a population of a given species will increase when there are no limits of any sort on its rate of growth. In mathematical terms, this is defined by the following formula:

$$\frac{dN}{dt} = r_m N$$

where N is the number of individuals in the population, dN/dt is the rate of change of its numbers over time, and r_m is the biotic potential of that population—its innate capacity for increase.

In practice, the biotic potential of a given population is very difficult to calculate, since the assumption of unrestricted growth is so unrealistic for most organisms. However, for varying sets of conditions, r, **the actual rate of population increase,** defined as the difference between the birth rate and the death rate per given number of individuals per unit of time, can be calculated. The actual rate of growth obviously may also be affected by net emi-

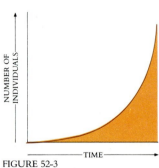

FIGURE 52-3

Exponential growth.

gration (movement out of the area) or net **immigration** (movement into the area). For example, the actual rate of increase for the human population of the United States during the closing decades of the twentieth century is more than half made up of immigrants to the country, less than half from the growth in numbers of the people already living in it.

The innate capacity of growth for any population is exponential and can be expressed by a curve like that shown in Figure 52-3. The *rate* of increase remains constant, but actual increase in the number of individuals accelerates rapidly as the size of the population grows. In practice such patterns of growth occur only for short periods, usually when an organism reaches a new habitat in which it has abundant resources. In this connection, one may think of such examples as dandelions reaching the fields, lawns, and meadows of North America from Europe for the first time; of algae colonizing a newly formed pond; or of the first terrestrial immigrants that arrive on an island recently thrust up from the sea (see boxed essay, pp. 1122-1123).

Carrying Capacity

No matter how rapidly populations may grow under such circumstances, however, they eventually reach some environmental limit imposed by shortages of an important factor such as space, light, water, or nutrients. A population ultimately stabilizes at a certain size, called the **carrying capacity** of the particular place where it lives. The carrying capacity is the number of individuals that can be supported at that place indefinitely.

The size at which a population stabilizes in a particular place is defined as the carrying capacity of that place for that species.

This kind of pattern of limitation by one or more factors in the environment can be approximated by the following equation:

$$\frac{dN}{dt} = rN\left(\frac{K - N}{K}\right)$$

In other words, the growth rate of the population under consideration *(dN/dt)* equals its rate of increase *(r)* multiplied by N, the number of individuals present at any one time, and then multiplied by an expression equal to K, the carrying capacity of the environment, minus N divided by K. As N increases (the population grows in size), the fraction by which r is multiplied becomes smaller and smaller, and the rate of increase of the population declines. In practical terms, this amounts to increasing competition among more and more individuals for a given set of resources present in a particular system. Graphically, this relationship is the S-shaped **sigmoid growth curve** characteristic of biological populations (Figure 52-4). As the size of a population stabilizes, its rate of growth slows down greatly, and it does not increase further.

Processes such as competition for resources, emigration, and accumulation of toxic waste products all increase as a population approaches its carrying capacity for a particular habitat. The resources for which the members of the population are competing may be food, shelter, light, mating sites, or any other factor necessary for the species to carry out its life cycle and reproduce itself. Among animals, territoriality, which is the result of competition for food and other resources between the members of the species in which it occurs, limits the size of the population.

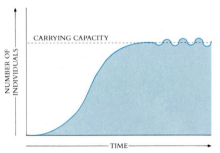

FIGURE 52-4

The sigmoid growth curve, characteristic of biological populations. Such a curve begins with a period of exponential growth like that shown in Figure 52-3; when the population approaches its environmental limits, the growth begins to slow down, and it finally stabilizes.

Density-Dependent and Density-Independent Effects

Effects such as those we have just discussed regulate the growth of populations and are called **density-dependent effects.** Among animals they are often accompanied by hormonal changes that may bring about alterations in behavior that directly affect the ultimate size of the population. One striking example occurs in migratory locusts ("short-horned" grasshoppers), which, when they are crowded, produce hormones that cause them to enter a new, migratory phase, in which the locusts take off as a swarm and fly long distances to new habitats (Figure 52-5). In contrast, **density-independent effects** are caused by factors, such as the weather and physical disruption of the habitat, that operate regardless of population size.

> **Density-dependent effects are controlled by factors that come into play particularly when the population size is larger; density-independent effects are controlled by factors that operate regardless of population size.**

Agriculture depends in part on the characteristics of the sigmoid growth curve, which we can apply to those of ecosystems at the beginning stages of succession. At such times, populations and individuals are growing rapidly, and net productivity is highest. Commercial fisheries exploit populations during their rapid growth phase, and they attempt to operate so that they are always harvesting populations in the steep, rapidly growing parts of the curve. The point of **optimal yield**—maximum sustainable catch from the population—lies partway up the sigmoid curve. Harvesting the population of an economically desirable species near this point will result in much better yields than can be obtained either when the population approaches the carrying capacity of its habitat or when it is very small (Figure 52-6). Overharvesting a population that is smaller than this critical size can destroy its productiveness for many years, as evidently happened to the Peruvian anchovy fishery after the populations had been depressed by the 1972 El Niño (see Chapter 50). It is difficult to determine the population levels of commercially valuable species that are most suitable for long-term, productive harvesting, but this is now the subject of much study.

> **In natural systems that are exploited by human beings, such as agricultural systems and fisheries, the aim is to exploit the population of the early, most productive part of the rising portion of the sigmoid growth curve.**

r Strategists and K Strategists

Many species, such as annual plants, a number of insects, and bacteria, have very fast rates of population growth. This growth cannot be controlled effectively by reducing their population sizes. In such species, small surviving populations will soon enter an exponential pattern of growth and regain their original sizes. In contrast, a comparable reduction in population size among relatively slow-breeding organisms, such as whales, rhinoceroses, California redwoods, or most tropical rain forest trees, can lead directly to extinction.

The populations of organisms that have sigmoid population growth curves are limited in number by the carrying capacity of the environment, or K. As you saw in Chapter 40, such organisms, including the relatively slow-breeding ones just mentioned, tend to live in habitats that are fairly stable and predictable. By contrast, species whose populations are characterized by exponential growth followed by sudden crashes in population size tend to live in unpredictable and rapidly changing environments. These organisms include the fugitive species, which, as you saw in the last chapter, are able to live in habitats that are available for only a short time. They have a high intrinsic rate of increase, or r, and are called r strategists. Using the terms we presented in Chapter 40, they have an r-selected reproductive strategy. The group for which an S-shaped growth curve is characteristic is called K strategists; they have a K-selected reproductive strategy. Many organisms are neither "pure" r strategists nor "pure" K strategists. Rather, their reproductive strategies lie somewhere between these two extremes or change from one extreme to the other under certain environmental circumstances.

In general, r strategists have many offspring (Figure 52-7). These offspring are small, mature rapidly, and receive little or no parental care. Many offspring are produced as a result of each reproductive event. Examples of such organisms include dandelions, aphids, and mice.

FIGURE 52-5

Migratory locusts, Locusta migratoria, *a legendary plague of large areas of Africa and Eurasia. When the population density reaches a certain level, the locusts change their characteristics and take off as a swarm.*

FIGURE 52-6

Fishing on Georges Bank, off the coast of New England. Intensive fishing here has driven the populations of commercially valuable fish species such as cod and halibut below the point of optimal yield. As increasingly sophisticated methods are used for fishing, the danger of such consequences increases.

FIGURE 52-7

The dandelion (Taraxacum officinale) is an r strategist; it produces numerous seeds by means of which the plants reproduce themselves rapidly, as every gardener knows. In the dandelion, the seeds are produced asexually, so that all of the daughter individuals are genetically identical to their parents.

FIGURE 52-8

The right whale (Eubalaena glacialis), a K strategist. Most whales have one calf at a time. Roger Payne, who took this photograph off Peninsula Valdes in Argentina, has written: "In our long sad history, we have brought hundreds of species to extinction. Unless we change our priorities, we will soon eradicate hundreds of thousands more. There is a curious fact, however: among all of the species we have destroyed, there is not one that occurred worldwide . . . the closest that we have ever come to taking that fatally insane step is with the right whale—a species that once occurred off the western and eastern shores of every continent. We came so close to destroying that species that it really is something of a miracle that it survived."

In contrast, the *K* strategists tend to have few offspring. These offspring are large, mature slowly, and often receive intensive parental care. This group consists of such organisms as coconut palms, whooping cranes, and whales (Figure 52-8). Many of them are in danger of extinction (Figure 52-1).

Human Populations

Human beings use their culture to avoid or postpone the effects of most limitations on their population size, which is currently increasing at the rate of about 1.7% per year. Some 85 million people are added annually to a total population that already numbers about 5 billion. At this rate the world population will double in just over 40 years. As we will discuss in Chapter 54, such a rate of growth has potential consequences for our future that are extremely grave. By reducing the population sizes of most other species, as our own populations grow so rapidly, we are driving to extinction those relatively large and slow-breeding species that we view as desirable (Figure 52-1), while favoring those weed and pest species that compete directly with our own enormous populations for food and other critical resources.

> **The rapidly growing human population is tending to exterminate *K* strategists, such as whales, lions, and forest trees, while favoring *r* strategists, including houseflies, cockroaches, and dandelions.**

MORTALITY AND SURVIVORSHIP

A population's intrinsic rate of increase depends on the ages of the organisms in it and the reproductive performance of the individuals in the various age groups. When a population lives in a constant environment for a few generations, its age distribution becomes stable. This distribution, however, differs greatly from species to species and even, to some extent, from place to place within a given species.

One way to express the characteristics of populations with respect to age distribution is the survivorship curve. **Survivorship** is defined as the percentage of an original population that survives to a given age. Samples of different kinds of survivorship curves are shown in Figure 52-9. In the hydra, individuals are equally likely to die at any age, as indicated by the straight survivorship curve. Oysters, on the other hand, produce vast numbers of offspring, only a few of which live to reproduce. Once they become established and grow into reproductive individuals, their rate of death, or **mortality,** is extremely low. Even though human babies are susceptible to death at relatively high rates, the highest mortality in people occurs later in life, in their postreproductive years.

By convention, life cycles are designated as type I—the type found in human beings, where a large proportion of the individuals appear to reach their physiologically determined maximum age; type II—the situation found in hydra, in which the mortality rate remains more or less constant at all ages; and type III—the kind of life cycle found in oysters, in which mortality is high in the early stages but then declines. Many animal and protist populations in nature probably have survivorship curves that lie somewhere between those characteristic of type II and type III, and many plant populations, with high mortality at the seed and seedling stages, are probably closer to type III. Humans have probably approached type I more and more closely through the years, with the birth rate remaining relatively constant or declining somewhat, but the death rate dropping markedly.

> **In type I survivorship curves, a large number of individuals reach their maximum theoretical age. In type II curves, mortality is largely independent of age. In type III curves, mortality is highest during the young stages. These curves are convenient ways to describe the different kinds of life histories.**

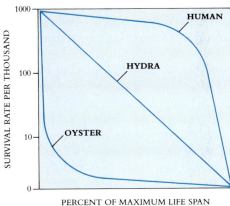

FIGURE 52-9

Survivorship curves for the oyster, hydra, and humans. Each population is assumed to have started with 1000 individuals. The shapes of the respective curves are determined by the mortality rates of these individuals at different ages.

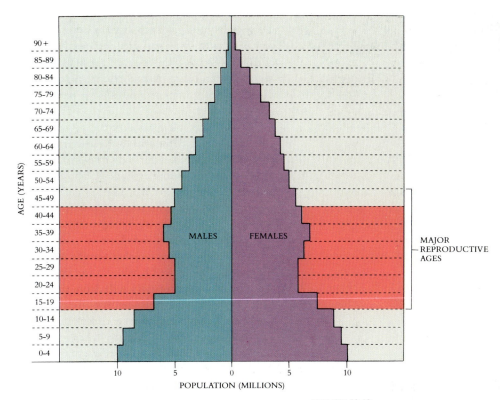

FIGURE 52-10

The age distribution of human males and females in the population of 1960.

DEMOGRAPHY

Demography is the statistical study of populations. The term comes from two Greek words, *demos,* "the people," (the same root we see in the word "democracy") and *graphos,* "measurement." It therefore means measurement of people, or, by extension, of the characteristics of populations. Demography is the science that enables us to predict the ways in which the sizes of populations will alter in the future. It takes into account the age distribution of the population and its changing size through time.

A population whose size remains the same through time is called a **stationary population.** In such a population, births plus immigration must exactly balance deaths plus emigration. In a stable population, the size not only remains constant, but so does the age structure.

An example of the age distribution of males and females in a population is shown in Figure 52-10. We can predict the future sizes of a population by multiplying the number of females in each age class by the average number of female offspring produced by a female alive at that age and then adding these totals for each age group to see whether the new number will exceed, equal, or be less than the number of females in the sample under consideration. By such means the future growth trends of the human population as a whole and of individual countries and regions can be determined. Such calculations form much of the basis for the considerations that we will present in Chapter 54.

INTERSPECIFIC INTERACTIONS THAT LIMIT POPULATION SIZE

Competition between individuals of two or more species for the same limiting resources—in other words, the direct effects of predation or parasitism of one species on another—is one of several kinds of interspecific interaction that may limit population size. We will discuss them in this context here, and in the next chapter we will return to them in relation to their role in the structure of the community.

Competition

Two kinds of organisms that are both using a resource that is in short supply are said to compete with one another. Interspecific competition is greatest between organisms that occupy the same trophic level: green plants compete mainly with other green plants, herbivores with other herbivores, carnivores with carnivores. In addition, competition is more acute between organisms that resemble one another closely than between ones that are less similar.

More than 50 years ago, the Soviet ecologist G.F. Gause formulated what he called the **principle of competitive exclusion.** This principle states that if two species are competing with one other for the same limited resource, then one of the species will be able to use that resource more efficiently than the other, and the former will therefore eventually eliminate the latter locally. The results of one of Gause's most important series of experiments are shown in Figure 52-11.

> **The principle of competitive exclusion states that if two species are competing with one another for the same limited resource in the same place, one of them will be able to use that resource more efficiently than the other and eventually will drive that second species to extinction locally.**

In subsequent experiments the results of growing individuals of two species together in the laboratory have not always been easily predictable. For example, when Thomas Park and his colleagues at the University of Chicago grew two species of flour beetle, *Tribolium* (Figure 52-12), together in the same container of flour, one always became extinct. The species that survived, however, varied. *Tribolium castaneum* usually won under relatively hot and damp conditions, while *Tribolium confusum* won under cooler, drier conditions. Subsequent experiments with these species demonstrated that a genetic component was also involved in the unpredictability of the outcome. Some strains of one species would win over some—but not all—strains of the other, under a given set of conditions.

Similarly, John Harper and his colleagues at the University College of North Wales grew two species of clover together, *Trifolium repens* and *Trifolium fragiferum*, white clover and strawberry clover. Each was sown at two densities, 36 and 64 plants per square foot (30.5 centimeters), using all the possible combinations. Although white clover initially

FIGURE 52-11

Results of Gause's experiment with two species of Paramecium, P. caudatum *and* P. aurelia. *These experiments illustrate that, when the two species are grown together in mixed culture,* P. aurelia *will drive* P. caudatum *to extinction. This indicates that* P. aurelia *uses the available food sources under these conditions more efficiently than* P. caudatum *does, probably because of its more rapid reproduction.*

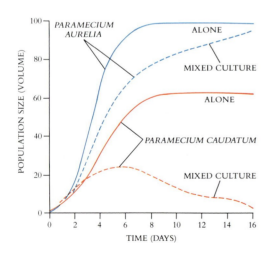

FIGURE 52-12

The flour beetle Tribolium confusum, *showing adults, larvae, and pupae. This species, and the closely similar* T. castaneum, *were used in experiments that demonstrated the unpredictability of results when two species of organisms are competing with one another.*

formed a dense canopy of leaves, the slower-growing strawberry clover, whose leaf stalks are longer, eventually produced enough leaves that overtopped the lower white clover and overcame it. Strawberry clover did this by competing more effectively for light; the outcome was the same regardless of the initial densities at which the plants were sown.

Competition In Nature

Examples demonstrating the principle of competitive exclusion can be found and studied, both by observation and by experiment, in nature. Such examples are directly related to the structure of the communities in which they occur, and they will be treated more extensively in the next chapter. However, we include a few examples here, as we consider the contribution of competitive interactions to the structure of natural communities.

The competitive interactions between two species of barnacles that grew together on the same rocks along the coast of Scotland have been investigated by J.H. Connell of the University of California, Santa Barbara. Barnacles, as you saw in Chapter 37, have free-swimming larvae that settle down, cement themselves to rocks, and then remain permanently attached to that point. Of the two species Connell studied, *Chthamalus stellatus* lives in shallower water, where it is often exposed to air by tidal action, and *Balanus balanoides* occurs lower down, where it is rarely exposed to the atmosphere (Figure 52-13). In this deeper zone, *Balanus* could always outcompete *Chthamalus* by crowding it off the rocks, undercutting it, and replacing it even where it had begun to grow. When Connell removed *Balanus* from the area, however, *Chthamalus* was easily able to occupy the deeper zone, indicating that no physiological or other general obstacles prevented it from becoming established there. In contrast, *Balanus* cannot survive in the shallow-water habitats where *Chthamalus* normally occurs. It evidently does not have the special physiological and morphological adaptations that allow *Chthamalus* to occupy this zone.

Viewing the same situation in evolutionary terms, one might suppose that *Chthamalus* evolved later than *Balanus* or that it reached the Scottish coast later than its rival. Unable to compete with *Balanus* in its main area of distribution, it nevertheless could survive in the relatively less favorable shallow-water habits where *Balanus* did not occur. Conversely, it is possible that *Balanus* evolved more recently or reached the area more recently than *Chthamalus* and possibly eliminated *Chthamalus* from most of its former range. In doing so

FIGURE 52-13

A *Balanus (the smaller, smooth barnacles) and* Chthamalus *(the larger, ridged barnacles) growing together on a rock. Competing species belonging to these two genera were studied by J.H. Connell along the coast of Scotland. The microscopic larvae of barnacles are dispersed widely, but they settle down as adults.*

B *The distribution of the two species with respect to different water levels and tidal effects is shown at the right.*

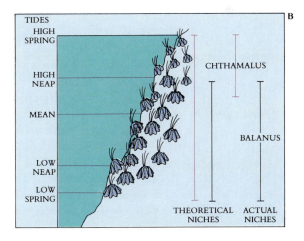

ISLAND BIOGEOGRAPHY

The number of species that live on an island is related to the size of that island, an important finding that was first expressed in quantitative terms in the 1950s. In 1967, Robert MacArthur of Princeton University and E.O. Wilson of Harvard published an important short book on this relationship, *The Theory of Island Biogeography*. They explained that species are constantly being dispersed to islands, so islands have a tendency to accumulate more and more species. At the same time that new species are being added, however, other species are being lost. Once the number of species in a particular group is high enough that it completely fills the capacity of that island, no more species can be established, on the average, unless one of the species that is already there becomes extinct. Every island of a given size, then, has a characteristic number of species that tends to be maintained through time, even though many of the individual species might change.

The initial stages in this process resemble those that occurred on Surtsey (see pp. 1122-1123). After such a new island is formed, if it is relatively close to a source area, colonization may be rapid. Eventually, however, an equilibrium number of species will be reached and subsequently maintained for each group of organisms.

The relationship, if expressed on a log/log basis, approximates a straight line. Thus Figure 52-A, modified from MacArthur and Wilson's book, expresses the relationship between the number of species of reptiles and amphibians on certain islands in the West Indies and the area of the islands.

The analysis of a great deal of data has shown that relationships such as this are best approximated by the formula

$$S = CA^z$$

In this formula S represents the number of species; C, the density of species, a factor that varies according to the group of organisms being considered and the region; A, the area of the island; and z the slope of the line in a graph like the one presented here. The number of species present on an individual island or set of islands varies somewhat with the height of the islands and their distance from the mainland or other source areas; more distant islands have fewer species, and higher ones have more species. Despite these variables, the relationship between species number and area is strikingly constant in a given region anywhere in the world. Thus useful predictions can be made about the expected numbers of species on islands or other areas when little or no information is available.

An experimental approach to understanding this relationship was carried out in small mangrove islands off the southern tip of Florida. In this region,

it would have restricted *Chthamalus* to the less favorable sites where it was subject to periodic drying.

In another set of field observations, the late Princeton ecologist Robert MacArthur studied five species of warblers, small, insect-eating birds that coexist during part of the year in the forests of the northeastern United States and adjacent Canada. Although they all appeared to be competing for the same resources, MacArthur found that each species actually spent most of its time feeding in different parts of the trees, so each ate different subsets of insects in those trees. Some of the species of warblers fed on insects near the ends of the branches, while others regularly penetrated well into the foliage; some stayed high on the trees, others fed on the lower branches; and these patterns were recombined in different ways characteristic of each of the warbler species (Figure 52-14). As a result of these different feeding habits, each species of warbler actually occupied a different niche; in other words, they had different ways of utilizing the resources of the environment and thus were not really in direct competition with one another for limited resources.

Robert MacArthur found that the five species of warblers he studied could coexist because the niches of the five species were different, and competitive exclusion between them did not occur.

islands about 12 meters across are inhabited by about 20 to 45 species of arthropods. Larger islands had more species; smaller ones had fewer species. When some of the islands were fumigated with an insecticide that killed all arthropods, they were recolonized within a year and rapidly attained a total number of species that roughly equaled the number before fumigation. Most interesting, however, was the fact that the new inhabitants of a given island mainly represented different species from those that lived there initially. The numbers of species that were able to coexist on a particular island tended to be set by the capacity of that island, which was presumably affected by the principles of competition, availability of habitat, and other factors we have discussed in this chapter. The exact species on an island, however, were determined largely by chance.

On islands far from source areas, like those in the central Pacific, the replacement of extinct species tends to be slow. This observation has important implications for conservation biology. The introduction of foreign species is apt to be especially disruptive in such ecologically fragile communities. The members of those groups that do disperse well, such as small, light insects or plants whose seeds drift in the sea or blow through the air, are apt to be better represented than other kinds of organisms with more limited powers of dispersal. Furthermore, if there is no source area, or if the source area is very distant—patterns that arise when the ecosystems of former source areas are destroyed by human activities or other activities—recolonization may not occur at all. As tropical vegetation is being cut up into small "islands," for example, the nature and size of these islands and the distance that separates them from the areas from which they might be recolonized become critical to species survival.

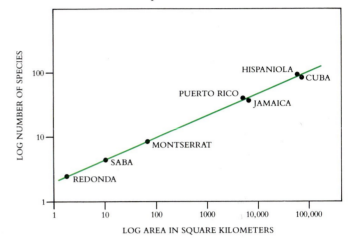

FIGURE 52-A

The relationship between the logarithm of species number (on the ordinate) and the logarithm of area (on the abscissa) for different islands approximates a straight line. The islands are all in the Caribbean.

A   B

FIGURE 52-14

Two of the warblers that Robert MacArthur studied, the Cape May warbler (A) and the Blackburnian warbler (B). Both species of warblers search for insects among the branches of trees, but the Cape May warbler spends at least half of its feeding time in the uppermost branches and crown of the trees, and the Blackburnian warbler searches mainly among the lower branches, occasionally passing up and down the trees. The three other species that coexisted with these two also differed sharply from one another, and from these, in their feeding habits.

In both the studies of Connell on barnacles and of MacArthur on warblers, the effects of competition on community structure are also apparent. Over a timescale long enough for evolution to occur, the characteristics of competing organisms may change as a result of natural selection to minimize the effects of the competition. This may occur actively, when populations of the competing species exist together; if this is the case, the process whereby they become more distinct from one another is known as **character displacement.** A similar process may also operate when species occur together that are only partly reproductively isolated. In such a case the factors may be strengthened that minimize the chances that individuals of the two species will breed together; for example, their breeding times may overlap initially and then become more distinct where the species occur together, because of the advantage of *not* producing hybrids. It is difficult to determine in a particular instance whether the critical differences that separate species and allow them to coexist have originated in isolated populations, or as a direct response to the challenges posed by their coexistence.

During the 1980s there has been a lively debate about the real importance of competition in nature in relation to community structure. Although the kinds of assumptions we have made in this section seem to be plausible explanations of natural situations of many kinds, they have not been tested as often as one might wish. Consider, for example, the herbs that grow under trees in eastern North America: do they actually compete with one another? What are the limiting factors for them? Is their distribution relative to one another set by competition, or is it set by chance, where a particular species happened to become established in the first place? Such considerations are of great importance when discussing whether character displacement plays a key role in the evolution of species and in determining the basic principles of island biogeography, which in turn are controlled by the number of species that can occur together in one place (see the boxed essay on pp. 1140-1141). As in any field of science, the development of appropriate tests of theory and the execution of these tests in a wide variety of natural situations will ultimately lead to a more comprehensive understanding of the way in which competition occurs in nature and of its consequences.

The Niche

Studies of the principle of competitive exclusion led to the development of the concept of the **niche.** A niche may be defined most simply as the role an organism plays in an ecosystem. It may be described in terms of space, food, temperature, appropriate conditions for mating, requirements for moisture, and so on. A full portrait of an organism's niche would also include its behavior and the ways in which this behavior changes at different seasons and different times of the day. In other words, *niche* is not synonymous with *habitat:* the habitat of an organism, the place where it lives, is defined by some, but by no means all, of the factors that make up its niche. The factors that make up an organism's niche determine whether it can exist in a given ecosystem, and also how many species can exist there together.

Niche, in the sense we have just been discussing, is sometimes called the **realized niche** of an organism—the role the organism actually plays in a particular ecosystem. It is thus distinguished from the niche that it might occupy if competitors were not present, called the **fundamental niche.** Thus the fundamental niche of the barnacle *Chthamalus* in Connell's experiments in Scotland included that of *Balanus,* but its realized niche was much narrowed because *Chthamalus* was outcompeted by *Balanus* in its own realized niche.

> **The fundamental niche of an organism is the niche that it might occupy if competitors were not present. The realized niche is the niche that it actually does occupy under natural circumstances.**

Another interesting example of some of the properties of a niche involves the flour beetles of genus *Tribolium* that we considered earlier. If *Tribolium* is grown in pure flour along with beetles of a second genus, *Oryzaephilus,* it will drive *Oryzaephilus* to extinction by means that are only partly understood. If, however, small pieces of glass tubing are added to the flour, providing refuges for *Oryzaephilus,* then both kinds of flour beetle will coexist indefinitely. In a sense this experiment suggests why so many kinds of organisms can coexist in a really complex ecosystem, such as a tropical rain forest, and why competition is more direct in an ecosystem with fewer species, such as the tundra.

When organisms reach habitats where no other plants, animals, or microorganisms have yet arrived, such as islands that have been raised above sea level or newly formed ponds, they are often able to occupy a much wider niche—play a wider variety of ecological roles—than their relatives in the communities from which they were derived. We considered this relationship in Chapter 20 when discussing the evolution of new species on islands. For example, Darwin's finches on the Galapagos Islands or the marsupials in Australia are groups of organisms that occupy many more different niches than do their relatives on the continent of South America. There the relatives of these animals compete with many other different kinds of organisms, some of which are their own close relatives. The Galapagos finches and Australian marsupials appear to have evolved from ancestors that reached their respective areas early and came to play a wide variety of ecological roles in them.

As species evolve, they become increasingly different from one another; eventually, they may be able to coexist, not then being subject to competitive exclusion. If two species become so different in isolation from one another that their niches differ sufficiently when they come into contact with each other, there will be no problem; if the niches do not, there will be selection so that they become more different, or the species cannot continue to coexist. In other words, the differences may evolve either before the species come into contact or after they do so.

Gause's principle of competitive exclusion can be restated, in terms of niches, as follows: *No two species can occupy the same niche*. Certainly species can and do coexist while competing for the same resources, and we have just seen some examples of such relationships. Nevertheless, Gause's theory predicts that when two species do coexist on a long-term basis, one or more features of their niches will always differ; otherwise the extinction of one species will inevitably result.

Many lines of native South American mammals were displaced by unrelated North American ones that moved in similar niches once the Isthmus of Panama was elevated above the sea between 3.1 and 3.6 million years ago (see Chapter 23). When this geological event occurred, animals and plants could move overland between these formerly isolated continents, creating an opportunity for competitive exclusion on a grand scale. The same process—competitive exclusion—occurs continuously in the woods, fields, and lakes near your home. Species that occupy the same niche (compete directly for the same limiting resource) cannot coexist indefinitely, whether in nature or under experimental conditions in the laboratory. Much of the research in animal ecology has been devoted to defining the ways in which the niches of coexisting organisms differ.

Niche is, of course, a complex concept, one that involves all facets of the environment that are important to individual species. No two species can coexist indefinitely if their niches overlap in too many ways. Within the past decade there has been a vigorous debate concerning the role of competitive exclusion, not only in determining the structure of communities, as we discussed earlier, but also in setting the course of evolution. Where one or more resources are obviously limiting, as in periods of drought, the role of competition becomes much more obvious than when they are not.

Predation

Predation is another factor that may limit the size of populations. In this sense predation includes everything from the situation that occurs when one kind of animal captures and eats another to parasitism (the condition of an organism living in or on another organism, at whose expense the parasite is maintained). Predation and parasitism are two ends of a biological spectrum between which there is no clearly marked distinction. They are governed by similar principles.

FIGURE 52-15

Oscillations of predator and prey species, in this case lynxes and snowshoe hares in northern Canada, as predicted by Gause's theory. The data are based on numbers of animal pelts from 1845 to 1935.

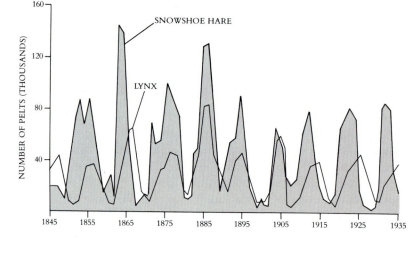

FIGURE 52-16

After an initial period in which prickly pear cacti (Opuntia), introduced from Latin America, choked many of the pastures of Australia with their rampant growth, they were controlled by the introduction of a cactus-feeding moth, Cactoblastis, from the areas where the cacti were native.

A *An infestation of prickly pear cacti in scrub in Queensland, Australia, in October, 1926.*
B *The same view in October, 1929, after the introduction of Cactoblastis.*

When experimental populations are set up under very simple circumstances in the laboratory—as Gause did, for example, with *Paramecium* and its predator protozoan *Didinium* (see Figure 9-6)—the predator often exterminates its prey and then becomes extinct itself, having nothing to eat. If refuges are provided for the prey, however, its populations will be driven to low levels but can recover; one would then expect the populations of predators and prey to follow a cyclical pattern. The low population levels of the prey species provide scant food for the predators; then the predators in turn become scarce. As this occurs the prey recover and again become abundant. Such interactions are expressed diagrammatically in Figure 52-15.

Relationships of this sort are of the greatest importance for biological control. If the prey becomes so rare that it is an infrequent event for the predator species to encounter it, the predator itself may become extinct. Ideally, in the case of deliberate biological control, the prey will survive in small numbers, thus making it possible for small populations of the predator to survive also and for the prey species to be controlled indefinitely.

In Australia, for example, prickly pear cactus *(Opuntia)* once overran the ranges and became so abundant that vast areas were effectively closed to cattle grazing (Figure 52-16). The situation was changed dramatically with the introduction of the moth *Cactoblastis*. The larvae of the moth feed on the pads of the cactus and rapidly destroy the plants. Within relatively few years, the moth had reduced the cactus to the status of a rare species in many

A

B

regions where it was formerly abundant. It is now exceptional to find an individual of *Cactoblastis,* but the moth is still present and evidently keeps the cactus in check.

The future is more problematic for the American chestnut *(Castanea americana),* a species of tree that has virtually been driven to extinction by the accidental introduction of an Asiatic ascomycete, *Endothia parasitica.* The chestnut used to be a dominant or codominant tree throughout much of the forests of eastern and central temperate North America. Chestnut blight, the disease caused by the fungus, was first seen in North America in New York State in 1904 and killed most of the trees in North America during the following 30 years. Today the American chestnut survives largely as sprouts that grow every year from the trunks of trees that were killed decades ago (Figure 52-17). Whether some populations are isolated from the chestnut blight in certain remote areas remains to be seen.

One factor in such situations is genetic variability. Asian chestnuts and other species of the genus *Castanea,* such as the chinquapin, *Castanea pumila,* of the southeastern United States, are partly or entirely resistant to the blight, but there is little evidence that resistant individuals of the American chestnut exist. This formerly widespread tree species may therefore actually be on the way to extinction, for reasons similar to, although less dramatic than, the ones we discussed in connection with the cheetah in Chapter 1.

Many additional examples of the ways in which predator-prey relationships operate are familiar. Diseases that completely kill their host species are not "successful," because once they have done so they have also eliminated their own source of food. Thus those strains of the disease-causing organism that are less virulent will be favored by natural selection and will survive.

A disease that always kills its host will die with it; one that produces sublethal effects will have the opportunity of spreading to another host.

The history of the viral disease myxomatosis, which was introduced into Australia and New Zealand to control rabbits, provides an instructive example of this principle. The rabbits were brought to these countries as a convenient source of meat, but they soon ran wild, with devastating effects on the countryside. When myxomatosis was introduced, most of the rabbits soon died. The most virulent strains of myxomatosis disappeared with the dead rabbits, and less lethal strains of the virus became apparent in the remaining rabbit populations. At the same time, strains of rabbits that were resistant to the disease began to appear. Now the populations of both kinds of organisms have achieved the kind of equilibrium relationship in which they can coexist indefinitely.

The relationships between large carnivores and grazing mammals are a subject of great interest and importance in many parts of the world. Appearances are sometimes deceiving. On Isle Royale in Lake Superior, for example, moose reached the island and multiplied freely there in isolation (Figure 52-18). When wolves later reached the island by crossing over the ice in winter as the moose had done earlier, it was widely assumed at first that they were playing the determining role in controlling the moose populations. More careful studies, however, have demonstrated that this is not the case. The moose that the wolves eat are old and diseased, for the most part, and would not survive long anyway. In general the moose are controlled by the amount of food available to them, their diseases, and many factors other than the wolves. Similar relationships are hotly discussed whenever conservationists enter into a dialogue with stockmen anywhere in the world.

On the plains of Africa, the balanced natural systems that involve a great deal of specialization among both the herbivores and the carnivores that prey on them have been greatly disrupted, not only because of selective hunting but also because the animals have been confined to areas much smaller than those they inhabited earlier. As human populations continue to grow, it will become more and more difficult for populations of many kinds of plants and animals to maintain themselves in nature. If we wish to preserve some of the more vulnerable species, such as the giant panda, the California condor, and the various species of rhinoceros, we will need to intervene actively by bringing these animals into zoos or protecting them stringently in natural reserves.

FIGURE 52-17

Although the chestnuts (Castanea) *have been killed over much of North America by the chestnut blight fungus,* Endothia parasitica, *the stumps of many individuals are still alive. Each season they send out new shoots, as seen here; a few of these survive long enough to produce fruits.*

FIGURE 52-18

Wolf-moose interactions on Isle Royale, Michigan.
A *A large pack of wolves in unsuccessful pursuit of a moose. This moose was chased for almost 2 kilometers; it then turned and faced the wolves, who by that time were exhausted from running through chest-deep snow. The wolves lay down and the moose walked away.*
B *Wolves beginning to feed on a freshly killed moose. A dominant female begins to open up the chest cavity, where the internal organs will be consumed first. Such dominant females, called alpha wolves, control the activities of the pack and generally produce the only offspring. Wolf predation selectively removes calves and old moose from the population— usually those more than 8 years old, and often with skeletal pathology such as arthritis. Isle Royale National Park provides an exceptional outdoor laboratory, the only site on earth where naturally regulated wolves and prey have been subjected to long-term study.*

A

B

The intricate interactions between predators and prey are an essential factor in the maintenance of groups of organisms occurring together that are rich and diverse in species. By controlling the levels of some species, the predators make possible the continued existence of others in that same community. In other words, by keeping the numbers of individuals of some of the competing species low, the predators prevent or greatly reduce competitive exclusion.

A given predator may very often feed on two, three, or more different kinds of plants or animals in a given community in a way that partly depends on their relative abundance. In other words, a predator may feed on species *A* when it is abundant and then switch to species *B* when *A* is rare. Similarly, a given prey species may be a primary source of food for increasing numbers of species as it becomes more abundant, which tends to limit the size of its population automatically. Such feedback systems are a key factor that determines the structure of many natural communities.

SUMMARY

1. Populations may be dispersed in an evenly spaced, clumped, or random manner. Clumped patterns are the most frequent.

2. The rate of growth of any population is defined as the difference between the birth rate and the death rate per individual per unit of time. The actual rate is affected both by emigration from the population and by immigration into it.

3. Most populations exhibit a sigmoid growth curve, which implies a relatively slow start in growth, a rapid increase, and then a leveling off when the carrying capacity of the species' environment is reached. Populations can be harvested most effectively when they are in the rapid growth phase.

4. *r* strategists have large broods and rapid rates of population growth; *K* strategists are limited in size by the carrying capacity of their environments. They tend to have fewer offspring, with slower rates of population growth.

5. Survivorship curves are used to describe the characteristics of growth in different kinds of populations. Type I populations are those in which a large proportion of the individuals approach their physiologically determined limits of age. Type II populations have a constant mortality throughout their lives. Type III populations have very high mortality in their early stages of growth, but an individual surviving beyond that point is likely to live a very long time.

6. Each species plays a specific role in its ecosystem; this role is called its niche.

7. Gause's principle of competitive exclusion states that if two species compete with one another for the same limited resources, then one will be able to utilize them more efficiently than the other and will drive its less efficient competitor to extinction. This principle can be restated in terms of the niche concept: No two species can occupy the same niche indefinitely. Organisms studied under field conditions display many different ways of avoiding direct competition for the same scarce resources.

8. An organism's fundamental niche is the total niche that the organism would occupy in the absence of competition. The realized niche of an organism is the actual niche that it occupies in nature.

9. Predator-prey relationships are of crucial importance in limiting population size in nature. They appear to maintain a kind of balance, because if the prey is completely eliminated, the predator will die also.

SELF-QUIZ

The following questions pinpoint important information in this chapter. You should be sure to know the answers to each of them.

1. The most common form of a population dispersion is a(n) _____ distribution.
   (a) evenly spaced
   (b) scattered
   (c) clumped

2. In the formula for calculating the biotic potential of a population, r stands for
   (a) predicted population increase
   (b) actual rate of population increase
   (c) predicted rate of population decrease
   (d) actual rate of population decrease
   (e) randomly fluctuating population size

3. The size at which a population stabilizes in a particular place is defined as the _____ _____ of that place for that species of organisms.

4. Biological populations tend to follow a sigmoid growth curve. True or false?

5. K strategists have
   (a) many offspring   (d) short generation time
   (b) large offspring   (e) none of the above
   (c) little parental care

6. The intrinsic rate of increase of a population depends on
   (a) the ages of the organisms making up the population
   (b) population density
   (c) the location of the organisms making up the population
   (d) the number of individuals in the population
   (e) density-independent factors

7. Match choices a, b, and c with the appropriate kind of survivorship curve below:
   (a) large numbers of individuals reach maximum theoretical age
   (b) mortality is highest during young stages
   (c) mortality is largely independent of age

   _____ type I survivorship curves

   _____ type II survivorship curves

   _____ type III survivorship curves

8. Interspecific competition is greatest between organisms that occupy the same trophic level. True or false?

9. The niche that an organism might occupy if competitors were not present is the
   (a) realized niche   (d) potential niche
   (b) actual niche   (e) fundamental niche
   (c) basic niche

10. Gause's principle of competitive exclusion states that two species can indefinitely occupy the same niche under special circumstances. True or false?

11. A "successful" disease will
    (a) evolve to the point where it is always able to kill its host   (d) always involve parasitism
    (b) produce sublethal effects   (e) always involve mutualism
    (c) evolve to the point where it has no effect on its host

12. Populations can be harvested most effectively when they are in the
    (a) rapid growth stage   (d) negative growth stage
    (b) slow growth stage   (e) equilibrium growth stage
    (c) zero growth stage

THOUGHT QUESTIONS

1. How is it possible for ecological generalists—those which, for example, feed on a wide range of foods—to coexist with specialists in natural communities, as they do throughout the world? Why do all species not simply become generalists or specialists? Which of these habits is a better strategy for survival? Why?

2. How can the principle of competitive exclusion be applied to the origin of large numbers of species of certain genera in the Hawaiian Islands? What would you expect to happen in the future to the hundreds of species of *Drosophila* that occur there, leaving aside the possibility of their extinction through the destruction of their habitats by human beings?

3. How do you think the intensity of competition would differ between an Arctic tundra and a tropical rain forest? How would you test your answer?

4. Which kinds of ecosystems are most resistant to introduced species? Why?

FOR FURTHER READING

BEDDINGTON, J.R., and R.M. MAY: "The Harvesting of Interacting Species in a Natural Ecosystem," *Scientific American,* November 1982, pages 62-69. An analysis of the changing populations of whales and other animals that feed on the krill (shrimp) populations in the Antarctic Ocean as an example of the problems of using a biological resource without destroying it.

COLINVAUX, P.A.: *Introduction to Ecology,* 2nd ed., John Wiley & Sons, Inc., New York, 1978. Excellent book that contains a particularly good treatment of the materials discussed in this chapter.

KREBS, C.J.: *Ecology: The Experimental Analysis of Distribution and Abundance,* Harper & Row, New York, 1972. Includes a strong, comprehensive treatment of ecology at the population level.

MacARTHUR, R.H., and E.O. WILSON: *The Theory of Island Biogeography,* Princeton University Press, Princeton, N.J., 1967. This short book introduced one of the more productive fields of modern ecology.

SCHOENER, T.W.: "The Controversy over Interspecific Competition," *American Scientist,* vol. 70, pages 586-595, 1982. Interesting discussion about the interplay between competition and predation in structuring natural communities.

TOWNSEND, C.R., and P. CALOW (eds.): *Physiological Ecology: An Evolutionary Approach to Resource Use,* Sinauer Associates, Inc., Sunderland, Mass., 1981. A series of excellent papers centering on the themes presented in this chapter.

WILSON, E.O., and W.H. BOSSERT: *A Primer of Population Biology,* Sinauer Associates, Inc., Sunderland, Mass., 1971. A good short introduction to the mathematical aspects of this field.

COMMUNITY ECOLOGY

Overview

Communities are bound together by the interactions of the organisms that make them up, but they are no more than the sum of the properties of those organisms. The species that occur together in a community have evolved together. The herbivores avoid feeding on certain plants and selectively use others; their actions are mediated by various characteristics of the plants. As a result, the herbivores themselves have become specialized. In some cases, they use the chemicals that they obtain from the plants they eat to protect them from their own predators; in others, they break down the chemicals and thus avoid their toxic effects. Symbiotic relationships generally act to order the flow of energy through ecosystems and to regulate the cycling of nutrients. Learning about these relationships individually is the key to learning about the functioning of ecosystems.

For Review   *Here are some important terms and concepts that you will encounter in this chapter. If you are not familiar with them, you should review them before proceeding.*

- **Biomes** (Chapter 50)

- **Energy flow** (Chapter 51)

- **Population dynamics** (Chapter 52)

- **Competition** (Chapter 52)

To what extent are the communities—associations of interacting populations of plants, animals, and microorganisms—that we observe in nature real, and to what extent are they simply arbitrary groupings that we devise for our own convenience? Ever since biologists began to describe the plant communities of central Europe in the nineteenth century, they have found this question difficult to answer. More recent research has shown that either alternative may be true, depending on one's particular point of view.

THE REDWOOD FOREST: AN EXAMPLE

Certainly, some kinds of organisms do tend to occur together in repeatable assemblages; nevertheless, difficulties arise when we attempt to define the limits of these assemblages in either space or time. Consider, for example, the redwood forest that extends along the coast

FIGURE 53-1

The redwood forest of coastal California and southwestern Oregon is dominated by the redwoods (Sequoia sempervirens) themselves.

A

B

C

FIGURE 53-2

Among the plants that occur regularly in the redwood community are the sword fern (Polystichum munitum) (A) and the redwood sorrel (Oxalis oregana) (B). One of the many characteristic animals is this ground beetle (family Carabidae), Scaphinotus velutinus, shown feeding on a slug on a leaf of sword fern (C). The ecological requirements of each of the organisms associated with the redwoods differ, but they overlap enough that they are able to occur together in what we perceive as the redwood community.

of central and northern California and into the southwestern corner of Oregon (Figure 53-1). This evergreen forest is dominated by redwoods, *Sequoia sempervirens,* the sole living survivors of a genus of gymnosperms that once ranged over much of the Northern Hemisphere. With these massive trees are regularly associated a number of other plants and animals (Figure 53-2). Thus we may be led to think that the redwood community is an easily defined and clearly delimited biological entity.

In this particular example, the redwoods themselves are the dominant species; they help to create the special circumstances under which the other plants, animals, and microorganisms in the community exist. Collectively, these organisms, together with the redwoods themselves, make up what we call the redwood community. The redwood trees always grow within the reach of coastal fogs, and moisture condensed from the fog drips in great quantities off their needles, especially in summer, helping to create a special habitat that is moist and whose conditions are quite constant all year round.

Problems arise, however, when we begin to chart the distributions of the other organisms, which seem to be so closely and predictably associated with the redwoods. Some have ranges that are less extensive than that of the redwood itself; others have ranges that are more extensive. Scarcely a single one has a distribution that even approximates that of the redwood. In other words, every plant and animal species that occurs in association with these majestic trees, and that helps to give this community its characteristic aspect, lives there because its own particular requirements—its ecological niche—overlap with those of the redwood. There is probably no other single kind of organism, however, whose ecological requirements coincide *exactly* with the redwood's.

The redwood trees themselves create the special atmosphere that distinguishes the redwood community. With them are regularly found a number of other plant and animal species that may also occur in other areas and whose ranges do not coincide exactly with the redwood's. The community, therefore, is an aggregation of organisms that occur together because their niches overlap. In this sense, it is not a superorganism; most of its properties are simply those of the organisms that make it up.

THE COMMUNITY AS A SUPERORGANISM

The studies of C.H. Muller, of the University of California, Santa Barbara, and his students have revealed some ways in which plant communities resemble superorganisms. In the coastal sage communities of the Santa Ynez Valley, near Santa Barbara, the investigators found that the hard-leaved, evergreen shrub *Ceanothus cuneatus* would eventually shade out the soft-leaved purple sage, *Salvia leucophylla*. Normally the *Ceanothus* is a member of the chaparral community, which grows in areas that receive more rainfall than those inhabited by the coastal sage community where the *Salvia* grows. Furthermore, nodules on the roots of the *Ceanothus* are inhabited by actinomycetes, which fix nitrogen and thus make it available to the plant. The *Salvia*, on the other hand, cannot fix nitrogen. Its leaves wither in the driest parts of the year, a strategy that conserves moisture. How does it compete in areas where the two plant species come together?

As anyone who has smelled its crushed leaves knows, *Salvia leucophylla*, like the other species of its genus, is highly aromatic. In fact, its leaves are rich in volatile terpenoids, which inhibit the establishment of the seedlings of other plant species near it; the suppression of one plant species by another by chemical means of this kind is known as **allelopathy.** If individual shrubs of the sage are growing in grassland, they are normally surrounded by a bare zone, in which no shrub seedlings grow (Figure 53-A). Only when the sage plants deteriorate are their colonies invaded by seedlings of *Ceanothus*, which eventually replace them. As long as the colonies of *Salvia* are healthy, they remain free of competition.

The full story is even more complicated, however—as is characteristic of field biology in general. Mice, birds, and insects live in the shelter of the sage shrubs and feed on seeds and seedlings that occur nearby. When these animals are excluded from

Little light reaches the ground under the evergreen redwood trees, so the plants that grow there must carry out photosynthesis efficiently at low light levels. Some of these plants, even though they may tolerate these low light levels and be able to grow in them, may actually grow better elsewhere, where more light is available. Similarly, some of the plant and animal species that grow under the redwoods may actually do better in drier places, although they can survive there. Thus the tolerances of the individual organisms overlap, and the different species are able to live together in what we perceive as a recognizable ecological unit.

We must also take into account the genetic variation within each species in the redwood community. All of them vary from place to place, as do virtually all kinds of organisms that have been studied in this respect. The redwoods themselves are not the same genetically at the southern limits of their range, near Big Sur along the rugged Santa Lucia Mountains on the central coast of California, as they are at their northern limits, in southwestern Oregon. The sword ferns that grow in clumps under the redwoods are not the same genetically as those that occur at middle elevations in the Cascade Mountains of Oregon and Washington, or the same as those that live on the exposed Pacific coastal bluffs. In a genetic sense, therefore, the redwood community exists because the species that make it up include ecotypes that can exist in that particular habitat.

The redwood community also has a historical dimension that we need to consider. The organisms that are now characteristic of this comunity have each had a complex and unique evolutionary history. At very different times in the past, they evolved and then came to be associated in part with the redwoods. Certainly the present-day redwood community has come into existence only during the past few million years. During this period, the range of the redwood was reduced as the tree was eliminated by climatic change from many of the areas in which it grew previously. In a historical sense, then, the redwood com-

the bare zone by cages, seedlings will become established in it. Their growth is often somewhat suppressed by the volatile compounds that the sage plants produce, however. How well such seedlings become established is also affected by the available moisture in that particular area, among other factors.

When one views the sage community as a whole, one can see several ways in which it might be viewed as a superorganism. By producing volatile chemicals, the sage shrubs inhibit the growth of other plants near them. By providing shelter for animals, which will eat nearby seeds and seedlings, they do the same thing in a different way. In a very loose sense, the animals free the individuals of *Salvia* from competition from other plants in exchange for shelter. Nevertheless, the whole community functions only because of the individual properties of the organisms that make it up. It may be considered a natural entity or superorganism of sorts, or as an aggregation of individual organisms, depending on one's point of view.

FIGURE 53-A

The bare zones around the colonies of Salvia leucophylla *are plainly visible in this aerial photograph taken in the mountains above Santa Barbara, California.*

munity that we recognize today represents the concordance of the individual histories of the plants, animals, and microogranisms that comprise it.

Why, then, do we recognize and speak of a redwood community? We do so largely because the redwoods are so dominant, so striking, and so memorable that our minds automatically create a category for them and all of the organisms that coexist with them. The association of these species at a particular place, or series of places, is certainly real, even though it changes in different parts of the range of the redwoods. In some ways, the community, once established, does function as a kind of ecological entity.

Many communities are very similar in their total species composition and aspect over wide areas. For example, the open savanna grassland that stretches across much of Africa includes many plant and animal species that coexist over thousands of square kilometers. Nonetheless, in the redwood community as in all others, the recognition of a particular "community" remains arbitrary. The particular assemblage of plants and animals differs in detail from place to place, and the categories themselves that we may choose to classify as communities and subcommunities can be determined only by statistical analysis. Indeed, the plant communities of Europe, some of which were recognized and described over a century ago, have now been subdivided and characterized more precisely by the use of sophisticated statistical methods. We can recognize as many or as few communities as we deem useful for the particular purpose we have in mind at a particular time.

FACTORS LIMITING DISTRIBUTIONS

The factors that limit the distributions of organisms can be grouped for convenience under three headings: climatic, biotic, and edaphic. Of these, climatic factors are perhaps the most important in determining the distributions of organisms. However, edaphic factors—

FIGURE 53-3

This aerial view of Jasper Ridge, in the Coast Ranges of California above Stanford University, clearly illustrates the difference between the vegetation on soil derived from serpentine rock, which is rich in magnesium and poor in calcium, and sandstone. The soil that formed from serpentine is dominated by the golden-flowered herb Lasthenia chrysostoma; *that derived from sandstone is dominated by grasses. Jasper Ridge was the locality for Paul Ehrlich's important studies of butterfly populations, illustrated in Figure 21-5.*

A

B

FIGURE 53-4

A *Mangroves at Cedar Keys National Wildlife Refuge, Levy County, Gulf Coast of Florida, in April 1984, that have been severely damaged by the hard freezes of the preceding two winters and especially by the severe freeze of December 1983. The effect of this damage near the northern end of the range of mangroves remains to be seen. Cedar Keys is an important area for wildlife as suggested by the nesting brown penguins seen in* **B.** *By processes such as the one shown here, the limits of organisms in communities are established.*

those pertaining to the soil—interact with climate in complex ways and may also play a key role in delineating distributions (Figure 53-3). Biotic, or biological, factors may likewise be significant; the kinds of biological communities that occur in particular places greatly influence the limits of distribution of organisms, both members of that community and others.

Climatic, biotic, and edaphic factors are often intricately linked. For example, redwoods directly affect the local climate that occurs around their bases and thus the distribution of many other kinds of organisms. In addition, certain communities may constitute barriers to the migration of many species of plants, animals, and microorganisms that do not occur in them. Plants and animals that need normal amounts of light cannot occur within, or even in shaded places at the edges of, the redwood forest; for them, the forest may be as inhospitable an environment as the sea.

Climatic, biotic, and edaphic factors interact to determine the limits of distribution of particular kinds of organisms.

The distribution of a particular species of organism at the margins of its range is generally determined by the continually changing climates in these areas, as well as by the genetic variability of its populations. Also of importance is the dispersibility of the organism. The conditions that occur in extremely hot, cold, dry, wet, or otherwise extraordinary years constantly act to eliminate populations of organisms at the limits of their distribution (Figure 53-4). On the other hand, if appropriate genetic combinations are represented in these marginal populations, or are achieved by recombination, and if the dispersal mechanisms of the species are adequate, the species may be able to evolve and to extend its range into new areas, reaching pockets of favorable habitat beyond their current ranges.

As noted in Chapter 34, the nutrients present in the soil play significant roles in determining the ranges of plant species. For example, the old, weathered soils of Australia are low in phosphates and in molybdenum, an essential element whose supply is generally adequate everywhere else in the world. The native plant communities of Australia are able to grow on these very poor soils; their characteristic plants often have leathery, evergreen leaves that appear to play an important role in conserving phosphorus and other nutrients. When we attempt to replace such plant communities with agricultural crops, however, great amounts of fertilizer often must be applied. Similar communities also occur on relatively infertile soils in South Africa (Figure 53-5).

Deciduous plants, whose leaves are shed at the end of each growing season, do not normally conserve their nutrients as effectively as evergreen plants do. Certain species of oaks and other deciduous trees and shrubs are exceptions to this generality. These plants hold their dead leaves through most or all of the winter; during this time, nutrients are extracted from these leaves for use mainly during the period of active growth in the spring. Because deciduous plants do not conserve nutrients as well as evergreen ones do, they are not as successful or well represented on the less fertile soils of Australia and similar regions. Not all plants with evergreen leaves function in this way, though, and we still have much to learn about the adaptive features of plants that enable them to grow successfully under such a wide array of circumstances throughout the world.

Biotic factors often limit the distribution of organisms in quite evident ways. For instance, a herbivore cannot spread beyond the limits of the plant on which it feeds unless new, genetically differentiated populations of individuals that can feed successfully on other kinds of plants evolve in it. Similarly, no carnivore can live apart from the animals that are its prey. A plant that is pollinated by a particular insect does not normally set seed beyond the range of that insect—although it may be physiologically able to do so, perhaps reproducing only vegetatively or perhaps setting seed only occasionally, if the other aspects of the environment are favorable. In all of these ways, organisms affect one another's ranges; they have done so ever since bacteria began to change the nature of the earth's atmosphere and its soils several billion years ago.

FIGURE 53-5

The fynbos plant community, shown here on the Cape Peninsula of South Africa, grows on highly infertile soils. Many of the plants that grow in the fynbos retain scarce nutrients such as phosphorus in their evergreen leaves.

COEVOLUTION

The primary producers, herbivores, and second- and higher-level predators in ecosystems have changed and adjusted to one another continually over millions of years. For example, many of the features of flowering plants, which we discussed in Chapter 30, have evolved in relation to the dispersal of the plant's gametes to other members of the same species by animals, especially insects. These animals, in turn, have evolved a number of special traits that enable them to obtain food or other resources efficiently from the plants they visit, often from their flowers. In addition, the seeds of many flowering plants have features that make them more likely to be dispersed to new areas of favorable habitat.

Such interactions, which involve the long-term, mutual evolutionary adjustment of the characteristics of the members of biological communities in relation to one another, are examples of **coevolution.** Many of the interspecific interactions we have covered earlier, for example in the last chapter, fall into this heading. Here we will consider some additional examples, grouping them under the general headings of predator-prey interactions and symbiosis.

> **Coevolution is a term that describes the long-term evolutionary adjustment of one group of organisms to another. It is an important factor in determining the structure of communities.**

PREDATOR-PREY INTERACTIONS
Plant-Herbivore Relationships

The plant–herbivore interface is the point at which the greatest transfer of energy takes place in the world's ecosystems, with the exception of the interface between organic matter and decomposers. In plant-herbivore interactions generally, some 300,000 species of autotrophic organisms, including about 250,000 species of plants, fix about 1% of the sunlight that falls on their leaves and green stems. Once these plants have carried out this process, at least 2 million species of herbivores feed on them. These herbivores integrate the materials in the plants, algae, or photosynthetic bacteria into their own bodies, actually converting about 10% of the stored energy in these organisms. The nature of life on earth has been largely determined by the ways in which plants avoid being eaten and by which herbivores succeed in eating them.

Plant Defenses

The most obvious ways in which terrestrial plants limit the activities of herbivores are morphological defenses. Thorns, spines, and prickles, for example, play an important role in discouraging browsers, although some of them have learned to overcome these obstacles (Figure 53-6). The defensive role of plant hairs, especially those that have a glandular, sticky tip (Figure 53-7), is less obvious. The enormous abundance of insects and the scale at which they operate as herbivores provide some indication of why hairs of various kinds are so important to the plants that possess them. Simple strengthening of the plant parts, too, as by the deposition of silica in the leaves, is an element in the protective system of some plants, such as grasses (see Figure 34-10). By depositing enough silica in their cells, the plants may simply become too tough to eat.

Significant as these morphological adaptations are, the chemical defenses that occur so widely in plants are even more crucial. As research in this area proceeds, we continue to be astonished by the number and diversity of these features. First, recall that most animals can manufacture only 8 of the 20 required amino acids; they must obtain all of the others from plants. Many plants have only small quantities, for example, of methionine, which animals do not synthesize. Consequently, these plants are not nutritious enough to support herbivores, unless the animals have another source of methionine. If plants or the individual parts of plants do not furnish adequate nutrients, those particular plants or plant parts will not support the growth of animals. As a result, they may be protected from herbivores.

Best known and perhaps most important in the defenses of plants against herbivores are the **secondary compounds,** so called to distinguish them from the **primary compounds,** or normal components of metabolism. Virtually all plants, and apparently many algae as well, contain secondary compounds that play a defensive role. For decades, botanists and chemists tended to regard these substances as waste products that the plants produced, but this notion has now been discredited.

Secondary compounds, chemicals that are not involved in primary metabolic processes, play the dominant role in protecting plants from being eaten by herbivores or predators.

FIGURE 53-6

Gerenuk, an antelope that grazes successfully on some of the spiny shrubs and trees of the African savanna that other browsers do not use as food.

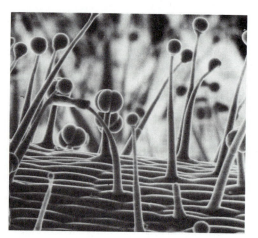

FIGURE 53-7

Glandular hairs of the wild potato Solanum berthaultii. *The hairs are defenses against aphids, which are the most important herbivores on these plants. They are of two types, type A and type B, both seen in this scanning electron micrograph. The shorter type A hairs have a four-lobed head that ruptures on contact, entrapping the aphid or other insect in a quick-setting sticky liquid. The type B hairs play an even more fascinating role in the plant's defenses. The sticky exudate that they secrete contains (E)-B farnesene, the main ingredient of the alarm pheromone in aphids. The extract from these hairs greatly disturbed aphids, driving many of them away, when it was tested experimentally by scientists at the Rothamsted Experimental Station in England. Together, the two types of hairs constitute a formidable barrier against the aphids.*

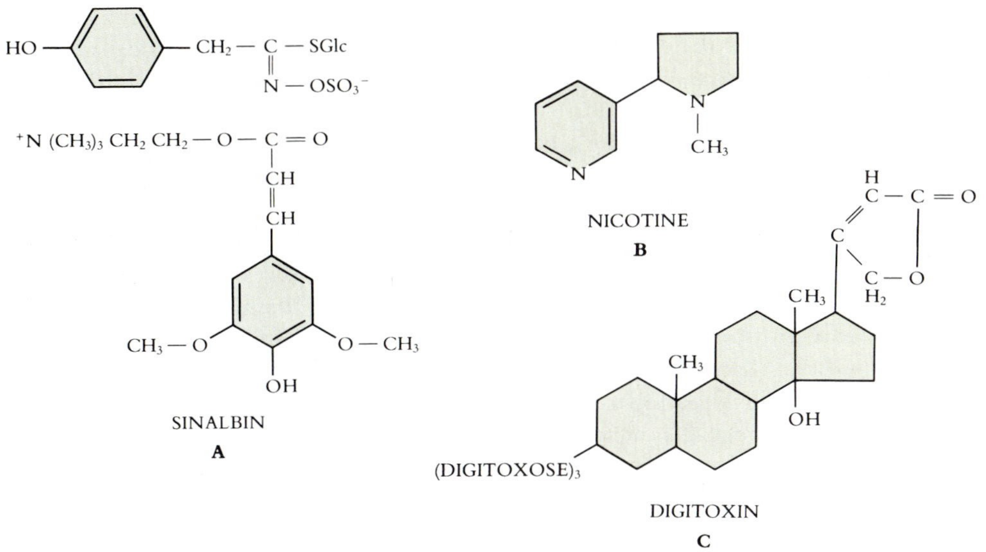

FIGURE 53-8

Plant secondary compounds differ greatly in structure; three are shown here.

A *Sinalbin, a mustard oil glycoside.*
B *Nicotine, an alkaloid that constitutes between 0.6% and 0.9% of the dry weight of tobacco* (Nicotiana tabacum) *leaves, has powerful effects on the human central nervous system and is highly addictive. Like digitoxin and the other cardiac glycosides in foxglove, nicotine and other alkaloids function in nature to prevent attacks by herbivores. Both foxglove and tobacco are members of the family Solanaceae, a family that is rich in secondary compounds.*
C *Digitoxin is one of the cardiac glycosides that occur in foxglove,* Digitalis. *These chemicals, either purified or as whole leaf extracts, are used to treat congestive heart failure. They increase the force of heart muscle contraction without increasing oxygen consumption, and thus cause the heart to pump more efficiently and to meet the needs of the circulatory system better. For the plant, these compounds are important deterrents to herbivores.*

We now know a great deal about the roles of secondary compounds in nature. As a rule, different kinds of compounds are characteristic of particular groups of plants. For example, the mustard family (Brassicaceae; also called Cruciferae) is characterized by mustard oils (Figure 53-8, *A*), the substances that give the pungent aromas and tastes to such plants as mustard, cabbage, watercress, radish, and horseradish. Mustard oils are also found in a few other families of plants—notably the mainly tropical caper family (Capparaceae), from which the temperate-zone mustard family probably originally evolved. The same tastes that we enjoy signal the presence of chemicals that are toxic to many groups of insects.

Another group of plants, the potato and tomato family (Solanaceae), is rich in alkaloids (Figure 53-8, *B*) and steroids, complex ring-forming molecules that exhibit an enormous range of structural diversity. Such compounds occur very widely among the flowering plants, but the particular kinds found in different groups differ greatly. The abundance of alkaloids in primitive flowering plants makes it likely that these compounds evolved early in the evolutionary history of the angiosperms and perhaps played an important role in their success. Plants of the milkweed family (Asclepiadaceae) and the closely related dogbane family (Apocynaceae) tend to produce a milky sap, which deters herbivores from eating them. In addition, these plants usually also contain cardiac glycosides, molecules named for their drastic effect on heart function in vertebrates.

These few examples demonstrate that angiosperms are provided with a rich and varied chemical arsenal. Mustard oils, alkaloids, steroids, and other classes of secondary compounds are either toxic to most herbivores or disturb their metabolism so greatly that they are unable to complete their development normally. As a consequence, most herbivores tend to avoid the plants that possess these compounds. The pattern of occurrence of such chemicals, therefore, has had major consequences on the evolution both of the plants themselves and of the herbivores, especially the insects, that feed on them.

Producing Defenses When They Are Needed

Some of the secondary compounds that plants use to defend themselves from herbivores are not normally present in their tissues. Rather, the plants produce them only when they are needed, in much the same way that immune systems operate in animals. When a leaf of tomato, potato, or alfalfa, for example, is injured, a chemical message travels rapidly through the plant that induces the synthesis and accumulation of proteins that inhibit digestion in an animal's gut. The signal is a fragment of the plant's cell wall, and the inhibitory enzymes are being identified in a number of laboratories at present. Similarly, the infection of a plant by a fungus or bacterium may cause the plant to synthesize a molecule that retards the spread of the infection. A familiar example of such a process is provided by the tannins that color apples brown when they are cut. These tannins interact with proteins and help to impede the activities of caterpillars, fungi, and other organisms, which produce symptoms similar to "cutting" when they attack an apple. Such defensive strategies are metabolically efficient: the plants are able to avoid the necessity of producing the chemical until it is actually needed. At other times, they are able to use the energy available to them for their growth and maintenance.

Recent studies by David Rhoades of the University of Washington have indicated that plants may react to attack by herbivores in a far more surprising way: it now appears likely that willows, alders, and other trees that are under attack by caterpillars may actually communicate with neighboring trees, and by doing so also induce them to alter their leaf chemistry—even though the caterpillars have not yet reached them! Tent caterpillars are colony-forming larvae of moths that attack trees in large numbers from the webbed nests they weave (Figure 53-9). The willow and alder trees that they besiege produce chemicals that make their leaves less and less nutritious to the caterpillars as the attack proceeds. In Rhoades' study, as the caterpillars attacked certain trees of these species, the leaves of nearby unattacked trees also became less nutritious, an effect that became pronounced after 3 days. Trees separated by a larger distance were unaffected. Since Rhoades and his coworkers had ruled out the possibility of underground connections between the trees, it seems likely that the plants were actually communicating their "distress" by some unknown, airborne substance—in effect, a pheromone—that produced the defensive reaction in trees that were likely to be attacked soon. This fascinating and previously unsuspected phenomenon obviously demands further investigation.

The Evolution of Herbivores

Associated with each family or other group of plants that is protected by a particular kind of secondary compound are certain groups of herbivores that are able to feed on these plants, often as their exclusive food source. How do these animals manage to avoid the chemical defenses of the plants, and what are the evolutionary antecedents and ecological consequences of such patterns of specialization?

As a starting point, we can offer the observation that some herbivore groups feed on plants of many different families, others on plants of just a few families. In general, those herbivores that feed on a restricted array of plant families, perhaps only one, feed on plant groups in which certain kinds of secondary compounds are well represented. For example, the larvae of cabbage butterflies (subfamily Pierinae) feed almost exclusively on plants of the mustard and caper families, as well as on a few other small families that are characterized by the presence of mustard oils (Figure 53-10). Similarly, the caterpillars of the monarch

FIGURE 53-9

Tent caterpillars form large colonies on willows, alders, wild cherries, and other plants. When they are attacked, the plants manufacture chemicals that make them less palatable to these herbivores. Surprisingly, the attacked plants also have the ability to transmit a signal through the air that causes other nearby plants of the same species to undergo a similar chemical transformation in advance of an actual attack by the insects. These tent caterpillars are feeding on the leaves of Sitka willow (Salix sitchensis) near Seattle, Washington.

FIGURE 53-10

A *The green caterpillars of the cabbage butterfly, Pieris rapae, are camouflaged on the leaves of cabbage and other plants on which they feed. Although these plants are protected by mustard oils, the caterpillars break down the compounds and are not protected by them.*

B *An adult cabbage butterfly.*

A

B

A   B

FIGURE 53-11

All stages of the life cycle of the monarch butterfly (Danaus plexippus) are protected from birds and other predators by the cardiac glycosides that occur in the milkweeds and dogbanes on which they feed as larvae. Both the caterpillars (A) and the adult butterflies (B) "advertise" their poisonous nature with warning coloration.

butterflies and their relatives (subfamily Danainae) feed on plants of the milkweed and dog-bane families (Figure 53-11). No other groups of butterflies have larvae that feed on these particular plants, and the members of other groups of insects that feed on them usually also belong to specialized subgroups whose feeding habits are associated exclusively with the members of those particular plant families.

We can explain the evolution of these particular patterns as follows. Once the ancestors of the caper and mustard families acquired the ability to manufacture mustard oils, they were protected for a time against most or all herbivores that were feeding on other plants in their area. The plants that contained the mustard oils apparently succeeded very well as a result of this particular defense, and they and their descendants thus evolved and migrated, eventually giving rise to the thousands of species of mustards and capers that now grow worldwide.

At some point, certain groups of insects—for example, the cabbage butterflies—developed the ability to break down the mustard oils and thus feed on these mustards and capers without harming themselves. Having acquired this ability, the butterflies themselves had registered an important evolutionary "breakthrough." They were able to use a new resource without having to face competition for it from other herbivores. The relationship that has formed between the cabbage butterflies and the plants of the mustard and caper families is an example of coevolution.

The members of many groups of plants are protected from most herbivores by their secondary compounds. Once the members of a particular herbivore group acquire the ability to feed on them, however, these herbivores have gained access to a new resource, which they can exploit without competition from other herbivores.

Although the role of secondary compounds in protecting plants from herbivores was first discovered in flowering plants, other groups of photosynthetic organisms have similar defenses. In a detailed study of the coral reef community at Carrie Bow Key, off the Caribbean coast of Panama, for example, scientists from the Smithsonian Institution and the Scripps Institution of Oceanography found that many red, brown, and green algae produce terpenoids and other compounds that deter feeding by herbivores living on the reef, such as damsel fishes, and also inhibit the growth of bacteria in culture. The green alga *Halimeda* (Figure 53-12), for example, which is often the most abundant alga on tropical reefs, produces the diterpenoid halimedatrial, a molecule with a unique structure, that was unknown before it was isolated from this alga. Halimedatrial was shown in the laboratory to be toxic to reef fishes, to reduce feeding significantly in algal grazers, and to be a powerful poison to cell and microbial cultures. Both marine and terrestrial animals, too, may manufacture toxic chemicals that deter their predators. Therefore, similar systems seem to operate in both the sea and on land that deter the dominant herbivores from consuming the primary producers of each community, and that deter carnivores from consuming herbivores.

FIGURE 53-12

The green alga Halimeda *deters marine algal grazers by means of its toxic compounds. It dominates many reefs throughout the tropics and is often the most abundant alga in terms of its biomass.*

FIGURE 53-13

A *A cage-reared jay, which had never seen a monarch butterfly before, eating one.*
B *The same jay a few minutes later regurgitating the butterfly. Such a bird is not likely to attempt to eat an orange-and-black insect again.*

A   B

Chemical Defenses in Animals

Some groups of animals that feed on plants rich in secondary compounds receive an extra benefit, one of great ecological importance. When the caterpillars of monarch butterflies feed on plants of the milkweed family, for example, they do not break down the cardiac glycosides that protect these plants from most herbivores. Instead, they store them in their fat bodies. As a result, these butterflies are themselves protected from predators! The cardiac glycosides in the leaves that the caterpillars eat are concentrated, stored, and passed through the chrysalis stage to the adult and even to the eggs; all stages are protected from predators. A bird that eats a monarch butterfly quickly regurgitates it and thenceforth avoids the kind of conspicuous orange and black pattern that characterizes the adult monarch (Figure 53-13). Locally, however, some birds have acquired the ability to tolerate the protective chemicals and eat the monarchs (see boxed essay, p. 1073).

Insects that do feed regularly on plants of the milkweed family are, in general, brightly colored (Figure 53-14). Among them are brightly colored cerambycid beetles, whose larvae feed on the roots of the milkweed plants; bright blue or green chrysomelid beetles; and bright red true bugs of the order Hemiptera. In some parts of the world, there are also bright red grasshoppers and other very obvious insects. These herbivores clearly are "advertising" their poisonous nature by their bright colors, using an ecological strategy known as **warning coloration.** Similar relationships occur in marine communities; a recent investigation revealed that of the exposed common coral reef invertebrates at Lizard Island, on the Great Barrier Reef off the northeast coast of Australia, three fourths were toxic to fish; of the protectively camouflaged ones, only one fourth were toxic.

The kinds of insects that eat plants whose chemical defenses are less obvious than those of the milkweeds are seldom brightly colored. In fact, many of these insects are **cryptically** colored—colored so as to blend in with their surroundings and thus to be hidden from predators (Figure 53-15). This is also true of insects such as the cabbage butterflies, which, although they feed on plants with well-marked chemical defenses, are able to do so because of their ability to break down the molecules involved, rather than store them.

A particularly fascinating example of camouflage, which we might define as "aggressive camouflage," was discovered recently in Costa Rica by Elizabeth McMahan of the University of North Carolina. Assassin bugs are predaceous members of the order Hemiptera, most of which hunt actively for other arthropods. A few of them, however, hide under piles of organic material that they attach to their exoskeleton, and pounce on their prey from this concealed position. McMahan found that one species of assassin bug, *Salyavata variegata*, uses the drained bodies of the termites on which it has fed to "fish" for additional prey in the openings of the termites' papery nests (Figure 53-16). Completely covered with a crust of nest material, the assassin bug itself is apparently regarded by the termite soldiers, which are blind, as a part of the nest. When the bug presents the soldiers with the body of its latest victim and then in turn catches and drains its new victims, it is ignored by the other termites that investigate the "scene of the crime"!

A

B

FIGURE 53-14

The herbivores that feed on milkweeds obtain cardiac glycosides, which are toxic, from the plants, and advertise their poisonous nature by displaying warning coloration to their potential predators. Examples of such herbivores are the longhorn beetle Tetraopes **(A)** *and the milkweed bug,* Oncopeltus fasciatus **(B)**. *The larvae of* Tetraopes *feed on the roots of the milkweed plants, and the adults mate on the flower clusters. All stages of the life cycle of* Oncopeltus *take place on the leaves and stems.*

FIGURE 53-15

A few striking examples of cryptic coloration. **A** *A young jackrabbit in the desert near Tucson displays both adaptive stillness and camouflage.* **B** *A nighthawk is almost invisible against a background of pebbles.* **C** *Protective coloration in a flounder on the sea floor.* **D** *An African katydid,* Gelatopoia bicolor, *that resembles a lichen.*

Some marine animals—such as certain nudibranchs (sea slugs; see Chapter 36)—acquire defensive chemicals or defensive cells for their prey. Hydroids often provide such stinging cells to animals that graze on them, for example. Off the coast of Panama, the large nudibranch *Aplysia* grazes selectively on red algae of the genus *Laurencia,* which is protected by elatol, a powerful inhibitor of cell division. Perhaps this is the reason that few fish feed on *Aplysia.* An intensive investigation of marine animals for new drugs against cancer and other diseases, or as sources of antibiotics, is now being carried out. It holds great promise, because of the enormous diversity of chemical compounds that occur in marine organisms.

Animals also manufacture many of the chemicals that they use in their defense. In fact, animals manufacture and use a startling array of substances to perform a quite incredible variety of defensive functions. Bees, wasps, predatory bugs, scorpions, spiders, and many other arthropods have chemicals that they use to defend themselves and to hunt for prey (Figure 53-17).

A variety of chemical defenses is found among the vertebrates, too. Frogs of the family Dendrobatidae, for example, secrete toxins known as batrachotoxins in the mucus that

FIGURE 53-16

The assassin bug Salyavata variegata, *which feeds on termites, uses the drained bodies of termites to "fish" in termite nests for additional prey and disguises itself under piles of organic material that it attaches to its own body.*

FIGURE 53-17

A bombardier beetle squirting a solution of quinones at a temperature of about 100° C. The phenolic precursors of these ring compounds are held, together with hydrogen peroxide, in one chamber of a gland in the beetle's abdomen; when threatened, the beetle mixes the contents of this chamber with those of the second chamber, oxidizing the phenols to quinones in a heat-releasing reaction. Meanwhile, oxygen is released from the hydrogen peroxide and propels the hot, stinging mixture out at high speed.

FIGURE 53-18

covers their skin (Figure 53-18). These toxins are so powerful that a few micrograms will kill a person if injected into the bloodstream. They are used by some of the native peoples of northern South America and Central America to poison the darts with which they hunt. Early explorers feared these poisons greatly, and some colonial settlements were actually destroyed by native Americans who used their poison darts against the European settlers.

Aposematic Coloration

Warning coloration, as defined in the preceding discussion, is also known as **aposematic** coloration. Such coloration is characteristic of animals that have effective defense systems, including not only poisons, but also stings, bites, and other means of repelling predators. An organism possessing such a system will benefit by advertising the fact clearly—for example, by showy colors that are not normally found in that particular habitat. Otherwise, the distasteful or poisonous individual runs the risk of being killed while protecting itself.

> **Aposematic, or warning, coloration serves to keep potential predators away from poisonous or otherwise dangerous prey.**

Of course, the animals that display aposematic coloration must remain together, too, if the system is to be effective. A lone individual that exhibits its warning coloration but is eaten will not deliver a message useful for the survival of other members of its species. If genetically related individuals are similarly colored, however, and live in the same vicinity, the selective advantage is obvious.

Some examples of animals with aposematic coloration are shown in Figure 53-19. Such animals tend to live together in family groups, unlike those that are cryptically colored. If camouflaged animals lived together in groups, one might be discovered by a potential predator, offering a valuable clue to the presence of others.

Mimicry

Many unprotected species have come, during the course of their evolution, to resemble distasteful ones that exhibit aposematic coloration. Provided that the unprotected animals are present in numbers that are low relative to those of the species that they resemble, they too will be avoided by predators. Such a pattern of resemblance is called **Batesian mimicry,** after the British naturalist H.W. Bates, who first brought it to general attention in the 1860s. Bates also carried out the first scientific studies of this phenomenon, which has been of great interest to naturalists ever since.

> **In Batesian mimicry, unprotected species resemble others that are distasteful. Both species exhibit aposematic coloration. The unprotected mimics will be avoided by predators if they are relatively scarce.**

Many of the best-known examples of Batesian mimicry occur among butterflies and moths. Obviously, predators in systems of this kind must use visual cues to hunt for their prey. Otherwise similar patterns of coloration would not matter to potential predators.

The groups of butterflies that provide the **models** in Batesian mimicry are, not surprisingly, members of groups whose larvae feed on only one or a few closely related plant families; the plants on which they feed are strongly protected chemically. The model butterflies take poisonous molecules from these plants and retain them in their own bodies. The **mimic** butterflies, in contrast, belong to groups in which the feeding habits of the larvae are not so restricted. As caterpillars, these butterflies feed on a number of different plant families, but not including those that are protected by toxic chemicals.

One well-known mimic among North American butterflies is the viceroy, *Limenitis archippus* (Figure 53-20). This butterfly, which resembles the poisonous monarch, ranges

FIGURE 53-19

Warning coloration is displayed by the spotted skunk, Spilogale putorius **(A);** *the poisonous gila monster,* Heleoderma suspectus, *a member of the only genus of poisonous lizards in the world* **(B);** *and the red-and-black African grasshopper* Phymatus morbillosus **(C),** *which feeds on succulent spurges,* Euphorbia, *plants that are highly poisonous.*

FIGURE 53-20

A *and* **B** *The viceroy butterfly,* Limenitis archippus, *a North American mimic of the poisonous monarch (see Figure 53-11). The larvae of the viceroy, which feed on willows or other nontoxic plants, are cryptically colored and thus concealed from birds and other predators.*

C *and* **D** *The red-spotted purple,* Limenitis arthemis astyanax, *is another member of the same genus as the viceroy. It is a Batesian mimic of another poisonous butterfly, the pipevine swallowtail,* Battus philenor, *which obtains its toxic chemicals from plants of the pipevine family (Aristolochiaceae), on which its larvae feed. The very different appearances of these two species illustrate vividly the way in which selection can drastically change the appearances of mimics.*

DECEPTION IN ORCHIDS

Like many other members of the orchid family, the European species *Cephalanthera rubra* produces no nectar, but bees visit it anyway. Investigations of this phenomenon on the island of Gotland, off the southern coast of Sweden, led botanist L. Anders Nilsson to conclude that the flowers of the orchid were mimics of those of a common species of bellflower, *Campanula rotundifolia* (Figure 53-B, *1*). These flowers resemble one another in size, general shape, and color, except for the amount of red that they display. Most insects, as you saw in Chapter 31, cannot distinguish red as a distinct color, so the flowers of the orchid and the bellflower appear much more similar to them than they do to us.

The flowers of the bellflower produce abundant nectar. If the orchid grows in the same vicinity and is not too abundant, bees and other insects that visit the flowers of the bellflower for nectar will also occasionally visit those of the orchid, too. When they do so, the orchid flowers may attach to them a pollinium, or pollen sac, as seen on the thorax of the bee shown in Figure 53-B, *2*. The bees then continue their search for nectar, visiting more flowers of the bellflower and thus maintaining themselves. As they do so, any orchid pollinia that they have picked up earlier remain attached to their backs. When they finally do make a mistake and arrive at another orchid flower, the pollinium may fit into a groove in the orchid stigma and be transferred to it. With thousands of pollen grains in each pollinium, the

from central Canada south through much of the United States east of the Sierra Nevada and Cascade Range into Mexico. The larvae feed on willows and cottonwoods, and neither they nor the adults are distasteful to birds. Interestingly, the larvae of the viceroy are hidden from predators on the leaves of their host plants because they resemble bird droppings, whereas the distasteful larvae of the monarch are very conspicuous.

For those animals that acquire their protective chemicals as a result of their feeding habits, the selective pressures to continue feeding on those particular plants become even stronger if the animals are brightly colored and conspicuous. Otherwise, nonprotected individuals may be eaten, a fact that would not form the kind of association for the predators that makes them avoid additional prey individuals with similar characteristics. In fact, different individuals of the generally protected plant groups may vary with respect to the exact kinds and amounts of the toxic chemicals that they contain. From the standpoint of the individual plant, it is less costly in terms of energy *not* to produce the secondary compound, provided that a plant that lacks them will not be eaten. This fact is presumably the reason, for example, that certain individuals in several genera of legumes, such as clovers and lotuses (the genus *Lotus,* not to be confused with the oriental lotus, a member of another and very different plant family belonging to the genus *Nelumbo*), do not produce cyanide in their leaves and stems, as most members of the same populations do. As long as there are enough protected individuals in a population, the nonprotected individuals can survive, since they are relatively unlikely to be "found out" by herbivores.

Because two individual plants may contain different amounts and kinds of chemicals, it is inevitable that individual herbivores will feed on plants that are less well protected. When they do so, the herbivores themselves may not actually be distasteful to their predators. Lincoln Brower, an American student of mimicry, has experimentally demonstrated the existence of exactly such situations in monarch butterflies: individuals that do not feed as larvae on milkweed or dogbane plants are not protected from birds. Brower has termed this phenomenon **automimicry**—literally, "self-mimicry." Provided that the unprotected individuals are not too frequent, the system will still work.

Not every member of a plant or animal species needs to be protected by a chemical defense if most are. The phenomenon by which the relatively rare, nonprotected individuals avoid being eaten is called automimicry.

pollination of an entire capsule, crammed with minute seeds, is assured.

At least 10% of all species of flowering plants are orchids, members of the family Orchidaceae. Fewer than half of these have nectar-producing flowers. Rather, they exist, in general, in widely scattered populations in tropical forests, where situations involving the kind of mimicry just described may be common. Chemicals certainly help to attract bees to the flowers of some orchids, as does the nectar that the flowers of some species produce, but systems of deception like that demonstrated by Nilsson for *Cephalanthera rubra* are probably much more common than we now realize.

FIGURE 53-B

1 *A male solitary bee,* Chelostoma fuliginosum, *on the labellum, or lip, of a flower of the orchid* Cephalanthera rubra. **2** *The bees visit flowers of* Campanula rotundifolia.

Another kind of mimicry, **Muellerian mimicry,** was named for the German biologist Fritz Mueller, who first described it in 1878. In Muellerian mimicry, several unrelated but protected animal species come to resemble one another. Thus a number of different kinds of stinging wasps have black-and-yellow striped abdomens, but they may not all be descended from a common ancestor that had similar coloration. In general, yellow-and-black and bright red tend to be common color patterns that presumably warn predators relying on vision that animals with such coloration are to be avoided. It is more difficult to prove that a resemblance between two protected animals actually represents Muellerian mimicry than it is to demonstrate Batesian mimicry experimentally.

In both Batesian and Muellerian mimicry, mimic and model must not only look alike but also act alike, if predators are to be deceived. For example, the members of several families of beetles that resemble wasps (Figure 53-21) behave surprisingly like the wasps they

A

B

C

D

FIGURE 53-21

The familiar yellow-and-black stripes of the yellowjackets and other wasps, such as Vespula arenaria **(A),** *form the basis for large Batesian and Muellerian mimicry complexes. Here we see Batesian mimics representing three separate orders of insects, all rarer than yellowjackets and with patterns of behavior similar to those of the dangerous wasps that they resemble:* **(B)** *a flower fly,* Chrysotoxum; **(C)** *a longhorn beetle; and* **(D)** *a sesiid moth,* Aegeria rutilans, *whose caterpillars are a pest on strawberries. None of these mimics stings or is poisonous, yet all are conspicuous members of the communities where they occur, flying about actively and remaining in full view at all times. They would be easy prey if they were not protected by their resemblance to yellowjacket wasps.*

mimic, flying often and actively from place to place. Mimics must also spend most of their time in the same habitats as do their models. If they did not, predators would discover that all of those conspicuous animals in one area are not only easily seen but also quite tasty! If the animals that resemble one another are all poisonous, or dangerous, they still gain an advantage by resembling one another, thus achieving collective protection.

Muellerian mimicry is a phenomenon in which two or more unrelated but protected species resemble one another, thus achieving a kind of group defense.

Predatory Behavior in Animals

For animals, behavior plays a leading role in predator-prey interactions. Some of the kinds of behavior that animals use to avoid being eaten, such as we discussed in Chapter 49, and a few examples of predatory behavior will be outlined here. Most vertebrates and many arthropods and other invertebrates search actively for their prey, and extremely complex adaptations that affect such relationships—for example, the "flooding" of the environment with bamboo seeds or with periodical cicadas (see Chapter 31)—have evolved.

Animals must always obtain sufficient energy to maintain their activities, a fact that becomes particularly important for active hunters. For example, hummingbirds must obtain enough sugar in the nectar of the flowers that they visit to maintain their high temperatures and rapid movements. They therefore cannot feed on flowers that do not produce much nectar or on flowers whose nectar does not contain much sugar (see Chapter 31). Similar considerations apply to chickadees, which feed on insects while moving constantly through the trees. They must find enough insects to support their activities, or else move to another area or switch to another kind of food.

Animals that are searching for prey form a **search image,** a kind of mental abstract of the major characteristics related to the particular kind of prey that they are seeking at that time (Figure 53-22). For example, many flowers change color once they have been pollinated and have stopped producing nectar. The flowers that do not contain food are excluded in this way from the search image of the insects or other animals that are visiting them. By forming a search image, the predator can make its activities more efficient. On the other hand, the predator's search image is often rather rigidly constructed, and avoiding falling into such a pattern is the basis for the use by potential prey of mimicry, aposematic coloration, cryptic coloration, camouflage, and similar deceptive techniques.

One of the most fascinating instances of predation that has been studied in detail involves the relationships between bats and the moths they catch on the wing at night. A bat navigates through the darkness by echolocation, uttering cries that are above the frequency of human hearing (see Figure 47-9). By monitoring the frequency of these sounds and the time required for them to return, the bat can gauge the distance of the object off which the sounds are bouncing and determine whether the object is moving towards or away from itself. This is a remarkable system, but perhaps still more remarkable are the adaptations that have evolved among the moths on which the bats frequently feed.

Very few animals can hear the sounds that bats produce, but in several different groups of moths, ears have evolved that are tuned to these particular frequencies; these ears may be located on the moth's thorax, abdomen, or mouthparts. Most important, the moth can detect the bat when the bat is still too far away to detect the moth! When it does detect the presence of a bat, a moth immediately takes evasive action, flying rapidly in the opposite direction or suddenly folding its wings and tumbling to the ground (Figure 53-23).

Bat-moth relationships involve still other remarkable aspects. The detection of the bat's sonar is critical for the survival of the moths. Mites live on the moths, and some of them inhabit the moth's ears. In almost all cases, however, a given moth will have a mite in one ear only. Since a moth that has mites in both ears will be unable to detect the presence of bats, it is very likely to be eaten. Under such strong selective pressure, there has ap-

A

B

FIGURE 53-22

The search image of predatory fishes—the characteristics of a small fish—is imitated by this anglerfish (Antennarius), *which thereby lures the other fishes close enough to capture them. Resting on the bottom in shallow waters of the Indian Ocean or western Pacific Ocean, the anglerfish moves its bait, which is at the end of a modified dorsal-fin spine projecting from its snout, in such a way as to increase the bait's resemblance to a small fish.*
A *The anglerfish in luring posture.*
B *Two-second time exposure showing pattern of movement of the luring apparatus.*

parently evolved in the mites the habit of leaving one of the moth's ears free. The mites that have evolved this behavioral pattern are more likely to survive than those that have not, along with the moths on which they are living.

Some moths have even evolved the ability to emit high-pitched clicks that closely resemble those of the bats. By doing this in the presence of a bat, they effectively jam its radar system and thus may be able to escape. The clicks that certain moths emit are closely tuned to the frequencies of the particular bats that are important predators in the vicinity where the moths live.

On a larger scale, patterns of predation play a key role in determining the structure of communities and the representation of species in them. For example, the chalk grasslands of southern England are unusually rich in perennial plants with conspicuous flowers. However, when the rabbits that normally grazed on these grasslands were largely eliminated by a severe outbreak of myxomatosis, a viral disease, the grasses grew much more strongly than they had previously, and the grassland became a much more monotonous place. Similarly, the presence of sea stars grazing in intertidal communities helps to maintain a greater level of diversity among the groups of organisms on which the sea stars feed than would occur in their absence. The sea stars presumably limit the numbers of the more frequent species more than they limit those of the rarer ones; when the once-frequent species become rare, the sea stars turn their attention to other prey.

The American ecologist J.H. Connell, whose work on barnacles along the Scottish coast we discussed in the last chapter, has hypothesized that diversity reaches high levels in rain forests and coral reefs in part because of the periodic disturbances to which these areas are subjected. He suggests that these disturbances may play a role analogous to that of the grazing rabbits and sea stars in the above examples.

Grazing and other similar pressures limit the number of individuals of a given species in a community but tend to increase the number of species represented in that community.

SYMBIOSIS

Symbiotic relationships are those in which two kinds of organisms live together; we have described a number of examples in this book. For instance, mitochondria are almost certainly the descendants of bacteria that became symbiotic in the cells that gave rise to the eukaryotes. Several different kinds of photosynthetic bacteria later became symbiotic in a number of different lines of unicellular eukaryotic organisms, conferring on these organisms the ability to photosynthesize and thus becoming their chloroplasts.

Other examples of symbiosis that we have described include lichens, which are associations of certain fungi and green algae or cyanobacteria; mycorrhizae, associations of fungi and the roots of plants; and the associations of bacteria and root nodules that occur in legumes and certain other plants, which enable the host plants to fix atmospheric nitrogen and the bacteria to obtain carbohydrates. A coral reef is a highly complex symbiotic system, involving not only the coral animals but also coralline algae and other autotrophic organisms that contribute to the net productivity of the reef. More broadly, a coral reef is an entire biological community, one in which symbiotic relationships are especially prominent.

In a general sense, the relationships between flowering plants and the animals that visit them are symbiotic, since they involve persistent, long-term interactions during the course of which both partners have become modified. These and the other examples reviewed above all fall into the class of symbiotic interactions known as **mutualism,** in which both participating species benefit. Other kinds of symbiotic relationships are **commensalism,** in which one species benefits while the other neither benefits nor suffers, and **parasitism,** in which one species benefits while the other is harmed. Parasitism, as mentioned in Chapter 52, is a form of predation.

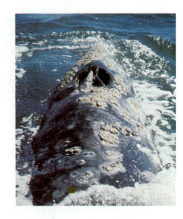

FIGURE 53-24

The barnacles that are evident on the back of this blowing gray whale, surfacing in San Ignacio Lagoon, Baja California, are carried from place to place with the whale, straining their food from water as they move about. Unlike most barnacles, which are anchored to fixed substrates, these animals have continuous access to fresh sources of the plankton on which they feed.

FIGURE 53-25

Leafcutter ants are enormously abundant in the tropics, and they often consume a high proportion of the available leaves at a particular spot to build their "fungus gardens."

FIGURE 53-26

These ants (Crematogaster) are tending willow aphids, obtaining the "honeydew" that the aphids excrete continuously, moving them from place to place, and protecting them from potential predators.

Symbiotic relationships are those in which two or more kinds of organisms live together. They include mutualism, in which both participating species benefit; commensalism, in which one of the species benefits while the other neither benefits nor is harmed; and parasitism, in which one species benefits while the other is harmed.

Commensalism

In nature, the individuals of one species often grow attached to those of another. For example, birds nest in trees, or epiphytes grow on the branches of other plants. In general, the host plant is unharmed, while the organism that grows or nests on it benefits. Similarly, various marine animals, such as barnacles or sea anemones, grow on other, often actively moving sea animals and thus are carried passively from place to place (Figure 53-24). These "passengers" presumably gain more protection from predation than they would if they were fixed in one place, and they also reach new sources of food. The increased water circulation that such animals receive as their host moves around may be of great importance, particularly if the passengers are filter feeders. The gametes of the passenger are also more widely dispersed than would be the case otherwise.

The best-known examples of commensalism—those that virtually define the concept—involve the relationships between certain small tropical fishes and sea anemones (see Figure 35-14, *C*). These fishes have evolved the ability to live among the tentacles of the sea anemones, even though these tentacles would quickly paralyze other fishes that touched them. The anemone fishes feed on the detritus left from the meals of the host anemone, remaining uninjured under remarkable circumstances. Other fishes live in similar associations with larger fishes or with other marine animals.

On land, an analogous relationship exists between certain birds and grazing animals. The birds spend most of their time clinging to the animals, picking off insects and other small bits of food, and carry out their entire life cycles in close association with the host animal.

In each of these instances, it is difficult to be certain whether the second partner receives a benefit or not, and there is no clear-cut boundary between commensalism and mutualism. For instance, it may be advantageous to the sea anemone to have particles of food removed from its tentacles; it may then be better able to catch other prey. The association of the grazing mammals and the birds, on the other hand, is quite clearly an example of mutualism. The mammal benefits by having parasites and other insects removed from its body, and the birds benefit by having access to a dependable source of food. The factors that are involved in a particular association can be learned only by patient observation and experimentation.

Mutualism

As you have seen, examples of mutualism abound in nature, and they are of fundamental importance in determining the structure of biological communities. In the tropics, for example, leafcutter ants (Figure 53-25) are often so abundant that they can remove a quarter or more of the total leaf surface of the plants in a given area. They do not eat these leaves directly; rather, they take them to their underground nests, where they chew them up and inoculate them with the spores of particular fungi. These fungi are cultivated by the ants and brought from one specially prepared bed to another, where they grow and reproduce. In turn, the fungi constitute the primary food of the ants and their larvae. The relationship between the leafcutter ants and these fungi, fueled by material cut from the leaves of plants, is an excellent example of mutualism.

Another relationship of this kind involves ants and aphids. Aphids, as you saw in Chapter 34, suck fluids from the phloem of living plants with their piercing mouthparts. They extract a certain amount of the sucrose and other nutrients from this fluid, but much runs out, in an altered form, through their anus. Certain ants have taken advantage of this habit—in effect domesticating the aphids—by carrying the aphids to new plants and using the "honeydew" that they excrete as food (Figure 53-26).

A particularly striking example of mutualism involving ants concerns certain Latin American species of the plant genus *Acacia*. In these species, the stipules are modified as paired, hollow thorns (Figure 53-27); consequently, these particular species are called "bull's horn acacias." It had been known since the mid-nineteenth century that these thorns were inhabited by stinging ants of the genus *Pseudomyrmex* (Figure 53-28, *A*). It was not until the early 1950s, however, that Daniel Janzen, now of the University of Pennsylvania, was able to demonstrate the full extent of the symbiotic association between these ants and their host plants.

At the tips of the leaflets of these acacias form unique, protein-rich bodies called Beltian bodies (Figure 53-28, *B*)—after Thomas Belt, a nineteenth-century British naturalist who first wrote about them based on his observations in Nicaragua. Beltian bodies do not occur in species of *Acacia* that are not inhabited by ants, and their role is clear: they serve as a primary food for the ants. In addition, the plants secrete nectar from glands near the bases of their leaves. The ants take this nectar also, eat it, and feed it and the Beltian bodies to their larvae as well.

Apparently this association is beneficial to the ants, and one can readily see why they inhabit acacias of this group. The ants and their larvae are protected within the swollen thorns, and the trees provide a ready source of a balanced diet, including the sugar-rich nectar and the protein-rich Beltian bodies. What, if anything, do the ants do for the plants? This had been the question that had fascinated observers for nearly a century, until it was answered by Janzen in a beautifully conceived and executed series of field experiments.

Whenever any herbivore lands on the branches or leaves of an acacia that is inhabited by *Pseudomyrmex* ants, the ants immediately fall on it and devour it. This protects the acacias from being eaten and also provides additional food for the ants, which continually patrol the branches. Related species of acacias that do not have the special features of the bull's horn acacias and are not protected by ants have bitter-tasting substances in their leaves that the bull's horn acacias lack. Evidently, these bitter-tasting substances protect the acacias in which they occur in much the same way that the ants protect the species that they inhabit.

The ants that live in the bull's horn acacias also help their hosts to compete with other plants. The ants cut away any branches of other plants that touch the bull's horn acacia in which they are living—creating, in effect, a tunnel of light through which the acacia can grow, even in the lush deciduous forests of lowland Central America (Figure 53-28, *C*).

FIGURE 53-27

The bull's horn acacias, which occur from Mexico to northern South America, have inflated, thornlike stipules.

FIGURE 53-28

A *Ants of the genus* Pseudomyrmex *live within the thorns.*
B *Beltian bodies are unique, protein-rich structures borne at the ends of the leaflets; the ants harvest them for food.*
C *The ants clear away the surrounding vegetation, allowing the acacias to survive in dense vegetation.*

A

B

C

Without the ants, as Janzen showed experimentally by poisoning the ant colonies that inhabited individual plants, the acacia is unable to compete successfully in this habitat.

Janzen determined these relationships by first making very careful observations of what was going on. Once he had formulated his hypotheses about the nature of the system, he poisoned the ant colonies in some of the plants and studied the fates of those particular plants, which were growing in the same habitats as others that had healthy ant colonies. Many similar experiments will be needed before we begin to understand the functioning of temperate ecosystems better, much less the vastly more complex ones of the tropics.

Parasitism

The concept of parasitism seems obvious, but individual instances are often surprisingly difficult to distinguish from predation (discussed in Chapter 52) and from various other kinds of symbiosis. Many instances of parasitism are well known. They include, for the vertebrates, members of many different phyla of animals and protists. Invertebrates also have series of parasites that live within their bodies. However, bacteria and viruses are often not considered parasites, even though they fit our definition precisely. Lice (see Figure 37-26, *F*), which live on the bodies of vertebrates—mainly birds and mammals—are normally considered parasites, but mosquitoes are not, even though they draw food from the same birds and mammals in a similar manner. Although the mosquitoes may fly away when they have finished feeding, they may be extremely closely associated with their hosts ecologically and in every other sense.

Internal parasitism is generally marked by much more extreme specialization than external parasitism, as shown by the many different examples of protist and invertebrate parasites that we covered in Chapters 27 and 35. The more closely the life of the parasite is linked with that of its host, the more its morphology and behavior are likely to have been modified during the course of its evolution. The same, of course, is true of symbiotic relationships of all sorts. Conditions inside the body of another organism are very different from those encountered outside and are apt to be much more constant in every way. Consequently, the structure of the parasite is often simplified, and unnecessary armaments and structures are lost as it evolves.

Despite these qualifications and inconsistencies in terminology, the meaning of the term *parasite* is clear. Parasitism may be regarded as a special form of predation in which the predator is much smaller than the prey and remains closely associated with it. Parasitism is harmful to the prey organism and beneficial to the parasite.

WHAT ARE COMMUNITIES?

As we said at the beginning of this chapter, communities of organisms exist because of the overlapping tolerances of the organisms that occur in them. This is far from saying, however, that communities are passive assemblages of organisms that do not interact with one another. Rather, the organisms in a given community have evolved in such a way that their relationships are extremely complex, doubtless to a far greater degree than we can even imagine. By their interrelationships, these organisms efficiently regulate the flow of energy that is ultimately derived from the sun and expend it in very precisely coordinated steps. At the same time, they control the cycling of nutrients within the ecosystem in such a way that these nutrients are available to a series of organisms before they return to their nonliving reservoirs. Our future, like our past, depends on our ability to understand and take advantage of the principles that govern the operation of natural ecosystems.

SUMMARY

1. Communities exist because the ecological requirements of the organisms that comprise them overlap in the areas where they occur together. Even when we examine a community such as the redwood community, which is dominated by one very obvious kind of organism that also affects its habitat directly, we realize that the different organisms function individually and that they display genetic variability from place to place.

2. Climatic, biotic, and edaphic factors limit the distributions of organisms. Among these, climatic factors are the most important, but the others are significant as well.

3. Coevolution is the process by which different kinds of organisms adjust to one another by genetic change over long periods of time. It is a stepwise process that ultimately involves adjustment of both groups of organisms.

4. Plants are often protected from herbivores, fungi, and other agents by chemicals that they manufacture. Such chemicals, which are not part of the primary metabolism of the plant, are called secondary compounds.

5. Particular classes of secondary compounds are usually characteristic of individual plant families or groups of closely related families. The herbivores that can feed on such plants either break the secondary compounds down or store them in their bodies. If they do the latter, the chemicals may in turn protect the herbivores from their predators.

6. Batesian mimicry is a situation in which a palatable or nontoxic organism resembles another kind of organism that is distasteful or toxic. Muellerian mimicry occurs when several toxic or dangerous kinds of organisms resemble one another.

7. In Batesian mimicry, the models are generally herbivores that feed on plants that contain toxic secondary compounds. The models retain such chemicals in their bodies. The mimics, in contrast, usually do not feed on such plants, but instead feed on others that are not protected in this way.

8. Predators construct a search image that enables them to recognize their prey and gather it efficiently. Camouflaged organisms remain concealed unless a predator is able to form a search image of them.

9. Symbiotic relationships are those in which two kinds of organisms live together. There are three principal sorts of symbiotic relationships: in commensalism, one benefits and the other is unaffected; in mutualism, each organism benefits; and in parasitism, one benefits and the other is harmed.

SELF-QUIZ

The following questions pinpoint important information in this chapter. You should be sure to know the answers to each of them.

1. Which of the following adjectives describes one of the factors that plays a role in determining the distributional limits of particular kinds of organisms?
   - (a) climatic
   - (b) biotic
   - (c) edaphic
   - (d) a and b
   - (e) all of the above

2. Leathery evergreen leaves serve to
   - (a) give a broad surface for gas exchange
   - (b) conserve phosphorus and other nutrients
   - (c) keep a plant healthy by rotating annually
   - (d) a and b

3. Interactions that involve the long-term evolutionary adjustment of the properties of the members of biological communities in relation to one another are grouped under the general heading of _____.

4. The flow of energy through the ecosystem of the world is such that the _____ interface is the point at which the greatest transfer of energy takes place.
   (a) plant-carnivore
   (b) plant-decomposer
   (c) carnivore-herbivore
   (d) plant-herbivore
   (e) herbivore-decomposer

5. Poisonous herbivores tend to be
   (a) brightly colored
   (b) cryptically colored

6. Another name for warning coloration is _____ coloration.

7. In _____ mimicry, unprotected species resemble others that are distasteful.

8. The larvae of a model butterfly feed on
   (a) nontoxic unprotected plants
   (b) chemically protected and often toxic plants
   (c) mimic butterfly larvae
   (d) themselves
   (e) none of the above

9. When two or more protected species resemble one another, thus achieving a kind of group protection, we refer to this phenomenon as
   (a) automimicry
   (b) Batesian mimicry
   (c) Muellerian mimicry
   (d) all of the above

10. Grazing tends to decrease the variety of species that make up a community. True or false?

11. _____ relationships are those in which two kinds of organisms live together.

12. If two species live together, with one benefiting and the other neither benefiting nor being harmed, their relationship is described as
    (a) mutualism
    (b) commensalism
    (c) parasitism
    (d) predation
    (e) competition

THOUGHT QUESTIONS

1. Why is it more difficult to demonstrate Muellerian mimicry than Batesian mimicry experimentally? How would you go about demonstrating each?

2. In the early decades of the twentieth century, the American ecologist F.E. Clements, who championed the concepts of succession and climax plant communities, held that communities were superorganisms. Another American ecologist, H.A. Gleason, held that communities consisted of nothing more than the sum of the properties of the organisms that coexisted in them. What experiments would you perform to distinguish between these two hypotheses?

3. In Batesian mimicry, why are the mimics rare?

4. Some kinds of insects are more resistant to insecticides than others. Can you think of a reason why, based on your reading in this chapter?

FOR FURTHER READING

CONNELL, J.H.: "Diversity in Tropical Rain Forests and Coral Reefs," *Science,* vol. 199, pages 1302-1309, 1979. A comparison of the two richest and most diverse ecosystems.

FUTUYMA, D.J., and M. SLATKIN (eds.): *Coevolution,* Sinauer Associates, Inc., Sunderland, Mass., 1983. A thorough review, comprising papers by many distinguished authors, that clearly demonstrates the scope of coevolutionary interactions in nature.

GILBERT, L.E., and P.H. RAVEN (eds.): *Coevolution,* 2nd ed., University of Texas Press, Austin, 1980. A series of papers by some of the leading experts on the full range of coevolutionary topics.

HARPER, J.L.: *Population Biology of Plants,* Academic Press, Inc., New York, 1977. The standard work in the field, by its leading student.

HUTCHINSON, G.E.: *The Ecological Theater and the Evolutionary Play,* Yale University Press, New Haven, Conn., 1965. This series of beautifully written essays, by one of the great masters of ecology, is an outstanding introduction to many of its most interesting concepts.

JACKSON, J.B.C., and T.P. HUGHES: "Adaptive Strategies of Coral-Reef Invertebrates," *American Scientist,* vol. 73, pages 265-274, 1985. Coral-reef environments, which are regularly disturbed by storms and predation, provide an excellent example of a complex, balanced marine community.

MOORE, D.M. (ed.): *Green Planet,* University of Cambridge Press, New York, 1982. A well-written account of many aspects of plant biology, including a number of topics pertinent to this chapter.

MORSE, D.H.: "Milkweeds and Their Visitors," *Scientific American,* July 1985, pages 112-119. Nectar feeders, herbivores, predators, and parasites gather on milkweed plants, forming a model ecological community, which the author has studied in detail.

OWEN, D.: *Camouflage and Mimicry,* University of Chicago Press, Chicago, Ill., 1980. A short, beautifully illustrated book; many of the photographs are based on the author's field experience in West Africa.

SMITH, R.E.: *Ecology and Field Biology,* 2nd ed., Harper & Row, Publishers, Inc., New York, 1974. Covers many aspects of ecology, including those emphasized in this chapter, from a field perspective.

Chapter 54

THE FUTURE OF THE BIOSPHERE

Overview

Biologists have an important contribution to make in feeding the rapidly growing world population by applying their knowledge about the characteristics of ecosystems and organisms to enhanced and stable food production. The development of agriculture by the roughly 5 million people who lived in the world about 11,000 years ago led through a series of stages to the very large human population—5 billion people—that exists now. More than half of these people live in the tropics and subtropics, where about 90% of human population growth over the next few decades will occur. Through the improvement of existing crops and domesticated animals by traditional plant and animal breeding methods and genetic engineering, as well as the selection of additional crops, biologists will play the key role in improving the quality of human life throughout the world. Extinction is a problem of great concern, with as much as 15% of the world's biological diversity likely to disappear over the next 3 or 4 decades, mostly in the tropics. Biological literacy is an essential component of human progress in the future.

For Review

Here are some important terms and concepts that you will encounter in this chapter. If you are not familiar with them, you should review them before proceeding.

- **Human evolution** (Chapter 24)
- **Characteristics of biomes** (Chapter 50)
- **Cycling of minerals and flow of energy** (Chapter 51)
- **Population growth characteristics** (Chapter 52)

We have now completed our review of the field of biology, and it remains to consider what the future holds for us. The human race is now growing so rapidly and affecting the global ecosystem to such a profound and unprecedented degree that we must apply our biological knowledge diligently and wisely to the management of the life-giving systems on which our collective prosperity—and even our survival—depends. In this chapter, we will consider our needs for food, water, energy, and other resources; the adequacy of the supplies of these commodities that are available to us now; and the ways in which we may be able to apply our biological knowledge to the solution of the problems presented to our rapidly expanding populations by the finite resources of the world on which we live.

To understand how we might deal with the world as it is now, we must first review briefly how we got to where we are. In Chapter 24, we traced the history of our own species and reviewed the evidence for the dating of the major steps that have marked the long process of our evolution. No less than 5 million years ago, the australopithecines had evolved as bipedal ground walkers and unquestionable hominids. Their brains had not increased in size over those of the other great apes. Nevertheless, it was clear that even then our ancestors were evolving along lines that were distinct from the other apes and were developing the cultural features that have become our primary characteristics.

After a period of perhaps 3 million years of evolution, *Australopithecus* gave rise to the first humans, members of the species *Homo habilis*—the first tool makers. This species, and therefore our genus, appeared in the fossil record about 2 million years ago. By about 1.7 million years ago, a second species, *Homo erectus,* had appeared. The bands of this species gathered food, scavenged, and probably hunted occasionally; they occurred widely over Africa, Europe, and Asia by 1 million years ago, developing the use of fire (no less than 1.4 million years ago) and many other traits that we consider human.

The earliest fossils that can clearly belong to *Homo sapiens,* our species, come from western Europe; they are about 500,000 years old. The Cro-Magnons, people of essentially modern character, suddenly replaced more archaic races, including the Neanderthals, in western Eurasia starting about 40,000 years ago. These people reached the New World, no less than 12,000 to 13,000 years ago. They were expert hunters, who apparently soon drove most of the remnants of the great Pleistocene herds of mammals (see Figure 23-2) to extinction. By about 11,000 years ago they had developed agriculture (Figure 54-1).

THE FUTURE OF AGRICULTURE

What are the prospects for increased agricultural productivity in the future, and how can we improve them? Viewing this problem in historical context, it is estimated that about 25,000 years ago there were perhaps 3 million human beings in the world. By the close of the Pleistocene Epoch, 10,000 years ago, there may have been about 5 million, distributed over all of the continents except Antarctica. When the continental ice sheets withdrew and the climate moderated throughout the world's temperate regions, the stage was set for the major advance that made possible the rapid increase in human population—the development of agriculture. This took place independently in the Middle East, in eastern Asia, and at one or more centers in North and South America, with different kinds of plants and animals domesticated in each area (see Chapter 24).

As a result of these early efforts, nearly all of the major crops in world commerce were discovered; most of them have been cultivated for hundreds or, more often, thousands of years; only a few, including rubber and oil palms (Figure 54-2), have entered widespread cultivation since 1800. One key feature for which nearly all of our important crops were

FIGURE 54-1

One of the most critical human inventions was that of agriculture. Here, in southern India, the people still thresh sorghum using primitive tools whose designs are probably thousands of years old. By producing abundant supplies of food, such agriculture made possible the growth of cities and the future development of human culture.

A

B

FIGURE 54-2

A *Oil palms (Eleais guineensis), which are native to Africa, were introduced into wide-scale cultivation in the early part of the twentieth century. They are now a multibillion-dollar crop, grown throughout the tropical regions of the world, where trees often are much more successful than the sorts of herbaceous crops that are the mainstay of temperate agriculture.* **B** *The roots of Casuarina, a fast-growing tree, are inhabited by nitrogen-fixing actinomycetes, which form nodules there. Casuarina is a different kind of crop, one that is receiving increasing attention as a possible source of firewood and for watershed protection in the tropics.*

THE POTENTIAL FOR NEW CROPS

With their vast and diverse arsenal of natural insecticides, plants offer a reservoir of useful products that we have barely tapped. A number of these plants have proved useful already, and many more will do so in the future, as we explore further.

One of the most fertile areas for biological exploration is the study of plant use by people who live in direct contact with natural plant communities. These people know a great deal about the plants with which they come into contact, and the screening that they and their ancestors have performed over many generations can, if its results are understood and catalogued, help us to recognize new useful plants. For example, the herbal curer in the jungles of southern Surinam has a profound knowledge of uses of the plants that grow in his region, but that knowledge is being lost as civilization and modern medicine move into the area.

Some recent triumphs in the search for economically important plants are the following:

Guayule *(Parthenium argenteum)* (Figure 54-A, *1*): This widespread shrub of northern Mexico and the southwestern United States is a rich source of rubber. The yield from some cultivated strains is up to 20% rubber. Guayule is now being cultivated in more than 30 countries, and efforts are being made to bring it into even wider cultivation.

Periwinkle (*Catharanthus roseus*) (Figure 54-A, *2*): Periwinkle is a garden plant, now widespread in cultivation; as a native plant, it is found only in Madagascar. Two drugs, vincablastine and vincacristine, which have been developed from periwinkle by Eli Lilly & Company, are effective in the treatment of certain forms of leukemia. For example, a child who develops Hodgkin's disease, a form of childhood leukemia, now has a 90% chance of survival if treated with these drugs. In 1950, such a child would have had only a 20% chance to live.

Jojoba (*Simmondsia chinensis*) (Figure 54-A, *3*): Despite its scientific name, which was conferred because the scientist who studied the first specimens incorrectly believed that they had come from China, jojoba is native to the deserts of northwestern Mexico and the southwestern United States. It is now an important crop, in production in several parts of the world since the late 1970s, because the liquid waxes in its seeds have characteristics similar

1

2

initially selected was ease of growth by relatively simple methods, often by the women who stayed behind in camps while the men were away hunting.

Fewer than 200 kinds of vascular plants, of the roughly 240,000 known, play an important role in agriculture and international trade at present. We obtain about 85% of our food from 20 of these, and about half—directly or indirectly—from just three: corn (maize), wheat, and rice. The traits for which we select new crops now, however, are often very different from those that appealed to our ancestors, living as they did in small groups around the foothills of the Near East or on the temperate slopes of the mountains in Mexico. Standards of cultivation have changed, and we use many products from plants—oils, drugs, and other chemicals, for example—that often would not have led to their cultivation

to those of sperm whale oil and are important for certain kinds of fine lubrication. Like guayule, jojoba can be cultivated in lands too arid for most other crops. The new availability of the liquid waxes from jojoba seeds has helped to make the hunting of sperm whales uneconomical, and these threatened animals may thus be saved from extinction.

Grain amaranths (*Amaranthus* species) (Figure 54-A, *4*): Important grain crops of the Latin American highlands in the days of the Incas and Aztecs, the grain amaranths are now minor crops in Latin America; their use was suppressed because they played a role in pagan ceremonies of which the Spaniards disapproved. The grain amaranths, fast-growing plants that produce abundant grain rich in lysine—an amino acid that is rare in most plant proteins but essential for animal nutrition—are now being investigated seriously for widespread development.

Winged bean (*Psophocarpus tetragonolobus*) (Figure 54-A, *5*): The winged bean is a nitrogen-fixing tropical vine that produces highly nutritious seeds, pods, and leaves. Its tubers are eaten like potatoes, and its seeds produce large quantities of edible oil. First cultivated locally in New Guinea and Southeast Asia, the winged bean has spread, since the 1970s, throughout the tropics, where it holds great promise as a productive source of food, especially for small farmers.

3

4

5

earlier. In a time when processes that we will describe in this chapter threaten to drive many of the world's plants and other organisms to extinction during our lifetimes and those of our children, we need to search systematically for new useful crops that fit the multiple needs of modern society in ways that would not have been considered earlier. We must initiate more diligent efforts to find and bring into cultivation more kinds of useful plants before they are gone forever.

Fewer than 200 kinds of plants, of the roughly 240,000 known, play an important role in agriculture and international trade at present. A careful search for new crops could yield many more.

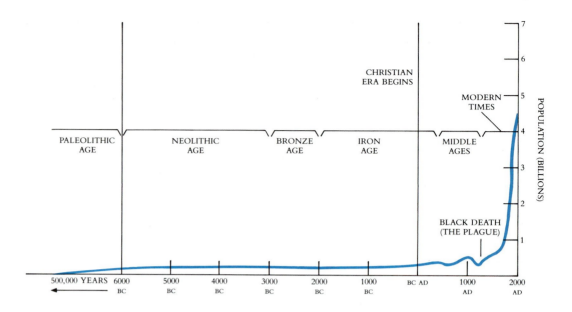

FIGURE 54-3

Growth curve of the human population.

FIGURE 54-4

Johann Gensfleisch Gutenberg, a German who lived from about 1400 to about 1468, introduced printing in Europe by means of movable type. Printing greatly accelerated the dissemination of knowledge.

THE POPULATION EXPLOSION

Only about 5 million people lived throughout the entire world when agriculture was first developed, but given the new and much more dependable sources of food that became available as a result of agriculture, human populations began to grow rapidly, climbing to an estimated 130 million people by the time of Christ—more than a 25-fold increase during the roughly 9000 years following the development of agriculture (Figure 54-3). The global population 2000 years ago was about half that of the United States and Canada at present. As human populations grew in the Middle East, their herds of domestic animals became tremendously destructive to the pastures and slopes on which they fed, ultimately often causing erosion and flooding. Similar effects were repeated later in the New World when the same animals were introduced there. Nonetheless, thanks to the increased production of food made possible by agriculture, large numbers of people were able to live together in permanent settlements for the first time. As a result, villages and towns developed by about 3000 BC in all areas where agriculture was practiced. By that time, some 5000 years ago, the plains of Europe were inhabited by people who lived in more or less permanent villages and were supported by agriculture.

The ability to live together on a long-term basis in relatively large settlements made possible, in turn, the specialization of professions in these centers. Such specialization was the necessary condition for the development of what we regard as modern culture. Such cultural specializations as the development of metallurgy, for example, could not originate until after the towns were settled.

The world population increased some 25-fold during the 9000 years from the development of agriculture to the time of Christ, when it amounted to an estimated 130 million people.

By 1650, the world population had multiplied four-fold and had reached 500 million, with many people living in substantial urban centers, in which many important innovations were taking place (Figure 54-4). The Renaissance in Europe, with its renewed interest in science, ultimately led to the establishment of industry in the seventeenth century, and to the Industrial Revolution of the late eighteenth and early nineteenth centuries. Centers of cultural innovation arose in Europe and elsewhere—making possible, in turn, still more increases in world population.

TABLE 54-1   WORLD POPULATION DISTRIBUTION

REGION OR COUNTRY	1985 POPULATION (MILLIONS)	POPULATION DOUBLING TIME (YR)	LIFE EXPECTANCY AT BIRTH (YR)	PER CAPITA GNP ($)
World	4845	41	62	2760
Developed	1174	118	73	9380
Less Developed	3671	34	58	700
Africa	551	24	50	750
Asia	2829	39	60	940
U.S. and Canada	264	99	75	13,820
Latin America	406	30	65	1890
Europe	492	240	73	8200
USSR	278	71	69	6350
Oceania	24	56	71	8570
Bangladesh	101.5	25	48	130
Brazil	138.4	30	63	1890
China	1042.0	65	65	290
France	55.0	198	75	10,390
Japan	120.8	110	77	10,180
Kenya	20.2	17	53	340
India	762.2	32	53	260
Mexico	79.7	27	66	2240
United Kingdom	56.4	630	73	9050

Data from 1985 World Population Data Sheet, Population Reference Bureau, Inc.

The Present Situation

For the past 300 years, and probably for much longer, the human birth rate (as a global average) has remained relatively constant or nearly so. It may be lower now than it has been historically, at about 27 births per year per 1000 people, but this difference is probably not significant. With the spread of better sanitation, coupled with modern medical techniques, however, the death rate has fallen steadily, to an estimated 1985 level of about 11 deaths per 1000 people per year. The difference between these two figures amounts to an annual, worldwide increase in human population of approximately 1.7%.

It is estimated that the world population will reach an unprecedented 5 billion people by mid-1987, with the *annual* increase in population amounting to about 85 million people, a number substantially larger than the total population of Britain, France, or Germany. At this rate, more than 232,000 people are added to the world population each day, or more than 160 every minute! The world population is expected to rise much higher—to well over 6 billion people by the end of the century, and then to add about 1.7 billion more during the following 20 years! In 1930, the total world population was only about 2 billion people. In view of the limited resources available to the human population and the necessity of our learning how to manage these resources well, the first and most necessary step toward global prosperity is to stabilize the population. If and when we develop technologies that would allow greater numbers of people to inhabit the earth in a stable condition, it would be possible to increase our numbers to whatever level might be appropriate at that time.

> **In the mid-1980s, the global human population of just under 5 billion people is growing at a rate of approximately 1.7% annually. At that rate, it will reach well over 6 billion people by 2000, and nearly 7.8 billion by 2020.**

The growth of human populations is much greater in the tropical and subtropical parts of the world than in the temperate ones, as Table 54-1 demonstrates.

FIGURE 54-5

With its human population growth outstripping its resources, India faces an uncertain future. At present rates of growth, India will pull even with China in population in about 2020, when each country will have about 1.2 billion people. This crowded scene was photographed in Ranchi, Bihar State, India, where it is increasingly difficult for the city administrators to collect taxes and pay for the basic services they provide.

By the year 2000, about 60% of the people in the world will be living in countries that are at least partly tropical or subtropical, 20% will be living in China, and the remaining 20%—one in five—will be in the so-called developed countries of Europe, the Soviet Union, Japan, the United States, Canada, Australia, and New Zealand together. While the populations of the developed countries are growing at an annual rate of only about 0.6%, those of the less developed, mostly tropical countries (excluding China) are growing at an annual rate estimated in 1985 to be about 2.4%.

In predicting the future growth patterns of populations, it is essential to know what proportion of the individuals making up those populations have not yet reached child-bearing age. In the developed countries, only about 23% of the people are less than 15 years old, whereas in the tropical countries, the corresponding figure is about 40%. Thus, even if the policies that most tropical and subtropical countries have established to limit their own population growth, so that their people can be more prosperous, are carried out consistently for decades, the populations of these countries will continue to grow well into the twenty-first century, and the developed countries will constitute a smaller and smaller proportion of the world's population. For example, if India, with a mid-1985 population of 762 million, of whom 39% are under 15 years old, managed to slow down to a replacement reproductive rate by the year 2000, its population still would not stop growing until the mid-twenty-first century (Figure 54-5). By the year 2020, India will have a population of more than 1.2 billion and will still be growing rapidly, judged from current patterns of growth. As a result of such worldwide trends, the proportion of the world's population living in the developed countries, a third in 1950, will have fallen to a sixth by 2020; the proportion living in the tropics will have risen from 45% to 64% in the same 70-year period.

The proportion of people living in the developed countries is falling—from a third of the total world population in 1950 to a projected sixth in 2020—while the proportion of people living in countries that are at least partly tropical or subtropical is rising from 45% to 64% during the same period.

Most countries are devoting considerable attention to slowing their rate of population growth, and there are genuine signs of progress. It is estimated, given the efforts that are under way and considering the direction of growth trends, that the world population may stabilize in the twenty-first century at about 12 to 18 billion people. No one knows whether the world can support so many people indefinitely, but we must assume that we will be able to find the ways for it to do so. Among the tasks that we must accomplish is the development of new ways to feed, clothe, and shelter adequately a population two to four times larger than the present one. The quality of life that will be available for our children and grandchildren in the twenty-first century will depend to a large extent on our ability to achieve this goal.

FOOD AND POPULATION

Even though experts estimate that enough food is produced in the world to provide an adequate diet for everyone in it, the distribution of this food is so unequal that large numbers of people live in hunger. In the United States, the Soviet Union, Japan, and Europe, there is, on the average, a quarter to a third more food available per person than the amount the United Nations Food and Agriculture Organization (FAO) set in 1978–1980 as the calorie intake necessary to maintain moderate physical activity. In contrast, countries such as Angola, Bangladesh, Bolivia, Ecuador, Haiti, India, Kenya, and Zimbabwe have 10% to 15% less than this minimum available per capita and, for the most part, few cash reserves with which to purchase more. Only the United States, Canada, Argentina, and one or two other temperate-zone countries have emerged as consistent exporters of food.

Of the estimated 2.5 billion people living in the tropics in the mid-1980s, the World Bank estimated that about a billion were living in absolute poverty. These people cannot reasonably expect to be able consistently to provide adequate food for themselves and their families. About 500 million people are estimated by the World Health Organization actually to be malnourished. In the developing world, about 10 million children under the age of 5 starve to death each year, mainly of malnutrition and the complications associated with it (Figure 54-6). Perhaps another 10 to 15 million older people die of starvation annually; it is difficult to be sure of the exact number because starvation brings on so many other complications. Thus about 60,000 to 80,000 people starve to death each day; many more people are on the verge of starvation, ill, or perhaps mentally retarded permanently as a result of a diet insufficient in protein when they were babies.

> **About 40% of the approximately 2.5 billion people living in or near the tropics in the mid-1980s existed in a state of absolute poverty; many of them were malnourished.**

In Africa, the production of food per capita has been dropping each year since at least 1961, while the population has continued to grow rapidly. For example, in East Africa, the estimated mid-1985 population of 159 million was increasing at an annual rate of 3.1%; thus it is expected to reach 258 million by the year 2000 and 452 million by 2020. These people have a life expectancy of 49 years at birth, nearly 30 years less than that in the United States. Some 47% of them are under 15 years of age, and the average annual income was about $300 in 1985. Available for their consumption was only about 85% of the U.N.-estimated minimum amount of necessary calories, and the food production per capita is declining rapidly. Few advances are even being planned seriously, let alone accomplished, that will substantially increase the region's ability to produce more food. In recent years, in fact, even the *absolute* food production in Africa has been dropping annually. Ethiopia and 13 other countries of sub-Saharan Africa were ravaged by famine in the early and mid-1980s, a consequence of drought added to a background of inadequate food supplies such as is characteristic of much of Africa even in "normal" years.

In the developed countries, about 70% of the population was living in urban centers by the mid-1970s—about twice the proportion that was then doing so in the less developed countries—and a far smaller proportion of the total population engaged in farming in the developed countries than did so in the developing countries. One of the most alarming

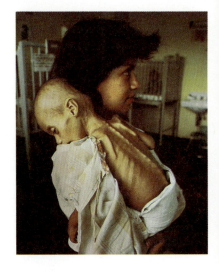

FIGURE 54-6

Consuelo Andena, 16, holding her 1-year-old baby, Carolina, in the malnutrition ward of the Women and Children's Hospital in Tegucigalpa, Honduras, in 1983. Such tragic scenes are common throughout the tropics, where more than 10 million babies perish each year from starvation.

trends that is taking place in tropical countries, however, is a massive movement to urban centers. For example, Mexico City, the largest city in the world, is plagued by smog, traffic, earthquakes, waste, and other problems, and is projected to have a population of more than 30 million by the end of the century (see Figure 1-9). The prospects of supplying adequate food, water, and sanitation to these people, in a country whose 1986 population (about 82 million) is expected to increase by an additional 30 million people by the year 2000 and then by about 38 million more by 2020, are almost unimaginable. It is no wonder that there is such massive emigration from Mexico and countries that are similarly troubled.

Consider what such rapid population growth means to a country. Brazil, which had about 142 million people in mid-1986 as well as the world's largest international debt and one of the highest rates of inflation anywhere, was then growing at about 2.3% per year, and it is estimated that its population will double in 30 years. Thus Brazil would need to double all of its food production, water supplies, fuel, housing, educational and medical facilities, and more during the next 30 years just to stay even (Figure 54-7). Even if Brazil does perform this unprecedented economic miracle, then its people will be only as well off in 30 years as they are now—no better off. If the miracle does not occur, then more than the currently estimated 40% of the people of Brazil who are living in absolute poverty will be in that condition.

> **The populations of the less developed countries are growing so rapidly that an unprecedented effort will be required to keep the proportion of hungry, poor people—now a fifth of the total world population—at its current level while their populations double during an average period of 29 years.**

The considerations about population growth that we have just presented indicate clearly why a stable population structure is as necessary for us as it is for the other organisms that share this planet with us. Although we have the ingenuity and ability to increase the supply of resources available to us, we urgently need to do so; meanwhile, we must find ways to ameliorate the poverty, malnutrition, disease, and starvation that haunt the lives of fully a quarter of our fellow human beings. That is the great task of biology.

WATER, ENERGY, AND MINERALS

Compounding the stresses placed on the earth's productive capacity for food are those associated with its limited supplies of water, energy, and minerals. We will briefly review some aspects of these problems to provide a suitable perspective within which to view the world's overall ecological situation.

Water

Fresh water is abundant in the world as a whole, but, like food, it is so unevenly distributed that it is often insufficient locally. Water is also very expensive to transport from place to place and easily polluted (see Chapter 51). About two thirds of the world's major rivers, many of them in the tropics and in seasonally dry areas, flow through two or more countries. The conflicts that have arisen over the use of their water, together with the demands of an increasing human population in the future, suggest that competition for water may be a major source of local aggression in the twenty-first century. The increasing pollution of groundwater and the economic and political difficulties in transporting water from place to place will loom ever larger in the future.

Conflicts over water use arise not only between countries but also between different parts of the same country and between different, competitive purposes within a region. In the United States, for example, agriculture currently consumes 83% of the water withdrawn, and an increase of 17% in need is projected by the year 2000. If major synthetic fuel (synfuel) programs involving the use of coal and oil shale are initiated, large quantities of

water currently not available would be needed for the synfuel projects. Most of the increased demand would take place in the West, where water is already often scarce.

In developed countries, vast amounts of water are used, but only a very small amount is used directly for human consumption; it is in its agricultural and industrial roles that water is most significant as a limiting resource. When it is used repeatedly for agriculture in a given region, it often becomes so salty that it can no longer be used, or the land itself becomes unsuitable for the cultivation of crops. As a result, and because noncultivated, naturally salty lands are available in some parts of the world, a strong effort is being made to develop strains of present crops and new crops that will tolerate and thrive in these salty conditions (Figure 54-8).

FIGURE 54-8

In these experiments at the University of Arizona, Tucson, scientists are attempting to select salt-tolerant crops that will be suited for the naturally saline soils of the world, as well as for those that have become saline through continued irrigation.

THE FIREWOOD CRISIS

In developed countries, the energy used to cook food accounts for less than 5% of all energy consumed. In the developing countries, however, it often accounts for between half and two thirds of total energy consumption, and the millions of often inefficient stoves that burn gas, kerosene, wood, or coal in these nations constitute one of the major expenditures of energy on a world scale. In sub-Saharan Africa, three fourths of the energy used comes from firewood, and in countries such as Tanzania, the proportion may be as high as 94%. Because of the relatively high cost of gas and kerosene, firewood is usually the most important fuel in third world households, and poor families in certain parts of the tropics may spend from a fifth to a third of their budget on this fuel, if they can afford to buy it at all.

In India, for example, firewood provides about half of the energy consumed for cooking in the cities and more than two thirds of that consumed in the villages. As the search for firewood is intensified, forests are being cut and whole areas are being denuded; the demand for firewood far outstrips the supply. In the villages, women may walk 10 kilometers or more to the nearest sources of wood and then return home with 25-kilogram loads; some of the firewood in India and other countries may be brought from more than 100 kilometers away to the places where it is consumed (Figure 54-B, *1*). Nearly all of this firewood is in the form of twigs and branches removed from living trees; the poor cannot afford to fell whole trees. For the truly poor, firewood is often nearly impossible to obtain, sometimes even more so than food itself. People in India are increasingly having to turn to burning cow dung as a source of energy, thus depleting the soil of nutrients that might otherwise enrich it.

Prospects for the future are disheartening. With firewood becoming increasingly scarce, people are less often able to cook their food properly; if the food is inadequately cooked, its digestibility decreases, and parasites in it are not as apt to be killed. Furthermore, the ceaseless search for firewood by the rural poor constitutes the major cause of deforestation in some areas. Haiti, for example, was once covered with majestic forests; the poor woodlands and shrublands that remain are constantly devastated and kept from regeneration (Figure 54-B, *2*). In Africa, all of the major urban areas are surrounded by zones, often 25 kilometers or more in diameter, that have been completely denuded of firewood, so that the people cannot even reach areas where they could gather more. In Nairobi, the poor spend as much as 30% of their income on firewood—sometimes more than they spend for the food they use it to cook. Watersheds throughout the

---

Energy

FIGURE 54-9

Solar collectors such as these harvest sunlight in much the same way that all photosynthetic organisms do. Sunlight constitutes a clean and inexhaustible source of energy.

The world's supplies of fossil fuels—oil, gas, and coal—however we measure them, are limited. They may last for decades, perhaps even for centuries, but eventually they will run out. Before they do, and better sooner than later, human beings will need to make the transition to renewable sources of energy—ones for which new supplies continually become available. Among these are wind and tidal forces, nuclear energy, and solar energy. Calculations suggest that the amounts of energy to be derived from wind and tidal forces are relatively limited, but these sources certainly need to be exploited fully.

Nuclear energy has also been proposed as an alternative to fossil fuels. However, in systems involving fission, the production of nuclear power also creates undesirable by-products that are difficult to dispose of satisfactorily. In addition, these by-products and the technology involved in the construction of the nuclear power plants can be used to construct military devices. The potential use of such devices by warring nations or by terrorists who might seize them could have disastrous consequences for the entire planet. Fusion technology, on the other hand, appears to offer considerable promise, and what appear to be the appropriate systems are under development.

In any consideration of energy, the sun stands out as the most promising renewable source in the long run (Figure 54-9). Whether this energy is captured directly by solar col-

tropics are being threatened by the ecological consequences of the destruction of trees and other plant cover—consequences that include erosion, nutrient loss, loss of agricultural potential, and the destruction of steady, year-round supplies of water.

As Anil Agarwal pointed out in the *New Scientist* of February 10, 1983, any third world government that wants to meet the energy needs of its people without destroying the environment will need to adopt an integrated national policy on energy for cooking. Such a strategy must include the use of the most promising trees for firewood (see Figure 54-2, *B*) under conditions where they could grow best and provide a regular harvest. Placing improved stoves in the hands of the rural poor is of great importance in conserving energy supplies and in making them accessible to the people involved; some notable successes have been achieved in doing this, especially in India. Plans formulated according to such policies need to take the truly poor into account, because otherwise these people will not be able to afford the benefits that they might attain.

In a sense, the firewood crisis represents a microcosm of the problem of the very poor in the tropics and subtropics. Unless appropriate ways are developed whereby they can find food, clothing, and shelter on a sustainable basis, their plight clearly will worsen and their impact on the environment will become even more serious—without any lasting gain for the people involved.

1

2

FIGURE 54-B

lectors or fixed by plants (Figure 54-10) and other photosynthetic organisms and then converted, it provides an inexhaustible supply, with comparatively few associated ecological problems. It is estimated that the world's total coal reserves—and coal is by far our most abundant source of fossil fuel—contain approximately as much energy as reaches the earth from the sun every 15 days.

Solar energy, whether collected directly or by plants, is clean and, by all practical measures, inexhaustible. In these respects it contrasts with most other actual or potential sources.

Energy is an especially severe problem in tropical countries, most of which lack both their own energy resources and hard currency with which to purchase oil or coal from abroad. In 1981 it was estimated that about 80% of the total world supply of industrial energy was consumed by the 25% of the world's people living in the developed countries; 7.3% by the nearly 22% of the total population who live in China; and only about 12% by the 54% of the people who live in the less developed regions of the world. For them, the development of renewable sources of energy is a matter of desperate urgency; only by doing this will they be able to stabilize their economic development and improve the lot of large numbers of their people. Solar energy on a small scale would appear to be emi-

A

B

C

FIGURE 54-10

Plants efficiently convert energy from the sun to biomass, and the process can be utilized in turn as a huge source of domestic or industrial energy.

A *Water hyacinth, despite its beautiful flowers, is generally regarded as a noxious weed, clogging waterways throughout the tropics. It is a native of South America.*

B *The same rapid growth that makes it such a harmful weed, however, allows it to produce large amounts of biomass rapidly. Here it is being harvested in India as animal feed; it can readily be dried and burned as a source of energy also.*

C *The milky sap of plants of the genus* Euphorbia *is rich in hydrocarbons and particularly promising as a future source of energy. Members of this genus (which also include the poinsettia) are being investigated intensively for their energy-producing potential. Shown here is* Euphorbia tirucelli, *cultivated for this purpose in Kenya, where the need for energy is especially severe and growing rapidly.*

nently suitable for the needs of tropical and subtropical areas, where sunlight is abundant and the need is so great. Here and elsewhere the development of solar energy, or other renewable energy sources, would enable us to avoid the consumption of our finite fossil fuel resources—and alleviate the global carbon dioxide problem at the same time. It is very much in the interests of the people of the developed nations, who depend so heavily on developing countries for trade and in many other ways, to help them to accomplish this development of renewable energy sources.

Minerals

The countries of the developed world, which will include only a fifth of the world's population at the end of this century, were using some 85% to 95% of most of the world's materials, both mineral and nonmineral, in the the mid-1980s. For example, they used 86% of the iron, 90% of the copper, and 92% of the aluminum that was consumed worldwide. Increasing expectations throughout the world, as well as increasing populations, make it extremely unlikely that the per capita supply of these materials in the developed countries can be maintained into the future. In addition, the countries of the developed world import many of the minerals they use from the tropics; future imports depend on the stability of the tropical countries involved.

THE PROSPECTS FOR MORE FOOD

The most rapid growth of human populations is taking place in the tropics and subtropics, and a majority of the world's people already lives there. If people everywhere are to be fed adequately, the increases in food production must take place primarily in these regions. Even though it is sometimes thought that the productive soils of temperate regions could simply be used more efficiently to produce sufficient food to feed everyone in the world, this is not the case, at least for the foreseeable future, as illustrated by the following relationship. Exports from the developed world provided about 8% of the total food consumed in the developing world in the mid-1980s; yet these exports constituted half of the total food exports from the United States and other major food-producing countries. The agricultural lands in the United States are being cultivated so intensively that a considerable amount of topsoil is already being lost there, and genuine concern exists for the future; drastic increases in productivity do not seem possible in the near future. About 70 million people per year are being added to the population of the tropics and subtropics, and these people must be fed also if global standards are to be raised. In addition, the developing countries had already accumulated a collective debt of more than $750 billion to the developed countries by the mid-1980s. This background makes it clear that substantial increases in food for the people who live in the tropics and subtropics must come from productivity within these countries themselves, although the United States and other countries should continue to assist as much as possible, especially at times when famine is particularly acute—as in sub-Saharan Africa in the 1980s.

During the 1950s and 1960s, the so-called Green Revolution took place. This revolution depended on the development of new, improved strains of wheat and rice at international centers that were organized by groups such as the Rockefeller and Ford Foundations, with the assistance of many other governments and agencies. As a result of the efforts made at these centers, the production of wheat in Mexico, for example, increased nearly 10-fold between 1950 and 1970, and Mexico temporarily became a wheat exporter, rather than an importer (Figure 54-11). During the same decades, food production in India was largely able to outpace even a population growth of approximately 2.1% annually, and China became self-sufficient in food.

Despite these and other similar success stories, the Green Revolution has had its limitations. The agricultural techniques of the developed world that the Green Revolution established in parts of the developing world require the expenditure of large amounts of energy, which often are not readily available in the tropics and subtropics. Because of the extensive use of fertilizers, pesticides, herbicides, and machinery, it requires about 10 times as much energy per calorie to produce rice in the United States as it does in the Philippines, and more than 1000 times as much energy to produce each calorie of energy in wheat in the United States as it does by traditional farming methods in India. However, energy prices are often held at artificially low levels in developing countries, so that it may actually be more expensive for a poor rural farmer to grow an equivalent amount of grain as for a large-scale farmer using developed-world technology. In this respect, the introduction of Green Revolution methods in some regions has actually *worsened* poverty for many of the people and lessened their access to food, fuel, and other essential commodities.

The Green Revolution has depended for its success on improved versions of current crops grown on lands that were already under cultivation; almost all of the land that can be cultivated by available methods is already cultivated. Each hectare of cultivated lands in the tropics that supported approximately three people in 1975 will need to support approximately twice as many in 2000—and it will need to do so year after year without the use of abundant fertilizers or other methods that are not economically feasible. Many of the trends that are taking place in tropical countries, such as the loss of agriculturally productive land to urban sprawl or the deterioration of soil quality because of intensive irrigation, actually run counter to the need to protect the currently cultivated land and to keep it in prime condition. Furthermore, the increased use of pesticides and herbicides may have undesirable side effects, and erosion tends to accompany the intensification of agriculture throughout the world.

The Green Revolution has resulted in increased food production in many parts of the world through the use of improved crop strains. These strains often depend on increased inputs of fertilizers, water, pesticides, and herbicides, as well as the greater use of machinery—means that are not normally available to the poor.

Certain nutritional problems have also arisen in connection with the Green Revolution. The overconcentration on cereal crops has tended to lower the production of other nutritionally important plants, including legumes, oilseeds, and vegetables of all kinds. Rectifying this imbalance remains a serious problem for third world planners. The linked sets of cereals and legumes—such as rice and beans—that our Stone Age ancestors domesticated and used are as nutritionally important to us as they were to them. The reason that legumes and cereals are often combined as food sources is that together they provide a balanced set of the amino acids required by humans for proper growth. In the modern world, we are tending to neglect such combinations. The varied strains of crops that are grown on small farms may also be driven out by many fewer kinds of modern strains, which produce a better yield if large-scale inputs of chemicals and the use of machinery are

FIGURE 54-11

Improved strains such as this dwarf wheat helped to make Mexico an exporter, rather than an importer, of wheat during some years in the 1960s and 1970s. The improved strains of wheat grown on them were developed by CIMMYT, the International Center for the Improvement of Maize and Wheat, an important research institution located in Mexico. CIMMYT is one of approximately 14 such institutions located throughout the world, each striving to improve some aspect of third world agriculture.

FIGURE 54-12

A traditional farm in Thailand, with coconuts, Leucaena *(a fast-growing, nitrogen-fixing legume that provides food and firewood), sugar palms, litchi nuts, and nitrogen-fixing* Azolla *growing in ponds. Land crabs are frequent in such farms in Thailand, and are harvested as a source of protein. Such farms, with their trees and mixed crops, are extremely suitable for many sites in the tropics, but the overall tendency is to try to replace them with large-scale agriculture modeled after that which is successful in temperate regions.*

possible (Figure 54-12). Despite these short-term advantages, the loss of the unique, traditional strains of crops currently cultivated by small, rural farmers throughout the world may ultimately prevent the particular crop plants from being able to grow in less favorable habitats or to withstand important diseases in the longer run (Figure 54-13). Biological diversity made possible the development of agriculture in the first place; genetic diversity still enables crops to adapt to new situations. The preservation of plant and other genetic material in so-called gene banks, either artificial—as in botanical gardens, zoos, or seed banks—or natural (in reserves) is a matter of high priority.

In improving the world food supply, the most promising strategy is to improve the productivity of crops that are already being grown on lands under cultivation. There are relatively few parts of the world where additional land exists that can be brought into cultivation with currently available technology; the land that is already cultivated and the crops that are already known to be highly productive are therefore our most important resources for the immediate future. In the development of these crops, biologists have a critical role to play. Traditional methods of plant breeding and selection must be applied fully and to many crops of importance in the tropics and subtropics, in addition to wheat, corn, and rice. The techniques of genetic engineering (see Chapter 18) are being applied extensively to crop plants, and to some extent animals, and they offer great promise for the future. For example, it will be possible to produce plants that are resistant to specific herbicides, which can therefore be applied much more effectively to crops for weed control. It will soon be possible through the methods of genetic engineering to produce new strains of plants that will grow successfully in areas where the particular crop plant could not grow before. It will eventually also be possible to introduce desirable characteristics, such as the ability to fix nitrogen, carry out C_4 photosynthesis, or produce substances that deter pests and diseases in abundance, into important crop plants. The ability to transfer genes between organisms, which became a practical possibility for the first time only in 1973, will be of great importance in the improvement of crop plants before the twentieth century ends.

In addition to the improvement of the crops that are now being cultivated, new crops—as discussed above—must also be developed for new areas. Ways must be developed also to harvest natural ecosystems, or to modify existing ones such as tropical rain forests so that they will produce products of importance for humans while remaining in

A

place. New technologies must be created to supplement those already available, and alternatives must be sought on a wide basis.

The oceans were once regarded as an inexhaustible source of food, but overexploitation of their resources is actually limiting the world catch from year to year, while these catches are costing more in terms of energy. It has been estimated that the mismanagement of fisheries, principally through overfishing, local pollution, and the destruction of fish breeding and feeding grounds, has already lowered the catch of fish in the sea by about 20% from its maximum levels. The decline in the numbers of whales in the oceans of the world is a tragic and well-known example of the way in which fisheries have been and are being destroyed.

The development of new kinds of food, such as microorganisms grown in culture in nutrient solutions, definitely should be pursued. We discussed the cyanobacterium *Spirulina*, which is being used in this way, in Chapter 26 (see Figure 26-10), and there are many similar efforts, often involving green algae and yeasts, under way throughout the world. Ultimately the enrichment of human food through protein-rich concentrates of microorganisms will provide important nutritional augmentation to our diets. There are, however, psychological barriers that must be overcome to persuade people to eat such foods, and the processes required to produce them tend to be energy expensive.

THE TROPICS

More than half of the world's population was living in the tropics and subtropics by the 1980s, and this percentage is increasing rapidly. The well-being of these people should be a matter of concern to everyone, including inhabitants of the developed countries, such as ourselves. Although this is clearly true on humanitarian grounds alone, it also can easily be justified economically and in terms of the self-interest of the people living in developed countries. Not only do we obtain many commodities, including foods, minerals, and other goods, *from* the tropics, but as many as one in six of the jobs in manufacturing in the developed world depends on exports *to* the tropics. In addition, as much as half of the food exports of developed countries go to the tropics. If this world commerce is to persist, the tropical countries must be able to develop sustainable agricultural systems and other kinds

FIGURE 54-13

Corn (Zea mays) and a wild relative.

A *The genetic uniformity of hybrid corn, enhanced by selection over decades, is clearly demonstrated by this crop in Illinois, one of the most productive areas for corn in the world.*

B *Corn is an annual grass, but this field in Mexico is dominated by a wild perennial relative,* Zea diploperennis, *which was discovered in 1977. The two species can be crossed to produce fertile hybrids, and Z.* diploperennis *is resistant to the seven main types of virus diseases that damage corn as a crop. It has also been used to develop a new kind of perennial corn, which shows promise for cultivation in some of the infertile soils of subtropical areas.*

C *Spikelets of* Zea diploperennis.

D *This commercially important and scientifically interesting relative of corn might never have been found at all because it occurs in only one field, which is about the size of a football field. Logging operations are widespread in the Sierra de Manantlán, Jalisco, Mexico, where it lives, and its single population could easily have been destroyed by the rapidly increasing human population pressures of the area.*

B

C

D

FIGURE 54-14

A family has cleared a small patch of forest in the northern Amazon Basin of Brazil; they will be able to grow crops on it for a few years and then will need to move on to another part of the original forest. Such activities can be sustained indefinitely if population levels are low enough, but that will rarely be the case in the closing years of the twentieth century.

of productive industries by means of which they can more adequately feed and otherwise provide for their rapidly growing populations.

Many people in the tropics engage in shifting agriculture—clearing and cultivating a patch of forest, growing crops for a few years, and then moving on (Figure 54-14). Such agricultural systems, more colorfully called "slash-and-burn" cultivation, are effective when the human populations that practice them are relatively small. Indeed, they still seem to be working well in areas such as Zaire, where about 35 million people are spread out over a very large area. In these systems, the minerals in the trees run out into the soil, and crops like manioc (tapioca, cassava; Figure 54-7) and beans can be cultivated successfully for a few years. The fertility of the soil is then exhausted, and the cultivator must move on and clear another patch of forest.

Provided that the forests have decades or even centuries to recover—depending on the nature of the forest being cut—shifting agriculture can be practiced indefinitely without permanent harmful effects. There is renewed interest in the ways in which people have traditionally cultivated these forests, and some of the discoveries that have been made—for example, the extensive irrigated terraces of the Mayas—certainly may have applicability for meeting contemporary needs. With the rapidly growing human populations of the tropics, however, the pressures on the forests have been intensified, and they often do not have sufficient amounts of time to recover; nor is there time for the development of new, more appropriate forest cultivation technologies or for the reintroduction of traditional ones. The greater the population of a given area, the more often a given patch of forest is cultivated and the less completely it recovers.

Shifting agriculture is the major factor in the destruction of tropical forests, but cattle farming on a large scale, especially in Latin America, and lumbering, especially in tropical Asia and usually with no effort being made to replace the forests being cut, are also playing significant roles. The beef and timber that are produced by such operations are mostly exported to the developed countries for cash, and the beef consumption and availability of timber in most third world countries have decreased substantially over the past several decades.

As a result of such overexploitation, experts predict, there will be little undisturbed tropical forest left anywhere in the world by the early twenty-first century. Many areas now occupied by forest will still be tree covered, but these trees will represent only a small percentage of those that now grow in these areas. Many species of tropical plants, animals, and microorganisms cannot reproduce outside of the conditions under which they live in the original forest. Consequently they are threatened with extinction or, at the very least, with banishment from large areas (see Figure 52-1). In fact, it is estimated that perhaps 15% or even 20% of the species on earth will become extinct during the next 30 or 40 years, amounting to perhaps 750,000 species. This catastrophic event will occur during the lifetime of a majority of people who are alive today. Put another way, one or two species per day are probably becoming extinct now—many of them on ecologically devastated islands such as St. Helena in the south Atlantic, and Rodrigues in the western Indian Ocean (Figure 54-15). By early in the twenty-first century, the rate could easily reach a few species an hour; and it would continue to climb for at least another 50 years. Overall, this would amount to an extinction event that has been unparalleled for at least 65 million years, since the end of the age of dinosaurs. The actual proportion of species becoming extinct over the course of the twenty-first century could easily be much greater than the proportion that became extinct at that time.

The prospect of such an extinction event is of major concern. In view of the fact that no more than one of every six tropical organisms has even been given a scientific name, it is clear that many species that are about to become extinct will never have been seen by any human being. As these species disappear, so does our opportunity to learn about them, not only scientifically, but also in terms of their possible benefits for ourselves. We have an intrinsic curiosity about the plants, animals, and microorganisms with which we share this planet, and we would like to understand them better individually. Each is unique, and with

A

B

its loss goes forever the chance to know any more about it, to understand it, or to use it for any purpose whatever. The fact that our entire supply of food is based primarily on 20 kinds of plants, out of the quarter million that are available, should give us pause to consider what it means to be living in a generation during which a high proportion of the remainder are being lost permanently. Many of them would surely be of great use to us if we knew about their properties (Figure 54-13; boxed essay, pp. 1176-1177).

What biologists can do in the face of this crisis is to help design intelligent plans for finding those organisms most likely to be of use and saving them from extinction. Biologists must also participate in sound, globally based schemes to preserve as much as possible of the biological diversity of life on earth so that our descendants may have as many options as possible. With the loss of tropical forest, and of biological communities throughout the world, we are permanently losing many opportunities not only for knowledge, but also for increased prosperity—whether we realize it or not. It is biologists who must understand this message and inform their fellow citizens of its importance to them.

> **Very little undisturbed tropical forest will be left anywhere in the world by the early twenty-first century. Primarily as a result of this destruction, 15% or more of the estimated 4 to 5 million species of plants, animals, and microorganisms on earth will become extinct during the next several decades, many of them without ever having been discovered or considered in terms of their potential usefulness to human beings.**

Viewing the consequences of uncontrolled deforestation in the tropics and subtropics in another way, tropical forests are complex, productive ecosystems that work well in the areas where they have evolved. The sad truth is that we do not know, for the most part, how to replace them with other productive ecosystems that will support human beings (Figure 54-16). When we clear a wood or open a prairie in the North Temperate Zone, we provide the basis for a farm that we know can be worked for generations. In the tropics, in general, we simply do not know how to engage in continuous agriculture in most areas that are not now under cultivation. When we clear a tropical forest, we engage in a one-time consumption of natural resources that will not be available again (Figure 54-17). The

FIGURE 54-15

A *This palm is the last individual of its species,* Hyophorbe verschaffeltii, *known in nature. It is shown here in forest remnants on the island of Rodrigues, in the western Indian Ocean. Who knows whether this palm, or one of the hundreds of other species of palms threatened with extinction around the world, holds the economic promise of the oil palm (Figure 54-2)?*

B Hyophorbe verschaffeltii *is not likely to become extinct, however, because it is preserved in cultivation on the campus of the University of Mauritius, on the island of Mauritius. The palms are the second most important family of plants in the world economically, surpassed only by the grasses.*

FIGURE 54-16

A One of the most difficult aspects to contend with in developing areas occupied by tropical rain forest is the soils of these regions, which are very thin. The roots of the trees occupy only the top few centimeters of soil.

B In an experiment shown here at San Carlos de Río Negro, in southern Venezuela, scientists working on a UNESCO Man and Biosphere ecosystem project are measuring the amounts of nutrients that seep through this thin soil layer—only a few percent of the total.

complex ecosystems that have been built up over billions of years are now being dismantled, in almost complete ignorance, by the human species—perhaps the most pervasive and dominant life form ever to have affected world ecology.

What biologists must do is learn more about the construction of sustainable agricultural ecosystems that will meet human needs in tropical and subtropical regions. The ecological principles that we have been reviewing in the past four chapters are universal principles. The undisturbed tropical rain forest has one of the highest rates of net primary productivity of any plant community on earth, and it is therefore logical to assume that it can be harvested for human purposes in a sustainable, intelligent way. Simply allowing it to be harvested—consumed—is something that biologists should attempt to avoid. Sound development of the tropics, an urgent matter in view of the very large numbers of humans who live in tropical countries, must be based on sound biology, as well as on achieving stable human population levels and alleviating the problems of poverty and widespread malnutrition.

FIGURE 54-17

A When tropical evergreen forests are removed, the ecological consequences can be disastrous. These fires are destroying rain forest in Brazil, which is being cleared for cattle pasture.

B The consequences of deforestation can be seen on these mid-elevation slopes in Ecuador, which now support only low-grade pastures where highly productive forest, which protected the watersheds of the area, grew in the 1970s.

C As time goes by, the consequences of tropical deforestation may become even more severe, as shown here by the extensive erosion in this area of Tanzania from which forest has been removed.

A

NUCLEAR WINTER: THE BIOLOGICAL CONSEQUENCES

For the immediate future, the destructive potentialities of a possible nuclear war must be considered an even more serious threat. The two atomic bombs dropped at Hiroshima and Nagasaki, Japan, in August 1945, had an explosive power of approximately 20,000 tons of TNT each (Figure 54-18); all of the weapons exploded in World War II had a combined explosive force of 2 million tons of TNT (2 megatons). In sharp contrast, the nuclear arsenals currently held by the United States and the Soviet Union total some *12,000 megatons* for strategic and theater weapons alone. Under these circumstances, an exchange of several thousand megatons is certainly possible.

After a nuclear exchange of this magnitude, many scientists believe, the fine dust raised by high-yield nuclear surface blasts and by smoke from fires ignited by aerial bursts would cause the sunlight reaching the earth's surface to diminish to a small percentage of present levels, producing a condition described as **nuclear winter.** This dust and smoke would encircle the world in about 1 to 2 weeks and would result in temperatures in continental interiors of about −15° to −25° C, regardless of season and persisting for several months. Compounded by radioactive fallout at lethal levels over much of the land in the Northern Hemisphere and by heavy ultraviolet radiation caused by destruction of much of the ozone layer, these persistent subfreezing temperatures would have extremely severe biological consequences (Figure 54-19, *A*). If, as seems highly possible, the sun-obscuring cloud were to spread into the tropics and the Southern Hemisphere, the consequences would be far worse (Figure 54-19, *B*).

The World Health Organization has calculated that after a major nuclear exchange, there would be about 1.1 billion immediate deaths and about as many additional injuries requiring medical attention, many of which could not be treated under the postwar conditions. The 2 to 3 billion people not killed more or less immediately would have to search for food and water in the dark in subfreezing temperatures under highly radioactive conditions. Even after sunlight began to return—in several months to a year—the re-establishment of agriculture would be extremely difficult, with many of the seed stocks destroyed. The kinds of agriculture that could be pursued would be far less sophisticated than

FIGURE 54-18

The detonation of an atomic bomb in southern Nevada in 1953. Such devices caused the deaths of hundreds of thousands of people at Hiroshima and Nagasaki, as well as widespread destruction. They are only a small fraction of the size of the weapons that are stockpiled in large numbers today.

B

C

A   B

FIGURE 54-19

A *In a spring or summer nuclear war, subfreezing temperatures would kill or damage virtually all crops in the Northern Hemisphere. The low light levels would inhibit photosynthesis, and the consequences would affect all food chains.*
B *If the cold and darkness following a nuclear war in the Northern Hemisphere extended into the tropics and subtropics, it would cause large-scale injury to plants and animals there and immediate, widespread extinction. People in places like Central America (shown here) would be forced to wander in search of food and shelter.*

those practiced now. If darkened skies and low temperatures did spread over the entire planet, a severe extinction event would immediately reduce the biological diversity of earth to probably less than half of what it is now, with the extinction of most tropical plants and animals especially likely.

> **A major nuclear war would probably produce a cloud that would lower temperatures drastically across the Northern Hemisphere, and possibly spread to the Southern Hemisphere as well. The subfreezing temperatures would be disastrous for agricultural systems and for many aspects of organized human society. Depending on the cloud's spread and duration, many species of organisms, especially in the tropics, could become extinct.**

Even a small-scale nuclear exchange of, say, 500 megatons, targeted on cities, would probably produce many of these negative effects. In a large-scale exchange, the extinction of the human species could not be ruled out as a possibility, even if only one side were to attack the other. Certainly, the small bands of people who might survive could not conceivably reconstitute social systems remotely resembling those whose members would have initiated the war in the first place. In view of these possible grim consequences, it is important for biologists to continue to model the biological effects of a nuclear exchange, and to make their findings known so that they can be taken into account by decision makers throughout the world.

WHAT BIOLOGISTS HAVE TO CONTRIBUTE

Our human population is at an unprecedented level and is straining the productive capacity of the global ecosystem more than it ever has. Of the estimated 5 billion people living in the mid-1980s, 1 in 5 is living in absolute poverty, 1 in 10 is malnourished, and 20 to 30 million are starving to death each year. The productive agricultural soils of temperate regions are rapidly being lost to erosion, our water supplies are becoming increasingly contaminated and restricted, and species of plants, animals, and microorganisms are being lost to extinction at a rate that has not occurred during the past 65 million years. The social and economic consequences of these biological problems are increasingly familiar, manifesting themselves as governmental instability, inflation, massive international debts, war, and huge numbers of displaced people and immigrants moving throughout the world.

The development of appropriate solutions to these massive problems must rest partly on the shoulders of politicians, economists, bankers, engineers—many different kinds of people. The basis for solving them permanently, however, and achieving a stable, productive world, must be biological (Figure 54-20). The energy that bombards the earth comes in essentially inexhaustible amounts from the sun; living systems capture that energy and fix it in molecules that can be utilized by all organisms to sustain their life processes. We and all other living beings depend on the proper utilization of that renewable energy, and of the other materials on which our civilization depends. Biologists, therefore, must come to understand better the processes involved, and develop new systems that will work efficiently in areas where they are not available now. That is our challenge for the future.

It is clear that a scientific education has become necessary for everyone, so that we may understand the basis for our continued existence on earth and the steps that we will need to take to improve the quality of our lives. Biology should play a major part in that education and is of critical importance in providing an improved standard of living for our fellow human beings. Biological literacy is no longer a luxury; for intelligent human beings who want to play a constructive role in improving the world, it has become a necessity.

FIGURE 54-20

Donald Perry, a biologist from the University of California, Los Angeles, investigating the canopy layers of the tropical rain forest.

SUMMARY

1. Human beings, who belong to the genus *Homo,* evolved from apelike animals of the genus *Australopithecus* about 2 million years ago. Our species, *Homo sapiens,* first appears in the fossil record about 500,000 years ago, and people of modern aspect replaced humans of more archaic aspect in western Eurasia starting about 40,000 years ago. They developed agriculture in several centers by no less than 11,000 years ago.

2. Herds of domesticated animals rapidly brought widespread ecological destruction in the Old World, and this phenomenon was exported to the Americas starting about 500 years ago.

3. About 5 million people lived throughout the world when agriculture was first developed. With the availability of this new steady source of food, the population grew rapidly, reaching about 130 million at the time of Christ and 500 million by 1650. The development of villages, towns, and finally cities was made possible as a result of agriculture. Cultural specialization occurred in these centers, resulting in the development of writing and, ultimately, of science.

4. By 1986, the world population was just under 5 billion, growing at a rate of 1.7% per year, and estimated to double within approximately 40 years. It is growing much more rapidly in the tropics and subtropics than in developed countries, so the relative proportions of people in these two areas are changing rapidly. By 2020, nearly two thirds of the world's people are expected to live in the tropics and subtropics, and only a sixth in developed countries—half the proportion that lived there in 1950.

5. In 1984, the World Bank estimated that about 1 billion of the roughly 2.5 billion people living in the tropics and subtropics existed in a state that it described as absolute poverty. Some 20 to 30 million people a year were starving to death, about 10 million of them children under 5 years old. With the populations of tropical countries doubling every 29 years, on the average, the task of feeding all of these extra people even as well as is being done now will be extraordinarily difficult.

6. In the last half of the twentieth century, some 80% to 90% of all the world's resources were being used by the dwindling minority of the world's population that lived in the developed countries. This uneven distribution of resources has led, and is likely to continue to lead, to global political and social instability, fostered also by the lack of attention to the needs of the rural poor in the tropics and subtropics and a lack of understanding of the ways in which tropical soils and climates can be managed to produced sustained agricultural yields.

7. More food must be produced for the inhabitants of the tropics and subtropics. It can be produced in appropriate quantity only in the countries where it is needed. As a result, a great deal of research into the properties of sustainable agricultural systems in the tropics and subtropics is required, and strict attention must be paid to the needs of the rural poor if stable systems are to be put in place.

8. The Green Revolution resulted from large-scale effort on the part of agricultural research institutes to increase the productivity of wheat and corn (maize) in the subtropics. It was enormously successful in increasing the food production of such countries as Mexico and India, although it had relatively little effect on the well-being of the rural poor and tended to be energy expensive.

9. The destruction or disturbance of virtually all of the forests of the tropics during the next several decades will bring about the extinction of as much as 15% of the species of plants, animals, and microorganisms on earth. Many of these will never have been seen by a scientist, and their possible contribution to human welfare will be lost.

10. A major nuclear war could result in the production of a cloud of smoke that would cut off most of our sunlight and result in the lowering of temperatures throughout the Northern Hemisphere to subfreezing levels. If this cloud were to spread into the tropics and the Southern Hemisphere, many kinds of organisms could become extinct. Even if it did not, the darkness and cold, coupled with the immediate and widespread death and destruction and the subsequent radiation, would result in the end of human society as we know it, in the Northern Hemisphere at least.

11. The application of the principles of biology to the human condition has never been more necessary than it is now. Only the full attention of society and of many talented individuals to the solution of these grave problems will make it possible for our children and grandchildren to enjoy the same benefits that we have.

SELF-QUIZ

The following questions pinpoint important information in this chapter. You should be sure to know the answers to each of them.

1. We derive about 85% of our food, directly or indirectly, from _____ kinds of plants of the approximately 240,000 that are available.

2. It is estimated that about 15% or more of the kinds of plants, animals, and microorganisms in the world may become extinct within the next 30 to 40 years. Is this a "normal" rate of extinction, one that has occurred constantly for many millions of years?

3. Nearly all of the major crops in world commerce have been in cultivation for thousands of years. True or false?

4. How many people were living in the world when agriculture was first being developed?
   (a) 50,000   (d) 5,000,000
   (b) 200,000   (e) 25,000,000
   (c) 1,000,000

5. The population of developed countries is expected to double in approximately
   (a) 10 years   (d) 40 years
   (b) 20 years   (e) 120 years
   (c) 30 years

6. For all practical purposes, _____ energy is inexhaustible.

7. In the latter half of the twentieth century, the developed countries used what percentage of all the world's resources?
   (a) 40-50   (d) 80-90
   (b) 50-60   (e) 90-100
   (c) 60-70

8. The Green Revolution
   (a) has resulted in increased food production
   (b) utilizes improved crop strains
   (c) greatly aided the very poor population
   (d) a and b
   (e) all of the above

9. The most necessary step in the attainment of peace and prosperity for everyone is
   (a) increasing the food supply
   (b) attaining stability in the world population
   (c) decreasing disease and other health problems
   (d) a and b
   (e) none of the above

10. What is the main problem with "shifting cultivation" practiced in the tropics?
   (a) forests are not being allowed sufficient recovery time
   (b) the crops drain the soil of its nutrients
   (c) once the land is used it is unable to support crops for many years
   (d) all of the above

11. Which of the following would be a consequence of a nuclear war?
   (a) cold temperatures
   (b) destruction of the ozone layer
   (c) increase in ultraviolet radiation
   (d) extinction of many species of organisms
   (e) all of the above

12. A 500-megaton exchange could cause a "nuclear winter." True or false?

THOUGHT QUESTIONS

1. About 8% of the food that was consumed in the world's less-developed countries in the early 1980s came from the developed countries; this included about half of all the food exported from the United States. Evaluate the likelihood of producing a food surplus in the United States and other temperate-zone countries sufficient to alleviate the problem of hunger among the rural poor in the tropics. List the factors that are involved.

2. What lessons can we learn from the high productivity of undisturbed tropical rain forests, and how can we apply them for human benefit?

3. Decreased birth rates have ultimately followed decreased death rates in many countries, such as Germany and Great Britain. Will this eventually occur worldwide and solve our population problems? Discuss the reasons for your answer.

4. Has the human population level already exceeded the long-term carrying capacity of the earth? What evidence can you bring to bear on this question?

FOR FURTHER READING

BATIE, S.S., and R.G. HEALY: "The Future of American Agriculture," *Scientific American,* February 1983, pages 45-53. An interesting analysis of the ecological and economic factors involved.

BROWN, L.R., W.U. CHANDLER, C. FLAVIN, C. POLLOCK, S. POSTEL, L. STARKE, and E.C. WOLF: *State of the World 1985: A Worldwatch Institute Report on Progress Toward a Sustainable Society,* W.W. Norton & Co., New York, 1985. An excellent overall assessment of ecological indicators throughout the world.

CHANG, T.T.: "Conservation of Rice Genetic Resources: Luxury or Necessity?" *Science,* vol. 224, pages 251-256, 1984. This article discusses the rich genetic diversity that is available in cultivated rice and its wild relatives, and suggests ways that it might be preserved.

EHRLICH, P.R., A.H. EHRLICH, and J.P. HOLDREN: *Ecoscience: Population, Resources, and Environment,* W.H. Freeman & Co., San Francisco, 1977. Offers ample coverage of all aspects of human population and resource questions.

GATES, D.M.: *Energy and Ecology,* Sinauer Associates, Inc., Sunderland, Mass., 1985. A practical appraisal of the ecological consequences of different forms of energy use.

HARWELL, M.A.: *Nuclear Winter: The Human and Environmental Consequences of Nuclear War,* Springer-Verlag, New York, 1984. A biologically oriented account of the nuclear winter phenomenon.

HUXLEY, A.: *Green Inheritance,* Anchor Press/ Doubleday, Garden City, N.Y., 1985. An outstanding account of the many ways in which we depend on plants and should therefore be concerned with their preservation.

MOORE, J.A.: "Science as a Way of Knowing—Human Ecology," *American Zoologist,* vol. 25, pages 1-155, 1985. Excellent essay on the evolution of human beings and their impact on the global environment.

MYERS, N.: *The Primary Source: Tropical Forests and Our Future,* W.W. Norton & Co., New York, 1984. A good account of tropical forests and the role they play in human welfare.

PIMENTEL, D., et al.: "Water Resources in Food and Energy Production," *BioScience,* vol. 32, pages 861-867, 1982. A discussion of the conflicts over water use that will play a major role in determining the future of the United States and other countries.

SOUTHWICK, C.H. (ed.): *Global Ecology,* Sinauer Associates, Inc., Publishers, Sunderland, Mass., 1985. An excellent collection of readings on human interactions with the environment.

VINING, D.R., JR.: "The Growth of Core Regions in the Third World," *Scientific American,* April 1985, pages 42-49. An analysis of the way in which rapid population gains registered in third world core urban regions pose grave social and economic problems.

The classification used in this book is that explained in Chapter 22. It recognizes a separate kingdom, Monera, for the bacteria (prokaryotes) and divides the eukaryotes into four kingdoms: the diverse and predominantly unicellular Protista, and three large, characteristic multicellular groups derived from them: Fungi, Plantae, and Animalia.

All of the phyla or divisions (equivalent terms used for different groups of organisms) discussed in this book are outlined in this appendix, together with the classes that we have mentioned. Viruses, which are considered nonliving, are not included in this appendix, but are treated in Chapter 25.

KINGDOM MONERA

Bacteria; the prokaryotes. Single-celled organisms, sometimes forming filaments or other forms of colonies. Bacteria lack a membrane-bound nucleus and chromosomes, sexual recombination, and internal compartmentalization of the cells; their flagella are simple, composed of a single fiber of protein. They are much more diverse metabolically than the eukaryotes. Their reproduction is predominantly asexual. About 2500 species are currently recognized, but that is probably only a small fraction of the actual number.

Phylum Methanocreatrices

Methane-producing bacteria; bacteria that form methane (CH_4) by reducing CO_2 and oxidizing hydrogen (H_2). Methane-producing bacteria are either nonmotile or flagellated rods, cocci, or spirilla. They are linked with other bacteria that reduce sulfur, fix nitrogen, or are photosynthetic by their unique rRNA base sequences, which differ sharply from those of all other bacteria; a separate kingdom, Archaebacteria, has been proposed for this group.

Phylum Omnibacteria

Mainly rigid, rod-shaped, heterotrophic, gram-negative bacteria; many important pathogens are included in this phylum, as well as the ubiquitous *Escherichia coli*. Another group, the vibrios, consists of comma-shaped bacteria with a single terminal flagellum; still another comprises the rickettsias, very small bacteria that are obligate parasites.

Phylum Cyanobacteria

Photosynthetic bacteria with chlorophyll *a* and phycobilins as accessory pigments. Cyanobacteria, formerly called "blue-green algae," have been prominent for at least 3 billion years and have played a major role in raising the earth's oxygen to its present levels. Many can fix atmospheric nitrogen in specialized cells called heterocysts. Some cyanobacteria form complex filaments or other colonies. A few hundred living species.

Phylum Chloroxybacteria

Prochloron, a single genus of nonmotile, marine cocci, symbiotic in colonial ascidians; contains chlorophylls *a* and *b*.

Phylum Aphragmabacteria

Mycoplasmas and spiroplasmas; nonmotile, heterotrophic, very small bacteria that lack cell walls. Some members of this phylum are less than 0.2 micrometer in diameter, the smallest cellular organisms. They cause important diseases of plants and animals.

Phylum Spirochaetae

Spirochetes; tightly coiled spirilla in which the flagella are located beneath the outer lipoprotein membrane of their gram-negative outer cell wall; there are 2 to more than 100 flagella, which propel the spirochetes through liquid.

Pseudomonads

Pseudomonads are straight or curved gram-negative rods that bear one to several flagella at one end; some are autotrophic, many are serious plant pathogens, and some prey on other bacteria.

Phylum Actinobacteria

The actinomycetes have a filamentous growth habit; they produce spores by the division of erect, terminal branches that divide into chains of small segments. Important pathogens, causing leprosy and tuberculosis, among other diseases. *Frankia* forms nitrogen-fixing nodules on the roots of certain plants. Actinomycetes are the source of more than half of the antibiotics in use today.

Phylum Myxobacteria

The gliding bacteria; unicellular rods less than 1.5 micrometers thick but up to 5 micrometers long. Myxobacteria, which are structurally the most complex prokaryotes, often aggregate into gliding masses and may form complex spore-bearing structures; chemotrophs or heterotrophs.

Nitrogen-fixing aerobic bacteria

Nitrogen-fixing aerobic bacteria are gram negative and mostly flagellated. Some genera are free living, but *Rhizobium,* the key genus in nitrogen fixation worldwide, forms nodules on the roots of legumes.

Chemoautotrophic Bacteria

Chemoautotrophic bacteria use the reduced gases ammonia, methane, and hydrogen sulfide as sources of energy; they are rods or cocci, some of them flagellated. Certain chemoautotrophic bacteria make possible specific steps in the nitrogen cycle, including the conversion of nitrites to nitrates.

KINGDOM PROTISTA

Eukaryotic organisms, including many evolutionary lines of primarily single-celled organisms. Eukaryotes have a membrane-bound nucleus and chromosomes, sexual recombination, and extensive internal compartmentalization of the cells; their flagella are complex, with 9 + 2 internal organization. Some protists have pseudopods, and some are sessile (immobile). They are diverse metabolically, but much less so than bacteria; protists are heterotrophic or autotrophic, and may capture prey, absorb their food, or photosynthesize. Reproduction in protists is either sexual, involving meiosis and syngamy, or asexual. These are general features of all eukaryotes; Fungi, Animalia, and Plantae differ in their specific characteristics. Although some protists would technically be regarded as belonging to divisions, we shall consider them all phyla here for simplicity, since the two categories are equivalent.

Phylum Caryoblastea

One species of amoebalike organism, *Pelomyxa palustris,* which lacks mitosis, mitochondria, and chloroplasts, but does have two kinds of bacterial symbionts. *Pelomyxa,* which occurs on the muddy bottoms of freshwater ponds, probably represents a very early stage in the evolution of eukaryotes.

Phylum Dinoflagellata

Dinoflagellates; unicellular, photosynthetic organisms, most of which are clad in stiff, cellulose plates and have two unequal flagella that beat in grooves encircling the body at right angles. Many are symbiotic in other organisms, and are then known as zooxanthellae. Chlorophylls *a* and *c*. About 1000 species.

Phylum Rhizopoda

Amoebas; heterotrophic, unicellular organisms that move from place to place by cellular extensions called pseudopods and reproduce only asexually, by fission. Hundreds of species.

Phylum Sporozoa

Sporozoans; unicellular, heterotrophic, nonmotile, spore-forming parasites of animals. Sporozoans have complex life cycles that involve both sexual and asexual phases, the former with an alternation of haploid and diploid generations. About 3900 species.

Phylum Acrasiomycota

Cellular slime molds; unicellular, amoebalike, heterotrophic organisms that aggregate in masses at certain stages of their life cycle and form compound sporangia. Meiosis occurs within cysts in these sporangia. About 65 species.

Phylum Myxomycota

Plasmodial slime molds; heterotrophic organisms that move from place to place as a multicellular, gelatinous mass, which forms sporangia at times. Diploid spores are produced within the sporangium, and meiosis occurs within them at the time of germination. About 450 species.

Phylum Zoomastigina

Zoomastigotes; a highly diverse phylum of mostly unicellular, heterotrophic, flagellated, free-living or parasitic protists (from one flagellum to thousands). Sexual reproduction rare. Thousands of species.

Phylum Euglenophyta

Euglenoids; mostly unicellular, photosynthetic or heterotrophic protists with two unequal flagella. Chlorophylls *a* and *b* in photosynthetic forms. Euglenoids closely resemble certain zoomastigotes. About 800 species.

Phylum Phaeophyta

Brown algae; multicellular, photosynthetic, mostly marine protists with chlorophylls *a* and *c* and an abundant carotenoid (fucoxanthin) that colors the organisms brownish. About 1500 species.

Phylum Bacillariophyta

Diatoms; mostly unicellular, photosynthetic organisms with chlorophylls *a* and *c* and fucoxanthin. Diatoms have a unique double shell made of opaline silica, resembling a small box with a lid. Sexual reproduction is apparently rare; flagella absent. About 11,500 living species.

Phylum Chlorophyta

Green algae; a large and diverse phylum of unicellular or multicellular, mostly aquatic organisms with chlorophylls *a* and *b* and carotenoids, and with starch accumulated within the plastids (as it is also in plants) as the food storage product. Many green algae have two equal whiplash flagella; others are nonflagellated. Some form cell plates like those of plants; many have cellulose prominent in their cell walls. About 7000 species.

Phylum Ciliophora

Ciliates; diverse, mostly unicellular, heterotrophic protists, characteristically with large numbers of cilia. Outer layer of cell proteinaceous, flexible. Most ciliates have a complex sexual process known as conjugation. About 8000 species.

Phylum Oomycota

Oomycetes; water molds, white rusts, and downy mildews. Aquatic or terrestrial unicellular or multicellular parasites or saprobes that absorb dead organic matter; cellulose is the primary constituent of the cell wall in many oomycetes. About 475 species.

Phylum Rhodophyta

Red algae; mostly marine, mostly multicellular protists with chloroplasts containing chlorophyll *a* and, like the cyanobacteria that presumably gave rise to their plastids, phycobilins; no flagellated cells present. About 4000 species.

KINGDOM FUNGI

Filamentous, multinucleate, heterotrophic eukaryotes with cell walls rich in chitin; no flagellated cells present. Mitosis in fungi takes place within the nuclei, the nuclear envelope never breaking down. The filaments of fungi grow through the substrate, secreting enzymes and digesting the products of their activity. Septa between the nuclei in the hyphae normally complete only when sexual or asexual reproductive structures are being cut off. Asexual reproduction frequent in some groups. The nuclei of fungi are haploid, with the zygote the only diploid stage in the life cycle. About 100,000 named species.

Division Zygomycota

Zygomycetes; bread molds and other microscopic fungi that occur on decaying organic matter. Zygomycetes are the fungal partners in endomycorrhizae. Hyphae aseptate except when forming sporangia or gametangia. Sexual reproduction by equal gametangia containing numerous nuclei; after fusion, zygotes are formed within a zygospore, a structure that often forms a characteristic thick wall. Meiosis occurs during the germination of the zygospore. About 600 species.

Division Ascomycota

Ascomycetes; yeasts, molds, many important plant pathogens, morels, cup fungi, and truffles. Hyphae divided by incomplete septa except when asci, the structures characteristic of sexual reproduction, are formed. Dikaryotic hyphae form after appropriate fusions of monokaryotic ones, and eventually differentiate asci, which are often club shaped, within an ascocarp. Syngamy within each ascus produces the zygote, which divides immediately by meiosis to produce the ascospores. About 30,000 named species.

Division Basidiomycota

Basidiomycetes; mushrooms, toadstools, bracket and shelf fungi, rusts, and smuts. Most ectomycorrhizae involve basidiomycetes; a few involve ascomycetes. Hyphae and life cycle similar to that of ascomycetes but differing in important details. Basidiospores elevated from the basidium on threadlike projections called sterigmata; basidiocarps may be large and elaborate or absent, depending on the group. About 25,000 named species.

Fungi Imperfecti

An artificial group of about 25,000 named species. Most are ascomycetes for which the sexual reproductive structures are not known; this may be because the stages are rare or do not occur, or because the fungus is poorly known. In either situation, the classification must be made according to the details that are known, which necessitates the recognition of the Fungi Imperfecti.

Lichens

Lichens are symbiotic associations between an ascomycete (a few basidiomycetes are involved also) and either a green alga or a cyanobacterium; the fungus provides protection and structure, and the photosynthetic organism enclosed in it manufactures carbohydrates. There are at least 25,000 species of lichens.

KINGDOM PLANTAE

Multicellular, photosynthetic, primarily terrestrial eukaryotes derived from the green algae (phylum Chlorophyta) and, like them, containing chlorophylls *a* and *b,* together with carotenoids, in chloroplasts and storing starch in chloroplasts. The cell walls of plants have a cellulose matrix and sometimes become lignified; cell division is by means of a cell plate that forms across the mitotic spindle. The vascular plants have an elaborate system of conducting cells, consisting of xylem (in which water and minerals are transported) and phloem (in which carbohydrates are transported); the mosses have a reduced vascular system, something the liverworts and hornworts—which may not be directly related to the mosses—lack. Plants have a waxy cuticle that helps them to retain water, and most have stomata, flanked by specialized guard cells, which allow water to escape and carbon dioxide to reach the chloroplast-containing cells within their leaves and stems. All plants have an alternation of generations with reduced gametophytes and multicellular gametangia. About 270,000 species.

Division Bryophyta

Mosses, hornworts, and liverworts, three groups of simple plants that may not be directly related, comprise the bryophytes. Bryophytes have green, photosynthetic gametophytes and usually brownish or yellowish sporophytes with little or no chlorophyll. Bryophytes have multicellular gametangia and biflagellated sperm. About 16,600 species.

Class Hepaticopsida

Liverworts; some "leafy," and some thallose, strap-shaped plants that creep along the soil or other substrate on which they grow. Sporophytes are usually enclosed within gametophyte tissue. About 6500 species.

Class Antheroceratopsida

Hornworts; thallose plants with green, creeping gametophytes. The photosynthetic sporophytes arise from the upper surface of the gametophyte. About 100 species.

Class Muscopsida

The mosses; usually brownish sporophytes, with complex sporangia, arise from the green, leafy gametophytes. Both gametophytes and sporophytes usually contain reduced vascular tissue, and mosses may be, in effect, reduced vascular plants. About 10,000 species.

Division Psilophyta

Whisk ferns. Vascular plants (like those in all the remaining plant divisions) with well-developed vascular tissue, consisting of xylem and phloem. Gametophytes complex, with multicellular gametangia. Homosporous. Motile sperm. No differentiation between root and shoot. Two genera and several species.

Division Lycophyta

Lycopods (including club mosses and quillworts). Vascular plants. Homosporous *(Lycopodium)* or homosporous *(Selaginella, Isoetes)*. Gametophytes complex, with multicellular gametangia. Motile sperm. Five genera and about 1000 species.

Division Sphenophyta

Horsetails. Vascular plants with characteristic ribbed, jointed stems. Homosporous. Gametophytes complex, with multicellular gametangia. Motile sperm. One genus *(Equisetum)*, 15 species.

Division Pterophyta

Ferns. Vascular plants, often with characteristic divided, feathery leaves (fronds). Mostly homosporous, a few genera heterosporous. Gametophytes complex, with multicellular gametangia. Motile sperm. About 12,000 species.

Division Coniferophyta

Conifers. Vascular plants, mostly trees, some shrubs, often with hard or leathery leaves. Heterosporous and seed-forming; the seeds are naked at the time of fertilization (therefore, the conifers are gymnosperms). Archegonia multicellular, antheridia lacking; sperm immotile, carried to the vicinity of the egg by the pollen tube. Gametophytes reduced, held within the ovule (megagametophytes) or pollen grain (microgametophytes). About 50 genera, with about 550 species.

Division Cycadophyta

Cycads; palmlike, heterosporous gymnosperms with large, pinnate leaves and terminal cones. Archegonia multicellular, antheridia lacking; sperm flagellated but carried to the vicinity of the egg by the pollen tube. Gametophytes greatly reduced. Ten genera, about 100 species.

Division Ginkgophyta

One species, the ginkgo or maidenhair tree; a tall, deciduous tree with fan-shaped leaves that have open dichotomous venation. Heterosporous gymnosperm with fleshy ovules. Details of gametes, gametangia, and gametophytes same as in cycads.

Division Gnetophyta

Gnetophytes, a very diverse group of three genera of gymnosperms. Some gnetophytes have archegonia, others do not; all lack antheridia. Sperm immotile; gametophytes reduced. About 70 species.

Division Anthophyta

Flowering plants, or angiosperms, the dominant group of plants. Angiosperms lack antheridia and archegonia, and have very reduced gametophytes consisting of a few cells; their sperm are immotile, carried to the vicinity of the ovule by the pollen tube. Angiosperms are characterized by their flowers, unique structures that may have as many as four whorls of appendages, and fruits, which enclose the seeds at maturity. Fertilization is indirect, and the ovules are enclosed at the time of pollination. Fertilization is double, one sperm fusing with the egg to produce a zygote, and the other fusing with the polar nuclei to form the primary endosperm nucleus, which gives rise to the endosperm—a distinctive tissue that nourishes the developing embryo. About 235,000 species.

Class Monocotyledons

The monocots; angiosperms with one seedling leaf (cotyledon). Monocots usually have parallel venation and flower parts in threes or multiples of threes. The vascular bundles in the stem are scattered, and true secondary growth is present in only a few genera. About 65,000 species.

Class Dicotyledones

The dicots; angiosperms with two seedling leaves (cotyledons). Dicots usually have reticulate (netlike) venation and flower parts in fours or fives, or multiples of fours or fives. The vascular bundles in the stem usually form a ring, and true secondary growth is characteristic. About 160,000 species.

KINGDOM ANIMALIA

Animals are multicellular eukaryotes that characteristically ingest their food. Their cells are usually flexible; in all of the approximately 35 phyla except sponges, these cells are organized into structural and functional units called tissues, which in turn make up organs in most animals. In animals, the cells move extensively during the development of the embryos; the blastula, a hollow ball of cells, forms early in this process and is characteristic of the group. Most animals reproduce sexually; their nonmotile eggs are much larger than their small, flagellated sperm. The gametes fuse directly to produce a zygote and do not divide by mitosis as in plants. More than a million species of animals have been described, and at least several times that many await discovery.

SUBKINGDOM PARAZOA

Animals that mostly lack definite symmetry, and possess neither tissues nor organs.

Phylum Porifera

Sponges. Larval sponges are free swimming, whereas the adults are sessile. Sponges feed by circulating water through pores and canals lined with flagellum-bearing choanocytes. The structure of these choanocytes indicates that sponges evolved from certain zoomastigotes. Many sponges have a skeleton composed of spicules (calcium carbonate or silica) or spongin (a protein), or both. About 10,000 marine species and 150 that live in fresh water.

Phylum Placozoa

A single marine species, *Trichoplax adherens,* which resembles a very large amoeba but is actually composed of thousands of cells, of which the surface ones are ciliated; lacks tissues, organs, and organ systems. Heterotrophic; both sexual and asexual reproduction are known.

SUBKINGDOM EUMETAZOA

Animals with definite symmetry, either radial or bilateral, and which have definite tissues and usually organs.

Phylum Cnidaria

Corals, jellyfish, hydras. Radially symmetrical animals that mostly have distinct tissues; two basically different body forms: polyps and medusae. Digestive cavity (gastrovascular cavity, or coelenteron) has only one opening. Cnidocysts, specialized stinging cells, occur only in the members of this phylum. Nearly all marine, a few in fresh water. About 10,100 species.

Class Hydrozoa

Hydroids; most have both polyp and medusa stages in their life cycle, with the polyp stage dominant. Most hydroids are marine and colonial. About 2700 species.

Class Scyphozoa

Jellyfish, medusa stage dominant, with the polyp stage small, inconspicuous, and simple in structure. About 200 species.

Class Anthozoa

Corals, sea anemones, and their relatives; only polyps occur in the members of this class, and they may be colonial or solitary. About 6200 species.

Phylum Ctenophora

Comb jellies and sea walnuts; translucent, gelatinous, free-swimming, spherical or ribbonlike, radially symmetrical animals with a gastrovascular cavity and an anal pore. Specialized tentacles with colloblasts in one group. About 90 species, now usually thought not to be closely related to the cnidarians.

Phylum Mesozoa

Bilaterally symmetrical acoelomates; mesozoans are minute, wormlike animals composed of very few cells, with an outer layer of 20 to 30 ciliated cells. A few dozen species.

Phylum Platyhelminthes

Flatworms; bilaterally symmetrical acoelomates. Flatworms are the simplest animals that have organs; they lack a circulatory system. Flatworm guts have only one opening; they are mostly hermaphroditic, with complex reproductive systems. About 13,000 species.

Class Turbellaria

Free-living flatworms; they move from place to place by means of their ciliated epidermis, and have eyespots, which are absent in their parasitic relatives. Abundant in fresh water and the sea.

Class Trematoda

Flukes; parasitic flatworms with a digestive tract, and often complex life cycles that involve two or more hosts.

Class Cestoda

Tapeworms; parasitic flatworms that lack a digestive tract and absorb food through their body walls; complex life cycles.

Phylum Rhynchocoela

Ribbon worms; bilaterally symmetrical, large, ribbon-shaped or thread-shaped aquatic, acoelomate worms, mainly marine, with a long, extensible proboscis and complete digestive systems. About 650 species.

Phylum Nematoda

Nematodes, eelworms, and roundworms; ubiquitous, bilaterally symmetrical, cylindrical, unsegmented, pseudocoelomate worms, including many important parasites of plants and animals. More than 12,000 described species, but the actual number is probably 500,000 or more species.

Phylum Nematomorpha

Horsehair worms; threadlike, bilaterally symmetrical, pseudocoelomate worms. Larval horsehair worms are parasitic in arthropods; the adults occur in fresh water or (one genus) in the sea. About 230 species.

Phylum Rotifera

Rotifers; small, wormlike or spherical, bilaterally symmetrical, pseudocoelomate animals with a crown of cilia. Most rotifers occur in fresh water, with a few in soil and a few in the sea. More than 1500 species.

Phylum Loricifera

Minute, bilaterally symmetrical, pseudocoelomate animals that live in the spaces between grains of sand in the sea; mouthparts consist of a unique flexible tube. Several species; the discovery of this phylum was announced in 1983.

Phylum Mollusca

Mollusks; bilaterally symmetrical, protostome coelomate animals that occur in marine, freshwater, and terrestrial habitats. Mollusks have a visceral mass and a muscular foot that is used in locomotion; many also have a head. Many mollusks form a shell, and all except for the bivalves have a radula, a rasping, tonguelike organ used for scraping, drilling, or capturing prey; the circulatory system consists of a heart and, usually, an open system through which the blood circulates freely. At least 110,000 species.

Class Polyplacophora

Chitons; marine mollusks with eight overlapping calcareous dorsal plates embedded in the mantle. The chitons are grazing herbivores. About 600 species.

Class Gastropoda

Snails and slugs; most gastropods have a spiral shell and a head with one or two pairs of tentacles. Slugs are snails that have lost their shell during the course of evolution. Of the approximately 80,000 species, nearly half are terrestrial.

Class Bivalvia

Bivalves—clams, scallops, oysters, mussels, and related mollusks—that have two shells, hinged together, and a wedge-shaped foot; bivalves lack distinct heads and radulas, and are usually sessile filter-feeders that disperse from place to place largely as larvae. About 10,000 species.

Class Cephalopoda

Octopuses, squids, and nautilus; active, intelligent marine predators in which the foot has evolved into a series of tentacles—8 in octopuses, 10 in squids, and 80 to 90 in the nautilus. The cephalopods have two horny jaws, highly developed eyes, and complex, efficient nervous systems. The shell is internal (squids), external (nautilus), or absent (octopuses). More than 600 species.

Phylum Annelida

Annelids; segmented, bilaterally symmetrical, protostome coelomates; the segments are divided internally by septa. Cerebral ganglion (brain) well developed; circulatory system closed; digestive tract complete, one-way. About 12,000 species.

Class Polychaeta

Polychetes; clamworms, plumed worms, scaleworms, and their relatives. Mainly marine worms with a distinct head and specialized sense organs, including eyes. Fleshy, paddlelike flaps called parapodia, used in locomotion, occur on most segments. The trochopore larvae are free swimming. About 8000 species.

Class Oligochaeta

Earthworms; terrestrial, freshwater, and marine annelids with fewer setae than the polychetes and no parapodia. Earthworms lack a distinct head, and their sense organs are not as specialized as those of the polychetes. About 3100 species.

Class Hirundinea

Leeches; dorsiventrally flattened external parasites, predators, or scavengers, with suckers at one or both ends of the body. About 300 species.

Phylum Arthropoda

Arthropods; bilaterally symmetrical protostome coelomates with a segmented body, chitinous exoskeleton, complete digestive tract, dorsal brain and paired nerve cord, and jointed appendages. Arthropods are the largest phylum of animals, with nearly a million species described and many more to be found.

Subphylum Chelicerata
Class Arachnida

Arachnids; spiders, mites, ticks, and scorpions. Largely terrestrial, carnivorous, air-breathing animals with chelicerae (pincers or fangs), pedipalps (usually sensory, but often used for catching and handling prey), and four pairs of walking legs. About 57,000 species.

Class Merostomata

Horseshoe crabs. Marine chelicerates with five pairs of walking legs, compound eyes, and book gills. Three genera and four species.

Class Pycnogonida

Sea spiders; small marine chelicerates that are mostly external parasites or predators on other animals. Sea spiders have a sucking proboscis and four to six pairs of legs. More than 1000 species.

Subphylum Crustacea
Class Crustacea

Crustaceans; lobsters, crayfish, shrimps, crabs, and many others. Mainly aquatic mandibulate animals with biramous (fundamentally two-branched) appendages; many crustaceans have a distinctive, free-swimming nauplius larva. Crustaceans have two pairs of antennae and legs on both their thorax and abdomen; many have compound eyes. About 35,000 species.

Subphylum Uniramia
Class Chilopoda

Centipedes; active, mandibulate predators with a head region followed by 15 to 177 body segments, each with a pair of legs. About 2500 species.

Class Diplopoda

Millipedes; mostly herbivorous, mandibulate animals with a head region followed by 20 to 200 segments, each with two pairs of legs. More than 10,000 named species.

Class Insecta

Insects; by far the largest group of organisms. Mostly terrestrial mandibulates, with a body divided into three regions: head, thorax, and abdomen. Insects have a complex series of mouthparts, which are highly specialized in different orders; they have compound eyes and one pair of antennae on the head. The six legs are located on the thorax, as are the wings, if present. Insects breathe by means of tracheae and excrete their wastes by a system of Malpighian tubules. Many have complex metamorphosis. More than 750,000 described species, but several times that many doubtless await discovery.

Phylum Onychophora

Wormlike protostome animals with a chitinous exoskeleton and tracheae; embryo development identical to that of the annelids and Uniramia. Onychophora appear to link annelids with Uniramia; the 70 living species appear to be evolutionary relics.

Phylum Echinodermata

Echinoderms; sea stars, brittle stars, sand dollars, sea cucumbers, and sea urchins. Complex deuterostome, coelomate, marine animals that are more or less radially symmetrical as adults; calcareous plates called ossicles are more or less abundant in the epidermis. Water vascular system extends through perforated plates as tube feet, and is a specialized feature of the phylum. About 6000 living species.

Class Crinoidea

Crinoids; sea lilies and feather stars. Filter-feeding echinoderms in which the mouth and anus are located in a disk on the upper surface; 5 to more than 200 feathery arms are located around the margins of this disk. About 600 species.

Class Asteroidea

Sea stars; star-shaped echinoderms, with the five to many arms set off more or less sharply from the disk. The mouth and anus are directed downward, and the animals move by means of rows of tube feet on the arms. About 1500 species.

Class Ophiuroidea

Brittle stars; star-shaped echinoderms with very slender, long, often spiny, highly flexible arms that are sharply set off from the disk. About 2000 species.

Class Echinoidea

Sea urchins and sand dollars; echinoderms that lack distinct arms and have a rigid external covering. Echinoids walk by means of their tube feet or jointed spines. About 950 species.

Class Holothuroidea

Sea cucumbers; soft, sluglike echinoderms that lie on their sides. About 1500 species.

Phylum Chaetognatha

Arrow worms; rapid-swimming, translucent, bilaterally symmetrical, marine, deuterostome, coelomate worms with large eyes, powerful jaws, and a head ringed with numerous sharp, movable hooks. About 70 species.

Phylum Hemichordata

Acorn worms; soft-bodied, bilaterally symmetrical, burrowing, marine, deuterostome, coelomate worms with three-segmented bodies (proboscis, collar, and trunk). About 90 species.

Phylum Chordata

Chordates; bilaterally symmetrical, deuterostome, coelomate animals that at some stage of their development have a notochord, pharyngeal slits, a hollow nerve cord on their dorsal side, and a tail. The best-known group of animals; about 45,000 species.

Subphylum Urochordata

Tunicates; larval tunicates are free-swimming and have a notochord and nerve cord, structures that are absent in the sessile, usually saclike adults, which also lack any sign of segmentation or of a body cavity. Tunicates are marine chordates that obtain their food by ciliary action. About 1250 species.

Subphylum Cephalochordata

Lancelets; marine fishlike animals with a permanent notochord, a nerve cord, a pharynx with gill slits, and no internal skeleton. Lancelets are filter-feeders that draw water through their pharyngeal slits by ciliary action. About 23 species.

Subphylum Vertebrata

Vertebrates; notochord replaced by cartilage or bone, forming a segmented vertebral column (the backbone). Vertebrates have a distinct head with a skull and brain. Their hollow dorsal nerve cord is usually protected within a U-shaped groove formed by paired projections from the vertebral column. About 43,700 species.

Class Agnatha

Agnathans; lampreys and hagfishes. Naked, eel-like, jawless fishes with a cartilaginous skeleton; the agnathans lack scales, bones, and fins. Parasites on other fishes, or scavengers. About 63 species.

Class Chondrichthyes

Sharks, skates, and rays. Almost entirely marine fishes with a cartilaginous skeleton, efficient fins, complex copulatory organs, and small, pointed scales (denticles); but lacking air bladders. About 850 species.

Class Osteichthyes

Bony fishes; the members of this class, which are abundant both in the sea and in fresh water, have bony skeletons, efficient fins, scales, and usually air bladders (by means of which they regulate their density and therefore their level in the water). More than 18,000 species.

Class Amphibia

Salamanders, frogs, and toads. Tetrapod, egg-laying, ectothermic vertebrates that lack scales; amphibians respire with gills as larvae and with lungs as adults. They have incomplete double circulation. Amphibians were the first terrestrial vertebrates; they still depend on a moist environment for at least a portion of their life cycles. About 2800 species.

Class Reptilia

The reptiles; lizards, snakes, turtles, and crocodiles. Tetrapod, ectothermic vertebrates with an amniotic egg; reptiles have lungs and are covered with scales; most are fully terrestrial, although some are aquatic. Reptiles have incomplete double circulation. The four legs are absent in snakes and some lizards. Nearly 6000 species.

Class Aves

Birds. Tetrapod, endothermic vertebrates in which the forelimbs are modified into wings; most are capable of flight, and all lay amniotic eggs. Birds have lungs and are fully terrestrial, although some live in water. They have complete double circulation; feathers are a characteristic feature of the class. About 9000 species.

Class Mammalia

Mammals. Tetrapod, endothermic vertebrates with complete double circulation and usually hairy skins. The forelimbs are modified into wings in bats, and all four limbs are modified into flippers in some aquatic mammals. Monotremes lay eggs; marsupials retain their young in a marsupium, or pouch, for a prolonged period; and the great majority of mammals—the placental mammals—nourish their young in the womb by means of a specialized structure, the placenta, modified from the amniotic egg characteristic of their ancestors. About 4500 species.

CHAPTER 2

1. isotopes
2. ions
3. d
4. 2
5. 8
6. b
7. e, d, b, a, c
8. 8
9. carbon
10. N, O, C, H
11. d
12. 18 grams

CHAPTER 3

1. c
2. c
3. reducing
4. e
5. coacervates
6. b
7. 2 billion years
8. ^{12}C
9. 2.5
10. oxygen gas
11. cyanobacteria
12. a

CHAPTER 4

1. c
2. false
3. c
4. false
5. d
6. two
7. 6
8. two
9. d
10. 64
11. UACG
12. RNA

CHAPTER 5

1. yes
2. c
3. autotroph
4. a
5. positive
6. true
7. nucleus
8. aerobic
9. no
10. human
11. one
12. swell

CHAPTER 6

1. e
2. microtubule
3. tubulin
4. a
5. false
6. c
7. nucleoli
8. e
9. b
10. b
11. chloroplast
12. two

CHAPTER 7

1. no
2. 9
3. c
4. b
5. no
6. a
7. myosin
8. true
9. d
10. 75 nm
11. actin
12. microfilaments

CHAPTER 8

1. binary fission
2. e
3. fungi; protists
4. d
5. centromeres
6. 1
7. 92
8. animal
9. b
10. telophase
11. anaphase
12. cytokinesis

CHAPTER 9

1. c
2. d
3. a
4. true
5. d
6. false
7. b
8. b
9. 3
10. true
11. a, d, e
12. a, e

CHAPTER 10

1. c
2. negative
3. true
4. a
5. anaerobic
6. f
7. 3
8. heterocysts
9. reduce
10. oxidize
11. e
12. true

CHAPTER 11

1. b
2. b
3. 2
4. b
5. true
6. pyruvate
7. c
8. 36
9. 3
10. oxygen
11. no
12. b

CHAPTER 12

1. short
2. violet
3. a
4. d
5. RuDP
6. NADPH2
7. d
8. H$_2$O
9. d
10. c
11. *b*
12. inside

CHAPTER 13

1. c
2. Koelreuter
3. b
4. d
5. dominant; recessive
6. 3:1
7. a
8. b
9. 25%
10. 25%
11. a
12. c
13. c
14. e
15. false
16. epistasis
17. true
18. false
19. linkage

CHAPTER 14

1. false
2. haploid
3. reduction meiosis
4. once, before prophase I
5. sister chromatids
6. prophase I
7. e
8. synaptonemal
9. chiasma
10. metaphase I
11. anaphase I
12. 4
13. no; must show subsequent segregation
14. yes
15. c
16. true
17. c
18. heterochromatin
19. chromomere
20. centromere

CHAPTER 15

1. a6, b9, c8, d2, e4, f3, g1, h5, i7
2. a9, b2, c10, d1, e6, f5, g7, h3, i8, j4
3. b
4. yes
5. S (protein); P (DNA)
6. d
7. thymine
8. helix
9. b
10. 1:3
11. b
12. two

CHAPTER 16

1. transcription: e, f, h; translation: a, b, c, d, g, i
2. e
3. true
4. ribosome
5. c
6. a
7. amino acid and tRNA
8. 20
9. no
10. three
11. no
12. 5′ to 3′

CHAPTER 17

1. b
2. thymine; cytosine
3. true
4. unequal crossing-over
5. true
6. insertional inactivation
7. c
8. p (a, b, e) e (c, d, f, g, h, i)
9. c, f, e, b, a, d
10. all but c
11. false
12. no

CHAPTER 18

1. restriction endonuclease
2. c
3. CTT
4. e
5. c
6. b
7. sarcoma
8. metastases
9. ⅓
10. smoking
11. tyrosine
12. 40

CHAPTER 19

1. a
2. natural selection
3. d
4. d
5. b
6. a
7. b
8. Homology
9. b
10. d
11. b
12. c

CHAPTER 20

1. macroevolution
2. b
3. Fitness
4. a
5. a
6. c
7. c
8. Genetic drift
9. c
10. a
11. e
12. true

CHAPTER 21

1. niche
2. false
3. many
4. Geographic isolation
5. seasonally
6. mechanical
7. Postzygotic isolating
8. clones
9. true
10. c
11. false
12. c

CHAPTER 22

1. c
2. a
3. true
4. type
5. e
6. d
7. c
8. family
9. c, d
10. phylogeny
11. artificial
12. Advanced; primitive

CHAPTER 23

1. b
2. Isotropic ratios
3. c
4. acritarchs
5. Cambrian
6. true
7. a
8. c
9. Pangaea
10. Mesozoic
11. b
12. true

CHAPTER 24

1. e
2. prosimians
3. true
4. false
5. a
6. false
7. c
8. Brachiation
9. true
10. true
11. *Australopithecus*
12. about 500,000 years

CHAPTER 25

1. c
2. c
3. c
4. c
5. e
6. c
7. d
8. Jenner
9. rearrangement of genes
10. Viroids
11. Poxviruses
12. d

CHAPTER 26

1. false
2. b
3. false
4. e
5. true
6. true
7. true
8. mutation
9. true
10. true
11. true
12. c

CHAPTER 27

1. true
2. e
3. cyanobacteria; *Prochloron; Heliobacterium*
4. yes
5. yes
6. e
7. false
8. Gametangia
9. zygote
10. schizogony
11. a
12. false

CHAPTER 28

1. false
2. b
3. d
4. d
5. a
6. false
7. meiosis
8. zygospores
9. ascus
10. true
11. false
12. d

CHAPTER 29

1. true
2. a, d, e
3. spore, gametes, zygote
4. c
5. xylem; phloem
6. alternation of generations
7. d
8. true
9. b
10. false
11. true
12. calyx and corolla

CHAPTER 30

1. b
2. true
3. leaves; stamens; sepals
4. a
5. false
6. c
7. c
8. b
9. false
10. cross
11. false
12. true

CHAPTER 31

1. apical meristems
2. 2
3. a
4. c
5. e
6. phelloderm; phellogen; cork (phellem)
7. girth; length
8. parallel
9. pericycle
10. e
11. true
12. true

CHAPTER 32

1. haploid
2. cell walls
3. false
4. false
5. a
6. divide
7. endosperm
8. b
9. stratification
10. true
11. leaf primordia
12. d

CHAPTER 33

1. darkened
2. a
3. adenine, buds
4. f
5. e
6. amyloplasts, apical meristem
7. e
8. d
9. true
10. true
11. carbon dioxide; ethylene
12. false

CHAPTER 34

1. c
2. adhesion; cohesion
3. water; solute; pressure
4. a
5. energy
6. true
7. b
8. c
9. guttation
10. translocation; source; roots
11. true
12. phosphorus; nitrogen

CHAPTER 35

1. true
2. a
3. Parazoa
4. nematocysts
5. bilateral
6. false
7. pseudocoelomates
8. a
9. pharynx
10. true
11. false
12. has body cavity

CHAPTER 36

1. mesoderm
2. c
3. b
4. protostomes; deuterostomes
5. bilateral
6. a
7. false
8. d
9. nephridia
10. true
11. d
12. clitellum

CHAPTER 37

1. e
2. Tagmata
3. molting, or ecdysis
4. false
5. c
6. e
7. Malpighian tubules
8. neoteny
9. 3
10. pheromones
11. yes
12. b

CHAPTER 38

1. a
2. true
3. crinoids
4. no
5. Chordata
6. e
7. true
8. true
9. tunic
10. false
11. 7
12. jaws

CHAPTER 39

1. d
2. c
3. swim bladders
4. e
5. amniotic
6. e
7. d
8. b
9. a
10. false
11. d
12. false

CHAPTER 40

1. a
2. true
3. no
4. a
5. a
6. c
7. hypothalamus
8. follicular; luteal
9. acrosome
10. e
11. d
12. c

CHAPTER 41

1. d
2. cleavage
3. a
4. c
5. false
6. b
7. b
8. neural crest
9. true
10. induction
11. false
12. 1b, 2a, 3d, 4a, 5a

CHAPTER 42

1. c
2. a
3. actin, myosin
4. myofibril
5. b
6. b
7. false
8. smooth
9. b
10. an axon
11. c
12. a2, b5, c1, d4, e3, f6

CHAPTER 43

1. proteases
2. c
3. a, 1 and 3; b, 1; c, 3
4. zymogens
5. b
6. b
7. a
8. bile salts
9. a
10. a
11. a
12. vitamins

CHAPTER 44

1. e
2. b
3. false
4. gills
5. d
6. countercurrent exchange
7. false
8. glottis
9. diaphragm
10. a
11. hemoglobin
12. d

CHAPTER 45

1. e
2. c
3. true
4. b
5. lymphatic
6. a
7. platelets
8. SA node
9. b
10. His
11. systolic
12. d

CHAPTER 46

1. axon
2. d
3. glial
4. negative
5. true
6. b
7. b
8. cerebellum
9. d
10. a
11. true
12. d

CHAPTER 47

1. nociceptors
2. d
3. statocysts
4. fluid
5. a
6. basilar
7. sonar
8. true
9. a
10. fovea
11. c
12. pit

CHAPTER 48

1. true
2. a
3. false
4. adrenalin
5. b
6. nephrons
7. Malpighian corpuscle
8. True
9. b
10. true
11. loop of Henle
12. d

CHAPTER 49

1. c
2. hypothalamus
3. no
4. c
5. a
6. classical
7. b
8. agonistic
9. yes
10. aggression; sociality
11. less
12. operant

CHAPTER 50

1. Biomes
2. c
3. b
4. true
5. e
6. thermocline
7. false
8. c
9. e
10. e
11. a
12. estivation

CHAPTER 51

1. b
2. e
3. Nitrogen
4. true
5. b
6. b
7. biomass
8. a
9. a
10. e
11. phosphates
12. d

CHAPTER 52

1. c
2. b
3. carrying capacity
4. true
5. b
6. a
7. a, c, b
8. true
9. e
10. false
11. b
12. a

CHAPTER 53

1. e
2. b
3. coevolution
4. d
5. a
6. aposematic
7. Batesian
8. b
9. c
10. false
11. Symbiotic
12. b

CHAPTER 54

1. 20
2. no
3. true
4. d
5. e
6. solar
7. d
8. d
9. b
10. a
11. e
12. true

GLOSSARY

abdomen The posterior portion of the body.

abortion The removal or induced elimination of the embryo from the uterus before the third trimester of pregnancy.

abscisic acid (ABA) (L. *ab*, away, off + *scissio*, dividing) A plant hormone with a variety of inhibitory effects; brings about dormancy in buds, maintains dormancy in seeds, effects stomatal closing, and is involved in the geotropism of roots; also known as the "stress hormone."

abscission (L. *ab*, away, off + *scissio*, dividing) In vascular plants, the dropping of leaves, flowers, fruits, or stems at the end of a growing season, as the result of formation of a layer of specialized cells (the abscission zone) and the action of a hormone (ethylene).

absorption (L. *absorbere*, to swallow down) The movement of water and of substances dissolved in water into a cell, tissue, or organism.

absorption spectrum The characteristic array of wavelengths (colors) of light that a particular substance absorbs.

accessory pigment In plants, a secondary pigment that captures light energy and transfers it to chlorophyll *a*.

acetylcholine The most important of the numerous chemical neurotransmitters responsible for the passing of nerve impulses across synaptic junctions; the neurotransmitter in neuromuscular (nerve-muscle) junctions.

acid A proton donor; a substance that dissociates in water, releasing hydrogen ions (H^+) and so causing a relative increase in the concentration of H^+ ions; having a pH in solution of less than 7; the opposite of "base."

acoelomate (Gr. *a*, not + *koiloma*, cavity) Without a coelom, as in flatworms and proboscis worms.

acquired immune deficiency syndrome (AIDS) An infectious and usually fatal human disease caused by a retrovirus, HTLV III, which attacks T cells. The virus multiplies within and kills individual T cells, releasing thousands of progeny that infect and kill other T cells, until no T cells remain, leaving the affected individual helpless in the face of microbial infections because his or her immune system is now incapable of marshaling a defense against them. *See* T cell.

actin (Gr. *actis*, a ray) One of the two major proteins that make up microfilaments (the other is myosin); the principal protein constituent in contractile tissues containing many microfilaments, such as muscle.

action potential A transient, all-or-none reversal of the electric potential across a membrane; its strength depends only on characteristics of the membrane (in nerves, the diameter) and is independent of the strength of the stimulus that triggers it; in neurons, an action potential initiates transmission of a nerve impulse.

action spectrum The characteristic array of wavelengths (colors) of light that elicit a particular reaction or response.

activation energy The energy that must be possessed by a molecule in order for it to undergo a specific chemical reaction.

active site The region of an enzyme surface to which a specific set of substrates binds, lowering the activation energy required for a particular chemical reaction and so facilitating it.

active transport The pumping of individual ions or other molecules across a cellular membrane from a region of lower concentration to one of higher concentration (that is, against a concentration gradient); because the ion or molecule is moved in a direction other than the one in which simple diffusion would take it, this transport process requires energy, which is typically supplied by the expenditure of ATP.

adaptation (L. *adaptare*, to fit) A peculiarity of structure, physiology, or behavior that promotes the likelihood of an organism's survival and reproduction in a particular environment.

adaptive radiation The evolution of several divergent forms from a primitive and unspecialized ancestor.

adductor (L. *ad*, to + *ducere*, to lead) A pulling muscle; in vertebrates, a muscle that draws a limb toward a median axis; in mollusks, a muscle that draws the two valves of the shell together.

adenine (Gr. *aden*, gland) A purine base; a nitrogen-containing ring compound present in nucleic acids (DNA and RNA), in adenosine phosphates (ADP and ATP), and in nucleotide coenzymes.

adenosine A nucleotide composed of adenine and a ribose sugar.

adenosine diphosphate (ADP) A nucleotide consisting of adenine, ribose sugar, and two phosphate groups; formed by the removal of one phosphate from an ATP molecule; the formation of ADP from ATP releases usable metabolic energy.

adenosine monophosphate (AMP) A nucleotide consisting of adenine, ribose sugar, and one phosphate group; can be formed by the removal of two phosphate groups from an ATP molecule, a reaction that releases energy and is often found at irreversible steps of metabolic pathways. The formation of AMP from ATP often renders a reaction irreversible because the two-phosphate ("pyrophosphate") reaction product is readily cleaved into individual phosphate units that are not easily rejoined.

adenosine triphosphate (ATP) A nucleotide consisting of adenine, ribose sugar, and three phosphate groups. ATP is the energy currency of cell metabolism in all organisms. ATP is formed from ADP + P in an enzymatic reaction that traps chemical energy released by metabolic processes or light energy captured in photosynthesis. Upon hydrolysis, ATP loses one phosphate and one hydrogen to become adenosine diphosphate (ADP), releasing energy in the process.

adrenal gland (L. *ad*, near + *renes*, kidney) A hormone-producing gland; a vertebrate endocrine gland whose cortex (outer surface) secretes cortisol, aldosterone, and other steroid hormones, and whose medulla (inner core) is the source of adrenaline and noradrenaline.

adrenaline (L. *ad*, to + *renalis*, pertaining to kidneys) A hormone produced by the medulla of the adrenal gland. Adrenaline is responsible for the physiological changes associated with the alarm response: it increases the concentration of sugar in the blood, raises blood pressure and heart rate, and increases muscular power and resistance to fatigue; also a neurotransmitter across synaptic junctions. Sometimes called "epinephrine."

adsorption (L. *ad*, to + *sorbere*, to suck in) The adhesion of molecules to a solid.

adventitious (L. *adventicius*, not properly belonging to) Referring to a structure arising from an unusual place, such as stems from roots or roots from stems.

aerobic (Gr. *aer*, air + *bios*, life) Requiring free oxygen; any biological process that can occur in the presence of gaseous oxygen (O_2).

aerobic pathway A metabolic pathway at least one step of which is an oxidation/reduction reaction that depends on oxygen gas as an electron acceptor; such a pathway will not proceed in the absence of oxygen.

afferent (L. *ad*, to + *ferre*, to carry) Adjective meaning leading inward or bearing toward some organ, for example, nerves conducting impulses toward the brain, or blood vessels carrying blood toward the heart; opposite of "efferent."

agonistic behavior (Gr. *agonistes*, combatant) Any behavior involving fighting; an aggressive act or threat, or retreat from such an act or threat.

aldosterone (Gr. *aldainein*, to nourish + *stereos*, solid) A hormone produced by the adrenal cortex that regulates the concentration of ions in the blood by stimulating the reabsorption of sodium and the excretion of potassium by the kidney.

alga, pl. **algae** A unicellular or simple multicellular photosynthetic organism lacking multicellular sex organs; the blue-green algae, or cyanobacteria, are photosynthetic bacteria.

alkaloids Nitrogen-containing ring compounds that are produced by plants and are physiologically active in vertebrates; examples are nicotine, quinine, and strychnine.

allantois (Gr. *allas*, sausage + *eidos*, form) A membrane of the amniotic egg that functions in respiration and excretion in birds and reptiles and plays an important role in the development of the placenta in most mammals.

allele (Gr. *allelon*, of one another) One of two or more alternative states of a gene.

allele frequency The relative proportion of a particular allele among the chromosomes carried by individuals of a population. Not equivalent to "gene frequency," although the two are sometimes confused.

allometry (Gr. *allos*, other + *metros*, measure) Relative growth of a part in relation to the whole organism.

allosteric interaction (Gr. *allos*, other + *stereos*, shape) A change in the shape of a protein resulting from the binding to the protein of a nonsubstrate molecule, called an effector. The new shape typically has different properties, becoming activated or inhibited. Cells control the activities of enzymes, transcription regulators, and other proteins by modulating the cellular concentration of particular allosteric effectors.

alpha helix (Gr. *alpha*, first + L. *helix*, spiral) A right-handed coil; in DNA, the usual form of biological DNA molecules (the one originally proposed in 1961 by Watson and Crick), although left-handed forms (called "Z DNA") also occur; in proteins, a secondary structure composed of a regular right-handed coil.

alternation of generations A reproductive cycle in which a haploid (1*n*) phase, the gametophyte, gives rise to gametes which, after fusion to form a zygote, germinate to produce a diploid (2*n*) phase, the sporophyte. Spores produced by meiotic division from the sporophyte give rise to new gametophytes, completing the cycle.

altruism Self-sacrifice for the benefit of others; in formal terms, a behavior that increases the fitness of the recipient while reducing the fitness of the altruistic individual.

alveolus, pl. **alveoli** (L., a small cavity) One of the many small, thin-walled air sacs within the lungs in which the bronchioles terminate.

amino acids (Gr. *Ammon*, referring to the Egyptian sun god, near whose temple ammonium salts were first prepared from camel dung) Nitrogen-containing organic molecules. Amino acids are the units, or "building blocks," from which protein molecules are assembled; all amino acids have the same underlying structure, $H_2N-RCH-COOH$, where R stands for a side group that is different for each kind of amino acid; 20 different amino acids combine to make proteins.

amniocentesis (Gr. *amnion*, membrane

around the fetus + *centes*, puncture) Examination of a fetus indirectly, by the carrying out of tests on cell cultures grown from fetal cells obtained from a sample of the amniotic fluid surrounding the developing embryo; a procedure often carried out in pregnant women over the age of 35, since the examination of fetal chromosomes readily reveals Down syndrome if it is present, permitting the choice of a therapeutic abortion.

amnion (Gr., membrane around the fetus) The innermost of the extraembryonic membranes; the amnion forms a fluid-filled sac around the embryo in amniotic eggs.

amniotic egg An egg that is isolated and protected from the environment by a more or less impervious shell during the period of its development and that is completely self-sufficient, requiring only oxygen from the outside.

amoebocyte (Gr. *amoibe*, change + *kytos*, hollow vessel) Any free body cell that can move by means of pseudopodia, as amoebas do.

amoeboid (Gr. *amoibe*, change) Moving or eating by means of pseudopodia (temporary cytoplasmic protrusions from the cell body).

amylase (L. *amylum*, starch + *-ase*, suffix meaning enzyme) An enzyme that breaks down (hydrolyzes) starch (amylose) into smaller units.

anabolism (Gr. *ana*, up + *bolein*, to throw) The biosynthetic or constructive part of metabolism; those chemical reactions involved in biosynthesis; opposite of "catabolism."

anaerobic (Gr. *an*, without + *aer*, air + *bios*, life) Any process that can occur without oxygen, such as anaerobic fermentation or H_2S photosynthesis; anaerobic organisms can live without free oxygen; obligate anaerobes cannot live in the presence of oxygen; facultative anaerobes can live with or without oxygen.

analogous (Gr. *analogos*, proportionate) Structures that are similar in function but different in evolutionary origin, such as the wing of a bat and the wing of a butterfly.

anatomy The study of the internal structure of organisms; opposed to morphology, the study of the external structure of organisms.

androecium (Gr. *andros*, man + *oilos*, house) The floral whorl that comprises the stamens.

androgen (Gr. *andros*, man + *genos*, origin, descent) Any of a group of vertebrate male sex hormones, such as testosterone.

aneuploidy (Gr. *an*, without + *eu*, good + *ploid*, multiple of) An organism whose cells have lost or gained a chromosome; cells that have a chromosome number of 2*n* − 1 (one fewer than normal) or 2*n* + 1 (one extra chromosome). Down syndrome, which results from an extra copy of human chromosome 21, is an example of aneuploidy in humans.

Ångström (after J. Ångström, a Swedish physicist, 1814-1874) An obsolete measure of length over atomic dimensions; one angstrom (Å) unit corresponds to 10^{-10} meter, or a tenth of a nanometer.

anhydrase (Gr. *an*, not + *hydor*, water + *-ase*, enzyme suffix) An enzyme that catalyzes a reaction that involves the removal of water from a substrate molecule. Carbonic anhydrase, for example, catalyzes the removal of water from carbonic acid, producing carbon dioxide.

anion (Gr. *anienae*, to go up) A negatively charged ion.

annual (L. *annus*, year) A plant that completes its life cycle (from seed germination to seed production) and dies within a single growing season.

annulus (L., ring) Any ringlike structure.

antenna (L., a long spike used to suport a particular kind of sail) Paired sensory appendages that occur on the head of many kinds of arthropods.

anterior (L. *ante*, before) Located before or towards the front; in animals, the head end of an organism.

anther (Gr. *anthos*, flower) In angiosperm flowers, the pollen-bearing portion of a stamen.

antheridium, pl. **antheridia** A sperm-producing organ; antheridia occur in some plants and protists.

anthropoid (Gr. *anthropos*, human) Closely related to humans; a higher primate (monkeys, apes, and humans).

antibiotic (Gr. *anti*, against + *biotikos*, pertaining to life) An organic molecule that is produced by a microorganism and kills or retards the growth of other microorganisms.

antibody (Gr. *anti*, against) A protein called an immunoglobulin that is produced by lymphocytes in response to a foreign substance (antigen) and released into the bloodstream.

anticodon The three-nucleotide sequence at the end of a transfer RNA molecule that is complementary to, and base pairs with, an amino-acid–specifying codon in messenger RNA.

antidiuretic hormone (ADH) (Gr. *anti*, against + *diurgos*, thoroughly wet + *hormaein*, to excite) A peptide hormone produced by the hypothalamus that promotes the reabsorption of water from the nephrons of the kidneys; also called vasopressin. Drinking excessive alcohol increases urine output because alcohol inhibits ADH.

antigen (Gr. *anti*, against + *genos*, origin) A foreign substance, usually a protein or polysaccharide, that stimulates one or more lymphocyte cells to begin to proliferate and secrete specific antibodies that bind to the foreign substances, labeling it as foreign and destined for destruction.

anus The terminal opening of the gut; the solid residues of digestion are eliminated through the anus.

aorta (Gr. *aeirein*, to lift) The major artery of vertebrate systemic blood circulation; in mammals, carries oxygenated blood away from the heart to all regions of the body except the lungs.

apical (L. *apex*, top) Pertaining to the tip or apex.

apical dominance (L. *apex*, top) In plants, the influence of a terminal bud in suppressing the growth of lateral buds by producing hormones.

apical meristem (L. *apex*, top + Gr. *meristos*, divided) In vascular plants, the growing point at the tip of the root or stem.

arboreal (L. *arbor*, tree) Living in trees.

archegonium, pl. **archegonia** (Gr. *archegonos*, first of a race) In bryophytes and some vascular plants, the multicellular egg-producing organ.

archenteron (Gr. *arche*, beginning + *enteron*, gut) The principal cavity of a vertebrate embryo in the gastrula stage; lined with endoderm, it opens to the outside and represents the future digestive cavity.

artery A thick-walled, elastic tube, lined with muscle, that carries blood from the heart to the tissues.

artificial selection The breeding of selected organisms for the purpose of producing descendants with desired traits.

ascospore A fungal spore produced within an ascus.

ascus pl. **asci** (Gr. *askos*, wineskin, bladder) A specialized cell, characteristic of the ascomycetes, in which two haploid nuclei fuse to produce a zygote that divides immediately by meiosis; at maturity, an ascus contains ascospores.

aseptate (Gr. *a-*, not + L. *septum*, fence) A term used to describe algal or fungal filaments lacking cross walls; an equivalent term is nonseptate.

asexual Without distinct sexual organs; an organism that reproduces without forming gametes—that is, without sex. Asexual reproduction, therefore, does not involve sex.

assimilation (L. *assimilatio*, bringing into conformity) In biology, the absorption by cells of nutrients and macromolecules, which are assembled into complex organic cell constituents.

atmospheric pressure (Gr. *atmos*, vapor + *sphaira*, globe) The weight of the earth's atmosphere over a unit area of the earth's surface; measured with a mercury barometer at sea level, this corresponds to the pressure required to lift a column of mercury 760 millimeters.

atom (Gr. *atomos*, indivisible) The smallest particle into which a chemical element can be divided and still retain the properties characteristic of the element; consists of a central core or nucleus composed of protons and neutrons, encircled by one or more electrons that move around the nucleus in characteristic orbits whose distance from the nucleus depends on their energy.

atomic number The number of protons in the nucleus of an atom; in an atom that does not bear an electric charge (i.e., one that is not an ion), the atomic number is also equal to the number of electrons.

atomic weight The weight of a representative atom of an element (estimated as the average weight of all its isotopes) relative to the weight of an atom of the most common isotope of carbon (which is by convention assigned the integral value of 12); the atomic weight of an atom is approximately equal to the number of protons plus neutrons in its nucleus.

ATP *See* adenosine triphosphate.

atrium (L., vestibule or courtyard) An antechamber; in the heart, a thin-walled chamber that receives venous blood and passes it on to the thick-walled ventricle; in the ear, the tympanic cavity.

auricle (L. *auricula*, a little ear) Any earlike lobe or process; the external ear, or pinna; one of the less muscular chambers of a heart.

autonomic nervous system (Gr. *autos*, self + *nomos*, law) The involuntary neurons and ganglia of the peripheral nervous system of vertebrates; regulates the heart, glands, visceral organs, and smooth muscle; subdivided into the sympathetic and parasympathetic divisions, whose effects oppose one another (one stimulating, the other inhibiting).

autoradiograph A photographic recording of the positions on a film image where radioactive decay of isotopes has occurred; if one attaches radioactive label to a specific molecule, an autoradiograph permits an investigator to locate precisely where in a cell the molecule becomes situated.

autosome (Gr. *autos*, self + *soma*, body) Any eukaryotic chromosome that is not a sex chromosome; autosomes are present in the same number and kind in both males and females of the species.

autotroph (Gr. *autos*, self + *trophos*, feeder) An organism able to build all the complex organic molecules that it requires as its own food source, using only simple inorganic compounds. Plants, algae, and some bacteria are autotrophs. Contrasts with *heterotroph*.

auxin (Gr. *auxein*, to increase) A plant hormone that controls cell elongation, among other effects.

auxotroph (L. *auxillium*, help + Gr. *trophos*, feeder) An organism that is nutritionally defective because of one or more mutations which destroy its ability to synthesize a particular molecule that must therefore be supplied for normal growth.

axon (Gr., axle) A process extending out from a neuron that conducts impulses away from the cell body.

axoneme (L. *axis,* axle + Gr. *nema,* thread) The microtubules in a cilium or flagellum, usually arranged as a circlet of nine pairs enclosing one central pair.

B cell A type of lymphocyte which, when confronted with a suitable antigen, is capable of secreting a specific antibody protein; each individual lymphocyte B cell is capable of secreting only one form of antibody, but few if any B lymphocytes secrete the same form of antibody.

bacillus, pl. **bacilli** (L. *baculum,* rod) A rod-shaped bacterium.

backcross The crossing of a hybrid individual with one of its parents or with a genetically equivalent individual; the most common type of backcross is a "testcross," a cross between an individual with a dominant phenotype (which might in principle be either homozygous for a dominant allele or heterozygous for that allele and a recessive one) with an individual that is homozygous for the recessive allele.

bacteriophage (Gr. *bakterion,* little rod + *phagein,* to eat) A virus that infects bacterial cells; also called a *phage.*

bark In plants, all tissues outside the vascular cambium in a woody stem.

basal body A self-reproducing cylinder-shaped cytoplasmic organelle composed of nine triplets of microtubules from which the flagella or cilia arise; identical in structure to the centriole, which is involved in mitosis and meiosis in most protists and animals and in some plants.

base A substance that dissociates in water, causing a decrease in the relative number of hydrogen ions (H^+), often by producing hydroxide ions (OH^-), which combine with hydrogen ions to form water and thus reduce the concentration of free hydrogen ions; having a pH greater than 7; the opposite of acid. *See* alkaline.

basidiospore A spore of the basidiomycetes; produced within and borne on a basidium after nuclear fusion and meiosis.

basidium, pl. **basidia** (L., a little pedestal) A specialized reproductive cell of the basidiomycetes, often club-shaped, in which nuclear fusion and meiosis occur.

behavior A coordinated neuromotor response to changes in external or internal conditions; a product of the integration of sensory, neural, and hormonal factors.

biennial A plant that normally requires two growing seasons to complete its life cycle. Biennials flower in the second year of their lives.

bilateral symmetry (L. *bi,* two + *lateris,* side; Gr. *symmetria,* symmetry) A body form in which the right and left halves of an organism are approximate mirror images of each other.

bile A solution of organic salts that is secreted by the vertebrate liver and temporarily stored in the gallbladder; emulsifies fats in the small intestine.

binary fission (L. *binarius,* consisting of two things or parts + *fissus,* split) Asexual reproduction by division of one cell or body into two equal, or nearly equal, parts.

binomial system (L. *bi,* twice, two + Gr. *nomos,* usage, law) A system of nomenclature in which the name of a species consists of two parts, the first of which designates the genus.

biodegradable Capable of being broken down by living organisms into simpler molecules; plastics are an example of stable materials that are not biodegradable and remain in the environment for a long time.

biogeochemical cycle (Gr. *bios,* life + *geo,* earth + *chemeia,* alchemy; *kyklos,* circle, wheel). The cyclic path of an inorganic substance, such as carbon or nitrogen, through an ecosystem.

biological clock Internal timing mechanism of organisms; many of these mechanisms are innate rhythms that are modulated by cyclic changes in the environment, in the same sense that an alarm clock is driven internally by a spring but the time it indicates is set by its user; internal timing mechanisms govern reproductive, physiological, and behavioral changes in almost all organisms.

bioluminescence Light production by living organisms; in bioluminescent organisms, proteins called luciferins, in the presence of oxygen, are converted by an enzyme called luciferase to oxyluciferins, with the liberation of light.

biomass (Gr. *bios,* life + *maza,* lump or mass) The weight of all the living organisms in a given population, area, or other unit being measured.

biome One of the major terrestrial ecosystems, characterized by climatic and soil conditions; the largest ecological unit.

biosphere (Gr. *bios,* life + *sphaira,* globe) That part of earth containing living organisms.

biosynthesis (Gr. *bios,* life + *synthesis,* a putting together) Formation by living organisms of complex organic molecules from simpler molecules.

biotic Pertaining to life.

bivalent (L. *bis,* twice + *valere,* to be strong) A pair of synapsed homologous chromosomes; in the first meiotic division, a tetrad.

blade The broad, expanded part of a leaf; also called the lamina.

blastocoel (Gr. *blastos,* a sprout + *koilos,* hollow) The central cavity of the blastula stage of vertebrate embryos.

blastocyst (Gr. *blastos,* a sprout + *cystos,* a cavity) In mammalian development, a modified blastula stage consisting of a hollow ball of cells, called a trophoblast, with an inner cell mass at one end that gives rise to the embryo.

blastodisc (Gr. *blastos,* sprout + *discos,* a round plate) In the development of birds, a disklike area on the surface of a large, yolky egg that undergoes cleavage and gives rise to the embryo.

blastomere (Gr. *blastos,* sprout + *meros,* part) One of the cells of a blastula; a product of the early cleavage divisions in the development of a vertebrate embryo.

blastopore (Gr. *blastos,* sprout + *poros,* a path or passage) In vertebrate development, the opening that connects the archenteron cavity of a gastrula stage embryo with the outside; represents the future mouth in some animals (protostomes), the future anus in others (deuterostomes).

blastula (Gr., a little sprout) In vertebrates, an early embryonic stage consisting of a hollow, fluid-filled ball of cells one layer thick; a vertebrate embryo after cleavage and before gastrulation.

blood pressure In vertebrates, the hydrostatic or fluid pressure within the circulatory system, generated by heart contractions; usually measured at large arteries of the systemic circulation, such as in an arm.

blood type In humans, the type of cell surface antigens present on the red blood cells of an individual; genetically determined, alternative alleles yield different surface antigens. When two different blood types are mixed, the cell surfaces often interact, leading to agglutination. One genetic locus encodes the ABO blood group, another the Rh blood group, and still others encode other surface antigens.

bond strength The amount of energy required to break a chemical bond; conventionally measured in kilocalories per mole.

botany (Gr. *botane,* a plant) The study of plants. One who studies plants is a *botanist.*

Bowman's capsule In the vertebrate kidney, the bulbous unit of the nephron, which surrounds the glomerulus. The kidney works by forced filtration, blood pressure driving blood plasma from the glomerular capillaries into the Bowman's capsule, after which it passes through the nephron, where most water and ions are reabsorbed into the bloodstream and the residue is excreted as urine.

brainstem The most posterior portion of the vertebrate brain; includes the medulla, pons, and midbrain.

branchial (Gr. *branchia,* gills) Referring to gills.

bronchus, pl. **bronchi** (Gr. *bronchos,* windpipe) One of a pair of respiratory tubes branching from the lower end of the trachea (windpipe) into either lung;

the respiratory path further subdivides into progressively finer passageways, the bronchioles, culminating in the alveoli.

brown fat Mitochondria-rich, heat-generating adipose tissue of endothermic vertebrates.

buccal (L. *bucca*, cheek) Referring to the mouth cavity.

bud An asexually produced outgrowth that develops into a new individual. In plants, an embryonic shoot, often protected by young leaves; buds may give rise to branch shoots.

buffer Any substance or chemical compound that tends to keep pH levels constant when acids or bases are added; a combination of a weak acid and a weak base that counteracts sudden swings in pH by combining with hydrogen ions (H^+) when the H^+ concentration rises, and by releasing H^+ ions when the H^+ concentration falls.

bulk flow In cells, the overall movement of water or some other liquid across the cell membrane, induced by gravity, pressure, or an interplay of both.

bundle sheath In plants, a layer or layers of cells surrounding a vascular bundle, especially characteristic of C_4 plants.

C_4 pathway A form of photosynthesis that avoids photorespiration in hot climates by increasing the internal CO_2 concentration of cells within which the dark reactions of photosynthesis occur. C_4 plants initially fix carbon dioxide to a compound known as phosphoenopyruvate (PEP) to yield oxaloacetate, a four-carbon compound—hence the name C_4. Also known as the Hatch-Slack pathway.

callose A complex branched carbohydrate that is a common wall constituent associated with the sieve areas of sieve elements; callose develops in reaction to the injury of the sieve elements.

callus (L. *callos*, hard skin) Undifferentiated tissue; a term used in tissue culture, grafting, and wound healing.

calorie (L. *calor*, heat) The amount of energy in the form of heat required to raise the temperature of 1 gram of water 1° C.

Calvin cycle The dark reactions of photosynthesis; a series of enzymatically mediated photosynthetic reactions in which carbon dioxide is bound to ribulose 1,5-diphosphate and then reduced to 3-phosphoglyceraldehyde, and the carbon-dioxide–accepting ribulose 1,5-diphosphate is regenerated. For every six molecules of carbon dioxide entering the cycle, a net gain of two molecules of glyceraldehyde 3-phosphate results. Also called the Calvin-Benson cycle.

calyx (Gr. *kalyx*, a husk, cup) The sepals collectively; the outermost flower whorl.

CAM *See* Crassulacean acid metabolism.

cambium, pl. **cambia** In vascular plants, embryonic tissue zones (meristems) that run parallel to the sides of roots and stems; consists of the vascular cambium and the cork cambium.

cancer Unrestrained invasive cell growth; a tumor or cell mass resulting from uncontrollable cell division; a result of mutational damage that destroys the normal control of cell division. A cancer cell is said to be "malignant," and in many (although by no means all) tissues, malignant cells can migrate to other body regions (metastasize), where they form secondary tumors.

capillaries (L. *capillaris*, hair-like) The smallest blood vessels; the very thin walls of capillaries are permeable to many molecules, and exchanges between blood and the tissues occur across them; the vessels that connect arteries with veins.

capillary action The movement of a liquid along a surface as a result of the combined effects of cohesion and adhesion.

capsule (L. *capsula*, a little chest) (1) In flowering plants, a dehiscent, dry fruit that develops from two or more carpels. (2) In bacteria, a slimy layer that forms around the cells. (3) In bryophytes, the sporangium.

carapace (Fr., from Sp., *carapacho*, shell) Shieldlike plate covering the cephalothorax of decapod crustaceans; the dorsal part of the shell of a turtle.

carbohydrate (L. *carbo*, charcoal + *hydro*, water) An organic compound consisting of a chain or ring of carbon atoms to which hydrogen and oxygen atoms are attached in a ratio of approximately 2:1; a compound of carbon, hydrogen, and oxygen having the generalized formula $(CH_2O)^n$; carbohydrates include sugars, starch, glycogen, and cellulose.

carbon cycle The worldwide circulation and reutilization of carbon atoms.

carbon fixation The conversion of CO_2 into organic compounds during photosynthesis; the first stage of the dark reactions of photosynthesis, in which carbon dioxide from the air is combined with ribulose 1,5-diphosphate.

carboxyl (*carbon* + *oxygen* + *-yl*, a chemical suffix) The acid group of organic molecules, —COOH.

carcinogen Any agent capable of inducing cancer; because cancer occurs as a result of mutation, most carcinogens are also potent mutagens. Programs that screen chemicals to detect potential carcinogens employ tests which measure the propensity of a chemical to cause mutations.

cardiac cycle In vertebrates, the pumping cycle of the heart, an alternation of contraction (systole) and relaxation (diastole).

cardiovascular system The blood circulatory system and the heart that pumps it; collectively, the blood, heart, and blood vessels.

carnivore (L. *carnis*, flesh + *voro*, to devour) Any organism that eats animals; a meat eater, as opposed to a plant eater, or herbivore.

carotene (L. *carota*, carrot + *-ene*, a prefix used for unsaturated straight-chain hydrocarbons) A red, yellow, or orange pigment belonging to the group of carotenoids; precursor of vitamin A. Carotenoids function as accessory pigments in photosynthesis.

carpel (Gr. *karpos*, fruit) A leaflike organ in angiosperms that encloses one or more ovules; one of the members of the gynoecium.

cartilage (L. *cartilago*, gristle) A connective tissue in skeletons of vertebrates. Cartilage forms much of the skeleton of embryos, very young vertebrates, and some adult vertebrates, such as sharks and their relatives. In most adult vertebrates much of it is converted into bone.

Casparian strip (Robert Caspary, a German botanist) In plants, a thickened, waxy strip that extends around and seals the walls of endodermal root cells, thus restricting the diffusion of solutes across the endodermis into the vascular tissues of the root.

catabolism (Gr. *katabole*, throwing down) In a cell, those metabolic reactions that result in the breakdown of complex molecules into simpler compounds, often with the release of energy.

catalyst (Gr. *kata*, down + *lysis*, a loosening) A substance that accelerates the rate of a chemical reaction by lowering the activation energy, without being used up in the reaction; enzymes are the catalysts of cells.

cation (Gr. *katienai*, to go down) A positively charged ion.

cecum, also **caecum** (L. *caecus*, blind) In vertebrates, a blind pouch at the beginning of the large intestine; any similar pouch.

cell (L. *cella*, a chamber or small room) The structural unit of organisms; the smallest unit that can be considered living; a cell consists of cytoplasm encased within a membrane.

cell cycle The repeating sequence of growth and division through which cells pass each generation.

cell division The division of a cell and its contents into two roughly equal parts. Bacteria divide by fission, eukaryotes by mitosis.

cell membrane The outermost membrane of the cell; also called the plasma membrane.

cell plate The structure that forms at the equator of the spindle during early telophase in the dividing cells of plants and a few green algae. The cell plate becomes the middle lamella during the course of development.

cellular respiration The metabolic harvesting of energy by oxidation; carried out by the citric acid cycle and oxidative phosphorylation, in which considerable energy is extracted from sugar molecule fragments left over from glycolysis.

cellulase An enzyme that hydrolyzes cellulose.

cellulose (L. *cellula*, a little cell) The chief constituent of the cell wall in all green plants, some algae, and a few other organisms. Cellulose is an insoluble complex carbohydrate $(C_6H_{10}O_5)^n$ formed of microfibrils of glucose molecules.

cell wall The rigid, outermost layer of the cells of plants, some protists, and most bacteria; the cell wall surrounds the cell (plasma) membrane.

central nervous system That portion of the nervous system where most association occurs; in vertebrates, it is composed of the brain and spinal cord; in invertebrates it usually consists of one or more cords of nervous tissue, together with their associated ganglia.

centriole (Gr. *kentron*, center of a circle + L. *olus*, little one) A cytoplasmic organelle located outside the nuclear membrane, identical in structure to a basal body; found in animal cells and in the flagellated cells of other groups. The centriole divides and organizes spindle fibers during mitosis and meiosis.

centromere (Gr. *kentron*, center + *meros*, a part) That position on a eukaryotic chromosome to which the spindle fibers are attached during cell division; composed of highly repeated DNA sequences (satellite DNA). Also called the kinetochore.

cephalization (Gr. *kephale*, head) In animals, the process by which neural organization and specialization of the sensory organs become localized in the anterior (head) end.

cephalothorax (Gr. *kephale*, head + *thorax*) A body division found in many arachnids and crustaceans, in which the head is fused with some or all of the thoracic segments.

cerebellum (L., little brain) The hindbrain region of the vertebrate brain, which lies above the medulla (brainstem) and behind the forebrain; it integrates information about body position and motion, coordinates muscular activities, and maintains equilibrium.

cerebral cortex The thin surface layer of neurons and glial cells covering the cerebrum; well developed only in mammals, and particularly prominent in humans. The cerebral cortex is the seat of conscious sensations and voluntary muscular activity.

cerebrum (L., brain) The portion of the vertebrate brain (the forebrain) that occupies the upper part of the skull, consisting of two cerebral hemispheres united by the corpus callosum. The cerebrum, which consists of paired hemispheres occupying the upper part of the skull overlying the thalamus, hypothalamus, and pituitary, is the primary association center of the brain. It coordinates and processes sensory input and coordinates motor responses.

chelicera, pl. **chelicerae** (Gr. *chele*, claw + *keras*, horn) The first pair of appendages in horseshoe crabs, sea spiders, and arachnids—the chelicerates, a group of arthropods. Chelicerae usually take the form of pincers or fangs.

chemical reaction The making or breaking of chemical bonds between atoms or molecules.

chemiosmosis The mechanism by which ATP is generated in mitochondria and chloroplasts. Energetic electrons excited by light (chloroplasts) or extracted by oxidation in the citric acid cycle (mitochondria) are used to drive proton pumps, creating a higher external proton concentration; when protons subsequently flow back in, it is through channels that couple their passage to the synthesis of ATP.

chemoautotroph An autotrophic bacterium that uses chemical energy released by specific inorganic reactions to power its life processes, including the synthesis of organic molecules.

chemoreceptor A sensory cell or organ that responds to the presence of a specific chemical stimulus by initiating a nerve impulse; includes taste and smell receptors.

chemotactic An organism that responds to a chemical stimulus by moving toward or away from it.

chiasma, pl. **chiasmata** (Gr., a cross) During meiosis, the region of contact between homologous chromatids where crossing-over has occurred during synapsis; a chiasma has the appearance of the letter X.

chitin (Gr. *chiton*, tunic) A tough, resistant, nitrogen-containing polysaccharide that forms the cell walls of certain fungi, the exoskeleton of arthropods, and the epidermal cuticle of other surface structures of certain other invertebrates.

chlorinated hydrocarbon Synthetic, chlorine-containing compound; often such compunds are not readily degradable in the environment. DDT is an example.

chlorophyll (Gr. *chloros*, green + *phyllon*, leaf) The green steroid pigment of plant cells, which is the receptor of light energy in photosynthesis; chlorophyll is also found in some protists and bacteria.

chloroplast (Gr. *chloros*, green + *plastos*, molded) A cell-like organelle present in algae and plants that contains chlorophyll (and usually other pigments) and carries out photosynthesis.

cholinergic (Gr. *chole*, bile + *ergon*, work) Type of neuron that releases acetylcholine from axon terminal.

chordate (L. *chorda*, chord, string) A member of the animal phylum Chordata, of which all members possess a notochord, dorsal nerve cord, pharyngeal gill pouches, and a tail, at least at some stage of the life cycle.

chorion (Gr., skin) The outer member of the double membrane that surrounds the embryo of reptiles, birds, and mammals; in placental mammals, it contributes to the structure of the placenta.

chromatid (Gr. *chroma*, color + L. *-id*, daughters of) One of the two daughter strands of a duplicated chromosome which are joined by a single centromere; separates and becomes a daughter chromosome at anaphase of mitosis or anaphase of the second meiotic division.

chromatin (Gr. *chroma*, color) The complex of DNA and proteins of which eukaryotic chromosomes are composed.

chromatophore (Gr. *chroma*, color + *phorus*, a bearer) A pigment-containing cell. In vertebrates, chromatophores usually occur in the dermis. In some bacteria, chromatophores occur as discrete vesicles delimited by a single membrane and containing photosynthetic pigments.

chromomere (Gr. *chroma*, color + *meros*, part) One of the chromatin bands on the chromosome; individual chromomeres may be identical to a gene or a cluster of genes.

chromonema, pl. **chromonemata** (Gr. *chroma*, color + *nema*, thread) Eukaryotic chromosomes are convoluted threads called chromonemata when they first become visible in prophase of

chromosome (Gr. *chroma*, color + *soma*, body) the vehicle by which hereditary information is physically transmitted from one generation to the next; the organelle that carries the genes. In bacteria, the chromosomes consist of a single naked circle of DNA; in eukaryotes, they consist of a single linear DNA molecule and associated proteins.

chromosome map A diagram of the linear order of the genes on a chromosome; chromosome maps are determined from the frequency of recombination between pairs of genes.

chrysalis (L., from Gr. *chrysos*, gold) The pupal stage of a butterfly, the resting stage during which the form of the adult butterfly differentiates.

cilium, pl. **cilia** (L., eyelash) A short, hairlike flagellum, especially used to refer to such structures when they are numerous. Cilia may be used in locomotion; in animals, they aid the movement of substances across surfaces.

circadian rhythms (L. *circa*, about + *dies*, day) Regular rhythms of growth or activity that occur on an approximately 24-hour cycle.

cisterna, pl. **cisternae** (L., a reservoir) In cells, a flattened or saclike space between membranes of the endoplasmic reticulum or Golgi body.

cistron (L. *cista*, box) A sequential series of codons in DNA that codes for an entire polypeptide chain.

citric acid cycle The cyclic series of reactions in which pyruvate, the product of glycolysis, enters the cycle to form citric acid and is then oxidized to carbon dioxide. Also called the Krebs cycle (after its discoverer) and the tricarboxylic acid (TCA) cycle (citric acid possesses three carboxyl groups).

cladistics (Gr. *cladus*, branch) A system of arranging taxa by the analysis of primitive and derived characteristics so that the arrangement will reflect phylogenetic relationships.

class A taxonomic category between phyla (divisions) and orders. A class contains one or more orders, and belongs to a particular phylum or division.

cleavage In vertebrates, a rapid series of successive cell divisions of a fertilized egg, forming a hollow sphere of cells, the blastula.

climax (Gr. *klimax*, staircase) Orgasm.

climax community In ecology, the final stage in a successional series; the climax stage is determined primarily by the climate and soil type of the area.

cloaca (L., sewer) In some animals, the common exit chamber from the digestive, reproductive, and urinary systems; in others, the cloaca may also serve as a respiratory duct.

clone (Gr. *klon*, twig) A line of cells, all of which have arisen from the same single cell by mitotic division; one of a population of individuals derived by asexual reproduction from a single ancestor; one of a population of genetically identical individuals.

cloning Producing a cell line or culture all of whose members contain identical copies of a particular nucleotide sequence; an essential element in genetic engineering, cloning is usually carried out by inserting the desired gene into a virus or plasmid, infecting a cell culture with the hybrid virus, and selecting for culture a cell that has taken up the gene.

cnidocyte (Gr. *knide*, nettle + *kytos*, vessel) Specialized cell that holds a nematocyst; cnidocytes are characteristic of cnidarians.

coacervate (L. *coacervatus*, heaped up) A spherical aggregation of lipid molecules in water, held together by hydrophobic forces.

coccus, pl. **cocci** (Gr. *kokkos*, a berry) A spherical bacterium.

cochlea (Gr. *kochlios*, a snail) In terrestrial vertebrates, a tubular cavity of the inner ear containing the essential organs of hearing; occurs in crocodiles, birds, and mammals; spirally coiled in mammals.

cocoon (Fr. *cocon*, shell) Protective covering of a resting or developmental stage, sometimes used to refer to both the covering and its contents; for example, the cocoon of a moth or the protective covering for the developing embryos in earthworms and some other annelids.

codominance In genetics, a situation in which the effects of both alleles at a particular locus are apparent in the phenotype of the heterozygote.

codon (L., code) The basic unit ("letter") of the genetic code; a sequence of three adjacent nucleotides in DNA or mRNA that code for one amino acid, or for polypeptide chain termination. *See* anticodon.

coelenteron (Gr. *koilos*, a hollow + *enteron*, gut) A digestive cavity with only one opening, characteristic of the cnidarians and ctenophores.

coelom (Gr. *koilos*, a hollow) A body cavity formed between layers of mesoderm and in which the digestive tract and other internal organs are suspended.

coenocytic (Gr. *koinos*, shared in common + *kytos*, a hollow vessel) An organism or tissue in which the cells are multinucleate, but the nuclei are not separated from one another by cell membranes; also referred to as syncytial.

coenzyme (L. *co-*, together + Gr. *en*, in + *zyme*, leaven) A nonprotein organic molecule that plays an accessory role in enzyme-catalyzed processes, often by acting as a donor or acceptor of electrons. NAD^+, FAD, and coenzyme A are common coenzymes.

coevolution (L. *co-*, together + *e-*, out + *volvere*, to fill) The simultaneous development of adaptations in two or more populations, species, or other categories that interact so closely that each is a strong selective force on the other.

cofactor One or more nonprotein components required by enzymes in order to function; many cofactors are metal ions, others are coenzymes. *See* coenzyme.

cohesion (L. *cohaerere*, to stick together) The mutual attraction (coherence) of molecules of the same substance.

collagen (Gr. *kolla*, glue + *genos*, descent) A tough, fibrous protein occurring in vertebrates as the chief constituent of bones, tendons, and other connective tissue. Collagen also occurs in invertebrates—for example, in the cuticle of nematodes.

collenchyma (Gr. *kolla*, glue + *en*, in + *chymein*, to pour) In plants, a supporting tissue composed of collenchyma cells; often found in regions of primary growth in stems and in some leaves. Collenchyma cells are elongated living cells with irregularly thickened primary cell walls.

colloid (Gr. *kolla*, glue + *eidos*, form) A two-phase system in which fine particles of one phase are stably suspended in the second phase.

colony A group of organisms living together in close association.

commensalism (L. *cum*, together with + *mensa*, table) A relationship in which one individual lives close to or on another and benefits, and the host is unaffected; a kind of symbiosis.

community (L. *communitas*, community, fellowship) All of the organisms inhabiting a common environment and interacting with one another.

companion cell In the phloem of flowering plants, a specialized kind of elongate parenchyma cell associated with a sieve-tube member.

competition Interaction between members of the same population or of two or more populations in order to obtain a mutually required resource available in limited supply; in competition, one species interferes with another enough to keep it from gaining access to the resource.

competitive exclusion The hypothesis that two species with identical ecological requirements cannot exist in the same locality indefinitely, and that the more efficient of the two in utilizing the available scarce resources will exclude the other; also known as Gause's principle, after the Russian biologist G.F. Gause.

compound (L. *componere*, to put together) A molecule composed of two or more kinds of atoms in definite ratios, held together by chemical bonds.

compound eye In arthropods, a complex eye composed of many separate elements, each element composed of light-sensitive cells and a lens that can form an image; all of the individual images are merged in the arthropod central nervous system to form a detailed, three-dimensional image.

compound leaf A leaf whose blade is divided into several distinct leaflets.

concentration gradient The concentration difference of a substance as a function of distance; in a cell, a greater concentration of its molecules in one region than in another.

condensation reaction (L. *co-*, together + *densare*, to make dense) A type of chemical reaction in which two molecules join to form one larger molecule, simultaneously splitting out a molecule of water. One molecule is stripped of a hydrogen atom and another is stripped of a hydroxyl group (—OH), resulting in the joining of the two molecules. The H^+ and —OH released may combine to form a water molecule. The biosynthetic reactions in which monomers (e.g., monosaccharides, amino acids) are joined to form polymers (e.g., polysaccharides, polypeptides) are condensation reactions. Condensation is also called dehydration synthesis.

conditioning A form of learning in which a behavioral response becomes associated (by means of a reinforcing stimulus) with a new stimulus not previously capable of invoking the response.

cone (1) In plants, the reproductive structure of a conifer. (2) In vertebrates, a type of light-sensitive neuron in the retina, concerned with the perception of color and with the most acute discrimination of detail.

conjugation (L. *conjugare*, to yoke together) Temporary union of two unicellular organisms, during which genetic material is transferred from one cell to the other; occurs in bacteria, protists, and certain algae and fungi.

conjugation tube A cytoplasmic tube through which genetic material passes during conjugation.

connective tissues A collection of vertebrate tissues derived from the mesoderm. Some kinds of connective tissue (lymphocytes and macrophages) are mobile hunters of invading bacteria, others secrete a matrix (cartilage and bone) that provides the body with structural support, and still others provide sites of fat storage (adipose tissue) or a site for hemoglobin (red blood cells).

consumer In ecology, a heterotroph that derives its energy from living or freshly killed organisms or parts thereof. Primary consumers are herbivores; higher-level consumers are carnivores.

continuous variation Variation in traits to which many different genes make a contribution; in such traits, a gradation of small differences is observed; often exhibiting a "normal" or bell-shaped distribution.

contractile vacuole In protists and some animals, a clear fluid-filled cell vacuole that takes up water from within the cell and then contracts, releasing it to the outside through a pore in a cyclical manner. Contractile vacuoles function primarily in osmoregulation and excretion.

control element A region of DNA that influences the transcription of structural genes; in maize, a transposon.

convergent evolution The independent development of similar structures in organisms that are not directly related; often found in organisms living in similar environments.

copulation (Fr., from L. *copulare*, to couple) Sexual union of opposite sexes, in which sperm are introduced into the body of the female to facilitate internal fertilization.

copy DNA (cDNA) DNA "copied" from an mRNA molecule, using the viral enzyme reverse transcriptase.

cork (L. *cortex*, bark) A secondary tissue produced by a cork cambium; made up of flattened cells, nonliving at maturity, with walls infiltrated with suberin, a waxy or fatty material that is resistant to the passage of gases and water vapor; the outer part of the periderm. Also called phellem.

cork cambium The lateral meristem that forms the periderm, producing cork (phellem) toward the surface (outside) of the plant and phelloderm toward the inside; common in stems and roots of gymnosperms and dicots. Also called phellogen.

cornea (L. *corneus*, horny) The transparent outer layer of the vertebrate eye.

corneum (L. *corneus*, horny) In human skin, the outer epithelial layer of dead, keratinized cells; cornified cells; also referred to as the stratum corneum.

corolla (L. *corona*, crown) The petals, collectively; usually the conspicuously colored flower whorl. Petals may be free or fused with one another or with the members of other floral whorls.

corpus luteum (L., yellowish body) A structure that develops from a ruptured follicle in the ovary after ovulation; the corpus luteum secretes estrogens and progesterone, which maintain the uterus during pregnancy.

cortex (L., bark) The outer layer of a structure; in animals, the outer, as opposed to the inner, part of an organ, as in the adrenal, kidney, and cerebral cortexes; in vascular plants, the primary ground tissue of a stem or root, bounded externally by the epidermis and internally by the central cylinder of vascular tissue.

cotyledon (Gr. *kotyledon*, a cup-shaped hollow) Seed leaf; cotyledons generally store food in dicotyledons and absorb it in monocotyledons. The food is used during the course of seed germination.

countercurrent exchange In organisms, the passage of heat or of molecules (such as oxygen, water, or sodium ions) from one circulation path to another moving in the opposite direction; because the flow of the two paths is in opposite directions, there is always a concentration difference between the two channels, facilitating transfer.

countercurrent multiplier A ∪-shaped countercurrent exchange arrangement in which a molecule such a sodium ion is actively transported out of one arm, to re-enter the other arm by diffusion. In such an arrangement, an ion concentration

gradient develops, highest at the bend of the ∪. It was once thought that a countercurrent multiplier was responsible for water retention in the vertebrate kidney, although this idea has since been abandoned.

coupled reactions In cells, the linking of an endergonic (energy-requiring) reaction to an exergonic (energy-releasing) reaction. In coupled reactions, the decrease in free energy associated with the exergonic reaction is typically greater than the increase in free energy associated with the endergonic reaction, so that the net change in free energy is negative and the two-reaction process proceeds— even though the first step would not have proceeded alone. In metabolism, coupling reactions to ATP hydrolysis (a highly exergonic reaction) is a common strategy employed to drive endergonic reactions.

covalent bond (L. *co-*, together + *valere*, to be strong) A chemical bond formed between atoms as a result of the sharing of one or more pairs of electrons.

Crassulacean acid metabolism (CAM metabolism) A variant of the C_4 pathway; phosphoenolpyruvate fixes CO_2 in C_4 compounds at night and then, during the day, the fixed CO_2 is transferred to the ribulose bisphosphate of the Calvin cycle within the same cell. CAM metabolism is characteristic of such succulent plants as cacti.

creatine phosphate High-energy phosphate compound found in the muscle of vertebrates and some invertebrates, used to regenerate stores of ATP.

crista, pl. **cristae** (L., crest) In mitochondria, the enfoldings of the inner mitochondrial membrane, which form a series of "shelves" containing the electron-transport chains involved in ATP formation.

cross-fertilization Fusion of gametes formed by different individuals; the opposite of self-fertilization.

crossing-over In meiosis, the exchange of corresponding chromatid segments between homologous chromosomes; responsible for genetic recombination between homologous chromosomes.

cross-pollination The transfer of pollen from the anther of one plant to the stigma of a flower of another plant.

cuticle (L. *cuticula*, little skin) A waxy or fatty, noncellular layer on the outer wall of epidermal cells. The cuticle is formed of a substance called cutin.

cyclic AMP (cAMP) A form of adenosine monophosphate (AMP) in which the atoms of the phosphate group form a ring; found in almost all organisms. cAMP functions as an intracellular hormone that regulates a diverse array of metabolic activities.

cyclic photophosphorylation In photosynthesis, the path taken by a light-excited electron from a photosystem through a membrane-embedded transport chain to proton pumps, then back to the photosystem; more generally, the chemiosmotic synthesis of ATP in photosynthesis.

cytochrome (Gr. *kytos*, hollow vessel + *chroma*, color) Any of several iron-containing protein pigments that serve as electron carriers in transport chains of photosynthesis and cellular respiration.

cytokinesis (Gr. *kytos*, hollow vessel + *kinesis*, movement) Division of the cytoplasm of a cell after nuclear division.

cytokinin (Gr. *kytos*, hollow vessel + *kinesis*, motion) A class of plant hormones that promote cell division, among other effects.

cytology (Gr. *kytos*, hollow vessel + *logos*, word) The study of cell structure.

cytoplasm (Gr. *kytos*, hollow vessel + *plasma*, anything molded) The living matter within a cell, excluding the nucleus; the protoplasm.

cytosine A pyrimidine; one of the four nitrogen-containing bases in the nucleic acids DNA and RNA; in a DNA duplex molecule, cytosine pairs with guanine, forming three hydrogen bonds.

cytoskeleton A network of protein microfilaments and microtubules within the cytoplasm of a eukaryotic cell that maintains the shape of the cell, anchors its organelles, and is involved in animal cell motility.

day-neutral plants Plants that flower without regard to day length.

deciduous (L. *decidere*, to fall off) In vascular plants, shedding all the leaves at a certain season.

decomposers Organisms (bacteria, fungi, heterotrophic protists) that break down organic material into smaller molecules, which are then recirculated.

dehiscence (L. *de*, down + *hiscere*, split open) The opening of an anther, fruit, or other structure, which permits the escape of reproductive bodies contained within.

demography (Gr. *demos*, people + *graphein*, to draw) The properties of the rate of growth and the age structure of populations.

denaturation The loss of the native configuration of a protein or nucleic acid as a result of excessive heat, extremes of pH, chemical modification, or changes in solvent ionic strength or polarity that disrupt hydrophobic interactions; usually accompanied by loss of biological activity.

dendrite (Gr. *dendron*, tree) A process extending from the cell body of a neuron, typically branched, that conducts impulses inward toward the cell body; although they may be long,

most dendrites are short, and a single neuron may possess many of them.

denitrification The conversion of nitrate to gaseous nitrogen; carried out by certain soil bacteria.

deoxyribonucleic acid (DNA) The genetic material of all organisms; composed of two complementary chains of nucleotides wound in a double helix; local unwinding of the helix by disruption of hydrogen bonds between strands permits RNA polymerase molecules to transcribe mRNA copies of genes, and permits DNA polymerase molecules to replicate copies of the duplex molecule. *See* alpha helix.

deoxyribose (L. *deoxy*, loss of oxygen + *ribose*, a pentose sugar) A five-carbon sugar having one oxygen atom less than ribose; a component of deoxyribonucleic acid (DNA).

dermis (Gr. *derma*, skin) The inner, sensitive mesodermal layer of skin, beneath the epidermis; corium.

determinate cleavage The type of cleavage, usually spiral, in which the fate of the blastomeres is determined very early in development; characteristic of protostomes.

determinate growth Growth of limited duration, as is characteristic of floral meristems and of leaves.

detritivores (L. *detritus*, worn down + *vorare*, to devour) Organisms that live on dead organic matter; included are large scavengers, smaller animals such as earthworms and some insects, and decomposers (fungi and bacteria).

deuterostome (Gr. *deuteros*, second + *stoma*, mouth) An animal in whose embryonic development the anus forms at or near the blastopore and the mouth forms secondarily elsewhere. Deuterostomes are also characterized by radial cleavage during the earliest stages of development and by enterocoelous formation of the coelom. *See* protostome.

diapause (Gr. *diapausis*, pause) A period of arrested development in the life cycle of insects and certain other animals in which physiological activity is very low and the animal is vulnerable to unfavorable external conditions.

diaphragm (Gr. *diaphrassein*, to barricade) (1) In mammals, a sheet of muscle tissue that separates the abdominal and thoracic cavities and functions in breathing. (2) A contraceptive device used to block the entrance to the uterus temporarily and thus prevent sperm from entering during sexual intercourse.

dicot Short for dicotyledon; a class of flowering plants generally characterized as having two cotyledons, net-veined leaves, and flower parts usually in fours or fives.

differentially permeable membrane A membrane through which some

substances can diffuse and others cannot.

differentiation A developmental process by which a relatively unspecialized cell undergoes a progressive change to a more specialized form or function; differentiation in plants may be reversed under suitable conditions, but it is rarely reversible in animals.

diffusion (L. *diffundere*, to pour out) The net movement of dissolved molecules or other particles from a region where they are more concentrated to a region where they are less concentrated, as a result of the random movement of individual molecules; the process tends to distribute molecules uniformly.

digestion (L. *digestio*, separating out, dividing) The breakdown of complex, usually insoluble foods into molecules that can be absorbed into cells and there degraded to yield energy and the raw materials for synthetic processes.

dihybrid (Gr. *dis*, twice + L. *hibrida*, mixed offspring) An individual heterozygous at two different loci; for example, A/a B/b.

dikaryotic (Gr. *di*, two + *karyon*, kernel) In fungi, having pairs of nuclei within each cells.

dioecious (Gr. *di*, two + *oikos*, house) Having the male and female elements on different individuals.

diploblastic (Gr. *diploos*, double + *blastos*, bud) In animals, having two germ layers, endoderm and ectoderm; all animals are now thought to be fundamentally triploblastic.

diploid (Gr. *diploos*, double + *eidos*, form) Having two sets of chromosomes ($2n$); in animals, twice the number characteristic of gametes; in plants, the chromosome number characteristic of the sporophyte generation; in contrast to haploid ($1n$).

disaccharide A carbohydrate formed of two simple sugar molecules bonded covalently; sucrose is a disaccharide composed of two glucose molecules linked together.

distal Situated away from or far from a point of reference (usually the main part of the body); opposite of proximal.

division A major taxonomic group; kingdoms are divided into divisions (or phyla, which are equivalent), and divisions are divided into classes.

dominant allele One allele is said to be dominant with respect to an alternative allele if a heterozygous individual bearing the two alleles is indistinguishable from an individual homozygous for the dominant allele; the alternative not detected in the heterozygote is said to be recessive with respect to the dominant allele. An allele that is dominant with respect to one alternative allele may be recessive with respect to another.

dormancy (L. *dormire*, to sleep) A

period during which growth ceases and is resumed only if certain requirements, as of temperature or day length, have been fulfilled.

dorsal (L. *dorsum,* the back) Toward the back, or upper surface; opposite of ventral.

double fertilization The fusion of the egg and sperm (resulting in a 2*n* fertilized egg, the zygote) and the simultaneous fusion of the second male gamete with the polar nuclei (resulting in a primary endosperm nucleus, which is often triploid, 3*n*); a unique characteristic of all angiosperms.

Down syndrome A congenital syndrome, whose manifestations include mental retardation, caused by the cells in a person's body having an extra copy of a segment of chromosome 21; also called trisomy 21.

duodenum (L. *duodeni,* 12 each—from its length, about 12 fingers' breadth) In vertebrates, the upper portion of the small intestine; the principal site of food digestion, where carbohydrates, proteins, and fats are broken down into sugars, amino acids, and fatty acids.

ecdysis (Gr. *ekdysis,* stripping off) Shedding of outer cuticular layer; molting, as in insects or crustaceans.

ecdysone (Gr. *ekdysis,* stripping off) Molting hormone of arthropods, which stimulates growth and ecdysis.

ecology (Gr. *oikos,* house + *logos,* word) The study of the interactions of organisms with one another and with their physical environment.

ecosystem (Gr. *oikos,* house + *systema,* that which is put together) A major interacting system that involves both organisms and their nonliving environment.

ecotype (Gr. *oikos,* house + L. *typus,* image) A locally adapted variant of an organism, differing genetically from other ecotypes.

ectoderm (Gr. *ecto,* outside + *derma,* skin) One of the three embryonic germ layers of early vertebrate embryos; ectoderm gives rise to the outer epithelium of the body (skin, hair, nails) and to the nerve tissue, including the sense organs, brain, and spinal cord.

ectotherm (Gr. *ectos,* outside + *therme,* heat) An organism, such as a reptile, that regulates its body temperature by taking in heat from the environment or giving it off to the environment; contrasts with endothermic; equivalent to *poikilotherm.*

edaphic (Gr. *adaphos,* ground, soil) Pertaining to the soil.

effector (L. *efficere,* to bring to pass) In animals, an organ, tissue, or cell that becomes active in response to stimulation; within cells, a small molecule that induces allosteric changes in the shapes of proteins. *See* allosteric interaction.

efferent (L. *ex,* out of + *ferre,* to bear) Leading or conveying away from some origin—for example, nerve impulses conducted away from the brain, or blood conveyed away from the heart; contrasts with afferent.

egg A female gamete; nonmotile and usually containing abundant cytoplasm; often larger than a male gamete.

electrolyte A substance that dissociates into ions in aqueous solution.

electron A subatomic particle with a negative electric charge equal in magnitude to the positive charge of the proton but with a much smaller mass; electrons orbit the atom's positively charged nucleus and determine its chemical properties. *See* atom.

electron acceptor A substance that accepts or receives electrons in an oxidation-reduction reaction, becoming reduced in the process.

electron carrier A molecule that can lose and gain electrons reversibly, alternately becoming oxidized and reduced; examples are NAD^+/NADH and cytochromes.

electron donor A substance that donates or gives up electrons in an oxidation-reduction reaction, becoming oxidized in the process.

electron transport The passage of energetic electrons through a series of membrane-associated electron-carrier molecules to proton pumps embedded within mitochondrial or chloroplast membranes; as the electrons arrive at the proton pumping channel, their energy drives the transport of protons out across the membrane, leading to the chemiosmotic synthesis of ATP. *See* chemiosmosis.

element A substance composed only of atoms of the same atomic number, which cannot be decomposed by ordinary chemical means; one of more than 100 distinct natural or synthetic types of matter that, singly or in combination, compose all materials of the universe.

embryo (Gr. *en,* in + *bryein,* to swell) The early developmental stage of an organism produced from a fertilized egg; in plants, a young sporophyte, before its initial period of rapid growth; in animals, a young organism before it emerges from the egg, or from the body of its mother; in humans, refers to the first 2 months of intrauterine life.

embryonic induction Processes by which one group of embryonic cells affects an adjacent group, thereby inducing those cells to differentiate in a manner they otherwise would not have.

embryo sac The female gametophyte of angiosperms, generally an eight-nucleate, seven-celled structure; the seven cells are the egg cell, two synergids and three antiopodals (each with a single nucleus), and the central cell (with two nuclei).

endergonic (Gr. *endon,* within + *ergon,* work) Describing a chemical reaction that requires energy; energy from an outside source must be added before the reaction proceeds; a thermodynamically "uphill" process; opposite of exergonic.

endocrine gland (Gr. *endon,* within + *krinein,* to separate) Ductless gland that secretes hormones into the extracellular spaces, from which they diffuse into the circulatory system; in vertebrates, includes the pituitary, sex glands, adrenal, thyroid, and others.

endocytosis (Gr. *endon,* within + *kytos,* hollow vessel) The uptake of material into cells by inclusion within an invagination of the plasma membrane; the material becomes trapped within a vacuole when the edges of the invagination fuse together. If solid material is included, the uptake is called phagocytosis; if dissolved material, it is called pinocytosis.

endoderm (Gr. *endon,* within + *derma,* skin) One of the three embryonic germ layers of early vertebrate embryos, destined to give rise to the epithelium that lines certain internal structures, such as most of the digestive tract and its outgrowths, most of the respiratory tract, and the urinary bladder, liver, pancreas, and some endocrine glands.

endodermis (Gr. *endon,* within + *derma,* skin) In vascular plants, a layer of cells forming the innermost layer of the cortex in roots and some stems. The endodermis is characterized by a Casparian strip within radial and transverse walls.

endometrium (Gr. *endon,* within + *metrios,* of the womb) The lining of the uterus in mammals; thickens in response to secretion of estrogens and progesterone and is sloughed off in menstruation.

endoplasmic reticulum (Gr. *endon,* within + *plasma,* from cytoplasm; L. *reticulum,* network) An extensive system of membranes present in most eukaryotic cells, dividing the cytoplasm into compartments and channels; those portions containing a dense array of ribosomes are called "rough ER," and other portions with fewer ribosomes are called "smooth ER."

endorphin One of a group of small neuropeptides produced by the vertebrate brain; like morphine, endorphins modulate pain perception; they are also implicated in many other functions.

endoskeleton (Gr. *endon,* within + *skeletos,* hard) A skeleton or supporting framework within the living tissues of an organism; in chordates, the internal framework of bone or cartilage or both; contrasts with exoskeleton

endosperm (Gr. *endon,* within + *sperma,* seed) A storage tissue characteristic of the seeds of angiosperms, which develops from the union of a male nucleus and the polar nuclei of the embryo sac. The endosperm is digested by the growing sporophyte either before the maturation of the seed or during its germination.

endotherm (Gr. *endon*, within + *therme*, heat) An organism that regulates its body temperature internally through metabolic processes, as do birds and mammals; *see also* homeotherm; contrasts with ectothermic.

energy Capacity to do work.

enhancer A nucleotide sequence that increases the utilization of eukaryotic promoters, presumably by altering chromatin configurations to render the promoter site more accessible to the RNA polymerase.

enteron (Gr. *intestine*) The digestive cavity.

entropy (Gr. *en*, in + *tropos*, change in manner) A measure of the randomness or disorder of a system; a measure of how much energy in a system has become so dispersed (usually as evenly distributed heat) that it is no longer available to do work.

enzyme (Gr. *enzymos*, leavened, from *en*, in + *zyme*, leaven) A protein that is capable of speeding up specific chemical reactions by lowering the required activation energy, but is unaltered itself in the process; a biological catalyst.

epidermis (Gr. *epi*, on or over + *derma*, skin) The outermost layers of cells; in plants, the exterior primary tissue of leaves, young stems, and roots; in vertebrates, the nonvascular external layer of skin, of ectodermal origin; in invertebrates, a single layer of ectodermal epithelium.

epididymis (Gr. *epi*, on + *didymos*, testicle) A sperm storage vessel; a coiled part of the sperm duct that lies near the testis.

epinephrine *See* adrenaline.

epiphyte (Gr. *epi*, on + *phyton*, plant) A plant that grows on another organism but is not parasitic on it.

episome A plasmid that is able to integrate into chromosomal DNA.

epistasis (Gr. *epi*, on + *stasis*, a standing still) Interaction between two nonallelic genes in which one of them modifies the phenotypic expression of the other; the masking or prevention of the expression of one gene by another gene at another locus.

epithelium (Gr. *epi*, on + *thele*, nipple) In animals, a type of tissue that covers an exposed surface or lines a tube or cavity.

equilibrium (L. *aequus*, equal + *libra*, balance) A stable condition; a system in which no further net change is occurring; the point at which a chemical reaction proceeds as rapidly in the reverse direction as it does in the forward direction, so that there is no further net change in the concentrations of products or reactants.

erythrocyte (Gr. *erythros*, red + *kytos*, hollow vessel) Red blood cell, the carrier of hemoglobin. In mammals, erythrocytes lose their nuclei, whereas in other vertebrates the nuclei are retained.

estrogens (Gr. *oestros*, frenzy + *genos*, origin) A group of steroid hormones, such as estradiol, that affect female secondary sex characteristics, estrus, and the human menstrual cycle.

estrus (L. *oestrus*, frenzy) The period of maximum female sexual receptivity, associated with ovulation of the egg; being "in heat."

ethology (Gr. *ethos*, habit or custom + *logos*, discourse) The study of patterns of animal behavior in nature.

ethylene A simple hydrocarbon that is a plant hormone involved in the ripening of fruit; $H_2C{=}CH_2$.

etiolation (Fr. *etioler*, to blanch) A condition that develops when plants are grown with insufficient or no light; it involves increased stem elongation, poor leaf development, and lack of chlorophyll.

euchromatin (Gr. *eu*, good + *chroma*, color) That portion of eukaryotic chromosomes that is transcribed into mRNA; contains active genes.

eukaryote (Gr. *eu*, good + *karyon*, kernel) A cell characterized by membrane-bound organelles, most notably the nucleus, and one that possesses chromosomes whose DNA is associated with proteins; an organism composed of such cells; contrasts with prokaryote.

eutrophication Process whereby a body of fresh water becomes enriched with nutrients, increases in productivity, and accumulates organic debris.

evaporation The escape of water molecules from the liquid to the gas phase at the surface of a body of water.

evolution (L. *evolvere*, to unfold) Genetic change in a population of organisms; in general, evolution leads to progressive change from simple to complex. Darwin proposed that natural selection was the mechanism of evolution.

excision repair The mechanism by which cells repair single strand damage to their DNA, such as dimerized pyrimidines; the damaged portion is removed, and a new stretch of nucleotides is synthesized in its place, with the other strand used as a template.

exergonic (L. *ex*, out + Gr. *ergon*, work) An energy-yielding process or chemical reaction; energy is released from the reactants, so that the products contain less chemical potential energy than the reactants; a "downhill" process that will proceed spontaneously.

exocrine glands (Gr. *ex*, out of + *krinein*, to separate) A type of gland that releases its secretion through a duct, such as digestive glands and sweat glands; contrasts with endocrine.

exocytosis (Gr. *ex*, out of + *kytos*, vessel) A type of bulk transport out of cells; cytoplasmic particles are encased within membranes, forming a vacuole that is transported to the cell surface; there, the vacuole membrane fuses with the cell membrane, discharging the vacuole's contents to the outside.

exon (Gr. *exo*, outside) A segment of DNA that is both transcribed into RNA and translated into protein, specifying the amino acid sequence of part of a polypeptide; contrasts with intron. Exons are characteristic of eukaryotes.

exoskeleton (Gr. *exo*, outside + *skeletos*, hard) An external skeleton, as in arthropods.

exteroceptor (L. *exter*, outward + *capere*, to take) A sense organ excited by stimuli from the external world.

F₁ (first filial generation) The offspring resulting from a cross; the parents of the cross are referred to as the parental generation.

F₂ (second filial generation) The offspring resulting from a cross between members of the F_1 generation; if these F_2 offspring were to mate and produce progeny, the progeny would be the F_3 generation, and so forth.

facilitated diffusion Carrier-assisted diffusion; the transport of molecules across a cellular membrane through specific channels (carrier molecules embedded in the membrane) from a region of high concentration to a region of low concentration; the process is driven by the concentration difference and does not require energy. The chief difference from free diffusion is that the membrane is impermeable to the molecule except for passage through the carrier channels; contrasts with active transport, which is in the direction of higher concentration and requires energy.

FAD Flavine adenine dinucleotide, an electron acceptor in the respiratory chain.

family A taxonomic group made up of one or more genera; one of the subdivisions of an order. The ending of family names in animals and heterotrophic protists is *-idae*; in all other organisms it is *-aceae*.

fat A molecule composed of glycerol and three fatty acid molecules; the proportion of oxygen to carbon is much less in fats than it is in carbohydrates; fats in the liquid state are called oils.

fatty acid A long hydrocarbon chain ending with a —COOH group; fatty acids are components of fats, oils, phospholipids, and waxes.

feedback inhibition Control

mechanism whereby an increase in the concentration of some molecule inhibits the synthesis of that molecule; more generally, the regulation of the level of any factor sensitive to its own magnitude; important in the regulation of enzyme and hormone levels, ion concentrations, temperature, and many other factors.

fermentation (L. *fermentum,* ferment) The enzyme-catalyzed extraction of energy from organic compounds without the involvement of oxygen; the enzymatic conversion, without oxygen, of carbohydrates to alcohols, acids, and carbon dioxide; the conversion of pyruvate to ethanol or lactic acid.

ferredoxin One of a class of electron-transferring proteins characterized by high iron content; some are involved in photosynthetic photophosphorylation.

fertilization The fusion of two haploid gamete nuclei to form a diploid zygote nucleus.

fetus (L., pregnant) An unborn or unhatched vertebrate that has passed through the earliest developmental stages; in humans, a developing individual is a fetus from about the second month of gestation until birth.

fiber (L. *fibra,* thread) An elongated, tapering, generally thick-walled sclerenchyma cell of vascular plants; its walls may or may not be lignified; it may or may not have a living protoplast at maturity.

fibril (L. *fibrila,* a small thread) Any minute, threadlike organelle within a cell.

filter feeding Any feeding process by which particulate food is filtered from water in which it is suspended.

filtration The filtering of blood in the kidney; blood plasma is forced, under pressure, out of the glomerular capillaries into Bowman's capsule, through which it enters the renal tubule; the filtrate contains water and ions (which are recovered) and metabolic wastes (which are eliminated as urine), but not red blood cells or large proteins, which are too large to pass through the glomerular capillary wall.

fission (L., a splitting) Asexual reproduction by a division of the cell or body into two or more parts of roughly equal size. *See* asexual reproduction, clone.

fitness The genetic contribution of an individual to succeeding generations, relative to the contributions of other individuals in the population.

flagellum, pl. **flagella** (L. *flagellum,* whip) A fine, long, threadlike organelle protruding from the surface of a cell; in bacteria, a single protein fiber, capable of rotary motion, that propels the cell through the water; in eukaryotes, an array of microtubules with a characteristic internal 9 + 2

microtubule structure, capable of vibratory but not rotary motion; used in locomotion and feeding; common in protists and motile gametes. A cilium is a small flagellum.

flower The reproductive structure of an angiosperm, which contains at least one stamen or one carpel, and may contain both kinds of structures and sepals and petals as well.

FMN Flavin mononucleotide, a carrier in the electron transport chain in respiration.

follicle (L. *folliculus,* small ball) In a mammalian ovary, one of the spherical chambers containing an oocyte.

fossil fuels The altered remains of once-living organisms that are burned to release energy. Examples are coal, oil, and natural gas.

fovea (L., a small pit) A small depression in the center of the retina with a high concentration of cones; the area of sharpest vision.

free energy Energy available to do work.

free energy change The total change in usable energy that results from a chemical reaction or other process; equal to the change in total energy (the heat content or enthalpy) minus the change in unavailable energy (the disorder or entropy times temperature). *See* entropy.

frond The leaf of a fern; any large, divided leaf.

fruit In angiosperms, a mature, ripened ovary (or group of ovaries), containing the seeds; also applied informally to the reproductive structures of some other kinds of organisms.

gametangium, pl. **gametangia** (Gr. *gamein,* to marry + L. *tangere,* to touch) A cell or organ in which gametes are formed.

gamete (Gr., wife) A haploid reproductive cell; upon fertilization, its nucleus fuses with that of another gamete of the opposite sex; the resulting diploid cell (zygote) may develop into a new diploid individual, or, in some protists and fungi, may undergo meiosis to form haploid somatic cells.

gametophyte In plants, the haploid (1*n*), gamete-producing generation, which alternates with the diploid (2*n*) sporophyte.

ganglion, pl. **ganglia** (Gr., a swelling) An aggregation of nerve cell bodies; in invertebrates, ganglia are the integrative centers; in vertebrates, the term is restricted to aggregations of nerve cell bodies located outside the central nervous system.

gap junction A junction between adjacent animal cells that allows the passage of materials between the cells; a system of pipes joining the cytoplasms of two adjacent cells.

gastrula (Gr., little stomach) In vertebrates, the embryonic stage in

which the blastula with its single layer of cells turns into a three-layered embryo made up of ectoderm, mesoderm, and endoderm, surrounding a cavity (archenteron) with one opening (blastopore).

gel (gelatin; L. *gelare,* to freeze) A semisolid substance; a colloidal system in which the solid particles form the continuous phase and the fluid medium forms the discontinuous phase.

gene (Gr. *genos,* birth, race) The basic unit of heredity; a sequence of DNA nucleotides on a chromosome that encodes a protein, tRNA, or rRNA molecule, or regulates the transcription of such a sequence.

gene frequency The relative occurrence of a particular allele in a population.

genetic code The "language" of the genes, dictating the correspondence between nucleotide sequence in DNA and amino acid sequence in proteins; a series of 64 different three-nucleotide sequences, or triplets (codons); except for three "stop" signals, each codon corresponds to one of the 20 amino acids.

genetic drift Random fluctuation in allele frequencies over time by chance.

genome (Gr. *genos,* offspring + L. *oma,* abstract group) The total genetic constitution of an organism; in bacteria, all the genes in the main circular chromosome or in associated plasmids; in a eukaryote, all the genes in a haploid set of chromosomes.

genotype (Gr. *genos,* offspring + *typos,* form) The total set of genes present in the cells of an organism, as contrasted with the phenotype, which is the realized expression of these genes; often also used to refer to the genetic constitution underlying a single trait or set of traits.

genus, pl. **genera** (L., race) A taxonomic group that includes species; families are divided into genera.

geotropism *See* gravitropism.

germ cells Gametes, or cells that give rise directly to gametes.

germination (L. *germinare,* to sprout) The resumption of growth and development by a spore or seed.

germ layer A layer of distinctive cells in an animal embryo; each germ layer is fated to give rise to certain tissues or structures as the organism develops. The majority of multicellular animals have three embryonic layers: ectoderm, mesoderm, and endoderm.

gestation (L. *gestare,* to bear) The period in which the offspring of placental mammals are carried in the uterus.

gibberellins (*Gibberella,* a genus of fungi) A group of plant growth hormones, the best-known effect of which is on the elongation of plant stems.

gill A respiratory organ of aquatic animals, usually a thin-walled projection from some part of the external body surface, endowed with a rich capillary bed and having a large surface area; in basidiomycete fungi, the plates on the underside of the cap.

gland (L. *glans*, acorn) A cluster of secretory cells; in animals, typically an organ composed of modified epithelial cells specialized to produce one or more secretions.

glomerulus (L., a little ball) In the vertebrate kidney, a cluster of capillaries enclosed by Bowman's capsule; also, a small spongy mass of tissue in the proboscis of hemichordates, presumed to have an excretory function. Also, a concentration of nerve fibers situated in the olfactory bulb. *See* Bowman's capsule, filtration.

glucagon (Gr. *glukus*, sweet + *ago*, to lead toward) A vertebrate hormone produced in the pancreas that acts to initiate the breakdown of glycogen to glucose subunits and so raise the concentration of blood sugar.

glucose A common six-carbon sugar ($C_6H_2O_6$); the most common monosaccharide in most organisms.

glycerol Three-carbon molecule with three hydroxyl groups attached; combines with fatty acids to form fat or oil.

glycogen (Gr. *glykys*, sweet + *gen*, of a kind) Animal starch; a complex branched polysaccharide that serves as a food reserve in animals, bacteria, and fungi; can be broken down readily into glucose subunits.

glycolysis (Gr. *glykys*, sweet + *lyein*, to loosen) The anaerobic breakdown of glucose; the enzyme-catalyzed breakdown of glucose to two molecules of pyruvic acid, with the net liberation of two molecules of ATP.

glycoxysome A small cellular organelle or microbody containing enzymes necessary for the conversion of fats into carbohydrates; glycoxysomes play an important role during seed germination in plants.

Golgi body (after Camillo Golgi, Italian histologist) An organelle present in many eukaryotic cells; consisting of flat, disk-shaped sacs, tubules, and vesicles, it functions as a collecting and packaging center for substances that the cell manufactures for export; also called dictyosome in plants. The terms "Golgi apparatus" and "Golgi complex" are used to refer collectively to all of the Golgi bodies of a given cell.

gonad (L. *gonas*, a primary sex organ) An organ that produces gametes (ovary in the female and testis in the male).

graded potential Short-range nerve impulses that show variations in intensity (amplitude); the summed effect of all the graded potentials arriving at a neuron cell body from its many dendrites may together be strong enough to initiate an action potential and so generate a nerve impulse that travels down the axon, or it may be too weak to engender a message.

grana, sing. **granum** (L., grain or seed) In chloroplasts, stacks of membrane-bound disks (thylakoids); the thylakoids contain the chlorophylls and carotenoids and are the sites of the light reactions of photosynthesis.

gravitropism (L. *gravis*, heavy + *tropes*, turning) Growth response to gravity in plants; formerly called geotropism.

ground meristem (Gr. *meristos*, divisible) The primary meristem, or meristematic tissue, that gives rise to the plant body (except for the epidermis and vascular tissues).

guanine (Sp. from Quechua, *huanu*, dung) A purine base found in DNA and RNA; its name derives from the fact that it occurs in high concentration as a white crystalline base, $C_5H_5N_5O$, in guano and other animal excrements.

guard cells Pairs of specialized epidermal cells that surround a stoma; when the guard cells are turgid, the stoma is open, and when they are flaccid, it is closed.

guttation (L. *gutta*, a drop) The exudation of liquid water from leaves due to root pressure.

gymnosperm (Gr. *gymnos*, naked + *sperma*, seed) A seed plant with seeds not enclosed in an ovary; the conifers are the most familiar group.

gynoecium (Gr. *gyne*, woman + *oikos*, house) The aggregate of carpels in the flower of a seed plant.

HTLV Human T cell lymphotropic virus, a group of single plus-strand RNA retroviruses responsible for several very serious human diseases, including a common form of leukemia and AIDS.

habitat (L. *habitare*, to inhabit) The environment of an organism; the place where it is usually found.

habituation (L. *habitus*, condition) A form of learning; a diminishing response to a repeated stimulus; the ignoring of an often-repeated stimulus.

haploid (Gr. *haploos*, single + *ploion*, vessel) Having only one set of chromosomes (n), in contrast to diploid $(2n)$; characteristic of eukaryotic gametes, of gametophytes in plants, and of some protists and fungi.

Hardy-Weinberg equilibrium A mathematical description of the fact that the relative frequencies of two or more alleles in a population do not change because of Mendelian segregation; allele and genotype frequencies remain constant in a random-mating population in the absence of inbreeding, selection, or other evolutionary forces; usually stated: if the frequency of allele *a* is *p* and the frequency of allele *b* is *q*, then the genotype frequencies after one generation of random mating will always be $p^2(a) + 2pq(ab) + q^2(b)$.

haustorium, pl. **haustoria** (L. *haustus*, from *haurire*, to drink, draw) A projection of fungal hypha, stem, or plant part that acts as a penetrating and absorbing organ.

heart In animals, a muscular organ that pumps the blood through a closed circulatory system.

heme (Gr. *haima*, blood) An iron-containing porphyrin group in proteins such as hemoglobin and the cytochromes.

hemoglobin (Gr. *haima*, blood + L. *globus*, a ball) A globular protein in vertebrate red blood cells and in the plasma of many invertebrates that carries oxygen and carbon dioxide; an essential part of each molecule is an iron-containing heme group, which both binds O_2 and CO_2 and gives blood its red color.

hemophilia (Gr. *haima*, blood + *philios*, friendly) A group of hereditary diseases characterized by failure of the blood to clot and consequent excessive bleeding from even minor wounds; a mutation in a gene encoding one of the protein factors involved in blood clotting.

hepatic (Gr. *hepatikos*, of the liver) Pertaining to the liver.

herb (L. *herba*, grass) A nonwoody seed plant with a relatively short-lived aerial portion. A herb is said to be *herbaceous*.

herbivore (L. *herba*, grass + *vorare*, to devour) Any organism subsisting on plants. Adj., *herbivorous*.

heredity (L. *heredis*, heir) The transmission of characteristics from parent to offspring through the gametes.

hermaphrodite (Gr. *hermaphroditos*, containing both sexes; from Greek mythology, Hermaphroditos, son of Hermes and Aphrodite) An organism with both male and female functional reproductive organs; many plants and some animals such as deepsea fishes are hermaphroditic; hermaphrodites may or may not be self-fertilizing.

heterochromatin (Gr. *heteros*, different + *chroma*, color) That portion of eukaryotic chromosomes that is not transcribed into RNA; stains intensely in histological preparations; characteristic of centromeres.

heterocyst (Gr. *heteros*, different + *kystis*, bladder) A large, transparent, thick-walled cell that forms under appropriate conditions in the filaments of certain cyanobacteria; nitrogen fixation takes place in heterocysts.

heterokaryotic (Gr. *heteros*, other + *karyon*, kernel) In fungi, having two or more genetically distinct types of nuclei within the same mycelium.

heterosporous (Gr. *heteros*, other + *sporos*, seed) In vascular plants, having spores of two kinds, namely, microspores and megaspores.

heterotroph (Gr. *heteros*, other +

trophos, feeder) An organism that cannot derive energy from photosynthesis or inorganic chemicals, and so must feed on other plants and animals, obtaining chemical energy by degrading their organic molecules; animals, fungi, and many unicellular organisms are heterotrophs; *see also* autotroph.

heterozygote (Gr. *heteros,* other + *zygotos,* a pair) A diploid individual that carries two different alleles on homologous chromosomes at one or more genetic loci. Adj., *heterozygous.* Opposite of homozygote.

heterozygote superiority The greater fitness of an organism heterozygous at a given genetic locus as compared with either homozygote.

hibernation (L. *hiberna,* winter) In mammals, the passing of winter in a torpid state in which the body temperature drops nearly to freezing and the metabolism drops close to zero.

histone (Gr. *histos,* tissue) A group of relatively small, very basic polypeptides, rich in arginine and lysine. An essential component of eukaryotic chromosomes, histones form the core of nucleosomes around which DNA is wrapped in the first stage of chromosome condensation.

holoblastic cleavage (Gr. *holos,* whole + *blastos,* germ) Process in vertebrate embryos in which the cleavage divisions all occur at the same rate, yielding a uniform cell size in the blastula; found in mammals, lancelets, and many aquatic invertebrates that have eggs with a small amount of yolk.

homeostasis (Gr. *homeos,* similar + *stasis,* standing) The maintaining of a relatively stable internal physiological environment in an organism, or steady-state equilibrium in a population or ecosystem; usually involves some form of feedback self-regulation.

homeotherm (Gr. *homoios,* same or similar + *therme,* heat) An organism, such as a bird or mammal, capable of maintaining a stable body temperature independent of the environmental temperature; at lower environmental temperatures this involves the metabolic generation of heat at considerable expense in terms of ATP utilized; "warm blooded." *See* endotherm. Contrasts with poikilotherm.

hominid (L. *homo,* man) Any primate in the human family, Hominidae. *Homo sapiens* is the only living representative; *Australopithecus* is an extinct genus.

hominoid (L. *homo,* man) Collectively, hominids and apes; together with the monkeys, hominoids constitute the anthropoid primates.

homokaryotic (Gr. *homos,* same + *karyon,* kernel) In fungi, having nuclei with the same genetic makeup within a mycelium.

homologous chromosome (Gr. *homologia,* agreement) In diploid cells, one chromosome of a pair that carry equivalent genes; chromosomes that associate in pairs in the first stage of meiosis; also called "homologues."

homology (Gr. *homologia,* agreement) A condition in which the similarity between two structures or functions is indicative of a common evolutionary origin. Adj., *homologous.*

homospory (Gr. *homos,* same or similar + *sporos,* seed) In some plants, production of only one type of spore rather than differentiated types. Compare heterospory. Adj., *homosporous.*

homozygote (Gr. *homos,* same or similar + *zygotos,* a pair) A diploid individual that carries identical alleles at one or more genetic loci on its two homologous chromosomes; opposite of heterozygote.

homozygous Being a homozygote; the term is usually applied to one or more specific loci, as in "homozygous with respect to the *w* locus" (that is, the genotype is *w/w*).

hormone (Gr. *hormaein,* to excite) A chemical messenger; a molecule, usually a peptide or steroid, that is produced in one part of an organism and triggers a specific cellular reaction in target tissues and organs some distance away.

host An organism on or in which a parasite lives.

hybrid (L. *hybrida,* the offspring of a tame sow and a wild boar) Offspring of two different varieties or of two different species; alternatively, offspring of two parents that differ in one or more heritable characteristics.

hybridization The mating of unlike parents.

hybridoma (contraction of *hybrid* + *myeloma*) A fast-growing cell line produced by fusing a cancer cell (myeloma) to some other cell, such as an antibody-producing cell. *See* monoclonal antibody.

hydrocarbon (Gr. *hydor,* water + L. *carbo,* charcoal) An organic compound consisting only of carbon and hydrogen atoms.

hydrogen bond A weak and very directional molecular attraction involving hydrogen atoms; produced by the interaction of the partial positive charge of a polar hydrogen atom (typically, hydrogen atoms covalently linked to oxygen or nitrogen, which more strongly attract the shared electron and so render the hydrogen nucleus partially positive) with the partial negative charge of another polar atom (typically an oxygen or nitrogen atom without a covalently bound hydrogen). *See* polar.

hydroid The polyp form of a cnidarian,

as distinguished from the medusa form. Any cnidarian of the class Hydrozoa, order Hydroida.

hydrolysis (Gr. *hydoe,* water + *lysis,* loosening) Splitting of one molecule into two parts by addition of H^+ and OH^- ions, derived from water. *See opposite terms* condensation reaction, dehydration synthesis.

hydrophilic (Gr. *hydor,* water + *philios,* friendly) Having an affinity for water; applied to polar molecules, which readily form hydrogen bonds with water and so readily dissolve in water (are water soluble). *See* polar molecule.

hydrophobic (Gr. *hydor,* water + *phobos,* hating) Repelled by water; refers to nonpolar molecules, which do not form hydrogen bonds with water and so are not soluble in water.

hydrophobic interaction (L. *hydrophobos,* hated by water) The propensity for water molecules to exclude nonpolar molecules, as oil is excluded from water; water tends to form the maximum number of hydrogen bonds, and more hydrogen bonds are possible if nonpolar molecules (which do not form hydrogen bonds) are not present to interfere with the hydrogen bonds between water molecules. Hydrophobic interactions are responsible for much of the three-dimensional structure of proteins, which is why the addition of nonpolar solvents denatures (unfolds) proteins.

hydroxyl group (hydrogen + oxygen + *-yl*) An OH^- group; a negatively charged ion formed by the disassociation of a water molecule.

hyperosmotic (Gr. *hyper,* over + *osmos,* impulse) Refers to a hypertonic solution whose osmotic pressure is greater than that of another solution with which it is compared; contrasts with hypoosmotic.

hypertonic (Gr. *hyper,* above + *tonos,* tension) Refers to a solution that contains a higher concentration of solute particles; water moves across a semipermeable membrane into hypertonic solution.

hypha, pl. **hyphae** (Gr. *hyphe,* web) A filament of a fungus or oomycete; collectively, the hyphae comprise the mycelium.

hypoosmotic (Gr. *hypo,* under + *osmos,* impulse) Refers to a hypotonic solution whose osmotic pressure is less than that of another solution with which it is compared; contrasts with hyperosmotic.

hypophysis (Gr. *hypo,* under + *physis,* growth) A portion of the pituitary gland.

hypothalamus (Gr. *hypo,* under + *thalamos,* inner room) A region of the vertebrate brain just below the cerebral hemispheres, under the thalamus; a center of the autonomic nervous system, responsible for the integration and correlation of many neural and endocrine functions.

hypothesis (Gr. *hypo*, under + *tithenai*, to put) A guess as to what might be; a postulated explanation of a phenomenon consistent with available information; no hypothesis is ever *proved* to be correct—all hypotheses are provisional, working ideas that are accepted for the time being but may be rejected in the future if not consistent with data generated by further experiments; a hypothesis that survives many tests and is very unlikely to be discarded is referred to as a "theory."

hypotonic (Gr. *hypo*, under + *tonos*, tension) Refers to a solution that contains a lower concentration of solute particles; water moves across a semipermeable membrane out of a hypotonic solution.

IAA *See* indoleacetic acid.

immune response In vertebrates, a defensive reaction of the body to invasion by a foreign substance or organism; the invader is recognized as foreign or "not-self" by antibodies produced by B cell lymphocytes, and then is eliminated by macrophages. T cells are responsible for protecting an individual from its own antibodies (cell-mediated immunity). *See* antibody, B cell.

immunoglobulin (L. *immunis*, free + *globus*, globe) An antibody.

inbreeding The breeding of genetically related plants or animals. In plants, inbreeding results from self-pollination; in animals, inbreeding results from matings between relatives; inbreeding tends to increase homozygosity.

incomplete dominance The ability of two alleles to produce a heterozygous phenotype that is different from either homozygous phenotype.

independent assortment Mendel's second law; the principle that segregation of alternative alleles at one locus into gametes is independent of the segregation of alleles at other loci; only true for gene loci located on different chromosomes, or so far apart on one chromosome that crossing-over is very frequent between the loci. *See* Mendel's second law.

indeterminate cleavage In animals, a type of embryonic development in which determination (the irreversible commitment of individual blastomeres to particular developmental fates) is delayed until later in the development of the embryo; for example, in echinoderms and vertebrates.

indeterminate growth In plants, unrestricted or unlimited growth, as with a vegetative apical meristem that produces an unrestricted number of lateral organs indefinitely.

indoleacetic acid (IAA) A naturally occurring auxin, one of the plant hormones.

indusium, pl. **indusia** (L., a woman's undergarment) Membranous growth of the epidermis of a fern leaf that covers a sorus.

inferior ovary An ovary that is completely or partially attached to the calyx; the other floral whorls appear to arise from its top.

inflammation (L. *inflammare*, from *flamma*, flame) The mobilization of body defenses against foreign substances and infectious agents, and the repair of damage from such agents; involves phagocytosis by macrophages and is often accompanied by an increase in the local temperature.

inflorescence A flower cluster; many definite kinds of floral arrangement occur in inflorescences.

innate (L. *innatus*, inborn) Describing a characteristic based partly or wholly on inherited gene differences.

insertion sequences Relatively short sequences of DNA that can move from one chromosomal location to other places in the same chromosome or in other chromosomes. Insertion sequences are short sequences of nucleotides that do not include any other genes; they often function as the ends of such larger transposable elements as transposons and retroviruses.

instar (L., form) Stage in the life of an insect or other arthropod between molts.

instinct (L. *instinctus*, impelled) Stereotyped, predictable, genetically programmed behavior. Learning may or may not be involved.

insulin A peptide hormone produced by the vertebrate pancreas which acts to promote glycogen formation and thus to lower the concentration of sugar in the blood.

integration, neural The summation of the depolarizing and repolarizing effects contributed by all excitatory and inhibitory synapses acting on a neuron. *See* graded potential.

integument (L. *integumentum*, covering) In plants, the outermost layer or layers of tissue enveloping the nucellus of the ovule; develops into the seed coat.

interferon In vertebrates, a protein produced in virus-infected cells that inhibits viral multiplication.

internode In plants, the region of a stem between two successive nodes.

interneuron Neuron that transmits nerve impulses from one neuron to another within the central nervous system; an individual may receive impulses from and transmit impulses to many different neurons.

interphase The period between two mitotic or meiotic divisions in which a cell grows and its DNA replicates; includes G1, S, and G2 phases.

intervening sequence An intron.

intron (L. *intra*, within) Portion of mRNA as transcribed from eukaryotic DNA that is removed by enzymes before the mature mRNA is translated into protein. These untranscribed regions comprise the bulk of most eukaryotic genes; typically, the transcribed portion of the gene exists as numerous short segments called "exons" that are scattered in no particular order within a much longer stretch of nontranscribed DNA. Those segments of the background nontranscribed DNA that fall between two exons are called introns. *See* exon.

invagination (L. *in*, in + *vagina*, sheath) The local infolding of a layer of tissue, especially in animal embryos, so as to form a depression or pocket opening to the outside.

inversion (L. *invertere*, to turn upside down) A reversal in order of a segment of a chromosome; also, to turn inside out, as in embryogenesis of sponges or discharge of a nematocyst.

ion Any atom or molecule containing an unequal number of electrons and protons and therefore carrying a net positive or net negative charge; gain of an extra electron produces a negatively charged cation, and loss of an electron produces a positively charged anion.

ionic bond A chemical bond formed as a result of the mutual attraction of ions of opposite charge; ionic bonds are nondirectional and form between one ion and nearly all ions of opposite charge.

irritability (L. *irritare*, to provoke) The ability to respond to stimuli; a general property of all organisms.

isolating mechanisms Mechanisms that prevent genetic exchange between individuals of different populations or species; may be behavioral, morphological, or physiological.

isomer (Gr. *isos*, equal + *meros*, part) One of a group of molecules identical in atomic composition but differing in structural arrangement, for example, glucose and fructose.

isomorphic (Gr. *isos*, equal + *morphe*, form) A term used to describe alternation of generations in some kinds of organisms, in which the gametophyte and sporophyte generations are similar in form.

isotonic (Gr. *isos*, equal + *tonos*, tension) Refers to two solutions that have equal concentrations of solute particles; if two isotonic solutions are separated by a semipermeable membrane, there will be no net flow of water across the membrane.

isotope (Gr. *isos*, equal + *topos*, place) An alternative form of a chemical element; differs from other atoms of the same element in the number of neutrons in the nucleus; isotopes thus differ in atomic weight. All isotopes have the same chemical behavior, as all contain the same number of protons and electrons. Some isotopes are unstable and emit radiation.

juvenile hormone Hormone produced by the corpora allata of insects; among its effects are maintenance of larval or nymphal characteristics during development.

karyotype (Gr. *karyon*, kernel + *typos*, stamp or print) The morphology of the chromosomes of an organism as viewed with a light microscope.

kb (kilobase) An abbreviation for 1000 base pairs of DNA.

keratin (Gr. *kera*, horn + *in*, suffix used for proteins) A tough, fibrous protein formed in epidermal tissues and modified into skin, feathers, hair, and hard structures such as horns and nails.

kidney In vertebrates, the organ that filters the blood to remove nitrogenous wastes and regulates the balance of water and solutes in blood plasma. *See* filtration, Bowman's capsule, glomerulus.

kinetic energy Energy of motion.

kinetin (Gr. *kinetikos*, causing motion) A synthetically produced purine that probably does not occur in nature but that acts as a cytokinin in plants.

kinetochore (Gr. *kinetikos*, putting in motion + *choros*, chorus) Disk-shaped protein structure within the centromere to which the spindle fibers attach during mitosis or meiosis. *See* centromere.

kinetosome (Gr. *kinetos*, moving + *soma*, body) A basal body; the self-duplicating organelle at the base of a flagellum or cilium; similar in structure to a centriole. *See* basal body.

kingdom The chief taxonomic category, for example, Monera or Plantae. In this book we recognize five kingdoms.

kin selection Selection favoring relatives; an increase in the frequency of related individuals (kin) in a population, leading to an increase in the relative frequency in the population of those alleles shared by members of the kin group.

Krebs cycle Another name for the citric acid cycle; also called the tricarboxylic acid (TCA) cycle.

K-selection (from the *K* term in the logistic equation) Natural selection under conditions that favor survival when populations are controlled primarily by density-dependent factors.

labrum (L., a lip) The upper lip of insects and crustaceans situated above or in front of the mandibles.

lamella (L., a little plate) A thin, platelike structure; in chloroplasts, a layer of chlorophyll-containing membranes; in bivalve mollusks, one of the two plates forming a gill; in vertebrates, one of the thin layers of bone laid concentrically around an osteon (Haversian) canal.

laminarin The principal storage product of the brown algae; a polymer of glucose.

larva, pl. **larvae** (L., a ghost) Immature form of an animal that is quite different from the adult and undergoes metamorphosis in reaching the adult form; examples are caterpillars and tadpoles.

larynx The voice box; a cartilaginous organ that lies between the pharynx and trachea and is responsible for sound production in vertebrates.

lateral meristems (L. *latus*, side + Gr. *meristos*, divided) In vascular plants, the meristems that give rise to secondary tissue; the vascular cambium and cork cambium.

leaf gap In plants, the region of parenchyma tissue in the primary vascular cylinder above the point of departure of the leaf trace or traces.

leaf primordium (L. *primordium*, beginning) A lateral outgrowth from the apical meristem that will eventually become a leaf.

learning The modification of behavior by experience.

lenticels (L. *lenticella*, a small window) Spongy areas in the cork surfaces of stem, roots, and other plant parts that allow interchange of gases between internal tissues and the atmosphere through the periderm.

leucoplast (Gr. *leukos*, white + *plasein*, to form) In plant cells, a colorless plastid in which starch grains are stored; usually found in cells not exposed to light, such as roots and internal stem tissue.

leukocyte (Gr. *leukos*, white + *kytos*, hollow vessel) A white blood cell; a diverse array of non-hemoglobin–containing blood cells, including phagocytic macrophages and antibody-producing lymphocytes.

life cycle The sequence of phases in the growth and development of an organism, from zygote formation to gamete formation.

light reactions First stage of photosynthesis; the absorption by a pigment of a photon of light from sunlight, the transport of the resulting excited electron to a proton pump to drive the chemiosmotic synthesis of ATP or the reduction of $NADP^+$, and the return of the electron to the pigment.

lignin A stiffening substance that is the most abundant polymer in plant cell walls after cellulose.

linkage Lack of independent segregation; the tendency for two or more genes to segregate together in a cross owing to the fact that they are located on the same chromosome.

lipase (Gr. *lipos*, fat + *-ase*, enzyme suffix) An enzyme that catalyzes the hydrolysis of fats.

lipid (Gr. *lipos*, fat) a nonpolar organic molecules; fatlike; one of a large variety of nonpolar hydrophobic molecules that are insoluble in water (which is polar) but that dissolve readily in nonpolar organic solvents; includes fats, oils, waxes, steroids, phospholipids, and carotenes.

lithosphere (Gr. *lithos*, rock + *sphaira*, ball) The rocky component of the earth's surface layers.

locus, pl. **loci** (L., place) The location of a gene on a chromosome; more precisely, the position of a particular transcription unit on a chromosome.

loop of Henle (after F.G.J. Henle, German pathologist) In the kidney of birds and mammals, a hairpin-shaped portion of the renal tubule in which water and salt are reabsorbed from the glomerular filtrate by diffusion.

lymph (L. *lympha*, water) In animals, a colorless fluid derived from blood by filtration through capillary walls in the tissues.

lymphatic system In animals, an open vascular system that reclaims water which has entered interstitial regions from the bloodstream (lymph); also transports fats from small intestine to the bloodstream; consists of lymph capillaries, which begin blindly in the tissues and lead to a network of progressively larger vessels that empty into the vena cava; also includes the lymph nodes, spleen, thymus, and tonsils.

lymph node In animals, a mass of spongy tissues that serve to filter the lymphatic system. Located throughout the lymphatic system, lymph nodes remove dead cells, debris, and foreign particles from the circulation.

lymphocyte (L. *lympha*, water + Gr. *kytos*, hollow vessel) A type of white blood cell. Lymphocytes are responsible for the immune response; there are two principal classes—B cells (which differentiate into antibody-producing plasma cells) and T cells (which interact directly with the foreign invader and are responsible for cell-mediated immunity).

lymphoid system In mammals, a collective term referring to those tissues and organs that function as cell-producing centers—bone marrow, thymus, lymph nodes, spleen, appendix, tonsils, adenoids, and patches of small intestine.

lysis (Gr., a loosening) Disintegration of a cell by rupture of its cell membrane.

lysogenic bacteria (Gr. *lysis*, a loosening + *genos*, race or descent) Bacteria carrying a silent virus integrated into the chromosome as if it were a bacterial gene; such bacteria carry the seeds of their own destruction (are "lyso-genic") because at some future time the virus may initiate active reproduction, causing lysis of the bacterial cells.

lysosome (Gr. *lysis*, a loosening + *soma*, body) A membrane-bound cell organelle containing hydrolytic (digestive) enzymes that are released when the lysosome ruptures; important in recycling worn-out mitochondria and other cellular debris.

macromolecule (Gr. *makros*, large + L. *moliculus*, a little mass) An extremely large molecule; a molecule of very high molecular weight; refers specifically to proteins, nucleic acids, polysaccharides, and complexes of these.

macronutrients (Gr. *makros*, large + L. *nutrire*, to nourish) Inorganic chemical elements required in large amounts for plant growth, such as nitrogen, potassium, calcium, phosphorus, magnesium, and sulfur.

Malpighian tubules (Marcello Malpighi, Italian anatomist, 1628-1694) Blind tubules opening into the hindgut of terrestrial arthropods; they function as excretory organs.

mandibles (L. *mandibula*, jaw) In crustaceans, insects, and myriapods, the appendages immediately posterior to the antennae; used to seize, hold, bite, or chew food.

mantle The soft, outermost layer of the body wall in mollusks; the mantle secretes the shell.

mass In chemistry, the total number of protons and neutrons in the nucleus of an atom. Approximately equal to the atomic weight.

mating type A strain of organisms incapable of sexual reproduction with one another but capable of such reproduction with members of other strains of the same organism.

matrix (L. *mater*, mother) In mitochondria, the solution in the interior of a mitochondrion, surrounding the cristae; contains the enzymes and other molecules involved in oxidative respiration; more generally, the intercellular substance of a tissue, or that part of a tissue within which an organ or process is embedded.

mechanoreceptor A sensory cell or organ that responds to mechanical stimuli associated with gravity and pressure; includes hearing, touch, and balance.

medulla (L., marrow) The inner portion of an organ, in contrast to the cortex or outer portion, as in the kidney or adrenal gland. Also, the most posterior region of the vertebrate brain, the hindbrain.

medusa (Gr. mythology, a female monster with snake-entwined hair) A jellyfish, or the free-swimming stage in the life cycle of cnidarians in general; medusae are free swimming and bell or umbrella shaped.

megagametophyte (Gr. *megas*, large + *gamos*, marriage + *phyton*, plant) In heterosporous plants, the female gametophyte; located within the ovule of seed plants.

megasporangium, pl. **megasporangia** A sporangium in which megaspores are produced in heterosporous plants.

megaspore (Gr. *megas*, large + *sporos*, seed) In heterosporous plants, a haploid (1*n*) spore that develops into a female gametophyte; megaspores are usually larger than microspores.

meiosis (Gr. *meioun*, to make smaller) Reduction division; the two successive nuclear divisions in which a single diploid (2*n*) cell forms four haploid (1*n*) nuclei, halving the chromosome number; segregation, crossing-over, and reassortment all occur during meiosis; in animals, meiosis usually occurs in the last two divisions in the formation of the mature egg or sperm; in plants, spores—which divide by mitosis—are produced as a result of meiosis. Compare with mitosis.

Mendel's first law The law of allele segregation: the factors specifying a pair of alternative characteristics (alleles) are separate, and only one may be carried in a particular gamete; gametes combine randomly in forming progeny. Chromosomes had not been observed in Mendel's time, and meiosis was unknown. Modern form: alleles segregate as chromosomes do.

Mendel's second law The law of independent assortment: The inheritance of alternative characteristics (alleles) of one trait is independent of the simultaneous inheritance of other traits—different traits (genes) assort independently. Mendel never tested a pair of traits that were located close together on a chromosome (although two of the seven genes he studied in fact were), and so never discovered that only genes that are not close together assort independently. Modern form: unlinked genes assort independently.

menopause (Gr. *men*, month + *pauein*, to cease) In the human female, the time when ovulation ceases; cessation of the menstrual cycle.

menstruation (L. *mens*, month) Periodic sloughing off of the blood-enriched lining of the uterus when pregnancy does not occur. The menstrual cycle in primates is the cycle of hormone-regulated changes in the condition of the uterine lining, which is marked by the periodic discharge of blood and disintegrated uterine lining through the vagina (menstruation).

meristem (Gr. *merizein*, to divide) Undifferentiated plant tissue from which new cells arise.

meroblastic (Gr. *meros*, part + *blastos*, germ) In vertebrates, a pattern of embryonic cleavage divisions in fertilized eggs having a large amount of yolk at the vegetal pole; cleavage divisions occur more rapidly at the animal pole and in extreme cases are restricted to a small area at the animal pole.

mesenchyme (Gr. *mesos*, middle + *enchyma*, infusion) Embryonic connective tissue; irregular or amoebocytic cells, often embedded in a gelatinous matrix.

mesenteries (Gr. *mesos*, middle + *enteron*, gut) Double layers of mesoderm that suspend the digestive tract and other internal organs within the coelom.

mesoderm (Gr. *mesos*, middle + *derma*, skin) One of the three embryonic germ layers that form in the gastrula; gives rise to muscle, bone and other connective tissue, the peritoneum, the circulatory system, and most of the excretory and reproductive systems.

mesoglea (Gr. *mesos*, middle + *glia*, glue) The layer of jellylike or cement material between the epidermis and gastrodermis in cnidarians; may also be used to refer to the jellylike matrix between the epithelial layers in sponges.

mesophyll (Gr. *mesos*, middle + *phyllon*, leaf) The photosynthetic parenchyma of a leaf, located within the epidermis. The vascular strands (veins) run through the mesophyll.

messenger RNA (mRNA) The RNA transcribed from structural genes; RNA molecules complementary to one strand of DNA, which are translated by the ribosomes into protein.

metabolic pathways In all living organisms, sequences of enzyme-catalyzed reactions in which the product of one reaction serves as the starting substance for the next.

metabolism (Gr. *metabole*, change) The sum of all chemical processes occurring within a living cell or organism; includes carbon fixation (photosynthesis), digestion (catabolism), extraction of chemical energy (respiration), and synthesis of organic molecules (anabolism).

metamorphosis (Gr. *meta*, after + *morphe*, form + *osis*, state of) Process in which there is a marked change in form during postembryonic development, for example, tadpole to frog or larval insect to adult.

metaphase (Gr. *meta*, middle + *phasis*, form) The stage of mitosis or meiosis during which microtubules become organized into a spindle and the chromosomes come to lie in the spindle's equatorial plane.

microbody A cellular organelle bounded by a single membrane and containing a variety of enzymes; generally derived from endoplasmic reticulum; includes peroxisomes and glyoxysomes.

microfibril A threadlike component of the plant cell wall, composed of cellulose molecules, visible only with an electron microscope.

microfilament (Gr. *mikros*, small + L.

filum, a thread) In cells, a fine protein thread composed of actin and myosin; capable of contraction when supplied with ATP. Contraction of microfilaments is the basic mechanism underlying changes in shape of eukaryotic cells, including cell motility and muscle contraction.

microgametophyte (Gr. *mikros,* small + *gamos,* marriage + *phyton,* plant) In heterosporous plants, the male gametophyte.

micrometer A unit of microscopic measurement convenient for describing cellular dimensions; 10^{-6} meter (about 1/25,000 of an inch); its symbol is μm. Replaces the now obsolete term "micron."

micronutrient (Gr. *mikros,* small + L. *nutrire,* to nourish) A mineral required in only minute amounts for plant growth, such as iron, chlorine, copper, manganese, zinc, molybdenum, and boron.

micropyle In the ovules of seed plants, an opening in the integuments through which the pollen tube usually enters.

microspore (Gr. *mikros,* small + *sporos,* seed) In plants, a spore that develops into a male gametophyte; in seed plants, it develops into a pollen grain.

microtrabecula, pl. **microtrabeculae** (Gr. *mikros,* small + L. *trabecula,* little beam) In eukaryotic cells, wisplike fibers that interconnect all of the other structures within the cell; visible in the cytoplasm under a high-voltage electron microscope.

microtubule (Gr. *mikros,* small + L. *tubulus,* little pipe) In eukaryotic cells, a long, hollow protein cylinder, about 25 nanometers in diameter, composed of the protein tubulin. Microtubules influence cell shape, move the chromosomes in cell division, and provide the functional internal structure of cilia and flagella.

microvillus (Gr. *mikros,* small + L. *villus,* shaggy hair) Cytoplasmic projection from epithelial cells, usually containing microtubules; microvilli greatly increase the surface area of the small intestine.

middle lamella The layer of intercellular material, rich in pectic compounds, that cements together the primary walls of adjacent plant cells.

mimicry (Gr. *mimos,* mime) The resemblance in form, color, or behavior of certain organisms (mimics) to other more powerful or more protected ones (models), which results in the mimics being protected in some way.

mineral A naturally occurring element or inorganic compound.

miscarriage Spontaneous expulsion of an animal embryo from the uterus before the third trimester of pregnancy.

mitochondrion, pl. **mitochondria** (Gr. *mitos,* thread + *chondrion,* small grain) The site of oxidative respiration in eukaryotes; a bacterium-like organelle found within the cells of all but one species of eukaryote. Mitochondria contain the enzymes catalyzing the citric acid cycle, which generates electrons that drive proton pumps within the mitochondrial membrane and so fosters the chemiosmotic synthesis of ATP. Almost all of the ATP of nonphotosynthetic eukaryotic cells is produced in mitochondria.

mitosis (Gr. *mitos,* thread) Somatic cell division; nuclear division in which the duplicated chromosomes separate to form two genetically identical daughter nuclei; usually accompanied by cytokinesis, producing two daughter cells. After mitosis, the chromosome number in each daughter nucleus is the same as it was in the original dividing cell. Mitosis is the basis of reproduction of single-celled eukaryotes, and of the physical growth of multicellular eukaryotes.

mitotic spindle During nuclear division, a microtubular structure that separates sister chromatids of a chromosome from each other.

mole (L. *moles,* mass) The atomic weight of a substance, expressed in grams; 1 mole is defined as the mass of 6.022×10^{23} atoms (Avogadro's number), the number of atoms in 22.4 liters.

molecular weight The sum of the atomic weights of the constituent atoms in a molecule.

molecule (L. *moliculus,* a small mass) A collection of two or more atoms held together by chemical bonds; the smallest unit of a compound that displays the properties of the compound.

molting Shedding of all or part of an organism's outer covering; in arthropods, periodic shedding of the exoskeleton to permit an increase in size.

monoclonal antibody An antibody of a single type that is produced by genetically identical plasma cells (clones); a monoclonal antibody is typically produced from a cell culture derived from the fusion product of a cancer cell and an antibody-producing cell. *See* hybridoma.

monocot Short for monocotyledon; a flowering plant in which the embryos have only one cotyledon, the floral parts are generally in threes, and the leaves typically are parallel-veined. Compare dicot.

monocyte (Gr. *monos,* single + *kytos,* hollow vessel) A type of leukocyte that becomes a phagocytic cell (macrophage) after moving into tissues.

monoecious (Gr. *monos,* single + *oecos,* house) A plant in which the staminate and pistillate flowers are separate, but borne on the same individual.

monohybrid (Gr. *monos,* single + L. *hybrida,* mongrel) A heterozygous (hybrid) offspring of parents differing in one specified character.

monomer (Gr. *monos,* single + *meros,* part) A subunit; a molecule capable of linking with other like molecules to form polymers.

monosaccharide (Gr. *monos,* one + *sakcharon,* sugar, from Sanskrit *sarkara,* gravel, sugar) A simple sugar that cannot be decomposed into smaller sugar molecules; the most common are five-carbon pentoses (such as ribose) and six-carbon hexoses (such as glucose).

morphogenesis (Gr. *morphe,* form + *genesis,* origin) The development of form; construction of the architectural features of organisms; the formation and differentiation of tissues and organs.

morphological (Gr. *morphe,* form + *logos,* discourse) Pertaining to form and structure

morphology (Gr. *morphe,* form + *logos,* discourse) The study of form and its development; includes cytology (the study of cell structure), histology (the study of tissue structure), and anatomy (the study of gross structure).

morula (L., a little mulberry) Solid ball of cells in the early stage of embryonic development.

motor neuron Neuron that transmits nerve impulses from the central nervous system to an effector, which is typically a muscle or a gland; an efferent neuron.

mRNA *See* messenger RNA.

multigene family A collection of related genes on a chromosome; a family of genes duplicated by unequal crossing-over whose members have since diverged in nucleotide sequence but are clearly related to one another. The majority of eukaryotic genes appear to be members of multigene families.

muscle fiber Muscle cell; a long, cylindrical, multinucleated cell containing numerous myofibrils, which is capable of contraction when stimulated.

mutagen (L. *mutare,* to change + Gr. *genaio,* to produce) An agent that induces changes in DNA (mutations); includes physical agents that damage DNA and chemicals that alter one or more DNA bases, link them together, or delete one or more of them. An X ray is an example of a physical mutagen that breaks both strands of a DNA molecule; cigarette tar is an example of a chemical that produces cancer-inducing mutations in lung cells.

mutant (L. *mutare,* to change) A mutated gene; alternatively, an organism carrying a gene that has undergone a mutation.

mutation A permanent change in a cell's DNA; includes changes in nucleotide sequence, alteration of gene position, gene loss or duplication, and insertion of foreign sequences.

mutalism (L. *mutuus*, lent, borrowed) The living together of two or more organisms in a symbiotic association in which both members benefit.

mycelium (Gr. *mykes*, fungus) In fungi or oomycetes, a mass of hyphae.

mycology The study of fungi. One who studies fungi is called a *mycologist*.

mycorrhiza, pl. **mycorrhizae** (Gr. *mykes*, fungus + *rhiza*, root) A symbiotic association between fungi and the roots of a plant.

myelin sheath (Gr. *myelinos*, full of marrow) A fatty layer surrounding the long axons of motor neurons in the peripheral nervous system of vertebrates; made up of the membranes of Schwann cells.

myofibril (Gr. *myos*, muscle + L. *fibrilla*, little fiber) A contractile microfilament within muscle, composed of myosin and actin.

myoglobin (Gr. *mys*, muscle + L. *globus*, a ball) An oxygen-binding, heme-containing globular protein found in vertebrate muscles.

myosin (Gr. *mys*, muscle + *in*, belonging to) One of the two protein components of microfilaments (the other is actin); a principal component of vertebrate muscle.

NAD (nicotinamide adenine dinucleotide) A coenzyme that functions as an electron acceptor in many of the oxidation reactions of respiration. NAD^+ is the oxidized form of NAD; NADH is the reduced form.

NADP (nicotinamide adenine dinucleotide phosphate) A coenzyme that functions as an electron donor in many of the reduction reactions of biosynthesis. $NADP^+$ is the oxidized form; $NADPH_2$, the reduced form of NADP.

natural selection The differential reproduction of genotypes; caused by factors in the environment; leads to evolutionary change.

nectar (Gr. *nektar*, the drink of the gods) A sugary fluid that attracts insects to plants. Nectar is produced in structures called nectaries.

negative feedback A homeostatic control mechanism whereby an increase in some substance or activity inhibits the process leading to the increase; also known as feedback inhibition.

nematocyst (Gr. *nema*, thread + *kystos*, bladder) In cnidarians, a specialized cellular capsule containing a tiny barb with a poisonous, paralyzing substance, which can be discharged against predator or prey.

neoteny (Gr. *neos*, new + *teinein*, to extend) The attainment of sexual maturity in the larval condition. Also, the retention of larval characters into adulthood.

nephridium, pl. **nephridia** (Gr. *nephros*, kidney) In invertebrates such as earthworms, a tubular excretory structure.

nephron (Gr. *nephros*, kidney) Functional unit of the vertebrate kidney; one of numerous tubules (a human kidney contains about 1 million) involved in filtration and selective reabsorption of blood; each nephron consists of a Bowman's capsule, an enclosed glomerulus, and a long attached tubule; in humans, called a renal tubule.

nerve A group or bundle of nerve fibers (axons) with accompanying neuroglial cells, held together by connective tissue; located in the peripheral nervous system (a bundle of nerve fibers within the central nervous system is known as a tract).

nerve fiber An axon.

nerve impulse An action potential; a rapid, transient, self-propagating reversal in electric potential that travels along the membrane of a neuron.

nerve net In some invertebrates, neurons dispersed through epithelium yet functionally linked to sensory cells, to each other, and to muscle tissue. Permits diffuse response to stimuli; unlike nervous system, has little orientation to information flow (the neurons can carry signals in either direction).

nervous system All the nerve cells of an animal. In humans, the nervous system consists of sensory receptors, efferent sensory nerves that carry sensory information to the brain, interneurons within the brain and spinal cord, and afferent motor neurons that carry commands to muscles and glands.

neuroglia (Gr. *neuron*, nerve + *glia*, glue) Nonconducting nerve cells that are intimately associated with neurons and appear to provide nutritional support; in vertebrates they represent at least half of the volume of the nervous system, yet the function of most is not well understood.

neuron (Gr., nerve) A nerve cell specialized for signal transmission; includes cell body, dendrites, and axon.

neurosecretory cell Any cell of the nervous system that produces a hormone.

neurotransmitter (Gr. *neuron*, nerve + L. *trans*, across + *mittere*, to send) A chemical released at the axon terminal of a neuron that travels across the synaptic cleft, binds a specific receptor on the far side, and, depending on the nature of the receptor, depolarizes or hyperpolarizes a second neuron or a muscle or gland cell.

neutron (L. *neuter*, neither) An uncharged subatomic particle of about the same size and mass as a proton but having no electric charge.

niche The role played by a particular species in its environment.

nicotinamide adenine dinucleotide See NAD.

nicotinamide adenine dinucleotide phosphate See NADP.

nitrification The oxidation of ammonia or ammonium to nitrites, carried out by certain soil bacteria.

nitrogen cycle Worldwide circulation and reutilization of nitrogen atoms, chiefly caused by metabolic processes of living organisms. Plants take up inorganic nitrogen and convert it into organic compounds (chiefly proteins), which are assimilated into the bodies of one or more animals; excretion and bacterial and fungal action on dead organisms return nitrogen atoms to the inorganic state.

nitrogen fixation The incorporation of atmospheric nitrogen into nitrogen compounds, a process that can be carried out only by certain microorganisms.

nitrogenous base A nitrogen-containing molecule having basic properties; a purine or pyrimidine; one of the building blocks of nucleic acids.

node (L. *nodus*, knot) The part of a plant stem where one or more leaves are attached; *see* internode.

nodules Enlargements or swellings on the roots of legumes and certain other plants inhabited by symbiotic nitrogen-fixing bacteria.

noncyclic photophosphorylation The photosynthetic light reactions of green plants; electrons derived from water molecules flow through two photosystems and two transport chains, producing ATP and $NADPH_2$.

noradrenaline A hormone produced by the medulla of the adrenal gland which increases the concentration of sugar in the blood, raises blood pressure and heart rate, and increases muscular power and resistance to fatigue; also one of the principal neurotransmitters of the parasympathetic nervous system; sometimes called *norepinephrine*.

notochord (Gr. *noto*, back + L. *chorda*, cord) In chordates, a dorsal rod of cartilage that runs the length of the body and forms the primitive axial skeleton in the embryos of all chordates; in most adult chordates the notochord is replaced by a vertebral column that forms around (but not from) the notochord.

nucellus (L. *nucella*, a small nut) Tissue composing the chief part of the young ovule, in which the embryo sac develops; equivalent to a megasporangium.

nuclear envelope (L. *nucleus*, a kernel) The double membrane that surrounds the nucleus within a eukaryotic cell; the outer membrane is often continuous with endoplasmic reticulum.

nucleic acid A nucleotide polymer; a

long chain of nucleotides; chief types are deoxyribonucleic acid (DNA), which is double stranded, and ribonucleic acid (RNA), which is typically single stranded.

nucleolar organizer A special area on certain chromosomes associated with the formation of the nucleolus.

nucleolus (L., a small nucleus) In eukaryotes, the site of rRNA synthesis; a spherical body composed chiefly of rRNA in the process of being transcribed from multiple copies of rRNA genes.

nucleoplasm (L. *nucleus*, kernel + Gr. *plasma*, mold) Protoplasm of nucleus, as distinguished from cytoplasm.

nucleoprotein A molecule composed of nucleic acid and protein.

nucleosome (L. *nucleus*, kernel + *soma*, body) The fundamental packaging unit of eukaryotic chromosomes; a complex of DNA and histone proteins in which one-and-three-quarters turns of the double-helical DNA are wound around eight molecules of histone. Chromatin is composed of long strings of nucleosomes, like beads on a string.

nucleotide A single unit of nucleic acid, composed of a phosphate, a five-carbon sugar (either ribose or deoxyribose), and a purine or a pyrimidine.

nucleus In atoms, the central core, containing positively charged protons and (in all but hydrogen) electrically neutral neutrons. In eukaryotic cells, the membranous organelle that houses the chromosomal DNA; in the central nervous system, a cluster of nerve cell bodies.

nymph (L. *nympha*, nymph, bride) An immature stage (following hatching) of a hemimetabolous insect that lacks a pupal stage.

obligate anaerobe An organism that is metabolically active only in the absence of oxygen.

ocellus, pl. **ocelli** (L., little eye) A simple light receptor common among invertebrates.

olfactory (L. *olfacere*, to smell) Pertaining to smell.

ommatidium, pl. **ommatidia** (Gr., little eye) The visual unit in the compound eye of arthropods; contains light-sensitive cells and a lens able to form an image.

oncogene (Gr. *oncos*, cancer + *genos*, birth) A cancer-causing gene; a mutant form of a growth-regulating gene that is inappropriately "on," causing unrestrained cell growth and division. Cancer appears to result only when several such controls have been abrogated.

ontogeny (Gr. *ontos*, being + *geneia*, act of being born, from genes, born) The course of development of an individual from egg to senescence.

oocyte (Gr. *oion*, egg + *kytos*, vessel) A cell that gives rise to an ovum by meiosis.

operator A site of gene regulation; a sequence of nucleotides overlapping the promoter site and recognized by a repressor protein. Binding of the repressor prevents binding of the polymerase to the promoter site and so blocks transcription of the structural gene.

operon (L. *operis*, work) A cluster of adjacent structural genes transcribed as a unit into a single mRNA molecule; transcription of all the genes of an operon is regulated coordinately by controlling binding of RNA polymerase to the single promoter site, typically via an adjacent and overlapping operator site. A common mode of gene organization in bacteria, but rare in eukaryotes.

ophthalmic (Gr. *ophthalmos*, an eye) Pertaining to the eye.

orbital (L. *orbis*, circle) The volume of space surrounding the atomic nucleus in which an electron will be found most of the time.

order A category of classification above the level of family and below that of class; orders are composed of one or more families.

organ (Gr. *organon*, tool) A body structure composed of several different tissues grouped together in a structural and functional unit.

organelle (Gr. *organella*, little tool) Specialized part of a cell; literally, a small organ analogous to the organs of multicellular animals.

organic Pertaining to living organisms in general, to compounds formed by living organisms, and to the chemistry of compounds containing carbon.

organism Any individual living creature, either unicellular or multicellular.

osmole Molecular weight of a solute, in grams, divided by the number of ions or particles into which it dissociates in solution. Adj., osmolar.

osmoregulation Maintenance of constant internal salt and water concentrations in an organism; the active regulation of internal osmotic pressure.

osmosis (Gr. *osmos*, act of pushing, thrust) The diffusion of water across a selectively permeable membrane (a membrane that permits the free passage of water but prevents or retards the passage of a solute). In the absence of differences in pressure or volume, the net movement of water is from the side containing a lower concentration of solute to the side containing a higher concentration.

osmotic pressure The potential pressure developed by a solution separated from pure water by a differentially permeable membrane. Measured as the pressure required to stop the osmotic movement of water

into a solution, it is an index of the solute concentration of the solution; the higher the solute concentration, the greater the osmotic potential of the solution. *Osmotic potential* is a synonym.

osteoblast (Gr. *osteon*, bone + *blastos*, bud) A bone-forming cell.

ovary (L. *ovum*, egg) (1) In animals, the organ in which eggs are produced. (2) In flowering plants, the enlarged basal portion of a carpel, which contains the ovule(s); the ovary matures to become the fruit.

oviduct (L. *ovum*, egg + *ductus*, duct) In vertebrates, the passageway through which ova (eggs) travel from the ovary to the uterus.

oviparity (L. *ovum*, egg + *parere*, to bring forth) Reproduction in which unfertilized eggs are released by the female; fertilization and development of offspring occur outside the maternal body. Adj., oviparous.

ovoviviparity (L. *ovum*, egg + *vivere*, to live + *parere*, to bring forth) Reproduction in which fertilized eggs develop within the maternal body without obtaining additional nourishment from it, and hatch there or immediately after laying. Adj., ovoviviparous.

ovulation In animals, the release of an egg or eggs from the ovary.

ovule (L. *ovulum*, a little egg) A structure in seed plants that contains the female gametophyte and is surrounded by the nucellus and one or two integuments; when mature, an ovule becomes a seed.

ovum, pl. **ova** (L., egg) The egg cell; female gamete.

oxidation (Fr. *oxider*, to oxidize) Loss of an electron by an atom or molecule. In metabolism, often associated with a gain of oxygen or loss of hydrogen. Oxidation (loss of an electron) and reduction (gain of an electron) take place simultaneously, because an electron that is lost by one atom is accepted by another. Oxidation-reduction reactions are an important means of energy transfer within living systems.

oxidative phosphorylation Using electrons derived from oxidative repiration to make ATP; the energetic electrons are delivered by NADH to a chain of carriers within the mitochondrial membrane that deliver them to proton pumps, driving the chemiosmotic synthesis of ATP. *See* chemiosmosis.

pacemaker A patch of excitatory tissue in the vertebrate heart that initiates the hearbeat; the evolutionary remnant of the first chamber of the fish heart, it is called the sinoatrial node and is located in the same position, where the superior vena cava enters the right atrium. Mechanical pacemakers, inserted near the heart of certain human patients, have the same function.

paleontology (Gr. *palaios*, old + *onta*, things that exist + *logos*, discourse) The study of ancient life, mainly carried out by means of fossils. *Paleobotany* is the study of ancient plant life.

palisade cells (L. *palus*, stake) In plant leaves, the columnar, chloroplast-containing parenchyma cells of the mesophyll. Also called *palisade parenchyma*.

pancreas (Gr. *pan*, all + *kreas*, meat, flesh) In vertebrates, the principal digestive gland; a small gland located between the stomach and the duodenum which produces digestive enzymes and the hormones insulin and glucagon.

parallel evolution In two or more different evolutionary lines, the development of similar structures having similar functions as a result of the same kinds of selection pressures.

parapodium (Gr. *para*, beside + *pous*, foot) One of the paired lateral processes on each side of most segments in polychete annelids; variously modified for locomotion, respiration, or feeding.

parasexual cycle In certain fungi, the fusion and segregation of heterokaryotic haploid nuclei to produce recombinant nuclei.

parasite (Gr. *para*, beside + *sitos*, food) An organism that lives on or in an organism of a different species and derives nutrients from it.

parasympathetic nervous system (Gr. *para*, beside + *syn*, with + *pathos*, feeling) In vertebrates, one of two subdivisions of the autonomic nervous system (the other is the sympathetic). The two subdivisions operate antagonistically, the parasympathetic system stimulating resting activities such as digestion and restoring the body to normal after emergencies by inhibiting alarm functions initiated by the sympathetic system.

parenchyma (Gr. *para*, beside + *en*, in + *chein*, to pour) A plant tissue composed of *parenchyma* cells; such cells are living, thin walled, and randomly arranged, and have large vacuoles; usually photosynthetic or storage tissue.

parthenogenesis (Gr. *parthenos*, virgin + *genesis*, birth) The development of an egg without fertilization, as in aphids, bees, ants, and some lizards.

passive transport The movement of a molecule across a cell membrane without the expenditure of energy by the cell. Diffusion, osmosis, and bulk flow are passive transport mechanisms.

pathogen (Gr. *pathos*, suffering + *genesis*, beginning) A disease-causing organism.

pelagic (Gr. *pelagikos*, of the open sea) Pertaining to the open ocean.

penis In reptiles and mammals, the male reproductive organ through which sperm are delivered into the female reproductive tract during sexual intercourse.

peptidase (Gr. *peptein*, to digest + *-ase*, enzyme suffix) An enzyme that breaks down short chains of amino acids (peptides) to individual amino acids.

peptide Two or more amino acids linked by peptide bonds.

peptide bond (Gr. *peptein*, to soften, digest) The type of bond that links amino acids together in proteins; formed by removing an OH from the carboxy (—COOH) group of one amino acid and an H from the amino (—NH$_2$) group of another to form an amide group CO≡NH≡.

perennial (L. *per*, through + *annus*, a year) A plant that lives for more than a year and produces flowers on more than one occasion.

perianth (Gr. *peri*, around + *anthos*, flower) In flowering plants, the petals and sepals taken together.

periclinal Parallel with the surface.

pericycle (Gr. *peri*, around + *kykos*, circle) In vascular plants, one or more cell layers surrounding the vascular tissues of the root, bounded externally by the endodermis and internally by the phloem.

periderm (Gr. *peri*, around + *derma*, skin) Outer protective tissue in vascular plants that is produced by the cork cambium and functionally replaces epidermis when it is destroyed during secondary growth; the periderm includes the cork, cork cambium, and phelloderm.

peripheral nervous system (Gr. *peripherein*, to carry around) All of the neurons and nerve fibers outside the central nervous system, including motor neurons, sensory neurons, and the autonomic nervous system.

peristalsis (Gr. *peristaltikos*, compressing around) Pumping by waves of contraction; in animals, a series of alternating contracting and relaxing muscle movements along the length of a tube such as the oviduct or alimentary canal that tend to force material such as an egg cell or food through the tube.

peritoneum (Gr. *peritonios*, stretched over) A membrane that lines the body cavity of coelomate animals and forms the external covering of the organs within it.

permeable (L. *permeare*, to pass through) Refers to a membrane through which a specified molecule, ion, or other solute can pass freely.

peroxisome A microbody that plays an important role in glycolic acid metabolism associated with photosynthesis; the site of photorespiration.

petal A flower part, usually conspicuously colored; one of the units of the corolla.

petiole (L. *petiolus*, a little foot) The stalk of a leaf.

pH A measure of the relative concentration of hydrogen ions in a solution; equal to the negative logarithm of the hydrogen ion concentration. pH values range from 0 to 14; the lower the value, the more hydrogen ions it contains (the more acidic it is); pH 7 is neutral, less than 7 is acidic, more than 7 is alkaline.

phage *See* bacteriophage.

phagocyte (Gr. *phagein*, to eat + *kytos*, hollow vessel) Any cell that engulfs and devours microorganisms or other particles.

phagocytosis (Gr., cell-eating) Endocytosis of solid particles; the cell membrane folds inward around the particle (which may be another cell) and then discharges the contents into the cell interior as a vacuole; characteristic of protists, amoebas, the digestive cells of some invertebrates, and vertebrate white blood cells.

pharynx (Gr., gullet) In vertebrates, a muscular tube that connects the mouth cavity and the esophagus; it serves as the gateway to the digestive tract and to the windpipe (trachea).

phenotype (Gr. *phainein*, to show + *typos*, stamp or print) The realized expression of the genotype; the physical appearance or functional expression of a trait; the result of the biological activity of proteins or RNA molecules transcribed from the DNA.

pheromone (Gr. *pherein*, to carry + *hormonos*, exciting, stirring up) Chemical substance released by the exocrine glands of one organism that influences the behavior or physiological processes of another organism of the same species. Some pheromones serve as sex attractants, as trail markers, and as alarm signals.

phloem (Gr. *phloos*, bark) In vascular plants, a food-conducting tissue basically composed of sieve elements, various kinds of parenchyma cells, fibers, and sclereids.

phosphagen A term for creatine phosphate and arginine phosphate, which in animals store high-energy phosphate bonds.

phosphate group —PO$_4$; a chemical group commonly involved in high-energy bonds.

phospholipid A phosphorylated lipid; similar in structure to a fat, but only two fatty acids are attached to the glycerol backbone, with the third space linked to a phosphorylated molecule. Phospholipid molecules have a polar hydrophilic "head" end (contributed by the phosphate group) and a nonpolar hydrophobic "tail" end (contributed by the fatty acids); they orient spontaneously in water to form bimolecular membranes in which the nonpolar tails of the molecules are oriented inward towards one another, away from the polar water environment. Phospholipids are the foundation for all cell membranes. *See* lipid, fatty acid.

phosphorylation (Gr. *phosphoros*, bringing light) A reaction in which a phosphate group is added to a compound; phosphorylation often results in the formation of a high-energy bond (as in the formation of ATP from ADP and inorganic phosphate) and also plays a key role in preventing cellular loss of metabolites by diffusion. (By phosphorylating glucose and other sugars, for example, the cell prevents them from diffusing outward to where the concentration is lower because the sugar's phosphate group is repelled by the interior of the cell membrane.)

photon (Gr. *photos*, light) The elementary particle of electromagnetic energy; light.

photoperiodism (Gr. *photos*, light + *periodos*, a period) The tendency of biological reactions to respond to the duration and timing of day and night; a mechanism for measuring seasonal time.

photophosphorylation (Gr. *photos*, light + *phosphoros*, bringing light) The formation of ATP in the chloroplast during photosynthesis.

photoreceptor (Gr. *photos*, light) A light-sensitive sensory cell.

photorespiration The light-dependent production of glycolic acid in chloroplasts and its subsequent oxidation in peroxisomes; this process tends to short-circuit photosynthesis, and becomes progressively more of a drain to plants at higher temperatures. C_4 and CAM photosynthesis are evolutionary responses to this dilemma.

photosynthesis (Gr. *photos*, light + *syn*, together + *tithenai*, to place) The utilization of light energy to create chemical bonds; the synthesis of organic compounds from carbon dioxide and water, using chemical energy (ATP) and reducing power (NADPH) generated by photosynthesis.

photosystem A functional light-trapping unit; an organized collection of chlorophyll and other pigment molecules embedded in the thylakoids of chloroplasts which trap photon energy and channel it in the form of energetic electrons to the thylakoid membrane.

phototaxis (Gr. *photos*, light + *taxis*, order) A response, either positive or negative, to light.

phototropism (Gr. *photos*, light + *trope*, turning) In plants, a growth response to a light stimulus.

phycobilins A group of water-soluble accessory pigments, including phycocyanins and phycoerythrins, which occur in red algae and cyanobacteria.

phycology (Gr. *phykos*, seaweed) The study of algae.

phylogeny (Gr. *phylon*, race, tribe) The evolutionary relationships among any group of organisms.

phylum, pl. **phyla** (Gr. *phylon*, race, tribe) A major category, between kingdom and class, of taxonomic classifications. *Division* is an equivalent term used in all groups except animals and heterotrophic protists.

physiology (Gr. *physis*, nature + *logos*, a discourse) The study of life's functions; how cells, organs, or entire organisms function.

phytochrome (Gr. *phyton*, plant + *chroma*, color) A plant pigment that is associated with the absorption of light; photoreceptor for red to far-red light; involved in a number of timing processes, such as flowering, dormancy, leaf formation, and seed germination.

phytoplankton (Gr. *phyton*, plant + *planktos*, wandering) Autotrophic plankton.

pigment (L. *pigmentum*, paint) A molecule that absorbs light.

pinna (L., feather, sharp point) The external ear. Also a feather, wing, or fin or similar part; in plants, the primary division of a compound leaf (a leaflet).

pinocytosis (Gr. *pinein*, to drink + *kytos*, hollow vessel + *osis*, condition) The taking up of fluid by endocytosis; refers to cells.

pistil (L. *pistillum*, pestle) Central organ of flowers, typically consisting of ovary, style, and stigma; a pistil may consist of one or more fused carpels, and is more technically and better known as the gynoecium. A flower with carpels but no functional stamens is called *pistillate*.

pit In plant cells, a recessed cavity in a cell wall where the secondary wall does not form.

pith The ground tissue occupying the center of the stem or root within the vascular cylinder; usually consists of parenchyma.

pituitary (L. *pituita*, phlegm) Perhaps the most important of the endocrine glands in vertebrates; under the hormonal control of the hypothalamus, which directs it via releasing hormones; the anterior lobe secretes tropic hormones, growth hormone, and prolactin; the posterior lobe stores and releases oxytocin and ADH produced by the hypothalamus.

placenta, pl. **placentae** (L., a flat cake) (1) In flowering plants, the part of the ovary wall to which the ovules or seeds are attached. (2) In mammals, a tissue formed in part from the inner lining of the uterus and in part from other membranes, through which the embryo (later the fetus) is nourished while in the uterus and wastes are carried away.

plankton (Gr. *planktos*, wandering) Free-floating, mostly microscopic, aquatic organisms.

planula (L. *planulus*, a little wanderer) The ciliated, free-swimming type of larva formed by many cnidarians.

plaque Clear area in a sheet of bacterial cells or eukaryotic cells growing in culture, resulting from the killing (lysis) of contiguous cells by viruses.

plasma (Gr., form) The fluid of vertebrate blood; contains dissolved salts, metabolic wastes, hormones, and a variety of proteins, including antibodies and albumin; blood minus the blood cells.

plasma cell An antibody-producing cell resulting from the multiplication and differentiation of a B lymphocyte that has interacted with an antigen; a mature plasma cell can produce from 3000 to 30,000 antibody molecules per second.

plasma membrane The membrane surrounding the cytoplasm of an animal cell; the outermost membrane of a cell; consists of a single bilayer of membrane; also called cell membrane and plasmalemma.

plasmid (Gr. *plasma*, a form or mold) A small fragment of extrachromosomal DNA, usually circular, that replicates independently of the main chromosome, although it may have been derived from it. Plasmids make up about 5% of the DNA of many bacteria, but are rare in eukaryotic cells.

plasmodesma, pl. **plasmodesmata** (Gr. *plassein*, to mold + *desmos*, a bond) Minute cytoplasmic connections that connect adjacent cells; in plants they extend through pores in the cell walls.

plasmodium (Gr. *plasma*, a form, mold + *eidos*, form) Stage in the life cycle of myxomycetes (plasmodial slime molds); a multinucleate mass of protoplasm surrounded by a membrane.

plastid (Gr. *plastos*, formed or molded) An organelle in the cells of photosynthetic eukaryotes (plants and algae) that is the site of photosynthesis and, in plants and green algae, of starch storage; plastids are bounded by a double membrane.

platelet (Gr. dim. of *plattus*, flat) In mammals, a fragment of a white blood cell that circulates in the blood and functions in the formation of blood clots at sites of injury.

pleiotropic (Gr. *pleros*, more + *trope*, a turning) Describing a gene that produces more than one phenotypic effect.

pleura (Gr., side, rib) In vertebrate animals, the membrane that lines each half of the thorax and covers the lungs.

poikilotherm (Gr. *poikilos*, changeable + *therme*, heat) An animal with a body temperature that fluctuates with that of the environment; cold blooded. *See* ectothermic.

point mutation An alteration of one nucleotide in a chromosomal DNA molecules; includes addition, deletion, or substitution of individual nucleotides.

polar (L. *polus*, end of axis) Having parts or areas with opposed or contrasting properties, such as positive and negative charges, head and tail.

polar body Minute, nonfunctioning cell produced during the meiotic divisions leading to gamete formation in vertebrates; contains a nucleus but very little cytoplasm.

polar molecule A molecule with positively and negatively charged ends; one portion of a polar molecule attracts electrons more strongly (is more electronegative) than another portion, with the result that the electron-rich portion carries a partial negative charge contributed by the electron excess, and the electron-poor portion carries a partial positive charge because of the electron deficit.

polar nuclei In flowering plants, two nuclei (usually), one derived from each end (pole) of the embryo sac, which become centrally located; they fuse with a male nucleus to form the primary (3*n*) endosperm nucleus.

pollen (L., fine dust) A collective term for pollen grains. In seed plants, each pollen grain contains an immature male gametophyte enclosed within a protective outer covering; pollen grains may be two-celled or three-celled when shed.

pollen tube A tube formed after germination of the pollen grain; carries the male gametes into the ovule.

pollination The transfer of pollen from an anther to a stigma.

polygamy (Gr. *polys*, many + *gamos*, marriate) Condition of having more than one mate at one time.

polygenic inheritance The pattern of heredity characteristic of traits (called "metric" traits) determined by the additive contributions of many genes; also called quantitative inheritance.

polymer (Gr. *polus*, many + *meris*, part) A molecule composed of many similar or identical molecular subunits. Starch is a polymer of glucose.

polymorphism (Gr. *polys*, many + *morphe*, form) The presence of more than one allele of a gene at a frequency greater than that of newly arising mutations; operationally, a population in which the most common allele at a locus has a frequency of less than 99%.

polynucleotide A single-stranded DNA or RNA molecule.

polyp (Fr. *polype*, octopus, from L. *polypus*, many footed) The vase-shaped, sessile, sedentary stage in the life cycle of cnidarians.

polypeptide (Gr. *polys*, many + *peptein*, to digest) A molecule consisting of many joined amino acids; not as complex as a protein.

polyploid (Gr. *polys*, many + *ploin*, vessel) An organism, tissue, or cell with more than two complete sets of homologous chromosomes.

polysaccharide (Gr. *polys*, many + *sakcharon*, sugar, from Sanskrit *sarkara*, gravel, sugar) A sugar polymer; a carbohydrate composed of many monosaccharide sugar subunits linked together in a long chain; examples are glycogen, starch, and cellulose.

polysome (Gr. *polys*, many + *soma*, body) In bacteria, a tandem cluster of ribosomes all engaged in translating the same messenger RNA molecule, one after the other. Also called polyribosome.

polytene chromosomes (Gr. *polys*, many + *tainia*, band) Chromosomes in the somatic cells of flies and some other insects in which the chromatin replicates repeatedly without undergoing mitosis.

population (L. *populus*, the people) Any group of individuals, usually of a single species, occupying a given area at the same time.

portal system (L. *porta*, gate) Any system of large veins beginning and ending with a bed of capillaries; for example, the hepatic (liver) portal and renal (kidney) portal systems in vertebrates.

posterior Of or pertaining to the rear end. In humans, the "rump" is often referred to as the posterior.

potential energy Energy that is not being used, but could be; energy in a potentially usable form; often called "energy of position."

predaceous (L. *praeda*, prey) Describing an animal that lives by killing and consuming other animals; predatory.

prehensile (L. *prehendere*, to seize) Adapted for grasping.

prey (L. *prehendere*, to grasp, seize) An organism eaten by another organism.

primary endosperm nucleus In flowering plants, the result of the fusion of a sperm nucleus and the (usually) two polar nuclei.

primary growth In vascular plants, growth originating in the apical meristems of shoots and roots, as contrasted with secondary growth; results in an increase in length.

primary meristematic tissue In vascular plants, the tissues derived from the apical meristem; of three kinds: protoderm, procambium, and ground meristem.

primary plant body The part of the vascular plant body arising from the apical meristems and their derivative meristematic tissues; composed entirely of primary tissues. The primary plant body is said to be made up of *primary tissues*.

primary structure of a protein The amino acid sequence of a protein.

primary wall In plants, the wall layer deposited during the period of cell expansion.

primitive streak (L. *primus*, first) In the early embryos of birds, reptiles, and mammals, a dorsal, longitudinal strip of ectoderm and mesoderm that is equivalent to the blastopore in other forms.

primordium, pl. **primordia** (L. *primus*, first + *ordiri*, to begin to weave) A cell or organ in its earliest stage of differentiation.

proboscis (Gr. *pro*, before + *boskein*, feed) A snout or trunk. Also, tubular sucking or feeding organ with the mouth at the end, as in planarians, leeches, and insects. Also, the sensory and defensive organ at the anterior end of certain invertebrates.

procambium (L. *pro*, before + *cambiare*, to exchange) In vascular plants, a primary meristematic tissue that gives rise to primary vascular tissues.

producers (L. *producere*, to bring forth) In ecology, organisms, such as plants, that are able to produce their own food from inorganic substances.

productivity A measure of the rate at which energy is assimilated by an organism or group of organisms.

progesterone (L. *progerere*, to carry forth or out + *steiras*, barren) In mammals, a steroid hormone secreted by the corpus luteum that prepares the uterus for implantation of the fertilized egg (ovum) and maintains the uterus during pregnancy.

prokaryote (Gr. *pro*, before + *karyon*, kernel) A bacterium; a cell lacking a membrane-bound nucleus or membrane-bound organelles. Prokaryotic cells are more primitive than eukaryotic cells, which evolved from them.

promoter A specific nucleotide sequence on a chromosome to which RNA polymerase attaches to initiate transcription of mRNA from a gene.

prophage Noninfectious bacteriophage units linked with the bacterial chromosome that multiply with the growing and dividing bacteria but do not bring about lysis of the bacteria. Prophage is a stage in the life cycle of a temperate phage. *See* lysogenic.

prophase (Gr. *pro*, before + *phasis*, form) An early stage in nuclear division, characterized by the formation of a microtubule spindle along the future axis of division, the shortening and thickening of the chromosomes, and their movement toward the equator of the spindle (the "metaphase plate").

proprioceptor (L. *proprius*, one's own) In vertebrates, a sensory receptor that senses the body's position and movements; located deep within the tissues, especially muscles, tendons, and joints.

prosimian (Gr. *pro*, before + L. *simia*, ape) Non-anthropoid primates, including lemurs, tarsiers, and lorises.

prostaglandins (Gr. *prostas*, a porch or vestibule + L. *glans*, acorn) A group of modified fatty acids, originally discovered in semen, that function as chemical messengers and have powerful effects on smooth muscle, nerves, circulation, and reproductive organs; synthesized in most, possibly all, cells of the body.

prostate gland (Gr. *prostas*, a porch or vestibule) In male mammals, a mass of glandular tissue at the base of the urethra that secretes an alkaline fluid which has a stimulating effect on the sperm as they are released.

protease (Gr. *proteios*, primary + *ase*, enzyme ending) An enzyme that digests proteins by breaking peptide bonds. Also called peptidases.

protein (Gr. *proteios*, primary) A chain of amino acids joined by peptide bonds; a protein typically contains over 100 amino acids and may be composed of more than one polypeptide.

protist (Gr. *protos*, first) A member of the kingdom Protista, which includes the unicellular eukaryotic organisms and some multicellular lines derived from them.

protoderm (Gr. *protos*, first + *derma*, skin) A primary meristematic tissue in vascular plants that gives rise to epidermis.

proton A subatomic, or elementary, particle with a single positive charge equal in magnitude to the charge of an electron and a mass of 1, very close to that of a neutron; the nucleus of a hydrogen atom is composed of a single proton.

protoplasm (Gr. *protos*, first + *plasma*, form) A vague term referring to the matrix of living cells (but not the organelles); the cytoplasm and nucleoplasm of the cell.

protoplast The protoplasm of an individual cell.

protostome (Gr. *protos*, first + *stoma*, mouth) An animal in whose embryonic development the mouth forms at or near the blastopore. Protostomes are also characterized by spiral cleavage during the earliest stages of development and by schizocoelous formation of the coelom.

proximal (L. *proximus*, near) Situated near the point of reference, usually the main part of a body or the point of attachment; opposite of distal.

pseudocoel, or **pseudocoelom** (Gr. *pseudos*, false + *koiloma*, cavity) A body cavity not lined with peritoneum and not a part of the blood or digestive systems, embryonically derived from the blastocoel.

pseudogene (Gr. *pseudos*, false + *genos*, birth) A silent gene; a copy of a gene that is not transcribed.

pseudopod, or **pseudopodium** (Gr. *pseudes*, false + *pous*, foot) "False foot"; a nonpermanent cytoplasmic extension of the cell body.

pulmonary circulation In terrestrial vertebrates, the pathway of blood circulation leading to and from the lungs.

pulse pressure The force generated by the heart's contraction; measured as the difference between systolic blood pressure (the maximum pressure exerted on arteries when blood is forced from heart) and diastolic blood pressure (the lowest pressure, which occurs just before blood is pumped out); generally measured at a main artery.

punctuated equilibrium A model of the mechanism of evolutionary change which proposes that long periods of little or no change are punctuated by periods of rapid evolution.

pupa (L., girl, doll) A developmental stage of some insects in which the organism is nonfeeding, immotile, and sometimes encapsulated or in a cocoon; the pupal stage occurs between the larval and adult phases. In butterflies, it is called a chrysalis.

purine (Gr. *purinos*, fiery, sparkling) The larger of the two general kinds of nucleotide base found in DNA and RNA; a nitrogenous base with a double-ring structure, such as adenine or guanine.

pyrimidine (alt. of pyridine, from Gr. *pyr*, fire) The smaller of the two general kinds of nucleotide base found in DNA and RNA; a nitrogenous base with a single-ring structure, such as cytosine, thymine, or uracil.

pyruvate The three-carbon compound that is the end product of glycolysis and the starting material of the citric acid cycle.

quaternary structure of a protein The level of aggregation of a globular protein molecule that consists of two or more polypeptide chains; a monomer has one subunit only, a dimer has two, a trimer three, and a tetramer four.

radial cleavage In animals, a type of embryological development in which early cleavage planes are symmetrical to the polar axis, each blastomere of one tier lying directly above the corresponding blastomere of the next layer; indeterminate cleavage.

radial symmetry (L. *radius*, a spoke of a wheel + Gr. *summetros*, symmetry) The regular arrangement of parts around a central axis such that any plane passing through the central axis divides the organism into halves that are approximate mirror images.

radioisotope An unstable isotope of an element that decays or disintegrates spontaneously, emitting radiation.

radula (L., scraper) Rasping tongue found in most mollusks.

recessive allele (L. *recedere*, to recede) An allele whose phenotypic effect is masked in the heterozygote by that of another, dominant allele; heterozygotes, although they contain a copy of the recessive allele, are phenotypically indistinguishable from dominant homozygotes.

reciprocal altruism Performance of an altruistic act with the expectation that the favor will be returned. A key and very controversial assumption of many theories dealing with the evolution of social behavior. *See* altruism.

recombinant DNA Fragments of DNA from two different species, such as a bacterium and a mammal, spliced together in the laboratory into a single molecule; a key component of genetic engineering technology, in which, for example, a gene that makes a certain bacterium resistant to a chemical weed killer is transferred to a crop plant.

recombination The formation of new gene combinations; in bacteria, it is accomplished by the transfer of genes into cells, often in association with viruses; in eukaryotes, it is accomplished by reassortment of chromosomes during meiosis, and by crossing-over.

reduction (L. *reductio*, a bringing back; originally "bringing back" a metal from its oxide) The gain of an electron by an atom; takes place simultaneously with oxidation (loss of an electron by an atom), because an electron that is lost by one atom is accepted by another.

reflex (L. *reflectere*, to bend back) In the nervous system, an "automatic" reponse to a stimulus; a motor response subject to little associative modification; among the simplest neural pathways, involving only a sensory neuron, sometimes (but not always) an interneuron, and one or more motor neurons.

releaser (L. *relaxare*, to unloose) A relatively simple environmental stimulus that triggers an innate behavior pattern.

releasing hormone A peptide hormone produced by the hypothalamus that stimulates or inhibits the secretion of specific hormones by the anterior pituitary.

renal (L. *renes*, kidneys) Pertaining to the kidney.

replication (L. *replicatio*, a folding back) The production of a second "daughter" molecule of DNA exactly like the first "parent" molecule; the parent molecule is used as a template.

repressor (L. *reprimere*, to press back, keep back) A protein that regulates DNA transcription by preventing RNA polymerase from attaching to the promoter and transcribing the structural gene. *See* operator.

respiration (L. *respirare*, to breathe) The utilization of oxygen; in terrestrial vertebrates, the inhalation of oxygen and the exhalation of carbon dioxide; in cells, the oxidation (electron removal) of food molecules, particularly pyruvate in the citric acid cycle, to obtain energy.

resting membrane potential The charge difference (difference in electric potential) that exists across a neuron at rest (about 70 millivolts).

restriction endonuclease An enzyme that cleaves a DNA duplex molecule at a particular base sequence.

restriction map A diagram of a segment of DNA, showing the relative locations of sites ("restriction sites") cleaved by particular restriction endonucleases ("restriction enzymes").

reticular (L., *reticulum*, small net) Resembling a net in appearance or structure.

retina (L., a small net) The photosensitive layer of the vertebrate eye; contains several layers of neurons and light receptors (rods and cones); receives the image formed by the lens and transmits it to the brain via the optic nerve.

retrovirus (L. *retro*, turning back) An RNA virus; a virus whose genetic material is RNA. When a retrovirus enters a cell, the cell's machinery "reads" the virus RNA, which contains a gene encoding an enzyme—reverse transcriptase—that then transcribes the virus RNA into duplex DNA, which the cell's machinery replicates as if it were its own; the DNA copies of many retroviruses appear able to enter eukaryotic chromosomes, where they act much like transposons. Retroviruses are associated with many human diseases, including cancer and AIDS.

reverse transcriptase An enzyme that transcribes RNA into DNA; found only in association with retroviruses.

rhizoid (Gr. *rhiza*, root) Slender, rootlike anchoring structure in some fungi, algae, and plant gametophytes.

rhizome (Gr. *rhizoma*, mass of roots) In vascular plants, a usually more or less horizontal underground stem; may be enlarged for storage, or may function in vegetative reproduction.

ribonucleic acid (RNA) A class of nucleic acids characterized by the presence of the sugar ribose (DNA contains deoxyribose instead) and the pyrimidine uracil (DNA contains thymine instead); includes mRNA, tRNA, and rRNA.

ribose A five-carbon sugar.

ribosomal RNA (rRNA) A class of RNA molecules found, together with characteristic proteins, in ribosomes; transcribed from the DNA of the nucleolus.

ribosome The molecular machine that carries out protein synthesis; the most complicated aggregation of proteins in a cell, also containing three different rRNA molecules; may be free in the cytoplasm or in eukaryotes sometimes attached to the membranes of the endoplasmic reticulum.

RNA *See* ribonucleic acid.

rod Light-sensitive nerve cell found in the vertebrate retina; sensitive to very dim light; responsible for "night vision."

root The usually descending axis of a plant, normally below ground, which anchors the plant and serves as the major point of entry for water and minerals.

root cap In vascular plants, a thimble-like mass of cells covering and protecting the growing tip of a root; the cells of the root cap are sloughed off as the root grows through the soil.

root hairs In vascular plants, tubular outgrowths of the epidermal cells of the root just back of its apex; most water enters through the root hairs.

root pressure In vascular plants, the pressure that develops in roots as the result of osmosis, which causes guttation of water from leaves and exudation from cut stumps.

r-selection (from the *r* term in the logistic equation) Natural selection under conditions that favor survival when populations are controlled primarily by density-independent factors; contrast with *K*-selection.

salt Ionic substance containing neither H^+ nor OH^-.

sarcolemma (Gr. *sarx*, flesh + *lemma*, rind) The thin, noncellular sheath that encloses a striated muscle fiber.

sarcomere (Gr. *sarx*, flesh + *meris*, part of) Fundamental unit of contraction in skeletal muscle; repeating bands of actin and myosin that appear between two Z lines.

sarcoplasm (Gr. *sarx*, flesh + *plasma*, mold) The cytoplasm of a vertebrate striated muscle cell; clear, semifluid substance between fibrils of muscle tissue.

satellite DNA A nontranscribed region of the chromosome with a distinctive base composition; a short nucleotide sequence repeated tandemly many thousands of times.

sclereid (Gr. *skleros*, hard) In vascular plants, a sclerenchyma cell with a thick, lignified, secondary wall having many pits; not elongate like a fiber, the other principal kind of sclerenchyma cell.

sclerenchyma cell (Gr. *skleros*, hard + *en*, in + *chymein*, to pour) A cell of variable form and size with more or less thick, often lignified, secondary walls; may or may not be living at maturity; includes fibers and sclereids. Collectively, sclerenchyma cells may make up a kind of tissue called sclerenchyma.

scrotum (L., bag) The pouch that contains the testes in most mammals.

secondary growth In vascular plants, an increase in stem and root diameter made possible by cell division of the lateral meristems. Secondary growth produces the secondary plant body.

secondary sex characteristics External differences between male and female animals; not directly involved in reproduction.

secondary structure of a protein The

twisting or folding of a polypeptide chain; results from the formation of hydrogen bonds between different amino acid side groups of a chain; the most common structures that form are a single-stranded helix, an extended sheet, or a cable containing three (as in collagen) or more strands.

secondary tissues In vascular plants, tissues produced by the lateral meristems—the vascular cambium and cork cambium.

secondary wall In plants, the innermost layer of the cell wall, formed in certain cells after cell elongation has ceased; secondary walls have a highly organized microfibrillar structure, and are often impregnated with lignin.

secretion (L. *secerere*, to sever, separate) Any molecule that is exported from the cell that produces it; a molecule or collection of molecules that is released to perform its function outside the gland that produced it.

seed A structure that develops from the mature ovule of a seed plant; seeds generally consist of seed coat, embryo, and a food reserve.

seed coat The outer layer of a seed, developed from the integuments of the ovule.

segregation of alleles *See* Mendel's first law.

selectively permeable membrane (L. *seligere*, to gather apart + *permeare*, to go through) A membrane that permits passage of water and some solutes but blocks passage of one or more solutes. Same as differentially permeable. Used to be referred to as "semipermeable."

self-fertilization The union of egg and sperm produced by a single hermaphroditic organism.

self-pollination The transfer of pollen from another to stigma in the same flower or to another flower of the same plant, leading to self-fertilization.

semen (L., seed) In reptiles and mammals, sperm-bearing fluid expelled from the penis during male orgasm; contains millions of sperm in each milliliter of seminal fluid, an alkaline, fructose-containing liquid that nourishes the sperm cells suspended in it.

semipermeable membrane *See* selectively permeable membrane.

sensory neuron A neuron that transmits nerve impulses from a sensory receptor to the central nervous system or central ganglion; an afferent neuron.

sensory receptor A cell, tissue, or organ that responds to internal or external stimuli by initiating a nerve impulse.

sepal (L. *sepalum*, a covering) A member of the outermost floral whorl of a flowering plant; collectively, the sepals constitute the calyx.

septate (L. *septum*, fence) Divided by cross-walls into cells or compartments.

septum, pl. **septa** (L., fence) A wall between two cavities.

serotonin (L. *serum*, serum) 5-Hydroxytryptamine; a phenolic amine found in many animal tissues that has profound but poorly understood metabolic and vascular effects, particularly on the central nervous system.

serum (L., serum) Vertebrate blood plasma minus all circulating cells and the protein fibrinogen; the liquid that separates from blood after coagulation.

sessile (L. *sessilis*, of or fit for sitting, low, dwarfed) Attached; not free to move about. In vascular plants, a leaf lacking a petiole is said to be sessile.

sex chromosomes Chromosomes that are different in the two sexes and that are involved in sex determination. All other chromosomes are called autosomes. *See* autosome.

sex-linked characteristic A genetic characteristic, such as color blindness in humans or white eye in fruit flies, that is determined by a gene located on a sex chromosome and that therefore shows a different pattern of inheritance in males than in females.

sexual reproduction The fusion of gametes followed by meiosis and recombination of some point in the life cycle.

shoot In vascular plants, the aboveground portions, such as the stem and leaves.

short-day plants Plants that must be exposed to light periods shorter than some critical length for flowering to occur; they flower primarily in autumn.

sieve cell In the phloem (food-conducting tissue) of vascular plants, a long, slender sieve element with relatively unspecialized sieve areas and with tapering end walls that lack sieve plates; found in all vascular plants except angiosperms, which have sieve-tube members.

sieve tube In the phloem of angiosperms, a series of sieve-tube members arranged end-to-end and interconnected by sieve plates.

sieve-tube member One of the component cells of a sieve tube.

simple leaf An undivided leaf, as opposed to a compound leaf.

sinistral (L. *sinister*, left) To the left; dextral is to the right.

sinoatrial node *See* pacemaker.

sinus (L., curve) A cavity or space in tissues or in bone.

smooth muscle Nonstriated muscle; lines the walls of internal organs and arteries and is under involuntary control.

sodium-potassium pump Transmembrane channels engaged in the active (ATP-driven) transport of sodium ions, exchanging them for potassium ions; maintains the resting membrane potential of neurons.

solute A molecule dissolved in some solution. As a general rule, solutes dissolve only in solutions of similar polarity—for example, glucose (polar) dissolves in (forms hydrogen bonds with) water (also polar), but not in vegetable oil (nonpolar).

solute potential The change in free energy or chemical potential of water produced by solutes; also called osmotic potential. *See* osmotic pressure.

solution A homogeneous mixture of the molecules of two or more substances; the substance present in the greatest amount (usually a liquid) is called the solvent, and the substances present in lesser amounts are called solutes.

somatic cells (Gr. *soma*, body) The differentiated cells composing body tissues of multicellular plants and animals; all body cells except those giving rise to gametes.

somatic nervous system (Gr. *soma*, body) In vertebrates, the neurons of the peripheral nervous system that control skeletal muscle; the "voluntary" system, as contrasted with the "involuntary," or autonomic, nervous system.

somite (Gr. *soma*, body) One of the blocks, or segments, of tissue into which the mesoderm is divided during differentiation of the vertebrate embryo.

speciation (L. *species*, kind) The process by which new species are formed during the course of evolution.

species, pl. **species** (L., kind, sort) A kind of organism; species are designated by binomial names written in italics.

specific heat The amount of heat (in calories) required to raise the temperature of 1 gram of a substance 1° C. The specific heat of water is 1 calorie per gram.

specificity Selectivity; as in the highly specific choice of substrate by an enzyme.

sperm (Gr. *sperma*, seed) A mature male gamete, usually motile and smaller than the female gamete.

spermatid (Gr. *sperma*, seed) In animals, each of four haploid ($1n$) cells that result from the meiotic divisions of a spermatocyte; each spermatid differentiates into a sperm cell.

spermatocytes (Gr. *sperma*, seed + *kytos*, vessel) In animals, the diploid ($2n$) cells formed by the enlargement of the spermatogonia; they give rise by meiotic division to the spermatids.

spermatogenesis (Gr. *sperma*, seed + *genesis*, origin) In animals, the process by which spermatogonia develop into sperm.

spermatogonia (Gr. *sperma*, seed + *gonos*, a child) In animals, the unspecialized diploid ($2n$) cells on the walls of the testes that, by meiotic

division, become spermatocytes, then spermatids, then sperm cells.

spermatogonium (Gr. *sperma*, seed + *gone*, offspring) In animals, the precursor of a mature male reproductive cell; gives rise directly to a spermatocyte.

spermatozoon, pl. **spermatozoa** (Gr. *sperma*, seed + *zoos*, living) A sperm cell.

sphincter (Gr. *sphinkter*, band, from *sphingein*, to bind tight) In vertebrate animals, a ring-shaped muscle capable of closing a tubular opening by constriction (such as the one between stomach and small intestine, or between anus and exterior).

spindle The motive assembly that carries out the separation of chromosomes during cell division; composed of microtubules and assembled during prophase at the equator of the dividing cell.

spindle fibers A group of microtubules that together make up the spindle.

spiracle (L. *spiraculum*, from *spirare*, to breathe) External opening of a trachea in arthropods.

spiral cleavage In animals, a type of early embryonic cleavage in which cleavage planes are diagonal to the polar axis and unequal cells are produced by the alternate clockwise and counterclockwise cleavage around the axis of polarity; determinate cleavage.

spirillum, pl. **spirilla** (L. *spira*, coil) A long coiled or spiral bacterium.

spongy parenchyma A leaf tissue composed of loosely arranged, chloroplast-bearing cells. *See* palisade parenchyma.

sporangium, pl. **sporangia** (Gr. *spora*, seed + *angeion*, a vessel) A structure in which spores are produced.

spore A haploid reproductive cell, usually unicellular, capable of developing into an adult without fusion with another cell. Spores result from meiosis, as do gametes, but gametes fuse immediately to produce a new diploid cell.

sporophyte (Gr. *spora*, seed + *phyton*, plant) The spore-producing, diploid ($2n$) phase in the life cycle of a plant having alternation of generations.

stamen (L., thread) The organ of a flower that produces the pollen; usually consists of another and filament; collectively, the stamens make up the androecium.

starch (Mid. Eng. *sterchen*, to stiffen) An insoluble polymer of glucose; the chief food storage substance of plants; typically composed of 1000 or more glucose units.

statocyst (Gr. *statos*, standing + *kystis*, sac) A sensory receptor sensitive to gravity and motion; consists of a vesicle containing granules of sand (statoliths) or some other material that stimulates surrounding tufts of cilia when the organism moves.

statolith (Gr. *statos,* standing + *lithos,* stone) Small calcareous body resting on tufts of cilia in the statocyst.

stem The aboveground axis of vascular plants; stems are sometimes below ground (as in rhizomes and corms).

stereoscopic vision (Gr. *stereos,* solid + *optikos,* pertaining to the eye) Ability to perceive a single, three-dimensional image from the simultaneous but slightly divergent two-dimensional images delivered to the brain by each eye.

steroid (Gr. *stereos,* solid + L. *ol,* from *oleum,* oil) One of a group of lipids having a molecular skeleton of four fused carbon rings and, often, a hydrocarbon tail. Cholesterol, sex hormones, and the hormones of the adrenal cortex are steroids.

stigma (Gr., mark, tattoo mark) (1) In angiosperm flowers, the region of a carpel that serves as a receptive surface for pollen grains. (2) Light-sensitive eyespot of some algae.

stimulus (L., goad, incentive) Any internal or external change or signal that can be detected by an organism; every stimulus is some form of energy change (for example, a change in heat, sound wave, chemical, or light energy).

stolon (L. *stolo,* shoot) (1) A stem that grows horizontally along the ground surface and may form adventitious roots, such as runners of the strawberry plant. (2) A functionally similar structure that occurs in some colonial cnidarians and ascidians.

stoma, pl. **stomata** (Gr., mouth) In plants, a minute opening bordered by guard cells in the epidermis of leaves and stems; water passes out of a plant mainly through the stomata, and CO_2 passes in chiefly by the same pathway.

striated muscle (L., from *striare,* to groove) Skeletal voluntary muscle and cardiac muscle. The name derives from its striped appearance, which reflects the arrangement of contractile elements. *See* voluntary muscle.

stroma (Gr., anything spread out) (1) The ground substance of plastids. (2) The supporting connective tissue framework of an animal organ; filmy framework of red blood corpuscles and certain cells.

structural gene A transcription unit; any sequence of nucleotides that encodes a protein, tRNA, or rRNA molecules, in distinction to other sequences that have regulatory functions but are not transcribed.

style (Gr. *stylos,* column) In flowers, the slender column of tissue which arises from the top of the ovary and through which the pollen tube grows.

subspecies A subdivision of a species, often a geographically distinct race.

substrate (L. *substratus,* strewn under)

The foundation to which an organism is attached; a molecule upon which an enzyme acts.

substrate level phosphorylation Formation of ATP that takes place via a coupled reaction during glycolysis.

succession In ecology, the slow, orderly progression of changes in community composition that takes place through time. *Primary succession* occurs in nature over long periods of time; *secondary succession* occurs when a climax community has been disturbed.

sucrase An enzyme that hydrolyzes sucrose into glucose and fructose; also called invertase.

sucrose Cane sugar; a common disaccharide found in many plants; a molecule of glucose linked to a molecule of fructose.

sugar Any monosaccharide or disaccharide.

superior ovary In flowers, a condition in which the outer floral whorls (androecium, corolla, and calyx) arise at the bottom of the ovary. *See* inferior ovary.

surface tension A tautness of the surface of a liquid, caused by the cohesion of the molecules of liquid. Water has an extremely high surface tension. *See* cohesion.

suspension A solid dispersed in a liquid; refers to solid particles sufficiently large that they will settle out of the fluid under the influence of gravity.

symbiosis (Gr. *syn,* together with + *bios,* life) The living together in close association of two or more dissimilar organisms; includes parasitism (in which the association is harmful to one of the organisms), commensalism (in which it is beneficial to one, of no significance to the other), and mutualism (in which the association is advantageous to both).

sympathetic nervous system A subdivision of the autonomic nervous system of vertebrates that functions as an alarm response; increases heartbeat and dilates blood vessels while putting the body's everyday functions, such as digestion, on hold; usually operates antagonistically with parasympathetic nerves; in times of stress, danger, or excitement, it mobilizes the body for rapid response.

synapse (Gr. *synapsis,* a union) A junction between a neuron and another neuron or muscle cell; the two cells do not touch, the gap being bridged by neurotransmitter molecules.

synapse, excitatory A synapse in which the receiving cell's receptors respond to the arrival of neurotransmitter molecules by increasing the receptor cell membrane's permeability to potassium or chloride or both; this drives its membrane potential toward threshold; excitatory synapses increase the

likelihood that a receiving cell will fire an action potential.

synapse, inhibitory A synapse in which the receiving cell's receptors respond to the arrival of neurotransmitter molecules by decreasing the receptor cell membrane's permeability to sodium and potassium; this drives its membrane potential away from threshold. Inhibitory synapses decrease the likelihood that a receiving cell will fire an action potential.

synapsis (Gr., contact, union) The point-by-point alignment (pairing) of homologous chromosomes that occurs before the first meiotic division; crossing-over occurs during synapsis.

syndrome (Gr. *syn,* with + *dramein,* to run) A group of symptoms characteristic of a particular disease or abnormality.

syngamy (Gr. *syn,* together with + *gamos,* marriage) The process by which two haploid cells fuse to form a diploid zygote; fertilization.

synthesis (Gr. *syntheke,* a putting together) The formation of a more complex molecule from simpler ones.

systematics (Gr. *systema,* that which is put together) Scientific study of the kinds and diversity of organisms and of the relationships between them; equivalent to *taxonomy.*

systemic circulation The circulation path of blood leading to and from all body parts except the lungs.

T cell A type of lymphocyte involved in cell-mediated immunity and interactions with B cells; the "T" refers to the fact that T cells are produced in the thymus; also called a T lymphocyte.

tactile (L. *tactilis,* able to be touched, from *tangere,* to touch) Pertaining to touch.

tagma pl. **tagmata** (Gr., arrangement, order, row) A compound body section of an arthropod resulting from embryonic fusion of two or more segments; for example, head, thorax, abdomen. The process of fusion is called *tagmatization* or *tagmosis.*

tandem repeats Multiple copies of the same gene lying side by side in series.

taxis, pl. **taxes** (Gr., arrangement) An orientation movement by a (usually) simple organism in response to an environmental stimulus.

taxon, pl. **taxa** A general term for any taxonomic category; therefore species, families, and phyla are all taxa.

taxonomy (Gr. *taxis,* arrangement + *nomos,* law) The science of the classification of organisms; equivalent to *systematics.*

telencephalon (Gr. *telos,* end + *encephalon,* brain) The most anterior portion of the brain, including the cerebrum and associated structures.

teleology (Gr. *telos,* end + *logos,* word)

The philosophical view that natural events are goal directed and preordained, as opposed to the scientific view of mechanical determinism.

telophase (Gr. *telos*, end + *phasis*, form) The last stage of the nuclear division of mitosis and meiosis, during which the chromosomes become reorganized into two nuclei.

template A pattern guiding the formation of a negative or complementary image; with reference to DNA duplication, each strand serves as a template on which a complementary strand is assembled.

tendril (L. *tendere*, to extend) A slender coiling structure, usually a modified leaf or stem, that aids in the support of a plant.

tentacles (L. *tentare*, to touch) Long, flexible protrusions located about the mouth in many invertebrates; usually prehensile or tactile.

territory (L. *territorium*, from *terra*, earth) An area or space occupied and defended by an individual or a group of animals.

tertiary structure of a protein The three-dimensional shape of a protein; primarily the result of hydrophobic interactions of amino acid side groups and, to a lesser extent, of hydrogen bonds between them; forms spontaneously.

testcross A mating between a phenotypically dominant individual of unknown genotype (it could in principle be either homozygous or heterozygous) and a homozygous recessive "tester," done to determine whether the phenotypically dominant individual is homozygous or heterozygous for the relevant gene. *See* backcross.

testis, pl. **testes** (L., *witness*) In mammals, the sperm-producing organ; also the source of male sex hormone.

testosterone (Gr. *testis*, testicle + *steiras*, barren) The male sex hormone; a steroid hormone secreted by the testes of mammals that stimulates the development and maintenance of male sex characteristics and the production of sperm.

tetrapods (Gr. *tetras*, four + *pous*, foot) Four-footed (actually, four-limbed) vertebrates; the group includes amphibians, reptiles, birds, and mammals.

thalamus (Gr. *thalamos*, chamber) That part of the vertebrate forebrain just posterior to the cerebrum; governs the flow of information from all other parts of the nervous system to the cerebrum.

theory (Gr. *theorein*, to look at) A well-tested hypothesis, one unlikely to be rejected by future tests.

thermodynamics (Gr. *therme*, heat + *dynamis*, power) The study of transformations of energy, using heat as the most convenient form of measurement of energy. The first law of thermodynamics states that the total energy of the universe remains constant. The second law of thermodynamics states that the entropy, or degree of disorder, tends to increase.

thigmotropism In plants, unequal growth in some structure that comes about as a result of physical contact with an object. The unequal growth that causes a vine to curl around a fence post is an example.

thorax (Gr., a breastplate) (1) In vertebrates, that portion of the trunk containing the heart and lungs. (2) In crustaceans and insects, the fused, leg-bearing segments between head and abdomen.

threshold value In neurons, the minimum change in membrane potential necessary to produce an action potential.

thylakoid (Gr. *thylakos*, sac + *-oides*, like) A saclike membranous structure in cyanobacteria and the chloroplasts of eukaryotic organisms; in chloroplasts, stacks of thylakoids form the grana; chlorophyll is found in the thylakoids.

thymine A pyrimidine occurring in DNA but not in RNA; *see also* uracil.

tight junction Region of actual fusion of cell membranes between two adjacent animal cells that prevents materials from leaking through the tissue; for example, intestinal epithelial cells are surrounded by tight junctions.

tissue (L. *texere*, to weave) A group of similar cells organized into a structural and functional unit.

trachea, pl. **tracheae** (L., windpipe) A tube for breathing; in terrestrial vertebrates, the windpipe that carries air between the larynx and bronchi (which leads to the lungs); in insects and some other terrestrial arthropods, a system of chitin-lined air ducts.

tracheid (Gr. *tracheia*, rough) In vascular plants, an elongated, thick-walled conducting and supporting cell of xylem. Tracheids, which are dead when functional, have tapering ends and pitted walls without perforations. Compare with *vessel member*.

transcription (L. *trans*, across + *scribere*, to write) The enzyme-catalyzed assembly of an RNA molecule complementary to a strand of DNA; may be either mRNA (for protein-encoding genes), tRNA, or rRNA.

transcription unit The portion of a gene that is actually transcribed into mRNA; a sequence of DNA that begins with a promoter and encompasses the leader region (where the ribosome binds to mRNA) and one or more structural genes. A transcription unit does not include the regulatory portions of a gene such as operators, enhancers, or repressor-encoding sequences.

transduction The transfer of genes from one organism to another, with a virus used as a vector (carrier); "piggyback" recombination; occurs naturally among bacteria, when viruses mistakenly incorporate bacterial DNA into themselves and so transfer it to other cells they infect.

transfection The transformation of eukaryotic cells in culture.

transfer RNA (tRNA) (L. *trans*, across + *ferre*, to bear or carry) A class of small RNAs (about 80 nucleotides) with two functional sites; to one site an "activating enzyme" adds a specific amino acid; the other site carries the nucleotide triplet (anticodon) for that amino acid. Each type of tRNA transfers a specific amino acid to a growing polypeptide chain as specified by the nucleotide sequence of the messenger RNA being translated. *See* activating enzyme.

transformation (L. *trans*, across + *formare*, to shape) The transfer of naked DNA from one organism to another; also called simply "gene transfer." First observed as uptake of DNA fragments among pneumococcal bacteria, transformation is now routinely carried out in fruitflies and other eukaryotes by microinjection of eggs, with transposons often used as vectors.

translation (L. *trans*, across + *latus*, that which is carried) The assembly of a protein on the ribosomes, using mRNA to direct the order of amino acids.

translocation (L. *trans*, across + *locare*, to put or place) (1) In plants, the long-distance transport of soluble food molecules (mostly sucrose), which occurs primarily in the sieve tubes of phloem tissue. (2) In genetics, the interchange of chromosome segments between nonhomologous chromosomes.

transpiration (L. *trans*, across + *spirare*, to breathe) The loss of water vapor by plant parts; most transpiration occurs through the stomata.

transposon (L. *transponere*, to change the position of) A DNA sequence carrying one or more genes and flanked by insertion sequences that confer the ability to move from one DNA molecule to another; an element capable of transposition, the changing of chromosomal location.

tricarboxylic acid cycle or **TCA cycle** *See* Krebs cycle.

trichome (Gr. *trichos*, hair) In plants, an outgrowth of the epidermis, such as a hair, scale, or water vesicle.

triple fusion In angiosperms, the fusion of the second male gamete, or sperm, with the polar nuclei, resulting in formation of a primary endosperm nucleus, which is most often triploid (3*n*).

triploblastic (Gr. *triploos*, triple + *blastos*, germ) Pertaining to metazoans in which the embryo has three primary germ layers—ectoderm, mesoderm, and endoderm. All metazoans are probably triploblastic, although some were formerly interpreted as *diploblastic*—having two primary germ layers in the embryo.

triploid (Gr. *triploos*, triple) Having three complete chromosome sets per cell (3*n*).

trophic level (Gr. *trophos*, feeder) A step in the movement of energy through an ecosystem.

trophoblast (Gr. *trephein*, to nourish + *blastos*, germ) In vertebrate embryos, the outer ectodermal layer of the blastodermic vesicle; in mammals it is part of the chorion and attaches to the uterine wall.

tropism (Gr. *trope*, a turning) A response to an external stimulus.

tropomyosin (Gr. *tropos*, turn + *myos*, muscle) Low-molecular-weight protein surrounding the actin filaments of striated muscle.

troponin Complex of globular proteins positioned at intervals along the actin filament of skeletal muscle; thought to serve as a calcium-dependent "switch" in muscle contraction.

tube feet Numerous small, muscular, fluid-filled tubes projected from the body of echinoderms; part of the water-vascular system; used in locomotion, clinging, food handling, and respiration.

tubulin (L. *tubulus*, small tube + *in*, belonging to) Globular protein subunit forming the hollow cylinder of microtubules.

turgor (L. *turgere*, to swell) The pressure exerted on the inside of a plant cell wall by the fluid contents of the cell; the interior of the cell is hypertonic in relation to the fluids surrounding it and so gains water by osmosis.

turgor pressure (L. *turgor*, a swelling) The pressure within a cell resulting from the movement of water into the cell. A cell with high turgor pressure is said to be *turgid*. *See* osmotic pressure.

umbilical (L. *umbilicus*, navel) In humans, refers to the navel, or umbilical cord.

unequal crossing-over A crossover between two identical sequences lying at different locations on homologous chromosomes.

unicellular Composed of a single cell.

uracil A pyrimidine found in RNA but not in DNA; *see also* thymine.

urea (Gr. *ouron*, urine) An organic molecule formed in the vertebrate liver; the principal form of disposal of nitrogenous wastes by mammals.

urethra (Gr. from *ourein*, to urinate)

The tube carrying urine from the bladder to the exterior of mammals.

urine (Gr. *ouron*, urine) The liquid waste filtered from the blood by the kidney and stored in the bladder pending elimination through the urethra; the principal excretory product in birds, reptiles, and insects.

uterus (L., womb) In mammals, a chamber in which the developing embryo is contained and nurtured during pregnancy.

vacuole (L. *vacuus*, empty) A space or cavity within the cytoplasm of a cell; typically filled with a watery fluid, the cell sap; part of the lysosomal compartment of the cell.

vagina Female accessory reproductive organ that receives sperm from the male penis; forms part of the birth canal, and acts as a channel to the exterior for menstrual flow.

vagus nerve (L. *vagus*, wandering) A parasympathetic nerve that innervates the heart and visceral organs.

variables Different but related conditions that might affect the outcome of a scientific test.

variety A group of plants of less than species rank; some botanists view the term as equivalent to subspecies in animals, whereas others view it as a taxon of lower rank.

vascular (L. *vasculum*, a small vessel) Containing or concerning vessels that conduct fluid.

vascular bundle In vascular plants, a strand of tissue containing primary xylem and primary phloem (and procambium if still present) and frequently enclosed by a bundle sheath of parenchyma or fibers.

vascular cambium In vascular plants, a cylindrical sheath of meristematic cells, the division of which produces secondary phloem outwardly and secondary xylem inwardly; the activity of the vascular cambium increases stem or root diameter.

vas deferens (L. *vas*, a vessel + *deferre*, to carry down) In mammals, the tube carrying sperm from the testes to the urethra.

vector (L., a bearer, carrier, from *vehere*, to carry) In genetics, any virus or plasmid DNA into which a gene is integrated and subsequently transferred into a cell; in disease, any agent that carries and transmits pathogenic microorganisms from one host to another host.

vegetative Nonsexual; in flowering plants, pertaining to parts of the shoot other than flowers or fruits.

vein (L. *vena*, a blood vessel) (1) In plants, a vascular bundle forming a part of the framework of the conducting and supporting tissue of a stem or leaf. (2) In animals, a blood

vessel carrying blood from the tissues to the heart.

ventral (L. *venter*, belly) Pertaining to the undersurface of an animal that moves on all fours, and to the front surface of an animal that holds its body erect.

ventricle (L. *ventriculus*, the stomach—i.e., the belly of the heart) A muscular chamber of the heart that receives blood from an atrium and pumps blood out to either the lungs or the body tissues.

vertebrate An animal having a backbone made of bony segments called vertebrae.

vesicle (L. *vesicula*, a little bladder) A small, intracellular, membrane-bound sac in which various substances are transported or stored.

vessel (L. *vas*, a vessel) A tubelike element in the xylem of angiosperms, composed of dead cells (vessel elements) arranged end to end. Its function is to conduct water and minerals from the soil.

vessel element In vascular plants, a typically elongated cell, dead at maturity, which conducts water and solutes in the xylem; vessel elements make up vessels. *Tracheids* are less specialized conducting cells.

viable (L. *vita*, life) Able to live.

villus, pl. **villi** (L., a tuft of hair) In vertebrates, one of the minute, fingerlike projections lining the small intestine that serve to increase the absorptive surface area of the intestine.

viscera (L., internal organs) Internal organs in the body cavity of an animal.

vitamin (L. *vita*, life + *amine*, of chemical origin) An organic substance that cannot be synthesized by a particular organism but is required in small amounts for normal metabolic function; must be supplied in the diet or synthesized by bacteria living in the intestine.

viviparity (L. *vivus*, alive + *parere*, to bring forth) Reproduction in which eggs develop within the mother's body, with her nutritional aid; characteristic of mammals, many reptiles, and some fishes; offspring are born as juveniles. Adj., viviparous.

water cycle Worldwide circulation of water molecules, powered by the sun.

water potential The potential energy of water molecules. Regardless of the reason (e.g., gravity, pressure, concentration of solute particles) for the water potential, water moves from a region where water potential is greater to a region where water potential is lower.

water-vascular system In echinoderms, a unique system of fluid-filled closed tubes and ducts; used to move tentacles and tube feet, which function in clinging, food handling, locomotion, and respiration.

whorl A circle of leaves or of flower parts borne at a single node.

wild type In genetics, the phenotype or genotype that is characteristic of the majority of individuals of a species in a natural environment.

wood Accumulated secondary xylem.

xylem (Gr. *xylon,* wood) In vascular plants, a specialized tissue, composed primarily of elongate, thick-walled conducting cells, which transports water and solutes through the plant body.

yolk The stored food in egg cells; it nourishes the embryo.

zoology (Gr. *zoe,* life + *logos,* word) The study of animals. A *zoologist* is a student of animals.

zooplankton (Gr. *zoe,* life + *plankton,* wanderer) The nonphotosynthetic organisms present in plankton.

zoospore A motile spore.

zygote (Gr. *zygotos,* paired together) The diploid (2*n*) cell resulting from the fusion of male and female gametes (fertilization). A zygote may either develop into a diploid individual by mitotic divisions or undergo meiosis to form haploid (*n*) individuals that divide mitotically to form a population of cells.

ILLUSTRATION ACKNOWLEDGMENTS

Page vi L. West

Page ix © Rod Planck/Tom Stack & Associates

Page xiv © T.E. Adams, 1984/Visuals Unlimited

Page xv Peter Parks/Oxford Scientific Films

Page xvi L. West

Page xvii © David M. Phillips, 1984/Visuals Unlimited

Page xviii L. West

Page xix © Leonard Lee Rue III, Blairstown, NJ 07825

Page xx Sylvia Duran Sharnoff, 1984/Visuals Unlimited

Page xxi Keith H. Murakami/Tom Stack & Associates

Page xxii © Ann Duncan/Tom Stack & Associates

Page xxii L. West

Page xxiii © Ed Robinson/Tom Stack & Associates

Page xxiv E.S. Ross

Page xxv Brian Parker/Tom Stack & Associates

Page xxvi M. Philip Kahl, Jr./Tom Stack & Associates

Page xxvii © Alan G. Nelson/Tom Stack & Associates

Page xxvii Brian Parker/Tom Stack & Associates

Page xxviii Brian Parker/Tom Stack & Associates

Page xxix Jeff Rotman

Page xxx L. West

Page xxxi John Gerlach/Tom Stack & Associates

I-A S. Chester/Tom Stack & Associates

I-B NASA

I-C John Reader

1-1 Leonard Lee Rue III

1-2 Robert Noonan

1-3 Maureen Hanson

1-4 American Cancer Society

1-5 Centers for Disease Control, Atlanta, Ga.

1-6 R. Rowan/Photo Researchers, Inc.

1-7 John Gerlach/Tom Stack & Associates

1-8 John Cunningham/Visuals Unlimited

1-9 Stephanie Maze © National Geographic Society

1-10 Thomas McAvoy, LIFE Magazine © Time, Inc.

1-11 Matthew A. Gonda, National Cancer Institute

1-13 Ron Kimball, "Wildlife Safari," Science, vol. 221, pages 459-462 and cover, July 29, 1983.

1-A John Gerard

1-15 E.S. Ross

1-16 E.S. Ross

1-17, A John Reader

1-17, B NASA

2-1 Illustration by William C. Ober

2-6, C Michael Gadomski/Tom Stack & Associates

2-7, B National Archives/Picture Research

2-10, A E.S. Ross

2-10, B Frank Awbrey/Visuals Unlimited

2-10, C Kjell Sandved

2-10, D Gerald Corsi/Tom Stack & Associates

2-11 Suzi Barnes/Tom Stack & Associates

2-13 E.S. Ross

2-15 Illustration by W. Ober

3-1 Bob McKeever/Tom Stack & Associates

3-5, A Mary Field

3-5, B Alan Briere/Tom Stack & Associates

3-5, C E.S. Ross

3-5, D John Cunningham/Visuals Unlimited

3-6 Y. Arthus-Bertrand/Peter Arnold, Inc.

3-7 Peter Parks/Oxford Scientific Films/Animals Animals

3-8 Runk-Schoenberger/Grant Heilman Photography

3-9 Bob Gossington/Tom Stack & Associates

3-10 Sidney Fox

3-11 John Cunningham/Visuals Unlimited

3-A-C NASA

3-D © California Institute of Technology, 1959

3-E National Optical Astronomy Observatories

4-6 S. Meola/Visuals Unlimited

4-7, B Barry King, University of California, Davis; BPS/Tom Stack & Associates

4-8, B J.D. Litvay/Visuals Unlimited

4-9, A Kjell B. Sandved

4-15, A David M. Phillips/Visuals Unlimited

4-15, B J. David Robertson, from Charles Flickinger, Medical Cell Biology, W.B. Saunders Co., Philadelphia, 1979.

4-15, C Illustration by W. Ober

4-16 Illustration by W. Ober

4-24, B Illustration by Ron Ervin

4-25 Oscar L. Miller, Science, vol. 169, pages 392-395, July, 1970.

II-A Manfred Kage/Peter Arnold, Inc.

II-B Manfred Kage/Peter Arnold, Inc.

II-C A. Paseika/Science Photo Library International

II-D David Scharf/Peter Arnold, Inc.

5-1 R. Hooke, Micrographia . . ., London 1665/Picture Research

5-2 E.S. Ross

5-3 S.M. Siegel

5-6 Frank Sloop

5-8 Paul Johnson, University of Rhode Island; BPS/Tom Stack & Associates

5-9 J.W. Schopf, Journal of Paleontology, vol. 45, pages 925-960, 1971.

5-10 J.W. Schopf, *Journal of Paleontology,* vol. 42, pages 651-688, 1968.
5-11, A Illustration by W. Ober
5-11, B Richard Rodewald
5-12 Ed Reschke
5-13 Illustration by Ron Ervin
5-16 T.E. Adams/Visuals Unlimited
5-17 Illustration by W. Ober
5-19 Manfred Kage/Peter Arnold, Inc.

6-1 James A. Spudich
6-2 Illustration by W. Ober
6-3 A K.G. Murti/Visuals Unlimited
6-3 B V.M. Kersey and Norman K. Wessels, *Journal of Cell Biology,* vol. 68, page 264, 1976.
6-4, A Illustration by Ron Ervin
6-4, B Richard Rodewald
6-5, A Illustration by Ron Ervin
6-5, B T.W. Tillack
6-5, C Charles Flickinger, *Medical Cell Biology,* W.B. Saunders Co., Philadelphia, 1979.
6-6, A Charles Flickinger, *Medical Cell Biology,* W.B. Saunders Co., Philadelphia, 1979.
6-6, B Illustration by Ron Ervin
6-8 K.G. Murti/Visuals Unlimited
6-9, A Illustration by Ron Ervin
6-9, B Charles Flickinger, *Medical Cell Biology,* W.B. Saunders Co., Philadelphia, 1979.
6-10 Micrograph by Kenneth Miller; illustration by W. Ober
6-11, A Jim and Cathy Church
6-11, B Steinhart Aquarium, San Francisco
6-12, A Joan Nowicke
6-12, B Manfred Kage/Peter Arnold, Inc.
6-12, C Richard Rodewald
6-12, D David M. Phillips/Visuals Unlimited

7-3, A L.M. Pope, University of Texas at Austin; BPS/Tom Stack & Associates
7-4 Illustration by W. Ober
7-5 Illustration by W. Ober
7-6 Illustration by W. Ober
7-7 K.G. Murti/Visuals Unlimited
7-8 Klaus Hausmann, *Scientific American,* August 1980, page 66.
7-9, A Phillip Harrington/Peter Arnold, Inc.
7-9, B L.E. Roth, University of Tennessee; BPS/Tom Stack & Associates
7-10 Micrographs by William Dentler; illustration by W. Ober
7-11 K.G. Murti/Visuals Unlimited
7-14 W.L. Dentler, University of Kansas; BPS/Tom Stack & Associates
7-15 Guenter Albrecht-Buehler
7-17 Susan Lowey

7-19 Illustration by W. Ober
7-20 Susumi Ito, from Charles Flickinger, *Medical Cell Biology,* W.B. Saunders Co., Philadelphia, 1979.

8-1 Brian Parker/Tom Stack & Associates
8-2 Illustration by W. Ober
8-3 Walter Engler
8-4 L. Maziarski/Visuals Unlimited
8-6 Jan de May, Janssen Pharmaceutica, *Scientific American,* vol. 243, August 1980, page 76.
8-7, A-F Ed Reschke
8-9, A and B Lionel I. Rebhun
8-A Paul W. Johnson; BPS/Tom Stack & Associates
8-B M.S. Fuller
8-11, A David M. Phillips/Visuals Unlimited
8-11, B and C Elias Lazarides, *Scientific American,* vol. 240, May 1979, page 107.
8-12 David M. Phillips/Visuals Unlimited
8-13, A B.A. Palevitz and E.H. Newcomb; BPS/Tom Stack & Associates

9-3, A Gary Grimes/ TAURUS Photos, Inc.
9-3, B Thomas Eisner
9-6, A and B Gary Grimes/TAURUS Photos, Inc.
9-7 Birgit Satir
9-8, A Camillo Peracchia, from Charles Flickinger, *Medical Cell Biology,* W.B. Saunders Co., Philadelphia, 1979.
9-8, C Elliot Hertzberg, *Journal of Biological Chemistry,* vol. 254, page 2143, 1979.
9-9 Richard Rodewald
9-10 E.H. Newcomb, University of Wisconsin; BPS/Tom Stack & Associates
9-12 Illustration by Ron Ervin
9-13 Illustration by W. Ober
9-16 Illustration by W. Ober

10-1 Robert Caputo © National Geographic Society
10-10 Paul W. Johnson, University of Rhode Island; BPS/Tom Stack & Associates

11-5 Grant Heilman/Grant Heilman Photography
11-7 Lester J. Reed
11-A E.S. Ross
11-12 Illustration by W. Ober
11-13 Efraim Racker

12-5 Walther Stoeckenius
12-9, A Sherman Thomson/Visuals Unlimited

12-9, B David Dennis/Tom Stack & Associates
12-11 Manfred Kage/Peter Arnold, Inc.
12-12 Illustration by Ron Ervin
12-13 Lewis K. Shumway
12-15 Carolina Biological Supply

III-A Alfred Paseika/Science Photo Library International
III-B John T. Lis
III-C John Cowell/Grant Heilman Photography
13-1 Schmidecker/Alpha
13-2 Richard Gross
13-3 National Library of Medicine/ Picture Research
13-4 Runk-Schoenberger/Grant Heilman Photography
13-5, A Grant Heilman/Grant Heilman Photography
13-5, B Illustration by Ron Ervin
13-6 Illustration by Ron Ervin
13-8 W.F. Bodmer and L.L. Cavalli-Sforza, *Genetics, Evolution, and Man* W.H. Freeman & Co., Publishers, New York, 1979.
13-9 Illustration by M.K. Ryan
13-12 Illustration by M.K. Ryan
13-13 Illustration by M.K. Ryan
13-14, B Albert Blakeslee, "Corn and Man," *Journal of Heredity,* vol. 5, page 511, 1914.
13-15 Leonard Lee Rue III

14-2 Diedrich von Wettstein; Reproduced with permission from *Annual Review of Genetics,* vol. 6, © 1972 by Annual Reviews, Inc.
14-3 James Kezer
14-4 Illustration by Ron Ervin
14-6, A-L Walter Plaut
14-7 Carolina Biological Supply Company
14-8 Illustration by M.K. Ryan
14-13 Oscar L. Miller
14-14 Illustration by W. Ober
14-16 William Wilson
14-17 Ed Reschke

15-2 L.L. Sims/Visuals Unlimited
15-4, A and B Karl Illmensee and Peter Hoppe, *Cell,* vol. 23, pages 9-18, January 1981; © MIT 1981.
15-5 Illustration by W. Ober
15-6 Illustration by Ron Ervin
15-7, A A.K. Kleinschmidt
15-7, B and C Lee D. Simon/Photo Researchers, Inc.
15-9 Photographer A.C. Barrington Brown, from J.D. Watson, *The Double Helix,* page 215, Atheneum, New York, 1968. © 1968 by J.D. Watson.
15-14 Illustration by Ron Ervin

16-1 O.L. Miller and Barbara R. Beatty, *Science,* vol. 164, cover, May 1969.

16-4 Jack Griffith

16-B Illustrations by Ron Ervin

16-11 Oscar L. Miller

16-C, 1 and 2 Bert O'Malley, *Proceedings of the National Academy of Sciences,* vol. 76, page 2256, 1979.

16-16 David R. Wolstenholme, *Chromosoma,* vol. 43, 1973.

16-17 Ulrich Laemmli

17-7 Micrograph by Charles C. Brinton and Judith Carnahan

17-9, A Jack Griffith

17-11 Stanley N. Cohen; reprinted by permission from *Nature,* vol. 221, no. 5226, pages 1273-1277. © 1969 Macmillan Journals, Ltd.

17-12 Illustration by M.K. Ryan

17-16 M.L. Pardue, *Chromosomes Today,* vol. 3, pages 47-52.

17-17, A O.L. Miller and Barbara R. Beatty, *Science,* vol. 164, cover, May 1969.

17-17, B Donald Brown, *Proceedings of the National Academy of Sciences,* vol. 68, page 3175, 1975.

18-1 Michael Wotton (Weyerhauser Company)

18-4 Stanley N. Cohen

18-6 R.L. Brinster and R.E. Hammer

18-8 Guenter Albrecht-Buehler

18-9 Illustration by W. Ober

18-13 Illustration by Ron Ervin

18-A Frank Sloop

IV-A Nadine Orabona/Tom Stack & Associates

IV-B Larry West

IV-C Cleveland P. Hickman, Jr.

IV-D Kjell Sandved

19-1, A and B E.S. Ross

19-2 E.S. Ross

19-3 BBC Copyright Photographs

19-4 Illustration by Ron Ervin

19-5 National Library of Medicine/ Picture Research

19-6, A and B Cleveland P. Hickman, Jr.

19-7, A and C Gary Romig

19-7, A Los Angeles Pigeon Club

19-8 National Library of Medicine/ Picture Research

19-9 Illustration by Ron Ervin

19-A Jeffrey Mitton

19-11 Illustration by Ron Ervin

19-12 Illustration courtesy of C.P. Hickman, Jr. *Integrated Principles of Zoology,* 7th ed., Times Mirror/ Mosby College Publishing, St. Louis, 1984.

20-1 Heather Angel

20-2 Thaddeus Kelly and William Wilson

20-3 Frank Sloop

20-5, A Library of Congress/Picture Research

20-5, B National Library of Medicine/ Picture Research

20-6, A The Bettmann Archive

20-7 E.S. Ross

20-8 Gerald D. Carr

20-9 E.S. Ross

20-10 John Cunningham/Visuals Unlimited

20-12, A and B J. Antonovics/Visuals Unlimited

20-13 From the Collection of the University of Wisconsin-Madison Archives/Picture Research

21-1 C.R. Bantock

21-2 Grant Heilman/Grant Heilman Photography

21-3, A-F E.S. Ross

21-4 Diane Pierce © National Geographic Society

21-5 P.R. Ehrlich

21-6 Illustration by M.K. Ryan

21-7, A and B E.S. Ross

21-8, A-C Kenneth Y. Kaneshiro

21-8, D Raymond Mendez/Animals Animals

21-A © Walt Disney Productions

21-9 Steen Willadsen; reprinted by permission from *Nature,* vol. 307, no. 5952, © 1984, Macmillan Journals Inc.

21-10, A John D. Cunningham/ Visuals Unlimited

21-10, B E.S. Ross

21-11, A and B E.S. Ross

21-B Olov Bjorkman/Carnegie Institute of Washington, Plant Biology Station

21-12, A and B Nita van der Werff

21-13 Illustration by George Venable

21-14 John S. Dunning

21-15 John D. Cunningham/Visuals Unlimited

21-16 Tom Stack/Tom Stack & Associates

22-1 Courtesy of Hunt Institute for Botanical Documentation, Carnegie-Mellon University, Pittsburgh.

22-2 Library of Congress/Picture Research

22-3 Illustration by M.K. Ryan

22-4 Missouri Botanical Garden

22-5, A Leonard Lee Rue III

22-5, B Brian Parker/Tom Stack & Associates

23-5 E.S. Ross

23-6 Tom Stack & Associates

22-7, A Rod Planck/Tom Stack & Associates

22-7, B E.S. Ross

22-7, C Brian Parker/Tom Stack & Associates

22-8, A and B E.S. Ross

22-8, C W.A. Derby/E.S. Ross

22-9, A John D. Cunningham/Visuals Unlimited

22-9, B Rod Planck/Tom Stack & Associates

22-10 Abraham and Beachey; BPS/ Tom Stack & Associates

22-12, A and C E.S. Ross

22-12, B Kjell Sandved

22-13 A. Friedman/Visuals Unlimited

22-14, A J. Robert Waaland; BPS/ Tom Stack & Associates

22-14, B T.E. Adams/Visuals Unlimited

22-14, C Marty Snyderman

23-1 Illustration by John Gurche

23-2 Natural History Museum/Los Angeles County

23-3, A Gail Gibson

23-3, B Reproduced by the courtesy of the South Australian Museum

23-5 E.S. Ross

23-6 Tom Stack & Associates

23-7 M.R. Walter

23-8 Andy Knoll

23-A Illustration by W. Ober

23-9 Illustration by John Gurche

23-10, A S. Conway Morris, *Phil. Trans. R. Soc. Lond.,* vol. B307, pages 507-586, 1985.

23-10, B S. Conway Morris, *Paleontology,* vol. 20, pages 623-640, 1977.

23-10, C S. Conway Morris, *Phil. Trans. R. Soc. Lond.,* vol. B285, pages 227-274, 1979.

23-10, D and E David L. Bruton

23-11, A Karl Nicklas

23-11, B William A. Shear, in *Science,* vol. 224, pages 492-494 © 1984 by the American Association for the Advancement of Science.

23-12 Illustration by Ron Ervin

23-13 John D. Cunningham/Visuals Unlimited

23-14 Rudolph Zallinger/Yale Peabody Museum

23-15 Illustration by Ron Ervin

23-16 Illustration by Ron Ervin, modified from material in *Science,* vol. 220, pages 712-715, 1983.

23-17 Field Museum of Natural History

23-18 E.S. Ross

23-19, A-C A. Montanari

24-1 R. Malenky

24-2 Illustration by Ron Ervin

24-3 Warren Garst/Tom Stack & Associates

24-4 Illustration by Ron Ervin

24-5, A Russell Mittermeier

24-5, B J.I. Pollock

24-5, C Gary Milburn/Tom Stack & Associates

24-6, A Leonard Lee Rue III
24-6, B and C E.S. Ross
24-6, D Nadine Orabona/Tom Stack & Associates
24-6, E E.S. Ross
24-6, F Russell Mittermeier
24-7 Cris Crowley/Tom Stack & Associates
24-8 Russell Mittermeier and Andy Young
24-9, A and B Warren and Genny Garst/Tom Stack & Associates
24-9, C Russell Mittermeier
24-10, A Russell Mittermeier and Andy Young
24-10, B Russell Mittermeier
24-11, A-C Russell Mittermeier
24-13 Illustration by Ron Ervin
24-A, 1 Russell Mittermeier
24-A, 2 Andy Young
24-A, 3 Jean-Jacques Petter
24-A, 4 and 5 Russell Mittermeier
24-14 Illustration by Ron Ervin
24-15 John D. Cunningham/Visuals Unlimited
24-16 Illustration by Ron Ervin
24-17, A Transvaal Museum, Pretoria, South Africa
24-17, B John Reader
24-18, A and B R.D. Leakey/National Museums of Kenya
24-19 Illustration by Ron Ervin
24-20 Photograph by Chester Tarka; courtesy of Ian Tattersall and Eric Delson
24-21 Illustration by Ron Ervin
24-22 Photograph by Chester Tarka; courtesy of Ian Tattersall and Eric Delson
24-23 John D. Cunningham/Visuals Unlimited
24-24, A and B E.S. Ross
24-25 Courtesy of Hunt Institute for Botanical Documentation, Carnegie-Mellon University, Pittsburgh.

V-A Manfred Kage/Peter Arnold, Inc.
V-B David Scharf/Peter Arnold, Inc.
V-C Brian Parker/Tom Stack & Associates
V-D Lee D. Simon/Photo Researchers
25-1 National Library of Medicine/Visuals Unlimited
25-2 K.G. Murti/Visuals Unlimited
25-3 Illustration by Ron Ervin
25-4 K.G. Murti/Visuals Unlimited
25-5 T.J. Beveridge, University of Guelph; BPS/Tom Stack & Associates; illustration by W. Ober
25-6 Illustration by W. Ober
25-7 Centers for Disease Control, Atlanta, Ga.
25-8, A E.S. Ross
25-8, B Centers for Disease Control, Atlanta, Ga.

25-9 K.G. Murti/Visuals Unlimited
25-10, A U.S. Department of Agriculture
25-10, B T.O. Diener, U.S. Department of Agriculture
25-11 Centers for Disease Control, Atlanta, Ga.
25-12, A-C W. Ober
25-13 Centers for Disease Control, Atlanta, Ga.
25-A Centers for Disease Control/World Health Organization

26-1, A J.J. Cardamone, Jr. and B.K. Pugashetti; BPS/Tom Stack & Associates
26-1, B D.M. Phillips/Visuals Unlimited
26-1, C Ed Reschke
26-1, D David White
26-1, E J. Robert Waaland; BPS/Tom Stack & Associates
26-2 Eli Lilly and Company
26-3 Ed Reschke
26-4 Abraham & Beachey; BPS/Tom Stack & Associates
26-5 T.J. Beveridge, University of Guelph; BPS/Tom Stack & Associates
26-A D. Balkwill and D. Maratea, Scientific American, December 1981, page 63.
26-7, A Purdue University
26-7, B B.J. Jacobsen
26-8 Centers for Disease Control, Atlanta, Ga.
26-9, A and B Karl O. Stetter; reprinted by permission from Nature, vol. 300, pages 258-259. © 1982, Macmillan Journals
26-9, C W. Stoeckenius
26-10, A-C David Hall
26-B Photo by F.R. Turner, courtesy of Howard Gest
26-11 Eldon H. Newcomb and T.D. Pugh
26-12, B David M. Phillips/Visuals Unlimited
26-13, A Richard Gross
26-13, B R.E. Davis, U.S. Department of Agriculture
26-14 Hubert A. Lechevalier
26-15, A U.S. Department of Agriculture
26-15 B and C R.R. Hebert, R.D. Holstein, and R.W.F. Hardy, E.I. du Pont de Nemours & Company

27-1, A Brian Parker/Tom Stack & Associates
27-1, B Tracy I. Borlund
27-1, C E.S. Ross
27-2 John D. Cunningham/Visuals Unlimited
27-3, A Marty Snyderman
27-3, B John D. Cunningham/Visuals Unlimited
27-3, C Jeff Rotman

27-4 E.W. Daniels, from K.W. Jeon, The Biology of the Amoebae, Academic Press, Inc., New York, 1973.
27-5 Terry Ashley/Tom Stack & Associates
27-7 Illustration by W. Ober
27-8, A E.S. Ross
27-8, B Robert Sisson © National Geographic Society
27-8, C and D James A. Dykens
27-9 Florida Department of Natural Resources
27-10 D. Davidson/Tom Stack & Associates
27-11 Centers for Disease Control, Atlanta, Ga.
27-12, A and B Jere H. Lipps
27-13 Kjell Sandved
27-A Illustration by Ron Ervin
27-15, A-C John Shaw/Tom Stack & Associates
27-17, A Ed Reschke
27-17, B E.S. Ross
27-18 Illustration by W. Ober
27-19 Illustration by W. Ober
27-20 Howard Hall/Tom Stack & Associates
27-21, A David M. Phillips/Visuals Unlimited
27-21, B Ed Reschke
27-21, C William Patterson/Tom Stack & Associates
27-22 Illustration by M.K. Ryan
27-23 Ed Reschke
27-24, A and B Bart J. Baca
27-25, A and B Manfred Kage/Peter Arnold, Inc.
27-26 Illustration by W. Ober
27-27 Manfred Kage/Peter Arnold, Inc.
27-29 Rick Harbo
27-30 W.H. Hodge/Peter Arnold, Inc.

28-1, A Carolina Biological Supply Company
28-1, B Kjell Sandved
28-1, C Robert Simpson/Tom Stack & Associates
28-2 Barbara Pfeffer/Peter Arnold, Inc.
28-3 Harvey C. Hoch
28-A, 1 R. Wenger, Sandoz, Ltd.
28-A, 2 R. Wenger, Sandoz, Ltd.
28-4 Kjell Sandved
28-6 Ed Pembleton
28-7 Herman Thompson/Visuals Unlimited
28-9 Ed Reschke
28-10 D.M. Phillips/Visuals Unlimited
28-11, A Walt Anderson/Tom Stack & Associates
28-11, B John D. Cunningham/Visuals Unlimited
28-11, C E.S. Ross
28-12 Illustration by George Venable

28-13 Terry Ashley/Tom Stack & Associates
28-14 H.C. Huang/Visuals Unlimited
28-15, **A** David M. Phillips/Visuals Unlimited
28-15, **B** E.S. Ross
28-16, **A** E.S. Ross
28-16, **B** Sylvia Duran Sharnoff/ Visuals Unlimited
28-16, **C** Stephen Sharnoff/Visuals Unlimited
28-B R.W. Kerrigan/Visuals Unlimited
28-17 Ed Reschke
28-C George Barron and Greg Thorn, *Science '84,* June, page 7.
28-18, **A and B** E. Imre Friedman
28-19 Lorene Segal and Thomas H. Nash III, *Ecology,* vol. 64, no. 6, pages 1343-1354, 1983.
28-20 R. Ronacordi/Visuals Unlimited
28-21, **A-C** Mark Brundrett
28-22, **A** N.C. Schenck/Visuals Unlimited
28-22, **B** D.H. Marx/Visuals Unlimited

VI-A Keith H. Murakami/Tom Stack & Associates
VI-B Larry West
VI-C John Shaw/Tom Stack & Associates
VI-D E.J. Cable/Tom Stack & Associates
29-1 Warren Garst/Tom Stack & Associates
29-2, **A** Ken Davis/Tom Stack & Associates
29-2, **B** E.S. Ross
29-2, **C** Rod Planck/Tom Stack & Associates
29-2, **D** E.S. Ross
29-4 Terry Ashley/Tom Stack & Associates
29-6, **A** Richard Gross
29-6, **B** Ed Reschke
29-8, **A** E.S. Ross
29-8, **B** Kirtley Perkins/Visuals Unlimited
29-9, **A** Ed Reschke
29-9, **B** Richard Gross
29-A T.S. Elias, *Science,* vol. 219, cover, January 7, 1983.
29-B-E D.R. Given
29-10 Illustration by W. Ober
29-11, **A** John Shaw/Tom Stack & Associates
29-11, **B** John D. Cunningham/ Visuals Unlimited
29-13 **A,** Rod Planck/Tom Stack & Associates
29-13 **B,** Ed Pembleton
29-13, **C** Kjell Sandved
29-13, **D,** Runk-Schoenberger/Grant Heilman Photography
29-14 Brian Parker/Tom Stack & Associates

29-16, **A and B** Joan Nowicke
29-19 Illustration by Ron Ervin
29-20 E.S. Ross
29-21 Illustration by W. Obert
29-22, **A** E.S. Ross
29-22, **B** E.S. Ross
29-24 Photo by Ed Reschke

30-1, **A** Rod Planck/Tom Stack & Associates
30-1, **B** Ed Pembleton
30-1, **C** William J. Weber/Visuals Unlimited
30-1, **D** C.P. Hickman, Jr.
30-1, **E** John D. Cunningham/Visuals Unlimited
30-1, **F** E.S. Ross
30-2 Sherwin Carlquist/Rancho Santa Ana Botanic Garden, Claremont, Calif.
30-3 David Dilcher, *Science,* vol. 224, May 4, 1984, page 511.
30-4 Ed Pembleton
30-5 Kjell Sandved
30-6 Sherwin Carlquist/Rancho Santa Ana Botanic Garden, Claremont, Calif.
30-7, **A** Kirtley-Perkins/Visuals Unlimited
30-7, **B** William J. Weber/Visuals Unlimited
30-8, **A and B** E. S. Ross; illustration by W. Ober
30-9 Illustration by W. Ober
30-10, **A and C** E.S. Ross
30-10, **B** Kjell Sandved
30-11 Kjell Sandved
30-12 John D. Cunningham/Visuals Unlimited
30-13, **A** Roy B. Clarkson
30-13 **B** John Skvarla
30-14 David W. Newman/Visuals Unlimited
30-15 Gary Milburn/Tom Stack & Associates
30-16 E.S. Ross
30-17, **A and B** E.S. Ross
30-18, **A** Steve Maslowski/Visuals Unlimited
30-18, **B** E.S. Ross
30-19 Thomas Eisner
30-20 E.S. Ross
30-21, **A-C** Peter Hoch
30-22, **A** Sal Giordano
30-22, **B** Ed Pembleton
30-22, **C** E.S. Ross
30-23, **A** Ed Pembleton
30-23, **B and C** John Cunningham/ Visuals Unlimited
30-24, **A** Ed Pembleton
30-24, **B** Heather Angel

31-2, **A and B** Photos by Dan Sindelar
31-3 John D. Cunningham/Visuals Unlimited
31-A and B George B. Schaller

31-4, **A and B** E.S. Ross
31-5 Peter Hoch
31-6 Terry Ashley/Tom Stack & Associates
31-7 Illustration by W. Ober
31-8 Randy Moore/Visuals Unlimited
31-9 T. Lawrence Mellichamp/Visuals Unlimited
31-11, **A and B** Randy Moore/Visuals Unlimited
31-12 E.J. Cable/Tom Stack & Associates
31-13, **A** Terry Ashley/Tom Stack & Associates
31-13, **B** Randy Moore/Visuals Unlimited
31-13, **C** James Ehleringer
31-14, **A and B** E.S. Ross
31-15, **A and B** Kjell Sandved
31-17 E.J. Cable/Tom Stack & Associates
31-18, **A** E.S. Ross
31-18, **B and C** W. and M. Ober
31-19 R.A. Gregory/Visuals Unlimited
31-20, **A** Ed Reschke
31-20, **B** E.J. Cable/Tom Stack & Associates
31-21 E.J. Cable/Tom Stack & Associates
31-22 R.F. Evert
31-23 R.F. Evert
31-24 John Shaw/Tom Stack & Associates
31-25, **A** E.S. Ross
31-25, **B** John D. Cunningham/ Visuals Unlimited
31-26 W. and M. Ober
31-27 R.A. Gregory/Visuals Unlimited
31-28 E.S. Ross
31-29 John D. Cunningham/Visuals Unlimited
31-30 Terry Ashley/Tom Stack & Associates
31-31 Randy Moore/Visuals Unlimited
31-32 E.J. Cable/Tom Stack & Associates
31-33, **A** John D. Cunningham/ Visuals Unlimited
31-33, **B** Sal Giordano
31-34 Tom Ulrich/Visuals Unlimited

31-1 John D. Cunningham/Visuals Unlimited
31-2 Courtesy Hunt Institute for Botanical Documentation, Carnegie-Mellon University, Pittsburgh.
32-3 Illustration by M.K. Ryan
32-4, **A-C** Maureen Hanson
32-5, **A and B** B.V. Conger et al. *Science,* vol. 221, August 1983, pages 850-851.
32-A, **1-4** Richard Branton
32-6 Maureen Hanson

32-7 W. and M. Ober
32-10 Barry L. Runk/Grant Heilman Photography
32-11 D.A. Priestley

33-1 Illustration by M.K. Ryan
33-2 Illustration by M.K. Ryan
33-3 Illustration by M.K. Ryan
33-5 *St. Louis Globe-Democrat*
33-8 Silvan H. Wittwer
33-9, A and B Silvan H. Wittwer
33-11 John D. Cunningham/Visuals Unlimited
33-12 John D. Cunningham/Visuals Unlimited
33-13, A and B W. and M. Ober
33-14, A and B Richard Gross
33-15 Alan P. Godlewski
33-17 Grant Heilman/Grant Heilman Photography
33-18 L.R. Heckard, Jepson Herbarium, University of California, Berkeley.

34-2 E.S. Ross
34-3 Illustration by M.K. Ryan
34-5 Laura Riley/Bruce Coleman, Inc.
34-6 Illustration by W. Ober
34-A, 1 Grant Heilman/Grant Heilman Photography
34-A, 2 Gerrit Davidse
34-A, 3 Thomas Croat
34-7 Grant Heilman/Grant Heilman Photography
34-9, A-F Emanuel Epstein
34-10 E.S. Ross
34-11 E.S. Ross
34-14 F. Stuart Chapin
34-15, A Gerald A. Peters, *Plant Physiology,* vol. 53, pages 813-819, 1974.
34-15, B BPS/Tom Stack & Associates
34-15, C Kjell Sandved
34-16, A Sal Giordano III
34-16, B R. Mitchell/Tom Stack & Associates
34-16, C John D. Cunningham/Visuals Unlimited
34-16, D Kjell Sandved
34-16, E H.A. Miller/Visuals Unlimited
34-16, F Thomas Croat
34-B, 1-3 Daniel H. Janzen

VII-A Larry West
VII-B Jeff Rotman
VII-C Rob McKenzie
35-1, A Marty Snyderman
35-1, B Rick Harbo
35-1, C Jim and Cathy Church
35-2, A Karl G. Grell
35-3 Illustration by W. Ober
35-4, A Jim and Cathy Church
35-4, B W. Ober
35-5 Illustration by W. Ober

35-6, A Bob Gossington/Tom Stack & Associates
35-6, B Manfred Kage/Peter Arnold, Inc.
35-7, A Marty Snyderman
35-7, B Gwen Fidler/Tom Stack & Associates
35-7, C Neil G. McDaniel/Tom Stack and Associates
35-8 Illustration by W. Ober
35-9 Tom Stack/Tom Stack and Associates
35-10 Illustration by W. Ober
35-11 Kjell Sandved
35-12 Illustration by W. Ober
35-13 Claudia Mills, *Nature,* February 1985, page 737.
35-14, A Marty Snyderman
35-14, B W. Ober
35-14, C R. Myers/Visuals Unlimited
35-15 W. Ober
35-16 Jeff Rotman
35-19, A Jim and Cathy Church
35-19, B E.J. Cable/Tom Stack & Associates
35-20 Illustration by W. Ober
35-21, A John Cunningham/Visuals Unlimited
35-21, B Centers for Disease Control, Atlanta, Ga.
35-22 Illustration by W. Ober
35-23 Kjell Sandved
35-24, A J.M. Ferris
35-24, B and C V.R. and J.M. Ferris
35-25 V.R. Ferris
35-26, A From V.R. Ferris and J.M. Ferris, "*Lindseyus costatus* n. sp. and notes on Rogueidae and Swangeriidae," *Proceedings of the Helminthological Society of Washington,* vol. 40, page 43, 1973.
35-26, B Illustration by V.R. Ferris
35-26, C From N.A. Cobb, Nematodes and their relationships: Yearbook of the Department of Agriculture, pages 457-490, 1914.
35-27 Illustration by M.K. Ryan
35-28 Ed Reschke
35-29 E.S. Ross
35-30 T.E. Adams/Visuals Unlimited
35-A Illustration by M.K. Ryan, after Robert Higgins/National Museum of Natural History, Smithsonian Institution

36-1, A Milton Rand/Tom Stack & Associates
36-1, B Kjell Sandved
36-3, A Richard Radtke
36-3, B Robert K. Trench
36-4 Illustration by W. Ober
36-5, A and B Kjell Sandved
36-6 Illustration by W. Ober
36-7 Kjell Sandved
36-8 Kjell Sandved
36-9, A, B, and D Kjell Sandved
36-9, C Milton Rand/Tom Stack & Associates

36-10 A and B Kjell Sandved
36-11, A A.E. Spreitzer
36-11, B Courtesy Dr. Steven Ahlstedt
36-12, A Marty Snyderman
36-12, B Alan Broder
36-12, C Alex Kerstitch
36-A Jack Donnelly, Woods Hole Oceanographic Institute
36-13 Alex Kerstitch
36-14 Illustration by George Venable
36-15 Illustration by M.K. Ryan
36-16, A Kjell Sandved
36-16, B Jeff Rotman
36-16, C Kjell Sandved
36-17 David M. Dennis/Tom Stack & Associates
36-18 Gwen Fidler/Tom Stack & Associates

37-1 Illustration by M.K. Ryan
37-2 Illustration by M.K. Ryan
37-A E.S. Ross
37-3 Illustration by M.K. Ryan
37-4 Illustration by M.K. Ryan
37-5, A Kjell Sandved
37-5, B C.P. Hickman, Jr.
37-6, A Kjell Sandved
37-6, B Illustration by W. Ober
37-7 Illustration by George Venable
37-8, A and B Illustrations by W. Ober
37-8, C Thomas Eisner
37-9 E.S. Ross
37-10 E.S. Ross
37-11, A John D. Cunningham/Visuals Unlimited
37-11, B and C Kjell Sandved
37-12 Kjell Sandved
37-B, 1 Kjell Sandved
37-B, 2 E.S. Ross
37-B, 3-5 Jonathan Coddington
37-13, A Rod Planck/Tom Stack & Associates
37-13, B Ann Moreton/Tom Stack & Associates
37-14, A E.S. Ross
37-14, B John Shaw/Tom Stack & Associates
37-14, C E.S. Ross
37-15 E.S. Ross
37-17 Rudolf Arndt/Visuals Unlimited
37-18 Heather Angel
37-19, A C.P. Hickman, Jr.
37-19, B Jeff Rotman
37-19, C C.P. Hickman, Jr.
37-20 Illustration by George Venable
37-21, A E.S. Ross
37-21, B Kjell Sandved
37-21, C T.E. Adams/Visuals Unlimited
37-22, A and B T.E. Adams/Visuals Unlimited
37-23 Kjell Sandved
37-24, A Alex Kerstitch
37-24, B E.S. Ross
37-25, A, C, D, and F E.S. Ross

37-25, **B** Kjell Sandved
37-25, **E** J.A. Adcock/Visuals Unlimited
37-26, **A** C.P. Hickman, Jr.
37-26, **B** Kjell Sandved
37-26, **C** Don and Pat Valenti/Tom Stack & Associates
37-26, **D** Kjell Sandved
37-26, **E and F** E.S. Ross
37-27 Illustration by George Venable
37-28 Illustration by George Venable
37-29 John Shaw/Tom Stack & Associates
37-C Grant Heilman/Grant Heilman Photography
37-30 Kjell Sandved
37-31 Illustration by W. Ober
37-32 **A** John Shaw/Tom Stack & Associates
37-32, **B-D** Gwen Fidler/Tom Stack & Associates
37-32, **E** John Shaw/Tom Stack & Assocites
37-32, **F and G** K.A. Blanchard/ Visuals Unlimited
37-32, **H and I** Sal Giordano III
37-33 Illustration by W. Ober

38-1, **A** Alex Kerstitch
38-1, **B and C** Carl Roessler/Tom Stack & Associates
38-1, **D** W. Ober
38-1, **E** Daniel W. Gotshall
38-1, **F** Jeff Rotman
38-2 Illustration by George Venable
38-3 Kjell Sandved
38-4 Photo by Harbor Branch Foundation, Inc., courtesy of David L. Pawson
38-5 Illustration by W. Ober
38-6 David L. Meyer
38-7 Photo by Harbor Branch Foundation, Inc., courtesy of David L. Pawson
38-8 Peter Moran/Australian Institute of Marine Science
38-9 Gary Milburn/Tom Stack & Associates
38-10 John D. Cunningham/Visuals Unlimited
38-11, **A and B** D.P. Wilson/Eric and David Hosking
38-12 D.P. Wilson/Eric and David Hosking
38-13 Illustration by W. Ober
38-14 Illustration by Ron Ervin
38-15, **A** Rick Harbo
38-15, **B** Jim and Cathy Church
38-15, **C** Illustration by W. Ober
38-16 Illustration by W. Ober
38-17, **A-D** Richard R. Olson
38-18, **A** Heather Angel
38-18, **B** Illustration by W. Ober

VIII-A Lennart Nilsson
VIII-B Jeff Rotman

VIII-C Gary Milburn/Tom Stack & Associates
VIII-D Brian Parker/Tom Stack & Associates
39-1 Illustration by M.K. Ryan
39-2 Illustration courtesy of C.P. Hickman, Jr., *Integrated Principles of Zoology*, 7th ed., Times Mirror/ Mosby College Publishing, St. Louis, 1984.
39-3 Steve Martin/Tom Stack & Associates
39-4 Illustration by W. Ober
39-5, **A and B** Marty Snyderman
39-5, **C** Carl Roessler/Tom Stack & Associates
39-6 Courtesy of the Natural History Museum of Los Angeles County
39-A Peter Scoones/Contact Stock Images
39-8, **A** Alex Kerstitch
39-8, **B** Marty Snyderman
39-8, **C** Rob McKenzie
39-8, **D** John D. Cunningham/Visuals Unlimited
39-8, **E** Brian Parker/Tom Stack & Associates
39-9 Fukuda Toyfumi © DISCOVER Magazine, Time Inc.
39-10, **A and B** John Shaw/Tom Stack & Associates
39-10, **C** C.P. Hickman, Jr.
39-10, **D** John D. Cunningham/ Visuals Unlimited
39-10, **E** David M. Dennis/Tom Stack & Associates
39-11, **A** Brian Parker/Tom Stack & Associates
39-11, **B** William J. Weber/Visuals Unlimited
39-11, **C** John Cancalosi/Tom Stack & Associates
39-11, **D** Rod Planck/Tom Stack & Associates
39-12 Tamra Raven
39-13 Illustration by W. Ober
39-14, **A** E.S. Ross
39-14, **B** John D. Cunningham/ Visuals Unlimited
39-14, **C** William J. Weber/Visuals Unlimited
39-14, **D and E** E.S. Ross
39-14, **F** John D. Cunningham/ Visuals Unlimited
39-14, **G** Alan Nelson/Tom Stack & Associates
39-B Illustration by George Venable
39-15 Illustration by M.K. Ryan
39-16 John D. Cunningham/Visuals Unlimited
39-17, **A** J.E. Wapstra, National Photo Index of Australian Wildlife
39-17, **B** J. Adcock/Visuals Unlimited
39-18 Tom Stack/Tom Stack & Associates
39-19 Illustration by W. Ober
39-20, **A** Brian Parker/Tom Stack & Associates

39-20, **B** Rod Planck/Tom Stack & Associates
39-20, **C** E.S. Ross
39-20, **D** C.P. Hickman, Jr.
39-20, **E** Marty Snyderman
39-21 Tim W. Clark
39-C1-4 Merlin D. Tuttle
39-22, **A and B** Mrs. Rudolph A. Peterson
39-22, **C** Flip Nicklin/Nicklin & Associates

40-1, **A** Ed Reschke
40-1, **B and C** E.S. Ross
40-A and B Steven G. Hoffman
40-2 David and Anne Doubilet © DISCOVER magazine, Time Inc.
40-3 Hans Pfletschinger/Peter Arnold, Inc.
40-4 C.P. Hickman, Jr.
40-9 Illustration by Ron Ervin
40-10 Lennart Nilsson
40-11 Illustration by W. Ober
40-12 Illustration by Ron Ervin
40-13 Lennart Nilsson
40-14 Illustration by Ron Ervin
40-15 Ed Reschke
40-17 Illustration by Ron Ervin
40-19 © Joel Gordon Photography 1982
40-20 Illustration by Ron Ervin

41-1, **A and B** David M. Phillips/ Visuals Unlimited
41-2, **A** Everett Anderson
41-4 Illustration after Hickman and Roberts, in Integrated Principles of Zoology, 7th ed., Times Mirror/ Mosby College Publishing, St. Louis, 1984.
41-5 David M. Phillips/Visuals Unlimited
41-6, **A-D** Carolina Biological Supply Company
41-7 David M. Phillips/Visuals Unlimited
41-9 Illustration by Ron Ervin
41-11 Illustration by Ron Ervin
41-13 Illustration by W. Ober
41-15 Heather Angel
41-16 Illustration by W. Ober
41-17, **B** Courtesy of Norman K. Wessels and Kathryn W. Tosney, from *Tissue Interactions and Development*, Benjamin Cummings, Menlo Park, Calif., 1977.
41-A Richard Hutchings/Photo Researchers, Inc.
41-18, **A-D** Lennart Nilsson
41-19 Illustration by Ron Ervin
41-20 Illustration by Ron Ervin

42-1 Illustration by M.K. Ryan
42-2, **A-C** Ed Reschke
42-3 Illustration by W. Ober
42-4 Emma Shelton
42-A Illustration by Ron Ervin
42-5 Jerome Gross

42-6 Illustration by Ron Ervin
42-7 Illustration by Ron Ervin
42-8 David M. Phillips/Visuals Unlimited
42-9, A-C Ed Reschke
42-10 Illustration by Ron Ervin
42-11 Illustration by Ron Ervin
42-12 Lennart Nilsson
42-13 Illustration by M.K. Ryan
42-14 John D. Cunningham/Visuals Unlimited

43-2 Illustration by M.K. Ryan
43-3 Illustration by Ron Ervin
43-4 Illustration by M.K. Ryan
43-5 Illustration by Ron Ervin
43-6 Illustration by Ron Ervin
43-7 Illustration by Ron Ervin
43-8 Illustration by Ron Ervin
43-9 Illustration by W. Ober
43-A Mary Field
43-B W.J. Weber/Visuals Unlimited
43-C Steve Maslowski/Visuals Unlimited
43-D John Cunningham/Visuals Unlimited
43-E Robert A. Tyrrell
43-10 Illustration by Ron Ervin and W. Ober
43-11, A Illustration by Ron Ervin
43-11, B D.M. Phillips © 1984/ Visuals Unlimited
43-12, A Courtesy of Richard Kessell, from Richard G. Kessel and Randy H. Kardon, *Tissues and Organs: a Text–Atlas of Scanning Electron Microscopy,* W.H. Freeman & Co., Publishers, New York, 1979.
43-12, B Barry King, University of California, Davis; BPS/Tom Stack & Associates
43-13 BPS/Tom Stack & Associates
43-14 Illustration by M.K. Ryan

44-4 Russ Kinne/National Audubon Collection/Photo Researchers
44-5 Illustration by M.K. Ryan
44-6 Illustration by M.K. Ryan
44-8 Courtesy C.P. Hickman, Jr., from *Integrated Principles of Zoology,* 7th ed., Times Mirror/Mosby College Publishing, St. Louis, 1984.
44-9 Illustration by George Venable
44-10 H.R. Dunker
44-11 Illustration by Ron Ervin
44-12 Illustration by Ron Ervin
44-13 Ellen Dirksen/Visuals Unlimited
44-14 Illustration by Ron Ervin

45-2 W. Rosenberg/BPS/Tom Stack & Associates
45-3 Illustration by Ron Ervin
45-5 Ed Reschke
45-6 Illustration by Ron Ervin
45-7 Ed Reschke
45-8 Illustration by Ron Ervin

45-9 Ed Reschke
45-A David M. Phillips/Visuals Unlimited
45-B Gernsheim Collection, Harry Ranson Humanities Research Center, University of Texas at Austin
45-10 Illustration by Ron Ervin
45-11 Makio Murayama/BPS/Tom Stack and Associates
45-12 Illustration by W. Ober
45-15 Illustration by W. Ober
45-18 Illustration by Ron Ervin
45-19 Illustration by Ron Ervin
45-20 Illustration by Ron Ervin
45-D Illustration by Ron Ervin
45-E Frank Sloop
45-21 W. Ober

46-1 Lennart Nilsson
46-2 Illustration by M.K. Ryan
46-3 Illustration by W. Ober; micrograph by C.S. Raines
46-4 E.R. Lewis/BPS/Tom Stack & Associates
46-6 Illustration by M.K. Ryan
46-8 Courtesy of Lennart Heimer, from *The Human Brain and Spinal Cord,* Springer-Verlag, Inc., New York, 1983.
46-9 Illustration by W. Ober
46-10 Ed Reschke
46-11 Illustration courtesy of C.P. Hickman, Jr. from *Integrated Principles of Zoology,* 7th ed., Times Mirror/Mosby College Publishing, St. Louis, 1984.
46-14 Illustration by M.K. Ryan
46-15 Illustration by Ron Ervin
46-16 Illustration by W. Ober
46-17 Illustration by W. Ober
46-19 Illustration by Ron Ervin
46-20 Illustration by Ron Ervin

47-1, A Martha Swope
47-1, B Michael Fogden
47-2 Illustration by W. Ober
47-4 Illustration by W. Ober; micrograph by A.J. Hudspeth and R. Jacobs
47-5 Illustration by M.K. Ryan
47-6, A Illustration by Ron Ervin
47-6, B Ed Reschke
47-7 Illustration by Ron Ervin
47-8 Illustration by W. Ober
47-9 E.S. Ross
47-11, B E.R. Lewis, F.S. Werblin, and Y.Y. Zeevi, University of California
47-14, A Kjell Sandved
47-14, B William J. Weber/Visuals Unlimited
47-14, C-F Kjell Sandved
47-15 Illustration by Ron Ervin
47-17, A Martin M. Rotker/ TAURUS Photos, Inc.

47-17, B Illustration by W. Ober
47-18, A Leonard L. Rue III
47-18, B Illustration by M.K. Ryan
47-19 Steve Martin/Tom Stack & Associates

48-1 Lennart Nilsson
48-3 Illustration by Ron Ervin
48-4 Illustration by Ron Ervin
48-5 Illustration by Ron Ervin
48-7 Illustration by Ron Ervin
48-8 Illustration by Ron Ervin
48-10 Illustration by W. Ober
48-12 Ed Reschke
48-14 Courtesy of Richard Kessel, from Richard G. Kessel and Randy H. Kardon, *Tissues and Organs: a Text–Atlas of Scanning Electron Microscopy,* W.H. Freeman & Co., Publishers, New York, 1979.
48-16 Rob McKenzie
48-17 John Gerlach/Tom Stack & Associates
48-20 Illustration by W. Ober
48-21 Illustration by Ron Ervin

49-1 E.S. Ross
49-2, A Larry Ditto/Tom Stack & Associates
49-2, B Warren Garst/Tom Stack & Associates
49-3 Stephen Dalton/National Audubon Society Collection/Photo Researchers
49-4 Animals Animals/Oxford Scientific Films
49-5 James R. Fisher/National Audubon Society Collection/Photo Researchers
49-6 Osler Library, McGill University
49-7 Bridgewater College
49-8 Linda Koebner/Bruce Coleman, Inc.
49-9 E.S. Ross
49-A C.P. Hickman, Jr.
49-10 Fawcett/Tom Stack & Associates
49-11 C.P. Hickman, Jr.
49-12 Mark Newman/Tom Stack & Associates
49-13 E.S. Ross
49-B Christopher Springmann
49-14 John D. Cunningham/Visuals Unlimited
49-15 Jeff Foott/Bruce Coleman, Inc.
49-17 Arthur Holzman/Animals Animals
49-C, 1-4 L.P. Brower
49-19 Fran Allen/Animals Animals
49-20 George Johnson

IX-A Brian Parker/Tom Stack & Associates
IX-B Larry West
IX-C W.J. Weber/Visuals Unlimited

IX-D John Gerlach/Tom Stack & Associates
50-1 Charles A. Munn
50-2 Marty Snyderman
50-5 E.S. Ross
50-A, 1 and 2 C.P. Hickman, Jr.
50-7 Marty Snyderman
50-8 Rick Harbo
50-9 John D. Cunningham/Visuals Unlimited
50-10, B-D Chesapeake Bay Foundation
50-11 W. and M. Ober
50-13 John D. Cunningham/Visuals Unlimited
50-14 Daniel W. Gotshall
50-16, A E.S. Ross
50-16, B Richard Gross
50-16, C E.S. Ross
50-16, D Kjell Sandved
50-17 C.P. Hickman, Jr.
50-18, A and B E.S. Ross
50-19 J. Cancalosi/Tom Stack & Associates
50-B, 1, 2, 4, 5, and 6 William L. Hamilton
50-B, 3 E.S. Ross
50-20 John Gerard
50-21 Carol Sutherland
50-22 John D. Cunningham/Visuals Unlimited
50-23 John D. Cunningham/Visuals Unlimited
50-24 Photo by E.S. Ross
50-25 John Gerard
50-26 E.S. Ross
50-27, A C.P. Hickman, Jr.
50-27, B Leonard L. Rue III
50-28, A William J. Weber/Visuals Unlimited
50-28, B Rick McIntyre/Tom Stack & Associates
50-28, C C.P. Hickman, Jr.

51-1 E.S. Ross
51-2 E.S. Ross
51-3 E.S. Ross
51-6 Gene Likens
51-7 John D. Cunningham/Visuals Unlimited
51-8 John D. Cunningham/Visuals Unlimited
51-9, A-C E.S. Ross
51-13 E.S. Ross
51-14, A and B Peter Frenzen
51-15 E.S. Ross
51-16 Rod Planck/Tom Stack & Associates
51-17 Mickey Gibson/Tom Stack & Associates
51-18 L.L.T. Rhodes/Earth Scenes
51-A John D. Cunningham/Visuals Unlimited

51-B Kathleen Blanchard/Visuals Unlimited

52-1, A and B World Wildlife Fund/Helmut Diller
52-1, C World Wildlife Fund/Paul Barruel
52-2 W.C. Gagné, Department of Entomology, Bishop Museum; courtesy Warren Wagner
52-5 Mike Holmes/Animals Animals
52-6 Richard Howard
52-7 John D. Cunningham/Visuals Unlimited
52-8 Roger Payne
52-12 Heather Angel
52-13 Anne Wertheim/Animals Animals
52-14, A and B Steven Maslowski/Visuals Unlimited
52-16 Biological Branch, Department of Lands, Queensland, Australia
52-17 C.P. Hickman, Jr.
52-18, A and B Rolf O. Peterson

53-1 Dale Jorgenson/Tom Stack & Associates
53-2, A-C E.S. Ross
53-A Roger Del Moral
53-3 Charles E. Comfort
53-4, A and B U.S. Department of the Interior, Fish & Wildlife Service
53-5 E.S. Ross
53-6 Gerald Corsi/Tom Stack & Associates
53-7 R.W. Gibson
53-9 David F. Rhoades
53-10, A and B E.S. Ross
53-11 E.S. Ross
53-12 Valerie J. Paul
53-13 Lincoln P. Brower
53-14, A and B E.S. Ross
53-15, A E.S. Ross
53-15, B and C John D. Cunningham/Visuals Unlimited
53-15, D E.S. Ross
53-16 Raymond Mendez/Animals Animals
53-17 T. Eisner and D. Aneshansley
53-18 John Gerlach/Tom Stack & Associates
53-19, A William J. Weber/Visuals Unlimited
53-19, B Alex Kerstitch
53-19, C E.S. Ross
53-20, A-D Paul A. Opler
53-B, 1 P. Armstrong/Visuals Unlimited
53-B, 2 Anders Nilsson. Reprinted by permission from Nature, vol. 305, no. 5937, pages 799-800 (cover) © 1983, Macmillan Journals, Inc.
53-21, A-D E.S. Ross

53-22 David Grobecker, Science, vol. 201, no. 4353, page 370 and cover, July 28, 1978.
53-23 Stephen Dalton, from Borne on the Wind: The Extraordinary World of Insects in Flight, Reader's Digest Press. Distributed by E.P. Dutton & Co., Inc., New York, 1975.
53-24 J. Harvey/Visuals Unlimited
53-25 E.S. Ross
53-26 E.S. Ross
53-27 Daniel H. Janzen
53-28, A and C Daniel H. Janzen
53-28, B and C W. Rettenmeyer

54-1 E.S. Ross
54-2, A Ghillean Prance
54-2, B D.O. Hall, King's College, University of London
54-A, 1 George H. Huey/Earth Scenes
54-A, 2 L. Mellichamp/Visuals Unlimited
54-A, 3 George H. Huey/Earth Scenes
54-A, 4 Ronald Orenstein/Earth Scenes
54-A, 5 L. Mellichamp/Visuals Unlimited
54-4 Library of Congress/Picture Research
54-5 Earthscan/Glenn Edwards
54-6 J.B. Forbes, St. Louis Post-Dispatch
54-7, A and B E.S. Ross
54-7, C D.O. Hall, King's College, University of London
54-8 Alex Kerstitch
54-B, 1 and 2 Earthscan/Mark Edwards
54-9 Dan McCoy/Rainbow
54-10, A-C D.O. Hall, King's College, University of London
54-11 Nigel Smith/Earth Scenes
54-12 D.O. Hall, King's College, University of London
54-13, A Hugh Iltis
54-13, B Hugh Iltis
54-13, C Hugh Iltis
54-13, D Hugh Iltis
54-14 E.S. Ross
54-15 Harold Moore
54-16, A Monte Lloyd
54-16, B Carl F. Jordan
54-17, A Jim Blair, National Geographic Society
54-17, B C.W. Rettenmeyer
54-17, C Inga Hedberg
54-18 U.S. Navy/Picture Research
54-19, A and B Rob Wood, Stansbury Ronsaville, Wood, Inc., 1983. © Open Space Institute, Inc. The Center on the Consequences of Nuclear War.
54-20 Donald R. Perry

INDEX

Agriculture—cont'd
  nuclear winter and, 1193-1194
  pollution and, 1126-1127
  population density and, 1135
  systems of, 1189-1192
  in tropics, 12-13
  water use and, 1183
Agriope aurantia, 785
AIDS; *see* Acquired immune deficiency
      syndrome
Ailurus fulgens, 838
Air
  composition of, 941-942
    diffusion rate and, 943
  current, 1084
  heat and, 1027
  height of, 941-942
  pollution of
    bryophyte and, 590
    lichen and, 575
    succession and, 1125-1128
  pressure and altitude and, 941
Air sac, 950
  anterior, 950
  posterior, 950
Aixsponsa, 837
Alanine, 40
Albinism, 245
Albumin
  egg and, 836
  serum, 967
Alcohol
  ethanol, 196-197, 567
  ethyl, 1185
  fluid absorption and, 933
  phosphorylated, 64
Alder, 1158
Aleurone layer, 670
Alfalfa butterfly, 795
Alfalfa mosaic, 500
Alga, 95, 556-557
  brown, 549-550
  defensive reaction and, 1159, 1161
  genome size and, 340
  green
    defensive reaction of, 1159
    new food and, 1189
    reproduction and, 852
    symbiosis and, 1167
  photosynthesis and, 224
  Protista and, 439
  red, 555-556
    flagella and, 129
  sponge symbiosis and, 726
Alkaloid, 1157
Alkaptonuria, 292
Allantois, 836
Allele
  definition of, 244
  migration and, 391-392
  multiple, 252
  probability and, 248-250
Allelopathy, 1152
Alligator, 834
Allium sativum, 644
Allogona townsendiana, 752
Allometry in human evolution, 480-481
Allosteric enzyme, 178
Allozyme
  definition of, 401
  gel electrophoresis and, 376-377
  genetic variation and, 377-378
  maintenance of, 396
Aloe, false, 598
alpha-Actinin, 131, 132, 133-134, 135
alpha-Ketoglutaric acid, 202
Alzheimer's disease, 502
Alteration, genetic, 253
Alternation of generation, 539, 587-588
  brown alga and, 550
  foram and, 543
Alternative trait, 240

Altruistic behavior, 1068
ALU element, 341
Aluminum
  chemical characteristics of, 24
  depletion of, 1184
  human body and, 20
Alvarex, L., 460
Alveolus, 951
  lung with, 945
  reptile respiration and, 949
Amanita muscaria, 561
Amaranth, 1177
Amaranthus, 1177
American chestnut, 1145
Amino acid(s)
  abiotic synthesis and, 179
  balanced set of, 1187
  cell wall and, 86, 87, 88
  collagen and, 906-907
  conversion into glucose, 971-972
  deamination, 972
  digestion and, 920-921
  essential, 937
  hominoid evolution and, 476-478
  insulin and, 294
  nutrition and, 935, 936
  plant defense and, 1156
  primitive earth and, 40
  protein and, 68-70
  sequence, 71-72
    evolution and, 376
  transcription and, 302
  water and, 70
gamma-Aminobutyric acid, 997
Ammodramus maritimus, 406
Ammonia
  carbon dioxide and, 972
  conversion to nitrite, 184
  fertilizer and, 712-713
  in living organism, 33-34
  nitrogen cycle and, 1115
  nitrogen fixation and, 181
  primitive atmosphere and, 38
Ammonite, 763
Ammonium ion, 711
Amnion
  human embryo and, 894
  reptile and bird and, 857
  reptilian, 836
Amniotic egg, 835-836
Amoeba, 102, 542
  flagella and, 129
AMP; *see* Adenosine monophosphate
Amphibian (Amphibia), 820, 826, 833-
      834
  brain of, 1002, 1003
  cell cleavage pattern of, 881
  digestion and, 924
  dorsal lip transplant experiment in, 889
  earliest, 832
  gastrulation of, 884
  genome size and, 340
  globin gene and, 342
  heart, 972, 975
  kidney and, 1049
  larva, 945
  lung, 948
  origin of, 454, 457
  parthenogenesis, 851
  sex and reproduction and, 855-856
Amphioxus, 818, 881, 887
Amphipathic, definition of, 65
Amphipod (Amphipoda), 776, 790
Ampulla, 807-808
Ampullae of Lorenzini, 1028
Amylase, 920, 921, 925
Amylopectin, 60
Amyloplast, 115, 670
  gravitropism and, 685
Anabaena azollae, 711
Anabolic process, 191
Anaerobic bacteria, 179

Anaerobic environment, 178
Anaerobic photosynthesis, 47
Anal sphincter, 934
Anaphase
  I, 263-264
  mitosis and, 146-147
Anax junius, 777
Androecium, 603, 617
Andromeda, 51
Anemia, sickle cell, 396-397, 401
  adaptive change and, 405
  molecular basis of, 294, 295
Anemone, sea, 733
  commensalism and, 1168
Aneuploid chromosome, 334
Angelfish, 831
Angina pectoris, 980
Angiosperm, 599-600, 609-629; *see also*
      Flower
  cell differentiation and, 666
  embryo and, 668-670
  evolution of, 610-618
  floral specialization and, 618-620
  fruit and, 625-627
  genome size and, 340
  gibberellic acid and, 682
  gymnosperm and, 610-611
  life cycle of, 603-606
  pollination and, 620-624
  secondary compound and, 1157
Angular acceleration, 1014-1015
Animal (Animalia)
  angiosperm pollination and, 615
  cell of, 95, 130, 131
  chemical defense of, 1160-1162
  classification of, 723
  cytoplasmic cleavage in, 148
  development of, 658
  domestication of, 486-487
  ecological race in, 417
  ectothermic, 836-837
  endothermic, 837
  homeothermic, 837
  invertebrate; *see* Invertebrate
  mitosis in, 144
  poikilothermic, 837
  pollination and, 621
  temperature control in land, 836-837
  vertebrate; *see* Vertebrate
  virus and, 500
Anion channel, 161
Annelid (Annelida), 751, 764-768; *see also*
      Worm
  circulatory system and, 959-960
  classes of, 766
  digestion and, 923
  hemoglobin and, 953
Annual plant, 635
Annual ring, 648
Anopheles, 4
Anostraca, 776, 790, 791
Ant, 795, 851
  aphid and, 1168-1169
  behavior and, 1063
  bulldog, 774
  desert, 778
Ant plant, 714
Antagonistic control, 1033-1034
  muscle and, 1038
  pituitary hormone and, 1041-1042
Anteater, 841
Antelope, 1156
Antenna, 774
Anther, 664
Antheridium
  ascomycetes and, 566
  bryophyte, 592
  definition of, 588-589
  fern and, 600
  moss, 591
  oomycete and, 555
Anthocyanin, 254

Hormone
  circulation, 961
  flowering, 690
  juvenile, 801, 802
  molting, 801, 802
  nervous system and, 988
  pituitary, 1040-1042
  plant and, 658, 677-684, 692
  sexual reproduction and, 860-861
Hornwort, 590-591
Horse
  digestion and, 925, 934
  evolution of, 374
Horsefly, 795
Horsehair worm, 747
Horseradish, 1157
Horseshoe bat, 845
Horseshoe crab, 776, 788
Horsetail, 596
Housefly, 795, 798, 799
Hubbard Brook, 1126-1127
Human, 843
  birth rate and, 1179
  blood group and, 386-389
  brain and, 1004-1005
    size and, 475
  central nervous system, 1001
  chimpanzee, gorilla, and, 475-479
  classification of, 433
  digestion and, 924, 925-926
  element in, 20
  embryonic development and, 893-896
  evolution of, 466-489
    allometry in, 480-481
    animal and plant domestication and,
      486-487
    hominids and, 481-486
    hominoids and, 475-481
    mammals and, 467-468
    primates and, 468-475
  globin gene and, 342
  heart, 977-983
  intercourse and, 868-870
  liver fluke and, 737, 740-741
  male gamete, 863-864
  menstrual cycle, 862
  nutrition and, 936-937
  origin of, 485
  pair bonding and, 870
  population and, 1178-1181
  race and, 420
  respiration and, 951-952
  skeleton and, 909
  T cell leukemia virus, 7-8
Humboldt Current, 1086-1087
Hummingbird
  beak of, 928, 929
  pollination and, 623
  search image and, 1166
Humpbacked whale, 846
Humus, 713
Hunger, 1003
Hyacinth, 1186
Hybrid, 338; see also Hybridization
Hybridization, 338
  barrier to, 409-415
  DNA, 478
  evolution and, 421-423
  first, 239-240
  Mendel and, 240-248
  postzygotic mechanisms and, 413
  prezygotic mechanisms and, 409-412
  significance of, 423
Hydnophytum, 714
Hydra, 730, 731, 851
  nervous system and, 999
Hydracarina, 787
Hydrarch primary succession, 1122
Hydration shell, 29
Hydrocarbon, 179
  chlorinated, pollution and, 1127-1128
  definition of, 34

Hydrochloric acid, 928, 929, 930
Hydrogen
  in ammonia, 33
  atom of, 16
  bond, 174
    definition of, 27
    water and, 30
  chemical characteristics of, 24
  compounds of, 34
  diatomic molecule of, 22
  fermentation and, 196-197
  glycolysis and, 197
  human body and, 20
  in living organism, 24-25
  methane and, 23
  molecular origin and, 28-29
  original atmosphere and, 38
  plant and, 706-707, 708
  removal of, by bacteria, 183
  water ionization and, 31
Hydrogen cyanide, 39-40
Hydrogen peroxide, 112
Hydrogen sulfide, 220
  original atmosphere and, 38
  photosynthesis and, 180-181, 182, 218
  vent organisms and, 763
Hydroid, 731
  stinging cells and, 1161
Hydrolysis, 164
Hydrophobic bonding, 30
Hydrophobic compound, 30
Hydrophobic substrates, 175
Hydroponic culture, 706-707
Hydrostatic pressure, 155
Hydrostatic skeleton, 743
Hydroxyapatite, 908
Hydroxyl group, 56
Hylobatidae, 472
Hymenoptera, 794, 795, 851
Hyophorbe verschaffeltii, 1191
Hyoseris longiloba, 598
Hyperosmosis, 1045
Hyperpolarization, 1022-1024
Hypersensitivity, 905
Hypertonic solution, 154
  plant cell and, 156
Hypha, 562, 577
  ascomycetes and, 566
  basidiomycetes and, 569
  fungal reproduction and, 564
  zygomycetes and, 565-566
Hypocotyl, 632
Hypogymnia enteromorpha, 575
Hypolimnion, 1092
Hypoosmosis, 1045
Hypopharynx, 795
Hypophysis, 1039
Hypoplectrus, 853
Hypothalamus, 1002, 1003
  motivation state and, 1058
  neuroendocrine control and, 1039-1040
  regulation of body functions by, 860
Hypothesis, 10
  countercurrent multiplier, 1050
  definition of, 282
  one gene-one enzyme, 294
Hypotonic solution, 154
Hysterectomy, 872

I

Ichthyologist, 830
Ichthyostegia, 455
Icosahedral RNA virus, 504
Icosahedron, 496
Igneous rock, 696
Imaging, 1016
Immune deficiency syndrome, acquired,
  7-8
Immune system, 916
  AIDS infection and, 8
  macrophage and, 904

Implantation
  of embryo, 868
  prevention of, 872
Imprinting, 1061
Impulse, nerve, 992-994, 995
Inactivation, insertional, 332
Incisor teeth, 925
Inclusive fitness, 1068-1069
Incompatibility, genetic, 620
Incomplete dominance, 253
Independent assortment, 263
  genetic, 250-251
  occurrence of, 339
Indeterminate growth, 616
Indian pipe, 610
Indoleacetic acid oxidase, 679
Induced fit, 175
Induction, 890
  enzyme transcription and, 297
Inductive reasoning, 1062
Industrial melanism, 397-398
Infarction, myocardial, 980, 981
Infection
  bacteria and, 520-523
  feline peritonitis, 9-10
  flu virus and, 502
  macrophages and, 904-905
  Staphylococcus, 517
  viral; *see* Virus
Inferior vena cava, 979
Information, neuron and, 990-991
Infrared light, 214
Ingram, V., 294, 295
Inhalation, 952
Inheritance; *see also* Gene; Genetics
  of acquired characteristics, 368
  Mendelian, 265-268
  patterns of, 237-258
    early ideas of, 239-240
    gene concept, 252-254
    independent assortment, 250-251
    Mendel and, 240-248
    probability and allele distribution,
      248-250
    variation and, 237-239
Inhibitors, actin, 134
Inhibitory enzyme, 1157-1158
Inhibitory synapse, 997
Initial
  fusiform, 647
  ray, 648
Initiation complex, 313
Injury, leukocyte and, 971
Innate behavior, 1058-1060
Innate releasing mechanism, 1059
Inner ear, 1019
Inoculation, smallpox, 506
Inorganic molecule, 34
Insect (Insecta), 776, 793-801; *see also*
    Arthropod
  angiosperm and, 614
  commensalism and, 1168
  communication of, 798
  digestion and, 934
  ecological communities and, 1152; *see*
    also Community ecology
  genome size and, 340
  hemocyanin and, 953
  nutrition and, 937
  oxygen diffusion and, 945
  pheromone, 796-797
  pollination and, 615, 621, 622; *see also*
    Bee
  respiration and, 948
  social behavior and, 1063
  wing, 796
Insect eater, 843
Insecticide, 796-797
Insectivora, 842, 843
  primate evolution and, 468
Insertional inactivation, 332
Instar, 799

Metabolism—cont'd
  cell—cont'd
    chemical bond and, 192-193
    chemical energy and, 191-192
    citric acid cycle and, 202-203, 205
    fermentation and, 196-197
    glucose catabolism and, 206-207
    glycolysis and, 193-195
    heterotrophic metabolism and, 209
    mitochondria and, 208
    oxidative respiration and, 198-201
  chemical bond and, 171-172
  definition of, 170
  enzyme and, 173
    catalysis and, 174-175
  eukaryotic, 101
  evolution of, 178-186, 195
    aerobic respiration and, 185-186
    anaerobic photosynthesis in, 180-181
    chemoautotrophs and, 183-184
    degradation in, 179
    fermentation in, 179-180
    nitrogen fixation in, 181
    oxygen to drive redox processes and, 182-183
    oxygen-forming photosynthesis in, 182
  glucose
    efficiency of, 206-207
    energy released during, 201
  life and, 42, 43
  metabolic journey in, 186
  oxidative, 178
    coupled channel and, 166
    heterotroph and, 940
    nutrition and, 935
  plant, 707
  primary compound and, 1156
Metabolite
  blood plasma and, 967
  cell and, 157-158
  osmosis and, 97-98
Metamorphic rock, 697
Metamorphosis, 856
  arthropod and, 773
  complete, 798, 799-801
  simple, 798, 799-801
Metaphase, 144-145, 263
Metaphase I, 263
Metaphase plate, 144-145
Metastasis, 354; see also Cancer
Metazoan
  digestion and, 922
  movement of nutrients and, 959
Meteorite, 460
Methane, 34
  covalent bonds of, 23
  original atmosphere and, 38
Methane-producing bacteria, 85-86, 179, 523-524
Methanogen, 47, 523-524
Methionine, 1156
Methionine-carrying tRNA, 313
Method, scientific, 282-283
Methylase, 349
Micelle, 66
Micranthena, 783
Microbody, 110-112
Microevolution, 385, 401; see also Adaptation; Evolution
Microfilament, 124, 131-134
  contraction of, 102, 122, 130-135
  cytochalasin B and, 102
  eukaryotic cell and, 102
  protein and, 131
  structure of, 131-132
Microfossil, 46-47
  age of, 450
Microgametophyte
  angiosperm and, 603
  definition of, 589
  pine and, 602
  seed and, 598

Microgametophyte—cont'd
  seed plant and, 599
Microhabitat, 658-659
Micronucleus, 554
Micronutrient, 706, 715
Microplankton, 812
Micropteropus pusillus, 845
Micropyle, 666
Microsporangium
  angiosperm and, 617
  pine and, 602
  seed and, 597-598
  seed plant and, 589, 597-599
Microtrabecular lattice, 124-125, 126
Microtubular apparatus, 143-144
Microtubular movement, 122, 124-130
Microtubular organizing centers, 125
Microtubular sliding, 147
Microtubules, 124, 125-130, 133
  adaptations of, 125-126
  centromere and, 263
  colchicine and, 103
  eukaryotic cell and, 102-103, 117
  fungus and, 563-564
  structure of, 125-126
Microvilli, 135, 931
  actin and, 135
Midbrain, 1001, 1002
Middle ear, 1019
Middle lamella, 150
Migration
  innate behavior and, 1058
  population change and, 391-392
Migratory locust, 1134-1135
Migratory memory, 1071
Milk
  coconut, 664
  young and, 840-841
Milkweed
  African, 618
  butterfly and, 800, 1072-1073, 1159
  secondary compound and, 1157
Milkweed bug, 1160
Milky sap, 1186
Miller, S., 39, 40-41
Miller-Urey experiment, 39, 40-41
Millipede, 776, 790-794
Mimicry, 1162-1166
  Muellerian, 1165
Mimosa pudica, 687
Mineral
  future of biosphere and, 1182-1186
  nutrition and, 937
  soil and, 696
  in tropical soils, 12
Mink, 843
Minus-strand RNA virus, 501-502
Miracidium, 740
Mispairing
  repeated sequence and, 325
  spontaneous, 326-327
Mite, 776, 786-787
  moths and, 1166-1167
Mitochondria, 84, 102, 537
  ATP generation and, 208
  cytokinesis and, 148
  distribution of, 113
  DNA and, 113
  eukaryotic cell and, 112-113, 117
  function of, 113
  internal structure of, 205
  membrane and, 113
  origin of, 90, 91, 92
  organic molecule breakdown and, 185
  proton pump and, 166
  replacement of, 111
  replication of, 113
  support and, 124
  symbiosis and, 1167
Mitosis, 138, 141, 142-148
  anaphase in, 146-147
  animal versus plant, 144
  in eukaryote, 146-148

Mitosis—cont'd
  fungus and, 563-564
  interphase of, 142-143
  meiosis and, 261, 276
  metaphase in, 144-145
  prophase in, 143-144
  telophase in, 147-148
Mitral valve, 978
Model of heredity, 245-246
  Watson-Crick, 290-291
Molar, 925
Molar concentration of hydrogen in water, 31
Mold
  genome size and, 340
  slime, 535
  water, 555
Mole, 843
  definition of, 31
  starnosed, 842
Mole rat, naked, 1068-1069
Molecular analysis, 839
Molecular clock of evolution, 478
Molecular concentration, 943
Molecule
  atom and, 19
  complex, origin of, 44-45
  composite, 57
  diatomic, 22
  digestion and, 920
  energy-storing, 57-64
  nature of
    atom and, 16-19, 24-25
    chemical bond and, 19-23
    electron orbital and, 17-19
    living organism and, 24-25
    water and, 25-31
  organic, 34, 38-40
  origin of, 28-29
  phospholipid, 42-43
  polar, 33
  polysaccharide, 515
  transport of, 104, 105
Mollusk (Mollusca), 751, 755-763
  atmospheric air, 948
  bivalve, 756
    sea stars and, 811
  body plan of, 756-758
  classes of, 758
  digestion and, 923
  hemoglobin and, 953
  genome size and, 340
  oxygen diffusion and, 944
  reproduction of, 758
  respiration and, 945
Molting hormone, 801, 802
Molybdenum
  depletion of, in soil, 1154
  human body and, 20
  nutrition and, 937
  plant and, 706-707
Monarch butterfly, 800, 801
  chemical defense of, 1160
  defensive reaction of, 1158-1159
  migration and, 1072-1073
Monera, 437
Monkey, 843
  evolution of, 470, 471-472
  nutrition and, 936
  Old World versus New World, 487
  owl, 470
  red, 470
  velvet, 471
Monocot, 613
  embryo and, 631
Monocotyledon, 613
Monoecious plant, 618
Monogenea, 740
Monosaccharide, 57
Monosynaptic reflex arc, 1035
Monotreme, 841
  ovipary of, 857
Monotropa uniflora, 610

Monsoon, 1085
Moose, 1145
Moray eel, 831
Morgan, T., 266-267
Morphogenesis, 895
Mortality, 1136
    average juvenile-adult, 858
Morula, 868, 881
Mosaic disease, 500
Mosquito, 795
    filariasis and, 746
    larvae of, 796
    malaria and, 544-545
    sense receptor of, 798
Moss, 590-594; see also Bryophyte
    fern and, 600
    gameteophyte of, 585
    life cycle of, 592-594
    versatility of, 592-593
Moth, 795
    bats and, 1166-1167
    luna, 794
    peppered, 385
        industrial melanism and, 397-398
    scales on wing of, 797
    sense receptor of, 798
    silkworm, 802
    wing of, 796
Mother cell, spore, 587
Motile cell, 131
Motion; see Movement
Motivation state, 1058
Motor cell, 686
Motor cortex, 1006
Motor end plate, 916, 996
Motor nerve, 989, 1034
Motor neuron, 990
Motor pathway, 989
Mouse, 843
    ecological community and, 1152
Mouth, 924-925
Movement
    carbohydrate, 703-706
    cells and, 135-136
    cephalopod and, 762
    echinoderm and, 808
    life and, 42
    sensing of, 1014-1015
    turgor, 686-688
    water in plant and, 698-703
mRNA; see Messenger RNA
Mucosa, stomach, 926
Mucus, 927
Mud puppy, 833
Mudskipper, 832
Mueller, F., 1165
Muellerian mimicry, 1165
Muller, C., 1152
Multicellularity
    evolution and, 538-539
        motile cells and, 131
    prokaryote versus eukaryote and, 514
Multienzyme complex, 198
Multifaceted eye, 1025
Multigene family, 342
    copy number and, 343
Multiple allele, 252
Multiple sclerosis, 502
Multituberculate, 458, 841
Murmur, heart, 982
Musca domestica, 795, 798
Muscle, 122
    antagonistic control of, 1036
    blood vessel and, 963
    cell, 134
    ciliary, 1025
    contraction and
        acetylcholine release and, 996
        microfilament and, 134
        receptor and, 1012-1013
    cnidarian and, 729
    energy use and, 196-197

Muscle—cont'd
    nervous system and, 1034-1036
    respiratory, 952
    tissue and, 910-914
Musculoskeletal system, 916
Mushroom, 573
Mustard, 1157
Mustela nigripes, 843
Mutagen, 324
    cancer and, 356, 359-361
    chemical, 325
        DNA damage and, 326
    spontaneous, 325
Mutant; see also Mutation
    arginine, 293
    Drosophila melanogaster, 266-268
    strain and, 292
Mutation, 292, 324-327, 332
    bacteria and, 516-517
    cancer and, 358-359
    population change and, 391
    significance of, 377-378
    source and types of, 325
Mycelium, 562
    basidiomycetes and, 569
Mycobacterium, 529-530
Mycologist, 560
Mycoplasma, 528-529
    characteristics of, 523
    genome size and, 340
Mycorrhiza, 575-577
    fungus and, 561-562
    phosphorus cycle and, 708
    succession and, 1124
    symbiosis and, 1167
    terrestrial plant and, 586
    tropical rain forest and, 1094
Myelin sheath, 991
    development of, 994
    nerve impulse propagation and, 995
Myelinated fiber, 991
Myocardial infarction, 980, 981
Myofibril, 910, 911, 912, 913
Myosin, 102, 131-132, 134, 910, 912, 914
Myxobacteria, 530
    characteristics of, 523
Myxomatosis, 1145, 1167
Myxomycota, 546-547

N

Na; see Sodium
NAD$^+$; see Nicotinamide adenine
        dinucleotide
NADH, citric acid cycle and, 203
NADH dehydrogenase, 204
NADP$^+$; see Nicotine adenine
        dinucleotide phosphate
NADPH
    generation of, 223
    glycolytic reversal and, 225
Na$^+$-K$^+$ pump; see Sodium-potassium
        pump
Naked mole rat, 1068-1069
Naphthalanacetic acid, 679
Natural population, changes in, 396-399;
        see also Population
Natural selection, 14, 372, 381, 384-385
    behavior and, 1071-1073
    definition of, 386
    evolution and, 338, 394-395
Natural variation, 338
Nauplius, 775, 776, 789
Nautilus, 761-763
Nautilus pompilius, 761
Navigation, magnetic, 516-517
Neanderthal, 485, 1175
Necator, 746
Nectar, 615
    pollination and, 623
Negative ion, 20

Neisseria gonorrhoeae, 522
Nekton, 1090
Nelumbo nucifera, 668, 691
Nematocyst, 729-730
Nematode (Nematoda), 744
    digestion and, 923
    evolution of circulatory system and,
        959
Neon
    chemical characteristics of, 24
    energy level of, 18
Neospora experiment, 292-294
Neotenic amphibian larvae, 945
Neoteny, 818
Nepenthes, 712
Nephrid organ, 1045
Nephridia
    annelid and, 765
    mollusk and, 757
Nephron tube
    fish and, 1047-1048
    mammalian, 1050
Nephrostome, 757
Neritic zone, 1089-1090
Nerve; see also Nervous system
    blood vessel and, 963
    cell, 914, 989, 990-998
        baroreceptor and, 1013
    fiber and, 1000
    impulse, 915, 990, 992-993
        propagation of, 995
        retinal and, 214
        saltatory conduction and, 994
    motor, 989
    optic, 1026
    parasympathetic, 989, 1036, 1037
    sensory receptor and, 1011-1012
    sympathetic, 989, 1036, 1037
    tissue and, 914-916
    transferring information from
        to nerve, 997-998
        to tissue, 994-996
Nerve cord, 814, 999
    chordate and, 815
Nerve gas, 996
Nerve net, 999
    cnidarian and, 729
Nervous system, 916, 987-1009
    arthropod and, 781
    associative activities and, 999
    autonomic, 1036
        antagonistic control of, 1037-1038
    brain and, 1000-1007
    central, 989; see also Central nervous
        system
    elaboration of, 887-888
    evolution of, 998-1004
    flatworm, 739
    information transmission and, 988
    neuron in, 914, 989, 990-998
    neurosensory communication and,
        1011
    parasympathetic, 989, 1036, 1037
    peripheral, 989
        central nervous system and, 1000
        importance of, 1007
        nerve tissue and, 915
    sensory receptor and
        external environment and, 1016-1028
        internal environment and, 1011-1016
    simplest, 999
    sympathetic, 989, 1036, 1037
    vertebrate, 1010-1031
        body activity and, 1033-1034
        evolution and, 1000
        neuroendocrine control and, 1039-
            1044
        neuromuscular control and, 1034-
            1036
        neurovisceral control and, 1036-1038
        organization of, 988-989
Net primary productivity, 1117

Thyroid-stimulating hormone, 1041
T$_i$ plasmid, 352
Tick, 776, 786, 787
Tiger, 410
Tiger salamander, 833
Tiger shark, 1048
Time scale, geological, 46
Tin, 20
Tinbergen, 1074
Tissue
  animal, 722
    movement of cells of, 130
  connective; see Connective tissue
  plant, 660-663
    vascular, 586
Titanium, 24
TMV; see Tobacco mosaic virus
Toad, 833
  African clawed, 350
  sexual reproduction and, 855-856
Tobacco, 4, 239-240, 359-361
Tobacco mosaic virus
  experiment and, 289
  purification of, 495
Togaviruses, 500
Tolerance, ecological communities and,
  1152
Tolypocladium inflatum, 561
Tomato, 661
  hydroponic, 706
  secondary compound and, 1157
Tomato bushy stunt, 500
Tonicella lineata, 758
Tooth decay, 522
Tortoise, 834
  Galapagos, 371
Totipotent cell, 659
Touch, 1013-1014
  insect and, 798
Toxic shock syndrome, 521
Toxin, red tide and, 541
Trace element, 936, 937
Trachea, 926, 951
  arthropods and, 780, 945
  insect and, 797
  respiration and, 948
Tracheary element, 596
Tracheid
  definition of, 615
  element, 639
Tracheoles, 780
Trade wind, 1084
Trait
  adaptation and, 385-386
  alternative, segregation of, 240
  direct transmission of, 239
  dominant and recessive, 242
  enzyme and, 294
  sex-linked, 267
Transcript, primary, 311
Transcription
  control of, 308-309
  enzyme and, 297
  genetic, 301, 302
  signal and, 305
  site of, 333
  unit of, 319
Transcutaneous respiration, 944
Transducin, 1023
Transduction, 1011
Transfection, 357-358
Transfer, gene, 328-333
  bacteria and, 330
  transposition and, 331, 332
  transposition in, 332-333
Transfer RNA, 307
Transfusion, 387
Transition state, 173
Translation, genetic, 302, 305
Translocation, 335
  carbohydrate and, 703-706

Translocation—cont'd
  group, 162-163
Transmembrane channel, 158
  protein and, 311
  type of, 159
Transmembrane protein, 159
  Halobacterium and, 219
Transmembrane proton-pumping
  channel, 217
Transmission, 1011
  definition of, 1011
  genetic material and, 239
  nerve impulse and, 993
Transmitter
  chemical; see Chemical, transmitter and
  chemical structure of, 997
  nervous system and, 991, 997
  synapse and, 995
Transpiration, 700, 701-702
Transplant
  cyclosporin and, 562-563
  nuclear, 285
Transport
  active, 163; see Active transport
  carbon dioxide, 960
  channel and, 159-166
    importance of, 166-167
  disaccharide and, 58-59
  ion, neuronal membrane and, 992, 993
  membrane and, 67
  nutrient, 960
  oxygen, 960
  plant and, 695-717
    carbohydrate and, 703-706
    water and, 698-703
  system of, 159-166
    photosynthesis and, 221, 222
  waste, 960
  water and, 1182
Transposase, 331, 332
Transposition, 324, 325, 331, 332
  impact of, 332-333
  occurrence of, 339
Transposition element, 340-341
  copy number and, 343
Transposon, 328, 331, 332
  crossing-over and, 339
  eukaryotic genome and, 340
Trap-door spider, 784
Trauma, 905
Traveler's diarrhea, 518
Tree
  agricultural growth of, 1175
  cloning of, 348
  defensive reaction of, 1158
  elephant, 434
  firewood and, 1185-1185
  maidenhair, 596
  silk-cotton, 585
Tree finch, 418-419
Tree frog, 833
Trehalose, 59
Trematoda, 740
Treponema, 529
Treponema pallidum, 522
Triassic Period, 456
  mammals and, 458
Tribolium, niche and, 1142
Tribolium confusum, 1138
Tricarboxylic acid cycle, 199
Trichinella, 746
Trichinosis, 746
Trichogyne, 566
Trichome, 640, 641
Trichoplax adhaerens, 723
Trichoptera, 793
Tricuspid valve, 979
Tridacna maxima, 755
Trigger hair, 687
Triglyceride, 62, 63
Trillium ozarkanum, 613

Trilobite, 455, 781
Trimester of pregnancy, 893-896
Triphoblastic eumetazoan, 728
Tritium, 16
Trochophore, mollusk, 758
  larva and, 758
Trophic chain, 209
Trophic level, 1119-1121
Trophoblast, 883
  human embryo and, 893-894
Tropical fish, 1167
Tropical soil, 12-13
Tropics
  agriculture and, 12-13
  bat and, 844
  forest in
    rain, 1094-1095
    undisturbed, 1189-1192
  future of biosphere and, 1189-1192
  population of, 12-13
Tropism, 685-686, 692
Tropomyosin, 131, 916
Troponin, 131
Tryon, R., 1070
Trypanosome, 333
Trypsin, 920, 921
  specificity and, 176
Trypsinogen, 921
Tryptophan, 679
Tsetse fly, 547
Tuatara, 834
Tubal ligation, 872
Tube
  Eustachian, 1019
  nephron
    fish and, 1047-1048
    mammalian, 1050
  pollen, 598
  sieve, 640
    nutrition and, 705
Tube worm, 766
Tuber, 650
Tuberculosis, 521
Tubule
  Malpighian, 780-781, 790, 1045
  renal, 1050
Tubulin, 102
  a or b, 125
  subunit of, 125
    dimer and, 102
  synthesis of, 142, 143
Tufted puffin, 837
Tularemia, 787
Tumbleweed, 626
Tumor, 354; see also Cancer
Tundra, 1104, 1105
Tunicate, 815
Turbellaria, 739
Turesson, Gote, 416
Turgor movement, plant and, 686-688,
  692
Turgor pressure, 156
Turtle, 834
  sexual reproduction and, 856
Tuttle, M., 844
Two-cycle pump, heart, 972, 975-976
Twofold rotational symmetry, 349
Two-solute model of kidney function,
  1051-1053
Two-way channel, 161-162
Tympanic membrane, 1019
Tympanum, 798
Typhus, 521
  dehydration and, 933
Tyrosine kinase, 357, 358

U

Uakari, 470
Uercus suber, 648
Ulcer, 928
Ultraviolet energy, 39